METALLIC MATERIALS SPECIFICATION HANDBOOK

METALLIC MATERIALS SPECIFICATION HANDBOOK

FOURTH EDITION

Robert B. Ross

Ross Materials Technology Ltd, East Kilbride, Glasgow

CHAPMAN & HALL
London · New York · Tokyo · Melbourne · Madras

Published by Chapman & Hall, 2–6 Boundary Row, London SE1 8HN

Chapman & Hall, 2–6 Boundary Row, London SE1 8HN, UK

Chapman & Hall, 29 West 35th Street, New York, NY10001, USA

Chapman & Hall Japan, Thomson Publishing Japan, Hirakawacho Nemoto Building, 7F, 1-7-11 Hirakawa-cho, Chiyoda-ku, Tokyo 102, Japan

Chapman & Hall Australia, Thomas Nelson Australia, 102 Dodds Street, South Melbourne, Victoria 3205, Australia

Chapman & Hall India, R. Seshadri, 32 Second Main Road, CIT East, Madras 600 035, India

---

First edition published in 1968 as *Metallic Materials*
Second edition 1972
Third edition 1980
Fourth edition 1992

Typeset by MFK Typesetting Limited, Hitchin, Hertfordshire
Printed in Great Britain by Clays Ltd, St Ives

ISBN 0 412 36940 0

A catalogue record for this book is available from the British Library

Library of Congress Cataloging-in-Publication data available

Contents

Preface to the fourth edition		xi
How to use this book		xii
1	**Aluminium Al**	**1**
1.1	General notes on aluminium	1
1.2	Notes on specifications and trade names	1
1A	Aluminium – commercially pure – wrought and cast	4
1B	Aluminium–manganese wrought alloys	10
1C	Aluminium–silicon wrought and cast alloys	13
1D	Aluminium–magnesium wrought alloys	19
1E	Aluminium–magnesium–silicon wrought alloys	28
1F	Aluminium–copper wrought alloys	35
1G	Aluminium–zinc wrought alloys	44
1H	Aluminium alloys clad with aluminium	48
1J	Aluminium–magnesium cast alloys	51
1K	Aluminium–silicon–magnesium cast alloys	54
1L	Aluminium–copper cast alloys	57
1M	Aluminium–magnesium–zinc cast alloys	64
1N	Aluminium – miscellaneous alloys	66
2	**Antimony Sb**	**68**
2.1	General notes on antimony	68
3	**Arsenic As**	**70**
3.1	General notes on arsenic	70
4	**Barium Ba**	**71**
4.1	General notes on barium	71
5	**Beryllium Be**	**72**
5.1	General notes on beryllium	72
6	**Bismuth Bi**	**74**
6.1	General notes on bismuth	74
7	**Boron B**	**76**
7.1	General notes on boron	76
8	**Cadmium Cd**	**77**
8.1	General notes on cadmium	77
9	**Caesium Cs (cesium)**	**79**
9.1	General notes on caesium	79
10	**Calcium Ca**	**80**
10.1	General notes on calcium	80

11	**Cerium Ce**	**81**
11.1	General notes on cerium	81
12	**Chromium Cr**	**83**
12.1	General notes on chromium	83
13	**Cobalt Co**	**85**
13.1	General notes on cobalt	85
13A	Cobalt alloys – wrought and cast	86
	Columbium Cb – see Niobium Nb	
14	**Copper Cu**	**94**
14.1	General notes on copper	94
14.2	Notes on specifications and trade names	94
14A	Copper – commercially pure – wrought and cast	96
14B	Copper–zinc wrought alloys	106
14C	Copper–zinc wrought and cast alloys	118
14D	Copper–beryllium wrought and cast alloys	126
14E	Copper–nickel wrought and cast alloys	129
14F	Copper–nickel–zinc wrought and cast alloys	134
14G	Copper–aluminium wrought and cast alloys	139
14H	Copper alloys for brazing fillers and fuse materials	146
14J	Copper sintered alloys	150
14K	Copper–tin wrought and cast alloys	152
14L	Copper–chromium wrought and cast alloys	165
14M	Copper–silicon wrought and cast alloys	167
14N	Copper alloys – miscellaneous	169
15	**Gallium Ga**	**171**
15.1	General notes on gallium	171
16	**Germanium Ge**	**172**
16.1	General notes on germanium	172
17	**Gold Au**	**173**
17.1	General notes on gold	173
17A	**Hafnium**	**177**
17A.1	General notes on hafnium	177
18	**Indium In**	**178**
18.1	General notes on indium	178
19	**Iridium Ir**	**179**
19.1	General notes on iridium	179
20	**Iron Fe**	**181**
20.1	General notes on iron	181
20A	Iron – commercially pure	182
20B	Iron – cast	184
20C	Iron–nickel alloys – magnetic	197

21	**Lead Pb**	**202**
21.1	General notes on lead	202
22	**Lithium Li**	**209**
22.1	General notes on lithium	209
23	**Magnesium Mg**	**211**
23.1	General notes on magnesium	211
23A	Magnesium – commercially pure – wrought and cast	213
23B	Magnesium alloys	214
24	**Manganese Mn**	**223**
24.1	General notes on manganese	223
25	**Mercury Hg**	**225**
25.1	General notes on mercury	225
26	**Molybdenum Mo**	**226**
26.1	General notes on molybdenum	226
26A	Molybdenum – all alloys	227
27	**Nickel Ni**	**229**
27.1	General notes on nickel	229
27.2	Nickel – notes on specifications	230
27A	Nickel – commercially pure – wrought and cast	231
27B	Nickel–chromium wrought and cast non-ageing alloys	234
27C	Nickel–chromium wrought and cast ageing alloys	242
27D	Nickel–copper wrought and cast alloys	253
27E	Nickel–iron alloys	256
27F	Nickel – miscellaneous alloys	258
28	**Niobium Nb**	**264**
28.1	General notes on niobium (also called columbium, Cb)	264
29	**Osmium Os**	**266**
29.1	General notes on osmium	266
30	**Palladium**	**267**
30.1	General notes on palladium	267
31	**Platinum Pt**	**269**
31.1	General notes on platinum	269
32	**Plutonium Pu**	**271**
32.1	General notes on plutonium	271
33	**Potassium K**	**272**
33.1	General notes on potassium	272
34	**Radium Ra**	**273**
34.1	General notes on radium	273
35	**The rare earth elements**	**274**

36	**Rhenium Re**	**276**
36.1	General notes on rhenium	276
37	**Rhodium Rh**	**278**
37.1	General notes on rhodium	278
38	**Rubidium Rb**	**280**
38.1	General notes on rubidium	280
39	**Ruthenium Ru**	**281**
39.1	General notes on ruthenium	281
40	**Selenium Se**	**282**
40.1	General notes on selenium	282
41	**Silicon Si**	**284**
41.1	General notes on silicon	284
42	**Silver Ag**	**286**
42.1	General notes on silver	286
43	**Sodium Na**	**293**
43.1	General notes on sodium	293
44	**Steel**	**295**
44.1	The thermal treatment of steel	295
44.2	Notes on specifications and trade names	297
44A1	Steel – plain carbon 0.05–0.2 per cent	300
44A2	Steel – plain carbon 0.25–0.45 per cent	338
44A3	Steel – plain carbon 0.5–0.8 per cent	356
44A4	Steel – plain carbon 0.8 per cent minimum	364
44B1	Steel – low carbon with silicon	369
44B2	Steel – medium carbon with silicon	371
44C	Steel – low and medium carbon with boron	372
44D	Steel–manganese alloys	373
44E1	Steel – low carbon with chromium	375
44E2	Steel – medium carbon with chromium	380
44E3	Steel – high carbon with chromium	384
44F1	Steel – low carbon with nickel	388
44F2	Steel – medium carbon with nickel	393
44F3	Steel – low carbon, high nickel – maraging	395
44G1	Steel – low carbon with molybdenum	397
44G2	Steel – medium and high carbon with molybdenum	403
44H	Steel – all carbon contents with tungsten	404
44J	Steel – all carbon contents with vanadium	406
44K1	Steel – low carbon alloys	409
44K2	Steel – medium carbon alloys	452
44K3	Steel – high carbon alloys	471
44L	Steel–cobalt alloys	485

44M1	Steel – low and medium carbon, high chromium	496
44M2	Steel – high carbon, high chromium	518
44N1	Steel – chromium–nickel – austenitic wrought and cast	524
44N2	Steel – chromium–nickel – wrought and cast, age hardening	569
44N3	Steel – chromium–nickel	573
44P	Steel – high manganese austenitic	578
45	**Strontium Sr**	**582**
45.1	General notes on strontium	582
46	**Tantalum Ta**	**583**
46.1	General notes on tantalum	583
47	**Tellurium Te**	**585**
47.1	General notes on tellurium	585
48	**Thallium Tl**	**586**
48.1	General notes on thallium	586
49	**Thorium Th**	**587**
49.1	General notes on thorium	587
50	**Tin Sn**	**588**
50.1	General notes on tin	588
50A	Tin alloys	590
51	**Titanium Ti**	**594**
51.1	General notes on titanium	594
51A	Titanium – commercially pure	595
51B	Titanium alloys – not solution treated	598
51C	Titanium alloys – solution treated	602
52	**Tungsten W**	**606**
52.1	General notes on tungsten	606
53	**Uranium U**	**610**
53.1	General notes on uranium	610
54	**Vanadium V**	**612**
54.1	General notes on vanadium	612
55	**Zinc Zn**	**614**
55.1	General notes on zinc	614
55A	Zinc alloys	615
56	**Zirconium Zr**	**618**
56.1	General notes on zirconium	618
	Appendix I Names and addresses of firms and organizations	621
	Appendix II Conversion tables	626
	Appendix IIA Temperature conversion table	626
	Appendix IIB Strength conversion table	628
	Appendix IIC Impact data conversion table	630

Appendix IID Hardness and tensile values approximate conversion table for steel 631
Appendix IIE Thickness conversion table – approximate 633
Appendix III List of elements with their symbols 634
Index 635

Preface to the fourth edition

It is now ten years since the third edition of Metallic Materials, and over twenty since the first edition. Over this period the work has been extensively used as a comprehensive sourcebook by all those concerned with the use of engineering materials.

Since the third edition, there have been a number of basic changes in the methods of specifying materials. Probably the most important of which is the introduction of the Unified Numbering System (UNS). These codes are issued in agreement between the Society of Automotive Engineers (SAE) and the American Society for Testing and Materials (ASTM), and reflect the acknowledgement of a need for a numbering system with some logic. The UNS codes have been issued to cover all the trade names, codes and specifications used in the US, and cognizance has been taken of codes used in other areas. The codes have a letter prefix – A for aluminium, C for copper, etc. – followed by a 5 figure code. This refers to a basic analysis and lists US specifications and trade names covered by this analysis. This edition of *Metallic Materials* lists approximately 4,000 UNS codes and refers the reader either to the basic analysis for this code, or to one of the existing specifications or trade names covered by the code.

In this edition steps have been taken to update the trade names and specifications, to eliminate mistakes, and to supply as much contemporary information as possible. The extent of the additions has made it necessary to examine the way that some of the specifications are known under various designations, and the implications of this examination are explained in the 'How to use this book' section.

The information in this book should not be used as an authority to use any material in lieu of a designer's specification. The information supplied is basic, and may constitute only the first step in identifying a possible, similar material. Further information will very often be required.

In any book of this type there will be errors of omission and commission. I hope that readers who find any will contact either myself or the publishers, in order that the next edition may be corrected.

Robert B. Ross

How to use this book

There are three reasons for using this book:

1. to identify the chemical constituents and other properties of a trade name, specification or symbol applied to a metal or alloy;
2. to obtain information on similar materials;
3. to obtain a specification, trade name or symbol, for a material for which mechanical and other properties have been identified.

Set out below is an explanation of each of these three methods of access.

How to find a metal or alloy

The book is divided into 118 sections according to the primary constituent of the material concerned. At the back of the book the index lists the names of materials in numero-alphabetical order, i.e. with numbers taking precedence over letters. Thus, entry 0.0671 appears before entry 15 Cr Mo 6, and this before entry A 32 STEEL, etc.

If the material is included in the book it will be found in the index. Where there is more than one specification identification in common use, there will only be one entry in the appropriate section, with two or more entries in the index.

To save space in this edition, use has been made of the basic 'designation' used by the British Standard (BS) for a specific steel. The full title of a material might, therefore, be BS 970 817 M40. Under BS 970 the index will say 'see designation', and the reader should then refer to index entry 817 M40 (the designation), which, in this case, will refer to Section 44K2. Once within the relevant section, the reader will find the composition and properties of the material in question, listed against the material's trade name or symbol.

With ASTM specifications the reader who has information on the designation alone will be referred to the full specification and the section identity. Thus for a material of full specification ASTM A757 A1Q, the reader with the designation A1Q alone will find themselves referred to ASTM A757 and then to Section 44A2.

Similarly, with DIN (German) and other national specifications such as JIS (Japanese) and GOST (Russian), the designations alone may be found in the index referring the reader to the full specification and the section number.

With the UNS (Unified Numbering System) codes the reader will be referred to a 5-figure number with letter prefix. Having found this in the index, they will be referred to a designation and then a section.

Thus the following types of entry will be found:

UNS S 30300	see designation
S 30300	see 303
303	Section 44N1

How to obtain information on similar materials

The 118 sections group materials by their chemical constituents. Thus, once a given material has been found, the reader will be able to identify other metallic materials having similar composition and mechanical properties, by looking at other entries within the same section. In this edition the constituent elements have been sorted to appear in the same order.

It must be emphasized that this book does not attempt to give equivalent specifications, but allows the reader to identify trade names and specifications where descriptive information can be obtained. It must be appreciated that where alloys exist, there may be an overlap between sections, and the reader should also look at adjacent sections. This occurs particularly with Sections 44P, 44N2 and 44N3, where the overlap in analysis and the information from the supplier can result in apparently similar materials appearing in different sections.

How to obtain a specification

Each of the 118 sections begins with a list of information on the physical properties of the main constituent of the materials in the group (specific gravity, conductivity, expansion, etc.).

Along with the trade names and symbols, listed by section there is information on the mechanical properties and chemical analysis of alloys in the groups. This information enables the reader to identify a trade name, symbol or specification in line with the properties required. Conversion tables and factors for all the common physical properties listed can be found in the Appendices.

It should be noted that many trade names and specifications in the book are no longer current, and the reader should therefore contact a reputable material supplier to ensure that the specification or trade name chosen is still in fact available. Appendix I lists the names and addresses of suppliers and associations mentioned in the book.

1. Aluminium Al

Physical properties		
Atomic number		13
Atomic weight		26.97
Crystal structure		Face centred cubic
Colour		Tin-white
Specific gravity		2.7
Density		2700 kg/m^3
Melting point		658 °C
Boiling point		2270 °C
Specific heat		0.90 J/g °C
Thermal conductivity		210 W/m °C
Coefficient of linear expansion (20–100 °C)		24 × 10^{-6}/ °C
Latent heat of fusion		386.9 J/g
Latent heat of vaporization		9462 J/g
Thermal neutron absorption cross-section		0.215 barns/atom
Electrical conductivity		62–62.9% IACS (copper 100%)
Specific resistance		27.8–27.4 microhm mm
Temperature coefficient of electrical resistance		0.0041/ °C
Electrochemical equivalent		0.3354 g/A/h
Electrode potential		−1.69 V
Magnetic susceptibility		0.6 × 10^{-6}
Young's modulus of elasticity		68.3 × 10^9 N/m^2
Tensile strength	annealed	45 N/mm^2
Hardness	annealed	15 DPN

1.1 General notes on aluminium

Aluminium is available in commercial grades from 99.0% to 99.99% purity, with a well-defined series of alloys based on the metal, which are found in all the normal forms of castings and wrought products, and also as a fine powder used in paints and metal spraying.

The metal and its alloys are used because of their lightness, corrosion resistance and strength.

Aluminium alloys weigh approximately one-third as much as steel, have a much better corrosion resistance to atmospheric conditions and depending on the alloys can be of a comparable strength to low alloy steel.

Following are listed all known specifications and trade names covering aluminium and its alloys. These are grouped in such a manner that all specifications for each well-defined alloy appear under the one heading, with a general note covering their properties, processing and uses.

1.2 Notes on specifications and trade names

An explanation of the different classifications and systems used by the various bodies and companies whose alloys are listed in the following pages may be of help in rapidly finding any desired material and in understanding the symbols.

British Standards Institution (BS)

This body issues individual specifications covering a particular alloy.

The present policy however, is to have a master specification related to the form of the material and in the main these replaced individual specifications. Thus:

BS 1470	Sheet strip and plate – all alloys
BS 1471	Drawn tube – all alloys
BS 1472	Forgings and forging stock – all alloys
BS 1473	Rivet, bolt and screw stock for forging – all alloys
BS 1474	Bar, extruded tube and sections – all alloys
BS 1475	Wire – all alloys
BS 1490	Castings – all alloys

BS 1476 and BS 1477 have now been withdrawn and included in BS 1470 and BS 1474 respectively.

The latest British Standard retains the same 'master specifications', but has replaced the detailed 'designation' with a four-digit code in line with American and Australian practice.

The first figure of four digits indicates the alloy group according to the major alloying element. The simplified code is below:

1xxx	aluminium 99% and better
2xxx	copper aluminium alloys
3xxx	manganese aluminium alloys
4xxx	silicon aluminium alloys
5xxx	magnesium aluminium alloys
6xxx	magnesium and silicon heat treatable aluminium alloys
7xxx	zinc aluminium alloys
8xxx	other elements

There is a coding system for casting using three digits but this is rather complex and requires interpretation. The reader is advised to refer to the Aluminium Federation Handbook or the relevant specification for detailed information.

The replaced "designations" are still in common use and are included in this book.

With this the specification number was followed by the letters N or H, denoting non-heat-treatable (N) alloys, or heat-treatable (H) alloys. Commercially pure aluminiums have no letter.

The form was next identified by the following code letters:

B	Bolt and screw stock for forging
C	Clad sheet and strip
E	Bars and sections
F	Forgings and forging stock
G	Wire
LM	Casting
P	Plate
PC	Clad plate
R	Rivet stock
S	Sheet up to and including 0.252 in. thick, and strip up to and including 0.192 in. thick
T	Drawn tube
V	Extruded round tube and hollow section

This was followed by a number with which the actual alloy is coded. Thus 1 is always 99.99%; aluminium 1A is always 99.98%; aluminium 6 is always 5% magnesium–aluminium alloy.

In the index alloys are listed under their full title and also in the abbreviated form, but only the full form is shown in the classified sections. Thus high purity aluminium is shown under 1A 99.98% aluminium, and under BS 1470 S1A 99.98% aluminium sheet, BS 1471 T1A 99.98% aluminium tube, etc.; and 1.5% copper 1% magnesium 1% silicon aluminium alloy is shown under H 11, and under BS 1472 HF11 1.5% copper 1% magnesium 1% silicon aluminium alloy forging, etc.

The abbreviated alloy forms will be encountered under other specification numbers than those listed above. They are also sometimes incorporated into trade names.

The condition in which the material is supplied or used is often coded and then added to the specification as a suffix.

This practice has not been adhered to in the following pages, as it is felt that the code is too arbitrary to be easily memorized. Also the BS code is similar to, but differs in detail from, the American code. Thus the condition has been added to the description wherever possible.

The code used by the BS range is as follows:

H8	Fully hardened by cold working (previously H)
H2	Cold worked by approximately one-quarter of possible amount, or fully cold worked then partially annealed to remove approximately three-quarters of the cold work (previously H4)
H4	As H2 but cold worked to half possible (previously H½)
H6	As H2 but cold worked to three-quarters possible (previously H¾)
M	As cast; as rolled; as extruded
O	Annealed
TB	Solution treated only (may room age, depending on the alloy) (previously W)
TE	Precipitation treated (aged only) (previously P)
TH	Drawn after solution treatment, then aged (previously WDP)
TF	Solution treated and aged (previously WP)

In the above and in the following pages the term 'aged' indicates artificial ageing. Room ageing is always described as such.

In addition to the normal BS range, a series of aircraft alloys are listed. These are individual specifications which have the prefix L. The range also includes magnesium alloys. As these are commonly known as BSL and L alloys, both forms of identification are listed in the index, the L alloy being referred to the full BSL specification to avoid confusion.

American Aluminum Association (AA)

This body does not issue mandatory specifications, but has organized a four-figure system which is partly descriptive. It is often referred to as 'Commercial Designation'. In the following pages the system is referred to as 'designation used by AA'. This alloy classification is now used by ASTM and SAE but not under AMS. There are five groups of alloys, identified by the first digit thus:

1xxx	Pure aluminium
2xxx	Copper aluminium alloys
3xxx	Manganese aluminium alloys
4xxx	Silicon aluminium alloys
5xxx	Magnesium aluminium alloys
6xxx	Silicon magnesium aluminium alloys
7xxx	Zinc aluminium alloys
8xxx	Other elements

The remaining three digits signified by x are used to describe the alloying elements. This is the system being used by UK and other countries.

To date this system applies to wrought products and not to castings.

There is a code for castings but it is complex, and not readily summarized.

American Society for Testing and Materials (ASTM)

This body issues several volumes of specifications at frequent intervals. The non-ferrous alloys are prefixed by the letter B followed by a number which identifies the specification. In some instances this will cover one material only, but more commonly it will cover all alloys in one form. Thus B209 is the specification for sheet aluminium alloys and must be followed by the four figures (designation used by AA), identifying the actual alloys.

The full specification then has the date of issue and in many instances the letter T signifying that the specification is still tentative. An example of such a specification would be ASTM B209 3003/61. This is for manganese aluminium alloy sheet, the specification being issued in 1961. The full specification appears only in the index, the year of issue being omitted in the classified section of this book.

Previous to 1961 ASTM used their own alloy designation. This code is still used for casting alloys. Letters are given for up to two principal alloying elements, and then up to three figures code the amount of these elements. Letters following the figures denote slight variations of the same basic specification. The alphabetical code used is:

A	Al	H	Th	Q	Ag
B	Bi	K	Zr	R	Cr
C	Cu	L	Li	S	Si
D	Cd	M	Mn	T	Sn
E	Rare earths	N	Ni	Y	Sb
F	Fe	P	Pb	Z	Zn
G	Mg				

Thus ASTM B209 M1A58 is for manganese aluminium alloy sheet, now ASTM B209 3003/61. In the following pages only the alloy designation (i.e. M1A) is listed, and cross-referred to the present designation (i.e. 3003). The ASTM casting alloys still employ this system of classification.

American Society of Mechanical Engineers (ASME)

This body issues specifications which are identical as far as material content is concerned to ASTM. They use the prefix S in lieu of ASTM (e.g. SB 125 would be identical to ASTM B125).

Society of Automotive Engineers (SAE) of America

This body issues specifications for general purpose wrought and cast alloys which were identified by two or three figures, prefixed by SAE. The wrought alloys now use the AAA four-figure code prefixed AA. There is no attempt to code the casting alloys.

This society also issue a series of specifications for aircraft materials. Aerospace Material Specifications have four figures prefixed by the letters AMS. The suffix letter indicates the issue number.

Thus AMS 4006 is a manganese aluminium alloy. These four figures bear no relationship to the AA code.

Temper designations used by American societies

The condition in which the material is supplied or used has been coded and this is in general use as a suffix to specifications in America. This has not been used in the following pages, as the condition is described in full.

The code is as follows:

Type 1	Alloys strengthened only by cold work
Type 2	Alloys which respond to heat treatment
F	As fabricated
H	Strain hardened – (cold worked), sub-divided as follows:
H1	Strain hardened (cold worked) only. A second number defines the degree of cold work, up to 8, which is fully hard. Thus H14 is half hard, H16 three-quarters hard.
H2	Strain hardened (cold worked) and partly annealed. A second number indicates the degree of cold work remaining after softening, up to 8, which is fully hard. Thus, for example, H24 is half hard.
H3	Strain hardened (cold worked) and stabilized. A second number indicates the degree of cold work in the stabilized alloy, up to 8, which is fully hard. Thus H34 is half hard.
O	Annealed
T2	Annealed – castings only
T3	Solution treated and cold worked
T4	Solution treated – room aged
T5	Aged
T6	Solution treated and aged
T7	Solution treated and stabilized
T8	Solution treated, cold worked and aged
T9	Solution treated, aged and cold worked
T10	Aged and cold worked
W	Solution treated – unstable condition

In the above and in the following pages the term 'aged' indicates artificial ageing: room ageing is always described as such.

There are now a number of variations on the above and readers are advised to confirm their information by reference to the specification.

UNS – Unified Numbering System

This system is described in ASTM E527.

It covers only metals and alloys which have a commercial standing.

There are 18 series of numbers. These are listed at the start of Section 44 in this book.

There is a single letter prefix followed by five digits.

Any UNS number for aluminium or its alloys will be headed 'A' from A00001 to A99999.

In this book the the UNS number is given, in the index, and refers the reader to the 'designation' used.

Thus the reader who requires information on UNS A91035 will find in the index:

UNS	see designation for detail
1035	Section 1A
A91035	UNS designation for 1035

In Section 1A the reader will find that 1035 is 99.35% wrought – this being the designation used in the USA.

Trade names and symbols

These generally include some means of identifying the company, and often attempt to describe the alloy, either by percentage or using one of the national alloy designations.

Wherever possible all the forms under which the alloy might be used have been included in the following pages. It is possible, however, that some have been omitted. If, therefore, the desired alloy cannot be found in the index, attempts should be made to add or subtract the firm's name as a prefix to the symbol.

German National Standards (DIN)

This body is analogous to the British Standards Institution and like them has issued individual specifications covering certain alloys used under particular conditions.

There are now two distinct systems of designations, one using a descriptive code with chemical symbols and numbers, the second using numbers only, known as 'Werkstoff number'. Both systems of designation are contained in master specifications of which there are four principal ones relating to aluminium.

These are DIN 1712, high purity aluminium; DIN 1725, aluminium alloys; DIN 1732, welding wire and DIN 8512, brazing and soldering fillers.

Examples of the methods of designation are first the descriptive code:

Al Mg 7	7% magnesium aluminium alloy
Al Cu Mg 2	4.2% copper 1.6% magnesium 0.6% manganese aluminium alloy

This is in addition to a numeral code which is designed to suit modern computing machines. All aluminium alloys are included in the range 3.000 to 3.499; examples are:

3.3557	7% magnesium aluminium alloy
3.1355	4.2% copper 1.6% magnesium 0.7% manganese aluminium alloy

1A Aluminium – commercially pure – wrought and cast

Specific gravity		2.7
Density		2700 kg/m^3
Solidus/liquidus	wrought	646–660 °C
	cast	645–655 °C
Thermal conductivity		230 W/m °C
Coefficient of linear expansion		24×10^{-6}/ °C
Electrical conductivity	wrought, annealed	59–61% IACS (copper 100%)
	cast, annealed	57% IACS
Specific resistance	wrought, annealed	28 microhm mm
	cast, annealed	30 microhm mm
Young's modulus of elasticity		68.3×10^9 N/m^2
Impact	wrought	27 J
Fatigue strength (50×10^6 cycles)	wrought, annealed	± 40 N/mm^2
	wrought, ½-hard	± 55 N/mm^2
	wrought, hard	± 75 N/mm^2
	cast, annealed	± 30 N/mm^2

Hot strength

Temperature	Annealed		Half-hard	
°C	Tensile strength N/mm^2	Elongation %	Tensile strength N/mm^2	Elongation %
150	50	65	110	16
260	22	85	23	85
315	17	90	17	90
370	9.0	95	9.0	95

The above properties are typical of the following group, and may not apply exactly to any one specification. It is possible that with certain specifications some of the values may not be applicable.

General metallurgical characteristics

The specifications listed in this section in general are 99% aluminium upwards, to 99.999% aluminium. This covers wrought and cast materials. Very few castings are manufactured in high purity aluminium. Ingot material is covered by these specifications.

High purity aluminium has excellent corrosion resistance and is capable of accepting cold work to a very high degree to give very thin foils. This is 'silver paper'.

Also included are materials with up to 1% of alloying elements such as zinc for cladding and other minute quantities of alloying elements which have little effect on the mechanical properties but will invariably reduce corrosion resistance and the ability to accept cold work.

These materials are used as primary metals in ingot form for alloying with other elements to produce the range of alloys listed in other sections.

The wrought material is used for cladding and also for foils for food wrapping such as 'silver paper'.

Many domestic articles are now manufactured in high purity aluminium for decorative purposes where low strength and attractive appearance is required.

These materials can be anodized and colour anodized to give an attractive range of colours with excellent corrosion resistance.

No heat treatment is possible with these materials. Where some impurities exist the material will cold work and can be softened by annealing. Welding requires the use of inert gas shielded techniques with great care being taken as aluminium oxidizes readily.

These materials are almost invariably difficult to machine, being too soft to give a good finish.

Symbol	Nominal analysis, supplier, condition and remarks.
1A	99.99% Al: Designation used by BS
1B	99.5% Al: Designation used by BS
1C	99.0% Al: Designation used by BS
1E	0.05% Cu 99.5% Al: For electrical purposes; designation used by BS
1S	99.5% Al: Sheet and strip; Alcan
2S	99.0% Al: Sheet; Alcan
2S	99.0% Al: Sheet; Alcoa for BS alloy S1C
2S	99.0% Al: Previous AA designation; now 1100
3.0185	98% Al with Cu and Cr: German Standard DIN 1712
3.0200	99.0% Al: German Standard DIN 1712
3.0205	99.0% Al: German Standard DIN 1712
3.0250	99.5% Al: German Standard DIN 1712
3.0255	99.5% Al: Designation used in German Standard DIN 1712
3.0256	99.5% Al: Wrought; German Standard DIN 1712
3.0257	99.5% Al: German Standard DIN 1712
3.0270	99.7% Al: Wrought; German Standard DIN 1712
3.0275	99.7% Al: Designation used in German Standard DIN 1712
3.0280	99.8% Al: Wrought; German Standard DIN 1712
3.0285	99.8% Al: German Standard DIN 1712
3.0305	99.85% Al: German Standard DIN 1712
3.0385	99.99% Al: German Standard DIN 1712
3.040	99.99% Al: Wrought; German Standard DIN 1712
3.4415	1% Zn Al alloy: Used for cladding; German Standard DIN 1725
72S	1% Zn Al alloy: Used for cladding; previous AA designation; now 7072
91E	0.04% Cu 0.5% Mg 0.5% Si Al: Designation used by BS
99.8	99.8% Al: Bar, tube, etc.; British Aluminium Co. for BS alloy 1A
100.1	99.0% Al (min): Casting; designation used in the UK and USA
130.1	99.3% Al (min): Casting; designation used in the UK and USA
150.1	99.5% Al (min): Casting; designation used in the UK and USA
160.1	0.1% Si (max) 0.2% Fe (max) Al: Casting
170.1	99.7% Al: Casting; designation used in the UK and USA
990A	99.0% Al: Previous ASTM designation; now 1100
1030	99.3% Al (min): Wrought; designation used in the UK and USA
1035	99.35% Al (min): Wrought; designation used in the UK and USA
1040	99.4% Al (min): Wrought; designation used in the UK and USA
1045	99.45% Al (min): Wrought; designation used in the UK and USA
1050	99.5% Al (min): Wrought; designation used in the UK and USA
1055	99.5% Al (min): Wrought; obsolete
1060	99.6% Al (min): Wrought; designation used in the UK and USA
1065	99.65% Al (min): Wrought; designation used in the UK and USA

Note. The following abbreviations and units are used in the tables:

DPN	Hardness, diamond pyramid number
UTS	Ultimate tensile strength, N/mm^2
Elon	Elongation, %
Proof	0.1% proof strength, N/mm^2

1 N/mm^2=0.1 hbar=0.102 kgf/mm^2=0.06475 $tonf/in.^2$=145.04 $lbf/in.^2$=1 MPa

See Appendix II for other abbreviations and conversion tables.

Symbol	Nominal analysis, supplier, condition and remarks.
1070	99.7% Al (min): Wrought; designation used in the UK and USA
1075	99.75% Al (min): Wrought; obsolete
1080	99.8% Al (min): Wrought; designation used in the UK and USA
1085	99.85% Al (min): Wrought; designation used in the UK and USA
1090	99.9% Al (min): Wrought; designation used in the UK and USA
1095	99.95% Al (min): Wrought; obsolete
1100	99.0% Al (min): Wrought; designation used in the UK and USA
1130	99.0% Al: Wrought; designation used by AA; no ASTM specification
1135	99.35% Al (min): Wrought; designation used in the UK and USA
1145	0.55% Si 99.45% Al + Fe: Foil for cladding; AA designation; no ASTM alloy
1145	99.45% Al (min): Wrought; designation used in the UK and USA
1170	99.7% Al (min): Wrought; designation used in the UK and USA
1175	99.75% Al (min): Wrought; designation used in the UK and USA
1180	99.8% Al (min): Wrought; designation used in the UK and USA
1185	99.85% Al (min): Wrought; designation used in the UK and USA
1188	99.88% Al (min): Wrought; designation used in the UK and USA
1193	99.93% Al (min): Wrought; obsolete; designation used in the UK and USA
1199	99.99% Al (min): Wrought; designation used in the UK and USA
1200	99.0% Al (min): Wrought; designation used in the UK and USA
1230	0.1% Cu 0.1% Zn Al: Used for cladding; designation used by AA
1230	99.3% Al (min): Wrought; designation used in the UK and USA
1235	99.35% Al: Foil; AA designation; no ASTM alloy
1235	99.35% Al (min): Wrought; designation used in the UK and USA
1250	99.5% Al (min): Wrought; designation used in the UK and USA
1260	99.6% Al (min): Wrought; designation used in the UK and USA
1285	99.85% Al (min): Wrought; designation used in the UK and USA
1345	99.45% Al (min): Wrought; designation used in the UK and USA
1350	99.45% Al (min): Wrought; designation used in the UK and USA
8001	0.5% Fe 11% Ni Al: Wrought; designation used in the UK and USA
8013	0.3% Cr Al: Wrought; obsolete; designation used in the UK and USA
8017	0.15% Cu 0.04% Mg 0.6% Fe Al: Wrought; designation used in the UK and USA
8020	0.2% Cu 0.03% B 0.5% Fe Al: Wrought; designation used in the UK and USA
8040	0.2% Zr Al: Wrought; designation used in the UK and USA
8076	0.12% Mg Al: Wrought; designation used in the UK and USA
8077	0.2% Mg 0.2% Fe 0.05% Zr Al: Wrought; designation used in the UK and USA
8111	0.7% Si 0.6% Fe Al: Wrought; designation used in the UK and USA
8112	0.7% Mg (max) 1% Fe (max) 1% Zn (max) Al: Wrought; designation used in the UK and USA

Symbol	Nominal analysis, supplier, condition and remarks.
8130	0.1% Cu 0.6% Fe Al: Wrought; designation used in the UK and USA
8176	0.1% Si 0.6% Fe Al: Wrought; designation used in the UK and USA
8177	0.1% Si 0.3% Fe Al: Wrought; designation used in the UK and USA
A 2	99.8% Al: Obsolete; Southern Forge; replaced by Impalco 030
A 3	99.5% Al: Obsolete; Southern Forge; replaced by Impalco 050
A 4	99% Al: Sheet, tube, bar, etc.; Southern Forge; replaced by Impalco 102
A 4	99.00% Al: Ingot, billet, etc.; L'Aluminium Français
A 4	99.00% Al: Ingot; French Standard; part of NFA S7.101
A 4.5	0.1% Cu Al: Wrought; French Standard
A 5	99.5% Al: Ingot; French Standard; part of NFA S7.101
A 5	99.5% Al: Wrought; French Standard
A 5	99.5% Al: Ingot, billet, etc.; L'Aluminium Français
A 5B	99.5% Al: Low Si and Fe for foil production; L'Aluminium Français
A 5L	99.5% Al: For electrical conductors; L'Aluminium Français
A 5.3 (E91100)	99.0% Al: Welding electrode; AWS designation
A 7	99.70% Al: Ingot, billet, etc.; L'Aluminium Français
A 7	99.70% Al: Ingot; French Standard; part of NFA S7.101
A 8	99.80% Al: Ingot; French Standard; part of NFA S7.101
A 8	99.8% Al: Ingot, billet, etc.; L'Aluminium Français
A 9	99.99% Al: Super purity, ingot, billet, etc.; L'Aluminium Français
AGG 1	99.5% Al: Wrought; further information from Aluminium Zentrale
AGG 2	99.0% Al: Wrought; further information from Aluminium Zentrale
AL 3	99.0% Al: Sheet, tube, forgings, etc.; James Booth for BS alloy 1C
AL 5	99.5% Al: Sheet, tube, forgings and plate; James Booth for BS alloy 1B
AL 99/99R	99.99% Al: For high reflectivity; previous German Standard designation
ALCAN GB 1S	99.5% Al: Sheet, tube, red forgings, etc.; Alcan for BS alloy 1B
ALCAN GB 2S	99.25% Al: Wire and bar; Alcan for BS alloy 1C
ALCAN GB 99.7	99.7% Al: Sheet Alcan; annealed **UTS: 80 Elon: 50%**
ALCAN GB 99.8	99.8% Al: Alcan for BS alloy S1A
ALCAN GB 99.99	99.99% Al: Alcan for BS alloy S1
ALCAN GB 100	99.5% Al: Casting; Alcan; as cast **DPN: 24 UTS: 8 Elon: 55% Proof: 30**
ALCAN GB B1S	99.5% Al: Wire for electrical purposes; Alcan for BS alloy 1E
ALCAN GB C1S	99.5% Al: Wire for electrical purposes; Alcan for BS alloy 1E
ALCOA 050	99.5% Al: Wrought; Alcoa; annealed **UTS: 75 Elon: 30% Proof: 30**

Note. The following abbreviations and units are used in the tables:

DPN	Hardness, diamond pyramid number
UTS	Ultimate tensile strength, N/mm^2
Elon	Elongation, %
Proof	0.1% proof strength, N/mm^2

1 N/mm^2=0.1 hbar=0.102 kgf/mm^2=0.06475 tonf/in.^{2}=145.04 lbf/in.^{2}=1 MPa

See Appendix II for other abbreviations and conversion tables.

Symbol	Nominal analysis, supplier, condition and remarks.
ALCOA 051	99.5% Al 0.5% Cu+Si+Fe (max): Wrought; Alcoa; annealed **UTS: 60 Elon: 25%**
ALCOA 102	99.0% Al: Wrought; Alcoa; annealed **UTS: 80 Elon: 28%**
ALMINAL 3	0.1% Cu 0.5% Si 0.1% Mn Al: Wrought alloy; origin unknown
ALMINTRODE 96.10	0.2% Si 0.2% Fe 99.6% Al: For electrodes; ESAB
ALOLINE	Hg free Al for anodes: Wilson Walton
ALUM T	99.0% Al: Bar; E Kaye for BS alloy 1E
ALUMOWELD	Mild steel core wire coated with high conductivity aluminium; Copperweld Co.
Al Zn 1	1% Zn Al alloy: Previous German Standard designation
AMS 4000 B	99.7% Al: Sheet; annealed **UTS: 80**
AMS 4001 B	99.0% Al: Sheet; annealed **UTS: 80 Elon: 20%**
AMS 4003 B	99.0% Al: Sheet; ½-hard **UTS: 110 Elon: 2%**
AMS 4062 C	99.0% Al: Seamless tube; ½-hard; AMS for AA alloy 1100
AMS 4011	99.45% Al
AMS 4102 A	99.9% Al: Bar as rolled; AMS for AA alloy 1100
AMS 4180 B	99.0% Al: Wire for spraying; cold drawn; AMS for AA alloy 1100
AMS 4208	99.35% Al
AMS 4209	99.35% Al
AMTROD 18.01	0.15% Fe 0.06% Zn Al: Welding wire; ESAB: for inert gas; as welded **UTS: 70 Elon: 25%**
ASTM B37	Aluminium for use in steel manufacture: Purity denoted by grade symbol
ASTM B179/995 A	99.5% Al: Ingot; primary metal
ASTM B184/A12	99.5% Al (min): For electrode filler wire; may be flux coated
ASTM B209/1060	99.6% Al: Sheet; annealed **UTS: 45 Elon: 25% Proof: 7**
ASTM B209/1060	99.6% Al: Sheet; cold rolled ¼-hard **UTS: 60 Elon: 10% Proof: 45**
ASTM B209/1100	99.0% Al: Sheet; annealed **UTS: 80 Elon: 30% Proof: 7**
ASTM B209/1100	99.0% Al: Sheet; hard **UTS: 140 Elon: 4%**
ASTM B209/1230	0.7% Si+Fe 99.3% Al: Sheet and plate
ASTM B209/7072	0.7% Si+Fe 1.0% Zn Al alloy: Sheet and plate
ASTM B210/1060	99.6% Al: Tube; annealed **UTS: 45 Proof: 7**
ASTM B210/1060	99.6% Al: Tube; cold drawn **UTS: 90 Proof: 70**
ASTM B210/1100	99.0% Al: Tube; annealed **UTS: 90**
ASTM B210/1100	99.0% Al: Tube; cold drawn **UTS: 140**
ASTM B210/7072	0.7% Si+Fe 1.0% Zn Al alloy: Tube
ASTM B211/1060	99.6% Al: Bar; annealed **UTS: 45 Elon: 25% Proof: 7**
ASTM B211/1060	99.6% Al: Bar; cold drawn **UTS: 90 Proof: 70**
ASTM B211/1100	99.0% Al: Bar; annealed **UTS: 60 Elon: 25% Proof 7**
ASTM B211/1100	99.0% Al: Bar; cold drawn **UTS: 140**
ASTM B221/1060	99.6% Al: Extruded bar, tube, etc.
ASTM B221/1100	99.0% Al: Extruded bar; annealed **UTS: 60 Elon: 25% Proof: 7**
ASTM B221/7072	0.7% Si+Fe 1.0% Zn Al alloy: Extruded bar, etc.
ASTM B233 1350	99.5% Al: Rod for electrical purposes; 61.5% IACS
ASTM B234/1060	99.6% Al: Tube; cold drawn **UTS: 60 Proof: 45**

Symbol	Nominal analysis, supplier, condition and remarks.
ASTM B234/7072	1.0% Zn Al alloy: Seamless drawn tube
ASTM B235/1060	99.6% Al: Tube; extruded annealed **UTS: 60 Proof: 15**
ASTM B235/1100	99.0% Al: Tube; extruded annealed **UTS: 90**
ASTM B241/1060	99.6% Al: Pipe and tube
ASTM B241/1100	1.0% Si+Fe Al: Pipe and tube
ASTM B241/7072	0.7% Si+Fe 1.0% Zn Al: Pipe and tube
ASTM B247/1100	99.0% Al: Forging; as forged **DPN: 20 UTS: 60 Elon: 25% Proof: 15**
ASTM B285 ER1060	99.6% Al: Electric welding rod; obsolete
ASTM B285 ER1100	99.0% Al: Welding wire
ASTM B285 ER1260	99.6% Al: Welding wire
ASTM B307/1100	99.0% Al: Coiled tube; annealed **UTS: 60**
ASTM B313/1100	99.0% Al: Welded tube; cold rolled **UTS: 140 Elon: 4%**
ASTM B316/1100	99.0% Al: For rivets; annealed **UTS: 90**
ASTM B327 G1C	0.9% Mg 95.0% (min) Al alloy: Hardener for Zn based die castings
ASTM B345/1060	99.6% Al: Pipe; annealed **UTS: 45 Proof: 7**
ASTM B361	Weld fitments on Al and Al alloys; graded by AA code (see also other A sections)
ASTM B373/1145	99.45% (min) Al alloy: Foil for capacitors
ASTM B373/1235	99.35% (min) Al alloy: Foil for capacitors
ASTM B404/1100	99.6% (min) Al alloy: For condenser tubes; graded by AA system (see also other A sections)
ASTM B544	99.5% Al: Rod for electrical conductivity; 61.5% IACS
ASTM B547	Al alloy: Drawn and welded tube; alloys designated by AAA number
ASTM B566	Cu clad Al wire
ASTM B606	Zn coated steel-cored Al alloy: Conductors
AWCO 99.5	99.5% Al: Wire; Aluminium Wire Co.
AWCO 99.8	99.8% Al: Wire; Aluminium Wire Co.
AWCO EP	0.05% Cu 0.5% Al: For electrical purposes; Aluminium Wire and Cable Co. for BS alloy 1E
AWCO SP	99.9% Al: Wire; Aluminium Wire and Cable Co. for BS alloy 1A
BA	Hg free Al: For anodes; British Aluminium Co.
BA 99	99.0% Al: Sheet, bar, forging, etc.; British Aluminium Co. for BS alloy 1C
BA 99.5	99.5% Al: Bar, tube, etc.; British Aluminium Co. for BS alloy 1B
BA 99.8	99.8% Al: Sheet, bar, forging, etc.; British Aluminium Co. for BS alloy 1A
BA 177	0.7% Si 0.7% Fe Al alloy: Sheet and strip; British Aluminium Co.; $\frac{1}{2}$-hard
BA COMMERCIAL QUALITY	99.0% Al: Bar, tube, etc.; British Aluminium Co. for BS alloy 1C
BA ELECTRICAL PURITY	0.05% Cu 99.5% Al: For electrical purposes; British Aluminium Co. for BS alloy E1E **DPN: 22 UTS: 75 Elon: 47% Proof: 30**
BA SUPER PURITY	99.99% Al: Bar, tube, etc.; British Aluminium Co. **DPN: 15 UTS: 45 Proof: 30**
BIRMETAL 1	99.5% Al: In wrought form; Birmetal
BIRMETAL 2M	99.0% Al: In wrought form; Birmetal
BS 215	99.5% Al: Wire for conductors; material conforms to BS 2627 G1E
BS 359	98.0% Al: Ingot for remelting; replaced by BS 1490
BS 360	99.0% Al: Ingot for remelting; replaced by BS 1490
BS 385	Pure Al tubes. Replaced by BS 1471
BS 386	Pure Al bar and sections. Replaced by BS 1476
BS 1453 G1	99.9% Al: For filler rod for welding pure Al; BS alloy 1
BS 1453 G1A	99.8% Al: For filler rod for welding pure Al; BS alloy 1A
BS 1453 G1B	99.5% Al: For filler rod for welding pure Al; BS alloy 1B
BS 1453 G1C	99.0% Al: For filler rod for welding Al; BS alloy 1C

Symbol	Nominal analysis, supplier, condition and remarks.
BS 1470 1050A	99.5% Al: Replaces 1B designations; plate and sheet; H4 condition **UTS: 135 Elon: 6%**
BS 1470 1080A	99.8% Al: Replaces 1A designation; plate and sheet; H4 condition **UTS: 120 Elon: 7%**
BS 1470 1200	99.0% Al: Replaces 1C designation; plate and sheet; H4 condition **UTS: 130 Elon 5%**
BS 1470 S1	99.99% Al: Sheet and strip; annealed **UTS: 60 Elon: 45%**
BS 1470 S1	99.99% Al: Sheet and strip; $\frac{1}{2}$-hard **UTS: 90 Elon: 12%**
BS 1470 S1	99.99% Al: Sheet and strip; hard **UTS: 90 Elon: 6%**
BS 1470 S1A	99.8% Al: Sheet and strip; annealed **UTS: 70 Elon: 35%**
BS 1470 S1A	99.8% Al: Sheet and strip; $\frac{1}{2}$-hard **UTS: 110 Elon: 8%**
BS 1470 S1A	99.8% Al: Sheet and strip; hard **UTS: 120 Elon: 5%**
BS 1470 S1B	99.5% Al: Sheet and strip; annealed **UTS: 90 Elon: 30%**
BS 1470 S1B	99.5% Al: Sheet and strip; $\frac{1}{2}$-hard **UTS: 110 Elon: 8%**
BS 1470 S1B	99.5% Al: Sheet and strip; hard **UTS: 120 Elon: 5%**
BS 1470 S1C	99.0% Al: Sheet and strip; annealed **UTS: 90 Elon: 30%**
BS 1470 S1C	99.0% Al: Sheet and strip; $\frac{1}{4}$-hard **UTS: 110 Elon: 12%**
BS 1470 S1C	99.0% Al: Sheet and strip; $\frac{1}{2}$-hard **UTS: 120 Elon: 7%**
BS 1470 S1C	99.0% Al: Sheet and strip; $\frac{3}{4}$-hard **UTS: 140 Elon: 5%**
BS 1470 S1C	99.0% Al: Sheet and strip; hard **UTS: 140 Elon: 3%**
BS 1471 1050A	99.5% Al: Drawn tube; H4 condition **UTS: 95**
BS 1471 1200	99.0% Al: Drawn tube; H4 condition **UTS: 120**
BS 1471 T1A	99.8% Al: Tube; annealed **UTS: 70**
BS 1471 T1A	99.8% Al: Tube; hard **UTS: 90**
BS 1471 T1B	99.5% Al: Tube; annealed
BS 1471 T1B	99.5% Al: Tube; hard
BS 1471 T1C	99.0% Al: Tube; annealed
BS 1471 T1C	99.0% Al: Tube; hard
BS 1471 F1A	99.8% Al: For forging and bar; annealed **UTS: 45 Elon: 30%**
BS 1472 1050A	99.5% Al: Forging; as forged **UTS: 60 Elon 22%**
BS 1472 F1B	99.5% Al: For forging and bar; annealed **UTS: 60 Elon: 25%**
BS 1472 F1C	99.0% Al: For forging and bar; annealed **UTS: 60 Elon: 20%**
BS 1473 1050A	99.5% Al: Bolt and screw stock **UTS: 60**
BS 1473 R1B	0.5% Cu + Si + Fe + Mn + Zn (max) 99.5% Al: Roll and screw stock; cold rolled **UTS: 110**
BS 1473 R1C	99.0% Al: For rivets; $\frac{1}{2}$-hard **UTS: 110**
BS 1474 1050A	99.5% Al: Bar and extruded tube **UTS: 60 Elon 25%**
BS 1474 1200	99.0% Al: Bar and extruded tube **UTS: 65**
BS 1474 V1A	99.8% Al: For tube; as drawn **UTS: 45 Elon: 30%**

Symbol	Nominal analysis, supplier, condition and remarks.
BS 1474 V1B	99.5% Al: For tube; as drawn **UTS: 60** **Elon: 25%**
BS 1474 V1C	99.0% Al: For tube; as drawn **UTS: 60** **Elon: 20%**
BS 1475 1080A	99.8% Al: Wire; H8 condition **UTS: 125**
BS 1475 1080A	99.5% Al: Wire; H8 condition **UTS: 135**
BS 1475 G1	99.99% Al: Wire; hard drawn **UTS: 90**
BS 1475 G1A	99.8% Al: Wire; annealed **UTS: 70**
BS 1475 G1A	99.8% Al: Wire; hard drawn **UTS: 120**
BS 1475 G1B	99.5% Al: Wire; annealed **UTS: 90**
BS 1475 G1C	99.0% Al: Wire; annealed **UTS: 90**
BS 1475 G1C	99.0% Al: Wire; hard drawn **UTS: 140**
BS 1476 E1A	99.8% Al: Bar; as rolled **UTS: 45** **Elon: 30%**
BS 1476 E1B	99.5% Al: Bar; as rolled **UTS: 60** **Elon: 25%**
BS 1476 E1C	99.0% Al: Bar; as rolled **UTS: 60** **Elon: 20%**
BS 1477 P1A	99.8% Al: Plate; ½-hard **UTS: 110** **Elon: 8%**
BS 1477 P1B	99.5% Al: Plate; ½-hard **UTS: 110** **Elon: 8%**
BS 1477 P1C	99.0% Al: Plate; ½-hard **UTS: 110** **Elon: 7%**
BS 1490 LMO	99.5% Al: Ingot
BS 1490 LMOM	99.5% Al: Castings; as cast
BS 1616 A	99.5% Al: Welding rod; flux coated; wire to BS alloy 1B
BS 1683	99.0% Al: Sheet or strip for wrapping cheese with a protective coating
BS 2627 G1E	0.05% Cu 99.5% Al (max): Wire for electrical purposes; annealed; electrical conductivity 58% IACS
BS 2627 G1E	0.05% Cu 99.5% Al (max): Wire for electrical purposes; ¾-hard; electrical conductivity 56% IACS
BS 2627 G1E	0.05% Cu 99.5% Al (max): Wire for electrical purposes; hard; electrical conductivity 56% IACS
BS 2791	99.5% Al: Wire for electrical conductors as BS 2627 G1E
BS 2897 1350	99.5% Al: replaces 1E designation; plate and sheet; H4 condition **UTS: 120** **Elon: 8%**
BS 2897 D1E	0.05% Cu 99.5% Al (max): Strip for electrical purposes; annealed **UTS: 90** **Elon: 25%**
BS 2897 D1E	0.05% Cu 99.5% Al (max): Strip for electrical purposes; ½-hard **UTS: 110** **Elon: 38%**
BS 2897 D1E	0.05% Cu 99.5% Al (max): Strip for electrical purposes; hard **UTS: 140** **Elon: 3%**
BS 2898 1350	99.5% Al: Bar and extruded tube; H2 condition **UTS: 85** **Elon: 15%**
BS 2898 E1E	0.05% Cu 99.5% Al (max): Bar for electrical purposes; ¼-hard **UTS: 70** **Elon: 15%**
BS 2898 E91E	0.04% Cu 0.5% Mg 0.5% Si Al alloy: For electrical purposes
BS 2901 G1	99.99% Al: Rod; for all types of welding
BS 2901 G1A	99.8% Al: Rod; for all types of welding
BS 2901 G1B	99.5% Al: Rod; for all types of welding
BS 2901 G1C	99.0% Al: Rod; for all types of welding
BS 3313 S1C	99.0% Al: Foil for dairy product containers; ½-hard **UTS: 120**
BS 3988 C1E	0.5% Cu (max) or 0.5% Cu+Si+Fe (max) Al: Wrought; for electrical purposes **UTS: 80** **Elon: 25%**
BS 43002 BTRS1	0.5% Mg Al alloy: For bright trim; annealed **UTS: 120**
BS 4300/7 NS41	0.9% Mg Al alloy: Sheet and strip; cold rolled **UTS: 180** **Elon: 3%** **Proof: 160**
BS 43009 NG41	0.9% Mg Al alloy: Wire; cold drawn **UTS: 180**
BS 6791	0.05% Cu 99.5% Al (max): For insulated cables order as BS 2627 N68
BE T9	Aluminium tube; obsolete **UTS: 110**
CINDAL	0.2% Mg 0.3% Cr 0.12% Zn Al alloy: Origin unknown
DIN 1712	Chemical composition of high purity, aluminium products; German Standard
DIN 1725	Chemical composition of aluminium cast and wrought alloys; German Standard
DURALBRITE 1	99.8% (min) Al extrusions: J Booth; hard drawn **UTS: 120**
DURALBRITE 2	1.0% Mg 0.3% Mn Al extrusion: J Booth; hard drawn **UTS: 210**
DURALBRITE 3	0.7% Mg Al extrusion: J Booth; hard drawn **UTS: 210**
DURALCOTE	Pre-painted aluminium sheet: Alcan Booth
DURCILIUM T	0.05% Cu 99.5% Al: For electrical purposes; E Kaye for BS alloy 1E
E Al	99.9% Al: Wrought; previous German Standard designation
E Al 99.5	99.5% Al: Swiss designation
ERMAL 99.0	99.0% Al: Sheet and strip; Enfield spec for BS 1470 S1C
ERMAL 99.5	99.5% Al: Sheet and strip; Enfield spec for BS 1470 S1B
ERMAL 99.7	99.7% Al: Sheet and strip; Enfield; annealed **UTS: 75** **Elon: 50%**
HA 2.9900	99.00% Al: Ingot for re-melting; Canadian Standard
HA 2.9950	99.50% Al: Ingot for re-melting; Canadian Standard
HA 2.9970	99.70% Al: Ingot for re-melting; Canadian Standard
HA 2.9980	99.80% Al: Ingot for re-melting; Canadian Standard
HA 2.9990	99.90% Al: Ingot for re-melting; Canadian Standard
HA 2.9999	99.99% Al: Ingot for re-melting; Canadian Standard
HA 4.990	99.0% Al: Plate or sheet; Canadian Standard
HA 5.990C	99.0% Al: Bar etc.; Canadian Standard
HA 6.990	99.0% Al: For rivets and brazing wire; Canadian Standard
HA 7.995	99.5% Al: Tube; Canadian Standard
H Al 1	99.9999% Al: Ingot; Koch Light Ltd; high purity metal
H Al 6A	99.999% Al: Ingot; Koch Light Ltd; high purity metal
H Al 9	99.999% Al: Sheet; 2 mm thick; Koch Light Ltd; high purity metal
HIDUMINIUM 1A	99.8% Al: Sheet, bar, forgings, etc.; High Duty Alloys for BS alloy 1A
HIDUMINIUM 1B	99.5% Al: Sheet, bar, forgings, etc.; High Duty Alloys for BS alloy 1B
HIDUMINIUM 1C	99.0% Al: Sheet, bar, forgings, etc.; High Duty Alloys for BS alloy 1C

Note. The following abbreviations and units are used in the tables:

DPN	Hardness, diamond pyramid number
UTS	Ultimate tensile strength, N/mm^2
Elon	Elongation, %
Proof	0.1% proof strength, N/mm^2

1 N/mm^2=0.1 hbar=0.102 kgf/mm^2=0.06475 tonf/in.^{2}=145.04 lbf/in.^{2}=1 MPa

See Appendix II for other abbreviations and conversion tables.

Symbol	Nominal analysis, supplier, condition and remarks.
HIDUMINIUM 100	Commercially pure Al: Sintered powder, bar and sheet; High Duty Alloys **UTS: 350 Elon: 7% Proof: 210**
IMPALCO 030	99.8% Al: Sheet, tube, bar, etc.; Imperial Aluminium Co. for BS alloy 1A
IMPALCO 050	99.5% Al: Sheet, tube, wire, bar, etc.; Imperial Aluminium Co. for BS alloy 1B
IMPALCO 051	99.5% Al: Wire and bar for electrical purposes; Imperial Aluminium Co. for BS alloy 1E
IMPALCO 102	99.0% Al: Sheet, tube bar, etc.; Imperial Aluminium Co. for BS alloy 1C
IMPALCO 700	0.7% Zn Al alloy: Imperial Aluminium Co.
IMPALCO P3	99.8% Al: Sheet, strip and plate; Imperial Aluminium Co. for BS alloy 1A
IMPALCO P5	99.5% Al: Sheet, strip, tube, bar, forgings and plate; Imperial Aluminium Co. for BS alloy 1B
IMPALCO P5E	99.5% Al: For wire and rod; high conductivity aluminium; Imperial Aluminium Co. for BS alloy 1E
IMPALCO P10	99.0% Al: For sheet, strip, tube, bar, forgings and plate; Imperial Aluminium Co. for BS alloy 1C
KYNAL P5	99.5% Al: ICI obsolete; now Imperial Aluminium Co.
KYNAL P10	99.0% Al: ICI obsolete; now Imperial Alluminium Co.
L4	High purity Al: Sheet; obsolete
L16	99.0% Al: Sheet; ½-hard **UTS: 120 Proof: 110**
L17	99.0% Al: Sheet; soft **UTS: 90**
L30	98.0% Al: Casting
L31	99.0% Al: Ingots; virgin metal for re-melting
L32	Aluminium bars; obsolete
L34	99.0% Al: Bar or section **UTS: 70**
L36	99.0% Al: Wire or tube for rivets **UTS: 110**
L48	99.7% Al: Ingots; virgin metal for re-melting
L49	99.0% Al: Ingots; secondary metal for re-melting
L54	99.0% Al: Tube; hydraulically tested cold drawn
L67	99.0% Al: Tube; hydraulically tested cold drawn **UTS: 110**
L116	99.0% Al: Tube
Metco 54NS	Al metal spray deposit
Metco 105	Al oxide: Spray deposit with traces of other oxides; Metco **DPN: 550**
Metco 105 NS	Al oxide: Spray deposit with traces of other oxides; Metco **DPN: 550**
Metco 105 NS.1	Al oxide: Spray deposit with traces of other oxides; Metco **DPN: 550**
Metco 105 SF	Al oxide: Spray deposit; hard dense deposit; Metco **DPN: 950**
Metco Aluminium	Al: Spray deposit; Metco
MG11A	1% Zn Al alloy: Previous ASTM designation; now 7072
NF A57-101	Classification of pure Al: French Standard
NF A57-101-A4	99.0% Al: Ingot; French National Standard
NF A57-101-A5	99.5% Al: Ingot; French National Standard
NF A57-101-A7	99.7% Al: Ingot
NF A57-101-A8	99.8% Al: Ingot
NORAL S	99.0% Al: Sheet and tube; Northern Aluminium Co. for BS alloy 1C
QQ A 411b	1.0% Fe + Si 99% Al: Bar, US Federal; annealed **UTS: 90 Elon: 25%**
QQ A 561	99.0% Al: Sheet and strip; US Federal; annealed **UTS: 70 Elon: 20%**
RAFFINAL 990	99.99% Al: Wrought; Anglo-Swiss Aluminium Co.
REINALUMINIUM 99.0	99.0% Al: Wrought; Anglo-Swiss Aluminium Co.
REINALUMINIUM 99.5	99.5% Al: Wrought; Anglo-Swiss Aluminium Co.
ROTOR ALUMINIUM 04	99.5% Al: Ingot; Anglo-Swiss Aluminium Co.
SAE 25	99.0% Al: Wrought annealed; now AA 1100 **UTS: 90 Elon: 30%**
SAE 25	99.0% Al: Wrought hard; now AA 1100 **UTS: 140 Elon: 4%**
SAE 28	0.1% Cu 99.3% Al: Used only for cladding; analysis applies before cladding; now AA 1230
SAE 214	1% Zn Al alloy: Used for cladding; now 7072
Soudalu Al 100	0.3% Si 0.2% Fe Al: Welding electrode; Soudometal for welding Al **UTS: 70 Elon 25% Proof 30**
Soudoteg Al 99.5	0.05% Si 0.2% Fe Al: Welding electrode; Soudometal for inert gas welding **UTS: 80 Elon: 35% Proof: 30**
Soudor G Al 99.5	0.05% Si 0.2% Fe Al: Welding electrode; Soudometal **UTS: 80 Elon: 35% Proof: 30**
SS 40-20	99.8% Al: Billet ingot or casting; Swedish Standard
SS 40-21	99.70% Al: Billet ingot or casting; Swedish Standard
SS 40-22	99.50% Al: Billet ingot or casting; Swedish Standard
SS 40-24	99.00% Al: Billet ingot or casting; Swedish Standard
SS 4005	99.7% Al: Sheet and strip; Swedish Standard; annealed **DPN: 25 UTS: 70 Elon: 35% Proof: 30**
SS 4005	99.7% Al: Sheet and strip; Swedish Standard; cold worked **DPN: 45 UTS: 150 Elon: 3% Proof: 110**
SS 4007	99.5% Al: Sheet; strip and tube; Swedish Standard; annealed **DPN: 25 UTS: 90 Elon: 30% Proof: 30**
SS 4007	99.5% Al: Sheet, strip and tube; Swedish Standard; cold worked **DPN: 45 UTS: 150 Elon: 3% Proof: 120**
SS 4008	99.5% Al: Bar and tube; for electrical purposes; Swedish Standard; annealed **DPN: 25 UTS: 70 Elon: 30% Proof: 30**
SS 4008	99.5% Al: Bar and tube; for electrical purposes; Swedish Standard; cold worked **DPN: 35 UTS: 120 Elon: 5% Proof: 70**
SS 4020	99.8% Al: Ingot; Swedish Standard; primary metal
SS 4021	99.7% Al: Ingot; Swedish Standard; primary metal
SS 4022	99.5% Al: Ingot; Swedish Standard; primary metal
SS 4023	99.3% Al: Ingot; Swedish Standard; primary metal
SS 4024	99.0% Al: Ingot; Swedish Standard; primary metal
STA 7 A2	99.8% Al: Sheet; replaced by BS 1470 S1A
STA 7 A3	99.5% Al: Bar; replaced by BS 1476 E1B
STA 7 A4B	99.0% Al: Bar; replaced by BS 1476 E1C
STA 7 A4C	99.0% Al: Tube; BS 1471 T1C
STA 7 A4D	99.0% Al: Sheet; replaced by BS 1470 S1C
STA 7 A4E	99.5% Al alloy: Wire; replaced by BS 14735 R1C
Star 99.2	99.0% Al; Star
Star 99.5	99.5% Al; Star
Star 99.5E	99.5% Al; Star
Star 99.8	99.8% Al; Star
Stub G1B	99.5% Al: For electric arc welding; Stubs
T 9	Al: Tube; obsolete; as BS T9
UNI 3021 APR04	0.4% Cr Al alloy: Ingot; primary metal; Italian Standard
UNI 3021 APT04	0.4% Ti Al alloy: Ingot; primary metal; Italian Standard
UNI 3567/66	99.0% Al: Italian Standard
UNI 3950 AP0	99.00% Al: Ingot; Italian Standard
UNI 3950 AP3	99.3% Al: Ingot; Italian Standard
UNI 3950 AP5	99.5% Al: Ingot; Italian Standard
UNI 3950 AP7	99.70% Al: Ingot; Italian Standard
UNI 3950 AP8	99.80% Al: Ingot; Italian Standard
UNI 4507	99.5% Al: Italian Standard
UNI 4508	99.7% Al: Italian Standard
UNI 4509	99.8% Al: Italian Standard
UNI 820	98–99.5% Al: Ingot; Italian Standard
W Al 8	99.999% Al: Powder; Light Ltd; high purity metal
WW T 783a	1.0% Fe + Si 99.0% Al (max): Tube; US Federal

1B Aluminium–manganese wrought alloys

Specific gravity		2.73
Density		2730 kg/m^3
Solidus/liquidus		643–654 °C
Thermal conductivity	annealed	193 W/m °C
	½-hard	176 W/m °C
	hard	155 W/m °C
Coefficient of linear expansion		23.2×10^{-6}/ °C
Electrical conductivity	annealed	51% IACS (copper 100%)
	¾-hard	41% IACS
	hard	40% IACS
Specific resistance	annealed	33.5 microhm mm
	hard	43.1 microhm mm
Young's modulus of elasticity		66.9×10^9 N/m^2
Impact	annealed	35 J
	½-hard	28 J
Fatigue strength (50×10^6 cycles)	annealed	± 55 N/mm^2
	½-hard	± 70 N/mm^2
	hard	± 85 N/mm^2

Hot strength

Temperature	Annealed		Hard	
°C	Tensile strength N/mm^2	Elongation %	Tensile strength N/mm^2	Elongation %
150	75	47	156	12
200	55	50	114	15
260	36	60	69	25
315	25	60	35	55

The above properties are typical of the following group, and may not apply exactly to any one specification. It is possible that with certain specifications some of the values may not be applicable.

This section has alloys where approximately 1% manganese is added to aluminium. This increases the mechanical strength to a limited extent.

It has the considerable advantage, however, that these materials will accept cold work and can thus show a marked increase in mechanical strength with only a limited decrease in their corrosion resistance.

The materials are not heat treatable but can be cold worked and annealed where necessary, and then cold worked again.

The specifications in this section are commonly used to produce sheets corrugated or embossed for cladding other structures. Holloware where limited deep drawing is required can be produced from these materials.

Probably the most common use for this material is for bodies of vehicles such as buses, trucks and underground trains. Because of the good corrosion resistance these materials are very often used in the unpainted condition.

Materials can be welded using inert gas shielding under much the same conditions as pure aluminium.

Symbol	Nominal analysis, supplier, condition and remarks.
3S 1	1.2% Mn Al alloy: Previous AA designation; now 3003
3S	1.25% Mn Al alloy: Alcan for BS alloy N3
3.0515	1.25% Mn Al alloy: German Standard DIN 1725
3.0570	0.5% Si 1% Mn 0.8% Fe Al alloy: German Standard
4S	1% Mg 1.2% Mn Al alloy: Previous AA designation; now 3004
3002	0.1% Mg 0.2% Mn Al alloy: Wrought; designation used in the UK and USA
3003	0.1% Cu 1.2% Mn Al alloy: Wrought; designation used in the UK and USA
3004	1% Mg 1.2% Mn Al alloy: Wrought; designation used in the UK and USA
3005	0.4% Mg 1.2% Mn Al alloy: Wrought; designation used in the UK and USA
3006	0.2% Cu 0.5% Mg 0.6% Mn Al alloy: Wrought; designation used in the UK and USA
3007	0.2% Cu 0.6% Mn Al alloy: Wrought; designation used in the UK and USA
3009	1.6% Mn Al alloy: Wrought; designation used in the UK and USA

Note. The following abbreviations and units are used in the tables:

DPN	Hardness, diamond pyramid number
UTS	Ultimate tensile strength, N/mm^2
Elon	Elongation, %
Proof	0.1% proof strength, N/mm^2

1 N/mm^2=0.1 hbar=0.102 kgf/mm^2=0.06475 tonf/in.^{2}=145.04 lbf/in.^{2}=1 MPa

See Appendix II for other abbreviations and conversion tables.

Symbol	Nominal analysis, supplier, condition and remarks.
3010	0.2% Cu 0.7% Mn Al alloy: Wrought; designation used in the UK and USA
3011	0.2% Cu 1% Mn Al alloy: Wrought; designation used in the UK and USA
3015	0.5% Mg 0.7% Mn Al alloy: Wrought; designation used in the UK and USA
3016	0.6% Mg 0.7% Mn Al alloy: Wrought; designation used in the UK and USA
3102	0.2% Mn Al alloy: Wrought; designation used in the UK and USA
3104	0.1% Cu 1% Mg 1.1% Mn Al alloy: Wrought; designation used in the UK and USA
3105	0.5% Mg 0.7% Mn Al alloy: Wrought; designation used in the UK and USA
3107	0.1% Cu 0.7% Mn Al alloy: Wrought; designation used in the UK and USA
3303	0.1% Cu 1.2% Mn Al alloy: Wrought; designation used in the UK and USA
3307	0.7% Mn Al alloy: Wrought; designation used in the UK and USA
8006	0.7% Mn 1.5% Fe Al alloy: Wrought; designation used in the UK and USA
8007	0.7% Mn 1.6% Fe Al alloy: Wrought; designation used in the UK and USA
8010	0.2% Cu 0.3% Mg 0.5% Mn 0.5% Fe Al alloy: Wrought; designation used in the UK and USA
8014	0.4% Mn 1.4% Fe Al alloy: Wrought; designation used in the UK and USA
A5.3 (E93003)	1.2% Mn Al alloy electrode: AWS designation
AA 3003	1.2% Mn Al alloy: Wrought; SAE designation; formerly SAE 29
AA 3004	1% Mg 1.2% Mn Al alloy: Wrought; SAE designation; formerly SAE 20
ALCAN GB 3S	1.2% Mn Al alloy: Alcan for BS alloy N3
ALCOA 190	1.2% Mn Al alloy: Wrought; Alcoa; ½-hard H4 condition **UTS: 150 Elon: 6%**
ALM	1% Mn Al alloy: Sheet; J Booth for BS alloy N3
ALMINTRODE 96.20	1.30% Mn 0.3% Fe 98.0% Al: For electrodes; ESAB
Al Mn	0.2% Mg 1.2% Mn Al alloy: Wrought; previous German Standard designation
ALUMAL	1.25% Mn Al alloy: Origin unknown
ALUMAN 100	1.2% Mn Al alloy: Wrought; Anglo-Swiss Aluminium Co.
ALUMAN 100	1.2% Mn All alloy: Star
AM 1	1.2% Mn All alloy: Wrought; L'Aluminium Français
AMG05	0.4% Mg 1.2% Mn Al alloy: Sheet; French Standard
AMI	1.2% Mn Al alloy: Wrought; French Standard
AMIG	1.0% Mg 1.2% Mn Al alloy: Sheet; French Standard
AMS 4006 B	1.2% Mn Al: Sheet; annealed; AMS for AA alloy 3003 **UTS: 120 Elon: 17%**
AMS 4008 C	1.2% Mn Al: Sheet; ½-hard; AMS for AA alloy 3003 **UTS: 160 Elon: 3%**
AMS 4010	1.2% Mn Al: Foil; hard rolled; AMS for AA alloy 3003 **UTS: 180**
AMS 4065 B	1.2% Mn Al alloy: Tube; annealed; AMS for AA alloy 3003 **UTS: 110**
AMS 4067 B	1.2% Mn Al alloy: Tube; ½-hard; AMS for AA alloy 3003 **UTS: 120**
ASTM B209/3003	1.2% Mn Al alloy: Sheet; annealed **UTS: 110 Elon: 23% Proof: 7**
ASTM B209/3003	1.2% Mn Al alloy: Sheet; cold rolled; ½-hard **UTS: 180 Elon: 12% Proof: 30**
ASTM B209/3004	1% Mg 1.2% Mn Al alloy: Sheet and plate
ASTM B209/3005	0.4% Mg 1.2% Mn Al alloy: Sheet and plate
ASTM B210/3003	1.2% Mn Al alloy: Tube; annealed **UTS: 110 Proof: 25**

Symbol	Nominal analysis, supplier, condition and remarks.
ASTM B210/3003	1.2% Mn Al alloy: Tube; cold drawn **UTS: 180 Proof: 150**
ASTM B210/3004	1% Mg 1.2% Mn Al alloy: Tube; annealed **UTS: 190 Proof: 45**
ASTM B210/3004	1% Mg 1.2% Mn Al alloy: Tube; cold drawn **UTS: 250 Proof: 190**
ASTM B211/3003	1.2% Mn Al alloy: Bar; cold drawn **UTS: 180**
ASTM B221/3003	1.2% Mn Al alloy: Extruded bar; annealed **UTS: 70 Elon: 25% Proof: 15**
ASTM B221/3004	1.2% Mn Al alloy: Extruded bar, etc.
ASTM B234/3003	1.2% Mn Al alloy: Tube; cold drawn **UTS: 150 Proof: 120**
ASTM B235/3003	1.2% Mn Al alloy: Extruded tube; annealed **UTS: 120 Elon: 25% Proof: 30**
ASTM B235/3004	1% Mg 1.2% Mn Al alloy: Extruded tube; annealed **UTS: 190 Proof: 45**
ASTM B241/3003	1.2% Mn Al alloy: Pipe; cold drawn **UTS: 180 Proof: 150**
ASTM B247/3003	1.2% Mn Al alloy: Forging: as forged **DPN: 25 UTS: 70 Elon: 25% Proof: 15**
ASTM B285 ER3004	1% Mg 1.2% Mn Al alloy: Welding rod; obsolete
ASTM B307/3003	1.2% Mn Al alloy: Coiled tube; annealed **UTS: 110**
ASTM B307/3004	1.0% Mg 1.3% Mn Al alloy: Drawn tube
ASTM B307/5005	0.8% Mn Al alloy: Drawn tube
ASTM B313/3003	1.2% Mn Al alloy: Welded tube; cold rolled **UTS: 180 Elon: 3%**
ASTM B318/3003	1.2% Mn Al alloy: Drawn coiled tubes; annealed **UTS: 120**
ASTM B318/3004	1% Mg 1.2% Mn Al alloy: Drawn coiled tube; annealed **UTS: 180**
ASTM B345/3003	1.2% Mn Al alloy: Pipe; annealed **UTS: 70 Proof: 15**
ASTM B345/3004	1% Mg 1.2% Mn Al alloy: Pipe; annealed **UTS: 150 Proof: 30**
ASTM B404/3003	1.25% Mn Al alloy: Condenser tubes
AWCO 60	1.25% Mn Al alloy: Wire; Aluminium Wire Co. for BS alloy N3
BA 60	1.25% Mn Al: Sheet; British Aluminium Co. for BS alloy N3
BIRMETAL 3	1.0% Mn Al alloy: In wrought form; Birmetal
BS 1453 N3	1.2% Mn Al alloy: Welding rod for welding Al alloy N3 type
BS 1470 3103	1.2% Mn Al alloy: Replaces N3 designation; plate and sheet; H4 condition **UTS: 150 Elon 5%**
BS 1470 NS3	1% Mn Al alloy: For sheet and strip; ¾-hard **UTS: 170 Elon: 5%**
BS 1470 NS3	1% Mn Al alloy: For sheet and strip; hard **UTS: 170 Elon: 3%**
BS 1470 NS3	1% Mn Al alloy: For sheet and strip; annealed **UTS: 110 Elon: 30%**
BS 1470 NS3	1% Mn Al alloy: For sheet and strip; ¼-hard **UTS: 120 Elon: 12%**
BS 1470 NS3	1% Mn Al alloy: For sheet and strip; ½-hard **UTS: 150 Elon: 7%**
BS 1471 5083	1.2% Mn 0.1% Cr Al alloy: Drawn tube; H2 condition **UTS: 310 Elon: 5% Proof: 235**
BS 1472 NF3	1% Mn Al alloy: For forgings and bar; annealed **UTS: 90 Elon: 20% Proof: 45**
BS 1475 3103	1.2% Mn Al alloy: Wire; H8 condition **UTS: 175**
BS 1475 NG3	1% Mn Al alloy: Wire; annealed **UTS: 120**
BS 1475 NG3	1% Mn Al alloy: Wire; hard drawn **UTS: 170**
BS 1477 NP3	1% Mn Al: Plate; ½-hard **UTS: 150 Elon: 7%**

Symbol	Nominal analysis, supplier, condition and remarks.
BS 2901 NG3	1.2% Mn Al alloy: Rod for inert gas shielded arc welding
BS 3313 NS3	1% Mn Al alloy: Strip for dairy product containers; ½-hard **UTS: 150**
BS 4300/1 NJ3	1.2% Mn Al alloy: Welded tube; as manufactured **UTS: 180 Elon: 3%**
BS 4300/6 NS31	0.4% Mg 0.7% Mn Al alloy: Sheet and strip; cold rolled **UTS: 210 Elon: 2% Proof: 190**
DIN 1732 S Al Mn	1.5% Mn Al alloy: Welding electrode
DTD 213	Mn Al alloy: Replaced by BSL59
DTD 653	Mn Al alloy: Replaced by BSL60
ELM 15	1% Mn Al alloy: Strip; Elm Engineering
HA4 MC10	0.15% Cu 1.2% Mn Al alloy: Plate; Canadian Standard
HA7 MC10	0.1% Cu 1.2% Mn Al alloy: Tube; Canadian Standard
HIDUMINIUM 11	1.25% Mn Al: For extrusions and tube; as rolled; High Duty Alloys **DPN: 30 UTS: 110 Elon: 30% Proof: 70**
HIDUMINIUM 11	1.25% Mn Al: For extrusions and tube; ½-hard; High Duty Alloys **DPN: 45 UTS: 150 Elon: 7% Proof: 110**
IMPALCO 190	0.15% Cu 0.6% Si 1.2% Mn Al alloy: Sheet and wire; Imperial Aluminium Co. for BS alloy N3
IMPALCO PA19	1% Mn Al alloy: For sheet, strip and wire; Imperial Aluminium Co. for BS alloy N3
KYNAL PA19	1.25% Mn Al alloy: ICI; obsolete; now Imperial Aluminium Co.
L59	1% Mn Al alloy: For sheet; ¾-hard **UTS: 170 Elon: 5%**
L60	1% Mn Al alloy: Sheet and strip; ¼-hard **UTS: 120 Elon: 12%**
L61	1% Mn Al alloy: For sheet and strip; annealed **UTS: 90 Elon: 30%**
M1A	1.2% Mn Al alloy: Previous ASTM designation; now 3003
MANGALAL	1.2% Mn Al alloy: Sheet; J Booth for BS alloy N3
MG 11A	1% Mg 1.2% Mn Al alloy: Previous ASTM designation; now 3004
N 3	1.2% Mn Al alloy: Designation used by BS
NG 3	1.2% Mn Al alloy: Designation used by BS
NJ 3	1.2% Mn Al alloy: Designation used by BS
NP 3	1.2% Mn Al alloy: Designation used by BS
NS 3	1.2% Mn Al alloy: Designation used by BS
NS 31	0.4% Mg 0.7% Mn Al alloy: Sheet; designation used by BS
NORAL 3S	1.25% Mn Al; North Aluminium Co. for BS 1470 NS3
QQ A 356b	1.2% Mn Al alloy: Bar; US Federal; annealed **UTS: 110 Elon: 25% Proof: 30**
QQ A 359a	1.2% Mn Al alloy: Sheet; US Federal; annealed **UTS: 110 Elon: 22%**
SAE 20	1% Mg 1.2% Mn Al alloy: Wrought hard; now AA 3004 **UTS: 250 Elon: 3%**
SAE 20	1% Mg 1.2% Mn Al alloy: Wrought annealed; now AA 3004
SAE 29	1.3% Mn Al alloy: Sheet; annealed; now AA 3003 **UTS: 120 Elon: 25%**
SAE 29	1.3% Mn Al alloy: Sheet; hard; now AA 3003 **UTS: 180 Elon: 4%**
SOUDALU Mn 2	0.4% Si 1.3% Mn 0.3% Fe Al alloy: Welding electrode; Soudometal **UTS: 150 Elon: 10% Proof: 80**
SS 4054	1.25% Mn Al alloy: Sheet and strip; Swedish Standard; annealed **DPN: 30 UTS: 120 Elon: 25% Proof: 45**
SS 4054	1.25% Mn Al alloy: Sheet and strip; Swedish Standard; cold rolled **DPN: 60 UTS: 170 Elon: 3% Proof: 140**
STA7 AW 3C	1.2% Mn Al alloy: Sheet; replaced by BS 1470 NS3
STUBS 49	1.5% Mn Al alloy: For welding; Stubs **DPN: 60 UTS: 175 Elon: 25% Proof: 100**
TI 11	1.25% Mn Al: Sheet; British Aluminium Co. spec for BS 1470 NS3
UNI 3568	1.2% Mn Al alloy: For forging; Italian Standard
UNI 6361	1.0% Mg 1.2% Mn Al alloy: For forging; Italian Standard
V Al Mn 1	0.5% Si 1% Mn Al alloy: Previous German Standard designation
W 3	1.2% Mn Al alloy: Sheet; replaced by IMPALCO 190; Southern Forge
WW T 788A	1.2% Mn Al alloy: Tube; US Federal

Note. The following abbreviations and units are used in the tables:

DPN	Hardness, diamond pyramid number
UTS	Ultimate tensile strength, N/mm^2
Elon	Elongation, %
Proof	0.1% proof strength, N/mm^2

1 N/mm^2=0.1 hbar=0.102 kgf/mm^2=0.06475 $tonf/in.^2$=145.04 $lbf/in.^2$=1 MPa

See Appendix II for other abbreviations and conversion tables.

1C Aluminium–silicon and wrought cast alloys

Specific gravity		2.6–2.7
Density		2600–2700 kg/m^3
Solidus/liquidus		525–625 °C
Thermal conductivity		101–126 W/m °C
Coefficient of linear expansion		20–23 $\times$ 10^{-6}/ °C
Electrical conductivity	annealed	26–39% IACS (copper 100%)
Specific resistance		40–50 microhm mm
Young's modulus of elasticity		71.0 $\times$ 10^9 N/m^2
Impact		7–8.5 J
Fatigue strength (50 $\times$ 10^6 cycles)	sand cast	$\pm$ 45 N/mm^2
Hot strength		

Temperature °C	Tensile strength N/mm^2
100	240
150	210
200	180
250	90
300	60
350	37
400	34

The above properties are chosen to show the typical values for the alloys listed, and may not apply exactly to any one specification. It is possible that with certain specifications some of the values may not be applicable.

General metallurical characteristics

The materials in this section cover castings and wrought material. Silicon reduces the melting point of aluminium. When approximately 12% silicon is added to aluminium the melting point is reduced to approximately 500 °C. Most of these alloys are therefore used for their casting characteristics. They have excellent fluidity at casting temperatures and thus give good reproducibility of any patterns.

The straightforward silicon–aluminium alloys have relatively low strength but this can be increased to some extent by alloying with copper and other elements.

Very few of these alloys can be heat treated to increase the mechanical properties and the alloys have relatively low mechanical strengths. Their corrosion resistance is also markedly less than that of pure aluminium or alloys with magnesium additions.

A limited number of wrought alloys are included with these specifications, the most common being for welding wire, brazing wire, etc.

Thin walled castings can be produced and items such as fuse boxes, electrical gear, etc. are commonly manufactured from these materials.

Welding requires the use of inert gas.

The corrosion resistance can be increased either by anodizing or, to a lesser extent, by chromate treatment.

Symbol	Nominal analysis, supplier, condition and remarks
3.2131	1.2% Cu 0.4% Mg 5.5% Si Al alloy: Casting; German Standard; solution treated and aged **DPN: 95 UTS: 230 Elon: 1% Proof: 200**
3.2152	4% Cu 7% Si Al alloy: German Standard DIN 1725
3.2153	3% Cu 0.2% Mg 7% Si 0.4% Mn 1.2% Zn Al alloy: Casting; German Standard DIN 1725
3.2245	0.2% Mg 5.25% Si Al alloy: German Standard DIN 1725
3.2285	7.5% Si Al alloy: German Standard DIN 1725
3.2341	5% Si Al alloy: German Standard DIN 1725
3.2371	0.4% Mg 7% Si Al alloy: Casting; German Standard
3.2381	0.4% Mg 11.5% Si 0.5% Mn Al alloy: Casting; German Standard

Note. The following abbreviations and units are used in the tables:

DPN	Hardness, diamond pyramid number
UTS	Ultimate tensile strength, N/mm^2
Elon	Elongation, %
Proof	0.1% proof strength, N/mm^2

1 N/mm^2=0.1 hbar=0.102 kgf/mm^2=0.06475 tonf/in.^{2}=145.04 lbf/in.^{2}=1 MPa

See Appendix II for other abbreviations and conversion tables.

Symbol	Nominal analysis, supplier, condition and remarks.
3.2383	0.3% Mg 10% Si 0.2% Al alloy: Casting; German Standard
3.2385	0.3% Cu 12% Si Al alloy: German Standard DIN 1725
3.2572	0.6% Cu 12% Si 0.3% Mn Al alloy: German Standard DIN 1725
3.2573	3.7% Cu 9% Si Al alloy: German Standard DIN 1725
3.2581	11.5% Si Al alloy: Casting; German Standard
3.2582	0.1% Cu 12% Si Al alloy: German Standard DIN 1725
3.2583	2% Cu 12.5% Si 0.3% Mn Al alloy: Casting; German Standard DIN number not known
3.2585	0.3% Cu 1% Si Al alloy: German Standard DIN 1725
13	11% Si Al alloy: Casting; Alcoa for BS alloy LM 20
32S	1% Cu 1% Mg 12% Si Al alloy: Previous AA designation; now 4032 for pistons
43	5% Si Al alloy: Casting; AA designation for ASTM alloy 85A or 85B
43	5% Si Al alloy: Casting; Alcoa for BS alloy LM 18
43S	5% Si Al alloy: Wire; Alcoa for BS alloy NW 21
43S	5% Si Al alloy: Previous AA designation; now 4043
123	5% Si Al alloy: Casting; Comalco for BS alloy LM 18
160	12% Si Al alloy: Casting; Comalco for BS alloy LM 6
160S	12% Si 0.3% Mg Al alloy: Casting; Comalco for BS alloy LM 6
220	0.3% Mg 7% Si Al alloy: Casting; AA designation for ASTM alloy G 10A

Symbol	Nominal analysis, supplier, condition and remarks.
305.0	1.2% Cu 5% Si Al alloy: Casting; designation used in the UK and USA
305.2	1.2% Cu 5% Si Al alloy: Ingot; designation used in the UK and USA
324.0	0.5% Cu 0.5% Mg 7.5% Si Al alloy: Casting; designation used in the UK and USA
324.1	0.5% Cu 0.5% Mg 7.5% Si Al alloy: Ingot; designation used in the UK and USA
324.2	0.5% Cu 0.5% Mg 7.5% Si Al alloy: Ingot; designation used in the UK and USA
328.0	1.5% Cu 0.4% Mg 8% Si Al alloy: Casting; designation used in the UK and USA
328.1	1.5% Cu 0.4% Mg 8% Si Al alloy: Ingot; designation used in the UK and USA
336.0	1% Cu 1% Mg 12% Si 2.5% Ni Al alloy: Casting; designation used in the UK and USA
336.1	1% Cu 1% Mg 12% Si 2.5% Ni Al alloy: Ingot; designation used in the UK and USA
336.2	1% Cu 1% Mg 12% Si 2.5% Ni Al alloy: Ingot; designation used in the UK and USA
339.0	2.2% Cu 1% Mg 12% Si 1% Ni Al alloy: Casting; designation used in the UK and USA
339.1	2.2% Cu 1% Mg 12% Si 1% Ni Al alloy: Ingot; designation used in the UK and USA
343.0	0.7% Cu 7.2% Si 1.7% Zn Al alloy: Casting; designation used in the UK and USA
343.1	0.7% Cu 7.2% Si 1.7% Zn Al alloy: Ingot; designation used in the UK and USA
354.0	1.8% Cu 0.5% Mg 9% Si Al alloy: Casting; designation used in the UK and USA
354.1	1.8% Cu 0.5% Mg 9% Si Al alloy: Ingot; designation used in the UK and USA
355.0	1.2% Cu 0.5% Mg 5% Si Al alloy: Casting; designation used in the UK and USA
355.1	1.2% Cu 0.5% Mg 5% Si Al alloy: Ingot; designation used in the UK and USA
355.2	1.2% Cu 0.5% Mg 5% Si Al alloy: Ingot; designation used in the UK and USA
356.0	0.3% Mg 7% Si Al alloy: Casting; designation used in the UK and USA
356.1	0.3% Mg 7% Si Al alloy: Ingot; designation used in the UK and USA
356.2	0.3% Mg 7% Si Al alloy; Ingot; designation used in the UK and USA
357.0	0.5% Mg 5% Si Al alloy: Casting; designation used in the UK and USA
359.0	0.6% Mg 9% Si Al alloy: Casting; designation used in the UK and USA
360.0	0.5% Mg 9.5% Si Al alloy: Casting; designation used in the UK and USA
360.2	0.5% Mg 9.5% Si Al alloy: Ingot; designation used in the UK and USA
361.0	0.25% Cu 0.5% Mg 10% Si 0.25% Ni Al alloy: Casting; designation used in the UK and USA
361.1	0.25% Cu 0.5% Mg 10% Si 0.25% Ni Al alloy: Ingot; designation used in the UK and USA
363.0	3% Cu 0.25% Mg 5.2% Si 3.7% Zn Al alloy: Casting; designation used in the UK and USA
363.1	3% Cu 0.25% Mg 5.2% Si 3.7% Zn Al alloy: Ingot; designation used in the UK and USA
364.0	0.3% Mg 8.5% Si 0.4% Cr 0.03% Be Al alloy: Casting; designation used in the UK and USA
364.2	0.3% Mg 8.5% Si 0.4% Cr 0.03% Be Al alloy: Ingot; designation used in the UK and USA
369.0	0.35% Cu 0.35% Mg 11.5% Si Al alloy: Casting; designation used in the UK and USA
369.1	0.35% Cu 0.35% Mg 11.5% Si Al alloy: Ingot; designation used in the UK and USA
380.0	3.5% Cu 8.5% Si 3% Zn (max) Al alloy: Casting; designation used in the UK and USA
380.2	3.5% Cu 8.5% Si Al alloy: Ingot; designation used in the UK and USA
383.0	2.5% Cu 10.5% Si 3% Zn (max) Al alloy: Casting; designation used in the UK and USA
383.1	2.5% Cu 10.5% Si 3% Zn (max) Al alloy: Ingot; designation used in the UK and USA
383.2	2.5% Cu 10.5% Si 0.1% Zn (max) Al alloy: Ingot; designation used in the UK and USA
384.0	3.7% Cu 11% Si 3% Zn (max) Al alloy: Casting; designation used in the UK and USA
384.1	3.7% Cu 11% Si 3% Zn (max) Al alloy: Ingot; designation used in the UK and USA
384.2	3.7% Cu 11% Si 0.1% Zn (max) Al alloy: Ingot; designation used in the UK and USA
385.0	3% Cu 12% Si 3% Zn (max) Al alloy: Casting; designation used in the UK and USA
385.1	3% Cu 12% Si 3% Zn (max) Al alloy: Ingot; designation used in the UK and USA
390.0	4.5% Cu 0.5% Mg 17% Si Al alloy: Casting; designation used in the UK and USA
390.2	4.5% Cu 0.5% Mg 17% Si Al alloy: Ingot; designation used in the UK and USA
392.0	0.6% Cu 1% Mg 19% Si Al alloy: Casting; designation used in the UK and USA
392.1	0.6% Cu 1% Mg 19% Si Al alloy: Ingot; designation used in the UK and USA
393.0	0.9% Cu 1% Mg 22% Si 0.2% Be 2.2% Ni 0.15% Ti Al alloy: Casting; designation used in the UK and USA
393.1	0.9% Cu 1% Mg 22% Si 0.2% Be 2.2% Ni 0.15% Ti Al alloy: Ingot; designation used in the UK and USA
393.2	0.9% Cu 1% Mg 22% Si 0.2% Be 2.2% Ni 0.15% Ti Al alloy: Ingot; designation used in the UK and USA
408.2	9% Si 1% Fe Al alloy: Ingot for steel coating; designation used in the UK and USA
409.2	9.5% Si 1% Fe Al alloy: Ingot for steel coating; designation used in the UK and USA
411.2	11% Si 1% Fe Al alloy: Ingot for steel coating; designation used in the UK and USA
413.0	12% Si Al alloy: Casting; designation used in the UK and USA
413.2	12% Si Al alloy: Ingot; designation used in the UK and USA
435.2	3.5% Si Al alloy: Ingot; designation used in the UK and USA
443.0	4.7% Si Al alloy: Casting; designation used in the UK and USA
443.1	4.7% Si Al alloy: Ingot; designation used in the UK and USA
443.2	4.7% Si Al alloy: Ingot; designation used in the UK and USA
444.0	7% Si Al alloy: Casting; designation used in the UK and USA
444.2	7% Si Al alloy: Ingot; designation used in the UK and USA
445.2	7% Si 1% Fe Al alloy: Ingot; designation used in the UK and USA
4008	0.4% Mg 7% Si Al alloy: Wrought; designation used in the UK and USA

Note. The following abbreviations and units are used in the tables:

DPN	Hardness, diamond pyramid number
UTS	Ultimate tensile strength, N/mm^2
Elon	Elongation, %
Proof	0.1% proof strength, N/mm^2

$1\ N/mm^2 = 0.1\ hbar = 0.102\ kgf/mm^2 = 0.06475\ tonf/in.^2 = 145.04\ lbf/in.^2 = 1\ MPa$

See Appendix II for other abbreviations and conversion tables.

Symbol	Nominal analysis, supplier, condition and remarks.
4043	5.3% Si Al alloy: Brazing rod; designation used in the UK and USA
4044	8.5% Si Al alloy: Brazing rod; designation used in the UK and USA
4045	10% Si Al alloy: Wrought; designation used in the UK and USA
4047	12% Si Al alloy: Brazing rod; designation used in the UK and USA
4104	1.5% Mg 10% Si 0.1% Bi Al alloy: Wrought; designation used in the UK and USA
4145	4% Cu 10% Si Al alloy: Brazing rod; designation used in the UK and USA
4343	7.8% Si Al alloy: Brazing rod; designation used in the UK and USA
4543	0.2% Mg 6% Si Al alloy: Wrought; designation used in the UK and USA
4643	0.2% Mg 4% Si Al alloy: Welding rod; designation used in the UK and USA
A5.3 (E94043)	5.2% Si Al alloy: Welding electrode; AWS designation
A5.8 (BA1 Si 2)	7.5% Si Al alloy: Welding electrode; AWS designation
A5.8 (BAl Si 3)	10.2% Si Al alloy: Welding electrode; AWS designation
A5.8 (BAl Si 4)	12% Si Al alloy: Welding electrode; AWS designation
A5.8 (BAl Si 5)	10% Si Al alloy: Welding electrode; AWS designation
A5.8 (BAl Si 7)	9.7% Si Al alloy: Welding electrode; AWS designation
A5.9 (BAl Si 11)	7.5% Si Al alloy: Welding electrode; AWS designation
AC 5	0.7% Mg 5% Si 0.2% Mn 0.1% Ti Al alloy: Anglo-Swiss
AL 2	12% Si Al alloy: Casting; Russian designation
AL 4	0.4% Mg 11.5% Si 0.5% Mn Al alloy: Casting; Russian designation
AL 5	1.2% Cu 0.5% Mg 5% Si Al alloy: Casting; Russian designation
AL 9	0.4% Mg 7% Si Al alloy: Casting; Russian designation
AL 26	1% Cu 1% Mg 24% Si 1% Ni Al alloy: Casting; Russian designation
AL 30	1.2% Cu 1.2% Mg 11% Si Al alloy: Casting; Russian designation
ALAR 005	5% Si Al alloy: Casting; American proprietary alloy for BS alloy LM 18
ALAR 0012	12% Si Al alloy: Casting; American proprietary alloy for BS alloy LM 20
ALCAN GB 38S	1% Cu 1% Mg 11% Si 1% Ni Al alloy: Forging; solution treated and aged; Alcan **DPN: 105 UTS: 350 Elon: 9% Proof: 270**
ALCAN GB 160	11.5% Si Al alloy: Casting; Alcan for BS alloy LM 6
ALCAN GB 33S	5% Si Al alloy: Wire; Alcan; annealed **UTS: 110 Elon: 40% Proof: 30**
ALCAN GB 33S	5% Si Al alloy: Wire; Alcan; hard **UTS: 210 Elon: 5% Proof: 170**
ALCAN GB B33S	0.2% Mg 5% Si Al alloy: Wire; Alcan; solution treated and aged **UTS: 180 Elon: 10% Proof: 150**
ALMINAL 2	1% Cu 10% Si Al alloy: Casting; origin unknown
ALMINAL 6	12% Si Al alloy: Casting; origin unknown
ALMINAL 8	0.4% Mg 5% Si 0.5% Mn Al alloy: Casting; origin unknown
ALMINAL 9	0.4% Mg 12.5% Si 0.5% Mn Al alloy: Casting; origin unknown
ALMINAL 10	10.5% Si Al alloy: Casting; origin unknown
ALMINAL 20	3% Cu 0.2% Mg 9% Si 1% Fe 0.5% Zn Al alloy: Casting; origin unknown
ALMINTRODE 96.0	11.0% Si 0.5% Fe Al: For electrodes; ESAB **UTS: 170 Elon: 13**
ALPAX	13% Si Al alloy: Casting; Light Alloy Ltd for BS alloy LM 6
AMS 4145 E	1% Cu 1% Mg 11% Si 1% Ni Al alloy: Forging; solution treated and aged; AMS for AA 4032
AMS 4181	0.4% Mg 7% Si Al alloy: Wrought
AMS 4184 B	4% Cu 10% Si Al alloy: Wire for brazing; AMS for AA alloy 4145

Symbol	Nominal analysis, supplier, condition and remarks.
AMS 4185 A	12% Si Al alloy: Wire for brazing; AMS for AA alloy 4047
AMS 4189	0.2% Mg 4.1% Si Al alloy: Wrought
AMS 4190 A	5% Si Al alloy: Wire for brazing; AMS for Al alloy 4043
AMS 4219	0.5% Mg 7% Si 0.05% Be 0.15% Ti Al alloy: Casting
AMS 4241	0.5% Mg 7% Si Al alloy: Casting
AMS 4248	0.3% Mg 7% Si Al alloy: Casting
AMS 4261	0.3% Mg 7% Si Al alloy: Casting
ANQQ A-405	5.0% Si Al alloy: Casting; US Service; as cast **UTS: 90 Elon: 3%**
ANTICORODAL 5S1	0.7% Mg 5% Si 0.2% Mn 0.1% Ti Al alloy: Casting; Anglo-Swiss; solution treated and aged; obsolete **DPN: 95 UTS: 280 Elon: 2% Proof: 250**
ANTICORODAL 34	0.4% Mg 0.2% Mn 0.15% Ti Si Al: Primary casting metal; Anglo-Swiss Aluminium Co.
AP 403	5.2% Si Al alloy: Casting; Australian specification
A-S4G	0.6% Mg 4.2% Si 45% Mn Al alloy: Casting; L'Aluminium Français
A-S4G	0.4% Mg 5.5% Si Al alloy: Casting; French Standard
A-S5	5.2% Si Al alloy: Wire; French Standard
A-S7G	0.3% Mg 7% Si Al alloy: Casting; French Standard
A-S7G	0.3% Mg 7.0% Si Al alloy: Casting; L'Aluminium Français
A-S9U3Y4	1.2% Cu 10.2% Si Al alloy: Casting; French Standard
A-S10G	0.25% Mg 10% Si Al alloy: Casting; L'Aluminium Français
A-S10G	0.3% Mg 11.5 % Si 0.4% Mn Al alloy: Casting; French Standard
A-S10UG	2.3% Cu 1.2% Mg 10% Si 0.5% Mn 1.0% Fe Al alloy: Casting; L'Aluminium Français
A-S12	12% Si Al alloy: Wire; French Standard
A-S12N2G	1.0% Cu 1.0% Mg 12% Si 2.2% Ni Al alloy: Casting; L'Aluminium Français
A-S12UN	1.0% Cu 1.0% Mg 12% Si 1.0% Ni Al alloy: Wrought and cast; L'Aluminium Français
A-S12UN	1.1% Cu 1.2% Mg 11% Si Al alloy: Casting; French Standard
A-S13	12.5% Si Al alloy: Casting; L'Aluminium Français
A-S22	22% Si Al master alloy; L'Aluminium Français
A-S22UNK	1.5% Cu 1.0% Mg 21% Si 0.5% Mn 1.1% Ni 1.0% Co Al alloy: Casting; L'Aluminium Français
AS 401	11.5% Si Al alloy: Casting; Australian specification
AS 601	0.4% Mg 7% Si Al alloy: Casting; Australian specification
ASTM B26 S5A	5% Si Al alloy: Sand casting; as cast **UTS: 110 Elon: 3% Proof: 22**
ASTM B26 S5B	5% Si 0.25% Cr Al alloy: Sand casting; as cast **UTS: 110 Elon: 3% Proof: 30**
ASTM B85 S5C	5% Si Al alloy: Die casting; as cast **UTS: 220 Elon: 9% Proof: 70**
ASTM B85 S12A	12% Si Al alloy: Die casting; as cast **UTS: 280 Elon: 3.5% Proof: 120**
ASTM B85 S12B	12% Si 2% Fe Al alloy: Die casting; as cast **UTS: 300 Elon: 2.5% Proof: 140**
ASTM B108 S5A	5% Si Al alloy: Casting; as cast **UTS: 140 Elon: 2.5% Proof: 22**
ASTM B108 S5B	5% Si 0.2% Cr Al alloy: Casting; as cast **UTS: 140 Elon: 2% Proof: 30**
ASTM B108 SC41	1.5% Cu 0.7% Mg 12% Si Al alloy: Chill cast; solution treated and aged; obsolete **DPN: 100 UTS: 300 Elon: 1% Proof: 250**
ASTM B108 SC9IAI	0.6% Mg 9.0% Si Al alloy: Die casting
ASTM B179 57A	7.0% Si Al alloy: Ingot
ASTM B179 S5A	5% Si Al alloy: Ingot; primary metal
ASTM B179 S5B	5% Si Al alloy: Ingot; primary metal
ASTM B179 S5C	5% Si Al alloy: Ingot; primary metal
ASTM B179 S12A–B	12% Si Al alloy: Ingot; Primary metal
ASTM B179 SC94A	3.5% Cu 0.3% Mg 9% Si Al alloy: Ingot
ASTM B179 SC71A	0.5% Mg 7.0% Si Al alloy: Ingot
ASTM B179 SC91A	0.6% Mg 9% Si Al alloy: Ingot

Symbol	Nominal analysis, supplier, condition and remarks.
ASTM B184 A143	5% Si Al alloy: For electrode filler wire; may be flux coated
ASTM B247 4032	1% Cu 1% Mg 12% Si 1% Ni Al alloy: Forging; solution treated and aged **DPN: 115 UTS: 370 Elon: 5% Proof: 300**
ASTM B260 B Al Si 1	5% Si Al alloy: Brazing filler; melting range 577–630 °C
ASTM B260 B Al Si 2	7% Si Al alloy: Brazing filler; melting range 577–613 °C
ASTM B260 B Al Si 3	4% Cu 10% Si Al alloy: Brazing filler; melting range 521–584 °C
ASTM B260 B Al Si 4	12% Si Al alloy: Brazing filler; melting range 577–582 °C
ASTM B260 B Al Si 5	7.5% Si Al alloy: Braze metal; brazing range 588–604 °C
ASTM B285 ER4043	5.2% Si Al alloy: Welding wire
ASTM B285 RS5B	5% Si Al alloy: Welding rod; obsolete
ASTM B285 RSC51A	1.2% Cu 0.5% Mg 5% Si Al alloy: Welding wire
ASTM B285 RSG70A	0.3% Mg 7% Si Al alloy: Welding wire
AWCO 40	12% Si Al alloy: For welding wire; Aluminium Wire Co. for BS alloy N2
AWCO 45	5% Si Al alloy: For welding wire; Aluminium Wire Co. for BS alloy N21
AWCO 46	3.5% Cu 12% Si Al alloy: Wire; Aluminium Wire Co. for welding
B 443.0	5.2% Si Al alloy: Casting; ANSI specification
BA 40	11% Si Al alloy: Casting; British Aluminium Co. for BS LM 6
BA 45	5% Si Al alloy: Casting; British Aluminium Co. for BS alloy LM 18
BIRMASIL SPEC	11% Si 3% Ni Al alloy: Casting; Birmingham Aluminium Co. for BS alloy LM17
BIRMASTIC	12% Si 3% Ni Al alloy: Origin unknown
BIRMETAL 005	5% Si Al alloy: Wire; Birmabright for BS alloy N21
BS 1354 NG2	12% Si Al alloy: Welding rod; for welding BS alloys H9, H10, H20, H30 and cast alloys
BS 1453 NG21	5% Si Al alloy: Welding rod; for welding BS alloys H9, H10, H20, H30 and cast alloys
BS 1475 NG2	12% Si Al alloy: Wire
BS 1475 NG21	5% Si Al alloy: Wire
BS 1475 4043A	5.2% Si Al alloy: Wire; welders electrode
BS 1475 4047A	12% Si Al alloy: Wire; welders electrode
BS 1490 LM2	1.5% Cu 10% Si Al alloy: Ingot; chill cast **UTS: 150**
BS 1490 LM2M	1.5% Cu 10% Si Al alloy: For die castings; chill cast **UTS: 150**
BS 1490 LM6	11% Si Al alloy: Ingots; chill cast **UTS: 180 Elon: 7%**
BS 1490 LM6M	11% Si Al alloy: For sand and die castings; chill cast **UTS: 180 Elon: 7%**
BS 1490 LM17	11% Si 3% Ni Al alloy: Casting; obsolete
BS 1490 LM18	5% Si Al alloy: Ingot; chill cast **UTS: 140 Elon: 4%**
BS 1490 LM18M	5% Si Al alloy: Sand or die casting; chill cast **UTS: 140 Elon: 4%**
BS 1490 LM20	12% Si Al alloy: Ingot; chill cast **UTS: 180 Elon: 5%**
BS 1490 LM20M	12% Si Al alloy: Die casting; chill cast **UTS: 180 Elon: 5%**
BS 1490 LM25	0.3% Mg 7.0% Si Al alloy: Casting; sand cast **UTS: 230 Proof: 200**
BS 1616 B	5% Si Al alloy: Welding rod flux coated wire to BS alloy NG 21
BS 1942/1	3% Cu 12% Si Al alloy: For brazing; melting range 550–570 °C
BS 1942/2	12% Si Al alloy: Brazing filler BS alloy N2; melting range 565–575 °C
BS 1942/3	8% Si Al alloy: Brazing filler; melting range 565–600 °C
BS 1942/4	5% Si Al alloy: Brazing filler BS alloy N21; melting range 565–625 °C
BS 2901 NG2	12% Si Al alloy: Rod for inert gas shielded arc welding
BS 2901 NG21	5% Si Al alloy: Rod for inert gas shielded arc welding
BS 4300/16-8011	0.7% Si 0.8% Fe Al alloy: ½-hard; H4 condition **UTS: 150 Elon: 4%**
CHROMET	10% Si Al alloy: Bearings for hard shafts; origin unknown
COMALCO 160	12% Si 0.3% Fe Al alloy: Casting; Comalco for BS alloy LM6
COMALCO 160S	12% Si 0.2% Fe 0.03% Na Al alloy: Casting; Comalco for BS alloy LM6
CP 601	0.2% Mg 7% Si Al alloy: Casting; Australian specification
DEF 30	Si Al alloy: For die-casting; covers BS alloys LM2 to LM6 and LM24
DIN 1732 S Al Si12	12% Si Al alloy: Welding electrode
DS 15000 Al501	11.5% Si Al alloy: Casting; Danish specification
DS 15000	11.5% Si Al alloy: Casting; Danish specification
DTD 135	Si Al casting alloy: Cancelled; obsolete
DTD 324A	1% Cu 1% Mg 11% Si 1% Ni Al alloy: Forging; solution treated and aged **UTS: 300 Elon: 3%**
DTD 5028	0.3% Mg 7.0% Si Al alloy: Sand casting **UTS: 220 Elon: 3% Proof: 180**
G Al Si 5 Cu 1	1% Cu 0.4% Mg 5% Si Al alloy: Casting; previous German Standard designation
G Al Si 5 Mg	0.6% Mg 5% Si Al alloy: Casting; previous German Standard designation
G Al Si 6 Cu 4	4% Cu 0.2% Mg 6% Si Al alloy: Casting; previous German Standard designation
G Al Si 6 Cu 6	3% Cu 5% Si 0.5% Mn Al alloy: Casting; previous German Standard designation
G Al Si 7 Cu 3	3% Cu 0.2% Mg 7% Si Al alloy: Casting; previous German Standard designation
G Al Si 9 (Cu)	1.5% Cu 9% Si Al alloy: Casting; previous German Standard designation
G Al Si 10 Mg	0.2% Mg 9.5% Si Al alloy: Casting; previous German Standard designation
G Al Si 10 Mg (Cu)	0.2% Cu 0.3% Mg 10% Si Al alloy: Casting; previous German Standard designation
G Al Si 12	11% Si Al alloy: Casting; previous German Standard designation
G Al Si 12 (Cu)	2% Cu 12.5% Si 0.3% Mn Al alloy: Casting; previous German Standard designation
GD Al Si 6 Cu 3	3% Cu 6% Si 0.4% Mn Al alloy: Casting; previous German Standard designation
GD Al Si 10 (Cu)	0.8% Cu 0.2% Mg 10% Si Al alloy: Casting; previous German Standard designation
GD Al Si 12.	12% Si Al alloy: Casting; previous German Standard designation
GD Al Si 12 (Cu)	0.6% Cu 12% Si 0.3% Mn Al alloy: Casting; previous German Standard designation
GOST 2685-63 Al2	11.5% Si Al alloy: Casting; Russian specification
GOST 2685-63 Al4	0.4% Mg 11.5% Si 0.5% Mn Al alloy: Casting; Russian specification
GOST 2685-63 Al4B	0.4% Mg 11.5% Si 0.5% Mn Al alloy: Casting; Russian specification
GOST 2685-63 Al8	10% Si Al alloy: Casting; Russian specification
GOST 2685-63 Al9	0.5% Mg 5% Si Al alloy: Casting; Russian specification

Note. The following abbreviations and units are used in the tables:

DPN	Hardness, diamond pyramid number
UTS	Ultimate tensile strength, N/mm^2
Elon	Elongation, %
Proof	0.1% proof strength, N/mm^2

1 N/mm^2=0.1 hbar=0.102 kgf/mm^2=0.06475 $tonf/in.^2$=145.04 $lbf/in.^2$=1 MPa

See Appendix II for other abbreviations and conversion tables.

Symbol	Nominal analysis, supplier, condition and remarks.
GOST 2685-63 Al9	0.5% Mg 5% Si Al alloy: Casting; Russian specification
GOST 2685-63 Al9B	0.5% Mg 5% Si Al alloy: Casting; Russian specification
GOST 2685-63 Al25	0.9% Cu 1.2% Mg 12% Si 1.6% Ni Al alloy: Casting; Russian specification
GOST 2685-63 Al27	10% Si Al alloy: Casting; Russian specification
GOST 2685-63 Al 30	0.9% Cu 1.2% Mg 12% Si 1.6% Ni Al alloy: Casting; Russian specification
GRINATAL	0.3% Mg 4.5% Si Al alloy: Wrought; Anglo-Swiss Aluminium Co.
GRINATAL 350	0.3% Mg 4.0% Si Al alloy: Wrought; Anglo-Swiss Aluminium Co.
H 49-2	1.6% Cu 10% Si Al casting alloy: Australian specification
H 49-5	12% Si Al casting alloy: Australian specification
H 49-6	0.5% Mg 4.7% Si Al casting alloy: Australian specification
H 49-7	0.4% Mg 11.5% Si 0.5% Mn Al alloy: Casting; Australian specification
H 49-11	0.8% Cu 1.2% Mg 12% Si 1.5% Ni Al alloy: Casting; Australian specification
H 49-14	5.2% Si Al alloy: Casting; Australian specification
H 49-15	11.5% Si Al alloy: Casting; Australian specification
HA 3 S5	5.2% Si Al alloy: Ingot; Canadian Standard
HA 3 S12N	12.5% Si Al alloy: Ingot; Canadian Standard
HA 3 S12P	12.0% Si 0.8% Fe Al alloy: Ingot; Canadian Standard
HA 3 SG70N	0.3% Mg 7.0% Si 0.2% Fe Al alloy: Ingot; Canadian Standard
HA 3 SG70P	0.3% Mg 7.0% Si Al alloy: Ingot; Canadian Standard
HA 3 SG71	0.5% Mg 7.0% Si Al alloy: Ingot; Canadian Standard
HA 3 SN122	1.0% Mg 12% Si 2.5% Ni Al alloy: Ingot; Canadian Standard
HA 8 SG121	0.9% Cu 1.0% Mg 12% Si 0.9% Ni Al alloy: Forging; Canadian Standard
HA 9 S5	5.2% Si Al alloy: Sand casting; Canadian Standard
HA 9 S12N	12.0% Si Al alloy; Sand casting; Canadian Standard
HA 9 SG70N	0.3% Mg 7.0% Si Al alloy: Sand casting; Canadian Standard
HA 9 SG71	0.5% Mg 7.0% Si Al alloy: Sand casting; Canadian Standard
HA 10 S5	5.2% Si Al alloy: Die casting; Canadian Standard
HIDUMINIUM 00	1.5% Cu 10% Si Al alloy: Die castings; High Duty Alloys for BS alloy LM2
HIDUMINIUM 08	1% Cu 1% Mg 11% Si 1% Ni Al alloy: Forging; solution treated and aged; High Duty Alloys
HIDUMINIUM 10	11% Si Al alloy: For casting thin sections; High Duty Alloys for BS alloy LM6
HIDUMINIUM 512	1% Cu 1% Mg 11% Si 1% Ni Al alloy: Forging; solution treated and aged; High Duty Alloys
IMPALCO 450	12% Si Al alloy: Wire for brazing; Imperial Aluminium Co. for BS alloy N2
IMPALCO 460	5% Si Al alloy: Wire for brazing; Imperial Aluminium Co. for BS alloy N21
IMPALCO 470	7.5% Si Al alloy: Wire for furnace brazing; Imperial Aluminium Co.
IMPALCO PA15	12% Si Al alloy: Wire for brazing; Imperial Aluminium Co. for BS alloy N2
IMPALCO PA16	5% Si Al alloy: Wire for brazing; Imperial Aluminium Co. for BS alloy N21
IMPALCO PA17	7.5% Si Al alloy: For brazing; Imperial Aluminium Co.
KYNAL PA15	12% Si Al alloy: Wire for brazing; ICI; obsolete; now Imperial Aluminium Co.
KYNAL PA16	5% Si Al alloy: Wire for brazing; ICI; obsolete; now Imperial Aluminium Co.
KYNAL PA17	5% Si Al alloy: Wire for brazing; ICI; obsolete; now Imperial Aluminium Co.
L8	12% Si Al alloy: Casting; obsolete
L33	11% Si Al alloy: Casting; chill cast **UTS: 180 Elon: 7% Proof: 70**
L99	0.3% Mg 7.0% Si Al alloy: Casting; sand cast **UTS: 230 Elon: 2% Proof: 180**
L 252	11.5% Si Al alloy: Casting; Spanish specification
L 253	11.5% Si Al alloy: Casting; Spanish specification
L 254	0.4% Mg 11.5% Si 0.5% Mn Al alloy: Casting; Spanish specification
L 255	0.9% Cu 1.2% Mg 12% Si 1.6% Ni Al alloy: Casting; Spanish specification
L 256	0.4% Mg 11.5% Si 0.5% Mn Al alloy: Casting; Spanish specification
L 257	0.55% Mg 5% Si Al alloy: Casting; Spanish specification
LAC 112 A	1.5% Cu 10% Si Al alloy: Casting; obsolete; Air Ministry specification similar to BS alloy LM2
L Al Si 12	0.3% Cu 12% Si Al alloy: German Standard; DIN number not known
LM 2	1.5% Cu 10% Si Al alloy: Casting; Intal for BS alloy LM2
LM2	1.5% Cu 10% Si Al alloy: Casting; designation used by BS
LM 6	11% Si Al alloy: Casting; Intal for BS alloy LM6
LM 6	11% Si Al alloy: Casting
LM 17	11% Si 3% Ni Al alloy: Casting
LM 18	5% Si Al alloy: Casting
LM 18	5% Si Al alloy: Casting; Intal for BS alloys LM18
LM 20	12% Si Al alloy: Casting
LM 25	0.3% Mg 7% Si Al alloy: Casting
METCO 52C NS	12% Si Al alloy: Spray deposit; Metco
METCO 601 NS	Si Al with plastic: Spray deposit; Metco
METCO SF Al	Si Al alloy: Spray deposit; Metco
MVC	11% Si Al: Casting; AEI; chill cast **UTS: 180**
N 2	12% Si Al alloy: Designation used by BS
N 21	5% Si Al alloy: Designation used by BS
NG 2	12% Si Al alloy: Weld rod; designation by BS
NG 21	5% Si Al alloy: Weld rod; designation used by BS
NBN436 Al Si 5 Mg	0.6% Mg 5% Si Al alloy: Casting; Belgium specification
NBN 436 Al Si 10 Mg	0.4% Mg 11.5% Si 0.5% Mn Al alloy: Casting; Belgium specification
NBN 436 Al Si 12	12% Si Al alloy: Casting; Belgium specification
NORAL 33S	5% Si Al alloy: Wire; Northern Aluminium Co. for BS alloy N21
NORAL 38S	1% Cu 1% Mg 11% Si 1% Ni Al alloy: Forging; solution treated and aged; Northern Aluminium Co. **UTS: 300 Elon: 3%**
NORAL 123	5% Si Al alloy: Casting; Northern Aluminium Co. for BS alloy LM 18
NORAL 158	11% Si 3% Ni Al alloy: Casting; Northern Aluminium Co. for BS alloy LM 17
NORAL 160	11% Si Al alloy: Casting; Northern Aluminium Co. for BS alloy LM20
NURAL 1761	1.0% Cu 1.0% Mg 17% Si 0.5% Cr 3.4% Ni Al alloy: Casting; Alcan (Germany) for pistons
NURAL 1761P	1.0% Cu 1.0% Mg 16.8% Si 1.0% Ni Al alloy: Forgings; Alcan (Germany) for pistons
NURAL 1762	1.0% Cu 1.0% Mg 17.2% Si Al alloy: Alcan (Germany)
NURAL 2361	1.0% Cu 1.0% Mg 23% Si 0.5% Cr 1.0% Ni Al alloy: Castings; Alcan (Germany) for pistons
NURAL 3210	1.1% Cu 1.0% Mg 12% Si 1.0% Ni 0.1% Ti Al alloy: Casting; Alcan (Germany) for pistons
QQ A 591/1	12% Si Al alloy: Casting; US Federal
QQ A 591/2	12.0% Si Al alloy: Casting; US Federal
QQ A 591/3	5.0% Si Al alloy: Casting; US Federal
QQ A 591/8	5.0% Si Al alloy: Casting; US Federal
QQ A 596/7	5.0% Si Al alloy: Casting; US Federal; as cast **UTS: 120 Elon: 2.5%**
QQ A 601/2	5.0% Si Al alloy: Casting; US Federal; as cast **UTS: 90 Elon: 3%**
SAE 35	5% Si Al alloy: Casting; sand cast **UTS: 110 Elon: 3% Proof: 30**
SAE 205	5% Si Al alloy: Welding wire; now AA4043

Symbol	Nominal analysis, supplier, condition and remarks.
SAE 206	7% Si Al alloy: Used only for cladding; analysis applies before cladding; now AA4343
SAE 290	1% Cu 1% Mg 12% Si Al alloy: Solution treated and aged; now AA 4043
	DPN: 115 UTS; 360 Elon: 5% Proof: 280
SAE 304	5% Si Al alloy: Casting; die cast; as cast
	UTS: 200 Elon: 5%
SAE 305	12% Si Al alloy: Casting; die cast; as cast
	UTS: 240 Elon: 3.5%
S Al Si 5	5% Si Al alloy: Wrought; German Standard
S Al Si 12	8% Si Al alloy: Wrought; German Standard
SC 92A	1.8% Cu 0.5% Mg 9% Si Al alloy: Casting; ASTM designation
Sf 1	13% Si 0.35% Mn Al alloy: Casting; Anglo-Swiss code for Silafont 1
Sf 2	0.45% Mg 11% Si 0.3% Mn Al alloy: Casting; Anglo-Swiss code for Silafont 2
Sf 3	0.4% Mg 9% Si 0.3% Mn Al alloy: Casting; Anglo-Swiss code for Silafont 3
Sf 4	1.2% Cu 0.65% Mg 11% Si 0.3% Mn Al alloy: Casting; Anglo-Swiss code for Silafont 4
Sf 5	0.3% Mg 9.5% Si 0.5% Co Al alloy: Casting; Anglo-Swiss code for Silafont 5
Sf 6	0.25% Mg 10% Si Al alloy: Casting; Anglo-Swiss code for Silafont 6
Sf 7	1.2% Cu 1% Mg 13% Si 1% Ni Al alloy: Casting; Anglo-Swiss code for Silafont 7
SF Aluminium	6% Si Al: Wire for metal spraying; as sprayed; Metco
	UTS: 240
SILAFONT 1	13% Si 0.35% Mn Al alloy: Casting; Anglo-Swiss; as cast; obsolete
	DPN: 55 UTS: 210 Elon: 7% Proof: 110
SILAFONT 2	0.45% Mg 11% Si 0.3% Mn Al alloy: Casting; Anglo-Swiss; solution treated and aged; obsolete
	DPN: 90 UTS: 300 Elon: 2% Proof: 240
SILAFONT 3	0.4% Mg 9% Si 0.3% Mn Al alloy: Casting; Anglo-Swiss; solution treated and aged; obsolete
	DPN: 90 UTS: 270 Elon: 2% Proof: 220
SILAFONT 4	1.2% Cu 0.65% Mg 11% Si 0.3% Mn Al alloy: Casting; Anglo-Swiss; as cast; obsolete
	DPN: 90 UTS: 210 Elon: 2% Proof: 150
SILAFONT 5	0.3% Mg 9.4% Si 0.5% Co Al alloy: Casting; Anglo-Swiss; solution treated and aged; obsolete
	DPN: 70 UTS: 270 Elon: 4% Proof: 220
SILAFONT 6	0.25% Mg 10% Si Al alloy: Casting; Anglo-Swiss as cast; obsolete
	DPN: 60 UTS: 210 Elon: 2% Proof: 150
SILAFONT 7	1.2% Cu 1% Mg 13% Si 1% Ni Al alloy: Casting; Anglo-Swiss; solution treated and aged; obsolete
	DPN: 120 UTS: 250 Elon: 1% Proof: 210
SILAFONT 14	13% Si 0.3% Mn Al alloy: Casting; Anglo-Swiss Aluminium Co.
SILAFONT 74	1.2% Cu 1.0% Mg 12% Si 1.0% Ni Al alloy: Ingot; Anglo-Swiss Aluminium Co.
SILAFONT 84	10% Si 0.3% Mn Al alloy: Casting; Anglo-Swiss Aluminium Co.
SONDALU Si 5	5% Si 0.8% Fe Al alloy: Welding electrode; Soudometal
	UTS: 120 Elon: 15% Proof: 30
SONDALU Si 12	11% Si 0.5% Fe Al alloy: Welding electrode; Soudometal
	UTS: 180 Elon: 5% Proof: 30
SONDOR Cr Al Si 5	5% Si 0.2% Fe Al alloy: Welding electrode; Soudometal
	UTS: 145 Elon: 14% Proof: 60
SONDOTIG Al Si 5	5% Si 0.2% Fe Al alloy: Welding electrode; for inert gas welding; Soudometal
	UTS: 145 Elon: 14% Proof: 60
SS 4244	0.3% Mg 7% Si Al alloy: Casting; Swedish Standard; as cast
	DPN: 90 UTS; 220 Elon: 2% Proof: 200
SS 4247	9% Si 0.5% Mn 1% Fe Al alloy: Casting; Swedish Standard; as cast
	DPN: 70
SS 4253	0.3% Mg 10% Si Al alloy: Casting; Swedish Standard; as cast
	DPN: 90 UTS; 220 Elon: 2% Proof: 200
SS 4255	10% Si Al alloy: Casting; Swedish Standard; as cast
	DPN: 60
SS 4260	12% Si 0.4% Mn Al alloy: Casting; Swedish Standard; as cast
	DPN: 60
SS 4261	12% Si Al alloy: Casting; Swedish Standard; as cast
	DPN: 60
STA 7 AC2	1.5% Cu 10% Si Al alloy: Casting; replaced by BS 1490 LM2
STA 7 AC6	11% Si Al alloy: Casting; replaced by BS 1490; LM6
STA 7 AW1	5.0% Si Al alloy: Wire; replaced by BS 1473/5 NW21
STA 7 AW2C	12% Si Al alloy: Wire; replaced by BS 1473/5 NW2
STAR Si 050	5.2% Si Al alloy; Star
STAR Si 120	12% Si Al alloy; Star
STUBS NG 21	5% Si Al alloy: For electric arc welding; Stubs
STUBS 4	High Si Al alloy: For brazing; melting range 580 °C; Stubs
STUBS 48	12% Si Al alloy: For welding and brazing; Stubs
	DPN: 60 UTS: 210 Elon: 30% Proof: 90
UNI 3022 MAS6	6% Si Al alloy: Ingot; primary metal; Italian Standard
UNI 3022 MA625	25% Si Al alloy: Ingot; primary metal; Italian Standard
UNI 3048	0.8% Cu 13% Si 0.3% Mn Al alloy: Italian Standard
UNI 3049	0.3% Mg 12% Si 0.5% Mn Al alloy: Italian Standard
UNI 3572	1.0% Cu 1.0% Mg 12% Si 0.8% Ni Al alloy: For forging; Italian Standard
UNI 4513	2.0% Cu 12% Si 0.3% Mn Al alloy: Casting; Italian Standard
UNI 4514	13% Si Al alloy: Casting; Italian Standard
UNI 5074	0.4% Mg 9.0% Si 0.6% Fe Al alloy: Casting; Italian Standard
UNI 5076	2.0% Cu 12% Si 0.7% Fe Al alloy: Casting Italian Standard
UNI 5077	5.0% Si 0.8% Fe Al alloy: Casting; Italian Standard
UNI 5078	9.0% Si 0.7% Fe Al alloy: Casting; Italian Standard
UNI 5079	13% Si 0.7% Al alloy: Casting; Italian Standard
UNI 6250	0.8% Cu 1.0% Mg 12.7% Si 2.2% Ni Al alloy: Casting; Italian Standard
UNI 6251	1.6% Cu 0.6% Mg 21% Si 0.7% Mn 1.5% Ni 0.8% Co Al alloy: Casting; Italian Standard
UNI 7257	0.3% Mg 7% Si Al alloy: Italian specification
V 1036 A/B BVI	5.5% Si Al alloy: Casting; Dutch specification
V 1036 A/B BVII	11.5% Si Al alloy: Casting; Dutch specification
V 1036 A/B BIX	0.5% Mg 5% Si Al alloy: Casting; Dutch specification
V 1036 A/B BXI	0.4% Cu 11.5% Si 0.5% Mn Al alloy: Casting: Dutch specification
WILMIL	13% Si Al alloy: Casting; W Mills Ltd for BS alloy LM6

Note. The following abbreviations and units are used in the tables:

DPN	Hardness, diamond pyramid number
UTS	Ultimate tensile strength, N/mm^2
Elon	Elongation, %
Proof	0.1% proof strength, N/mm^2

1 N/mm^2=0.1 hbar=0.102 kgf/mm^2=0.06475 $tonf/in.^2$=145.04 $lbf/in.^2$=1 MPa

See Appendix II for other abbreviations and conversion tables.

1D Aluminium–magnesium wrought alloys

Specific gravity		2.65–2.7
Density		2650–2700 kg/m^3
Solidus/liquidus		580–650 °C
Thermal conductivity	annealed	120–190 W/m °C
	½-hard	138–147 W/m °C
Coefficient of linear expansion		23.8–24.3 × 10^{-6}/ °C
Electrical conductivity	annealed	30–50% IACS (copper 100%)
Specific resistance	annealed	34–59 microhm mm
	½-hard	40–55 microhm mm
	hard	64 microhm mm
Young's modulus of elasticity		66.9–72.4 × 10^9 N/m^2
Impact	annealed	38–49 J
	½-hard	28 J
Fatigue strength (50 × 10^6 cycles)	annealed	± 90–140 N/mm^2
Hot strength		

Temperature °C	Tensile strength N/mm^2	Elongation %
150	150–180	40–45
200	145–160	50–55
250	110–125	60–65
300	48–54	

The above properties are included as a guide to the group of specifications as a whole. There is considerable variation within the group, and as would be expected these properties vary in general with the magnesium percentage.

General metallurgical properties

Magnesium in small quantities of less than 1%, added to aluminium, gives excellent hot working properties and therefore is useful in producing extrusions and forgings where intricate shapes are required.

Alloys with 1% magnesium and up to 1% silicon have a specific characteristic and are listed and described in Section 1E. These alloys are heat treatable.

The alloys in this section, i.e. the aluminium–magnesium alloys, are not heat treatable except to remove cold work.

Additions of magnesium up to 7% or 8% give alloys with excellent corrosion resistance, particularly under marine conditions. The addition of copper, manganese and chromium strengthens the alloys and increases their ability to accept cold work, but reduces their corrosion resistance.

Welding requires the use of inert gas techniques.

These alloys are commonly used for architectural purposes where moderate strength but good corrosion resistance is required. The alloys can be anodized and colour anodized to give a range of colours with an attractive appearance. These alloys, particularly the higher magnesium alloys, have the best marine corrosion resistance of any aluminium alloy.

Many alloys are now used on motor car and commercial vehicle trims and some use has been found in the chemical industry.

Because of the cold work characteristic of the lower magnesium alloys, they are used for extrusions and the most common and intricate aluminium alloy extrusions are manufactured from these alloys.

Symbol	Nominal analysis, supplier, condition and remarks.
3.3308	0.5% Mg Al alloy: German Standard DIN 1725
3.3309	0.5% Mg Al alloy: German Standard DIN 1725
3.3318	0.8% Mg Al alloy: German Standard DIN 1725

Note. The following abbreviations and units are used in the tables:

DPN	Hardness, diamond pyramid number
UTS	Ultimate tensile strength, N/mm^2
Elon	Elongation, %
Proof	0.1% proof strength, N/mm^2

1 N/mm^2=0.1 hbar=0.102 kgf/mm^2=0.06475 tonf/in.^{2}=145.04 lbf/in.^{2}=1 MPa

See Appendix II for other abbreviations and conversion tables.

Symbol	Nominal analysis, supplier, condition and remarks.
3.3319	1% Mg Al alloy: German Standard DIN 1725
3.3329	2% Mg Al alloy: German Standard DIN 1725
3.3350	10% Mg Al alloy: Wrought; German Standard DIN 1725
3.3315	0.9% Mg 0.2% Mn 0.2% Cr Al alloy: Wrought; German Standard DIN 1725; annealed **DPN: 30 UTS: 90 Elon: 20%**
3.3525	2.0% Mg 0.3% Mn 0.2% Cr Al alloy: Wrought; German Standard DIN 1725
3.3527	2% Mg 1% Mn 0.2% Cr Al alloy: Wrought; German Standard DIN 1725
3.3535	3% Mg 0.2% Cr Al alloy: Wrought; German Standard DIN 1725
3.3537	7% Mg 0.45% Mn Al alloy: Wrought; German Standard DIN 1725

Symbol	Nominal analysis, supplier, condition and remarks.
3.3547	4.4% Mg 0.6% Mn 0.1% Cr Al alloy: German Standard
3.3555	5% Mg 0.2% Cr Al alloy: Wrought; German Standard DIN 1725; annealed **DPN: 55 UTS: 220 Elon: 17% Proof: 90**
3.3557	7% Mg Al alloy: German Standard DIN 1725
3.3575	6.8% Mg 0.2% Cr Al alloy: Wrought; German Standard DIN 1725
50S	1.2% Mg Al alloy: Previous AA designation now 5050
52S	2% Mg Al: Sheet and tube; Alcoa for BS alloy N4
52S	2.5% Mg 0.2% Cr Al alloy: Previous AA designation now 5052
56S	5% Mg Al alloy: Sheet and strip; Alcoa for BS alloy N6
56S	5% Mg 0.15% Cr Al alloy: Previous AA designation now 5056
1435	0.4% Mg Al alloy: Wrought; designation used in the UK and USA
5005	0.8% Mg Al alloy: Wrought; designation used in the UK and USA
5006	1% Mg 0.6% Mn Al alloy: Wrought; designation used in the UK and USA
5010	0.4% Mg 0.2% Mn Al alloy: Wrought; designation used in the UK and USA
5016	1.6% Mg 0.5% Mn Al alloy: Wrought; designation used in the UK and USA
5017	0.2% Cu 2% Mg 0.7% Mn Al alloy: Wrought; designation used in the UK and USA
5034	1% Mg 0.3% Mn 0.5% Fe Al alloy: Wrought; obsolete; designation used in the UK and USA
5039	0.15% Cu 3.8% Mg 0.4% Mn Al alloy: Wrought; obsolete; designation used in the UK and USA
5040	0.2% Cu 1.2% Mg 1% Mn Al alloy: Wrought; designation used in the UK and USA
5042	3.5% Mg 0.3% Mn Al alloy: Wrought; designation used in the UK and USA
5043	0.2% Cu 1% Mg 0.9% Mn Al alloy: Wrought; designation used in the UK and USA
5050	1.6% Mg Al alloy: Wrought; designation used in the UK and USA
5051	2% Mg Al alloy: Wrought; designation used in the UK and USA
5052	2.6% Mg 0.2% Cr Al alloy: Wrought; designation used in the UK and USA
5056	5% Mg 0.1% Mn 0.15% Cr Al alloy: Wrought; designation used in the UK and USA
5082	4.5% Mg Al alloy: Wrought; designation used in the UK and USA
5083	4.5% Mg 0.6% Mn 0.2% Cr Al alloy: Wrought; designation used in the UK and USA
5086	4% Mg 0.5% Mn 0.2% Cr Al alloy: Wrought; designation used in the UK and USA
5151	1.8% Mg Al alloy: Wrought; designation used in the UK and USA
5154	3.5% Mg 0.2% Cr Al alloy: Wrought; designation used in the UK and USA
5182	4.5% Mg Al alloy: Wrought; designation used in the UK and USA
5183	4.8% Mg 0.7% Mn 0.2% Cr Al alloy: Wrought; designation used in the UK and USA

Note. The following abbreviations and units are used in the tables:

DPN	Hardness, diamond pyramid number
UTS	Ultimate tensile strength, N/mm^2
Elon	Elongation, %
Proof	0.1% proof strength, N/mm^2

$1\ N/mm^2 = 0.1\ hbar = 0.102\ kgf/mm^2 = 0.06475\ tonf/in.^2 = 145.04\ lbf/in.^2 = 1\ MPa$

See Appendix II for other abbreviations and conversion tables.

Symbol	Nominal analysis, supplier, condition and remarks.
5205	0.8% Mg 0.08% Cr Al alloy: Wrought; designation used in the UK and USA
5250	1.5% Mg 0.1% Mn Al alloy: Wrought; designation used in the UK and USA
5252	2.6% Mg Al alloy: Wrought; designation used in the UK and USA
5254	3.5% Mg 0.2% Cr Al alloy: Wrought; designation used in the UK and USA
5351	1.9% Mg Al alloy: Wrought; designation used in the UK and USA
5352	2.6% Mg Al alloy: Wrought; designation used in the UK and USA
5356	5% Mg 0.15% Mn 0.15% Cr Al alloy: Wrought; designation used in the UK and USA
5357	1% Mg 0.3% Mn Al alloy: Wrought; designation used in the UK and USA
5451	2% Mg 0.2% Cr Al alloy: Wrought; designation used in the UK and USA
5454	2.7% Mg 0.7% Mn 0.15% Cr Al alloy: Wrought; designation used in the UK and USA
5456	5% Mg 0.7% Mn 0.15% Cr Al alloy: Wrought; designation used in the UK and USA
5457	1% Mg 0.3% Mn Al alloy: Wrought; designation used in the UK and USA
5552	2.6% Mg Al alloy: Wrought; designation used in the UK and USA
5554	2.7% Mg 0.15% Cr Al alloy: Wrought; designation used in the UK and USA
5556	5% Mg 0.15% Cr 0.1% Ti Al alloy: Wrought; designation used in the UK and USA
5557	0.6% Mg 0.2% Mn Al alloy: Wrought; designation used in the UK and USA
5652	2.6% Mg 0.2 Cr Al alloy: Wrought; designation used in the UK and USA
5654	3.5% Mg 0.2% Cr Al alloy: Wrought; designation used in the UK and USA
5657	0.8% Cr Al alloy: Wrought; designation used in the UK and USA
AG 0.5	0.5% Mg Al alloy: Wrought; L'Aluminium Français
AG 0.6	0.8% Mg Al alloy: French Standard; Wrought
AG 0.6	0.8% Mg Al alloy: Wrought; French Standard
AG 1	1.0% Mg Al alloy: Wrought; L'Aluminium Français
AG 1	1.4% Mg Al alloy: Wrought; French Standard
AG 2	2.2% Mg Al alloy: Wrought; French Standard
AG 2	2.1% Mg Al alloy: Wrought; L'Aluminium Français
AG 2M	2% Mg Al alloy: French Standard; Wrought
AG 2.5MC	2.7% Mg 0.7% Mn 0.1% Cr Al alloy: French Standard; Wrought
AG 3	3.1% Mg 0.8% Mn Al alloy: Wrought; L'Aluminium Français
AG 3	3.2% Mg Al alloy: Wrought; French Standard
AG 4	4.2% Mg 0.8% Mn Al alloy: Wrought; L'Aluminium Français
AG 4 MC	4.2% Mg Al alloy: Wrought; French Standard
AG 4.5 MC	4.4% Mg 0.7% Mn 0.1% Cr Al alloy: French Standard; Wrought
AG 5	5.0% Mg 0.6% Mn Al alloy: Sheet; French Standard
AG 5	5.2% Mg 0.8% Mn Al alloy: Wrought; L'Aluminium Français
AG 7	7.0% Mg 0.8% Mn Al alloy: Wrought; L'Aluminium Français
AGG 4	Mg Mn Al alloy: Wrought; further information from Aluminium Zentrale
AGG 50	1% Mg 0.2% Mn 0.2% Cr Al alloy: Wrought; further information from Aluminium Zentrale
AGG 54	4% Mg Al alloy: Wrought; further information from Aluminium Zentrale
AGG 57	3% Mg 0.2% Cr Al alloy: Wrought; further information from Aluminium Zentrale
AL 8	10.2% Mg Al alloy: Casting; Russian designation
AL 27	10% Mg Al alloy: Casting; Russian designation

Symbol	Nominal analysis, supplier, condition and remarks.
ALCAN GB 54S	3.5% Mg 0.3% Mn Al alloy: Alcan for BS alloy N5
ALCAN GB 58S	7.2% Mg 0.3% Mn Al alloy: Alcan for BS alloy N7
ALCAN GBA 56S	5% Mg 0.2% Mn Al alloy: Alcan for BS alloy N6
ALCAN GBB 53S	2.7% Mg 0.7% Mn 0.1% Cr Al alloy: Alcan **UTS: 250 Elon: 6% Proof: 200**
ALCAN GBD 54S	4.2% Mg 0.7% Mn 0.12% Cr Al alloy: Alcan for BS alloy N5
ALCAN GBL 57S	1% Mg Al alloy: Sheet and strip; Alcan; for bright anodizing trim
ALCAN GBM 57S	2% Mg 0.3% Mn Al alloy: Alcan for BS alloy N4
ALCAN GBS 57S	0.6% Mg Al alloy: Bar and forging for bright anodizing; Alcan for BT 1
ALCOA 510	2.0% Mg Al alloy; Wrought; Alcoa; ½-hard; H4 condition **UTS: 225 Elon: 5% Proof: 170**
ALCOA 520	3.5% Mg Al alloy: Wrought; Alcoa; ½-hard; H4 condition **UTS: 300 Elon: 6% Proof: 220**
ALCOA 540	1.0% Mg Al alloy: Wrought; Alcoa; ½-hard; H4 condition **UTS: 155**
ALCOA 550	0.5% Mg Al alloy: Wrought; Alcoa; ½-hard; H4 condition; for anodizing **UTS: 150**
Al Mg 1	1% Mg 0.2% Mn 0.2% Cr Al alloy: Wrought; previous German Standard designation
Al Mg 2	2% Mg 0.2% Mn 0.2% Cr A1 alloy: Wrought; previous German Standard designation
Al Mg 3	3% Mg 0.2% Mn 0.2% Cr Al alloy: Wrought; previous German Standard designation
Al Mg 3 Si	3% Mg 0.6% Si 0.5% Mn 0.2% Cr Al alloy: Wrought; previous German Standard designation
Al Mg 5	5% Mg 0.4% Mn 0.2% Cr Al alloy: Wrought; previous German Standard designation
A1 Mg 7	7% Mg 0.4% Mn 0.2% Cr Al alloy: Wrought; previous German Standard designation
Al Mg Mn	2% Mg 1% Mn 0.2% Cr Al alloy: Wrought; previous German Standard designation
ALMINAL 4	0.15% Cu 1.8% Mg 0.6% Si 0.5% Mn 0.5% Cr Al alloy: Wrought; origin unknown
ALMINAL 5	0.15% Cu 3.0% Mg 0.6% Si 1.0% Mn 0.5% Cr Al: Wrought; origin unknown
ALMINAL 6	0.15% Cu 4.5% Mg 0.6% Si 1.0% Mn Al alloy: Wrought; origin unknown
Al R Mg 0.5	0.5% Mg Al alloy: For reflective uses; previous German Standard designation
Al R Mg 1	1% Mg Al alloy: For reflective uses; previous German Standard designation
Al R Mg 2	2% Mg Al alloy: For reflective uses; previous German Standard designation
ALUMAGNESE 10 C	1% Mg Al alloy: E Kaye; hard drawn **UTS: 180 Elon: 5%**
ALUMAGNESE 20 V	2% Mg Al alloy: E Kaye for BS alloy N4
ALUMAGNESE 35 B	3.5% Mg Al alloy: E Kaye for BS alloy N5
ALUMAGNESE 45	4.5% Mg 0.8% Mn Al alloy: E Kaye for BS alloy N8
ALUMAGNESE 50J	5% Mg Al alloy: E Kaye for BS alloy H6
ALUMAGNESE CS	1% Mg Al alloy: Tube; E Kaye; ½-hard **UTS: 210 Elon: 9%**
AMG 3	3.5% Mg Al alloy: Russian designation
AMS 4004	2.5% Mg 0.25% Cr Al alloy: Foil
AMS 4005	5% Mg 0.1% Cr Al alloy
AMS 4015 E	2.5% Mg 0.25% Cr Al alloy: Sheet and plate; annealed; AMS for AA alloy 5052 **UTS: 180 Elon: 17%**
AMS 4016 E	2.5% Mg 0.25% Cr Al alloy: Sheet and plate; ¼-hard; AMS for AA alloy 5052 **UTS: 220 Elon: 6%**
AMS 4017 E	2.5% Mg 0.25% Cr Al alloy: Sheet and plate; ½-hard; AMS for AA alloy 5052 **UTS: 250 Elon: 4%**
AMS 4018	3.5% Mg 0.25% Cr Al alloy: Sheet and plate; annealed; AMS for AA alloy 5154 **UTS: 240 Elon: 14% Proof: 70(0.2%)**
AMS 4019	3.5% Mg 0.25% Cr Al alloy: Sheet plate; cold rolled; AMS for AA alloy 5154 **UTS: 240 Elon: 8% Proof: 170**
AMS 4056 B	4.5% Mg 0.6% Mn 0.15% Cr Al alloy: Sheet; annealed; AMS for AA alloy 5083
AMS 4057 A	4.5% Mg 0.6% Mn 0.15% Cr Al alloy: Sheet; ¼-hard; AMS for AA alloy 5083
AMS 4058 A	4.5% Mg 0.6% Mn 0.15% Cr Al alloy: Sheet; ½-hard; AMS for AA alloy 5083
AMS 4059 B	4.5% Mg 0.6% Mn 0.15% Cr Al alloy: Sheet; ⅛-hard; AMS for AA alloy 5083
AMS 4069	2.5% Mg 0.25% Cr Al alloy: Seamless tube; annealed; AMS for AA alloy 5052
AMS 4070 F	2.5% Mg 0.25% Cr Al alloy: Seamless tube; annealed; AMS for AA alloy 5052
AMS 4071 F	2.5% Mg 0.25% Cr Al alloy: Seamless tube; annealed and drawn; AMS for AA alloy 5052
AMS 4114 B	2.5% Mg 0.25% Cr Al alloy: Bar as rolled; AMS for AA alloy 5052
AMS 4175	2.4% Mg 0.2% Cr Al alloy: Wrought
AMS 4176	5% Mg 0.15% Cr Al alloy: Wrought
AMS 4177	5% Mg 0.15% Cr Al alloy: Wrought
AMS 4178	2.4% Mg 0.2% Cr Al alloy: Wrought
AMS 4182 A	5% Mg 0.12% Mn 0.12% Cr Al alloy: Wire; annealed; AMS for AA alloy 5056
AMS 4348	2.4% Mg 0.2% Cr Al alloy: Wrought
AMS 4349	5% Mg 0.15% Cr Al alloy: Wrought
AN WW C 561a/1	5.2% Mg Al alloy: Wire; US Service
ASTM B199 AZ81A	7.5% Mg 0.7% Zn Al alloy: Die casting
ASTM B209/5005	1% Mg Al alloy: Sheet; annealed **UTS: 110 Elon: 22%**
ASTM B209/5005	1% Mg Al alloys: Sheet; cold rolled **UTS: 170 Elon: 3%**
ASTM B209/5050	1.5% Mg Al alloy: Sheet, annealed **UTS: 12; Elon: 20% Proof: 22**
ASTM B209/5050	1.5% Mg Al alloy: Sheet; cold rolled **UTS: 200 Elon: 3%**
ASTM B209/5052	2.5% Mg 0.2% Cr Al alloy: Sheet; annealed **UTS: 180 Elon: 18% Proof: 45**
ASTM B209/5052	2.5% Mg 0.2% Cr Al alloy: Sheet; cold rolled **UTS: 280 Elon: 4%**
ASTM B209/5083	4.5% Mg 0.7% Mn 0.2% Cr Al alloy: Sheet; annealed **UTS: 310 Elon: 16% Proof: 100**
ASTM B209/5083	4.5% Mg 0.7% Mn 0.2% Cr Al alloy: Sheet; cold rolled **UTS: 370 Elon: 6% Proof: 270**
ASTM B209/5086	4% Mg 0.5% Mn 0.2% Cr Al alloy: Sheet; annealed **UTS: 250 Elon: 18% Proof: 70**
ASTM B209/5086	4% Mg 0.5% Mn 0.2% Cr Al alloy: Sheet; cold rolled **UTS: 340 Elon: 4% Proof: 250**
ASTM B209/5154	3.5% Mg 0.25% Cr 0.2% Ti Al alloy: Sheet; annealed **UTS: 240 Elon: 18% Proof: 45**
ASTM B209/5154	3.5% Mg 0.25% Cr 0.2 Ti Al alloy: Sheet; cold rolled **UTS: 310 Elon: 5% Proof: 240**
ASTM B209/5155	4.2% Mg 0.4% Mn 0.15% Cr Al alloy: Sheet and plate
ASTM B209/5252	2.5% Mg Al alloy: Sheet and plate
ASTM B209/5254	3.5% Mg 0.25% Cr Al alloy: Sheet; annealed **UTS: 240 Elon: 18% Proof: 45**
ASTM B209/5254	3.5% Mg 0.25% Cr Al alloy: Sheet; cold rolled **UTS: 300 Elon: 5% Proof: 240**
ASTM B209/5257	0.4% Mg Al alloy: Sheet and plate
ASTM B209/5454	2.7% Mg 0.7% Mn 0.15% Cr 0.2% Ti Al alloy: Sheet; annealed **UTS: 240 Elon: 18% Proof: 60**

Symbol	Nominal analysis, supplier, condition and remarks.
ASTM B209/5454	2.7% Mg 0.7% Mn 0.15% Cr 0.2% Ti Al alloy: Sheet; cold rolled **UTS: 310 Elon: 10% Proof: 180**
ASTM B209/5456	5% Mg 0.7% Mn 0.15% Cr 0.2% Ti Al alloy: Sheet; annealed **UTS: 340 Elon: 16% Proof: 180**
ASTM B209/5456	5% Mg 0.7% Mn 0.15% Cr 0.2% Ti Al alloy: Sheet; ½-hard **UTS: 400 Elon: 9% Proof: 270**
ASTM B209/5457	1.0% Mg 0.3% Mn Al alloy: Sheet and plate
ASTM B209/5557	0.6% Mg Al alloy: Sheet and plate
ASTM B209/5652	2.5% Mg 0.25% Cr Al alloy: Sheet; annealed **UTS: 200 Elon: 20%**
ASTM B209/5652	2.5% Mg 0.25% Cr Al alloy: Sheet; cold rolled **UTS: 280 Elon: 4%**
ASTM B209/5657	0.8% Mg Al alloy: Sheet and plate
ASTM B210/5050	1.5% Mg Al alloy: Tube; annealed **UTS: 140 Proof: 15**
ASTM B210/5050	1.5% Mg Al alloy: Tube; cold drawn **UTS: 210 Proof: 150**
ASTM B210/5052	2.5% Mg 0.2% Cr Al alloy: Tube; annealed **UTS: 170 Proof: 45**
ASTM B210/5052	2.5% Mg 0.2% Cr Al alloy: Tube; cold drawn **UTS: 270 Proof: 210**
ASTM B210/5083	4.4% Mg 0.7% Mn 0.2% Cr Al alloy: Tube
ASTM B210/5086	4% Mg 0.5% Mn 0.15% Cr Al alloy: Tube
ASTM B210/5154	3.5% Mg 0.2% Cr 0.2% Ti Al alloy: Tube; annealed **UTS: 200 Proof: 60**
ASTM B210/5154	3.5% Mg 0.2% Cr 0.2% Ti Al alloy: Tube; cold drawn **UTS: 310 Proof: 220**
ASTM B210/5254	3.5% Mg 0.2% Cr Al alloy: Tube; annealed **UTS: 200 Proof: 60**
ASTM B210/5254	3.5% Mg 0.2% Cr Al alloy: Tube; cold drawn **UTS: 310 Proof: 220**
ASTM B210/5652	2.5% Mg 0.2% Cr Al alloy: Tube; annealed **UTS: 200 Proof: 45**
ASTM B210/5652	2.5% Mg 0.2% Cr Al alloy: Tube; cold drawn **UTS: 280 Proof: 210**
ASTM B211/5052	2.5% Mg 0.2% Cr Al alloy: Bar; cold drawn **UTS: 270**
ASTM B211/5056	5% Mg 0.1% Cr Al alloy: Bar; cold rolled **UTS: 390**
ASTM B211/5154	3.5% Mg 0.2% Cr 0.2% Ti Al alloy: Bar; cold drawn **UTS: 310**
ASTM B211/5254	3.5% Mg 0.2% Cr Al alloy: Bar; cold drawn **UTS: 310**
ASTM B211/5652	2.5% Mg 0.2% Cr Al alloy: Bar; cold drawn **UTS: 270**
ASTM B221/5052	2.4% Mg 0.2% Cr Al alloy: Extruded bar, etc.
ASTM B221/5083	4.5% Mg 0.8% Mn 0.2% Cr Al alloy: Extruded bar; cold drawn **UTS: 270 Elon: 12% Proof: 150**
ASTM B221/5086	4% Mg 0.6% Mn 0.2% Cr Al alloy: Extruded bar; cold drawn **UTS: 240 Elon: 12% Proof: 110**
ASTM B221/5154	3.5% Mg 0.2% Cr Al alloy: Extruded bar; cold drawn **UTS: 200 Proof: 75**

Note. The following abbreviations and units are used in the tables:

DPN	Hardness, diamond pyramid number
UTS	Ultimate tensile strength, N/mm^2
Elon	Elongation, %
Proof	0.1% proof strength, N/mm^2

1 N/mm^2=0.1 hbar=0.102 kgf/mm^2=0.06475 $tonf/in.^2$=145.04 $lbf/in.^2$=1 MPa

See Appendix II for other abbreviations and conversion tables.

Symbol	Nominal analysis, supplier, condition and remarks.
ASTM B221/5454	2.8% Mg 0.8% Mn 0.1% Cr Al alloy: Extruded bar; cold drawn **UTS: 210 Elon: 12% Proof: 60**
ASTM B221/5456	5% Mg 0.8% Mn 0.1% Cr Al alloy: Extruded bar; cold drawn **UTS: 280 Elon: 12% Proof: 170**
ASTM B221/5652	2.4% Mg 0.2% Cr Al alloy: Extruded bar, etc.
ASTM B234/5052	2.5% Mg 0.2% Cr Al alloy: Tube; cold drawn **UTS: 220 Proof: 170**
ASTM B234/5454	2.8% Mg 0.7% Mn 0.1% Cr Al alloy: Tube; cold drawn **UTS: 270 Proof: 200**
ASTM B235/5052	2.5% Mg 0.2% Cr Al alloy: Tube; extruded **UTS: 220 Proof: 60**
ASTM B235/5154	3.5% Mg 0.2% Cr Al alloy: Tube; annealed **UTS: 220 Proof: 75**
ASTM B235/5254	3.5% Mg 0.2% Cr Al alloy: Extruded tube; annealed; free of copper **UTS: 220 Proof: 75**
ASTM B235/5454	2.6% Mg 0.7% Mn 0.1% Cr Al alloy: Extruded tube; annealed **UTS: 240 Elon: 14% Proof: 75**
ASTM B235/5456	5% Mg 0.7% Mn 0.1% Cr Al alloy: Extruded tube; hard drawn **UTS: 310 Elon: 12% Proof: 120**
ASTM B235/5652	2.5% Mg 0.2% Cr Al alloy: Extruded tube; annealed **UTS: 220 Proof: 60**
ASTM B241/5052	2.5% Mg 0.45% Si+Fe 0.25% Cr Al: Pipe and tube
ASTM B241/5083	4.5% Mg 0.7% Mn 0.15% Cr Al: Pipe and tube
ASTM B241/5154	3.5% Mg 0.2% Cr Al alloy: Pipe; cold drawn **UTS: 310 Proof: 220**
ASTM B241/5254	3.5% Mg 0.2% Cr Al alloy: Pipe; low copper; cold drawn **UTS: 310 Proof: 220**
ASTM B241/5454	2.7% Mg 0.7% Mn 0.1% Cr Al alloy: Pipe; annealed **UTS: 210 Elon: 12% Proof: 60**
ASTM B241/5456	5% Mg 0.7% Mn 0.1% Cr Al alloy: Pipe; annealed **UTS: 280 Elon: 12% Proof: 120**
ASTM B241/5652	2.5% Mg 0.40% Si+Fe 0.25% Cr Al alloy: Pipe and tube
ASTM B247/5083	4.5% Mg 0.7% Mn 0.15% Cr Al alloy: Die and hand forgings
ASTM B247/5456	5.1% Mg 0.8% Mn 0.15% Cr Al alloy: Die and hand forgings
ASTM B285 ER5050	1.5% Mg Al alloy: Electric welding rod; obsolete
ASTM B285 ER5052	2.5% Mg 0.2% Cr Al alloy: Welding rod; obsolete
ASTM B285 ER5154	3.5% Mg 0.2% Cr Al alloy: Welding wire
ASTM B285 ER5183	4.8% Mg 0.7% Mn 0.15% Cr Al alloy: Welding wire
ASTM B285 ER5254	3.5% Mg 0.2% Cr Al alloy: Welding wire; low Cu
ASTM B285 ER5356	5% Mg 0.1% Mn 0.1% Cr 0.1% Ti Al alloy: Welding wire
ASTM B285 ER5554	2.8% Mg 0.8% Mn 0.1% Cr 0.1% Ti Al alloy: Welding wire
ASTM B285 ER5556	5.1% Mg 0.7% Mn 0.1% Cr 0.1% Ti Al alloy: Welding wire
ASTM B285 ER5652	2.5% Mg 0.2% Cr Al alloy: Welding wire
ASTM B307/5050	1.5% Mg Al alloy: Coiled tube; annealed **UTS: 120**
ASTM B307/5052	2.5% Mg 0.25% Cr Al alloy: Drawn tube
ASTM B308/5083	4.5% Mg 0.7% Mn 0.1% Cr Al alloy: Section; annealed **UTS: 250 Elon: 16% Proof: 90**
ASTM B308/5086	4% Mg 0.5% Mn 0.1% Cr Al alloy: Section; as rolled or extruded **UTS: 240 Elon: 12% Proof: 110**
ASTM B308/5454	2.7% Mg 0.7% Mn 0.1% Cr Al alloy: Section; annealed **UTS: 210 Elon: 14% Proof: 60**
ASTM B308/5456	5% Mg 0.7% Mn 0.1% Cr Al alloy: Section; annealed **UTS: 280 Elon: 16% Proof: 120**

Symbol	Nominal analysis, supplier, condition and remarks.
ASTM B313/5050	1.5% Mg Al alloy: Welded tube; cold rolled **UTS: 270 Elon: 4%**
ASTM B313/5052	2.5% Mg 0.2% Cr Al alloy: Welded tube; cold rolled **UTS: 270 Elon: 4%**
ASTM B313/5086	4% Mg 0.5% Mn Al alloy: Welded tube; cold rolled **UTS: 330 Elon: 5% Proof: 250**
ASTM B313/5154	3.5% Mg 0.2% Cr Al alloy: Welded tube; cold rolled **UTS: 4% Proof: 240**
ASTM B316/5052	2.5% Mg 0.2% Cr Al alloy: Rivets; annealed **UTS: 210**
ASTM B316/5056	5% Mg 0.1% Mn 0.1% Cr Al alloy: Rivets; annealed **UTS: 310**
ASTM B316/5652	2.5% Mg 0.2% Cr Al alloy: Rivets; annealed **UTS: 210**
ASTM B318/5005	0.8% Mg Al alloy: Drawn coiled tubes; annealed **UTS: 120**
ASTM B318/5050	1.5% Mg Al alloy: Drawn coiled tube; annealed **UTS: 140**
ASTM B318/5052	2.5% Mg 0.2% Cr Al alloy: Drawn coiled tube; annealed **UTS: 220**
ASTM B345/5050	1.4% Mg Al alloy: Pipe; annealed **UTS: 110 Proof: 15**
ASTM B345/5052	2.5% Mg 0.2% Cr Al alloy: Pipe; annealed **UTS: 170 Proof: 45**
ASTM B345/5083	4.5% Mg 0.8% Mn 0.1% Cr Al alloy: Pipe; as drawn **UTS: 270 Elon: 12%**
ASTM B345/5086	4% Mg 0.5% Mn Al alloy: Pipe; annealed **UTS: 240 Elon: 14% Proof: 75**
ASTM B345/5154	3.5% Mg 0.2% Cr Al alloy: Pipe; annealed **UTS: 200 Elon: 12% Proof: 60**
ASTM B345/5456	5.1% Mg 0.8% Mn 0.1% Cr Al alloy: Pipe; annealed **UTS: 280 Elon: 16% Proof: 120**
ASTM B396/5805	0.8% Mg Al alloy: Electrical; wire
ASTM B404/5052	2.5% Mg 0.25% Cr Al alloy: Condenser tubes
ASTM B404/5454	2.7% Mg 0.75% Mn 0.12% Cr Al alloy: Condenser tubes
ASTM B531	0.7% Mg Al alloy: Rod for electrical purposes; 54% IACS conductivity
AWCO 07	7% Mg Al alloy: Wire; Aluminium Wire Co. for BS alloy N7
AWCO 21	2% Mg Al alloy: Wire; Aluminium Wire Co. for BS alloy N4
AWCO 27	3% Mg Al alloy: For wire; Aluminium Wire Co. for BS N5
AWCO 28	5% Mg Al alloy: Wire; Aluminium Wire Co. for BS alloy N6
AWCO 282	4.8% Mg 0.8% Mn 0.15% Cr Al alloy: Wire; Aluminium Wire Co.; for welding
AWCO 283	5.2% Mg 0.8% Mn 0.15% Cr 0.15% Ti Al alloy: Wire; Aluminium Wire Co.; for welding
AWCO 284	2.7% Mg 0.8% Mn 0.15% Cr 0.15% Ti Al alloy: Wire; Aluminium Wire Co.; for welding
AWCO 285	5.0% Mg 1.0% Mn 0.15% Cr 0.15% Ti Al alloy: Aluminium Wire Co.; for welding
BA 21	2% Mg Al alloy: Sheet and tube; British Aluminium Co. for BS alloy N4
BA 27	3% Mg Al alloy: Sheet, strip and tube; British Aluminium Co. for BS alloy N5
BA 28	5% Mg Al alloy: Sheet, strip and tube; British Aluminium Co. for BS alloy N6
BA 211	1.1% Mg Al alloy: Sheet and strip for bright anodizing; British Aluminium Co. for BT2
BA 212	0.6% Mg Al alloy: Sheet and strip for bright anodizing; British Aluminium Co. for BT1
BA 213	1.1% Mg Al alloy: Sheet and strip for bright anodizing; British Aluminium Co. for BT2
BA 218	2.0% Mg Al alloy: Sheet; British Aluminium Co.
BA 226	1.0% Mg Al alloy: Tube; British Aluminium Co.
BA 227	7% Mg Al alloy: Forgings; British Aluminium Co. for BS alloy N7

Symbol	Nominal analysis, supplier, condition and remarks.
BA 271	2.7% Mg 0.8% Mn 0.1% Cr Al alloy: Sheet and strip; British Aluminium Co.
BA 281	4.5% Mg 0.7% Mn Al alloy: British Aluminium Co. for BS alloy N8
BA 284	4.0% Mg 0.5% Mn 0.1% Cr Al alloy: Sheet and strip; British Aluminium Co.; ½-hard **UTS: 375**
BA 5052	2.5% Mg 0.25% Cr Al alloy: Sheet; British Aluminium Co.; ¼-hard; for transportation containers **Elon: 5.5% Proof: 240**
BA SP11	0.5% Mg Al alloy: Sheet and strip for bright anodizing; British Aluminium Co. for BT4
BA SP12	1% Mg Al alloy: Sheet and strip for bright anodizing; British Aluminium Co. for BT5
BB 1	1% Mg Al alloy: Tube; Birmabright; annealed **DPN: 40 UTS: 150 Elon: 25% Proof: 60**
BB 1	1% Mg Al alloy: Tube; Birmabright; hard **DPN: 60 UTS: 220 Elon: 7% Proof: 180**
BB 1 - X	1% Mg Al alloy: Sheet; Birmabright; annealed **DPN: 30 UTS: 110 Elon: 27% Proof: 50**
BB 1 - X	1% Mg Al alloy: Sheet; Birmabright; ¾-hard **DPN: 45 UTS: 200 Elon: 4% Proof: 180**
BB 2	2% Mg Al alloy: Sheet and tube; Birmabright for BS alloy N4
BB 3	3% Mg Al alloy: Sheet, strip and tube; Birmabright for BS alloy N5
BB 4	4.5% Mg 0.75% Mn Al alloy: Wrought form; Birmetal for BS alloy N8
BB 5	5% Mg Al alloy: Sheet, strip and tube; Birmabright for BS alloy N6
BB 5 - X	4.5% Mg 0.7% Mn Al alloy: Birmabright for BS alloy N8
BB 7	7% Mg Al alloy: Sheet, strip, tube and forging; Birmabright for BS alloy N7
BB 17	0.6% Mg Al alloy: Sheet and strip for bright anodizing; Birmabright for BT 1
BB 127	1.1% Mg Al alloy: Sheet and strip for bright anodizing; Birmabright for BT2
BS 477	Al alloy: Bar; replaced by BS 1476
BS 532	Al alloy: Forgings; replaced by BS 1472
BS 1453 NG5	3.5% Mg Al alloy: Welding rod; for welding Al alloys N4 and N5 type
BS 1453 NG6	5% Mg Al alloy: Welding rod; for welding Al alloys, cast and wrought with up to 5% Mg
BS 1470 5083	4.4% Mg 0.7% Mn 0.1% Cr Al alloy: Plate and sheet; H4 condition; replaces N8 designation **UTS: 160 Elon: 4% Proof: 100**
BS 1470 5154A	3.5% Mg 0.2% Cr + Mn Al alloy: Plate and sheet; H4 condition; replaces N5 designation **UTS: 365 Elon: 6% Proof: 270**
BS 1470 5251	2% Mg 0.3% Mn Al alloy: Plate and sheet; H6 condition; replaces N4 designation **UTS: 250 Elon: 5% Proof: 175**
BS 1470 NS4	2% Mg Al alloy: For sheet and strip; ½-hard **UTS: 220 Elon: 5% Proof: 170**
BS 1470 NS4	2% Mg Al alloy: For sheet and strip; ¼-hard **UTS: 200 Elon: 8% Proof: 120**
BS 1470 NS4	2% Mg Al alloy: For sheet and strip; annealed **UTS: 180 Elon: 18%**
BS 1470 NS5	3% Mg Al alloy: For sheet and strip; annealed **UTS: 210 Elon: 18%**
BS 1470 NS5	3% Mg Al alloy: For sheet and strip; ¼-hard **UTS: 240 Elon: 8% Proof: 150**
BS 1470 NS5	3% Mg Al alloy: For sheet and strip; ½-hard **UTS: 270 Elon: 5% Proof: 210**
BS 1470 NS6	5% Mg Al alloy: For sheet and strip; annealed **UTS: 250 Elon: 18% Proof: 120**
BS 1470 NS6	5% Mg Al alloy: For sheet and strip; ¼-hard **UTS: 280 Elon: 8% Proof: 210**
BS 1471 5154A	3.6% Mg Al alloy: drawn tube; H4 condition **UTS: 245 Elon: 4% Proof: 200**

Symbol	Nominal analysis, supplier, condition and remarks.
BS 1471 NT4	2% Mg Al alloy: Tube; ½-hard **UTS: 220 Elon: 5% Proof: 170**
BS 1471 NT4	2% Mg Al alloy: Tube; annealed **UTS: 170 Elon: 18%**
BS 1471 NT5	3% Mg Al alloy: Tube; annealed **UTS: 210 Elon: 18% Proof: 110**
BS 1471 NT5	3% Mg Al alloy: Tube; ½-hard **UTS: 240 Elon: 5% Proof: 180**
BS 1471 NT6	5% Mg Al alloy: Tube; annealed **UTS: 250 Elon: 18% Proof: 120**
BS 1471 NT6	5% Mg Al alloy: Tube; ½-hard **UTS: 260 Elon: 5% Proof: 210**
BS 1471 NT8	4.5% Mg 0.75% Mn Al alloy: Tube; ½-hard H4 condition **UTS: 310 Elon: 5% Proof: 230**
BS 1472 5154A	3.5% Mg Al alloy: forging; as forged **UTS: 215 Elon: 16% Proof: 100**
BS 1472 5251	2% Mg 0.3% Mn Al alloy: forging; as forged **UTS: 170 Elon: 16% Proof: 60**
BS 1472 NF4	2% Mg Al alloy: For forgings and bar; annealed **UTS: 170 Elon: 18% Proof: 75**
BS 1472 NF5	3% Mg Al alloy: For forgings and bar; annealed **UTS: 210 Elon: 18% Proof: 90**
BS 1472 NF6	5% Mg Al alloy: For forgings and bar; annealed **UTS: 250 Elon: 18% Proof: 120**
BS 1472 NF7	7% Mg Al alloy: For forgings and bar; annealed **UTS: 300 Elon: 18% Proof: 120**
BS 1472 NF8	4.5% Mg 0.75% Mn Al alloy: For forgings and bar; annealed **UTS: 270 Elon: 15% Proof: 120**
BS 1473 5056A	5% Mg 0.7% Mn Al alloy: Bolt and screw stock; H4 condition **UTS: 340 Proof: 240**
BS 1473 5154A	3.5% Mg Al alloy: Bolt and screw stock; annealed **UTS: 200**
BS 1473 NB6	5.0% Mg 0.3% Mn+Cr Al alloy: Rivet and screw stock; cold rolled **UTS: 350 Proof: 240**
BS 1473 NG7	7% Mg Al alloy: Welding rod; for welding Al cast and wrought alloys with high Mg
BS 1473 NR5	3% Mg Al alloy: For rivets; annealed **UTS: 210**
BS 1473 NR6	5% Mg Al alloy: For rivets; annealed **UTS: 240**
BS 1473 NR6	5% Mg Al alloy: For bolt stock; ½-hard **UTS: 310 Proof: 220**
BS 1474 5083	4.5% Mg 0.8% Mn 0.1% Cr Al alloy: Bar and extruded tube; as drawn **UTS: 280 Elon: 12% Proof: 130**
BS 1474 5154A	3.5% Mg Al alloy: Bar and extruded tube; as drawn **UTS: 215 Elon: 16% Proof: 100**
BS 1474 5251	2.1% Mg 0.3% Mn Al alloy: Bar and extruded tube; as drawn **UTS: 170 Elon: 16% Proof: 60**
BS 1474 NV4	2% Mg Al alloy: For tube; as drawn **UTS: 170 Elon: 18%**
BS 1474 NV5	3% Mg Al alloy: For tube; as drawn **UTS: 210 Elon: 18% Proof: 90**
BS 1474 NV6	5% Mg Al alloy: For tube; as drawn **UTS: 250 Elon: 18% Proof: 120**
BS 1474 NV8	4.5% Mg 0.75% Mn Al alloy: As drawn **UTS: 260 Elon: 16% Proof: 120**
BS 1475 5056	5% Mg 0.4% Mn Al alloy: Wire; condition H4; welding electrode **UTS: 340 Elon: 10%**
BS 1475 5154A	3.6% Mg Al alloy: Wire; welding electrode
BS 1475 5251	2% Mg 0.3% Mn Al alloy: Wire; condition H8 **UTS: 260**
BS 1475 5556A	5.5% Mg 0.9% Mn Al alloy: Wire; welding electrode
BS 1475 NG4	2% Mg Al alloy: Wire; annealed **UTS: 180**
BS 1475 NG4	2% Mg Al alloy: Wire; hard drawn **UTS: 250**
BS 1475 NG5	3.5% Mg Al alloy: Wire
BS 1475 NG6	5% Mg Al alloy: Wire; annealed **UTS: 260**
BS 1475 NG6	5% Mg Al alloy: Wire; ½-hard **UTS: 330**
BS 1475 NG6	5% Mg Al alloy: Wire; hard **UTS: 370**
BS 1476 NE4	2% Mg Al alloy: Bar; as rolled **UTS: 170 Elon: 18%**
BS 1476 NE5	3.5% Mg Al alloy: Bar; as rolled **UTS: 200 Elon: 18% Proof: 90**
BS 1476 NE6	5% Mg Al alloy: Bar; as rolled **UTS: 250 Elon: 18% Proof: 120**
BS 1476 NE8	4.5% Mg 0.75% Mn Al alloy: Bar; annealed **UTS: 250 Elon: 16% Proof: 120**
BS 1477 NP4	2% Mg Al alloy: Plate; as rolled **UTS: 180 Elon: 12%**
BS 1477 NP6	5% Mg Al alloy: Plate; annealed **UTS: 250 Elon: 20% Proof: 90**
BS 1477 NP8	4.5% Mg 0.75% Mn Al alloy: Plate; annealed **UTS: 160 Elon: 16% Proof: 120**
BS 1616 C	5% Mg Al alloy: Welding rod flux coated wire to BS alloy N6
BS 2901 NG4	2.2% Mg Al alloy: Rod for inert gas shielded arc welding
BS 2901 NG5	3.5% Mg Al alloy: Rod for all welding
BS 2901 NG6	5.0% Mg Al alloy: Rod for all welding
BS 2901 NG7	7.0% Mg Al alloy: Rod for inert gas shielded arc welding
BS 2901/5056 A	5.0% Mg Al alloy: Rod for gas shielded arc welding
BS 2901/5154 A	3.5% Mg Al alloy: Rod for gas shielded arc welding
BS 2901/5183	4.8% Mg Al alloy: Rod for gas shielded arc welding
BS 2901/5356	5% Mg Al alloy: Rod for gas shielded arc welding
BS 2901/5554	2.5% Mg Al alloy: Rod for gas shielded arc welding
BS 2901/5556 A	5.5% Mg Al alloy: Rod for gas shielded arc welding
BS 4300/1 5251	2% Mg 0.4% Mn Al alloy: Longitudinal welded tube **UTS: 245 Elon: 4% Proof: 220**
BS 4300/1 NJ4	2.2% Mg Al alloy: Longitudinal welded tube; as welded **UTS: 240 Elon: 3% Proof: 220**
BS 4300/1 NJ5	3.4% Mg Al alloy: Longitudinal welded tube; as welded **UTS: 290 Elon: 5% Proof: 250**
BS 4300/2 BTRS1	0.5% Mg Al alloy: For bright trim; ½-hard condition **UTS: 150**
BS 4300/2 BTRS2	0.09% Mg Al alloy: For bright trim; ½-hard condition **UTS: 170**
BS 4300/2 BTRS2	1.0% Mg Al alloy: For bright trim; annealed **UTS: 150**
BS 4300/6 3105	0.5% Mg 0.5% Mn Al alloy: replaces N31 designation; Plate and sheet; H4 condition **UTS: 180 Elon: 3% Proof: 145**
BS 4300/7 5005	0.8 Mg Al alloy: replaces N41 designation; Plate and sheet; H4 condition **UTS: 165 Elon: 4% Proof: 100**

Note. The following abbreviations and units are used in the tables:

DPN	Hardness, diamond pyramid number
UTS	Ultimate tensile strength, N/mm^2
Elon	Elongation, %
Proof	0.1% proof strength, N/mm^2

$1\ N/mm^2 = 0.1\ hbar = 0.102\ kgf/mm^2 = 0.06475\ tonf/in.^2 = 145.04\ lbf/in.^2 = 1\ MPa$

See Appendix II for other abbreviations and conversion tables.

Symbol	Nominal analysis, supplier, condition and remarks.
BS 4300/7 NS41	0.8% Mg Al alloy: Sheet and strip; ½-hard H4 condition **UTS: 160 Elon: 4% Proof: 100**
BS 4300/8 5454	2.7% Mg 0.8% Mn 0.1% Cr Al alloy: Plate and sheet; H4 condition **UTS: 300 Elon: 5% Proof: 200**
BS 4300/8 NS51	2.7% Mg 0.7% Mn 0.12% Cr Al alloy: Plate and sheet; ½-hard H4 condition **UTS: 300 Elon: 5% Proof: 200**
BS 4300/9 NG41	0.8% Mg Al alloy: Wire; hard H8 condition **UTS: 180**
BS 4300/10 NT51	2.7% Mg 0.7% Mn 0.12% Cr Al alloy: Drawn tube; ½-hard H4 condition **UTS: 300 Elon: 4% Proof: 220**
BS 4300/11 5454	2.7% Mg 0.7% Mn 0.1% Cr Al alloy: Forging; as forged **UTS: 270 Elon: 16% Proof: 90**
BS 4300/11 NF51	2.7% Mg 0.7% Mn 0.12% Cr Al alloy: Forging; as forged **UTS: 270 Elon: 18% Proof: 80**
BS 4300/12 5454	2.7% Mg 0.7% Mn 0.1% Cr Al alloy: Bar and extruded tube; as drawn **UTS: 215 Elon: 16% Proof: 100**
BS 4300/12 NE51	2.8% Mg 0.7% Mn 0.12% Cr Al alloy: Bar, tube, etc.; as drawn **UTS: 250 Elon: 17% Proof: 90**
BS 4300/13 5554	2.7% Mg 0.7% Mn Al alloy: Wire; welding electrode
BS 4300/13 NG52	2.8% Mg 0.7% Mn 0.12% Cr Al alloy: Welding wire
BS 4300/13 NG52	2.7% Mg 0.7% Mn 0.7% Cr 0.1% Ti Al alloy: Wire; for welding
BT 1	0.6% Mg Al alloy: For bright anodizing; annealed; designation given by Aluminium Federation **UTS: 120**
BT 1	0.6% Mg Al alloy: For bright anodizing; ¼-hard; designation given by Aluminium Federation **UTS: 140**
BT 1	0.6% Mg Al alloy: For bright anodizing; ½-hard; designation given by Aluminium Federation **UTS: 150**
BT 1	0.6% Mg Al alloy: For bright anodizing; hard; designation given by Aluminium Federation **UTS: 170**
BT 2	1.1% Mg Al alloy: For bright anodizing; annealed; designation given by Aluminium Federation **UTS: 170**
BT 2	1.1% Mg Al alloy: For bright anodizing; ½-hard; designation given by Aluminium Federation **UTS: 180**
BT 2	1.1% Mg Al alloy: For bright anodizing; hard; designation given by Aluminium Federation **UTS: 200**
BT 4	0.5% Mg Al alloy: For bright anodizing; annealed; designation given by Aluminium Federation **UTS: 90**
BT 4	0.5% Mg Al alloy: For bright anodizing; ¼-hard; designation given by Aluminium Federation **UTS: 110**
BT 4	0.5% Mg Al alloy: For bright anodizing; ½-hard; designation given by Aluminium Federation **UTS: 120**
BT 4	0.5% Mg Al alloy: For bright anodizing; hard; designation given by Aluminium Federation **UTS: 15;**
BT 5	1% Mg Al alloy: For bright anodizing; annealed; designation given by Aluminium Federation **UTS: 120**
BT 5	1% Mg Al alloy: For bright anodizing; ¼-hard; designation given by Aluminium Federation **UTS: 140**
BT 5	1% Mg Al alloy: For bright anodizing; ½-hard; designation given by Aluminium Federation **UTS: 170**
BT 5	1% Mg Al alloy: For bright anodizing; hard; designation given by Aluminium Federation **UTS: 180**
DTD 175	Mg Al alloy: Obsolete
DTD 180B	3% Mg Al alloy: Sheet; ¼-hard **UTS: 250 Elon: 8% Proof: 170**
DTD 182B	7% Mg Al alloy: Sheet and strip; annealed **UTS: 330 Elon: 18%**
DTD 186B	7% Mg Al alloy: Tube; ½-hard **UTS: 370 Elon: 5% Proof: 260**
DTD 190	Obsolete. No information available
DTD 297A	7% Mg Al alloy: Bar and forging; annealed **UTS: 300 Elon: 18% Proof: 140**
DTD 303	5% Mg Al alloy: Wire as drawn; replaced by L 58 **UTS: 260**
DTD 310	2% Mg Al alloy: Replaced by L 56
DTD 440	2% Mg Al alloy: Tube; ½-hard; cold drawn **UTS: 220**
DTD 606A	2% Mg Al alloy: Sheet; ½-hard; cold rolled **UTS: 220 Elon: 15% Proof: 180**
DTD 634A	2% Mg Al alloy: Sheet and strip; annealed **UTS: 180 Elon: 18%**
DURALBRITE 4	2.5% Mg Al alloy: Extrusion; Alcan; hard drawn **UTS: 255**
ELM 8	2% Mg Al alloy: Strip and welded tube; Elm Engineering **UTS: 220 Elon: 5% Proof: 170**
ELM 10	2% Mg Al alloy: Strip and welded tube; Elm Engineering **UTS: 260 Elon: 4% Proof: 220**
ERMAL NS4	2% Mg Al alloy: Sheet and strip; Enfield specification for BS 1470 NS4
G 1A	1.5% Mg Al alloy: Previous ASTM designation, now 5050
G 1B	0.8% Mg Al alloy: Previous ASTM designation, now 5005
GM 31A	2.8% Mg 0.8% Mn 0.1% Cr Al alloy: Previous ASTM designation, now 5454
GM 40A	4% Mg 0.5% Mn 0.2% Cr Al alloy: Previous ASTM designation, now 5086
GM 41A	4.5% Mg 0.7% Mn 0.15% Cr Al alloy: Previous ASTM designation, now 5083
GM 50A	5% Mg 0.15% Mn 0.1% Cr Al alloy: Previous ASTM designation, now 5056
GM 51A	5% Mg 0.7% Mn 0.15% Cr Al alloy: Previous ASTM designation, now 5456
GR 20A	0.1% Cu 2.5% Mg 0.3% Cr Al alloy: Previous ASTM designation, now 5052
GR 20B	2.5% Mg 0.3% Cr Al alloy: Previous ASTM designation, now 5652
GR 40A	3.5% Mg 0.2% Cr Al alloy: Previous ASTM designation, now 5154
GR 40B	3.5% Mg 0.2% Cr Al alloy: Previous ASTM designation, now 5254
HA 4 GM 31N	2.7% Mg 0.8% Mn 0.1% Cr Al alloy: Plate; Canadian Standard
HA 4 GM41	4.5% Mg 0.7% Mn 0.1% Cr Al alloy: Plate; Canadian Standard
HA 4 GR20	2.6% Mg 0.25% Cr Al alloy: Plate; Canadian Standard
HA 5 GM41	4.5% Mg 0.7% Mn 0.1% Cr Al alloy: Bar, etc.; Canadian Standard
HA 5 GR20	2.6% Mg 0.2% Cr Al alloy: Bar, etc.; Canadian Standard
HA 6 GM50R	5.0% Mg 0.1% Mn 0.1% Cr Al alloy: For rivets and brazing wire; Canadian Standard
HA 6 GS11P	1.2% Mg 0.2% Cr Al alloy: For rivets and brazing wire; Canadian Standard
HA 7 GR20	2.5% Mg Al alloy: Tube; Canadian Standard
HIDUMINIUM 05	5% Mg Al alloy: For tubes and extrusions; High Duty Alloys for BS alloy N6

Symbol	Nominal analysis, supplier, condition and remarks.
HIDUMINIUM 07	7% Mg Al alloy: For forgings and tubes; High Duty Alloys for BS alloy N7
HIDUMINIUM 12	1.3% Mg Al alloy: Sheet; High Duty Alloys; annealed **UTS: 150 Elon: 25% Proof: 50**
HIDUMINIUM 14	1% Mg 0.5% Mn Al alloy: Tube; High Duty Alloys; hard drawn **UTS: 210 Elon: 6% Proof: 170**
HIDUMINIUM 16	0.6% Mg Al alloy: Sheet and strip for bright anodizing; High Duty Alloys for BT 1
HIDUMINIUM 17	1.1% Mg Al alloy: Sheet and strip for bright anodizing; High Duty Alloys for BT 2
HIDUMINIUM 22	2% Mg Al alloy: For extrusions and tubes; High Duty Alloys for BS alloy N4
HIDUMINIUM 24	2.5% Mg Al alloy: Sheet; High Duty Alloys; annealed **UTS: 200 Elon: 18%**
HIDUMINIUM 33	3% Mg Al alloy: For extrusions and tubes; High Duty Alloys for BS alloy N5
HIDUMINIUM 35	4.5% Mg 0.8% Mn Al alloy: High Duty Alloys for BS alloy N8
IMPALCO 510	2% Mg Al alloy: Sheet, tube, bar, etc.; Imperial Aluminium Co. for BS alloy N4
IMPALCO 520	3.5% Mg 1% Mn Al alloy: Sheet, tube, bar, etc.; Imperial Aluminium Co. for BS alloy N5
IMPALCO 530	4.5% Mg 1% Mn Al alloy: Plate; Imperial Aluminium Co. for BS alloy N5/6
IMPALCO 531	4.4% Mg Al alloy: Plate, bar and forgings; Imperial Aluminium Co. for BS alloy N8
IMPALCO 540	1% Mg Al alloy: Sheet and bar; Imperial Aluminium Co.; annealed **UTS: 150**
IMPALCO 540	1% Mg Al alloy: Sheet and bar; Imperial Aluminium Co.; ½-hard **UTS: 180**
IMPALCO 540	1% Mg Al alloy: Sheet and bar; Imperial Aluminium Co.; ½-hard **UTS: 200**
IMPALCO 550	0.5% Mg Al alloy: Sheet; Imperial Aluminium Co.; annealed **UTS: 120**
IMPALCO 550	0.5% Mg Al alloy: Sheet; Imperial Aluminium Co.; ½-hard **UTS: 140**
IMPALCO 560	5% Mg 0.5% Cr Al alloy: Sheet and bar; Imperial Aluminium Co. for BS alloy N6
IMPALCO 570	7% Mg Al alloy: Sheet and bar; Imperial Aluminium Co. for BS alloy N7
IMPALCO 900	1% Mg 0.2% Mn Al alloy: Tube; Imperial Aluminium Co.; annealed **UTS: 120 Elon: 20%**
IMPALCO 900	1% Mg 0.2% Mn Al alloy: Tube; Imperial Aluminium Co.; ½-hard **UTS: 200 Elon: 6% Proof: 150**
IMPALCO 900	1% Mg 0.2% Mn Al alloy: Tube; Imperial Aluminium Co.; hard **UTS: 220 Elon: 4% Proof: 200**
IMPALCO 901	1% Mg Al alloy: Tube; Imperial Aluminium Co.; as drawn **UTS: 140 Elon: 25% Proof: 75**
IMPALCO M31	0.6% Mg Al alloy: Sheet and strip for bright anodizing; Imperial Aluminium Co. for BT1
IMPALCO M32	10% Mg Al alloy: For sheet and strip; ¼-hard; Imperial Aluminium Co. **UTS: 140 Elon: 12% Proof: 110**
IMPALCO M32	1% Mg Al alloy: Sheet and strip; annealed; Imperial Aluminium Co. for high reflectivity **UTS: 140 Elon: 20%**
IMPALCO M32	1% Mg Al alloy: For sheet and strip; hard; Imperial Aluminium Co. **UTS: 180 Elon: 4% Proof: 140**
IMPALCO M32X	1% Mg Al alloy: For sheet and strip; annealed; Imperial Aluminium Co. Similar to M32 **UTS: 140 Elon: 20%**
IMPALCO M32X	1% Mg Al alloy: For sheet and strip; ¼-hard; Imperial Aluminium Co. **UTS: 140 Elon: 12% Proof: 120**
IMPALCO M32X	1% Mg Al alloy: For sheet and strip; hard; Imperial Aluminium Co. **UTS: 180 Elon: 4% Proof: 140**
IMPALCO M34	0.05% Cu 1% Mg Al alloy: For tube; hard; Imperial Aluminium Co. **DPN: 70 UTS: 210 Elon: 6% Proof: 180**
IMPALCO M34	0.05% Cu 1% Mg Al alloy: For tube; annealed; Imperial Aluminium Co. **DPN: 36 UTS: 130 Elon: 22%**
IMPALCO M35/1	2% Mg Al alloy: For sheet, strip, tube, wire, plate and sections; Imperial Aluminium Co. for BS alloy N4
IMPALCO M35/2	0.1% Cu 3.5% Mg Al alloy: For sheet, strip, tube, wire, bar and sections; Imperial Aluminium Co. for BS alloy N5
IMPALCO M35/2	4% Mg 1% Mn Al alloy: For plate; Imperial Aluminium Co. for BS alloy N5/6
IMPALCO M36	5% Mg 1% Mn Al alloy: For sheet, strip, wire, bar, rod and sections; Imperial Aluminium Co. for BS alloy N6
KYNAL M35/1	2% Mg Al alloy: ICI; obsolete; now Imperial Aluminium Co.
KYNAL M35/2	3% Mg Al alloy: ICI; obsolete; now Imperial Aluminium Co.
KYNAL M36	5% Mg Al alloy: ICI; obsolete; now Imperial Aluminium Co.
KYNAL M37	7% Mg Al alloy: ICI; obsolete; now Imperial Aluminium Co.
L44	2% Mg Al alloy: Bar; as rolled **UTS: 170 Elon: 18%**
L46	Mg Al alloy: Obsolete
L55	2% Mg Al alloy: Tube; ½-hard; cold drawn **UTS: 220**
L56	2% Mg Al alloy: Tube; annealed **UTS: 210**
L58	5% Mg Al alloy: Wire; as drawn **UTS: 260**
L80	2.25% Mg Al alloy: Sheet; annealed **UTS: 180 Elon: 18%**
L81	2.25% Mg Al alloy: Sheet; ½-hard **UTS: 220 Elon: 5% Proof: 170**
L82	3.5% Mg Al alloy: Sheet; annealed **UTS: 210 Elon: 18%**
MG 2	2% Mg Al alloy: Sheet and tube; J Booth for BS alloy N4
MG 3	3% Mg Al alloy: Sheet, strip and tube; J Booth for BS alloy N5
MG 5	5% Mg Al alloy: Sheet, strip and tube; J Booth for BS alloy N6
MG 5S	4.5% Mg 0.7% Mn 0.15% Cr Al alloy: Tube forgings; J Booth; annealed **UTS: 260 Elon: 16% Proof: 120**
MG 7	7% Mg Al alloy: Sheet, strip and tube forging; J Booth for BS alloy N7
N4	2% Mg Al alloy: Wrought; designation used by BS
N5	3.5% Mg Al alloy: Wrought; designation used by BS
N6	5% Mg Al alloy: Wrought; designation used by BS
N7	7% Mg Al alloy: Wrought; designation used by BS

Note. The following abbreviations and units are used in the tables:

DPN	Hardness, diamond pyramid number
UTS	Ultimate tensile strength, N/mm^2
Elon	Elongation, %
Proof	0.1% proof strength, N/mm^2

1 N/mm^2=0.1 hbar=0.102 kgf/mm^2=0.06475 tonf/in.^{2}=145.04 lbf/in.^{2}=1 MPa

See Appendix II for other abbreviations and conversion tables.

Symbol	Nominal analysis, supplier, condition and remarks.
N8	4.5% Mg 0.75% Mn Al alloy: Wrought; designation used by BS
N41	0.8% Mg Al alloy: Wrought; designation used by BS
N51	2.7% Mg 0.7% Mn 0.12% Cr Al alloy: Wrought; designation used by BS
N52	2.8% Mg 0.7% Mn 0.12% Cr Al alloy: Wrought; designation used by BS
NORAL 54S	3% Mg Al alloy: Sheet, strip and tube; Northern Aluminium Co. for BS alloy N5
NORAL 58S	7% Mg Al alloy: Sheet, strip and tube forging; Northern Aluminium Co. for BS alloy N7
NORAL A56S	5% Mg Al alloy: Sheet, strip and tube; Northern Aluminium Co. for BS alloy N6
NORAL M57S	2% Mg Al alloy: Sheet and tube; Northern Aluminium Co. for BS alloy N4
P 3G	3.1% Mg 0.3% Si 0.2% Mn 0.1% Ti Al alloy: Casting; Anglo-Swiss code for Peraluman 3G
P 5G	5.2% Mg 0.9% Si 0.3% Mn 0.1% Ti Al alloy: Casting; Anglo-Swiss code for Peraluman 5G
P 9G	0.2% Mg 0.3% Mn 0.025% Be Al alloy: Casting; Anglo-Swiss code for Peraluman 9G
Pe 15	1.5% Mg 0.2% Mn Al alloy: Wrought; Anglo-Swiss code for Peraluman 15
Pe 30	2.8% Mg 0.4% Mn Al alloy: Wrought; Anglo-Swiss code for Peraluman 30
Pe 40	4% Mg 0.4% Mn Al alloy: Wrought; Anglo-Swiss code for Peraluman 40
Pe 50	0.4% Cu 5.2% Mg 0.4% Mn Al alloy: Wrought; Anglo-Swiss code for Peraluman 50
PERALUMAN 3G	3.1% Mg 0.3% Si 0.2% Mn 0.1% Ti Al alloy: Casting; Anglo-Swiss solution treated and aged; obsolete **DPN: 60 UTS: 210 Elon: 6% Proof: 140**
PERALUMAN 5G	5.2% Mg 0.9% Si 0.3% Mn 0.1% Ti Al alloy: Casting; Anglo-Swiss as die cast; obsolete **DPN: 70 UTS: 210 Elon: 5% Proof: 110**
PERALUMAN 9G	9.2% Mg 0.3% Mn 0.025% Be Al alloy: Casting; Anglo-Swiss as die cast; obsolete **DPN: 70 UTS: 220 Elon: 2% Proof: 140**
PERALUMAN 15	1.5% Mg 0.2% Mn Al alloy: Wrought; Anglo-Swiss; cold drawn; obsolete **DPN: 70 UTS: 220 Elon: 6% Proof: 210**
PERALUMAN 30	2.8% Mg 0.4% Mn Al alloy: Wrought; Anglo-Swiss; hard drawn; obsolete **DPN: 85 UTS: 280 Elon: 9% Proof: 250**
PERALUMAN 40	4% Mg 0.4% Mn Al alloy: Wrought; Anglo-Swiss; cold drawn; obsolete **DPN: 90 UTS: 310 Elon: 9% Proof: 280**
PERALUMAN 50	0.4% Cu 5.2% Mg 0.4% Mn Al alloy: Wrought; Anglo-Swiss; hard drawn; obsolete **DPN: 100 UTS: 360 Elon: 9% Proof: 300**
PERALUMAN 100	0.8% Mg 0.2% Mn 0.2% Cr Al alloy: Wrought; Anglo-Swiss Aluminium Co.
PERALUMAN 150	1.6% Mg 0.2% Mn Al alloy: Wrought; Anglo-Swiss Aluminium Co.
PERALUMAN 200	2% Mg 0.3% Mn Al alloy: Wrought; Anglo-Swiss Aluminium Co.
PERALUMAN 260	2.7% Mg 0.8% Mn 0.1% Cr Al alloy: Wrought; Anglo-Swiss Aluminium Co.
PERALUMAN 260	2.6% Mg 0.8% Mn 0.15% Cr Al alloy: Wrought; Anglo-Swiss Aluminium Co.
PERALUMAN 300	2.9% Mg 0.3% Mn Al alloy: Wrought; Anglo-Swiss Aluminium Co.
PERALUMAN 350	3.5% Mg Al alloy: Wrought; Anglo-Swiss Aluminium Co.
PERALUMAN 400	4.0% Mg 0.3% Mn Al alloy: Wrought; Anglo-Swiss Aluminium Co.
PERALUMAN 460	4.3% Mg 0.7% Mn 0.15% Cr Al alloy: Wrought; Anglo-Swiss Aluminium Co.
PERALUMAN 460	4.5% Mg 0.8% Mn 0.2% Cr Al alloy: Wrought; Anglo-Swiss Aluminium Co.
PERALUMAN 500	5% Mg 0.3% Mn Al alloy: Wrought; Anglo-Swiss Aluminium Co.
QQ A 315	2.4% Mg 0.2% Cr Al alloy: Bar; US Federal; annealed **UTS: 180 Elon: 25% Proof: 75**
QQ A 318	2.5% Mg 0.3% Cr Al alloy: Plate; US Federal; cold rolled **UTS: 250 Elon: 2% Proof: 180**
REFLECTAL 050	0.5% Mg Al alloy: For anodizing; Anglo-Swiss Aluminium Co.
REFLECTAL 100	1.0% Mg Al alloy: For anodizing; Anglo-Swiss Aluminium Co.
SAE 201	2.5% Mg 0.25% Cr Al alloy: Wrought; annealed; now AA 5652 **UTS: 210 Elon: 19%**
SAE 201	2.5% Mg 0.25% Cr Al alloy: Wrought; hard; now AA 5052 **UTS: 250 Elon: 4%**
SAE 207	1.5% Mg Al alloy: Cold rolled; now AA 5050 **DPN: 63 UTS: 210 Elon: 6% Proof: 180**
SAE 208	3.5% Mg 0.2% Cr Al alloy: Cold rolled; now AA 5154 **DPN: 80 UTS: 330 Elon: 10% Proof: 280**
SAE 209	1% Mg 0.3% Mn Al alloy: Wrought; annealed; now AA 5357 **UTS: 110 Elon: 20%**
SAE 251	2.7% Mg 0.7% Mn 0.1% Cr Al alloy: Wrought; annealed; now AA 5454 **UTS: 260 Elon: 18% Proof: 60**
SAE 252	1% Mg 0.2% Mn Al alloy: Wrought; annealed; now AA 5457 **UTS: 140 Elon: 20%**
SOUDOTEG Al Mg	4.7% Mg 0.8% Mn Al alloy: Welding electrode for inert gas welding; Soudometal **UTS: 280 Elon: 20% Proof: 120**
SOUDOTEG Al Mg 5	5% Mg 0.1% Ti 0.2% Fe Al alloy: Welding electrode for inert gas welding; Soudometal **UTS: 240 Elon: 18% Proof: 90**
SONDOR G Al Mg	4.7% Mg 0.5% Mn Al alloy: Welding electrode; Soudometal **UTS: 280 Elon: 20% Proof: 120**
SONDOR G Al Mg 5	5% Mg 0.2% Fe Al alloy: Welding electrode; Soudometal **UTS: 240 Elon: 18% Proof: 90**
SS 4106	0.9% Mg Al alloy: Sheet and strip; Swedish Standard; annealed **DPN: 30 UTS: 110 Elon: 25% Proof: 45**
SS 4106	0.9% Mg Al alloy: Sheet and strip; Swedish Standard; cold rolled **DPN: 60 UTS: 150 Elon: 3% Proof: 120**
SS 4120	2.3% Mg Al alloy: Bar, sheet and strip; Swedish Standard; annealed **DPN: 55 UTS: 180 Elon: 18% Proof: 45**
SS 4120	2.3% Mg Al alloy: Bar, sheet and strip; Swedish Standard; cold worked **DPN: 85 UTS: 250 Elon: 3% Proof: 210**
SS 4163	5% Mg 1% Si Al alloy: Casting; Swedish Standard; as cast **DPN: 60**
STA 7 AW4A	2% Mg Al alloy: Bar; replaced by BS 1476 NE4
STA 7 AW4B	2% Mg Al alloy: Tube, replaced by BS 1471 NT4
STA 7 AW4C	2% Mg Al alloy: Sheet; replaced by BS 1470 NS4
STA 7 AW5A	3% Mg Al alloy: Bar; replaced by BS 1476 NE5
STA 7 AW5B	3% Mg Al alloy: Tube; replaced by BS 1471 NT5
STA 7 AW5C	3% Mg Al alloy: Sheet; replaced by BS 1470 NS5
STA 7 AW6A	5% Mg Al alloy: Bar; replaced by BS 1476 NE6
STA 7 AW6B	5% Mg Al alloy: Tube; replaced by BS 1471 NT6
STA 7 AW6C	5% Mg Al alloy: Sheet; replaced by BS 1470 NS6
STA 7 AW6D	5% Mg Al alloy: Wire; replaced by BS 1473–5 NR6 NV6 NG6
STA 7 AW7A	7% Mg Al alloy: Forging and bar; replaced by BS 1472 NF7
STA 7 AW7B	7% Mg Al alloy: Tube; replaced by BS 1471 NT7

Symbol	Nominal analysis, supplier, condition and remarks.
STA 7 AW7C	7% Mg Al alloy: Replaced by BS 1470 NS7
STUBS NG 6	5% Mg Al alloy: For electric arc welding; Stubs
STUBS NG 61	5.3% Mg Al alloy: For electric arc welding; Stubs
TI 05	5% Mg Al alloy: For tube; British Aluminium Co. for BS alloy N6
TI 07	7% Mg Al alloy: Tube and forging; British Aluminium Co. for BS alloy N7
TI 22	2% Mg Al alloy: Sheet and tube; British Aluminium Co. for BS alloy N4
TI 33	3% Mg Al alloy: Sheet, strip and tube; British Aluminium Co. for BS alloy N5
UNI 3573	1.5% Mg Al alloy: For forging; Italian Standard
UNI 3574	2.5% Mg Al alloy: For forging; Italian Standard
UNI 3575	3.5% Mg Al alloy: For forging; Italian Standard
UNI 3576	5.0% Mg Al alloy: For forging; Italian Standard
UNI 4510	0.9% Mg Al alloy: Sheet for decorative use; Italian Standard
UNI 4511	2.0% Mg Al alloy: For forging; Italian Standard
UNI 4512	0.5% Mg Al alloy: For decorative use; Italian Standard
UNI 5452	4.4% Mg Al alloy: For forging; Italian Standard
UNI 5764	0.8% Mg Al alloy: Italian Standard
UNI 5784	0.8% Mg Al alloy: For forging; Italian Standard
UNI 6360	0.1% Cu 0.9% Mg Al alloy: For decorative purposes; Italian Standard
UNI 7789	2.7% Mg 0.7% Mn 0.1% Cr Al alloy: Italian Standard
UNI 7790	4.4% Mg 0.7% Mn 0.1% Cr Al alloy: Italian Standard
UV Al Mg	10% Mg Al alloy: Wrought; previous German Standard designation

Symbol	Nominal analysis, supplier, condition and remarks.
W 4	2.2% Mg Al alloy: Sheet, bar, wire, etc.; Southern Forge; replaced by IMPALCO 510
W 5	3.5% Mg Al alloy: Sheet, bar, etc.; Southern Forge; replaced by IMPALCO 520
W 6	5% Mg 1% Mn Al alloy: Sheet, bar and wire; Southern Forge; replaced by IMPALCO 560
W 010	1% Mg Al alloy: Sheet, bar and forging; Southern Forge; replaced by IMPALCO 540
W 011	1% Mg Al alloy: Southern Forge; replaced by IMPALCO 900
W 012	1% Mg Al alloy: Southern Forge; replaced by IMPALCO 901
WWT 787	2.4% Mg 0.2% Cr Al alloy: Tube; US Federal; annealed **UTS: 220** **Proof: 120**

Note. The following abbreviations and units are used in the tables:

DPN	Hardness, diamond pyramid number
UTS	Ultimate tensile strength, N/mm^2
Elon	Elongation, %
Proof	0.1% proof strength, N/mm^2

1 N/mm^2=0.1 hbar=0.102 kgf/mm^2=0.06475 tonf/in.^{2}=145.04 lbf/in.^{2}=1 MPa

See Appendix II for other abbreviations and conversion tables.

1E Aluminium–magnesium–silicon wrought alloys

Specific gravity		2.7
Density		2700 kg/m^3
Solidus/liquidus		552–652 °C
Thermal conductivity	annealed	190–210 W/m °C
	solution treated and aged	170–200 W/m °C
Coefficient of linear expansion		23–25 × 10^{-6}/ °C
Electrical conductivity	solution treated and aged	40–55% IACS (copper 100%)
Specific resistance	annealed	32–38 microhm mm
	solution treated and aged	36–45 microhm mm
Young's modulus of elasticity		65.5–68.9 × 10^9 N/m^2
Impact	solution treated and aged	27–41 J
Fatigue strength (10^8 cycles)	solution treated and aged	±80–120 N/mm^2
Hot strength		

Temperature °C	Tensile strength* N/mm^2	Elongation* %
100	300	21
150	250	24
200	205	21
250	120	23
315	47	38
350	37	48
400	30	57

*The values shown apply to the 1% Mg 1% Si 0.7% Mn alloys. Other alloys have lower values above 150 °C.

The above properties are typical of the following specifications, and may not apply exactly to any one. It is possible that some of the values are inapplicable.

General metallurgical characteristics

After the addition of between 0.5% and 1.0% magnesium and 0.5% and 1.0% silicon, with or without copper but generally with some chromium, the mechanical properties of aluminium alloys can be improved by heat treatment.

This heat treatment is a solution treat and age (precipitation hardening). The alloys listed all have good working properties and can be forged, extruded and drawn. When heat treated there is a marked increase in tensile strength and yield strength with some reduction in elongation but there is less reduction in corrosion resistance with these alloys in relation to the increase in mechanical strength than for any other range of aluminium alloys.

With the addition of small amounts of copper, chromium and manganese the mechanical properties can be enhanced.

These materials can be welded using the inert gas shielded method. This will affect the mechanical properties if the alloys have been heat treated but can be carried out prior to heat treatment when little or no decrease in mechanical properties will be found.

These alloys find use for architectural purposes, for window glazing bars and supports. The structural members of many transport vehicles such as railways, trucks, etc. are manufactured in these alloys as are tubular furniture and similar items.

Their corrosion resistance can be improved dramatically using anodizing and, to a lesser extent, by chromate treatment.

Symbol	Nominal analysis, supplier, condition and remarks.
2EC	0.5% Mg 0.5% Si Al alloy: For electrical purposes; AA designation
3.0615	Mg Si Pb Al alloy: German Standard DIN 1725
3.2305	0.6% Mg 0.5% Si Al alloy: German Standard DIN 1725
3.2315	0.7% Mg 1% Si 0.6% Mn Al alloy: German Standard DIN 1725
3.3205	0.7% Mg 0.4% Si Al alloy: German Standard DIN 1725
3.3206	0.7% Mg 0.6% Si Al alloy: Wrought; German Standard DIN 1725
3.3241	3% Mg 1% Si 0.3% Mn 0.1% Ti Al alloy: Casting; German Standard DIN 1725
3.3243	3% Mg 0.7% Si 0.3% Mn 0.1% Ti Al alloy: Casting; German Standard DIN 1725 **DPN: 60 UTS: 170 Elon: 5% Proof: 75**
3.3245	3% Mg 0.6% Si 0.6% Mn 0.2% Cr Al alloy: Wrought; German Standard DIN 1725 **DPN: 45 UTS: 170 Elon: 17% Proof: 75**
62S	0.2% Cu 1% Mg 0.6% Si 0.1% Cr Al alloy: Previous AA designation now 6062
63S	0.75% Mg 0.5% Si Al alloy: Wire; Alcoa for BS alloy H9
63S	0.8% Mg 0.4% Si Al alloy: Previous AA designation now 6063
4009	1.2% Cu 0.5% Mg 5% Si Al alloy: Wrought; designation used in the UK and USA
4011	0.5% Mg 7% Si Al alloy: Wrought; designation used in the UK and USA
4013	0.1% Cu 0.1% Mg 4% Si 1% Bi Al alloy: Wrought; designation used in the UK and USA
4032	1% Cu 1% Mg 12% Si 1% Ni Al alloy: Wrought; designation used in the UK and USA
6003	1.2% Mg 0.6% Si Al alloy: Wrought; designation used in the UK and USA
6004	0.5% Mg 0.5% Si 0.4% Mn Al alloy: Wrought; designation used in the UK and USA
6005	0.5% Mg 0.8% Si Al alloy: Wrought; designation used in the UK and USA
6006	0.2% Cu 0.5% Mg 0.4% Si 0.1% Mn Al alloy: Wrought; designation used in the UK and USA
6007	0.7% Mg 1% Si 0.1% Mn 0.1% Cr Al alloy: Wrought; designation used in the UK and USA
6009	0.4% Cu 0.6% Mg 0.8% Si 0.6% Mn Al alloy: Wrought; designation used in the UK and USA
6010	0.4% Cu 0.8% Mg 1% Si 0.6% Mn Al alloy: Wrought; designation used in the UK and USA
6011	0.6% Cu 0.9% Mg 0.9% Si Al alloy: Wrought; designation used in the UK and USA
6013	0.8% Cu 0.9% Mg 0.8% Si 0.6% Mn Al alloy: Wrought; designation used in the UK and USA
6017	0.1% Cu 0.5% Mg 0.6% Si 0.2% Fe Al alloy: Wrought; designation used in the UK and USA
6053	1.2% Mg 0.5% Si 0.2% Cr Al alloy: Wrought; designation used in the UK and USA
6060	0.5% Mg 0.5% Si Al alloy: Wrought; designation used in the UK and USA
6061	0.3% Cu 1% Mg 0.6% Si 0.2% Cr Al alloy: Wrought; designation used in the UK and USA
6101	0.5% Mg 0.5% Si Al alloy: Wrought; designation used by AA; no ASTM alloy listed
6105	0.6% Mg 0.8% Si Al alloy: Wrought; designation used in the UK and USA
6110	0.5% Cu 0.8% Mg 1.2% Si 0.2% Cr Al alloy: Wrought; designation used in the UK and USA
6111	0.7% Mg 0.9% Si Al alloy: Wrought; designation used in the UK and USA
6151	0.6% Mg 0.9% Si 0.2% Cr Al alloy: Wrought; designation used in the UK and USA
6162	0.9% Mg 0.6% Si Al alloy: Wrought; designation used in the UK and USA
6201	0.8% Mg 0.7% Si Al alloy: Wrought; designation used in the UK and USA
6205	0.5% Mg 0.7% Si 1% Mn 0.1% Cr Al alloy: Wrought; designation used in the UK and USA
6253	1.2% Mg 0.6% Si 0.2% Cr Al: Cladding alloy; designation used in the UK and USA
6261	0.2% Cu 0.9% Mg 0.5% Si Al alloy; Obsolete; designation used in the UK and USA
6262	1% Mg 0.6% Si 0.1% Cr 0.5% Bi 0.5% Pb Al alloy: Wrought; free machining; designation used in the UK and USA
6301	0.7% Mg 0.7% Si Al alloy: Wrought; designation used in the UK and USA
6351	0.6% Mg 1% Si 0.6% Mn Al alloy: Wrought; designation used in the UK and USA
6463	0.1% Cu 0.6% Mg 0.4% Si Al alloy: Wrought; designation used in the UK and USA
6563	0.2% Cu 0.7% Mg 0.4% Si Al alloy: Designation used by AA; no ASTM alloy listed
6763	0.1% Cu 0.6% Mg 0.4% Si Al alloy: Wrought; designation used in the UK and USA
6961	0.6% Mg 0.35% Si 0.2% Cr Al alloy: Wrought; designation used in the UK and USA
A 51S	0.75% Mg 1% Si Al alloy: Forgings; Alcoa
AC	0.8% Mg 1% Si Al alloy: Anglo-Swiss code for Anticorodal

Note. The following abbreviations and units are used in the tables:

DPN	Hardness, diamond pyramid number
UTS	Ultimate tensile strength, N/mm^2
Elon	Elongation, %
Proof	0.1% proof strength, N/mm^2

1 N/mm^2=0.1 hbar=0.102 kgf/mm^2=0.06475 tonf/in.^{2}=145.04 lbf/in.^{2}=1 MPa

See Appendix II for other abbreviations and conversion tables.

Symbol	Nominal analysis, supplier, condition and remarks.
AD 3	0.2% Cu 1% Mg 0.6% Si 0.2% Cr Al alloy: Russian designation
AD 31	0.6% Mg 0.4% Si Al alloy: Russian designation
AGG 51	0.4% Mg 0.55% Si Al alloy: Wrought; further information from Aluminium Zentrale
AGS	0.8% Mg 0.8% Si Al alloy: Wrought; French Standard; part of NFA. 57. 350
AGS 50	0.5% Mg 0.4% Si Al alloy: Wrought; L'Aluminium Français
AGS 65	0.65% Mg 0.5% Si Al alloy: Wrought; L'Aluminium Français
AGS/L	0.7% Mg 0.6% Si Al alloy: Wrought; L'Aluminium Français
AGSUC	0.2% Cu 1% Mg 0.6% Si 0.15% Cr Al alloy: Wrought; French Standard
ALCAN GB 50S	0.7% Mg 0.4% Si Al alloy: Alcan for BS alloy HE9
ALCAN GB 51S	0.7% Mg 1% Si Al alloy: Alcan for BS alloy HE19
ALCAN GB 65S	1% Mg 0.7% Si 0.2% Cr Al alloy: Alcan for BS alloy H20
ALCAN GB B51S	0.7% Mg 1% Si 0.5% Mn Al alloy: Alcan for BS alloy H30
ALCAN GB C50S	0.7% Mg 0.4% Si Al alloy: Bar and forging; Alcan for alloy BT3
ALCAN GB D50S	0.7% Mg 0.5% Si Al alloy: Alcan for BS 2898. E91E
ALCAN GB D57S	0.3% Cu 0.7% Mg 0.4% Si Al alloy: Bar and forging; Alcan for BT6
ALCOA 910	0.6% Mg 0.5% Si Al alloy: Wrought; Alcoa; solution treated and aged **UTS: 250 Elon: 13% Proof: 230**
ALCOA 912	0.6% Mg 0.4% Si Al alloy: Wrought; Alcoa; solution treated and aged **UTS: 240 Elon: 15% Proof: 210**
ALCOA 918	0.05% Cu 0.6% Mg 0.5% Si (max) Al alloy: Wrought; Alcoa; solution treated and aged 55% IACS; for electrical purposes **UTS: 240 Elon: 16% Proof: 210**
ALCOA 920	0.8% Mg 1.0% Si Al alloy: Wrought; Alcoa; solution treated and aged **UTS: 340 Elon: 11% Proof: 300**
ALCOA 940	0.25% Cu 1.0% Mg 0.6% Si Al alloy: Wrought; Alcoa; solution treated and aged **UTS: 310 Elon: 12% Proof: 270**
ALCOA 945	0.25% Cu 1.0% Mg 0.6% Si Al alloy: Wrought; Alcoa; solution treated and aged **UTS: 310 Elon: 12% Proof: 280**
ALCOA 946	0.25% Cu 1.0% Mg 0.6% Si 0.5% Pb 0.5% Bi Al alloy: Wrought; Alcoa; solution treated and aged; free machining **UTS: 310 Elon: 14% Proof: 280**
ALDREY	0.4% Mg 0.5% Si Al alloy: Anglo-Swiss Aluminium Co., for high electrical conductivity
ALDREY 051	0.4% Mg 0.6% Si Al alloy: Wrought; Anglo-Swiss Aluminium Co.
ALDREY 051	0.6% Mg 0.5% Si Al alloy; Star
ALMASILIUM	1% Mg 2% Si Al alloy: Heat treatable; origin unknown
Al Mg Si 0.5	0.7% Mg 0.5% Si Al alloy: Wrought; previous German Standard designation
Al Mg Si 1	1% Mg 1.4% Si 0.4% Mn 0.2% Cr Al alloy: Wrought; previous German Standard designation
Al Mg Si Pb	1% Mg 1.2% Si 0.8% Mn 0.2% Cr + Pb Al alloy: Wrought; previous German Standard designation; free machining
ALMINAL 9	0.15% Cu 0.4% Mg 0.3% Si Al alloy: Wrought; origin unknown
ALMINAL 10	0.15% Cu 0.4% Mg 0.3% Si 1.0% Mn 0.5% Cr Al alloy: Wrought; origin unknown
ALMINAL 11	1% Cu 0.5% Mg 0.8% Si 1% Mn Al alloy: Wrought; origin unknown
AMS 4009	0.2% Cu 1% Mg 0.6% Si 0.2% Cr Al alloy: Wrought
AMS 4025 D	0.25% Cu 1% Mg 0.6% Si 0.25% Cr Al alloy: Sheet; annealed. AMS for AA alloy 6061
AMS 4026 D	0.25% Cu 1% Mg 0.6% Si 0.25% Cr Al alloy: Sheet; AMS for AA alloy 6061; solution treated
AMS 4027 E	0.25% Cu 1% Mg 0.6% Si 0.25% Cr Al alloy: Sheet; AMS for AA alloy 6061; solution treated and aged
AMS 4043	0.25% Cu 1% Mg 0.6% Si 0.25% Cr Al alloy: Plate; AMS for AA alloy 6061
AMS 4053	0.25% Cu 1% Mg 0.6% Si 0.25% Cr Al alloy: Solution treated, stretched and aged; AMS for AA alloy 6061
AMS 4079	0.25% Cu 1% Mg 0.6% Si 0.25% Cr Al alloy: Seamless tube; annealed; AMS for AA alloy 6061
AMS 4080 E	0.25% Cu 1% Mg 0.6% Si 0.25% Cr Al alloy: Seamless tube; annealed; AMS for AA alloy 6061
AMS 4081 A	0.25% Cu 1% Mg 0.5% Si 0.25% Cr Al alloy: Seamless tube; solution treated; AMS for AA alloy 6061
AMS 4082 E	0.25% Cu 1% Mg 0.6% Si 0.25% Cr Al alloy: Seamless tube; solution treated and aged; AMS for AA alloy 6061
AMS 4083 D	0.25% Cu 1% Mg 0.6% Si 0.25% Cr Al alloy: Seamless tube; solution treated and aged; AMS for AA alloy 6061
AMS 4091	0.25% Cu 1% Mg 0.6% Si 0.1% Cr Al alloy: Tube; solution treated; AMS for AA alloy 6062
AMS 4092	0.25% Cu 1% Mg 0.6% Si 0.1% Cr Al alloy: Tube; solution treated and aged; AMS for AA alloy 6062
AMS 4093	0.25% Cu 1% Mg 0.6% Si 0.1% Cr Al alloy: Tube; solution treated and aged; AMS for AA alloy 6062
AMS 4113	0.2% Cu 1% Mg 0.6% Si 0.2% Cr Al alloy
AMS 4115	0.25% Cu 1% Mg 0.6% Si 0.25% Cr Al alloy: Bar; as rolled; AMS for AA alloy 6061
AMS 4116 A	0.3% Cu 1% Mg 0.6% Si 0.25% Cr Al alloy: Bar; solution treated; AMS for AA alloy 6061
AMS 4117 A	0.3% Cu 1% Mg 0.6% Si 0.25% Cr Al alloy: Bar; solution treated and aged; AMS for AA alloy 6061
AMS 4125 E	0.6% Mg 1% Si 0.25% Cr Al alloy: Forgings; solution treated and aged; AMS for AA alloy 6151
AMS 4127 B	0.3% Cu 1% Mg 0.6% Si 0.25% Cr Al alloy: Forging; solution treated and aged; AMS for AA alloy 6061
AMS 4128	0.2% Cu 1% Mg 0.6% Si 0.2% Cr Al alloy
AMS 4146	0.3% Cu 1% Mg 0.6% Si 0.25% Cr Al alloy: Forging; solution treated; AMS for AA alloy 6061
AMS 4150 C	0.25% Cu 1% Mg 0.6% Si 0.25% Cr Al alloy: Extrusion; solution treated and aged; AMS for AA alloy 6061
AMS 4155 A	0.25% Cu 1% Mg 0.6% Si 0.1% Cr Al alloy: Extrusion; solution treated and aged; AMS for AA alloy 6062
AMS 4156 C	0.6% Mg 0.4% Si Al alloy: Extrusion; solution treated and aged; AMS for AA alloy 6063
AMS 4160	0.25% Cu 1% Mg 0.6% Si 0.25% Cr Al alloy: Extrusion; annealed; AMS for AA alloy 6061
AMS 4161	0.25% Cu 1% Mg 0.6% Si 0.25% Cr Al alloy: Extrusion; solution treated; AMS for AA alloy 6061
AMS 4172	0.2% Cu 1% Mg 0.6% Si 0.2% Cr Al alloy: Wrought
AMS 4173	0.2% Cu 1% Mg 0.6% Si 0.2% Cr Al alloy: Wrought
AMS 4312	0.2% Cu 1% Mg 0.6% Si 0.2% Cr Al alloy: Wrought
AMS 4347	0.8% Cu 1% Mg 0.8% Si Al alloy: Wrought
ANTICORODAL	0.8% Mg 1% Si Al alloy: Anglo-Swiss Aluminium Co.; cold rolled; solution treated and aged; obsolete **DPN: 110 UTS: 340 Elon: 16% Proof: 300**

Note. The following abbreviations and units are used in the tables:

DPN	Hardness, diamond pyramid number
UTS	Ultimate tensile strength, N/mm^2
Elon	Elongation, %
Proof	0.1% proof strength, N/mm^2

1 N/mm^2=0.1 hbar=0.102 kgf/mm^2=0.06475 tonf/in.^{2}=145.04 lbf/in.^{2}=1 MPa

See Appendix II for other abbreviations and conversion tables.

Symbol	Nominal analysis, supplier, condition and remarks.
ANTICORODAL 041	0.35% Mg 0.4% Si Al alloy: Wrought; Anglo-Swiss Aluminium Co.
ANTICORODAL 090	0.9% Mg 1% Si 0.7% Mn Al alloy; Star
ANTICORODAL 110	0.8% Mg 1.0% Si 0.7% Mn Al alloy: Wrought; Anglo-Swiss Aluminium Co.
ANTICORODAL 850	0.6% Mg 0.4% Si Al alloy; Star
ANTICORODAL 990	0.45% Mg 0.55% Si Al alloy: Wrought; Anglo-Swiss Aluminium Co.
ANTICORODAL Pb 108	0.8% Mg 1.0% Si 1.2% Pb 0.3% Co Al alloy: Wrought; Anglo-Swiss Aluminium Co.; free machining
A SG	1.0% Mg 1.1% Si Al alloy: Wrought; L'Aluminium Français
A SG	1.0% Mg 1.2% Si Al alloy: Wrought; French Standard
A SGM 0.7	0.8% Mg 1% Si 0.6% Mn Al alloy; Wrought; French Standard
A SGM	1.0% Mg 1.1% Si 0.5% Mn 0.2% Cr Al alloy: Wrought; French Standard
ASTM B209/6003	1.2% Mg 0.5% Si Al alloy: Sheet and plate
ASTM B209/6061	0.2% Cu 1% Mg 0.6% Si 0.2% Cr Al alloy: Sheet; annealed **UTS: 140 Elon: 18% Proof: 60**
ASTM B209/6061	0.2% Cu 1% Mg 0.6% Si 0.2% Cr Al alloy: Sheet; solution treated and aged **UTS: 280 Elon: 9% Proof: 220**
ASTM B210/6061	0.2% Cu 1% Mg 0.6% Si 0.2% Cr Al alloy: Tube; solution treated and aged **UTS: 280 Proof: 240**
ASTM B210/5062	0.2% Cu 1% Mg 0.6% Si 0.1% Cr Al alloy: Tube; solution treated and aged **UTS: 280 Proof: 240**
ASTM B210/6063	0.7% Mg 0.5% Si Al alloy: Tube; solution treated and aged **UTS: 220 Proof: 180**
ASTM B211/6061	0.2% Cu 1% Mg 0.6% Si 0.2% Cr Al alloy: Bar; solution treated and aged **UTS: 280**
ASTM B211/6262	1.0% Mg 0.6% Si 0.1% Cr 0.5% Bi 0.5% Pb Al alloy: Bar; free machining
ASTM B221/6061	0.2% Cu 1% Mg 0.6% Si 0.2% Cr Al alloy: Extruded bar; solution treated and aged **UTS: 250 Elon: 10% Proof: 240**
ASTM B221/6062	0.2% Cu 1% Mg 0.6% Si 0.1% Cr Al alloy: Extruded bar; solution treated and aged **UTS: 250 Elon: 10% Proof: 240**
ASTM B221/6063	0.5% Mg 0.4% Si Al alloy: Extruded bar; solution treated and aged **UTS: 200 Elon: 10% Proof: 170**
ASTM B221/6351	0.6% Mg 1% Si 0.6% Mn Al alloy: Extruded bar; solution treated and aged **UTS: 280 Elon: 10% Proof: 250**
ASTM B234/6061	1% Mg 0.6% Si 0.2% Cr Al alloy: Tube; solution treated and aged **UTS: 280 Elon: 11% Proof: 240**
ASTM B234/6062	0.2% Cu 1% Mg 0.6% Si 0.1% Cr Al alloy: Tube; solution treated and aged **UTS: 280 Elon: 11% Proof: 240**
ASTM B235/6061	0.2% Cu 1% Mg 0.6% Si 0.2% Cr Al alloy: Extruded tube; solution treated and aged **UTS: 250 Elon: 10% Proof: 240**
ASTM B235/6062	0.2% Cu 1% Mg 0.6% Si 0.1% Cr Al alloy: Extruded tube; solution treated and aged **UTS: 250 Elon: 10% Proof: 240**
ASTM B235/6063	0.7% Mg 0.3% Si Al alloy: Extruded tube; solution treated and aged **UTS: 200 Elon: 10% Proof: 170**
ASTM B235/6351	1% Cu 0.6% Mg 0.6% Mn Al alloy: Extruded tube; solution treated and aged **UTS: 280 Elon: 10% Proof: 250**
ASTM B241/6061	0.2% Cu 1% Mg 0.6% Si 0.2% Cr Al alloy: Pipe; solution treated and aged **UTS: 260 Elon: 10% Proof: 240**
ASTM B241/6062	0.2% Cu 1% Mg 0.6% Si 0.1% Cr Al alloy: Pipe; solution treated and aged **UTS: 260 Elon: 10% Proof: 240**
ASTM B241/6063	0.7% Mg 0.4% Si Al alloy: Pipe; solution treated and aged **UTS: 200 Elon: 8% Proof: 170**
ASTM B241/6351	0.6% Mg 1% Si 0.6% Mn Al alloy: Pipe; solution treated and aged **UTS: 280 Elon: 10% Proof: 250**
ASTM B247/6053	1.2% Mg 0.5% Si 0.2% Cr Al alloy: Forging; solution treated and aged **DPN: 75 UTS: 240 Elon: 16% Proof: 200**
ASTM B247/6061	0.2% Cu 1% Mg 0.6% Si 0.2% Cr Al alloy: Forging; solution treated and aged **DPN: 80 UTS: 260 Elon: 10% Proof: 240**
ASTM B247/6151	0.6% Mg 0.9% Si 0.2% Cr Al alloy: Forging; solution treated and aged **DPN: 90 UTS: 300 Elon: 14% Proof: 250**
ASTM B308/6061	0.2% Cu 1% Mg 0.6% Si 0.2% Cr Al alloy: Section; solution treated and aged **UTS: 250 Elon: 10% Proof: 240**
ASTM B308/6062	0.2% Cu 1% Mg 0.6% Si 0.1% Cr Al alloy: Section; solution treated and aged **UTS: 250 Elon: 10% Proof: 240**
ASTM B308/6066	1% Cu 1.2% Mg 1.5% Si 0.8% Mn Al alloy: Section; solution treated and aged **UTS: 350 Elon: 8% Proof: 310**
ASTM B308/6351	0.6% Mg 1% Si 0.6% Mn Al alloy: Section; solution treated and aged **UTS: 280 Elon: 10% Proof: 250**
ASTM B313/6061	0.2% Cu 1% Mg 0.6% Si 0.2% Cr Al alloy: Welded tube; solution treated and aged **UTS: 280 Elon: 10% Proof: 240**
ASTM B316/6053	1.2% Mg 0.7% Si 0.2% Cr Al alloy: Rivets; annealed
ASTM B316/6061	0.2% Cu 1% Mg 0.6% Si 0.2% Cr Al alloy: Rivets; annealed
ASTM B317	0.6% Mg 0.5% Si Al alloy: For electrical purposes; solution treated and aged 55% IACS **UTS: 200 Proof: 170**
ASTM B317/6101	0.55% Mg 0.5% Si Al alloy: Bar
ASTM B345/6061	0.2% Cu 1% Mg 0.6% Si 0.2% Cr Al alloy: Pipe; solution treated and aged **UTS: 250 Elon: 8% Proof: 220**
ASTM B345/6062	0.2% Cu 1% Mg 0.6% Si 0.1% Cr Al alloy: Pipe; solution treated and aged **UTS: 250 Elon: 8% Proof: 220**
ASTM B345/6063	0.7% Mg 0.4% Si Al alloy: Pipe; solution treated and aged **UTS: 200 Elon: 8% Proof: 170**
ASTM B345/6351	0.6% Mg 1% Si 0.6% Mn Al alloy: Pipe; solution treated and aged **UTS: 280 Elon: 10% Proof: 250**
ASTM B398/6201	0.75% Mg 0.7% Si Al alloy: Wire
ASTM B404/6061	0.3% Cu 1.0% Mg 0.6% Si 0.2% Cr Al alloy: Condenser tube
ASTM B429/6061	0.30% Cu 1.0% Mg 0.6% Si 0.2% Cr Al alloy: Extruded pipe and tube
ASTM B429/6063	0.7% Mg 0.4% Si Al alloy: Extruded pipe and tube
AWCO 22	0.3% Cu 1% Mg 0.6% Si 0.6% Mn Al alloy: Aluminium Wire Co. for BS alloy H 20
AWCO 24	0.75% Mg 0.5% Si Al alloy: Wire; Aluminium Wire Co. for BS alloy H9
AWCO 25	0.75% Mg 1% Si Al alloy: Wire; Aluminium Wire Co. for BS alloy H10
AWCO SILMALEC	0.04% Cu 0.6% Mg 0.5% Si Al alloy: For electrical purposes; Aluminium Wire Co. for BS alloy E91E
BA 22	0.25% Cu 1% Mg 0.6% Si 0.6% Mn Al alloy: British Aluminium Co. for BS alloy H20

Symbol	Nominal analysis, supplier, condition and remarks.
BA 24	0.75% Mg 0.5% Si Al alloy: Extrusion; British Aluminium Co. for BS H9
BA 25	0.75% Mg 1% Si Al alloy: Sheet, strip, tube and extrusion; British Aluminium Co.
BA 241	0.7% Mg 0.4% Si Al alloy: Bar and forging for bright anodizing; British Aluminium Co. for BT 3
BA 251	1% Mg 1% Si Al alloy: British Aluminium Co. for BS alloy H19
BA SILMALEC	0.04% Cu 0.6% Mg 0.5% Si Al alloy: For electrical purposes; British Aluminium Co. for BS alloy E91E
BA SP 16	0.3% Cu 0.7% Mg 0.4% Si Al alloy: Bar and forging; British Aluminium Co. for BT 6
BB 019	0.7% Mg 1% Si Al alloy: Sheet, strip and tube; Birmabright
BIRMETAL 016	0.3% Cu 1% Mg 0.7% Mn Al alloy: Birmabright for BS alloy H20
BIRMETAL 055	0.75% Mg 0.5% Si Al alloy: Birmabright for BS alloy H9
BIRMETAL 065	0.7% Mg 0.4% Si Al alloy: Bar and forging for bright anodizing; Birmabright for BT3
BIRMETAL 069	1% Mg 1% Si alloy: Birmabright for BS alloy H19
BIRMETAL 071	1% Mg 1% Si 0.7% Mn Al alloy: Birmabright for BS alloy H30
BS 414	Al alloy: Sheet and strip; heat treated; replaced by BS 1470
BS 1470 6082	1% Mg 1.0% Si 0.8% Mn Al alloy: Plate-sheet; solution treated and aged; British Aluminium Co. **UTS: 295 Elon: 8% Proof: 255**
BS 1470 HS 20	0.3% Cu 1% Mg 0.7% Si 0.6% Mn Al alloy: Sheet and strip; solution treated and aged **UTS: 260 Elon: 8% Proof: 220**
BS 1470 HS 30	1% Mg 1% Si 0.75% Mn Al alloy: Sheet and strip; solution treated and aged **UTS: 280 Elon: 8% Proof: 240**
BS 1471 6061	0.2% Cu 0.7% Mg 0.4% Si Al alloy: Drawn tube; solution treated and aged; British Aluminium Co. **UTS: 200 Elon: 8% Proof: 180**
BS 1471 6063	0.7% Mg 0.4% Si Al alloy: Drawn tube; solution treated and aged; British Aluminium Co. **UTS: 200 Elon: 8% Proof: 180**
BS 1471 6082	0.9% Mg 1% Si 0.6% Mn Al alloy: Drawn tube; solution treated and aged; British Aluminium Co. **UTS: 310 Elon: 9% Proof: 240**
BS 1471 HT 9	0.7% Mg 0.5% Si Al alloy: Tube; solution treated and aged **UTS: 200 Elon: 8% Proof: 170**
BS 1471 HT19	1% Mg 1% Si Al alloy: For tube; solution treated and aged **UTS: 280 Elon: 8% Proof: 210**
BS 1471 HT20	0.3% Cu 1% Mg 0.5% Si 0.6% Mn Al alloy: Tube; solution treated and aged **UTS: 280 Elon: 8% Proof: 210**
BS 1471 HT30	1% Mg 1% Si 0.75% Mn Al alloy: Tube; solution treated and aged **UTS: 300 Elon: 8% Proof: 220**
BS 1472 2031	2.3% Cu 0.9% Mg 0.9% Si 0.9% Fe 1% Ni Al alloy: Extruded; solution treated and aged; British Aluminium Co. **UTS: 385 Elon: 6% Proof: 200**
BS 1472 6063	0.6% Mg 0.4% Si Al alloy: Forging; solution treated and aged; British Aluminium Co. **UTS: 185 Elon: 10% Proof: 130**
BS 1472 6082	1% Mg 1% Si 0.8% Mn Al alloy: Forging; solution treated and aged; British Aluminium Co. **UTS: 295 Elon: 8% Proof: 260**
BS 1472 HF9	0.75% Mg 0.5% Si Al alloy: For forging and bar; solution treated and aged **UTS: 180 Elon: 10% Proof: 140**
BS 1472 HF30	1% Mg 1% Si 0.75% Mn Al alloy: For forging and bar; solution treated and aged **UTS: 280 Elon: 10% Proof: 240**
BS 1473 6061	0.3% Cu 1% Mg 0.6% Si 0.2% Cr Al alloy: Bolt and screw stock; solution treated; cold work and aged; British Aluminium Co. **UTS: 310 Proof: 245**
BS 1473 6082	1% Mg 1% Si 0.7% Mn Al alloy: Bolt and screw stock; solution treated and aged; British Aluminium Co. **UTS: 300 Proof: 270**
BS 1473 HB30	1% Mg 1% Si 0.75% Mn Al alloy: For bolt stock; solution treated and aged **UTS: 280 Proof: 240**
BS 1473 HR30	1% Mg 1% Si 0.75% Mn Al alloy: For rivets; solution treated room aged **UTS: 200**
BS 1474 6061	0.2% Cu 1% Mg 0.6% Si 0.2% Cr Al alloy: Bar and extruded tube; solution treated and aged; British Aluminium Co. **UTS: 280 Elon: 8% Proof: 240**
BS 1474 6063	0.7% Mg 0.4% Si Al alloy: Bar and extruded tube; solution treated and aged; British Aluminium Co. **UTS: 150 Elon: 6% Proof: 130**
BS 1474 6063A	0.8% Mg 0.5% Si Al alloy: Bar and extruded tube; solution treated and aged; British Aluminium Co. **UTS: 230 Elon: 8% Proof: 190**
BS 1474 6082	0.9% Mg 1% Si 0.6% Mn Al alloy: Bar and extruded tube; solution treated and aged; British Aluminium Co. **UTS: 280 Elon: 5% Proof: 240**
BS 1474 HV9	0.75% Mg 0.7% Si Al alloy: For tube; solution treated and aged **UTS: 200 Elon: 12% Proof: 150**
BS 1474 HV19	1% Mg 1% Si Al alloy: For tube; solution treated and aged **UTS: 260 Elon: 10% Proof: 220**
BS 1474 HV30	1% Mg 1% Si 0.75% Mn Al alloy: For tube; solution treated and aged **UTS: 280 Elon: 10% Proof: 240**
BS 1475 6061	0.2% Cu 1% Mg 0.6% Si 0.2% Cr Al alloy: Wire; solution treated; cold work and aged; British Aluminium Co. **UTS: 355**
BS 1475 6063	0.7% Mg 0.4% Si Al alloy: Wire; solution treated and aged; British Aluminium Co. **UTS: 185**
BS 1475 HG9	0.75% Mg 0.5% Si Al alloy: Wire; solution treated and drawn **UTS: 260**
BS 1475 HG20	0.3% Cu 1% Mg 0.75% Si 0.5% Mn Al alloy: Wire; solution treated and drawn **UTS: 350**
BS 1475 HG30	1% Mg 1% Si 0.75% Mn Al alloy: Wire; solution treated and aged **UTS: 280**
BS 1476 HE9	0.75% Mg 0.5% Si Al alloy: Bar; solution treated and aged **UTS: 180 Elon: 12% Proof: 150**
BS 1476 HE19	1% Mg 1% Si Al alloy: Bar; solution treated and aged **UTS: 260 Elon: 10% Proof: 220**
BS 1476 HE20	0.3% Cu 1% Mg 0.7% Si 0.6% Mn Al alloy: Bar; solution treated and aged **UTS: 260 Elon: 10% Proof: 220**

Note. The following abbreviations and units are used in the tables:

DPN	Hardness, diamond pyramid number
UTS	Ultimate tensile strength, N/mm^2
Elon	Elongation, %
Proof	0.1% proof strength, N/mm^2

1 N/mm^2=0.1 hbar=0.102 kgf/mm^2=0.06475 $tonf/in.^2$=145.04 $lbf/in.^2$=1 MPa

See Appendix II for other abbreviations and conversion tables.

Symbol	Nominal analysis, supplier, condition and remarks.
BS 1476 HE30	1% Mg 1% Si 0.7% Mn Al alloy; Bar; solution treated and aged **UTS: 280 Elon: 10% Proof: 240**
BS 1477 HP20	0.3% Cu 1% Mg 0.7% Mn Al alloy: Plate; solution treated and aged **UTS: 260 Elon: 8% Proof: 210**
BS 1477 HP30	1% Mg 1% Si 0.7% Mn Al alloy: Plate; solution treated and aged **UTS: 280 Elon: 8% Proof: 220**
BS 2898 6101A	0.6% Mg 0.5% Si Al alloy: Bar and extruded tube; solution treated and aged; British Aluminium Co. **UTS: 200 Elon: 10% Proof: 170**
BS 3242	Mg Si Al wrought alloy: For electrical conductors; electrical conductivity 52% IACS
BS 4300/4	0.6% Mg 0.3% Si Al alloy: For bright trim; solution treated and aged; bars and section **UTS: 185 Elon: 10% Proof: 160**
BS 4300/4 6463	0.7% Mg 0.4% Si Al alloy: Bar and extruded tube; solution treated and aged; British Aluminium Co. **UTS: 185 Elon: 10% Proof: 160**
BS 4300/3	0.6% Mg 0.4% Si Al alloy: For bright trim; solution treated and aged; forging **UTS: 185 Elon: 10% Proof: 160**
BT 3	0.7% Mg 0.4% Si Al alloy: For bright anodizing; solution treated and aged; designation given by Al industry **UTS: 180 Elon: 12% Proof: 150**
BT 6	0.3% Cu 0.7% Mg 0.4% Si Al alloy: For bright anodizing; solution treated and aged; designation given by Al industry **UTS: 220 Elon: 10%**
DTD 346 A	1% Mg 1% Si 0.7% Mn Al alloy: Sheet and strip; annealed **UTS: 170 Elon: 18%**
DTD 372 B	0.6% Mg 0.5% Si Al alloy: Bar; solution treated; suitable for welding **UTS: 140 Elon: 17%**
DTD 5080	1% Mg 1% Si 0.7% Mn Al alloy: Sheet; solution treated and aged; weldable **UTS: 280 Elon: 8% Proof: 240**
DURALBRITE 6	0.7% Mg 0.4% Si Al alloy: Extrusion; Alcan; solution treated and aged **UTS: 190 Elon: 8%**
DURAL F	0.25% Cu 1% Mg 0.6% Si 0.25% Cr Al alloy: Sheet and tube forging; J Booth for BS alloy H20
DURAL H	0.75% Mg 1% Si Al alloy: Sheet and tube forging; Alcan for BS alloy H30
DURAL X	0.8% Mg 1% Si 0.15% Mn Al alloy: Tube; Alcan for BS alloy H19
DURCILIUM E	0.6% Mg 0.5% Si Al alloy: For electrical purposes; E Kaye; solution treated and aged; electrical conductivity 54% IACS
DURCILIUM Q	0.3% Cu 1% Mg 0.7% Si 0.6% Mn Al alloy: E Kaye for BS alloy H20
DURCILIUM R	1% Cu 1% Mg 1% Si Al alloy: E Kaye for BS alloy H19
DURCILIUM S	1% Mg 1% Si 0.7% Mn Al alloy: E Kaye for BS alloy H30
DURCILIUM W	0.75% Mg 0.5% Si Al alloy: E Kaye for BS alloy H9
E Al Mg Si	0.4% Mg 0.5% Si Al alloy: Wrought; previous German Standard designation
ED	0.7% Mg 0.7% Si Al alloy: Anglo-Swiss code for Extrudal
EXTRUDAL	0.7% Mg 0.7% Si Al alloy: Anglo-Swiss Aluminium Co.; cold rolled; solution treated and aged **DPN: 85 UTS: 260 Elon: 14% Proof: 210**
EXTRUDAL 050	0.6% Mg 0.6% Si Al alloy: Wrought; Anglo-Swiss Aluminium Co.
EXTRUDAL 050	0.7% Mg 0.5% Si Al alloy: Star
GS 10A	0.7% Mg 0.4% Si Al alloy: Previous ASTM designation; now 6063
GS 11A	0.2% Cu 1% Mg 0.5% Si 0.2% Cr Al alloy: Previous ASTM designation; now 6061
GS 11B	1.2% Mg 0.7% Si 0.2% Cr Al alloy: Previous ASTM designation; now 6053
GS 11C	0.2% Cu 1% Mg 0.6% Si 0.1% Cr Al alloy: Previous ASTM designation; now 6062
H 9	0.7% Mg 0.5% Si Al alloy: Wrought; designation used by BS
H 19	1% Mg 1% Si Al alloy: Wrought; designation used by BS
H 20	0.3% Cu 1% Mg 0.5% Si 0.6% Mn Al alloy: Wrought; designation used by BS
H 30	1% Mg 1% Si 0.75% Mn Al alloy: Wrought; designation used by BS
HA 4 GS11N	0.25% Cu 1.0% Mg 0.6% Si 0.2% Cr Al alloy: Plate; Canadian Standard
HA 5 GS10	0.7% Mg 0.5% Si Al alloy: Bar, etc.; Canadian Standard
HA 5 GS11N	0.3% Cu 1.0% Mg 0.6% Si 0.2% Cr Al alloy: Bar, etc.; Canadian Standard
HA 5 GS11T	0.2% Cu 0.9% Mg 0.5% Si 0.25% Mn Al alloy: Bar, etc.; Canadian Standard
HA 5 GS11R	0.6% Mg 1.0% Si 0.6% Mn Al alloy: Bar, etc.; Canadian Standard
HA 6 GS11N	0.3% Cu 1.0% Mg 0.6% Si 0.2% Cr Al alloy: For rivets and brazing wire; Canadian Standard
HA 7 GS10	0.6% Mg 0.4% Si Al alloy: Tube; Canadian Standard
HA 7 GS11N	0.2% Cu 1.0% Mg 0.6% Si 0.2% Cr Al alloy: Tube; Canadian Standard
HA 7 GS11T	0.2% Cu 0.8% Mg 0.5% Si 0.3% Mn Al alloy: Tube; Canadian Standard
HA 7 GS11R	0.6% Mg 1.0% Si 0.6% Mn Al alloy: Tube; Canadian Standard
HA 8 GS11N	0.2% Cu 1.0% Mg 0.6% Si 0.2% Cr Al alloy: Forging; Canadian Standard
HA 8 GS11P	0.7% Mg 0.8% Si 0.2% Cr Al alloy: Forging; Canadian Standard
HIDUMINIUM 18	0.7% Mg 0.4% Si Al alloy: Bar and forging for bright anodizing; High Duty Alloys for BT 3
HIDUMINIUM 42	1% Mg 1% Si Al alloy: High Duty Alloys for BS alloy H19
HIDUMINIUM 43	0.2% Cu 1% Mg 0.6% Si 0.6% Mn 0.2% Cr Al alloy: High Duty Alloys for BS alloy H20
HIDUMINIUM 44	0.75% Mg 1% Si Al alloy: For forgings, extrusions and tubes; High Duty Alloys
HIDUMINIUM 46	0.6% Mg 0.5% Si Al alloy: High Duty Alloys for BS alloy H9
IMPALCO 910	0.5% Mg 0.5% Si Al alloy: Tube, bar and forging; Imperial Aluminium Co. for BS alloy H9
IMPALCO 912	0.6% Mg 0.5% Si Al alloy: Bar and forgings; high purity 910; Imperial Aluminium Co.; solution treated and aged **UTS: 180 Elon: 12% Proof: 150**
IMPALCO 918	0.5% Mg 0.5% Si Al alloy: Bar; Imperial Aluminium Co. for BS 2898 E91E
IMPALCO 920	1% Mg 1% Si 0.7% Mn 0.2% Cr Al alloy: Sheet and bar, etc.; Imperial Aluminium Co. for BS alloy H30
IMPALCO 940	1% Mg 0.6% Si 0.2% Cr Al alloy: Sheet and bar, etc.; Imperial Aluminium Co. for BS alloy H20
IMPALCO 945	1% Mg 0.6% Si 0.1% Cr (max) Al alloy: Bar; Imperial Aluminium Co.; solution treated and aged **UTS: 280 Elon: 10% Proof: 240**
IMPALCO 946	1% Mg 0.6% Si 0.5% Pb 0.5% Bi Al alloy: Bars and forgings; Imperial Aluminium Co.; solutions treated and aged **DPN: 110 UTS: 330 Proof: 300**
IMPALCO 950	0.9% Mg 1% Si 0.1% Cr (max) Al: Bar and tube; Imperial Aluminium Co. for BS alloy H19
IMPALCO M38	0.7% Mg 0.5% Si Al alloy: Bar for electrical purposes; Imperial Aluminium Co. for BS 2898 E91E

Symbol	Nominal analysis, supplier, condition and remarks.
IMPALCO M39/1	0.7% Mg 0.5% Si Al alloy: Imperial Aluminium Co. for BS alloy H9
IMPALCO M39/2	1% Mg 1% Si 0.7% Mn 0.3% Cr Al alloy: Imperial Aluminium Co. for BS alloy H30
IMPALCO M40	0.2% Cu 1% Mg 0.7% Si 0.6% Mn Al alloy: Imperial Aluminium Co. for BS alloy H20
IMPALCO M41	1% Mg 1% Si 0.4% Cr Al alloy: Strip, tube and bar; Imperial Aluminium Co.; solution treated and aged **UTS: 280 Elon: 9% Proof: 220**
IMPALCO M42	1% Mg 1% Si Al alloy: Imperial Aluminium Co. for BS alloy H19
IMPALCO SUPERSPEED 946	0.25% Cu 1.0% Mg 0.6% Si 0.5% Pb 0.5% Bi Al alloy: Imperial Aluminium Co.; free machining bar **UTS: 270 Elon: 10% Proof: 225**
KYNAL C69	1% Cu 1% Mg Al alloy: ICI; obsolete; now Imperial Aluminium Co.
KYNAL M39/1	0.7% Mg 0.5% Si Al alloy: ICI; obsolete; now Imperial Aluminium Co.
KYNAL M39/2	0.7% Mg 1% Si Al alloy: ICI; obsolete; now Imperial Aluminium Co.
L111	0.8% Mg 1% Si 0.6% Mn Al alloy: Bar
L112	0.8% Mg 1% Si 0.6% Mn Al alloy: Forging
L113	0.8% Mg 1% Si 0.6% Mn Al alloy: Sheet and strip
L114	0.8% Mg 1% Si 0.7% Mn Al alloy: Tube
L117	0.2% Cu 1% Mg 0.6% Si 0.2% Cr Al alloy: Tube
L118	0.2% Cu 1% Mg 0.6% Si 0.2% Cr Al alloy: Tube as L117; hydraulically tested
NORAL 50S	0.75% Mg 0.5% Si Al alloy: Wire; Northern Aluminium Co. for BS alloy H9
NORAL 51S	0.75% Mg 1% Si Al alloy: Sheet and tube forging; Northern Aluminium Co.
NORAL 62S	1% Cu 1% Mg 1% Si Al alloy: For tube forging; Northern Aluminium Co.; solution treated and aged **UTS: 370 Elon: 6% Proof: 310**
QQ A 327	0.3% Cu 1.0% Mg 0.6% Si 0.2% Cr Al alloy: Sheet; US Federal; solution treated and aged **UTS: 260 Elon: 9% Proof: 200**
QQ A331B	1.3% Mg 0.7% Si 0.2% Cr Al alloy: Bar; US Federal; Solution treated and aged **UTS: 210 Elon: 14% Proof: 150**
QQ A 334	1.3% Mg 0.7% Si 0.25% Cr Al alloy: Sheet; US Federal: solution treated and aged **UTS: 220 Elon: 9% Proof: 170**
SAE 210	1.2% Mg 0.8% Si Al alloy: Wrought; used only for cladding; now 6003
SAE 211	0.2% Cu 1% Mg 0.6% Si 0.1% Cr Al alloy: Solution treated and aged; now 6062 **DPN: 95 UTS: 300 Elon: 17% Proof: 270**
SAE 212	0.7% Mg 0.4% Si Al alloy: Solution treated and aged; now 6062 **DPN: 73 UTS: 220 Elon: 12% Proof: 200**
SAE 213	0.2% Cu 0.6% Mg 0.3% Si Al alloy: Now 6951
SAE 253	0.7% Mg 0.4% Si Al alloy: Wrought; solution treated and aged; now 6463 **UTS: 150 Elon: 10% Proof: 110**
SAE 280	0.6% Mg 1% Si 0.25% Cr Al alloy: Forgings; solution treated and aged; now 6151 **UTS: 300 Elon: 14% Proof: 250**
SAE 281	0.25% Cu 1% Mg 0.6% Si 0.25% Cr Al alloy: Tube and sheet forging; solution treated and aged; now 6061 **Elon: 8% Proof: 240**
SAE 282	1.5% Mg 0.7% Si 0.25% Cr Al alloy: Wrought; solution treated and aged **UTS: 210 Elon: 10% Proof: 170**
SG 11 B	1% Cu 1.1% Mg 1.5% Si 0.8% Mn Al alloy: Previous ASTM designation; now 6066
SIMALEC	0.04% Cu 0.5% Mg 0.5% Si Al alloy: For electrical purposes; Aluminium Wire Co. and British Aluminium Co. for BS alloy E91E
SIMGAL	0.75% Mg 0.5% Si Al alloy: For wire; Alcan for BS alloy H9
SS 4104	0.7% Mg 0.5% Si Al alloy: Bar and tube; Swedish Standard; as rolled **DPN: 40 UTS: 140 Elon: 17% Proof: 110**
SS 4212	0.9% Mg 1% Si 0.8% Mn Al alloy: Bar and forging; Swedish Standard; as forged **DPN: 70 UTS: 200 Elon: 15% Proof: 90**
STA 7 AW9A	0.7% Mg 0.5% Si Al alloy: Bar; replaced by BS 1476 HE9
STA 7 AW9B	0.7% Mg 0.5% Si Al alloy: Bar; replaced by BS 1476 HE9
STA 7 AW10A	0.7% Mg 1% Si Al alloy: Forging and bar; replaced by BS 1472
STA 7 AW 10B	0.7% Mg 1% Si Al alloy: Forging and bar; replaced by BS 1472
STA 7 AW10C	0.7% Mg 1% Si Al alloy: Tube; replaced by BS 1471
STA 7 AW10D	0.7% Mg 1% Si Al alloy: Sheet; replaced by BS 1470
STA 7 AW10E	0.7% Mg 1% Si Al alloy: Sheet; replaced by BS 1470
STA 7 AW10F	0.7% Mg 1% Si Al alloy: Wire; replaced by BS 1473-5
TI 03	1% Cu 1% Mg 1% Si Al alloy: Forging; TI Aluminium; solution treated and aged **UTS: 370 Elon: 10% Proof: 300**
TI 40	0.75% Mg 0.5% Si Al alloy: TI Aluminium for BS alloy H9
TI 44	0.75% Mg 1% Si Al alloy: Sheet, strip and tube; TI Aluminium
UNI 3501	0.5% Mg 0.7% Si 0.7% Mn Al alloy: Italian Standard
UNI 3569	0.7% Mg 0.4% Si Al alloy: For forging; Italian Standard
UNI 3570	0.5% Mg 0.5% Si Al alloy: For forging; Italian Standard
UNI 3571	0.6% Mg 1.0% Si 0.3% Mn Al alloy: For forging; Italian Standard
UNI 6170	0.25% Cu 1.0% Mg 0.6% Si 0.25% Cr Al alloy: For forging; Italian Standard
UNI 6359	0.1% Cu 0.5% Mg 0.45% Si Al alloy: For forging; Italian Standard
VIVAL	1.0% Mg 0.5% Si 0.5% Mn Al rolling alloy: Origin unknown
W 9	0.6% Mg 0.5% Si Al alloy: Southern Forge; replaced by IMPALCO 910
W 10	1% Mg 1% Si Al alloy: Southern Forge; replaced by IMPALCO 950
W 20	0.2% Cu 1% Mg 0.6% Si 0.2% Cr Al alloy: Southern Forge; replaced by IMPALCO 940
W 30	1% Mg 1% Si 0.8% Mn Al alloy: Southern Forge; replaced by IMPALCO 920
WO 90	0.5% Mg 0.4% Si Al alloy: Southern Forge; replaced by IMPALCO 912
WWT 789	1.0% Mg 0.6% Si 0.2% Cr 0.1% Ti Al alloy: Tube; US Federal; solution treated and aged **UTS: 270 Elon: 9% Proof: 220**
WWT 790	1.2% Mg 0.7% Si 0.2% Cr Al alloy: Tube; US Federal; solution treated and aged

Note. The following abbreviations and units are used in the tables:

DPN	Hardness, diamond pyramid number
UTS	Ultimate tensile strength, N/mm^2
Elon	Elongation, %
Proof	0.1% proof strength, N/mm^2

$1\ N/mm^2 = 0.1\ hbar = 0.102\ kgf/mm^2 = 0.06475\ tonf/in.^2 = 145.04\ lbf/in.^2 = 1\ MPa$

See Appendix II for other abbreviations and conversion tables.

1F Aluminium–copper wrought alloys
With magnesium, silicon, titanium, nickel and iron additions

Specific gravity		2.7–2.75
Density		2700–2750 kg/m^3
Solidus/liquidus		530–640 °C
Thermal conductivity		147–180 W/m °C
Coefficient of linear expansion		22–24 × 10^{-6}/ °C
Electrical conductivity	solution treated and aged	34–41% IACS (copper 100%)
Specific resistance	solution treated and aged	41–51 microhm mm
Young's modulus of elasticity		68.95 × 10^9 N/m^2
Impact	solution treated and aged	8–11 J
Fatigue strength (50 × 10^6 cycles)	solution treated and aged	± 120–150 N/mm^2

Hot strength

Temperature °C	Tensile strength* N/mm^2	Elongation* %
100	410	18
150	370	20
200	330	20
250	270	17
300	150	18
350	75	45
400	45	60

*These are approximate values which might be expected from the medium alloyed specifications listed. Some of the 4% copper highly alloyed specifications give better results.

These properties are chosen to be typical of the group of specifications as a whole. There is considerable variation within the group and in general the properties vary in proportion to the copper content.

General metallurgical characteristics

The specifications in this section have up to 5% copper. This is the original aluminium alloy where it was found that mechanical properties would be enhanced to a marked degree by solution treating and ageing (precipitation treatment).

Many of these alloys come under the term 'Dural'.

Additions of silicon and manganese improve the mechanical strength with chromium being commonly added.

Many of these alloys are precipitation treated at room temperature and this is where the term 'ageing' originated. After solution treating, which is quenching from about 500 °C such as could be involved with a casting technique, it was found that after a period of time the mechanical properties of the alloys had improved thus the term 'ageing' was used.

In order to prevent this happening, it requires that if solution treatment and ageing cannot be carried out immediately then the solution treatment is carried out and the components are held at below 0 °C where cold working can be carried out and then they can be allowed to age harden. This is very useful when riveting is required.

Additions of iron, nickel and titanium and grain refining elements stabilize the materials, giving some high temperature properties in some instances.

None of the alloys exhibit good corrosion resistance compared with the manganese, magnesium alloys, or pure aluminium.

These alloys are used where high strength, low weight, with reasonable corrosion resistance is required. They are relatively expensive and generally it is found that the alloys listed in Section 1E find more commercial use than the alloys listed in this section. Where the strength to weight ratio is important, however, as in aircraft or in trucks which are used over a long period of time, then these alloys can be economical.

Welding requires the use of the inert gas shielded process.

The corrosion resistance can be improved by anodizing or to a lesser extent by chromate treatment.

Brazing, using special fluxes and Al/Si filler rods, is readily accomplished, but the product will not have the strength of the parent metal.

Post-treatment. Where possible the parts should be re-solution treated and re-aged. This will seldom be possible owing to the danger of distortion and incipient melting of the weld area where the melting point has been reduced. The parts must always be re-aged at the correct temperature after welding or brazing. This re-age, as well as restoring the mechanical properties as much as possible, reduces the stresses caused by welding.

Symbol	Nominal analysis, supplier, condition and remarks.
3.1190	5% Cu Al alloy: German Standard DIN number not known
3.1255	4.5% Cu 0.6% Mg 0.8% Si 0.9% Mn Al alloy: German Standard DIN 1725
3.1305	2.5% Cu 0.4% Mg Al alloy: Wrought; German Standard DIN 1725
3.1325	4% Cu 0.7% Mg 0.6% Si 0.7% Mn Al alloy: Wrought; German Standard DIN 1725
3.1335	4% Cu 0.7% Mg 0.7% Si 0.5% Mn Al alloy: Wrought; German Standard DIN 1725
3.1355	4.3% Cu 1.3% Mg 0.7% Mn Al alloy: Wrought; German Standard DIN 1725
3.1365	4.3% Cu 1.5% Mg Al alloy: German Standard DIN 1725
3.1590	3.3% Cu Al alloy: German Standard; DIN number not known
3.1645	4.2% Cu 1% Mg 0.7% Mn 2% (Pb+Sn+Bi+Sb+Cd) Al alloy: Wrought; German Standard DIN 1725
3.1655	5.5 Cu Al alloy: German Standard
3.2131	2% Cu 3% Si Al alloy: German Standard DIN 1725
11S	5.5% Cu Al alloy: Previous AA designation; now 2011
14S	4.4% Cu 0.6% Mg 0.7% Si 0.6% Mn Al alloy: Sheet and forging; Alcoa for BS alloy H15
14S	4.5% Cu 0.6% Mg 1% Si 1% Mn Al alloy: Previous AA designation; now 2014
17S	4% Cu 0.6% Mg 0.5% Mn Al alloy: Forgings; Alcoa for BS alloy H14
17S	4% Cu 0.6% Mg 0.6% Mn Al alloy: Previous AA designation; now 2017
18S	4% Cu 1.5% Mg 2% Ni Al alloy: Forgings; Alcoa for BS alloy H17
18S	4% Cu 0.8% Mg 2% Ni Al alloy: Previous AA designation; now 2018
24S	4.5% Cu 1.5% Mg 0.6% Mn Al alloy: Previous AA designation; now 2024
25S	4.5% Cu 0.8% Si 0.9% Mn Al alloy: Previous AA designation; now 2025
2004	6% Cu Al alloy: Wrought; designation used in the UK and USA
2008	1% Cu 0.3% Mg 0.6% Si Al alloy: Wrought; designation used in the UK and USA
2011	5.5% Cu 0.4% Pb 0.4% Bi Al alloy: Wrought; designation used in the UK and USA
2014	4.5% Cu 0.6% Mg 0.7% Si Al alloy: Wrought; designation used in the UK and USA
2017	4% Cu 0.6% Mg 2% Ni Al alloy: Wrought; designation used in the UK and USA
2018	4% Cu 0.7% Mg 2% Ni Al alloy: Wrought; designation used in the UK and USA
2020	6.5% Cu 0.1% Cd 0.15% Ti 0.1% V Al alloy: Wrought; obsolete; designation used in the UK and USA
2021	6.5% Cu 0.1% Cd 0.15% Ti 0.1% V Al alloy: Obsolete; designation used in the UK and USA
2024	4.2% Cu 1.6% Mg Al alloy: Wrought; designation used in the UK and USA
2025	4.5% Cu 0.7% Si Al alloy: Wrought; designation used in the UK and USA

Note. The following abbreviations and units are used in the tables:

DPN	Hardness, diamond pyramid number
UTS	Ultimate tensile strength, N/mm^2
Elon	Elongation, %
Proof	0.1% proof strength, N/mm^2

$1\ N/mm^2 = 0.1\ hbar = 0.102\ kgf/mm^2 = 0.06475\ tonf/in.^2 = 145.04\ lbf/in.^2 = 1\ MPa$

See Appendix II for other abbreviations and conversion tables.

Symbol	Nominal analysis, supplier, condition and remarks.
2034	4.6% Cu 1.6% Mg 0.1% Zr Al alloy: Wrought; designation used in the UK and USA
2036	2.6% Cu 0.5% Mg Al alloy; Wrought; designation used in the UK and USA
2037	1.8% Cu 0.5% Mg Al alloy: Wrought; designation used in the UK and USA
2038	1.3% Cu 0.6% Mg 0.7% Si Al alloy: Wrought; designation used in the UK and USA
2048	3.2% Cu 1.6% Mg Al alloy: Wrought; designation used in the UK and USA
2090	2.7% Cu 2.2% Li Al alloy: Wrought; designation used in the UK and USA
2091	2% Cu 1.6% Mg 2% Li Al alloy: Wrought; designation used in the UK and USA
2117	2.7% Cu 0.3% Mg Al alloy: Wrought; designation used in the UK and USA
2124	4.2% Cu 1.6% Mg Al alloy: Wrought; designation used in the UK and USA
2214	4.5% Cu 0.6% Mg 0.8% Si Al alloy: Wrought; designation used in the UK and USA
2218	4% Cu 1.6% Mg 2% Ni Al alloy: Wrought; designation used in the UK and USA
2219	6.3% Cu 0.15% Ti 0.1% V 0.15% Zr Al alloy: Wrought; designation used in the UK and USA
2224	4.1% Cu 1.6% Mg Al alloy: Wrought; designation used in the UK and USA
2319	6.2% Cu 0.15% Ti 0.1% V 0.15% Zr Al alloy: Wrought; designation used in the UK and USA
2324	4% Cu 1.6% Mg Al alloy: Wrought; designation used in the UK and USA
2419	6.2% Cu 0.08% Ti 0.1% V 0.15% Zr Al alloy: Wrought; designation used in the UK and USA
2519	5.8% Cu 0.2% Mg 0.08% Ti 0.1% V 0.15% Zr Al alloy: Wrought; designation used in the UK and USA
2618	2.2% Cu 1.5% Mg 1.1% Fe 1% Ni 0.08% Ti Al alloy: Wrought; designation used in the UK and USA
A 17S	2.8% Cu 0.3% Mn Al alloy: Previous AA designation; now 2117
A 17S	2.2% Cu 03% Mg Al alloy: Wire for rivets; Alcoa for BS alloy H13
AK 4-1	1.2% Cu 1% Mg 0.2% Si 1.1% Ni Al alloy: Russian designation
AK 8	4.2% Cu 0.6% Mg 0.7% Si 0.8% Mn Al alloy: Russian designation
ALCAN GB 16S	2.2% Cu 0.3% Mg Al alloy: Alcan; solution treated; room aged **UTS: 300 Elon: 20% Proof: 150**
ALCAN GB 17S	4.2% Cu 0.7% Mg 0.7% Mn Al alloy: Alcan for BS alloy H14
ALCAN GB 19S	4.25% Cu 1.5% Mg 2% Ni Al alloy: Forging; Alcan; solution treated; room aged **UTS: 360 Elon: 15% Proof: 210**
ALCAN GB 24S	4.5% Cu 1.5% Mg 0.7% Mn Al alloy: Alcan; solution treated; room aged **DPN: 120 UTS: 450 Elon: 19% Proof: 300**
ALCAN GB 26S	4% Cu 0.7% Mg 0.7% Si 0.7% Mn Al alloy: Sheet and plate; Alcan
ALCAN GB 28S	5.2% Cu 0.5% Bi 0.5% Pb Al alloy: Free machining; Alcan; solution treated and aged **UTS: 330 Elon: 17% Proof: 220**
ALCAN GB 42S	2.2% Cu 1.5% Mg 1.2% Ni 1% Fe Al alloy: Forging; Alcan for BS alloy H18
ALCAN GB 62S	1.5% Cu 1% Mg 1% Si 0.7% Mn Al alloy: Alcan for BS alloy H11
ALCAN GB B19S	2.5% Cu 1.5% Mg 1.2% Ni 1.1% Fe 0.1% Ti Al alloy: Alcan; solution treated and aged **UTS: 400 Elon: 11% Proof: 390**
ALCAN GB B26S	4.2% Cu 0.7% Mg 0.7% Si 0.7% Mn Al alloy: Alcan; solution treated and aged **DPN: 145 UTS: 460 Elon: 10% Proof: 400**
ALCAN GB D19S	2.5% Cu 1.5% Mg 1.3% Ni Al alloy: Plate; Alcan

Symbol	Nominal analysis, supplier, condition and remarks.
ALCOA 660	4.5% Cu 0.5% Mg 0.7% Si 0.8% Mn Al alloy: Wrought; Alcoa; solution treated and aged **UTS: 480 Elon: 7% Proof: 400**
ALCOA 800	5.5% Cu 0.5% Pb 0.5% Bi Al alloy: Wrought; Alcoa; solution treated and aged; free machining **UTS: 380 Elon: 13% Proof: 300**
Al Cu Mg 0.5	2.5% Cu 0.4% Mg Al alloy: Wrought; previous German Standard designation
Al Cu Mg 1	4% Cu 0.6% Mg 0.6% Si 0.8% Mn Al alloy: Wrought; previous German Standard designation
Al Cu Mg 2	4.2% Cu 1.6% Mg 0.8% Mn Al alloy: Wrought; previous German Standard designation
Al Cu 4 Mg 1.5	4.2% Cu 1.6% Mg 0.5% Mn Al alloy: Swiss designation
Al Cu Mg Pb	4.5% Cu 1% Mg 0.7% Mn + Pb Al alloy: Wrought; previous German Standard designation; free machining
Al Cu Si Mn	4.5% Cu 0.6% Mg 1% Si 0.8% Mn Al alloy: Wrought; previous German Standard designation
Al Cu4 Si Mn	4.2% Cu 0.5% Mg 0.7% Si 0.7% Mn Al alloy: Swiss designation
Al Cu6 Bi Pb	5.5% Cu Al alloy: Swiss designation
ALDURAL Q	4.4% Cu 0.6% Mg 0.7% Si 0.6% Mn Al alloy: Sheet and strip; J Booth for BS alloy H15
ALMINAL 11	4.5% Cu Al alloy: Origin unknown
ALMINAL 15	4.0% Cu 0.6% Mg 1.5% Si 1.2% Mn 1.0% Fe Al alloy: Wrought; origin unknown
AMS 4007	4.2% Cu 1.6% Mg 0.6% Mn Al alloy
AMS 4014	4.5% Cu 0.5% Mg 0.8% Si 0.8% Mn Al alloy: Plate; solution treated; cold rolled and aged **UTS: 450 Elon: 5:**
AMS 4028 A	4.5% Cu 0.5% Mg 0.8% Si 0.8% Mn Al alloy: Sheet; AMS for alloy 2014; annealed
AMS 4029 A	4.5% Cu 0.5% Mg 0.8% Si 0.8% Mn Al alloy: Sheet; AMS for alloy 2014; solution treated and aged
AMS 4031	6.3% Cu 0.3% Mn 0.18% Zr 0.1% V 0.06% Ti Al alloy: Sheet; AMS for alloy 2219; annealed **UTS: 200 Elon: 12% Proof: 120**
AMS 4033 A	4.5% Cu 1.5% Mg 0.6% Mn Al alloy: Plate; AMS for alloy 2024; solution treated and stretched
AMS 4035 E	4.5% Cu 1.5% Mg 0.6% Mn Al alloy: Sheet; AMS for alloy 2024; annealed
AMS 4037 F	4.5% Cu 1.5% Mg 0.6% Mn Al alloy: Sheet; AMS for 2024; solution treated
AMS 4066	6.2% Cu 0.3% Mn 0.08% Ti 0.1% V Al alloy: Wrought
AMS 4086 F	4.5% Cu 1.5% Mg 0.6% Mn Al alloy: Seamless tube; solution treated; AMS for alloy 2024
AMS 4087 C	4.5% Cu 1.5% Mg 0.6% Mn Al alloy: Seamless tube; annealed; AMS for alloy 2024
AMS 4088 E	4.5% Cu 1.5% Mg 0.6% Mn Al alloy: Seamless tube; solution treated; AMS for alloy 2024
AMS 4101	4.2% Cu 1.6% Mg 0.6% Mn Al alloy
AMS 4112	4.2% Cu 1.6% Mg 0.6% Mn Al alloy
AMS 4118 C	4% Cu 0.5% Mg 0.7% Mn Al alloy: Bar; solution treated; AMS for alloy 2017
AMS 4119 A	4.5% Cu 1.5% Mg 0.6% Mn Al alloy: Bar; solution treated and cold rolled; AMS for alloy 2024
AMS 4120 F	4.5% Cu 1.5% Mg 0.6% Mn Al alloy: Bar; cold rolled; solution treated; AMS for alloy 2024
AMS 4121 C	4.5% Cu 0.5% Mg 0.9% Si 0.8% Mn Al alloy: Bar; cold rolled; solution treated and aged; AMS for alloy 2014
AMS 4130 G	4.5% Cu 0.8% Si 0.8% Mn Al alloy: Forging; solution treated and aged; AMS for alloy 2025
AMS 4132 A	2.3% Cu 1.6% Mg 1% Fe 1% Ni 1% Ti Al alloy: Forging; solution treated and aged; AMS for alloy 2618
AMS 4133	4.2% Cu 0.6% Mg 0.8% Si Al alloy
AMS 4134 A	4.4% Cu 0.4% Mg 0.8% Si 0.8% Mn Al alloy: Forging; solution treated: AMS for alloy 2014
AMS 4135 J	4.5% Cu 0.5% Mg 0.9% Si 0.8% Mn Al alloy: Forging; solution treated and aged; AMS for alloy 2014
AMS 4140 D	4% Cu 0.7% Mg 2% Ni Al alloy: Forging; solution treated and aged; AMS for alloy 2018
AMS 4142 D	4% Cu 1.5% Mg 0.7% Si 2% Ni Al alloy: Forging; as forged; AMS for alloy 2218
AMS 4143	6.1% Cu 0.3% Mn 0.1% V Al alloy
AMS 4152 G	4.5% Cu 1.5% Mg 0.6% Mn Al alloy: Extrusions; solution treated; AMS for alloy 2024
AMS 4153 B	4.5% Cu 0.5% Mg 1% Si 0.8% Mn Al alloy: Extrusion; solution treated and aged; AMS for alloy 2014
AMS 4162	6.2% Cu 0.3% Mn 0.15% Ti 0.1% V Al alloy: Wrought
AMS 4163	6.2% Cu 0.3% Mn 0.15% Ti 0.1% V Al alloy: Wrought
AMS 4164 C	4.4% Cu 1.5% Mg 0.6% Mn Al alloy: Extrusion; AMS for alloy 2024
AMS 4165 C	4.5% Cu 1.5% Mg 0.6% Mn Al alloy: Extrusion; AMS for alloy 2024
AMS 4191 A	6.3% Cu 0.3% Mn 0.2% Zn 0.15% Ti 0.1% V Al alloy: Welding rod; AMS for alloy 2319
AMS 4192	4.2% Cu 1.6% Mg 0.5% Mn Al alloy: Wrought
AMS 4193	4.2% Cu 1.6% Mg 0.5% Mn Al alloy: Wrought
AMS 4194	4.2% Cu 1.6% Mg 0.5% Mn Al alloy: Wrought
AMS 4195	4.2% Cu 1.6% Mg 0.5% Mn Al alloy: Wrought
AMS 4221	4.2% Cu 1.6% Mg 0.5% Mn Al alloy
AMS 4313	6.2% Cu 0.3% Mn 0.05% Ti 0.1% V Al alloy: Wrought
AMS 4314	4.3% Cu 0.6% Mg 1% Si Al alloy: Wrought
AMS 4346	2.8% Cu 2.2% Li 0.1% Zr Al alloy: Wrought
AMS 4351	2.8% Cu 2.2% Li 0.1% Zr Al alloy: Wrought
ANA 8	4.2% Cu 1.0% Si 1.0% Mn Al alloy: Forging; US Service; solution treated; room aged **UTS: 330 Elon: 12% Proof: 210**
ANA 12.1	4.2% Cu 1.5% Mg 0.2% Cr Al alloy: Sheet; US Service; solution treated and aged **UTS: 440 Elon: 14% Proof: 250**
ANA 13	4.3% Cu 1.5% Mg 0.2% Cr Al alloy: Sheet; US Service; solution treated and aged **UTS: 420 Elon: 12% Proof: 240**
AN QQ W 298/2	4.2% Cu 1.5% Mg 0.5% Mn Al alloy: Wire; US Service; solution treated and aged **UTS: 420**
ARDAL	2% Cu 1.5% Fe 0.6% Ni Al alloy: Origin unknown
ASTM B209/2024	4% Cu 1.5% Mg Al alloy: Sheet; annealed **UTS: 210 Elon: 10% Proof: 75**
ASTM B209/2024	4% Cu 1.5% Mg Al alloy: Sheet; solution treated; room aged **UTS: 420 Elon: 6% Proof: 270**
ASTM B209/2219	6.3% Cu 0.3% Mn 0.06% Ti Al alloy: Sheet
ASTM B210/2024	4% Cu 1.5% Mg 0.5% Mn Al alloy: Tube; solution treated and cold worked **UTS: 450 Proof: 280**
ASTM B211/2011	5.5% Cu 0.4% Bi 0.4% Pb Al alloy: Bar; free machining
ASTM B211/2014	4.5% Cu 0.6% Mg 1% Si 1% Mn Al alloy: Bar; solution treated and aged **UTS: 460**
ASTM B211/2017	4% Cu 0.6% Mg 0.7% Mn Al alloy: Bar; solution treated; room aged **UTS: 390**
ASTM B211/2024	4% Cu 1.5% Mg 0.6% Mn Al alloy: Bar; solution treated, room aged **UTS: 440**
ASTM B211/2219	6.2% Cu 0.3% Mn 0.05% Ti Al alloy: Bar
ASTM B221/2014	4.5% Cu 0.6% Mg 1% Si 1% Mn Al alloy: Extruded bar; solution treated and aged **UTS: 420 Elon: 7% Proof: 370**
ASTM B221/2024	4% Cu 1.5% Mg 0.6% Mn Al alloy: Extruded bar; solution treated; room aged **UTS: 400 Elon: 12% Proof: 280**
ASTM B221/2219	6.2% Cu 0.3% Mn 0.05% Ti Al alloy: Extruded bar, etc.

Symbol	Nominal analysis, supplier, condition and remarks.
ASTM B235/2014	4.5% Cu 0.6% Mg 1% Si 1% Mn Al alloy: Extruded tube; solution treated and aged **UTS: 480 Elon: 7% Proof: 420**
ASTM B235/2024	4.5% Cu 1.5% Mg 0.7% Mn Al alloy: Extruded tube; solution treated; room aged **UTS: 450 Elon: 10% Proof: 350**
ASTM B241/2014	4.5% Cu 0.5% Mg 0.8% Si 0.8% Mn Al alloy: Pipe and tube
ASTM B241/2024	4.3% Cu 1.5% Mg 0.6% Mn Al alloy: Pipe and tube
ASTM B241/2219	6.3% Cu 0.3% Mn 0.6% Ti Al alloy: Pipe and tube
ASTM B247/2014	4.5% Cu 0.6% Mg 1% Si 0.8% Mn Al alloy: Forging; solution treated and aged **DPN: 125 UTS: 460 Elon: 10% Proof: 390**
ASTM B247/2018	4% Cu 0.8% Mg 2% Ni Al alloy: Forging; solution treated and aged **DPN: 100 UTS; 390 Elon: 10% Proof: 270**
ASTM B247/2025	4.5% Cu 1% Si 0.7% Mn Al alloy: Forging; solution treated and aged **DPN: 100 UTS: 420 Elon: 16% Proof: 220**
ASTM B247/2218	4% Cu 1.5% Mg 2% Ni Al alloy: Forging: solution treated and aged **DPN: 100 UTS: 390 Elon: 10% Proof: 270**
ASTM B285 ER2014	4.5% Cu 0.6% Mg 1% Mn Al alloy: Electric welding rod; obsolete
ASTM B285 RC4A	4.5% Cu Al alloy: Welding wire
ASTM B285 RCN42A	4% Cu 1.5% Mg 2% Ni Al alloy: Welding wire
ASTM B308/2014	4.5% Cu 0.6% Mg 0.9% Si 0.7% Mn Al alloy: Section; solution treated and aged **UTS: 420 Elon: 7% Proof: 370**
ASTM B316/2017	4% Cu 0.6% Mg 0.8% Mn Al alloy: Rivets; annealed **UTS: 240**
ASTM B316/2024	4.5% Cu 1.5% Mg 0.6% Mn Al alloy: Rivets; annealed **UTS: 240**
ASTM B316/2117	2.7% Cu 0.4% Mg Al alloy: Rivert; annealed **UTS: 170**
A U2G	2.4% Cu 0.4% Mg 0.8% Si Al alloy: Wrought; L'Aluminium Français
A U2G	2.5% Cu 0.3% Mg Al alloy: Wrought; French Standard
A U2GN	2.2% Cu 1.6% Mg 1.2% Fe 1.1% Ni Al alloy: Wrought; L'Aluminium Français
A U2GN	2.2% Cu 1.6% Mg 1.1% Ni 1.1% Fe Al alloy: Wrought; French Standard
A U2N	2.2% Cu 0.9% Mg 0.9% Si 1.2% Ni 1.1% Fe Al alloy: Tube; French Standard
A U2N	2.1% Cu 1.0% Mg 0.8% Si 1.0% Fe 0.1% Ti 1.0% Ni Al alloy: Wrought; L'Aluminium Français
A U4G	4% Cu 0.7% Mg 0.5% Si 0.5% Mn Al alloy: Wrought; L'Aluminium Français
A U4G	4% Cu 0.6% Mg 0.7% Mn Al alloy: Wrought; French Standard
A U4G	4.1% Cu 0.7% Mg 0.6% Si 0.6% Mn Al alloy: Wrought; French Standard
A U4G1	4.2% Cu 1.6% Mg 0.6% Mn Al alloy: Wrought; French Standard
A U4G1	4.2% Cu 1.4% Mg 0.7% Mn Al alloy: Wrought; L'Aluminium Français
A U4N	4.0% Cu 1.5% Mg 2.0% Ni Al alloy: Wrought; L'Aluminium Français

Note. The following abbreviations and units are used in the tables:

DPN	Hardness, diamond pyramid number
UTS	Ultimate tensile strength, N/mm^2
Elon	Elongation, %
Proof	0.1% proof strength, N/mm^2

$1\ N/mm^2 = 0.1\ hbar = 0.102\ kgf/mm^2 = 0.06475\ tonf/in.^2 = 145.04\ lbf/in.^2 = 1\ MPa$

See Appendix II for other abbreviations and conversion tables.

Symbol	Nominal analysis, supplier, condition and remarks.
A U4PB	4.3% Cu 1.0% Mg 1.2% Pb Al alloy: Wrought; free machining; French Standard
A U4SG	4.5% Cu 0.5% Mg 0.7% Si 0.6% Mn Al alloy: Wrought; French Standard
A U4SG	4.4% Cu 0.5% Mg 0.9% Si 0.8% Mn Al alloy: Wrought; French Standard
A U4SG	4.4% Cu 0.6% Mg 0.8% Si 0.8% Mn Al alloy: Wrought; L'Aluminium Francais
A U5 Pb Bi	5.5% Cu Al alloy: Wire; French Standard
Av 22	4% Cu 0.7% Mg 0.5% Mn Al alloy: Anglo-Swiss code for Avional 22
Av 24	4.3% Cu 1.1% Mg 0.8% Mn 0.1% Cr Al alloy: Anglo-Swiss code for Avional 24
AVIONAL 22	4% Cu 0.7% Mg 0.5% Mn Al alloy: Anglo-Swiss Aluminium Co.; solution treated and aged; obsolete **DPN: 110 UTS: 390 Elon: 23% Proof: 270**
AVIONAL 24	4.3% Cu 1.1% Mg 0.8% Mn 0.1% Cr Al alloy: Anglo-Swiss Aluminium Co.; solution treated and aged; obsolete **DPN: 130 UTS: 480 Elon: 16% Proof: 330**
AVIONAL 050	2.5% Cu 0.3% Mg Al alloy: Wrought; Anglo-Swiss Aluminium Co.
AVIONAL 100	4.0% Cu 0.6% Mg 0.5% Mn Al alloy: Wrought; Anglo-Swiss Aluminium Co.
AVIONAL Pb 118	4.7% Cu 1.0% Mg 0.7% Mn 1.0% Pb Al alloy: Wrought; Anglo-Swiss Aluminium Co.; free machining
AVIONAL 150	4.5% Cu 1.5% Mg 0.8% Mn 1.0% Cr Al alloy: Wrought; Anglo-Swiss Aluminium Co.
AVIONAL 660	4.5% Cu 0.5% Mg 0.7% Si 0.8% Mn Al alloy: Star
AVIONAL 662	4.5% Cu 0.5% Mg 0.7% Si 0.8% Mn Al alloy: Star
AWCO 31	5% Cu Al alloy: Welding wire; Aluminium Wire Co.; as drawn
AWCO 301	4% Cu 0.6% Mg 0.5% Mn Al alloy: Wire; Aluminium Wire Co. for BS alloy N14
AWCO 304	2.2% Cu 0.3% Mg Al alloy: Wire for rivets; Aluminium Wire Co. for BS alloy N13
AWCO 305	4.4% Cu 0.6% Mg 0.7% Si 0.6% Mn Al alloy: Wire; Aluminium Wire Co. for BS alloy N15
AWCO 308	5.5% Cu 0.5% Pb 0.5% Bi Al alloy: Bar etc.; Aluminium Wire Co.; free machining
AWCO 309	4.5% Cu 1.6% Mg 0.6% Mn Al alloy: Bar; Aluminium Wire Co.
B 18S	4% Cu 1.5% Mg 2% Ni Al alloy: Previous AA designation; now 2218
BA 35	3.5% Cu 0.7% Mg 0.5% Sb 0.2% Sn Al alloy: Extrusion; British Aluminium Co.; as rolled **UTS: 240 Elon: 10% Proof: 110**
BA 301	4% Cu 0.6% Mg 0.5% Mn Al alloy: Sheet, strip and tube; British Aluminium Co. for BS alloy N14
BA 303	4.4% Cu 0.6% Mg 0.7% Si 0.6% Mn Al alloy: Sheet and tube; British Aluminium Co. for BS alloy N15
BA 305	4.6% Cu 0.7% Mg 0.8% Si 0.7% Mn Al alloy: British Aluminium Co. for BS alloy N15
BA 306	1.7% Cu 0.7% Mg 1% Si 0.6% Mn Al alloy: British Aluminium Co. for BS alloy N11
BA 307	2% Cu 0.9% Mg 0.8% Si 1% Ni 0.1% Ti Al alloy: British Aluminium Co.; solution treated and aged **UTS: 370 Elon: 8% Proof: 280**
BA 308	5.5% Cu 0.5% Pb 0.5% Bi Al alloy: Bar; British Aluminium Co.; solution treated and aged; free machining **UTS: 300 Elon: 8% Proof: 240**
BA 309	4.3% Cu 1.6% Mg 0.5% Mn 0.05% Pb Al alloy: Plate; British Aluminium Co.; solution treated and aged **UTS: 420 Elon: 10% Proof: 250**
BA 352	4% Cu 0.6% Mg 0.5% Mn Al alloy: Sheet and strip; British Aluminium Co. for BS alloy N14
BA 353	4.4% Cu 0.6% Mg 0.7% Si 0.6% Mn Al alloy: Sheet and strip; British Aluminium Co. for BS alloy N15
BIRMETAL 161	1.5% Cu 1% Mg 1% Si Al alloy: Birmabright for BS alloy H11

Symbol	Nominal analysis, supplier, condition and remarks.
BIRMETAL 230	Cu Mg Al alloy: For rivets; Birmabright **UTS: 270**
BIRMETAL 477	4% Cu 1% Mg 0.5% Si 0.7% Mn Al alloy: Birmabright for BS alloy H14
BIRMETAL 478	4% Cu 0.5% Mg 0.5% Si 1% Mn Al alloy: Birmabright for BS alloy H15
BMB 473	4% Cu 0.6% Mg 0.5% Mn Al alloy: Sheet and strip; Birmabright for BS alloy H14
BMB 551	4.4% Cu 0.6% Mg 0.7% Si 0.6% Mn Al alloy: Forging; Birmabright for BS alloy H15
BMB 761	1.5% Cu 1% Mg 1% Si Al alloy: Forging; Birmabright; solution treated and aged **UTS: 370 Elon: 10% Proof: 300**
BRILLUM	1.5% Cu 2% Ni Al alloy: ICI; obsolete; now Imperial Aluminium Co.
BS 395	Al alloy (Duralumin): Sheet and strip; replaced by BS 1470
BS 396	Al alloy (Duralumin): Tube; replaced by BS 1471
BS 478	Al alloy bar (Y alloy): Replaced by BS 1476
BS 533	Al alloy forging (Y alloy): Replaced by BS 1472
BS 1080	3% Cu 0.5% Mg Al alloy: Bar **UTS: 240 Elon: 10% Proof: 110**
BS 1470 HS14	4% Cu 1% Mg 0.75% Mn Al alloy: For sheet and strip; solution treated; room aged **UTS: 370 Elon: 15% Proof: 120**
BS 1470 HS15	4% Cu 0.5% Mg 0.7% Si 0.75% Mn Al alloy: For sheet and strip; solution treated; room aged **UTS: 370 Elon: 15% Proof: 220**
BS 1470 2014A Clad	4.2% Cu 0.5% Mg 0.7% Si 0.8% Mn Al alloy: Plate and sheet; solution treated and aged; replaces H15 designation; clad **UTS: 430 Elon: 7% Proof: 375**
BS 1471 HT14	4% Cu 1% Mg 0.75% Mn Al alloy: Tube **UTS: 390 Elon: 10% Proof: 270**
BS 1471 HT15	4% Cu 0.5% Mg 0.7% Si 0.75% Mn Al alloy: For tube **UTS: 420 Elon: 8% Proof: 350**
BS 1471 2014A	4.5% Cu 0.4% Mg 0.8% Mn Al alloy: Drawn tube; solution treat and aged; British Aluminium Co. **UTS: 450 Elon: 6% Proof: 370**
BS 1472 HF11	1.5% Cu 1% Mg 1% Si Al alloy: For forging and bar; solution treated and aged **UTS: 360 Elon: 8% Proof: 280**
BS 1472 HF12	2% Cu 1% Mg 1% Si 1% Fe 1% Ni Al alloy: For forging and bar; solution treated and aged **UTS: 370 Elon: 8% Proof: 280**
BS 1472 HF14	4% Cu 0.6% Mg 0.5% Mn Al alloy: For forging; solution treated; room aged **UTS: 360 Elon: 15% Proof: 200**
BS 1472 HF15	4% Cu 0.75% Mg 0.75% Si 1% Mn Al alloy: For forging and bar; solution treated and aged **UTS: 420 Elon: 8% Proof: 370**
BS 1472 HF16	2.2% Cu 1.5% Mg 1.2% Si 1.1% Ni Al alloy: Forging; solution treated and aged **UTS: 430 Elon: 5% Proof: 340**
BS 1472 HF17	4% Cu 1.5% Mg 2% Ni Al alloy: For forging; solution treated and aged **UTS: 330 Elon: 8%**
BS 1472 HF18	2% Cu 1.5% Mg 1% Ni 1% Fe Al alloy: For forging and bar; solution treated and aged **UTS: 360 Elon: 6% Proof: 250**
BS 1472 2014A	4.5% Cu 0.6% Mg 0.7% Si 0.8% Mn Al alloy: Extrusion; solution treated and aged; British Aluminium Co. **UTS: 480 Elon: 7% Proof: 430**
BS 1472 2618A	2.3% Cu 1.5% Mg 0.2% Si 1% Ni Al alloy: Forging; solution treated and aged; British Aluminium Co. **UTS: 430 Elon: 5% Proof: 340**
BS 1472 5083	4.5% Mg 0.7% Mn 0.1% Cr Al alloy: Forging; as forged **UTS: 280 Elon: 12% Proof: 130**
BS 1473 HB15	4% Cu 0.75% Mg 0.75% Si 1% Mn Al alloy: For bolt stock; solution treated and aged **UTS: 420 Proof: 360**
BS 1473 HR15	4% Cu 0.7% Mg 0.7% Si 1% Mn Al alloy: For rivets; solution treated; room aged **UTS: 370**
BS 1473 2014A	4.2% Cu 0.6% Mg 0.7% Si 1% Mn Al alloy: Bolt and screw stock; solution treated; room aged **UTS: 385**
BS 1474 HV11	1.5% Cu 1% Mg 1% Si Al alloy: For tubes; solution treated and aged **UTS: 330 Elon: 8% Proof: 270**
BS 1474 HV14	4% Cu 1% Mg 0.5% Si 0.75% Mn Al alloy: For tubes; solution treated; room aged **UTS: 360 Elon: 12% Proof: 210**
BS 1474 HV15	4% Cu 0.75% Mg 0.75% Si 1% Mn Al alloy: For tubes; solution treated and aged **UTS: 420 Elon: 8% Proof: 360**
BS 1474 2014A	4.4% Cu 0.6% Mg 0.8% Si 0.8% Mn Al alloy: Bar and extruded tube; solution treated and aged **UTS: 380 Elon: 10% Proof: 250**
BS 1475 HG15	4% Cu 0.6% Mg 0.7% Si 1% Mn Al alloy: Wire; solution treated and aged **UTS: 420**
BS 1475 2014A	4.5% Cu 0.6% Mg 0.7% Si 1% Mn Al alloy: Wire; solution treated and aged; TF **UTS: 430**
BS 1476 HE11	1.5% Cu 1% Mg 1% Si Al alloy: Bar; solution treated and aged **UTS: 370 Elon: 8% Proof: 300**
BS 1476 HE14	4% Cu 1% Mg 0.5% Si 0.7% Mn Al alloy: Bar; solution treated; room aged **UTS: 370 Elon: 10% Proof: 220**
BS 1476 HE15	4% Cu 0.5% Mg 0.5% Si 1% Mn Al alloy: Bar; solution treated and aged **UTS: 450 Elon: 8% Proof: 390**
BS 1477 HP14	4% Cu 1% Mg 0.5% Si 0.7% Mn Al alloy: Plate; solution treated and aged **UTS: 370 Elon: 10% Proof: 220**
BS 1477 HP15	4% Cu 0.5% Mg 0.7% Si 1% Mn Al alloy: Plate; solution treated and aged **UTS: 410 Elon: 8% Proof: 310**
BS 4300/5 FC 1	5.5% Cu 0.5% Bi 0.5% Pb Al alloy: Bar; solution treated and aged; free machining **UTS: 300 Elon: 6% Proof: 200**
BS 4300/5 201	5.5% Cu Al alloy: Bar and extruded tube; solution treated and aged; British Aluminium Co. **UTS: 300 Elon: 6% Proof: 230**
BS T4	Aluminium alloy: Tube; obsolete; see L 62
CB 60A	5.5% Cu Al alloy: Previous ASTM designation; now 2011
CG 30A	2.5% Cu 0.3% Mg Al alloy: Previous ASTM designation; now 2117
CG 42A	4.2% Cu 1% Mg 0.7% Mn Al alloy: Previous ASTM designation; now 2024
CM 41A	4% Cu 0.6% Mg 0.8% Mn Al alloy: Previous ASTM designation; now 2017
CS 41A	4.5% Cu 0.6% Mg 1% Si 0.8% Mn Al alloy: Previous ASTM designation; now 2014
CONLOY	4.5% Cu Al alloy: Tube; Constrictor Ltd; obsolete **DPN: 75 UTS: 220 Proof: 170**
D 16	4.2% Cu 1.6% Mg 0.5% Mn Al alloy: Russian designation
D 18	2.5% Cu 0.3% Mg Al alloy: Russian designation
DECOLTAL 500	5.5% Cu 0.4% Si 0.4% Pb Al alloy: Wrought; Anglo-Swiss Aluminium Co.; free machining
DTD 130 B	2% Cu 1% Mg 0.15% Ti Al alloy: Bar and forgings; solution treated and aged **UTS: 370 Elon: 9% Proof: 300**

Symbol	Nominal analysis, supplier, condition and remarks.
DTD 147	4% Cu 0.6% Mg 0.5% Mn Al alloy: Forgings; solution treated **UTS: 360 Elon: 15% Proof: 200**
DTD 150 A	4% Cu 0.6% Mg 0.5% Mn Al alloy: Forgings; solution treated and room aged **UTS: 360 Elon: 15% Proof: 200**
DTD 246 B	2% Cu 1% Mg 1% Si 1% Ni 0.1% Ti Al alloy: Forgings; solution treated and aged **DPN: 85 UTS: 240 Elon: 16% Proof: 105**
DTD 273	4% Cu 1.4% Mg 0.4% Mn Al alloy: Tube; solution treated **UTS: 440 Elon: 7% Proof: 330**
DTD 327	2.2% Cu 0.3% Mg Al alloy: Wire; solution treated **UTS: 250**
DTD 364	4% Cu 0.7% Mg 0.7% Si Al alloy: Forging
DTD 410	2% Cu Mg Si Fe Ni Al alloy: Forging **DPN: 100 UTS: 340 Elon: 6% Proof: 220**
DTD 423 C	1.5% Cu 1% Mg 1% Si Al alloy: Bar and forging; solution treated and aged **UTS: 370 Elon: 10% Proof: 280**
DTD 443 A	1.5% Cu 1% Mg 1% Si Al alloy: Bar and forging; solution treated; room aged **UTS: 270 Elon: 15% Proof: 170**
DTD 450 A	1.5% Cu 1% Mg 1% Si Al alloy: Tube; solution treated; room aged **UTS: 250 Elon: 12% Proof: 170**
DTD 460 A	1.5% Cu 1% Mg 1% Si Al alloy: Tube; solution treated and aged **UTS: 370 Elon: 6% Proof: 310**
DTD 464	Cu Mn Al alloy: Replaced by BS alloy L63
DTD 520	Cu Si Al alloy: Obsolete
DTD 546 B	4.4% Cu 0.6% Mg 0.7% Si 0.6% Mn Al alloy: Sheet; solution treated and aged **UTS: 400 Elon: 8% Proof: 310**
DTD 603 B	4.4% Cu 0.6% Mg 0.7% Si 0.6% Mn Al alloy: Sheet; solution treated; room aged **UTS: 390 Elon: 15% Proof: 240**
DTD 646 B	4.4% Cu 0.6% Mg 0.7% Si 0.6% Mn Al alloy: Sheet; solution treated and aged **UTS: 420 Elon: 8% Proof: 340**
DTD 706	4.4% Cu 0.7% Mg 0.7% Si 0.6% Mn Al alloy: Sheet; solution treated and aged **UTS: 420 Elon: 8% Proof: 340**
DTD 717	2% Cu 1.5% Mg 1% Fe 1% Ni Al alloy: Forgings; solution treated and aged; for high temperature use **UTS: 390 Elon: 12% Proof: 240**
DTD 724	2.2% Cu 1.5% Mg 1.0% Fe 1.2% Ni 0.1% Ti Al alloy: Forging; solution treated and aged **UTS: 390 Elon: 10% Proof: 250**
DTD 731 A	2.2% Cu 1.5% Mg 1% Ni 1% Fe Al alloy: Forging; solution treated and aged **UTS: 420 Elon: 6% Proof: 310**
DTD 745	2.2% Cu 1.5% Mg 1% Fe 1% Ni Al alloy: Forging; solution treated and aged **UTS: 400 Elon: 6% Proof: 300**
DTD 746	4.2% Cu 0.7% Mg 0.7% Si 1% Mn Al alloy: Sheet; solution treated and aged **UTS: 420 Elon: 8% Proof: 330**
DTD 5004	6% Cu 0.3% Mn Al alloy: Forging; solution treated and aged; for high temperature use **UTS: 360 Elon: 6% Proof: 210**
DTD 5010	4.2% Cu 0.7% Mg 0.7% Si 1% Mn Al alloy: Plate; solution treated and aged **UTS: 390 Elon: 12% Proof: 240**
DTD 5014	2.2% Cu 1.5% Mg 1% Fe 1% Ni Al alloy: Bar; solution treated and aged **UTS: 390 Elon: 7% Proof: 300**
DTD 5020 A	4.4% Cu 0.7% Mg 0.7% Si 0.6% Mn Al alloy: Plate; solution treated and aged **UTS: 390 Elon: 7% Proof: 330**
DTD 5030	4.4% Cu 0.7% Mg 0.7% Si 0.6% Mn Al alloy: Plate; solution treated and room aged **UTS: 370 Elon: 12% Proof: 220**
DTD 5084	2.2% Cu 1.6% Mg 1.2% Fe 1.2% Ni Al alloy: Forging; solution treated and aged; for high temperature use **UTS: 370 Elon: 8% Proof: 280**
DTD 5090	4.4% Cu 1.5% Mg 0.6% Mn Al alloy: Plate; solution treated; room aged **UTS: 400 Elon: 12% Proof: 270**
DU BRAND	4% Cu 0.6% Mg 0.5% Mn Al alloy: Forging and tube; High Duty Alloys; see Hiduminium 01
DURAL B	4% Cu 0.6% Mg 0.5% Mn Al alloy: Sheet, tube and forging; Alcan for BS alloy H14
DURAL E	1.8% Cu 0.8% Mg 1% Si 0.7% Mn Al alloy: Tube and forging: Alcan for BS alloy H11
DURAL G	4.2% Cu 1.5% Mg 0.2% Si 0.6% Mn Al alloy: Sheet, tube and forging; Alcan; solution treated; room aged **UTS: 420 Elon: 12% Proof: 270**
DURAL J	2.2% Cu 1.5% Mg 0.8% Si 1% Ni 0.1% Ti Al alloy: Forging; Alcan for BS alloy H18
DURAL JJ	2.2% Cu 1.5% Mg 0.15% Si 1% Fe 1.1% Ni Al alloy: Tube and forging; Alcan: solution treated and aged **UTS: 390 Elon: 6% Proof: 300**
DURAL M	2.2% Cu 0.3% Mg Al alloy: Wire for rivets; Alcan
DURAL Q	4% Cu 1.4% Mg 0.4% Mn Al alloy: Tube; Alcan; solution treated **UTS: 440 Elon: 7% Proof: 330**
DURAL S	4.4% Cu 0.6% Mg 0.7% Si 0.6% Mn Al alloy: Sheet, tube and forging; Alcan for BS alloy H15
DURAL T	2% Cu 1% Mg 0.8% Si 1% Ni 0.1% Ti Al alloy: Forging; Alcan; solution treated and aged **UTS: 370 Elon: 8% Proof: 280**
DURICILIUM F	4% Cu 0.7% Mg 0.7% Si 1% Mn Al alloy: Tube; E Kaye for BS alloy H14/15
DURICILIUM H	4% Cu 1% Mg 0.5% Si 0.7% Mn Al alloy: E Kaye for BS alloy H14
DURICILIUM K	4% Cu 0.5% Mg 0.5% Si 1% Mn Al alloy: E Kaye for BS alloy H15
DURICILIUM M	1.5% Cu 1% Mg 1% Si Al alloy: E Kaye for BS alloy H11
FMA	5.5% Cu 0.5% Bi 0.5% Pb Al alloy: Extrusions; free machining; McKechnie; solution treated and aged **DPN: 100 UTS: 330 Elon: 10% Proof: 250**
H 11	1.5% Cu 1% Mg 1% Si Al alloy: Wrought; designation used by BS
H 12	2% Cu 1% Mg 1% Si 1% Fe 1% Ni Al alloy: Wrought; designation used by BS
H 14	4% Cu 1% Mg 0.7% Mn Al alloy: Wrought; designation used by BS
H 15	4% Cu 0.5% Mg 0.7% Si 0.7% Mn Al alloy: Wrought; designation used by BS
H 16	2.2% Cu 1.5% Mg 1.2% Si 1.1% Ni Al alloy: Wrought; designation used by BS
H 17	4% Cu 1.5% Mg 2% Ni Al alloy: Wrought; designation used by BS
H 18	2% Cu 1.5% Mg 1% Ni 1% Fe Al alloy: Wrought; designation used by BS
HA 4 CG42	4.3% Cu 1.6% Mg 0.6% Mn Al alloy: Sheet and plate; Canadian Standard

Note. The following abbreviations and units are used in the tables:

DPN	Hardness, diamond pyramid number
UTS	Ultimate tensile strength, N/mm^2
Elon	Elongation, %
Proof	0.1% proof strength, N/mm^2

$1\ N/mm^2 = 0.1\ hbar = 0.102\ kgf/mm^2 = 0.06475\ tonf/in.^2 = 145.04\ lbf/in.^2 = 1\ MPa$

See Appendix II for other abbreviations and conversion tables.

Symbol	Nominal analysis, supplier, condition and remarks.
HA 5 CB60	5.5% Cu Al alloy: Bar, etc.; Canadian Standard
HA 5 CG42	4.3% Cu 1.6% Mg 0.6% Mn Al alloy: Bar, etc.; Canadian Standard
HA 5 CS41N	4.5% Cu 0.6% Mg 0.8% Si 0.8% Mn Al alloy: Bar, etc.; Canadian Standard
HA 6 CG30	2.6% Cu 0.4% Mg Al alloy: For rivets and brazing wire; Canadian Standard
HA 6 CM41	4.0% Cu 0.6% Mg 0.6% Si 0.7% Mn Al alloy: For rivets and brazing wire; Canadian Standard
HA 7 CG42	4.3% Cu 1.6% Mg 0.6% Mn Al alloy: Tube; Canadian Standard
HA 8 CN42	4.0% Cu 0.7% Mg 2.0% Ni Al alloy: Forging; Canadian Standard
HA 8 CS41N	4.5% Cu 0.6% Mg 0.8% Si 0.8% Mn Al alloy: Forging; Canadian Standard
HA 8 CS41P	4.5% Cu 0.6% Mg 0.8% Mn Al alloy: Forging; Canadian Standard
HA 8 ZG62	1.6% Cu 2.5% Mg 0.2% Cr 5.5% Zn Al alloy: Forging; Canadian Standard
HIDUMINIUM 01	4% Cu 0.6% Mg 0.5% Mn Al alloy: For forgings and tube. High Duty Alloys for BS alloy H14 (DU BRAND)
HIDUMINIUM 02	4% Cu 1.5% Mg 2% Ni Al alloy: For pistons and cylinder heads; High Duty Alloys for BS alloy H17 (Y alloy)
HIDUMINIUM 03	1% Cu 1% Mg 1% Si Al alloy: For forgings and extrusions; High Duty Alloys for BS alloy H11
HIDUMINIUM 55	2.5% Cu 1% Mg 1% Si 0.3% Mn 0.7% Ni Al alloy: For forgings; High Duty Alloys for BS alloy H12
HIDUMINIUM 66	4.5% Cu 0.6% Mg 0.7% Si 0.6% Mn Al alloy: For forgings; High Duty Alloys for BS alloy H15
HIDUMINIUM 72	4.5% Cu 1.6% Mg 0.6% Mn Al alloy: Plate; High Duty Alloys; solution treated and aged **UTS: 420 Elon: 10% Proof: 250**
IMPALCO 660	4.1% Cu 1% Mn Al alloy: Bar and forgings, etc.; Imperial Aluminium Co. for BS alloy H15
IMPALCO 690	1.5% Cu 1% Mg Al alloy: Bar and forgings; Imperial Aluminium Co. for BS alloy H11
IMPALCO 800	5.5% Cu 0.4% Pb 0.4% Bi Al alloy: Bar and tube; Imperial Aluminium Co.; solution treated and aged; free machining **UTS: 220 Elon: 11% Proof: 300**
IMPALCO 810	15% Cu 0.7% Mg 1% Si 1% Pb Al alloy: Bar and forgings; Imperial Aluminium Co.; solution treated and aged; free machining **UTS: 270 Elon: 10% Proof: 220**
IMPALCO 830	2.3% Cu 1% Mg 1% Si 1% Ni Al alloy: Bar and forging; Imperial Aluminium Co. for BS alloy H12
IMPALCO 840	2.2% Cu 1.5% Mg 1.3% Si 1% Ni Al alloy: Forging; Imperial Aluminium Co. for BS alloy H18
IMPALCO 860	4% Cu 1.5% Mg 2% Ni Al alloy: Forging; Imperial Aluminium Co.; solution treated and aged **UTS: 340 Elon: 15% Proof: 210**
IMPALCO 2024	4.2% Cu 1.6% Mg 0.6% Mn 0.2% Cr Al alloy: Plate; Imperial Aluminium Co.; solution treated and aged; temporary designation **UTS: 420 Elon: 10% Proof: 250**
IMPALCO C66	4% Cu 0.7% Mg 1% Mn 1% Fe Al alloy: Sheet, strip and rod; Imperial Aluminium Co. for BS alloy H15
IMPALCO C69	1.5% Cu 1% Mg 1% Si 1% Mn Al alloy: Imperial Aluminium Co. for BS alloy H11
IMPALCO C80	5.5% Cu 0.5% Pb Al alloy: For bar and section; Imperial Aluminium Co.; free machining **UTS: 280 Elon: 12% Proof: 210**
KYNAL 90	2.2% Cu 0.3% Mg Al alloy: ICI: obsolete; now Imperial Aluminium Co.
KYNAL C65	4% Cu 0.6% Mg 0.5% Mn Al alloy: ICI: obsolete; now Imperial Aluminium Co.
KYNAL C66	4.4% Cu 0.6% Mg 0.7% Si 0.6% Mn Al alloy: ICI: obsolete; now Imperial Aluminium Co.
KYNAL C67	4.4% Cu 0.6% Mg 0.7% Si 0.6% Mn Al alloy: ICI; obsolete; now Imperial Aluminium Co.
KYNAL Y88	2.6% Cu 1% Mg 1% Si 1% Ni 0.1% Ti Al alloy: ICI: obsolete; now Imperial Aluminium Co.
KYNAL Y92	4% Cu 1.5% Mg 2% Ni Al alloy: ICI: obsolete; now Imperial Aluminium Co.
L1	4% Cu Mg Mn Al alloy: For forgings; obsolete; replaced by BSL 64
L3	Al alloy: Sheet and coils; obsolete; no information available
L24	4% Cu 1.5% Mg 2% Ni Al alloy: Y alloy; obsolete
L25	4% Cu 1.5% Mg 2% Ni Al alloy: Forging; solution treated and aged **UTS: 360 Elon: 15% Proof: 210**
L37	4% Cu 0.75% Si 1% Mn Al alloy: For wires and rivets; cold drawn; can be solution treated and aged **UTS: 370**
L39	4.4% Cu 0.6% Mg 0.7% Si 0.7% Mn Al alloy: Forging; solution treated; room aged **UTS: 370 Elon: 15% Proof: 220**
L40	Cu Mg Si Al alloy: Obsolete; replaced by L 65
L42	2.2% Cu 1.5% Mg 0.8% Si 1.2% Ni 1% Fe 0.1% Ti Al alloy: Forging; solution treated and aged **UTS: 370 Elon: 8% Proof: 250**
L43	Cu Mg Al alloy: Forgings; obsolete; replaced by L 65
L45	Cu Al alloy: Forgings; obsolete; replaced by L 65
L57	2.2% Cu 0.3% Mg Al alloy: Wire; solution treated **UTS: 250**
L62	4% Cu 0.75% Mg 0.75% Si 0.8% Mn Al alloy: For tubes; solution treated; room aged **UTS: 390 Proof: 270**
L63	4% Cu 0.75% Mg 0.75% Si 0.8% Mn Al alloy: For tubes; solution treated and aged **UTS: 440 Proof: 340**
L64	4% Cu 0.75% Mg 0.75% Si 0.8% Mn Al alloy: For bars and forgings; solution treated; room aged **UTS: 340 Elon: 13% Proof: 210**
L65	4% Cu 0.75% Mg 0.75% Si 0.8% Mn Al alloy: For bars and forgings; solution treated and aged **UTS: 440 Elon: 8% Proof: 370**
L69	Cu Mg Al alloy: Wire; obsolete; replaced by L 86
L70	4% Cu 0.75% Mg 0.75% Si 0.8% Mn Al alloy: For sheet and strip; solution treated; room aged **UTS: 370 Elon: 15% Proof: 240**
L71	4% Cu 0.75% Mg 0.75% Si 0.8% Mn Al alloy: For sheet and strip; solution treated and aged **UTS: 420 Elon: 8% Proof: 340**
L76	4.4% Cu 0.7% Mg 0.7% Si 0.6% Mn Al alloy: Bar and forging; solution treated; room aged **UTS: 360 Elon: 15% Proof: 200**
L77	4.4% Cu 0.7% Mg 0.7% Si 0.6% Mn Al alloy: Bar and forging; solution treated and aged **UTS: 440 Elon: 8% Proof: 370**
L83	2% Cu 1% Mg 0.9% Si 1% Ni 0.9% Fe Al alloy: Bar and forging; solution treated and aged **UTS: 390 Elon: 10% Proof: 270**
L84	1.5% Cu 0.8% Mg 1% Si Al alloy: Bar; solution treated and aged **UTS: 300 Elon: 15% Proof: 180**
L85	1.5% Cu 0.8% Mg 1% Si Al alloy: Bar and forging; solution treated and aged **UTS: 360 Elon: 10% Proof: 270**
L86	2.5% Cu 0.4% Mg Al alloy: For gold forged rivets; cold drawn; can be solution treated and aged **UTS: 250**
L87	4.4% Cu 0.7% Mg 0.7% Si 0.6% Mn Al alloy: Bar; solution treated and aged **UTS: 420 Elon: 8% Proof: 360**
L102	4.5% Cu 0.6% Mg 0.7% Si 0.7% Mn Al alloy: Bar
L103	4.3% Cu 0.6% Mg 0.7% Si 0.8% Mn Al alloy: Forging
L105	4.2% Cu 0.6% Mg 0.7% Si 0.6% Mn Al alloy: Tube

Symbol	Nominal analysis, supplier, condition and remarks.
L109	4.2% Cu 1.6% Mg 0.5% Mn Al alloy: Clad sheet and strip; clad with type 1050A
L110	4.2% Cu 1.6% Mg 0.5% Mn Al alloy: Clad sheet and strip; clad with type 1050A
L119	5% Cu 0.25% Mn 1.5% Ni Al alloy: Casting; solution treated and aged **UTS: 215 Elon: 1% Proof: 190**
L154	4.2% Cu 1.5% Si Al alloy: Chill cast; solution treated and aged **UTS: 265 Elon: 13% Proof: 165**
L155	4.2% Cu 1.25% Si Al alloy: Chill cast; Solution treated and aged **UTS: 310 Elon: 9% Proof: 200**
L156	4.4% Cu 0.6% Mg 0.7% Si 0.8% Mn Al alloy: Sheet and strip
L157	4.2% Cu 0.6% Mg 0.7% Si 0.7% Mn Al alloy: Sheet and strip
L158	4.2% Cu 0.6% Mg 0.7% Si 0.6% Mn Al alloy: Sheet and strip; close tolerance as L157
L159	4.2% Cu 0.6% Mg 0.7% Si 0.6% Mn Al alloy: Sheet and strip; close tolerance as L157
L163	4.2% Cu 0.6% Mg 0.7% Si 0.6% Mn Al alloy: Clad; sheet and strip; clad with 1050A
L164	4.2% Cu 0.6% Mg 0.7% Si 0.6% Mn Al alloy: Clad; sheet and strip; clad with 1050A
L165	4.2% Cu 0.6% Mg 0.7% Si 0.6% Mn Al alloy: Clad; sheet and strip; clad with 1050A
L166	4.2% Cu 0.6% Mg 0.7% Si 0.6% Mn Al alloy: Clad; sheet and strip; as L164; close tolerance
L167	4.2% Cu 0.6% Mg 0.7% Si 0.6% Mn Al alloy: Clad; sheet and strip; as L165; close tolerance
L168	4.3% Cu 0.6% Mg 0.7% Si 0.6% Mn Al alloy: Bar
ML 2	5.5% Cu 0.4% Si 0.4% Pb 0.4% Bi Al alloy: Bar; AALCO; free machining; American equivalent 2011T3 **DPN: 95 UTS: 300 Elon: 10% Proof: 240**
NORAL 16S	2.2% Cu 0.3% Mg Al alloy: Wire for rivets; Northern Aluminium Co.
NORAL 17S	4% Cu 0.6% Mg 0.5% Mn Al alloy: Sheet, strip and tube; Northern Aluminium Co. for BS alloy H14
NORAL 19S	4% Cu 1.5% Mg 2% Ni Al alloy: Forging; Northern Aluminium Co. for BS alloy H17
NORAL 24S	4% Cu 1.4% Mg 0.4% Mn Al alloy: Tube; Northern Aluminium Co.; solution treated **UTS: 440 Elon: 7% Proof: 330**
NORAL 26S	4.4% Cu 0.6% Mg 0.7% Si 0.6% Mn Al alloy: Sheet tube and forging; Northern Aluminium Co. for BS alloy H15
NORAL 42S	2.2% Cu 1.5% Mg 0.8% Si 1% Ni 0.1% Ti Al alloy: Forgings; Northern Aluminium Co. for BS alloy H18
QQ A 35lb	4.0% Cu 0.4% Mg Al alloy: Bar and forging; US Federal; solution treated; room aged **UTS: 360 Elon: 12% Proof: 210**
QQ A 353a	4.0% Cu 0.6% Mg 0.6% Mn Al alloy: Sheet; US Federal **UTS: 370 Elon: 12% Proof: 200**
QQ A 354a	4.1% Cu 1.2% Mg 0.5% Mn Al alloy: Forging; US Federal; solution treated and aged **UTS: 420 Elon: 12% Proof: 270**
QQ A 355A	4.2% Cu 1.5% Mg 0.2% Cr Al alloy: Sheet; US Federal; Solution treated and aged **UTS: 440 Elon: 15% Proof: 250**
QQ A 361	4.0% Cu 0.6% Mg 0.2% Cr Al alloy: Sheet; US Federal; solution treated and aged **UTS: 340 Elon: 14% Proof: 180**
QQ A 362/1	4.2% Cu 1.5% Mg 0.2% Cr Al alloy: Sheet; US Federal **UTS: 370 Elon: 13% Proof: 180**
QQ A 367b/1	4.5% Cu 0.5% Mg 1.0% Mn Al alloy: Forging; US Federal; solution treated and aged **DPN: 125 UTS: 450 Elon: 10% Proof: 300**
RR 56	2% Cu 0.9% Mg 0.8% Si 1% Ni 0.1% Ti Al alloy: For forgings and extrusions; High Duty Alloys specification for DTD 130B DTD 410
RR 57	6% Cu 0.25% Mn 0.1% Ti Al alloy: For forging; High Duty Alloys; solution treated and aged **DPN: 110 UTS: 370 Elon: 8% Proof: 220**
RR 58	2% Cu 1.5% Mg 1.2% Ni 0.1% Ti Al alloy: For forging; High Duty Alloys for DTD 717 DTD 724
RR 59	2% Cu 1.5% Mg 0.8% Si 1.2% Ni 0.1% Ti Al alloy: For forging; High Duty Alloys for BS alloy H18
RR 257	6% Cu 1% Ni 0.25% Co 0.25% Sb 0.2% Ti Al alloy: High Duty Alloys; solution treated and aged **UTS: 340 Elon: 8% Proof: 220**
RR S R	2.5% Cu 0.1% Mg 0.75% Ni 0.1% Ti Al alloy: For forgings; High Duty Alloys; used for sealing rings **DPN: 70**
SAE 24	4.5% Cu 1.5% Mg 0.6% Mn Al alloy: Wrought; solution treated; now 2024 **UTS: 440 Elon: 12% Proof: 270**
SAE 26	4% Cu 0.5% Mg 0.5% Mn Al alloy: Bar, wire, etc.; solution treated; now 2017 **UTS: 390 Elon: 16% Proof: 210**
SAE 202	5.5% Cu 0.4% Pb 0.4% Bi Al alloy: For free machining; solution treated, cold worked and aged; now 2011 **DPN: 100 UTS: 420 Elon: 12% Proof: 300**
SAE 203	4% Cu 1.5% Mg 1% Ni Al alloy: Solution treated and stabilized; now 2218 **DPN: 95 UTS: 330 Elon: 11% Proof: 240**
SAE 204	2.6% Cu 0.4% Mg Al alloy: Wrought; now 2117
SAE 260	4.5% Cu 0.4% Mg 0.8% Si 0.8% Mn Al alloy: Bar and forging; solution treated and aged; now 2014 **UTS: 460 Elon: 10% Proof: 390**
SAE 270	4% Cu 0.5% Mg 2% Ni Al alloy: Forgings; solution treated and aged; now 2018 **UTS: 390 Elon: 10% Proof: 270**
SS 4335	4.2% Cu 1% Mg 0.8% Mn 1.2% Pb Al alloy: Bar; Swedish Standard **DPN: 120 UTS: 420 Elon: 6% Proof: 300**
SS 4338	4.5% Cu 0.6% Mg 0.9% Si 0.8% Mn Al alloy: Bar, forging, etc.; Swedish Standard; as rolled **DPN: 130 UTS: 450 Elon: 8% Proof: 340**
STA 7 AW8	3.4% Cu 0.7% Mg 0.5% Sb 0.2% Sn Al alloy: Bar (free cutting); as rolled; free cutting; obsolete **UTS: 240 Elon: 10% Proof: 105**
STA 7 AW11A	1% Cu 1% Mg 1% Si Al alloy: Bar; replaced by BS 1476 HE11
STA AW11B	1% Cu 1% Mg 1% Si Al alloy: Bar; replaced by BS 1476 HE 11
STA 7 AW12	2.6% Cu 1% Mg 1% Si 0.3% Mn 1% Fe 0.7% Ni Al alloy: Forging; replaced by BS 1472 HF 12
STA 7 AW13	2.2% Cu 0.3% Mg Al alloy: Wire; replaced by BS 1473-5
STA 7 AW14	4% Cu 0.6% Mg 0.5% Mn Al alloy: Wire; replaced by BS 1473-5 HR15
STA 7 AW15A	4% Cu 0.6% Mg 0.5% Mn Al alloy: Forging and bar; replaced by BS alloy H14
STA 7 AW15B	4.4% Cu 0.6% Mg 0.7% Si 0.6% Mn Al alloy: Forging; replaced by BS alloy H15

Note. The following abbreviations and units are used in the tables:

DPN	Hardness, diamond pyramid number
UTS	Ultimate tensile strength, N/mm^2
Elon	Elongation, %
Proof	0.1% proof strength, N/mm^2

$1\ N/mm^2 = 0.1\ hbar = 0.102\ kgf/mm^2 = 0.06475\ tonf/in.^2 = 145.04\ lbf/in.^2 = 1\ MPa$

See Appendix II for other abbreviations and conversion tables.

Symbol	Nominal analysis, supplier, condition and remarks.
STA 7 AW15C	4.4% Cu 0.6% Mg 0.7% Si 0.6% Mn Al alloy: Tube; replaced by BS 1471 HT15
STA 7 AW15D	4.4% Cu 0.6% Mg 0.7% Si 0.6% Mn Al alloy: Tube; replaced by BS 1471 HT15
STA 7 AW15E	4% Cu 0.6% Mg 0.5% Mn Al alloy: Sheet; replaced by BS 1470 HS14
STA 7 AW15G	4.4% Cu 0.6% Mg 0.7% Si 0.6% Mn Al alloy: Sheet; replaced by BS 1470 HS15
STA 7 AW15H	4.4% Cu 0.6% Mg 0.7% Si 0.6% Mn Al alloy: Sheet; replaced by BS 1470 HS15
STA 7 AW17B	4% Cu 1.5% Mg 2% Ni Al alloy: Forgings; replaced by BS 1472 HF17
STA 7 AW18	2.2% Cu 1.5% Mg 0.8% Si 1% Fe 1.2% Ni 0.1% Ti Al alloy: Forging; replaced by BS 1472 HF18
TI 01	4% Cu 0.6% Mg 0.5% Mn Al alloy: Tube; British Aluminium Co. for BS alloy H14
TI 04	4% Cu 0.6% Mg 0.5% Mn Al alloy: Sheet and strip; British Aluminium Co. for BS alloy H14
TI 53	3.5% Cu 0.7% Mg 0.5% Sb 0.2% Sn Al alloy: Extrusions; British Aluminium Co.; as rolled **UTS: 240 Elon: 10% Proof: 105**
TI 55	2.6% Cu 1% Mg 1% Si 0.3% Mn 0.7% Ni Al alloy: Forgings; British Aluminium Co. for BS alloy H12
TI 56	2% Cu 1% Mg 0.8% Si 1% Ni 0.1% Ti Al alloy: Forging; British Aluminium Co.; solution treated and aged **UTS: 400 Elon: 10% Proof: 320**
TI 66	4.4% Cu 0.6% Mg 0.7% Si 0.6% Mn Al alloy: Sheet and tube; British Aluminium Co. for BS alloy H15
UNI 3022 MAC33	33% Cu Al alloy: Ingot; primary metal; Italian Standard
UNI 3577	2.5% Cu 0.3% Mg 0.3% Si Al alloy: For forging; Italian Standard
UNI 3578	3.5% Cu 0.6% Mg 0.6% Si 1.5% Fe 0.6% Ni Al alloy: For forging; Italian Standard
UNI 3579	4.0% Cu 0.5% Mg 0.5% Mn Al alloy: For forging; Italian Standard
UNI 3581	4.4% Cu 0.4% Mg 0.8% Si 0.8% Mn Al alloy: For forging; Italian Standard
UNI 3583	4.5% Cu 1.5% Mg 0.6% Mn Al alloy: For forging; Italian Standard
UNI 6362	5.5% Cu 0.5% Pb 0.5% Bi Al alloy: For forging; Italian Standard; free machining
UNI 7256	4.5% Cu Al alloy: Italian specification
V Al Cu	5% Cu Al alloy: Previous German Standard designation
V Al Cu Mn 1	3.3% Cu Al alloy: Previous German Standard designation
W 11	1.5% Cu 0.8% Mg 1% Si 1% Mn Al alloy: Southern Forge; replaced by IMPALCO 690
W 12	2.2% Cu 1% Mg 1% Si 1% Fe 1% Ni Al alloy: Southern Forge; replaced by IMPALCO 830
W 17	4% Cu 1.5% Mg 2% Ni Al alloy: Southern Forge; replaced by IMPALCO 860
W 18	2.2% Cu 1.5% Mg 1% Si 1% Fe 1% Ni Al alloy: Southern Forge; replaced by IMPALCO 840
W 150	4% Cu 0.8% Mg 1% Si 1% Mn Al alloy: Southern Forge; replaced by IMPALCO 660
W 152	1.5% Cu 1% Mg 1% Si 1% Pb Al alloy: Free machining; Southern Forge; replaced by IMPALCO 810
WW T 785	4.2% Cu 1.5% Mg 0.6% Mn 0.2% Cr Al alloy: Tube; US Federal; solution treated and aged **UTS: 420 Elon: 13% Proof: 270**
WW T 786a	4.0% Cu 0.6% Mg 0.6% Mn Al alloy: Tube; US Federal; solution treated and aged **UTS: 370 Elon: 13% Proof: 270**
Y ALLOY	4% Cu 5% Mg 2% Ni Al alloy: For pistons and cylinder heads; High Duty Alloys; HIDUMINIUM 02
ZIRKONAL	15% Cu 0.5% Si 0.8% Mn Al alloy: Origin unknown

Note. The following abbreviations and units are used in the tables:

DPN	Hardness, diamond pyramid number
UTS	Ultimate tensile strength, N/mm^2
Elon	Elongation, %
Proof	0.1% proof strength, N/mm^2

$1\ N/mm^2 = 0.1\ hbar = 0.102\ kgf/mm^2 = 0.06475\ tonf/in.^2 = 145.04\ lbf/in.^2 = 1\ MPa$

See Appendix II for other abbreviations and conversion tables.

1G Aluminium–zinc wrought alloys
With copper, manganese, magnesium and chromium additions

Specific gravity		2.01–2.84
Density		2010–2840 kg/in.^{3}
Solidus/liquidus		477–629 °C
Thermal conductivity	solution treated and aged	134 W/m °C
Coefficient of linear expansion		20–23.2 × 10^{-6}/ °C
Electrical conductivity		30% IACS (copper 100%)
Specific resistance	annealed	30–37 microhm mm
	solution treated and aged	50–57 microhm mm
Young's modulus of elasticity		71.02 × 10^9 N/m^2
Impact	solution treated and aged	7 J
Fatigue strength	solution treated and aged	± 150 N/mm^2
Hot strength		

Temperature °C	Tensile strength N/mm^2
100	480–550
150	330–370
200	150–200
250	90
300	60

The above properties are chosen to be typical of the group of specifications as a whole. There is some variation within the group, and certain properties may not apply to some specifications.

General metallurgical properties

The alloys listed in this section have up to 7% zinc, seldom having less than 4% zinc. They also have approximately 3% magnesium and can have reasonable quantities of copper with titanium, manganese and chromium also present.

These have the highest tensile strength of any aluminium alloys, the strength being obtained by thermal treatment, i.e. by solution treating and ageing (precipitation hardening).

The alloys have relatively poor corrosion resistance and this limits, to a large extent, their use in normal industry.

These alloys do not accept cold work without becoming brittle and this also limits their usefulness.

In the correctly heat treated condition these materials have a yield and tensile strength comparable with that of mild steel, with approximately one-third the weight of the latter.

Welding is not advised on these materials without specialist advice.

These alloys are used in the main in the aircraft industry where considerable control is necessary regarding their poor corrosion resistance. Items such as landing gear and many major structural items on aircraft are manufactured from these materials which are then carefully corrosion protected and subjected to frequent inspection.

The materials should be considered where the strength-to-weight ratio is important, and where corrosion is not a significant problem.

If corrosion protection is required, this will generally be by anodizing, often followed by painting.

Symbol	Nominal analysis, supplier, condition and remarks.
3.0001	5% Cu 2.5% Mg 5.5% Zn Al alloy: German Standard DIN 1725
3.0002	4.5% Cu 2% Mg 3% Zn Al alloy: German Standard DIN 1725
3.0120	4% Cu 1% Mg 1.5% Zn Al alloy: German Standard DIN 1725
3.4335	1.2% Mg 0.2% Mn 0.2% Cr 4.5% Zn Al alloy: German Standard
3.4345	0.7% Cu 3.1% Mg 0.4% Mn 0.2% Cr 4.5% Zn Al alloy: German Standard DIN 1725
3.4355	2.7% Mg 0.5% Mn 0.2% Cr 4.8% Zn Al alloy: German Standard DIN 1725
3.4365	1.6% Cu 2.5% Mg 0.2% Cr 5.6% Zn Al alloy: German Standard DIN 1725
3.4375	1% Cu 2.7% Mg 0.5% Cr 5% Zn Al alloy: German Standard DIN 1725
75S	0.4% Cu 2.7% Mg 0.5% Mn 5% Zn Al alloy: Tube; Alcoa; solution treated and aged **UTS: 450 Elon: 8% Proof: 400**
75S	1.5% Cu 2.5% Mg 0.3% Cr 5.5% Zn Al alloy: Previous AA designation – now 7075
76S	1% Cu 2.7% Mg 5% Zn Al alloy: Bar; Alcoa; solution treated and aged **UTS: 500 Elon: 460**
7001	0.2% Cu 3% Mg 7.5% Zn Al alloy: Wrought; designation used in the UK and USA
7004	1.5% Mg 0.5% Mn 4.5% Zn Al alloy: Wrought; designation used in the UK and USA

Note. The following abbreviations and units are used in the tables:

DPN	Hardness, diamond pyramid number
UTS	Ultimate tensile strength, N/mm^2
Elon	Elongation, %
Proof	0.1% proof strength, N/mm^2

1 N/mm^2=0.1 hbar=0.102 kgf/mm^2=0.06475 tonf/in.^{2}=145.04 lbf/in.^{2}=1 MPa

See Appendix II for other abbreviations and conversion tables.

Symbol	Nominal analysis, supplier, condition and remarks.
7005	1.4% Mg 0.5% Mn 0.1% Cr 4.5% Zn Al alloy: Wrought; designation used in the UK and USA
7008	1% Mg 0.2% Cr 5% Zn Al alloy: Wrought; designation used in the UK and USA
7010	1.7% Cu 2.4% Mg 6.3% Zn Al alloy: Wrought; designation used in the UK and USA
7011	1.3% Mg 0.2% Mn 0.1% Cr 4.7% Zn Al alloy: Wrought; designation used in the UK and USA
7013	1.2% Mn 1.7% Zn Al alloy: Wrought; designation used in the UK and USA
7016	0.6% Cu 1.1% Mg 4.5% Zn Al alloy: Wrought; designation used in the UK and USA
7021	1.6% Mg 5.5% Zn Al alloy: Wrought; designation used in the UK and USA
7029	0.7% Cu 1.7% Mg 4.7% Zn Al alloy: Wrought; designation used in the UK and USA
7039	2.8% Mg 0.2% Cr 4% Zn Al alloy: Wrought; designation used in the UK and USA
7046	1.3% Mg 7.1% Zn Al alloy: Wrought; designation used in the UK and USA
7049	2% Cu 2.4% Mg 0.18% Cr 7.7% Zn Al alloy: Wrought; designation used in the UK and USA
7050	2.3% Cu 2.3% Mg 6.2% Zn Al alloy: Wrought; designation used in the UK and USA
7070	1.5% Zn Al alloy: Wrought; designation used in the UK and USA
7072	1% Zn Al alloy: Wrought; designation used in the UK and USA
7075	1.8% Cu 2.5% Mg 0.23% Cr 5.6% Zn Al alloy: Wrought; designation used in the UK and USA
7076	0.7% Cu 1.8% Mg 0.5% Mn 7.5% Zn Al alloy: Wrought; designation used in the UK and USA
7079	0.6% Cu 3.4% Mg 0.2% Cr 4.2% Zn Al alloy: Wrought; designation used in the UK and USA
7090	1% Cu 2.5% Mg 1.5% Co 8% Zn Al alloy: Wrought; designation used in the UK and USA
7091	1.4% Cu 2.5% Mg 0.4% Co 6.4% Zn Al alloy: Wrought; designation used in the UK and USA
7104	0.7% Mg 4% Zn Al alloy: Wrought; obsolete
7108	5% Zn 0.2% Zr Al alloy: For cladding
7116	0.8% Cu 4.7% Zn Al alloy: Wrought; designation used in the UK and USA
7129	0.7% Cu 1.7% Mg 4.7% Zn Al alloy: Wrought; designation used in the UK and USA
7146	1.5% Cu 2.5% Mg 0.17% Cr 7.7% Zn Al alloy: Wrought; designation used in the UK and USA
7149	1.5% Cu 2.5% Mg 0.17% Cr 7.7% Zn Al alloy: Wrought; designation used in the UK and USA
7150	2.2% Cu 2.3% Mg 6.4% Zn Al alloy: Wrought; designation used in the UK and USA
7175	1.6% Cu 2.5% Mg 0.22% Cr 5.6% Zn Al alloy: Wrought; designation used in the UK and USA
7178	2% Cu 2.8% Mg 0.25% Cr 6.8% Zn Al alloy: Wrought; designation used in the UK and USA
7179	0.6% Cu 3.2% Mg 0.2% Mn 0.15% Cr 4.2% Zn Al alloy: Wrought; designation used in the UK and USA
7229	0.7% Cu 1.7% Mg 4.7% Zn Al alloy: Wrought; designation used in the UK and USA
7277	1.2% Cu 2% Mg 0.2% Cr 4% Zn Al alloy: Wrought; designation used in the UK and USA
7472	1.3% Mg 1.6% Zn Al alloy: Wrought; designation used in the UK and USA
7475	1.7% Cu 2.4% Mg 0.2% Cr 5.7% Zn Al alloy: Wrought; designation used in the UK and USA
ALCOA 740	0.5% Cu 2.7% Mg 0.5% Mn 5.7% Zn Al alloy: Wrought; Alcoa; solution treated and aged **UTS: 550 Elon: 8% Proof: 510**
ALCOA 760	1.0% Cu 2.7% Mg 0.5% Mn 5.5% Zn Al alloy: Wrought; Alcoa; solution treated and aged **UTS: 630 Elon: 7% Proof: 580**
ALCAN 74S	1.8% Mg 4.3% Zn Al alloy: Sheet, tube, bar, etc.; Alcan; solution treated and aged; weldable **UTS: 370 Elon: 15% Proof: 300**
ALCAN GB B75S	0.7% Cu 2.5% Mg 0.2% Mn 0.12% Cr 5.7% Zn Al alloy: Alcan; solution treated and aged **DPN: 170 UTS: 600 Elon: 10% Proof: 550**
ALCAN GB C75S	0.7% Cu 2.7% Mg 0.7% Mn 5.5% Zn Al alloy: Alcan; solution treated and aged **UTS: 590 Elon: 11% Proof: 540**
ALCAN GB C77S	1.7% Cu 2.2% Mg 0.12% Mn 0.12% Cr 7.2% Zn Al alloy: Alcan; solution treated and aged **DPN: 175 UTS: 610 Elon: 10% Proof: 570**
ALCAN GB M75S	1.3% Cu 2.5% Mg 0.2% Mn 0.12% Cr 6% Zn Al alloy: Alcan **DPN: 110 UTS: 250 Elon: 12% Proof: 150**
Al Zn Mg 3	2.7% Mg 0.2% Cr 5% Zn Al alloy: Wrought; previous German Standard designation
Al Zn Mg Cu 0.5	0.8% Cu 3% Mg 0.2% Cr 4.5% Zn Al alloy: Wrought; previous German Standard designation
Al Zn Mg Cu 1.5	1.8% Cu 2.7% Mg 0.3% Cr 5.5% Zn Al alloy: Wrought; previous German Standard designation
AMS 4024 A	0.6% Cu 3% Mg 0.2% Mn 0.2% Cr 4.3% Zn Al alloy: Sheet; AMS for AA alloy 7079; solution treated and aged
AMS 4038	1.6% Cu 2.5% Mg 0.3% Cr 5.6% Zn Al alloy: Plate; AMS for AA alloy 7075; solution treated; cold worked and aged
AMS 4044 C	1.6% Cu 2% Mg 0.25% Cr 5.6% Zn Al alloy: Plate; AMS for AA alloy 7075; annealed
AMS 4045 C	1.6% Cu 2.5% Mg 0.25% Cr 5.6% Zn Al alloy: Sheet; AMS for AA alloy 7075; solution treated and aged
AMS 4050	2.2% Cu 2.2% Mg 6.2% Zn Al alloy
AMS 4078	1.6% Cu 2.5% Mg 0.2% Cr 5.6% Zn Al alloy
AMS 4084	1.7% Cu 2.2% Mg 0.2% Cr 5.7% Zn Al alloy
AMS 4083	1.7% Cu 2.2% Mg 0.2% Cr 5.7% Zn Al alloy
AMS 4089	1.5% Cu 2.2% Mg 0.2% Cr 5.7% Zn Al alloy: Wrought
AMS 4090	1.5% Cu 2.2% Mg 0.2% Cr 5.7% Zn Al alloy: Wrought
AMS 4107	2.3% Cu 2.2% Mg 6.1% Zn Al alloy
AMS 4108	2.3% Cu 2.2% Mg 6.1% Zn Al alloy
AMS 4111	1.5% Cu 2.5% Mg 0.2% Cr 7.7% Zn Al alloy
AMS 4122 C	1.6% Cu 2.5% Mg 0.3% Cr 5.6% Zn Al alloy: Bar; AMS for AA alloy 7075; solution treated and aged
AMS 4123 A	1.5% Cu 2.5% Mg 0.3% Cr 5.6% Zn Al alloy: Bar; AMS for AA alloy 7075; cold drawn
AMS 4124	1.8% Cu 2.5% Mg 0.22% Cr 5.6% Zn Al alloy
AMS 4126	1.8% Cu 2.5% Mg 0.22% Cr 5.6% Zn Al alloy
AMS 4131	1.8% Cu 2.5% Mg 0.22% Cr 5.6% Zn Al alloy
AMS 4136	0.6% Cu 3.3% Mg 0.2% Cr 4.3% Zn Al alloy: Forging; AMS for AA alloy 7079; solution treated and aged
AMS 4137 A	0.7% Cu 1.6% Mg 0.5% Mn 7.3% Zn Al alloy: Forging; AMS for AA alloy 7076; solution treated and aged
AMS 4138	0.6% Cu 3.3% Mg 0.2% Mn 0.2% Cr 4.3% Zn Al alloy: Forging; AMS for AA alloy 7079; solution treated and aged
AMS 4139 F	1.6% Cu 2.5% Mg 0.25% Cr 5.6% Zn Al alloy: Forging; AMS for AA alloy 7075; solution treated and aged
AMS 4141	1.8% Cu 2.5% Mg 0.22% Cr 5.6% Zn Al alloy
AMS 4147	1.7% Cu 2.5% Mg 0.2% Cr 5.6% Zn Al alloy: Wrought
AMS 4148	1.7% Cu 2.5% Mg 0.2% Cr 5.6% Zn Al alloy: Wrought
AMS 4149	1.7% Cu 2.5% Mg 0.2% Cr 5.6% Zn Al alloy: Wrought
AMS 4154 F	1.6% Cu 2.5% Mg 0.3% Cr 5.6% Zn Al alloy: Extrusion; AMS for AA alloy 7075; solution treated and aged

Symbol	Nominal analysis, supplier, condition and remarks.
AMS 4158 A	2% Cu 2.7% Mg 0.3% Cr 6.8% Zn Al alloy: Extrusion; AMS for AA alloy 7178; solution treated and aged
AMS 4166	1.7% Cu 2.5% Mg 0.2% Cr 5.6% Zn Al alloy: Wrought
AMS 4167	1.7% Cu 2.5% Mg 0.2% Cr 5.6% Zn Al alloy: Wrought
AMS 4168 A	1.6% Cu 2.5% Mg 0.3% Cr 5.6% Zn Al alloy: Extrusion; AMS for AA alloy 7075; solution treated and aged
AMS 4169 B	1.6% Cu 2.5% Mg 0.3% Cr 5.6% Zn Al alloy: Extrusion; AMS for AA alloy 7075; solution treated and aged
AMS 4170	1.6% Cu 2.5% Mg 0.25% Cr 5.6% Zn Al alloy: Extrusion; AMS for AA alloy 7075; solution treated and aged
AMS 4171 A	0.6% Cu 3.3% Mg 0.2% Mn 0.2% Cr 4.3% Zn Al alloy: Extrusion; AMS for AA alloy 7079; solution treated and aged
AMS 4174	1.7% Cu 2.5% Mg 0.2% Cr 5.6% Zn Al alloy: Wrought
AMS 4179	1.7% Cu 2.5% Mg 5.6% Zn 0.2% Cr Al alloy: Wrought
AMS 4186	1.7% Cu 2.4% Mg 0.2% Cr 5.6% Zn Al alloy: Wrought
AMS 4187	1.7% Cu 2.4% Mg 0.2% Cr 5.6% Zn Al alloy: Wrought
AMS 4200	1.7% Cu 2.4% Mg 0.2% Cr 7.7% Zn Al alloy: Wrought
AMS 4201	2.3% Cu 2.3% Mg 6.1% Zn Al alloy: Wrought
AMS 4202	1.7% Cu 2.3% Mg 0.2% Cr 5.7% Zn Al alloy: Wrought
AMS 4203	1.7% Cu 2.3% Mg 6.2% Zn Al alloy: Wrought
AMS 4204	1.7% Cu 2.3% Mg 6.2% Zn Al alloy: Wrought
AMS 4205	1.7% Cu 2.3% Mg 6.2% Zn Al alloy: Wrought
AMS 4207	1.5% Cu 2.2% Mg 0.2% Cr 5.7% Zn Al alloy: Wrought
AMS 4247	1.7% Cu 2.4% Mg 0.2% Cr 7.7% Zn Al alloy: Wrought
AMS 4306	2.2% Cu 2.3% Mg 6.4% Zn Al alloy: Wrought
AMS 4307	2.2% Cu 2.3% Mg 6.4% Zn Al alloy: Wrought
AMS 4310	1.7% Cu 2.4% Mg 0.2% Cr 5.6% Zn Al alloy: Wrought
AMS 4311	1.7% Cu 2.4% Mg 0.2% Cr 5.6% Zn Al alloy: Wrought
AMS 4320	1.5% Cu 2.4% Mg 0.2% Cr 7.7% Zn Al alloy: Wrought
AMS 4321	1.5% Cu 2.4% Mg 0.2% Cr 7.7% Zn Al alloy: Wrought
AMS 4323	1.7% Cu 2.5% Mg 0.2% Cr 5.6% Zn Al alloy: Wrought
AMS 4333	2.3% Cu 2.3% Mg 6.1% Zn Al alloy: Wrought
AMS 4340	2.3% Cu 2.2% Mg 6.2% Zn Al alloy: Wrought
AMS 4341	2.3% Cu 2.2% Mg 6.2% Zn Al alloy: Wrought
AMS 4342	2.3% Cu 2.2% Mg 6.2% Zn Al alloy: Wrought
AMS 4343	1.5% Cu 2.4% Mg 0.2% Cr 7.7% Zn Al alloy: Wrought
AMS 4344	1.7% Cu 2.5% Mg 0.2% Cr 5.6% Zn Al alloy: Wrought

Note. The following abbreviations and units are used in the tables:

DPN	Hardness, diamond pyramid number
UTS	Ultimate tensile strength, N/mm^2
Elon	Elongation, %
Proof	0.1% proof strength, N/mm^2

$1\ N/mm^2 = 0.1\ hbar = 0.102\ kgf/mm^2 = 0.06475\ tonf/in.^2 = 145.04\ lbf/in.^2 = 1\ MPa$

See Appendix II for other abbreviations and conversion tables.

Symbol	Nominal analysis, supplier, condition and remarks.
ASTM B209/7039	2.8% Mg 0.3% Mn 0.2% Cr 4% Zn Al alloy: Plate and sheet
ASTM B209/7075	1.7% Cu 2.5% Mg 0.3% Cr 5.5% Zn Al alloy: Sheet; annealed **UTS: 270 Elon: 10% Proof: 140**
ASTM B209/7075	1.7% Cu 2.5% Mg 0.3% Cr 5.5% Zn Al alloy: Sheet; solution treated and aged **UTS: 530 Elon: 6% Proof: 470**
ASTM B210/7075	1.6% Cu 2.4% Mg 0.3% Cr 5.6% Zn Al alloy: Tube
ASTM B211/7075	1.5% Cu 2.5% Mg 0.3% Cr 5.5% Zn Al alloy: Bar; solution treated and aged **UTS: 530**
ASTM B221/7075	1.5% Cu 2.5% Mg 0.3% Cr 5.5% Zn Al alloy: Extruded bar; solution treated and aged **UTS: 590 Elon: 7% Proof: 510**
ASTM B221/7079	0.6% Cu 3% Mg 0.2% Mn 0.2% Cr 4% Zn Al alloy: Extruded bar; solution treated and aged **UTS: 540 Elon: 7% Proof: 480**
ASTM B221/7178	2% Cu 2.8% Mg 0.3% Cr 7% Zn Al alloy: Extruded bar; solution treated and aged **UTS: 600 Elon: 5% Proof: 540**
ASTM B235/7075	1.7% Cu 2.5% Mg 0.3% Cr 5.5% Zn Al alloy: Extruded tube; solution treated and aged **UTS: 590 Elon: 7% Proof: 510**
ASTM B241/7075	1.6% Cu 2.5% Mg 0.3% Cr 5.6% Zn Al alloy: Pipe and tube
ASTM B241/7079	0.6% Cu 3.3% Mg 0.2% Mn 0.15% Cr 4.3% Zn Al alloy: Pipe and tube
ASTM B241/7178	2.0% Cu 2.8% Mg 0.3% Cr 6.8% Zn Al alloy: Pipe and tube
ASTM B247/7075	1.6% Cu 2.5% Mg 0.2% Cr 5.5% Zn Al alloy: Forging; solution treated and aged **DPN: 135 UTS: 540 Elon: 10% Proof: 460**
ASTM B285/R ZG61A	0.6% Mg 0.5% Cr 5.7% Zn 0.2% Ti Al alloy: Welding wire
ASTM B316/7075	1.7% Cu 2.5% Mg 0.2% Cr 5.5% Zn Al alloy: Rivets; annealed **UTS: 270**
ASTM B327 ZG71A	0.85% Mg 0.7% Zn Al alloy: Hardener for Zn base die casting
AWCO 701	1% Cu 3% Mg 6% Zn Al alloy: Wire; Aluminium Wire Co.; solution treated and aged **UTS: 490 Elon: 7% Proof: 420**
A Z3G2	2.2% Mg 0.2% Cr 3.5% Zn Al alloy: Wrought; French Standard
A Z4G	2.0% Mg 0.5% Mn 0.3% Cr 3.0% Zn Al alloy: Wrought; L'Aluminium Français
A Z5G	1.2% Mg 0.8% Mn 0.15% Cr 4.5% Zn Al alloy: Wrought; French Standard
A Z5GU	1.0% Cu 2.8% Mg 5.7% Zn Al alloy: Wrought; L'Aluminium Français
A Z5GU	1.8% Cu 2.4% Mg 0.2% Cr 5.7% Zn Al alloy: Wrought; French Standard
A Z6G	2.0% Mg 0.5% Mn 0.3% Cr 6.0% Zn Al alloy: Wrought; L'Aluminium Français
A Z8GU	2.0% Cu 2.0% Mg 0.3% Cr 7.2% Zn Al alloy: Wrought: L'Aluminium Français
BA 703	0.3% Cu 2.5% Mg 0.2% Mn 0.1% Cr 5.7% Zn Al alloy: British Aluminium Co.; solution treated and aged **UTS: 490 Elon: 6% Proof: 420**
BA 704	1.5% Cu 2.2% Mg 0.2% Mn 0.1% Cr 7% Zn Al alloy: British Aluminium Co.; solution treated and aged **UTS: 540 Elon: 5% Proof: 480**
BA 705	0.5% Cu 2.5% Mg 0.5% Mn 5.7% Zn Al alloy: British Aluminium Co.; solution treated and aged **UTS: 510 Elon: 6% Proof: 440**
BA 706	1% Cu 2.79% Mg 0.5% Mn 5.7% Zn Al alloy: British Aluminium Co.; solution treated and aged **UTS: 540 Elon: 5% Proof: 460**

Symbol	Nominal analysis, supplier, condition and remarks.
BA 707	1.3% Cu 2.9% Mg 0.4% Mn 0.1% Cr 5.7% Zn Al alloy: British Aluminium Co.; solution treated and aged
	UTS: 480 Elon: 6% Proof: 400
BBZ 36	1% Cu 2.7% Mg 5% Zn Al alloy: Bar; Birmabright; solution treated and aged
	UTS: 510 Elon: 5% Proof: 460
BIRMETAL 212	1.6% Cu 2.5% Mg 0.15% Cr 0.2% Zn Al alloy: In wrought form; Birmetal for DTD 5074
BIRMETAL Z36	3% Cu 5% Zn Mg Al alloy: Birmabright
	DPN: 170 UTS: 600 Proof: 520
BMB 1306	0.4% Cu 2.7% Mg 0.5% Mn 5% Zn Al alloy: Birmabright; solution treated and aged
	UTS: 480 Elon: 5% Proof: 400
BMB 2308	1% Cu 2.7% Mg 5% Zn Al alloy: Bar; Birmabright; solution treated and aged
	UTS: 530 Elon: 5% Proof: 460
BS 918	Cu Zn Al alloy: Bars; obsolete
BS 4300/14–7020	1.2% Mg 0.3% Mn 4.5% Zn Al alloy: Sheet and plate; solution treated and aged; TF
	UTS: 320 Elon: 10% Proof: 270
BS 4300/14 HS17	1.2% Mg 0.5% Mn 4.2% Zn Al alloy: Sheet and plate; solution treated and aged; TF condition
	UTS: 320 Elon: 10% Proof: 270
BS 4300/15.7020	1.2% Mg 0.3% Mn 0.2% Cr 4.5% Zn Al alloy: Bar and extruded tube; solution treated and aged TF
	UTS: 340 Elon: 10% Proof: 280
BS 4300/15 HE17	1.2% Mg 0.3% Mn 0.2% Cr 4.5% Zn Al alloy: Bar and tube
DTD 363 A	3% Cu 5% Zn Mg Al alloy: Extrusions
	UTS: 510 Elon: 5% Proof: 450
DTD 683 A	Cu Mg Zn Mn Al alloy: For forgings; replaced by DTD 5024
DTD 693	0.4% Cu 2.7% Mg 0.5% Mn 5.3% Zn Al alloy: Tube; solution treated and aged
	UTS: 450 Elon: 8% Proof: 400
DTD 5024	0.5% Cu 2.5% Mg 0.5% Mn 5.7% Zn Al alloy: Forgings; solution treated and aged
	UTS: 470 Elon: 7% Proof: 400
DTD 5034	0.5% Cu 2.7% Mg 0.2% Cr 5.5% Zn Al alloy: Forgings; solution treated and aged
	UTS: 470 Elon: 7% Proof: 400
DTD 5044	0.5% Cu 2.7% Mg 0.5% Mn 5.5% Zn Al alloy: Bar; solution treated and aged
	UTS: 480 Elon: 7% Proof: 450
DTD 5050 B	0.9% Cu 2.7% Mg 0.16% Cr 5.7% Zn Al alloy: Plate; solution treated and aged
	UTS: 480 Elon: 7% Proof: 450
DTD 5054	0.5% Cu 2.7% Mg 0.16% Cr 5.7% Zn Al alloy: Bar; solution treated and aged
	UTS: 520 Elon: 7% Proof: 450
DTD 5064	1.1% Cu 2.8% Mg 0.5% Mn 5.5% Zn Al alloy: Bar; solution treated and aged
	UTS: 540 Elon: 5% Proof: 450
DTD 5074	1.6% Cu 2.5% Mg 0.16% Cr 6.2% Zn Al alloy: Bar; solution treated and aged
	UTS: 540 Elon: 5% Proof: 470
DTD 5094	0.5% Cu 2.5% Mg 0.5% Mn 5.5% Zn Al alloy: Forgings; solution treated and aged
	UTS: 460 Elon: 8% Proof: 400
DTD 5104 A	0.5% Cu 2.8% Mg 0.5% Mn 5.8% Zn Al alloy: Forgings
DTD 5110	1.8% Cu 2.6% Mg 0.15% Cr 5.8% Zn Al alloy: Clad as designation 7075; clad with 7072
DTD 5114	0.5% Cu 2.8% Mg 0.2% Cr 5.6% Zn Al alloy: Bar
DTD 5120 B	1.7% Cu 2.5% Mg 6.2% Zn Al alloy: as designation 7010
DTD 5124	1.8% Cu 2.6% Mg 0.2% Cr 5.6 Zn Al alloy: Bar
DTD 5124	2.5% Cu 0.5% Mn 5.5% Zn Al alloy: Forging; similar to DTD 5024 with improved stress corrosion resistance
DTD 5130	1.7% Cu 2.5% Mg 6.2% Zn 0.15% Zr Al alloy
DTD 5636	1.7% Cu 2.5% Mg 6.2% Zn 0.15% Zr Al alloy
DURALUMIN KC	1.8% Cu 2.5% Mg 6% Zn 0.05% Pb Al alloy: Bar and forging; Alcan; solution treated and aged
	UTS: 530 Elon: 5% Proof: 450
DURALUMIN LC	0.5% Cu 2.8% Mg 0.1% Cr 5.5% Zn 0.05% Pb Al alloy: Forgings; Alcan; solution treated and aged
	UTS: 520 Elon: 7% Proof: 440
DURAL K	0.4% Cu 2.7% Mg 0.5% Mn 5% Zn Al alloy: Forgings and tube; Alcan; solution treated and aged
	UTS: 530 Elon: 5% Proof: 450
DURAL L	0.6% Cu 3% Mg 0.4% Mn 5.8% Zn Al alloy: Tube, forging and plate; Alcan; solution treated and aged
	UTS: 480 Elon: 5% Proof: 400
DURICILIUM N	Zn-containing Al alloy: For welded structure; E Kaye; solution treated
	UTS: 330 Elon: 14% Proof: 220
DURICILIUM P	1.6% Cu 2.5% Mg 0.2% Cr 6.2% Zn Al alloy: E Kaye; solution treated and aged
	UTS: 600 Elon: 6% Proof: 530
DURICILIUM X	1% Cu 2.8% Mg 0.5% Mn 5.5% Zn Al alloy: E Kaye; solution treated and aged
	UTS: 600 Elon: 6% Proof: 530
DURICILIUM Z	0.5% Cu 2.7% Mg 0.16% Cr 5.7% Zn Al alloy: E Kaye; solution treated and aged
	UTS: 540 Elon: 8% Proof: 480
HA 4 ZG62	1.6% Cu 2.5% Mg 0.2% Cr 5.6% Zn Al alloy: Plate; Canadian Standard
HA4 ZG62 (ALCLAD)	1.6% Cu 2.5% Mg 0.2% Cr 5.6% Zn Al alloy: Plate clad with Al; Canadian Standard
HA 5 ZG62	1.6% Cu 2.5% Mg 0.2% Cr 5.6% Zn Al alloy: Bar, etc.; Canadian Standard
HA 7 ZG62	1.6% Cu 2.5% Mg 0.2% Cr 5.5% Zn Al alloy: Tube; Canadian Standard
HIDUMINIUM 48	Zn Mg Al alloy: Wrought; High Duty Alloys
	UTS: 495 Elon: 12% Proof: 450
HIDUMINIUM 78	0.5% Cu 2.8% Mg 0.1% Mn 0.15% Cr 6% Zn Al alloy: Bar and plate; High Duty Alloys; solution treated and aged
	UTS: 400 Elon: 6% Proof: 440
HIDUMINIUM 89	1.4% Cu 2.5% Mg 0.1% Mn 0.2% Cr 5.8% Zn Al alloy: Bar; High Duty Alloys; solution treated and aged
	UTS: 550 Elon: 5% Proof: 460
IMPALCO 740	0.4% Cu 3% Mg 5.7% Zn Al alloy: Bar and forgings; Imperial Aluminium Co.; solution treated and aged
	UTS: 500 Elon: 7% Proof: 440
IMPALCO 750	0.5% Cu 2.8% Mg 0.2% Cr 5.8% Zn Al alloy: Bar and forgings; Imperial Aluminium Co.; solution treated and aged
	UTS: 520 Elon: 7% Proof: 440
IMPALCO 760	1% Cu 2.8% Mg 5.5% Zn Al alloy: Bar; Imperial Aluminium Co.; solution treated and aged
	UTS: 540 Elon: 5% Proof: 460
IMPALCO 770	1.7% Cu 2.5% Mg 6% Zn Al alloy: Bar; Imperial Aluminium Co.; solution treated and aged
	UTS: 540 Elon: 5% Proof: 460
KYNAL Z93	0.4% Cu 2.7% Mg 5.3% Zn Al alloy: ICI; obsolete; now Imperial Aluminium Co.
L 88	1.8% Cu 2.6% Mg 0.2% Cr 5.8% Zn Al alloy: Sheet and strip; solution treated and aged; TF
	UTS: 480 Elon: 8% Proof: 420
L 160	1.8% Cu 2.5% Mg 0.2% Cr 5.6% Zn Al alloy: Bar
L 161	1.8% Cu 2.5% Mg 0.2% Cr 5.6% Zn Al alloy: Forging
L 162	1.8% Cu 2.5% Mg 0.2% Cr 5.6% Zn Al alloy: Forging
NORAL C77S	1% Cu 2.7% Mg 5% Zn Al alloy: Bar; Northern Aluminium Co.; solution treated and aged
	UTS: 560 Elon: 5% Proof: 460
NORAL M75S	0.4% Cu 2.7% Mg 0.5% Mn 5% Zn Al alloy: Forgings; Northern Aluminium Co.; solution treated and aged
	UTS: 490 Elon: 7% Proof: 420

Symbol	Nominal analysis, supplier, condition and remarks.
PERUNAL	1.5% Cu 2.3% Mg 0.2% Mn 0.2% Cr 6.2% Zn 0.1% Ti Al alloy: Anglo-Swiss Aluminium Co.; solution treated and aged **DPN: 180 UTS: 600 Proof: 530**
PERUNAL 215	1.5% Cu 2.2% Mg 0.2% Mn 0.2% Cr 5.7% Zn 0.1% Ti Al alloy: Wrought; Anglo-Swiss Aluminium Co.
Pu	1.5% Cu 2.3% Mg 0.2% Mn 0.2% Cr 6.2% Zn 0.1% Ti Al alloy: Anglo-Swiss Aluminium Co.; code for Perunal
RR 77	0.4% Cu 2.7% Mg 0.5% Mn 5% Zn Al alloy: For forgings and tubes; High Duty Alloys for DTD 683, DTD 693
RR 88	1% Cu 2.7% Mg 0.5% Mn 5% Zn Al alloy: For extrusions; High Duty Alloys for DTD 363A
SAE 215	1.8% Cu 2.5% Mg 0.2% Cr 5.5% Zn Al alloy: Solution treated and aged; now 7075 **UTS: 570 Elon: 7% Proof: 510**
STA 7 AW16A	1.0% Cu 2.7% Mg 0.5% Mn 5.3% Zn Al alloy: Bar; solution treated and aged; obsolete **UTS: 540 Elon: 5% Proof: 460**
TI 77	0.4% Cu 2.7% Mg 0.5% Mn 5% Zn Al alloy: Tube; British Aluminium Co.; solution treated and aged **UTS: 450 Elon: 8% Proof: 400**
TI 88	1% Cu 2.7% Mg 5% Zn Al alloy: Bar; British Aluminium Co.; solution treated and aged **UTS: 560 Elon: 5% Proof: 460**
UNI 3735	1.6% Cu 2.5% Mg 5.8% Zn Al alloy: For forging; Italian Standard
UNI 3737	1.6% Cu 2.5% Mg 7.8% Zn Al alloy: For forging; Italian Standard
UNI 3738	1.6% Cu 2.5% Mg 7.8% Zn Al alloy: For forging; Italian Standard
UNI 7791	1.2% Mg 0.3% Mn 4.5% Zn Al alloy: Italian Standard
UNIDOR 080	0.8% Mg 4.3% Zn Al alloy: Wrought; Anglo-Swiss Aluminium Co.
UNIDOR 100	1.2% Mg 0.4% Mn 0.2% Cr 4.2% Zn Al alloy: Wrought; Anglo-Swiss Aluminium Co.
URAL 1	4% Cu 1% Mg 1.5% Zn Al alloy: German Standard
URAL III	4.5% Cu 2% Mg 3% Zn Al alloy: German Standard
URAL IV	5% Cu 2.5% Mg 5.5% Zn Al alloy: German Standard
V 95	1.6% Cu 2.5% Mg 0.2% Cr 5.6% Zn Al alloy: Russian designation
W 16	1.8% Cu 2.5% Mg 0.2% Cr 6% Zn Al alloy: Southern Forge; replaced by IMPALCO 770
W 26	1% Cu 2.8% Mg 0.5% Mn 5.5% Zn Al alloy: Southern Forge; replaced by IMPALCO 760
W 160	0.5% Cu 2.8% Mg 0.2% Cr 5.8% Zn Al alloy: Southern Forge; replaced by IMPALCO 750
W 260	0.5% Cu 2.8% Mg 5.8% Zn Al alloy: Southern Forge; replaced by IMPALCO 740
ZG 62A	1.5% Cu 2.5% Mg 0.3% Cr 5.5% Zn Al alloy: Previous ASTM designation; now 7075
ZISIUM	2% Cu 15% Zn 0.5% Sn Al alloy: Forging; origin unknown

Note. The following abbreviations and units are used in the tables:

DPN	Hardness, diamond pyramid number
UTS	Ultimate tensile strength, N/mm^2
Elon	Elongation, %
Proof	0.1% proof strength, N/mm^2

1 N/mm^2=0.1 hbar=0.102 kgf/mm^2=0.06475 tonf/in.^{2}=145.04 lbf/in.^{2}=1 MPa

See Appendix II for other abbreviations and conversion tables.

1H Aluminium alloys clad with aluminium

This section covers the series of specifications, trade names, etc. where aluminium alloys are clad with pure aluminium or aluminium alloy.

There are two reasons for cladding, the first and most important being to improve the corrosion resistance of the alloy itself. In general, where the mechanical properties are improved by the addition of alloy elements, the corrosion resistance is decreased. This decrease in some instances, for example the zinc–aluminium alloys, can be quite dramatic.

It will be obvious that only relatively simple shapes can have the cladding carried out. Cladding is produced by forming the base material and then taking thin sheets of the material for cladding and rolling the two materials together. The shapes are therefore generally in sheet form, but extrusions can be produced as can some sophisticated forgings.

The materials will have the mechanical properties of the base material with the corrosion resistance at the surface improved.

In general the material used is high purity aluminium, either 99.99% or 99.95% aluminium.

The clad skin is very thin and thus will be susceptible to damage. Great care must be taken when working with clad materials, either during production or in service, that this thin skin is not destroyed.

It must always be appreciated that corrosion can be more serious when it is local than when it is general and where stressed areas are involved then the problem of stress corrosion might exist.

Some 1% zinc alloys are used to clad the wrought materials listed in Section 1G where high strength aluminium alloys exist with zinc as the major alloying element.

The second reason for cladding is to produce a film which will braze to a similar film. The high strength material therefore has a cladding of an aluminium–silicon alloy which reduces the melting point and thus allows it to be brazed without affecting the heat treatment or mechanical properties of the base material.

Considerable skill is necessary, firstly in the design, and also in the application of brazing on these components.

The materials in this section are used for architectural purposes, very often on components used in transport where the pure aluminium skin obviates the need to paint. High production components where brazing is required are also included in this section.

Symbol	Nominal analysis, supplier, condition and remarks.
11	1.2% Mn Al sheet clad one side with 7.5% Al alloy: For brazing; SAE designation; formerly SAE 247
12	1.2% Mn Al sheet clad both sides with 7.5% Si Al alloy: For brazing; SAE designation; formerly SAE 248
21	0.6% Mg 0.4% Si Al alloy sheet clad one side with 7.5% Si Al alloy: SAE designation; formerly SAE 249 for brazing
22	0.6% Mg 0.4% Si Al alloy clad both sides with 7.5% Si Al alloy: For brazing; SAE designation; formerly SAE 250
23	0.2% Cu 0.6% Mg 0.3% Si Al alloy sheet clad one side with 9% Si Al alloy: For brazing; SAE designation
24	0.2% Cu 0.6% Mg 0.3% Si Al alloy sheet clad both sides with 9% Si Al alloy: For brazing; SAE designation
ALCAN GB24S ALCLAD	4.5% Cu 1.5% Mg 0.7% Mn Al alloy clad with Al: Alcan; solution treated and room aged **UTS: 420 Elon: 18% Proof: 280**
ALCAN GB26S ALCLAD	4.2% Cu 0.7% Mg 0.7% Si 0.7% Mn Al alloy clad with Al: Alcan; solution treated and aged **UTS: 440 Elon: 10% Proof: 360**
ALCAN GBM75S ALCLAD	1.3% Cu 2.5% Mg 0.12% Cr 6% Zn Al alloy clad with Al: Alcan; solution treated and aged **UTS: 510 Elon: 11% Proof: 420**
ALCOA 460	5.0% Si Al alloy: Clad on Alcoa 190 sheet; for brazing
ALCOA 470	7.5% Si Al alloy: Clad on Alcoa 190 sheet; for brazing
ALDURAL B	4.2% Cu 0.7% Mg 0.6% Mn Al alloy clad with Al: J Booth; clad version of DURAL B
ALDURAL G	4.2% Cu 1.5% Mg 0.6% Mn 0.2% Cr Al alloy clad with Al: Sheet; J Booth; clad version of DURAL G
ALDURAL JJ	2.3% Cu 1.5% Mg 0.15% Si 1% Fe 1.1% Ni Al alloy clad with Al: Sheet; J Booth; clad version of DURAL JJ
ALDURAL K	1.1% Cu 3% Mg 5.7% Zn Al alloy clad with 1% Zn Al alloy: J Booth; clad version of DURAL K clad with 1% Zn Al
ALDURAL L	0.6% Cu 3% Mg 5.8% Zn Al alloy sheet clad with 1% Zn Al alloy: J Booth; clad version of DURAL L clad with 1% Zn Al alloy
ALDURAL S	4.2% Cu 0.7% Mg 0.7% Si 0.7% Mn Al alloy clad with Al: Sheet; J Booth; clad version of DURAL S
AMS 4020	0.25% Cu 1% Mg 0.6% Si 0.25% Cr Al alloy clad with Al/Zn: AMS for alloy 6061 clad with 7072
AMS 4021 B	0.25% Cu 1% Mg 0.6% Si 0.25% Cr Al alloy clad with Al/Zn: AMS for alloy 6061 clad with 7072 annealed
AMS 4022 C	0.25% Cu 1% Mg 0.6% Si 0.25% Cr Al alloy clad with Al/Zn: AMS for alloy 6061 clad with 7072 solution treated
AMS 4023 C	0.25% Cu 1% Mg 0.6% Si 0.25% Cr Al alloy clad with Al/Zn: AMS for alloy 6061 with 7072 solution treated and aged
AMS 4034 A	4.5% Cu 1.5% Mg 0.6% Mn Al alloy clad with Al: AMS for AA alloy 2024 clad with 1230
AMS 4036	4.5% Cu 1.5% Mg 0.6% Mn Al alloy clad with Al one side: AMS for AA alloy 2024 clad with Al
AMS 4039	1.6% Cu 2.5% Mg 0.3% Cr 5.6% Zn Al alloy clad with Al/Zn: AMS for AA alloy 7075 clad with Al/Zn
AMS 4040 E	4.5% Cu 1.5% Mg 0.6% Mn Al alloy clad with Al: AMS for AA alloy 2024 clad with 1230; annealed
AMS 4041 G	4.5% Cu 1.5% Mg 0.6% Mn Al alloy clad with Al: AMS for AA alloy 2024 clad with 1230
AMS 4042 E	4.5% Cu 1.5% Mg 0.6% Mn Al alloy clad with Al: AMS for AA alloy 2024 clad with Al
AMS 4046	1.6% Cu 2.5% Mg 0.25% Cr 5.6% Zn Al alloy clad with Al one side: AMS for AA 7075 clad with Al/Zn
AMS 4047 B	1.6% Cu 2.5% Mg 0.25% Cr 5.6% Zn Al alloy Al clad Al/Zn roll tapered: AMS for AA alloy 7075 clad with Al/Zn
AMS 4048 C	1.6% Cu 2.5% Mg 0.25% Cr 5.6% Zn Al alloy clad with Al, annealed with Al/Zn: AMS for AA alloy 7075 clad with Al/Zn
AMS 4049 C	1.6% Cu 2.5% Mg 0.25% Cr 5.6% Zn Al alloy clad with Al/Zn: AMS for AA alloy 7075 clad with Al/Zn
AMS 4051 A	2% Cu 2.7% Mg 0.3% Cr 6.8% Zn Al alloy clad with Al/Zn: Annealed **UTS: 240 Elon: 10%**
AMS 4052 A	2% Cu 2.7% Mg 0.3% Cr 6.8% Zn Al alloy clad with Al/Zn: Solution treated and aged **UTS: 570 Elon: 6% Proof: 490**
AMS 4054	0.3% Cu 0.6% Mg 0.35% Si Al alloy clad with braze metal: Annealed and clad one side with Al alloy braze metal
AMS 4055	0.3% Cu 0.6% Mg 0.3% Si Al alloy clad with braze metal: Annealed; clad both sides with Al alloy braze metal
AMS 4063	3003 alloy clad with 4343; Braze alloy
AMS 4064	3003 alloy clad with 4343: Braze alloy
AMS 4077	2024 alloy clad with 99.3% Al
AMS 4094	Designation 2219 clad with alloy 7072
AMS 4095	Alloy designation 2219 clad with alloy 7072
AMS 4096	Alloy designation 2219 clad with alloy 7072
AMS 4100	7475 alloy clad with 7072
AMS 4196	1% Zn alloy 7075 clad with Al alloy
AMS 4197	1% Zn alloy 7075 clad with Al alloy
AMS 4198	1% Zn alloy 7079 clad with Al alloy
AMS 4199	1% Zn alloy 7079 clad with Al alloy
AMS 4243	1% Zn alloy 7050 clad with Al alloy
ASTM B209/2014 CLAD	4.5% Cu 0.5% Mg 1% Si Al alloy sheet clad with Al: Annealed and clad with ASTM 6003 **UTS: 210 Elon: 16% Proof: 75**
ASTM B209/2014 CLAD	4.5% Cu 0.5% Mg 1% Si Al alloy sheet clad with Al: Clad with ASTM 6003; solution treated and aged **UTS: 450 Elon: 4% Proof: 400**
ASTM B209/2024 CLAD	4% Cu 1.5% Mg Al alloy sheet clad with Al: Annealed, clad with ASTM 1230 **UTS: 210 Elon: 10% Proof: 75**
ASTM B209/2024 CLAD	4% Cu 1.5% Mg Al alloy sheet clad with Al: Solution treated; room aged; clad with ASTM 1230 **UTS: 420 Elon: 6% Proof: 270**
ASTM B209/3003 CLAD	1.2% Mn Al alloy sheet clad with Al/Zn: Annealed, clad with ASTM 7072 **UTS: 110 Elon: 23% Proof: 75**
ASTM B209/3003 CLAD	1.2% Mn Al alloy sheet clad with Al/Zn: Cold rolled; ½-hard **UTS: 150 Elon: 8% Proof: 90**
ASTM B209/3004 CLAD	3004 alloy sheet clad with 7072 alloy
ASTM B209/5155 CLAD	5155 alloy clad with alloy 7072
ASTM B209/6061 CLAD	0.2% Cu 1% Mg 0.6% Si 0.2% Cr Al alloy sheet clad with Al/Zn: Annealed; clad with ASTM 7072 **UTS: 120 Elon: 18% Proof: 60**
ASTM B209/6061 CLAD	0.2% Cu 1% Mg 0.6% Si 0.2% Cr Al alloy sheet clad with Al/Zn: Solution treated and aged; clad with ASTM 7072 **UTS: 280 Elon: 8% Proof: 240**
ASTM B209/7075	1.7% Cu 2.5% Mg 0.3% Cr 5.5% Zn Al alloy sheet clad with Al/Zn

Note. The following abbreviations and units are used in the tables:

DPN	Hardness, diamond pyramid number
UTS	Ultimate tensile strength, N/mm^2
Elon	Elongation, %
Proof	0.1% proof strength, N/mm^2

1 N/mm^2=0.1 hbar=0.102 kgf/mm^2=0.06475 tonf/in.^{2}=145.04 lbf/in.^{2}=1 MPa

See Appendix II for other abbreviations and conversion tables.

Symbol	Nominal analysis, supplier, condition and remarks.
ASTM B210/3003 CLAD	1.2% Mn Al alloy tube clad with Al/Zn: Annealed; clad with ASTM 7072 **UTS: 75 Proof: 15**
ASTM B210/3003 CLAD	1.2% Mn Al alloy tube clad with Al/Zn: Cold drawn; clad with ASTM 7072 **UTS: 170 Proof: 150**
ASTM B211/5056 CLAD	5% Mg 0.1% Cr Al alloy bar clad with Al: Cold drawn; clad with ASTM 6253 **UTS: 340**
ASTM B234/3003 ALCLAD	1.2% Mn Al tube coated with Al/Zn: Cold drawn; clad with 7072 **UTS: 140 Proof: 110**
ASTM B235/3003 ALCLAD	1.2% Mn Al alloy tube clad with Al/Zn: Annealed; clad with 7072 and extruded **UTS: 100 Proof: 22**
ASTM B241/3003 ALCLAD	1.3% Mn Al pipe and tube clad with ASTM B241/7072
ASTM B307/3003 ALCLAD	1.2% Mn Al alloy coiled tube clad with Al/Zn: Annealed; clad with 7072 **UTS: 90**
ASTM B307/5050 ALCLAD	1.4% Mg Al alloy clad with 1.2% Mg 1.6% Zn Al alloy: Drawn tube
ASTM B313/3004 ALCLAD	1% Mg 1.2% Mn Al alloy welded tube clad with Al: Cold rolled; clad with 7072 **UTS: 250 Elon: 4%**
ASTM B345/3003 ALCLAD	1.2% Mn Al alloy pipe clad with Al/Zn: As drawn; clad with 7072 **UTS: 750 Proof: 30**
ASTM B345/6061 ALCLAD	0.2% Cu 1% Mg 0.6% Si 0.2% Cr Al alloy clad with Al/Zn: Pipe; solution treated and aged; clad with 7072 **UTS: 220 Proof: 200**
ASTM B404/3003 ALCLAD	1.25% Mn 1.0% Zn Al alloy clad with Al alloy: Condenser tubes
BA 351	4.1% Cu 0.8% Mg 0.5% Si 0.7% Mn Al alloy clad with Al: Plate; British Aluminium Co. for BS alloy HC14
BA 355	4.6% Cu 0.7% Mg 0.8% Si 0.7% Mn Al alloy clad with Al: British Aluminium Co. for BS alloy HC15
BA 359	4.2% Cu 1.6% Mg 0.6% Mn 0.05% Pb Al alloy clad with Al: Plate; British Aluminium Co.; solution treated and aged **UTS: 400 Elon: 10% Proof: 240**
BA 751	1.3% Cu 2.9% Mg 0.4% Mn 0.1% Cr 5.7% Zn Al alloy: Clad; British Aluminium Co.; solution treated and aged; clad with 1% Zn alloy **UTS: 460 Elon: 8% Proof: 390**
BA 757	1.3% Cu 2.9% Mg 0.4% Mn 0.1% Cr 5.7% Zn Al alloy: Clad; British Aluminium Co.; solution treated and aged; clad with 1% Zn alloy **UTS: 460 Elon: 6% Proof: 390**
BIRMETAL 477 CLAD	4% Cu 1% Mg 0.5% Si 0.7% Mn Al alloy clad with Al: Birmabright for BS alloy HC 14
BIRMETAL 478 CLAD	4% Cu 0.5% Mg 0.5% Si 1% Mn Al alloy clad with Al: Birmabright for BS alloy HC15
BIRMETAL Z36 CLAD	3% Cu 5% Zn Mg Al alloy clad with Al: Birmabright **UTS: 480 Elon: 11% Proof: 420**
BS 1470 HC14	4% Cu 1% Mg 0.75% Mn Al alloy: Al coated sheet and strip; solution treated; room aged **UTS: 360 Elon: 15% Proof: 210**
BS 1470 HC15	4% Cu 0.7% Mg 0.6% Si 0.8% Mn Al alloy: Sheet and strip coated with Al: solution treated and aged **UTS: 390 Elon: 8% Proof: 300**
BS 1477 HPC14	4% Cu 1% Mg 0.5% Si 0.7% Mn Al alloy plate clad with Al: Solution treated; room aged; coated with Al **UTS: 360 Elon: 10% Proof: 210**
BS 1477 HPC15	4% Cu 0.5% Mg 0.7% Si 1% Mn Al alloy: Al clad plate; solution treated and aged; coated with Al **UTS: 390 Elon: 8% Proof: 300**
CLAD CG 42A	4.2% Cu 1.5% Mg Al alloy clad with Al: Previous ASTM designation; now 2024; clad with 1230
CLAD CS 41A	4.5% Cu 0.6% Mg 1% Si Al alloy clad with Si/Mg Al: Previous ASTM designation; now 2014; clad with 6003
CLAD GM 50A	5% Mg 0.1% Cr Al alloy clad with Al/Zn alloy: Previous ASTM designation; now 5056; clad with 6253
CLAD GS 11A	0.2% Cu 1% Mg 0.6% Si 0.2% Cr Al alloy clad with Al/Zn: Previous ASTM designation; now 6061; clad with 7072
CLAD MIA	1.2% Mn Al alloy clad with Al/Zn alloy: Previous ASTM designation; now 3003; clad with 7072
CLAD MG 11A	1% Mg 1.2% Mn Al alloy clad with Al/Zn alloy: Previous ASTM designation; now 3004; clad with 7072
CLAD ZG 62A	1.7% Cu 2.5% Mg 0.3% Cr 5.5% Zn Al alloy clad with Al/Zn alloy: Previous ASTM designation; now 7075; clad with 7072
DTD 390	Al clad sheet: Replaced by DTD 610
DTD 610	4% Cu 1.5% Si 1% Fe Al alloy clad with Al
DTD 687B	1% Cu 3% Mg 0.2% Cr 5% Zn Al alloy sheet coated with Al: Solution treated and aged; coated with 1% Zn Al alloy **UTS: 450 Elon: 8% Proof: 390**
DTD 710B	4% Cu 0.7% Mg 0.7% Si 1% Mn Al sheet coated with Al: Solution treated; room aged; coated with Al **UTS: 370 Elon: 15% Proof: 220**
DTD 5040	4.4% Cu 0.7% Mg 0.7% Si 0.6% Mn Al alloy plate clad with Al: Solution treated and aged; coated with Al **UTS: 420 Elon: 8% Proof: 330**
DTD 5060	1.5% Cu 2.7% Mg 0.8% Mn 5% Zn Al alloy coated with Al: Solution treated and aged; coated with 1% Zn Al alloy **UTS: 480 Elon: 6% Proof: 400**
DTD 5070A	2.5% Cu 1.5% Mg 1.2% Ni 1% Fe Al alloy coated with Al: Sheet; solution treated and aged; coated with 1% Zn Al alloy **UTS: 370 Elon: 6% Proof: 300**
DTD 5100	4.4% Cu 1.5% Mg 0.6% Mn Al alloy clad with Al: Plate; solution treated; room aged; coated with Al **UTS: 420 Elon: 10% Proof: 250**
HA 4 CG42 ALCLAD	4.3% Cu 1.6% Mg 0.6% Mn Al plate clad with 99.7% Al: Canadian Standard
HA 4 CS41N ALCLAD	4.5% Cu 0.6% Mg 0.8% Si 0.8% Mn Al alloy plate clad with 1% Mg Al: Canadian Standard
HIDUMINIUM 04	4.2% Cu 0.8% Mg 0.5% Si 0.8% Mn Al alloy clad with Al: Plate; High Duty Alloys for BS alloy HC14
IMPALCO C66A	4% Cu 0.7% Mg 1% Mn Al alloy clad with Al: Sheet; Imperial Aluminium Co. for BS alloy HC15
KYNAL-CORE C65A	4% Cu 0.6% Mg 0.5% Mn Al alloy clad with Al: ICI: obsolete; now Imperial Aluminium Co.
KYNAL-CORE C66A	Al clad form of KYNAL C66: ICI: obsolete; now Imperial Aluminium Co.
KYNAL-CORE C67A	Al clad version of KYNAL C67: ICI: obsolete; now Imperial Aluminium Co.
KYNAL-CORE C68A	Al clad version of KYNAL C67: ICI; obsolete; now Imperial Aluminium Co.
KYNAL-CORE Z93A	Al clad form of KYNAL Z93; ICI; obsolete; now Imperial Aluminium Co.
L38	Cu Al alloy sheet clad with Al: Obsolete
L47	Cu Al alloy sheet clad with Al: Obsolete; replaced by L73

Note. The following abbreviations and units are used in the tables:

DPN	Hardness, diamond pyramid number
UTS	Ultimate tensile strength, N/mm^2
Elon	Elongation, %
Proof	0.1% proof strength, N/mm^2

1 N/mm^2=0.1 hbar=0.102 kgf/mm^2=0.06475 $tonf/in.^2$=145.04 $lbf/in.^2$=1 MPa

See Appendix II for other abbreviations and conversion tables.

Symbol	Nominal analysis, supplier, condition and remarks.
L72	4% Cu 0.75% Mg 0.75% Si 0.8% Mn Al alloy sheet coated with Al: Solution treated; room aged **UTS: 370 Elon: 15% Proof: 220**
L73	4% Cu 0.75% Mg 0.75% Si 0.8% Mn Al alloy sheet coated with Al: Solution treated and aged **UTS: 400 Elon: 8% Proof: 310**
SAE 240	4.5% Cu 1.5% Mg 0.6% Mn Al alloy clad with Al: Sheet; solution treated; SAE 24 clad with Al **UTS: 420 Elon: 12% Proof: 250**
SAE 242	1% Mg 1.2% Mn Al alloy clad with Al/Zn alloy: Cold rolled; AA alloy 3004 clad with 7072 **UTS: 270 Elon: 5% Proof: 240**
SAE 244	1% Mg 0.60% Si Al alloy clad with Al/Zn alloy: AA alloy 6061; clad with 7072
SAE 245	1.2% Mn Al alloy clad with Al/Zn alloy: Cold rolled; AA alloy 3003 clad with 7072 **UTS: 200 Elon: 4% Proof: 170**
SAE 246	4.5% Cu 0.6% Mg 1% Si 1% Mn 0.1% Cr Al alloy clad with Al: AA alloy 2014 clad with 6003
SAE 247	7.5% Si 1.2% Mn Al sheet clad one side with Al: For brazing; alloy 3003 clad with 4343; now AA11
SAE 248	1.2% Mn Al sheet clad both sides with 7.5% Si Al alloy: For brazing; alloy 3003 clad with 4343; now AA12
SAE 249	0.6% Mg 0.4% Si Al alloy clad one side with 7.5% Si Al alloy: Alloy 6951 clad with 4343; now AA21
SAE 250	0.6% Mg 0.4% Si Al alloy sheet clad both sides with 7.5% Si Al alloy: Alloy 6951 clad with 4343; now AA22
UNI 3736	1.6% Cu 2.5% Mg 0.2% Mn 0.15% Cr 5.8% Zn 0.1% Ti Al alloy clad with 1.0% Zn Al: Italian Standard
UNI 3580	4.0% Cu 0.5% Mg 0.5% Mn Al alloy clad with Al: Italian Standard
UNI 3582	4.4% Cu 0.4% Mg 0.8% Si 0.8% Mn Al alloy clad with 1% Si Mg Mn alloy: UNI 3571; Italian Standard

Note. The following abbreviations and units are used in the tables:

DPN	Hardness, diamond pyramid number
UTS	Ultimate tensile strength, N/mm^2
Elon	Elongation, %
Proof	0.1% proof strength, N/mm^2

1 N/mm^2=0.1 hbar=0.102 kgf/mm^2=0.06475 tonf/in.^{2}=145.04 lbf/in.^{2}=1 MPa

See Appendix II for other abbreviations and conversion tables.

1J Aluminium–magnesium cast alloys

Specific gravity		2.55–2.65
Density		2550–2650 kg/m^3
Solidus/liquidus		450–642 °C
Thermal conductivity		86–138 W/m °C
Coefficient of linear expansion		24–25 × 10^{-6}/ °C
Electrical conductivity		20–30% IACS (copper 100%)
Specific resistance		60–80 microhm mm
Young's modulus of elasticity		71.0 × 10^9 N/m^2
Impact	sand cast	8 J
	chill cast	12–15 J
Fatigue strength (10^6 cycles)	sand cast	± 60–75 N/mm^2
	chill cast	± 90–120 N/mm^2
Hot strength		No detailed figures available. Strength falls off rapidly above 100–125 °C

The above properties are typical of the following group, and may not apply exactly to any one specification. It is possible that with certain specifications some of the values may not be applicable.

The materials in this section contain aluminium with magnesium from 4 to 10%, very often with some silicon and other minor elements added.

In general these materials are not heat treatable. As the magnesium content increases so does the corrosion resistance.

There are some problems in casting the materials, particularly as the magnesium content is increased, and these materials should not be designed for use where thin wall castings are required.

Many of the materials in this section give the best compromise of the aluminium alloys between strength and corrosion resistance. They are difficult to cast, however, and problems can be encountered.

Materials are used for marine purposes and also where decorative castings are required with reasonably intricate shapes difficult or expensive to produce by forging or machining. Many of these castings are used in builders' hardware such as rain-water castings, etc., and other architectural features.

Welding requires considerable care and skill, using the inert gas shielded technique.

Symbol	Nominal analysis, supplier, condition and remarks.
2.3561	4.5% Mg 0.5% Mn Al alloy: Casting; German Standard
3.3261	4.8% Mg 1% Si 0.1% Ti Al alloy: Casting; German Standard DIN 1725 **DPN: 70 UTS: 220 Elon: 5% Proof: 75**
3.3282	7.5% Mg 0.7% Si 0.4% Mn Al alloy: Casting; German Standard DIN 1725 **DPN: 70 UTS: 200 Elon: 2% Proof: 140**
3.3292	9% Mg 0.5% Si 0.4% Mn Al alloy: Casting; German Standard DIN 1725 **DPN: 70 UTS: 220 Elon: 2% Proof: 140**
3.3591	10% Mg 0.2% Mn 0.1% Ti Al alloy: Casting; German Standard DIN 1725
220	10% Mg Al alloy: Casting; Alcoa for BS alloy LM 10
320	4% Mg Al alloy: Casting; Comalco; sand cast **DPN: 60 UTS: 150 Elon: 4% Proof: 60**
340	8% Mg Al alloy: Casting; Comalco; sand cast **DPN: 70 UTS: 250 Elon: 7% Proof: 110**
350	10% Mg Al alloy: Casting; Comalco for BS alloy LM 10
511.0	4% Mg 0.5% Si Al alloy: Casting; designation used in the UK and USA
511.1	4% Mg 0.5% Si Al alloy: Ingot; designation used in the UK and USA
511.2	4% Mg 0.5% Si Al alloy: Ingot; designation used in the UK and USA
512.0	4% Mg 1.8% Si Al alloy: Casting; obsolete; designation used in the UK and USA
512.2	4% Mg 1.8% Si Al alloy: Ingot; designation used in the UK and USA
513.0	4% Mg 1.8% Zn Al alloy: Casting; designation used in the UK and USA
513.2	4% Mg 1.8% Zn Al alloy: Ingot; designation used in the UK and USA
514.0	4% Mg Al alloy: Casting; designation used in the UK and USA
514.1	4% Mg Al alloy: Ingot; designation used in the UK and USA
514.2	4% Mg Al alloy: Ingot; designation used in the UK and USA
515.0	3.2% Mg 0.7% Si 0.5% Mn Al alloy: Casting; designation used in the UK and USA
515.2	3.2% Mg 0.7% Si 0.5% Mn Al alloy: Ingot; designation used in the UK and USA
516.0	3.2% Mg 0.7% Si 0.3% Mn 0.3% Ni Al alloy: Casting; designation used in the UK and USA
516.1	3.2% Mg 0.7% Si 0.3% Mn 0.3% Ni Al alloy: Ingot; designation used in the UK and USA
518.0	8% Mg Al alloy: Casting; designation used in the UK and USA
518.1	8% Mg Al alloy: Ingot; designation used in the UK and USA
518.2	8% Mg Al alloy: Ingot; designation used in the UK and USA
520.0	10% Mg Al alloy: Casting; designation used in the UK and USA
520.2	10% Mg Al alloy: Ingot; designation used in the UK and USA

Note. The following abbreviations and units are used in the tables:

DPN	Hardness, diamond pyramid number
UTS	Ultimate tensile strength, N/mm^2
Elon	Elongation, %
Proof	0.1% proof strength, N/mm^2

$1\ N/mm^2 = 0.1\ hbar = 0.102\ kgf/mm^2 = 0.06475\ tonf/in.^2 = 145.04\ lbf/in.^2 = 1\ MPa$

See Appendix II for other abbreviations and conversion tables.

Symbol	Nominal analysis, supplier, condition and remarks.
535.0	6.8% Mg 0.1% Mn 0.15% Ti Al alloy: Casting; designation used in the UK and USA
535.2	6.8% Mg 0.1% Mn 0.15% Ti Al alloy: Ingot; designation used in the UK and USA
A 10	10% Mg Al master alloy: L'Aluminium Français
A G3T	3.0% Mg Al alloy: Casting; L'Aluminium Français
A G4Z	4% Mg 1.2% Zn Al alloy: Casting; L'Aluminium Français
A G6	4.5% Mg 0.4% Mn Al alloy: Casting; French Standard
A G6	6.0% Mg Al alloy: Casting; L'Aluminium Français
A G10Y4	10.2% Mg Al alloy: Casting; French Standard
Al 28	4.5% Mg 0.5% Mn Al alloy: Casting; Russian designation
ALCAN GB 350	10% Mg Al alloy: Casting; Alcan for BS alloy LM 10
ALCAN GB B320	4% Mg 0.5% Mn Al alloy: Casting; Alcan for BS alloy LM 5
AMS 4238 A	6.8% Mg 0.2% Mn 0.2% Ti Al alloy: Casting; as cast
AMS 4239	6.8% Mg 0.2% Mn 0.2% Ti Al alloy: Casting; annealed
AMS 4240 C	10% Mg Al alloy: Casting; solution treated
AN QQ A 366	8.0% Mg Al alloy: Die casting; US Service; as cast **UTS: 210 Elon: 5% Proof: 120**
AN QQ A 392	10% Mg Al alloy: Casting; US Service; aged **UTS: 280 Elon: 12% Proof: 140**
AN QQ A 402	4.0% Mg Al alloy: Casting; US Service; as cast **UTS: 110**
AP 501	4.5% Mg 0.5% Mn Al alloy: Casting; Australian specification
AP 505	10.2% Mg Al alloy: Casting; Australian specification
ASTM B26 G4A	4% Mg 0.35% Si Al alloy: Sand casting; as cast **UTS: 150 Elon: 6% Proof: 45**
ASTM B26 G10A	10% Mg Al alloy: Sand casting; solution treated **UTS: 280 Elon: 12% Proof: 140**
ASTM B26 GM70B	7% Mg Al alloy: Sand casting **UTS: 240 Elon: 9% Proof: 110**
ASTM B26 GS42A	4% Mg 2% Si Al alloy: Sand casting; as cast **UTS: 110 Proof: 50**
ASTM B85 G8A	8% Mg Al alloy: Die casting; as cast **UTS: 320 Elon: 5% Proof: 180**
ASTM B108 GM42A	4% Mg 0.2% Cr Al alloy: Casting; as cast **UTS: 120 Elon: 1.5%**
ASTM B108 GM70B	7% Mg 0.2% Mn 0.2% Ti Al alloy: Casting; as cast **UTS: 240 Elon: 8% Proof: 110**
ASTM B108 GZ42A	4% Mg 2% Zn Al alloy: Casting; as cast **UTS: 140 Elon: 2.5% Proof: 75**
ASTM B179 G4A	4% Mg Al alloy: Ingot
ASTM B179 G8A	8% Mg Al alloy: Ingot
ASTM B179 G10A	10% Mg Al alloy: Ingot
ASTM B179 GM70B	7% Mg 0.2% Ti Al alloy: Ingot
ASTM B179 GS42A	4% Mg 2% Si Al alloy: Ingot
B 514.0	4.0% Mg 1.8% Si Al alloy: Investment casting; American National Standard
BA 28	5% Mg 0.5% Mn Al alloy: Casting; British Aluminium Co. for BS alloy LM 5
BA 29	10% Mg Al alloy: Casting; British Aluminium Co. for BS alloy LM 10
BB 5	5% Mg 0.5% Mn Al alloy: Casting; Birmabright for BS alloy LM 5
BIRMABRIGHT	5% Mg 0.5% Mn Al alloy: Casting; Birmabright for BS alloy LM 5
BS 1490 LM5	4.5% Mg 0.5% Mn Al alloy: Ingot; chill cast **UTS: 170 Elon: 5%**
BS 1490 LM5M	4.5% Mg 0.5% Mn Al alloy: For sand and die castings; chill cast **UTS: 170 Elon: 5%**
BS 1490 LM10	10% Mg Al alloy: Ingots; chill cast **UTS: 300 Elon: 12%**
BS 1490 LM10W	10% Mg Al alloy: Sand or die castings; chill cast; solution treated **UTS: 300 Elon: 12%**
DSI 5000 A1252	4.5% Mg 0.5% Mn Al alloy: Casting; Danish specification

Symbol	Nominal analysis, supplier, condition and remarks.
DTD 165 A	4.5% Mg 0.5% Mn Al alloy: Castings; chill cast; as cast **UTS: 170 Elon: 5% Proof: 75**
DTD 300 A	10.5% Mg Al alloy: Casting; chill cast; solution treated **UTS: 300 Elon: 12% Proof: 170**
DTD 5018	7.5% Mg 0.2% Mn 1.2% Zn Al alloy: Sand casting **UTS: 280 Elon: 5% Proof: 170**
G Al Mg 3	3% Mg Al alloy: Casting; previous German Standard designation
G Al Mg 3(Cu)	0.4% Cu 3% Mg Al alloy: Casting; previous German Standard designation
G Al Mg 5	0.2% Cu 5% Mg 1% Si Al alloy: Casting; previous German Standard designation
G Al Mg 10	0.2% Cu 10% Mg Al alloy: Casting; previous German Standard designation
G Al Mg(Cu)	0.4% Cu 4% Mg Al alloy: Casting; previous German Standard designation
GD Al Mg 8(Cu)	0.2% Cu 8% Mg 1% Si 0.4% Mn Al alloy: Casting: previous German Standard designation
GD Al Mg 9	9% Mg 0.4% Mn Al alloy: Casting; previous German Standard designation
GOST 2685-63 AL13	4.5% Mg 0.5% Mn Al alloy: Casting; Russian specification
H 49 4	4.5% Mg 0.5% Mn Al alloy: Casting; Australian specification
H 49 8	10% Mg Al alloy: Casting; Australian specification
HA 3 G8	0.1% Cu 8% Mg Al alloy: Ingot; Canadian Standard
HA 3 G10	10% Mg Al alloy: Ingot; Canadian Standard
HA 3 GS40	4% Mg 0.5% Si Al alloy: Ingot; Canadian Standard
HA 9 G10	10% Mg Al alloy: Sand casting; Canadian Standard
HA 9 GS40	4.0% Mg 0.5% Si Al alloy: Sand casting; Canadian Standard
HIDUMINIUM 90	10.5% Mg Al alloy: Casting; High Duty Alloys for BS alloy LM 10
L53	10% Mg Al alloy: Casting; chill cast; solution treated **UTS: 300 Elon: 12% Proof: 170**
L231	10% Mg Al alloy: Casting; Spanish specification
LM 10	10% Mg Al alloy: Casting; INTAL for BS alloy LM 10
MG 7	7% Mg 1.5% Fe Al alloy: For die casting; Birmabright; as cast **UTS: 220 Elon: 3.5%**
NBN 436 Al Mg 6	5% Mg 0.5% Mn Al alloy: Casting; Belgium specification
NBN 436 Al Mg 11	10.2% Mg Al alloy: Casting; Belgium specification
NORAL 350	10% Mg Al alloy: Casting; Northern Aluminium Co. for BS alloy LM 10
PERALUMAN 34	3.4% Mg 0.2% Mn 0.15% Ti Al alloy: Casting; Anglo-Swiss Aluminium Co.
PERALUMAN 75	7.0% Mg 1.78% Si 0.4% Mn 0.4% Fe 0.15% Ti Al alloy: Ingot; Anglo-Swiss Aluminium Co.
QQ A 59/7	8.0% Mg Al alloy: Die casting; US Federal; as cast **UTS: 210 Elon: 4% Proof: 120**
QQ A 591/9	3.2% Mg Al alloy: die casting; US Federal; as cast **UTS: 270 Elon: 10% Proof: 120**
QQ A 601/5	3.8% Mg Al alloy: Casting; US Federal; as cast **UTS: 140**
SAE 320	4% Mg Al alloy: Casting; as cast; sand cast **UTS: 150 Elon: 6% Proof: 45**
SAE 324	10% Mg Al alloy: Casting; sand cast; solution treated **UTS: 280 Elon: 12% Proof: 150**
STA 7 AC5	5.0% Mg 0.5% Mn Al alloy: Casting; replaced by BS 1490 LM 5
STA 7 AC10	10.5% Mg Al alloy: Casting; replaced by BS 1490 LM 10
UNI 3056	10% Mg Al alloy: Casting; Italian Standard
UNI 3057	7.0% Mg Al alloy: Casting; Italian Standard
UNI 3058	5.0% Mg Al alloy: Casting; Italian Standard
UNI 3059	3.0% Mg Al alloy: Casting; Italian Standard
UNI 5080	7.5% Mg 0.8% Fe Al alloy: Casting; Italian Standard
V1036 A/B BXIV	10% Mg Al alloy: Casting; Dutch specification

Note. The following abbreviations and units are used in the tables:

DPN	Hardness, diamond pyramid number
UTS	Ultimate tensile strength, N/mm^2
Elon	Elongation, %
Proof	0.1% proof strength, N/mm^2

1 N/mm^2=0.1 hbar=0.102 kgf/mm^2=0.06475 $tonf/in.^2$=145.04 $lbf/in.^2$=1 MPa

See Appendix II for other abbreviations and conversion tables.

1K Aluminium–silicon–magnesium cast alloys

Specific gravity	2.65–2.7
Density	2650–2700 kg/m^3
Solidus/liquidus	525–650 °C
Thermal conductivity	117–170 W/m °C
Coefficient of linear expansion	19–23 × 10^{-6} °C
Electrical conductivity	29–43% IACS (copper 100%) The electrical resistance increases with silicon content and in the heat treated condition.
Specific resistance	41–64 microhm mm
Young's modulus of elasticity	71.0 × 10^9 N/m^2
Impact	1–2 J
Fatigue strength (5 × 10^6 cycles)	
sand cast, solution treated and aged	50–80 N/mm^2
chill cast, solution treated and aged	80–100 N/mm^2

Hot strength

Temperature °C	Tensile strength N/mm^2
100	320
150	300
200	260
250	150
300	90
350	50
400	37

The above values are chosen to show the typical properties of this group of alloys. It may be that none of the values applies to certain of the specifications listed.

General metallurgical characteristics

To some extent the materials in this section are comparable as castings to the specifications, trade names, etc. found in Section 1E. The materials in this section can generally be solution treated and aged to improve the mechanical properties but being castings will never have the same ductility as the wrought materials.

The addition of higher quantities of silicon increases the fluidity of the castings and thus gives good reproducibility. By adding copper, nickel and other minor elements mechanical strength can be improved.

The heat treatment of these materials is similar to that of other heat treatable aluminium alloys, i.e. solution treatment and ageing, and requires considerable control and care.

These materials are used for items such as pump casings, crank cases, gear box cases, brackets and other components which are reasonably highly stressed, because they have good mechanical strength with reasonable corrosion resistance. The castability of these is not as good as the materials in Section 1J but is better than many other aluminium alloys.

Included in this group are the alloys which are used to manufacture internal combustion engine piston castings.

Symbol	Nominal analysis, supplier, condition and remarks.
135	0.3% Mg 7% Si 0.1% Ti Al alloy: Casting; Comalco; solution treated and aged **DPN: 65 UTS: 210 Elon: 5% Proof: 110**
162	1% Cu 1.1% Mg 12% Si 2.5% Ni Al alloy: Casting; Comalco for BS alloy LM13
354	1.8% Cu 0.5% Mg 9% Si Al alloy: Casting; designation used in the USA
355	1.1% Cu 0.5% Mg 5% Si Al alloy: Casting: Alcoa for BS alloy LM16
355	1.2% Cu 0.5% Mg 5% Si Al alloy: Casting; AA designation for ASTM alloy SC51A
356	0.3% Mg 5.5% Si 0.4% Mn Al alloy: Casting; Alcoa for BS alloy LM8

Note. The following abbreviations and units are used in the tables:

DPN	Hardness, diamond pyramid number
UTS	Ultimate tensile strength, N/mm^2
Elon	Elongation, %
Proof	0.1% proof strength, N/mm^2

1 N/mm^2=0.1 hbar=0.102 kgf/mm^2=0.06475 tonf/in.^{2}=145.04 lbf/in.^{2}=1 MPa

See Appendix II for other abbreviations and conversion tables.

Symbol	Nominal analysis, supplier, condition and remarks.
356	0.3% Mg 7% Si 0.2% Fe Al alloy: Casting; Birmabright; chill cast; solution treated and aged **DPN: 80 UTS: 220 Elon: 5% Proof: 150**
356	0.3% Mg 7% Si Al alloy: Casting; AA designation for ASTM alloy SG 70A
356.0	0.3% Mg 7.0% Si Al alloy: Investment casting; American National Standard; ANSI
360	0.5% Mg 9.5% Si Al alloy: Casting; Comalco; as cast **DPN: 85 UTS: 270 Elon: 3% Proof: 150**
357.0	0.5% Mg 7% Si Al alloy: Casting; designation used in the USA
357.1	0.5% Mg 7% Si Al alloy: Ingot; designation used in the USA
358.0	0.5% Mg 8% Si 0.15% Ti Al alloy: Casting; designation used in the USA
358.2	0.5% Mg 8% Si 0.15% Ti Al alloy: Ingot; designation used in the USA
359.0	0.6% Mg 9% Si Al alloy: Casting; designation used in the USA
359.2	0.6% Mg 9% Si Al alloy: Casting; designation used in the USA
A 132	1% Cu 1% Mg 12% Si 2.5% Ni Al alloy: Casting; AA designation for ASTM alloy SN 122A
A 132	1% Cu 1% Mg 12% Si 2.5% Ni Al alloy: Casting; Alcoa for BS alloy LM 13

Symbol	Nominal analysis, supplier, condition and remarks.
A 356	0.3% Mg 7% Si 0.1% Fe (max) Al alloy: Casting; Birmabright; chill cast; solution treated and aged **DPN: 85 UTS: 250 Elon: 9% Proof: 180**
A 356.0	0.3% Mg 7.0% Si Al alloy: Investment casting; American National Standard; ANSI
A 360	0.5% Mg 9.5% Si Al alloy: Casting; AA designation for ASTM alloy SG 100A
A 360	0.5% Mg 9.5% Si Al alloy: Casting; Comalco for BS alloy LM 9
A 03560	Unified number for alloy 356.0; ASTM system
A 13560	Unified number for alloy A 356.0; ASTM system
ALCAN GB 125	1.2% Cu 0.5% Mg 5% Si Al alloy: Casting; Alcan for BS alloy LM 16
ALCAN GB 162	1% Cu 1% Mg 11.5% Si 2% Ni Al alloy: Casting; Alcan for BS alloy LM 13
ALCAN GB B116	0.5% Mg 4.5% Si Al alloy: Casting; Alcan for BS alloy LM 8
ALCAN GB C125	1.5% Cu 0.5% Mg 5% Si Al alloy: Casting; Alcan for BS alloy LM 16
ALCAN GB D135	7% Si Mg Al alloy: Casting; Alcan for BS alloy LM 25
ALPAX BETA	0.4% Mg 11% Si 0.5% Mn Al alloy: Casting; Light Alloys Ltd for BS alloy LM 9
ALPAX GAMMA	0.4% Mg 11% Si 0.5% Mn Al alloy: Casting; Light Alloys Ltd for BS alloy LM 9
ALMELEC	0.7% Mg 0.5% Si Al alloy: Source unknown **UTS: 300**
AMS 4210 F	1.2% Cu 1.5% Mg 5% Si Al alloy: Casting; AMS for AA alloy 355
AMS 4212 E	1.2% Cu 0.5% Mg 5% Si Al alloy: Casting; solution treated and aged; AMS for AA alloy 355
AMS 4214 D	1.2% Cu 0.5% Mg 5% Si Al alloy: Casting; solution treated and stabilized; AMS for AA alloy 355
AMS 4215 B	1.2% Cu 0.5% Mg 5% Si Al alloy: Casting; premium grade; AMS for AA alloy 355
AMS 4217 D	0.3% Mg 7% Si Al alloy: Casting; solution treated and aged; AMS for AA alloy 356
AMS 4218 B	0.3% Mg 7% Si Al alloy: Casting; premium grade AMS for AA alloy 356
AMS 4260	0.3% Mg 7% Si Al alloy: Casting; investment; solution treated and aged; AMS for AA alloy 356
AMS 4280 E	1.2% Cu 0.5% Mg 5% Si Al alloy: Casting; solution treated and over-aged; AMS for AA alloy 355
AMS 4281 C	1.2% Cu 0.5% Mg 5% Si Al alloy: Casting; solution treated and aged; AMS for AA alloy 355
AMS 4284 D	0.3% Mg 7% Si Al alloy: Casting; solution treated and aged; AMS for AA alloy 356
AMS 4285	0.3% Mg 7% Si Al alloy: Centri-cast; solution treated and aged; AMS for AA alloy 356
AMS 4286 A	0.3% Mg 7% Si Al alloy: Casting; aged; AMS for AA alloy 356
AMS 4290	0.5% Mg 9.5% Si Al alloy: Casting; die cast **UTS: 280 Elon: 2% Proof: 150**
AN QQ A 376	1.2% Cu 0.5% Mg 5.0% Si Al alloy: Casting; US Service **UTS: 210 Elon: 2% Proof: 120**
AN QQ A 386	1.0% Cu 1.0% Mg 12% Si 2.5% Ni Al alloy: Casting; US Service; as cast **UTS: 210**
AN QQ A 394	0.3% Mg 7.0% Si Al alloy: Casting; US Service; solution treated and aged **UTS: 170 Elon: 3% Proof: 110**
ANTICORODAL	0.7% Mg 6.5% Si 0.3% Mn 0.15% Ti Al alloy: Casting; Anglo-Swiss Aluminium Co.
ANTICORODAL 70	0.3% Mg 7.0% Si 0.1% Ti Al alloy: Casting; Anglo-Swiss Aluminium Co.
ANTICORODAL 71	0.45% Mg 7.2% Si Al alloy: Primary casting ingot; Anglo-Swiss Aluminium Co.
ASTM B26 SC82A	1.5% Cu 0.5% Mg 8% Si Al alloy: Sand casting; solution treated and aged **UTS: 220 Elon: 1% Proof: 140**
ASTM B26 SG70A	0.3% Mg 7% Si Al alloy: Sand casting; solution treated and aged **UTS: 200 Elon: 3% Proof: 120**
ASTM B85 SG100A	0.5% Mg 10% Si Al alloy: Die casting; as cast **UTS: 320 Elon: 3.5% Proof: 150**
ASTM B85 SG100B	0.5% Mg 10% Si 2% Fe Al alloy: Die casting; as cast **Elon 2.5% Proof: 170**
ASTM B108 SG51B	1.2% Cu 0.5% Mg 5% Si 0.2% Ti Al alloy: Casting; solution treated and aged **UTS: 270 Elon: 3% Proof: 200**
ASTM B108 SG70A	0.3% Mg 7% Si 0.2% Ti Al alloy: Casting; solution treated and aged **UTS: 220 Elon: 3% Proof: 140**
ASTM B108 SG70B	0.3% Mg 7% Si 0.2% Ti Al alloy: Casting; solution treated and aged **UTS: 250 Elon: 5% Proof: 170**
ASTM B108 SN122A	1% Cu 1% Mg 12% Si 0.2% Ti Al alloy: Casting; solution treated and aged **UTS: 270**
ASTM B179 SC51B	1% Cu 0.5% Mg 5% Si Al alloy: Ingot; primary metal
ASTM B179 SC82A	1.5% Cu 0.4% Mg 8% Si 0.5% Mn Al alloy: Ingot; primary metal
ASTM B179 SG70A	0.3% Mg 7% Si Al alloy: Ingot; primary metal
ASTM B179 SG70B	0.3% Mg 7% Si Al alloy: Ingot; primary metal
ASTM B179 SG100A-B	0.5% Mg 9.5% Si Al alloy: Ingot; primary metal
ASTM B179 CG100A	10% Cu 0.3% Mg 2% Si Al alloy: Ingot; primary metal
BA 41	0.4% Mg 11% Si 0.5% Mn Al alloy: Casting; British Aluminium Co. for BS alloy LM 9
BA 42	1% Cu 1% Mg 12% Si 2.5% Ni Al alloy: Casting; British Aluminium Co. for BS alloy LM 13
BA 451	0.5% Mg 5.0% Si Al alloy: Bar; British Aluminium Co.; solution treated and aged **UTS: 180 Elon: 12% Proof: 150**
BIRMIDAL	0.3% Mg 5.5% Si 0.4% Mn Al alloy: Casting; Birmabright for BS alloy LM 8
BS 1490 LM8	0.5% Mg 4.5% Si Al alloy: Ingot; chill cast **UTS: 270 Elon: 2%**
BS 1490 LM8M	0.5% Mg 4.5% Si Al alloy: Sand and die casting; chill cast **UTS: 220 Elon: 3%**
BS 1490 LM8P	0.5% Mg 4.5% Si Al alloy: Sand and die casting; chill cast and aged **UTS: 180 Elon: 2%**
BS 1490 LM8W	0.5% Mg 4.5% Si Al alloy: Sand and die casting; chill cast; solution treated and room aged **UTS: 220 Elon: 5%**
BS 1490 LM8WP	0.5% Mg 4.5% Si Al alloy: Sand and die casting; chill cast; solution treated and aged **UTS: 270 Elon: 2%**
BS 1490 LM9	0.4% Mg 12% Si 0.5% Mn Al alloy: Ingot; chill cast **UTS: 280**
BS 1490 LM9WP	0.4% Mg 12% Si 0.5% Mn Al alloy: Sand and die casting; chill cast; solution treated and aged **UTS: 280**
BS 1490 LM13	1% Cu 1% Mg 12% Si 2% Ni Al alloy: Ingot; chill cast **UTS: 270**
BS 1490 LM13WP	1% Cu 1% Mg 12% Si 2% Ni Al alloy: Sand or die piston casting; chill cast; solution treated and aged **UTS: 270**
BS 1490 LM16	1% Cu 0.5% Mg 5% Si Al alloy: Ingot; chill cast **UTS: 270**
BS 1490 LM16WP	1% Cu 0.5% Mg 5% Si Al alloy: Sand and die casting; chill cast; solution treated and aged **UTS: 270**
COMALCO 135	0.3% Mg 7% Si 0.2% Fe 0.1% Ti Al alloy: Casting; Comalco alloy for SG 70A
COMALCO 162	1% Cu 1.2% Mg 12% Si 0.3% Fe 2.5% Ni Al alloy: Casting; Comalco alloy for LM 13

Symbol	Nominal analysis, supplier, condition and remarks.
COMALCO 360	0.5% Mg 9.5% Si 0.9% Fe Al alloy: casting; Comalco alloy for SG 100B
D 12	7% Cu 0.3% Mg 5.5% Si Al alloy: Casting; Birmabright; solution treated and aged; pistons **DPN: 132 UTS: 330 Proof: 300**
DTD 245 A	0.5% Mg 11.5% Si 0.5% Mn Al alloy: Casting; chill cast; replaced by L75 **UTS: 280 Proof: 240**
DTD 716 A	0.5% Mg 5% Si Al alloy: Casting; chill cast **UTS: 160 Elon: 3% Proof: 80**
DTD 722 A	0.5% Mg 5% Si Al alloy: Casting; chill cast and aged **UTS: 180 Elon: 2% Proof: 130**
DTD 727 A	0.5% Mg 5% Si Al alloy: Casting; chill cast; solution treated **UTS: 220 Elon: 5% Proof: 100**
DTD 735 A	0.5% Mg 5% Si Al alloy: Casting; chill cast; solution treated and aged **UTS: 270 Elon: 2% Proof: 210**
DTD 5028	0.3% Mg 7.0% Si Al alloy: Casting
HA 9 SC51N	1.2% Cu 0.5% Mg 5.0% Si Al alloy: Sand casting; Canadian Standard
HA 9 SC51P	1.2% Cu 0.5% Mg 5.0% Si Al alloy: Sand casting; Canadian Standard
HA 10 SC51N	1.2% Cu 0.5% Mg 5.0% Si Al alloy: For die casting; Canadian Standard
HA 10 SC51P	1.2% Cu 0.5% Mg 5.0% Si Al alloy: For die casting; Canadian Standard
HA 10 SG70N	0.3% Mg 7.0% Si Al alloy: For die casting; Canadian Standard
HA 10 SG70P	0.3% Mg 7.0% Si Al alloy: For die casting; Canadian Standard
HA 10 SG71	0.5% Mg 7.0% Si Al alloy: For die casting; Canadian Standard
HA 10 SN122	1.0% Cu 1.0% Mg 12% Si 2.5% Ni Al alloy: For die casting; Canadian Standard
HIDUMINIUM 40	0.3% Mg 5% Si 0.4% Mn Al alloy: For sand and die casting; High Duty Alloys for BS alloy LM 8
L75	0.4% Mg 12% Si 0.5% Mn Al alloy: Casting; chill cast; solution treated and aged **UTS: 280 Proof: 240**
LM 5	4.5% Mg 0.5% Mn Al alloy: Casting; designation used by BS
LM 8	0.5% Mg 4.5% Si Al alloy: Casting; INTAL for BS alloy LM 8
LM 8	0.5% Mg 4.5% Si Al alloy: Casting; designation used by BS
LM 9	0.4% Mg 12% Si 0.5% Mn Al alloy: Casting; designation used by BS
LM 10	10% Mg Al alloy: Casting; designation used by BS
LM 13	1% Cu 1% Mg 12% Si 2% Ni Al alloy: Casting; designation used by BS
LM 13	1% Cu 1% Mg 12% Si 2% Ni Al alloy: Casting; INTAL for BS alloy LM 13
LM 16	1% Cu 5% Mg 5% Si Al alloy: Casting; designation used by BS
LM 16	1% Cu 0.5% Mg 5% Si Al alloy: Casting; INTAL for BS alloy LM 16
LO-EX	1% Cu 1% Mg 12% Si 2% Ni Al alloy: Casting; BKL for BS alloy LM 13
NORAL 125	1.1% Cu 0.5% Mg 5% Si Al alloy: Casting; Northern Aluminium Co. for BS alloy LM 16
NORAL 161	0.4% Mg 11% Si 0.5% Mn Al alloy: Casting; Northern Aluminium Co. for BS alloy LM 9
NORAL 162	1% Cu 1% Mg 12% Si 2.5% Ni Al alloy: Casting; Northern Aluminium Co. for BS alloy LM 13
NURAL 25	1.5% Mg 5.5% Si 0.2% Mn Al alloy: Casting; Alcan (Germany); general purpose alloy
QQ A 596/6	1.2% Cu 0.5% Mg 5.0% Si Al alloy: Casting; US Federal; aged **UTS: 210 Elon: 1.5%**
QQ A 596/8	0.3% Mg 7.0% Si Al alloy: Casting; US Federal; solution treated and aged **UTS: 200 Elon: 3%**
QQ A 596/9	1.0% Cu 1.0% Mg 12% Si 2.5% Ni Al alloy: Casting; US Federal; aged **UTS: 200**
QQ A 596/10	1.5% Cu 0.5% Mg 12% Si Al alloy: Casting; US Federal; aged **UTS: 180**
QQ A 601/3	0.3% Mg 7.0% Si Al alloy: Casting; US Federal; solution treated and aged **UTS: 200 Elon: 3%**
QQ A 601/10	12% Cu 0.5% Mg 5.0% Si Al alloy: Casting; US Federal; aged **UTS: 200 Elon: 2%**
QQ A 601/11	1.5% Cu 0.5% Mg 5.0% Si 0.8% Ni Al alloy: Casting; US Federal; aged **UTS: 150**
QQ A 601/12	0.7% Cu 1.0% Mg 12% Si 2.5% Ni Al alloy: Casting; US Federal; aged **UTS: 200**
SAE 309	0.5% Mg 10% Si Al alloy: Casting; die cast; as cast **UTS: 300 Elon: 5%**
SAE 321	1% Cu 1% Mg 12% Si 2.5% Ni Al alloy: Casting; solution treated and aged **UTS: 270**
SAE 322	1.25% Cu 0.5% Mg 5% Si Al alloy: Casting; sand cast; solution treated and aged **UTS: 210 Elon: 2% Proof: 130**
SAE 323	0.3% Mg 7% Si Al alloy: Casting; sand cast; solution treated and aged **UTS: 200 Elon: 3% Proof: 130**
SAE 327	1.5% Cu 0.5% Mg 8% Si 0.5% Mn Al alloy: Casting; solution treated and aged **UTS: 220 Elon: 1% Proof: 140**
SAE 328	1.5% Cu 0.8% Mg 12% Si 0.8% Mn Al alloy: Casting; solution treated and aged **UTS: 280 Proof: 250**
SAE 336	0.3% Mg 7% Si Al alloy: Casting; solution treated and aged **UTS: 250 Elon: 5% Proof: 180**
SILAFONT 30	0.3% Mg 9.5% Si Al alloy: Casting; Anglo-Swiss Aluminium Co.
SILAFONT 54	0.3% Mg 9.5% Si Al alloy: Primary ingot; Anglo-Swiss Aluminium Co.
SILAFONT 64	0.2% Mg 10% Si Al alloy: Casting; Anglo-Swiss Aluminium Co.
SILAFONT 66	0.3% Mg 10% Si Al alloy: Casting; Anglo-Swiss Aluminium Co.
STA 7 AC8	0.3% Mg 5.5% Si 0.4% Mn Al alloy: Casting; replaced by BS 1490 LM8
STA 7 AC9	0.4% Mg 11% Si 0.5% Mn Al alloy: Casting; replaced by BS 1490 LM9
STA 7 AC13	1.0% Cu 1.0% Mg 12% Si 2.5% Ni Al alloy: Casting; replaced by BS 1490 LM13
UNI 3051	0.35% Mg 9% Si 0.5% Mn Al alloy: Casting; Italian Standard
UNI 3054	0.65% Mg 4.5% Si 0.7% Mn Al alloy: Casting; Italian Standard

Note. The following abbreviations and units are used in the tables:

DPN	Hardness, diamond pyramid number
UTS	Ultimate tensile strength, N/mm^2
Elon	Elongation, %
Proof	0.1% proof strength, N/mm^2

1 N/mm^2=0.1 hbar=0.102 kgf/mm^2=0.06475 tonf/in.^{2}=145.04 lbf/in.^{2}=1 MPa

See Appendix II for other abbreviations and conversion tables.

Symbol	Nominal analysis, supplier, condition and remarks.
UNI 3055	0.65% Mg 2.0% Si 0.7% Mn Al alloy: Casting; Italian Standard
UNI 3599	0.3% Mg 7.0% Si 0.5% Mn Al alloy: Casting; Italian Standard
UNI 3600	1.3% Cu 0.5% Mg 5.0% Si Al alloy: Casting; Italian Standard
WILMIL M	0.4% Mg 11% Si 0.5% Mn Al alloy: Casting; W Mills Ltd for BS alloy LM 9; obsolete

Note. The following abbreviations and units are used in the tables:

DPN	Hardness, diamond pyramid number
UTS	Ultimate tensile strength, N/mm^2
Elon	Elongation, %
Proof	0.1% proof strength, N/mm^2

1 N/mm^2=0.1 hbar=0.102 kgf/mm^2=0.06475 tonf/in.^{2}=145.04 lbf/in.^{2}=1 MPa
See Appendix II for other abbreviations and conversion tables.

1L Aluminium–copper cast alloys
With silicon, zinc, nickel etc.

Specific gravity	2.7–2.9
Density	2700–2900 kg/m^3
Solidus/liquidus	510–650 °C
Thermal conductivity*	121–176 W/m °C
Coefficient of linear expansion	21–23 × 10^{-6}/ °C
Electrical conductivity*	26–35% IACS (copper 100%)
Specific resistance	36–49 microhm mm
Young's modulus of elasticity	69.6–71.0 N/m^2
Impact	0.5–4 J
Fatigue strength (10^6 cycles)	± 50–140 N/mm^2
Hot strength	

Temperature °C	Tensile strength N/mm^2	Tensile strength † N/mm^2
100	300	
150	280	220
200	240	200
250	170	180
300	90	150
350	50	80

* Alloys with low percentages of alloying elements have higher values.
† The values shown on the right apply to the alloys developed for hot strength by addition of titanium and nickel. Alloys with low percentages of alloying elements have higher values.

The above properties are chosen to be typical of the group as a whole. Certain alloys listed may have specific properties which do not apply.

The materials in this group are quite complex with the addition of alloying elements which are designed firstly to improve the castability and secondly to improve the mechanical properties and in certain instances some physical properties.

The mechanical properties of the majority of alloys in this section can be increased by heat treatment but the improvement is not as dramatic as in other heat treatable aluminium alloys.

The materials tend to be more brittle than other lower strength materials. The corrosion resistance is also relatively low in many of these alloys and corrosion protection is therefore advised.

Welding requires care and the use of the inert gas shielded process.

The castings can be anodized.

These materials are used similarly to those in Section 1K, i.e. for gear boxes, sumps, electrical fittings, brackets etc. Some of the alloys in this section have reasonably good high temperature properties and therefore can be used at elevated temperatures where heat is involved, i.e. up to approximately 200–300 °C.

Symbol	Nominal analysis, supplier, condition and remarks.
3.1263	4.5% Cu 3% Si 2.5% Zn 0.5% Ni Al alloy: Casting; German Standard DIN 1725
3.1371	4.5% Cu 0.2% Mg 0.2% Ti Al alloy: Casting; German Standard DIN 1725
3.1841	4.5% Cu 0.2% Ti Al alloy: Casting; German Standard DIN 1725
3.2151	4% Cu 0.2% Mg 6% Si 2% Zn Al alloy: Casting; German Standard DIN 1725
3.2161	3.5% Cu 8.5% Si Al alloy: Casting; German Standard
113	7% Cu 3% Si 3% Zn Al alloy: Casting; Alcoa for BS alloy LM 1
113	7% Cu 3% Si Al alloy: Casting; AA designation for ASTM alloy CS 72A
117	3% Cu 5% Si 0.5% Mn Al alloy: Casting; Comalco for BS alloy LM 4
122	10% Cu 0.3% Mg Al alloy: Casting; Alcoa for BS alloy LM 12
122	10% Cu 0.2% Mg Al alloy: Casting; AA designation for ASTM alloy CG 100A
127	4.5% Cu 5% Si 0.3% Fe Al alloy: Casting; Comalco for BS alloy LM 22
138	10% Cu 4% Si 0.3% Fe Al alloy: Casting; as cast **DPN: 100 UTS: 200 Elon: 1.5% Proof: 170**
142	4% Cu 1.5% Mg 2% Ni Al alloy: Casting; Alcoa for BS alloy LM 14
142	4% Cu 1.5% Mg 2% Ni Al alloy: Casting; AA designation for ASTM alloy CN 42A
143	3.5% Cu 8.5% Si 0.8% Fe Al alloy: Casting; Comalco for BS alloy LM 24
152	6% Cu 0.4% Mg 5.5% Si Al alloy: Casting; AA designation for ASTM alloy CS 66A
195	4.5% Cu Al alloy: Casting; Alcoa for BS alloy LM 11
201.0	4.6% Cu 0.3% Mg 0.8% Ag Al alloy: Casting; designation used in the UK and USA
201.2	4.6% Cu 0.3% Mg 0.8% Ag Al alloy: Casting; designation used in the UK and USA
202.0	4.6% Cu 0.3% Mg 0.4% Cr 0.8% Ag Al alloy: Casting; obsolete; designation used in the UK and USA
202.2	4.6% Cu 0.3% Mg 0.4% Cr 0.8% Ag Al alloy: Casting; obsolete; designation used in the UK and USA
203.0	5% Cu 1.5% Ni 0.2% Ti Al alloy: Casting; designation used in the UK and USA
203.2	5% Cu 1.5% Ni 0.2% Ti Al alloy: Casting; designation used in the UK and USA
204.0	4.8% Cu 0.2% Mg 0.2% Ti Al alloy: Casting; designation used in the UK and USA
204.2	4.5% Cu 0.2% Mg 0.2% Ti Al alloy: Casting; designation used in the UK and USA
206.0	4.6% Cu 0.25% Mg Al alloy: Casting; designation used in the UK and USA
206.2	4.6% Cu 0.25% Mg Al alloy: Ingot; designation used in the UK and USA
208.0	4% Cu 3% Si Al alloy: Casting; designation used in the UK and USA
208.1	4% Cu 3% Si Al alloy: Ingot; designation used in the UK and USA
208.1	4% Cu 3% Si Al alloy: Ingot; designation used in the UK and USA
208.2	4% Cu 3% Si Al alloy: Ingot; designation used in the UK and USA
213.0	7% Cu 2% Si 2% Zn Al alloy: Ingot; designation used in the UK and USA
213.1	7% Cu 2% Si 2% Zn Al alloy: Ingot; designation used in the UK and USA
220.0	10% Cu 0.2% Mg Al alloy: Casting; designation used in the UK and USA
222.1	10% Cu 0.25% Mg Al alloy: Ingot; designation used in the UK and USA
224.0	5% Cu 0.3% Mn 0.1% V 0.1% Zr Al alloy: Casting; designation used in the UK and USA
224.2	5% Cu 0.3% Mn 0.1% V 0.1% Zr Al alloy: Ingot; designation used in the UK and USA
226	4.8% Cu 0.2% Ti Al alloy: Casting; Comalco for BS alloy LM 11
238.0	10% Cu 4% Si 0.2% Mn 1% Zn Al alloy: Obsolete; designation used in the UK and USA
238.1	10% Cu 4% Si 0.2% Mn 1% Zn Al alloy: Ingot; obsolete; designation used in the UK and USA
238.2	10% Cu 4% Si 0.2% Mn Al alloy: Ingot; obsolete; designation used in the UK and USA
240.0	8% Cu 6% Mg 0.5% Si 0.5% Mn 0.5% Ni Al alloy: Casting; designation used in the UK and USA
240.1	8% Cu 6% Mg 0.5% Si 0.5% Mn 0.5% Ni Al alloy: Ingot; designation used in the UK and USA
242.0	4% Cu 1.6% Mn 2% Ni Al alloy: Casting; designation used in the UK and USA
242.1	4% Cu 1.5% Mn 2% Ni Al alloy: Ingot; designation used in the UK and USA
242.2	4% Cu 1.5% Mn 2% Ni Al alloy: Ingot; designation used in the UK and USA
243.0	4% Cu 2% Mg 0.3% Cr 2% Ni Al alloy: Casting; designation used in the UK and USA
243.1	4% Cu 2% Mg 0.3% Cr 2% Ni Al alloy: Ingot; designation used in the UK and USA
295.0	4.5% Cu 1% Ti Al alloy: Casting; designation used in the UK and USA
295.1	4.5% Cu 1% Ti Al alloy: Ingot; designation used in the UK and USA
295.2	4.5% Cu 1% Ti Al alloy: Ingot; designation used in the UK and USA
296.0	4.5% Cu 2.5% Si Al alloy: Casting; designation used in the UK and USA
296.1	4.5% Cu 2.5% Si Al alloy: Ingot; designation used in the UK and USA
296.2	4.5% Cu 2.5% Si Al alloy: Ingot; designation used in the UK and USA
308.0	4.5% Cu 5.5% Si Al alloy: Casting; designation used in the UK and USA
308.1	4.5% Cu 5.5% Si Al alloy: Ingot; designation used in the UK and USA
308.2	4.5% Cu 5.5% Si Al alloy: Ingot; designation used in the UK and USA
319	3% Cu 4.5% Si 0.5% Mn Al alloy: Casting; Alcoa for BS alloy LM 4
319	4% Cu 6% Si Al alloy: Casting; AA designation for ASTM alloy SC 64B or SC 64C
319.0	3.5% Cu 6% Si Al alloy: Casting; designation used in the UK and USA
319.1	3.5% Cu 6% Si Al alloy: Ingot; designation used in the UK and USA
319.2	3.5% Cu 6% Si Al alloy: Ingot; designation used in the UK and USA
320	3% Cu 0.3% Mg 6.5% Si Al alloy: Casting; designation used in the UK and USA
320.1	3% Cu 0.3% Mg 6.5% Si Al alloy: Ingot; designation used in the UK and USA
328.0	1.5% Cu 0.4% Mg 8.0% Si 0.4% Mn Al alloy: Investment casting; American National Standard; ANSI

Note. The following abbreviations and units are used in the tables:

DPN	Hardness, diamond pyramid number
UTS	Ultimate tensile strength, N/mm^2
Elon	Elongation, %
Proof	0.1% proof strength, N/mm^2

$1\ N/mm^2 = 0.1\ hbar = 0.102\ kgf/mm^2 = 0.06475\ tonf/in.^2 = 145.04\ lbf/in.^2 = 1\ MPa$

See Appendix II for other abbreviations and conversion tables.

Symbol	Nominal analysis, supplier, condition and remarks.
332.0	3% Cu 1% Mg 9.5% Si Al alloy: Casting; designation used in the UK and USA
332.1	3% Cu 1% Mg 9.5% Si Al alloy: Ingot; designation used in the UK and USA
332.2	3% Cu 1% Mg 9.5% Si Al alloy: Ingot; designation used in the UK and USA
333.0	3.5% Cu 1% Mg 9% Si Al alloy: Casting; designation used in the UK and USA
333.1	3.5% Cu 1% Mg 9% Si Al alloy: Ingot; designation used in the UK and USA
339.0	2.2% Cu 1% Mg 12% Si 1% Ni Al alloy: Casting; designation used in the UK and USA
355.0	1.2% Cu 0.5% Mg 5.0% Si Al alloy: Investment casting; American National Standard; ANSI
380	3.5% Cu 8% Si Al alloy: Casting; AA designation for ASTM alloy SC 84B
384	3.5% Cu 11% Si Al alloy: Casting; AA designation for ASTM alloy SC 114A
645	3.5% Cu 11% Zn Al alloy: Casting; Alcoa for BS alloy LM 3
6063	0.6% Mg 0.4% Si Al alloy: Wrought; designation used in the UK and USA
6066	1% Mg 1.3% Si 1% Mn Al alloy: Wrought; designation used in the UK and USA
6070	0.25% Cu 0.8% Mg 1.3% Si 0.8% Mn Al alloy: Wrought; designation used in the UK and USA
6101	0.5% Mg 0.5% Si Al alloy: Wrought; designation used in the UK and USA
A 108	4.5% Cu 5.5% Si Al alloy: Casting; AA designation for ASTM alloy SC 64A
A 143	3% Cu 1% Mg 10% Si 0.4% Fe 1% Ni Al alloy: Casting; Comalco; aged
	DPN: 105 UTS: 240 Elon: 1% Proof: 180
A 380	3.5% Cu 8% Si Al alloy: Casting; AA designation for ASTM alloy SC 84A
A 33550	Unified number for alloy C355.0
AERAL	3.5% Cu 1.8% Mg 0.6% Si 2.25% Cd Al alloy: Source unknown
	UTS: 450
AEROLITE	1.15% Cu 0.4% Mg 0.5% Si 0.1% Zn 1.0% Fe Al alloy: Source unknown
AERON	4% Cu 1.0% Si Al alloy: Solution treated and aged; source unknown
	UTS: 370
Af 3	4.5% Cu 0.2% Ti Al alloy: Casting; Anglo-Swiss code for Alufont 3
Al 16 V	4% Cu 0.2% Mg 6% Si 0.4% Mn Al alloy: Casting; Russian designation
Al 18 V	10% Cu 0.3% Mg 2.5% Si Al alloy: Casting; Russian designation
Al 75	3% Cu 5% Si 0.5% Mn Al alloy: Casting; American name for BS alloy LM 22
ALAR 308	3.5% Cu 8% Si Al alloy: Casting; American name for BS alloy LM 24
ALCAN GB 117	3% Cu 5% Si 0.5% Mn Al alloy: Casting; Alcan for BS alloy LM 4
ALCAN GB 218	4% Cu 1.5% Mg 2% Ni Al alloy: Casting; Alcan for BS alloy LM 14
ALCAN GB 226	4.5% Cu Al alloy: Casting; Alcan for BS alloy LM 11
ALCAN GB A143	3% Cu 1% Mg 9% Si Al alloy: Casting for pistons; Alcan for BS alloy LM 26
ALCAN GB B117	3% Cu 5% Si 0.5% Mn Al alloy: Casting; Alcan for BS alloy LM 21
ALCAN GB C117	3.5% Cu 5% Si 0.5% Mn Al alloy: Casting; Alcan for BS alloy LM 22
ALDAL	4% Cu 0.5% Mg 0.6% Si 0.5% Mn Al alloy; Source unknown
	UTS: 450
ALFERIUM	2.5% Cu 0.6% Mg 0.3% Si 0.5% Mn Al alloy: Source unknown
	UTS: 450

Symbol	Nominal analysis, supplier, condition and remarks.
ALUFONT 3	4.5% Cu 0.2% Ti Al alloy: Casting; Anglo-Swiss Aluminium Co.; solution treated and aged
	DPN: 105 UTS: 340 Elon: 6% Proof: 220
ALUFONT 42	4.7% Cu 0.25% Ti Al alloy: Casting; Anglo-Swiss Aluminium Co.
ALUFONT 47	4.7% Cu 0.2% Mg 0.25% Ti Al alloy: Casting; Anglo-Swiss Aluminium Co.
ALUMINAL 7	2.0% Cu 0.2% Mg 2.5% Si 1.0% Ni Al alloy: Casting: origin unknown
AMS 4220 D	4% Cu 1.5% Mg 0.2% Cr 2% Ni Al alloy: Casting; solution treated and over-aged; AMS for AA alloy 142
AMS 4222 D	4% Cu 1.5% Mg Al alloy: Casting; solution treated and over-aged; AMS for AA alloy 142
AMS 4223	4.5% Cu 0.2% Mg 0.3% Mn 0.6% Ag Al alloy: Casting
AMS 4224	4% Cu 2% Mg 0.3% Mn 0.3% Cr 2% Ni 0.1% Ti 0.1% V Al Alloy: Casting; stabilized
AMS 4225	5% Cu 0.25% Mn 1.5% Ni Al alloy: Casting
AMS 4226	5% Cu 0.3% Mn 0.2% Zr Al alloy: Casting
AMS 4227 A	8% Cu 6% Mg 0.5% Mn 0.5% Ni Al alloy: Casting
AMS 4228	4.5% Cu 0.2% Mg 0.3% Mn Al alloy: Casting
AMS 4229	4.5% Cu 0.2% Mg 0.3% Mn Al alloy: Casting
AMS 4230 C	4.5% Cu Al alloy: Casting; solution treated; AMS for AA alloy 195
AMS 4231 C	4.5% Cu Al alloy: Casting; solution treated and aged; AMS for AA alloy 195
AMS 4235	4.7% Cu 0.25% Mg 0.3% Mn 0.2% Ti Al alloy: Casting
AMS 4236	4.7% Cu 0.25% Mg 0.3% Mn 0.2% Ti Al alloy: Casting
AMS 4237	4.6% Cu 0.25% Mg Al alloy: Casting
AMS 4242	4.3% Cu 0.2% Mg 0.3% Mn Al alloy: Casting
AMS 4282 E	4.5% Cu 2.5% Si Al alloy: Casting; solution treated and aged
AMS 4283 D	4.5% Cu 2.5% Si Al alloy: Casting; solution treated
AMS 4291 B	3.5% Cu 5% Si or 8.5% Si Al alloy: Die casting; as cast
AN A4	3.5% Cu 3.5% Si 1.0% Fe Al alloy: Casting; US Federal
	UTS: 110 Elon: 1.5%
AN A5	4.0% Cu 2.5% Si Al alloy: Casting; US Service; aged
	UTS: 170
AN QQ A 379	4.0% Cu 1.5% Mg 2.0% Ni Al alloy: Casting; US Service; solution treated and aged
	UTS: 200
AN QQ A383	4.5% Cu 2.5% Si Al alloy: Casting; US Service; aged
	UTS: 200
AN QQ A390	4.5% Cu 1.5% Si Al alloy: Casting; US Service
	UTS: 180
AN QQ A397	4.0% Cu 3.0% Si Al alloy: Casting; US Service
	UTS: 110 Elon: 1.5%
AN QQ A399	8.0% Cu 1.2% Si 1.5% Fe Al alloy: Casting; US Service; as cast
	UTS: 110
AP 303	3% Cu 5% Si 0.4% Mn Al alloy: Casting; Australian specification
AP 309	1.2% Cu 0.5% Mg 5% Si Al alloy: Casting; Australian specification
A S2U	1.5% Cu 0.1% Mg 2.1% Si 1.1% Ni 1.1% Fe Al alloy: Casting; French Standard
A S2U	1.3% Cu 0.1% Mg 2.3% Si 1.0% Fe Al alloy: Casting; L'Aluminium Français
A S5U	3.2% Cu 5% Si 0.4% Mn Al alloy: Casting; French Standard
A S5U2	4% Cu 0.2% Mg 6% Si 0.4% Mn Al alloy: Casting; French Standard
A S5U3	3.2% Cu 0.1% Mg 5.0% Si Al alloy: Casting; L'Aluminium Français
A S9U3Y4	3.5% Cu 8.2% Si Al alloy: Casting; French Standard
AS 307	1.5% Cu 10.2% Si Al alloy: Casting; Australian specification
AS 313	3.5% Cu 8.5% Si Al alloy: Casting; Australian specification

Symbol	Nominal analysis, supplier, condition and remarks.
ASTM B26 C4A	4.5% Cu 1.5% Si 1% Fe 0.25% Ti Al alloy: Sand casting; solution treated and aged **UTS: 210 Elon: 3% Proof: 120**
ASTM B26 CG100A	10% Cu 2% Si 1.5% Fe Al alloy: Sand casting; solution treated and aged **UTS: 200**
ASTM B26 CN42A	4% Cu 1.5% Mg 1% Fe 2% Ni Al alloy: Sand casting; solution treated and aged **UTS: 210 Proof: 120**
ASTM B26 CS43A	4% Cu 3% Si 1.2% Fe Al alloy: Sand casting; as cast **UTS: 110 Elon: 1.5% Proof: 60**
ASTM B26 CS72A	7% Cu 3% Si 1.4% Fe 2.5% Zn Al alloy: Sand casting; as cast; obsolete **UTS: 120 Elon: 1% Proof: 60**
ASTM B26 SC8	3% Cu 5% Si Al alloy: Casting; chill cast; solution treated and aged **UTS: 300 Elon: 3% Proof: 150**
ASTM B26 SC51A	1.2% Cu 0.5% Mg 5% Si Al alloy: Sand casting; solution treated and aged **UTS: 210 Elon: 2% Proof: 120**
ASTM B26 SC64D	4% Cu 6% Si Al alloy: Sand casting; solution treated and aged **UTS: 210 Elon: 2.5% Proof: 120**
ASTM B26 CS74A	7.0% Cu 3.5% Si 1.4% Fe 0.2% Ti Al alloy: Casting; replaces CS72A
ASTM B85 SC84A	3.5% Cu 8% Si Al alloy: Die casting; as cast **UTS: 330 Elon: 3.5% Proof: 150**
ASTM B85 SC84B	3.5% Cu 8% Si 2% Fe Al alloy: Die casting; as cast **UTS: 320 Elon: 2.5% Proof: 150**
ASTM B85 SC114A	3.5% Cu 11% Si Al alloy: Die casting; as cast **UTS: 330 Elon: 2.5% Proof: 150**
ASTM B108 CG100A	10% Cu 0.2% Mg 2% Si Al alloy: Casting; solution treated and aged **UTS: 270**
ASTM B108 CN42A	4% Cu 1.5% Mg 0.7% Si 2% Ni Al alloy: Casting; solution treated and aged **UTS: 270**
ASTM B108 CS42A	4.5% Cu 2.5% Si Al alloy: Casting; solution treated and aged **UTS: 240 Elon: 2% Proof: 140**
ASTM B108 CS66A	6.5% Cu 0.4% Mg 5.5% Si Al alloy: Casting; aged **UTS: 210 Proof: 170**
ASTM B108 CS72A	7% Cu 2.5% Si Al alloy: Casting; as cast; obsolete **UTS: 150 Proof: 90**
ASTM B108 CS104A	10% Cu 0.2% Mg 4% Si Al alloy: Casting; as cast
ASTM B108 SC51A	1.2% Cu 0.8% Mg 5% Si Al alloy: Casting; solution treated and aged **UTS: 250 Elon: 1.5% Proof: 150**
ASTM B108 SC64D	4% Cu 6% Si 0.2% Ti 0.3% Ni Al alloy: Casting; solution treated and aged **UTS: 270 Elon: 2% Proof: 150**
ASTM B108 SC92A	2.0% Cu 0.5% Mg 9% Si Al alloy: Die casting
ASTM B108 SC94A	3.5% Cu 0.25% Mg 9% Si Al alloy: Die casting
ASTM B108 SC103A	3% Cu 1% Mg 9% Si Al alloy: Casting; aged **UTS: 210**
ASTM B108 SC122A	1.5% Cu 0.7% Mg 12% Si 0.2% Ti Al alloy: Casting; solution treated and aged **UTS: 280 Proof: 250**

Note. The following abbreviations and units are used in the tables:

DPN	Hardness, diamond pyramid number
UTS	Ultimate tensile strength, N/mm^2
Elon	Elongation, %
Proof	0.1% proof strength, N/mm^2

1 N/mm^2=0.1 hbar=0.102 kgf/mm^2=0.06475 tonf/in.^{2}=145.04 lbf/in.^{2}=1 MPa

See Appendix II for other abbreviations and conversion tables.

Symbol	Nominal analysis, supplier, condition and remarks.
ASTM B179 C4A	4.5% Cu 1% Si Al alloy: Ingot; primary metal
ASTM B179 CG100A	10% Cu 0.3% Mg 2% Si Al alloy: Ingot; primary metal
ASTM B179 CN42A	4% Cu 1.5% Mg 2% Ni Al alloy: Ingot; primary metal
ASTM B179 CS43A	4% Cu 3% Si Al alloy: Ingot; primary metal
ASTM B179 CS66A	6% Cu 0.3% Mg 5.5% Si Al alloy: Ingot; primary metal
ASTM B179 CS72A	7% Cu 3% Si Al alloy: Ingot; primary metal
ASTM B179 CS76A	7% Cu 35% Si Al alloy: Ingot
ASTM B179 CS104A	10% Cu 0.3% Mg 4% Si Al alloy: Ingot; primary metal
ASTM B179 SC51A	1% Cu 0.5% Mg 5% Si 0.25% Cr Al alloy: Ingot; primary metal
ASTM B179 SC64C	4% Cu 6% Si Al alloy: Ingot; primary metal
ASTM B179 SC84A–B	3.5% Cu 8% Si 2.9% Zn (max) Al alloy: Ingot; primary metal
ASTM B179 SC103A	2.5% Cu 1% Mg 9.5% Si Al alloy: Ingot; primary metal
ASTM B179 SC114A	4% Cu 11% Si Al alloy: Ingot; primary metal
ASTM B179 SC122A	1.5% Cu 0.7% Mg 12% Si Al alloy: Ingot; primary metal
ASTM B179 SC649	4% Cu 6% Si Al alloy: Ingot; primary metal
A U4NT	4.0% Cu 1.4% Mg Al alloy: Casting; L'Aluminium Français
A U5GT	4.8% Cu 0.25% Mg Al alloy: Casting; L'Aluminium Français
A U10G	10% Cu 0.25% Mg Al alloy: Casting; L'Aluminium Français
A U4NT	4% Cu 1.4% Mg 2% Ni Al alloy: Casting; French Standard
A U40T5	55% Cu 5% Ti Al master alloy: L'Aluminium Français
A U5GT	4.5% Cu Al alloy: Casting; French Standard
A U50	49% Cu Al master alloy: L'Aluminium Français
A U1054	10% Cu 0.25% Mg 4.2% Si Al alloy: Casting; L'Aluminium Français
B 132	1.5% Cu 1.0% Mg 12% Si 0.7% Mn Al alloy: Casting; AA designation for ASTM alloy SC122A
B 143	3% Cu 1.2% Mg 9.5% Si Al alloy: Casting; Comalco; aged **DPN: 103 UTS: 240 Elon: 1% Proof: 170**
B 195	4.5% Cu 2.5% Si Al alloy: Casting; AA designation for ASTM alloy CS42A
BA 32	4.5% Cu Al alloy: Casting; British Aluminium Co. for BS alloy LM 11
BA 33	4% Cu 1.5% Mg 2% Ni Al alloy: Casting; British Aluminium Co. for BS alloy LM 14
BIRMALITE	10% Cu 0.3% Mg Al alloy: Castings; Birmabright for BS alloy LM 12
BKL 305	3% Cu 5% Si 0.5% Mn Al alloy: Castings; BKL chill cast similar to BS alloy LM 4 with wider limits **DPN: 85 UTS: 170 Elon: 1.5% Proof: 75**
BKL 308	3.5% Cu 8.5% Si Al alloy: Casting; BKL for BS alloy LM 24
Bohn L4	3% Cu 1% Mg 9% Si 1% Ni Al alloy: Casting; American proprietary alloy **DPN: 105 UTS: 210 Elon: 0% Proof: 200**
BS 361	7% Cu Al alloy: Casting; replaced by BS 1490
BS 362	12% Cu Al alloy: Casting; replaced by BS alloy LM12
BS 363	Zn Cu Al alloy: Casting for crankcases; replaced by BS alloy LM 3
BS 702	Si Cu Al alloy: Casting; replaced by BS 1490
BS 703	4% Cu 1.5% Mg 2% Ni 0.2% Ti Al alloy: Castings; Y alloy; replaced
BS 1490 LM1	7% Cu 3% Si 3% Zn Al alloy: Ingot; chill cast **UTS: 150**
BS 1490 LM1M	7% Cu 3% Si 3% Zn Al alloy: Casting; as cast; chill cast **UTS: 150**
BS 1490 LM3	3% Cu 10% Zn Al alloy: Ingot; chill cast **UTS: 140**
BS 1490 LM3M	3% Cu 10% Zn Al alloy: For sand casting; chill cast **UTS: 140**

Symbol	Nominal analysis, supplier, condition and remarks.
BS 1490 LM4	3% Cu 5% Si 0.5% Mn Al alloy: Ingot; chill cast **UTS: 150**
BS 1490 LM4M	3% Cu 5% Si 0.5% Mn Al alloy: For sand and die casting; chill cast **UTS: 150**
BS 1490 LM4WP	3% Cu 5% Si 0.5% Mn Al alloy: Sand and die casting; solution treated and aged **UTS: 220**
BS 1490 LM7	1.75% Cu 2% Si 1% Ni 1% Fe Al alloy: Casting; chill cast **UTS: 180 Elon: 3%**
BS 1490 LM11	4.5% Cu 0.3% Ti Al alloy: Ingot; chill cast **UTS: 300 Elon: 9%**
BS 1490 LM11WP	4.5% Cu 0.3% Ti Al alloy: For sand casting; chill cast; solution treated and aged **UTS: 300 Elon: 9%**
BS 1490 LM12	10% Cu 0.3% Mg Al alloy: Ingot; chill cast **UTS: 170**
BS 1490 LM12WP	10% Cu 0.3% Mg Al alloy: Die castings; chill cast; solution treated and aged **UTS: 270**
BS 1490 LM14	4% Cu 1.5% Mg 2% Ni Al alloy: Ingot; chill cast **UTS: 270**
BS 1490 LM14WP	4% Cu 1.5% Mg 2% Ni Al alloy: Sand or die castings; chill cast; solution treated and aged **DPN: 120 UTS: 270**
BS 1490 LM15	2% Cu 1% Mg 1.5% Si 1.5% Ni 1% Fe Al alloy: Casting; chill cast; solution treated and aged **DPN: 125 UTS: 330**
BS 1490 LM21	4% Cu 6% Si 0.5% Mn Al alloy: Ingot; chill cast **UTS: 170 Elon: 1%**
BS 1490 LM21M	4% Cu 6% Si 0.5% Mn Al alloy: Sand or die casting; chill cast **UTS: 170 Elon: 1%**
BS 1490 LM22	3% Cu 5% Si 0.5% Mn Al alloy: Ingot; chill cast **UTS: 240 Elon: 8%**
BS 1490 LM22W	3% Cu 5% Si 0.5% Mn Al alloy: Die casting; chill cast; solution treated, room aged **UTS: 240 Elon: 8%**
BS 1490 LM23	1.5% Cu 2% Si 1% Fe 1% Ni 0.2% Ti Al alloy: Ingot; chill cast **UTS: 180 Elon: 3%**
BS 1490 LM23P	1.5% Cu 2% Si 1% Fe 1% Ni 0.2% Ti Al alloy: Sand or die casting; chill cast and aged **UTS: 180 Elon: 3%**
BS 1490 LM24	3.5% Cu 8% Si Al alloy: Ingots; chill cast **UTS: 170 Elon: 1.5%**
BS 1490 LM24M	3.5% Cu 8% Si Al alloy: Die castings; as cast; chill cast **UTS: 170 Elon: 1.5%**
BS 1490 LM26	3.0% Cu 1.0% Mg 9.5% Si Al alloy: Casting; chill cast **DPN: 100 UTS: 210 Proof: 160**
BS 1490 LM27	2.0% Cu 7.0% Si 0.4% Mn Al alloy: Casting; chill cast **UTS: 160 Elon: 2% Proof: 90**
BS 1490 LM28	1.5% Cu 1.2% Mg 19% Si 1.2% Ni Al alloy: Casting; chill cast **UTS: 190 Proof: 160**
BS 1490 LM29	1.1% Cu 1.1% Mg 23% Si 1.1% Ni Al alloy: Casting; chill cast **UTS: 190 Proof: 170**
BS 1490 LM30	4.5% Cu 0.5% Mg 17% Si Al alloy: Casting; chill cast **UTS: 160 Proof: 160**
CERALUMIN ASM	1.5% Cu 0.6% Mg 1% Si 0.8% Ni 0.1% Nb Al alloy: Casting; J Stone; chill cast; solution treated and aged **UTS: 300 Elon: 12% Proof: 170**
C 355.0	1.25% Cu 0.5% Mg 5.0% Si Al alloy: Investment casting; American National Standard; ANSI
CERALUMIN B	1% Cu 0.1% Mg 2.5% Si 1% Ni 0.2% Ti Al alloy: Casting; J Stone for BS alloy LM 7
CERALUMIN C	1.5% Cu 0.8% Mg 0.75% Si 1% Ni 0.2% Ti Al alloy: Casting; J Stone for BS alloy LM 15
COMALCO 117	3% Cu 5% Si 0.5% Mn 0.3% Fe Al alloy: Casting; Comalco for BS alloy LM 4
COMALCO 127	4.5% Cu 5% Si 0.3% Fe Al alloy: Casting; Comalco for BS alloy LM 22
COMALCO 138	10% Cu 4% Si 0.3% Fe Al alloy: Casting; Comalco for ASTM alloy CS 104A
COMALCO 143	3.5% Cu 8% Si 0.8% Fe Al alloy: Casting; Comalco for ASTM alloy LM 24
COMALCO A143	3% Cu 1% Mg 9.5% Si 0.5% Fe 1% Ni Al alloy: Casting; Comalco
COMALCO B143	3% Cu 1% Mg 9.5% Si 0.3% Fe Al alloy: Casting; Comalco for ASTM alloy SC 103A
CQ 51A	4.6% Cu 0.3% Mg 0.35% Mn Al alloy: Investment casting; designation used by ASTM
D 7	3% Cu 4.5% Si 0.5% Mn Al alloy: Casting; Birmabright for BS alloy LM4
D 8	3% Cu 4.5% Si 0.5% Mn Al alloy: Casting; Birmabright for BS alloy LM 4
DS 15000 Al 101	10% Cu 0.3% Mg Al alloy: Casting; Danish specification
DS 15000 Al 511	3% Cu 5% Si Al alloy: Casting; Danish specification
DS 15000 Al 512	3.5% Cu 8.5% Si Al alloy: Casting; Danish specification
DTD 131	Cu Al alloy: Casting; obsolete; replaced by L52
DTD 133C	1% Cu 2.5% Si 1% Ni 1% Fe 0.2% Ti Al alloy: Casting; chill cast and aged **UTS: 190 Elon: 3% Proof: 120**
DTD 276A	1.2% Cu 0.5% Mg 5% Si Al alloy: Casting; solution treated and aged; chill cast; replaced by L78 **UTS: 250 Proof: 180**
DTD 287	1% Cu 0.5% Si 1% Ni 1% Fe 0.2% Ti Al alloy: Casting; chill cast and aged **UTS: 190 Elon: 3% Proof: 120**
DTD 298B	4.5% Cu 0.2% Ti Al alloy: Casting; solution treated; chill cast **UTS: 250 Elon: 13% Proof: 130**
DTD 304B	4.5% Cu 0.2% Ti Al alloy: Casting; solution treated and aged; chill cast **UTS: 300 Elon: 9% Proof: 180**
DTD 361B	4.5% Cu Al alloy: Casting; solution treated and aged; chill cast **UTS: 390 Elon: 4% Proof: 340**
DTD 424A	3% Cu 5% Si 0.5% Mn Al alloy: Casting; as cast; replaced by L79 **UTS: 150 Elon: 2%**
DTD 741A	4% Cu 2% Mg 0.7% Co 0.2% Nb Al alloy: Casting; chill cast; solution treated and aged **UTS: 330 Proof: 240**
DTD 5018	7.7% Mg 1.2% Zn Al alloy: Casting
E QQ A 601/13	4.0% Cu 2.5% Si Al alloy: Casting; US Federal; aged **UTS: 170**
E QQ A 601/14	3.5% Cu 3.0% Si 1.0% Fe Al alloy: Casting; US Federal **UTS: 110 Elon: 1.5%**
G Al Cu 4 Ti	4.5% Cu 0.2% Ti Al alloy: Casting; previous German Standard designation
G Al Cu 4 Ti Mg	4.5% Cu 0.2% Mg 0.2% Ti Al alloy: Casting; previous German Standard designation
G Al Cu 5 Si 3	5% Cu 3% Si Al alloy: Casting; previous German Standard designation
GMNA 100	6% Cu 6% Si Al alloy: Casting; Comalco; as cast **DPN: 65 UTS: 150 Elon: 3% Proof: 75**
GOST 2685/63 Al 1	4% Cu 1.5% Mg 2.0% Ni Al alloy: Casting; Russian specification
GOST 2685/63 Al 3	1.3% Cu 0.5% Mg 5.0% Si Al alloy: Casting; Russian specification
GOST 2685/63 Al 3B	1.3% Cu 0.5% Mg 5.0% Si Al alloy: Casting; Russian specification
GOST 2685/63 Al 6	Cu Mg Si Al alloy: Casting; Russian specification
GOST 2685/63 Al 7	4.5% Cu Al alloy: Casting; Russian specification
GOST 2685/63 Al7B	4.5% Cu Al alloy: Casting; Russian specification

Symbol	Nominal analysis, supplier, condition and remarks.
GOST 2685/63 Al 10B	7% Cu 3% Si 3% Zn Al alloy: Casting; Russian specification
GOST 2685/63 Al 12	10% Cu 0.3% Mg Al alloy: Casting; Russian specification
GOST 2685/63 Al 14B	3.5% Cu 8.5% Si Al alloy: Casting; Russian specification
GOST 2685/63 Al 15	7% Cu 3% Si 3% Zn Al alloy: Casting; Russian specification
GOST 2685/63 Al 16B	3% Cu 5% Si 0.5% Mn Al alloy: Casting; Russian specification
GOST 2685/63 Al 18B	Cu Si Zn Al alloy: Casting; Russian specification
GOST 2685/63 Al 19	4.5% Cu Al alloy: Casting; Russian specification
GOST 2685/63 Al 20B	7% Cu 3% Si 3% Zn Al alloy: Casting; Russian specification
GOST 2685/63 Al 21	4% Cu 1.5% Mg 2% Ni Al alloy: Casting; Russian specification
H 49–1	7% Cu 3% Si 3% Zn Al alloy: Casting; Australian specification
H 49–3	3% Cu 5% Si Al alloy: Casting; Australian specification
H 49–9	4.5% Cu Al alloy: Casting; Australian specification
H 49–10	10% Cu 0.3% Mg Al alloy: Casting; Australian specification
H 49–12	4.0% Cu 1.5% Mg 2.0% Ni Al alloy: Casting; Australian specification
H 49–13	1.2% Cu 0.5% Mg 5.0% Si Al alloy: Casting; Australian specification
H 49–16	3.5% Cu 8.5% Si Al alloy: Casting; Australian specification
HA 3 C4	4.5% Cu 0.25% Si 0.1% Ti Al alloy: Ingot; Canadian Standard
HA 3 CG50	4.8% Cu 0.2% Mg 0.2% Ti Al alloy: Ingot; Canadian Standard
HA 3 CS42	4.5% Cu 2.5% Si 0.1% Ti Al alloy: Ingot; Canadian Standard
HA 3 CS72	7.2% Cu 2.0% Si 1% Fe Al alloy: Ingot; Canadian Standard
HA 3 SC51N	1.2% Cu 0.5% Mg 5.0% Si 0.2% Fe Al alloy: Ingot; Canadian Standard
HA 3 SC51P	1.2% Cu 0.5% Mg 5.0% Si Al alloy: Ingot; Canadian Standard
HA 3 SC53	3.0% Cu 5.0% Si 0.5% Mn Al alloy: Ingot; Canadian Standard
HA 3 SC84N	3.5% Cu 8.5% Si Al alloy: Ingot; Canadian Standard
HA 3 SC84P	3.5% Cu 8.5% Si 0.8% Fe Al alloy: Ingot; Canadian Standard
HA 3 SC84R	3.5% Cu 0.6% Mg 8.5% Si 1.0% Fe 1.0% Zn Al alloy: Ingot; Canadian Standard
HA 9	4.5% Cu Al alloy: Sand casting; Canadian Standard
HA 9 CG50	4.8% Cu 0.2% Mg Al alloy: Sand casting; Canadian Standard
HA 9 CS42	4.5% Cu 2.5% Si Al alloy: Sand casting; Canadian Standard
HA 9 CS72	7.2% Cu 2.0% Si 1.2% Fe Al alloy: Sand casting; Canadian Standard
HA 9 SC53	3.0% Cu 5.0% Si 0.5% Mn Al alloy: Sand casting; Canadian Standard

Note. The following abbreviations and units are used in the tables:

DPN	Hardness, diamond pyramid number
UTS	Ultimate tensile strength, N/mm^2
Elon	Elongation, %
Proof	0.1% proof strength, N/mm^2

1 N/mm^2=0.1 hbar=0.102 kgf/mm^2=0.06475 $tonf/in.^2$=145.04 $lbf/in.^2$=1 MPa
See Appendix II for other abbreviations and conversion tables.

Symbol	Nominal analysis, supplier, condition and remarks.
HA 9 ZG61N	0.6% Mg 0.5% Cr 0.2% Ti 5.8% Zn Al alloy: Sand casting; Canadian Standard
HA 9 ZG61P	0.5% Cu 0.7% Mg 6.5% Zn Al alloy: Sand casting; Canadian Standard
HA 10 CS42	4.5% Cu 2.5% Si Al alloy: Die casting; Canadian Standard
HA 10 SC53	3.0% Cu 5.0% Si 0.5% Mn Al alloy: For die casting; Canadian Standard
HIDUMINIUM 20	3% Cu 4.5% Si 0.5% Mn Al alloy: For sand and die casting; High Duty Alloys for BS alloy LM 4
HIDUMINIUM 80	4.5% Cu 0.15% Ti Al alloy: High strength sand casting; High Duty Alloys for BS alloy LM 11
INTAL 7Q5	7.2% Cu 5.0% Si 5.5% Mn Al alloy: For die casting; Canadian Standard
L 4	3% Cu 1% Mg 9% Si 1% Ni Al alloy: Casting; Birmabright; chill cast for pistons **DPN: 105 UTS: 210 Proof: 200**
L35	4% Cu 1.5% Mg 2% Ni 0.2% Ti Al alloy: Ingot and casting; chill cast; Y alloy **UTS: 270 Proof: 210**
L50	4% Cu 0.75% Mg 0.6% Mn Al alloy: For ingots; secondary metal for re-melting
L51	1% Cu 2.5% Si 0.9% Ni 1% Fe 0.2% Ti Al alloy: Casting; chill cast and aged **UTS: 190 Elon: 3% Proof: 120**
L52	2.5% Cu 1% Mg Ni Si Fe Al alloy: For sand or die castings; chill cast **UTS: 300 Proof: 250**
L78	1.25% Cu 0.5% Mg 5% Si Al alloy: Casting; chill cast **UTS: 290 Proof: 170**
L79	Cu Si Al alloy: Casting; obsolete
L91	4.5% Cu Al alloy: Casting; sand cast **UTS: 220 Elon: 7% Proof: 160**
L92	4.5% Cu Al alloy: Casting; sand cast **UTS: 280 Elon: 4% Proof: 200**
L 211	10% Cu 0.3% Mg Al alloy: Casting; Spanish specification
L 212	10% Cu 0.3% Mg Al alloy: Casting; Spanish specification
L 213	7% Cu 3% Si 3% Zn Al alloy: Casting; Spanish specification
L 214	4.5% Cu Al alloy: Casting; Spanish specification
L 215	4% Cu 1.5% Mg 2% Ni Al alloy: Casting; Spanish specification
L 216	4% Cu 1.5% Mg 2% Ni Al alloy: Casting; Spanish specification
L 232	4.5% Cu 0.5% Mn Al alloy: Casting; Spanish specification
L 258	1.4% Cu 0.13% Mg 2.2% Si 1.2% Fe 1.3% Ni 0.12% Ti Al alloy: Casting; Spanish specification
L 411	7% Cu 3% Si 3% Zn Al alloy: Casting; Spanish specification
L 451	3% Cu 5% Si 0.5% Mn Al alloy: Casting: Spanish specification
L 452	4% Cu 6% Si 0.35% Mn Al alloy: Casting; Spanish specification
L 453	3.5% Cu 8.5% Si Al alloy: Casting; Spanish specification
LAC 10	10% Cu 0.3% Mg Al alloy: Casting; obsolete; Air Ministry for BS 1490 LM 12
LAC 113B	3% Cu 10% Zn Al alloy: Casting; obsolete; Air Ministry for BS 1490 LM 3
LM 1	7% Cu 3% Si 3% Zn Al alloy: Casting; INTAL for BS alloy LM 1
LM 2	10% Cu 0.3% Mg Al alloy: Casting; designation used by BS
LM 3	3% Cu 10% Zn Al alloy: Casting; INTAL for BS alloy LM 3
LM 4	3% Cu 5% Si 0.5% Mn Al alloy: Casting; designation used by BS

Symbol	Nominal analysis, supplier, condition and remarks.
LM 4	3% Cu 5% Si 0.5% Mn Al alloy: Casting; INTAL for BS alloy LM 4
LM 7	1.75% Cu 2% Si 1% Ni 1% Fe Al alloy: Casting; designation used by BS
LM 11	4.5% Cu 0.3% Ti Al alloy: Casting; designation used by BS
LM 11	4.5% Cu 0.3% Ti Al alloy: Casting; INTAL for BS alloy LM 11
LM 12	10% Cu 0.3% Mg Al alloy: Casting; designation used by BS
LM 14	4% Cu 1.5% Mg 2% Ni Al alloy: Casting; designation used by BS
LM 14	4% Cu 1.5% Mg 2% Ni Al alloy: Casting; INTAL for BS alloy LM 14
LM 15	2% Cu 1% Mg 1.5% Si 1.5% Ni Al alloy: Casting; designation used by BS
LM 21	4% Cu 6% Si 0.5% Mn Al alloy: Casting; designation used by BS
LM 22	3% Cu 5% Si 0.5% Mn Al alloy: Casting; designation used by BS
LM 23	1.5% Cu 2% Si 1% Fe 1% Ni 0.2% Ti Al alloy: Casting; designation used by BS
LM 23	1.5% Cu 2% Si 1% Fe 1% Ni 0.2% Ti Al alloy: Casting; INTAL for BS alloy LM 23
LM 24	3.5% Cu 8% Si Al alloy: Casting; designation used by BS
LM 24	3.5% Cu 8% Si Al alloy: Casting; INTAL for BS alloy LM 24
LM 26	3% Cu 1% Mg 9.5% Si Al alloy: Casting; designation used by BS
LM 27	2% Cu 0.4% Mg 7% Si Al alloy: Casting; designation used by BS
LM 28	1.5% Cu 1.2% Mg 19% Si 1.2% Ni Al alloy: Casting; designation used by BS
LM 29	1.1% Cu 1.1% Mg 23% Si 1.1% Ni Al alloy: Casting; designation used by BS
LM 30	4.5% Cu 0.5% Mg 17% Si Al alloy: Casting; designation used by BS
MIRALITE	0.3% Si 4% Ni 0.4% Fe 0.04% Na 0.05% Pb Al alloy: Casting and wire; MIRALITE
NBN 436 AlCu2NiMg	2% Cu 1.5% Mg 2% Ni Al alloy: Casting; Belgium specification
NBN 436 AlCu4Ni2Mg	4% Cu 1.5% Mg 2.0% Ni Al alloy: Casting; Belgium specification
NBN 436 AlCu5MgTi	4.5% Cu 0.3% Ti Al alloy: Casting; Belgium specification
NBN 436 AlSi2Cu	1.4% Cu 0.1% Mg 2.2% Si 1.1% Fe 1.3% Ni Al alloy: Casting; Belgium specification
NBN 436 AlSi5Cu	3% Cu 5% Si Al alloy: Casting; Belgium specification
NBN 436 AlSi9Cu3	3.5% Cu 8.5% Si Al alloy: Casting; Belgium specification
NORAL 117	3% Cu 4.5% Si 0.5% Mn Al alloy: Casting; Northern Aluminium Co. for BS alloy LM 4
NORAL 218	4% Cu 1.5% Mg 2% Ni Al alloy: Casting; Northern Aluminium Co. for BS alloy LM 14
NORAL 226	4.5% Cu Al alloy: Casting; Northern Aluminium Co. for BS alloy LM 11
NORAL 237	7% Cu 3% Si 3% Zn Al alloy: Casting; Northern Aluminium Co. for BS alloy LM 1
NORAL 252	10% Cu 0.3% Mg Al alloy: Casting; Northern Aluminium Co. for BS alloy LM 12
NORAL A111	1% Cu 0.1% Mg 2.5% Si 1% Ni 0.2% Ti Al alloy: Casting; Northern Aluminium Co. for BS alloy LM 7
QQ A 591/4	7.0% Cu 3.5% Si 2.0% Fe 1.5% Zn Al alloy: Casting; US Federal
QQ A 591/5	4.0% Cu 5.0% Si 2.0% Fe 1.0% Zn Al alloy: Casting; US Federal
QQ A 591/10	3.5% Cu 6.5% Si 1.2% Fe Al alloy: Casting; US Federal
QQ A 591/11	3.5% Cu 6.5% Si 2.0% Fe 1.0% Zn Al alloy: Casting; US Federal
QQ A 596/1	7.0% Cu 4.0% Si 1.2% Fe 2.5% Zn Al alloy: Casting; US Federal **UTS: 140**
QQ A 596/3	4.0% Cu 1.5% Mg 2.0% Ni Al alloy: Casting; US Federal; aged **UTS: 200**
QQ A 596/4	4.5% Cu 2.5% Si Al alloy: Casting; US Federal; aged **UTS: 200**
QQ A 596/5	4.5% Cu 5.5% Si Al alloy: Casting; US Federal; as cast **UTS: 150**
QQ A 601/4	4.5% Cu 1.2% Si Al alloy: Casting; US Federal; aged **UTS: 180**
QQ A 601/6	4.0% Cu 1.5% Mg 2.0% Ni Al alloy: Casting; US Federal; as cast **UTS: 150**
QQ A 601/8	4.0% Cu 3.0% Si Al alloy: Casting; US Federal **UTS: 110 Elon: 1.5%**
QQ A 601/9	8.0% Cu 1.2% Si 1.5% Fe Al alloy: Casting; US Federal; as cast **UTS: 110**
RR 50	1% Cu 2.5% Si 1% Ni 0.2% Ti Al alloy: Sand and die casting; High Duty Alloys for BS alloy LM 7
RR 53B	1.5% Cu 0.8% Mg 0.75% Si 1% Ni 0.2% Ti Al alloy: Sand and die casting; High Duty Alloys for BS alloy LM 15
RR 250	5% Cu 0.25% Mn 1% Ni 0.2% Ti 0.25% Co 0.25% Sb Al alloy: For sand casting; High Duty Alloys; for high temperature use **DPN: 90 UTS: 210 Elon: 1% Proof: 150**
RR 350	5% Cu 1.5% Ni 0.25% Co 0.25% Zr 0.25% Sb 0.2% Ti Al alloy: Casting; High Duty Alloys; solution treated and aged; for high temperature use **UTS: 240 Elon: 2% Proof: 150**
RRAC 9 A	1.2% Cu 0.5% Mg 1.3% Si 1.6% Ni 5% Sn Al alloy: Casting; aged; High Duty Alloys; used as a bearing alloy **DPN: 80 Proof: 150**
SAE 33	7% Cu 3% Si Al alloy: Casting; sand cast **UTS: 120 Proof: 70**
SAE 34	10% Cu 0.3% Mg Al alloy: Casting; sand cast **UTS: 200**
SAE 38	4.5% Cu 0.8% Si Al alloy: Casting; solution treated and aged **UTS: 240 Proof: 180**
SAE 39	4% Cu 1.5% Mg 2% Ni Al alloy: Casting; solution treated and aged **UTS: 210 Proof: 120**
SAE 300	6% Cu 0.5% Mg 6% Si Al alloy: Casting; die cast; aged **DPN: 95 UTS: 210 Proof: 170**
SAE 303	3.5% Cu 11% Si Al alloy: Casting; die cast; as cast **UTS: 280 Elon: 2%**
SAE 306	3.5% Cu 8% Si Al alloy: Casting; die cast; as cast **UTS: 280 Elon: 3%**
SAE 307	3.5% Cu 5% Si Al alloy: Casting; die cast; as cast **UTS: 250 Elon: 2.5%**
SAE 308	3.5% Cu 8% Si Al alloy: Casting; die cast; as cast **UTS: 300 Elon: 2%**
SAE 326	4% Cu 6% Si Al alloy: Casting; solution treated and aged **UTS: 200 Elon: 2.5% Proof: 130**
SAE 329	4% Cu 6% Si Al alloy: Casting; solution treated and aged **UTS: 210 Elon: 1.5% Proof: 130**
SAE 330	4.5% Cu 5.5% Si Al alloy: Casting; die cast; as cast **UTS: 150 Proof: 70**
SAE 331	4% Cu 9% Si Al alloy: Casting; solution treated and aged **UTS: 220**
SAE 332	3% Cu 1% Mg 9% Si 1% Ni Al alloy: Casting: aged **UTS: 210**

Symbol	Nominal analysis, supplier, condition and remarks.
SAE 334	2.3% Cu 1% Mg 12% Si Al alloy: Casting; aged **UTS: 210**
SAE 335	1.2% Cu 0.4% Mg 5% Si Al alloy: Casting; solution treated and aged **UTS: 270 Elon: 3% Proof: 200**
SAE 380	4.5% Cu 2.5% Si Al alloy: Casting; solution treated and aged **UTS: 220 Elon: 2% Proof: 140**
SAE 382	4.6% Cu 0.3% Mg 0.3% Mn 0.2% Ti Al alloy: Casting
SAE 383	2.5% Cu 10% Si Al alloy: Casting
SS 4230	3% Cu 6% Si 0.4% Mn Al alloy: Casting; Swedish Standard; as cast **DPN: 70**
SS 4231	3.2% Cu 5% Si 0.5% Mn Al alloy: Casting; Swedish Standard; as cast **DPN: 70**
SS 4251	3.5% Cu 8% Si 0.5% Mn Al alloy: Casting; Swedish Standard; as cast **DPN: 70**
SS 4252	3.5% Cu 8% Si 0.5% Mn 1% Fe Al alloy: Casting; Swedish Standard; as cast **DPN: 80**
SS 4254	3.0% Cu 9.0% Si 2.0% Zn Al alloy: Casting; Swedish Standard **DPN: 85 UTS: 250 Elon: 1% Proof: 180**
STA 7 AC1	7% Cu 3% Si 3% Zn Al alloy: Casting; replaced by BS 1490 LM 1
STA 7 AC3	3.5% Cu 11% Zn Al alloy: Casting; replaced by BS 1490 LM 3
STA 7 AC4	3% Cu 4.5% Si 0.5% Mn Al alloy: Casting; replaced by BS 1490 LM 4
STA 7 AC7	1% Cu 0.1% Mg 2.5% Si 0.9% Ni 0.2% Ti Al alloy: Casting; replaced by BS 1490 LM 7
STA 7 AC11	4.5% Cu Al alloy: Casting; replaced by BS 1490 LM 11
STA 7 AC12	10% Cu 0.3% Mg 1% Fe Al alloy: Casting; replaced by BS 1490 LM 12
STA 7 AC14	4% Cu 1.5% Mg 2% Ni Al alloy: Casting; replaced by BS 1490 LM 14
STA 7 AC15	1.5% Cu 0.8% Mg 0.7% Si 1% Fe 1% Ni 0.2% Ti Al alloy: Casting; replaced by BS 1490 LM 15
UNI 3040	12% Cu Al alloy: Primary metal; Italian Standard
UNI 3041	10% Cu 0.3% Mg 1.0% Fe Al alloy: Primary metal; Italian Standard
UNI 3042	10% Cu 0.25% Mg 1.0% Si 1.5% Ni 0.15% Ti Al alloy: Casting; Italian Standard
UNI 3043	8% Cu Al alloy: Casting; Italian Standard
UNI 3044	4.5% Cu Al alloy: Casting; Italian Standard
UNI 3045	4% Cu 1.5% Mg 2.0% Ni Al alloy: Casting; Italian Standard
UNI 3046	3% Cu 0.6% Mg 0.7% Si 1.5% Fe 0.6% Ni 0.15% Ti Al alloy: Casting; Italian Standard
UNI 3050	2.2% Cu 1.4% Mg 10% Si 1.0% Ni Al alloy: Italian Standard
UNI 3052	4.0% Cu 5.5% Si Al alloy: Casting; Italian Standard
UNI 3601	3.5% Cu 8.5% Si Al alloy: Casting; Italian Standard
UNI 3602	0.6% Mg 1.0% Fe 0.2% Ti 5.0% Zn Al alloy: Casting; Italian Standard
UNI 5075	3.5% Cu 8.5% Si 0.7% Fe Al alloy: Casting; Italian Standard
V 1036 A/B BXIII	4.5% Cu 0.5% Mn Al alloy: Casting; Dutch specification
V 1036 A/B BXV	1.3% Cu 0.5% Mg 5% Si Al alloy: Casting; Dutch specification
V 1036 A/B BXVI	3% Cu 5% Si 0.5% Mn Al alloy: Casting; Dutch specification
VANALUIM	5% Cu 14% Zn 0.75% Fe 0.25% V Al alloy: Origin unknown
VITAL	1% Cu 0.9% Si 1.2% Zn Al alloy: Origin unknown
X-ALLOY	3.5% Cu 0.6% Mg 0.6% Si 1.25% Fe 0.6% Ni Al alloy: Origin unknown
Y-ALLOY	4% Cu 1.5% Mg 2% Ni Al alloy: Sand and die castings; High Duty Alloys for BS alloy LM 14
Z 3	7% Cu 3% Si 3% Zn Al alloy: Casting; Birmabright for BS alloy LM 1

Note. The following abbreviations and units are used in the tables:

DPN	Hardness, diamond pyramid number
UTS	Ultimate tensile strength, N/mm^2
Elon	Elongation, %
Proof	0.1% proof strength, N/mm^2

1 N/mm^2=0.1 hbar=0.102 kgf/mm^2=0.06475 $tonf/in.^2$=145.04 $lbf/in.^2$=1 MPa

See Appendix II for other abbreviations and conversion tables.

1M Aluminium–magnesium–zinc cast alloys With and without titanium, copper and chromium

Specific gravity		2.76–2.8
Density		2760–2800 kg/m^3
Solidus/liquidus		570–760 °C
Thermal conductivity		138 W/m °C
Coefficient of linear expansion		21×10^{-6}/ °C
Electrical conductivity		35% IACS (copper 100%)
Specific resistance		49 microhm mm
Young's modulus of elasticity		71.0×10^9 N/m^2
Impact		–
Fatigue strength (5×10^6 cycles)	sand cast	± 50 N/mm^2
	chill cast	± 75 N/mm^2

The above properties are typical of the following group, and may not apply exactly to any one specification. It is possible that with certain specifications some of the values may not be applicable.

General metallurgical properties

This is a comparatively small range of alloys which are based on 5–7% zinc, always with magnesium present. These are medium to high strength materials, reasonably low cost with poor corrosion resistance.

They can be subjected to heat treatment to increase the mechanical properties but this is not very dramatic and generally the casting is used in the solution treatment condition with an ageing operation added to improve the properties under many circumstances.

Welding is difficult and should not be attempted on high integrity components. It requires the use of the inert gas shielded process.

There are very few alloys in this section which are used in any great quantity in the United Kingdom. They are used in the United States of America as general purpose aluminium castings with reasonably good mechanical properties and low price with low, or not very good, corrosion resistance.

Symbol	Nominal analysis, supplier, condition and remarks.
3.1263	4.5% Cu 3% Si 2.5% Zn 0.5% Ni Al alloy: Casting; German Standard DIN 1725
40 E	0.6% Mg 0.5% Cr 5.5% Zn 0.2% Ti Al alloy: Casting; AA designation for ASTM alloy ZG 61A
249.0	4.2% Cu 0.3% Mg 3% Zn Al alloy: Casting; obsolete; designation used in the UK and USA
249.2	4.2% Cu 0.3% Mg 3% Zn Al alloy: Casting; obsolete; designation used in the UK and USA
705.0	0.3% Cu 1.6% Mg 3% Zn Al alloy: Casting; designation used in the UK and USA
705.2	0.3% Cu 1.6% Mg 3% Zn Al alloy: Ingot; designation used in the UK and USA
707.0	03% Cu 2% Mg 4.2% Zn Al alloy: Casting; designation used in the UK and USA
707.1	0.3% Cu 2% Mg 4.2% Zn Al alloy: Ingot; designation used in the UK and USA
710.0	0.5% Cu 0.7% Mg 6.5% Zn Al alloy: Casting; designation used in the UK and USA
710.1	0.5% Cu 0.7% Mg 6.5% Zn Al alloy: Ingot; designation used in the UK and USA
711.0	0.5% Cu 0.3% Mg 6.5% Zn Al alloy: Casting; designation used in the UK and USA
711.1	0.5% Cu 0.3% Mg 6.5% Zn Al alloy: Ingot; designation used in the UK and USA
712.0	0.5% Cu 0.6% Mg 0.2% Ti 5.7% Zn Al alloy: Casting; designation used in the UK and USA
712.2	0.5% Cu 0.6% Mg 0.2% Ti 5.7% Zn Al alloy: Ingot; designation used in the UK and USA
713.0	0.7% Cu 0.4% Mg 7.5% Zn Al alloy: Casting; designation used in the UK and USA
713.1	0.7% Cu 0.4% Mg 7.5% Zn Al alloy: Ingot; designation used in the UK and USA
771.0	0.9% Mg 0.1% Cr 0.15% Ti 7% Zn Al alloy: Casting; designation used in the UK and USA
771.2	0.9% Mg 0.1% Cr 0.15% Ti 7% Zn Al alloy: Ingot; designation used in the UK and USA
772.0	0.7% Mg 0.1% Cr 0.15% Ti 6.5% Zn Al alloy: Casting; designation used in the UK and USA
772.2	0.7% Mg 0.1% Cr 0.15% Ti 6.5% Zn Al alloy: Casting; designation used in the UK and USA
Al 100 (79)	1% Cu 3% Mg 5% Zn Al alloy: Wrought; further information from Aluminium Zentrale
Al B42	1% Cu 3% Mg 5% Zn Al alloy: Wrought; further information from Aluminium Zentrale

Note. The following abbreviations and units are used in the tables:

DPN	Hardness, diamond pyramid number
UTS	Ultimate tensile strength, N/mm^2
Elon	Elongation, %
Proof	0.1% proof strength, N/mm^2

1 N/mm^2=0.1 hbar=0.102 kgf/mm^2=0.06475 tonf/in.^{2}=145.04 lbf/in.^{2}=1 MPa

See Appendix II for other abbreviations and conversion tables.

Symbol	Nominal analysis, supplier, condition and remarks.
Al B133	3% Mg 1% Si 1% Mn 0.2% Cr 5% Zn Al alloy: Wrought; further information from Aluminium Zentrale
Al Zn 4.5 Mg 1	1.2% Mg 0.2% Cr 4.5% Zn Al alloy: Casting; Swiss designation
AN A17	0.5% Mg 0.5% Cr 5.0% Zn 0.2% Ti Al alloy: Casting; US Service; as cast **UTS: 210 Elon: 3% Proof: 120**
AP 701	0.7% Mg 5.2% Zn Al alloy: Casting; Australian specification
ASTM B26 ZC81A	0.7% Cu 0.3% Mg 7.5% Zn Al alloy: Sand casting; aged **UTS: 200 Elon: 3% Proof: 150**
ASTM B26 ZC32A	1.5% Mg 0.5% Mn 0.3% Cr 3% Zn Al alloy: Casting; room aged **UTS: 200 Elon: 5% Proof: 110**
ASTM B26 ZG42A	2% Mg 0.5% Mn 0.3% Cr 4% Zn Al alloy: Casting; solution treated and aged **UTS: 250 Elon: 1% Proof: 200**
ASTM B26 ZG61A	0.6% Mg 0.5% Cr 5.5% Zn Al alloy: Casting; room aged **UTS: 210 Elon: 3% Proof: 150**
ASTM B26 ZG61B	0.5% Cu 0.7% Mg 6.5% Zn Al alloy: Casting; room aged **UTS: 210 Elon: 2% Proof: 120**
ASTM B108 ZC60A	0.5% Cu 0.3% Mg 6.5% Zn Al alloy: Casting; aged **Elon: 7% Proof: 110**
ASTM B108 ZC81B	0.6% Cu 0.4% Mg 0.3% Cr 7.5% Zn Al alloy: Casting; aged **UTS: 210 Elon: 4% Proof: 140**
ASTM B108 ZG32A	1.5% Mg 0.5% Mn 3% Zn 0.3 Cr Al alloy: Casting; aged **UTS: 250 Elon: 10% Proof: 110**
ASTM B108 ZG42A	2% Mg 0.5% Mn 0.3% Cr 4.2% Zn Al alloy: Casting; aged **UTS: 300 Elon: 4% Proof: 180**
ASTM B179 ZC60A	0.5% Cu 0.4% Mg 6.5% Zn Al alloy: Ingot; primary metal
ASTM B179 ZC81A	0.7% Cu 0.4% Mg 7.5% Zn Al alloy: Ingot; primary metal
ASTM B179 ZC81B	0.6% Cu 0.4% Mg 7.5% Zn Al alloy: Ingot; primary metal
ASTM B179 ZG32A	1.6% Mg 0.5% Mn 3% Zn Al alloy: Ingot; primary metal
ASTM B179 ZG42A	2% Mg 0.3% Mn 4% Zn Al alloy: Ingot; primary metal
ASTM B179 ZG61A	0.6% Mg 0.5% Cr 5.5% Zn Al alloy: Ingot; primary metal
ASTM B179 ZG61B	0.5% Cu 0.7% Mg 6.5% Zn Al alloy: Ingot; primary metal
AZ 5G	0.2% Cu 0.5% Mg 0.2% Cr 0.2% Ti 5.0% Zn Al alloy: Casting; L'Aluminium Français
AZ 5G	0.6% Mg 0.5% Cr 5.2% Zn Al alloy: Casting; French Standard
C 612	0.5% Cu 0.3% Mg 6.5% Zn Al alloy: Casting; AA designation for ASTM alloy ZC60A
D 712.0	0.6% Mg 0.5% Cr 5.7% Zn Al alloy: Investment casting; American National Standard; ANSI

Symbol	Nominal analysis, supplier, condition and remarks.
DSI 5000 A1621	0.6% Mg 5.2% Zn Al alloy: Casting; Danish specification
DTD 5008A	0.6% Mg 0.5% Cr 5% Zn 0.2% Ti Al alloy: Casting; chill cast and aged **UTS: 220 Elon: 5% Proof: 180**
FRONTIER 40E	0.60% Mg 0.5% Cr 5% Zn 0.2% Ti Al alloy: Casting; chill cast; also known as 40E; American alloy **UTS: 240 Elon: 7% Proof: 170**
GOST 2685/63 A124	0.6% Cu 0.5% Cr 5.3% Zn 0.2% Ti Al alloy: Casting; Russian specification
H 49-17	0.6% Mg 5.2% Zn Al alloy: Casting; Australian specification
HA 3 ZG61N	0.55% Mg 0.5% Cr 0.2% Ti 5.7% Zn Al alloy: Ingot; Canadian Standard
HA 3 ZG61P	0.5% Cu 0.75% Mg 6.5% Zn Al alloy: Ingot; Canadian Standard
L5	Cn Zn Al alloy: Casting; obsolete
SAE 310	0.5% Mg 0.5% Cr 5.5% Zn Al alloy: Casting; sand cast and aged **UTS: 210 Elon: 3% Proof: 150**
SAE 311	1.5% Mg 0.5% Mn 0.3% Cr 3% Zn Al alloy: Casting; sand cast and aged **UTS: 200 Elon: 5% Proof: 110**
SAE 312	2% Mg 0.5% Mn 0.3% Cr 4% Zn Al alloy: Casting; solution treated and aged **UTS: 220 Elon: 1%**
SAE 313	0.5% Cu 0.7% Mg 6.5% Zn 0.2% Ti Al alloy: Casting; sand cast and aged **UTS: 210 Elon: 2% Proof: 130**
SAE 314	0.5% Cu 0.3% Mg 6.5% Zn 0.2% Ti Al alloy: Casting; die cast; aged **UTS: 180 Elon: 7%**
SAE 315	0.8% Cu 0.3% Mg 0.3% Cr 7.5% Zn Al alloy: Casting; aged **UTS: 200 Elon: 5% Proof: 150**
SS 4438	0.6% Mg 0.4% Cr 5% Zn 0.2% Ti Al alloy: Casting: Swedish Standard **DPN: 80**
TENZALOY	0.8% Cu 0.3% Mg 7.5% Zn Al alloy: Casting; AA designation for ASTM alloy ZC81A
TERNALLOY 5	1.5% Mg 0.3% Cr 3% Zn Al alloy: Casting; AA designation for ASTM alloy ZG32A
TERNALLOY 7	2% Mg 0.5% Mn 0.3% Cr 4.2% Zn Al alloy: Casting; AA designation for ASTM alloy ZG42A
U15	1% Mg 5% Zn Al alloy: Casting; Anglo-Swiss Aluminium Co.
UNIFONT 5	1% Mg 5% Zn Al alloy: Casting; Anglo-Swiss Aluminium Co.; room aged; obsolete **DPN: 75 UTS: 240 Elon: 6% Proof: 170**
UNIFONT 54	0.9% Mg 0.2% Cr 5.2% Zn Al alloy: Casting; Anglo-Swiss Aluminium Co.
ZG 71B	0.9% Mg 0.1% Cr 7.0% Zn Al alloy: Investment casting; designation used by ASTM
ZG 81A	0.7% Cu 0.35% Mg 7.5% Zn Al alloy: Investment casting; designation used by ASTM
ZIMALUIM	7.5% Mg 11.5% Zn Al alloy: German origin

Note. The following abbreviations and units are used in the tables:

DPN	Hardness, diamond pyramid number
UTS	Ultimate tensile strength, N/mm^2
Elon	Elongation, %
Proof	0.1% proof strength, N/mm^2

1 N/mm^2=0.1 hbar=0.102 kgf/mm^2=0.06475 tonf/in.^{2}=145.04 lbf/in.^{2}=1 MPa

See Appendix II for other abbreviations and conversion tables.

1N Aluminium – miscellaneous alloys

Following is listed a small number of aluminium alloys which have special purposes and do not fit into any of the preceding groups.

Very little information is available on these materials apart from that appearing with each alloy.

The alloys in this section do not fit into any of the previous sections. In many cases they are master alloys used to make castings or wrought material by the addition of specific alloying elements.

Some of them have very specific purposes and where possible the information is given alongside the alloy itself.

Symbol	Nominal analysis, supplier, condition and remarks.
3.2685	3% Cu 11% Si 10% Sn + Cd + Ni Al alloy: German Standard
850.0	1% Cu 6.2% Sn Al alloy: Casting; designation used in the USA
850.1	1% Cu 6.2% Sn Al alloy: Ingot; designation used in the UK and USA
851.0	1% Cu 2.5% Si 0.5% Ni 6.2% Sn Al alloy: Casting; designation used in the UK and USA
851.1	1% Cu 2.5% Si 0.5% Ni 6.2% Sn Al alloy: Ingot; designation used in the UK and USA
852.0	2% Cu 0.8% Mg 1.2% Ni 6.2% Sn Al alloy: Casting; designation used in the UK and USA
852.1	2% Cu 0.8% Mg 1.2% Ni 6.2% Sn Al alloy: Ingot; designation used in the UK and USA
853.0	3.5% Cu 6% Si 6.2% Sn Al alloy: Casting; designation used in the UK and USA
853.2	3.5% Cu 6% Si 6.2% Sn Al alloy: Ingot; designation used in the UK and USA
4002	0.1% Cu 1% Mg 4% Si 1% Cd Al alloy: Wrought; designation used in the UK and USA
4004	1.5% Mg 9.7% Si Al alloy: Wrought brazing alloy; designation used in the UK and USA
8081	1% Cu 20% Sn Al alloy: Wrought; designation used in the UK and USA
8280	1.5% Si 1% Fe 6.2% Sn Al alloy: Wrought; designation used in the UK and USA
A C4	3.7% Cr Al master alloy: L'Aluminium Français
A Fe 10	10% Fe Al master alloy: L'Aluminium Français
ALCAN DURALCOTE	Trade name for pre-coated Al sheet; Alcan
ALCAN GB 730	1% Cu 2.5% Si 0.5% Ni 6.5% Sn Al alloy: Casting; Alcan; aged; for bearings **DPN: 50 UTS: 150 Elon: 5% Proof: 75**
Al-Zr	47.5% Zr Al alloy: Origin unknown
A M4	4.0% Mn Al alloy: Casting; L'Aluminium Français

Note. The following abbreviations and units are used in the tables:

DPN	Hardness, diamond pyramid number
UTS	Ultimate tensile strength, N/mm^2
Elon	Elongation, %
Proof	0.1% proof strength, N/mm^2

1 N/mm^2=0.1 hbar=0.102 kgf/mm^2=0.06475 tonf/in.^{2}=145.04 lbf/in.^{2}=1 MPa

See Appendix II for other abbreviations and conversion tables.

Symbol	Nominal analysis, supplier, condition and remarks.
A M10	10% Mn Al master alloy: L'Aluminium Français
AMS 4275B	1% Cu 1% Ni 6% Sn Al alloy: Casting for bearings; AMS for SAE 770
A N20	20% Ni Al master alloy: L'Aluminium Français
A S9K7	9.5% Si 7% Co Al master alloy: L'Aluminium Français
ASTM B199 HK31A	3.2% Th 0.7% Zr Al alloy: Die casting
ASTM B199 QE22A	2.2% Rare earth 0.7% Zr Al alloy: Die casting
ASTM B327 CG181A	18.0% Cu 0.8% Mg Al alloy: Hardener for Zn base; die casting
A T4	4% Ti Al master alloy: L'Aluminium Français
A Zr5	5.5% Zr Al master alloy: L'Aluminium Français
B 850.0	2.0% Cu 0.7% Mg 1.2% Ni 6.2% Sn Al alloy: Investment casting; American National Standard; ANSI
FERRO ALUMINIUM	0.1% Cu 0.2% Si 0.04% C 50% Fe Al: Metal Alloys; primary metal
GALVALUME	1.5% Si 43% Zn Al alloy: For galvanizing sheet; Zinc Development Association
HIDUMINIUM 29	1% Cu 6.5% Sn 0.8% Ni Al alloy: Casting; High Duty Alloys; chill cast; aged for bearings **DPN: 40 UTS: 140 Elon: 15% Proof: 50**
L11	Al alloy: Casting; obsolete; no information available
L27	Al alloy: Obsolete; no information available
L28	Al alloy: Obsolete; no information available
L29	Al alloy: Obsolete; no information available
L Al Si Sn	3% Cu 11% Si 10% Sn + Cd + Ni Al alloy: Designation used by German Standards
MB 7	1% Cu 1% Mg 0.6% Bi 1.7% Ni 7% Sn Al alloy: Casting; Birmabright; aged for bearings **UTS: 210 Elon: 1.5% Proof: 170**
METCO 101B NS	Ti Al oxides: Spray deposit; mixed oxides; Metco **DPN: 540**
METCO 101 NS	Ti Al oxides: Spray deposit; mixed oxides; Metco **DPN: 600**
METCO 101 SF	Ti Al oxides: Spray deposit; Metco **DPN: 950**
NF A57 350	Classifications of Al alloys: Wrought; French Standard
NF A57 650	Classification of Al alloy: Sheets; French Standard
SAE 770	1% Cu 1% Ni 6% Sn Al alloy: For bearings
SAE 780	1% Cu 1.5% Si 6% Sn Al alloy: Bearing metal; cast on steel backing
SAE 781	4% Si 1% Cd Al alloy: Bearing metal; cast on steel backing
SAE 782	1.0% Cu 1.0% Ni 3.0% Cd Al alloy: For bearings bonded to steel backing
UNI 3022 MAF5	5% Fe Al alloy: Ingot; primary metal; Italian Standard
UNI 3022 MAF10	10% Fe Al alloy: Ingot; primary metal; Italian Standard
UNI 3022 MAK10	10% Co Al alloy: Ingot; primary metal; Italian Standard
UNI 3022 MAM10	10% Mn Al alloy: primary metal; Italian Standard
UNI 3022 MAN 25	25% Ni Al alloy: Ingot; primary metal; Italian Standard
UNI 3022 MAR4	4% Cr Al alloy: Ingot; primary metal; Italian Standard
UNI 3022 MAT3	3% Ti Al alloy: Ingot; primary metal; Italian Standard
UNI 6252	1.0% Cu 6% Si 1.0% Ni 0.12% Ti Al alloy: Casting; Italian Standard
UNI 6253	2% Mn 2% Ni 0.15% Ti Al alloy: Casting; Italian Standard
WOLFRAMIUM	0.3% Cu 1.4% Sb 0.2% Fe 0.1% Sn 0.4% W Al alloy: Good corrosion resistance; origin unknown
X 8081	1.0% Cu 20% Sn Al alloy: For bearings; Alcoa; available in coil form
Z - ALLOY	6.5% Ni 0.5% Ti Al alloy: Used for hard shafts and medium bearings; origin unknown
ZICRAL	1.5% Cu 2.25% Mg 0.7% Si 0.4% Mn 0.4% Cr 8.0% Zn Al alloy: Wrought; origin unknown

Note. The following abbreviations and units are used in the tables:

DPN	Hardness, diamond pyramid number
UTS	Ultimate tensile strength, N/mm^2
Elon	Elongation, %
Proof	0.1% proof strength, N/mm^2

$1\ N/mm^2 = 0.1\ hbar = 0.102\ kgf/mm^2 = 0.06475\ tonf/in.^2 = 145.04\ lbf/in.^2 = 1\ MPa$

See Appendix II for other abbreviations and conversion tables.

2. Antimony Sb

Physical properties	
Atomic number	51
Atomic weight	121.76
Crystal structure	Hexagonal
Colour	Silver-white with bluish tinge
Specific gravity	6.62
Density	6620 kg/m^3
Melting point	630.5 °C
Boiling point	1380 °C
Specific heat	0.206 J/g °C
Thermal conductivity	18.6 W/m °C
Coefficient of linear expansion (20–100 °C)	10.9×10^{-6}/ °C
Latent heat of fusion	168.3 J/g
Latent heat of vaporization	1604 J/g
Thermal neutron absorption cross-section	5.7 barns/atom
Electrical conductivity	4.5% IACS (copper 100%)
Specific resistance	410 microhm mm
Temperature coefficient of electrical resistance	0.0039/ °C
Electrochemical equivalent	1.494 g/A/h
Electrode potential	+ 0.1 V
Magnetic susceptibility	0.82×10^{-6}
Young's modulus of elasticity	77.9×10^9 N/m^2
Tensile strength	108 N/mm^2
Hardness	30–60 DPN

2.1 General notes on antimony

Antimony does not occur in the metallic state. It is generally purified from the sulphide ore (stibnite) by reduction with wrought iron, using common salt as the flux in a reverberatory furnace.

The pure metal is brittle with little strength and is a poor conductor of heat and electricity but a good light reflector.

Antimony, like arsenic and bismuth, is a metalloid having many characteristics of the non-metals although not to quite the same degree as arsenic.

No commercial use is made of the primary metal and there are no alloys with antimony as the basic ingredient. It is however purified from its ore and is available under several names as a primary metal.

As an alloying element, antimony finds use as follows:

Lead alloys. 0.25–0.5% antimony has little effect on the physical and mechanical properties of lead but improves its castability.

5–12% antimony added to lead gives a range of alloys with increased strength, corrosion resistance and hardness. These are used in electrical storage batteries, as electroplating anodes, and vat linings. The name 'Regulus Metal' is applied to these alloys, as well as to the commercially pure form of antimony. The alloys have a better creep strength than pure lead.

Bearings. 3–20% antimony is added to both lead based and tin based bearing alloys.

Antimony forms intermetallic compounds with tin, which are cubic in shape and give hardness and strength to an otherwise soft matrix.

Type metal. Antimony in small quantities added to certain lead based alloys lowers the melting point, and reduces the solidification contraction. This increases the casting fluidity ensuring complete filling of intricate moulds while the lack of contraction gives excellent reproduction of the pattern being cast. Considerable quantities of these alloys were continuously used for newspaper type.

Brittanic metal. This is basically 80% tin, 18% antimony, 1.5% copper. It has good castability, is sonorous and retains its polish; hence it is used for bell castings and household ornaments.

Pewter. Previously this was 80% tin, 20% lead, with antimony additions. Now it resembles Brittanic metal.

Ammunition. Up to 15% antimony has been added to lead for bullets. This hardens the lead, making it brittle.

Following are listed the specifications, trade names and symbols used for the primary metal.

Symbol	Nominal analysis, supplier, condition and remarks.
ANTIMONY	High purity Sb metal: Impurities less than 1 p.p.m.; Johnson Matthey; available as cast bar for semiconductors
ANTIMONY	Sb metal: Blackwells; pure and commercial quality metal
ARSENIC-ANTIMONY	10% As Sb alloy: Blackwells; primary metal
ASTM B237A	0.05% As 0.15% Pb Sb metal: Primary metal
ASTM B237 B	0.1% As 0.2% Pb Sb metal: Primary metal
BRIOUDE	1.3% As 0.25% Pb 98.3% Sb metal: Brand name for antimony regulus; further information from The Lead Development Association
COOKSON'S C	99.6% Sb metal: Brand name for antimony metal; further information from The Lead Development Association
HALLETTS	98.5% Sb metal: Brand name for antimony regulus; further information from The Lead Development Association
h Sb 1	99.9999% Sb metal: Ingot; Light Ltd; high purity metal
h Sb 96	95.99% Sb 121: Stable isotope 121; Light Ltd
h Sb 97	95.99% Sb 123: Stable isotope 123; Light Ltd
LA LUCETTE	99.35% Sb metal: Brand name for antimony metal; further information from The Lead Development Association
LONESTAR	99.7% Sb metal: US brand name for antimony regulus
MIL A 10841	99.5% Sb metal: US Federal specification
REGULUS	Name given to commercially pure antimony: The same name is used for Pb/Sb alloys
RMM	99.3% Sb metal: US brand name for antimony regulus
TYNE	99.0% Mn Sb metal: Brand name for antimony regulus; further information from The Lead Development Association
WCC	99.848% Sb metal: Brand name for antimony regulus; further information from The Lead Development Association

Note. The following abbreviations and units are used in the tables:

DPN	Hardness, diamond pyramid number
UTS	Ultimate tensile strength, N/mm^2
Elon	Elongation, %
Proof	0.1% proof strength, N/mm^2

1 N/mm^2=0.1 hbar=0.102 kgf/mm^2=0.06475 $tonf/in.^2$=145.04 $lbf/in.^2$=1 MPa

See Appendix II for other abbreviations and conversion tables.

3. Arsenic As

Physical properties

Atomic number		33
Atomic weight		74.9
Crystal structure		Rhombohedral
Colour		Steel-grey
Specific gravity	stable form	5.7
	yellow	3.7
	black	4.7
Density	stable	5700 kg/m^3
	yellow	3700 kg/m^3
	black	4700 kg/m^3
Melting point	under pressure	814 °C
Boiling point		Sublimes above 450 °C at normal pressure
Specific heat		0.33 J/g °C
Thermal conductivity		–
Coefficient of linear expansion		4.7×10^{-6}/ °C
Latent heat of fusion		370.5 J/g
Latent heat of vaporization		416.9 J/g
Thermal neutron absorption cross-section		4.2 barns/atom
Electrical conductivity		3.8% IACS (copper 100%)
Specific resistance		460 microhm mm
Temperature coefficient of electrical resistance		0.0043/ °C
Electrochemical equivalent		0.932 g/A/h
Electrode potential		–
Magnetic susceptibility		-0.31×10^{-6}
Young's modulus of elasticity		–
Tensile strength		–
Hardness		–

3.1 General notes on arsenic

Arsenic is found with the ores of copper, lead, tin, zinc and gold and recovered as a by-product when these ores are treated.

The metal is obtained by reduction with carbon when crude arsenic is obtained. It is purified by subliming over charcoal under vacuum.

The stable form is steel grey, brittle and crystalline. Arsenic exists in two other less stable forms (allotropes), yellow arsenic, specific gravity 3.7, and black arsenic, specific gravity 4.7.

Arsenic forms no useful alloys where it is the main constituent but is used as an alloying element to improve the heat resistance of copper, to harden lead and also to harden certain white metal bearing alloys. In small amounts it is used to de-oxide copper.

The metal itself is not toxic but many of the soluble compounds, notably the oxides, are poisonous. They are used in insecticides, sheep dips, and fungicides. Arsenic is a metalloid, in that it has the characteristics of a non-metal and a metal, and along with boron and silicon has the least metallic nature of the elements listed here as metals.

Symbol	Nominal analysis, supplier, condition and remarks.
ARSENIC	High purity metal: Impurities less than 4 p.p.m.; Johnson Matthey; supplied sealed under vacuum
ARSENIC	98–99% commercially pure metal: Blackwells; primary metal
ARSENIC COPPER	50% Cu As alloy: Blackwells; primary metal
ARSENIC-IRON	45% Fe As alloy: Blackwells; primary metal
MISPICKEL	FeS As ore: The most common arsenic ore
ORPIMENT	As_2S_3 ore: Generally associated with lead and copper ore
P ASLA	99.9999% As metallic lumps: Each 100 g; Light Ltd; high purity metal
REALGAR	As_2S_2 ore: Generally associated with lead and copper ores
WHITE ARSENIC	As_2O_3: Arsenious oxide

Note. The following abbreviations and units are used in the tables:

DPN	Hardness, diamond pyramid number
UTS	Ultimate tensile strength, N/mm^2
Elon	Elongation, %
Proof	0.1% proof strength, N/mm^2

1 N/mm^2=0.1 hbar=0.102 kgf/mm^2=0.06475 tonf/in.^{2}=145.04 lbf/in.^{2}=1 MPa

See Appendix II for other abbreviations and conversion tables.

4. Barium Ba

Physical properties

Atomic number	56
Atomic weight	137.36
Crystal structure	–
Colour	Yellowish white
Specific gravity	3.66
Density	3660 kg/m^3
Melting point	704 °C
Boiling point	1620 °C
Specific heat	0.285 J/g °C
Thermal conductivity	–
Coefficient of linear expansion (20–100 °C)	18 × 10^{-6}/ °C
Latent heat of fusion	55.7 J/g
Latent heat of vaporization	1290 J/g
Thermal neutron absorption cross-section	1.17 barns/atom
Electrical conductivity	3.5% IACS (copper 100%)
Specific resistance	500 microhm mm
Temperature coefficient of electrical resistance	–
Electrochemical equivalent	2.56 g/A/h
Electrode potential	−2.9 V
Magnetic susceptibility	0.9 × 10^{-6}
Young's modulus of elasticity	–
Tensile strength	–
Hardness	–

4.1 General notes on barium

Barium metal is still a laboratory curiosity, as it is very similar to calcium which is much cheaper. One limited use of barium is as an alloy with magnesium and aluminium as a 'getter' for high vacuum valves, when it is allowed to combine with the last traces of oxygen inside the finished valve. It is claimed that the thin film of barium which is produced in the process acts as a high temperature lubricant to any moving electrode systems.

There are records of barium used in conjunction with lead and calcium in a bearing alloy, but no current information is available.

Symbol	Nominal analysis, supplier, condition and remarks.
BARIUM w Ba 21	Commercially pure metal: Blackwells; primary metal 99.5% Ba rod: 20 mm in diameter; Light Ltd; high purity metal

Note. The following abbreviations and units are used in the tables:

DPN	Hardness, diamond pyramid number
UTS	Ultimate tensile strength, N/mm^2
Elon	Elongation, %
Proof	0.1% proof strength, N/mm^2

1 N/mm^2=0.1 hbar=0.102 kgf/mm^2=0.06475 tonf/in.^{2}=145.04 lbf/in.^{2}=1 MPa

See Appendix II for other abbreviations and conversion tables.

5. Beryllium Be

Physical properties	
Atomic number	4
Atomic weight	9.02
Crystal structure	–
Colour	Steel grey
Specific gravity	1.844
Density	1844 kg/m^3
Melting point	1285 °C
Boiling point	2507 °C
Specific heat	1.78 J/g °C
Thermal conductivity	147 W/m °C
Coefficient of linear expansion (20–100 °C)	13×10^{-6}/ °C
Latent heat of fusion	1047–1130 J/g
Latent heat of vaporization	24773 J/g
Thermal neutron absorption cross-section	0.009 barns/atom
Electrical conductivity	38.9–43.1% IACS (copper 100%)
Specific resistance	122 microhm mm
Temperature coefficient of electrical resistance	0.006/ °C
Electrochemical equivalent	–
Electrode potential	−1.7 V
Magnetic susceptibility	-1.0×10^{-6}
Young's modulus of elasticity	276×10^9 N/m^2
Tensile strength	370 N/mm^2
Hardness	55–60 DPN

5.1 General notes on beryllium

Pure beryllium metal can be worked at about 400 °C and is available as extrusions, castings (vacuum cast) and powder compacts, but it is not possible to cold work beryllium which is relatively brittle.

The metal has a low neutron capture cross-section which renders it useful in atomic energy equipment, where the high melting point, chemical stability and relatively high strength enhance this usefulness. The metal has a tensile strength of 300–450 N/mm^2 at room temperature and about 150 N/mm^2 at 600 °C. The elongation of the wrought material is about 20% in the direction of working, but is very low in the transverse direction. The pure metal finds use as electrodes in neon tubes as a component in neutron sources, and the powder has been used to surface harden copper, nickel and iron alloys.

The metal and its oxide are dangerous to health and very great care must be taken when handling. In addition many soluble compounds can cause severe dermatitis. Because of this, the use of the metal and many of its compounds is restricted to those establishments which comply with the necessary precautions to safeguard operators' health. These do not apply, however, to metallic alloys where beryllium is present in relatively small amounts in a stable form.

The largest amount of beryllium is used as an alloy with copper, when it is present up to a maximum of 2.8% and does not present any health hazards. These are the strongest copper alloys known and form the subject of a separate group, 'Copper–beryllium', Section 14D.

Alloys of beryllium and nickel, or iron, have age hardening properties, but to date have found no commercial application. When traces of beryllium are added to platinum there is a measurable increase in hardness.

Some use has been found for beryllium as an alloy with aluminium for coating steel, when there is a reduction in the interface steel/aluminium layer, with a consequent increase in the coating ductility.

There has also been a report that traces added to magnesium improve the castability of the metal and refine the grain structure.

Apart from the uses in nuclear engineering, where all concerned are familiar with the discipline necessary for handling potentially dangerous materials, no practical use is made of beryllium because of the health hazard. As stated above, this does not exist with the beryllium–copper alloys.

Symbol	Nominal analysis, supplier, condition and remarks.
AMS 7900	Be metal: Bar; rod and shapes; primary metal
AMS 7901	Be metal: Bar; rod and shapes
AMS 7902	Be metal: Plate and sheet
BERYLLIUM	Commercially pure metal: Blackwells; primary metal
I 400	4.2% BeO 92% (min) Be billet: QMV grade; Brush Beryllium Co.; for instruments **UTS: 330**
N 50A	1.0% BeO Be block: Brush Beryllium Co.
N 100A	1.2% BeO Be block: Brush Beryllium Co.
N 200A	2.0% BeO Be block: Brush Beryllium Co.
P Be 12	99.99% Be flakes: Light Ltd; high purity metal
QMV	Trade name for commercial beryllium: Brush Beryllium Co.; available in various forms and grades
S 100C	1.2% BeO 98.5% (min) Be billet: QMV grade; Brush Beryllium Co. **UTS: 220** **Elon: 1%** **Proof: 180**
S 200C	2.0% BeO 98.0% (min) Be billet: QMV grade; Brush Beryllium Co. **UTS: 270** **Elon: 1%** **Proof: 200**
S 300C	3.0% BeO 97.4% (min) Be billet: QMV grade; Brush Beryllium Co. **UTS: 270** **Elon: 1%** **Proof: 200**
SP 100C	1.2% BeO Be powder: Brush Beryllium Co.
SP 200C	2.0% BeO Be powder: Brush Beryllium Co.
SP 300C	3.0% BeO Be powder: Brush Beryllium Co.
SR 200	2.0% BeO Be sheet: QMV grade; Brush Beryllium Co.; hot rolled **UTS: 480** **Elon: 5%** **Proof: 330**

Note. The following abbreviations and units are used in the tables:

DPN	Hardness, diamond pyramid number
UTS	Ultimate tensile strength, N/mm^2
Elon	Elongation, %
Proof	0.1% proof strength, N/mm^2

1 N/mm^2=0.1 hbar=0.102 kgf/mm^2=0.06475 $tonf/in.^2$=145.04 $lbf/in.^2$=1 MPa

See Appendix II for other abbreviations and conversion tables.

6. Bismuth Bi

Physical properties

Atomic number	83
Atomic weight	209.0
Crystal structure	Rhombohedral
Colour	White with reddish tinge
Specific gravity	9.8
Density	9800 kg/m^3
Melting point	271.3 °C
Boiling point	1420 °C
Specific heat	0.13 J/g °C
Thermal conductivity	10 W/m °C
Coefficient of linear expansion (20–100 °C)	13.3×10^{-6}/ °C
Latent heat of fusion	54.4 J/g
Latent heat of vaporization	854 J/g
Thermal neutron absorption cross-section	0.032 barns/atom
Electrical conductivity	1.7% IACS (copper 100%)
Specific resistance	1120 microhm mm
Temperature coefficient of electrical resistance	0.0043/ °C
Electrochemical equivalent	2.59 g/A/h
Electrode potential	+0.2 V
Magnetic susceptibility	-1.35×10^{-6}
Young's modulus of elasticity	31.7×10^9 N/m^2
Tensile strength	–
Hardness	–

6.1 General notes on bismuth

Bismuth is soft but quite brittle, being similar in many respects to antimony, arsenic and zinc. It has considerably more metallic properties than either antimony or arsenic but is a very poor heat conductor and, like antimony, expands on solidification, this property being transferred to its alloys.

The pure metal can be extruded as a very fine wire which is used in high quality pyrometer and galvanometer suspensions, as well as hair lines in optical instruments.

The bismuth alloys are characterized by their low melting point, low contraction rate and high fluidity. These low melting point alloys are used as fusible plugs or links which melt when heat is applied, and are found in fire sprinkler systems, holding open fire doors, and other safety devices which must automatically operate when the temperature rises above a certain limit. Alloys melting from 60 °C upwards are available.

Bismuth alloys exist which expand on solidification, remain dimensionally constant or contract very slightly. All of these are used in engineering to fill pipes while bending, to hold intricate parts during machining, or as patterns to simulate certain shrinkage characteristics.

Molten bismuth does not readily oxidize and it is this property which enhances the fluidity of some alloys. This, with the low contraction rate, gives castings of a very high definition, and it is thus used as type metal in good quality printing.

Small quantities of bismuth added to austenitic stainless steel are reported to increase the machinability without detracting from the corrosion resisting or mechanical properties, while some zinc alloys have also been produced with bismuth added for the same purpose.

Note. The following abbreviations and units are used in the tables:

DPN	Hardness, diamond pyramid number
UTS	Ultimate tensile strength, N/mm^2
Elon	Elongation, %
Proof	0.1% proof strength, N/mm^2

1 N/mm^2=0.1 hbar=0.102 kgf/mm^2=0.06475 tonf/in.^{2}=145.04 lbf/in.^{2}=1 MPa
See Appendix II for other abbreviations and conversion tables.

Symbol	Nominal analysis, supplier, condition and remarks.
AAF 11078A/1	27% Pb 13% Sn 11% Cd Bi alloy: US Service; melting point 71 °C
AAF 11078A/2	42% Pb 7.0% Cd Bi alloy: US Service; melting range 80–91 °C
ANATOMICAL ALLOY	17% Pb 19% Sn 10% Hg Bi alloy: Melting point 60 °C; used for making bone replicas
ASTM B774-117	22.5% Pb 8.2% Sn 5.2% Cd 19% In Bi alloy: Melting point 47 °C
ASTM B774-136	18% Pb 12% Sn 21% In Bi alloy: Melting point 58 °C

Symbol	Nominal analysis, supplier, condition and remarks.
ASTM B774-158	38% Pb 13% Sn 10% Cd Bi alloy: Melting point 70 °C
ASTM B774-158/190	38% Pb 11% Sn 9.5% Cd Bi alloy: Melt range 70–88 °C
ASTM B774-174	17% Sn 26% In Bi alloy: Melting point 79 °C
ASTM B774-203	32% Pb 15.5% Sn Bi alloy: Melting point 95 °C
ASTM B774-255	44.5% Pb Bi alloy: Melting point 124 °C
ASTM B774-281	45% Sn Bi alloy: Melting point 138 °C
BISMANOL	20% Mn Bi alloy: Pressed powder for magnets; consult PMA for further information
BISMUTH	Commercially pure Bi metal: Blackwells; primary metal
BISMUTH	High purity Bi metal: Impurities 10 p.p.m.; Johnson Matthey; supplied as ingot or wire
CERROBASE	44.5% Pb Bi alloy: Mining and Chemical Co. Co.; melting point 124 °C
CERRO-CAST	Bi base alloy: Melting range 140–170 °C; Mining & Chemical Co. Co.
CERRO-LOW 117	Bi base alloy: Melting point 47.5 °C; Mining & Chemical Co. Co.
CERRO-LOW 136	Bi base alloy: Melting point 58 °C; Mining & Chemical Co. Co.
CERRO-LOW 147	Bi base alloy: Melting range 61–65 °C; Mining & Chemical Co. Co.
CERROSAFE	37.5% Pb 11.3% Sn 8.5% Cd Bi alloy: Melting range 70–90 °C; Mining & Chemical Co. Products Ltd
CERROTRIC	42% Sn Bi alloy: Melting point 138.5 °C; Mining & Chemical Co. Products Ltd
CERROTRU	42% Sn 1% Pb Bi alloy: Mining & Chemical Co. Co; melting range 134–135 °C; zero contraction on solidification
D'ARCET	25% Sn 25% Pb Bi alloy: Information from Tin Research Institute; melting range 96–98 °C
GLANCE	Bismuth sulphide ore
hBi1	99.9999% Bi rod, 12 mm diameter: Light Ltd; high purity metal
LICHTENBERG	20% Sn 30% Pb Bi alloy: Information from Tin Research Institute; melting range 96–100 °C
LIPOWITZ	13.3% Sn 26.7% Pb 10% Cd Bi alloy: Information from Tin Research Institute; melting range 70–73 °C
MALOTTE	34% Sn 20% Pb Bi alloy: Information from Tin Research Institute; melting range 96–123 °C
MCP 47	Sn Pb Cd In Bi alloy: Mining & Chemical Co.; melting point 47 °C **DPN: 16 UTS: 3 Elon: 25% Proof: 2**
MCP 58	Sn Pb In Bi alloy: Mining & Chemical Co.; melting point 58 °C **DPN: 15 UTS: 4 Elon: 17% Proof: 2.5**
MCP 70	Sn Pb Cd Bi alloy: Mining & Chemical Co.; melting point 70 °C **DPN: 13 UTS: 2 Elon: 200% Proof: 1**
MCP 78	Sn In Bi alloy: Mining & Chemical Co.; melting point 78 °C
MCP 92	Pb Cd Bi alloy: Mining & Chemical Co.; melting point 92 °C
MCP 96	Sn Pb Bi alloy: Mining & Chemical Co.; melting point 96 °C **DPN: 14 UTS: 2.5 Elon: 100% Proof: 1.5**
MCP 100	Sn Pb Bi alloy: Mining & Chemical Co.; melting point 100 °C
MCP 103	Sn Cd Bi alloy: Mining & Chemical Co.; melting point 103 °C
MCP 124	Pb Bi alloy: Mining & Chemical Co.; melting point 124 °C **DPN: 14 UTS: 4 Elon: 80% Proof: 2**
MCP 137	Sn Bi alloy: Mining & Chemical Co.; melting point 137 °C **DPN: 21 UTS: 6 Elon: 70% Proof: 3.5**
NEWTON	18.8% Sn 31.2% Pb Bi alloy: Information from Tin Research Institute; melting range 96–97 °C
ONION	20% Sn 30% Pb Bi alloy: Information from Tin Research Institute; melting range 96–100 °C
ROSE	22% Sn 28% Pb Bi alloy: Information from Tin Research Institute; melting range 96–100 °C
WOODS	12.5% Sn 25% Pb 12.5% Cd Bi alloy: Information from Tin Research Institute; melting range 70–72 °C

Note. The following abbreviations and units are used in the tables:

DPN	Hardness, diamond pyramid number
UTS	Ultimate tensile strength, N/mm^2
Elon	Elongation, %
Proof	0.1% proof strength, N/mm^2

1 N/mm^2=0.1 hbar=0.102 kgf/mm^2=0.06475 $tonf/in.^2$=145.04 $lbf/in.^2$=1 MPa

See Appendix II for other abbreviations and conversion tables.

7. Boron B

Physical properties

Atomic number	5
Atomic weight	10.82
Crystal structure	Monoclinic
Colour	Yellow
Specific gravity	2.6
Density	2600 kg/m^3
Melting point	2100 °C
Boiling point	3550 °C
Specific heat	1.30 J/g °C
Thermal conductivity	–
Coefficient of linear expansion (20–100 °C)	8.3×10^{-6}/ °C
Latent heat of fusion	–
Latent heat of vaporization	3140 J/g
Thermal neutron absorption cross-section	–
Electrical conductivity	–
Specific resistance	–
Temperature coefficient of electrical resistance	–
Electrochemical equivalent	–
Electrode potential	–
Magnetic susceptibility	0.69×10^{-6}
Young's modulus of elasticity	–
Tensile strength	–
Hardness	–

7.1 General notes on boron

At present pure boron metal is a laboratory curiosity, although there are suggestions that it may find use in the nuclear industry as a neutron absorber.

Traces of boron in steel (less than 0.003%) increase the hardenability by an amount equal to 1% nickel 0.5% chromium or 0.25% molybdenum. Boron-containing steels are now commonly used where high strength, high quality welds are required.

Small amounts of boron, up to 0.1%, added to cast iron inhibit graphite formation, thus increasing the possibility of obtaining local surface hardening when desired.

Boron is added in amounts up to 3.0% to certain hard facing alloys as a fluxing agent. This melts above a certain temperature and acts as a bonding agent for the hard infusible particles.

Like arsenic and silicon, boron is a metalloid with only feeble metallic properties. There are no alloys based on boron.

Note. The following abbreviations and units are used in the tables:

DPN	Hardness, diamond pyramid number
UTS	Ultimate tensile strength, N/mm^2
Elon	Elongation, %
Proof	0.1% proof strength, N/mm^2

1 N/mm^2=0.1 hbar=0.102 kgf/mm^2=0.06475 tonf/in.^{2}=145.04 lbf/in.^{2}=1 MPa

See Appendix II for other abbreviations and conversion tables.

Symbol	Nominal analysis, supplier, condition and remarks.
BORON	Pure amorphous powder: Blackwells; primary metal
BORON CARBIDE	Blackwells
HB 6	99.9995% B rod: 5 mm diameter; Light Ltd; high purity metal

8. Cadmium Cd

Physical properties	
Atomic number	48
Atomic weight	112.4
Crystal structure	Hexagonal
Colour	White with blue tinge
Specific gravity	8.64
Density	8640 kg/m^3
Melting point	321 °C
Boiling point	767 °C
Specific heat	0.230 J/g °C
Thermal conductivity	92 W/m °C
Coefficient of linear expansion (20–100 °C)	30×10^{-6}/ °C
Latent heat of fusion	55.3 J/g
Latent heat of vaporization	1197 J/g
Thermal neutron absorption cross-section	2.400 barns/atom
Electrical conductivity	22.5% IACS (copper 100%)
Specific resistance	77 microhm mm
Temperature coefficient of electrical resistance	0.0042/ °C
Electrochemical equivalent	2.09 g/A/h
Electrode potential	−0.403 V
Magnetic susceptibility	0.18×10^{-6}
Young's modulus of elasticity	55.2×10^9 N/m^2
Tensile strength	75 N/mm^2
Hardness	22 DPN

8.1 General notes on cadmium

Approximately 60% of the world's supply of cadmium is used as electroplating anodes for deposition of a corrosion protection coating. Although much dearer than zinc, cadmium plating is commonly used and is specified for aircraft use wherever steel has to be protected by plating. Zinc reacts in contact with aluminium to produce a bubbly corrosive product. This does not occur with cadmium. It must not be used in contact with food as it can be poisonous if ingested. Cadmium oxide fumes are toxic and are a hazard if allowed to accumulate during melting, welding or brazing.

Metallic cadmium is a neutron absorber and is used in rod form, or in electroplated graphite rods as fission moderators in atomic piles.

There are very few alloys based on cadmium, but a few brazing materials have been developed which have a high shear strength. These are zinc/cadmium or silver/cadmium, and are used for tipping high speed tools. There are also a few cadmium based alloys used for bearing materials.

Cadmium is used as an alloying element with copper, when up to 1% cadmium is added to increase the tensile strength with very little reduction in electrical conductivity. These alloys are used for traction conductor wires and for long span transmission conductor wires.

With lead, cadmium forms an alloy which does not work harden, and has been used for cable sheathing purposes, where a harder coating than pure lead is required.

Low melting point alloys used for fusible links generally have some cadmium content. Where alloys with cadmium are tested above their melting point for any reason, great care is necessary to prevent ingestion of cadmium fumes, which are now known to be extremely toxic.

Note. The following abbreviations and units are used in the tables:

DPN	Hardness, diamond pyramid number
UTS	Ultimate tensile strength, N/mm^2
Elon	Elongation, %
Proof	0.1% proof strength, N/mm^2

1 N/mm^2=0.1 hbar=0.102 kgf/mm^2=0.06475 tonf/in.^{2}=145.04 lbf/in.^{2}=1 MPa

See Appendix II for other abbreviations and conversion tables.

Symbol	Nominal analysis, supplier, condition and remarks.
2.2480	20% Zn Cd brazing alloy: Working temperature 280 °C; German Standard
BS 2686/1	99.95% Cd (Sb + As + Th 0.01%(max)) anodes: Cast; rolled or extruded
CADMIUM	Cd metal: In form of powder, stick, sheet, etc.; Blackwells; primary metal
CAZIN	17.4% Zn Cd alloy: For soldering steel cables; origin unknown; melting point 263 °C
DIN 8512 L Cd Zn 20	20% Zn Cd brazing alloy: Working temperature 280 °C

Symbol	Nominal analysis, supplier, condition and remarks.
ELMET SILNO	Ag + Ni with Ag and Cd oxides: Sintered material; Metro Cutanit Ltd; range of materials
h Cd 1	99.9999% Cd rod: 13 mm diameter; Light Ltd; high purity metal
HT 5	16.5% Zn 5% Ag Cd alloy: Braze metal; melting range 270–285 °C; Sheffield Smelting Co. **DPN: 60 UTS: 200 Elon: 8% Proof: 150**
L 01900	99.9% Cd: UNS designation
L 01950	99.95% Cd: UNS designation
L Cd Zn 20	20% Zn Cd brazing alloy: Designation used by German Standard
LM 5	Ag Cd base soft solder: Melting range 338–390 °C; Johnson Matthey; electrical conductivity 22.4% IACS **UTS: 130 Elon: 25%**
QQ A671	99.9% Cd: US Federal specification
SAE 18	1.3% Ni Cd alloy: For bearings
SAE 180	0.6% Cu 0.7% Ag Cd alloy: For bearings

Note. The following abbreviations and units are used in the tables:

DPN	Hardness, diamond pyramid number
UTS	Ultimate tensile strength, N/mm^2
Elon	Elongation, %
Proof	0.1% proof strength, N/mm^2

1 N/mm^2=0.1 hbar=0.102 kgf/mm^2=0.06475 tonf/in.^{2}=145.04 lbf/in.^{2}=1 MPa

See Appendix II for other abbreviations and conversion tables.

9. Caesium Cs (cesium)

Physical properties	
Atomic number	55
Atomic weight	132.91
Crystal structure	Body-centred cubic
Colour	Silvery white
Specific gravity	1.89
Density	1890 kg/m^3
Melting point	28.4 °C
Boiling point	705 °C
Specific heat	0.209 J/g °C
Thermal conductivity	36 W/cm °C
Coefficient of linear expansion (20–100 °C)	97 × 10^{-6}/ °C
Latent heat of fusion	15.78 J/g
Latent heat of vaporization	611 J/g
Thermal neutron absorption cross-section	29 barns/atom
Electrical conductivity	9% IACS (copper 100%)
Specific resistance	190 microhm mm
Temperature coefficient of electrical resistance	0.0044/ °C
Electrochemical equivalent	–
Electrode potential	−3.0 V
Magnetic susceptibility	−0.22 × 10^{-6}
Young's modulus of elasticity	–
Tensile strength	–
Hardness	–

9.1 General notes on caesium

Caesium salts used in the manufacture of electronic high vacuum valves result in the reduction of caesium from the chloride salt by molten barium metal. This is a by-product of the high vacuum process required by modern electronic equipment and is not a real source of the metal.

Caesium is of interest in that it is the most basic element known, and could be termed the most metallic element. The metal reacts explosively with oxygen or water and thus finds very little commercial use.

When subjected to light, gamma radiation or X-rays it emits electrons in proportion to the amount falling on it. The sensitivity to light rays is very similar to the human eye. Thus, in spite of the difficulty of using the metal, photo-electric cell mechanisms are manufactured with caesium.

The metal is not radio-active.

As would be expected, there are no alloys based on caesium. To date no use has been made of the metal as an alloying element, but it has found some use as a catalyst in the resin industry.

No specifications or proprietary names could be found for caesium metal.

10. Calcium Ca

Physical properties		
Atomic number		20
Atomic weight		40.08
Crystal structure		Face-centred cubic
Colour		Yellowish white
Specific gravity		1.54
Density		1540 kg/m^3
Melting point		851 °C
Boiling point		1482 °C
Specific heat		0.63 J/g °C
Thermal conductivity		126 W/m °C
Coefficient of linear expansion (20–100 °C)		23×10^{-6}/ °C
Latent heat of fusion		216.4 J/g
Latent heat of vaporization		4660 J/g
Thermal neutron absorption cross-section		0.43 barns/atom
Electrical conductivity		48.7% IACS (copper 100%)
Specific resistance		46 microhm mm
Temperature coefficient of electrical resistance		0.00457/ °C
Electrochemical equivalent		0.747 g/A/h
Electrode potential		−2.8 V
Magnetic susceptibility		1.1×10^{-6}
Young's modulus of elasticity		23.4×10^9 N/m^2
Tensile strength	rolled	110 N/mm^2
	annealed	40 N/mm^2
Hardness		17 DPN

10.1 General notes on calcium

Calcium resembles aluminium and magnesium in many of its properties but is more difficult to cast as it oxidizes much more rapidly. It is also much softer and more ductile, being readily cut with a knife and bent by hand. Although it burns readily in air it is not subject to spontaneous combustion and can be safely handled.

Calcium, either as the metal or as a compound, finds several uses as a de-oxidizer in the preparation of some of the rarer metals such as beryllium, titanium, vanadium, uranium and zirconium.

There are no commercial calcium based alloys, but the metal is used as an addition to aluminium and magnesium alloys, where it acts as a grain refiner. There is reported to be an improvement in surface finish when machining these alloys.

Lead/calcium alloys have improved fatigue strength and have been used for bearing alloys and cable sheathing, as well as electrical storage battery grids.

Calcium silicide is used as a graphite controller in 'inoculated' cast irons.

The metal in particle form is now finding considerable use as a catalyst in chemical engineering.

Symbol	Nominal analysis, supplier, condition and remarks.
CALCIUM	Ca metal: In form of lumps and filings; Blackwells; primary metal
DOMAL 2	99.9% Ca metal: Dominion Magnesium
DOMAL 3	0.3% Al 0.75% Mg Ca metal: Dominion Magnesium
DOMAL 4	0.35% Al 1% Mg 98.5% Ca metal: Dominion Magnesium
NELCO	0.25% Al 0.2% Mg 99% Ca metal: Nelco Metal Corp
W Ca 17	99.9% Ca metal: In grain form; Light Ltd; high purity metal

Note. The following abbreviations and units are used in the tables:

DPN	Hardness, diamond pyramid number
UTS	Ultimate tensile strength, N/mm^2
Elon	Elongation, %
Proof	0.1% proof strength, N/mm^2

1 N/mm^2=0.1 hbar=0.102 kgf/mm^2=0.06475 tonf/in.^{2}=145.04 lbf/in.^{2}=1 MPa

See Appendix II for other abbreviations and conversion tables.

11. Cerium Ce

Physical properties	
Atomic number	58
Atomic weight	140.13
Crystal structure	Face-centred cubic
Colour	Steel grey
Specific gravity	6.7
Density	6700 kg/m^3
Melting point	795 °C
Boiling point	3470 °C
Specific heat	0.188 J/g °C
Thermal conductivity	109 W/m °C
Coefficient of linear expansion (20–100 °C)	8.4 × 10^{-6}/ °C
Latent heat of fusion	35.6 J/g
Latent heat of vaporization	–
Thermal neutron absorption cross-section	0.73 barns/atom
Electrical conductivity	2.2% IACS (copper 100%)
Specific resistance	780 microhm mm
Temperature coefficient of electrical resistance	0.00087/ °C
Electrochemical equivalent	–
Electrode potential	–
Magnetic susceptibility	15 × 10^{-6}
Young's modulus of elasticity	41.4 × 10^9 N/m^2
Tensile strength	150 N/mm^2
Hardness	24 DPN

11.1 General notes on cerium

This is one of the elements forming the group known as 'rare earth metals', which are not as rare as was at one time thought, but are by no means plentiful. In general they are found together, and great difficulty is experienced in obtaining any one element free from the remainder. The history of the group gives some indication of the difficulty in purification, in that what at one time was thought to be one element turned out in fact to be a mixture of four of the rare earth metals.

Cerium is one of the less rare metals in the group and probably the most widely used, and for this reason has been separated from the other 'rare earths' which are treated as one group in Section 35 of this book.

Cerium metal is soft, being readily cut with a knife, oxidizes rapidly in air, but does not burn spontaneously. It reacts with water, releasing hydrogen but not violently. It burns more intensely in air than does magnesium and this affinity for oxygen is utilized as a 'getter' when cerium combines with the last traces of oxygen in vacuum electronic equipment.

Cerium alloyed with other rare earth metals is used as a powerful reducing agent and catalyst in chemical engineering.

An interesting use of cerium is as an alloy with iron in proportions of approximately 50% each, when it has peculiar pyrophoric qualities, giving off copious sparks when rubbed with a file. These alloys are used as 'flints' in cigarette lighters and similar equipment. This use is now diminishing with the advent of the 'catalytic reaction' type equipment.

Cerium is used as an alloying element with aluminium and magnesium, when it enhances the high temperature properties, fatigue and creep strength.

The alloy mischmetall, 50% cerium, is used as a deoxidizer in the steel industry, and has been found to give some grain refinement. In small amounts cerium gives an improvement in the hot working properties of austenitic steels and nickel alloys.

Symbol	Nominal analysis, supplier, condition and remarks.
AUER	35% Fe 24% La 4% Yb 2% Er Ce alloy: Common name; pyrophoric material
CERIUM	99.5% Ce metal: Ingot, sheet, rod and wire; Johnson Matthey
CERIUM	Ce metal: Blackwells, primary metal
h Ce 16	99.9% Ce metal: Ingot; Light Ltd; high purity metal
h Ce 19	99.9% Ce metal: Foil; Light Ltd; high purity metal
h Ce 20a	99.9% Ce metal: Wire; 1 mm in diameter; Light Ltd; high purity metal
h Ce 96	99.5% Ce 140 isotope: Light Ltd; stable isotopic form
HUBER	15% Mg Ce alloy: Common name; pyrophoric material
MISCHMETALL	40% La + Sa + Nd + Pr 10% Fe + Yb Ce alloy: Obtained by electrolysis; common name
SENSITIVE	25% Pt Ce alloy: Common name; pyrophoric material
WELSBACH	30% Fe 10% rare earth Ce alloy: Common name; pyrophoric material

Note. The following abbreviations and units are used in the tables:

DPN	Hardness, diamond pyramid number
UTS	Ultimate tensile strength, N/mm^2
Elon	Elongation, %
Proof	0.1% proof strength, N/mm^2

1 N/mm^2=0.1 hbar=0.102 kgf/mm^2=0.06475 $tonf/in.^2$=145.04 $lbf/in.^2$=1 MPa

See Appendix II for other abbreviations and conversion tables.

12. Chromium Cr

Physical properties

Atomic number	24
Atomic weight	52.01
Crystal structure	Rhombohedral
Colour	Greyish white
Specific gravity	7.19
Density	7190 kg/m^3
Melting point	1860 °C
Boiling point	2200 °C
Specific heat	0.461 J/g °C
Thermal conductivity	69.1 W/m °C
Coefficient of linear expansion (20–100 °C)	6.2×10^{-6}/ °C
Latent heat of fusion	293.1 J/g
Latent heat of vaporization	5945 J/g
Thermal neutron absorption cross-section	2.9 barns/atom
Electrical conductivity	14% IACS (copper 100%)
Specific resistance	130 microhm mm
Temperature coefficient of electrical resistance	0.00588/ °C
Electrochemical equivalent	0.971 g/A/h
Electrode potential	−0.56 V
Magnetic susceptibility	3.08×10^{-6}
Young's modulus of elasticity	248.2×10^9 N/m^2
Tensile strength	890 N/mm^2
Hardness annealed	90 DPN

12.1 General notes on chromium

Metallic chromium of 99% purity is available but is too brittle for any practical purpose. Higher purity grades are generally only laboratory materials, but there are indications that a more ductile and therefore useful metal may be available, which could have many potential uses. At present there is no alloy system based on chromium but the metal is used for chromium plating and as an alloy addition, often with startling results.

In electroplating, chromium is used for two distinct purposes. First, as a decorative finish it is always plated on top of nickel in the form of a very thin cracked deposit. As this is known to be porous, recent work would appear to indicate that the microcracking produces the necessary small anode to large cathode ratio known to result in galvanic protection.

The coating is so thin that, unless the undercoat is highly polished, the chromium will appear dull. The hard blue-white colour of the deposited chromium is retained under all normal atmospheric conditions. The chrome is seldom or never the source of failure, which is caused by corrosion of the base metal from a too thin, badly prepared, or porous nickel deposit.

Deposits of hard chrome for wear resistance purposes are now common. Chromium has a low coefficient of friction and this is aided by the electroplated deposit having a micro-cracked surface which retains a film of oil even under adverse conditions. The deposit is plated directly on the base metal, and should not be less than 0.1 mm thick after grinding. Electroplated chromium is very brittle and must not be subjected to shock loads. Considerable care must also be taken when grinding the deposit as this is very readily cracked and softened. Because of this, hard chromium plating is being replaced by metal deposit techniques using other alloyed materials which have good wear resistance and can better withstand shock loading. Chromium deposits, however, have certain high temperature wear resisting properties which appear to be unique.

Details of the use of chromium as an alloying element appear with the appropriate section on the base material. Briefly these are as follows:

Aluminium. Additions of chromium in the order of 0.3% are made as grain refiners and to improve the corrosion resistance of some higher strength aluminium wrought alloys.

Copper. A chromium content of 0.5% in copper produces an age hardening alloy used for welding electrodes.

Cobalt. Up to 25% chromium is found in the cobalt based alloys, where it forms a solid solution. If carbon is present some of the chromium will form hard carbides, and there is evidence to show that it is the surface layer of chromium oxide which gives these alloys their excellent high temperature oxidation and corrosion resistance.

Steel. Chromium has two attributes when alloyed in steel. First, it combines with carbon to form very hard chromium carbides. These considerably increase the hardenability of the steel. The chromium content varies from as low as 0.5% to about 5%, and is often found in conjunction with other carbide-forming elements such as vanadium, molybdenum and tungsten. Most high speed tools have some chromium added.

The second virtue of chromium is that it enhances corrosion resistance and resistance to oxidation at temperature. Chromium alone in quantities from 10% up to 20% gives steels with very good corrosion resistance. When over 7% nickel is added to steel with 17% and more chromium an austenitic stainless steel is formed. Detailed information on the effects of chromium on steels appears in the sections concerned.

Nickel. Additions of up to 20% chromium to nickel give a range of materials with excellent high temperature properties. The basic 80/20 nickel/chromium alloys now include many complex elements. Details of these are given in the specific sections.

Symbol	Nominal analysis, supplier, condition and remarks.
608	15% Ni 2.0% W chromium carbide: General Electric Co. **UTS: 240 Elon: 1.1% Proof: 60**
ASTM A560/50/50	0.1% C 50% Ni Cr alloy
ASTM A560/60/40	0.1% C 40% Ni Cr alloy
CHROMIUM	Cr metal: Fused powder; Blackwells; pure metal
CHROMIUM 99	0.01% C 0.2% Fe 0.1% Al 0.1% Si 0.01% Cu Cr metal: Metal Alloys Ltd
CHROMIUM 99 LOW GAS	0.08% C 0.03% N 0.001% H Cr metal: Metal Alloys Ltd
CRONITE 50/50	0.1% C 48% Ni 2.0% W Cr alloy: Casting; Cronite Ltd; as cast; heat resistant
CRONITE 60/40	0.1% C 40% Ni 2.0% W Cr alloy: Casting; Cronite Ltd; as cast; heat resistant **DPN: 350 UTS: 520 Elon: 1.0% Proof: 400**
h Cr 7a	99.999% Cr flake: Light Ltd; high purity metal
h Cr 7b	99.999% Cr pellets: Light Ltd; high purity metal
h Cr 8	99.999% Cr crystals (ex iodide): Light Ltd; high purity metal
h Cr 12	99.99% Cr flake: Light Ltd; high purity metal
METCO 81NS	Ni Cr matrix Cr C: Spray deposit; Metco **DPN: 350**
METCO 83VF	Ni Cr matrix Cr C: Spray deposit; Metco **DPN: 540**
METCO 106	Cr oxide: Spray deposit; Metco **DPN: 975**
METCO 136CP	Si and Cr oxides: Spray deposit; Metco **DPN: 800**
METCO 136F	Si and Cr oxides: Spray deposit; Metco **DPN: 1300**
METCO 430 NS	Ni Al matrix Cr C: Spray deposit; Metco **DPN: 460**
p Cr 23	99.0% Cr crystalline powder: Light Ltd
TOLOY 60/40	0.1% C 40% Ni Cr alloy: Casting; Wellman

Note. The following abbreviations and units are used in the tables:

DPN	Hardness, diamond pyramid number
UTS	Ultimate tensile strength, N/mm^2
Elon	Elongation, %
Proof	0.1% proof strength, N/mm^2

1 N/mm^2=0.1 hbar=0.102 kgf/mm^2=0.06475 $tonf/in.^2$=145.04 $lbf/in.^2$=1 MPa

See Appendix II for other abbreviations and conversion tables.

13. Cobalt Co

Physical properties	
Atomic number	27
Atomic weight	58.94
Crystal structure	Close-packed hexagonal up to 417 °C; face-centred cubic up to melting point
Colour	Silver-white with bluish tinge
Specific gravity	8.8
Density	8800 kg/m^3
Melting point	1493 °C
Boiling point	3100 °C
Specific heat	0.440 J/g °C
Thermal conductivity	69.21 W/cm °C
Coefficient of linear expansion (20–100 °C)	12.5×10^{-6}/ °C
Latent heat of fusion	259.6 J/g
Latent heat of vaporization	3280 J/g
Thermal neutron absorption cross-section	34.8 barns/atom
Electrical conductivity	27.6% IACS (copper 100%)
Specific resistance	62.4 microhm mm
Temperature coefficient of electrical resistance	0.0053/ °C
Electrochemical equivalent	1.1 g/A/h
Electrode potential	−0.28 V
Magnetic susceptibility	Ferromagnetic
Young's modulus of elasticity	211×10^9 N/m^2
Tensile strength	240 N/mm^2
Hardness	230 DPN

13.1 General notes on cobalt

Cobalt is commonly produced up to 99.5% purity, with purities up to 99.9% available.

Cobalt is a silver-white metal with a slight bluish tinge normally stable in the atmosphere but tending to oxidize more easily than nickel.

The pure metal has few uses but because of recent developments is now commercially available as bar sheet and expanded metal, and is therefore likely to find more widespread applications. At present cobalt is used because of its good magnetic properties which are retained up to 1120 °C, considerably higher than any iron based alloy. It is also used as an electroplated deposit, and as polished sheet for reflective purposes under arduous conditions.

Cobalt base alloys and cobalt-containing alloys, of which there are a great number, find important commercial applications as magnetic alloys, creep and heat resistant alloys, ultra-high tensile steels, dental and surgical alloys and in a wide range of non-ferrous alloys where its addition contributes either directly or indirectly to the required properties.

The advances made in engineering materials to meet the demands of rapidly expanding technologies have resulted in a dramatic growth in the consumption of cobalt to today's high level.

Symbol	Nominal analysis, supplier, condition and remarks.
COBALT	High purity metal: Impurities 10 p.p.m.; Johnson Matthey; supplied as sponge, rod, sheet and wire
COBALT	Co metal: In form of powder, briquettes, strip and wire; British Metal Corporation Ltd
COBALT	Co metal: In all forms; Brandhurst Co. Ltd
COBALT	Co metal: In form of powder, sponge, bar, sheet; New Metals & Chemicals Ltd
COBALT	Co metal: In form of powder, cubes, sheets, etc.; Blackwells; pure metal
L Co 8	99.999% Co metal: In form of sponge; Light Ltd; high purity metal
L Co 11	99.999% Co metal: In form of rod 5 mm diameter; Light Ltd; high purity metal
L Co 73	99.99% Co metal: In form of single crystals 6.3 × 12 mm; Light Ltd
W Co 23	99.5% Co metal: In form of powder; Light Ltd

Note. The following abbreviations and units are used in the tables:

DPN	Hardness, diamond pyramid number
UTS	Ultimate tensile strength, N/mm^2
Elon	Elongation, %
Proof	0.1% proof strength, N/mm^2

1 N/mm^2=0.1 hbar=0.102 kgf/mm^2=0.06475 tonf/in.^{2}=145.04 lbf/in.^{2}=1 MPa

See Appendix II for other abbreviations and conversion tables.

13A Cobalt alloys – wrought and cast

Specific gravity	8.2–9
Density	8200–9000 kg/m^3
Solidus/liquidus	14.7 W/m °C
Coefficient of linear expansion (20–100 °C)	9.5–11 × 10^{-6}/ °C
Electrical conductivity	1–2% IACS (copper 100%)
Specific resistance	850–17000 microhm mm
Young's modulus of elasticity	179–248 × 10^9 N/m^2
Impact	Will vary considerably
Fatigue strength (10^6 cycles)	± 250 N/mm^2 (applies to Stellite X40)

Hot strength

Temperature °C	Tensile strength N/mm^2	Elongation %
650	300 (after 1500 h)	5.6
750	200 (after 1000 h)	6.5
815	120 (after 750 h)	1.5

The above properties are typical of the following group, and may not apply exactly to any one specification. It is possible that with certain specifications some of the values may not be applicable.

General metallurgical characteristics

These alloys are based on the solid solution of cobalt and chromium. They have additions of nickel to improve the ductility, while carbon and carbide formers such as tungsten and molybdenum are added to increase the hardness.

Some of the more complex alloys are capable of being age hardened to a limited extent in a manner analogous to the nickel chromium alloys. To date none of the cobalt materials produced has as marked age hardening properties as those in the nickel series.

The ductility of all the alloys is very low; hence little or no cold work can be carried out without causing undue brittleness often resulting in crumbling.

Cobalt alloys all have the ability to retain their hardness at elevated temperatures and this with their corrosion resistance is the principal reason for their existence. This extreme difficulty in hot working results in few wrought products, although there are notable exceptions. The alloys are generally used as castings, or as welded or sprayed coatings on more ductile materials.

The cobalt alloys have excellent damping characteristics and this property can be of considerable value in high duty rotating assemblies operating at elevated temperatures. The magnetic properties of some cobalt alloys arise from the high anisotropy constant of cobalt which leads to magnetic properties such as coercivity and magnetic energy. Cobalt also has the highest curie point. Thus, the alloys retain their magnetism at a higher temperature than any other magnetic materials. Alloyed with iron, cobalt increases the saturation induction. Only a few of the magnetic alloys appear in the following pages as cobalt is seldom the principal constituent. These alloys appear under 20C and 44L.

The effect of alloying elements are briefly as follows:

Chromium. This is the principal element, forming a stable solid solution, analogous in many ways to the nickel chrome series. There is evidence that it is the chromium which imparts the hot oxidation resistance.

Nickel. The presence of nickel increases the malleability with the minimum effect on the remaining properties. Nickel also helps to take some of the refractory metals into solution, thus aiding the hot strength, and is an essential element in certain precipitation hardening alloys as it is a constituent in the necessary intermetallic compounds.

Molybdenum. This strengthens the solid solution matrix, increasing the hot strength and also taking part in precipitation age hardening.

Tungsten, niobium and tantalum. These act in a similar manner to molybdenum, generally being more active and in some cases more refractory.

Titanium. This is one of the elements which forms intermetallic compounds with nickel, allowing age hardening to take place.

Aluminium. This acts in a similar manner to titanium, again forming intermetallic compounds with nickel.

Zirconium and other refractory materials are added to increase the hot strength or corrosion resistance under specific conditions.

Carbon is present with carbide forming elements to give the very hard carbides which give wear resistance. Depending on the metal used, these can have a very high hardness even at elevated temperatures. With the correct thermal treatment, some age hardening can be obtained with these carbides but this is not as marked as with the intermetallic compounds.

Boron. This is used in small amounts to improve the hot strength, and in large amounts (above 1%) to act as a fluxing agent in the sprayed or welded alloys to improve the bond to the base metal.

Many cobalt alloys are applied by metal spray or weld deposition. This allows the use of a higher ductile base material with the relevant surfaces being coated with corrosion, abrasion resistant materials generally capable of withstanding high temperatures.

Techniques are now available for castings, sometimes with very fine dimensional limits using the complete range of cobalt alloys.

In general the cobalt alloys are more or less unmachinable except with specialist techniques and cooling. The materials, either as a casting or a surface coating, are not heat treatable except to remove cold work. This requires elevated temperatures in the region of 700 °C.

The materials can be welded using sophisticated techniques and specialist advice is recommended.

Cobalt alloys find use where high temperatures are involved with corrosive atmospheres. These include gas turbine components, equipment in the petro-chemical industry and other chemical engineering industries. Some alloys have specific uses in the nuclear industry.

These are expensive materials which are difficult to handle and it is only where high temperatures with corrosive atmospheres are involved that their cost and difficulty of handling are justified.

Symbol	Nominal analysis, supplier, condition and remarks.
19	0.45% C 26% Cr 15% Ni 6% Mo 1% Ti Co alloy: Casting; American proprietary alloy listed in SAE year book
25 Ni	0.17% C 19% Cr 25% Ni 10% W 1% Fe 1.5% Nb 42.5% Co alloy: General Electric Co.
25 Ni & V	0.2% C 19% Cr 25% Ni 11% W 1.2% Nb 3% V 2% Fe Co alloy: General Electric Co.
422	0.45% C 26% Cr 15% Ni 6% Mo 1% Fe Co alloy: American proprietary alloy listed in SAE year book
422/19	0.45% C 26% Cr 15% Ni 6% Mo 1% Fe Co alloy: Information from Cobalt Information Centre
ABRASODUR 83	5.5% C 25.0% Cr 6.8% Nb 1.5% Ti 6.0% Fe Co alloy: Welding electrode; Soudometal; for abrasive resistance at high temperatures
AF 94	0.12% C 15% Cr 10% Ni 5% Mo 10% W 1% Nb 2% Fe Co alloy: Allegheny Ludlum
AIR RESIST 13	0.45% C 1.0% Ni 11.0% W 2.0% Nb 3.5% Al 2.5% Fe Co alloy: Origin unknown
AIR RESIST 213	Modified Air Resist 13
AIR RESIST 215	Modified Air Resist 13
AISI 670	0.12% C 19.8% Cr 9.9% Ni 15.2% W 1.6% Fe Co alloy: Annealed **UTS: 1050 Elon: 64% Proof: 450**
AISI 671	0.4% C 19.6% Cr 20.3% Ni 10% Fe 4% Mo 4% W 4% Nb Co alloy: Solution treated and aged **UTS: 1000 Elon: 35% Proof: 370**
ALLOY 6	1% C 30% Cr 1% Si 5% W Co alloy: Casting; Dewrance **DPN: 394 UTS: 670**
ALLOY 9	1.35% C 31% Cr 1% Si 8% W Co alloy: Casting; Dewrance **DPN: 521 UTS: 470**
ALLOY 63	26% Cr 2% C 20% W 1% V Co alloy: Casting; Dewrance **DPN: 500 UTS: 540**
ALTEMP S 816	0.4% C 20% Cr 20% Ni 4% Mo 4% W 4% Nb Co alloy: Allegheny Ludlum
AMS 4783	0.2% C 20% Cr 13% Ni 2.4% Mo 2.8% W 20% Fe Co alloy
AMS 5373	1.1% C 29% Cr 1.5% Ni 0.7% Mo 4% W 1.5% Fe Co alloy: Castings
AMS 5375 B	25% Cr 1.8% Ni 5% W Co alloy: Investment casting
AMS 5378 B	25% Cr 32% Ni 5.5% Mo Co alloy: Investment casting
AMS 5380 C	26% Cr 15% Ni 6% Mo Co alloy: Investment casting
AMS 5382	0.5% C 25.5% Co 10% Ni 7% W 1.5% Fe Co alloy: Castings
AMS 5385	0.2% C 27% Cr 2.5% Ni 5.5% Mo 1.7% Fe Co alloy: Castings
AMS 5385 C	27% Cr 2.8% Ni 5.5% Mo Co base: Investment casting
AMS 5387	28% Cr 5% W Co alloy: Investment casting
AMS 5530 C	15.5% Cr 16% Mo 5.5% Fe 3.8% W Co alloy: Sheet
AMS 5534	0.37% C 20% Cr 20% Ni 4% Mo 4% W 4% Nb 3.5% Fe Co alloy: Sheet
AMS 5537	0.1% C 20% Cr 10% Ni 15% W 1.5% Fe Co alloy: Sheet
AMS 5608	0.1% C 22% Cr 22% Ni 14.5% W 0.1% La Co alloy
AMS 5758	0.025% C (max) 20% Cr 35% Ni 9.7% Mo Co alloy
AMS 5759	0.1% C 20% Cr 10% Ni 15% W 1.6% Fe Co alloy: Forgings
AMS 5759 B	20% Cr 10% Ni 15% W Co alloy: Bar and forging
AMS 5765	0.37% C 20% Cr 20% Ni 4% Mo 4% W 4% Nb 3.4% Fe Co alloy: Forgings
AMS 5772	0.1% C 22% Cr 22% Ni 14.5% W 0.1% La Co alloy
AMS 5788	28% Cr 4.5% W Co alloy: For coating
AMS 5789	0.5% C 25.5% Cr 10.5% Ni 7.5% W Co alloy
AMS 5796	20% Cr 10% Ni 15% W Co alloy: Wire
AMS 5797	0.1% C (max) 20% Cr 10% Ni 15% W 1.6% Fe Co alloy: Wire
AMS 5801	0.1% C 22% Cr 22% Ni 14.5% W 0.1% La Co alloy
AMS 5833	0.15% C (max) 20% Cr 15.5% Ni 7% Mo 15% Fe Co alloy
AMS 5834	0.15% C (max) 20% Cr 15.5% Ni 7% Mo 15% Fe Co alloy
AMS 5841	0.04% C (max) 19% Cr 33% Ni 7% Mo 3% Ti 0.5% Nb 9% Fe Co alloy
AMS 5842	0.04% C (max) 19% Cr 33% Ni 7% Mo 3% Ti 0.5% Nb 9% Fe Co alloy
AMS 5843	0.04% C (max) 19% Cr 33% Ni 7% Mo 3% Ti 0.5% Nb 9% Fe Co alloy
AMS 5844	0.025% C (max) 20% Cr 35% Ni 9.7% Mo Co alloy
AMS 5845	0.025% C (max) 20% Cr 35% Ni 9.7% Mo Co alloy
AMS 5875	0.15% C (max) 20% Cr 15.5% Ni 7% Mo 15% Fe Co alloy
AMS 5876	0.15% C (max) 20% Cr 15.5% Ni 7% Mo 15% Fe Co alloy
AMS 7238	1.4% C 30% Cr 1.5% Ni 8.3% W Co alloy
AMS 7468	0.4% C 18.5% Cr 10% Ni 15% W Co alloy
AMS 7475	0.04% C (max) 19% Cr 33% Ni 3% Ti 9% Fe Co alloy
ANC 13	0.5% C 25% Cr 10% Ni 7% W Co alloy: Investment casting; designation used by BS
ANC 14	0.2% C 27% Cr 2.5% Ni 5.5% Mo Co alloy: Investment casting; designation used by BS
ARNAVAR	0.15% C 20% Cr 13% Ni 2% Mo 2.8% W 28% Fe Co alloy: Origin unknown
AS 25	0.1% C 20% Cr 10% Ni 1.5% Mn 15% W 3% Fe (max) Co alloy: Origin unknown

Note. The following abbreviations and units are used in the tables:

DPN	Hardness, diamond pyramid number
UTS	Ultimate tensile strength, N/mm^2
Elon	Elongation, %
Proof	0.1% proof strength, N/mm^2

$1\ N/mm^2 = 0.1\ hbar = 0.102\ kgf/mm^2 = 0.06475\ tonf/in.^2 = 145.04\ lbf/in.^2 = 1\ MPa$

See Appendix II for other abbreviations and conversion tables.

Symbol	Nominal analysis, supplier, condition and remarks.
ASTM A461/671	0.37% C 20% Cr 20% Ni 4.0% Mo 4.0% W 4.0% Nb + Ta 5.0% Fe Co alloy: Bar
ASTM A567/1	0.25% C 27% Cr 2.7% Ni 5.5% Mo 3% Fe Co alloy: Casting
ASTM A567/2	0.5% C 25.5% Cr 10.5% Ni 7.5% W 2% Fe Co alloy: Casting
ASTM A639/671	0.37% C 20.0% Cr 20% Ni 4.0% Mo 4.0% W 4.0% Nb + Ta 5.0% Fe (max) Co alloy
ATG Z	0.4% C 20% Cr 20% Ni 4.0% Mo 5.0% Fe 4.0% W 4.0% Nb Co alloy: Imphy
BCO 1	0.8% C 19% Cr 17% Ni 8% Si 4% W 0.8% B Co alloy: Welding electrode
BS 3146 ANC13	0.5% C 25% Cr 10% Ni 7% W Co alloy: Investment casting; as cast **UTS: 630 Elon: 5% Proof: 400**
BS 3146 ANC14	0.2% C 27% Cr 2.5% Ni 5.5% Mo Co alloy: Investment casting; as cast **UTS: 600 Elon: 5% Proof: 420**
BS 3531/1–4	0.1% C 20% Cr 15% W 10% Ni Co alloy: For implant and surgery tools; wrought **UTS: 880 Elon: 39%**
BS 3531/13	29% Cr 6.0% Mo Co alloy: For implant and surgery tools; cast **UTS: 650 Elon: 8% Proof: 450**
CM 6	1% C 30% Cr 1% Si 5% W Co alloy: For hard facing; Dewrance; gas welding
CM 9	1.3% C 31% Cr 1% Si 8% W Co alloy: For hard facing; Dewrance; gas welding
CM 10	4.2% C 15% Cr 2% Ni 8% Mo 1.5% Si 1% V Co alloy: For hard facing; Dewrance; gas welding
CM 63	2% C 28% Cr 20% W 1% V Co alloy: For hard facing; Dewrance; gas welding
CM 106	1% C 30% Cr 1% Si 5% W Co alloy: For hard facing; Dewrance; electric welding
CM 110	4.25% C 16% Cr 2% Ni 8% Mo 1.5% Si 1% V Co alloy: For hard facing; Dewrance; electric welding
COBSTEL 1	2.5% C 30% Cr 13% W 1% Fe Co alloy: Electrode; Metrode **DPN: 600**
COBSTEL 6	1% C 30% Cr 5.5% W 1% Fe Co alloy: Electrode; Metrode **DPN: 500**
COBSTEL 8	0.2% C 27% Cr 3% Ni 5% Mo 1% Fe Co alloy: Electrode; Metrode **DPN: 450**
COBSTEL 12	1.4% C 30% Cr 8% W 1% Fe Co alloy: Electrode; Metrode **DPN: 550**
COBSTEL 25	0.1% C 20% Cr 10% Ni 15% W 1% Fe Co alloy: Electrode; Metrode **DPN: 400**
COBSTEL 50	0.2% C 28% Cr 0.5% Nb 20% Fe Co alloy: Electrode; Metrode **DPN: 350**
CO-ELINVAR	10% Cr 30% Fe Co alloy: Details from Cobalt Information Centre
CROFORM	27% Cr 5% Mo Co alloy: For dental uses; details from Cobalt Information Centre **DPN: 390 UTS: 700 Elon: 4% Proof: 510**

Note. The following abbreviations and units are used in the tables:

DPN	Hardness, diamond pyramid number
UTS	Ultimate tensile strength, N/mm^2
Elon	Elongation, %
Proof	0.1% proof strength, N/mm^2

1 N/mm^2=0.1 hbar=0.102 kgf/mm^2=0.06475 $tonf/in.^2$=145.04 $lbf/in.^2$=1 MPa
See Appendix II for other abbreviations and conversion tables.

Symbol	Nominal analysis, supplier, condition and remarks.
CUNICO 11	24% Ni 35% Cu Co alloy: For magnets; US material listed by PMA
DELCRO 1300K	0.1% C 28% Cr 20% Fe 0.7% Si Co alloy: Deloro; for furnace equipment; origin unknown
DIN 17410	German Standard covering magnetic alloy
DURATHERM 477	17.9% Cr 23% Ni 4% Mo Co alloy: Origin unknown
DURATHERM 600	0.05% C (max) 12% Cr 4% Mo 2% Ti 8.7% Fe 4% W 30% Ni Co alloy: Origin unknown
DURATHERM 700	0.05% C (max) 12% Cr 4% Mo 3.5% Ti 6.8% Fe 4% W 30% Ni Co alloy: Origin unknown
DURATHERM 2602	0.05% C (max) 12% Cr 4% Mo 1% Ti 10% Fe 4% W 30% Ni Co alloy: Origin unknown
DYNAMET	20% Cr 13% Ni 2.5% W 17% Fe Co alloy: Wrought; American proprietary alloy for springs
DYNAVAR	20% Cr 13% Ni 2.5% W 17% Fe Co alloy: Wrought; American proprietary alloy for springs
E Co Cr A	1.2% C 28.5% Cr 4.5% W Co alloy: Weld electrode; designation used by AWS
E Co Cr B	1.3% C 28.5% Cr 8% W Co alloy: Weld electrode
E Co Cr C	2.3% C 28.5% Cr 12.5% W Co alloy: Weld electrode
E Co Cr E	0.25% C 26.5% Cr 3% Ni 5.5% Mo Co alloy: Weld electrode
ELGILOY	0.15% C 20% Cr 15% Ni 7% Mo 16% Fe 0.04% Be Co alloy: American proprietary alloy listed in SAE year book **DPN: 702 UTS: 2460 Proof: 1510**
ELGILOY	0.15% C 20% Cr 15% Ni 15% Fe 7.0% Mo 0.04% Be Co alloy: Strip; Gilby-Brunton; aged **DPN: 700 UTS: 2550 Proof: 1600**
EMS 242	1.2% C 30% Cr 2.25% Ni 4.5% W 3% Fe Co alloy: Origin unknown; for hard facing **DPN: 440**
ESCO ALLOY 72	0.1% C 28% Cr 20% Fe 0.7% Si Co alloy: Eastern Stainless Steel Co.; for furnace equipment
ESSCALLOY L 605	0.1% C 1.5% Mn 20% Cr 10% Ni 15% W 3% Fe Co alloy: Eastern Stainless Steel Co.
F 484	0.4% C 25% Cr 5% Ni 4% Mo 4% W 4% Nb 1% Be 2% Fe Co alloy: Allegheny Ludlum
G 32	0.3% C 19% Cr 12% Ni 2% Mo 1.2% Nb 2.8% V 13% Fe Co alloy: Bar and sheet; Jessop Saville; solution treated and aged **DPN: 320 UTS: 1000 Elon: 8% Proof: 720**
G 34	0.8% C 19% Cr 12.5% Ni 2% Mo 1.3% Nb 2.8% V Co alloy: Casting; Jessop Saville; as cast **DPN: 305 UTS: 600 Elon: 2.5% Proof: 400**
G 34C	0.8% C 19% Cr 12% Ni 15% Fe 2% Mo 2.5% V 1.3% Nb Co alloy: Casting; Jessop Saville
G 60	0.25% C 27% Cr 2.8% Ni 5.5% Mo Co alloy: Casting; Jessop Saville
G 65	0.45% C 25% Cr 10% Ni 7% W 2% Fe Co alloy: Casting; Jessop Saville; as cast **UTS: 700 Elon: 17% Proof: 400**
G 75	0.45% C 21% Cr 11% W 2% Nb Co alloy: Casting; Jessop Saville
G 87	0.10% C 21.0% Cr 10.0% Ni 15.0% W Co alloy: Jessop Saville; scale or creep resistant
G 88	0.10% C 16.0% Cr 11.0% Ni 1.5% Mo 1.5% Nb Co alloy: Jessop Saville; scale or creep resistant
GAUSSIT 180	14% V Co alloy: Rolled strip; German alloy listed by PMA
HA 3	2.5% C 30% Cr 3% Ni 12% W 3% Fe Co alloy: Code for Haynes 3
HA 4	0.6% C 30% Cr 14% W 3% Fe Co alloy: Code for Haynes 4
HA 6	1.0% C 30% Cr 3% Ni 4% W 3% Fe Co alloy: Code for Haynes 6
HA 12	1.3% C 30% Cr 8% Ni 8% W 3% Fe Co alloy: Code for Haynes 12
HA 19	1.7% C 31% Cr 10% W 3% Fe Co alloy: Code for Haynes 19

Symbol	Nominal analysis, supplier, condition and remarks.
HA 25	0.1% C 20% Cr 10% Ni 15% W 3% Fe Co alloy: Code for Haynes 25
HA 36	0.4% C 18.5% Cr 10% Ni 15% W 0.03% B Co alloy: Code for Haynes 36
HA 98 M2	2.0% C 30% Cr 3.5% Ni 18% W 1% Mo 2% Fe Co alloy: Code for Haynes 98 M2
HA 150	0.1% C 28% Cr 20% Fe 0.7% Si Co alloy: Code for Haynes 150
HA 151	0.5% C 20% Cr 12.5% W 0.05% B Co alloy: Code for Haynes 151
HA 152	0.4% C 20% Cr 1% Ni 11% W 2% Nb & Ta Co alloy: Code for Haynes 152
HA 302	0.85% C 21.5% Cr 10% W 9% Ta 0.2% Zr Co alloy: Code for Haynes 302
HAVAR	0.2% C 20% Cr 13% Ni 2% Mo 2.8% W 0.04% Be Co alloy: American proprietary alloy; consult SAE **DPN: 650 UTS: 310 Proof: 1620**
HAYNES 3	2.5% C 30% Cr 3% Ni 12% W 3% Fe Co alloy: Casting; S Osborn; cutting tools; wear resistant parts **DPN: 670 UTS: 540 Elon: 1%**
HAYNES 4	0.8% C 30% Cr 14% W 8% Fe Co alloy: Casting; S Osborn; abrasion and corrosion resistant **DPN: 460**
HAYNES 6	1% C 30% Cr 3% Ni 4% W 3% Fe Co alloy: Casting; S Osborn; tough grades abrasion resistant **DPN: 460 UTS: 820 Elon: 1%**
HAYNES 6B	1% C 30% Cr 3% Ni 4% W 3% Fe Co alloy: Bar; S Osborn; abrasion and corrosion resistant **DPN: 450 UTS: 1040 Elon: 7% Proof: 600**
HAYNES 6K	1.6% C 31% Cr 3% Ni 4% W 3% Fe Co alloy: Bar; S Osborn; abrasion and corrosion resistant **DPN: 500 UTS: 1280 Elon: 3.5%**
HAYNES 12	1.3% C 30% Cr 3% Ni 8% W 3% Fe Co alloy: Casting; S Osborn; tough grade; abrasion resistant **DPN: 520 UTS: 740**
HAYNES 19	1.7% C 31% Cr 10% W 3% Fe Co alloy: Casting; S Osborn; cutting alloy; wear resistant **DPN: 570 UTS: 750 Elon: 1%**
HAYNES 25	0.1% C 20% Cr 10% Ni 15% W 3% Fe Co alloy: Wrought; S Osborn; solution treated; heat resistant; wear resistant **UTS: 1000 Elon: 64% Proof: 450**
HAYNES 36	0.4% C 18.5% Cr 10% Ni 15% W 0.03% B Co alloy: Union Carbide casting
HAYNES 98 M2	2% C 30% Cr 3.5% Ni 18% W 1% Mo 2% Fe Co alloy: Casting; S Osborn; cutting tools; wear resistant parts **DPN: 760 UTS: 660**
HAYNES 150	0.1% C 28% Cr 20% Fe 0.7% Si Co alloy: Union Carbide; for furnace equipment **DPN: 250 UTS: 520 Elon: 8% Proof: 300**
HAYNES 151	0.5% C 20% Cr 12.5% W 0.05% B Co alloy: Casting; Union Carbide
HAYNES 152	0.4% C 20% Cr 1% Ni 11% W 2% Nb & Ta Co alloy: Union Carbide **DPN: 375 UTS: 820 Elon: 5% Proof: 600**
HAYNES 302	0.85% C 21.5% Cr 10% W 9% Ta 0.2% Zr Co alloy: Union Carbide; turbine blade alloy
HAYNES STAR J	2.5% C 32% Cr 2.5% Ni 17% W 2% Fe Co alloy: Union Carbide for metal cutting **DPN: 750 UTS: 480**
HDA 8151	0.5% C 20% Cr 12.5% W 0.05% B Co alloy: Casting; Union Carbide
HE 1049	0.4% C 26% Cr 10% Ni 15% W 3% Fe 0.4% B Co alloy: Casting; American proprietary alloy listed in SAE yearbook
HEV 4	0.1% C 20% Cr 10% Ni 15% W Co alloy: For valves; annealed; Designation used by SAE **UTS: 1100 Elon: 65% Proof: 470**
HIPERCO 50	0.01% C 0.05% Si 1.9% V 49% Fe Co alloy: Carpenter; for magnetic purposes
HONIAL 2	Al Ni Co alloy: For magnets; French alloy listed by PMA
HONIAL 5	Al Ni Co alloy: For magnets; French alloy listed by PMA
HONIAL 12	Al Ni Co alloy: For magnets; French alloy listed by PMA
HR 40	0.1% C 20% Cr 10% Ni 1.5% Mn 15% W 3% Fe (max) Co alloy: BS Aerospace designation
HS 21	0.25% C 27% Cr 3% Ni 5% Mo 1% Fe Co alloy: Casting; American proprietary alloy; listed in SAE yearbook
HS 23	0.4% C 24% Cr 2% Ni 6% W Co alloy: Casting; listed by Cobalt Information Centre
HS 23	0.4% C 24.0% Cr 2.0% Ni 5.0% W 1.0% Fe Co alloy: Origin unknown
HS 25	0.1% C 20% Cr 10% Ni 15% W Co alloy: American proprietary alloy listed in SAE yearbook **DPN: 220 UTS: 1020 Elon: 65% Proof: 470**
HS 27	0.4% C 25.0% Cr 3.2% Ni 6% W Co alloy: Casting; listed by Cobalt Information Centre
HS 30	0.45% C 26% Cr 15% Ni 6% Mo 1% Fe Co alloy: American proprietary alloy listed in SAE yearbook
HS 31	0.5% C 25% Cr 10% Ni 7.5% Mo 1.5% Fe Co alloy: Casting; American proprietary alloy listed in SAE yearbook **UTS: 700 Elon: 11% Proof: 500**
HS 36	0.4% C 19% Cr 10% Ni 15% W 1% Fe 0.03% B Co alloy: American proprietary alloy listed in SAE yearbook **DPN: 300 UTS: 700 Elon: 5% Proof: 600**
HS 188	0.1% C 22% Cr 22% Ni 14.5% W 0.1% La Co alloy: Origin unknown
I 336	0.2% C 20% Cr 15% Ni 12% W 1% Nb Co alloy: Listed by Cobalt Information Centre
ILLIUM D	0.2% C 27% Cr 4.5% Mo 1% W 1% Fe Co alloy: American proprietary alloy listed in SAE yearbook
ILLIUM H	0.1% C 28% Cr 20% Fe 0.7% Si Co alloy: Listed in SAE yearbook
ILLIUM X	0.8% C 28% Cr 1% Ni 15% W 2% Fe Co alloy: American proprietary alloy listed in SAE yearbook
J 1570	0.2% C 20% Cr 28% Ni 6% W 4% Ti Co alloy: Universal Cyclop **DPN: 385**
J 1650	0.2% C 19% Cr 27% Ni 12% W 3.8% Ti 2% Ta 0.02% B Co alloy: Universal Cyclop
JETALLOY 209	0.02% C 20% Cr 10% Ni 15% W 2% Ti Co alloy: Quebec Metallurgical
JETALLOY 249	0.03% C 25% Cr 10% Ni 7% W Co alloy: Quebec Metallurgical
JETALLOY 1570	0.2% C 20% Cr 30% Ni 6.5% W 4% Ti Co alloy: General Electric
KOERFLEX 30	8% V 39% Fe Co alloy: Strip for magnets; German alloy listed by PMA
KOERFLEX 200	8% V 39% Fe Co alloy: Wire and strip for magnets; German alloy listed by PMA
KOERFLEX 300	8% V 4% Cr 36% Fe Co alloy: Strip and wire for magnets; German alloy listed by PMA
KOERZIT H	8% V 39% Fe Co alloy: Strip and wire for magnets; German alloy listed by PMA
KOERZIT T	4% Cr 36% Fe 0.8% V Co alloy: Permanent alloy; details from Cobalt Information Centre
L 202	1.5% C 28.5% Cr 6.5% Ni 3.7% Mo 4.0% Nb Co alloy: VEW
L 210	0.9% C 31% Cr 13.5% W Co alloy: VEW
L 215	1.9% C 31% Cr 13.5% W Co alloy: VEW
L 216	2.5% C 31% Cr 13.5% W Co alloy: VEW
L 218	1.4% C 31% Cr 0.2% Mo 8.5% W Co alloy: VEW
L 219	1.2% C 25% Cr 4.5% W Co alloy: VEW
L 225	0.25% C 27% Cr 6% Mo Co alloy: VEW

Symbol	Nominal analysis, supplier, condition and remarks.
L 251	0.4% C 19% Cr 10% Ni 15% W 1% Fe 0.03% B Co alloy: Casting; American proprietary alloy listed in SAE yearbook
L 605	0.1% C 20% Cr 10% Ni 15% W 3% Fe Co alloy: Used by American Companies; consult SAE yearbook
LODEX	Fe Co sintered alloy: With lead for magnets; further information from PMA
M 203	0.07% C 20% Cr 25% Ni 12% W 2% Ti 0.7% Al 1.5% Nb 1.6% Fe Co alloy: General Electric
M 204	0.07% C 18% Cr 25% Ni 12% W 1.6% Fe 1.2% Nb Co alloy: General Electric
M 205	0.07% C 18% Cr 25% Ni 12% W 2.75% Al 1.2% Nb 1.6% Fe Co alloy: General Electric
MAGNETOFLEX 20	20% Ni Co alloy: Rolled strip; German alloy listed by PMA
MALCALLOY	13.5% Al Co alloy: For magnets; Japanese alloy listed by PMA
MAR – M 918	0.05% C 20.0% Ni 7.5% Ta 0.1% Zr Co alloy: Martin **UTS: 880 Elon: 48% Proof: 380**
METCO 18C	Co alloy: Spray deposit; Metco **DPN: 550**
METCO 45C NS	Cr Ni W Co alloy: Sprayed deposit; Metco **DPN: 260**
METCO 45VF NS	Cr Ni W Co alloy: Sprayed deposit; Metco **DPN: 260**
METCO 68F NSI	17% Cr Mo Co alloy: Powder coating; Metco **DPN: 460**
METCO 439	50% W C Co: Spray deposit; Metco **DPN: 460**
ML 1700	0.2% C 25% Cr 15% W 0.4% B Co alloy: Casting; American proprietary alloy; listed in SAE yearbook
MP 35N ALLOY	0.02% C (max) 20% Cr 35% Ni 10% Mo Co alloy: Carpenters **UTS: 1560 Elon: 14% Proof: 1500**
NFW 1	Al Ni Co alloy: Casting for magnets; Japanese alloy listed by PMA
NFW 2	Al Ni Co alloy: Casting for magnets; Japanese alloy listed by PMA
NFW 3	Al Ni Co alloy: Casting for magnets; Japanese alloy listed by PMA
NFW 4	Al Ni Co alloy: Casting for magnets; Japanese alloy listed by PMA
NFW 5	Al Ni Co alloy: Casting for magnets; Japanese alloy listed by PMA
NFW 6	Al Ni Co alloy: Casting for magnets; Japanese alloy listed by PMA
NICKELVAC L. 605	0.12% C 20% Cr 10% Ni 15% W 1.7% Mn Co alloy: Vanadium alloy; double vacuum cast
NICROBRAZ 210	0.4% C 19.0% Cr 0.8% B 17% Ni 8.0% Si 4.0% W Co alloy: For brazing; Wall Colmonoy; melting range 1110–1150 °C
NICROBRAZ 300	0.8% C 21.0% Cr 3.25% B 10.0% W 17.0% Ni Co alloy: For brazing; Wall Colmonoy; melting range 1040–1120 °C
NIVAFLEX	0.03% C 18% Cr 21% Ni 4% Mo 4% W 1% Ti 0.3% Be Co alloy: Information from Cobalt Information Centre
NIVCO 10	0.05% C 22.5% Ni 1.8% Ti 0.2% Al 1% Fe 1% Zr Co alloy: American proprietary alloy listed in SAE yearbook
ONERAL M 47	1% C 28% Cr 6% Ni 10% Mo 0.3% Ti 3% Fe Co alloy: Casting; listed by Cobalt Information Centre
ONERAL S 90	0.3% C 27% Cr 17% Ni 5% Mo Co alloy: Information from Cobalt Information Centre
P 6	6.0% Ni 4.0% V 45% Fe Co alloy: Rolled strip; US alloy listed by PMA for hysteresis motors
PERMENDUR 24	49% Fe 2% V Co alloy: More ductile than Permendur; Telcon
PERMENDUR V	2% V 49% Fe Co alloy: Telcon; high permeability for high flux densities
R 30004	0.2% C 20% Cr 13% Ni 2.4% Mo 2.8% W 23% Fe Co alloy: Designation used by UNS
R 30816	0.36% C 20% Cr 20% Ni 4% Mo 5% Fe 4% Nb + Ta 4% W Co alloy: UNS designation
R 30816	UNS designation for ASTM A639/671
R Co Cr A	1.1% C 29% Cr Co alloy: Weld electrode
R Co Cr B	1.4% C 30% Cr 8.3% W Co alloy: Weld electrode
R Co Cr C	2.5% C 30% Cr 1.5% Ni 13% W Co alloy: Weld electrode; designation used by AW
REFRACTALOY 70	0.05% C 20% Cr 20% Ni 8% Mo 4% W 15% Fe Co alloy: American proprietary alloy; consult SAE
REMENDUR	2.5% V Fe Co alloy: For magnets; American alloy listed by PMA
REXWELD 33W	2.2% C 3% Cr 18% W Co alloy: For hard facing; Crucible Steel Co. **DPN: 700**
REXWELD A	1.0% C 28% Cr 4.5% W Co alloy: For hard facing; Crucible Steel Co. **DPN: 400**
REXWELD B	1.5% C 28% Cr 7.5% W Co alloy: For hard facing; Crucible Steel Co. **DPN: 420**
REXWELD C	2.5% C 30% Cr 12.5% W Co alloy: For hard facing; Crucible Steel Co. **DPN: 550**
REXWELD VT	0.3% C 27% Cr 3% Ni 5.5% Mo Co alloy: For hard facing; Crucible Steel Co. **DPN: 350**
S–816	0.38% C 20% Cr 20% Ni 4% Mo 4% Nb 4% W Co alloy: Origin unknown
S 816	0.38% C 20% Cr 20% Ni 3% Fe 4% Mo 4% W 4% Nb Co alloy: Allegheny Ludlum
S 844	0.25% C 25% Cr 20% Ni 2% Mo 2% W 2% Nb 3% Fe Co alloy: Allegheny Ludlum
SAE VF 5	1.7% C 25% Cr 22% Ni 12% W 2% Fe Co alloy: For hard facing
SL 959	High C High Cr Co alloy: Casting; Swift Levrick **DPN: 450**
SL 970	0.4% C 24.5% Cr 5% W Co alloy: Swift Levrick
SM 302	0.85% C 21.5% Cr 10% W 9% Ta 0.2% Ti 0.005% B 1% Fe Co alloy: Martin Metal Co.
SM 322	1.0% C 21.5% Cr 9% W 4.5% Ta 0.7% Ti 2.2% Zr 0.7% Fe Co alloy: Martin Metal Co.
SOUDOKAY STELKAY 1-G	2.7% C 28% Cr 13% W 5% Fe Co alloy: Welding electrode; Soudometal **DPN: 580**
SOUDOKAY STELKAY 6-G	1.1% C 29% Cr 5% W 5% Fe Co alloy: Welding electrode; Soudometal **DPN: 400**
SOUDOKAY STELKAY 12-G	1.6% C 29% Cr 7.7% W 5% Fe Co alloy: Welding electrode; Soudometal **DPN: 470**
SOUDOKAY STELKAY 21-G	0.2% C 26% Cr 2.5% Ni 4.9% Mo 5% Fe Co alloy: Welding electrode; Soudometal **DPN: 330**
SOUDOSPRAY 162-5	C Cr W Co alloy: Spray powder; Soudometal; grade 12 Stellite **DPN: 470**

Note. The following abbreviations and units are used in the tables:

DPN	Hardness, diamond pyramid number
UTS	Ultimate tensile strength, N/mm^2
Elon	Elongation, %
Proof	0.1% proof strength, N/mm^2

$1\ N/mm^2 = 0.1\ hbar = 0.102\ kgf/mm^2 = 0.06475\ tonf/in.^2 = 145.04\ lbf/in.^2 = 1\ MPa$

See Appendix II for other abbreviations and conversion tables.

Symbol	Nominal analysis, supplier, condition and remarks.
SOUDOSPRAY 166-1	C Cr W Co alloy: Spray metal powder; Soudometal; deposits grade 6 Stellite **DPN: 470**
SOUDOSPRAY 166-2	C Cr W Co alloy: Spray metal powder; Soudometal; deposits grade 6 Stellite **DPN: 470**
SOUDOSPRAY 166-3	C Cr W Co alloy: Spray metal powder; Soudometal; deposits grade 6 Stellite **DPN: 470**
SOUDOSPRAY 166-5	C Cr W Co alloy: Spray metal powder; Soudometal; deposits grade 6 Stellite **DPN: 470**
SOUDOSPRAY 166-6	C Cr W Co alloy: Spray metal powder; Soudometal; deposits grade 6 Stellite **DPN: 470**
SOUDOSPRAY 167-0	C Cr Ni W Co alloy: Spray metal; Soudometal; deposits grade F cobalt alloy **DPN: 430**
SOUDOSPRAY 167-1	C Cr Ni W Co alloy: Spray metal; Soudometal; deposits grade F cobalt alloy **DPN: 430**
SOUDOSPRAY 167-2	C Cr Ni W Co alloy: Spray metal; Soudometal; deposits grade F cobalt alloy **DPN: 430**
SOUDOSPRAY 167-3	C Cr Ni W Co alloy: Spray metal; Soudometal; deposits grade F cobalt alloy **DPN: 430**
SOUDOSPRAY 167-5	C Cr Ni W Co alloy: Spray metal; Soudometal; deposits grade F cobalt alloy **DPN: 430**
SOUDOSPRAY 167-6	C Cr Ni W Co alloy: Spray metal; Soudometal; deposits grade F cobalt alloy **DPN: 430**
SOUDOSPRAY 168-5	C Cr Ni Mo Co alloy: Spray metal; Soudometal; deposits type 21 Co alloy **DPN: 320**
SOUDOSPRAY 327-0	C Cr Ni B Co alloy: Powder with W_2C; Soudometal; for powder deposits – variable W_2C **DPN: 700**
SOUDOSPRAY 327-1	C Cr Ni B Co alloy: Powder with W_2C; Soudometal; for powder deposits – variable W_2C **DPN: 700**
SOUDOSPRAY 327-2	C Cr Ni B Co alloy: Powder with W_2C; Soudometal; for powder deposits – variable W_2C **DPN: 700**
SOUDOSPRAY 327-3	C Cr Ni B Co alloy: Powder with W_2C; Soudometal; for powder deposits – variable W_2C **DPN: 700**
SOUDOSPRAY 327-5	C Cr Ni B Co alloy: Powder with W_2C; Soudometal; for powder deposits – variable W_2C **DPN: 700**
SOUDOSPRAY 327-6	C Cr Ni B Co alloy: Powder with W_2C; Soudometal; for powder deposits – variable W_2C **DPN: 700**
SOUDOSPRAY 328-0	C Cr Ni B Co alloy: Powder with W_2C; Soudometal; for powder deposits – high W_2C content
SOUDOSPRAY 328-1	C Cr Ni B Co alloy: Powder with W_2C; Soudometal; for powder deposits – high W_2C content
SOUDOSPRAY 328-2	C Cr Ni B Co alloy: Powder with W_2C; Soudometal; for powder deposits – high W_2C content
SOUDOSPRAY 328-3	C Cr Ni B Co alloy: Powder with W_2C; Soudometal; for powder deposits – high W_2C content
SOUDOSPRAY 328-5	C Cr Ni B Co alloy: Powder with W_2C; Soudometal; for powder deposits – high W_2C content
SOUDOSPRAY 328-6	C Cr Ni B Co alloy: Powder with W_2C; Soudometal; for powder deposits – high W_2C content
SOUDOSTEL 1 AP	2.3% C 32.4% Cr 2.2% Ni 0.1% Mo 12.7% W 3.0% Fe Co alloy: Electrode; Soudometal; for abrasion resistance at high temperature **DPN: 600**
SOUDOSTEL 1 C	2.4% C 31.9% Cr 2.2% Ni 0.2% Mo 12.2% W 2.3% Fe Co alloy: Welding electrode; Soudometal; for abrasion resistance at high temperature **DPN: 560**
SOUDOSTEL 6 AP	1.3% C 27.5% Cr 2.7% Ni 4.4% W 2.7% Fe Co alloy: Welding electrode; Soudometal; for abrasion resistance at high temperature **DPN: 450**
SOUDOSTEL 6 C	1.2% C 27.2% Cr 2.7% Ni 4.6% W 2.4% Fe Co alloy: Welding electrode; Soudometal; for abrasion resistance at high temperature **DPN: 450**
SOUDOSTEL 12 AP	1.6% C 30.7% Cr 2.7% Ni 8.6% W 2.8% Fe Co alloy: Welding electrode; Soudometal; for wear resistance **DPN: 500**
SOUDOSTEL 12 C	1.6% C 30.8% Cr 2.4% Ni 8.5% W 2.3% Fe Co alloy: Welding electrode; Soudometal; for wear resistance **DPN: 500**
SOUDOSTEL 12 HC	1.8% C 28.5% Cr 2.5% Ni 8.5% W 2.3% Fe Co alloy: Welding electrode; Soudometal; for wear resistance **DPN: 520**
SOUDOSTEL 21 AP	0.27% C 27.9% Cr 2.8% Ni 5.3% Mo 2.4% Fe Co alloy: Welding electrode; Soudometal; for wear resistance at high temperature **DPN: 350**
SOUDOSTEL 21 C	0.25% C 27.2% Cr 2.5% Ni 5.4% Mo 1.7% Fe Co alloy: Welding electrode; Soudometal; for wear resistance at high temperature **DPN: 350**
SOUDOSTEL HR 1	2.2% C 27.8% Cr 0.1% Ni 0.1% Mo 12% W 3% Fe Co alloy: Welding electrode; Soudometal; for abrasion resistance at high temperature **DPN: 600**
SOUDOSTEL HR 6	1.0% C 29.4% Cr 4.7% W 1.8% Fe Co alloy: Welding electrode; Soudometal; for abrasion resistance at high temperature **DPN: 460**
SOUDOSTEL HR 12	1.3% C 29.3% Cr 7.9% W 2.6% Fe Co alloy: Welding electrode; Soudometal; for wear resistance **DPN: 500**
SOUDOSTEL HR 21	0.27% C 27.0% Cr 2.6% Ni 4.6% Mo 2.4% Fe Co alloy: Welding electrode; Soudometal; for wear resistance at high temperature **DPN: 350**
SOUDOSTEL HR 25	0.05% C 20.5% Cr 9.8% Ni 16% W 2% Fe Co alloy: Welding electrode; Soudometal; for abrasion and impact resistance **DPN: 230**
STAR J	2.5% C 32% Cr 2% N 17% W 3% Fe Co alloy: Casting; S Osborn; cutting tools; wear resistant parts **DPN: 760**
STELLITE 1	2.5% C 33% Cr 13% W Co alloy: Casting for hard facing rods (oxyacetylene) and electrodes; Deloro **DPN: 620 UTS: 600 Elon: 1%**
STELLITE 3	2.4% C 30% Cr 13% W Co alloy: Casting; Deloro **DPN: 620 UTS: 600 Elon: 1%**
STELLITE 4	1% C 31% Cr 14% W Co alloy: Casting; Deloro **DPN: 490 UTS: 1000 Elon: 1% Proof: 600**
STELLITE 6	1% C 26% Cr 5% W Co alloy: Casting for hard facing rods (oxyacetylene) and electrodes; Deloro **DPN: 400 UTS: 880 Elon: 1% Proof: 530**
STELLITE 7	0.4% C 26% Cr 6% W Co alloy: Casting; Deloro **DPN: 325 UTS: 900 Elon: 8% Proof: 450**
STELLITE 8	0.2% C 30% Cr 6% Mo Co alloy: Casting; Deloro for BS 3146/ANC 14 **DPN: 325 UTS: 900 Elon: 9% Proof: 470**
STELLITE 12	1.8% C 29% Cr 9% W Co alloy: Casting for hard facing rods (oxyacetylene) and electrodes; Deloro **DPN: 510 UTS: 820 Elon: 1% Proof: 630**

Symbol	Nominal analysis, supplier, condition and remarks.
STELLITE 12P	1.4% C 31.0% Cr 9.0% W Co alloy: Melting range 1260–1300 °C; Deloro Stellite **DPN: 470 UTS: 880 Elon: 1%**
STELLITE 20	2.5% C 3% Cr 18% W Co alloy: Casting for hard facing rods (oxyacetylene) and electrodes; Deloro **DPN: 660 UTS: 600 Elon: 1%**
STELLITE 21	0.25% C 27% Cr 2.5% Ni 4.5% Mo 2% Fe 0.007% B Co alloy: Casting; Deloro
STELLITE 23	0.4% C 24% Cr 2% Ni 5% W 2% Fe Co alloy: Deloro
STELLITE 27	0.4% C 25% Cr 32% Ni 5.5% Mo Co alloy: Deloro
STELLITE 30	0.45% C 26% Cr 15% Ni 6% Mo Co alloy: Deloro
STELLITE 31	0.5% C 25.5% Cr 10.5% Ni 7.5% W Co alloy: Casting; Deloro
STELLITE 100	2% C 34% Cr 19% W Co alloy: Casting; Deloro **DPN: 850 UTS: 690 Elon: 1%**
STELLITE 208	26.0% Cr 3.0% Mo 20% Fe alloy: Melting range 1400-1415 °C; Deloro Stellite **DPN: 280 UTS: 571 Elon: 12%**
STELLITE 306	1.4% C 25.0% Cr 5.0% Ni 2.0% W 6.0% Nb Co alloy: Melting range 1225-1305 °C; Deloro Stellite **DPN: 360**
STELLITE 506	1.6% C 35.0% Cr 7.5% W 3.0% Fe Co alloy: Deloro Stellite **DPN: 400**
STELLITE F	1.75% C 29% Cr 22% Ni 12% W Co alloy: Deloro
STELLITE X 40	0.3% C 25% Cr 10% Ni 7% W Co alloy: Casting; Deloro for BS 3146/ANC 13 **DPN: 325 UTS: 1120 Elon: 10% Proof: 430**
STUBS 701	C Cr Ni W Co alloy: Electrode; Stubs; for high temperature application **DPN: 600**
STUBS 701 A	C Cr Ni W Co alloy: Electrode; Stubs; for high temperature application **DPN: 600**
STUBS 706	C Cr W Co alloy: Welding electrode; Stubs; for tool stud welding; high temperature application **DPN: 450**
STUBS 706 A	C Cr W Co alloy: Welding electrode; Stubs; for tool stud welding; high temperature application **DPN: 450**
STUBS 712	C Si W Co alloy: Welding electrode; Stubs; for tool steels; for corrosive and high temperature application **DPN: 550**
STUBS 712 A	C Si W Co alloy: Welding electrode; Stubs; for tool steels; for corrosive and high temperature application **DPN: 550**
SUPERMENDUR	49% Fe 2% V Co alloy: For magnets; Telcon; rectangular hysteresis loop
TANTUNG	3% C 30% Cr 16% W 4.5% Nb or Ta 2% Mn 3% Fe Co alloy: Vasoloy; harder than Tantung G **DPN: 830**
TANTUNG G	3.0% C 30% Cr 16% W 4.5% Nb or Ta 2% Mn 3% Fe Co alloy: Vasoloy; Ramet; cutting tools **DPN: 790**
THERMELAST 2602	0.03% C 12% Cr 26% Ni 13% Fe 4% W 4% Mo 1% Ti 0.2% Be Co alloy: Details from Cobalt Information Centre
TRIBALOY 100	35% Mo 10% Si Co alloy: Intermetallic material for wear resistant purposes; Du Pont
TRIBALOY 400	28% Mo 8% Cr 2% Si Co alloy: Intermetallic material for wear resistant purposes; suitable for hard facing; Du Pont
TRIBALOY 800	28% Mo 17% Cr 3% Si Co alloy: Intermetallic material for wear resistant purposes; suitable for hard facing; Du Pont
TRIBALOY T400	0.8% C 8% Cr 28% Mo 3% Ni + Fe Co alloy: For valve facings; Du Pont
UCAR 25	0.07% C 20.0% Cr 10.0% Ni 0.5% Mo 1.5% Fe Co alloy: Casting; Union Carbide
UCAR FSX414	0.25% C 29.0% Cr 10.0% Ni 7.0% W Co alloy: Union Carbide
UCAR M509	0.6% C 21.5% Cr 10.0% Ni 7.0% W Co alloy: Union Carbide
UCAR X45	0.25% C 25.5% Cr 10.5% Ni 7.0% W Co alloy: Union Carbide
UM Co 50	0.1% C 28% Cr 20% Fe 0.7% Si Co alloy: American proprietary alloy; consult SAE yearbook **UTS: 540 Elon: 8% 4310**
UM Co 51	0.27% C 28% Cr 20% Fe 2% Nb 0.7% Si Co alloy: American proprietary alloy; consult SAE yearbook **UTS: 630 Elon: 4% Proof: 460**
UNITEMP L 605	0.1% C 1.5% Mn 20% Cr 10% Ni 15% W 3% Fe Co alloy: Universal Cyclops
UNITEMP S 816	0.38% C 1.5% Mn 20% Cr 20% Ni 4% Mo 4% W 4% Nb Co alloy: Universal Cyclops
UNS R30001-R39999	Co metal and alloys: American Standard system; unified number
V 36	0.3% C 25% Cr 20% Ni 4% Mo 2% W 2.2% Nb 3% Fe Co alloy: Allegheny Ludlum
VELINVAR	31% Fe 8.5% V Co alloy: Details from Cobalt Information Centre
VF 2	1.2% C 28% Cr 3% Ni 4.5% W 0.5% Mo 3% Fe Co alloy: For hard facing; designation used by SAE
VF 5	1.7% C 25% Cr 22% Ni 12% W 2% Fe Co alloy: For hard facing; designation used by SAE
VF 6	2.5% C 30% Cr 1.5% Ni 0.5% Mo 13% W 3% Fe Co alloy: For valve facing; SAE designation
VF 7	1.4% C 30% Cr 1.5% Ni 8.3% W 3% Fe Co alloy: For valve facing; SAE designation
VF 8	0.8% C 8% Cr 28% Mo 3% Fe + Ni Co alloy: For valve facing; SAE designation
VICALLOY	38% Fe 10% V Co alloy: Magnet alloy strip; Telcon; annealed **DPN: 200 UTS: 760**
VICALLOY 1	38% Fe 10% V Co alloy: Permanent magnet; Telcon
VICALLOY 11	35% Fe 13% V Co alloy: Permanent magnet; Telcon
VIMETAL	0.05% C 9% Cr 17% Ni 5.5% Fe 4% W 4% Mo 3% Nb 1.5% Ti Co alloy: Details from Cobalt Information Centre
VITALLIUM	35% Cr 5% Mo Co alloy: American proprietary alloy **DPN: 388 UTS: 280 Elon: 4% Proof: 420**
VMA 4	23.5% Cr 10% Ni 7% W 3.5% Ti Co alloy: Casting; BS designation; obsolete
WAI-MET 50	0.1% C 28% Cr 20% Fe 0.7% Si Co alloy: Wai-Met alloys for furnace equipment
WALLEX 1	2.2% C 30% Cr 12.5% W Co alloy: Castings and rods for facing; Wall Colmonoy; melting point 1270 °C; good corrosion resistance; low friction **DPN: 570**
WALLEX 1NE	2.5% C 3.0% Fe 30% Cr 12.0% W Co alloy: Bronze alloy metal; Wall Colmonoy; melting point 1255 °C
WALLEX 4	1.0% C 1.5% Si (max) 30.5% Cr 14.0% W Co alloy: Bronze metal; Wall Colmonoy; melting point 1260 °C
WALLEX 6	1% C 29% Cr 4.5% W Co alloy: Castings and rods for spraying; Wall Colmonoy; melting point 1275 °C; heat resistant; low friction **DPN: 430 UTS: 810**

Note. The following abbreviations and units are used in the tables:

DPN	Hardness, diamond pyramid number
UTS	Ultimate tensile strength, N/mm^2
Elon	Elongation, %
Proof	0.1% proof strength, N/mm^2

$1\ N/mm^2 = 0.1\ hbar = 0.102\ kgf/mm^2 = 0.06475\ tonf/in.^2 = 145.04\ lbf/in.^2 = 1\ MPa$

See Appendix II for other abbreviations and conversion tables.

Symbol	Nominal analysis, supplier, condition and remarks.
WALLEX 6NE	1.0% C 2.0% Fe (max) 31% Cr 5.5% N Co alloy: Bronze metal; Wall Colmonoy; melting point 1285 °C
WALLEX 7	0.2% C 27% Cr 5% Mo Co alloy: Welding electrode; Wall Colmonoy
WALLEX 7NE	0.3% C 3.0% Fe (max) 30% Cr 3.0% Ni 5.5% Mo Co alloy: Bronze metal; Wall Colmonoy; melting point 1260 °C
WALLEX 12	1.0% C 1.25% Si 9.0% W 29% Cr Co alloy: Bronze metal; Wall Colmonoy; melting point 1280 °C
WALLEX 12NE	1.5% C 3.0% Fe (max) 30% Cr 8.5% W Co alloy: Bronze metal; Wall Colmonoy; melting point 1280 °C
WALLEX 42	Cr W Co alloy: For powder coating; Wall Colmonoy **DPN: 460**
WALLEX 50	0.7% C 25% Cr 11% Ni 1% Fe 10% W 3% B Co alloy: Powder; Wall Colmonoy; melting point 1110 °C; similar to Wallex 1 **DPN: 550**
WALLEX F	1.7% C 26% Cr 23% Ni 12% W Co alloy: Welding electrode; Wall Colmonoy
WF 11	0.1% C 1.5% Mn 20% Cr 10% Ni 15% W 3% Fe Co alloy: American proprietary alloy; consult SAE yearbook.
WF 31	0.15% C 20% Cr 10% Ni 2.6% Mo 10% W 1% Ti Co alloy: American proprietary alloy listed in SAE yearbook
WI 52	0.4% C 20% Cr 1% Ni 11% W 2% Nb + Ta Co alloy: Wai-Met Alloy Co.
WIPTAM	0.1% C 28.3% Cr 24.4% Ni Co alloy: For dental use; details from Cobalt Information Centre **UTS: 620 Elon: 52% Proof: 830**
X 40	0.5% C 25% Cr 10% Ni 7.5% W 1.5% Fe Co alloy: American proprietary alloy listed in SAE yearbook
X 40	0.45% C 25% Cr 10% Ni 7% W Co alloy: Casting; Jessop Saville as G 65 **UTS: 730 Elon: 17% Proof: 400**
X 50	0.75% C 22% Cr 20% Ni 12% W 2.5% Fe Co alloy: Casting: listed by Cobalt Information Centre
X 63	0.4% C 23% Cr 10% Ni 6% Mo 1% Fe Co alloy: Casting; American proprietary alloy listed in SAE yearbook
XSH	20.0% Cr 10.0% Ni 15.0% W Co alloy: Origin unknown

Note. The following abbreviations and units are used in the tables:

DPN Hardness, diamond pyramid number
UTS Ultimate tensile strength, N/mm^2
Elon Elongation, %
Proof 0.1% proof strength, N/mm^2

1 N/mm^2=0.1 hbar=0.102 kgf/mm^2=0.06475 tonf/in.^{2}=145.04 lbf/in.^{2}=1 MPa
See Appendix II for other abbreviations and conversion tables.

14. Copper Cu

Physical properties

Atomic number		27
Atomic weight		63.54
Crystal structure		Face-centred cubic
Colour		Reddish orange
Specific gravity		8.96
Density		8960 kg/m^3
Melting point		1083 °C
Boiling point		2582 °C
Specific heat		0.385 J/g °C
Thermal conductivity		385 W/m °C
Coefficient of linear expansion (20–100 °C)		16.4×10^{-6}/ °C
Latent heat of fusion		204.8 J/g
Latent heat of vaporization		5234 J/g
Thermal neutron absorption cross-section		3.8 barns/atom
Electrical conductivity		100% IACS (copper 100%)
Specific resistance		17 microhm mm
Temperature coefficient of electrical resistance		0.0039/ °C
Electrochemical equivalent	monovalent	2.380 g/A/h
	divalent	1.185 g/A/h
Electrode potential	monovalent	–0.52 V
	divalent	–0.34 V
Magnetic susceptibility		-0.080×10^{-6}
Young's modulus of elasticity		110×10^9 N/m^2
Tensile strength	annealed	210 N/mm^2
Hardness	annealed	50 DPN
	cold worked	100 DPN

14.1 General notes on copper

Copper and its alloys have been in the service of mankind for longer than any other metal, and until the second half of this century it had the largest use of any metal after iron. Aluminium has now taken second place and copper is third in the table for usefulness.

Articles made from bronze – a tin copper alloy – have been found which were made by man in his very early days as a creative animal. The history of copper illustrates the technical development and ability of men through the ages. These attributes continue today, with highly complex copper alloys soaring into space.

Relatively large amounts of native copper metal have been found widely dispersed throughout the world. The largest deposit is probably that on the shores of Lake Superior, USA. These deposits probably account for the early use of copper, but the ease with which it can be reduced from its ore must also be taken into account.

Copper is used as high purity metal and has a range of alloys all of which are described and listed in the following sections.

Copper is also used as an alloying element, the largest proportion being with nickel where it forms a series of heat and corrosion resistant alloys.

It is also commonly alloyed with zinc to form the brasses.

In small quantities of up to 5% it forms intermetallic compounds with aluminium, giving the well-known age hardening series of aluminium alloys.

Many tin and lead based bearing metals have up to 10% additions. These strengthen the matrix and also in some cases help in the formation of intermetallic compounds.

In quantities up to 0.5%, copper is sometimes added to steel where it gives a degree of atmospheric corrosion resistance and some increase in yield. It is also a constituent in the new maraging steels.

Between 1 and 2% copper added to some types of cast iron ensures uniformity of properties in individual castings and can enhance the mechanical properties.

Copper has had many years in the service of man but recently has met with competition from the younger metals. There is no danger whatever of copper being completely supplanted by these materials as the versatility of copper ensures that new uses are found to replace those more economically carried out by aluminium and stainless steels.

14.2 Notes on specifications and trade names

An explanation of the different methods of classification and designation used with copper-based materials may be of assistance for understanding and rapidly finding the different specifications and symbols.

British Standards Institution (BS)

This body issues specifications covering individual alloys, and until recently their policy was to have an individual specification for each material, and often for each form of the material.

The present policy is to have a master specification number covering the form, with alloy designations covering the chemical composition. These on the whole replace the individual specification. Copper-based materials tend to be used in traditional industries, however, where the old specifications die hard.

The current specifications are:

BS 1400 Copper and alloy casting
BS 2870 Copper and alloy sheet and strip
BS 2871 Copper and alloy tube
BS 2872 Copper and alloy forgings and forging stock
BS 2873 Copper and alloy wire
BS 2874 Copper and alloy bar and section
BS 2875 Copper and alloy plate

These are followed by the alloy designation which always has a prefix letter or letters denoting the main alloying constituent.

Thus CA designates aluminium–copper, CB beryllium–copper, CN nickel–copper, CZ zinc–copper (brass), etc. A three figure code then identifies the actual copper or alloy, for example CZ101 is 10% zinc–copper, CB101 1.8% beryllium–copper.

Suffix letters and figures can be used to indicate the heat treatment condition, or degree of cold work carried out. These are as follows:

H¼	Cold worked to approximately one-quarter the possible amount, or fully cold worked and then partially annealed to remove about three-quarters of the cold work.
H½	As H¼ but cold worked to half possible.
H¾	As H¼ but cold worked to three-quarters possible.
M	As cast – as rolled – as extruded
O	Annealed
W	Solution treated
P	Precipitation treated – aged.

Examples of these specifications would be:

BS 2874 CZ101	10% zinc–copper alloy in bar form
BS 2870 CB101	1.8% beryllium–copper alloy in sheet form

Full specifications covering the mechanical and physical properties are not available for the complete range of alloy designations of all the forms. In the following pages the nominal analysis has been supplied for the full range for identification purposes, but only where the British Standards Institution supply mechanical or other properties are these quoted, and only the latter specifications are available. The alloy designations are often quoted without reference to the BS number and have been indexed under both the full specification and the alloy designation, thus BS 2874 CZ101 and CZ101 appear as separate entries.

In addition there are a limited number of aircraft specifications with the prefix BS B or T followed by a number. These are individual specifications covering one analysis and generally in one form.

American Society for Testing and Materials (ASTM)

This body issues several volumes of specifications at intervals. All the non-ferrous alloys are prefixed with the letter B followed by a number for the actual alloy. The year of issue follows, with the letter T for tentative issues of a specification. Typical examples would be:

ASTM B36/4	Copper alloy sheet containing 20% zinc
ASTM B291	Copper alloy sheet containing 1% manganese and 29% zinc

This book does not show the year of issue.

Although some of these specifications are subdivided to group together similar alloys, there is no real classification system at present with ASTM, and no attempt to code the form or condition of the material.

This body devised a designation code for the various forms of pure copper and these are listed in ASTM B224. In general the code is the initial letter of the description, thus OFC is oxygen-free copper. This method is now finding general use and is used for some proprietary alloys in this country.

Society of Automotive Engineers (SAE)

This body issues a handbook at intervals included in which are their own alloys with the prefix SAE. Previously these had a two or three digit number prefixed by SAE, sometimes with a letter suffix indicating the issue number. The present system uses an alloy designation agreed by the American Copper Association, with the prefix CA followed by three figures. This alloy designation is used with and without the SAE prefix and without any prefix; thus all forms are listed.

This society also issues a series of aerospace material specifications with the prefix AMS followed by four figures for the actual material condition and form. The suffix letter indicates the issue number, and is not shown in this book.

Unified Numbering System (UNS)

This system is described in ASTM E527.

It covers only metals and alloys which have a commercial standing.

There are 18 series of numbers. These are listed at the start of Section 44 in this book.

There is a single letter prefix followed by five digits.

Any UNS number for copper or its alloys will be headed C from C00001 to C99999. These will be found listed in the index and will refer the reader to the commercial alloy involved. The reader should then look for the alloy in the index and will be guided to the section in which it is listed.

Trade names and symbols

These generally incorporate some code or symbol indicating the supplier followed by letters or figures to identify the alloy. Sometimes the national designations are incorporated in the trade names.

Whenever possible all forms of the names and specifications have been included in the following lists, but it is possible that some have been omitted and it is suggested that if the name is not found at the first attempt then attempts are made to add or subtract the maker's name.

14A Copper – commercially pure – wrought and cast

Specific gravity		8.9
Density		8900 kg/m^3
Solidus/liquidus		1083 °C
Thermal conductivity		314–402 W/m °C
Coefficient of linear expansion		17.3×10^{-6}/ °C
Electrical conductivity	annealed	80–101% IACS (copper 100%)
Specific resistance	annealed	21.5–17.241 microhm mm
Young's modulus of elasticity		124×10^9 N/m^2
Impact	annealed	60 J
	cold worked	30 J
Fatigue strength	annealed	± 60 N/mm^2
	cadmium copper	± 150 N/mm^2

Hot strength

Temperature °C	Tensile strength N/mm^2	Elongation %
100	205	50
200	172	48
300	150–151.5	46
400	108–130	42–48

The above are applicable to the specifications listed, but may not be exactly applicable to any one material.

General metallurgical characteristics

Depending on its method of manufacture and in some instances the source of the ore, there are several varieties of commercially pure copper. Briefly these are as follows.

Cathode copper. This is the product of electrolytic refining and is the raw material for many of the other forms listed below. It can be remelted and cast, but apart from this finds little use. It has relatively poor conductivity owing to the presence of minute quantities of impurities.

Tough pitch copper. This is copper with a controlled amount of oxygen present which improves the workability and conductivity of the metal. There are various forms of tough pitch copper.

Electrolytic tough pitch copper. This is cathode copper remelted and allowed to pick up oxygen in the process. The product is used for all forms of electrical equipment, and also where good thermal conductivity is important.

With oxygen controlled to 0.04% max the annealed wrought product has a conductivity of 100% IACS.

Fire refined tough pitch copper. This is copper refined in a furnace and not electrolytically. As this is generally more difficult and does not recover any precious metals which may be present in the raw material, the type of ore and local conditions largely control the use of the method.

The final product is comparable with electrolytic tough pitch copper with a conductivity in the wrought annealed condition of 100% IACS or better.

Ordinary tough pitch copper. The oxygen content and other impurities are not as carefully controlled as with the above two types. The electrical and thermal conductivities are lower – thus it is not used for electrical purposes, but finds many general engineering uses. This copper is also cast into ingots and used as the primary metal for alloys.

Tough pitch arsenical copper. This has approximately 0.4% arsenic in solid solution. The electrical and thermal conductivities are reduced, but the strength and corrosion resistance are improved, and these improvements are maintained up to approximately 300 °C.

Oxygen-free copper. There are various forms of copper that are either produced oxygen free or are subsequently de-oxidized, as follows.

Oxygen-free high conductivity copper. This is difficult and expensive to produce. In America and Finland it is made by remelting and pouring cathode copper under an inert atmosphere resulting in high conductivity copper which can be brazed and welded. This material will accept a higher degree of cold work than any other form of copper.

Phosphorus de-oxidized copper. Phosphorus up to 0.05% but not generally above 0.04% is used to de-oxidize copper. This eases casting, working, brazing and welding problems, but reduces the thermal and electrical conductivity, which falls to about 70% IACS. Higher phosphorus contents are used in brazing filler rods.

De-oxidized arsenical copper. This is phosphorus de-oxidized copper with small quantities of arsenic added which has little effect on the other properties but increases the corrosion resistance and hot strength.

Pure copper will always be one of the above when found in industry and commerce. Also classified as copper are several alloys where the alloying element is less than 1%.

These elements are as follows.

Silver. This has no effect on the conductivity but does raise the annealing temperature required to remove cold work. Thus cold worked copper with approximately 0.1% silver has to be heated to above 300 °C before it softens, whereas ordinary copper tends to soften at 200 °C. This silver copper is used where cold worked material is to be soldered. Silver also improves the creep strength of copper.

Cadmium. This has a similar effect to silver and in addition toughens the material and improves the fatigue strength. Approximately 0.8% cadmium reduces the conductivity in the annealed state to 95% IACS. This material is used for overhead contact wires for traction and for long span conductors.

Tellurium and selenium. These also increase the softening temperature, in a similar manner to silver and cadmium. Their main use in copper, however, is to improve machinability. A good surface finish and dimensional stability is obtained with annealed tellurium or selenium copper which has a conductivity above 90% IACS.

Sulphur. This has a similar effect to tellurium and is now preferred by many users.

All the above coppers are capable of accepting a considerable amount of cold work, with the de-oxidized varieties superior in this context to the tough pitch varieties.

Commercially pure copper cannot be heat treated except to remove cold work, which requires temperatures of between 300 and 400 °C before all evidence of cold work has been removed.

No problems occur where copper has to be joined by soldering or brazing but the type of commercially pure copper will determine whether or not it can be welded. High conductivity coppers on the whole cannot be welded under normal circumstances as they have a very high affinity for oxygen.

The use of gas shielded welding is advisable when welding is essential but advice should be taken before attempting to weld any copper.

Copper has a high corrosion resistance under most circumstances. In normal atmospheres it forms a green patina which protects it from further corrosion. This is a complex compound which forms on the surface and is quite attractive. It is now common practice if the copper colour is required that the copper is cleaned chemically and then treated with some clear lacquer. Again advice is advised.

These materials are all difficult to machine in the soft or less than fully hard condition. Some materials have additions such as sulphur or tellurium which improve the machinability but on the whole these are soft materials which tend to tear rather than machine.

Copper finds many uses because it is one of the two coloured metals, the other being gold. With the advent of anodizing of aluminium, however, where it is possible to produce colours of a very attractive nature similar to, or identical to, gold or copper, the use of copper has become less in this area.

To take each of the types of copper in turn, cathode copper is a primary metal used to produce the copper alloys.

Tough pitch copper is used for electrical purposes, conductor wires, etc. It is also used for roofing and similar areas where welding is not required.

Oxygen-free high conductivity copper is more expensive but is one of the few coppers which can be welded without the use of gas shielding.

De-oxidized copper is used for chemical purposes such as water pipes and for industrial and decorative purposes. It is used in food containers and for architectural purposes where appearance is important.

Copper is now being largely replaced as an electrical conductor because of its cost except where joints are essential. Thus, where electricity is conducted at high voltages over considerable distances such as the grid system and in industry where conductors can be assembled by professionals, then aluminium has replaced copper almost completely.

Under normal industrial conditions, however, and in domestic circumstances copper remains the principal conductor. The reason for this is that joints can be easily made whereas with aluminium very specialized techniques are required in order to prevent high resistance and thus loss of conductivity locally.

The use of copper in the food industry for cookers, containers, etc. has largely been replaced by austenitic stainless steel.

Symbol	Nominal analysis, supplier, condition and remarks.
1C	High conductivity Cu: Langley Alloys **UTS: 180 Elon: 40% Proof: 30**
2.0050	99.9% Cu: Electrolytic; German Standard
2.0060	0.03% Ag 99.95% Cu: German Standard
2.0070	99.9% Cu: German Standard
2.0080	99.9% Cu: German Standard
2.0090	99.9% Cu: German Standard
2.0100	99.75% Cu: German Standard
2.0110	99.8% Cu: German Standard
2.0120	99.5% Cu: German Standard
2.0150	0.4% As 99.25% Cu: German Standard
2.0170	0.4% As 99.0% Cu: German Standard
2.0181	99.0% Cu: German Standard
2.1006	0.8% Sn Cu: German Standard
2.1211	1.0% Ag Cu: German Standard
101	99.96% Cu: Oxygen-free; designation agreed by CDAA; not used
104	99.95% Cu: Ag always present; oxygen-free; designation agreed by CDAA; not used
105	99.95% Cu: Ag always present; oxygen-free; designation agreed by CDAA; not used
107	99.95% Cu: Ag always present; oxygen-free; designation agreed by CDAA; not used
111	0.01% P 0.4% S Cu bar: McKechnie; 95% IACS **DPN: 90 UTS: 320 Elon: 25%**
121	0.01% P 99.9% Cu: With Ag; designation agreed by CDAA; not used
123	0.02% P 99.9% Cu: With Ag; designation agreed by CDAA; not used
125	99.88% Cu: Fire refined; tough pitch; designation agreed by CDAA; not used
127	99.88% Cu: Fire refined; tough pitch; with Ag; designation agreed by CDAA; not used
128	99.88% Cu: Fire refined; tough pitch; with Ag; designation agreed by CDAA; not used
130	99.88% Cu: Fire refined; tough pitch; with Ag; designation agreed by CDAA; not used

Note. The following abbreviations and units are used in the tables:

DPN Hardness, diamond pyramid number
UTS Ultimate tensile strength, N/mm^2
Elon Elongation, %
Proof 0.1% proof strength, N/mm^2

1 N/mm^2=0.1 hbar=0.102 kgf/mm^2=0.06475 $tonf/in.^2$=145.04 $lbf/in.^2$=1 MPa
See Appendix II for other abbreviations and conversion tables.

Symbol	Nominal analysis, supplier, condition and remarks.
141	0.4% As 99.4% Cu: Arsenical tough pitch; designation agreed by CDAA; not used
142	0.03% P 0.4% As 99.4% Cu: Designation agreed by CDAA; not used
164	0.3% Sn 0.7% Cd Cu alloy: Wrought; designation agreed by CDAA; not used
165	0.6% Sn 0.8% Cd Cu alloy: Wrought; designation agreed by CDAA; not used
189	0.7% Sn 0.3% Si Cu alloy: Wrought; designation agreed by CDAA; not used
ABS/B42	0.04% P (max) 99.9% Cu: Pipe; American Bureau of Shipping
AL 0 1	Low Zn Cu: Thomas Bolton; for singe plate
AMPCOLAY 90	Cu: 92% IACS; Ampco
AMPCOLAY 900	99.9% Cu High conductivity: Ampco
AMPCOLAY 901	99.9% Cu with Ag: 93% IACS; Ampco
AMPCOLAY 910	Cu: 90% IACS; Ampco
AMS 4500 D	Cu: Sheet and strip; annealed
AMS 4501	99.95% Cu: Oxygen-free
AMS 4602	99.95% Cu: Oxygen-free
AMS 4700	99.99% Cu: Oxygen-free; electrolytic
AMS 4701 B	Cu: Wire; annealed
ANACOS	99.90% Cu: Wire; may be tinned; Fred Smith; hard drawn or annealed
ANACOS CADMIUM	Trace P 1% Cd Cu: Wire; may be tinned; Fred Smith; hard drawn
ANACOS SILVER	Trace P 0.1% Ag Cu: Wire; Fred Smith; hard drawn
ASTM B1	Cu: Wire; hard drawn; mechanical properties vary with diameter
ASTM B2	Cu: Wire; medium hard drawn; mechanical properties vary with diameter
ASTM B3	Cu: Wire; annealed; 100% IACS **Elon: 25%**
ASTM B4	99.9% Cu: Wire, bar, ingots, etc.; Lake Copper; covers copper of low resistivity and high resistivity
ASTM B5	99.5% Cu: Wire, bar, ingots, etc.; electrolytic refined high conductivity copper
ASTM B11/110	99.88% Cu: Plate for fireboxes
ASTM B11/122	0.02% P 99.90% Cu: Plate for fireboxes
ASTM B11/125	99.88% Cu: Plate for fireboxes
ASTM B11/141	0.35% As 99.4% Cu: Plate for fireboxes
ASTM B11/142	0.03% P 0.3% As Cu: Plate for fireboxes
ASTM B11 ATP	0.2% As 99.4% Cu: Plate; tough pitch arsenical copper; hot rolled **UTS: 210 Elon: 35%**
ASTM B11 DHP	0.02% P 99.9% Cu: Phosphorized; hot rolled **UTS: 210 Elon: 30%**
ASTM B11 DPA	0.02% P 0.3% As 99.4% Cu: Phosphorized arsenical copper; hot rolled **UTS: 210 Elon: 35%**
ASTM B11 ETP	99.88% Cu: Plate; tough pitch copper **UTS: 200 Elon: 30%**
ASTM B11 FRTP	99.88% Cu: Plate; tough pitch copper **UTS: 200 Elon: 30%**
ASTM B12	Cu: Rods for locomotive staybolts
ASTM B12/102	99.9% Cu: Rods for staybolts; oxygen-free; previously OF
ASTM B12/110	99.9% Cu: Rods for staybolts; electrolytic tough pitch; previously ETP

Note. The following abbreviations and units are used in the tables:

DPN	Hardness, diamond pyramid number
UTS	Ultimate tensile strength, N/mm^2
Elon	Elongation, %
Proof	0.1% proof strength, N/mm^2

$1\ N/mm^2 = 0.1\ hbar = 0.102\ kgf/mm^2 = 0.06475\ tonf/in.^2 = 145.04\ lbf/in.^2 = 1\ MPa$

See Appendix II for other abbreviations and conversion tables.

Symbol	Nominal analysis, supplier, condition and remarks.
ASTM B12/120	99.9% Cu: Rods for staybolts; phosphorus de-oxidized; previously DLP
ASTM B12/122	99.9% Cu: Rods for staybolts; phosphorus de-oxidized; previously DHP
ASTM B12/125	99.9% Cu: Rods for staybolts; fire refined tough pitch; previously FRTP
ASTM B12/141	0.3% As 99.88% Cu: Rods for staybolts; previously ATP
ASTM B12/142	0.3% As 99.88% Cu: Rods for staybolts; previously DPA
ASTM B42	0.04% P 99.9% Cu: Seamless pipe; as drawn **UTS: 260 Proof: 240**
ASTM B42/102	99.9% Cu: Seamless pipe; previously OF
ASTM B42/120	99.9% Cu: Seamless pipe; phosphorus de-oxidized; previously DLP
ASTM B42/122	99.9% Cu: Seamless pipe; phosphorus de-oxidized; previously DHP
ASTM B47	Cu: Wire for overhead trolleys **UTS: 310 Elon: 4%**
ASTM B48	Cu: Rod; rectangular section for conductors **Elon: 30%**
ASTM B49	Cu: Rod; hot rolled for electrical purposes; conductivity 100% IACS
ASTM B52A	14.0% P (min) Cu: Ingot
ASTM B52B	10% P (min) Cu: Ingot
ASTM B68 DHP	0.02% P 99.9% Cu: Seamless tube; annealed **UTS: 200 Elon: 40%**
ASTM B68 DLP	0.01% P 99.9% Cu: Seamless tube; annealed **UTS: 200 Elon: 40%**
ASTM B68 OF	99.92% Cu: Seamless tube; annealed **UTS: 200 Elon: 40%**
ASTM B72 A	99.75% Cu: For castings; high purity copper; fire refined
ASTM B72 B	0.3% Pb 0.1% As Cu: For casting; fire refined copper
ASTM B75/102	99.95% Cu: Seamless tube; oxygen-free; previously OF
ASTM B75/120	0.01% P 99.90% Cu: Seamless tube; previously DLP
ASTM B75/122	0.02% P 99.9% Cu: Seamless tube; previously DHP
ASTM B75/142	0.02% P 99.4% Cu: Seamless tube; previously DPA
ASTM B75 DHP	0.02% P 99.9% Cu: Seamless tube; as drawn **UTS: 240**
ASTM B75 DHP	0.02% P 99.9% Cu: Seamless tube; cold drawn **UTS: 320**
ASTM B75 DLP	0.01% P 99.9% Cu: Seamless tube; as drawn **UTS: 240**
ASTM B75 DLP	0.01% P 99.9% Cu: Seamless tube; cold drawn **UTS: 330**
ASTM B75 DPA	0.02% P 0.3% As 99.4% Cu: Seamless tube; as drawn **UTS: 240**
ASTM B75 DPA	0.02% P 0.3% As 99.4% Cu: Seamless tube; cold drawn **UTS: 330**
ASTM B75 OF	99.2% Cu: Oxygen-free seamless tube; as drawn **UTS: 240**
ASTM B75 OF	99.2% Cu: Oxygen-free seamless tube; cold drawn **UTS: 330**
ASTM B88	0.04% P 99.9% Cu: Seamless tube; annealed **UTS: 200**
ASTM B111	0.4% As 99.4% Cu: Tube
ASTM B111	0.04% P (max) 99.9% Cu: Tube
ASTM B111/102	99.5% Cu: Seamless tube; oxygen-free; previously OF
ASTM B111/120	0.01% P 99.90% Cu: Seamless tube; previously DLP
ASTM B111/122	99.9% Cu: Seamless tube; phosphorus de-oxidized; previously DHP
ASTM B111/142	0.2% As 99.4% Cu: Seamless tube; previously DPA
ASTM B115	99.9% Cu: Electrolyic cathode copper
ASTM B124/12	99.9% Cu: Forging and bar; mechanical properties not quoted
ASTM B152/102	99.95% Cu: Sheet and bar; oxygen-free
ASTM B152/104	99.95% Cu: Sheet and bar; oxygen-free; silver bearing
ASTM B152/105	99.95% Cu: Sheet and bar; oxygen-free; silver bearing
ASTM B152/107	99.95% Cu: Sheet and bar; oxygen-free

Symbol	Nominal analysis, supplier, condition and remarks.
ASTM B152/122	0.03% P 99.90% Cu: Sheet and bar; phosphorized; high residual phosphorus
ASTM B152/123	99.90% Cu: Sheet and bar; phosphorized silver bearing
ASTM B152 ATP	0.25% As 99.4% Cu: Sheet, strip, plate and bar; annealed; arsenical tough pitch **UTS: 270**
ASTM B152 DHP	0.02% P 99.9% Cu: Strip, sheet, plate and bar; annealed; phosphorized copper **UTS: 270**
ASTM B152 DPS	0.04% P (max) 99.9% Cu: Sheet, strip, plate and bar; silver bearing; phosphorized **UTS: 270**
ASTM B152 ETP	99.9% Cu: Sheet, strip, plate and bar; annealed; electrolytic; tough pitch **UTS: 270**
ASTM B152 FRTP	99.88% Cu: Sheet, strip, plate and bar; annealed; fire refined; tough pitch **UTS: 270**
ASTM B152 OF	99.92% Cu: Sheet, strip, plate and bar; annealed; oxygen-free **UTS: 270**
ASTM B152 OFS	99.9% Cu: Sheet, strip, plate and bar; annealed; silver bearing; oxygen-free **UTS: 270**
ASTM B152 STP	99.9% Cu: Sheet, strip, plate and bar; annealed; silver bearing copper; tough pitch **UTS: 270**
ASTM B187	Cu: Bus bar rods and shapes for electrical use; conductivity IACS 97.5% (min)
ASTM B188	Cu: Pipe and tube; seamless for electrical use; conductivity IACS 97.5% (min)
ASTM B224 ATP	Arsenical tough pitch Cu: Classification of Cu
ASTM B224 CAST	Casting Cu: Classification of Cu
ASTM B224 CATH	Electrolytic cathode Cu: Classification of Cu
ASTM B224 DHP	Phosphorized Cu: High residual phosphorus; classification of Cu
ASTM B224 DLP	Phosphorized Cu: Low residual phosphorus; classification of Cu
ASTM B224 DPA	Phosphorized arsenical Cu: Classification of Cu
ASTM B224 DPS	Phosphorized silver bearing Cu: Classification of Cu
ASTM B224 DPTE	Phosphorized tellurium bearing Cu: Classification of Cu
ASTM B224 ETP	Electrolytic tough pitch Cu: Classification of Cu
ASTM B224 FRHC	Fire refined high conductivity Cu: Classification of Cu
ASTM B224 FRTP	Fire refined tough pitch Cu: Classification of Cu
ASTM B224 OF	Oxygen-free Cu without residual de-oxidants: Classification of Cu
ASTM B224 OFP	Oxygen-free phosphorus bearing Cu: Classification of Cu
ASTM B224 OFPTE	Oxygen-free phosphorus and tellurium bearing Cu: Classification of Cu
ASTM B224 OFS	Oxygen-free silver bearing Cu: Classification of Cu
ASTM B224 OFTE	Oxygen-free tellurium bearing Cu: Classification of Cu
ASTM B224 SATP	Silver bearing arsenical tough pitch Cu: Classification of Cu
ASTM B224 STP	Silver bearing tough pitch Cu: Classification of Cu
ASTM B229	Standard copper and copper-clad steel wire composite conductors
ASTM B246	Cu: Wire; hard and medium; hard drawn; tin coated **Elon: 1%**
ASTM B260 B Cu	0.07% P 99.9% Cu: Braze filler; melting point 1080 °C
ASTM B260 B Cu 1	99.90% (min) Cu: Brazing filler metal; braze range 1093–1149 °C
ASTM B260 B Cu 1a	99.0% (min) Cu: Brazing filler metal; braze range 1093–1149 °C
ASTM B280	99.92% Cu: Seamless tube
ASTM B280 DHP	0.02% P 99.2% Cu: Seamless tube
ASTM B280 DLP	0.01% P 99.9% Cu: Seamless tube
ASTM B301	0.5% Te Cu alloy: Bar; ½-hard **UTS: 260 Elon: 12% Proof: 200**
ASTM B301	0.5% Te Cu alloy: Bar; hard **UTS: 300 Elon: 10% Proof: 260**
ASTM B359/102	99.95% Cu: Firmed tube
ASTM B359/120	99.9% Cu: Firmed tube
ASTM B359/122	99.9% Cu: Firmed tube
ASTM B359/142	99.4% Cu: Firmed tube
ASTM B360	0.02% P 99.9% Cu: Capillary tube **UTS: 450**
ASTM B370	99.88% (min) Cu: Sheet and strip for buildings
ASTM B379	Cu billets: P de-oxidized; classified by P content
ASTM B442	99.90% Cu: Chemically refined; conductivity 100% IACS
ASTM B447	99.4% Cu (min): Welded tube; graded by Cu content
ASTM B451	99.50% Cu (min): Foil strip, etc. for printed circuits
ASTM B465	0.01% Zn 0.03% P 2.4% Fe Cu alloy: Plate and bar; conductivity 62% IACS
ASTM B469	0.02% P 1.0% Fe 98.7% Cu (min): Tube for pressure application **UTS: 280 Elon: 18% Proof: 180**
ASTM B623	Tough pitch fire refined copper. See designation
ASTM B640	Specification for welded Cu and Cu alloy tubes: Designation by UNS
ASTM B641	Seamless and welded copper tube: See CDA designation
ASTM B707	Copper pipe – seamless for water tubes: See UNS designation
ASTM B716	Copper tube – welded for water: See UNS designation
ASTM B743	Seamless copper tube: See UNS designation
ASTM B747 C15100	0.1% Zr Cu: Sheet and strip
B 37	Copper casting: High conductivity; Anti-attrition Co. Ltd
BOLTOMET 103	Oxygen-free high conductivity copper: Thomas Bolton
BOLTOMET 105	Electrolytically refined copper: Thomas Bolton for BS 1036
BOLTOMET 107	Fire refined copper: Thomas Bolton for BS 1037
BOLTOMET 108	Anode copper: Thomas Bolton for BS 1038; withdrawn
BOLTOMET 112	0.1% Ag oxygen-free high conductivity copper: Thomas Bolton
BOLTOMET 113	0.1% Ag Cu: For electrical purposes; Thomas Bolton for BS 1434
BOLTOMET 115	0.15% Ag Cu: For electrical conductors; hollow section; Thomas Bolton
BOLTOMET 117	Calcium boride de-oxidized copper: Thomas Bolton; suitable for brazing; withdrawn
BOLTOMET 121	Phosphorus de-oxidized electrolytically refined copper: Thomas Bolton for BS 1172; low bismuth
BOLTOMET 123	Phosphorus de-oxidized fire refined copper: Thomas Bolton for BS 1172
BOLTOMET 152	Phosphorus de-oxidized arsenical copper: Thomas Bolton for BS 1174
BOLTOMET 154	Low arsenical copper: For printing rollers; Thomas Bolton
BOLTOMET 156	Arsenical Cu: Thomas Bolton for BS 1173; withdrawn
BOLTOMET 160	High phosphorus de-oxidized non-arsenical copper: Wire; Thomas Bolton
BOLTOMET 162	0.03% P de-oxidized copper: Thomas Bolton
BOLTOMET 170	Low Zn Cu: Thomas Bolton for singe plates; withdrawn
BOLTOMET 175	0.05% P de-oxidized copper: Thomas Bolton
BOLTOMET 206	0.6% Cd Cu: Thomas Bolton for BS 672
BOLTOMET 208	0.8% Cd Cu: Thomas Bolton for BS 23
BOLTOMET 210	1% Cd Cu: For welding electrodes; Thomas Bolton
BOLTOMET 302	0.4% Sn Cu: Wire; Thomas Bolton for armature wires
BOLTOMET 304	1% Sn Cu: Wire; Thomas Bolton for armature wires
BOLTOMET 366	Mn Si de-oxidized Cu: Thomas Bolton for BS 2901 C7
BOLTOMET 820	Zr Cu alloy: For commutator bar; Thomas Bolton; withdrawn
BOLTOMET 909	Te de-oxidized high conductivity Cu: Thomas Bolton; withdrawn
BOLTOMET 952	Te tough pitch high conductivity Cu: Thomas Bolton; free machining
BOROFIL	Boron Cu alloy: Welding rod; IMI melting point 1083 °C

Symbol	Nominal analysis, supplier, condition and remarks.
BS 23	Copper of high conductivity for trolley wires can be 99.9% Cu or 0.7% Cd Cu: As BS 2873 C108
BS 24/5	99.2% Cu: For railway fireboxes, tubes, etc.; BS 1173 or BS 1174
BS 125	Hard drawn copper 0.7% Cd Cu: Wire for overhead transmission
BS 198	Electrolytic copper: Wire, bar, cakes, slabs, etc.; replaced by BS 1035-40
BS 198	Electrolytic copper: Ingots and ingot bars; replaced by BS 1035-40
BS 200	Tough copper: Cakes and billets; replaced by BS 1035-40
BS 201	Fire copper: Cakes; replaced by BS 1035-40
BS 202	Electrolytic cathode copper: Replaced by BS 1035-40
BS 203	'Best Select' copper: Replaced by BS 1035-40
BS 378 C106	0.05% P Cu: Phosphorus de-oxidized copper tube; as drawn for condenser tubes, etc.
BS 378 C107	0.05% P 0.4% As Cu: Phosphorus de-oxidized arsenical copper tube; as drawn for condenser tubes, etc. **DPN: 100**
BS 518	Medium hard copper: Strip, bar and rod for electrical purposes; replaced by BS 1432-3
BS 659	P or P and As de-oxidized Cu: Tube; light drawn
BS 672	Cd Cu alloy: For electrical conductor wires; hard drawn **UTS: 530 Elon: 2%**
BS 899 C101	Electrolytic tough pitch high conductivity copper: Sheet and strip; annealed; electrical conductivity 99.25% IACS **UTS: 210 Elon: 35%**
BS 899 C101	Electrolytic tough pitch high conductivity copper: Sheet and strip; hard; electrical conductivity 97% IACS **UTS: 300**
BS 899 C102	Fire refined tough pitch high conductivity copper: Sheet and strip; annealed; electrical conductivity 99.25% IACS **UTS: 210 Elon: 35%**
BS 899 C102	Fire refined tough pitch high conductivity copper: Sheet and strip; hard; electrical conductivity 97% IACS **UTS: 300**
BS 899 C103	Oxygen-free high conductivity copper: Sheet and strip; annealed; electrical conductivity 99.25% IACS **UTS: 210 Elon: 35%**
BS 899 C103	Oxygen-free high conductivity copper: Sheet and strip; hard; electrical conductivity 97% IACS **UTS: 300**
BS 899 C104	99.85% Cu: Non-arsenical tough pitch; sheet and strip; annealed **UTS: 210 Elon: 35%**
BS 899 C104	99.85% Cu: Non-arsenical tough pitch; sheet and strip; hard **UTS: 300**
BS 899 C105	Tough pitch arsenical copper: Annealed **UTS: 210 Elon: 35%**
BS 899 C105	Tough pitch arsenical copper: Hard **UTS: 300**
BS 899 C106	Non-arsenical copper: Phosphorus de-oxidized; sheet; annealed **UTS: 210 Elon: 35%**
BS 899 C106	Non-arsenical copper: Phosphorus de-oxidized; sheet; hard **UTS: 300**
BS 899 C107	Arsenical copper: Phosphorus de-oxidized; sheet and strip; annealed **UTS: 210 Elon: 35%**
BS 899 C107	Arsenical copper: Phosphorus de-oxidized; sheet and strip; hard **UTS: 300**
BS 1035	99.9% Cu: Produced as cathodes by electrolysis; Ag is counted as copper
BS 1036	99.9% Cu: Electrolytic refined; tough pitch high conductivity copper
BS 1037	99.9% Cu: Fire refined; tough pitch high conductivity copper
BS 1038	99.85% Cu: Tough pitch copper; electrical conductivity not specified
BS 1039	99.75% Cu: Tough pitch copper; electrical conductivity not specified
BS 1040	99.5% Cu: Tough pitch copper; electrical conductivity not specified
BS 1110	Cu: Sheet and strip; electrical; hard; replaced by BS 1432
BS 1172	99.85% Cu: Phosphorus de-oxidized non-arsenical; electrical conductivity not specified
BS 1173	0.5% As 99.2% Cu: Tough pitch arsenical copper
BS 1174	0.5% As 99.2% Cu: Phosphorus de-oxidized arsenical copper
BS 1400 HCC1C	99.9% Cu: Castings; high conductivity; electrical conductivity 95% IACS **UTS: 150 Elon: 25%**
BS 1400 HCC2C	Cu: Castings; analysis not specified; electrical conductivity 75% IACS **UTS: 150 Elon 25%**
BS 1401	99.9% Cu: Phosphorus or arsenical de-oxidized; tube; solid drawn tube; analysis to BS 1172 or BS 1174
BS 1432 C101	99.9% Cu: Electrolytic refined tough pitch copper; sheet; annealed; electrical conductivity 99.25% IACS **UTS: 210 Elon: 35%**
BS 1432 C101	99.9% Cu: Electrolytic refined tough pitch copper; sheet; $\frac{3}{4}$-hard; electrical conductivity 97% IACS **UTS: 240 Elon: 12%**
BS 1432 C101	99.9% Cu: Electrolytic refined tough pitch copper; Sheet; hard; electrical conductivity 97% IACS **UTS: 300**
BS 1432 C102	99.9% Cu: Fire refined tough pitch copper; sheet; annealed; electrical conductivity 99.25% IACS **UTS: 210 Elon: 35%**
BS 1432 C102	99.9% Cu: Fire refined tough pitch copper; sheet; $\frac{1}{2}$-hard; electrical conductivity 99.25% IACS **UTS: 240 Elon: 12%**
BS 1432 C102	99.9% Cu: Fire refined tough pitch copper; sheet; hard; electrical conductivity 97% IACS **UTS: 300**
BS 1432 C103	99.9% Cu: Oxygen-free copper; sheet; annealed; 99.25% IACS **UTS: 210 Elon: 35%**
BS 1432 C103	99.9% Cu: Oxygen-free copper; sheet; $\frac{1}{2}$-hard; 97% IACS **UTS: 240 Elon: 12%**
BS 1432 C103	99.9% Cu: Oxygen-free copper; sheet; hard; 97% IACS **UTS: 300**
BS 1433	99.9% Cu: Tough pitch copper; bar; annealed **UTS: 210 Elon: 50%**
BS 1433	99.9% Cu: Tough pitch copper; bar; medium hard **UTS: 240 Elon: 22%**
BS 1433	99.9% Cu: Tough pitch copper; bar; hard **UTS: 280 Elon: 9%**

Note. The following abbreviations and units are used in the tables:

DPN	Hardness, diamond pyramid number
UTS	Ultimate tensile strength, N/mm^2
Elon	Elongation, %
Proof	0.1% proof strength, N/mm^2

1 N/mm^2=0.1 hbar=0.102 kgf/mm^2=0.06475 $tonf/in.^2$=145.04 $lbf/in.^2$=1 MPa

See Appendix II for other abbreviations and conversion tables.

Symbol	Nominal analysis, supplier, condition and remarks.
BS 1434	99.9% Cu: Tough pitch copper; for electrical purposes; hard drawn; 97% IACS; for commutator bars **DPN: 82 UTS: 280**
BS 1453 C1	1% Ag Cu alloy: Weld filler rod; for welding copper
BS 1541 C106	0.06% P 99.85% Cu: Plate; annealed **UTS: 210 Elon: 35%**
BS 1845 CU1	99.9% Cu: For brazing; total impurities 0.03%; melting point 1085 °C
BS 1845 CU2	99.9% Cu: For brazing; total impurities 0.04%; melting point 1085 °C
BS 1845 CU3	99.95% Cu: For brazing; total impurities 0.03%; melting point 1085 °C
BS 1845 CU4	99.85% Cu: For brazing; total impurities 0.05%; melting point 1085 °C
BS 1845 CU5	99.5% Cu: For brazing; total impurities 0.1%; melting point 1085 °C
BS 1845 CU6	99.85% Cu: For brazing; melting point 1085 °C
BS 1861	99.95% Cu: Oxygen-free high conductivity copper
BS 1977 C101	99.9% Cu: Electrolytic tough pitch Cu; tube; annealed Cu to BS 1036; 99.25% IACS **UTS: 200 Elon: 40%**
BS 1977 C102	99.9% Cu: Fire refined tough pitch Cu; tube; annealed Cu to BS 1037; 99.25% IACS **UTS: 200 Elon: 40%**
BS 1977 C102	99.9% Cu: Oxygen-free Cu; tube; annealed Cu to BS 1861; 99.25% IACS **UTS: 200 Elon: 40%**
BS 2027	99.8% Cu: Plate for general purposes; annealed analysis to BS 1038, BS 1172–4 **UTS: 200 Elon: 35%**
BS 2755/2	Cd Cu alloy: For overhead wires
BS 2755 a	High conductivity Cu: For overhead wires
BS 2870 C101	99.9% Cu: Electrolyic tough pitch Cu; sheet; annealed **UTS: 200 Elon: 35%**
BS 2870 C101	99.9% Cu: Electrolyic tough pitch Cu; sheet; hard **UTS: 280**
BS 2870 C102	99.9% Cu: Fire refined tough pitch Cu; sheet; annealed **UTS: 200 Elon: 35%**
BS 2870 C102	99.9% Cu: Fire refined tough pitch Cu; sheet; hard **UTS: 280**
BS 2870 C103	99.95% Cu: Oxygen-free Cu; sheet; annealed **UTS: 200 Elon: 35%**
BS 2870 C103	99.95% Cu: Oxygen-free Cu; sheet; hard **UTS: 270**
BS 2870 C104	99.85% Cu: Tough pitch non-arsenical; sheet; annealed **UTS: 200 Elon: 35%**
BS 2870 C104	99.85% Cu: Tough pitch non-arsenical; sheet; ½-hard **UTS: 240 Elon: 10%**
BS 2870 C104	99.85% Cu: Tough pitch non-arsenical; sheet; hard **UTS: 280**
BS 2870 C105	0.4% As 99.2% Cu: Tough pitch arsenical; sheet; annealed **UTS: 210 Elon: 35%**
BS 2870 C105	0.4% As 99.2% Cu: Tough pitch arsenical; sheet; ½-hard **UTS: 240 Elon: 10%**
BS 2870 C105	0.4% As 99.2% Cu: Tough pitch arsenical; sheet; hard **UTS: 280**
BS 2870 C106	0.04% P 99.85% Cu: Phosphorus de-oxidized Cu; sheet; annealed **UTS: 200 Elon: 35%**
BS 2870 C106	0.04% P 99.85% Cu: Phosphorus de-oxidized Cu: sheet; hard **UTS: 280**
BS 2870 C107	0.1% P 0.4% As 99.2% Cu: Phosphorus de-oxidized arsenical Cu; sheet; annealed **UTS: 200 Elon: 35%**
BS 2870 C107	0.1% P 0.4% As 99.2% Cu: Phosphorus de-oxidized Cu; sheet; hard **UTS: 280**
BS 2870 C108	1% Cd Cu alloy: Sheet; specification not issued
BS 2870 C109	0.5% Te Cu alloy: Sheet; specification not issued
BS 2871 C101	99.9% Cu: Electrolytic refined tough pitch; tube; annealed **UTS: 210**
BS 2871 C102	99.9% Cu: Fire refined tough pitch; tube; annealed **UTS: 210**
BS 2871 C103	99.95% Cu: Oxygen-free copper; tube; annealed **UTS: 210**
BS 2871 C104	99.8% Cu: Non-arsenical tough pitch; tube; specification not issued
BS 2871 C105	0.4% As 99.2% Cu: Tough pitch copper; tube; specification not issued
BS 2871 C106	0.04% P 99.85% Cu: Phosphorus de-oxidized; tube; annealed **UTS: 220**
BS 2871 C107	0.05% P 0.4% As 99.2% Cu: Phosphorus arsenical; tube; annealed **UTS: 220**
BS 2871 C108	Cd Cu: Tube; specification not issued
BS 2871 C109	0.5% Te Cu alloy: Tube; specification not issued
BS 2872 C101	99.9% Cu: Forging electrolytic tough pitch Cu; specification not issued
BS 2872 C102	99.9% Cu: Fire refined tough pitch; forging; specification not issued
BS 2872 C103	99.95% Cu: Forging; oxygen-free; high conductivity; specification not issued
BS 2872 C104	99.85% Cu: Forging; tough pitch; specification not issued
BS 2872 C105	0.4% As 99.2% Cu: Forging; tough pitch; specification not issued
BS 2872 C106	0.04% P 99.85% Cu: Forging; specification not issued
BS 2872 C107	0.06% P 0.4% As 99.2% Cu: Forging; specification not issued
BS 2872 C108	1% Cd Cu alloy: Forging; specification not issued
BS 2872 C109	0.5% Te Cu alloy: Forging; specification not issued
BS 2873 C101	99.9% Cu: Electrolytic refined tough pitch wire; annealed; electrical conductivity 97% IACS **Elon: 15%**
BS 2873 C102	99.9% Cu: Fire refined tough pitch; wire; annealed; electrical conductivity 100% IACS **Elon: 25%**
BS 2873 C103	99.95% Cu: Oxygen-free high conductivity; wire; mechanical properties not quoted; annealed; electrical conductivity 100% IACS
BS 2873 C104	99.85% Cu: Tough pitch; wire; specification not issued
BS 2873 C105	99.2% Cu: Tough pitch; wire; specification not issued
BS 2873 C106	99.85% Cu: Phosphorus de-oxidized; wire; specification not issued
BS 2873 C107	99.2% Cu: Phosphorus de-oxidized; wire; specification not issued
BS 2873 C108	1% Cd Cu alloy: Wire; mechanical properties not quoted; hard; electrical conductivity 80% IACS
BS 2873 C109	0.5% Te Cu alloy: Wire; specification not issued
BS 2874 C101	99.9% Cu: Electrolytic tough pitch high conductivity; bar; annealed **UTS: 220 Elon: 40%**
BS 2874 C101	99.9% Cu: Electrolytic tough pitch high conductivity; bar; hard **UTS: 320 Elon: 9%**
BS 2874 C102	99.9% Cu: Fire refined tough pitch high conductivity; bar; annealed **UTS: 220 Elon: 40%**
BS 2874 C102	99.9% Cu: Fire refined tough pitch high conductivity; bar; hard **UTS: 320 Elon: 9%**
BS 2874 C103	99.95% Cu: Oxygen-free high conductivity; bar; annealed **UTS: 220 Elon: 40%**
BS 2874 C103	99.95% Cu: Oxygen-free high conductivity; bar; hard **UTS: 320 Elon: 9%**

Symbol	Nominal analysis, supplier, condition and remarks.
BS 2874 C104	99.85% Cu: Tough pitch non-arsenical; bar; specification not issued
BS 2874 C105	0.4% As 99.2% Cu: Tough pitch arsenical; bar; annealed **UTS: 210 Elon: 40%**
BS 2874 C106	0.04% P 99.85% Cu: Bar; annealed **UTS: 210 Elon: 40%**
BS 2874 C106	0.04% P 99.85% Cu: Bar; hard **UTS: 220 Elon: 15%**
BS 2874 C107	0.4% As 99.2% Cu: Bar; annealed **UTS: 210 Elon: 40%**
BS 2874 C108	1% Cd Cu alloy: Bar; specification not issued
BS 2874 C109	0.4% Te Cu alloy: Bar; annealed **UTS: 210 Elon: 35%**
BS 2875 C101	99.9% Cu: Electrolytic tough pitch; high conductivity; plate; annealed **UTS: 210 Elon: 35%**
BS 2875 C101	99.9% Cu: Electrolytic tough pitch; high conductivity; plate; hard **UTS: 250 Elon: 17%**
BS 2875 C102	99.9% Cu: Fire refined tough pitch; high conductivity; plate; annealed **UTS: 210 Elon: 35%**
BS 2875 C102	99.9% Cu: Fire refined tough pitch; high conductivity; plate; hard **UTS: 250 Elon: 17%**
BS 2875 C103	99.95% Cu: Oxygen-free high conductivity; plate; annealed **UTS: 210 Elon: 35%**
BS 2875 C103	99.95% Cu: Oxygen-free high conductivity; plate; hard **UTS: 250 Elon: 17%**
BS 2875 C104	99.85% Cu: Tough pitch non-arsenical; plate; annealed **UTS: 210 Elon: 35%**
BS 2875 C104	99.85% Cu: Tough pitch non-arsenical; plate; hard **UTS: 250 Elon: 17%**
BS 2875 C105	0.4% As 99.2% Cu: Tough pitch arsenical; plate; annealed **UTS: 210 Elon: 35%**
BS 2875 C105	0.4% As 99.2% Cu: Tough pitch arsenical; plate; hard **UTS: 250 Elon: 17%**
BS 2875 C106	0.04% P 99.85% Cu: Phosphorus de-oxidized non-arsenical; plate; annealed **UTS: 210 Elon: 35%**
BS 2875 C106	0.04% P 99.85% Cu: Phosphorus de-oxidized non-arsenical; plate; annealed **UTS: 250 Elon: 17%**
BS 2875 C107	0.04% P 0.4% As 99.2% Cu: Arsenical; plate; annealed **UTS: 210 Elon 35%**
BS 2875 C107	0.04% P 0.4% As 99.2% Cu: Arsenical; plate; hard **UTS: 270 Elon: 17%**
BS 2875 C108	1% Cd Cu alloy: Plate; electrical conductivity 80% IACS **UTS: 280 Elon: 17%**
BS 2875 C109	0.5% Te Cu: Plate; specification not issued
BS 2901 C7	0.2% Mn 0.3% Si 98.5% Cu: Rod for all welding types
BS 2901 C8	0.2% Al 0.2% Ti 99.4% Cu: Rod for all welding types
BS 4109 C101	High purity copper: Electrolytic tough pitch as BS 1036 C101

Note. The following abbreviations and units are used in the tables:

DPN	Hardness, diamond pyramid number
UTS	Ultimate tensile strength, N/mm^2
Elon	Elongation, %
Proof	0.1% proof strength, N/mm^2

1 N/mm^2=0.1 hbar=0.102 kgf/mm^2=0.06475 tonf/in.^{2}=145.04 lbf/in.^{2}=1 MPa

See Appendix II for other abbreviations and conversion tables.

Symbol	Nominal analysis, supplier, condition and remarks.
BS 4109 C102	High purity copper: Fire refined tough pitch as BS 1036 C102
BS 4393	High purity copper: Tin or tin/lead coated
BS 4577 A1/1	99.93% Cu: For resistance welding electrodes; as drawn; conductivity 98% IACS **DPN: 90 UTS: 260 Elon: 25%**
BS 4577 A1/2	0.1% Ni 1% Te Cu: For resistance welding electrodes; as drawn; conductivity 85% IACS **DPN: 90 UTS: 250 Elon: 14%**
BS 4577 A1/3	1% Cd Cu: For resistance welding electrodes; as drawn; conductivity 80% IACS **DPN: 90 UTS: 300 Elon: 14%**
BS 4577 A4/3	6% Ag Cu: For resistance welding electrodes; wrought; conductivity 80% IACS **DPN: 150 UTS: 490 Elon: 9%**
BS 4608	99.9% Cu: For electrical purposes; sheet and strip
BS 6360	99.9% Cu: Tough pitch high conductivity copper; as BS 4109; as BS 1036 and BS 1037
C 7	98.5% Cu: Rod for gas shielded arc welding
C 8	99.4% Cu: Rod for gas shielded arc welding
C 24	97.9% Cu: Rod for gas shielded arc welding
C 80	Copper sintered material: Sintered Products Ltd; 8–11% porosity **UTS: 750 Elon: 3%**
C 80	Sintered pure Cu: 9% porosity; specific gravity 8.1; Durasint **DPN: 40 UTS: 100 Elon: 3%**
C10100	99.99% Cu: Oxygen-free; electronic; CDA and UNS designation; previous OFE
C10200	99.95% Cu: Oxygen-free; CDA and UNS designation; previous OF
C10300	0.003% P 99.95% Cu: CDA and UNS designation; previous OFXLP
C10400	0.027% Ag 99.95% Cu: CDA and UNS designation; previous OFS
C10500	0.034% Ag 99.95% Cu: CDA and UNS designation; previous OFS
C10700	0.085% Ag 99.95% Cu: CDA and UNS designation; previous OFS
C10800	0.01% P 99.95% Cu: CDA and UNS designation; previous OFLP
C10920	99.9% Cu: CDA and UNS designation
C10930	0.04% Ag 99.9% Cu: CDA and UNS designation
C10940	0.08% Ag 99.9% Cu: CDA and UNS designation
C11000	99.9% Cu: Tough pitch; CDA and UNS designation; previous ETP
C11010	99.9% Cu: Remelted; high conductivity; CDA and UNS designation; previous RHC
C11020	99.9% Cu: Fire refined; high conductivity; CDA and UNS designation; previous FRHC
C11030	99.9% Cu: Chemically refined tough pitch; CDA and UNS designation; previous CRTP
C11100	99.9% Cu: Electrolytic; tough pitch; anneal resistant; CDA and UNS designation
C11300	0.03% Ag 99.9% Cu: Tough pitch with Ag; CDA and UNS designation; previous STP
C11400	0.034% Ag 99.9% Cu: CDA and UNS designation; previous STP
C11500	0.055% Ag 99.9% Cu: CDA and UNS designation; previous STP
C11600	0.085% Ag 99.9% Cu: CDA and UNS designation; previous STP
C11700	0.04% P 0.01% B 99.9% Cu: CDA and UNS designation
C11904	0.03% Ag 99.9% Cu: CDA and UNS designation
C11905	0.034% Ag 99.9% Cu: CDA and UNS designation
C11906	0.05% Ag 99.9% Cu: CDA and UNS designation
C11907	0.085% Ag 99.9% Cu: CDA and UNS designation
C12000	0.008% P 99.9% Cu: Low recorded P; CDA and UNS designation; previous DLP

Symbol	Nominal analysis, supplier, condition and remarks.
C12100	0.008% P 0.015% Ag 99.9% Cu: CDA and UNS designation; previous DLP
C12200	0.03% P 99.9% Cu: High recorded P; CDA and UNS designation; previous DHP
C12210	0.02% P 99.9% Cu: CDA and UNS designation
C12220	0.05% P 99.9% Cu: CDA and UNS designation
C12300	0.02% P 99.9% Cu: CDA and UNS designation
C12500	0.012% Ag 0.003% Sb 99.88% Cu: Fire refined tough pitch; CDA and UNS designation; previous FRTP
C12700	0.027% Ag 0.012% As 0.003% Sb 99.88% Cu: Fire refined tough pitch with Ag; CDA and UNS designation; previous FRSTP
C12800	0.034% Ag 0.012% As 0.003% Sb 99.88% Cu: Fire refined tough pitch with Ag; CDA and UNS designation; previous FRSTP
C12900	0.054% Ag 0.012% As 0.003% Sb 99.88% Cu: Fire refined tough pitch with Ag; CDA and UNS designation; previous FRSTP
C13000	0.085% Ag 0.012% As 0.003% Sb 99.88% Cu: Fire refined tough pitch with Ag; CDA and UNS designation; previous FRSTP
C14100	0.3% As 99.4% Cu: Tough pitch; CDA and UNS designation; obsolete
C14180	0.075% P 99.9% Cu: CDA and UNS designation
C14181	0.002% P 99.9% Cu: CDA and UNS designation
C14200	0.02% P 0.3% As 99.4% Cu: CDA and UNS designation; previous DPA
C14210	0.03% P 0.4% As 99.2% Cu: CDA and UNS designation
C14300	0.01% Cd 99.9% Cu: Cd de-oxidized; CDA and UNS designation
C14310	0.02% Cd 99.9% Cu: Cd de-oxidized; CDA and UNS designation
C14400	0.15% Sn 0.02% P 0.003% Sb Cu: CDA and UNS designation
C14410	0.12% Sn 0.05% Pb 0.01% P 99.9% Cu: CDA and UNS designation
C14420	0.1% Sn 0.04% Te 99.9% Cu: CDA and UNS designation
C14430	0.3% Sn Cu: CDA and UNS designation
C14440	0.01% Sn 99.96% Cu: CDA and UNS designation
C14500	0.01% P 0.5% Te Cu: CDA and UNS designation
C14510	0.05% Pb 0.02% P 0.5% Te Cu: CDA and UNS designation
C14520	0.01% P 0.5% Te Cu: CDA and UNS designation; previous DPTE
C14530	0.01% Sn 0.004% P 0.02% Te Cu: CDA and UNS designation
C14700	0.004% P 0.3% S Cu: CDA and UNS designation
C14710	0.02% Sb 0.1% S Cu: CDA and UNS designation
C14720	0.35% S 99.5% Cu: CDA and UNS designation
C14730	99.8% Cu: Sulphur bearing; CDA and UNS designation; obsolete
C15000	0.15% Zr 99.8% Cu: Zr copper; CDA and UNS designation
C15100	0.005% Al 99.82% Cu: CDA and UNS designation
C15150	0.02% Zr 99.9% Cu: Zr copper; CDA and UNS designation
C15500	0.06% P 0.027% Ag 0.1% Mg Cu: CDA and UNS designation
C15600	0.07% P 0.25% Co 0.02% Mg Cu: CDA and UNS designation
C15710	0.1% Al 0.1% O 99.71% Cu: CDA and UNS designation
C15715	0.15% Al 0.15% O 99.6% Cu: CDA and UNS designation
C15720	0.2% Al 0.2% O 99.5% Cu: CDA and UNS designation
C15725	0.25% Al 0.24% O 99.4% Cu: CDA and UNS designation

Symbol	Nominal analysis, supplier, condition and remarks.
C15735	0.32% Al 0.32% O 99.25% Cu: CDA and UNS designation
C15760	0.6% Al 0.52% O 98.77% Cu: CDA and UNS designation
C16200	1% Cd Cu: CDA and UNS designation; previous cadmium copper
C16210	1% Cd Cu: CDA and UNS designation; previous cadmium copper
C16400	0.75% Cd 99.8% Cu: UNS designation; obsolete
C16500	0.6% Sn 0.8% Cd Cu: CDA and UNS designation
C18700	1.1% Pb Cu: CDA and UNS designation
C18900	0.7% Sn 0.2% Si Cu: CDA and UNS designation
C18980	1% Sn max Cu: CDA and UNS designation
C19200	0.02% P 1% Fe Cu: CDA and UNS designation
C19210	0.03% P 0.1% Fe Cu: CDA and UNS designation
C19220	0.07% Sn 0.05% P 0.2% Fe Cu: CDA and UNS designation
C19250	0.3% Ni 0.3% Fe Cu: UNS designation; obsolete
C19260	0.6% Fe Cu: CDA and UNS designation
C19280	0.5% Sn 0.5% Zn 0.01% P 1% Fe Cu: CDA and UNS designation
C19400	0.15% Sn 0.1% P 2% Fe Cu: CDA and UNS designation
C19410	0.7% Sn 0.15% Zn 0.03% P 2% Fe Cu: CDA and UNS designation
C19450	0.8% Sn 0.02% P 2.2% Fe Cu: CDA and UNS designation
C19500	0.4% Sn 2% P 1.5% Fe 1% Co Cu: CDA and UNS designation
C19520	2% Pb 1% Fe Cu: CDA and UNS designation
C19600	0.3% P 1% Fe Cu: CDA and UNS designation
C19700	0.3% P 1% Fe Cu: CDA and UNS designation
C19750	0.2% Sn 0.3% P 1% Fe Cu: CDA and UNS designation
C55180	5% P Cu: CDA and UNS designation
C55181	7.2% P Cu: CDA and UNS designation
C55280	7% P 2% Ag Cu: CDA and UNS designation
C55281	6% P 5% Ag Cu: CDA and UNS designation
C55282	6.8% P 5% Ag Cu: CDA and UNS designation
C55283	7.2% P 6% Ag Cu: CDA and UNS designation
C55284	5% P 15% Ag Cu
C80100	99.95% Cu: Cast; CDA and UNS designation
C80300	0.03% Ag (min) 99.95% Cu: Cast; CDA and UNS designation
C80500	0.03% Ag (min) 99.75% Cu: Cast; CDA and UNS designation
C80700	0.02% B 99.75% Cu: Cast; CDA and UNS designation
C80900	0.03% Ag (min) 99.7% Cu: Cast; CDA and UNS designation
C81100	99.7% Cu: Cast; CDA and UNS designation
C81200	0.06% P 99.9% Cu: Cast; CDA and UNS designation
C81540	2.5% Ni 0.6% Si 0.8% Cu alloy: Cast; CDA and UNS designation
CADMIUM COPPER	0.75% Cd Cu alloy: High tensile strength; used for telephone wires; origin unknown
CATHODE COPPER	Copper produced by electrolysis; low conductivity; common name
CHILI BAR	1% S + usual impurities Cu: Crude; origin unknown
COMBARLOY	0.1% Ag Cu: For electrical purposes; Thomas Bolton
COPPER	Cu metal: Ingots; granulated, powder, foil etc.; Blackwell; primary metal
COPPER	High purity metal: Impurities 10 p.p.m.; Johnson Matthey; supplied as ingot, powder, wire and sheet
COPPERWELD HM	Mild steel copper-coated wire: Hard drawn steel; 40% of wire is copper for electrical leads; Copperweld Steel Co.
COPPERWELD LC	Mild steel copper-coated wire: Annealed steel; 40% of wire is copper for electrical leads; Copperweld Steel Co.
Cu/a1	99.9% Cu: Oxygen-free; French Standard; part of NFA 53-100

Symbol	Nominal analysis, supplier, condition and remarks.
Cu/a2	99.9% Cu: French Standard; part of NFA 53-100
Cu/a3	99.75% Cu: French Standard; part of NFA 53-100
Cu/b	99.9% Cu: 0.03% O: French Standard; part of NFA 53-100
Cu/c1	99.92% Cu: For conductors; French Standard; part of NFA 53-100
Cu/c2	0.0003% O (max) 99.96% Cu: For conductors; French Standard; part of NFA 53-100
Cu/d	99.9% Cu cathode copper: French Standard; part of NFA 53-100
CUPRALITH	Series of copper–lithium alloys containing 1–10% Li
CUPROMET 100	99.5% Cu electrode: Metrode
Cu-Zr MASTER ALLOY	12.5% Zr Cu alloy: Origin unknown
DHP	0.05% P 99.9% Cu (min) de-oxidized
DIN 1708	Covers high purity coppers
DIN 1733 S Cu Ag	1.0% Ag Cu
DIN 1733 S Cu Sn	0.6% Sn Cu
DIN 1785 S B Cu	0.03% P 0.4% As Cu: Wrought
DIN 1785 S D Cu	0.03% P Cu: Wrought
DO	0.03% P 99.95% Cu: Extrusions; phosphorus de-oxidized; McKechnie Bros for BS alloy C106
DONA	0.04% P 0.02% As 99.9% Cu: Non-arsenical; de-oxidized; Birmingham Battery for BS 1172
DONA	Phosphorus de-oxidized non-arsenical copper: IMI 131
DOXA	0.04% P 0.35% As 99.5% Cu: Arsenical de-oxidized; Birmingham Battery for BS 1174
DTD 607	99.9% Cu: Strip; annealed; 0.05% Ag is optional **DPN: 55**
DTD 607	99.9% Cu: Strip; cold rolled and tempered; 0.05% Ag is optional **DPN: 95**
DTD 631	99.8% Cu: Silver coated strip suitable for brazing
E Cu	1% Al (max) 1% Si (max) Cu: weld electrode; designation used by AWS
EC	99.9% Cu: Extrusions; electrolytic tough pitch; McKechnie Bros for BS alloy C101
ELKALOY A	Cd Cu alloy: For resistance welding electrode; Johnson Matthey; electrical conductivity 85% IACS **DPN: 90 UTS: 450 Elon: 20% Proof: 370**
ERM ALW	0.1% Ni 0.6% Te Cu alloy: Drawn bar; Enfield Rolling Mills Ltd; electrode material for resistance welding **DPN: 100 UTS: 270 Elon: 16% Proof: 240**
ERM HSM	0.6% Te 0.03% O Cu alloy: Drawn bar; Enfield Rolling Mills Ltd; commutator bars etc. **DPN: 90 UTS: 260 Elon: 20% Proof: 220**
ETP	99.9% Cu (min): Electrolytic tough pitch Cu
FRSTP	99.88% Cu (min): Fire refined tough pitch Cu with Ag
FRTP	99.88% Cu (min): Fire refined tough pitch Cu
h Cu 6 a	99.998% Cu: Bars; Light Ltd; high purity metal
HC	99.95% Cu: Extrusions; fire refined tough pitch; McKechnie Bros for BS alloy C102
HCOKOF	Oxygen-free high conductivity copper: Thomas Bolton; withdrawn
HCOKOF	High purity, high conductivity, oxygen-free Cu: Outokumpu
HSM/S	0.4% S Cu alloy: Rod for free machining; Enfield specification for BS 2874 C111
IMI 100	Oxygen-free high conductivity copper: IMI
IMI 103	Tough pitch high conductivity copper: Electrolytic; IMI for BS 125, BS 128, BS 1432, BS 1433
IMI 121	99.8% Cu: Fire refined for rivets; IMI; withdrawn
IMI 131	Phosphorus de-oxidized non-arsenical copper; IMI for BS 1172
IMI 134	99.85% Cu: Tough pitch non-arsenical; IMI for BS 1038, BS 1039
IMI 138	0.4% As Cu: Tough pitch; IMI
IMI 146	0.4% As Cu: De-oxidized arsenical; IMI
IMI 153	Ag Cu alloy: Brazing rod; IMI for BS alloy C1
IMI 161	Boron Cu: Welding rod; IMI Borofil
IMI 161	0.2% Al 0.2% Ti Cu alloy: Welding rod; IMI Nitrofil
IMI 166	1% Cd Cu alloy: Wire and bar; IMI for BS alloy C108
IMI 181	0.5% Te Cu: For free machining; IMI KUTERN; electrical conductivity 90% IACS
IMI 651	1% Sn Cu alloy: Wire and bar; IMI for PO bronze **DPN: 170 UTS: 540 Elon: 2% Proof: 480**
IMI 981	High conductivity copper wire: Tinned; IMI wire for either IMI 103 or IMI 100
KUFIL	1% Ag Cu alloy: Welding rod; IMI BS alloy C1; melting range 1073–1078 °C
KUTERN	0.5% Te Cu alloy: Rod and sections; IMI free machining copper; IMI 181
LAKE COPPER	Copper which originates on north peninsula of Michigan, USA: See ASTM B4
MANGANIN	13% Mn Cu alloy: For resistors, etc.; Driver
METCO 55	Cu with some oxide: Spray deposit; Metco
METCO COPPER	Low oxide Cu: Spray deposit; Metco
NFA 53-100	Classification of copper: French Standard
NITROFIL	0.3% A1 0.3% Ti Cu alloy: Filler rod; IMI; melting point 1080 °C; for arcon arc welding Cu
OFE	99.99% Cu min: Oxygen-free electronic copper
OFLP	0.01% P 99.95% Cu min: Oxygen-free low P
OFXLP	0.007% P 99.95% Cu: Oxygen-free extra low P
PO Bronze	1% Sn Cu alloy: Traditional name; originally used for telephone line wire
QQ C 501	99.9% Cu: Plate or sheet; US Federal **Elon: 20%**
QQ W 336	High conductivity Cu: Wire; US Federal; hard drawn **UTS: 430 Elon: 3%**
QQ W 341	High conductivity Cu: Wire; US Federal; annealed **UTS: 240 Elon: 28%**
RWMA Class 1	Cd Cu alloy: Mallory Metallurgical code for Elkaloy A
SAE 71	99.9% Cu: Sheet and strip; ½-hard; now SAE CA 116 **UTS: 280**
SAE 71	99.9% Cu: Sheet and strip; extra hard; now SAE CA 116 **UTS: 370**
SAE 75	99.9% Cu: Tube; light drawn; now SAE CA 122 **UTS: 240**
SAE 75	99.9% Cu: Tube; hard drawn; now SAE CA 122 **UTS: 330**
SAE 83	99.9% Cu: Wire; annealed; electrical conductivity 100% IACS; now SAE CA 110 **UTS: 240 Elon: 30%**
SAE CA 102	99.95% Cu: Oxygen-free; suitable for welding
SAE CA 110	99.9% Cu: Electrolytic tough pitch; formerly SAE 71 and 83
SAE CA 111	99.9% Cu: Electrolytic tough pitch with trace Cd; formerly SAE 71
SAE CA 113	Trace Ag 99.9% Cu: Tough pitch; formerly SAE 71; ASTM alloy STP
SAE CA 114	Trace Ag 99.9% Cu: Tough pitch; formerly SAE 71; ASTM alloy STP; higher Ag than CA 113
SAE CA 115	0.06% Ag Cu alloy: Wrought
SAE CA 116	Trace Ag 99.9% Cu: Tough pitch; formerly SAE 71; ASTM alloy STP; higher Ag than CA 113 and CA 114
SAE CA 120	0.008% P 99.9% Cu: De-oxidized; formerly SAE 75; ASTM alloy DLP
SAE CA 122	0.02% P 99.9% Cu: De-oxidized; formerly SAE 75; ASTM alloy DHP

Note. The following abbreviations and units are used in the tables:

DPN	Hardness, diamond pyramid number
UTS	Ultimate tensile strength, N/mm^2
Elon	Elongation, %
Proof	0.1% proof strength, N/mm^2

1 N/mm^2=0.1 hbar=0.102 kgf/mm^2=0.06475 $tonf/in.^2$=145.04 $lbf/in.^2$=1 MPa

See Appendix II for other abbreviations and conversion tables.

Symbol	Nominal analysis, supplier, condition and remarks.
SAE CA 145	0.008% P 0.5% Te 99.5% Cu: Free machining; electrical conductivity IACS 93%
SAE CA 147	0.3% S 99.7% Cu: Free machining; IACS 93%
SAE CA 150	0.12% Zr 99.88% Cu: For welding electrode
SAE CA 162	1% Cd 99% Cu alloy
SAE CA 187	1% Pb Cu alloy: Free machining
SAE CA 192	0.03% P 1% Fe Cu alloy: Wrought
SAE CA 962	10% Ni 1.5% Fe Cu alloy: Casting
SC	0.3% S 99.7% Cu: Extrusions; free machining; McKechnie; as drawn; electrical conductivity IACS 95%
	DPN: 90 UTS: 240 Elon: 20% Proof: 150
For SIS specifications see SS	
SOUDOCUIVRE	0.8% Sn 2.5% Mn Cu alloy: Welding electrode; Soudometal; for welding pure copper
	DPN: 50 UTS: 250 Elon: 35%
SOUDOGEN BE50	6% P Cu: Bronze metal; Soudometal
	UTS: 250
SOUDOR G Cu	0.8% Sn 0.2% Mn Cu: Welding electrode; Soudometal
	UTS: 220 Elon: 30%
SOUDOTIG Cu	0.8% Sn 0.2% Mn Cu: Welding electrode; Soudometal; for inert gas welder
	UTS: 220 Elon: 30%
SS 5010	99.9% Cu: Bar, sheet, forging, etc.; Swedish Standard; annealed; electrical conductivity 100% IACS
	DPN: 50 UTS: 240 Elon: 50% Proof: 45
SS 5010	99.9% Cu: Bar, sheet, forging, etc.; Swedish Standard; cold worked; electrical conductivity 100% IACS
	DPN: 70 UTS: 420 Elon: 30% Proof: 210
SS 5011	99.5% Cu: Bar and sheet; oxygen-free high conductivity; Swedish Standard; annealed; electrical conductivity 100% IACS
	DPN: 50 UTS: 200 Elon: 50% Proof: 45
SS 5011	99.5% Cu: Bar and sheet; oxygen-free high conductivity; Swedish Standard; cold rolled; electrical conductivity 100% IACS
	DPN: 100 UTS: 300 Elon: 12% Proof: 240
SS 5013	99.8% Cu: Plate; fire refined; Swedish Standard; annealed
	DPN: 65 UTS: 210 Elon: 50% Proof: 45
SS 5013	99.8% Cu: Plate; fire refined; Swedish Standard; cold rolled
	DPN: 100 UTS: 330 Elon: 12% Proof: 300
SS 5015	0.03% P 99.8% Cu: Plate and bar; fire refined; Swedish Standard; annealed
	DPN: 50 UTS: 200 Elon: 45% Proof: 45
SS 5015	0.03% P 99.8% Cu: Plate and bar; fire refined; Swedish Standard; cold rolled
	DPN: 85 UTS: 240 Elon: 30% Proof: 200
SS 5030	0.08% Ag 99.9% Cu: Extrusions; Swedish Standard; cold drawn
	DPN: 100 UTS: 280 Elon: 10% Proof: 240
SS 5053	Cd bearing Cu: Bar; Swedish Standard; cold drawn
	UTS: 580 Proof: 420
STA 7 C1	99.75% Cu: Tough pitch for driving bands; annealed
	DPN: 50 UTS: 180 Elon: 40%
STA 7 C2	0.1% P 99.75% Cu: Phosphorus de-oxidized for driving bands; annealed
	DPN: 50 UTS: 180 Elon: 40%
STA 7 C3	99.9% Cu: Tough pitch high conductivity; sheet; annealed BS 899
	UTS: 200 Elon: 35%
STA 7 C3	99.9% Cu: Tough pitch high conductivity; sheet; ½-hard
	UTS: 270
STA 7 C4	99.9% Cu: Tough pitch; tubes; annealed
	UTS: 210 Elon: 40%
STA 7 C5	99.9% Cu: Tough pitch; bars; annealed
	UTS: 210 Elon: 40%
STA 7 C5	99.9% Cu: Tough pitch; bars; ½-hard
	DPN: 65 UTS: 270 Elon: 12%
STA 7 C6	99.9% Cu cathode copper: BS 1035
STA 7 C7A	99.9% Cu: Electrolytic high conductivity; BS 1036
STA 7 C7B	99.9% Cu: Fire refined high conductivity; BS 1037
STA 7 C8A	99.85% Cu: Tough pitch non-arsenical; BS 1038
STA 7 C8B	99.75% Cu: Tough pitch non-arsenical; BS 1039
STA 7 C8C	99.5% Cu: Tough pitch non-arsenical; BS 1040
STA 7 C9	0.4% As 99.2% Cu: Tough pitch; now BS 1173
STA 7 C10	0.1% P 99.85% Cu: Phosphorus de-oxidized; now BS 1172
STA 7 C11	0.1% P 0.4% As 99.2% Cu: Now BS 1174
STA 13 C1	99.75% Cu: Tough pitch for projectile driving bands
STA 13 C2	0.05% P 99.75% Cu: Phosphorus de-oxidized for projectile driving bands
STP	99.9% Cu min: Tough pitch Cu with Ag
T7	0.4% As 99.2% Cu: Seamless tube; ½-hard
	UTS: 300
T51	0.5% As 99.2% Cu: Seamless tube; as drawn
	UTS: 240
TC	0.5% Te 99.5% Cu: Extrusions; free machining; McKechnie; electrical conductivity IACS 90%
	DPN: 100 UTS: 280 Elon: 15% Proof: 220
TOUGH PITCH	Pure copper with O: High conductance; common name
VDM Mn Bz 2	0.3% Nimoy 1.9% Mn Cu alloy: VDM
WHITE PINE LAKE COPPER	99.9% Cu Ag: American trade name
WW T 799a	99.9% Cu: Seamless tube; US Federal
WW T 799a/N	99.9% Cu: Seamless tube; US Federal

Note. The following abbreviations and units are used in the tables:

DPN	Hardness, diamond pyramid number
UTS	Ultimate tensile strength, N/mm^2
Elon	Elongation, %
Proof	0.1% proof strength, N/mm^2

1 N/mm^2=0.1 hbar=0.102 kgf/mm^2=0.06475 $tonf/in.^2$=145.04 $lbf/in.^2$=1 MPa

See Appendix II for other abbreviations and conversion tables.

14B Copper–zinc wrought alloys
Less than 38 per cent zinc – alpha brass

Specific gravity		8.4–8.8
Density		8400–8800 kg/m^3
Solidus/liquidus	5% Zn	1044 °C
	30% Zn	900 °C
Thermal conductivity	5% Zn	188 W/m °C
	30% Zn	109 W/m °C
Coefficient of linear expansion		(18.2–20.5) × 10^{-6}/ °C
Electrical conductivity	5% Zn alloys	44% IACS (copper 100%)
	30% Zn alloys	27% IACS (copper 100%)
Specific resistance	5% Zn	50 microhm mm
	30% Zn	65 microhm mm
Young's modulus of elasticity		(105–117) × 10^9 N/m^2
Impact		60 J
Fatigue strength (10^8 cycles)	annealed	80–110 N/mm^2
	cold worked	120–140 N/mm^2
		Fatigue strength varies directly with the zinc content.

Hot strength

Temperature °C	Tensile strength N/mm^2		Elongation %	
	5 % Zn	30 % Zn	5 % Zn	30 % Zn
100	660	570	8	7
200	670	580	8	7
300	450	570	28	8
400	420	380	50	42
500	340	300	60	45
600	320	270	65	47

The above properties are typical of the following group, and may not apply exactly to any one specification. It is possible that with certain specifications some of the values may not be applicable.

General metallurgical characteristics

The majority of copper alloys are based on the copper zinc series. There are three groups, alpha, alpha plus beta, and beta, depending on the zinc content. Each of these groups has its own characteristics, which are modified first by the zinc content and then by the addition of alloying elements to give increased strength and corrosion resistance. The alpha brasses are listed in this section, while the alpha plus beta and beta alloys are found in 14C.

The alpha group has up to 38% zinc present. At the higher zinc contents depending on the heat treatment and other alloying elements, some beta phase may be present.

These alpha brasses are similar in many respects to copper and are often used in lieu of copper where economy is necessary or corrosion resistance and electrical properties are not of paramount importance, for example, in decorative articles and household goods. There are several groups of these alloys, as follows.

Cap copper. Zinc content 2–5%. This, in effect, is copper de-oxidized with zinc. The result is a ductile readily worked material of low mechanical strength. The electrical properties are not much affected, the conductivity being 75–95% IACS. The alloy can be strengthened by cold work to a limited extent.

Gilding metal. Zinc content 5–20%. These alloys are used for their pleasing reddish copper colour which becomes less coppery and more brassy as the zinc content increases. They are all capable of being cold worked, the hardness increasing with zinc content.

These alloys are not prone to the defect known as 'season cracking' which affects most of the higher zinc brasses. This cracking is caused by local corrosion attack at grain boundaries when cold worked brasses are stored in certain atmospheres.

The electrical conductivity of 10% zinc gilding metal is as low as 43% IACS.

Cartridge brass. Zinc content 28–32%. This is the most ductile and easily worked of the zinc copper alloys, finding use where deep drawing operations are required. Interstage annealing is necessary between heavy deep draws and time and temperature must be carefully controlled to prevent grain growth resulting in an orange peel surface finish. This material is subject to season cracking and should always be stress relieved after the final forming operation. The electrical conductivity of these alloys is 27% IACS.

Basis or common brass. Zinc content is 35–38%. This is the cheapest zinc copper alloy, and is generally found only as sheet. The ductility is reduced as the zinc content is increased, owing to the formation of some beta phase. The material is generally used for shallow pressings for low duty purposes. The corrosion resistance and electrical conductivity are relatively low.

Effect of alloying elements

All of the above materials can have certain properties modified by the addition of alloying elements with little effect on the remaining properties.

Tin. The addition of up to 1% improves the corrosion resistance. 'Admiralty Brass' is 'Cartridge Brass' with 1% tin added. This differs from 'Naval Brass' in the zinc content.

Lead. Small amounts of lead (up to 2%) are added to the brasses to improve their machinability. The lead does not go into solution, but acts as a discontinuity giving built in lubrication and smaller chips. There is a slight reduction in corrosion resistance and ductility.

Aluminium. This improves the resistance of brasses to sea water, up to 2% aluminium being used. One part by weight of aluminium has the equivalent effect of six parts zinc, as far as other properties are concerned. Thus aluminium brasses have approximately 22% zinc, 2% aluminium, but are equivalent for all practical purposes to the 30% zinc series.

In general it can be said that corrosion resistance, electrical and heat conductivity and price all fall as the zinc content rises. The ability to accept cold work rises to the 30% zinc alloy, then drops with further increase in zinc.

None of the alloys listed can have their mechanical properties improved by heat treatment. The effects of cold work can be eliminated at temperatures between 300 and 600 °C. Some care is required with heating these alloys as in an oxidizing atmosphere de-zincification can occur which can seriously reduce the fatigue properties.

It is important, however, to remove the effect of cold work on these materials in order to reduce the danger of season cracking. This is an alternative name for stress corrosion and can be serious in certain atmospheres, particularly where any cold work on the surface exists.

Welding of these alloys is seldom required and special techniques and specialist advice are necessary.

Brazing and soldering present no problems.

These alloys in general have reasonable corrosion resistance although they do form a greenish patina under normal atmospheric conditions. This can be readily removed.

They are not suitable, however, for use in marine atmospheres or where corrosive conditions exist. They are particularly prone to de-zincification under oxidizing conditions and, as stated above, this can reduce the fatigue resistance and ability to withstand bending in addition to being unsightly.

The materials on the whole have excellent deep drawing properties. Considerable competition is being experienced from austenitic stainless steels which generally have better corrosion resistance and do not require cleaning to the same extent.

These alloys, however, have an attractive colour and remain popular for many purposes.

Symbol	Nominal analysis, supplier, condition and remarks.
2.0290	2% Pb 35% Zn Cu alloy: Casting; German Standard
2.0291	2% Pb 35% Zn Cu alloy: Casting; German Standard
2.0293	2% Pb 35% Zn Cu alloy: Casting; German Standard
2.0295	2% Pb 35% Zn Cu alloy: Casting; German Standard
2.0470	1.0% Sn 28% Zn Cu alloy: German Standard
8M	1% Sn 0.2% Pb 35% Zn 4% Al 1% Fe Cu alloy: Wrought; Langley Alloys **UTS: 540 Elon: 15% Proof: 270**
20	2% Pb 35.5% Zn Cu alloy: Extrusion; McKechnie Bros; as drawn **DPN: 120 UTS: 370 Elon: 30% Proof: 200**
21	35% Zn Cu alloy: Extrusion; McKechnie Bros; annealed **DPN: 90 UTS: 330 Elon: 45% Proof: 100**
65	29% Zn 4.25% Al 0.75% Fe 0.75% Mn Cu alloy: Extrusion; McKechnie Bros; as extruded **DPN: 130 UTS: 600 Elon: 15% Proof: 300**
193	5% Zn 2.2% Fe Cu alloy: Wrought; designation agreed by CDAA; not used
205	2% Zn Cu alloy: Wrought; designation agreed by CDAA; not used
226	12.5% Zn Cu alloy: Wrought; designation used by CDAA
234	15% Zn Cu alloy: Wrought; designation agreed by CDAA; not used
250	25% Zn Cu alloy: Wrought; designation agreed by CDAA; not used
261	30% Zn 0.04% P Cu alloy: Wrought; designation agreed by CDAA; not used
262	32% Zn Cu alloy: Wrought; designation agreed by CDAA; not used
274	37% Zn Cu alloy: Wrought; designation agreed by CDAA; not used
305	0.8% Sn 1.2% Pb 37% Zn 1.3% Mn Cu alloy: McKechnie **DPN: 130 UTS: 500 Elon: 30% Proof: 250**
306	0.5% Sn 1.2% Pb 37% Zn 1.3% Mn Cu alloy: McKechnie **DPN: 140 UTS: 520 Elon: 25% Proof: 280**
310	0.5% Pb 8% Zn Cu alloy: Wrought; designation agreed by CDAA; not used
310	0.2% Pb 20% Zn Cu alloy: Extrusions; McKechnie Bros; annealed **DPN: 90 UTS: 280 Elon: 50% Proof: 100**
311	30% Zn Cu alloy: Extrusion; McKechnie Bros; annealed **DPN: 90 UTS: 280 Elon: 60% Proof: 100**
314	1.7% Pb 9.25% Zn Cu alloy: Wrought; designation used by CDAA

Note. The following abbreviations and units are used in the tables:

DPN	Hardness, diamond pyramid number
UTS	Ultimate tensile strength, N/mm^2
Elon	Elongation, %
Proof	0.1% proof strength, N/mm^2

1 N/mm^2=0.1 hbar=0.102 kgf/mm^2=0.06475 tonf/in.^{2}=145.04 lbf/in.^{2}=1 MPa

See Appendix II for other abbreviations and conversion tables.

Symbol	Nominal analysis, supplier, condition and remarks.
316	1.8% Pb 9% Zn 1.0% Ni Cu alloy: Wrought; designation agreed by CDAA; not used
320	0.5% Sn 0.9% Pb 37% Zn 0.8% Al 1% Mn Cu alloy: McKechnie; as drawn
	DPN: 160 UTS: 540 Elon: 25% Proof: 300
320	2% Pb 12% Zn Cu alloy: Wrought; designation agreed by CDAA; not used
323	0.6% Sn 1% Pb 37% Zn 1% Al 1% Mn Cu alloy: McKechnie; extruded
	DPN: 160 UTS: 540 Elon: 25% Proof: 270
325	0.6% Pb 37% Zn 1.5% Al 2% Mn Cu alloy: McKechnie; as drawn
	DPN: 190 UTS: 650 Elon: 20% Proof: 400
325	2.8% Pb 25% Zn Cu alloy: Wrought; designation agreed by CDAA; not used
328	37% Zn 1.7% Al 3% Mn 1% Si Cu alloy: McKechnie; as drawn
	DPN: 190 UTS: 610 Elon: 20% Proof: 380
330	28% Zn 4% Al 0.7% Mn Cu alloy: McKechnie; extruded
	DPN: 190 UTS: 690 Elon: 20% Proof: 380
332	1.7% Pb 35% Zn Cu alloy: Tube; designation used by CDAA
335	0.5% Pb 37% Zn Cu alloy: Wrought; designation used by CDAA
340	1% Pb 35% Zn Cu alloy: Wrought; designation used by CDAA
342	1.2% Sn 0.4% Pb 36% Zn Cu alloy: McKechnie; Naval brass
	DPN: 120 UTS: 380 Elon: 35%
344	0.7% Pb 35% Zn Cu alloy: Wrought; designation agreed by CDAA; not used
347	1.5% Pb 35% Zn Cu alloy: Wrought; designation agreed by CDAA; not used
348	0.6% Pb 38% Zn Cu alloy: Wrought; designation agreed by CDAA: not used
353	2% Pb 38% Zn Cu alloy: Wrought; designation used by CDAA
356	2.5% Pb 38% Zn Cu alloy: Wrought; designation not used by CDAA
362	4.0% Pb 37% Zn Cu alloy: Wrought; designation agreed by CDAA; not used
365	0.8% Pb 38% Zn Cu alloy: Wrought; designation used by CDAA
366	0.7% Pb 38% Zn 0.07% As Cu alloy: Wrought; designation used by CDAA
367	0.7% Pb 38% Zn 0.07% Sb Cu alloy: Wrought; designation used by CDAA
368	0.7% Pb 38% Zn 0.06% P Cu alloy: Wrought; designation used by CDAA
370	1.1% Pb 38% Zn Cu alloy: Wrought; designation used by CDAA
371	1.0% Pb 38% Zn Cu alloy: Wrought; designation agreed by CDAA; not used
377	2.0% Pb 38% Zn Cu alloy: For forging; designation used by CDAA
405	1.0% Sn 5% Zn Cu alloy: Wrought; designation agreed by CDAA; not used
408	2.0% Sn 5% Zn Cu alloy: Wrought; designation agreed by CDAA; not used
409	0.7% Sn 5% Zn Cu alloy: Wrought; designation agreed by CDAA; not used
410	2.5% Sn 5% Zn Cu alloy: Wrought; designation agreed by CDAA; not used
411	0.5% Sn 7% Zn Cu alloy: Wrought; designation agreed by CDAA; not used
413	1.0% Sn 7% Zn Cu alloy: Wrought; designation agreed by CDAA; not used
415	2% Sn 6% Zn Cu alloy: Wrought; designation agreed by CDAA; not used
420	1.7% Sn 8% Zn 0.2% P Cu alloy: Wrought; designation agreed by CDAA; not used
421	2.6% Sn 8% Zn 0.3% P Cu alloy: Wrought; designation agreed by CDAA; not used
422	1% Sn 10% Zn 0.3% P Cu alloy: Wrought; designation agreed by CDAA; not used
422	2% Pb 36% Zn 0.1% As Cu alloy: McKechnie; resists de-zincification
	DPN: 130 UTS: 380 Elon: 30%
425	2% Sn 10% Zn 0.3% P Cu alloy: Wrought; designation agreed by CDAA; not used
430	2.2% Sn 12% Zn Cu alloy: Wrought; designation agreed by CDAA; not used
432	0.5% Sn 14% Zn 0.3% P Cu alloy: Wrought; designation agreed by CDAA; not used
434	0.8% Sn 15% Zn Cu alloy: Wrought; designation agreed by CDAA; not used
435	1.0% Sn 18% Zn Cu alloy: Wrought; designation agreed by CDAA; not used
436	0.4% Sn 18% Zn Cu alloy: Wrought; designation agreed by CDAA; not used
438	1.2% Sn 18% Zn Cu alloy: Wrought; designation agreed by CDAA; not used
440	2.75% Pb 36% Zn Cu alloy; McKechnie
441	3.1% Pb 36% Zn Cu alloy; McKechnie
442	1.0% Sn 28% Zn Cu alloy: Wrought; designation agreed by CDAA; not used
443	1.0% Sn 30% Zn 0.07% As Cu alloy: Wrought; designation used by CDAA
444	1.0% Sn 30% Zn 0.07% Sb Cu alloy: Wrought; designation used by CDAA
445	1.0% Sn 30% Zn 0.06% P Cu alloy: Wrought; designation used by CDAA
448	2% Pb 37% Zn Cu alloy; McKechnie
452	0.5% Pb 38% Zn Cu alloy; McKechnie
462	0.7% Sn 38% Zn Cu alloy: Wrought; designation agreed by CDAA; not used
476	2% Sn 2% Pb 10% Zn 0.05% P Cu alloy: Wrought; designation agreed by CDAA; not used
482	0.7% Sn 0.8% Pb 38% Zn Cu alloy: Wrought; designation agreed by CDAA; not used
485	0.7% Sn 1.8% Pb 38% Zn Cu alloy: Wrought; designation used by CDAA
665	20% Zn 1.0% Mn Cu alloy: Wrought; designation agreed by CDAA; not used
667	30% Zn 1.2% Mn Cu alloy: Wrought; designation agreed by CDAA; not used
674	0.2% Sn 0.4% Pb 38% Zn 1.5% Al 3.0% Mn 1.0% Si Cu alloy: Wrought; designation agreed by CDAA; not used
675	1.0% Sn 38% Zn 0.3% Mn Cu alloy: Wrought; designation used by CDAA
676	1.0% Sn 0.7% Pb 38% Zn 0.4% Mn Cu alloy: Wrought; designation agreed by CDAA; not used
685	12% Zn 4.0% Al 2.0% Fe Cu alloy: Wrought; designation agreed by CDAA; not used
687	20% Zn 2.1% Al 0.7% As Cu alloy: Wrought; designation used by CDAA

Note. The following abbreviations and units are used in the tables:

DPN	Hardness, diamond pyramid number
UTS	Ultimate tensile strength, N/mm^2
Elon	Elongation, %
Proof	0.1% proof strength, N/mm^2

$1\ N/mm^2 = 0.1\ hbar = 0.102\ kgf/mm^2 = 0.06475\ tonf/in.^2 = 145.04\ lbf/in.^2 = 1\ MPa$

See Appendix II for other abbreviations and conversion tables.

Symbol	Nominal analysis, supplier, condition and remarks.
862	23% Zn 5% Al 3% Fe 3% Mn Cu alloy: Casting; designation used by CDAA
863	26% Zn 6% Al 3% Fe 3% Mn Cu alloy: Casting; designation used by CDAA
864	39% Zn 1% Al 1% Fe 1% Mn Cu alloy: Casting; designation used by CDAA
ABS/B43	15% Zn Cu alloy: Piping; American Bureau of Shipping
ABS TYPE 2	1% Sn (max) 0.4% Pb (max) 39% Zn 0.7% Al 1.2% Fe Cu alloy: Casting; American Bureau of Shipping **UTS: 460 Elon: 20% Proof: 175**
ADMIRALTY BRASS	1% Sn 28% Zn Cu alloy: Common name; further information from Copper Development Association
AICHS METAL	38% Zn 1.8% Fe Cu alloy: Casting; source unknown
ALDURBRA	22% Zn 2% Al Cu alloy: Wrought; C Clifford
ALDURBRA	22% Zn 2% Al Cu alloy: Casting; C Clifford **DPN: 130 UTS: 590 Elon: 16% Proof: 300**
ALUMBRO	22% Zn 2% Al 0.04% As Cu alloy: Strip and sheet; IMI; annealed **DPN: 100 UTS: 400 Elon: 60% Proof: 100**
ALUMBRO	22% Zn 2% Al 0.04% As Cu alloy: Strip and sheet; IMI; ¼-hard **DPN: 160 UTS: 500 Elon: 25% Proof: 420**
ALUMINIUM BRASS	22% Zn 2% Al Cu alloy: Common name; further information from the Copper Development Association
AMERICAN BRASS	3% Pb 35% Zn Cu alloy: Bar; free cutting; common name for BS alloy CZ124 type; further information from the Copper Development Association
AMS 4505 D	30% Zn Cu alloy: Sheet and strip; annealed
AMS 4507 C	30% Zn Cu alloy: Sheet and strip; ½-hard
AMS 4508 B	30% Zn Cu alloy: Laminated sheet
AMS 4555 C	31% Zn Cu alloy: Seamless tube; light annealed; similar to SAE 74C
AMS 4558 B	1.6% Pb 32% Zn Cu alloy: Seamless tube; as drawn
AMS 4610 H	3% Pb 35.5% Zn Cu alloy: Bar; free cutting; ½-hard
AMS 4611 C	0.8% Sn 38.5% Zn Cu alloy: Bar; Naval brass; ½-hard
AMS 4612 D	0.8% Sn 38.5% Zn Cu alloy: Bar; Naval brass; hard drawn
AMS 4614 D	2% Pb 38% Zn Cu alloy: Forging; free machining
AMS 4710	35% Zn Cu alloy: Wire; tinned; annealed
AMS 4710	35% Zn Cu alloy: Wrought
AMS 4712 A	35% Zn Cu alloy: Wire; annealed
AMS 4713 A	35% Zn Cu alloy: Wire; 1/8-hard
AMS 4862 B	24% Zn 5.2% Al 3% Fe 4% Mn Cu alloy: Casting
ASTM B16	3% Pb 35% Zn Cu alloy: Rod; annealed **UTS: 330 Elon: 20% Proof: 10**
ASTM B16	3% Pb 35% Zn Cu alloy: Rod; hard drawn **UTS: 500 Elon: 4% Proof: 24**
ASTM B19	30% Zn Cu alloy: Cartridge brass; annealed **UTS: 330 Elon: 45%**
ASTM B19	30% Zn Cu alloy: Cartridge brass; hard **UTS: 530**
ASTM B19	30% Zn Cu alloy: Cartridge brass; extra spring hard **UTS: 760**
ASTM B22 E	25% Zn 5% Al 3% Fe 4% Mn Cu alloy: Casting; as cast **DPN: 223 UTS: 800 Elon: 12% Proof: 42**
ASTM B36/1	5% Zn Cu alloy: Sheet and bar; ¼-hard **UTS: 300**
ASTM B36/1	5% Zn Cu alloy: Sheet and bar; extra spring hard **UTS: 460**
ASTM B36/2	10% Zn Cu alloy: Sheet and bar; ¼-hard **UTS: 340**
ASTM B36/2	10% Zn Cu alloy: Sheet and bar; extra spring hard **UTS: 530**
ASTM B36/3	15% Zn Cu alloy: Sheet and bar; ¼-hard **UTS: 340**
ASTM B36/3	15% Zn Cu alloy: Sheet and bar; extra spring hard **UTS: 620**
ASTM B36/4	20% Zn Cu alloy: Sheet and bar; ¼-hard **UTS: 370**
ASTM B36/4	20% Zn Cu alloy: Sheet and bar; extra spring hard **UTS: 680**
ASTM B36/6	30% Zn Cu alloy: Sheet and bar; ¼-hard **UTS: 370**
ASTM B36/6	30% Zn Cu alloy: Sheet and bar; extra spring hard **UTS: 730**
ASTM B36/8	35% Zn Cu alloy: Sheet and bar; ¼-hard **UTS: 380**
ASTM B36/8	35% Zn Cu alloy: Sheet and bar; extra spring hard **UTS: 660**
ASTM B43	15% Zn Cu alloy: Seamless pipe; annealed
ASTM B105	5% Sn 10% Zn 3.5% Al 3% Si 1.5% Cd (max) Cu: Wire; mechanical and electrical properties depend on size, etc.
ASTM B111/230	15% Zn Cu alloy: For condenser tubes
ASTM B111/442	1% Sn 28% Zn Cu alloy: Seamless tube
ASTM B111/443	1% Sn 28% Zn 0.5% As Cu alloy: Seamless tube
ASTM B111/444	1% Sn 28% Zn 0.05% Sb Cu alloy: Seamless tube
ASTM B111/445	1% Sn 28% Zn 0.05% P Cu alloy: Seamless tube
ASTM B111/687	20% Zn 2.1% Al 0.05% As Cu alloy: For condenser tubes
ASTM B111 A	1% Sn 30% Zn Cu alloy: Tube
ASTM B111 A1 BRASS	22% Zn 2% Al Cu alloy: Tube
ASTM B111 B	1% Sn 30% Zn 0.1% As Cu alloy: Tube
ASTM B111 C	1% Sn 30% Zn 0.1% Sb Cu alloy: Tube
ASTM B111 D	1% Sn 30% Zn 0.1% P Cu alloy: Tube
ASTM B111 MUNTZ	0.3% Pb 38% Zn Cu alloy: Tube
ASTM B111 RED BRASS	15% Zn Cu alloy: Tube
ASTM B119	Cast Cu based alloys: Replaced by individual specifications
ASTM B121	2.5% Pb 35% Zn Cu alloy: Sheet and bar; extra spring hard; available in all degrees of hardness **UTS: 620**
ASTM B121/1	0.5% Pb 10% Zn Cu alloy: Sheet and bar; extra hard; available in all degrees of hardness **UTS: 490**
ASTM B121/2	0.5% Pb 35% Zn Cu alloy: Sheet and bar; extra spring hard; available in all degrees of hardness **UTS: 620**
ASTM B121/3	1.1% Pb 35% Zn Cu alloy: Sheet and bar; extra spring hard; available in all degrees of hardness **UTS: 620**
ASTM B121/4	1.8% Pb 37% Zn Cu alloy: Sheet and bar; extra spring hard; available in all degrees of hardness **UTS: 620**
ASTM B121/5	2% Pb 35% Zn Cu alloy: Sheet and bar; extra spring hard; available in all degrees of hardness **UTS: 620**
ASTM B124/13	0.7% Sn 1.5% Pb 37% Zn Cu alloy: Forging and bar; mechanical properties not quoted
ASTM B129	30% Zn Cu alloy: For cartridge cups; mechanical properties not quoted
ASTM B130	10% Zn Cu alloy: Strip; ¼-hard **UTS: 320**
ASTM B130	10% Zn Cu alloy: Strip; extra spring hard **UTS: 530**
ASTM B131	10% Zn Cu alloy: For pressings; for bullet jacket cups
ASTM B134/1	5% Zn Cu alloy: Wire; ¼-hard **UTS: 300**
ASTM B134/1	5% Zn Cu alloy: Wire; spring hard **UTS: 510**
ASTM B134/2	10% Zn Cu alloy: Wire; ¼-hard **UTS: 370**
ASTM B134/2	10% Zn Cu alloy: Wire; spring hard **UTS: 600**
ASTM B134/3	15% Zn Cu alloy: Wire; ¼-hard **UTS: 420**

Symbol	Nominal analysis, supplier, condition and remarks.
ASTM B134/3	15% Zn Cu alloy: Wire; spring hard **UTS: 730**
ASTM B134/4	20% Zn Cu alloy: Wire; ¼-hard **UTS: 500**
ASTM B134/4	20% Zn Cu alloy: Wire; spring hard **UTS: 870**
ASTM B134/6	30% Zn Cu alloy: Wire; ¼-hard **UTS: 530**
ASTM B134/6	30% Zn Cu alloy: Wire; spring hard **UTS: 890**
ASTM B134/7	35% Zn Cu alloy: Wire; ¼-hard **UTS: 530**
ASTM B134/7	35% Zn Cu alloy: Wire; spring hard **UTS: 890**
ASTM B134/8	38% Zn Cu alloy: Wire; ¼-hard **UTS: 530**
ASTM B134/8	38% Zn Cu alloy: Wire; spring hard **UTS: 890**
ASTM B135/1	15% Zn Cu alloy: Tube; hard drawn **UTS: 400**
ASTM B135/2	30% Zn Cu alloy: Tube; hard drawn **UTS: 460**
ASTM B135/3	0.5% Pb 33% Zn Cu alloy: Tube; hard drawn **UTS: 460**
ASTM B135/4	1.5% Pb 32% Zn Cu alloy: Tube; hard drawn **UTS: 460**
ASTM B135/7	10% Zn Cu alloy: Tube; hard drawn **UTS: 360**
ASTM B138 A	1% Sn 39% Zn 1.5% Fe 0.1% Mn Cu alloy: Bar; annealed **UTS: 390 Elon: 20% Proof: 140**
ASTM B138 A	1% Sn 39% Zn 1.5% Fe 0.1% Mn Cu alloy: Bar; hard **UTS: 530 Elon: 10% Proof: 360**
ASTM B138 B	23% Zn 4.5% Al 3% Fe 3.5% Mn Cu alloy: Bar; annealed; manganese bronze **UTS: 620 Elon: 10% Proof: 330**
ASTM B138 B	23% Zn 4.5% Al 3% Fe 3.5% Mn Cu alloy: Bar; hard; manganese bronze **UTS: 820 Elon: 5% Proof: 480**
ASTM B140 A	2% Pb 15% Zn Cu alloy: Bar; annealed **UTS: 220 Elon: 15% Proof: 450**
ASTM B140 B	2% Pb 10% Zn 1% Ni Cu alloy: Bar; annealed **UTS: 220 Elon: 15% Proof: 450**
ASTM B171	Condenser tube plates in copper alloys; covers various types; see also Section 14G
ASTM B176 ZS331A	35% Zn 1% Si Cu alloy: Casting; die cast **UTS: 400 Elon: 15% Proof: 130**
ASTM B283	Cu alloy: Die forgings; various types; see also Section 14G
ASTM B 291	29% Zn 1% Mn Cu alloy: Sheet and strip; ¼-hard **UTS: 420**
ASTM B 291	29% Zn 1% Mn Cu alloy: Sheet and strip; extra spring hard **UTS: 750**
ASTM B 359/230	15% Zn Cu alloy: Finned tube
ASTM B 359/442	1.0% Sn 30% Zn Cu alloy: Finned tube
ASTM B 359/443	1.0% Sn 30% Zn Cu alloy: Finned tube
ASTM B 359/444	1.0% Sn 30% Zn Cu alloy: Finned tube
ASTM B 359/445	1.0% Sn 30% Zn Cu alloy: Finned tube
ASTM B453	Pb Zn Cu alloy: Rods; graded by CDAA system
ASTM B505	Cu alloys: For continuous casting; graded by CDAA alloys; see also Sections 14G and 14K
ASTM B508/411	0.5% Sn 9.5% Zn Cu alloy: Strip for flexible piping
ASTM B 642	5% Zn Cu alloy: Welded tube
ASTM B 706	15% Zn 1% Ni 1% Si Cu alloy: Pipe and tube
B 42 HC	3% Pb 35% Zn Cu alloy: Bar and forging; Manganese Bronze Ltd for BS alloy CZ124
B 76	1.5% Pb 35% Zn Cu alloy: Extrusion; Delta Metal Ltd
BARRONIA	4.0% Sn 0.5% Pb 12.5% Zn Cu alloy: Common name
BASIS BRASS	27% Zn Cu alloy: Sheet; origin unknown
BASIS QUALITY BRASS	23.5% Zn Cu alloy: Origin unknown
BATALBRA	22% Zn 2% Al 0.03% As Cu alloy: Tube; Birmingham Battery for BS alloy CZ110
BATNAVAL	1.2% Sn 38% Zn Cu alloy: Sheet and plate; Birmingham Battery for BS alloy CZ112
BATURNAL	2% Pb 35% Zn Cu alloy: Tube; Birmingham Battery for BS alloy CZ119
BINDING BRASS	1.5% Pb 35% Zn Cu alloy: Origin unknown
BOBIERE'S METAL	37% Zn Cu alloy: Origin unknown
BOLTOMET 506	3% Zn Cu alloy: T. Bolton; cap copper
BOLTOMET 510	10% Zn Cu alloy: T. Bolton for BS 713
BOLTOMET 514	15% Zn Cu alloy: T. Bolton for BS 712
BOLTOMET 516	20% Zn Cu alloy: T. Bolton for BS 711
BOLTOMET 518	30% Zn Cu alloy: T. Bolton for BS 267
BOLTOMET 520	35% Zn Cu alloy: T. Bolton for BS 266 and BS 2786
BOLTOMET 522	37% Zn Cu alloy: T. Bolton for BS 265
BOLTOMET 570	30% Zn Cu alloy: Wire; T. Bolton
BOLTOMET 607	1.25% Pb 27% Zn Cu alloy: Rod; T. Bolton
BOLTOMET 710	1.25% Sn 35% Zn Cu alloy: Naval brass; T. Bolton for BS 251, BS 409, BS 1541
BOLTOMET 731	10% Zn Cu alloy: T. Bolton; commercial bronze
BROWN METAL	15% Zn Cu alloy: Common name
BS 251 CZ112	1.25% Sn 35% Zn Cu alloy: Naval brass and forgings; as rolled **UTS: 360 Elon: 20%**
BS 265 CZ108	37% Zn Cu alloy: Strip; cold rolled **UTS: 360 Elon: 15%**
BS 266 CZ107	35% Zn Cu alloy: Sheet and strip; annealed **DPN: 80 UTS: 270 Elon: 45%**
BS 266 CZ107	35% Zn Cu alloy: Sheet and strip ¼-hard **DPN: 75(min)UTS: 330 Elon: 35%**
BS 266 CZ107	35% Zn Cu alloy: Sheet and strip; ½-hard **DPN: 110 UTS: 360 Elon: 20%**
BS 266 CZ107	35% Zn Cu alloy: Sheet and strip; hard **DPN: 135 UTS: 450 Elon: 3%**
BS 267 CZ106	30% Zn Cu alloy: Sheet and strip; cold rolled **DPN: 75 UTS: 270 Elon: 50%**
BS 378 CZ105	30% Zn 0.04% As Cu alloy: Arsenical brass tube; as drawn; for condenser tubes, etc. **DPN: 150**
BS 378 CZ110	20% Zn 2% Al 0.04% As Cu alloy: Aluminium brass; as drawn; for condenser tubes, etc. **DPN: 150**
BS 378 CZ111	1.2% Sn 28% Zn 0.04% As Cu alloy: Tube; as drawn; Admiralty brass for condenser tubes, etc. **DPN: 150**
BS 711 CZ103	20% Zn Cu alloy: Sheet and strip; annealed 80/20 brass **DPN: 80 UTS: 250 Elon: 40%**
BS 711 CZ103	20% Zn Cu alloy: Sheet and strip; ½-hard 80/20 brass **DPN: 95 UTS: 330 Elon: 10%**
BS 711 CZ103	20% Zn Cu alloy: Sheet and strip; hard 80/20 brass **DPN: 120 UTS: 390 Elon: 5%**
BS 712 CZ102	15% Zn Cu alloy: Sheet and strip; annealed 85/15 brass **DPN: 80 UTS: 240 Elon: 35%**
BS 712 CZ102	15% Zn Cu alloy: Sheet and strip; ½-hard 85/15 brass **DPN: 95 UTS: 330 Elon: 7%**
BS 712 CZ102	15% Zn Cu alloy: Sheet and strip; hard 85/15 brass **DPN: 110 UTS: 360 Elon: 3%**

Note. The following abbreviations and units are used in the tables:

DPN	Hardness, diamond pyramid number
UTS	Ultimate tensile strength, N/mm^2
Elon	Elongation, %
Proof	0.1% proof strength, N/mm^2

1 N/mm^2=0.1 hbar=0.102 kgf/mm^2=0.06475 $tonf/in.^2$=145.04 $lbf/in.^2$=1 MPa

See Appendix II for other abbreviations and conversion tables.

Symbol	Nominal analysis, supplier, condition and remarks.
BS 713 CZ101	10% Zn Cu alloy: Sheet and strip; annealed 90/10 brass **DPN: 80 UTS: 240 Elon: 35%**
BS 713 CZ101	10% Zn Cu alloy: Sheet and strip; ½-hard 90/10 brass **DPN: 95 UTS: 300 Elon: 7%**
BS 713 CZ101	10% Zn Cu alloy: Sheet and strip; hard 90/10 brass **DPN: 110 UTS: 350 Elon: 3%**
BS 885 CZ105	30% Zn 0.04% As Cu alloy: Tube; annealed 70/30 arsenical brass **UTS: 270**
BS 885 CZ110	20% Zn 2% Al 0.04% As Cu alloy: Tube; annealed aluminium brass **UTS: 330**
BS 920	Naval brass die casting; replaced by BS 1400
BS 1400 SCB1-C	2% Sn 3% Pb 21% Zn Cu alloy: Casting; sand cast **DPN: 50 UTS: 180 Elon: 30% Proof: 60**
BS 1400 SCB2-1	3.5% Pb 30% Zn Cu alloy: Brass ingots; sand cast **UTS: 170 Elon: 12%**
BS 1400 SCB2-C	3.5% Pb 30% Zn Cu alloy: Brass castings; sand cast **UTS: 170 Elon: 12%**
BS 1400 SCB3-1	2.5% Pb 30% Zn Cu alloy: Brass ingots; sand cast **UTS: 180 Elon: 12%**
BS 1400 SCB3-C	2.5% Pb 30% Zn Cu alloy: Brass castings; sand cast **UTS: 180 Elon: 12%**
BS 1400 SCB5-C	1% Sn 12% Zn Cu alloy: Casting; brazeable sand cast **DPN: 50 UTS: 200 Elon: 30% Proof: 60**
BS 1400 SCB6-1	15% Zn Cu alloy: Brass ingots; sand cast, brazeable **UTS: 170 Elon: 20%**
BS 1400 SCB6-C	15% Zn Cu alloy: Brass castings; sand cast, brazeable **UTS: 170 Elon: 20%**
BS 1402	30% Zn Cu alloy: Tube; solid drawn and annealed; for gas installation work
BS 1403	30% Zn Cu alloy: Tube; annealed and cold drawn; for gas lighting
BS 1464 CZ105	30% Zn 0.04% As Cu alloy: Tube; arsenical brass **DPN: 95**
BS 1464 CZ110	20% Zn 2% Al Cu alloy: Tube; aluminium brass **DPN: 95**
BS 1464 CZ111	1.2% Sn 28% Zn Cu alloy: Tube; Admiralty brass **DPN: 95**
BS 1541 CZ110	20% Zn 1.8% Al 0.04% As Cu alloy: Plate; aluminium brass, as rolled **UTS: 270 Elon: 40%**
BS 1541 CZ112	1% Sn 35% Zn Cu alloy: Plate; Naval brass; as rolled **UTS: 340 Elon: 20%**
BS 1845 CZ1	49% Zn Cu alloy: For brazing; melting range 860–870 °C
BS 1845 CZ2	46% Zn Cu alloy: For brazing; melting range 870–880 °C
BS 1845 CZ3	40% Zn Cu alloy: For brazing; melting range 885–890 °C
BS 1845 CZ4	1.0% Sn 46% Zn Cu alloy: For brazing; melting range 860–870 °C
BS 1845 CZ5	1.0% Sn 40% Zn Cu alloy: For brazing; melting range 880–890 °C
BS 1845 CZ6	40% Zn 0.3% Si Cu alloy: For brazing; melting range 875–895 °C
BS 1845 CZ7	40% Zn 0.1% Mn 0.2% Si Cu alloy: For brazing; melting range 870–900 °C
BS 2785 CZ118	1% Pb 35% Zn Cu alloy: Strip; ½-hard; for clocks and watches **DPN: 130 UTS: 370 Elon: 10%**
BS 2785 CZ118	1% Pb 35% Zn Cu alloy: Strip; extra hard; for clocks and watches **DPN: 180 UTS: 510 Elon: 3%**
BS 2785 CZ119	2% Pb 35% Zn Cu alloy: Strip; ½-hard; for clocks and watches **DPN: 130 UTS: 390 Elon: 10%**
BS 2785 CZ119	2% Pb 35% Zn Cu alloy: Strip; extra hard; for clocks and watches **DPN: 180 UTS: 510 Elon: 3%**
BS 2785 CZ120	2% Pb 38% Zn Cu alloy: Strip; ½-hard; for clocks and watches **DPN: 130 UTS: 510 Elon: 10%**
BS 2785 CZ120	2% Pb 38% Zn Cu alloy: Strip; extra hard; for clocks and watches **DPN: 180 UTS: 560 Elon: 5%**
BS 2786 CZ107	35% Zn Cu alloy: Wire; hard drawn **UTS: 750**
BS 2870 CZ101	10% Zn Cu alloy: Sheet; annealed **DPN: 80 UTS: 240 Elon: 35%**
BS 2870 CZ101	10% Zn Cu alloy: Sheet; ½-hard **UTS: 900 Elon: 7%**
BS 2870 CZ101	10% Zn Cu alloy: Sheet; hard **DPN: 105 UTS: 330 Elon: 3%**
BS 2870 CZ102	15% Zn Cu alloy: Sheet; annealed **DPN: 80 UTS: 240 Elon: 35%**
BS 2870 CZ102	15% Zn Cu alloy: Sheet; ½-hard **DPN: 90 UTS: 300 Elon: 7%**
BS 2870 CZ102	15% Zn Cu alloy: Sheet; hard **DPN: 105 UTS: 340 Elon: 3%**
BS 2870 CZ103	20% Zn Cu alloy: Sheet; annealed **DPN: 80 UTS: 250 Elon: 40%**
BS 2870 CZ103	20% Zn Cu alloy: Sheet; ½-hard **DPN: 90 UTS: 330 Elon: 10%**
BS 2870 CZ103	20% Zn Cu alloy: Sheet; hard **DPN: 115 UTS: 370 Elon: 5%**
BS 2870 CZ104	1% Pb 20% Zn Cu alloy: Sheet; specification not issued
BS 2870 CZ105	30% Zn 0.05% As Cu alloy: Sheet; specification not issued
BS 2870 CZ106	30% Zn Cu alloy: Sheet; annealed **DPN: 80 UTS: 270 Elon: 50%**
BS 2870 CZ106	30% Zn Cu alloy: Sheet; ¼-hard **DPN: 75 UTS: 330 Elon: 35%**
BS 2870 CZ106	30% Zn Cu alloy: Sheet; ¼-hard **DPN: 100 UTS: 340 Elon: 20%**
BS 2870 CZ106	30% Zn Cu alloy: Sheet; hard **DPN: 120 UTS: 390 Elon: 3%**
BS 2870 CZ107	35% Zn Cu alloy: Sheet; annealed **DPN: 80 UTS: 270 Elon: 45%**
BS 2870 CZ107	35% Zn Cu alloy: Sheet; ¼-hard **DPN: 75 UTS: 320 Elon: 35%**
BS 2870 CZ107	35% Zn Cu alloy: Sheet; ½-hard **DPN: 100 UTS: 360 Elon: 20%**
BS 2870 CZ107	35% Zn Cu alloy: Sheet; hard **DPN: 130 UTS: 420 Elon: 3%**
BS 2870 CZ107	35% Zn Cu alloy: Sheet; extra hard **DPN: 165 UTS: 510**
BS 2870 CZ108	38% Zn Cu alloy: Sheet; common brass; annealed **DPN: 80 UTS: 270 Elon: 40%**
BS 2870 CZ108	38% Zn Cu alloy: Sheet; ¼-hard **DPN: 75(min)UTS: 330 Elon: 30%**
BS 2870 CZ108	38% Zn Cu alloy: Sheet; ½-hard **DPN: 100 UTS: 360 Elon: 15%**
BS 2870 CZ108	38% Zn Cu alloy: Sheet; hard **DPN: 130 UTS: 420 Elon: 3%**
BS 2870 CZ108	38% Zn Cu alloy: Sheet; common brass; extra hard **DPN: 165 UTS: 510**
BS 2870 CZ110	20% Zn 2% Al 0.05% As Cu alloy: Sheet; aluminium brass
BS 2870 CZ111	1.2% Sn 28% Zn 0.05% As Cu alloy: Sheet; Admiralty brass
BS 2870 CZ112	1.2% Sn 35% Zn Cu alloy: Sheet; annealed Naval brass **UTS: 330 Elon: 25%**
BS 2870 CZ112	1.2% Sn 35% Zn Cu alloy: Sheet; hard; Naval brass **UTS: 390 Elon: 20%**
BS 2870 CZ116	30% Zn 4.5% Al 1% Fe 1.5% Mn Cu alloy: Sheet; high tensile brass; specification not issued
BS 2870 CZ118	1% Pb 35% Zn Cu alloy: Sheet; ½-hard **DPN: 120 UTS: 360 Elon: 10%**
BS 2870 CZ118	1% Pb 35% Zn Cu alloy: Sheet; hard **DPN: 150 UTS: 420 Elon: 5%**

Symbol	Nominal analysis, supplier, condition and remarks.
BS 2870 CZ118	1% Pb 35% Zn Cu alloy: Sheet; extra hard
	DPN: 180 UTS: 510 Elon: 3%
BS 2870 CZ119	2% Pb 35% Zn Cu alloy: Sheet; $\frac{1}{2}$-hard
	DPN: 120 UTS: 360 Elon: 10%
BS 2870 CZ119	2% Pb 35% Zn Cu alloy: Sheet; hard
	DPN: 150 UTS: 420 Elon: 5%
BS 2870 CZ119	2% Pb 35% Zn Cu alloy: Sheet; extra hard
	DPN: 180 UTS: 510 Elon: 3%
BS 2870 CZ124	3% Pb 35% Zn Cu alloy: Sheet; specification not issued
BS 2870 CZ125	4.0% Zn Cu alloy: Sheet; etc.; cap copper
	DPN: 75 UTS: 100
BS 2871 CZ101	10% Zn Cu alloy: Tube; specification not issued
BS 2871 CZ102	15% Zn Cu alloy: Tube; specification not issued
BS 2871 CZ103	20% Zn Cu alloy: Tube; specification not issued
BS 2871 CZ104	20% Zn Cu alloy: Tube; specification not issued
BS 2871 CZ105	30% Zn 0.04% As Cu alloy: Tube; arsenical brass; annealed
	UTS: 300
BS 2871 CZ106	30% Zn Cu alloy: Tube; specification not issued
BS 2871 CZ107	35% Zn Cu alloy: Tube; specification not issued
BS 2871 CZ108	38% Zn Cu alloy: Tube; common brass; specification not issued
BS 2871 CZ110	20% Zn 2% Al 0.03% As Cu alloy: Tube; aluminium brass; annealed
	UTS: 370
BS 2871 CZ111	1% Sn 28% Zn 0.03% As Cu alloy: Tube; Admiralty brass
BS 2871 CZ116	35% Zn 4.5% Al 1% Mn Cu alloy: Tube; specification not issued
BS 2871 CZ118	1% Pb 35% Zn Cu alloy: Tube; specification not issued
BS 2871 CZ119	2% Pb 35% Zn Cu alloy: Tube; annealed
	UTS: 330
BS 2871 CZ124	3% Pb 35% Zn Cu alloy: Tube; specification not issued
BS 2871 CZ126	30% Zn 0.04% As Cu alloy
BS 2871 CZ127	12% Zn 1% Al 1% Ni 1% Si Cu alloy
BS 2872 CZ101	10% Zn Cu alloy: Forging; specification not issued
BS 2872 CZ102	15% Zn Cu alloy: Forging; specification not issued
BS 2872 CZ103	20% Zn Cu alloy: Forging; specification not issued
BS 2872 CZ104	1% Pb 20% Zn Cu alloy: Forging; specification not issued
BS 2872 CZ105	30% Zn 0.04% As Cu alloy: Forging; specification not issued
BS 2872 CZ106	30% Zn Cu alloy: Forging; specification not issued
BS 2872 CZ107	35% Zn Cu alloy: Forging; specification not issued
BS 2872 CZ108	37% Zn Cu alloy: Forging; specification not issued
BS 2872 CZ110	20% Zn 2% Al 0.04% As Cu alloy: Forging; specification not issued
BS 2872 CZ111	1.2% Sn 28% Zn 0.04% As Cu alloy: Forging; specification not issued
BS 2872 CZ112	1.2% Sn 35% Zn Cu alloy: Forging; as forged
	UTS: 330 Elon: 20%
BS 2872 CZ116	1% Pb 30% Zn 4.5% Al 1% Mn Cu alloy: Forging; as forged
	UTS: 450 Elon: 15% Proof: 270
BS 2872 CZ118	1% Pb 35% Zn Cu alloy: Forging; specification not issued
BS 2872 CZ119	2% Pb 35% Zn Cu alloy: Forging; specification not issued
BS 2872 CZ124	3% Pb 35% Zn Cu alloy: Forging; specification not issued
BS 2873 CZ101	10% Zn Cu alloy: Wire; specification not issued
BS 2873 CZ102	15% Zn Cu alloy: Wire; annealed
	UTS: 280 Elon: 30%
BS 2873 CZ102	15% Zn Cu alloy: Wire; hard
	UTS: 580
BS 2873 CZ103	20% Zn Cu alloy: Wire; annealed
	UTS: 300 Elon: 35%
BS 2873 CZ103	20% Zn Cu alloy: Wire; hard
	UTS: 600
BS 2873 CZ104	0.8% Pb 20% Zn Cu alloy: Wire; specification not issued
BS 2873 CZ105	30% Zn 0.04% As Cu alloy: Wire; specification not issued
BS 2873 CZ106	30% Zn Cu alloy: Wire; annealed
	UTS: 300 Elon: 45%
BS 2873 CZ106	30% Zn Cu alloy: Wire; hard
	UTS: 600
BS 2873 CZ107	35% Zn Cu alloy: Wire; annealed
	UTS: 320 Elon: 40%
BS 2873 CZ107	35% Zn Cu alloy: Wire; extra hard
	UTS: 700
BS 2873 CZ108	38% Zn Cu alloy: Wire; annealed
	UTS: 320 Elon: 40%
BS 2873 CZ108	38% Zn Cu alloy: Wire; extra hard
	UTS: 680
BS 2873 CZ110	20% Zn 2% Al 0.04% As Cu alloy: Wire; specification not issued
BS 2873 CZ111	1.2% Sn 28% Zn 0.04% As Cu alloy: Wire; specification not issued
BS 2873 CZ112	1.2% Sn 35% Zn Cu alloy: Wire; specification not issued
BS 2873 CZ116	30% Zn 4.5% Al 1% Fe 1% Mn Cu alloy: Wire; specification not issued
BS 2873 CZ118	1% Pb 35% Zn Cu alloy: Wire; specification not issued
BS 2873 CZ119	2% Pb 35% Zn Cu alloy: Wire; specification not issued
BS 2873 CZ124	3% Pb 35% Zn Cu alloy: Wire; specification not issued
BS 2874 CZ101	10% Zn Cu alloy: Bar; specification not issued
BS 2874 CZ102	15% Zn Cu alloy: Bar; specification not issued
BS 2874 CZ103	20% Zn Cu alloy: Bar; as rolled
	UTS: 300 Elon: 25%
BS 2874 CZ104	0.7% Pb 20% Zn Cu alloy: Bar; as rolled
	UTS: 300 Elon: 25%
BS 2874 CZ105	30% Zn 0.04% As Cu alloy: Bar; specification not issued
BS 2874 CZ106	30% Zn Cu alloy: Bar; annealed
	UTS: 270 Elon: 50%
BS 2874 CZ107	35% Zn Cu alloy: Bar; specification not issued
BS 2874 CZ108	0.3% Pb 38% Zn Cu alloy: Bar; specification not issued
BS 2874 CZ110	20% Zn 2% Al 0.04% As Cu alloy: Bar; specification not issued
BS 2874 CZ111	1.2% Sn 28% Zn 0.04% As Cu alloy: Bar
BS 2874 CZ116	30% Zn 4.5% Al 1% Fe 1% Mn Cu alloy: Bar; as rolled
	UTS: 530 Elon: 15% Proof: 270
BS 2874 CZ118	1% Pb 35% Zn Cu alloy: Bar; specification not issued
BS 2874 CZ119	2% Pb 35% Zn Cu alloy: Bar; as rolled
	UTS: 330 Elon: 25%
BS 2874 CZ124	3% Pb 35% Zn Cu alloy: Bar; annealed
	UTS: 320 Elon: 15% Proof: 120
BS 2874 CZ124	3% Pb 35% Zn Cu alloy: Bar; hard drawn
	UTS: 530 Elon: 4% Proof: 220
BS 2874 CZ132	2.3% Pb 36% Zn 0.1% As Cu alloy
BS 2875 CZ101	10% Zn Cu alloy: Plate; specification not issued
BS 2875 CZ102	15% Zn Cu alloy: Plate; specification not issued
BS 2875 CZ103	20% Zn Cu alloy: Plate; specification not issued
BS 2875 CZ104	1% Pb 20% Zn Cu alloy: Plate; specification not issued
BS 2875 CZ105	30% Zn 0.04% As Cu alloy: Plate; annealed
	UTS: 270 Elon: 45%

Note. The following abbreviations and units are used in the tables:

DPN	Hardness, diamond pyramid number
UTS	Ultimate tensile strength, N/mm^2
Elon	Elongation, %
Proof	0.1% proof strength, N/mm^2

$1\ N/mm^2 = 0.1\ hbar = 0.102\ kgf/mm^2 = 0.06475\ tonf/in.^2 = 145.04\ lbf/in.^2 = 1\ MPa$

See Appendix II for other abbreviations and conversion tables.

Symbol	Nominal analysis, supplier, condition and remarks.
BS 2875 CZ105	30% Zn 0.04% As Cu alloy: Plate; hard **UTS: 340 Elon: 22%**
BS 2875 CZ106	30% Zn Cu alloy: Plate; annealed **UTS: 270 Elon: 45%**
BS 2875 CZ106	30% Zn Cu alloy: Plate; hard **UTS: 340 Elon: 22%**
BS 2875 CZ107	35% Zn Cu alloy: Plate; specification not issued
BS 2875 CZ108	38% Zn Cu alloy: Plate; hard **UTS: 370 Elon: 20%**
BS 2875 CZ110	20% Zn 2% Al 0.04% As Cu alloy: Plate; annealed **UTS: 340 Elon: 20%**
BS 2875 CZ110	20% Zn 2% Al 0.04% As Cu alloy: Plate; hard **UTS: 390 Elon: 20%**
BS 2875 CZ111	1.2% Sn 28% Zn 0.04% As Cu alloy: Plate; specification not issued
BS 2875 CZ112	1.2% Sn 35% Zn Cu alloy: Plate
BS 2875 CZ116	30% Zn 4.5% Al 1% Fe 1% Mn Cu alloy: Plate; specification not issued
BS 2875 CZ124	3% Pb 35% Zn Cu alloy: Plate; specification not issued
BS 2901 C14	1.2% Sn 30% Zn 0.04% As Cu alloy: Rod for inert gas shielded arc welding
BS 2901 C15	30% Zn 2% Al 0.04% As Cu alloy: Rod for inert gas shielded arc welding
BS 3127/3	35% Zn Cu alloy: Tubing; ¼-hard drawn for bourdon tubes **DPN: 120**
BS B11	20% Zn Cu alloy: Bar; used as brazing filler wire **UTS: 300 Elon: 25% Proof: 60**
C 14	1.2% Sn 30% Zn 0.04% As Cu alloy: Welding rod; designation used by BS
C 15	30% Zn 2% Al 0.04% As Cu alloy: Welding rod; designation used by BS
C 20500	1.5% Zn Cu alloy: CDA and UNS designation
C 21000	5% Zn Cu alloy: Gilding metal; CDA and UNS designation
C 22000	10% Zn Cu alloy: Commercial bronze; CDA and UNS designation
C 22600	12% Zn Cu alloy: Jewellery bronze; CDA and UNS designation
C 23000	15% Zn Cu alloy: Red brass; CDA and UNS designation
C 23030	16% Zn 0.2% Si Cu alloy: CDA and UNS designation
C 23400	18% Zn Cu alloy: CDA and UNS designation
C 24000	20% Zn Cu alloy: CDA and UNS designation
C 24080	21% Zn Cu alloy: CDA and UNS designation
C 25000	25% Zn Cu alloy: CDA and UNS designation
C 26000	30% Zn Cu alloy: Cartridge brass; CDA and UNS designation
C 26100	32% Zn 0.03% P Cu alloy: CDA and UNS designation
C 26130	32% Zn 0.06% As Cu alloy: CDA and UNS designation
C 26200	32% Zn Cu alloy: CDA and UNS designation
C 26380	30% Zn Cu alloy: CDA and UNS designation
C 26800	34% Zn Cu alloy: Yellow brass; CDA and UNS designation
C 27000	35% Zn Cu alloy: Yellow brass; CDA and UNS designation
C 27200	36% Zn Cu alloy: Yellow brass; CDA and UNS designation
C 27400	37% Zn Cu alloy: Yellow brass; CDA and UNS designation
C 28000	38% Zn Cu alloy: Muntz metal; CDA and UNS designation
C 29800	50% Zn Cu alloy: Obsolete; UNS designation
C 31000	0.5% Pb 10% Zn Cu alloy: UNS designation; obsolete
C 31200	1% Pb 10% Zn Cu alloy: CDA and UNS designation
C 31400	2% Pb 9% Zn Cu alloy: CDA and UNS designation
C 31600	2% Pb 9% Zn 1% Ni Cu alloy: CDA and UNS designation
C 32000	1.8% Pb 10% Zn Cu alloy: CDA and UNS designation
C 32500	2.7% Pb 36% Zn Cu alloy: UNS designation; obsolete
C 32510	0.5% Pb 32% Zn Cu alloy: CDA and UNS designation
C 33000	0.5% Pb 32% Zn Cu alloy: CDA and UNS designation
C 33100	1% Pb 32% Zn Cu alloy: CDA and UNS designation
C 33200	2% Pb 32% Zn Cu alloy
C 33500	0.5% Pb 35% Zn Cu alloy
C 33530	0.5% Pb 35% Zn Cu alloy
C 34000	1% Pb 35% Zn Cu alloy
C 34200	2% Pb 34% Zn Cu alloy
C 34400	0.7% Pb 35% Zn Cu alloy: UNS designation; obsolete
C 34500	2% Pb 34% Zn Cu alloy: CDA and UNS designation
C 34700	1.4% Pb 35% Zn Cu alloy: UNS designation; obsolete
C 34800	0.6% Pb 37% Zn Cu alloy: UNS designation; obsolete
C 34900	0.4% Pb 37% Zn Cu alloy: UNS designation; obsolete
C 35000	1.8% Pb 36% Zn Cu alloy
C 35300	2% Pb 37% Zn Cu alloy
C 35330	2.5% Pb 37% Zn Cu alloy
C 35340	2% Pb 37% Zn 0.2% Fe Cu alloy: CDA and UNS designation
C 35600	2.5% Pb 37% Zn Cu alloy: CDA and UNS designation
C 36000	3% Pb 37% Zn Cu alloy: CDA and UNS designation
C 36200	4% Pb 37% Zn Cu alloy
C 36500	0.5% Pb 38% Zn Cu alloy
C 36600	0.5% Pb 38% Zn Cu alloy: Arsenical
C 36700	0.5% Pb 38% Zn Cu alloy: Antimonial; CDA and UNS designation
C 36800	0.5% Pb 38% Zn Cu alloy: Phosphorized; CDA and UNS designation
C 37000	1% Pb 38% Zn Cu alloy: CDA and UNS designation
C 37100	1% Pb 38% Zn Cu alloy
C 37700	2% Pb 38% Zn Cu alloy: CDA and UNS designation
C 40900	0.7% Sn 5% Zn Cu alloy: UNS designation; obsolete
C 43800	1.2% Sn 18% Zn Cu alloy: UNS designation; obsolete
C 45450	0.3% Sn 0.2% Pb 35% Zn 0.3% Al Cu alloy: CDA and UNS designation
C 46200	0.7% Sn 35% Zn Cu alloy: Naval brass
C 46210	37% Zn Cu alloy
C 47200	3.5% Sn 37% Zn Cu alloy: Brazing; UNS designation; obsolete
C 47600	2% Sn 2% Pb 10% Zn 0.1% Mn Cu alloy: CDA and UNS designation
C 47940	1.7% Sn 1.5% Pb 32% Zn 0.4% Ni Cu alloy: CDA and UNS designation
C 48510	1.2% Sn 1.7% Pb 38% Zn Cu alloy: UNS designation; obsolete
C 66400	11.5% Zn 1.5% Fe Cu alloy: CDA and UNS designation
C 66410	11.5% Zn 2% Fe Cu alloy: CDA and UNS designation
C 66700	30% Zn 1% Mn Cu alloy
C 66800	35% Zn 2.7% Mn 1% Si Cu alloy
C 66900	26% Zn 12% Mn Cu alloy
C 68700	22% Zn 2% Al Cu alloy: CDA and UNS designation
C 68800	22.5% Zn 3.4% Al Cu alloy: CDA and UNS designation
C 69000	22.5% Zn 3.4% Al Cu alloy: CDA and UNS designation
C 69100	22% Zn 1% Al 1% Ni Cu alloy: CDA and UNS designation
C 69800	32% Zn 1% Si Cu alloy: CDA and UNS designation
C 83300	1.5% Sn 1.5% Pb 4% Zn Cu alloy: Cast; CDA and UNS designation
C 83400	10% Zn Cu alloy: Cast; CDA and UNS designation
C 83410	1.5% Sn 9% Zn Cu alloy: Cast; CDA and UNS designation
C 83420	3% Sn 9% Zn Cu alloy: Cast; CDA and UNS designation
C 83450	3% Sn 2.5% Pb 6% Zn Cu alloy: Cast; CDA and UNS designation
C 85200	1.2% Sn 2.7% Pb 23% Zn Cu alloy: Cast; CDA and UNS designation
C 85210	2% Sn 3.5% Pb 22% Zn Cu alloy: Cast; CDA and UNS designation

Symbol	Nominal analysis, supplier, condition and remarks.
C 85300	30% Zn Cu alloy: Casting; UNS designation; obsolete
C 85310	1% Sn 3.5% Pb 23% Zn Cu alloy: Cast; CDA and UNS designation
C 85400	1.2% Sn 2.2% Pb 28% Zn Cu alloy: Cast; CDA and UNS designation
C 85500	39% Zn Cu alloy
C 85600	38% Zn Cu alloy: Casting; UNS designation; obsolete
C 85700	1.2% Sn 1.2% Pb 36% Zn Cu alloy: Cast
C 85710	1.7% Pb 38% Zn Cu alloy: Cast
C 85800	1% Sn 1% Pb 36% Zn Cu alloy: Cast
C 86100	23% Zn 4% Al 3% Fe 3.2% Mn Cu alloy: Cast; CDA and UNS designation
C 86200	23.5% Zn 4% Al 3% Fe 3.2% Mn Cu alloy: Cast; CDA and UNS designation
C 86300	23.5% Zn 6% Al 3% Fe 3.2% Mn Cu alloy: Cast; CDA and UNS designation
C 86400	1% Sn 38% Zn 1% Al 1.2% Fe 0.7% Mn Cu alloy: Cast; CDA and UNS designation
C 86500	38% Zn 1% Al 1% Fe 1% Mn Cu alloy: Cast; CDA and UNS designation
C 86700	1% Sn 34% Zn 2% Al 2% Fe 2% Mn Cu alloy: Cast; CDA and UNS designation
C 86800	1% Sn 36% Zn 2% Al 2% Fe 3% Mn Cu alloy: Cast; CDA and UNS designation
C 87900	33% Zn 1% Si Cu alloy: Cast; CDA and UNS designation
CAP COPPER	3.5% Zn Cu alloy: Strip; common name; further information from Copper Development Association
CARO-BUHLER	Pb Zn P Cu alloy: For bar and tube; Carobronze; free machining
CARTRIDGE BRASS	30% Zn Cu alloy: Sheet; common name
CHINESE ART METAL	1% Sn 17.5% Pb 10% Zn Cu alloy: Origin unknown
CLOCK BRASS	2% Pb 37% Zn Cu alloy: Bar and sheet; American trade name
COMMERCIAL BRONZE	10% Zn Cu alloy: Common American name
CON	0.5% Sn 0.5% Pb 34% Zn 1.5% Al 1.75% Ni 1% Fe 2.75% Mn Cu alloy: Extrusions; McKechnie Bros; as drawn
	DPN: 150 UTS: 550 Elon: 20% Proof: 270
CORRONIUM	5% Sn 15% Zn Cu alloy: Origin unknown
CSC STAR	37.5% Zn Cu alloy: Extrusion; McKechnie Bros; annealed
	DPN: 90 UTS: 320 Elon: 45% Proof: 100
DIN 1709 G Ms 65	2% Pb 35% Zn Cu alloy
	DPN: 60 UTS: 200 Elon: 20% Proof: 75
DIN 1785 K Ms 63	35% Zn Cu alloy
DIN 1785 K Ms 72	30% Zn Cu alloy: Wrought
DIN 1785 So Ms 71	1% Sn 30% Zn Cu alloy: Wrought
DIN 1785 So Ms 76	20% Zn 2% Al Cu alloy: Wrought
DIN 8513 L Cu Zn 40	Zn Si Cu alloy: Welders rod
DIN 17656 GBMs 65 A	2% Pb 35% Zn Cu alloy: Casting
DIN 17660	Specification covering a series of copper zinc alloys; designations prefixed Ms are included

Note. The following abbreviations and units are used in the tables:

DPN Hardness, diamond pyramid number
UTS Ultimate tensile strength, N/mm^2
Elon Elongation, %
Proof 0.1% proof strength, N/mm^2

1 N/mm^2=0.1 hbar=0.102 kgf/mm^2=0.06475 tonf/in.^{2}=145.04 lbf/in.^{2}=1 MPa

See Appendix II for other abbreviations and conversion tables.

Symbol	Nominal analysis, supplier, condition and remarks.
DIN 17661	Copper zinc alloys; with additions; designations with prefix So Ms
DTD 253 A	10% Zn 1% Al 1% Ni 1% Si Cu alloy: Tube
	UTS: 450
DTD 283 A	15% Zn 1% Al 1% Ni 1% Si Cu alloy: Sheet; annealed
	UTS: 370
DTD 307	10% Zn 1.5% Mn 2% Si Cu alloy: Annealed; solid drawn tube
	UTS: 370
DTD 318 A	4% Sn 15% Zn 1% Fe Cu alloy: Tube; annealed; solid drawn
	UTS: 450
DTD 319	12% Zn 1% Al 1% Ni 1% Si Cu alloy: Bars; as rolled
	UTS: 450 Elon: 20% Proof: 250
DTD 323 B	12% Zn 1% Al 1% Ni 1% Si Cu alloy: Tube; annealed
	UTS: 400
DTD 337	4.0% Zn Cu alloy: Tube for radiators
DTD 387	4.0% Zn Cu alloy: Tube for radiators
DTD 604	1% Sn 30% Zn Cu alloy: Tubes (Sn optional); annealed; solid drawn
	UTS: 300 Proof: 60
DTD 627	1.8% Pb 35% Zn Cu alloy: Bar; not suitable for rivets
DTD 5009	15% Zn 1% Al 1% Ni 1% Si Cu alloy: Bar and wire; cold drawn for springs
DTD 5019	15% Zn 1.0% Al 1.2% Ni 1.0% Si Cu alloy: Tube; annealed
	UTS: 450 Elon: 40%
E 48	Pb 30% Zn Cu alloy: Extrusions; McKechnie Bros; leaded variety of 311
E 56	30% Zn As (trace) Cu alloy: Extrusions; McKechnie Bros; resists de-zincification
E 78	10% Zn Cu alloy: Extrusions; McKechnie Bros; annealed
	DPN: 90 UTS: 250 Elon: 50% Proof: 90
F 7S	1.2% Sn 0.5% Pb 37% Zn Cu alloy: Extrusions; McKechnie Bros; as drawn
	DPN: 115 UTS: 39 Elon: 30% Proof: 14
F 7S STAR	1.2% Sn 37% Zn Cu alloy: Extrusions; McKechnie Bros; lead-free variety of F 7S
G Cu 65 Zn	2% Pb 35% Zn Cu alloy: Casting; designation used by German Standards
G Ms 65	2% Pb 35% Zn Cu alloy: Casting; designation used by German Standards
GB Cu 65 Zn	2% Pb 35% Zn Cu alloy: Casting; designation used by German Standards
GB Ms 65 A	2% Pb 35% Zn Cu alloy: Casting; designation used by German Standards
GB Ms 65 B	2% Pb 35% Zn Cu alloy: Casting; designation used by German Standards
GB Ms 65 C	2% Pb 35% Zn Cu alloy: Casting; designation used by German Standards
GILDING METAL	10% Zn Cu alloy: Common name; further information from Copper Development Association
HARDWARE BRONZE	2% Pb 13% Zn Cu alloy: Rod; American trade name
HECORROS 76	20% Zn 2% Al 0.03% As Cu alloy: VDM
IMI 210	10% Zn Cu alloy: Gilding metal; IMI for BS alloy CZ101
IMI 212	12% Zn Cu alloy: Gilding metal; IMI
IMI 215	15% Zn Cu alloy: Gilding metal; IMI for BS alloy CZ102
IMI 220	20% Zn Cu alloy: Gilding metal; IMI for BS alloy CZ103
IMI 230	30% Zn Cu alloy: Bar; IMI for BS alloy CZ106
IMI 237	37% Zn Cu alloy: Bar and wire; IMI for BS alloy CZ108
IMI 239	37% Pb Zn Cu alloy: Wire; IMI
IMI 276	30% Zn As Cu alloy: Bar
IMI 303	22% Zn 2% Al Cu alloy: Welding wire; IMI 'Alumbro'; withdrawn
IMI 345	1% Sn 11% Zn Cu alloy: Wire; IMI

Symbol	Nominal analysis, supplier, condition and remarks.
IMI 432	2% Pb 35% Zn Cu alloy: Bar; IMI for BS alloy CZ119; nipple wire
IMI 433	2% Pb 35% Zn Cu alloy: Wire; IMI for BS alloy CZ119
IMI 443	0.25% Pb 37% Zn Cu alloy: Free machining; IMI
IMMADIUM II	30% Zn Al Fe Mn Cu alloy: Manganese Bronze Ltd
IMMADIUM V	1% Pb 30% Zn 4.5% Al 1% Mn Cu alloy: Bar and forging; Manganese Bronze Ltd for BS alloy CZ116
IMMADIUM VI	28% Zn Al Fe Mn Cu alloy: Manganese Bronze Ltd
JEWELRY BRONZE	12.5% Zn Cu alloy: Wrought; used for Au plated parts; traditional name
K Ms 63	35% Zn Cu alloy: Wrought; designation used by German Standards
K Ms 72	30% Zn Cu alloy: Wrought; designation used by German Standards
KUNIFORM	Zn/Cu strip; fine grained; IMI; covers 60/40 to 90/10 Cu/Zn alloys
LBZ	1% Sn 2.4% Pb 35% Zn 0.4% Al 0.5% Fe 1.75% Mn Cu alloy: Extrusions; McKechnie Bros; as drawn **DPN: 130 UTS: 480 Elon: 20% Proof: 210**
LED-O-LOY	2% Sn 2% Pb 9% Zn 0.05% P Cu alloy: Strip; American trade name
LOW BRASS	20% Zn Cu alloy: Common American name
MANGANESE BRONZE	30% Zn 4% Al 3% Fe 3% Mn Cu alloy: Common name; analysis varies considerably
METER BRONZE	0.6% Sn 7% Zn Cu alloy: American trade name
Ms 56	2% Pb 40% Zn Cu alloy: Wrought; designation used by German Standards
Ms 58	2% Pb 40% Zn Cu alloy: Wrought; designation used by German Standards
Ms 60	40% Zn Cu alloy: Wrought; designation used by German Standards
Ms 60 Pb	2.0% Pb 38% Zn Cu alloy: Wrought; designation used by German Standards
Ms 63	37% Zn Cu alloy: Wrought; designation used by German Standards
Ms 63 Pb	2.0% Pb 3.6% Zn Cu alloy: Wrought; designation used by German Standards
Ms 67	33% Zn Cu alloy: Wrought; designation used by German Standards
Ms 72	30% Zn Cu alloy: Wrought; designation used by German Standards
Ms 80	20% Zn Cu alloy: Wrought; designation used by German Standards
Ms 85	15% Zn Cu alloy: Wrought; designation used by German Standards
Ms 90	10% Zn Cu alloy: Wrought; designation used by German Standards
MUNTZ	0.3% Pb 38% Zn Cu alloy: Common name; lead not always present; further information from the Copper Development Association
NEBALOY	37% Zn Cu alloy: Strip; American trade name
NIPPLE	2% Pb 35% Zn Cu alloy: Wire; common name
QQ B 601/A	0.3% Pb 15% Zn Cu alloy: Casting; US Federal
QQ B 601/B	0.5% Pb 15% Zn Cu alloy: Casting; US Federal
QQ B 611a/E	30% Zn Cu alloy: Plate or sheet; US Federal
QQ B 621/B	0.5% Sn 2.0% Pb 30% Zn Cu alloy: Casting; US Federal **UTS: 200 Elon: 20%**
QQ B 621/C	1.5% Sn 3.0% Pb 20% Zn Cu alloy: Casting; US Federal **UTS: 220 Elon: 25%**
QQ B 721a/B	30% Zn 4.0% Al 3.0% Fe 4% Mn Cu alloy: Bar and forging; US Federal **UTS: 600 Elon: 20% Proof: 270**
QQ B 726/B	0.2% Pb 30% Zn 5% Al 3% Fe 3% Mn Cu alloy: Casting; US Federal 'manganese bronze' **UTS: 620 Elon: 20% Proof: 300**
QQ B 726/C	0.2% Pb 30% Zn 5% Al 3% Fe 3% Mn Cu alloy: Casting; US Federal **UTS: 750 Elon: 12% Proof: 420**
QQ C 591/A	1.5% Sn 4.2% Zn 1.5% Mn 3.0% Si Cu alloy: Sheet; US Federal **UTS: 370 Elon: 50% Proof: 120**
QQ C 591a/D	2.2% Sn 4.0% Zn 2.0% Si Cu alloy: Bar; US Federal **UTS: 600 Elon: 20% Proof: 300**
QQ S 551/C	0.3% Pb 30% Zn Cu alloy: Braze filler; US Federal; melting range 899–960 °C
QQ S 551/D	0.2% Pb 20% Zn Cu alloy: Braze filler; US Federal; melting range 968–996 °C
QQ W 321/A	0.1% Pb 35% Zn Cu alloy: Wire; US Federal; annealed **UTS: 300**
QQ W 321/C	28% Zn Cu alloy: Wire; US Federal; hard drawn **UTS: 820**
RED BRASS	15% Zn Cu alloy: Common name
S1B	35% Zn 1.75% Al 3% Mn 1% Si Cu alloy: Extrusions; McKechnie Bros; as extruded **DPN: 130 UTS: 580 Elon: 15% Proof: 270**
SAE 70A	30% Zn Cu alloy: Sheet and strip; ¼-hard **UTS: 370**
SAE 70A	30% Zn Cu alloy: Sheet and strip; extra hard **UTS: 620**
SAE 70B	30% Zn Cu alloy: Sheet and strip; ¼-hard **UTS: 370**
SAE 70B	30% Zn Cu alloy: Sheet and strip; extra hard **UTS: 620**
SAE 70C	35% Zn Cu alloy: Sheet and strip; ¼-hard **UTS: 370**
SAE 70C	35% Zn Cu alloy: Sheet and strip; extra hard **UTS: 600**
SAE 72	3% Pb 35% Zn Cu alloy: Bar; annealed **UTS: 300 Elon: 25% Proof: 100**
SAE 72	3% Pb 35% Zn Cu alloy: Bar; ½-hard **UTS: 360 Elon: 20% Proof: 270**
SAE 74B	0.8% Pb 35% Zn Cu alloy: Tube; as drawn **UTS: 370**
SAE 74B	0.8% Pb 35% Zn Cu alloy: Tube; hard drawn **UTS: 460**
SAE 74C	30% Zn Cu alloy: Tube; as drawn **UTS: 370**
SAE 74C	30% Zn Cu alloy: Tube: hard drawn **UTS: 460**
SAE 74D	15% Zn Cu alloy: Tube: Light drawn **UTS: 370**
SAE 74D	15% Zn Cu alloy: Tube; hard drawn **UTS: 400**
SAE 79A	15% Zn Cu alloy: Sheet and strip; ¼-hard **UTS: 340**
SAE 79A	15% Zn Cu alloy: Sheet and strip; extra hard **UTS: 530**
SAE 79B	20% Zn Cu alloy: Sheet and strip; ¼-hard **UTS: 370**
SAE 79B	20% Zn Cu alloy: Sheet and strip; extra hard **UTS: 600**
SAE 80A	30% Zn Cu alloy: Wire; ¼-hard **UTS: 460**
SAE 80A	30% Zn Cu alloy: Wire; extra hard **UTS: 800**
SAE 80B	35% Zn Cu alloy: Wire; ¼-hard **UTS: 460**
SAE 80B	35% Zn Cu alloy: Wire; extra hard **UTS: 800**
SAE 795	0.5% Sn 10% Zn Cu alloy: For bearings; wrought
SAE 890	1.5% Pb 20% Zn Cu alloy: Sintering; density 750 kg/m^3
SAE 891	1.5% Pb 20% Zn Cu alloy: Sintering; density 770 kg/m^3
SAE CA 210	5% Zn Cu alloy: Gilding metal
SAE CA 220	10% Zn Cu alloy: Commercial bronze
SAE CA 230	15% Zn Cu alloy: Red brass; formerly SAE 74D and SAE 79A
SAE CA 240	20% Zn Cu alloy: Low brass; formerly SAE 79B
SAE CA 260	30% Zn Cu alloy: Cartridge brass; formerly SAE 70A, SAE 74C, SAE 80A

Symbol	Nominal analysis, supplier, condition and remarks.
SAE CA 268	33.7% Zn Cu alloy: Yellow brass sheet, formerly SAE 70C
SAE CA 270	35% Zn Cu alloy: Rod and wire; formerly SAE 80B
SAE CA 330	0.5% Pb 33.5% Zn Cu alloy: Tube; formerly SAE 74B
SAE CA 331	1% Pb 32.5% Zn Cu alloy: Wrought
SAE CA 342	2% Pb 33.5% Zn Cu alloy: Wrought
SAE CA 345	2% Pb 35% Zn Cu alloy: Wrought
SAE CA 350	1% Pb 36.5% Zn Cu alloy: Wrought
SAE CA 360	3% Pb 35.5% Zn Cu alloy: Formerly SAE 72
SAE CA 670	24.2% Zn 4.5% Al 3% Fe 3.7% Mn Cu alloy: Wrought
SAE CA 673	2% Pb 33.7% Zn 2.7% Mn 1% Si Cu alloy: Wrought
SAE CA 852	1% Sn 2% Pb 23% Zn Cu alloy: Casting
SAE CA 854	1% Sn 2% Pb 28% Zn Cu alloy: Casting
SAE CA 858	1% Sn 1% Pb 32% Zn Cu alloy: Casting
SAE CA 862	25% Zn 3% Fe 3.2% Mn 4% Al
SAE CA 863	25% Zn 6% Al 3% Fe 3.2% Mn Cu alloy: Casting
SAE CA 865	38% Zn 1% Al 1% Fe 1% Mn Cu alloy: Casting
SAE CA 872	1% Sn 5% Zn 2.5% Fe Cu alloy: Casting
SAE CA 874	1% Pb 14% Zn Cu alloy: Casting
SAE CA 875	14% Zn Cu alloy: Casting
SAE CA 878	14% Zn Cu alloy: Casting
SAE CA 879	35% Zn Cu alloy: Casting
SEVA	30% Zn 0.04% As Cu alloy: For tube; Yorkshire Imp Metals
For SIS specifications see SS	
So Ms 57 Al 2	37% Zn 2.0% Al 1.0% Ni 2.0% Mn Cu alloy: Wrought; designation used by German Standards
So Ms 58	0.2% Sn 37% Zn 1.5% Ni 1% Fe 0.2% Mn Cu alloy: Wrought; designation used by German Standards
So Ms 58 Al 1	38% Zn 1.0% Al 1.0% Ni 2.0% Mn Cu alloy: Wrought; designation used by German Standards
So Ms 58 Pb	1.5% Pb 37% Zn 0.5% Al 0.2% Ni 0.4% Fe 1.0% Mn Cu alloy: Wrought; designation used by German Standards
So Ms 59	37% Zn 1.0% Al 2.5% Ni 2.0% Mn Cu alloy: Wrought; designation used by German Standards
So Ms 60	0.8% Sn 39% Zn Cu alloy: Wrought
So Ms 71	1% Sn 30% Zn Cu alloy: Wrought; designation used by German Standards
So Ms 76	20% Zn 2% Al Cu alloy: Wrought; designation used by German Standards
SPINNING BRASS	1% Pb 37% Zn Cu alloy: Bar and sheet; American trade name
SPRABRASSY	34% Zn Cu alloy: Wire for spraying; Metco Ltd
SPRABRASSY	Zn Cu alloy: Spray deposit; Metco
SPRABRONZE C	10% Zn Cu alloy: Wire for spraying; Metco Ltd
SS 5112	15% Zn Cu alloy: Plate and bar; Swedish Standard; annealed **DPN: 70 UTS: 300 Elon: 50% Proof: 75**
SS 5112	15% Zn Cu alloy: Plate and bar; Swedish Standard; cold rolled **DPN: 130 UTS: 400 Elon: 12% Proof: 300**
SS 5114	20% Zn Cu alloy: Plate; Swedish Standard; annealed **DPN: 60 UTS: 300 Elon: 50% Proof: 75**
SS 5114	20% Zn Cu alloy: Plate; Swedish Standard; cold rolled **DPN: 110 UTS: 420 Elon: 20% Proof: 340**
SS 5122	30% Zn Cu alloy: Plate; Swedish Standard; annealed **DPN: 70 UTS: 340 Elon: 55% Proof: 100**
SS 5122	30% Zn Cu alloy: Plate; Swedish Standard; cold rolled **DPN: 140 UTS: 450 Elon: 15% Proof: 360**
SS 5124	35% Zn Cu alloy: Bar; Swedish Standard; annealed **DPN: 75 UTS: 300 Elon: 45% Proof: 90**
SS 5124	35% Zn Cu alloy: Bar; Swedish Standard; cold rolled **DPN: 140 UTS: 420 Elon: 10% Proof: 340**
SS 5140	1% Pb 35% Zn Cu alloy: Bar; Swedish Standard; annealed **DPN: 75 UTS: 320 Elon: 45% Proof: 90**
SS 5140	1% Pb 35% Zn Cu alloy: Bar; Swedish Standard; cold rolled **DPN: 150 UTS: 440 Elon: 10% Proof: 360**
SS 5144	1% Sn 2% Pb 30% Zn Cu alloy: Casting; Swedish Standard; as cast **DPN: 50 UTS: 200 Elon: 20% Proof: 90**
SS 5150	36% Zn Cu alloy: Plate and bar; Swedish Standard; annealed **DPN: 75 UTS: 360 Elon: 45% Proof: 100**
SS 5150	36% Zn Cu alloy: Plate and bar; Swedish Standard; cold rolled **DPN: 160 UTS: 600 Elon: 10% Proof: 580**
SS 5202	3.5% Sn 3.5% Pb 9% Zn Cu alloy: Casting; Swedish Standard; as cast **DPN: 70 UTS: 180 Elon: 15% Proof: 100**
SS 5217	31% Zn 2% Al 0.04% As Cu alloy: Bar; Swedish Standard; annealed **DPN: 90 UTS: 390 Elon: 45% Proof: 170**
SS 5220	1.1% Sn 28% Zn 0.04% As Cu alloy: Bar; Swedish Standard; annealed **DPN: 100 UTS: 360 Elon: 45% Proof: 170**
SS 5234	25% Zn 4.5% Al 2% Fe 3.5% Mn Cu alloy: Bar and forging; Swedish Standard; hot rolled **DPN: 200 UTS: 650 Elon: 10% Proof: 330**
SS 5236	2% Pb 30% Zn 2% Al 0.08% Mn Cu alloy: Bar; Swedish Standard; cold rolled **DPN: 140 UTS: 480 Elon: 20% Proof: 330**
SS 5238	1% Sn 0.8% Pb 34% Zn 0.8% Al 1% Fe 2% Mn Cu alloy: Bar and forging; Swedish Standard; cold rolled **DPN: 160 UTS: 510 Elon: 10% Proof: 340**
STA 7 CM4B	35% Zn 4.5% Al 0.7% Fe Cu alloy: Bar and forging **UTS: 530 Elon: 15% Proof: 270**
STA 7 CM4C	1% Sn 0.2% Pb 35% Zn 4% Al 1% Fe Cu alloy: Bar and forging **UTS: 530 Elon: 15% Proof: 270**
STA 7 CS1A	5% Zn 2% Fe 1% Mn 4% Si Cu alloy: Bar; annealed **UTS: 340 Elon: 45%**
STA 7 CS1A	5% Zn 2% Fe 1% Mn 4% Si Cu alloy: Bar; cold worked **UTS: 530 Elon: 13%**
STA 7 CS1B	5% Zn 2% Fe 1% Mn 4% Si Cu alloy: Sheet; cold worked **UTS: 530 Elon: 20%**
STA 7 CS1D	5% Zn 2% Fe 1% Mn 4% Si Cu alloy: Sheet; annealed **UTS: 340 Elon: 45%**
STA 7 CS2	5% Zn 1% Mn 2.5% Si Cu alloy: Tube; hard drawn **UTS: 530**
STA 7 CX1	15% Zn Cu alloy: Casting; gilding metal
STA 7 CX2	20% Zn Cu alloy: Casting
STA 7 CZ1	3.5% Zn Cu alloy: Strip; cap copper **DPN: 70**
STA 7 CZ2	10% Zn Cu alloy: Sheet; gilding metal **DPN: 70**
STA 7 CZ3	22% Zn 2% Al Cu alloy: Tube; for condenser tubes
STA 7 CZ4A	30% Zn Cu alloy: Tube
STA 7 CZ4B	1.2% Sn 29% Zn Cu alloy: Tube; annealed **UTS: 340**
STA 7 CZ4B	1.2% Sn 29% Zn Cu alloy: Tube; hard drawn **UTS: 450**
STA 7 CZ5	30% Zn Cu alloy: Sheet; annealed; cartridge brass **DPN: 70**

Note. The following abbreviations and units are used in the tables:

DPN	Hardness, diamond pyramid number
UTS	Ultimate tensile strength, N/mm^2
Elon	Elongation, %
Proof	0.1% proof strength, N/mm^2

$1\ N/mm^2 = 0.1\ hbar = 0.102\ kgf/mm^2 = 0.06475\ tonf/in.^2 = 145.04\ lbf/in.^2 = 1\ MPa$

See Appendix II for other abbreviations and conversion tables.

Symbol	Nominal analysis, supplier, condition and remarks.
STA 7 CZ5	30% Zn Cu alloy: Sheet; cold rolled; cartridge brass **DPN: 130**
STA 7 CZ6	35% Zn Cu alloy: Sheet; annealed **DPN: 75 UTS: 270 Elon: 45%**
STA 7 CZ6	35% Zn Cu alloy: Sheet; cold rolled **DPN: 110 UTS: 420 Elon: 17%**
STA 7 CZ7	37% Zn Cu alloy: Sheet; annealed **DPN: 75 UTS: 270 Elon: 40%**
STA 7 CZ7	37% Zn Cu alloy: Sheet; cold rolled **DPN: 110 UTS: 420 Elon: 17%**
STA 7 CZ8A	1.2% Sn 38% Zn Cu alloy: Bar **UTS: 360 Elon: 20%**
STA 7 CZ8B	1.2% Sn 38% Zn Cu alloy: Plate; annealed **UTS: 340 Elon: 25%**
STA 7 CZ8B	1.2% Sn 38% Zn Cu alloy: Plate; cold rolled **UTS: 390 Elon: 20%**
STA 7 CZ11A	2.5% Pb 35% Zn Cu alloy: Bar; annealed **UTS: 300 Elon: 25% Proof: 120**
STA 17	3.5% Zn Cu alloy: Strip for shells and caps; cap copper alloy **DPN: 70**
STA 19	10% Zn Cu alloy: Strip for bullet envelopes; gilding metal **DPN: 70**
STA 51	30% Zn Cu alloy: Bar for vent tubes
STUBS 1MS	Zn Si Cu alloy: For brazing and welding; Stubs; melting range 880–900 °C **UTS: 480 Elon: 26%**
STUBS 1S	Zn Si Cu alloy: Electrodes for brazing and welding; Stubs; melting range 860–880 °C **DPN: 180 UTS: 520 Elon: 17%**
STUBS 2H	Zn Ni Cu alloy: For hard surfacing; Stubs; good coefficient of friction **DPN: 260**
STUBS 2H M	Zn Ni Cu alloy: For hard surfacing; Stubs; good coefficient of friction **DPN: 260**
T8	Seamless brass tubes; annealed
T18	Brass tube; hard drawn, seamless; obsolete; see BS 885
T47	30% Zn Cu alloy: Tube for radiators; as drawn **UTS: 290**
T48	Brass tube for radiators; obsolete
TAURUS XXII	28% Zn 5% Al 1.5% Fe 1% Mn Cu alloy: Casting; D Brown; sand cast **DPN: 150 UTS: 600 Elon: 17% Proof: 300**
TAURUS XXIII	30% Zn 5.5% Al 1.5% Fe 1.5% Mn Cu alloy: Casting; D Brown; sand cast **DPN: 190 UTS: 760 Elon: 13% Proof: 420**
TAURUS XXIV	5% Sn 5% Pb 17% Zn Cu alloy: Casting; D Brown; centri-cast **DPN: 120 UTS: 340 Elon: 2% Proof: 270**
TAURUS XXX	2% Sn 10% Zn Cu alloy: Casting; D Brown; sand cast **DPN: 65 UTS: 200 Elon: 32%**
TUNGUM	10% Zn 1% Al 1.2% Ni 1% Si Cu alloy: Casting; Tungum Co.
TUNGUM	10% Zn 1% Al 1.2% Ni 1% Si Cu alloy: Sheet, bar and tube; Tungum Co.; as rolled **UTS: 600 Elon: 20% Proof: 330**
UZ 29E1	1.0% Sn 28.0% Zn Cu alloy: Origin unknown
VDM NAVAL BRASS	0.7% Sn 37% Zn Cu alloy: VDM
VIALBRA	22% Zn 2% Al Cu alloy: Origin unknown
WAS	Cu alloy: Welding rod; Delta metal for BS 1724A
WHEEL BRASS	2% Pb 30% Zn Cu alloy: Origin unknown
WW T 791/1	15% Zn Cu alloy: Seamless tube; US Federal
WW T 791/2	0.8% Pb 35% Zn Cu alloy: Seamless tube; US Federal
YALE BRONZE	1% Sn 1% Pb 7.75% Zn Cu bronze; Used for nuts and bolts; origin unknown
YELLOW BRASS	40% Zn Cu alloy: Common name; further information from Copper Development Association
YELLOW INGOT METAL	1% Sn 2% Pb 32% Zn Cu alloy: Used for plumber's fitting; common name
YORCALBRO	22% Zn 2% Al 0.04% As Cu alloy: Tube; Yorkshire Imperial Metals Ltd
YR	1.5% Pb 37% Zn Cu alloy: Extrusions; McKechnie Bros; as drawn **DPN: 125 UTS: 390 Elon: 27% Proof: 170**
YS	2.75% Pb 36% Zn Cu alloy: Extrusions; McKechnie Bros; as BS alloy CZ 124

Note. The following abbreviations and units are used in the tables:

DPN	Hardness, diamond pyramid number
UTS	Ultimate tensile strength, N/mm^2
Elon	Elongation, %
Proof	0.1% proof strength, N/mm^2

1 N/mm^2=0.1 hbar=0.102 kgf/mm^2=0.06475 $tonf/in.^2$=145.04 $lbf/in.^2$=1 MPa

See Appendix II for other abbreviations and conversion tables.

14C Copper–zinc wrought and cast alloys 37–50 per cent zinc – alpha + beta and beta alloys

Specific gravity	8.4
Density	8400 kg/m^3
Solidus/liquidus	890–920 °C
Thermal conductivity	109 W/m °C
Coefficient of linear expansion	21×10^{-6}/ °C
Electrical conductivity	25–28% IACS (copper 100%)
Specific resistance	62–68 microhm mm
Young's modulus of elasticity	103×10^9 N/m^2
Impact	40 J
Fatigue strength (10^7 cycles)	±150 N/mm^2
Hot strength	–

The above properties are typical of the following group, and may not apply exactly to any one specification. It is possible that with certain specifications some of the values may not be applicable.

General metallurgical characteristics

The alloys listed all have sufficient zinc – above 37% – to ensure the presence of some beta phase in the alpha matrix. This causes the alloys to be stronger than plain alpha brass (up to 38% zinc) (Section 14B) but means that ductility is reduced to such an extent that little or no cold forming or drawing can be carried out. The strength and lack of ductility increase proportionately up to about 46% zinc, at which point beta only is present and then a third phase – gamma – makes its appearance.

There will inevitably be some overlap between what are classed as true alpha brasses and the alpha–beta group. The alpha brasses are listed in Section 14B. With copper at 55–60% and alloying elements in addition to zinc, there is often sufficient overlap in the individual analysis for one specification to be alpha or alpha–beta depending on the actual analysis.

At 50% Zn only gamma is present and the alloys are so brittle as to be virtually useless. There is also a considerable drop in the corrosion resistance.

By far the largest and most important group is based on the 60% copper, 40% zinc analysis – the 60/40 brasses. This is a true alpha–beta brass, in that it has poor cold working properties but excellent hot working characteristics. Thus good quality forgings, stampings, plates, bars and hot extrusions can be produced from this relatively inexpensive material which has good corrosion resistance with acceptable strength, ductility and thermal and electrical properties.

Considerable modifications can be made to the basic alloy by additions of lead, tin, aluminium, silicon, manganese and nickel. In small quantities, generally replacing zinc, these metals do not affect the general metallurgy of the alloys but have the following effects.

Lead. Up to 3.5% lead additions are made to improve machinability. Bars with this amount of lead find considerable use in automachining when the lead acts as a lubricant and also produces small chips which do not clog the tool. There is a reduction in the impact value and corrosion resistance proportional to the lead content.

Tin. Up to 1% is added to improve the corrosion resistance. This gives a slight increase in tensile strength with no reduction in ductility. The alloy is termed Naval brass and is distinct from Admiralty, which is an alpha brass.

Iron. This strengthens the alpha phase, giving a higher tensile strength with little reduction in ductility. It is generally added with aluminium and manganese.

Aluminium. 1% aluminium has the same effect as about 6% zinc on the structure of copper–zinc alloys. In addition it considerably enhances the corrosion resistance, particularly to sea water and high temperature oxidation.

Silicon. Small amounts of silicon are sometimes added to 60/40 brass to improve the initial castability prior to forging. This also enhances the corrosion resistance. Other alloying elements are always present.

These alloys must not be confused with the silicon bronze series which have different and distinct properties.

Manganese. The term 'manganese bronze' was for many years applied to 60/40 brasses with any other alloying elements, if manganese was present. This is a misnomer as they are not true bronzes and manganese probably has the least effect of any of the alloying elements. It does give some strength to the alpha matrix but its main contribution is to form an oxide during forging which in addition to enhancing the corrosion resistance has a pleasing brown colour.

Nickel. This strengthens the alpha phase and considerably improves the corrosion resistance.

High tensile brasses. These are essentially 60/40 brasses with up to 2% of aluminium, tin and nickel, and about 1% of iron and manganese. Some have small amounts of lead added to improve machinability. Apart from the increase in tensile strength and corrosion resistance these materials have all the characteristics of alpha–beta brasses and are often incorrectly termed 'manganese bronze'.

High zinc alloys. These are beta alloys and are generally too brittle for most practical purposes. They can be readily hot worked but are prone to de-zincification and corrosion. Some use is made of them as brazing alloys owing to their relatively low melting range.

The mechanical properties of these materials cannot be improved by any form of heat treatment. They can be cold worked but become quite brittle with a limited amount of cold work and therefore require stress releasing or annealing at frequent intervals if cold work is applied. Care must be taken when softening as de-zincification can occur on some alloys at a relatively low temperature.

At higher temperatures grain growth can occur along with de-zincification which seriously affects the corrosion resistance and fatigue strength of the materials.

These materials can only be welded under very specialized conditions and expert advice is necessary.

Brazing and soldering can be readily carried out with care being taken to remove the fluxes which can be very corrosive.

Almost all these materials, particularly those containing some lead, have excellent machinability.

The materials in this group are used for hot stampings where they probably can produce the most economical high production materials where low strength is required. They are also excellent for extrusions and general engineering purposes, nuts, bolts, pipe fittings, etc.

Some of the materials are used as brazing rods.

Symbol	Nominal analysis, supplier, condition and remarks.
2.0340	40% Zn 0.8% Al Cu alloy: Casting; German Standard
2.0341	40% Zn 0.3% Al Cu alloy: Casting; German Standard
2.0343	40% Zn 0.3% Al Cu alloy: Casting; German Standard
2.0366	0.2% Sn 40% Zn 0.3% Si Cu alloy: Wrought; German Standard
2.0531	40% Zn Cu alloy: Wrought; German Standard
2.0590	1% Sn 40% Zn 1% Fe 2% Mn Cu alloy: Casting; German Standard
2.0592	40% Zn 2% Al 1.5% Ni 1.5% Fe 2.5% Mn Cu alloy: Casting; German Standard
2.0596	40% Zn 4% Al 1.5% Ni 2% Fe 3% Mn Cu alloy: Casting; German Standard
2.0598	40% Zn 7% Al 1.5% Ni 3% Fe 4% Mn Cu alloy: Casting; designation used by German Standards
4	3% Pb 39% Zn Cu alloy: Extrusions; McKechnie Bros; as drawn **DPN: 140 UTS: 400 Elon: 20% Proof: 200**
4M	40% Zn 5% Al 2% Ni 1% Fe 3% Mn Cu alloy: Casting; Langley Alloys
4VH	3% Pb 39% Zn Cu alloy: Extrusion code for Hidurit 11; McKechnie Bros; variant of 4
5	1% Pb 38% Zn Cu alloy: Extrusion; Delta Metal Ltd
5V	1% Pb 38% Zn Cu alloy: Extrusion; Delta Metal Ltd
7M	0.7% Sn 1% Pb 40% Zn 1% Al 1% Fe 1.5% Mn Cu alloy: Wrought; Langley Alloys; as Hidurit 15
9M	5% Sn 40% Zn 2.5% Al 1% Ni 3% Mn Cu alloy: Casting; Langley Alloys for Hidurit 12
10	1% Sn 1% Pb 37% Zn Cu alloy: Extrusion; Delta Metal Ltd
11M	40% Zn 5% Al 1% Ni 4% Mn Cu alloy: Casting; Langley Alloys; Hidurit 10
31	40% Zn Sn Pb Fe Al Ni Cu alloy: Extrusion; Delta Metal Ltd; alloying elements varied to suit requirements
37	Sn Pb 40% Zn Al Ni Fe Mn Cu alloy: Extrusion; Delta Metal Ltd; alloying elements varied to suit requirements
51A	50% Zn Cu alloy: Bronze filler; melting range 860–870 °C; Langley Alloys
51A	50% Zn Cu alloy: Brazing spelter; Delta Metal for BS 1845C
60L	1% Pb 38% Zn Cu alloy: Extrusion; Delta Metal Ltd
115	3% Pb 41% Zn Cu alloy: Extrusion; McKechnie Bros; as drawn **DPN: 100 UTS: 360 Elon: 22% Proof: 170**
210	2.5% Pb 40% Zn Cu alloy: Extrusion; McKechnie Bros; as drawn **DPN: 130 UTS: 400 Elon: 22% Proof: 200**
218	Sn Pb 40% Zn Al Ni Fe Mn Cu alloy: Extrusion; Delta Metal Ltd; alloying elements varied to suit requirements
280	40% Zn Cu alloy: Wrought; designation used by CDAA
298	50% Zn Cu alloy: For brazing; designation agreed by CDAA; not used
301	1% Pb 40% Zn Cu alloy: Extrusion; Delta Metal
312	40% Zn Cu alloy: Extrusion; McKechnie Bros; annealed **DPN: 100 UTS: 330 Elon: 40% Proof: 140**
380	2.0% Pb 40% Zn Cu alloy: Wrought; designation agreed by CDAA; not used
385	3% Pb 40% Zn Cu alloy: Wrought; designation used by CDAA
457	1.7% Pb 39% Zn Cu alloy: McKechnie Bros
462	2% Pb 39% Zn Cu alloy: McKechnie Bros
463	2.2% Pb 40% Zn Cu alloy: McKechnie Bros
465	0.7% Sn 40% Zn 0.07% As Cu alloy: Wrought; designation used by CDAA
466	0.7% Sn 40% Zn 0.07% Sb Cu alloy: Wrought; designation used by CDAA
466	4% Pb 39% Zn Cu alloy: McKechnie Bros
467	0.7% Sn 40% Zn 0.06% P Cu alloy: Wrought; designation used by CDAA
467	2.2% Pb 40% Zn Cu alloy: McKechnie Bros
470	0.8% Sn 40% Zn 0.01% Al Cu alloy: Brazing rod; designation agreed by CDAA; not used
472	3.5% Sn 50% Zn Cu brazing alloy: Designation agreed by CDAA; not used
473	3% Pb 39% Zn Cu alloy: McKechnie Bros
540	0.75% Sn 1.25% Pb 38.5% Zn 0.75% Fe 1.25% Mn Cu alloy: Extrusions; McKechnie Bros; as drawn **DPN: 130 UTS: 480 Elon: 20% Proof: 210**
671	0.7% Sn 0.2% Pb 40% Zn 0.2% Mn Cu alloy: Wrought; designation agreed by CDAA; not used
677	0.8% Pb 40% Zn 2.0% Ni 1.0% Fe 0.2% Mn 0.4% As Cu alloy: Wrought; designation agreed by CDAA; not used
680	0.9% Sn 40% Zn 0.6% Ni 0.4% Mn Cu alloy: Wrought; designation agreed by CDAA; not used
681	0.9% Sn 40% Zn 0.4% Mn Cu alloy: Wrought; designation agreed by CDAA; not used
796	1.0% Pb 40% Zn 10% Ni 2.0% Mn Cu alloy: Wrought; designation agreed by CDAA; not used
798	2.0% Pb 40% Zn 10% Ni 2.0% Mn Cu alloy: Wrought; designation agreed by CDAA; not used
850	41% Zn 0.2% Si Cu alloy: Braze filler rod; McKechnie Bros; melting range 890–900 °C
912	0.5% Pb 40% Zn 1.5% Al 2.5% Mn 0.7% Si Cu alloy: Bar and tube; Manganese Bronze Co.
A 0	40% Zn Cu alloy: Bar and forging; Manganese Bronze Co.; for BS alloy CZ109
A 71 NB	1.2% Sn 40% Zn Cu alloy: Bar and forging; Manganese Bronze Co.; for BS alloy CZ112
ABS TYPE 3	1.0% Sn (max) 43% Zn 3.2% Ni 1.2% Fe 2.0% (max) Al Cu alloy: Casting; American Bureau of Shipping **UTS: 530 Elon: 18% Proof: 225**
AITCH METAL	38% Zn 1.5% Fe Cu alloy: Obsolete; Blackwell

Note. The following abbreviations and units are used in the tables:

DPN	Hardness, diamond pyramid number
UTS	Ultimate tensile strength, N/mm^2
Elon	Elongation, %
Proof	0.1% proof strength, N/mm^2

$1\ N/mm^2 = 0.1\ hbar = 0.102\ kgf/mm^2 = 0.06475\ tonf/in.^2 = 145.04\ lbf/in.^2 = 1\ MPa$

See Appendix II for other abbreviations and conversion tables.

Symbol	Nominal analysis, supplier, condition and remarks.
AM	Sn Pb 40% Zn Al Ni Fe Mn Cu alloy: Extrusion; Delta Metal Ltd; alloying elements varied to suit requirements
AMPCOLOY 62	40% Zn 1% Al 1% Fe 0.2% Mn Cu alloy: Casting; Ampco
AMPCOLOY 64	0.2% Pb 30% Zn 5% Al 3% Fe 4% Mn Cu alloy: Casting; Ampco
AMPCOLOY 66	0.2% Pb 30% Zn 5% Al 3% Fe 4% Mn Cu alloy: Casting; Ampco
AMPCOLOY 666	Zn Mn Cu alloy: Manganese bronze; Ampco
AMS 4619 B	40% Zn 1% Al 1% Fe 0.2% Mn Cu alloy: Forging
AMS 4860 A	39% Zn 1% Al 1% Fe 0.3% Mn Cu alloy: Casting
AN QQ B 646	1.0% Sn 40% Zn Cu alloy: Bar; US Service **UTS: 420 Elon: 30% Proof: 200**
ARCHITECTURAL BRONZE C94	44% Zn Fe Mn Cu alloy: Manganese Bronze Co.
ASTM B21 A	0.7% Sn 40% Zn Cu alloy: Bar; Naval brass; annealed **UTS: 370 Elon: 30% Proof: 120**
ASTM B21 A	0.7% Sn 40% Zn Cu alloy: Bar; Naval brass; hard drawn **UTS: 460 Elon: 13% Proof: 330**
ASTM B21 B	0.7% Sn 0.7% Pb 40% Zn Cu alloy: Bar; Naval brass; annealed **UTS: 370 Elon: 25% Proof: 120**
ASTM B21 B	0.7% Sn 0.7% Pb 40% Zn Cu alloy: Bar; Naval brass; hard drawn **UTS: 460 Elon: 11% Proof: 330**
ASTM B21 C	0.7% Sn 1.8% Pb 40% Zn Cu alloy: Bar; Naval brass; annealed **UTS: 370 Elon: 20% Proof: 120**
ASTM B21 C	0.7% Sn 1.8% Pb 40% Zn Cu alloy: Bar; Naval brass; hard drawn **UTS: 460 Elon: 10% Proof: 330**
ASTM B30/6 C	1% Sn 1% Pb 36.5% Zn 0.3% Al Cu alloy: Ingot; as cast **UTS: 270 Elon: 15%**
ASTM B30/7 A	0.7% Sn 0.7% Pb 37% Zn 0.7% Al 1.2% Fe 0.5% Mn Cu alloy: Ingot; as cast **UTS: 420 Elon: 15%**
ASTM B30/8 A	38% Zn 1.25% Al 1.25% Fe 1% Mn Cu alloy: Ingot; as cast **UTS: 460 Elon: 20%**
ASTM B30/8 B	25% Zn 5% Al 3% Fe 4% Mn Cu alloy: Ingot; as cast **UTS: 450 Elon: 18%**
ASTM B30/8 C	25% Zn 5% Al 3% Fe 4% Mn Cu alloy: Ingot; as cast **UTS: 800 Elon: 12%**
ASTM B111/280	39% Zn Cu alloy: For condenser tubes
ASTM B124/2	2% Pb 39% Zn Cu alloy: Forging and bar; mechanical properties not quoted
ASTM B124/3	0.8% Sn 39% Zn Cu alloy: Forging and bar; mechanical properties not quoted
ASTM B124/4	1% Sn 39% Zn 0.1% Mn Cu alloy: Forging and bar; mechanical properties not quoted
ASTM B132 A	1.5% Sn 40% Zn 1.5% Al 1% Mn Cu alloy: Casting; as cast **UTS: 420 Elon: 15% Proof: 120**
ASTM B132 B	1.5% Sn 38% Zn 3% Al 3% Fe 3% Mn Cu alloy: Casting; as cast **UTS: 580 Elon: 15% Proof: 210**

Note. The following abbreviations and units are used in the tables:

DPN	Hardness, diamond pyramid number
UTS	Ultimate tensile strength, N/mm^2
Elon	Elongation, %
Proof	0.1% proof strength, N/mm^2

1 N/mm^2=0.1 hbar=0.102 kgf/mm^2=0.06475 $tonf/in.^2$=145.04 $lbf/in.^2$=1 MPa

See Appendix II for other abbreviations and conversion tables.

Symbol	Nominal analysis, supplier, condition and remarks.
ASTM B135/5	40% Zn Cu alloy: Tube; as drawn **UTS: 370**
ASTM B135/6	1% Pb 39% Zn Cu alloy: Tube; as drawn **UTS: 370**
ASTM B147/8 A	40% Zn 1% Al 1% Fe 0.2% Mn Cu alloy: Casting **UTS: 460 Elon: 20% Proof: 160**
ASTM B147/8 B	0.2% Pb 30% Zn 5% Al 3% Fe 4% Mn Cu alloy: Casting **UTS: 660 Elon: 18% Proof: 320**
ASTM B147/8 C	0.2% Pb 30% Zn 5% Al 3% Fe 4% Mn Cu alloy: Casting **UTS: 810 Elon: 12% Proof: 420**
ASTM B259 RB Cu Zn A	0.8% Sn 40% Zn Cu alloy: Welding rod **UTS: 340**
ASTM B259 RB Cu Zn B	1% Sn 40% Zn 0.5% Ni Cu alloy: Welding rod **UTS: 400**
ASTM B259 RB Cu Zn C	1% Sn 40% Zn Cu alloy: Welding rod **UTS: 400**
ASTM B259 RB Cu Zn D	40% Zn 10% Ni Cu alloy: Welding rod
ASTM B260 RB Cu Zn A	0.8% Sn 40% Zn Cu alloy: Braze filler; melting range 887–898 °C
ASTM B455/380	2.0% Pb 40.5% Zn Cu alloy: Extrusion
ASTM B455/385	2.9% Pb 39.6% Zn Cu alloy: Extrusion
B 31	Sn Pb 40% Zn Al Ni Fe Mn Cu alloy: Extrusion; Delta Metal Ltd; alloying elements varied to suit requirements
B 41	1% Sn 38% Zn Cu alloy: Casting; Anti-Attrition Ltd for BS alloy SCB4; Naval brass
B 42F	2% Pb 40% Zn Cu alloy: Bar; Manganese Bronze Co.; hard drawn for BS alloy CZ122
B 42HS	3% Pb 40% Zn Cu alloy: Bar; Manganese Bronze Co.; for BS alloy CZ121
B 50	0.5% Pb 40% Zn Cu alloy: Bar and forging; Manganese Bronze Co.; for BS alloy CZ123
B STAR H	41.5% Zn Cu alloy: Extrusion; McKechnie Bros; annealed **DPN: 100 UTS: 330 Elon: 40% Proof: 140**
BATH METAL	45% Zn Cu alloy: Origin unknown
BETA BRASS	Sn Zn Al Ni Fe Mn Cu alloy: Casting; common name for BS alloy HTB
BOLTOMET 524	39% Zn Cu alloy: T Bolton for BS 1949/A; withdrawn
BOLTOMET 611	1.75% Pb 39% Zn Cu alloy: T Bolton for BS 218
BOLTOMET 613	2.25% Pb 40% Zn Cu alloy: T Bolton for BS 249
BOLTOMET 619	0.4% Pb 40% Zn Cu alloy: T Bolton for BS 1949/B
BOLTOMET 715	0.7% Sn 1% Pb 35% Zn 1% Al 1% Fe 1% Mn Cu alloy: T Bolton for BS 250 CZ 114; high tensile brass
BOLTOMET 721	1% Sn 1% Pb 40% Zn 1% Fe Cu alloy: T Bolton for BS 1001; high tensile brass
BS 31	Copper alloy bars for automatic machinery; obsolete
BS 207	Brass ingot and casting; replaced by BS 1400
BS 208	Brass castings; replaced by BS 1400
BS 218	2% Pb 40% Zn Cu alloy: Bars and forgings **UTS: 300 Elon: 25%**
BS 249 CZ121	2.5% Pb 40% Zn Cu alloy: Rods and sections; free machining **UTS: 370 Elon: 15%**
BS 250 CZ114	0.7% Sn 1% Pb 40% Zn 1% Al 1% Fe 1.5% Mn Cu alloy: Bars; as drawn **UTS: 450 Elon: 20% Proof: 220**
BS 250 CZ122	2% Pb 40% Zn Cu alloy: Bar; hard drawn **UTS: 450 Elon: 20% Proof: 220**
BS 252 CZ113	1% Sn 40% Zn Cu alloy: Naval brass bars; as rolled **UTS: 400 Elon: 20%**
BS 264	Hot rolled yellow metal plate and sheet; obsolete
BS 409	1% Sn 40% Zn Cu alloy: Plate, sheet and strip; annealed; Naval brass **UTS: 330 Elon: 25%**
BS 409	1% Sn 40% Zn Cu alloy: Plate, sheet and strip; cold rolled; Naval brass **UTS: 400 Elon: 20%**

Symbol	Nominal analysis, supplier, condition and remarks.
BS 932	Brass gravity die casting; replaced by BS 1400
BS 944/1	2% Pb 39% Zn Cu alloy: Bars and forgings **UTS: 300 Elon: 25%**
BS 1001 CZ115	1% Sn 1% Pb 40% Zn 1% Fe 1.5% Mn Cu alloy: Forgings; soldering quality; high tensile brass; as forged **UTS: 450 Elon: 20% Proof: 180**
BS 1002	High tensile brass; replaced by BS 1001
BS 1400 DCB1/1	40% Zn 0.4% Al Cu alloy: Brass ingots; chill cast; for die casting **UTS: 270 Elon: 25%**
BS 1400 DCB1 C	40% Zn Cu alloy: Brass castings; chill cast; for die casting **UTS: 270 Elon: 25%**
BS 1400 DCB2/1	1% Sn 40% Zn 0.4% Al Cu alloy: Brass ingots; chill cast; for die casting **UTS: 300 Elon: 20%**
BS 1400 DCB2 C	1% Sn 40% Zn Cu alloy: Naval brass castings; chill cast; for die casting **UTS: 300 Elon: 20% Proof: 100**
BS 1400 DCB3/1	1.5% Pb 40% Zn 0.7% Al Cu alloy: Brass ingots; chill cast; for die casting **UTS: 290 Elon: 15%**
BS 1400 DCB3 C	1.5% Pb 40% Zn Cu alloy: Brass castings; chill cast; for die casting **UTS: 290 Elon: 15%**
BS 1400 HTB1/1	1.5% Sn 40% Zn 2.5% Al 1% Ni 1.2% Fe 3% Mn Cu alloy: Ingots; chill cast **UTS: 480 Elon: 20% Proof: 210**
BS 1400 HTB1 C	1.5% Sn 40% Zn 2.5% Al 1% Ni 1.2% Fe 3% Mn Cu alloy: Castings; chill cast **UTS: 480 Elon: 20% Proof: 210**
BS 1400 HTB2/1	0.5% Sn 40% Zn 5% Al 2% Ni 1.2% Fe 3% Mn Cu alloy: Ingots; sand cast **UTS: 530 Elon: 15% Proof: 270**
BS 1400 HTB2 C	0.5% Sn 40% Zn 5% Al 2% Ni 1.2% Fe 3% Mn Cu alloy: Castings; sand cast **UTS: 530 Elon: 15% Proof: 270**
BS 1400 HTB3/1	0.2% Sn 40% Zn 5% Al 1% Ni 1.2% Fe 4% Mn Cu alloy: Ingots; sand cast; beta brass **UTS: 730 Elon: 12% Proof: 390**
BS 1400 HTB3 C	0.2% Sn 40% Zn 5% Al 1% Ni 1.2% Fe 4% Mn Cu alloy: Castings; sand cast; beta brass **UTS: 730 Elon: 12% Proof: 390**
BS 1453 C2	40% Zn 0.3% Si Cu alloy: Weld filler rod for brazing steel to copper or welding brasses
BS 1453 C4	40% Zn 0.2% Mn 0.2% Si Cu alloy: Weld filler rod for brazing of copper to cast iron steel and for welding brasses
BS 1541 CZ123	0.5% Pb 40% Zn Cu alloy: Plate; as rolled **UTS: 330 Elon: 20%**
BS 1845/8	50% Zn Cu alloy: Braze filler; melting range 860–870 °C
BS 1845/9	45% Zn Cu alloy: Braze filler; melting range 870–880 °C
BS 1845/10	40% Zn Cu alloy: Braze filler; melting range 885–890 °C
BS 1845/11	1% Sn 44% Zn Cu alloy: Braze filler; melting range 860–870 °C
BS 1845/12	1% Sn 39% Zn Cu alloy: Braze filler; melting range 880–890 °C
BS 1949 CZ109	40% Zn Cu alloy: Brass rods and forgings; as rolled **UTS: 310 Elon: 30% Proof: 150**
BS 1949 CZ123	0.5% Pb 40% Zn Cu alloy: Brass rod and forgings; as rolled **UTS: 310 Elon: 30%**
BS 2870 CZ109	40% Zn Cu alloy: Sheet; specification not issued
BS 2870 CZ113	1.0% Sn 40% Zn Cu alloy: Sheet; specification not issued
BS 2870 CZ114	0.7% Sn 1% Pb 40% Zn 1% Al 1% Fe 1.5% Mn Cu alloy: Sheet; high tensile brass; specification not issued
BS 2870 CZ115	1% Sn 1.2% Pb 40% Zn 1% Fe 1.5% Mn Cu alloy: Sheet; high tensile brass soldering quality; specification not issued
BS 2870 CZ120	2% Pb 40% Zn Cu alloy: Sheet; $\frac{1}{2}$-hard **DPN: 120 Elon: 10%**
BS 2870 CZ120	2% Pb 40% Zn Cu alloy: Sheet; hard **DPN: 150 UTS: 490 Elon: 5%**
BS 2870 CZ120	2% Pb 40% Zn Cu alloy: Sheet; extra hard **DPN: 180 UTS: 560 Elon: 3%**
BS 2870 CZ121	3% Pb 40% Zn Cu alloy: Sheet; specification not issued
BS 2870 CZ122	2% Pb 40% Zn Cu alloy: Sheet; specification not issued
BS 2870 CZ123	0.5% Pb 40% Zn Cu alloy: Sheet; as rolled **UTS: 360 Elon: 20%**
BS 2871 CZ109	40% Zn Cu alloy: Lead-free brass; specification not issued
BS 2871 CZ112	1.2% Sn 38% Zn Cu alloy: Tube; Naval brass; specification not issued
BS 2871 CZ113	1% Sn 40% Zn Cu alloy: Tube; Naval brass; specification not issued
BS 2871 CZ114	0.7% Sn 1% Pb 40% Zn 1.0% Al 1% Fe 1.5% Mn Cu alloy: Tube; specification not issued
BS 2871 CZ115	1% Sn 1% Pb 40% Zn 1% Fe 1.5% Mn Cu alloy: Tube; specification not issued
BS 2871 CZ117	0.7% Sn 40% Zn 2.5% Al 1% Fe 1% Mn Cu alloy: Tube; specification not issued
BS 2871 CZ120	2% Pb 40% Zn Cu alloy: Tube; specification not issued
BS 2871 CZ121	3% Pb 40% Zn Cu alloy: Tube; specification not issued
BS 2871 CZ122	2% Pb 40% Zn Cu alloy: Tube; specification not issued
BS 2871 CZ123	0.5% Pb 40% Zn Cu alloy: Tube; specification not issued
BS 2872 CZ109	40% Zn Cu alloy: Forging; as forged **UTS: 300 Elon: 30%**
BS 2872 CZ113	1% Sn 40% Zn Cu alloy: Forging; specification not issued
BS 2872 CZ114	0.7% Sn 1% Pb 40% Zn 1% Fe 1.5% Mn Cu alloy: Forging **UTS: 450 Elon: 20%**
BS 2872 CZ115	1.0% Sn 1.0% Pb 35% Zn 1.0% Fe 1.0% Mn Cu alloy: Forging; as forged **UTS: 450 Elon: 20% Proof: 180**
BS 2872 CZ120	2% Pb 40% Zn Cu alloy: Forging; specification not issued
BS 2872 CZ121	3% Pb 40% Zn Cu alloy: Forging; specification not issued
BS 2872 CZ122	2% Pb 40% Zn Cu alloy: Forging; as forged **UTS: 300 Elon: 25%**
BS 2872 CZ123	0.5% Pb 40% Zn Cu alloy: Forging; as forged **UTS: 300 Elon: 30%**
BS 2872 CZ128	2% Pb 38% Zn Cu alloy
BS 2872 CZ129	1.3% Pb 38% Zn Cu alloy
BS 2873 CZ109	40% Zn Cu alloy: Wire; specification not issued
BS 2873 CZ113	1% Sn 40% Zn Cu alloy: Wire; specification not issued
BS 2873 CZ114	0.8% Sn 1% Pb 40% Zn 1% Fe 1.5% Mn Cu alloy: Wire; specification not issued
BS 2873 CZ115	1.0% Sn 1.0% Pb 35% Zn 1.0% Fe 1.0% Mn Cu alloy: Wire; specification not issued
BS 2873 CZ120	2% Pb 40% Zn Cu alloy: Wire; specification not issued
BS 2873 CZ121	3% Pb 40% Zn Cu alloy: Wire; specification not issued
BS 2873 CZ122	2% Pb 40% Zn Cu alloy: Wire; specification not issued
BS 2873 CZ123	0.5% Pb 40% Zn Cu alloy: Wire; specification not issued
BS 2874 CZ109	40% Zn Cu alloy: Bar; as rolled **UTS: 310 Elon: 30% Proof: 150**
BS 2874 CZ112	1.2% Sn 38% Zn Cu alloy: Bar; as rolled **UTS: 360 Elon: 20%**
BS 2874 CZ113	1% Sn 40% Zn Cu alloy: Bar; as rolled **UTS: 360 Elon: 20%**
BS 2874 CZ114	1% Sn 1% Pb 40% Zn 1% Al 1% Fe 1.5% Mn Cu alloy: Bar; as rolled **UTS: 480 Elon: 18% Proof: 240**

Symbol	Nominal analysis, supplier, condition and remarks.
BS 2874 CZ115	1% Sn 1% Pb 40% Zn 1% Fe 1.5% Mn Cu alloy: Bar; cold worked and stress relieved **UTS: 480 Elon: 15% Proof: 240**
BS 2874 CZ120	2% Pb 40% Zn Cu alloy: Bar; specification not issued
BS 2874 CZ121	3% Pb 40% Zn Cu alloy: Bar; as rolled **UTS: 340 Elon: 16%**
BS 2874 CZ122	2% Pb 40% Zn Cu alloy: Bar; hard drawn **UTS: 400 Elon: 22% Proof: 180**
BS 2874 CZ123	0.5% Pb 40% Zn Cu alloy: Bar; as rolled **UTS: 320 Elon: 30%**
BS 2874 CZ128	2% Pb 38% Zn Cu alloy
BS 2874 CZ129	1.3% Pb 38% Zn Cu alloy
BS 2874 CZ130	3% Pb 40% Zn Cu alloy
BS 2874 CZ131	1.7% Pb 38% Zn Cu alloy
BS 2874 CZ133	0.7% Sn 39% Zn Cu alloy
BS 2874 CZ134	1.2% Sn 2% Pb 38% Zn Cu alloy
BS 2874 CZ135	38% Zn 1.5% Al 2% Mn Cu alloy
BS 2874 CZ136	3% Pb 39% Zn Cu alloy
BS 2874 CZ137	1.7% Pb 39% Zn Cu alloy
BS 2875 CZ109	40% Zn Cu alloy: Plate; specification not issued
BS 2875 CZ113	1% Sn 40% Zn Cu alloy: Plate; specification not issued
BS 2875 CZ114	0.7% Sn 1% Pb 40% Zn 1.0% Al 1% Fe 1.5% Mn Cu alloy: Plate; specification not issued
BS 2875 CZ115	1% Sn 1% Pb 40% Zn 1% Fe 1.5% Mn Cu alloy: Plate; specification not issued
BS 2875 CZ121	3% Pb 40% Zn Cu alloy: Plate; specification not issued
BS 2875 CZ122	2% Pb 40% Zn Cu alloy: Plate; specification not issued
BS 2875 CZ123	0.5% Pb 40% Zn Cu alloy: Plate; as rolled **UTS: 320 Elon: 20%**
BS 2901 C4	40% Zn 0.2% Mn 0.2% Si Cu alloy: Welding rod for inert gas shielded arc welding
BS B1	High tensile brass bar; obsolete
BS B12	40% Zn Cu alloy: Sheet; annealed
BS B13	40% Zn Cu alloy: Bar; free machining
BS B20	Brass bar for hot stamping; obsolete
C 2B	1% Sn 40% Zn 0.5% Ni 0.5% Fe Cu alloy: Rod for welding; designation for BS 1453
C 2C	1.0% Sn 40% Zn 0.5% Fe Cu alloy: Rod for welding; designation for BS 1453
C 4	40% Zn 0.2% Mn 0.2% Si Cu alloy: Welding rod; designation used by BS
C 50	2% Pb 40% Zn Cu alloy: Extrusion; Delta Metal for BS 249
C 90	1% Sn 1% Pb 40% Zn 1% Fe 1% Mn Cu alloy: Bar; Manganese Bronze Co.; Parsons C 90
C 92	Sn 40% Zn Al Fe Mn Cu alloy: Manganese Bronze Co.; Parsons C 92
C 94	44% Zn Fe Mn Cu alloy: Manganese Bronze Co.; architectural bronze
C 28200	39% Zn 0.2% P Cu alloy: Obsolete; UNS designation
C 28580	50% Zn Cu alloy: CDA and UNS designation
C 37710	2% Pb 40% Zn Cu alloy: CDA and UNS designation
C 37800	2% Pb 40% Zn Cu alloy: CDA and UNS designation
C 38000	2% Pb 40% Zn Cu alloy: CDA and UNS designation
C 38010	2.2% Pb 40% Zn 0.5% Al Cu alloy: CDA and UNS designation
C 38500	3% Pb 41% Zn Cu alloy: CDA and UNS designation
C 38510	3.5% Pb 39% Zn Cu alloy: CDA and UNS designation
C 38590	2.7% Pb 40% Zn Cu alloy: CDA and UNS designation

Note. The following abbreviations and units are used in the tables:

DPN	Hardness, diamond pyramid number
UTS	Ultimate tensile strength, N/mm^2
Elon	Elongation, %
Proof	0.1% proof strength, N/mm^2

1 N/mm^2=0.1 hbar=0.102 kgf/mm^2=0.06475 tonf/in.^{2}=145.04 lbf/in.^{2}=1 MPa

See Appendix II for other abbreviations and conversion tables.

Symbol	Nominal analysis, supplier, condition and remarks.
C 38600	3% Pb 40% Zn Cu alloy: Obsolete; UNS designation
C 46400	0.8% Sn 40% Zn Cu alloy: CDA and UNS designation
C 46420	1.2% Sn 39% Zn Cu alloy: CDA and UNS designation
C 46500	0.7% Sn 39% Zn Cu alloy: Arsenical; CDA and UNS designation
C 46600	0.7% Sn 39% Zn Cu alloy: Antimonial; CDA and UNS designation
C 46700	0.7% Sn 39% Zn Cu alloy: Phosphorized; CDA and UNS designation
C 47000	0.7% Sn 40% Zn Cu alloy: Braze and weld rod; CDA and UNS designation
C 48200	0.7% Sn 0.7% Pb 39% Zn Cu alloy: CDA and UNS designation
C 48500	0.7% Sn 1.8% Pb 39% Zn Cu alloy: CDA and UNS designation
C 48600	1.2% Sn 1.7% Pb 39% Zn Cu alloy: CDA and UNS designation
C 49080	3.5% Sn 47% Zn Cu alloy: CDA and UNS designation
C 67000	27% Zn 4.5% Al 3% Fe 3.2% Mn Cu alloy: CDA and UNS designation
C 67130	1% Sn 1% Pb 31% Zn 0.7% Al 1% Ni 1% Mn Cu alloy: CDA and UNS designation
C 67300	2.0% Pb 35% Zn 2.7% Mn 1% Si Cu alloy: CDA and UNS designation
C 67400	36% Zn 1.2% Al 2.7% Mn 1% Si Cu alloy: CDA and UNS designation
C 67410	40% Zn 1.8% Al 1.8% Mn 1% Si Cu alloy: CDA and UNS designation
C 67500	39% Zn 1.2% Fe 0.3% Mn Cu alloy: CDA and UNS designation
C 67620	1% Pb 40% Zn 0.8% Fe 1.5% Mn Cu alloy: CDA and UNS designation
C 67700	0.7% Pb 38% Zn 2% Ni 1% Fe 0.2% Mn 0.6% As Cu alloy: CDA and UNS designation
C 67800	39% Zn 1% Al 1% Fe 0.4% Mn Cu alloy: CDA and UNS designation
C 67810	40% Zn 1% Al 1% Mn Cu alloy: CDA and UNS designation
C 67820	0.7% Sn 38% Zn 1% Al 0.8% Fe 1.2% Mn Cu alloy: CDA and UNS designation
C 68000	1% Sn 39% Zn 0.6% Ni 1% Fe Cu alloy: CDA and UNS designation
C 68100	1% Sn 39% Zn 1% Fe Cu alloy: CDA and UNS designation
C 68200	39% Zn 0.8% Fe Cu alloy: CDA and UNS designation
C 68600	0.8% Sn 38% Zn 1% Al 1% Fe Cu alloy: CDA and UNS designation
C 79820	2.7% Sn 42% Zn 9.5% Ni Cu alloy: Obsolete; UNS designation
CHANNEL BRONZE	1% Sn 39% Zn Cu alloy: Origin unknown
CM 2	Sn Pb 40% Zn Al Ni Fe Mn Cu alloy: Extrusion; Delta Metal for BS 1002
CX 1	2% Pb 40% Zn Cu alloy: Extrusion; Delta Metal
D 1	1% Sn 1% Pb 40% Zn 1% Fe 1% Mn Cu alloy: Bar; Manganese Bronze Co.; for BS alloy CZ114
D 7	40% Zn Cu alloy: Lead-free bars; Manganese Bronze Co.; annealed **UTS: 330 Elon: 50%**
D 9	40% Zn Cu alloy: Free machining bar; Manganese Bronze Co.; for small sections **UTS: 370 Elon: 25%**
DIL	1% Sn 1% Pb 40% Zn 1% Fe 1% Mn Cu alloy: Bar; Manganese Bronze Co.; for BS alloy CZ 114
DIN 1709 GD Ms 60	40% Zn 0.8% Al Cu alloy: Casting; die cast **DPN: 100 UTS: 330 Elon: 4% Proof: 150**
DIN 1709 GK Ms 60	40% Zn 0.8% Al Cu alloy: Casting
DIN 1709 G So Ms F 30	1% Sn 40% Zn 1% Fe 2% Mn Cu alloy: Casting **DPN: 85 UTS: 330 Elon: 25% Proof: 150**

Symbol	Nominal analysis, supplier, condition and remarks.
DIN 1709 G So Ms F 45	40% Zn 2% Al 1.5% Ni 1.5% Fe 2.5% Mn Cu alloy: Casting **DPN: 130 UTS: 530 Elon: 25% Proof: 180**
DIN 1709 G So Ms F 60	40% Zn 4% Al 1.5% Ni 2% Fe 3% Mn Cu alloy: Casting **DPN: 160 UTS: 600 Elon: 20% Proof: 300**
DIN 1709 G So Ms F 75	40% Zn 7% Al 1.5% Ni 3% Fe 4% Mn Cu alloy: Casting **DPN: 100 UTS: 360 Elon: 35% Proof: 90**
DIN 1733 Ms 60	0.2% Sn 40% Zn 0.3% Si Cu alloy: Wrought
DIN 1733 S So Ms	40% Zn Cu alloy: Wrought
DIN 17656 GB Ms 60 A	40% Zn 0.3% Al Cu alloy: Casting
E	1.5% Pb 39% Zn Cu alloy: Extrusion; Delta Metal Ltd
E 22	1% Sn 1% Pb 40% Zn 1% Fe Cu alloy: T Bolton for BS 1001
E D	Sn Pb 40% Zn Al Ni Fe Mn Cu alloy: Extrusion; Delta Metal Ltd; alloying elements varied to suit requirements
EMBEESH	Sn 40% Zn Al Fe Mn Cu alloy: Manganese Bronze Co.
ETS	0.75% Sn 0.1% Pb 39% Zn Cu alloy: Extrusion; McKechnie Bros; as drawn **DPN: 120 UTS: 370 Elon: 25% Proof: 140**
ETW	0.3% Sn 40% Zn 0.2% Si Cu alloy: Brazing filler rod; McKechnie Bros; melting range 890–900 °C
F 58	1% Sn 0.5% Pb 40% Zn Cu alloy: Extrusion; McKechnie Bros; as drawn **DPN: 120 UTS: 370 Elon: 25% Proof: 140**
FXY	1% Sn 2.25% Pb 38% Zn Cu alloy: Extrusion; McKechnie Bros; as drawn **DPN: 120 UTS: 370 Elon: 22% Proof: 140**
G Cu 55 Zn Al	40% Zn 2% Al 1.5% Ni 1.5% Fe 2.5% Mn Cu alloy: Casting; designation used by German Standards
G Cu 55 Zn Al 2	40% Zn 4% Al 1.5% Ni 2% Fe 3% Mn Cu alloy: Casting; designation used by German Standards
G Cu 55 Zn Al 4	40% Zn 7% Al 1.5% Ni 3% Fe 4% Mn Cu alloy: Casting; designation used by German Standards
G Cu 55 Zn Mn	1% Sn 40% Zn 1% Fe 2% Mn Cu alloy: Casting; designation used by German Standards
G So Ms F 30	1% Sn 40% Zn 1% Fe 2% Mn Cu alloy: Casting; designation used by German Standards
G So Ms F 45	40% Zn 2% Al 1.5% Ni 1.5% Fe 2.5% Mn Cu alloy: Casting; designation used by German Standards
G So Ms F 60	40% Zn 4% Al 1.5% Ni 2% Fe 3% Mn Cu alloy: Casting; designation used by German Standards
G So Ms F 75	40% Zn 7% Al 1.5% Ni 3% Fe 4% Mn Cu alloy: Casting; designation used by German Standards
GB	3.75% Pb 38% Zn Cu alloy: Extrusion; McKechnie Bros; as drawn **DPN: 110 UTS: 370 Elon: 20% Proof: 170**
GB Cu 60 Zn	40% Zn 0.3% Al Cu alloy: Casting; designation used by German Standards
GB Ms 60A	40% Zn 0.3% Al Cu alloy: Casting; designation used by German Standards
GB Ms 60B	40% Zn 0.3% Al Cu alloy: Casting; designation used by German Standards
GD Cu 60 Zn	40% Zn 0.8% Al Cu alloy: Casting; die cast
GD Ms 60	40% Zn 0.8% Al Cu alloy: Casting; die cast
GK Cu 60 Zn	40% Zn 0.8% Al Cu alloy: Casting; designation used by German Standards
GK Ms 60	40% Zn 0.8% Al Cu alloy: Casting; designation used by German Standards
H 15	1.5% Pb 39% Zn Cu alloy: Extrusion; Delta Metal Co.
HECORROS 76	20% Zn 2% Al Cu alloy: Tube and pipe; VDM
HIDURIT 10	40% Zn 5% Al 1% Ni 1% Fe 4% Mn Cu alloy: Casting; Langley Alloys for BS alloy HTB3
HIDURIT 11	40% Zn 5% Al 2% Ni 1% Fe 3% Mn Cu alloy: Casting; Langley Alloys for BS alloy HTB2
HIDURIT 12	1.5% Sn 40% Zn 2.5% Al 1% Ni 1% Fe 3% Mn Cu alloy: Casting; Langley Alloys for BS alloy HTB1
HIDURIT 15	0.7% Sn 1% Pb 40% Zn 1% Al 1% Fe 1.5% Mn Cu alloy: Wrought; Langley Alloys for BS alloy CZ114
HIGH TENSILE BRASS	40% Zn with Sn and Al Fe Mn Cu alloy: BS alloy CZ114; further information from Copper Development Association
IAT	1% Sn 0.5% Pb 40% Zn 1% Fe 1% Mn Cu alloy: Bar; Manganese Bronze Co. for BS alloy CZ114; low Pb
IMI 246	45% Zn Cu alloy: Brazing rod; IMI for BS 1845/9; melting range 870–880 °C
IMI 312	37% Zn 3% Al 1% Fe 2% Mn Cu alloy: Bar; IMI **DPN: 180 UTS: 630 Elon: 20% Proof: 330**
IMI 330	1.2% Sn 0.3% Pb 39% Zn 0.5% Al 1% Fe 1.5% Mn Cu alloy: Bar; IMI **DPN: 160 UTS: 540 Elon: 30% Proof: 240**
IMI 333	0.7% Sn 1.2% Pb 40% Zn 1.5% Fe 1.5% Mn Cu alloy: Bar; IMI and Manganese Bronze Co. **DPN: 140 UTS: 480 Elon: 25% Proof: 240**
IMI 334	1.2% Pb 38% Zn 1.2% Al 1% Fe 1.5% Mn Cu alloy: Bar; IMI **DPN: 170 UTS: 560 Elon: 20% Proof: 300**
IMI 360	1% Sn 37% Zn Cu alloy: Bar; IMI Naval brass: annealed **DPN: 100 UTS: 330 Elon: 56% Proof: 100**
IMI 365	1% Sn 39% Zn Cu alloy: Brazing rod; IMI for BS 1845/12; melting range 880–890 °C
IMI 366	40% Zn Cu alloy: Small amount Sn; IMI 'Tobin Brass' for brazing high Cu brasses
IMI 388	40% Zn 0.3% Si Cu alloy: Brazing metal; silicon brass for brazing steel and Cu alloys; melting range 880–890 °C; IMI
IMI 442	2.6% Pb 39% Zn Cu alloy: Bar; IMI
IMI 452	0.5% Pb 40% Zn Cu alloy: IMI for BS 2873 CZ123; melting range 883–890 °C
IMI 453	0.7% Pb (max) 40% Zn Cu alloy: Bar; IMI
IMI 456	0.7% Pb 40% Zn Cu alloy: Bar; IMI
IMI 467	2.2% Pb 40% Zn Cu alloy: Free machining bar for hot stamping; IMI
IMI 469	3% Pb 40% Zn Cu alloy: Free machining bar; IMI for BS alloy CZ121
IMMADIUM IV	1% Sn 1% Pb 40% Zn 1% Fe 1% Mn Cu alloy: Bar; Manganese Bronze Co. for BS alloy CZ115; soldering quality
ITA	Sn 40% Zn Al Fe Mn and Cu alloy: Manganese Bronze Co.; Parsons ITA
K	0.75% Sn 1.25% Pb 38% Zn 0.25% Al 0.75% Fe 1.25% Mn Cu alloy: Extrusion; McKechnie Bros; as drawn **DPN: 140 UTS: 500 Elon: 20% Proof: 240**
KP	0.75% Sn 0.75% Pb 38% Zn 0.25% Al 0.25% Fe 1.25% Mn Cu alloy: Extrusion; McKechnie Bros; as drawn **DPN: 130 UTS: 450 Elon: 25% Proof: 210**
KS	0.75% Pb 40% Zn 0.25% Fe 1.75% Mn Cu alloy: Extrusion; McKechnie Bros; as drawn **DPN: 100 UTS: 420 Elon: 30% Proof: 180**
KT	2% Pb 40% Zn Cu alloy: Extrusion; Delta Metal Ltd
KZ	0.5% Sn 0.75% Pb 38.5% Zn 0.75% Al 0.75% Fe 0.75% Mn Cu alloy: Extrusion; McKechnie Bros; as drawn **DPN: 140 UTS: 530 Elon: 15% Proof: 270**
M 1	Sn Pb 40% Zn Al Ni Fe Mn Cu alloy: Extrusion; Delta Metal Ltd for BS 1001
MANGANESE BRONZE DI	1% Sn 1% Pb 40% Zn 1% Fe 1% Mn Cu alloy: Bar; Manganese Bronze Co. for BS alloy CZ114
MANGANESE BRONZE DIL	1% Sn 1% Pb 40% Zn 1% Fe 1% Mn Cu alloy: Bar; Manganese Bronze Co. for BS alloy CZ114
MANGANESE BRONZE IAT	1% Sn 0.5% Pb 40% Zn 1% Fe 1% Mn Cu alloy: Bar; Manganese Bronze Co. for BS alloy CZ114; low Pb
MANGANESE BRONZE IX	1% Sn 1% Pb 40% Zn 1% Fe 1% Mn Cu alloy: Bar; Manganese Bronze Co. for BS alloy CZ114

Symbol	Nominal analysis, supplier, condition and remarks.
MKDC 1	0.5% Sn 2.5% Pb 37% Zn Cu alloy: Casting; McKechnie Bros; chill cast **DPN: 105 UTS: 340 Elon: 17% Proof: 150**
NAVAL BRASS	1.25% Sn 38% Zn Cu alloy: Common name; further information from Copper Development Association
NF 2	Sn Pb 40% Zn Al Ni Fe Mn Cu alloy: Extrusion; Delta Metal Ltd; alloying elements varied to suit requirements
NIL	1% Sn 1% Pb 37% Zn Cu alloy: Extrusion; Delta Metal Ltd
No. 1 HIGH TENSILE BRASS	35% Zn 4.5% Al 1% Ni 2% Fe 4% Mn Cu alloy: Casting; Stone Manganese Marine for BS alloy HTB3
No. 3 HIGH TENSILE BRASS	35% Zn 5% Al 2% Ni 2% Fe 3% Mn Cu alloy: Casting; Stone Manganese Marine for BS alloy HTB2
No. 4 HIGH TENSILE BRASS	0.5% Sn 35% Zn 2% Al 1% Fe 1.5% Mn Cu alloy: Casting; Stone Manganese Marine
No. 6 HIGH TENSILE BRASS	1.5% Sn 2.5% Al 1.5% Fe 3% Mn Cu alloy: Casting; Stone Manganese Marine for BS alloy HTB1
NS	44% Zn 10% Ni 0.2% Si Cu alloy: Brazing filler rod; McKechnie Bros; melting range 905–915 °C
ORM	Sn Pb 40% Zn Al Ni Fe Mn Cu alloy: Extrusion; Delta Metal Ltd; alloying elements varied to suit requirements
PARSONS C 90	1% Sn 1% Pb 40% Zn 1% Fe 1% Mn Cu alloy: Bar; Manganese Bronze Co. for BS alloy CZ115
PARSONS C 92	Sn 40% Zn Al Fe Mn Cu alloy: Manganese Bronze Co.
PARSONS ITA	Sn 40% Zn Al Fe Mn Cu alloy: Manganese Bronze Co.
PARSONS No 1 STANDARD	1% Sn 1% Pb 40% Zn 1% Fe 1% Mn Cu alloy; Bar; Manganese Bronze Co. for BS alloy CZ114; made from virgin metal
PARSONS SB 6	46% Zn 6% Ni Cu alloy: Manganese Bronze Co.
PL	2% Pb 40% Zn Cu alloy: Extrusion; Delta Metal Ltd
PROPELLER BRASS	1% Sn 40% Zn 0.2% Al 1% Fe 0.2% Mn Cu alloy: Casting; further information from Copper Development Association
PW	0.75% Sn 40% Zn 0.3% Fe Cu alloy: Brazing filler rod; McKechnie Bros; melting range 885–895 °C
QQ B 611a/A	2.0% Pb 40% Zn Cu alloy: Bar and forging; US Federal
QQ B 611a/B	3.0% Pb 40% Zn Cu alloy: Bar and forging; US Federal **UTS: 340 Elon: 15% Proof: 120**
QQ B 636	0.8% Sn 40% Zn Cu alloy: Bar; US Federal **UTS: 400 Elon: 28% Proof: 200**
QQ B 721a/A	1.0% Sn 40% Zn Cu alloy: Plate; US Federal **UTS: 370 Elon: 22% Proof: 140**
QQ S 551/A	0.5% Pb 48% Zn Cu alloy: Brazing filler; US Federal
QQ S 551/B	3.5% Sn 0.5% Pb 45% Zn Cu alloy: Brazing filler; US Federal; melting point 871 °C
S	3% Pb 40% Zn 0.3% Al Cu alloy: Extrusion; McKechnie Bros; as drawn **DPN: 120 UTS: 400 Elon: 22% Proof: 180**
S Ms 60	0.2% Sn 40% Zn 0.3% Si Cu alloy: Wrought; designation used by German Standards
S So Ms	40% Zn Cu alloy: Wrought; designation used by German Standards
SAE 43	40% Zn 1% Al 1.5% Mn Cu alloy: Casting; as cast **UTS: 460 Elon: 20% Proof: 170**
SAE 73	1% Sn 40% Zn Cu alloy: Bar and forging; annealed; ASTM B 21 A **UTS: 370 Elon: 40% Proof: 120**
SAE 73	1% Sn 40% Zn Cu alloy: Bar and forging; hard; ASTM B 21 A **UTS: 450 Elon: 25% Proof: 240**
SAE 74 A	0.8% Pb 40% Zn Cu alloy: Tube; as drawn; ASTM B 135/5 **UTS: 370**
SAE 88	2% Pb 40% Zn Cu alloy: Bar and forging; as forged; ASTM B 124/2 **UTS: 370 Elon: 30%**
SAE 430 A	30% Zn 5% Al 3% Fe 3% Mn Cu alloy: Casting **UTS: 650 Elon: 20% Proof: 330**
SAE 430 B	30% Zn 5% Al 3% Fe 3% Mn Cu alloy: Casting **UTS: 810 Elon: 12% Proof: 420**
SAE CA 377	2% Pb 39% Zn Cu alloy: Formerly SAE 88
SAE CA 464	0.75% Sn 39.25% Zn Cu alloy: Naval brass; Formerly SAE 73
SAE CA 465	40% Zn 0.5% Al Cu alloy: Wrought
SAE CA 466	40% Zn 0.5% Sb Cu alloy: Wrought
SAE CA 467	40% Zn 0.5% P Cu alloy: Wrought
SAE CA 674	36% Zn 1.2% Al 2.7% Mn 1% Si Cu alloy: Wrought
SAE CA 675	1% Sn 38.8% Zn 1.4% Fe 0.2% Mn Cu alloy: Wrought
SB	Sn Pb 40% Zn Al Ni Fe Mn Cu alloy: Extrusion; Delta Metal Ltd; alloying elements varied to suit requirements
SB 6	46% Zn 6% Ni Cu alloy: Manganese Bronze Co.
SCB 4	1.2% Sn 40% Zn Cu alloy: Designation used by BS
SIFBRONZE	39% Zn 0.25% Si Cu alloy: Brazing rod; Suffolk Iron Foundry; melting point 850 °C
SILICON BRASS	40% Zn 0.3% Si Cu alloy: Braze metal; IMI 388
For SIS specifications see SS	
SPRABRONZE TM	0.8% Sn 40% Zn 0.7% Fe 0.2% Mn Cu alloy: Wire for spraying; Metco Ltd; as sprayed **UTS: 180**
SPRABRONZE TM	1% Sn 0.1% Pb 38% Zn Cu alloy: Spray deposit; Metco **DPN: 100**
SS	1.2% Pb 40% Zn Cu alloy: Extrusion; McKechnie Bros; as drawn **DPN: 135 UTS: 390 Elon: 30% Proof: 180**
SS 1	0.5% Pb 39.5% Zn Cu alloy: Extrusion; McKechnie Bros; annealed **DPN: 100 UTS: 370 Elon: 35% Proof: 220**
SS 5163	0.5% Pb 40% Zn Cu alloy: Plate; Swedish Standard; annealed **DPN: 90 UTS: 330 Elon: 40% Proof: 120**
SS 5163	0.5% Pb 40% Zn Cu alloy: Plate; Swedish Standard; cold rolled **DPN: 125 UTS: 370 Elon: 20% Proof: 240**
SS 5165	1.2% Pb 40% Zn Cu alloy: Plate and bar; Swedish Standard; annealed **DPN: 80 UTS: 330 Elon: 35% Proof: 120**
SS 5165	1.2% Pb 40% Zn Cu alloy: Plate and bar; Swedish Standard; cold rolled **DPN: 135 UTS: 370 Elon: 15% Proof: 200**
SS 5168	2% Pb 40% Zn Cu alloy: Plate; Swedish Standard; cold rolled **DPN: 140 UTS: 390 Elon: 12% Proof: 280**
SS 5170	3% Pb 38% Zn Cu alloy: Bar and forging; Swedish Standard; annealed **DPN: 95 UTS: 390 Elon: 35% Proof: 140**
SS 5170	3% Pb 38% Zn Cu alloy: Bar and forging; Swedish Standard; cold rolled **DPN: 140 UTS: 450 Elon: 15% Proof: 330**
SS 5173	1% Pb 40% Zn Cu alloy: Extrusion; Swedish Standard; as rolled **DPN: 95 UTS: 400 Elon: 30% Proof: 150**

Note. The following abbreviations and units are used in the tables:

DPN	Hardness, diamond pyramid number
UTS	Ultimate tensile strength, N/mm^2
Elon	Elongation, %
Proof	0.1% proof strength, N/mm^2

1 N/mm^2=0.1 hbar=0.102 kgf/mm^2=0.06475 $tonf/in.^2$=145.04 $lbf/in.^2$=1 MPa

See Appendix II for other abbreviations and conversion tables.

Symbol	Nominal analysis, supplier, condition and remarks.
SS 5272	0.8% Pb 40% Zn 0.8% Al Cu alloy: Extrusion; Swedish Standard; as rolled **DPN: 130 UTS: 480 Elon: 20% Proof: 170**
STA 7 CM1	1% Sn 38% Zn 1% Fe 2% Mn Cu alloy: Bar; as rolled BS 250, BS 1001 **UTS: 450 Elon: 20% Proof: 220**
STA 7 CM1	1% Sn 38% Zn 1% Fe 2% Mn Cu alloy: Bar; cold drawn; stress relieved **UTS: 530 Elon: 15% Proof: 240**
STA 7 CM2	0.5% Sn 1% Pb 40% Zn 1% Al 1% Fe 1% Mn Cu alloy: Bar and forging **UTS: 450 Elon: 20% Proof: 220**
STA 7 CM3	0.5% Sn 1% Pb 40% Zn 2.5% Al 1% Fe 1.5% Mn Cu alloy: Bar and forging **UTS: 530 Elon: 15% Proof: 240**
STA 7 CM4A	1% Sn 0.5% Pb 38% Zn 1.5% Al 1% Fe 1% Mn Cu alloy: Bar and forging **UTS: 530 Elon: 15% Proof: 240**
STA 7 CM5	0.5% Sn 2% Pb 40% Zn 0.2% Al 0.7% Fe 1% Mn Cu alloy: Casting; as cast **UTS: 300 Elon: 12% Proof: 120**
STA 7 CM6 A and B	0.5% Sn 1% Pb 40% Zn 1% Al 0.7% Fe 2% Mn Cu alloy: Casting **UTS: 440 Elon: 17% Proof: 200**
STA 7 CM7 A and B	0.2% Sn 0.2% Pb 40% Zn 4% Al 1.5% Fe 2% Mn Cu alloy: Casting **UTS: 600 Elon: 14% Proof: 330**
STA 7 CX3	1% Sn 40% Zn 0.2% Al 1% Fe 0.2% Mn Cu alloy: Casting; Admiralty Propeller brass **UTS: 500 Elon: 15% Proof: 240**
STA 7 CX4	1% Sn 40% Zn 0.8% Al 1% Fe 1.5% Mn Cu alloy: Casting **UTS: 500 Elon: 15% Proof: 240**
STA 7 CX4	1% Sn 40% Zn 0.8% Al 1% Fe 1.5% Mn Cu alloy: Bar and forging **UTS: 500 Elon: 25% Proof: 240**
STA 7 CX5	40% Zn 3% Al 1% Fe 2% Mn Cu alloy: Section; base fuse alloy **UTS: 600 Elon: 15% Proof: 300**
STA 7 CX6	High tensile brass; no specific composition; as drawn **UTS: 450 Elon: 28% Proof: 220**
STA 7 CX13	1% Pb 40% Zn 3% Al 2% Mn Cu alloy: Section; leaded base fuse metal **UTS: 600 Elon: 15% Proof: 300**
STA 7 CZ9A	2% Pb 40% Zn Cu alloy: Forging **UTS: 300 Elon: 25%**
STA 7 CZ9B	2% Pb 40% Zn Cu alloy: Bar **UTS: 300 Elon: 25%**
STA 7 CZ9B1	1.5% Pb (max) 40% Zn Cu alloy: Bar and forging **UTS: 300 Elon: 25%**
STA 7 CZ9C	2% Pb 40% Zn Cu alloy: Bar; drawn and reeled **UTS: 300 Elon: 30% Proof: 180**
STA 7 CZ9D	2% Pb 40% Zn Cu alloy: Bar; drawn and reeled up to 40 mm diameter **UTS: 400 Elon: 20% Proof: 220**
STA 7 CZ9E	2% Pb 40% Zn Cu alloy: Bar; drawn and reeled up to 25 mm diameter **UTS: 450 Elon: 15% Proof: 260**
STA 7 CZ10	3% Pb 40% Zn Cu alloy: Bar; free machining **UTS: 370 Elon: 20%**
STA 7 CZ11B	2% Pb 40% Zn Cu alloy: Sheet and strip **UTS: 370**
STA 7 CZ14	40% Zn Cu alloy: Die casting; as cast **UTS: 260 Elon: 25%**
STA 7 CZ15	1% Sn 38% Zn Cu alloy: Die casting; as cast **UTS: 300 Elon: 20%**
STEREO	38% Zn 2% Fe Cu alloy: Blackwell; obsolete
SUPERSTON 50	0.2% Pb 30% Zn 5% Al 1.5% Fe 2.2% Mn Cu alloy: Casting; Stone Manganese for BS 207
TAURUS XXI	1% Sn 40% Zn 0.5% Al 1% Fe 1% Mn Cu alloy: Casting; D Brown; sand cast **DPN: 110 UTS: 480 Elon: 24% Proof: 200**
TAURUS XXXI	1% Pb 40% Zn Cu alloy: Casting; D Brown; sand cast **DPN: 70 UTS: 290 Elon: 15%**
TOBIN BRASS	40% Zn Cu alloy: With small addition Sn; IMI 366
TOBIN BRONZE	0.75% Sn 0.1% Pb 39% Zn Cu alloy: Extrusions; common name for this alloy; further information from Copper Development Association
TONUM	35% Zn 1.7% Al 1% Fe 1.5% Mn Cu alloy: Casting; Stone Manganese **UTS: 520 Elon: 20% Proof: 240**
TURBISTON 3	1% Sn 41% Zn 0.25% Al 1% Fe 0.2% Mn Cu alloy: Casting; Stone Manganese **DPN: 150 UTS: 510 Elon: 18% Proof: 240**
TURBISTON M	35% Zn 1.7% Al 1% Fe 1.5% Mn Cu alloy: Casting; Stone Manganese **UTS: 52 Elon: 20% Proof: 24(0.15%)**
UNI 6896	39% Zn 1.0% Al 1.0% Fe 1.0% Mn Cu alloy: Wrought; Italian Standard; high tensile brass **DPN: 140 UTS: 480 Elon: 15% Proof: 200**
VICTOR BRONZE	39% Zn 1.5% Al 1.0% Fe 0.05% V Cu alloy: Origin unknown
W	2% Pb 40% Zn Cu alloy: Extrusion; McKechnie Bros; as drawn **DPN: 130 UTS: 40 Elon: 22% Proof: 20**
WS	1.75% Pb 39% Zn Cu alloy: Extrusion; McKechnie Bros; as drawn **DPN: 125 UTS: 39 Elon: 30% Proof: 18**
XX	50% Zn Cu alloy: Brazing filler rod; McKechnie Bros; melting range 875–885 °C

Note. The following abbreviations and units are used in the tables:

DPN	Hardness, diamond pyramid number
UTS	Ultimate tensile strength, N/mm^2
Elon	Elongation, %
Proof	0.1% proof strength, N/mm^2

$1\ N/mm^2 = 0.1\ hbar = 0.102\ kgf/mm^2 = 0.06475\ tonf/in.^2 = 145.04\ lbf/in.^2 = 1\ MPa$

See Appendix II for other abbreviations and conversion tables.

14D Copper–beryllium wrought and cast alloys

Specific gravity		8.25
Density		8250 kg/m^3
Solidus/liquidus		890–990 °C
Thermal conductivity	solution treated	84 W/m °C
	solution treated and aged	105 W/m °C
Coefficient of linear expansion		17×10^{-6}/ °C
Electrical conductivity	solution treated	16–18% IACS
	aged maximum hardness	20–25% IACS (copper 100%)
	aged to maximum conductivity	32–38% IACS
Specific resistance	solution treated	96–108 microhm mm
	aged to maximum hardness	69–86 microhm mm
	aged to maximum conductivity	46–54 microhm mm
Young's modulus of elasticity		$(124–131) \times 10^9$ N/m^2
Impact – solution treated and aged		100 J
Fatigue strength (10^8 cycles)		±220–270 N/mm^2
Hot strength		–

The above properties are typical of the following group, and may not apply exactly to any one specification. It is possible that with certain specifications some of the values may not be applicable.

General metallurgical characteristics

The alloys listed have up to 2% beryllium, with additions of cobalt or nickel. They all exhibit age hardening characteristics and give the best mechanical properties obtainable from any copper alloy.

In the annealed condition two phases, alpha and gamma, are present and the alloy is reasonably ductile. On heating to above about 750 °C, depending on the beryllium content, the gamma phase dissolves in the alpha and if quenched in water will remain in solution – 'solution treated'. In this condition it resembles pure copper, being soft, ductile and capable of accepting a considerable amount of cold work. If, however, the alloy is then heated to 315–470 °C, the gamma phase agglomerates and precipitates out of solution. This causes disturbance of the crystal lattice making the material less ductile but giving a considerable increase in hardness. Moreover this increase in hardness can be additive to that obtained from cold work. The electrical and thermal conductivity is reduced as the hardness increases. Corrosion resistance is also affected but not to such a marked degree.

There are two basic ranges of alloys: those with less than 1% beryllium, and those with about 2% beryllium. The cobalt or nickel content generally varies inversely with the beryllium content, their presence increasing the response to thermal treatment.

The mechanical properties of beryllium coppers improve at sub-zero temperatures as low as −200 °C. This improvement is about 10% and includes ductility and impact strength.

In order to obtain the maximum mechanical properties from these materials, they must first be solution treated, which requires a temperature of approximately 900 °C, and then aged. The solution treatment will result in the material being soft and ductile and the ageing will increase the mechanical properties depending on the ageing temperature chosen.

For maximum mechanical strength the ageing is in the region of 450 °C but, where a higher ductility or maximum electrical conductivity is required, ageing will be at the higher end, 450–480 °C.

The ageing operation is time and temperature dependent, i.e. by holding the material at the same temperature for a longer time the material will first increase in tensile strength and hardness and then begin to soften. If the material is correctly solution treated and is then held at the same temperature anywhere above approximately 300–350 °C, then it will gradually increase in tensile properties and decrease in ductility, and having reached a maximum will then decrease in tensile strength and improve in ductility.

The maximum electrical conductivity and tensile strength is achieved by slightly over-ageing.

Beryllium and beryllium oxide are toxic but as long as they are contained as an alloy in beryllium copper then no problems exist. There are some health hazards where heat treatment or welding is involved in that beryllium oxide will form.

Welding does not present any technical problems which do not exist with other copper alloys. That is, because of the high conductivity difficulty can exist in achieving a satisfactory weld without pre-heating and normally fluxes are needed where gas welding, electric arc welding or brazing is required. These fluxes will invariably be aggressive and can cause corrosion problems if not removed.

It is recommended that anyone brazing or welding these alloys should wear a mask or use good air extraction.

There can also be a limited health hazard from dust when these components are machined or ground; therefore any grinding, etc., must be carried out with good extraction and the collected dust must be disposed of under controlled conditions.

These materials probably have the highest mechanical strength of any non-ferrous alloy and in addition they have much better electrical conductivity than the ferrous alloys of comparable strength.

They have the added advantage that they do not produce sparks when ground or hit with a hammer and thus have considerable advantages for safety where tools are used under hazardous conditions. Spanners, hammers, etc., used where ammunition, petrol or gas are involved are generally manufactured from beryllium copper alloys.

They also find considerable use as springs where electrical conductivity is involved, and draught excluders where doors open and shut are commonly manufactured from beryllium copper alloys. They are also used for high strength corrosion resistant tubing, dies for deep drawing of metals, and occasionally ornaments where the attractive bronze finish with strength is required.

As these materials can be produced in a very soft condition, allowing extruding, forming, etc., and can then be subjected to controlled hardening to allow accurate machining in an efficient manner, and finally can be hardened, again by ageing, means that they have a considerable range of possibilities in engineering.

Symbol	Nominal analysis, supplier, condition and remarks.
3	1.8% Ni 0.4% Be Cu alloy: Brush Wellman
10	0.6% Be 2.5% Co Cu alloy: Wrought; Brush Beryllium Co.
10C	0.6% Be 2.5% Co Cu alloy: Casting; Brush Beryllium Co.
10CR	0.6% Be 2.5% Co Cu alloy: Casting; Beryllium Corporation
11CR	1% Ni 0.6% Be 2% Co Cu alloy: Casting; Beryllium Corporation
20C	2.1% Be 0.5% Co Cu alloy: Casting; Brush Beryllium Co.
20CR	2.2% Be 0.5% Co Cu alloy: Casting; Beryllium Corporation
25	2% Be 0.25% Co Cu alloy: Wrought; Brush Beryllium Co.
30CR	1.8% Ni 0.6% Be Cu alloy: Casting; Beryllium Corporation
35	1.5% Ni 0.3% Be Cu alloy: Wrought; Brush Beryllium Co.
35C	1.5% Ni 0.35% Be Cu alloy: Casting; Brush Beryllium Co.
50	0.4% Be 1.5% Co 1% Ag Cu alloy: Wrought; Beryllium Corporation
50C	0.5% Be 1.1% Ag Cu alloy: Casting; Brush Beryllium Co.
50CR	0.5% Be 1.5% Co 1% Ag Cu alloy: Casting; Beryllium Corporation
50P	0.5% Be 2.5% Co Cu alloy: Powder; Brush Beryllium Co.; for sintering
70CR	0.9% Cr 0.15% Be Cu alloy: Casting; Beryllium Corporation; for welding electrodes
150P	1.5% Be 0.25% Co Cu alloy: Powder; Brush Beryllium Co.; for sintering
165	1.7% Be 0.25% Co Cu alloy: Wrought; Brush Beryllium Co.
165C	1% Ni 0.6% Be 2% Co Cu alloy: Casting; Brush Beryllium Co.
165CR	1.7% Be 0.2% Co Cu alloy: Casting; Brush Beryllium Co.
174	0.3% Be 0.5% Co Cu alloy: Brush Wellman
190	1.9% Be 0.4% Co Cu alloy: Brush Wellman
200P	2.0% Be Cu alloy: Powder; Brush Beryllium Co.; for sintering
245C	2.5% Be 0.5% Co Cu alloy: Casting; Brush Beryllium Co.
245Cr	2.5% Be 0.5% Co Cu alloy: Casting; Beryllium Corporation
275C	2.75% Be 0.5% Co Cu alloy: Casting; Brush Beryllium Co.
275Cr	2.7% Be 0.5% Co Cu alloy: Casting; Beryllium Corporation
290	1.9% Be 0.4% Co Cu alloy: Brush Wellman
AMPCOLOY 83	1.7% Be 0.25% Co Cu alloy: Wrought; Ampco Metal
AMPCOLOY 84	Be Cu alloy: Ampco
AMPCOLOY 96	1.5% Ni 0.3% Be Cu alloy: Casting; Ampco Metal
AMS 4530 B	1.9% Be Cu alloy: Sheet and strip; solution treated
AMS 4532 A	1.9% Be Cu alloy: Sheet and strip; ½-hard
AMS 4650 D	1.9% Be Cu alloy: Bar and strip; solution treated
AMS 4651	0.2% (min) Ni 1.9% Be and Co Cu alloy
AMS 4725 A	1.9% Be Cu alloy: Wire; solution treated
AMS 4890 A	0.3% Si 2% Be 0.4% Co Cu alloy: Investment casting; heat treated
ASTM B194	0.6% Ni 1.9% Be + Co + Fe (max) Cu alloy: Sheet and bar; annealed **UTS: 460**
ASTM B194	0.6% Ni 1.9% Be + Co + Fe (max) Cu alloy: Sheet and bar; hard **UTS: 820**
ASTM B196	0.6% Ni 1.9% Be + Co + Fe (max) Cu alloy: Bar; annealed **UTS: 460**
ASTM B196	0.6% Ni 1.9% Be + Co + Fe (max) Cu alloy: Bar; solution treated; hard drawn and aged **UTS: 1370**
ASTM B197	0.6% Ni 1.9% Be + Co + Fe (max) Cu alloy: Wire; annealed **UTS: 460**
ASTM B197	0.6% Ni 1.9% Be + Co + Fe (max) Cu alloy: Wire; cold drawn; solution treated; cold drawn and aged **UTS: 1540**
ASTM B441	0.55% Be 2.5% Co Cu alloy: Bar; solution treated and aged; conductivity 20% IACS **UTS: 800**
ASTM B643	0.2% Ni + Co (max) 1.9% Be Cu alloy: Tube; seamless
ASTM B768 C 17400	0.3% Be 0.2% Co Cu alloy: Strip **UTS: 700 Elon: 8% Proof: 640**
ASTM B768 C 17410	0.3% Be 0.2% Co Cu alloy: Strip **UTS: 810 Elon: 9% Proof: 730**
BERALOY A	2% Be 0.5% Co Cu alloy: Driver
BERYDUR	1% Be 0.25% Co Cu alloy: Wrought; Beryllium Corporation
BERYLCO	4% Be Cu alloy: Ingot; Beryllium Corporation
BERYLCO	0.55% Be 2.5% Co Cu alloy: Wrought; Beryllium Corporation; solution treated and aged **UTS: 1070 Proof: 750**
BERYLCO 25	1.9% Be 0.2% Co Cu alloy: Wrought; Beryllium Corporation; cold worked; solution treated and aged **UTS: 1540 Proof: 1370**
BERYLCO 33/25	2% Be 0.25% Co Cu alloy: Beryllium Corporation; same as alloy 25; free machining
BERYLCO 50	0.4% Be 1.5% Co 1.0% Ag Cu alloy: Bar and forging; Beryllium Corporation; solution treated and aged **UTS: 1070 Proof: 750**
BERYLCO 165	1.7% Be 0.2% Co Cu alloy: Strip and bar; Beryllium Corporation; cold worked; solution treated and aged **UTS: 1370 Proof: 1140**
BERYLLIUM BRONZE	2.5% Be Cu alloy: Common name
BERYLLOY	2.7% Be 0.5% Co Cu alloy: Telcon Metals
BRUSH 190	1.9% Be 0.2% Co or Ni Cu alloy: Wrought; Brush Beryllium Co.

Note. The following abbreviations and units are used in the tables:

DPN	Hardness, diamond pyramid number
UTS	Ultimate tensile strength, N/mm^2
Elon	Elongation, %
Proof	0.1% proof strength, N/mm^2

1 N/mm^2=0.1 hbar=0.102 kgf/mm^2=0.06475 tonf/in.^{2}=145.04 lbf/in.^{2}=1 MPa

See Appendix II for other abbreviations and conversion tables.

Symbol	Nominal analysis, supplier, condition and remarks.
BS 2870 CB101	1.8% Be 0.3% Co + Ni Cu alloy: Sheet; solution treated and aged **DPN: 350**
BS 2870 CB101	1.8% Be 0.3% Co + Ni Cu alloy: Sheet; solution treated and cold rolled; hard **DPN: 225 UTS: 680 Elon: 4%**
BS 2870 CB101	1.8% Be 0.3% Co + Ni Cu alloy: Sheet; solution treated; cold rolled and aged **DPN: 350**
BS 2870 CB101	1.8% Be 0.3% Co + Ni Cu alloy: Sheet; solution treated; cold rolled and aged **DPN: 350**
BS 2872 CB101	0.2% Ni 1.8% Be or Co Cu alloy: Forging; specification not issued
BS 2873 CB 101	0.2% Ni 1.8% Be + Co Cu alloy: Wire; solution treated and aged **UTS: 1050**
BS 2873 CB101	0.2% Ni 1.8% Be + Co Cu alloy: Wire; solution treated; cold drawn and aged **UTS: 1200**
BS 2874 CB101	1.8% Be 0.2% Co + Ni Cu alloy: Bar; specification not issued
BS 2875 CB101	1.8% Be 0.2% Co + Ni Cu alloy: Plate; specification not issued
BS 3127/2	1.8% Be 0.3% Co Cu alloy: Tube; solution treated for bourdon tube **DPN: 90**
BS 3127/2	1.8% Be 0.3% Co Cu alloy: Tube; solution treated; $\frac{1}{4}$-hard **DPN: 160**
BS 4577 A3/1	0.4% Be 2.5% Co Cu alloy: For resistance welding electrodes; conductivity 45% IACS; wrought **DPN: 180 UTS: 620 Elon: 9%**
BS 4577 A4/2	2.0% Be Cu alloy: For resistance welding electrodes; wrought; conductivity 23% IACS **DPN: 350 UTS: 1050 Elon: 2%**
C 17000	0.2% Ni 1.7% Be + Co Cu alloy: CDA and UNS designation
C 17200	0.2% Ni 1.9% Be + Co Cu alloy: CDA and UNS designation
C 17300	0.2% Ni 1.9% Be + Co Cu alloy: CDA and UNS designation
C 17400	0.35% Be 0.3% Co Cu alloy: CDA and UNS designation
C 17410	0.35% Be 0.5% Co Cu alloy: CDA and UNS designation
C 17420	0.2% Be 0.3% Co Cu alloy: CDA and UNS designation
C 17500	0.6% Be 2.5% Co Cu alloy: CDA and UNS designation
C 17510	1.8% Ni 0.4% Be 0.3% Co Cu alloy: CDA and UNS designation
C 17520	1% Ni 0.2% Be Cu alloy: CDA and UNS designation
C 17600	0.35% Be 1.5% Co 1% Ag Cu alloy: CDA and UNS designation
C 17700	0.6% Be 2.5% Co 0.5% Te Cu alloy: CDA and UNS designation

Note. The following abbreviations and units are used in the tables:

DPN	Hardness, diamond pyramid number
UTS	Ultimate tensile strength, N/mm^2
Elon	Elongation, %
Proof	0.1% proof strength, N/mm^2

1 N/mm^2=0.1 hbar=0.102 kgf/mm^2=0.06475 tonf/in.^{2}=145.04 lbf/in.^{2}=1 MPa

See Appendix II for other abbreviations and conversion tables.

Symbol	Nominal analysis, supplier, condition and remarks.
C 81300	0.08% Be 0.8% Co Cu alloy: Cast; CDA and UNS designation
C 81400	0.8% Cr 0.08% Be Cu alloy: Cast; CDA and UNS designation
C 81700	1% Ni 0.45% Be 1% Co 1% Ag Cu alloy: Cast; CDA and UNS designation
C 81800	0.45% Be 1.5% Co 1% Ag Cu alloy: Cast; CDA and UNS designation
C 82000	0.6% Be 2.5% Co Cu alloy: Cast; CDA and UNS designation
C 82100	1% Ni 0.6% Be 1% Co Cu alloy: Cast; CDA and UNS designation
C 82200	1.5% Ni 0.6% Be Cu alloy: Cast; CDA and UNS designation
C 82400	1.7% Be 0.3% Co Cu alloy: Cast; CDA and UNS designation
C 82500	0.3% Si 2% Be 0.5% Co Cu alloy: Cast; CDA and UNS designation
C 82510	0.3% Si 2% Be 1% Co Cu alloy: Cast; CDA and UNS designation
C 82600	0.3% Si 2.35% Be 0.5% Co Cu alloy: Cast; CDA and UNS designation
C 82700	1.25% Ni 2.5% Be Cu alloy: Cast; CDA and UNS designation
C 82800	0.3% Si 2.6% Be 0.6% Co Cu alloy: Cast; CDA and UNS designation
CABRA 170	1.7% Be 0.2% Co Cu alloy: Casting; tentative SAE designation
CDA 170	1.7% Be 0.25% Co Cu alloy: Wrought; Copper Development Association of America; tentative
CDA 172	2% Be 0.25% Co Cu alloy: Wrought; Copper Development Association of America
CDA 175	0.4% Be 1.5% Co 1% Ag Cu alloy: Wrought; Copper Development Association of America; tentative
CDA 176	0.5% Be 2.5% Co Cu alloy: Wrought; Copper Development Association of America; tentative
CU BE 50	0.5% Be Cu alloy: Strip and sheet; Telcon Metals; solution treated and aged; electrical conductivity 50% IACS **DPN: 200 UTS: 760 Elon: 10%**
CU BE 50	0.5% Be Cu alloy: Strip and sheet; Telcon Metals; solution treated; cold rolled and aged; electrical conductivity 50% IACS **DPN: 250 UTS: 800 Elon: 8%**
CU BE 250	1.8% Be 0.25% Co Cu alloy: Sheet and strip; Telcon Metals; annealed **UTS: 460 Elon: 45% Proof: 170**
CU BE 250	1.8% Be 0.25% Co Cu alloy: Sheet and strip; Telcon Metals; solution treated and cold rolled **UTS: 900 Elon: 5% Proof: 720**
CU BE 250	1.8% Be 0.25% Co Cu alloy: Sheet and strip; Telcon Metals; solution treated; cold rolled and aged **UTS: 1300 Elon: 2% Proof: 1100**
CU BE 275	2.5% Be 0.5% Co Cu alloy: Casting; Telcon; for high strength moulds and dies
CUBELLOY	4.0% Be Cu master alloy: Imperial Smelting Corporation; primary material
ERM N S	0.4% Be 2.25% Co Cu alloy: Bar or forging; Enfield Rolling Mills Ltd; resistance weld electrodes; electrical conductivity 50% IACS **DPN: 250 UTS: 810 Elon: 10% Proof: 700**
IMI 185	1.8% Be 0.3% Co + Ni Cu alloy: Strip and wire; IMI; previously KYCUBE 25
KYCUBE 25	1.8% Be 0.3% Co + Ni Cu alloy: Strip and wire; IMI for BS alloy CB101
KYCUBE 25/1	2% Be 0.6% Co + Fe + Ni Cu alloy: Strip; IMI for BS alloy CB101; withdrawn
M 25	0.4% Pb 1.9% Be 0.4% Co Cu alloy: Brush Wellman
MALLORY 53B	1.7% Be 0.25% Co Cu alloy: Casting; Mallory Metallurgical; American alloy

Symbol	Nominal analysis, supplier, condition and remarks.
MALLORY 73	1.8% Be 0.1% Co Cu alloy: Castings; Mallory Metallurgical; electrical conductivity 20% IACS **DPN: 370 UTS: 750 Elon: 2% Proof: 670**
MALLORY 73	Be Co Cu alloy: For flash welding dies; Mallory Metallurgical for BS alloy CB101; electrical conductivity 23% IACS **DPN: 370 UTS: 1120 Elon: 4% Proof: 1020**
MALLORY 100	Be Co Cu alloy: Casting; Mallory Metallurgical; electrical conductivity 45% IACS **DPN: 210 UTS: 600 Elon: 6% Proof: 530**
MALLORY 100	Be Co Cu alloy: For resistance welding electrodes for stainless steel; Mallory Metallurgical; electrical conductivity **DPN: 200 UTS: 680 Elon: 10% Proof: 600**
MALLORY 150	0.5% Be 1.5% Co 1.1% Ag Cu alloy: Casting; Mallory Metallurgical; welding electrodes; American alloy
MATTHEY 100	0.4% Be 2.3% Co Cu alloy: For resistance electrodes; Johnson Matthey; conductivity 45% IACS **DPN: 180 UTS: 690 Elon: 10%**
RWMA Class 3	Be Co Cu alloy: Mallory Metallurgical code for Mallory 100
RWMA Class 4	1.8% Be 0.1% Co Cu alloy: Mallory Metallurgical code for Mallory 73
SAE CA170	1.7% Be 0.3% Co Cu alloy: Strip
SAE CA 172	1.9% Be Cu alloy: Strip
SAE CA 175	0.5% Be 2.5% Co Cu alloy: Strip
SAE CA 176	0.35% Be 1.5% Co Cu alloy: Bar

Note. The following abbreviations and units are used in the tables:

DPN	Hardness, diamond pyramid number
UTS	Ultimate tensile strength, N/mm^2
Elon	Elongation, %
Proof	0.1% proof strength, N/mm^2

1 N/mm^2=0.1 hbar=0.102 kgf/mm^2=0.06475 tonf/in.^{2}=145.04 lbf/in.^{2}=1 MPa

See Appendix II for other abbreviations and conversion tables.

14E Copper–nickel wrought and cast alloys
Up to 50 per cent nickel with additions of iron and manganese

Specific gravity		8.9–8.97
Density		8900–8970 kg/m^3
Solidus/liquidus		1199 °C
Thermal conductivity		25–59 W/m °C
Coefficient of linear expansion		(14.5-17) × 10^{-6}/ °C
Electrical conductivity		6.5% IACS (copper 100%)
Specific resistance		265 microhm mm
Young's modulus of elasticity		(124–159) × 10^9 N/m^2
Impact		107 J
Fatigue strength	cold drawn	±230 N/mm^2
	annealed	±150 N/mm^2
Hot strength		–

Temperature °C	Tensile strength N/mm^2
205	320
425	240
645	140
875	37

The above properties are typical of the following group, and may not apply exactly to any one specification. It is possible that with certain specifications some of the values may not be applicable.

General metallurgical characteristics

Copper and nickel form a continuous series of solid solutions. That is, copper and nickel are soluble in each other without forming intermetallic compounds, the characteristics of copper gradually changing to those of nickel from 0% nickel 100% copper, to 100% nickel 0% copper.

Nickel increases the strength and electrical and corrosion resistance (particularly marine) in direct proportion to the alloy content. Iron is added to increase the mechanical strength and to provide resistance to impingement attack.

Manganese is added to de-sulphurize, thus preventing the brittleness and reduction in corrosion resistance which results when sulphur is present.

Titanium is added to welding rods and alloys commonly welded to reduce porosity at the weld.

Silicon additions made to some of the alloys with about 2.5% nickel result in a degree of age hardening caused by the precipitation of intermetallic compounds from solution. The effect is not very marked and the alloys are as often used in the annealed or cold worked condition as the heat treated condition.

The ability to accept cold work without requiring an anneal varies inversely with the nickel content. Alloys with less than 10% nickel can be subjected to considerable cold work without trouble and all the alloys can be readily hot worked. Alloys with more than 50% Ni are listed in Section 27D as nickel–copper alloys and include 'Monel' metals which accept very little cold work without becoming brittle. The copper–nickel alloys are seldom used as castings. Some copper–nickel alloys have high electrical resistance, with very low temperature coefficient of resistance.

With the exception of a limited number of alloys containing some nickel and silicon, none of these alloys can be thermally treated to improve their mechanical properties. The low nickel silicon alloys have some age hardening properties but this requires specialist advice.

As stated above, these alloys on the whole can accept a considerable quantity of cold work, and when annealing or softening is required they must be heated to 700 °C.

Welding and brazing can be carried out using normal techniques with fluxed rods or electric or gas welding. Brazing presents no problem but wherever fluxes are used care must be taken to ensure that all evidence of fluxes is removed to prevent corrosion problems.

The copper–nickel alloys were traditionally used for marine conditions where sea-water was involved. In many of these uses the alloys are now being replaced with austenitic stainless steel but the copper nickel alloys still find specific uses in components such as condensers and condenser tubes. Where water is handled hot these alloys probably have better corrosion resistance than the austenitic stainless steels.

The higher nickel alloys in the Monel range are finding use offshore for cladding where the dual purpose of corrosion protection and the inhibition of marine growth has made these alloys very useful.

The higher nickel alloys produce a range of materials where constant electrical resistance over a wide range of temperatures is required.

Symbol	Nominal analysis, supplier, condition and remarks.
2.0837	30% Ni 1% Fe 0.2% Ti Cu alloy: German Standard
2.0882	30.0% Ni Cu alloy: German Standard
4F	2.7% Ni 0.5% Si Cu alloy: Bar and forging; Langley Alloys
30 ALLOY	2.25% Ni Cu alloy: British Driver-Harris; electrical conductivity 30% IACS **UTS: 300**
60 ALLOY	6% Ni Cu alloy: For resistors; Driver
90 ALLOY	12% Ni Cu alloy: For resistors; Driver
95 ALLOY	11% Ni Cu alloy: Driver-Harris **UTS: 360**
111 ALLOY	Ni Cu alloy: Driver-Harris; electrical conductivity 55% IACS **UTS: 300**
180 ALLOY	22% Ni Cu alloy: For resistors; Driver
190	0.2% P 1.1% Ni Cu alloy: Wrought; designation agreed by CDAA; not used
191	0.2% P 1.1% Ni Cu alloy: Wrought; designation agreed by CDAA; not used
647	2.0% Ni 0.6% Si Cu alloy: Wrought; designation agreed by CDAA; not used
702	2.5% Ni 0.4% Mn Cu alloy: Wrought; designation agreed by CDAA; not used
703	5.2% Ni 0.5% Mn Cu alloy: Wrought; designation agreed by CDAA; not used
704	1.0% Zn 5.5% Ni 1.5% Fe 0.5% Mn Cu alloy: Wrought; designation agreed by CDAA; not used
707	10% Ni 0.5% Mn Cu alloy: Wrought; designation agreed by CDAA; not used
708	11.0% Ni 0.15% Mn Cu alloy: Wrought; designation agreed by CDAA; not used
709	1.0% Zn 15% Ni 0.6% Mn Cu alloy: Wrought; designation agreed by CDAA; not used
720	42% Ni 2.0% Fe 1.2% Mn Cu alloy: Wrought; designation agreed by CDAA; not used
964	30% Ni Cu alloy: Designation used by CDAA
ADNIC	1% Sn 29% Ni 1.3% Mn Cu alloy: Origin unknown
ADVANCE	44% Ni 1.5% Mn Cu alloy: British Driver-Harris **UTS: 450**

Note. The following abbreviations and units are used in the tables:

DPN	Hardness, diamond pyramid number
UTS	Ultimate tensile strength, N/mm^2
Elon	Elongation, %
Proof	0.1% proof strength, N/mm^2

1 N/mm^2=0.1 hbar=0.102 kgf/mm^2=0.06475 tonf/in.^{2}=145.04 lbf/in.^{2}=1 MPa

See Appendix II for other abbreviations and conversion tables.

Symbol	Nominal analysis, supplier, condition and remarks.
ALLOY 400	Ni 2.5% Fe 0.2% C 30% Cu alloy: Carpenter; annealed **UTS: 500 Elon: 53% Proof: 280**
ALLOY Cu Ni 90/10	10% Ni 1.4% Fe 0.7% Mn Cu alloy: VDM
ALLOY Cu Ni 70/30	31% Ni 0.7% Fe 1% Mn Cu alloy: VDM
AMPCOLOY 521	30% Ni Cu alloy: Casting; Ampco
AMPCOLOY 522	30% Ni Cu alloy: Ampco
AMPCOLOY 525	10% Ni 1.2% Fe Cu alloy: Casting; Ampco
AMPCOLOY 526	10% Ni 1% Fe Cu alloy: Ampco
AMS 4732	22.5% Ni Cu alloy
ANTIMONY BRONZE	2% Ni 7.5% Sb Sn free Cu alloy: Common name
ASTM B111 70/30 Cu Ni	31% Ni Cu alloy: Tube
ASTM B111 80/20 Cu Ni	21% Ni Cu alloy: Tube
ASTM B111 90/10 Cu Ni	10% Ni Cu alloy: Tube
ASTM B111/704	5.2% Ni 1.5% Fe 0.5% Mn Cu alloy: Seamless tube
ASTM B111/706	10% Ni 1.4% Fe Cu alloy: Seamless tube
ASTM B111/710	21% Ni 0.8% Fe Cu alloy: Seamless tube
ASTM B111/715	31% Ni 0.5% Fe Cu alloy: Seamless tube
ASTM B111/716	31% Ni 5.2% Fe Cu alloy: Seamless tube
ASTM B111/720	41% Ni 2.0% Fe 1.2% Mn Cu alloy: Seamless tube
ASTM B122/5	30% Ni Cu alloy: Sheet; ¼-hard **UTS: 390**
ASTM B122/5	30% Ni Cu alloy: Sheet; spring hard **UTS: 500**
ASTM B122/6	20% Ni Cu alloy: Sheet; ¼-hard **UTS: 390**
ASTM B122/6	20% Ni Cu alloy: Sheet; spring hard **UTS: 590**
ASTM B171/706	10% Ni 1.2% Fe Cu alloy: Rolled plate
ASTM B171/715	31% Ni 0.55% Fe Cu alloy: Rolled plate
ASTM B225 E Cu Ni	29% Ni Cu alloy: Welding electrode **UTS: 340**
ASTM B259 R Cu Ni	1% Sn 1% Zn 30% Ni Cu alloy: Welding rod **UTS: 340**
ASTM B358 A	30% Ni Cu alloy: Ingot; primary metal
ASTM B358 B	50% Ni Cu alloy: Ingot; primary metal
ASTM B360 B	30% Ni 0.9% Fe Cu alloy: Casting; as cast **UTS: 420 Elon: 20% Proof: 210**
ASTM B369 A	10% Ni 1.2% Fe Cu alloy: Casting; as cast **UTS: 300 Elon: 20% Proof: 170**
ASTM B402/70/30	31% Ni 0.6% Fe Cu alloy: Plate and sheet
ASTM B402/90/10	10% Ni 1.5% Fe Cu alloy: Plate and sheet
ASTM B411	1.9% Ni 0.6% Si Cu alloy: Rod and bar
ASTM B412	1.9% Ni 0.6% Si Cu alloy: Wire
ASTM B422	1.9% Ni 0.6% Si Cu alloy: Sheet and strip
ASTM B433/70/30	30% Ni 0.7% Fe 1.2% Mn 0.5% Si 1.0% Nb Cu alloy: Casting
ASTM B433/90/10	10% Ni 1.3% Fe 1.2% Mn Cu alloy: Casting
ASTM B466	Ni Cu alloy: Pipe and tube; graded by CDAA system

Symbol	Nominal analysis, supplier, condition and remarks.
ASTM B467	Ni Cu alloy: Welded pipe and tube; graded by CDAA system
ASTM B492	20% Ni 1.0% Fe 0.8% Mn 0.4% Si Cu alloy: Casting for tailshaft sleeves
ASTM B740 Cu 9 Ni 6 Sn	6% Sn 9% Ni Cu alloy: Stock
ASTM B740 Cu 15 Ni 8 Sn	8% Sn 15% Ni Cu alloy: Stock
B 47	Ni Cu alloy: Casting; Anti Attrition Ltd for electrical slip rings
	UTS: 240 Elon: 25%
B 68	Ni Cu alloy: Casting for food containers; Anti Attrition Ltd
	DPN: 60 UTS: 210 Elon: 10%
BATNICKON 5	5% Ni Fe Mn Cu alloy: Tube, sheet, etc.; Birmingham Battery
BATNICKON 10	10% Ni Fe Cu alloy: Tube, sheet, etc.; Birmingham Battery
BATNICKON 30	30% Ni Fe Cu alloy: Sheet, tube, etc.; Birmingham Battery
BS 374 CN103	15% Ni 0.2% Mn Cu alloy: Sheet, annealed
	UTS: 260 Elon: 36%
BS 374 CN104	20% Ni 0.2% Mn Cu alloy: Sheet, annealed
	UTS: 300 Elon: 36%
BS 374 CN105	25.0% Ni 0.25% Mn Cu alloy: Sheet, annealed
	UTS: 330 Elon: 33%
BS 374 CN106	30.0% Ni Cu alloy: Sheet, annealed
	UTS: 360 Elon: 33%
BS 378 CN102	10% Ni 1.5% Fe 0.75% Mn Cu alloy: Tube; as drawn for condenser tubes, etc.
	DPN: 150
BS 378 CN107	30% Ni 0.75% Fe 1% Mn Cu alloy: Tube; as drawn for condenser tubes, etc.
	DPN: 150
BS 1400 CN1	31% Ni 0.7% Fe 1.8% Cr Cu alloy: Casting
BS 1400 CN2	30% Ni 1.2% Fe 1.3% Nb + Ta Cu alloy: Casting
BS 1464 CN107	30% Ni 0.7% Fe 1% Mn Cu alloy: Tube
	DPN: 120
BS 1541 CN101	5.5% Ni 1.2% Fe 0.6% Mn Cu alloy: Plate; as rolled
	UTS: 220 Elon: 30%
BS 1541 CN102	10.5% Ni 1.5% Fe 0.75% Mn Cu alloy: Plate; as rolled
	UTS: 260 Elon: 30%
BS 1541 CN104	20% Ni 0.2% Mn Cu alloy: Plate; as rolled
	UTS: 250 Elon 30%
BS 1541 CN106	30% Ni 0.5% Mn Cu alloy: Plate; as rolled
	UTS: 300 Elon 30%
BS 1541 CN107	30% Ni 0.6% Fe 1% Mn Cu alloy: Plate; as rolled
	UTS: 300 Elon 30%
BS 1845 C28	9.5% Ni 0.3% Si Cu alloy: For brazing; melting range 920–980 °C
BS 2870 CN101	5.5% Ni 1.2% Fe 0.5% Mn Cu alloy: Sheet; as rolled
	UTS: 220 Elon 35%
BS 2870 CN102	10% Ni 1.5% Fe 0.7% Mn Cu alloy: Sheet
BS 2870 CN103	15% Ni 0.2% Mn Cu alloy: Sheet; annealed
	UTS: 260 Elon 35%
BS 2870 CN104	20% Ni 0.2% Mn Cu alloy: Sheet; annealed
	UTS: 300 Elon 35%
BS 2870 CN105	25% Ni 0.3% Mn Cu alloy: Sheet; annealed
	UTS: 330 Elon 30%
BS 2870 CN106	30% Ni Cu alloy: Sheet; annealed
	UTS: 360 Elon 30%
BS 2870 CN107	30% Ni 0.7% Fe 1% Mn Cu alloy: Sheet; specification not issued
BS 2871 CN101	5.5% Ni 1.2% Fe 0.5% Mn Cu alloy: Tube; annealed
	UTS: 300
BS 2871 CN102	10% Ni 1.5% Fe 0.7% Mn Cu alloy: Tube; annealed
	UTS: 330
BS 2871 CN103	15% Ni 0.2% Mn Cu alloy: Tube; specification not issued
BS 2871 CN104	20% Ni 0.2% Mn Cu alloy: Tube; specification not issued
BS 2871 CN105	25% Ni 0.4% Mn Cu alloy: Tube; specification not issued
BS 2871 CN106	30% Ni Cu alloy: Tube; specification not issued
BS 2871 CN107	30% Ni 0.75% Fe 1% Mn Cu alloy: Tube; annealed
	UTS: 420
BS 2871 CN108	30% Ni 2% Fe 2% Mn Cu alloy
BS 2872 CN101	5.5% Ni 1.2% Fe 0.5% Mn Cu alloy: Forging; specification not issued
BS 2872 CN102	10% Ni 1.5% Fe 0.7% Mn Cu alloy: Forging; specification not issued
BS 2872 CN103	15% Ni 0.3% Mn Cu alloy: Forging; specification not issued
BS 2872 CN104	20% Ni 0.3% Mn Cu alloy: Forging; specification not issued
BS 2872 CN105	25% Ni 0.4% Mn Cu alloy: Forging; specification not issued
BS 2872 CN106	30% Ni Cu alloy: Forging; specification not issued
BS 2872 CN107	30% Ni 1% Mn Cu alloy: Forging; specification not issued
BS 2873 CN101	5% Ni 1.2% Fe 0.6% Mn Cu alloy: Wire; specification not issued
BS 2873 CN102	10% Ni 1.5% Fe 0.8% Mn Cu alloy: Wire; specification not issued
BS 2873 CN103	15% Ni 0.3% Mn Cu alloy: Wire; specification not issued
BS 2873 CN104	20% Ni 0.2% Mn Cu alloy: Wire; specification not issued
BS 2873 CN105	25% Ni 0.4% Mn Cu alloy: Wire; specification not issued
BS 2873 CN106	30% Ni Cu alloy: Wire; specification not issued
BS 2873 CN107	31% Ni 0.8% Fe 1% Mn Cu alloy: Wire; specification not issued
BS 2874 CN101	5.5% Ni 1.2% Fe 0.6% Mn Cu alloy: Bar; specification not issued
BS 2874 CN102	10% Ni 1.5% Fe 0.7% Mn Cu alloy: Bar; specification not issued
BS 2874 CN103	15% Ni 0.3% Mn Cu alloy: Bar; specification not issued
BS 2874 CN104	20% Ni 0.3% Mn Cu alloy: Bar; specification not issued
BS 2874 CN105	25% Ni 0.4% Mn Cu alloy: Bar; specification not issued
BS 2874 CN106	30% Ni Cu alloy: Bar; specification not issued
BS 2874 CN107	30% Ni 0.7% Fe 1% Mn Cu alloy: Bar; specification not issued
BS 2875 CN101	5.5% Ni 1% Fe 0.5% Mn Cu alloy: Plate; as rolled
	UTS: 220 Elon: 30%
BS 2875 CN102	10% Ni 1.5% Fe 0.7% Mn Cu alloy: Plate; as rolled
	UTS: 260 Elon: 30%
BS 2875 CN103	15% Ni 0.2% Mn Cu alloy: Plate; specification not issued
BS 2875 CN104	20% Ni 0.3% Mn Cu alloy: Sheet; as rolled
	DPN: 250 Elon: 30%
BS 2875 CN105	25% Ni 0.25% Mn Cu alloy: Plate; specification not issued
BS 2875 CN106	30% Ni Cu alloy: Plate; as rolled
	UTS: 300 Elon: 30%
BS 2875 CN107	30% Ni 0.8% Fe 1% Mn Cu alloy: Plate; as rolled
	UTS: 300 Elon: 30%
BS 2901 C16	0.3% Al 10% Ni Cu alloy: Rod for inert gas shielded arc welding; obsolete
BS 2901 C17	20% Ni 0.3% Ti Cu alloy: Rod for all welding
BS 2901 C18	0.3% Al 30% Ni Cu alloy: Rod for all welding
BS 2901 C19	0.3% Al 5.5% Ni 1.2% Fe 0.5% Cu alloy: Rod for all welding
BS 4577 A3/2	2.5% Ni 0.5% Si Cu alloy: For resistance welding electrodes; wrought; conductivity 38% IACS
	DPN: 200 UTS: 650 Elon: 14%
BS 4577 A4/1	0.2% P 0.1% Ni Cu alloy: For resistance welding electrodes; wrought; conductivity 50% IACS
	DPN: 130 UTS: 400 Elon: 20%

Symbol	Nominal analysis, supplier, condition and remarks.
C 16	11.0% Ni 1.6% Fe Cu alloy: Rod for gas shielded arc welding
C 18	30% Ni 0.7% Fe Cu alloy: Rod for gas shielded arc welding
C 19000	1.1% Ni Cu: CDA and UNS designation
C 19010	1.4% Ni Cu: CDA and UNS designation
C 19020	0.5% Sn 2.2% Ni Cu: CDA and UNS designation
C 19100	0.2% P 1.2% Ni Cu: CDA and UNS designation
C 64700	1.8% Ni 0.6% Si Cu alloy: CDA and UNS designation
C 64710	0.4% Zn 3.2% Ni 0.8% Si Cu alloy: CDA and UNS designation
C 64720	0.3% Zn 2% Ni 0.5% Si Cu alloy: CDA and UNS designation
C 70100	3.5% Ni Cu alloy: CDA and UNS designation
C 70200	2.5% Ni Cu alloy: CDA and UNS designation
C 70250	3.2% Ni Cu alloy: CDA and UNS designation
C 70300	5.2% Ni Cu alloy: Obsolete; UNS designation
C 70320	3.2% Ni Cu alloy: CDA and UNS designation
C 70400	5.5% Ni 1.5% Fe Cu alloy: CDA and UNS designation
C 70440	5.2% Ni 1.4% Fe Cu alloy: CDA and UNS designation
C 70500	6.8% Ni Cu alloy: CDA and UNS designation
C 70600	10% Ni 1.4% Fe Cu alloy: CDA and UNS designation
C 70610	10.5% Ni 1.5% Fe Cu alloy: CDA and UNS designation
C 70690	10% Ni Cu alloy: CDA and UNS designation
C 70700	10% Ni Cu alloy: CDA and UNS designation
C 70800	11.5% Ni Cu alloy: CDA and UNS designation
C 70900	15% Ni Cu alloy: CDA and UNS designation
C 71000	21% Ni Cu alloy: CDA and UNS designation
C 71100	23% Ni Cu alloy: CDA and UNS designation
C 71110	22.5% Ni Cu alloy: CDA and UNS designation
C 71300	25% Ni Cu alloy: CDA and UNS designation
C 71500	31% Ni 0.8% Fe Cu alloy: CDA and UNS designation
C 71580	31% Ni Cu alloy: CDA and UNS designation
C 71581	31% Ni 0.6% Fe Cu alloy: CDA and UNS designation
C 71590	31% Ni Cu alloy: CDA and UNS designation
C 71630	31% Ni 0.8% Fe Cu alloy: CDA and UNS designation
C 71640	30.5% Ni 2% Fe Cu alloy: CDA and UNS designation
C 71700	30.5% Ni 0.8% Fe Cu alloy: CDA and UNS designation
C 71900	30.5% Ni Cu alloy: CDA and UNS designation
C 72150	44.5% Ni Cu alloy: CDA and UNS designation
C 72200	16.5% Ni 0.7% Fe Cu alloy: CDA and UNS designation
C 72400	2% Al 13% Ni Cu alloy: CDA and UNS designation
C 72420	1.5% Al 15% Ni 1% Fe 4.5% Mn Cu alloy: CDA and UNS designation
C 72500	2% Sn 9.5% Ni Cu alloy: CDA and UNS designation
C 72600	4% Sn 4% Ni Cu alloy: CDA and UNS designation
C 72700	6% Sn 9% Ni Cu alloy: CDA and UNS designation
C 72800	8% Sn 10% Ni Cu alloy: CDA and UNS designation
C 72900	8% Sn 15% Ni Cu alloy: CDA and UNS designation
C 72950	5% Sn 21% Ni Cu alloy: CDA and UNS designation
C 96200	10% Ni 1.4% Fe 0.7% Nb Cu alloy: CDA and UNS designation
C 96300	20% Ni 1.2% Fe 1% Mn 1% Nb Cu alloy: CDA and UNS designation
C 96400	30% Ni 1.2% Fe 1% Nb Cu alloy: CDA and UNS designation

Note. The following abbreviations and units are used in the tables:

DPN	Hardness, diamond pyramid number
UTS	Ultimate tensile strength, N/mm^2
Elon	Elongation, %
Proof	0.1% proof strength, N/mm^2

1 N/mm^2=0.1 hbar=0.102 kgf/mm^2=0.06475 tonf/in.^{2}=145.04 lbf/in.^{2}=1 MPa

See Appendix II for other abbreviations and conversion tables.

Symbol	Nominal analysis, supplier, condition and remarks.
C 96600	31% Ni 1% Fe 0.7% Be Cu alloy: CDA and UNS designation
C 96700	31% Ni 0.8% Fe 1% Be Cu alloy: CDA and UNS designation
C 96800	10% Ni 0.4% Mn 0.2% Nb Cu alloy: CDA and UNS designation
C 99400	3% Zn 1.5% Al 2.2% Ni 2% Fe 1.2% Si Cu alloy: Cast; CDA and UNS designation
C 99500	1.5% Zn 1.5% Al 4.5% Ni 4% Fe 1.5% Si Cu alloy: Cast; CDA and UNS designation
CB CUFRON	44% Ni 3% Mn Cu alloy: Carpenter; annealed; constant resistance alloy **UTS: 410 Elon: 25%**
CB CUFRON	45% Ni Cu alloy: Carpenter; cold worked constant resistance alloy **UTS: 790 Elon: 25%**
CBX CUFRON	44% Ni 3% Mn Cu alloy: Carpenter; annealed; constant resistance alloy **UTS: 415 Elon: 25%**
CONSTANTAN	45% Ni 55% Cu alloy: Electrical conductivity 3.5% IACS; A Heckford Ltd
Cu Ni 10 Fe	10% Ni Cu alloy: Wrought and cast; Inco; annealed **DPN: 90 UTS: 310 Elon: 40% Proof: 140**
Cu Ni 10 Fe Mn	10% Ni with Fe Mn Cu alloy: Wrought and cast; Inco; annealed **DPN: 100 UTS: 310 Elon: 30% Proof: 140**
Cu Ni 30 Fe	30% Ni with Fe Cu alloy: Wrought and cast; Inco; annealed **DPN: 100 UTS: 300 Elon: 45% Proof: 150**
Cu Ni 30 Fe 2 Mn 2	30% Ni with Fe and Mn Cu alloy: Wrought and cast; Inco; annealed **DPN: 110 UTS: 460 Elon: 45% Proof: 190**
Cu Ni 30 Fe Mn	30% Ni with Fe and Mn Cu alloy: Wrought and cast; Inco; annealed **DPN: 100 UTS: 430 Elon: 35% Proof: 220**
CUNICO 1	21% Ni 29% Co Cu alloy: For magnets; American material listed by PMA
CUNIFE 1	20% Ni 20% Fe Cu alloy: For magnets; American material listed by PMA
CUNIFE 11	20% Ni 27.5% Fe 2.5% Co Cu alloy: For magnets; American material listed by PMA
CUNIFER 10	10% Ni 1.4% Fe 0.7% Mn Cu alloy: VDM
CUNIFER 30	31% Ni 0.7% Fe 1% Mn Cu alloy: VDM
CUNIFER 302	31% Ni 1.8% Fe 1.7% Mn Cu alloy: VDM
CUNIFER B 7030	30.5% Ni 1% Fe 0.05% C (max) Cu alloy: Strip; VDM
CUNIFER S 7030	30.5% Ni 1% Fe 0.05% C (max) 0.3% Ti Cu alloy: Welding rod; VDM
CUNIFER S 9010	10% Ni 1% Fe 0.05% C (max) Cu alloy: Welding rod; VDM
CUPRO NICKEL	Series of copper–nickel alloys containing 15–17% Ni: Common name
CUPRON	45% Ni Cu alloy: Brunton for use up to 500 °C; electrical conductivity 4% IACS
CUPRONET N10	11% Ni 1.7% Fe Cu alloy: Electrode; Metrode
CUPRONET N30	31% Ni 0.6% Fe Cu alloy: Electrode; Metrode
DGS 320	30% Ni Cu alloy: Admiralty specification for alloy CN3
DGS 8454	30% Ni Cu alloy: Admiralty specification for alloy CN3
DIN 1733 El Cu Ni 10 Mn	10.8% Ni 1.3% Fe 0.004% C Cu alloy: Welding electrode
DIN 1733 S Cu Ni 30 Fe	30% Ni 1% Fe 0.2% Ti Cu alloy: Wrought
DIN 1785 Cu Ni 10 Fe	16% Ni 1% Fe 0.5% Mn Cu alloy: Wrought
DIN 1785 Cu Ni 20 Fe	21% Ni 0.8% Fe 1% Mn Cu alloy: Wrought
DIN 1785 Cu Ni 30 Fe	31% Ni 0.8% Fe 1% Mn Cu alloy: Wrought
DTD 498	2.7% Ni 0.5% Si Cu alloy: Bar and forging; as rolled or forged **UTS: 570 Elon: 15% Proof: 400**

Symbol	Nominal analysis, supplier, condition and remarks.
DTD 504	2.7% Ni 0.5% Si Cu alloy: Bar; cold rolled; solution treated and aged **DPN: 200 UTS: 630 Elon: 15% Proof: 560**
E Cu Ni	31% Ni 0.5% Fe 1.7% Mn Cu alloy: Welding electrode; designation used by AWS
ERM 3A	0.2% P 1% Ni Cu alloy: Bar, forging and casting; Enfield; as drawn; electrical conductivity 60% IACS **DPN: 160 UTS: 320 Elon: 22% Proof: 270**
FERRY	45% Ni Cu alloy: For electrical resistance; H Wiggin; annealed; electrical resistance 49 microhm/cm; non-magnetic **DPN: 156 UTS: 480 Elon: 47% Proof: 370**
FIVEOHM	Ni Mn Cu alloy: Specific electrical resistance 5 microhm/cm/cm
HAI 30 ALLOY	2% Ni Cu alloy: Harrison Alloys
HAI 60 ALLOY	6% Ni Cu alloy: Harrison Alloys
HAI 90 ALLOY	11% Ni Cu alloy: Harrison Alloys
HAI 180 ALLOY	24% Ni Cu alloy: Harrison Alloys
HAI-EN	45% Ni Cu alloy: For thermocouples; Harrison Alloys; specific resistance 49.9 microhm/cm
HAI-II ALLOY	1% Ni Cu alloy: For thermocouples; Harrison Alloys; specific resistance 4.2 microhm/cm
HAI-JN	45% Ni Cu alloy: For thermocouples; Harrison Alloys; specific resistance 49.9 microhm/cm
HAI-TN	45% Ni Cu alloy: For thermocouples; Harrison Alloys; specific resistance 49.9 microhm/cm
HECNUM	45% Ni Cu alloy: Electrical conductivity 5.5% IACS; A Heckford Ltd; low temperature coefficient of electrical resistance **DPN: 155 UTS: 480 Elon: 38% Proof: 220**
HIDURAL 5	2.7% Ni 0.5% Si Cu alloy: Bar and forging; Langley Alloys; annealed **UTS: 600 Elon: 20% Proof: 450**
HIDURAL 5	2.7% Ni 0.5% Si Cu alloy: Forging; Langley Alloys; cold drawn **UTS: 730 Elon: 20% Proof: 640**
HIDURAL 5	2.5% Ni 0.5% Si Cu alloy: Langley Alloys
HIDURAX SPECIAL 13A	3.0% Al 14.5% Ni 1.7% Fe 0.3% Mn Cu alloy: Langley; aged **DPN: 250 UTS: 860 Elon: 14% Proof: 700**
HIDURAX SPECIAL 19A	1.7% Al 17.5% Ni 4.5% Mn Cu alloy: Langley; aged; improved properties can be obtained by cold working **DPN: 240 UTS: 790 Elon: 25% Proof: 490**
HIDURON 102	10% Ni Cu alloy: Langley Bronze
HIDURON 102	10% Ni 1.5% Fe Cu alloy: Langley Alloys
HIDURON 107	30% Ni Cu alloy: Langley Bronze
HIDURON 107	30% Ni 0.7% Fe 1% Mn Cu alloy: Langley Alloys
HIDURON 501	12% Al Ni Fe Mn Cu alloy: Cast; Langley Alloys **UTS: 450 Elon: 35% Proof: 220**
IMI 842	20% Ni Cu alloy: Strip and wire; IMI; annealed **DPN: 90 UTS: 330 Elon: 50% Proof: 110**
IMI 842	20% Ni Cu alloy: Strip and wire; IMI; cold drawn **DPN: 160 UTS: 530 Elon: 8% Proof: 480**
IMI 849	30% Ni Cu alloy: Strip and wire; IMI
IS 2283NS10	10% Ni 1.5% Fe 0.7% Mn Cu alloy: Sheet; Indian Standard; annealed **DPN: 85 UTS: 320 Elon: 42% Proof: 120**
KUNIFER 5	5.5% Ni 1.2% Fe 0.5% Mn Cu alloy: Tube, plate and sheet; ICI and Yorkshire Imperial Metals; resistant to corrosion and erosion **DPN: 70 UTS: 290 Elon: 47% Proof: 110**
KUNIFER 5T	5% Ni 1.2% Fe 0.5% Mn 0.3% Ti Cu alloy: Welding rod; IMI
KUNIFER 10	10% Ni 2% Fe 1% Mn Cu alloy: Sheet and tube; ICI and Yorkshire Imperial Metals; corrosion resistant **DPN: 100 UTS: 320 Elon: 47% Proof: 150**
KUNIFER 10T	10% Ni 1.5% Fe 0.7% Mn 0.3% Ti Cu alloy: Welding rod; IMI
KUNIFER 30	31% Ni 0.6% Fe 0.8% Mn Cu alloy: Tube; Yorkshire Imperial Metals
KUNIFER 30	30% Ni 0.7% Fe 0.8% Mn Cu alloy: Wrought; IMI alloy obtained from Yorkshire Imperial Metals **DPN: 100 UTS: 440 Elon: 45% Proof: 140**
KUNIFER 30A	30% Ni 2% Fe 2% Mn Cu alloy: Wrought; IMI **DPN: 110 UTS: 650 Elon: 45% Proof: 180**
KUNIFER 30T	30% Ni 0.5% Fe 0.7% Mn 0.4% Ti Cu alloy: Welding rod; IMI
LOHM	6% Ni Cu alloy: British Driver-Harris; electrical conductivity 18% IACS **UTS: 520**
MAGNETOFLEX 1 2	20% Ni Cu alloy: Strip for magnets; German alloy listed by PMA
MANGANIN	4% Ni 12% Mn Cu alloy: Wire; British Driver-Harris **UTS: 450**
MIDOHM	23% Ni Cu alloy: British Driver-Harris **UTS: 520**
MONEL 450	31% Ni Cu alloy: INCA **UTS: 380 Elon: 46% Proof: 160**
NEN 6033/Cu Ni 10	10% Ni 1.5% Fe 0.7% Mn Cu alloy: Sheet; Dutch specification; annealed **DPN: 85 UTS: 320 Elon: 42% Proof: 120**
NEWLOY	1% Sn 35% Ni Cu alloy: Harrison Fischer
NF A51 102 Cu Ni 10 Fe 1 Mn	10% Ni 1.5% Fe 0.7% Mn Cu alloy: Tube; French Standard; annealed **DPN: 95 UTS: 320 Elon: 40% Proof: 150**
NF A51 102 Cu Ni 30 Mn 1 Fe	30% Ni 0.7% Fe 1% Mn Cu alloy: Tube; French Standard; annealed **DPN: 105 UTS: 420 Elon: 42% Proof: 180**
PERMET	21% Ni 29% Co Cu alloy: For magnets; American alloy listed by PMA
REGENT	20% Ni Cu alloy: A Heckford; electrical conductivity 6.5% IACS **DPN: 128 UTS: 330 Elon: 39%**
SAE CA 706	10% Ni 1.3% Fe Cu alloy: Wrought
SAE CA 710	20% Ni 1% Fe 1% Mn Cu alloy: Wrought
SAE CA 715	1% Zn (max) 30% Ni 1% Mn (max) Cu alloy: Wrought

For SIS specifications see SS

Symbol	Nominal analysis, supplier, condition and remarks.
SPECIAL ADVANCE	42% Ni Cu alloy: British Driver-Harris; thermocouples
SS 5667	10% Ni 1.5% Fe 1% Mn Cu alloy: Bar; Swedish Standard; annealed **DPN: 80 UTS: 300 Elon: 35% Proof: 150**
SS 5682	30% Ni 1% Fe 1% Mn Cu alloy: Bar; Swedish Standard; annealed **DPN: 100 UTS: 320 Elon: 32% Proof: 170**
STA 7 CX11	2.5% Ni 0.5% Si Cu alloy: Bar; solution treated; cold worked and aged **UTS: 630 Elon: 15% Proof: 560**
STA CX11	2.5% Ni 0.5% Si Cu alloy: Bar; solution treated and aged **UTS: 570 Elon: 15% Proof: 400**
STABILOHM 43	Ni Mn Cu alloy: Wire for resistors; Johnson Matthey
STUBS 387	30.8% Ni 0.6% Fe 0.08% C 0.3% Ti Cu alloy: Welders electrode; Stubs **UTS: 390 Elon: 30% Proof: 240**
STUBS 389	10.8% Ni 1.3% Fe 0.004% C Cu alloy: Welders electrode; Stubs **UTS: 320 Elon: 20% Proof: 240**
TAURUS	1.5% Sn 4% Ni Cu alloy: Casting; D Brown; centri-cast **DPN: 80 UTS: 290 Elon: 35% Proof: 150**
TAURUS XXVI	11% Sn 7% Pb 38% Ni Cu alloy: Casting; D Brown; centri-cast **DPN: 170 UTS: 340 Elon: 3% Proof: 180**
TELCALLOY 1	5% Ni Cu alloy: For electrical resistance; Telcon; annealed; electrical conductivity 18% IACS **UTS: 240**
TELCALLOY 1.5	10% Ni Cu alloy: For electrical resistance; Telcon; annealed; electrical conductivity 12% IACS **UTS: 260**

Symbol	Nominal analysis, supplier, condition and remarks.
TELCALLOY 2	15% Ni Cu alloy: For electrical resistance; Telcon; annealed; electrical conductivity 8.5% IACS **UTS: 300**
TELCALLOY 2	Ni Cu alloy: Wire for heaters; specific resistance 20 microhm/cm; Telcon Ltd
TELCALLOY 3	25% Ni Cu alloy: For electrical resistance; Telcon; annealed; electrical conductivity 5.5% IACS **UTS: 370**
TELCALLOY 3	Ni Cu alloy: For resistance wire; Telcon; specific resistance 30 microhm/cm
TELCALLOY 4	Ni Cu alloy: For electrical resistance; Telcon
TELCONSTAN	Ni Cu alloy: For electrical resistance; Telcon; electrical conductivity 3.5% IACS; constant electrical resistivity over normal temperature range; annealed **UTS: 400**
TENOHM	Ni Mn Cu alloy: Specific electrical resistance 10 microhm/cm/cm; origin unknown
UN 30	31.0% Ni 0.5% Fe Cu alloy: Origin unknown
UNI 6785	10% Ni 1.5% Fe 0.7% Mn Cu alloy: Tube; Italian Standard; annealed **DPN: 85 UTS: 320 Elon: 40% Proof: 150**
UNI 6786	30% Ni 0.7% Fe 1% Mn Cu alloy: Tube; Italian specification; annealed **DPN: 105 UTS: 420 Elon: 42% Proof: 180**
VSM 10803/Cu Ni 10 Fe Mn	10% Ni 1.5% Fe 0.7% Mn Cu alloy: Sheet; Swiss Standard; annealed **DPN: 85 UTS: 320 Elon: 42% Proof: 120**
VSM 10803/Cu Ni 30 Fe Mn	30% Ni 0.7% Fe 1% Mn Cu alloy: Tube; Swiss Standard; annealed **DPN: 105 UTS: 420 Elon: 42% Proof: 180**
WYMDALOY	20% Ni 20% Mn Cu alloy: Good corrosion resistance and high resistance to wear; origin unknown
YORCORON	30% Ni 2% Fe 2% Mn Cu alloy: Wrought; Yorkshire Imperial Metals **DPN: 70 UTS: 290 Elon: 47% Proof: 110**
YORCORON 3/3	30% Ni 3% Fe 3% Mn Cu alloy: Yorkshire Imperial Metals; annealed **DPN: 125 UTS: 450 Elon: 35% Proof: 190**

Note. The following abbreviations and units are used in the tables:

DPN	Hardness, diamond pyramid number
UTS	Ultimate tensile strength, N/mm^2
Elon	Elongation, %
Proof	0.1% proof strength, N/mm^2

1 N/mm^2=0.1 hbar=0.102 kgf/mm^2=0.06475 $tonf/in.^2$=145.04 $lbf/in.^2$=1 MPa

See Appendix II for other abbreviations and conversion tables.

14F Copper–nickel–zinc wrought and cast alloys
With additions of lead, tin, manganese and silicon – 'nickel silver' and 'nickel brass'

Specific gravity		8.46–8.60
Density		8460–8600 kg/m^3
Solidus/liquidus		930–1020 °C
Thermal conductivity		37.3–46 W/m °C
Coefficient of linear expansion		(15.0–16.7) $\times 10^{-6}$/ °C
Electrical conductivity		6–15% IACS (copper 100%)
Specific resistance		110–120 microhm mm
Young's modulus of elasticity		120×10^9 N/m^2
Impact	on 30% Ni 15% Zn annealed alloy	108 J
		118 J
Fatigue strength (10×10^7 cycles)	20% Ni casting	±100 N/mm^2

Hot strength

Temperature °C	Tensile strength N/mm^2	Elongation %
180	420	16.5
290	390	10.5
340	320	1
400	320	2
480	290	3.5

The above properties are typical of the following group, and may not apply exactly to any one specification. It is possible that with certain specifications some of the values may not be applicable.

General metallurgical characteristics

These alloys all have 40–60% copper with from 10% to 30% nickel, the remainder zinc. The copper content generally remains constant, with the nickel and the zinc varying in inverse proportion. Lead is added to increase machinability, tin to improve corrosion resistance, manganese acts as a de-sulphurizer, while silicon strengthens the alpha matrix.

These alloys are all white in colour, with reasonably good corrosion resistance and the ability to take and retain a high polish.

They can all be cold worked and only require interstage annealing when very deep drawing has been carried out, although this is less marked with the high zinc alloys – nickel brass.

All the lead-free alloys can be hot worked but best results are obtained with low nickel, high zinc which produces an alpha–beta structure.

The structure resembles that of the alpha brasses, except for the higher zinc materials where some beta phase appears. This reduces somewhat their ability to accept cold work.

All the alloys can be used to make castings, but these have no significant use except for ornament.

None of the materials in this section can have their mechanical properties improved by heat treatment. When cold worked they will have their tensile strength increased and ductility reduced. If the cold work must be removed this will require annealing at a maximum temperature of 700 °C.

Welding presents no problems apart from a limited health hazard where the higher zinc alloys are involved.

Brazing is regularly accomplished as is soldering.

Where fluxes are involved then these must be removed after joining in order to prevent corrosion.

These alloys are used in the main for their appearance. In many of their uses they have been replaced with austenitic stainless steel.

These alloys were used for components such as cutlery, and they were used as the base material in EPNS (electro-plated nickel silver) which was then silver plated. Many household ornaments are still found using these materials, also silver plated.

Architectural use remains quite common, as does plumbing such as taps, etc.

Some of the high nickel materials in this group are used as resistance wires and where electrical spring contacts are required to have a constant electrical resistance.

Symbol	Nominal analysis, supplier, condition and remarks.
2A	Zn Ni Cu alloy: Nickel silver; welding rod; Delta Metal for BS1724C
370	2% Pb 42% Zn 9.5% Ni Cu alloy: McKechnie **DPN: 140 UTS: 580 Elon: 25%**
732	4.5% Zn 21% Ni 1.0% Mn Cu alloy: Wrought; designation agreed by CDAA; not used
735	10% Zn 18% Ni 0.5% Mn Cu alloy: Wrought; designation agreed by CDAA; not used
736	12% Zn 15% Ni 0.5% Mn Cu alloy: Wrought; designation agreed by CDAA; not used
740	18% Zn 10% Ni 0.5% Mn Cu alloy: Wrought; designation agreed by CDAA; not used
745	25% Zn 10% Ni 0.5% Mn Cu alloy: Wrought; designation agreed by CDAA
754	21% Zn 15% Ni 0.5% Mn Cu alloy: Wrought; designation agreed by CDAA
757	23% Zn 12% Ni 0.5% Mn Cu alloy: Wrought; designation agreed by CDAA
762	18% Zn 12% Ni 0.5% Mn Cu alloy: Wrought; designation agreed by CDAA; not used
764	12% Zn 17% Ni 0.5% Mn Cu alloy: Wrought; designation agreed by CDAA; not used
766	30% Zn 12% Ni 0.5% Mn Cu alloy: Wrought; designation agreed by CDAA; not used
767	28% Zn 15% Ni 0.5% Mn Cu alloy: Wrought; designation agreed by CDAA; not used
773	40% Zn 0.2% P 10% Ni 0.2% Si Cu alloy: Wrought; designation agreed by CDAA; not used
774	40% Zn 10% Ni Cu alloy: Wrought; designation agreed by CDAA; not used
776	0.2% Pb 42% Zn 13% Ni 0.2% Mn Cu alloy: Wrought; designation agreed by CDAA; not used
782	2.0% Pb 23% Zn 8% Ni 0.5% Mn Cu alloy: Wrought; designation agreed by CDAA; not used
784	1.2% Pb 30% Zn 10% Ni 0.5% Mn Cu alloy: Wrought; designation agreed by CDAA; not used
786	1.5% Pb 30% Zn 10% Ni 0.5% Mn Cu alloy: Wrought; designation agreed by CDAA; not used
788	1.7% Pb 24% Zn 10% Ni 0.5% Mn Cu alloy: Wrought; designation agreed by CDAA; not used
790	2.0% Pb 20% Zn 12% Ni 0.5% Mn Cu alloy: Wrought; designation agreed by CDAA; not used
792	1.0% Pb 22% Zn 12% Ni 0.5% Mn Cu alloy: Wrought; designation agreed by CDAA; not used
794	1.0% Pb 15% Zn 18.5% Ni 0.5% Mn Cu alloy: Wrought; designation agreed by CDAA; not used
ALBATRA METAL	20% Zn 20% Ni Cu alloy: May contain 1.25% Pb
ALFENIDE METAL	30% Zn 10% Ni Cu alloy: Origin unknown
AMBRAC (1)	5% Zn 20% Ni Cu alloy: For corrosion resistance; origin unknown
AMBRAC (2)	5% Zn 30% Ni Cu alloy: Tube; origin unknown
ASTM B30/10 A	2% Sn 9% Pb 20% Zn 12% Ni Cu alloy: Ingot; as cast; leaded nickel silver **UTS: 200 Elon: 8%**
ASTM B30/11 A	4% Sn 4% Pb 8% Zn 20% Ni Cu alloy: Ingot; as cast; leaded nickel silver **UTS: 260 Elon: 10%**
ASTM B30/11 B	5% Sn 1.5% Pb 2% Zn 25% Ni Cu alloy: Ingot; as cast; leaded nickel silver **UTS: 340 Elon: 12%**
ASTM B122/1	10% Zn 18% Ni Cu alloy: Sheet; ¼-hard **UTS: 450**
ASTM B122/1	10% Zn 18% Ni Cu alloy: Sheet; extra hard **UTS: 600**
ASTM B122/2	7% Zn 18% Ni Cu alloy: Sheet; ¼-hard **UTS: 450**
ASTM B122/2	17% Zn 18% Ni Cu alloy: Sheet; spring hard **UTS: 720**
ASTM B122/3	24% Zn 10% Ni Cu alloy: Sheet; ¼-hard **UTS: 460**
ASTM B122/3	24% Zn 10% Ni Cu alloy: Sheet; spring hard **UTS: 740**
ASTM B122/4	27% Zn 18% Ni Cu alloy: Sheet; ¼-hard **UTS: 560**
ASTM B122/4	27% Zn 18% Ni Cu alloy: Sheet; spring hard **UTS: 830**
ASTM B122/7	5% Zn 20% Ni Cu alloy: Sheet; ¼-hard **UTS: 420**

Note. The following abbreviations and units are used in the tables:

DPN	Hardness, diamond pyramid number
UTS	Ultimate tensile strength, N/mm^2
Elon	Elongation, %
Proof	0.1% proof strength, N/mm^2

1 N/mm^2=0.1 hbar=0.102 kgf/mm^2=0.06475 tonf/$in.^2$=145.04 lbf/$in.^2$=1 MPa

See Appendix II for other abbreviations and conversion tables.

Symbol	Nominal analysis, supplier, condition and remarks.
ASTM B122/7	5% Zn 20% Ni Cu alloy: Sheet; spring hard **UTS: 630**
ASTM B122/8	29% Zn 12% Ni Cu alloy: Sheet; ¼-hard **UTS: 530**
ASTM B122/8	29% Zn 12% Ni Cu alloy: Sheet; spring hard **UTS: 840**
ASTM B122/9	20% Zn 10% Ni Cu alloy: Sheet; ¼-hard **UTS: 440**
ASTM B122/9	20% Zn 10% Ni Cu alloy: Sheet; extra hard **UTS: 620**
ASTM B124/14	45% Zn 10% Ni Cu alloy: Forging and bar; mechanical properties not part of specification
ASTM B149/10 A	2% Sn 9% Pb 20% Zn 12% Ni Cu alloy: Casting **UTS: 200 Elon: 8% Proof: 75**
ASTM B149/11 A	4% Sn 4% Pb 8% Zn 20% Ni Cu alloy: Casting **UTS: 200 Elon: 8% Proof: 110**
ASTM B149/11 B	5% Sn 1.5% Pb 2% Zn 25% Ni Cu alloy: Casting **UTS: 330 Elon: 15% Proof: 140**
ASTM B151 A	17% Zn 18% Ni Cu alloy: Bar; hard **UTS: 560**
ASTM B151 B	27% Zn 18% Ni Cu alloy: Bar; hard **UTS: 600**
ASTM B151 B1	22% Zn 18% Ni Cu alloy: Bar; hard **UTS: 600**
ASTM B151 C	1% Pb 19% Zn 18% Ni Cu alloy: Bar; hard **UTS: 560**
ASTM B151 D	23% Zn 12% Ni Cu alloy: Bar; hard **UTS: 600**
ASTM B151 E	25% Zn 10% Ni Cu alloy: Bar; hard **UTS: 600**
ASTM B206 A	17% Zn 18% Ni Cu alloy: Wire; hard **UTS: 750**
ASTM B206 B	27% Zn 18% Ni Cu alloy: Wire; spring hard **UTS: 950**
ASTM B206 B1	22% Zn 18% Ni Cu alloy: Wire; spring hard **UTS: 950**
ASTM B206 C	19% Zn 18% Ni Cu alloy: Wire; ½-hard **UTS: 570**
ASTM B206 D	23% Zn 12% Ni Cu alloy: Wire; spring hard **UTS: 950**
ASTM B206 E	25% Zn 10% Ni Cu alloy: Wire; spring hard **UTS: 950**
ASTM B260 RB Cu Zn D	40% Zn 0.2% P 10% Ni 0.2% Si Cu alloy: Braze filler; melting range 920–931 °C
ASTM B292 A	5% Sn 2% Zn 5% Ni Cu alloy: Casting; as cast **UTS: 320 Elon: 25% Proof: 120**
ASTM B292 A	5% Sn 2% Zn 5% Ni Cu alloy: Casting; heat treated **UTS: 540 Elon: 5% Proof: 340**
ASTM B292 B	5% Sn 1% Pb 2% Zn 5% Ni Cu alloy: Casting; as cast **UTS: 180 Elon: 20% Proof: 120**
AWA	Zn Ni Cu alloy: Nickel silver welding rod; Delta
BS 790 NS103	28% Zn 10% Ni Cu alloy: Sheet; annealed **DPN: 100**
BS 790 NS103	28% Zn 10% Ni Cu alloy: Sheet; extra hard **DPN: 185**
BS 790 NS104	26% Zn 12.0% Ni Cu alloy: Sheet; annealed **DPN: 100**
BS 790 NS104	26% Zn 12.0% Ni Cu alloy: Sheet; extra hard **DPN: 190**

Note. The following abbreviations and units are used in the tables:

DPN	Hardness, diamond pyramid number
UTS	Ultimate tensile strength, N/mm^2
Elon	Elongation, %
Proof	0.1% proof strength, N/mm^2

1 N/mm^2=0.1 hbar=0.102 kgf/mm^2=0.06475 tonf/in.^{2}=145.04 lbf/in.^{2}=1 MPa

See Appendix II for other abbreviations and conversion tables.

Symbol	Nominal analysis, supplier, condition and remarks.
BS 790 NS105	25% Zn 15% Ni Cu alloy: Sheet; annealed **DPN: 105**
BS 790 NS105	25% Zn 15% Ni Cu alloy: Sheet; extra hard rolled **DPN: 195**
BS 790 NS106	20% Zn 18% Ni Cu alloy: Sheet; annealed **DPN: 110**
BS 790 NS106	20% Zn 18% Ni Cu alloy: Sheet; extra hard rolled **DPN: 200**
BS 790 NS108	20% Zn 20% Ni Cu alloy: Sheet; annealed **DPN: 110**
BS 790 NS108	20% Zn 20% Ni Cu alloy: Sheet; extra hard rolled **DPN: 205**
BS 790 NS109	17% Zn 25% Ni Cu alloy: Sheet; annealed **DPN: 115**
BS 790 NS109	17% Zn 25% Ni Cu alloy: Sheet; extra hard rolled **DPN: 210**
BS 790 NS110	12% Zn 30% Ni Cu alloy: Strip; annealed **DPN: 115**
BS 790 NS110	12% Zn 30% Ni Cu alloy: Strip; extra hard rolled **DPN: 210**
BS 1453 C5	40% Zn 10% Ni 0.4% Si Cu alloy: Filler rod; for brazing mild steel cast iron
BS 1453 C6	40% Zn 15% Ni 0.3% Si Cu alloy: Filler rod; for brazing cast iron
BS 1824 NS104	25% Zn 12% Ni Cu alloy: Strip; annealed **DPN: 115**
BS 1824 NS104	25% Zn 12% Ni Cu alloy: Strip; cold rolled **DPN: 220**
BS 1824 NS107	25% Zn 18% Ni Cu alloy: Strip; annealed **DPN: 115**
BS 1824 NS107	25% Zn 18% Ni Cu alloy: Strip; cold rolled **DPN: 220**
BS 2870 NS101	2% Pb 45% Zn 10% Ni 0.5% Mn Cu alloy: Sheet; specification not issued
BS 2870 NS102	2% Pb 42% Zn 14% Ni 2% Mn Cu alloy: Sheet; specification not issued
BS 2870 NS103	25% Zn 10% Ni Cu alloy: Sheet; annealed **DPN: 100**
BS 2870 NS103	25% Zn 10% Ni Cu alloy: Sheet; extra hard **DPN: 185**
BS 2870 NS104	26% Zn 12% Ni Cu alloy: Sheet; annealed **DPN: 100**
BS 2870 NS104	26% Zn 12% Ni Cu alloy: Sheet; extra hard **DPN: 190**
BS 2870 NS105	25% Zn 15% Ni Cu alloy: Sheet; annealed **DPN: 105**
BS 2870 NS105	25% Zn 15% Ni Cu alloy: Sheet; extra hard **DPN: 195**
BS 2870 NS106	20% Zn 18% Ni Cu alloy: Sheet; annealed **DPN: 110**
BS 2870 NS106	20% Zn 18% Ni Cu alloy: Sheet; extra hard **DPN: 200**
BS 2870 NS107	25% Zn 18% Ni Cu alloy: Sheet; soft **DPN: 100**
BS 2870 NS107	24% Zn 18% Ni Cu alloy: Sheet; extra hard **DPN: 220**
BS 2870 NS108	20% Zn 20% Ni Cu alloy: Sheet; annealed **DPN: 110**
BS 2870 NS108	16% Zn 20% Ni Cu alloy: Sheet; ½-hard **DPN: 140**
BS 2870 NS108	16% Zn 20% Ni Cu alloy: Sheet; hard **DPN: 175**
BS 2870 NS108	16% Zn 20% Ni Cu alloy: Sheet; extra hard **DPN: 205**
BS 2870 NS109	17% Zn 25% Ni Cu alloy: Sheet; annealed **DPN: 115**
BS 2870 NS109	17% Zn 25% Ni Cu alloy: Sheet; ½-hard **DPN: 150**
BS 2870 NS109	17% Zn 25% Ni Cu alloy: Sheet; hard **DPN: 180**

Symbol	Nominal analysis, supplier, condition and remarks.
BS 2870 NS109	17% Zn 25% Ni Cu alloy: Sheet; extra hard **DPN: 210**
BS 2870 NS110	12% Zn 30% Ni Cu alloy: Sheet
BS 2870 NS111	1.5% Pb 30% Zn 10% Ni Cu alloy: Sheet; specification not issued
BS 2870 NS112	0.7% Pb 24% Zn 15% Ni Cu alloy: Sheet
BS 2870 NS113	0.6% Pb 20% Zn 18% Ni Cu alloy: Sheet
BS 2871 NS101	2.0% Pb 40% Zn 10% Ni 0.5% Mn Cu alloy: Tube; specification not issued
BS 2871 NS102	2.0% Pb 40% Zn 14% Ni 2% Mn Cu alloy: Tube; specification not issued
BS 2871 NS103	28% Zn 10% Ni Cu alloy: Tube; specification not issued
BS 2871 NS104	26% Zn 12% Ni Cu alloy: Tube; specification not issued
BS 2871 NS105	25% Zn 15% Ni Cu alloy: Tube; specification not issued
BS 2871 NS106	20% Zn 18% Ni Cu alloy: Tube; specification not issued
BS 2871 NS107	25% Zn 18% Ni Cu alloy: Tube; specification not issued
BS 2871 NS108	20% Zn 20% Ni Cu alloy: Tube; specification not issued
BS 2871 NS109	17% Zn 25% Ni Cu alloy: Tube; specification not issued
BS 2871 NS110	12% Zn 30% Ni Cu alloy: Tube; specification not issued
BS 2872 NS101	2.0% Pb 40% Zn 10% Ni Cu alloy: Forging; as forged **UTS: 450 Elon: 10%**
BS 2872 NS102	2.0% Pb 40% Zn 14% Ni 2% Mn Cu alloy: Forging; specification not issued
BS 2872 NS103	28% Zn 10% Ni Cu alloy: Forging; specification not issued
BS 2872 NS104	26% Zn 12% Ni Cu alloy: Forging; specification not issued
BS 2872 NS105	25% Zn 15% Ni Cu alloy: Forging; specification not issued
BS 2872 NS106	20% Zn 18% Ni Cu alloy: Forging; specification not issued
BS 2872 NS107	25% Zn 18% Ni Cu alloy: Forging; specification not issued
BS 2872 NS108	20% Zn 20% Ni Cu alloy: Forging; specification not issued
BS 2872 NS109	17% Zn 25% Ni Cu alloy: Forging; specification not issued
BS 2872 NS110	12% Zn 30% Ni Cu alloy: Forging; specification not issued
BS 2872 NS111	1.5% Pb 30% Zn 10% Ni Cu alloy: Forging; specification not issued
BS 2872 NS112	0.7% Pb 24% Zn 15% Ni Cu alloy: Forging; specification not issued
BS 2872 NS113	0.5% Pb 20% Zn 18% Ni Cu alloy: Forging; specification not issued
BS 2873 NS101	2% Pb 40% Zn 10% Ni 0.5% Mn Cu alloy: Wire; specification not issued
BS 2873 NS102	2% Pb 40% Zn 14% Ni 2% Mn Cu alloy: Wire; specification not issued
BS 2873 NS103	28% Zn 10% Ni Cu alloy: Wire; mechanical properties not quoted
BS 2873 NS104	26% Zn 12% Ni Cu alloy: Wire; mechanical properties not quoted
BS 2873 NS105	25% Zn 15% Ni Cu alloy: Wire; mechanical properties not quoted
BS 2873 NS106	20% Zn 18% Ni Cu alloy: Wire; specification not issued
BS 2873 NS107	25% Zn 18% Ni Cu alloy: Wire; mechanical properties not quoted
BS 2873 NS108	20% Zn 20% Ni Cu alloy: Wire; mechanical properties not quoted
BS 2873 NS109	17% Zn 25% Ni Cu alloy: Wire; mechanical properties not quoted
BS 2873 NS110	12% Zn 30% Ni Cu alloy: Wire; mechanical properties not quoted
BS 2873 NS111	1.5% Pb 30% Zn 10% Ni Cu alloy: Wire; specification not issued
BS 2873 NS112	0.8% Pb 24% Zn 15% Ni Cu alloy: Wire; mechanical properties not quoted
BS 2873 NS113	0.6% Pb 20% Zn 18% Ni Cu alloy: Wire; mechanical properties not quoted
BS 2874 NS101	2% Pb 40% Zn 10% Ni 0.5% Mn Cu alloy: Bar; as rolled **UTS: 450 Elon: 10%**
BS 2874 NS102	2% Pb 40% Zn 14% Ni 2% Mn Cu alloy: Bar; as rolled **UTS: 490 Elon: 10%**
BS 2874 NS103	38% Zn 10% Ni Cu alloy: Bar; specification not issued
BS 2874 NS104	26% Zn 12% Ni Cu alloy: Bar; specification not issued
BS 2874 NS105	25% Zn 15% Ni Cu alloy: Bar; specification not issued
BS 2874 NS106	20% Zn 18% Ni Cu alloy: Bar; specification not issued
BS 2874 NS107	25% Zn 18% Ni Cu alloy: Bar; specification not issued
BS 2874 NS108	20% Zn 20% Ni Cu alloy: Bar; specification not issued
BS 2874 NS109	17% Zn 30% Ni Cu alloy: Bar; specification not issued
BS 2874 NS110	12% Zn 30% Ni Cu alloy: Bar; specification not issued
BS 2874 NS111	1.5% Pb 30% Zn 10% Ni Cu alloy: Bar
BS 2874 NS112	0.8% Pb 24% Zn 15% Ni Cu alloy: Bar
BS 2874 NS113	0.6% Pb 20% Zn 18% Ni Cu alloy: Bar
BS 2875 NS101	2% Pb 40% Zn 10% Ni 0.5% Mn Cu alloy: Plate; specification not issued
BS 2875 NS102	2% Pb 40% Zn 14% Ni 2% Mn Cu alloy: Plate; specification not issued
BS 2875 NS103	28% Zn 10% Ni Cu alloy: Plate; specification not issued
BS 2875 NS104	26% Zn 12% Ni Cu alloy: Plate; specification not issued
BS 2875 NS105	25% Zn 15% Ni Cu alloy: Plate; specification not issued
BS 2875 NS106	20% Zn 18% Ni Cu alloy: Plate; specification not issued
BS 2875 NS107	25% Zn 18% Ni Cu alloy: Plate; specification not issued
BS 2875 NS108	20% Zn 20% Ni Cu alloy: Plate; specification not issued
BS 2875 NS109	17% Zn 25% Ni Cu alloy: Plate; specification not issued
BS 2875 NS110	12% Zn 30% Ni Cu alloy: Plate; specification not issued
BS 2875 NS111	1.5% Pb 30% Zn 10% Ni 0.2% Mn Cu alloy: Plate; specification not issued
BS 2875 NS112	0.7% Pb 24% Zn 15% Ni Cu alloy: Plate; specification not issued
BS 2875 NS113	0.6% Pb 20% Zn 18% Ni Cu alloy: Plate; specification not issued
BS T3	Zn Ni Cu alloy: Seamless Ni brass tube; obsolete
C 63	2% Pb 40% Zn 10% Ni Cu alloy: Bar and forging; Manganese Bronze for BS alloy NS 101; Parsons C 63
C73150	12% Zn 5.5% Ni Cu alloy: CDA and UNS designation
C73200	4.5% Zn 21% Ni Cu alloy: CDA and UNS designation
C73500	12% Zn 17% Ni Cu alloy: CDA and UNS designation
C73800	18% Zn 12% Ni Cu alloy: CDA and UNS designation
C74000	19% Zn 10% Ni Cu alloy: CDA and UNS designation
C74300	27% Zn 8% Ni Cu alloy: CDA and UNS designation
C74500	25% Zn 10% Ni Cu alloy: CDA and UNS designation
C75200	17% Zn 18% Ni Cu alloy: CDA and UNS designation
C75400	20% Zn 15% Ni Cu alloy: CDA and UNS designation
C75700	23% Zn 12% Ni Cu alloy: CDA and UNS designation
C75720	26% Zn 12% Ni Cu alloy: CDA and UNS designation
C75900	21% Zn 18% Ni Cu alloy: CDA and UNS designation
C76000	30% Zn 8% Ni Cu alloy: CDA and UNS designation
C76100	30% Zn 10% Ni Cu alloy: CDA and UNS designation
C76200	30% Zn 12.2% Ni Cu alloy: CDA and UNS designation
C76300	20% Zn 18% Ni Cu alloy: CDA and UNS designation

Symbol	Nominal analysis, supplier, condition and remarks.
C76390	13.5% Zn 24.5% Ni Cu alloy: CDA and UNS designation
C76400	23% Zn 18% Ni Cu alloy: CDA and UNS designation
C76600	32% Zn 12% Ni Cu alloy: CDA and UNS designation
C76700	29% Zn 15% Ni Cu alloy: CDA and UNS designation
C77000	27% Zn 18% Ni Cu alloy: CDA and UNS designation
C77010	27% Zn 18% Ni Cu alloy: CDA and UNS designation
C77300	42% Zn 10% Ni Cu alloy: CDA and UNS designation
C77310	42% Zn 10% Ni Cu alloy: CDA and UNS designation
C77400	44% Zn 10% Ni Cu alloy: CDA and UNS designation
C77600	44% Zn 13% Ni Cu alloy: CDA and UNS designation
C78200	27% Zn 8% Ni Cu alloy: CDA and UNS designation
C78800	26% Zn 10% Ni Cu alloy: CDA and UNS designation
C79000	24% Zn 12% Ni Cu alloy: CDA and UNS designation
C79200	25% Zn 12% Ni Cu alloy: CDA and UNS designation
C79300	30% Zn 12% Ni Cu alloy: CDA and UNS designation
C79600	1% Pb 43% Zn 10% Ni Cu alloy: CDA and UNS designation
C79620	1.5% Pb 42% Zn 9.5% Ni Cu alloy: CDA and UNS designation
C79800	2% Pb 40% Zn 10% Ni Cu alloy: CDA and UNS designation
C79810	2.7% Pb 40% Zn 9.5% Ni Cu alloy: CDA and UNS designation
C79900	1.2% Pb 43% Zn 7.5% Ni Cu alloy: CDA and UNS designation
C97300	2.2% Sn 9.5% Pb 19% Zn 12.5% Ni Cu alloy: Cast; CDA and UNS designation
C97400	3% Sn 5% Pb 16% Zn 16.2% Ni Cu alloy: Cast; CDA and UNS designation
C97600	4% Sn 4% Pb 6% Zn 20% Ni Cu alloy: Cast; CDA and UNS designation
C97800	4.7% Sn 1.7% Pb 2.5% Zn 25.5% Ni Cu alloy: Cast; CDA and UNS designation
CHROMAX BRONZE	12% Zn 3% Al 15.2% Ni 3% Cr Cu alloy: Origin unknown
DIN 8513 L L Cu Ni 10 Zn 42	Zn Ni Cu alloy: Welders rod
IMI 511	28% Zn 10% Ni Cu alloy: Bar and wire; IMI
IMI 512	26% Zn 12% Ni Cu alloy: Bar and wire; IMI
IMI 513	23% Zn 15% Ni Cu alloy: Bar and wire; IMI
IMI 514	20% Zn 18% Ni Cu alloy: Bar and wire; IMI
IMI 515	18% Zn 20% Ni Cu alloy: Bar and wire; IMI
IMI 525	40% Zn 10% Ni Cu alloy: Bar; IMI; annealed **DPN: 120 UTS: 450 Elon: 25% Proof: 200**
IMI 530	20% Zn 1.7% Al 6% Ni Cu alloy: Bar; IMI; heat treated **DPN: 200 UTS: 560 Elon: 20% Proof: 330**
IMI 551	2% Pb 38% Zn 10% Ni Cu alloy: Bar; IMI; nickel brass; free machining **DPN: 140 UTS: 480 Elon: 20% Proof: 220**
IN 732X	30% Ni 2.8% Cr Cu alloy: International Nickel Co.; annealed **DPN: 150 Proof: 340**
KUNIAL BRASS	20% Zn 2% Al 6% Ni Cu alloy: Bar; sheet and wire; IMI; solution treated; cold rolled and tempered for springs **DPN: 270 UTS: 810**
NICKEL BRASS	38% Zn 10% Ni Cu alloy: Common name; further information from Copper Development Association
NICKEL SILVER	25% Zn 10% Ni Cu alloy: Common name for BS alloys NS 103 to NS 110; further information from Copper Development Association
NICKELOID	Trade name for nickel silver: Barker and Allen Ltd
PARSONS C 63	2% Pb 40% Zn 10% Ni Cu alloy: Bar and forging; Manganese Bronze for BS alloy NS 101
PARSONS SS 15	2.2% Pb 40% Zn 14% Ni Cu alloy: Bar and forging; Manganese Bronze for BS alloy NS 102
QQ N 321/A	20% Zn 18% Ni Cu alloy: Bar and plate; US Federal
QQ N 321/B	1.0% Pb 20% Zn 18% Ni Cu alloy: Bar and sheet; US Federal
SAE CA 752	17% Zn 18% Ni Cu alloy: Wrought
SAE CA 770	27% Zn 18% Ni Cu alloy: Wrought
SCIMITARS	18% Zn Ni Cu alloy: Barker and Allen Ltd.
SILMET	Trade name for nickel silver: Barker and Allen Ltd
For SIS specifications see SS	
SLIDEX	12% Zn Cu alloy: Wire and strip; Jones and Rooke
SPEDEX	Trade name for nickel silver: Barker and Allen Ltd
SS 15	2.2% Pb 40% Zn 14% Ni Cu alloy: Bar and forging; Manganese Bronze for BS alloy NS 102; Parsons SS 15
SS 5243	24% Zn 12% Ni 0.2% Mn Cu alloy: Plate and bar; Swedish Standard; annealed **DPN: 90 UTS: 370 Elon: 40% Proof: 150**
SS 5243	24% Zn 12% Ni 0.2% Mn Cu alloy: Plate and bar; Swedish Standard; cold rolled **DPN: 190 UTS: 690 Elon: 5% Proof: 480**
SS 5246	16% Zn 18% Ni 0.5% Mn Cu alloy: Plate; Swedish Standard; annealed **DPN: 95 UTS: 400 Elon: 35% Proof: 170**
SS 5246	16% Zn 18% Ni 0.5% Mn Cu alloy: Plate; Swedish Standard; cold rolled **DPN: 150 UTS: 630 Elon: 17% Proof: 550**
SWM	1.5% Pb 44% Zn 8.5% Ni Cu alloy: Extrusions; McKechnie; as extruded **DPN: 130 UTS: 530 Elon: 20% Proof: 240**
VICTOR METAL	35% Zn 15% Ni Cu alloy: Good machinability; origin unknown
WESSELS ALLOY	15% Zn 25% Ni 1% Ag Cu alloy: Origin unknown
WHITE BENEDICT METAL	1% Sn 45% Pb 18% Zn 16.5% Ni Cu alloy: Origin unknown
WINNS BRONZE	0.75% Pb 32% Zn 2.2% Ni 0.25% Fe Cu alloy: Origin unknown
WM	2% Pb 41% Zn 9.5% Ni Cu alloy: Extrusion; McKechnie; as drawn **DPN: 150 UTS: 530 Elon: 20% Proof: 260**
WM 11	2% Pb 37% Zn 11% Ni Cu alloy: Extrusion; McKechnie; as extruded **DPN: 130 UTS: 530 Elon: 20% Proof: 240**
WMW	1.5% Pb 43% Zn 8% Ni Cu alloy: Extrusion; McKechnie; as drawn **DPN: 150 UTS: 530 Elon: 20% Proof: 260**
WOLFRAM BRASS	22% Zn 14% Ni 4% W Cu alloy: Origin unknown
ZODIAC	16% Zn 20% Ni Cu alloy: Electrical resistance alloy; origin unknown

Note. The following abbreviations and units are used in the tables:

DPN	Hardness, diamond pyramid number
UTS	Ultimate tensile strength, N/mm^2
Elon	Elongation, %
Proof	0.1% proof strength, N/mm^2

1 N/mm^2=0.1 hbar=0.102 kgf/mm^2=0.06475 tonf/in.^{2}=145.04 lbf/in.^{2}=1 MPa

See Appendix II for other abbreviations and conversion tables.

14G Copper–aluminium wrought and cast alloys
With iron, manganese, nickel, tin and lead – aluminium bronze

Specific gravity		8.34
Density		8340 kg/m^3
Solidus/liquidus		990–1100 °C
Thermal conductivity		42–88 W/m °C
Coefficient of linear expansion		(16–19) × 10^{-6}/ °C
Electrical conductivity		6–15% IACS (copper 100%)
Specific resistance		120–240 microhm mm
Young's modulus of elasticity		(103–110) × 10^9 N/m^2
Impact	annealed	34 J
	heat treated	20 J
Fatigue strength (30 × 10^7 cycles)	hard drawn	±150 N/mm^2

Hot strength

Temperature °C	Tensile strength N/mm^2
95	600
205	500
315	500
425	260
540	140

The above properties are typical of the following group, and may not apply exactly to any one specification. It is possible that with certain specifications some of the values may not be applicable.

General metallurgical characteristics

These alloys with up to 11% aluminium all have excellent corrosion resistance, including good oxidation resistance at elevated temperature as a result of the formation of a thin surface layer of aluminium oxide.

There are two basic series of alloys.

Up to 5% aluminium. These have an alpha structure.

With 7–11% aluminium. These have an alpha structure, with a second phase, beta, present at room temperature. The latter increases the strength but decreases the ductility. A third phase, 'gamma two' may be present if the alloys are cooled very slowly and this further decreases the ductility.

All the alloys generally have one or more of the following elements.

Iron. This strengthens the alpha phase with little effect on the corrosion resistance up to 2% iron. It also helps to refine the structure of high aluminium alloys.

Manganese. Small quantities are used to de-sulphurize during casting. Larger quantities help to strengthen the alloy, and also may improve the corrosion resistance.

Nickel. This strengthens the alpha phase, improves the corrosion resistance and like iron inhibits the formation of the undesirable 'gamma two' phase.

Lead. This is also added to improve machinability. It reduces to some extent the ductility and corrosion resistance.

All these alloys maintain their properties at relatively high temperatures – up to 300–400 °C – being the best copper alloys for this purpose.

The corrosion resistance is also excellent at these temperatures.

The lower aluminium alloys can be readily cold and hot worked, while alloys with 9% or more of aluminium are not always suitable for cold working but are readily hot worked. The structure of these high aluminium alloys above approximately 550 °C is alpha plus beta. The beta, however, may break down to the brittle 'gamma two' phase below 550 °C. This can be prevented by rapid cooling from above 550 °C, when the beta phase will remain.

A limited number of the alloys in this group, those with between 7 and 11% aluminium, have a duplex structure and can have their properties improved by heat treatment. The heat treatment involved is quite complex and technical advice should always be sought before specifying these materials in the heat treated condition. The increase in mechanical properties is quite slight and requires very careful control.

The rest of the materials in this group cannot accept very much cold work without becoming brittle and thus require interstage annealing which will be carried out between 300 and 650 °C.

Casting of these alloys is extremely difficult because of the formation of aluminium oxide which results in the formation of 'hot shuts'. This precludes the use of any normal casting technique where splashing can occur and requires special techniques.

The same difficulties apply when welding in that the aluminium oxide tends to form a film and prevents good fusion unless great care is taken, either using aggressive fluxes or inert gas.

Where fluxes are used it is essential that they are properly removed following brazing or welding to prevent corrosion in service.

Because of the aluminium oxide (alumina) which exists on the surface of these materials, they can present considerable problems in that they are abrasive and can cause considerable wear if any rubbing or moving contact exists between them and other materials. There is evidence that they can cause wear on carburized and nitrided surfaces.

They have excellent corrosion resistance under marine conditions and are probably the most flexible materials where strength, wear resistance and corrosion resistance are involved under marine conditions.

They therefore find use in gears, pumps, impellors, control rods, shafts, nuts and bolts, turbine blades, compressor blades, acid handling equipment – particularly where marine conditions and reasonably high strength are involved.

The materials in this group are competing to a large extent with austenitic stainless steels but under marine conditions and where stress corrosion is involved they probably have considerable advantages. However there are more difficulties in producing and working them than with austenitic stainless steel.

Symbol	Nominal analysis, supplier, condition and remarks.
1A	9% Al 3% Fe + Ni (max) Cu alloy: Bar and forging; Langley Alloy for BS alloy CA103
2.0917	6% Al 0.4% Ni 1.0% Mn Cu alloy: Wrought; German Standard
2.0928	9.5% Al Cu alloy: Casting; German Standard
2.0929	9% Al Cu alloy: Casting; German Standard
2.0940	10% Al 3% Fe Cu alloy: Casting; German Standard
2.0941	10% Al 3.0% Fe Cu alloy: Casting; German Standard
2.0962	8% Al 1.5% Ni 6% Mn Cu alloy: Casting; German Standard
2.0966	9.5% Al 5.5% Ni 2.0% Fe 1.0% Mn Cu alloy: German Standard
2.0970	8% Al 5% Ni 5% Fe Cu alloy: Casting; German Standard
2.0971	9% Al 5% Ni 5% Fe Cu alloy: Casting; German Standard
2.0975	9% Al 5% Ni 5% Fe Cu alloy: Casting; German Standard
2.0980	10% Al 6% Ni 6% Fe Cu alloy: Casting; German Standard
2A	9% Al Cu alloy: Bar and forging; Langley Alloys code for Hidurax 5
3A	9.5% Al 2% Ni 2% Fe Cu alloy: Forging; Langley Alloys code for Hidurax 2
4A	9.5% Al 2% Fe Cu alloy: Casting; Langley Alloys code for Hidurax 2
6A	10% Al 5% Ni 5% Fe Cu alloy: Bar and forging; Langley Alloys code for Hidurax 1
7A	10% Al 4.5% Ni 4.5% Fe Cu alloy: Casting; Langley Alloys code for Hidurax 1
10A	10% Al 5% Ni 5% Fe 2% Mn Cu alloy: Bar and forging; Langley Alloys **UTS: 750 Elon: 20% Proof: 450**
12A	10% Al 4.5% Ni 4.5% Fe Cu alloy: Casting; Langley Alloys code for Hidurax
17A	Al Cu alloy: Langley Alloys code for Hidurax
18A	11% Al 4% Ni 4% Fe Cu alloy: Casting; Langley Alloys **UTS: 740 Elon: 15% Proof: 400**
21A	10% Al 5% Ni 3% Fe 1% Mn Cu alloy: Bar; Langley Alloys **UTS: 600 Elon: 10% Proof: 300**
24A	5% Al Cu alloy: Bar and forging; Langley Alloys code for Hidurax 6
26A	10% Al 5% Ni 2% Fe 1% Mn Cu alloy: Wrought; Langley Alloys
36A	10% Al 4.5% Ni 4.5% Fe Cu alloy: Langley Alloys code for Hidurax

Note. The following abbreviations and units are used in the tables:

DPN	Hardness, diamond pyramid number
UTS	Ultimate tensile strength, N/mm^2
Elon	Elongation, %
Proof	0.1% proof strength, N/mm^2

1 N/mm^2=0.1 hbar=0.102 kgf/mm^2=0.06475 tonf/in.^{2}=145.04 lbf/in.^{2}=1 MPa

See Appendix II for other abbreviations and conversion tables.

Symbol	Nominal analysis, supplier, condition and remarks.
41A	7% Al 2% Si Cu alloy: Wrought; Langley Alloys code for Hidurax 7
42A	7% Al 2% Si Cu alloy: Langley Alloy code for Hidurax 7
75	11.5% Al 5% Ni 5% Fe 1% Mn Cu alloy: Extrusion; McKechnie Bros; as extruded **DPN: 250 UTS: 860 Elon: 8% Proof: 450**
160	9.5% Al Cu alloy: Extrusions; McKechnie Bros; as drawn **DPN: 150 UTS: 530 Elon: 25% Proof: 240**
164	9.3% Al 1.5% Ni 1.5% Fe Cu alloy: Extrusions; McKechnie Bros; as drawn **DPN: 170 UTS: 580 Elon: 18% Proof: 300**
164	9.5% Al 2% Ni 2% Fe Cu alloy: Bar and forging; Manganese Bronze Ltd; as Al Bronze 184
197	9.5% Al 5% Ni 5% Fe Cu alloy: Bar and forging; Manganese Bronze Ltd; as Al Bronze 197
197	10% Al 4.5% Ni 4.5% Fe Cu alloy: Extrusion; McKechnie Bros; as drawn **DPN: 220 UTS: 750 Elon: 15% Proof: 420**
606	5.5% Al Cu alloy: Wrought; designation agreed by CDAA; not used
607	1.8% Zn 2.6% Al Cu alloy: Wrought; designation agreed by CDAA; not used
608	5.8% Al 0.3% As Cu alloy: Wrought; designation agreed by CDAA; not used
610	8% Al Cu alloy: Wrought; designation agreed by CDAA; not used
612	8% Al Cu alloy: Wrought; designation agreed by CDAA; not used
613	0.4% Sn 7% Al 3.5% Fe Cu alloy: Wrought; designation agreed by CDAA; not used
614	7% Al 3.0% Fe 1.0% Mn Cu alloy: Wrought; designation used by CDAA
616	1.0% Zn 9% Al 1.0% Ni 4.0% Fe 1.5% Mn Cu alloy: Wrought; designation agreed by CDAA; not used
618	10% Al 1.5% Fe Cu alloy: Wrought; designation agreed by CDAA; not used
620	10% Al 3.5% Fe Cu alloy: Wrought; designation agreed by CDAA; not used
622	11% Al 4.0% Fe Cu alloy: Wrought; designation agreed by CDAA; not used
626	10.3% Al 3.8% Ni 3.0% Fe 1.5% Mn Cu alloy: Wrought; designation agreed by CDAA; not used
628	9.5% Al 5.5% Ni 2.2% Fe Cu alloy: Wrought; designation agreed by CDAA; not used
634	8.8% Al Cu alloy: Extrusion; McKechnie Bros; as drawn **DPN: 150 UTS: 530 Elon: 40% Proof: 220**
639	7.2% Al 0.2% Ni 2.2% Si Cu alloy: Wrought; designation agreed by CDAA; not used
642	1.0% Zn 8% Al 4.0% Fe 2% Si Cu alloy: Wrought; designation agreed by CDAA; not used
705	7.0% Al 0.15% Mn Cu alloy: Wrought; designation agreed by CDAA; not used
764	9% Al 1.2% Ni 1% Fe Cu alloy: McKechnie; as drawn **DPN: 180 UTS: 510 Elon: 25% Proof: 310**
768	9.8% Al 4.5% Ni 4.5% Fe Cu alloy: McKechnie; as drawn **DPN: 240 UTS: 785 Elon: 15% Proof: 560**

Symbol	Nominal analysis, supplier, condition and remarks.
775	6% Al 2.2% Si Cu alloy: McKechnie; as drawn **DPN: 210 UTS: 520 Elon: 25% Proof: 425**
778	6.8% Al 1.7% Si Cu alloy: McKechnie; as drawn **DPN: 190 UTS: 640 Elon: 25% Proof: 420**
781	9% Al 4.2% Ni 4.2% Fe Cu alloy: McKechnie; as drawn **DPN: 240 UTS: 740 Elon: 21% Proof: 490**
952	9% Al 3% Fe Cu alloy: Casting; designation used by CDAA
953	10% Al 1% Fe Cu alloy: Casting; designation used by CDAA
954	11% Al 4% Fe Cu alloy: Casting; designation used by CDAA
955	11% Al 4% Ni 4% Fe Cu alloy: Casting; designation used by CDAA
AB	1.5% Pb 9.8% Al Cu alloy: Extrusion; McKechnie Bros; as drawn **DPN: 200 UTS: 550 Elon: 15% Proof: 260**
ABS TYPE 4	10% Al 4.2% Ni 4.0% Fe Cu alloy: Casting; American Bureau of Shipping **UTS: 600 Elon: 15% Proof: 245**
ABS TYPE 5	7.8% Al 2.2% Ni 3.0% Fe 12.5% Mn Cu alloy: Casting; American Bureau of Shipping **UTS: 635 Elon: 20% Proof: 280**
AETERNA 614	7.2% Al 2.6% Fe (max) 0.4% Mn Cu alloy VDM
AETERNA 630	9.8% Al 4.7% Ni 2.8% Fe 1% Mn Cu alloy VDM
AL 1	9.4% Al Cu alloy: Wrought; Delta Metal
AL BRONZE 184	9.5% Al 2% Ni 2% Fe Cu alloy: Bar and forging; Manganese Bronze Ltd; as DTD 164 **DPN: 170 UTS: 550 Elon: 17% Proof: 260**
AL BRONZE 197	9.5% Al 5% Ni 5% Fe Cu alloy: Bar and forging; Manganese Bronze Ltd; as DTD 197 **DPN: 220 UTS: 720 Elon: 15% Proof: 400**
Al Bz 5	5% Al 0.3% As Cu alloy: Wrought; designation used by German Standards
Alloy D	7.2% Al 2.8% Fe 0.4% Mn (max) Cu alloy VDM
Alloy E	9.8% Al 4.7% Ni 2.7% Fe 1% Mn Cu alloy VDM
AMPCO 8	0.25% Sn 6.5% Al 2.5% Fe Cu alloy: Wrought; Ampco **UTS: 570 Elon: 35% Proof: 310**
AMPCO 12	8.9% Al 2.9% Fe Cu alloy: Casting; Ampco **DPN: 130 UTS: 550 Elon: 37% Proof: 220**
AMPCO 15	9.3% Al 3.1% Fe Cu alloy: Extrusion; Ampco **DPN: 180 UTS: 670 Elon: 25% Proof: 330**
AMPCO 16	10.1% Al 3.3% Fe Cu alloy: Casting; Ampco **DPN: 160 UTS: 650 Elon: 23% Proof: 230**
AMPCO 18	10.5% Al 3.5% Fe Cu alloy: Wrought and cast; Ampco **DPN: 190 UTS: 720 Elon: 15% Proof: 270**
AMPCO 18/13	10.5% Al 3.5% Fe Cu alloy: Casting; Ampco **DPN: 175 UTS: 700 Elon: 20% Proof: 250**
AMPCO 18/22	10.5% Al 3.5% Fe Cu alloy: Casting; Ampco **DPN: 225 UTS: 750 Elon: 9% Proof: 380**
AMPCO 18/23	10.5% Al 3.5% Fe Cu alloy: Casting; Ampco **DPN: 205 UTS: 750 Elon: 15% Proof: 380**
AMPCO 20	11.3% Al 3.8% Fe Cu alloy: Casting; Ampco **DPN: 220 UTS: 630 Elon: 6% Proof: 270**
AMPCO 20/13	11.3% Al 3.8% Fe Cu alloy: Casting; Ampco **DPN: 205 UTS: 610 Elon: 8% Proof: 240**
AMPCO 21	13.1% Al 4.4% Fe Cu alloy: Wrought and cast; Ampco **DPN: 290 UTS: 610 Elon: 2% Proof: 390**
AMPCO 22	14.1% Al 4.7% Fe Cu alloy: Casting; Ampco **DPN: 330 UTS: 580 Elon: 0.5% Proof: 490**
AMPCO 24	Al Cu alloy: For dies; Ampco **DPN: 350**
AMPCO 25	Al Cu alloy: For dies; Ampco **DPN: 360**
AMPCOLAY 45	10% Al 5% Ni 3% Fe 1% Mn Cu alloy: Wrought; Ampco
AMPCOLAY 405	8% Al 1% Ni 4% Fe 1% Mn 2% Si Cu alloy: Wrought; Ampco
AMPCOLAY 483	Al Ni Fe Cu alloy: Ampco
AMPCOLAY 495	8% Al 2% Ni 3% Fe 12% Mn Cu alloy: Forging; Ampco
AMPCOLAY 570	10% Al 15% Ni 0.7% Fe 1.5% Co Cu alloy: Casting; Ampco
AMPCOLAY A1	9% Al 3% Fe Cu alloy: Casting; Ampco
AMPCOLAY B2	10% Al 1% Fe Cu alloy: Casting; Ampco
AMPCOLAY B2 (wrought)	8% Al 1% Ni 4% Fe 1% Mn 2% Si Cu alloy: Wrought; Ampco
AMPCOLAY C3	11% Al 4% Fe Cu alloy: Casting; Ampco
AMPCOLAY D4	11% Al 4% Ni 4% Fe Cu alloy: Casting; Ampco
AMPCOLAY E5	8% Al 1% Ni 4% Fe 1% Mn 2% Si Cu alloy: Wrought; Ampco
AMS 4630 E	8.5% Al Cu alloy: Tube; annealed
AMS 4631 C	7.5% Al 2% Si Cu alloy: Bar and forging
AMS 4632 C	8.5% Al Cu alloy: Bar; hard drawn
AMS 4635 B	10% Al 3% Fe Cu alloy: Bar and forging
AMS 4640 C	10.5% Al 5% Ni 2.5% Fe Cu alloy: Bar and forging
AMS 4870 B	11% Al 3.6% Fe Cu alloy: Casting; chill cast; as cast
AMS 4871 B	11% Al 3.6% Fe Cu alloy: Casting; chill cast; heat treated
AMS 4872 B	11% Al 3.6% Fe Cu alloy: Sand casting; as cast
AMS 4873 A	11% Al 3.6% Fe Cu alloy: Sand casting; heat treated
AMS 4880	10.5% Al 5% Ni 2.5% Fe Cu alloy: Centri-casting; heat treated
AMS 4881	11% Al 5% Ni 5.2% Fe Cu alloy casting
AN B 16	8.0% Al 5.0% Ni 2.2% Si Cu alloy: Bar; US Service **UTS: 600 Elon: 17% Proof: 200**
AN QQ B 672	11% Al 5.0% Ni 5.0% Mn Cu alloy: Casting; US Service **UTS: 580 Elon: 3%**
ASTM B30/9 A	9% Al 3% Fe Cu alloy: Ingot; as cast **UTS: 460 Elon: 20%**
ASTM B30/9 B	10% Al 1% Fe Cu alloy: Ingot; heat treated **UTS: 580 Elon: 12%**
ASTM B30/9 C	11% Al 4% Fe Cu alloy: Ingot; heat treated **UTS: 660 Elon: 6%**
ASTM B30/9 D	11% Al 4% Ni 4% Fe Cu alloy: Ingot; heat treated **UTS: 800 Elon: 5%**
ASTM B111/608	5.5% Al Cu alloy: Seamless tube
ASTM B111/687	2.2% Al Cu alloy: Seamless tube
ASTM B111 A1 BRONZE	6% Al Cu alloy: Tube
ASTM B124/11 A	8% Al 1% Ni 1.5% Mn Cu alloy: Forging and bar; mechanical properties not quoted
ASTM B124/11 B	10% Al 5% Ni 3% Fe 1.5% Mn Cu alloy: Forging and bar; mechanical properties not quoted
ASTM B148/9 A	9% Al 3% Fe Cu alloy: Casting; as cast **DPN: 110 UTS: 460 Elon: 20% Proof: 170**
ASTM B148/9 B	10% Al 1% Fe Cu alloy: Casting; heat treated **DPN: 160 UTS: 580 Elon: 12% Proof: 260**
ASTM B148/9 C	11% Al 4% Fe Cu alloy: Casting; heat treated **DPN: 190 UTS: 660 Elon: 6% Proof: 330**
ASTM B148/9 D	11% Al 4% Ni 4% Fe Cu alloy: Casting; heat treated **DPN: 200 UTS: 800 Elon: 5% Proof: 420**
ASTM B150/1	8% Al 1% Ni 4% Fe 1% Mn 2% Si Cu alloy: Bar **UTS: 530 Elon: 10% Proof: 240**
ASTM B150/2	10% Al 5% Ni 3% Fe 1% Mn Cu alloy: Bar **UTS: 690 Elon: 6% Proof: 330**
ASTM B150/3	7% Al 2% Fe Cu alloy: Bar **UTS: 530 Elon: 30% Proof: 240**
ASTM B150/614	7.0% Al 2.5% Fe Cu alloy: Rod; previously alloy No 3
ASTM B150/630	10.0% Al 4.7% Ni 3.0% Fe Cu alloy: Rod; previously alloy No 2
ASTM B150/642	8.5% Al 1.0% Ni (max) 4.0% Fe (max) Cu alloy: Rod; previous alloy No 1
ASTM B169/612	8% Al Cu alloy: Al bronze; sheet and bar
ASTM B169/612	7% Al 2.5% Fe Cu alloy: Al bronze; sheet and bar
ASTM B169 A	5.5% Al Cu alloy: Sheet and bar; annealed **UTS: 330 Elon: 40% Proof: 110**
ASTM B169 A	5.5% Al Cu alloy: Sheet and bar; hard drawn **UTS: 360 Elon: 30% Proof: 120**

Symbol	Nominal analysis, supplier, condition and remarks.
ASTM B169 C	8% Al Cu alloy: Sheet and bar; annealed **UTS: 390 Elon: 25% Proof: 150**
ASTM B169 C	8% Al Cu alloy: Sheet and bar; hard drawn **UTS: 400 Elon: 22% Proof: 170**
ASTM B169 D	7% Al 2% Fe Cu alloy: Sheet and bar; annealed **UTS: 510 Elon: 33% Proof: 240**
ASTM B169 D	7% Al 2% Fe Cu alloy: Sheet and bar; hard drawn **UTS: 580 Elon: 32% Proof: 330**
ASTM B171/464	0.75% Sn 40% Zn Cu alloy: Rolled plate
ASTM B171/614	2.5% Sn 7% Al 2.5% Fe Cu alloy: Rolled plate
ASTM B171/628	10.5% Al 5.5% Ni 2.5% Fe 1.2% Mn Cu alloy: plate
ASTM B225 E Cu Al A 1	7% Al Cu alloy: Welding electrode **UTS: 390**
ASTM B225 E Cu Al A 2	10% Al Cu alloy: Welding electrode **UTS: 420**
ASTM B225 E Cu Al B	11% Al 4% Fe Cu alloy: Welding electrode **UTS: 460**
ASTM B259 R Cu Al A 2	10% Al Cu alloy: Welding rod **UTS: 450**
ASTM B259 R Cu Al B	11% Al 4% Fe Cu alloy: Welding rod **UTS: 470**
ASTM B359/608	5.5% Al Cu alloy: Finned tube
ASTM B359/687	2.2% Al Cu alloy: Finned tube
B 29	9.5% Al 2.5% Fe 1% Mn Cu alloy: Casting; Anti-Attrition for BS alloy AB1
B 31	0.5% Zn 9.5% Al 5.5% Ni 4.5% Fe 1.5% Mn Cu alloy: Casting; Anti-Attrition for BS alloy AB2
BATTERIUM	9% Al 1% Ni Cu alloy: Batterium Metal Ltd **DPN: 168 UTS: 730 Elon: 48%**
BOLTOMET 803	9.25% Al Cu alloy: T Bolton for BS alloy CA 103
BOLTOMET 807	9.5% Al 5% Ni 5% Fe Cu alloy: T Bolton for BS alloy CA104
BS 378 CA102	7% Al 2.5% Fe + Mn Cu alloy: Tubes; as drawn for condenser tubes 150
BS 1031	Aluminium bronze ingots and castings; replaced by BS 1400
BS 1032	Aluminium bronze ingots and castings; replaced by BS 1400
BS 1072	Al bronze ingots and castings; high tensile; replaced by BS 1400
BS 1073	Al bronze ingots and castings; high tensile; replaced by BS 1400
BS 1400 AB1 C	9.5% Al 2% Fe Cu alloy: Aluminium bronze castings; chill cast **UTS: 530 Elon: 20% Proof: 200**
BS 1400 AB2 1	9.5% Al 5% Ni 4.5% Fe Cu alloy: Aluminium bronze ingots; chill cast **UTS: 630 Elon: 15% Proof: 240**
BS 1400 AB2 C	9.5% Al 5% Ni 4.5% Fe Cu alloy: Aluminium bronze castings; chill cast **UTS: 630 Elon: 15% Proof: 240**
BS 1400 AB3	6.2% Al 0.6% Fe 2.2% Si Cu alloy casting
BS 1400 CMA1	1% Sn 8% Al 3% Ni 3% Fe 12% Mn Cu alloy: Ingots; chill cast **UTS: 650 Elon: 30% Proof: 300**
BS 1400 CMA1C	1% Sn 8% Al 3% Ni 3% Fe 12% Mn Cu alloy: Castings; chill cast **UTS: 650 Elon: 30% Proof: 300**
BS 1400 CMA2 1	1% Sn 8.75% Al 3% Ni 3% Fe 12% Mn Cu alloy: Ingots; chill cast **UTS: 730 Elon: 10% Proof: 370**
BS 1400 CMA2 C	1% Sn 8.75% Al 3% Ni 3% Fe 12% Mn Cu alloy: Castings; sand cast **UTS: 730 Elon: 10% Proof: 370**
BS 1464 CA 102	7% Al Cu tube; aluminium bronze; Ni Fe and Mn may be present to a total of 2.5% **DPN: 120**
BS 1541 CA102	7% Al Cu alloy: Plate; aluminium bronze; as rolled; Ni Fe and Mn may be present up to 2.5% **UTS: 450 Elon: 35%**
BS 1541 CA105	10.0% Al 5.0% Ni 2.5% Fe 1.0% Mn Cu alloy: Plate; as rolled **UTS: 580 Elon: 12%**
BS 1541 CA106	7% Al 3% Fe Cu alloy: Plate; as rolled **UTS: 450 Elon: 35%**
BS 1867	7% Al 2% Ni + Fe + Mn (max) Cu alloy: Tube; annealed **UTS: 400 Elon: 50%**
BS 2032 CA103	9% Al 3% Fe + Ni (max) Cu alloy: Bar and forging; annealed **UTS: 510 Elon: 15% Proof: 180**
BS 2033 CA104	10% Al 5% Ni 5% Fe Cu alloy: Bar and forging; as rolled or forged **UTS: 690 Elon: 15% Proof: 370**
BS 2870 CA101	5% Al Cu alloy: Sheet; as rolled; aluminium bronze **UTS: 330 Elon: 40%**
BS 2870 CA102	7% Al Cu alloy: Sheet; Ni, Fe and Mn up to 2.5%; aluminium bronze; specification not issued
BS 2870 CA103	9% Al 3% Fe + Ni Cu alloy: Sheet; specification not issued
BS 2870 CA104	10% Al 5% Ni 5% Fe Cu alloy: Sheet; specification not issued
BS 2870 CA105	10% Al 5% Ni 2.5% Fe 1.0% Mn Cu alloy: Sheet; specification not issued
BS 2870 CA106	7% Al 3% Fe Cu alloy: Sheet; specification not issued
BS 2871 CA101	5% Al Cu alloy: Tube; aluminium bronze; annealed **UTS: 390**
BS 2871 CA102	7% Al 2.5% Ni + Fe + Mn Cu alloy: Tube; aluminium bronze; annealed **UTS: 420**
BS 2871 CA103	9% Al 3% Fe + Ni Cu alloy: Tube; aluminium bronze; specification not issued
BS 2871 CA104	10% Al 5% Ni 5% Fe Cu alloy: Tube; aluminium bronze; specification not issued
BS 2872 CA101	5% Al Cu alloy: Forging; specification not issued
BS 2872 CA102	7% Al 2.5% Ni + Fe + Mn (max) Cu alloy: Forging; specification not issued
BS 2872 CA103	9% Al 3% Fe + Ni Cu alloy: Forging; as forged **UTS: 510 Elon: 25% Proof: 220**
BS 2872 CA104	10% Al 5% Ni 5% Fe Cu alloy: Forging; as forged **UTS: 690 Elon: 15% Proof: 370**
BS 2872 CA105	10% Al 5% Ni 2.5% Fe 1% Mn Cu alloy: Forging; specification not issued
BS 2872 CA106	7% Al 3.0% Fe Cu alloy: Forging; as forged **UTS: 520 Elon: 35% Proof: 210**
BS 2873 CA101	5% Al Cu alloy: Wire; specification not issued
BS 2873 CA102	7% Al 2.5% Ni + Fe + Mn (max) Cu alloy: Wire; specification not issued
BS 2873 CA103	9% Al 3% Fe + Ni Cu alloy: Wire; specification not issued
BS 2873 CA104	10% Al 5% Ni 5% Fe Cu alloy: Wire; specification not issued
BS 2873 CA105	10% Al 5.0% Ni 2.5% Fe 1.0% Mn Cu alloy: Wire; specification not issued
BS 2873 CA106	7% Al 3% Fe Cu alloy: Wire; specification not issued
BS 2874 CA101	5% Al Cu alloy: Bar; specification not issued
BS 2874 CA102	7% Al 2.5% Ni + Fe + Mn (max) Cu alloy: Bar; specification not issued

Note. The following abbreviations and units are used in the tables:

DPN	Hardness, diamond pyramid number
UTS	Ultimate tensile strength, N/mm^2
Elon	Elongation, %
Proof	0.1% proof strength, N/mm^2

$1\ N/mm^2 = 0.1\ hbar = 0.102\ kgf/mm^2 = 0.06475\ tonf/in.^2 = 145.04\ lbf/in.^2 = 1\ MPa$

See Appendix II for other abbreviations and conversion tables.

Symbol	Nominal analysis, supplier, condition and remarks.
BS 2874 CA103	9.0% Al 3% Fe + Ni Cu alloy: bar; annealed **UTS: 510 Elon: 15% Proof: 180**
BS 2874 CA104	10% Al 5% Ni 5% Fe Cu alloy: Bar; as rolled **UTS: 660 Elon: 15% Proof: 330**
BS 2874 CA105	10% Al 5% Ni 2.5% Fe 1% Mn Cu alloy: Bar; specification not issued
BS 2874 CA106	7% Al 3% Fe Cu alloy: Bar; annealed **UTS: 510 Elon: 35% Proof: 210**
BS 2874 CA107	6.2% Al 2.2% Si Cu alloy
BS 2875 CA101	5% Al Cu alloy: Plate; specification not issued
BS 2875 CA102	7% Al 2.5% Fe + Mn + Ni Cu alloy: Plate; as rolled **UTS: 450 Elon: 35%**
BS 2875 CA103	9% Al 3% Fe + Ni Cu alloy: Plate; specification not issued
BS 2875 CA104	10% Al 5% Ni 5% Fe Cu alloy: Plate; specification not issued
BS 2875 CA105	10% Al 5% Ni 2.5% Fe 1% Mn Cu alloy: Plate; as rolled **UTS: 580 Elon: 12%**
BS 2875 CA106	7% Al 3% Fe Cu alloy: Plate; as rolled **UTS: 450 Elon: 35%**
BS 2901 C12	7% Al 2% Fe + Ni + Mn Cu alloy: Rod for all welding
BS 2901 C13	10% Al Cu alloy: Rod for all welding
BS 2901 C20	9% Al 5% Ni 2% Fe 1% Mn Cu alloy: Rod for gas and electric welding
BS 4577 A4/4	10% Al 5% Ni 5% Fe Cu alloy: For resistance welding electrodes; wrought; conductivity 10% IACS **DPN: 190 UTS: 700 Elon: 13%**
C12	7.0% Al Cu alloy: rod for gas shielded arc welding
C12 Fe	7.5% Al 3.0% Fe Cu alloy: Rod for gas shielded arc welding
C13	10% Al 1% Ni 1.2% Fe Cu alloy: Rod for gas shielded arc welding
C20	9.0% Al 4.5% Ni 2.5% Fe Cu alloy: Rod for gas shielded arc welding
C22	8.0% Al 2.0% Ni 3.0% Fe 13.0% Mn Cu alloy: Rod for gas shielded arc welding
C23	6.2% Al 0.6% Fe 2.0% Si Cu alloy: Rod for gas shielded arc welding
C26	9.0% Al 5.0% Ni 4.0% Fe Cu alloy: Rod for gas shielded arc welding
C60600	5.5% Al Cu alloy: CDA and UNS designation
C60700	1.8% Sn 2.5% Al Cu alloy: CDA and UNS designation
C60800	5.7% Al Cu alloy: CDA and UNS designation
C61000	7.2% Al Cu alloy: CDA and UNS designation
C61300	0.3% Sn 6.7% Al 2.5% Fe Cu alloy: CDA and UNS designation
C61400	7% Al 2.2% Fe Cu alloy: CDA and UNS designation
C61500	7% Al 2% Ni Cu alloy: CDA and UNS designation
C61550	6% Al 2% Ni Cu alloy: CDA and UNS designation
C61800	10.2% Al 1% Fe Cu alloy: CDA and UNS designation
C61900	9.2% Al 3.7% Fe Cu alloy: CDA and UNS designation
C62200	11.5% Al 3.6% Fe Cu alloy: CDA and UNS designation
C62300	9.2% Al 3% Fe Cu alloy: CDA and UNS designation
C62400	10.7% Al 3.2% Fe Cu alloy: CDA and UNS designation
C62500	13% Al 4.5% Fe Cu alloy: CDA and UNS designation
C62580	12.5% Al 4% Fe Cu alloy: CDA and UNS designation
C62581	13.5% Al 4% Fe Cu alloy: CDA and UNS designation
C62582	14.5% Al 4% Fe Cu alloy: CDA and UNS designation
C62730	9.2% Al 5% Ni 5% Fe Cu alloy: CDA and UNS designation
C63000	10% Al 4.7% Ni 3% Fe Cu alloy: CDA and UNS designation
C63010	10.2% Al 5% Ni 2.7% Fe Cu alloy: CDA and UNS designation
C63020	11% Al 5% Ni 4.7% Fe Cu alloy: CDA and UNS designation
C63200	9.2% Al 4.4% Ni 4% Fe 1.8% Mn Cu alloy: CDA and UNS designation
C63230	9% Al 4.7% Ni 4% Fe Cu alloy: CDA and UNS designation
C63280	9% Al 4.7% Ni 4% Fe 2.5% Mn Cu alloy: CDA and UNS designation
C63300	6.2% Al 1.7% Ni 4% Fe 12% Mn Cu alloy: CDA and UNS designation
C63380	7.7% Al 2.2% Ni 3% Fe 12% Mn Cu alloy: CDA and UNS designation
C63400	3% Al 0.3% Si Cu alloy: CDA and UNS designation
C63600	3.5% Al 1% Si Cu alloy: CDA and UNS designation
C63800	2.8% Al 1.8% Si Cu alloy: CDA and UNS designation
C64110	1.5% Pb 9.5% Al Cu alloy: CDA and UNS designation
C64200	7% Al 2% Si Cu alloy: CDA and UNS designation
C64210	6.8% Al 1.7% Si Cu alloy: CDA and UNS designation
C64250	6.5% Al 1.8% Si Cu alloy: CDA and UNS designation
C64400	4% Al 4.8% Ni 1% Si Cu alloy: CDA and UNS designation
C95200	9% Al 3.2% Fe Cu alloy: Cast; CDA and UNS designation
C95210	9% Al 0.5% Ni 3.2% Fe Cu alloy: Cast; CDA and UNS designation
C95220	10% Al 2% Ni 3.2% Fe Cu alloy: Cast; CDA and UNS designation
C95300	10% Al 1% Fe Cu alloy: Cast; CDA and UNS designation
C95400	10.7% Al 1% Ni 4% Fe Cu alloy: Cast; CDA and UNS designation
C95410	10.7% Al 2% Ni 4% Fe Cu alloy: Cast; CDA and UNS designation
C95420	11.2% Al 3.7% Fe Cu alloy: Cast; CDA and UNS designation
C95500	10.7% Al 4.2% Ni 4% Fe Cu alloy: Cast; CDA and UNS designation
C95510	10% Al 5% Ni 2.7% Fe Cu alloy: Cast; CDA and UNS designation
C95520	11% Al 5.1% Ni 5% Fe Cu alloy: Cast; CDA and UNS designation
C95600	7% Al 0.2% Ni 2% Si Cu alloy: Cast; CDA and UNS designation
C95700	7.7% Al 2.2% Ni 3% Fe 12% Mn Cu alloy: Cast; CDA and UNS designation
C95710	7.7% Al 2.2% Ni 3% Fe 12% Mn Cu alloy: Cast; CDA and UNS designation
C95800	9% Al 4.5% Ni 4% Fe 1% Mn Cu alloy: Cast; CDA and UNS designation
C95810	9% Al 4.5% Ni 4% Fe 1% Mn Cu alloy: Cast; CDA and UNS designation
C95900	12.7% Al 4% Fe Cu alloy: Cast; CDA and UNS designation
C99300	11% Al 15.5% Ni 0.8% Fe 1.5% Co Cu alloy: Cast; CDA and UNS designation
C99350	8.5% Zn 10% Al 15.2% Ni Cu alloy: Cast; CDA and UNS designation
CROTORITE	Al bronze casting; Manganese Bronze Ltd for DTD 160 DTD 174
CROTORITE IV	10% Al + Ni + Mn Cu alloy: Manganese Bronze Ltd
CROTORITE V	9% Al 3% Fe + Ni Cu alloy: Bar and forging; Manganese Bronze Ltd for BS alloy CA103; low Fe
CROTORITE Z	9% Al 3% Fe + Ni Cu alloy: Bar and forging; Manganese Bronze Ltd for BS alloy CZ103
Cu Al 5	5% Al Cu alloy: German designation
Cu Al 8	8% Al Cu alloy: German designation
Cu Al 8 Fe	7.7% Al 2.8% Fe Cu alloy: German Standard
Cu Al 9 Mn	8.7% Al 2.2% Mn Cu alloy: German designation
Cu Al 10 Ni	9.5% Al 5% Ni 4.5% Fe Cu alloy: German designation
Cu Al 10 Ni 5 Fe 4	9.8% Al 4.5% Ni 4.5% Fe Cu alloy: German Standard
Cu Al 11 Ni	11.5% Al 6% Ni 6% Fe Cu alloy: German designation
Cu Al 15	5% Al Cu alloy: German Standard
Cu Al 17	7% Al Cu alloy: German Standard
Cu Al 18	8% Al Cu alloy: German Standard
DGS 129	6.2% Al 0.6% Fe Cu alloy: Admiralty specification

Symbol	Nominal analysis, supplier, condition and remarks.
DGS 348	9.1% Al 5% Ni 4.2% Fe Cu alloy: Admiralty specification
DGS 357	Al Fe Cu alloy: Admiralty specification for alloy
DGS 361A	9.4% Al 4.7% Ni 4.7% Fe Cu alloy: Admiralty specification
DGS 1043	9.2% Al 5% Ni 5% Fe Cu alloy: Admiralty specification
DGS 1043	9.2% Al 4.7% Ni 4.7% Fe Cu alloy: Admiralty specification
DGS 1043	10% Al 5% Ni 5% Fe Cu alloy: Admiralty specification
DGS 1044	6.2% Al 0.1% Ni 0.6% Fe Cu alloy: Admiralty specification
DGS 1044	6.2% Al 0.6% Fe 2.2% Si Cu alloy: Admiralty specification
DGS 8451	9% Al 3% Fe + Ni Cu alloy: Bar and forging; Admiralty for BS alloy CA103; heat treated; further information from the Copper Development Association **DPN: 150 UTS: 550 Elon: 35 Proof: 280**
DGS 8452A	10% Al 5% Ni 5% Fe Cu alloy: Admiralty specification
DGS 8453	7% Al 2% Si Cu alloy: Admiralty specification
DIN 1714 Cu Al 10 Fe	9.5% Al 2.5% Fe Cu alloy: Wrought
DIN 1714 Cu Al 10 Ni	9.5% Al 5.0% Ni 4.5% Fe Cu alloy: Wrought
DIN 1714 G Al Bz 9	9.5% Al Cu alloy: Casting **DPN: 110 UTS: 420 Elon: 25% Proof: 170**
DIN 1714 G Fe Al Bz F 50	10% Al 3% Fe Cu alloy: Casting **DPN: 135 UTS: 530 Elon: 20% Proof: 210**
DIN 1714 G Mn Al Bz F 42	8% Al 6% Mn 1.5% Cu alloy: Casting **DPN: 120 UTS: 510 Elon: 26% Proof: 220**
DIN 1714 G Ni Al Bz F 50	8% Al 5% Ni 5% Fe Cu alloy: Casting **DPN: 150 UTS: 580 Elon: 25% Proof: 240**
DIN 1714 G Ni Al Bz F 60	9% Al 5% Ni 5% Fe Cu alloy: Casting **DPN: 170 UTS: 630 Elon: 18% Proof: 330**
DIN 1714 G Ni Al Bz F 68	10% Al 6% Ni 6% Fe Cu alloy: Casting **DPN: 190 UTS: 750 Elon: 8% Proof: 370**
DIN 1714 GZ Ni Al Bz F 70	9% Al 5% Ni 5% Fe Cu alloy: Casting **DPN: 180 UTS: 750 Elon: 16% Proof: 360**
DIN 1733 S Al Bz 6	6% Al 0.4% Ni 1.0% Mn Cu alloy: Wrought
DIN 1733 S Al Bz 8	8% Al 0.4% Ni 1.0% Mn Cu alloy: Wrought
DIN 1785 Al Bz 5	5% Al 0.3% As Cu alloy: Wrought
DIN 8555 E31 300	Al Fe Cu welding electrode
DIN 17656 GB Al Bz 9	9% Al Cu alloy: Casting
DIN 17656 GB Fe Al Bz	10% Al 3.0% Fe Cu alloy: Casting
DIN 17656 GB Ni Al Bz	9% Al 5% Ni 5% Fe Cu alloy: Casting
DIN 17665 Cu Al 5	5% Al Cu alloy: Wrought
DIN 17665 Cu Al 8	8% Al Cu alloy: Wrought
DIN 17665 Cu Al 8 Fe	7.2% Al 2.7% Fe Cu alloy: Wrought
DIN 17665 Cu Al 9 Mn	9% Al 2% Mn Cu alloy: Wrought
DIN 17665 Cu Al 10 Fe	9.5% Al 4% Ni + Fe Mn Cu alloy: Wrought
DIN 17665 Cu Al 10 Ni	9.7% Al 5% Ni 5% Fe Cu alloy: Wrought

Symbol	Nominal analysis, supplier, condition and remarks.
DIN 17665 Cu Al 11 Ni	11.5% Al 6% Ni 6% Fe Cu alloy: Wrought
DNC M 38	9% Al 5% Ni 4.5% Fe 1.5% Mn Cu alloy: Casting; Admiralty specification for BS alloy AB2 C; further information from the Copper Development Association
DTD 160	9.5% Al Cu alloy: Bar; tempered and quenched; for valve seats **DPN: 150 UTS: 530**
DTD 164 A	9.5% Al 2% Ni 2% Fe Cu alloy: Forgings; tempered **UTS: 560 Elon: 17% Proof: 270**
DTD 174 A	8% Al 3% Ni 2% Fe 3% Mn Cu alloy: Casting **UTS: 480 Elon: 20%**
DTD 197 A	9.5% Al 5% Ni 5% Fe Cu alloy: Bar and forging; as rolled or forged **UTS: 650 Elon: 15% Proof: 330**
DTD 412	10% Al 4.5% Ni 4.5% Fe Cu alloy: Casting; as cast **UTS: 600 Elon: 12% Proof: 240**
E 86	7% Al 2% Si Cu alloy: Extrusion; McKechnie Bros; as drawn **DPN: 150 UTS: 580 Elon: 35% Proof: 300**
E Cu Al A2	7.5% Al Cu alloy: Weld electrode; designation used by AWS
E Cu Al C	10% Al 4.2% Fe Cu alloy: Weld electrode; designation used by AWS
E Cu Al D	11.5% Al 5% Fe Cu alloy: Weld electrode; designation used by AWS
E Cu Al E	13% Al 5% Fe Cu alloy: Weld electrode; designation used by AWS
E Cu Mn Ni Al	6.5% Al 1.7% Ni 4% Fe 12% Mn Cu alloy: Weld electrode
E Cu Ni Al	7.5% Al 5% Ni 4.5% Fe Cu alloy: Weld electrode; designation used by AWS
E in C No. C 106	9% Al 5% Ni 4.5% Fe 1.5% Mn Cu alloy: Casting; Admiralty for BS alloy AB2 C; further information from the Copper Development Association
G Cu Al 8 Mn	8% Al 1.5% Ni 6% Mn Cu Alloy: Casting; designation used by German Standards
G Cu Al 9	9.5% Al Cu alloy: Casting; designation used by German Standards
G Cu Al 9 Ni	8% Al 5% Ni 5% Fe Cu alloy: Casting; designation used by German Standards
G Cu Al 10 Fe	9.5% Al 3% Fe Cu alloy: German Standard
G Cu Al 10 Fe	10% Al 3% Fe Cu alloy: Casting; designation used by German Standards
G Cu Al 10 Ni	9% Al 5% Ni 5% Fe Cu alloy: Casting; designation used by German Standards
G Cu Al 10 Ni	10.8% Al 4.2% Ni 4% Fe Cu alloy: German Standard
G Cu Al 11 Ni	10% Al 6% Ni 6% Fe Cu alloy: Casting; designation used by German Standards
G Fe Al Bz F 50	10% Al 3% Fe Cu alloy: Casting; designation used by German Standards
GB Cu Al 9	9% Al Cu alloy: Casting; designation used by German Standards
GB Cu Al 9 Ni	9% Al 5% Ni 5% Fe Cu alloy: Casting; designation used by German Standards
GB Cu Al 10 Fe	10% Al 3.0% Fe Cu alloy: Casting; designation used by German Standards
GZ Cu Al 10 Ni	9% Al 5% Ni 5% Fe Cu alloy: Casting; designation used by German Standards
HEUSLER ALLOY	15% Al 30% Mn Cu alloy: Magnetic alloy; further information from Permanent Magnet Association
HIDURAL 7	Zn Al Cu alloy: Wrought; composition not supplied; Langley Alloys **DPN: 145 UTS: 570 Elon: 20% Proof: 260**
HIDURAX 1	10% Al 4.5% Ni 4.5% Fe Cu alloy: Casting; Langley Alloys for BS alloy AB 2
HIDURAX 1	10% Al 5% Ni 5% Fe Cu alloy: Bar and forging; Langley Alloys for BS alloy CA104
HIDURAX 2	9.5% Al 2% Fe Cu alloy: Casting; Langley Alloys for BS alloy AB 1

Note. The following abbreviations and units are used in the tables:

DPN	Hardness, diamond pyramid number
UTS	Ultimate tensile strength, N/mm^2
Elon	Elongation, %
Proof	0.1% proof strength, N/mm^2

1 N/mm^2=0.1 hbar=0.102 kgf/mm^2=0.06475 tonf/in.^{2}=145.04 lbf/in.^{2}=1 MPa

See Appendix II for other abbreviations and conversion tables.

Symbol	Nominal analysis, supplier, condition and remarks.
HIDURAX 2	9.5% Al 2% Ni 2% Fe Cu alloy: Forging; Langley Alloys
	UTS: 690 Elon: 22% Proof: 330
HIDURAX 3	10% Al Cu alloy: For casting; Langley Alloys; for corrosion resistance
	DPN: 130 UTS: 540 Elon: 25% Proof: 230
HIDURAX 4	Al Cu alloy: Composition not supplied; Langley Alloys; cast
	DPN: 250 UTS: 820 Elon: 8% Proof: 540
HIDURAX 4	Al Cu alloy: Composition not supplied; Langley Alloys; wrought
	DPN: 270 UTS: 800 Elon: 7% Proof: 490
HIDURAX 5	9% Al Cu alloy: Bar and forging; Langley Alloys
	UTS: 530 Elon: 20% Proof: 250
HIDURAX 6	5% Al Cu alloy: Bar and forging; Langley Alloys
	UTS: 390 Elon: 60% Proof: 150
HIDURAX 7	7% Al 2% Si Cu alloy: Wrought; Langley Alloys
HIDURON 130	14.5% Ni with Al Fe and Mn Cu alloy: Langley Alloy
	DPN: 230 UTS: 880 Elon: 12% Proof: 680
HIDURON 191	Al Fe Cu alloy: Wrought; Langley Alloys; similar to BS 2033
	UTS: 490 Elon: 32% Proof: 300
HOLFOS AB1	9.5% Al 2.5% Fe Cu alloy: Casting; J Holroyd; as cast
	DPN: 130 UTS: 550 Elon: 20% Proof: 200
HOLFOS AB2	10% Al 5.5% Ni 4.5% Fe Cu alloy: Casting; J Holroyd; as cast
	DPN: 165 UTS: 660 Elon: 15% Proof: 250
HOLFOS HTB1	36% Zn 1% Al 1% Fe 2% Mn Cu alloy: Casting; J Holroyd; as cast
	DPN: 115 UTS: 500 Elon: 20% Proof: 220
HOLFOS HTB2	33% Zn 3.5% Al 1.5% Fe 2% Mn Cu alloy: Casting; J Holroyd; as cast
	DPN: 135 UTS: 590 Elon: 15% Proof: 280
HOLFOS HTB3	30% Zn 5% Al 2% Fe 3% Mn Cu alloy: Casting; J Holroyd; as cast
	DPN: 185 UTS: 750 Elon: 12% Proof: 420
IMI 581	Al Mn Cu alloy: Resistance wire; IMI; Kuthern 41
IMI 756	5% Al Cu alloy: Welding rod and wire; IMI
IMI 757	7% Al Cu alloy: Welding rod and wire; IMI
IMI 764	10% Al Cu alloy: Bar; as drawn; IMI
	DPN: 160 UTS: 560 Elon: 20% Proof: 330
INCRAMET 800	10% Al 15% Ni 0.7% Fe 1.5% Co Cu alloy: Casting; Copper Development Association; for bottle moulds
	UTS: 660
INOXYDA	10% Al 4.5% Ni 4% Fe Cu alloy: Casting; information from International Nickel
KUMANAL	2% Al 10% Mn Cu alloy: Hard drawn wire and strip; IMI; electrical conductivity 4.5% IACS; heating elements
	UTS: 670 Elon: 10%
KUTHERM 41	Al Mn Cu alloy: Wire; IMI; electrical conductivity 4.5% IACS; withdrawn
Marinel	14.5% Ni 4.5% Mn with Al Fe + Cr Cu alloy: Langley Alloys
	DPN: 280 UTS: 900 Elon: 12% Proof: 750
MEIGH METAL	9% Al 5% Ni 4% Fe Cu alloy: Casting; Meigh Casting Co. for BS alloy AB2 C
Met Bronze A6MN	6.5% Al 2% Ni 4% Fe 12% Mn Cu alloy: Electrode Metrode
Met Bronze A8N	7.5% Al 5% Ni 4% Fe Cu alloy: Electrode Metrode
Met Bronze A8N	7.5% Al 5% Ni 4% Fe Cu alloy: Electrode Metrode
Metco 51	10% Al Cu spray deposit Metco
	DPN: 120
Metco 51F NS	10% Al Cu spray deposit Metco
	DPN: 160
Metco 605 NS	10% Al Cu with plastic spray deposit Metco
Metco 610 NS	10% Al Cu with plastic spray deposit Metco
Metco Sprabronze AA	10% Al Cu spray deposit Metco
	DPN: 160
NARITE	14% Al 1% Ni 5% Fe Cu alloy: Casting for deep drawing dies; N C Ashton; 120 tons/in.^{2} compression strength
NIKALIUM	9% Al 5% Ni 4.5% Fe Cu alloy: Casting for marine turbines; Manganese Bronze Ltd for BS alloy AB2 C
No. 4 ALUMINIUM BRONZE	9.5% Al 5% Ni 4.5% Fe Cu alloy: Casting; Stone Manganese Marine for BS alloy AB2
No. 6 ALUMINIUM BRONZE	9% Al 1% Ni 2% Fe 1% Mn Cu alloy: Casting; Stone Manganese Marine for BS alloy AB1
NOVOSTON	8% Al 2% Ni 3% Fe 12% Mn Cu alloy: Forging; Stone Manganese Marine; as forged
	DPN: 200 UTS: 750 Elon: 28% Proof: 400
QQ B 666 A	5.0% Al Cu alloy: Sheet; US Federal
	UTS: 370 Elon: 27% Proof: 140
QQ B 666 B	8% Al 5.0% Ni 2.2% Si Cu alloy: Bar and forging; US Federal
	UTS: 530 Elon: 17% Proof: 210
QQ B 671 A	0.5% Sn 10% Al 3.5% Fe 3% Mn Cu alloy: Casting; US Federal
	UTS: 450 Elon: 22% Proof: 150
QQ B 671 B	10% Al 1% Fe Cu alloy: Casting; US Federal
	DPN: 170 UTS: 530 Elon: 12% Proof 240
QQ B 671 C	0.5% Sn 10% Al 5% Ni 3% Mn Cu alloy: Casting; US Federal
	DPN: 180 UTS: 530 Elon: 12% Proof: 220
QQ B 671 D	5% Ni 10% 3% Mn Cu alloy: Casting; US Federal
	UTS: 600 Elon: 6% Proof: 300
RESISCO	7% Al Cu alloy: Tube; Yorkshire Imperial Metals Ltd
S Al Bz 6	6% Al 0.4% Ni 1.0% Mn Cu alloy: Wrought; designation used by German Standards
S Al Bz 8	8% Al 0.4% Ni 1.0% Mn Cu alloy: Wrought; designation used by German Standards
SAE 68 A	9% Al 3% Fe Cu alloy: Casting; as cast
	UTS: 460 Elon: 20% Proof: 170
SAE 68 B	10% Al 1% Fe Cu alloy: Casting; as cast
	UTS: 460 Elon: 20% Proof: 170
SAE 68 B	10% Al 1% Fe Cu alloy: Casting; solution treated and aged
	UTS: 580 Elon: 12% Proof: 260
SAE 71D	7% Al 2% Fe Cu alloy: Wrought
SAE 701 A	5% Al Cu alloy: Bar and forging; properties not specified
SAE 701 B	10% Al 1% Ni 4% Fe 1.5% Mn Cu alloy: Bar and forging
	UTS: 530 Elon: 15% Proof: 240
SAE 701 C	10% Al 5% Ni 3% Fe 1.5% Mn Cu alloy: Bar and forging
	UTS: 690 Elon: 12% Proof: 330
SAE CA 608	5% Al Cu alloy: Wrought
SAE CA 614	7% Al 2% Fe Cu alloy: Wrought; formerly SAE 701 D
SAE CA 617	8.5% Al 1.5% Fe Cu alloy: Wrought; formerly SAE 701 B
SAE CA 618	10% Al 1% Fe Cu alloy: Wrought
SAE CA 623	9% Al 3% Fe Cu alloy: Wrought; formerly SAE 701 B
SAE CA 624	10.7% Al 3% Fe Cu alloy: Wrought; formerly SAE 701 B
SAE CA 630	10% Al 5% Ni 3% Fe Cu alloy: Wrought; formerly SAE 701 C
SAE CA 637	7.5% Al 1.7% Si Cu alloy: Wrought; formerly SAE 701 B
SAE CA 642	7% Al 2% Si Cu alloy: Wrought
SAE CA 952	9% Al 3.2% Fe Cu alloy: Casting
SAE CA 953	10% Al 1.2% Fe Cu alloy: Casting
SAE CA 954	10.7% Al 2.5% Ni 4% Fe Cu alloy: Casting
SAE CA 955	10.7% Al 4% Ni 4% Fe Cu alloy: Casting
SAE CA 958	9% Al 4.5% Ni 4% Fe Cu alloy: Casting
SILVA BRONZE	Al Cu wrought alloy: IMI; for fire back boilers
	DPN: 100 UTS: 400 Elon: 75% Proof: 90

Symbol	Nominal analysis, supplier, condition and remarks.
Soudobronze 101	7.5% Al 0.5% Fe 0.5% Mn 0.5% Si Cu alloy: Welding electrode; Soudometal for surface deposits **DPN: 125**
Soudobronze 201	9.6% Al 0.9% Mn 1.6% Si Cu alloy: Welding electrode; Soudometal for surface deposits **DPN: 225**
Soudobronze 301	11% Al 1% Mn 1.6% Si Cu alloy: Welding electrode; Soudometal for surface deposits **DPN: 280**
Soudobronze MnS	5.6% Al 2.6% Ni 3.2% Fe 12.5% Mn Cu alloy: Welding electrode; Soudometal for welding Al bronze **DPN: 155**
Soudor G Cu Al 8	8% Al 0.4% Fe Cu alloy: Welding electrode; Soudometal **UTS: 430 Elon: 40% Proof: 200**
Soudoteg Cu Al 8	8% Al 0.4% Fe Cu alloy: Welders electrical; Soudometal for inert gas welding **UTS: 430 Elon: 40% Proof: 200**
SPRABRONZE AA	9% Al 1% Fe Cu alloy: Wire for spraying; Metco Ltd; as sprayed **DPN: 146 UTS: 200**
STA 7 CA1	5% Al Cu alloy: Bar and sheet **UTS: 300 Elon: 25%**
STA 7 CA2	0.1% Pb 9.5% Al Cu alloy: Bar **DPN: 150 UTS: 530**
STA 7 CA3	10% Al 2% Ni 1% Mn 1% Fe Cu alloy: Casting; as cast **UTS: 480 Elon: 20%**
STA 7 CA4	10% Al 4.5% Ni 4% Fe 3% Mn Cu alloy: Casting; sand and die cast **UTS: 600 Elon: 12%**
STA 7 CA5	10% Al 5% Ni 5% Fe 2% Mn Cu alloy: Bar and forging **DPN: 200 UTS: 660 Elon: 15%**
STA 7 CX10	10% Al 3% Ni 3% Fe Cu alloy: Casting; die cast **DPN: 190**
Stubs 34	9% Al Cu alloy: Welding rod; Stubs for facing and welding bronze, cast iron **DPN: 130 UTS: 450 Elon: 30%**
Stubs 34 N	Al bronze welding rod; Stubs for welding and facing bronze, cast iron **DPN: 230 UTS: 680 Elon: 20%**
Stubs 343	Al Fe Cu alloy: Welding electrode; Stubs for wear resistance **DPN: 300**
Stubs A34	9.0% Al Cu alloy: Welding electrode; Stubs for joining Al bronze and for low friction coating **DPN: 130 UTS: 490 Elon: 30% Proof: 185**
Stubs A3444	9.0% Al 4.5% Ni 3.5% Fe 1.0% Mn Cu alloy: Welding rod; Stubs for joining Al bronzes containing N **DPN: 200 UTS: 620 Elon: 12% Proof: 400**
SUPERSTON	1% Sn 8% Al 3% Ni 3% Fe 12% Mn Cu alloy: Casting; Phosphor Bronze Ltd for BS alloys CMA 1 C and CMA 2 C
SUPERSTON 10	4% Al 2% Ni 3% Fe 12% Mn Cu alloy: Sheet and plate; Stone Manganese Marine; as rolled **UTS: 420 Elon: 62% Proof: 200**
SUPERSTON 40	8% Al 2% Ni 3% Fe 12% Mn Cu alloy: Casting; Stone Manganese Marine; as cast **DPN: 180 UTS: 690 Elon: 27% Proof: 300**
SUPERSTON 40	8% Al 2% Ni 3% Fe 12% Mn Cu alloy: Forging; Stone Manganese Marine; as forged **DPN: 200 UTS: 750 Elon: 28% Proof: 400**
SUPERSTON 60	9% Al 2% Ni 3% Fe 12% Mn Cu alloy: Casting; Stone Manganese Marine; as cast **DPN: 220 UTS: 750 Elon: 14% Proof: 400**
SUPERSTON 60	9% Al 2% Ni 3% Fe 12% Mn Cu alloy: Casting; Stone Manganese Marine; as forged **DPN: 250 UTS: 830 Elon: 15% Proof: 450**
SUPERSTON 70	7.5% Al 2.5% Ni 3% Fe 15% Mn Cu alloy: Casting; Stone Manganese Marine; as cast **DPN: 200 UTS: 660 Elon: 20% Proof: 300**
SUPERTRODE	8% Al 2% Ni 3% Fe 13% Mn Cu alloy: Welding rod; Stone Manganese Marine; as welded **DPN: 180 UTS: 680 Elon: 25% Proof: 370**
TAURUS XX	9.5% Al 5% Ni 4% Fe 1% Mn Cu alloy: Casting; D Brown; centri-cast **DPN: 155 UTS: 630 Elon: 14% Proof: 330**
U-A 10NM	9.5% Al 5.5% Ni 2.2% Fe 1.2% Mn Cu alloy: Origin unknown
VULCAN METAL	0.4% Sn 1% Al 1.5% Ni 4.4% Fe 1% Si 0.7% Cr Cu alloy: Origin unknown
XANTAL	0.05% Zn 9.5% Al 0.2% Ni 0.2% Fe Cu alloy: Cast and wrought; origin unknown

Note. The following abbreviations and units are used in the tables:

DPN	Hardness, diamond pyramid number
UTS	Ultimate tensile strength, N/mm^2
Elon	Elongation, %
Proof	0.1% proof strength, N/mm^2

1 N/mm^2=0.1 hbar=0.102 kgf/mm^2=0.06475 $tonf/in.^2$=145.04 $lbf/in.^2$=1 MPa

See Appendix II for other abbreviations and conversion tables.

14H Copper alloys for brazing fillers and fuse materials
Containing silver, gold, palladium, phosphorus, cadmium or zinc

General metallurgical characteristics

These alloys have been developed in the first instance for the temperature range over which they are molten, as this very often is the deciding factor regarding their use. It is also necessary that a braze metal is capable of wetting the materials being joined and gives off no noxious fumes.

Probably the most important aspect, however, is the workability of the molten metal. This must be as fluid as possible for joints relying on capillary action, but should be less so for manipulating hand brazes. The casting qualities of the braze metal should be of the first order.

The various alloying elements added to copper firstly lower the melting point and then, depending on the element, have one of several effects.

Silver. Lowers the melting point, increases the fluidity, improves the strength and soundness of the braze including the fatigue properties and has good electrical conductivity and corrosion resistance. It is sometimes present alone, or with any of the listed alloying elements except gold.

Zinc. Lowers the melting point and acts as a de-oxidizer. It is usually added as the most economical method of achieving this but has adverse effects on the strength and corrosion resistance. It is never present alone in the listed alloys but brazing alloys will also be found in the copper zinc groups 14B and 14C.

Cadmium. Reduces the melting point to the lowest commonly used when present with silver and some zinc, without seriously affecting the electrical or corrosion properties. It is never found as the sole alloying element in brazing materials. Cadmium coppers for electrical purposes are found in the pure copper group 14A. These cadmium alloys present a health hazard when used for brazing or casting.

Phosporus. This is a powerful de-oxidizer, but can have serious deleterious effects on the electrical resistance and ductility. It also increases the fluidity of the molten metal. The straight copper phosphorus alloys are generally used only when brazing copper.

Gold. This is used on a limited scale where excellent corrosion resistance brazing is necessary, with high melting points. These alloys are also oxidation resistant up to their melting range, which is quite narrow. Gold is always the sole alloying element.

Palladium. This usually, but not always, has silver also present. The alloys melt over a narrow range and result in a hard, very corrosion resistant braze with stable electrical properties. The higher palladium silver alloys are white in colour.

None of the materials in this group can have their properties improved by heat treatment. Using considerable skill and expertise it is possible to improve the mechanical properties of a braze by 'peening'. In this technique a skilled man uses either a small hammer or a round-nosed punch and the mechanical properties of the weld are improved by work hardening. This technique is not advised unless considerable skill and advice are available.

Where, for any reason, cold work is induced into the material by machining or for some other reason then this can be removed by softening at a temperature of about 300 °C.

These materials in the main are used as welding and brazing filler rods and therefore are seldom themselves welded or brazed.

There could be occasions where brazing or soldering is required and no problems will be found provided that the parts are correctly cleaned and any fluxes used are cleaned off after the operation.

The materials have the same corrosion resistance as other copper alloys of the same type, i.e. markedly better than that of steel and comparable with, sometimes better than and sometimes not as good as, that of austenitic stainless steel depending on the corrosion resistance involved.

Some alloys in this section can be placed in sub-groups as follows.

Silver alloys. These are principally used as braze metals where the melting point is important. Cadmium is now present in many of these alloys and they therefore present a health hazard in use.

Phosphorus alloys. These are used as brazing metals where simple joins are required and electrical resistance is not a problem. These are probably the most commonly used materials in this section.

Palladium alloys. These have high strengths and give high melting points with good corrosion resistance. They are therefore used for jewellery type work where a silver cover is desirable.

Gold alloys. These have specific electrical properties and also have the ability to withstand considerable corrosion at high temperatures. The largest use of these materials is in the field of high temperature fuses to protect furnace heat treatment equipment. They do not oxidize and thus have a permanent electrical resistivity but melt at a sharp point when a specific temperature is achieved, thus cutting off the current and protecting the equipment.

Symbol	Nominal analysis, supplier, condition and remarks.
22 MS	Zn Ni Cu alloy electrodes for brazing and welding stubs for use on Zn Ni Cu alloy: melting range 890–920 °C **UTS: 660** **Elon: 27%**
279	Ni Mn Pd Cu brazing alloy: Melting range 1060–1105 °C; Engelhard Industries; electrical conductivity 3% IACS **DPN: 135** **UTS: 600** **Elon: 35%**
ARGO BRAZE 25	33% Zn 2% Ni 2% Mn 25% Ag Cu alloy: Braze metal; Johnson Matthey for tungsten carbide
ARGO BRAZE 27	20% Zn 5% Ni 9% Mn 27% Ag Cu alloy: Braze metal; Johnson Matthey for tungsten carbide

Note. The following abbreviations and units are used in the tables:

DPN	Hardness, diamond pyramid number
UTS	Ultimate tensile strength, N/mm^2
Elon	Elongation, %
Proof	0.1% proof strength, N/mm^2

1 N/mm^2=0.1 hbar=0.102 kgf/mm^2=0.06475 tonf/in.^{2}=145.04 lbf/in.^{2}=1 MPa

See Appendix II for other abbreviations and conversion tables.

Symbol	Nominal analysis, supplier, condition and remarks.
ARGO-BOND	23% Zn Ag Cd Cu alloy: Braze metal; Johnson Matthey; melting range 616–735 °C
ARGO-SWIFT	29% Zn Ag Cd Cu alloy: Braze metal; Johnson Matthey; melting range 607–700 °C
ASTM B260 B Au 1	37% Au Cu alloy: Braze metal; temperature range 1016–1093 °C
ASTM B260 B Au 3	3% Ni 34.5% Au Cu alloy: Braze metal; temperature range 1030–1090 °C
ASTM B260 B Cu 1	99.9% (min) Cu: Braze metal; temperature range 1093–1149 °C
ASTM B260 B Cu 1 A	99.0% (min) Cu: Braze metal; temperature range 1093–1149 °C
ASTM B260 B Cu 2	86.5% (min) Cu: Braze metal; brazing range 1090–1150 °C
ASTM B260 B Cu Au 1	37.5% Au Cu alloy: Braze filler; melting range 955–990 °C
ASTM B260 B Cu P 1	5% P Cu alloy: Brazing filler; melting range 706–890 °C
ASTM B260 B Cu P 2	7% P Cu alloy: Brazing filler; melting range 706–805 °C
ASTM B260 B Cu P 3	6% P 6% Ag Cu alloy: Brazing filler; melting range 640–800 °C

Symbol	Nominal analysis, supplier, condition and remarks.
ASTM B260 B Cu P 4	7% P 6% Ag Cu alloy: Brazing filler; melting range 640–719 °C
ASTM B260 B Cu P 5	5% P 15% Ag Cu alloy: Brazing filler; melting range 640–815 °C
B 6	16% Zn Ag Cu alloy: Braze metal; Johnson Matthey; melting range 790–830 °C
B-BRONZE	3% Ni 0.03% B Cu alloy: Braze metal; Johnson Matthey; melting range 1080–1100 °C
BS 1845/6	5% P 14% Ag Cu alloy: Braze filler; melting range 625–780 °C
BS 1845/7	7% P Cu alloy: Braze filler; melting range 705–800 °C
BS 1845 Au 3	37.5% Au Cu alloy: For brazing metal; melting range 980–1000 °C
BS 1845 Au 4	30% Au Cu alloy: For braze metal; melting range 995–1020 °C
BS 1845 Cp 1	4.6% P 14.5% Ag Cu alloy: Braze metal; melting range 645–700 °C
BS 1845 Cp 2	6.5% P 2.0% Ag Cu alloy: Braze metal; melting range 645–740 °C
BS 1845 Cp 3	7.7% P Cu alloy: Braze metal; melting range 705–800 °C
BS 1845 Pd 8	18% Pd Cu alloy: For brazing metal; melting range 1080–1090 °C
BS 1845 Pd 12	15% Ni 10% Mn 20% Pd Cu alloy: For brazing metal; melting range 1060–1110 °C
BS 1845 Pd 13	20% Ni 15% Mn 30% Pd Cu alloy: For brazing metal; melting range 1070–1090 °C
C 4	24% Zn Ag Cu alloy: Braze metal; Johnson Matthey; melting range 740–780 °C
C BRONZE	2.5% Ni 11% Mn Cu alloy: Braze metal; Johnson Matthey; melting range 965–995 °C
COLMONOY 15	7.0% P Cu alloy: Bronze metal; Wall Colmonoy; melting point 795 °C
COPPER-FLO	7.4% P Cu alloy: Braze metal; Johnson Matthey; melting range 714–810 °C
CPNM 2	Ni Mn Pd Cu brazing alloy: Melting range 1060–1105 °C; Englehard Industries; electrical conductivity 3% IACS **DPN: 135 UTS: 600 Elon: 35%**
D 3	33% Zn Ag Cu alloy: Braze metal; Johnson Matthey; melting range 700–740 °C
D BRONZE	10% Mn 4% Co Cu alloy: Braze metal; Johnson Matthey; melting range 980–1030 °C
DIN 1734 L Ag 8	37% Zn 8% Ag Cu alloy: Braze metal; brazing temperature 860 °C
DIN 1734 L Ag 12	35% Zn 12% Ag Cu alloy: Brazing metal; working temperature 830 °C
DIN 1734 L Ag 12 Cd	30% Zn 12% Ag 7% Cd Cu alloy: Braze metal; working temperature 800 °C
DIN 1734 L Ag 15	35% Zn 15% Ag 10% Cd Cu alloy: Braze metal; working temperature 770 °C
DIN 1734 L Ag 15 P	3% P 15% Ag Cu alloy: Braze metal; working temperature 710 °C
DIN 1734 L Ag 20	30% Zn 20% Ag 15% Cd Cu alloy: Braze metal; working temperature 750 °C
DIN 1734 L Ag 25	30% Zn 25% Ag Cu alloy: Braze metal; working temperature 780 °C
DIN 1734 L Ag 25 Cd	20% Zn 25% Ag 15% Cd Cu alloy: Braze metal; working temperature 730 °C

Note. The following abbreviations and units are used in the tables:

DPN	Hardness, diamond pyramid number
UTS	Ultimate tensile strength, N/mm^2
Elon	Elongation, %
Proof	0.1% proof strength, N/mm^2

$1\ N/mm^2 = 0.1\ hbar = 0.102\ kgf/mm^2 = 0.06475\ tonf/in.^2 = 145.04\ lbf/in.^2 = 1\ MPa$

See Appendix II for other abbreviations and conversion tables.

Symbol	Nominal analysis, supplier, condition and remarks.
DIN 1734 L Ag 27	20% Zn 5% Ni 8% Mn 27% Ag Cu alloy: Braze metal; working temperature 840 °C
DIN 1734 L Ag 30 Cd 5	20% Zn 30% Ag 5% Cd Cu alloy: Braze metal; working temperature 770 °C
DIN 1734 L Ag 30 Cd 12	20% Zn 30% Ag 12% Cd Cu alloy: Braze metal; working temperature 700 °C
DIN 1734 L Ag 38	3% Sn 25% Zn 38% Ag Cu alloy: Braze metal; working temperature 800 °C
ERM SC 65	6% Ag 0.03% O Cu alloy: Forgings; Enfield Rolling Mills Ltd; resistance weld electrodes **DPN: 140 UTS: 440 Elon: 12% Proof: 340**
F BRONZE	38.5% Zn 2% Mn 2% Co Cu alloy: Braze metal; Johnson Matthey; melting range 890–910 °C
G BRONZE	2.5% Ni 0.6% Si Cu alloy: Braze metal; Johnson Matthey; melting range 1090–1100 °C
H 12	Zn Ag Cu alloy: For brazing; silver solder; Sheffield Smelting Co.; melting point 780 °C; electrical conductivity 8.9% IACS **DPN: 106 UTS: 260 Elon: 30% Proof: 190**
H BRONZE	9% Ni 39% Mn Cu alloy: Braze metal; Johnson Matthey; melting range 880–920 °C
JMMB BRONZE	Cu alloy: For brazing stainless steel; molybdenum or tungsten; Johnson Matthey
L 3	38% Zn 10% Ag Cu alloy: Braze metal; Sheffield Smelting; melting range 840–855 °C **DPN: 178 UTS: 530 Elon: 7% Proof: 480**
L 7	35% Zn 14% Ag Cu alloy: Braze metal; Sheffield Smelting; melting range 810–835 °C **DPN: 174 UTS: 550 Elon: 3% Proof: 500**
LX 13	30% Zn 20% Ag 10% Cd Cu alloy: Braze metal; Sheffield Smelting; melting range 718–773 °C **DPN: 111 UTS: 370 Elon: 9% Proof: 300**
LX 18	27.5% Zn 25% Ag 17.5% Cd Cu alloy: Braze metal: Sheffield Smelting; melting range 605–710 °C **DPN: 116 UTS: 370 Elon: 14% Proof: 300**
M 1	34% Zn 31% Ag Cu alloy: Braze metal; Sheffield Smelting; melting range 714–755 °C **DPN: 112 UTS: 480 Elon: 5% Proof: 450**
MATTABRAZE 12	31% Zn 12% Ag 7% Cd Cu alloy: Braze metal; Johnson Matthey; melting range 620–820 °C
MATTABRAZE 45	18% Zn 20% Cd 17% Cu Ag alloy: Braze metal; Johnson Matthey; melting range 620–635 °C
METALFLO	25% Zn 25% Ag 17% Cd Cu alloy: Braze metal; Johnson Matthey; melting range 605–720 °C
METASIL	25% Zn 20% Ag 15% Cd Cu alloy: Braze metal; Johnson Matthey; melting range 605–760 °C
MICROBRAZ 5025	5% P 38% Ni 7% Cr Cu alloy: Braze filler; Wall Colmonoy; melting range 1065–1150 °C
MICROBRAZ 5027	26.5% P 4.9% Cr Cu alloy: Braze metal; Wall Colmonoy; melting range 1065–1150 °C
MX 4	Zn Ag Cu alloy: For brazing; silver solder; Sheffield Smelting; melting range 608–665 °C; electrical conductivity 26% IACS **DPN: 103 UTS: 240 Elon: 15% Proof: 150**
NICORO	3.0% Ni 35% Au Cu alloy: For brazing; melting range 1000–1030 °C, Wesgo
OROBRAZE 998	37.5% Ag Cu alloy: Braze metal; Johnson Matthey; melting range 980–998 °C
OROBRAZE 1018	30% Au Cu alloy: Braze metal; Johnson Matthey; melting range 996–1018 °C
PALLABRAZE	15% Pb Ag Cu alloy: Braze metal; Johnson Matthey; melting range 856–880 °C
PALLABRAZE 810	5% Pd Ag Cu alloy: Braze metal; Johnson Matthey; melting range 807–810 °C
PALLABRAZE 840	10% Pd Ag Cu alloy: Braze metal; Johnson Matthey; melting range 830–840 °C
PALLABRAZE 850	10% Pd Ag Cu alloy: Braze metal; Johnson Matthey; melting range 824-850 °C
PALLABRAZE 880	15% Pd Ag Cu alloy: For brazing; Johnson Matthey; melting range 856–880 °C

Symbol	Nominal analysis, supplier, condition and remarks.
PALLABRAZE 900	20% Pd Ag Cu alloy: Braze metal; Johnson Matthey; melting range 876–900 °C
PALLABRAZE 950	25% Pd Ag Cu alloy: Braze metal; Johnson Matthey; melting range 901–950 °C
PALLABRAZE 1090	18% Pd Cu alloy: Braze metal; Johnson Matthey; melting range 1080–1090 °C
PHOSPHALLOY No 1	5% P 15% Ag Cu alloy: Braze metal; Sheffield Smelting; melting range 625–780 °C
PHOSPHALLOY No 2	6.5% P 2% Ag Cu alloy: Braze metal; Sheffield Smelting; melting range 690–835 °C **DPN: 130 UTS: 420 Elon: 5% Proof: 240**
QQ S 561/3	5.0% P 15% Ag Cu alloy: Solder; US Federal
SB 260 B Cu 2	See ASTM B260 B Cu 2
SILBERLOT 8	37% Zn 8% Ag Cu alloy: Braze metal; designation used by German Standards
SILBERLOT 12	35% Zn 12% Ag Cu alloy: Braze metal; designation used by German Standards
SILBERLOT 12 Cd	30% Zn 12% Ag 7% Cd Cu alloy: Braze metal; designation used by German Standards
SILBERLOT 15	35% Zn 15% Ag 10% Cd Cu alloy: Braze metal; designation used by German Standards
SILBERLOT 15 P	3% P 15% Ag Cu alloy: Braze metal; designation used by German Standards
SILBERLOT 20	30% Zn 20% Ag 15% Cd Cu alloy: Braze metal; designation used by German Standards
SILBERLOT 25	30% Zn 25% Ag Cu alloy: Braze metal; designation used by German Standards
SILBERLOT 25 Cd	20% Zn 25% Ag 15% Cd Cu alloy: Braze metal; designation used by German Standards
SILBERLOT 27	20% Zn 5% Ni 8% Mn 27% Ag Cu alloy: Braze metal; designation used by German Standards
SILBERLOT 30 Cd 5	20% Zn 30% Ag 5% Cd Cu alloy: Braze metal; designation used by German Standards
SILBERLOT 30 Cd 12	20% Zn 30% Ag 12% Cd Cu alloy: Braze metal; designation used by German Standards
SILBERLOT 38	3% Sn 25% Zn 38% Ag Cu alloy: Braze metal; designation used by German Standards
SILBRALLOY	2% P Ag Cu alloy: Braze metal; Johnson Matthey; melting range 638–694 °C; for brazing copper and alloys **DPN: 195 UTS: 530 Elon: 5%**
SILBRAZE	41% Zn 0.3% Si 1% Ag Cu alloy: Braze metal; Sheffield Smelting; melting range 886–893 °C **DPN: 115 UTS: 480 Elon: 43% Proof: 210**
SILFLO 0	7.0% P Cu alloy: Brazing rod for high electrical conductivity joints; 98% IACS; All-State; brazing range 710–790 °C
SILFLO 5	6.0% P 5.0% Ag Cu alloy: Brazing rod; All-State; high electrical conductivity; 98% IACS; brazing range 700–850 °C
SILFLO 15	5.0% P 15.0% Ag Cu alloy: Brazing rod; All-State; brazing range 700–830 °C; high electrical conductivity; 98% IACS
SIL-FOS	4.6% P 14.5% Ag Cu alloy: Braze metal; Johnson Matthey; melting range 644–800 °C
SIL-FOS	15% P Ag Cu alloy: Braze metal for copper and its alloys; Johnson Matthey; melting range 625–780 °C **DPN: 187 UTS: 690 Elon: 10%**
SIL-FOS 5	6% P 5% Ag Cu alloy: Braze metal; Johnson Matthey; melting range 644–815 °C
SIL-FOS 6	7.2% P 6% Ag Cu alloy: Braze metal; Johnson Matthey; melting range 644–718 °C
SILVER-FLO 5	4.0% Zn 5% Ag Cu alloy: Braze metal; Johnson Matthey; melting range 830–870 °C
SILVER-FLO 12	4.0% Zn 12% Ag Cu alloy: Braze metal; Johnson Matthey; melting range 800–830 °C
SILVER-FLO 20	36% Zn 20% Ag Cu alloy: Braze metal; Johnson Matthey; melting range 776–815 °C
SILVER-FLO 25	34% Zn 25% Ag Cu alloy: Braze metal; Johnson Matthey; melting range 700–800 °C
SOUDEN BE 55	6% P 5% Ag Cu alloy: Braze metal; Soudometal **UTS: 250**
SOUDOGEN BE 5	6.5% P 2% Ag Cu alloy: Bronze metal; Soudometal **UTS: 250 Elon 5%**
SOUDOGEN BE 300	25% Zn 15% Co 20% Ag Cu alloy: Bronze metal; Soudometal
STAN-FOS	7% Sn 6.7% P Cu alloy: Braze metal; Johnson Matthey; melting range 640–680 °C
STUBS 6	Zn Ni Ag Cu alloy: Brazing electrode; Stubs for joining steel SG iron and Ni alloy; melting range 890–920 °C **UTS: 880 Elon: 29%**
STUBS 6M	Zn Ni Ag Cu alloy: Electrode for brazing; Stubs for brazing steel SG iron, Ni alloys; melting range 890–920 °C **UTS: 880 Elon: 29%**
STUBS 7	Zn 18% Ag 44% Cu alloy: For brazing; Stubs; high zinc silver solder melting range 784–815 °C **DPN: 130 UTS: 470**
STUBS 7M	Zn 18% Ag 44% Cu alloy: For brazing; Stubs; high zinc silver solder melting range 784–815 °C **DPN: 130 UTS: 470**
STUBS 22	Zn Ni Cu alloy: For brazing and welding; Stubs for joining Zn Ni Cu alloys **UTS: 660 Elon: 29%**
STUBS 22M	Zn Ni Cu alloy: For brazing and welding; Stubs for Zn Ni Cu alloy **UTS: 660 Elon: 29%**
STUBS 35R	7.0% P 0.5% Ag Cu alloy: Brazing rod; Stubs for electrical joints; 10% IACS; melting range 650–710 °C
STUBS 35S	7.0% P 0.5% Ag Cu alloy: Brazing rod; Stubs for electrical joints; 10% IACS; melting range 650–710 °C
STUBS 36	6.0% P 2.0% Ag Cu alloy: Brazing rod; Stubs for electrical joints; 9% IACS; melting range 645–740 °C
STUBS 38	Ag Cu alloy: Welding electrode; Stubs; MP 1070 for welding de-oxidized copper
THESSCO	Trade name for brazing and soldering material; Sheffield

Note. The following abbreviations and units are used in the tables:

DPN	Hardness, diamond pyramid number
UTS	Ultimate tensile strength, N/mm^2
Elon	Elongation, %
Proof	0.1% proof strength, N/mm^2

1 N/mm^2=0.1 hbar=0.102 kgf/mm^2=0.06475 $tonf/in.^2$=145.04 $lbf/in.^2$=1 MPa

See Appendix II for other abbreviations and conversion tables.

14J Copper sintered alloys
With iron, tungsten and tungsten carbide

General metallurgical characteristics

These alloys are produced for two basic reasons, firstly as porous metals and secondly to give artificial hardness. In both cases the excellent properties of copper regarding conductivity, corrosion resistance and as a bearing metal are combined with other properties by physical, not metallurgical, means.

The series of iron–copper sintered alloys are harder than most other copper alloys and their porosity can be controlled.

The tungsten–copper and tungsten carbide sintered alloys have these very hard particles set in the copper matrix with its excellent electrical conductivity.

The sintering process employs very fine particles, the various constituents being thoroughly mixed and then compacted to the desired shape. The pressure employed at this stage decides the percentage porosity of the finished article. The shapes are then heated when the copper melts and surrounds each particle of the alloying material. With some processes pressure and heat are applied at the same time.

During production of these alloys there are two stages, the first of which is to use pressure to produce a shape with the material involved. The second stage is to increase the temperature where one of the components in the sinter will melt or fuse. This then becomes the bond or the liquid glue which holds the remaining components together. Generally this is the copper alloy component within the matrix.

No other form of heat treatment can be carried out on these materials.

Welding or brazing should not be attempted on these materials without specific advice being obtained from the producers of the alloys. The materials can be joined by welding and brazing but this will generally result in the closing up of the pores and the material adjacent to the weld or braze will no longer have the properties of the sinter material.

Corrosion should not present a problem provided that the materials are not subjected to corrosive atmospheres with changes in temperature or pressure. The corrosive atmosphere could be absorbed into the pores with corrosion resulting in a dramatic manner.

Electroplating of these components must never be attempted without expert advice.

These materials are used principally as bearings but many uses are now being found for intricate shapes which can be produced using the sinter process and do not require any machining.

Tolerances of a very tight limit can be produced provided that the correct dies are manufactured and the control of the process is to the requisite standard.

Many of the materials have controlled porosity which allows them to absorb liquids such as oil and thus to act as constant lubricated bearings or similar devices.

Symbol	Nominal analysis, supplier, condition and remarks.
AMS 4805 B	10% Sn Cu alloy sinter: Oil impregnated
B 10	Iron–copper sintered material: Sintered Products Ltd; 17–22% porosity; withdrawn **UTS: 220 Elon: 1.5%**
B 10	Bronze sinter: Firth Cleveland; specific gravity 6.8–7.2; as sintered **DPN: 115 UTS: 130 Elon: 4%**
B 15	Iron–copper sintered material: Sintered Products Ltd; 17–22% porosity; withdrawn **UTS: 240 Elon: 1.5%**
B 20	Bronze sinter: Firth Cleveland; specific gravity 8.0–8.4; as sintered **DPN: 90 UTS: 320 Elon: 40%**
BM 78	Iron–copper–carbon sintered material: Sintered Products Ltd; 18–22% porosity **UTS: 250 Elon: 10%**
C 10	Cu sinter: Firth Cleveland; specific gravity 7.0–7.4; as sintered **DPN: 115 UTS: 160 Elon: 13%**

Note. The following abbreviations and units are used in the tables:

DPN	Hardness, diamond pyramid number
UTS	Ultimate tensile strength, N/mm^2
Elon	Elongation, %
Proof	0.1% proof strength, N/mm^2

1 N/mm^2=0.1 hbar=0.102 kgf/mm^2=0.06475 $tonf/in.^2$=145.04 $lbf/in.^2$=1 MPa

See Appendix II for other abbreviations and conversion tables.

Symbol	Nominal analysis, supplier, condition and remarks.
CT 0010-N	10% Sn 0.25% C (max) Cu sinter alloy: USA designation
CT 0010-R	10% Sn 0.25% C (max) Cu sinter alloy: USA designation
DURASINT	Sintered metal for friction; origin unknown
ELKONITE 1 W 3	W Cu alloy: For arc resistant contacts; Mallory Metallurgical; conductivity 41% IACS **DPN: 140**
ELKONITE 3 W 3	W Cu alloy: For arc resistant contacts; Mallory Metallurgical; conductivity 33% IACS **DPN: 160**
ELKONITE 10 W 3	W Cu alloy: For welding electrodes; Mallory Metallurgical
ELKONITE 20 K 3	W C Cu alloy: For riveting and welding electrodes: Mallory Metallurgical; electrical conductivity 22% IACS **DPN: 300 UTS: 530**
ELKONITE 20 W 3	W Cu alloy: For resistance welding electrodes; Mallory Metallurgical; electrical conductivity 28% IACS **DPN: 230 UTS: 630**
ELKONITE 30 W 3	W Cu alloy: For resistance welding electrodes; Mallory Metallurgical; electrical conductivity 28% IACS **DPN: 240 UTS: 65**
ELKONITE 100 M	W Cu sintered alloy: For hot riveting electrodes; Mallory Metallurgical; conductivity 30% IACS **DPN: 170 UTS: 35**
ELKONITE 100 W	W Cu sintered alloy: For resistance brazing electrodes; Mallory Metallurgical; conductivity 30% IACS **DPN: 550 UTS: 22**
FC 3/75	Iron–copper sintered material: Sintered Products Ltd; 22–28% porosity **UTS: 20 Elon: 1.5%**

Symbol	Nominal analysis, supplier, condition and remarks.
FC 75	Iron–copper sintered material: Sintered Products Ltd; 22–28% porosity **UTS: 21 Elon: 1.5%**
FC 75/1	Iron–copper–carbon sintered material: Sintered Products Ltd; 22–23% porosity **UTS: 24 Elon: 1.5%**
FC 80	Iron–copper sintered material: Sintered Products Ltd; 18–22% porosity **UTS: 24 Elon: 1.5%**
FC 85/1	Iron–copper–carbon sintered material: Sintered Products Ltd; 15–20% porosity **UTS: 28 Elon: 1.5%**
FEC 90	Iron–copper sintered material: Sintered Products Ltd; 4–8% porosity; withdrawn **UTS: 34 Elon: 5.5%**
FGC 75	Iron–copper–carbon sintered material: Sintered Products Ltd; 22–23% porosity; now FC75/1 **UTS: 24 Elon: 1.5%**
FGC 85	Iron–copper–carbon sintered material: Sintered Products Ltd; 15–20% porosity; now FC85/1 **UTS: 28 Elon: 1.5%**
FU E10-58	10% Sn 0.25% C (max) Cu sintered alloy: French designation
FU E10-62	10% Sn 0.25% C (max) Cu sintered alloy: French designation
FX 10	Fe Cu sinter: Firth Cleveland; specific gravity 7.1–7.6; as sintered **DPN: 140 UTS: 400 Elon: 4%**
H 75	Iron–copper–carbon sintered material: Sintered Products Ltd; 15–20% porosity; withdrawn **UTS: 22 Elon: 1%**
MALLORY 1000	W Cu alloy: Sintered; Mallory Metallurgical; electrical conductivity 14% IACS; 'No Chat' **DPN: 270 UTS: 75 Elon: 2.5%**
MALLORY NO-CHAT	W Cu alloy: Sintered; Mallory Metallurgical; electrical conductivity 14% IACS; Mallory 1000 **DPN: 270 UTS: 75 Elon: 2.5%**
M BOL-1	10% Sn 0.25% C (max) Cu alloy: Sintered bearing; Gilite; 27% porosity min
M BOL-2	10% Sn 0.25% C (max) Cu alloy: Sintered bearing; Gilite; 22% porosity min
NO-CHAT	W Cu alloy: Sintered; Mallory Metallurgical; electrical conductivity 14% IACS **DPN: 270 UTS: 75 Elon: 2.5%**
POROSINT	Bronze sintered metal filters: Sintered Products Ltd
RIGIDMESH	Sintered metal for filters: origin unknown
RWMA CLASS 12	W Cu alloy: For riveting and welding electrodes; Mallory Metallurgical code for Elkonite 20 W 3
RWMA CLASS 13	Mallory Metallurgical code for Elkonite 100 W
S 60	Iron–carbon sintered material: Bearings; Sintered Products Ltd; 25–30% porosity; withdrawn **UTS: 9**
SAE 890	0.1% Sn 1.5% Pb 0.2% Fe Cu sintered material: Density 7.2–7.7 **UTS: 9**
SAE 891	0.1% Sn 1.5% Pb 0.2% Fe Cu sintered material: Density 7.7 (min) **UTS: 10**
SILVER-FLO 302	2% Sn 32% Zn 30% Ag Cu alloy: Braze metal; Johnson Matthey; melting range 665–755 °C
SINT A50	10% Sn 0.25% C (max) Cu sintered alloy: German designation
SPARKOMITE 10	W Cu sintered tube: For spark erosion electrodes; as Elkonite 10 W 3; Mallory Metallurgical
SX 20	Fe Cu with C sinter: Firth Cleveland; specific gravity 7.1–7.6; hardened and tempered; file hard **UTS: 700 Elon: 5%**
U 3	W Cu contact material: Sheffield Smelting; electrical conductivity 28% IACS **DPN: 240**
U 4	W Cu contact material: Sheffield Smelting; electrical conductivity 35% IACS **DPN: 220**
U 8	W Cu contact material: Sheffield Smelting; electrical conductivity 40% IACS **DPN: 180**
WOLFRAM BRONZE	10% W Cu alloy: Origin unknown
Z 10	Brass sinter: Firth Cleveland; specific gravity 7.2–7.6; as sintered **DPN: 115 UTS: 160 Elon: 11%**

Note. The following abbreviations and units are used in the tables:

DPN	Hardness, diamond pyramid number
UTS	Ultimate tensile strength, N/mm^2
Elon	Elongation, %
Proof	0.1% proof strength, N/mm^2

1 N/mm^2=0.1 hbar=0.102 kgf/mm^2=0.06475 $tonf/in.^2$=145.04 $lbf/in.^2$=1 MPa

See Appendix II for other abbreviations and conversion tables.

14K Copper–tin wrought and cast alloys
With additions of phosphorus, lead and zinc – bronze and 'phosphor bronze'

Specific gravity	8.92
Density	8920 kg/m^3
Solidus/liquidus	1050 °C
Thermal conductivity	75.4 W/m °C
Coefficient of linear expansion	18 × 10^{-6}/ °C
Electrical conductivity	16–17% IACS (copper 100%)
Specific resistance	110–115 microhm mm
Young's modulus of elasticity	110 × 10^9 N/m^2
Impact 6% Sn 10% Pb cast alloy	26 J
Fatigue strength (10 × 10^8 cycles) 6% Sn 10% Pb alloy	±90 N/mm^2

Hot strength

Temperature °C	Tensile strength N/mm^2
95	240
150	210
205	200

The above properties are typical of the following group and may not apply exactly to any one specification. It is possible that with certain specifications some of the values may not be applicable.

General metallurgical characteristics

The term 'bronze' was originally confined to the series of copper–tin alloys, but has gradually been widened to include additions of other elements, such as phosphorus and lead.

The range of alloys covered by the terms 'phosphor bronze' and 'lead bronze' all contain up to approximately 10% tin and retain the characteristics of the true 'bronze' with additional properties supplied by these supplementary elements. This is not the case with aluminium, manganese or silicon bronzes, which never have appreciable additions of tin and have properties entirely different from tin bronze. Each of these groups appears under its own heading. To some extent the word 'bronze' is now taken to mean a copper alloy which is not based on the copper zinc series. When it is preceded by appreciable quantities of manganese, beryllium, aluminium, etc., none of these alloys contain tin.

In the true wrought bronzes, i.e. copper–tin alloys, the tin is usually present to between 5 and 8% and is soluble in the copper matrix, giving an alpha solid solution. This results in a stronger material than copper, with the ability to accept a considerable degree of cold work, thus enhancing the mechanical properties. Corrosion resistance is improved but there is a considerable drop in electrical and thermal conductivities.

Phosphorus acts as a de-oxidizer during casting and is invariably present up to about 0.5% when it increases the tensile strength and improves the cold working properties, with little effect on the ductility. With the higher phosphorus contents there is a hard constituent present which gives cast phosphor bronze, containing up to about 10% tin, good bearing properties. Lead may be added in small amounts to improve the machinability of the wrought alloys and a series of copper–tin–lead casting materials is used as high-duty bearings, particularly for reciprocating loads.

Additions of up to 10% zinc are made as an economical de-oxidizer, particularly in the cast alloys, where the zinc has no effect on the mechanical properties. As zinc and phosphorus both act as de-oxidizers they are not generally found together.

The copper–tin–zinc series are known as 'gunmetals', and find considerable use for their excellent casting properties and corrosion resistance, particularly in marine conditions. When lead up to 5% is added, a range of leaded gunmetals is formed which have better machining characteristics than the straight gunmetals and are cheaper to cast.

Alloys with more than 10% tin are hard and brittle, and find little industrial use apart from 'bell' and 'speculum' metal.

Bell metal has 20% tin, generally with antimony present. It has good casting but poor mechanical properties, with the advantage of a very sonorous note when struck. Copper alloys containing about 45% tin have a white lustrous colour and are capable of taking and retaining a high polish. These are the speculum metals, which have very poor mechanical properties but find some use because of their good reflectivity and are now electroplated successfully. These materials, after brasses, are the most popular of the copper alloys.

None of the materials can have their mechanical properties improved by any form of heat treatment.

Cold working achieved either by machining or hammering, etc., or misuse during service, can be removed normally at 300 °C but may require a temperature as high as 450–600 °C.

These materials are not normally welded although specific techniques are now available under highly controlled conditions where welding can be carried out. Very specialist advice and control is essential. Brazing and soldering can be carried out on these materials with no problems. Care must be taken to remove all evidence of flux following any brazing or soldering operations.

Very often the term 'weld' is used as a synonym for 'braze'.

These materials in the main, with some exceptions, have excellent corrosion resistance particularly against marine atmospheres. The presence of tin will almost invariably be an advantage with tin being the 'good fairy' against lead and zinc the 'bad fairies'.

Electroplating of these materials is unusual but should present no problems.

Their uses are diminishing because of the advent of austenitic stainless steel. Bronzes still find use as these bronzes are the easiest materials to cast and thus the high material cost can be offset against the ease of castability and thus the low cost of rejects. This very often eliminates or considerably reduces the cost of machining the finished product.

Where marine atmosphere is involved the higher the tin content the better. Tin bronzes are still commonly specified where high quality corrosion protection is required under marine conditions.

Many of these materials make excellent bearings, the phosphor bronzes and lead bronzes being very popular and quite cheap.

Speculum metal which is white is now a general replacement for pewter.

Speculum metal can be electroplated, thus producing an attractive non-toxic surface on components which has a reasonable resistance to tarnishing and can be readily tarnish protected. Speculum metal can also be used for reflectors.

Within this group are the materials which have unique properties for bells. These are easy to cast, have an attractive appearance with the correct shape and are sonorous. They can be readily tuned by machining.

Symbol	Nominal analysis, supplier, condition and remarks.
2.0921	8% Sn 0.4% Ni 1.0% Mn Cu alloy: Wrought; German Standard
2.1021	7% Sn 0.2% Pb Cu alloy: German Standard
2.1050	10% Sn 1.0% Pb 0.5% Zn Cu alloy: Casting; German Standard
2.1051	10% Sn Cu alloy: Casting; German Standard
2.1052	12% Sn 1.0% Pb 0.5% Zn Cu alloy: Casting; German Standard
2.1053	12% Sn 0.2% P Cu alloy: Wrought; German Standard
2.1056	14% Sn 1.0% Pb 0.5% Zn Cu alloy: Casting; German Standard
2.1057	14% Sn Cu alloy: Casting; German Standard
2.1086	10.5% Sn 1.0% Pb 2% Zn Cu alloy: Casting; German Standard
2.1087	10% Sn 2% Zn Cu alloy: Casting; German Standard
2.1090	7% Sn 6% Pb 4% Zn Cu alloy: Casting; German Standard **DPN: 80 UTS: 260 Elon: 19% Proof: 120**
2.1091	7% Sn 6% Pb 4.5% Zn Cu alloy: Casting; German Standard
2.1096	5% Sn 5% Pb 5% Zn Cu alloy: Casting; German Standard **DPN: 70 UTS: 250 Elon: 18% Proof: 110**
2.1097	5% Sn 5% Pb 5% Zn Cu alloy: Casting; German Standard
2.1166	3% Sn 26% Pb 2% Ni Cu alloy: Casting; German Standard
2.1170	10% Sn 5% Pb 1% Ni Cu alloy: Casting; German Standard
2.1171	10% Sn 5% Pb Cu alloy: Casting; German Standard
2.1176	10% Sn 10% Pb 1% Ni Cu alloy: Casting; German Standard
2.1177	10% Sn 10% Pb Cu alloy: Casting; German Standard
2.1182	8% Sn 15% Pb 1.5% Ni Cu alloy: Casting; German Standard
2.1183	8% Sn 15% Pb Cu alloy: Casting; German Standard
2.1188	5% Sn 20% Pb 2% Ni Cu alloy: Casting; German Standard
2.1189	4.5% Sn 21% Pb Cu alloy: Casting; German Standard
419	5% Sn 4% Zn Cu alloy: Wrought; designation agreed by CDAA; not used
502	1.2% Sn 0.04% P Cu alloy: Wrought; designation used by CDAA
505	1.5% Sn 0.3% P Cu alloy: Wrought; designation agreed by CDAA; not used
507	1.8% Sn 0.04% P Cu alloy: Wrought; designation agreed by CDAA; not used
508	3.0% Sn 0.04% P Cu alloy: Wrought; designation agreed by CDAA; not used
509	3.1% Sn 0.2% P Cu alloy: Wrought; designation agreed by CDAA; not used
518	5% Sn 0.3% P 0.01% Al Cu alloy: Wrought; designation agreed by CDAA; not used
524	10% Sn 0.2% P Cu alloy: Wrought; designation used by CDAA
532	4.7% Sn 3.2% Pb 0.3% P Cu alloy: Wrought; designation agreed by CDAA; not used
534	5% Sn 1.0% Pb 0.2% P Cu alloy: Wrought; designation agreed by CDAA; not used
546	4.0% Sn 4% Pb 3.0% Zn 0.4% P Cu alloy: Wrought; designation agreed by CDAA; not used
836	5% Sn 5% Pb 5% Zn Cu alloy: Casting; designation used by CDAA
838	4% Sn 6% Pb 7% Zn Cu alloy: Casting; designation used by CDAA
842	5% Sn 2% Pb 13% Zn Cu alloy: Casting; designation used by CDAA
844	3% Sn 7% Pb 9% Zn Cu alloy: Casting; designation used by CDAA
848	3% Sn 6% Pb 15% Zn Cu alloy: Casting designation used by CDAA
903	8% Sn 4% Zn Cu alloy: Casting; designation used by CDAA
905	10% Sn 2% Zn Cu alloy: Casting; designation used by CDAA
907	11% Sn 1.5% P Cu alloy: Casting; designation used by CDAA
910	15% Sn Cu alloy: Casting; designation used by CDAA
913	19% Sn Cu alloy: Casting; designation used by CDAA
915	10% Sn 2.5% Pb 3.5% Ni Cu alloy: Casting; designation used by CDAA
922	6% Sn 2% Pb 4% Zn Cu alloy: Casting; designation used by CDAA
923	8% Sn 1% Pb 4% Zn Cu alloy: Casting; designation used by CDAA
925	11% Sn 1% Pb 1.5% P 1.5% Ni Cu alloy: Casting; designation used by CDAA
927	10% Sn 2% Pb Cu alloy: Casting; designation used by CDAA
928	16% Sn 5% Pb Cu alloy: Casting; designation used by CDAA
932	7% Sn 7% Pb 3% Zn Cu alloy: Casting; designation used by CDAA
934	8% Sn 8% Pb Cu alloy: Casting; designation used by CDAA
935	5% Sn 9% Pb 1% Zn Cu alloy: Casting; designation used by CDAA
937	10% Sn 10% Pb Cu alloy: Casting; designation used by CDAA

Note. The following abbreviations and units are used in the tables:

DPN	Hardness, diamond pyramid number
UTS	Ultimate tensile strength, N/mm^2
Elon	Elongation, %
Proof	0.1% proof strength, N/mm^2

1 N/mm^2=0.1 hbar=0.102 kgf/mm^2=0.06475 $tonf/in.^2$=145.04 $lbf/in.^2$=1 MPa

See Appendix II for other abbreviations and conversion tables.

Symbol	Nominal analysis, supplier, condition and remarks.
938	7% Sn 15% Pb Cu alloy: Casting; designation used by CDAA
939	6% Sn 16% Pb Cu alloy: Casting; designation used by CDAA
940	13% Sn 15% Pb Cu alloy: Casting; designation used by CDAA
941	5% Sn 20% Pb Cu alloy: Casting; designation used by CDAA
943	5% Sn 25% Pb Cu alloy: Casting; designation used by CDAA
947	5% Sn 2% Zn 5% Ni Cu alloy: Casting; designation used by CDAA
948	5% Sn 1% Pb 2% Zn 5% Ni Cu alloy: Casting; designation used by CDAA
A 25	Bronze welding rod; high tensile; low fuming; Delta Metals
AAPHOS	10% Sn 0.25% Pb 0.5% P Cu alloy: Casting; Anti Attrition for BS alloy PB 1
ACID BRONZE	Range of leaded nickel-copper-tin bronzes 9% Sn 9.5% Pb 0.75% Ni Cu alloy: Common name
ADMIRALTY GUNMETAL	10% Sn 2% Zn Cu alloy: Common name
ALLEN'S METAL	5% Sn 40% Pb Cu alloy
AMOLCTC 2	10% Sn 0.25% Pb 0.5% P Cu alloy: Continuous cast; Eyre Smelting for BS alloy PB 1; free machining **DPN: 75 UTS: 280 Elon: 25%**
AMPCOLOY 31	8% Sn 4% Zn Cu alloy: Casting; Ampco
AMPCOLOY 32	10% Sn 10% Pb Cu alloy: Casting; Ampco
AMPCOLOY 35	7% Sn 7% Pb 3% Zn Cu alloy: Casting; Ampco
AMPCOLOY 38	8% Sn 4% Zn Cu alloy: Casting; Ampco
AMPCOLOY 54	Sn P Cu alloy: For gears; Ampco
AMPCOLOY 71	6% Sn 1.5% Pb 4.5% Zn Cu alloy: Casting
AMPCOLOY 72	8% Sn 4% Zn Cu alloy: Casting; Ampco
AMPCOLOY 74	5% Sn 5% Pb 5% Zn Cu alloy: Casting; Ampco
AMPCOLOY 79	10% Sn 2% Zn Cu alloy: Casting; Ampco
AMPCOLOY 711	11% Sn 0.5% Pb 0.2% P Cu alloy: Casting; Ampco
AMPCOLOY 712	Sn P Cu alloy: For gears; Ampco
AMPCOLOY 715	Sn P Cu alloy: For gears; Ampco
AMS 4510 C	5% Sn P Cu alloy: Sheet and strip; hard drawn
AMS 4520 E	4% Sn 3% Zn Cu alloy: Strip
AMS 4625 D	5% Sn P Cu alloy: Bar and tube; ½-hard
AMS 4720 B	5% Sn P Cu alloy: Wire; spring hard
AMS 4824	1.2% Sn 24% Pb Cu alloy: Casting
AMS 4827	10% Sn 9.5% Pb Cu alloy: Casting
AMS 4840 A	5.5% Sn 24.5% Pb Cu alloy: Casting
AMS 4842 A	10% Sn 10% Pb Cu alloy: Casting
AMS 4845 D	10% Sn 2% Zn Cu alloy: Casting
AMS 4846 A	11% Sn 2% Zn Cu alloy: Casting
AMS 4855 B	5% Sn 5% Pb 5% Zn Cu alloy: Casting
AMS 7320	16% Sn 5% Pb Cu alloy: Casting
AMS 7322	19% Sn 1% P (max) Cu alloy: Casting
ARIEL BRONZE	10% Sn 0.25% Pb 0.5% P Cu alloy: Casting; Eyre Smelting for BS alloy PB 1
ARTIC BRONZE	Series of leaded bearing bronzes; grain structure produced by rapid cooling in (chill cast) metal moulds; origin unknown
ASTM B22 A	19% Sn 0.25% Pb Cu alloy: Casting; deformation limit of 170 N/mm^2; minimum specified
ASTM B22 B	16% Sn 0.25% Pb 1% P Cu alloy: Casting; deformation limit of 120 N/mm^2; minimum specified
ASTM B22 C	10% Sn 9.5% Pb 0.1% P 1% Ni Cu alloy: Casting
ASTM B22 D	10% Sn 0.3% Pb 2.5% Zn 1% Ni Cu alloy: Casting; as cast **UTS: 260 Elon: 20% Proof: 100**
ASTM B30/1 A	10% Sn 2% Zn Cu alloy: Ingot 'G' Bronze; as cast **UTS: 260 Elon: 20%**
ASTM B30/1 B	8% Sn 4% Zn Cu alloy: Ingot 'G' Bronze; as cast **UTS: 260 Elon: 20%**
ASTM B30/2 A	6% Sn 1.5% Pb 4.5% Zn Cu alloy: Ingot Navy 'M'; as cast **UTS: 240 Elon: 30%**
ASTM B30/2 B	8.5% Sn 0.5% Pb 4% Zn Cu alloy: Ingot 'Navy PC'; as cast **UTS: 240 Elon: 18%**
ASTM B30/3 A	10% Sn 10% Pb Cu alloy: Ingot; as cast **UTS: 220 Elon: 22%**
ASTM B30/3 B	7% Sn 7% Pb 3% Zn Cu alloy: Ingot; as cast **UTS: 200 Elon: 12%**
ASTM B30/3 C	5% Sn 9% Pb 1% Zn Cu alloy: Ingot; as cast **UTS: 170 Elon: 8%**
ASTM B30/3 D	7% Sn 15% Pb Cu alloy: Ingot; as cast **UTS: 170 Elon: 10%**
ASTM B30/4 A	5% Sn 5% Pb 5% Zn Cu alloy: Ingot; as cast **UTS: 210 Elon: 22%**
ASTM B30/4 B	4% Sn 6% Pb 7% Zn Cu alloy: Ingot; as cast **UTS: 200 Elon: 15%**
ASTM B30/5 A	3% Sn 7% Pb 9% Zn Cu alloy: Ingot; as cast **UTS: 200 Elon: 18%**
ASTM B30/5 B	2.5% Sn 6.5% Pb 15% Zn Cu alloy: Ingot; as cast **UTS: 170 Elon: 15%**
ASTM B30/6 A	1% Sn 3% Pb 24% Zn Cu alloy: Ingot; as cast **UTS: 240 Elon: 25%**
ASTM B30/6 B	1% Sn 3% Pb 29% Zn Cu alloy: Ingot; as cast **UTS: 200 Elon: 20%**
ASTM B61	6% Sn 1.5% Pb 4% Zn 1% Ni (max) Cu alloy: Casting; as cast **UTS: 220 Elon: 22% Proof: 90**
ASTM B62	5% Sn 5% Pb 5% Zn 1% Ni (max) Cu alloy: Casting; as cast **UTS: 200 Elon: 20% Proof: 75**
ASTM B66/HARD BRONZE	7.5% Sn 13% Pb Cu alloy: Casting for locomotives
ASTM B66/ MEDIUM BRONZE	7% Sn 19% Pb Cu alloy: Casting for locomotives
ASTM B66/PHOS BRONZE	8% Sn 10.5% Pb 0.2% P Cu alloy: Casting for locomotives
ASTM B66/SOFT BRONZE	5% Sn 25% Pb Cu alloy: Casting for locomotives
ASTM B100/1	4.6% Sn 0.2% P Cu alloy: Plate and sheet; as rolled or annealed **DPN: 130 UTS: 420 Elon: 10%**
ASTM B103/A	4.5% Sn 0.2% P Cu alloy: Sheet and bar; annealed **UTS: 360**
ASTM B103 A	4.5% Sn 0.2% P Cu alloy: Sheet and bar; extra spring hard **UTS: 760**
ASTM B103 B	4.5% Sn 3% Pb 0.2% P Cu alloy: Sheet and bar; annealed **UTS: 370**
ASTM B103 B	4.5% Sn 3% Pb 0.2% P Cu alloy: Sheet and bar; extra spring hard **UTS: 750**
ASTM B103 B1	4.2% Sn 1% Pb 0.2% P Cu alloy: Sheet and bar annealed **UTS: 370**
ASTM B103 B1	4.2% Sn 1% Pb 0.2% P Cu alloy: Sheet and bar; extra spring hard **UTS: 750**

Note. The following abbreviations and units are used in the tables:

DPN	Hardness, diamond pyramid number
UTS	Ultimate tensile strength, N/mm^2
Elon	Elongation, %
Proof	0.1% proof strength, N/mm^2

1 N/mm^2=0.1 hbar=0.102 kgf/mm^2=0.06475 tonf/in.^{2}=145.04 lbf/in.^{2}=1 MPa

See Appendix II for other abbreviations and conversion tables.

Symbol	Nominal analysis, supplier, condition and remarks.
ASTM B103 C	8% Sn 0.2% P Cu alloy: Sheet and bar; annealed **UTS: 420**
ASTM B103 C	8% Sn 0.2% P Cu alloy: Sheet and bar; extra spring hard **UTS: 810**
ASTM B103 D	10% Sn 0.2% P Cu alloy: Sheet and bar; annealed **UTS: 390**
ASTM B103 D	10% Sn 0.2% P Cu alloy: Sheet and bar; extra spring hard **UTS: 930**
ASTM B139 A	4% Sn 0.3% P Cu alloy: Bar; annealed **UTS: 370**
ASTM B139 A	4% Sn 0.3% P Cu alloy: Bar; spring hard **UTS: 730 Elon: 5%**
ASTM B139 B1	4% Sn 1% Pb 0.3% P Cu alloy: Bar; hard **UTS: 370 Elon: 12%**
ASTM B139 B2	4% Sn 4% Pb 3% Zn 0.4% P Cu alloy: Bar; hard **UTS: 370 Elon: 12%**
ASTM B139 C	8% Sn 0.3% P Cu alloy: Bar; annealed **UTS: 420**
ASTM B139 C	8% Sn 0.3% P Cu alloy: Bar; hard **UTS: 460 Elon: 10%**
ASTM B139 D	10% Sn 0.3% P Cu alloy: Bar; annealed **UTS: 480**
ASTM B139 D	10% Sn 0.3% P Cu alloy: Bar; hard **UTS: 560 Elon: 10%**
ASTM B143/1 A	10% Sn 2% Zn Cu alloy: Casting; as cast **UTS: 260 Elon: 20% Proof: 100**
ASTM B143/1 B	8% Sn 4% Zn Cu alloy: Casting; as cast **UTS: 260 Elon: 20% Proof: 100**
ASTM B143/2 A	6% Sn 1.5% Pb 4.5% Zn Cu alloy: Casting; as cast **UTS: 220 Elon: 22% Proof: 90**
ASTM B143/2 B	8% Sn 1% Pb 4% Zn Cu alloy: Casting; as cast **UTS: 240 Elon: 18% Proof: 90**
ASTM B144/3 A	10% Sn 10% Pb Cu alloy: Casting; as cast **UTS: 170 Elon: 8% Proof: 60**
ASTM B144/3 B	7% Sn 7% Pb 3% Zn Cu alloy: Casting; as cast **UTS: 200 Elon: 12% Proof: 75**
ASTM B144/3 C	5% Sn 9% Pb 1% Zn Cu alloy: Casting; as cast **UTS: 170 Elon: 8% Proof: 60**
ASTM B144/3 D	7% Sn 15% Pb Cu alloy: Casting; as cast **UTS: 170 Elon: 10% Proof: 75**
ASTM B144/3 E	5% Sn 25% Pb Cu alloy: Casting; as cast **UTS: 140 Elon: 7%**
ASTM B145/4 A	5% Sn 5% Pb 5% Zn Cu alloy: Casting **UTS: 200 Elon: 20% Proof: 75**
ASTM B145/4 B	4% Sn 6% Pb 7% Zn Cu alloy: Casting **UTS: 200 Elon: 15% Proof: 60**
ASTM B145/5 A	3% Sn 7% Pb 9% Zn Cu alloy: Casting **UTS: 200 Elon: 18% Proof: 60**
ASTM B145/5 B	3% Sn 6% Pb 15% Zn Cu alloy: Casting **UTS: 170 Elon: 15% Proof: 60**
ASTM B146/6 A	1% Sn 3% Pb 24% Zn Cu alloy: Casting **UTS: 240 Elon: 25% Proof: 60**
ASTM B146/6 B	1% Sn 3% Pb 29% Zn Cu alloy: Casting **UTS: 200 Elon: 20% Proof: 60**
ASTM B146/6 C	1% Sn 1% Pb 37% Zn Cu alloy: Casting **UTS: 260 Elon: 15% Proof: 75**
ASTM B147/7 A	0.7% Sn 0.7% Pb 35% Zn 0.7% Al 1% Fe 0.2% Mn Cu alloy: Casting **UTS: 420 Elon: 15% Proof: 120**
ASTM B159 A	5% Sn 0.3% P Cu alloy: Wire; annealed **UTS: 330**
ASTM B159 A	5% Sn 0.3% P Cu alloy: Wire; hard **UTS: 880**
ASTM B159 C	8% Sn 0.3% P Cu alloy: Wire; annealed **UTS: 400**
ASTM B159 C	8% Sn 0.3% P Cu alloy: Wire; hard **UTS: 1000**
ASTM B159 D	10% Sn 0.3% P Cu alloy: Wire; annealed **UTS: 480**
ASTM B159 D	10% Sn 0.3% P Cu alloy: Wire; hard **UTS: 1080**
ASTM B176 Z 30 A	1.5% Sn 1.5% Pb 30% Zn Cu alloy: Casting; die cast **UTS: 330 Elon: 10% Proof: 150**
ASTM B225 E Cu	1% Sn 98% Cu alloy: Welding electrode **UTS: 170**
ASTM B225 E Cu Si	1.5% Sn 3% Si Cu alloy: Welding electrode **UTS: 340**
ASTM B225 E Cu Sn A	5% Sn 0.3% P Cu alloy: Welding electrode **UTS: 240**
ASTM B225 E Cu Sn C	8% Sn 0.3% P Cu alloy: Welding electrode **UTS: 260**
ASTM B259 R Cu	1% Sn 98% Cu alloy: Welding rod **UTS: 150**
ASTM B259 R Cu Sn A	5% Sn 0.3% P Cu alloy: Welding rod **UTS: 220**
ASTM B427/908	12% Sn Cu alloy: Casting for gears **DPN: 95 UTS: 350 Elon: 12%**
ASTM B427/915	10% Sn 2.7% Pb 3.2% Ni Cu alloy: Castings for gears **DPN: 75 UTS: 310 Elon: 8% Proof: 170**
ASTM B427/916	10.2% Sn 1.75% Ni Cu alloy: Castings for gears **DPN: 65 UTS: 320 Elon: 10% Proof: 180**
ASTM B427/917	12% Sn 1.75% Ni Cu alloy: Casting for gears **DPN: 85 UTS: 320 Elon: 12% Proof: 280**
ASTM B508/505	1.3% Sn Cu alloy: Strip for flexible piping
ATLAS CTC 1	10% Sn 0.25% Pb 0.5% P Cu alloy: Continuous cast; Eyre Smelting for BS alloy PB 1 **DPN: 95 UTS: 360 Elon: 17%**
B 1	High Sn Cu alloy: Casting for high duty bearings; Anti-Attrition **DPN: 103 UTS: 240 Elon: 2%**
B 2	Sn Cu alloy: Casting for thrust bearings; Anti-Attrition **DPN: 120 UTS: 240 Elon: 1%**
B 6	Sn Cu alloy: Casting for unlined bearings; Anti-Attrition **DPN: 95 UTS: 220 Elon: 10%**
B 9	10% Sn 0.25% Pb 0.5% P Cu alloy: Casting; Anti-Attrition for BS alloy PB 1
B 13	Sn P Cu alloy: Casting for rolling mill bearings; Anti-Attrition **DPN: 92 UTS: 220 Elon: 6%**
B 14	12% Sn 0.15% P Cu alloy: Casting for worm wheels; Anti-Attrition for BS alloy PB 2
B 18	Sn Cu alloy: Casting; electrical bronze; Anti-Attrition; for good conductivity fittings **DPN: 75 UTS: 260 Elon: 30%**
B 20	5% Sn 5% Pb 5% Zn 2% Ni Cu alloy: Casting; Anti-Attrition for BS alloy LG 2
B 32	5% Sn 20% Pb 1% Zn 0.1% P 2% Ni Cu alloy: Casting; Anti-Attrition for BS alloy LB 5
B 38	Sn Cu alloy: Brazing metal; Anti-Attrition
B 39	10% Sn 1.5% Pb 2% Zn Cu alloy: Casting; Anti-Attrition for BS alloy G 1
B 40	8% Sn 1.5% Pb 4% Zn Cu alloy: Casting; Anti-Attrition for BS alloy G2
B 45	Sn Pb Cu alloy: Casting; gunmetal; Anti-Attrition **DPN: 60 UTS: 120 Elon: 14%**
B 60	Copper–tin sintered material: Sintered Products Ltd; 30–33% porosity **UTS: 70 Elon: 2.5%**
B 65	Copper–tin sintered material: Sintered Products Ltd; 25–30% porosity **UTS: 750 Elon: 2.5%**
B 70	Copper–tin sintered material: Sintered Products Ltd; 22–25% porosity **UTS: 90 Elon: 3%**
B 80	Copper–tin sintered material: Sintered Products Ltd; 10–15% porosity **UTS: 100 Elon: 4%**

Symbol	Nominal analysis, supplier, condition and remarks.
B 80	Sintered bronze: 22% porosity; specific gravity 7.1; Durasint **DPN: 50 UTS: 140 Elon: 5%**
B 80/1	Sintered graphite bronze: 18% porosity; specific gravity 710; Durasint **DPN: 60 UTS: 140 Elon: 6%**
B 85	Sintered bronze: 7% porosity; specific gravity 8.2; Durasint **DPN: 70 UTS: 350 Elon: 8%**
B 85/P	Copper–tin sintered material: Sintered Products Ltd; 4–8% porosity; withdrawn **UTS: 180 Elon: 20%**
B 90/P	Copper–tin sintered material: Sintered Products Ltd; 1–3% porosity; withdrawn **UTS: 300 Elon: 35%**
B O H T BRONZE	11.5% Sn 0.7% P Cu alloy: Casting; spuncast; I Holroyd; heat treated version of BO Bronze **DPN: 95 UTS: 320 Elon: 20% Proof: 170**
BEARING BRONZE	11% Sn 0.2% Pb Cu alloy: Cast; origin unknown
BELL METAL	22.5% Sn Cu alloy: Used for bells or as bearing bronze; origin unknown
BO BRONZE	11.5% Sn 0.7% P Cu alloy: Casting; spuncast; I Holroyd for BS alloy PB 1 and PB 2; as cast **DPN: 100 UTS: 290 Elon: 3% Proof: 210**
BOLTOMET 10	5% Sn Cu alloy: Previous name for Bolt 317; Thomas Bolton; withdrawn
BOLTOMET 11	6% Sn Cu alloy: Previous name for Bolt 320; Thomas Bolton; withdrawn
BOLTOMET 12	3% Sn Cu alloy: Previous name for Bolt 312; Thomas Bolton; withdrawn
BOLTOMET 15	8% Sn Cu alloy: Previous name for Bolt 338; Thomas Bolton; withdrawn
BOLTOMET 16	7% Sn Cu alloy: Previous name for Bolt 327; Thomas Bolton; withdrawn
BOLTOMET 58	Sn Pb Cu alloy: Previous name for Bolt 445; Thomas Bolton; withdrawn
BOLTOMET 305	1% Sn Cu alloy: Strip; hard; for flexible tubing; Thomas Bolton **DPN: 145 UTS: 48 Elon: 8%**
BOLTOMET 307	1.2% Sn 0.3% P Cu alloy: Rod and strip; annealed; Thomas Bolton **DPN: 65 UTS: 260 Elon: 50% Proof: 75**
BOLTOMET 307	1.2% Sn 0.3% P Cu alloy: Rod; hard; Thomas Bolton **DPN: 140 UTS: 400 Elon: 22% Proof: 360**
BOLTOMET 307	1.2% Sn 0.03% P Cu alloy: Wire and strip; Thomas Bolton **DPN: 182 UTS: 560 Elon: 3% Proof: 480**
BOLTOMET 309	2.5% Sn 0.04% P Cu alloy: Rod and strip: annealed; Thomas Bolton **DPN: 75 UTS: 300 Elon: 50% Proof: 90**
BOLTOMET 309	2.5% Sn 0.04% P Cu alloy: Hard; Thomas Bolton **DPN: 157 UTS: 450 Elon: 21% Proof: 390**
BOLTOMET 309	2.5% Sn 0.04% P Cu alloy: Strip; hard; Thomas Bolton **DPN: 200 UTS: 600 Elon: 4% Proof: 550**
BOLTOMET 312	3.5% Sn 0.07% P Cu alloy: Rod and strip; annealed; Thomas Bolton **DPN: 70 UTS: 300 Elon: 55% Proof: 90**
BOLTOMET 312	3.5% Sn 0.07% P Cu alloy: Rod; hard; Thomas Bolton **DPN: 180 UTS: 530 Elon: 18% Proof: 460**
BOLTOMET 312	3.5% Sn 0.07% P Cu alloy: Strip; hard; Thomas Bolton **DPN: 220 UTS: 700 Elon: 3% Proof: 600**
BOLTOMET 317	5% Sn 0.15% P Cu alloy: Rod and strip; annealed; Thomas Bolton **DPN: 80 UTS: 330 Elon: 60% Proof: 120**
BOLTOMET 317	5% Sn 0.15% P Cu alloy: Rod; hard; Thomas Bolton **DPN: 210 UTS: 590 Elon: 20% Proof: 460**
BOLTOMET 317	5% Sn 0.15% P Cu alloy: Strip; hard; Thomas Bolton **DPN: 225 UTS: 730 Elon: 3% Proof: 630**
BOLTOMET 319	6.25% Sn 0.1% P Cu alloy: Bar; hard drawn; Thomas Bolton **DPN: 170 UTS: 550 Elon: 19% Proof: 470**
BOLTOMET 320	6.3% Sn 0.25% P Cu alloy: Strip; annealed; Thomas Bolton **DPN: 82 UTS: 340 Elon: 68% Proof: 120**
BOLTOMET 320	6.3% Sn 0.25% P Cu alloy: Strip; hard; used for springs; Thomas Bolton **DPN: 241 UTS: 780 Elon: 3% Proof: 690**
BOLTOMET 327	6.8% Sn 0.25% P Cu alloy: Wire; hard; Thomas Bolton for BS 384 **UTS: 830**
BOLTOMET 338	7.8% Sn 0.3% P Cu alloy: For rod and strip; annealed; Thomas Bolton **DPN: 90 UTS: 370 Elon: 75% Proof: 140**
BOLTOMET 338	7.8% Sn 0.3% P Cu alloy: For rod; hard; Thomas Bolton **DPN: 232 UTS: 660 Elon: 20% Proof: 510**
BOLTOMET 338	7.8% Sn 0.3% P Cu alloy: For strip; hard; Thomas Bolton **DPN: 243 UTS: 800 Elon: 4% Proof: 700**
BOLTOMET 349	10% Sn 0.15% P Cu alloy: Strip; annealed; Thomas Bolton **DPN: 97 UTS: 420 Elon: 70% Proof: 150**
BOLTOMET 349	10% Sn 0.15% P Cu alloy: Strip; hard; Thomas Bolton **DPN: 248 UTS: 820 Elon: 4% Proof: 730**
BOLTOMET 356	2.0% Sn 0.05% P Cu alloy: Strip; hard; Thomas Bolton **DPN: 167 UTS: 500 Elon: 4% Proof: 470**
BOLTOMET 445	5% Sn 1% Pb 0.15% P Cu alloy: Rod; annealed; free machining; Thomas Bolton **DPN: 85 UTS: 330 Elon: 55% Proof: 140**
BOLTOMET 445	5% Sn 1% Pb 0.15% P Cu alloy: Rod; hard; free machining; Thomas Bolton **DPN: 160 UTS: 500 Elon: 17% Proof: 360**
BRONZE 44	8% Sn 15% P Cu alloy: Welding electrode; Murex Ltd
BRONZE 66	8% Sn 0.15% P Cu alloy: Welding electrode; Murex Ltd
BRONZETRODE 94.15	7.40% Sn 0.20% P 0.50% Mn Cu alloy: Electrodes; ESAB **UTS: 463 Elon: 28%**
BS 99	5% Sn 2% Pb 8% Zn Cu alloy: Covers all forms of screwed fittings **UTS: 180 Elon: 15%**
BS 352	Phosphor bronze turbine blading; obsolete
BS 369 PB102	5% Sn 0.25% P Cu alloy: Phosphor bronze bar; as rolled **UTS: 450 Elon: 20% Proof: 330**
BS 382/3	Bronze; gunmetal; ingots and castings; replaced by BS 1400
BS 384	5.7% Sn 0.02% Pb (max) 0.2% P Cu alloy: Phosphor bronze wire; cold drawn for general engineering **UTS: 830**
BS 407 PB101	4% Sn 0.3% P Cu alloy: Sheet and strip; annealed **DPN: 80 UTS: 290 Elon: 40%**
BS 407 PB101	4% Sn 0.2% P Cu alloy: Sheet and strip; ¼–hard **DPN: 100 UTS: 330 Elon: 30% Proof: 100**
BS 407 PB101	4% Sn 0.2% P Cu alloy: Sheet and strip; ½–hard **DPN: 150 UTS: 450 Elon: 8% Proof: 290**
BS 407 PB101	4% Sn 0.3% P Cu alloy: Sheet and strip; cold rolled **DPN: 180 UTS: 530 Elon: 4% Proof: 370**

Note. The following abbreviations and units are used in the tables:

DPN	Hardness, diamond pyramid number
UTS	Ultimate tensile strength, N/mm^2
Elon	Elongation, %
Proof	0.1% proof strength, N/mm^2

$1\ N/mm^2 = 0.1\ hbar = 0.102\ kgf/mm^2 = 0.06475\ tonf/in.^2 = 145.04\ lbf/in.^2 = 1\ MPa$

See Appendix II for other abbreviations and conversion tables.

Symbol	Nominal analysis, supplier, condition and remarks.
BS 407 PB101	4% Sn 0.2% P Cu alloy: Sheet and strip; extra hard **DPN: 190 UTS: 600 Proof: 450**
BS 407 PB102	5% Sn 0.2% P Cu alloy: Sheet and strip; annealed **DPN: 85 UTS: 300 Elon: 45%**
BS 407 PB102	5% Sn 0.2% P Cu alloy: Sheet and strip; ¼-hard **DPN: 110 UTS: 340 Elon: 35% Proof: 120**
BS 407 PB102	5% Sn 0.2% P Cu alloy: Sheet and strip; ½-hard **DPN: 160 UTS: 480 Elon: 10% Proof: 330**
BS 407 PB102	5% Sn 0.2% P Cu alloy: Sheet and strip; hard **DPN: 180 UTS: 580 Elon: 4% Proof: 420**
BS 407 PB102	5% Sn 0.2% P Cu alloy: Sheet and strip; extra hard **DPN: 210 UTS: 640 Proof: 480**
BS 407 PB103	7% Sn 0.2% P Cu alloy: Sheet and strip; spring hard; up to 15 cm wide 0.056 cm thick **DPN: 220**
BS 407 PB103	7% Sn 0.2% P Cu alloy: Sheet and strip; extra spring hard; up to 15 cm wide 0.056 cm thick **DPN: 240**
BS 407 PB103	7% Sn 0.2% P Cu alloy: Sheet and strip; hard **DPN: 200 UTS: 600 Elon: 6% Proof: 450**
BS 407 PB103	7% Sn 0.2% P Cu alloy: Sheet and strip; extra hard **DPN: 215 UTS: 690 Proof: 530**
BS 407 PB103	7% Sn 0.2% P Cu alloy: Sheet and strip; annealed **DPN: 90 UTS: 330 Elon: 50%**
BS 407 PB103	7% Sn 0.2% P Cu alloy: Sheet and strip; ¼-hard **DPN: 115 UTS: 370 Elon: 40% Proof: 180**
BS 407 PB103	7% Sn 0.2% P Cu alloy: Sheet and strip; ½-hard **DPN: 170 UTS: 520 Elon: 12% Proof: 360**
BS 421	Phosphor bronze castings; replaced by BS 1400
BS 897	Leaded gunmetal casting; replaced by BS 1400
BS 898	Leaded gunmetal casting; replaced by BS 1400
BS 960	Leaded bronze casting and ingot; replaced by BS 1400
BS 961	Leaded bronze casting and ingot; replaced by BS 1400
BS 962	Leaded bronze casting and ingot; replaced by BS 1400
BS 963	Leaded bronze casting and ingot; replaced by BS 1400
BS 964	Leaded bronze casting and ingot; replaced by BS 1400
BS 965	Leaded bronze casting and ingot; replaced by BS 1400
BS 1021	Copper alloy: Ingot and casting; replaced by BS 1400
BS 1022	Copper alloy: Ingot and casting; replaced by BS 1400
BS 1023	Copper alloy: Ingot and casting; replaced by BS 1400
BS 1024	Copper alloy: Ingot and casting; replaced by BS 1400
BS 1025	Copper alloy: Ingot and casting; replaced by BS 1400
BS 1026	Copper alloy: Ingot and casting; replaced by BS 1400
BS 1027	Copper alloy: Ingot and casting; replaced by BS 1400
BS 1028	Copper alloy: Ingot and casting; replaced by BS 1400
BS 1058	Phosphor bronze: Ingot and casting; replaced by BS 1400
BS 1059	Phosphor bronze: Ingot and casting; replaced by BS 1400
BS 1060	Phosphor bronze: Ingot and casting; replaced by BS 1400
BS 1061	Phosphor bronze: Ingot and casting; replaced by BS 1400
BS 1158	Copper alloy: Ingot and casting; replaced by BS 1400
BS 1159	Copper alloy: Ingot and casting; replaced by BS 1400
BS 1400 G1/1	10% Sn 2.25% Zn Cu alloy: Gunmetal ingot; sand cast **UTS: 240 Elon: 15%**
BS 1400 G1 C	10% Sn 2% Zn Cu alloy: Gunmetal casting; chill cast **UTS: 220 Elon: 3% Proof: 120**
BS 1400 G2/1	8% Sn 4.25% Zn Cu alloy: Gunmetal ingot; sand cast **UTS: 240 Elon: 15%**
BS 1400 G2 C	8% Sn 4% Zn Cu alloy: Gunmetal casting; chill cast **UTS: 210 Elon: 3% Proof: 120**
BS 1400 G3/1	7% Sn 2.5% Zn 5.5% Ni Cu alloy: Ni gunmetal ingot; sand cast **UTS: 270 Elon: 18%**
BS 1400 G3 C	7% Sn 0.4% Pb 2.2% Zn 5.5% Ni Cu alloy: Casting; sand cast; as cast **UTS: 270 Elon: 18% Proof: 140**
BS 1400 G3 WP	7% Sn 0.4% Pb 2.2% Zn 5.5% Ni Cu alloy: Casting **DPN: 160 UTS: 420 Elon: 3% Proof: 270**
BS 1400 LB1/1	9% Sn 15% Pb Cu alloy: Lead bronze ingot; sand cast **UTS: 170 Elon: 4%**
BS 1400 LB1 C	9% Sn 15% Pb Cu alloy: Lead bronze casting; chill cast **UTS: 200 Elon: 3% Proof: 120**
BS 1400 LB2/1	10% Sn 10% Pb Cu alloy: Lead bronze ingot; sand cast **UTS: 180 Elon: 5%**
BS 1400 LB2 C	10% Sn 10% Pb Cu alloy: Lead bronze casting; chill cast **UTS: 210 Elon: 3% Proof: 140**
BS 1400 LB3/1	10% Sn 5% Pb Cu alloy: Lead bronze ingot; sand cast **UTS: 180 Elon: 5%**
BS 1400 LB3 C	10% Sn 5% Pb Cu alloy: Lead bronze casting; chill cast **UTS: 200 Elon: 3% Proof: 140**
BS 1400 LB4/1	5% Sn 10% Pb Cu alloy: Lead bronze ingot; sand cast **UTS: 150 Elon: 8%**
BS 1400 LB4 C	5% Sn 10% Pb Cu alloy: Lead bronze casting; chill cast **UTS: 200 Elon: 5% Proof: 75**
BS 1400 LB5/1	5% Sn 21% Pb Cu alloy: Lead bronze ingot; sand cast **UTS: 150 Elon: 6%**
BS 1400 LB5 C	5% Sn 21% Pb Cu alloy: Lead bronze casting; chill cast **UTS: 170 Elon: 6% Proof: 75**
BS 1400 LG1/1	3% Sn 5% Pb 9% Zn Cu alloy: Lead gunmetal ingot; sand cast **UTS: 170 Elon: 12%**
BS 1400 LG1 C	3% Sn 4.5% Pb 8.5% Zn Cu alloy: Lead gunmetal casting; sand cast **UTS: 170 Elon: 12%**
BS 1400 LG2/1	5% Sn 5% Pb 5.5% Zn Cu alloy: Lead gunmetal ingot; sand cast **UTS: 200 Elon: 15%**
BS 1400 LG2 C	5% Sn 5% Pb 5% Zn Cu alloy: Lead gunmetal casting; chill cast **UTS: 200 Elon: 7% Proof: 100**
BS 1400 LG3/1	7% Sn 2% Pb 4.5% Zn Cu alloy: Lead gunmetal bronze ingot; sand cast **UTS: 210 Elon: 12%**
BS 1400 LG3 C	7% Sn 2% Pb 4% Zn Cu alloy: Lead gunmetal bronze casting; chill cast **UTS: 220 Elon: 5% Proof: 110**
BS 1400 LG4/1	7% Sn 3% Pb 2% Zn Cu alloy: Lead gunmetal bronze ingot; sand cast **UTS: 240 Elon: 18%**
BS 1400 LPB1/1	7.5% Sn 3.5% Pb 0.4% P Cu alloy: Ingot; sand cast; leaded phosphor bronze **UTS: 180 Elon: 3%**
BS 1400 LPB1 C	7.5% Sn 3.5% Pb 0.3% P Cu alloy: Casting; chill cast; leaded phosphor bronze **UTS: 210 Elon: 2% Proof: 120**
BS 1400 PB1/1	10% Sn 0.6% P Cu alloy: Phosphor bronze casting; sand cast **UTS: 210 Elon: 3%**
BS 1400 PB1 C	10% Sn 0.5% P Cu alloy: Phosphor bronze casting; chill cast **UTS: 300 Elon: 2% Proof: 170**
BS 1400 PB2/1	12% Sn 0.25% P Cu alloy: Phosphor bronze ingot; sand cast **UTS: 210 Elon: 5%**
BS 1400 PB2 C	12% Sn 0.15% P Cu alloy: Phosphor bronze casting; chill cast **UTS: 240 Elon: 3% Proof: 170**
BS 1400 PB3/1	9.5% Sn 0.25% P Cu alloy: Phosphor bronze ingot; sand cast **UTS: 220 Elon: 7%**
BS 1400 PB3 C	9.5% Sn 0.25% P Cu alloy: Phosphor bronze casting; chill cast **UTS: 220 Elon: 7% Proof: 120**
BS 1400 PB4/1	9.5% Sn 0.6% P Cu alloy: Phosphor bronze ingot; sand cast **UTS: 180 Elon: 3%**

Symbol	Nominal analysis, supplier, condition and remarks.
BS 1400 PB4 C	9.5% Sn 0.5% P Cu alloy: Phosphor bronze casting; chill cast **UTS: 240 Elon: 2% Proof: 140**
BS 1400 SCB1/1	2.5% Sn 3.5% Pb 25% Zn Cu alloy: Brass ingot; sand cast **UTS: 170 Elon: 20%**
BS 1400 SCB1 C	2.5% Sn 3.5% Pb 25% Zn Cu alloy: Brass casting; sand cast **UTS: 170 Elon: 20%**
BS 1400 SCB4/1	1.2% Sn 40% Zn Cu alloy: Naval brass ingot; sand cast **UTS: 160 Elon: 20%**
BS 1400 SCB4 C	1.2% Sn 40% Zn Cu alloy: Naval brass casting; sand cast **UTS: 160 Elon: 20% Proof: 450**
BS 1400 SCB5/1	1.5% Sn 10% Zn Cu alloy: Brass ingot; sand cast for brazed parts in contact with salt water **UTS: 180 Elon: 20%**
BS 1400 SCB5 C	1.5% Sn 10% Zn Cu alloy: Brass casting; sand cast for brazed parts in contact with salt water **UTS: 180 Elon: 20%**
BS 2061	5% Sn 0.3% P Cu alloy: Spring washers; specification is for spring washers **DPN: 200**
BS 2870 PB101	4% Sn 0.3% P Cu alloy: Sheet; hard **DPN: 170 UTS: 520 Elon: 4% Proof: 340**
BS 2870 PB101	4% Sn 0.3% P Cu alloy: Sheet; extra hard **DPN: 190 UTS: 600 Proof: 450**
BS 2870 PB101	4% Sn 0.3% P Cu alloy: Sheet; annealed **DPN: 80 UTS: 290 Elon: 40%**
BS 2870 PB101	4% Sn 0.3% P Cu alloy: Sheet; ¼-hard **DPN: 100 UTS: 330 Elon: 30% Proof: 100**
BS 2870 PB101	4% Sn 0.3% P Cu alloy: Sheet; ½-hard **DPN: 140 UTS: 440 Elon: 8% Proof: 240**
BS 2870 PB102	5% Sn 0.2% P Cu alloy: Sheet; annealed **DPN: 85 UTS: 300 Elon: 45%**
BS 2870 PB102	5% Sn 0.2% P Cu alloy: Sheet; ¼-hard **DPN: 110 UTS: 340 Elon: 35% Proof: 120**
BS 2870 PB102	5% Sn 0.2% P Cu alloy: Sheet; ½-hard **DPN: 150 UTS: 460 Elon: 10% Proof: 300**
BS 2870 PB102	5% Sn 0.2% P Cu alloy: Sheet; hard **DPN: 170 UTS: 530 Elon: 4% Proof: 370**
BS 2870 PB102	5% Sn 0.2% P Cu alloy: Sheet; extra hard **DPN: 210 UTS: 630**
BS 2870 PB103	7% Sn 0.2% P Cu alloy: Sheet; extra hard **DPN: 215 UTS: 680 Proof: 530**
BS 2870 PB103	7% Sn 0.2% P Cu alloy: Sheet; spring hard **DPN: 230**
BS 2870 PB103	7% Sn 0.2% P Cu alloy: Sheet; extra hard **DPN: 240 (min)**
BS 2870 PB103	7% Sn 0.2% P Cu alloy: Sheet; annealed **DPN: 90 UTS: 340 Elon: 50%**
BS 2870 PB103	7% Sn 0.2% P Cu alloy: Sheet; ¼-hard **DPN: 115 UTS: 370 Elon: 40% Proof: 180**
BS 2870 PB103	7% Sn 0.2% P Cu alloy: Sheet; ½-hard **DPN: 160 UTS: 480 Elon: 12% Proof: 300**
BS 2870 PB103	7% Sn 0.2% P Cu alloy: Sheet; hard **DPN: 180 UTS: 580 Elon: 6% Proof: 400**
BS 2870 PB104	8% Sn 0.3% P Cu alloy: Sheet; specification not issued
BS 2871 PB101	3.5% Sn 0.3% P Cu alloy: Tube; specification not issued
BS 2871 PB102	5% Sn 0.2% P Cu alloy: Tube; annealed **UTS: 320**
BS 2871 PB103	7% Sn 0.2% P Cu alloy: Tube; specification not issued
BS 2871 PB104	8% Sn 0.3% P Cu alloy: Tube; annealed **UTS: 360**
BS 2872 PB101	3.5% Sn 0.3% P Cu alloy: Forging; specification not issued
BS 2872 PB102	5% Sn 0.2% P Cu alloy: Forging; specification not issued
BS 2872 PB103	7% Sn 0.2% P Cu alloy: Forging; specification not issued
BS 2872 PB104	8% Sn 0.3% P Cu alloy: Forging; specification not issued
BS 2873 PB101	3.5% Sn 0.3% P Cu alloy: Wire; specification not issued
BS 2873 PB102	5% Sn 0.2% P Cu alloy: Wire; annealed **UTS: 330 Elon: 40%**
BS 2873 PB102	5% Sn 0.2% P Cu alloy: Wire; extra hard **UTS: 830**
BS 2873 PB103	7% Sn 0.2% P Cu alloy: Wire; annealed **UTS: 260 Elon: 45%**
BS 2873 PB103	7% Sn 0.2% P Cu alloy: Wire; extra hard **UTS: 880**
BS 2873 PB104	8% Sn 0.3% P Cu alloy: Wire; specification not issued
BS 2874 PB101	3.5% Sn 0.3% P Cu alloy: Bar; specification not issued
BS 2874 PB102	5% Sn 0.2% P Cu alloy: Bar; as rolled **UTS: 370 Elon: 20% Proof: 300**
BS 2874 PB103	7% Sn 0.2% P Cu alloy: Bar; specification not issued
BS 2874 PB104	8% Sn 0.3% P Cu alloy: Bar; as rolled **UTS: 450 Elon: 20%**
BS 2875 PB101	3.5% Sn 0.3% P Cu alloy: Plate; as rolled **UTS: 290 Elon: 35%**
BS 2875 PB101	3.5% Sn 0.3% P Cu alloy: Plate; hard **UTS: 400 Elon: 15%**
BS 2875 PB102	5% Sn 0.2% P Cu alloy: Plate; annealed **UTS: 300 Elon: 40%**
BS 2875 PB102	5% Sn 0.2% P Cu alloy: Plate; hard **UTS: 400 Elon: 15%**
BS 2875 PB103	7% Sn 0.2% P Cu alloy: Plate; specification not issued
BS 2875 PB104	8% Sn 0.3% P Cu alloy: Plate; specification not issued
BS 2901 C10	5% Sn 0.3% P Cu alloy: Rod for all forms of welding
BS 2901 C11	7% Sn 0.3% P Cu alloy: Rod for all forms of welding
BS 3127/5	5% Sn 0.15% P Cu alloy: Tube; ¼-hard; drawn for bourdon tubes **DPN: 130**
BS B8	10% Sn 0.5% P Cu alloy: Casting; chill cast for bearings **UTS: 240 Elon: 3%**
BS T52	5.5% Sn 0.2% P Cu alloy: Tube; hard drawn **UTS: 360**
C10	5.0% Sn 0.2% P Cu alloy: Rod for gas shielded arc welding
C11	6.5% Sn 0.2% P Cu alloy: Rod for gas shielded arc welding
C25	5.0% Sn 1.4% Ni 0.6% Si Cu alloy: Rod for gas shielded arc welding.
C18990	2% Sn 0.15% Cr Cu: CDA and UNS designation
C40400	0.5% Sn 2.5% Zn Cu alloy: CDA and UNS designation
C40500	1% Sn 4% Zn Cu alloy: CDA and UNS designation
C40800	2% Sn 3% Zn Cu alloy: CDA and UNS designation
C41000	2.4% Sn 4% Zn Cu alloy: CDA and UNS designation
C41100	0.5% Sn 10% Zn Cu alloy: CDA and UNS designation
C41300	1% Sn 8% Zn Cu alloy: CDA and UNS designation
C41500	2% Sn 9% Zn Cu alloy: CDA and UNS designation
C41900	5% Sn 3% Zn Cu alloy: absolute UNS designation
C42000	1.7% Sn 10% Zn Cu alloy: CDA and UNS designation
C42100	2.7% Sn 10% Zn 0.2% Mn Cu alloy: CDA and UNS designation
C42200	1% Sn 10% Zn Cu alloy: CDA and UNS designation

Note. The following abbreviations and units are used in the tables:

DPN	Hardness, diamond pyramid number
UTS	Ultimate tensile strength, N/mm^2
Elon	Elongation, %
Proof	0.1% proof strength, N/mm^2

1 N/mm^2=0.1 hbar=0.102 kgf/mm^2=0.06475 tonf/in.^{2}=145.04 lbf/in.^{2}=1 MPa

See Appendix II for other abbreviations and conversion tables.

Symbol	Nominal analysis, supplier, condition and remarks.
C42500	2.2% Sn 10% Zn Cu alloy: CDA and UNS designation
C43000	2.2% Sn 10% Zn Cu alloy: CDA and UNS designation
C43200	0.5% Sn 12% Zn Cu alloy: CDA and UNS designation
C43400	0.8% Sn 12% Zn Cu alloy: CDA and UNS designation
C43500	1% Sn 18% Zn Cu alloy: CDA and UNS designation
C43600	0.3% Sn 18% Zn Cu alloy: CDA and UNS designation
C44300	1% Sn 27% Zn Cu alloy: Arsenical; CDA and UNS designation
C44400	1% Sn 27% Zn Cu alloy: Antimonial; CDA and UNS designation
C44500	1% Sn 27% Zn Cu alloy: Phosphorized; CDA and UNS designation
C50100	0.7% Sn 0.03 P Cu alloy: CDA and UNS designation
C50200	1.2% Sn Cu alloy: CDA and UNS designation
C50500	1.2% Sn 0.3% P Cu alloy: CDA and UNS designation
C50700	1.7% Sn Cu alloy: CDA and UNS designation
C50710	2% Sn 0.3% Ni Cu alloy: CDA and UNS designation
C50715	2% Sn 0.03% P 0.1% Fe Cu alloy: CDA and UNS designation
C50800	3% Sn 0.05% P Cu alloy: CDA and UNS designation
C50900	3.2% Sn 0.2% P Cu alloy: CDA and UNS designation
C51000	5% Sn 0.2% P Cu alloy: CDA and UNS designation
C51100	4.2% Sn 0.2% P Cu alloy: CDA and UNS designation
C51800	5% Sn 0.2% P Cu alloy: CDA and UNS designation
C51900	6% Sn 0.2% P Cu alloy: CDA and UNS designation
C52100	8% Sn 0.2% P Cu alloy: CDA and UNS designation
C52400	10% Sn 0.2% P Cu alloy: CDA and UNS designation
C52600	2.7% Sn 0.25% P Cu alloy: Obsolete; UNS designation
C52900	8% Sn 0.25% P 1.5% Mn Cu alloy: Obsolete; UNS designation
C53200	4.7% Sn 3.2% Pb 0.2% P Cu alloy: CDA and UNS designation
C53400	4.5% Sn 1% Pb 0.2% P Cu alloy: CDA and UNS designation
C54400	4% Sn 4% Pb 3.2% Zn 0.3% P Cu alloy: CDA and UNS designation
C54600	4% Sn 4% Pb 3% Zn Cu alloy: Obsolete; UNS designation
C54800	5% Sn 5% Pb 0.2% P Cu alloy: CDA and UNS designation
C64900	1.4% Sn 1% Si Cu alloy: CDA and UNS designation
C83500	6% Sn 4.5% Pb 1.7% Zn Cu alloy: Cast; CDA and UNS designation
C83520	4% Sn 4% Pb 2.5% Zn Cu alloy: Cast; CDA and UNS designation
C83600	5% Sn 5% Pb 5% Zn Cu alloy: Cast; CDA and UNS designation
C83800	4% Sn 6% Pb 7% Zn Cu alloy: Cast; CDA and UNS designation
C83810	2.7% Sn 5% Pb 8.5% Zn Cu alloy: Cast; CDA and UNS designation
C84200	5% Sn 2.5% Pb 13% Zn Cu alloy: Cast; CDA and UNS designation
C84400	3% Sn 7% Pb 8.5% Zn Cu alloy: Cast; CDA and UNS designation
C84410	3.7% Sn 8% Pb 8.5% Zn Cu alloy: Cast; CDA and UNS designation
C84500	3% Sn 6.7% Pb 12% Zn Cu alloy: Cast; CDA and UNS designation
C84800	2.5% Sn 6.2% Pb 15% Zn Cu alloy: Cast; CDA and UNS designation
C90200	7% Sn Cu alloy: Cast; CDA and UNS designation
C90250	10% Sn Cu alloy: Cast; CDA and UNS designation
C90300	8.2% Sn 4% Zn Cu alloy: Cast; CDA and UNS designation
C90500	10% Sn 2% Zn Cu alloy: Cast; CDA and UNS designation
C90700	11% Sn Cu alloy: Cast; CDA and UNS designation
C90710	11% Sn 1% P Cu alloy: Cast; CDA and UNS designation
C90800	12% Sn Cu alloy: Cast; CDA and UNS designation

Symbol	Nominal analysis, supplier, condition and remarks.
C90810	12% Sn Cu alloy: Cast; CDA and UNS designation
C90900	13% Sn Cu alloy: Cast; CDA and UNS designation
C91000	15% Sn Cu alloy: Cast; CDA and UNS designation
C91100	16% Sn Cu alloy: Cast; CDA and UNS designation
C91300	19% Sn Cu alloy; Cast; CDA and UNS designation
C91600	10% Sn 1.8% Ni Cu alloy: Cast; CDA and UNS designation
C91700	12% Sn 1.8% Ni Cu alloy: Cast; CDA and UNS designation
C92200	6% Sn 1.5% Pb 4% Zn Cu alloy: Cast; CDA and UNS designation
C92300	8.2% Sn 0.7% Pb 3.5% Zn Cu alloy: Cast; CDA and UNS designation
C92310	8% Sn 1% Pb 4% Zn Cu alloy: Cast; CDA and UNS designation
C92400	10% Sn 1.7% Pb 2% Zn Cu alloy: Cast; CDA and UNS designation
C92410	7% Sn 3% Pb 2.2% Zn Cu alloy: Cast; CDA and UNS designation
C92500	11% Sn 1.2% Pb 1% Ni Cu alloy: Cast; CDA and UNS designation
C92600	10% Sn 1.2% Pb 2% Zn Cu alloy: Cast; CDA and UNS designation
C92610	10% Sn 1% Pb 2.2% Zn Cu alloy: Cast; CDA and UNS designation
C92700	10% Sn 1.7% Pb Cu alloy: Cast; CDA and UNS designation
C92710	10% Sn 5% Pb Cu alloy: Cast; CDA and UNS designation
C92800	16% Sn 5% Pb Cu alloy: Cast; CDA and UNS designation
C92810	13% Sn 5% Pb 1% Ni Cu alloy: Cast; CDA and UNS designation
C92900	10% Sn 2.6% Pb 3.2% Ni Cu alloy: Cast; CDA and UNS designation
C93100	7.5% Sn 3.5% Pb 2% Zn Cu alloy: Cast; CDA and UNS designation
C93200	7% Sn 7% Pb 2.5% Zn Cu alloy: Cast; CDA and UNS designation
C93400	8% Sn 8% Pb Cu alloy: Cast; CDA and UNS designation
C93500	5.2% Sn 9% Pb 2% Zn Cu alloy: Cast; CDA and UNS designation
C93600	7% Sn 12% Pb Cu alloy: Cast; CDA and UNS designation
C93700	10% Sn 9.5% Pb Cu alloy: Cast; CDA and UNS designation
C93720	4% Sn 8% Pb 4% Zn Cu alloy: Cast; CDA and UNS designation
C93800	7% Sn 14.5% Pb Cu alloy: Cast; CDA and UNS designation
C93900	6% Sn 10% Pb 1.5% Zn Cu alloy: Cast; CDA and UNS designation
C94000	13% Sn 15% Pb 0.7% Ni Cu alloy: Cast; CDA and UNS designation
C94100	5.5% Sn 20% Pb Cu alloy: Cast; CDA and UNS designation
C94200	3.5% Sn 3.5% Pb 3% Zn (max) Cu alloy: casting; obsolete; UNS designation
C94300	5.2% Sn 23% Pb Cu alloy: Cast; CDA and UNS designation
C94310	2.2% Sn 30% Pb 0.7% Ni Cu alloy: Cast; CDA and UNS designation
C94320	5.5% Sn 28% Pb Cu alloy: Cast; CDA and UNS designation
C94330	3.5% Sn 23% Pb 3% Zn Cu alloy: Cast; CDA and UNS designation
C94400	8% Sn 10.5% Pb Cu alloy: Cast; CDA and UNS designation
C94500	7% Sn 19% Pb 1% Zn Cu alloy: Cast; CDA and UNS designation

Symbol	Nominal analysis, supplier, condition and remarks.
C94700	5.2% Sn 1.7% Zn 5.2% Ni Cu alloy: Cast; CDA and UNS designation
C94800	5.2% Sn 0.7% Pb 1.7% Zn 5.2% Ni Cu alloy: Cast; CDA and UNS designation
C94900	5% Sn 5% Pb 5% Zn 5% Ni Cu alloy: Cast; CDA and UNS designation
C98200	1.4% Sn 24% Pb Cu alloy: Cast; CDA and UNS designation
C98400	0.5% Sn 30% Pb 1.5% Ag Cu alloy: Cast; CDA and UNS designation
C98600	0.5% Sn 35% Pb 1.5% Ag Cu alloy: Cast; CDA and UNS designation
C98800	0.25% Sn 40% Pb 5.5% Ag Cu alloy: Cast; CDA and UNS designation
C98820	1.25% Sn 42% Pb Cu alloy: Cast; CDA and UNS designation
C98840	1.25% Sn 46% Pb Cu alloy: Cast; CDA and UNS designation
CAROBRONZE	8.5% Sn 0.3% P Cu alloy: Bar; Carobronze; cold drawn **DPN: 145 UTS: 530 Elon: 25% Proof: 370**
CHINESE BRONZE	22% Sn Cu alloy: Origin unknown
CITOBRONZE A5	6.5% Sn 0.1% P 1.4% Mn Cu alloy: Welding electrode Soudometal for welding bronze and cladding steel **DPN: 105**
CITOBRONZE A8	8.5% Sn 0.1% P 1.2% Mn Cu alloy: Welding electrode Soudometal for welding bronze
CITOBRONZE AA	7.9% Sn 0.1% P Cu alloy: Welding electrode; Soudometal for welding bronze **DPN: 105**
COGWHEEL 11	Sn P Cu alloy: Casting; Phosphor Bronze Ltd; chill cast **DPN: 100 UTS: 290 Elon: 7% Proof: 150**
COGWHEEL VII	Sn 1.5% P Cu alloy: Casting; Phosphor Bronze Ltd; chill cast **DPN: 125 UTS: 340 Elon: 3% Proof: 180**
COGWHEEL VIII	High Sn P Cu alloy: Casting; Phosphor Bronze Ltd; chill cast; for bushes **DPN: 140 UTS: 340 Elon: 2% Proof: 220**
COGWHEEL XI	Sn 8% Pb P Cu alloy: Casting; Phosphor Bronze Ltd; chill cast; bearings **DPN: 100 UTS: 270 Elon: 1.5% Proof: 170**
COINAGE BRONZE	4% Sn 1% Zn Cu alloy: Wrought tin bronze used for coins; origin unknown
CUSILAY	1.5% Sn 0.85% Fe 2% Si Cu alloy: Origin unknown
DIN 1705 G Sn Bz 10	10% Sn 1.0% Pb 0.5% Zn Cu alloy: Casting
DIN 1705 G Sn Bz 14	14% Sn 1.0% Pb 0.5% Zn Cu alloy: Casting
DIN 1705 Rg 5	5% Sn 5% Pb 5% Zn Cu alloy: Casting
DIN 1705 Rg 7	7% Sn 6% Pb 4% Zn Cu alloy: Casting
DIN 1705 Rg 10	10.5% Sn 1.0% Pb 2% Zn Cu alloy: Casting
DIN 1716 G Pb Bz 25	3% Sn 26% Pb 2% Ni Cu alloy: Casting **DPN: 30**
DIN 1716 G Sn Pb Bz 5	10% Sn 5% Pb 1% Ni Cu alloy: Casting **DPN: 85 UTS: 220 Elon: 18% Proof: 120**
DIN 1716 G Sn Pb Bz 10	10% Sn 10% Pb 1% Ni Cu alloy: Casting **DPN: 75 UTS: 220 Elon: 14% Proof: 100**
DIN 1716 G Sn Pb Bz 15	8% Sn 15% Pb 1.5% Ni Cu alloy: Casting **DPN: 70 UTS: 210 Elon: 12% Proof: 100**

Note. The following abbreviations and units are used in the tables:

DPN	Hardness, diamond pyramid number
UTS	Ultimate tensile strength, N/mm^2
Elon	Elongation, %
Proof	0.1% proof strength, N/mm^2

$1\ N/mm^2 = 0.1\ hbar = 0.102\ kgf/mm^2 = 0.06475\ tonf/in.^2 = 145.04\ lbf/in.^2 = 1\ MPa$

See Appendix II for other abbreviations and conversion tables.

Symbol	Nominal analysis, supplier, condition and remarks.
DIN 1716 G Sn Pb Bz 20	5% Sn 20% Pb 2% Ni Cu alloy: Casting **DPN: 55 UTS: 200 Elon: 10% Proof: 90**
DIN 1733 S Cu Al 8	9% Al Cu alloy: Welding electrode
DIN 1733 S Cu Sn 13	12% Sn 0.25% P (max) Cu alloy: Welding electrode
DIN 1733 S Cu Sn 13	Sn Cu welding electrode
DIN 1733 S Sn Bz 6	7% Sn 0.2% Pb Cu alloy
DIN 1733 S Sn Bz 12	12% Sn 0.2% P Cu alloy: Wrought
DIN 1785 K Sn Bz 2	1.7% Sn Cu alloy: Wrought
DIN 17656 GB Rg 10	10% Sn 2% Zn Cu alloy: Casting
DIN 17656 GB Sn Bz 14	14% Sn Cu alloy: Casting
DIN 17656 GB Sn Pb Bz 5	10% Sn 5% Pb Cu alloy: Casting
DIN 17656 GB Sn Pb Bz 10	10% Sn 10% Pb Cu alloy: Casting
DTD 265 A	8% Sn 0.3% P Cu alloy: Bar and tube; hard drawn; for bushes **DPN: 120 UTS: 450 Elon: 20%**
DTD 318 A	4.0% Sn 15% Zn 1.0% Fe Cu alloy: Tube; annealed **UTS: 450**
DTD 459	15% Sn 5% Pb Cu alloy: Casting lead bronze; sand or chill cast; for bearings
E Cu Sn A	5% Sn 0.2% P Cu alloy: Weld electrode; designation used by AWS
E Cu Sn C	8% Sn 0.2% P Cu alloy: Weld electrode; designation used by AWS
ENCON	Sn Cu alloy: Continuously cast; Enfield alloys for BS 1400
ESCO TCC1	10% Sn 0.25% Pb 0.5% P Cu alloy: Casting; Eyre Smelting for BS alloy PB 1
ESCO TCC2	7% Sn 3.5% Pb 2% Zn 0.3% P 1% Ni Cu alloy: Casting; Eyre Smelting Co. for BS alloy LPB 1
G Cu Pb 5 Sn	10% Sn 5% Pb 1% Ni Cu alloy: Casting; designation used by German Standards
G Cu Pb 10 Sn	10% Sn 10% Pb 1% Ni Cu alloy: Casting; designation used by German Standards
G Cu Pb 15 Sn	8% Sn 15% Pb 1.5% Ni Cu alloy: Casting; designation used by German Standards
G Cu Pb 20 Sn	5% Sn 20% Pb 2% Ni Cu alloy: Casting; designation used by German Standards
G Cu Pb 25	3% Sn 26% Pb 2% Ni Cu alloy: Casting; designation used by German Standards
G Cu Sn 5 Zn Pb	5% Sn 5% Pb 5% Zn Cu alloy: Casting; designation used by German Standards
G Cu Sn 7 Zn Pb	7% Sn 6% Pb 4% Zn Cu alloy: Casting; designation used by German Standards
G Cu Sn 10	10% Sn 1.0% Pb 0.5% Zn Cu alloy: Casting; designation used by German Standards
G Cu Sn 10 Zn	10.5% Sn 1.0% Pb 2% Zn Cu alloy: Casting; designation used by German Standards
G Cu Sn 12	12% Sn 1.0% Pb 0.5% Zn Cu alloy: Casting; designation used by German Standards
G Cu Sn 14	14% Sn 1.0% Pb 0.5% Zn Cu alloy: Casting; designation used by German Standards
G I BRONZE	10% Sn 2% Zn Cu alloy: Casting continuous cast; J Holroyd for BS alloy G1 **DPN: 105 UTS: 360 Elon: 25% Proof: 170**
G Pb Bz 25	3% Sn 26% Pb 2% Ni Cu alloy: Casting; designation used by German Standards
G Sn Bz 10	10% Sn 1.0% Pb 0.5% Zn Cu alloy: Casting; designation used by German Standards
G Sn Bz 12	12% Sn 1.0% Pb 0.5% Zn Cu alloy: Casting; designation used by German Standards
G Sn Bz 14	14% Sn 1.0% Pb 0.5% Zn Cu alloy: Casting; designation used by German Standards
G Sn Pb Bz 5	10% Sn 5% Pb 1% Ni Cu alloy: Casting; designation used by German Standards

Symbol	Nominal analysis, supplier, condition and remarks.
G Sn Pb Bz 10	10% Sn 10% Pb 1% Ni Cu alloy: Casting; designation used by German Standards
G Sn Pb Bz 15	8% Sn 15% Pb 1.5% Ni Cu alloy: Casting; designation used by German Standards
G Sn Pb Bz 20	5% Sn 20% Pb 2% Ni Cu alloy: Casting; designation used by German Standards
GB Cu Pb 5 Sn	10% Sn 5% Pb Cu alloy: Casting; designation used by German Standards
GB Cu Pb 10 Sn	10% Sn 10% Pb Cu alloy: Casting; designation used by German Standards
GB Cu Pb 15 Sn	8% Sn 15% Pb Cu alloy: Casting; designation used by German Standards
GB Cu Pb 20 Sn	4.5% Sn 21% Pb Cu alloy: Casting; designation used by German Standards
GB Cu Sn 5 Zn Pb	5% Sn 5% Pb 5% Zn Cu alloy: Casting; designation used by German Standards
GB Cu Sn 7 Zn Pb	7% Sn 6% Pb 4.5% Zn Cu alloy: Casting; designation used by German Standards
GB Cu Sn 10	10% Sn Cu alloy: Casting; designation used by German Standards
GB Cu Sn 10 Zn	10% Sn 2% Zn Cu alloy: Casting; designation used by German Standards
GB Cu Sn 12	12% Sn Cu alloy: Casting; designation used by German Standards
GB Cu Sn 14	14% Sn Cu alloy: Casting; designation used by German Standards
GB Rg 5	5% Sn 5% Pb 5% Zn Cu alloy: Casting; designation used by German Standards
GB Rg 7	7% Sn 6% Pb 4.5% Zn Cu alloy: Casting; designation used by German Standards
GB Rg 10	10% Sn 2% Zn Cu alloy: Casting; designation used by German Standards
GB Sn Bz 10	10% Sn Cu alloy: Casting; designation used by German Standards
GB Sn Bz 12	12% Sn Cu alloy: Casting; designation used by German Standards
GB Sn Bz 14	14% Sn Cu alloy: Casting; designation used by German Standards
GB Sn Pb Bz 5	10% Sn 5% Pb Cu alloy: Casting; designation used by German Standards
GB Sn Pb Bz 10	10% Sn 10% Pb Cu alloy: Casting; designation used by German Standards
GB Sn Pb Bz 15	8% Sn 15% Pb Cu alloy: Casting; designation used by German Standards
GB Sn Pb Bz 20	4.5% Sn 21% Pb Cu alloy: Casting; designation used by German Standards
GC Cu Sn 12	12% Sn 1.0% Pb 0.5% Zn Cu alloy: Casting; designation used by German Standards
GC Cu Sn 5 Zn Pb	5% Sn 5% Pb 5% Zn Cu alloy: Casting; designation used by German Standards
GC Cu Sn 7 Zn Pb	7% Sn 6% Pb 4% Zn Cu alloy: Casting; designation used by German Standards
GC Cu Sn 10 Zn	10.5% Sn 1.0% Pb 2.0% Zn Cu alloy: Casting; designation used by German Standards
GC Rg 5	5% Sn 5% Pb 5% Zn Cu alloy: Casting; designation used by German Standards
GC Rg 7	7% Sn 6% Pb 4% Zn Cu alloy: Casting; designation used by German Standards
GC Rg 10	10.5% Sn 1.0% Pb 2.0% Zn Cu alloy: Casting; designation used by German Standards
GC Sn Bz 12	12% Sn 1.0% Pb 0.5% Zn Cu alloy: Casting; designation used by German Standards
GZ Cu Sn 5 Zn Pb	5% Sn 5% Pb 5% Zn Cu alloy: Casting; designation used by German Standards
GZ Cu Sn 10 Zn	10.5% Sn 1.0% Pb 2% Zn Cu alloy: Casting; designation used by German Standards
GZ Cu Sn 12	12% Sn 1.0% Pb 0.5% Zn Cu alloy: Casting; designation used by German Standards
GZ Rg 5	5% Sn 5% Pb 5% Zn Cu alloy: Casting; designation used by German Standards
GZ Rg 7	7% Sn 6% Pb 4% Zn Cu alloy: Casting; designation used by German Standards
GZ Rg 10	10.5% Sn 1.0% Pb 2% Zn Cu alloy: Casting; designation used by German Standards
GZ Sn Bz 12	12% Sn 1.0% Pb 0.5% Zn Cu alloy: Casting; designation used by German Standards
HOLFOS BO	11.5% Sn 0.7% P Cu alloy: Casting; Holroyd **DPN: 100 UTS: 300 Elon: 5% Proof: 210**
HOLFOS BOHT	11.5% Sn 0.7% P Cu alloy: Casting; Holroyd; heat treated **DPN: 90 UTS: 330 Elon: 20% Proof: 170**
HOLFOS G1	10% Sn 2% Zn Cu alloy: Casting; Holroyd; centrifugally cast **DPN: 95 UTS: 300 Elon: 10% Proof: 170**
HOLFOS G2	8% Sn 4% Zn Cu alloy: Casting; Holroyd; centrifugally cast **DPN: 90 UTS: 240 Elon: 9% Proof: 140**
HOLFOS G3	7% Sn 2.2% Zn 0.5% P (max) 5.5% Ni Cu alloy: Casting; Holroyd; centrifugally cast; as cast **DPN: 95 UTS: 300 Elon: 14% Proof: 150**
HOLFOS G3WP	7% Sn 2.2% Zn 0.5% P (max) 5.5% Ni Cu alloy: Holroyd; centrifugally cast; heat treated **DPN: 180 UTS: 420 Elon: 3% Proof: 270**
HOLFOS JH17	14% Sn 1.3% P Cu alloy: Casting; Holroyd **DPN: 140 UTS: 340 Elon: 1.0% Proof: 270**
HOLFOS LB1	9% Sn 15% Pb Cu alloy: Casting; Holroyd; centrifugally cast **DPN: 70 UTS: 210 Elon: 7% Proof: 140**
HOLFOS LB2	10% Sn 10% Pb Cu alloy: Casting; Holroyd; centrifugally cast **DPN: 90 UTS: 240 Elon: 8% Proof: 170**
HOLFOS LB3	10% Sn 5% Pb Cu alloy: Casting; Holroyd; centrifugally cast **DPN: 95 UTS: 250 Elon: 12% Proof: 210**
HOLFOS LB4	5% Sn 10% Pb Cu alloy: Casting; Holroyd; centrifugally cast **DPN: 70 UTS: 220 Elon: 10% Proof: 140**
HOLFOS LB5	5% Sn 20% Pb Cu alloy: Casting; Holroyd; centrifugally cast **DPN: 70 UTS: 170 Elon: 7% Proof: 140**
HOLFOS LG1	3% Sn 5% Pb 9% Zn Cu alloy: Casting; Holroyd; centrifugally cast **DPN: 70 UTS: 220 Elon: 35% Proof: 120**
HOLFOS LG2	5% Sn 5% Pb 5% Zn Cu alloy: Casting; Holroyd; centrifugally cast **DPN: 75 UTS: 210 Elon: 15% Proof: 140**
HOLFOS LG3	7% Sn 2% Pb 5% Zn Cu alloy: Casting; Holroyd; centrifugally cast **DPN: 80 UTS: 240 Elon: 12% Proof: 170**
HOLFOS LG4	7% Sn 3% Pb 3% Zn Cu alloy: Casting; Holroyd; centrifugally cast **DPN: 70 UTS: 250 Elon: 15% Proof: 140**
HOLFOS LG773	7% Sn 7% Pb 3% Zn Cu alloy: Casting; Holroyd; centrifugally cast **DPN: 85 UTS: 300 Elon: 15% Proof: 140**
HOLFOS LPB1	7.5% Sn 4% Pb 0.3% P (min) Cu alloy: Casting; J Holroyd; centrifugally cast **DPN: 85 UTS: 270 Elon: 20% Proof: 170**
HOLFOS PB1	10% Sn 0.5% P (min) Cu alloy: Casting; Holroyd; centrifugally cast **DPN: 100 UTS: 300 Elon: 10% Proof: 210**
HOLFOS PB2	11.5% Sn 0.15% P (min) Cu alloy: Casting; Holroyd; centrifugally cast **DPN: 100 UTS: 300 Elon: 8% Proof: 180**
HOLFOS PB3	9.5% Sn 0.3% P Cu alloy: Casting; Holroyd; centrifugally cast **DPN: 95 UTS: 270 Elon: 9% Proof: 180**
HOLFOS SPUNCAST	11.5% Sn 0.7% P Cu alloy: Casting; spuncast; J Holroyd; alternative name for BO BRONZE

Symbol	Nominal analysis, supplier, condition and remarks.
HOLFOS WW	11.5% Sn 0.2% P Cu alloy: Casting; chill cast; J Holroyd for BS alloy PB 2 **DPN: 90 UTS: 270 Elon: 5% Proof: 170**
IMI 657	5% Sn 0.2% P Cu alloy: Strip and wire; IMI for BS alloy PB 102
IMI 659	7% Sn 0.2% P Cu alloy: Strip and wire; IMI for BS alloy PB 103
IMI 660	8% Sn 0.3% P Cu alloy: Bar and wire; IMI
IMI 661	8% Sn 0.3% P Cu alloy: Strip and wire; IMI for BS alloy PB 104; withdrawn
JH 17	14% Sn 1.3% P Cu alloy: Casting; spuncast; J Holroyd **DPN: 140 UTS: 320 Elon: 1% Proof: 270**
K Sn Bz 2	1.7% Sn Cu alloy: Wrought; designation used by German Standards
MET-BRONZE PT8	8% Sn 0.1% P Cu alloy: Electrode; Metrode
NBN 267/21/22 Br Sn 5 Zn 5 Pb 5	5% Sn 5% Pb 5% Zn Cu alloy: Casting; Belgian specification
NBN 267/21/22 Br Sn 7 Zn 3 Pb 4	7% Sn 4% Pb 3% Zn Cu alloy: Casting; Belgian specification
NBN 267/21/22 Br Sn 10 Zn 2	10% Sn 2% Zn Cu alloy: Casting; Belgian specification
NBN 267/21/22 Br Sn 10	11.5% Sn 0.7% P Cu alloy: Casting; Belgian specification
NBN 267/21/22 Br Sn 10	10% Sn 0.5% P Cu alloy: Casting; Belgian specification
NBN 267/21/22 Br Sn 12	11.5% Sn 0.15% P Cu alloy: Casting; Belgian specification
NBN 267/21/22 Br Sn 12	11.5% Sn 0.2% P Cu alloy: Casting; Belgian specification
NBN 267/21/23 Br Sn 9 P	9.5% Sn 0.3% P Cu alloy: Casting; Belgian specification
NBN 267/23 Br Pb 15 Sn 8	9% Sn 1.5% Pb Cu alloy: Casting; Belgian specification
NBN 267/23 Br Pb 24 Sn 5	5% Sn 20% Pb Cu alloy: Casting; Belgian specification
NBN 267/23 Br Sn 10 Pb 10	10% Sn 10% Pb Cu alloy: Casting; Belgian specification
NF A53 012	Classification of bronze alloy; French Standard
NF A53 607	General specification covering range of bronze alloys; French Standard
NF A53 707	Classification of bronze alloys; French Standard
No. 1 GUNMETAL	10% Sn 1.5% Pb 2% Zn 1% Ni Cu alloy: Casting; Stone Manganese Marine for BS alloy G1
No. 1 LEADED BRONZE	10% Sn 10% Pb Cu alloy: Casting; Stone Manganese Marine; as AMS 4842
No. 1 PHOSPHOR BRONZE	12% Sn 0.5% Pb 0.15% P Cu alloy: Casting; Stone Manganese Marine; for BS alloy PB2
No. 2 GUNMETAL	7% Sn 2% Pb 4% Zn 2% Ni Cu alloy: Casting; Stone Manganese Marine; as BS alloy LG3
No. 2 LEADED BRONZE	9% Sn 15% Pb 1% Zn 0.1% P 2% Ni Cu alloy: Casting; Stone Manganese Marine for BS alloy LB1
No. 2 PHOSPHOR BRONZE	10% Sn 0.25% Pb 0.5% P Cu alloy: Casting; Stone Manganese Marine for BS alloy PB1
No. 3 GUNMETAL	5% Sn 5% Pb 5% Zn 2% Ni Cu alloy: Casting; Stone Manganese Marine for BS alloy LG2
No. 3 PHOSPHOR BRONZE	9% Sn 0.25% Pb 0.3% P Cu alloy: Casting; Stone Manganese Marine for BS alloy PB3
No. 4 LEADED BRONZE	5% Sn 9% Pb 0.5% Ni Cu alloy: Casting; Stone Manganese Marine; as SAE 66

Note. The following abbreviations and units are used in the tables:

DPN	Hardness, diamond pyramid number
UTS	Ultimate tensile strength, N/mm^2
Elon	Elongation, %
Proof	0.1% proof strength, N/mm^2

$1\ N/mm^2 = 0.1\ hbar = 0.102\ kgf/mm^2 = 0.06475\ tonf/in.^2 = 145.04\ lbf/in.^2 = 1\ MPa$

See Appendix II for other abbreviations and conversion tables.

Symbol	Nominal analysis, supplier, condition and remarks.
No. 4 PHOSPHOR BRONZE	7% Sn 3% Pb 2% Zn 0.3% P 1% Ni Cu alloy: Casting; Stone Manganese Marine for BS alloy LPB1
NOIL	20% Sn Cu alloy: Obsolete; consult the Copper Development Association **DPN: 158**
OUNCE METAL	5% Sn 5% Pb 5% Zn 1% Ni (max) Cu alloy: Casting; common name; refer to Copper Development Association for further information
PHOSPHOR BRONZE	10% Sn with up to 0.5% P Cu alloy: Covered by BS alloys PB 1–4
PROMET 6	5.5% Sn 14% Pb 1.5% Ni Cu alloy: Bearings mould; origin unknown **DPN: 60**
QQ B 691/1	6.0% Sn 1.5% Pb 4.0% Zn Cu alloy: Casting; US Federal **UTS: 220 Elon: 22%**
QQ B 691/2	5% Sn 5% Pb 5% Zn Cu alloy: Casting; US Federal **UTS: 200 Elon: 20%**
QQ B 691/3	5% Sn 2.5% Pb 13% Zn 0.7% Ni Cu alloy: Casting; US Federal **UTS: 170 Elon: 15%**
QQ B 691/6	8% Sn 4.0% Zn Cu alloy: Casting; US Federal **UTS: 220 Elon: 22%**
QQ B 691/11	3.0% Sn 7% Pb 8% Zn Cu alloy: Casting; US Federal **UTS: 200 Elon: 13%**
QQ B 691b/5	8% Sn 4.0% Zn Cu alloy: Casting; US Federal **UTS: 270 Elon: 20%**
QQ B 746/A	4.0% Sn Cu alloy: Bar and forging; US Federal **UTS: 450 Elon: 15% Proof: 270**
QQ C 591a/B	2.2% Sn 1.2% Zn 1.5% Si Cu alloy: Rod; US Federal **UTS: 520 Elon: 10% Proof: 370**
QQ W 401	3.5% Sn 0.3% P Cu alloy: Wire; US Federal; hard drawn for springs **UTS: 750 Elon: 5%**
SAE 40	5% Sn 5% Pb 5% Zn Cu alloy: Casting; as cast **UTS: 200 Elon: 20%**
SAE 41	1.5% Sn 2.5% Pb 30% Zn Cu alloy: Casting; as cast **UTS: 200 Elon: 20%**
SAE 48	30% Pb Cu alloy: For bearings; as cast
SAE 49	24% Pb Cu alloy: For bearings; as cast
SAE 62	10% Sn 0.3% Pb 2% Zn 1% Ni Cu alloy: Casting; as cast **UTS: 270 Elon: 20%**
SAE 63	10% Sn 2% Pb 0.2% P 1% Ni Cu alloy: Casting; as cast **UTS: 240 Elon: 10%**
SAE 64	10% Sn 9% Pb 0.25% P 0.5% Ni Cu alloy: Casting; as cast **UTS: 170 Elon: 8%**
SAE 65	11% Sn 0.5% Pb 0.2% P Cu alloy: Casting; as cast **UTS: 240 Elon: 10%**
SAE 66	5% Sn 9% Pb 0.5% Ni Cu alloy: Casting; as cast **UTS: 170 Elon: 8%**
SAE 67	6% Sn 16% Pb 0.7% Ni Cu alloy: Casting; semi-plastic bronze; as cast **UTS: 120 Elon: 10%**
SAE 77 A	4% Sn 0.3% P Cu alloy: Sheet and strip, annealed **UTS: 340**
SAE 77 A	4% Sn 0.3% P Cu alloy: Sheet and strip; extra hard **UTS: 600**
SAE 77 C	8% Sn 0.2% P Cu alloy: Sheet and strip; annealed **UTS: 420**
SAE 77 C	8% Sn 0.2% P Cu alloy: Sheet and strip; extra hard **UTS: 750**
SAE 81	4% Sn 0.25% P Cu alloy: Bar and wire; hard drawn **UTS: 830 Elon: 3.5%**
SAE 480	35% Pb Cu alloy: For bearings; as cast
SAE 481	40% Pb Cu alloy: For bearings; as cast
SAE 482	5.0% Sn 28% Pb Cu alloy: For bearings
SAE 484	3.0% Sn 42% Pb Cu alloy: For bearings

Symbol	Nominal analysis, supplier, condition and remarks.
SAE 620	8% Sn 0.3% Pb 4% Zn 1% Ni Cu alloy: Casting; as cast **UTS: 270 Elon: 20%**
SAE 621	8% Sn 1% Pb 4% Zn 1% Ni Cu alloy: Casting; as cast **UTS: 240 Elon: 18%**
SAE 622	6% Sn 1.5% Pb 4% Zn 1% Ni Cu alloy: Casting; as cast **UTS: 240 Elon: 22%**
SAE 640	11% Sn 1.2% Pb 0.25% P 1% Ni Cu alloy: Casting; as cast **UTS: 240 Elon: 10%**
SAE 660	7% Sn 7% Pb 3% Zn 0.5% Ni Cu alloy: Casting; as cast **UTS: 200 Elon: 12%**
SAE 791	4% Sn 4% Pb 4% Zn Cu alloy: For bearings; wrought
SAE 792	10% Sn 10% Pb Cu alloy: For bearings; as cast
SAE 793	4% Sn 8% Pb 4% Zn Cu alloy: For bearings; cast
SAE 794	3.5% Sn 23% Pb Cu alloy: For bearings; as cast
SAE 797	10% Sn 10% Pb Cu alloy: For bearings; sintered
SAE 798	4% Sn 8% Pb 4% Zn Cu alloy: For bearings; sintered
SAE 799	3.5% Sn 23% Pb Cu alloy: For bearings; sintered
SAE 840	10% Sn 1.7% C Cu alloy: Sinter for bearings; density 6.0 g/cm^3
SAE 841	10% Sn 1.7% C Cu alloy: Sinter for bearings; density 6.6 g/cm^3
SAE 842	10% Sn 1.7% C Cu alloy: Sinter for bearings; density 7000 kg/m^3
SAE 843	2.5% Pb 1.5% C Cu alloy: Sinter for bearings; density 6700 kg/m^3
SAE CA 510	5% Sn 0.2% P Cu alloy: Wrought; formerly SAE 77A and SAE 81
SAE CA 511	4% Sn 0.2% P Cu alloy: Wrought
SAE CA 521	8% Sn 0.2% P Cu alloy: Wrought; formerly SAE 77C
SAE CA 524	10% Sn 0.2% P Cu alloy: Wrought
SAE CA 544	4% Sn 4% Pb 4% Zn Cu alloy: Wrought
SAE CA 836	5% Sn 5% Pb 5% Zn Cu alloy: Casting
SAE CA 838	3.8% Sn 6% Pb 6.5% Zn Cu alloy: Casting
SAE CA 903	8.2% Sn 4% Zn Cu alloy: Casting
SAE CA 905	10% Sn 2% Zn Cu alloy: Casting
SAE CA 909	11% Sn Cu alloy: Casting
SAE CA 922	6% Sn 1.5% Pb 4% Zn Cu alloy: Casting
SAE CA 923	8.2% Sn 0.7% Pb 3.7% Zn Cu alloy: Casting
SAE CA 925	11% Sn 1.2% Pb 1.2% Ni Cu alloy: Casting
SAE CA 927	10% Sn 1.7% Pb Cu alloy: Casting
SAE CA 929	10% Sn 2.6% Pb 3.1% Cu alloy: Casting
SAE CA 932	7% Sn 7% Pb 3% Zn Cu alloy: Casting
SAE CA 935	5.2% Sn 9.5% Pb 2% Zn Cu alloy: Casting
SAE CA 937	10% Sn 9.5% Pb Cu alloy: Casting
SAE CA 938	7% Sn 14% Pb Cu alloy: Casting
SAE CA 943	5.2% Sn 23.5% Pb Cu alloy: Casting
SAE CA 947	5.2% Sn 1.7% Zn 5.2% Ni Cu alloy: Casting
SAE CA 948	5.2% Sn 0.7% Pb 1.7% Zn 5.2% Ni Cu alloy: Casting
SFG	5% or 7% Sn 0.2% P Cu alloy: Strip; IMI grade with improved properties over BS alloys PB 102–103
For SIS specifications see SS	
SOUDOR G Cu Sn 6	6.3% Sn 0.2% P Cu alloy: Welding electrode; Soudometal **UTS: 300 Elon: 40% Proof: 130**
SOUDOTEG Cu Sn 6	6.3% Sn 0.2% P Cu alloy: Welders electrode; Soudometal; for inert gas welding **UTS: 300 Elon: 40% Proof: 130**
SPECULUM	35% Sn Cu alloy: For electroplating; used for plating cutlery and household goods
SPOOLARC GRADE C	8.0% Sn 0.2% P Cu alloy: Brazing rod; All-State for welding phosphorus bronze **DPN: 90 UTS: 420 Elon: 55% Proof: 200**
SPRABRONZE P	5% Sn Cu alloy: Wire for spraying; Metco Ltd **UTS: 100**
SS 5203	5.5% Sn 3% Pb 7% Zn Cu alloy: Casting; Swedish Standard; as cast **DPN: 70 UTS: 200 Elon: 20% Proof: 100**
SS 5204	5% Sn 5% Pb 5% Zn Cu alloy: Casting; Swedish Standard; as cast **DPN: 70 UTS: 200 Elon: 20% Proof: 100**
SS 5420	1% Sn 0.2% P Cu alloy: Plate; Swedish Standard; annealed **DPN: 65 UTS: 270 Elon: 50% Proof: 60**
SS 5420	1% Sn 0.2% P Cu alloy: Plate; Swedish Standard; cold drawn **DPN: 100 UTS: 340 Elon: 15% Proof: 290**
SS 5428	7% Sn 0.3% P Cu alloy: Plate and bar; Swedish Standard; annealed **DPN: 85 UTS: 390 Elon: 55% Proof: 150**
SS 5428	7% Sn 0.3% P Cu alloy: Plate and bar; Swedish Standard; cold rolled **DPN: 210 UTS: 720 Elon: 10% Proof: 630**
SS 5431	9% Sn 0.3% P Cu alloy: Plate and bar; Swedish Standard; cold rolled **DPN: 200 UTS: 530 Elon: 25% Proof: 390**
SS 5443	10% Sn Cu alloy: Casting; Swedish Standard; as cast **DPN: 75 UTS: 250 Elon: 18% Proof: 150**
SS 5444	9% Sn 3% Pb 2% Zn Cu alloy: Casting; Swedish Standard; as cast **DPN: 75 UTS: 240 Elon: 16% Proof: 150**
SS 5465	12% Sn Cu alloy: Casting; Swedish Standard; as cast **DPN: 90 UTS: 240 Elon: 12% Proof: 150**
SS 5475	14% Sn Cu alloy: Casting; Swedish Standard; as cast **DPN: 100 UTS: 210 Elon: 4% Proof: 170**
SS 5640	10% Sn 10% Pb Cu alloy: Casting; Swedish Standard; as cast **DPN: 70 UTS: 180 Elon: 12% Proof: 90**
SS 8300	0.1% C (max) sintered Cu **UTS: 463 Elon: 28%**
SS 8320	10% Sn 0.5% C Cu alloy: Sinter
SS 8321	10% Sn 1.5% C Cu alloy: Sinter
SS 8330	10% Sn 10% P 0.5% C and Cu alloy: Sinter
STA 7 CG1	10% Sn 2% Zn Cu alloy: Bar; hard rolled **UTS: 530 Elon: 18%**
STA 7 CG2	10% Sn 0.5% Pb 2.0% Zn Cu alloy: Casting; sand cast **UTS: 240 Elon: 10%**
STA 7 CG3	8.0% Sn 0.5% Pb 4.0% Zn Cu alloy: Casting; sand cast **UTS: 240 Elon: 10%**
STA 7 CG4	7% Sn 2.0% Pb 5.0% Zn Cu alloy: Casting; sand cast **UTS: 210 Elon: 10%**
STA 7 CG5	5.0% Sn 5.0% Pb 5.0% Zn Cu alloy: Casting; sand cast **UTS: 181 Elon: 10%**
STA 7 CG6	3.0% Sn 5.0% Pb 9.0% Zn Cu alloy: Casting; sand cast **UTS: 170 Elon: 10%**
STA 7 CP1	5.0% Sn 0.3% P Cu alloy: Sheet, tube and wire; annealed **DPN: 80 UTS: 300 Elon: 45%**
STA 7 CP1	5.0% Sn 0.3% P Cu alloy: Sheet, tube and wire; hard **UTS: 860**
STA 7 CP2	4.0% Sn 0.3% P Cu alloy: Bar; as drawn **UTS: 370 Elon: 20%**
STA 7 CP3	7.0% Sn 3.0% Pb 2.0% Zn 0.3% P 1.0% Ni Cu alloy: Casting; chill cast **UTS: 210 Elon: 1.5%**
STA 7 CP4	8.0% Sn 0.3% P Cu alloy: Casting; as cast; for bearings **UTS: 220 Elon: 10%**
STA 7 CP5	10% Sn 0.2% Pb 0.5% P Cu alloy: Casting; chill cast **UTS: 240 Elon: 1.5%**
STA 7 CP6	12% Sn 0.5% Pb 0.15% P 0.5% Ni Cu alloy: Casting; chill cast: **DPN: 90 UTS: 250 Elon: 3%**
STA 7 CX7	8.0% Sn 0.3% P Cu alloy: Bar and tube; hard drawn **DPN: 120 UTS: 450 Elon: 20%**
STA 7 CX8	9.0% Sn 2.0% Zn Cu alloy **UTS: 210 Elon: 3.5%**
STA 7 CX9A	15% Sn 1.0% Zn 0.2 Pb Cu alloy
STA 7 CX9B	18% Sn 0.2% Pb 0.7% Zn Cu alloy

Symbol	Nominal analysis, supplier, condition and remarks.
STA 7 CX12	6.0% Sn 2.0% Zn 0.3% P 2.0% Ni Cu alloy: Casting; as cast **UTS: 300 Elon: 20% Proof: 150**
STA 7 CZ12	2.0% Sn 3.0% Pb 35% Zn 1.0% Ni Cu alloy: Casting; as cast **UTS: 170 Elon: 20%**
STA 7 CZ13	1.0% Sn 3.0% Pb 35% Zn 1.0% Ni 1.0% Fe Cu alloy: Casting; as cast **UTS: 180 Elon: 12%**
STUBS 32	Sn Cu alloy: Welding electrode; Stubs; for welding bronze and surface coating steel/cast iron **DPN: 75 UTS: 340 Elon: 34%**
STUBS 32W	Sn Cu alloy: Welding electrode; Stubs; for welding bronze and surface coating steel/cast iron **DPN: 100 UTS: 290 Elon: 20%**
STUBS 39	Sn Mn Si Cu alloy: Welding electrode; Stubs; for welding pure copper **DPN: 60 UTS: 200 Elon: 35%**
STUBS 39W	Sn Mn Si Cu alloy: Welding rod; Stubs; for welding pure copper **DPN: 60 UTS: 200 Elon: 35%**
STUBS 320	Sn Cu alloy: Welding electrode; Stubs; welding bronzes, surfacing steel and cast iron **DPN: 100 UTS: 260 Elon: 25%**
STUBS A 320	12% Sn Cu alloy: Welding electrode; Stubs; for joining bronze and facing steel, cast iron **DPN: 100 UTS: 295 Elon: 25% Proof: 160**
SUPER-HOLFOS WW	11.5% Sn 0.2% P Cu alloy: Casting; centri-cast; J Holroyd for BS alloy PB 2 **DPN: 100 UTS: 300 Elon: 6% Proof: 180**
T 52	5.5% Sn 0.2% P Cu alloy: Tube; see BS T52
TAURUS	10.5% Sn 0.25% P 0.7% Ni Cu alloy: Casting; D Brown; centri-cast **DPN: 75 UTS: 270 Elon: 22% Proof: 170**
TAURUS I	12% Sn 0.3% P Cu alloy: Casting; D Brown; centri-cast **DPN: 90 UTS: 280 Elon: 5% Proof: 210**
TAURUS II	10% Sn 0.2% P Cu alloy: Casting; D Brown; sand cast **DPN: 70 UTS: 270 Elon: 15% Proof: 120**
TAURUS III	14% Sn 0.5% P Cu alloy: Casting; D Brown; sand cast **DPN: 100 UTS: 210 Elon: 2% Proof: 180**
TAURUS IV	10.5% Sn 0.25% P 1% Ni Cu alloy: Casting; D Brown; sand cast **DPN: 80 UTS: 270 Elon: 20% Proof: 170**
TAURUS VI	10% Sn 2% Zn Cu alloy: Casting; D Brown; sand cast **DPN: 60 UTS: 270 Elon: 15% Proof: 140**
TAURUS VII	12% Sn 5% Pb 0.3% P 1% Ni Cu alloy: Casting; D Brown; sand cast **DPN: 75 UTS: 220 Elon: 6% Proof: 180**
TAURUS VIII	10% Sn 3% Pb 0.25% P 1% Ni Cu alloy: Casting; D Brown; sand cast **DPN: 70 UTS: 240 Elon: 9% Proof: 170**
TAURUS IX	10% Sn 11.5% Pb 0.25% P 1% Ni Cu alloy: Casting; D Brown; sand cast **DPN: 70 UTS: 220 Elon: 8% Proof: 170**
TAURUS XI	5% Sn 5% Pb 5% Zn Cu alloy: Casting; D Brown; sand cast **DPN: 50 UTS: 220 Elon: 20% Proof: 90**
TAURUS XIV	7% Sn 2% Pb 5% Zn Cu alloy: Casting; D Brown; sand cast **DPN: 60 UTS: 240 Elon: 10% Proof: 100**
TAURUS XVI	9% Sn 0.25% P 0.5% Ni Cu alloy: Casting; D Brown; sand cast **DPN: 75 UTS: 330 Elon: 25% Proof: 180**
TAURUS XXXIII	9.5% Sn 2% Zn 1.5% Ni Cu alloy: Casting; D Brown; sand cast **DPN: 85 UTS: 300 Elon: 12% Proof: 180**
TELFOS	5% Sn Cu alloy: Bar; free machining quality; C Clifford
UE 3Z9	3.0% Sn 8.2% Zn 0.2% P (max) Cu alloy: Designation for French Standards
UE 5P	5.0% Sn 0.3% P (max) Cu alloy: Designation used by French Standards
UE 5Pb 5Z5	5.0% Sn 5.0% Pb 5.0% Zn Cu alloy: Casting; designation used by French Standards
UE 5Z4	4.0% Sn 4.0% Zn 0.2% P (max) Cu alloy: Designation used by French Standards
UE 7Z5Pb4	7.2% Sn 3.9% Pb 5.0% Zn Cu alloy: Casting; designation used by French Standards
UE 8Z2	8.5% Sn 2.0% Pb (max) 2.5% Zn Cu alloy: Casting; designation for French Standards
UE 9P	8.2% Sn 0.35% P (max) Cu alloy: Designation used by French Standards
UE 10Z1	9.5% Sn 0.2% Pb (max) 0.2% Zn (max) Cu alloy: Casting; designation used by French Standards
UE 12P	11.5% Sn 1.0% Pb (max) 0.2% Zn (max) Cu alloy: Casting; designation used by French Standards
UE 12Z1	11.5% Sn 1.5% Pb (max) 0.2% Zn (max) Cu alloy: Casting; designation used by French Standards
UE 14	13.5% Sn 1.0% Pb (max) 1.0% Zn (max) Cu alloy: Casting; designation used by French Standards
UE 16	16.0% Sn 1.0% Pb (max) 1.0% Zn (max) Cu alloy: Casting; designation used by French Standards
UE 18	18.0% Sn 1.0% Pb (max) 0.5% Zn (max) Cu alloy: Casting; designation used by French Standards
UE Pb Z	4.5% Sn (max) 5.0% Pb 5.0% Zn Cu alloy: Casting; designation used by French Standards
VALVE BRONZE	6% Sn 4.5% Pb 6% Zn Cu alloy: Origin unknown
VALVE METAL	3% Sn 7% Pb 9% Zn Cu alloy: Origin unknown
VALVIT	9% Sn Cu alloy: For bearings; origin unknown
YORCASTON	12% Sn Cu alloy: Tube; Yorkshire Imperial Metals Ltd

Note. The following abbreviations and units are used in the tables:

DPN	Hardness, diamond pyramid number
UTS	Ultimate tensile strength, N/mm^2
Elon	Elongation, %
Proof	0.1% proof strength, N/mm^2

$1\ N/mm^2 = 0.1\ hbar = 0.102\ kgf/mm^2 = 0.06475\ tonf/in.^2 = 145.04\ lbf/in.^2 = 1\ MPa$

See Appendix II for other abbreviations and conversion tables.

14L Copper–chromium wrought and cast alloys

Specific gravity		8.94
Density		8940 kg/m^3
Solidus/liquidus		1073–1081 °C
Thermal conductivity	solution treated and aged	306 W/m °C
Coefficient of linear expansion		17.7 × 10^{-6}/ °C
Electrical conductivity	aged	80% IACS (copper 100%)
	solution treated	45% IACS (copper 100%)
Specific resistance	aged	22 microhm mm
	solution treated	39 microhm mm
Young's modulus of elasticity		110 × 10^9 N/m^2
Impact		Can be expected to be low
Fatigue strength (10^8 cycles)		±190 N/mm^2
Hot strength		The alloys retain their strength up to 350 °C

The above properties are typical of the following group and may not apply exactly to any one specification. It is possible that with certain specifications some of the values may not be applicable.

General metallurgical characteristics

The specifications listed have approximately 0.5% chromium added to pure copper. The resulting intermetallics can be dissolved in the copper by heating to the solution treatment temperature and then quenching, when the material is very soft and ductile but suffers a drop in conductivity. After ageing at a low temperature, there is an increase in hardness and conductivity with a decrease in ductility. These properties are maintained in service up to 350 °C at least, which gives chromium–copper an advantage over work hardened copper alloys when hardness at high temperatures is necessary.

These materials are difficult to cold work and therefore only a limited amount of annealing is required.

The majority of materials in this group can be solution treated whereby the chromium is taken into solution and with a subsequent ageing operation some hardening is achieved.

It is doubtful whether this relatively expensive technique is used on many components but it will certainly enhance the mechanical strength of the material and, where maximum hardness or strength is required with the other properties, then this is an advantage.

It is unusual to have these materials welded although brazing can be carried out quite simply provided the surface oxide is removed mechanically and active fluxes are removed. They will need considerable pre-heating as they have reasonably good thermal conductivity and good electrical conductivity.

No corrosion protection will be required on these alloys under normal circumstances. Painting or electroplating can be used where necessary.

The principal use of these materials is in resistance welding electrodes. Their high electrical conductivity along with their good heat conductivity and reasonably high strength makes them ideal for this purpose. Resistance electrodes are required to conduct electricity into the component involved, to withstand reasonably high temperatures when the weld takes place and to have sufficient strength not to 'mushroom' when the pressure is applied.

They also find use as switch contacts, rotors in electric motors, and in some cases cylinder heads where high strength and good corrosion resistance at high temperature are required.

Symbol	Nominal analysis, supplier, condition and remarks.
2H	1% Cr 0.1% Zr Cu alloy: Langley Alloys; hard rolled; bar
4H	1% Cr 0.1% Zr Cu alloy: Langley Alloys; cold formed
182	1.0% Cr Cu alloy: Wrought; designation agreed by CDAA; not used
185	0.7% Cr Cu alloy: Wrought; designation agreed by CDAA; not used
BOLTOMET 814	Cr Cu alloy: For commutator bars; T Bolton
BOLTOMET 818	Cr Mg Cu alloy: For machined products; Thomas Bolton
BS 4577 A2/1	1% Cr Cu alloy: For resistance welding electrodes; wrought; conductivity 78% IACS **DPN: 120 UTS: 380 Elon: 15%**
BS 4577 A2/2	1% Cr 0.1% Zr Cu alloy: Resistance welding electrodes; wrought; conductivity 75% IACS **DPN: 130 UTS: 380 Elon: 15%**
C18000	2.5% Ni 0.5% Cr Cu alloy: CDA and UNS designation
C18030	0.1% Sn 0.15% Cr Cu alloy: CDA and UNS designation
C18040	0.25% Sn 0.3% Cr Cu alloy: CDA and UNS designation
C18050	0.1% Cr Cu alloy: CDA and UNS designation
C18070	0.05% Si 0.25% Cr Cu alloy: CDA and UNS designation
C18090	1% Sn 1% Ni 0.7% Cr Cu alloy: CDA and UNS designation
C18100	0.6% Cr Cu alloy: CDA and UNS designation
C18135	0.4% Cr Cu alloy: CDA and UNS designation

Note. The following abbreviations and units are used in the tables:

DPN	Hardness, diamond pyramid number
UTS	Ultimate tensile strength, N/mm^2
Elon	Elongation, %
Proof	0.1% proof strength, N/mm^2

1 N/mm^2=0.1 hbar=0.102 kgf/mm^2=0.06475 tonf/in.^{2}=145.04 lbf/in.^{2}=1 MPa

See Appendix II for other abbreviations and conversion tables.

Symbol	Nominal analysis, supplier, condition and remarks.
C18150	1% Cr Cu alloy: CDA and UNS designation
C18200	1% Cr Cu alloy: CDA and UNS designation
C18400	0.8% Cr Cu alloy: CDA and UNS designation
C18500	0.7% Cr Cu alloy: CDA and UNS designation
C18550	0.12% Sn 0.8% Cr Cu alloy: CDA and UNS designation
C81500	1% Cr Cu alloy: Cast; CDA and UNS designation
CHROMIUM COPPER	0.5% Cr Cu alloy: Solution treated; used for cylinder heads and resistance welding electrodes; common name
CUPALOY	0.5% Cr 0.1% Ag Cu alloy: Origin unknown
CUPROCHROM	1.8% Cr Cu alloy: For electronic grids. Driver
DTD 354	2% Sn 2% Fe 1.5% Cr Cu alloy: Bar and tube; chromium bronze; as drawn for valve guides **UTS: 330 Elon: 25%**
ERM CCS	0.6% Cr Cu alloy: Drawn bar; Enfield Rolling Mills Ltd; welding electrodes, switchgear, etc.; electrical conductivity 85% IACS **DPN: 150 UTS: 480 Elon: 25% Proof: 400**
ERM CCS/Mg	0.6% Cr 0.03% Mg Cu alloy: Bar and forging; Enfield; 80% IACS; as drawn **DPN: 150 UTS: 480 Elon: 25% Proof: 400**
ERM CCS/Z	0.6% Cr 0.15% Zr Cu alloy: Bar and forging; Enfield for switchgear; IACS 80%; as drawn
HIDURAL 6	Cr Cu alloy: For resistance electrodes; Langley Alloys; 80% IACS **DPN: 100 UTS: 490 Proof: 330**
HIDURAL 640	Cr Zr Cu alloy: For resistance electrodes; Langley Alloys; 80% IACS **DPN: 100 UTS: 490 Proof: 330**
IMI 171	0.5% Cr Cu alloy: For electrode tips; IMI; KUMIUM
KUMIUM	0.7% Cr Cu alloy: Bar; solution treated and cold rolled; IMI; resistance weld electrodes switch contacts **UTS: 490 Elon: 12% Proof: 400**
MALLORY 3	Cr Cu alloy: Casting; Mallory Metallurgical; conductivity 75% IACS **DPN: 120 UTS: 330 Elon: 15% Proof: 270**
MALLORY 3	Cr Cu alloy: For resistance welding electrodes for mild steel; Mallory Metallurgical; electrical conductivity 80% IACS **DPN: 140 UTS: 510 Elon: 15% Proof: 460**
MALLORY 328	Cr Zr Cu alloy: For resistance welding electrodes for light alloys; Mallory Metallurgical; electrical conductivity 80% IACS **DPN: 140 UTS: 510 Elon: 15% Proof: 460**
MALLORY 328	Cr Zr Cu alloy: Casting; Mallory Metallurgical; electrical conductivity 75% IACS **DPN: 120 UTS: 330 Elon: 15% Proof: 240**
MATTHEY 3	1% Cr Cu alloy: For resistance welding electrodes; Johnson Matthey; conductivity 80% IACS **DPN: 120 UTS: 520 Elon: 15%**
MATTHEY 328	1% Cr 0.1% Zr Cu alloy: For resistance welding electrodes; Johnson Matthey; conductivity 75% IACS **DPN: 140 UTS: 520 Elon: 15%**
RWMA CLASS 2	Cr Cu alloy: Mallory Metallurgical code for Mallory 3 and 328
SAE CA184	0.8% Cr Cu alloy: For welding electrodes
VERILITE	1.5% Ni 1.5% Cr 1.0% Co Cu alloy: Origin unknown

Note. The following abbreviations and units are used in the tables:

DPN	Hardness, diamond pyramid number
UTS	Ultimate tensile strength, N/mm^2
Elon	Elongation, %
Proof	0.1% proof strength, N/mm^2

1 N/mm^2=0.1 hbar=0.102 kgf/mm^2=0.06475 $tonf/in.^2$=145.04 $lbf/in.^2$=1 MPa

See Appendix II for other abbreviations and conversion tables.

14M Copper–silicon wrought and cast alloys
With manganese, zinc and iron additions – silicon brass and bronze

Specific gravity		8.52
Density		8520 kg/m^3
Solidus/liquidus		1020 °C
Thermal conductivity		32.7 W/m °C
Coefficient of linear expansion		18 × 10^{-6} °C
Electrical conductivity	annealed	6.6% IACS (copper 100%)
Specific resistance	annealed	257 microhm mm
Young's modulus of elasticity		103 × 10^9 N/m^2
Impact	20 °C	89 J
	30 °C	100 J
	50 °C	98 J
	80 °C	94 J
	115 °C	87 J
Fatigue strength (30 × 10^7 cycles)	cold drawn	±220 N/mm^2
	annealed	±120 N/mm^2

Hot strength

Temperature °C	Tensile strength N/mm^2	Elongation %
20	360	66
100	330	59
200	300	54
300	270	52

The above properties are typical of the following group and may not apply exactly to any one specification. It is possible that with certain specifications some of the values may not be applicable.

General metallurgical characteristics

Silicon added to copper acts as a de-oxidizer and also enhances the strength and corrosion resistance, especially to acids. The fluidity of the molten metal is increased, which helps to give sound castings and makes welding easier. There is a large increase in the electrical resistance, allowing the use of these alloys for resistance welding purposes where good corrosion resistance is an advantage.

Manganese up to about 1% is generally present with all the straight silicon bronzes, acting as a de-sulphurizer and aiding the corrosion resistance. When zinc is present up to about 15% the alloys are known as silicon brasses. These alloys have excellent casting characteristics and can be readily cold worked; they are not affected by season cracking which can be troublesome with high copper–zinc alloys, the actual cracking taking place during storage. The zinc-free alloys maintain their mechanical properties including ductility at temperatures down to −100 °C.

The mechanical properties of these materials cannot be improved by thermal treatment. They are difficult to cold work but, where cold working does exist without any problems, then this can be removed at temperatures as low as 300 °C but may require temperatures up to 550 °C.

Like the aluminium copper alloys these materials form a silicon oxide at the surface when heat treated and this makes them difficult to braze or weld without the use of very active fluxes. At the same time this is one of their advantages in corrosion resistance.

It is not advised to weld or braze these materials and if it is essential then specific advice should be taken.

Corrosion protection is not normally required on these materials.

These alloys find many uses in the marine industry for components such as nails, bolts, nuts, etc. Paper-making equipment also makes use of silicon bronzes and brasses and in some cases they are used as bearings. Some bells are made from these materials but this is not as common as with the standard bronzes.

Engineering makes use of the materials for acid storage tanks, piping, pickling crates and general handling equipment.

Many of these uses are being replaced by type 316 austenitic stainless steel.

Symbol	Nominal analysis, supplier, condition and remarks.
2.1461	3.5% Si Cu alloy: German Standard
651	1.5% Zn 0.7% Mn 1.5% Si Cu alloy: Wrought; designation used by CDAA
653	2.3% Si Cu alloy: Wrought; designation agreed by CDAA; not used
656	1.5% Sn 1.5% Zn 1.5% Mn 3.4% Si Cu alloy: Wrought; designation agreed by CDAA; not used
658	1.5% Mn 3.2% Si Cu alloy: Wrought; designation agreed by CDAA; not used

Note. The following abbreviations and units are used in the tables:

DPN	Hardness, diamond pyramid number
UTS	Ultimate tensile strength, N/mm^2
Elon	Elongation, %
Proof	0.1% proof strength, N/mm^2

1 N/mm^2=0.1 hbar=0.102 kgf/mm^2=0.06475 tonf/in.^{2}=145.04 lbf/in.^{2}=1 MPa

See Appendix II for other abbreviations and conversion tables.

Symbol	Nominal analysis, supplier, condition and remarks.
661	0.6% Pb 1.5% Zn 1.5% Mn 3.2% Si Cu alloy: Wrought; designation agreed by CDAA; not used
692	10% Zn 1.2% Si Cu alloy: Wrought; designation agreed by CDAA; not used
694	0.2% Pb 20% Zn 4.0% Si Cu alloy: Wrought; designation agreed by CDAA; not used
697	1.0% Pb 20% Zn 3.0% Si Cu alloy: Wrought; designation agreed by CDAA; not used
AMS 4615 C	3.2% Si Cu alloy: Bar; hard drawn
AMS 4616 B	28% Zn 1.5% Fe 3.2% Si Cu alloy: Bar; forging and tube
AMS 4665 A	3.2% Si Cu alloy: Seamless tube; annealed
ARGOFIL	0.25% Mn 0.25% Si Cu alloy: Wire; IMI; filler rod for welding copper
ASTM B 30/12 A	0.5% Sn 3% Zn 1% Al 1.5% Fe 1% Mn 3% Si Cu alloy: Ingot; as cast **UTS: 320 Elon: 20%**
ASTM B 30/13 A	0.5% Pb 14% Zn 3.5% Si Cu alloy: Ingot; as cast **UTS: 340 Elon: 18%**
ASTM B 30/13 B	14% Zn 4% Si Cu alloy: Ingot; as cast **UTS: 420 Elon: 16%**
ASTM B53 A	11% Si Cu alloy: Ingot
ASTM B53 B	20% Si Cu alloy: Ingot
ASTM B53 C	30% Si Cu alloy: Ingot
ASTM B96 A	3.2% Si Cu alloy: Plate and sheet; annealed **UTS: 400 Elon: 40% Proof: 100**
ASTM B96 C	2.5% Si Cu alloy: Plate and sheet; annealed **UTS: 400 Elon: 40% Proof: 100**
ASTM B97	3.2% Si Cu alloy: Sheet, strip and bar; annealed **UTS: 400**
ASTM B97	3.2% Si Cu alloy: Sheet; strip and bar; spring hard **UTS: 810**
ASTM B97 B	1.4% Si Cu alloy: Sheet and bar; annealed **UTS: 290**
ASTM B97 B	1.4% Si Cu alloy: Sheet and bar; spring hard **UTS: 550**
ASTM B97 C	2.5% Si Cu alloy: Sheet and bar; annealed **UTS: 400**
ASTM B97 C	2.5% Si Cu alloy: Sheet and bar; spring hard **UTS: 810**
ASTM B98/651	1.5% Zn (max) 1.4% Si Cu alloy: Rod; previously alloy B
ASTM B98/655	1.5% Mn (max) 3.3% Si Cu alloy: Rod; previously alloy A
ASTM B98/661	0.4% Pb 1.5% Mn (max) 3.1% Si Cu alloy: Rod; previously alloy D
ASTM B98 A	3.2% Si Cu alloy: Bar; annealed **UTS: 360 Elon: 35% Proof: 90**
ASTM B98 A	3.2% Si Cu alloy: Bar; extra hard **UTS: 730 Elon: 7% Proof: 390**
ASTM B98 B	1.4% Si Cu alloy: Bar; annealed **UTS: 270 Elon: 30% Proof: 75**
ASTM B98 B	1.4% Si Cu alloy: Bar; extra hard **UTS: 580 Elon: 8% Proof: 330**
ASTM B98 D	0.6% Pb 3.1% Si Cu alloy: Bar; annealed **UTS: 360 Elon: 35% Proof: 90**
ASTM B98 D	0.6% Pb 3.1% Si Cu alloy: Bar; extra hard **UTS: 730 Elon: 7% Proof: 390**

Note. The following abbreviations and units are used in the tables:

DPN	Hardness, diamond pyramid number
UTS	Ultimate tensile strength, N/mm^2
Elon	Elongation, %
Proof	0.1% proof strength, N/mm^2

$1\ N/mm^2 = 0.1\ hbar = 0.102\ kgf/mm^2 = 0.06475\ tonf/in.^2 = 145.04\ lbf/in.^2 = 1\ MPa$

See Appendix II for other abbreviations and conversion tables.

Symbol	Nominal analysis, supplier, condition and remarks.
ASTM B99 A	3.2% Si Cu alloy: Wire; annealed **UTS: 440 Elon: 47%**
ASTM B99 A	3.2% Si Cu alloy: Wire; spring hard **UTS: 950 Elon: 4%**
ASTM B99 B	1.4% Si Cu alloy: Wire; annealed **UTS: 290 Elon: 40%**
ASTM B99 B	1.4% Si Cu alloy: Wire; spring hard **UTS: 750 Elon: 6%**
ASTM B124/7	3% Si Cu alloy: Forging and bar; mechanical properties not quoted
ASTM B176 ZS 144 A	15% Zn 4% Si Cu alloy: Casting; die cast **UTS: 630 Elon: 25% Proof: 340**
ASTM B198/12 A	1% Sn 5% Zn 1.5% Al 4% Si Cu alloy: Casting; sand cast **UTS: 300 Elon: 20% Proof: 100**
ASTM B198/13 A	1% Pb 14% Zn 3% Si Cu alloy: Casting; sand cast **UTS: 330 Elon: 18% Proof: 120**
ASTM B198/13 B	14% Zn 4% Si Cu alloy: Casting; sand cast **UTS: 420 Elon: 16% Proof: 150**
ASTM B259 R Cu Si A	1% Sn 1% Zn 3% Si Cu alloy: Welding rod **UTS: 340**
ASTM B259 R Cu Si B	1% Sn 1% Zn 1.5% Si Cu alloy: Welding rod **UTS: 220**
ASTM B315/615	1.4% Si Cu alloy: Pipe
ASTM B315/655	3.2% Si Cu alloy: Pipe
ASTM B315/658	0.9% Mn 3.2% Si Cu alloy: Pipe
ASTM B315 A	1% Mn 3.2% Si Cu alloy: Pipe
ASTM B315 A7	1.2% Mn 3.1% Si Cu alloy: Pipe
ASTM B371 A	15% Zn 4% Si Cu alloy: Rod; annealed **UTS: 560 Elon: 15% Proof: 270**
ASTM B371 B	20% Zn 3% Si Cu alloy: Rod; annealed **UTS: 420 Elon: 22% Proof: 210**
BOLTOMET 968	3% Si Cu alloy: T Bolton
BS 1029	Silicon bronze ingots and casting; replaced within BS 1400
BS 1030	Silicon bronze ingots and casting; replaced within BS 1400
BS 1541 CS101	1% Mn 3% Si Cu alloy: Plate; as rolled **UTS: 340 Elon: 45%**
BS 1866	1% Mn 3% Si Cu alloy: Tube; solid drawn; annealed **UTS: 370 Elon: 50%**
BS 1948 CS101	1% Mn 3% Si Cu alloy: Rods and sections; annealed **UTS: 360 Elon: 50%**
BS 2870 CS101	1% Mn 3% Si Cu alloy: Sheet; as rolled; silicon bronze **UTS: 360 Elon: 50%**
BS 2871 CS101	1% Mn 3% Si Cu alloy: Tube; annealed **UTS: 420**
BS 2872 CS101	1% Mn 3% Si Cu alloy: Forging; as forged **UTS: 330 Elon: 50%**
BS 2873 CS101	1% Mn 3% Si Cu alloy: Wire; mechanical properties not quoted
BS 2874 CS101	1% Mn 3% Si Cu alloy: Bar; annealed **UTS: 360 Elon: 50%**
BS 2875 CS101	1% Mn 3% Si Cu alloy: Plate; as rolled **UTS: 340 Elon: 45%**
BS 2901 C9	1% Mn 3% Si Cu alloy: Rod for all forms of welding
C9	3.0% Si Cu alloy: Rod for gas shielded arc welding
C65100	1.5% Zn 1.8% Si Cu alloy: CDA and UNS designation
C65300	2.3% Si Cu alloy: CDA and UNS designation
C65400	1.5% Sn 3% Si 0.1% Cr Cu alloy: CDA and UNS designation
C65500	3.2% Si Cu alloy: CDA and UNS designation
C65600	3.2% Si Cu alloy: CDA and UNS designation
C65620	2% Zn 1.5% Fe 3.2% Si Cu alloy: CDA and UNS designation
C65800	3.2% Si Cu alloy: CDA and UNS designation
C66100	0.6% Pb 3.2% Si Cu alloy: CDA and UNS designation
C69400	15% Zn 4% Si Cu alloy: CDA and UNS designation
C69430	15% Zn 4% Si Cu alloy: CDA and UNS designation
C69440	15% Zn 4% Si Cu alloy: CDA and UNS designation
C69450	15% Zn 4% Si Cu alloy: CDA and UNS designation

Symbol	Nominal analysis, supplier, condition and remarks.
C69700	1% Pb 18% Zn 3% Si Cu alloy: CDA and UNS designation
C69710	1% Pb 18% Zn 3% Si Cu alloy: CDA and UNS designation
C69720	1% Pb 18% Zn 3% Si Cu alloy: CDA and UNS designation
C69730	1% Pb 18% Zn 3% Si Cu alloy: CDA and UNS designation
C87200	5% Zn 3% Si Cu alloy: Obsolete; UNS designation
C87300	1.2% Mn 4% Si Cu alloy: Cast; CDA and UNS designation
C87400	14% Zn 3.2% Si Cu alloy: Cast; CDA and UNS designation
C87410	14% Zn 3.2% Si Cu alloy: Cast; CDA and UNS designation
C87420	14% Zn 3.2% Si Cu alloy: Cast; CDA and UNS designation
C87430	14% Zn 3.2% Si Cu alloy: Cast; CDA and UNS designation
C87500	14% Zn 4% Si Cu alloy: Cast; CDA and UNS designation
C87510	14% Zn 4% Si Cu alloy: Cast; CDA and UNS designation
C87520	14% Zn 4% Si Cu alloy: Cast; CDA and UNS designation
C87530	14% Zn 4% Si Cu alloy: Cast; CDA and UNS designation
C87600	5.5% Zn 4.5% Si Cu alloy: Cast; CDA and UNS designation
C87610	4% Zn 4% Si Cu alloy: Cast; CDA and UNS designation
C87800	14% Zn 4% Si Cu alloy: Cast; CDA and UNS designation
DIN 1733 S Cu Si	3.5% Si Cu alloy: Wrought
DTD 263	10% Zn 1.5% Si Cu alloy: Silicon brass; annealed **UTS: 420 Elon: 30%**
DTD 267	10% Zn 1.5% Si Cu alloy: Sheet; ½-hard; cold rolled **UTS: 360 Elon: 20%**
DTD 355	2% Zn 1.7% Fe 3.5% Si Cu alloy: Casting **UTS: 300 Elon: 12%**
E Cu Si	3% Si Cu alloy: Weld electrode; designation used by AWS
EVERDUR	1% Mn 4% Si Cu alloy: Casting; IMI; corrosion resistant; withdrawn
EVERDUR A	1% Mn 3% Si Cu alloy: Bar, sheet, wire and rod; IMI; corrosion resistant with high ductility **UTS: 340 Elon: 70%**
EVERDUR D	1% Mn 4% Si Cu alloy: Ingot; IMI
HIDUREL 6	1% Cr 0.1% Zr Cu alloy: Langley Alloys
HIDUREL 640	1% Cr 0.1% Zr Cu alloy: Langley Alloys
HOLFOS JHR42	2% Zn 1.5% Fe 0.5% Mn 3.5% Si Cu alloy: Casting; J Holroyd; as cast **DPN: 110 UTS: 340 Elon: 18% Proof: 200**
IMI 176	0.25% Mn 0.25% Si Cu alloy: Welding rod; IMI; Argofil
IMI 705	1% Mn 3% Si Cu alloy: Sheet and wire; IMI; Everdur A
JB 33 A	10% Zn 1.5% Si Cu alloy: Proprietary name; consult Copper Development Association for further details
JHR 42	2% Zn 1.7% Fe 3.5% Si Cu alloy: Casting; J Holroyd for DTD 355 **DPN: 105 UTS: 340 Elon: 15% Proof: 200**
MALLORY 53	Ni Si Cu alloy: For welding electrodes; Mallory Metallurgical
PMG	2% Zn 3.5% Si Cu alloy: Casting; Vickers Armstrong; DTD 355
QQ B 746/A	0.3% P 3.5% Si Cu alloy: Plate and sheet; US Federal **UTS: 690 Elon: 2.5%**
QQ C 593	0.5% Zn 2.5% Fe 3.0% Si Cu alloy: Casting; US Federal **UTS: 300 Elon: 15%**
S Cu Si	3.5% Si Cu alloy: Wrought; designation used by German Standards
SA 4	8% (Si + Al) Cu alloy: Manganese Bronze Ltd
SAE CA655	3% Si Cu alloy: Wrought
SAE CA 655	1.5% Fe 3% Si + Mn + Zn (max) Cu alloy: Wrought
SILICON BRONZE SA 4	8% (Si + Al) Cu alloy: Manganese Bronze Ltd
STA 7 CSIC	5.0% Zn 2.0% Fe 1.0% Mn 4.0% Si Cu alloy: Casting; as cast **UTS: 300 Elon: 15%**

Note. The following abbreviations and units are used in the tables:

DPN	Hardness, diamond pyramid number
UTS	Ultimate tensile strength, N/mm^2
Elon	Elongation, %
Proof	0.1% proof strength, N/mm^2

1 N/mm^2=0.1 hbar=0.102 kgf/mm^2=0.06475 tonf/in.^{2}=145.04 lbf/in.^{2}=1 MPa

See Appendix II for other abbreviations and conversion tables.

14N Copper alloys – miscellaneous

In this group are a relatively small number of alloys which have been found which do not fit into any of the existing alloy listings in this book. These alloys have been produced for specific purposes. In some cases the purposes for which they have been produced are known and will be indicated alongside the alloy itself. In other instances a very limited amount of information is available on the material but it was felt advisable to inform readers of the maximum information which was available rather than to ignore the fact that an alloy with a specific reference existed.

Any information which readers can supply on these materials will be acknowledged and used in any future edition.

Symbol	Nominal analysis, supplier, condition and remarks.
AMS 4764	9.5% Ni 38% Mn Cu alloy
C69900	2% Al 44% Mn Cu alloy: Incramute; CDA and UNS designation
C69900	2% Al 44% Mn Cu alloy: CDA and UNS designation
C69910	4% Zn 1.2% Fe 30% Mn Cu alloy: Incramute; CDA and UNS designation
C69910	0.6% Al 1.2% Fe 30% Mn Cu alloy: CDA and UNS designation
C69950	9.5% Ni 38% Mn Cu alloy: Incramute; CDA and UNS designation
C69950	9.5% Ni 38% Mn Cu alloy: CDA and UNS designation
C72400	2% Al 13% Ni Cu alloy: CDA and UNS designation
C72420	1.5% Al 15% Ni 4.5% Mn Cu alloy: CDA and UNS designation
C72500	2.2% Sn 9.5% Ni Cu alloy: CDA and UNS designation
C72600	4% Sn 4% Ni Cu alloy: CDA and UNS designation

Symbol	Nominal analysis, supplier, condition and remarks.
C99600	1.8% Al 42% Mn Cu alloy: Cast; CDA and UNS designation
C99600	1.7% Al 42% Mn Cu alloy: CDA and UNS designation
C99700	22% Zn 2% Al 5% Ni 14% Mn 5% Nb Cu alloy: Casting; CDA and UNS designation
C99700	22% Zn 2% Al 5% Ni 13% Mn Cu alloy: Cast; CDA and UNS designation
C99750	1.7% Sn 1.7% Pb 20% Zn 20% Mn 2% Ag Cu alloy: Casting; CDA and UNS designation
C99750	1.7% Sn 20% Zn 1.8% Al 5% Ni 20% Mn Cu alloy: Cast; CDA and UNS designation
INCRAMUTE 1	1.9% Al 42% Mn Cu alloy: Cast; origin unknown

Note. The following abbreviations and units are used in the tables:

DPN	Hardness, diamond pyramid number
UTS	Ultimate tensile strength, N/mm^2
Elon	Elongation, %
Proof	0.1% proof strength, N/mm^2

1 N/mm^2=0.1 hbar=0.102 kgf/mm^2=0.06475 $tonf/in.^2$=145.04 $lbf/in.^2$=1 MPa

See Appendix II for other abbreviations and conversion tables.

15. Gallium Ga

Physical properties	
Atomic number	31
Atomic weight	69.72
Crystal structure	Orthorhombic
Colour	Silver white
Specific gravity	5.91
Density	5910 kg/m^3
Melting point	29.8 °C
Boiling point	2403 °C
Specific heat	0.33 J/g °C
Thermal conductivity	41 W/m °C
Coefficient of linear expansion (20–30 °C)	58 × 10^{-6}/ °C (thermal expansion varies in different directions)
Latent heat of fusion	80.4 J/g
Latent heat of vaporization	3894 J/g
Thermal neutron absorption cross-section	2.8 barns/atom
Electrical conductivity (a-axis)	10% IACS (copper 100%)
Specific resistance (resistivity) (a-axis)	174 microhm mm
Temperature coefficient of electrical resistance	0.0040/ °C
Electrochemical equivalent	2.45 g/A/h
Electrode potential	−0.52 V
Magnetic susceptibility	−0.24 × 10^{-6}
Young's modulus of elasticity	–
Tensile strength	less than 150 N/mm^2
Hardness	50 DPN

15.1 General notes on gallium

Gallium is one of those elements which appear very sparingly in nature, and yet whose occurrence is widespread. Most coal has a minute percentage of the element which can be recovered from the flue dust when large quantities of coal are burned.

The metal is soft and readily cut, but is also reasonably brittle. It has a brilliant lustre when first cut, then oxidizes rapidly. It resembles aluminium and indium in many respects.

The long temperature range of the liquid phase, from 30–2000 °C, makes this a potentially interesting material. Unfortunately, liquid gallium, even at reasonably low temperatures, is corrosive to most other metals. The metals tantalum and tungsten resist this attack, as do quartz, graphite and alumina. Using quartz capillary tubes gallium thermometers have been produced that are capable of measuring accurately temperatures up to 1200 °C.

Gallium alloys are available with very low melting points. Two eutectic alloys exist with sharp melting points.

82% Ga 12% Sn 6% Zn	Melting point 17 °C
76% Ga 24% In	Melting point 15.7 °C

Again, these materials attack other metals at elevated temperatures; otherwise they would probably find considerable use as transfer liquids.

Like antimony and bismuth, gallium expands on solidification. With the low melting points of the element and some of its alloys, this can be a serious disadvantage, causing strain on the container in the same manner as water freezing and fracturing its pipe.

Some use has been made of gallium in vapour arc lamps, usually in conjunction with cadmium, but also with mercury.

There are reports that additions of gallium to magnesium and some magnesium–tin alloys markedly increase the corrosion resistance.

The high cost of extraction and purification and the corrosive attack on other metals will probably prevent any wider use of this metal and its alloys.

Note. The following abbreviations and units are used in the tables:

DPN	Hardness, diamond pyramid number
UTS	Ultimate tensile strength, N/mm^2
Elon	Elongation, %
Proof	0.1% proof strength, N/mm^2

1 N/mm^2=0.1 hbar=0.102 kgf/mm^2=0.06475 tonf/in.^{2}=145.04 lbf/in.^{2}=1 MPa

See Appendix II for other abbreviations and conversion tables.

Symbol	Nominal analysis, supplier, condition and remarks.
GALLIUM	High purity metal: Impurities less than 4 p.p.m.; Mallory Metallurgical; supplied as rod for semiconductors
h Ga 1	99.9999% Ga rod: Light Ltd; high purity gallium

16. Germanium Ge

Physical properties	
Atomic number	32
Atomic weight	72.6
Crystal structure	Cubic
Colour	Greyish white
Specific gravity	5.32
Density	5320 kg/m^3
Melting point	958 °C
Boiling point	2700 °C
Specific heat	0.301 J/g °C
Thermal conductivity	6.28 W/m °C
Coefficient of linear expansion (20–100 °C)	6×10^{-6}/ °C
Latent heat of fusion	427 J/g
Latent heat of vaporization	3345 J/g
Thermal neutron absorption cross-section	2.5 barns/atom
Electrical conductivity	3% IACS (copper 100%)
Specific resistance	500 microhm mm
Temperature coefficient of electrical resistance	–
Electrochemical equivalent	–
Electrode potential	–
Magnetic susceptibility	-0.12×10^{-6}
Young's modulus of elasticity	–
Tensile strength	–
Hardness	–

16.1 General notes on germanium

Germanium is a metalloid, having some non-metallic characteristics, but not as marked as in silicon, or boron. It is a very brittle material, quite stable in air up to red heat.

Until the unusual electrical properties became known germanium had very little use. There are no commercial alloys based on the metal, but additions to aluminium copper alloys ('Duralumin') increases their strength and hot working properties.

When added to tin an alloy with considerably greater hardness and some reduction in ductility is obtained. Gold with 13% germanium forms a eutectic alloy melting at 356 °C which finds some use as a special purpose solder.

The high cost and scarcity of germanium tend to persuade potential users to alternative materials.

By far the largest use of germanium is in the electronics industry where its electrical rectification properties have been found to be extremely useful.

The voltage range of germanium covers the gap between silicon and selenium rectifiers, but by far the largest use of the metal is as 'transistors'. These are germanium crystals, which rectify radio signals using considerably less power in the process than conventional valves. In addition they are very much smaller and more robust, and constitute the core of modern electronic devices which can operate for long periods on minute power supplies, the transistor radio set being a good example.

Note. The following abbreviations and units are used in the tables:

DPN	Hardness, diamond pyramid number
UTS	Ultimate tensile strength, N/mm^2
Elon	Elongation, %
Proof	0.1% proof strength, N/mm^2

1 N/mm^2=0.1 hbar=0.102 kgf/mm^2=0.06475 tonf/in.^{2}=145.04 lbf/in.^{2}=1 MPa

See Appendix II for other abbreviations and conversion tables.

Symbol	Nominal analysis, supplier, condition and remarks.
GERMANIUM	99.999% Ge ingot: Koch Light
GERMANIUM I	Ge metal: Impurities below spectographic detection; Mallory Metallurgical; ingot; specific resistance 0.02 microhm mm
GERMANIUM II	High purity metal Ge: Impurities less than 1 p.p.m.; Mallory Metallurgical; ingot; specific resistance 0.5 microhm mm
h Ge 6	99.999% Ge: Ingot; Light Ltd; high purity germanium
h Ge 8	99.999% Ge: Powder; Light Ltd; high purity germanium

17. Gold Au

Physical properties	
Atomic number	79
Atomic weight	197.2
Crystal structure	Face-centred cubic
Colour	Yellow
Specific gravity	19.32
Density	19320 kg/m^3
Melting point	1063 °C
Boiling point	2960 °C
Specific heat	0.1323 J/g °C
Thermal conductivity	301 W/m °C
Coefficient of linear expansion (20–100 °C)	14.4 × 10^{-6}/ °C
Latent heat of fusion	66.2 J/g
Latent heat of vaporization	1738 J/g
Thermal neutron absorption cross-section	99 barns/atom
Electrical conductivity	80% IACS (copper 100%)
Specific resistance	22 microhm mm
Temperature coefficient of electrical resistance	0.0034/ °C
Electrochemical equivalent	2.45 g/A/h
Electrode potential	+1.68 V
Magnetic susceptibility	−0.15 × 10^{-6}
Young's modulus of elasticity	80 × 10^9 N/m^2
Tensile strength annealed	100 N/mm^2
Hardness annealed	30 DPN

17.1 General notes on gold

Along with copper, gold is the only truly coloured metal, being pale yellow in the massive state, and red or even purple when finely divided.

The ability of gold to resist oxidation and chemical attack, together with its attractive colour, has made it popular as a means of decoration and a standard of value since the dawn of history.

One major use of gold is in international currency and as such most of it is securely locked in underground vaults in the financial centres of the world.

Considerable quantities are still used for ornamental purposes, often being added as fine powder to pottery glazes and glass ware.

Industrially, gold has only limited uses owing to its artificially high cost and low strength.

The ability to resist oxidation over a large range of temperatures and its high electrical conductivity are now finding use in the electrical and electronic industries. For these purposes very thin electroplated deposits are used. A thin deposit of nickel between the gold and copper base metal prevents diffusion of the gold into copper. Gold plating for purely decorative purposes is also increasing. Modern techniques are increasing the hardness of the plated deposit, which, as it does not oxidize or readily corrode, has no need for abrasive cleaning and can therefore be very thin.

There are also a limited number of chemical engineering and laboratory uses for gold, notably in the man-made fibre industry. Some gold alloy brazing materials are used where excellent joint conductivity under oxidizing conditions is necessary. Gold leaf is still used on articles such as presentation leather books and for outdoor decorations, such as name signs. The high malleability and almost complete absence of cold working, coupled with the high density allows gold to be rolled or beaten into intricate shapes, or very thin foils – leaf gold.

Addition of platinum and palladium to gold can give the alloys age hardening characteristics. The temperatures required are in the region of 300–400 °C, the most popular uses being for personal ornaments and dental fillings.

The 'carat' applied to gold is not the weight measure used for precious stones, but a proportional measure giving the percentage of gold present. Twenty-four carat gold is pure metal; 12 carat gold is 50% gold alloyed with other metals, generally copper with palladium, nickel or silver to give some increase in hardness.

Following are listed the known specifications and trade names with gold as the main constituent.

Symbol	Nominal analysis, supplier, condition and remarks.
9ct No. 90	Gold alloy: Reddish yellow colour; Sheffield Smelting Co.; melting range 880–895 °C **DPN: 110**
9ct No. 91	Gold alloy: Yellow colour; Sheffield Smelting Co.; melting range 855–870 °C **DPN: 110**
9ct No. 92	Gold alloy: Reddish yellow colour; Sheffield Smelting Co.; melting range 920–940 °C **DPN: 110**
9ct No. 93	Gold alloy: Yellow colour; Sheffield Smelting Co.; melting range 870–890 °C **DPN: 190**
9ct No. 94	Gold alloy: Yellow colour; Sheffield Smelting Co.; melting range 835–860 °C **DPN: 110**
9ct No. 95	Gold alloy: Yellow colour; Sheffield Smelting Co.; melting range 860–880 °C; age hardenable **DPN: 110**
9ct No. 97	Gold alloy: Red colour; Sheffield Smelting Co.; melting range 940–955 °C **DPN: 110**
9ct No. 98	Gold alloy: Low zinc; red colour; Sheffield Smelting Co.; melting range 925–945 °C **DPN: 90**
9ct No. 99	Gold alloy: Yellow colour; Sheffield Smelting Co.; melting range 820–840 °C **DPN: 190**
9ct No. 910	Gold alloy: Greenish yellow colour; Sheffield Smelting Co.; melting range 895–920 °C; age hardenable **DPN: 90**
9ct No. 911	Gold alloy: Reddish yellow colour; Sheffield Smelting Co.; melting range 845–870 °C; age hardenable **DPN: 110**
9ct No. 912	Gold alloy: Reddish yellow colour; Sheffield Smelting Co.; melting range 910–935 °C **DPN: 110**
9ct S 91 (easy)	37.5% Au alloy: For soldering; Sheffield Smelting Co.; melting range 610–660 °C
9ct S 92 (medium)	37.5% Au alloy: For soldering; Sheffield Smelting Co.; melting range 700–710 °C
9ct S 93 (hard)	37.5% Au alloy: For soldering; Sheffield Smelting Co.; melting range 740–760 °C
9ct W 91	Gold alloy: White; Sheffield Smelting Co. **DPN: 140**
9ct W 93	Gold alloy: White; Sheffield Smelting Co. **DPN: 85**
9ct W 94	Gold alloy: White; Sheffield Smelting Co. **DPN: 60**
9ct W 95	Gold alloy: White; Sheffield Smelting Co. **DPN: 85**
9ct W 96	Gold alloy: White; Sheffield Smelting Co. **DPN: 85**
9ct W 97	Gold alloy: White; Sheffield Smelting Co. **DPN: 85**
9ct W 98	Gold alloy: White; Sheffield Smelting Co. **DPN: 60**
9ct W 99	Gold alloy: White; Sheffield Smelting Co. **DPN: 85**

Note. The following abbreviations and units are used in the tables:

DPN	Hardness, diamond pyramid number
UTS	Ultimate tensile strength, N/mm^2
Elon	Elongation, %
Proof	0.1% proof strength, N/mm^2

1 N/mm^2=0.1 hbar=0.102 kgf/mm^2=0.06475 $tonf/in.^2$=145.04 $ibf/in.^2$=1 MPa

See Appendix II for other abbreviations and conversion tables.

Symbol	Nominal analysis, supplier, condition and remarks.
14/450	Au based solder for dental use: Engelhard Industries; melting range 620–755 °C
14ct No. 144	Gold alloy: Green colour; Sheffield Smelting Co.; melting range 950–975 °C **DPN: 90**
14ct No. 145	Gold alloy: Red colour; Sheffield Smelting Co.; melting range 885–895 °C **DPN: 110**
14ct No. 146	Gold alloy: Reddish yellow colour; Sheffield Smelting Co.; melting range 840–880 °C **DPN: 90**
14ct No. 147	Gold alloy: Green colour; Sheffield Smelting Co.; melting range 920–935 °C **DPN: 110**
14ct No. 148	Gold alloy: Yellow colour; Sheffield Smelting Co.; melting range 830–855 °C; age hardenable **DPN: 190**
14ct No. 149	Gold alloy: Reddish yellow colour; Sheffield Smelting Co.; melting range 810–838 °C **DPN: 90**
14ct S 143 (medium)	58.5% Au alloy: For soldering; Sheffield Smelting Co.; melting range 830–835 °C
14ct S 145 (hard)	58.5% Au alloy: For soldering; Sheffield Smelting Co.; melting range 835–870 °C
14ct S 146 (easy)	58.5% Au alloy: For soldering; Sheffield Smelting Co.; melting range 720–740 °C
14ct S 147 (easy)	58.5% Au alloy: For soldering; Sheffield Smelting Co.; melting range 700–720 °C
14ct W 142	Gold alloy: White; Sheffield Smelting Co. **DPN: 140**
14ct W 143	Gold alloy: White; Sheffield Smelting Co. **DPN: 60**
15ct S 151 (easy)	62.5% Au alloy: For soldering; Sheffield Smelting Co.; melting range 720–750 °C
16/550	Au based solder for dental use: Engelhard Industries; melting range 767–783 °C
18/650	Au based solder for dental use: Engelhard Industries; melting range 785–798 °C
18ct No. 185	Gold alloy: Reddish yellow colour; Sheffield Smelting Co.; melting range 890–915 °C; age hardenable **DPN: 150**
18ct No. 186	Gold alloy: Reddish yellow colour; Sheffield Smelting Co.; melting range 900–920 °C **DPN: 150**
18ct No. 187	Gold alloy: Yellow colour; Sheffield Smelting Co.; melting range 885–905 °C **DPN: 110**
18ct No. 188	Gold alloy: Reddish yellow colour; Sheffield Smelting Co.; melting range 890–910 °C **DPN: 150**
18ct S 181 (hard)	75% Au alloy: For soldering; Sheffield Smelting Co.; melting range 885–905 °C
18ct S 185 (easy)	75% Au alloy: For soldering; Sheffield Smelting Co.; melting range 660–765 °C
18ct S 186 (medium)	75% Au alloy: For soldering, Sheffield Smelting Co.; melting range 650–800 °C
18ct W 181	Gold alloy: White; Sheffield Smelting Co. **DPN: 85**
18ct W 182	Gold alloy: White; Sheffield Smelting Co. **DPN: 226**
18ct W 183	Gold alloy: White; Sheffield Smelting Co. **DPN: 140**
18ct W 184	Gold alloy: Reddish yellow; Sheffield Smelting Co.; melting range 920–935 °C **DPN: 150**
18ct W 184	Gold alloy: White; Sheffield Smelting Co. **DPN: 140**
18ct W 185	Gold alloy: White; Sheffield Smelting Co. **DPN: 60**
20/750	Au based solder for dental use: Engelhard Industries; melting range 707–822 °C

Symbol	Nominal analysis, supplier, condition and remarks.
22/800	Au based solder for dental use: Engelhard Industries; melting range 722–845 °C
22ct No. 223	Gold alloy: Deep yellow; Sheffield Smelting Co.; melting range 890–910 °C **DPN: 90**
424	Cu Au brazing alloy: Melting point 889 °C; Engelhard Industries; electrical conductivity 15% IACS **DPN: 250 UTS: 450 Elon: 17%**
429	Cu Au brazing alloy: Melting range 905–915 °C; Engelhard Industries; electrical conductivity 15% IACS **DPN: 112 UTS: 400 Elon: 38%**
441	Ni Au brazing alloy: Melting point 950 °C; Engelhard Industries **DPN: 245 UTS: 690 Elon: 5%**
450 (fine)	Au based solder for dental use: Engelhard Industries; melting range 620–755 °C
550 (fine)	Au based solder for dental use: Engelhard Industries; melting range 767–783 °C
625R	Cu Ag Au alloy: For sliding contacts; conductivity 14% IACS; Johnson Matthey **DPN: 95**
650 (fine)	Au based solder for dental use: Engelhard Industries; melting range 785–798 °C
750 (fine)	Au based solder for dental use: Engelhard Industries; melting range 707–822 °C
800 (fine)	Au based solder for dental use: Engelhard Industries; melting range 722–845 °C
AMERICAN GOLD	90% Au 10% Cu standard coinage alloy of the USA
AMS 4784	25% Ni 25% Pd Au alloy: Braze metal
AMS 4785	36% Ni 34% Pd Au alloy: Braze metal
AMS 4786	22% Ni 8% Pd Au: Braze alloy
AMS 4787	28% Ni Au alloy: Braze metal
AMS 7731	99.95% Au: Refined gold
ANORMAT	78.3% Au + Pt for dental purposes: Mallory Metallurgical; melting range 915–917 °C; hardened **DPN: 165 Elon: 26%**
ASTM B260/B Au2	20% Cu Au brazing filler metal: Brazing temperature range 890–982 °C
ASTM B260/B Au4	18.5% Ni Au brazing filler metal: Brazing temperature range 919–1004 °C
ASTM B260 B Cu Au2	20% Cu Au alloy: Braze filler; melting range 880–887 °C
AUBEL	92% (min) Au for dental purposes: Mallory Metallurgical; melting range 970–995 °C **DPN: 67 Elon: 50%**
AUDENWIRE	75% Au + Pt alloy: Wire for dental purposes; Mallory Metallurgical; melting range 965–1000 °C; hardened **DPN: 339**
AUMET	83.3% Au + Pt for dental purposes: Mallory Metallurgical; melting range 1000–1085 °C **DPN: 71 Elon: 47%**
BAKER 4	Gold alloy used in dentistry: With Pt and Pd; Engelhard Industries; annealed; melting range 850–910 °C **DPN: 130 UTS: 480 Elon: 25% Proof: 370**
BAKER 4	Gold alloy used in dentistry: With Pt and Pd; Engelhard Industries; age hardened **DPN: 228 UTS: 850 Elon: 4% Proof: 600**
BINORMAT	78.3% Au + Pt for dental purposes: Mallory Metallurgical; melting range 925–945 °C **DPN: 100 Elon: 42%**
BORAWIRE	61% Au + Pt alloy: Wire for dental purposes; Mallory Metallurgical; melting range 1080–1180 °C; hardened **DPN: 296**
BS 1845 Au 1	19% Cu 1% Fe Au alloy: For braze metal; melting range 905–910 °C
BS 1845 Au 2	37.5% Cu Au alloy: For braze metal; melting range 930–940 °C
BS 1845 Au 5	17.5% Ni Au alloy: For braze metal; melting point 950 °C
BS 1845 Au 6	25% Ni Au alloy: For braze metal; melting point 950 °C
BS 3384 A	65% Au (min) dental solder
BS 3384 B	53% Au (min) dental solder
BS 4425	Au alloy used for dental casting graded by mechanical properties
CHICAGO 4	Gold alloy used in dentistry: High Pt; Engelhard Industries; annealed; melting range 890–940 °C **DPN: 150 UTS: 550 Elon: 30% Proof: 340**
CHICAGO 4	Gold alloy used in dentistry: High Pt; Engelhard Industries; age hardened; melting range 890–940 °C **DPN: 235 UTS: 780 Elon: 5% Proof: 660**
COINAGE GOLD	0.15% Ag Au alloy: 22 carat; balance Cu
CORODENT	71% Au + Pt for dental purposes: Mallory Metallurgical; melting range 935–995 °C; hardened **DPN: 192 Elon: 17%**
DENTECON	63% Au + Pt for dental purposes: Mallory Metallurgical; melting range 885–945 °C; hardened **DPN: 266 Elon: 12%**
DENTORMAT	76% Au + Pt for dental purposes: Mallory Metallurgical; melting range 880–925 °C; hardened **DPN: 252 Elon: 8%**
DORDENT	72.5% Au + Pt for dental purposes: Mallory Metallurgical; melting range 870–920 °C; hardened **DPN: 298 Elon: 2.5%**
GOLD	High purity metal: Impurities less than 5 p.p.m.; Mallory Metallurgical; supplied as sponge rod sheet and wire
h Au 3	99.9999% Au sponge: Light Lab Ltd; high purity metal
h Au 8	99.999% Au powder: Light Lab Ltd; high purity metal
JMC 625	Cu Ag Au alloy: For wiping contacts; Mallory Metallurgical
JMC 625 R	Cu Ag alloy: For wiping contacts; Mallory Metallurgical
JMM 625	Cu Ag Au alloy: For sliding contacts; conductivity 12% IACS; Johnson Matthey **DPN: 175**
NIORO	18% Ni Au alloy: For brazing; melting point 950 °C; Wesgo
NICORO 80	2.0% Ni 16% Cu Au alloy: For brazing; melting range 910–925 °C; Wesgo
ORO BRAZE	Cu Fe 80% Au alloy: Brazing metal; Mallory Metallurgical; melting range 908–910 °C
ORO BRAZE 910	Cu Fe 80% Au alloy: For brazing; Mallory Metallurgical; melting range 908–910 °C
ORO BRAZE 940	Cu 62.5% Au alloy: Brazing metal; Mallory Metallurgical; melting range 930–940 °C
ORO BRAZE 950	17.5% Ni Au alloy: Bearing metal; Mallory Metallurgical; melting point 950 °C
ORO BRAZE 990	25% Ni Au alloy: Bearing metal; Mallory Metallurgical; melting range 950–990 °C
ORO BRAZE 1040	30% Ag Au alloy: For brazing; Mallory Metallurgical; melting range 1030–1040 °C
ORO CAST	Gold alloy: Used in dentistry; with Pt metals; Engelhard Industries; annealed; melting range 850–910 °C **DPN: 142 UTS: 510 Elon: 25% Proof: 330**
ORO CAST	Gold alloy: Used in dentistry; with Pt metals; Engelhard Industries; age hardened; melting range 850–910 °C **DPN: 229 UTS: 800 Elon: 6% Proof: 770**
PALNIRO 1	25% Pd 25% Ni Au alloy: For brazing; melting range 1102–1121 °C; Wesgo
PALORO	8.0% Pd Au alloy: For brazing; melting range 1200–1240 °C; Wesgo
PLATINUM COLOUR SOLDER	Au based solder with Pd and Pt for dental use: Engelhard Industries; melting range 832–860 °C
POO560	4.2% Ag 33% Cu 6.7% Zn Au alloy: UNS designation
POO580	3.5% Ag 31% Cu 7% Zn Au alloy: UNS designation
QA Wire	Gold based Pt and Pd alloy: Engelhard Industries
SILCORO 60	20% Cu 20% Ag Au alloy: For brazing; melting range 835–845 °C; Wesgo

Symbol	Nominal analysis, supplier, condition and remarks.
SILCORO 75	20% Cu 5.0% Ag Au alloy: For brazing; melting range 885–895 °C; Wesgo
SUPER ORALIUM	Gold alloy used in dentistry: With Pt and Pd; white colour; Engelhard Industries; annealed; melting range 923–975 °C **DPN: 175 UTS: 530 Elon: 27% Proof: 390**
SUPER ORALIUM	Gold alloy used in dentistry: With Pt and Pd; white colour; Engelhard Industries; age hardened; melting range 923–975 °C **DPN: 220 UTS: 830 Elon: 2% Proof: 790**
TRUCAST HARD	Gold alloy used in industry: Contains Pt; Engelhard Industries; annealed; melting range 900–950 °C **DPN: 127 UTS: 400 Elon: 24% Proof: 210**
TRUCAST HARD	Gold alloy used in dentistry: Contains Pt; Engelhard Industries; age hardened; melting range 900–950 °C **DPN: 159 UTS: 480 Elon: 8% Proof: 320**
TRUCAST MEDIUM	Gold alloy used in dentistry: Pt free; Engelhard Industries; melting range 920–955 °C **DPN: 90 UTS: 340 Elon: 31% Proof: 170**
TRUCAST SOFT	Gold alloy used in dentistry: Pt free; Engelhard Industries; melting range 940–970 °C **DPN: 59 UTS: 250 Elon: 33% Proof: 100**
UNIGOLD	Gold alloy used in dentistry: With Pt and Pd; Engelhard Industries; annealed; melting range 850–890 °C **DPN: 142 UTS: 460 Elon: 8% Proof: 320**
UNIGOLD	Gold alloy used in dentistry: With Pt and Pd; Engelhard Industries; age hardened **DPN: 229 UTS: 890 Elon: 1% Proof: 780**
WHITE GOLD SOLDER	12% Ni 15% Zn Au alloy: Origin unknown

Note. The following abbreviations and units are used in the tables:

DPN	Hardness, diamond pyramid number
UTS	Ultimate tensile strength, N/mm^2
Elon	Elongation, %
Proof	0.1% proof strength, N/mm^2

1 N/mm^2=0.1 hbar=0.102 kgf/mm^2=0.06475 $tonf/in.^2$=145.04 $lbf/in.^2$=1 MPa

See Appendix II for other abbreviations and conversion tables.

17A. Hafnium Ha

Physical properties	
Atomic number	72
Atomic weight	178.5
Crystal structure	–
Colour	Silver
Specific gravity	13.31
Density	13310 kg/m^3
Melting point	2150 °C
Boiling point	5400 °C
Specific heat	0.144 J/g °C
Thermal conductivity	22–29 W/m °C
Coefficient of linear expansion (20–100 °C)	6×10^{-6}/ °C
Latent heat of fusion	–
Latent heat of vaporization	–
Thermal neutron absorption cross-section	High
Electrical conductivity	48% IACS (copper 100%)
Specific resistance	35.5 microhm mm
Temperature coefficient of electrical resistance	–
Electrochemical equivalent	–
Electrode potential	–
Magnetic susceptibility	–
Young's modulus of elasticity	–
Tensile strength	–
Hardness	–

17A.1 General notes on hafnium

Hafnium is a metal very similar to zirconium and is invariably found with zirconium. It is found in the USA, principally in Florida, and in Australia and Brazil. Some is found also in Africa and India.

The metal is quite hard, silver in colour, looking like stainless steel. It is used to fabricate nuclear reactor control rods but apart from this has no other use.

Hafnium has good mechanical properties and high corrosion resistance. It can be successfully alloyed with other metals such as iron, titanium, niobium and tantalum. It is an extremely rare element, however, and because of its scarcity finds very few uses.

A mixture of 4 parts of tantalum carbide to 1 part hafnium carbide is the most refractory substance known, having a melting point of 4215 K (3940 °C).

There are no hafnium alloys.

Note. The following abbreviations and units are used in the tables:

DPN	Hardness, diamond pyramid number
UTS	Ultimate tensile strength, N/mm^2
Elon	Elongation, %
Proof	0.1% proof strength, N/mm^2

1 N/mm^2=0.1 hbar=0.102 kgf/mm^2=0.06475 tonf/in.^{2}=145.04 lbf/in.^{2}=1 MPa

See Appendix II for other abbreviations and conversion tables.

Symbol	Nominal analysis, Supplier, condition and remarks.
ASTM B737 R1	High purity Hf rod and wire nuclear grade
ASTM B737 R2	High purity Hf rod and wire nuclear grade
ASTM B737 R3	Hf rod and wire alloying grade
ASTM B776 R1	High purity Hf
ASTM B776 R2	High purity Hf
ASTM B776 R3	Alloy grade Hf

18. Indium In

Physical properties	
Atomic number	49
Atomic weight	114.82
Crystal structure	Face-centred tetragonal
Colour	Silver-white
Specific gravity	7.31
Density	7310 kg/m^3
Melting point	156.4 °C
Boiling point	2100 °C
Specific heat	0.239 J/g °C
Thermal conductivity	24.7 W/m °C
Coefficient of linear expansion (20–100 °C)	33×10^{-6}/ °C
Latent heat of fusion	28.5 J/g
Latent heat of vaporization	2022 J/g
Thermal neutron absorption cross-section	196 barns/atom
Electrical conductivity	20% IACS (copper 100%)
Specific resistance	90 microhm mm
Temperature coefficient of electrical resistance	0.0045/ °C
Electrochemical equivalent	1.4 g/A/h
Electrode potential	−0.34 V
Magnetic susceptibility	0.11×10^{-6}
Young's modulus of elasticity	10.8×10^9 N/m^2
Tensile strength annealed	4.5 N/mm^2
Hardness	less than 10 DPN

18.1 General notes on indium

The compounds of indium occur in small quantities widely scattered throughout the world. To date no deposits large enough to warrant extraction of the metal are known, but certain tin, lead and zinc ores contain appreciable quantities.

The purified metal is very soft and ductile, being readily cut with a knife. The surface does not oxidize in air at room temperature and indium can be easily welded to itself without heat and with very little pressure.

The only known use for the metal is as a sealing material for high vacuum equipment under sterile conditions and no alloys are known with indium as the base metal.

The largest quantity is electrodeposited, some for its attractive lustre and good wearing properties. However, it is more generally applied as a very thin top coat on lead, silver or cadmium bearings. This is then diffused into the substrate coating by a heat treatment and imparts corrosion resistance and improved frictional properties. Some modern cars use thin steel shell bearings coated with lead bronze, indium plated.

When indium is added to the lead–tin–bismuth alloy – Woods Metal – the melting point can be reduced to as low as 48 °C. This alloy is used in dentistry and plastic surgery. Small additions of indium strengthen gold and platinum alloys without detracting too much from their other properties.

Indium and its compounds are now finding some use as semiconductors in the electronics industry.

Following are listed the known suppliers.

Note. The following abbreviations and units are used in the tables:

DPN	Hardness, diamond pyramid number
UTS	Ultimate tensile strength, N/mm^2
Elon	Elongation, %
Proof	0.1% proof strength, N/mm^2

1 N/mm^2=0.1 hbar=0.102 kgf/mm^2=0.06475 tonf/in.^{2}=145.04 lbf/in.^{2}=1 MPa

See Appendix II for other abbreviations and conversion tables.

Symbol	Nominal analysis, supplier, condition and remarks.
ASTM B774 244	48% Sn In alloy: Melting point 118 °C
ASTM B774 296	3% Ag In alloy: Melting point 147 °C
ASTM B774 300/302	15% Pb 5% Ag In alloy: Melting range 149–150 °C
ASTM B774 320/345	30% Pb In alloy: Melting range 160–174 °C
h In 1	99.9999% In rod 12 mm diameter: Light Ltd; high purity metal
h In 96	99.8% In 115 stable isotope: Light Ltd; isotope 115
INDIUM	High purity metal: Impurities 1 p.p.m.; Mallory Metallurgical; supplied as ingot, sheet, wire
INDIUM	In metal: Blackwells; pure metal

19. Iridium Ir

Physical properties

Atomic number		77
Atomic weight		192.1
Crystal structure		Face-centred cubic
Colour		Grey
Specific gravity		22.65
Density		22 650 kg/m^3
Melting point		2443 °C
Boiling point		4500 °C
Specific heat		0.135 J/g °C
Thermal conductivity		147 W/m °C
Coefficient of linear expansion (20–100 °C)		6.5×10^{-6}/ °C
Latent heat of fusion		27 600 J/g
Latent heat of vaporization		636 000 J/g
Thermal neutron absorption cross-section		440 barns/atom
Electrical conductivity		32% IACS (copper 100%)
Specific resistance		47 microhm mm
Temperature coefficient of electrical resistance		0.0041/ °C
Electrochemical equivalent		–
Electrode potential		+ 1.0 V
Magnetic susceptibility		0.15×10^{-6}
Young's modulus of elasticity		524×10^9 N/m^2
Tensile strength	cold drawn	2000 N/mm^2
	annealed	1000 N/mm^2
Hardness	cold drawn	650 DPN
	annealed	370 DPN

19.1 General notes on iridium

Iridium invariably occurs in association with platinum, often being found as the native platinum/osmium/iridium alloy 'osmiridium'.

Metal of 99.9% purity is obtained as a by-product in the preparation of platinum. This can be hot worked at white heat to bars or hollow shapes required for laboratory use. At room temperature it is hard and brittle, readily crumbling to a coarse powder if cold worked to any great extent. Iridium is the densest known element.

The pure metal has excellent chemical resistance to the extent of not being attacked by aquaregia. This property accounts for the limited use made of the metal which would be increased but for the extreme difficulty of working and shaping.

Platinum is hardened by the addition of iridium, with little reduction in ductility up to 20% alloy content. Above this figure the material loses its ductility. These platinum–iridium alloys are used for laboratory ware and electrical contacts where hammering action at high speeds occurs. Under certain conditions of alternating current these materials have superior electrical performance to tungsten contacts. The natively occurring alloy 'osmiridium' is still used for tipping pen nibs. Until recently this alloy was used exclusively for this purpose, a typical analysis being 35% osmium, 30% iridium, remainder platinum, rhodium and ruthenium. It is now largely superseded by ruthenium sintered powder. The alloys of iridium and platinum, with and without osmium and rhodium, are used to a great extent in jewellery as mounts for precious stones. The 10% iridium/platinum alloy was used to manufacture the standard metre, the standard pound weight and the standard kilogram; the majority of sub-standards use a similar type of alloy. This is indicative of the high degree of stability attributed to these alloys.

One of the isotopes of iridium – Ir 192 – is radioactive and increasing use is being made of this as a source of non-destructive testing.

In general this is a scarcely used, expensive material with a number of potentially useful applications limited by price and the difficulty of working the metal.

Following are listed the known iridium suppliers with their identifying symbols.

Symbol	Nominal analysis, supplier, condition and remarks.
ASTM B671 – 99.8	99.8% Ir
ASTM B671 – 99.9	99.9% Ir
h Ir 8	99.999% Ir sponge: Light Ltd; high purity metal
IRIDIUM	99.9% Ir wire: International Nickel; cold drawn **DPN: 625 UTS: 2000 Elon: 2%**
IRIDIUM	99.9% Ir wire and rod: International Nickel; annealed at 1000 °C for 30 min **DPN: 500 UTS: 1020 Elon: 15%**
IRIDIUM	99.9% Ir wire and rod: International Nickel; annealed at 1200 °C for 30 min **DPN: 350**
IRIDIUM	High purity metal: Impurities less than 10 p.p.m.; Mallory Metallurgical; supplied as sponge
IRIDOSIUM	10% Pt 1.5% Rh 27% Os 6% Ru Ir alloy: Naturally occurring
OSMIRIDIUM	30% Os Ir alloy: Naturally occurring; details from International Nickel; for pen nib tips, etc.
PLATINIRIDIUM	Pt Os Rh Pd Ir alloy: Up to 75% Ir; rare naturally occurring alloy

Note. The following abbreviations and units are used in the tables:

DPN	Hardness, diamond pyramid number
UTS	Ultimate tensile strength, N/mm^2
Elon	Elongation, %
Proof	0.1% proof strength, N/mm^2

1 N/mm^2=0.1 hbar=0.102 kgf/mm^2=0.06475 $tonf/in.^2$=145.04 $lbf/in.^2$=1 MPa

See Appendix II for other abbreviations and conversion tables.

20. Iron Fe

Physical properties	
Atomic number	26
Atomic weight	55.85
Crystal structure	Body-centred cubic up to 950 °C Face-centred cubic up to 1425 °C
Colour	Greyish white
Specific gravity	7.9
Density	7900 kg/m^3
Melting point	1535 °C
Boiling point	2800 °C
Specific heat	0.44 J/g °C
Thermal conductivity	76.2 W/m °C
Coefficient of linear expansion (20–100 °C)	12.2×10^{6}/ °C
Latent heat of fusion	272 J/g
Latent heat of vaporization	6615 J/g
Thermal neutron absorption cross-section	2.53 barns/atom
Electrical conductivity	19% IACS (copper 100%)
Specific resistance	89 microhm mm
Temperature coefficient of electrical resistance	0.0062/ °C
Electrochemical equivalent	1.042 g/A/h – divalent 0.695 g/A/h – trivalent
Electrode potential	−0.04 V
Magnetic susceptibility	Ferromagnetic
Young's modulus of elasticity	$193–200 \times 10^9$ N/m^2
Tensile strength	540 N/mm^2
Hardness	150 DPN

20.1 General notes on iron

In this book iron, iron alloys and cast iron appear in this section, divorced from the carbon/iron alloys – steel – which appear later in Section 44.

Iron followed copper in the service of man and in certain parts of the world, notably India, it is possible that iron tools and ornaments were in fact in use before bronze.

The ores of iron are found in many areas, often in highly concentrated pockets, and there is little doubt that the first iron was accidentally produced when ore was used as fire-bricks or fireplace material. It is this ease of reduction from the ore coupled with large ore deposits which has made iron the principal tool in man's struggle to civilize and improve his lot.

All the industrial countries have or had large iron ore deposits, many of which are becoming exhausted. Large deposits are available in some of the world's hinterlands, such as within Alaska, Australia and North Canada, which can be mined and shipped economically using modern machinery and techniques.

A limited amount of native iron is found, notably one twenty-five ton lump in Greenland, but no commercial quantities have been discovered. Meteorites almost always contain native iron alloyed with nickel, cobalt and other metals.

Until the end of the last century a considerable quantity of wrought iron was made. Iron of a very high purity, albeit in contact with stringers of slag, was produced by this method. This material has now almost all been replaced by mild steel which is made using considerably less human sweat and muscle. For some uses, such as chains, wrought iron was specified until quite recently.

High purity iron is very much a laboratory material and has no real commercial applications. When highly purified, iron has considerable resistance to corrosion. A metallurgical enigma is the 'Delhi Pillar' which is a large structure of pure iron standing unprotected for centuries in the open air with no visible evidence of corrosion.

The 'pure' iron normally used tarnishes in the atmosphere, rapidly forming a scale of complex iron oxides which is only loosely adherent and is then readily removed, allowing the fresh surface to form new scale and repeat the process.

Industrially iron in the cast form finds considerable use.

Further details of wrought and cast iron are given in the subsequent sections.

Steel is an alloy of iron and carbon, often with other alloying elements. These materials are dealt with in Section 44.

There is also a range of highly alloyed heat resistant iron alloys. These are dealt with under nickel–iron, nickel, chrome–iron, etc., in Section 44, but Section 20C is devoted to the iron–nickel alloys which have been developed for their specific magnetic properties.

20A Iron – commercially pure

Specific gravity	7.5–7.8
Density	7500–7800 kg/m^3
Solidus/liquidus	1520–1530 °C
Thermal conductivity	17–38 W/m °C
Coefficient of linear expansion (20–100 °C)	10.2–11.9 × 10^{-6}/ °C
Electrical conductivity	3.5–9% IACS (copper 100%)
Specific resistance	180–520 microhm mm
Young's modulus of elasticity	211.7 × 10^9 N/m^2
Impact	–
Fatigue strength	–
Hot strength	–

The above properties are typical of the following group, and may not apply exactly to any one specification. It is possible that with certain specifications some of the values may not be applicable.

General metallurgical characteristics

Iron is a relatively soft ductile material which has few industrial applications as it cannot compete economically with materials such as mild steel.

Very high purity iron, which is exceedingly difficult to produce and of academic interest only, has excellent corrosion resistance. All other grades tarnish and corrode rapidly. Apart from wrought iron, the specifications listed are generally the primary materials and laboratory use.

The majority of materials in this section cannot have their mechanical properties improved either by thermal treatment or cold working.

High quality pure iron does not harden during cold work at room temperature and thus is capable of accepting considerable amounts of cold work without any increase in hardness. Where minute quantities of impurities are present, however, hardening will occur and annealing or softening will require a temperature of at least 300 °C and probably as high as 500–600 °C.

The materials in this section can be welded and brazed without much problem.

It should be noted, however, that wrought iron is very similar to a book. If the outer cover of the book has a fillet weld attached to it, then this will only attach itself to the cover of the book and any load applied to the cover will tend to open the book at a very low load. Fillet welding should therefore not be attempted on any of the materials in this group without specific advice and trials being carried out.

The same comments apply to brazing because of the lamella nature of wrought iron.

Corrosion protection is not normally required on these components or on any components made from this material as wrought iron has an excellent history of good corrosion resistance.

Very little wrought iron is now used in industry. Its principal use was in the form of chains and similar material where the lack of cold work meant that it did not need annealing after a period of time whereas normal mild steel chain and similar materials cold work during use when they are overloaded and require to be softened periodically and then re-proof tested.

Most of the materials in this group have been replaced either with austenitic type cast iron which is in many cases an alternative substitute or austenitic stainless steel, copper or copper alloys.

Symbol	Nominal analysis, supplier, condition and remarks.
00 Iron	0.025% C 0.1% Si 0.12% Mn steel: Casting; Edgar Allen; Code 687
A GRADE	Low carbon iron: Low Moor for DTD 5092; DTD 5102
ALFENOL	Fe alloy containing 16% Al
ALFER	Fe alloy containing 11/13% Al
ALSIFER	Master alloy: 20% Al 40% Fe 40% Si
ALSIMIN	Ferrosilicon aluminium alloy: 45% Si 15% Al iron
AMS 7706	Fe: Commercially pure plate, sheet, bar, etc.; annealed
AMS 7707	Fe: Commercially pure plate, sheet, bar, etc.; as rolled
ARMCO IRON	Purest form of commercial iron: Soft iron containing less than 0.1% impurities; trade name
ARMICO	0.03% C 0.005% Si 0.15% Ni 0.2% Cu steel: Strip; Bairds
ASTM A42	0.09% (max) Mn wrought iron plate
ASTM A84	0.06% Mn wrought iron for staybolts **UTS: 340 Elon: 30%**
ASTM A207	Wrought iron: Bars and shapes; 0.09% Mn (max)
ASTM A382	Wrought iron: For heat exchanger and condenser tubes
BEST YORKSHIRE	0.06% Mn 0.16% P wrought iron: General name for BS 858
B GRADE	Low carbon iron: Low Moor for DTD 5102
BS 48	Wrought iron: Replaced by BS 51
BS 51 A	0.1% Mn (max) wrought iron: Bar and plate; as rolled **UTS: 330 Elon: 25%**
BS 51 B	0.15% Mn (max) wrought iron: Bar and plate; as rolled **UTS: 330 Elon: 23%**
BS 51 C	Wrought iron: Analysis not specified; as rolled **UTS: 330 Elon: 19%**
BS 601/1	Low C steel: Strip for magnetic purposes; non-oriented; graded by magnetic properties **UTS: 330 Elon: 25% Proof: 170**

Note. The following abbreviations and units are used in the tables:

DPN	Hardness, diamond pyramid number
UTS	Ultimate tensile strength, N/mm^2
Elon	Elongation, %
Proof	0.1% proof strength, N/mm^2

1 N/mm^2=0.1 hbar=0.102 kgf/mm^2=0.06475 tonf/in.^{2}=145.04 lbf/in.^{2}=1 MPa

See Appendix II for other abbreviations and conversion tables.

Symbol	Nominal analysis, supplier, condition and remarks.
BS 601/2	Low C steel: Strip for magnetic purposes; oriented; graded by magnetic properties
BS 762	0.075% Mn wrought iron bars **UTS: 330 Elon: 25% Proof: 170**
BS 858	0.06% Mn 0.16% P wrought iron: 'Best Yorkshire' wrought iron **UTS: 330 Elon: 21% Proof: 170**
CODE 687	0.025% C 0.1% Si 0.12% Mn steel: Casting; Edgar Allen; 00 Iron
DTD 330	Soft iron: Replaced by DTD 5092 and DTD 5102
DTD 5092	Soft iron: Sheet, bar and forging; annealed; magnetic properties quoted in specification
DTD 5102	Soft iron: Sheet, bar and forging; annealed; magnetic properties included in specification
E 10	Fe high purity sinter: Firth Cleveland; specific gravity 6.6–7.0; as sintered **DPN: 100 UTS: 190 Elon: 7%**
E 20	Fe high purity sinter: Firth Cleveland; specific gravity 7.1–7.5; as sintered **DPN: 130 UTS: 250 Elon: 12%**
F 10	Fe sintered products: Specific gravity 5.7–6.1; Firth Cleveland; as sintered **UTS: 120 Elon: 2%**
F 20	Fe sintered products: Specific gravity 6.1–6.5; Firth Cleveland; as sintered **UTS: 160 Elon: 6%**
F 30	3% Cu Fe sintered products: Specific gravity 6.1–6.5; Firth Cleveland; as sintered **UTS: 240 Elon: 1%**
F 40	5.0% Cu Fe sintered products: Specific gravity 6.1–6.5; Firth Cleveland; as sintered **UTS: 250 Elon: 1%**
F 50	10% Cu Fe sintered products: Specific gravity 6.1–6.5; Firth Cleveland; as sintered **UTS: 270 Elon: 1%**
F 70	Iron: Sintered material; bearings; Sintered Products Ltd; 23–28% porosity **UTS: 90 Elon: 1.5%**
F 80	Iron: Sintered material; bearings; Sintered Products Ltd, 15–22% porosity **UTS: 100 Elon: 2.5%**
FE 85	High purity iron: Sintered material; for high magnetic permeability; Sintered Products Ltd; 12–14% porosity **UTS: 200 Elon: 14%**
FE 90	High purity iron: Sintered material for high magnetic permeability; Sintered Products Ltd; 4–8% porosity **UTS: 220 Elon: 26%**
FERROCOR 216	Electrical steel: For fractional horse power motors; Richard Thomas & Baldwins
FERROCOR 253	Electrical steel: For fractional horse power motors; Richard Thomas & Baldwins
FERROCOR 320	Electrical steel: For fractional horse power motors; Richard Thomas & Baldwins
FERROSIL 100	Low C iron: Strip for magnetic purposes; Richard Thomas & Baldwins for BS 601/1
FERROSIL 107	Low C iron: Strip for magnetic purposes; Richard Thomas & Baldwins for BS 601/1
FERROSIL 146	Low C iron: Strip for magnetic purposes; Richard Thomas & Baldwins for BS 601/1
FERROSIL 170	Low C iron: Strip for magnetic purposes; Richard Thomas & Baldwins for BS 601/1
FERROSIL 187	Low C iron: Strip for magnetic purposes; Richard Thomas & Baldwins for BS 601/1
FERROSIL 216	Low C iron: Strip for magnetic purposes; Richard Thomas & Baldwins for BS 601/1
FERROSIL CR253	Low C iron: Strip for magnetic purposes; Richard Thomas & Baldwins for BS 601/1
FERROVAC E	0.007% C Fe high purity iron: Crucible Steel Co.; for magnetic purposes
HAI – JP	99.5% Fe for thermocouples: Harrison Alloys; specific resistance 10 microhm cm
h Fe 11a	99.998% Fe rod 5 mm diameter: Light Ltd; high purity metal
HIPERM (SUPER)	Low carbon iron: Obsolete; Low Moor for DTD 5092 and DTD 5201
IRON	Fe powder: Electrolytic; Blackwells; pure metal
IRON	High purity metal: Impurities less than 10 p.p.m.; Mallory Metallurgical; supplied as sponge, rod, wire and sheet
KB 90	Sintered Fe: High impact strength; specific gravity 7.25; Durasint **DPN: 120 UTS: 430 Elon: 12%**
KB 90/3/4	Sintered Fe: High impact strength; specific gravity 7.25; Durasint **DPN: 240 UTS: 620 Elon: 2%**
PERMET PF1	100% Fe: For magnets; origin unknown
S 10	5% Cu Fe with C sintered products: Firth Cleveland; hardened and tempered **UTS: 500 Elon: 0.5%**
S 20	Fe sinter with C: Firth Cleveland; hardened and tempered **UTS: 470 Elon: 0.5%**
S 30	3.0% Cu Fe with C sinter: Firth Cleveland; hardened and tempered **UTS: 500 Elon: 0.5%**
S 40	Cu P Fe with C sinter: Firth Cleveland; hardened and tempered; file hard **UTS: 500 Elon: 0.8%**
SKF REMKO	0.02% C 0.12% Mn iron: Origin unknown
TRAN-COR A 5	Low carbon steel: Strip for audio transformers; Armco; hot rolled **DPN: 180 UTS: 480 Elon: 10% Proof: 370**
TRAN-COR A 6	Low carbon steel: Strip for audio transformers; Armco; hot rolled **DPN: 180 UTS: 480 Elon: 10% Proof: 370**
TRAN-COR M 14	Low carbon steel: Strip for transformers; Armco; hot rolled **DPN: 200 UTS: 490 Elon: 2% Proof: 450**
TRAN-COR M 15	Low carbon steel: Strip for transformers; Armco; hot rolled **DPN: 200 UTS: 490 Elon: 2% Proof: 450**
TRAN-COR M 17	Low carbon steel: Strip for transformers; Armco; hot rolled **DPN: 190 UTS: 490 Elon: 3% Proof: 440**
TRAN-COR M 19	Low carbon steel: Strip for transformers; Armco; hot rolled **DPN: 175 UTS: 460 Elon: 11% Proof: 390**
TRAN-COR M 22	Low carbon steel: Strip for transformers; Armco; hot rolled **DPN: 171 UTS: 460 Elon: 12% Proof: 370**
TRAN-COR M 27	Low carbon steel: Strip for transformers; Armco; hot rolled **DPN: 160 UTS: 420 Elon: 18% Proof: 330**
TRAN-COR M 36	Low carbon steel: Strip for magnetos; Armco; hot rolled **UTS: 330 Elon: 26% Proof: 210**
TRAN-COR M 43	Low carbon steel: Strip for electric motors; Armco; hot rolled **UTS: 270 Elon: 29% Proof: 150**
W Fe 18	99.9% Fe powder: Light Ltd; made from iron carbonyl

Note. The following abbreviations and units are used in the tables:

DPN	Hardness, diamond pyramid number
UTS	Ultimate tensile strength, N/mm^2
Elon	Elongation, %
Proof	0.1% proof strength, N/mm^2

1 N/mm^2=0.1 hbar=0.102 kgf/mm^2=0.06475 tonf/in.^{2}=145.04 lbf/in.^{2}=1 MPa

See Appendix II for other abbreviations and conversion tables.

20B Iron – cast

Specific gravity	7.2–7.4
Density	7200–7400 kg/m^3
Solidus/liquidus	1150–1450 °C
Thermal conductivity	46–63 W/m °C
Coefficient of linear expansion	10–17 × 10^{-6}/ °C
Electrical conductivity	5–6.5% IACS (copper 100%)
Specific resistance	270–370 microhm cm
Young's modulus of elasticity	138–152 × 10^9 N/m^2
Impact	5.2–22 J
Fatigue strength (10 × 10^6 cycles)	± 15
Hot strength	Not applicable owing to extremely wide variations

The above figures are intended to show typical values and do not apply exactly to any one specification. As with all casting materials wide variations will be found, many of which are considerably outside the above listed ranges.

General metallurgical characteristics

Cast iron can be described as an alloy of iron and carbon.

The normal accepted minimum carbon content is 1.5%; however, most cast irons have carbon in the range 2.5–3.5%. Commercial cast irons are not simply alloys of iron and carbon, but invariably have some silicon, manganese, phosphorus and sulphur present. These elements affect the carbon solubility and give rise to various types of cast iron dependent on the form in which the carbon is present.

Alloying elements may be added to give special effects and various thermal treatments can be applied. The types, alloying contents and thermal treatments overlap to such an extent that it is not considered possible or desirable to split the various cast irons into separate groups.

The fact that these materials are cast in small batches, varying in some degree from each other, makes cast iron rather unique among metallic materials. Control of the low alloy irons is usually by mechanical properties only, the founder using his skills to vary the composition according to size, sectional thickness and surface required on the finished casting.

Briefly the types of cast iron are as follows:

Grey cast iron. This is the most common type, being used for general purpose castings where low cost is the prime factor. This material is easily produced, gives excellent reproduction and allows considerable flexibility in casting design. It is the cheapest metallic material used. The structure is usually a mixture of pearlite and ferrite with flakes of carbon in the form of graphite. The proportion of pearlite and ferrite present is controlled by alloying elements, notably silicon, and also by the rate of cooling from liquid to solid state, a fully pearlitic structure generally giving the best tensile and transverse properties of the unalloyed grey cast irons. The ferritic grades have slightly better ductility.

White cast irons. These contain no free carbon as graphite in the 'as cast' condition and are therefore very hard and brittle with limited use. When alloyed with nickel, chromium or molybdenum a tough, hard abrasion resistant iron is obtained.

Whiteheart malleable cast iron. In the Whiteheart process white iron castings are first produced and then annealed in an oxidizing atmosphere. This results in a gradual migration of the carbon from the centre of the casting to the skin where it is oxidized and removed from the metal. In this process therefore the annealed casting has a low carbon content – sometimes less than 1% with thin sections. The rest of the carbon is present partly in nodular form as free graphite and partly combined, the latter appearing as pearlite in the matrix. The final structure and properties depend on the time and temperature of the heat treatment and also on the original composition of the metal. This material is seldom used now as the long term anneal makes it expensive.

Blackheart malleable cast iron. The Blackheart process differs from the Whiteheart process in that the anneal is carried out in a neutral or slightly carburizing atmosphere; consequently the carbon content of the metal is not appreciably reduced. The carbon in the final casting approaches that of grey cast iron. In the annealing process however, it is deposited in a finely divided form (temper carbon) giving a product with properties more resembling steel than normal grey cast iron.

Modern malleabilizing techniques have substantially improved these processes. Malleable cast iron is used for many engineering components and is particularly suitable when something stronger and tougher than grey cast iron is required, but where cheapness is essential. Compared with other irons malleable cast iron shows less corrosion resistance but its other properties make it suitable for components subject to shock.

There is a certain restriction in the size of malleable castings – generally above 100 lb the process becomes uneconomical as the time at temperature to convert the carbon is excessive. The relationship between surface area and mass is the important factor in the economics of the process.

It should be noted that the term 'malleable' is now being used for spheroidal graphite irons, particularly in America.

Nodular or spheroidal graphite cast irons. These cover a range of cast irons in which the graphite present is in spheroidal form instead of the flake in ordinary cast iron. The change can be caused by various treatments (usually patented) of molten cast iron in the ladle prior to pouring. This controls the manner in which the carbon forms during solidification. The brittleness of normal grey cast iron is partly due to the edges of the graphite flakes forming easy cleavage paths, thus giving areas of weakness. When the graphite is in the nodular form, this weakness is eliminated and a material is obtained of greater strength plus ductility. There can also be a dramatic improvement in impact strength. The structure of nodular iron can be pearlitic or ferritic or combinations of both. The pearlitic type as cast has high strength with relatively low elongation but can be heat treated to produce a ferritic type with somewhat lower strength but much greater elongation. Under certain conditions the ferritic type can be produced 'as cast' but this is not usual.

Acicular cast irons. The term 'acicular' means 'needle-like' and describes the ferritic structure which forms in grey cast iron during the change from austenite into pearlite and graphite on cooling. This can only be produced on unalloyed iron by quenching from between 480 and 260 °C. This is impracticable with castings made in sand and it is necessary to retard the rate of transformation by the addition of alloying elements such as nickel and molybdenum.

Tensile strengths in the range 380–540 N/mm^2 are obtained from acicular iron with flake graphite. By over-alloying and tempering, tensile strength of 630 N/mm^2 can be obtained. Impact resistance of acicular iron is about twice that of high duty iron with a pearlite matrix and resistance to wear is also better owing to higher initial hardness and the fact that acicular irons have work hardening properties.

Most cast irons are covered by the preceding types. The properties of each are dependent on various elements, some of which are normally present because of their presence in the pig iron used to produce the metal. Other alloying elements may be added to produce specific results. Briefly the effects of these elements are as follows.

Silicon. This is a graphite former, helping to convert the iron carbide–cementite in white cast iron to graphite or aiding the formation of ferrite in grey cast iron. It also helps the fluidity at casting and very few irons are produced with less than 2% silicon. Irons with up to 5% silicon generally with nickel and chromium have good oxidation resistance, while higher silicon irons – 15–20% – although brittle and virtually unmachinably hard, have excellent resistance to sulphuric and nitric acids of all concentrations and temperatures.

Chromium. This is a carbide former and thus tends to reduce the amount of graphite present. The carbides formed are very hard and brittle, increasing the hardness and hardenability of cast iron. Chromium also forms an adherent oxide film and when present above about 5% has an appreciable effect on the corrosion resistance.

As with steel, 18% chromium when present with 8% nickel gives an austenite structure which has excellent corrosion resistance. Cast irons never have a truly single phase structure owing to their high carbon content; thus these austenitic irons are not stainless in the sense that applies to 18/8 austenitic steels.

Nickel. This is a ferrite strengthener which encourages the formation of graphite and more than any other single element improves the toughness of cast irons. As stated above, when present in sufficient quantity with chromium, austenitic corrosion resistant irons are formed.

Molybdenum. This is a carbide former which considerably increases the hardenability of cast iron. All hardenable irons with any appreciable sectional thickness will have some molybdenum present, generally with nickel to improve the ductility.

Phosphorus. This forms a hard brittle phase which for most purposes is highly undesirable. It does aid the fluidity of the molten metal however. Phosphorus is present in most pig irons and remains as a residual element unless special fluxing techniques are used for its removal.

Depending on the matrix of the cast iron involved this can have its properties improved in a similar manner to steel by heat treatment. It is strongly advised, however, that no attempt is made to have any heat treatment carried out on any cast iron without the advice of a competent metallurgist who is knowledgeable in the problems which can be involved with cast iron.

It is extremely easy, by the misuse of heat treatment or carrying out heat treatment without knowing exactly the composition of the material being heat treated, to produce a glass-like component. This can have disastrous results in service.

It is also possible to take a material with a reasonable mechanical strength and reduce this mechanical strength but improve the ductility and impact strength; however, this also could result in failure in service.

Surface hardening of cast iron can be accomplished but this again requires skilled advice and very tight controls. This can be achieved either by flame hardening which is difficult to control, or by induction hardening which can be scientifically controlled. This can produce a component with a very hard glass-like surface, for example, on a lathe bed with the underlying material having the properties of cast iron.

It must be appreciated that the underlying materials will themselves be brittle and therefore considerable care is required.

There are a limited number of techniques where welding can be carried out but this requires very specialist equipment, high skills and highly controlled conditions.

Much of the welding carried out on cast iron is in fact a form of brazing. Provided the temperature of about 700 °C is not exceeded then a join can be satisfactorily obtained with cast iron. There are many welding rods which are in fact brazing rods and with the correct flux high quality joins can be achieved. Again, however, it must be appreciated that the temperature of approximately 700 °C maximum should never be exceeded without the considerable danger that a very brittle structure is produced.

Brazing with copper alloys is commonly carried out and again it must be appreciated that a maximum temperature of 700–750 °C is all that is permitted to prevent the formation of a very hard brittle heat affected zone. Pre-heating is generally necessary with most high quality brazes.

Cast irons in the 'as cast' state have an excellent corrosion resistance. The oxide produced at casting is adherent and is almost non-corrosive. Once this is machined however, cast iron will have a corrosion resistance typical of that of mild steel.

The austenitic cast irons have better corrosion resistance but this is not comparable to those of austenitic stainless steel.

Uses of cast iron can be divided into their types.

Grey cast iron is a general purpose material. It is used for drain pipes, machine tool castings and low load components such as brackets where strength is not important. A limited number of gears and shafts are made in this material because of the excellent surface lubrication given by the graphite. These gears and shafts, however, will not withstand impact or high mechanical loads.

Grey cast iron has an almost unique property in that it will damp vibrations. The graphite has the ability to absorb mechanical vibrations and grey cast iron is therefore a silent material. It has this property along with certain magnesium castings and is used on machine tools, etc., very often for this reason.

Malleable cast iron is still used to a limited extent where high production small, almost minute, components can be cast in grey iron and with a relatively simple heat treatment can be transformed into a ductile material. These components will generally have an intricate shape which would be expensive to machine.

White cast iron is an extremely brittle material and is used for wear resistant parts such as on dredger buckets, excavator grabs, crusher jaws, etc. It must invariably be supported with a more ductile material, generally a steel.

Nodular or spheroidal graphite cast irons are finding an ever increasing use in engineering. They are used for crankshafts, camshafts and many other rotating parts. They are used for high duty gears where the strength of the tooth is essential rather than the self-lubricating properties. They are competing with carburized steel components under many circumstances.

Austenitic cast irons are used for marine purposes, pump bodies, general engineering where corrosion is present. They are finding increasing use in the petro-chemical industry where complicated shapes can be manufactured in a casting much more economically than in austenitic stainless steel which then requires to be machined.

It must, however, be appreciated that the corrosion resistance of austenitic cast iron will never be comparable to that of austenitic steel such as Type 316.

Symbol	Nominal analysis, supplier, condition and remarks.
0707	Nodular graphite cast iron: Pearlitic/ferritic structure; designation used by Danish Standards **DPN: 230 UTS: 600 Elon: 3% Proof: 350**
0708	Nodular graphite cast iron: Pearlitic structure; designation used by Danish Standards **DPN: 265 UTS: 700 Elon: 2% Proof: 400**
0715	Nodular graphite cast iron: Pearlitic/ferritic matrix; designation used by Danish Standards **DPN: 165 UTS: 400 Elon: 17% Proof: 250**
0716	Nodular graphite cast iron: Pearlitic/ferritic matrix; designation used by Danish Standards **DPN: 165 UTS: 400 Elon: 17% Proof: 250**
0727	Nodular graphite cast iron: Pearlitic/ferritic structure; designation used by Danish Standards **DPN: 205 UTS: 500 Elon: 7% Proof: 310**
380–17	Spheroidal graphite cast iron: Bulgaria Standard
400–12	Spheroidal graphite cast iron: Bulgaria Standard
450–5	Spheroidal graphite cast iron: Bulgaria Standard
500–2	Spheroidal graphite cast iron: Bulgaria Standard
600–2	Spheroidal graphite cast iron: Bulgaria Standard
700–2	Spheroidal graphite cast iron: Bulgaria Standard
800–2	Spheroidal graphite cast iron: Bulgaria Standard
900–2	Spheroidal graphite cast iron: Bulgaria Standard
A 32/101	General standard for flake cast iron: Grey; French Standard; see designation for details
A 32/101 F40D	Fine ground flake graphite cast iron: French Standard **DPN: 250 UTS: 400**
A 32/201	General standard for nodular graphite cast iron: French Standard; see designation for details
AMS 5310 B	Pearlitic malleable cast iron: Hardened and tempered
AMS 5315	Nodular ductile cast iron **UTS: 420**
AMS 5316	Nodular ductile cast iron **UTS: 580**
AMS 5328	0.3% C 0.8% Cr 1.8% Ni 0.35% Mo: Investment cast iron
AMS 5329	0.3% C 0.8% Cr 1.8% Ni 0.35% Mo: Sand cast iron
AMS 5330	0.42% C 0.8% Cr 1.8% Ni 0.35% Mo: Investment cast iron
AMS 5331	0.42% C 0.8% Cr 1.8% Ni 0.35% Mo: Sand cast iron
AMS 5333	0.15% C 0.5% Cr 0.5% Ni 0.2% Mo: Investment cast iron
AMS 5334A	0.3% C 0.5% Cr 0.5% Ni 0.2% Mo: Investment cast iron
AMS 5335A	0.3% C 0.5% Cr 0.5% Ni 0.2% Mo: Sand cast iron
AMS 5336	0.3% C 0.95% Cr 0.2% Mo: Investment cast iron
AMS 5338	0.4% C 0.95% Cr 0.2% Mo: Investment cast iron
ASTM A43	Pig iron for foundry use; various grades by analysis
ASTM A47	Malleable cast iron: Composition not specified **UTS: 370 Elon: 18% Proof: 240**
ASTM A48	Grey iron castings: Suffix number indicates tensile strength in p.s.i. × 1000
ASTM A53 T/T 60/40/18	Spheroidal cast iron with ferritic matrix **DPN: 160 UTS: 36 Elon: 17% Proof: 22**
ASTM A53 T/T 65/45/12	Spheroidal cast iron with mainly ferritic matrix **DPN: 170 UTS: 40 Elon: 12% Proof: 27**
ASTM A53 T/T 80/55/06	Spheroidal cast iron with ferritic/pearlitic matrix **UTS: 48 Elon: 7% Proof: 33**
ASTM A53 T/T 100/70/03	Spheroidal cast iron with pearlitic structure **UTS: 65 Elon: 2% Proof: 42**
ASTM A53 T/T 120/90/02	Spheroidal cast iron with pearlitic matrix **UTS: 72 Elon: 2% Proof: 45**
ASTM A126 A	Grey cast iron: Min tensile 145 N/mm^2
ASTM A126 B	Grey cast iron: Min tensile 215 N/mm^2
ASTM A159	Grey cast iron: For automotive use; graded in same manner as SAE alloys
ASTM A197	Malleable iron: Free of primary graphite **UTS: 280 Elon: 5% Proof: 200**
ASTM A220	Malleable pearlitic cast iron: Graded in same manner as SAE alloys
ASTM A276	Grey iron castings: For temperatures up to 300 °C; classified by tensile properties
ASTM A278	Grey cast iron: Graded by strength
ASTM A319/1	4.1% C grey cast iron
ASTM A319/11	3.8% C grey cast iron
ASTM A319/111	3.5% C grey cast iron
ASTM A398	Welding rods for cast iron: Classified according to analysis; ASTM system
ASTM A436/1	3.0% C (max) 15% Ni 6.5% Cu 2.2% Cr cast iron: Austenitic grey
ASTM A436/1b	3.0% C (max) 15% Ni 6.5% Cu 3.0% Cr cast iron: Austenitic grey
ASTM A436/2	3.0% C (max) 20% Ni 32.2% Cr cast iron: Austenitic grey
ASTM A436/2b	3.0% C 20% Ni 4.5% Cr cast iron: Austenitic grey
ASTM A436/3	2.6% C (max) 30% Ni 3.0% Cr cast iron: Austenitic grey
ASTM A436/4	2.6% C (max) 5.5% Si 30.5% Ni 5.0% Cr cast iron: Austenitic grey
ASTM A436/5	2.4% C (max) 35% Ni cast iron: Austenitic grey

Note. The following abbreviations and units are used in the tables:

DPN	Hardness, diamond pyramid number
UTS	Ultimate tensile strength, N/mm^2
Elon	Elongation, %
Proof	0.1% proof strength, N/mm^2

1 N/mm^2=0.1 hbar=0.102 kgf/mm^2=0.06475 tonf/in.^{2}=145.04 lbf/in.^{2}=1 MPa

See Appendix II for other abbreviations and conversion tables.

Symbol	Nominal analysis, supplier, condition and remarks.
ASTM A436/6	3.0% C (max) 20% Ni 4.5% Cu 1.5% Cr cast iron: Austenitic grey
ASTM A439 D2	3.0% C (max) 20% Ni 2.2% Cr iron: Casting austenitic; ductile
ASTM A439 D2B	3.0% C (max) 20% Ni 3.3% Cr iron: Casting austenitic; ductile
ASTM A439 D2C	2.9% C (max) 22.5% Ni iron: Casting austenitic; ductile
ASTM A439 D3	2.6% C (max) 30% Ni 3.0% Cr iron: Casting austenitic; ductile
ASTM A439 D3A	2.6% C (max) 30% Ni 1.2% Cr iron: Casting austenitic; ductile
ASTM A439 D4	2.6% C (max) 30% Ni 5.0% Cr iron: Casting austenitic; ductile
ASTM A439 D5	2.4% C (max) 35% Ni iron: Casting austenitic; ductile
ASTM A439 D5B	2.4% C (max) 35% Ni 2.5% Cr iron: Casting austenitic; ductile
ASTM A445	3.0% C (max) 2.5% Si 0.08% P cast iron: Ferritic; ductile
ASTM A447	0.2% C 12% Ni 25.5% Cr iron: Casting for high temperature use
ASTM A448	0.55% C 35% Ni 15.5% Cr iron: Casting for high temperature use; previously ASTM B207
ASTM A476	3.0% C (min) 3.0% Si 0.08% P 0.05% S cast iron: Ductile
ASTM A518	0.9% C 14.5% Si cast iron: Corrosion resistant
ASTM A532/1	3% C (total) 4% Ni 1.5% Cr white cast iron: Abrasion resistant; three grades varying slightly in analysis
ASTM A532(1D)	8.5% Cr 5.5% Ni white cast iron
ASTM A532/11	3% C (total) 16% Cr 3% Mo white cast iron: Abrasion resistant; two grades varying slightly in analysis
ASTM A532(11A)	12% Cr 0.75% Mo white cast iron
ASTM A532(11B)	16% Cr 2% Mo white cast iron
ASTM A532(11C)	16% Cr 3% Mo white cast iron
ASTM A532(11D)	20.5% Cr white cast iron
ASTM A532(11E)	20.5% Cr 1.5% Mo white cast iron
ASTM A532/111	2.8% C (total) 26% C white cast iron: Abrasion resistant; two grades varying slightly in analysis
ASTM A532(111A)	25.5% Cr white cast iron
ASTM A536 60/40/18	Nodular graphite cast iron **UTS: 414 Elon: 18% Proof: 276**
ASTM A536 65/45/12	Nodular graphite cast iron **UTS: 450 Elon: 12% Proof: 310**
ASTM A536 80/55/06	Nodular graphite cast iron **UTS: 550 Elon: 6% Proof: 380**
ASTM A536 80/60/03	Nodular graphite cast iron **UTS: 550 Elon: 3% Proof: 410**
ASTM A536 100/70/03	Nodular graphite cast iron **UTS: 690 Elon: 3% Proof: 480**
ASTM A536 120/90/02	Nodular graphite cast iron **UTS: 825 Elon: 2% Proof: 620**
ASTM A567 HH90	0.9% C 26% Cr 12.5% Ni Fe alloy: Casting
ASTM A567 HI50C	0.5% C 28% Cr 16% Ni 1.0% Nb + Ta 0.1% Fe alloy: Casting
ASTM A567 HT50C	0.5% C 14% Cr 35% Ni 1.0% Nb + Ta Fe alloy: Casting
ASTM A571	2.5% C 2.0% Si 4.2% Mn 23% Ni Fe alloy: Casting; ductile iron
ASTM A602 M3210	Spheroidal graphite malleable iron: Casting; annealed **DPN: 156 UTS: 35 Elon: 10% Proof: 22**
ASTM A602 M4504	Spheroidal graphite malleable iron: Casting; hardened and tempered **DPN: 185 UTS: 45 Elon: 4% Proof: 32**
ASTM A602 M5003	Spheroidal graphite malleable iron: Casting; hardened and tempered **DPN: 210 UTS: 53 Elon: 3% Proof: 38**
ASTM A602 M5503	Spheroidal graphite malleable iron: Casting; hardened and tempered **DPN: 210 UTS: 53 Elon: 3% Proof: 39**
ASTM A602 M7002	Spheroidal graphite malleable iron: Casting; hardened and tempered **DPN: 250 UTS: 63 Elon: 2% Proof: 50**
ASTM A602 M8501	Spheroidal graphite malleable iron: Casting; hardened and tempered **DPN: 285 UTS: 74 Elon: 1% Proof: 60**
ASTM A667	Specification covering dual metal grey and white cast iron cylinder
ASTM A716	Ductile iron: Pipe for culverts **UTS: 410 Elon: 10% Proof: 290**
ASTM A746	Spheroidal graphite (ductile) iron for sewer pipes
ASTM A746	Ductile iron for gravity sewer pipes; grade 60–42–10 **UTS: 415 Elon: 10% Proof: 290**
ASTM A748	Chilled white-grey cast iron for rolls: These rolls have a chilled surface; grey core
ASTM A823 ACA	Flake graphite cast iron: Ferritic matrix **DPN: 175 UTS: 207**
ASTM A823 ACB	Flake graphite cast iron: Ferritic matrix **DPN: 175 UTS: 175**
ASTM A823 ACC	Flake graphite cast iron: Ferritic matrix **DPN: 175 UTS: 140**
ASTM A823 ASA	Flake graphite cast iron: Ferritic matrix **DPN: 180 UTS: 205**
ASTM A823 ASB	Flake graphite cast iron: Ferritic matrix **DPN: 180 UTS: 175**
ASTM A823 ASC	Flake graphite cast iron: Ferritic matrix **DPN: 180 UTS: 138**
ASTM A823 ASS	Flake graphite cast iron: Ferritic matrix **DPN: 170 UTS: 124**
ASTM A823 N-CA	Flake graphite cast iron: 15% Pearlite **DPN: 200 UTS: 205**
ASTM A823 N-CB	Flake graphite cast iron: 15% Pearlite at centre **DPN: 200 UTS: 170**
ASTM A823 N-CC	Flake graphite cast iron: 40% Pearlite/ferritic matrix **DPN: 200 UTS: 140**
ASTM A823 N-SA	Flake graphite cast iron: 20% Pearlite **DPN: 195 UTS: 205**
ASTM A823 N-SB	Flake graphite cast iron: 20% Pearlite **DPN: 195 UTS: 175**
ASTM A823 N-SC	Flake graphite cast iron: 50% Pearlite at centre **DPN: 195 UTS: 140**
ASTM A823 N-SS	Flake graphite cast iron: 50% Pearlite at centre **DPN: 170 UTS: 125**
AWS A5-15(RC1)	3.3% C 2.8% Si 0.5% P Fe: Welding rod
AWS A5-15(RC1·A)	3.3% C 0.3% Ni 2.2% Si 0.3% P Fe: Welding rod
AWS A5-15(RC1·B)	3.5% C 3.5% Si 0.5% Mg Fe: Welding rod
BS 309 W22/4	Whiteheart malleable cast iron **UTS: 300 Elon: 4%**
BS 309 W24/8	Whiteheart malleable cast iron **UTS: 360 Elon: 8%**
BS 310 B18/6	Malleable Blackheart iron: Casting **UTS: 270 Elon: 6% Proof: 170**
BS 310 B20/10	Malleable Blackheart iron: Casting **UTS: 300 Elon: 10% Proof: 80**
BS 310 B22/14	Malleable Blackheart iron: Casting **UTS: 330 Elon: 14% Proof: 200**
BS 321 A	Grey cast iron: General purpose; as cast **UTS: 320**
BS 321 C	Grey cast iron: General purpose; as cast **UTS: 140**
BS 821 (High)	Iron castings: For gears and gear blanks; as cast **DPN: 220 UTS: 300**
BS 821 (Medium)	Iron castings: For gears and gear blanks; stress relieved **DPN: 200 UTS: 220**
BS 821 (Ordinary)	Iron castings: For gears and gear blanks; as cast **DPN: 160 UTS: 170**
BS 1452/10	Grey iron castings **DPN: 200 UTS: 150**
BS 1452/12	Grey iron castings **DPN: 200 UTS: 180**

Symbol	Nominal analysis, supplier, condition and remarks.
BS 1452/14	Grey iron castings **DPN: 200 UTS: 210**
BS 1452/17	Grey iron castings **DPN: 210 UTS: 240**
BS 1452/20	Grey iron castings **DPN: 210 UTS: 300**
BS 1452/23	Grey iron castings **DPN: 210 UTS: 340**
BS 1452/26	Grey iron castings **DPN: 260 UTS: 390**
BS 1453 B1	3.3% C 3.2% Si 1.5% P (max) low S: Weld filler rod; for welding cast iron
BS 1453 B2	3.3% C 2.2% Si 1.5% P low S: Weld filler rod; for welding cast iron; harder deposit than B1
BS 1453 B3	3.2% C 2.2% Si 1.5% Ni: Weld filler rod; for welding Ni bearing and high strength cast irons
BS 1591	0.8% C 15% Si 1.0% Mn 1.0% P 0.1% S cast iron: Annealed; for acid resistant purposes
BS 2789/1	Cast iron: Replaced by BS alloys SNG 32/2 and SNG 32/7
BS 2789 2 A	Cast iron: Replaced by BS alloy SNG 27/12
BS 2789/2 B	Cast iron: Replaced by BS alloy SNG 24/17
BS 2789 SNG24/17	Spheroidal cast iron: With ferritic matrix **DPN: 160 UTS: 360 Elon: 17% Proof: 220**
BS 2789 SNG27/12	Spheroidal cast iron: With mainly ferritic matrix **DPN: 170 UTS: 400 Elon: 12% Proof: 270**
BS 2789 SNG32/7	Spheroidal cast iron: With ferritic/pearlitic matrix **UTS: 480 Elon: 7% Proof: 330**
BS 2789 SNG37/2	Spheroidal cast iron: With pearlitic matrix **UTS: 560 Elon: 2% Proof: 370**
BS 2789 SNG42/2	Spheroidal cast iron: With pearlitic structure **UTS: 650 Elon: 2% Proof: 420**
BS 2789 SNG47/2	Spheroidal cast iron: With pearlitic matrix **UTS: 720 Elon: 2% Proof: 450**
BS 3333 P28/6	Pearlitic malleable cast iron **UTS: 420 Elon: 6% Proof: 270**
BS 3333 P33/4	Pearlitic malleable cast iron **UTS: 500 Elon: 4% Proof: 300**
BS 3468 AUS101 A	3% C 2% Si 2% Cr 15% Ni 6% Cu cast iron: Austenitic structure; high expansion **DPN: 212 UTS: 140 Elon: 2%**
BS 3468 AUS101 B	3% C 2% Si 3% Cr 15% Ni 6% Cu cast iron: Austenitic structure; high expansion **DPN: 248 UTS: 180**
BS 3468 AUS102 A	3% C 2% Si 2% Cr 20% Ni cast iron: Austenitic structure; high expansion **DPN: 212 UTS: 140 Elon: 2%**
BS 3468 AUS102 B	3% C 2% Si 3% Cr 20% Ni cast iron: Austenitic structure; high expansion **DPN: 248 UTS: 180**
BS 3468 AUS104	2% C 5% Si 3.5% Cr 10% Ni cast iron: Austenitic structure; for heat resistance **DPN: 248 UTS: 180 Elon: 2%**
BS 3468 AUS105	2.5% C 1.5% Si 3% Cr 30% Ni cast iron: Austenitic structure; corrosion resistance **DPN: 212 UTS: 170**
BS 3468 AUS202 A	3% C 2.5% Si 2% Cr 20% Ni cast iron: Austenitic structure with spheroidized graphite **DPN: 201 UTS: 360 Elon: 8% Proof: 220**
BS 3468 AUS202 B	3% C 2.5% Si 3% Cr 20% Ni cast iron: Austenitic structure with spheroidized graphite **DPN: 255 UTS: 360 Elon: 6% Proof: 220**
BS 3468 AUS203	3% C 2.5% Si 0.5% Cr 22% Ni cast iron: Austenitic structure with spheroidized graphite; good impact strength **DPN: 170 UTS: 360 Elon: 20% Proof: 220**
BS 3468 AUS204	3% C 5% Si 2% Cr 20% Ni cast iron: Austenitic structure; spheroidized graphite **DPN: 230 UTS: 360 Elon: 10% Proof: 220**
BS 3468 AUS205	2.5% C 2.0% Si 3% Cr 30% Ni cast iron: Austenitic structure; spheroidized graphite **DPN: 201 UTS: 360 Elon: 7% Proof: 220**
BS 5001	3.5% C 1.8% Si 1.0% Mn 1.0% P 0.12% S cast iron: Piston rings **UTS: 240**
BS 5022	Malleable iron: Castings for automobiles; withdrawn
BS 5024	Cast iron: For air cooled and jacketed cylinders for automobiles; withdrawn
BS 5025	Cast iron: For sand cast pistons and valve guides for automobiles; withdrawn
BS 5026	Cast iron: For flywheels for automobiles; withdrawn
BS K6	3.5% C (max) 1.8% Si 0.12% S 1.0% P cast iron: Combined carbon 0.7%; for piston rings
BS K11	3.5% C 2.0% Si 1.0% Mn 0.12% S 1.0% P cast iron: Combined carbon 0.7% **UTS: 200**
C1 MATCH COMP Fe C	2.5% C 0.5% Mn 4% Si Fe alloy electrode: Metrode; for welding cast iron
C1 MATCH Sg 20 Ni	2% C 2% Cr 2% Si 20% Ni Fe alloy electrode: Metrode; for welding cast iron
CAUSAL METAL	Austenitic grey cast iron: Corrosion resistance comparable to nickel; origin unknown
CB	Cast iron: For chemical corrosion resistance; Meehanite; as cast for salt water and acid resistance **DPN: 190 UTS: 300**
CB 3	Cast iron: For use with concentrated sulphuric acid; Meehanite; as cast; suitable for up to 100% at 95 °C **DPN: 200 UTS: 300**
CC	Cast iron: Excellent corrosion resistance; Meehanite; as cast for salt water resistance, etc. **DPN: 190 UTS: 300**
CR1	Flake graphite austenitic cast iron: Meehanite
CR2	Flake graphite austenitic cast iron: Meehanite
CR3	Flake graphite austenitic cast iron: Meehanite
CR4	Flake graphite austenitic cast iron: Meehanite
CR5	Flake graphite austenitic cast iron: Meehanite
CR6	Flake graphite austenitic cast iron: Meehanite
CR7	Flake graphite austenitic cast iron: Meehanite
CR8	Flake graphite austenitic cast iron: Meehanite
CR9	Flake graphite austenitic cast iron: Meehanite
CRS	Austenitic nodular graphite cast iron: Corrosion resistance; Meehanite **DPN: 250 UTS: 380 Elon: 20%**
CRS 1	Nodular graphite austenitic cast iron: Meehanite
CRS 2	Nodular graphite austenitic cast iron: Meehanite
CRS 3	Nodular graphite austenitic cast iron: Meehanite
CRS 4	Nodular graphite austenitic cast iron: Meehanite
CRS 5	Nodular graphite austenitic cast iron: Meehanite
CRS 6	Nodular graphite austenitic cast iron: Meehanite
CRS 7	Nodular graphite austenitic cast iron: Meehanite
CRS 8	Nodular graphite austenitic cast iron: Meehanite
CRS 9	Nodular graphite austenitic cast iron: Meehanite
CRS 10	Nodular graphite austenitic cast iron: Meehanite
CRS 11	Nodular graphite austenitic cast iron: Meehanite
D	14.5% Si Fe casting: Duriron Co.; symbol for Duriron
D 51	14.5% Si 4.5% Cr Fe casting: Duriron Co.; symbol for Durichlor 51
D 4018	Ductile SG cast iron: Ferritic matrix; SAE J434 **DPN: 170**

Note. The following abbreviations and units are used in the tables:

DPN	Hardness, diamond pyramid number
UTS	Ultimate tensile strength, N/mm^2
Elon	Elongation, %
Proof	0.1% proof strength, N/mm^2

1 N/mm^2=0.1 hbar=0.102 kgf/mm^2=0.06475 $tonf/in.^2$=145.04 $ibf/in.^2$=1 MPa

See Appendix II for other abbreviations and conversion tables.

Symbol	Nominal analysis, supplier, condition and remarks.
D 4512	Ductile SG cast iron: Ferritic pearlitic matrix; SAE J434 **DPN: 185**
D 5506	Ductile SG cast iron: Ferritic pearlitic matrix; SAE J434 **DPN: 220**
D 7003	Ductile SG cast iron: Pearlitic matrix; SAE J434 **DPN: 270**
DIN 1691	General standard for flake cast iron: Grey; German Standard; see designation for details
DIN 1691 GG40	Fine ground flake graphite cast iron: German Standard **DPN: 250 UTS: 400**
DIN 1693	General standard for nodular cast iron: German Standard; see designation for details
DIN 1693 GGG38	Spheroidal cast iron with ferritic matrix **DPN: 160 UTS: 360 Elon: 17% Proof: 220**
DIN 1693 GGG42	Spheroidal cast iron with mainly ferritic matrix **DPN: 170 UTS: 400 Elon: 12% Proof: 270**
DIN 1693 GGG45	Spheroidal cast iron with ferritic/pearlitic matrix **UTS: 480 Elon: 7% Proof: 330**
DIN 1693 GGG50	Spheroidal cast iron with ferritic/pearlitic matrix **UTS: 480 Elon: 7% Proof: 330**
DIN 1693 GGG60	Spheroidal cast iron with pearlitic matrix **UTS: 560 Elon: 2% Proof: 370**
DIN 1693 GGG70	Spheroidal cast iron with pearlitic matrix **UTS: 270 Elon: 2% Proof: 450**
DQ & T	Ductile SG cast iron: Quenched and tempered; martensitic matrix; SAE J434
DRICO	3.2% C 1.5% Si cast iron: Origin unknown
DS 11/301	General standard for flake cast iron: Grey; Danish Standard; see designation for details
DS 11/301 GG40	Fine ground flake graphite cast iron: Danish Standard **DPN: 250 UTS: 400**
DS 11/303	General Standard for nodular graphite cast iron: Danish Standard; see designation for details
DTD 233 A	3.5% C total 0.7% C combined 2% Si 0.5% Cr 0.7% Mo cast iron: For piston rings; centri-cast pots **DPN: 275**
DTD 413	3.4% C total 0.7% C combined 2% Si 1% Mo cast iron: For piston rings
DTD 462	1.7% C total 30% Cr 1.5% Si 1% Mn cast iron: Centri-cast; for piston rings; annealed **DPN: 300**
DTD 485 A	2.9% C total 2% C combined 1% Cr 0.9% Mn 8% Mo cast iron: Centri-cast; for piston rings; annealed **DPN: 280**
DTD 614	1.8% C total 1.8% Si 16% Cr 0.1% S and P cast iron **DPN: 300**
DTD 719	1.7% C total 14% Cr 0.4% Mo cast iron: Centri-cast; for rings and liners; annealed
DURATLAS GM	3.3% C 0.5% Si 0.5% Mn 2% Cr 4.8% Ni 0.5% Mo cast iron: Firth Brown **DPN: 500**
DURICHLOR 51	0.9% C 14.5% Si 4.5% Mn iron alloy: Duriron Co.; excellent corrosion resistance
DURIRON	0.9% C 1.5% Mn 14.5% Si iron alloy: Duriron Co.; excellent corrosion resistance
ELVERITE CG	2.5% Cr 4% Ni 1% Mo white cast iron: Origin unknown; see ASTM A 532–1
ELVERITE K	2.5% Cr 4% Ni 1% Mo white cast iron: Origin unknown; see ASTM A 532–1
ELVERITE I	9% Cr 6% Ni 1% Mo white cast iron: Origin unknown; see ASTM A 532–1
F 30000	UNS for ductile cast iron: Heat treated; classified by hardness
F 47001	2.2% Cr 6% Cu 15% Ni cast iron; UNS designation
F 47002	2.2% Cr 20% Cu 15% Ni cast iron; UNS designation
FC 10 GRADE 1	Fine grained grey cast iron: Designation used by Japanese Standards **UTS: 150**

Symbol	Nominal analysis, supplier, condition and remarks.
FC 15	Fine grained flake graphite cast iron: Designation used by Portuguese Standards **DPN: 170 UTS: 150**
FC 15 GRADE 2	Fine grained grey cast iron: Designation used by Japanese Standards **UTS: 150**
FC 20	Fine grained flake graphite cast iron: Designation used by Portuguese Standards **DPN: 195 UTS: 225**
FC 20 GRADE 3	Fine grained grey cast iron: Designation used by Japanese Standards **UTS: 200**
FC 25	Fine grained flake graphite cast iron: Designation used by Portuguese Standards **DPN: 200 UTS: 250**
FC 25 GRADE 4	Fine grained grey cast iron: Designation used by Japanese Standards **UTS: 250**
FC 30	Fine grained flake graphite cast iron: Designation used by Portuguese Standards **DPN: 210 UTS: 300**
FC 30 GRADE 5	Fine grained grey cast iron: Designation used by Japanese Standards **UTS: 300**
FC 35	Fine grained flake graphite cast iron: Designation used by Portuguese Standards **DPN: 220 UTS: 350**
FC 35 GRADE 6	Fine grained grey cast iron: Designation used by Japanese Standards **UTS: 350**
FC 40	Fine grained flake graphite cast iron: Designation used by Portuguese Standards **DPN: 250 UTS: 400**
FC 150	Fine grained flake cast iron: Romania Standard
FC 200	Fine grained flake cast iron: Romania Standard
FC 250	Fine grained flake cast iron: Romania Standard
FC 275	Compacted graphite cast iron: Meehanite; for high temperature use **DPN: 150 UTS: 275 Elon: 2% Proof: 220**
FC 300	Fine grained flake cast iron: Romania Standard
FC 350	Fine grained flake cast iron: Romania Standard
FC 400	Fine grained flake cast iron: Romania Standard
FCD 40	Spheroidal graphite cast iron: Japanese Standard
FCD 45	Spheroidal graphite cast iron: Japanese Standard
FCD 50	Spheroidal graphite cast iron: Japanese Standard
FCD 60	Spheroidal graphite cast iron: Japanese Standard
FCD 70	Spheroidal graphite cast iron: Japanese Standard
FE 15 Si	0.8% C 15% Si Fe alloy electrode: Metrode for welding cast iron
FG 10	Fine grained flake graphite cast iron: Designation used by Spanish Standards **DPN: 170 UTS: 150**
FG 15	Fine grained flake graphite cast iron: Designation used by Spanish Standards **DPN: 170 UTS: 150**
FG 15D	Fine grained flake graphite cast iron: Designation used by French Standards **DPN: 170 UTS: 150**
FG 20	Fine grained flake graphite cast iron: Designation used by Spanish Standards **DPN: 195 UTS: 225**
FG 20D	Fine grained flake graphite cast iron: Designation used by French Standards **DPN: 190 UTS: 200**
FG 20D	Fine grained flake graphite cast iron: Designation used by French Standards **DPN: 195 UTS: 225**
FG 25	Fine grained flake graphite cast iron: Designation used by Spanish Standards **DPN: 200 UTS: 250**

Symbol	Nominal analysis, supplier, condition and remarks.
FG 25D	Fine grained flake graphite cast iron: Designation used by French Standards **DPN: 205 UTS: 275**
FG 30	Fine grained flake graphite cast iron: Designation used by Spanish Standards **DPN: 210 UTS: 300**
FG 30D	Fine grained flake graphite cast iron: Designation used by French Standards **DPN: 210 UTS: 300**
FG 35	Fine grained flake graphite cast iron: Designation used by Spanish Standards **DPN: 220 UTS: 350**
FG 35	Fine grained flake graphite cast iron: Designation used by Spanish Standards **DPN: 250 UTS: 400**
FG 35D	Fine grained flake graphite cast iron: Designation used by French Standards **DPN: 220 UTS: 350**
FG 40D	Fine grained flake graphite cast iron: French Standard **DPN: 250 UTS: 400**
FGE 38.17	Nodular graphite cast iron: Mainly ferritic structure; designation used by Spanish Standard **DPN: 180 UTS: 380 Elon: 17% Proof: 230**
FGE 42.12	Nodular graphite cast iron: Ferritic structure; designation used by Spanish Standard **DPN: 200 UTS: 420 Elon: 12% Proof: 250**
FGE 50.7	Nodular graphite cast iron: Pearlitic/ferritic structure; designation used by Spanish Standard **DPN: 205 UTS: 500 Elon: 7% Proof: 310**
FGE 60.2	Nodular graphite cast iron: Pearlitic/ferritic structure; designation used by Spanish Standard **DPN: 235 UTS: 600 Elon: 3% Proof: 350**
FGE 70.2	Nodular graphite cast iron: Pearlitic structure; designation used by Spanish Standard **DPN: 265 UTS: 700 Elon: 2% Proof: 400**
FGE 80.2	Nodular graphite cast iron: Pearlitic structure; designation used by Spanish Standard **DPN: 300 UTS: 800 Elon: 2% Proof: 460**
FGE 370.17	Spheroidal graphite cast iron: Portuguese Standard
FGE 400.12	Spheroidal graphite cast iron: Portuguese Standard
FGE 500.7	Spheroidal graphite cast iron: Portuguese Standard
FGE 600.3	Spheroidal graphite cast iron: Portuguese Standard
FGE 700.2	Spheroidal graphite cast iron: Portuguese Standard
FGE 800.2	Spheroidal graphite cast iron: Portuguese Standard
FGG	Fine grained flake graphite cast iron: Belgian Standard **DPN: 250 UTS: 400**
FGG 10	Fine grained flake graphite cast iron: Designation used by Belgian Standard
FGG 15	Fine grained flake graphite cast iron: Designation used by Belgian Standard **DPN: 170 UTS: 150**
FGG 20	Fine grained flake graphite cast iron: Designation used by Belgian Standard **DPN: 190 UTS: 200**
FGG 20	Fine grained flake graphite cast iron: Designation used by Belgian Standard **DPN: 195 UTS: 225**

Note. The following abbreviations and units are used in the tables:

DPN	Hardness, diamond pyramid number
UTS	Ultimate tensile strength, N/mm^2
Elon	Elongation, %
Proof	0.1% proof strength, N/mm^2

1 N/mm^2=0.1 hbar=0.102 kgf/mm^2=0.06475 tonf/in.^{2}=145.04 ibf/in.^{2}=1 MPa

See Appendix II for other abbreviations and conversion tables.

Symbol	Nominal analysis, supplier, condition and remarks.
FGG 25	Fine grained flake graphite cast iron: Designation used by Belgian Standard **DPN: 205 UTS: 275**
FGG 30	Fine grained flake graphite cast iron: Designation used by Belgian Standard **DPN: 210 UTS: 300**
FGG 35	Fine grained flake graphite cast iron: Designation used by Belgian Standard **DPN: 220 UTS: 350**
FGG 40	Fine grained cast iron: Belgian Standard
FGN 70.3	Spheroidal graphite cast iron: Romanian Standard
FGS 38.15	Nodular graphite cast iron: Pearlitic/ferritic matrix; designation used by French Standard **DPN: 165 UTS: 400 Elon: 17% Proof: 250**
FGS 42.12	Nodular graphite cast iron: Mainly ferritic matrix; designation used by French Standard **DPN: 200 UTS: 420 Elon: 12% Proof: 250**
FGS 50.7	Nodular graphite cast iron: Pearlitic/ferritic structure; designation used by French Standard **DPN: 205 UTS: 500 Elon: 7% Proof: 310**
FGS 60.2	Nodular graphite cast iron: Pearlitic/ferritic structure; designation used by French Standard **DPN: 230 UTS: 600 Elon: 3% Proof: 350**
FGS 70.2	Nodular graphite cast iron: Pearlitic structure; designation used by French Standard **DPN: 265 UTS: 700 Elon: 2% Proof: 400**
FGS 370.17	Spheroidal graphite cast iron: French Standard
FGS 400.12	Spheroidal graphite cast iron: French Standard
FGS 500.7	Spheroidal graphite cast iron: French Standard
FGS 600.3	Spheroidal graphite cast iron: French Standard
FGS 700.2	Spheroidal graphite cast iron: French Standard
FGS 800.2	Spheroidal graphite cast iron: French Standard
FNG 1	Nodular graphite cast iron: Pearlitic/ferritic structure; designation used by Belgian Standard **DPN: 205 UTS: 500 Elon: 7% Proof: 310**
FNG 38.17	Nodular graphite cast iron: Pearlitic/ferritic matrix; designation used by Belgian Standard **DPN: 165 UTS: 400 Elon: 17% Proof: 250**
FNG 42.12	Nodular graphite cast iron: Mainly ferritic matrix; designation used by Belgian Standard **DPN: 200 UTS: 420 Elon: 12% Proof: 250**
FNG 50.7	Nodular graphite cast iron: Pearlitic/ferritic structure; designation used by Belgian Standard **DPN: 205 UTS: 500 Elon: 7% Proof: 310**
FNG 60.2	Nodular graphite cast iron: Pearlitic/ferritic structure; designation used by Belgian Standard **DPN: 230 UTS: 600 Elon: 3% Proof: 350**
FNG 70.2	Nodular graphite cast iron: Pearlitic structure; designation used by Belgian Standard **DPN: 265 UTS: 700 Elon: 2% Proof: 400**
FNG 80.2	Nodular graphite cast iron: Pearlitic structure; designation used by Belgian Standard **DPN: 300 UTS: 800 Elon: 2% Proof: 460**
FT 15 D	Fine grained flake cast iron: French Standard
FT 20 D	Fine grained flake cast iron: French Standard
FT 25 D	Fine grained flake cast iron: French Standard
FT 30 D	Fine grained flake cast iron: French Standard
FT 35 D	Fine grained flake cast iron: French Standard
FT 40 D	Fine grained flake cast iron: French Standard
G 10	Fine grained flake graphite cast iron: Designation used by Italian Standard **DPN: 170 UTS: 150**
G 15	Fine grained flake graphite cast iron: Designation used by Italian Standard **DPN: 170 UTS: 150**
G 20	Fine grained flake graphite cast iron: Designation used by Italian Standard **DPN: 195 UTS: 225**
G 25	Fine grained flake graphite cast iron: Designation used by Italian Standard **DPN: 200 UTS: 250**

Symbol	Nominal analysis, supplier, condition and remarks.
G 30	Fine grained flake graphite cast iron: Designation used by Italian Standard **DPN: 210 UTS: 300**
G 35	Fine grained flake graphite cast iron: Designation used by Italian Standard **DPN: 250 UTS: 400**
G 1800	Flake graphite grey cast iron: SAE designation in J 859 **DPN: 190 UTS: 124**
G 2000	Grey cast iron: SAE designation; as cast; formerly SAE 110 **DPN: 187 UTS: 120**
G 2500	Flake graphite grey cast iron: SAE designation in J 859 **DPN: 200 UTS: 173**
G 3000	3.4% total C 2% Si grey cast iron: SAE designation; as cast; formerly SAE 111 **DPN: 200 UTS: 200**
G 3500	3.5% total C 1.5% Si grey cast iron: SAE designation; as cast; formerly SAE 120 **DPN: 210 UTS: 220**
G 4000	3.4% total C 1.5% Si grey cast iron: SAE designation; as cast; formerly SAE 121 **DPN: 230 UTS: 270**
G 4500	Grey cast iron: SAE designation; as cast; formerly SAE 122 **DPN: 235 UTS: 300**
GA	Cast iron: Can be hardened and tempered; Meehanite; as cast **DPN: 220 UTS: 330**
GA 350	Fine grained flake graphite cast iron: Meehanite; previously grade GA **DPN: 220 UTS: 350**
GB	Cast iron: Meehanite; as cast **DPN: 210 UTS: 300**
GB 300	Fine grained flake graphite cast iron: Meehanite; previously grade GB **DPN: 210 UTS: 300**
GC	Cast iron: Meehanite; as cast **DPN: 195 UTS: 270**
GC 275	Fine grained flake graphite cast iron: Previously grade GC **DPN: 205 UTS: 275**
GD	Cast iron: Meehanite; as cast **DPN: 185 UTS: 240**
GD 250	Fine grained flake graphite cast iron: Previously grade GD **DPN: 200 UTS: 250**
GE	Cast iron: Meehanite; as cast **DPN: 170 UTS: 200**
GE 200	Fine grained flake graphite cast iron: Meehanite; previously grade GE **DPN: 190 UTS: 200**
GE 225	Fine grained flake graphite cast iron: Meehanite; previously grade GE **DPN: 195 UTS: 225**
GF	Whiteheart cast iron: Britannia; spheroidized **UTS: 480 Elon: 10% Proof: 300**
GF 150	Fine grained flake graphite cast iron: Meehanite **DPN: 170 UTS: 150**
GG 10	Fine grained flake graphite cast iron: Designation used by Austrian Standards
GG 10	Fine grained flake graphite cast iron: Designation used by Danish Standards
GG 10	Fine grained flake graphite cast iron: Designation used by German Standards
GG 15	Fine grained flake graphite cast iron: Designation used by Austrian Standards **DPN: 170 UTS: 150**
GG 15	Fine grained flake graphite cast iron: Designation used by Danish Standards **DPN: 170 UTS: 150**
GG 15	Fine grained flake graphite cast iron: Designation used by German Standards **DPN: 170 UTS: 150**
GG 15	Fine grained flake graphite cast iron: Designation used by Dutch Standards **DPN: 170 UTS: 150**
GG 20	Fine grained flake graphite cast iron: Designation used by Danish Standards **DPN: 190 UTS: 200**
GG 20	Fine grained flake graphite cast iron: Designation used by German Standards **DPN: 190 UTS: 200**
GG 20	Fine grained flake graphite cast iron: Designation used by Austrian Standards **DPN: 195 UTS: 225**
GG 20	Fine grained flake graphite cast iron: Designation used by Danish Standards **DPN: 195 UTS: 225**
GG 20	Fine grained flake graphite cast iron: Designation used by German Standards **DPN: 195 UTS: 225**
GG 20	Fine grained flake graphite cast iron: Designation used by Dutch Standards **DPN: 195 UTS: 225**
GG 25	Fine grained flake graphite cast iron: Designation used by Austrian Standards **DPN: 205 UTS: 275**
GG 25	Fine grained flake graphite cast iron: Designation used by Danish Standards **DPN: 205 UTS: 275**
GG 25	Fine grained flake graphite cast iron: Designation used by Dutch Standards **DPN: 205 UTS: 275**
GG 25	Fine grained flake graphite cast iron: Designation used by German Standards **DPN: 205 UTS: 275**
GG 30	Fine grained flake graphite cast iron: Designation used by Dutch Standards **DPN: 210 UTS: 300**
GG 30	Fine grained flake graphite cast iron: Designation used by German Standards **DPN: 210 UTS: 300**
GG 30	Fine grained flake graphite cast iron: Designation used by Austrian Standards **DPN: 210 UTS: 300**
GG 30	Fine grained flake graphite cast iron: Designation used by Danish Standards **DPN: 210 UTS: 300**
GG 35	Fine grained flake graphite cast iron: Designation used by Dutch Standards **DPN: 220 UTS: 350**
GG 35	Fine grained flake graphite cast iron: Designation used by Austrian Standards **DPN: 220 UTS: 350**
GG 35	Fine grained flake graphite cast iron: Designation used by Danish Standards **DPN: 220 UTS: 350**
GG 35	Fine grained flake graphite cast iron: Designation used by German Standards **DPN: 220 UTS: 350**
GG 40	Fine grained flake graphite cast iron: Danish Standard **DPN: 250 UTS: 400**
GG Ft 15	Fine grain flake graphite cast iron: Swiss Standard
GG Ft 20	Fine grain flake graphite cast iron: Swiss Standard
GG Ft 25	Fine grain flake graphite cast iron: Swiss Standard
GG Ft 30	Fine grain flake graphite cast iron: Swiss Standard
GGG 35.3	Nodular graphite cast iron: Mainly ferritic; designation used by German Standards **DPN: 145 UTS: 350 Elon: 24% Proof: 220**
GGG 38	Spheroidal graphite cast iron: Ferritic matrix; German Standard

Symbol	Nominal analysis, supplier, condition and remarks.
GGG 40	Nodular graphite cast iron: Pearlitic/ferritic graphite; designation used by German Standards **DPN: 165 UTS: 400 Elon: 17% Proof: 250**
GGG 40.3	Nodular graphite cast iron: Mainly ferritic; designation used by German Standards **DPN: 150 UTS: 400 Elon: 20% Proof: 250**
GGG 42	Spheroidal graphite cast iron: Mainly ferritic matrix; designation used by German Standards
GGG 45	Spheroidal graphite cast iron: Ferritic/pearlitic matrix; designation used by German Standards
GGG 50	Nodular graphite cast iron: Pearlitic/ferritic structure; designation used by German Standards **DPN: 205 UTS: 500 Elon: 7% Proof: 310**
GGG 60	Nodular graphite cast iron: Pearlitic/ferritic structure; designation used by German Standards **DPN: 230 UTS: 600 Elon: 3% Proof: 310**
GGG 70	Nodular graphite cast iron: Pearlitic structure; designation used by German Standards **DPN: 265 UTS: 700 Elon: 2% Proof: 400**
GGG 80	Nodular graphite cast iron: Pearlitic structure; designation used by German Standards **DPN: 300 UTS: 800 Elon: 2% Proof: 460**
GGG 400	Spheroidal graphite cast iron: Austrian Standard
GGG 400 K	Spheroidal graphite cast iron: Austrian Standard
GGG 500	Spheroidal graphite cast iron: Austrian Standard
GGG 600	Spheroidal graphite cast iron: Austrian Standard
GGG 700	Spheroidal graphite cast iron: Austrian Standard
GGG FGS 38	Spheroidal graphite cast iron: Swiss Standard
GGG FGS 42	Spheroidal graphite cast iron: Swiss Standard
GGG FGS 50	Spheroidal graphite cast iron: Swiss Standard
GGG FGS 60	Spheroidal graphite cast iron: Swiss Standard
GGG FGS 70	Spheroidal graphite cast iron: Swiss Standard
GGL	Ni Cu Cr flake cast iron: Austenitic; German Standard
GM	Cast iron: Can be hardened and tempered; Meehanite; as cast **DPN: 230 UTS: 360**
GM 400	Fine grained grey cast iron: Meehanite; previously grade GM **DPN: 250 UTS: 400**
GN 38	Nodular graphite cast iron: Pearlitic/ferritic matrix; designation used by Dutch Standards **DPN: 165 UTS: 400 Elon: 17% Proof: 250**
GN 42	Nodular graphite cast iron: Mainly ferritic matrix; designation used by Dutch Standards **DPN: 200 UTS: 420 Elon: 12% Proof: 250**
GN 50	Nodular graphite cast iron: Pearlitic/ferritic structure; designation used by Dutch Standards **DPN: 205 UTS: 500 Elon: 7% Proof: 310**
GN 60	Nodular graphite cast iron: Pearlitic/ferritic structure; designation used by Dutch Standards **DPN: 230 UTS: 600 Elon: 3% Proof: 310**
GN 70	Nodular graphite cast iron: Pearlitic structure; designation used by Dutch Standards **DPN: 265 UTS: 700 Elon: 2% Proof: 400**
GOST 1412	General standard for grey cast iron: Russian standard; see designation for details
GOST 7293/54	Nodular graphite cast iron: Specification for Russian Standard; details under designation
GOV 38	Spheroidal graphite case iron: Hungarian Standard
GOV 40	Spheroidal graphite case iron: Hungarian Standard
GOV 45	Spheroidal graphite case iron: Hungarian Standard
GOV 50	Spheroidal graphite case iron: Hungarian Standard
GOV 60	Spheroidal graphite case iron: Hungarian Standard
GOV 70	Spheroidal graphite case iron: Hungarian Standard
GRAIN 21.80	0.08% C 1.3% Mn 0.4% Si Fe: Powder for welding; Esab; for submerged arc; properties of deposit varies with flux used
GRAIN 21.81	0.03% C 0.8% Mn 25% Ni Fe: Powder for welding; Esab; for submerged arc; properties of deposit vary with flux
GRP 38	Nodular graphite cast iron: Pearlitic/ferritic matrix; designation used by Finnish Standards **DPN: 165 UTS: 400 Elon: 17% Proof: 250**
GRP 40	Nodular graphite cast iron: Mainly ferritic matrix; designation used by Finnish Standards **DPN: 200 UTS: 420 Elon: 12% Proof: 250**
GRP 50	Nodular graphite cast iron: Pearlitic/ferritic structure; designation used by Finnish Standards **DPN: 205 UTS: 500 Elon: 7% Proof: 310**
GRP 60	Nodular graphite cast iron: Pearlitic/ferritic structure; designation used by Finnish Standards **DPN: 230 UTS: 600 Elon: 3% Proof: 350**
GRP 70	Nodular graphite cast iron: Pearlitic structure; designation used by Finnish Standards **DPN: 265 UTS: 700 Elon: 2% Proof: 400**
GRP 340	Spheroidal graphite: Cast iron; Finnish Standard
GRP 370	Spheroidal graphite: Cast iron; Finnish Standard
GRP 400	Spheroidal graphite: Cast iron; Finnish Standard
GRP 500	Spheroidal graphite: Cast iron; Finnish Standard
GRP 600	Spheroidal graphite: Cast iron; Finnish Standard
GRP 700	Spheroidal graphite: Cast iron; Finnish Standard
GRP 800	Spheroidal graphite: Cast iron; Finnish Standard
GRS 10	Fine grained flake graphite cast iron: Designation used by Finnish Standards
GRS 15	Fine grained flake graphite cast iron: Designation used by Finnish Standards **DPN: 170 UTS: 150**
GRS 20	Fine grained flake graphite cast iron: Designation used by Finnish Standards **DPN: 195 UTS: 225**
GRS 25	Fine grained flake cast iron: Finnish Standard
GRS 30	Fine grained flake graphite cast iron: Designation used by Finnish Standards **DPN: 210 UTS: 300**
GRS 35	Fine grained flake graphite cast iron: Designation used by Finnish Standards **DPN: 220 UTS: 350**
GRS 40	Fine grained flake cast iron: Finnish Standard
GS 370.17	Nodular graphite cast iron: Mainly ferritic structure; designation used by Italian Standards **DPN: 180 UTS: 370 Elon: 17% Proof: 230**
GS 400.12	Nodular graphite cast iron: Mainly ferritic structure; designation used by Italian Standards **DPN: 200 UTS: 400 Elon: 12% Proof: 250**
GS 500.7	Nodular graphite cast iron: Pearlitic/ferritic structure; designation used by Italian Standards **DPN: 205 UTS: 500 Elon: 7% Proof: 310**
GS 600.2	Nodular graphite cast iron: Pearlitic/ferritic structure; designation used by Italian Standards **DPN: 230 UTS: 600 Elon: 3% Proof: 350**
GS 700.2	Nodular graphite cast iron: Pearlitic structure; designation used by Italian Standards **DPN: 265 UTS: 700 Elon: 2% Proof: 400**
GS 800	Nodular graphite cast iron: Pearlitic structure; designation used by Italian Standards **DPN: 300 UTS: 800 Elon: 2% Proof: 460**
H 11-52	General standard for flake cast iron: Grey; Finnish Standards; see designation for details
H 11-52 GRS40	Fine grained flake graphite cast iron: Finnish Standard **DPN: 250 UTS: 400**

Note. The following abbreviations and units are used in the tables:

DPN	Hardness, diamond pyramid number
UTS	Ultimate tensile strength, N/mm^2
Elon	Elongation, %
Proof	0.1% proof strength, N/mm^2

$1\ N/mm^2 = 0.1\ hbar = 0.102\ kgf/mm^2 = 0.06475\ tonf/in.^2 = 145.04\ lbf/in.^2 = 1\ MPa$

See Appendix II for other abbreviations and conversion tables.

Symbol	Nominal analysis, supplier, condition and remarks.
HA	Cast iron: For use up to 650 °C for burner parts, etc.; Meehanite; as cast **DPN: 220 UTS: 330**
HAYES COMPACTED IRON	Modified grey cast iron: Hayes **UTS: 420 Elon: 3%**
HAYES HYDRAULIC IRON	Grey cast iron to BS 1452/20 Hayes **UTS: 312 Elon: 0%**
HAYES HYDRAULIC NODULAR	Spheroidal graphite cast iron to BS 2789 **UTS: 500 Elon: 7%**
HB	Cast iron: For use up to 700 °C; abrasion resistance for furnace equipment trays, etc.; Meehanite; as cast **DPN: 300 UTS: 240**
HC	Cast iron: For use up to 700 °C; abrasion resistance for oil refinery trays, valves, etc.; Meehanite; as cast **DPN: 300 UTS: 240**
HD	Cast iron: For use up to 620 °C; for oil refinery equipment; Meehanite; as cast **DPN: 200 UTS: 220**
HE	Cast iron: For intermittent temperature service for ingot moulds, etc.; Meehanite; as cast **DPN: 170 UTS: 210**
HR	Cast iron: For use up to 850 °C; abrasion resisting for furnace and burner equipment; Meehanite; as cast **DPN: 450 UTS: 240**
HR 1	Cast iron: For use up to 790 °C; abrasion resisting for oil refinery parts; Meehanite; as cast; similar to HR **DPN: 450 UTS: 240**
HS	Nodular graphite cast iron: Heat resistant for use up to 900 °C; Meehanite **DPN: 200 UTS: 400**
HSV	Modular graphite cast iron: Meehanite for high temperature cast **DPN: 240 UTS: 650 Elon 2% Proof: 345**
INCANITE 1	Cast iron: Incanite **DPN: 190 UTS: 210**
INCANITE 2	Cast iron: Incanite **DPN: 210 UTS: 300**
INCANITE 3	Cast iron: Incanite for BS alloys 20 and 23 (BS 1452) **DPN: 225 UTS: 330**
INCANITE 4	Cast iron: Close grained; Incanite **DPN: 225 UTS: 300**
INCANITE 5	Cast iron: Incanite **DPN: 200 UTS: 340**
INCANITE 6	Cast iron: Incanite **DPN: 225 UTS: 390**
INCANITE HRA	Haematite type cast iron: Incanite; for Al and Zn melting pots, etc. **DPN: 190 UTS: 210**
INCANITE HRB	Heat resistant cast iron: Superior to grade HRA; Incanite **DPN: 210 UTS: 330**
JIS G5501	General standard for grey cast iron: Japanese Standard; see designation for details
JIS G5502	General standard for spheroidal graphite cast iron: Japanese Standard
JUS CJ 2020	General standard for fine grain flake cast iron: Yugoslavian Standard; see designation for details
JUS CJ 2022	General standard for spheroidal graphite cast iron: Yugoslavian Standard; see designation for details
KC	Cast iron: For use with strong alkalis; Meehanite; as cast
L NC 20.2	Flake graphite austenitic cast iron: French Standard **DPN: 215 UTS: 170 Elon: 2%**
L NC 30.3	Flake graphite austenitic cast iron: French Standard **DPN: 215 UTS: 190 Elon: 2%**
L NUC 15.6.2	Flake graphite austenitic cast iron: French Standard **DPN: 210 UTS: 170 Elon: 2%**
M 3191	Flake graphite cast iron: Austrian Standard; covers various grades
M 3191 GG35	Fine grained cast iron: Flake graphite; Austrian Standard **DPN: 250 UTS: 400**
M 3193	General standard for nodular graphite cast iron: Austrian Standard; see designation for details
M 3210	Malleable cast iron: Nodular iron; SAE designation in J 158 **DPN: 160 UTS: 345 Elon: 10% Proof: 225**
M 4504	Malleable cast iron: Nodular iron; SAE designation in J 158 **DPN: 190 UTS: 450 Elon: 4% Proof: 310**
M 5003	Malleable cast iron: Nodular iron; SAE designation in J 158 **DPN: 205 UTS: 520 Elon: 3% Proof: 345**
M 5503	Malleable cast iron: Nodular iron; SAE designation in J 158 **DPN: 205 UTS: 520 Elon: 3% Proof: 380**
M 7002	Malleable cast iron: Nodular iron; SAE designation in J 158 **DPN: 250 UTS: 620 Elon: 2% Proof: 480**
M 8501	Malleable cast iron: Nodular iron; SAE designation in J 158 **DPN: 285 UTS: 725 Elon: 1% Proof: 590**
MIL Spec 1·24137(B)	2% Cr 20% Ni cast iron
MIL Spec 1·24137(C)	0.5% Cr (max) 21.5% Ni cast iron
MNC 706	Swedish Standard; summary of spheroidal graphite cast irons
MNC 707 E	Specification covering several grades of Blackheart malleable cast iron; Swedish Standard
MNC 1205	General standard for flake graphite cast iron: Grey; Swedish Standard; see SIS for details
MSZ 8280 66 C 81	General standard for flake cast iron: Hungarian Standard; see designation for details
MSZ 8277	General standard for spheroidal graphite cast iron: Hungarian Standard
N 3	3% C (max) 0.85% Mn 2.2% Si 2.1% Cr 20% Ni cast iron: For furnace fittings; Wellman
NBN 830.01	General specification for flake cast iron: Grey; Belgian Standard; see designation for details
NBN 830.01 FGG40	Fine grained flake graphite cast iron: Belgian Standard **DPN: 250 UTS: 400**
NBN 830.02	General standard for nodular graphite cast iron: Belgian Standard; see designation for details
NEN 6002 A	General standard used for flake cast iron: Grey; Dutch Standard; see designation for details
NEN 6002 D	General standard for nodular graphite cast iron: Dutch Standard; see designation for details
NFA 32.101	General standard for fine grain flake cast iron: French Standard; see designation for details
NFA 32.201	General standard for spheroidal graphite cast iron: French Standard
Ni HARD	3.3% C 0.7% Si 2.5% Cr 4.1% Ni 0.4% P white cast iron: Impact 30 J; International Nickel **UTS: 550**
Ni HARD 1	2.5% Cr 4% Ni 1% Mn (max) white cast iron: International Nickel; ASTM A532/1
Ni HARD 2	2.5% Cr 4% Ni 1% Mn (max) white cast iron: International Nickel; ASTM A532/1
Ni HARD 3	1.2% Cr 3% Ni 1% Mn (max) white cast iron: International Nickel; ASTM A532/1
Ni HARD 4	9% Cr 6% Ni 1% Mn (max) white cast iron: International Nickel; ASTM A532/1
Ni HARD 4	3% C 6% Ni 8% Cr martensitic white iron: High impact strength; International Nickel **DPN: 560**
Ni-RESIST 1	2.0% C 1.7% Si 15% Ni 6% Cu 2% Cr austenitic cast iron: High corrosion resistance; International Nickel **DPN: 150 UTS: 190 Elon: 2%**

Symbol	Nominal analysis, supplier, condition and remarks.
Ni-RESIST 1B	2.0% C 1.9% Si 15% Ni 6% Cu 3% Cr austenitic cast iron: High corrosion resistance; International Nickel **DPN: 180 UTS: 220**
Ni-RESIST 2	2.0% C 1.9% Si 20% Ni 2% Cr austenitic cast iron: For steam service; International Nickel **DPN: 145 UTS: 190**
Ni-RESIST 2B	2.0% C 1.9% Si 20% Ni 5% Cr austenitic cast iron: For high temperature use; International Nickel **DPN: 210 UTS: 220**
Ni-RESIST 3	2.0% C 1.5% Si 30% Ni 3% Cr austenitic cast iron: Low expansion; International Nickel **DPN: 140 UTS: 22**
Ni-RESIST 4	2.0% C 5.0% Si 31% Ni 5% Cr austenitic cast iron: Use with foodstuff; International Nickel **DPN: 180 UTS: 220**
Ni-RESIST 5	2.0% C 1% Si 35% Ni austenitic cast iron: High shock resistance; International Nickel **DPN: 115 UTS: 160**
Ni-RESIST D 2	2.0% C 20% Ni 2% Cr austenitic cast iron: Spheroidal graphite; high expansion; International Nickel **DPN: 170 UTS: 450 Elon: 14% Proof: 240**
Ni-RESIST D 2B	2.0% C 20% Ni 3% Cr austenitic cast iron: Spheroidal graphite; high corrosion resistance; International Nickel **DPN: 180 UTS: 470 Elon: 11% Proof: 240**
Ni-RESIST D 2C	2.0% C 22% Ni austenitic cast iron: Spheroidal graphite; high ductility **DPN: 150 UTS: 440 Elon: 30% Proof: 230**
Ni-RESIST D 2M	2.0% C 4% Mn 22% Ni austenitic cast iron: Spheroidal graphite; low temperature ductility; International Nickel **DPN: 145 UTS: 460 Elon: 40% Proof: 210**
Ni-RESIST D 3	2.0% C 30% Ni 3% Cr austenitic cast iron: Spheroidal graphite; shock resistant; International Nickel **DPN: 170 UTS: 420 Elon: 11% Proof: 240**
Ni-RESIST D 3A	2.0% C 30% Ni 5% Cr austenitic cast iron: Spheroidal graphite; wear resistant; International Nickel **DPN: 160 UTS: 410 Elon: 15% Proof: 240**
Ni-RESIST D 4	2.0% C 5% Si 30% Ni 5% Cr austenitic cast iron: Spheroidal graphite; high corrosion resistance; International Nickel **DPN: 205 UTS: 470 Elon: 3% Proof: 280**
Ni-RESIST D 5	2.0% C 35% Ni austenitic cast iron: Spheroidal graphite; low expansion; International Nickel **DPN: 155 UTS: 400 Elon: 30% Proof: 230**
Ni-RESIST D 5B	2.0% C 35% Ni 3% Cr austenitic cast iron: Spheroidal graphite; low expansion; International Nickel **DPN: 165 UTS: 410 Elon: 8% Proof: 270**
NITRON	3.2% C 1.5% Si 0.12% S 0.03% P (max) cast iron: Origin unknown
NL 38	Spheroidal graphite cast iron: Yugoslavian Standard
NL 42	Spheroidal graphite cast iron: Yugoslavian Standard
NL 50	Spheroidal graphite cast iron: Yugoslavian Standard
NL 60	Spheroidal graphite cast iron: Yugoslavian Standard
NL 70	Spheroidal graphite cast iron: Yugoslavian Standard
NMX	0.5% C 5% Mr 5% Cr 13% Ni 0.3% V 2% Cu Fe electrode: Metrode; for welding cast iron
NP 1759	General standard for spheroidal graphite cast iron: Portuguese Standard

Note. The following abbreviations and units are used in the tables:

DPN	Hardness, diamond pyramid number
UTS	Ultimate tensile strength, N/mm^2
Elon	Elongation, %
Proof	0.1% proof strength, N/mm^2

1 N/mm^2=0.1 hbar=0.102 kgf/mm^2=0.06475 tonf/in.^{2}=145.04 ibf/in.^{2}=1 MPa

See Appendix II for other abbreviations and conversion tables.

Symbol	Nominal analysis, supplier, condition and remarks.
NS 722	General standard for flake graphite cast iron: Grey; Norwegian Standard; see designation for details
NS 11301	Norwegian Standard for nodular graphite cast iron: See designation for details
NS 11338	Nodular graphite cast iron: Mainly ferritic structure; designation used by Norwegian Standards **DPN: 180 UTS: 370 Elon: 17% Proof: 230**
NS 11342	Nodular graphite cast iron: Mainly ferritic structure; designation used by Norwegian Standards **DPN: 180 UTS: 380 Elon: 15% Proof: 230**
NS 11350	Nodular graphite cast iron: Pearlitic/ferritic structure; designation used by Norwegian Standards **DPN: 205 UTS: 500 Elon: 7% Proof: 310**
NS 11360	Nodular graphite cast iron: Pearlitic/ferritic structure; designation used by Norwegian Standards **DPN: 235 UTS: 600 Elon: 3% Proof: 350**
NS 11370	Nodular graphite cast iron: Pearlitic structure; designation used by Norwegian Standards **DPN: 265 UTS: 700 Elon: 2% Proof: 400**
OV 15	Fine grained flake cast iron: Hungarian Standard
OV 20	Fine grained flake cast iron: Hungarian Standard
OV 25	Fine grained flake cast iron: Hungarian Standard
OV 30	Fine grained flake cast iron: Hungarian Standard
OV 35	Fine grained flake cast iron: Hungarian Standard
OV 40	Fine grained flake cast iron: Hungarian Standard
PC 400	Compacted graphite cast iron: Meehanite; for high temperature use **DPN: 225 UTS: 400 Elon: 1% Proof: 330**
PNH 82123	General standard for spheroidal graphite cast iron: Polish Standard; see designation for details
PNH 83101	General standard for fine grain flake cast iron: Polish Standard; see designation for details
R 001	General standard for flake graphite cast iron: Grey; Portuguese Standard; see designation for details
SAE 110	3.5% C 0.7% Mn 2.5% Si 0.15% S 0.25% P grey cast iron: As cast **DPN: 187 UTS: 120**
SAE 111	3.3% C 0.8% Mn 2.2% Si 0.15% S 0.2% P grey cast iron: As cast **DPN: 200 UTS: 180**
SAE 113	3.4% C 0.8% Mn 1.5% Si 0.14% S 0.2% P grey cast iron: Alloying elements permitted **DPN: 200 UTS: 180**
SAE 114	3.4% C 0.8% Mn 1.5% Si 0.14% S 0.2% P grey cast iron: Alloying elements permitted **DPN: 230 UTS: 180**
SAE 120	3.3% C 0.8% Mn 2% Si 0.15% S 0.15% P grey cast iron: As cast **DPN: 210 UTS: 240**
SAE 121	3.2% C 0.8% Mn 2% Si 0.15% S 0.12% P grey cast iron: As cast **DPN: 220 UTS: 270**
SAE 122	3.1% C 0.9% Mn 2% Si 0.15% S 0.1% P grey cast iron: As cast **DPN: 230 UTS: 300**
SAE 32510	Primary graphite free malleable cast iron: Annealed **UTS: 340 Elon: 10% Proof: 210**
SAE 35013	Primary graphite free malleable cast iron: Annealed **UTS: 370 Elon: 18% Proof: 220**
SAE 35018	Malleable cast iron **UTS: 370 Elon: 18% Proof: 220**
SAE 43010	Pearlitic malleable cast iron: Hardened and tempered **DPN: 180 UTS: 420 Elon: 10% Proof: 290**
SAE 48005	Pearlitic malleable cast iron: Hardened and tempered **DPN: 200 UTS: 500 Elon: 5% Proof: 320**
SAE 53004	Pearlitic malleable iron: Casting; hardened and tempered **DPN: 200 UTS: 580 Elon: 4% Proof: 370**
SAE 60003	Pearlitic malleable iron: Casting; hardened and tempered **DPN: 220 UTS: 580 Elon: 3% Proof: 420**

Symbol	Nominal analysis, supplier, condition and remarks.
SAE 70002	Pearlitic malleable iron: Casting; hardened and tempered **DPN: 260 UTS: 600 Elon: 2% Proof: 500**
SAE D4018	Nodular graphite cast iron: Ferritic structure **DPN: 170 UTS: 415 Elon: 18% Proof: 275**
SAE D4512	Nodular graphite cast iron: Pearlitic/ferritic structure **DPN: 185 UTS: 450 Elon: 12% Proof: 310**
SAE D5506	Nodular graphite cast iron: Pearlitic/ferritic structure **DPN: 225 UTS: 550 Elon: 6% Proof: 380**
SAE D7003	Nodular graphite cast iron: Pearlitic structure **DPN: 270 UTS: 690 Elon: 3% Proof: 480**
SAE G 2000	Grey cast iron: As cast **DPN: 187 UTS: 120**
SAE G 3000	3.4% C (total) 2.0% Si 0.12% S 0.15% P grey cast iron: As cast **DPN: 200 UTS: 200**
SAE G 3500	3.5% C (total) 1.6% Si 0.12% S 0.15% P grey cast iron: As cast **DPN: 220 UTS: 220**
SAE G 4000	3.4% C (total) 1.6% Si 0.12% S 0.15% P grey cast iron: As cast **DPN: 220 UTS: 270**
SAE G 4500	Grey cast iron: As cast **DPN: 250 UTS: 290**
SAE J158	Specification for malleable cast iron: Graded by strength
SAE J434	Specification for ductile cast iron: Spheroidal graphite SG iron; graded by strength
SAE J434 B	Nodular graphite cast iron: Covers various grades
SAE J859	Specification for flake graphite grey cast iron: Graded by strength
SANDHOLME GA	Cast iron: Hardened and tempered after machining; Sandholme
SC	Cast iron: For use up to 900 °C; Meehanite; as cast; furnace equipment **DPN: 300 UTS: 180**
SC 12–28	Fine grained grey cast iron: Designation used by Russian Standards **UTS: 150**
SC 15–32	Fine grained grey cast iron: Designation used by Russian Standards **UTS: 150**
SC 21–40	Fine grained grey cast iron: Designation used by Russian Standards **UTS: 225**
SC 28–48	Fine grained grey cast iron: Designation used by Russian Standards **UTS: 275**
SC 32–52	Fine grained grey cast iron: Designation used by Russian Standards **UTS: 300**
SC 36–56	Fine grained grey cast iron: Designation used by Russian Standards **UTS: 350**
SC 40–60	Fine grained grey cast iron: Designation used by Russian Standards **UTS: 400**
SD 51	14.5% Si 4.5% Cr Fe casting: Duriron Co.; symbol for Superchlor
SF	Cast iron: Ferritic matrix structure; Meehanite; as cast for BS alloy 27/12 (BS 2789) **DPN: 170 UTS: 390 Elon: 15% Proof: 290**
SF 400	Nodular graphite cast iron: Pearlitic/ferritic matrix; Meehanite **DPN: 165 UTS: 400 Elon: 17% Proof: 250**
SF 420.12	Nodular graphite cast iron: Meehanite **DPN: 160 UTS: 420 Elon: 12% Proof: 260**
SFA 420–12	Nodular graphite cast iron: Mainly ferritic matrix; Meehanite; previously grade SF **DPN: 200 UTS: 420 Elon: 12% Proof: 250**
SFF	Cast iron: Ferritic matrix structure; Meehanite; as cast for BS alloy 24/17 (BS 2789) **DPN: 160 UTS: 370 Elon: 20% Proof: 240**
SFF 350	Nodular graphite cast iron: Mainly ferritic; Meehanite; previously SFF **DPN: 145 UTS: 350 Elon: 24% Proof: 220**
SFF 400	Nodular graphite cast iron: Mainly ferritic; Meehanite; previously SFF **DPN: 150 UTS: 400 Elon: 20% Proof: 250**
SFP 500	Nodular graphite cast iron: Pearlitic/ferritic structure; Meehanite; previously grade SPF **DPN: 205 UTS: 500 Elon: 7% Proof: 310**
SFSH 1152	General standard for fine grained flake cast iron: Finnish Standard; see designation for details
SFT	Cast iron: Ferritic; nodular graphite structure; Meehanite; as cast; for use at temperatures up to 900 °C **DPN: 235 UTS: 370 Elon: 2% Proof: 300**
SG 38	Nodular graphite cast iron: Pearlitic/ferritic matrix; designation used by Austrian Standards **DPN: 165 UTS: 400 Elon: 17% Proof: 250**
SG 42	Nodular graphite cast iron: Mainly ferritic matrix; designation used by Austrian Standards **DPN: 200 UTS: 420 Elon: 12% Proof: 350**
SG 50	Nodular graphite cast iron: Pearlitic/ferritic structure: designation used by Austrian Standards **DPN: 205 UTS: 500 Elon: 7% Proof: 310**
SG 60	Nodular graphite cast iron: Pearlitic/ferritic structure; designation used by Austrian Standards **DPN: 230 UTS: 600 Elon: 3% Proof: 350**
SG 70	Nodular graphite cast iron: Pearlitic structure; designation used by Austrian Standards **DPN: 265 UTS: 700 Elon: 2% Proof: 400**
SH	Cast iron: Nodular graphite type; Meehanite; hardened and tempered for rotating parts, etc. **DPN: 350 UTS: 910 Elon: 2% Proof: 600**
SH 800	Nodular graphite cast iron: Pearlitic structure; Meehanite; previously grade SM **DPN: 300 UTS: 800 Elon: 2% Proof: 460**
SH 1000	Nodular graphite cast iron: Pearlitic structure; Meehanite **UTS: 1000 Elon: 1%**
SHH	Nodular graphite cast iron: For abrasive resistance; hardened and tempered; Meehanite **DPN: 320 UTS: 1000**
SH N	Nodular graphite cast iron: For crankshafts, etc.; normalized; Meehanite **DPN: 270 UTS: 700 Elon: 3% Proof: 500**
SJ G15	Fine grained flake graphite cast iron: Designation used by Norwegian Standards **DPN: 170 UTS: 150**
SJ G20	Fine grained flake graphite cast iron: Designation used by Norwegian Standards **DPN: 195 UTS: 225**
SJ G25	Fine grained flake graphite cast iron: Designation used by Norwegian Standards **DPN: 195 UTS: 225**
SJ G30	Fine grained flake graphite cast iron: Designation used by Norwegian Standards **DPN: 210 UTS: 300**
SJ G35	Fine grained flake graphite cast iron: Designation used by Norwegian Standards **DPN: 220 UTS: 350**
SJ G40	Fine grained flake graphite cast iron: Designation used by Norwegian Standards **DPN: 250 UTS: 400**
SL 15	Fine grained flake cast iron: Yugoslavian Standard
SL 20	Fine grained flake cast iron: Yugoslavian Standard
SL 25	Fine grained flake cast iron: Yugoslavian Standard
SL 30	Fine grained flake cast iron: Yugoslavian Standard
SL 35	Fine grained flake cast iron: Yugoslavian Standard

Symbol	Nominal analysis, supplier, condition and remarks.
SL 40	Fine grained flake cast iron: Yugoslavian Standard
SN 22	Spheroidal graphite SG cast iron: Austenitic; French Standard; impact 20 J **DPN: 170 UTS: 370 Elon: 20% Proof: 170**
S NC 20.2	Spheroidal graphite SG cast iron: Austenitic; French Standard; impact 13 J **DPN: 200 UTS: 370 Elon: 7% Proof: 210**
S NC 30.3	Spheroidal graphite SG cast iron: Austenitic; French Standard **DPN: 200 UTS: 370 Elon: 7% Proof: 210**
S NM 23.4	Spheroidal graphite SG cast iron: Austenitic; French Standard; impact 24 J **DPN: 180 UTS: 440 Elon: 25% Proof: 210**
SOUDOFONTE GS	3.5% C 3.5% Si iron alloy: Soudometal; for welding cast iron **DPN: 200**
SP	Pearlitic nodular cast iron: Meehanite; as cast for BS alloy 37/2 (BS 2789) **DPN: 220 UTS: 580 Elon: 2% Proof: 390**
SP 700	Nodular graphite cast iron: Pearlitic structure; Meehanite **DPN: 265 UTS: 700 Elon: 2% Proof: 400**
SPF	Pearlitic nodular cast iron: Meehanite; as cast for BS alloy 32/7 (BS 2789) **DPN: 200 UTS: 500 Elon: 7% Proof: 330**
SPF 600	Nodular graphite cast iron: Pearlitic/ferritic structure; Meehanite; previously SP grade **DPN: 230 UTS: 600 Elon: 3% Proof: 350**
SS 08–1002	Blackheart malleable cast iron: Swedish Standard; replaced by SS 08–1500
SS 08–1400	Blackheart malleable cast iron: Swedish Standard **DPN: 149 UTS: 300 Elon: 6%**
SS 08–1500	Blackheart malleable cast iron: Swedish Standard **DPN: 130 UTS: 320 Elon: 12% Proof: 190**
SS 08–5200	Blackheart malleable cast iron: Swedish Standard **DPN: 160 UTS: 400 Elon: 7% Proof: 240**
SS 08–5400	Blackheart malleable cast iron: Swedish Standard **DPN: 190 UTS: 500 Elon: 5% Proof: 300**
SS 08–5403	Blackheart malleable cast iron: Swedish Standard; replaced by SS 08–5400
SS 08–5600	Blackheart malleable cast iron: Swedish Standard **DPN: 220 UTS: 600 Elon: 4% Proof: 380**
SS 08–5603	Blackheart malleable cast iron: Swedish Standard; replaced by SS 08–5600
SS 08–6203	Blackheart malleable cast iron: Swedish Standard **DPN: 260 UTS: 700 Elon: 3% Proof: 530**
SS 08–6403	Blackheart malleable cast iron: Swedish Standard **DPN: 280 UTS: 800 Elon: 2% Proof: 600**
SS 14–0814	Blackheart malleable iron: Swedish Standard **DPN: 149 UTS: 300 Elon: 5%**
SS 14–0815	Blackheart malleable iron: Swedish Standard **DPN: 127 UTS: 320 Elon: 12% Proof: 190**
SS 14–0852	Pearlitic malleable iron: Swedish Standard **DPN: 155 UTS: 400 Elon: 7% Proof: 240**
SS 14–0854	Pearlitic malleable iron: Swedish Standard **DPN: 190 UTS: 500 Elon: 5% Proof: 300**
SS 14–0856	Pearlitic malleable iron: Swedish Standard **DPN: 220 UTS: 600 Elon: 4% Proof: 380**

Note. The following abbreviations and units are used in the tables:

DPN	Hardness, diamond pyramid number
UTS	Ultimate tensile strength, N/mm^2
Elon	Elongation, %
Proof	0.1% proof strength, N/mm^2

1 N/mm^2=0.1 hbar=0.102 kgf/mm^2=0.06475 $tonf/in.^2$=145.04 $lbf/in.^2$=1 MPa

See Appendix II for other abbreviations and conversion tables.

Symbol	Nominal analysis, supplier, condition and remarks.
SS 14–0862	Pearlitic malleable iron: Swedish Standard **DPN: 260 UTS: 700 Elon: 3% Proof: 530**
SS 14–0864	Pearlitic malleable iron: Swedish Standard **DPN: 290 UTS: 800 Elon: 2% Proof: 600**
SS 0100	Grey cast iron: As cast; Swedish Standard
SS 0110	Grey cast iron: As cast; Swedish Standard **DPN: 180**
SS 0115	Grey cast iron: As cast; Swedish Standard **DPN: 190**
SS 0120	Grey cast iron: As cast; Swedish Standard **DPN: 210**
SS 0125	Grey cast iron: As cast; Swedish Standard **DPN: 230**
SS 0130	Grey cast iron: As cast; Swedish Standard **DPN: 250**
SS 0135	Grey cast iron: As cast; Swedish Standard **DPN: 260**
SS 0140	Grey cast iron: As cast; Swedish Standard
SS 0212	Grey cast iron: Swedish Standard **DPN: 170**
SS 0215	Grey cast iron: Swedish Standard **DPN: 175**
SS 0217	Grey cast iron: Swedish Standard **DPN: 190**
SS 0219	Grey cast iron: Swedish Standard **DPN: 210**
SS 0221	Grey cast iron: Swedish Standard **DPN: 230**
SS 0223	Grey cast iron: Swedish Standard **DPN: 250**
SS 0707	Spheroidal graphite cast iron: As cast; Swedish Standard **DPN: 260**
SS 0717	Spheroidal graphite cast iron: Swedish Standard **DPN: 150 UTS: 400 Elon: 18% Proof: 250**
SS 0727	Spheroidal graphite cast iron: Swedish Standard **DPN: 200 UTS: 500 Elon: 7% Proof: 350**
SS 0732	Spheroidal graphite cast iron: Swedish Standard **DPN: 230 UTS: 600 Elon: 5% Proof: 400**
SS 0737	Spheroidal graphite cast iron: Swedish Standard **DPN: 245 UTS: 700 Elon: 3% Proof: 450**
SS 0810	Blackheart malleable cast iron: Annealed; Swedish Standard **DPN: 140**
SS 0854	Cast iron: Pearlitic malleable iron; Swedish Standard **DPN: 135 UTS: 500 Elon: 5% Proof: 300**
SS 0856	Cast iron: Pearlitic malleable iron; Swedish Standard **DPN: 220 UTS: 600 Elon: 4% Proof: 360**
STA 8	Grey iron casting: Replaced by BS 1452/17
STA 9	Grey iron casting: Replaced by BS 1452/20
STA 25	0.8% C 15% Si cast iron: Replaced by BS 1591
STAS 568	General standard for fine grain flake cast iron: Romanian Standard; see designation for details
STAS 6071	General standard for spheroidal graphite cast iron: Romanian Standard
SUPERCHLOR	0.9% C 1.5% Mn 14.5% Si 4.5% Cr iron alloy: Duriron Co.; excellent corrosion resistance
TENSOLA I	Pearlitic cast iron: Gloucester Foundry **DPN: 180 UTS: 450 Elon: 8% Proof: 300**
TENSOLA II	Pearlitic cast iron: Gloucester Foundry **DPN: 200 UTS: 530 Elon: 6% Proof: 330**
TENSOLA III	Pearlitic cast iron: Gloucester Foundry **DPN: 245 UTS: 690 Elon: 3% Proof: 420**
UNE 36–111	General standard for flake graphite cast irons: Grey; Spanish Standard; see designation for details
UNE 36–118–73	Nodular graphite cast iron: Grey; Spanish Standard; details given under designation
UNI 5007	General standard for flake graphite cast iron: Grey; Italian Standard; see designation for details
UNI 14544	Nodular graphite cast iron: Italian Standard; details given under designation

Symbol	Nominal analysis, supplier, condition and remarks.
VAM 20	20% Cr 1.5% Ni (max) 1.5% Mo 1.2% Cu (max) white cast iron: See ASTM A 532/11
VAM 27	24% Cr 1.5% Ni (max) 1.5% Mo (max) 1.2% Cu (max) white cast iron: See ASTM A 532/11
VCh 15	Fine grained flake cast iron: Bulgarian Standard
VCh 20	Fine grained flake cast iron: Bulgarian Standard
VCh 25	Fine grained flake cast iron: Bulgarian Standard
VCh 30	Fine grained flake cast iron: Bulgarian Standard
VCh 35	Fine grained flake cast iron: Bulgarian Standard
VCh 38.17	Spheroidal graphite cast iron: Russian Standard
VCh 42.12	Spheroidal graphite cast iron: Russian Standard
VCh 45.0	Nodular graphite cast iron: Pearlitic/ferritic structure; designation used by Russian Standards **DPN: 225 UTS: 440 Proof: 350**
VCh 45.5	Nodular graphite cast iron: Pearlitic/ferritic structure; designation used by Russian Standards **DPN: 185 UTS: 440 Elon: 5% Proof: 320**
VCh 50.1.5	Nodular graphite cast iron: Pearlitic/ferritic structure; designation used by Russian Standards **DPN: 225 UTS: 490 Elon: 1.5% Proof: 372**
VCh 50.2	Spheroidal graphite cast iron: Russian Standard
VCh 50.7	Spheroidal graphite cast iron: Russian Standard
VCh 60.2	Nodular graphite cast iron: Pearlitic/ferritic structure; designation used by Russian Standards **DPN: 235 UTS: 585 Elon: 2% Proof: 410**
VCh 60.2	Spheroidal graphite cast iron: Russian Standard
VCh 70.2	Spheroidal graphite cast iron: Russian Standard
VCh 80.2	Spheroidal graphite cast iron: Russian Standard
VCh 100.2	Spheroidal graphite cast iron: Russian Standard
VCR 40–10	Nodular graphite cast iron: Pearlitic/ferritic structure; designation used by Russian Standards **DPN: 180 UTS: 390 Elon: 10% Proof: 290**
VSM 10693	General standard for spheroidal graphite cast iron: Swiss Standard
VSM 10691	General standard for fine grain flake cast iron: Swiss Standard; see designation for details
W 22/4	Whiteheart malleable cast iron: Designation used by BS
W 24/8	Whiteheart malleable cast iron: Designation used by BS
WA	Cast iron with good wear resistance: Meehanite; as cast; for gears, brake drums, etc. **DPN: 300 UTS: 340**
WB	Cast iron: Wear resisting; Meehanite; as cast; for mill liners, sand blast nozzles, etc. **DPN: 450 UTS: 300**
WBC	Cast iron: Suitable for chill casting; Meehanite; as cast; for crusher jaws, mill liners, etc. **DPN: 550 UTS: 300**
WBC	Cast iron: Suitable for chill casting; Meehanite; stress relieved; for rope and wire guides; truck wheels, etc. **DPN: 550 UTS: 210**
WH	Cast iron: Wear resisting; Meehanite; as cast; for gravel pipes, sand blast tables, etc. **DPN: 600 UTS: 270**
WSH	Nodular graphite cast iron: For abrasive resistance; hardened and tempered to give martensitic structure; Meehanite **DPN: 600 UTS: 1200**
WSH 1	Nodular graphite cast iron: Meehanite; for wear resistance **DPN: 500 UTS: 600**
WSH 2	Nodular graphite cast iron: Meehanite; for wear resistance **DPN: 670 UTS: 300**
Z 115	Fine grained flake cast iron: Polish Standard
Z 120	Fine grained flake cast iron: Polish Standard
Z 125	Fine grained flake cast iron: Polish Standard
Z 130	Fine grained flake cast iron: Polish Standard
Z 135	Fine grained flake cast iron: Polish Standard
Z 140	Fine grained flake cast iron: Polish Standard
Zs 3817	Spheroidal graphite cast iron: Polish Standard
Zs 4012	Spheroidal graphite cast iron: Polish Standard
Zs 4505	Spheroidal graphite cast iron: Polish Standard
Zs 5002	Spheroidal graphite cast iron: Polish Standard
Zs 6002	Spheroidal graphite cast iron: Polish Standard
Zs 8002	Spheroidal graphite cast iron: Polish Standard
Zs 9002	Spheroidal graphite cast iron: Polish Standard

Note. The following abbreviations and units are used in the tables:

DPN	Hardness, diamond pyramid number
UTS	Ultimate tensile strength, N/mm^2
Elon	Elongation, %
Proof	0.1% proof strength, N/mm^2

1 N/mm^2=0.1 hbar=0.102 kgf/mm^2=0.06475 $tonf/in.^2$=145.04 $lbf/in.^2$=1 MPa

See Appendix II for other abbreviations and conversion tables.

20C Iron–nickel alloys – magnetic With chromium, titanium or cobalt

Specific gravity	8.1
Density	8100 kg/m^3
Solidus/liquidus	1450 °C
Thermal conductivity	34–45 W/m °C
Coefficient of linear expansion (20–100 °C)	$1-10 \times 10^{-6}$/ °C (These are low and controlled expansion alloys)
Electrical conductivity	2.1–3.5% IACS (copper 100%)
Specific resistance	500–800 microhm mm
Young's modulus of elasticity	145×10^9 N/m^2
Impact	Will vary considerably
Fatigue strength	–
Hot strength	–

The above properties have been chosen to show typical values for the specifications listed and may not apply exactly to any one specification. It is possible that with some specifications the values may not be applicable.

General metallurgical characteristics

Iron of a very high purity has been shown to have magnetic properties considerably better than commercially pure iron. This is seldom economically desirable and is briefly discussed in Section 20A.

Iron–silicon alloys with a low carbon content also have useful magnetic properties; they are treated as steels and are discussed in Section 44B1.

With additions of nickel the magnetic properties of iron first fall, until at about 30% nickel the alloys are completely non-magnetic at room temperature. With further additions of nickel, a series of magnetically useful alloys is available at 50% iron/nickel. With 85% nickel a third series of useful alloys exist which are discussed briefly in Section 27E.

Apart from their magnetic and low expansion properties, the alloys listed have no significance. It is of interest that this property is sometimes useful in the negative sense, being absent, and some of the alloys listed can be made either magnetic or non-magnetic by a relatively small temperature alteration.

The magnetic properties are also considerably influenced by the direction of rolling or cold working, this being known as 'grain orienting', and by the addition of chromium up to about 6%.

Any thermal treatment carried out on these materials will be to alter the magnetic properties of the material as cast.

This is a very complex technique and should never be attempted without specific advice from physicists and metallurgists.

Advice can be obtained from the Permanent Magnet Association when necessary.

Certain of these materials can change from being magnetic to being completely non-magnetic at specific temperatures and this is one reason for their use. By heat treatment the temperature at which the change takes place can be altered and thus the use of the materials can be changed dramatically. It is most important that the correct advice is obtained before any attempt is made to heat treat for magnetic purposes. No other reason for heat treating could ever be involved with these materials.

These materials should not be welded under any circumstances. Any joining should be achieved by the use of adhesives or low temperature brazing or soldering, or pinning.

These materials are invariably used for magnetic and electronic purposes. They are used as chokes, inductances, transformer cores, and as screening materials to prevent high frequency interference. Many of the alloys are highly specialized and it is recommended that further information is obtained from the suppliers or, if in doubt, from the Permanent Magnet Association.

Symbol	Nominal analysis, supplier, condition and remarks.
1.3921	0.8% Cr 48% Ni Fe alloy: German Standard
1.3926	46% Ni Fe alloy: German Standard
1.3927	46% Ni Fe alloy: German Standard
2.4472	45% Fe Ni alloy: German Standard
2.4480	0.8% Cr 47% Fe Ni alloy: German Standard
2.4486	6.0% Cr 46% Fe Ni alloy: German Standard
22.3 ALLOY	0.1% C 3.1% Cr 22% Ni Fe alloy: Carpenter; high expansion coefficient
42 ALLOY	0.05% C (max) 41% Ni Fe alloy: Carpenter; low expansion coefficient
142 ALLOY	41% Ni Fe alloy: Driver Harris; expansion coefficient 5.3×10^{-6}; electrical conductivity 2.5% IACS **UTS: 800**
146 ALLOY	46% Ni Fe alloy: Driver Harris **UTS: 800**
152 ALLOY	49% Fe Ni alloy: Driver Harris; glass metal seals **UTS: 480**
193 ALLOY	2.7% Cr 32% Ni Fe alloy: Driver Harris; heavy duty rheostats; withdrawn
426 ALLOY	0.08% C (max) 6% Cr 42% Ni Fe alloy: Darwin
AC HT	0.5% C 15.5% Cr 35% Ni Fe alloy: Origin unknown
ADR	40% Ni Fe alloy: Low expansion; Imphy
AFNOR 23.15 NK 38	38% Ni 1.5% Co 1.5% Ti 3% Nb Fe alloy: French specification
ALCHROME 750	15% Cr 4% Al Fe alloy: Carpenter **UTS: 620 Elon: 20% Proof: 480**
ALCHROME D	15% Cr 5.5% Al Fe alloy: For heating elements; Driver
ALCHROME DK	21% Cr 4.5% Al Fe alloy: Carpenter; for heating elements **UTS: 630 Elon: 15% Proof: 440**
ALNI (high coercivity)	30.0% Ni 12.5% Al 4.0% Cu 0.5% Ti Fe magnetic alloy: Coercivity 680 oersteds; information from PMA
ALNICO (high remanence)	16.4% Ni 9.2% Al 5.0% Cu 12.3% Co Fe alloy: For magnets; remanence 8000 gauss; information from PMA
ALNICO 3	26% Ni 12% Al Fe alloy: For magnets; Simonds
AMS 5221	0.06% C (max) 0.5% Al 5.5% Cr 2.5% Ti 42% Ni Fe alloy: Precipitation hardening
AMS 5223 A	5.2% Cr 42% Ni 2.3% Ti 0.5% Al Fe alloy: Strip; 10% cold reduced
AMS 5225	0.06% C (max) 0.5% Al 5.5% Cr 2.5% Ti 42% Ni Fe alloy: Precipitation hardening
AMS 5225 A	5.2% Cr 42% Ni 2.3% Ti 0.5% Al Fe alloy: Strip; 50% cold reduced
AMS 5392 G	2.1% Cr 15% Ni 6% Cu Fe alloy: Sand casting
AMS 5393 B	2% Cr 20% Ni Fe alloy: Sand casting
AMS 5394	2% Cr 20% Ni Fe alloy: Sand casting
AMS 5395	22% Ni Fe alloy: Nodular casting
AMS 7701	Ni Fe alloy: Sheet and strip; annealed
AMS 7702	Ni Fe alloy: Sheet and strip; ½-hard
AMS 7705	Ni Fe alloy: Bar and forging
AMS 7717	Ni Fe alloy: Sheet and strip; for magnetic purposes; forming quality
AMS 7718	Ni Fe alloy: Bar, forging, etc.; magnetic alloy
AMS 7719	Ni Fe alloy: Sheet and strip; for magnetic purposes; stamping quality
AMS 7726 A	29% Ni 17% Co Fe alloy: Wire; low expansion alloy; for glass sealing
AMS 7727	29% Ni 17% Co Fe alloy: Bar and forging; low expansion; for glass sealing
AMS 7728 A	29% Ni 17% Co Fe alloy: Sheet and strip; low expansion; for glass sealing
AMS 7734	0.1% C (max) 42.5% Ni Fe alloy
ASC	35% Ni 10% Co Fe alloy: Low expansion; Imphy
ASC A	28% Ni 18% Co Fe alloy: Low expansion; Imphy
AUDIOLLOY	48% Ni Fe alloy: Origin unknown

Note. The following abbreviations and units are used in the tables:

DPN	Hardness, diamond pyramid number
UTS	Ultimate tensile strength, N/mm^2
Elon	Elongation, %
Proof	0.1% proof strength, N/mm^2

$1\ N/mm^2 = 0.1\ hbar = 0.102\ kgf/mm^2 = 0.06475\ tonf/in.^2 = 145.04\ lbf/in.^2 = 1\ MPa$

See Appendix II for other abbreviations and conversion tables.

Symbol	Nominal analysis, supplier, condition and remarks.
BS 933/4	Ni Fe alloy: Sheet for magnetic purposes; composition not part of specification
BS 2857	36% Ni Fe alloy: Strip for laminations
BS 2857 B	50% Fe Ni alloy: Strip for laminations
BS 2857 C	35% Ni Fe alloy: For transformer laminations
BS 2857 D	36% Ni Fe alloy: For transformer laminations
BS 3127/9	0.06% C 5.2% Cr 42% Ni 0.5% Al 2.3% Ti Fe alloy: Tubing; annealed; for bourdon tubes **DPN: 160**
BS 3127/9	0.06% C 5.2% Cr 42% Ni 0.5% Al 2.3% Ti Fe alloy: Tubing; lightly cold worked; for bourdon tubes **DPN: 200**
CASTEES SG	Ni Fe welding electrode: Phillips; for cast iron
CERAMIC SEALING ALLOY	25% Co 27% Ni Fe alloy: For ceramic metal seal
COLUMNAR HYCOMAX 11	14.5% Ni 7.0% Al 28.5% Co 3.0% Cu 4.0% Ti 2.0% Nb 0.2% S Fe alloy: For magnets; Swift Levick
COMET	2.7% Cr 32% Ni Fe alloy: Driver Harris; heavy duty rheostats; withdrawn **UTS: 600**
DILVER P0	29% Ni 21.8% Co Fe alloy: Low expansion; Imphy
DILVER P1	29% Ni 18% Co Fe alloy: Low expansion; Imphy
DIN 17745 Ni 48	46% Ni Fe alloy
DIN 17745 Ni 49	0.8% Cr 48% Ni Fe alloy
DIN 17745 Ni Fe 45	45% Fe Ni alloy
DIN 17745 Ni Fe 48 Cr	0.8% Cr 47% Fe Ni alloy
DUMET	0.1% C (max) 42% Ni Fe alloy: Sealing
E 10	0.06% C 0.2% Si 0.2% Mn 3.6% Ni Fe alloy: Jessop; controlled expansion alloy
EHC ALLOY	14.0% Ni 7.3% Al 34.0% Co 4.5% Cu 5.5% Ti Fe alloy: For magnets; origin unknown
ELINVAR	9% Cr 35% Ni Fe alloy: Telcon; annealed; expansion 6 $\times 10^{-6}$ **UTS: 690**
ELINVAR	9% Cr 35% Ni Fe alloy: Telcon; cold drawn **DPN: 1540**
ER Ni Fe Mn Ci	0.5% C (max) 12% Mn 40% Ni Fe welding electrode: Designation used by AWS
GILGRID	45% Fe 10% Mo Ni alloy: Wire for electronic use; Gilbey-Brunton
GLASS SEALING 27	0.05% C 28% Cr 0.5% Ni Fe alloy: Carpenter; low expansion coefficient
GLASS SEALING 42	0.05% C (max) 4.1% Ni alloy: Carpenter; low expansion coefficient
GLASS SEALING 42.6	0.05% C (max) 5.7% Cr 42% Ni Fe alloy: Carpenter; low expansion coefficient
HAI 36 ALLOY	36% Ni Fe alloy: Harrison Alloy
HAI 42 ALLOY	42% Ni Fe alloy: Harrison Alloy
HAI 49 ALLOY	49% Ni Fe alloy: Harrison Alloy
HAI 373 ALLOY	29% Ni 27% Co Fe alloy: Harrison Alloy
HAI P6 ALLOY	6% Ni 45% Co Fe alloy: Harrison Alloy
HCR	50% Fe Ni alloy: Strip; Telcon; annealed; has rectangular hysteresis loop **DPN: 110 UTS: 370**
HIGH PERMEABILITY 49	0.02% C 0.35% Si 48% Ni Fe alloy: Carpenter; annealed **UTS: 500 Elon: 45% Proof: 150**
HIPERNIK	50% Ni Fe alloy: Strip; Westinghouse Electric; for magnetic laminations
HYREM RADIOMETAL	50% Ni Fe alloy: for magnetic purposes; Telcon
HYRHO RADIOMETAL	Ni Fe alloy: High resistivity with high saturation flux density; Telcon; cold rolled **DPN: 250 UTS: 700**
HYRHO RADIOMETAL	50% Ni Fe alloy: With higher resistivity than Radiometal; Telcon
IN 568	2.5% Ni 2.5% Si Fe alloy: International Nickel Co.
INCOLOY 805	0.12% C 7.5% Cr 0.5% Mo 36% Ni Fe alloy: For springs; Huntington; low temperature coefficient of modulus of elasticity
INCOLOY 903	38% Ni 15% Co 1.4% Ti 3% Nb Fe alloy: Inco; solution treated and aged **UTS: 1300 Elon: 14% Proof: 1100**
INCOLOY 907	38% Ni 13% Co 4.7% Nb 1.5% Ti Fe alloy: Inco
INCOLOY 909	0.01% C 38% Ni 13% Co 4.7% Nb 1.5% Ti Fe alloy: Inco
INVAR	0.1% C 0.2% Si 0.5% Mn 36% Ni Fe alloy: Latrobe Steel Co.; annealed; low expansion; also Telcon **DPN: 131 UTS: 450 Elon: 40% Proof: 220**
INVAR 36	36% Ni Fe alloy: Low expansion; Telcon; annealed; expansion 2.0 $\times 15^{-6}$/ °C up to 100 °C **DPN: 140 UTS: 480 Elon: 30% Proof: 260**
INVAR 42	42% Ni Fe alloy: Low expansion; Telcon; annealed; expansion 4.5 $\times 10^{-6}$/ °C up to 300 °C **DPN: 145 UTS: 520 Elon: 30% Proof: 240**
IRWAR FM	0.1% Co 36% Ni 0.15% Se Fe alloy: Latrobe; low expansion alloy; free machining
IRWAR STANDARD	36% Ni Fe alloy: Low expansion; Imphy
K 94100	0.05% C (max) 41% Ni Fe alloy: Designation used by UNS
K 94600	0.05% C (max) 46% Ni Fe alloy: Designation used by UNS
K 94760	0.07% C (max) 5.6% Cr 42% Ni Fe alloy: Designation used by UNS
K 94800	0.05% C (max) 48% Ni Fe alloy: Designation used by UNS
KOERZIT 120K	26.0% Ni 13.0% Al 4.0% Co 3.0% Cu 1.0% Ti Fe alloy: For magnets; German alloy
KOERZIT 120R	22.0% Ni 12.0% Al 3.0% Co 3.0% Cu 0.5% Ti Fe alloy: For magnets; German alloy
KOERZIT 160	21.0% Ni 10.0% Al 17.0% Co 3.0% Cu 1.0% Ti Fe alloy: For magnets; German alloy
KOERZIT 190	21.0% Ni 10.0% Al 17.0% Co 1.0% Ti 1.0% Ti Fe alloy: For magnets; German alloy
KOERZIT 220	15.0% Ni 7.0% Al 28.0% Co 5.0% Cu 8.0% Ti Fe alloy: For magnets; German alloy
KOERZIT 350	15.0% Ni 7.0% Al 30.0% Co 4.0% Cu 5.0% Ti Fe alloy: For magnets; German alloy
KOERZIT 400K	15.0% Ni 9.0% Al 24.0% Co 3.0% Cu 1.0% Ti Fe alloy: For magnets; German alloy
KOERZIT 450	15.0% Ni 7.0% Al 30.0% Co 4.0% Cu 5.0% Ti Fe alloy: For magnets; German alloy
KOERZIT 500	15.0% Ni 9.0% Al 24.0% Co 3.0% Cu Fe alloy: For magnets; German alloy
KOERZIT 600	15.0% Ni 9.0% Al 24.0% Co 3.0% Cu Fe alloy: For magnets; German alloy
LEGA 42	42% Ni Fe alloy: Low expansion; Driver Harris
LEGA 46	46% Ni Fe alloy: Low expansion; Driver Harris
MAGLOY 1	14.0% Ni 8.0% Al 24.0% Co 3.0% Cu Fe alloy: For magnets; origin unknown
MAGLOY 2	14.0% Ni 8.0% Al 24.0% Co 3.0% Cu 1.2% Ti Fe alloy: For magnets; origin unknown
MAGLOY 5	18.0% Ni 10.0% Al 13.0% Co 6.0% Cu Fe alloy: For magnets; origin unknown
MAGLOY 7	21.0% Ni 9.0% Al 20.0% Co Fe alloy: For magnets; origin unknown
MAGLOY 8	17.0% Ni 7.0% Al 30.0% Co 5.0% Ti Fe alloy: For magnets; origin unknown
MAGLOY 10	14.0% Ni 8.0% Al 24.0% Co 3.0% Cu Fe alloy: For magnets; origin unknown
MAGLOY 100	14.0% Ni 8.0% Al 24.0% Co 3.0% Cu Fe alloy: For magnets; origin unknown
N 42	42% Ni Fe alloy: Low expansion; Imphy
N 501	50% Ni 1% Cr. Fe alloy: Low expansion; Imphy
NIALCO I	25.0% Ni 12.0% Al 2.0% Co 4.0% Cu Fe alloy: For magnets; French alloy
NIALCO IF	25.0% Ni 12.0% Al 2.0% Co 4.0% Cu Fe alloy: For magnets; French alloy
NIALCO II	25.0% Ni 10.0% Al 4.0% Co 4.0% Cu Fe alloy: For magnets; French alloy

Symbol	Nominal analysis, supplier, condition and remarks.
NIALCO III	20.0% Ni 10.0% Al 12.0% Co 4.0% Cu Fe alloy: For magnets; French alloy
NIALCO IIIF	20.0% Ni 10.0% Al 12.0% Co 4.0% Cu Fe alloy: For magnets; French alloy
NIALCO IV	20.0% Ni 8.0% Al 10.0% Co 6.0% Cu Fe alloy: For magnets; French alloy
NIALCO IVF	20.0% Ni 8.0% Al 10.0% Co 6.0% Cu Fe alloy: For magnets; French alloy
NIALCO V	21.0% Ni 10.0% Al 17.0% Co 3.0% Cu 1.0% Ti Fe alloy: For magnets; French alloy
NICOSEAL	29% Ni 17% Co Fe alloy: Low expansion; Carpenter
NICOSEL	29% Ni 17% Co Fe alloy: Low expansion; Firth Brown
NIKO	Ni Co Fe alloy: Swift Levick; low expansion **DPN: 170 UTS: 510 Elon 41% Proof: 360**
NILGRO 42	42% Ni Fe alloy: Low expansion; Darwins; coefficient of expansion 6×10^{-6}/ °C up to 350 °C **UTS: 550 Elon: 42% Proof: 390**
NILO 36	36% Ni Fe alloy: Wrought; expansion 1.5×10^{-6}; H Wiggin; electrical conductivity 2% IACS
NILO 40	40% Ni Fe alloy: Wrought; expansion 4.1×10^{-6}; H Wiggin; specification Res. 680 microhm/mm; electrical conductivity 2.5% IACS
NILO 42	42% Ni Fe alloy: Wrought; expansion 5.3×10^{-6}; H Wiggin; electrical conductivity 2.75% IACS
NILO 48	48% Ni Fe alloy: Wrought; expansion 8.5×10^{-6}; H Wiggin; electrical conductivity 3.5% IACS
NILO 50	50% Ni Fe alloy: Wrought; expansion 9.3×10^{-6}; H Wiggin; electrical conductivity 3.5% IACS
NILO 475	47% Ni 5% Cr Fe alloy: Wrought; expansion 8.2×10^{-6}; H Wiggin; electrical conductivity 2% IACS
NILO K	29% Ni 17% Co Fe alloy: Wrought; expansion 6×10^{-6}; H Wiggin; electrical conductivity 3.5% IACS
NILO K45	32.5% Ni 13.0% Co Fe alloy: For glass seals; International Nickel Co.
NILO P59	50% Ni Fe alloy: For glass seals; International Nickel Co.
NILOMAG 471	47.0% Ni 3.0% Mo Fe alloy: Strip made by powder metallurgy; International Nickel Co.
NILVAR	36% Ni Fe alloy: Expansion 1×10^{-6}; Driver Harris; electrical conductivity 2% IACS **UTS: 530**
NIPERMAG	12.0% Al 30.0% Ni 0.4% Ti Fe alloy: For magnets; origin unknown **DPN: 450**
NI-ROD 55	1.5% C 45% Fe Ni alloy: Welding rod; Huntington; for welding high phosphorus cast irons
NIROMET 42	42% Ni Fe alloy: For glass/metal seal; Driver
NIROMET 46	46% Ni Fe alloy: For glass/metal seal; Driver
NI-SPAN-C	0.02% C 5.4% Cr 2.4% Ti 42% Ni Fe alloy: Bar and sheet; Huntington; now NI-SPAN-C 902
NI-SPAN-C 902	0.02% C 5.4% Cr 2.4% Ti 42% Ni Fe alloy: Bar and sheet; Huntington; age hardening
NI-SPAN-C 902	0.06% C (max) 5.5% Cr 42% Ni 0.5% Al Fe alloy: Inco; annealed **UTS 690 Elon: 45% Proof: 310**
No. 4 ALLOY	5.5% Cr 42% Ni Fe alloy: British Driver Harris; glass metal seals for soft glasses
NS 36	36% Ni Fe alloy: Swift Levick; low expansion **DPN: 135 UTS: 450 Elon: 42% Proof: 240**
NS 42	42% Ni Fe alloy: Swift Levick; low expansion **DPN: 143 UTS: 520 Elon: 46% Proof: 210**
NS 49	49% Ni Fe alloy: Swift Levick; low expansion **DPN: 143 UTS: 530 Elon: 45% Proof: 270**
OERSTIT 120R	22.0% Ni 12.0% Al 3.0% Co 3.0% Cu 0.5% Ti Fe alloys: For magnets; German alloy
ORTHOMUMETAL	Soft magnetic material: Telcon
PERMALLOY 'B'	46% Ni Fe alloy: Sheet, bar, forging, etc.; Standard Telephones; high values of flux density
PERMALLOY 'D'	36% Ni Fe alloy: Standard Telephones; electrical conductivity 1.8% IACS
R 2800	Ni Fe alloy: Telcon; hard rolled; Curie point 150 °C **DPN: 170**
RADIO METAL 36	36% Ni Fe alloy: Induction melted castings; Telcon; annealed; similar to Mumetal, but cheaper; see Section 27C **DPN: 110 UTS: 520**
RADIO METAL 36	36% Ni Fe alloy: Induction melted castings; Telcon; hard rolled **DPN: 290 UTS: 970**
RADIO METAL 50	50% Fe Ni alloy: Sheet and bar; Telcon; annealed; high initial permeability **DPN: 100 UTS: 400**
RADIO METAL 50	50% Fe Ni alloy: Sheet and bar; Telcon; hard rolled **DPN: 250 UTS: 750**
RECO 100	24.0% Ni 14.0% Al Fe alloy: For magnets; Dutch alloy
RECO 120	26.0% Ni 13.0% Al 4.0% Co 3.0% Cu 1.0% Ti Fe alloy: For magnets; Dutch alloy
RECO 140	24.0% Ni 10.0% Al 5.0% Co 7.0% Cu 0.8% Ti Fe alloy: For magnets; Dutch alloy
RECO 160	18.5% Ni 10.0% Al 13.0% Co 7.5% Cu 1.9% Ti Fe alloy: For magnets; Dutch alloy
RECO 170	24.0% Ni 9.5% Al 10.0% Co 6.0% Cu 5.0% Ti Fe alloy: For magnets; Dutch alloy
RECO 220	15.0% Ni 7.0% Al 26.0% Co 5.0% Cu 7.0% Ti Fe alloy: For magnets; Dutch alloy
RODAR	29% Ni 17% Co Fe alloy: For glass/metal seal; Driver
SANBOLD NA 35	35% Ni Fe alloy: Strip; Sanderson; electrical conductivity 1.8% IACS
SANBOLD NA 47	47% Ni Fe alloy: Strip for transformers, etc.; Sanderson; electrical conductivity 3% IACS
SATMUMETAL	Soft magnetic alloy: 0.025 oersteds coercivity; Telcon
SUPER RADIO METAL 50	50% Fe Ni alloy: Similar to Radio metal 50 Ni alloy; Telcon; annealed; initial permeability 3 times Radio metal 50 **DPN: 110 UTS: 420**
SUPER RADIO METAL 50	50% Fe Ni alloy: Similar to Radio metal 50 Ni alloy; Telcon; hard rolled **DPN: 250 UTS: 750**
SUPER UGIMAX 600	14.0% Ni 8.0% Al 24.0% Co 3.0% Cu Fe alloy: For magnets; French alloy
SUPER UGIMAX 800	14.0% Ni 8.0% Al 24.0% Co 3.0% Cu Fe alloy: For magnets; French alloy
TELCOSEAL I	29% Ni 17% Co Fe alloy: For use with borosilicate glass; Telcon; expansion 5.3×10^{-6}; electrical conductivity 3.75% IACS; annealed **DPN: 150 UTS: 480 Elon: 30%**
TELCOSEAL II	42% Ni Fe alloy: For use with lead borosilicate glass; Telcon; expansion 5.8×10^{-6}; electrical conductivity 3% IACS; annealed **DPN: 125 UTS: 530 Elon: 27%**
TELCOSEAL III	6% Cr 42% Ni Fe alloy: For use with lead glass; Telcon; expansion 8.7×10^{-6}; electrical conductivity 1.8% IACS; annealed **DPN: 120 UTS: 550 Elon: 35%**
TELCOSEAL V	25% Co Fe alloy: For use with soft lead glass; Telcon; expansion 10.6×10^{-6}; electrical conductivity 2.5% IACS; annealed **DPN: 150**

Note. The following abbreviations and units are used in the tables:

DPN	Hardness, diamond pyramid number
UTS	Ultimate tensile strength, N/mm^2
Elon	Elongation, %
Proof	0.1% proof strength, N/mm^2

1 N/mm^2=0.1 hbar=0.102 kgf/mm^2=0.06475 tonf/in.^{2}=145.04 lbf/in.^{2}=1 MPa

See Appendix II for other abbreviations and conversion tables.

Symbol	Nominal analysis, supplier, condition and remarks.
TELCOSEAL VI	49% Ni Fe alloy: For use with soft lead/lime glass; Telcon; expansion 9.1 × 10^{-6}; electrical conductivity 4.2% IACS; annealed **DPN: 120 UTS: 480 Elon: 30%**
THERLO	29% Ni 17% Co Fe alloy: Driver Harris; glass/metal seals **UTS: 540**
TICONAL 190	21.0% Ni 12.0% Al 14.0% Co 3.0% Cu Fe alloy: For magnets; Dutch alloy
TICONAL 360	15.0% Ni 8.5% Al 24.0% Co 3.0% Cu 1.5% Ti Fe alloy: For magnets; Dutch alloy
TICONAL 400	14.0% Ni 8.5% Al 24.0% Co 3.0% Cu Fe alloy: For magnets; Dutch alloy
TICONAL 450	14.5% Ni 7.5% Al 34.0% Co 4.5% Cu 5.0% Ti Fe alloy: For magnets; Dutch alloy
TICONAL 500	14.0% Ni 8.5% Al 24.0% Co 3.0% Cu Fe alloy: For magnets; Dutch alloy
TICONAL 600	14.0% Ni 8.5% Al 24.0% Co 3.0% Cu Fe alloy: For magnets; Dutch alloy
TICONAL 600	14.0% Ni 8.0% Al 24.0% Co 3.0% Cu Fe alloy: For magnets; French alloy
TICONAL 650	14.0% Ni 8.5% Al 24.0% Co 3.0% Cu Fe alloy: For magnets; Dutch alloy
TICONAL 700	14.0% Ni 8.0% Al 24.0% Co 3.0% Cu Fe alloy: For magnets; French alloy
TICONAL 750	14.0% Ni 8.0% Al 24.0% Co 3.0% Cu Fe alloy: For magnets; French alloy
TICONAL 750	14.0% Ni 8.5% Al 24.0% Co 3.0% Cu Fe alloy: For magnets; Dutch alloy
TICONAL 800	14.0% Ni 8.0% Al 24.0% Co 3.0% Cu Fe alloy: For magnets; French alloy
TICONAL 1500	15.0% Ni 7.0% Al 35.0% Co 4.0% Cu 5.0% Ti Fe alloy: For magnets; French alloy
TICONAL 1500F	15.0% Ni 7.0% Al 35.0% Co 4.0% Cu 5.0% Ti Fe alloy: For magnets; French alloy
TICONAL 1800	15.0% Ni 7.0% Al 35.0% Co 4.0% Cu 5.0% Ti Fe alloy: For magnets; French alloy
TICONAL C	13.9% Ni 8.0% Al 24.0% Co 3.0% Cu 0.8% Nb Fe alloy: For magnets; origin unknown
TICONAL FRITTE	14.0% Ni 8.0% Al 24.0% Co 3.0% Cu Fe alloy: For magnets; French alloy

Symbol	Nominal analysis, supplier, condition and remarks.
TOPHET D	20% Cr 36% Ni Fe alloy: For heating elements; Driver
UGIMAX 600	14.0% Ni 8.0% Al 24.0% Co 3.0% Cu Fe alloy: For magnets; French alloy
UGIMAX 800	14.0% Ni 8.0% Al 24.0% Co 3.0% Cu Fe alloy: For magnets; French alloy
UNS L05001–L05999	Lead based alloys: American Standard system; unified numbers
VACODIL 36	36% Ni Fe alloy: Low expansion; German proprietary alloy
VACODIL 42	42% Ni Fe alloy: Low expansion; German proprietary alloy
VACODIL 46	46% Ni Fe alloy: Low expansion; German proprietary alloy
VACON 10	28% Ni 18% Co Fe alloy: Low expansion; German proprietary alloy
VACON 12	28% Ni 18% Co Fe alloy: Low expansion; German proprietary alloy
VACON 20	28% Ni 21% Co Fe alloy: Low expansion; German proprietary alloy
VACON 70	28% Ni 23% Co Fe alloy: Low expansion; German proprietary alloy
VACOVIT 485	48% Ni 5% Cr Fe alloy: Low expansion; German proprietary alloy
VACOVIT 501	49% Ni 1% Cr Fe alloy: Low expansion; German proprietary alloy
ZERALOY	31% Ni 5% Co Fe alloy: Low expansion; Swift Levick

Note. The following abbreviations and units are used in the tables:

DPN	Hardness, diamond pyramid number
UTS	Ultimate tensile strength, N/mm^2
Elon	Elongation, %
Proof	0.1% proof strength, N/mm^2

1 N/mm^2=0.1 hbar=0.102 kgf/mm^2=0.06475 $tonf/in.^2$=145.04 $ibf/in.^2$=1 MPa

See Appendix II for other abbreviations and conversion tables.

21. Lead Pb

Physical properties	
Atomic number	82
Atomic weight	207.21
Crystal structure	Face-centred cubic
Colour	Blue-grey
Specific gravity	11.34
Density	11340 kg/m^3
Melting point	327 °C
Boiling point	1750 °C
Specific heat	0.129 J/g °C
Thermal conductivity	17.6 W/m °C
Coefficient of linear expansion (20–100 °C)	29.1×10^{-6}/ °C
Latent heat of fusion	24.1 J/g
Latent heat of vaporization	850 J/g
Thermal neutron absorption cross-section	0.17 barns/atom
Electrical conductivity	9% IACS (copper 100%)
Specific resistance	208 microhm mm
Temperature coefficient of electrical resistance	0.0043/ °C
Electrochemical equivalent	3.858 g/A/h
Electrode potential	−0.126 V
Magnetic susceptibility	-0.12×10^{-6}
Young's modulus of elasticity	14×10^9 N/m^2
Tensile strength	18 N/mm^2
Hardness	5 DPN

21.1 General notes on lead

Lead occurs in deposits widespread throughout the world and it is of note that this metal, which has been used for plumbing since Roman times at least, makes up only 0.002% of the earth's crust. Lithium, which is looked on as a rare metal, is present as 0.004%. Lead ores, however, tend to be found in concentrated pockets which makes their mining more economical than lithium, which is evenly distributed.

The excellent ductility of lead which has a complete absence of cold working properties has made this a favourite material. This is further enhanced by its excellent corrosion resistance and the ease with which it can be cold forge welded.

The Romans had lead water pipes and used lead as a roofing material. Lead pipes are now being replaced by copper, aluminium and plastic tubing, as these have thinner wall sections for the same strength and do not suffer creep failure at normal temperatures. The use of lead for sheathing electric cables is also largely being replaced with aluminium and plastic. Lead will be attacked corrosively by soft or neutral water which contains carbon dioxide. This can cause a health hazard when lead pipe is used.

The largest modern use of lead and its alloys is in the lead/sulphuric acid electrical storage battery or 'accumulator'. No other material can supply the necessary electrical, chemical and corrosion resistant properties as economically as lead.

The chemical industry still uses quantities of lead in contact with acids and chromium electroplating equipment. This is being reduced by the use of titanium.

Some of the newer plastics, stainless steel and titanium are now threatening other uses of lead, which suffers from the disadvantage of being the heaviest metal in common use. This added to its low strength and very low creep properties means that it almost always has to be supported by other metals. Additions of antimony to lead increase the strength without affecting its chemical resistance to any great extent, but make it more difficult to weld or solder.

A range of bearing metals exists with lead as the main constituent. Here the lead is used as a soft matrix supporting the hard intermetallic compounds of antimony, copper, tin and cadmium. These alloys are used for high bearing loads, and can stand reasonably high temperatures. Lead bearings do not however have good fatigue strength and must not be used where reciprocating loads are involved unless special precautions are taken. A number of different alloys have been developed which cater for conditions varying from lightly loaded slow moving, up to such purposes as bearings on stone crushing equipment.

It is strongly advised that use is made of the detailed information available from the sources listed for these very special purpose alloys.

Lead is also used as a base for soldering and some fusible alloys.

As an alloying element lead has many uses. It is more economical than tin and has many similar properties, being present in many tin based bearing alloys and almost all tin based solders and fusible alloys.

Lead is added to steel, aluminium and copper alloys up to 0.3% to improve machinability. It does not dissolve in any of these materials and thus has little effect on their properties, but does form a discontinuity which prevents long curling cuttings. It will reduce the ductility, however. The soft lead also acts to some extent as a built-in solid lubricant, but there remains doubt as to the extent of the practical advantage of this property.

Lead sheeting has been the standard protection until recent years against X-rays and radioactivity.

Lead and its alloys have no natural resonant frequencies, and it is thus a 'silent' metal.

The corrosion resistant and bearing properties of lead when applied as an electroplated deposit are now being exploited.

Lead oxide as a base for paints has been used for many years, and there are modern 'paints' chemically producing an adherent deposit of metallic lead on steel which show considerable promise. Metallic lead and most of its compounds are toxic, and thus cannot be used in contact with foodstuffs or on such articles as children's toys.

Lead also gives off toxic fumes when heated. This results in a health hazard to welders, scrap merchants, demolishers etc.

Symbol	Nominal analysis, supplier, condition and remarks.
2.3010	99.99% Pb: German Standard
2.3020	99.985% Pb: German Standard
2.3021	0.06% Cu 99.9% Pb: Chemical grade; German Standard
2.3030	99.94% Pb: German Standard
2.3040	99.9% Pb: For chemical use; German Standard
2.3075	0.1% Cu 99.75% Pb: German Standard
2.3085	98.5% Pb: German Standard
2.3131	0.005% Sn and Sb Pb: German Standard
2.3132	0.12% Sb 0.005% Sn Pb: German Standard
2.3137	0.8% Sb 0.005% Sn Pb: German Standard
2.3138	2.5% Sn Pb alloy: German Standard
2.3139	0.04% Te 0.005% Sn and Sb Pb: German Standard
2.3201	0.7% Sb 1.2% As Pb alloy: German Standard
2.3202	0.25% Sb Pb alloy: German Standard
2.3203	3% Sb 1.5% As Pb alloy: German Standard
2.3205	6% Sb Pb alloy: German Standard
2.3208	8% Sb Pb alloy: German Standard
2.3212	12% Sb Pb alloy: German Standard
2.3229	8.8% Sb Pb alloy: German Standard
2.3299	8.8% Sb Pb alloy: German Standard
A 1 BABBITT	Pb base bearing metal: Medium duty; Magnolia Anti-Friction Metal Co. **DPN: 26 UTS: 66**
A 5	Ag Pb base soft solder: Mallory Metallurgical; electrical conductivity 7.2% IACS; melting range 304–370 °C **UTS: 37 Elon: 35%**
A 25	Ag Pb base soft solder: Mallory Metallurgical; electrical conductivity 7.2% IACS; melting point 304 °C **UTS: 37 Elon: 35%**
A 36	Lead base white metal: For railway waggons; Anti-Attrition **DPN: 25**
A 105	Lead base white metal: For gland packing; soft; Anti-Attrition
A 110	Lead base white metal: For heavy duty and eccentric bearings; Anti-Attrition
A 122	Lead base white metal: Capping metal; Anti-Attrition
A 205	Lead base white metal: For gland packing; hard; Anti-Attrition
A 210	Lead base white metal: For railway waggons; Anti-Attrition **DPN: 32**
ALLULOY No. 1	Lead base white metal: For heavy duty and eccentric bearings; Anti-Attrition; as A 110
ALLULOY No. 2	Lead base white metal: Capping metal; Anti-Attrition; as A 122
AMS 4750A	45% Sn Pb alloy: Solder
AMS 4755A	5.5% Ag Pb alloy: Solder
AMS 4756	1.5% Ag 1% Sn Pb alloy: Solder
AMS 7720	6.5% Sb 0.5% Sn Pb alloy: Casting
AMS 7721	6.5% Sb 0.5% Sn Pb alloy: Sheet and extrusion
ANTIMONIAL LEAD	30% Sb Pb: Used in chemical plants battery plates; common name
ARSENIC-LEAD	25% As Pb alloy: Blackwells; primary metal
ASTM B23/7	10% Sn 15% Sb 0.5% Cu Pb alloy: Bearing metal; melting point 240 °C; compressive strength 100 N/mm^2 **DPN: 22**
ASTM B23/8	5% Sn 15% Sb 0.5% Cu Pb alloy: Bearing metal; melting point 237 °C; compressive strength 100 N/mm^2 **DPN: 20**
ASTM B23/13	6% Sn 10% Sb 0.5% Cu Pb alloy: Bearing metal
ASTM B23/15	1% Sn 16% Sb 0.6% Cu Pb alloy: Bearing metal; melting point 248 °C **DPN: 21**
ASTM B29	99.85–99.94% Pb pig: Specification covers 4 grades
ASTM B32/1/5 S	1% Sn 1.5% Ag Pb solder: Melting point 309 °C
ASTM B32/2A	2% Sn Pb solder
ASTM B32/2 B	2% Sn 0.3% Sb Pb solder
ASTM B32/2/5 S	2.5% Ag Pb solder: Melting point 304 °C
ASTM B32/5 A	5% Sn Pb solder: Melting range 270–312 °C
ASTM B32/5 B	5% Sn 0.3% Sb Pb: Solder; melting range 270–312 °C
ASTM B32/10 B	10% Sn 0.3% Sb Pb: Solder; melting range 268–299 °C
ASTM B32/15 B	15% Sn 0.3% Sb Pb: Solder; melting range 227–288 °C
ASTM B32/20 B	20% Sn 0.3% Sb Pb: Solder; melting range 183–277 °C
ASTM B32/20 C	20% Sn 1% Sb Pb: Solder; melting range 183–277 °C
ASTM B32/25 A	25% Sn Pb: Solder; melting range 183–266 °C
ASTM B32/25 B	25% Sn 0.3% Sb Pb: Solder; melting range 183–266 °C
ASTM B32/25 C	25% Sn 1.3% Sb Pb: Solder; melting range 183–266 °C
ASTM B32/30 A	30% Sn Pb: Solder; melting range 183–255 °C
ASTM B32/30 B	30% Sn 0.3% Sb Pb: Solder; melting range 183–255 °C
ASTM B32/30 C	30% Sn 1.6% Sb Pb: Solder; melting range 183–255 °C
ASTM B32/35 A	35% Sn Pb: Solder; melting range 183–247 °C
ASTM B32/35 B	35% Sn 0.3% Sb Pb: Solder; melting range 183–247 °C
ASTM B32/35 C	35% Sn 1.8% Sb Pb: Solder; melting range 183–247 °C
ASTM B32/40 A	40% Sn Pb: Solder; melting range 183–238 °C
ASTM B32/40 B	40% Sn 0.3% Sb Pb: Solder; melting range 183–238 °C
ASTM B32/40 C	40% Sn 2% Sb Pb: Solder; melting range 183–238 °C
ASTM B32/45 A	45% Sn Pb: Solder; melting range 183–227 °C
ASTM B32/45 B	45% Sn 0.3% Sb Pb: Solder; melting range 183–227 °C
ASTM B102 Y10A	10% Sb Pb alloy: Bearing metal
ASTM B102 YT155A	5% Sn 15% Sb Pb alloy: Bearing metal
ASTM B325	For refined secondary Pb; two listed; soft; Cu free and Cu bearing with 0.04% Cu
AUTO A	16% Sb 1% Cu Pb alloy; Bearing metal; Stone Manganese
BABBITT	Sn Cu Sb Pb white metal bearing alloys: Further information from Lead Development Association
BABBITT No. 6	Pb base bearing metal: For general engineering; Eyre Smelting; liquidus 272 °C **DPN: 20 UTS: 45 Elon: 3%**
BAHN METAL	0.6% Na 0.7% Ca 0.05% Ni Pb: Bearing metal; German origin

Note. The following abbreviations and units are used in the tables:

DPN	Hardness, diamond pyramid number
UTS	Ultimate tensile strength, N/mm^2
Elon	Elongation, %
Proof	0.1% proof strength, N/mm^2

1 N/mm^2=0.1 hbar=0.102 kgf/mm^2=0.06475 tonf/in.^{2}=145.04 lbf/in.^{2}=1 MPa

See Appendix II for other abbreviations and conversion tables.

Symbol	Nominal analysis, supplier, condition and remarks.
BS 219 C	40% Sn 2.2% Pb alloy: Solder; melting range 185– 227 °C
BS 219 D	30% Sn 1.5% Sb Pb alloy: Solder; melting range 185–248 °C
BS 219 G	40% Sn 0.4% Sb (max) Pb alloy: Solder; melting range 183–234 °C
BS 219 H	35% Sn 0.3% Sb (max) Pb alloy: Solder; melting range 183–255 °C
BS 219 J	30% Sb 0.3% Sb (max) Pb alloy: Solder; melting range 183–255 °C
BS 219 L	32% Sn 1.8% Sb Pb alloy: Solder, melting range 185–243 °C
BS 219 M	45% Sn 2.5% Sb Pb alloy: Solder; melting range 185–215 °C
BS 219 N	18% Sn 1% Sb Pb alloy: Solder; melting range 185–275 °C
BS 219 R	45% Sn 0.4% Sb (max) Pb alloy: Solder, melting range 183–224 °C
BS 219 V	20% Sn 0.2% Sb (max) Pb alloy: Solder; melting range 183–276 °C
BS 219/1S	1.2% Sn 0.1% Sb (max) 1.5% Ag Pb alloy: Solder; melting range 309–310 °C
BS 219/5S	5% Sn 0.1% Sb (max) 1.5% Ag Pb alloy: Solder; melting range 296–301 °C
BS 334 A	99.99% Pb chemical lead: Primary material
BS 334 B	Chemical Pb containing protective elements: Primary material 0.005% Bi (max)
BS 335/2	7% Sb Pb alloy: Casting; corrosion resistant for lining tanks; 'Regulus Metal' **DPN: 15 UTS: 40**
BS 335/3	9% Sb Pb alloy: Casting; 'Regulus Metal'; corrosion resistant; wear resistant **DPN: 16 UTS: 45**
BS 335/4	11% Sb Pb alloy: Casting; 'Regulus metal'; corrosion resistant; machinable **DPN: 17 UTS: 57**
BS 335/5	12% Sb (min) Pb alloy: Casting; 'Regulus Metal'; corrosion resistant; machinable **DNP: 19 UTS: 60**
BS 602/1	99.8% Pb: Pipe
BS 602/2	99.5% Pb: Pipe
BS 602/3	0.05% Te Pb: Pipe
BS 643	5% Sn 15% Sb Pb alloy: Used for capping wire ropes
BS 801	99.8% Pb: For cable sheathing
BS 801 B	0.9% Sb Pb alloy: For sheathing
BS 801 D	0.25% Cd 0.5% Sb Pb alloy: For sheathing
BS 801 E	0.4% Sn 0.2% Sb Pb alloy: For sheathing
BS 1085	0.004% Ag 0.004% Cu Pb: Pipe
BS 1178	99.9% Pb alloy: Sheet and strip for building
BS 3332/7	12% Sn 13% Sb 1% Cu Pb alloy: Bearing metal
BS 3332/8	5% Sn 15% Sb Pb alloy: Bearing metal
BULLET ALLOY	0.75% Sb 99.2% Pb: Origin unknown
CAPSULE METAL	8% Sn Pb alloy: Origin unknown
CERROBASE	Bi Pb alloy: Casting for pattern metal; Mining & Chemical Ltd; can be cast into paper moulds etc.; melting point 124 °C
COMSOL	Ag Sn Pb base soft solder: Mallory Metallurgical; electrical conductivity 8% IACS; melting point 296 °C **UTS: 37 Elon: 40%**

Note. The following abbreviations and units are used in the tables:

DPN	Hardness, diamond pyramid number
UTS	Ultimate tensile strength, N/mm^2
Elon	Elongation, %
Proof	0.1% proof strength, N/mm^2

1 N/mm^2=0.1 hbar=0.102 kgf/mm^2=0.06475 tonf/in.^{2}=145.04 ibf/in.^{2}=1 MPa

See Appendix II for other abbreviations and conversion tables.

Symbol	Nominal analysis, supplier, condition and remarks.
CRUSHER	Pb base hardened bearing metal for dusty conditions: Magnolia Anti-Friction Metal Co.
DIN 1707 LPB 98.5	98.5% Pb alloy
DIN 1707 LSn 8	8% Sn 0.5% Sb Pb alloy
DIN 1707 LSn 25	25% Sn 1.7% Sb Pb alloy
DIN 1707 LSn 30	30% Sn 2.0% Sb Pb alloy
DIN 1707 LSn 33	33% Sn 2.2% Sb Pb alloy
DIN 1707 LSn 35	35% Sn 2.3% Sb Pb alloy
DIN 1707 LSn 40	40% Sn 2.7% Sb Pb alloy
DIN 1719	Pb specification covers various grades pure lead
DIN 1741 Sb Pb 46	2% Cu 40% Sn 12% Sb Pb alloy **DPN: 17 UTS: 60 Elon: 4%**
DIN 1741 Sb Pb 59	3% Cu 25% Sn 13% Sb Pb alloy **DPN: 18 UTS: 60 Elon: 3%**
DIN 1741 Sb Pb 85	5% Sn 10% Sb Pb alloy **DPN: 18 UTS: 60 Elon: 8%**
DIN 1741 Sb Pb 87	13% Sb Pb alloy **DPN: 14 UTS: 45 Elon: 10%**
DIN 1741 Sb Pb 97	3% Sb Pb alloy **DPN: 9 UTS: 45 Elon: 20%**
DIN 16512 Pb Sn 3 Sb 4	3.0% Sn 4% Sb Pb alloy
DIN 16512 Pb Sn 3 Sb 12	3% Sn 12% Sb Pb alloy: Typemetal
DIN 16512 Pb Sn 4 Sb 15	4% Sn 15% Sb Pb alloy
DIN 16512 Pb Sn 5 Sb 12	5% Sn 12% Sb Pb alloy
DIN 16512 Pb Sn 5 Sb 28	5.5% Sn 29% Sb Pb alloy
DIN 16512 Pb Sn 9 Sb 17	9% Sn 17% Sb Pb alloy
DIN 16512 Pb Sn 15 Sb 4	15% Sn 4.5% Sb Pb alloy
DIN 16512 V Pb Sn 5 Sb 28	5% Sn 28% Sb Pb alloy
DIN 16512 V Pb Sn 30 Sb 6	30% Sn 6% Sb Pb alloy
DIN 17640	Specification covering grades of lead
DIN 17641	Pb and Pb alloys: Specification covers several analyses
DISPLAY MONOTYPE ALLOY	17% Sb 8% Sr Pb alloy: Origin unknown
DTD 685	1.5% Ag 1.2% Sn Pb alloy: Solder; melting range 309–313 °C
DURASTIC A 1	Cu Pb Sn base bearing for reciprocating engines: Eyre Smelting; melting range 186–350 °C **DPN: 31 UTS: 90 Elon: 3%**
ELECTRO TYPE	3% Sn 3% Sb Pb alloy: White metal; further information from Lead Development Association
E QQ S 571/R	2.5% Ag Pb alloy solder: US Federal; melting point 304 °C
ESD 32	10.0% Co 18.0% Fe Pb alloy: Origin unknown
ESD 42	10.0% Co 18.0% Fe Pb alloy: Origin unknown
FLOWER BRAND	Pb base graphite impregnated bearing metal: Magnolia Anti-Friction Metal Co.; constant load and speed **DPN: 25.5 UTS: 66**
FRARY METAL	1% Ba 0.8% Ca Pb alloy: Origin unknown
HOYT 3 M	Pb base bearing metal: Hoyt
HOYT 4 A	Pb base bearing metal: Hoyt
HOYT 30	30% Sn Pb solder: Contains up to 1.7% Sb; melting range 183–255 °C; Hoyt
HOYT 32	32% Sn 1.8% Sb Pb solder: Melting range 185–243 °C; Hoyt
HOYT 35	35% Sn Pb solder: Melting range 183–255 °C; Hoyt
HOYT 40	40% Sn Pb solder: Slow setting; may contain Sb; melting range 183–234 °C; Hoyt
HOYT 142	Sn Pb base bearing metal: Hoyt
HOYT 155	Sn Pb base bearing metal: Hoyt
HOYT STAR	Pb base bearing metal: Hoyt
h Pb 1	99.9999% Pb: Rod; Light Ltd; high purity metal

Symbol	Nominal analysis, supplier, condition and remarks.
h Pb 2	99.9999% Pb: Shot; Light Ltd; high purity metal
h Pb 4b	99.9999% Pb: Sheet 1 mm thick; Light Ltd; high purity metal
h Pb 73a	99.99% Pb single crystal 6.35 mm diameter × 25 mm: Light Ltd
IBIS	Sn Pb base bearing metal for general purpose: Phosphor Bronze Ltd; available in various grades
INTERTYPE METAL	3% Sn 12% Pb alloy: White metal; further information from Lead Development Association
Kb Pb	0.005% Sb and Sn Pb: Designation used by German Standard
Kb Pb Sb	0.12% Sn 0.005% Sn Pb: Designation used by German Standard
Kb Pb Sn 2.5	2.5% Sn Pb alloy: Designation used by German Standard
Kb Pb Te 0.04	0.04% Te 0.005% Sn and Sb Pb: Designation used by German Standard
L 05120	5% Sn 9% Sb Pb alloy: Babbit Metal; UNS designation
L 50001	99.9999% Pb: Zone refined; UNS designation
L 50005	99.999% Pb: Refined soft; UNS designation
L 50010	99.99% Pb: Refined soft; UNS designation
L 50020	99.985% Pb: Refined soft; UNS designation
L 50025	99.97% Pb: UNS designation
L 50035	99.95% Pb (min): UNS designation
L 50040	99.94% Pb (min): UNS designation
L 50042	99.94% Pb: Corroding lead; UNS designation
L 50045	99.94% Pb: Common lead; UNS designation
L 50050	99.9% Pb: Grade A lead; UNS designation
L 50060	99.85% Pb (min): Type III lead; UNS designation
L 50065	99.7% Pb (min): Grade C lead; UNS designation
L 50070	99.5% Pb (min): Grade B lead; UNS designation
L 50080	95% Pb (min): Grade B lead; UNS designation
L 50090	Pb not specified: Grade B lead; UNS designation
L 50101	99.8% Pb 0.2% Ag: Cable sheathing; UNS designation
L 50110	99.5% Pb 0.5% Ag: UNS designation
L 50113	2.5% Sn 0.5% Ag Pb alloy: Solder; UNS designation
L 50115	0.75% Ag Pb: Anode alloy; UNS designation
L 50120	1% Ag Pb: Anode alloy; UNS designation
L 50121	1% Sn 1% Ag Pb: Solder; UNS designation
L 50122	1% As 1% Ag Pb: Anode alloy; UNS designation
L 50131	1% Sn 1.5% Ag Pb: Solder; UNS designation
L 50132	1% Sn 1.5% Ag Pb alloy: Solder; UNS designation
L 50134	5% Sn 1.5% Ag Pb alloy: UNS designation
L 50140	2% Ag Pb alloy: For cathodic protection; UNS designation
L 50150	2.5% Ag Pb alloy: UNS designation
L 50151	2.5% Ag Pb alloy: UNS designation
L 50152	2% Sn 2.5% Ag Pb alloy: Solder; UNS designation
L 50170	5% Ag Pb alloy: Solder; UNS designation
L 50172	5% Ag 5% In Pb alloy: UNS designation
L 50180	5.5% Ag Pb alloy: UNS designation
L 50310	0.1% Sn 0.15% As 0.1% Bi Pb alloy: For sheathing; UNS designation
L 50510	0.05% Ba Pb alloy: UNS designation
L 50520	1% Sn 0.05% Ba Pb alloy: UNS designation
L 50521	1.5% Sn 0.05% Ba Pb alloy: UNS designation
L 50522	2% Sn 0.05% Ba Pb alloy: UNS designation
L 50530	1% Sn 0.1% Ba Pb alloy: UNS designation
L 50535	2% Sn 0.1% Ba Pb alloy: UNS designation
L 50540	0.4% Ba 0.8% Ca Pb alloy: Frary metal; UNS designation
L 50541	1% Ba 0.5% Ca Pb alloy: UNS designation
L 50542	1.2% Ba 0.8% Ca Pb alloy: UNS designation
L 50543	2% Ba 0.8% Ca Pb alloy: UNS designation
L 50710	0.008% Ca Pb metal: UNS designation
L 50712	0.025% Ca Pb metal: UNS designation
L 50713	0.3% Sn 0.025% Ca Pb alloy: Sheathing; UNS designation
L 50720	0.03% Ca Pb metal: UNS designation
L 50722	0.06% Cu 0.03% Ca Pb alloy: UNS designation
L 50725	0.05% Cu 0.04% Ca Pb alloy: Sheathing; UNS designation
L 50728	0.5% Sn 0.04% Ca Pb alloy: Battery grids; UNS designation
L 50730	0.5% Ag 0.06% Ca Pb alloy: For anodes; UNS designation
L 50735	0.06% Ca Pb alloy: For battery grids; UNS designation
L 50736	0.2% Sn 0.065% Ca Pb alloy: For battery grids; UNS designation
L 50737	0.5% Sn 0.065% Ca Pb alloy: For battery grids; UNS designation
L 50740	0.7% Sn 0.065% Ca Pb alloy: For battery grids; UNS designation
L 50745	1% Sn 0.065% Ca Pb alloy: For battery grids; UNS designation
L 50750	1.3% Sn 0.065% Ca Pb alloy: For battery grids; UNS designation
L 50755	1.5% Sn 0.065% Ca Pb alloy: For battery grids; UNS designation
L 50760	0.07% Ca Pb metal: For battery grids; UNS designation
L 50765	0.7% Sn 0.07% Ca Pb alloy: For battery grids; UNS designation
L 50770	0.1% Ca Pb metal: For battery grids; UNS designation
L 50775	0.3% Sn 0.1% Ca Pb alloy: For battery grids; UNS designation
L 50780	0.5% Sn 0.1% Ca Pb alloy: For battery grids; UNS designation
L 50790	1% Sn 0.1% Ca Pb alloy: For battery grids; UNS designation
L 50795	0.3% Sn 0.12% Ca Pb alloy: For battery grids; UNS designation
L 50800	0.3% Sn 0.21% Ca Pb alloy: For battery grids; UNS designation
L 50810	0.02% Al 0.04% Li 0.7% Ca 0.6% Na Pb alloy: Bearing metal; UNS designation
L 50820	0.02% Al 0.04% Li 0.7% Ca 0.2% Na 0.4% Ba Pb alloy: Bearing metal; UNS designation
L 50840	1% Ca Pb metal: UNS designation
L 50850	2% Ca Pb metal: UNS designation
L 50880	6% Ca Pb alloy: UNS designation
L 50940	17% Cd Pb metal: Eutectic alloy; UNS designation
L 51110	0.05% Cu Pb metal: Copperized lead; UNS designation
L 51121	99.9% Pb: UNS designation
L 51123	99.85% Pb (min): Grade D; UNS designation
L 51124	99.82% Pb (min): UNS designation
L 51125	99.9% Pb: UNS designation
L 51180	3% Sn 45% Cu Pb alloy: Bearing; UNS designation
L 51510	2.3% Ag 4.8% In Pb alloy: Solder; UNS designation
L 51511	5% In Pb: Solder; UNS designation
L 51512	2.5% Ag 5% In Pb: Solder; UNS designation
L 51530	19% In Pb alloy: UNS designation
L 51532	20% In Pb alloy: UNS designation
L 51535	25% In Pb alloy: UNS designation
L 51540	40% In Pb alloy: UNS designation
L 51550	50% In Pb alloy: UNS designation
L 51705	0.01% Li 99.9% Pb: UNS designation
L 51708	0.02% Li 99.9% Pb: UNS designation
L 51710	0.03% Li 99.9% Pb: UNS designation
L 51720	0.06% Li 99.9% Pb: UNS designation
L 51740	0.02% Li 0.35% Sn 99.6% Pb: UNS designation
L 51748	0.04% Li 0.7% Sn 99.2% Pb: UNS designation
L 52515	0.015% As 0.2% Sb 99.8% Pb: For cable sheathing; UNS designation
L 52525	0.03% As 0.3% Sb 0.03% Te Pb: For cable sheathing; UNS designation
L 52540	0.25% Cd 0.5% Sb 99.2% Pb: For cable sheathing; UNS designation
L 52550	0.04% Cu 0.6% Sb 0.04% Te Pb: For cable sheathing; UNS designation
L 52560	0.75% Sb 99.2% Pb: Bullet alloy; UNS designation
L 52605	1% Sb 99% Pb: UNS designation

Symbol	Nominal analysis, supplier, condition and remarks.
L 52615	0.1% As 1% Sb 0.3% Sn Pb alloy: Die casting; UNS designation
L 52620	1.4% Cd 1.5% Sb 97% Pb: Battery alloy; UNS designation
L 52625	0.45% As 1.65% Sb 98% Pb: Shot alloy; UNS designation
L 52725	2.5% Sb 97.5% Pb: Bullet alloy; UNS designation
L 52760	0.18% As 0.07% Cu 2.7% Sb 0.2% Sn Pb alloy: UNS designation
L 52765	0.3% As 0.07% Cu 2.7% Sb 0.3% Sn Pb alloy: UNS designation
L 52805	3% Sb 97% Pb: UNS designation
L 52810	0.15% As 3% Sb 0.3% Sn Pb: Battery alloy; UNS designation
L 52860	3.5% Sb 4.1% Sn Pb: Bearing alloy; UNS designation
L 52901	4% Sb Pb alloy: UNS designation
L 52910	4% Sb 1% Sn Pb: UNS designation
L 52915	4% Sb 3% Sn Pb: Type metal; UNS designation
L 53020	5% Sb 5% Sn Pb: Bullet alloy; UNS designation
L 53105	6% Sb Pb alloy: UNS designation
L 53120	0.4% As 6% Sb Pb: Anode alloy; UNS designation
L 53122	0.4% As 6% Sb Pb: Sheet; UNS designation
L 53125	0.6% As 6% Sb Pb: Pipe; heat resistant; UNS designation
L 53130	0.6% As 6% Sb 0.6% Sn Pb alloy: UNS designation
L 53235	1.25% As 8% Sb Pb: Hard shot alloy; UNS designation
L 53238	2% As 8% Sb Pb: Hard shot alloy; UNS designation
L 53260	8% Sb 3% Sn Pb: For spin casting; UNS designation
L 53265	8% Sb 4% Sn Pb: Typemetal; UNS designation
L 53305	9% Sb Pb alloy: UNS designation
L 53320	9% Sb 5% Sn Pb: White metal bearings; UNS designation
L 53340	10% Sb Pb alloy: For die casting; UNS designation
L 53343	10% Sb 6% Sn Pb alloy: Bearings; UNS designation
L 53345	10% Sb 6% Sn Pb alloy: UNS designation
L 53346	10% Sb 6% Sn Pb: Beams alloy; UNS designation
L 53405	11% Sb Pb alloy: UNS designation
L 53420	11% Sb 3% Sn Pb alloy: UNS designation
L 53425	11% Sb 5% Sn Pb alloy: UNS designation
L 53452	12% Sb 3% Sn Pb alloy: UNS designation
L 53455	12% Sb 4% Sn Pb alloy: Eutectic; UNS designation
L 53456	12% Sb 5% Sn Pb alloy: UNS designation
L 53460	12% Sb 6% Sn Pb alloy: UNS designation
L 53465	12.5% Sb 10% Sn Pb alloy: UNS designation
L 53470	0.4% As 12.7% Sb 0.7% Sn Pb alloy: CT Metal; UNS designation
L 53480	3% As 12.7% Sb 0.7% Sn Pb alloy: Babbitt; UNS designation
L 53505	1% As 13% Sb 1% Sn Pb alloy: UNS designation
L 53510	13% Sb 6.5% Sn Pb alloy: UNS designation
L 53530	14% Sb 6% Sn Pb alloy: UNS designation
L 53550	15% Sb Pb alloy: UNS designation
L 53555	15% Sb 2% Sn Pb alloy: UNS designation
L 53558	15% Sb 4% Sn Pb alloy: UNS designation
L 53560	15% Sb 5% Sn Pb alloy: Die casting; UNS designation
L 53565	15% Sb 5% Sn Pb alloy: Bearing metal; UNS designation
L 53570	15% Sb 7% Sn Pb alloy: Monotype alloy; UNS designation

Note. The following abbreviations and units are used in the tables:

DPN Hardness, diamond pyramid number
UTS Ultimate tensile strength, N/mm^2
Elon Elongation, %
Proof 0.1% proof strength, N/mm^2

1 N/mm^2=0.1 hbar=0.102 kgf/mm^2=0.06475 tonf/in.^{2}=145.04 ibf/in.^{2}=1 MPa
See Appendix II for other abbreviations and conversion tables.

Symbol	Nominal analysis, supplier, condition and remarks.
L 53575	15% Sb 8% Sn Pb alloy: Stereotype alloy; UNS designation
L 53580	15% Sb 10% Sn Pb alloy: UNS designation
L 53585	0.4% As 15% Sb 10% Sn Pb alloy: UNS designation
L 53620	1% As 16% Sb 1% Sn Pb alloy: UNS designation
L 53650	17% Sb 8% Sn Pb alloy: UNS designation
L 53655	17% Sb 9% Sn Pb alloy: UNS designation
L 53685	19% Sb 9% Sn Pb alloy: UNS designation
L 53710	1.5% Cu 20% Sb 20% Sn Pb alloy: UNS designation
L 53730	2.5% Sn 2.5% Sb 95% Pb: Electrotype metal; UNS designation
L 53740	24% Sb 9% Sn Pb alloy: UNS designation
L 53750	24% Sb 12% Sn Pb alloy: UNS designation
L 53780	1.5% Cu 25% Sb 13% Sn Pb alloy: Typemetal; UNS designation
L 53790	28% Sb 5% Sn Pb alloy: Typemetal; UNS designation
L 53795	29% Sb 5.5% Sn Pb alloy: Typemetal; UNS designation
L 54030	0.07% Cd 0.2% Sn Pb: For cable sheaths; UNS designation
L 54050	0.15% Cd 0.4% Sn Pb: For cable sheaths; UNS designation
L 54210	2% Sn Pb: Solder; UNS designation
L 54211	0.3% Sb 2% Sn Pb: Solder; UNS designation
L 54250	5% Sb 2.6% Sn Pb: Solder; UNS designation
L 54280	5% Sb 3% Sn Pb: Solder; UNS designation
L 54310	3% Sb 4% Sn Pb alloy: UNS designation
L 54320	5% Sn Pb: Solder; UNS designation
L 54321	0.3% Sb 5% Sn Pb: Solder; UNS designation
L 54322	5% Sn Pb: Solder; UNS designation
L 54360	0.5% As 4% Sb 5% Sn Pb: Solder; UNS designation
L 54370	7% Sn Pb alloy: For plating bearings; UNS designation
L 54410	0.3% Sb 8% Sn Pb alloy: Solder; UNS designation
L 54510	10% Sn Pb alloy: For plating bearings; UNS designation
L 54520	0.3% Sb 10% Sn Pb: Solder; UNS designation
L 54525	2% Ag 10% Sn Pb: Solder; UNS designation
L 54530	0.5% As 4% Sb 10% Sn Pb: Solder; UNS designation
L 54540	0.4% Sb 12% Sn Pb: Solder; UNS designation
L 54555	2.5% Sb 14.5% Sn Pb: Solder; UNS designation
L 54560	0.3% Sb 15% Sn Pb: Solder; UNS designation
L 54570	2.5% Sb 15% Sn Pb: Solder; UNS designation
L 54580	4.5% Sb 15% Sn Pb: Solder; UNS designation
L 54610	1.5% Sb 19.5% Sn Pb: Solder; UNS designation
L 54710	20% Sn Pb: Solder; UNS designation
L 54711	0.3% Sb 20% Sn Pb: Solder; UNS designation
L 54712	1% Sb 20% Sn Pb: Solder; UNS designation
L 54713	1.2% Sb 22% Sn Pb: Solder; UNS designation
L 54715	3% Sb 23% Sn Pb: Solder; UNS designation
L 54720	25% Sn Pb: Solder; UNS designation
L 54721	0.3% Sb 25% Sn Pb: Solder; UNS designation
L 54722	1.2% Sb 25% Sn Pb alloy: Solder; UNS designation
L 54727	3% Cu 13% Sb 25% Sn Pb: Bearing alloy; UNS designation
L 54750	3% Ag 27% Sn Pb alloy: Solder; UNS designation
L 54755	21% Bi 27% Sn Pb: Fusible alloy
L 54805	1.5 Sb 28% Sn Pb: Solder; UNS designation
L 54810	28.5% Bi 28.5% Sn Pb: Fusible alloy; UNS designation
L 54815	1% Sb 29.5% Sn Pb alloy: Solder; UNS designation
L 54820	30% Sn Pb; Solder; UNS designation
L 54821	0.3% Sb 30% Sn Pb: Solder; UNS designation
L 54822	1.6% Sb 30% Sn Pb: Solder; UNS designation
L 54827	6% Sb 30% Sn Pb: Typemetal; UNS designation
L 54829	20% Bi 30% Sn Pb alloy: Fusible metal; UNS designation
L 54830	30.8% Bi 31% Sn Pb: Fusible metal; UNS designation
L 54832	1.8% Sb 31.5% Sn Pb alloy: Solder; UNS designation
L 54833	3% Sb 32% Sn Pb alloy: Solder; UNS designation
L 54835	33.3% Bi 33.3% Sn Pb: Fusible metal; UNS designation
L 54840	32% Bi 34% Sn Pb: Fusible metal; UNS designation

Symbol	Nominal analysis, supplier, condition and remarks.
L 54850	35% Sn Pb alloy: Solder; UNS designation
L 54851	0.3% Sb 35% Sn Pb alloy: Solder; UNS designation
L 54852	1.8% Sb 35% Sn Pb alloy: Solder; UNS designation
L 54855	3% Ag 35.5% Sn Pb: Solder; UNS designation
L 54860	21% Bi 36% Sn Pb: Fusible metal; UNS designation
L 54865	21% Bi 37% Sn Pb: Fusible metal; UNS designation
L 54905	1.7% Sb 38.2% Sn Pb: Solder; UNS designation
L 54910	12.6% Bi 40% Sn Pb: Fusible metal; UNS designation
L 54915	40% Sn Pb: Solder; UNS designation
L 54916	0.3% Sb 40% Sn Pb: Solder; UNS designation
L 54918	2% Sb 40% Sn Pb: Solder; UNS designation
L 54925	2% Cu 12% Sb 40% Sn Pb: Bearing metal; UNS designation
L 54930	4% Bi 40% Sn Pb: Fusible metal; UNS designation
L 54935	14% Bi 43% Sn Pb: Fusible metal; UNS designation
L 54940	1.7% Sb 43.2% Sn Pb: Solder; UNS designation
L 54945	1% Ag 44% Sn Pb: Solder; UNS designation
L 54950	45% Sn Pb: Solder; UNS designation
L 54951	0.3% Sb 45% Sn Pb: Solder; UNS designation
L 54955	2.5% Sb 45% Sn Pb: Solder; UNS designation
L 55005	48% Sn Pb: Solder; UNS designation
L 55030	50% Sn Pb: Solder; UNS designation
L 55031	0.3% Sb 50% Sn Pb: Solder; UNS designation
L 55210	0.3% Sn 0.06% Sr Pb: Battery alloy; UNS designation
L 55260	0.03% Al 0.2% Sr Pb: Battery alloy; UNS designation
L 55290	2% Sr Pb alloy: UNS designation
LANSTON STANDARD	19% Sb 9% Sn Pb alloy: Origin unknown
LEAD	Pb: Powder, stick, sheet, wire, foil, etc.; Blackwell; pure metal
LEAD 6 A	6% Sb Pb alloy: Wire for spraying; Metco Ltd
LINOTYPE METAL	3% Sn 12% Sb Pb alloy: White metal; further information from Lead Development Association
LINOTYPE METAL	11% Sb 4% Sn Pb alloy: Common name
LOCO	Pb base hardened bearing metal for dusty conditions: Magnolia Anti-Friction Metal Co. **DPN: 36**
LYMANS ALLOY	5% Sb 5% Sn Pb alloy: Origin unknown
MAGNOLIA METAL	Pb base bearing metal: Further information from the Lead Development Association
MALTEX X No 1	8.7% Sn 16.7% Sb Pb alloy: Bearing metal; Stone Manganese
MALTEX X No 1 A	4% Sn 14% Sb 1% Cu Pb alloy: Bearing metal; Stone Manganese; as BS 3332/8
MALTEX X No 7	40% Sn 16% Sb 1% Cu Pb alloy: Bearing metal; Stone Manganese
MARINE I	4% Sn 14% Sb 1% Cu Pb alloy: Bearing metal; Stone Manganese; as BS 3332/8
MARINE II	22.5% Sn 7.7% Sb Pb alloy: Bearing metal; Stone Manganese
MASCOT	Sn Pb base melting metal for general use: Eyre Smelting; melting range 240–300 °C **DPN: 23 UTS: 66 Elon: 2%**
MONOTYPE ALLOY	15% Sb 7% Sn Pb alloy: Common name
MONOTYPE METAL	9% Sn 17% Sb Pb alloy: White metal; further information from Lead Development Association
Pb 5	Ag Sn Pb alloy: For brazing; Johnson Matthey
Pb Sb 5	6% Sb Pb alloy: Designation used by German Standards
Pb Sb 8	8% Sb Pb alloy: Designation used by German Standards
Pb Sb 9	8.8% Sb Pb alloy: Designation used by German Standards
Pb Sb 9X	8.8% Sb Pb alloy: Designation used by German Standards
Pb Sb 12	12.5% Sb Pb alloy: Designation used by German Standards
PLASTIC METAL	Pb base bearing metal: Further information from Lead Development Association

Symbol	Nominal analysis, supplier, condition and remarks.
PLUMBERS SOLDER	33.3% Sn Pb alloy: Further information from Lead Development Association
P M	99.97% Pb: Refined lead; Platt Metals Ltd
QQ C 40	99.7% Pb: US Federal specification
QQ L 171	99.9% Pb: US Federal specification
QQ L 171 (Grade C)	99.9% Pb: US Federal specification
QQ L 201	99.5% Pb: Remelted lead; US Federal specification
QQ L 201 A	99.9% Pb: Sheet; US Federal specification
QQ L 201 B	99.5% Pb: Sheet; US Federal specification
QQ S 571 (Ag 2.5)	2.5% Ag Pb alloy: Solder; US Federal specification
QQ S 571 (Ag 5.5)	5.5% Ag Pb alloy: Solder; US Federal specification
QQ S 571 (Pb 65)	0.3% Sb 35% Sn Pb: Solder; US Federal specification
QQ S 571 (Pb 70)	0.3% Sb 30% Sn Pb: Solder; US Federal specification
QQ S 571 (Pb 80)	0.3% Sb 20% Sn Pb: Solder; US Federal specification
QQ S 571 (Sn 5)	5% Sn Pb: Solder; US Federal specification
QQ S 571 (Sn 20)	1% Sb 20% Sn Pb: Solder; US Federal specification
QQ S 571 (Sn 30)	1.6% Sb 30% Sn Pb: Solder; US Federal specification
QQ S 571 (Sn 35)	1.8% Sb 35% Sn Pb: Solder; US Federal specification
QQ S 571 (Sn 40)	0.3% Sb 40% Sn Pb: Solder; US Federal specification
QQ S 571 (Sn 50)	0.3% Sb 50% Sn Pb: Solder; US Federal specification
QQ S 571 B	40% Sn Pb alloy: Solder; US Federal specification
QQ S 571 D	0.45% Sb 40% Sn Pb: Solder; US Federal specification; melting range 182–238 °C
QQ S 571 D Ag 1.5	1.5% Ag 1.0% Sn Pb: Solder; US Federal specification; melting point 309 °C
QQ S 571 D Ag 2.5	2.5% Ag Pb: Solder; US Federal specification; melting point 304 °C
QQ S 571 D Ag 5.5	5.5% Ag Pb: Solder; US Federal specification; melting range 304–366 °C
QQ S 571 D Pb 65	0.45% Sb 37% Sn Pb: Solder; US Federal specification; melting range 182–246 °C
QQ S 571d Pb 70	0.45% Sb 31% Sn 70% Pb: Solder; US Federal specification; melting range 182–254 °C
QQ S 571 D Pb 80	0.45% Sb 20% Sn Pb: Solder; US Federal specification; melting range 182–277 °C
QQ S 571 D Sn 5	5.0% Sn Pb: Solder; US Federal specification; melting range 270–313 °C
QQ S 571 D Sn 10	10% Sn Pb: Solder; US Federal specification; melting range 269–299 °C
QQ S 571 D Sn 20	1.0% Sb 20.5% Sn Pb: Solder; US Federal specification; melting range 182–277 °C
QQ S 571 D Sn 30	1.6% Sb 30.5% Sn Pb: Solder; US Federal specification; melting range 182–254 °C
QQ S 571 D Sn 35	1.8% Sb 35.5% Sn Pb: Solder; US Federal specification; melting range 182–246 °C
QQ T 390	5% Sn 9% Sb Pb alloy: US Federal specification
RAILWAY A	40% Sn 16% Sb 1% Cu Pb alloy: Bearing metal; Stone Manganese
REGULUS METAL	6–12% Sb Pb alloy: Casting; see BS 335
RM 1 METAL	Pb base hardened bearing metal for heavy duty conditions: Magnolia Anti-Friction Metal Co.
R Pb	0.7% Sb 1.2% As Pb alloy: Designation used by German Standards
RULES MONOTYPE ALLOY	15% Sb 10% Sn Pb alloy: Origin unknown
SAE 1 A	45% Sn 0.4% Sb (max) Pb alloy: Solder; melting range 182–227 °C
SAE 1 B	43% Sn 1.7% Sb Pb alloy: Solder; melting range 185–223 °C
SAE 2 A	40% Sn 0.4% Sb (max) Pb alloy: Solder; melting range 182–223 °C
SAE 2 B	38% Sn 1.7% Sb Pb alloy: Solder; melting range 185–232 °C
SAE 3 A	30% Sn 0.5% Sb (max) Pb alloy: Solder; melting range 182–254 °C
SAE 3 B	30% Sn 1.0% Sb Pb alloy: Solder; melting range 185–250 °C
SAE 4 A	25% Sn 0.4% Sb (max) Pb alloy: Solder; melting range 182–266 °C

Symbol	Nominal analysis, supplier, condition and remarks.
SAE 4 B	25% Sn 1.5% Sb Pb alloy: Solder; melting range 185–260 °C
SAE 5 A	20% Sn 0.4% Sb (max) Pb alloy: Solder; melting range 182–279 °C
SAE 5 B	20% Sn 1.5% Sb Pb alloy: Solder; melting range 185–266 °C
SAE 6 A	15% Sn 0.4% Sb (max) Pb alloy: Solder; melting range 224–290 °C
SAE 6 B	15% Sn 2.7% Sb (max) Pb alloy: Solder; melting range 224–290 °C
SAE 7 A	50% Sn 0.4% Sb (max) Pb alloy: Solder; melting range 182–216 °C
SAE 8 A	35% Sn 0.4% Sb (max) Pb alloy: Solder; melting range 182–245 °C
SAE 9 B	2.7% Sn 5.2% Sb Pb alloy: Solder; melting range 240–290 °C
SAE 13	6% Sn 10% Sb Pb alloy: For bearings
SAE 14	10% Sn 15% Sb Pb alloy: For bearings
SAE 15	1% Sn 15% Sb Pb alloy: For bearings
SAE 16	4.5% Sn 3.5% Sb Pb alloy: For bearings; cast on a sintered matrix
SAE 19	10% Sn Pb alloy: For bearings; this alloy is electroplated onto Cu or Ag based bearings
SAE 190	7% Sn Pb alloy: For bearings; this alloy is electroplated onto Cu or Ag based bearings
SAE 485	3% Sn 46% Cu Pb alloy: For bearings; cast onto steel or sintered backing
SHOT ALLOY	0.5% As 1.6% Sb Pb alloy: Origin unknown
SPRABABBITT L	0.25% Cu 10% Sn 13% Sb Pb alloy: Wire for spraying; Metco Ltd; obsolete
STA 7 LB1	Pb base bearing metal: Obsolete
STA 7 LB2	Pb base bearing metal: Obsolete; 'Magnolia Metal'
STA 7 TB2A	Pb base bearing metal: Obsolete; 'Babbit'
STEREOTYPE	6% Sn 16% Sb Pb alloy: White metal; further information from Lead Development Association
STEREOTYPE ALLOY	15% Sb 8% Sn Pb alloy: Common name
SX 25	0.25% Cu 2.5% Ag Pb alloy: Soft solder; Sheffield Smelting; melting range 302–305 °C **DPN: 8.0 UTS: 37 Elon: 45% Proof: 23**
TANDEM HDL	Pb base bearing metal: For shock loads; Eyre Smelting; liquidus 350 °C **DPN: 34 UTS: 90 Elon: 1%**
TANDEM RM	Pb base bearing metal: For shock loads; Eyre Smelting; liquidus 350 °C **DPN: 31 UTS: 85 Elon: 3%**
TANDEM SC	Pb base bearing metal: For low speed high duty; Eyre Smelting; liquidus 300 °C **DPN: 27 UTS: 75 Elon: 2%**
UNS L5001–L5999	Lead based alloys: American Standard System
V Pb Sn 5 Sb 28	5% Sn 28% Sb Pb alloy: Designation used by German Standards
V Pb Sn 30 Sb 6	30% Sn 6% Sb Pb alloy: Designation used by German Standards
W E WATSON BRAND	Sn Pb base bearing metal: For heavy duty purposes; Eyre Smelting; melting range 240–370 °C **DPN: 30 UTS: 67 Elon: 1%**

Note. The following abbreviations and units are used in the tables:

DPN	Hardness, diamond pyramid number
UTS	Ultimate tensile strength, N/mm^2
Elon	Elongation, %
Proof	0.1% proof strength, N/mm^2

1 N/mm^2=0.1 hbar=0.102 kgf/mm^2=0.06475 $tonf/in.^2$=145.04 $lbf/in.^2$=1 MPa

See Appendix II for other abbreviations and conversion tables.

22. Lithium Li

Physical properties	
Atomic number	3
Atomic weight	6.94
Crystal structure	Body-centred cubic
Colour	Silver-white
Specific gravity	0.53
Density	530 kg/m^3
Melting point	186 °C
Boiling point	1371 °C
Specific heat	4.02 J/g °C
Thermal conductivity	71.2 W/m °C
Coefficient of linear expansion (20–100 °C)	56×10^{-6}/ °C
Latent heat of fusion	137 J/g
Latent heat of vaporization	19 460 J/g
Thermal neutron absorption cross-section	70 barns/atom
Electrical conductivity	20% IACS (copper 100%)
Specific resistance	84 microhm mm
Temperature coefficient of electrical resistance	0.0045/ °C
Electrochemical equivalent	0.262 g/A/h
Electrode potential	−3.02 V
Magnetic susceptibility	0.5×10^{-6}
Young's modulus of elasticity	–
Tensile strength	Less than 15 N/mm^2
Hardness	Less than 5 DPN

22.1 General notes on lithium

Lithium occurs generally in combination with aluminium as the complex double oxide or fluoride. Approximately 0.004% of the earth's crust is composed of lithium, which is twice the amount of lead, but as it is more evenly distributed mining is less economical.

The metal is softer than lead, can be cut with a knife and tarnishes rapidly in moist air. It reacts vigorously with water, liberating hydrogen, and thus must be contained out of contact with air.

Lithium is the lightest known metal, having a specific gravity one fifth that of aluminium and half that of water.

The low melting point and high specific heat – almost that of water – and high boiling point make lithium an obvious choice as a heat exchange medium. Some use is being made of lithium in preference to sodium and potassium in nuclear submarine reactors, but the corrosive nature of the liquid metal necessitates expensive containers, such as tantalum. Liquid lithium at 200 °C attacks most common metals and glass or porcelain.

Alloys of lithium with silver or copper are used as deoxidizers for copper and silver and their alloys, while the alloy with calcium is a powerful scavenging agent for oxygen. The affinity of lithium for oxygen has also been used in heat treatment atmospheres, where the metal is added to the burnt town gas. Any free oxygen is converted to lithium oxide, but even more important is the ability of lithium to combine with oxygen in water vapour. This is one of the more troublesome components of atmospheric gas at heat treatment temperatures and the reaction has the secondary effect of releasing hydrogen to enrich the reducing properties of the resultant gas. The more efficient use of hydrogen and cracked ammonia furnace atmospheres has to a lesser extent reduced the need for lithium enriched atmospheres.

Lithium metal is used in chemical engineering as a catalyst in the production of some synthetic rubbers and forms the base for a series of metallic greases which have exceptional viscosity stability over a range of temperatures.

There are no commercial alloys based on lithium but in small amounts up to 1% lithium has a wide field of application. When added to lead based bearing metals and certain zinc base sheet metal alloys there is a considerable increase in hardness. The aluminium–zinc cast and wrought alloys are strengthened by 0.5% lithium and magnesium alloys have their corrosion resistance increased with no drop in strength.

When added to solder and braze materials lithium improves the wetting properties with in some instances improvement in shear strength. This increase in the ability to wet means enhanced brazing on materials such as chromium and tungsten.

The metal lithium has many useful and interesting properties, although its lack of stability in air precludes its use as the main constituent in normal engineering alloys.

Symbol	Nominal analysis, supplier, condition and remarks.
ASTM B357	99.9% Li metal in ingot form
L 06990	99.99% Li: UNS designation
LITHIUM	Li metal: Blackwells; pure metal
W Li 16	99.98% Li: Ingot; Light Ltd; high purity metal
W Li 20	99.9% Li: Wire 3 mm diameter; Light Ltd

Note. The following abbreviations and units are used in the tables:

DPN	Hardness, diamond pyramid number
UTS	Ultimate tensile strength, N/mm^2
Elon	Elongation, %
Proof	0.1% proof strength, N/mm^2

1 N/mm^2=0.1 hbar=0.102 kgf/mm^2=0.06475 $tonf/in.^2$=145.04 $lbf/in.^2$=1 MPa

See Appendix II for other abbreviations and conversion tables.

23. Magnesium Mg

Physical properties

Atomic number		12
Atomic weight		24.32
Crystal structure		Close-packed hexagonal
Colour		Silver-white
Specific gravity		1.74
Density		1740 kg/m^3
Melting point		650 °C
Boiling point		1103 °C
Specific heat		1.03 J/g °C
Thermal conductivity		159 W/m °C
Coefficient of linear expansion (20–100 °C)		26.1×10^{-6}/ °C
Latent heat of fusion		371.8 J/g
Latent heat of vaporization		4760 J/g
Thermal neutron absorption cross-section		0.06 barns/atom
Electrical conductivity		37% IACS (copper 100%)
Specific resistance		46 microhm mm
Temperature coefficient of electrical resistance		0.0040/ °C
Electrochemical equivalent		0.454 g/A/h
Electrode potential		−2.37 V
Magnetic susceptibility		0.55×10^{-6}
Young's modulus of elasticity		44×10^9 N/m^2
Tensile strength	annealed	150 N/mm^2
Hardness		35 DPN

23.1 General notes on magnesium

Magnesium is found generally as the carbonate, the double carbonate with calcium, the sulphate, chloride and silicate, and is the third most common metal. The ores are widespread throughout the world and are all more or less soluble in water. There is therefore a tendency for them to be dissolved by rivers and streams, and to concentrate in the oceans where they form the salt in sea water.

Magnesium is a modern metal which owes its place in the table of usefulness almost exclusively to the aircraft industry. The use of higher temperatures and stresses in this industry is to some extent lessening its popularity. The high tonnage requirements, however, caused modern plants to be built and reduced the price of magnesium to that of a competitive material where weight-saving is of prime importance.

The affinity of magnesium for oxygen is a considerable disadvantage from the corrosion point of view, necessitating complex corrosion resisting procedures and first quality paint films. Because of this property the metal is used for such purposes as metal de-oxidizing and as an oxygen 'getter' in thermionic valves. The well known property of burning in air to give a very white bright light is still used for flares, fireworks and photographic flashes. It forms the base of most incendiary devices.

Magnesium has gained the reputation that it corrodes dangerously rapidly, which is unfortunate. While there is no doubt that severe corrosion can occur with magnesium and its alloys this is not as bad as with steel but can be more spectacular and it is probably easier to protect magnesium permanently than steel. Alkalis and hydrofluoric acid have no effect on magnesium.

Pure magnesium is not often used commercially but there is a well defined range of alloys, some of which can have their mechanical properties enhanced by thermal treatment. Details of these materials are given in the following sections.

Magnesium is also used as an alloying element, particularly with aluminium where it has the effect of increasing strength when cold worked, without detracting from the corrosion resistance. It has a similar though less marked effect on zinc alloys.

There are also heat treatable aluminium–magnesium alloys, particularly with silicon, where the magnesium forms intermetallic compounds which take part in the hardening process.

23.2 Magnesium – note on specifications

An explanation of the different methods of classification and designation used with magnesium base materials may be of assistance in identifying these alloys rapidly. This being a comparatively modern metal, there are as yet relatively few specifications and trade names in existence. There are also very few basic suppliers and these are almost all American or Canadian, with the result that the same alloy appears more or less under the same name or symbol to a much greater extent than with the older metals, or even aluminium.

British Standards Institution (BS)

The specifications issued to date are almost all based on the system of classifying by form – sheet, bar, etc. – with the different alloys given a designation, which identifies the analysis. A letter is used to confirm the form.

These specifications are:

BS 2970	Castings – no identifying letter used
BS 3370	Strip – identifying letter 'S'
BS 3371	Tube – identifying letter 'T'
BS 3372	Forging – identifying letter 'F'
BS 3373	Bar – identifying letter 'E'
BS 3374	Plate – identifying letter 'F'

The alloy designation is prefixed by the letters MAG followed by figures which identify the alloy. Each alloy appears three times in the index of this book, for example:

(a) The full title	BS 3370 MAG S111 – Aluminium–zinc–magnesium alloy strip
(b) also	MAG 111 – Aluminium–zinc–magnesium alloy
(c) and	111 – Aluminium–zinc–magnesium alloy

The two last entries are always referred to the BS range, where the necessary information will be found under the full title. The British Standard designation for the alloys uses the letters preceding the figures and this is the correct form which appears in other specifications than those listed above.

One exception is the aircraft material specifications which use the letter 'L' as prefix followed by three digits. Aluminium and its alloys use the same prefix followed by two digits.

American Society for Testing and Materials (ASTM)

This body issues separate specifications covering different forms and materials. Each specification is identified by the letters ASTM followed by 'B' indicating a non-ferrous specification, then a number. The specification is then subdivided by analysis, each of which is identified by a letter and figure code which is almost universally used for magnesium alloys.

This uses two or more letters to indicate the alloys content, for example: 'A' for aluminium, 'Z' for zinc, 'M' for magnesium, followed by a two or three digit number identifying the actual alloy. There is no attempt to make the numbers give the percentage alloying elements or other information.

In this book the alloys are always given their full specification number and the designation is also listed in the index as these alloys tend to be called by this rather than their full title. The index entry refers to the appropriate section where the full title appears.

Examples would be:

ASTM B93 AZ 63 A/61	6% Al 3% Zn Mg alloy ingot – issued in 1961
AZ 63A	6% Al 3% Zn Mg alloy – this is the commonly used designation.

The full title should always be followed by two figures, giving the year of issue, for example, 61 indicates 1961. The letter 'T' following indicates that the specification was tentative at that time. The issue year and other suffix information is not always shown in this book.

Society of Automotive Engineers (SAE)

This body issues a series of specifications with the prefix SAE followed by two or three digits, the first of which is always 5. There has been no attempt to make these figures describe or code the alloy content.

This body also issues a series of Aerospace Material Specifictions prefixed with the letters AMS using four digits. A suffix letter denotes the issue number. Again, no effort has been made to code the analysis or form using these figures.

Unified Numbering System (UNS)

This system is described in ASTM E527.

It covers only metals and alloys which have a commercial standing, with a few exceptions where the UNS specification is unique.

There are 18 series of numbers. These are listed at the start of Section 44 in this book.

There is a single letter prefix followed by five digits.

Any UNS number for magnesium or its alloys will be headed 'M' from M00001 to M20000.

German Standards (DIN)

This uses two distinct systems, the first of which has a letter and figure code descriptive of the form and analysis of the alloy. These are contained in a master specification generally covering the form – i.e. sheet, bar, etc. – and prefixed by the letters DIN.

The second and newer system uses numbers only, the first of which is always 3 (in common with aluminium alloys) followed by a period and then four other digits, the first two classifying the alloy, the second two indicating the quantity of alloying element. This system was devised to suit modern accounting machines and computers and is included in the DIN specification, along with original descriptive method.

In this book the DIN specification is given and the new figure code. The descriptive designation is indexed and referred to the DIN range in the appropriate section.

Examples are:

DIN 1729 Mg Mn 2	1.7% Mn Mg alloy
Mg Mn 2	1.7% Mn Mg alloy: Designation used by German Standards
3.5200	1.7% Mn Mg alloy: German Standard

Trade names and specifications

These very often use the same code or one based on the same system as that attributed to ASTM.

While every effort has been made to include all possible alternative names, it is suggested that if the material cannot be identified immediately, the manufacturer's name or initials should be added or substituted to or from the symbol requiring identification.

23A Magnesium – commercially pure – wrought and cast

Specific gravity		1.738
Density		1738 kg/m^3
Solidus/liquidus		650 °C
Thermal conductivity		157 W/m °C
Coefficient of linear expansion (20–100 °C)		26.1 × 10^{-6}/ °C
Electrical conductivity		38% IACS (copper 100%)
Specific resistance		44.61 microhm mm
Young's modulus of elasticity		44 × 10^9 N/m^2
Impact	wrought	16–24 J
Fatigue strength		± 74 N/mm^2
Hot strength		–

The above properties have been chosen to show typical values for the specifications listed and may not apply exactly to any one specification. It is possible that with some specifications the values may not be applicable.

General metallurgical characteristics

Magnesium is the lightest metal commonly used in the sevice of mankind, the comparative figures being magnesium 1, aluminium 1.6, steel 4.4, copper 4.9 and titanium 2.5. Commercially pure magnesium however has few properties which make it an attractive engineering material. It has poor casting characteristics, low strength and does not work harden to any appreciable extent, while the corrosion resisting properties of the pure metal are only slightly superior to that of the alloys.

Commercially pure magnesium has very limited uses in industry. Where it is used it will require corrosion protection and specific advice should be obtained in this area.

Under dry conditions the corrosion resistance is reasonably satisfactory. Where any humidity at temperature is involved, however, or any corrosive atmosphere, then corrosion will occur very rapidly and quite dramatically under some circumstances.

Machinability of pure magnesium is very good and it is one of the few advantages of the metal. Considerable care, however, is required in that the metal in fine form will ignite very readily, this being either in the form of ribbon or dust.

The materials in this section are generally confined to primary metal but one specific use is as flares which are still used for lighting purposes under outdoor conditions, or for emergencies. Magnesium lights readily and burns with a fierce very white intense light.

Symbol	Nominal analysis, supplier, condition and remarks.
3.5002	99.95% Mg: German Standard
3.5003	99.8% Mg: German Standard
ASTM B92/9980 A	99.8% Mg: Ingot
ASTM B92/9990 A	99.9% Mg: Ingot
ASTM B92/9990 B	99.9% Mg: Ingot; low iron
ASTM B92/9995 A	99.95% Mg: Ingot
BS L120	99.5% Mg: Ingot; primary metal
DIN 17800 H Mg 99.8	99.8% Mg
DIN 17800 H Mg 99.95	99.95% Mg
HG 2-9990	99.9% Mg: Primary metal; Domal
HG 2-9995	99.95% Mg: Primary metal; Domal
HG 2-9997	99.97% Mg: Primary metal; Domal
HG 2-9999	99.99% Mg: High purity metal; Domal
H Mg 11a	99.99% Mg: Ingot; Light Ltd; high purity metal
MAGNESIUM	Mg: Ingot, stick, cubes, ribbon, wire, etc.; Blackwells; primary metal
MELPURE	99.9% Mg: Ingot; MEL primary metal
SS 4602	99.95% Mg: Ingot; Swedish Standard; primary metal
SS 4604	99.8% Mg: Ingot; Swedish Standard; primary metal

Note. The following abbreviations and units are used in the tables:

DPN	Hardness, diamond pyramid number
UTS	Ultimate tensile strength, N/mm^2
Elon	Elongation, %
Proof	0.1% proof strength, N/mm^2

1 N/mm^2=0.1 hbar=0.102 kgf/mm^2=0.06475 tonf/in.^{2}=145.04 lbf/in.^{2}=1 MPa

See Appendix II for other abbreviations and conversion tables.

23B Magnesium alloys
With zinc, aluminium, manganese, zirconium, rare earths or thorium

Specific gravity		1.8
Density		1800 kg/m^3
Solidus/liquidus		600–630 °C
Thermal conductivity		63–142 W/m °C
Coefficient of linear expansion (20–100 °C)		26–27 × 10^{-6}/ °C
Electrical conductivity		10–30% IACS (copper 100%)
Specific resistance		70–160 microhm mm
Young's modulus of elasticity		44.8 × 10^9 N/m^2
Impact		28 J
Fatigue strength (50 × 10^6 cycles)	wrought	± 100–120 N/mm^2
	cast	± 60–90 N/mm^2

Hot strength

Temperature °C	Tensile strength N/mm^2
20	210
100	180
200	140
300	75
400	45

The above properties have been chosen to show typical values for the specifications listed and may not apply exactly to any one specification. It is possible that with some specifications the values may not be applicable.

General metallurgical characteristics

Magnesium is the lightest metal in general engineering use, being approximately two-thirds the weight of aluminium. The low strength coupled with its low corrosion resistance means that very few uses are found for the pure metal.

The addition of certain alloying elements has the effect of considerably increasing the strength with some improvement in corrosion resistance. Briefly these elements and their effects are as follows:

Zinc. This was the earliest used alloying element and acts as a hardening agent helping to refine the grain. Magnesium alloys containing more than about 1% zinc are however very prone to weld cracking. Zinc is never the sole alloying element, generally being used with aluminium and zirconium. Up to 6% zinc is commonly used in magnesium alloys.

Aluminium. This hardens the alloy with some grain refinement but gives a long freezing range leading to casting porosity in many cases. There are some intermetallic compounds formed which can give age hardening properties. Up to 10% aluminium is commonly used.

Manganese. This has little effect on the strength of the alloys, but improves the corrosion resistance. There is a tendency to coarsen the grain size thus reducing the fatigue strength. Up to 2% manganese is used alone, with considerably less in conjunction with aluminium and zinc.

Zirconium. This has a profound effect, grain refining the cast structure with consequent increase in strength. Zirconium is only slightly soluble in magnesium and is thus always present in very small quantities.

Aluminium and manganese form compounds with zirconium which are insoluble in magnesium and can thus be used to harden by precipitation treatment. Zirconium improves the workability of wrought products. Up to about 0.7% zirconium is generally used.

Rare earth metals. These strengthen magnesium and improve the high temperature properties. The rare earths also reduce the tendency of cracking in the zirconium–zinc–magnesium alloys.

There is also an improvement in the castability, notably in freedom from porosity, when the rare earth elements, especially cerium, are present. This is sufficient to warrant the use of these expensive alloys in some cases where the hot strength is not required.

Thorium. This has a very similar effect to the rare earths in that it improves the creep strength and other high temperature properties, improves the castability and, in addition, enhances the fatigue strength. Less than 2% thorium is used for alloying properties.

Silver. This has recently been used added to rare earth and zirconium alloys of magnesium and results in considerable age hardening properties of castings. Although to date silver is not used in wrought alloys there are indications that this series of silver–rare earth–zirconium alloys could be the basis of magnesium alloys of comparable strength to some aluminium alloys.

All magnesium alloys have good damping characteristics, but are rather more notch sensitive than other common low strength engineering materials. This, coupled with their high galvanic corrosion characteristics, makes the correct design of magnesium articles of prime importance. This should always include generous radii, an absence of pockets to collect moisture and no intimate contact with other metals. Joins must always be carefully sited away from areas of possible stress concentration.

There are a limited number of magnesium alloys which can be solution treated and aged.

Where magnesium is cold worked for any reason it must be softened using a temperature of approximately 300 °C.

Specialist advice should always be obtained.

Welding of magnesium can be carried out but generally

requires the use of inert gas and specialist advice should be obtained.

Magnesium has a history of having serious problems with corrosion and to a large extent this poor corrosion resistance is inhibiting the expansion of the use of magnesium alloys in industry.

There are techniques for protecting the materials but these tend to be quite complex, difficult to control and therefore are expensive.

There is also the problem that magnesium and its alloys can be quite flammable. In finely divided form they can in fact be explosive.

Any machining of magnesium requires particular care to make certain that the turnings, drillings and dust are carefully stored and precautions taken at a high standard. Any grinding or polishing requires specialist equipment.

Where magnesium fires exist water must not be used as the fire can result in the production of hydrogen which will obviously feed the fire. The fire must therefore be extinguished using dry powder or inert gases.

The advantages of magnesium alloys are their excellent machinability, their lightness and their ability to absorb vibrations and thus be silent metals. These are largely outweighed by the disadvantages of poor corrosion resistance and accident prone-ness regarding fire hazards.

Magnesium is still used to some extent in aircraft but is being replaced in many instances by titanium. The motor industry still makes some use of magnesium but there is no indication that this will increase.

Considerable quantities of magnesium are still used as anodes in the corrosion protection of steel.

Symbol	Nominal analysis, supplier, condition and remarks.
3.5101	4.5% Zn 0.7% Zr 1.2% rare earth Mg alloy: Casting; German Standard
3.5103	2.3% Zn 0.6% Zr 3% rare earth Mg alloy: Casting; German Standard
3.5104	4.5% Zn 0.7% Zr Mg alloy: Casting; German Standard
3.5161	5.5% Zn 0.5% Zr Mg alloy: Wrought; German Standard **DPN: 60 UTS: 250 Elon: 5% Proof: 180**
3.5200	1.8% Mn Mg alloy: Wrought; German Standard **DPN: 40 UTS: 180 Elon: 1.5% Proof: 140**
3.5312	3% Al 1% Zn 0.2% Mn Mg alloy: Wrought; German Standard **DPN: 45 UTS: 240 Elon: 10% Proof: 150**
3.5612	6% Al 1% Zn 0.2% Mn Mg alloy: Wrought; German Standard **DPN: 55 UTS: 240 Elon: 6% Proof: 170**
3.5632	6% Al 3% Zn 0.2% Mn Mg alloy: Casting; German Standard **DPN: 60 UTS: 150 Elon: 5% Proof: 90**
3.5812	8% Al 0.8% Zn 0.2% Mn Mg alloy: Casting; German Standard **DPN: 60 UTS: 210 Elon: 5% Proof: 120**
3.5812	8% Al 0.7% Zn 0.3% Mn Mg alloy: Casting; German Standard
3.5912	9% Al 0.8% Zn 0.2% Mn Mg alloy: Casting; German Standard **DPN: 65 UTS: 210 Elon: 3% Proof: 140**
3.5912	8% Al 0.9% Zn 0.3% Mn Mg alloy: Casting; German Standard
3.5922	8.5% Al 1.5% Zn 0.2% Mn Mg alloy: Casting; German Standard **DPN: 65 UTS: 200 Elon: 2% Proof: 120**
3.6104	4.5% Zn 0.7% Zr 1.2% rare earth Mg alloy: Casting; German Standard
3.6204	2.3% Zn 0.6% Zr 3% rare earth Mg alloy: Casting; German Standard
A 8	8% Al 0.5% Zn 0.3% Mn Mg alloy: Casting; MEL; solution treated; ductile; shock resistant **UTS: 220 Elon: 8% Proof: 75**

Note. The following abbreviations and units are used in the tables:

DPN	Hardness, diamond pyramid number
UTS	Ultimate tensile strength, N/mm^2
Elon	Elongation, %
Proof	0.1% proof strength, N/mm^2

1 N/mm^2=0.1 hbar=0.102 kgf/mm^2=0.06475 $tonf/in.^2$=145.04 $ibf/in.^2$=1 MPa

See Appendix II for other abbreviations and conversion tables.

Symbol	Nominal analysis, supplier, condition and remarks.
A 2855	8% Al 0.4% Zn 0.3% Mn Mg alloy: Forging; MEL **DPN: 70 UTS: 300 Elon: 11% Proof: 180**
AM 60 A	6% Al Mg alloy: Casting
AM 503	1.5% Mn Mg alloy: Sheet, bar and tube; MEL; weldable; good corrosion resistance **UTS: 240 Elon: 7% Proof: 120**
AM 503	1.5% Mn Mg alloy: Sheet and tube; Birmabright for BS alloy 101
AM 503 S	0.7% Mn Mg alloy: MEL for nuclear engineering
AMS 4350 F	6% Al 1% Zn Mg alloy: Extrusions; as extruded; AMS for ASTM alloy AZ 61 A
AMS 4352 A	5.5% Zn Zr Mg alloy: Extrusion; aged; AMS for ASTM alloy ZK 60 A
AMS 4358 A	6% Al 1% Zn Mg alloy: Forging; as forged; AMS for ASTM alloy AZ 61 X
AMS 4360 C	8.5% Al 0.5% Zn Mg alloy: Forging; aged; AMS for ASTM alloy AZ 80 A
AMS 4362	5% Zn Mg alloy: Forging; aged; AMS for ASTM alloy ZK 60 A
AMS 4363	0.8% Mn 2% Th Mg alloy: Wrought
AMS 4375	3% Al 1% Zn Mg alloy: Casting
AMS 4375 D	3% Al 1% Zn Mg alloy: Sheet; annealed; AMS for ASTM alloy AZ 31 B
AMS 4376 A	3% Al 1% Zn Mg alloy: Plate; AMS for ASTM alloy AZ 31 B
AMS 4377 B	3% Al 1% Zn Mg alloy: Sheet; AMS specification for ASTM alloy AZ 31 B
AMS 4382	3% Al 1% Zn Mg alloy: Casting
AMS 4383	0.8% Mn 2% Th Mg alloy: Wrought
AMS 4384 B	3.2% Th 0.7% Zr Mg alloy: Sheet; annealed; AMS for ASTM alloy HK 31 A
AMS 4385 C	3.2% Th 0.7% Zr Mg alloy: Sheet; AMS for ASTM alloy HK 31 A
AMS 4386	1.2% Al 14% Li Mg alloy: Casting
AMS 4387	2.3% Zn 0.6% Zr Mg alloy: Extrusion; as extruded
AMS 4388	3% Th 1.5% Mn Mg alloy: Extrusion; as extruded
AMS 4389 A	3% Th 1.5% Mn Mg alloy: Extrusion; aged
AMS 4390 B	2% Th 0.8% Mn Mg alloy: Sheet; cold rolled; AMS for ASTM alloy HM 21 A
AMS 4395 A	9% Al 2% Zn Mg alloy: Welding wire; AMS for ASTM alloy AZ 92 A
AMS 4396	2.5% Zn 0.7% Zr 3.3% Ce Mg alloy: Welding wire; AMS for ASTM alloy EZ 33 A
AMS 4397	1.2% Al 14% Li Mg alloy: Casting
AMS 4418 A	2.5% Ag 2% Dy 1% Zr Mg alloy: Sand casting; solution treated and aged; AMS for ASTM alloy QE 222 A
AMS 4419	2.5% Ag 1.2% Th 0.8% Zr 1% rare earth Mg alloy: Casting
AMS 4420 H	6% Al 3% Zn Mg alloy: Casting; sand cast; AMS for ASTM alloy AZ 63 A

Symbol	Nominal analysis, supplier, condition and remarks.
AMS 4422 J	6% Al 3% Zn Mg alloy: Sand casting; solution treated; AMS for ASTM alloy AZ 63 A
AMS 4424 F	6% Al 3% Zn Mg alloy: Sand casting; solution treated and aged; AMS for ASTM alloy AZ 63 A
AMS 4425	6% Zn 0.8% Zr 2.7% rare earth Mg alloy: Casting
AMS 4434 F	9% Al 2% Zn Mg alloy: Sand casting; solution treated and aged; AMS for ASTM alloy AZ 92 A
AMS 4437	8.7% Al 0.1% Zn Mg alloy: Sand casting; solution treated and aged; AMS for ASTM alloy AZ 91 C
AMS 4437 B	10% Al 0.7% Zn 0.3% Mn Mg alloy: Casting
AMS 4438	5.7% Zn 1.8% Th Mg alloy: Sand casting; aged; AMS for ASTM alloy ZH 62 A
AMS 4439	4.2% Zn 0.8% Zr 1.2% rare earth Mg alloy: Casting
AMS 4440 A	0.7% Zr 3.5% Ce Mg alloy: Sand casting; aged; AMS for ASTM alloy EK 41 A
AMS 4441 A	0.7% Zr 3.5% Ce Mg alloy: Sand casting; solution treated and aged; AMS for ASTM alloy EK 41 A
AMS 4442 A	2.5% Zn 0.7% Zr 3.3% Ce Mg alloy: Sand casting; aged; AMS for ASTM alloy EZ 33 A
AMS 4442 B	2.3% Zn 0.6% Zr 3% rare earth Mg alloy: Casting
AMS 4443 A	4.5% Zn 0.7% Zr Mg alloy: Sand casting; aged; AMS for ASTM alloy ZK 51 A
AMS 4443 B	4.5% Zn 0.7% Zr Mg alloy: Casting
AMS 4444	6% Zn 1% Zr Mg alloy: Sand casting; aged; AMS for ASTM alloy ZK 61 A
AMS 4445 A	3.3% Th 0.8% Zr Mg alloy: Sand casting; solution treated and aged; AMS for ASTM alloy HK 31 A
AMS 4446	8.8% Al 0.2% Mn 0.6% Zr Mg alloy: Casting
AMS 4447	2.1% Zn 3.3% Th 0.8% Zr Mg alloy: Sand casting; aged; AMS for ASTM alloy HZ 32 A
AMS 4452	8.6% Al Mg alloy: Casting
AMS 4453	9% Al 2% Zn Mg alloy: Investment casting; solution treated and aged; AMS for ASTM alloy AZ 92 A
AMS 4455	10% Al Mg alloy: Investment casting; solution treated and aged; AMS for ASTM alloy AM 100 A
AMS 4483	10% Al Mg alloy: Permanent mould castings; solution treated and aged; AMS for ASTM alloy AM 100 A
AMS 4484	9% Al 2% Zr Mg alloy: Casting
AMS 4484 E	9% Al 2% Zn Mg alloy: Permanent mould casting; solution treated and aged; AMS for ASTM alloy AZ 92 A
AMS 4490 D	9% Al 0.7% Zn Mg alloy: Die casting; as cast; AMS for ASTM alloy AZ 91
AN M 16	9% Al 1.0% Zn Mg alloy: Casting; US Service
AN QQ M 56/1A	9.0% Al 2.0% Zn Mg alloy: Casting; US Service; as cast **UTS: 140 Elon: 4% Proof: 35**
AN QQ M 56/1C	9.0% Al 2.0% Zn Mg alloy: Casting; US Service; as cast **UTS: 120 Proof: 35**
AS 41 A	4.5% Al 0.35% Mn Mg alloy: Casting
ASTM B80 AM 100 A	10% Al Mg alloy: Sand casting; solution treated and aged **UTS: 240 Proof: 100**
ASTM B80 AZ 63 A	6% Al 3% Zn Mg alloy: Sand casting; solution treated and aged **UTS: 220 Elon: 3% Proof: 90**
ASTM B80 AZ 81 A	7.5% Al 0.7% Zn Mg alloy: Sand casting; solution treated **UTS: 220 Elon: 7% Proof: 60**
ASTM B80 AZ 91 C	9% Al 0.6% Zn Mg alloy: Sand casting; solution treated and aged **UTS: 220 Elon: 3% Proof: 90**
ASTM B80 AZ 92 A	9% Al 2% Zn Mg alloy: Sand casting; solution treated and aged **UTS: 220 Elon: 1% Proof: 100**
ASTM B80 EK 30 A	3% rare earth 0.2% Zr Mg alloy: Sand casting; solution treated and aged **UTS: 120 Elon: 2% Proof: 75**
ASTM B80 EK 41 A	4% rare earth 0.7% Zr Mg alloy: Sand casting; solution treated and aged **UTS: 140 Elon: 1% Proof: 90**
ASTM B80 EZ 33 A	2.6% Zn 3.2% rare earth 0.7% Zr Mg alloy: Sand casting; aged **UTS: 120 Elon: 2% Proof: 75**
ASTM B80 HK 31 A	3.2% Th 0.6% Zr Mg alloy: Sand casting; solution treated and aged **UTS: 180 Elon: 4% Proof: 75**
ASTM B80 HZ 32 A	2% Zn 3.2% Th 0.7% Zr Mg alloy: Sand casting; aged **UTS: 180 Elon: 4% Proof: 75**
ASTM B80 K 1 A	0.6% Zr Mg alloy: Sand casting; as cast **UTS: 150 Elon: 14%**
ASTM B80 QE 22 A	2.2% rare earth 0.6% Zr Mg alloy: Sand casting; solution treated and aged **UTS: 240 Elon: 2% Proof: 170**
ASTM B80 ZE 41 A	4.2% Zn 1.2% rare earth 0.6% Zr Mg alloy: Sand casting; aged **UTS: 180 Elon: 2.5% Proof: 120**
ASTM B80 ZH 62 A	5.8% Zn 2% Th 0.7% Zr Mg alloy: Sand casting; aged **UTS: 240 Elon: 5% Proof: 150**
ASTM B80 ZK 51 A	5% Zn 0.7% Zr Mg alloy: Sand casting; aged **UTS: 220 Elon: 5% Proof: 120**
ASTM B80 ZK 61 A	6% Zn 0.8% Zr Mg alloy: Sand casting; solution treated and aged **UTS: 270 Elon: 5% Proof: 170**
ASTM B90 AZ 31 B	3% Al 1% Zn Mg alloy: Sheet; annealed **UTS: 200 Elon: 10%**
ASTM B90 AZ 31 B	3% Al 1% Zn Mg alloy: Sheet; cold rolled **UTS: 240 Elon: 6% Proof: 140**
ASTM B90 AZ 31 B	3% Al 1% Zn Mg alloy: Forging; as forged **UTS: 220 Elon: 6% Proof: 120**
ASTM B90 HK 31 A	3.2% Th 0.6% Zr Mg alloy: Sheet; annealed **UTS: 200 Elon: 12% Proof: 90**
ASTM B90 HK 31 A	3.2% Th 0.6% Zr Mg alloy: Sheet; cold rolled **UTS: 220 Elon: 4% Proof: 170**
ASTM B90 HM 21 A	0.8% Al 2% Th Mg alloy: Sheet; solution treated; cold worked and aged **UTS: 200 Elon: 6% Proof: 100**
ASTM B90 ZE 10 A	0.2% rare earth 1% Zn Mg alloy: Sheet; annealed **UTS: 200 Elon: 14% Proof: 100**
ASTM B90 ZE 10 A	0.2% rare earth 1% Zn Mg alloy: Sheet; cold rolled **UTS: 220 Elon: 4% Proof: 140**
ASTM B91 AZ 61 A	6.5% Al 1% Zn Mg alloy: Forging; as forged **UTS: 250 Elon: 6% Proof: 140**
ASTM B91 AZ 80 A	8.5% Al 0.6% Zn Mg alloy: Forging; aged **UTS: 290 Elon: 2% Proof: 180**
ASTM B91 TA 54 A	3.5% Al Mg alloy: Forging; as forged **UTS: 280 Elon: 7% Proof: 140**
ASTM B91 ZK 60 A	5.5% Zn Mg alloy: Forging; aged **UTS: 290 Elon: 7% Proof: 170**
ASTM B93 AM 100 A	10% Al Mg alloy: Ingot
ASTM B93 AZ 63 A	6% Al 3% Zn Mg alloy: Ingot
ASTM B93 AZ 81 A	7.6% Al 0.7% Zn Mg alloy: Ingot
ASTM B93 AZ 91 A	9% Al 0.6% Zn Mg alloy: Ingot
ASTM B93 AZ 91 B	9% Al 0.6% Zn Mg alloy: Ingot
ASTM B93 AZ 91 C	9% Al 0.6% Zn Mg alloy: Ingot
ASTM B93 AZ 92 A	9% Al 2% Zn Mg alloy: Ingot

Note. The following abbreviations and units are used in the tables:

DPN	Hardness, diamond pyramid number
UTS	Ultimate tensile strength, N/mm^2
Elon	Elongation, %
Proof	0.1% proof strength, N/mm^2

1 N/mm^2=0.1 hbar=0.102 kgf/mm^2=0.06475 tonf/in.^{2}=145.04 ibf/in.^{2}=1 MPa

See Appendix II for other abbreviations and conversion tables.

Symbol	Nominal analysis, supplier, condition and remarks.
ASTM B94 AZ 91 A	9% Al 0.6% Zn 0.1% Cu Mg alloy: Die casting; as cast **DPN: 63 UTS: 240 Elon: 3% Proof: 170**
ASTM B94 AZ 91 B	9% Al 0.6% Zn 0.3% Cu Mg alloy: Die casting; as cast **DPN: 63 UTS: 240 Elon: 3% Proof: 170**
ASTM B107 AZ 31 B	3% Al 1% Zn Mg alloy: Bar; as drawn **UTS: 220 Elon: 7% Proof: 150**
ASTM B107 AZ 31 C	3% Al 1% Zn Mg alloy: Bar; as drawn **UTS: 220 Elon: 7% Proof: 150**
ASTM B107 AZ 61 A	6.5% Al 1% Zn Mg alloy: Bar; as drawn **UTS: 280 Elon: 8% Proof: 150**
ASTM B107 AZ 80 A	8.5% Al 0.6% Zn Mg alloy: Bar; solution treated **UTS: 330 Elon: 4% Proof: 220**
ASTM B107 M 1 A	1.2% Mn Mg alloy: Bar; as drawn **UTS: 210 Elon: 2%**
ASTM B107 ZK 60 A	5.5% Zn 0.45% Zr Mg alloy: Bar; solution treated **UTS: 320 Elon: 4% Proof: 250**
ASTM B199 AM 91 C	9% Al 0.6% Zn Mg alloy: Casting; solution treated and aged **UTS: 220 Elon: 3% Proof: 90**
ASTM B199 AM 92 A	9% Al 2% Zn Mg alloy: Casting; solution treated and aged **UTS: 220 Proof: 100**
ASTM B199 AM 100 A	10% Al Mg alloy: Casting; solution treated and aged **UTS: 220 Elon: 2% Proof: 100**
ASTM B199 EK 41 A	4% rare earth 0.6% Zr Mg alloy: Casting; solution treated and aged **UTS: 150 Elon: 1% Proof: 90**
ASTM B199 EZ 33 A	2.5% Zn 3% rare earth 0.7% Zr Mg alloy: Casting; aged **UTS: 140 Elon: 2% Proof: 90**
ASTM B217 AZ 31 B	3% Al 1% Zn Mg alloy: Tube; as drawn **UTS: 210 Elon: 8% Proof: 90**
ASTM B217 AZ 31 C	3% Al 1% Zn Mg alloy: Tube; as drawn **UTS: 210 Elon: 8% Proof: 90**
ASTM B217 AZ 61 A	6.5% Al 1% Zn Mg alloy: Tube; as drawn **UTS: 240 Elon: 7% Proof: 90**
ASTM B217 M 1 A	1.2% Mn Mg alloy: Tube; as drawn **UTS: 180 Elon: 2%**
ASTM B217 ZK 60 A	5% Zn 0.4% Zr Mg alloy: Tube; solution treated and aged **UTS: 320 Elon: 4% Proof: 250**
ASTM B260 B Mg 1	9% Al 2% Zn Mg alloy: Braze filler
ASTM B260 B Mg 2	12% Al 5% Zn Mg alloy: Brazing filler metal; braze temperature range 582–610 °C
ASTM B260 B Mg 2a	12% Al 5% Zn 0.005% Be Mg alloy: Brazing filler metal; braze temperature range 582–610 °C
ASTM B403 AM 100 A	10% Al 0.10% (min) Mn Mg alloy: Investment casting
ASTM B403 AZ 81 A	7.5% Al 0.13% (min) Mn 0.7% Zn Mg alloy: Investment casting
ASTM B403 AZ 91 C	8.8% Al 0.13% (min) Mn 0.7% Zn Mg alloy: Investment casting
ASTM B403 AZ 92 A	8.5% Al 0.10% (min) Mn 2.0% Zn Mg alloy: Investment casting
ASTM B403 EZ 33 A	2.5% Zn 0.75% Zr 3.5% rare earth Mg alloy: Investment casting
ASTM B403 HK 31 A	0.7% Zr 3.5% Th Mg alloy: Investment casting
ASTM B403 K 1 A	0.7% Zr Mg alloy: Investment casting
ASTM B403 QE 22 A	0.7% Zr 2.5% Ag 2.2% rare earth Mg alloy: Investment casting
ASTM B403 ZK 61 A	6.0% Zn 0.8% Zr Mg alloy: Investment casting
AZ 21 X	1.7% Al 0.6% Zn Mg alloy: Rod, bar and forging; Domal **UTS: 210 Elon: 5% Proof: 90**
AZ 31	3% Al 1% Zn 0.2% Mn Mg alloy: Wrought; previous German designation
AZ 31	3% Al 1% Zn 0.3% Mn Mg alloy: Sheet and tube; MEL and Birmetal for BS alloy 111
AZ 31 X	3% Al 1% Zn Mg alloy: Rod, tube and forging; Domal **UTS: 220 Elon: 8% Proof: 120**
AZ 61	6% Al 1% Zn 0.3% Mn Mg alloy: Wrought; previous German designation
AZ 61 X	6.2% Al 1.1% Zn Mg alloy: Rod, tube and forging; Domal **UTS: 250 Elon: 7% Proof: 120**
AZ 63	6% Al 3% Zn 0.2% Mn Mg alloy: Casting; previous German designation
AZ 80 X	9% Al 6% Zn Mg alloy: Casting; Domal; as cast **UTS: 150 Elon: 3%**
AZ 80 X	9% Al 0.6% Zn Mg alloy: Casting; Domal; solution treated, room aged **UTS: 220 Elon: 7%**
AZ 80 X	9% Al 0.6% Zn Mg alloy: Bar and forging; Domal; as forged **UTS: 300 Elon: 6% Proof: 180**
AZ 80 X	9% Al 0.6% Zn Mg alloy: Bar and forging; Domal; aged **UTS: 330 Elon: 3% Proof: 220**
AZ 81	8% Al 0.8% Zn 0.2% Mn Mg alloy: Casting; previous German designation
AZ 91	9% Al 0.8% Zn 0.2% Mn Mg alloy: Casting; previous German designation
AZ 91	9.5% Al 0.5% Zn 0.3% Mn Mg alloy: Casting; MEL; solution treated and aged; for pressure die casting **UTS: 220 Elon: 2% Proof: 100**
AZ 91 E	8.7% Al 0.2% Mn 0.6% Zn Mg alloy: Casting; designation used in the UK and USA
AZ 91 X	9.5% Al 0.4% Zn 0.3% Mn 0.0015% Be Mg alloy: Casting; MEL; as cast; for die casting **UTS: 210 Elon: 2% Proof: 100**
AZ 91 X	8.7% Al 0.8% Zn 0.3% Mn Mg alloy: Casting; Domal; as cast; die cast **UTS: 140**
AZ 91 X	8.7% Al 0.8% Zn 0.3% Mn Mg alloy: Casting; Domal; solution treated and aged; die cast **UTS: 220 Elon: 3% Proof: 90**
AZ 92	8.5% Al 1.5% Zn 0.2% Mn Mg alloy: Casting; previous German designation
AZ 855	8% Al 0.4% Zn 0.3% Mn Mg alloy: Forging; MEL; for highly stressed forgings **UTS: 300 Elon: 10% Proof: 180**
AZM	6% Al 1% Zn 0.3% Mn Mg alloy: Bar and forging; MEL and Birmetal for BS alloy 121
BIRMAG	Free machining magnesium alloy with properties similar to Birmetal alloy AZM
B Mg 1	9% Al 1% Mn 2% Zn 0.0006% Be Mg: Weld electrode; designation used by AWS
BS 1272/80	Mg alloy: Ingot and casting; replaced BS 2970
BS 1453 D1	10% Al 0.2% Mn Mg alloy: Weld filler rod for welding Mg alloys
BS 1453 D2	1.5% Mn Mg alloy: Weld filler rod for welding Mg alloys
BS 2901 D1	9.5% Al Mg alloy: Welding rod for inert gas shielded arc welding
BS 2901 D2	1.2% Mn Mg alloy: Rod for inert gas shielded arc welding
BS 2901 D3	7% Al 0.25% Mn Mg alloy: Rod for inert gas shielded arc welding
BS 2901 D4	3% Al 1% Zn 0.4% Mn Mg alloy: Rod for inert gas shielded arc welding
BS 2901 D5	1% Zn 0.8% Zr Mg alloy: Rod for inert gas shielded arc welding
BS 2901 D6	2.5% Zn 2.0% Cd 0.8% Zr Mg alloy: Rod for inert gas shielded arc welding
BS 2901 D7	4.0% Zn 1.2% rare earth 0.6% Zr Mg alloy: Rod for inert gas shielded arc welding
BS 2901 D8	3.2% rare earth 2% Zn 0.8% Zr Mg alloy: Rod for inert gas shielded arc welding

Symbol	Nominal analysis, supplier, condition and remarks.
BS 2970 MAG 1 M	8% Al 0.7% Zn 0.3% Mn Mg alloy: Casting; chill cast; as cast **UTS: 180 Elon: 4% Proof: 75**
BS 2970 MAG 1 W	8% Al 0.7% Zn 0.3% Mn Mg alloy: Casting; chill cast; solution treated **UTS: 220 Elon: 10% Proof: 70**
BS 2970 MAG 2 M	8% Al 0.7% Zn 0.4% Mn Mg alloy: Casting; chill cast; high purity **UTS: 180 Elon: 4% Proof: 75**
BS 2970 MAG 2 W	8% Al 0.7% Zn 0.4% Mn Mg alloy: Casting; chill cast; solution treated; high purity **UTS: 220 Elon: 10% Proof: 70**
BS 2970 MAG 3 M	10% Al 0.7% Zn 0.3% Mn Mg alloy: Casting; chill cast; as cast **UTS: 170 Elon: 2% Proof: 90**
BS 2970 MAG 3 W	10% Al 0.7% Zn 0.3% Mn Mg alloy: Casting; chill cast; solution treated **UTS: 210 Elon: 5% Proof: 75**
BS 2970 MAG 3 WP	10% Al 0.7% Zn 0.3% Mn Mg alloy: Casting; chill cast; solution treated and aged **UTS: 210 Elon: 2% Proof: 110**
BS 2970 MAG 4 M	4.5% Zn 0.7% Zr Mg alloy: Casting; chill cast; as cast **UTS: 220 Elon: 10% Proof: 100**
BS 2970 MAG 4 P	4.5% Zn 0.7% Zr Mg alloy: Casting; chill cast; aged **UTS: 240 Elon: 7% Proof: 130**
BS 2970 MAG 5 M	4.5% Zn 0.7% Zr 1.2% rare earth Mg alloy: Casting; chill cast; as cast **UTS: 180 Elon: 4% Proof: 100**
BS 2970 MAG 5 P	4.5% Zn 0.7% Zr 1.2% rare earth Mg alloy: Casting; chill cast; aged **UTS: 210 Elon: 4% Proof: 120**
BS 2970 MAG 6	2.3% Zn 0.6% Zr 3% rare earth Mg alloy: Casting; chill cast; as cast or aged **UTS: 150 Elon: 3% Proof: 90**
BS 2970 MAG 7 M	8% Al 0.9% Zn 0.3% Mn Mg alloy: Casting; chill cast; as cast **UTS: 170 Elon: 2% Proof: 75**
BS 2970 MAG 7 W	8% Al 0.9% Zn 0.3% Mn Mg alloy: Casting; chill cast; solution treated **UTS: 210 Elon: 5% Proof: 70**
BS 2970 MAG 7 WP	8% Al 0.9% Zn 0.3% Mn Mg alloy: Casting; chill cast; solution treated and aged **UTS: 210 Elon: 2% Proof: 90**
BS 3370 MAG 131	2.2% Zn 0.2% Mn Mg alloy: Casting
BS 3370 MAG S101	1.5% Mn Mg alloy: Strip; as rolled **UTS: 200 Elon: 5% Proof: 60**
BS 3370 MAG S111	3% Al 1% Zn 0.5% Mn Mg alloy: Strip; annealed **UTS: 220 Elon: 12% Proof: 100**
BS 3370 MAG S141	1% Zn 0.5% Zr Mg alloy: Strip; as rolled **UTS: 240 Elon: 8% Proof: 150**
BS 3370 MAG S151	3% Zn 0.5% Zr Mg alloy: Strip; as rolled **UTS: 250 Elon: 8% Proof: 170**
BS 3371 MAG T101	1.5% Mn Mg alloy: Tube; as drawn **UTS: 220 Elon: 4% Proof: 100**
BS 3371 MAG T111	3% Al 1% Zn 0.5% Mn Mg alloy: Tube; as drawn **UTS: 220 Elon: 10% Proof: 140**
BS 3371 MAG T121	6% Al 1% Zn 0.3% Mn Mg alloy: Tube; as drawn **UTS: 250 Elon: 7% Proof: 150**
BS 3371 MAG T141	1% Zn 0.5% Zr Mg alloy: Tube; as drawn **UTS: 240 Elon: 4% Proof: 150**
BS 3372 MAG F101	1.5% Mn Mg alloy: Forgings; as forged **UTS: 200 Elon: 5% Proof: 90**
BS 3372 MAG F121	6% Al 1.2% Zn 0.2% Mn Mg alloy: Forging; as forged **UTS: 280 Elon: 8% Proof: 150**
BS 3372 MAG F151	3.25% Zn 0.6% Zr Mg alloy: Forging; as forged **UTS: 250 Elon: 8% Proof: 170**
BS 3373 MAG E101	1.5% Mn Mg alloy: Bar; as extruded **UTS: 210 Elon: 3% Proof: 100**
BS 3373 MAG E111	3% Al 1% Zn 0.3% Mn Mg alloy: Bar; as extruded **UTS: 220 Elon: 10% Proof: 140**
BS 3373 MAG E121	6% Al 1% Zn 0.25% Mn Mg alloy: Bar; as extruded **UTS: 240 Elon: 9% Proof: 150**
BS 3373 MAG E141	1% Zn 0.6% Zr Mg alloy: Bar; as extruded **UTS: 240 Elon: 10% Proof: 150**
BS 3373 MAG E151	3% Zn 0.6% Zr Mg alloy: Bar; as extruded **UTS: 290 Elon: 10% Proof: 200**
BS 3373 MAG E161	5% Zn 0.6% Zr Mg alloy: Bar; as extruded, straightened and heat treated **UTS: 300 Elon: 8% Proof: 210**
BS 3374 MAG P101	1.5% Mn Mg alloy: Plate; as rolled and flattened **UTS: 180 Elon: 5% Proof: 60**
BS 3374 MAG P111	3% Al 1% Zn 0.3% Mn Mg alloy: Plate; as rolled and flattened **UTS: 210 Elon: 10% Proof: 100**
BS 3374 MAG P141	1% Zn 0.6% Zr Mg alloy: Plate; as rolled and flattened **UTS: 210 Elon: 8% Proof: 100**
BS 3374 MAG P151	3% Zn 0.6% Zr Mg alloy: Plate; as rolled and flattened **UTS: 220 Elon: 8% Proof: 120**
BS L	See L121–L515 for details
C	8% Al 1% Zn 0.15% Mn Mg alloy: Casting; MEL; solution treated and aged **UTS: 210 Elon: 2% Proof: 90**
DIN 1729 Mg Al 3 Zn	3% Al 1% Zn 0.2% Mn Mg alloy: Wrought **DPN: 45 UTS: 240 Elon: 10% Proof: 150**
DIN 1729 Mg Al 6 Zn	6% Al 1% Zn 0.2% Mn Mg alloy: Wrought **DPN: 55 UTS: 240 Elon: 6% Proof: 170**
DIN 1729 Mg Al 8 Zn	8.5% Al 0.6% Zn 0.2% Mn Mg alloy: Wrought **DPN: 60 UTS: 250 Elon: 6% Proof: 180**
DIN 1729 Mg Mn 2	1.8% Mn Mg alloy: Wrought **DPN: 40 UTS: 180 Elon: 1.5% Proof: 140**
DIN 1729 Mg Zn 6 Zr	5.5% Zn 0.5% Zr Mg alloy: Wrought **DPN: 60 UTS: 250 Elon: 5% Proof: 180**
DIN 1729G Mg Al 6 Zn 3	6% Al 3% Zn 0.2% Mn Mg alloy: Casting **DPN: 60 UTS: 150 Elon: 5% Proof: 90**
DIN 1729G Mg Al 8 Zn 1	8% Al 0.8% Zn 0.2% Mn Mg alloy: Casting **DPN: 60 UTS: 210 Elon: 5% Proof: 120**
DIN 1729G Mg Al 9 Zn 1	9% Al 0.8% Zn 0.2% Mn Mg alloy: Casting **DPN: 65 UTS: 210 Elon: 3% Proof: 140**
DIN 1729G Mg Al 9 Zn 2	8.5% Al 1.2% Zn 0.2% Mn Mg alloy: Casting **DPN: 65 UTS: 200 Elon: 2% Proof: 120**
DOMAL AZ 21X	1.7% Al 0.6% Zn Mg alloy: Rod, bar and forging; Domal **UTS: 210 Elon: 5% Proof: 90**
DOMAL AZ 31X	3% Al 1% Zn Mg alloy: Rod, tube and forging; Domal **UTS: 220 Elon: 8% Proof: 120**
DOMAL AZ 61X	6.2% Al 1.1% Zn Mg alloy: Rod, tube and forging; Domal **UTS: 250 Elon: 7% Proof: 120**
DOMAL AZ 80X	9% Al 0.6% Zn Mg alloy: Bar and forging; Domal; as forged **UTS: 300 Elon: 6% Proof: 180**
DOMAL AZ 80X	9% Al 0.6% Zn Mg alloy: Bar and forging; Domal; aged **UTS: 330 Elon: 3% Proof: 220**
DOMAL K 1	0.8% Zr Mg alloy: Casting; Domal; as cast; high damping capacity **UTS: 140 Elon: 15% Proof: 450**

Note. The following abbreviations and units are used in the tables:

DPN	Hardness, diamond pyramid number
UTS	Ultimate tensile strength, N/mm^2
Elon	Elongation, %
Proof	0.1% proof strength, N/mm^2

$1\ N/mm^2 = 0.1\ hbar = 0.102\ kgf/mm^2 = 0.06475\ tonf/in.^2 = 145.04\ lbf/in.^2 = 1\ MPa$

See Appendix II for other abbreviations and conversion tables.

Symbol	Nominal analysis, supplier, condition and remarks.
DTD 88 C	6% Al 1% Zn 0.2% Mn Mg alloy: Forgings **UTS: 270 Elon: 8% Proof: 150**
DTD 118 B	1.5% Mn Mg alloy: Sheet and strip; annealed **UTS: 200 Elon: 5% Proof: 60**
DTD 140 C	1.5% Mn Mg alloy: Casting; as cast **UTS: 90 Elon: 3%**
DTD 142 A	1.5% Mn Mg alloy: Bar; as rolled **UTS: 220 Elon: 4% Proof: 120**
DTD 259 A	7% Al 1.5% Zn 0.3% Mn Mg alloy: Bar **UTS: 250 Elon: 10% Proof: 170**
DTD 348 A	6% Al 1% Zn 0.3% Mn Mg alloy: Tube **UTS: 250 Elon: 8%**
DTD 619	3% Zn 0.7% Zr Mg alloy: Forging; as forged **UTS: 290 Elon: 8% Proof: 200**
DTD 622 A	3% Zn 0.7% Zr Mg alloy: Bar; as rolled and stress released **UTS: 290 Elon: 10% Proof: 200**
DTD 626 B	3% Zn 0.5% Zr Mg alloy: Sheet; as rolled **UTS: 250 Elon: 8% Proof: 170**
DTD 684	8% Al 0.8% Zn 0.4% Mn Mg alloy: Casting; annealed for corrosion resistance purposes **UTS: 180 Elon: 4% Proof: 75**
DTD 690	8% Al 0.7% Zn 0.4% Mn Mg alloy: Casting; solution treated; chill cast **UTS: 220 Elon: 10% Proof: 70**
DTD 708	2.5% Zn 3% rare earth 1% Zr Mg alloy: Casting; chill cast; aged **UTS: 150 Elon: 3% Proof: 100**
DTD 711 A	4.5% Zn 0.7% Zr Mg alloy: Casting; chill cast **UTS: 220 Elon: 7%**
DTD 718	0.5% Zn 3% rare earth 0.7% Zr Mg alloy: Casting; chill cast and aged **UTS: 150 Elon: 3% Proof: 100**
DTD 721 A	4% Zn 0.7% Zr Mg alloy: Casting; chill cast; stress released **UTS: 240 Elon: 7% Proof: 200**
DTD 728	0.7% Zr 3% rare earth Mg alloy: Casting; chill cast and aged **UTS: 150 Elon: 3% Proof: 100**
DTD 729	3% Zn 0.7% Zr Mg alloy: Forging **UTS: 250 Elon: 8% Proof: 170**
DTD 732 A	3% Al 1% Zn 0.3% Mn Mg alloy: Sheet; annealed **UTS: 220 Elon: 12% Proof: 100**
DTD 737	1.5% Mn Mg alloy: Tube **UTS: 220 Elon: 4%**
DTD 738	4.2% Zn 1.5% rare earth 0.6% Zr Mg alloy: Casting; chill cast **UTS: 180 Elon: 4% Proof: 100**
DTD 742 A	3% Al 1% Zn 0.5% Mn Mg alloy: Sheet; cold rolled; ½-hard **UTS: 240 Elon: 8% Proof: 150**
DTD 748	4.2% Zn 0.6% Zr 1.5% rare earth Mg alloy: Casting; chill cast and stress released **UTS: 210 Elon: 4% Proof: 120**
DTD 749	7% Al 1.5% Zn 0.3% Mn Mg alloy: Bar **UTS: 210 Elon: 8% Proof: 140**
DTD 5001 A	1.2% Zn 0.6% Zr Mg alloy: Sheet; as rolled **UTS: 240 Elon: 8% Proof: 150**
DTD 5005	2.2% Zn 3% Th 0.7% Zr Mg alloy: Casting; chill cast and stress released **UTS: 180 Elon: 5% Proof: 75**
DTD 5011	1.3% Zn 0.7% Zr Mg alloy: Bar; stress released **UTS: 250 Elon: 10% Proof: 170**
DTD 5015	5.5% Zn 1.8% Th 0.7% Zr Mg alloy: Casting; chill cast; stress released **UTS: 250 Elon: 5% Proof: 140**
DTD 5021	1.3% Zn 0.7% Zr Mg alloy: Tube; stress released **UTS: 240 Elon: 5% Proof: 170**
DTD 5025	2.5% Ag 1.6% rare earth 0.6% Zr Mg alloy: Casting
DTD 5031	5.5% Zn 0.7% Zr Mg alloy: Bar; stress released **UTS: 290 Elon: 10% Proof: 200**
DTD 5035	2.5% Ag 2.5% rare earth 0.6% Zr Mg alloy: Casting; chill cast; solution treated and aged **UTS: 240 Elon: 2% Proof: 170**
DTD 5041	5.5% Zn 0.7% Zr Mg alloy: Bar; stress released **UTS: 300 Elon: 8% Proof: 210**
DTD 5051	1.5% Mn Mg alloy: Plate; annealed **UTS: 200 Elon: 5% Proof: 60**
DTD 5055	0.8% Zr 2.2% rare earth 2.5% Ag Mg alloy: Casting
DTD 5061	3% Al 1% Zn 0.3% Mn Mg alloy: Plate; as rolled **UTS: 220 Elon: 10% Proof: 120**
DTD 5071	1.3% Zn 0.6% Zr Mg alloy: Plate; as rolled; weldable **UTS: 220 Elon: 12% Proof: 120**
DTD 5081	3% Zn 0.6% Zr Mg alloy: Plate; as rolled **UTS: 250 Elon: 10% Proof: 170**
DTD 5091	2.0% Mg 1.0% Mn Mg alloy: Sheet and strip; annealed
DTD 5101	0.2% Zn 1.0% Mn Mg alloy: Sheet and strip; ½-hard
DTD 5111	0.6% Zn 0.7% Zr 0.8% Th Mg alloy: Wrought
ELEKTRON MCZ	0.7% Zr 3% rare earth Mg alloy: Origin unknown
ELEKTRON Z5Z	0.7% Zr 4.5% Zn Mg alloy: Origin unknown
ELEKTRON ZCM	0.5% Mn 2.8% Cr 6% Zn Mg alloy: Casting; origin unknown
ELEKTRON ZZ	0.7% Zr 3.0% Zn Mg alloy: Origin unknown
ER AZ61A	6.5% Al 1% Zn 0.0005% Be Mg alloy: Weld electrode; designation used by AWS
F3	3% Al 1% Zn 0.3% Mn Mg alloy: Wrought; commercial; French designation
F10	10% Al 0.7% Zn 0.3% Mn Mg alloy: Casting; commercial; French designation
FT	8% Al 0.7% Zn 0.3% Mn Mg alloy: Casting; commercial; French designation
G A 3 Z1	3% Al 1% Zn 0.5% Mn Mg alloy: Wrought; French specification
G A 6 Z1	6% Al 1.2% Zn 0.2% Mn Mg alloy: Wrought; French specification
G A 9	8% Al 0.7% Zn 0.3% Mn Mg alloy: Casting; French specification
G A 9 Z1	10% Al 0.7% Zn 0.3% Mn Mg alloy: Casting; French specification
G Ag 25 TR	2.5% Ag 2.5% rare earth 0.6% Zr Mg alloy: Casting; French specification
G M2	1.5% Mn Mg alloy: Wrought; French specification
G Mg Al 6 Zn 3	6% Al 3% Zn 0.2% Mn Mg alloy: Casting; designation used by German Standards
G Z 4 TR	4.5% Zn 0.7% Zr 1.2% rare earth Mg alloy: Casting; French specification
G Z 5 Zr	4.5% Zn 0.7% Zr Mg alloy: Casting; French specification
GD Mg Al Zn 1	8% Al 0.8% Zn 0.2% Mn Mg alloy: Casting; designation used by German Standards
GD Mg Al 9 Zn 1	9% Al 0.8% Zr 0.2% Mn Mg alloy: Casting; designation used by German Standards
GD Mg Al 9 Zn 2	8.5% Al 1.5% Zn 0.2% Mn Mg alloy: Casting; designation used by German Standards
GK Mg Al 8 Zn 1	8% Al 0.8% Zn 0.2% Mn Mg alloy: Casting; designation used by German Standards
GK Mg Al 9 Zn 1	9% Al 0.8% Zn 0.2% Mn Mg alloy: Casting; designation used by German Standards
GK Mg Al 9 Zn 2	8.5% Al 1.5% Zn 0.2% Mn Mg alloy: Casting; designation used by German Standards
HK 31	3% Th 0.7% Zr Mg alloy: Casting; Domal; solution treated and aged **UTS: 18 Elon: 4% Proof: 75**
HZ 32	2% Zn 3.2% Th 0.7% Zr Mg alloy: Casting; Domal; aged **UTS: 180 Elon: 4% Proof: 75**
HZ 32 A	2% Zn 3.2% Th 0.7% Zr Mg alloy: Casting
K 1	0.8% Zr Mg alloy: Casting; Domal; as cast; high damping capacity **UTS: 140 Elon: 15% Proof: 45**
KIA	0.7% Zr Mg alloy: Casting

Symbol	Nominal analysis, supplier, condition and remarks.
L121	8% Al 0.7% Zn 0.3% Mn Mg alloy: Casting; chill cast; as cast; designation used by British Standards **UTS: 180 Elon: 4% Proof: 7.5**
L122	8% Al 0.7% Zn 0.3% Mn Mg alloy: Casting; chill cast; solution treated; designation used by British Standards **UTS: 220 Elon: 10% Proof: 60**
L123	10% Al 0.7% Zn 0.3% Mn Mg alloy: Casting; chill cast; as cast; designation used by British Standards **UTS: 170 Elon: 2% Proof: 90**
L124	10% Al 0.7% Zn 0.3% Mn Mg alloy: Casting; chill cast; solution treated; designation used by British Standards **UTS: 210 Elon: 5% Proof: 75**
L125	10% Al 0.7% Zn 0.3% Mn Mg alloy: Casting; chill cast; solution treated and aged; designation used by British Standards **UTS: 210 Elon: 2% Proof: 100**
L126	2% Zn 0.8% Zr 3% rare earth Mg alloy: Casting; designation used by British Standards
L127	4.5% Zn 0.8% Zr Mg alloy: Casting; designation used by British Standards
L128	4.2% Zn 0.8% Zr 1.2% rare earth Mg alloy: Casting; designation used by British Standards
L503	6% Al 1% Zn Mg alloy: Wrought; designation used by British Standards
L504	3.2% Zn 0.6% Zr Mg alloy: Wrought; designation used by British Standards
L505	3.2% Zn 0.6% Zr Mg alloy: Wrought; designation used by British Standards
L508	1.2% Zn 0.6% Zr Mg alloy: Wrought; designation used by British Standards
L512	6% Al 1.2% Zn Mg alloy: Wrought; designation used by British Standards
L513	6% Al 1% Zn Mg alloy: Wrought; designation used by British Standards
L514	3.2% Zn 0.7% Zr Mg alloy: Wrought; designation used by British Standards
L515	1.2% Zn 0.6% Zr Mg alloy: Wrought; designation used by British Standards
M 1	6% Al 1% Zn 0.25% Mn Mg alloy: Wrought; commercial designation; French
M 2	1.8% Mn Mg alloy: Wrought; previous German designation
MAG 1	8% Al Mg alloy: Rod; for gas shielded arc welding; designation for BS 2901
MAG 3	10% Al 0.5% Zn Mg alloy: Rod; for gas shielded arc welding; designation for BS 2901
MAG 5	4% Zn Mg alloy: Rod; for gas shielded arc welding; designation for BS 2901
MAG 6	2% Zn 3% rare earth Mg alloy: Rod; for gas shielded arc welding; designation for BS 2901
MAG 8	2% Zn 3% Th Mg alloy: Rod; for gas shielded arc welding; designation for BS 2901
MAG 9	5.5% Zn 2% Th Mg alloy: Rod; for gas shielded arc welding; designation for BS 2901
MAG 111	3% Al 1% Zn Mg alloy: Rod; for gas shielded arc welding; designation for BS 2901
MAG 141	1% Zn Mg alloy: Rod; for gas shielded arc welding; designation for BS 2901

Note. The following abbreviations and units are used in the tables:

DPN	Hardness, diamond pyramid number
UTS	Ultimate tensile strength, N/mm^2
Elon	Elongation, %
Proof	0.1% proof strength, N/mm^2

$1\ N/mm^2 = 0.1\ hbar = 0.102\ kgf/mm^2 = 0.06475\ tonf/in.^2 = 145.04\ lbf/in.^2 = 1\ MPa$

See Appendix II for other abbreviations and conversion tables.

Symbol	Nominal analysis, supplier, condition and remarks.
MAGNOX A 12	1% Al 0.01% Be Mg alloy: MEL; for nuclear purposes
MAGNUMINIUM 133	1.5% Mn Mg alloy: Tube, forgings and extrusions; High Duty Alloys for BS alloy MAG 101 **UTS: 210 Elon: 4% Proof: 120**
MAGNUMINIUM 133 X	0.65% Mn Mg alloy: Bar and forgings; High Duty Alloys; as drawn or forged **UTS: 240 Elon: 5% Proof: 120**
MAGNUMINIUM 266	6% Al 1% Zn 0.3% Mn Mg alloy: Bar and forging; High Duty Alloys for BS alloy MAG 121 **UTS: 270 Elon: 10% Proof: 140**
MG C 42	4.5% Zn 0.7% Zr Mg alloy: Casting; AECMA
MG C 43	4.5% Zn 0.7% Zr 1.2% rare earth Mg alloy: Casting; AECMA
MG C 51	2.5% Ag 2.5% rare earth 0.6% Zr Mg alloy: Casting; AECMA
MG C 61	8% Al 0.7% Zr 0.3% Mn Mg alloy: Casting; AECMA
MG C 91	2.3% Zn 0.6% Zr 3% rare earth Mg alloy: Casting; origin unknown
MG P 43	3% Zn 0.6% Zr Mg alloy: Wrought; AECMA
MG P 62	8% Al 0.7% Zn 0.3% Mn Mg alloy: Casting; AECMA
MG P 63	6% Al 1% Zn 0.3% Mn Mg alloy: Casting; AECMA
MIL M 8916	3% Th Mg alloy: Wrought; military specification USA
MIL M 8917	0.8% Mn 2% Th Mg alloy: Wrought; military specification USA
MIL M 26075	0.6% Zr 3.2% Th Mg alloy: Wrought; military specification USA
MIL M 45207	0.6% Zr Mg alloy: Casting; military specification USA
MIL M 46037	1.2% Zn 0.2% rare earth Mg alloy: Wrought; military specification USA
MIL M 46062	9% Al 0.6% Zn Mg alloy: Casting; military specification USA
MIL M 46062 B	10% Al 0.7% Zn 0.3% Mn Mg alloy: Casting; military specification USA
MIL R 6944	6.5% Al 1.2% Zn Mg alloy: Wrought; military specification USA
MNC 46E	Swedish Standard; summary of magnesium and magnesium alloy ingots and castings
MSR A	0.6% Zr 2.5% Ag 1.7% rare earth Mg alloy: Casting; MEL; solution treated and aged **UTS: 220 Elon: 4% Proof: 150**
MSR B	0.6% Zr 2.5% Ag 2.5% rare earth Mg alloy: Casting; MEL; solution treated and aged; weldable; used up to 250 °C **UTS: 220 Elon: 2% Proof: 170**
MTZ	0.7% Zr 3% Th Mg alloy: Casting; MEL; solution treated and aged; weldable; used up to 350 °C **UTS: 200 Elon: 5% Proof: 75**
QE 22	0.5% Cu 1.5% Ag 0.7% Zr 2.2% rare earth Mg alloy: Casting; designation used in USA
QE 22A	2.2% rare earth 0.8% Zr Mg alloy: Casting
QH 21	0.25% Ag 0.8% Zr 1% rare earth Mg alloy: Casting; designation used in USA
QH 21A	2.5% Ag 1.2% Th 0.8% Zr 1% rare earth Mg alloy: Casting; designation used in USA
QQ M 31	3% Al 1% Zn Mg alloy: Wrought; military specification USA
QQ M 31 B	1.5% Mn Mg alloy: Wrought; US Federal specification
QQ M 38	9% Al 0.7% Zn Mg alloy: Casting; military specification USA
QQ M 40	3% Al 1% Zn Mg alloy: Wrought; military specification
QQ M 40 B	1.5% Mn Mg alloy: Wrought; US Federal specification
QQ M 44	3% Al 1% Zn 0.5% Mn Mg alloy: Wrought; US military specification
QQ M 54	1.5% Mn Mg alloy: Wrought; US Federal specification
QQ M 55	10% Al Mg alloy: Casting; military specification
QQ M 56	6% Al 3% Zn Mg alloy: Casting; military specification
QQ M 56 B	8% Al 0.7% Zn 0.3% Mn Mg alloy: Casting; US Federal specification

Symbol	Nominal analysis, supplier, condition and remarks.
RZ 5	4% Zn 0.7% Zr 1.2% rare earth Mg alloy: Casting; MEL; weldable **UTS: 200 Elon: 4% Proof: 120**
SAE 50	6% Al 3% Zn Mg alloy: Casting; ASTM alloy AZ 63 A; solution treated and aged **UTS: 220 Elon: 3% Proof: 90**
SAE 51	1.2% Mn Mg alloy: Sheet; annealed; ASTM alloy M 1 A **UTS: 250 Elon: 12%**
SAE 52	3% Al 1% Zn Mg alloy: Extrusion; ASTM alloy AZ 31 B; as extruded **UTS: 210 Elon: 7% Proof: 120**
SAE 53	3.5% Al 5% Sn Mg alloy: As forged; ASTM alloy TA 54 A **UTS: 240 Elon: 7% Proof: 140**
SAE 500	9% Al 2% Zn Mg alloy: Casting; ASTM alloy AZ 92 A; solution treated and aged **UTS: 220 Elon: 1% Proof: 100**
SAE 501	9% Al 0.7% Zn Mg alloy: Die casting; low Cu content; as cast; ASTM alloy AZ 91 A **UTS: 220 Elon: 3% Proof: 140**
SAE 501A	9% Al 0.7% Zn 0.3% Cu Mg alloy: Die casting; as cast; ASTM alloy AZ 91 B **UTS: 220 Elon: 3% Proof: 140**
SAE 502	10% Al Mg alloy: Casting; solution treated and aged; ASTM alloy AM 100 A **UTS: 220 Proof: 100**
SAE 503	9% Al 2% Zn Mg alloy: Casting; solution treated and aged; ASTM alloy AZ 92 A **UTS: 220 Proof: 100**
SAE 504	8.7% Al 0.8% Zn Mg alloy: Sand casting; solution treated and aged; ASTM alloy AZ 91 C **UTS: 220 Elon: 3% Proof: 90**
SAE 505	7.5% Al 0.7% Zn Mg alloy: Casting; solution treated **UTS: 220 Elon: 7% Proof: 45**
SAE 506	2.5% Zn 3% rare earth 0.7% Zr Mg alloy: Casting; aged **UTS: 120 Elon: 2% Proof: 75**
SAE 507	3.2% Th 0.7% Zr Mg alloy: Casting; solution treated and aged **UTS: 180 Elon: 4% Proof: 75**
SAE 507	3.2% Th 0.7% Zr Mg alloy: Wrought; annealed **UTS: 200 Elon: 12% Proof: 90**
SAE 508	5.7% Zn 2% Th 0.7% Zr Mg alloy: Casting; aged **UTS: 220 Elon: 4% Proof: 140**
SAE 509	4.5% Zn 0.7% Zr Mg alloy: Casting; aged **UTS: 220 Elon: 5% Proof: 120**
SAE 510	3% Al 1% Zn Mg alloy: Sheet; annealed **UTS: 270 Elon: 12%**
SAE 510	3% Al 1% Zn Mg alloy: Sheet; cold rolled **UTS: 270 Elon: 4% Proof: 200**
SAE 513	6% Zn 0.8% Zr Mg alloy: Casting; solution treated and aged **UTS: 290 Elon: 5% Proof: 170**
SAE 520	7% Al 1% Zn Mg alloy: Extrusion; ASTM alloy AZ 61 A; as extruded **UTS: 250 Elon: 8% Proof: 170**
SAE 522	1.2% Mn Mg alloy: Extrusion; ASTM alloy M 1 A; as extruded **UTS: 200 Elon: 2%**
SAE 523	9% Al 0.6% Zn Mg alloy: Extrusion; ASTM alloy AZ 80 A; aged **UTS: 320 Elon: 4% Proof: 200**
SAE 524	5% Zn Mg alloy: Extrusion; ASTM alloy ZK 60 A; as extruded **UTS: 320 Elon: 4% Proof: 250**
SAE 531	6.5% Al 1% Zn Mg alloy: As forged; ASTM alloy AZ 61 A **UTS: 250 Elon: 6% Proof: 140**
SAE 532	8.5% Al 0.6% Zn Mg alloy: As forged; ASTM alloy AZ 80 A **UTS: 290 Elon: 5% Proof: 170**
SAE 533	1.2% Mn Mg alloy: Forging; as forged; ASTM alloy M 1 A **UTS: 200 Elon: 3% Proof: 90**
SAE 534	1.2% Zn 0.19% rare earth Mg alloy: Wrought; annealed **UTS: 210 Elon: 5%**
SAE 534	1.2% Zn 0.19% rare earth Mg alloy: Wrought; ½-hard **UTS: 220 Elon: 4%**
SS 4635	9% Al 0.4% Mn 0.5% Zn 0.001% Be Mg alloy: Casting; Swedish Standard; as cast
SS 4637	8% Al 0.4% Mn 0.5% Zn Mg alloy: Casting; Swedish Standard
SS 4640	8% Al 0.2% Mn 1.2% Zn Mg alloy: Casting; Swedish Standard
T2	1.5% Mn Mg alloy: Wrought; commercial designation; French
TZ 6	5.5% Zn 0.7% Zr 1.3% Th Mg alloy: Casting; MEL; aged; weldable **UTS: 270 Elon: 8% Proof: 150**
WW T 825	3% Al 1% Zn 0.5% Mn Mg alloy: Wrought; US military specification
WW T 825	3% Al 1% Zn Mg alloy: Wrought; US military specification
WW T 825 B	1.5% Mn Mg alloy: Wrought; US Federal specification
Z 52	4.5% Zn 0.7% Zr Mg alloy: Casting; MEL; aged **UTS: 240 Elon: 7% Proof: 140**
ZA	0.55% Zr Mg alloy: MEL for nuclear engineering
ZC 63	2.7% Cu 0.5% Mn 6% Zn Mg alloy: Casting; designation used in USA
ZCM 630	0.5% Mn 2.8% Cu 6% Zn Mg alloy: Casting; origin unknown
ZE 41 A	4.2% Zn 1.2% rare earth 0.8% Zr Mg alloy: Casting
ZE 63 A	6% Zn 2.6% rare earth 0.8% Zr Mg alloy: Casting
ZH 62	5.8% Zn 1.8% Th 0.7% Zr Mg alloy: Casting; Domal; aged **UTS: 240 Elon: 4% Proof: 150**
ZK 60	5.5% Zn 0.5% Zr Mg alloy: Bar and forging; Domal; aged **UTS: 320 Elon: 4% Proof: 240**
ZK 60	5.5% Zn 0.5% Zr Mg alloy: Casting; Domal; solution treated and aged **UTS: 270 Elon: 5% Proof: 170**
ZK 60	5.5% Zn 0.5% Zr Mg alloy: Previous German designation
ZK 61	5.5% Zn 0.5% Zr Mg alloy: Bar and forging; Domal; aged **UTS: 320 Elon: 4% Proof: 240**
ZK 61	5.5% Zn 0.5% Zr Mg alloy: Casting; Domal; solution treated and aged **UTS: 270 Elon: 5% Proof: 170**
ZRE 1	2.2% Zn 0.6% Zr 2.7% rare earth Mg alloy: Casting; MEL; annealed; weldable; creep resistant to 250 °C **UTS: 150 Elon: 4% Proof: 75**
ZT 1	2.2% Zn 0.7% Zr 3% Th Mg alloy: Casting; MEL; aged; weldable; used up to 350 °C **UTS: 200 Elon: 7% Proof: 70**
ZTY	0.5% Zn 0.6% Zr 0.75% Th Mg alloy: Sheet and forgings; MEL; as wrought; weldable **UTS: 200 Elon: 9% Proof: 100**
ZW 1	1.2% Zn 0.6% Zr Mg alloy: Sheet, bar and tube; Birmabright for BS alloy 141
ZW 1	1.3% Zn 0.6% Zr Mg alloy: Sheet, plate and tube; MEL; as wrought; weldable **UTS: 220 Elon: 8% Proof: 140**
ZW 3	3% Zn 0.75% Zr Mg alloy: Bar and forging; High Duty Alloys for BS alloy MAG 151
ZW 3	3% Zn 0.6% Zr Mg alloy: Sheet, bar and forging; Birmabright for BS alloy 151

Symbol	Nominal analysis, supplier, condition and remarks.
ZW 3	3% Zn 0.6% Zr Mg alloy: Sheet, plate and forging; MEL; as wrought; weldable **UTS: 270 Elon: 8% Proof: 180**
ZW 6	5.5% Zn 0.6% Zr Mg alloy: Bar; Birmabright for BS alloy 161
ZW 6	5.5% Zn 0.6% Zr Mg alloy: Extrusion; MEL; aged; not weldable **UTS: 300 Elon: 8% Proof: 210**

Note. The following abbreviations and units are used in the tables:

DPN	Hardness, diamond pyramid number
UTS	Ultimate tensile strength, N/mm^2
Elon	Elongation, %
Proof	0.1% proof strength, N/mm^2

1 N/mm^2=0.1 hbar=0.102 kgf/mm^2=0.06475 $tonf/in.^2$=145.04 $lbf/in.^2$=1 MPa

See Appendix II for other abbreviations and conversion tables.

24. Manganese Mn

Physical properties	
Atomic number	25
Atomic weight	54.93
Crystal structure	Cubic – complex – other structures also known
Colour	White-grey
Specific gravity	7.44
Density	7440 kg/m^3
Melting point	1244 °C
Boiling point	2150 °C
Specific heat	0.448 J/g °C
Thermal conductivity	8 W/m °C
Coefficient of linear expansion (20–100 °C)	22.8×10^{-6}/ °C
Latent heat of fusion	271 J/g
Latent heat of vaporization	4091 J/g
Thermal neutron absorption cross-section	12.6 barns/atom
Electrical conductivity	5.8% IACS (copper 100%)
Specific resistance	280 microhm mm
Temperature coefficient of electrical resistance	0.0039/ °C
Electrochemical equivalent	1.025 g/A/h – divalent 0.684 g/A/h – trivalent
Electrode potential	−1.05 V
Magnetic susceptibility	11.8×10^{-6}
Young's modulus of elasticity	159×10^9 N/m^2
Tensile strength	480 N/mm^2
Hardness	500 DPN

24.1 General notes on manganese

Native manganese does not exist, but several oxides and hydrated oxides are found in various parts of the world, particularly in the USSR.

High purity manganese metal is hard and brittle, and under certain circumstances has a yellowish tinge. The freshly fractured or cut surfaces tarnish slowly when left in air.

Manganese has no known use, nor does it form the base of many useful alloys. Two which have been produced are a braze metal with cobalt and nickel which has been used to braze stainless steel under certain circumstances and an alloy with copper, nickel and aluminium produced because of its high damping characteristics.

Neither of these alloys is in common use and very little information is available.

Manganese, however, is widely used as an alloying element. The most common use is as a de-oxidizer and de-sulphurizer in the manufacture of steel. Iron oxide and sulphide are brittle refractory materials which tear the plastic steel during rolling or forging. With manganese present these are converted to manganese oxide and sulphide, both of which are plastic at the working temperature of steel.

Up to approximately 1% manganese can be present in steel without affecting its properties, except as described above. With austenitic steels there is generally 3–5% manganese present. Additional manganese acts first as a pearlite former in carbon steels and with small amounts of molybdenum present gives air hardening steels.

When above about 12% manganese is present the steel becomes austenitic, but is unstable and readily converted to martensite by cold working.

Manganese also de-sulphurizes copper alloys and when added to brasses imparts a pleasant colour and improves the corrosion resistance with some increase in strength. Higher manganese–copper alloys can have a very low coefficient of expansion.

Additions of manganese to nickel give alloys with a special temperature coefficient of resistance.

Aluminium with approximately 2% manganese has a corrosion resistance almost equal to that of the pure metal, with an increase in strength which can be considerably increased by cold work. Manganese also adds to the strength of the more complex age hardening alloys.

Magnesium has its corrosion resistance improved with additions of small quantities of manganese.

Manganese oxide dust can be a health hazard under certain conditions. There is no evidence that any of the alloys with manganese has ever been dangerous, even when being ground or emery dressed.

Manganese is one of the more useful metallurgical elements, which makes itself felt over a wide range of alloys without forming any useful alloys in its own right. Following are listed the specifications and trade names of manganese and its alloys.

Symbol	Nominal analysis, supplier, condition and remarks.
AMS 4780	16% Ni 16% Co 0.8% B Mn alloy: Braze metal
ASTM A601 A	99.9% Mn metal: Regular grade; electrolytic
ASTM A601 B	99.9% Mn metal: Intermediate hydrogen; electrolytic
ASTM A601 C	99.9% Mn metal: Low hydrogen; electrolytic
ASTM A601 D	94.5% Mn metal 4.5% N: Electrolytic
ASTM A601 E	93.5% Mn metal 6.0% N: Electrolytic
ASTM A601 F	99.9% Mn metal: Powder
ASTM A701	0.08% C (max) 30.0% Si 0.05% P (max) 5% Fe Mn: Ferro-manganese silicon
FERRO-MANGANESE	0.1% C 3% Si 0.4% P 20% Fe Mn: Metal Alloys; primary metal
FERRO-MANGANESE	0.08% C 2% Si 0.08% P 20% Fe Mn low P: Metal Alloys; primary metal
h Mn 11	99.99% Mn flake: Light Ltd; high purity metal
MANGANESE	Mn metal: Carbon free; Blackwells; primary metal
MANGANESE 94–95	0.1% C 2.8% Fe 1% Si 1% Al 0.1% Cu Mn metal: Metal Alloys; primary metal
MANGANESE 96–97	0.1% C 1.2% Fe 0.8% Si 0.8% Al 0.1% Cu Mn metal: Metal Alloys; primary metal
SONOSTON	4% Al 2.5% Ni 30% Cu Mn alloy: Casting; Stone Manganese Marine; high damping capacity **UTS: 530** **Elon: 20%** **Proof: 220**

Note. The following abbreviations and units are used in the tables:

DPN	Hardness, diamond pyramid number
UTS	Ultimate tensile strength, N/mm^2
Elon	Elongation, %
Proof	0.1% proof strength, N/mm^2

1 N/mm^2=0.1 hbar=0.102 kgf/mm^2=0.06475 $tonf/in.^2$=145.04 $lbf/in.^2$=1 MPa

See Appendix II for other abbreviations and conversion tables.

25. Mercury Hg

Physical properties	
Atomic number	80
Atomic weight	200.61
Crystal structure	Rhombohedral
Colour	White
Specific gravity	13.55
Density	13 550 kg/m^3
Melting point	−38.87 °C
Boiling point	356.58 °C
Specific heat	0.1386 J/g °C
Thermal conductivity	8.8 W/m °C
Coefficient of linear expansion (20–100 °C)	61 × 10^{-6}/ °C
Latent heat of fusion	11.56 J/g
Latent heat of vaporization	292 J/g
Thermal neutron absorption cross-section	380 barns/atom
Electrical conductivity	1.9% IACS (copper 100%)
Specific resistance	940 microhm mm
Temperature coefficient of electrical resistance	0.0010/ °C
Electrochemical equivalent	7.45 g/A/h – monovalent 3.73 g/A/h – divalent
Electrode potential	+ 0.789 V
Magnetic susceptibility	−0.168 × 10^{-6}
Young's modulus of elasticity	This is not applicable to mercury

25.1 General notes on mercury

A limited amount of native mercury occurs trapped in the ore, which is composed of sulphides called cinnabar, found in areas of extinct volcanoes. The most famous deposit is in Spain where mercury has been mined and refined since 400 BC. Deposits are found all down the western spine of the American continent, but none are as rich or extensive as in the Spanish mines.

It is the only metal remaining liquid at normal temperatures, solidifying at −39 °C, when it is ductile, malleable and soft enough to cut with a knife.

The pure metal is used in laboratory type apparatus such as thermometers, and barometers, and has some industrial use as a gas sealing medium. It has always had a certain fascination for its 'quicksilver' qualities. Note should be taken that the metal itself and some of its compounds can be a serious health hazard.

Mercury can be used as a solvent for many metals including all the precious metal elements, nickel, cobalt and iron being the only common metals which are not dissolved. At one time use was made of this as a standard method of purification for gold, but it has been almost completely replaced by the cyanide process.

Tin amalgam has been used for silvering mirrors, and amalgams of gold, copper and zinc are still used in dentistry.

Zinc amalgam is used in dry batteries. The action only takes place when a circuit is completed; thus they find use in devices which operate at infrequent intervals where normal batteries would require changing to prevent corrosion.

Mercury is also used in vacuum rectifier equipment, and some use is now being made of the metal in lieu of steam to drive power turbines.

Most mercury is still used in chemical compounds for drugs, although a considerable amount is made into mercury fulminate for explosive detonators and railway fog warning devices.

Some mercury is used as the electrolyte for the production of chlorine and caustic soda.

Note. The following abbreviations and units are used in the tables:

DPN	Hardness, diamond pyramid number
UTS	Ultimate tensile strength, N/mm^2
Elon	Elongation, %
Proof	0.1% proof strength, N/mm^2

1 N/mm^2=0.1 hbar=0.102 kgf/mm^2=0.06475 tonf/in.^{2}=145.04 lbf/in.^{2}=1 MPa

See Appendix II for other abbreviations and conversion tables.

Symbol	Nominal analysis, supplier, condition and remarks.
AMALGAM	Hg and other metal: Name given to solution of metal in mercury
ARISTALOY	Sn and other metals Hg amalgam: Engelhard Industries; for dental fillings
p Hg 1	99.9999% Hg: Light Ltd; high purity metal
QUICKSILVER	Hg metal: Common name for mercury

26. Molybdenum Mo

Physical properties		
Atomic number		42
Atomic weight		95.95
Crystal structure		Body-centred cubic
Colour		Dull silver
Specific gravity		10.3
Density		10 300 kg/m^3
Melting point		2620 °C
Boiling point		5560 °C
Specific heat		0.255 J/g °C
Thermal conductivity		145 W/m °C
Coefficient of linear expansion (20–100 °C)		4.9×10^{-6}/ °C
Latent heat of fusion		293 J/g
Latent heat of vaporization		5610 J/g
Thermal neutron absorption cross-section		2.5 barns/atom
Electrical conductivity		34% IACS (copper 100%)
Specific resistance		517 microhm mm
Temperature coefficient of electrical resistance		0.0039/ °C
Electrochemical equivalent		1.790 g/A/h
Electrode potential		−0.2 V
Magnetic susceptibility		0.93×10^{-6} cm/g
Young's modulus of elasticity		324×10^9 N/m^2
Tensile strength	annealed	324×10^9 N/m^2
Hardness	annealed	230 DPN

26.1 General notes on molybdenum

Molybdenum is generally found as the sulphide ore molybdenite which occurs in local concentrations but is one of the rarer metal elements in the earth's crust. The largest and best known deposit is in Colorado, USA, where a mountain is gradually being removed to recover molybdenum ore containing approximately ten pounds metallic material per ton of mountainside.

Until recently molybdenum was available only as powder and powder compacts. It is now possible to cast the metal using the electric arc principle under high vacuum or inert gas. Ingots up to one ton are produced capable of being forged and rolled.

When heated above 500 °C it is necessary to exclude air from molybdenum to prevent formation of the oxide which does not form a protective coating. The rate of oxide formation is proportional to temperature and is very rapid above 1000 °C. Hot working can be carried out in air after heating in vacuum or inert atmosphere at about 1000 °C. Nitrogen rich atmospheres cause a brittle nitrided layer to form, but salt baths have been successfully used.

At present the largest use of molybdenum is as an alloying element in steel. The hardenability of plain carbon and low alloy steel is increased with as little as 0.2% of the metal. This means that larger sections can be hardened or the same size of section can be hardened with a more gentle quenching medium, for example oil in place of water or air instead of oil.

Molybdenum also reduces the danger of 'temper' or 'blue' brittleness which is a condition affecting plain carbon or nickel and chromium low alloy steels tempered in the range 300–500 °C.

Molybdenum in higher quantities up to 5% is present in almost all high speed tool steels, where hot hardness is necessary. The fact that molybdenum forms stable hard carbides with available carbon accounts for these qualities and the increased creep strength of nickel chromium molybdenum steel. Molybdenum has a similar effect to tungsten and vanadium but is generally more economic.

When added to austenitic stainless steel molybdenum considerably enhances the resistance of the metal to corrosive acids and marine atmospheres. The martensitic stainless steels have their hot strength and corrosion resistance improved.

Molybdenum is added to cast iron to increase the strength and heat resistance.

Molybdenum itself has a very low coefficient of friction which can be a problem at forging but is useful for preventing hot frettage where this is a problem. This property is also present in the disulphide compound, and considerable use is made now of molybdenum disulphide as an additive to lubricating oils and greases.

26A Molybdenum – all alloys

Specific gravity		10
Density		10 000 kg/m^3
Solidus/liquidus		2500–2600 °C
Thermal conductivity		Variable depending on alloy
Coefficient of linear expansion		Variable depending on alloy
Electrical conductivity		30% IACS (copper 100%)
Specific resistance		53 microhm mm
Young's modulus of elasticity		–
Impact		Variable
Fatigue strength	tensile	± 300 N/mm^2
Hot strength		300 N/mm^2 at 1200 °C in inert atmosphere

The above properties are typical of the following group, and may not apply exactly to any one specification. It is possible that with certain specifications some of the values may not be applicable.

General metallurgical characteristics

Molybdenum and its alloys are characterized by the ease of working and fabrication at room temperature with excellent mechanical properties at elevated temperatures, but very poor resistance to hot oxidation. The material is now available in most forms and in several alloys although the pure metal is still the most common.

Additions of small percentages of titanium and zirconium increase the creep strength, and raise the temperature required to remove cold work.

Molybdenum metal has a very low coefficient of friction which causes considerable difficulty when forging. The property, however, can be useful to prevent frettage, or excessive wear.

The properties of molybdenum alloys cannot be improved by heat treatment. Their surfaces can be carburized or nitrided, however, in a similar manner to steels.

Any cold work will be removed at a temperature of about 800 °C. It is possible to add sulphur to the surface of molybdenum giving a very low friction molybdenum disulphide.

Welding is difficult and requires either very careful control with inert gas shielding or preferably electron beam welding under high vacuum.

Machining of the alloys should present no problem.

These alloys are extremely expensive and as they are not oxidation resistant at high temperatures they find little use, although they have excellent hot strength.

Where the use is under controlled atmospheres or vacuum such as in valves or in controlled atmosphere furnaces, then these alloys find uses as support grids, furnace elements and general furniture for inside furnaces.

Some of the alloys find use as high temperature parts in jet engines and space rocket motors.

If a satisfactory technique can be devised to ensure the surface protection against oxidation at high temperatures, then use of molybdenum alloys will undoubtedly increase.

Symbol	Nominal analysis, supplier, condition and remarks.
AMS 78 CD	0.5% Ti 0.08% Zr Mo arc cast alloy: Forging; stress relieved; interim AMS
AMS 5662 B	19% Cr 5% Nb + Ta 3% Mo 1% Ti 19% Fe Mo alloy **DPN: 330 UTS: 1450 Proof: 1110**
AMS 5663 B	19% Cr 5% Nb + Ta 3% Mo 1% Ti 19% Fe Mo alloy **DPN: 330 UTS: 1450 Proof: 1110**
AMS 5664 A	19% Cr 5% Nb + Ta 3% Mo 1% Ti 19% Fe Mo alloy **DPN: 330 UTS: 1450 Proof: 1110**
AMS 7800	Mo-sintered alloy: Sheet and strip; stress relieved
AMS 7801	Mo arc cast alloy: Sheet and strip; stress relieved
AMS 7805	Mo arc cast alloy: Rods; stress relieved
AMS 7806	Mo-sintered alloy: Powder and forgings; stress relieved
AMS 7807	Mo arc cast alloy: Forgings; stress relieved
AMS 7811 A	0.5% Ti Mo arc cast alloy: Sheet and strip; stress relieved
AMS 7813	0.5% Ti Mo arc cast alloy: Bar; stress relieved
AMS 7817	0.03% C 0.47% Ti 0.09% Zr Mo alloy
AMS 7819	0.5% Ti 0.09% Zr Mo arc cast alloy: Bar; stress relieved
AMS 7819	0.02% C 0.47% Ti 0.09% Zr Mo alloy
CMX FB 30W 1	0.02% C 30% W Mo billets for forging: Climax **DPN: 200**
CMX FB TZM 1	0.03% C 0.5% Ti 0.08% Zr Mo alloy: Forging; Climax; annealed **DPN: 200**
CMX S 1	0.03% C Mo alloy: Sheet; produced by vacuum arc casting; Climax; stress relieved **UTS: 740 Elon: 11% Proof: 640**
CMX S TZM 1	0.02% C 0.5% Ti 0.1% Zr 99.25% Mo metal: Sheet; Climax **UTS: 1000 Elon: 7% Proof: 830**
CMX WB LC 1	0.005% C Mo alloy: Bar; produced by vacuum arc casting; Climax; stress relieved **DPN: 230 UTS: 420 Elon: 30% Proof: 210**
CMX WB T 1	0.03% C 0.5% Ti Mo alloy: Bar; wrought; Climax; stress relieved **DPN: 260 UTS: 480 Elon: 12% Proof: 240**
CMX WB TZM 1	0.06% C 0.5% Ti 0.08% Zr Mo alloy: Bar; Climax; stress relieved **DPN: 270 UTS: 560 Elon: 15% Proof: 320**

Note. The following abbreviations and units are used in the tables:

DPN	Hardness, diamond pyramid number
UTS	Ultimate tensile strength, N/mm^2
Elon	Elongation, %
Proof	0.1% proof strength, N/mm^2

1 N/mm^2=0.1 hbar=0.102 kgf/mm^2=0.06475 tonf/in.^{2}=145.04 lbf/in.^{2}=1 MPa

See Appendix II for other abbreviations and conversion tables.

Symbol	Nominal analysis, supplier, condition and remarks.
CMX WB TZM 2	0.02% C 0.5% Ti 0.1% Zr Mo alloy: Bar; Climax; as rolled and stress relieved **DPN: 270 UTS: 760 Elon: 10% Proof: 600**
ELKONITE 100 M	Mo for rivetting and welding electrodes: Mallory Metallurgical; electrical conductivity 30% IACS **DPN: 190 UTS: 530**
L MOLLA	99.99% Mo rod: 5 mm diameter; Light Ltd; high purity metal
METCO 63 NS	Pure Mo: Spray deposit; Metco
METCO 64	Mo: Spray deposit; Metco
METCO 350	C Fe Mo alloy: Casting deposit; Metco **DPN: 460**
METCO 350 C	C Fe Mo alloy: Casting deposit; Metco **DPN: 460**
METCO 501	Fe Ni Co Mo alloy: Spray deposit; Metco **DPN: 570**
METCO 505	Mo alloy: Deposit; self fluxing; Metco **DPN: 420**
MOLYBDENUM	Mo powder: Wire, rod, sheet, etc.; Blackwells; pure metal
MTC	0.5% Ti Mo alloy: Climax
R03601	0.04% C (max) 99.9% Mo sealing alloy: Designation used by UNS
R03602	0.005% C (max) 99.9% Mo sealing alloy
S Mo 1000	Mo metal: Fansteel Metallurgical
SPRABOND	99% Mo (min): Wire for spraying; Metco; as sprayed **DPN: 382 UTS: 45**
TZM 1	0.02% C 0.5% Ti 0.1% Zr 99.25% Mo metal: Climax; designation used for this specific analysis
UNS R3001–R3999	Mo metal and alloys: American Standards system

Note. The following abbreviations and units are used in the tables:

DPN	Hardness, diamond pyramid number
UTS	Ultimate tensile strength, N/mm^2
Elon	Elongation, %
Proof	0.1% proof strength, N/mm^2

1 N/mm^2=0.1 hbar=0.102 kgf/mm^2=0.06475 $tonf/in.^2$=145.04 $ibf/in.^2$=1 MPa

See Appendix II for other abbreviations and conversion tables.

27. Nickel Ni

Physical properties

Atomic number		28
Atomic weight		58.69
Crystal structure		Face-centred cubic
Colour		White
Specific gravity		8.88
Density		8880 kg/m^3
Melting point		1455 °C
Boiling point		3380 °C
Specific heat		0.46 J/g °C
Thermal conductivity		60.7 W/m °C
Coefficient of linear expansion (20–100 °C)		13.1×10^{-6}/ °C
Latent heat of fusion		305.6 J/g
Latent heat of vaporization		5862 J/g
Thermal neutron absorption cross-section		4.8 barns/atom
Electrical conductivity		26% IACS (copper 100%)
Specific resistance		64 microhm mm
Temperature coefficient of electrical resistance		0.0066/ °C
Electrochemical equivalent		1.0945 g/A/h
Electrode potential		−0.25 V
Magnetic susceptibility		Ferromagnetic
Young's modulus of elasticity		207×10^9 N/m^2
Tensile strength	annealed	45 N/mm^2
Hardness	annealed	75 DPN

27.1 General notes on nickel

Nickel is found as the sulphide, often associated with arsenic and copper and always with cobalt. Meteorites almost always contain some nickel, with iron and cobalt, but apart from this it is not found as the native metal.

A large deposit occurs in Sudbury, Canada, and others in Australia.

Nickel of 99.9% purity, almost free from cobalt and sulphur, is obtained by the Mond process, and this is the material used for the nickel base alloys.

The product however requires careful melting and deoxidizing before it can be rolled, forged or subjected to any degree of hot or cold work.

Information on commercially pure nickel and nickel alloys is given in the following sections.

Nickel is also used as an alloying element in the following fields.

Low alloy steels. Nickel is a ferrite strengthener and thus imparts toughness to steels, increasing the ductility without decreasing the tensile properties.

Engineering steels with 4% nickel were fairly commonplace until nickel scarcity and cost forced metallurgists to look for alternative alloying materials. It was found that almost comparable results could be obtained more economically with chromium, molybdenum and 1–2% nickel.

A range of steels with 9–10% nickel has now been developed which maintain ductility at very low temperatures, and a further series of high nickel, low carbon steels, known as maraging alloys, exhibit very high tensile properties with excellent ductility, weldability and ease of thermal treatment.

Austenitic stainless steels. It was found that by adding 8–10% nickel to the 12–18% chromium steels a completely stainless, austenitic steel was formed.

This now accounts for the largest single use of nickel. Details of these steels are given in Section 44N.

Cupro-nickels. Nickel and copper form a series of solid solutions over the complete range of analysis.

Details of these alloys will be found in Sections 14E and 27D.

Nickel also strengthens the alpha phase in copper–zinc alloys and almost all high strength, corrosion resistant brasses and bronzes contain nickel.

The series of 'copper–nickel–zinc' alloys and 'nickel silver' are discussed in Section 14F.

Aluminium alloys. Nickel enhances the high temperature properties of aluminium alloys, particularly the aluminium–copper series.

Cobalt alloys. Many of these complex heat and wear resistant alloys have nickel present, sometimes as the ductile matrix, in others to add malleability without detracting too much from other properties.

Cast iron. Nickel strengthens the ferrite in cast iron and considerably improves the heat resistance. Many special purpose nickel cast irons are now produced.

In general nickel can be said to be one of the most useful and versatile of all alloying elements, quite apart from the range of nickel based high temperature alloys and the many and varied applications found for the pure metal.

27.2 Nickel – notes on specifications

The following information may be of use in helping to find unknown specifications. Although it is only in comparatively modern times that nickel alloys have been used in large quantities in everyday engineering, many nickel based materials have had specialized use for many years. This may be the cause of the greater tendency to use the trade names of nickel alloys rather than national specifications, probably more than with any of the other modern materials. This has been helped by the relatively few producers using a small number of trade names which have the added advantage of being to some extent descriptive.

British Standards Institution (BS)

This body originally issued a series of individual specifications covering separate analysis and form. These have been largely withdrawn in favour of the system where a master specification is used to cover one form – bar, sheet, etc. – within which each different alloy is identified by its own designation. The master specifications are:

BS 3071	Nickel and nickel alloy castings
BS 3072	Nickel and nickel alloy sheet
BS 3073	Nickel and nickel alloy strip
BS 3074	Nickel and nickel alloy tube
BS 3075	Nickel and nickel alloy wire
BS 3076	Nickel and nickel alloy rod and section

To conserve space only the full title is listed in the classified section. The designation is included in the index, and referred to the BS range in the appropriate section.

American Society for Testing and Materials (ASTM)

This body issues a series of separate specifications, each prefixed by the letters ASTM followed by the letter B signifying a non-ferrous specification, the actual alloy being identified by a number sometimes qualified by letters. The final two numbers, for example 61, are for the year of issue, while the letter T shows the specification to be tentative. In the interests of clarity the year of issue is not shown in the classified section but is given in the index.

Society of Automotive Engineers (SAE)

The specifications issued by this body use the prefix SAE followed by two or three figures, sometimes having letters in addition to the numbers. There is no obvious system or code to indicate alloy content form.

This society also issues Aerospace Material Specifications which carry the prefix AMS followed by four figures. The suffix letter which sometimes appears is used to indicate the issue number. No attempt has been made to code or classify the alloys.

Unified Numbering System (UNS)

Information can be obtained from Society of Automotive Engineers, 400 Commonwealth Drive, Warrendale, PA 15096-0001, USA.

This system is described in ASTM E527. It covers only metals and alloys which have a commercial standing. Any organization can contact the UNS through the SAE at the above address to have new alloys included in UNS designations.

There are 18 series of numbers, each with a single prefix followed by five digits. The letter is intended to be suggestive of the family of metals identified.

Those underlined may be found in Section 44 – steel.

A	aluminium
C	copper
D	steels with specified mechanical properties
E	rare earths
F	cast irons and steels
G	AISI and SAE designated steels
H	AISI and SAE designated steels
J	AISI and SAE designated steels
K	AISI and SAE designated steels
L	low melting alloys
M	miscellaneous alloys
N	nickel alloys
P	precious metals and alloys
R	reactive and refractory metals and alloys
S	stainless and heat resistant steels
T	tool steels, AISI designated H steels, cast and wrought tool steels
W	weld filler materials
Z	zinc and zinc alloys

The UNS number is seldom a specification, but will almost always refer to an existing specification.

German Standard Specification (DIN)

These are issued under master specifications using the prefix DIN with a number, generally covering the form – bar, sheet, etc. The alloy analysis is then identified using two systems, the earliest having letters and figures to describe and code the analysis and form. The latest method uses the numbers 2.4000 to 2.4999. The first two figures after the point signify the alloy content and form, while the last two figures attempt to give the quantity of the principal alloying elements.

This system was developed for use with modern accounting machines and computers.

Trade names

These very often attempt to describe either the content or the use of the alloy. The trade names of the principal producers, Nimonic and Inconel, cover specific types of alloys, the number being used to identify the analysis of form, the higher numbers indicating more complex alloys.

27A Nickel – commercially pure – wrought and cast

Specific gravity		8.88
Density		8880 kg/m^3
Solidus/liquidus		1435–1445 °C
Thermal conductivity		60.7 W/m °C
Coefficient of linear expansion		13.3×10^{-6}/ °C
Electrical conductivity		19% IACS (copper 100%)
Specific resistance		91 microhm mm
Young's modulus of elasticity		207×10 N/m^2
Impact	cold rolled	160 J
Fatigue strength	cold rolled	±290 N/mm^2
Hot strength		

Temperature °C	Tensile strength N/mm^2	Elongation %
20	450	47
150	440	44.5
250	440	45
370	360	61.5
480	220	66

The above properties have been chosen to show typical values for the specification listed and may not apply exactly to any one specification. It is possible that with some specifications the values may not be applicable.

General metallurgical characteristics

The largest single use for pure nickel is as anodes for electroplating. These are generally cast with a controlled amount of oxide to produce a fine grain and are known as 'de-polarized anodes'. These dissolve more evenly, and thus more economically, than the high purity wrought nickel. Some use is now made of electrolytic nickel which has a sulphur content giving the same effect at lower cost.

Some of the materials listed are primary metals used in many industries as their source of nickel. Depending on the method of manufacture the nickels listed will have no contaminants or will have oxygen, carbon, sulphur, or cobalt present in small amounts.

With oxygen and carbon the cold working properties and weldability of nickel are adversely affected, while sulphur affects the corrosion resistance and weldability. Cobalt in small amounts has no serious effect on the properties of nickel.

The properties of nickel cannot be improved by any form of heat treatment. Nickel does cold work and where this must be removed it is necessary to go above 1000 °C. With oxygen and carbon absent nickel can accept a considerable amount of cold work, which increases the hardness without it becoming too brittle. Thin sheets can be produced with interstage annealing.

Nickel has excellent corrosion resistance and is not attacked by many acids even at elevated temperatures. An adherent oxide film is formed on the surface which successfully protects the metal from further oxidation even at very high temperatures.

Nickel has special electrical and magnetic properties caused by the peculiar configuration of its outer shell of electrons and resembles iron and cobalt in this respect.

These three metals, iron, nickel and cobalt, are the only ferromagnetic metals.

The adherent oxide which is self healing on nickel makes brazing and certain forms of welding difficult. Specialist advice should therefore be sought.

Nickel has excellent corrosion resistance, many acids having little or no effect upon it. This property makes it useful over a wide range of activities but its high cost precludes it from becoming a very popular metal.

Austenitic stainless steels are largely replacing nickel in many of the uses which it at one time had such as for laboratory equipment, spatulas, crucibles, balance weights, etc.

Much of the nickel used is as an electro-deposit beneath chromium. All chromium plating for decorative purposes has an underlay of nickel which is very much thicker than the chromium.

Electroless nickel is now being commonly used because it can be hardened and still retains good corrosion resistance, thus giving the rather unusual combination of excellent corrosion resistance and good abrasion resistance.

Nickel in massive form sometimes finds use for its magnetic properties but it is an unusual metal to be found in industry.

Symbol	Nominal analysis, supplier, condition and remarks.
1.4876	0.4% Ti Ni: German Standard
2.4050	99.8% Ni: German Standard
2.4051	0.02% Mg 99.8% Ni anode: German Standard
2.4052	0.05% Mg 99.7% Ni anode: German Standard
2.4053	0.07% Mg 99.7% Ni anode: German Standard
2.4056	0.2% Si 99.6% Ni anode: German Standard
2.4060	99.6% Ni: German Standard
2.4062	0.4% Fe 99.4% Ni: German Standard
2.4066	99.2% Ni: German Standard
2.4068	0.02% C 99.0% Ni: German Standard
2.4106	0.8% Mn 98% Ni: German Standard
99 ALLOY	99.8% Ni: Driver-Harris **UTS: 530**
200 ALLOY	0.25% Fe (max) 0.35% Mn (max) 0.07% C (max) Ni and Co: Carpenter; high points Ni
201 ALLOY	0.25% Fe (max) 0.35% Mn (max) 0.02% C (max) Ni and Co: Carpenter; high points Ni
205 ALLOY	0.03% Ti 0.35% Mn (max) 0.07% C (max) 0.06% Mg Ni: Carpenter; for magnetic purposes
A Nickel	0.06% C 99.5% Ni: Bar, sheet, etc.; Huntington; now Nickel 200
A Nickel	99.0% Ni (min): For electrical leads; H Wiggin; annealed; electrical conductivity 20% IACS **DPN: 95 UTS: 370 Elon: 45% Proof: 100**
ACTIVE Ni	0.2% Si Mg Ni: For thermionic valves; British Driver-Harris
ALLOY 200	0.1% C (max) 99.2% Ni: VDM
ALLOY 201	0.02% C (max) 99.0% Ni: VDM
ALLOY 205	0.02% C (max) 99.6% Ni: VDM
AMPCO 501	Nickel: Cast; Ampco **DPN: 130**
AMPCOLAY 502	Nickel: Wrought; Ampco for A Nickel
AMS 5865	0.02% C (max) 2.2% Th oxide Ni
AMS 5890	0.02% C (max) 2.2% Th oxide Ni
ANS A5 15 E Ni Ci	0.6% C Ni: Welding electrode
ASTM A636	0.5% Fe (max) 0.9% Cu (max) 1.3% Co (max) 75% Ni: Sinter for alloying
ASTM B39 A Shot	0.9% Fe 0.75% C 0.07% S 97.7% Ni: Shot; virgin metal
ASTM B39 Electrolytic	0.6% Fe 0.1% C 0.02% S 99.5% Ni: Virgin metal
ASTM B39 Ingot	98.5% Ni: Virgin metal
ASTM B39 X Shot	0.6% Fe 0.25% C 0.05% S 98.9% Ni: Shot; virgin metal
ASTM B160	0.4% Fe 0.2% Cu 0.3% Mn 0.15% C Ni: Bar; annealed **UTS: 370 Elon: 40% Proof: 75**
ASTM B161	0.4% Fe 0.25% Cu 0.3% Mn 0.15% C Ni: Tube; annealed **UTS: 370 Elon: 40% Proof: 90**
ASTM B162	0.4% Fe 0.25% Cu 0.3% Mn 0.15% C Ni: Sheet and strip; hard **UTS: 670 Elon: 2% Proof: 480**
ASTM B304 ER Ni 3	0.15% C 93.0% Ni (min): Welding electrode; previous designation ERN 61
ASTM B304 R Ni 2	0.15% C 97.0% Ni (min): Welding electrode; previous designation RN 41
AT NICKEL	0.4% Fe 0.25% Cu 0.15% C 0.01% S 99% Ni: Wrought; H Wiggin; annealed; sheet, bar and plate **DPN: 100 UTS: 370 Elon: 40% Proof: 20**

Note. The following abbreviations and units are used in the tables:

DPN Hardness, diamond pyramid number
UTS Ultimate tensile strength, N/mm^2
Elon Elongation, %
Proof 0.1% proof strength, N/mm^2

1 N/mm^2=0.1 hbar=0.102 kgf/mm^2=0.06475 tonf/in.^{2}=145.04 lbf/in.^{2}=1 MPa

See Appendix II for other abbreviations and conversion tables.

Symbol	Nominal analysis, supplier, condition and remarks.
AT NICKEL	0.4% Fe 0.25% Cu 0.15% C 0.01% S 99% Ni: Wrought; H Wiggin; cold rolled; sheet, bar and plate **DPN: 200 Elon: 10% Proof: 30**
BALLAST NICKEL	99.8% Ni (min): For current limiting controls; Driver-Harris
BS 375 A	0.1% C 99.5% Ni: Refined metal; primary metal
BS 558	98.5% Ni: For anodes; anodes for nickel plating
BS 1525	Nickel and nickel alloy: Sheet; replaced by BS 3072 and BS 3073
BS 1526	Nickel and nickel alloy: Sheet; replaced by BS 3072 and BS 3073
BS 1527	Nickel and nickel alloy: Sheet; replaced by BS 3072 and BS 3073
BS 1528	Nickel and nickel alloy: Bar; replaced by BS 3076
BS 1529	Nickel and nickel alloy: Bar; replaced by BS 3076
BS 1530	Nickel and nickel alloy: Bar; replaced by BS 3076
BS 1531	Nickel and nickel alloy: Tube; replaced by BS 3074
BS 1532	Nickel and nickel alloy: Tube; replaced by BS 3074
BS 1533	Nickel and nickel alloy: Tube; replaced by BS 3074
BS 1534	Nickel and nickel alloy: Wire; replaced by BS 3075
BS 1535	Nickel and nickel alloy: Wire; replaced by BS 3075
BS 1536	Nickel and nickel alloy: Wire; replaced by BS 3075
BS 2901 NA32	3% Ti Ni alloy: Rod; for gas shielded arc welding
BS 2901 NA32	3% Ti Ni alloy: Rod for all forms of welding
BS 3072 NA11	0.15% C (max) 99% Ni (min): Sheet; annealed **UTS: 390 Elon: 35% Proof: 90**
BS 3072 NA12	0.02% C (max) 99% Ni (min): Sheet; annealed **UTS: 360 Elon: 35% Proof: 90**
BS 3073 NA11	0.15% C (max) 99% Ni (min): Strip; annealed **DPN: 100 UTS: 390 Elon: 35% Proof: 90**
BS 3074 NA11	0.15% C (max) 99% Ni (min): Tube; annealed **UTS: 360 Elon: 40% Proof: 90**
BS 3074 NA11	0.15% C (max) 99% Ni (min): Tube; cold drawn **UTS: 390 Elon: 15% Proof: 270**
BS 3074 NA12	0.02% C (max) 99% Ni: Tube; annealed **UTS: 340 Elon: 40% Proof: 75**
BS 3074 NA12	0.02% C (max) 99% Ni: Tube; cold drawn **UTS: 400 Elon: 15% Proof: 210**
BS 3075 NA11	0.15% C (max) 99% Ni (min): Wire; annealed **UTS: 340 Elon: 25%**
BS 3076 NA11	0.15% C (max) 99% Ni (min): Rod and section; annealed **UTS: 390 Elon: 35% Proof: 90**
BS 3076 NA11	0.15% C (max) 99% Ni (min): Rod and section; cold drawn **UTS: 530 Elon: 15% Proof: 340**
BS 3076 NA12	0.02% C (max) 99% Ni: Bar and section; annealed **UTS: 330 Elon: 35% Proof: 60**
BS 3076 NA12	0.02% C (max) 99% Ni: Bar and section; cold drawn **UTS: 450 Elon: 15% Proof: 300**
BS 3975 NA11	0.15% C (max) 99% Ni (min): Wire; cold drawn **UTS: 480**
CARBONIZED NICKEL	99% Ni (min): For anode plates; Driver-Harris
CASTEES N	Ni core welding electrode; Philips; for welding cast iron
CATHODE NICKEL	99.5% Ni (min): For electronic valve cathodes; Driver-Harris
CI CAVITY FILL Ni	2% Fe 0.8% C Ni alloy: Electrode; Metrode; for cast iron
CI LOW CON Ni	2% Fe 0.8% C Ni alloy: Electrode; Metrode; for cast iron
CI SOFT FLOW Ni	2% Fe 0.8% C Ni alloy: Welding electrode; Metrode; for cast iron
CI-NI	1.0% C 97.0% Ni: Electrode; Metrode; for general repair welding
CORNES 11	99% Ni (min): Philip Cornes
CORNES 12	0.02% C (max) 99% Ni (min): Philip Cornes
CZ 100	3% Fe (max) 1.5% Mn (max) 1% C (max) Ni: Casting; US designation
CZ 100	1.25% Cu (max) 1% C (max) Ni: Casting; US designation

Symbol	Nominal analysis, supplier, condition and remarks.
DIN 1701 C Ni 98.5	0.5% C 0.01% S 98.5% Ni
DIN 1701 C Ni 99.5	0.13% C 0.001% S 99.5% Ni
DIN 1701 C Ni 99.8	0.08% C 0.0005% S 99.8% Ni
DIN 1701 E Ni 99.5	0.1% C 0.02% S 99.5% Ni
DIN 1701 E Ni 99.8	0.1% C 0.01% S 99.8% Ni
DIN 1701 M Ni 99.5	0.1% C 0.02% S 99.5% Ni
DIN 1701 W Ni 99	0.1% C 0.015% S 99.0% Ni
DIN 1702	High purity Ni for anodes
DIN 17740 LC Ni 99	0.02% C 99.0% Ni
DIN 17740 Ni 99.8	99.8% Ni
DIN 17741 Ni Mn 1	0.8% Mn 98% Ni
DURANICKEL	0.5% Ti 4.5% Al Ni alloy: High carbon bar; H Wiggin; for springs and dies; acid resistant **DPN: 295 Elon: 28%**
DURANICKEL	Ni with additions: Driver-Harris; melting point 1435 °C
E Ni 1	0.2% Ti 0.1% C (max) Ni: Weld electrode; designation used by AWS
GFA NICKEL	Low carbon 99.0% Ni (min): For anodes; H Wiggin
GRADE A	High purity nickel: For thermionic valves; British Driver-Harris
h Ni 8	99.999% Ni: Sponge; Light Ltd; high purity metal
h Ni 15	99.995% Ni: Wire 1 mm diameter; Light Ltd
HAI Ni 200	99% Ni (min): Harrison Alloy
HAI Ni 201	99% Ni (min): Harrison Alloy
HAI Ni 205	99% Ni (min): Harrison Alloy
HAI Ni 211	4.5% Mn Ni alloys: Harrison Alloy
HAI Ni 233	99% Ni (min): Harrison Alloy
HAI Ni 233 C	99% Ni: Carbonized; for anode plates; Harrison Alloy
HAI Ni 499 C	99.8% Ni: Carbonized; for anode plates; Harrison Alloy
HG NICKEL	0.06% C 99.0% Ni (min): Wire; H Wiggin; for electronic use
HPM	0.02% C 99.97% Ni: Bar and sheet; H Wiggin; now Nickel 270
METCO 56 C NS	High purity Ni: Powder deposit; casting; Metco **DPN: 85**
METCO 56 F NS	High purity Ni: For coating deposit; Metco **DPN: 85**
METCO NICKEL	99% Ni (min): Sprayed deposit; Metco **DPN: 110**
MNS NICKEL	High carbon 2.0% Mn Ni alloy: Rod and wire; H Wiggin; improved machinability
N 220 C	0.04% Mg 99% Ni (min): Commercially pure Ni
Ni 99.8 Mg	0.02% Mg 99.8% Ni: Anode; designation used by German Standards
NICKEL	High purity metal: Impurities less than 15 p.p.m.; Mallory Metallurgical; supplied as sponge, rod, wire and sheet
NICKEL	Ni: Cubes, sticks, powder and wire; Blackwells; pure metal
NICKEL 2	0.05% C 97% Ni: Welding electrode; Metrode
NICKEL 41	0.06% C 99.5% Ni: Filler rod; Huntington; for oxyacetylene welding of nickel
NICKEL 61	3% Ti 0.06% C 96% Ni: Filler rod; Huntington; inert gas shield welding of nickel
NICKEL 131	3% Ti 0.5% Al 0.25% C Ni alloy: Filler rod; Huntington; metal arc welding of nickel
NICKEL 141	2.2% Ti 0.25% Al 0.05% C Ni alloy: Filler rod; Huntington; for welding nickel and steel
NICKEL 200	0.06% C 99.5% Ni: Bar, sheet, forgings, etc.; Huntington; annealed **UTS: 450 Elon: 47% Proof: 120**
NICKEL 201	0.01% C 99.5% Ni: Bar, sheet, forging, etc.; Huntington
NICKEL 205	0.02% Ti 0.06% C 0.04% Mg 99.5% Ni: Bar, sheet, etc.; Huntington
NICKEL 213	High carbon 2.0% Mn Ni alloy: Bar and wire; H Wiggin; improved machinability
NICKEL 220	0.02% Ti 0.06% C 0.04% Mg 99.5% Ni: Bar and tube; Huntington
NICKEL 222	0.07% Mg 99.5% Ni (min): Tube and strip; H Wiggin; for electronic use
NICKEL 222	0.05% Mg Ni: Inco; for electronic use
NICKEL 225	0.02% Ti 0.06% C 0.04% Mg 99.5% Ni: Bar and tube; Huntington
NICKEL 230	0.003% Ti 0.09% C 0.06% Mg 99.5% Ni: Sheet; Huntington
NICKEL 233	0.003% Ti 0.09% C 0.07% Mg 99.5% Ni: Sheet and tube; Huntington
NICKEL 270	0.02% C 99.97% Ni: Bar, sheet and tube; Huntington
NI-ROD	3% Fe 1% C Ni alloy welding electrode: Huntington; welding cast iron, etc.
NIVAC P	0.007% C Ni; high purity metal: Crucible Steel; annealed; for electrical purposes **UTS: 270 Elon: 40% Proof: 210**
NIVAC P	0.007% C Ni; high purity metal: Crucible Steel; cold rolled **UTS: 600 Elon: 5% Proof: 580**
NORMAL Ni	High purity Ni for thermionic valves: British Driver-Harris; purity between that of passive and active Ni
PA 22 NICKEL	0.07% Mg 99.5% Ni (min): Tube and strip; H Wiggin; electronic use; now Nickel 222
PA 23 NICKEL	0.05% Mg 99.5% Ni (min): Tube and strip; H Wiggin; electronic use
PASSIVE Ni	High purity Ni for thermionic valves: British Driver-Harris
PERMANICKEL	Ni with additions: Driver-Harris
PERMANICKEL 300	0.5% Ti 0.25% C 0.35% Mg 98.6% Ni strip: Huntington; annealed **UTS: 800 Elon: 41% Proof: 300**
SOUDOFONTE B1	0.6% C Ni: Welding electrode; Soudometal; for welding cast iron **DTN: 160**
SOUDOFONTE B12	1.6% C Ni: Welding electrode; Soudometal; for welding cast iron **DTN: 160**
SOUDOFONTE B24	0.6% C Ni: Welding electrode; Soudometal; for welding cast iron **DTN: 155**
SOUDOR G Ni Ti	3.2% Ti 0.01% C Ni: Welding electrode; Soudometal **UTS: 420 Elon: 20% Proof: 200**
SOUDOTIG Ni Ti	3.2% Ti 0.01% C Ni: Welding electrode; Soudometal; for inert gas welding **UTS: 420 Elon: 20% Proof: 200**
STUBS 8	98% Ni: Welding electrode; Stubs; for welding cast iron **DTN: 175 UTS: 490 Elon: 30%**
STUBS 8 NC	98% Ni: Welding electrode; Stubs; for welding cast iron **DTN: 175 UTS: 490 Elon: 30%**
STUBS 80 Ni	2% Ti 0.014% C Ni: Welding electrode; Stubs; for welding nickel **UTS: 450 Elon: 30% Proof: 300**
VDM LC NICKEL 99.2	0.02% C (max) 99.2% Ni: Tube, sheet, etc.; VDM
VDM LC NICKEL 99.6	0.02% C (max) 99.6% Ni: Tube, sheet, etc.; VDM
VDM NICKEL 99.2	0.1% C 99.2% Ni: Tube, sheet, etc.; VDM
VDM NICKEL B 9604	2.5% Ti 0.05% C (max) Ni: Strip for welding; VDM
VDM NICKEL S 9604	2.5% Ti 0.05% C (max) Ni: Filler weld rod; VDM
X 10 Ni Cr	0.35% Al + Ti Ni: German Standard
Z-NICKEL	0.2% Fe 0.2% Si 0.02% Cu 0.2% Mn age hardening Ni alloy: Rod or strip; electrical conductivity 12% IACS; origin unknown **UTS: 690**

27B Nickel–chromium wrought and cast non-ageing alloys With iron and other non-ageing elements present

Specific gravity	8.37
Density	8370 kg/m^3
Solidus/liquidus	1380–1430 °C
Thermal conductivity	20.9 W/m °C
Coefficient of linear expansion (20–100 °C)	12.2 × 10^{-6}/ °C
Electrical conductivity	1.6% IACS (copper 100%)
Specific resistance	1100 microhm mm
Young's modulus of elasticity	186 × 10^9 N/m^2
Impact	108 J
Fatigue strength (10 × 10^6 cycles)	±180 N/mm^2
Hot strength	

Temperature °C	Tensile strength N/mm^2	Elongation %
20	780	44
200	740	40
300	740	40
500	720	37
600	520	29
700	340	29
800	200	85
900	100	80
1000	75	100

The above properties are included as a guide to the group of specifications as a whole. There is considerable variation within the group and some of these properties may not be applicable to all the specifications.

General metallurgical characteristics

Chromium strengthens the nickel matrix and by forming a stable, very adherent oxide considerably increases the hot corrosion resistance.

The chromium dissolves in the nickel forming a solid solution stable at room, as well as elevated temperatures. There are no phase changes and no intermetallic compounds to cause precipitation hardening to any measurable extent in the alloys listed.

Iron reduces the corrosion resistance and hot strength, but has some desirable electrical and magnetic properties, and is, of course, much cheaper than nickel.

The presence of carbon increases the strength but can cause trouble, as some of the chromium carbides precipitate under certain conditions, resulting in local brittleness. Additions of many elements cause formation of intermetallic compounds which can be precipitation age hardened. These alloys are described and listed in the groups following this section. The elements boron and silicon are often present as fluxing agents in those alloys which are applied by metal spraying or used as brazing materials. Tungsten carbide may also be present to strengthen and harden the alloy. Like all nickel alloys those listed are readily hardened by cold work and this is accompanied by a reduction in ductility.

The alloys listed all have excellent corrosion and acid resistance, are relatively easily welded, and maintain their strength and corrosion resistance at high temperatures. The creep strength of these materials is slightly superior to that of the austenitic steels but does not compare with that of the age hardening nickel chromium materials.

The properties of the alloys in this section cannot be improved by thermal treatment. They do, however, accept a considerable amount of cold work with an increase in tensile strength and a reduction in ductility. This cold work can be removed by holding at a temperature of above 1000 °C for quite a short time.

Joining by welding requires the use of inert gas shielding or very active corrosive fluxes which are not recommended.

The materials are difficult to machine, partly because they are inherently tough but also because of their cold working properties.

Specialist advice is usually required but in general it is necessary to make certain that the tool remains below the surface, cuts and does not rub.

The principal use for these materials is where hot oxide resistance is required. Their hot strength is not particularly high nor is their creep resistance but they have this excellent ability to withstand oxidation at temperature. This makes them useful for furnace elements, heating elements in electric fires, jigs and fixtures associated with heat treatment and combustion equipment.

Springs for high temperature use where the load is low are also commonly made in these materials.

Quite often the materials are sprayed or welded onto base materials which have a higher mechanical strength but less oxidation resistance.

The materials are expensive, however.

Symbol	Nominal analysis, supplier, condition and remarks.
2.4640	15% Cr 8% Fe Ni alloy: German Standard
2.4816	15.5% Cr 8% Fe 0.15% C Ni alloy: German Standard
2.4816	16% Cr 10% Fe 0.03% C (max) Ni alloy: German Standard
2.4816	16% Cr 10% Fe 0.05% C Ni alloy: German Standard
2.4858	21.5% Cr 30% Fe 3% Mo 2.2% Cu 0.05% C (max) Ni alloy: German Standard
2.4867	15% Cr 21% Fe Ni alloy: German Standard
2.4869	20% Cr 1.0% Si Ni alloy: German Standard
2.4870	10% Cr Ni alloy: German Standard
2.4951	19.5% Cr 5% Fe (max) 0.1% C 5% Co (max) Ni alloy: German Standard
14.75 Mn Nb	15.0% Cr 8.0% Fe 2.0% Nb 6.0% Mn 0.06% C Ni alloy: Electrode; Metrode; for welding austenitic to ferritic steel
14.75 Mo Nb	15.0% Cr 8.0% Fe 2.0% Mo 2.0% Nb 0.06% C Ni alloy: Electrode; Metrode; for welding Incoloy 800
14.75 Nb	15.0% Cr 8.0% Fe 2% Nb 0.06% C Ni alloy: Welding electrode; Metrode
15.60 Mn WNb	15.0% Cr 2.0% Nb 6.0% Mn 0.06% C 5.0% W Ni alloy: Electrode; Metrode
15.60 Nb	15.0% Cr 30% Fe 0.06% C Ni alloy: Welding electrode; Metrode
20.55 4Si Cu W	20.0% Cr 3.5% Mo 4.0% Si 4.0% Cu 0.06% C 2.0% W Ni alloy: Electrode; Metrode
22H	26% Cr 20% Fe 0.5% C 5% W Ni alloy: Casting; Thompson L'Hospied; annealed **DPN: 180 UTS: 460 Elon: 3%**
25.55.6 Cu W	25.0% Cr 6.0% Mo 6.0% Cu 0.06% C 1.5% W Ni alloy: Electrode; Metrode
50.50	50% Cr 0.08% C Ni alloy: Electrode; Metrode
50.50 Nb	50% Cr 1.5% Nb 0.08% C Ni alloy: Electrode; Metrode
55HSW	17% Cr 30% Fe 0.6% C 1.35% W Ni alloy: For furnace furniture, etc.; Wellman
80.20 ALLOY	20% Cr 0.2% C Ni alloy: Carpenter **DPN: 210 UTS: 830 Elon: 35% Proof: 410**
A FNDR 81.347 EF 2070 Ni Cr Mn Fe B 14020 BH	15.5% Cr 8% Fe 1.9% Nb 8% Mn 0.05% C Ni alloy: Welding electrode; French Standard
ACCOLAY	15% Cr Ni alloy: Source unknown
ACI HW	12% Cr 25% Fe 0.5% C Ni alloy: Origin unknown
ACI HX	17% Cr 13% Fe 0.4% C Ni alloy
ALLOY 75	20.5% Cr 5% Fe (max) 0.4% Ti 0.1% C Ni alloy: VDM
ALLOY 600	16% Cr 8.5% Fe 0.07% C Ni alloy: VDM
ALLOY 600 H	16% Cr 8% Fe 0.2% Al 0.3% C Ni alloy: VDM
ALLOY 601 H	23% Cr 14% Fe 0.4% Ti 1.5% Al 0.4% C Ni alloy: VDM
ALLOY 690	28.5% Cr 9% Fe 0.02% C (max) Ni alloy: VDM
ALL-STATE 8.60	57% Ni alloy: Welding electrode for repairing cast iron; All State
ALUMEL	5% Cr Ni alloy: Wire for thermocouples; source unknown
AMPCOLAY 531	15% Cr 7% Fe 0.15% C Ni alloy: Ampco
AMS 4775 A	16.5% Cr 4% Fe 3.8% B 4% Si Ni alloy: Bronze metal
AMS 4776	Low carbon 16.5% Cr 4% Fe 3.8% B 4% Si Ni alloy: Braze metal

Note. The following abbreviations and units are used in the tables:

DPN	Hardness, diamond pyramid number
UTS	Ultimate tensile strength, N/mm^2
Elon	Elongation, %
Proof	0.1% proof strength, N/mm^2

$1\ N/mm^2 = 0.1\ hbar = 0.102\ kgf/mm^2 = 0.06475\ tonf/in.^2 = 145.04\ lbf/in.^2 = 1\ MPa$

See Appendix II for other abbreviations and conversion tables.

Symbol	Nominal analysis, supplier, condition and remarks.
AMS 4777	7% Cr 3% Fe 3% B 4% Si Ni alloy: Braze metal
AMS 5540 F	15.5% Cr 8% Fe Ni alloy: Sheet and strip
AMS 5580 C	15.5% Cr 8% Fe Ni alloy: Seamless tube
AMS 5665 F	15.5% Cr 8% Fe Ni alloy: Bar and forging
AMS 5676	20% Cr Ni alloy: Welding wire
AMS 5677	20% Cr 1.6% Nb + Ta Ni alloy: Electrode; coated wire
AMS 5679 B	15.5% Cr 8% Fe 2% Nb + Ta Ni alloy: Welding wire; cold drawn
AMS 5682 A	20% Cr Ni alloy: Rod or wire coated electrode
AMS 5683 B	15.5% Cr 8% Fe Ni alloy: Welding wire
AMS 5684 B	15% Cr 9% Fe 2% Nb + Ta Ni alloy: Coated electrode
AMS 5687 C	15.5% Cr 8% Fe Ni alloy: Wire; annealed
AMS 5716	18.5% Cr 33% Fe 0.08% C (max) Ni alloy
AMS 5870	23% Cr 16% Fe 1.5% Al 0.1% C (max) Ni alloy
AMS 7232	15.5% Cr 8% Fe 0.15% C (max) Ni alloy
AN 1	21% Cr 40% Fe 0.4% Al 0.1% C (max) Ni alloy: Vallruna
AN 2	21.5% Cr 30% Fe 3% Mo 1% Ti 2.2% Cu 0.05% C (max) Ni alloy: Vallruna
AN N4	14% Cr 9% Fe 0.15% C Ni alloy: Welding rod; US Service
AN QQ N 268	13% Cr 9% Fe Ni alloy: Bar and forging; US Service
AN QQ N 271	13% Cr 9% Fe Ni alloy: Sheet; US Service
AN WW T 831	13% Cr 9% Fe Ni alloy: Seamless tube; US Service
AN WW T 833	13% Cr 9% Fe Ni alloy: Welded tube; US Service **UTS: 720 Elon: 35% Proof: 300**
ANC 5C	15% Cr 25% Fe Ni alloy: Investment casting; designation used by BS
ANC 8	20% Cr 0.4% Ti 0.8% Al 0.1% C Ni alloy: Investment casting; designation used by BS
ASTM A296 CW 12M	17.5% Cr 7% Fe 18% Mo 0.12% C 5% W 2.5% Co Ni alloy: Casting
ASTM A296 CY40	15.5% Cr 11% Fe 0.4% C (max) Ni alloy: Casting
ASTM A297 HW	12% Cr 38% Fe 0.35% C Ni alloy: Casting; as cast **UTS: 340**
ASTM A297 HX	17% Cr 17% Fe 0.35% C Ni alloy: Casting; as cast **UTS: 400**
ASTM A494 CW12M2B	19% Cr 3.0% Fe (max) 19% Mo 0.07% C (max) Ni alloy: Casting
ASTM A494 N12M2B	1.0% Cr (max) 31.5% Mo 0.07% C (max) Ni alloy: Casting
ASTM A608 HW50	12.0% Cr 28.0% Fe 0.5% C Ni alloy: Tube; centrifugally cast
ASTM A608 HX50	17.0% Cr 17.0% Fe 0.5% C Ni alloy: Tube; centrifugally cast
ASTM B166	15% Cr 7% Fe 0.15% C Ni alloy: Rod and bar; annealed **UTS: 580 Elon: 30% Proof: 240**
ASTM B166	15% Cr 7% Fe 0.15% C Ni alloy: Rod and bar; cold drawn **UTS: 780 Elon: 8% Proof: 650**
ASTM B167	16% Cr 7% Fe 0.15% C Ni alloy: Tube; annealed **UTS: 580 Elon: 35% Proof: 210**
ASTM B168	16% Cr 7% Fe 0.15% C Ni alloy: Sheet and strip; annealed **UTS: 580 Elon: 30% Proof: 220**
ASTM B168	16% Cr 7% Fe 0.15% C Ni alloy: Sheet and strip; hard **UTS: 910 Elon: 2% Proof: 650**
ASTM B260 B Ni 1	14% Cr 4.5% Fe 3.3% B 4% Si 0.75% C Ni alloy: Braze metal; temperature range 1066–1204 °C
ASTM B260 B Ni 2	7% Cr 3% Fe 3.15% B 4.5% Si Ni alloy: Braze metal; temperature range 1010–1177 °C
ASTM B260 B Ni 5	19% Cr 10.2% Si Ni alloy: Braze metal; temperature range 1149–1204 °C
ASTM B260 B Ni 6	11% P Ni alloy: Braze metal; temperature range 927–1024 °C
ASTM B260 B Ni 7	13% Cr 10% P Ni alloy: Braze metal; temperature range 927–1038 °C
ASTM B260 B Ni Cr	17% Cr 3% B Ni alloy: Braze filler

Symbol	Nominal analysis, supplier, condition and remarks.
ASTM B295 E 3 N 12	15% Cr 11% Fe 3% Nb 0.1% C Ni alloy: Welding electrode
	UTS: 580
ASTM B295 E 4 N 12	17% Cr 4% Fe 3% Nb 1% Mn 0.15% C Ni alloy: Welding electrode
ASTM B295 E Ni Cr 1	17.5% Cr (min) 3% Nb Ni alloy: Welding rod; previous designation E 4 N 12
ASTM B295 E Ni Cr Fe 1	15% Cr 10% Fe 3% Nb Ni alloy: Welding rod; previous designation E 3 N 12
ASTM B295 E Ni Cr Fe 2	15% Cr 9% Fe 1.7% Mo 2% Nb 2% Mn Ni alloy: Welding rod
ASTM B295 E Ni Cr Fe 3	15% Cr 8% Fe 1.5% Nb 7% Mn Ni alloy: Welding rod
ASTM B304 ER Ni Cr 3	20% Cr 1.5% Nb 3.0% Mn Ni alloy: Welding rod
ASTM B304 ER Ni Cr Fe 5	17% Cr 8% Fe 2.5% Nb Ni alloy: Welding rod; previous designation ERN 62
ASTM B304 ER Ni Cr Fe 6	16% Cr 8% Fe 3% Ti 2.3% Mn Ni alloy: Welding rod
ASTM B304 ERN 6N	20% Cr 2% Fe 0.3% Ti 0.1% C Ni alloy: Welding electrode
ASTM B304 ERN 62	16% Cr 8% Fe 2% Nb 0.1% C Ni alloy: Welding electrode
ASTM B304 RN 42	16% Cr 8% Fe 0.1% C Ni alloy: Welding electrode
ASTM B304 R Ni Cr Fe 4	17% Cr 8% Fe Ni alloy: Welding rod; previous designation RN 42
ASTM B344	19% Cr 45% Fe 1% Mn Ni alloy: For resistance wire
ASTM B344/1	20% Cr 1% Fe 2.5% Mn Ni alloy: For resistance wire
ASTM B344/2	16% Cr 23% Fe 1% Mn Ni alloy: For resistance wire
ATGF	15.0% Cr 7.0% Fe 2.5% Ti 0.8% Al 1.0% Nb Ni alloy: Origin unknown
AURIGA XXXII	20% Cr 35% Fe 0.5% C Ni alloy: Casting; D Brown
BLAW KNOX 22H	26.5% Cr 22% Fe 0.47% C 5% W Ni alloy: Casting; for furnace fittings, etc.; Wellman
BRIGHTRAY 35	20.0% Cr 0.3% Al 0.1% C (max) 0.1% Ti 35.0% Ni Fe alloy: H Wiggin; electrical resistance alloy
	UTS: 600
BRIGHTRAY B	16% Cr 25% Fe 0.2% Cu 0.1% C Ni alloy: Wrought; H Wiggin; annealed; for use up to 950 °C
	DPN: 186 UTS: 680
BRIGHTRAY C	19% Cr 1.6% Si 0.1% C Ni alloy: Wrought; H Wiggin; annealed; for use up to 1150 °C
	DPN: 196 UTS: 740
BRIGHTRAY H	19% Cr 0.5% Fe 1% Si 3.5% Al 0.1% C Ni alloy: Wrought; H Wiggin; annealed; for use up to 1200 °C
	DPN: 200 UTS: 760
BRIGHTRAY S	20% Cr 1% Fe 0.4% Ti 0.1% C Ni alloy: Wrought; H Wiggin; annealed; for use up to 1150 °C
	DPN: 196 UTS: 730
BS 1648 H	20% Cr 35% Fe 3.0% Si 0.5% C Ni alloy: Casting; as cast; contained in BS 3100
BS 1648 K	15% Cr 20% Fe 3.0% Si 0.75% C Ni alloy: Casting; as cast; contained in BS 3100
BS 1845 N13	7% Cr 3% Fe 3% B 4% Si Ni alloy: For brazing; melting range 955–1000 °C
BS 1845 N16	15% Cr 4% Fe 3.2% B 4% Si 0.75% C Ni alloy: For braze metal; melting range 1000–1060 °C
BS 1845 N17	15% Cr 4% Fe 3.2% B 4% Si 0.06% C (max) Ni alloy: For braze metal; melting range 1010–1070 °C

Note. The following abbreviations and units are used in the tables:

DPN	Hardness, diamond pyramid number
UTS	Ultimate tensile strength, N/mm^2
Elon	Elongation, %
Proof	0.1% proof strength, N/mm^2

1 N/mm^2=0.1 hbar=0.102 kgf/mm^2=0.06475 tonf/in.^{2}=145.04 lbf/in.^{2}=1 MPa

See Appendix II for other abbreviations and conversion tables.

Symbol	Nominal analysis, supplier, condition and remarks.
BS 1845 N18	19% Cr 10.2% Si 0.15% C (max) Ni alloy: For braze metal; melting range 1080–1135 °C
BS 2901 NA34	20% Cr Ni alloy: Rod for all forms of welding
BS 2901 NA35	20% Cr 3% Fe Ni alloy: Rod for gas shielded arc welding
BS 2901 NA40	21% Cr 18% Fe 9% Mo 2% Co Ni alloy: Rod for gas shielded arc welding
BS 2901 NA43	21% Cr 5% Fe 9% Mo Ni alloy: Rod for gas shielded arc welding
BS 2901 NA45	16% Cr 3% Fe 16% Mo 2% Co Ni alloy: Rod for gas shielded arc welding
BS 3072 NA14	15% Cr 8% Fe Ni alloy: Sheet; annealed
	DPN: 180 UTS: 530 Elon: 30% Proof: 170
BS 3073 NA14	15% Cr 8% Fe Ni alloy: Strip; annealed
	DPN: 180 UTS: 530 Elon: 30% Proof: 180
BS 3074 NA14	15% Cr 8% Fe Ni alloy: Tube; annealed
	UTS: 530 Elon: 35% Proof: 200
BS 3075 NA14	15% Cr 8% Fe Ni alloy: Wire; annealed
	UTS: 530 Elon: 22%
BS 3075 NA14	15% Cr 8% Fe Ni alloy: Wire; cold drawn
	UTS: 770
BS 3076 NA14	15% Cr 8% Fe Ni alloy: Bar and section; annealed
	UTS: 530 Elon: 32% Proof: 150
BS 3076 NA14	15% Cr 8% Fe Ni alloy: Bar and section; cold drawn
	UTS: 700 Elon: 7% Proof: 530
BS 3076 NA17	18% Cr 40% Fe 2.2% Si 1.2% Mn Ni alloy: Rod
BS 3146 ANC5C	15% Cr 25% Fe Ni alloy: Investment casting; as cast
BS 3146 ANC8	20% Cr 0.4% Ti 0.8% Al 0.1% C Ni alloy: Investment casting; as cast
	UTS: 400 Elon: 5%
CALOMIC	15% Cr 20% Fe Ni alloy: Casting; electrical resistance and heating purposes; Telcon Metals
	UTS: 760
CALOMIC	15% Cr 20% Fe Ni alloy: Wrought; electrical resistance up to 1000 °C; Telcon Metals; annealed; specific resistance 1100 microhm mm
	UTS: 760
CALORITE	12% Cr 15% Fe 8% Mn Ni alloy: Heat resistant series; origin unknown
CHLORIMET 2	3.0% Fe 31.5% Mo 0.07% C Ni alloy: Casting; Duriron Co.
	UTS: 506 Elon: 20% Proof: 323
CHLORIMET 3	19% Cr 19% Mo 0.07% C (max) Ni alloy: Casting; Duriron Co.
	UTS: 500 Elon: 25% Proof: 320
CHLORIMET 3	13% Cr 5% Fe 3% B 0.7% C WC Ni alloy: Origin unknown
CHROMEL A	20% Cr Ni alloy: Wire for thermocouples; source unknown
CHROMEL P	10% Cr Ni alloy: Wire for thermocouples; source unknown
CHRONITE	13.5% Cr 10% Fe 0.4% Si 1% Al 1% Mn Ni alloy: Heat and corrosion resistant; used for burners and certain types of valves; origin unknown
CM 53	7% Cr 3% Fe 3% B 4.5% Si Ni alloy: For hard facing; Dewrance; gas welding
CM 55	11% Cr 4% Fe 2% B 4% Si Ni alloy: For hard facing; Dewrance; gas welding
CM 56	16% Cr 4% Fe 3.5% B 4.5% Si Ni alloy: For hard facing; Dewrance; gas welding
COLMONOY 4	10% Cr 2.5% Fe 2% B 0.4% C Ni alloy: Rod for spraying; Wall Colmonoy; melting point 1110 °C; low friction
	DPN: 370
COLMONOY 4	10% Cr 2.5% Fe 2% B 0.4% C Ni alloy: Casting; Wall Colmonoy; melting point 1110 °C; low friction
	DPN: 370
COLMONOY 5	11.5% Cr 4.2% Fe 2.5% B 0.6% C Ni alloy: Casting; Wall Colmonoy; melting point 1070 °C; low friction
	DPN: 480

Symbol	Nominal analysis, supplier, condition and remarks.
COLMONOY 5	11.5% Cr 4.2% Fe 2.5% B 0.6% C Ni alloy: Rod for hard facing; Wall Colmonoy; melting point 1070 °C; low friction **DPN: 480**
COLMONOY 6	13% Cr 5% Fe 3% B 0.7% C Ni alloy: Rod for hard facing; Wall Colmonoy; melting point 1040 °C; low friction **DPN: 670**
COLMONOY 6	13% Cr 5% Fe 3% B 0.7% C Ni alloy: Castings; Wall Colmonoy; melting point 1040 °C; low friction **DPN: 670**
COLMONOY 8	26% Cr 1% Fe 3.5% B 0.9% C Ni alloy: Powder for spraying; Wall Colmonoy; melting point 1080 °C; low friction **DPN: 640**
COLMONOY 20	5% Cr 3.5% Fe 1% B 0.25% C Ni alloy: Casting; Wall Colmonoy; melting point 1220 °C; good wear resistance **DPN: 220**
COLMONOY 21	5% Cr 2% Fe 1.2% B 0.25% C Ni alloy: Powder for spraying; Wall Colmonoy; for repairing moulds **DPN: 285**
COLMONOY 29	Ni base alloy for powder coating: Colmonoy **DPN: 270**
COLMONOY 41	10.0% Cr 2.7% Fe 2.0% B 2.7% Si 0.45% C Ni alloy: Bronze powder; Wall Colmonoy; melting point 1105 °C
COLMONOY 45	10% Cr 3.5% Fe 2.2% B 0.5% C Ni alloy: Powder for spraying; Wall Colmonoy; melting point 1085 °C **DPN: 420**
COLMONOY 56	13% Cr 4.5% Fe 2.7% B 0.7% C Ni alloy: Rods for metal spraying; Wall Colmonoy; melting point 1050 °C; low friction corrosion resistance **DPN: 580**
COLMONOY 56	13% Cr 1.5% Fe 2.7% B 0.7% C Ni alloy: Casting; Wall Colmonoy; melting point 1050 °C; low friction **DPN: 580**
COLMONOY 60	14.0% Cr 4.3% Fe 1.75% Mo 3.0% B 3.71% Si 1.7% Cu 0.7% C Ni alloy: Bronze powder; Wall Colmonoy; melting point 1040 °C
COLMONOY 61	14.0% Cr 4.6% Fe 3.0% B 3.7% Si 0.7% C Ni alloy: Bronze powder; Wall Colmonoy; melting point 1025 °C
COLMONOY 62	14.0% Cr 4.6% Fe 3.0% B 3.7% Si 0.7% C Ni alloy: Bronze powder; Wall Colmonoy; melting point 1025 °C
COLMONOY 72	13.0% Cr 3.5% Fe 3.5% Si 0.7% C 12.0% W Ni alloy: Bronze metal; Wall Colmonoy; melting point 1060 °C
COLMONOY 75	13% Cr 5% Fe 3% B 0.7% C WC Ni alloy: Powder for spraying; Wall Colmonoy; melting point 1040 °C; wear resistant **DPN: 660**
CORNES 14	15.5% Cr 8% Fe Ni alloy: Philip Cornes
CORNES 15 H	21% Cr Fe + Al + Ti Ni alloy: Philip Cornes
CORNES 16	21.5% Cr Mo + Ti + Cu + Fe Ni alloy: Philip Cornes
CORNES 17	18% Cr Fe + Si Ni alloy: Philip Cornes
CORNES 80/20	20% Cr Ni alloy with Ti: Philip Cornes
CORNES 601	23% Cr 14% Fe Si + Al + Ti + C Ni alloy: Philip Cornes
CORNES 625	21.5% Cr 9% Mo Ni alloy: Philip Cornes
CORRONEL 230	35% Cr Ni alloy: Bar and sheet; H Wiggin; pickling and chemical plant
CRONIX 70 E	30% Cr 1.2% Si 0.07% C (max) Ni alloy: VDM
CRONIX 80 E	20% Cr 1.2% Si 0.07% C (max) Ni alloy: VDM
CUSTOM AGE 625	20% Cr 5% Fe 8.5% Mo 1.3% Ti 3.3% Nb 0.01% C Ni alloy: Carpenter
CY 40	15.5% Cr 1.0% Fe 0.4% C (max) Ni alloy: Casting; US designation
DELORO SF 40	7.5% Cr 1.5% Fe 1.5% B 4% Si Ni alloy: Casting; Deloro **DPN: 380**
DELORO SF 50	10% Cr 4% Fe 1.5% B 4% Si Ni alloy: Casting; Deloro **DPN: 550**

Symbol	Nominal analysis, supplier, condition and remarks.
DELORO SF 60	15% Cr 4.5% Fe 3% B 4.5% Si Ni alloy: Casting; Deloro **DPN: 750**
DIN 1736 S Ni Cr 19 ND	20% Cr 3% Fe 2% Mo 2.3% Nb 0.05% C (max) Ni alloy: Welding electrode
DIN 17470 Ni Cr 80/20	20% Cr Ni alloy
DIN 17742 Ni Cr 10	10% Cr Ni alloy
DIN 17742 Ni Cr 15 Fe Mo	16% Cr 15% Fe 7% Mo Ni alloy
DTD 328	15% Cr 10% Fe 1% Mn 0.2% C (max) Ni alloy: Sheet **UTS: 450**
DTD 703B	20% Cr 0.5% Ti 0.1% C Ni alloy: Sheet; annealed **UTS: 750 Elon: 30% Proof: 300**
DTD 5047	16.5% Cr 35% Fe 3.2% Mo Ni alloy: Sheet and strip; weldable; heat resistant
DTD 5057	18% Cr 7.0% Mo 14% Co Ni alloy: Sheet and strip; weldable; heat resistant
DURCO CY40	15.5% Cr 11.0% Fe (max) 0.4% C (max) Ni alloy: Casting; Duriron Co. **UTS: 490 Elon: 30% Proof: 195**
E 320	20% Cr 40% Fe 2.5% Mo 3.5% Cu 0.07% C (max) Ni alloy: Weld electrode; designation used by AWS
E 320 LR	20% Cr 38% Fe 2.5% Mo 3.5% Cu 0.03% C (max) Ni alloy: Weld electrode; designation used by AWS
E Ni Cr 3T	20% Cr 2.5% Nb 0.1% C (max) Ni alloy: Weld electrode; designation used by AWS
E Ni Cr A	12% Cr 3% Fe 2.5% B 0.45% C Ni alloy: Weld electrode; designation used by AWS
E Ni Cr B	14% Cr 4% Fe 3% B 0.6% C Ni alloy: Weld electrode; designation used by AWS
E Ni Cr C	15% Cr 4.5% Fe 3.5% B 0.75% C Ni alloy: Weld electrode; designation used by AWS
E Ni Cr Fe 1	15% Cr 3.2% Nb 0.08% C (max) Ni alloy: Weld electrode; designation used by AWS
E Ni Cr Fe 2	15% Cr 1.5% Mo 2.2% Nb 0.1% C (max) Ni alloy: Weld electrode; designation used by AWS
E Ni Cr Fe 3	15% Cr 1.7% Nb 0.1% C (max) Ni alloy: Weld electrode; designation used by AWS
E Ni Cr Fe 4	15% Cr 2.2% Mo 1.2% Nb 0.2% C (max) Ni alloy: Weld electrode; designation used by AWS
E Ni Cr Mo 1	23% Cr 9% Mo 0.1% C Ni alloy: Weld electrode; designation used by AWS
E Ni Cr Mo 4	15.5% Cr 5.5% Fe 16% Mo 0.02% C (max) 3.7% W Ni alloy: Weld electrode; designation used by AWS
E Ni Cr Mo 6	14% Cr 7% Mo 1.5% Nb 0.1% C (max) Ni alloy: Weld electrode; designation used by AWS
E Ni Cr Mo 7	16% Cr 15.5% Mo 0.015% C (max) Ni alloy: Weld electrode; designation used by AWS
E Ni Cr Mo 9	22% Cr 7% Mo 2% Cu 0.02% C (max) Ni alloy: Weld electrode; designation used by AWS
E Ni Cr Mo 12	21.5% Cr 9.3% Mo 2% Nb 0.03% C (max) Ni alloy: Weld electrode; designation used by AWS
E Ni Fe Cr 2T	19% Cr 25% Fe 3% Mo 0.08% C (max) Ni alloy: Weld electrode; designation used by AWS
E Ni Mo 3	7% Cr 5.5% Fe 25% Mo 0.12% C (max) Ni alloy: Weld electrode; designation used by AWS
ER Ni 1	2.7% Ti 0.15% C (max) 93.0% Ni (min): Weld electrode; designation used by AWS
ER Ni C 1	4% Fe (max) 4% Cu (max) 1% C (max) Ni: Weld electrode; designation used by AWS
ER Ni Cr 3	20% Cr 2.5% Nb 0.1% C (max) Ni alloy: Weld electrode; designation used by AWS
ER Ni Cr Fe 6	15.5% Cr 3% Ti 0.08% C (max) Ni alloy: Weld electrode; designation used by AWS
ER Ni Cr Fe 7	15.5% Cr 0.7% Al 1% Nb 0.08% C (max) Ni alloy: Weld electrode; designation used by AWS
ER Ni Cr Mo 1	22% Cr 9% Mo 0.1% C (max) 12.5% Co Ni alloy: Weld electrode; designation used by AWS

Symbol	Nominal analysis, supplier, condition and remarks.
ER Ni Cr Mo 1	22% Cr 19.5% Fe 6.5% Mo 2% Cu 0.05% C (max) Ni alloy: Weld electrode; designation used by AWS
ER Ni Cr Mo 3	22% Cr 9% Mo 3.5% Nb 0.1% C (max) Ni alloy: Weld electrode; designation used by AWS
ER Ni Cr Mo 7	15.5% Cr 15% Mo 0.015% C (max) Ni alloy: Weld electrode; designation used by AWS
ER Ni Cr Mo 8	24.5% Cr 6% Mo 1.2% Ti 1% Cu 0.03% C (max) Ni alloy: Weld electrode; designation used by AWS
ER Ni Fe Cr 1	21.5% Cr 30% Fe 3% Mo 2.2% Cu 0.05% C (max) Ni alloy: Weld electrode; designation used by AWS
ER Ni Fe Cr 2	19% Cr 20% Fe 3% Mo 0.08% C (max) Ni alloy: Weld electrode; designation used by AWS
ERA HR 5	20% Cr 20% Fe Ni alloy: Casting; Hadfields specification for BS 1648 K **UTS: 740 Elon: 12% Proof: 330**
FOREMOST HR3	22% Cr 15% Fe 0.2% C Ni alloy: Casting; Swift Levick; annealed **DPN: 350 UTS: 810 Elon: 20% Proof: 610**
FOREMOST HR4	20% Cr 3% Ti 1.0% Al 0.1% C Ni alloy: Casting; Swift Levick; annealed **DPN: 300 UTS: 780 Elon: 44% Proof: 330**
G 63	15.5% Cr 0.1% C Ni alloy: Jessop; scale or creep resistant
GL 1	15% Cr 0.15% C (max) Ni alloy: Vallruna
HAI 64 ALLOY	2% Cr 2% Mn Ni alloy: For spark plugs; Harrison Alloy
HAI 66 ALLOY	16% Cr 2% Fe Ni alloy: For spark plugs; Harrison Alloy
HAI 600 ALLOY	16% Cr 8% Fe Ni alloy: Harrison Alloy
HAI EP	10% Cr Ni alloy: For thermocouples; Harrison Alloy; specific resistance 70.6 microhm/cm
HAI KP	10% Cr Ni alloy: For thermocouples; Harrison Alloy; specific resistance 70.5 microhm/cm
HAI Ni Cr 60	16% Cr 24% Fe Ni alloy: For resistance heating wear; Harrison Alloy
HAI Ni Cr 70	30% Cr Ni alloy: For resistance heating wear; Harrison Alloy
HAI Ni Cr 80	20% Cr Ni alloy: For resistance heating wear; Harrison Alloy
HAI NP	14% Cr 1.5% Si Ni alloy: For thermocouples; Harrison Alloy; specific resistance 97.2 microhm/cm
HAS B	5% Fe 28% Mo 0.04% C Ni alloy: Electrode; Metrode; for welding Hastellas B
HAS B 2L	1% Fe 28% Mo 0.02% C Ni alloy: Electrode; Metrode; for welding Hastellas B
HAS C 276	15% Cr 5% Fe 16% Mo 0.02% C 4% W Ni: Electrode; Metrode; for welding C276 alloy
HAS CH	15% Cr 5% Fe 16% Mo 0.2% C 4% W Ni alloy: Electrode; Metrode
HAS G	22% Cr 10% Fe 6% Mo 2% Nb 2% Cu 0.04% C 0.5% W Ni alloy: Electrode; Metrode; for welding Hastellas G
HAS G3	22% Cr 10% Fe 7% Mo 2% Cu 0.02% C 1% W 4% Co Ni alloy: Electrode; Metrode; for welding Hastellas G3
HAS N	7% Cr 4% Fe 16% Mo 0.04% C Ni alloy: Electrode; Metrode; for welding Hastellas N
HAS W	5% Cr 5% Fe 25% Mo 0.04% C Ni alloy: Electrode; Metrode; for welding Hastellas W

Note. The following abbreviations and units are used in the tables:

DPN	Hardness, diamond pyramid number
UTS	Ultimate tensile strength, N/mm^2
Elon	Elongation, %
Proof	0.1% proof strength, N/mm^2

1 N/mm^2=0.1 hbar=0.102 kgf/mm^2=0.06475 tonf/in.^{2}=145.04 lbf/in.^{2}=1 MPa

See Appendix II for other abbreviations and conversion tables.

Symbol	Nominal analysis, supplier, condition and remarks.
HAS X	22% Cr 21% Fe 9% Mo 0.08% C 0.7% W 1.5% Co Ni alloy: Electrode; Metrode; for welding Hastellas X
HASTELLOY D	1% Fe 10% Si 4% Cu 0.1% C Ni alloy: Casting; Haynes Stellite
HEATING ELEMENTS	14% Cr 23% Fe 0.2% Cu 0.5% Zr Ni alloy: Origin unknown
HR 5	19.5% Cr 5% Fe (max) 0.12% C 5% Co (max) Ni alloy: BS Aerospace designation
HS 28W	26% Cr 20% Fe 2% Mo 1% Al 0.45% C 3% W Ni alloy: Thomson L'Hospied **DPN: 180 UTS: 570 Elon: 5% Proof: 370**
HW	12% Cr 23% Fe 0.55% C Ni alloy: Casting; Huntington; aged **UTS: 580 Elon: 4% Proof: 340**
HX	17% Cr 20% Fe 0.55% C Ni alloy: Casting; Thomson L'Hospied **DPN: 172 UTS: 450 Elon: 10% Proof: 250**
INCOLOY 65	21.0% Cr 30% Fe 1.7% Cu 0.03% C Ni alloy: H Wiggin; for gas shielded arc welding of Incoloy 825
INCOLOY 135	29.0% Cr 26.0% Fe 3.2% Mo 1.8% Cu 0.05% C Ni alloy: H Wiggin; electrode for metal arc welding of Incoloy 825
INCONEL	15% Cr 7% Fe 0.04% C Ni alloy: International Nickel
INCONEL	15% Cr 10% Fe Ni alloy: H Wiggin; furnace equipment; acid resistant
INCONEL	15.8% Cr 7.2% Fe 0.04% C Ni alloy: Huntington; now Inconel 600; Inconel used as trade name
INCONEL 42	16% Cr 7.2% Fe 0.04% C Ni alloy: Filler; Huntington; for gas welding Inconel 600 and Incoloy 800
INCONEL 62	16% Cr 7.5% Fe 2.3% Nb 0.04% C Ni alloy: Filler; Huntington; electric welding of Inconel 600 and Incoloy 800
INCONEL 112	21.5% Cr 4.0% Fe 9.0% Mo 3.6% Nb + Ta 0.05% C Ni alloy: H Wiggin; electrode for metal arc welding of Inconel 625
INCONEL 132	14% Cr 8.5% Fe 2% Nb 0.05% C 0.1% Co Ni alloy: Filler; Huntington; electric welding of Inconel 600
INCONEL 182	14% Cr 7.5% Fe 2% Nb 7.5% Mn 0.05% C 0.1% Co Ni alloy: Filler; Huntington; welding Inconel 600 to steel
INCONEL 600	16% Cr 8% Fe 0.7% Cu 0.2% C Ni alloy: Wrought; H Wiggin; annealed **DPN: 180 UTS: 620**
INCONEL 600	15% Cr 8% Fe Ni alloy: Wrought; H Wiggin; annealed; tensile strength at 800 °C 10 tons/in.^{2} **DPN: 180 UTS: 600 Elon: 42% Proof: 240**
INCONEL 600	15% Cr 8% Fe Ni alloy: Wrought; H Wiggin; cold drawn **DPN: 280 UTS: 810 Elon: 12% Proof: 760**
INCONEL 600	15.8% Cr 7.2% Fe 0.04% C Ni alloy: Bar, sheet, etc.; Huntington; annealed; Inconel 602 was Inconel 600 **DPN: 150 UTS: 660 Elon: 45% Proof: 350**
INCONEL 600	15.8% Cr 7.2% Fe 0.04% C Ni alloy: Bar, sheet, etc.; Huntington; cold drawn; Inconel 604 was Inconel 600 **DPN: 250 UTS: 910 Elon: 25% Proof: 680**
INCONEL 601	22.0% Cr 15% Fe 1.5% Al 0.1% C (max) Ni alloy: H Wiggin; heat and corrosion resistance **DPN: 70 UTS: 690 Elon: 50% Proof: 560**
INCONEL 601	23.0% Cr 17.0% Fe 1.2% Al Ni alloy: For hot oxidation resistance; Huntington **DPN: 120 UTS: 600 Elon: 50% Proof: 250**
INCONEL 602	See Inconel 600
INCONEL 604	15.8% Cr 7.2% Fe 2% Nb 0.04% C Ni alloy: Bar, sheet, etc.; Huntington; previously Inconel 600
INCONEL 610	15.5% Cr 9.0% Fe 1.0% Nb 0.5% Cu 0.2% C Ni alloy: Huntington
INCONEL 690	29% Cr 9% Fe 0.05% C (max) Ni alloy
INCONEL 705	15.5% Cr 8.0% Fe 5.5% Si 0.5% Cu 0.3% C Ni alloy: Huntington
INCONEL 706	16% Cr 38% Fe 1.7% Ti 3% Nb 0.06% C (max) Ni alloy

Symbol	Nominal analysis, supplier, condition and remarks.
INCONEL FILLER 82	20.0% Cr 3.0% Fe (max) 2.5% Nb 3.0% Mn Ni alloy: Filler rod; Inco
INCONEL FILLER 92	15.5% Cr 8.0% Fe 2.2% Mn 0.08% C (max) Ni alloy: Filler rod; Inco
INCONEL FM62	15.5% Cr 8% Fe 2.2% Nb 0.08% C (max) Ni alloy
INCONEL FM82	20% Cr 2.5% Nb 3% Mn 0.1% C (max) Ni alloy
INCONEL FM92	15.5% Cr 8% Fe (max) 3% Ti 0.08% C (max) Ni alloy
KH 70	29.5% Cr 5.0% Fe 0.07% C Ni alloy: Russian Standard designation
KH 701U	29.5% Cr 1.0% Fe (max) 0.1% C (max) Ni alloy: Russian Standard designation
KROMORE	15% Cr Ni alloy: Driver-Harris; as Nichrome III; withdrawn
LESCALLOY 600	15% Cr 7% Fe 0.07% C Ni alloy: Latrobe; for nuclear reactor parts
MANAURITE 50W	28% Cr 5% W Ni alloy: Pompey **UTS: 550** **Proof: 260**
MANAURITE 60	15% Cr 25% Fe Ni alloy: Pompey **UTS: 450** **Elon: 22%** **Proof: 240**
MAXHETE 5	14.5% Cr 20% Fe 0.2% C Ni alloy: Edgar Allen; heat treatment equipment
MAXHETE 8	19.5% Cr 0.1% C Ni alloy: Edgar Allen; electrical resistance elements
METCO 12C	10% Cr 2.5% Fe 2.5% B 2.5% Si 0.15% C Ni alloy: Powder for spraying; Metco Ltd **DPN: 310**
METCO 14E	Cr B Ni alloy: Spray deposit; Metco **DPN: 550**
METCO 15C	17% Cr 4% Fe 3.5% B 4% Si 1% C Ni alloy: Powder for spraying; Metco Ltd **DPN: 802**
METCO 15E	Cr B Ni alloy: Spray deposit; Metco **DPN: 630**
METCO 15F	Cr B Ni alloy: Spray deposit; fine grained; Metco **DPN: 670**
METCO 19E	Cr B Ni alloy: Spray deposit; Metco **DPN: 670**
METCO 31C	11% Cr 2.5% Fe 2.5% B 2.5% Si 0.5% C 35% WC Ni alloy: Powder for spraying; Metco Ltd; tungsten carbide in Ni alloy matrix **DPN: 865**
METCO 43C	20% Cr Ni alloy: Powder for spraying; Metco Ltd **DPN: 189**
METCO 44	Cr Ni alloy: Powder casting; Metco **DPN: 170**
METCO XP 1150	14.5% Cr 3.5% Fe 3.0% B 3.5% Si 0.5% C Ni alloy: For metal spray; Metco Ltd
METCOLOY 33	15% Cr 25% Fe Ni alloy: Wire for spraying; Metco Ltd
MIL-N-6840	15.8% Cr 7.2% Fe 0.04% C Ni alloy
N 75	20% Cr 5% Fe 0.4% Ti Ni alloy: H Wiggin
N 06008	30% Cr 1.2% Si 0.15% C (max) Ni alloy: Designation used by UNS
N 06009	20% Cr 1.2% Si 1.2% Nb 0.15% C (max) Ni alloy: Designation used by UNS
N 06333	25.5% Cr 17% Fe 3.2% Mo 0.1% C 3.2% W 3.2% Co Ni alloy: UNS designation
N 06621	19.5% Cr 5% Fe (max) 0.1% C 5% Co (max) Ni alloy: Designation used by UNS
N 06625	21.5% Cr 5% Fe (max) 9% Mo 0.1% C (max) Ni alloy: UNS designation
N 08421	21% Cr 30% Fe 5.7% Mo 1.7% Cu 0.025% C (max) Ni alloy: designation used by UNS
N 08825	21% Cr 3% Mo 1% Ti 2.2% Cu 0.05% C (max) Ni alloy: UNS designation
N 09911	12.5% Cr 38% Fe 5.7% Mo 0.015% B 0.04% C Ni alloy: designation used by UNS
N 09925	21.5% Cr 37% Fe 3% Mo 2% Ti 2.2% Cu 0.03% C (max) Ni alloy: designation used by UNS
Na 22 H	27% Cr 17% Fe 0.5% C Ni alloy: American proprietary alloy listed in SAE year book

Symbol	Nominal analysis, supplier, condition and remarks.
NACE MR 01.75	21.5% Cr 9% Mo 5% Nb 0.05% C (max) Ni alloy: NACE specification
NC 20T	20% Cr Ni alloy with Ti: French Standard
NCFI	15.5% Cr 8.0% Fe 0.5% Cu (max) 0.15% C (max) Ni alloy: Japanese Standard designation; Inconel 600
NH	17% Cr 35% Fe 1% Si 1% Mn Ni: Ingot; primary metal; H Wiggin; for addition to cast iron
Ni Cr 15 Fe	15% Cr 8% Fe Ni alloy: Designation used by German Standards
Ni Cr 60/15	15% Cr 21% Fe Ni alloy: Designation used by German Standards
Ni Cr 70/30	30% Cr 1.2% Si 0.07% C (max) Ni alloy: VDM
Ni Cr 80/20	20% Cr 1.2% Si 0.07% C (max) Ni alloy: VDM
NICHROME	Low C 15% Cr 20% Fe Ni alloy: For resistance elements; British Driver-Harris; annealed **UTS: 170**
NICHROME II	18% Cr 14% Fe Ni alloy: British Driver-Harris; withdrawn
NICHROME III	15% Cr Ni alloy: British Driver-Harris; as Kromore; withdrawn
NICHROME IV	20% Cr Ni alloy: British Driver-Harris; withdrawn
NICHROME V	20% Cr Ni alloy: British Driver-Harris; use up to 1150 °C **UTS: 870**
NICHROME V	20% Cr Ni alloy: For electrical resistance; British Driver-Harris; annealed **UTS: 170**
NICKEL FILLER 61	2.7% Ti 1.5% Al 1.0% Mn 0.15% C (max) 0.03% P (max) Ni: Filler wire for overlaying steel; H Wiggin
NICKELVAC N	15.5% Cr 8% Fe 0.07% C Ni alloy: Vanadium Alloys Ltd; double vacuum cast
NICRAL Z	16% Cr 2% Fe Ni alloy: Imphy **DPN: 145** **UTS: 630** **Elon: 35%** **Proof: 270**
NICREX 3	15% Cr 25% Fe 0.8% Nb 0.07% C Ni alloy: Welding electrode; Murex **UTS: 600** **Elon: 35%**
NICREX 4	18% Cr 1.8% Nb 0.2% C Ni alloy: Welding electrode; Murex **UTS: 650** **Elon: 30%**
NICROBODY 5007	11.2% Cr 0.06% Co 8% P Ni alloy: Braze filler; Wall Colmonoy; melting range 1010–1120 °C
NICROBRAZ 30	19% Cr 10% Si 0.1% C Ni alloy: Braze metal; Wall Colmonoy; melting range 1080–1130 °C
NICROBRAZ 35	9.5% Si 9.5% Mn 19.5% C Ni alloy: For brazing; Wall Colmonoy; melting range 1135–1160 °C
NICROBRAZ 50	13% Cr 0.1% C 10% P Ni alloy: Braze metal; Wall Colmonoy; melting point 890 °C
NICROBRAZ 51	25% Cr 10% P Ni alloy: Bronze metal; Wall Colmonoy; melting range 980–1095 °C
NICROBRAZ 120	13.5% Cr 4.5% Fe 3.5% B 4.5% Si 0.8% C Ni alloy: Braze metal; Wall Colmonoy; as 'Standard' but coarser mesh powder
NICROBRAZ 125	14.0% Cr 4.5% Fe 3.0% B 0.7% C Ni alloy: For brazing; Wall Colmonoy; melting range 970–1040 °C
NICROBRAZ 150	15% Cr 2.5% B 0.1% C Ni alloy: Braze metal; Wall Colmonoy; melting point 1050 °C
NICROBRAZ 160	10% Cr 2.5% Fe 2% B 2.5% Si 0.45% C Ni alloy: Braze metal; Wall Colmonoy; melting range 970–1160 °C
NICROBRAZ 170	11.5% Cr 3.7% Fe 2.5% B 3.25% Si 0.55% C 16% W Ni alloy: Braze metal; Wall Colmonoy; melting range 980–1100 °C
NICROBRAZ 180	5% Cr 3.5% Fe 1% B 3% Si 0.2% C Ni alloy: Braze metal; Wall Colmonoy; melting range 970–1180 °C
NICROBRAZ 200	3% Fe 3.2% B 4.5% Si 7% C 6% W Ni alloy: Braze metal; Wall Colmonoy; melting range 975–1040 °C
NICROBRAZ 3001	11.5% Cr 6.0% Si Ni alloy: For brazing; Wall Colmonoy; melting range 1220–1235 °C
NICROBRAZ 3002	15.0% Cr 8.0% Si Ni alloy: For brazing; Wall Colmonoy; melting range 1180–1200 °C

Symbol	Nominal analysis, supplier, condition and remarks.
NICROBRAZ 3003	17% Cr 0.1% B 9.5% Si Ni alloy: For brazing; Wall Colmonoy; melting range 1150–1180 °C
NICROBRAZ 3004	11.5% Cr 0.4% B 7.0% Si Ni alloy: For brazing; Wall Colmonoy; melting range 1160–1180 °C
NICROBRAZ 3005	13.0% Cr 0.35% B 8.0% Si Ni alloy: For brazing; Wall Colmonoy; melting range 1160–1180 °C
NICROBRAZ 5007	11.2% Cr 0.6% Co 8% P Ni alloy: Braze filler; Wall Colmonoy; melting range 1010–1120 °C
NICROBRAZ LC	13% Cr 4.5% Fe 3.5% B 4.5% Si 0.06% C Ni alloy: Braze metal; Wall Colmonoy; melting range 970–1080 °C
NICROBRAZ LM	6.5% Cr 2.5% Fe 3% B 4.5% Si 0.06% C Ni alloy: Braze metal; Wall Colmonoy; melting range 970–1000 °C
NICROBRAZ LMO1	3.0% Cr 1.3% Fe 1.9% B 3.0% Si Ni alloy: For brazing; Wall Colmonoy; melting range 1090–1155 °C
NICROBRAZ LMO2	2.25% Cr 1.0% Fe 1.7% B 2.75% Si Ni alloy: For brazing; Wall Colmonoy; melting range 1090–1155 °C
NICROBRAZ LMO3	2.5% Cr 1.1% Fe 1.8% B 3.0% Si Ni alloy: For brazing; Wall Colmonoy; melting range 1090–1155 °C
NICROBRAZ LMO4	2.3% Cr 1.2% Fe 1.8% B 3.0% Si Ni alloy: For brazing; Wall Colmonoy; melting range 1090–1160 °C
NICROBRAZ LMO5	2.7% Cr 1.2% Fe 1.8% B 3.0% Si Ni alloy: For brazing; Wall Colmonoy; melting range 1090–1160 °C
NICROBRAZ LMO6	2.3% Cr 1.0% Fe 1.5% B 2.4% Si Ni alloy: For brazing; Wall Colmonoy; melting range 1090–1160 °C
NICROBRAZ LMO7	5.6% Cr 2.4% Fe 2.5% B 3.5% Si Ni alloy: For brazing; Wall Colmonoy; melting range 1120–1160 °C
NICROBRAZ LMO8	3.8% Cr 1.7% Fe 2.2% B 3.5% Si Ni alloy: For brazing; Wall Colmonoy; melting range 1090–1155 °C
NICROBRAZ (STANDARD)	13.5% Cr 4.5% Fe 3.5% B 4.5% Si 0.8% C Ni alloy: Braze metal; Wall Colmonoy; melting range 980–1040 °C
NICROBRAZ WG	11.5% Cr 3.5% Fe 3% B 3.5% Si 0.06% C Ni alloy: Braze metal; Wall Colmonoy; melting range 970–1090 °C; obsolete
NICROCOAT	7% Cr 3% Fe Ni alloy: Powder; with additive; Wall Colmonoy; for coating
NICROCOAT 1	13.5% Cr Ni alloy: Powder; with additive; Wall Colmonoy; for coating
NICROCOAT 3	13% Cr Ni alloy: Powder; with additive; Wall Colmonoy; for coating
NICROCOAT 4	15% Cr Ni alloy: Powder; with additive; Wall Colmonoy; for coating
NICROCOAT 6	7% Cr 6% W Ni alloy: Powder; with additive; Wall Colmonoy; for coating
NICROCOAT 8	10% Cr Ni alloy: Powder; with additive; Wall Colmonoy; for coating
NICROCOAT 9	19% Cr 10% Si Ni alloy: Powder; with additive; Wall Colmonoy; for coating
NICROCOAT 610	13.5% Cr 10% Ti boride Ni alloy: Powder; Wall Colmonoy; for coating
NICROCOAT 620	18% Cr 5% Si 8.5% Ti boride Ni alloy: Powder; Wall Colmonoy; for coating
NICROCOAT 621	18.5% Cr 7.3% Si 4.5% Ti silicide Ni alloy: Powder; Wall Colmonoy; for coating
NICROCOAT 630	17% Cr 5.3% Si 6.5% Ti 6.5% TiN silicide Ni alloy: Powder; Wall Colmonoy; for coating

Note. The following abbreviations and units are used in the tables:

DPN	Hardness, diamond pyramid number
UTS	Ultimate tensile strength, N/mm^2
Elon	Elongation, %
Proof	0.1% proof strength, N/mm^2

1 N/mm^2=0.1 hbar=0.102 kgf/mm^2=0.06475 tonf/in.^{2}=145.04 lbf/in.^{2}=1 MPa

See Appendix II for other abbreviations and conversion tables.

Symbol	Nominal analysis, supplier, condition and remarks.
NICROCOAT 700	13% Cr 5% Al 10% boride oxide Ni alloy: Powder; Wall Colmonoy; for coating
NICROFER 3220 H	21% Cr Fe + Al + Ti Ni alloy: Philip Cornes
NICROFER 3718	18% Cr Fe + Si Ni alloy: Philip Cornes
NICROFER 4221	21% Cr Fe + Mo + Ti + Cu Ni alloy: Philip Cornes
NICROFER 6020 h Mo	21.5% Cr 9% Mo Ni alloy: Philip Cornes
NICROFER 6023	23% Cr 14% Fe + Si + Al + Ti Ni alloy: Philip Cornes
NICROFER 6023 H	23% Cr 14% Fe 0.4% Ti 11.3% Al 0.1% C (max) Ni alloy: VDM
NICROFER 6030	28.5% Cr 9% Fe 0.02% C (max) Ni alloy: VDM
NICROFER 7216	15.5% Cr 8% Fe Ni alloy: Philip Cornes
NICROFER 7216	16% Cr 8.5% Fe 0.07% C (max) Ni alloy: VDM
NICROFER 7216 H	16% Cr 8.5% Fe 0.2% Ti 0.3% C Ni alloy: VDM
NICROFER 7520	20% Cr Ni + Ti alloy: Philip Cornes
NICROFER 7520	20.5% Cr 0.5% Fe (max) 0.4% Ti 0.1% C Ni alloy: VDM
NICROFER B 7020	20% Cr 3% Fe (max) 2.5% Nb 3.2% Mn 0.05% C (max) Ni alloy: Strip; VDM
NICROFER S 7020	20% Cr 3% Fe (max) 0.05% C (max) Ni alloy: Weld; red; VDM
NICROTUNG	12.0% Cr 4.0% Ti 4.0% Al 0.1% C 10.0% Co 8.0% W Ni alloy: Origin unknown
NIKROTHAL 60	15% Cr 25% Fe Ni alloy: Kanthal A B; for heating elements
NIKROTHAL 80	20% Cr Ni alloy: Kanthal A B; for heating elements
NIMOCAST 75	20.0% Cr 50% Fe 0.4% Ti 0.1% C Ni alloy: Casting; H Wiggin
NIMOCAST 257	20.0% Cr 5.0% Fe 1.6% Ti 0.08% C 16.0% Co Ni alloy: Casting; H Wiggin
NIMOCAST 258	10.0% Cr 2.0% Fe 5.0% Mo 3.7% Ti 4.8% Al 0.2% C 20.0% Co Ni alloy: Casting; H Wiggin
NIMOCAST 258	10.0% Cr 2.0% Fe 5.0% Mo 3.7% Ti 4.8% Al 0.2% C 20.0% Co Ni alloy: Casting; H Wiggin
NIMOCAST PD16	6.0% Cr 2.0% Mo 6.0% Al 0.13% C Ni alloy: Casting; H Wiggin; for turbine blades
NIMOCAST PE10	20.0% Cr 6.0% Mo 6.5% Nb 0.02% C 1.0% Co 2.5% W Ni alloy: Casting; H Wiggin
NIMONIC 75	20.0% Cr 5% Fe 0.4% Ti Ni alloy: H Wiggin; annealed; tensile strength at 750 °C 220 N/mm^2 **DPN: 200 UTS: 730 Elon: 44% Proof: 340**
NIMONIC 75	20% Cr 5% Fe 0.4% Ti Ni alloy: H Wiggin; cold rolled **DPN: 300**
NIMROD 182	15% Cr 8% Fe 2% Nb 8% Mn 0.05% C Ni alloy: Electrode; Metrode
NIMROD 182 KS	15% Cr 8% Fe 2% Nb 8% Mn 0.05% C Ni alloy: Electrode; Metrode; for positional welding
NIMROD 625	21% Cr 1% Fe 9% Mo 4% Nb 0.04% C Ni alloy: Electrode; Metrode
NIMROD 625 KS	21% Cr 1% Fe 9% Mo 4% Nb 0.04% C Ni alloy: Electrode; Metrode; for positional welding
NIMROD A KS	15% Cr 8% Fe 2% Mo 2% Nb 3% Mn 0.05% C Ni alloy: Electrode; Metrode; for positional welding
NIMROD AB	15% Cr 8% Fe 2% Mo 2% Nb 3% Mn 0.05% C Ni alloy: Electrode; Metrode
PER 1	20.0% Cr 0.4% Ti 5.0% Co Ni alloy: Origin unknown
PER 2	22.0% Cr 4.0% Ti 5.0% Co Ni alloy: Origin unknown
PER 2B	20.0% Cr 2.0% Ti 20.0% Co Ni alloy: Origin unknown
PER 2U	20.0% Cr 5.0% Fe 2.0% Ti 1.0% Al 20.0% Co Ni alloy: Origin unknown
PER 2V	19.0% Cr 2.0% Ti 1.5% Al 2.0% Co Ni alloy: Origin unknown
PER 13	13.0% Cr 4.0% Mo 6.0% Al Ni alloy: Origin unknown
PIREKS 60	20% Cr 20% Fe Ni alloy: Casting; Darwins; heat treatment equipment
PIREKS 60/13	20% Cr 20% Fe Ni alloy: Rod or wire; Darwins; electrical resistors; heating elements
PYROMET 88	16% Cr 3% Ti 0.03% C Ni alloy: Carpenter **UTS: 1120 Elon: 12% Proof: 720**
PYROMET 600	15.5% Cr 7% Fe 0.1% C (max) Ni alloy: Carpenter

Symbol	Nominal analysis, supplier, condition and remarks.
PYROMET ALLOY X 750	15% Cr 0.7% Fe 2.5% Ti 1% Nb + Ta 0.08% C (max) Ni alloy: Carpenter **UTS: 1110 Elon: 22% Proof: 630**
PYROMIC	20% Cr Ni alloy: For electrical resistance up to 1150 °C; Telcon Metals; annealed **UTS: 770**
QQ C.551	3% Fe 23% Cu 3% Mn Ni alloy: Casting; 2.0% C + Si; US Federal **UTS: 450 Elon: 25% Proof: 180**
R 870G	20.0% Cr 15% Fe 0.2% C Ni alloy: Casting; scaling temperature 1050 °C; Fagerston
RA 333	25% Cr 20% Fe 3.0% Mo 0.08% C 3.5% W 3.5% Co Ni alloy: Simonds
RNC 30	25% Cr 30% Fe Ni alloy: For electrical resistance; Imphy
RNC-CARBIMPHY	20% Cr 35% Fe Ni alloy: For electrical resistance; Imphy; for use in carburizing atmospheres
RNC-SUPERIMPHY	20% Cr Ni alloy: For electrical resistance; Imphy
SAE 70334	12% Cr 25% Fe 0.5% Mo 0.5% C Ni alloy: Casting
SAE 70335	17% Cr 22% Fe 0.5% Mo 0.5% C Ni alloy: Casting
SAE VF 1	20% Cr 1% Fe 0.2% C Ni alloy: For hard facing
SANICRO 70	16% Cr 10% Fe 0.03% C (max) Ni alloy: Sandvik
SANICRO 71	16% Cr 9% Fe 0.04% C Ni alloy: For tube; Sandvik **DPN: 160 UTS: 560 Elon: 30% Proof: 230**
SANICRO 71	16% Cr 0.05% C 10% Co Ni alloy: Sandvik
SF 40	7.5% Cr 1.5% Fe 1.5% B 4% Si Ni alloy: Deloro
SF 50	10% Cr 4% Fe 1.5% B 4% Si Ni alloy: Deloro
SF 60	15% Cr 4.5% Fe 3% B 4.5% Si Ni alloy: Deloro
SIMALLOY 600	16% Cr 8% Fe 0.15% C Ni alloy: Simonds
SOUDONEL BS	15.5% Cr 8% Fe 1.7% Mo 1.7% Nb 2.7% Mn 0.08% C Ni alloy: Electrode; Soudometal; for welding high Ni steel; impact 50 J at −200 °C **UTS: 650 Elon: 32% Proof: 410**
SOUDOR G Ni Cr 3	20% Cr 2.7% B 3% Mn 0.01% C 0.4% TiN Ni alloy: Welding electrode; Soudometal **UTS: 560 Elon: 30% Proof: 400**
SOUDOTIG 625	23.3% Cr 0.8% Fe 9.1% Mo 0.2% Ti 3.4% Nb 0.01% C Ni alloy: Welding electrode; Soudometal; for inert gas welding **UTS: 830 Elon: 30% Proof: 415**
SOUDOTIG Ni Cr 3	20.5% Cr 1% Fe 0.4% Ti 2.7% Nb 0.015% C Ni alloy: Welding electrode; Soudometal; for inert gas welding; impact at −200 °C 80 J **UTS: 560 Elon: 30% Proof: 400**
STUBS 068 HH	20% Cr 3.0% Fe 2% Mo 2.3% Nb + Ta 5.5% Mn 0.05% C (max) Ni alloy: Welding electrode; Stubs; nuclear; chemical industry; welding **UTS: 650 Elon: 30% Proof: 390**
STUBS 4221	25.8% Cr 20% Fe 3.5% Mo 1.6% Cu 0.04% C Ni alloy: Welding electrode; Stubs **DPN: 150 UTS: 540 Elon: 30% Proof: 340**
STUBS 6222 Mo	21.5% Cr 2.0% Fe 8.8% Mo 3.5% Nb + Ta 0.02% C Ni alloy: Welding electrode; Stubs; for heat resistant properties **UTS: 760 Elon: 30% Proof: 450**
STUBS 7015	16% Cr 6.5% Fe 2.5% Nb + Ta 6% Mn 0.03% C Ni alloy: Welding electrode; Stubs **DPN: 170 UTS: 620 Elon: 35% Proof: 380**
STUBS 7015 Mo	15.7% Cr 6.2% Fe 1% Mo 1.7% Nb + Ta 3% Mn 0.02% C Ni alloy: Welding electrode; Stubs **DPN: 170 UTS: 620 Elon: 35% Proof: 380**
T 1	10% Cr Ni alloy: Thermocouple; British Driver-Harris **UTS: 650**
TOLOY 50	26% Cr 20% Fe 2% Mo 1% Al 0.45% C 3.0% W Ni alloy: Thomson L'Hospied **DPN: 180 UTS: 570 Elon: 5% Proof: 370**
TOLOY 50/50	50% Cr 0.1% C Ni alloy: Casting; Wellman; superheater tube supports, etc.
TOLOY 65	17% Cr 20% Fe 0.55% C Ni alloy: Casting; Thomson L'Hospied **DPN: 172 UTS: 450 Elon: 10% Proof: 250**
TOLOY 65 HX	17% Cr 18% Fe 0.55% C Ni alloy: Casting for furnace fittings, etc.; Wellman
TOPHET 30	30% Cr Ni alloy: For heating elements; Driver-Harris
TOPHET A	20% Cr Ni alloy: For heating elements and resistance wire; Driver-Harris
TOPHET A	20% Cr Ni alloy: Strip for furnace elements; Gilbey-Brunton
TOPHET C	16% Cr 24% Fe Ni alloy: Strip for furnace elements; Gilbey-Brunton
TOPHET C	15% Cr Ni alloy: For heating elements and resistance wire; Driver-Harris
TPM	16% Cr 3.1% Ti 0.05% C Ni alloy: Origin unknown
UCAR 75	19.5% Cr 0.1% C Ni alloy: Union Carbide
UCAR 600	15.5% Cr 2.2% Nb 0.1% C (max) Ni alloy: Union Carbide
UCAR 601	23.0% Cr 1.35% Al 0.05% C Ni alloy: Union Carbide
VALRAY 1	20% Cr Ni alloy: With C Mn and Si for facing valves; H Wiggin
VF 1	20% Cr 1% Fe 0.2% C Ni alloy: For hard facing; designation used by SAE
VIKRO 1	13.5% Cr 24.5% Fe 0.7% C Ni alloy: Casting; Firth Vickers
VIKRO 11	20.0% Cr 35% Fe 0.4% C Ni alloy: Casting; Firth Vickers
VMA 5	20% Cr 6% Mo 2% Ti 20% Co Ni alloy: Investment casting; BS designation; obsolete
VMA 6A	13% Cr 4% Mo 6% Al 2% Nb Ni alloy: Investment casting; BS designation; obsolete
VMA 6B	13% Cr 4% Mo 6% Al 2% Nb Ni alloy: Investment casting; BS designation; obsolete
VMA 6C	13% Cr 4% Mo 6% Al 2% Nb Ni alloy: Investment casting; BS designation; obsolete
VMA 7A	15% Cr 10% Fe 5% Mo 2% Ti 3.5% Al Ni alloy: Investment casting; BS designation; obsolete
VMA 7B	15% Cr 10% Fe 5% Mo 2% Ti 3.5% Al Ni alloy: Investment casting; BS designation; obsolete
VMA 8	15% Cr 8% Mo 3.6% Ti 4.2% Al Ni alloy: Investment casting; BS designation; obsolete
VMA 9	18% Cr 3% Ti 3% Al 18% Co Ni alloy: Investment casting; BS designation; obsolete
VMA 10	14.5% Cr 4% Mo 3.5% Ti 4.5% Al 15% Co Ni alloy: Investment casting; BS designation; obsolete
VMA 11	15% Cr 4.5% Mo 2.4% Ti 4.4% Al 22% Co Ni alloy: Investment casting; BS designation; obsolete
VMA 12	9.5% Cr 3% Mo 4.7% Ti 5.5% Al 1% V 15% Co Ni alloy: Investment casting; BS designation; obsolete
VMA 13	19% Cr 18.7% Fe 3% Mo 1% Ti 0.6% Al 5% Nb Ni alloy: Investment casting; BS designation; obsolete
VMA 14	9% Cr 2.5% Mo 1.5% Ti 5.5% Al 1.5% Ta 10% Co 1.0% W Ni alloy: Investment casting; BS designation; obsolete
VMA 15	9% Cr 1.5% Ti 5.5% Al 2.5% Ta 10% Co 1.5% Ha Ni alloy: Investment casting; BS designation; obsolete
VMA 16A	16% Cr 1.7% Mo 3.5% Ti 3.5% Al 1% Nb 1.7% Ta 2.5% W 8.5% Co Ni alloy: Investment casting; BS designation; obsolete
VMA 16B	16% Cr 1.7% Mo 3.5% Ti 3.5% Al 1% Nb 1.7% Ta 2.5% W 8.5% Co Ni alloy: Investment casting; BS designation; obsolete
WAUKESHA 88	12.5% Cr 2% Fe (max) 2.7% Mo 4% Bi 0.05% C (max) 4% Sn Ni alloy: Casting; Dewramet; non-galling **DPN: 150 UTS: 310 Elon: 7% Proof: 262**
WE 132	14.5% Cr 10% Fe (max) 3.2% Nb 0.08% C (max) Ni alloy

27C Nickel–chromium wrought and cast ageing alloys With titanium, molybdenum, aluminium and cobalt

Specific gravity	8.1
Density	8100 kg/m^3
Solidus/liquidus	1340–1390 °C
Thermal conductivity	113 W/m °C
Coefficient of linear expansion (20–100 °C)	11.7×10^{-6}/ °C
Electrical conductivity	1.4% IACS (copper 100%)
Specific resistance	1200 microhm mm
Young's modulus of elasticity	221×10^9 N/m^2
Impact	42–100 J
Fatigue strength (15×10^6 cycles)	±370 N/mm^2 at 750 °C
Hot strength	

Temperature °C	Tensile strength N/mm^2	Elongation %
20	1210	33
200	1170	32
400	1080	32
500	1050	32
600	1030	18
700	810	10
800	550	15
900	220	31
1000	60	87

* The hot strength figures quoted are average. Tensile strengths of 170 N/mm^2 at 1000 °C are possible with some of these alloys.

All the above properties are included as a guide to the group of specifications as a whole. There is considerable variation within the group and some of these properties may not be applicable to all the specifications.

General metallurgical characteristics

These alloys are based on the 80/20 nickel/chromium materials described and listed in Section 27B with the addition of elements which form intermetallic compounds with each other and chromium. These take part in the metallurgical reactions necessary to impart age hardening. The principal effect is to increase tensile properties at high temperatures, the most important of which is the creep strength. Most of the alloys listed have excellent high temperature oxidation and corrosion resistance, although some of the more highly alloyed materials with the better creep strength have poorer hot corrosion resistance.

They are all expensive, on the whole difficult to work and are only used when their high temperature, high strength properties are essential.

These materials are all capable of being solution treated and aged. In addition they can be cold worked to increase the mechanical strength to a considerable extent with a reduction in ductility.

To remove the cold working the material must be annealed to above 1000 °C.

Solution treatment and ageing requires skill and information should be sought before attempting these treatments.

The materials can be welded using the inert gas techniques. Specialist advice should be obtained, and it should be noted that any welding will affect the mechanical properties of components which have been aged.

The materials listed are difficult to machine as they are extremely tough, and with their cold working facility this can result in considerable tool wear. It is important that the tools are made to cut below the surface and remain below the surface. Surface rubbing results in severe cold work and lack of machinability.

These materials are used for turbine blades in gas and steam turbines, combustion equipment which requires high strength, internal combustion engine exhaust valves and chemical plant where high temperatures are used.

The principal characteristic used with these materials is the high creep strength.

Note. The following abbreviations and units are used in the tables:

DPN	Hardness, diamond pyramid number
UTS	Ultimate tensile strength, N/mm^2
Elon	Elongation, %
Proof	0.1% proof strength, N/mm^2

1 N/mm^2=0.1 hbar=0.102 kgf/mm^2=0.06475 tonf/in.^{2}=145.04 lbf/in.^{2}=1 MPa

See Appendix II for other abbreviations and conversion tables.

Symbol	Nominal analysis, supplier, condition and remarks.
1/360	10% Cr 4.5% Fe 5% Mo 0.3% B 6% Al 0.1% C 2% Nb Ni alloy: Casting; American proprietary alloy listed in SAE yearbook
1C 901	13% Cr 40% Fe 6% Mo 3% Ti 0.07% C alloy: Bofors; age hardened **DPN: 340 UTS: 1020 Elon: 12% Proof: 670**
2.4537	15.5% Cr 6.0% Fe 16.0% Mo 0.02% C (max) 3.7% W 2.5% Co 0.35% V Ni alloy

Symbol	Nominal analysis, supplier, condition and remarks.
2.4602	16% Cr 6% Fe 17% Mo 4% W Ni alloy: German Standard
2.4605	22% Cr 25% Fe 7% Mo 2.0% Nb Ni alloy: German Standard
2.4634	14.7% Cr 5% Mo 0.008% B 1.2% Ti 4.7% Al 0.15% C 20% Co Ni alloy: German Standard
2.4662	16.5% Cr 35% Fe 3.2% Mo 1.7% Al 0.06% C 2% Co (max) Ni alloy: German Standard
2.4665	21.5% Cr 18% Fe 9% Mo 0.1% C 0.7% W 1.5% Co Ni alloy: German Standard
2.4858	22% Cr 35% Fe 3% Mo 2% Cu 0.05% C Ni alloy: German Standard
2.4889	15% Cr 15% Fe 7% Mo Ni alloy: German Standard
2.4952	19.5% Cr 2.2% Ti 1.4% Al 0.06% C Ni alloy: German Standard
2.4969	19.5% Cr 2.5% Ti 1.5% Al 0.13% C (max) 18% Co Ni alloy: German Standard
69 INCONEL X	15% Cr 6.8% Fe 2.4% Ti 0.8% Al 0.8% Nb 0.04% C Ni alloy: Filler; Huntington; now Inconel 69
713C	12.5% Cr 4.2% Mo 0.8% Ti 6.1% Al 2.2% Nb + Ta 0.12% C Ni alloy: International Nickel Co.; vacuum melted **UTS: 860 Elon: 8% Proof: 750**
713C	12.5% Cr 4.2% Mo 0.12% B 0.8% Ti 6.1% Al 2.2% Nb 0.12% C 0.1% Zr Ni alloy: Union Carbide
713LC	12% Cr 4.5% Mo 0.01% B 0.7% Ti 6.0% Al 2.3% Nb 0.05% C 0.1% Zr Ni alloy: Union Carbide
713LC	12% Cr 4.5% Mo 0.6% Ti 5.9% Al 2.0% Nb + Ta 0.05% C Ni alloy: International Nickel Co.; as cast **UTS: 910 Elon: 15% Proof: 760**
718	19% Cr 18% Fe 3% Mo 1% Ti 0.6% Al 5% Nb + Ta 0.08% C Ni alloy: Crucible Steel Co.; solution treated and aged **UTS: 1380 Proof: 1100**
738	16.0% Cr 1.75% Mo 0.01% B 3.4% Ti 3.4% Al 0.9% Nb 0.17% C 8.5% Co 2.6% W 1.75% Ta 0.1% Zr Ni alloy: Union Carbide
AF 1753	16.2% Cr 9.5% Fe 1.6% Mo 3.2% Ti 1.9% Al 0.25% C 7.2% Co 8.4% W 0.06% Zr Ni alloy: Information from SAE handbook
AF2 – 1DA	12% Cr 3% Mo 0.015% B 4.5% Al 0.32% C Ni alloy: Origin unknown
AISI 664	15% Cr 30% Fe 4% Mo 0.01% B 3% Ti 1% Al 0.06% C 3.6% W Ni alloy: Solution treated and aged **UTS: 1470 Elon: 15% Proof: 1020**
AISI 680	21.5% Cr 18.5% Fe 9% Mo 0.1% C 0.6% W 1.5% Co Ni alloy: Solution treated; has slight ageing properties **UTS: 830 Elon: 52% Proof: 270**
AISI 681	12.5% Cr 37% Fe 6% Mo 0.015% B 2.5% Ti 0.2% Al 0.05% C Ni alloy: Solution treated; double aged **UTS: 1220 Elon: 17% Proof: 740**
AISI 682	12.5% Cr 37% Fe 5.7% Mo 0.015% B 2.8% Ti 0.2% Al 0.05% C Ni alloy: Solution treated; double aged **UTS: 1220 Elon: 17% Proof: 740**
AISI 683	19% Cr 1.8% Fe 10% Mo 3.1% Ti 1.5% Al 0.09% C 11% Co Ni alloy: Solution treated and aged **UTS: 1470 Elon: 14% Proof: 1020**
AISI 684	17.5% Cr 0.5% Fe 4.2% Mo 0.005% B 3% Ti 3% Al 0.1% C 18.5% Co Ni alloy: Solution treated; double aged **UTS: 1250 Elon: 16% Proof: 800**
AISI 685	19.7% Cr 0.7% Fe 4.45% Mo 0.005% B 3% Ti 1.4% Al 0.07% C 13.5% Co Ni alloy: Solution treated; double aged **UTS: 1380 Elon: 25% Proof: 800**
AISI 686	15% Cr 10% Fe 5% Mo 2.5% Ti 2% Al 0.12% C Ni alloy: Solution treated and aged **UTS: 1220 Elon: 21% Proof: 830**
AISI 687	15% Cr 0.5% Fe 5.25% Mo 0.03% B 3.5% Ti 4.2% Al 0.07% C 18.5% Co Ni alloy: Solution treated; double aged **UTS: 1470 Elon: 17% Proof: 760**
AISI 688	15% Cr 6.7% Fe 2.5% Ti 0.8% Al 0.8% Nb Ni alloy: Solution treated; double aged **UTS: 1130 Elon: 24% Proof: 650**
AISI 689	20% Cr 10% Mo 0.005% B 2.6% Ti 1% Al 0.15% C 10% Co Ni alloy: Solution treated and aged **UTS: 1220 Elon: 16% Proof: 880**
AISI 690	18% Cr 18% Fe 3.2% Mo 2.7% Ti 0.2% Al 0.03% C 20% Cr Ni alloy: Solution treated; double aged **UTS: 1020 Elon: 19% Proof: 650**
ALLOY 49	30% Cr 5% Fe 2.5% C 14% W Ni alloy: Casting; Dewrance **DPN: 395 UTS: 770**
ALLOY 80	16% Cr 5% Fe 17% Mo 0.15% C 5% W Ni alloy: Casting; Dewrance **DPN: 265 UTS: 540**
ALLOY 81	1% Cr 5% Fe 28% Mo 0.12% C Ni alloy: Casting; Dewrance **DPN: 235 UTS: 540**
ALLOY 333	25% Cr 17% Fe 3% Mo 0.04% C 3% W 3% Co Ni alloy: VDM
ALLOY 617	21.5% Cr 0.4% Ti 1% Al 0.07% C 11.5% Co Ni alloy: VDM
ALLOY 625 H	22% Cr 4% Fe (max) 9% Mo 3.8% Nb 0.07% C Ni alloy: VDM
ALLOY 713 C	12.5% Cr 1.0% Fe 4.5% Mo 0.012% B 0.6% Ti 6.0% Al 2.0% Nb 0.12% C 0.1% Zr Ni alloy: Information from SAE handbook
ALLOY X	21.5% Cr 9% Fe 9% Mo 0.07% C 0.8% W 0.7% Co Ni alloy: VDM
ALLVAC 500ZB	19% Cr 4% Mo 0.006% B 3% Ti 3% Al 0.08% C 17% Co 0.06% Zr Ni alloy: Vanadium Alloys; double vacuum cast
ALLVAC 718	19% Cr 20% Fe 3% Mo 0.004% B 1% Ti 0.5% Al 5.1% Nb + Ta 0.05% C Ni alloy: Vanadium Alloys; double vacuum cast
ALLVAC GMR 235	15.5% Cr 10% Fe 5.5% Mo 2.5% Ti 2% Al 0.12% C Ni alloy: Vanadium Alloys; double vacuum cast
ALLVAC M 252	19% Cr 9.7% Mo 0.06% B 2.6% Ti 1.1% Al 0.14% C 10.2% Co Ni alloy: Vanadium Alloys; double vacuum cast
ALLVAC R 41	19% Cr 9.7% Mo 0.006% B 3.2% Ti 1.5% Al 0.08% C 11% Co Ni alloy: Vanadium Alloys; double vacuum cast
ALLVAC WASPALOY	19.5% Cr 4.25% Mo 0.006% B 3% Ti 1.4% Al 0.08% C 14% Co 0.06% Zr Ni alloy: Vanadium Alloys; double vacuum cast
AMD 57 BX	16.5% Cr 9% Fe 1.5% Mo 3% Ti 2% Al 7.5% Co 8.5% W Ni alloy: Bar and forging; interim AMS
AMPCOLAY 561	20% Cr 30% Fe 3% Mo Ni alloy: Ampco for Ni O NEL
AMS 5383	19% Cr 3% Mo 1% Ti 0.6% Al 5% Nb 0.08% C (max) Ni alloy
AMS 5384	18% Cr 2% Fe 4% Mo 3% Ti 3% Al 18% Co Ni alloy: Casting; solution treated and aged; vacuum melted
AMS 5388	16% Cr 5.7% Fe 17% Mo 0.07% C 1.2% Co 4% W 0.3% V Ni alloy: Castings
AMS 5389	15% Cr 6% Fe 16% Mo 3.5% W 2.5% Co 0.3% V Ni alloy: Castings
AMS 5390	22% Cr 18% Fe 9% Mo 0.1% C 1.5% Co 0.6% W Ni alloy: Castings
AMS 5391	12.5% Cr 1% Fe 4.5% Mo 6% Al 2% Nb 0.12% C Ni alloy: Castings
AMS 5396	1% Cr 2% Fe 28% Mo 2.5% Co 0.3% V Ni alloy: Castings
AMS 5397	10% Cr 3% Mo 4.7% Ti 5.5% Al 0.18% C 15% Co Ni alloy: Castings

Symbol	Nominal analysis, supplier, condition and remarks.
AMS 5399	19% Cr 9.7% Mo 3.2% Ti 1.6% Al 0.12% C (max) Ni alloy
AMS 5401	21.5% Cr 9% Mo 3.5% Nb 0.1% C (max) Ni alloy
AMS 5402	21.5% Cr 9% Mo 3.5% Nb 0.1% C (max) Ni alloy
AMS 5404	12.7% Cr 2% Mo 4% Ti 3.5% Al 0.12% C 9% Co 4% Ta 4.2% W 0.1% Zr Ni alloy: Casting
AMS 5405	9% Cr 2% Ti 1% Nb 0.15% C 10% Co 12.5% W 0.05% Zr Ni alloy: Casting
AMS 5406	9% Cr 2% Ti 1% Nb 0.15% C 10% Co 12.5% W 0.05% Zr Ni alloy: Casting
AMS 5407	9% Cr 2% Ti 1% Nb 0.15% C 10% Co 12.5% W 0.05% Zr Ni alloy: Casting
AMS 5509	15% Cr 28% Fe 4% Mo 3% Ti 1% Al 4% W Ni alloy: Sheet and strip; vacuum melted
AMS 5530	15% Cr 6% Fe 16% Mo 3.5% W 2.5% Co 0.3% V Ni alloy: Sheet
AMS 5536	22% Cr 18% Fe 9% Mo 0.1% C 1.5% Co 0.6% W Ni alloy: Sheet
AMS 5536 C	22% Cr 18% Fe 9% Mo 1.5% Co 0.6% W Ni alloy: Sheet
AMS 5541	15% Cr 6% Fe 2.5% Ti 0.6% Al 0.04% C Ni alloy: Sheet
AMS 5541 A	15.5% Cr 2.5% Ti 0.7% Al Fe Ni alloy: Sheet
AMS 5542 C	15.5% Cr 2.5% Ti 0.7% Al 1% Nb + Ta Fe Ni alloy: Sheet
AMS 5544	19.5% Cr 4.3% Mo 3% Ti 1.4% Al 13.5% Co Ni alloy: Sheet; annealed; vacuum melted
AMS 5545	19% Cr 10% Mo 3% Ti 1.5% Al 11% Co Ni alloy: Sheet; solution treated; vacuum melted
AMS 5550	15.6% Cr 0.7% Ti 3.4% Al 0.1% Cu 0.04% C Ni alloy: Sheet
AMS 5550A	15.5% Cr 0.7% Ti 3.25% Al Ni alloy: Sheet
AMS 5551	19% Cr 10% Mo 2.5% Ti 1% Al 10% Co Ni alloy: Sheet; solution treated; vacuum melted
AMS 5581	21.5% Cr 9% Mo 3.6% Nb 0.1% C (max) Ni alloy
AMS 5582	15.5% Cr 7% Fe 2.5% Ti 0.7% Al 1% Nb + Ta Ni alloy: Seamless tube
AMS 5583	15.5% Cr 6.5% Fe 2.5% Ti 1% Nb 0.08% C (max) Ni alloy
AMS 5586	19.5% Cr 4.2% Mo 3% Ti 0.06% C 0.08% Zr 13.5% Co Ni alloy
AMS 5587	21.7% Cr 18.5% Fe 9% Mo 0.1% C 0.8% W 1.5% Co Ni alloy
AMS 5588	21.7% Cr 18.5% Fe 9% Mo 0.1% C 0.8% W 1.5% Co Ni alloy
AMS 5589	19% Cr 10% Fe 3% Mo 0.9% Ti 0.6% Al 5.2% Nb 0.08% C (max) Ni alloy
AMS 5590	19% Cr 10% Fe 3% Mo 0.9% Ti 0.6% Al 5.2% Nb 0.08% C (max) Ni alloy
AMS 5593	25% Cr 17% Fe 0.04% C (max) 3% Co 3% W Ni alloy: Sheet
AMS 5596	19% Cr 3% Mo 1% Ti 0.6% Al 5% Nb + Ta Ni alloy: Sheet; annealed; vacuum melted
AMS 5597	19% Cr 10% Fe 3% Mo 1% Ti 0.6% Al 5.2% Nb 0.08% C (max) Ni alloy
AMS 5598	15% Cr 7% Fe 2.5% Ti 0.8% Al 0.8% Nb + Ta 0.04% C Ni alloy
AMS 5599	21.5% Cr 9% Mo 3.6% Nb 0.1% C (max) Ni alloy
AMS 5660	12.5% Cr 34% Fe 6% Mo 2.7% Ti 0.15% Al 0.05% C (max) Ni alloy: Forgings
AMS 5660 A	12.5% Cr 34% Fe 6% Mo 2.5% Ti Ni alloy: Bar and forging; vacuum melted
AMS 5661	12.5% Cr 34% Fe 5.8% Mo 3% Ti Ni alloy: Bar and forging; vacuum melted
AMS 5661	12.5% Cr 34% Fe 5.8% Mo 2.9% Ti 0.05% C (max) Ni alloy: Forgings
AMS 5666	21.5% Cr 9% Mo 3.6% Nb 0.1% C (max) Ni alloy
AMS 5667	15% Cr 6.7% Fe 2.5% Ti 0.8% Al 0.8% Nb 0.04% C Ni alloy: Forgings
AMS 5668 D	15.5% Cr 7% Fe 2.5% Ti 0.7% Al 1% Nb + Ta Ni alloy: Bar
AMS 5669	15% Cr 7% Fe 2.5% Ti 0.8% Al 0.8% Nb + Ta 0.04% C Ni alloy
AMS 5671	15.5% Cr 2.5% Ti 0.7% Al 1% Nb 0.08% C (max) Ni alloy
AMS 5675	15.5% Cr 7% Fe 3% Ti 2.3% Mn Ni alloy: Welding wire; cold drawn
AMS 5698	15% Cr 6.7% Fe 2.5% Ti 0.8% Nb 0.04% C (max) Ni alloy: Wire
AMS 5699 B	15.5% Cr 7% Fe 2.5% Ti 0.7% Al 1% Nb + Ta Ni alloy: Wire; spring temper
AMS 5704	19.5% Cr 4.2% Mo 3% Ti 1.4% Al 0.06% C 0.07% Zr 13.5% Co Ni alloy
AMS 5706	19.5% Cr 4.5% Mo 3% Ti 1.5% Al 13.5% Co Ni alloy: Bar and forging; solution treated; vacuum melted
AMS 5707 A	19.5% Cr 4.5% Mo 3% Ti 1.5% Al 13.5% Co Ni alloy: Bar and forging; stabilized and aged
AMS 5708	19.5% Cr 4.5% Mo 3% Ti 1.5% Al 13.5% Co Ni alloy: Bar and forging; solution treated
AMS 5709	19.5% Cr 4.5% Mo 3% Ti 1.5% Al 13.5% Co Ni alloy: Bar and forging; solution treated, stabilized and aged
AMS 5711	15.7% Cr 15.5% Mo 0.3% Al 0.02% C (max) 0.05% La Ni alloy
AMS 5712	19% Cr 10% Mo 3% Ti 1.5% Al 11% Co Ni alloy: Bar and forging; solution treated; vacuum melted
AMS 5713	19% Cr 10% Mo 3% Ti 1.5% Al 11% Co Ni alloy: Bar and forging; solution treated and aged; vacuum melted
AMS 5714	15.5% Cr 7% Fe 2.3% Ti 0.7% Al 0.08% C (max) Ni alloy
AMS 5715	23% Cr 16% Fe 1.5% Al 0.1% C (max) Ni alloy
AMS 5717	25% Cr 30% Fe 3.2% Mo 3.2% Co 3.2% W Ni alloy: Bar and forging
AMS 5746	15% Cr 30% Fe 4% Mo 1% Al 4% W Ni alloy: Bar and forging; vacuum melted
AMS 5747	15.5% Cr 2.5% Ti 0.7% Al 1% Nb 0.08% C (max) Ni alloy
AMS 5750	15% Cr 6% Fe 16% Mo 3.5% W 2.5% Co 0.3% V Ni alloy: Forgings
AMS 5751	18% Cr 4% Fe 4% Mo 3% Ti 3% Al 17% Co Ni alloy: Bar and forging; solution treated; stabilized and aged
AMS 5753	18% Cr 4% Mo 3% Ti 3% Al 17% Co Ni alloy: Bar and forging; solution treated; vacuum melted
AMS 5754D	22% Cr 18% Fe 9% Mo 1.5% Co 0.6% W Ni alloy: Bar and forging
AMS 5755	5% Cr 5% Fe 24% Mo 0.06% C 1.2% Co 0.6% V Ni alloy: Forgings
AMS 5756	19% Cr 10% Mo 2.5% Ti 1% Al 10% Co Ni alloy: Bar and forging; solution treated; vacuum melted
AMS 5757	19% Cr 10% Mo 2.5% Ti 1% Al 10% Co Ni alloy: Bar and forging; solution treated and aged; vacuum melted
AMS 5778	15.5% Cr 7% Fe 2.5% Ti 0.7% Al 1% Nb + Ta Ni alloy: Electrode
AMS 5779	15% Cr 2% Ti 0.6% Al Nb + Ta Ni alloy: Coated electrode
AMS 5786	5% Cr 24.5% Mo 0.06% C (max) 1.25% Co Ni alloy: Wire
AMS 5787	4.5% Cr 5.5% Fe 24% Mo Ni alloy: Coated electrode

Note. The following abbreviations and units are used in the tables:

DPN	Hardness, diamond pyramid number
UTS	Ultimate tensile strength, N/mm^2
Elon	Elongation, %
Proof	0.1% proof strength, N/mm^2

1 N/mm^2=0.1 hbar=0.102 kgf/mm^2=0.06475 tonf/in.^{2}=145.04 lbf/in.^{2}=1 MPa

See Appendix II for other abbreviations and conversion tables.

Symbol	Nominal analysis, supplier, condition and remarks.
AMS 5798	22% Cr 18.5% Fe 9% Mo 1.5% Co 0.6% W Ni alloy: Wire
AMS 5799	22% Cr 18.5% Fe 9% Mo 1.5% Co 0.6% W Ni alloy: Coated electrode
AMS 5800 A	19% Cr 10% Mo 3.2% Ti 1.5% Al 11% Co Ni alloy: Welding wire; vacuum melted
AMS 5828	19.5% Cr 4.3% Mo 3% Ti 4% Al 13.5% Co Ni alloy: Welding wire; vacuum melted
AMS 5829	19.5% Cr 2.4% Ti 1.4% Al 0.13% C (max) 18% Co Ni alloy
AMS 5832	19% Cr 10% Fe 3% Mo 1% Ti 0.6% Al 5.2% Nb 0.08% C (max) Ni alloy
AMS 5837	21.5% Cr 9% Mo 3.6% Nb 0.1% C (max) Ni alloy
AMS 5838	15.7% Cr 15.5% Mo 0.3% Al 0.02% C (max) 0.05% La Ni alloy
AMS 5846	15% Cr 5% Mo 3.2% Ti 4% Al 0.06% C 18.5% Co Ni alloy
AMS 5851	15% Cr 5% Mo 3.5% Ti 4% Al 0.04% C 17% Co Ni alloy
AMS 5852	15% Cr 5% Mo 3.5% Ti 4% Al 0.04% C 17% Co Ni alloy
AMS 5855	12% Cr 3% Mo 3% Ti 4.6% Al 0.32% C 6% W 1.5% Ta 0.1% Zr 10% Co Ni alloy
AMS 5856	12% Cr 3% Mo 3% Ti 4.6% Al 0.32% C 6% W 1.5% Ta 0.1% Zr 10% Co Ni alloy
AMS 5872	20% Cr 5.8% Mo 2.6% Al + Ti 0.06% C 20% Co Ni alloy
AMS 5881	12% Cr 3% Mo 3% Ti 4.6% Al 0.32% C 6% W 1.5% Ta 0.1% Zr 10% Co Ni alloy
AMS 5882	15% Cr 5% Mo 3.5% Ti 4% Al 0.04% C 17% Co Ni alloy
AMS 5886	20% Cr 5.8% Mo 2.6% Al + Ti 0.06% C 20% Co Ni alloy
AMS 7237	22% Cr 18.5% Fe 9% Mo 0.1% C 0.6% W 1.7% Co Ni alloy
AMS 7246	7% Fe 2.5% Ti 0.7% Al 1% Nb 0.08% C 15.5% Co Ni alloy
AMS 7469	19% Cr 9.7% Mo 3.1% Ti 1.4% Al 0.12% C (max) 11% Co Ni alloy
AMS 7471	19.5% Cr 4.2% Mo 3% Ti 0.07% C 0.08% Zr 13.5% Co Ni alloy
AMS 7490	21.5% Cr 9% Mo 3.6% Nb 0.1% C (max) Ni alloy
ASTM A461/684	17.5% Cr 4.0% Fe 4.0% Mo 0.008% B 2.7% Ti 2.8% Al 0.15% C (max) 16.5% Co Ni alloy: Bar
ASTM A461/685	19.5% Cr 2% Fe 4.2% Mo 0.008% B 3.0% Ti 1.4% Al 0.06% C 13.5% Co Ni alloy: Bar
ASTM A461/688	15.5% Cr 7.0% Fe 2.5% Ti 0.7% Al 1.0% Nb + Ta 0.08% C (max) 1.0% Co (max) Ni alloy: Bar
ASTM A461/689	19% Cr 5% Fe 9.7% Mo 0.008% B 2.5% Ti 1.0% Al 0.15% C 10% Co Ni alloy: Bar
ASTM A567/4	16.5% Cr 5.7% Fe 0.3% Nb + Ta 0.12% C (max) 2.0% Co 17% W 4.5% V Ni alloy: Casting
ASTM A567/5	21.7% Cr 18.5% Fe 9% Mo 0.2% C (max) 1.7% Co 0.7% W Ni alloy: Casting; as AISI 680
ASTM A567/6 V	19% Cr 2% Fe 4.0% Mo 3% Ti 3% Al 0.1% C (max) 18% Co Ni alloy: Casting; as AISI 684
ASTM A567/7 V	13% Cr 4.5% Mo 0.7% Ti 6% Al 2.5% Nb + Ta 0.14% C 0.1% Zr Ni alloy: Casting
ASTM A567/8	15.5% Cr 10% Fe 5.2% Mo 2.0% Ti 3% Al 0.15% C Ni alloy: Casting
ASTM A567/9 V	15.5% Cr 10% Fe 5.2% Mo 2.0% Ti 3.0% Al 0.15% C Ni alloy: Casting
ASTM A567/10 V	15.5% Cr 4.2% Fe 5.2% Mo 2.5% Ti 3.6% Al 0.15% C Ni alloy: Casting
ASTM A670	18% Cr 20% Fe 3% Mo 1% Ti 0.6% Al 5% Nb + Ta 0.08% C Ni alloy: Precipitation hardened **UTS: 1240 Elon: 12% Proof: 1030**
ASTM B260 B Ni 7	13% Cr 10% P Ni alloy: Brazing filler metal; braze temperature range 927–1038 °C
ASTM B295 E3N1B	1% Cr 5% Fe 28% Mo 0.1% C 2% Co Ni alloy: Welding rod; as welded; as E4N1B **UTS: 830**
ASTM B295 E3N1C	15% Cr 5% Fe 17% Mo 0.12% C 2% Co 4% W Ni alloy: Welding electrode; as welded; as E4N1C **UTS: 650**
ASTM B295 E3N19	15% Cr 11% Fe 2% Ti 3% Nb 0.2% C Ni alloy: Welding electrode; age hardened **UTS: 830**
ASTM B295 E4N1B	1% Cr 5% Fe 28% Mo 0.1% C 2% Co Ni alloy: Welding rod; as welded; as E3N1B **UTS: 830**
ASTM B295 E4N1C	15% Cr 5% Fe 17% Mo 0.12% C 2% Co 4% W Ni alloy: Welding electrode; as welded; as E3N1C **UTS: 650**
ASTM B304 ERN7B	1% Cr 6% Fe 28% Mo 0.08% C 2.5% Co 0.4% V Ni alloy: Welding electrode
ASTM B304 ERN7C	15% Cr 6% Fe 16% Mo 0.08% C 2.5% Co 3.5% W Ni alloy: Welding electrode
ASTM B304 ERN7W	5% Cr 6% Fe 24% Mo 0.12% C 2.5% Co 0.6% V Ni alloy: Welding electrode
ASTM B304 ERN69	16% Cr 7% Fe 2.5% Ti 1% Al 1% Nb 0.08% C Ni alloy: Welding electrode
ASTM B322 Ni Mo	1% Cr 5% Fe 28% Mo 2.5% Co 0.4% V Ni alloy: Casting; annealed **UTS: 530 Elon: 6% Proof: 320**
ASTM B322 Ni Mo Cr	16% Cr 6% Fe 17% Mo 4.5% W 2.5% Co 0.3% V Ni alloy: Casting; annealed **UTS: 530 Elon: 4% Proof: 320**
ASTM B333	1% Cr 5% Fe 28% Mo 2.5% Co 0.3% V Ni alloy: Sheet; hot rolled and annealed **UTS: 810 Elon: 45% Proof: 340**
ASTM B334	15% Cr 6% Fe 16% Mo 4% W 2.5% Co 0.3% V Ni alloy: Sheet; hot rolled and annealed **UTS: 810 Elon: 40% Proof: 340**
ASTM B335	1% Cr 5% Fe 28% Mo 2.5% Co 0.3% V Ni alloy: Rod; annealed **UTS: 810 Elon: 32% Proof: 320**
ASTM B336	15% Cr 6% Fe 16% Mo 3.5% W 2.5% Co 0.3% V Ni alloy: Rod; annealed **UTS: 790 Elon: 22% Proof: 320**
ASTM B637	Ni alloy: Bar and forging; graded by UNS
ASTM B637/80 A	19.5% Cr 3.0% Fe (max) 2.0% Ti 1.2% Al 0.1% C (max) Ni alloy
ASTM B637/684	17.5% Cr 4.0% Fe 4.0% Mo 0.008% B 2.8% Ti 3.0% Al 0.15% C (max) 17.5% Co Ni alloy
ASTM B637/685	19.0% Cr 2.0% Fe 4.2% Mo 0.008% B 3.0% Ti 1.4% Al 0.07% C 13.0% Co 0.09% Zn Ni alloy
ASTM B637/688	15.5% Cr 7.0% Fe 2.5% Ti 0.8% Al 0.08% C (max) 1.0% Co (max) 1.0% Ta Ni alloy
ASTM B637/689	19.0% Cr 5.0% Fe 9.7% Mo 2.5% Ti 1.0% Al 0.5% Mn (max) 0.15% C 10.0% Co Ni alloy
ASTM B637/718	19.0% Cr 20% Fe 3.0% Mo 0.8% Ti 0.5% Al 5.1% Nb + Ta 0.08% C (max) 1.0% Co (max) Ni alloy **UTS: 965 Elon: 30% Proof: 550**
ASTM B670	19% Cr 3% Mo 0.8% Ti 0.6% Al 5.2% Nb 0.08% C (max) Ni alloy: Plate and sheet; Precipitation and aged **UTS: 1240 Elon: 10% Proof: 1035**
ASTROLOY	15% Cr 5.2% Mo 0.03% B 3.5% Ti 4.4% Al 0.06% C 15% Co Ni alloy: American proprietary alloy listed in SAE yearbook
ASTROLOY M	15% Cr 5.2% Mo 3.5% Ti 4% Al 0.04% C Ni alloy: Origin unknown
ATG R	20% Cr 4% Fe 0.4% Ti 0.11% C 4% Co Ni alloy: Imphy **UTS: 770 Elon: 40% Proof: 330**
ATG S3	20% Cr 5.0% Fe 2.0% Ti 1.2% Al 0.1% C Ni alloy: Imphy; solution treated and aged **UTS: 1000 Elon: 15% Proof: 590**

Symbol	Nominal analysis, supplier, condition and remarks.
ATG S4	20% Cr 2.0% Fe 2.2% Ti 1.2% Al 0.12% C 20% Co Ni alloy: Imphy; solution treated and aged **UTS: 1060 Elon: 15% Proof: 700**
BP 87	20% Cr 0.5% Ti 2.5% Nb 3% Mn 0.02% C Ni alloy: Filler; Huntington; now Inconel 82
BS 2901 NA 36	20% Cr 3% Ti 18% Co Ni alloy: Rod; for gas shielded arc welding
BS 2901 NA 37	18% Cr 7% Mo 2% Ti 14% Co Ni alloy: Rod; for gas shielded arc welding
BS 2901 NA 38	20% Cr 6% Mo 20% Co Ni alloy: Rod; for gas shielded arc welding
BS 2901 NA 39	16% Cr 8% Fe 3% Ti Ni alloy: Rod; for gas shielded arc welding
BS 2901 NA 42	16.5% Cr 44% Fe 3% Mo 2% Co Ni alloy: Rod; for gas shielded arc welding
BS 3076 NA 16	21% Cr 32% Fe 3.0% Mo 0.9% Ti 2.2% Cu Ni alloy: Rod
BS 3146 ANC9	20% Cr 2.8% Ti 1.4% Al 0.8% C Ni alloy: Investment casting; solution treated and aged **UTS: 570 Elon: 5% Proof: 400**
BS 3146 ANC10	20% Cr 1.2% Al 0.1% C 17% Co Ni alloy: Investment casting; solution treated and aged **UTS: 570 Elon: 5% Proof: 400**
BS 3146 ANC11	21% Cr 10% Mo 0.32% C 10% Co Ni alloy: Investment casting; as cast **UTS: 400 Elon: 5%**
BS 3146 ANC12	21% Cr 10% Mo 2.5% Ti 0.8% Al 0.1% C (max) 1.0% Co Ni alloy: Investment casting; as cast **UTS: 500 Elon: 5%**
BS 3146 ANC16	17% Cr 6% Fe 17% Mo 0.1% C 4.5% W Ni alloy: Investment casting; as cast **UTS: 420 Elon: 5% Proof: 220**
BS 3976 NA19	19% Cr 2.2% Ti 1.3% Al 18% Co Ni alloy: Rod
C 242	21.5% Cr 10.5% Mo 10% Co Ni alloy: Casting; H Wiggin; see Nimonic 242
C 263	20% Cr 5.9% Mo 2.2% Ti 0.5% Al 20% Co Ni alloy: Forging; H Wiggin; for use up to 850 °C
C 276	15.0% Cr 5.5% Fe 16.0% Mo 4.0% W 2.0% Co Ni alloy: Climax Moly
C 1023	15.5% Cr 8.3% Mo 0.006% B 3.6% Ti 4.1% Al 0.15% C 9.75% Co Ni alloy: Union Carbide
CG 27	13.2% Cr 35% Fe 5.5% Mo 2.5% Ti 1.5% Al 0.04% C Ni alloy: Origin unknown
CM 40	29% Cr 2.5% C 10% Co 14% W Ni alloy: For hard facing; Dewrance; gas welding
COLMONOY 70	11.5% Cr 3.7% Fe 2.5% B 0.5% C 16% W Ni alloy: Casting; Wall Colmonoy; melting point 1120 °C; resists high temperature frettage and corrosion **DPN: 580**
COLMONOY 70	11.5% Cr 3.7% Fe 2.5% B 0.5% C 16% W Ni alloy: Rod for hard facing; Wall Colmonoy; melting point 1120 °C; resists high temperature frettage and corrosion **DPN: 580**
CORNES C276	16% Cr 5.5% Fe 16% Mo 0.02% C 0.35% V 4% W Ni alloy: Philip Cornes
CORRONEL 220	3% Fe 29% Mo 0.2% Cu 0.8% Mn 0.05% C 2% V Ni alloy: H Wiggin; acid resistant **UTS: 1060 Elon: 37% Proof: 550**
CORRONEL 230	36% Cr 5% Fe 1% Ti 0.5% Al 1% Cu 0.08% C Ni alloy: Wrought; H Wiggin; annealed **DPN: 250 UTS: 800**
CRONITE	17% Cr 25% Fe 0.3% Mo 0.7% C 1.0% W 0.2% Co Ni alloy: Casting; Cronite; as cast **DPN: 195 UTS: 580 Elon: 5% Proof: 300**
CRUCIBLE 718	19% Cr 18% Fe 3% Mo 1% Ti 0.06% Al 5% Nb + Ta 0.08% C Ni alloy: Crucible Steel Co.; solution treated and aged **UTS: 1300 Proof: 1120**
CRUCIBLE M252	19% Cr 10% Mo 2.5% Ti 1% Al 0.15% C 10% Co Ni alloy: Crucible Steel Co.; solution treated and aged; AISI 689 **UTS: 1150 Elon: 16% Proof: 740**
CRUCIBLE X750	15% Cr 7% Fe 2.5% Ti 0.8% Al 0.8% Nb + Ta 0.04% C Ni alloy: Crucible Steel Co.; double solution treated and aged; AISI 688 **UTS: 1150 Elon: 23% Proof: 800**
D 979	15% Cr 27% Fe 4% Mo 0.01% B 3% Ti 1% Al 0.05% C 4% W Ni alloy: American proprietary alloy listed in SAE yearbook
DARWIN 55	23% Cr 10% Fe 4% Mo 5% Cu 0.2% C 2% W Ni alloy: Casting; Darwin; acid resistant
DARWIN 654A	20% Fe 20% Mo 0.1% C (max) Ni alloy: Casting; Darwin; acid resistant; not oxidizing acids
DARWIN 655B	Low carbon 8% Fe 27% Mo Ni alloy: Casting; Darwin; acid resistant; not oxidizing acids
DARWIN 656C	Low carbon 14% Cr 6% Fe 17% Mo 5% W Ni alloy: Casting; Darwin; acid resistant
DCM	15% Cr 5% Fe 5.2% Mo 0.08% B 3.5% Ti 4.6% Al 0.08% C Ni alloy: Casting; American proprietary alloy listed in SAE yearbook
DELORO B	5% Fe 29% Mo 0.1% C Ni alloy: Casting; Deloro **DPN: 270 UTS: 600 Elon: 10% Proof: 360**
DELORO C	17% Cr 6% Fe 17% Mo 0.1% C 5% W Ni alloy: Casting; Deloro **DPN: 275 UTS: 600 Elon: 10% Proof: 370**
DIN 17742 Ni Cr 22 Mo	22% Cr 25% Fe 7% Mo 2.0% Nb Ni alloy
DIN 17742 Ni Mo 16 Cr	16% Cr 6% Fe 16% Mo 4% W Ni alloy
DTD 725	19% Cr 2% Ti 1% Al 0.1% C (max) 2% Co (max) Ni alloy: Forging; solution treated and aged; N80 material **DPN: 250**
DTD 736	19% Cr 5% Fe 2.2% Ti 1% Al 0.1% C 2% Co (max) Ni alloy: Forging; solution treated and aged; N80A material **DPN: 250**
DTD 747A	20% Cr 5% Fe 2.2% Ti 1.3% Al 0.1% C 17% Co Ni alloy: Forging; solution treated and aged; N90 **DPN: 250**
DTD 5007	14.5% Cr 5.0% Mo 1.2% Ti 4.7% Al 0.2% C 20% Co Ni alloy: Tensile strength 110 N/mm² at 940 °C; N115 **DPN: 300**
DTD 5017	15% Cr 4.0% Mo 4.0% Ti 5.0% Al 0.2% C 15% Co Ni alloy: Tensile strength 110 N/mm² at 980 °C; N115 **DPN: 330**
DTD 5027	20% Cr 2.2% Ti 1.3% Al 0.13% C 17% Co Ni alloy sheet: Tensile strength 140 N/mm² at 870 °C; solution treated and aged; N90 **DPN: 230 UTS: 1070 Elon: 20% Proof: 630**
DTD 5067	15% Cr 3.5% Mo 3.8% Ti 4.8% Al 15% Co Ni alloy: Vacuum melted
DTD 5077	19.5% Cr 2.25% Ti 1.4% Al Ni alloy: Wrought; for studs, nuts, etc.
DTD 5087	19.5% Cr 1.5% Al 0.13% C (max) 18% Co Ni alloy: Wire
E Ni Cr Mo 1	22% Cr 20% Fe 6.5% Mo 2% Cu 0.05% C (max) Ni alloy: Weld electrode; designation used by AWS
E Ni Cr Mo 2	21.5% Cr 18% Fe 9% Mo 0.11% C 1.7% Co 0.8% W Ni alloy: Weld electrode; designation used by AWS

Note. The following abbreviations and units are used in the tables:

DPN	Hardness, diamond pyramid number
UTS	Ultimate tensile strength, N/mm²
Elon	Elongation, %
Proof	0.1% proof strength, N/mm²

1 N/mm²=0.1 hbar=0.102 kgf/mm²=0.06475 tonf/in.²=145.04 lbf/in.²=1 MPa

See Appendix II for other abbreviations and conversion tables.

Symbol	Nominal analysis, supplier, condition and remarks.
EPK 24	See Nimocast PK24
ER Ni Cr Fe 5	15.5% Cr 7% Fe 2.2% Nb 0.08% C (max) Ni alloy: Weld electrode; designation used by AWS
ER Ni Cr Mo 2	21% Cr 9% Mo 0.1% C 1.5% Co 0.7% W Ni alloy: Weld electrode; designation used by AWS
ER Ni Cr Mo 11	29% Cr 5% Mo 1% Nb 0.03% C (max) 2.5% W Ni alloy: Weld electrode; designation used by AWS
EVANOHM	20% Cr 2.7% Al 2.7% Cu Ni alloy: Wire; Brunton; specific resistance 1340 microhm mm; for use up to 300 °C
FS 718	18.5% Cr 20% Fe 3.0% Mo 0.9% Ti 0.6% Al 5.0% Nb 0.05% C Ta Ni alloy: Firth Sterling; solution treated and aged **UTS: 1500 Elon: 21% Proof: 1350**
FS 901	12.5% Cr 37% Fe 5.7% Mo 2.9% Ti 0.2% Al 0.05% C Ni alloy: Firth Sterling; solution treated and aged **UTS: 1180 Elon: 21% Proof: 850**
FS X 750	15.25% Cr 6.7% Fe 2.5% Ti 0.8% Al 0.9% Nb 0.04% C Ta Ni alloy: Firth Sterling; solution treated and aged **UTS: 1200 Elon: 28% Proof: 770**
G 39	19.5% Cr 5% Fe 3% Mo 1.5% Nb 0.5% C 3% W Ni alloy: Casting; Jessops; used as cast **UTS: 480 Elon: 5%**
G 44	20% Cr 6% Mo 1.6% Ti 1% Al 0.08% C 8% Co Ni alloy: Casting; Jessops; solution treated and aged **UTS: 780 Elon: 40% Proof: 400**
G 54	19% Cr 3% Mo 1.5% Nb 0.5% C 3% W 1.5% Ta Ni alloy: Casting; Jessops
G 55	15% Cr 4% Mo 2.5% Ti 2.5% Al 0.15% C 2% W Ni alloy: Casting; Jessops; as cast **UTS: 760 Elon: 8.5% Proof: 580**
G 62	15% Cr 6% Mo 1.2% Ti 5% Al 0.25% C 19% Co Ni alloy: Bar and forgings; Jessops; solution treated and aged; vacuum melted **UTS: 1260 Elon: 15% Proof: 800**
G 64	11% Cr 4% Fe 3% Mo 6% Al 2% Nb 0.1% C 4% W Ni alloy: Casting; Jessops; solution treated; vacuum cast **UTS: 810 Elon: 3.5% Proof: 650**
G 66	13% Cr 1.7% Mo 2.2% Ti 0.05% C Ni alloy: Bar and forging; Jessop **UTS: 910 Elon: 12% Proof: 620**
G 67	16% Cr 4% Fe 3% Mo 1% Ti 6% Al 0.1% C 4% W Ni alloy: Casting; Jessops; solution treated; vacuum cast **UTS: 900 Elon: 1.6% Proof: 760**
G 69	20% Cr 3% Mo 3% Nb 0.3% C 3% W 10% Co Ni alloy: Casting; Jessops; as cast **UTS: 630 Elon: 20% Proof: 310**
G 70	15% Cr 6% Mo 3% Ti 4% Al 0.15% C 18% Co Ni alloy: Jessops; solution treated and aged **UTS: 1270 Elon: 14% Proof: 940**
G 73	15% Cr 4.5% Mo 2.3% Ti 4.3% Al 0.08% C 26% Co Ni alloy: Casting; Jessops
G 74	Cr Co W Ti Al Ni alloy: Casting; Jessops; solution treated; vacuum cast **UTS: 970 Elon: 3% Proof: 830**
G 76	21% Cr 10% Mo 2.5% Ti 0.7% Al 0.06% C Ni alloy: Casting; Jessops
G 77	20% Cr 6% Mo 6.5% Nb 0.04% C 2.5% W Ni alloy: Casting; Jessops
G 79	14% Cr 4% Mo 0.8% Ti 6% Al 2% Nb 0.14% C Ni alloy: Casting; Jessops
G 80	20% Cr 2.5% Ti 1.5% Al 0.05% C Ni alloy: Jessops; solution treated and aged **UTS: 1100 Elon: 28% Proof: 630**
G 81	20% Cr 2.3% Ti 1.3% Al 0.05% C 18% Co Ni alloy: Wrought; Jessops
G 82	16.0% Cr 3.0% Ti 2.0% Al 0.10% C Ni alloy: Jessops; scale or creep resistant
G 83	10% Cr 10% Mo 3% Ti 2% Al 0.07% C 10% Co Ni alloy: Wrought; Jessops
G 84	10% Cr 3% Mo 5% Ti 5% Al 0.17% C 2% W 15% Co Ni alloy: Casting; Jessops; vacuum cast **UTS: 990 Elon: 4% Proof: 760**
G 94	9% Cr 3.5% Mo 6% Al 4% Nb 0.06% C 4.5% W Ni alloy: Casting; Jessops
G 95	15.0% Cr 5.0% Mo 2.25% Ti 2.75% Al 0.15% C Ni alloy: Jessops; scale or creep resistant
G 100	10% Cr 3% Mo 5% Ti 5.5% Al 0.18% C 0.75% V 15% Co Ni alloy: Casting; Jessops
G 101	13% Cr 35% Fe 5.5% Mo 2.7% Ti 0.2% Al 0.06% C Ni alloy: Wrought; Jessops
G 103	5.0% Cr 4.0% Mo 0.10% C 8.0% W 15.0% Co Ni alloy: Jessops; scale or creep resistant
G 157	27% Cr 6% Fe 1.5% Mo 2% Ti 0.7% Al 0.06% C 1.5% W Ni alloy: American proprietary alloy listed in SAE yearbook
G 267	13% Cr 40% Fe 5.5% Mo 0.05% C with Ti + Al Ni alloy: For shear blades; Jessops **DPN: 320**
GMR 235	15% Cr 10% Fe 5.2% Mo 0.06% B 2% Ti 3% Al 0.15% C Ni alloy: Casting; American proprietary alloy listed in SAE yearbook
GMR 235D	15% Cr 4.5% Fe 5% Mo 0.05% B 2.5% Ti 3.5% Al 0.15% C Ni alloy: Casting; American proprietary alloy listed in SAE yearbook
HAS C	15% Cr 5% Fe 16% Mo 0.03% C 4% W Ni alloy: Electrode; Metrode; for welding Hastelloy C
HAS G	22% Cr 30% Fe 6% Mo 2% Nb 2% Cu 0.04% C 0.5% W Ni alloy: Electrode; Metrode; for welding Hastelloy C
HAS G-3	22% Cr 20% Fe 7% Mo 2% Cu 0.02% C 1% W 4% Co Ni alloy: Electrode; Metrode; for welding Hastelloy C
HASTELLOY A	20% Fe 20% Mo 0.1% C Ni alloy: Haynes Stellite
HASTELLOY B	1% Cr 5% Fe 28% Mo 0.1% C Ni alloy: Casting; Haynes Stellite
HASTELLOY B	1% Cr 5% Fe 28% Mo 0.12% C 0.5% V 2.5% Co Ni alloy: Casting; Osborn; resists hydrochloric and phosphoric acid; solution treated and aged **UTS: 590 Elon: 12% Proof: 330**
HASTELLOY B282	0.6% Cr 5.0% Fe 28.0% Mo 0.02% C 2.0% V Ni alloy: Langley Alloys
HASTELLOY C	16% Cr 5% Fe 17% Mo 0.1% C 4% W Ni alloy: Casting; Haynes Stellite
HASTELLOY C	16% Cr 5% Fe 17% Mo 0.12% C 4% W 2.5% Co Ni alloy: Casting; Osborn; solution treated; acid and thermal shock resistant **UTS: 550 Elon: 8% Proof: 330**
HASTELLOY C276	Low carbon 15.5% Cr 5.5% Fe 16.0% Mo 3.7% W Ni alloy: Langley Alloys
HASTELLOY F	22% Cr 21% Fe 6.5% Mo 2.1% Nb 0.02% C 1.2% Co Ni alloy: Haynes Stellite
HASTELLOY N	7% Cr 5% Fe 16% Mo 0.01% B 0.06% C 0.5% Co Ni alloy: Haynes Stellite
HASTELLOY R235	15% Cr 10% Fe 5.5% Mo 2.5% Ti 2% Al 0.15% C 2.5% Co Ni alloy: Haynes Stellite
HASTELLOY W	5% Cr 5% Fe 24% Mo 0.12% C 2.5% Co 0.5% V Ni alloy: Haynes Stellite
HASTELLOY X	22% Cr 24% Fe 9% Mo 0.15% C Ni alloy: Haynes Stellite
HASTELLOY X	21% Cr 18% Fe 9% Mo 0.2% C 0.7% W 2% Co Ni alloy: Casting; Osborn; as cast; resists oxidation up to 1200 °C **UTS: 440 Elon: 11% Proof: 290**
HASTELLOY X	21% Cr 18% Fe 9% Mo 0.1% C 0.7% W 2% Co Ni alloy: Wrought; Osborn; solution treated; resists oxidation to 1200 °C **UTS: 760 Elon: 41% Proof: 360**
HEV 2	16% Cr 3% Ti 0.05% C 0.5% Co Ni alloy: For exhaust valves; designation used by SAE; annealed **UTS: 1100 Elon: 33% Proof: 630**

Symbol	Nominal analysis, supplier, condition and remarks.
HEV 3	15% Cr 2.5% Ti 0.7% Al 0.05% C 0.7% Co Ni alloy: For valves; designation used by SAE; annealed **UTS: 1110 Elon: 25% Proof: 630**
HEV 5	20% Cr 5% Fe 2.5% Ti 1.1% Al 0.05% C 2% Co Ni alloy: For valves; designation used by SAE; annealed **UTS: 1140 Elon: 39% Proof: 630**
HEV 6	20% Cr 2.5% Ti 1.1% Al 0.05% C 18% Co Ni alloy: For valves; designation used by SAE; annealed **UTS: 1220 Elon: 40% Proof: 740**
HEV 8	22.7% Cr 13% Fe 2% Mo 2.3% Ti 1.3% Al 0.04% C Ni alloy for valves; SAE designation
HR 1	19.5% Cr 2.2% Ti 0.07% C Ni alloy: BS Aerospace designation
HR 2	19.5% Cr 2.5% Ti 1.5% Al 0.13% C (max) 18% Co Ni alloy: BS Aerospace designation
HR 3	14.7% Cr 0.006% B 1.2% Ti 4.7% Al 0.15% C 20% Co Ni alloy: BS Aerospace designation
HR 4	15% Cr 4% Mo 0.06% B 4% Ti 0.16% C 14% Co Ni alloy: BS Aerospace designation
HR 6	21.5% Cr 18% Fe 9% Mo 0.1% C 0.7% W 1.5% Co Ni alloy: BS Aerospace designation
HR 10	20% Cr 5.8% Mo 2.6% Al + Ti 0.06% C 20% Co Ni alloy: BS Aerospace designation
HR 53	12.5% Cr 5.7% Mo 0.015% B 3% Ti 0.04% C Ni alloy: BS Aerospace designation
HR 55	16.5% Cr 35% Fe 3.2% Mo 1.7% Al 0.06% C 2% Co (max) Ni alloy: BS Aerospace designation
HX	17% Cr 11% Fe 0.5% Mo 0.55% C Ni alloy: Casting; Huntington; aged **UTS: 510 Elon: 9% Proof: 290**
HYC	15% Cr 5% Fe 16% Mo 0.8% Mn 0.1% C 4% W Ni alloy: Electrode; Murex; for hard facing **DPN: 300**
IC 901	13% Cr 37% Fe 6% Mo 3.0% Ti 0.05% C Ni alloy: Bofors; solution treated and aged **DPN: 320 UTS: 1050 Elon: 12% Proof: 700**
ILLIUM 98	28% Cr 1% Fe 8.5% Mo 5.5% Cu 0.05% C Ni alloy: American proprietary alloy listed in SAE yearbook
ILLIUM B	28% Cr 8.5% Mo 0.2% B 4.5% Si 5% Cu 0.05% C Ni alloy: American proprietary alloy listed in SAE yearbook
ILLIUM G	22.5% Cr 6.5% Fe 6.5% Mo 6.5% Cu 0.2% C Ni alloy: American proprietary alloy listed in SAE yearbook
ILLIUM R	22% Cr 1% Fe 4% Mo 3% Cu 0.07% C Ni alloy: American proprietary alloy listed in SAE yearbook
IN 100	10% Cr 1% Fe 3% Mo 0.015% B 5% Ti 5.5% Al 0.18% C 15% Co 0.05% Zr Ni alloy: Casting; American proprietary alloy listed in SAE yearbook
IN 102	15% Cr 7.0% Fe 2.9% Mo 0.5% Ti 0.5% Al 2.9% Nb + Ta 0.06% C 3.0% W Ni alloy: International Nickel Co.; hot worked **UTS: 910 Elon: 50% Proof: 460**
IN 162	10.0% Cr 4.0% Mo 0.9% Ti 6.4% Al 1.0% Nb 0.12% C 2.0% W 2.0% Ta Ni alloy: International Nickel Co.; vacuum cast **UTS: 950 Elon: 6% Proof: 190**
IN 713	13% Cr 4.2% Mo 0.7% Ti 6% Al 2.2% Nb 0.1% Zr 0.012 C Ni alloy: Origin unknown

Note. The following abbreviations and units are used in the tables:

DPN	Hardness, diamond pyramid number
UTS	Ultimate tensile strength, N/mm^2
Elon	Elongation, %
Proof	0.1% proof strength, N/mm^2

1 N/mm^2=0.1 hbar=0.102 kgf/mm^2=0.06475 tonf/in.^{2}=145.04 lbf/in.^{2}=1 MPa

See Appendix II for other abbreviations and conversion tables.

Symbol	Nominal analysis, supplier, condition and remarks.
IN 722	15% Cr 6.5% Fe 2.4% Ti 0.6% Al 0.04% C Ni alloy: Union Carbide
IN 738	16.0% Cr 1.7% Mo 3.4% Ti 3.4% Al 0.9% Nb 0.17% C 8.5% Co 2.6% W 1.7% Ta Ni alloy: International Nickel Co.; vacuum cast **UTS: 1120 Elon: 5% Proof: 970**
INCO	See also Incoloy
INCO 276	15% Cr 16% Mo 0.01% C (max) 3.7% W 2.5% Co (max) Ni alloy: Inco **UTS: 790 Elon: 50% Proof: 415**
INCO 700	15% Cr 4% Fe 3% Mo 2.2% Ti 3% Al 2% Mn 0.1% C 29% Co Ni alloy: Inco
INCO 739	15% Cr 0.5% Fe 1.7% Ti 2.7% Al 0.07% C Ni alloy: Inco
INCO 901	12.8% Cr 35% Fe 5.6% Mo 2.5% Ti 0.05% C Ni alloy: Inco
INCO G3	22% Cr 19% Fe 7% Mo 2% Cu 0.015% C (max) 1.5% W (max) 5% Co (max) Ni alloy: Inco; annealed **UTS: 690 Elon: 50% Proof: 320**
INCO HX	22% Cr 19% Fe 9% Mo 0.1% C 1.5% Co 0.8% W Ni alloy: Inco
INCO WELD A	16% Cr 6.7% Fe 3% Ti 2.2% Mn 0.03% C Ni alloy: Filler; Huntington; now Inconel 92
INCOLOY	See also Inco
INCOLOY 804	29.3% Cr 25% Fe 0.4% Ti 0.25% Al 0.06% C Ni alloy: Wrought; Huntington; resists carburization and sulphur penetration
INCOLOY 807	20% Cr 30% Fe 0.4% Ti 0.3% Al 0.08% C 8.0% Co Ni alloy: H Wiggin; for high temperature use **DPN: 225 UTS: 660**
INCOLOY 825	21.5% Cr 30% Fe 3% Mo 0.9% Ti 0.03% C Ni alloy: Bar and sheet; Huntington; annealed **UTS: 650 Elon: 50% Proof: 220**
INCOLOY 825	22% Cr 35% Fe 3.0% Mo 2.0% Cu 0.05% C Ni alloy: Wrought; H Wiggin; annealed **DPN: 150 UTS: 640**
INCOLOY 901	13% Cr 37% Fe 5.5% Mo 2.9% Ti 0.1% C Ni alloy: H Wiggin; solution treated and aged **DPN: 302 UTS: 1120 Elon: 12% Proof: 740**
INCOLOY 901	13.5% Cr 34% Fe 6.2% Mo 2.5% Ti 0.05% C Ni alloy: Bar and sheet; Huntington; age hardened
INCOLOY 901Mod	12.5% Cr 34.0% Fe 5.8% Mo 0.015% B 2.9% Ti 0.05% C Ni alloy
INCOLOY 925	21% Cr 28% Fe 3% Mo 2% Ti 0.3% Al 0.01% C Ni alloy: Inco; solution treated and aged **UTS: 1200 Elon: 24% Proof: 810**
INCONEL 69	15% Cr 6.8% Fe 2.4% Ti 0.8% Al 0.8% Nb 0.04% C Ni alloy: Filler; Huntington; inert gas arc welding of Inconel X750
INCONEL 82	20% Cr 1% Fe 0.5% Ti 2.5% Nb 3% Mn 0.02% C Ni alloy: Filler; Huntington; inert gas welding of Inconel and steel
INCONEL 92	16.5% Cr 6.7% Fe 3.1% Ti 2.2% Mn 0.03% C Ni alloy: Filler; Huntington; inert gas arc welding of austenitic steels
INCONEL 617	22% Cr 9% Mo 1.2% Al 0.1% C 12.5% Co Ni alloy: Inco
INCONEL 625	22% Cr 3% Fe 9% Mo 4% Nb 0.05% C Ni alloy: Bar, sheet, etc.; Huntington; annealed **UTS: 1000 Elon: 42% Proof: 600**
INCONEL 700	15% Cr 3.7% Mo 2.2% Ti 3% Al 0.12% C 28% Co Ni alloy: Bar; Huntington; age hardened **UTS: 1210 Elon: 25% Proof: 760**
INCONEL 702	15.6% Cr 0.7% Ti 3.4% Al 0.04% C Ni alloy: Bar, sheet, etc.; Huntington
INCONEL 718	19% Cr 18% Fe 3% Mo 0.8% Ti 0.6% Al 5% Nb 0.04% C Ni alloy: Bar, sheet, etc.; H Wiggin; age hardened **UTS: 1520 Elon: 23% Proof: 1300**
INCONEL 721	16% Cr 7.2% Fe 3% Ti 2.2% Mn 0.04% C Ni alloy: Bar; Huntington

Symbol	Nominal analysis, supplier, condition and remarks.
INCONEL 722	15% Cr 6.5% Fe 2.4% Ti 0.6% Al 0.04% C Ni alloy: Bar, sheet, etc.; Huntington; age hardened **UTS: 1210 Elon: 31% Proof: 690**
INCONEL 751	15% Cr 6.7% Fe 2.5% Ti 1.2% Al 1% Nb 0.04% C Ni alloy: Bar; Huntington
INCONEL ALLOY 725	21% Cr 1.0% Fe 8.7% Mo 1.3% Ti 3.2% Nb 0.03% C (max) Ni alloy
INCONEL FM69	15.5% Cr 6.5% Fe 2.3% Ti 1% Nb 0.08% C (max) Ni alloy
INCONEL M	16% Cr 7.2% Fe 3% Ti 2.2% Mn 0.04% C Ni alloy: Huntington; now Inconel 721
INCONEL M	15% Cr 6.5% Fe 2.4% Ti 0.6% Al 0.04% C Ni alloy: Bar, sheet, etc.; Huntington; now Inconel 722
INCONEL MA 754	20% Cr 1% Fe 0.3% Al 0.05% C 0.6% Mo Ni alloy: Inco
INCONEL W	15% Cr 7% Fe 2.5% Ti 0.6% Al 0.04% C Ni alloy: Inco
INCONEL X	15% Cr 7% Fe 2.5% Ti 1% Al 1% Nb 0.04% C Ni alloy: Inco
INCONEL X	15% Cr 6.7% Fe 2.5% Ti 0.8% Al 0.8% Nb 0.04% C Ni alloy: Bar, sheet, etc.; Huntington; now Inconel X750
INCONEL X550	15% Cr 7% Fe 2.5% Ti 1% Al 1% Nb 0.04% C Ni alloy: Inco
INCONEL X550	15% Cr 6.7% Fe 2.5% Ti 1.2% Al 1% Nb 0.04% C Ni alloy: Bar; Huntington; now Inconel 751
INCONEL X750	15% Cr 6.7% Fe 2.5% Ti 0.8% Al 0.8% Nb 0.04% C Ni alloy: Bar, sheet, etc.; Huntington; age hardened **UTS: 1170 Elon: 22% Proof: 650**
J 1500	20% Cr 10% Mo 3% Ti 1% Al 0.15% C 10% Co Ni alloy: American proprietary alloy listed in SAE yearbook
K 42B	18% Cr 13% Fe 2.5% Ti 0.2% Al 0.05% C 22% Co Ni alloy: Westinghouse
KARMA	20% Cr 5% Fe + Al Ni alloy: For electrical resistors; British Driver-Harris
KH 70MVTIUB	17.5% Cr 5.0% Fe (max) 5.0% Mo 2.5% Ti 1.3% Al 0.9% Nb 0.12% C (max) 2.7% W Ni alloy: Russian Standard designation
KH 70VMTIU	18.5% Cr 4.0% Fe (max) 4.5% Mo 2.6% Ti 1.2% Al 0.08% C (max) 4.5% W Ni alloy: Russian Standard designation
KH 77TIUR	20.5% Cr 4.0% Fe (max) 2.5% Ti 0.7% Al 0.06% C (max) Ni alloy: Russian Standard designation
KH 80TBIU	16.5% Cr 3.0% Fe (max) 2.1% Ti 0.7% Al 1.2% Nb 0.08% C (max) Ni alloy: Russian Standard designation
KHN 60V	25% Cr 4.0% Fe (max) 0.45% Ti 0.5% Al (max) 0.1% C (max) Ni alloy: Russian Standard designation
L 306	20% Cr 3% Fe 2.5% Ti 1.5% Al 0.1% C (max) Ni alloy: VEW
L 335	15% Cr 7% Fe 2.5% Ti 0.8% Al 1% Nb 0.08% C (max) Ni alloy: VEW
LANGALLOY 4R	1% Cr 5% Fe 28% Mo 0.12% C 0.5% V 2.5% Co Ni alloy: Casting; Langley; Hastelloy B type **DPN: 220 UTS: 550 Elon: 8% Proof: 360**
LANGALLOY 5R	16% Cr 5% Fe 17% Mo 0.12% C 4% W 2.5% Co Ni alloy: Casting; Langley; Hastelloy C type **DPN: 200 UTS: 480 Elon: 9% Proof: 340**
LANGALLOY 7R	23% Cr 5% Fe 6% Mo 3% Si 6% Cu 0.08% C 2% W Ni alloy: Langley **DPN: 200 UTS: 440 Elon: 8% Proof: 330**
LESCALLOY 718 VAC–ARC	19% Cr 3.0% Mo 0.9% Ti 5.2% Nb + Ta 0.8% Al Ni alloy: Latrobe Steel; solution treated and aged **DPN: 370 UTS: 800 Elon: 54% Proof: 450**
LESCALLOY D979	15% Cr 30% Fe 4% Mo 3% Ti 1.0% Al 0.04% C 4% W Ni alloy: Latrobe; for turbine discs
LESCALLOY X750	15.5% Cr 7% Fe 2.5% Ti 0.9% Nb + Ta 0.04% C Ni alloy: Latrobe
M 21	5.7% Cr 2.0% Mo 6.0% Al 1.5% Nb 0.13% C 11.0% W 0.1% Zr Ni alloy: International Nickel Co.; vacuum cast **UTS: 800 Elon: 5.5% Proof: 760**
M 22	5.7% Cr 2.0% Mo 6.3% Al 0.13% C 11.0% W 3.0% Ta 0.6% Zr Ni alloy: International Nickel Co.; vacuum cast **UTS: 740 Elon: 5.5% Proof: 690**
M 22B	5.7% Cr 2.0% Mo 6.3% Al 0.13% C 11.0% W 3.0% Ta 0.5% Zr Ni alloy: International Nickel Co.; vacuum cast **UTS: 800 Elon: 4% Proof: 720**
M 252	19% Cr 10% Mo 2.5% Ti 1% Al 0.15% C 10% Co Ni alloy: Crucible Steel Co.; solution treated and aged – AISI 689 **UTS: 1140 Elon: 16% Proof: 740**
M 252	19% Cr 9.7% Mo 2.5% Ti 1% Al 0.12% C 10% Co Ni alloy: Origin unknown
M 313	30% Cr 1.7% Ti 0.9% Al 0.06% C (max) 0.05% Zr Ni alloy: International Nickel Co.; solution treated and aged **DPN: 245 UTS: 1130 Elon: 30% Proof: 420**
M 600	19% Cr 13% Fe 7% Mo 2.3% Ti 1.1% Al 0.08% C Ni alloy: American proprietary alloy listed in SAE yearbook
MACHINETRODE 92.78	30.0% Cu 0.8% Mn 0.6% C 2.4% Mn Ni alloy: Electrode; Esab **DPN: 150**
MAR M 200	9% Cr 2% Ti 5% Al 1% Nb 0.15% C 12.5% W 10% Co Ni alloy: Casting; Martin Metals Ltd **UTS: 1000 Elon: 6% Proof: 830**
MAR M 246	9% Cr 2.5% Mo 0.02% B 1.5% Ti 5.5% Al 0.15% C 10% Co 10% W 1.5% Ta Ni alloy: Martin Metals Ltd **UTS: 900 Elon: 4% Proof: 780**
MAR M 421	Cr Ni alloy: For creep resistance; Martin Metals Ltd; for turbine blades
MC 102	20% Cr 6.0% Mo 6.6% Nb + Ta 0.04% C 2.5% W Ni alloy: International Nickel Co.; solution treated and aged **UTS: 690 Elon: 5% Proof: 620**
MC 102	20.0% Cr 3.0% Fe 6.0% Mo 0.25% Si 6.6% Nb 0.3% Mn 0.04% C 2.5% W Ni alloy: Union Carbide
METCO 16C	16% Cr 2.5% Fe 3% Mo 4% B 4% Si 3% Cu 0.5% C Ni alloy: Powder for spraying; Metco Ltd **DPN: 765**
METCO 17F	17% Cr 3% Fe 2% Mo 3% B 4% Si 2% Cu 0.5% C Ni alloy: Powder for spraying; Metco Ltd; as sprayed **DPN: 425**
METCO 440	Cr Fe Mo Al Ni alloy: Sprayed deposit; Metco **DPN: 160**
MIL-N-7786	15% Cr 6.7% Fe 2.5% Ti 0.8% Al 0.8% Nb 0.04% C Ni alloy
MIL-N-18088	15% Cr 6% Fe 16% Mo 3.5% W 2.5% Co 0.3% V Ni alloy
MIL-R-5031	1% Cr 5% Fe 28% Mo 2.5% Co 0.3% V Ni alloy
MONEL R405	2.5% Fe (max) 30% Cu 0.3% C (max) Ni alloy: Inco; annealed **UTS: 550 Elon: 40% Proof: 240**
MP 35N	20% Cr 10% Mo 35% Co Ni alloy: Climax Molybdenum
MUMETAL	14% Fe 4% Mo 5% Cu Ni alloy: Sheet and rod; Telcon; annealed; high permeability low hysteresis **DPN: 110 UTS: 530**
MUMETAL	14% Fe 4% Mo 5% Cu Ni alloy: Sheet and rod; Telcon; hard rolled; magnetic properties are reduced **DPN: 290 UTS: 900**
MUMETAL 40	14% Fe 4% Mo 5% Cu Ni alloy: Sheet and rod; Telcon; annealed; high permeability **DPN: 110 UTS: 540**

Symbol	Nominal analysis, supplier, condition and remarks.
MUMETAL 40	14% Fe 4% Mo 5% Cu Ni alloy: Sheet and rod; Telcon; hard rolled; permeability reduced **DPN: 290 UTS: 900**
MUMETAL 40CT	Similar to Mumetal 40; vacuum melted material; Telcon
MUMETAL 60	77% Ni alloy with minimum initial permeability of 60 000; Telcon
N 07718	19% Cr 18% Fe 3% Mo 1% Ti 0.6% Al 5% Nb + Ta 0.08% C (max) Ni alloy: UNS designation
NA 19	19.5% Cr 2.5% Ti 1.5% Al 0.13% C (max) 18% Co Ni alloy: BS 3072-6
NA 20	19.5% Cr 2.5% Ti 1.4% Al 0.07% C Ni alloy
NA 21	21.5% Cr 9% Mo 3.5% Nb 0.1% C (max) Ni alloy
NA 22H (c)	27% Cr 16% Fe 0.5% C 6% W Ni alloy: Blaw-Knox
NB	5% Fe 28% Mo 0.04% C Ni alloy: Wrought; Japanese alloy
NC	15% Cr 5% Fe 13% Mo 0.05% C 4% W Ni alloy: Wrought; Japanese alloy
NI 3021	14.8% Cr 5% Mo 4.8% Al 0.15% C 20% Co Ni alloy: Designation used by UNS
Ni Cr Mo 4	15.5% Cr 6.5% Fe 16% Mo 0.02% C (max) Ni alloy: Weld electrode; designation used by ANS
NI O NEL 65	21% Cr 30% Fe 3% Mo 1% Ti 1.7% Cu 0.03% C Ni alloy: Filler; Huntington; inert gas arc welding of Incoloy 825
NI O NEL 135	19% Cr 31% Fe 5.5% Mo 1% Nb 1.8% Cu 0.05% C Ni alloy: Filler; Huntington; electric welding of Incoloy 825
NI O NEL 825	21% Cr 34% Fe 3% Mo 2% Cu Ni alloy: Wrought; H Wiggin **UTS: 640 Elon: 30% Proof: 220**
NI O NEL 825	21.5% Cr 30% Fe 3% Mo 0.9% Ti 0.03% C Ni alloy: Bar and sheet; Huntington; now Incoloy 825
NICKELVAC 700	15% Cr 3.7% Mo 2.5% Ti 3.2% Al 0.13% C 29% Co Ni alloy: Vanadium Alloys; double vacuum melted
NICKELVAC 901	12.5% Cr 35% Fe 6.1% Mo 0.15% B 3% Ti 0.06% C Ni alloy: Vanadium Alloys; double vacuum melted
NICKELVAC W	15.5% Cr 8% Fe 2.5% Ti 0.75% Al 0.06% C Ni alloy: Vanadium Alloys; double vacuum cast
NICKELVAC X	15.5% Cr 6.7% Fe 2.5% Ti 0.7% Al 1% Nb 0.06% C Ni alloy: Vanadium Alloys; double vacuum cast
NICROFER 4626 Mo W	25% Cr 17% Fe 3% Mo 0.04% C 3% W 3% Co Ni alloy: VDM
NICROFER 4722 Co	21.5% Cr 9% Fe 9% Mo 0.07% C 0.8% W 0.7% Co Ni alloy: VDM
NICROFER 5120 Co Ti	20% Cr 5.8% Mo 1.8% Ti 0.5% Al 0.06% C 20% Co Ni alloy: VDM
NICROFER 5219 Nb	18.5% Cr 18.5% Fe 3% Mo 1% Ti 0.5% Al 5.1% Nb 0.05% C Ni alloy: VDM
NICROFER 5520 Co	21.5% Cr 0.4% Ti 1% Al 0.07% C 11.5% Co Ni alloy: VDM
NICROFER 5716 h Mo W	16% Cr 5.5% Fe 16% Mo 0.02% C 0.35% V 4% W Ni alloy: Philip Cornes
NICROFER 6022 h Mo	22% Cr 4% Fe (max) 9% Mo 3.8% Nb 0.07% C Ni alloy: VDM
NICROFER 7016 Ti Nb	15.5% Cr 2.6% Ti 0.7% Al 1% Nb 0.05% C Ni alloy: VDM
NICROFER 7520 Ti	20% Cr 2.3% Ti 1.4% Al 0.06% C Ni alloy: VDM
NICROFER B 6020	21.5% Cr 9% Mo 3.5% Nb 0.05% C (max) Ni alloy: VDM
NICROFER S 4225	25% Cr 30% Fe (max) 5.5% Mo 2.2% Cu 0.02% C (max) Ni alloy: Weld rod; VDM
NICROFER S 4722	22% Cr 18% Fe 9% Mo 0.01% C 0.7% W 2.2% Co Ni alloy: Weld rod; VDM
NICROFER S 5520	22% Cr 9% Mo 1.2% Al 0.1% C (max) 12% Co Ni alloy: Weld rod; VDM
NICROFER S 5621	21% Cr 3% Fe 13.5% Mo 0.008% C (max) 0.2% V 3% W Ni alloy: Weld rod; VDM
NICROFER S 5716	15.5% Cr 5.5% Fe 16% Mo 0.02% C (max) 4.2% W Ni alloy: Weld rod
NICROFER S 6020	21.5% Cr 9% Mo 0.05% C (max) Ni alloy: Weld rod; VDM
NICROFER S 6616	16% Cr 15.5% Mo 0.5% Ti 0.015% C (max) Ni alloy: Weld rod; VDM
NICROTUNG	12% Cr 0.05% B 4% Ti 4% Al 0.1% C 10% Co 8% W 0.05% Zr Ni alloy: Casting; American proprietary alloy listed in SAE yearbook
NIMOCAST 80	20% Cr 5.0% Fe 2.0% Ti 1.0% Al 0.1% C 2.0% Co Ni alloy: Casting; H Wiggin; cast variety of N80A
NIMOCAST 90	20% Cr 5.0% Fe 2.5% Ti 1.5% Al 0.13% C 18% Co Ni alloy: Casting; H Wiggin; cast variety of N90
NIMOCAST 258	10.0% Cr 5.0% Mo 3.5% Ti 4.8% Al 0.2% C 20.0% Co Ni alloy: Casting; H Wiggin
NIMOCAST 713	13.4% Cr 4.5% Mo 1.0% Ti 6.2% Al 2.3% Nb Ni alloy: Casting; H Wiggin; for cast blading used up to 1000 °C
NIMOCAST 713LC	As 713 with 0.05% C **DPN: 370 UTS: 820**
NIMOCAST PE10	20% Cr 6.0% Mo 6.5% Nb 2.5% W Ni alloy: Casting; H Wiggin; for use up to 850 °C
NIMOCAST PK24	10% Cr 3.0% Mo 5.2% Ti 5.5% Al 15% Co Ni alloy: Casting; H Wiggin; vacuum melted; for use up to 1000 °C
NIMOCAST PK24	9.5% Cr 3.0% Mo 4.7% Ti 5.5% Al 0.17% C 15.0% Co 1.0% V Ni alloy: H Wiggin; for use up to 1000 °C
NIMOLOY PK37	18.5% Cr 4.0% Fe 2.5% Ti 1.0% Al 0.1% C 18.5% Co Ni alloy: H Wiggin; abrasion and thermal shock resistant
NIMONIC 80 A	20% Cr 5% Fe 2% Ti 1% Al 0.1% C 2% Co Ni alloy: H Wiggin; solution treated and aged; tensile strength 600 N/mm^2 at 750 °C **DPN: 300 UTS: 1070 Elon: 40% Proof: 600**
NIMONIC 81	30% Cr 1.8% Ti 0.9% Al 0.05% C (max) 1.0% Co Ni alloy: H Wiggin; for turbine components **DPN: 230 UTS: 1050**
NIMONIC 90	20% Cr 5% Fe 2.5% Ti 1.5% Al 0.13% C 18% Co Ni alloy: H Wiggin; solution treated and aged; tensile strength 700 N/mm^2 at 750 °C **DPN: 300 UTS: 1210 Elon: 33% Proof: 750**
NIMONIC 93	19.5% Cr 2% Fe 0.3% Mo 3% Ti 1.5% Al 17% Co Ni alloy: For turbine blades; H Wiggin
NIMONIC 95	19.5% Cr 5.0% Fe 0.15% C 18.0% Co Ni alloy: H Wiggin
NIMONIC 100	11.0% Cr 2.0% Fe 5.0% Mo 1.5% Ti 5.0% Al 0.3% C 20.5% Co Ni alloy: H Wiggin
NIMONIC 100	19.0% Cr 5.0% Mo 1.5% Ti 5.0% Al 0.3% C 20% Co Ni alloy: H Wiggin
NIMONIC 105	14% Cr 1% Fe 1.2% Ti 4.5% Al 0.2% C 20% Co Ni alloy: H Wiggin; double solution treated and aged **DPN: 370 UTS: 990 Elon: 7% Proof: 760**
NIMONIC 115	15% Cr 3.5% Mo 4% Ti 5% Al 0.15% C 15% Co Ni alloy: H Wiggin; double solution treated; tensile strength 1000 N/mm^2 at 800 °C **DPN: 400 UTS: 1210 Elon: 27% Proof: 800**
NIMONIC 118	15% Cr 3.5% Mo 4% Ti 5% Al 0.15% C 15% Co Ni alloy: Wrought; vacuum cast; H Wiggin; improved high temperature strength over N115 **DPN: 400 UTS: 1210 Elon: 27% Proof: 800**

Note. The following abbreviations and units are used in the tables:

DPN	Hardness, diamond pyramid number
UTS	Ultimate tensile strength, N/mm^2
Elon	Elongation, %
Proof	0.1% proof strength, N/mm^2

1 N/mm^2=0.1 hbar=0.102 kgf/mm^2=0.06475 $tonf/in.^2$=145.04 $lbf/in.^2$=1 MPa

See Appendix II for other abbreviations and conversion tables.

Symbol	Nominal analysis, supplier, condition and remarks.
NIMONIC 120	12.5% Cr 5.7% Mo 2.5% Ti 4.5% Al 0.04% C 10.0% Co Ni alloy: H Wiggin; for gas turbine blades **UTS: 1150**
NIMONIC 242	21.5% Cr 10.5% Mo 10% Co Ni alloy: Casting; H Wiggin; for cast turbine engine blades
NIMONIC 263	20% Cr 6% Mo 2.2% Ti 0.5% Al 20% Co Ni alloy: H Wiggin **DPN: 320 UTS: 650 Elon: 45% Proof: 400**
NIMONIC 901	12.5% Cr 42% Fe 2.7% Ti Ni alloy: H Wiggin; for turbine discs and blades **DPN: 360 UTS: 780 Elon: 20% Proof: 580**
NIMONIC 942	12.5% Cr 6.0% Mo 3.7% Ti 0.6% Al 0.03% C 1.0% Co Ni alloy: H Wiggin; for compressor blades, etc.
NIMONIC AP1	15% Cr 5% Mo 0.025% B 3.5% Ti 4% Al 0.02% C 17% Co Ni alloy: Inco
NIMONIC PE 11	18% Cr 35% Fe 5% Mo 2.5% Ti 0.8% Al Ni alloy: Forging; H Wiggin; vacuum cast material for use up to 600 °C
NIMONIC PE 13	22% Cr 18% Fe 9.0% Mo 1.5% Co 0.6% W Ni alloy: Sheet; H Wiggin; for furnace parts
NIMONIC PE 16	16.5% Cr 35% Fe 3.2% Mo 1.2% Ti 1.2% Al Ni alloy: Forging; H Wiggin; vacuum cast material for use up to 750 °C
NIMONIC PK 25	18.0% Cr 4.0% Mo 2.9% Ti 0.07% C 2.7% Cl 17.5% Co Ni alloy: H Wiggin; creep resistant **UTS: 1360**
NIMONIC PK 31	20% Cr 4.5% Mo 2.3% Ti 0.4% Al 5.0% Nb 14% Co Ni alloy: Forging; H Wiggin; gas turbine discs
NIMONIC PK 33	19% Cr 7.0% Mo 2.0% Ti 2.0% Al 14% Co Ni alloy: Forging; H Wiggin; vacuum processed for use up to 950 °C
PE 10	20% Cr 6.0% Mo 6.5% Nb 2.5% W Ni alloy: Casting; H Wiggin; see Nimocast PE10
PE 11	18% Cr 35% Fe 5.0% Mo 2.5% Ti 0.8% Al Ni alloy: Forging; H Wiggin; see Nimonic PE 11
PE 13	22% Cr 18% Fe 9.0% Mo 0.6% W 1.5% Co Ni alloy: Sheet; H Wiggin; for furnace parts
PE 16	16.5% Cr 34.15% Fe 3.25% Mo 0.003% B 1.25% Ti 1.25% Al 0.06% C 0.04% Zr Ni alloy: Union Carbide
PERMALLOY C	13% Fe 4% Mo 5% Cu Ni alloy: Standard Telephones; where high initial permeability required
PK 24	10% Cr 3.0% Mo 5.2% Ti 5.5% Al 15% Co Ni alloy: Casting; H Wiggin; see Nimocast PK24
PK 31	20% Cr 4.5% Mo 2.3% Ti 0.4% Al 5.0% Nb 14% Co Ni alloy: Forging; H Wiggin; see Nimonic PK 31
PK 33	19% Cr 7.0% Mo 2.0% Ti 2.0% Al 14% Co Ni alloy: Forging; H Wiggin; see Nimonic PK 33
PK 42	15% Cr 6% Mo 3.2% Ti 4% Al 0.15% C 19% Co Ni alloy: Forging; H Wiggin
PYROMET 31	22.5% Cr 15% Fe 2% Mo 2.5% Ti 1.3% Al 0.9% Cu 0.45% C Ni alloy
PYROMET 80A	20% Cr 0.7% Fe 2.3% Ti 1.2% Al 0.06% C 1% Co Ni alloy: Carpenter **UTS: 1000 Elon: 39% Proof: 620**
PYROMET 751	15% Cr 6.7% Fe 2.5% Ti 1.2% Al 1% Nb 0.04% C Ni alloy: Carpenter **UTS: 1200 Elon: 20% Proof: 750**
PYROMET ALLOY 31	22.5% Cr 15% Fe 2% Mo 2.5% Ti 1.5% Al 1% Nb 0.04% C Ni alloy: Carpenter **UTS: 1250 Elon: 25% Proof: 770**
PYROMET ALLOY 41	19% Cr 5% Fe 9.5% Mo 3.2% Ti 1.5% Al 0.1% C 11% Co Ni alloy: Carpenter **UTS: 1420 Elon: 14% Proof: 1050**
PYROMET ALLOY 88	16.4% Cr 6.5% Fe 3% Ti 0.03% C Ni alloy: Carpenter
PYROMET ALLOY 90	19.5% Cr 2.4% Ti 1.4% Al 0.07% C 18% Co Ni alloy: Carpenter **UTS: 1200 Elon: 33% Proof: 800**
PYROMET ALLOY 860	14% Cr 0.0% Fe 6% Mo 3% Ti 1% Al 0.1% C (max) 4% Co Ni alloy: Carpenter **UTS: 1240 Elon: 21% Proof: 800**
PYROMET ALLOY M252	19% Cr 5% Fe 9.5% Mo 2.5% Ti 1% Al 0.15% C 10% Co Ni alloy: Carpenter **UTS: 1200 Elon: 25% Proof: 760**
PYROTOOL 7	19% Cr 20% Fe 3% Mo 0.6% Al 0.05% C + Ni + Co alloy: Carpenter
PYROTOOL EX	14% Cr 37% Fe 6% Mo 3% Ti 0.05% C 4% Co Ni alloy: Carpenter
PYROTOOL M	19% Cr 2% Fe 10% Mo 2.5% Ti 1% Al 0.12% C 10% Co Ni alloy: Carpenter
PYROTOOL W	19.5% Cr 1% Fe 4.2% Mo 3% Ti 1.2% Al 0.05% C 13% Co Ni alloy: Carpenter
R 235	15% Cr 10.0% Fe 5.5% Mo 2.5% Ti 2.0% Al 0.12% C 1.1% Co Ni alloy: Information from SAE handbook
REFRACTALOY 26	18% Cr 18% Fe 3% Mo 2.8% Ti 0.2% Al 0.05% C 20% Co Ni alloy: Westinghouse
RENE 41	19% Cr 10% Mo 0.01% B 3% Ti 1.5% Al 0.09% C 11% Co Ni alloy: American proprietary alloy listed in SAE yearbook
RENE 41	19% Cr 9.7% Mo 3.2% Ti 1.6% Al 0.12% C (max) 11% Co Ni alloy
REXWELD 66	16% Cr 6% Fe 17% Mo 0.15% C 4.5% W Ni alloy: For hard facing; Crucible Steel Co.; work hardened and aged
SAE HEV 2	16% Cr 3% Ti 0.05% C 0.5% Co Ni alloy: For exhaust valves; annealed **UTS: 1110 Elon: 25% Proof: 640**
SAE HEV 3	15% Cr 2.5% Ti 0.7% Al 0.05% C 0.7% Co Ni alloy: For valves; annealed **UTS: 1180 Elon: 25% Proof: 640**
SAE HEV 5	20% Cr 5% Fe 2.5% Ti 1.1% Al 0.05% C 2% Co Ni alloy: For valves; annealed **UTS: 1140 Elon: 39% Proof: 640**
SAE HEV 6	20% Cr 2.5% Ti 1.1% Al 0.06% C 18% Co Ni alloy: For valves; annealed **UTS: 1210 Elon: 40% Proof: 740**
SAE VF 3	29% Cr 6% Fe 2.4% C 10% Co 15% W Ni alloy: For hard facing
SAE VF 4	26% Cr 4% Fe 2.0% C 0.5% Co 8.5% W Ni alloy: For hard facing
SEL 1	15.0% Cr 4.4% Mo 0.015% B 2.4% Ti 4.5% Al 0.08% C 22.0% Co Ni alloy: Union Carbide
SIMALLOY 750	16% Cr 7.0% Fe 2.5% Ti 0.7% Al 1.0% Nb 0.08% C Ni alloy: Simonds
SUPER MUMETAL 50	14% Fe 4% Mo 4% Cu Ni alloy: Telcon; annealed; better permeability than Mumetal **DPN: 110 UTS: 540**
SUPER MUMETAL 50	14% Fe 4% Mo 4% Cu Ni alloy: Telcon; hard rolled **DPN: 290 UTS: 910**
SUPER MUMETAL 100	14% Fe 4% Mo 5% Co Ni alloy: Telcon; annealed; better permeability than Mumetal **DPN: 110 UTS: 540**
SUPER MUMETAL 100	14% Fe 4% Mo 5% Co Ni alloy: Telcon; hard rolled **DPN: 290 UTS: 910**
SUPER MUMETAL Co	Similar to Mumetal; now referred to as Mumetal 60; Telcon
TICONIUM	23% Cr 6% Fe 6% Mo 0.01% C 31% Co Ni alloy: Westinghouse
TRW 1800	13% Cr 0.07% B 0.6% Ti 6% Al 0.09% C 9% W 0.07% Zr 1% V Ni alloy: Casting; American proprietary alloy listed in SAE yearbook
UCAR 16	16.5% Cr 3.0% Fe 3.2% Mo 1.2% Ti 1.2% Al 0.05% C Ni alloy: Union Carbide
UCAR 80	19% Cr 0.4% Mo 2.5% Ti 1.2% Al 0.08% C 2% Co Ni alloy: Union Carbide; solution treated and aged **UTS: 760 Elon: 15% Proof: 540**
UCAR 90	20% Cr 0.4% Mo 2.8% Ti 1.2% Al 0.08% C 17% Co Ni alloy: Union Carbide; solution treated and aged **UTS: 740 Proof: 540**
UCAR 625	22.0% Cr 3.0% Fe 9.0% Mo 0.2% Ti 0.2% Al 4.0% Nb 0.05% C Ni alloy: Union Carbide

Symbol	Nominal analysis, supplier, condition and remarks.
UCAR 700	15.0% Cr 0.7% Fe 3.7% Mo 2.1% Ti 3.0% Al 0.12% C 28.5% Co Ni alloy: Union Carbide
UCAR 702	15.5% Cr 0.4% Fe 0.7% Ti 3.5% Al 0.04% C Ni alloy: Union Carbide
UCAR 713C	13% Cr 4.7% Mo 0.7% Ti 6% Al 2% Nb + Ta 0.12% C 1% Co 0.1% Zr Ni alloy: Union Carbide; as cast; AMS 5391 **UTS: 910 Elon: 5% Proof: 770**
UCAR 713LC	12.0% Cr 4.5% Mo 0.6% Ti 6.0% Al 2.0% Nb 0.05% C Ni alloy: Union Carbide
UCAR 718	19% Cr 3% Mo 1% Ti 0.6% Al 5.2% Nb + Ta 0.08% C 1% Co Ni alloy: Union Carbide; wrought; solution treated and aged **UTS: 1300 Elon: 15% Proof: 1070**
UCAR 718	19% Cr 3% Mo 1% Ti 0.6% Al 5.2% Nb + Ta 0.08% C 1% Co Ni alloy: Union Carbide; vacuum cast; solution treated and aged **UTS: 1070 Elon: 25% Proof: 770**
UCAR 901	13.5% Cr 6.2% Mo 2.5% Ti 0.2% Al 0.05% C 35% Fe Ni alloy: Union Carbide
UCAR C130	21.5% Cr 0.5% Fe 6.0% Mo 2.1% Ti 0.05% C 20.0% Co Ni alloy: Union Carbide
UCAR C1023	15.5% Cr 8.3% Mo 3.6% Ti 4.1% Al 0.15% C 9.7% Co Ni alloy: Union Carbide
UCAR GMR 235	15% Cr 5.2% Mo 0.07% B 2% Ti 3% Al 0.15% C Ni alloy: Union Carbide; as cast **UTS: 760 Elon: 4% Proof: 660**
UCAR GMR 235D	16% Cr 5% Mo 0.07% B 2.5% Ti 3.7% Al 0.15% C Ni alloy: Union Carbide; vacuum cast **UTS: 760 Elon: 3% Proof: 720**
UCAR HASTELLOY R 235	15% Cr 5% Mo 2.5% Ti 2% Al 0.16% C 2.5% Co Ni alloy: Sheet; Union Carbide; solution treated and aged **UTS: 1210 Elon: 21% Proof: 820**
UCAR IN100	10% Cr 3% Mo 5% Ti 5.5% Al 0.17% C 15% Co Ni alloy: Union Carbide; vacuum cast; as cast **UTS: 1070 Elon: 9% Proof: 870**
UCAR IN731	9.5% Cr 2.5% Mo 4.6% Ti 5.5% Al 0.18% C 10.0% Co Ni alloy: Union Carbide
UCAR IN792	12.7% Cr 2.0% Mo 4.2% Ti 3.2% Al 0.2% C 9.0% Co 4.0% W Ni alloy: Union Carbide
UCAR M21	5.7% Cr 2.0% Mo 0.02% B 6.0% Al 1.5% Nb 0.21% C 11.0% W 0.12% Zr Ni alloy: Vacuum cast; Union Carbide
UCAR R 41	19% Cr 10% Mo 0.008% B 3.1% Ti 1.5% Al 0.1% C 11% Co Ni alloy: Sheet and bar; Union Carbide; solution treated and aged **UTS: 1540 Elon: 14% Proof: 1110**
UCAR U500	19.0% Cr 4.0% Fe 4.0% Mo 3.0% Ti 3.0% Al 0.08% C 18.0% Co Ni alloy: Union Carbide
UCAR U520	19.0% Cr 6.0% Mo 3.0% Ti 2.0% Al 0.05% C 12.0% Co 1.0% W Ni alloy: Union Carbide
UCAR U700	15.0% Cr 5.2% Mo 3.5% Ti 4.2% Al 0.15% C 18.5% Cr Ni alloy: Union Carbide
UCAR U710	18.0% Cr 3.2% Mo 5.0% Ti 2.5% Al 0.07% C 15.0% Co 1.5% W Ni alloy: Union Carbide
UCAR U722	15.5% Cr 2.4% Ti 0.7% Al 0.08% C (max) Ni alloy: Union Carbide
UCAR WASPALLOY	19% Cr 4.2% Mo 3% Ti 1.2% Al 0.08% C 14% Co 0.1% Zr Ni alloy: Sheet; Union Carbide; solution treated **UTS: 1070 Elon: 25% Proof: 580**
UCAR X	22.0% Cr 9.0% Mo 0.1% C 1.5% Co 0.6% W Ni alloy: Union Carbide
UDIMET 200	12.5% Cr 42% Fe 2.7% Ti Ni alloy: For turbine discs and blades; American proprietary alloy **DPN: 360 UTS: 780 Elon: 20% Proof: 580**
UDIMET 500	19% Cr 4% Fe 4% Mo 0.01% B 3% Ti 3% Al 0.08% C 19% Co Ni alloy: American proprietary alloy listed in SAE yearbook
UDIMET 500	17.5% Cr 4% Mo 2.8% Ti 3% Al 0.15% C (max) 14.5% Co Ni alloy
UDIMET 600	17.5% Cr 4% Fe 4% Mo 0.04% B 2.8% Ti 4.2% Al 0.1% C 16.5% Co Ni alloy: American proprietary alloy listed in SAE yearbook
UDIMET 700	15% Cr 1% Fe 5% Mo 0.1% B 3.5% Ti 4.2% Al 0.15% C 18% Co Ni alloy: American proprietary alloy listed in SAE yearbook
UNITEMP AF 1753	16% Cr 10% Fe 1.6% Mo 0.008% B 3.2% Ti 2% Al 0.2% C 7% Co 8.4% W 0.06% Zr Ni alloy: American proprietary alloy listed in SAE yearbook
UNS N07718	UNS designation for ASTM A670/718
VACUMELTROL 41	19% Cr 5% Fe 9.7% Mo 0.007% B 3.15% Ti 1.5% Al 0.1% C 11% Co Ni alloy: Carpenter
VF 3	29% Cr 6% Fe 2.4% C 10% Co 15% W Ni alloy: For hard facing; designation used by SAE
VF 4	26% Cr 4% Fe 2% C 0.5% Co 8.5% W Ni alloy: For hard facing; designation used by SAE
VMA 2	5.7% Cr 2% Mo 6% Al 11% W Ni alloy: Investoral casting; BS designation; obsolete
VMA 3	21% Cr 10% Mo 2.5% Ti 1% Al Ni alloy: Investoral casting; BS designation; obsolete
WASPALOY	19% Cr 1% Fe 4.3% Mo 0.005% B 3% Ti 1.3% Al 0.7% C 14% Co Ni alloy: American proprietary alloy listed in SAE yearbook; contact H Wiggin
WASPALOY MOD	19% Cr 1% Fe 7% Mo 2.5% Ti 1.2% Al 0.05% C 11.5% Co Ni alloy: American proprietary alloy listed in SAE yearbook; contact H Wiggin
X 750	15% Cr 7% Fe 2.5% Ti 0.8% Al 0.8% Nb + Ta 0.04% C Ni alloy: Crucible Steel Co.; double solution treated and aged; AISI 688 **UTS: 1140 Elon: 23% Proof: 890**
Z 15 CNS 25-20	16% Cr 5.5% Fe 16% Mo 0.02% C 4% W 0.35% V Ni alloy: French Standard

Note. The following abbreviations and units are used in the tables:

DPN	Hardness, diamond pyramid number
UTS	Ultimate tensile strength, N/mm^2
Elon	Elongation, %
Proof	0.1% proof strength, N/mm^2

$1\ N/mm^2 = 0.1\ hbar = 0.102\ kgf/mm^2 = 0.06475\ tonf/in.^2 = 145.04\ lbf/in.^2 = 1\ MPa$

See Appendix II for other abbreviations and conversion tables.

27D Nickel–copper wrought and cast alloys

Specific gravity		8.83
Density		8830 kg/m^3
Solidus/liquidus		1300–1350 °C
Thermal conductivity		26 W/m °C
Coefficient of linear expansion (20–100 °C)		13.6×10^{-6}/ °C
Electrical conductivity		3.6% IACS (copper 100%)
Specific resistance		480 microhm mm
Young's modulus of elasticity		165.5×10^9 N/m^2
Impact		136 J
Fatigue strength (10×10^7 cycles)	annealed	±240 N/mm^2
	cold worked	±330 N/mm^2

Hot strength

Temperature °C	Tensile strength N/mm^2	Elongation %
20	530	48
100	530	48
300	520	49
400	450	50
500	340	30
600	240	30
700	140	40
800	75	50
900	50	54
980	34	60

The above properties have been chosen to show typical values for the specifications listed and may not apply exactly to any one specification. It is possible that with some specifications the values may not be applicable.

General metallurgical characteristics

Nickel and copper form a complete series of solid solutions. In the alloys listed the copper is dissolved in the nickel matrix, and this solution persists at room temperature. Thus one phase only is present, resulting in excellent corrosion resistance over a wide range of temperatures. The copper improves the electrical and thermal conductivity over other nickel alloys and also increases the ease of brazing.

Some of these alloys have been in existence for many years, as there are naturally occurring nickel/copper ores which on reduction give a 70/30 nickel/copper alloy. The fact that this is readily worked, easily machined and has good heat and corrosion resistance ensured it a ready market. This alloy was probably the first material which allowed designers to start using the power available when high temperatures were harnessed. Iron may also be present in many of the alloys listed, either as an impurity with adverse affects on corrosion resistance, or as an additive along with manganese and silicon to improve mechanical properties.

With titanium and aluminium additions intermetallic compounds are formed which impart some age hardening properties at the expense of corrosion resistance.

Like all nickel alloys these materials harden considerably when cold worked, increasing the tensile properties and decreasing the ductility. It is possible to apply a controlled amount of cold work to these alloys and then to use them in service. It is also possible to apply excessive cold work and then a controlled amount of annealing but in both instances specialist advice is advised.

A limited number of the alloys can be solution treated and aged and this is indicated in the specification.

These alloys have magnificent marine corrosion resistance and more than reasonably good temperature, particularly oxidation, resistance and therefore find use on all types of marine equipment where temperatures are involved.

They are therefore used with combustion equipment, direct heated boilers, fire-boxes, etc.

In many of the applications, however, they are now being replaced with some of the more sophisticated austenitic stainless steels.

Note. The following abbreviations and units are used in the tables:

DPN	Hardness, diamond pyramid number
UTS	Ultimate tensile strength, N/mm^2
Elon	Elongation, %
Proof	0.1% proof strength, N/mm^2

1 N/mm^2=0.1 hbar=0.102 kgf/mm^2=0.06475 tonf/in.^{2}=145.04 lbf/in.^{2}=1 MPa

See Appendix II for other abbreviations and conversion tables.

Symbol	Nominal analysis, supplier, condition and remarks.
2.4360	2.0% Fe 31% Cu Ni alloy: German Standard
2.4374	1.5% Fe 0.8% Ti 3.0% Al 30% Cu Ni alloy: German Standard
2.4375	2% Fe 1% Ti 3% Al 30% Cu 0.25% C Ni alloy
44 K MONEL	0.5% Ti 2.8% Al 29.5% Cu 0.15% C Ni alloy: Filler; Huntington; now Monel 44
64 K MONEL	0.5% Ti 2.8% Al 29.5% Cu 0.15% C Ni alloy: Filler; Huntington; now Monel 64

Symbol	Nominal analysis, supplier, condition and remarks.
134 K MONEL	0.3% Ti 2% Al 27% Cu 2.5% Mn 0.25% C Ni alloy: Filler; Huntington; now Monel 134
400	2% Fe 31% Cu 2.0% Mn Ni alloy: International Nickel Co.; cold drawn rod **DPN: 220 UTS: 790 Elon: 20% Proof: 570**
ALLOY 82	9% Si 3% Cu 0.12% C Ni alloy: Casting; Dewrance **DPN: 270 UTS: 680**
ALLOY 400	1.2% Fe 31% Cu 0.15% C (max) Ni alloy: VDM; Nicorros
ALLOY K 500	1.2% Fe 0.4% Ti 3% Al 30% Cu 0.15% C (max) Ni alloy: VDM; Nicorros Al
ALLOY R 405	31% Cu 0.3% C (max) Ni alloy: VDM; Nicorros S
AMPCOLAY 551	1% Fe 0.5% Ti 2.8% Al 29.5% Cu 0.15% C Ni alloy: Wrought; Ampco for K Monel
AMPCOLAY 552	30% Cu Ni alloy: Wrought; Ampco for Monel
AMPCOLAY 553	3% Fe 3.7% Si 32% Cu 1% Mn 0.12% C Ni alloy: Casting; Ampco for S Monel
AMPCOLAY 557	3% Fe 2.7% Si 32% Cu 1% Mn 0.12% C Ni alloy: Casting; Ampco for H Monel
AMPCOLAY 558	30% Cu Ni alloy: Weldable; Ampco for Monel
AMPCOLAY 559	30% Cu Ni alloy: Cast; Ampco for Monel
AMS 4544 B	30% Cu Ni alloy: Sheet; annealed
AMS 4574 B	30% Cu Ni alloy: Seamless tube; annealed
AMS 4575	44% Cu 0.3% C (max) Ni alloy
AMS 4674 C	30% Cu 0.04% S Ni alloy: Bar and forging; free machining
AMS 4675	30% Cu Ni alloy: Bar and forging
AMS 4675 B	30% Cu Ni alloy: Brazed tube; annealed
AMS 4676	0.6% Ti 3% Al 30% Cu Ni alloy: Bar and forging; K Monel
AMS 4677	3% Al 30% Cu 0.1% C (max) Ni alloy
AMS 4730C	30% Cu Ni alloy: Wire; annealed; AMS for 400 Monel
AMS 4731	2.5% Fe (max) 31% Cu Ni alloy
AMS 4892	4% Si 30% Cu Ni alloy: Casting; as cast
AMS 4893	4% Si 30% Cu Ni alloy: Casting; heat treated
AMS 7233	37% Cu 0.3% C (max) Ni alloy: Monel 400
AMS 7234	34% Cu 0.3% C (max) Ni alloy: Monel K405
ASTM B127	2% Fe 30% Cu Ni alloy: Plate, sheet and strip; annealed **UTS: 580 Elon: 35% Proof: 180**
ASTM B127	2% Fe 30% Cu Ni alloy: Plate, sheet and strip; hard **UTS: 740 Elon: 2% Proof: 640**
ASTM B164 A	2.5% Fe 33% Cu 2% Mn Ni alloy: Rod and bar; annealed; low sulphur **UTS: 480 Elon: 35% Proof: 150**
ASTM B164 A	2.5% Fe 33% Cu 2% Mn Ni alloy: Rod and bar; cold drawn; low sulphur **UTS: 760 Elon: 8% Proof: 610**
ASTM B164 B	2.5% Fe 33% Cu 2% Mn Ni alloy: Bar and rod; annealed (0.02–0.06% S) **UTS: 480 Elon: 35% Proof: 150**
ASTM B164 B	2.5% Fe 33% Cu 2% Mn Ni alloy: Bar and rod; cold drawn (0.02–0.06% S) **UTS: 680 Elon: 8% Proof: 340**
ASTM B165	2.5% Fe 33% Cu 2% Mn 0.024% S (max) Ni alloy: Pipe; annealed **UTS: 540 Elon: 35% Proof: 180**

Note. The following abbreviations and units are used in the tables:

DPN	Hardness, diamond pyramid number
UTS	Ultimate tensile strength, N/mm^2
Elon	Elongation, %
Proof	0.1% proof strength, N/mm^2

1 N/mm^2=0.1 hbar=0.102 kgf/mm^2=0.06475 tonf/in.^{2}=145.04 lbf/in.^{2}=1 MPa

See Appendix II for other abbreviations and conversion tables.

Symbol	Nominal analysis, supplier, condition and remarks.
ASTM B295 E3N10	1% Ti 1.5% Al 30% Cu 4% Mn 0.4% C Ni alloy: Welding electrode **UTS: 540**
ASTM B295 E3N14	2% Fe 3% Al 35% Cu 4% Mn 0.4% C Ni alloy: Welding electrode; age hardened **UTS: 740**
ASTM B295 E4N10	2% Fe 1% Ti 3% Nb 30% Cu 2% Mn 0.15% C Ni alloy: Welding electrode
ASTM B295 E Ni Cu 1	3% Nb (max) 30% Cu 4% Mn 0.15% C (max) Ni alloy: Welding rod; previous designation E4N10
ASTM B295 E Ni Cu 2	2.5% Nb (max) 30% Cu 6% Mn 0.15% C (max) Ni alloy: Welding rod
ASTM B295 E Ni Cu 4	30% Cu 4.0% Mn 0.4% C (max) Ni alloy: Welding rod; previous designation E3N10
ASTM B304 ER Ni Cu 7	28% Cu Ni alloy: Welding electrode; previous designation ERN60
ASTM B304 ERN60	2% Fe 1% Si 2% Ti 30% Cu 1% Mn 0.15% C Ni alloy: Welding electrode
ASTM B304 ERN64	2% Fe 1% Si 7% Ti 3% Al 30% Cu 0.25% C Ni alloy: Welding electrode
ASTM B304 R Ni Cu 5	30% Cu Ni alloy: Welding electrode; previous designation RN40
ASTM B304 RN40	2% Fe 30% Cu 2% Mn 0.3% C Ni alloy: Welding electrode
ASTM B304 RN43	1% Si 30% Cu 0.3% C Ni alloy: Welding electrode
AWS A 515 E Ni Cu B	30% Cu Ni alloy: Welding electrode
BS 1537	Cu Ni alloy: Casting; replaced by BS 3071
BS 2901 NA33	2% Fe 2.2% Ti 1.2% Al 25% Cu Ni alloy: Rod for all forms of welding
BS 3071 NA1	1% Si 30% Cu 1% Mn Ni alloy: Casting; as cast **UTS: 420 Elon: 16% Proof: 150**
BS 3071 NA2	2.7% Si 30% Cu 1% Mn Ni alloy: Casting; as cast **UTS: 540 Elon: 10% Proof: 220**
BS 3071 NA3	4% Si 30% Cu 1% Mn Ni alloy: Casting; as cast **DPN: 250 UTS: 610**
BS 3072 NA13	32% Cu Ni alloy: Annealed **UTS: 450 Elon: 33% Proof: 150**
BS 3073 NA13	32% Cu Ni alloy: Strip; annealed **DPN: 130 UTS: 450 Elon: 33% Proof: 150**
BS 3074 NA13	32% Cu Ni alloy: Annealed **UTS: 460 Elon: 35% Proof: 180**
BS 3074 NA13	32% Cu Ni alloy: Tube; cold drawn **UTS: 580 Elon: 10% Proof: 370**
BS 3075 NA13	32% Cu Ni alloy: Wire; annealed **UTS: 450 Elon: 22%**
BS 3075 NA13	32% Cu Ni alloy: Wire; cold drawn **UTS: 730**
BS 3076 NA13	32% Cu Ni alloy: Bar and section; annealed **UTS: 450 Elon: 35% Proof: 100**
BS 3076 NA13	32% Cu Ni alloy: Bar and section; cold drawn **UTS: 600 Elon: 14% Proof: 450**
BS 3076 NA18	0.7% Ti 3% Al 30% Cu Ni alloy: Rod
BS 3127/4	1% Ti 3% Al 30% Cu Ni alloy: Tube; solution treated for bourdon tubes **DPN: 190**
BS 3127/4	1% Ti 3% Al 30% Cu Ni alloy: Tube; solution treated and cold worked for bourdon tubes **DPN: 250**
BS 3146 ANC18A	1.0% Si 30% Cu 0.2% C Ni alloy: Investment casting; as cast **UTS: 300 Elon: 15% Proof: 100**
BS 3146 ANC18B	2.7% Si 30% Cu 0.1% C Ni alloy: Investment casting; as cast **UTS: 420 Elon: 10% Proof: 220**
BS 3146 ANC18C	4% Si 30% Cu 0.1% C Ni alloy: Investment casting; as cast **UTS: 580 Proof: 480**
CALMALLOY	2% Fe 29% Cu Ni alloy: Used for temperature compensation shunts; origin unknown
CASTEES M	Cu Ni alloy: Welding electrode; Philip; for cast iron

Symbol	Nominal analysis, supplier, condition and remarks.
CI SOFT FLOW Ni Cu	2% Fe 32% Cu 0.6% C Ni alloy: Electrode; Metrode; for cast iron
COPEL	45% Cu Ni alloy: Origin unknown
CORNES 13	27% Cu Ni alloy: Philip Cornes
CORNES 18	37% Cu + Al + Ti Ni alloy: Philip Cornes
CORRONEL	26% Cu 4% Mn Ni alloy: Origin unknown
DIN 17743 Ni Cu 30 Al	1.5% Fe 0.8% Ti 3% Al 30% Cu Ni alloy
DIN 17743 Ni Cu 30 Fe	2.0% Fe 31% Cu Ni alloy
DTD 10 B	2.5% Fe 35% Cu Ni alloy: Sheet **UTS: 450 Proof: 100**
DTD 192	2.5% Fe 35% Cu Ni alloy: Bar and forging; annealed **UTS: 540 Elon: 30%**
DTD 196	32% Cu 1.5% Mn Ni alloy: Bar; annealed **UTS: 540 Elon: 35%**
DTD 200 A	2.5% Fe 35% Cu Ni alloy: Bar and strip; cold rolled and tempered **UTS: 680 Elon: 14% Proof: 450**
DTD 204 A	2% Fe 31% Cu Ni alloy: Bar and tube; annealed **UTS: 560**
DTD 232	1.5% Fe 28% Cu 1.5% Mn 23% Zn Ni alloy: Sheet; cold rolled **UTS: 870 Elon: 12% Proof: 690**
DTD 237	30% Cu 23% Zn Ni alloy: Sheet and strip; softened **UTS: 540 Elon: 25% Proof: 220**
DTD 268	2% Fe 30% Cu 2% Mn 23% Zn Ni alloy: Bar and wire; annealed **UTS: 540**
DTD 477	2% Fe 30% Cu 1.5% Mn Ni alloy: Tube; annealed; solid drawn **UTS: 490**
DTD 487	2% Fe 3% Al 30% Cu 2% Mn Ni alloy: Bar; annealed and cold drawn to size; for cold headed bolts **UTS: 810 Elon: 20% Proof: 690**
E Ni Cu 7	35% Cu 0.15% C (max) Ni alloy: Weld electrode; designation used by ANS
E Ni Cu A	4.5% Fe 40% Cu 0.45% C Ni alloy: Weld electrode; designation used by ANS
E Ni Cu B	4.5% Fe 30% Cu 0.4% C Ni alloy: Weld electrode; designation used by ANS
EG 1	31% Cu 0.3% C (max) Ni alloy: Vallruna
EG 2	0.5% Ti 3% Al 30% Cu Ni alloy: Vallruna
ER Ni Cu 7	35% Cu 0.15% C (max) Ni alloy: Weld electrode; designation used by ANS
H MONEL	3% Fe 2.7% Si 32% Cu 1% Mn 0.12% C Ni alloy: Casting; H Wiggin; now Monel 506
HAI 400 ALLOY	32% Cu Ni alloy: Harrison Alloys
HAI Cu Ni 102	45% Cu Ni alloy: Harrison Alloys
HAI Cu Ni 103	45% Cu Ni alloy: Harrison Alloys
HAI Cu Ni 104	45% Cu Ni alloy: Harrison Alloys
JAE	30% Cu Ni alloy with high temperature coefficient of magnetism: H Wiggin; Curie point 70–75 °C **DPN: 110 UTS: 420 Elon: 52% Proof: 140**
K 500	2% Fe 1% Si 1% Ti 3% Al 30% Cu 1.5% Mn 0.25% C Ni alloy: Wrought; H Wiggin; annealed; can be age hardened **DPN: 170 UTS: 680**
K MONEL	1% Fe 0.5% Ti 2.8% Al 29.5% Cu 0.15% C Ni alloy: Bar, sheet, etc.; Huntington; now Monel K500
KR MONEL	0.5% Ti 2.8% Al 29.5% Cu 0.23% C Ni alloy: Bar; Huntington; now Monel 501
LANGALLOY 1N	1.5% Fe 1.8% Si 28.5% Cu 1% Mn Ni alloy: Langley **DPN: 120 UTS: 420 Elon: 25% Proof: 180**
LANGALLOY 2N	3% Fe 3% Si 30% Cu Ni alloy: Casting; Langley for BS alloy NA2
LANGALLOY 5N	3% Fe 4% Si 30% Cu Ni alloy: Casting; Langley for BS alloy NA3
LC MONEL	1.3% Fe 13% Cu 0.12% C Ni alloy: Sheet, bar, etc.; Huntington; now Monel 506
M 35–1	3.5% Fe (max) 30% Cu 0.35% C (max) Ni alloy: Casting; US designation
M 35–2	3.5% Fe (max) 30.5% Cu 0.35% C (max) Ni alloy: Casting; US designation
METCO 57 NS	Cu Ni alloy: For coating; Metco **DPN: 125**
METCO 58 NS	Fe Cu In Ni alloy: For coating; Metco **DPN: 125**
METCO MOREL	Cu Ni alloy: Sprayed deposit; Metco **DPN: 100**
MONEL 40	1.3% Fe 31% Cu 0.1% C Ni alloy: Filler rod; Huntington; oxyacetylene welding of Monel 400
MONEL 43	40% Cu 0.1% C Ni alloy: Filler rod; Huntington; oxyacetylene welding of Monel 402 and 403
MONEL 44	0.5% Ti 2.8% Al 29.5% Cu 0.15% C Ni alloy: Filler rod; Huntington; oxyacetylene welding of Monel K 500
MONEL 60	2.2% Ti 30.5% Cu 0.03% C Ni alloy: Filler; Huntington; inert gas shielded welding of Monel 400
MONEL 64	0.5% Ti 2.8% Al 29.5% Cu 0.15% C Ni alloy: Filler; Huntington; inert gas shielded welding of Monel K 500
MONEL 130	0.3% Ti 1% Al 27% Cu 2.5% Mn 0.15% C Ni alloy: Filler; Huntington; for welding Monel 400
MONEL 134	0.3% Ti 2% Al 27% Cu 2.5% Mn 0.25% C Ni alloy: Filler; Huntington; for welding Monel K 500
MONEL 140	0.7% Ti 0.3% Al 26% Cu 1.2% Mn 0.05% C 1.5% Nb Ni alloy: Filler; Huntington; for welding Monel 400 to steel
MONEL 180	0.7% Ti 0.3% Al 28% Cu 5% Mn 0.03% C 1.5% Nb Ni alloy: Filler; Huntington; for welding Monel alloys
MONEL 400	1.3% Fe 31.5% Cu 0.12% C Ni alloy: Bar, sheet, tube, etc.; Huntington; annealed **UTS: 550 Elon: 48% Proof: 200**
MONEL 400	3% Fe 1% Si 32% Cu 1% Mn 0.3% C Ni alloy: Casting; H Wiggins; as cast **DPN: 120 UTS: 420 Elon: 25% Proof: 200**
MONEL 400	2.5% Fe 32% Cu 2% Mn 0.3% C Ni alloy: Wrought; H Wiggin; annealed **DPN: 140 UTS: 550 Elon: 40% Proof: 220**
MONEL 400	2.5% Fe 32% Cu 2% Mn 0.3% C Ni alloy: Wrought; H Wiggin; cold rolled **DPN: 200 UTS: 760 Elon: 7% Proof: 690**
MONEL 401	53% Cu 0.03% C 0.5% Co Ni alloy: Strip; Huntington
MONEL 402	40% Cu 0.12% C Ni alloy: Sheet, bar, etc.; Huntington
MONEL 403	40% Cu 1.8% Mn 0.12% C Ni alloy: Bar, sheet, etc.; Huntington
MONEL 404	0.02% Al 44% Cu 0.06% C Ni alloy: Bar, sheet, etc.; Huntington
MONEL 406	1.3% Fe 13% Cu 0.12% C Ni alloy: Sheet, bar, etc.; Huntington
MONEL 414	2.5% Fe 32% Cu 1.0% Mn 0.4% C Ni alloy: Rod and wire; H Wiggin; free machining variety of Monel 400
MONEL 501	0.5% Ti 2.8% Al 29.5% Cu 0.23% C Ni alloy: Bar; Huntington
MONEL 502	3% Al 34% Cu Ni alloy: Monel 502
MONEL 505	3% Fe 3.7% Si 32% Cu 1% Mn 0.12% C Ni alloy: Casting; H Wiggin; as cast **DPN: 240 UTS: 630 Proof: 630**
MONEL 506	3% Fe 2.7% Si 32% Cu 1% Mn 0.12% C Ni alloy: Casting; H Wiggin; as cast **DPN: 200 UTS: 56 Elon: 12% Proof: 32**
MONEL C and C	2.5% Fe 32% Cu 1.0% Mn 0.4% C Ni alloy: For machining; H Wiggin; as Monel 400; higher carbon
MONEL FM60	35.5% Cu Ni alloy: Weld filler metal
MONEL K405	1.3% Fe 31.5% Cu 0.18% C Ni alloy: Bar; Huntington
MONEL K500	2% Fe 1% Si 1% Ti 3% Al 30% Cu 1.5% Mn 0.25% C Ni alloy: Wrought; H Wiggin; annealed; can be age hardened **DPN: 170 UTS: 69**
MONEL K500	1% Fe 0.5% Ti 2.8% Al 19.5% Cu 0.15% C Ni alloy: Bar, sheet, etc.; Huntington; age hardened **DPN: 275 UTS: 107 Elon: 25% Proof: 76**

Symbol	Nominal analysis, supplier, condition and remarks.
MONEL K500	1% Fe 0.5% Ti 2.8% Al 29.5% Cu 0.15% C Ni alloy: Huntington; cold drawn and aged **DPN: 300 UTS: 121 Elon: 20%**
NCC PIG	9% Cr 8% Fe 0.5% Si 25% Cu 0.5% Mn Ni alloy: Ingot; H Wiggin; for addition to cast iron
NICORROS	1.7% Fe 31% Cu 1.2% Mn 0.15% C (max) Ni alloy: VDM **UTS: 470 Elon: 30% Proof: 180**
NICORROS	27% Cu Ni alloy: Philip Cornes
NICORROS Al	1.2% Fe 0.4% Ti 3% Al 30% Cu 0.15% C (max) Ni alloy: VDM **UTS: 1120 Elon: 25% Proof: 790**
NICORROS Al	27% Cu + Al + Ti Ni alloy: Philip Cornes
NICORROS B 6530	2.2% Ti 31% Cu 3% Mn 0.15% C (max) Ni alloy: Strip; VDM
NICORROS S	31% Cu 0.3% C (max) Ni alloy: VDM **UTS: 582 Elon: 28% Proof: 520**
NICORROS S 6530	2.2% Ti 31% Cu 0.15% C (max) Ni alloy: Weld rod; VDM
NO 4402	31% Cu 0.04% C (max) Ni alloy: Designation used by UNS
PLATNAM	0.5% Fe 0.3% Al 33% Cu 13% Sn Ni alloy: Hopkinsons; for valve discs and seats up to 600 °C
QQ N 281	2.5% Fe 31% Cu Ni alloy: US Federal specification
R MONEL	1.3% Fe 31.5% Cu 0.18% C Ni alloy: Bar; Huntington; Monel 405
S MONEL	3% Fe 3.7% Si 32% Cu 1% Mn 0.12% C Ni alloy: Casting; H Wiggin; now Monel 505
SOUDOFONTE K	46% Fe 7.0% Cu 1.1% C Ni: Welding electrode; Soudometal **DPN: 200**
SOUDOFONTE Mo	30% Cu Ni: Welding electrode; Soudometal; for welding cast iron **DPN: 160**
SOUDONEL F	1.5% Fe 0.6% Ti 31% Cu 3.5% Mn 0.06% C Ni alloy: Welding electrode; Soudometal; for welding Cu/Ni alloys **UTS: 510 Elon: 43% Proof: 320**
SOUDOR Cr Ni Cu 7	2.2% Ti 28.5% Cu 3% Mn 0.02% C Ni alloy: Welding electrode; Soudometal **UTS: 490 Elon: 30% Proof: 210**
SOUDOTIG Ni Cu 7	0.2% Fe 2.2% Ti 28.5% Cu 0.02% C Ni alloy: Electrode; Soudometal; for inert gas welding **UTS: 490 Elon: 30% Proof: 210**
STUBS 80 MON	1% Fe 29.5% Cu 3% Mn 0.025% C Ni: Welding electrode; Stubs; welding Ni/Cu materials **UTS: 480 Elon: 35% Proof: 280**

Note. The following abbreviations and units are used in the tables:

DPN	Hardness, diamond pyramid number
UTS	Ultimate tensile strength, N/mm^2
Elon	Elongation, %
Proof	0.1% proof strength, N/mm^2

1 N/mm^2=0.1 hbar=0.102 kgf/mm^2=0.06475 $tonf/in.^2$=145.04 $lbf/in.^2$=1 MPa
See Appendix II for other abbreviations and conversion tables.

27E Nickel–iron alloys
With chromium, copper or molybdenum

Specific gravity	8.5
Density	8500 kg/m^3
Solidus/liquidus	1425 °C
Thermal conductivity	110 W/m °C
Coefficient of linear expansion (20–100 °C)	11.5×10^{-6}/ °C
Electrical conductivity	8.5% IACS (copper 100%)
Specific resistance	200 microhm mm
Impact	Variable
Fatigue strength	±210 N/mm^2
Hot strength	

Temperature °C	Tensile strength N/mm^2
95	580
205	550
310	550
485	520
540	500

The above properties have been chosen to show typical values for the specifications listed and may not apply exactly to any specification. It is possible that with some specifications the values may not be applicable.

General metallurgical characteristics

The alloys of nickel and iron have special magnetic properties and, on the whole, this is the sole reason for their existence. As nickel is added to iron the magnetic properties first increase and then decrease; the iron/nickel alloys and properties are discussed in Section 20C. Included in Section 20C are the series of alloys varying between 45% and 55% nickel. For clarity no attempt has been made to include any of these alloys in the present group although all those above 50% nickel are technically members of Section 27E.

Most of the alloys listed here contain between 70% and 80% nickel, the remainder being principally iron.

Chromium improves the hot corrosion resistance and strength and also increases the electrical resistance. The copper and molybdenum additions are usually to stabilize some magnetic property.

For none of these alloys are mechanical properties improved by any form of heat treatment.

Like all nickel alloys they work harden considerably and this work hardening can be removed by annealing at a temperature above 1000 °C.

There are some very specialist heat treatment operations which alter the magnetic properties of these alloys but they require highly specialist advice and usually specialist equipment.

The principal reason for using these materials is for their magnetic properties where they are used for magnetic shields, electrical resistors and other electronic, magnetic and telecommunication purposes.

Some are used for their oxidation resistance at high temperatures.

Symbol	Nominal analysis, supplier, condition and remarks.
2.4500–2.4519	2.0% Cr 16% Fe 5% Cu Ni alloy: German Standard numbers
2.4540–2.4559	16% Fe 4% Mo Ni alloy: German Standard numbers
2.4867	15% Cr 20% Fe Ni alloy: German Standard
141 ALLOY	30% Fe Ni alloy: British Driver-Harris; ballast resistance; withdrawn
ACI HU	19% Cr 38% Fe 0.5% C Ni alloy: Origin unknown
ACI HU 50	19% Cr 38% Fe 0.5% C Ni alloy: Origin unknown
ACI HW 50	12% Cr 26% Fe 0.5% C Ni alloy: Origin unknown
AMS 5605	16% Cr 39% Fe 1.7% Ti 2.9% Nb 0.06% C (max) Ni alloy
AMS 5606	16% Cr 39% Fe 1.7% Ti 2.9% Nb 0.06% C (max) Ni alloy
AMS 5633	13.2% Cr 30% Fe 5.5% Mo 2.5% Ti 1.5% Al 0.8% Nb 0.06% C Ni alloy
AMS 5634	13.2% Cr 30% Fe 5.5% Mo 2.5% Ti 1.5% Al 0.8% Nb 0.06% C Ni alloy
AMS 5701	16% Cr 39% Fe 1.7% Ti 2.9% Nb 0.06% C (max) Ni alloy
AMS 5702	16% Cr 39% Fe 1.7% Ti 2.9% Nb 0.06% C (max) Ni alloy
AMS 5703	16% Cr 39% Fe 1.7% Ti 2.9% Nb 0.06% C (max) Ni alloy
AWS A 515 E Ni Fe CI	46% Fe 0.8% C Ni alloy: Welding electrode
BALCO	Fe Ni alloy: Gilbey Brunton; high temperature coefficient of resistance
BS 2857 A	24% Fe Ni alloy: Strip for laminations
BS 2901 NA41	22% Cr 27% Fe 3% Mo 2% Cu Ni alloy: Rod for gas shielded arc welding
CI MET Ni Fe	45% Fe 1.3% C Ni alloy: Electrode; Metrode; for SG cast iron
CI Ni Fe	45% Fe 1.0% C Ni alloy: Electrode; Metrode; for high strength welding
DEF 5192	Cr 14% Fe Mo Cu Mn Ni alloy: For high permeability magnetic alloys
DIN 17470 Ni Cr 60/15	15% Cr 20% Fe Ni alloy
DIN 17745 Ni Fe 15 Mo	16% Fe 4% Mo Ni alloy
DIN 17745 Ni Fe 16 Cu Cr	2.0% Cr 16% Fe 5% Cu Ni alloy
DIN 17745 Ni Fe 16 Cu Mo	16% Fe 4% Mo 5% Cu Ni alloy
DIN 17745 Ni Fe 46 Cr	6% Cr 46% Fe Ni alloy

Note. The following abbreviations and units are used in the tables:

DPN	Hardness, diamond pyramid number
UTS	Ultimate tensile strength, N/mm²
Elon	Elongation, %
Proof	0.1% proof strength, N/mm²

1 N/mm²=0.1 hbar=0.102 kgf/mm²=0.06475 tonf/in.²=145.04 ibf/in.²=1 MPa

See Appendix II for other abbreviations and conversion tables.

Symbol	Nominal analysis, supplier, condition and remarks.
E Ni Cr Mo B	27% Cr 22% Fe 8.5% Mo 2.7% C 12% Co 3% W: Ni alloy; weld electrode; designation used by ANS
E Ni Fe CI	37% Fe 2% C (max) Ni alloy: Weld electrode; designation used by ANS
E Ni Fe CI A	8% Fe (max) 2.5% Al 2% C (max) Ni alloy: Weld electrode; designation used by ANS
E Ni Fe Mn CI	45% Fe 12% Mn 2% C (max) Ni alloy: Weld electrode; designation used by ANS
E Ni Fe T3 CI	42% Fe 4% Mn 2% C (max) Ni alloy: Weld electrode; designation used by ANS
FIXAMPER	28% Fe Ni alloy: Imphy
GLASS SEALING 52	49.5% Fe 0.01% C Ni alloy: Carpenter; low expansion coefficient
HAI 52 ALLOY	48% Fe Ni alloy: Harrison Alloys
HAI 380 ALLOY	30% Fe Ni alloy: Harrison Alloys
HIPERNOM	16% Fe 4.5% Mo 0.05% C Ni alloy: Westinghouse; for magnetic shielding
HIPERNOM ALLOY	15% Fe 4.2% Mo 0.35% Si 0.02% C Ni alloy: Carpenter; for magnetic properties
HY Mu 80	25% Fe 4.2% Mo 0.35% Si 0.02% C Ni alloy: Carpenter; annealed; high magnetic permeability **UTS: 550 Elon: 60% Proof: 170**
HY Mu 800	15% Fe 5% Mo 0.15% Si 0.01% C Ni alloy: Carpenter
HY Mu 800A	15% Fe 5% Mo 0.3% Al_2O_3 0.005% C (max) Ni alloy: Carpenter; magnetic properties with good wear resistance
HYTEMCO	30% Fe Ni alloy: British Driver-Harris; ballast resistance **UTS: 520**
KONEL	10% Fe 17% Co Ti Ni alloy: Tensile strength 450 N/mm² at 600 °C; source unknown
LEGA 52	49% Fe Ni alloy: Low expansion; Driver-Harris
N 52	48% Fe Ni alloy: Low expansion; Imphy
N 54	46% Fe Ni alloy: Low expansion; Imphy
N 58	42% Fe Ni alloy: Low expansion; Imphy
N 501	1.0% Cr 49% Fe Ni alloy: Low expansion; Imphy
NILGRO 36	36% Fe Ni alloy: Low expansion; Darwins; coefficient of expansion 1.5×10^{-6} up to 150 °C **UTS: 480 Elon: 31% Proof: 260**
NILO 51	49% Fe 0.15% C (max) Ni alloy: H Wiggin; for sealing to soft glasses **DPN: 140 UTS: 520**
NILO 51	49% Fe Ni alloy: For glass seals; International Nickel Co.
NILO 55	42% Fe (max) 0.15% C (max) Ni alloy: H Wiggin; for cast iron welding electrodes
NILOMAG 77	13.5% Fe 4.0% Mo 5.0% Cu Ni alloy: H Wiggin; laminations for transducers and switches
NILOMAG 77	13.5% Fe 4.2% Mo 5% Cu 0.02% C Ni alloy: Inco **DPN: 125 UTS: 540**
NIRON 52	49% Fe Ni alloy: For glass/metal seal; Driver-Harris
NY–RA 80	15% Fe 4.2% Mo 0.35% Si 0.02% C Ni alloy: Carpenter; for magnetic properties
PERMALLOY F	35% Fe Ni alloy: Standard Telephones; low coercive force
PLATINITE	48% Fe Ni alloy: Low expansion; Imphy
R 2799	Fe Ni alloy: Telcon; hard rolled; Curie point approximately 60 °C **DPN: 190**

Symbol	Nominal analysis, supplier, condition and remarks.
SANBOLD NA76	24% Fe Ni alloy: Strip for magnetic shielding, etc.; Sanderson; electrical conductivity 2.8% IACS
SOUDOFONTE C	46% Fe 0.8% C Ni alloy: Welding electrode; Soudometal; for welding cast iron **DPN: 190**
SOUDOFONTE D	46% Fe 1.9% C Ni alloy: Welding electrode; Soudometal; for welding cast iron **DPN: 180**
SOUDOFONTE F	49% Fe 1.7% C Ni alloy: Welding electrode; Soudometal; for welding cast iron **DPN: 200**
STUBS 8 FNN	Fe Ni alloy: Welding rod; Stubs; for welding SG cast iron **DPN: 180 UTS: 420 Elon: 20%**
STUBS 8 Nifer	Fe Ni alloy: Welding electrode with Si Mn and Cu; Stubs; for welding nodular and ductile cast iron **DPN: 220 UTS: 540 Elon: 23%**
STUBS 8 S	Fe Si Cu Mn Ni alloy: Welding electrode; Stubs; surface of grey cast iron **DPN: 175 UTS: 490 Elon: 20%**
STUBS 86 FN	Fe Ni alloy: Welding rod; Stubs; for welding grey nodular and SG cast iron **DPN: 180 UTS: 500 Elon: 18% Proof: 340**
STUBS 88	Fe Cu Mn Ni alloy: Welding electrode; Stubs; for build up of grey cast iron **DPN: 175 UTS: 490 Elon: 20%**
VACOVIT 511	1% Cr 49% Fe Ni alloy: Low expansion; German proprietary alloy
VACOVIT 540	46% Fe Ni alloy: Low expansion; German proprietary alloy
VACUMET HY Mu 800	15% Fe 5% Mo 0.15% Si 0.0% C Ni alloy: Carpenter; for magnetic properties

Note. The following abbreviations and units are used in the tables:

DPN	Hardness, diamond pyramid number
UTS	Ultimate tensile strength, N/mm^2
Elon	Elongation, %
Proof	0.1% proof strength, N/mm^2

1 N/mm^2=0.1 hbar=0.102 kgf/mm^2=0.06475 $tonf/in.^2$=145.04 $lbf/in.^2$=1 MPa

See Appendix II for other abbreviations and conversion tables.

27F Nickel – miscellaneous alloys

General metallurgical characteristics

The alloys listed have all a specialized use, based on one or other of the peculiar properties of nickel, modified to a greater or lesser extent by one or more alloying elements. The alloys do not fit into any of the preceding groups and have not sufficient variation in properties to warrant separate groups. The principal effects of the commonest alloying elements are given below, but it should be noted that the effect of any two elements together may result in an alloy of completely different properties.

It must be emphasized that the information available on many of these proprietary materials is rather meagre and the reader is strongly advised to obtain further details from the source if necessary. It was not possible to list the general physical and mechanical properties, as the variations within the group are too great for this to be meaningful.

Aluminium. This can form intermetallic compounds with other alloying elements which may be used to strengthen the nickel matrix. It is also used as a de-oxidizing metal, particularly with nickel alloy welding rods.

Beryllium. This forms intermetallic compounds in the nickel matrix in a manner analogous to the copper/beryllium series – Section 14D. These can be solution treated and aged but to date have not found the same number of applications as the copper/beryllium alloys.

Boron. This is used as a fluxing agent in metal spray alloys and some brazing materials. It lowers the melting point and acts as a vehicle for refractory particles. The boron also helps the material to wet the base metal, thus considerably aiding adhesion.

Copper. This forms a solid solution with nickel and is never the major constituent in the alloys listed. The nickel/copper alloys are discussed in Section 27D.

Iron. This may be added for some electrical purpose, but in general is an economical alloying element which has the minimum effect on other properties.

Manganese. This improves the hot corrosion resistance but the principal use of nickel manganese alloys is for their special electrical properties.

Molybdenum. This improves the resistance of nickel alloys to acids and also increases the hot strength properties. The alloys with a high molybdenum content – above 20% – have very special uses.

Palladium. This is added to certain alloys to give desirable brazing properties such as wetting, without detracting from the excellent corrosion resistance and hot strength of nickel alloys. There is a series of alloys designed for joining nickel/chromium and nickel/copper alloys by brazing.

Silicon. This is a de-oxidizing element which also has considerable fluxing properties and improves the corrosion resistance under certain circumstances.

Titanium. This forms intermetallic compounds which harden the nickel matrix. It is generally used along with aluminium, but can also be used alone as a grain refiner.

Like all nickel alloys the materials in this group will show an increase in mechanical strength when cold worked but a decrease in ductility. Annealing at temperatures above 1000 °C will soften the material.

Some of the materials in this group may be capable of being age hardened but specialist advice will be required.

The materials in this section are used for a wide range of specialist purposes, many of which are listed beside the specification. These include electrical resistors, low expansion materials and materials with controlled electrical or magnetic properties, or very specialist expansion characteristics.

Symbol	Nominal analysis, supplier, condition and remarks.
2.4108	1.0% Mn 0.2% C Ni alloy: German Standard
2.4110	2.0% Mn Ni alloy: German Standard
2.4116	5.0% Mn Ni alloy: German Standard
2.4122	1.0% Si 1.5% Al 3% Mn Ni alloy: German Standard
2.4128	0.5% Ti 4.5% Al Ni alloy: German Standard
2.4132	2.0% Be Ni alloy: German Standard
2.4400–2.4419	11% Fe 3.0% Mo 14% Cu Ni alloy: German Standard
2.4520–2.4539	16% Fe 4.0% Mo 5.0% Cu Ni alloy: German Standard
2.4600	5% Fe 28% Mo Ni alloy: German Standard
25.35.4 CW Co	25.0% Cr 15.0% Fe 0.5% C 5.0% W 15.0% Co Ni alloy: Electrode; Metrode
25.50 COW Nb Zr	25.0% Cr 3.0% Fe 2.0% Nb 0.5% C 9.0% W 12.0% Co 0.1% Zr Ni alloy: Electrode; Metrode
26.35 COW Nb	26.0% Cr 12.0% Fe 1% Nb 0.5% C 5.0% W 15.0% Co Ni alloy: Electrode; Metrode
28.37.4 Cu B	27.0% Cr 31.0% Fe 4.0% Mo 0.5% C 2.0% Co Ni alloy: Electrode; Metrode; for welding Incoloy 825
50.50 Nb	48% Cr 1.5% Nb 0.06% C Ni alloy: Electrode; Metrode
	DPN: 350
133 ALLOY	3.0% Si Ni alloy: Driver-Harris; withdrawn
	UTS: 910
200C	2.0% Be Ni alloy: Casting; Brush Beryllium Co.
211 ALLOY	0.75% Fe (max) 4.7% Mn 0.2% C (max) Ni alloy: Carpenter
220C	2.2% Be Ni alloy: Casting; Brush Beryllium Co.
260C	2.6% Be Ni alloy: Casting; Brush Beryllium Co.
288	Pd Ni brazing alloy: Melting point 1237 °C; Engelhard Industries
	DPN: 224 UTS: 760 Elon: 30%
318	Pd Mn Ni brazing alloy: Melting point 1120 °C; Engelhard Industries; electrical conductivity 9% IACS
	DPN: 330 UTS: 800 Elon: 31%
420CR	1.8% Be 1.7% Ti Ni alloy: Casting; Beryllium Corporation
2307	20% Cr 3% Fe 2.5% Ti 1.5% Al 0.3% C 18% Co Ni alloy: VEW
ABRASODUR 53	29.0% Cr 8.5% Fe 4.2% Mo 0.8% Ti 6.9% Nb 3.6% C 8.3% W Ni alloy: Welding electrode; Soudometal; abrasion resistance with good impact
	DPN: 600
AFNOR 81-347 EF2070 Ni Cr Fe B110 20BH	15% Cr 7.5% Fe 2% Nb 0.03% C Ni alloy: Welding electrode; French Standard
ALLOY 1	15% Mg Ni alloy: Ingot; H Wiggin; for cast iron additions
ALLOY 1M	15% Mg 0.8% Mischmetal Ni alloy: Ingot; H Wiggin; for cast iron additions
ALLOY 2	27% Si 18% Mg Ni alloy: Ingot; H Wiggin; for cast iron additions
ALLOY 600L	16% Cr 8% Fe 0.2% Ti 0.025% C (max) Ni alloy: Origin unknown
ALLOY 601	23% Cr 14% Fe 0.2% Ti 1.3% Al 0.05% C Ni alloy
ALLOY 625	3% Fe 22% Mo 3.5% Nb 0.025% C (max) 9% Mo Ni alloy: Origin unknown
ALLOY 690	28% Cr 9% Fe 0.02% C (max) Ni alloy: Origin unknown

Note. The following abbreviations and units are used in the tables:

DPN	Hardness, diamond pyramid number
UTS	Ultimate tensile strength, N/mm^2
Elon	Elongation, %
Proof	0.1% proof strength, N/mm^2

1 N/mm^2=0.1 hbar=0.102 kgf/mm^2=0.06475 tonf/in.^{2}=145.04 lbf/in.^{2}=1 MPa

See Appendix II for other abbreviations and conversion tables.

Symbol	Nominal analysis, supplier, condition and remarks.
ALLOY 718	18.5% Cr 18.5% Fe 3% Mo 0.8% Ti 0.6% Al 5% Nb 0.05% C Ni alloy: VDM; Nicrofer 5219 Nb
ALLOY 825	21% Cr 32% Fe 2.7% Mo 0.8% Ti 2.2% Cu 0.02% C (max) Ni alloy: VDM; Nicrofer 4221
ALLOY 825 h Mo	21% Cr 29.5% Fe 6% Mo 0.8% Ti 2.2% Cu 0.02% C (max) Ni alloy: VDM; Nicrofer 4221 L Mo
ALLOY B2	1% Cr (max) 28% Mo 0.01% C (max) Ni alloy: Origin unknown
ALLOY C4	16% Cr 2% Fe (max) 16% Mo 0.3% Ti 0.008% C (max) Ni alloy: Origin unknown
ALLOY C22	21% Cr 3% Fe 13.5% Mo 0.008% C (max) 0.2% V 3% W Ni alloy: VDM; Nicrofer 5621 L Mu W
ALLOY G	22% Cr 19.5% Fe 2% Nb 2% Cu 0.03% C (max) 6.5% Cr Ni alloy: VDM; Nicrofer 4520 L Mo
ALLOY G3	22.5% Cr 19% Fe 7% Mo 0.3% Nb 2% Cu 0.01% C (max) Ni alloy: VDM; Nicrofer 4823 L Mo
AMS 4778 A	3% B 4.5% Si Ni alloy: Braze metal
AMS 4779	1.8% B 3.5% Si Ni alloy: Braze metal
AMS 4782	19% Cr 10.2% Si 0.1% C (max) Ni alloy: Braze metal
AMS 5607	7% Cr 16.5% Mo 0.06% C Ni alloy
AMS 5771	7% Cr 16.5% Mo 0.06% C Ni alloy
AMS 5873	15.7% Cr 15.5% Mo 0.3% Al 0.02% C (max) 0.05% La Ni alloy
AMS 7232	15.5% Cr 8% Fe 0.15% C (max) Ni alloy
ARC 164	22% Fe 18% Mo Ni alloy: Imphy
ARSENIC-NICKEL	40% As Ni alloy: Blackwells; Primary metal
ASTM A494 Ni Mo	5.0% Fe 28% Mo 0.12% C (max) 2.5% Co 0.5% V Ni alloy: Casting
ASTM A494 Ni Mo Cr	16% Cr 6.0% Fe 17% Mo 0.12% C (max) 4.5% W 2.5% Co Ni alloy: Casting
ASTM B260 B Ni 3	3.15% B 4.5% Si Ni alloy: Brazing metal; temperature range 1010–1177 °C
ASTM B260 B Ni 4	1.6% B 3.5% Si Ni alloy: Brazing metal; temperature range 1010–1177 °C
ASTM B260 B Ni 6	11% P Ni alloy: Brazing metal; braze temperature range 927–1030 °C
ASTM B295 E Ni 1	3% Ti 0.1% C (max) Ni alloy: Welding rod; previous designation E4N11
ASTM B295 E Ni Mo 1	5.5% Fe 28% Mo 2% Co Ni alloy: Welding rod; previous designation E3N1B and E4N1B
ASTM B295 E Ni Mo 2	15.5% Cr 3.5% Fe 16.5% Mo 2% Co 3.7% W Ni alloy: Welding rod; previous designation E3N1C and E4N1C
ASTM B295 E Ni Mo 3	4% Cr 5.5% Fe 25% Mo 2% Co Ni alloy: Welding rod
ASTM B295 E3N11	1.0% Si 3.0% Ti 1.0% Al 0.7% C Ni alloy: Welding electrode
	UTS: 370
ASTM B295 E4N11	1.0% Si 3.0% Ti 0.1% C Ni alloy: Welding electrode
ASTM B304 ER Ni Mo 4	5% Fe 28% Mo 2% Co 0.4% V Ni alloy: Welding electrode; previous designation ERN7B
ASTM B304 ER Ni Mo 5	5% Fe 16% Mo 2% Co 3.5% W 0.2% V Ni alloy: Welding electrode; previous designation ERN7C
ASTM B304 ERN61	1.0% Fe 2.5% Ti 1.5% Al 1.0% Mn 0.15% C Ni alloy: Welding electrode
ASTM B304 RN41	1.0% Si 0.2% Cu 0.15% C Ni alloy: Welding electrode
B Ni 1	14% Cr 4.5% Fe 3.2% B 4.5% Si 0.75% C Ni alloy: Braze metal; designation used by ANS
B Ni 1A	14% Cr 4.5% Fe 3.2% B 4.5% Si 0.06% C (max) Ni alloy: Braze metal; designation used by ANS
B Ni 2	7% Cr 3% Fe 3.2% B 4.5% Si 0.06% C Ni alloy: Braze metal; designation used by ANS
B Ni 3	3.2% B 4.5% Si 0.06% C (max) Ni alloy: Braze metal; designation used by ANS
B Ni 4	1.8% B 3.5% Si 0.06% C (max) Ni alloy: Braze metal; designation used by ANS
B Ni 5	19% Cr 10.2% Si 0.1% C (max) Ni alloy: Braze metal; designation used by ANS
B Ni 6	0.01% C (max) 11% P Ni alloy: Braze metal; designation used by ANS
B Ni 7	14% Cr 0.08% C (max) 10% P Ni alloy: Braze metal; designation used by ANS

Symbol	Nominal analysis, supplier, condition and remarks.
B Ni 8	7% Si 22% Mn 0.1% C (max) Ni alloy: Braze metal; designation used by ANS
B Ni 9	15% Cr 3.7% B 0.06% C (max) Ni alloy: Braze metal; designation used by ANS
B Ni 10	11.5% Cr 3.5% Fe 2.5% B 3.5% Si 0.5% C 16% W Ni alloy: Braze metal; designation used by ANS
B Ni 11	10.2% Cr 3.2% Fe 3.75% Si 0.4% C 12% W Ni alloy: Braze metal; designation used by ANS
BS 1845 N11	11% P Ni alloy: For brazing; melting point 875 °C
BS 1845 N12	13% Cr 10% P Ni alloy: For brazing; melting point 890 °C
BS 1845 N14	3% B 4.5% Si Ni alloy: For braze metal; melting range 960–1040 °C
BS 1845 N15	1.7% B 3.5% Si Ni alloy: For braze metal; melting range 980–1070 °C
BS 1845 PD11	31% Mn 21% Pd Ni alloy: For braze metal; melting point 1120 °C
BS 2901 NA44	1% Cr 4% Fe 28% Mo 1% Co Ni alloy: Rod for gas shielded arc welding
BS 3146 ANC 15	5% Fe 27% Mo 0.1% C Ni alloy: Investment casting; as cast **UTS: 460 Elon: 10% Proof: 220**
BS 3146 ANC 17	9.5% Si 3.0% Cu 0.1% C Ni alloy: Investment casting; as cast; sulphuric acid resistant
CA NICKEL	4.0% Co Ni alloy: Strip; H Wiggin; for magnetostriction
CATHODE ALLOY	2% and 4% W Ni alloy: For electronic valve cathodes; Driver-Harris
CINEX	2.4% Fe 0.1% Mn 0.7% C Ni alloy: Welding rod; Murex; for cast iron
CM 52	3.0% B 4.5% Si Ni alloy: For hard facing; Dewrance; gas welding
COBANIC	45% Co Ni alloy: For valve filaments; Driver-Harris
COLMONOY 22	1.0% Fe 1.35% B 3.5% Si 0.15% C Ni alloy: Powder; Wall Colmonoy; for metal spraying; melting point 105 °C
COLMONOY 23	1.0% Fe 1.25% B 2.3% Si 0.1% C Ni alloy: Powder; Wall Colmonoy; for metal spraying; melting point 1065 °C
COLMONOY 23A	1.0% Fe 1.2% B 2.3% Si 0.1% C Ni alloy: Bronze metal; Wall Colmonoy; melting point 1065 °C
COLMONOY 24	1.0% Fe 1.2% B 2.3% Si 0.1% C Ni alloy: Bronze metal; Wall Colmonoy; melting point 1065 °C
COLMONOY 25	5.0% Cr 2.0% Fe 1.25% B 4.0% Si 0.25% C Ni alloy: Powder; Wall Colmonoy; for metal spraying; melting point 1220 °C
COLMONOY 26	1.0% Fe 1.35% B 3.5% Si 0.12% C Ni alloy: Powder; Wall Colmonoy; for metal spraying; melting point 1050 °C
COLMONOY 27	1.0% Fe 1.25% B 2.3% Si 0.1% C Ni alloy: Powder; Wall Colmonoy; for metal spraying; melting point 1065 °C
COLMONOY 28	1.25% B 2.3% Si 0.08% C Ni alloy: Powder; Wall Colmonoy; for metal spraying; melting point 1065 °C
COLMONOY 42	Cr B Si Ni alloy: Powder coating; Colmonoy **DPN: 420**
COLMONOY 43	Cr B Si Ni alloy: Powder coating; Colmonoy **DPN: 420**
COLMONOY 44	12% Cr 2.7% Fe 2.5% B 2.5% Si 0.45% C Ni alloy
COLMONOY 46	15% Cr 4.2% Fe 3.2% B 4.5% Si 0.7% C Ni alloy
COLMONOY 52	Cr B Si Ni alloy: Powder coating; Colmonoy **DPN: 520**
COLMONOY 53	Cr B Si Ni alloy: Powder coating; Colmonoy **DPN: 520**
COLMONOY 62	Cr B Si Ni alloy: Powder; for spray coating; Colmonoy **DPN: 600**
COLMONOY 63	Cr B Si Ni alloy: Powder for coating; Colmonoy **DPN: 600**
COLMONOY 69	Cr Mo B Si Cu Ni alloy: Powder for spray coating; Colmonoy
COMET 95	15.7% Cr 5.5% Fe 15.5% Mo 0.03% C 3.7% W Ni alloy: Welding electrode; Soudometal; oxidation corrosion resistance at high temperatures **DPN: 200**
COMET 95 Co	15.3% Cr 2.2% Fe 15.6% Mo 0.03% C 3.9% W 4.0% Co Ni alloy: Welding electrode; Soudometal; for oxidation resistance at high temperatures **DPN: 200**
COMET 97	4.6% Fe 27% Mo 0.015% C 0.2% V Ni alloy: Welding electrode; Soudometal; for severe corrosion conditions **DPN: 200**
CR1	0.6% Cr 2.75% Be 0.5% C Ni alloy: Casting; Beryllium Corporation
CR 2	0.6% Cr 2.7% Be 0.9% C Ni alloy: Casting; Beryllium Corporation
CW 12 M	17.5% Cr 7% Fe 18% Mo 0.12% C (max) 5% W 2% Co Ni alloy: US designation
CW 12M	17.7% Cr 7% Fe 18% Mo 0.12% C (max) 5% W Ni alloy: Casting; US designation
D 979	15% Cr 4.2% Mo 3% Ti 1% Al 0.08% C (max) 4.2% W Ni alloy: Origin unknown
D NICKEL	4.5% Mn 0.1% C Ni alloy: Bar and sheet; Huntington; now Nickel 211
DELORO 40G	7.5% Cr 5.0% Fe 1.2% B 4.0% Si Ni alloy: Melting range 985–1180 °C; Deloro Stellite **UTS: 400 Elon: 1%**
DELORO 45	7.5% Cr 1.5% Fe 1.5% B 4.0% Si Ni alloy: Melting range 995–1150 °C; Deloro Stellite **DPN: 380**
DELORO 50	10% Cr 4.0% Fe 1.5% B 4.0% Si Ni alloy: Melting range 975–1065 °C; Deloro Stellite **DPN: 550**
DELORO 60	15% Cr 4.5% Fe 3.0% B 4.5% Si Ni alloy: Melting range 965–1005 °C; Deloro Stellite **DPN: 750**
DELORO PW22	1.2% Cr 1.3% B 2.5% Si Ni alloy: Melting range 940–1260 °C; Deloro Stellite **DPN: 225**
DIN 17741 Ni Al 4 Ti	0.5% Ti 4.5% Al Ni alloy
DIN 17741 Ni Mn 3 Al	1.0% Si 1.5% Al 3.0% Mn Ni alloy
DIN 17745 Ni Cu 14 Fe Mo	11% Fe 3.0% Mo 14% Cu Ni alloy
DURANICKEL	0.5% Ti 4.5% Al Ni alloy: Bar (high carbon); H Wiggin; for springs, dies; acid resistant **DPN: 295 UTS: 1250 Elon: 28%**
DURANICKEL 301	0.5% Ti 4.5% Al 0.15% C Ni alloy: Bar and sheet; Huntington; age hardened **UTS: 1350 Elon: 28% Proof: 920**
E Ni CI	8% Fe (max) 2.5% Cu (max) 2% C (max) Ni alloy: Weld electrode; designation used by ANS
E Ni CI A	8% Fe (max) 2% Al 2% C (max) Ni alloy: Weld electrode; designation used by ANS
E Ni Mo 7	28% Mo 0.02% C (max) Ni alloy: Weld electrode; designation used by ANS
E NICKEL	2.0% Mn Ni alloy: Driver-Harris; International Nickel Co. **UTS: 610**

Note. The following abbreviations and units are used in the tables:

DPN	Hardness, diamond pyramid number
UTS	Ultimate tensile strength, N/mm^2
Elon	Elongation, %
Proof	0.1% proof strength, N/mm^2

1 N/mm^2=0.1 hbar=0.102 kgf/mm^2=0.06475 $tonf/in.^2$=145.04 $lbf/in.^2$=1 MPa

See Appendix II for other abbreviations and conversion tables.

Symbol	Nominal analysis, supplier, condition and remarks.
E NICKEL	2% Mn 0.1% N Ni alloy: Tube and sheet; Huntington; now Nickel 212
EATONITE	29% Cr 6.5% Fe 2.4% C 15% W 10% Co Ni alloy: For valves; origin unknown
EATONITE 3	29% Cr 8% Fe 5.5% Mo 1.8% C Ni alloy: For valve facing; origin unknown
EATONITE 5	29% Cr 8% Fe 8.5% Mo 2.2% C Ni alloy: For valve facing; origin unknown
ER Ni Mo 1	6% Fe (max) 30% Mo 0.12% C (max) Ni alloy: Weld electrode; designation used by ANS
ER Ni Mo 2	7% Cr 5% Fe (max) 17.5% Mo 0.06% C Ni alloy: Weld electrode; designation used by ANS
ER Ni Mo 3	5% Cr 5.5% Fe 25.5% Mo 0.12% C (max) Ni alloy: Weld electrode; designation used by ANS
ER Ni Mo 7	28% Mo 0.02% C (max) Ni alloy: Weld electrode; designation used by ANS
F NICKEL	1.5% Fe 5.5% Si 0.2% Cu 0.5% C 0.6% S Ni: Primary metal; H Wiggin; supplied as shot or ingot
HAI 63 ALLOY	1% Si 4% Mn Ni alloy: For spark plugs; Harrison Alloys
HAI Ni 634	4% W Ni alloy: Harrison Alloys
HAI-KN	5% Mn + Si + Al + Co Ni alloy: For thermocouples; Harrison Alloys; specific resistance 29.4 microhm/cm
HAI-NN	4.5% Si Ni alloy: For thermocouples; Harrison Alloys; specific resistance 35.7 microhm/cm
HAS CH	15% Cr 5% Fe 16% Mo 0.2% C 4% W Ni alloy: Electrode; Metrode
HASB	4.5% Fe 28.0% Mo 0.04% C 0.4% V Ni alloy: Electrode; Metrode; for welding Hastelloy B type alloys
HASB	1% Fe 28% Mo 0.04% C Ni alloy: Electrode; Metrode; work hardens; resists acid attack
HASC	15.0% Cr 4.5% Fe 16.0% Mo 0.09% C 4.0% W 0.1% V Ni alloy: Electrode; Metrode; for welding Hastelloy C type alloys
HASC	15% Cr 5% Fe 16% Mo 0.02% C 4% W Ni alloy: Electrode; Metrode **DPN: 350**
HASC4	17% Cr 2% Fe 16% Mo 0.01% C Ni alloy: Electrode; Metrode; for welding Hastelloy C4
HASD	9% Si 3% Cu 0.05% C Ni alloy: Electrode; Metrode; for Hastelloy D welding
HASF	22.0% Cr 22.2% Fe 6.0% Mo 2.0% Nb 0.07% C Ni alloy: Electrode; Metrode; for welding Hastelloy F type alloys
HASG	22.0% Cr 19% Fe 6.0% Mo 2.0% Cu 0.06% C 0.5% W Ni alloy: Electrode; Metrode; for welding Hastelloy E type alloys
HASN	7.0% Cr 4.0% Fe 16.0% Mo 0.04% C Ni alloy: Electrode; Metrode; for welding Hastelloy N type alloys
HASS	15.5% Cr 15.5% Mo 0.04% C Ni alloy: Electrode; Metrode; for welding Hastelloy S type alloys
HASTELLOY 5	16.2% Cr 14.7% Mo 0.4% Al 0.02% C (max) Ni alloy
HASTELLOY C4	16% Cr 16% Mo 0.015% C (max) Ni alloy
HASTELLOY C276	15.5% Cr 5.5% Fe 16.0% Mo 0.02% C 3.8% W Ni alloy: Casting; Inco
HASTELLOY G-2	24.5% Cr 15% Fe 6% Mo 1% Cu 0.03% C (max) Ni alloy
HASTELLOY G-3	19.5% Cr 4.2% Mo 3% Ti 0.07% C 13.5% Co Ni alloy
HASW	5.0% Cr 5.0% Fe 25.0% Mo 0.06% C 0.4% V Ni alloy: Electrode; Metrode; for welding Hastelloy W type alloys
HPW NICKEL	3.5% W Ni alloy: Made by powder metallurgy
IN 504	3.0% Mo 10% Si 5.7% Ti 2.5% Cu Ni alloy: International Nickel Co.
INCOCAL 10	0.6% Si 0.3% C 5.5% Ca Ni alloy: Primary metal
INCOLOY 904	50.0% Fe 1.6% Ti 0.02% C 14.5% Co Ni alloy: H Wiggin; low coefficient of expansion **UTS: 920**
INCOLOY FM65	21% Cr 22% Fe (min) 3% Mo 1% Ti 2.2% Cu 0.05% C (max) Ni alloy: Weld rod
INCOMAG 1	15% Mg Ni alloy: For cast iron additions; International Nickel Co.
INCOMAG 1LC	16% Mg low carbon Ni alloy: For de-oxidation purposes; International Nickel Co.
INCOMAG 1M	15% Mg 0.8% Mischmetall Ni alloy: For cast iron additions; International Nickel Co. (formerly alloy IM)
INCOMAG 2M	15% Mg 30% Se 0.8% Mischmetall Ni alloy: For cast iron additions; International Nickel Co.
INCOMAG 3	4.5% Mg Ni alloy: For cast iron additions; fumeless; International Nickel Co.
INCOMAG 4	3.4% Fe 4.2% Mg Ni alloy: For cast iron additions; reduced fume; International Nickel Co.
INCOMAG Z	15% Mg 30% Se Ni alloy: For cast iron additions; International Nickel Co. (formerly Alloy Z)
INCONEL MA754	21% Cr 0.5% Ti 0.3% Al 0.05% C 0.6% Y oxide Ni alloy
INOR 8	7% Cr 5.0% Fe 16% Mo 0.01% B 0.06% C 0.5% Co Ni alloy: Alternative name for Hastelloy N
KH 75MBTIU	20.5% Cr 8.0% Fe (max) 19.5% Mo 0.5% Ti 0.5% Al 1.1% Nb 0.1% C (max) Ni alloy: Russian Standard designation
KH 78T	20.5% Cr 6.0% Fe (max) 0.2% Ti 0.12% C (max) Ni alloy: Russian Standard designation
KH 601U	16.5% Cr 26% Fe 3.2% Al 0.1% C (max) Ni alloy: Russian Standard designation
L307	20% Cr 3% Fe 2.5% Ti 1.5% Al 0.3% C (max) 18% Co Ni alloy: VEW
L311	12.5% Cr 36% Fe 5.7% Mo 2.9% Ti 0.1% C (max) Ni alloy: VEW
L317	15% Cr 1% Fe 5% Mo 1.2% Ti 4.5% Al 0.2% C (max) 20% Co Ni alloy: VEW
L326	19% Cr 1% Fe 6% Mo 3% Ti 2% Al 0.05% C 1% W 12% Co Ni alloy: VEW
LANGALLOY 2R	1.0% Fe 1.0% Si 1.0% Mn 0.7% C Ni alloy: Casting; Langley Alloys **DPN: 110 UTS: 390 Elon: 25% Proof: 180**
M21	5.7% Cr 2.0% Mo 0.02% B 6.0% Al 1.5% Nb 0.13% C 11% W 0.12% Zr Ni alloy: Union Carbide
M22	5.7% Cr 0.75% Fe 2.0% Mo 0.6% Si 6.3% Al 0.13% C 11.0% W 3.0% Ta 0.6% Zr Ni alloy: Union Carbide
M1628	10% Fe 25% Mo Ni alloy: Imphy
MAGNO	4.5% Mn Ni alloy: Driver-Harris **UTS: 670**
MANGNOL	5% Mn Ni alloy: For valve grid wire; Driver-Harris
MANGONIC 2	2.0% Mn Ni alloy: H Wiggin; for filament wires
MANGONIC 3	3.0% Mn Ni alloy: For electrical leads; magnetic; H Wiggin; annealed; electrical conductivity 12% IACS **DPN: 140 UTS: 540 Elon: 56% Proof: 150**
MANGONIC 5	5.0% Mn Ni alloy: H Wiggin; filament wire
MAR M432	Ni alloy: Casting for creep property; Martin
MAR-M-ALLOY Hf Modified	9% Cr 2.5% Mo 1.5% Ti 5.5% Al 0.15% C 9.5% Co 1.5% Ta 10% W 1.7% Hf Ni alloy: Martin
METCO 31 CNS	35% WC Cr Co Ni alloy: Sprayed deposit; Metco **DPN: 1500**
METCO 36 C	35% WC Cr Ni alloy: Sprayed deposit; Metco **DPN: 1500**
METCO 404 NS	Al Ni alloy: Sprayed deposit; Metco **DPN: 140**
METCO 405	Al Ni alloy: Sprayed deposit; Metco; good bond strength **DPN: 245**
METCO 405 NS	Al Ni alloy: Sprayed deposit; Metco **DPN: 140**
METCO 443 NS	Cr Al Ni alloy: Comfort sprayed deposit; Metco **DPN: 180**
METCO 451	Ni Al intermetallic Ni alloy: For spray deposition; Metco **DPN: 280**

Symbol	Nominal analysis, supplier, condition and remarks.
METCO 470 AW	Cr Fe Ni alloy: Sprayed deposit; Metco **DPN: 190**
MODIFIED HILO	20% Co Ni alloy: For valve filaments; Driver-Harris
MP 35N MULTIPHASE	20% Cr 10% Mo 35% Co Ni alloy: Latrobe
MP–159	19% Cr 7% Mo 3% Ti 0.2% Al 0.5% Nb 0.04% C (max) 36% Co Ni alloy: Origin unknown
N 12M	6% Fe (max) 30.5% Mo 0.12% C (max) Ni alloy: Casting; US designation
N 19907	40% Fe 1.6% Ti 4.8% Nb 0.06% C (max) 14% Co Ni alloy: Designation used by UNS
N 19909	40% Fe 4.8% Nb 0.06% C (max) 14% Co Ni alloy: Designation used by UNS
Ni Mo 30	5% Fe 28% Mo Ni alloy: Designation used by German Standards
NICKEL 204	0.06% C 4.5% Co Ni alloy: Bar, sheet, etc.; Huntington
NICKEL 211	4.5% Mn 0.1% C Ni alloy: Bar, sheet, etc.; Huntington
NICKEL 211	5.0% Mn Ni alloy: H Wiggin; previously Mangonic 5
NICKEL 212	2.0% Mn 0.1% C Ni alloy: Tube and strip; Huntington
NICKEL 212	2.0% Mn Ni alloy: H Wiggin; previously Mangonic 2
NICKEL 215	5.0% Mn Ni alloy: Wire; International Nickel Co.
NICKEL 229	0.03% Al 2.0% W 0.05% Mg Ni: Made by powder metallurgy; International Nickel Co.
NICROBRAZ 10	0.1% C 1% P Ni alloy: Braze metal; Wall Colmonoy; melting point 880 °C
NICROBRAZ 60	8% Si 17% Mn 0.1% C Ni alloy: Braze metal; Wall Colmonoy; melting range 1010–1030 °C
NICROBRAZ 65	7.0% Si 4.5% Cu 23% Mn Ni alloy: For brazing; Wall Colmonoy; melting range 980–1010 °C
NICROBRAZ 130	3.0% B 4.5% Si 0.06% C Ni alloy: Braze metal; Wall Colmonoy; melting range 980–1040 °C
NICROBRAZ 135	1.8% B 3.5% Si Ni alloy: Braze metal; Wall Colmonoy; melting range 990–1057 °C
NICROBRAZ 171	10.0% Cr 3.5% Fe 2.5% B 3.5% Si 0.4% C 12.0% W Ni alloy: For brazing; Wall Colmonoy; melting range 970–1095°C
NICROBRAZ 220	4% Cr 1.0% B 45% Mn Ni alloy: Braze metal; Wall Colmonoy; melting range 995–1080 °C
NICROBRAZ 230	3.5% Cr 10% B 2.5% Si 35% Mn Ni alloy: Braze metal; Wall Colmonoy; melting range 980–1065 °C
NICROBRAZ 1351	1.5% B 3.0% Si Ni alloy: For brazing; Wall Colmonoy; melting range 1090–1180 °C
NICROBRAZ 5040	5.0% Cr 0.6% Fe 2.1% Si 4.0% P 1.2% N Ni alloy: For brazing; Wall Colmonoy; melting point 1010–1150 °C
NICROBRAZ 5060	8.0% Cr 0.4% Fe 0.8% B 1.4% Si 6.0% P Ni alloy: For brazing; Wall Colmonoy; melting point 1010–1120 °C
NICROBRAZ 5075	10.0% Cr 0.25% Fe 0.5% B 0.9% Si 7.5% P Ni alloy: For brazing; Wall Colmonoy; melting point 980–1090 °C
NICROCOAT 2	4.5% Si Ni alloy: Powder with additions; Wall Colmonoy; for coating
NICROCOAT 7	3.5% Si Ni alloy: Powder with additions; Wall Colmonoy; for coating
NICROFER 4221	21% Cr 33% Fe 2.7% Mo 0.8% Ti 2.2% Cu 0.02% C (max) Ni alloy: VDM **UTS: 590 Elon: 30% Proof: 250**
NICROFER 4221 h Mo	21% Cr 29% Fe 6% Mo 0.8% Ti 2.2% Cu 0.02% C (max) Ni alloy: VDM **UTS: 590 Elon: 30% Proof: 240**
NICROFER 4520 h Mo	22% Cr 19.5% Fe 6.5% Mo 2% Nb 2% Cu 0.03% C (max) Ni alloy: VDM **UTS: 590 Elon: 30% Proof: 210**
NICROFER 4823 h Mo	22.5% Cr 19% Fe 7% Mo 0.3% Nb 2% Cu 0.01% C (max) Ni alloy: VDM **UTS: 640 Elon: 55% Proof: 290**
NICROFER 5219 Nb	18.5% Cr 18.5% Fe 3% Mo 0.8% Ti 0.6% Al 5% Nb 0.05% C Ni alloy: VDM **UTS: 1240 Elon: 12% Proof: 1030**
NICROFER 5621 h Mo W	21% Cr 3% Fe 13.5% Mo 0.008% C (max) 0.2% V 3% W Ni alloy: VDM **UTS: 780 Elon: 55% Proof: 380**
NICROFER 5716 h Mo W	15.7% Cr 16% Mo 0.01% C (max) 0.2% V 3.7% W 2% Co Ni alloy: VDM
NICROFER 6020 h Mo	22% Cr 9% Mo 3.5% Nb 0.025% C (max) Ni alloy: VDM
NICROFER 6023	23% Cr 14% Fe 0.2% Ti 1.3% Al 0.05% C Ni alloy: VDM
NICROFER 6030	28.5% Cr 9% Fe 0.02% C (max) Ni alloy: VDM
NICROFER 6616 h Mo	16% Cr 2% Fe 16% Mo 0.3% Ti 0.008% C (max) Ni alloy: VDM
NICROFER 7216 LC	16% Cr 8.5% Fe 0.2% Ti 0.025% C (max) Ni alloy: VDM
NILOMAG 771	14.0% Fe 4.0% Mo 5.0% Cu Ni alloy: Wire or strip made by powder metallurgy; International Nickel Co.
NIMAX C	15% Cr 5% Fe 16% Mo 0.02% C 4% W Ni alloy: Electrode; Metrode **DPN: 350**
NIMAX CH	15% Cr 5% Fe 16% Mo 0.2% C 4% W Ni alloy: Electrode; Metrode
NIMOCAST 242	22.0% Cr 10.5% Mo 0.35% C 10.0% Co Ni alloy: H Wiggin; high thermal shock resistance
NIMOCAST 263	20.0% Cr 5.8% Mo 2.2% Ti 0.06% C 20.0% Co Ni alloy **DPN: 195 UTS: 1000**
NIMOFER	1% Cr (max) 28% Mo 0.02% C (max) Ni alloy: Weld rod; VDM
NIMOFER 6928	1% Cr (max) 2% Fe (max) 28.5% Mo 0.01% C (max) Ni alloy: VDM
NIMROD 625	21% Cr 1% Fe 9% Mo 4% Nb 0.04% C Ni alloy: Electrode; Metrode
Ni-Zr MASTER ALLOY	5% Fe 7.5% Si 10% Al 27.5% Zr Ni alloy: Origin unknown
NMP	Pd Mn Ni brazing alloy: Melting point 1120 °C; Engelhard Industries; electrical conductivity 9% IACS **DPN: 330 UTS: 780 Elon: 7%**
PA 10	4.0% W 0.05% Mg Ni alloy: Tube and strip; H Wiggin obsolete
PERMAGRID	4% Ti 0.3% Mg Ni alloy: For valve grid wire; Driver
PYROMET ALLOY 102	15% Cr 7.5% Fe 3% Mo 0.5% Ti 0.4% Al 3% Nb 0.08% C (max) 3% W Ni alloy: Carpenter
PYROMET ALLOY 625	22% Cr 5% Fe 9% Mo 3.5% Nb + Ta 0.1% C (max) Ni alloy: Carpenter **UTS: 1000 Elon: 35% Proof: 550**
PYROMET ALLOY 680	22% Cr 18% Fe 9% Mo 0.1% C 0.8% W 1.5% Co Ni alloy: Carpenter **UTS: 790 Elon: 50% Proof: 320**
PYROMET ALLOY 718	19% Cr 22% Fe 3% Mo 0.004% B 1% Ti 0.5% Al 0.1% C (max) Ni alloy: Carpenter **UTS: 1450 Elon: 22% Proof: 1200**
R 63	1.0% Si 4.0% Mn Ni alloy: Driver-Harris; spark plug electrodes **UTS: 490**
R Ni Cr A	11% Cr 2.25% Fe 2.5% B 2.2% Si 0.45% C Ni alloy: Weld metal; designation used by ANS
R Ni Cr B	13% Cr 4% Fe 3% B 4% Si 0.6% C Ni alloy: Weld metal; designation used by ANS

Note. The following abbreviations and units are used in the tables:

DPN	Hardness, diamond pyramid number
UTS	Ultimate tensile strength, N/mm^2
Elon	Elongation, %
Proof	0.1% proof strength, N/mm^2

1 N/mm^2=0.1 hbar=0.102 kgf/mm^2=0.06475 tonf/in.^{2}=145.04 lbf/in.^{2}=1 MPa

See Appendix II for other abbreviations and conversion tables.

Symbol	Nominal analysis, supplier, condition and remarks.
R Ni Cr C	15% Cr 4.5% Fe 3.5% B 4.5% Si 0.7% C Ni alloy: Weld metal; designation used by ANS
RA 333	25% Cr 18% Fe 0.05% C 3% W 3% Co Ni alloy: Origin unknown
REXWELD 64	13% Cr 3% B 3.0% Si 0.6% C Ni alloy: For hard facing; Crucible Steel Co. **DPN: 620**
SOUDER G625	22.3% Cr 0.8% Fe 9.1% Mo 0.2% Ti 3.4% Nb 0.01% C Ni alloy: Welding electrode; Soudometal **UTS: 830 Elon: 30% Proof: 415**
SOUDOKAY TOOL ALLOY CO	16% Cr 2.5% Fe 16.5% Mo 0.03% C 4.5% W Ni alloy: Welding electrode; Soudometal **DPN: 190**
SOUDONEL	15% Cr 7.5% Fe 2% Nb 7% Mn 0.03% C Ni alloy: Welding electrode; Soudometal; for welding high Ni steel; impact at −200 °C 80 J **UTS: 635 Elon: 40% Proof: 375**
SOUDONEL 625	22% Cr 4.5% Fe 9% Mo 3.5% Nb 0.06% C Ni alloy: Welding electrode; Soudometal; impact at −200 °C 45 J **UTS: 760 Elon: 32% Proof: 490**
SOUDONEL D	15% Cr 9% Fe 6.5% Mo 1.1% Nb 3.7% Mn 0.07% C 1.2% W Ni alloy: Electrode; impact at −200 °C 60 J **UTS: 680 Elon: 40% Proof: 410**
SOUDONEL G	22% Cr 1.6% Fe 9% Mo 3.5% Nb 0.03% C Ni alloy: Welding electrode; Soudometal; impact at−200 °C 55 J **UTS: 730 Elon: 40% Proof: 500**
SOUDONEL S4648	20% Cr 3% Fe 2.4% Nb 3.5% Mn 0.03% C Ni alloy: Welding electrode; Soudometal; impact at −200 °C 120 J **UTS: 675 Elon: 45% Proof: 400**
SOUDOSTEL 60	35.5% Cr 0.5% Fe 0.8% B 0.9% C 10.5% W Ni alloy: Welding electrode; Soudometal **DPN: 400**
SYLVALOY	3% Si Ni alloy: For valve filament; Driver-Harris
T 2	5% Al + Mn + Si Ni alloy: Thermocouple; Driver-Harris **UTS: 560**

Symbol	Nominal analysis, supplier, condition and remarks.
TRIBALOY 700	15% Cr 32% Mo 3% Si Ni alloy: Intermetallic material for wear resistant purposes; suitable for hard facing; Du pont
VF 9	29% Cr 8% Fe 5.5% Mo 1.8% C Ni alloy: For valve facing; SAE designation
VF 10	29% Cr 8% Fe 8.5% Mo 2.2% C Ni alloy: For valve facing; SAE designation
W 5	4.0% Si 0.3% Mn Ni alloy: Wire and strip; H Wiggin; for spark plug electrodes **DPN: 130 UTS: 550 Elon: 56% Proof: 150**
W 6	2.0% Si 0.5% Mn Ni alloy: Wire and strip; H Wiggin; for spark plug electrodes **DPN: 120 UTS: 530 Elon: 57% Proof: 140**
W 7	1.0% Si 2.75% Mn Ni alloy: Wire and strip; H Wiggin; for spark plug electrodes **DPN: 120 UTS: 530 Elon: 57% Proof: 140**
W 9	1.0% Si 4.5% Mn Ni alloy: Wire and strip; H Wiggin; for spark plug electrodes
W 9	1.0% Si 4.5% Mn with Zr additions Ni alloy: For spark plug electrodes; H Wiggin
X-782	26% Cr 2% C 8.7% W Ni alloy
XEV H	19% Cr 2% Fe 4.3% Mo 3% Ti 0.05% C 13% Co Ni alloy: For valves; SAE
XEV I	19% Cr 19% Fe 1% Ti 5% Nb + Ta 0.04% C Ni alloy: For valves; SAE

Note. The following abbreviations and units are used in the tables:

DPN	Hardness, diamond pyramid number
UTS	Ultimate tensile strength, N/mm^2
Elon	Elongation, %
Proof	0.1% proof strength, N/mm^2

1 N/mm^2=0.1 hbar=0.102 kgf/mm^2=0.06475 $tonf/in.^2$=145.04 $lbf/in.^2$=1 MPa

See Appendix II for other abbreviations and conversion tables.

28. **Niobium** Nb (also called columbium, Cb)

Physical properties

Atomic number		41
Atomic weight		92.91
Crystal structure		Body-centred cubic
Colour		Steel grey
Specific gravity		8.6
Density		8600 kg/m^3
Melting point		2468 °C
Boiling point		4927 °C
Specific heat		0.272 J/g °C
Thermal conductivity		52.3 W/m °C
Coefficient of linear expansion (20–100 °C)		7.1 × 10^{-6}/ °C
Latent heat of fusion		–
Latent heat of vaporization		7704 J/g
Thermal neutron absorption cross-section		1.1 barns/atom
Electrical conductivity		13.3% IACS (copper 100%)
Specific resistance		151 microhm mm
Temperature coefficient of electrical resistance		0.0039/ °C
Electrochemical equivalent		0.45 g/A/h
Electrode potential		−1.1 V
Magnetic susceptibility		2.28 × 10^{-6} cgs
Young's modulus of elasticity		103 × 10^9 N/m^2
Tensile strength	annealed	300 N/mm^2
	cold worked	600 N/mm^2
Hardness	annealed	80 DPN
	cold worked	130 DPN

28.1 General notes on niobium (also called columbium, Cb)

Niobium occurs in nature combined with iron as a complex oxide, generally associated with the similar tantalum–iron oxide. The ore is very scarce and as niobium is generally used as a low percentage alloy very little scrap is recovered.

The metal is quite ductile, capable of being rolled to sheet and bar. It can be welded using the inert gas shielded technique with copious gas flow before and after the weld on both sides.

Niobium resists attack from almost all chemicals, but is in no way superior to tantalum which is slightly easier to produce and more plentiful. The ability to withstand oxidation, especially at elevated temperatures, makes niobium potentially useful for gas turbine and rocket motor applications. It has comparable hot strength properties to molybdenum with much superior oxidation resistance. The present scarcity and the high cost prevent its widespread use. It has been used as a canning material in atomic reactors, but has few advantages over tantalum.

By far the best known use for niobium is as a carbide former in austenitic steels and nickel alloys. The carbide is insoluble and being very stable and evenly distributed prevents the formation of other soluble carbides which would be dissolved during welding, and precipitated from solution in the weld area, giving rise to the condition known as 'weld decay'. It is now added in small amounts – 0.01% – to weldable mild steel, as an optional element to aid grain refinement, thus preventing grain growth during welding. This enhances the impact properties.

In the nickel alloys niobium can be present up to 4% when it enhances the creep strength and hot oxidation resistance.

When added to low alloy chrome steels there is some improvement in impact strength and an increase in the rate of nitride penetration.

Some aluminium age-hardening alloys have been produced with up to 0.2% niobium added to refine the grain and strengthen the matrix.

The more complex magnet alloys have niobium added, principally to ensure the presence of very stable carbides.

There are a few alloys based on niobium which have potentially very many high temperature applications. Until a more plentiful supply of the metal is available these applications will continue to be filled by molybdenum and tantalum alloys.

This is the only element on which there has been no agreement regarding the chemical symbol: in Europe it is Nb for niobium; in the Americas it is Cb for columbium. Throughout this book it is referred to as Nb.

Symbol	Nominal analysis, supplier, condition and remarks.
AMS 7850	Nb: Sheet, strip and foil; Columbium
AMS 7851	0.03% C (max) 10% W 2.5% Zr Nb alloy
AMS 7852	0.03% C (max) 10% W 2.5% Zr Nb alloy
AMS 7855	0.03% C (max) 10% W 2.5% Zr Nb alloy
AMS 7857	0.015% C (max) 10% Hf Nb alloy
COLUMBIUM	Ferro columbium, free of carbon: Blackwells; primary metal; also called niobium
n Nb 16	99.9% Nb: Rod; 500 p.p.m. Ta; Light Ltd; high purity metal
n Nb 73	99.99% Nb metal: Single crystal; Light Ltd
NIOBIUM	99.6% Nb: Powder, granules and electrodes; Kennametal; columbium
PLATINUM SUBSTITUTE	70% Zr 39.5% Ta Nb alloy: Origin unknown
R04295	0.015% C (max) 10% Hf 1% Ti Nb alloy: Ingot; UNS designation
SU 13	0.12% C 17% W 3.5% Hf Nb alloy: IMI; heat treated **UTS: 460 Elon: 30% Proof: 340**
SU 16	0.08% C 11% W 3% Mo 2% Hf Nb alloy: IMI; solution treated; cold worked and aged **UTS: 530 Elon: 22% Proof: 440**
SU 31	0.12% C 17% W 3.5% Hf Nb alloy: IMI; heat treated **UTS: 460 Elon: 30% Proof: 340**
W Nb 18	99.9% Nb: Powder; Light Ltd

Note. The following abbreviations and units are used in the tables:

DPN	Hardness, diamond pyramid number
UTS	Ultimate tensile strength, N/mm^2
Elon	Elongation, %
Proof	0.1% proof strength, N/mm^2

$1\ N/mm^2 = 0.1\ hbar = 0.102\ kgf/mm^2 = 0.06475\ tonf/in.^2 = 145.04\ lbf/in.^2 = 1\ MPa$

See Appendix II for other abbreviations and conversion tables.

29. Osmium Os

Physical properties		
Atomic number		76
Atomic weight		190.2
Crystal structure		Close-packed hexagonal
Colour		Blue/white
Specific gravity		22.5
Density		22 500 kg/m^3
Melting point		3050 °C
Boiling point		4600 °C
Specific heat		1.31 J/g °C
Thermal conductivity		91.67 W/m °C
Coefficient of linear expansion (20–100 °C)		6.6×10^{-6}/ °C
Latent heat of fusion		26.8 J/g
Latent heat of vaporization		678 J/g
Thermal neutron absorption cross-section		15 barns/atom
Electrical conductivity		19% IACS (copper 100%)
Specific resistance		19 microhm mm
Temperature coefficient of electrical resistance		0.0066/ °C
Electrochemical equivalent		–
Electrode potential		+0.7 V
Magnetic susceptibility		0.04×10^{-6}
Young's modulus of elasticity		–
Tensile strength	annealed	1000 N/mm^2
Hardness	annealed	300 DPN
	cold worked	670 DPN

29.1 General notes on osmium

This is one of the platinum group of metals, always being found in association with its fellow members. Much of the metal occurs native in Africa, Alaska and Canada as an alloy with iridium, platinum, rhodium and ruthenium, known as osmiridium. This is often found alongside ores containing gold. Previously this native alloy was used to tip pen nibs and is now used as a source for the platinum group of metals. Pens are now tipped with a powder compact of ruthenium.

The metal, with a density of 22.5, is the heaviest known, and like all members of the platinum group has a high melting point, bluish white colour and excellent resistance to chemical attack. The pure metal is extremely brittle and crumbles to a powder when subjected to cold work, with only a very slight increase in ductility when hot worked. The metal forms an oxide in air at 1000 °C which is very poisonous and once formed is volatile above 200 °C.

Osmium has no properties which make it an attractive alternative to iridium or other more ductile members of its group, and its scarcity, brittleness and readily formed poisonous oxide make it an undesirable first choice.

A limited amount is used as sintered powder wear resistant points for delicate instruments and electrical contacts subject to severe hammering.

The chemical engineering industry has found a limited use for finely divided osmium as a catalyst.

Of all the metals in the platinum group (platinum, iridium, osmium, rhodium, ruthenium and palladium) osmium has by far the least attractive properties and the lowest potential usefulness.

Note. The following abbreviations and units are used in the tables:

DPN	Hardness, diamond pyramid number
UTS	Ultimate tensile strength, N/mm^2
Elon	Elongation, %
Proof	0.1% proof strength, N/mm^2

1 N/mm^2=0.1 hbar=0.102 kgf/mm^2=0.06475 tonf/in.^{2}=145.04 lbf/in.^{2}=1 MPa

See Appendix II for other abbreviations and conversion tables.

Symbol	Nominal analysis, supplier, condition and remarks.
h Os 18	99.97% Os: Powder; Light Ltd; high purity metal
OSMIUM	High purity metal: Impurities 10 p.p.m.; Mallory Metallurgical; supplied as sponge; platinum group metal

30. Palladium Pd

Physical properties		
Atomic number		46
Atomic weight		106.4
Crystal structure		Face-centred cubic
Colour		White
Specific gravity		12.02
Density		12 020 kg/m^3
Melting point		1552 °C
Boiling point		2900 °C
Specific heat		0.247 J/g °C
Thermal conductivity		71.2 W/m °C
Coefficient of linear expansion (20–100 °C)		11.1×10^{-6}/ °C
Latent heat of fusion		17.6 J/g
Latent heat of vaporization		373 J/g
Thermal neutron absorption cross-section		8 barns/atom
Electrical conductivity		16% IACS (copper 100%)
Specific resistance		100 microhm mm
Temperature coefficient of electrical resistance		0.038/ °C
Electrochemical equivalent		–
Electrode potential		−0.57 V
Magnetic susceptibility		5.8×10^{-6}
Young's modulus of elasticity		100×10^9 N/m^2
Tensile strength	annealed	180 N/mm^2
Hardness	annealed	37 DPN

30.1 General notes on palladium

Palladium is a member of the platinum group of metals, the others being iridium, osmium, rhodium and ruthenium.

It is always found associated with these, but being the least chemically resistant member of the group is generally combined with a non-metal rather than as native alloy. The largest deposits are found in Africa, Canada and Russia.

Being the most plentiful and most easily purified member of the group, palladium is the cheapest of the platinum metals.

Like all members of the group it has a high melting point, a bluish colour, and resists chemical attack, although in this last respect it has the least resistance of the six platinum metals.

Pure palladium is very ductile and can be rolled into thin sheets, or drawn to fine wire. It does work harden to some extent and must be annealed between stages of severe cold work. This should be at 1000 °C at least, in an atmosphere free from oxygen, hydrogen, or carbon-containing gases, all of which can be absorbed to give a hard brittle product. Suitable atmospheres are nitrogen, argon, or vacuum.

The fact that palladium oxidizes at temperatures above about 500 °C, coupled with its relatively low chemical resistance, precludes its use as a laboratory tool.

Palladium and its alloys are used in the main as substitutes for platinum. It is much cheaper and more plentiful than platinum and most of its properties are similar to, but not quite as good as, the corresponding properties of platinum. It is the lightest of the metals in the group, a density of 12 compared with 21 for platinum and 22.5 for osmium, and this makes it attractive in the jewellery field.

Most palladium metal is used for this purpose, either as the pure metal or alloyed with silver, molybdenum and other members of its own group.

The same type of alloys are used for electrical purposes, such as potentiometer rubbing contacts and light duty impact contactors. It is not as sensitive or wear resistant as platinum for these purposes, but much cheaper.

Alloyed with gold, palladium is used to protect furnaces. These alloys have good resistance to oxidation up to 1200 °C and melting points from 1000 up to 1200 °C. Having the added advantage of short melting ranges, they are used as fuses inside the furnace that melt at the pre-determined temperature, thus cutting off the electrical heating. The oxidation resistance of these alloys at these temperatures is sufficient to ensure stability of electrical resistance and melting point.

The same alloys are used as thermocouple materials in conjunction with other platinum group metals and alloys, as they generate a high e.m.f. over the range 500–1000 °C.

Alloys of palladium with copper, silver and gold are used in the manufacture of false teeth where good corrosion resistance and silver colour are added advantages to the good casting characteristics.

A series of palladium alloys has recently been produced for use as high temperature brazing materials. These are used for joining nickel alloys and other high temperature, corrosion resistant alloys which for some reason cannot be welded.

The largest single use for palladium is as a catalyst, either in the finely divided state, or sometimes as a fine wire gauze.

Symbol	Nominal analysis, supplier, condition and remarks.
226 ALLOY	30% Ag 10% Au 14% Cu 10% Pt Pd alloy: Origin unknown
267	40% Ag Pd alloy: Engelhard; annealed; electrical conductivity 4.1% IACS **DPN: 95**
278	40% Cu Pd alloy: Engelhard; annealed; electrical conductivity 4.9% IACS **DPN: 145**
AMS 7735	30% Ag 10% Au 14% Cu 10% Pt Pd alloy
BS 1845 PD14	40% Ni Pd alloy: For brazing metal; melting point 1235 °C
BV Pd 1	0.005% C (max) 35% Co Pd braze alloy: designation used by AWS
h Pd 15	99.9% Pd: Wire; 0.5 mm diameter, Light Ltd
h Pd 73	99.99% Pd single crystals 12 mm × 25 mm: Light Ltd
JMC 77	40% Cu Pd alloy: For wiping contacts; Mallory Metallurgical; age hardened **DPN: 350**
JMM 77	Cu Pd alloy: For sliding contacts; 4.6% IACS conductivity; Johnson Matthey **DPN: 210**
P 03590	40% Cu Pd alloy: Obsolete; designation used by UNS
P 03990	99.07% Pd (min) alloy: Obsolete; designation used by UNS
PALAURAL	34% Au + Pt Pd alloy: For dental purposes; Mallory Metallurgical; melting range 945–990 °C; hardened **DPN: 258**
PALCO	35% Co Pd alloy: For brazing; melting range 1230–1235 °C; Wesgo
PALINEY No. 7	30% Ag 10% Au 14% Cu 10% Pt Pd alloy: Origin unknown
PALLACAST	Palladium alloy: Used in dentistry; with Au and Ag; Engelhard Industries; annealed, melting range 920–1025 °C **DPN: 172 UTS: 460 Elon: 15% Proof: 330**
PALLACAST	Palladium alloy: Used in dentistry; with Au and Ag; Engelhard Industries; age hardened, melting range 920–1025 °C **DPN: 249 UTS: 850 Elon: 2% Proof: 770**
PALLADIUM	Chemically pure Pd: Engelhard Industries
PALLADIUM	High purity metal: Impurities 10 p.p.m.; Mallory Metallurgical; supplied as sponge and wire
PN 1	40% Ni Pd brazing alloy: Melting point 1237 °C; Engelhard Industries
W ALLOY	30% Ag 10% Au 14% Cu 10% Pt Pd alloy: Origin unknown

Note. The following abbreviations and units are used in the tables:

DPN	Hardness, diamond pyramid number
UTS	Ultimate tensile strength, N/mm^2
Elon	Elongation, %
Proof	0.1% proof strength, N/mm^2

1 N/mm^2=0.1 hbar=0.102 kgf/mm^2=0.06475 $tonf/in.^2$=145.04 $lbf/in.^2$=1 MPa

See Appendix II for other abbreviations and conversion tables.

31. Platinum Pt

Physical properties		
Atomic number		78
Atomic weight		195.09
Crystal structure		Face-centred cubic
Colour		White
Specific gravity		21.45
Density		21 450 kg/m^3
Melting point		1769 °C
Boiling point		3800 °C
Specific heat		0.134 J/g °C
Thermal conductivity		69.1 W/m °C
Coefficient of linear expansion (20–100 °C)		9×10^{-6}/ °C
Latent heat of fusion		113 J/g
Latent heat of vaporization		2407 J/g
Thermal neutron absorption cross-section		9 barns/atom
Electrical conductivity		18% IACS (copper 100%)
Specific resistance		98 microhm mm
Temperature coefficient of electrical resistance		0.0039/ °C
Electrochemical equivalent		1.8160 g/A/h
Electrode potential		+1.2 V
Magnetic susceptibility		1.1×10^{-6}
Young's modulus of elasticity		152×10^9 N/m^2
Tensile strength	annealed	140 N/mm^2
Hardness	annealed	37 DPN

31.1 General notes on platinum

This is the most important metal in the group of six to which it gives its name, the remaining members being iridium, osmium, palladium, rhodium and ruthenium. The metals are always found in association with each other, are all resistant to chemical attack and have high melting points.

Platinum is outstanding in the group in that it has corrosion resistance equal to the best, coupled with excellent workability. Wire finer than 0.01 mm in diameter is available, capable of withstanding nitric and sulphuric acids at high temperatures.

Platinum and its fellow group members occur in the native state, generally alloyed with each other. The known alluvial deposits are rapidly becoming exhausted, but native alloys have been discovered in Africa and sulphide ores in Canada.

The African deposit is in the form of nuggets, associated with gold, while the Canadian sulphide ores form a valuable byproduct in nickel extraction.

Platinum has been found to be a valuable crucible material for laboratory and chemical engineering use. It withstands repeated rapid heating and cooling, is not readily attacked chemically, thus does not contaminate the material being treated and being soft and ductile withstands the abuse normal to industrial process equipment. Examples of the industries using these crucibles are glass and metal manufacturing when high purity products are necessary.

Platinum is also used for general laboratory ware, where its constant weight under the most severe conditions makes it an invaluable material for items such as electrodes and crucibles.

Platinum and its alloys are the favourite setting materials for diamonds in jewellery. The brilliant permanent lustre and ability to be formed into intricate shapes is responsible for the ever increasing use of platinum alloys as personal jewellery and ornaments.

The dimensional stability of platinum alloys under all normal conditions is responsible for their use as the material for standards, and master instruments. Being capable of forming very thin sheets which can be shaped and readily forge welded, platinum is often the most economical chemical resistant material for sheathing instruments and equipment operating in corrosive, high temperatures. In this field competition from rhodium electroplating is now being felt.

The expansion coefficient of platinum and platinum–rhodium alloys is comparable to that of good quality glass and the corrosion resistant qualities of these materials makes them useful as glass metal seals, or on apparatus where good temperature resistance and ductility are required in conjunction with glass.

Platinum and its alloys are a favourable material for thermocouples, as the e.m.f. generated between, for example, a platinum/platinum–rhodium couple is comparatively high and the couple is robust and withstands hot corrosion and oxidation.

Platinum–rhodium alloys are used as the electrical resistance heating medium in small laboratory type furnaces where high temperatures are required and vacuum or gas atmospheres to protect the elements cannot be used.

Platinum is used as an electrical contact material where a constant resistance over a range of operating conditions and temperatures and under intermittent use is important. It would not be used where high conductivity is important as platinum has a reasonably high resistance.

A large amount of platinum is now used as a catalyst in the chemical industry and oil refining. Almost all high octane modern petrols owe their economical production to platinum catalysts.

Some platinum alloys are used as high temperature brazing materials for joining nickel alloy and other high temperature metals where for some reason welding is undesirable.

Platinum, with its excellent corrosion resistance, high melting point and temperature resistance coupled with good ductility and weldability, makes this one of the more useful metals in the service of mankind. Only its scarcity, difficulty of purification and resultant high price keep this material from general use.

Symbol	Nominal analysis, supplier, condition and remarks.
51	30% Ir Pt alloy: Engelhard; annealed; electrical conductivity 5.3% IACS **DPN: 285**
52	25% Ir Pt alloy: Engelhard; annealed; electrical conductivity 5.4% IACS **DPN: 240**
53	20% Ir Pt alloy: Engelhard; annealed; electrical conductivity 5.7% IACS **DPN: 200**
55	10% Ir Pt alloy: Engelhard; annealed; electrical conductivity 7.0% IACS **DPN: 120**
102	10% Ru Pt alloy: Engelhard; annealed; electrical conductivity 4.1% IACS **DPN: 200**
103	5% Ru Pt alloy: Engelhard; annealed; electrical conductivity 5.5% IACS **DPN: 125**

Note. The following abbreviations and units are used in the tables:

DPN	Hardness, diamond pyramid number
UTS	Ultimate tensile strength, N/mm^2
Elon	Elongation, %
Proof	0.1% proof strength, N/mm^2

1 N/mm^2=0.1 hbar=0.102 kgf/mm^2=0.06475 tonf/in.^{2}=145.04 lbf/in.^{2}=1 MPa
See Appendix II for other abbreviations and conversion tables.

Symbol	Nominal analysis, supplier, condition and remarks.
115	Platinum: Commercially pure Pt; Engelhard; electrical conductivity 15% IACS **DPN: 60**
450	Au Ag Pt alloy: Engelhard; annealed; electrical conductivity 10.2% IACS **DPN: 60**
h Pt 10	99.999% Pt: Wire; 0.5 mm diameter; Light Ltd; high purity metal
h Pt 71	99.9999% Pt single crystals 3 mm × 25 mm: Light Ltd
IRRU	Ir Ru Pt alloy: For sliding contacts; electrical conductivity 4.4% IACS; Johnson Matthey **DPN: 310**
OERSTIT 900 CP	22% Co Pt alloy: For magnets; German alloy listed by PMA
OSMIRIDIUM	30% Ir 35% Os Rh Pt alloy: Natively occurring material
P 04840	15% Ir Pt alloy: Obsolete; designation used by UNS
PALLABRAZE 1237	40% Ni 60% Pd alloy: Brazing metal; Johnson Matthey; melting point 1237 °C
PERMANIT 900 CP	22% Co Pt alloy: For magnets; Austrian alloy listed by PMA
PGS	Au Ag Pt alloy: Engelhard
PLACO	23% Co Pt alloy: For magnets; German alloy; listed by PMA
PLACOVAR	23% Co Pt alloy: For magnets; American proprietary alloy; consult SAE
PLATINAX 11	Co Pt alloy: For extremely powerful magnets; Johnson Matthey; coercive force 4800 oersted
PLATINUM	High purity metal: Impurities less than 5 p.p.m.; Johnson Matthey; supplied as sponge, wire and sheet

32. Plutonium Pu

Physical properties		
Atomic number		94
Atomic weight		239
Crystal structure		Monoclinic
Colour		Yellow/white
Specific gravity		19
Density		19 000 kg/m^3
Melting point		640 °C
Boiling point		3235 °C
Specific heat		Varies with structure
Thermal conductivity		8.4 W/m °C
Coefficient of linear expansion (20–100 °C)		55 × 10^{-6}/ °C
Latent heat of fusion		–
Latent heat of vaporization		–
Thermal neutron absorption cross-section		–
Electrical conductivity		1.2% IACS (copper 100%)
Specific resistance		1500 microhm mm
Temperature coefficient of electrical resistance		–
Electrochemical equivalent		2.8 g/A/h
Electrode potential		–
Magnetic susceptibility		2.2 × 10^{-6}
Young's modulus of elasticity	cast	96.5 × 10^9 N/m^2
Tensile strength	cast	400 N/mm^2
Hardness		250 DPN

32.1 General notes on plutonium

This is an artificial element formed from uranium by the capture of a neutron when nuclear energy is harnessed in an 'atomic pile'. It is, in fact, a member of the second group of 'rare earth elements'.

Plutonium can be separated from the uranium used as the source material by chemical means. The purified plutonium can then be used as the primary material in another type of nuclear reactor.

Although it does not release as much energy as uranium 235 – the higher reactive isotope contained in uranium – it is much more reactive than normal uranium.

Plutonium, therefore, is used as a means of producing nuclear energy.

There are no commercially available materials containing plutonium.

Plutonium occurs in minute quantities in nature, having been found in pitchblende in a concentration of 1 part in 10^{12}.

This material is extremely toxic.

Note. The following abbreviations and units are used in the tables:

DPN	Hardness, diamond pyramid number
UTS	Ultimate tensile strength, N/mm^2
Elon	Elongation, %
Proof	0.1% proof strength, N/mm^2

1 N/mm^2=0.1 hbar=0.102 kgf/mm^2=0.06475 tonf/in.^{2}=145.04 ibf/in.^{2}=1 MPa

See Appendix II for other abbreviations and conversion tables.

Symbol	Nominal analysis, supplier, condition and remarks.
M 05995	99.5% (min) Pu: Nuclear grade; obsolete; UNS designation

33. Potassium K

Physical properties	
Atomic number	19
Atomic weight	39.096
Crystal structure	Body-centred cubic
Colour	Silver white
Specific gravity	0.86
Density	860 kg/m^3
Melting point	63.7 °C
Boiling point	760 °C
Specific heat	0.724 J/g °C
Thermal conductivity	99.2 W/m °C
Coefficient of linear expansion (20–100 °C)	83 × 10^{-6}/ °C
Latent heat of fusion	61.55 J/g
Latent heat of vaporization	2077 J/g
Thermal neutron absorption cross-section	2.7 barns/atom
Electrical conductivity	25% IACS (copper 100%)
Specific resistance	66.4 microhm mm
Temperature coefficient of electrical resistance	0.0052/ °C
Electrochemical equivalent	1.46 g/A/h
Electrode potential	−2.92 V
Magnetic susceptibility	0.52 × 10^{-6}
Young's modulus of elasticity	–
Tensile strength	–
Hardness	–

33.1 General notes on potassium

Potassium is too reactive an element ever to be found as native metal, but is common in the combined state. It is an essential constituent of all plant life but in quantities far too small to be a source for the metal. The compounds of potassium are all water soluble, yet seawater contains very little of this metal. This would appear to be caused by the greater affinity of soil and plant life for potassium than sodium and magnesium, both of which can be recovered from seawater in economic quantities.

Potassium is a silvery white metal, soft enough to cut with a knife, but tarnishing immediately. If left for any time in air potassium will combine with the moisture in the air to form potassium hydroxide – caustic potash. Potassium reacts violently if brought into contact with water and the metal must always be stored under liquid paraffin in an air tight container.

Potassium is very feebly radioactive but not enough to be either dangerous or useful. This radioactivity is due to the presence of the isotope potassium 40.

An alloy of sodium and potassium is liquid from room temperature to about 400 °C and finds use in thermometers where a wider temperature range than that covered by mercury is required.

There are also some lead alloys with small amounts of potassium added as a hardener.

No alloys exist based on potassium and as most of the properties are almost identical to those of sodium this metal is the useful first choice. Sodium is more readily obtained than potassium and thus is more economical.

Note. The following abbreviations and units are used in the tables:

DPN	Hardness, diamond pyramid number
UTS	Ultimate tensile strength, N/mm^2
Elon	Elongation, %
Proof	0.1% proof strength, N/mm^2

1 N/mm^2=0.1 hbar=0.102 kgf/mm^2=0.06475 tonf/in.^{2}=145.04 lbf/in.^{2}=1 MPa

See Appendix II for other abbreviations and conversion tables.

Symbol	Nominal analysis, supplier, condition and remarks.
WK 16	99.97% K metal: Light Ltd; high purity metal

34. Radium Ra

Physical properties	
Atomic number	88
Atomic weight	226.05
Crystal structure	–
Colour	Brilliant white
Specific gravity	5.0
Density	5000 kg/m^3
Melting point	700 °C
Boiling point	1500 °C
Specific heat	–
Thermal conductivity	–
Coefficient of linear expansion (20–100 °C)	–
Latent heat of fusion	–
Latent heat of vaporization	–
Thermal neutron absorption cross-section	–
Electrical conductivity	–
Specific resistance	–
Temperature coefficient of electrical resistance	–
Electrochemical equivalent	–
Electrode potential	–
Magnetic susceptibility	–
Young's modulus of elasticity	–
Tensile strength	–
Hardness	–

34.1 General notes on radium

Radium is one of the very rare metal elements which would probably still be undiscovered but for the fact that it is very radioactive. It is almost always found in association with uranium, deposits of which occur in Canada, the Congo and USA. The principal source is the Canadian pitchblende deposits which also contain silver and copper with some nickel. The extraction and purification of the metal is a long drawn out complex process resulting in very little metallic radium from many hundred tons of ore.

Radium is a brilliant white, which rapidly blackens owing to the formation of the nitride. It reacts vigorously with water forming the hydroxide. The best known property of radium is the radioactivity, it being one of the most active, earliest investigated of the radioactive elements. Considerable romance is associated with the early work on these elements and the name of Madame Curie will always be linked with radium.

The metal itself is very chemically active and attacks glass and many types of metal containers. There is still some use made of the metal for radiography, industrially and medicinally, and much of the available metal is used for radiotherapy. The most popular use at present for the metal is in the manufacture of luminous paints, but as these present considerable and serious health hazards the uses are vey limited.

The modern radioactive isotopes of cobalt, tantalum and iridium are more easily handled than radium and in many cases give superior X-ray radiographs.

Some of the radiotherapy uses of radium are also being replaced by these materials. There are no industrial uses for any alloys of radium.

The fact that radium is radioactive constitutes a health hazard and users of the metal must comply with stringent regulations regarding hygiene and methods of handling.

Radium metal is an interesting rarity with no important industrial uses and a considerable health hazard.

35. The rare earth elements

Cerium Ce, Dysprosium Dy, Erbium Er, Europium Eu, Gadolinium Gd, Holmium Ho, Lutetium Lu, Neodymium Nd, Praseodymium Pr, Promethium Pm, Samarium Sm, Terbium Tb, Thulium Tm, Ytterbium Yb.

Americium Am, Berkelium Bk, Californium Cf, Curium Cm, Einsteinium Es, Fermium Fm, Mendelevium Md, Neptunium Np, Plutonium Pu, Protactinium Pa, Thorium Th, Uranium U.

Lanthanum La, Scandium Sc, Yttrium Y.

They fall into two well-defined groups, the first occurring naturally in reasonably large quantities, the second being radioactive and very largely man-made short life elements. A third group resemble the rare earths to such an extent that they are always classified together.

Four of the metals listed as rare earths are discussed and listed separately as some use has been found for them in industry; these are, cerium in the first group and thorium, plutonium and uranium in the second group.

The metals in the first group all occur in monazite sand and are so much alike that it is only in recent years that it has been appreciated that more than two or three elements were present. Monazite sand also contains thorium compounds and is a principal source of this metal, the rare earths generally being a byproduct in its extraction. The extraction and reduction is a complex process, resulting in an alloy 'Mischmetal' which is about 50% cerium, the remainder being other rare earth metals of the first group with metallic impurities.

The pure metals can only be separated from each other by very complicated and delicate chemical and physical means and apart from cerium there appears to be very little requirement for them in the pure state.

They are very useful laboratory research materials in that they can be used to illustrate the relationships of elements to each other within the atomic table of elements.

The uses of cerium have been discussed elsewhere, the uses of the remaining elements being few.

Probably the most important use is as grain refiners and strengtheners of magnesium alloys, where no attempt is made to separate the metals and they are always referred to as 'rare earths'.

Other uses are as cores for carbon arcs in the preparation of special optical glass, and some uses are now being found in nuclear engineering as control rod elements.

The metals in the second group are all radioactive and three, plutonium, thorium and uranium, are the subject of separate sections in this book. The remaining elements in this group are mostly man-made, or discovered during the recent researches into nuclear fission. They are all unstable elements as the result of the radioactivity, and apart from nuclear engineering they serve no useful purpose.

The three elements in the last group are generally found with the first group of rare earth metals in monazite sand. They have no practical significance.

Symbol	Nominal analysis, supplier, condition and remarks.
E 00000	99.0% Ac: Commercial grade actinium; UNS designation
E 01000	99.0% Ce: Commercial grade actinium; UNS designation
E 01401	30% Co Ce alloy: UNS designation
E 21000	38% La 12% Nd 4% Pr Ce alloy: Mischmetal; UNS designation
E 21100	2.5% Mg 32% La 11.7% Nd 3.9% Pr Ce alloy: Mischmetal; UNS designation
E 21101	5% Mg 31% La 11.4% Nd 3.9% Pr Ce alloy: Mischmetal; UNS designation
E 21102	10% Mg 29% La 11% Nd 3.5% Pr Ce alloy: Mischmetal; UNS designation
E 21103	20% Mg 26% La 9.5% Nd 3.5% Pr Ce alloy: Mischmetal; UNS designation
E 21200	1% Al mixed rare earth alloy: UNS designation
E 21201	20% Al mixed rare earth alloy: UNS designation
E 21300	1% Fe 32.5% La 12% Nd 4% Pr Ce alloy: Mischmetal; UNS designation
E 21301	5% Fe mixed rare earth alloy: UNS designation
E 21302	25% Fe mixed rare earth alloy: UNS designation

Note. The following abbreviations and units are used in the tables:

DPN	Hardness, diamond pyramid number
UTS	Ultimate tensile strength, N/mm^2
Elon	Elongation, %
Proof	0.1% proof strength, N/mm^2

1 N/mm^2=0.1 hbar=0.102 kgf/mm^2=0.06475 tonf/in.^{2}=145.04 ibf/in.^{2}=1 MPa

See Appendix II for other abbreviations and conversion tables.

Symbol	Nominal analysis, supplier, condition and remarks.
E 21500	1% Au mixed rare earth alloy: UNS designation
E 23130	2.5% Mg 1% Fe 31% La 11.5% Nd 4% Pr Ce alloy: UNS designation
E 23310	2.5% Mg 10% Fe mixed rare earth alloy: UNS designation
E 23311	2% Mg 22% Fe mixed rare earth alloy: UNS designation
E 23312	6% Mg 23% Fe mixed rare earth alloy: UNS designation
E 28531	5% Ca 14% Fe 36% Si 14% La 5.5% Nd 2% Pr Ce alloy: UNS designation
E 31000	20% La 19% Nd 6% Pr 4% Sm 2% Gd 2% Y Ce alloy: UNS designation
E 46006	99% Dy (min) commercial grade dysprosium: UNS designation
E 48000	99% Er (min) commercial grade erbium: UNS designation
E 52000	99% Gd (min) commercial grade gadolinium: UNS designation
E 56000	99% Ho (min) commercial grade holmium: UNS designation
E 58000	99% La (min) commercial grade lanthanum: UNS designation
E 58401	25% Co La alloy: UNS designation
E 68000	99% Lu (min) commercial grade lutetium: UNS designation
E 69006	99% Nd (min) commercial grade neodymium: UNS designation
E 74000	99% Pr (min) commercial grade praseodymium: UNS designation
E 74401	28% Co Pr alloy: UNS designation
E 78000	99% Pm (min) commercial grade promethium: UNS designation

Symbol	Nominal analysis, supplier, condition and remarks.
E 79000	99% Sm (min) commercial grade samarium: UNS designation
E 79401	40% Co Sm alloy: UNS designation
E 83000	99% Sc commercial grade scandium: UNS designation
E 85000	99% Tb commercial grade terbium: UNS designation
E 87000	99% Tm commercial grade thulium: UNS designation
E 88000	99% Yb commercial grade ytterbium: UNS designation
E 90000	99% Y commercial grade yttrium: UNS designation
E 90401	36% Co Y alloy: UNS designation
E 90501	5% Cr Y alloy: UNS designation
h Dy 16	99.9% Dy: Ingot; Light Ltd; high purity metal dysprosium
h Dy 20a	99.9% Dy: Wire 1 mm diameter; Light Ltd; high purity metal dysprosium
h Dy 74	99.9% Dy: Single crystals 5 mm × 5 mm × 5 mm; Light Ltd; dysprosium
h Er 16	99.9% Er: Ingot; Light Ltd; high purity erbium
h Er 20a	99.9% Er: Wire 1 mm diameter; Light Ltd; high purity metal erbium
h Eu 16	99.9% Eu: Ingot; Light Ltd; high purity europium
h Eu 20b	99.9% Eu: Wire 1 mm thick; Light Ltd; high purity europium
h Gd 16	99.9% Gd: Ingot; Light Ltd; high purity gadolinium
h Gd 20a	99.9% Gd: Wire 1 mm diameter; Light Ltd; gadolinium
h Gd 74	99.9% Gd: Single crystals 5 mm × 5 mm × 5 mm each; Light Ltd: gadolinium
h Lu 16	99.9% Lu: Ingot; Light Ltd; lutetium
h Nd 16	99.9% Nd in lumps: Light Ltd; high purity metal
h Nd 20	99.9% Nd: Wire; 1 mm diameter; Light Ltd
h Pr 16	99.9% Pr: Ingot; Light Ltd; high purity metal
h Pr 20	99.9% Pr: Wire; 1 mm diameter; Light Ltd; high purity metal
h Sc 16	99.95% Sc: Ingot; Light Ltd; high purity metal
h Sc 20	99.9% Sc: Wire; 1 mm diameter; Light Ltd
h Sm 16	99.7% Sm: Ingot; Light Ltd
h Sm 19	99.9% Sm: Foil; Light Ltd
h Tb 16	99.6% Tb: Ingot; Light Ltd; high purity metal
h Tb 74	99.9% Tb: Single crystal 5 mm × 5 mm × 5 mm; Light Ltd
h Tm 16	99.9% Tm: Ingot; Light Ltd; high purity metal
h Tm 20	99.9% Tm: Wire 1 mm diameter; Light Ltd
h Y 16	99.9% Y: Ingot; Light Ltd; high purity metal
h Y 20	99.9% Y: Wire; 1 mm diameter; Light Ltd
h Yb 16	99.9% Yb: Ingot; Light Ltd; high purity metal
h Yb 20	99.9% Yb: Wire; 1 mm diameter; Light Ltd; high purity metal
NEODYMIUM	Nd metal: Blackwells; pure metal

Note. The following abbreviations and units are used in the tables:

DPN Hardness, diamond pyramid number
UTS Ultimate tensile strength, N/mm^2
Elon Elongation, %
Proof 0.1% proof strength, N/mm^2

1 N/mm^2=0.1 hbar=0.102 kgf/mm^2=0.06475 $tonf/in.^2$=145.04 $lbf/in.^2$=1 MPa

See Appendix II for other abbreviations and conversion tables.

36. Rhenium Re

Physical properties		
Atomic number		75
Atomic weight		186.22
Crystal structure		Close-packed hexagonal
Colour		Silver-white
Specific gravity		21.2
Density		21 200 kg/m^3
Melting point		3167 °C
Boiling point		5900 °C
Specific heat		0.138 J/g °C
Thermal conductivity		71.2 W/m °C
Coefficient of linear expansion (20–100 °C)		6.7×10^{-6}/ °C
Latent heat of fusion		–
Latent heat of vaporization		–
Thermal neutron absorption cross-section		86 barns/atom
Electrical conductivity		8.5% IACS (copper 100%)
Specific resistance		200 microhm mm
Temperature coefficient of electrical resistance		0.0031/ °C
Electrochemical equivalent		–
Electrode potential		–
Magnetic susceptibility		–
Young's modulus of elasticity		460×10^9 N/m^2
Tensile strength	annealed	114 N/mm^2
Hardness	annealed	170 DPN
	15% cold worked	550 DPN

36.1 General notes on rhenium

Rhenium was one of the unknown elements when the periodic table was first produced. Not until 1925 was the metal prepared and its properties shown to be in general agreement with those predicted about 1870. It is now known that rhenium is widely distributed, generally as the sulphide associated with molybdenum ores. The first deposits to be worked commercially were in the Rhine area of Germany and this river has given its name to the metal.

Rhenium is silvery white, and can stand very little cold work without intermediate annealing. If the anneal which must be above 1500 °C is carried out in air an oxide forms which diffuses in from the surface causing brittleness. The formation of this oxide also precludes all normal hot working methods. Some success has been achieved by hot forging under hydrogen. The metal has good corrosion resistant properties for all except oxidizing liquids or atmospheres. No serious atmospheric oxidation occurs below red heat and the metal retains its tensile properties at these temperatures.

The high hardness achieved on cold worked rhenium has been utilized in pen nibs and as pivot bearing points in instruments.

Rhenium is readily applied by electroplating and is used to coat the inside of tanks for transporting and storing acids. It also has excellent resistance to molten metals such as tin, lead, zinc and silver. Heater elements of rhenium can be used to evaporate these metals.

For some purposes rhenium metal is preferred to tungsten for electronic tube electrodes. Rhenium or rhenium-plated electrical contacts have been shown to give excellent results in marine engine magnetos. When used to interrupt high current d.c. circuits rhenium contacts regularly give better results with less arcing than other contact materials, although it does not have particularly good electrical conductivity.

Rhenium when added to molybdenum considerably increases the ductility of the cast metal. Molybdenum–rhenium alloys have therefore been developed as weld filler rods for molybdenum alloys. Sintered alloys with between 50% and 90% rhenium, tungsten and molybdenum are successfully used for all ball point pens and other purposes where a smooth action is necessary over a protracted period.

Rhenium is one of the very modern metals which is scarce and expensive at present, but which has potential high temperature and corrosion resistant properties, particularly when alloyed with other refractory metals.

Symbol	Nominal analysis, supplier, condition and remarks.
RHENIUM	High purity Re metal: Impurities 10 p.p.m.; Johnson Matthey; supplied as powder and button
RHENIUM	Re pure metal: Blackwells

Note. The following abbreviations and units are used in the tables:

DPN	Hardness, diamond pyramid number
UTS	Ultimate tensile strength, N/mm^2
Elon	Elongation, %
Proof	0.1% proof strength, N/mm^2

1 N/mm^2=0.1 hbar=0.102 kgf/mm^2=0.06475 $tonf/in.^2$=145.04 $lbf/in.^2$=1 MPa

See Appendix II for other abbreviations and conversion tables.

37. Rhodium Rh

Physical properties		
Atomic number		45
Atomic weight		102.91
Crystal structure		Face-centred cubic
Colour		Bluish white
Specific gravity		12.4
Density		12 400 kg/m^3
Melting point		1960 °C
Boiling point		4500 °C
Specific heat		0.243 J/g °C
Thermal conductivity		151 W/m °C
Coefficient of linear expansion (20–100 °C)		8.5×10^{-6}/ °C
Latent heat of fusion		21.8 J/g
Latent heat of vaporization		532 J/g
Thermal neutron absorption cross-section		2.6 barns/atom
Electrical conductivity		40% IACS (copper 100%)
Specific resistance		43 microhm mm
Temperature coefficient of electrical resistance		0.0046/ °C
Electrochemical equivalent		–
Electrode potential		+0.8 V
Magnetic susceptibility		1.14×10^{-6}
Young's modulus of elasticity		359×10^9 N/m^2
Tensile strength		700 N/mm^2
Hardness	annealed	100 DPN
	as plated	800 DPN

37.1 General notes on rhodium

Rhodium is one of the platinum metals, always being found in association with the remaining members – osmium, iridium, palladium and ruthenium. It is present in small proportions in the natively occurring alloys iridosmium and osmiridium, but also occurs in the palladium rich sulphide ores.

The resultant high purity metal is relatively soft in the annealed condition but work hardens very readily, accepting very little work before crumbling to a powder. Slight impurities also increase the hardness and make cold working impossible. As hot work has to be carried out at high temperatures under reducing atmospheres or vacuum, industry has found very few uses for this rather intractable material in the pure form.

Electroplating techniques are now possible with rhodium. The as plated metal has a high hardness, excellent light reflectivity, resistance to oxidation up to and above red heat, is unaffected by all known acids, and has stable electrical resistance. The metal can only be deposited from acid solutions thus undercoatings of silver or nickel are generally required to prevent attack and faulty adhesion of the base metal – copper or steel.

Rhodium plating is now used in jewellery where its high lustre and hardness make it an attractive economical alternative to platinum. Many silver or silver plated trophies and 'plate' are now flash rhodium plated to prevent or at least considerably reduce tarnishing. The very thin coat allows the warmer silver colour to shine through, yet prevents scratching of the silver when it is necessary to re-polish.

The good light and heat reflectivity find use as searchlight reflectors and to concentrate heat in infra-red ovens. For best results light reflectors should be polished silver, rhodium plated.

Much chemical equipment and apparatus is now rhodium plated, as this has comparable chemical resistance to platinum and is much harder.

Electrical contact points with both surfaces rhodium plated have a zero electrochemical potential and will stand severe hammering under high corrosive conditions up to red heat. The high hardness is also being used for 'plug-in' type sockets where softer deposits would be removed by mechanical abrasion.

Molybdenum and its alloys, rhodium plated, are finding some use at the intermediate temperatures where molybdenum is oxidized but rhodium is not. At present this is rather limited, as rhodium starts to diffuse rapidly into the molybdenum matrix at about 900 °C. However, there is a sufficient increase in the electrical contact conductivity and surface hardness for some use to be found for rhodium plated molybdenum.

Rhodium used as an alloy hardens platinum, the alloys being workable up to about 40% rhodium with no effect on the resistance to chemical attack. The lower rhodium–platinum alloys are used in the man-made fibre field for such purposes as extrusion nozzles. The same alloys are used for high temperature furnace windings, potentiometers and the gauze and powders which find applications as catalysts in chemical engineering plants.

Although rhodium is for all practical purposes unworkable, the fact that it can be readily electroplated allows the attractive hard wearing and chemical resistant properties to be fully utilized. The use of rhodium electroplating will increase as more economical methods of producing the metal are developed.

Symbol	Nominal analysis, supplier, condition and remarks.
ASTM B616	Specification for refined Rh: Graded by purity
RHODIUM	High purity metal: Impurities 10 p.p.m.; Johnson Matthey; supplied as sponge

Note. The following abbreviations and units are used in the tables:

DPN	Hardness, diamond pyramid number
UTS	Ultimate tensile strength, N/mm^2
Elon	Elongation, %
Proof	0.1% proof strength, N/mm^2

1 N/mm^2=0.1 hbar=0.102 kgf/mm^2=0.06475 $tonf/in.^2$=145.04 $lbf/in.^2$=1 MPa

See Appendix II for other abbreviations and conversion tables.

38. Rubidium Rb

Physical properties	
Atomic number	37
Atomic weight	85.48
Crystal structure	Body-centred cubic
Colour	White
Specific gravity	1.525
Density	1525 kg/m^3
Melting point	39.0 °C
Boiling point	685 °C
Specific heat	0.336 J/g °C
Thermal conductivity	58 W/m °C
Coefficient of linear expansion (20–100 °C)	90×10^{-6}/ °C
Latent heat of fusion	25.5 J/g
Latent heat of vaporization	888 J/g
Thermal neutron absorption cross-section	0.7 barns/atom
Electrical conductivity	15% IACS (copper 100%)
Specific resistance	125 microhm mm
Temperature coefficient of electrical resistance	0.0052/ °C
Electrochemical equivalent	–
Electrode potential	−2.9 V
Magnetic susceptibility	0.2×10^{-6}
Young's modulus of elasticity	–
Tensile strength	–
Hardness	–

38.1 General notes on rubidium

The compounds of rubidium occur sparingly but widespread throughout the world, generally associated with potassium and always with caesium.

The pure metal is soft, silvery white, tarnishing in air and reacting vigorously in water liberating hydrogen. The metal is a laboratory curiosity with no industrial use although it has weak radioactivity and has replaced caesium in photo-electric cells for certain purposes.

Rubidium is one of the alkali metals, in the same group as sodium, lithium, potassium, magnesium and caesium.

Although it is comparatively plentiful in nature, the fact that rubidium has no commercial value ensures that it remains a rare metal. No alloys based on the metal, or uses for the metal as an alloying element, have been found to date.

Symbol	Nominal analysis, supplier, condition and remarks.
f Rb 16	99.9% Rb metal: In ampoules; Light Ltd
RUBIDIUM	Pure metal: Blackwells.

Note. The following abbreviations and units are used in the tables:

DPN	Hardness, diamond pyramid number
UTS	Ultimate tensile strength, N/mm^2
Elon	Elongation, %
Proof	0.1% proof strength, N/mm^2

1 N/mm^2=0.1 hbar=0.102 kgf/mm^2=0.06475 tonf/in.^{2}=145.04 lbf/in.^{2}=1 MPa

See Appendix II for other abbreviations and conversion tables.

39. Ruthenium Ru

Physical properties

Atomic number	44
Atomic weight	101.7
Crystal structure	Close-packed hexagonal
Colour	White
Specific gravity	12.3
Density	12 300 kg/m^3
Melting point	2250 °C
Boiling point	4110 °C
Specific heat	0.243 J/g °C
Thermal conductivity	116 W/m °C
Coefficient of linear expansion (20–100 °C)	9.6×10^{-6}/ °C
Latent heat of fusion	193 J/g
Latent heat of vaporization	6196 J/g
Thermal neutron absorption cross-section	2.6 barns/atom
Electrical conductivity	24% IACS (copper 100%)
Specific resistance	72 microhm mm
Temperature coefficient of electrical resistance	0.0010/ °C
Electrochemical equivalent	–
Electrode potential	+0.45 V
Magnetic susceptibility	0.56×10^{-6}
Young's modulus of elasticity	414×10^9 N/m^2
Tensile strength	annealed 370 N/mm^2
Hardness	annealed 220 DPN

39.1 General notes on ruthenium

Ruthenium is a member of the platinum group of metals and is always found associated with iridium, osmium, palladium and rhodium, as well as platinum. Like them it has a high melting point, bluish white colour and excellent resistance to chemical attack. Another similarity is the difficulty with which it is separated from its fellow members and obtained in a pure state.

Ruthenium metal itself has no practical use as it is unworkable, and other members of the platinum group have similar or better properties. To date ruthenium has not found much use as an electroplated deposit. There have been suggestions that it could be used as a very high temperature braze metal for joining refractory metals but there is no evidence of any applications to date.

Many fountain pen nibs are now tipped with a sintered ruthenium alloy containing 13% tungsten 10% molybdenum 3% platinum and cobalt. This is replacing the naturally occurring osmiridium alloy previously exclusively used.

The majority of ruthenium is used alloyed with platinum or palladium both of which are hardened with no reduction in chemical resistance. Above 15% ruthenium these alloys are unworkable.

The platinum–ruthenium alloys are used for electrical contacts where they have similar properties to platinum–iridium. They are also generally used as substitutes for platinum jewellery, the alloy being harder and more wear resistant than palladium or platinum but not so readily formed into fine or intricate designs.

Some work has been carried out with ruthenium–molybdenum alloys which show an increase in ductility over pure ruthenium. The resultant alloys have some advantage over molybdenum for high temperature oxidation resistance, but not sufficient to overcome the price advantage of molybdenum. Alloys of ruthenium with molybdenum have also been developed which can be used for brazing refractory materials at temperatures of above 200 °C.

Ruthenium is a close second to osmium in having the least practical use of the platinum metals and for the same reason, in that the metal cannot be readily fabricated.

There are some uses for ruthenium in nuclear reactors but very little information is available.

Note. The following abbreviations and units are used in the tables:

DPN	Hardness, diamond pyramid number
UTS	Ultimate tensile strength, N/mm^2
Elon	Elongation, %
Proof	0.1% proof strength, N/mm^2

1 N/mm^2=0.1 hbar=0.102 kgf/mm^2=0.06475 tonf/in.^{2}=145.04 lbf/in.^{2}=1 MPa
See Appendix II for other abbreviations and conversion tables.

Symbol	Nominal analysis, supplier, condition and remarks.
ASTM B717 99.8	99.8% Ru: High purity
ASTM B717 99.9	99.9% Ru: High purity
FISSIUM	26% Mo 23% Pd Rh Zr 40% Ru alloy: Alloy of varying content in spent nuclear fuel elements
Fs	26% Mo 23% Pd Rh Zr 40% Ru alloy: Symbol used for Fissium
RUTHENIUM	High purity metal: Impurities 10 p.p.m.; Johnson Matthey; supplied as sponge

40. Selenium Se

Physical properties	
Atomic number	34
Atomic weight	78.96
Crystal structure	Two allotropic forms
Colour	Steel grey
Specific gravity	4.81
Density	4810 kg/m^3
Melting point	220 °C
Boiling point	685 °C
Specific heat	0.352 J/g °C
Thermal conductivity	2.9 W/m °C
Coefficient of linear expansion (20–100 °C)	37×10^{-6}/ °C
Latent heat of fusion	84 J/g
Latent heat of vaporization	3140 J/g
Thermal neutron absorption cross-section	11.8 barns/atom
Electrical conductivity	15% IACS (copper 100%)
Specific resistance	120 microhm mm
Temperature coefficient of electrical resistance	–
Electrochemical equivalent	0.5 g/A/h
Electrode potential	–
Magnetic susceptibility	-0.32×10^{-6}
Young's modulus of elasticity	57.9×10^9 N/m^2
Tensile strength	–
Hardness	–

40.1 General notes on selenium

Selenium has the characteristics of non-metals and metals evenly divided, being a metalloid or semi-metal. The non-metallic characteristics strongly resemble those of sulphur, the range of compounds known as selenides being similar in many respects to the sulphides.

It occurs in nature as metallic selenides generally associated with pyrites – metallic sulphides.

Like the non-metal sulphur, selenium exists in several allotropes the stable form being silvery grey, produced when any form of selenium is heated at 200–230 °C, and allowed to cool. A yellow allotropic form known as 'flowers of selenium' is produced when metallic selenium is allowed to boil.

Most high purity selenium is used as a material for rectifying electric current. Selenium metal rectifiers are now used almost exclusively when large current rectification is necessary. Rotary convertors were previous first choice for this purpose. Germanium and silicon join selenium as rectification metals for electronic and low current use.

Almost all commercial photo-electric devices have selenium as the operating mechanism. This uses a remarkable property in that light falling on the metal generates a small electric current proportional to the amount of light. The exact reason for this is as yet unknown but is believed to be associated with the fact that light causes emission of electrons from the high purity selenium. Photo-electric devices are now used for many purposes including trip mechanisms which operate when a light beam is interrupted, or to switch on artificial light when sunlight is reduced. The same devices convert light into sound in modern films, the sound track running alongside the video film. They also take part in the modern methods of colorimetric analysis whereby accurate quantitative estimation is rapidly carried out.

There have been recent advances in the use of selenium as a corrosion preventative coating on steel. Selenium additions of up to 0.5% are now made to copper alloys and austenitic stainless steels to improve machinability when there is no reduction in the hot working properties and very little in ductility.

The largest single application of selenium makes use of low grade selenium compounds for decolourizing glass. Small quantities added to molten glass eliminate the blue/green tinge caused by traces of iron common to most glass making sands.

Selenious acid is used in the local corrosion protection of magnesium. Selenium and the majority of its compounds are toxic and care must always be taken to prevent ingestion particularly of small quantities over a long period.

Of all the semi-metals or metalloids, selenium is probably the most useful from the purely engineering and metallurgical point of view.

Symbol	Nominal analysis, supplier, condition and remarks.
AR Q	99.99% Se: In shot form; common name for selenium for rectifiers
DD Q	99.95% Se: In shot form; common name for selenium for rectifiers
h Se 2	99.999% Se pellets: Light Ltd; high purity metal
SELENIUM	Se metal; Blackwells; pure metal

Note. The following abbreviations and units are used in the tables:

DPN	Hardness, diamond pyramid number
UTS	Ultimate tensile strength, N/mm^2
Elon	Elongation, %
Proof	0.1% proof strength, N/mm^2

1 N/mm^2=0.1 hbar=0.102 kgf/mm^2=0.06475 $tonf/in.^2$=145.04 $lbf/in.^2$=1 MPa

See Appendix II for other abbreviations and conversion tables.

41. Silicon Si

Physical properties	
Atomic number	14
Atomic weight	28.06
Crystal structure	Tetrahedral cubic–amorphous
Colour	–
Specific gravity	2.49
Density	2490 kg/m^3
Melting point	1414 °C
Boiling point	2600 °C
Specific heat	0.670 J/g °C
Thermal conductivity	84 W/m °C
Coefficient of linear expansion (20–100 °C)	2.5×10^{-6}/ °C
Latent heat of fusion	1809 J/g
Latent heat of vaporization	10 614 J/g
Thermal neutron absorption cross-section	0.13 barns/atom
Electrical conductivity	Less than 1% IACS (copper 100%)
Specific resistance	100 000 microhm mm
Temperature coefficient of electrical resistance	–
Electrochemical equivalent	–
Electrode potential	–0.45 V
Magnetic susceptibility	0.13×10^{-6}
Young's modulus of elasticity	112.4×10^9 N/m^2
Tensile strength	–
Hardness	–

41.1 General notes on silicon

Silicon is the second most common element occurring on earth, the first being oxygen. Combined with oxygen, silicon forms sand, quartz and flint, all of which are forms of the silicon oxide, silica. It is also the main constituent of granite and many of the metals occur as single or complex silicates.

Silicon is not a true metal and has even less metallic characteristics than most metalloids or semi-metals. It occurs in two forms: amorphous silicon, a brown powder, and crystalline silicon, a black lustrous solid.

The only known industrial use of silicon is to rectify electrical current. For this purpose silicon is superior to selenium for higher temperatures and voltages and thus finds some special purpose applications in the electronic industry. Pure silicon is more difficult to produce, thus being more expensive than either selenium or germanium.

It has no other property resembling that of the true metals and no other industrial or commercial use.

The element has found considerable use, however, in the metallurgy of several commonly used metals. These sometimes combine directly with the element, or make use of the abundant oxide, silica, but this being refractory is not usual.

There are now being developed a host of compounds with very special properties, all based on the element silicon, in a manner not unlike the organic compounds based on carbon. Silicone greases, rubbers, cements and so on have all been manufactured, many of which have special low and high temperature applications.

The metallurgical applications of silicon and silica are as follows.

Copper alloys. Silicon acts as a de-oxidizer, but is not generally added for this purpose but to strengthen the matrix and improve the corrosion resistance.

The copper–silicon alloys with up to 3.5% silicon are sometimes used as substitutes for tin bronze where strength and corrosion resistance are of prime importance.

When added to the copper–zinc series silicon imparts some corrosion resistance and strength to the matrix, and these alloys are sometimes known as 'silicon brass'.

Aluminium alloys. Silicon acts as a powerful de-oxidizer but is not generally used for this purpose. It lowers the melting point, thus increasing the fluidity of castings and allowing more intricate patterns to be successfully cast. With about 12–13% silicon additions, the lowest melting point of any aluminium alloy is achieved and as these still make sound castings a series of brazing alloys has been developed allowing other aluminium alloys to be brazed.

Silicon also forms intermetallic compounds, particularly with magnesium, which dissolve in the aluminium matrix with correct thermal treatment and then precipitate out, causing 'age hardening'.

Iron. Silicon added to cast iron helps to reduce the carbon present as carbides and increases that present as graphite and is therefore termed a graphitizer. As only carbides harden this has the effect of giving softer, more ductile cast irons. As a secondary effect it increases the fluidity of molten cast iron, thus giving castings of better definition.

High silicon irons have excellent acid resistance, particularly to sulphuric acid.

Steel. Silicon is present in most steels up to 0.35% as it is one of the most commonly used de-oxidizing agents. In these proportions silicon has no effect on the physical or mechanical properties of steel. Between 1% and 2% silicon steels have excellent hardenability and fatigue strength, thus finding use as leaf and coil springs, very often alloyed with manganese and chromium.

Very low carbon steels with 2–4% silicon have special hysteresis properties which make them suitable for transformer cores and other electrical and electronic uses.

Silicon in higher quantities imparts considerable acid resistance, particularly to the oxidizing acids. A process analogous to carburizing has been developed – known as siliconizing or 'Ihrigizing'. With this the parts are packed in silicon carbide and chlorine gas is passed over them at 1000 °C. This gives a silicon 'case' of up to 0.075% deep if desired, which has good acid resistance with reasonably good wear. The process is best suited to plain carbon steel or malleable cast iron.

Athough silicon cannot truly be classed as a metal it has enough use to metallurgists to warrant the brief note given.

Symbol	Nominal analysis, supplier, condition and remarks.
CMSZ	5% Zr 30% Fe 32.5% Cr 32.5% Si alloy: Origin unknown
L Si 2	99.9999% Si granules: Light Ltd; high purity
L Si 71	Single crystal Si: Light Ltd
SILVAISE	6% Zr 10% Ti 10% V 6% Al 0.5% B Si alloy: Origin unknown

Note. The following abbreviations and units are used in the tables:

DPN	Hardness, diamond pyramid number
UTS	Ultimate tensile strength, N/mm^2
Elon	Elongation, %
Proof	0.1% proof strength, N/mm^2

1 N/mm^2=0.1 hbar=0.102 kgf/mm^2=0.06475 $tonf/in.^2$=145.04 $lbf/in.^2$=1 MPa

See Appendix II for other abbreviations and conversion tables.

42. Silver Ag

Physical properties		
Atomic number		47
Atomic weight		107.88
Crystal structure		Face-centred cubic
Colour		White
Specific gravity		10.5
Density		10 500 kg/m^3
Melting point		960.8 °C
Boiling point		2193 °C
Specific heat		0.234 J/g °C
Thermal conductivity		419 W/m °C
Coefficient of linear expansion (20–100 °C)		19.6×10^{-6}/ °C
Latent heat of fusion		105 J/g
Latent heat of vaporization		2332 J/g
Thermal neutron absorption cross-section		63 barns/atom
Electrical conductivity		105% IACS (copper 100%)
Specific resistance		15.5 microhm mm
Temperature coefficient of electrical resistance		0.0041/ °C
Electrochemical equivalent		4.025 g/A/h
Electrode potential		+0.799 V
Magnetic susceptibility		0.2×10^{-6}
Young's modulus of elasticity		76×10^9 N/m^2
Tensile strength	annealed	140 N/mm^2
Hardness	annealed	25 DPN

42.1 General notes on silver

Native silver is still found on occasions generally associated with native copper. The most common ore is the sulphide 'galena', found in South America, Western USA, Australia and Norway.

Silver ores are very often found with lead and copper, the lead mines in this country producing silver as a byproduct as long as enough lead was present to justify extraction.

Some silver has been mined in its own right in Scotland and Cornwall.

Metallic silver is pure white in colour, relatively soft and capable of taking a high polish. In the pure state it is the best conductor of heat and electricity and when polished one of the best reflectors of light.

The metal has excellent resistance to oxidation even at comparatively high temperatures, but tarnishes readily in sulphur-bearing atmospheres.

Silver has resistance to most acids, except nitric acid and concentrated hydrochloric acid, having particularly good resistance to organic materials. This accounts for the long use of silver for cutlery and plate and the modern use for silver plating for the same purpose, and for many utensils and equipment in food manufacture. The high polish which still retains a degree of warmth, not apparent with the platinum metals or chromium, coupled with the ease of manipulation and fabrication makes silver a favourite jewellery material.

Silver is now being used as a ladle material in the production of the alkali metals. A considerable quantity of pure silver in the form of very thin sheet is used to line or coat copper alloy pipes and vessels which handle chemicals during manufacture and storage. Silver plating is also used for these applications, some of which are being replaced by the modern acid resistant stainless steels or titanium.

After copper, silver is the most commonly used electrical contact material, having an advantage in that the contact resistance is not increased by oxidation at normal temperatures or by arcing.

Silver should not be used in direct contact with steel as severe corrosion can take place in the steel under certain conditions. Steel parts must be copper- or nickel-plated prior to silver. Stainless steel and nickel alloy parts operating at high temperatures are often silver-plated to prevent seizure of nuts and bolts, thus ensuring easy removal at overhaul.

Nickel silvers contain no silver, the name being applied to the range of copper–zinc–nickel alloys which resemble it when polished. This also applies to German silver.

Several alloys are based on silver, one reason for alloying being to increase the hardness. Coins now have less than 50% silver, the remainder being copper, nickel and zinc.

High quality trophies and ornamental household articles are made from similar types of alloys, sterling silver being 7.5% copper–silver alloy. Modern alloys often have nickel added.

Industrially there is little use for these alloys as the chemical and oxidation resistance is reduced. Silver 'solders' or brazing alloys make up the bulk of the alloys based on silver. These have low melting points, thus reducing the risk of surface oxidation and distortion of the parts being brazed. The resultant join is stronger than copper–zinc brazing as the silver solders have superior wetting and penetrating qualities in addition to their higher intrinsic strength.

These alloys have cadmium, copper, zinc and phosphorus additions to vary the melting range, increase fluidity and for economy reasons. The presence of cadmium is now known to constitute a health hazard at brazing.

Silver is used as an alloying element to pure copper where in small quantities it raises the temperature at which softening occurs after cold work.

Some copper bearing alloys have silver additions, the resultant material standing higher temperatures than the copper–lead series at the expense of frictional properties.

Considerable quantities of silver are used in photographic film and most modern mirrors are silver backed. Recently it has been found that silver can be used catalytically for several important chemical processes.

Silver is probably the most industrially useful of the so-called precious metals. Some of these applications are now finding stainless steels to be a suitable alternative but the ease of manipulation, fabrication and deposition ensure that silver will continue to be used commercially and industrially for as long as it is economically possible.

Symbol	Nominal analysis, supplier, condition and remarks.
15% MANGANESE SILVER	15% Mn Ag alloy: Braze metal; Johnson Matthey; melting range 951–960 °C
120	20% Ni Ag alloy: Engelhard; annealed; electrical conductivity 72% IACS **DPN: 50**
179	28% Cu Ag alloy: Melting point 779 °C; Engelhard; annealed; electrical conductivity 80% IACS **DPN: 88**
718	Cu Pd Ag brazing alloy: Melting range 970–1010 °C; Engelhard; electrical conductivity 36% IACS **DPN: 50 UTS: 210 Elon: 28%**
719	Cu Pd Ag brazing alloy: Melting range 1080–1090 °C; Engelhard; electrical conductivity 11% IACS **DPN: 85 UTS: 330 Elon: 30%**
722	Cu Pd Ag brazing alloy: Melting range 807–810 °C; Engelhard; electrical conductivity 45% IACS **DPN: 90 UTS: 440 Elon: 23%**
723	Cu Pd Ag brazing alloy: Melting range 824–852 °C; Engelhard; electrical conductivity 30% IACS **DPN: 75 UTS: 460 Elon: 22%**
724	Cu Pd Ag brazing alloy: Melting range 850–900 °C; Engelhard; electrical conductivity 18% IACS **DPN: 120 UTS: 450 Elon: 20%**
725	Cu Pd Ag brazing alloy: Melting range 901–950 °C; Engelhard; electrical conductivity 12% IACS **DPN: 185 UTS: 500 Elon: 15%**
748	Cu Pd Ag brazing alloy: Melting range 876–898 °C; Engelhard; electrical conductivity 15% IACS **DPN: 140 UTS: 490 Elon: 25%**
1103	Cd Ag alloy: Engelhard; annealed; electrical conductivity 35% IACS **DPN: 40**
1144	Fe Ag alloy: Engelhard; annealed; electrical conductivity 90% IACS **DPN: 65**
1322	5% Pd Ag alloy: Engelhard; annealed; electrical conductivity 45% IACS **DPN: 33**
1323	10% Pd Ag alloy: Engelhard; annealed; electrical conductivity 30% IACS **DPN: 40**
1325	20% Pd Ag alloy: Engelhard; annealed; electrical conductivity 17% IACS **DPN: 55**
1340	10% Au Ag alloy: Engelhard; annealed; electrical conductivity 48% IACS **DPN: 29**
1379	Cu Ag alloy: Standard silver; Engelhard; annealed; electrical conductivity 90% IACS **DPN: 56**
1380	Cu Ag alloy: Coin silver; Engelhard; annealed; electrical conductivity 86% IACS **DPN: 62**
1386	20% Cu Ag alloy: Engelhard; annealed; electrical conductivity 82% IACS **DPN: 85**
1396	50% Cu Ag alloy: Engelhard; annealed; electrical conductivity 82% IACS **DPN: 95**
1438	30% Ni Ag alloy: Engelhard; annealed; electrical conductivity 62% IACS **DPN: 55**
1439	40% Ni Ag alloy: Engelhard; annealed; electrical conductivity 55% IACS **DPN: 60**
1464	Cd Ag alloy: Engelhard; annealed; electrical conductivity 82% IACS **DPN: 60**
1465	Cd Ag alloy: Engelhard; annealed; electrical conductivity 72.0% IACS **DPN: 70**
1845/4	Cu Zn 61% Ag alloy: Braze metal; Johnson Matthey; BS 18454; melting range 690–737 °C
1845/5	Cu Zn 43% Ag alloy: Braze metal; Johnson Matthey; BS 1845-5; melting range 698–788 °C
2601	Pd Mn Ag brazing alloy: Melting range 1000–1120 °C; Engelhard; electrical conductivity 19% IACS **DPN: 95 UTS: 270 Elon: 11%**
2602	Pd Mn Ag brazing alloy: Melting range 1180–1200 °C; Engelhard; electrical conductivity 10% IACS **DPN: 130 UTS: 480 Elon: 25%**
2949	Cd Ag alloy: Engelhard; annealed; electrical conductivity 69.0% IACS **DPN: 70**
AB 49 TRI-FOIL	20% Zn 0.5% Ni 2% Mn 28% Cu Ag alloy: Braze metal; Johnson Matthey; for brazing tungsten carbide
Ag/Cu EUTECTIC	29% Cu Ag alloy: Braze metal; Johnson Matthey; melting point 778 °C
AMS 4765	2% Ni 42% Ni Ag alloy: Braze metal
AMS 4766 A	15% Mn Ag alloy: Braze metal
AMS 4767	7.2% Cu 0.2% Li Ag alloy: Braze metal
AMS 4768	26% Cu 21% Zn 18% Cd Ag alloy: Braze metal
AMS 4769	15% Cu 16% Zn 24% Cd Ag alloy: Braze metal
AMS 4770 B	15.5% Cu 16.5% Zn 18% Cd Ag alloy: Braze metal
AMS 4771	15.5% Cu 15.5% Zn 3% Ni 16% Cd Ag alloy: Braze metal
AMS 4772 A	40% Cu 5% Zn 1% Ni Ag alloy: Braze metal
AMS 4773	30% Cu 10% Sn Ag alloy: Braze metal
AMS 4774	28.5% Cu 6% Sn 2.5% Ni Ag alloy: Braze metal
ARGO-BRAZE 40	28% Zn 2% Ni 30% Cu Ag alloy: Braze metal; Johnson Matthey; for tungsten carbide

Note. The following abbreviations and units are used in the tables:

DPN	Hardness, diamond pyramid number
UTS	Ultimate tensile strength, N/mm^2
Elon	Elongation, %
Proof	0.1% proof strength, N/mm^2

1 N/mm^2=0.1 hbar=0.102 kgf/mm^2=0.06475 tonf/in.^{2}=145.04 ibf/in.^{2}=1 MPa

See Appendix II for other abbreviations and conversion tables.

Symbol	Nominal analysis, supplier, condition and remarks.
ARGO-BRAZE 49 H	23% Zn 7% Mn 4% Ni 16% Cu Ag alloy: Braze metal; Johnson Matthey; for tungsten carbide
ARGO-BRAZE 50	Cu Zn Cd Ni Mn 50% Ag alloy: For brazing; Johnson Matthey; melting range 639–688 °C
ARGO-BRAZE 56	14.5% In 2.5% Ni 27% Cu Ag alloy: Braze metal; Johnson Matthey; melting range 600–711 °C
ARGO-FLO	Cu Zn Cd 39% Ag alloy: Brazing metal; Johnson Matthey; melting range 605–651 °C
ASTM B260 Ag 7	22% Cu 16% Zn 5% Sn Ag alloy: Braze filler; melting range 617–650 °C
ASTM B260 Ag 18	30% Cu 10% Sn Ag alloy: Braze metal; melting range 718–843 °C
ASTM B260 Ag 19	8.7% Cu 0.22% Li Ag alloy: Braze metal; melting range 877–982 °C
ASTM B260 B Ag 1	15% Cu 17% Zn 24% Cd Ag alloy: Braze filler; melting range 604–617 °C
ASTM B260 B Ag 1 A	15% Cu 16% Zn 18% Cd Ag alloy: Braze filler; melting range 626–635 °C
ASTM B260 B Ag 2	26% Cu 20% Zn 18% Cd Ag alloy: Braze filler; melting range 605–700 °C
ASTM B260 B Ag 3	15% Cu 15% Zn 16% Cd 3% Ni Ag alloy: Braze filler; melting range 631–686 °C
ASTM B260 B Ag 4	30% Cu 29% Zn 2% Ni Ag alloy: Braze filler; melting range 670–776 °C
ASTM B260 B Ag 5	30% Cu 25% Zn Ag alloy: Braze filler; melting range 675–741 °C
ASTM B260 B Ag 6	32% Cu 15% Zn Ag alloy: Braze filler; melting range 686–775 °C
ASTM B260 B Ag 8	28% Cu Ag alloy: Braze filler; melting point 780 °C
ASTM B260 B Ag 8 A	27.7% Cu 0.22% Li Ag alloy: Braze metal; melting range 760–870 °C
ASTM B260 B Ag 9	20% Cu 14% Zn Ag alloy: Braze filler; melting range 692–717 °C
ASTM B260 B Ag 10	20% Cu 10% Zn Ag alloy: Braze filler; melting range 720–750 °C
ASTM B260 B Ag 11	22% Cu 3% Zn Ag alloy: Braze filler; melting range 737–787 °C
ASTM B260 B Ag 13	40% Cu 5% Zn 1% Ni Ag alloy: Braze metal; melting range 855–970 °C
ASTM B260 B Ag Mn	15% Mn Ag alloy: Braze filler; melting range 960–970 °C
ASTM F106 TB Ag	99.95% Ag (min)
B A Ag 27	13.5% Cd 35% Cu 26% Zn Ag: Braze metal; designation used by AWS
B Ag 1	24% Cd 15% Cu 16% Zn Ag: Braze metal; designation used by AWS
B Ag 1a	18% Cd 15.5% Cu 23% Zn Ag: Braze metal; designation used by AWS
B Ag 2	18% Cd 26% Cu 21% Zn Ag: Braze metal; designation used by AWS
B Ag 2a	20% Cd 27% Cu 23.5% Zn Ag: Braze metal; designation used by AWS
B Ag 3	16% Cd 15.5% Cu 15.5% Zn 3% Ni Ag: Braze metal; designation used by AWS
B Ag 4	30% Cu 28% Zn 2% Ni Ag: Braze metal; designation used by AWS
B Ag 5	30% Cu 25% Zn Ag: Braze metal; designation used by AWS
B Ag 6	34% Cu 16% Zn Ag: Braze metal; designation used by AWS
B Ag 7	22% Cu 17% Zn 5% Sn Ag: Braze metal; designation used by AWS
B Ag 8	37% Cu Ag: Braze metal; designation used by AWS
B Ag 8a	37% Cu 0.4% Li Ag: Braze metal; designation used by AWS
B Ag 9	20% Cu 15% Zn Ag: Braze metal; designation used by AWS
B Ag 10	20% Cu 10% Zn Ag: Braze metal; designation used by AWS
B Ag 13	41% Cu 5% Zn 1% Ni Ag: Braze metal; designation used by AWS
B Ag 18	30% Cu 10% Sn Ag: Braze metal; designation used by AWS
B Ag 19	7% Cu 0.2% Li Ag: Braze metal; designation used by AWS
B Ag 20	38% Cu 32% Zn Ag: Braze metal; designation used by AWS
B Ag 21	28.5% Cu 6% Sn 2.5% Ni Ag: Braze metal; designation used by AWS
B Ag 22	16% Cu 22% Zn 7.5% Mn 4.5% Ni Ag: Braze metal; designation used by AWS
B Ag 23	15% Mn Ag: Braze metal; designation used by AWS
B Ag 24	20% Cu 28% Zn 2% Ni Ag: Braze metal; designation used by AWS
B Ag 25	40% Cu 35% Zn 5% Mn Ag: Braze metal; designation used by AWS
B Ag 26	38% Cu 32.5% Zn 2% Mn 2% Ni Ag: Braze metal; designation used by AWS
B Ag 28	30% Cd 28% Zn 2% Sn Ag: Braze metal; designation used by AWS
B Ag 33	17.5% Cd 30% Cu 27% Zn Ag: Braze metal; designation used by AWS
B Ag 34	32% Cu 28% Zn 2% Sn Ag: Braze metal; designation used by AWS
BAZAR METAL	9% Ni Ag alloy: Sheet and wire; origin unknown
BS 1561	99.99% Ag: For anodes; cast or rolled; annealed
BS 1845/3	15% Cu 16% Zn 19% Cd Ag alloy: Braze filler; melting range 620–640 °C
BS 1845/4	28% Cu 10% Zn Ag alloy: Braze filler; melting range 690–735 °C
BS 1845/5	37% Cu 19% Zn Ag alloy: Braze filler; melting range 700–775 °C
BS 1845 Ag 1	15% Cu 16% Zn 19% Cd Ag alloy: Braze metal; melting range 620–640 °C
BS 1845 Ag 2	17% Cu 16% Zn 25% Cd Ag alloy: Braze metal; melting range 610–620 °C
BS 1845 Ag 3	20% Cu 21% Zn 20% Cd Ag alloy: Braze metal; melting range 605–650 °C
BS 1845 Ag 4	28.5% Cu 10% Zn Ag alloy: Braze metal; melting range 690–735 °C
BS 1845 Ag 5	37% Cu 19.5% Zn Ag alloy: Braze metal; melting range 700–775 °C
BS 1845 Ag 6	30% Cu Ag alloy: Braze metal; melting range 600–720 °C
BS 1845 Ag 7	28% Cu Ag alloy: Braze metal; melting point 780 °C
BS 1845 Ag 8	99.99% Ag: For brazing; melting range 960 °C
BS 1845 Pd 1	26.5% Cu 5.0% Pd Ag alloy: For braze metal; melting range 805–810 °C
BS 1845 Pd 2	31.5% Cu 10% Pd Ag alloy: For braze metal; melting range 825–850 °C
BS 1845 Pd 3	22.5% Cu 10% Pd Ag alloy: For braze metal; melting range 830–860 °C
BS 1845 Pd 4	20% Cu 15% Pd Ag alloy: For braze metal; melting range 875–900 °C
BS 1845 Pd 5	28% Cu 20% Pd Ag alloy: For braze metal; melting range 875–900 °C

Note. The following abbreviations and units are used in the tables:

DPN	Hardness, diamond pyramid number
UTS	Ultimate tensile strength, N/mm^2
Elon	Elongation, %
Proof	0.1% proof strength, N/mm^2

1 N/mm^2=0.1 hbar=0.102 kgf/mm^2=0.06475 tonf/in.^{2}=145.04 lbf/in.^{2}=1 MPa

See Appendix II for other abbreviations and conversion tables.

Symbol	Nominal analysis, supplier, condition and remarks.
BS 1845 Pd 6	21% Cu 25% Pd Ag alloy: For braze metal; melting range 900–950 °C
BS 1845 Pd 7	5% Pd Ag alloy: For braze metal; melting range 970–1010 °C
BS 1845 Pd 9	20% Pd 5% Mn Ag alloy: For brazing metal; melting range 1000–1120 °C
BS 1845 Pd 10	33% Pd 3% Mn Ag alloy: For brazing metal; melting range 1180–1200 °C
B V Ag 0	99.95% Ag (min): Braze rod; designation used by AWS
B V Ag 8	28% Cu Ag: Braze metal; designation used by AWS
B V Ag 8a	29% Cu 0.5% Ni Ag: Braze metal; designation used by AWS
B V Ag 18	30% Cu 10% Sn Ag: Braze metal; designation used by AWS
B V Ag 29	25% Cu 14.5% Zn Ag: Braze metal; designation used by AWS
B V Ag 30	27% Cu 5% Pd Ag: Braze metal; designation used by AWS
B V Ag 31	32% Cu 10% Pd Ag: Braze metal; designation used by AWS
B V Ag 32	21% Cu 25% Pd Ag: Braze metal; designation used by AWS
B V Ag C	49% Cu Ag: Braze metal; designation used by AWS
COIN SILVER	10% Cu Ag alloy: American origin
CUSIL	28% Cu Ag alloy: For brazing; melting point 780 °C; Wesgo
CUSIL DECARBONIZED	20% Cu Ag alloy: Free of trapped carbon; melting point 780 °C; brazing alloy; Wesgo
CUSIL VT	20% Cu Ag alloy: For brazing; melting point 780 °C; Wesgo
DIMPALLOY	Silver solder sheet with flux: Sheffield Smelting; available as MX 18, MX 12 and MX 8
DIN 1734 L Ag 44	25% Zn 32% Cu Ag alloy: Braze metal; working temperature 730 °C
DIN 1734 L Ag 45	15% Zn 20% Cd 19% Cu Ag alloy: Bronze metal; working temperature 620 °C
DIN 1734 L Ag 49	25% Zn 18% Cu 5% Mn 3% Ni Ag alloy: Braze metal; working temperature 690 °C
DIN 1734 L Ag 50	15% Zn 5% Cd 32% Cu Ag alloy: Braze metal; working temperature 700 °C
DIN 1735 L Ag 50 Cd	15% Zn 20% Cu 15% Cd Ag alloy: Braze metal; working temperature 650 °C
DIN 1735 L Ag 60	12% Zn 28% Cu Ag alloy: Braze metal; working temperature 710 °C
DIN 1735 L Ag 60 Cd	10% Zn 25% Cu 2% Sn 2% Cd Ag alloy: Braze metal; working temperature 680 °C
DIN 1735 L Ag 67	8% Zn 25% Cu Ag alloy: Braze metal; working temperature 730 °C
DIN 1735 L Ag 67 Cd	10% Zn 12% Cu 10% Cd Ag alloy: Braze metal; working temperature 710 °C
DIN 1735 L Ag 75	2% Zn 24% Cu Ag alloy: Braze metal; working temperature 770 °C
DIN 1735 L Ag 83	2% Zn 15% Cu Ag alloy: Braze metal; working temperature 830 °C
DIN ARGO-FLO	21% Zn 20% Cd 19% Cu Ag alloy: Braze metal
EASY-FLO	16% Zn 19% Cd 15% Cu Ag alloy: Braze metal; Johnson Matthey; melting range 620–630 °C
EASY-FLO 1	Zn Cd Cu 50% Ag alloy: Brazing metal; Johnson Matthey; melting range 620–630 °C; electrical conductivity 22% IACS **DPN: 131 UTS: 450 Elon: 35%**
EASY-FLO 2	Zn Cd Cu 42% Ag alloy: Brazing metal; Johnson Matthey; melting range 608–617 °C; electrical conductivity 20% IACS **DPN: 135 UTS: 450 Elon: 30%**
EASY-FLO 3	Zn Cd 3% Ni Cu 50% Ag alloy: Braze metal; Johnson Matthey; melting range 634–656 °C; for brazing carbide tips **DPN: 135 UTS: 470**
EASY-FLO 3	15% Zn 16% Cd 18% Cu 3% Ni Ag alloy: Braze metal; Johnson Matthey; melting range 634–656 °C
EASY-FLO TRI-FOIL C	16% Zn 19% Cd 15% Cu Ag alloy: Braze metal; Johnson Matthey: for tungsten carbide
ELKONITE 20S	W Ag alloy: For arc resistant contacts; Johnson Matthey; conductivity 43% IACS **DPN: 220**
ELKONITE 35S	W Ag alloy: For arc resistant contacts; Johnson Matthey; conductivity 52% IACS **DPN: 140**
ELKONITE 50S	W Ag alloy: For arc resistant contacts; Johnson Matthey; conductivity 61% IACS **DPN: 115**
ELKONITE D54	Cd Ag alloy: For contacts; Johnson Matthey; conductivity 82% IACS
ELKONITE D54L	Cd Ag alloy: For contacts; Johnson Matthey; conductivity 82% IACS
ELKONITE D54X	Cd Ag alloy: For contacts; Johnson Matthey; conductivity 82% IACS
ELKONITE D55X	Cd Ag alloy: For contacts; Johnson Matthey; conductivity 75% IACS
ELKONITE D56	Ni Ag alloy: For contacts; Johnson Matthey; conductivity 72% IACS
ELKONITE D58/1	1% graphite Ag alloy: For sliding contacts; Johnson Matthey; conductivity 96% IACS
ELKONITE D58/2	2% graphite Ag alloy: For sliding contacts; Johnson Matthey; conductivity 87% IACS
ELKONITE D510	Ni Ag alloy: For contacts; Johnson Matthey; conductivity 87% IACS
ELKONITE D520	Ni Ag alloy: For contacts; Johnson Matthey; conductivity 57% IACS
ELKONITE G13	W Ag alloy: For contacts; Johnson Matthey; conductivity 57% IACS **DPN: 110**
ELKONITE G14	W Ag alloy: For contacts; Johnson Matthey; conductivity 48% IACS **DPN: 200**
ELKONITE G17	Mo Ag alloy: For contacts; Johnson Matthey; conductivity 48% IACS **DPN: 180**
ELKONITE G18	Mo Ag alloy: For contacts; Johnson Matthey; conductivity 52% IACS **DPN: 150**
EUTECTIC ALLOY	29% Cu Ag alloy: Braze metal; Johnson Matthey; melting point 778 °C
G4	W Ag contact material: Sheffield Smelting; electrical conductivity 43% IACS **DPN: 225**
G 6 Cu Zn	67% Ag alloy: Braze metal for silver; Johnson Matthey; melting range 705–723 °C; white colour
G7	W Ag contact material: Sheffield Smelting; electrical conductivity 50% IACS **DPN: 165**
G9	W Ag contact material: Sheffield Smelting; electrical conductivity 55% IACS **DPN: 140**
GC15	Graphite Ag contact material: Sheffield Smelting; electrical conductivity 85% IACS **DPN: 64**
GD10	Cadmium oxide Ag sintered contact material: Sheffield Smelting; electrical conductivity 82% IACS **DPN: 60**
GD25	Cadmium oxide Ag oxidized contact material: Sheffield Smelting; electrical conductivity 82% IACS **DPN: 60**
GD35	Cadmium oxide Ag contact material: Sheffield Smelting; electrical conductivity 75% IACS **DPN: 65**
GN1	Ni Ag contact material: Sheffield Smelting; electrical conductivity 66% IACS **DPN: 90**

Symbol	Nominal analysis, supplier, condition and remarks.
H 12	28% Cu Ag alloy: Braze metal; Sheffield Smelting; melting point 778 °C **DPN: 106 UTS: 390 Elon: 30% Proof: 290**
h Ag 1	99.9999% Ag rod: 12 mm diameter; Light Ltd; high purity metal
h Ag 96	99.0% Ag: Isotope 109; Light Ltd
HARDSILVER	3% Cu Ag alloy: Used for contacts; common name
HW 10C	WC Ag contact material: Sheffield Smelting; electrical conductivity 60% IACS **DPN: 130**
IN 15	15% In 24% Cu Ag alloy: Braze metal; Johnson Matthey; melting range 630–705 °C
IRCUSIL 10	10% In 27% Cu Ag alloy: For brazing; melting range 685–730 °C; Wesgo
IRCUSIL 15	14.5% In 24% Cu Ag alloy: For brazing; melting range 630–705 °C; Wesgo
JMC 1715 Mg Ni	Ag alloy: For electrical spring contact; Johnson Matthey; electrical conductivity 55% IACS **DPN: 145 UTS: 400 Elon: 14% Proof: 340**
JMM 77	Pd Pt Au Ag alloy: For contacts; Johnson Matthey; age hardened **DPN: 320 UTS: 1100 Elon: 7% Proof: 1000**
JMM 1715	Ni Mg Ag alloy: For contacts; 60% IACS conductivity; Johnson Matthey **DPN: 145**
MATTHEY 50S	50% W Ag alloy: Sinter for contacts; 61% IACS conductivity; Johnson Matthey **DPN: 115**
MATTHEY 55X	15% Cd O Ag alloy: For contacts; 75% IACS conductivity; Johnson Matthey **DPN: 60**
MATTHEY 3045	45% WC Ag alloy: Sinter for contacts; 62% IACS conductivity; Johnson Matthey **DPN: 95**
MATTHEY D54	10% Cd O Ag alloy: For contacts; 82% IACS conductivity; Johnson Matthey **DPN: 50**
MATTHEY D54X	10% Cd O Ag alloy: For contacts; 82% IACS conductivity; Johnson Matthey **DPN: 58**
MATTHEY D55	15% Cd O Ag alloy: For contacts; 72% IACS conductivity; Johnson Matthey **DPN: 55**
MATTHEY D56	30% Ni Ag alloy: For contacts; 72% IACS conductivity; Johnson Matthey **DPN: 68**
MATTHEY D58	1% C (graphite) Ag: For contacts; 96% IACS conductivity; Johnson Matthey **DPN: 40**
MATTHEY D58	2% C (graphite) Ag: For contacts; 86% IACS conductivity; Johnson Matthey **DPN: 40**
MATTHEY D510	10% Ni Ag alloy: For contacts; 87% IACS conductivity; Johnson Matthey **DPN: 40**
MATTHEY D520	20% Ni Ag alloy: For contacts; 82% IACS; Johnson Matthey **DPN: 48**
MATTHEY G13	50% WC Ag alloy: Sinter for contacts; 57% IACS conductivity; Johnson Matthey **DPN: 110**
MATTIBRAZE 34	Cu Zn Cd 34% Ag alloy: For brazing; Johnson Matthey; melting range 612–668 °C
MELT-ESI MX 12	16.5% Cu 18% Zn 23.5% Cd Ag alloy: Braze metal; Sheffield Smelting; melting range 608–621 °C; electrical conductivity 26% IACS; now THESSCO **DPN: 135 UTS: 420 Elon: 27% Proof: 250**
MELT-ESI MX 18	15% Cu 15% Zn 22% Cd Ag alloy: Braze filler; Sheffield Smelting; melting range 630–640 °C; electrical conductivity 23.8% IACS; now THESSCO **DPN: 115 UTS: 420 Elon: 43% Proof: 300**
MELT-ESI MX 18 PLUS	15.5% Cu 15.5% Zn 18% Cd 3% Ni Ag alloy: Braze metal; Sheffield Smelting; melting range 640–660 °C; electrical conductivity 17.2% IACS; now THESSCO **DPN: 145 UTS: 450 Elon: 35% Proof: 300**
MELT-ESI No 2	16.5% Cu 18% Zn 23.5% Cd Ag alloy: Filler metal; Sheffield Smelting; melting range 608–621 °C **DPN: 135 UTS: 420 Elon: 10% Proof: 250**
MX 0	25% Cu 25% Zn 20% Cd Ag alloy: Braze metal; Sheffield Smelting; melting range 605–680 °C; electrical conductivity 27.2% IACS **DPN: 114 UTS: 340 Elon: 10% Proof: 220**
MX 5	26% Cu 21% Zn 18% Cd Ag alloy: Braze metal; Sheffield Smelting; melting range 608–665 °C **DPN: 103 UTS: 360 Elon: 15% Proof: 220**
MX 8	20% Cu 19% Zn 23% Cd Ag alloy: Braze metal; Sheffield Smelting; melting range 606–648 °C; electrical conductivity 25.9% IACS **DPN: 121 UTS: 420 Elon: 24% Proof: 240**
MX 12	16.5% Cu 18% Zn 23.5% Cd Ag alloy: Braze metal; Sheffield Smelting; see MELT-ESI MX 12; electrical conductivity 26.0% IACS
MX 18	15% Cu 15% Zn 22% Cd Ag alloy: Braze metal; Sheffield Smelting; see MELT-ESI MX 18
MX 18 PLUS	15.5% Cu 15.5% Zn 18% Cd 3% Ni Ag alloy: Braze metal; Sheffield Smelting; see MELT-ESI MX 18 PLUS
NICUSIL 3	0.7% Ni 28% Cu Ag alloy: For brazing; melting range 780–795 °C; Wesgo
P 07505	UNS for AWS designation B Ag 24
P 07507	UNS for AWS designation B V Ag C
P 07540	UNS for AWS designation B Ag 13
P 07547	UNS for AWS designation B V Ag 32
P 07560	UNS for AWS designation B Ag 13
P 07563	UNS for AWS designation B Ag 7
P 07587	UNS for AWS designation B V Ag 31
P 07600	UNS for AWS designation B Ag 18
P 07607	UNS for AWS designation B V Ag 18
P 07627	UNS for AWS designation B V Ag 29
P 07630	UNS for AWS designation B Ag 21
P 07650	UNS for AWS designation B Ag 9
P 07687	UNS for AWS designation B V Ag 30
P 07700	UNS for AWS designation B Ag 10
P 07720	UNS for AWS designation B Ag 8
P 07723	UNS for AWS designation B Ag 8a
P 07727	UNS for AWS designation B V Ag 8
P 07728	UNS for AWS designation B V Ag 8a
P 07850	UNS for AWS designation B Ag 23
P 07900	UNS for ASTM B 617
P 07925	UNS for AWS designation B Ag 19
P 07931	7.2% Cu Ag alloy: Sterling Standard; designation used by UNS
P 07932	7% Cu Ag alloy: Sterling Silversmith
PALCUSIL 5	5.0% Pd 27% Cu Ag alloy: For brazing; melting range 807–810 °C; Wesgo
PALCUSIL 10	10% Pd 32% Cu Ag alloy: For brazing; melting range 824–852 °C; Wesgo
PALCUSIL 15	15% Pd 20% Cu Ag alloy: For brazing; melting range 850–900 °C; Wesgo

Note. The following abbreviations and units are used in the tables:

DPN	Hardness, diamond pyramid number
UTS	Ultimate tensile strength, N/mm^2
Elon	Elongation, %
Proof	0.1% proof strength, N/mm^2

1 N/mm^2=0.1 hbar=0.102 kgf/mm^2=0.06475 $tonf/in.^2$=145.04 $lbf/in.^2$=1 MPa

See Appendix II for other abbreviations and conversion tables.

Symbol	Nominal analysis, supplier, condition and remarks.
PALCUSIL 25	25% Pd 21% Cu Ag alloy: For brazing; melting range 900–950 °C; Wesgo
PALLABRAZE 1010	5% Pd Ag alloy: Brazing metal; Johnson Matthey; melting range 970–1010 °C
PALLABRAZE 1225	30% Pd Ag alloy: For brazing; Johnson Matthey; melting range 1150–1225 °C
QQ 561/4	15% Cu 18% Cd 15% Zn Ag alloy: Solder; US Federal; melting range 600–634 °C
SCP 1	Cu Pd Ag brazing alloy: Melting range 807–810 °C; Engelhard; electrical conductivity 45% IACS **DPN: 90 UTS: 440 Elon: 23%**
SCP 2	Cu Pd Ag brazing alloy: Melting range 824–852 °C; Engelhard; electrical conductivity 30% IACS **DPN: 75 UTS: 460 Elon: 22%**
SCP 3	Cu Pd Ag brazing alloy: Melting range 850–900 °C; Engelhard; electrical conductivity 18% IACS **DPN: 120 UTS: 450 Elon: 20%**
SCP 4	Cu Pd Ag brazing alloy: Melting range 901–950 °C; Engelhard; electrical conductivity 12% IACS **DPN: 185 UTS: 440 Elon: 15%**
SCP 5	Cu Pd Ag brazing alloy: Melting range 970–1010 °C; Engelhard; electrical conductivity 36% IACS **DPN: 50 UTS: 210 Elon: 28%**
SCP 6	Cu Pd Ag brazing alloy: Melting range 1080–1090 °C; Engelhard; electrical conductivity 11% IACS **DPN: 85 UTS: 330 Elon: 30%**
SCP 7	Cu Pd Ag brazing alloy: Melting range 876–890 °C; Engelhard; electrical conductivity 15% IACS **DPN: 140 UTS: 480 Elon: 25%**
SILBERLOT 44	25% Zn 32% Cu Ag alloy: Braze metal; designation used by German Standards
SILBERLOT 45	15% Zn 20% Cd 19% Cu Ag alloy: Braze metal; designation used by German Standards
SILBERLOT 49	25% Zn 18% Cu 5% Mn 3% Ni Ag alloy: Braze metal; designation used by German Standards
SILBERLOT 50	15% Zn 5% Cd 32% Cu Ag alloy: Braze metal; designation used by German Standards
SILBERLOT 50 Cd	15% Zn 20% Cu 15% Cd Ag alloy: Braze metal; designation used by German Standards
SILBERLOT 60	12% Zn 28% Cu Ag alloy: Braze metal; designation used by German Standards
SILBERLOT 60 Cd	10% Zn 25% Cu 2% Sn 2% Cd Ag alloy: Braze metal; designation used by German Standards
SILBERLOT 67	8% Zn 25% Cu Ag alloy: Braze metal; designation used by German Standards
SILBERLOT 67 Cd	10% Zn 12% Cu 10% Cd Ag alloy: Braze metal; designation used by German Standards
SILBERLOT 75	2% Zn 24% Cu Ag alloy: Braze metal; designation used by German Standards
SILBERLOT 83	2% Zn 15% Cu Ag alloy: Braze metal; designation used by German Standards
SILMANAL	9.0% Mn 4.0% Al Ag alloy: For magnets; American alloy listed by PMA; cold rolled
SILVER	Ag powder: Blackwells; pure metal
SILVER	High purity metal: Impurities 10 p.p.m.; Johnson Matthey; supplied as crystals
SILVER COPPER EUTECTIC	28% Cu Ag alloy: Braze metal; Johnson Matthey; melting point 778 °C
SILVER FLO 33	33.5% Zn 33% Cu Ag alloy: Braze metal; Johnson Matthey; melting range 700–740 °C
SILVER FLO 34	27% Zn 37% Cu 2.2% Sn Ag alloy: Braze metal; Johnson Matthey; melting range 630–730 °C
SILVER FLO 38	29% Zn 31% Cu 2% Sn Ag alloy: Braze metal; Johnson Matthey; melting range 660–720 °C
SILVER FLO 40	28% Zn 30% Cu 2% Sn Ag alloy: Braze metal; Johnson Matthey; melting range 650–710 °C
SILVER FLO 43	20% Zn 37% Cu Ag alloy: Braze metal; Johnson Matthey; melting range 690–775 °C
SILVER FLO 44	26% Zn 30% Cu Ag alloy: Braze metal; Johnson Matthey; melting range 675–735 °C
SILVER FLO 55	22% Zn 21% Cu 2% Sn Ag alloy: Braze metal; Johnson Matthey; melting range 630–660 °C
SILVER FLO 56	17% Zn 22% Cu 5% Sn Ag alloy: Braze metal; Johnson Matthey; melting range 630–660 °C
SILVER FLO 60	14% Zn 26% Cu Ag alloy: Braze metal; Johnson Matthey; melting range 695–730 °C
SILVER FLO 452	25% Zn 2.3% Sn 27.7% Cu Ag alloy: Braze metal; Johnson Matthey; melting range 640–680 °C
SILVER SOLDER	Cu Zn Ag brazing alloys: Covered by BS 1845/3/4 and 5; common name
SILVERSMITHS G RANGE	7% Cu Ag alloy: Sterling Silver; common name
SIX EIGHTY ALLOY	65% Ag alloy: For dental amalgams; Johnson Matthey
SOUDOGEN BE 1	20% Cu 20% Cd 20% Zn Ag alloy: Braze metal; Soudometal
SOUDOGEN BE 2	28% Cu 21% Cd 20% Zn Ag alloy: Braze metal; Soudometal
SOUDOGEN BE 3	40% Cu 15% Cd 25% Zn Ag alloy: Braze metal; Soudometal
SOUDOGEN BE 15	23.5% Cu 4.5% Sn 17% Zn Ag alloy: Braze metal; Soudometal **UTS: 440**
SOUDOGEN BE 100	19% Cu 20% Cd 21% Zn Ag alloy: Braze metal; Soudometal
SOUDOGEN BE 200	28% Cu 21% Cd 21% Zn Ag alloy: Braze metal; Soudometal
SPM 1	Pd Mn Ag brazing alloy: Melting range 1000–1120 °C: Engelhard; electrical conductivity 19% IACS **DPN: 95 UTS: 270 Elon: 11%**
SPM 1	33% Pd 5% Mn Ag alloy: For brazing; International Nickel; melting range 1000–1115 °C
SPM 2	33% Pd 3% Mn Ag alloy: For brazing; International Nickel; melting range 1180–1200 °C
SPM 2	Pd Mn Ag brazing alloy: Melting range 1180–1200 °C: Engelhard; electrical conductivity 10% IACS **DPN: 130 UTS: 480 Elon: 25%**
STANDARD SILVER	Alternative name for Sterling Silver
STERLING	7.5% Cu Ag alloy: Hall marked alloy of British Commonwealth and USA; common name
STUBS 3	21% Cd 19% Cu 20% Zn Ag alloy: For brazing; Stubs; silver solder; melting range 595–630 °C **DPN: 140 UTS: 545**
STUBS 3M	21% Cd 19% Cu 20% Zn Ag alloy: For brazing; Stubs; silver solder; melting range 595–630 °C **DPN: 140 UTS: 545**
STUBS 31	28% Cu 22% Cd 20% Cd Ag alloy: For brazing; Stubs; silver solder; melting range 607–685 °C **DPN: 140 UTS: 545**
STUBS 31M	28% Cu 22% Cd 20% Cd Ag alloy: For brazing; Stubs; silver solder; melting range 607–685 °C **DPN: 140 UTS: 545**
STUBS 306	21.0% Cu 2.5% Sn 25% Zn Ag alloy: For brazing; Stubs; Cd free silver solder; melting range 630–660 °C **DPN: 130 UTS: 420**
STUBS 306M	21.0% Cu 2.5% Sn 25% Zn Ag alloy: For brazing; Stubs; Cd free; silver solder; melting range 630–660 °C **DPN: 130 UTS: 420**
STUBS 570 K	Sn Ag alloy: Brazing rod; Stubs; used for food purposes; melting range 220 °C
STUBS 3033	33% Cu 33% Zn Ag alloy: For brazing; Stubs; silver solder; melting range 700–740 °C **DPN: 130 UTS: 542**
STUBS 3033 M	33% Cu 33% Zn Ag alloy: For brazing; Stubs; silver solder; melting range 700–740 °C **DPN: 130 UTS: 542**
STUBS 3040	30% Cu 28% Zn 2% Sn Ag alloy: For brazing; Stubs; Cd free silver solder, melting range 650–710 °C **DPN: 130 UTS: 400**

Symbol	Nominal analysis, supplier, condition and remarks.
STUBS 3040 M	30% Cu 28% Zn 2% Sn Ag alloy: For brazing; Stubs; Cd free silver solder, melting range 650–710 °C **DPN: 130 UTS: 400**
THESSCO D	Sn 70% Ag: Dental amalgam alloy; Sheffield Smelting
THESSCO M	Sn 68% Ag: Dental amalgam alloy; Sheffield Smelting
THESSCO MX12	16.5% Cu 18% Zn 23.5% Cd Ag alloy: Braze metal; Sheffield Smelting; melting range 608–621 °C; previously MELT-ESI **DPN: 135 UTS: 430 Elon: 27% Proof: 240**
THESSCO MX18	15% Cu 15% Zn 22% Cd Ag alloy: Braze filler; Sheffield Smelting; melting range 630–640 °C; previously MELT-ESI **DPN: 115 UTS: 430 Elon: 43% Proof: 300**
THESSCO MX18 PLUS	15.5% Cu 15.5% Zn 18% Cd 3% Ni Ag alloy: Braze metal; Sheffield Smelting; melting range 640–660 °C; previously MELT-ESI **DPN: 145 UTS: 450 Elon: 35% Proof: 300**

Note. The following abbreviations and units are used in the tables:

DPN	Hardness, diamond pyramid number
UTS	Ultimate tensile strength, N/mm^2
Elon	Elongation, %
Proof	0.1% proof strength, N/mm^2

1 N/mm^2=0.1 hbar=0.102 kgf/mm^2=0.06475 $tonf/in.^2$=145.04 $lbf/in.^2$=1 MPa

See Appendix II for other abbreviations and conversion tables.

43. Sodium Na

Physical properties	
Atomic number	11
Atomic weight	22.997
Crystal structure	Body-centred cubic
Colour	Silver-white
Specific gravity	0.971
Density	971 kg/m^3
Melting point	97.8 °C
Boiling point	883 °C
Specific heat	1.235 J/g °C
Thermal conductivity	135 W/m °C
Coefficient of linear expansion (20–100 °C)	71×10^{-6}/ °C
Latent heat of fusion	114 J/g
Latent heat of vaporization	4208 J/g
Thermal neutron absorption cross-section	0.54 barns/atom
Electrical conductivity	42% IACS (copper 100%)
Specific resistance	41 microhm mm
Temperature coefficient of electrical resistance	0.0055/ °C
Electrochemical equivalent	0.86 g/A/h
Electrode potential	−2.71 V
Magnetic susceptibility	0.51×10^{-6}
Young's modulus of elasticity	–
Tensile strength	Under 10 N/mm^2
Hardness	0.1 DPN

43.1 General notes on sodium

As would be expected of this very chemically active metal, it is never found in its native state. Combined with chlorine and as the carbonate, vast quantities exist spread over the earth's surface. Sodium chloride – common salt – is found in the sea, and there are enormous deposits of rock salt found all over the world as the result of the evaporation of prehistoric seas. One of these deposits in Poland is said to be 800 km long, 32 km wide, 370 m thick and has been operated as a salt mine for 600 years.

Sodium is a silvery white, soft metal which can be cut with a knife and can be moulded easily. It tarnishes rapidly in air, converting in time to sodium hydroxide by combining with the moisture in the atmosphere. It reacts violently in water with evolution of hydrogen. Sodium must therefore be stored at all times out of contact with air, generally under paraffin in air tight containers. This property reduces the usefulness of the metal which has a low specific gravity, with good heat and electrical conductivity and light reflection.

One use is as a heat transfer medium in high duty internal combustion engine exhaust valves. Here the stem is hollow and filled with sodium which melts at engine running temperature and transfers the heat from the valve head through the stem and valve guides.

The relatively high specific heat and low melting point, coupled with the comparatively non-corrosive nature with metals make sodium a potentially attractive heat transfer medium in other fields, and alloyed with potassium it is used for this purpose in certain types of nuclear power generators, and in thermometers where the liquid range from 400 °C to room temperature is wider than for most liquids.

Large quantities of the pure metal are used in the chemical industry in the preparation of such items as sodium cyanide, tetraethyl lead and sodium peroxide.

Some success has been achieved with sodium filled thin wall copper tube for electrical conductors. These can be used where weight is a vital factor, as the final conductor can be almost one-third the weight of the equivalent pure copper wire.

Sodium is used in arc lamps, giving the well-known intense orange light.

The affinity for oxygen leads to its use as a de-oxidizer in the purification of some of the refractory metals such as titanium and zirconium. Small quantities added to lead increase the hardness.

Some aluminium–silicon alloys have up to 0.1% sodium added as a grain refiner, giving a considerable improvement in mechanical properties.

Sodium has some photo-electric effects, but not sufficient to make it competitive with selenium.

Symbol	Nominal analysis, supplier, condition and remarks.
SODIUM	Na metal: Blackwells; pure metal
w Na 11	99.95% Na: In ampoules; Light Ltd; high purity metal
w Na 16	99.98% Na: In lump form; Light Ltd; high purity metal

Note. The following abbreviations and units are used in the tables:

DPN	Hardness, diamond pyramid number
UTS	Ultimate tensile strength, N/mm^2
Elon	Elongation, %
Proof	0.1% proof strength, N/mm^2

1 N/mm^2=0.1 hbar=0.102 kgf/mm^2=0.06475 $tonf/in.^2$=145.04 $lbf/in.^2$=1 MPa

See Appendix II for other abbreviations and conversion tables.

44. Steel

Steel is an alloy of carbon and iron with between 0.05% and 2.0% carbon content; some additions are made to all types of steel, for example manganese. Alloy steels contain one or more additional elements to give properties not obtainable with plain carbon steels.

The old Open Hearth and Bessemer practices of steel making are now obsolete and the bulk of world steel production is by the basic oxygen steel method, BOS, or variations on this technique. These processes are extremely rapid and use for their feed liquid iron and steel scrap blown with oxygen. The liquid steel made by this process can be further treated in various ways to produce high quality steels, for example by vacuum treatments.

Modern aircraft and tool steels are generally manufactured by the electric arc process, or under vacuum. Some of these start with pig iron and scrap, but generally scrap alloy steel with conventional steel ingot is used as the raw material.

Electric melting is confined to the production of high quality alloy steels with the alloying elements added on completion of the fluxing cycle, either to the melt itself during tapping, or to the ladle after tapping. Electric melted steels are used for quality engineering and aircraft steels, tool steels and most of the stainless varieties of steel.

A refinement of this method is to carry out melting, fluxing and casting under vacuum. Some electric melted steels, double vacuum re-melted, are now produced which after forging have tensile properties with the transverse strength almost equal to the longitudinal strength. The use of these steels is confined at present to high duty aircraft parts and ball races.

Of all the metals, indeed of all materials used by man, steel is by far the most versatile. Few objects or articles exist, useful, essential or ornamental, which have not at one time been made of steel in some of the myriad forms in which it exists, from wire a millionth of an inch thick to castings weighing hundreds of tons, from massive ships to delicate jewellery, from teaspoons to tanks.

44.1 The thermal treatment of steel

Several excellent books have been written with this as their sole subject and the information in this section should be taken as no more than brief notes to clear up some common misunderstandings. It has been written in the form of extended definitions, using the same terms as appear in the text. It should be noted, however, that there are considerable variations in the use of the terms, dependent on the industry involved and very often the geographical location.

Probably the most abused term is 'heat treated', which tends to be taken as synonymous with the most common thermal treatment carried out in any specific location. It can mean hardened and tempered in one factory and normalized in a second. This book therefore does not make reference to any part being 'heat treated', but prefers the term 'thermal treatment' when making reference to the process in general.

The term covers the processes where the steel is heated by any means whatever for the purpose of removing mechanical stresses or causing a metallurgical change to the structure, and includes the method of cooling.

The whole basis of the thermal treatment of steel relies on the fact that iron – also called ferrite – can dissolve iron carbide – also called cementite – and other metallic carbides while remaining in the solid state.

This requires that the steel is heated to above the 'upper critical' point when there is a change in the lattice structure. Below the 'lower critical' point the structure consists of cubes with iron atoms at each corner and in the centre, being known as 'body-centred cubic'. Also present is the compound of iron and carbon called cementite, or iron carbide. This is insoluble in the iron or ferrite and appears as a separate phase. Thus, below the 'lower critical' point, steel consists of body-centred cubic iron and iron carbide. Above the 'upper critical' point the structure is still cubic with the iron atoms at each corner but the centre atoms have migrated to the faces of the cubes – this being known as a 'face-centred cubic' structure – and this can dissolve the iron carbide, resulting in a single phase structure. The structure is termed austenite and has the appearance under the microscope of a pure metal. For steels which can be hardened by thermal treatment the 'upper critical' point varies from 900 to 700 °C depending on the carbon content and alloying elements, while the 'lower critical' point is at 700–720 °C, varied only by alloying elements.

The thermal treatments used to affect the metallurgical condition all make use of the change in the position of the carbon atoms at these critical points which, incidentally, results in a volume change of about 3%.

It must be emphasized that the steel remains solid throughout all the changes and the solution (dissolving) and precipitation of the iron carbides is the result of the migration of the atoms within the iron molecule.

Thermal treatments are applied to steel for the purpose of increasing the hardness or for reducing the hardness and increasing the ductility. The treatments commonly applied for each of these are as follows.

Increase in hardness

Age, carburize, carbo-nitride, cyanide, flame harden, harden, induction harden, nitride, precipitation treat.

Reduce hardness – increase ductility

Anneal, cyclic anneal, normalize, solution treat, spheroidize, stress relief, temper.

As stated above some of these terms can have more than one definition, but the following are the meanings ascribed in the text of this book, listed in alphabetical order.

Age. This applies to a limited number of complex ferrous alloys where the upper critical point has been depressed to below room temperature by the addition of alloying elements. These steels are thus austenitic at ambient temperatures. This austenite can be made to dissolve certain other elements by a 'solution treatment' and the steel can be hardened by precipitating these intermetallic compounds from solution. The ageing process always requires a previous solution treatment and is often called 'precipitation treatment'. This is one of the more complex forms of thermal treatment.

The term 'age' has been used to describe the long term treatment sometimes given to castings by leaving them in the open air. This is in fact a form of 'stress relieving'.

Anneal. This term is often used to denote any method of thermal treatment involving softening but should be applied only when recrystallization has occurred and the material is fully softened. This can only happen when the steel has been taken above the 'upper critical' point by about 30–50 °C for long enough to dissolve all carbides and then cooled very slowly.

The classical anneal requires that the parts remain in the furnace during cooling, resulting in complete equilibrium and the softest possible condition. This is seldom carried out as it is not an economical treatment, the cooling generally being in an insulated pot, in a box packed in sand or in some other refractory material.

The ferritic structure cannot dissolve carbides, whereas these are taken into solution in austenite and at the same time there is a variation in volume in the change from ferrite to austenite. This 'recrystallization', as it is called, completely eliminates all trace of previous thermal treatment, giving a fresh start. The slow cool from the critical temperature maintains equilibrium conditions at all times, allowing the iron carbide to return to its natural position, resulting in a completely stress-free structure.

For many purposes the subcritical anneal described later results in a satisfactory softness, but it must be noted does not cause recrystallization of the grains which is generally necessary if further heavy cold work is to be carried out. The rate of cooling is less critical for low carbon low alloy steels but must be very slow for higher carbon and alloyed steels.

Carburize. This process makes use of the fact that steel held at a temperature in an atmosphere containing carbon will absorb this at the surface until 0.8% carbon is attained. The carbon also diffuses in from the surface giving in effect a higher carbon steel around a lower carbon core.

As the percentage of carbon virtually decides the hardness and ductility of steel it is possible using this process to combine high surface hardness – and low ductility, with low core hardness – and high ductility. The process uses two basic methods.

First in 'pack carburizing' the parts are surrounded by activated charcoal inside closed boxes and heated to about 920–950 °C. This requires little special equipment but it is slow and dirty.

The second method is 'gas carburizing' where a carbon-rich atmosphere is used inside special furnaces. The temperature used is in the region of 920–950 °C, the time required being much shorter than with the pack method, as there is no box and large mass of charcoal to be heated. Special furnaces and equipment are essential, but this process is more economical, rapid and cleaner than 'pack carburizing'. For thin or shallow case depths carbo-nitride is commonly used.

Carbo-nitride. This is an alternative method of surface hardening to carburizing or to cyanide hardening. It makes use of a blended gas which is basically very similar to that in gas carburizing, with ammonia being added to give the necessary active nitrogen.

Carbo-nitriding can be carried out over a range of temperatures which coincide with the hardening temperatures of low alloy, low carbon steels. This can therefore be used for direct hardening without the danger of grain growth which occurs with carburizing.

Carbo-nitriding is generally found to be more economical than carburizing for shallow case depths and is probably the most economical of the diffusion-type surface hardening techniques.

Cyanide. This is an alternative method of surface hardening to carburizing and makes use of molten sodium cyanide salt as the heating medium and source of carbon and nitrogen. Cyanides are compounds of carbon and nitrogen and, depending on the temperatures used, can carburize or nitride. With temperatures in the region of 880–920 °C a carburized case results, whereas temperatures of 500–550 °C give a case which is mostly nitride. Intermediate temperatures result in a true cyanide case partly carburized, partly nitrided.

The cyanide salt bath method is probably the cheapest means of surface hardening, but suffers from the disadvantage that all soluble cyanides are extremely poisonous. Apart from the obvious health hazard for operators there can be considerable trouble and expense with effluent disposal, and this generally makes the carbo-nitride process more economical.

Cyclic anneal. This is a combination of full anneal and sub-critical anneal, in that the parts are heated to above the 'upper critical' temperature to dissolve the carbides, and then cooled to just above the 'lower critical' temperature, when the carbides are precipitated in the form of spheres. With many of the highly alloyed steels more than one cycle is required to ensure that all carbides are first dissolved, then precipitated as small spheres. This process is used because the resultant structure is the best for machining high carbon steels and is sometimes called spheroidize or spheroidal anneal.

Flame hardening. This makes use of the high temperature of the oxy-acetylene flame. This, when played on the surface of steel, will rapidly raise the surface to above the critical temperature and thus can be used as a means of locally hardening the surface. In all respects, this technique is identical to induction hardening.

It suffers from the disadvantage that it is difficult to control accurately the temperature of the oxy-acetylene flame.

Harden. This requires that parts are taken to just above the 'upper critical' point for a sufficient length of time to dissolve all the carbides. The changes in structure at the 'critical points' take the carbon into solution and result in a volume change with the atoms migrating from the centre of each iron molecule to the faces of the cube.

In the hardening process the steel is 'quenched' using brine, water, oil or air as the quenching medium, dependent on the 'hardenability' of the steel. As the steel cools the structure attempts to revert to that of ferrite, but the iron carbide is held in a form of supersaturated solution. This results in an unstable, strained structure called 'martensite' which is the hardest structure obtainable with steel. The hardness is directly proportional to the carbon content, while alloying elements decide the rate of cooling required to give full hardness in addition to having a considerable effect on tempering. Full hardening can only be accomplished when all the carbides have been taken into solution and the cooling rate is sufficient to give the fully martensitic structure. Too rapid a quench gives 100% martensite but results in distortion and cracking. It is also possible with certain steels to retain austenite and thus prevent full hardening. In addition to distortion at the quench there can also be dimensional changes in service as the austenite changes to martensite. As a general rule martensite is too brittle for practical purposes and must be 'tempered'.

There are means of obtaining the desired tempered structure other than hardening and tempering such as 'martempering', 'austempering' or 'delayed quench hardening', all of which make use of the same principle as outlined above, but require very critical control of times and temperatures and are related to the mass of the part being treated. Technical advice should always be obtained before attempting any of these treatments.

Induction hardening. This makes use of the induction heating technique to raise the surface temperature of steel rapidly to above the critical temperature necessary for hardening. This heating is so rapid that the surface will be at the required temperature while the core or centre remains at ambient. On removing the source of heat, quenching can take place and this will be extremely efficient as the heat will dissipate towards the core as well as from the outer surface.

Induction hardening is very similar to flame hardening and is generally confined to medium carbon steels such as 0.4% carbon. With alloy steels of high hardenability, there can be problems in the demarcation between case and core being too rapid for good results.

The advantage of induction hardening over flame hardening is that the temperature and the area to be heated can be controlled accurately, with reproductibility of results.

This is the most economical method of producing surface hardened components.

Nitride. The compound of iron and nitrogen, iron nitride, is very hard and retains its hardness to relatively high temperatures, although its formation takes place at about 500 °C, with an actual increase in volume.

The process requires special furnaces and equipment using ammonia, which is a compound of hydrogen and nitrogen, as the source of nitrogen. It is a slow process requiring special steels and is the most expensive of the commonly applied surface hardening treatments. It has the considerable advantage, however, that no quenching is involved and the process temperature causes no distortion. There is a growth of about 0.01 mm during a 60-hour nitriding cycle which gives a case depth of about 0.3 mm. It is difficult to measure the nitrided case.

Normalize. This is similar in many respects to the anneal in that the temperature used is about 30–50 °C above the 'upper critical' for sufficient time to take all the carbon into solution and allow recrystallization to take place. The parts are then cooled in air, which can result in a degree of hardness proportional to the carbon content and alloying elements, as alloying has the effect of slowing the speed with which the iron carbide can return to equilibrium.

Normalizing results in complete recrystallization, thus eliminating any heterogeneous grain structure from previous hot or cold work. All unequal stresses are removed although the result is seldom completely strain free for there will always be some air hardening except in low carbon mild steels.

Precipitation treat. See age.

Solution treat. With ferrous metallurgy this term applies only to a limited number of materials which are so highly alloyed that they remain austenitic at room temperature, but have intermetallic compounds present which are soluble in the austenitic matrix.

The solution treatment generally requires a temperature in excess of 1000 °C which dissolves the intermetallic compounds, these being kept in solution by quenching when the steel is in the softest condition.

The steel is then hardened by ageing, when some precipitation of the intermetallic compounds takes place, causing strain.

Spheroidize – or Spheroidal anneal. See cyclic anneal.

Stress relieve. This is another term which can mean anything from a full anneal to a 200 °C short term de-embrittlement treatment after electroplating, depending on the circumstances.

Stress relieving treatment should not alter the metallurgical condition of the material and ideally should be at as high a temperature as possible without causing any metallurgical change. Because of the range involved the actual temperature and time required should always be specified, or the reason for stress relieving clearly stated.

The degree of stress relieving achieved is directly proportional to the temperature used and can often result in distortion; thus it should be carried out before machining if fine limits are involved. This distortion is caused by the locked-up stresses becoming greater than the yield strength of the steel, as this drops with rise in temperature. It can sometimes be minimized by reducing the rate of heating.

Subcritical anneal. For this process the steel is heated to just below the 'lower critical' point. At this temperature none of the carbides are taken into solution and no change from ferrite to austenite takes place. Most cold working stresses are released at the same time, but as the steel has not been recrystallized it is not in a suitable condition to accept further heavy cold work.

This subcritical anneal is very often referred to as 'anneal', 'soften', 'stress relief' or 'fully temper'. It is used after normalizing air hardening steels and for softening all steels after carburizing or cold working when they have to be machined prior to further thermal treatment.

Temper. After hardening, steel should be 100% martensite and depending on the carbon content will be hard and brittle. By tempering at 120 °C some increase in ductility with no reduction in hardness is obtained. No hardened steel should be used without being tempered at 120 °C minimum.

As the tempering temperature is increased from this figure so the hardness drops and the ductility increases up to the 'lower critical' point when maximum ductility with almost full softening is achieved. This is caused by the iron carbide, which was trapped in solution in iron and thus held in an unnatural state, migrating to its natural position as a separate phase as the increasing temperature increases the mobility of the iron atoms and releases the iron carbide.

The steel changes in structure from α martensite to a fully tempered spheroidized structure which is soft and ductile at all times. Microscopic examination clearly shows the presence of these constituents and tells the metallurgist whether or not the steel has been correctly treated.

Only steels which are transformed to 100% martensite and then tempered back to the desired structure give the full mechanical properties. 'Slack' quenching followed by a low temperature temper may give the same hardness figures as a correctly quenched and tempered steel, but not the same proof strength and ductility. The same comments apply to steel which has not been taken above the 'upper critical' temperature prior to quenching.

44.2 Notes on specifications and trade names

As would be expected, there have been a large number of

attempts to evolve a standard method of steel classification. These fall into three categories: (a) by analysis, (b) by mechanical properties, (c) by use. In general, the methods based on properties and use are related to specific industries and as such are difficult to describe, but there are many based on analysis. The following short notes may be of assistance in identifying more rapidly specifications which fall into this category.

British Standards Institution (BS)

This body has produced a large number of different specifications, ranging from the single analysis special purpose to complex methods of classification involving analysis and use. Many of the older single purpose specifications have now been replaced by these classified specifications.

For engineering purposes the designation used was En followed by a number which made no attempt to be descriptive, although in general the lower numbers denoted low carbon mild steels. The number may be followed by a letter indicating a slight difference in analysis. Each En designation detailed analysis range, mechanical properties, condition and other necessary information. BS 971 has been issued as a commentary on the steels in BS 970 and is recommended to those who require further details on why specific steels are used. The En series has now been replaced with a code and BS 970 has been re-issued. Readers are advised to consult British Standard PD 6431 'New designation system for alloy steel' for further information.

There is also a three number designation system used by BS for steels employed in the chemical industry. These are classified according to the form within one specification and include mechanical properties and condition.

For example:

BS 1501	Steel plate section and bar for the chemical, petroleum and allied industries
BS 1503	Steel forgings for these industries
BS 1504	Steel castings for these industries
BS 1506	Steel bolting materials for these industries

Within these specifications covering form appears the analysis designated by the three digits. A few examples of these are:

620	0.15% carbon 0.6% chromium 0.6% molybdenum steel
621	0.15% carbon 1.0% chromium 0.5% molybdenum steel
845	18% chromium 10% nickel 2.5% molybdenum steel

Full details of this number system are given in the appendix to BS 1501.

These designations are used within other specifications covering materials for other industries or different material forms.

BS 4360 Structural Steels – Weldable

This specification covers a series of steels, all of which are weldable.

The strength and impact properties are designated by two figures and a letter. The two figures give the ultimate tensile strength in hectobars. 1 hbar is equal to 10 N/mm^2; thus a 40 hbar steel has a nominal UTS of 400 N/mm^2. This is followed by a letter A, B, C, D or E.

A	indicates that no impact properties are specified.
B	indicates that impact properties are specified at room temperature.
C	indicates that impact properties are specified at 0 °C.
D	indicates that impact properties are specified at a temperature below zero, which is given in the specification.
E	indicates that this is a low temperature impact steel with through thickness tensile properties.

BS 970 Engineering Steels

The specification BS 970 was amended in 1970 to eliminate the use of En— as a designation and replace this with a designation for plain and low alloy steels, and a separate type of designation for stainless steels.

This was updated in 1983 using the same type of designation. Briefly, this is for plain carbon and low alloy steels. There are three figures which identify the specification. This is followed by a letter – A, H or M – and then two figures. The first three figures are unique for a specification and are based on the chemical analysis. The letters are as follows:

A	indicates that the material is specified by the chemical analysis.
H	indicates that the material is specified by its hardenability.
M	indicates that it is specified by its mechanical properties.

The final two figures indicate the carbon content. Thus, 15 indicates 0.15% carbon, 30 indicates 0.3% carbon.

The stainless steels are designated by three figures, either a 300 series or a 400 series. The 300 series is used for austenitic type stainless steels and follows the general US specification, i.e. Type 316 in the US is identical to BS 970 316. After the three figures there is a letter with two figures. This will give specific requirements within the general specification.

The 400 series are designations used for 12% chromium ferritic/martensitic type stainless steels.

In this book the index for BS 970 states: see designation for details. All the designations are listed in the index, with the section where they will be found.

The En specifications show the designation now used wherever possible.

The examples listed are not intended to be a complete list the reader is advised to contact the British Standards Institution for further details.

The British Standards Institution have also issued designations headed CLA which are attempts to classify the different materials according to analysis for specific purposes. These are not always material specifications as such, but have all been included in this book for reference purposes.

There is a series of aircraft steel specifications also issued, which should correctly be called British Standards S . . . but are very commonly referred to as BS S . . . and have been included in this book under both titles – BS S . . . and S . . ., the latter always being referred to the former in the interests of clarity.

American Iron and Steel Institute (AISI) Society of Automotive Engineers (SAE)

These bodies have devised a number code which classifies and describes the steel alloys. In general the same code is used by both bodies sometimes with minor variations, for example, in the prefix letter indicating the method of manufacture used by AISI but not SAE.

The code is based on four digits for plain carbon and low alloy steels, the first two describing the steel, the second two indicating the percentage of carbon present.

Briefly the code is as follows:

10xx	Plain carbon steel – thus 1050 would be a 0.5% carbon steel. B 1025 would be a 0.2% plain carbon steel made by the acid Bessemer process.
11xx	Free machining plain carbon steel – sulphur additions
12xx	Free machining plain carbon steel – sulphur and phosphorus additions
12Lxx	Leaded steels with 0.3% sulphur – free cutting
13xx	Manganese 1.75%
40xx	Molybdenum 0.20 or 0.25%
41xx	Chromium 0.50, 0.80 or 0.95%; molybdenum 0.12, 0.20 or 0.30%
43xx	Nickel 1.83%, chromium 0.50 or 0.80%; molybdenum 0.25%
44xx	Molybdenum 0.53%
46xx	Nickel 0.85 or 1.83%, molybdenum 0.20 or 0.25%
47xx	Nickel 1.05%, chromium 0.45%, molybdenum 0.20 or 0.35%
48xx	Nickel 3.50%, molybdenum 0.25%
50xx	Chromium 0.40%
51xx	Chromium 0.80, 0.88, 0.93, 0.95 or 1.00%
52xx	Carbon 1.04%, chromium 1.03 or 1.45%
61xx	Chromium 0.60 or 0.95%, vanadium 0.13 or 0.15% (min)
86xx	Nickel 0.55%, chromium 0.50%, molybdenum 0.20%
87xx	Nickel 0.55%, chromium 0.50%, molybdenum 0.25%
88xx	Nickel 0.55%, chromium 0.50%, molybdenum 0.35%
92xx	Silicon 2.00%
50Bxx	Chromium 0.28 or 0.50% with boron
51Bxx	Chromium 0.80%
81Bxx	Nickel 0.30%, chromium 0.45%, molybdenum 0.12% with boron
94Bxx	Nickel 0.45%, chromium 0.40%, molybdenum 0.12% with boron

With the stainless varieties of steel the AISI use a three digit code which is also used by SAE but in their case preceded by the figures 50.

The three digit code is as follows:

2xx	Chromium nickel manganese – non-hardenable, austenitic
3xx	Chromium nickel – non-hardenable, austenitic
4xx	Chromium – hardenable and non-hardenable martensitic and ferritic, depending on carbon content
5xx	Chromium – low chromium heat-resisting. This last group are not stainless and appear with the low chromium steels in this book.

There is also an AISI code for tool steels, which has a letter prefix followed by two digits. The letter identifies the type of steel, examples being

A	Air hardening steel
M	Molybdenum steel
T	Tungsten steel
H	Hot work steel
D	Cold work steel – high carbon and chrome
O	Cold work steel – oil hardening
W	Water hardening steel

These codes are very often included in the trade name of US proprietary tool steel.

In this book the steels are indexed and classified under the numerical code, as it has been found that the steels are now known by this form. These codes are becoming accepted in the UK. Many UK tool steel manufacturers now use the tool steel classification system.

American Society for Testing and Materials (ASTM)

This body issues individual specifications which always have ASTM as the prefix, followed by the letter A, denoting a ferrous material, then a number identifying the actual specification, which may be followed by letters or numbers subdividing the material by analysis. The AISI code is sometimes used for this purpose. Finally the year of origin is given. A letter T after this denotes a tentative specification. In general each specification covers a steel in a specific form or for a special purpose rather than by analysis.

In this book, year and suffix figures are omitted.

German Standards (DIN)

There are two systems used at present by the German Specifications. Both systems have a general specification, usually by form carrying a DIN number which is then subdivided. The original method was descriptive, using the chemical symbols and numbers in an attempt to describe the alloy concerned. The latest method consists of numbers only and was devised to suit modern machinery, particularly computers. There is one significant figure which classifies the metal and after the decimal point there are four figures, the first two of which are used to identify the alloy, the last two the quantity. These are called Werkstoff numbers. Most steels are covered by the significant figure 1, but some have no significant figure before the decimal point. Examples are:

DIN 17221 67SiCr5	0.65% carbon, 1.3% silicon, 0.5% chromium spring steel
DIN 17221-7103	0.65% carbon, 1.3% silicon, 0.5% chromium spring steel
DIN 17155-15Mo3	0.18% carbon, 0.3% molybdenum steel for boiler plate
DIN 17155-1.5415	0.18% carbon, 0.3% molybdenum steel for boiler plate

In this book the full DIN specification is given once, using the descriptive form, and the descriptive designation and the Werkstoff number are indexed and classified separately. Examples of the entries would be:

DIN 17221 67SiCr5	0.65% carbon, 1.3% silicon, 0.5% chromium spring steel
67 Si Cr 5	0.65% carbon, 1.3% silicon, 0.5% chromium steel – designation used by German Standards
0.7103	0.65% carbon, 1.3% silicon, 0.5% chromium steel – German Standard

Thus three separate entries appear for each steel.

Classification system used in this book

All steels appear in Section 44 divided into groups based on analysis, with each basic material given a letter of the alphabet. It is obvious that all carbon contents cannot sensibly be included in one group, and this has been sub-divided and identified by a number. Thus plain carbon steels are divided into four sections 44A1, 44A2, 44A3 and 44A4, according to the carbon content. A slight overlap has been allowed to cope with variations in tolerance, which might otherwise tend to separate materials which should be grouped together.

The decision to include steels with a single alloying element as separate groups was taken because it is believed that in general an alloying element alone has a more marked effect on a steel than when a number of alloying elements are present. It also helps to indicate clearly what each alloy element does to steel.

The problem of residual elements has not been resolved, principally because there is often no clear indication given by the supplier whether or not certain elements are desirable, or will merely be accepted up to a certain limit. Every effort has been made to be consistent, but examples will no doubt be found of low alloy steels wrongly grouped.

44A1 Steel – plain carbon 0.05–0.2 per cent

Specific gravity	7.86
Density	7860 kg/m^3
Solidus/liquidus	1500–1520 °C
Thermal conductivity	63 W/m °C
Coefficient of linear expansion	11 × 10^{-6}/ °C
Electrical conductivity	8.4% IACS (copper 100%)
Specific resistance	200 microhm mm
Young's modulus of elasticity	206 × 10^9 N/m^2
Impact	40–200 J
Fatigue strength	–
Hot strength	

Temperature °C	Tensile strength N/mm^2	Elongation %
100	400	34
200	450	27
300	480	23
400	370	37
500	300	38
600	190	48
700	100	56
800	60	65
900	35	75

The above properties are typical of the following group and may not apply exactly to any one specification. It is possible that with certain specifications some of the values may not be applicable.

General metallurgical characteristics

These steels have too little carbon for any great increase in hardness to be produced by thermal treatment. Depending on the method of manufacture and the care with which undesirable impurities have been removed, they will exhibit a high degree of ductility with little work hardening at the lower carbon range. The cheaper steels, with varying carbon and higher phosphorus, sulphur and oxygen content, can become relatively brittle and unsuitable for cold working.

The steels are seldom used in the hardened and tempered condition but are commonly normalized. They can, however, be carburized to give a high carbon surface which is then hardenable. Without alloying elements this case hardened layer can be rather brittle and there may be a tendency for it to separate from the soft ductile core and to fail by spalling or exfoliation. This can be caused by bending movement or by heat stresses during hardening or subsequent grinding if the carburizing is not correctly controlled.

Some of the materials listed have elements added to improve machinability. The most common are sulphur, lead and tellurium, all of which act as chip breakers and are reputed to supply some lubricating properties.

This group includes the steels which are the first choice for weldability because of the low hardenability.

These steels are not capable of being hardened to any appreciable extent but the mechanical properties of many of the steels in this group can be improved to some extent by water quenching. This requires that the steels are taken to a temperature not much below 900 °C and then quenched in water. The quench must be efficient – either the water must be agitated or the components mechanically moved. It has been found that brine is more efficient than water.

It will be obvious that the cross or ruling section will be critical with regard to the ability of the quench to remove the heat from the component. Thus, it is seldom possible for components with sections larger than 10–15 mm.

While this is a disadvantage if good mechanical properties are required, it is a considerable advantage with welding where undesirable hardenability results in brittleness in heat affected zones.

It must be appreciated that 'tramp' elements such as chromium, and to a lesser extent vanadium, molybdenum, etc., will increase the hardenability of these steels. Chromium as low as 0.1% and vanadium, or molybdenum, as low as 0.05% can have quite dramatic effects on the hardenability.

Where controlled welding is essential then readers are advised to make use of the carbon equivalent. The formula is

carbon equivalent (CE) =

$$C + \frac{Mn}{6} + \frac{Cr + Mo + V}{5} + \frac{Ni + Cu}{15}$$

The critical value for welding is in the region of 0.7%.

Brazing and soldering do not present problems but any fluxes used must be carefully removed to prevent corrosion.

Oxides known as rust form on the surface and are loosely adherent. This results in pitting and scaling. There are now proven techniques for preventing this corrosion. The extent of the corrosion resistance will depend on the procedure chosen and ensuring that this is correctly carried out.

The machinability of these materials is relatively poor, resulting in a torn surface. This can be improved by firstly keeping high speeds and constant tool pressure. Where necessary free machining steels containing sulphur, selenium, tellurium or lead can be specified.

Low quality steels with high quantities of sulphur and phosphorus will have better machinability than good quality steels which are clean and free of oxides and slag inclusions.

The materials in this group are by far the most common used in engineering. The uses include car bodies, ships, modern domestic appliances such as fridges, etc. The vast majority of structural steels used in civil engineering are included in this group although some will also be in 44A2.

These are by far the most economical materials used in engineering. They will therefore invariably be the first choice and only their disadvantages in low strength, poor corrosion resistance, etc. will preclude their use. Many of these steels can be carburized.

They are now finding competition from stainless steels and aluminium alloys.

Symbol	Nominal analysis, supplier, condition and remarks.
0 STEEL	0.1% C (max) 0.15% Mn 0.15% Si steel: Castings; Edgar Allen; Code No 602 Steel
0.0040	0.12% C 0.3% Mn 0.06% S 0.08% P steel: German Standard
0.0041	0.12% C (max) 0.3% Mn 0.1% Si 0.06% S 0.07% P steel: German Standard
0.0042	0.1% C (max) 0.3% Mn 0.1% Si 0.04% S and P steel: German Standard
0.0043	0.1% C (max) 0.3% Mn 0.1% Si 0.03% S and P steel: German Standard
0.0044	0.1% C (max) 0.4% Mn 0.08% Si 0.05% S and P steel: German Standard
0.0105	0.1% C (max) 0.3% Mn 0.06% S 0.07% P steel: German Standard
0.0115	0.12% C (max) 0.3% Mn 0.06% S 0.08% P steel: German Standard
0.0133	0.12% C (max) 0.3% Mn 0.05% S and P steel: German Standard
0.0402	0.1% C (max) 0.3% Mn 0.15% Si 0.05% S and P steel: German Standard
0.0403	0.1% C (max) 0.35% Mn 0.15% Si 0.05% S and P steel: German Standard
0.0404	0.1% C (max) 0.35% Mn 0.15% Si 0.05% S and P steel: German Standard
0.0415	0.12% C (max) 0.35% Mn 0.06% S 0.08% P steel: German Standard
0.0433	0.12% C (max) 0.35% Mn 0.15% Si 0.04% S and P steel: German Standard
0.0510	0.1% C (max) 0.35% Mn 0.1% Si 0.04% S and P steel: German Standard
0.0511	0.1% C (max) 0.4% Mn 0.1% Si 0.03% S and P steel: German Standard

Note. The following abbreviations and units are used in the tables:

DPN	Hardness, diamond pyramid number
UTS	Ultimate tensile strength, N/mm^2
Elon	Elongation, %
Proof	0.1% proof strength, N/mm^2

1 N/mm^2=0.1 hbar=0.102 kgf/mm^2=0.06475 tonf/in.^{2}=145.04 lbf/in.^{2}=1 MPa

See Appendix II for other abbreviations and conversion tables.

Symbol	Nominal analysis, supplier, condition and remarks.
0.0515	0.1% C (max) 0.3% Mn 0.1% Si 0.03% S and P steel: German Standard
0.0516	0.1% C (max) 0.3% Mn 0.08% Si 0.04% S and P steel: German Standard
0.0517	0.1% C (max) 0.3% Mn 0.04% S and P steel: German Standard
0.0518	0.1% C (max) 0.3% Mn 0.1% Si 0.03% S and P steel: German Standard
0.0524	0.1% C (max) 0.3% Mn 0.08% Si 0.03% S and P steel: German Standard
0.0525	0.1% C (max) 0.3% Mn 0.08% Si 0.04% S and P steel: German Standard
0.0572	0.16% C 0.3% Mn 0.2% Si 0.04% S and P steel: German Standard
0.0612	0.21% C 0.45% Mn 0.2% Si 0.04% S and P steel: German Standard
1.0022	0.13% C 0.3% Mn 0.06% S and P steel: German Standard
1.0061	0.17% C (max) 0.4% Mn 0.3% Si 0.05% S and P steel: German Standard
1.0100	0.17% C (max) 0.05% S 0.08% P steel: German Standard
1.0102	0.17% C (max) 0.05% S and P steel: German Standard
1.0103	0.17% C 0.05% S and P steel: German Standard
1.0104	0.17% C (max) 0.05% S and P steel: German Standard
1.0106	0.17% C (max) 0.05% S and P steel: German Standard
1.0110	0.2% C (max) 0.05% S 0.08% P steel: German Standard
1.0112	0.2% C (max) 0.05% S 0.06% P steel: German Standard
1.0116	0.2% C (max) 0.05% S and P steel: German Standard
1.0122	0.2% C (max) 0.05% S 0.06% P steel: German Standard
1.0123	0.2% C (max) 0.05% S 0.08% P steel: German Standard
1.0301	0.09% C 0.35% Mn 0.2% Si 0.045% S and P steel: German Standard
1.0305	0.17% C (max) 0.4% Mn 0.3% Si 0.05% S and P steel: German Standard
1.0308	0.18% C (max) 0.05% S and P steel: German Standard
1.0309	0.17% C (max) 0.4% Mn 0.2% Si 0.05% S and P steel: German Standard
1.0318	0.1% C (max) 0.3% Mn 0.04% S and P steel: German Standard
1.0322	0.07% C 0.45% Mn 0.03% S and P steel: German Standard

Symbol	Nominal analysis, supplier, condition and remarks.
1.0323	0.09% C 0.45% Mn 0.25% S and P steel: German Standard
1.0325	0.1% C 0.03% S and P steel: German Standard
1.0328	0.1% C 0.06% Mn 0.03% S and P steel: German Standard
1.0329	0.1% C 0.6% Mn 0.05% Si 0.025% S and P steel: German Standard
1.0330	0.1% C (max) 0.3% Mn 0.05% S and P steel: German Standard
1.0331	0.1% C (max) 0.45% Mn 0.1% Si 0.05% S and P steel: German Standard
1.0333	0.1% C (max) 0.35% Mn 0.1% Si 0.04% S and P steel: German Standard
1.0334	0.1% C (max) 0.45% Mn 0.1% Si 0.04% S and P steel: German Standard
1.0336	0.1% C (max) 0.3% Mn 0.08% Si 0.03% S and P steel: German Standard
1.0337	0.1% C (max) 0.45% Mn 0.08% Si 0.03% S and P steel: German Standard
1.0338	0.1% C (max) 0.3% Mn 0.1% Si 0.03% S and P steel: German Standard
1.0356	0.16% C (max) 0.5% Mn 0.2% Si 0.045% S and P steel: German Standard
1.0401	0.16% C 0.4% Mn 0.2% Si 0.045% S and P steel: German Standard
1.0402	0.2% C 0.45% Mn 0.25% Si 0.045% S and P steel: German Standard
1.0405	0.22% C (max) 0.45% Mn 0.2% Si 0.05% S and P steel: German Standard
1.0418	0.22% C (max) 0.4% Mn 0.2% Si 0.05% S and P steel: German Standard
1.0437	0.2% C (max) 0.6% Mn 0.2% Si 0.05% S and P steel: German Standard
1.0439	0.13% C 0.6% Mn 0.05% Si 0.025% S and P steel: German Standard
1.0448	0.13% C 0.6% Mn 0.05% Si 0.03% S and P steel: German Standard
1.0456	0.22% C (max) 0.5% Mn 0.2% Si 0.05% S and P steel: German Standard
1.0580	0.18% C 1.4% Mn steel: German Standard
1.0581	0.18% C 1.4% Mn steel: German Standard
1.0611	0.21% C 0.45% Mn 0.2% Si 0.045% S and P steel: German Standard
1.0711	0.12% C (max) 0.7% Mn 0.25% S 0.05% P steel: German Standard; free machining
1.0713	0.13% C (max) 1.1% Mn 0.23% S 0.07% P steel: German Standard; free machining
1.0716	0.12% C (max) 0.7% Mn 0.25% S 0.05% P 0.2% Pb steel: German Standard; free machining
1.0719	0.13% C (max) 1.1% Mn 0.23% S 0.03% P steel: German Standard; free machining
1.0721	0.09% C 0.7% Mn 0.2% Si 0.22% S 0.07% P steel: German Standard; free machining
1.0723	0.16% C 0.7% Mn 0.3% Si 0.22% S 0.07% P steel: German Standard; free machining
1.0724	0.2% C 0.7% Mn 0.2% Si 0.2% S 0.07% P steel: German Standard; free machining

Note. The following abbreviations and units are used in the tables:

DPN	Hardness, diamond pyramid number
UTS	Ultimate tensile strength, N/mm^2
Elon	Elongation, %
Proof	0.1% proof strength, N/mm^2

1 N/mm^2=0.1 hbar=0.102 kgf/mm^2=0.06475 $tonf/in.^2$=145.04 $lbf/in.^2$=1 MPa

See Appendix II for other abbreviations and conversion tables.

Symbol	Nominal analysis, supplier, condition and remarks.
1.0736	0.08% C 0.35% S steel: German Standard; free machining
1.0737	0.08% C 0.35% S 0.2% Pb steel: German Standard; free machining
1.0831	0.2% C 1.5% Mn 0.5% Si 0.05% S and P steel: German Standard
1.0832	0.2% C (max) 1.5% Mn 0.3% Si 0.05% S and P steel: German Standard
1.0833	0.2% C 0.05% S and P steel: German Standard
1.0841	0.2% C (max) 0.05% S and P steel: German Standard
1.1121	0.09% C 0.4% Mn 0.25% Si 0.035% S and P steel: German Standard
1.1141	0.16% C 0.4% Mn 0.2% Si 0.03% S and P steel: German Standard
1.1144	0.16% C 0.4% Mn 0.2% Si 0.035% S and P steel: German Standard
1.1151	0.21% C 0.5% Mn 0.25% Si 0.035% S and P steel: German Standard
1.5034	0.1% C 1.0% Mn 0.03% S and P steel: Welding rod; German Standard
1.5035	0.1% C 1.0% Mn 0.03% S and P steel: Welding rod; German Standard
1.5063	0.13% C 1.5% Mn 0.03% S and P steel: Welding rod; German Standard
1.5064	0.13% C 1.6% Mn 0.03% S and P steel: Welding rod; German Standard
1.5074	0.16% C 2.0% Mn 0.02% S and P steel: Sheet and strip; German Standard
1.5083	0.21% C 1.7% Mn 0.03% S and P steel: Welding rod; German Standard
1.5086	0.13% C 2.0% Mn 0.03% S and P steel: Welding rod; German Standard
1.5098	0.13% C 3.0% Mn 0.03% S and P steel: Welding rod; German Standard
1.5331	0.17% C 1.4% Mn 0.2% Ti 0.1% Al steel: Welding rod; German Standard
1.5340	0.17% C 1.1% Mn 0.1% Al 0.25% Zr steel: Welding rod; German Standard
1.5417	0.2% C 1.1% Mn steel: Seamless tube; German Standard
1B	0.08% C 0.25% Mn steel: Sandvik for SS 1160
1CR	0.08% C (max) 0.4% Mn 0.03% S 0.025% P steel: Designation in BS 1449
1CS	0.08% C (max) 0.4% Mn 0.03% S 0.025% P steel: Designation in BS 1449
1DT	0.06% C 0.15% Mn 0.04% Si steel: Sandvik for SS 1150
1DTR	0.04% C 0.15% Mn 0.04% Si steel: Sandvik
1HR	0.08% C (max) 0.4% Mn 0.030% S 0.025% P steel: Designation in BS 1449
1HS	0.08% C (max) 0.4% Mn 0.030% S 0.025% P steel: Designation in BS 1449
2CR	0.08% C (max) 0.4% Mn 0.035% S 0.03% P steel: Designation in BS 1449
2CS	0.08% C (max) 0.4% Mn 0.035% S 0.03% P steel: Designation in BS 1449
2HR	0.08% C (max) 0.45% Mn 0.035% S 0.03% P steel: Designation in BS 1449
2HS	0.08% C (max) 0.45% Mn 0.035% S 0.03% P steel: Designation in BS 1449
2LS	0.1% C 0.4% Mn 0.1% Si steel: Sandvik for SS 1265, 1350
2S	0.1% C 0.5% Mn 0.15% Si steel: Sandvik for SS 1233/34
2SE	0.1% C 0.4% Mn 0.15% Si 0.03% Cu (max) steel: Sandvik
3BS	0.1% C 0.45% Mn 0.15% Si steel: Sandvik for SS 1300
3BSX	0.1% C 0.4% Mn 0.15% Si steel (S and P lower than 3BS): Sandvik
3CR	0.1% C (max) 0.5% Mn 0.04% S 0.04% P steel: Designation in BS 1449

Symbol	Nominal analysis, supplier, condition and remarks.
3CS	0.1% C (max) 0.5% Mn 0.04% S 0.04% P steel: Designation in BS 1449
3HR	0.1% C (max) 0.5% Mn 0.04% S 0.04% P steel: Designation in BS 1449
3HS	0.1% C (max) 0.5% Mn 0.04% S 0.04% P steel: Designation in BS 1449
3L7	0.1% C 0.7% Mn 0.2% Si steel: Sandvik
3LS	0.15% C 0.45% Mn 0.25% Si steel: Sandvik for SS 1345
4CR	0.12% C (max) 0.6% Mn 0.05% S and P steel: Designation in BS 1449
4CS	0.12% C (max) 0.6% Mn 0.05% S and P steel: Designation in BS 1449
4HR	0.12% C (max) 0.6% Mn 0.05% S and P steel: Designation in BS 1449
4HS	0.12% C (max) 0.6% Mn 0.05% S and P steel: Designation in BS 1449
4L7	0.2% C 0.6% Mn 0.2% Si steel: Sandvik for SS 1434/35
4LM	0.2% C 1.4% Mn 0.2% Si steel: Sandvik
4LS	0.2% C 0.4% Mn 0.2% Si steel: Sandvik for SS 1357, 1410
9 S 20	0.12% C (max) 0.7% Mn 0.2% S 0.1% P steel: Designation used by German Standards; free machining
9 S 27	0.12% C (max) 0.8% Mn 0.25% S 0.1% P steel: Designation used by German Standards; free machining
9 S Mn 23	0.13% C (max) 1.1% Mn 0.23% S steel: Designation used by German Standards; free machining
9 S Mn 28	0.14% C (max) 1.1% Mn 0.28% S steel: Designation used by German Standards; free machining
9 S Mn 36	0.15% C (max) 1.2% Mn 0.38% S steel: Designation used by German Standards; free machining
9 S Mn Pb 23	0.13% C (max) 1.1% Mn 0.23% S 0.2% Pb steel: Designation used by German Standards; free machining
9 S Mn Pb 28	0.14% C (max) 1.1% Mn 0.28% S 0.2% Pb steel: Designation used by German Standards; free machining
9 S Mn Pb 36	0.15% C (max) 1.2% Mn 0.38% S 0.2% Pb steel: Designation used by German Standards; free machining
9 S Pb 23	0.12% C (max) 0.7% Mn 0.22% S 0.05% P 0.2% Pb steel: German Standard; free machining
10 M1	0.1% C 0.9% Mn steel: Pompey **UTS: 390 Elon: 25% Proof: 210**
10 S 20	0.1% C 0.7% Mn 0.3% Si 0.22% S 0.07% P steel: German Standard; free machining
11 Mn 4	0.11% C 1.1% Mn 0.2% Si steel: Designation used by German Standards
11 Mn 4 A1	0.11% C 1.0% Mn 0.1% Si steel: Designation used by German Standards
12.40	0.1% C 2.0% Mn steel: Wire; copper coated; Esab; for use with submerged arc welding; as welded **DPN: 170 UTS: 600 Elon: 24% Proof: 450**
12 Mn 6	0.12% C 1.5% Mn steel: Designation used by German Standards
12 Mn 6 A1	0.12% C 1.6% Mn 0.18% Si steel: Designation used by German Standards
12 Mn 8 A1	0.12% C 2.0% Mn 0.1% Si steel: Designation used by German Standards
12L13	0.13% C (max) 0.3% S 0.1% P 0.25% Pb steel: Designation used in the UK and USA
12L14	0.15% C (max) 0.3% S 0.08% P 0.25% Pb steel: Designation used in the UK and USA
13 Mn 12	0.13% C 3.0% Mn 0.22% Si steel: Designation used by German Standards
14 Mn 4	0.14% C 1.0% Mn 0.05% S and P steel: Designation used by German Standards
14HR	0.15% C (max) 0.6% Mn 0.05% S and P steel: Designation in BS 1449
14HS	0.15% C (max) 0.6% Mn 0.05% S and P steel: Designation in BS 1449
15HR	0.2% C (max) 0.9% Mn 0.05% S 0.06% P steel: Designation in BS 1449
15HS	0.2% C (max) 0.9% Mn 0.05% S 0.06% P steel: Designation in BS 1449
15 S 20	0.15% C 0.7% Mn 0.2% S 0.07% P steel: Designation used by German Standards; free machining; carburizing
17 Mn Ti 6	0.17% C 1.4% Mn 0.22% Ti 0.25% Si 0.1% Al steel: Designation used by German Standards
17 Mn Zr 4	0.17% C 1.1% Mn 0.5% Si 0.1% Al 0.25% Zr steel: Designation used by German Standards
21 Mn 6	0.22% C 1.7% Mn 0.2% Si steel: Designation used by German Standards
22 S 20	0.21% C 0.5% Mn 0.2% S steel: Designation used by German Standards; free machining; carburizing
25CB	0.25% C 0.75% Mn 0.6% Si steel: Electrode; Metrode; weld deposit hardenable **DPN: 375**
25MS	0.25% C 0.8% Mn steel: Electrode; Metrode
30 DAK	0.22% C (max) 1.6% Mn + Cr (max) steel: High yield for construction; Redheugh for BS 4360-55C **UTS: 600 Elon: 18% Proof: 430**
34/20	0.15% C (max) 1.2% Mn steel: Sheet and plate; designation used by British Standards
34/20 CR	0.5% C (max) 1.2% Mn 0.05% S and P steel: Designation in BS 1449
34/20 CS	0.5% C (max) 1.2% Mn 0.05% S and P steel: Designation in BS 1449
34/20 HR	0.5% C (max) 1.2% Mn 0.05% S and P steel: Designation in BS 1449
34/20 HS	0.5% C (max) 1.2% Mn 0.05% S and P steel: Designation in BS 1449
035 AHK	0.25% C steel: Sheet coated; SAE J1392 grade; strength given by figures; type by letters
37/20	0.2% C (max) 1.2% Mn steel: Sheet and plate; designation used by British Standards
37/23 CR	0.2% C (max) 1.2% Mn 0.05% S and P steel: Designation in BS 1449
37/23 CS	0.2% C (max) 1.2% Mn 0.05% S and P steel: Designation in BS 1449
37/23 HR	0.2% C (max) 1.2% Mn 0.05% S and P steel: Designation in BS 1449
37/23 HS	0.2% C (max) 1.2% Mn 0.05% S and P steel: Designation in BS 1449
40/30	0.15% C (max) 1.2% Mn steel: Sheet and plate; designation used by British Standards
040 A 04	0.08% C (max) 0.4% Mn steel: Designation in BS 970; replaces En 2
040 A 10	0.1% C 0.4% Mn steel: Designation in BS 970; replaces En 2
040 A 12	0.12% C 0.4% Mn steel: Designation in BS 970; replaces En 2
040 A 22	0.22% C steel: Designation in BS 970; obsolete
040 AHK	0.25% C steel: Sheet-coated; SAE T1392 grade; strength given by figures; type by letters
40 EE	0.16% C (max) steel: Normalized; weldable; designation for BS 4360 **UTS: 420 Elon: 25%**
40F30	0.12% C (max) 1.2% Mn steel: Sheet and plate; designation used by British Standards
42 A	0.22% C 1.3% Mn steel: High strength; SAE J 1442
43/25	0.25% C (max) 1.2% Mn steel: Sheet and plate; designation used by British Standards
43/25 HR	0.25% C (max) 1.2% Mn 0.05% S and P steel: Designation in BS 1449
43/25 HS	0.25% C (max) 1.2% Mn 0.05% S and P steel: Designation in BS 1449
43/35	0.15% C (max) 1.2% Mn steel: Sheet and plate; designation used by British Standards
43 EE	0.16% C (max) steel: Normalized; weldable; designation for BS 4360 **UTS: 500 Elon: 23%**
43F35	0.12% C (max) 1.2% Mn steel: Sheet and plate; designation used by British Standards

Symbol	Nominal analysis, supplier, condition and remarks.
045 A10	0.1% C 0.8% Mn steel: For case hardeners; designation in BS 970; replaces En 2
045 AHK	0.25% C steel: Sheet-coated; SAE T1392 grade; strength given by figures; type by letters
45FG	0.18% C (max) 1.1% Mn 0.04% S and P (max) steel: HOAG; Al killed
045 M10	0.1% C 0.45% Mn steel: For case hardeners; designation in BS 970; replaces En 2
46/40	0.15% C (max) 1.2% Mn steel: Sheet and plate; designation used by British Standards
46F40	0.12% C (max) 1.2% Mn steel: Sheet and plate; designation used by British Standards
48.04	0.07% C 1.2% Mn 0.4% Si steel: For electrodes; Esab **UTS: 550 Elon: 27% Proof: 460**
48.14	0.07% C 0.85% Mn 0.4% Si steel: For electrodes; Esab **UTS: 520 Elon: 28% Proof: 420**
50/35	0.2% C (max) 1.5% Mn steel: Sheet and plate; designation used by British Standards
50/35 HR	0.2% C (max) 1.5% Mn 0.05% S and P steel: Designation in BS 1449
50/35 HS	0.2% C (max) 1.5% Mn 0.05% S and P steel: Designation in BS 1449
50 A	0.22% C steel: High strength; SAE J 1442
050 A04	0.04% C 0.5% Mn steel: Designation in BS 970; obsolete
050 A10	0.1% C 0.5% Mn steel: Designation in BS 970; replaces En 2
050 A12	0.12% C 0.5% Mn steel: Designation in BS 970; replaces En 2
050 A15	0.15% C 0.5% Mn steel: Designation in BS 970; replaces En 2
050 A20	0.2% C 0.5% Mn steel: Designation in BS 970; replaces En 2
050 A22	0.22% C 0.5% Mn steel: Designation in BS 970; replaces En 2
50 DD	0.18% C (max) steel: Normalized; weldable; with Nb and V; designation for BS 4360 **UTS: 560 Elon: 20%**
50 EE	0.18% C (max) steel: Normalized; weldable; with Nb and V; designation for BS 4360 **UTS: 560 Elon: 20%**
50 F	0.22% C steel: High strength; SAE J 1442
50 F	0.16% C (max) steel: Quench and tempered; weldable with Nb and V; designation for BS 4360 **UTS: 560 Elon: 20%**
50F45	0.12% C (max) 1.2% Mn steel: Sheet and plate; designation used by British Standards
50FK	0.2% C (max) 1.3% Mn 0.04% S and P (max) steel: HOAG; Al killed
50 W	0.22% C steel: High strength; SAE J 1442
53.05	0.08% C 0.8% Mn 0.4% Si steel: For electrodes; Esab **UTS: 570 Elon: 27% Proof: 470**
53.35	0.08% C 1.0% Mn 0.5% Si steel: For electrodes; Esab **UTS: 540 Elon: 30% Proof: 432**
55 F	0.16% C (max) steel: Quench and tempered; weldable with Nb and V; designation for BS 4360 **UTS: 610 Elon: 19%**
55FK	0.2% C (max) 1.35% Mn 0.04% S and P (max) steel: HOAG; Al killed

Note. The following abbreviations and units are used in the tables:

DPN	Hardness, diamond pyramid number
UTS	Ultimate tensile strength, N/mm^2
Elon	Elongation, %
Proof	0.1% proof strength, N/mm^2

1 N/mm^2=0.1 hbar=0.102 kgf/mm^2=0.06475 tonf/in.^{2}=145.04 ibf/in.^{2}=1 MPa

See Appendix II for other abbreviations and conversion tables.

Symbol	Nominal analysis, supplier, condition and remarks.
055 M15	0.2% C (max) 0.8% Mn (max) steel: Designation in BS 970
60 A	0.25% C steel: High strength; SAE J 1442
060 A10	0.1% C 0.6% Mn steel: Designation in BS 970; obsolete
060 A12	0.12% C 0.6% Mn steel: Designation in BS 970; replaces En 2
060 A15	0.15% C 0.5% Mn steel: Designation in BS 970; replaces En 2
060 A17	0.17% C 0.6% Mn steel: Designation in BS 970; replaces En 2
060 A20	0.2% C 0.6% Mn steel: Designation in BS 970; replaces En 2
060 A22	0.22% C 0.6% Mn steel: Designation in BS 970; obselete
60 F	0.25% C steel: High strength; SAE J 1442
60F55	0.12% C (max) 1.5% Mn steel: Sheet and plate; designation used by British Standards
68F62	0.12% C (max) 1.5% Mn steel: Sheet and plate; designation used by British Standards
70 F	0.18% C steel: High strength; SAE J 1442
73.08	0.08% C 0.8% Mn 0.6% Ni 0.4% Cu steel: For electrodes; Esab standard **UTS: 590 Elon: 28% Proof: 490**
73.68	0.06% C 0.7% Mn 2.4% Ni 0.3% Si 0.01% S 0.01% P steel: For electrodes; Esab **UTS: 600 Elon: 28% Proof: 500**
75F70	0.12% C (max) 1.5% Mn steel: Sheet and plate; designation used by British Standards
080 A15	0.15% C 0.8% Mn steel: For case hardening; designation in BS 970
080 A17	0.17% C 0.8% Mn steel: Designation in BS 970; obsolete
080 A20	0.2% C 0.8% Mn steel: For case hardening; designation in BS 970
080 A22	0.22% C 0.8% Mn steel: Designation in BS 970; replaces En 2
080 A25	0.25% C 0.8% Mn steel: Designation in BS 970; replaces En 2
80 F	0.18% C steel: High strength; SAE J 1442
080 M15	0.15% C 0.8% Mn steel: For case hardening; designation in BS 970
120 M19	0.19% C 1.2% Mn steel: Designation in BS 970; replaces En 14
125 A15	0.15% C 1.2% Mn steel: For carburizing; designation in BS 970
130 M15	0.15% C 1.2% Mn steel: For case hardening; designation in BS 970
150 M19	0.19% C 1.5% Mn steel: Designation in BS 970; replaces En 14
170 H20	0.2% C 1% Mn steel: Designation in BS 970
170M	0.2% C 1.2% Mn 0.2% Si steel: Pompey
210 A15	0.15% C 1.1% Mn steel: For case hardening; designation in BS 970
214 A15	0.15% C 1.3% Mn steel: For case hardening; designation in BS 970
214 M15	0.15% C 1.4% Mn steel: For case hardening; designation in BS 970
220 M07	0.15% C (max) 1.1% Mn 0.25% S steel: Free machining; designation in BS 970; replaces En 1
230 M07	0.15% C (max) 1.1% Mn 0.3% S steel: Free machining; designation in BS 970; replaces En 1
240 M07	0.07% C steel: Designation in BS 970; obsolete
240 M07	0.15% C (max) 1.2% Mn 0.4% S steel: Designation in BS 970; obsolete
338	0.05% C 0.1% Mn steel: For carburizing; S Osborn **DPN: 130**
1005	0.06% C (max) 0.04% S and P (max) steel: Designation used in the UK and USA
1006	0.08% C (max) 0.04% S and P (max) steel: Designation used in the UK and USA

Symbol	Nominal analysis, supplier, condition and remarks.
1008	0.1% C (max) 0.04% S and P (max) steel: Designation used in the UK and USA
1009	0.15% C (max) 0.04% S and P (max) steel: Designation used in the UK and USA
1010	0.1% C 0.04% S and P (max) steel: Designation used in the UK and USA; obsolete
1011	0.11% C 0.04% S and P (max) steel: Designation used in the UK and USA; obsolete
1012	0.12% C 0.04% S and P (max) steel: Designation used in the UK and USA
1013	0.13% C 0.04% S and P (max) steel: Designation used in the UK and USA
1015	0.15% C 0.04% S and P (max) steel: Designation used in the UK and USA
1016	0.16% C 0.04% S and P (max) steel: Designation used in the UK and USA
1017	0.17% C 0.04% S and P (max) steel: Designation used in the UK and USA
1018	0.18% C 0.04% S and P (max) steel: Designation used in the UK and USA
1019	0.19% C 0.04% S and P (max) steel: Designation used in the UK and USA
1020	0.2% C 0.04% S and P (max) steel: Designation used in the UK and USA
1021	0.21% C 0.04% S and P (max) steel: Designation used in the UK and USA
1022	0.22% C 0.04% S and P (max) steel: Designation used in the UK and USA
1023	0.23% C 0.04% S and P (max) steel: Designation used in the UK and USA
1108	0.1% C 0.1% S 0.04% P (max) steel: Designation used in the UK and USA
1109	0.1% C 0.1% S 0.04% P (max) steel: Designation used in the UK and USA
1110	0.1% C 0.1% S 0.04% P (max) steel: Designation used in the UK and USA
1116	0.16% C 0.2% S 0.04% P (max) steel: Designation used in the UK and USA
1117	0.17% C 0.1% S 0.04% P (max) steel: Designation used in the UK and USA
1118	0.18% C 0.1% S 0.04% P (max) steel: Designation used in the UK and USA
1119	0.19% C 0.3% S 0.04% P (max) steel: Designation used in the UK and USA
1123	0.23% C 0.8% S 0.04% P (max) steel: Designation used in the UK and USA
1211	0.13% C (max) 0.2% S 0.1% P steel: Designation used in the UK and USA
1212	0.13% C (max) 0.2% S 0.1% P steel: Designation used in the UK and USA
1213	0.13% C (max) 0.28% S 0.1% P steel: Designation used in the UK and USA
1215	0.09% C (max) 0.3% S 0.07% P steel: Designation used in the UK and USA
1309	0.07% C (max) 0.25% Mn 0.025% S (max) 0.035% P (max) steel: French Standard designation
1513	0.13% C 1.2% Mn steel: Designation used in the UK and USA
1518	0.18% C 1.2% Mn steel: Designation used in the UK and USA
1522	0.22% C 1.2% Mn steel: Designation used in the UK and USA
1524	0.22% C 1.5% Mn steel: Designation used in the UK and USA
A 0	0.12% C 0.5% Mn steel: Cogne
A 1	0.15% C (max) 0.5% Mn (max) steel: Cogne
A 1	0.19% C 0.8% Mn 0.25% Si steel: Jessop
A 1	0.1% C (max) steel: Designation for BS 1453
A 2	0.21% C (max) 0.8% Mn (max) steel: Cogne
A 2	0.15% C (max) steel: Designation for BS 1453
A 7	0.12% C 0.3% Mn 0.25% Si steel: Jessop

Symbol	Nominal analysis, supplier, condition and remarks.
A 7	C not specified 0.04% S and P steel: Plate; Armco; for structural use **UTS: 500 Elon: 24% Proof: 220**
A 7	0.12% C (max) steel: Designation for BS 1453
A 15	0.1% C 1.2% Mn 0.6% Si steel: Welding rod; designation used by British Standards
A 15	0.12% C (max) 0.4% Cu (max) steel: Welding electrode; designation for BS 2901
A 17	0.15% C 1.1% Mn 0.5% Si 0.03% S and P steel: Welding rod; designation used by British Standards
A 17	0.12% C (max) steel: Welding electrode; designation for BS 2901
A 18	0.12% C (max) steel: Welding electrode; designation for BS 2901
A 19	0.12% C (max) 0.5% Al steel: For welding electrodes; designation for BS 2901
A 36	0.26% C 1.0% Mn 0.2% Si 0.04% S and P steel: Plate; Armco; for structural use **UTS: 540 Elon: 23% Proof: 220**
A 50/1	C not specified 0.05% S (max) 0.07% P (max) steel: Plate; French Standard designation **UTS: 550 Proof: 290**
A 50/2	C not quoted 0.045% S and P (max) steel: Plate; French Standard designation
A 52	0.2% C 1.5% Mn 0.05% S and P (max) steel: Belgian specification
A 52 HS	0.2% C 1.5% Mn 0.05% S and P (max) steel: Belgian specification
A 113	C not specified 0.04% S and P steel: Plate; Armco; covers three grades A, B and C **UTS: 450 Elon: 27% Proof: 180**
A 131	0.22% C 1.0% Mn 0.2% Si 0.04% S and P steel: Plate; Armco; cold flanging quality **UTS: 500 Elon: 24% Proof: 210**
A 283	C not specified 0.04% S and P steel: Plate; Armco; in four grades A, B, C and D **UTS: 500 Elon: 26% Proof: 210**
A 299	0.3% C 1.2% Mn 0.2% Si 0.035% S and P steel: Plate; Armco; boilers and pressure vessels **UTS: 600 Elon: 20% Proof: 270**
A 373	0.26% C 0.8% Mn 0.2% Si 0.04% S and P steel: Plate; Armco; for welded structures **UTS: 500 Elon: 24% Proof: 210**
A 442	0.23% C 1.0% Mn 0.2% Si 0.04% S and P steel: Plate; Armco; boilers and pressure vessels **UTS: 490 Elon: 26% Proof: 210**
A St 2	0.1% C (max) 0.3% Mn 0.15% Si 0.05% S and P steel: German Standard
A St 3	0.1% C (max) 0.4% Mn 0.1% Si 0.04% S and P steel: German Standard
A St 4	0.1% C (max) 0.35% Mn 0.08% Si 0.035% S and P steel: German Standard
A St 35/8	0.17% C (max) 0.05% S and P steel: German Standard
A St 45/8	0.22% C (max) 0.05% S and P steel: Designation used by German Standards
ABS	0.22% C 1.0% Mn 0.2% Si 0.04% S and P steel: Plate; Armco; cold flanging quality **UTS: 500 Elon: 24% Proof: 210**
ABS/A	0.25% C (max) 0.6% Mn (min) 0.04% S and P (max) steel: For ships' hulls; American Bureau of Shipping **UTS: 450 Elon: 21% Proof: 240**
ABS/AH32	0.18% C (max) 1.3% Mn 0.04% Ni (max) 0.25% Cr (max) 0.08% Mo (max) 0.1% V (max) 0.05% Nb (max) 0.35% Cu (max) steel: For ships' hulls; American Bureau of Shipping **UTS: 530 Elon: 19% Proof: 320**
ABS/AH36	0.18% C (max) 1.3% Mn 0.04% Ni (max) 0.25% Cr 0.08% Mo (max) 0.1% V (max) 0.05% Nb (max) 0.35% Cu (max) steel: For ships' hulls; American Bureau of Shipping **UTS: 560 Elon: 20% Proof: 360**

Symbol	Nominal analysis, supplier, condition and remarks.
ABS/B	0.21% C (max) 0.85% Mn 0.04% S and P (max) steel: For ships' hulls; American Bureau of Shipping **UTS: 450 Elon: 21% Proof: 240**
ABS/CS	0.16% C (max) 1.2% Mn 0.04% S and P (max) steel: For ships' hulls; American Bureau of Shipping **UTS: 450 Elon: 21% Proof: 250**
ABS/D	0.23% C (max) 0.9% Mn (max) steel: Plate; American Bureau of Shipping **UTS: 440 Elon: 23% Proof: 210**
ABS/D	0.21% C (max) 1.05% Mn 0.04% S and P (max) steel: For ships' hulls; American Bureau of Shipping; impact 27 J at −20 °C **UTS: 450 Elon: 21% Proof: 250**
ABS/D	0.12% C 0.42% Mn steel: Tube; American Bureau of Shipping
ABS/DH32	0.18% C (max) 1.4% Mn 0.04% Ni (max) 0.25% Cr (max) 0.08% Mo (max) 0.1% V (max) 0.05% Nb (max) 0.35% Cu (max) steel: For ships' hulls; American Bureau of Shipping; impact 34 J at −20 °C **UTS: 530 Elon: 20% Proof: 320**
ABS/DH36	0.18% C (max) 1.3% Mn 0.4% Ni (max) 0.25% Cr (max) 0.08% Mo (max) 0.1% V (max) 0.05% Nb (max) 0.35% Cu (max) steel: For ships' hulls; American Bureau of Shipping; impact 34 J at −20 °C **UTS: 560 Elon: 20% Proof: 360**
ABS/DS	0.16% C 1.15% Mn 0.04% S and P (max) steel: For ships' hulls; American Bureau of Shipping **UTS: 450 Elon: 21% Proof: 240**
ABS/E	0.18% C (max) 1.05% Mn 0.04% S and P (max) steel: For ships' hulls; American Bureau of Shipping; impact 27 J at −40 °C **UTS: 450 Elon: 21% Proof: 240**
ABS/EH32	0.18% C (max) 1.3% Mn 0.04% Ni (max) 0.25% Cr (max) 0.08% Mo (max) 0.1% V (max) 0.05% Nb (max) 0.35% Cu (max) steel: For ships' hulls; American Bureau of Shipping; impact 34 J at −40 °C **UTS: 550 Elon: 20% Proof: 320**
ABS/EH36	0.18% C (max) 1.3% Mn 0.4% Ni (max) 0.25% Cr (max) 0.08% Mo (max) 0.1% V (max) 0.05% Nb (max) 0.35% Cu (max) steel: For ships' hulls; American Bureau of Shipping; impact 34 J at −40 °C **UTS: 560 Elon: 20% Proof: 360**
ABS/G	0.12% C 0.42% Mn steel: Tube; American Bureau of Shipping
ABS/G2A	Flash butt welded or drop forged steel chain: American Bureau of Shipping **UTS: 570 Elon: 22%**
ABS/G2B	Cast steel chain: American Bureau of Shipping **UTS: 500 Elon: 22%**
ABS/G3A	Flash butt welded or drop forged steel chain: American Bureau of Shipping; impact 8 J at 0 °C **UTS: 700 Elon: 17%**
ABS/G3B	Cast steel chain: American Bureau of Shipping; impact 8 J at 0 °C **UTS: 700 Elon: 17%**
ABS/GA	0.17% C (max) 0.9% Mn (max) 0.25% Cu steel: Plate; for intermediate temperature **UTS: 330 Elon: 27% Proof: 170**

Note. The following abbreviations and units are used in the tables:

DPN	Hardness, diamond pyramid number
UTS	Ultimate tensile strength, N/mm^2
Elon	Elongation, %
Proof	0.1% proof strength, N/mm^2

1 N/mm^2=0.1 hbar=0.102 kgf/mm^2=0.06475 tonf/in.^{2}=145.04 lbf/in.^{2}=1 MPa

See Appendix II for other abbreviations and conversion tables.

Symbol	Nominal analysis, supplier, condition and remarks.
ABS/GB	0.22% C (max) 0.9% Mn (max) 0.27% Cu steel: Plate; for intermediate temperature service **UTS: 380 Elon: 25% Proof: 190**
ABS/GC	0.28% C (max) 0.9% Mn (max) 0.27% Cu steel: Plate; for intermediate temperature service **UTS: 430 Elon: 23% Proof: 210**
ABS/GI	Flash butt welded steel chain: American Bureau of Shipping **UTS: 360 Elon: 30%**
ABS/H	0.12% C 0.42% Mn steel: Tube; American Bureau of Shipping
ABS/J	0.27% C (max) 0.93% Mn (max) steel: Tube; American Bureau of Shipping
ABS/K	0.18% C (max) 0.7% Mn steel: Plate; American Bureau of Shipping **UTS: 430 Elon: 23% Proof: 210**
ABS/L	0.23% C (max) 0.75% Mn steel: Plate; American Bureau of Shipping **UTS: 460 Elon: 24% Proof: 225**
ABS/M	0.27% C (max) 1.0% Mn steel: Plate; American Bureau of Shipping **UTS: 500 Elon: 20% Proof: 245**
ABS/N	0.29% C (max) 1.0% Mn steel: Plate; American Bureau of Shipping **UTS: 550 Elon: 18% Proof: 270**
AC1 BAR	0.1% C 1% Mn 0.06% Si 0.3% S 0.04% P steel: Workington Iron & Steel; free machining; hot rolled **UTS: 420 Elon: 40%**
AC1 BAR	0.1% C 1% Mn 0.06% Si 0.3% S 0.04% P steel: Workington Iron & Steel; free machining; bright drawn **UTS: 600 Elon: 16%**
AC1 BEL	0.1% C 1% Mn 0.06% Si 0.3% S 0.04% P 0.2% Pb steel: Workington Iron & Steel; free machining; hot rolled **UTS: 420 Elon: 40%**
AC1 BEL	0.1% C 1% Mn 0.06% Si 0.03% S 0.04% P 0.2% Pb steel: Workington Iron & Steel; free machining; bright drawn **UTS: 600 Elon: 16%**
ACIBOND	0.16% C 1.5% Mn 0.05% Al 0.25% Cu 0.016% B steel: Workington Iron & Steel
AF No 3 81.309 E43 1/0 RR130 32	0.06% C 0.5% Mn 0.4% Si steel: Welders electrode; French Standard
AF No 3 81.309 E51 3/2 AR160 31	0.09% C 0.8% Mn 0.25% Si steel: Welders electrode; French specification
AF No 3 E43 2/2 RII	0.07% C 0.5% Mn 0.3% Si steel: Welders electrode; French specification
AFNOR 81.309 E43 2/2 R12	0.08% C 0.65% Mn 0.35% Si steel: Welders electrode; French Standard
AFNOR 81.309 E43 2/2 RR2	0.05% C 0.5% Mn 0.35% Si steel: Welders electrode; French specification
AFNOR 81.309 E43 2/2 RR 150 32	0.07% C 0.65% Mn 0.35% Si steel: Welders electrode; French specification
AFNOR 81.309 E43 3/2 C10	0.12% C 0.55% Mn 0.2% Si steel: Welders electrode; French specification
AFNOR 81.309 E43 3/3 AR150 32	0.1% C 0.55% Mn 0.2% Si steel: Welders electrode; French specification
AGP V 9	0.12% C 0.85% Mn 0.25% S steel: Pompey; free machining
AISI 12 L14	0.15% C (max) 1.0% Mn 0.3% S 0.07% P 0.3% Pb steel: Free machining
AISI 1006	0.08% C 0.3% Mn 0.04% S and P steel
AISI 1008	0.1% C 0.4% Mn 0.04% S and P steel
AISI 1009	0.15% C 0.6% Mn 0.04% S and P steel
AISI 1010	0.1% C 0.4% Mn 0.04% S and P steel
AISI 1012	0.12% C 0.4% Mn 0.04% S and P steel
AISI 1015	0.15% C 0.4% Mn 0.04% S and P steel
AISI 1016	0.15% C 0.8% Mn 0.04% S and P steel
AISI 1017	0.18% C 0.4% Mn 0.04% S and P steel
AISI 1018	0.18% C 0.8% Mn 0.04% S and P steel
AISI 1019	0.18% C 0.9% Mn 0.04% S and P steel

Symbol	Nominal analysis, supplier, condition and remarks.
AISI 1020	0.2% C 0.4% Mn 0.04% S and P steel
AISI 1021	0.2% C 0.8% Mn 0.04% S and P steel
AISI 1022	0.2% C 0.9% Mn 0.04% S and P steel
AISI 1108	0.1% C 0.7% Mn 0.1% S 0.04% P steel: Free cutting
AISI 1109	0.1% C 0.8% Mn 0.1% S 0.04% P steel: Free cutting
AISI 1110	0.1% C 0.4% Mn 0.1% S 0.04% P steel: Free machining
AISI 1111	0.13% C (max) 0.8% Mn 0.12% S 0.1% P steel: Acid Bessemer
AISI 1112	0.13% C (max) 0.9% Mn 0.2% S 0.1% P steel: Free machining
AISI 1113	0.13% C 0.9% Mn 0.2% S 0.1% P steel: Free cutting
AISI 1115	0.15% C 0.8% Mn 0.1% S 0.04% P steel: Free cutting
AISI 1116	0.16% C 1.2% Mn 0.2% S 0.04% P steel: Free machining
AISI 1117	0.17% C 1.2% Mn 0.1% S 0.04% P steel: Free cutting
AISI 1118	0.17% C 1.5% Mn 0.1% S 0.04% P steel: Free cutting
AISI 1119	0.17% C 1.2% Mn 0.3% S 0.04% P steel: Free cutting
AISI 1120	0.2% C 0.9% Mn 0.1% S 0.04% P steel: Free cutting
AISI 1211	0.13% C (max) 0.8% Mn 0.12% S 0.1% P steel: Free machining
AISI 1212	0.13% C (max) 0.8% Mn 0.2% S 0.1% P steel: Free machining
AISI 1213	0.13% C (max) 0.8% Mn 0.3% S 0.1% P steel: Free machining
AISI 1215	0.09% C (max) 0.9% Mn 0.3% S 0.07% P steel: Free machining
AISI 1320	0.2% C 1.7% Mn 0.04% S and P steel: Obsolete
ALDUR 45/60	0.21% C 1.3% Mn 0.04% S and P (max) steel: Voest
ALDUR 50	0.21% C 1.3% Mn 0.04% S and P (max) steel: Voest
ALDUR 50/65	0.23% C 1.5% Mn 0.04% S and P (max) steel: Voest
ALDUR 55	0.22% C 1.4% Mn 0.04% S and P (max) steel: Voest
ALDUR 55/63	0.23% C 1.7% Mn 0.04% S and P (max) steel: Voest
ALDUR 58	0.23% C 1.5% Mn 0.04% S and P (max) 0.02‰ Cu steel: Voest
ALDUR 58/72	0.23% C 1.7% Mn 0.04% S and P (max) steel: Voest
ALFORT	0.22% C 1.3% Mn 0.04% S and P steel: Austrian specification
ALFREEZE I	0.17% C (max) 1.5% Mn 0.05% S and P (max) steel: Appleby-Frodingham
ALFREEZE II	0.17% C (max) 1.5% Mn 0.04% S and P (max) with N additions steel: Appleby-Frodingham
ALFRIG 5UV	0.18% C 1.1% Mn 0.04% S and P steel: Austrian specification
ALGO-LOY 1315	0.15% C 1.3% Mn 0.4% Cu steel: Algoma
ALUMOWELD	Aluminium covered steel wire: Copperweld steel; used for overhead lines
AMS 5010 D	0.13% C 0.2% S 0.1% P steel: Bar; AMS for SAE 1112
AMS 5022 F	0.17% C 1.2% Mn steel: Bar and forging; free cutting; AMS specification for SAE 1117
AMS 5030 A	0.06% C (max) steel: Wire for welding
AMS 5031	0.12% C steel-coated welding electrode
AMS 5032 A	0.2% C steel: Wire; annealed; AMS for SAE 1020
AMS 5036 D	0.1% C (max) steel: Sheet and strip; Al coated
AMS 5040 E	0.15% C (max) steel: Sheet and strip; for deep forming; cold rolled; AMS for SAE 1010
AMS 5041	0.08% C (max) steel: Sheet and strip; for deep drawing; cold rolled; AMS for SAE 1006
AMS 5042 E	0.15% C (max) steel: For deep forming; cold rolled; AMS for SAE 1010
AMS 5044 C	0.15% C (max) steel: ½-hard; AMS for SAE 1010
AMS 5045 B	0.20% C steel: ½-hard; AMS for SAE 1020
AMS 5046	0.2% C steel: Type 1020
AMS 5047	Low C steel Al killed: For forming; AMS for SAE 1010
AMS 5050 E	0.15% C (max) steel: Annealed; AMS for SAE 1010
AMS 5053 B	0.13% C (max) steel: Tube; welded; annealed; AMS for SAE 1010
AMS 5060 B	0.15% C steel: Bar, forgings and tube; AMS for SAE 1015
AMS 5061 A	0.18% C steel: For bars and wire
AMS 5069	0.18% C steel: Bars, forging and tube; AMS for SAE 1018
AMS 5681	0.05% C (max) 11% Ni 18.5% Cr 0.7% Nb steel: Wire
AMS 7225	0.1% C 0.45% Mn steel
AMS 7496	0.25% C (max) 1% Mn steel
AN QQ S 646 1	0.2% C 0.4% Mn 0.05% S and P steel: US Service; as rolled **UTS: 300 Elon: 22% Proof: 220**
AN QQ W 435/3	0.1% C 0.4% Mn 0.05% S and P steel: Wire; US Service; zinc coated; annealed **UTS: 370**
AN S 11	0.2% C 0.4% Mn 0.05% S and P steel: Wire; US Service **UTS: 300 Elon: 22% Proof: 220**
ANSCOL 50	0.18% C (max) 0.7% Mn 0.015% Nb steel: Interlake Steel Co.
API 5A H40	C not specified 0.06% S (max) 0.04% P (max) steel: For casing, tube and drill pipe **UTS: 420 Elon: 20% Proof: 280**
API 5A J55	C not specified 0.06% S (max) 0.04% P (max) steel: For casing, tube and drill pipe **UTS: 520 Elon: 20% Proof: 380**
API 5A K55	C not specified 0.06% S (max) 0.04% P (max) steel: For casing, tube and drill pipe **UTS: 650 Elon: 20% Proof: 380**
API 5A N80	C not specified 0.06% S (max) 0.04% P (max) steel: For casing, tube and drill pipe **UTS: 700 Elon: 20% Proof: 550**
API 5L A	0.22% C (max) 0.9% Mn (max) 0.05% S (max) 0.04% P (max) steel: Line pipe; seamless or welded **UTS: 335 Elon: 25% Proof: 245**
API 5L A25/1	0.21% C (max) 0.45% Mn 0.06% S (max) 0.045% P (max) steel: Line pipe; seamless or welded **UTS: 310 Elon: 25*MB Proof: 175**
API 5L A25/2	0.21% C (max) 0.45% Mn 0.06% S (max) 0.08% P (max) steel: Line pipe; seamless or welded **UTS: 310 Elon: 25% Proof: 175**
API 5L B	0.27% C (max) 1.15% Mn (max) 0.05% S (max) 0.04% P (max) steel: Line pipe; seamless or welded **UTS: 410 Elon: 25% Proof: 245**
API 5L SA	0.21% C (max) 0.9% Mn (max) 0.05% S (max) 0.04% P (max) steel: Spiral weld line pipe **UTS: 330 Elon: 20% Proof: 210**
API 5L SB	0.26% C (max) 1.15% Mn (max) 0.05% S (max) 0.04% P (max) steel: Spiral weld line pipe **UTS: 330 Elon: 20% Proof: 242**
API 5L X70	0.23% C (max) 1.6% Mn (max) steel: High test line pipe; seamless or welded **UTS: 565 Elon: 20% Proof: 483**
AQUACIDOX	0.11% C 0.4% Mn 0.3% Si (max) 0.05% S and P 0.3% Cu steel: Tube; STD Services (see Resistco)
ARCTIC A	0.15% C 1.5% Mn 0.05% S and P steel: South Durham
ARCTIC B	0.22% C 1.6% Mn 0.05% S and P (max) steel: South Durham
ARCTIC C	0.22% C (max) 1.6% Mn 0.04% S and P (max) steel: South Durham
ARCTIC D	0.14% C (max) 1.5% Mn 0.1% Nb 0.045% S (max) 0.035% P (max) steel: South Durham
ARMCO HS No 4R	0.25% C (max) 1.4% Mn 0.2% Cu steel: Armco
ARMCO HS No 4S	0.25% C (max) 1.4% Mn 0.2% Cu steel: Armco
ARMCO HS No 7	0.13% C 0.7% Mn 0.02% Nb steel: Armco
ASTM A53 E GA	0.25% C steel pipe: Resistance welder **UTS: 330 Elon: 18% Proof: 200**
ASTM A53 E GB	0.3% C steel pipe: Resistance welder **UTS: 330 Elon: 18% Proof: 200**
ASTM A53 F	Mild steel pipe: Furnace welder **UTS: 310 Proof: 190**
ASTM A53 S GA	0.25% C steel pipe: Seamless **UTS: 330 Proof: 200**
ASTM A53 S GB	0.3% C steel pipe: Seamless **UTS: 330 Proof: 200**

Symbol	Nominal analysis, supplier, condition and remarks.
ASTM A67	0.08% (min) C 0.11% (max) P 0.20% (min) Cu: When specified steel for tie plates
ASTM A109/4	0.15% C 0.6% Mn 0.2% Cu (optional) cold rolled steel: Strip; annealed and final light rolled **UTS: 330 Elon: 32%**
ASTM A109/5	0.15% C 0.6% Mn 0.2% Cu (optional) cold rolled steel: Strip; annealed **UTS: 300 Elon: 39%**
ASTM A113	C not specified 0.05% S 0.07% P steel: For locomotives **UTS: 370 Elon: 26% Proof: 180**
ASTM A120	Mild steel pipe: Welded or seamless; galvanized; for gardeners' use
ASTM A135	C not specified 0.06% S 0.05% P electric resistance welded steel: Pipe
ASTM A161 (low carbon)	0.15% C 0.5% Mn 0.5% S and P steel: Steel tubes
ASTM A178 A	0.12% C 0.4% Mn 0.06% S 0.05% P electric resistance welded steel: Boiler tubes **UTS: 420 Elon: 30% Proof: 260**
ASTM A179	0.12% C 0.35% Mn 0.05% S and P steel: Cold drawn tube for condenser
ASTM A192	0.12% C 0.45% Mn steel: For boiler tubes
ASTM A194/1	0.15% C (min) steel: For bolts
ASTM A214	0.18% C (max) 0.45% Mn steel: Welded tubes
ASTM A226	0.12% C 0.45% Mn 0.2% Si steel: Tube; electric resistance welded
ASTM A233	Mild steel welding electrodes classified by coating
ASTM A235	C not specified 0.9% Mn 0.05% S and P steel: Forgings; normalized and tempered **UTS: 580 Elon: 21% Proof: 270**
ASTM A237	C not specified 0.05% S and P steel: Forgings; normalized and tempered **UTS: 630 Elon: 21% Proof: 420**
ASTM A251	Mild steel welding rods classified by strength of deposit
ASTM A254	0.1% C 0.45% Mn 0.05% S and P copper brazed steel: Tube
ASTM A263	Stainless steel clad mild steel sheet
ASTM A264	Austenitic stainless clad mild steel sheet
ASTM A265	Nickel and nickel alloy clad mild steel sheet
ASTM A283 A	Low C 0.05% S and P steel: Plate for structures; as rolled **UTS: 340 Elon: 27% Proof: 170**
ASTM A283 B	Low C 0.05% S and P steel: Plate for structures; as rolled **UTS: 370 Elon: 25% Proof: 180**
ASTM A283 C	Low C 0.05% S and P steel: Plate for structures; as rolled **UTS: 420 Elon: 23% Proof: 200**
ASTM A283 D	Low C 0.05% S and P steel: Plate for structures; as rolled **UTS: 450 Elon: 21% Proof: 220**
ASTM A285 A	0.15% C (max) 0.8% Mn 0.05% S and P steel: Plate
ASTM A285 B	0.2% C (max) 0.8% Mn 0.04% S and P steel: Plate
ASTM A285 C	0.25% C (max) 0.8% Mn 0.04% S and P steel: Plate
ASTM A366	0.15% C (max) 0.5% Mn 0.04% S and P steel: Sheet; cold rolled
ASTM A415	0.15% C 0.4% Mn steel: Sheet; may contain Cu; hot rolled
ASTM A424	0.04% C 0.2% Mn 0.04% S 0.015% P steel: Sheet for porcelain enamelling; graded by C content
ASTM A425	0.15% C (max) 0.45% Mn 0.04% S and P steel: Strip; Cu may be present; commercial quality
ASTM A441	0.25% C (max) 1.3% Mn 0.02% V 0.2% Cu steel: For structural purposes
ASTM A442	0.23% C 0.85% Mn 0.2% Si 0.04% S and P steel: Plate for pressure vessels
ASTM A489	C not specified 0.13% Si 0.05% S and P steel: For eyebolts **UTS: 450 Elon: 30% Proof: 210**
ASTM A502/1	0.19% C 0.6% Mn 0.2% Cu (optional) steel: For structural rivets
ASTM A502/2	0.24% C 1.4% Mn 0.2% Cu (optional) steel: For structural rivets
ASTM A512	Cold drawn plain carbon steel: Tubing; butt welded; graded according to AISI system
ASTM A513	Resistance welded steel tubing; plain carbon and low alloy; graded according to AISI system
ASTM A523 A	0.22% C (max) 0.9% Mn 0.04% S and P steel: Pipe for high pressure; seamless or electric resistance welded
ASTM A524	0.21% C (max) 1.0% Mn 0.05% S and P steel: Seamless pipe for process piping; in grades I and II
ASTM A529	0.27% C 1.2% Mn 0.04% S and P 0.2% Cu (optional) structural steel **UTS: 410 Elon: 19% Proof: 300**
ASTM A537	0.24% C (max) 1.0% Mn 0.04% Si steel: Plate
ASTM A539	0.15% C (max) 0.6% Mn 0.05% S and P steel: Electric resistance welded coiled tube
ASTM A548	Mild steel wire for screws; graded by AISI system
ASTM A549	Mild steel wire for wood screws; graded by AISI system
ASTM A556 A2	0.18% C (max) 0.5% Mn 0.05% S and P steel: Tube; seamless; cold drawn **UTS: 330 Elon: 35% Proof: 180**
ASTM A558 EL8	0.1% C 0.4% Mn 0.05% Si steel: Electrode
ASTM A558 EL8K	0.1% C 0.4% Mn 0.15% Si steel: Electrode
ASTM A558 EL12	0.12% C 0.5% Mn 0.05% Si steel: Electrode
ASTM A558 EM5K	0.06% C 1.2% Mn 0.6% Si steel: Electrode
ASTM A558 EM12	0.11% C 1.0% Mn 0.05% Si steel: Electrode
ASTM A558 EM12K	0.11% C 1.0% Mn 0.2% Si steel: Electrode
ASTM A558 EM13K	0.13% C 1.2% Mn 0.55% Si steel: Electrode
ASTM A558 EM15K	0.15% C 1.0% Mn 0.2% Si steel: Electrode
ASTM A558 EN14	0.14% C 2.0% Mn 0.05% Si steel: Electrode
ASTM A559 E60S1	0.13% C 0.35% Si steel: Electrode
ASTM A559 E60S2	0.06% C 0.2% Ti 0.55% Si 0.1% Zr steel: Electrode
ASTM A559 E60S3	0.12% C 1.2% Mn 0.55% Si steel: Electrode
ASTM A559 E70S4	0.11% C 0.75% Si steel: Electrode
ASTM A559 E70S5	0.13% C 0.45% Si steel: Electrode
ASTM A559 E70S6	0.11% C 1.6% Mn 1.0% Si steel: Electrode
ASTM A559 E70T	Low carbon welding electrodes; composite; graded by composition
ASTM A562	0.12% C (max) 1.2% Mn 0.45% Ti 0.3% Si 0.15% Cu (max) steel: Plate for glass or metallic coating
ASTM A569	0.15% C (max) 0.4% Mn 0.04% S and P steel: Strip; Cu may be present
ASTM A572	0.21% C (max) steel: For structural use; may contain V or Nb
ASTM A573	0.24% C (max) 1.0% Mn 0.2% Si steel: Plate of improved toughness
ASTM A575	Mild and medium carbon hot rolled steel bar; graded by AISI system
ASTM A576	Plain carbon hot rolled steel bars; graded by AISI system (see also 44A2 and 44A3)
ASTM A587	0.15% C (max) 0.45% Mn 0.05% S and P (max) 0.02% Al (min) steel: Pipe; electric welded **UTS: 331 Elon: 40% Proof: 210**
ASTM A589	Low C 0.06% S (max) 0.05% P (max) steel: Pipe; seamless for water pipes **UTS: 330 Elon: 35% Proof: 210**
ASTM A589 B	Low C 0.06% S (max) 0.05% P (max) steel: Pipe **UTS: 310 Elon: 30% Proof: 170**

Note. The following abbreviations and units are used in the tables:

DPN	Hardness, diamond pyramid number
UTS	Ultimate tensile strength, N/mm^2
Elon	Elongation, %
Proof	0.1% proof strength, N/mm^2

$1\ N/mm^2 = 0.1\ hbar = 0.102\ kgf/mm^2 = 0.06475\ tonf/in.^2 = 145.04\ lbf/in.^2 = 1\ MPa$

See Appendix II for other abbreviations and conversion tables.

Symbol	Nominal analysis, supplier, condition and remarks.
ASTM A591	Mild steel zinc coated sheet; cold rolled
ASTM A594, 594/1	0.15% C (max) 0.5% Mn (max) 0.02% Al (max) steel: For magnetic purposes
ASTM A594/2	0.1% C (max) 0.6% Mn (max) 0.02% Al (max) steel: For magnetic purposes
ASTM A594/3	0.08% C (max) 0.4% Mn (max) 0.02% Al (max) steel: For magnetic purposes
ASTM A594/4	0.06% C (max) 0.4% Mn (max) 0.015% Al (max) steel: For magnetic purposes
ASTM A595 A	0.2% C 0.4% Mn steel: Tube tapered for structural purposes **UTS: 450 Proof: 380**
ASTM A595 B	0.2% C 1.0% Mn 0.05% S and P (max) steel: Tapered tube **UTS: 480 Proof: 410**
ASTM A599	Mild steel electro tin plated sheet; cold rolled
ASTM A603	Mild steel wire rope; zinc coated
ASTM A606/2	0.2% C (max) 1.3% Mn (max) 0.06% S (max) alloy steel: Sheet with twice corrosion resistance of plain carbon steel; normalized **UTS: 450 Elon: 22% Proof: 310**
ASTM A606/4	0.26% C (max) 1.3% Mn (max) 0.06% S (max) alloy steel: Sheet with four times corrosion resistance of plain carbon steel; normalized **UTS: 450 Elon: 22% Proof: 310**
ASTM A607/45	0.26% C (max) 1.4% Mn (max) 0.005% V or 0.004% Nb steel: Sheet **UTS: 410 Elon: 23% Proof: 320**
ASTM A607/50	0.27% C (max) 1.4% Mn (max) 0.005% V or 0.004% Nb steel: Sheet **UTS: 450 Elon: 21% Proof: 345**
ASTM A607/55	0.29% C (max) 1.4% Mn (max) 0.005% V or 0.004% Nb steel: Sheet **UTS: 480 Elon: 19% Proof: 380**
ASTM A607/60	0.3% C (max) 1.55% Mn (max) 0.005% V or 0.004% Nb steel: Sheet **UTS: 520 Elon: 17% Proof: 415**
ASTM A607/65	0.3% C (max) 1.55% Mn (max) 0.005% V or 0.004% Nb steel: Sheet **UTS: 550 Elon: 15% Proof: 450**
ASTM A607/70	0.3% C (max) 1.7% Mn (max) 0.005% V or 0.004% Nb steel: Sheet **UTS: 590 Elon: 14% Proof: 485**
ASTM A611 A	0.2% C (max) 0.6% Mn (max) steel: Cold rolled sheet **UTS: 290 Elon: 26% Proof: 170**
ASTM A611 A, B, C and E	0.2% C (max) 0.04% S and P (max) 0.2% B and Cu (max) steel: Cold rolled sheet
ASTM A611 B	0.2% C (max) 0.6% Mn (max) steel: Cold rolled sheet **UTS: 310 Elon: 24% Proof: 205**
ASTM A611 C	0.2% C (max) 0.6% Mn (max) steel: Cold rolled sheet **UTS: 330 Elon: 22% Proof: 230**
ASTM A611 D	0.2% C (max) 0.9% Mn (max) steel: Cold rolled sheet **UTS: 360 Elon: 20% Proof: 275**
ASTM A611 D	0.2% C (max) 0.9% Mn (max) 0.04% S and P (max) 0.2% Cu (max) steel: Cold rolled sheet
ASTM A611 E	0.2% C (max) 0.6% Mn (max) steel: Cold rolled sheet **UTS: 570 Proof: 550**
ASTM A615/40	Plain carbon 0.05% P (max) steel: For concrete reinforcement; deformed **UTS: 483 Elon: 10% Proof: 276**
ASTM A615/60	Plain carbon 0.05% P (max) steel: For concrete reinforcement; deformed **UTS: 621 Elon: 8% Proof: 414**
ASTM A619	0.1% C (max) 0.5% Mn (max) 0.035% S (max) 0.025% P (max) steel: Sheet for deep drawing
ASTM A620	0.1% C (max) 0.5% Mn (max) 0.035% S (max) 0.025% P (max) steel: Sheet for deep drawing
ASTM A621	0.1% C (max) 0.5% Mn (max) 0.035% S (max) 0.025% P (max) steel: Sheet for deep drawing
ASTM A622	0.1% C (max) 0.5% Mn (max) 0.035% S (max) 0.025% P (max) steel: Sheet for deep drawing
ASTM A623 D	0.12% C (max) 0.6% Mn (max) steel: Sheet; tin plated
ASTM A623 L	0.12% C (max) 0.6% Mn (max) 0.12% N + Cr + Mo (max) steel: Sheet: tin plated
ASTM A623 MC	0.12% C (max) 0.7% Mn (max) steel: Sheet; tin plated
ASTM A623 MR	0.13% C (max) 0.6% Mn (max) steel: Sheet; tin plated
ASTM A633 A	0.18% C (max) 1.2% Mn steel: For structures; normalized **UTS: 500 Elon: 19% Proof: 290**
ASTM A633 B	0.18% C (max) 1.15% Mn 0.1% V (max) steel: For structures; normalized **UTS: 500 Elon: 20% Proof: 290**
ASTM A633 C	0.2% C (max) 1.2% Mn 0.03% Nb steel: For structures; normalized **UTS: 520 Elon: 20% Proof: 320**
ASTM A633 D	0.2% C (max) 1.0% Mn 0.25% Ni (max) 0.2% Cr (max) steel: For structures; normalized **UTS: 520 Elon: 20% Proof: 320**
ASTM A635	0.15% C (max) 0.5% Mn 0.18% Cu (optional) steel: Sheet
ASTM A640	Steel wire zinc coated for support cables
ASTM A641	Steel wire hot dip zinc coated; annealed **UTS: 500**
ASTM A641	Steel wire hot dip zinc coated; medium drawn **UTS: 580**
ASTM A641	Steel wire hot dip zinc coated; hard drawn **UTS: 650**
ASTM A643 A	0.25% C (max) 1.2% Mn steel: Castings for pressure vessels **UTS: 485 Elon: 22% Proof: 275**
ASTM A650	0.1% C (max) 0.035% S (max) 0.025% P (max) 0.5% (max) steel: Sheet; galvanized for deep drawing
ASTM A650	Low carbon steel plate; double reduced
ASTM A656/2	0.15% C (max) 0.9% Mn (max) 0.2% Ti 0.01% Al (min) steel: For structures **UTS: 720 Elon: 12% Proof: 550**
ASTM A657	Steel plate or strip; cold rolled and chromium plated
ASTM A659	0.2% C steel: Strip and sheet; hot rolled; graded by AISI number
ASTM A660 WCA	0.25% C (max) 0.7% Mn steel: Pipe; centrifugally cast; high temperature use **UTS: 414 Elon: 24% Proof: 207**
ASTM A660 WCB	0.3% C (max) 1.0% Mn steel: Pipe; centrifugally cast; high temperature use **UTS: 483 Elon: 22% Proof: 207**
ASTM A660 WCC	0.25% C (max) 1.2% Mn steel: Pipe; centrifugally cast; high temperature use **UTS: 483 Elon: 22% Proof: 276**
ASTM A662 A	0.17% C (max) 1.1% Mn steel: Plate for low temperature use; 27 J impact at −75 °C **UTS: 460 Elon: 20% Proof: 275**
ASTM A662 B	0.22% C (max) 1.2% Mn steel: Plate for low temperature use; 20 J impact at −50 °C **UTS: 520 Elon: 20% Proof: 275**
ASTM A663	C not specified 0.04% S and P (max) steel: Hot rolled bars; graded by tensile strength
ASTM A668 A-F	C not specified 1.1% Mn (max) 0.05% S and P (max) steel: For general use
ASTM A668 AH-FH	C not specified 1.1% Mn (max) 0.05% S and P (max) steel: For general use
ASTM A668 G-N	C and Mn not specified 0.04% S and P (max) steel: For general use
ASTM A668 GH-NH	C and Mn not specified 0.04% S and P (max) steel: For general use
ASTM A675	C not specified 0.04% S and P (max) steel: Bars; graded by tensile strength
ASTM A678 A	0.16% C (max) 1.2% Mn steel: Plate for structures; hardened and tempered **UTS: 520 Elon: 22% Proof: 345**
ASTM A678 B	0.2% C (max) 1.0% Mn steel: Plate for structures; hardened and tempered **UTS: 600 Elon: 22% Proof: 414**

Symbol	Nominal analysis, supplier, condition and remarks.
ASTM A678 C	0.22% C 1.3% Mn steel: Plate for structures; hardened and tempered **UTS: 700 Elon: 19% Proof: 480**
ASTM A707 L1/1	0.23% C (max) 1.5% Mn steel: Flanges for low temperature use; impact 41 J at −30 °C **UTS: 290 Elon: 22%**
ASTM A707 L1/2	0.23% C (max) 1.5% Mn (max) steel: Flanges for low temperature use; impact 54 J at −30 °C **UTS: 360 Elon: 25%**
ASTM A707 L1/3	0.23% C (max) 1.5% Mn (max) steel: Flanges for low temperature use; impact 68 J at −30 °C **UTS: 415 Elon: 23%**
ASTM A707 L1/4	0.23% C (max) 1.5% Mn (max) steel: Flanges for low temperature use; impact 68 J at −30 °C **UTS: 515 Elon: 20%**
ASTM A709/50 W	0.2% C (max) 1.35% Mn (max) with grain refining elements steel: For bridges **UTS: 485 Elon: 20% Proof: 345**
ASTM A709/100	0.15% C 1.0% Mn with grain refining elements steel: For bridges **UTS: 800 Elon: 17% Proof: 635**
ASTM A709/100 W	0.15% C 1.0% Mn with grain refining elements steel: For bridges **UTS: 800 Elon: 17% Proof: 635**
ASTM A714/1	0.26% C (max) 1.3% Mn (max) 0.18% Cu (min) steel: Pipe; welded or seamless **UTS: 463 Elon: 22% Proof: 345**
ASTM A715	0.15% C (max) 1.65% Mn (max) steel: Sheet; low alloy; high formability; fine grained; graded by alloy content and tensile properties
ASTM A724 A	0.22% C (max) 1.2% Mn steel: For welded pressure vessels **UTS: 700 Elon: 19% Proof: 485**
ASTM A727	0.28% C (max) 1.1% Mn steel: Forging for pipe fittings for notch toughness **UTS: 500 Elon: 22% Proof: 250**
ASTM A730 A	0.15% C (max) 0.45% Mn steel: Forging for railway use
ASTM A730 B	0.2% C 0.45% Mn steel: Forging for railway use
ASTM A732/1 A	0.2% C 0.4% Mn steel: Investment casting; annealed **UTS: 410 Elon: 35% Proof: 280**
ASTM A733 A53	Mild steel; welded or seamless; pipe nipples
ASTM A733 A120	Welded or seamless; pipe nipples
ASTM A737 B	0.22% C (max) 1.3% Mn 0.05% Nb (max) steel: Plate for pressure vessels
ASTM A741	Mild steel wire rope; zinc coated for guard rails
ASTM A749	General requirements for hot rolled strip: May have some alloy content
ASTM A766	0.17% C 1.2% Mn 0.1% S 0.04% P 0.2% Pb steel: For pressure containers; free machining **UTS: 415 Elon: 18% Proof: 200**
ASTM A769	0.2% C (max) 0.8% Mn steel: For welded steel strength; resistance welded
ASTM A769-36	0.2% C (max) 0.04% S and P (max) steel: Shapes for structural use **UTS: 365 Elon: 22% Proof: 250**
ASTM A769-40	0.2% C (max) 0.04% S and P (max) steel: Shapes for structural use **UTS: 380 Elon: 21% Proof: 275**

Note. The following abbreviations and units are used in the tables:

DPN	Hardness, diamond pyramid number
UTS	Ultimate tensile strength, N/mm^2
Elon	Elongation, %
Proof	0.1% proof strength, N/mm^2

1 N/mm^2=0.1 hbar=0.102 kgf/mm^2=0.06475 tonf/in.^{2}=145.04 lbf/in.^{2}=1 MPa

See Appendix II for other abbreviations and conversion tables.

Symbol	Nominal analysis, supplier, condition and remarks.
ASTM A769-45	0.2% C (max) 0.04% S and P (max) steel: Shapes for structural use **UTS: 415 Elon: 19% Proof: 310**
ASTM A769-45 W	0.2% C (max) 0.04% S and P (max) steel: Shapes for structural use **UTS: 450 Elon: 19% Proof: 310**
ASTM A769-50	0.2% C (max) 0.04% S and P (max) with trace element steel: Shapes for structural use **UTS: 450 Elon: 17% Proof: 345**
ASTM A769-50 W	0.2% C (max) 0.04% S and P (max) with trace element steel: Shapes for structural use **UTS: 485 Elon: 17% Proof: 345**
ASTM A769-60	0.2% C (max) 0.04% S and P (max) with trace element steel: Shapes for structural use **UTS: 520 Elon: 17% Proof: 415**
ASTM A769-80	0.2% C (max) 0.04% S and P (max) with trace element steel: Shapes for structural use **UTS: 620 Elon: 17% Proof: 550**
ASTM A782/1	0.2% C 0.7% Cr 0.4% Mo 0.1% Zr steel plate: For pressure vessels; quenched and tempered **UTS: 730 Elon: 18% Proof: 550**
ASTM A782/2	0.2% C 0.7% Cr 0.4% Mo 0.1% Zr steel plate: For pressure vessels; quenched and tempered **UTS: 800 Elon: 17% Proof: 620**
ASTM A782/3	0.2% C 0.7% Cr 0.4% Mo 0.1% Zr steel plate: For pressure vessels; quenched and tempered **UTS: 860 Elon: 16% Proof: 690**
ASTM A787 MT 1010	0.1% C steel: Resistance welded tube; Al or Zn coated
ASTM A787 MT 1015	0.15% C steel: Resistance welded tube; Al or Zn coated
ASTM A787 MT 1020	0.2% C steel: Resistance welded tube; Al or Zn coated
ASTM A787 MTX 1015	0.15% C steel: Resistance welded tube; Al or Zn coated
ASTM A787 MTX 1020	0.2% C steel: Resistance welded tube; Al or Zn coated
ASTM A794	0.2% C steel: Cold rolled; graded by AISI system 1015 to 1023
ASTM A795 A	0.25% C 1% Mn steel pipe: Galvanized
ASTM A795 B	0.3% C 1.2% Mn steel pipe: Galvanized
ASTM A795 E	Analysis not specified; electric resistance welded galvanized steel pipe
ASTM A795 F	C not specified; furnace welded galvanized steel pipe
ASTM A795 S	Analysis not specified; seamless galvanized steel pipe
ASTM A805	0.25% C (max) steel: Flat wire; cold rolled
ASTM A808	0.12% C (max) 0.1% V (max) 0.15% Nb + V (max) steel: 60 J impact at −45 °C **UTS: 450 Elon: 22% Proof: 320**
ASTM A812/65	0.23% C (max) 1.4% Mn 0.1% Nb + V steel: For pressure vessels **UTS: 585 Proof: 450**
ASTM A812/80	0.23% C (max) 1.5% Mn 0.1% Nb + V steel: For pressure vessels **UTS: 690 Proof: 550**
ASTM A812 G 65	0.23% C (max) 0.1% Nb steel: Sheet; for pressure vessels; weldable **UTS: 450 Proof: 585**
ASTM A812 G 65	0.23% C (max) 0.1% Nb steel: Sheet; for welded pressure vessels **UTS: 620 Elon: 10% Proof: 450**
ASTM A812 G 65	0.23% C (max) 0.3% Cr (max) 0.1% Nb + V steel: Sheet; hot rolled
ASTM A812 G 80	0.23% C (max) 0.1% Nb steel: Sheet; for pressure vessels; weldable **UTS: 550 Proof: 690**
ASTM A812 G 80	0.23% C (max) 0.1% Nb steel: Sheet; for welded pressure vessels **UTS: 670 Elon: 9% Proof: 550**
ASTM A812 G 80	0.23% C (max) 0.3% Cr (max) 0.1% Nb + V steel: Sheet; hot rolled

Symbol	Nominal analysis, supplier, condition and remarks.
ASTM A822	0.18% C (max) 0.45% Mn steel: Cold drawn tube **UTS: 310 Elon: 35% Proof: 170**
ASTM A827	Carbon steel for forgings; graded by AISI system
ASTM A836	0.2% C (max) 0.9% Mn (max) 0.9% Ti steel: Forgings; for glass lined vessels
ASTM A841	0.2% C (max) 0.25% Ni (max) 0.25% Cr (max) 0.08% Mo (max) steel: Plate; for pressure vessels
ASTM A847	0.2% C (max) 1.35% Mn (max) 0.2% Cu (min) steel: Tube; cold formed; for enhanced corrosion resistance
ASTM A850-1	0.27% C (max) 0.2% Cu (max) steel: Bar **UTS: 520 Elon: 18% Proof: 345**
ASTM A850-2	0.27% C (max) steel: Bar **UTS: 520 Elon: 18% Proof: 345**
ASTM A851	Austenitic steel: Induction welded tubes for condensers; see designation for steel type
ASTM A853	Carbon steel wire; analysis not specified; supplied in various conditions
ASTM A853	0.1% C steel: Wire for general use; steel grades AISI 1006 to 1020
ASTM A858	0.2% C (max) 1.35% Mn (max) 1% Ni + Cr + Mo + Cu (max) steel: Tube for low temperature use; 27 J impact at −45 °C **UTS: 485 Elon: 22% Proof: 250**
ASTM A860	0.2% C (max) 1% Ni + Cr + Mo + Cu (max) steel: For butt weld fittings; graded by yield strength and production method
ASTM A865	C not specified; 0.25% S 0.14% P steel: For pipe joints
ASTM B227	Hard drawn copper clad steel wire for electrical purposes
ASTM B228	Copper clad steel conductor wire; concentric lag
AURIGA 1A	0.12% C 0.5% Mn 0.4% Si 0.05% S and P steel: Casting; D Brown; high magnetic permeability **UTS: 39 Elon: 22%**
AURIGA 1B	0.17% C 0.5% Mn 0.4% Si 0.05% S and P steel: Casting; D Brown; high magnetic permeability **UTS: 44 Elon: 22%**
AURIGA V1A	0.22% C 1.2% Mn 0.05% S and P steel: Casting; D Brown; normalized and tempered **UTS: 60 Elon: 18%**
AUTROD 12.10	0.1% C 1.0% Mn steel: Weld metal; Esab; copper coated for submerged arc welding; flux can be used to modify analysis of deposit
AUTROD 12.30	0.1% C 1.3% Mn steel: Copper coated welding rod; Esab; for submerged arc welding; properties of the weld metal can be modified with different fluxes
AUTROD 12.51	0.1% C 1.0% Mn 0.6% Si steel: Copper coated wire; Esab; for carbon dioxide or gas shielded welding; as welded **UTS: 580 Elon: 25% Proof: 430**
AUTROD 13.42	0.12% C 1.4% Mn 0.3% Si 0.01% S and P (max) steel: Esab; welding rod; weld deposit; 50 J impact at −20 °C **UTS: 550 Elon: 27% Proof: 440**
AW 10	0.18% C 0.75% Mn 0.02% Nb 0.25% Cu steel: Alan Wood
AW 55	0.16% C 0.55% Mn 0.02% Nb steel: Alan Wood
AW AC	Mild steel Al coated conductivity wire: Copperweld Co
AWS A5-20 (E6XT1)	Mild steel: Weld electrode; obsolete; AWS designation
AWS A5-20 (E6XT4)	Mild steel: Weld electrode; obsolete; AWS designation
AWS A5-20 (E6XT5)	Mild steel: Weld electrode; obsolete; AWS designation
AWS A5-20 (E6XT6)	Mild steel: Weld electrode; obsolete; AWS designation
AWS A5-20 (E6XT7)	Mild steel: Weld electrode; obsolete; AWS designation
AWS A5-20 (E6XT8)	Mild steel: Weld electrode; obsolete; AWS designation
AWS A5-20 (E6XT11)	Mild steel: Weld electrode; obsolete; AWS designation
AWS A5-20 (E6XT13)	Mild steel: Weld electrode; flux cured; AWS designation
AWX 45	0.12% C 0.45% Mn 0.02% Nb steel: Alan Wood
AWX 50	0.5% Mn 0.02% Nb steel: Alan Wood
AWX 55	0.16% C 0.55% Mn 0.02% Nb steel: Alan Wood
B 0	0.1% C (max) steel: Breda
B 1	0.15% C (max) steel: Breda
B 2	0.2% C (max) steel: Breda
B 4	0.2% C steel: As rolled; Bofors **DPN: 130 UTS: 45 Elon: 24% Proof: 24**
B 4 V	0.15% C 0.7% Mn steel: For case hardening; Bofors; as rolled **DPN: 160**
B 303	0.15% C 0.7% Mn steel: Fagersta
BH 36	0.2% C (max) 1.2% Mn 0.04% S (max) 0.03% P (max) steel: Rheinstahl
BS 14	Low C 0.05% S (max) 0.05% P steel: Structural; pressure parts for marine boilers **UTS: 42 Elon: 24%**
BS 24/1/1	Low C 0.05% S and P steel: For loco axles; hardened and tempered **UTS: 50 Elon: 25% Proof: 24**
BS 24/1/2	Low C 0.05% S and P steel: For loco axles; hardened and tempered **UTS: 56 Proof: 27**
BS 24/1/3	Low C 0.06% S and P steel: For wagon axles; hardened and tempered **UTS: 56 Proof: 27**
BS 24/3 A 9	Low C 0.06% S and P steel: For railway spring buckles; as rolled **UTS: 39 Elon: 22%**
BS 24/3 A 10	Low C 0.06% S and P steel: For railway spring buckles; as rolled **UTS: 42 Elon: 22%B**
BS 24/3 A 11	Low C 0.06% S and P steel: For railway spring buckles; as rolled **UTS: 53 Elon: 20%**
BS 24/4 A	0.15% C 0.6% Mn 0.05% S and P steel: Forging for railway use; carburized parts **UTS: 39 Elon: 28%**
BS 24/4 B	0.2% C 0.6% Mn 0.05% S and P steel: Forging for railway use; boiler forgings **UTS: 46 Elon: 26%**
BS 24/4 E	0.16% C 0.6% Mn 0.06% S and P steel: Forging for railway use **UTS: 40 Elon: 28%**
BS 24/4 F	0.21% C 0.6% Mn 0.06% S and P steel: Forging for railway use **UTS: 45 Elon: 26%**
BS 24/5/611	0.16% C (max) 0.04% S and P steel: Plate; as BS alloy 611; for railway use
BS 24/5/612	Low C 0.05% S and P steel: Plate; as BS alloy 612; for railway use
BS 24/5/613	Low C 0.05% S and P steel: Bar; as BS alloy 613; for railway use
BS 29/22/26	Low C 0.05% S 0.05% P steel: Forging; marine forgings; welding quality **UTS: 36 Elon: 33%**
BS 29/28/32	Low C 0.05% S 0.05% P steel: Forging; marine forgings **UTS: 45 Elon: 28%**
BS 29/32/36	Low C 0.05% S 0.05% P steel: Forging; marine forgings **UTS: 51 Elon: 25%**
BS 29/36/40	Low C 0.05% S 0.05% P steel: Forging; marine forgings **UTS: 58 Elon: 22%**
BS 32/3	Low C free machining steel; withdrawn owing to low resistance to shock
BS 32/4	0.1% C 1% Mn 0.25% S 0.07% P steel: Free machining bar; replaced by BS 970 En 1A

Symbol	Nominal analysis, supplier, condition and remarks.
BS 32/5	0.1% C 1.2% Mn 0.45% S 0.06% P steel: Free machining bar; replaced by BS 970 En 1B
BS 75	Wrought steels for automobiles; replaced by BS 970
BS 400	0.2% C 0.6% Mn 0.3% Si 0.045% S and P steel: For gas cylinders
BS 401	0.2% C 0.7% Mn 0.3% Si 0.045% S and P steel: For gas cylinders
BS 640/2	0.12% C 0.05% Mn 0.04% Si 0.05% S and P steel: Electrode
BS 640/3	0.1% C 0.04% Mn 0.04% Si 0.04% S and P steel: Electrode
BS 725	Hot rolled mild steel strip; replaced by BS 1449
BS 782	Electrodes for metal arc welding; replaced by BS 639 and BS 2549
BS 806 A	Low C 0.05% S and P steel: Seamless pipe; cold drawn and annealed **UTS: 390**
BS 806 B	Low C 0.05% S and P steel: Seamless pipe; as drawn **UTS: 390**
BS 806 C	Low C 0.06% S and P steel: Seamless pipe; as drawn **UTS: 370**
BS 806 D	Low C 0.06% S and P steel: Hydraulic lap welded; as welded **UTS: 370**
BS 806 E	Low C 0.06% S and P steel: Roll lap welded pipe; as welded **UTS: 370**
BS 806 F	Low C steel: Welded pipe
BS 847	Cold rolled mild strip; replaced by BS 1449
BS 970	See designation for details
BS 971	Commentary on steels contained in BS 970 (En series)
BS 980 CDS1	0.2% C 0.05% S 0.05% P steel: Cold drawn tube; annealed **UTS: 370** **Proof: 170**
BS 980 CDS2	0.2% C 0.05% S 0.05% P steel: Cold drawn tube; as drawn and tempered **UTS: 420** **Proof: 360**
BS 980 CDS3	0.18% C 0.6% Mn steel: Cold drawn tube; as drawn and tempered **UTS: 420** **Proof: 360**
BS 980 CDS4	0.15% C 0.8% Mn 0.07% S steel: Cold drawn tube; as drawn and tempered **UTS: 420** **Proof: 360**
BS 1052	Low C 0.06% S 0.06% P steel: Wire; annealed **UTS: 390**
BS 1052	Low C 0.06% S 0.06% P steel: Wire; hard drawn **UTS: 690**
BS 1387	Low C 0.06% S and P steel: For steel tubes and sockets **UTS: 390**
BS 1449 CS1	0.07% C (max) 0.4% Mn 0.03% S and P steel: Strip; extra deep drawing; quality as BS alloy En 2A/1
BS 1449 CS2	0.08% C (max) 0.4% Mn 0.03% S and P steel: Strip; extra deep drawing; quality as BS alloy En 2A/1
BS 1449 CS3	0.1% C (max) 0.5% Mn 0.04% S and P steel: Strip; deep drawing; quality as BS alloy En 2A
BS 1449 CS4	0.12% C (max) 0.5% Mn 0.05% S and P steel: Strip; as BS alloy En 2
BS 1449 CS12	0.12% C 0.5% Mn 0.05% S and P steel: Strip; as BS alloy En 2B
BS 1449 CS17	0.17% C 0.5% Mn 0.05% S and P steel: Strip; as BS alloy En 2C
BS 1449 CS22	0.22% C 0.05% Mn 0.05% S and P steel: Strip; as BS alloy En 2C
BS 1449 HR1	0.09% C 0.45% Mn 0.03% S and P steel: Extra deep drawing (killed) sheet; as BS 970 En 2A/1 **UTS: 320** **Elon: 34%**
BS 1449 HR2	0.09% C 0.45% Mn 0.03% S and P steel: Extra deep drawing sheet; as BS 970 En 2A/1 **UTS: 320** **Elon: 34%**
BS 1449 HR3	0.1% C 0.5% Mn 0.04% S and P steel: Deep drawing sheet; as BS 970 En 2A **UTS: 330** **Elon: 31%**
BS 1449 HR4	0.15% C 0.6% Mn 0.05% S and P steel: Drawing quality sheet; as BS 970 En 2 **UTS: 290**
BS 1449 HR11	0.08% C 0.4% Mn 0.03% S and P steel: Extra deep drawing (killed) plate; as BS 970 En 2A/1 **UTS: 320** **Elon: 34%**
BS 1449 HR12	0.09% C 0.4% Mn 0.03% S and P steel: Extra deep drawing plate; as BS 970 En 2A/1 **UTS: 320** **Elon: 34%**
BS 1449 HR13	0.1% C 0.5% Mn 0.04% S and P steel: Deep drawing quality plate; as BS 970 En 2A **UTS: 320** **Elon: 31%**
BS 1449 HR14	0.15% C 0.6% Mn 0.05% S and P steel: Flanging or drawing quality plate; as BS 970 En 2A **UTS: 270**
BS 1449 HR15	0.2% C 0.9% Mn 0.06% S and P steel: Commercial quality plate; as BS 970 En 2 **UTS: 270**
BS 1449 HS1	0.07% C 0.4% Mn 0.03% S and P steel: Sheet and plate **UTS: 270** **Elon: 34%**
BS 1449 HS2	0.08% C 0.4% Mn 0.03% S and P steel: Sheet and plate **UTS: 270** **Elon: 34%**
BS 1449 HS3	0.1% C 0.5% Mn 0.04% S and P steel: Sheet and plate **UTS: 270** **Elon: 30%**
BS 1449 HS4A	0.12% C 0.5% Mn 0.05% S and P steel: Sheet and plate **UTS: 270** **Elon: 25%**
BS 1449 HS4B	0.13% C 0.6% Mn 0.06% S and P steel: Sheet and plate **UTS: 270**
BS 1449 HS12	0.12% C 0.5% Mn 0.05% S and P steel: Sheet and plate **UTS: 300** **Elon: 25%** **Proof: 170**
BS 1449 HS17	0.17% C 0.5% Mn 0.05% S and P steel: Sheet and plate **UTS: 340** **Elon: 25%** **Proof: 200**
BS 1449 HS20	0.2% C 1.5% Mn 0.2% Si 0.06% S and P steel: Sheet **UTS: 370** **Elon: 18%** **Proof: 340**
BS 1449 HS22	0.22% C 0.5% Mn 0.05% S and P steel: Sheet and plate **UTS: 390** **Elon: 20%** **Proof: 220**
BS 1449 HS23	0.22% C 0.6% Mn 0.05% S and P steel: Sheet and plate **UTS: 450** **Elon: 20%** **Proof: 240**
BS 1449 NHR12	0.1% C 0.5% Mn 0.04% S and P steel: Extra deep drawing quality plate **UTS: 290** **Elon: 26%**
BS 1449 NHR13	0.12% C 0.5% Mn 0.05% S and P steel: Deep drawing quality plate **UTS: 300** **Elon: 23%**
BS 1449 NHR14	0.15% C 0.5% Mn 0.05% S and P steel: Flanging quality plate **UTS: 300** **Elon: 22%**
BS 1449 NHR15	0.2% C 0.9% Mn 0.06% S and P steel: Commercial quality plate

Note. The following abbreviations and units are used in the tables:

DPN	Hardness, diamond pyramid number
UTS	Ultimate tensile strength, N/mm^2
Elon	Elongation, %
Proof	0.1% proof strength, N/mm^2

1 N/mm^2=0.1 hbar=0.102 kgf/mm^2=0.06475 $tonf/in.^2$=145.04 $lbf/in.^2$=1 MPa

See Appendix II for other abbreviations and conversion tables.

Symbol	Nominal analysis, supplier, condition and remarks.
BS 1449 NHR21	0.18% C 0.6% Mn 0.05% S and P steel: Plate **UTS: 390 Elon: 20% Proof: 210**
BS 1449 NHR22	0.18% C 0.6% Mn 0.05% S and P steel: Plate **UTS: 440 Elon: 18% Proof: 210**
BS 1449 NHR23	0.18% C 0.6% Mn 0.05% S and P steel: Plate **UTS: 450 Elon: 25% Proof: 240**
BS 1449 NHR25	0.2% C 1.2% Mn 0.06% S and P steel: Plate **UTS: 530 Elon: 14% Proof: 340**
BS 1453 A1	0.1% C 0.6% Mn 0.25% Ni steel: For welding filler rod to give butt weld of 309 N/mm^2 tensile
BS 1453 A2	0.15% C 1.2% Mn 0.2% Si steel: For welding filler rod to give butt weld of 420 N/mm^2 tensile
BS 1456 A	0.21% C 1.5% Mn 0.3% Si 0.05% S and P steel: Castings; hardened and tempered; contained in BS 3100 **UTS: 610 Elon: 18%**
BS 1501/101	C not specified 0.06% S and P steel: Plate and bar **UTS: 420 Elon: 20%**
BS 1501/151	0.2% C 0.05% S and P steel: Plate and bar; normalized **UTS: 420 Elon: 20% Proof: 220**
BS 1501/151 A	0.2% C 0.05% S and P steel: Plate and bar; normalized **UTS: 390 Elon: 25% Proof: 180**
BS 1501/151 B	0.2% C 0.05% S and P steel: Plate and bar; normalized **UTS: 390 Elon: 23% Proof: 210**
BS 1501/161 A	0.2% C 0.50% Mn 0.25% Si 0.05% S and P steel (silicon killed): Plate and bar; normalized **UTS: 360 Elon: 25% Proof: 180**
BS 1501/161 B	0.2% C 0.50% Mn 0.25% Si 0.05% S and P steel (silicon killed): Plate and bar; normalized **UTS: 390 Elon: 23% Proof: 200**
BS 1501/161 C	0.2% C 0.50% Mn 0.25% Si 0.05% S and P steel (silicon killed): Plate and bar; normalized **UTS: 450 Elon: 25% Proof: 220**
BS 1501/221	0.2% C 1.50% Mn 0.25% Si 0.05% S and P steel: Plate and bar; normalized **UTS: 500 Elon: 18% Proof: 240**
BS 1506/111	0.2% C 0.7% Mn 0.3% Si 0.06% S and P steel: Bar; as rolled **UTS: 450 Elon: 25% Proof: 220**
BS 1507/101	C not specified 0.06% S and P steel: Seamless pipes; as rolled **UTS: 360 Elon: 30%**
BS 1507/131	C not specified 0.06% S and P steel: Lap welded pipes **UTS: 370 Elon: 30%**
BS 1507/151	C not specified 0.05% S and P steel: Seamless pipes; as rolled **UTS: 370 Elon: 30%**
BS 1507/171	C not specified 0.05% S and P steel: Cold drawn seamless pipes; annealed after drawing **UTS: 370 Elon: 31%**
BS 1507/181	C not specified 0.05% S and P steel: Lap welded pipes **UTS: 360**
BS 1508/151	C not specified 0.05% S and P steel: Seamless tube; as rolled **UTS: 360 Elon: 29%**
BS 1508/171	C not specified 0.05% S and P steel: Cold drawn tube; annealed after drawing **UTS: 370 Elon: 30%**
BS 1617 A	0.15% C (max) 0.5% Mn 0.6% Si 0.06% S and P steel: Casting; annealed; high magnetic permeability **UTS: 420 Elon: 22%**
BS 1617 B	0.25% C (max) 0.5% Mn 0.6% Si 0.06% S and P steel: Casting; annealed; high magnetic permeability **UTS: 480 Elon: 20% Proof: 200**
BS 1627	0.12% C 0.5% Mn 0.05% S and P steel: Tube
BS 1627	0.15% C 0.50% Mn 0.3% Si 0.05% S and P steel: Cold drawn tubes; annealed **UTS: 300 Elon: 32%**
BS 1633 L	0.18% C (max) 0.2% Si 0.05% S and P steel: For land boilers **UTS: 580 Elon: 21%**
BS 1654	Mild steel tubes; replaced by BS 3059
BS 1678	0.15% C (max) 0.4% Mn 0.05% S and P steel: Tube
BS 1730	0.2% C (max) 0.5% Mn 0.05% S and P steel: Tube
BS 1775 CDS11	Low C 0.06% S and P steel: Seamless tube; cold drawn and annealed **UTS: 300 Proof: 170**
BS 1775 CDS13	Low C 0.06% S and P seamless tube; cold drawn and annealed **UTS: 330 Proof: 200**
BS 1775 CDS16	Low C 0.06% S and P steel: Seamless tube; cold drawn and annealed **UTS: 400 Proof: 240**
BS 1775 CDS20	Low C 0.06% S and P steel: Seamless tube; cold drawn and annealed **UTS: 530 Proof: 300**
BS 1775 CDS23	Low C 0.06% S and P steel: Seamless tube; cold drawn and annealed **UTS: 480 Proof: 340**
BS 1775 CDS24	Low C 0.06% S and P steel: Seamless tube; cold drawn or cold drawn and tempered **UTS: 420 Proof: 360**
BS 1775 CDS28	Low C 0.06% S and P steel: Seamless tube; cold drawn or cold drawn and tempered **UTS: 530 Proof: 420**
BS 1775 CDS35	Low C 0.06% S and P steel: Seamless tube; cold drawn or cold drawn and tempered **UTS: 630 Proof: 530**
BS 1775 CEW11	Low C 0.06% S and P steel: Electrically welded tube; cold drawn after welding; then annealed **UTS: 300 Proof: 170**
BS 1775 CEW16	Low C 0.06% S and P steel: Electrically welded tube; cold drawn after welding; then annealed **UTS: 370 Proof: 240**
BS 1775 CEW23	Low C 0.06% S and P steel: Electrically welded tube; cold drawn after welding; then annealed **UTS: 480 Proof: 340**
BS 1775 CEW24	Low C 0.06% S and P steel: Electrically welded tube; cold drawn after welding **UTS: 420 Proof: 360**
BS 1775 CEW28	Low C 0.06% S and P steel: Electrically welded tube; cold drawn after welding **UTS: 530 Proof: 420**
BS 1775 EFW16	Low C 0.06% S and P steel: Electric fusion welded tube; as welded **UTS: 390 Proof: 240**
BS 1775 ERW11	Low C 0.06% S and P steel: Electrically welded tube; as welded **UTS: 300 Proof: 170**
BS 1775 ERW16	Low C 0.06% S and P steel: Electrically welded tube; as welded **UTS: 370 Proof: 240**
BS 1775 ERW20	Low C 0.06% S and P steel: Electrically welded tube; as welded **UTS: 450 Proof: 300**
BS 1775 ERW23	Low C 0.06% S and P steel: Electrically welded tube; as welded **UTS: 480 Proof: 340**
BS 1775 HFS11	Low C 0.06% S and P steel: Seamless tube; as drawn **UTS: 300 Proof: 170**
BS 1775 HFS13	Low C 0.06% S and P steel: Seamless tube; as drawn **UTS: 330 Proof: 200**
BS 1775 HFS16	Low C 0.06% S and P steel: Seamless tube; as drawn **UTS: 400 Proof: 240**
BS 1775 HFS20	Low C 0.06% S and P steel: Seamless tube; as drawn **UTS: 550 Proof: 300**
BS 1775 HFS23	Low C 0.06% S and P steel: Seamless tube; as drawn **UTS: 480 Proof: 340**
BS 1775 HFW13	Low C 0.06% S and P steel: Welded tube; as welded **UTS: 330 Proof: 200**
BS 1775 HFW16	Low C 0.06% S and P steel: Welded tube; as welded **UTS: 400 Proof: 240**

Symbol	Nominal analysis, supplier, condition and remarks.
BS 1775 HFW23	Low C 0.06% S and P steel: Welded tube; as welded **UTS: 480** **Proof: 340**
BS 1775 HLW16	Low C 0.06% S and P steel: Hydraulically lap welded tube; as welded **UTS: 390** **Proof: 240**
BS 1775 OAW11	Low C 0.06% S and P steel: Oxy-acetylene welded tube; as welded **UTS: 300** **Proof: 170**
BS 1882	99.0% Ni mild steel to BS alloy 151–154 or 157 sheet
BS 2762 NDI	0.2% C (max) 1.5% Mn 0.06% S 0.05% P steel: Plate; notch ductile for bridges, etc. **UTS: 450** **Elon: 22%** **Proof: 220**
BS 2762 NDII	0.2% C 1.5% Mn 0.06% S 0.05% P steel: Plate; notch ductile for bridges, etc. **UTS: 450** **Elon: 22%** **Proof: 220**
BS 2762 NDIII	0.17% C (max) 1.5% Mn 0.3% Si 0.05% S and P steel: Plate; notch ductile for bridges, etc. **UTS: 450** **Elon: 22%** **Proof: 210**
BS 2762 NDIV	0.17% C (max) 1.5% Mn 0.3% Si 0.05% S and P steel: Plate; notch ductile for bridges, etc. **UTS: 450** **Elon: 22%** **Proof: 210**
BS 2772/3	0.2% C 1.5% Mn 0.05% S and P steel: Castings for colliery winding gear; hardened and tempered **DPN: 150** **UTS: 480** **Elon: 25%** **Proof: 300**
BS 2858	0.15% C (max) 0.5% Mn 0.05% Si 0.05% S and P steel: Plate for galvanizing pots
BS 2901 A15	0.1% C 1.2% Mn 0.6% Si steel: Rod for gas and electric welding
BS 2901 A17	0.15% C 1.1% Mn 0.5% Si 0.03% S and P steel: Rod for gas and electric welding
BS 3059/1	Low C 0.05% S and P steel: Tube for boilers; seamless; hot finished **UTS: 360**
BS 3059/2	Low C 0.05% S and P steel: Tube for boilers; cold drawn; seamless **UTS: 360**
BS 3059/3	Low C 0.05% S and P steel: Tube for boilers; electrically welded and normalized **UTS: 360**
BS 3059/4	Low C 0.05% S and P steel: Tube for boilers; electrically welded from strip, then cold drawn **UTS: 360**
BS 3100 A4	0.2% C 1.4% Mn 0.05% S and P (max) steel: Casting **UTS: 600** **Elon: 16%** **Proof: 320**
BS 3100 AL1	0.2% C (max) 1.1% Mn 0.04% S and P (max) steel: Casting for low temperature use; 20 J impact at −40 °C **UTS: 430** **Elon: 22%** **Proof: 230**
BS 3100 AM1	0.15% C (max) 0.5% Mn (max) 0.8% Cr + Mo + Ni + Cu (max) steel: Casting with high magnetic permeability **UTS: 400** **Elon: 22%** **Proof: 185**
BS 3100 AM2	0.25% C (max) 0.5% Mn (max) 0.8% Cr + Mo + Ni + Cu (max) steel: Casting with high magnetic permeability **UTS: 450** **Elon: 22%** **Proof: 215**
BS 3100 AW1	0.14% C 0.8% Mn 0.8% Cr + Mo + Ni + Cu (max) steel: Casting for case hardening **UTS: 460** **Elon: 12%**

Note. The following abbreviations and units are used in the tables:

DPN	Hardness, diamond pyramid number
UTS	Ultimate tensile strength, N/mm^2
Elon	Elongation, %
Proof	0.1% proof strength, N/mm^2

1 N/mm^2=0.1 hbar=0.102 kgf/mm^2=0.06475 tonf/in.^{2}=145.04 ibf/in.^{2}=1 MPa

See Appendix II for other abbreviations and conversion tables.

Symbol	Nominal analysis, supplier, condition and remarks.
BS 3141 C1A1	0.1% C 0.5% Mn 0.04% S and P steel: Classification system used by British Standards; discontinued
BS 3141 C2A1	0.12% C (max) 0.5% Mn 0.05% S and P steel: Classification system used by British Standards; discontinued
BS 3141 C2B1	0.15% C 0.4% Mn 0.05% S and P steel: Classification system used by British Standards; discontinued
BS 3141 C2E1	0.12% C 0.5% Mn 0.05% S and P steel: Classification system used by British Standards; discontinued
BS 3141 C2F1	0.12% C 0.8% Mn 0.05% S and P steel: Classification system used by British Standards; discontinued
BS 3141 C2G1	0.12% C 1.3% Mn 0.05% S and P steel: Classification system used by British Standards; discontinued
BS 3141 C2K1	0.15% C 0.4% Mn 0.05% S and P steel: Classification system used by British Standards; discontinued
BS 3141 C2L1	0.15% C 1.1% Mn 0.05% S and P steel: Classification system used by British Standards; discontinued
BS 3141 C3A1	0.2% C 0.5% Mn 0.05% S and P steel: Classification system used by British Standards; discontinued
BS 3141 C3B1	0.2% C 0.7% Mn 0.05% S and P steel: Classification system used by British Standards; discontinued
BS 3141 CF2C1	0.1% C 1% Mn 0.25% S 0.07% P steel: Free cutting; classification system used by British Standards; discontinued
BS 3141 CF2D1	0.1% C 1.2% Mn 0.5% S 0.06% P steel: Free cutting; classification system used by British Standards; discontinued
BS 3141 CF2H1	0.14% C 1.1% Mn 0.12% S 0.05% P steel: Free cutting; classification system used by British Standards; discontinued
BS 3141 CF2J1	0.16% C 1.2% Mn 0.14% S 0.06% P steel: Free cutting; classification system used by British Standards; discontinued
BS 3141 CF3D1	0.2% C 1.1% Mn 0.14% S 0.06% P steel: Free cutting; classification system used by British Standards; discontinued
BS 3141 CF3E1	0.2% C 1.1% Mn 0.14% S 0.05% P steel: Free cutting; classification system used by British Standards; discontinued
BS 3458	0.2% C 0.5% Ni 0.5% Cr 0.2% Mo steel: For chain slings; hardened and tempered **DPN: 250**
BS 3601 BW	C not specified 0.06% S and P steel: Butt welded pipe; available in grade 22 **UTS: 390**
BS 3601 CDS	C not specified 0.06% S and P steel: Seamless pipe; cold drawn; available in grades 22, 27 and 35 **UTS: 400**
BS 3601 EFW	C not specified 0.06% S and P steel: Electric fusion welded pipe; available in grade 26 **UTS: 420**
BS 3601 ERW	C not specified 0.06% S and P steel: Electrical resistance welded pipes; available in grades 22 and 27 **UTS: 370**
BS 3601 HFS	C not specified 0.06% S and P steel: Seamless pipe; hot finished; available in grades 22, 27 and 35 **UTS: 400**
BS 3601 HLW	C not specified 0.06% S and P steel: Hydraulic welded pipes; available in grade 26 **UTS: 420**
BS 3601 SFW	C not specified 0.06% S and P steel: Spiral seam fusion welded pipe available in grade 26 **UTS: 420**
BS 3602 CDS	0.22% C 0.5% Mn 0.2% Si 0.05% S and P steel: Seamless pipe; cold drawn; available in grades 23, 27 and 35 **UTS: 420** **Proof: 240**
BS 3602 EFW	Analysis not quoted; electric fusion welded steel pipes
BS 3602 ERW	0.22% C 0.05% S and P steel: Electric resistance welded pipes; available in grades 23 and 27 **UTS: 420** **Proof: 210**

Symbol	Nominal analysis, supplier, condition and remarks.
BS 3602 HFS	0.22% C 0.5% Mn 0.2% Si 0.05% S and P steel: Seamless tube; hot finished; available in grades 23, 27 and 35
	UTS: 420 **Proof: 240**
BS 4360/40 A	0.27% C 0.06% S and P steel: As rolled; weldable
	UTS: 420 **Elon: 22%**
BS 4360/40 B	0.25% C 1.6% Mn 0.06% S and P steel: As rolled; weldable
	UTS: 420 **Elon: 22%**
BS 4360/40 C	0.22% C 1.6% Mn 0.06% S and P steel: As rolled; weldable
	UTS: 420 **Elon: 22%**
BS 4360/40 D	0.19% C 1.6% Mn 0.06% S and P steel: Normalized; weldable; with Nb
	UTS: 420 **Elon: 22%**
BS 4360/40 E	0.19% C 1.6% Mn 0.3% Si 0.05% S and P steel: Normalized; weldable
	UTS: 420 **Elon: 22%**
BS 4360/43 D	0.19% C 1.6% Mn 0.05% S and P steel: Normalized; weldable; with Nb
	UTS: 500 **Elon: 23%**
BS 4360/43 E	0.19% C 1.6% Mn 0.3% Si 0.05% S and P steel: Normalized; weldable
	UTS: 500 **Elon: 23%**
BS 4449	0.25% C (max) 0.05% S and P steel: For concrete reinforcement; as rolled; high yield
	UTS: 410 **Elon: 14%**
BS 4449	0.25% C (max) 0.06% S and P steel: For concrete reinforcement; as rolled; mild steel
	UTS: 250 **Elon: 22%**
BS 4461	0.25% C (max) 0.06% S and P (max) steel: For concrete reinforcement; cold rolled
	UTS: 440 **Elon: 13%**
BS S3	Low carbon steel: Sheet; 28 ton range suitable for welding; replaced by BS S510
BS S13	0.1% C steel: For carburizing; replaced
BS S14	0.14% C 0.7% Mn 0.2% Si 0.05% S and P steel: Bar and forging; blank carburized; hardened and tempered
	UTS: 480 **Elon: 20%**
BS S32	Low carbon mild steel: Wire for welding; obsolete
BS S84	Low carbon steel: Sheet; for welding; obsolete (see BS S511)
BS S91	0.15% C (max) 0.5% Mn 0.2% Si 0.04% S and P steel: Normalized; for bearing shells
	DPN: 140
BS S92	0.22% C 1.5% Mn 0.2% Si 0.04% S and P steel: Hardened and tempered; suitable for welding
	DPN: 210 **UTS: 760** **Elon: 17%** **Proof: 460**
BS S112	0.2% C 1.0% Mn 0.3% Si 0.14% S 0.05% P 0.2% Pb steel: Cold drawn; free machining
	UTS: 690 **Elon: 15%**
BS S510	0.21% C 0.6% Mn 0.2% Si 0.05% S and P steel: Sheet; normalized; suitable for welding
	UTS: 420 **Elon: 20%** **Proof: 240**
BS S511	0.12% C (max) 0.4% Mn 0.2% Si 0.05% S and P steel: Sheet; annealed; suitable for welding
	UTS: 330 **Elon: 25%**
BS S512	0.12% C (max) 0.5% Mn 0.2% Si 0.05% S and P steel: Tinned sheet; tin to be 99.5% pure and not less than 0.015 mm thick
BS S514	0.2% C 1.5% Mn 0.2% Si 0.05% S and P steel: Sheet; hardened and tempered
	UTS: 840 **Elon: 12%** **Proof: 600**
BS S515	0.2% C 1.5% Mn 0.2% Si 0.05% S and P steel: Sheet; annealed; suitable for welding
	DPN: 170 **UTS: 450** **Elon: 20%**
BS T6	0.18% C (max) 1.0% Mn 0.2% Si 0.05% S and P steel: Tube; as drawn and tempered, weldable
	UTS: 450 **Proof: 420**
BS T26	0.2% C 0.05% S and P steel: Tube; ½-hard

Symbol	Nominal analysis, supplier, condition and remarks.
BST 4KP2	0.22% C 0.6% Mn 0.3% Cr + Ni + Cu (max) 0.05% S (max) 0.04% P (max) steel: Plate; Russian Standard
C 1	0.15% C (max) 0.5% Mn steel: Siau
C 2	0.2% C (max) 0.8% Mn steel: Siau
C 8 UNI 4365	0.08% C 0.4% Mn 0.04% S and P steel: Italian Standard
C 9/15	0.15% C 0.4% Mn 0.04% S and P steel: Designation used by German Standards
C 9/22	0.21% C 0.4% Mn 0.04% S and P steel: Designation used by German Standards
C 10	0.1% C 0.3% Mn 0.045% S and P steel: Designation used by German Standards
C 10 UNI 4365	0.09% C 0.5% Mn 0.3% Si 0.035% S and P steel: Italian Standard
C 15	0.15% C 0.4% Mn 0.045% S and P steel: For carburizing; designation used by German Standards
C 15 UNI 4365	0.15% C 0.5% Mn 0.3% Si 0.035% S and P steel: Italian Standard; for bolts
C 20	0.2% C 0.6% Mn steel: Italian Standards designation; normalized
	DPN: 180 **UTS: 520** **Elon: 24%** **Proof: 240**
C 20 UNI 4365	0.2% C 0.8% Mn 0.4% Si 0.035% S and P steel: Italian Standard; for bolts
C 21 UNI 4365	0.2% C 0.45% Mn 0.1% Si 0.04% S and P steel: Italian Standard; for bolts
C 22	0.21% C 0.4% Mn 0.045% S and P steel: Designation used by German Standards
C 23	0.1% C 0.7% Mn steel: Welding electrode; Philips
CASONA	0.1% C (max) 0.25% Mn 0.25% Si steel: For carburizing; Hall and Pickles; obsolete
CD 28	0.12% C 0.5% Mn 0.2% Si steel: Fagersta
	UTS: 390 **Elon: 30%**
CD 32	0.14% C 0.7% Mn 0.3% Si steel: Fagersta
	UTS: 440 **Elon: 28%**
CD 40	0.17% C 0.8% Mn 0.3% Si steel: Fagersta
	UTS: 460 **Elon: 25%**
CELLOCITO PL2	0.12% C 0.55% Mn 0.2% Si: Soudametal; mild steel; welding for pipelines and ducts
	UTS: 530 **Elon: 26%** **Proof: 435**
CHAR-PAC	0.2% C (max) 1.0% Mn 0.04% S and P (max) steel: US Steel Corp
CHMS	0.14% C 0.7% Mn steel: For case hardening; ESC for BS alloy En 32
CODE 602	0.1% C (max) 0.15% Mn 0.15% Si steel: Casting; Edgar Allen; O steel
CODE 605	0.15% C 0.45% Mn 0.25% Si steel: Casting; Edgar Allen; T steel
COILEX	0.08% C 0.8% Mn 0.2% Si steel: Welding electrode; Murex
	UTS: 420 **Elon: 38%** **Proof: 330**
COILEX F	0.08% C 0.4% Mn 0.15% Si steel: Welding electrode; Murex
	UTS: 500 **Elon: 25%** **Proof: 440**
COLCLAD 12 Cr	0.08% C (max) 0.5% Ni 12% Cr stainless steel: Colvilles Stainless to BS alloy En56A
COLCLAD 13/Cr/Al	0.08% C (max) 0.5% Ni 12% Cr 0.2% Al stainless steel on mild steel: Colvilles stainless BS alloy 713
COLCLAD 18/8	0.08% C (max) 10% Ni 19% Cr stainless steel on mild steel: Colvilles stainless to BS alloy En58A
COLCLAD 18/8 ELC	0.03% C (max) 10% Ni 19% Cr stainless steel on mild steel: Colvilles stainless to BS alloy 801C
COLCLAD 18/8 Nb	0.08% C (max) 9% Ni 18% Cr 0.9% Nb stainless steel on mild steel: Colvilles stainless to BS alloy En58
COLCLAD 18/8 Ti	0.08% C (max) 9% Ni 18% Cr 0.5% Ti stainless steel on mild steel: Colvilles stainless to BS alloy En58B
COLCLAD 18/10/2 ELC	0.03% C (max) 12% Ni 18% Cr 2.5% Mo stainless steel on mild steel: Colvilles
COLCLAD 18/10/2 Nb	0.08% C (max) 12% Ni 18% Cr 2.5% Mo 0.9% Nb stainless steel on mild steel: Colvilles stainless to BS alloy 845

Symbol	Nominal analysis, supplier, condition and remarks.
COLCLAD 18/10/2 Ti	0.08% C (max) 12% Ni 18% Cr 2.5% Mo 0.5% Ti stainless steel on mild steel: Colvilles stainless to BS alloy En 58H
COLCLAD 18/10/3 ELC	0.03% C (max) 12% Ni 18% Cr 3% Mo stainless steel on mild steel: Colvilles
COLCLAD 18/10/3 Nb	0.08% C 12% Ni 18% Cr 3% Mo 0.9% Nb stainless steel on mild steel: Colvilles stainless to BS alloy En 58J
COLCLAD INCONEL	0.2% C 16% Cr 0.07% Cu 8% Fe Ni alloy on mild steel: Colvilles; Inconel on mild steel
COLCLAD NICKEL	0.02% C (max) 0.1% Ti (max) 0.2% Cu Ni on mild steel: Colvilles; Ni to BS alloy 1872 – low carbon
COLCLAD NICKEL	0.15% C (max) 0.1% Ti (max) 0.2% Cu Ni on mild steel; Colvilles; Ni to BS alloy 1872
COLORCOAT	Strip steel (also stainless, etc.) with a range of plastic or paint coatings; British Steel
COLTUF 26	0.15% C (max) 1.2% Mn 0.2% Si 0.05% S and P steel: Plate; Colvilles
	UTS: 390 Elon: 26% Proof: 240
COLTUF 28	0.16% C (max) 1.5% Mn 0.2% Si 0.05% S and P steel: Plate; Colvilles
	UTS: 420 Elon: 25% Proof: 240
COLTUF 32	0.22% C (max) 1.6% Mn 0.5% Ni 0.3% Si steel: Plate; Colvilles
	UTS: 540 Elon: 22% Proof: 330
COMET B	0.07% C 0.65% Mn 0.35% Si mild steel: Welding rod; Soudametal; horizontal welding of plate
	UTS: 550 Elon: 26% Proof: 480
COMET J 50	0.06% C 1.0% Mn 0.4% Si mild steel: Welders rod; Soudametal; for repair welding
	UTS: 550 Elon: 30% Proof: 460
COMET J 50 N	0.06% C 1.0% Mn 0.6% Si mild steel: Welding rod; Soudametal
	UTS: 550 Elon: 32% Proof: 440
COMET R 12	Analysis not supplied: Mild steel; welding electrode; Soudametal; for deep penetration welds
COMET T	0.06% C 0.5% Mn 0.4% Si mild steel: Welding rod; Soudametal; flexible position welding
	UTS: 550 Elon: 22% Proof: 495
COMET V	0.08% C 0.6% Mn 0.35% Si mild steel: Welding rod; Soudametal; welding of pipework, etc.
	UTS: 540 Elon: 26% Proof: 48
COMET W	0.01% C 0.55% Mn 0.2% Si mild steel: Welding rod; Soudametal; horizontal welding of sheet
	UTS: 500 Elon: 30% Proof: 420
CONLO 1	0.16% C 1.5% Mn 0.2% Si 0.05% S and P steel: Plate; Consett; normalized; suitable for welding; Al killed
	UTS: 440 Elon: 25% Proof: 240
CONLO 1	0.16% C (max) 1.2% Mn 0.05% S and P (max) steel: Consett
CONLO 11	0.16% C 1.2% Mn 0.05% S and P (max) steel: Consett
CONLO 11	0.16% C 1.2% Mn 0.2% Si 0.05% S and P steel: Plate; Consett; normalized; suitable for welding; Si killed
	UTS: 440 Elon: 25% Proof: 240
CON-PAC	0.2% C (max) 1.25% Mn 0.25% Cu steel: US Steel Corp.
COPPERWELD	Copper covered steel wire: Copperweld Steel Co.; for electronic uses

Note. The following abbreviations and units are used in the tables:

DPN	Hardness, diamond pyramid number
UTS	Ultimate tensile strength, N/mm^2
Elon	Elongation, %
Proof	0.1% proof strength, N/mm^2

1 N/mm^2=0.1 hbar=0.102 kgf/mm^2=0.06475 tonf/in.^{2}=145.04 lbf/in.^{2}=1 MPa

See Appendix II for other abbreviations and conversion tables.

Symbol	Nominal analysis, supplier, condition and remarks.
CORROSTITE	0.11% C 0.4% Mn 0.3% Si (max) 0.05% S and P 0.3% Cu steel: Tube; STD Service (see Resistco)
Cq 15	0.15% C 0.4% Mn 0.2% Si 0.04% S and P steel: German Standards
Cq 22	0.22% C 0.5% Mn 0.25% Si 0.04% S and P steel: German Standards
CREUSELSO 38	0.18% C (max) 1.2% Mn (max) Al killed steel: Creusot
CREUSELSO 42	0.24% C (max) 1.4% Mn (max) 0.03% S and P (max) steel: Creusot
CRO 42	0.2% C 0.6% Ni 0.2% Mo steel: For case hardening; Bofors
	DPN: 220
CRO 53	0.2% C 1.2% Ni 1.0% Cr 0.2% Mo steel: For case hardening; Bofors; annealed
	DPN: 220
Cu 1	0.1% C (max) 0.5% Mn steel: Siau
CUPREX	0.07% C 0.35% Mn 0.5% Cu steel: Welding electrode; Murex
	UTS: 460 Elon: 30% Proof: 440
D 1	0.12% C 0.5% Mn 0.2% Si steel: Pompey
D 1S	0.1% C 0.7% Mn 0.2% Si 0.2% S steel: Pompey; free machining
D 1SS	0.1% C 1.0% Mn 0.2% Si 0.15% S steel: Pompey; free machining
D 2	0.2% C 0.6% Mn steel: Pompey
D 2K	0.2% C 0.9% Mn 0.3% Si 0.35% S steel: Pompey; free machining
D 2Mn	0.2% C 1.3% Mn 0.2% Si steel: Pompey
D 2S	0.2% C 0.75% Mn 0.25% Si 0.15% S steel: Pompey; free machining
D 2SS	0.2% C 1.0% Mn 0.25% Si steel: Pompey
DC 1	0.1% C 0.5% Mn 0.3% Si steel: Pompey
DEEPEX	0.15% C 0.35% Mn steel: Welding electrode; Murex
DEF 13 B1	Mild steel: No properties or specifications listed
DEF 13 B2	Mild steel: Covers BS 970 En 3A 3C 4, etc.
	UTS: 420 Elon: 20% Proof: 220
DEF 13 B6	Carbon case hardening steel: Covers BS 970 En 32, etc.
	UTS 480 Elon: 20%
DEF 13 B6 B	Carbon manganese case hardening steel: Covers BS 970 En 201
	UTS: 600 Elon: 20%
DILLINAL 54	0.2% C (max) 1.3% Mn 0.04% S and P (max) steel: German proprietary specification; Al killed
DILLINAL 54T	0.2% C (max) 1.3% Mn 0.04% S and P (max) steel: German proprietary specification; Al killed
DILLINAL 56	0.2% C (max) 1.3% Mn 0.5% Ni 0.04% S and P (max) steel: German proprietary specification; Al killed
DILLINAL 56T	0.2% C (max) 1.3% Mn 0.5% Ni 0.12% V steel: German proprietary specification; Al killed
DIN 1613 St 35 13K	0.09% C 0.4% Mn steel: For chains
DIN 1624 St 0	0.12% C (max) 0.4% Mn 0.06% S 0.08% P steel: Cold rolled strip
DIN 1624 St 1	0.12% C (max) 0.4% Mn 0.06% S 0.07% P steel: Cold rolled strip
DIN 1624 St 2	0.1% C (max) 0.3% Mn 0.05% S and P steel: Cold rolled strip
DIN 1624 St 3	0.1% C 0.4% Mn 0.04% S and P steel: Cold rolled strip
DIN 1624 St 4	0.1% C (max) 0.4% Mn 0.035% S 0.03% P steel: Cold rolled strip
DIN 1626 42/2	0.25% C (max) 0.05% S (max) 0.06% P (max) steel: Pipe; German Standard
	UTS: 460 Proof: 260
DIN 1626 52/3	0.22% C (max) 1.5% Mn (max) 0.05% S and P (max) steel: Pipe; German Standard
	UTS: 560 Proof: 360
DIN 1626 St 33	Analysis not specified: Steel pipe; German Standard
	UTS: 410
DIN 1626 St 34/2	0.17% C (max) 0.05% S and P (max) steel: Pipe; German Standard
	UTS: 400 Proof: 240

Symbol	Nominal analysis, supplier, condition and remarks.
DIN 1626 St 37	0.2% C (max) 0.05% S (max) 0.08% P (max) steel: Pipe; German Standard **UTS: 400** **Proof: 240**
DIN 1626 St 37/2	0.2% C (max) 0.05% S (max) 0.06% P (max) steel: Pipe; German Standard **UTS: 410** **Proof: 260**
DIN 1626 St 42	0.25% C (max) 0.05% S (max) 0.08% P (max) steel: Pipe; German Standard **UTS: 460** **Proof: 260**
DIN 1629 St 00	Analysis not specified: Steel tube; German Standard
DIN 1629 St 35	0.18% C (max) 0.05% S and P (max) steel: Pipe; German Standard **UTS: 400** **Proof: 240**
DIN 1629 St 35/4	0.17% C (max) 0.4% Mn (max) 0.05% S and P (max) steel: Pipe; German Standard **UTS: 400** **Proof: 240**
DIN 1629 St 45	0.25% C (max) 0.05% S and P (max) steel: Pipe; German Standard **UTS: 500** **Proof: 260**
DIN 1629 St 45/4	0.22% C (max) 0.4% Mn (max) 0.04% S and P (max) steel: Pipe; German Standard **UTS: 500** **Proof: 260**
DIN 1629 St 52	0.2% C (max) 1.5% Mn (max) 0.05% S and P (max) steel: Pipe; German Standard **UTS: 570** **Proof: 360**
DIN 1629 St 52/4	0.2% C (max) 1.5% Mn (max) 0.05% S and P (max) steel: Pipe; German Standard **UTS: 570** **Proof: 360**
DIN 1651 9 S 20	0.12% C (max) 0.7% Mn 0.2% S 0.1% P steel: Free machining
DIN 1651 9 S 27	0.12% C (max) 0.8% Mn 0.25% S 0.1% P steel: Free machining
DIN 1651 9 S Mn 23	0.13% C (max) 1.1% Mn 0.23% S steel: Free machining
DIN 1651 9 S Mn 28	0.14% C (max) 1.1% Mn 0.28% S steel: Free machining **DPN: 160** **UTS: 460**
DIN 1651 9 S Mn 36	0.15% C (max) 1.2% Mn 0.38% S steel: Free machining; German Standard **DPN: 170** **UTS: 480**
DIN 1651 9 S Mn Pb 23	0.13% C (max) 1.1% Mn 0.23% S 0.2% Pb steel: Free machining
DIN 1651 9 S Mn Pb 28	0.14% C (max) 1.1% Mn 0.28% S 0.2% Pb steel: Free machining **DPN: 160** **UTS: 460**
DIN 1651 9 S Mn Pb 36	0.15% C (max) 1.2% Mn 0.38% S 0.2% Pb steel: Free machining **DPN: 170** **UTS: 480**
DIN 1651 10 S 20	0.09% C 0.7% Mn 0.2% S 0.07% P steel: Free cutting; carburizing
DIN 1651 15 S 20	0.15% C 0.7% Mn 0.2% S 0.07% P steel: Free machining; carburizing
DIN 1651 22 S 20	0.21% C 0.5% Mn 0.2% S steel: Free machining; for carburizing
DIN 1654 C 9 15	0.15% C 0.4% Mn 0.04% S and P steel: For screws
DIN 1654 C 9 22	0.21% C 0.4% Mn 0.04% S and P steel: For screws
DIN 2440	Plain carbon steel: Tubes for screwing; German Standard
DIN 2441	Plain carbon steel: Tubes for screwing; German Standard
DIN 2442	Plain carbon steel: Tubes for screwing; German Standard
DIN 17100 RST 34/1	0.17% C (max) 0.05% S (max) 0.08% P (max) steel: Plate; German specification **UTS: 380** **Proof: 200**
DIN 17100 RST 34/2	0.15% C (max) 0.05% S (max) 0.007% P (max) steel: Plate; German specification **UTS: 380** **Proof: 200**
DIN 17100 RST 37/1	0.2% C (max) 0.05% S (max) 0.07% P (max) steel: Plate; German specification **UTS: 410** **Proof: 230**
DIN 17100 RST 37/2	0.17% C (max) 0.05% S and P (max) 0.007% N (max) steel: Plate; German specification **UTS: 410** **Proof: 230**
DIN 17100 RST 46/7	0.2% C (max) 0.05% S (max) 0.05% P (max) steel: Plate; German specification **UTS: 490** **Proof: 280**
DIN 17100 St 33/1	Chemical analysis not specified: Steel for plate; German specification **UTS: 410** **Proof: 190**
DIN 17100 St 33/2	C not specified 0.05% S (max) 0.06% P (max) 0.007% N (max) steel: Plate; German specification **UTS: 410** **Proof: 190**
DIN 17100 St 37/3	0.17% C (max) 0.045% S and P (max) 0.009% N (max) steel: Plate; German specification **UTS: 410** **Proof: 230**
DIN 17100 St 46/3	2% C (max) 0.045% S (max) 0.045% P (max) steel: Plate; German specification **UTS: 490** **Proof: 280**
DIN 17100 St 50/1	0.2% C 0.05% S (max) 0.08% P (max) steel: Plate; German standard **UTS: 550** **Proof: 290**
DIN 17100 St 52/3	0.21% C (max) 1.5% Mn (max) 0.045% P (max) 0.045% (max) steel: Plate; German specification **UTS: 570** **Proof: 350**
DIN 17100 USt 34/1	0.17% C (max) 0.05% S (max) 0.08% P (max) steel: Plate; German specification **UTS: 380** **Proof: 200**
DIN 17100 USt 34/2	0.15% C (max) 0.05% S and P (max) 0.007% N (max) steel: Plate; German specification **UTS: 380** **Proof: 200**
DIN 17100 USt 37/1	0.2% C (max) 0.05% S (max) 0.07% P (max) steel: Plate; German specification **UTS: 410** **Proof: 230**
DIN 17100 USt 37/2	0.19% C (max) 0.05% S and P (max) 0.007% N (max) steel: Plate; German specification **UTS: 410** **Proof: 230**
DIN 17100 USt 42/1	0.25% C (max) 0.05% S (max) 0.08% P (max) steel: Plate; German specification **UTS: 460** **Proof: 250**
DIN 17110 MR St 34	0.1% C 0.3% Mn 0.05% S and P steel: For rivets
DIN 17110 MR St 44	0.15% C 0.6% Mn 0.06% S and P steel: For rivets
DIN 17110 MU St 34	0.1% C (max) 0.25% Mn 0.05% S and P steel: For rivets
DIN 17110 Q St 34	0.1% C (max) 0.3% Mn 0.04% S and P steel: For rivets
DIN 17110 TU St 34	0.08% C (max) 0.06% S 0.08% P steel: For rivets
DIN 17155 17 Mn 4	0.17% C 0.1% Mn 0.3% Cr (max) 0.05% S and P (max) steel: Plate; German Standard **UTS: 520** **Proof: 280**
DIN 17155 19 Mn 5	0.2% C 1.1% Mn 0.3% Cr (max) 0.05% S and P (max) steel: Plate; German Standard **UTS: 570** **Proof: 320**
DIN 17155 Gd 1	0.17% C (max) 0.05% S and P steel: Plate; for boiler
DIN 17155 Gd 11	0.23% C (max) 0.05% S and P steel: For boiler plate
DIN 17155 H11	0.2% C (max) 0.5% Mn (max) 0.3% Cr (max) 0.05% S and P (max): German Standard **UTS: 440** **Proof: 250**
DIN 17155 HI	0.16% C (max) 0.4% Mn (max) 0.3% Cr (max) 0.05% S and P (max) steel: Plate; German Standard **UTS: 400** **Proof: 220**
DIN 17155 HM	0.22% C (max) 0.55% Mn (max) 0.3% Cr (max) 0.05% S and P (max) steel: Plate; German Standard
DIN 17172 RRSt 34/7	0.17% C 0.35% Mn (max) 0.045% S and P steel: Pipe; German Standard **UTS: 390** **Proof: 210**
DIN 17172 RRSt 38/7	0.22% C (max) 0.4% Mn (max) 0.04% S and P (max) steel: Pipe; German Standard **UTS: 420** **Proof: 250**
DIN 17172 RSt 34/7	0.17% C 0.35% Mn (max) 0.04% S and P steel: Pipe; German Standard **UTS: 390** **Proof: 210**

Symbol	Nominal analysis, supplier, condition and remarks.
DIN 17172 RSt 38/7	0.22% C (max) 0.4% Mn (max) 0.04% S and P (max) steel: Pipe; German Standard **UTS: 430** **Proof: 250**
DIN 17172 St 43/7	0.22% C (max) 0.8% Mn 0.04% S and P (max) steel: Pipe; German Standard **UTS: 490** **Proof: 300**
DIN 17172 St 47/7	0.22% C (max) 1.0% Mn 0.04% S and P (max) steel: Pipe; German Standard **UTS: 490** **Proof: 300**
DIN 17172 St 53/7	0.22% C (max) 1.2% Mn 0.04% S and P (max) steel: Pipe; German Standard **UTS: 580** **Proof: 370**
DIN 17172 USt 34/7	0.17% C 0.35% Mn (max) 0.04% S and P steel: Pipe; German Standard **UTS: 390** **Proof: 210**
DIN 17172 USt 38/7	0.2% C (max) 0.4% Mn (max) 0.04% S and P (max) steel: Pipe; German Standard **UTS: 430** **Proof: 250**
DIN 17175 14 Mn 4	0.14% C 1.0% Mn 0.05% S and P steel: Tube
DIN 17175 St 35/8	0.17% C (max) 0.05% S and P steel: Tube
DIN 17175 St 35/8	0.17% C 0.4% Mn (max) 0.05% S and P (max) steel: Pipe; Japanese Standard; for high temperature **UTS: 400** **Proof: 240**
DIN 17175 St 45/8	0.22% C (max) 0.05% S and P steel: For seamless tube
DIN 17200 C10	0.1% C 0.45% Mn 0.045% S and P (max) steel: For structures; German Standard
DIN 17200 C15	0.15% C 0.45% Mn 0.045% S and P (max) steel: For structures; German Standard
DIN 17200 C22	0.21% C 0.45% Mn 0.045% S and P (max) steel: For structures; German Standard
DIN 17200 CK 10	0.1% C 0.45% Mn 0.035% S and P (max) steel: For structures; German Standard
DIN 17200 CK 15	0.15% C 0.45% Mn 0.035% S and P (max) steel: For structures; German Standard
DIN 17200 CK 22	0.22% C 0.45% Mn 0.035% S and P (max) steel: For structures; German Standard
DIN 17200 CM 15	0.15% C 0.45% Mn 0.035% S and P (max) steel: For structures; German Standard
DIN 17210 C 10	0.1% C 0.3% Mn 0.045% S and P steel: For carburizing
DIN 17210 C 15	0.15% C 0.4% Mn 0.045% S and P steel: Bar; for carburizing
DIN 17210 CK 10	0.1% C 0.3% Mn 0.035% S and P steel: For carburizing
DIN 17210 CK 15	0.15% C 0.3% Mn 0.035% S and P steel: Bar; for carburizing
DISCUS	Mild steel corrugated sheet: Galvanized; RTB for BS 3083
DOFASCOLOY 45 W	0.25% C 1.2% Mn 0.01% Nb steel: Dominion
DOFASCOLOY 50 W	0.25% C 1.25% Mn 0.01% Nb steel: Dominion
DOFASCOLOY 55 W	0.25% C 1.25% Mn 0.01% Nb steel: Dominion
DOFASCOLOY 60 W	0.25% C 1.25% Mn 0.01% Nb steel: Dominion
DOFASCOLOY 100 W	0.25% C 1.25% Mn 0.01% Nb steel: Dominion

Note. The following abbreviations and units are used in the tables:

DPN	Hardness, diamond pyramid number
UTS	Ultimate tensile strength, N/mm^2
Elon	Elongation, %
Proof	0.1% proof strength, N/mm^2

1 N/mm^2=0.1 hbar=0.102 kgf/mm^2=0.06475 $tonf/in.^2$=145.04 $lbf/in.^2$=1 MPa

See Appendix II for other abbreviations and conversion tables.

Symbol	Nominal analysis, supplier, condition and remarks.
DOFASCOLOY M	0.25% C 1.5% Mn 0.2% Cu steel: Dominion
DRAGONITE	Low carbon electric-galvanized sheet and plate: Steel Co. of Wales for BS alloy En 2
DTD 720A	0.15% C 0.7% Mn 0.04% S 0.04% P steel: Softened and cold drawn; for rivets **UTS: 540** **Elon: 27%**
DTD 5199	C steel: Casting; investment cast **UTS: 500**
DTD 5209	C Mn steel: Casting; investment cast **UTS: 620**
DUCOL QT 122	0.17% C (max) 1.4% Mn 0.05% S and P steel: Colvilles
DUNELT 3	0.1% C 0.06% S and P steel: Dunford; normalized **UTS: 340** **Elon: 30%**
DUNELT 4	0.3% C 1% Mn 0.15% S 0.06% P steel: Free machining; Dunford; hardened and tempered **UTS: 600** **Elon: 15%**
DUNELT 5	0.15% C 1.2% Mn 0.15% S 0.06% P steel: Free machining; Dunford; hardened and tempered **UTS: 560** **Elon: 15%**
DUNELT 7	0.15% C 1.2% Mn 0.3% S 0.04% P steel: Free machining; Dunford; cold drawn **UTS: 600** **Elon: 15%**
DUNELT 14	0.14% C 0.06% S and P steel: For case hardening; Dunford for BS alloy En32
DUNELT 21	0.2% C 1.0% Mn 0.06% S and P steel: Dunford; cold drawn **UTS: 420** **Elon: 17%**
DUNELT 39	0.22% C 1.5% Mn 0.06% S and P steel: Dunford; for BS alloy En 14A
E 7 X T1	0.17% C (max) steel: Weld electrode; designation used by AWS
E 7 X T2	Mild steel: Weld electrode; designation used by AWS
E 7 X T3	Mild steel;Weld electrode; designation used by AWS
E 7 X T4	1.8% Al (max) mild steel: Weld electrode; designation used by AWS
E 7 X T5	0.18% C (max) steel: Weld electrode; designation used by AWS
E 7 X T6	1.8% Al mild steel: Electrode; designation used by AWS
E 7 X T7	1.8% Al mild steel: Electrode; designation used by AWS
E 7 X T8	1.8% Al mild steel: Electrode; designation used by AWS
E 7 X T8 Ni 1	0.12% C (max) 1% Ni 1.8% Al steel: Weld electrode; designation used by AWS
E 7 X T9	1.18% (max) mild steel: Electrode; designation used by AWS
E 7 X T10	Mild steel: Electrode; designation used by AWS
E 7 X T11	1.8% Al mild steel: Electrode; designation used by AWS
E 7 X T12	1.15% C (max) mild steel: Electrode; designation used by AWS
E 7 X T14	Mild steel: Electrode; designation used by AWS
E 900	0.1% C 0.4% Mn steel: VEN
E 920	0.15% C 0.4% Mn steel: VEN
E 7014	0.2% Cr (max) 0.3% Ni (max) 0.3% Mo (max) steel: Weld electrode; designation used by AWS
E 7015	0.2% Cr (max) 0.3% Ni (max) 0.3% Mo (max) steel: Weld electrode; designation used by AWS
E 7016	0.2% Cr (max) 0.3% Ni (max) 0.3% Mo (max) steel: Weld electrode; designation used by AWS
E 7018	0.2% Cr (max) 0.3% Ni (max) 0.3% Mo (max) steel: Weld electrode; designation used by AWS
E 7024	0.2% Cr (max) 0.3% Ni (max) 0.3% Mo (max) steel: Weld electrode; designation used by AWS
E 7027	0.2% Cr (max) 0.3% Ni (max) 0.3% Mo (max) steel: Weld electrode; designation used by AWS
E 7028	0.2% Cr (max) 0.3% Ni (max) 0.3% Mo (max) steel: Weld electrode; designation used by AWS

Symbol	Nominal analysis, supplier, condition and remarks.
E 7048	0.2% Cr (max) 0.3% Ni (max) 0.3% Mo (max) steel: Weld electrode; designation used by AWS
EF 206	0.2% C 0.7% Mn 0.035% S and P (max) steel: German proprietary specification; Al killed
EG 7 X T1	0.2% Cr (max) 0.3% Ni (max) 0.35% Mo (max) steel: Weld electrode; designation used by AWS
EG 7 X T2	1.4% Mn 0.8% Ni steel: Weld electrode; designation used by AWS
EG XX S5	0.13% C 0.9% Al steel: Weld electrode; designation used by AWS
EH 10 K EW	0.1% C steel: Weld electrode; designation used by AWS
EH 14	0.15% C steel: Weld electrode; designation used by AWS
EL 8	0.1% C (max) steel: Weld electrode: designation used by AWS
EL 8K	0.1% C (max) 0.2% Si steel: Weld electrode; designation used by AWS
EL 12	0.12% C steel: Weld electrode: designation used by AWS
EL 12N	0.1% C steel: Weld electrode: designation used by AWS
EL GA 31	0.06% C 0.07% Mn steel: Electrode: Metrode
EL GA MAYETA II	0.06% C 0.085% Mn steel: Electrode: Metrode
EL GA P 43	0.06% C 0.5% Mn steel: Electrode: Metrode
EL GA P 44	0.08% C 0.5% Mn steel: Electrode: Metrode
EL GA P 47	0.07% C 1.4% Mn steel: Electrode: Metrode
EL GA P 51	0.06% C 0.9% Mn steel: Electrode: Metrode
ELECTROID IRON	0.02% C 0.08% Ni 0.2% Cr 0.05% V 0.12% Si 0.03% Al Fe alloy: Carpenter; for magnetic purposes
EM 12 EW	0.1% C steel: Weld electrode; designation used by AWS
EM 12 K EW	0.1% C 0.2% Si steel: Weld electrode; designation used by AWS
EM 12 KN	0.1% C 0.2% Si steel: Weld electrode; designation used by AWS
EM 13 K EW	0.1% C 0.5% Si steel: Weld electrode; designation used by AWS
EM 14 K	0.12% C 0.1% Ti 0.5% Si steel: Weld electrode; designation used by AWS
EM 15 K EW	0.15% C 0.2% Si steel: Weld electrode; designation used by AWS
En 1A	0.1% C 1% Mn 0.25% S 0.07% P steel: Free machining; designation in B 9970; replaced by 220 M07 **UTS: 420 Elon: 14%**
En 1B	0.1% C 1.2% Mn 0.4% S 0.06% P steel: Designation used in BS 970; replaced by 240 M07 **UTS: 370 Elon: 12%**
En 2	0.2% C (max) 0.06% S and P steel: Cold forming; designation used in BS 970; annealed; replaced by 050 A20 **UTS: 300 Elon: 28%**
En 2A	0.12% C (max) 0.05% S and P steel: Cold forming; designation used in BS 970; annealed; replaced by 040 A12 **UTS: 300 Elon: 28%**
En 2A/1	0.1% C (max) 0.04% S and P steel: Cold forming; designation used in BS 970; annealed; replaced by 040 A12 **UTS: 300 Elon: 28%**
En 2B	0.15% C (max) 0.05% S and P steel: Cold forming; designation used in BS 970; annealed; replaced by 040 A12 **UTS: 300 Elon: 28%**
En 2C	0.2% C 0.05% S and P steel: Cold forming; designation used in BS 970; annealed; replaced by 040 A12 **UTS: 300 Elon: 28%**
En 2D	0.22% C 0.05% S and P steel: Cold forming; designation used in BS 970; annealed; replaced by 050 A20 **UTS: 300 Elon: 28%**
En 7	0.2% C 0.15% S 0.06% P steel: Free cutting; designation used in BS 970; cold drawn; obsolete **UTS: 600 Elon: 15%**
En 32A	0.15% C (max) 0.05% S and P steel: For carburizing; designation used in BS 970; obsolete
En 32B	0.15% C 0.07% S 0.05% P steel: For carburizing; designation used in BS 970; obsolete
En 32C	0.14% C 0.05% S and P steel: For carburizing; designation used in BS 970; obsolete
En 32M	0.15% C 0.12% S 0.05% P steel: For carburizing; designation used in BS 970; obsolete
En 201	0.18% C (max) 0.05% S and P steel: For case hardening; designation used in BS 970; obsolete
En 202	0.18% C (max) 0.15% S 0.05% P steel: For case hardening; designation used in BS 970; obsolete
ENELEC SK	Mild steel: Copper coated; welding wire; Philip **UTS: 530 Elon: 27% Proof: 425**
ENROX	0.08% C 0.6% Mn steel: Welding electrode; Philip
ER 70 S4	0.13% C 0.75% Si steel: Weld electrode; designation used by AWS
ER 70 S6	0.13% C 1% Si steel: Weld electrode; designation used by AWS
ER 70 S7	0.13% C 0.65% Si steel: Weld electrode; designation used by AWS
ERA 60	0.18% C 1.2% Mn 0.05% S and P carburizing steel: Hadfields for BS alloy En 201
ERA 60M	0.18% C 1.2% Mn 0.15% S 0.05% P steel: Free machining; carburizing; Hadfields for BS alloy En 202
EX-TEN 42	0.21% C 0.9% Mn 0.02% V 0.01% Nb optional steel: US Steel Corp.
FAGERSTA 1150	0.07% C (max) 0.25% Mn steel: For deep drawing; Fagersta
FAGERSTA 1160	0.08% C (max) 0.25% Mn steel: For deep drawing; Fagersta
FAGERSTA 1265	0.1% C 0.35% Mn steel: For carburizing; Fagersta; weldable
FAGERSTA 1357	0.2% C 0.6% Mn steel: For carburizing; Fagersta; weldable
FASTEX 5	0.06% C 0.6% Mn steel: Welding electrode; Murex **UTS: 540 Elon: 29% Proof: 460**
FASTEX 7	0.08% C 0.06% Mn 0.4% Si steel: Welding electrode; Murex **UTS: 540 Elon: 25% Proof: 480**
FASTEX 100	0.07% C 0.4% Mn steel: Welding electrode; Murex **UTS: 420 Elon: 35% Proof: 330**
FB 50	0.2% C (max) 1.2% Mn 0.04% S and P (max) steel: German proprietary specification; Al killed
FCCH	0.14% C 0.6% Mn 0.25% Si steel: For carburizing; Firth Brown; for BS 970 En32A
Fe 33	C not specified 0.07% S (max) 0.07% P (max) steel: Designation used by Italian Standards
Fe 34A	C not specified 0.065% S and P steel: Designation used by Italian Standards
Fe 34B	0.2% C (max) 0.045% S and P (max) steel: Designation used by Italian Standards
Fe 34C	0.2% C (max) 0.05% S and P (max) steel: Designation used by Italian Standards
Fe 37A	0.25% C (max) 0.075% S (max) 0.065% P (max) steel: Designation used by Italian Standards
Fe 37B	0.25% C (max) 0.055% P (max) 0.055% (max) steel: Designation used by Italian Standards
Fe 37C	0.23% C 0.055% S (max) 0.05% P (max) steel: Designation used by Italian Standards
Fe 37D	0.22% C 0.045% S (max) 0.045% P (max) steel: Designation used by Italian Standards
Fe 42A	0.28% C (max) 0.075% S (max) 0.065% P (max) steel: Designation used by Italian Standards **UTS: 460 Elon: 23% Proof: 245**

Symbol	Nominal analysis, supplier, condition and remarks.
Fe 42B	0.24% C (max) 0.055% S (max) 0.055% P (max) 0.011% N (max) steel: Designation used by Italian Standards **UTS: 460 Elon: 23% Proof: 245**
Fe 42C	0.22% C (max) 0.055% S (max) 0.05% P (max) 0.01% N (max) steel: Designation used by Italian Standards **UTS: 460 Elon: 23% Proof: 245**
Fe 42D	0.22% C (max) 0.045% S and P (max) steel: Designation used by Italian Standards **UTS: 460 Elon: 23% Proof: 245**
Fe 44A	0.28% C (max) 0.065% S and P (max) steel: Designation used by Italian Standards
Fe 44B	0.24% C (max) 0.055% S and P (max) steel: Designation used by Italian Standards
Fe 44C	0.24% C (max) 0.045% S and P (max) steel: Designation used by Italian Standards
Fe 44D	0.22% C (max) 0.045% S and P (max) steel: Designation used by Italian Standards
Fe 50	C not specified 0.06% S 0.045% P steel: Designation used by Italian Standards **UTS: 550 Elon: 18% Proof: 295**
Fe 52B	0.26% C (max) 0.055% S and P steel: Designation used by Italian Standards
Fe 52C	0.24% C (max) 0.04% S and P (max) steel: Designation used by Italian Standards
Fe 52D	0.23% C (max) 0.045% S and P (max) steel: Designation used by Italian Standards
Fe 60	C not specified 0.06% S (max) 0.045% P (max) steel: Designation used by Italian Standards **UTS: 640 Elon: 15% Proof: 335**
Fe 70	C not specified 0.06% S (max) 0.055% P (max) steel: Designation used by Italian Standards **UTS: 750 Elon: 10% Proof: 365**
FEMAX 33.80	0.080% C 0.80% Mn 0.50% Sc steel: For electrodes; Esab **UTS: 556 Elon: 26% Proof: 479**
FERALSIN 52	0.2% C 1.3% Mn 0.03% S (max) 0.04% P (max) steel: Belgian specification; Al killed
FERALSIN 58	0.25% C 1.5% Mn 0.03% S (max) 0.04% P (max) steel: Belgian specification; Al killed
FERROARC R155	0.09% C 1.0% Mn 0.25% Si mild steel: Welding rod; Soudametal; horizontal welding of plate **UTS: 540 Elon: 27% Proof: 460**
FERROARC R160	0.05% C 1.1% Mn 0.3% Si mild steel: Welding rod; Soudametal **UTS: 580 Elon: 28% Proof: 500**
FILLETRODE 43.33	0.060% C 0.7% Mn 0.40% Si steel: For electrodes; Esab **UTS: 556 Elon: 27% Proof: 479**
FLT 2A	0.11% C 1.06% Mn 0.12% Cu steel: Fuji; Al killed
FLT 2B	0.11% C 1.16% Mn 0.01% P 0.12% Cu and S steel: Fuji; Al killed
FNB 36A	0.23% C (max) 1.4% Mn 0.1% Nb 0.035% S and P (max) steel: Fuji
FNB 36B	0.23% C (max) 1.4% Mn 0.1% Nb 0.035% S and P (max) steel: Fuji
FNB 40A	0.26% C (max) 1.4% Mn 0.1% Nb 0.035% S and P (max) steel: Fuji

Note. The following abbreviations and units are used in the tables:

DPN	Hardness, diamond pyramid number
UTS	Ultimate tensile strength, N/mm^2
Elon	Elongation, %
Proof	0.1% proof strength, N/mm^2

1 N/mm^2=0.1 hbar=0.102 kgf/mm^2=0.06475 $tonf/in.^2$=145.04 $lbf/in.^2$=1 MPa

See Appendix II for other abbreviations and conversion tables.

Symbol	Nominal analysis, supplier, condition and remarks.
FNB 40B	0.26% C (max) 1.4% Mn 0.1% Nb 0.035% S and P (max) steel: Fuji
FORTREX 30	0.07% C 0.7% Mn 0.4% Si steel: Welding electrode; Murex **UTS: 500 Elon: 35% Proof: 440**
FORTREX 35	0.08% C 0.7% Mn 0.5% Si steel: Welding electrode; Murex **UTS: 510 Elon: 34% Proof: 460**
FORTREX 35A	0.075% C 0.7% Mn 0.4% Si steel: Welding electrode; Murex **UTS: 480 Elon: 35% Proof: 440**
FTW 52	0.18% C (max) 1.5% Mn 0.03% S and P (max) steel: Fuji
FTW 55	0.18% C (max) 1.5% Mn 0.03% S and P (max) steel: Fuji
FTW 58	0.18% C (max) 1.5% Mn 0.03% S and P (max) steel: Fuji
G 3101/73/5534	C not specified 0.05% S and P (max) steel: For plate; Japanese specification **UTS: 390 Proof: 200**
GALVALUNE SHEET	Mild steel sheet coated with Galvalune
GALVAMATT	0.2% C steel: Baldwins Ltd
GENEX	0.02% C 0.1% Mn steel: Welding electrode; Murex
GLX 45W	0.16% C 0.68% Mn 0.025% Nb steel: National Steel Co.
GLX 50W	0.16% C 0.68% Mn 0.025% Nb steel: National Steel Co.
GLX 55	0.16% C 0.68% Mn 0.025% Nb steel: National Steel Co.
GLX 55	0.22% C (max) 1.35% Mn 0.01% Nb steel: National Steel Co.
GLX 60W	0.16% C 0.68% Mn 0.025% Nb steel: National Steel Co.
GLX 65	0.24% C 1.35% Mn 0.01% Nb steel: National Steel Co.
GLX 70	0.26% C 1.35% Mn 0.01% Nb steel: National Steel Co.
GOST 380 2SP2	0.12% C 0.35% Mn 0.05% S (max) 0.04% P (max) 0.3% Cu + Cr + Ni (max) steel: Plate; Russian National specification
GOST 380 BST 25 P	0.18% C (max) 0.45% Mn 0.05% S (max) 0.04% P (max) steel: Plate; Russian National specification
GOST 380 BST2 PS	0.12% C 0.35% Mn 0.05% S (max) 0.04% P (max) steel: Plate; Russian National specification
GOST 380 BST2 PS2	0.12% C 0.35% Mn 0.05% S (max) 0.04% P (max) 0.3% Cu + Cr + Ni (max) steel: Plate; Russian National specification
GOST 380 BST3 GPS	0.22% C (max) 0.9% Mn 0.05% S (max) 0.04% P (max) steel: Plate; Russian National specification **UTS: 440**
GOST 380 BST3 GPS2	0.22% C (max) 0.9% Mn 0.05% S (max) 0.04% P (max) 0.3% Cr + Ni + Cu (max) steel: Plate; Russian National specification **UTS: 420 Proof: 240**
GOST 380 BST3 KP2	0.18% C 0.45% Mn 0.05% S (max) 0.04% P (max) 0.3% Cu + Cr + Ni (max) steel: Plate; Russian National specification
GOST 380 BST3 PS	0.18% C 0.35% Mn 0.05% S (max) 0.04% P (max) steel: Plate; Russian National specification
GOST 380 BST3 SP	0.18% C 0.6% Mn 0.05% S (max) 0.04% P (max) steel: Plate; Russian National specification
GOST 380 BST3 SP2	0.18% C 0.5% Mn 0.05% S (max) 0.3% Cr + Ni + Cu (max) steel: Plate; Russian National specification
GOST 380 BST4 KP	0.22% C 0.6% Mn 0.05% S (max) 0.4% P (max) steel: Plate; Russian Standard
GOST 380 BST4 PS	0.22% C 0.5% Mn 0.05% S (max) 0.04% P (max) steel: Plate; Russian Standard
GOST 380 BST4 PS2	0.22% C 0.55% Mn 0.3% Cr + Cu (max) 0.05% S (max) 0.04% P (max) steel: Plate; Russian specification

Symbol	Nominal analysis, supplier, condition and remarks.
GOST 380 BST4 SP2	0.22% C 0.5% Mn 0.3% Cr + Ni + Cu (max) 0.05% S (max) 0.04% P (max) steel: Plate; Russian Standard
GOST 380 BST5 GPS	0.26% C 1.0% Mn 0.05% S (max) 0.4% P (max) steel: Plate; Russian Standard
GOST 380 ST0	0.23% C (max) 0.06% S (max) 0.07% P (max) steel: Plate; Russian National specification **UTS: 310**
GOST 380 ST1 KP	0.09% C 0.35% Mn 0.05% S 0.04% P steel: Plate; Russian National specification **UTS: 370**
GOST 380 ST1 KP2	0.09% C 0.35% P (max) 0.3% Cu + Cr + Ni (max) steel: Plate; Russian National specification **UTS: 370**
GOST 380 ST1 PS	0.09% C 0.35% Mn 0.05% S 0.04% P (max) steel: Plate; Russian National specification **UTS: 370**
GOST 380 ST1 PS2	0.09% C 0.35% Mn 0.05% S 0.04% P 0.3% Cu Cr Ni steel: Plate; Russian National specification **UTS: 370**
GOST 380 ST1 SP	0.09% C 0.35% Mn 0.05% S (max) 0.04% P (max) steel: Plate; Russian National specification **UTS: 370**
GOST 380 ST2 KP	0.12% C 0.35% Mn 0.05% S (max) 0.04% P (max) steel: Plate
GOST 380 ST2 KP2	0.12% C 0.35% Mn 0.05% S (max) 0.04% P (max) 0.03% Cu Cr Ni (max) steel: Plate; Russian National specification **UTS: 370**
GOST 380 VST2 KP	0.15% C (max) 0.4% Mn 0.05% S (max) 0.04% P (max) steel: Plate; Russian National specification **UTS: 350**
GOST 380 VST2 PS	0.15% C (max) 0.45% Mn 0.05% S 0.04% P (max) steel: Plate; Russian National specification **UTS: 390**
GOST 380 VST2 PS2	0.15% C (max) 0.45% Mn 0.05% S (max) 0.04% P (max) 0.3% Cu Cr Ni (max) steel: Plate; Russian National specification **UTS: 390** **Proof: 220**
GOST 380 VST2 SP2	0.15% C (max) 0.35% Mn 0.05% S (max) 0.04% P (max) 0.3% Cu Cr Ni (max) steel: Plate; Russian National specification **UTS: 390** **Proof: 220**
GOST 380 VST3 CPS3	0.22% C (max) 0.9% Mn 0.3% Cr Ni Cu (max) 0.05% S (max) 0.04% P (max) steel: Plate; Russian National specification **UTS: 440** **Proof: 240**
GOST 380 VST3 EPS4	0.22% C (max) 0.9% Mn 0.3% Cr Ni Cu (max) 0.05% S (max) 0.04% P (max) steel: Plate; Russian National specification **UTS: 420** **Proof: 240**
GOST 380 VST3 EPS5	0.22% C (max) 0.9% Mn 0.3% Cr Ni Cu (max) 0.05% S (max) 0:04% P (max) steel: Plate; Russian National specification **UTS: 420** **Proof: 240**
GOST 380 VST3 EPS6	0.22% C (max) 0.9% Mn 0.3% Cr Ni Cu (max) 0.05% S (max) 0.04% P (max) steel: Plate; Russian National specification **UTS: 420** **Proof: 240**
GOST 380 VST3 PS	0.22% C (max) 0.6% Mn 0.05% S (max) 0.04% P (max) steel: Plate; Russian National specification **UTS: 430**
GOST 380 VST3 PS2	0.22% C (max) 0.6% Mn 0.03% Cr Ni Cu (max) 0.05% S 0.04% P steel: Plate; Russian National specification **UTS: 430** **Proof: 240**
GOST 380 VST3 PS3	0.22% C (max) 0.6% Mn 0.3% Cr Ni Cu (max) 0.05% S (max) 0.04% P (max) steel: Plate; Russian National specification **UTS: 430** **Proof: 250**
GOST 380 VST3 PS4	0.22% C (max) 0.65% Mn 0.3% Cr Ni Cu (max) 0.05% S 0.04% P steel: Plate; Russian National specification **UTS: 420** **Proof: 240**
GOST 380 VST3 PS5	0.22% C (max) 0.65% Mn 0.3% Cr Ni Cu (max) 0.05% S (max) 0.04% P (max) steel: Plate; Russian National specification **UTS: 420** **Proof: 250**
GOST 380 VST3 PS6	0.22% C (max) 0.6% Mn 0.3% Cr Ni Cu (max) 0.05% S (max) 0.04% P (max) steel: Plate; Russian National specification **UTS: 430** **Proof: 250**
GOST 380 VST3 SP	0.22% C (max) 0.55% Mn 0.05% S (max) 0.04% P (max) steel: Plate; Russian National specification **UTS: 420**
GOST 380 VST3 SP2	0.22% C (max) 0.55% Mn 0.3% Cr Ni Cu (max) 0.05% S (max) 0.04% P (max) steel: Plate; Russian National specification **UTS: 420** **Proof: 240**
GOST 380 VST3 SP3	0.22% C (max) 0.55% Mn 0.3% Cr Ni Cu (max) 0.05% S (max) 0.04% P (max) steel: Plate; Russian National specification
GOST 380 VST3 SP4	0.22% C (max) 0.55% Mn 0.3% Cr Ni Cu (max) 0.05% S (max) 0.04% P (max) steel: Plate; Russian National specification **UTS: 420** **Proof: 240**
GOST 380 VST3 SP5	0.22% C (max) 0.55% Mn 0.05% S (max) 0.04% P (max) 0.3% Cr Ni Cu (max) steel: Plate; Russian National specification **UTS: 420** **Proof: 240**
GOST 380 VST3 SP6	0.22% C (max) 0.55% Mn 0.3% Cr Ni Cu (max) 0.05% S (max) 0.04% P (max) steel: Plate; Russian National specification **UTS: 420** **Proof: 250**
GOST 380 VST4 PS2	0.27% C (max) 0.6% Mn 0.3% Cr Ni Cu (max) 0.05% S (max) 0.04% P (max) steel: Plate; Russian Standard **UTS: 480** **Proof: 260**
GOST 380 VST4 PS3	0.27% C (max) 0.6% Mn 0.3% Cr Ni Cu (max) 0.05% S (max) 0.04% P (max) steel: Plate; Russian Standard **UTS: 480** **Proof: 270**
GOST 380 VST4 SP	0.27% C (max) 0.5% Mn 0.05% S (max) 0.04% P (max) steel: Plate; Russian Standard **UTS: 480** **Proof: 270**
GOST 380 VST4 SP3	0.27% C (max) 0.5% Mn 0.3% Cr Ni Cu (max) 0.05% S (max) 0.04% P (max) steel: Plate; Russian specification **UTS: 480** **Proof: 270**
GOST 380 VST25 P	0.15% C (max) 0.35% Mn 0.05% S (max) 0.04% P (max) steel: Plate; Russian National specification **UTS: 390**
GOST 550 10	0.1% C 0.5% Mn 0.25% Cr + Ni (max) 0.035% S and P (max) steel: Pipe; Russian Standard **UTS: 360** **Proof: 220**
GOST 550 10G2	0.12% C 1.4% Mn 0.025% Cr + Ni (max) 0.035% S and P (max) steel: Pipe; Russian Standard **UTS: 480** **Proof: 270**
GOST 550 20	0.2% C 0.5% Mn 0.25% Cr + Ni (max) 0.04% S and P (max) steel: Pipe; Russian Standard **UTS: 440** **Proof: 260**
GOST 613 E	C not specified 0.045% S and P (max) steel: Pipe; Russian Standard **UTS: 750** **Proof: 550**
GOST 631 D	C not specified 0.045% S and P (max) steel: Pipes; Russian Standard **UTS: 650** **Proof: 380**
GOST 631 K	C not specified 0.045% S and P (max) steel: Pipe; Russian Standard **UTS: 700** **Proof: 500**
GOST 631 L	C not specified 0.045% S and P (max) steel: Pipe; Russian Standard **UTS: 800** **Proof: 650**

Symbol	Nominal analysis, supplier, condition and remarks.
GOST 631 M	C not specified 0.045% S and P (max) steel: Pipe; Russian Standard **UTS: 900** **Proof: 750**
GOST 632 S	C not specified 0.045% S and P (max) steel: Pipe; Russian Standard **UTS: 550** **Proof: 320**
GOST 633 D	C not specified 0.045% S and P steel: Pipe; Russian Standard **UTS: 650** **Proof: 380**
GOST 1050/05 KP	0.06% C (max) 0.4% Mn (max) 0.25% Ni 0.1% Cr 0.035% S and P (max) steel: For structures; Russian Standard
GOST 1050/08	0.08% C 0.5% Mn 0.2% Ni (max) 0.0% Cr (max) 0.035% S and P (max) steel: For structures; Russian Standard
GOST 1050/08 KP	0.08% C 0.4% Mn 0.25% Ni (max) 0.1% Cr (max) 0.04% S and P (max) steel: For structures; Russian Standard
GOST 1050/10	0.1% C 0.5% Mn 0.2% Ni (max) 0.15% Cr (max) 0.035% S and P (max) steel: For structures; Russian Standard
GOST 1050/10 KP	0.1% C 0.4% Mn 0.2% Ni (max) 0.15% Cr (max) 0.04% S and P (max) steel: For structures; Russian Standard
GOST 1050/15	0.15% C 0.5% Mn 0.25% Cr and Ni (max) 0.04% S and P (max) steel: For structures; Russian Standard
GOST 1050/15 G	0.15% C 0.85% Mn 0.25% Cr and Ni (max) 0.04% S and P (max) steel: For structures; Russian Standard
GOST 1050/15 KP	0.15% C 0.4% Mn 0.025% Cr and Ni (max) 0.04% S and P (max) steel: For structures; Russian Standard
GOST 1050/20	0.2% C 0.45% Mn 0.25% Cr and Ni (max) 0.04% S and P (max) steel: For structures; Russian Standard
GOST 1050/20 G	0.2% C 0.85% Mn 0.25% Cr and Ni (max) 0.04% S and P (max) steel: For structures; Russian Standard
GOST 1050/20 KP	0.2% C 0.35% Mn 0.25% Cr and Ni (max) 0.04% S and P (max) steel: For structures; Russian Standard
GOST 1060/10	0.1% C 0.5% Mn 0.2% Ni (max) 0.15% Cr (max) 0.035% S and P (max) steel: Tube; Russian Standard **UTS: 350**
GOST 3262	Steel pipe for water and gas: Russian Standard
GOST 4543/09 C2S	0.12% C (max) 1.5% Mn 0.3% Cr and Ni (max) steel: Russian Standard
GOST 4543/09 G2	0.12% C (max) 1.6% Mn 0.3% Cr and Ni (max) steel: Russian Standard
GOST 4543/10 G2S1	0.12% C (max) 1.5% Mn 0.3% Cr and Ni (max) steel: Russian Standard
GOST 4543/12 GS	0.12% C 1.0% Mn 0.3% Cr and Ni (max) steel: Russian Standard
GOST 4543/14 G	0.14% C 0.85% Mn 0.3% Cr and Ni (max) steel: Russian Standard
GOST 4543/14 G2	0.14% C 1.4% Mn 0.3% Cr and Ni (max) steel: Russian Standard
GOST 4543/15 GF	0.15% C 1.05% Mn 0.3% Cr and Ni (max) steel: Russian Standard
GOST 4543/16 ES	0.16% C 1.05% Mn 0.3% Cr and Ni (max) steel: Russian Standard
GOST 4543/17 GS	0.17% C 1.2% Mn 0.3% Cr and Ni (max) steel: Russian Standard
GOST 4543/18 G2	0.18% C 1.4% Mn 0.3% Cr and Ni (max) steel: Russian Standard
GOST 4543/18 G2S	0.18% C 1.9% Mn 0.3% S and P (max) steel: Russian Standard
GOST 4543/19 G	0.19% C 0.95% Mn 0.3% Cr and Ni (max) steel: Russian Standard
GOST 4543/27 SG	0.27% C 1.2% Mn 0.25% Cr (max) steel: Russian Standard
GOST 5005/15	0.15% C 0.5% Mn 0.25% Cr and Ni (max) 0.04% S and P (max) steel: Pipe; Russian Standard; welded **UTS: 480** **Proof: 380**
GOST 5005/20	0.2% C 0.5% Mn 0.25% Cr and Ni (max) 0.04% S and P (max) steel: Pipe; Russian Standard; welded **UTS: 500** **Proof: 400**
GOST 5058/09 G2	0.12% C (max) 1.6% Mn 0.3% Cr Ni Cu (max) 0.04% S (max) 0.035% P (max) steel: Plate; Russian Standard
GOST 5058/09 G2S	0.12% C 1.5% Mn 0.3% Cr Ni Cu (max) 0.05% S (max) 0.035% P (max) steel: Plate; Russian Standard **UTS: 450** **Proof: 290**
GOST 5058/10 G2S1	0.12% C (max) 1.45% Mn 0.3% Cr Ni Cu (max) 0.05% S (max) 0.035% P (max) steel: Plate; Russian Standard **UTS: 500** **Proof: 350**
GOST 5058/12 GS	0.12% C 1.0% Mn 0.3% Cr Ni Cu (max) 0.04% S (max) 0.035% P (max) steel: Plate; Russian Standard
GOST 5058/14 G	0.15% C 0.8% Mn 0.03% Cr Ni Cu (max) 0.0035% P (max) 0.004% P (max) steel: Plate; Russian Standard **UTS: 400** **Proof: 290**
GOST 5058/14 G2	0.16% C 1.4% Mn 0.3% Cr Ni Cu (max) 0.04% S (max) 0.035% P (max) steel: Plate; Russian Standard
GOST 5058/14 KHGS	0.13% C 1.1% Mn 0.3% Cr Ni Cu (max) 0.04% S (max) 0.035% P (max) steel: Plate; Russian Standard **UTS: 500** **Proof: 350**
GOST 5058/16 GS	0.15% C 1.1% Mn 0.3% Cr Ni Cu (max) 0.05% S (max) 0.035% P (max) steel: Plate; Russian Standard **UTS: 440** **Proof: 270**
GOST 5058/17 GS	0.17% C 1.2% Mn 0.3% Cr Ni Cu (max) 0.04% S (max) 0.035% P (max) steel: Plate; Russian Standard **UTS: 500** **Proof: 340**
GOST 5058/18 G2	0.17% C 1.4% Mn 0.3% Cr Ni Cu (max) 0.04% S 0.035% P steel: Plate; Russian Standard **UTS: 520** **Proof: 360**
GOST 5058/19 G	0.19% C 0.95% Mn 0.3% Cr Ni Cu (max) 0.04% S (max) 0.55% P (max) steel: Plate; Russian Standard **UTS: 480** **Proof: 320**
GOST 5520/12 K	0.12% C 0.55% Mn 0.3% Cr Ni Cu (max) 0.04% S and P steel: Plate; Russian Standard **UTS: 410** **Proof: 220**
GOST 5520/15 K	0.16% C 0.5% Mn 0.3% Cr Ni Cu (max) 0.04% S and P (max) steel: Plate; Russian Standard **UTS: 440** **Proof: 220**
GOST 5520/18 K	0.18% C 0.65% Mn 0.3% Cr Ni Cu (max) 0.04% S and P steel: Plate; Russian Standard **UTS: 500** **Proof: 270**
GOST 5520/69/16 K	0.16% C 0.6% Mn 0.3% Cr Ni Cu (max) 0.3% S and P (max) steel: Plate; Russian Standard **UTS: 460** **Proof: 250**
GOST 5520/69/20 K	0.2% C 0.6% Mn 0.3% Cr Ni Cu (max) 0.04% S and P (max) steel: Plate; Russian Standard **UTS: 470** **Proof: 240**
GOST 5654/10	0.1% C 0.5% Mn 0.2% Ni (max) 0.15% Cr (max) 0.035% S and P (max) steel: Plate; Russian Standard; seamless **UTS: 360** **Proof: 220**
GOST 5654/20	0.12% C 0.5% Mn 0.2% Ni (max) 0.25% Cr (max) 0.04% S and P (max) steel: Plate; Russian Standard; seamless **UTS: 440** **Proof: 260**
GOST 8467 D	C not specified 0.045% S and P steel: Pipe; Russian Standard **UTS: 650** **Proof: 380**

Note. The following abbreviations and units are used in the tables:

DPN	Hardness, diamond pyramid number
UTS	Ultimate tensile strength, N/mm^2
Elon	Elongation, %
Proof	0.1% proof strength, N/mm^2

1 N/mm^2=0.1 hbar=0.102 kgf/mm^2=0.06475 $tonf/in.^2$=145.04 $lbf/in.^2$=1 MPa

See Appendix II for other abbreviations and conversion tables.

Symbol	Nominal analysis, supplier, condition and remarks.
GOST 8696/10/G2SD	0.12% C (max) 1.5% Mn 0.3% Cr and Ni (max) 0.04% S and P (max) steel: Pipe; Russian Standard; welded **UTS: 500** **Proof: 350**
GOST 8696 MST2KP	0.11% C 0.35% Mn 0.005% S (max) 0.045% P (max) steel: Pipe; Russian Standard; welded **UTS: 340** **Proof: 220**
GOST 8696 MST3	0.12% C (max) 0.4% Mn 0.05% S (max) 0.08% P (max) steel: Pipe; Russian Standard; welded **UTS: 380** **Proof: 240**
GOST 8696 MST3	0.18% C 0.5% Mn 0.055% S (max) 0.045% P (max) steel: Pipe; Russian Standard; welded **UTS: 380** **Proof: 240**
GOST 8696 MST3KP	0.18% C 0.45% Mn 0.055% S (max) 0.045% P (max) steel: Pipe; Russian Standard; welded **UTS: 380** **Proof: 240**
GOST 8731/10	0.1% C 0.5% Mn 0.2% Ni (max) 0.15% Cr (max) 0.035% S and P (max) steel: Pipe; Russian Standard; seamless **UTS: 340** **Proof: 210**
GOST 8731/10/G2	0.13% C 1.4% Mn 0.2% Ni (max) 0.25% Cr (max) 0.035% S and P (max) steel: Pipe; Russian Standard; seamless **UTS: 480** **Proof: 270**
GOST 8731/20	0.12% C 0.5% Mn 0.2% Ni (max) 0.25% Cr (max) 0.04% S and P (max) steel: Pipe; Russian Standard; seamless **UTS: 420** **Proof: 250**
GOST 8731 BMST4 SP	0.22% C 0.55% Mn 0.055% S (max) 0.045% P (max) steel: Pipe; Russian Standard; seamless **UTS: 420** **Proof: 250**
GOST 8733/10	0.12% C 0.55% Mn 0.2% Ni (max) 0.15% Cr (max) 0.035% S and P (max) steel: Pipe; Russian Standard; seamless **UTS: 340** **Proof: 210**
GOST 8733/10 G2	0.12% C 1.4% Mn 0.2% Ni (max) 0.15% Cr (max) 0.035% S and P (max) steel: Pipe; Russian Standard; seamless **UTS: 430** **Proof: 250**
GOST 8733/20	0.20% C 0.55% Mn 0.2% Ni (max) 0.15% Cr (max) 0.035% S and P (max) steel: Pipe; Russian Standard; seamless **UTS: 420** **Proof: 250**
GOST 10705/08	0.08% C 0.5% Mn 0.2% Ni (max) 0.1% Cr (max) 0.035% S and P (max) steel: Pipe; Russian Standard; welded **UTS: 340** **Proof: 320**
GOST 10705/10	0.1% C 0.5% Mn 0.2% Ni (max) 0.15% Cr (max) 0.04% S and P (max) steel: Pipe; Russian Standard; welded **UTS: 340**
GOST 10705/15	0.15% C 0.5% Mn 0.2% Ni (max) 0.25% Cr (max) 0.04% S and P (max) steel: Pipe; Russian Standard; welded **UTS: 380**
GOST 10705/20	0.2% C 0.5% Mn 0.2% Ni (max) 0.25% Cr (max) 0.04% S and P (max) steel: Pipe; Russian Standard; welded **UTS: 400**
GOST 10705 BKST3	0.18% C 0.55% Mn 0.055% S (max) 0.045% P (max) steel: Pipe; Russian Standard; welded **UTS: 380**
GOST 10705 BKST4	0.22% C 0.5% Mn 0.055% S (max) 0.045% P (max) steel: Pipe; Russian Standard; welded **UTS: 400**
GOST 10705 BST3	0.12% C (max) 0.4% Mn 0.06% S (max) 0.08% P (max) steel: Pipe; Russian Standard; welded **UTS: 380**
GOST 10705 BST4	0.14% C 0.45% Mn 0.06% S (max) 0.08% P (max) steel: Pipe; Russian Standard; welded **UTS: 440**
GOST 10705 MST2	0.13% C 0.4% Mn 0.055% S (max) 0.045% P (max) steel: Pipe; Russian Standard; welded
GOST 10705 MST3	0.18% C 0.45% Mn 0.055% S (max) 0.45% P (max) steel: Pipe; Russian Standard; welded
GOST 10705 MST4	0.22% C 0.55% Mn 0.055% S (max) 0.045% P (max) steel: Pipe; Russian Standard; welded
GOST 10706/09 G2	0.12% C (max) 1.6% Mn 0.3% Cr Ni and Cu (max) 0.045% S and P (max) steel: Pipe; Russian Standard; welded **UTS: 460** **Proof: 310**
GOST 10706/14 G	0.14% C 0.85% Mn 0.3% Cr Ni and Cu (max) 0.04% S and P (max) steel: Pipe; Russian Standard; welded **UTS: 460** **Proof: 290**
GOST 10706/14 G2	0.16% C 1.4% Mn 0.3% Cr Ni and Cu (max) 0.04% S and P (max) steel: Pipe; Russian Standard **UTS: 480** **Proof: 340**
GOST 10706/15 G2S	0.15% C 1.1% Mn 0.3% Cr Ni and Cu (max) 0.04% S and P (max) steel: Pipe; Russian Standard; welded **UTS: 500** **Proof: 350**
GOST 10706/18 G2S	0.18% C 1.4% Mn 0.3% Cr Ni and Cu (max) 0.04% S and P (max) steel: Pipe; Russian Standard; welded **UTS: 600** **Proof: 400**
GOST 10706/19 G	0.19% C 0.9% Mn 0.3% Cr Ni and Cu (max) 0.04% S and P (max) steel: Pipe; Russian Standard; welded **UTS: 470** **Proof: 300**
GOST 10706/102 SR	0.12% C (max) 1.45% Mn 0.3% Cr and Ni (max) 0.04% S and P (max) 0.22% Cu steel: Pipe; Russian Standard; welded **UTS: 500** **Proof: 350**
GOST 10802/15 GS	0.15% C 1.1% Mn 0.3% Cr and Ni (max) 0.04% S and P (max) steel: Pipe; Russian Standard **UTS: 500** **Proof: 300**
GOST 10802/20	0.13% C 0.5% Mn 0.25% Ni (max) 0.15% Cr (max) 0.04% S and P (max) steel: Pipe; Russian Standard **UTS: 410** **Proof: 250**
GOST 11017/20	0.2% C 0.5% Mn 0.2% Ni (max) 0.25% Cr (max) steel: Plate; Russian Standard; seamless **UTS: 400**
GOST 12132/08	0.08% C 0.5% Mn 0.25% Ni (max) 0.1% Cr (max) 0.035% S and P (max) steel: Pipe; Russian Standard; welded **UTS: 320**
GOST 12132/10	0.1% C 0.5% Mn 0.25% Ni (max) 0.15% Cr (max) 0.035% S and P (max) steel: Pipe; Russian Standard; welded **UTS: 340**
GOST 12132/20	0.2% C 0.5% Mn 0.25% Cr and Ni (max) 0.04% S and P (max) steel: Pipe; Russian Standard; welded **UTS: 420**
GOST 14162/08	0.08% C 0.5% Mn 0.25% Ni (max) 0.1% Cr (max) 0.035% S and P (max) steel: Capillary tube; Russian Standard **UTS: 320**
GOST 14162/10	0.1% C 0.5% Mn 0.25% Ni (max) 0.15% Cr (max) 0.035% S and P (max) steel: Capillary tube; Russian Standard **UTS: 340**
GOST VST4 PS	0.27% C (max) 0.6% Mn 0.05% S (max) 0.04% P (max) steel: Plate; Russian Standard **UTS: 480**
GOST VST4 SP2	0.27% C (max) 0.5% Mn 0.3% Cr Ni Cu (max) 0.05% S (max) 0.04% P (max) steel: Plate; Russian Standard **UTS: 460** **Proof: 260**
H 1	0.16% C (max) 0.4% Mn 0.3% Cr (max) 0.05% S and P steel: Designation used by German Standards
H1 V	0.26% C (max) 0.3% Cr steel: German Standards
HECLA 26	0.15% C steel: Casting; high permeability; Hadfields for BS 1617 A and B **UTS: 420** **Elon: 22%** **Proof: 200**

Symbol	Nominal analysis, supplier, condition and remarks.
HECLA 35	0.2% C 0.5% Mn steel: Hadfields for BS alloys En 2C, En 2D, En 3A, En 3C, En 4
HECLA 46	0.15% C 0.6% Mn 0.5% S and P: Carburizing steel; Hadfields for BS alloy En32A
HECLA CH 31	0.15% C 0.8% Mn 0.06% S and P steel: For carburizing; Hadfields for BS alloy En32B
HEDEX O	0.2% C 1% Mn 0.06% S 0.04% P steel: Wire for forging; Kiveton Park; annealed **UTS: 530**
HERMEES	0.09% C 0.8% Mn steel: Welders electrode; Philip
HIBIL 36	0.2% C (max) 1.5% Mn 0.1% Nb 0.04% S and P (max) steel: Nippon Kokan
HIBIL TYPE A	0.13% C (max) 1.05% Mn 0.05% Nb 0.35% S and P (max) steel: Nippon Kokan
HIBIL TYPE B	0.15% C (max) 1.18% Mn 0.05% Nb steel: Nippon Kokan
HIBIL TYPE C	0.15% C (max) 1.8% Mn 0.05% Nb steel: Nippon Kokan
HICON	0.16% C (max) 1.45% Mn 0.05% Nb 0.04% S and P (max) steel: Nippon Kokan
HICON 36	0.2% C (max) 1.5% Mn 0.1% Nb 0.04% S and P (max) steel: Nippon Kokan
HICON 40	0.2% C (max) 1.5% Mn 0.1% Nb 0.035% S and P (max) steel: Nippon Kokan
HI-F	0.12% C 0.75% Mn 0.015% Nb 0.2% Cu steel: McLouth Steel Corp.
HI-KILLED	0.25% C 1.3% Mn 0.2% Cu steel: Inland Steel Co.
HI-YIELD 45	0.22% C 1.25% Mn 0.025% Nb steel: Granite City Steel Co.
HI-YIELD 50	0.22% C 1.25% Mn 0.02% Nb steel: Granite City Steel Co.
HI-YIELD 55	0.25% C 1.35% Mn 0.02% Nb steel: Granite City Steel Co.
HI-YIELD 60	0.25% C 1.35% Mn 0.02% Nb steel: Granite City Steel Co.
HMS 35	0.1% C 0.4% Mn 0.25% Si 0.03% S and P steel: Krupp
HMS 35/Z	0.12% C 0.5% Mn 0.25% Si 0.05% S and P steel: Krupp
HMS 40	0.15% C 0.4% Mn 0.25% Si 0.03% S and P steel: Krupp
HMS 41/Z	0.15% C 0.6% Mn 0.25% Si 0.05% S and P steel: Krupp
HMS 45	0.22% C 0.45% Mn 0.25% Si 0.035% S and P steel: Krupp
HMS 45/Z	0.17% C 0.55% Mn 0.25% Si 0.05% S and P steel: Krupp
HSA 1	0.17% C 1.3% Mn 0.2% Cu steel: Toto Steel
HSB	0.23% C 1.3% Mn 0.045% S and P (max) 0.2% Cu steel: Yawata
HSB 40	0.18% C (max) 0.65% Mn 0.04% S and P (max) 0.02% Al steel: Phoenix Rheinrohr
HSB 40S	0.18% C (max) 0.65% Mn 0.04% S and P (max) 0.02% Al (max) steel: Phoenix Rheinrohr
HSB 45	0.18% C (max) 0.85% Mn 0.04% S and P (max) 0.02% Al (max) steel: Phoenix Rheinrohr
HSB 45S	0.18% C (max) 0.85% Mn 0.04% S and P (max) 0.02% Al (max) steel: Phoenix Rheinrohr
HSB 50	0.2% C (max) 1.2% Mn 0.04% S and P (max) 0.02% Al (max) steel: Phoenix Rheinrohr
HSB 50S	0.2% C (max) 1.2% Mn 0.04% S and P (max) 0.02% Al steel: Phoenix Rheinrohr
HTP 52W	0.18% C (max) 1.1% Mn 0.3% Cu steel: Kawasaki Iron and Steel Co.
HTP 57W	0.2% C (max) 1.3% Mn 0.1% Cr 0.03% Cu steel: Kawasaki Iron and Steel Co.
HYPLUS	0.2% C (max) 1% Mn 1.6% Cr + Mn (max): Appleby Frodingham; weldable **UTS: 540 Elon: 21% Proof: 330**
HYPLUS 23	0.2% C (max) 1.5% Mn 0.1% Nb 0.5% Si 0.05% S and P steel: Appleby Frodingham; weldable **UTS: 550 Elon: 20% Proof: 350**
HYPLUS 29	0.22% C (max) 1.6% Mn 0.2% V 0.3% Si 0.05% S and P steel: Appleby Frodingham; weldable **UTS: 1000 Elon: 20% Proof: 700**
HYPLUS 111	0.22% C (max) 1.5% Mn 0.05% Nb steel: Appleby Frodingham
HYPRESS 26-32	0.07% C steel: Origin unknown
IH 50	0.22% C 1.5% Mn 0.2% Cu steel: IH Wisconsin Steel
IN 1	0.17% C (max) 0.8% Mn 0.01% S and P steel: Yawata
IN 4	0.08% C (max) 0.68% Mn 0.005% S 0.01% P steel: Yawata
IN 5	0.09% C (max) 1.5% Mn 0.008% S 0.015% P steel: Yawata
IN 6	0.11% C 0.78% Mn 0.006% S 0.015% P steel: Yawata
IN 10	0.09% C (max) 0.8% Mn 0.07% Al steel: Yawata
IN A	0.1% C (max) 1.5% Mn 0.01% S and P steel: Yawata
IN B	0.13% C 1.4% Mn 0.015% S and P steel: Yawata
IRONEX 2	0.06% C 0.5% Mn steel: Welding electrode; Murex **UTS: 460 Elon: 30% Proof: 440**
IRONEX 5	0.06% C 0.5% Mn steel: Welding electrode; Murex **UTS: 480 Elon: 30% Proof: 450**
IRONEX 7	0.065% C 0.7% Mn 0.4% Si steel: Welding electrode; Murex **UTS: 510 Elon: 28% Proof: 480**
J SG4051 S20C	0.2% C 0.45% Mn 0.03% S and P (max) steel: For structures; Japanese Standard
J SG4106 S Mn 21	0.21% C 1.35% Mn 0.03% S and P (max) steel: Japanese Standard
JALTEN 2	0.15% C (max) 1.4% Mn 0.3% Cu steel: Jones and Laughlin
JALTEN 3	0.25% C (max) 1.6% Mn 0.2% Cu steel: Jones and Laughlin
JALTEN 3R	0.25% C (max) 1.6% Mn 0.2% Cu steel: Jones and Laughlin
JALTEN 3S	0.25% C (max) 1.6% Mn 0.2% Cu steel: Jones and Laughlin
JIS B2351 STPS1	0.2% C (max) 0.45% Mn 0.04% S and P steel: Pipe; Japanese Standard **UTS: 450 Proof: 180**
JIS B2351 STPS2	0.12% C 0.45% Mn 0.035% S and P (max) 0.2% Cu (max) steel: Pipe; Japanese Standard **UTS: 450 Proof: 180**
JIS C8305	0.1% C (max) 0.35% Mn 0.04% S and P (max) steel: Conduit tube; Japanese Standard **UTS: 280**
JIS G3101 SS41	C not specified 0.05% S and P (max) steel: Plate; National specification; Japanese **UTS: 455 Proof: 240**
JIS G3101 SS50	C not specified 0.05% S and P (max) steel: Plate; Japanese Standard **UTS: 560 Proof: 280**
JIS G3103 SB35	0.21% C 0.8% Mn (max) 0.04% S (max) 0.035% P (max) steel: Plate; Japanese Standard **UTS: 380 Proof: 190**
JIS G3103 SB42	0.25% C 0.8% Mn (max) 0.04% S (max) 0.035% P (max) steel: Plate; Japanese Standard **UTS: 460 Proof: 230**
JIS G3103 SB46	0.3% C (max) 0.9% Mn (max) 0.04% S (max) 0.035% P (max) steel: Plate; Japanese Standard **UTS: 500 Proof: 250**

Note. The following abbreviations and units are used in the tables:

DPN	Hardness, diamond pyramid number
UTS	Ultimate tensile strength, N/mm^2
Elon	Elongation, %
Proof	0.1% proof strength, N/mm^2

1 N/mm^2=0.1 hbar=0.102 kgf/mm^2=0.06475 $tonf/in.^2$=145.04 $lbf/in.^2$=1 MPa

See Appendix II for other abbreviations and conversion tables.

Symbol	Nominal analysis, supplier, condition and remarks.
JIS G3103 SG30	0.2% C (max) 1.0% Mn (max) 0.04% S and P steel: Plate; Japanese Standard **UTS: 450**
JIS G3103 SP49	0.35% C (max) 0.9% Mn (max) 0.04% S (max) 0.035% P (max) steel: Plate; Japanese Standard **UTS: 540** **Proof: 270**
JIS G3106 B	0.19% C (max) 1.5% Mn (max) 0.04% S and P steel: Plate; Japanese Standard **UTS: 560** **Proof: 320**
JIS G3106 SM41A	0.24% C 2.5% Mn 0.04% S and P (max) steel: Plate; Japanese National Standard **UTS: 460** **Proof: 250**
JIS G3106 SM41B	0.21% C 0.9% Mn 0.04% S and P (max) steel: Plate; Japanese National Standard **UTS: 460** **Proof: 240**
JIS G3106 SM41C	0.18% C (max) 1.4% Mn (max) 0.04% S and P (max) steel: Plate **UTS: 460** **Proof: 220**
JIS G3106 SM50C	0.18% C (max) 1.5% Mn (max) 0.04% S and P (max) steel: Plate; Japanese Standard **UTS: 560** **Proof: 320**
JIS G3106 SM50Y A and B	8.2% C (max) 1.5% Mn (max) 0.04% S and P (max) steel: Plate; Japanese Standard **UTS: 560** **Proof: 360**
JIS G3106 SM53 B and C	0.2% C (max) 1.5% Mn (max) 0.04% S and P (max) steel: Plate; Japanese Standard **UTS: 490** **Proof: 360**
JIS G3106 SM58	0.18% C (max) 1.5% Mn (max) 0.04% S and P steel: Plate; Japanese Standard; hardened; tempered **UTS: 650** **Proof: 460**
JIS G3114 SMA41 A, B and C	0.2% C (max) 1.4% Mn (max) 0.25% Cr 0.04% S and P 0.3% Cu steel: Plate; Japanese Standard **UTS: 460** **Proof: 240**
JIS G3115 SPV24	0.19% C 1.4% Mn (max) 0.04% S (max) 0.035% P (max) steel: Plate; Japanese Standard **UTS: 450** **Proof: 230**
JIS G3115 SPV32	0.18% C (max) 1.5% Mn (max) 0.04% S (max) 0.035% P (max) steel: Plate; Japanese Standard **UTS: 560** **Proof: 310**
JIS G3115 SPV36	0.2% C (max) 1.6% Mn 0.04% S (max) 0.035% P steel: Plate; Japanese Standard **UTS: 590** **Proof: 340**
JIS G3115 SPV46	0.18% C (max) 1.6% Mn (max) 0.04% S (max) 0.035% P (max) steel: Plate; Japanese Standard; hardened and tempered **UTS: 630** **Proof: 440**
JIS G3115 SPV50	0.18% C (max) 0.04% S (max) 0.025% P (max) steel: Plate; Japanese Standard; hardened and tempered **UTS: 670** **Proof: 420**
JIS G3116 SG26	0.2% C (max) 0.3% Mn (max) 0.04% S and P (max) steel: Plate; Japanese Standard **UTS: 410**
JIS G3116 SG30	0.2% C (max) 1.0% Mn (max) 0.04% S and P (max) steel: Plate **UTS: 450** **Proof: 300**
JIS G3116 SG33	0.2% C (max) 1.5% Mn (max) 0.04% S and P (max) steel: Plate; Japanese Standard **UTS: 500** **Proof: 270**
JIS G3116 SG37	0.2% C (max) 1.5% Mn (max) 0.04% S and P (max) steel: Plate; Japanese Standard **UTS: 550** **Proof: 370**
JIS G3118 SGV42	0.25% C 1.0% Mn 0.04% S (max) 0.035% P (max) steel: Plate; Japanese Standard **UTS: 460** **Proof: 230**
JIS G3118 SGV46	0.29% C (max) 1.0% Mn 0.04% S (max) 0.035% P (max) steel: Plate; Japanese Standard **UTS: 500** **Proof: 250**
JIS G3118 SGV49	0.31% C (max) 1.0% Mn 0.04% S (max) 0.035% P (max) steel: Plate; Japanese Standard **UTS: 540** **Proof: 270**
JIS G3126 SLA24A	0.15% C (max) 1.1% Mn 0.035% S and P (max) steel: Plate; Japanese Standard; normalized **UTS: 460** **Proof: 220**
JIS G3126 SLA24B	0.15% C (max) 1.1% Mn 0.035% S and P (max) steel: Plate; Japanese Standard; normalized **UTS: 450** **Proof: 220**
JIS G3126 SLA33A	0.16% C (max) 1.2% Mn 0.035% S and P (max) steel: Plate; Japanese Standard; normalized **UTS: 510** **Proof: 330**
JIS G3126 SLA33B	0.16% C (max) 1.2% Mn 0.035% S and P (max) steel: Plate; Japanese Standard; hardened and tempered **UTS: 510** **Proof: 330**
JIS G3126 SLA37	0.18% C (max) 1.2% Mn 0.035% S and P (max) steel: Plate; Japanese Standard; hardened and tempered **UTS: 560** **Proof: 370**
JIS G3444 STK30	C not specified 0.05% S and P steel: Tube; Japanese Standard **UTS: 300**
JIS G3444 STK41	0.25% C 0.04% S and P steel: Pipe; Japanese Standard **UTS: 410** **Proof: 240**
JIS G3444 STK50	0.18% C (max) 1.5% Mn (max) 0.04% S and P (max) steel: Pipe; Japanese Standard **UTS: 500** **Proof: 400**
JIS G3444 STK51	0.3% C (max) 0.7% Mn 0.04% S and P steel: Pipe; Japanese Standard **UTS: 510** **Proof: 360**
JIS G3444 STK55	0.23% C (max) 1.5% Mn (max) 0.04% S and P steel: Pipe; Japanese Standard **UTS: 550** **Proof: 400**
JIS G3445 STKM11A	0.12% C (max) 0.4% Mn 0.04% S and P (max) steel: Pipe; Japanese Standard **UTS: 300**
JIS G3445 STKM12A	0.2% C (max) 0.4% Mn 0.04% S and P (max) steel: Pipe; Japanese Standard **UTS: 400** **Proof: 280**
JIS G3445 STKM12B	0.2% C (max) 0.5% Mn 0.04% S and P (max) steel: Pipe; Japanese Standard **UTS: 400** **Proof: 280**
JIS G3445 STKM12C	0.2% C (max) 0.45% Mn 0.04% S and P steel: Pipe; Japanese Standard **UTS: 480** **Proof: 360**
JIS G3445 STKM13A	0.25% C (max) 0.6% Mn 0.04% S and P steel: Pipe; Japanese Standard **UTS: 380** **Proof: 220**
JIS G3445 STKM13B	0.25% C (max) 0.6% Mn 0.04% S and P (max) steel: Pipe; Japanese Standard **UTS: 450** **Proof: 310**
JIS G3445 STKM13C	0.25% C (max) 0.6% Mn 0.04% S and P (max) steel: Pipe; Japanese Standard **UTS: 520** **Proof: 390**
JIS G3445 STKM18A	0.18% C (max) 1.5% Mn (max) 0.04% S and P (max) steel: Pipe; Japanese Standard **UTS: 450** **Proof: 280**
JIS G3445 STKM18B	0.18% C (max) 1.5% Mn (max) 0.04% S and P (max) steel: Pipe; Japanese Standard **UTS: 500** **Proof: 320**
JIS G3445 STKM18C	0.18% C (max) 1.5% Mn (max) 0.04% S and P (max) steel: Pipe; Japanese Standard **UTS: 520** **Proof: 390**
JIS G3452 SGP	C not quoted 0.05% S and P (max) steel: Tube; Japanese Standard **UTS: 300**
JIS G3454 STPG38	0.25% C (max) 0.5% Mn 0.04% S and P (max) steel: Pipe; Japanese Standard **UTS: 380** **Proof: 220**
JIS G3455 STS35	0.11% C 0.45% Mn 0.035% S and P (max) 0.2% Cu (max) steel: Pipe; Japanese Standard **UTS: 350** **Proof: 180**
JIS G3455 STS38	0.25% C (max) 0.5% Mn steel: Pipe; Japanese Standard **UTS: 380** **Proof: 220**

Symbol	Nominal analysis, supplier, condition and remarks.
JIS G3456 STPT38	0.25% C (max) 0.6% Mn 0.035% S and P (max) 0.2% Cu (max) steel: Pipe; Japanese Standard **UTS: 380** **Proof: 220**
JIS G3457 STPY41	C not quoted 0.05% S and P (max) steel: Pipe; Japanese Standard **UTS: 410** **Proof: 230**
JIS G3460 STPL39	0.25% C (max) 1.35% Mn (max) 0.035% S and P 0.02% Cu (max) steel: Pipe; Japanese Standard **UTS: 390** **Proof: 210**
JIS G3461 STB30	0.2% C (max) 0.45% Mn 0.04% S and P steel: Pipe; Japanese Standard **UTS: 300**
JIS G3461 STB33	0.12% C 0.45% Mn 0.035% S and P 0.2% Cu (max) steel: Pipe; Japanese Standard **UTS: 330** **Proof: 180**
JIS G3461 STB35	0.12% C 0.45% Mn 0.035% S and P 0.2% Cu steel: Pipe; Japanese Standard **UTS: 330** **Proof: 180**
JIS G3461 STB42	0.32% C (max) 0.55% Mn 0.035% S and P (max) 0.2% Cu (max) steel: Pipe; Japanese Standard **UTS: 420** **Proof: 260**
JIS G3464 STBL39	0.25% C (max) 1.35% Mn (max) 0.035% S and P (max) 0.2% Cu (max) steel: Pipe; Japanese Standard **UTS: 390** **Proof: 210**
JIS G3465 STMC55	C not specified 0.04% S and P (max) steel: Tube for drilling; Japanese Standard **UTS: 550**
JIS G3465 STMC65	C not specified 0.04% S and P (max) steel: Tube for drilling; Japanese Standard **UTS: 650**
JIS G3465 STMR60	C not specified 0.04% S and P (max) steel: Tube for drilling; Japanese Standard **UTS: 600** **Proof: 380**
JIS G3465 STMR70	C not specified 0.04% S and P (max) steel: Tube for drilling; Japanese Standard **UTS: 700** **Proof: 450**
JIS G3465 STMR80	C not specified 0.04% S and P (max) steel: Tube for drilling; Japanese Standard **UTS: 800** **Proof: 530**
JIS G3466 STKR41	0.25% C (max) 0.04% S and P (max) steel: Pipe; Japanese Standard **UTS: 410** **Proof: 250**
JIS G3466 STKR50	0.18% C (max) 1.5% Mn (max) 0.04% S and P (max) steel: Pipe; Japanese Standard **UTS: 500** **Proof: 330**
JIS G4051 S9CK	0.1% C 0.45% Mn 0.025% S and P (max) steel: For structures; Japanese Standard
JIS G4051 S10C	0.1% C 0.45% Mn 0.03% S and P (max) steel: For structures; Japanese Standard
JIS G4051 S12C	0.12% C 0.45% Mn 0.03% S and P steel: For structures; Japanese Standard
JIS G4051 S15C	0.15% C 0.45% Mn 0.03% S and P (max) steel: For structures; Japanese Standard
JIS G4051 S17C	0.17% C 0.45% Mn 0.03% S and P (max) steel: For structures; Japanese Standard
JLX W 45	0.2% C (max) 0.75% Mn 0.01% Nb steel: Jones and Loughlin
JLX W 50	0.21% C (max) 0.75% Mn 0.01% Nb steel: Jones and Loughlin
JLX W 55	Jones and Loughlin
JLX W 60	0.2% C (max) 0.75% Mn 0.01% Nb steel: Jones and Loughlin
JOUVENCEL	Low carbon decarburized steel sheet; non-ageing; Esperance Longdoz **UTS: 310** **Elon: 41%** **Proof: 200**
JS 2	0.22% C 1.5% Mn 0.5% Si steel: Casting; Jopling **DPN: 180** **UTS: 620** **Elon: 18%**
K 02006	0.2% C (max) steel: Weld electrode; UNS designation
K 02802	0.28% C (max) 0.2% Si 0.04% S and P steel: UNS designation
K 03016	0.03% C (max) 0.75% Mn steel: Casting; designation used by UNS
KAISERALOY 45CB	0.2% C (max) 1.0% Mn 0.01% Nb steel: Kaiser Steel Corp.
KAISERALOY 50MM	0.27% C (max) 1.35% Mn 0.2% Cu steel: Kaiser Steel Corp.
KAISERALOY 55CB	0.24% C (max) 1.4% Mn 0.01% Nb steel: Kaiser Steel Corp.
KAISERALOY 60CB	0.26% C (max) 1.6% Mn 0.01% Nb steel: Kaiser Steel Corp.
KEA 581	Low C free cutting steel: Kayser Ellison
KF	0.18% C (max) 1.3% Mn 0.04% S and P (max): Alloy content controlled; DNV
KF 1	0.12% C 0.5% Mn 0.05% S and P case hardening steel: Kirkstall for BS alloy En32A
KF 1A	0.16% C 1.4% Mn 0.05% S and P case hardening steel: Kirkstall for BS alloy En201
KF 3	0.15% C 0.8% Mn 0.05% S and P case hardening steel: Kirkstall for BS alloy En32B
KF 4	0.2% C 0.8% Mn 0.06% S and P steel: Kirkstall for BS alloy En3
KF 23	0.1% C 0.8% Mn 0.1% Si 0.25% S 0.07% P steel: Free machining; Kirkstall for BS alloy En1A
KF 39	0.1% C 1% Mn 0.2% S 0.06% P case hardening steel: Free machining; Kirkstall; case hardened and tempered **UTS: 530** **Elon: 22%** **Proof: 370**
KF 39C	0.15% C 1% Mn 0.12% S 0.05% P case hardening steel: Free machining; Kirkstall for BS alloy En32M
KF 39E	0.15% C 1.4% Mn 0.23% S 0.05% P case hardening steel: Free machining; Kirkstall; case hardened and tempered **UTS: 580** **Elon: 20%**
KF 39E/1	0.14% C 1.4% Mn 0.14% S 0.05% P case hardening steel: Free machining; Kirkstall for BS alloy En202
KF 60	0.1% C 1.2% Mn 0.1% Si 0.5% S 0.06% P steel: Free machining; Kirkstall for BS alloy En1B
KHB 36	0.22% C (max) 1.55% Mn 0.05% S and P (max) steel: German proprietor's specification
KIRKASE	0.1% C 1% Mn 0.2% S 0.06% P case hardening steel: Free machining; Kirkstall for KF 39
KORA 2	0.16% C 0.8% Mn 0.12% Si 0.05% S and P steel: Sanderson for BS alloy En32B
KP 2	0.1% C 1% Mn 0.25% S 0.07% P free cutting steel: Kiveton Park for BS alloy En1
KP 3	0.15% C 0.8% Mn steel: For carburizing; Kiveton Park for BS alloy En32
KP 19	0.2% C 1% Mn 0.15% S 0.06% P steel: For case hardening; Kiveton Park for BS alloy En7
KUTTWELL	0.14% C 0.9% Mn 0.04% S and P steel: For case hardening; Swift Levick for BS alloy En32
L 1	0.22% C 1% Mn 0.6% Si steel: Casting; Jessop
L St 45/8	0.22% C (max) 0.05% S and P steel: Designation used by German Standards
LANTERN	0.15% C 0.8% Mn 0.05% S and P steel: For case hardening; Toledo for BS alloys En32A and En32B
LOWHEES	0.08% C 1.3% Mn steel: Welding electrode; Philips
LT 75N	0.2% C (max) 1.0% Mn steel: Lukens
LT 75QT	0.2% C (max) 1.0% Mn steel: Lukens
LUKENS 45	0.2% C 1.2% Mn 0.05% S and P (max) steel: Lukens

Note. The following abbreviations and units are used in the tables:

DPN	Hardness, diamond pyramid number
UTS	Ultimate tensile strength, N/mm^2
Elon	Elongation, %
Proof	0.1% proof strength, N/mm^2

1 N/mm^2=0.1 hbar=0.102 kgf/mm^2=0.06475 tonf/in.^{2}=145.04 lbf/in.^{2}=1 MPa

See Appendix II for other abbreviations and conversion tables.

Symbol	Nominal analysis, supplier, condition and remarks.
LUKENS 50	0.2% C (max) 1.35% Mn 0.05% S and P (max) steel: Lukens
LUKENS 55	0.22% C (max) 1.35% Mn 0.05% S and P (max) steel: Lukens
LUKENS 60	0.22% C (max) 1.6% Mn 0.05% S and P (max) steel: Lukens
LUKENS 80	0.2% C (max) 1.4% Mn 0.05% S and P (max) 0.3% Cu steel: Lukens
LUKENS 80	0.21% C 1.5% Mn 0.035% S and P 0.3% Cu steel: American proprietary steel; consult SAE for further information **Proof: 580**
LUKENS 440	0.28% C (max) 1.4% Mn 0.05% S and P (max) 0.2% Cu steel: Lukens
M 4	0.21% C 1.5% Mn 0.25% Si steel: Jessop
M 113	C not specified 0.04% S and P steel: Plate; Armco as A113 **UTS: 450 Elon: 27% Proof: 180**
M St 0	0.12% C (max) 0.3% Mn 0.06% S and P steel: German Standard
M St 34/2	0.17% C (max) 0.05% S and P steel: German Standard
M St 34/3	0.17% C 0.05% S steel: German Standard
M St 37/2	0.2% C (max) 0.05% S and P steel: German Standard
M St 37/3	0.2% C (max) 0.05% S and P steel: German Standard
M St 42/2	0.25% C (max) 0.05% S and P steel: German Standard
M St 42/3	0.25% C (max) 0.05% S and P steel: German Standard
M St 52/3	0.2% C (max) 0.05% S and P steel: German Standard
M1-F	0.12% C (max) 0.75% Mn 0.015% Nb 0.22% Cu steel: McLouth
MANGEAR	0.12% C 1.6% Mn 0.2% Si 0.04% S 0.04% P: Steel Peach & Tozer; also known as Phoenix Mangear **DPN: 160 UTS: 500 Elon: 37% Proof: 480**
MANITOBA	0.18% C steel: Casting; free machining; Bonds Foundry
MAN-TEN	0.25% C (max) 1.4% Mn 0.2% Cu steel: US Steel Corp.
MAN-TEN A440	0.28% C (max) 1.4% Mn 0.2% Cu steel: US Steel Corp.
MARINER	0.28% C 1.1% Mn 0.2% Cu steel: US Steel Corp.
MAX FINE GRAIN	0.2% C (max) 1.0% Mn 0.2% Cu 0.1% Zr steel: National Steel
MAX HIGH MARG	0.22% C 1.25% Mn 0.22% Cu steel: National Steel
MAXELOY	0.15% C (max) 1.2% Mn 0.2% Cu steel: Crucible Steel Co.
MBK 8	0.07% C 0.4% Mn 0.1% Si 0.03% S and P steel: German Standard
MBK 8A1	0.07% C 0.4% Mn 0.1% Si 0.03% S and PAl steel: German Standard
MBK 10	0.1% C 0.6% Mn 0.05% Si 0.025% S and P steel: German Standard
MBK 13	0.13% C 0.6% Mn 0.05% Si 0.025% S and P steel: German Standard
MED Mn A440	0.28% C (max) 1.5% Mn 0.3% Cu steel: Bethlehem Steel Co.
METCO 91	Mild steel: Powder coating; Metco **DPN: 280**
MIL S 645	0.15% C steel: Weld electrode; US Federal specifications
MIL S 780.9	0.2% C (max) 1.25% Mn steel: Bethlehem Steel Co.
MIL S 12505.3	0.14% C (max) 0.95% Mn with Ni, Mo and Cu steel: Bethlehem Steel Co.
MIL S 12505.6	0.15% C (max) 1.0% Mn 0.3% Cu steel: Bethlehem Steel Co.
MILDEES	0.07% C 0.5% Mn steel: Welding electrode; Philips
MILDTRODE 46.00	0.080% C 0.45% Mn 0.20% Si steel: For electrodes; Esab **UTS: 494 Elon: 25% Proof: 401**
MNC 810	0.19% C 0.17% Mn 0.25% Cr 0.35% Cu: Structural steels; Swedish Standard **UTS: 400 Elon: 24% Proof: 230**
MNC 811	Summary of structural steel tubing; Swedish Standard

Symbol	Nominal analysis, supplier, condition and remarks.
MNC 815	List of reinforcing steels: Swedish Standard; individual specification listed
MNC 815E	Specification covering steels for reinforcing: Swedish Standard
MNC 831	Summary of pressure vessel steel tubing; Swedish Standard
MNC 832	Summary of pressure vessel steel for gas cylinders; Swedish Standard
MNC 840	Summary of steels for screws, nuts and rivets; Swedish Standard
MNC 845	Summary of free cutting steels; Swedish Standard
MNC 851	Summary of case hardening steels; Swedish Standard
MNC 915	General specification for cold rolled sheet
MNC 916	General specification for structural steels: Swedish Standard
MNC 920	Summary of cold rolled steel strip; Swedish Standard
MONARC	0.09% C 0.45% Mn steel: Welders electrode; Philips
MONEMEL 1	Low carbon decarburized steel: Sheet; non-ageing; Esperance Longdoz; for direct one coat enamelling
MONEMEL 2	Low carbon decarburized steel: Sheet; non-ageing; Esperance Longdoz; for direct one coat enamelling and extra deep drawing
MR St 1	0.12% C (max) 0.3% Mn 0.15% Si 0.06% S and P steel: German Standard
MR St 2	0.1% C (max) 0.3% Mn 0.1% Si 0.05% S and P steel: German Standard
MR St 3	0.1% C (max) 0.3% Mn 0.1% Si 0.04% S and P steel: German Standard
MR St 3	0.1% C 0.3% Mn 0.05% S and P steel: Designation used by German Standards
MR St 4	0.1% C (max) 0.35% Mn 0.08% Si 0.035% S and P steel: German Standard
MR St 34/2	0.17% C (max) 0.05% S and P steel: German Standard
MR St 37/2	0.2% C 0.05% S and P steel: German Standard
MR St 42/2	0.25% C (max) 0.05% S and P steel: German Standard
MR St 44	0.15% C 0.6% Mn 0.06% S and P steel: Designation used by German Standards
MSV	0.1% C 0.5% Mn steel: Pompey
MSVB/W	0.2% C 1.1% Mn 0.5% Si 0.035% S and P steel: Krupp; up to 0.3% Cr may be present
MSVD	0.2% C 0.5% Mn steel: Pompey
MU St 1	0.12% C (max) 0.3% Mn 0.06% S and P steel: German Standard
MU St 2	0.1% C 0.3% Mn 0.05% S and P steel: German Standard
MU St 3	0.1% C (max) 0.3% Mn 0.04% S and P steel: German Standard
MU St 4	0.1% C (max) 0.3% Mn 0.03% S and P steel: German Standard
MU St 34	0.1% C (max) 0.2% Mn 0.05% S and P steel: Designation used by German Standards
MU St 34/2	0.17% C (max) 0.05% S and P steel: German Standard
MU St 37/2	0.2% C (max) 0.05% S and P steel: German Standard
MU St 42/2	0.25% C (max) 0.05% S and P steel: German Standard
MUK 8	0.07% C 0.4% Mn 0.03% S and P steel: German Standard
MUK 10	0.1% C 0.6% Mn 0.03% S and P steel: German Standard
MURAWIRE W1	0.06% C 1.5% Mn 0.6% Si 0.06% P steel: Welding electrode; Murex; copper coated **UTS: 480 Elon: 34% Proof: 390**
MURAWIRE W2	0.06% C 0.7% Mn 0.4% Si steel: Welding electrode; Murex **UTS: 440 Elon: 37% Proof: 330**
N54	0.17% C 1.5% Mn steel **DPN: 157 UTS: 550 Elon: 21%**
NF A35 501/A33	C not specified steel: Plate; French national specification **UTS: 370 Proof: 180**

Symbol	Nominal analysis, supplier, condition and remarks.	UTS	Elon	Proof
NF A35 501/A33/2	C not specified 0.05% S and P (max) steel: Plate; French national specification	370		170
NF A35 501/A34/1	C not specified 0.05% S (max) 0.06% P (max) steel: Plate; French national specification	370		160
NF A35 501/A34/2	C not specified 0.05% S and P (max) 0.007% N (max) steel: Plate; French national specification	380		150
NF A35 501/A37/1	0.18% C (max) 0.05% S (max) 0.06% P (max) steel: Plate; French national specification	410		240
NF A35 501/A37/2	0.18% C (max) 0.05% S and P 0.007% N (max) steel: Plate; French national specification			
NF A35 501/A37/3	0.16% C (max) 0.045% S and P (max) steel: Plate; French national specification	410		210
NF A35 501/A37/4	0.16% C (max) 0.04% S and P 0.02% Al (max) steel: Plate; French national specification	410		210
NF A35 501/A42/1	0.2% C (max) 0.05% S (max) 0.06% P (max) steel: Plate; French Standard	460		250
NF A35 501/A42/2	0.2% C (max) 0.05% Mn (max) 0.05% S and P (max) 0.007% N (max) steel: Plate; French Standard	460		240
NF A35 501/A42/3	0.18% C (max) 0.045% S and P steel: Plate; French Standard	460		230
NF A35 501/A42/4	0.18% C (max) 0.04% S and P (max) 0.02% Al steel: Plate; French Standard	460		230
NF A35 501/A47/2	0.24% C (max) 1.3% Mn (max) 0.05% S and P (max) steel: Plate; French Standard	520		250
NF A35 501/A47/3	0.2% C (max) 1.3% Mn (max) 0.04% S and P (max) steel: Plate; French Standard	520		270
NF A35 501/A47/4	0.2% C (max) 1.3% Mn (max) 0.04% S and P (max) 0.02% Al (max) steel: Plate; French Standard	520		270
NF A35 501/A52/2	0.24% C (max) 1.5% Mn (max) 0.05% S and P (max) steel: Plate; French Standard	570		330
NF A35 501/A52/3	0.2% C (max) 1.5% Mn (max) 0.045% S and P (max) steel: Plate; French specification	570		340
NF A35 501/A52/4	0.2% C (max) 1.5% Mn (max) 0.04% S and P (max) steel: Plate; French specification	570		340
NF A35 501/A60/1	C not specified 0.05% S (max) 0.07% P (max) steel: Plate; French Standard	660		340
NF A35 501/A60/2	C not specified 0.045% S and P (max) steel: Plate; French Standard	660		320
NF A35 501/E24/1	0.18% C (max) 0.05% S (max) 0.06% P (max) steel: Plate; French national specification	410		240
NF A35 501/E24/2	0.18% C (max) 0.05% S and P (max) 0.007% N (max) steel: Plate; French national specification	410		220
NF A35 501/E24/3	0.16% C (max) 0.045% S and P (max) steel: Plate; French national specification	410		210
NF A35 501/E24/4	0.16% C (max) 0.04% S and P (max) 0.02% Al (max) steel: Plate; French national specification	410		220
NF A35 501/E26/1	0.2% C (max) 0.05% S and P (max) 0.06% P (max) steel: Plate; French Standard	460		250
NF A35 501/E26/2	0.2% C (max) 0.05% S and P (max) 0.007% N (max) steel: Plate; French Standard	460		240
NF A35 501/E26/3	0.18% C (max) 0.045% S and P (max) steel: Plate; French Standard	460		230
NF A35 501/E26/4	0.18% C (max) 0.04% S and P (max) 0.02% Al steel: Plate; French Standard	460		240
NF A35 501/E30/2	0.24% C (max) 1.3% Mn (max) 0.05% S and P (max) steel: Plate; French Standard	520		250
NF A35 501/E30/3	0.2% C (max) 1.3% Mn (max) 0.045% S and P (max) steel: Plate; French Standard	520		270
NF A35 501/E30/4	0.2% C (max) 1.30% Mn (max) 0.04% S and P (max) 0.02% Al (max) steel: Plate; French Standard	520		270
NF A35 501/E36/2	0.24% C (max) 1.5% Mn (max) 0.05% S and P (max) steel: Plate; French specification	570		330
NF A35 501/E36/3	0.2% C (max) 1.5% Mn (max) 0.045% S and P (max) steel: Plate; French specification	570		340
NF A35 501/E36/4	0.2% C (max) 1.5% Mn (max) 0.04% S and P (max) steel: Plate; French specification	570		340
NF A35 501/E37/2	0.18% C (max) 0.05% S and P (max) 0.007% N (max) steel: Plate; French specification	410		220
NF A35 551/20M5	0.19% C 1.35% Mn steel: Plate; French Standard			
NF A35 551/X12	0.13% C 0.45% Mn steel: For structures; French Standard			
NF A35 551/XC10	0.09% C 0.045% Mn steel: For structures; French Standard			
NF A35 551/XC18	0.18% C 0.5% Mn steel: For structures; French Standard			
NF A35 561/12MF4	0.12% C (max) 1% Mn 0.2% S steel: Free machining; hardened and tempered; French Standard	880	11%	570
NF A35 561/S250	0.14% C (max) 1.2% Mn 0.28% S steel: Free machining; French Standard	400	23%	
NF A35 561/S250 Pb	0.14% C (max) 1.2% Mn 0.28% S 0.23% Pb steel: Free machining; French Standard	400	23%	
NF A35 561/S300	0.15% C (max) 0.3% Mn 0.35% S steel: Free machining; French Standard	420	21%	
NF A35 561/S300 Pb	0.15% C (max) 1.3% Mn 0.35% S 0.25% Pb steel: Free machining; French Standard	420	21%	
NF A35 562/10F2	0.11% C (max) 0.6% Mn 0.2% S steel: Free machining; hardened and tempered; French Standard	660	14%	370
NF A35 562/10 Pb F2	0.11% C 0.6% Mn 0.2% S 0.25% Pb steel: Free machining; hardened and tempered; French Standard	660	14%	390

Note. The following abbreviations and units are used in the tables:

DPN	Hardness, diamond pyramid number
UTS	Ultimate tensile strength, N/mm^2
Elon	Elongation, %
Proof	0.1% proof strength, N/mm^2

1 N/mm^2=0.1 hbar=0.102 kgf/mm^2=0.06475 $tonf/in.^2$=145.04 $lbf/in.^2$=1 MPa

See Appendix II for other abbreviations and conversion tables.

Symbol	Nominal analysis, supplier, condition and remarks.
NF A35 562/20F2	0.18% C (max) 0.7% Mn 0.2% S steel: Free machining; hardened and tempered; French Standard **UTS: 620 Elon: 17% Proof: 390**
NF A35 562/35MF4	0.35% C (max) 1.1% Mn 0.2% S steel: Free machining; hardened and tempered; French Standard **UTS: 890 Elon: 10% Proof: 690**
NF A36 205/A37C1	0.16% C (max) 0.4% Mn (max) 0.04% S and P (max) steel: Plate; French Standard **UTS: 410 Proof: 220**
NF A36 205/A37C2	0.15% C (max) 0.4% Mn (max) 0.035% S and P (max) steel: Plate; French Standard **UTS: 410 Proof: 220**
NF A36 205/A37P1	0.16% C (max) 0.4% Mn (max) 0.04% S and P (max) steel: Plate; French Standard **UTS: 410 Proof: 220**
NF A36 205/A37P2	0.15% C (max) 0.4% Mn 0.035% S and P (max) steel: Plate; French Standard **UTS: 410 Proof: 220**
NF A36 205/A42C1	0.18% C (max) 0.5% Mn (max) 0.04% S and P (max) steel: Plate; French Standard **UTS: 460 Proof: 240**
NF A36 205/A42C2	0.18% C (max) 0.6% Mn (max) 0.035% S and P (max) steel: Plate; French Standard **UTS: 460 Proof: 250**
NF A36 205/A42P1	0.18% C (max) 0.5% Mn (max) 0.04% S and P (max) steel: Plate; French Standard **UTS: 460 Proof: 240**
NF A36 205/A42P2	0.18% C (max) 0.6% Mn (max) 0.035% S and P (max) steel: Plate; French Standard **UTS: 460 Proof: 250**
NF A36 205/A48C1	0.22% C (max) 0.6% Mn (max) 0.04% S and P (max) steel: Plate; French Standard **UTS: 520 Proof: 280**
NF A36 205/A48C2	0.2% C (max) 0.8% Mn (max) 0.3% Cr Ni (max) 0.035% S and P steel: Plate; French Standard **UTS: 520 Proof: 280**
NF A36 205/A48P1	0.22% C (max) 0.6% Mn (max) 0.04% S and P (max) steel: Plate; French Standard **UTS: 520 Proof: 280**
NF A36 205/A48P2	0.2% C (max) 0.8% Mn (max) 0.3% Ni (max) 0.2% Cr (max) 0.035% S and P (max) steel: Plate; French Standard **UTS: 520 Proof: 280**
NF A36 205/A52C1	0.22% C (max) 0.9% Mn 0.04% S and P steel: Plate; French Standard **UTS: 570 Proof: 340**
NF A36 205/A52C2	0.2% C (max) 1.0% Mn (max) 0.3% Ni (max) 0.2% Cr (max) 0.035% S and P (max) steel: Plate; French Standard **UTS: 570 Proof: 340**
NF A36 205/A52P1	0.22% C (max) 0.9% Mn (max) 0.04% S and P steel: Plate; French Standard **UTS: 570 Proof: 340**
NF A36 205/A52P2	0.2% C (max) 1.0% Mn (max) 0.3% Ni (max) 0.2% Cr (max) 0.035% S and P (max) steel: Plate; French Standard **UTS: 570 Proof: 340**
NF A36 208/A42FP1	0.18% C (max) 0.5% Mn (max) 0.04% S and P (max) steel: Plate; French Standard; normalized **UTS: 440 Proof: 240**
NF A36 208/A42FP2	0.18% C (max) 0.6% Mn (max) 0.035% S and P (max) steel: Plate; French Standard; normalized **UTS: 460 Proof: 240**
NF A36 208/A48FP1	0.22% C (max) 0.6% Mn (max) 0.04% S and P (max) steel: Plate; French Standard; normalized **UTS: 530 Proof: 280**
NF A36 208/A48FP2	0.2% C (max) 0.8% Mn (max) 0.3% Ni (max) 0.2% Cr (max) 0.05 V (max) 0.1% Cu (max) steel: Plate; French Standard; normalized **UTS: 520 Proof: 280**

Symbol	Nominal analysis, supplier, condition and remarks.
NF A36 208/A52FP1	0.22% C (max) 0.9% Mn (max) 0.04% S and P (max) steel: Plate; French Standard; normalized **UTS: 570 Proof: 340**
NILSIL	0.06% C 0.5% Mn steel: Electrode; Metrode
NK Hiten 50	0.18% C (max) 1.4% Mn 0.035% S and P (max) steel: Nippon Kokan
NK Hiten 55	0.2% C (max) 1.5% Mn 0.035% S and P (max) steel: Nippon Kokan
NV 1-0	0.2% C (max) 0.4% Mn 0.05% S and P (max) 0.009% N (max) steel: Designation used by DNV **UTS: 410 Elon: 20% Proof: 240**
NV 1-1	0.16% C (max) 0.4% Mn 0.05% S and P (max) 0.009% N (max) steel: Designation used by DNV **UTS: 410 Elon: 20% Proof: 240**
NV 1-2	0.14% C (max) 0.4% Mn 0.04% S and P (max) 0.009% N (max) steel: Designation used by DNV **UTS: 410 Elon: 20% Proof: 240**
NV 2-0	0.2% C (max) 0.4% Mn 0.05% S and P (max) 0.009% N (max) steel: Designation used by DNV **UTS: 450 Elon: 20% Proof: 260**
NV 2-1	0.16% C (max) 0.4% Mn 0.05% S and P (max) 0.009% N (max) steel: Designation used by DNV **UTS: 450 Elon: 20% Proof: 260**
NV 2-2	0.16% C (max) 0.4% Mn 0.04% S and P (max) 0.009% N (max) steel: Designation used by DNV **UTS: 450 Elon: 20% Proof: 260**
NV 2-2	0.14% C (max) 1.1% Mn 0.04% S and P (max) 0.009% N (max) steel: Designation used by DNV **UTS: 450 Elon: 20% Proof: 260**
NV 2-4	0.14% C (max) 0.11% Mn 0.035% S and P (max) 0.009% N (max) steel: Designation used by DNV **UTS: 450 Elon: 20% Proof: 260**
NV 3-0	0.2% C (max) 0.4% Mn 0.05% S and P (max) 0.009% N (max) steel: Designation used by DNV **UTS: 480 Elon: 20% Proof: 260**
NV 3-1	0.16% C (max) 0.4% Mn 0.05% S and P (max) 0.009% N (max) steel: Designation used by DNV **UTS: 480 Elon: 20% Proof: 260**
NV 3-2	0.16% C (max) 0.4% Mn 0.04% S and P (max) 0.009% N (max) steel: Designation used by DNV **UTS: 480 Elon: 20% Proof: 260**
NV 4-0	0.2% C (max) 1.6% Mn (max) 0.05% S and P (max) 0.009% N (max) steel: Designation used by DNV **UTS: 540 Elon: 18% Proof: 330**
NV 4-1	0.16% C (max) 1.6% Mn (max) 0.04% S and P (max) 0.009% N (max) steel: Designation used by DNV **UTS: 540 Elon: 18% Proof: 330**
NV 4-2	0.16% C (max) 1.6% Mn (max) 0.04% S and P (max) 0.009% N (max) steel: Designation used by DNV **UTS: 540 Elon: 18% Proof: 330**
NV 4-3	0.15% C (max) 1.7% Mn (max) 0.04% S and P (max) 0.009% N (max) steel: Designation used by DNV **UTS: 540 Elon: 18% Proof: 330**
NV 4-4	0.15% C (max) 1.7% Mn (max) 0.035% S and P (max) 0.009% N (max) steel: Designation used by DNV **UTS: 540 Elon: 18% Proof: 330**
NV 5-0	0.2% C (max) 1.0% Mn 0.05% S and P (max) 0.009% N (max) steel: Designation used by DNV **UTS: 510 Elon: 19% Proof: 280**
NV 5-1	0.18% C (max) 1.1% Mn 0.04% S and P (max) 0.009% N (max) steel: Designation used by DNV **UTS: 510 Elon: 19% Proof: 280**
NV 5-2	0.16% C (max) 1.2% Mn 0.04% S and P (max) 0.009% N (max) steel: Designation used by DNV **UTS: 510 Elon: 19% Proof: 280**
NV 6-0	0.23% C (max) 1.15% Mn 0.05% S and P (max) 0.009% N (max) steel: Designation used by DNV **UTS: 570 Elon: 18% Proof: 320**
NV 6-1	0.2% C (max) 1.2% Mn 0.04% S and P (max) 0.009% N (max) steel: Designation used by DNV **UTS: 570 Elon: 18% Proof: 320**

Symbol	Nominal analysis, supplier, condition and remarks.
NV 6-2	0.18% C (max) 1.25% Mn 0.04% S and P (max) 0.009% N (max) steel: Designation used by DNV **UTS: 570 Elon: 18% Proof: 320**
NVA	C not specified 2.5% Mn (min) 0.05% S and P (max) steel: Plate for ships' hulls; DNV
NVA 27	Analysis as NVA; designation used by DNV **UTS: 470 Elon: 22% Proof: 270**
NVA 32	Analysis as NVA; designation used by DNV **UTS: 520 Elon: 22% Proof: 320**
NVA 36	Analysis as NVA; designation used by DNV **UTS: 550 Elon: 21% Proof: 360**
NVA 40	Analysis as NVA; designation used by DNV **UTS: 600 Elon: 20% Proof: 400**
NVC	0.21% C (max) 1.0% Mn 0.05% S and P (max) steel: Plate for ships' hulls; fine grained steel with Al; normalized; DNV
NVD	0.21% C (max) 1.0% Mn 0.05% S and P (max) steel: Plate for ships' hulls; DNV
NVD 27	Analysis as NVD; designation used by DNV **UTS: 470 Elon: 22% Proof: 270**
NVD 32	Analysis as NVD; designation used by DNV **UTS: 520 Elon: 22% Proof: 320**
NVD 36	Analysis as NVD; designation used by DNV **UTS: 550 Elon: 21% Proof: 360**
NVD 40	Analysis as NVD; designation used by DNV **UTS: 600 Elon: 20% Proof: 400**
NVE	0.18% C (max) 1.1% Mn 0.05% S and P (max) steel: Plate for ships' hulls; normalized; DNV
NVE 27	Analysis as NVE; designation used by DNV **UTS: 470 Elon: 22% Proof: 270**
NVE 32	Analysis as NVE; designation used by DNV **UTS: 520 Elon: 22% Proof: 320**
NVE 36	Analysis as NVE; designation used by DNV **UTS: 550 Elon: 21% Proof: 360**
NVE 40	Analysis as NVE; designation used by DNV **UTS: 600 Elon: 20% Proof: 400**
NVK1	C not specified 0.04% S and P (max) steel: For chains; normalized; designation used by DNV **UTS: 320 Elon: 30%**
NVK2	C not specified 0.04% S and P (max) steel: For chains; normalized; designation used by DNV **UTS: 550 Elon: 22% Proof: 300**
NVK3	C not specified 0.04% S and P (max) steel: For chains; hardened and tempered; designation used by DNV **UTS: 700 Elon: 17% Proof: 400**
NVR 1-1	0.17% C (max) 0.05% S and P (max) steel: Designation used by DNV
NVR 1-2	0.17% C (max) 0.4% Mn 0.2% Cr (max) 0.05% S and P (max) steel: Designation used by DNV
NVR 1-3	0.22% C (max) 0.6% Mn 0.2% Cr (max) 0.05% S and P (max) steel: Designation used by DNV
NVR 1-4	0.2% C (max) 1.1% Mn 0.2% Cr (max) 0.05% S and P (max) steel: Designation used by DNV
NVW	0.2% C (max) 0.5% Mn 0.05% S and P (max) steel: Plate for ships' hulls; DNV
ORELLOY 440	0.19% C 1.3% Mn 0.35% Cu steel: Oregon
OX 520A	0.18% C (max) 1.4% Mn 0.025% Nb 0.009% N steel: Oxelsund
OX 522	0.2% C (max) 1.6% Mn 0.04% Nb 0.015% Al steel: Plate; Srensk-Stal; graded by impact **UTS: 500 Elon: 22% Proof: 330**
OX 522A	0.18% C (max) 0.3% Mn 0.025% Nb 0.009% N steel: Oxelsund
OX 525B	0.18% C (max) 1.4% Mn 0.025% Nb 0.009% N steel: Oxelsund
OX 542	0.2% C (max) 1.8% Mn 0.05% Nb 0.015% Al steel: Plate; Srensk-Stal; graded by impact **UTS: 510 Elon: 22% Proof: 370**
P 30	0.06% C 0.5% Mn steel: Electrode; Metrode; replaced by ELGA P 43
P 151	0.15% C 0.8% Mn steel: For case hardening; Carrs
P 282	0.1% C 0.25% Mn 0.25% Si steel: For case hardening; Carrs
PAC	0.09% C 0.4% Mn steel: Pompey
PAC 1	0.12% C 0.4% Mn steel: Pompey
PAC 2	0.15% C 0.6% Mn steel: Pompey
PAR-TEN	0.12% C (max) 0.75% Mn 0.07% Cu steel: US Steel Corp.
PENETRODE 29.10	0.080% C 0.65% Mn 0.06% Si steel: For electrodes; Esab **UTS: 479 Elon: 28% Proof: 355**
PENETRODE 29.10	0.08% C 0.65% Mn 0.06% Si steel: For electrodes; Esab; as welded **UTS: 480 Elon: 28% Proof: 355**
PG 605	0.15% C 1.25% Mn 0.12% S 0.05% P steel: Free cutting; Park Gate; as rolled **UTS: 450 Elon: 30%**
PGF 1	0.1% C 1.2% Mn 0.35% S 0.09% P steel: Free cutting; Park Gate; as rolled **UTS: 390 Elon: 30%**
PGF 4	0.12% C 1% Mn 0.25% S 0.07% P steel: Free cutting; Park Gate; as rolled **UTS: 340 Elon: 26%**
PGF 5	0.12% C 1.2% Mn 0.5% S 0.06% P steel: Free cutting; Park Gate; as rolled **UTS: 340 Elon: 24%**
PHILIPS 23C	0.1% C 0.7% Mn steel: Welding electrode; Philips **UTS: 550 Elon: 28% Proof: 390**
PHILIPS 27	0.08% C 0.9% Mn steel: Welding electrode; Philips **UTS: 550 Elon: 26% Proof: 420**
PHILIPS 27P	0.07% C 1.2% Mn steel: Welding electrode; Philips; low hydrogen rod **UTS: 570 Elon: 28% Proof: 480**
PHILIPS 35S	0.07% C 0.9% Mn steel: Welding electrode; Philips **UTS: 540 Elon: 26% Proof: 420**
PHILIPS 36S	0.08% C 0.8% Mn steel: Welding electrode; Philips **UTS: 540 Elon: 26% Proof: 420**
PHILIPS 46	0.09% C 0.5% Mn steel: Welding electrode; Philips **UTS: 510 Elon: 26% Proof: 450**
PHILIPS 48	0.09% C 0.45% Mn steel: Welding electrode; Philips **UTS: 460 Elon: 30% Proof: 415**
PHILIPS 55C (55)	0.09% C 1% Mn steel: Welding electrode; Philips **UTS: 540 Elon: 28% Proof: 420**
PHILIPS 56S	0.08% C 1% Mn steel: Welding electrode; Philips **UTS: 550 Elon: 26% Proof: 420**
PHILIPS 68	0.09% C 0.42% Mn steel: Welding electrode; Philips **UTS: 520 Elon: 22% Proof: 440**
PHILIPS 78	0.08% C 0.6% Mn steel: Welding electrode; Philips **UTS: 500 Elon: 24% Proof: 420**
PHILIPS C16	0.1% C 1% Mn steel: Welding electrode; Philips **UTS: 500 Elon: 30% Proof: 420**
PHILIPS C33 M	0.09% C 0.7% Mn 0.4% Si steel: Welding electrode; Philips **UTS: 530 Elon: 28% Proof: 440**
PHILIPS ENROX	0.08% C 0.6% Mn steel: Welding electrode; Philips **UTS: 520 Elon: 27% Proof: 460**
PHILIPS HERMEES	0.09% C 0.8% Mn steel: Welding electrode; Philips **UTS: 530 Elon: 29% Proof: 470**

Note. The following abbreviations and units are used in the tables:

DPN Hardness, diamond pyramid number
UTS Ultimate tensile strength, N/mm^2
Elon Elongation, %
Proof 0.1% proof strength, N/mm^2

1 N/mm^2=0.1 hbar=0.102 kgf/mm^2=0.06475 tonf/in.^{2}=145.04 lbf/in.^{2}=1 MPa
See Appendix II for other abbreviations and conversion tables.

Symbol	Nominal analysis, supplier, condition and remarks.
PHILIPS LOWHEES	0.08% C 1.3% Mn steel: Welding electrode; Philips **UTS: 530 Elon: 29% Proof: 450**
PHILIPS MILDEES	0.07% C 0.5% Mn steel: Welding electrode; Philips **UTS: 470 Elon: 31% Proof: 410**
PHILIPS MONARC	0.09% C 0.45% Mn steel: Welding electrode; Philips **UTS: 480 Elon: 31% Proof: 400**
PHILIPS QUEENARC	0.08% C 0.4% Mn steel: Welding electrode; Philips **UTS: 450 Elon: 31% Proof: 380**
PHOENIX 1	0.2% C 0.7% Mn 0.06% S 0.06% P steel: Steel Peach and Tozer **UTS: 530 Elon: 20%**
PHOENIX 2	0.2% C 0.7% Mn 0.06% S 0.06% P steel: Steel Peach and Tozer **UTS: 420 Elon: 17%**
PHOENIX 4	0.2% C (max) 0.9% Mn 0.25% S 0.1% P: Free machining steel; Steel Peach and Tozer **UTS: 420 Elon: 14%**
PHOENIX MANGEAR	0.12% C 1.6% Mn 0.2% Si 0.04% S 0.04% P steel: Steel Peach and Tozer; for lifting chains **DPN: 160 UTS: 500 Elon: 37% Proof: 380**
PITT-TEN 2	0.15% C (max) 0.75% Mn 0.035% Cu steel: Pittsburgh Steel
PITT-TEN X45W	0.15% C 0.75% Mn 0.03% Nb steel: Pittsburgh Steel
PITT-TEN X50W	0.15% C 0.75% Mn 0.03% Nb steel: Pittsburgh Steel
PITT-TEN X55W	0.15% C 0.75% Mn 0.03% Nb steel: Pittsburgh Steel
PITT-TEN X60W	0.15% C 0.75% Mn 0.03% Nb steel: Pittsburgh Steel
PLANEMEL	Low carbon decarburized steel: Sheet; non-ageing; Esperance Longdoz; for vitreous enamel
PLASTICS HOBBING	0.1% C (max) 0.4% Mn 1% Si steel: Edgar Allen
PM 5	0.18% C 1.0% Mn 0.14% Si steel: Pompey
PM 6	0.2% C 1.5% Mn 0.15% Si steel: Pompey
POSITRODE 46.28	0.080% C 0.45% Mn 0.10% Si steel: For electrodes; Esab **UTS: 479 Elon: 28% Proof: 386**
PRIMAFIXE	0.07% C 0.5% Mn 0.35% Si mild steel: Welders rod; Soudametal; for contaminated sheet mild weld **UTS: 550 Elon: 26% Proof: 480**
PRIMEX	0.04% C 0.5% Mn steel: Welding electrode; Murex **UTS: 440 Elon: 33% Proof: 420**
PZ 6000	Mild steel: Wire welding electrode; Cu coated; Philips **UTS: 550 Elon: 27% Proof: 425**
PZ 6008	Mild steel: Welding wire; Philips **UTS: 530 Elon: 25% Proof: 425**
PZ 6101S	0.09% C 1.4% Mn 0.6% Si steel: Wire for welding; Philips; flux cored **UTS: 530 Elon: 28% Proof: 480**
PZ 6120S	0.08% C 1.4% Mn 0.4% Si steel: Wire for welding; Philips; flux cored **UTS: 520 Elon: 26% Proof: 420**
Q 9 100 A	0.2% C 1.5% Mn steel: SAE 368a
Q 9 100 B	0.2% C 1.5% Mn 0.0005% B steel: SAE 368a
Q 9 110 A	0.21% C 1.5% Mn steel: SAE 368a
Q 980	0.2% C 1.3% Mn steel: SAE J 368a
Q 980 A	0.2% C 1.3% Mn steel: SAE J 368a
Q 980 B	0.2% C 1.3% Mn 0.0005% B steel: J 368a
Q 990 A	0.2% C 1.3% Mn steel: SAE J 368a
Q 990 B	0.2% C 1.3% Mn 0.0005% B steel: SAE J 368a
Q St 34	0.1% C 0.3% Mn 0.04% S and P steel: Designation used by German Standards
Q St 34/2	0.17% C (max) 0.05% S and P steel: German Standard
Q St 34/3	0.17% C (max) 0.05% S and P steel: German Standard
Q St 37/2	0.2% C (max) 0.05% S and P steel: German Standard
Q St 37/3	0.2% C (max) 0.05% S and P steel: German Standard
Q St 42/2	0.25% C (max) 0.05% S and P steel: German Standard
Q St 42/3	0.25% C (max) 0.05% S and P steel: German Standard
Q St 52/3	0.2% C (max) 0.05% S and P steel: German Standard
QUEENARC	0.08% C 0.4% Mn steel: Welding electrode; Philips
R St 13	0.1% C (max) 0.3% Mn 0.1% Si 0.04% S and P steel: German Standard

Symbol	Nominal analysis, supplier, condition and remarks.
R St 13/03	0.1% C (max) 0.3% Mn 0.1% Si 0.04% S and P steel: German Standard
R St 13/04	0.1% C (max) 0.3% Mn 0.1% Si 0.04% S and P steel: German Standard
R St 13/05	0.1% C (max) 0.3% Mn 0.1% Si 0.04% S and P steel: German Standard
R St 14	0.1% C (max) 0.3% Mn 0.1% Si 0.03% S and P steel: German Standard
R St 14/04	0.1% C (max) 0.3% Mn 0.1% Si 0.03% S and P steel: German Standard
R St 14/05	0.1% C (max) 0.3% Mn 0.1% Si 0.03% S and P steel: German Standard
R St 34	0.17% C (max) 0.05% S and P steel: German Standard
R St 34/2	0.17% C (max) 0.05% S and P steel: German Standard
R St 37	0.2% C (max) 0.05% S 0.08% P steel: German Standard
R St 37/2	0.2% C (max) 0.05% S and P steel: German Standard
R St 37/202	0.2% C (max) 0.05% S and P steel: German Standard
R St 37/203	0.2% C (max) 0.05% S and P steel: German Standard
R St 37/204	0.2% C (max) 0.05% S and P steel: German Standard
R St 37/205	0.2% C (max) 0.05% S and P steel: German Standard
R St 42	0.25% C (max) 0.05% S 0.08% P steel: German Standard
R St 42/2	0.25% C (max) 0.05% S and P steel: German Standard
R St 42/202	0.25% C (max) 0.05% S and P steel: German Standard
R St 42/203	0.25% C (max) 0.05% S and P steel: German Standard
R St 42/204	0.25% C (max) 0.05% S and P steel: German Standard
R St 42/205	0.25% C (max) 0.05% S and P steel: German Standard
REPUBLIC M1	0.25% C (max) 1.4% Mn 0.2% Cu steel: Republic Steel Co.
REPUBLIC M2	0.25% C (max) 1.4% Mn 0.2% Cu steel: Republic Steel Co.
RESISTCO	0.11% C 0.4% Mn 0.3% Si (max) 0.05% S and P 0.3% Cu steel: Tube; STD Service; for boiler feed tubes **UTS: 390 Elon: 25% Proof: 180**
RG 45	0.08% C (max) steel: Weld electrode; designation used by ANS
RG 60	0.15% C (max) steel: Weld electrode; designation used by ANS
RG 65	C not specified steel: Weld electrode; designation used by ANS
RIO 214	0.08% C 5% Ni 26% Cr 1.5% Mo steel: Bofors **DPN: 260 UTS: 600 Elon: 20% Proof: 420**
RIVER ACE 60	0.18% C (max) 1.5% Mn 0.035% S and P (max) steel: Kawasaki Iron and Steel Co.
S 70	Iron–carbon sintered material: Sintered Products; 23–28% porosity **UTS: 120 Elon: 1%**
S St 2	0.1% C 0.35% Mn 0.15% Si 0.05% S and P steel: German Standard
S St 3	0.1% C (max) 0.4% Mn 0.1% Si 0.04% S and P steel: German Standard
S St 4	0.1% C (max) 0.4% Mn 0.08% Si 0.035% S and P steel: German Standard
SAE 12L14	0.15% C (max) 1% Mn 0.3% S 0.05% P steel: Free cutting
SAE 0022	0.2% C 0.7% Mn 0.6% Si 0.05% S and P cast steel: Suitable for carburizing
SAE 080	0.05% C (max) 0.05% S and P cast steel: Annealed **DPN: 163 UTS: 560 Elon: 18% Proof: 270**
SAE 850	0.2% C Fe: Sinter for bearings; density 5900 kg/m^3
SAE 853	0.2% C Fe: Sinter for mechanical parts; density 6300 kg/m^3
SAE 870	0.2% C 20% Cu Fe alloy: Sinter; density 7100 kg/m^3
SAE 950	0.2% C (max) 1.25% Mn 0.05% S 0.15% P steel: General specification covering weldable steels; as rolled **UTS: 480 Elon: 20% Proof: 320**
SAE 1005	0.06% C (max) 0.35% Mn 0.05% S 0.04% P steel: Bar, etc.

Symbol	Nominal analysis, supplier, condition and remarks.
SAE 1006	0.08% C 0.3% Mn 0.05% S 0.04% P steel: Hot rolled as AISI C1006
	DPN: 86 UTS: 300 Elon: 30% Proof: 150
SAE 1006	0.08% C 0.3% Mn 0.05% S 0.04% P steel: Cold drawn
	DPN: 95 UTS: 330 Elon: 20% Proof: 440
SAE 1008	0.1% C 0.4% Mn 0.05% S 0.04% P steel: Hot rolled as AISI C1008
	DPN: 86 UTS: 300 Elon: 30% Proof: 150
SAE 1008	0.1% C 0.4% Mn 0.05% S 0.04% P steel: Cold drawn
SAE 1009	0.15% C 0.6% Mn 0.05% S 0.04% P steel: Hot rolled as AISI C1009
	DPN: 86 UTS: 300 Elon: 30% Proof: 150
SAE 1009	0.15% C 0.6% Mn 0.05% S 0.04% P steel: Cold drawn
	DPN: 95 UTS: 330 Elon: 20% Proof: 440
SAE 1010	0.1% C 0.4% Mn 0.05% S 0.04% P steel: Hot rolled as AISI C1010
	DPN: 95 UTS: 330 Elon: 28% Proof: 170
SAE 1010	0.1% C 0.4% Mn 0.05% S 0.04% P steel: Cold drawn
	DPN: 105 UTS: 370 Elon: 20% Proof: 300
SAE 1012	0.12% C 0.4% Mn 0.05% S 0.04% P steel: Hot rolled as AISI C1012
	DPN: 95 UTS: 330 Elon: 28% Proof: 180
SAE 1012	0.12% C 0.4% Mn 0.05% S 0.04% P steel: Cold drawn
	DPN: 105 UTS: 370 Elon: 19% Proof: 320
SAE 1013	0.13% C 0.7% Mn 0.05% S 0.04% P steel: Bar, etc.
SAE 1015	0.15% C 0.4% Mn 0.05% S 0.04% P steel: Hot rolled as AISI C1015
	DPN: 101 UTS: 340 Elon: 28% Proof: 180
SAE 1015	0.15% C 0.4% Mn 0.05% S 0.04% P steel: Cold drawn
	DPN: 111 UTS: 390 Elon: 18% Proof: 330
SAE 1016	0.15% C 0.8% Mn 0.05% S 0.04% P steel: Bar; hot rolled as AISI C1016
	DPN: 111 UTS: 390 Elon: 25% Proof: 200
SAE 1016	0.15% C 0.8% Mn 0.05% S 0.04% P steel: Bar; cold drawn
	DPN: 121 UTS: 440 Elon: 18% Proof: 360
SAE 1017	0.18% C 0.4% Mn 0.05% S 0.04% P steel: Bar; hot rolled as AISI C1017
	DPN: 105 UTS: 370 Elon: 26% Proof: 200
SAE 1017	0.18% C 0.4% Mn 0.05% S 0.04% P steel: Bar; Cold drawn
	DPN: 116 UTS: 420 Elon: 18% Proof: 340
SAE 1018	0.18% C 0.8% Mn 0.05% S 0.04% P steel: Bar; hot rolled as AISI C1018
	DPN: 116 UTS: 400 Elon: 25% Proof: 210
SAE 1018	0.18% C 0.8% Mn 0.05% S 0.04% P steel: Bar; cold drawn
	DPN: 126 UTS: 450 Elon: 15% Proof: 370
SAE 1019	0.18% C 0.9% Mn 0.05% S 0.04% P steel: Bar; hot rolled as AISI C1019
	DPN: 116 UTS: 420 Elon: 25% Proof: 210
SAE 1019	0.18% C 0.9% Mn 0.05% S 0.04% P steel: Bar; cold drawn
	DPN: 131 UTS: 460 Elon: 15% Proof: 390
SAE 1020	0.2% C 0.4% Mn 0.05% S 0.04% P steel: Bar; hot rolled as AISI C1020
	DPN: 111 UTS: 390 Elon: 25% Proof: 200
SAE 1020	0.2% C 0.4% Mn 0.05% S 0.04% P steel: Bar; cold drawn
	DPN: 121 UTS: 440 Elon: 15% Proof: 360
SAE 1021	0.2% C 0.8% Mn 0.05% S 0.04% P steel: Bar
SAE 1022	0.2% C 0.9% Mn 0.05% S 0.04% P steel: Bar; hot rolled as AISI C1022
	DPN: 121 UTS: 440 Elon: 23% Proof: 220
SAE 1022	0.2% C 0.9% Mn 0.05% S 0.04% P steel: Bar; cold drawn
	DPN: 137 UTS: 500 Elon: 15% Proof: 400
SAE 1023	0.22% C 0.4% Mn 0.05% S 0.04% P steel: Bar; hot rolled as AISI C1023
	DPN: 111 UTS: 390 Elon: 25% Proof: 210
SAE 1023	0.22% C 0.4% Mn 0.05% S 0.04% P steel: Bar; cold drawn
	DPN: 121 UTS: 440 Elon: 15% Proof: 370
SAE 1108	0.1% C 0.7% Mn 0.1% S 0.04% P steel: Free cutting; hot rolled as AISI C1108
	DPN: 101 UTS: 340 Elon: 30% Proof: 180
SAE 1108	0.1% C 0.7% Mn 0.1% S 0.04% P steel: Free cutting; cold drawn
	DPN: 121 UTS: 390 Elon: 20% Proof: 330
SAE 1109	0.1% C 0.8% Mn 0.1% S 0.04% P steel: Free cutting; hot rolled as AISI B1109
	DPN: 101 UTS: 340 Elon: 30% Proof: 180
SAE 1109	0.1% C 0.8% Mn 0.1% S 0.04% P steel: Free cutting; cold drawn
	DPN: 121 UTS: 390 Elon: 20% Proof: 330
SAE 1110	0.1% C 0.4% Mn 0.1% S 0.04% P steel: Free machining
SAE 1111	0.13% C 0.8% Mn 0.12% S 0.1% P steel: Free cutting; hot rolled as AISI B1112
	DPN: 121 UTS: 420 Elon: 25% Proof: 220
SAE 1111	0.13% C 0.8% Mn 0.12% S 0.1% P steel: Free cutting; cold drawn
	DPN: 131 UTS: 480 Elon: 10% Proof: 390
SAE 1112	0.13% C 0.9% Mn 0.2% S 0.1% P steel: Free cutting; hot rolled as AISI B1111
	DPN: 121 UTS: 440 Elon: 25% Proof: 220
SAE 1112	0.13% C 0.9% Mn 0.2% S 0.1% P steel: Free cutting; cold drawn
	DPN: 137 UTS: 480 Elon: 10% Proof: 400
SAE 1113	0.13% C 0.9% Mn 0.3% S 0.1% P steel: Free cutting; hot rolled as AISI B1113
	DPN: 121 UTS: 440 Elon: 25% Proof: 220
SAE 1113	0.13% C 0.9% Mn 0.3% S 0.1% P steel: Free cutting; cold drawn
	DPN: 137 UTS: 480 Elon: 10% Proof: 400
SAE 1115	0.15% C 0.8% Mn 0.1% S 0.04% P steel: Free cutting; hot rolled as AISI C1115
	DPN: 111 UTS: 390 Elon: 25% Proof: 200
SAE 1115	0.15% C 0.8% Mn 0.1% S 0.04% P steel: Free cutting; cold drawn
	DPN: 121 UTS: 440 Elon: 20% Proof: 360
SAE 1116	0.16% C 1.2% Mn 0.2% S 0.04% P steel: Free machining
SAE 1117	0.17% C 1.2% Mn 0.1% S 0.04% P steel: Free cutting; hot rolled as AISI C1117
	DPN: 121 UTS: 440 Elon: 23% Proof: 220
SAE 1117	0.17% C 1.2% Mn 0.1% S 0.04% P steel: Free cutting; cold drawn
	DPN: 137 UTS: 520 Elon: 15% Proof: 400
SAE 1118	0.17% C 1.5% Mn 0.1% S 0.04% P steel: Free cutting; hot rolled as AISI C1118
	DPN: 131 UTS: 460 Elon: 23% Proof: 240
SAE 1118	0.17% C 1.5% Mn 0.1% S 0.04% P steel: Free cutting; cold drawn
	DPN: 143 UTS: 510 Elon: 15% Proof: 440
SAE 1119	0.17% C 1.2% Mn 0.3% S 0.04% P steel: Free cutting; hot rolled as AISI C1119
	DPN: 121 UTS: 440 Elon: 23% Proof: 220

Note. The following abbreviations and units are used in the tables:

DPN	Hardness, diamond pyramid number
UTS	Ultimate tensile strength, N/mm^2
Elon	Elongation, %
Proof	0.1% proof strength, N/mm^2

1 N/mm^2=0.1 hbar=0.102 kgf/mm^2=0.06475 $tonf/in.^2$=145.04 $lbf/in.^2$=1 MPa

See Appendix II for other abbreviations and conversion tables.

Symbol	Nominal analysis, supplier, condition and remarks.
SAE 1119	0.17% C 1.2% Mn 0.3% S 0.04% P steel: Free cutting; cold drawn **DPN: 137 UTS: 500 Elon: 15% Proof: 400**
SAE 1120	0.2% C 0.9% Mn 0.1% S 0.04% P steel: Free cutting; hot rolled as AISI C1120 **DPN: 121 UTS: 440 Elon: 23% Proof: 220**
SAE 1120	0.2% C 0.9% Mn 0.1% S 0.04% P steel: Free cutting; cold drawn **DPN: 137 UTS: 500 Elon: 15% Proof: 400**
SAE 1215	0.09% C 0.9% Mn 0.3% S 0.06% P steel: Free machining
SAE 1320	0.2% C 1.7% Mn 0.04% S and P steel: Obsolete
SAE 1513	0.13% C 1.2% Mn 0.05% S 0.04% P steel: For bar
SAE 1518	0.18% C 1.2% Mn 0.05% S 0.04% P steel: Bar, etc.
SAE 1522	0.22% C 1.2% Mn 0.05% S 0.04% P steel: Bar, etc.
SAE 12413	0.13% C 0.9% Mn 0.3% S 0.1% P steel: Free machining
SAE J 178	0.9% C 0.4% Mn steel: Cold drawn; music wire **UTS: 2350**
SAE J 435-0022	0.17% C 0.7% Mn 0.04% S and P steel: Casting
SAE J 435-0025	0.25% C (max) 0.75% Mn (max) 0.04% S and P steel: Casting
SAN BOLD 3	0.2% C 0.6% Mn 0.3% Si steel: Sanderson; for BS alloy En 3A
SB 36F	0.15% C 1.2% Mn 0.04% S and P (max) steel: German proprietor's specification; Al killed
SB 36FT	0.15% C 1.25% Mn 0.04% S and P (max) steel: German proprietor's specification; Al killed
SB 38F	0.2% C (max) 1.2% Mn 0.04% S and P (max) steel: German proprietor's specification; Al killed
SB 38FT	0.2% C (max) 1.2% Mn 0.04% S and P (max) steel: German proprietor's specification
SB 39FT	0.16% C 1.3% Mn 0.03% V 0.02% Al (max) 0.03% Tl steel: German proprietor's specification
SB 40F	0.2% C (max) 1.4% Mn 0.04% S and P (max) steel: German proprietor's specification; Al killed
SB 40FT	0.2% C (max) 1.3% Mn 0.04% S and P (max) steel: German proprietor's specification; Al killed
SB 42FT	0.16% C 1.3% Mn 0.1% V 0.035% S and P (max) steel: German proprietor's specification; Al killed
SCH 1	0.14% C 1.0% Mn 0.2% Si 0.06% S and P steel: Spencer; for BS alloy En32
SE 90	Iron–carbon sintered materials; Sintered Products; 4–8% porosity **UTS: 390 Elon: 15%**
SELCO 53	0.22% C (max) 1.5% Mn 0.04% S and P (max) steel: Italian proprietor's specification
SELCO 53V	0.2% C (max) 1.5% Mn 0.04% S and P (max) steel
SH 13-54	0.15% C 1.0% Mn 0.015% S 0.02% P 0.4% Cu steel: Fuji
SH 13-60	0.18% C 1.05% Mn steel: Fuji
SHEF-LO TEMP	0.2% C (max) 1.0% Mn steel: Armco
SHEF-SUPER LO TEMP	0.2% C (max) 1.1% Mn steel: Armco
SILICON CORE IRON A	0.03% C 1% Si Fe alloy: Carpenter; for magnetic purposes
SILICON CORE IRON A-FM	0.04% C (max) 1% Si Fe alloy: Carpenter; for magnetic purposes
For SIS specifications see SS	
SKF 017	0.17% C 0.7% Mn steel: Carburizing; SKF
SKF 21T2	0.18% C 1.4% Mn hollow steel: Bar **DPN: 170 UTS: 570 Elon: 23% Proof: 360**
SKF CEAX	0.5% C 0.7% Mn 0.05% S 0.05% P steel: For springs; SKF
SKF TERMEK	Steel: Free machining; bar; SKF
SMOOTHTRODE 43.23	0.07% C 0.68% Mn 0.4% Si steel: For electrodes; Esab **UTS: 556 Elon: 26% Proof: 279**
SMR St 2	0.1% C (max) 0.45% Mn 0.1% Si 0.05% S and P steel: German Standard
SMR St 3	0.1% C (max) 0.45% Mn 0.1% Si 0.04% S and P steel: German Standard
SMR St 4	0.1% C (max) 0.45% Mn 0.07% Si 0.03% S and P steel: German Standard
SMU St 2	0.1% C 0.45% Mn 0.05% S and P steel: German Standard
SMU St 3	0.1% C (max) 0.45% Mn 0.04% S and P steel: German Standard
SMU St 4	0.1% C (max) 0.45% Mn 0.03% S and P steel: German Standard
SOUDAIR	0.12% C 1.6% Mn 1.2% Al steel: Welding electrode; Soudometal **UTS: 590 Elon: 22% Proof: 490**
SOUDITIG G1	0.07% C 1.1% Mn steel: Welding electrode; Soudometal; for inert gas welding **UTS: 525 Elon: 30% Proof: 390**
SOUDITIG G3	0.08% C 1.6% Mn steel: Welding electrode; Soudometal; for inert gas welding **UTS: 620 Elon: 30% Proof: 490**
SOUDOFIL 1	0.07% C 1.5% Mn 0.9% Si steel: Welding electrode; Soudometal; for inert gas welding **UTS: 550 Elon: 22% Proof: 420**
SOUDOFIL 2	0.07% C 1.4% Mn 0.8% Si steel: Welding electrode; Soudometal; for inert gas welding **UTS: 550 Elon: 22% Proof: 420**
SOUDOKAY	0.09% C 0.9% Mn steel: Welding electrode; Soudometal **DPN: 170**
SOUDOKAY BU-CI	0.06% C 1% Mn 0.6% Si steel: Welding electrode; Soudometal; low hydrogen welding rod **DPN: 220 UTS: 560 Elon: 24% Proof: 450**
SOUDORECORD	0.05% C 0.5% Mn 0.35% Si steel: Soudometal; mild steel; welding rod **UTS: 550 Elon: 24% Proof: 480**
SOUNDOTENAX 52	0.2% C 1.3% Mn 0.32% Cr steel: Cockerill-Ougree
SOUNDOTENAX 56	0.2% C 1.3% Mn 0.42% Ni 0.32% Cr 0.22% Mo steel: Cockerill-Ougree
SOUNDOTENAX 66	0.2% C 1.3% Mn 0.9% Ni 0.3% Cr 0.4% Mo steel: Cockerill-Ougree
SP 25	Sintered Ni alloy steel similar to SP 24; Durasint **DPN: 210 UTS: 410 Elon: 2%**
SP 26	Sintered Ni alloy steel similar to SP 20; Durasint **DPN: 210 UTS: 550 Elon: 2%**
SPEAR B8	0.06% C 0.4% Mn 0.1% Si steel: Spear and Jackson; obsolete
SPELTAFAST	Mild steel: Sheet; galvanized; RTB for BS 2989
SPRASTEEL 10	0.1% C 0.5% Mn 0.04% S and P steel: Wire for spraying; Metco; as sprayed **DPN: 189 UTS: 200**
SPRASTEEL 10	0.1% C steel: Powder coating; Metco **DPN: 190**
SPRASTEEL 25	0.23% C 0.6% Mn 0.04% S and P steel: Wire for spraying; Metco; as sprayed **DPN: 189 UTS: 220**
SPRASTEEL 25	0.23% C steel: Powder coating; Metco **DPN: 205**
SS 1111	0.08% C 0.4% Mn 0.04% S and P steel: Swedish Standard; for drawn wire **UTS: 590**
SS 1140	0.15% C 0.5% Mn 0.06% S and P steel: Sheet; commercial grade; Swedish Standard
SS 1142	0.1% C 0.5% Mn 0.05% S and P steel: Sheet for pressing; Swedish Standard
SS 1142	0.1% C (max) 0.5% Mn (max) 0.04% S and P steel: Sheet; cold rolled
SS 1145	0.1% C 0.5% Mn 0.04% S and P steel: Sheet; for deep drawing; Swedish Standard
SS 1146	0.1% C 0.5% Mn 0.04% S and P steel: Sheet; for deep drawing; Swedish Standard
SS 1146	0.1% C (max) 0.5% Mn (max) 0.04% S and P steel: Sheet; cold rolled; Swedish Standard

Symbol	Nominal analysis, supplier, condition and remarks.
SS 1147	0.08% C (max) 0.45% Mn (max) 0.03% S and P steel: Sheet; cold rolled; Swedish Standard
SS 1147	0.1% C 0.5% Mn 0.03% S and P steel: Sheet; for extra deep drawing; Swedish Standard
SS 1148	0.1% C 0.5% Mn 0.03% S and P steel: Cold rolled; for extra deep drawing; Swedish Standard
SS 1150	0.07% C 0.3% Mn 0.06% Si 0.03% S and P 0.1% Cu steel: Strip; Swedish Standard; annealed **DPN: 95 UTS: 370**
SS 1150	0.07% C 0.3% Mn 0.06% Si 0.03% S and P 0.1% Cu steel: Strip; Swedish Standard; cold rolled **DPN: 130 UTS: 370**
SS 1151	0.12% C (max) 0.6% Mn (max) 0.04% S and P steel: Sheet; for hot dip galvanizing; Swedish Standard **UTS: 500**
SS 1151	0.12% C (max) 0.6% Mn 0.04% S and P steel: For cold rolled sheet; Swedish Standard **UTS: 500**
SS 1152	0.12% C (max) 0.5% Mn (max) 0.04% S and P steel: Sheet for galvanizing; Swedish Standard
SS 1157	0.08% C (max) 0.45% Mn (max) 0.03% S and P steel: Sheet for galvanizing; Swedish Standard
SS 1160	0.08% C (max) 0.3% Mn 0.04% S (max) 0.03% P (max) steel: Swedish Standard **DPN: 100 UTS: 300 Elon: 35%**
SS 1160	0.08% C (max) 0.35% Mn (max) 0.04% S 0.03% P steel: For cold rolled strip; Swedish Standard
SS 1211	0.12% C 0.4% Mn 0.06% S and P steel: Swedish Standard; for drawn wire **UTS: 500**
SS 1232	0.13% C 0.4% Mn 0.2% Si 0.05% S and P 0.009% N steel: Swedish Standard **UTS: 360 Elon: 25%**
SS 1232	0.13% C (max) 0.2% Cr (max) 0.03% V (max) 0.01% Nb (max) steel: Swedish Standard; tube
SS 1250	0.12% C 0.4% Mn 0.2% Si 0.05% S and P steel: Strip; Swedish Standard; for cold rolling
SS 1265	0.1% C 0.35% Mn 0.03% S and P steel: Strip; Swedish Standard; annealed **DPN: 115 UTS: 340**
SS 1265	0.1% C steel: Case hardening; possible Swedish Standard
SS 1265	0.1% C 0.35% Mn 0.03% S and P steel: Strip; Swedish Standard; cold rolled **DPN: 210 UTS: 640**
SS 1270	0.15% C (max) 1.0% Mn 0.05% S and P steel: Sheet; Swedish Standard; galvanized **UTS: 330 Elon: 18% Proof: 250**
SS 1305	0.18% C (max) 0.6% Mn 0.04% S and P steel: Swedish Standard; annealed **UTS: 390 Elon: 25% Proof: 210**
SS 1305	0.25% C (max) 0.3% Cu steel: Casting; Swedish Standard
SS 1306	0.18% C (max) 0.3% Cr (max) steel: Casting; Swedish Standard
SS 1311	0.14% C 0.5% Mn 0.06% S and P structural steel: Swedish Standard; as rolled **UTS: 390 Elon: 24% Proof: 220**
SS 1311	0.13% C 0.5% Mn 0.06% S 0.08% P steel: For reinforcing; Swedish Standard **Elon: 20% Proof: 220**
SS 1312	0.12% C 0.6% Mn 0.05% S and P 0.09% N structural steel: Swedish Standard; as rolled **UTS: 390 Elon: 24% Proof: 220**
SS 1313	0.12% C 0.7% Mn 0.05% S and P 0.009% N structural steel: Swedish Standard; as rolled **UTS: 390 Elon: 24% Proof: 220**
SS 1316	0.15% C (max) 0.7% Mn (max) 0.05% S and P steel: Plate; for cold rolling; Swedish Standard
SS 1331	0.15% C 0.2% Cr (max) 0.03% V (max) 0.01% Nb (max) 0.3% Cu (max) steel: Swedish Standard; pressurized
SS 1350	0.15% C 0.6% Mn 0.05% S and P steel: Swedish Steel; normalized **DPN: 120 UTS: 400 Elon: 27% Proof: 210**
SS 1350-6	0.15% steel: Bar; Swedish Standard; work hardened **DPN: 135 UTS: 420 Elon: 10% Proof: 360**
SS 1357	0.2% C 0.35% Mn 0.03% S and P steel: Strip; Swedish Standard; annealed **DPN: 135 UTS: 460**
SS 1357	0.2% C 0.35% Mn 0.03% S and P steel: Strip; Swedish Standard; cold rolled 225 740
SS 1360	0.15% C (max) 1.2% Mn 0.05% S and P steel: Sheet; Swedish Standard **UTS: 360 Elon: 18% Proof: 280**
SS 1370	0.15% C 0.7% Mn 0.03% S and P steel: Swedish Standard
SS 1386	Steel: Bar for reinforcing; Swedish Standard; included in MNC 815
SS 1387	Steel: For reinforcing; Swedish Standard included in MNC 815
SS 1411	0.17% C 0.7% Mn 0.06% S 0.08% P steel: For reinforcing; Swedish Standard **Elon: 20% Proof: 260**
SS 1412	0.2% C 0.3% Cr (max) 0.4% Cu steel: Structural; Swedish Standard
SS 1413	0.18% C 0.3% Cr (max) 0.4% Cu steel: Structural; Swedish Standard
SS 1414	0.18% C 0.3% Cr (max) 0.4% Cu steel: Structural; Swedish Standard
SS 1426	0.2% C (max) 1.0% Mn (max) 0.05% S and P steel: Sheet for cold rolling; Swedish Standard
SS 1430	0.18% C (max) 0.25% Cr (max) 0.03% V (max) 0.01% Nb (max) steel: Swedish Standard; tube
SS 1434	0.22% C (max) 0.25% Cr (max) 0.03% V (max) 0.01% Nb (max) steel: Swedish Standard; tube
SS 1435	0.22% C (max) 0.25% Cr (max) 0.03% V (max) 0.01% Nb (max) steel: Swedish Standard; tube
SS 1450	0.22% C 0.7% Mn steel: Swedish Standard; general engineering **DPN: 135 UTS: 480 Elon: 24% Proof: 230**
SS 1912	0.09% C 0.6% Mn 0.25% S 0.08% P steel: Free machining; Swedish Standard; cold rolled **UTS: 440**
SS 1914	0.11% C 1.1% Mn 0.3% S 0.25% Pb steel: Free machining; Swedish Standard
SS 1922	0.16% C 1% Mn 0.15% S 0.05% P steel: Free machining; Swedish Standard; cold rolled **UTS: 760**
SS 1926	0.15% C 1.0% Mn 0.2% S 0.25% Pb steel: Free machining; Swedish Standard
SS 2032	Steel: For reinforcing; Swedish Standard; included in MNC 815
SS 2101	0.2% C (max) 0.25% Cr (max) 0.03% V (max) 0.01% Nb (max) steel: Swedish Standard; tube
SS 2106	0.18% C 1.4% Mn 0.035% S (max) 0.035% P (max): For pressure vessels; Swedish Standard **UTS: 400 Elon: 21%**

Note. The following abbreviations and units are used in the tables:

DPN	Hardness, diamond pyramid number
UTS	Ultimate tensile strength, N/mm^2
Elon	Elongation, %
Proof	0.1% proof strength, N/mm^2

1 N/mm^2=0.1 hbar=0.102 kgf/mm^2=0.06475 $tonf/in.^2$=145.04 $lbf/in.^2$=1 MPa

See Appendix II for other abbreviations and conversion tables.

Symbol	Nominal analysis, supplier, condition and remarks.
SS 2106	0.2% C (max) 1.6% Mn (max) 0.02% N (max) steel: For pressure vessels; Swedish Standard; normalized **UTS: 510 Elon: 22% Proof: 350**
SS 2107	0.2% C (max) 1.4% Mn 0.035% S (max) 0.035% P (max): For pressure vessels; Swedish Standard **UTS: 400 Elon: 22%**
SS 2107	0.2% C (max) 1.6% Mn (max) 0.02% N (max) steel: For pressure vessels; Swedish Standard; normalized **UTS: 510 Elon: 22% Proof: 350**
SS 2116	0.2% C (max) 1.8% Mn (max) 0.03% S and P (max) 0.02% N (max) steel: For pressure vessels; Swedish Standard
SS 2117	0.2% C (max) 1.8% Mn (max) 0.035% S and P (max) 0.02% N (max) steel: For pressure vessels; Swedish Standard
SS 2121	0.2% C (max) 1.5% Mn 0.05% S and P (max) steel: Sheet; Swedish Standard; galvanized **UTS: 400 Elon: 16% Proof: 320**
SS 2122	0.2% C (max) 1.0% Mn 0.05% S 0.05% P steel: Plate; Swedish Standard **UTS: 390 Elon: 14%**
SS 2122	0.2% C (max) 1.5% Mn 0.05% S and P (max) steel: Plate; Swedish Standard; galvanized **UTS: 430 Elon: 14% Proof: 350**
SS 2122	0.2% C (max) 1.5% Mn 0.05% S and P steel: Sheet; Swedish Standard **UTS: 430 Elon: 14% Proof: 350**
SS 2127	0.15% C 1% Cr steel: Swedish Standard
SS 2132	0.2% C (max) 1.6% Mn 0.035% S and P (max) 0.02% N (max) steel: For structures; Swedish Standard
SS 2133	0.2% C (max) 1.6% Mn 0.035% S and P (max) 0.02% N steel: For structures; Swedish Standard
SS 2134	0.2% C (max) 1.6% Mn 0.035% S and P (max) 0.02% N steel: For structures; Swedish Standard
SS 2135	0.2% C (max) 1.6% Mn 0.035% S and P (max) 0.02% N steel: For structures; Swedish Standard
SS 2136	0.2% C (max) 1.5% Mn (max) 0.05% S and P steel: Sheet for cold rolling; Swedish Standard
SS 2136	0.2% C (max) 1.5% Mn 0.05% S and P steel: Sheet; Swedish Standard **UTS: 400 Elon: 16% Proof: 320**
SS 2137	Plain carbon steel: For reinforcing **UTS: 1000 Elon: 3% Proof: 800**
SS 2142	0.2% C (max) 1.8% Mn 0.035% S and P (max) 0.02% N steel: For structures; Swedish Standard
SS 2143	0.2% C (max) 1.8% Mn 0.035% S and P 0.02% N steel: For structures; Swedish Standard
SS 2144	0.2% C (max) 1.8% Mn 0.035% S and P (max) 0.02% N steel: For structures; Swedish Standard
SS 2145	0.2% C (max) 1.8% Mn 0.035% S and P (max) 0.02% N steel: For structures; Swedish Standard
SS 2146	0.2% C (max) 1.8% Mn steel: Swedish Standard
SS 2164	Steel: For reinforcing; Swedish Standard; included in MNC 815
SS 2165	0.24% C (max) 1.6% Mn (max) 0.05% S 0.06% P steel: For reinforcing; Swedish Standard **Elon: 15% Proof: 370**
SS 2167	Steel: For reinforcing; Swedish Standard; included in MNC 815
SS 2168	0.28% C (max) 1.6% Mn (max) 0.05% S 0.06% P steel: For reinforcing; Swedish Standard **Elon: 12% Proof: 590**
SS 2172-04	0.2% C (max) steel: For case hardening; Swedish Standard; normalized **UTS: 500 Elon: 21% Proof: 320**
SS 2172-21	0.2% C (max) 1.5% Mn (max) 0.05% Si steel: Casting; Swedish Standard; normalized **UTS: 500 Elon: 18% Proof: 300**
SS 3401	0.09% C 1.4% Mn steel: Weld electrode; Swedish Standard

Symbol	Nominal analysis, supplier, condition and remarks.
SS 3402	0.09% C 1.4% Mn steel: Weld electrode; Swedish Standard
SS 3403	0.09% C 1.4% Mn steel: Weld electrode; Swedish Standard
SS 3404	0.09% C 1.8% Mn steel: Weld electrode; Swedish Standard
SS 3405	0.09% C 1.8% Mn steel: Weld electrode; Swedish Standard
SS 3406	0.09% C 1.8% Mn steel: Weld electrode; Swedish Standard
SS 3421	0.09% C 1.4% Mn steel: Weld electrode; Swedish Standard
SS 3422	0.09% C 1.4% Mn steel: Weld electrode; Swedish Standard
SS 3423	0.09% C 1.4% Mn steel: Weld electrode; Swedish Standard
SS 3424	0.09% C 1.8% Mn steel: Weld electrode; Swedish Standard
SS 3425	0.09% C 1.8% Mn steel: Weld electrode; Swedish Standard
SS 3426	0.09% C 1.8% Mn steel: Weld electrode; Swedish Standard
SS 8001	0.2% C sintered iron: Swedish Standard
SS 8020	0.2% C (max) 0.35% P sintered steel: Swedish Standard
St 10	0.15% C (max) 0.35% Mn 0.06% S and P steel: German Standard
St 10/01	0.15% C (max) 0.35% Mn 0.06% S and P steel: German Standard
St 10/02	0.15% C (max) 0.35% Mn 0.06% S and P steel: German Standard
St 10/03	0.15% C (max) 0.35% Mn 0.06% S and P steel: German Standard
St 34/2	0.17% C (max) 0.05% S and P steel: German Standard
St 34/3	0.17% C 0.05% S and P 0.01% N steel: German Standard
St 35	0.18% C (max) 0.05% S and P steel: German Standard
St 35/4	0.17% C (max) 0.4% Mn 0.3% Si 0.05% S and P steel: German Standard
St 35/13 K	0.09% C 0.4% Mn steel: Designation used by German Standards
St 37	0.2% C (max) 0.05% S 0.08% P steel: German Standard
St 37/2	0.2% C 0.05% S 0.06% P steel: German Standard
St 37/3	0.2% C (max) 0.05% S and P steel: German Standard
St 42	0.25% C (max) 0.05% S 0.08% P steel: German Standard
St 42/2	0.25% C (max) 0.05% S and P steel: German Standard
St 42/3	0.25% C (max) 0.05% S and P steel: German Standard
St 45	0.25% C (max) 0.3% Cr 0.05% S and P steel: German Standard
St 45/4	0.22% C (max) 0.4% Mn 0.2% Si 0.05% S and P steel: German Standard
St 52	0.2% C (max) 1.5% Mn 0.5% Si 0.05% S and P steel: German Standard
St 52/3	0.2% C (max) 0.05% S and P steel: German Standard
St 52/4	0.2% C (max) 1.5% Mn 0.3% Si 0.05% S and P steel: German Standard
St 52/302	0.2% C (max) 0.05% S and P steel: German Standard
STA 5 V1A	0.1% C steel: Free machining; replaced by BS alloy En1A
STA 5 V1B	0.2% C steel: Semi-free cutting; replaced by BS alloy En7A
STA 5 V2	0.2% C steel: Cold drawn; replaced by BS alloy En2
STA 5 V2A	0.12% C steel: Replaced by BS alloy En2A
STA 5 V2A/1	0.1% C steel: Replaced by BS alloy En2A/1
STA 5 V2B	0.15% C steel: Replaced by BS alloy En2B
STA 5 V2C	0.2% C steel: Replaced by BS alloy En2C
STA 5 V2D	0.22% C steel: Replaced by BS alloy En2D
STA 5 V3	0.2% C steel: Replaced by BS alloy En3A
STA 5 V15	0.15% C (max) steel: Replaced by BS alloy En32C

Symbol	Nominal analysis, supplier, condition and remarks.
STA 5 V15/1	0.15% C (max) steel: Replaced by BS alloy En32C
STA 5 V15/A	0.18% C (max) steel: Replaced by BS alloy En201
STA 5 V15A/1	0.18% C steel: Obsolete
STA 5 V15AM	0.18% C (max) steel: Free cutting; replaced by BS alloy En202
STA 38 B	Low carbon killed steel: For drawn shell bodies
STC 1	0.2% C 0.8% Mn 0.2% Si 0.05% S and P steel: Spencer; normalized **DPN: 150 UTS: 450 Elon: 25%**
STELCO C845	0.2% C 1.2% Mn 0.005% Nb 0.2% Cu steel: Steel Co. of Canada
STELCO C850	0.2% C 1.2% Mn 0.005% Nb 0.2% Cu steel: Steel Co. of Canada
STELCO C855	0.2% C 1.2% Mn 0.005% Nb 0.2% Cu steel: Steel Co. of Canada
STELCO C860	0.2% C 1.2% Mn 0.005% Nb 0.2% Cu steel: Steel Co. of Canada
STELCOLOY G	0.12% C 0.75% Mn 0.45% Ni 0.45% Cr 0.45% Cu steel: Steel Co. of Canada
STELCOLOY S	0.15% C 1.35% Mn 0.3% Ni 0.32% Cr 0.02% V 0.3% Cu steel: Steel Co. of Canada
STELVETITE G	PVC on one side of hot dip galvanized steel: British Steel
STELVETITE R	PVC on both sides of zinc-plated steel: British Steel
STELVETITE Z	Coloured PVC on one side of zinc-plated steel: British Steel
STUBS 5	Low C Si Fe welders electrode; Stubs; for welding grey cast iron **DPN: 185 UTS: 340**
STUBS 610	Mild steel: Welding electrode; Stubs **UTS: 530 Elon: 25% Proof: 450**
STUBS 611	Mild steel: Welding electrode; Stubs **UTS: 510 Elon: 25% Proof: 450**
STUBS 612	Mild steel: Welding electrode; Stubs **UTS: 540 Elon: 25% Proof: 420**
STUBS 614	Mild steel: Welding electrode; Stubs; low hydrogen electrode **UTS: 550 Elon: 32% Proof: 410**
STUBS A1	Mild steel: Welding electrode; Stubs; general purpose
STUBS A15	0.12% C (max) 1.3% Mn 0.2% Al 0.4% Cu (max) steel: Welding electrode; Stubs; suitable for TIG welding
STUBS A18	0.12% C 1.2% Mn 0.03% S and P mild steel: Electrode; Stubs; for CO_2–argon welding **UTS: 524 Elon: 30% Proof: 400**
SUPAVIT	Mild steel: Sheet decarburized for vitreous enamelling; Richard Thomas and Baldwins
SUPER TUFCOR	0.15% C 0.9% Mn steel: Jonas; core properties; hardened and tempered **DPN: 170 UTS: 580 Elon: 20%**
SUPERCOUPE	Analysis not supplied: Mild steel; welding electrode; Soudametal; cutting and gouging
SUPERCOUPE A	Analysis not supplied: Mild steel; welding electrode; Soudametal; gouging and cutting
SUPERELSO 60	0.2% C (max) 1.2% Mn 0.03% S and P steel: Crusot
SUPERTOUGH B20	0.2% C 1.5% Mn steel: ESC for BS alloy En14A
T	0.15% C 0.45% Mn 0.3% Si steel: Casting; Edgar Allen; Code No 605 steel

Note. The following abbreviations and units are used in the tables:

DPN	Hardness, diamond pyramid number
UTS	Ultimate tensile strength, N/mm^2
Elon	Elongation, %
Proof	0.1% proof strength, N/mm^2

1 N/mm^2=0.1 hbar=0.102 kgf/mm^2=0.06475 $tonf/in.^2$=145.04 $lbf/in.^2$=1 MPa

See Appendix II for other abbreviations and conversion tables.

Symbol	Nominal analysis, supplier, condition and remarks.
T 708	0.09% C 1.0% Mn 0.25% S 0.05% P steel: Free machining; Fagersta **UTS: 390 Elon: 28%**
T 709	0.09% C 1.0% Mn 0.25% S 0.05% P steel: Free machining; Fagersta **UTS: 390 Elon: 28%**
T 715	0.15% C 0.9% Mn 0.25% Si 0.15% S 0.05% P steel: Free machining; Fagersta **UTS: 460 Elon: 25%**
T 725	0.25% C 0.9% Mn 0.25% Si 0.15% S 0.05% P steel: Free machining; Fagersta **UTS: 530 Elon: 22%**
T P W	0.06% C 0.5% Mn steel: Welding electrode; Murex
T St 0	0.12% C (max) 0.3% Mn 0.06% S 0.08% P steel: German Standards
T St 10	0.15% C (max) 0.35% Mn 0.06% S 0.08% P steel: German Standards
T St 10/01	0.15% C (max) 0.35% Mn 0.06% S 0.08% P steel: German Standards
T St 10/02	0.15% C (max) 0.35% Mn 0.06% S 0.08% P steel: German Standards
T St 10/03	0.15% C (max) 0.35% Mn 0.06% S 0.08% P steel: German Standards
TENAPSO	0.07% C 0.35% Mn 0.2% Si 0.35% Cu steel: Pompey
TENASOUDO 50	0.07% C 1.0% Mn 0.6% Si mild steel: Welding rod; Soudametal; for repair welds **UTS: 580 Elon: 26% Proof: 485**
TENSITRODE 48.30	0.08% C 1% Mn 0.6% Si steel: For electrodes; Esab **UTS: 589 Elon: 28% Proof: 494**
TENSITRODE 55.00	0.08% C 1.5% Mn 0.60% Si steel: For electrodes; Esab **UTS: 602 Elon: 28% Proof: 570**
TGS	0.2% C 1.3% Mn steel: Carpenter
THERMOFIXE	Analysis not supplied: Mild steel; welding rod; Soudametal; for free heating
TOLEDO 15	0.2% C 0.8% Mn 0.06% S and P steel: Toledo for BS alloy En2
TOLEDO 20	0.2% C 1% Mn 0.06% S and P steel: Toledo for BS alloy En3 En3A and En3B
TONCAN	0.03% C 0.07% Mo 0.45% Cu steel: Corrosion resistant; no further information
TT St 35	0.16% C (max) 0.5% Mn 0.2% Si 0.05% S and P steel: German Standard
TT St 41	0.2% C (max) 0.6% Mn 0.2% Si 0.05% S and P steel: German Standard
TT St 45	0.22% C (max) 0.5% Mn 0.2% Si 0.05% S and P steel: German Standard
TU St 1	0.12% C (max) 0.3% Mn 0.06% S and P steel: German Standard
TU St 2	0.1% C (max) 0.3% Mn 0.05% S and P steel: German Standard
TU St 34	0.08% C 0.25% Mn 0.06% S 0.08% P steel: Designation used by German Standards
TU St 37	0.2% C (max) 0.05% S 0.08% P steel: German Standard
TU St 37/02	0.2% C (max) 0.05% S 0.08% P steel: German Standard
TU St 37/03	0.2% C (max) 0.05% S 0.08% P steel: German Standard
TU St 37/04	0.2% C (max) 0.05% S 0.08% P steel: German Standard
TU St 37/05	0.2% C (max) 0.05% S 0.08% P steel: German Standard
TUBROD 15.10	0.06% C 1.7% Mn 0.9% Si steel: Tubular welding rod; Esab; as welded **UTS: 675 Elon: 24% Proof: 575**
TUFCOR	0.15% C 0.7% Mn steel: Jonas; core properties; hardened and tempered **DPN: 148 UTS: 480 Elon: 20%**
TYPE M	Mild steel: Multiwire Al-coated wire; Copperweld Co.
TYPE XX	0.02% C 0.1% Mn steel: Welding electrode; Murex **UTS: 400 Elon: 30% Proof: 370**

Symbol	Nominal analysis, supplier, condition and remarks.
U St 12	0.1% C (max) 0.3% Mn 0.05% S and P steel: German Standard
U St 12/03	0.1% C (max) 0.3% Mn 0.05% S and P steel: German Standard
U St 12/04	0.1% C (max) 0.3% Mn 0.05% S and P steel: German Standard
U St 12/05	0.1% C (max) 0.3% Mn 0.05% S and P steel: German Standard
U St 13	0.1% C (max) 0.3% Mn 0.04% S and P steel: German Standard
U St 13/03	0.1% C (max) 0.3% Mn 0.04% S and P steel: German Standard
U St 13/04	0.1% C 0.3% Mn 0.04% S and P steel: German Standard
U St 13/05	0.1% C (max) 0.3% Mn 0.04% S and P steel: German Standard
U St 14	0.1% C (max) 0.3% Mn 0.03% S and P steel: German Standard
U St 14/04	0.1% C (max) 0.3% Mn 0.03% S and P steel: German Standard
U St 14/05	0.1% C (max) 0.3% Mn 0.03% S and P steel: German Standard
U St 34/2	0.17% C (max) 0.05% S and P steel: German Standard
U St 37	0.2% C (max) 0.05% S 0.08% P steel: German Standard
U St 37/2	0.2% C (max) 0.05% S and P steel: German Standard
U St 37/202	0.2% C (max) 0.05% S and P steel: German Standard
U St 37/203	0.2% C (max) 0.05% S and P steel: German Standard
U St 37/204	0.2% C (max) 0.05% S and P steel: German Standard
U St 37/205	0.2% C (max) 0.05% S and P steel: German Standard
U St 42	0.25% C (max) 0.05% S and P steel: German Standard
U St 42/2	0.25% C (max) 0.05% S and P steel: German Standard
U St 42/202	0.25% C (max) 0.05% S and P steel: German Standard
U St 42/203	0.25% C 0.05% S and P steel: German Standard
U St 42/204	0.25% C (max) 0.05% S and P steel: German Standard
U St 42/205	0.25% C 0.05% S and P steel: German Standard
UHB 3	0.15% C (max) 0.5% Mn 0.15% Si 0.03% S and P steel: Tube; Uddelholm **UTS: 370 Elon: 30% Proof: 210**
UHB 4M10	0.22% C 0.8% Mn 0.2% Si 0.03% S and P steel: Tube; Uddelholm **UTS: 450 Elon: 30% Proof: 240**
ULTRAMILD	0.04% C 0.3% Si steel: Electrode; Metrode
UMS 45	0.22% C 0.45% Mn 0.25% Si 0.045% S and P steel: Krupp
UNBREAKABLE	0.14% C 0.9% Mn 0.04% S and P steel: For case hardening; Swift Levick for BS alloy En32
UNI 743	Replaced by UNI 5334
UNI 815	Replaced by UNI 5335
UNI 2952	Replaced by UNI 5331
UNI 2953	Replaced by UNI 5331
UNI 4365	Italian Standard covering: Plain carbon and low alloy steels for bolts
UNI 5331 C 10	0.1% C 0.5% Mn steel: For carburizing
UNI 5331 C 16	0.15% C 0.5% Mn steel: For carburizing
UNI 5334 Fe 33	C not specified: Hot rolled steel: Section **UTS: 40 Elon: 17%**
UNI 5334 Fe 34	0.17% C: Hot rolled steel: Section; graded by S and P content; Italian Standard **UTS: 380 Elon: 28% Proof: 180**
UNI 5334 Fe 37	0.2% C: Hot rolled steel: Section; graded by S and P content; Italian Standard **UTS: 400 Elon: 26% Proof: 210**
UNI 5334 Fe 42	0.2% C: Hot rolled steel: Section; graded by S and P content; Italian Standard **UTS: 460 Elon: 23% Proof: 220**
UNI 5334 Fe 50	C not specified: Hot rolled steel: Section; graded by S and P content; Italian Standard **UTS: 550 Elon: 19% Proof: 300**
UNI 5335 Fe 33	0.06% S 0.08% P steel: Plate; hot rolled; Italian Standard **UTS: 350 Elon: 17%**
UNI 5335 Fe 34	0.17% C steel: Plate; hot rolled; graded by S and P content; Italian Standard **UTS: 360 Elon: 27% Proof: 180**
UNI 5335 Fe 37	0.2% C steel: Plate; hot rolled; graded by S and P content; Italian Standard **UTS: 400 Elon: 25% Proof: 210**
UNI 5335 Fe 42	0.21% C steel: Plate; hot rolled; graded by S and P content; Italian Standard **UTS: 440 Elon: 22% Proof: 230**
UNI 5335 Fe 50	C not specified steel: Plate; hot rolled; graded by S and P content; Italian Standard **UTS: 540 Elon: 19% Proof: 280**
UNION 32	0.2% C 1.2% Mn 0.15% Cr 0.05% Al 0.25% Cu steel: Dortmund-Horder
UNION 36	0.22% C (max) 1.4% Mn 0.04% S and P steel: Dortmund-Horder; Al killed
UNION 40	0.24% C (max) 1.5% Mn 0.04% S and P (max) steel: Dortmund-Horder; Al killed
UNION 45	0.24% C (max) 1.6% Mn 0.15% Ti 0.04% S and P (max) steel: Dortmund-Horder; Al killed
UNION HB36	0.22% C (max) 1.4% Mn 0.05% S and P (max) steel: Dortmund-Horder; Al killed
UNION Q38	0.2% C (max) 1.1% Mn 0.04% S and P (max) steel: Dortmund-Horder
UNION Q42	0.22% C (max) 1.4% Mn 0.04% S and P (max) steel: Dortmund-Horder
UNITRODE 48.00	0.080% C 0.70% Mn 0.60% Si steel: For electrodes; Esab **UTS: 525 Elon: 30% Proof: 417**
VELVARC TWO	0.08% C 0.6% Mn steel: Welding electrode; Philips **UTS: 475 Elon: 24% Proof: 390**
VERSITRODE 46.58	0.08% C 0.045% Mn 0.15% Si steel: For electrodes; Esab **UTS: 510 Elon: 28% Proof: 401**
VODEX	0.06% C 0.4% Mn steel: Welding electrode; Murex **UTS: 500 Elon: 33% Proof: 450**
W 6	0.11% C steel: Welding electrode; Cu covered; Armco; as welded **UTS: 530 Elon: 27% Proof: 450**
W 61	0.11% C 1.0% Mn 0.2% Si steel: Welding electrode; Cu covered; Armco; as welded **UTS: 620 Elon: 25% Proof: 540**
WELCON 2H	0.18% C (max) 1.35% Mn 0.04% S and P (max) steel: Japan Steel Works
WELCON 50	0.18% C (max) 1.35% Mn 0.035% S and P (max) steel: Japan Steel Works
WEL-TEN 50	0.18% C 1.3% Mn 0.035% S and P (max) steel: Yawata
WEL-TEN 55	0.18% C (max) 1.35% Mn 0.035% S and P (max) steel: Yawata
WU St 12	0.1% C (max) 0.3% Mn 0.05% S and P steel: German Standard
WU St 12/03	0.1% C 0.3% Mn 0.05% S and P steel: German Standard
WU St 12/04	0.1% C 0.3% Mn (max) 0.05% S and P steel: German Standard
WU St 12/05	0.1% C 0.3% Mn (max) 0.05% S and P steel: German Standard
WU St 37/2	0.2% C 0.05% S 0.06% P steel: German Standard
WU St 37/202	0.2% C (max) 0.05% S 0.06% P steel: German Standard
WU St 37/203	0.2% C (max) 0.05% S 0.06% P steel: German Standard
WU St 37/204	0.2% C (max) 0.06% P 0.05% S steel: German Standard
WU St 37/205	0.2% C (max) 0.05% S 0.06% P steel: German Standard
XC 18 S	0.15% C 0.7% Mn steel: Carburizing; AFNOR

Symbol	Nominal analysis, supplier, condition and remarks.
XL CUT	0.09% C (max) 0.9% Mn 0.3% S steel: British Steel; free machining
XL CUT	0.09% C (max) 0.97% Mn 0.3% S 0.05% P (max) steel: British Steel; free cutting
XL CUT Pb	0.09% C (max) 0.9% Mn 0.3% S 0.15% Pb (min) steel: British Steel; free machining
XL CUT Pb	0.09% C (max) 0.97% Mn 0.3% S 0.06% P (max) 0.15% Pb (min) steel: British Steel; free machining
XL CUT SPb	0.08% C (max) 0.97% Mn 0.3% S 0.09% P (max) 0.2% Pb (min) 0.008% N steel: British Steel; free machining
XL CUT SPb	0.08% C (max) 0.9% Mn 0.3% S 0.2% Pb (min) steel: British Steel; free machining
Y 12	0.17% C (max) 0.25% Mn 0.025% S (max) 0.035% P (max) French Standard
YAW-TEN 50	0.16% C (max) 1.1% Mn 0.3% Ti 0.25% Cu steel: Yawata
YB TEN	0.12% C (max) 0.75% Mn 0.04% S and P steel: Youngstown Steel Co.
YEO 40	0.22% C 1.3% Mn 0.1% Ti steel: Yamata
YND 33	1.05% Mn 0.1% Cr 0.035% S and P (max) steel: Yamata
YOLOY 50W	0.17% C 0.7% Mn 0.05% S and P (max) steel: Youngstown Steel Co.
YOLOY M, G and A	0.25% C (max) 1.6% Mn 0.2% Cu steel: Youngstown Steel Co.
YO-MAN B	0.25% C (max) 1.6% Mn 0.2% Cu steel: Youngstown Steel Co.
Z 908	0.08% C 1.2% Mn 0.35% S 0.08% P steel: Free machining; VEW
Z 952	0.08% C 1.2% Mn 0.35% S 0.23% Pb steel: Free machining; VEW
ZINCGRIP	Mild steel: Sheet; Zn coated; Armco
ZINCROMETAL	Mild steel sheet coated with zinc-rich paint; weldable; Dacral
ZINTEX	Mild steel: Sheet; zinc coated; Origin unknown

Note. The following abbreviations and units are used in the tables:

DPN	Hardness, diamond pyramid number
UTS	Ultimate tensile strength, N/mm^2
Elon	Elongation, %
Proof	0.1% proof strength, N/mm^2

1 N/mm^2=0.1 hbar=0.102 kgf/mm^2=0.06475 tonf/in.^{2}=145.04 lbf/in.^{2}=1 MPa

See Appendix II for other abbreviations and conversion tables.

44A2 Steel – plain carbon 0.25–0.45 per cent

Specific gravity	7.86
Density	7860 kg/m^3
Solidus/liquidus	1470–1500 °C
Thermal conductivity	41.9 W/m °C
Coefficient of linear expansion (20–100 °C)	11 × 10^{-6}/ °C
Electrical conductivity	8–10% IACS (copper 100%)
Specific resistance	180–250 microhm mm
Young's modulus of elasticity	–
Impact	88–115 J
Fatigue strength	–
Hot strength	

Temperature °C	Tensile strength N/mm^2	Elongation %
200	450	37
300	450	39
400	420	38
500	200	38
650	210	45

The above properties are typical of the following group and may not apply to any one specification. It is possible that with certain specifications some of the values may not be applicable.

General metallurgical characteristics

The steels in this group are all hardenable by thermal treatment and cold work. The degree of hardenability is directly proportional to the carbon content, modified to some extent by the presence of impurities which tend to increase the hardness while decreasing the ductility by a disproportional amount. All the steels listed have some silicon and manganese, these being added as de-oxidizing and de-sulphurizing elements during manufacture. The quantities present have no significant effect on the properties.

The quality of the steels can be roughly assessed from the phosphorus and sulphur contents, as both these elements should be as low as possible, although some steels have high sulphur, lead or tellurium additions to aid machinability, but this is not normally required on these medium carbon materials.

Some of the lower carbon steels are used in the carburized condition.

Small additions of copper have been made to certain of the alloys listed to improve the atmospheric corrosion resistance and the tensile properties. Many of the steels will have residual elements present which could have an adverse effect on welding as they tend to increase the hardenability.

The properties of the steels in this group can be improved to a measurable degree by thermal treatment. This requires that they should be taken to almost 900 °C and quenched. The quench will almost invariably be with water but with some of the higher carbon steels, particularly those with residual elements, oil may be necessary to prevent cracking.

Where thicker sections, above 15–20 mm, are involved it will be essential to quench in water.

Along with the advantages in being able to harden by thermal treatment are the disadvantages that because of the increased hardenability these steels are prone to cracking at welding.

The presence of tramp elements, such as chromium, molybdenum and vanadium, can cause problems with welding. Where the material is to have its mechanical properties improved by thermal treatment then these tramp elements are a considerable advantage, whereas when the materials are to be welded they can cause problems resulting in heat affected zone cracking.

It is now common practice that the carbon equivalent (CE) is included in the material specification at a controlled level. This will seldom be above 0.7% for materials in this group. The carbon equivalent formula is

$$CE = C + \frac{Mn}{6} + \frac{Cr + Mo + V}{5} + \frac{Ni + Cu}{15}$$

It is recommended that materials in this section are received in the normalized condition prior to machining.

No technical problems exist with any of these steels regarding welding, brazing or soldering but as stated above, if the carbon equivalent is high, then pre-heating and controlled heat input is essential to prevent heat affected zone cracking.

These materials have similar corrosion resistance to those in Section 44A1. That is, they produce a loose oxide which falls off under very little stress allowing further corrosion to take place. The fact that the materials are sometimes used under reasonable stress means that corrosion pitting could have a serious effect on the fracture toughness and the fatigue resistance.

There are now well defined and proven techniques for carrying out corrosion protection on these steels. This requires expert advice.

The machinability of these materials can be reasonable but many have sulphur, tellurium, selenium or lead added to improve machinability. This will give some reduction in impact and ductility.

The material should always be machined in the normalized or hardened and tempered condition. Machining in the 'as forged' or 'as rolled' condition can result in considerable variations and local tearing and also dimensional changes during service.

These steels are general purpose engineering and the first choice for low price medium strength components such as boilers, sections for buildings under high stress, roof trusses, lifting equipment, gears, springs and low static and moving parts in all forms of engines particularly where the section is quite small. Pipelines, pylons, rails and sheeting which form part of stress components are all made using these steels.

The materials in this group find considerable competition from low alloy steels and from materials such as higher alloyed aluminium where stress is involved with corrosion and also, in many cases, stainless steel and some copper alloys.

Some materials in this group can be carburized.

Symbol	Nominal analysis, supplier, condition and remarks.
0.0652	0.36% C 0.6% Mn 0.2% Si 0.04% S and P steel: German Standard
0.0967	0.38% C 0.7% Mn 1.5% Si steel: For springs; German Standard
0.0968	0.46% C 0.7% Mn 1.7% Si steel: For springs; German Standard
1.0130	0.25% C (max) 0.05% S 0.08% P steel: German Standard
1.0132	0.25% C (max) 0.05% S 0.06% P steel: German Standard
1.0136	0.25% C (max) 0.05% S and P steel: German Standard
1.0142	0.25% C (max) 0.05% S 0.06% P steel: German Standard
1.0143	0.25% C (max) 0.05% S and P steel: German Standard

Note. The following abbreviations and units are used in the tables:

DPN	Hardness, diamond pyramid number
UTS	Ultimate tensile strength, N/mm^2
Elon	Elongation, %
Proof	0.1% proof strength, N/mm^2

1 N/mm^2=0.1 hbar=0.102 kgf/mm^2=0.06475 tonf/in.^{2}=145.04 lbf/in.^{2}=1 MPa

See Appendix II for other abbreviations and conversion tables.

Symbol	Nominal analysis, supplier, condition and remarks.
1.0507	0.36% C 0.6% Mn 0.05% S and P steel: German Standard
1.0509	0.36% C 0.4% Mn 0.2% Si 0.05% S and P steel: German Standard
1.0530	0.3% C 0.05% S 0.08% P steel: German Standard
1.0532	0.3% C 0.05% S 0.06% P steel: German Standard
1.0540	0.4% C 0.05% S 0.08% P steel: German Standard
1.0542	0.4% C 0.05% S 0.06% P steel: German Standard
1.0651	0.36% C 0.6% Mn 0.2% Si 0.045% S and P steel: German Standard
1.0726	0.38% C 0.7% Mn 0.2% Si 0.2% S 0.07% P steel: Free machining; German Standard
1.0727	0.46% C 0.7% Mn 0.2% Si 0.2% S 0.07% P steel: Free machining; German Standard
1.0903	0.51% C 0.65% Mn 0.045% S and P (max) steel: For springs; German Standard
1.0970	0.38% C 0.65% Mn 0.045% S and P (max) steel: For springs; German Standard
1.1165	0.31% C 1.4% Mn steel: German Standard
1.1174	0.36% C 0.5% Mn 0.2% Si 0.035% S and P steel: German Standard
1.1181	0.36% C 0.5% Mn 0.2% Si 0.035% S and P steel: German Standard
1.1183	0.36% C 0.5% Mn 0.25% Si 0.035% S and P steel: German Standard

Symbol	Nominal analysis, supplier, condition and remarks.
1.1191	0.46% C 0.6% Mn 0.25% Si 0.035% S and P steel: German Standard
1.1193	0.46% C 0.7% Mn 0.25% Si 0.035% S and P steel: German Standard
1.1194	0.46% C 0.7% Mn 0.25% Si 0.035% S and P steel: German Standard
1.1730	0.45% C 0.7% Mn 0.035% S and P (max) steel: German Standards designation
1.5038	0.4% C 1.0% Mn 0.035% S and P steel: Forging; German Standard
1.5066	0.3% C 1.3% Mn 0.3% Cr 0.03% S and P steel: Forging; German Standard
1.5067	0.36% C 1.3% Mn 0.035% S and P steel: Forging; German Standard
1.5120	0.38% C 1.1% Mn 0.8% Si steel: Forging; German Standard
1.5121	0.46% C 1.1% Mn 0.8% Si steel: Forging; German Standard
1.5122	0.37% C 1.2% Mn 1.2% Si steel: For springs; German Standard
6 LM	0.3% C 1.4% Mn 0.2% Si steel: Sandvik
7 L	0.35% C 0.5% Mn 0.25% Si steel: Sandvik for SS 1550
8 LM	0.4% C 1.3% Mn 0.2% Si steel: Sandvik for SS 2120
9 L	0.45% C 0.4% Mn 0.2% Si steel: Sandvik for SS 1650, 1660
9 L 7	0.45% C 0.75% Mn 0.25% Si steel: Sandvik
9 LM	0.45% C 1.4% Mn 0.2% Si steel: Sandvik
30 NCD 12 UNI 4365	0.3% C 2.9% Ni 0.8% Cr 0.45% Mo steel: Italian Standard; for bolts
35 S 20	0.35% C 0.6% Mn 0.2% S 0.05% P steel: Free machining; designation used by German Standards
36 Mn 5	0.36% C 1.4% Mn 0.25% Si steel: Designation used by German Standards
37 CB	0.37% C 0.8% Mn 0.4% Si steel: Electrode; Metrode
37 MS	0.37% C 0.8% Mn steel: Electrode; Metrode
38 Mn Si 4	0.38% C 1.1% Mn 0.8% Si steel: Designation used by German Standards
38 NCD 4 UNI 4365	0.38% C 0.8% Ni 0.7% Cr 0.2% Mo steel: Italian Standard; for bolts
38 Si 6	0.38% C 0.6% Mn 1.5% Si steel: Designation used by German Standards
38 Si 8	0.38% C 0.7% Mn 1.5% Si steel: Designation used by German Standards
40 NCD 7 UNI 4365	0.4% C 1.8% Ni 0.7% Cr 0.25% Mo steel: Italian Standard; for bolts
42 Mn V 7	0.42% C 1.8% Mn 0.1% V 0.25% Si steel: Designation used by German Standards
45 MS	0.45% C 0.8% Mn steel: Electrode; Metrode
45 S 20	0.46% C 0.7% Mn 0.2% S 0.05% P steel: Free machining; designation used by German Standards
46 Si 7	0.46% C 0.7% Mn 1.7% Si steel: Designation used by German Standards
50/45	0.2% C (max) 1.5% Mn steel: Sheet and plate; designation in BS 1449
50 Si 7	0.49% C 0.6% Mn 1.75% Si steel: Italian Standards designation **DPN: 250 UTS: 1450 Elon: 7% Proof: 1150**

Note. The following abbreviations and units are used in the tables:

DPN	Hardness, diamond pyramid number
UTS	Ultimate tensile strength, N/mm^2
Elon	Elongation, %
Proof	0.1% proof strength, N/mm^2

1 N/mm^2=0.1 hbar=0.102 kgf/mm^2=0.06475 tonf/in.^{2}=145.04 ibf/in.^{2}=1 MPa

See Appendix II for other abbreviations and conversion tables.

Symbol	Nominal analysis, supplier, condition and remarks.
55 EE	0.16% C (max) steel: Normalized, weldable with Nb and V; designation for BS 4360 **UTS: 610 Elon: 19%**
55 Si 8	0.55% C 0.85% Mn 2.0% Si steel: Italian Standards designation **DPN: 250 UTS: 1550 Elon: 5% Proof: 1250**
60/55	0.2% C (max) 1.5% Mn steel: Sheet and plate; designation in BS 1449
060A25	0.25% C 0.6% Mn steel: BS 970; obsolete
060A27	0.27% C 0.6% Mn steel: BS 970; obsolete
060A30	0.3% C 0.6% Mn steel: BS 970; obsolete
060A32	0.32% C 0.6% Mn steel: BS 970
060A35	0.35% C 0.6% Mn steel: BS 970; obsolete
060A37	0.37% C 0.6% Mn steel: BS 970; obsolete
060A40	0.4% C 0.6% Mn steel: BS 970
060A42	0.42% C 0.6% Mn steel: BS 970; obsolete
060A45	0.45% C 0.6% Mn steel: BS 970
060A47	0.47% C 0.6% Mn steel: BS 970
070M20	0.2% C 0.7% Mn steel: BS 970
070M26	0.26% C 0.7% Mn steel
080A27	0.27% C 0.8% Mn steel: BS 970; obsolete
080A30	0.3% C 0.8% Mn steel
080A32	0.32% C 0.8% Mn steel
080A35	0.35% C 0.8% Mn steel: BS 970
080A37	0.37% C 0.8% Mn steel: BS 970
080A40	0.4% C 0.8% Mn steel: BS 970
080A42	0.42% C 0.8% Mn steel: BS 970
080H36	0.36% C 0.8% Mn steel: BS 970
080H41	0.4% C 0.8% Mn steel: BS 970
080M30	0.3% C 0.8% Mn steel: BS 970
080M36	0.36% C 0.8% Mn steel: BS 970
080M40	0.4% C 0.8% Mn steel
120M28	0.28% C 1.2% Mn steel: BS 970
120M36	0.36% C 1.2% Mn steel: BS 970
135M44	0.44% C 1.3% Mn steel: BS 970
150M28	0.28% C 1.5% Mn steel: BS 970
150M36	0.30% C 1.5% Mn steel: BS 970
150M40	0.4% C 1.6% Mn steel: BS 970
212A37	0.37% C 1.2% Mn steel: BS 970
212A42	0.42% C 1.2% Mn steel: BS 970
212M36	0.36% C 1.2% Mn steel: BS 970
212M44	0.44% C 1.2% Mn 0.18% S steel: BS 970; obsolete
216A42	0.42% C 1.3% Mn steel: BS 970
216M28	0.28% C 1.3% Mn steel: BS 970
216M36	0.36% C 1.5% Mn 0.18% S steel: Free machining; BS 970
216M44	0.44% C 1.3% Mn steel: BS 970
225M36	0.36% C 1.2% Mn steel: BS 970
226M44	0.44% C 1.5% Mn 0.26% S steel: Free machining; BS 970
1022	0.22% C 0.9% Mn 0.04% S and P steel: Designation used by AISI
1025	0.25% C 0.04% S and P (max) steel: Designation used in the UK and USA
1026	0.26% C 0.04% S and P (max) steel: Designation used in the UK and USA
1029	0.29% C 0.8% Mn 0.05% S and P steel: Designation used by AISI
1030	0.3% C 0.04% S and P (max) steel: Designation used in the UK and USA
1033	0.33% C 0.04% S and P (max) steel: Designation used in the UK and USA
1034	0.34% C 0.04% S and P (max) steel: Designation used in the UK and USA
1035	0.35% C 0.04% S and P (max) steel: Designation used in the UK and USA
1037	0.37% C 0.04% S and P (max) steel: Designation used in the UK and USA
1038	0.38% C 0.04% S and P (max) steel: Designation used in the UK and USA

Symbol	Nominal analysis, supplier, condition and remarks.
1039	0.39% C 0.04% S and P (max) steel: Designation used in the UK and USA
1040	0.4% C 0.04% S and P (max) steel: Designation used in the UK and USA
1042	0.42% C 0.04% S and P (max) steel: Designation used in the UK and USA
1043	0.43% C 0.04% S and P (max) steel: Designation used in the UK and USA
1044	0.47% C 0.4% Mn 0.04% S and P steel: Designation used by AISI
1045	0.44% C 0.04% S and P (max) steel: Designation used in the UK and USA
1046	0.46% C 0.04% S and P (max) steel: Designation used in the UK and USA
1049	0.49% C 0.04% S and P (max) steel: Designation used in the UK and USA
1050	0.5% C 0.04% S and P (max) steel: Designation used in the UK and USA
1105	0.65% C 0.2% Mn 0.02% S and P (max) steel: French Standards designation
1132	0.32% C 0.1% S 0.04% P (max) steel: Designation used in the UK and USA
1137	0.37% C 0.1% S 0.04% P (max) steel: Designation used in the UK and USA
1139	0.39% C 1.5% Mn 0.17% S 0.04% P steel: Free machining; designation used by AISI
1140	0.4% C 0.1% S 0.04% P (max) steel: Designation used in the UK and USA
1141	0.41% C 0.1% S 0.04% P (max) steel: Designation used in the UK and USA
1144	0.44% C 0.3% S 0.04% P (max) steel: Designation used in the UK and USA
1145	0.45% C 0.1% S 0.04% P (max) steel: Designation used in the UK and USA
1146	0.46% C 0.1% S 0.04% P (max) steel: Designation used in the UK and USA
1151	0.51% C 0.1% S 0.04% P (max) steel: Designation used in the UK and USA
1306	0.55% C 0.25% Mn 0.025% S (max) 0.035% P (max) steel: French Standards designation
1307	0.45% C 0.25% Mn 0.025% S (max) 0.035% P (max) steel: French Standards designation
1308	0.38% C 0.25% Mn 0.025% S (max) 0.035% P (max) steel: French Standards designation
1330	0.3% C 1.7% Mn steel: Designation used in the UK and USA
1335	0.35% C 1.7% Mn steel: Designation used in the UK and USA
1340	0.4% C 1.7% Mn steel: Designation used in the UK and USA
1345	0.45% C 1.7% Mn steel: Designation used in the UK and USA
1524	0.24% C 1.5% Mn steel: Designation used in the UK and USA
1525	0.25% C 1% Mn steel: Designation used in the UK and USA
1526	0.26% C 1.2% Mn steel: Designation used in the UK and USA
1527	0.27% C 1.4% Mn steel: Designation used in the UK and USA
1533	0.33% C 1.4% Mn steel: Designation used in the UK and USA
1534	0.34% C 1.3% Mn steel: Designation used in the UK and USA
1536	0.36% C 1.3% Mn steel: Designation used in the UK and USA
1541	0.4% C 1.5% Mn steel: Designation used in the UK and USA
1547	0.47% C 1.5% Mn steel: Designation used in the UK and USA

Symbol	Nominal analysis, supplier, condition and remarks.
1548	0.48% C 1.2% Mn steel: Designation used in the UK and USA
1551	0.51% C 1% Mn steel: Designation used in the UK and USA
1552	0.52% C 1.3% Mn steel: Designation used in the UK and USA
2321	0.45% C steel: French Standards designation
9255	0.55% C 2% Si steel: Designation used in the UK and USA
9260	0.6% C 2% Si steel: Designation used in the UK and USA
9262	0.62% C 2% Si steel: Designation used in the UK and USA
A 2	0.23% C 0.8% Mn 0.25% Si steel: Jessop
A 3	0.32% C 0.8% Mn 0.25% Si steel: Jessop
A 3	0.32% C (max) 0.8% Mn steel: Cogne
A 4	0.42% C (max) 0.9% Mn steel: Cogne
A 4	0.35% C 0.8% Mn 0.25% Si steel: Jessop
A 5	0.4% C 0.8% Mn 0.25% Si steel: Jessop
A 5	0.52% C (max) 0.9% Mn steel: Cogne
A 6	0.45% C 0.8% Mn 0.25% Si steel: Jessop
A 16	0.27% C 1.4% Mn 0.4% Si 0.03% S and P steel: Welding rod; designation used by British Standards
A 16	0.27% C steel: Welding electrode; designation for BS 2901
A 201 A	0.25% C 0.8% Mn 0.2% Si 0.35% S and P steel: Plate; Armco; flange and firebox grades **UTS: 450 Elon: 28% Proof: 200**
A 201 B	0.3% C 0.8% Mn 0.2% Si 0.035% S and P steel: Plate; Armco; flange and firebox quality **UTS: 530 Elon: 25% Proof: 210**
A 212 A	0.3% C 0.9% Mn 0.2% Si 0.35% S and P steel: Plate; Armco; flange and firebox quality **UTS: 540 Elon: 24% Proof: 220**
A 212 B	0.32% C 0.9% Mn 0.2% Si 0.35% S and P steel: Plate; Armco; flange and firebox quality **UTS: 600 Elon: 22% Proof: 240**
A 285	0.24% C 0.8% Mn 0.035% S and P steel: Plate; Armco; flange and firebox quality in three grades A, B and C **UTS: 340 Elon: 30% Proof: 180**
A 455	0.33% C 1.0% Mn 0.1% Si 0.04% S and P steel: Plate; Armco **UTS: 580 Elon: 16% Proof: 220**
A Mn	0.25% C Mn steel: Casting; Firth Brown; normalized; tempered **DPN: 170 UTS: 550 Elon: 25% Proof: 340**
ABS/F	0.3% C (max) 0.9% Mn (max) steel: Plate; American Bureau of Shipping **UTS: 500 Elon: 19% Proof: 245**
ABS/F	0.35% C (max) 0.8% Mn (max) steel: Tube; American Bureau of Shipping
ABS/G	0.33% C (max) 0.9% Mn (max) steel: Plate; American Bureau of Shipping **UTS: 550 Elon: 17% Proof: 270**
AISI 1023	0.22% C 0.4% Mn 0.04% S and P steel
AISI 1024	0.23% C 1.5% Mn 0.04% S and P steel
AISI 1025	0.25% C 0.4% Mn 0.05% S 0.04% P steel
AISI 1026	0.26% C 0.7% Mn 0.04% S and P steel
AISI 1027	0.25% C 1.3% Mn 0.05% S 0.04% P steel
AISI 1029	0.29% C 0.8% Mn 0.05% S and P steel
AISI 1030	0.3% C 0.8% Mn 0.05% S 0.04% P steel
AISI 1033	0.33% C 0.9% Mn 0.05% S 0.04% P steel
AISI 1035	0.35% C 0.8% Mn 0.05% S 0.04% P steel
AISI 1036	0.34% C 1.2% Mn 0.05% S 0.04% P steel
AISI 1037	0.35% C 0.9% Mn 0.05% S 0.04% P steel
AISI 1038	0.38% C 0.8% Mn 0.05% S 0.04% P steel
AISI 1039	0.4% C 0.9% Mn 0.05% S 0.04% P steel
AISI 1040	0.4% C 0.8% Mn 0.05% S 0.04% P steel
AISI 1041	0.4% C 1.5% Mn 0.05% S 0.04% P steel
AISI 1042	0.43% C 0.8% Mn 0.05% S 0.04% P steel

Symbol	Nominal analysis, supplier, condition and remarks.
AISI 1043	0.43% C 0.9% Mn 0.05% S 0.04% P steel
AISI 1045	0.46% C 0.8% Mn 0.05% S 0.04% P steel
AISI 1046	0.46% C 0.9% Mn 0.05% S 0.04% P steel
AISI 1126	0.26% C 0.9% Mn 0.1% S 0.04% P steel: Free cutting
AISI 1132	0.31% C 1.5% Mn 0.1% S 0.04% P steel: Free cutting
AISI 1137	0.36% C 1.5% Mn 0.1% S 0.04% P steel: Free cutting
AISI 1138	0.37% C 0.9% Mn 0.1% S 0.04% P steel: Free cutting
AISI 1139	0.39% C 1.5% Mn 0.17% S 0.04% P steel: Free machining
AISI 1140	0.4% C 0.9% Mn 0.1% S 0.04% P steel: Free cutting
AISI 1141	0.41% C 1.5% Mn 0.1% S 0.04% P steel: Free cutting
AISI 1144	0.44% C 1.5% Mn 0.3% S 0.04% P steel: Free cutting
AISI 1145	0.45% C 0.9% Mn 0.05% S 0.04% P steel: Free cutting
AISI 1146	0.46% C 0.9% Mn 0.1% S 0.04% P steel: Free cutting
AISI 1330	0.3% C 1.8% Mn 0.3% Si steel
AISI 1335	0.35% C 1.8% Mn 0.3% Si steel
AISI 1340	0.4% C 1.8% Mn 0.3% Si steel
AISI 1345	0.4% C 1.8% Mn 0.3% Si steel
AM 4	0.35% C 1.5% Mn 0.25% Si steel: Jessop for BS alloy En 15
AMS 5020	0.35% C 0.1% S steel: Free machining
AMS 5024 C	0.37% C 1.5% Mn steel: Bar and forging; free cutting; AMS for SAE 1137
AMS 5062 A	0.25% C (max) steel: For bar, forging, tube, etc.
AMS 5070 B	0.22% C steel: Bar and forging; AMS for SAE 1022 **UTS: 370**
AMS 5075 A	0.25% C steel: Seamless tube; AMS for SAE 1025 **UTS: 370**
AMS 5077 A	0.25% C steel: Welded tube; AMS for SAE 1025 **UTS: 370**
AMS 5080 A	0.35% C steel: Bar, forging and tube; AMS for SAE 1035
AMS 5082	0.35% C steel: Seamless tube; AMS for SAE 1035 **UTS: 650**
AN QQ W 429/1	Plain carbon steel: Wire; US Service **UTS: 1830**
AN S 4	0.35% C 0.7% Mn 0.05% S and P steel: US Service
AN T 4/1	0.25% C 0.4% Mn 0.05% S and P steel: Tube; welded; US Service **UTS: 300 Elon: 22% Proof: 210**
AN WW T 846/1	0.25% C 0.4% Mn 0.05% S and P steel: Seamless tube; US Service **UTS: 300 Elon: 22% Proof: 210**
API 5A C75/1	0.5% C (max) 1.9% Mn (max) 0.5% Cr + Ni + Cu (max) 0.2% Mo steel: Casing and tubing; normalized and tempered **UTS: 655 Elon: 20% Proof: 515**
API 5A C75/2	0.4% C (max) 1.5% Mn steel: Casing and tubing; hardened and tempered **UTS: 655 Elon: 20% Proof: 515**
API 5A C75/3	0.42% C 0.85% Mn 1.0% Cr 0.2% Mo steel: Casing and tubing; normalized and tempered **UTS: 655 Elon: 20% Proof: 515**
API 5A C95	0.45% C 1.9% Mn steel: Casing and tubing; hardened and tempered **UTS: 730 Elon: 20% Proof: 655**
API 5L S	Spiral weld line pipe; See API designation for grades
API 5L X42	0.29% C (max) 1.25% Mn (max) 0.05% S (max) 0.04% P (max) steel: High test line pipe; seamless and welded **UTS: 415 Elon: 20% Proof: 290**
API 5L X46	0.30% C (max) 1.35% Mn (max) 0.05% S (max) 0.04% P (max) steel: High test line pipe; seamless and welded **UTS: 435 Elon: 20% Proof: 315**
API 5L X52	0.30% C (max) 1.35% Mn (max) 0.05% S (max) 0.04% P (max) steel: High test line pipe; seamless and welded **UTS: 470 Elon: 20% Proof: 360**
API 5L X56	0.26% C (max) 1.35% Mn (max) with Nb, V or Ti steel: High test line pipe; seamless or welded **UTS: 500 Elon: 20% Proof: 385**
API 5L X60	0.26% C (max) 1.35% Mn (max) with Nb, V or Ti steel: High test line pipe; seamless or welded **UTS: 530 Elon: 20% Proof: 415**
API 5L X65	0.26% C (max) 1.4% Mn with Nb and V steel: High test line pipe; seamless or welded **UTS: 540 Elon: 20% Proof: 445**
API 5LU X80	0.26% C (max) 1.4% Mn (max) with other elements as agreed steel: Ultra-high test line pipe **UTS: 690 Elon: 20% Proof: 550**
API 5LU X100	0.26% C (max) 1.4% Mn (max) with other elements as agreed steel: Ultra-high test line pipe **UTS: 830 Elon: 20% Proof: 690**
ASTM 122	0.25% C (max) steel: French specification
ASTM A27/60/30	0.3% C (max) 0.6% Mn 0.05% S and P steel: Casting
ASTM A27/65/35	0.3% C (max) 0.7% Mn 0.05% S and P steel: Casting
ASTM A27/70/36	0.35% C (max) 0.7% Mn 0.05% S and P steel: Casting
ASTM A27/70/40	0.25% C (max) 1.0% Mn 0.05% S and P steel: Casting
ASTM A27 N1	0.25% C (max) 0.7% Mn 0.05% S and P steel: Casting
ASTM A27 N2	0.35% C (max) 0.6% Mn 0.05% S and P steel: Casting
ASTM A27 U60/30	0.25% C (max) 0.7% Mn 0.05% S and P steel: Casting
ASTM A31 A	C not specified 0.4% Mn 0.06% S (max) 0.05% P (max) steel: For rivets **UTS: 350 Elon: 27% Proof: 160**
ASTM A31 B	0.3% C 0.5% Mn 0.06% S (max) 0.05% P (max) steel: For rivets **UTS: 440 Elon: 22% Proof: 200**
ASTM A36	0.30% C 1.0% Mn 0.10% Si steel: Plate and bar for structural use
ASTM A105/1	0.35% C 0.9% Mn 0.05% S 0.05% P steel: For flanges; normalized
ASTM A106 A	0.25% C (max) 0.5% Mn 0.05% S and P steel: Pipe
ASTM A106 B	0.3% C (max) 0.6% Mn 0.05% S and P steel: Pipe
ASTM A106 C	0.35% C (max) 0.6% Mn 0.05% S and P steel: Pipe
ASTM A108	Carbon steel: Bars and shafting; cold finished; graded to AISI code
ASTM A109/1	0.25% C 0.6% Mn 0.2% Cu (optional) cold rolled steel: Strip; hard **UTS: 640 Elon: 3%**
ASTM A109/2	0.25% C 0.6% Mn 0.2% Cu (optional) cold rolled steel: Strip; $\frac{1}{2}$-hard **UTS: 450 Elon: 10%**
ASTM A109/3	0.25% C 0.6% Mn 0.2% Cu (optional) cold rolled steel: Strip; $\frac{1}{4}$-hard **UTS: 370 Elon: 20%**
ASTM A131 A	C not specified 0.05% (max) S 0.04% (max) P steel: For ships
ASTM A131 B	0.21% C 0.95% Mn 0.05% (max) S 0.04% (max) P steel: For ships
ASTM A131 C	0.23% C 0.75% Mn 0.05% (max) S 0.04% (max) P steel: For ships
ASTM A139 A	C not specified 0.7% Mn 0.04% S and P steel: Arc welded; pipe **UTS: 330 Elon: 35% Proof: 200**
ASTM A139 B	0.3% C (max) 0.7% Mn 0.4% S and P steel: Arc welded; pipe **UTS: 400 Elon: 30% Proof: 240**

Note. The following abbreviations and units are used in the tables:

DPN	Hardness, diamond pyramid number
UTS	Ultimate tensile strength, N/mm^2
Elon	Elongation, %
Proof	0.1% proof strength, N/mm^2

1 N/mm^2=0.1 hbar=0.102 kgf/mm^2=0.06475 tonf/in.^{2}=145.04 lbf/in.^{2}=1 MPa

See Appendix II for other abbreviations and conversion tables.

Symbol	Nominal analysis, supplier, condition and remarks.
ASTM A148	Steel: Casting; high strength; graded by mechanical properties
ASTM A178 C	0.35% C 0.8% Mn 0.05% P 0.06% S steel: Electric resistance welded boiler tubes
ASTM A181/1	0.35% C (max) 0.9% Mn steel: For pipe flanges **UTS: 440 Elon: 22% Proof: 220**
ASTM A183	0.30% C (min) steel: For bolts
ASTM A194/2	0.4% C (min) steel: For bolts
ASTM A210 A1	0.27% C (max) 0.9% Mn 0.1% Si steel: Seamless tube
ASTM A210 C	0.35% C (max) 0.7% Mn 0.1% Si steel: Seamless tube
ASTM A216 WCA	0.25% C (max) 0.7% Mn 0.6% Si steel: Casting for high temperature service
ASTM A216 WCB	0.3% C (max) 1.0% Mn 0.6% Si steel: Casting for high temperature service
ASTM A216 WCC	0.25% C (max) 1.2% Mn 0.6% Si steel: Casting for high temperature service
ASTM A217 WC1	0.25% C (max) 0.65% Mn 0.5% Mo steel: Casting
ASTM A242	0.25% C 1.25% Mn 0.05% S steel: For structural use; as rolled **UTS: 480 Elon: 19% Proof: 340**
ASTM A266/1 and 2	0.35% C (max) 0.6% Mn steel: For drawn forging
ASTM A266/3	0.5% C (max) 0.65% Mn steel: For drawn forging
ASTM A284 A	0.22% C 0.9% Mn 0.2% Si 0.04% S and P steel: Plate; as rolled **UTS: 330 Elon: 28% Proof: 150**
ASTM A284 B	0.3% C 0.9% Mn 0.2% Si 0.04% S and P steel: Plate; as rolled **UTS: 370 Elon: 27% Proof: 180**
ASTM A284 C	0.33% C 0.9% Mn 0.2% Si 0.04% S and P steel: Plate; as rolled **UTS: 420 Elon: 25% Proof: 200**
ASTM A284 D	0.35% C 0.9% Mn 0.2% Si 0.04% S and P steel: Plate; as rolled **UTS: 420 Elon: 24% Proof: 220**
ASTM A290 A	0.3% C 0.8% Mn 0.05% S and P steel: Forging; hardened and tempered **DPN: 160 UTS: 580 Elon: 20% Proof: 270**
ASTM A290 B	0.35% C 0.8% Mn 0.05% S and P steel: Forging; hardened and tempered **DPN: 160 UTS: 580 Elon: 20% Proof: 270**
ASTM A290 C1	0.45% C 0.8% Mn 0.05% S and P steel: Forging; hardened and tempered **DPN: 170 UTS: 610 Elon: 20% Proof: 300**
ASTM A291/1	0.4% C 0.5% Mn 0.05% S and P steel: Forging; hardened and tempered **UTS: 610 Elon: 19% Proof: 340**
ASTM A291/2	0.42% C 0.05% S and P steel: Forging; hardened and tempered **UTS: 680 Elon: 20% Proof: 480**
ASTM A299	0.3% C 1.2% Mn 0.2% Si 0.04% S and P steel: Plate for pressure vessels
ASTM A306/45	Plain carbon steel: Bar; as rolled **UTS: 330 Elon: 33% Proof: 150**
ASTM A306/50	Plain carbon steel: Bar; as rolled **UTS: 370 Elon: 30% Proof: 170**
ASTM A306/55	Plain carbon steel: As rolled **UTS: 370 Elon: 26% Proof: 180**
ASTM A306/60	Plain carbon steel: As rolled **UTS: 440 Elon: 22% Proof: 200**
ASTM A306/65	Plain carbon steel: As rolled **UTS: 460 Elon: 20% Proof: 210**
ASTM A325	0.3% C 0.5% Mn steel: Hardened and tempered **UTS: 710 Elon: 14% Proof: 530**
ASTM A333/1	0.3% C 0.7% Mn steel: Pipe for low temperature use
ASTM A333/6	0.3% C 0.7% Mn steel: Pipe for low temperature use
ASTM A350 LF1	0.3% C (max) 1% Mn 0.04% S and P steel: For low temperature use **UTS: 400 Elon: 25% Proof: 200**
ASTM A350 LF2	0.3% C (max) 1.3% Mn 0.2% Si 0.04% S and P steel: For low temperature use **UTS: 460 Elon: 22% Proof: 240**

Symbol	Nominal analysis, supplier, condition and remarks.
ASTM A352 LCB	0.3% C 1.0% Mn 0.6% Si steel: Normalized and tempered for low temperature service **UTS: 440 Elon: 24% Proof: 220**
ASTM A356/1	0.35% C (max) steel: Casting for turbines
ASTM A372 Class I	0.30% C (max) 1.0% Mn (max) 0.27% Si steel: Forging for pressure vessels
ASTM A372 Class II	0.40% C (max) 1.29% Mn (max) 0.27% Si steel: Forging for pressure vessels
ASTM A372 Class III	0.48% C (max) 1.65% Mn (max) 0.27% Si steel: Forging for pressure vessels
ASTM A372 Class VII	0.4% C 0.7% Mn 1.8% Ni 0.8% Cr 0.2% Mo 0.3% Si steel: For pressure vessels
ASTM A374	0.22% C (max) 1.2% Mn steel: Sheet and strip; cold rolled
ASTM A375	0.22% C (max) 1.25% Mn steel: Sheet and strip; hot rolled
ASTM A381	0.32% C (max) 1.3% Mn 0.05% S and P steel: For high pressure pipe; mechanical properties vary with grades
ASTM A413	0.33% C (max) 1% Mn 0.05% S and P steel: For chains
ASTM A414	0.25% C steel: Sheet of firebox quality; various grades according to C content
ASTM A433	0.2% C 0.25% Pb free machining steel: For pressure vessels; graded according to C content
ASTM A440	0.3% C (max) 0.3% Si 0.06% S and P 0.2% Cu steel: For structural purposes
ASTM A449	0.42% C 0.6% Mn 0.04% S and P steel: For studs; hardened and tempered **DPN: 230**
ASTM A454	0.3% C 1% Mn steel: For conveyor chain; classified by tensile properties
ASTM A455	0.3% C 1.0% Mn 0.2% Si 0.04% S and P steel: Plates for pressure vessels; graded by carbon content
ASTM A465 LI	0.3% C (max) 0.9% Mn 0.05% S and P 0.2% Pb steel: For pipe flanges; as forged **UTS: 400 Elon: 22% Proof: 200**
ASTM A465 LII	0.3% C (max) 0.9% Mn 0.05% S and P 0.2% Pb steel: For pipe flanges; normalized **UTS: 400 Elon: 25% Proof: 200**
ASTM A465 LIII	0.35% C (max) 0.9% Mn 0.05% S and P 0.2% Pb steel: For pipe flanges; as forged **UTS: 460 Elon: 28% Proof: 240**
ASTM A465 LIV	0.35% C (max) 0.9% Mn 0.05% S and P 0.2% Pb steel: For pipe flanges; normalized **UTS: 460 Elon: 22% Proof: 240**
ASTM A466	0.25% C 0.05% S and P steel: Chain; weldless
ASTM A467	0.33% C (max) 1% Mn 0.05% S and P steel: Machine and coil chain
ASTM A470/1	0.45% C (max) 0.03% V steel: For rotors; vacuum treated
ASTM A486	Steel: Casting for bridges; classified by tensile properties
ASTM A487/3 N	0.35% C (max) 1.5% Mn 0.4% Mo steel: Casting for pressure; normalized; 3Q grade is hardened and tempered
ASTM A490	0.45% C 0.04% S and P steel: For bolts
ASTM A500	0.26% C (max) 0.04% S and P 0.2% Cu steel: For structural tubing; cold formed
ASTM A501	0.26% C (max) 0.04% S and P 0.2% Cu steel: For structural tubing; cold formed
ASTM A515	0.25% C steel: Plate for fusion welded pressure vessels; graded by tensile strength
ASTM A516	0.24% C steel: Plate for fusion welded pressure vessels; graded by tensile strength
ASTM A519	Seamless steel mechanical tubing: Low and medium carbon or alloy steel; graded according to AISI system
ASTM A523 B	0.26% C (max) 1.1% Mn 0.04% S and P steel: Pipe for high pressure; seamless or electric resistance welded
ASTM A541/1	0.35% C (max) 0.06% V steel: Forging
ASTM A556 B2	0.27% C (max) 0.6% Mn 0.05% S and P steel: Seamless tube; cold drawn **UTS: 420 Elon: 30% Proof: 260**

Symbol	Nominal analysis, supplier, condition and remarks.
ASTM A556 C2	0.3% C (max) 0.65% Mn 0.05% S and P steel: Seamless tube; cold drawn **UTS: 490 Elon: 30% Proof: 280**
ASTM A557	Electric welded steel tubes; see ASTM A556 for analysis etc. of grades
ASTM A563	0.35% C steel: For nuts; different grades indicate type of steel and heat treatment
ASTM A570	0.25% C steel: Sheet and strip; graded by Mn content
ASTM A573/70	0.27% C (max) 1.0% Mn 0.2% Si steel of improved toughness
ASTM A612	0.3% C (max) 1.2% Mn 0.04% S and P (max) steel: Plate for pressure vessels **UTS: 600 Elon: 20% Proof: 345**
ASTM A618/1	0.26% C (max) 1.3% Mn (max) steel: Tube; seamless or welded for structure **UTS: 20% Proof: 345**
ASTM A649/2	0.5% C (max) 0.7% Mn steel: For rolls; corrugated
ASTM A649/4	0.35% C (max) 0.8% Mn steel: For rolls; corrugated
ASTM A680	Cold rolled carbon steel: Strip; see ASTM A682 for analysis; see also 44A3 and 44A4
ASTM A694	0.3% C (max) 1.5% Mn (max) steel: For pipe flanges and fittings; graded by yield strength
ASTM A695 A	C not specified 0.04% S and P (max) steel: Bar; graded by yield strength
ASTM A695 B	0.35% C (max) 1.1% Mn 0.04% S and P (max) steel: Bar; graded by yield strength
ASTM A695 C	C not specified 0.04% S and P (max) steel: Bar; graded by yield strength
ASTM A695 D	C not specified 0.04% S and P (max) 0.2% Pb steel: Bar; graded by yield strength
ASTM A696 B	0.32% C (max) 1.04% Mn (max) 0.04% S and P (max) steel: Bar for pressure use **UTS: 415 Elon: 20% Proof: 240**
ASTM A696 C	0.32% C (max) 1.04% Mn 0.04% S and P (max) steel: Bar for pressure use **UTS: 485 Elon: 18% Proof: 275**
ASTM A707 L2/1	0.33% C (max) 1.4% Mn steel: Flanges for low temperature use; 41 J impact at −46 °C **UTS: 290 Elon: 22%**
ASTM A707 L2/2	0.33% C (max) 1.4% Mn steel: Flanges for low temperature use; 54 J impact at −46 °C **UTS: 360 Elon: 25%**
ASTM A707 L2/3	0.33% C (max) 1.4% Mn steel: Flanges for low temperature use; 68 J impact at −46 °C **UTS: 415 Elon: 23%**
ASTM A707 L2/4	0.33% C (max) 1.4% Mn steel: Flanges for low temperature use; 68 J impact at −46 °C **UTS: 515 Elon: 20%**
ASTM A709/36	0.26% C 1.0% Mn with grain refining elements steel: For bridges **UTS: 400 Elon: 20% Proof: 250**
ASTM A709/50	0.23% C (max) 1.35% Mn (max) with grain refining elements steel: For bridges **UTS: 450 Elon: 20% Proof: 345**
ASTM A711	Steel for forging; graded by AISI designation
ASTM A722	C not specified 0.04% S and P (max) steel: Bar; for prestressed concrete **UTS: 435 Proof: 810**
ASTM A732/2 A	0.3% C 0.8% Mn steel: Investment casting; annealed **UTS: 450 Elon: 25% Proof: 310**
ASTM A732/2 Q	0.3% C 0.8% Mn steel: Investment casting; hardened and tempered **UTS: 590 Elon: 10% Proof: 415**
ASTM A732/3 A	0.4% C 0.8% Mn steel: Investment casting; annealed **UTS: 515 Elon: 25% Proof: 330**
ASTM A732/3 Q	0.4% C 0.8% Mn steel: Investment casting; hardened and tempered **UTS: 690 Elon: 10% Proof: 620**
ASTM A738	0.24% C (max) 1.5% Mn (max) steel: Plate for pressure use; 27 J impact at −45 °C; hardened and tempered **UTS: 600 Elon: 20% Proof: 310**
ASTM A757-A1Q	0.3% C 1% Ni + Cr + Mo + V + Cu (max) steel: Casting; quench and temper; 17 J impact at −50 °C **UTS: 450 Elon: 24% Proof: 240**
ASTM A757-A2Q	0.25% C 1% Ni + Cr + Mo + V + Cu (max) steel: Casting; quench and temper; 20 J impact at −50 °C **UTS: 485 Elon: 22% Proof: 275**
ASTM A757-A1Q	0.3% C 1% Mn 0.025% S and P steel: Casting; for low temperature use
ASTM A757-A2Q	0.25% C 1.2% Mn 0.025% S and P steel: Casting; for low temperature use
ASTM A759	0.75% C 0.03% S and P steel: For crane rails
ASTM A767	Analysis not specified steel: Reinforcement bars for concrete; galvanized
ASTM A775	Analysis not specified steel: Reinforcement bars for concrete; epoxy-coated
ASTM A779	Analysis not specified steel: Seven wire for concrete; reinforcement for pre-stressed concrete
ASTM A821	0.7% C 0.04% S and P (max) steel: For pre-stressed concrete; hard drawn **UTS: 210**
ASTM A857	0.25% C (max) steel: For sheet piling
AURIGA I	0.25% C 0.7% Mn 0.4% Si 0.05% S and P steel: Casting; D Brown **UTS: 460 Elon: 22%**
AURIGA III	0.42% C 0.8% Mn 0.3% Si 0.05% S and P steel: Casting; D Brown **UTS: 630 Elon: 12%**
AURIGA IIIA	0.35% C 0.7% Mn 0.4% Si 0.05% S and P steel: Casting; D Brown **UTS: 580 Elon: 15%**
AURIGA VIB	0.29% C 1.5% Mn 0.04% S and P steel: Casting; D Brown; hardened and tempered **UTS: 760 Elon: 15%**
B 3	0.3% C (max) steel: Breda
B 3 Mn	0.32% C 1.2% Mn 0.3% Si steel: Pompey
B 4	0.4% C (max) steel: Breda
B 5	0.5% C (max) steel: Breda
B 7	0.35% C steel: As rolled; Bofors **DPN: 160 UTS: 540 Elon: 22% Proof: 250**
B 9 S	0.45% C steel: Bofors **DPN: 190 UTS: 660 Elon: 16%**
B 103	0.46% C 0.8% Mn steel: Fagersta; normalized **DPN: 200 UTS: 630 Elon: 18%**
B 805	0.4% C 1.3% Mn steel: Fagersta
B METAL	0.25% C 0.7% Mn 0.3% Si steel: Casting; Edgar Allen; code 603 steel
BS 15/1	0.25% C 0.06% S 0.06% P steel: General purpose structural steel **UTS: 450 Elon: 25% Proof: 240**
BS 15/2	0.25% C 0.06% S 0.06% P 0.3% Cu steel: General purpose structural steel **UTS: 450 Elon: 25% Proof: 240**
BS 15/3	0.25% C 0.06% S 0.06% P 0.4% Cu steel: General purpose structural steel **UTS: 450 Elon: 25% Proof: 240**

Note. The following abbreviations and units are used in the tables:

DPN	Hardness, diamond pyramid number
UTS	Ultimate tensile strength, N/mm^2
Elon	Elongation, %
Proof	0.1% proof strength, N/mm^2

1 N/mm^2=0.1 hbar=0.102 kgf/mm^2=0.06475 tonf/in.^{2}=145.04 ibf/in.^{2}=1 MPa

See Appendix II for other abbreviations and conversion tables.

Symbol	Nominal analysis, supplier, condition and remarks.
BS 24/2/4	Medium C 0.05% S and P steel: For loco tyres; hardened and tempered **UTS: 1120 Elon: 9%**
BS 24/2/5	Medium C 0.06% S and P steel: For wagon tyres; hardened and tempered **UTS: 1120 Elon: 8%**
BS 24/3A/3	0.4% C 0.8% Mn 1.7% Si steel: For wagon springs; as BS alloy En 46
BS 24/3A/4	0.4% C 0.8% Mn 1.7% Si steel: For wagon springs; as BS alloy En 46
BS 24/4C	0.32% C 0.7% Mn 0.1% Si 0.05% S and P steel: Forging; for railway use in couplings etc. **UTS: 540 Elon: 24%**
BS 24/4D	0.42% C 0.7% Mn 0.1% Si 0.05% S and P steel: Forging; for railway use in couplings etc. **UTS: 630 Elon: 18%**
BS 32/1	0.4% C 0.7% Mn 0.06% S 0.06% P steel: Bar; replaced by BS 970 En 6A
BS 32/2	0.25% C 1% Mn 0.06% S 0.06% P steel: Bar; replaced by BS 970 En 3B
BS 47A	0.25% C 0.8% Mn 0.15% Si 0.075% S and P steel: For railway fishplates
BS 47B	0.36% C 0.8% Mn 0.15% Si 0.075% S and P steel: For railway fishplates
BS 53	Cold drawn weldless steel tubes for boilers; replaced by BS 3059
BS 165	Hard drawn steel: Wire for reinforcement; replaced by BS 785
BS 399	0.44% C 0.7% Mn 0.3% Si 0.05% S and P steel: For gas cylinders
BS 494	Cold drawn mild steel: Tube for boilers; replaced by BS 3059
BS 548	0.3% C (max) 0.05% S 0.05% P structural steel: As rolled **UTS: 610 Elon: 18% Proof: 320**
BS 592 A	0.25% C (max) 1.0% Mn 0.6% Si 0.06% S and P steel: Casting; contained in BS 3100 **UTS: 420 Elon: 22% Proof: 210**
BS 592 B	0.35% C (max) 1.0% Mn 0.6% Si 0.06% S and P steel: Casting; contained in BS 3100 **UTS: 480 Elon: 20% Proof: 240**
BS 592 C	0.45% C (max) 1.0% Mn 0.6% Si 0.06% S and P steel: Casting; contained in BS 3100 **UTS: 530 Elon: 15% Proof: 240**
BS 640/1	0.37% C 0.9% Mn 0.1% Si 0.05% S and P steel: Electrodes
BS 785	0.3% C (max) 0.05% S 0.05% P steel: For concrete reinforcement; high tensile form (C content near top limit) **UTS: 610 Elon: 16% Proof: 330**
BS 785	0.3% C (max) 0.05% S 0.05% P steel: For concrete reinforcement; medium tensile form (higher C than mild steel) **UTS: 540 Elon: 16% Proof: 240**
BS 970	See designation for details
BS 980 CDS5	0.3% C 0.7% Mn steel: Tube; cold drawn; annealed **UTS: 390 Proof: 240**
BS 980 CDS6	0.3% C 0.6% Mn steel: Tube; cold drawn; as drawn and tempered **UTS: 530 Proof: 420**
BS 980 CDS7	0.45% C 0.6% Mn steel: Tube; cold drawn; annealed **UTS: 500 Proof: 330**
BS 980 CDS9	0.25% C 1.5% Mn steel: Tube; cold drawn; annealed **UTS: 450 Proof: 270**
BS 980 CDS10	0.25% C 1.5% Mn steel: Tube; cold drawn; as drawn and tempered **UTS: 610 Proof: 450**
BS 980 CEW3	0.3% C 0.06% S and P steel: Tube; cold drawn; electric welded
BS 980 CEW4	0.3% C 0.06% S and P steel: Tube; cold drawn; electric welded
BS 980 ERW2	0.3% C (max) 0.06% S and P steel: Tube
BS 980 ERW3	0.4% C (max) 0.06% S and P steel: Tube
BS 1045	0.4% C (max) 1.5% Mn 0.3% Si 0.05% S and P steel: For gas cylinders; normalized **UTS: 670 Elon: 15% Proof: 390**
BS 1144	Cold rolled and twisted mild steel: Bar; for concrete reinforcing **UTS: 450 Elon: 13% Proof: 370**
BS 1287	0.44% C 0.7% Mn 0.045% S and P steel: For gas cylinders
BS 1414B	0.35% C (max) 0.9% Mn 0.05% S and P steel: For valves
BS 1449-10	0.21% C 0.7% Mn steel: Sheet and plate
BS 1449-20	0.2% C 1.5% Mn steel: Sheet and plate
BS 1449-30	0.3% C 0.7 Mn steel: Sheet and plate
BS 1449 HR5A	0.25% C 0.8% Mn 0.05% S and P rimming steel: Sheet; as BS 970 En 2C/A **UTS: 400 Elon: 25% Proof: 210**
BS 1449 HR5B	0.25% C 0.8% Mn 0.05% S and P balanced steel: Sheet; as BS 970 En 2C/B **UTS: 400 Elon: 25% Proof: 210**
BS 1449 HR5C	0.25% C 0.8% Mn 0.05% S and P steel: Sheet; killed **UTS: 400 Elon: 25% Proof: 210**
BS 1449 HR6A	0.27% C 0.8% Mn 0.05% S and P rimming steel: Sheet **UTS: 450 Elon: 20% Proof: 240**
BS 1449 HR6B	0.27% C 0.8% Mn 0.05% S and P balanced steel: Sheet **UTS: 450 Elon: 20% Proof: 240**
BS 1449 HR6C	0.27% C 0.8% Mn 0.05% S and P steel: Sheet; killed **UTS: 450 Elon: 20% Proof: 240**
BS 1449 HR16A	0.25% C 0.8% Mn 0.05% S and P rimming steel: Plate **UTS: 400 Elon: 28% Proof: 220**
BS 1449 HR16B	0.25% C 0.8% Mn 0.05% S and P balanced steel: Plate **UTS: 400 Elon: 28% Proof: 220**
BS 1449 HR16C	0.25% C 0.8% Mn 0.05% S and P steel: Plate; killed **UTS: 400 Elon: 28% Proof: 220**
BS 1449 HR17A	0.27% C 0.8% Mn 0.05% S and P rimming steel: Plate
BS 1449 HR17B	0.27% C 0.8% Mn 0.05% S and P balanced steel: Plate
BS 1449 HR17C	0.27% C 0.8% Mn 0.05% S and P steel: Plate; killed
BS 1449 HS30	0.3% C 0.8% Mn 0.2% Si 0.06% S and P steel: Sheet **UTS: 480 Elon: 18% Proof: 270**
BS 1449 HS40	0.4% C 0.8% Mn 0.2% Si 0.06% S and P steel: Sheet **UTS: 530 Elon: 16% Proof: 280**
BS 1449 NHR24	0.3% C 0.8% Mn 0.06% S and P steel: Sheet **UTS: 520 Elon: 15% Proof: 270**
BS 1453 A3	0.27% C 1.5% Mn 0.4% Si steel: For welding filler rod; to give butt weld strength of 480 N/mm^2
BS 1456 B	0.3% C 1.5% Mn 0.4% Si 0.05% S and P steel: Casting; normalized or hardened and tempered; contained in BS 3100 **UTS: 610 Elon: 15%**
BS 1458	Composition to suit mechanical properties 0.05% S and P steel: Casting; normalized or hardened and tempered; contained in BS 3100 **DPN: 230 UTS: 610 Elon: 15% Proof: 480**
BS 1459	Composition to suit mechanical properties 0.05% S and P steel: Casting; normalized or hardened and tempered; contained in BS 3100 **DPN: 280 UTS: 610 Elon: 12% Proof: 580**
BS 1503/161 A and B	0.25% C 0.2% Si 0.05% S and P steel: Forging; normalized **UTS: 420 Proof: 200**
BS 1503/161 C	0.3% C 1.0% Mn 0.2% Si 0.05% S and P steel: Forging; normalized **UTS: 530 Proof: 240**
BS 1503/221	0.23% C 1.5% Mn 0.25% Si 0.05% S and P steel: Forging; normalized **UTS: 540 Proof: 240**
BS 1504/101 A	C not specified 0.06% S and P steel: Casting; normalized **UTS: 420 Elon: 20% Proof: 200**

Symbol	Nominal analysis, supplier, condition and remarks.
BS 1504/101 B	C not specified 0.06% S and P steel: Casting; normalized **UTS: 460 Elon: 20% Proof: 210**
BS 1504/101 C	C not specified 0.06% S and P steel: Casting; normalized **UTS: 560 Elon: 15% Proof: 240**
BS 1504/161 A	0.25% C (max) 0.9% Mn 0.5% Si 0.05% S and P steel: Casting; normalized **UTS: 420 Elon: 22% Proof: 210**
BS 1504/161 B	0.3% C (max) 0.9% Mn 0.5% Si 0.05% S and P steel: Casting; normalized **UTS: 460 Elon: 22% Proof: 240**
BS 1560 B	0.35% C (max) 0.05% S and P steel: For pipe flanges
BS 1570 B	0.35% C (max) 0.05% S and P steel: For valves
BS 1633 A	0.28% C (max) 0.6% Mn 0.2% Si 0.05% S and P steel: For land boilers; normalized **UTS: 390 Elon: 30%**
BS 1633 B	0.28% C (max) 0.6% Mn 0.2% Si 0.05% S and P steel: For land boilers; normalized **UTS: 420 Elon: 28%**
BS 1633 C	0.28% C (max) 0.6% Mn 0.2% Si 0.05% S and P steel: For land boilers; normalized **UTS: 450 Elon: 26%**
BS 1633 D	0.28% C (max) 0.6% Mn 0.2% Si 0.05% S and P steel: For land boilers; normalized **UTS: 480 Elon: 23%**
BS 1633 E	0.32% C (max) 0.6% Mn 0.2% Si 0.05% S and P steel: For land boilers; normalized **UTS: 550 Elon: 22%**
BS 1663	0.3% C (max) 0.05% S and P steel: For electrically welded chain
BS 1717 CDS 101	0.2% C 0.05% S and P steel: Seamless tube; annealed **UTS: 300 Proof: 160**
BS 1717 CDS101	0.2% C 0.05% S and P steel: Seamless tube; normalized **UTS: 360 Proof: 210**
BS 1717 CDS102	0.2% C 0.05% S and P steel: Seamless tube; as drawn **UTS: 420 Proof: 360**
BS 1717 CDS102	0.2% C 0.05% S and P steel: Seamless tube; after welding, brazing or annealing **UTS: 300 Proof: 170**
BS 1717 CDS103	0.25% C 0.6% Mn 0.05% S and P steel: Seamless tube; annealed **UTS: 360 Proof: 210**
BS 1717 CDS103	0.25% C 0.6% Mn 0.05% S and P steel: Seamless tube; normalized **UTS: 420 Proof: 270**
BS 1717 CDS104	0.25% C 0.6% Mn 0.05% S and P steel: Seamless tube; as drawn **UTS: 480 Proof: 390**
BS 1717 CDS104	0.25% C 0.6% Mn 0.05% S and P steel: Seamless tube; after welding, brazing or annealing **UTS: 360 Proof: 210**
BS 1717 CDS105	0.35% C 0.6% Mn 0.05% S and P steel: Seamless tube; annealed **UTS: 420 Proof: 270**
BS 1717 CDS105	0.35% C 0.6% Mn 0.05% S and P steel: Seamless tube; normalized **UTS: 480 Proof: 330**
BS 1717 CDS106	0.35% C 0.6% Mn 0.05% S and P steel: Seamless tube; as drawn **UTS: 530 Proof: 420**
BS 1717 CDS106	0.35% C 0.6% Mn 0.05% S and P steel: Seamless tube; after annealing or brazing **UTS: 420 Proof: 270**
BS 1717 CDS107	0.5% C 0.6% Mn 0.05% S and P steel: Seamless tube; annealed **UTS: 500 Proof: 330**
BS 1717 CDS107	0.5% C 0.6% Mn 0.05% S and P steel: Seamless tube; normalized **UTS: 580 Proof: 420**
BS 1717 CDS108	0.5% C 0.6% Mn 0.05% S and P steel: Seamless tube; as drawn **UTS: 690 Proof: 580**
BS 1717 CDS108	0.5% C 0.6% Mn 0.05% S and P steel: Seamless tube; after welding, brazing or annealing **UTS: 500 Proof: 330**
BS 1717 CEW101	0.2% C 0.6% Mn 0.06% S and P steel: Electrically welded tube; cold drawn strip; annealed after welding **UTS: 300 Proof: 170**
BS 1717 CEW102	0.2% C 0.6% Mn 0.06% S and P steel: Electrically welded tube; cold drawn strip prior to welding **UTS: 420 Proof: 360**
BS 1717 CEW102	0.2% C 0.6% Mn 0.06% S and P steel: Electrically welded tube; after annealing or brazing **UTS: 300 Proof: 170**
BS 1717 CEW103	0.25% C 0.6% Mn 0.06% S and P steel: Electrically welded tube; cold drawn strip, annealed after welding **UTS: 360 Proof: 210**
BS 1717 CEW104	0.25% C 0.6% Mn 0.06% S and P steel: Electrically welded tube; cold drawn strip prior to welding **UTS: 480 Proof: 390**
BS 1717 CEW104	0.25% C 0.6% Mn 0.06% S and P steel: Electrically welded tube; after annealing or brazing **UTS: 360 Proof: 210**
BS 1717 ERW101	0.1% C 0.6% Mn 0.06% S and P steel: Electrically welded tube; as welded **UTS: 300 Proof: 170**
BS 1717 ERW102	0.2% C 0.6% Mn 0.06% S and P steel: Electrically welded tube; as welded **UTS: 370 Proof: 220**
BS 1717 ERW102	0.2% C 0.6% Mn 0.06% S and P steel: Electrically welded tube; annealed or brazed **UTS: 300 Proof: 170**
BS 1717 ERW103	0.3% C 0.6% Mn 0.06% S and P steel: Electrically welded tube; as welded **UTS: 450 Proof: 300**
BS 1717 ERW103	0.3% C 0.6% Mn 0.06% S and P steel: Electrically welded tube; annealed or brazed **UTS: 370 Proof: 220**
BS 2763	C not specified 0.050% S and P steel: Wire for rope; specification covers bright and galvanized; tensile strength varies with diameter
BS 2772/2	0.27% C 0.2% Si 0.05% S and P steel: For colliery winding equipment; hardened and tempered
BS 2901 A 16	0.27% C 1.4% Mn 0.4% Si 0.03% S and P steel: Rod for gas and electric welding
BS 3059/5	0.25% C (max) 0.05% S and P steel: Tube for boilers; seamless; hot finished **UTS: 480**
BA 3059/6	0.25% C 0.05% S and P steel: Tube for boilers; seamless; cold drawn **UTS: 480**
BS 3083	Mild steel: Sheet; galvanized and corrugated
BS 3100	Steel: Casting for general engineering purposes, covering a range of BS specifications
BS 3100 A1	0.25% C (max) 0.9% Mn (max) 0.8% Cr + Mo + Ni + Cu (max) 0.06% S and P (max) steel: Casting **UTS: 430 Elon: 22% Proof: 230**

Note. The following abbreviations and units are used in the tables:

DPN	Hardness, diamond pyramid number
UTS	Ultimate tensile strength, N/mm^2
Elon	Elongation, %
Proof	0.1% proof strength, N/mm^2

1 N/mm^2=0.1 hbar=0.102 kgf/mm^2=0.06475 $tonf/in.^2$=145.04 $lbf/in.^2$=1 MPa

See Appendix II for other abbreviations and conversion tables.

Symbol	Nominal analysis, supplier, condition and remarks.
BS 3100 A2	0.35% C (max) 1.0% Mn (max) 0.06% S and P (max) steel: Casting **UTS: 490 Elon: 18% Proof: 260**
BS 3100 A3	0.45% C (max) 1.0% Mn (max) 0.06% S and P (max) steel: Casting **UTS: 540 Elon: 14% Proof: 295**
BS 3100 A5	0.3% C 1.4% Mn 0.015% S and P (max) steel: Casting **UTS: 700 Elon: 13% Proof: 370**
BS 3100 A6	0.3% C 1.4% Mn 0.05% S and P (max) steel: Casting **UTS: 800 Elon: 13% Proof: 495**
BS 3100 AW2	0.45% C 1.0% Mn (max) 0.8% Cr + Mo + Ni + Cu (max) steel: Casting for surface hardening **UTS: 620 Elon: 12% Proof: 325**
BS 3100 AW3	0.55% C 1.0% Mn (max) 0.8% Cr + Mo +Ni + Cu (max) steel: Casting for surface hardening **UTS: 690 Elon: 8% Proof: 370**
BS 3100 BT1	C not specified 0.05% S and P (max) steel: Casting for high tensile use **UTS: 750 Elon: 11% Proof: 495**
BS 3100 BT2	C not specified 0.04% S and P (max) steel: Casting **UTS: 925 Elon: 8% Proof: 585**
BS 3100 BT3	C not specified 0.03% S and P (max) steel: Casting **UTS: 1080 Elon: 6% Proof: 695**
BS 3127/8	0.35% C 1.2% Mn 0.05% S and P steel: Tube; lightly cold worked; for bourdon tubes **DPN: 170**
BS 3141 C3C1	0.28% C (max) 0.7% Mn 0.05% S and P steel: Classification system used by BS; discontinued
BS 3141 C3E1	0.25% C 0.5% Mn 0.06% S and P steel: Classification system used by BS; discontinued
BS 3141 C3G1	0.25% C 0.7% Mn 0.05% S and P steel: Classification system used by BS; discontinued
BS 3141 C3H1	0.28% C 0.8% Mn 0.06% S and P steel: Classification system used by BS; discontinued
BS 3141 C4A1	0.32% C (max) 0.7% Mn 0.05% S and P steel: Classification system used by BS; discontinued
BS 3141 C4B1	0.3% C 0.8% Mn 0.06% S and P steel: Classification system used by BS; discontinued
BS 3141 C4C1	0.3% C 0.5% Mn 0.06% S and P steel: Classification system used by BS; discontinued
BS 3141 C4D1	0.3% C 0.8% Mn 0.05% S and P steel: Classification system used by BS; discontinued
BS 3141 C4E1	0.32% C 0.8% Mn 0.06% S and P steel: Classification system used by BS; discontinued
BS 3141 C4F1	0.32% C 1% Mn 0.05% S and P steel: Classification used by BS; discontinued
BS 3141 C4G1	0.36% C 0.8% Mn 0.06% S and P steel: Classification system used by BS; discontinued
BS 3141 C4J1	0.35% C 0.8% Mn 0.05% S and P steel: Classification system used by BS; discontinued
BS 3141 C4K1	0.4% C 0.8% Mn 0.06% S and P steel: Classification system used by BS; discontinued
BS 3141 C4L1	0.37% C 1% Mn 0.05% S and P steel: Classification system used by BS; discontinued
BS 3141 C5A1	0.41% C 0.8% Mn 0.06% S and P steel: Classification system used by BS; discontinued
BS 3141 C5C1	0.4% C 0.8% Mn 0.05% S and P steel: Classification system used by BS; discontinued
BS 3141 C5D1	0.4% C 1% Mn 0.06% S and P steel: Classification system used by BS; discontinued
BS 3141 C5F1	0.42% C 0.8% Mn 0.05% S and P steel: Classification system used by BS; discontinued
BS 3141 C5G1	0.42% C 1% Mn 0.05% S and P steel: Classification system used by BS; discontinued
BS 3141 C5J1	0.45% C 0.6% Mn 0.05% S and P steel: Classification system used by BS; discontinued
BS 3141 C5K1	0.45% C 1.1% Mn 0.06% S and P steel: Classification system used by BS; discontinued
BS 3141 C5L1	0.47% C 0.8% Mn 0.06% S and P steel: Classification system used by BS; discontinued
BS 3141 CF4H1	0.35% C 1.1% Mn 0.18% S 0.06% P steel: Free cutting; classification system used by BS; discontinued
BS 3141 CF4M1	0.37% C 1.1% Mn 0.18% S 0.06% P steel: Free cutting; classification system used by BS; discontinued
BS 3141 CF5B1	0.4% C 1.1% Mn 0.18% S 0.06% P steel: Free cutting; classification system used by BS; discontinued
BS 3141 CF5E1	0.4% C 1.1% Mn 0.18% S 0.06% P steel: Free cutting; classification system used by BS; discontinued
BS 3141 CF5H1	0.42% C 1.2% Mn 0.18% S 0.06% P steel: Free cutting; classification system used by BS; discontinued
BS 3146 CLA3	C not specified 0.05% S and P steel: Investment casting; hardened and tempered **DPN: 220 UTS: 690 Elon: 15% Proof: 480**
BS 3146 CLA4	C not specified 0.05% S and P steel: Investment casting; hardened and tempered **DPN: 275 UTS: 920 Elon: 12% Proof: 580**
BS 3146 CLA5A	C not specified 0.02% S 0.025% P steel: Investment casting; hardened and tempered **DPN: 300 UTS: 920 Elon: 12% Proof: 670**
BS 3146 CLA5B	C not specified 0.02% S 0.025% P steel: Investment casting; hardened and tempered **DPN: 360 UTS: 1140 Elon: 7% Proof: 910**
BS 3706/1	0.25% C 0.06% S and P steel: As rolled **UTS: 420 Elon: 20%**
BS 3706/2	0.25% C 0.06% S and P 0.3% Cu steel: As rolled **UTS: 420 Elon: 20%**
BS 3706/3	0.25% C 0.06% S and P 0.45% Cu steel: As rolled **UTS: 420 Elon: 20%**
BS 3740	Mild steel plate clad with stainless may be 13% Cr or one of the 18/8 Cr/Ni type
BS 4360/43 A	0.3% C 0.06% S and P steel: As rolled; weldable **UTS: 500 Elon: 22%**
BS 4360/43A1	0.3% C 0.06% S and P steel: As rolled; weldable **UTS: 300 Elon: 22%**
BS 4360/43 B	0.26% C 1.6% Mn 0.06% S and P steel: As rolled; weldable **UTS: 500 Elon: 22%**
BS 4360/43 C	0.22% C 1.6% Mn 0.06% S and P steel: As rolled; weldable **UTS: 500 Elon: 22%**
BS 4360/50 A	0.27% C 1.7% Mn 0.06% S and P steel: Sheet; as rolled; weldable **UTS: 560 Elon: 20%**
BS 4360/50 B	0.24% C 1.6% Mn 0.06% S and P steel: Sheet; as rolled; weldable; with Nb **UTS: 560 Elon: 20%**
BS 4360/50 C	0.24% C 1.6% Mn 0.06% S and P steel: Weldable; with Nb **UTS: 560 Elon: 20%**
BS 4360/50 D	0.22% C 1.6% Mn 0.3% Si 0.05% S and P steel: Normalized; weldable; with Nb **UTS: 560 Elon: 20%**
BS 4360/55 C	0.26% C 1.7% Mn 0.05% S and P steel: Plate; as rolled; weldable; with Nb **UTS: 610 Elon: 17% Proof: 440**
BS 4360/55 E	0.26% C 1.7% Mn 0.05% S and P steel: Plate; normalized; with Nb **UTS: 610 Elon: 17% Proof: 440**
BS 5006	Cold worked steel: Bar and strip for automobiles; withdrawn; replaced by BS 1449 and BS 2453
BS 5009	Steel: Tube for automobiles; withdrawn; replaced by BS 980
BS 5010	Steel: For laminated springs for automobiles; withdrawn; replaced by BS 970
BS 5028	Steel: Castings (No 1 and 2 grades) for automobiles; withdrawn
For BS S specifications see S	
C 3	0.32% C (max) 0.8% Mn steel: Origin unknown
C 4	0.42% C (max) 0.8% Mn steel: Origin unknown
C 5	0.5% C (max) 0.8% Mn steel: Origin unknown

Symbol	Nominal analysis, supplier, condition and remarks.
C 30	0.3% C 0.7% Mn steel: Italian Standards designation; normalized
	DPN: 200 UTS: 620 Elon: 21% Proof: 280
C 30 UNI 4365	0.3% C 0.8% Mn 0.4% Si 0.035% S and P steel: Italian Standard; for bolts
C 35	0.35% C 0.7% Mn steel: Italian Standards designation; normalized
	DPN: 210 UTS: 620 Elon: 19% Proof: 300
C 35 UNI 4365	0.35% C 0.7% Mn 0.35% Si 0.04% S and P steel: Italian Standard; for bolts
C 40	0.4% C 0.7% Mn steel: Italian Standards designation; normalized
	DPN: 240 UTS: 670 Elon: 17% Proof: 340
C 40 UNI 4365	0.4% C 0.8% Mn 0.4% Si 0.035% S and P steel: Italian Standard; for bolts
C 45	0.45% C 8.7% Mn steel: Italian Standards designation; normalized
	DPN: 235 UTS: 720 Elon: 16% Proof: 370
C 45	0.46% C 0.7% Mn steel: Italian Standard
	DPN: 200 UTS: 1300 Elon: 8% Proof: 1000
C 45W3	0.45% C 0.7% Mn 0.035% S and P (max) steel: German Standards designation
C 50	0.5% C 0.75% Mn steel: Italian Standards designation
	DPN: 245 UTS: 750 Elon: 15% Proof: 400
C 140	0.4% C 1.5% Mn 0.25% Si 0.02% S and P steel: SHD drills, etc.
C METAL	0.32% C 0.7% Mn 0.3% Si steel: Casting; Edgar Allen; code 604
CASE 1280	0.28% C 1.1% Mn 0.2% Si 0.035% S and P steel: Plate; Armco; pressure vessels
	UTS: 560 Elon: 22% Proof: 240
CCU	0.32% C 0.7% Mn 0.3% Si 0.9% Cu steel: Casting; Edgar Allen; code 642
CEAX	0.45% C 0.6% Mn 0.035% S and P steel: Bars; SKF
Cf 35	0.36% C 0.6% Mn 0.2% Si 0.035% S and P steel: German Standard
Cf 43	0.46% C 0.7% Mn 0.2% Si 0.035% S and P steel: German Standard
Cf 53	0.36% C 0.7% Mn steel: German Standard
CMN 1	0.25% C 1.5% Mn 0.25% Si steel: Firth Brown
CMN 2	0.35% C 1.5% Mn 0.25% Si steel: Firth Brown for BS alloy En 15 and En 15A
CODE 603	0.25% C 0.7% Mn 0.3% Si steel: Casting; Edgar Allen; B metal
CODE 604	0.32% C 0.7% Mn 0.22% Si steel: Casting; Edgar Allen; C metal
CODE 606	0.45% C 0.8% Mn 0.3% Si steel: Casting; Edgar Allen; D metal
CODE 609	0.32% C 0.7% Mn 0.6% Si steel: Casting; Edgar Allen; 'Retort'
CODE 610	0.25% C 1.4% Mn steel: Casting; Edgar Allen; MT 'P' steel
CODE 611	0.37% C 0.7% Mn 0.3% Si steel: Casting; Edgar Allen; 'Matte Ladle'
CODE 626	0.3% C 1.4% Mn steel: Casting; Edgar Allen; MT 'R'
CODE 642	0.32% C 0.7% Mn 0.3% Si 0.9% Cu steel: Casting; Edgar Allen; code 642
CPV	0.2% V steel: Balfour Darwin; three ranges of carbon
Cq 35	0.35% C 0.6% Mn 0.25% Si 0.04% S and P steel: German Standard
Cq 45	0.45% C 0.65% Mn 0.25% Si 0.04% S and P steel: German Standard
D 3	0.32% C 0.7% Mn 0.3% Si steel: Pompey
D 3S	0.35% C 0.8% Mn 0.25% Si steel: Pompey
D 4	0.4% C 0.6% Mn 0.3% Si steel: Pompey
D 4Mn	0.4% C 1.3% Mn 0.3% Si steel: Pompey
D 48	0.48% C 0.6% Mn 0.3% Si steel: Pompey
D METAL	0.45% C 0.85% Mn 0.3% Si steel: Casting; Edgar Allen; code 606
DEF 13 B3	Carbon and carbon manganese steel: Covers BS alloys En 5, 6, 8 and includes En 14
	UTS: 550 Elon: 15% Proof: 240
DEF 13 B3A	Carbon and low alloy hardened and tempered steel: Covers BS alloy En 5, 8, 15 and includes En 12 and 14
	UTS: 610 Elon: 20% Proof: 370
DEF 13 B3B	Carbon and low alloy hardened and tempered steel: Covers BS alloys En 5 and 8 and includes En 15, 17, 29 and 100
	UTS: 670 Elon: 20% Proof: 450
DIN 1629 St 55	0.36% C (max) 0.05% S and P (max) steel: Pipe; German Standard
	UTS: 600 Proof: 300
DIN 1629 St 55/4	0.36% C (max) 0.4% Mn (max) 0.05% S and P (max) steel: Pipe; German Standard
	UTS: 600 Proof: 300
DIN 1651 35 S 20	0.35% C 0.6% Mn 0.2% S 0.05% P steel: Free machining
DIN 1651 45 S 20	0.46% C 0.7% Mn 0.2% S 0.05% P steel: Free machining
DIN 1654 C9/35	0.36% C 0.6% Mn 0.04% S and P steel: For screws
DIN 1654 C9/45	0.46% C 0.7% Mn 0.04% S and P steel: For screws
DIN 17100 RSt 42/1	0.25% C (max) 0.05% S (max) 0.08% P (max) steel: Plate; German specification
	UTS: 460 Proof: 250
DIN 17100 RSt 42/2	0.23% C (max) 0.05% S (max) 0.05% P (max) steel: Plate; German specification
	UTS: 460 Proof: 250
DIN 17100 St 50/2	0.3% C 0.05% S and P (max) steel: Plate
	UTS: 550 Proof: 290
DIN 17100 St 60/1	0.35% C 0.05% S (max) 0.08% P (max) steel: Plate; German Standard
	UTS: 660 Proof: 330
DIN 17100 St 60/2	0.4% C 0.05% S and P steel: Plate; German Standard
	UTS: 660 Proof: 330
DIN 17100 UST 42/2	0.25% C (max) 0.05% S (max) 0.05% P (max) steel: Plate; German specification
	UTS: 460 Proof: 250
DIN 17155 H1V	0.26% C (max) 0.05% S and P steel: For boiler plate
DIN 17155 H1V	0.26% C (max) 0.6% Mn (max) 0.3% Cr (max) 0.05% S and P (max) steel: Plate; German Standard
	UTS: 520 Proof: 280
DIN 17155 H1VA	0.26% C (max) 0.05% S and P steel: For boiler plate
DIN 17155 H1VL	0.26% C (max) 0.05% S and P steel: For boiler plate
DIN 17200 28 Mn 6	0.28% C 1.5% Mn 0.035% S and P (max) steel: German Standard
DIN 17200 30 Mn 5	0.3% C 1.4% Mn 0.035% S and P steel: Forging
DIN 17200 37 Mn Si 5	0.37% C 1.2% Mn 1.2% Si steel: Forging
DIN 17200 40 Mn 4	0.4% C 0.9% Mn 0.035% S and P steel
DIN 17200 40 Mn 4	0.4% C 0.95% Mn 0.035% S and P (max) steel: German Standard
DIN 17200 C 35	0.35% C 0.65% Mn 0.045% S and P (max) steel: For structures; German Standard
DIN 17200 C 45	0.45% C 0.65% Mn 0.045% S and P (max) steel: For structures; German Standard
DIN 17220 C 45	0.46% C 0.7% Mn 0.045% S and P steel
DIN 17200 CK 35	0.36% C 0.6% Mn 0.04% S and P steel
DIN 17200 CK 35	0.35% C 0.65% Mn 0.035% S and P (max) steel: For structures; German Standard

Note. The following abbreviations and units are used in the tables:

DPN	Hardness, diamond pyramid number
UTS	Ultimate tensile strength, N/mm^2
Elon	Elongation, %
Proof	0.1% proof strength, N/mm^2

1 N/mm^2=0.1 hbar=0.102 kgf/mm^2=0.06475 $tonf/in.^2$=145.04 $lbf/in.^2$=1 MPa

See Appendix II for other abbreviations and conversion tables.

Symbol	Nominal analysis, supplier, condition and remarks.
DIN 17200 CK 45	0.45% C 0.65% Mn 0.035% S and P (max) steel: For structures; German Standard
DIN 17200 CK 45	0.46% C 0.7% Mn 0.035% S and P steel
DIN 17200 CM 35	0.35% C 0.65% Mn 0.035% S and P (max) steel: For structures; German Standard
DIN 17200 CM 45	0.45% C 0.65% Mn 0.035% S and P (max) steel: For structures; German Standard
DIN 17221 38 Si 6	0.38% C 0.7% Mn 1.5% Si steel: For springs
DIN 17221 38 Si 7	0.38% C 0.65% Mn 0.045% S and P (max) steel: For springs; German Standard; heat treated
DIN 17221 46 Si 7	0.46% C 0.6% Mn 1.7% Si steel: For springs
DISQUE E	0.45% C 0.7% Mn 1.75% Si steel: Pompey
DISQUE H	0.55% C 0.7% Mn 1.75% Si steel: Pompey
DMS	0.35% C 1.2% Mn 1.2% Si steel: Pompey
DOUBLE EXTRA	1.0% C tool steel: Balfour Darwin
DTD 503 A	0.25% C 1.5% Mn 0.05% S 0.05% P steel: Tube; annealed **UTS: 370**
DTD 666	C not specified 0.02% S 0.025% P steel: Casting; casting to be X-rayed **DPN: 269 UTS: 910 Elon: 12% Proof: 630**
DTD 705	C not specified 0.02% S 0.025% P steel: Casting; casting to be X-rayed **DPN: 340 UTS: 1210 Elon: 7% Proof: 910**
DTD 5032	0.4% C 0.8% Mn 0.05% S 0.05% P steel: Swaged; blued if necessary at 350 °C **UTS: 910 Elon: 10%**
DTD 5072	C not specified 0.02% S 0.025% P steel: Casting; hardened, tempered and X-rayed **DPN: 370 UTS: 1210 Elon: 7% Proof: 910**
DUCOL HT	0.3% C (max) 1.6% Mn 0.05% S and P steel: Plate and section; Colvilles alloy for BS 548 **UTS: 560 Elon: 18% Proof: 330**
DUNELT	0.4% C 0.06% S and P steel: Dunford for BS alloy En 8
DUNELT 40	0.35% C 1.6% Mn 0.06% S and P steel: Dunford for BS alloy En 15A
DUNELT 41	0.38% C 1.2% Mn 0.15% S 0.06% P steel: Free machining; Dunford; hardened and tempered **DPN: 250 UTS: 720 Elon: 15%**
E 5M	0.5% C 0.6% Mn 0.3% Si steel: Pompey
En 3	0.25% C (max) 0.06% S and P steel: Designation used in BS 970; obsolete
En 3A	0.2% C 0.06% S and P steel: Designation used in BS 970; obsolete
En 3B	0.25% C (max) 0.06% S and P steel: Cold drawn; designation used in BS 970; obsolete
En 4	0.3% C (max) 0.06% S and P steel: Normalized; designation used in BS 970; obsolete
En 4A	0.3% C (max) 0.06% S and P steel: Cold drawn; designation used in BS 970; obsolete
En 5	0.3% C 0.06% S and P steel: Hardened and tempered; designation used in BS 970; replaced by 080A30
En 5A	0.27% C 0.06% S and P steel: Hardened and tempered; designation used in BS 970; replaced by 080A27
En 5C	0.32% C 0.06% S and P steel: Hardened and tempered; designation used in BS 970; replaced by 080A32
En 5D	0.3% C 0.06% S and P steel: Hardened and tempered; designation used in BS 970; replaced by 080A30
En 5K	0.3% C 0.05% S and P steel: Hardened and tempered; designation used in BS 970; replaced by 080A30
En 6	0.4% C (max) 0.06% S and P steel: Designation used in BS 970; obsolete
En 6A	0.4% C (max) 0.06% S and P steel: Designation used in BS 970; obsolete
En 6B	0.4% C (max) 0.06% S and P steel: Designation used in BS 970; obsolete
En 6K	0.35% C (max) 0.05% S and P steel: Designation used in BS 970; obsolete
En 8	0.4% C 0.06% S and P steel: Designation used in BS 970; replaced by 080M40
En 8A	0.36% C 0.06% S and P steel
En 8B	0.38% C 0.06% S and P steel: Designation used in BS 970; replaced by 080A35
En 8C	0.4% C 0.06% S and P steel: Designation used in BS 970; replaced by 080A40
En 8D	0.43% C 0.06% S and P steel: Designation used in BS 970; replaced by 080A42
En 8E	0.37% C 0.06% S and P steel: Designation used in BS 970; replaced by 080A37
En 8K	0.4% C 0.05% S and P steel: Designation used in BS 970; replaced by 080A40
En 8M	0.4% C 0.16% S 0.07% P steel: Free machining; designation used in BS 970; replaced by 212M36
En 15	0.35% C steel: Designation used in BS 970; replaced by 150M36
En 15A	0.35% C steel: Designation used in BS 970; replaced by 150M36
En 15B	0.37% C steel: Designation used in BS 970; replaced by 120M36
En 16A	0.27% C steel: Designation used in BS 970; replaced by 605A27
En 16B	0.32% C steel: Designation used in BS 970; replaced by 605A32
En 16C	0.37% C steel: Designation used in BS 970; replaced by 605A37
En 46	0.4% C 1.7% Si steel: For springs; designation used in BS 970; obsolete
ERA 51	0.35% C 1.5% Mn steel: Hadfields for BS alloy En 15 and En 15A
ERA 53	0.37% C 1.2% Mn steel: Hadfields for BS alloy En 15B
ETG	0.44% C 1.5% Mn 0.3% S steel: Free machining; Von Moos; cold roller **DPN: 280 UTS: 960 Elon: 7% Proof: 865**
ETG88	0.44% C 1.5% Mn 0.3% S steel: Free machining; Von Moos; low internal stress **DPN: 250 UTS: 810 Elon: 10% Proof: 685**
F114	0.46% C 0.7% Mn 1.7% Si steel: VEW
FAGERSTA B805	0.4% C 1.3% Mn steel: Fagersta
FAGERSTA 1660	0.45% C 0.6% Mn steel: Fagersta
FF 3	0.35% C 0.5% Mn 0.25% Si steel: Pompey **UTS: 780 Elon: 12% Proof: 530**
FF 3H	0.37% C 0.7% Mn 0.3% Si steel: Pompey **UTS: 740 Elon: 10% Proof: 450**
FF 4	0.42% C 0.7% Mn 0.3% Si steel: Pompey **UTS: 920 Elon: 9% Proof: 660**
FF 38	0.38% C 0.7% Mn 0.3% Si steel: Pompey
FUJI CON SUPER	0.33% C 1.2% Mn 0.015% S and P steel: Fuji
GOST 380 BST5 PS	0.33% C 0.65% Mn 0.05% S (max) 0.04% P (max) steel: Plate; Russian Standard
GOST 380 BST5 SP	0.33% C 0.65% Mn 0.05% S (max) 0.04% P (max) steel: Plate; Russian Standard
GOST 380 BST5 SP2	0.33% C 0.65% Mn 0.3% Cr Ni Cu (max) 0.05% S (max) 0.04% P (max) steel: Plate; Russian Standard
GOST 380 BST6 PS	0.42% C 0.65% Mn 0.05% S (max) 0.04% P (max) steel: Plate; Russian Standard
GOST 380 BST6 PS2	0.33% C 0.65% Mn 0.3% Cr Ni Cu (max) 0.05% S (max) 0.04% P (max) steel: Plate; Russian Standard
GOST 380 BST6 SP	0.42% C 0.65% Mn 0.05% S (max) 0.04% P (max) steel: Plate; Russian Standard
GOST 380 BST6 SP2	0.42% C 0.65% Mn 0.3% Cr Ni Cu (max) 0.05% S (max) 0.04% P (max) steel: Plate; Russian Standard
GOST 380 VST5 GPS	0.3% C (max) 1.0% Mn 0.05% S (max) 0.04% P steel: Plate; Russian Standard **UTS: 480**
GOST 380 VST5 GPS2	0.3% C (max) 1.0% Mn 0.3% Cr Ni Cu (max) 0.05% S (max) 0.04% P (max) steel: Plate; Russian Standard
GOST 380 VST5 PS	0.37% C (max) 0.75% Mn 0.05% S (max) 0.04% P steel: Plate; Russian Standard **UTS: 570**

Symbol	Nominal analysis, supplier, condition and remarks.
GOST 380 VST5 PS2	0.37% C (max) 0.75% Mn 0.3% Cr Ni Cu (max) 0.05% S 0.04% P steel: Plate; Russian Standard **UTS: 570** **Proof: 280**
GOST 380 VST5 SP	0.37% C (max) 0.75% Mn 0.04% S 0.05% P steel: Plate; Russian Standard **UTS: 570**
GOST 380 VST5 SP2	0.37% C (max) 0.75% Mn 0.3% Cr Ni Cu (max) 0.05% S (max) 0.04% P (max) steel: Plate; Russian Standard **UTS: 570** **Proof: 280**
GOST 1050/25	0.26% C 0.65% Mn 0.25% Cr and Ni (max) 0.04% S and P (max) steel: For structures; Russian Standard
GOST 1050/25 G	0.26% C 0.85% Mn 0.25% Cr and Ni (max) 0.04% S and P (max) steel: For structures; Russian Standard
GOST 1050/30	0.3% C 0.65% Mn 0.25% Cr and Ni (max) 0.04% S and P (max) steel: For structures; Russian Standard
GOST 1050/30 G	0.3% C 0.85% Mn 0.25% Cr and Ni (max) 0.04% S and P (max) steel: For structures; Russian Standard
GOST 1050/35	0.35% C 0.65% Mn 0.25% Cr and Ni (max) 0.04% S and P (max) steel: For structures; Russian Standard
GOST 1050/35 G	0.35% C 0.85% Mn 0.25% Cr and Ni (max) 0.04% S and P (max) steel: For structures; Russian Standard
GOST 1050/40	0.4% C 0.65% Mn 0.25% Cr and Ni (max) 0.04% S and P (max) steel: For structures; Russian Standard
GOST 1050/40 G	0.4% C 0.85% Mn 0.25% Cr and Ni (max) 0.04% S and P (max) steel: For structures; Russian Standard
GOST 1050/45	0.45% C 0.65% Mn 0.25% Cr and Ni (max) 0.04% S and P (max) steel: For structures; Russian Standard
GOST 1050/45 G	0.45% C 0.85% Mn 0.25% Cr and Ni (max) 0.04% S and P (max) steel: For structures; Russian Standard
GOST 2052/50 S2	0.51% C 0.75% Mn 0.3% Cr and Ni (max) 0.04% S and P (max) steel: Bar; Russian Standard
GOST 2052/55 SG	0.55% C 0.9% Mn 0.3% Cr and Ni (max) 0.04% S and P (max) steel: Bar; Russian Standard
GOST 2052/55 SG	0.55% C 0.9% Mn 0.4% Ni (max) 0.3% Cr (max) 0.04% S and P (max) 0.25% Cu (max) steel: For springs; Russian Standard
GOST 4543/25 G2S	0.25% C 1.4% Mn 0.3% Cr and Ni (max) steel: Russian Standard
GOST 4543/35 G2	0.35% C 1.6% Mn 0.25% Cr (max) steel: Russian Standard
GOST 4543/35 GS	0.34% C 1.0% Mn 0.3% Cr and Ni (max) steel: Russian Standard
GOST 4543/35 SG	0.35% C 1.25% Mn 0.25% Cr (max) steel: Russian Standard
GOST 4543/36 G2S	0.36% C 1.65% Mn 0.25% Cr (max) steel: Russian Standard
GOST 4543/40 G2	0.4% C 1.6% Mn 0.25% Cr (max) steel: Russian Standard
GOST 4543/45 G2	0.45% C 1.6% Mn 0.25% Cr (max) steel: Russian Standard
GOST 7909/36 Cr 2S	0.36% C 0.4% Ni (max) 0.3% Cr (max) 0.045% S (max) 0.04% P (max) steel: Pipe; Russian Standard **UTS: 700** **Proof: 500**
GOST 8467/36 G2S	0.36% C 1.65% Mn 0.4% Ni (max) 0.3% Cr (max) 0.045% S and P (max) steel: Pipe; Russian Standard **UTS: 700** **Proof: 500**
GOST 8731/35	0.35% C 0.65% Mn 0.2% Ni (max) 0.25% Cr (max) 0.04% S and P (max) steel: Pipe; Russian Standard; seamless **UTS: 520** **Proof: 300**
GOST 8731/45	0.45% C 0.65% Mn 0.2% Ni (max) 0.25% Cr (max) 0.04% S and P (max) steel: Pipe; Russian Standard; seamless **UTS: 600** **Proof: 330**
GOST 8731 BMST5 SP	0.32% C 0.65% Mn 0.055% S (max) 0.045% P (max) steel: Pipe; Russian Standard; seamless **UTS: 500** **Proof: 270**
GOST 8733/35	0.35% C 0.65% Mn 0.2% Ni (max) 0.15% Cr (max) 0.04% S and P (max) steel: Pipe; Russian Standard; seamless **UTS: 520** **Proof: 300**
GOST 8733/45	0.45% C 0.65% Mn 0.2% Ni (max) 0.15% Cr (max) 0.04% S and P (max) steel: Pipe; Russian Standard; seamless **UTS: 600** **Proof: 330**
GOST 10706/24 G	0.24% C 0.9% Mn 0.3% Cr + Ni and Cu (max) 0.04% S and P (max) steel: Pipe; Russian Standard; welded **UTS: 490** **Proof: 330**
GOST 10706/25 G2S	0.24% C 1.4% Mn 0.3% Cr + Ni and Cu (max) 0.04% S and P (max) steel: Pipe; Russian Standard; welded **UTS: 600** **Proof: 400**
GOST 12132/35	0.35% C 0.65% Mn 0.25% Cr and Ni (max) 0.04% S and P (max) steel: Pipe; Russian Standard; welded **UTS: 520**
GOST 12132/45	0.45% C 0.05% Mn 0.25% Cr and Ni (max) 0.04% S and P (max) steel: Pipe; Russian Standard; welded **UTS: 600**
HARDFLEX 9	0.45% C steel: Strip; Sandvik; supplied hardened and tempered **UTS: 900** **Proof: 750**
HARDFLEX II	0.55% C steel: Strip; Sandvik; supplied hardened and tempered **UTS: 1000** **Proof: 850**
HECLA 37	0.4% C 0.8% Mn steel: Hadfields for BS alloys En 8 and En 8A, B and C
HECLA 37M	0.4% C 1.1% Mn 0.15% S steel: Free machining; Hadfields for BS alloy En 8M
HECLA 40	0.3% C 0.8% Mn steel: Hadfields for BS alloys En 5 and En 5A, B, C and K
HEDEX 1	0.27% C 0.9% Mn 0.05% S 0.04% P steel: Wire for forging; Kiveton Park; annealed **UTS: 530**
HEDEX 1A	0.32% C 0.8% Mn 0.05% S 0.04% P steel: Wire for forging; Kiveton Park; annealed **UTS: 530**
HEDEX 1B	0.37% C 0.8% Mn 0.05% S 0.04% P steel: Wire for forging; Kiveton Park; annealed **UTS: 560**
HEDEX 2	0.42% C 0.8% Mn 0.05% S 0.04% P steel: Wire for forging; Kiveton Park; annealed **UTS: 610**
HEDEX 3	0.38% C 0.8% Mn 0.05% S 0.04% P steel: Wire for forging; Kiveton Park; annealed **UTS: 560**
HEDEX 12	0.22% C 1.5% Mn 0.05% S 0.04% P steel: Wire for forging; Kiveton Park; annealed **UTS: 550**
HEDEX 13	0.37% C 0.8% Mn 0.05% S 0.04% P steel: Wire for forging; Kiveton Park; annealed **UTS: 560**
HI STRENGTH C	0.24% C 1.3% Mn 0.05% Nb steel: Armco; weldable; in 4 grades **UTS: 530** **Elon: 15%** **Proof: 420**
HI-MAN	0.3% C 1.3% Mn 0.2% Cu steel: Inland Steel Co.
HI-STRESS	Medium carbon hot rolled deformed steel: Section; Park Gate; for concrete reinforcing **Proof: 280**

Note. The following abbreviations and units are used in the tables:

DPN	Hardness, diamond pyramid number
UTS	Ultimate tensile strength, N/mm^2
Elon	Elongation, %
Proof	0.1% proof strength, N/mm^2

$1\ N/mm^2 = 0.1\ hbar = 0.102\ kgf/mm^2 = 0.06475\ tonf/in.^2 = 145.04\ lbf/in.^2 = 1\ MPa$

See Appendix II for other abbreviations and conversion tables.

Symbol	Nominal analysis, supplier, condition and remarks.
HITENSPEED 45	0.38% C 1.3% Mn 0.26% S 0.04% P (max) steel: Hardened and tempered **UTS: 700 Elon: 12% Proof: 500**
HITENSPEED 45A	0.42% C 1.3% Mn 0.26% S 0.04% P (max) steel: Hardened and tempered **UTS: 700 Elon: 12% Proof: 540**
HMS 55	0.35% C 0.55% Mn 0.25% Si 0.035% S and P steel: Krupp
HMS 65	0.45% C 0.65% Mn 0.25% Si 0.035% S and P steel: Krupp
HTC	0.4% C 0.8% Mn steel: ESC for BS alloy En 8
JIS 4051/28 C	0.28% C 0.75% Mn 0.03% S and P (max) steel: For structures; Japanese Standard
JIS G3101 SS55	0.3% C 1.6% Mn (max) 0.04% S and P (max) steel: Plate; Japanese Standard **UTS: 550 Proof: 410**
JIS G3106 SM50A	0.21% C (max) 1.5% Mn (max) 0.04% S and P (max) steel: Plate; Japanese Standard **UTS: 560 Proof: 330**
JIS G3445 STKM14A	0.3% C (max) 0.7% Mn 0.04% S and P (max) steel: Pipe; Japanese Standard **UTS: 420 Proof: 250**
JIS G3445 STKM14B	0.3% C (max) 0.7% Mn 0.04% S and P steel: Pipe; Japanese Standard **UTS: 510 Proof: 360**
JIS G3445 STKM14C	0.3% C (max) 0.7% Mn 0.04% S and P (max) steel: Pipe; Japanese Standard **UTS: 560 Proof: 420**
JIS G3445 STKM15A	0.3% C 0.7% Mn 0.04% S and P steel: Pipe; Japanese Standard **UTS: 480 Proof: 280**
JIS G3445 STKM15C	0.3% C 0.7% Mn 0.04% S and P (max) steel: Pipe; Japanese Standard **UTS: 590 Proof: 440**
JIS G3445 STKM16A	0.4% C 0.7% Mn 0.04% S and P steel: Pipe; Japanese Standard **UTS: 520 Proof: 330**
JIS G3445 STKM16C	0.4% C 0.7% Mn 0.04% S and P (max) steel: Pipe; Japanese Standard **UTS: 630 Proof: 470**
JIS G3454 STPG42	0.3% C (max) 0.5% Mn 0.04% S and P steel: Pipe; Japanese Standard **UTS: 420 Proof: 250**
JIS G3455 STS42	0.3% C (max) 0.6% Mn 0.035% S and P 0.2% Cu (max) steel: Pipe; Japanese Standard
JIS G3455 STS49	0.33% C (max) 0.7% Mn 0.035% S and P 0.2% Cu steel: Pipe; Japanese Standard **UTS: 490 Proof: 280**
JIS G3456 STPT42	0.3% C (max) 0.65% Mn 0.035% S and P (max) 0.2% Cu (max) steel: Pipe; Japanese Standard **UTS: 420 Proof: 250**
JIS G3456 STPT49	0.33% C (max) 0.65% Mn 0.035% S and P (max) 0.2% Cu (max) steel: Pipe; Japanese Standard **UTS: 490 Proof: 280**
JIS G4051 S22C	0.22% C 0.45% Mn 0.03% S and P (max) steel: For structures; Japanese Standard
JIS G4051 S25C	0.25% C 0.45% Mn 0.03% S and P (max) steel: For structures; Japanese Standard
JIS G4051 S30C	0.3% C 0.75% Mn 0.03% S and P (max) steel: For structures; Japanese Standard
JIS G4051 S33C	0.33% C 0.75% Mn 0.03% S and P (max) steel: For structures; Japanese Standard
JIS G4051 S35C	0.35% C 0.75% Mn 0.03% S and P (max) steel: For structures; Japanese Standard
JIS G4051 S38C	0.38% C 0.75% Mn 0.03% S and P (max) steel: For structures; Japanese Standard
JIS G4051 S40C	0.4% C 0.75% Mn 0.03% S and P (max) steel: For structures; Japanese Standard
JIS G4051 S43C	0.43% C 0.75% Mn 0.03% S and P (max) steel: For structures; Japanese Standard
JIS G4051 S45C	0.45% C 0.75% Mn 0.03% S and P (max) steel: For structures; Japanese Standard
JIS G4106 S Mn 1	0.33% C 0.03% S and P (max) steel: Japanese Standard
JIS G4106 S Mn 2	0.38% C 1.5% Mn 0.03% S and P (max) steel: Japanese Standard
JIS G4106 S Mn 3	0.43% C 1.5% Mn 0.03% S and P (max) steel: Japanese Standard
JS 3	0.45% C (max) 1.0% Mn 0.6% Si steel: Casting; Jopling **DPN: 170 UTS: 530 Elon: 15% Proof: 240**
JS 12	0.3% C 1.5% Mn steel: Casting; Jopling **DPN: 230 UTS: 760 Elon: 15%**
JS 16	0.35% C (max) 1.0% Mn 0.6% Si steel: Casting; Jopling **DPN: 160 UTS: 480 Elon: 20% Proof: 240**
K945	0.45% C 0.7% Mn 0.035% S and P steel: VEW
K02603	0.26% C (max) 0.04% S and P steel: UNS designation
K03500	0.35% C (max) 0.04% S and P steel: Casting; designation used by UNS
K12810	0.28% C (max) 0.2% Cu (min) steel: Designation used by UNS
KF 4A	0.22% C 1.6% Mn 0.06% S and P steel: Kirkstall; hardened and tempered **UTS: 640 Elon: 20% Proof: 450**
KF 5	0.22% C 0.8% Mn 0.06% S and P steel: Kirkstall for BS alloy En 4A
KF 8	0.38% C 0.8% Mn 0.06% S and P steel: Kirkstall for BS alloy En 8
KF 9	0.42% C 1.2% Mn 0.06% S and P steel: Kirkstall for BS alloy En 8K
KF 10	0.44% C 0.7% Mn 0.06% S and P steel: Kirkstall for BS alloy En 8K
KF 10A	0.48% C 0.8% Mn 0.06% S and P steel: Kirkstall for BS alloy En 43A
KF 40	0.36% C 1.5% Mn 0.05% S and P steel: Kirkstall for BS alloy En 15
KF 40A	0.28% C 1.5% Mn 0.06% S and P steel: Kirkstall for BS alloy En 14B
KF 51	0.4% C 1.2% Mn 0.2% Si 0.17% S 0.06% P steel: Free machining; Kirkstall for BS alloy En 8M
KF 59	0.25% C 1.2% Mn 0.25% Si 0.14% S 0.06% P steel: Free machining; Kirkstall for BS alloy En 7
KF 63	0.38% C 1.5% Mn 0.2% Si 0.2% S 0.06% P steel: Free machining; Kirkstall for BS alloy En 15 AM
KF 63A	0.44% C 1.5% Mn 0.3% Si 0.3% S 0.05% P steel: Free machining; Kirkstall
KF 64	0.3% C 1.4% Mn 0.3% Si 0.16% S 0.05% P steel: Free machining; Kirkstall; hardened and tempered **UTS: 610 Elon: 17% Proof: 530**
KP 4	0.3% C 1% Mn 0.06% S and P steel: Kiveton Park for BS alloy En 4
KP 5	0.3% C 0.9% Mn 0.06% S and P steel: Kiveton Park for BS alloy En 5
KP 6	0.4% C 0.8% Mn 0.06% S and P steel: Kiveton Park for BS alloy En 8
KP 20	0.35% C 1.5% Mn 0.06% S and P steel: Kiveton Park for BS alloy En 15
KP 21	0.25% C 1.5% Mn 0.06% S and P steel: Kiveton Park for BS alloy En 14
L 2	0.28% C 1% Mn 0.6% Si steel: Casting; Jessop; grade B
L 3	0.37% C 1% Mn 0.6% Si steel: Casting; Jessop; grade C
M 3	0.46% C 1.6% Mn 0.22% Si steel: Jessop
M St 50/2	0.3% C 0.05% S 0.06% P steel: German Standard
M St 60/2	0.4% C 0.05% S 0.06% P steel: German Standard
MACHINERY 30	0.3% C 0.8% Mn 0.2% Si 0.04% S and P steel: Atlas; as AISI 1030
MACHINERY 40	0.4% C 0.8% Mn 0.2% Si 0.04% S and P steel: Atlas; as AISI 1040

Symbol	Nominal analysis, supplier, condition and remarks.
MATTE LADLE	0.37% C 0.7% Mn 0.3% Si steel: Casting; Edgar Allen; code 611
MBV	0.4% C 0.9% Mn 0.4% Si 0.035% S and P steel: Krupp
MBV/m	0.36% C 1.35% Mn 0.25% Si 0.035% S and P steel: Krupp; up to 0.3% Cr may be present
MBV/w	0.3% C 1.35% Mn 0.25% Si 0.035% S and P steel: Krupp; up to 0.3% Cr may be present
MEL-TROL TGS	0.2% C 1.3% Mn 0.2% Si steel: Carpenter
Mil S 43	0.32% C 1.3% Mn 0.15% Si steel: Casting; US Federal specification
Mil S 3289	0.3% C 0.04% S and P steel: US Federal specification
Mil S 15083-70-36	0.35% C 0.05% S and P (max) steel: Casting; US Federal specification
Mil S 15083-80-40	0.5% C 0.05% S and P (max) steel: Casting; US Federal specification
Mil S 18591-1040	0.4% C 0.04% S and P (max) steel: Casting; US Federal specification
Mil S 22141-1040	0.4% C 0.025% S and P (max) steel: Casting; US Federal specification
Mil S 22141-1050	0.5% C 0.025% S and P (max) steel: Casting; US Federal specification
Mil S 22698(C)	0.27% C 0.04% S and P steel: US Federal specification
Mil S 81591-1050	0.5% C 0.04% S and P (max) steel: Casting; US Federal specification
MNC 720	Swedish Standard; summary of steel castings
MNC 852	General standard for SS range of steels for hardening and tempering; Swedish Standard
MNC 854	Swedish Standard; summary of steel for induction and flame hardening
MNC 870	Covers spring steels of various types; Swedish Standard; includes detailed specifications
MS 34	0.38% C 1.05% Mn 0.8% Si 0.035% S and P steel: Krupp
MS 43	0.46% C 1.05% Mn 0.8% Si 0.035% S and P steel: Krupp
MS BV	0.37% C 1.25% Mn 1.25% Si 0.035% S and P steel: Krupp
MS V3	0.3% C 0.75% Mn steel: Pompey
MT 'P'	0.25% C 1.4% Mn steel: Casting; Edgar Allen; code 610
MT 'R'	0.3% C 1.4% Mn steel: Casting; Edgar Allen; code 626
N 82	0.4% C 1.3% Mn steel: Bofors; hardened and tempered **DPN: 230 UTS: 730 Elon: 17% Proof: 480**
N 91	0.45% C 0.8% Mn steel: For surface hardening; Bofors; normalized; hardened by flame or induction heating **DPN: 200 UTS: 670 Elon: 18% Proof: 300**
NF A35 551/35 M5	0.36% C 1.2% Mn steel: French Standard
NF A35 551/XC25	0.26% C 0.55% Mn steel: For structures; French Standard
NF A35 551/XC32	0.32% C 0.65% Mn steel: For structures; French Standard
NF A35 551/XC38	0.38% C 0.65% Mn steel: For structures; French Standard
NF A35 551/XC42	0.42% C 0.65% Mn steel: For structures; French Standard
NF A35 571/41S7	0.41% C 0.65% Mn 0.035% S and P (max) steel: Bar; French Standard
NF A35 571/46S7	0.46% C 0.65% Mn 0.035% S and P (max) steel: Bar; French Standard
NF A35 571/50S7	0.5% C 0.65% Mn 0.05% S and P (max) steel: Bar; French Standard
NF A35 571/51S7	0.51% C 0.65% Mn 0.3% Cr (max) 0.035% S and P (max) steel: Bar; French Standard
NV9	0.39% C steel: For valves; SAE designation
OCM 6	0.35% C 1.5% Mn steel: W Marrison for BS alloy En 15
Philips 250	0.3% C 1% Mn 0.8% Si steel: Welding electrode; Philips **DPN: 260**
Q BRAND	0.4% C 0.8% Mn steel: For cars; die inserts, wedges, etc. **UTS: 650**
QQ W 409a	Low and medium carbon steel: Wire for cold heating; US Federal specification covering a range of AISI alloys
QQ W 414a	Carbon steel: Wire; for book binding; US Federal specification
QQ W 418a	Carbon steel: Wire; for concrete reinforcing; US Federal specification; cold drawn
RETORT	0.32% C 0.7% Mn 0.6% Si steel: Casting; Edgar Allen; code 609
RS	0.45% C 0.6% Mn 1.7% Si steel: Pompey
RS 1	0.45% C 0.7% Mn 1.8% Si steel: Pompey
S1 A	0.2% C 0.7% Mn 0.2% Si 0.05% S and P steel: Bar; cold drawn; BS aircraft standard **UTS: 530 Elon: 12%**
S1 B	0.3% C 0.7% Mn 0.2% Si 0.05% S and P steel: Bar; cold drawn; BS aircraft standard **UTS: 530 Elon: 12%**
S1 C	0.35% C 0.7% Mn 0.2% Si 0.05% S and P steel: Bar; cold drawn; BS aircraft standard **UTS: 530 Elon: 12%**
S 3F	0.35% C 0.6% Mn 0.3% Si steel: Pompey
S 4F	0.4% C 0.6% Mn 0.3% Si steel: Pompey
S 4M	0.45% C 0.7% Mn 0.2% Si steel: Pompey
S 4S	0.4% C 0.6% Mn 0.3% Si steel: Pompey
S6	0.4% C steel: Normalized; replaced by BS S93
S20	Steel: Sheet; tinned; replaced by BS S512
S21	0.25% C (max) 0.8% Mn 0.2% Si 0.05% S and P steel: Cold drawn; BS aircraft standard **UTS: 420 Elon: 17%**
S23	Medium carbon steel: Replaced; BS aircraft standard
S24	Bright steel: Bar for keys; replaced; BS aircraft standard
S26	Medium carbon steel: Forgings; 600 N/mm^2; replaced; BS aircraft standard
S27	Medium carbon steel: Bar; 600 N/mm^2; replaced; BS aircraft standard
S31	Steel: Wire for forging bolts; replaced; BS aircraft standard
S71	0.3% C steel: Normalized; obsolete; BS aircraft standard
S76	0.4% C steel: Hardened and tempered; obsolete; BS aircraft standard
S77	0.3% C steel: Hardened and tempered; obsolete; BS aircraft standard
S93	0.4% C 0.7% Mn 0.2% Si 0.04% S and P steel: Normalized; BS aircraft standard **DPN: 180 UTS: 610 Elon: 20% Proof: 300**
S105	0.4% C 0.9% Mn 0.2% Si 0.04% S and P steel: Wire for forged bolts; hardened and tempered after forging; BS aircraft standard **UTS: 610 Elon: 18% Proof: 660**

Note. The following abbreviations and units are used in the tables:

DPN	Hardness, diamond pyramid number
UTS	Ultimate tensile strength, N/mm^2
Elon	Elongation, %
Proof	0.1% proof strength, N/mm^2

1 N/mm^2=0.1 hbar=0.102 kgf/mm^2=0.06475 tonf/in.^{2}=145.04 lbf/in.^{2}=1 MPa

See Appendix II for other abbreviations and conversion tables.

Symbol	Nominal analysis, supplier, condition and remarks.
S113	0.4% C 0.8% Mn 0.2% Si 0.04% S and P 0.2% Pb steel: Hardened and tempered; Pb content optional; BS aircraft standard **DPN: 230 UTS: 760 Elon: 15%**
S116	0.4% C 0.9% Mn 0.2% Si 0.04% S and P steel: Bar; hardened and tempered; BS aircraft standard **DPN: 280 UTS: 920 Elon: 18% Proof: 660**
S516	0.45% C 1.5% Mn 0.2% Si 0.05% S and P steel: Sheet; hardened and tempered; BS aircraft standard **UTS: 920 Elon: 8% Proof: 760**
SAE 003 C	0.3% C (max) 0.7% Mn 0.6% Si 0.05% S and P cast steel: Hardened and tempered; weldable **DPN: 131 UTS: 450 Elon: 24% Proof: 220**
SAE 0025	0.25% C 0.7% Mn 0.8% Si steel: Casting; as cast; suitable for welding **DPN: 187 UTS: 420 Elon: 22% Proof: 200**
SAE 0035	0.3% C 0.8% Mn 0.6% Si 0.05% S and P investment cast steel: Hardened and tempered **UTS: 480 Elon: 22% Proof: 210**
SAE 0050	0.45% C 0.7% Mn 0.4% Si 0.05% S and P cast steel: Hardened and tempered **DPN: 200 UTS: 730 Elon: 10% Proof: 480**
SAE 090	C not specified 0.05% S and P cast steel: Hardened and tempered **DPN: 190 UTS: 670 Elon: 20% Proof: 420**
SAE 0105	C not specified 0.05% S and P cast steel: Hardened and tempered **DPN: 220 UTS: 760 Elon: 17% Proof: 610**
SAE 0120	C not specified 0.05% S and P cast steel: Hardened and tempered **DPN: 250 UTS: 890 Elon: 14% Proof: 740**
SAE 0150	C not specified 0.05% S and P cast steel: Hardened and tempered **DPN: 310 UTS: 1070 Elon: 9% Proof: 910**
SAE 0175	C not specified 0.05% S and P cast steel: Hardened and tempered **DPN: 360 UTS: 1300 Elon: 6% Proof: 1070**
SAE 851	0.4% Fe sinter: For bearings; density 5900 kg/m^3
SAE 861	0.3% C 4.0% Cu Fe sintered metal: Specific density 5.8–6.2 **UTS: 150**
SAE 862	0.25% C 9.0% Cu Fe alloy sinter: For bearings; density 6000 kg/m^3
SAE 863	0.25% C 22% Cu Fe alloy sinter: For bearings; density 6000 kg/m^3
SAE 1024	0.23% C 1.5% Mn 0.05% S 0.04% P steel: Bar; hot rolled as AISI C1024 **DPN: 149 UTS: 530 Elon: 20% Proof: 300**
SAE 1024	0.23% C 1.5% Mn 0.05% S 0.04% P steel: Bar; cold drawn **DPN: 163 UTS: 600 Elon: 12% Proof: 580**
SAE 1025	0.25% C 0.4% Mn 0.05% S 0.04% P steel: Bar; hot rolled as AISI C1025 **DPN: 116 UTS: 500 Elon: 25% Proof: 210**
SAE 1025	0.25% C 0.4% Mn 0.05% S 0.04% P steel: Bar; cold drawn **DPN: 126 UTS: 450 Elon: 15% Proof: 370**
SAE 1026	0.26% C 0.7% Mn 0.04% S and P steel: As AISI C1026
SAE 1027	0.25% C 1.3% Mn 0.05% S 0.04% P steel: Bar; hot rolled **DPN: 149 UTS: 540 Elon: 18% Proof: 300**
SAE 1027	0.25% C 1.3% Mn 0.05% S 0.04% P steel: Bar; cold drawn **DPN: 163 UTS: 610 Elon: 12% Proof: 500**
SAE 1029	0.29% C 0.75% Mn 0.05% S 0.04% P steel: Bar, etc.
SAE 1030	0.3% C 0.8% Mn 0.05% S 0.04% P steel: Bar; hot rolled **DPN: 137 UTS: 480 Elon: 20% Proof: 240**
SAE 1030	0.3% C 0.8% Mn 0.05% S 0.04% P steel: Bar; cold drawn **DPN: 149 UTS: 550 Elon: 12% Proof: 450**
SAE 1033	0.33% C 0.9% Mn 0.05% S 0.04% P steel: Bar; hot rolled **DPN: 143 UTS: 530 Elon: 18% Proof: 240**
SAE 1033	0.33% C 0.9% Mn 0.05% S 0.04% P steel: Bar; cold drawn **DPN: 163 UTS: 580 Elon: 12% Proof: 480**
SAE 1035	0.35% C 0.8% Mn 0.05% S 0.04% P steel: Bar; hot rolled **DPN: 143 UTS: 530 Elon: 18% Proof: 240**
SAE 1035	0.35% C 0.8% Mn 0.05% S 0.04% P steel: Bar; cold drawn **DPN: 163 UTS: 580 Elon: 12% Proof: 480**
SAE 1036	0.34% C 1.2% Mn 0.05% S 0.04% P steel: Bar; hot rolled **DPN: 163 UTS: 610 Elon: 16% Proof: 320**
SAE 1036	0.34% C 1.2% Mn 0.05% S 0.04% P steel: Bar; cold drawn **DPN: 187 UTS: 660 Elon: 12% Proof: 540**
SAE 1037	0.35% C 0.9% Mn 0.05% S 0.04% P steel: Bar; hot rolled **DPN: 143 UTS: 540 Elon: 18% Proof: 240**
SAE 1037	0.35% C 0.9% Mn 0.05% S 0.04% P steel: Bar; hot rolled **DPN: 143 UTS: 540 Elon: 18% Proof: 240**
SAE 1037	0.35% C 0.9% Mn 0.05% S 0.04% P steel: Bar; cold drawn **DPN: 167 UTS: 600 Elon: 12% Proof: 480**
SAE 1038	0.38% C 0.8% Mn 0.05% S 0.04% P steel: Bar; hot rolled **DPN: 149 UTS: 550 Elon: 18% Proof: 300**
SAE 1038	0.38% C 0.8% Mn 0.05% S 0.04% P steel: Bar; cold drawn **DPN: 163 UTS: 610 Elon: 12% Proof: 500**
SAE 1039	0.4% C 0.9% Mn 0.05% S 0.04% P steel: Bar; hot rolled **DPN: 156 UTS: 580 Elon: 16% Proof: 300**
SAE 1039	0.4% C 0.9% Mn 0.05% S 0.04% P steel: Bar; cold drawn **DPN: 179 UTS: 630 Elon: 12% Proof: 530**
SAE 1040	0.4% C 0.8% Mn 0.05% S 0.04% P steel: Bar; hot rolled **DPN: 149 UTS: 550 Elon: 18% Proof: 300**
SAE 1040	0.4% C 0.8% Mn 0.05% S 0.04% P steel: Bar; cold drawn **DPN: 170 UTS: 630 Elon: 12% Proof: 550**
SAE 1041	0.4% C 1.5% Mn 0.05% S 0.04% P steel: Bar; hot rolled **DPN: 187 UTS: 660 Elon: 15% Proof: 360**
SAE 1041	0.4% C 1.5% Mn 0.05% S 0.04% P steel: Bar; cold drawn **DPN: 207 UTS: 750 Elon: 10% Proof: 630**
SAE 1042	0.43% C 0.8% Mn 0.05% S 0.04% P steel: Bar; hot rolled **DPN: 163 UTS: 580 Elon: 16% Proof: 300**
SAE 1042	0.43% C 0.8% Mn 0.05% S 0.04% P steel: Bar; cold drawn **DPN: 179 UTS: 650 Elon: 12% Proof: 550**
SAE 1043	0.43% C 0.9% Mn 0.05% S 0.04% P steel: Bar; hot rolled **DPN: 163 UTS: 600 Elon: 16% Proof: 320**
SAE 1043	0.43% C 0.9% Mn 0.05% S 0.04% P steel: Bar; cold drawn **DPN: 179 UTS: 660 Elon: 12% Proof: 560**
SAE 1045	0.46% C 0.8% Mn 0.05% S 0.04% P steel: Bar; hot rolled **DPN: 163 UTS: 600 Elon: 16% Proof: 320**
SAE 1045	0.46% C 0.8% Mn 0.05% S 0.04% P steel: Bar; cold drawn **DPN: 179 UTS: 660 Elon: 12% Proof: 560**
SAE 1046	0.46% C 0.9% Mn 0.05% S 0.04% P steel: Bar; hot rolled

Symbol	Nominal analysis, supplier, condition and remarks.
SAE 1046	0.46% C 0.9% Mn 0.05% S 0.04% P steel: Bar; cold drawn
SAE 1126	0.26% C 0.9% Mn 0.1% S 0.04% P steel: Free cutting; cold drawn
SAE 1126	0.26% C 0.9% Mn 0.1% S 0.04% P steel: Free cutting; hot rolled
SAE 1132	0.31% C 1.5% Mn 0.1% S 0.04% P steel: Free cutting; cold drawn
SAE 1132	0.31% C 1.5% Mn 0.1% S 0.04% P steel: Free cutting; hot rolled
SAE 1137	0.36% C 1.5% Mn 0.1% S 0.04% P steel: Free cutting; cold drawn
SAE 1137	0.36% C 1.5% Mn 0.1% S 0.04% P steel: Free cutting; hot rolled
SAE 1138	0.37% C 0.9% Mn 0.1% S 0.04% P steel: Free cutting; cold drawn
SAE 1138	0.37% C 0.9% Mn 0.1% S 0.04% P steel: Free cutting; hot rolled
SAE 1139	0.39% C 1.5% Mn 0.18% S 0.04% P steel: Free cutting; cold drawn
SAE 1140	0.4% C 0.9% Mn 0.1% S 0.04% P free cutting steel: Hot rolled **DPN: 156 UTS: 580 Elon: 16% Proof: 300**
SAE 1140	0.4% C 0.9% Mn 0.1% S 0.04% P free cutting steel: Cold drawn **DPN: 170 UTS: 640 Elon: 12% Proof: 530**
SAE 1141	0.41% C 1.5% Mn 0.1% S 0.04% P free cutting steel: Hot rolled **DPN: 187 UTS: 680 Elon: 15% Proof: 360**
SAE 1141	0.41% C 1.5% Mn 0.1% S 0.04% P free cutting steel: Cold drawn **DPN: 212 UTS: 770 Elon: 10% Proof: 630**
SAE 1144	0.44% C 1.5% Mn 0.3% S 0.4% P free cutting steel: Hot rolled **DPN: 197 UTS: 700 Elon: 15% Proof: 370**
SAE 1144	0.44% C 1.5% Mn 0.3% S 0.4% P free cutting steel: Cold drawn
SAE 1145	0.45% C 0.9% Mn 0.05% S 0.04% P free cutting steel: Hot rolled **DPN: 170 UTS: 630 Elon: 15% Proof: 320**
SAE 1145	0.45% C 0.9% Mn 0.05% S 0.04% P free cutting steel: Cold drawn **DPN: 187 UTS: 680 Elon: 12% Proof: 580**
SAE 1146	0.46% C 0.9% Mn 0.1% S 0.04% P free cutting steel: Hot rolled **DPN: 170 UTS: 630 Elon: 15% Proof: 320**
SAE 1146	0.46% C 0.9% Mn 0.1% S 0.04% P free cutting steel: Cold drawn **DPN: 187 UTS: 680 Elon: 12% Proof: 580**
SAE 1330	0.3% C 1.8% Mn 0.3% Si steel: Mechanical properties not specified
SAE 1335	0.35% C 1.8% Mn 0.3% Si steel: Mechanical properties not specified
SAE 1340	0.4% C 1.8% Mn 0.3% Si steel: Mechanical properties not specified
SAE 1345	0.45% C 1.8% Mn 0.3% Si steel: Mechanical properties not specified
SAE 1524	0.24% C 1.5% Mn 0.05% S 0.04% P steel: Bar, etc.
SAE 1525	0.25% C 1.0% Mn 0.05% S 0.04% P steel: Bar, etc.
SAE 1527	0.27% C 1.3% Mn 0.05% S 0.04% P steel: Bar, etc.
SAE 1541	0.41% C 1.5% Mn 0.05% S 0.04% P steel: Bar, etc.
SAE J 113-1	0.6% C 1% Mn steel: Wire for springs **UTS: 1720**
SAE J 113-2	0.62% C 1% Mn steel: Wire for springs **UTS: 1800**
SAE J 435-0050	0.45% C 0.04% S and P (max) steel: Casting
SAE NV1	0.4% C 1.5% Mn 0.25% Si steel: For hardening and tempering and for flame and induction hardening; as SAE 1041
SAE NV2	0.47% C 1.5% Mn 0.2% Si steel: For inlet valves; as SAE 1047
SANBOLD 8	0.4% C 0.8% Mn 0.2% Si steel: Sanderson for BS alloy En 8 **DPN: 200 UTS: 750 Elon: 14% Proof: 400**
SANBOLD 15	0.35% C 1.5% Mn 0.2% Si steel: Sanderson for BS alloy En 15
SANBOLD 52	0.35% C 1.5% Mn 0.2% Si steel: Sanderson; replaced by Sanbold 15
SANBOLD 3890	0.4% C 0.8% Mn 0.2% Si steel: Sanderson; replaced by Sanbold 8
SIMANCON	0.23% C 1.6% Mn 0.6% Si 0.045% S and P 0.04% Nb steel: Plate; for welding; normalized; Consett **UTS: 600 Elon: 22% Proof: 390**
For SIS specifications see SS	
SK 42	0.27% C (max) 1.2% Mn 0.04% S and P (max) steel: Phoenix
SPEAR 412	0.35% C 1.3% Mn 0.5% Ni steel: Spear and Jackson for BS alloy En 12
Sprasteel LS	Medium carbon steel: Sprayed deposit; Metco **DPN: 240**
SS 1410	0.25% C 0.9% Mn 0.02% Si structural steel: Swedish Standard **UTS: 460 Elon: 20% Proof: 240**
SS 1412	0.2% C 0.3% Cr (max) 0.4% Cu (max) steel: Structural purposes; Swedish Standard
SS 1450	0.25% C 0.6% Mn 0.05% S and P steel: Bar, forging, etc.; Swedish Standard; normalized **DPN: 130 UTS: 450 Elon: 24% Proof: 220**
SS 1450-0	0.25% C steel: Swedish Standard; as rolled; suffix indicates condition **DPN: 135 UTS: 420 Elon: 24% Proof: 220**
SS 1505	0.3% C 0.6% Mn 0.04% S and P steel: Swedish Standard; annealed **UTS: 460 Elon: 18% Proof: 240**
SS 1510	0.3% C 1.2% Mn structural steel: Swedish Standard; as rolled **UTS: 580 Elon: 20% Proof: 330**
SS 1550	0.35% C 0.6% Mn 0.05% S and P steel: Bar, forging, etc.; Swedish Standard; normalized **DPN: 160 UTS: 530 Elon: 22% Proof: 240**
SS 1550-06	0.35% C steel: Bar; Swedish Standard; work hardened **DPN: 165 UTS: 530 Elon: 8% Proof: 450**
SS 1572	0.35% C steel: For flame and induction hardening; mechanical properties not specified; Swedish Standard
SS 1572	0.35% C steel: Swedish Standard; hardened and tempered **UTS: 600 Elon: 19% Proof: 370**
SS 1606	0.23% C 1.6% Mn 0.6% Si 0.045% S and P steel: Swedish Standard; casting **DPN: 205 UTS: 650 Elon: 10% Proof: 300**
SS 1650	0.45% C 20.6% Mn 0.04% S and P steel: Plate, bar and forging; Swedish Standard; normalized **DPN: 190 UTS: 630 Elon: 18% Proof: 300**
SS 1655	0.52% C 0.7% Mn steel: Swedish Standard; general engineering **DPN: 230 UTS: 760 Elon: 12% Proof: 350**
SS 1660	0.44% C 0.4% Mn 0.03% S and P steel: Strip; Swedish Standard; annealed **DPN: 155 UTS: 650**

Note. The following abbreviations and units are used in the tables:

DPN	Hardness, diamond pyramid number
UTS	Ultimate tensile strength, N/mm^2
Elon	Elongation, %
Proof	0.1% proof strength, N/mm^2

1 N/mm^2=0.1 hbar=0.102 kgf/mm^2=0.06475 $tonf/in.^2$=145.04 $lbf/in.^2$=1 MPa

See Appendix II for other abbreviations and conversion tables.

Symbol	Nominal analysis, supplier, condition and remarks.
SS 1660	0.44% C 0.4% Mn 0.03% S and P steel: Strip; Swedish Standard; cold rolled **DPN: 255 UTS: 760**
SS 1672	0.45% C 0.8% Mn 0.03% S and P steel: For flame hardening; Swedish Standard; annealed **DPN: 200**
SS 1672	0.46% C steel: Swedish Standard; hardened and tempered **UTS: 700 Elon: 16% Proof: 420**
SS 1674	0.52% C steel: Swedish Standard **UTS: 740 Elon: 13% Proof: 510**
SS 1674	0.52% C steel: Swedish Standard; hardened and tempered **UTS: 740 Elon: 13% Proof: 510**
SS 1757	Reinforcing steel: Cold rolled **UTS: 1000 Elon: 2.5%**
SS 1940	0.26% C 1% Mn 0.15% S 0.05% P free machining steel: Swedish Standard; normalized **UTS: 480**
SS 1957	0.38% C 1% Mn 0.15% S 0.05% P free machining steel: Swedish Standard; normalized **UTS: 530**
SS 2120	0.41% C 1.2% Mn 0.03% S and P steel: Bar and forging; Swedish Standard; hardened and tempered **DPN: 290 UTS: 1030 Elon: 8% Proof: 680**
SS 2120	0.41% C 1.2% Mn steel: Swedish Standard; hardened and tempered **UTS: 700 Elon: 12% Proof: 450**
SS 2128	0.4% C 1.6% Mn 0.05% S and P steel: For gas cylinders; Swedish Standard; as forged **DPN: 220 UTS: 750 Elon: 15% Proof: 420**
SS 2165	0.25% C 1.6% Mn 0.05% S and P steel: Swedish Standard; included in MNC 815
SS 2168	0.25% C 1.6% Mn 0.05% S and P steel: Swedish Standard; reinforced steel; included in MNC 815
SS 2168	0.28% C (max) 1.6% Mn (max) 0.05% S and P steel: For reinforcing **Elon: 12% Proof: 590**
SS 8010	0.4% C sintered iron: Swedish Standard
SS 8051	0.4% C 2.0% Cu sintered steel
St 50	0.3% C 0.05% S 0.08% P steel: German Standard
St 50/2	0.3% C 0.05% S 0.06% P steel: German Standard
St 50/202	0.3% C 0.05% S 0.06% P steel: German Standard
St 50/203	0.3% C 0.05% S 0.06% P steel: German Standard
St 55	0.36% C 0.05% S and P steel: German Standard
St 55/4	0.35% C 0.3% Mn 0.2% Si 0.05% S and P steel: German Standard
St 60	0.4% C 0.05% S 0.08% P steel: German Standard
St 60/2	0.4% C 0.05% S 0.06% P steel: German Standard
St 60/202	0.4% C 0.05% S 0.06% P steel: German Standard
St 60/203	0.4% C 0.05% S 0.06% P steel: German Standard
STA 5 V4	0.3% C steel: Replaced by BS alloy En 5
STA 5 V4/1	0.3% C steel: Replaced by BS alloy En 5A
STA 5 V4/2	0.4% C steel: Replaced by BS alloy En 8A
STA 5 V4/3	0.3% C steel: Replaced by BS alloy En 5A
STA 5 V4/4	0.3% C steel: Replaced by BS alloy En 5C
STA 5 V4A	0.4% C steel: Replaced by BS alloy En 8K
STA 5 V4A/1	0.4% C steel: Replaced by BS alloy En 8B
STA 5 V4A/2	0.4% C steel: Replaced by BS alloy En 8D
STA 5 V4A/3	0.4% C steel: Replaced by BS alloy En 8C
STA 5 V4AM	0.35% C 1.0% Mn 0.25% Si 0.15% S 0.05% P En 8M
STA 5 V7	0.35% C steel: Replaced by BS alloy En 15
STA 28	Medium carbon steel: Strip; mainspring quality; hardened and tempered
STA 30	Medium carbon steel: For rifle barrels, etc.
STA 36 A	Medium carbon steel: Forging for shell bodies; hardened and tempered
STC 2	0.3% C 1.0% Mn 0.2% Si 0.06% S and P steel: Spencer; normalized **DPN: 150 UTS: 650 Elon: 25%**
STC 3	0.35% C 1.0% Mn 0.2% Si 0.05% S and P steel: Spencer; normalized **DPN: 150 UTS: 530 Elon: 25%**
STC 4	0.4% C 1.0% Mn 0.2% Si 0.05% S and P steel: Spencer; normalized **DPN: 170 UTS: 620 Elon: 20%**
STRENLITE 50	0.33% C 1.65% Mn steel: Steel Co. of Canada
STRENLITE 440	0.27% C 1.4% Mn 0.2% Cu steel: Steel Co. of Canada
SUPERTOUGH B 35	0.35% C 1.5% Mn steel: ESC for BS alloy En 15
T1	0.4% C (max) 1.5% Mn 0.3% Si 0.05% S and P steel: Tube; hardened and tempered **UTS: 530**
T5	Medium carbon steel: Tube; obsolete
T14	Medium carbon steel: Tube for axles; tempered; obsolete
T21	Carbon steel: Tube; annealed; obsolete
T35	0.25% C (max) 1.5% Mn 0.2% Si 0.05% S and P steel: Tube; hardened and tempered; weldable **UTS: 530 Proof: 450**
T45	0.25% C (max) 1.5% Mn 0.2% Si 0.05% S and P steel: Tube; hardened and tempered; weldable **UTS: 670 Proof: 610**
T54	0.25% C 0.6% Mn 0.2% Si 0.05% S and P steel: Tube; as drawn and tempered; weldable **UTS: 530 Proof: 450**
T 735	0.35% C 1.0% Mn 0.25% Si 0.15% S 0.05% P steel: Fagersta; free machining **UTS: 580 Elon: 20%**
TOLEDO 30	0.3% C 1% Mn 0.06% S and P steel: Toledo for BS alloy En 5
TOLEDO 40	0.4% C 0.8% Mn 0.06% S and P steel: Toledo for BS alloy En 6A
TOWER 50	0.33% C 1.2% Mn 0.04% S and P (max) steel: Algoma
TOWER 55	0.33% C 1.65% Mn 0.04% S and P (max) steel: Algoma
UMS 55	0.35% C 0.55% Mn 0.25% Si 0.045% S and P steel: Krupp
UMS 65	0.45% C 0.65% Mn 0.25% Si 0.045% S and P steel: Krupp
V 732	0.26% C 2.5% Mn steel: VEW
V 762	0.37% C 1.3% Mn steel: VEW
V 930	0.31% C 1.4% Mn steel: VEW
V 935	0.30% C 0.7% Mn steel: VEW
V 945	0.45% C 0.7% Mn steel: VEW
VIGILANT	0.55% C 0.7% Mn steel: Origin unknown
W1	C not specified 0.04% S and P galvanized steel: Wire; high tensile wire
W2	C not specified 0.04% S and P galvanized steel: Wire
W3	0.45% C 1.2% Mn 0.3% Si 0.05% S and P steel: Wire **UTS: 840 Elon: 10%**
W8	0.45% C 1.2% Mn 0.2% Si 0.05% S and P steel: Tie rods **UTS: 840 Elon: 10%**
W9	C not specified 0.04% S and P galvanized steel: Wire
WHC	0.58% C 0.5% Mn steel: Origin unknown
WLM WARRANTED	Plain carbon steel 0.15–0.55% C : W Marrison
Y 35	0.35% C 0.25% Mn 0.025% S (max) 0.035% P (max) steel: French Standards designation
Y 45	0.45% C 0.25% Mn 0.025% S (max) 0.035% P (max) steel: French Standards designation
Y 45S7	0.45% C 0.6% Mn steel: French Standard
Y 55	0.55% C 0.25% Mn 0.025% S (max) 0.035% P (max) steel: French Standards designation
Y 55S7	0.55% C 0.8% Mn 0.4% Cr (max) steel: French Standards designation
Z 908	0.08% C 1.2% Mn 0.35% S 0.08% P steel: Free machining; VEW
Z 912	0.3% C 1.6% Mn 0.2% S 0.05% P 0.15% S steel: Free machining; VEW

Symbol	Nominal analysis, supplier, condition and remarks.
Z 952	0.08% C 1.2% Mn 0.35% S 0.23% Pb steel: Free machining; VEW
Z 986	0.46% C 0.7% Mn steel; VEW

Note. The following abbreviations and units are used in the tables:

DPN	Hardness, diamond pyramid number
UTS	Ultimate tensile strength, N/mm^2
Elon	Elongation, %
Proof	0.1% proof strength, N/mm^2

1 N/mm^2=0.1 hbar=0.102 kgf/mm^2=0.06475 tonf/in.^{2}=145.04 lbf/in.^{2}=1 MPa

See Appendix II for other abbreviations and conversion tables.

44A3 Steel – plain carbon 0.5–0.8 per cent

Specific gravity	7.86
Density	7860 kg/m^3
Solidus/liquidus	1460–1490 °C
Thermal conductivity	46.1 W/m °C
Coefficient of linear expansion	11 × 10^{-6}/ °C
Electrical conductiivity	8–10% IACS (copper 100%)
Specific resistance	180–250 microhm mm
Young's modulus of elasticity	–
Impact	61–115 J
Fatigue strength (no. of cycles)	–
Hot strength	

Temperature °C	Tensile strength N/mm^2	Elongation %
200	800	21
300	840	17.5
400	660	27
500	460	35
600	300	41

The above properties are typical of the following group, and may not apply exactly to any one specification. It is possible that with certain specifications some of the values may not be applicable.

General metallurgical characteristics

This group is the medium high carbon range capable of 1500 N/mm^2 tensile, provided a low ductility can be accepted. The steels are available in all the normal forms of sheet, bar, tube, forgings and castings, but thin section sheet and tubing are not normally available as the materials cold work and require frequent anneals to prevent embrittlement. They are reasonably hardenable but not under still air conditions. Almost without exception the materials listed are used in the normalized, hardened and tempered, or cold drawn condition.

These steels are hardenable by thermal treatment. They require a temperature above 800 °C for hardening and, where sections are thin, i.e. less than 10 mm, may require oil quenching. Technical advice should be obtained where high integrity or high cost items are involved.

The steels in this group, and some of the steels in Group 44A2, are ideal for induction hardening or where necessary flame hardening. Readers are strongly advised to obtain specialist advice before any of these techniques is attempted.

The steels in this group should only be welded after obtaining technical advice. They will be prone to heat affected zone cracking, and will require pre-heating and careful control of the heat input.

These steels do not have any improved corrosion protection over steels with a lower carbon content. The fact that they can be hardened and thus be subjected to stresses in service means that corrosion pitting can be extremely dangerous in producing stress raisers to cause impact and fatigue failure.

The carbon equivalent CE can be used to evaluate potential welding problems. The formula is

$$CE = C + \frac{Mn}{6} + \frac{Cr + Mo + V}{5} + \frac{Ni + Cu}{15}$$

A carbon equivalent higher than 0.7% indicates the possibility of problems during welding.

Many of these steels will be hardened and tempered to above the critical strength of 60 tons per square inch which requires that following any electroplating operation they must be de-embrittled to prevent hydrogen embrittlement. This requires a temperature of 200–250 °C for a minimum of 2 hours.

In general these materials will machine well.

The materials in this group are used for springs, collets, fixtures, mandrils, shear blades of all kinds and general engineering purposes where medium high hardness is required but little or no shock resistance will exist.

These materials must never be used under conditions where high strength, high impact is involved. It must also be appreciated that as the thickness of section increases then the possibility of obtaining a correct thermal treatment becomes less and thus the yield or proof strength might be difficult to achieve.

Symbol	Nominal analysis, supplier, condition and remarks.
0.0073	0.75% C 0.7% Mn 0.35% Si 0.045% S and P steel: German Standard
0.0722	0.48% C 0.7% Mn 0.2% Si 0.045% S and P steel: German Standard
0.0740	0.53% C 0.5% Mn 0.035% Si 0.045% S and P steel: German Standard
0.0761	0.68% C 0.7% Mn 0.35% Si 0.045% S and P steel: German Standard
0.0763	0.74% C 0.45% Mn 0.2% Si 0.045% S and P steel: German Standard
0.0931	0.6% C 1.0% Mn 1.2% Si steel: For springs; German Standard
0.0969	0.51% C 0.3% Mn 1.7% Si steel: For springs; German Standard
0.0970	0.56% C 0.8% Mn 1.7% Si steel: For springs; German Standard
0.0971	0.64% C 0.8% Mn 1.7% Si steel: For springs; German Standard
0.1210	0.53% C 0.5% Mn 0.35% Si 0.035% S and P steel: German Standard
0.1221	0.61% C 7% Mn 0.35% Si 0.035% S and P steel: German Standard
0.1231	0.68% C 0.7% Mn 0.35% Si 0.035% S and P steel: German Standard
0.1248	0.75% C 0.5% Mn 0.2% Si 0.035% S and P steel: German Standard
0.5028	0.65% C 0.8% Mn 1.7% Si steel: For springs; German Standard
0.5029	0.71% C 0.7% Mn 1.7% Si steel: For springs; German Standard
0.70 Li Pja	0.7% C 0.5% Mn 0.25% Si steel: Fagersta
1.0505	0.53% C 0.55% Mn 0.4% Si 0.05% S and P steel: German Standard
1.0601	0.61% C 0.7% Mn 0.35% Si 0.045% S and P steel: German Standard
1.0603	0.68% C 0.7% Mn 0.4% Si 0.045% S and P steel: German Standard
1.0605	0.75% C 0.7% Mn 0.4% Si 0.045% S and P steel: German Standard
1.0632	0.5% C 0.05% S and P steel: German Standard
1.0728	0.62% C 0.7% Mn 0.3% Si 0.2% S 0.07% P steel: German Standard
1.0751	0.61% C 0.7% Mn 0.25% Si 0.045% S and P steel: German Standard
1.0902	0.45% C 0.7% Mn 1.7% Si spring steel: German Standard
1.0904	0.55% C 0.9% Mn 1.7% Si spring steel: German Standard
1.0906	0.65% C 0.9% Mn 1.7% Si spring steel: German Standard
1.0909	0.6% C 0.9% Mn 2% Si spring steel: German Standard
1.0913	0.5% C 1.8% Mn steel: German Standard
1.1210	0.53% C 0.5% Mn 0.35% Si 0.035% S and P steel: German Standard
1.1213	0.53% C 0.6% Mn 0.25% Si 0.035% S and P steel: German Standard
1.1221	0.61% C 0.7% Mn 0.35% Si 0.035% S and P steel: German Standard
1.1231	0.68% C 0.7% Mn 0.35% Si 0.035% S and P steel: German Standard
1.1248	0.75% C 0.5% Mn 0.2% Si 0.035% S and P steel: German Standard
1.1249	0.72% C 0.3% Mn 0.2% Si 0.03% S and P steel: German Standard
1.1740	0.6% C 0.7% Mn 0.035% S and P (max) steel: German Standards designation
46 Mn Si 4	0.48% C 1.1% Mn 0.8% Si: Designation used by German Standards
50 Mn 7	0.5% C 1.8% Mn steel: German Standard
50 Mn Si 4	0.5% C 1.1% Mn 0.8% Si steel: Designation used by German Standards
53 Mn Si 4	0.54% C 1.0% Mn 0.9% Si steel: Designation used by German Standards
55 Si 7	0.55% C 0.9% Mn 1.7% Si spring steel: German Standard
60 Si 7	0.6% C 0.9% Mn 2% Si spring steel: German Standard
060A52	0.52% C 0.6% Mn steel: BS 970; obsolete
060A57	0.57% C 0.6% Mn steel: BS 970
060A62	0.62% C 0.6% Mn steel: BS 970
060A67	0.67% C 0.6% Mn steel: BS 970
060A72	0.72% C 0.6% Mn steel: BS 970
060A78	0.78% C 0.6% Mn steel: BS 970
060A81	0.81% C 0.6% Mn steel: BS 970
070M55	0.55% C 0.7% Mn steel: BS 970
080A47	0.47% C 0.8% Mn steel: BS 970
080A52	0.52% C 0.8% Mn steel: BS 970
080A57	0.57% C 0.8% Mn steel: BS 970
080A62	0.62% C 0.8% Mn steel: BS 970; obsolete
080A67	0.69% C 0.8% Mn steel: BS 970
080A72	0.72% C 0.8% Mn steel: BS 970; obsolete
080A78	0.78% C 0.8% Mn steel: BS 970; obsolete
080M46	0.46% C 0.8% Mn steel: BS 970
080M50	0.5% C 0.8% Mn steel: BS 970
080M55	0.55% C 0.8% Mn steel: BS 970
1046	0.46% C 0.9% Mn 0.05% S 0.04% P steel: Designation used by AISI
1047	0.47% C 1.5% Mn 0.04% S and P steel: Designation used by AISI
1053	0.53% C 0.04% S and P (max) steel: Designation used in the UK and USA
1055	0.55% C 0.04% S and P (max) steel: Designation used in the UK and USA
1059	0.59% C 0.04% S and P (max) steel: Designation used in the UK and USA
1060	0.6% C 0.04% S and P (max) steel: Designation used in the UK and USA
1064	0.64% C 0.04% S and P (max) steel: Designation used in the UK and USA
1065	0.65% C 0.04% S and P (max) steel: Designation used in the UK and USA
1069	0.69% C 0.04% S and P (max) steel: Designation used in the UK and USA
1070	0.7% C 0.04% S and P (max) steel: Designation used in the UK and USA
1074	0.74% C 0.04% S and P (max) steel: Designation used in the UK and USA

Note. The following abbreviations and units are used in the tables:

DPN	Hardness, diamond pyramid number
UTS	Ultimate tensile strength, N/mm^2
Elon	Elongation, %
Proof	0.1% proof strength, N/mm^2

1 N/mm^2=0.1 hbar=0.102 kgf/mm^2=0.06475 $tonf/in.^2$=145.04 $lbf/in.^2$=1 MPa

See Appendix II for other abbreviations and conversion tables.

Symbol	Nominal analysis, supplier, condition and remarks.
1104	0.75% C 0.02% S and P (max) steel: French Standards designation
1164	0.75% C 0.2% Mn 0.02% S and P (max) steel: French Standards designation
1165	0.65% C 0.2% Mn 0.02% S and P (max) steel: French Standards designation
1204	0.75% C 0.25% Mn 0.025% S and P (max) steel: French Standards designation
1304	0.75% C 0.25% Mn 0.025% S (max) 0.035% P (max) steel: French Standards designation
1305	0.65% C 0.25% Mn 0.025% S (max) 0.035% P (max) steel: French Standards designation
1553	0.53% C 1% Mn steel: Designation used in the UK and USA
1561	0.61% C 0.9% Mn steel: Designation used in the UK and USA
1566	0.66% C 0.9% Mn steel: Designation used in the UK and USA
1570	0.7% C 1% Mn steel: Designation used in the UK and USA
1572	0.72% C 1.1% Mn steel: Designation used in the UK and USA
1580	0.8% C 1% Mn steel: Designation used in the UK and USA
A 6	0.62% C (max) 0.9% Mn steel: Cogne
A 32 STEEL	0.55% C 0.8% Mn steel: ESC for BS alloy En 9
A STEEL	0.5% C 0.8% Mn steel: ECS for BS alloy En 43A
ABRADUR 2	0.7% C 1.0% Mn 0.35% Si steel: Pompey
ACIBRADE	0.68% C 1.5% Mn 0.05% Si 0.06% S and P steel: Workington Iron & Steel; hot rolled
AISI 1048	0.48% C 1.3% Mn 0.04% S and P steel
AISI 1049	0.49% C 0.8% Mn 0.05% S 0.04% P steel
AISI 1050	0.51% C 0.8% Mn 0.05% S 0.04% P steel
AISI 1051	0.51% C 1.0% Mn 0.05% S and P steel
AISI 1052	0.52% C 1.3% Mn 0.05% S and P steel
AISI 1052	0.51% C 1.2% Mn 0.05% S 0.04% P steel
AISI 1053	0.53% C 0.8% Mn 0.05% S and P steel
AISI 1055	0.55% C 0.8% Mn 0.05% S 0.04% P steel
AISI 1060	0.6% C 0.8% Mn 0.05% S 0.04% P steel
AISI 1064	0.63% C 0.6% Mn 0.05% S 0.04% P steel
AISI 1065	0.65% C 0.8% Mn 0.05% S 0.04% P steel
AISI 1070	0.7% C 0.8% Mn 0.05% S 0.04% P steel
AISI 1074	0.75% C 0.6% Mn 0.05% S 0.04% P steel
AISI 1078	0.78% C 0.5% Mn 0.05% S 0.04% P steel
AISI 1080	0.82% C 0.8% Mn 0.05% S 0.04% P steel
AISI 1151	0.51% C 0.9% Mn 0.1% S 0.04% P steel: Free cutting
AMS 5085	0.50% C steel: Sheet and strip; AMS for SAE alloy 1050
AMS 5110 A	0.8% C steel: Music wire; commercial quality; AMS for SAE alloy 1080
AMS 5115 B	0.7% C steel: Spring wire; AMS for SAE alloy 1070
AMS 5120 D	0.75% C steel: Spring and strip; annealed; AMS for SAE alloy 1074
AMS 7240	0.6% C 0.75% Mn steel
AQUATOUGH 70	0.7% C 0.3% Mn steel: Clyde Alloy **DPN: 750**
ASTM A1	0.68% C 0.8% Mn 0.15% Si steel: Rails
ASTM A21	0.48% C 0.75% Mn steel: Axles for railway use

Note. The following abbreviations and units are used in the tables:

DPN	Hardness, diamond pyramid number
UTS	Ultimate tensile strength, N/mm^2
Elon	Elongation, %
Proof	0.1% proof strength, N/mm^2

1 N/mm^2=0.1 hbar=0.102 kgf/mm^2=0.06475 tonf/in.^{2}=145.04 lbf/in.^{2}=1 MPa

See Appendix II for other abbreviations and conversion tables.

Symbol	Nominal analysis, supplier, condition and remarks.
ASTM A25 Class A	0.57% C 0.67% Mn (max) steel: For steel wheels for electrical service
ASTM A25 Class B	0.62% C 0.67% Mn steel: For steel wheels for electrical service
ASTM A25 Class C	0.72% C 0.67% Mn steel: For steel wheels for electrical service
ASTM A25 Class V	0.72% C 0.67% Mn steel: for steel wheels for electrical service
ASTM A227	0.55% C 0.9% Mn 0.2% Si 0.045% S and P steel: Spring wire; hard drawn; actual tensile varies with diameter **UTS: 1550**
ASTM A228	0.85% C 0.4% Mn 0.2% Si 0.03% S and P steel: Spring wire; cold drawn; piano wire **UTS: 3400**
ASTM A229 A	0.6% C 1.0% Mn 0.2% Si 0.045% S and P steel: Spring wire; hardened and tempered **UTS: 1840**
ASTM A229B	0.6% C 0.8% Mn 0.2% Si 0.045% S and P steel: Spring wire; hardened and tempered **UTS: 1840**
ASTM A230	0.65% C 0.7% Mn 0.2% Si 0.03% S and P steel: Spring wire; hardened and tempered for valve springs **UTS: 1550**
ASTM A241	0.58% C steel: For tie plates
ASTM A290 C2	0.55% C 0.8% Mn 0.05% S and P steel: Forging; hardened and tempered **DPN: 170 UTS: 610 Elon: 20% Proof: 300**
ASTM A306/70	Plain carbon steel: As rolled **UTS: 540 Elon: 18% Proof: 220**
ASTM A306/75	Plain carbon steel: As rolled **UTS: 550 Elon: 18% Proof: 240**
ASTM A306/80	Plain carbon steel: As rolled **UTS: 550 Elon: 17% Proof: 270**
ASTM A321	0.55% C 0.8% Mn steel: Hardened and tempered **UTS: 630 Elon: 18% Proof: 450**
ASTM A407	0.57% C 0.9% Mn steel: Wire for coil springs; upholstery springs
ASTM A417	0.62% C 0.9% Mn 0.04% S and P steel: Wire for upholstery springs
ASTM A421	0.85% C 0.8% Mn 0.2% Si 0.04% S and P steel: Wire for pre-stressed concrete
ASTM A504	0.6% C 0.7% Mn 0.05% S and P steel: For wheels
ASTM A551	0.7% C 0.7% Mn 0.2% Si 0.05% S and P steel: Tyres graded by C content
ASTM A583	0.7% C steel: Castings for railway wheels; classified by use
ASTM A631	0.7% C 0.7% Mn: Cast steel wheels for railway service; classified by use
ASTM A648/1	0.55% C 0.8% Mn steel: Wire; hard drawn for reinforcing **UTS: 1300**
ASTM A648/11	0.65% C 0.8% Mn steel: Wire; hard drawn for reinforcing **UTS: 1500**
ASTM A648/111	0.7% C 0.8% Mn steel: Wire; hard drawn for reinforcing
ASTM A679	0.7% C 0.8% Mn steel: Wire; hard drawn **UTS: 2000**
ASTM A680	Medium high carbon steel: Strip; cold rolled for springs
ASTM A682	Cold rolled carbon steel: Strip; coded by AISI type; see also Sections 44A1, 44A2 and 44A4
ASTM A682	Medium high carbon steel: Strip; cold rolled; graded by AISI system
ASTM A684	Medium high carbon steel: Strip for springs
ASTM A713	High carbon steel: For springs; graded by AISI designation
ASTM A713	Medium high carbon steel: Wire for springs; graded to AISI system

Symbol	Nominal analysis, supplier, condition and remarks.
ASTM A729	0.6% C (max) 1.5% Mn steel: For railway axles; hardened and tempered **UTS: 690 Elon: 20% Proof: 450**
ASTM A730 C, D and E	0.47% C 0.75% Mn steel: Forgings for railway use
ASTM A730 F	0.52% C 0.75% Mn steel: Forgings for railway use
ASTM A730 G and H	C not specified 0.75% Mn steel: Forgings for railway use
ASTM A732/4 A	0.5% C 0.8% Mn steel: Investment casting; annealed **UTS: 620 Elon: 20% Proof: 345**
ASTM A732/4 Q	0.5% C 0.8% Mn steel: Investment casting; hardened and tempered **UTS: 860 Elon: 5% Proof: 680**
ASTM A759	0.75% C 0.8% Mn steel: For crane rails; may be supplied heat treated
ASTM A764	0.65% C steel: Wire for springs; drawn and galvanized
ASTM B606	0.65% C 1.0% Mn steel: Wire; galvanized; for core wire on A1 electrical conductors
ATERCLIFFE	Plain carbon tool steels 0.6%–1.5% C: Sanderson; carbon content specified by a temper no.
AURIGA V	0.55% C 0.7% Mn 0.4% Si 0.05% S and P steel: Casting; D Brown **UTS: 670 Elon: 10%**
AW 5	0.55% C 0.6% Mn steel: ESC; spanners, pliers, etc. **DPN: 690**
AW 8	0.8% C 0.8% Mn steel: ESC; picks, hammers, etc. **DPN: 690**
B 1	0.55% C 0.6% Mn 0.25% Si steel: Jessop for BS alloy En 9
B 6	0.6% C (max) steel: Breda
B 10	0.5% C steel: Normalized; Bofors **DPN: 190 UTS: 610 Elon: 18% Proof: 300**
B 12	0.6% C steel: Normalized; Bofors; obsolete **DPN: 220 UTS: 720 Elon: 15% Proof: 330**
B 14	0.7% C steel: Normalized; Bofors **DPN: 230 UTS: 770 Elon: 13% Proof: 370**
BS 9	0.55% C 1.0% Mn 0.06% S and P steel: For bull head rails
BS 11	0.55% C 1.0% Mn 0.06% S and P steel: For flat bottom rails
BS 24/3 A1	0.55% C 0.8% Mn 1.7% Si steel: For wagon springs; as BS alloy En 45
BS 24/3 A2	0.55% C 0.8% Mn 1.7% Si steel: For wagon springs; as BS alloy En 45
BS 24/3 A5	0.77% C 0.6% Mn 0.3% Si steel: For wagon springs; as BS alloy En 42
BS 24/3 A6	0.77% C 0.6% Mn 0.3% Si steel: For wagon springs; as BS alloy En 42
BS 24/3 A7	0.5% C 0.7% Mn 0.3% Si steel: For wagon springs; as BS alloy En 43
BS 24/3 A8	0.55% C 0.7% Mn 0.3% Si steel: For wagon springs; as BS alloy En 43
BS 24/3 BB	0.6% C 0.8% Mn 1.8% Si 0.05% S and P steel: For springs; as BS alloy En 45
BS 24/3 BC	0.6% C 0.6% Mn 1.9% Si 0.05% S and P steel: For springs; as BS alloy En 45A
BS 224/1	0.6% C 0.7% Mn 0.05% S 0.05% P steel: For die blocks; hardened and tempered **DPN: 250**
BS 970	See designation for details
BS 980 CDS 8	0.5% C 0.6% Mn steel: Cold drawn tube; as drawn and tempered **UTS: 690 Proof: 580**
BS 1408	0.7% C 1% Mn 0.3% Si steel: Cold drawn; composition complies with BS 970 En 49A, B, C and D **UTS: 1550**
BS 1429	0.6%–1.0% C steel: Wire for springs; annealed; covered by BS alloys En 42, En 44, En 45, En 47 and En 50
BS 1429 En 42C	0.75% C 0.7% Mn 0.05% S and P steel: For springs
BS 1449 HS50	0.5% C 0.8% Mn 0.2% Si 0.05% S and P steel: Sheet
BS 1449 HS60	0.6% C 0.7% Mn 0.2% Si 0.05% S and P steel: Sheet
BS 1449 HS70	0.7% C 0.7% Mn 0.2% Si 0.05% S and P steel: Sheet
BS 1449 HS80	0.8% C 0.7% Mn 0.2% Si 0.05% S and P steel: Sheet
BS 1449-40	0.4% C 0.7% Mn steel: Sheet and plate
BS 1449-50	0.5% C 0.7% Mn steel: Sheet and plate
BS 1449-60	0.6% C 0.7% Mn steel: Sheet and plate
BS 1449-70	0.7% C 0.7% Mn steel: Sheet and plate
BS 1449-80	0.8% C 0.7% Mn steel: Sheet and plate
BS 1506/162	0.6% C (max) 0.8% Mn 0.2% Si 0.05% S and P steel: Bar; normalized **UTS: 760 Elon: 18% Proof: 330**
BS 1760 A	0.45% C 1.0% Mn 0.6% Si 0.06% S and P steel: Casting for surface hardening; contained in BS 3100 **UTS: 610 Elon: 12% Proof: 300**
BS 1760 B	0.6% C 1.0% Mn 0.6% Si 0.06% S and P steel: Casting for surface hardening; contained in BS 3100 **UTS: 670 Elon: 10% Proof: 330**
BS 2453	0.4% C 0.8% Mn 0.25% Si 0.05% S and P steel: For spokes; may be zinc coated **UTS: 1070**
BS 2691	0.8% C 0.7% Mn 0.2% Si 0.05% S and P steel: Wire for concrete; cold drawn and stress released **UTS: 1680 Proof: 1380**
BS 2803	0.6% C 0.8% Mn 0.3% Si steel: Wire; ground; different grades define slight difference **UTS: 1540**
BS 3111/1	0.4% C 0.8% Mn 0.2% Si 0.05% S and P steel: Wire; annealed and drawn; for cold forged bolts **UTS: 610**
BS 3141 C6A1	0.5% C 0.6% Mn 0.05% S and P steel: Classification system used by BS; discontinued
BS 3141 C6B1	0.5% C 0.8% Mn 0.06% S and P steel: Classification system used by BS; discontinued
BS 3141 C6C1	0.5% C 1.1% Mn 0.06% S and P steel: Classification system used by BS; discontinued
BS 3141 C6E1	0.5% C 0.4% Mn 0.05% S and P steel: Classification system used by BS; discontinued
BS 3141 C6F1	0.5% C 0.7% Mn 0.05% S and P steel: Classification system used by BS; discontinued
BS 3141 C6G1	0.52% C 0.9% Mn 0.06% S and P steel: Classification system used by BS; discontinued
BS 3141 C6G1	0.55% C 0.8% Mn 0.06% S and P steel: Classification system used by BS; discontinued
BS 3141 C6H1	0.55% C 1.1% Mn 0.06% S and P steel: Classification system used by BS; discontinued
BS 3141 C7A1	0.6% C 0.8% Mn 0.05% S and P steel: Classification system used by BS; discontinued
BS 3141 C7B1	0.62% C 0.5% Mn 0.05% S and P steel: Classification system used by BS; discontinued
BS 3141 C7C1	0.65% C 0.8% Mn 0.05% S and P steel: Wire; classification system used by BS; discontinued
BS S25	High carbon steel: Bar; replaced
BS S70	0.55% C 0.7% Mn 0.2% Si 0.04% S and P steel: Normalized **DPN: 230 UTS: 760 Elon: 15%**
BS S79	0.55% C 0.7% Mn 0.2% Si 0.04% S and P steel: Hardened and tempered **DPN: 280 UTS: 920 Elon: 13%**
BS S513	0.8% C 0.7% Mn 0.3% Si 0.05% S and P steel: For springs; hardened and tempered **DPN: 450**
BS S517	0.47% C 1.5% Mn 0.2% Si 0.05% S and P steel: Sheet; hardened and tempered **UTS: 1210 Elon: 5% Proof: 1000**
BS W6	C not specified 0.04% S and P steel: Galvanized wire for balloon cables **UTS: 2310**
BULL HEAD RAIL	0.55% C 1% Mn 0.06% S and P steel: See BS 9
BULLS HEAD	0.6% C (min) steel: Marsh Brothers; applies to a range of plain C steels
C 6	0.6% C (max) 0.8% Mn steel: Siau

Symbol	Nominal analysis, supplier, condition and remarks.
C45W	0.51% C 0.6% Mn 0.035% S and P (max) steel: German Standard
C 53	0.53% C 0.5% Mn 0.35% Si 0.045% S and P steel: German Standard
C 60	0.6% C 0.75% Mn steel: Italian Standards designation; normalized **DPN: 260 UTS: 820 Elon: 12% Proof: 450**
C 60	0.61% C 0.7% Mn steel: Italian Standards designation **DPN: 230 UTS: 1450 Elon: 6% Proof: 1050**
C 60 W	0.6% C 0.7% Mn 0.035% S and P (max) steel: German Standard
C 60 W3	0.6% C 0.7% Mn 0.035% S and P (max) steel: German Standards designation
C 70	0.69% C 0.7% Mn steel: Italian Standards designation **DPN: 250 UTS: 1450 Elon: 5% Proof: 1050**
C 75	0.75% C 0.55% Mn steel: Italian Standards designation **DPN: 265 UTS: 1450 Elon: 5% Proof: 1050**
C 90	0.9% C 0.55% Mn steel: Italian Standards designation **DPN: 270 UTS: 1500 Elon: 4% Proof: 1100**
C 175	0.75% C 0.3% Mn 0.25% Si 0.02% S and P steel: Drills, etc.; SHD
C 255	0.55% C 0.7% Mn 0.25% Si 0.04% S and P steel: Drills, etc.; SHD
CD 60	0.5% C 0.8% Mn 0.4% Si steel: Fagersta **UTS: 760 Elon: 18%**
Cf 70	0.71% C 0.3% Mn 0.2% Si 0.03% S and P steel: German Standard
CK 53	0.53% C 0.5% Mn 0.35% Si 0.035% S and P steel: German Standard
CK 60	0.61% C 0.7% Mn 0.04% S and P steel: Designation used by German Standards
CK 67	0.67% C 0.7% Mn 0.35% Si 0.035% S and P steel: German Standard
COLORADO	Plain carbon tool steels 0.6%–1.5% C: Sanderson; carbon content specified by temper no.
COMMON HARDENING	0.55% C 0.7% Mn steel: Origin unknown
CRESCENT 6	0.65% C steel: Spencer
D 5	0.55% C 0.7% Mn 0.3% Si steel: Pompey
D 6	0.6% C 0.8% Mn 0.3% Si steel: Pompey
D 7	0.7% C 0.7% Mn 0.3% Si steel: Pompey
D 8	0.8% C 0.6% Mn 0.3% Si steel: Pompey
DIN 1651 60 S 20	0.6% C 0.7% Mn 0.2% S 0.05% P steel: Free machining
DIN 17100 St 70/2	0.5% C 0.05% S and P (max) 0.007% N (max) steel: Plate; German Standard **UTS: 770 Proof: 360**
DIN 17200 C 55	0.55% C 0.75% Mn 0.045% S and P steel: For structures; German Standard
DIN 17200 C 60	0.6% C 0.75% Mn 0.045% S and P steel: For structures; German Standard
DIN 17200 C60	0.61% C 0.7% Mn 0.045% S and P steel
DIN 17200 CK 55	0.55% C 0.75% Mn 0.035% S and P (max) steel: For structures; German Standard
DIN 17200 CK 60	0.6% C 0.75% Mn 0.035% S and P (max) steel: For structures; German Standard
DIN 17200 CK 60	0.61% C 0.7% Mn 0.4% S and P steel
DIN 17200 CM 55	0.55% C 0.75% Mn 0.035% S and P (max) steel: For structures; German Standard
DIN 17200 CM 60	0.6% C 0.75% Mn 0.035% S and P (max) steel: For structures; German Standard
DIN 17221 51 S 7	0.51% C 0.6% Mn 1.7% Si steel: For springs
DIN 17221 55 S 7	0.55% C 0.9% Mn 1.7% Si steel: For springs
DIN 17221 60 Si Mn 5	0.6% C 1.0% Mn 1.2% Si steel: For springs
DIN 17221 65 Si 7	0.65% C 0.8% Mn 1.7% Si steel: For springs
DIN 17221 66 Si 7	0.66% C 0.8% Mn 1.7% Si steel: For springs
DIN 17222 60 Si Mn 5	0.6% C 1.0% Mn 1.2% Si steel: For springs
DIN 17222 65 Si 7	0.65% C 0.8% Mn 1.7% Si steel: For springs
DIN 17222 66 Si 7	0.66% C 0.8% Mn 1.7% Si steel: For springs
DIN 17222 71 Si 7	0.71% C 0.7% Mn 1.7% Si steel: For springs
DIN 17222 C67	0.68% C 0.7% Mn 0.035% S and P steel: For springs
DIN 17222 C75	0.75% C 0.7% Mn 0.045% S and P steel: For springs
DIN 17222 M75	0.75% C 0.4% Mn 0.045% S and P steel: For springs
DIN 17222 MK75	0.75% C 0.5% Mn 0.035% S and P steel: For springs
DTD 5 A	0.75% C 1% Mn 0.04% S 0.04% P steel: For springs; hard drawn and tempered **UTS: 1680**
DTD 215 A	0.7% C 0.5% Mn 0.03% S 0.03% P steel: Wire; cold drawn and tempered; not for valve springs **UTS: 1550**
DTD 239 A	0.8% C 1% Mn 0.045% S 0.04% P steel: For springs; hardened and tempered; not for valve springs **UTS: 1380**
DUNELT 53	0.55% C 0.7% Mn 0.06% S and P steel: Dunford for BS alloy En 9
E 7	0.7% C 0.3% Mn 0.25% Si steel: Pompey
E 8 FX	0.8% C 0.3% Mn 0.12% Si steel: Pompey
En 9	0.55% C 0.06% S and P steel: Designation used in BS 970; obsolete
En 9K	0.55% C 0.05% S and P steel: Designation used in BS 970; obsolete
En 42	0.8% C 0.05% S and P steel: For springs; designation used in BS 970; obsolete
En 42A	0.85% C 0.05% S and P steel: For springs; designation used in BS 970; obsolete
En 42B	0.65% C 0.05% S and P steel: For springs; designation used in BS 970; obsolete
En 43	0.58% C 0.05% S and P steel: For springs; designation used in BS 970; replaced by 080 A 50
En 43A	0.5% C 0.06% S and P steel: For springs; designation used in BS 970; replaced by 080 A 50
En 43B	0.48% C 0.06% S and P steel: For springs; designation used in BS 970; replaced by 080 M 47
En 43C	0.52% C 0.06% S and P steel: For springs; designation used in BS 970; replaced by 080 M 52
En 43D	0.62% C 0.06% S and P steel: Designation used in BS 970; replaced by 080 A 62
En 43E	0.67% C 0.06% S and P steel: Designation used in BS 970; replaced by 080 A 67
En 43F	0.5% C 0.05% S and P steel: Obsolete
En 45	0.6% C 1.75% Si steel: For springs; designation used in BS 970; obsolete
En 45A	0.6% C 1.8% Si steel: For springs; designation used in BS 970; obsolete
En 49	0.6% C 0.05% S and P steel: For springs; designation used in BS 970
En 49A	0.6% C 0.05% S and P steel: For springs; designation used in BS 970
En 49B	0.65% C 0.05% S and P steel: For springs; designation used in BS 970
En 49C	0.7% C 0.04% S and P steel: For springs; designation used in BS 970
En 49D	0.75% C 0.04% S and P steel: For springs; designation used in BS 970
EXTRA	C as specified 0.25% Mn steel: For tools; Braeburn
F100	0.52% C 0.7% Mn 1.7% Si steel: VEW
F105	0.64% C 0.9% Mn 1.7% Si steel: VEW
F108	0.6% C 0.9% Mn 2.0% Si steel: VEW

Note. The following abbreviations and units are used in the tables:

DPN	Hardness, diamond pyramid number
UTS	Ultimate tensile strength, N/mm^2
Elon	Elongation, %
Proof	0.1% proof strength, N/mm^2

1 N/mm^2=0.1 hbar=0.102 kgf/mm^2=0.06475 $tonf/in.^2$=145.04 $lbf/in.^2$=1 MPa

See Appendix II for other abbreviations and conversion tables.

Symbol	Nominal analysis, supplier, condition and remarks.
F110	0.56% C 0.9% Mn 1.7% Si steel: VEW
F180	0.5% C 1.8% Mn 0.3% Si steel: VEW
F200	0.67% C 0.5% Mn 0.5% Cr steel: VEW
FABIS	Plain carbon tool steels 0.6%–1.5% C: Sanderson; carbon content specified by temper no.
FAGERSTA 1680	0.6% C 0.3% Mn steel: Fagersta
FAGERSTA 1773	0.75% C 0.3% Mn steel: Fagersta
FAGERSTA 1778	0.7% C 0.6% Mn steel: For springs; Fagersta
FAGERSTA 1779	0.85% C 0.5% Mn steel: For springs; Fagersta
FLAT BOTTOM RAILS	0.55% C 1% Mn 0.06% S and P steel: See BS 11
GOST 1050/50	0.5% C 0.75% Mn 0.03% S and P steel: For structures: Russian Standard
GOST 1050/50 G	0.5% C 0.85% Mn 0.25% Cr and Ni (max) 0.04% S and P (max) steel: For structures; Russian Standard
GOST 1050/55	0.55% C 0.65% Mn 0.25% Cr and Ni (max) 0.04% S and P (max) steel: For structures; Russian Standard
GOST 1050/60	0.6% C 0.75% Mn 0.25% Cr and Ni (max) 0.04% S and P (max) steel: For structures; Russian Standard
GOST 1050/60 G	0.6% C 0.85% Mn 0.25% Cr and Ni (max) 0.04% S and P (max) steel: For structures; Russian Standard
GOST 1050/65	0.65% C 0.75% Mn 0.25% Cr and Ni (max) 0.04% S and P (max) steel: For structures; Russian Standard
GOST 1050/65 G	0.65% C 1.05% Mn 0.25% Cr and Ni (max) 0.04% S and P (max) steel: For structures; Russian Standard
GOST 1050/70	0.7% C 0.65% Mn 0.25% Cr and Ni (max) 0.04% S and P (max) steel: For structures; Russian Standard
GOST 1050/70 G	0.7% C 1.05% Mn 0.25% Cr and Ni (max) 0.04% S and P (max) steel: For structures; Russian Standard
GOST 1050/75	0.75% C 0.65% Mn 0.25% Cr and Ni (max) 0.04% S and P (max) steel: For structures; Russian Standard
GOST 1050/80	0.8% C 0.75% Mn 0.04% S and P steel: For structures; Russian Standard
GOST 2052/55 S2	0.55% C 0.75% Mn 0.3% Cr and Ni (max) 0.04% S and P (max) steel: Russian Standard
GOST 2052/60 S2A	0.6% C 0.85% Mn 0.4% Ni (max) 0.3% Cr (max) 0.03% S and P (max) 0.25% Cu (max) steel: For springs; Russian Standard
GOST 2052/60 SG	0.6% C 0.9% Mn 0.4% Cr and Ni (max) 0.04% S and P (max) steel: Bar; Russian Standard
GOST 2052/60 SG	0.6% C 0.9% Mn 0.4% Ni 0.3% Cr 0.04% S and P (max) 0.25% Cu steel: For springs; Russian Standard
GOST 2052/60 SGA	0.6% C 0.9% Mn 0.3% Cr and Ni (max) 0.03% S and P (max) steel: Bar; Russian Standard
GOST 2052/60 SGA	0.6% C 0.9% Mn 0.4% Ni (max) 0.3% Cr (max) 0.03% S and P (max) 0.25% Cu (max) steel: For springs; Russian Standard
GOST 2052/62 S2A	0.62% C 0.85% Mn 0.4% Ni (max) 0.3% Cr (max) 0.03% S and P (max) 0.25% Cu (max) steel: For springs; Russian Standard
GOST 2052/65	0.65% C 0.25% Cr and Ni (max) 0.04% S and P (max) steel: Bar; Russian Standard
GOST 2052/65 G	0.65% C 1.05% Mn 0.4% Cr and Ni (max) 0.04% S and P (max) steel: Bar; Russian Standard
GOST 2052/70	0.7% C 0.65% Mn 0.25% Cr and Ni (max) 0.04% S and P steel: Bar; Russian Standard
GOST 2052/70 S3A	0.7% C 0.75% Mn 0.3% Cr and Ni (max) 0.03% S and P (max) steel: Bar; Russian Standard
GOST 2052/75	0.75% C 0.65% Mn 0.3% Cr and Ni (max) 0.045% S and P (max) 0.25% Cu (max) steel: For springs; Russian Standard
GOST 4543/50 G2	0.5% C 1.6% Mn 0.25% Cr (max) steel: Russian Standard
HARDFLEX 13M	0.75% C steel: Strip; Sandvik; supplied hardened and tempered **UTS: 1300** **Proof: 1100**
HDB 1	0.6% C 0.7% Mn 0.3% Si steel: Huntsman; large press dies
HECLA 18	0.65% C 0.4% Mn steel: Hadfields for BS alloy En 42E
HECLA 36	0.85% C 0.7% Mn steel: Hadfields for BS alloy En 42D
HECLA 41	0.55% C 0.7% Mn steel: Hadfields for BS alloy En 9

Symbol	Nominal analysis, supplier, condition and remarks.
HECLA 42	0.8% C 0.65% Mn steel: Hadfields for BS alloy En 42
HECLA D17	0.55% C 0.7% Mn steel: Hadfields for BS alloy En 9
HECLA S55	0.55% C 0.85% Mn steel: Hadfields for BS alloys En 45, En 45A and En 46
HEDEX 11	0.55% C 0.7% Mn 0.05% S 0.04% P steel: Wire for forging; Kiveton Park for BS alloy En 9
HEDEX 26	0.62% C 1% Mn steel: Wire for forging; Kiveton Park; annealed **UTS: 670**
HMS 75	0.53% C 0.65% Mn 0.25% Si 0.035% S and P steel: Krupp
HMS 80	0.6% C 0.65% Mn 0.25% Si 0.035% S and P steel: Krupp
ICSW 1	0.8% C steel: For castings; Inman
JEM No.5	0.8% C steel: Jonas
JEM No.6	0.65% C steel: Jonas
JIS G405 S48C	0.48% C 0.75% Mn 0.03% S and P (max) steel: For structures; Japanese Standard
JIS G3445 STKM 17A	0.5% C 0.7% Mn 0.04% S and P steel: Pipe; Japanese Standard **UTS: 560** **Proof: 350**
JIS G3445 STKM 17 C	0.5% C 0.7% Mn 0.04% S and P (max) steel: Pipe; Japanese Standard **UTS: 660** **Proof: 490**
JIS G4051 S50C	0.5% C 0.75% Mn 0.03% S and P (max) steel: For structures; Japanese Standard
JIS G4051 S53G	0.53% C 0.75% Mn 0.03% S and P (max) steel: For structures; Japanese Standard
JIS G4051 S55C	0.55% C 0.75% Mn 0.03% S and P (max) steel: For structures; Japanese Standard
JIS G4051 S58C	0.58% C 0.75% Mn 0.03% S and P (max) steel: For structures; Japanese Standard
JIS G4801 SUP7	0.6% C 0.85% Mn 0.035% S and P (max) steel: For springs; Japanese Standard
K950	0.51% C 0.6% Mn 0.035% S and P steel: VEW
K960	0.6% C 0.7% Mn 0.035% S and P steel: VEW
K05003	0.5% C (max) 0.05% S and P (max) steel: Casting; designation used by UNS
K05800	0.6% C 0.2% Cr (min) 0.05% P (max) steel: Casting; designation used by UNS
KF 10B	0.52% C 0.8% Mn 0.06% S and P steel: Kirkstall for BS alloy En 9
KF 10C	0.58% C 0.7% Mn 0.06% S and P steel: Kirkstall for BS alloy En 9
KF 10D	0.63% C 0.7% Mn 0.06% S and P steel: Kirkstall; hardened and tempered **UTS: 910** **Elon: 12%** **Proof: 760**
KP 7	0.55% C 0.7% Mn 0.06% S and P steel: Kiveton Park for BS alloy En 9
KP 8	0.75% C 0.6% Mn 0.05% S and P steel: Kiveton Park for BS alloy En 42
KP 25	0.5% C 0.8% Mn 1.7% Si steel: For springs; Kiveton Park for BS alloy En 45
L 7	0.7% C steel: Pompey
LC 60	0.65% C steel: Water quenched; Low Moor; obsolete **DPN: 790**
LC 80	0.8% C steel: Water quenched; Low Moor; obsolete **DPN: 790**
LRT	0.5% C 0.5% Mn 0.2% Si steel: Pompey
M 75	0.75% C 0.4% Mn 0.045% S and P steel: Designation used by German Standards
MC 7	0.75% C 0.3% Mn 0.25% Si steel: Pompey
MIL W 21425-1	0.65% C 1% Mn 0.04% S and P (max) steel: Casting; designation used by UNS
MK 75	0.75% C 0.5% Mn 0.035% S and P steel: Designation used by German Standards
MR 6	0.6% C 0.7% Mn 0.3% Si steel: Pompey
MS 45	0.5% C 1.05% Mn 0.8% Si 0.035% S and P steel: Krupp
MS 50	0.53% C 1.05% Mn 0.9% Si 0.035% S and P steel: Krupp

Symbol	Nominal analysis, supplier, condition and remarks.
NF A35 351/XC55	0.55% C 0.65% Mn steel: For structures; French Standard
NF A35 501/A70/2	C not specified 0.045% S and P (max) steel: Plate; French Standard
NF A35 551/XC48	0.48% C 0.65% Mn steel: For structures; French Standard
NF A35 551/XC65	0.65% C 0.75% Mn steel: For structures; French Standard
NF A35 551/XC70	0.7% C 0.75% Mn steel: For structures; French Standard
NF A35 551/XC80	0.8% C 0.65% Mn steel: For structures; French Standard
NF A35 571/56S7	0.56% C 0.75% Mn 0.45% Cr (max) 0.035% S and P (max) steel: For springs; French Standard
NF A35 571/60S7	0.6% C 0.85% Mn 0.05% S and P (max) steel: Bar; French Standard
NF A35 571/61S7	0.6% C 0.85% Mn 0.45% Cr (max) 0.035% S and P (max) steel: For springs; French Standard
No.7 TEMPER	0.7% C 0.2% Mn steel: Picks, screw-drivers, etc.; ESC **DPN: 630**
No.8 TEMPER	0.8% C 0.2% Mn steel: Rock drills, etc.; ESC **DPN: 630**
P 256	0.5% C 0.7% Mn steel: Chisels, drifts, etc.: Carr's **DPN: 580**
PEGASE S	0.5% C 0.5% Mn 0.08% Si steel: Pompey
QQ W 461f	Carbon steel: Wire; round; US Federal specification
RS 2	0.5% C 0.6% Mn 1.7% Si steel: Pompey
RS 3	0.55% C 0.8% Mn 1.7% Si steel: Pompey
RS 4	0.6% C 0.85% Mn 1.85% Si steel: Pompey
S 6F	0.6% C 0.7% Mn 0.35% Si steel: Pompey
S 7F	0.65% C 0.7% Mn 0.3% Si steel: Pompey
S 8	0.8% C 0.7% Mn 0.3% Si steel: Pompey
S 25	High carbon steel: BS aircraft standard
S 65	0.65% C 0.7% Mn 0.3% Si steel: Pompey
S 70	0.55% C 0.7% Mn 0.04% S and P steel: BS aircraft standard
S 75	0.75% C 0.5% Mn 0.3% Si steel: Pompey
S 79	0.55% C 0.7% Mn 0.04% S and P steel: BS aircraft standard
S 145	0.55% C 0.7% Mn 1.7% Si steel: For springs; Bofors; hardened and tempered **DPN: 450 UTS: 1550 Elon: 7% Proof: 1150**
S 310	0.55% C 30.8% Mn 21.7% Si steel: Fagersta **DPN: 610**
S 513	0.8% C 0.7% Mn 0.05% S and P steel: BS aircraft standard
S 517	0.47% C 1.5% Mn 0.05% S and P steel. BS aircraft standard
SAE 864 A	10.8% C 21.5% Cu Fe: Sintered material; porosity 18%; specific gravity 5.7–6.1 **UTS: 140**
SAE 864 B	0.8% C 1.5% Cu Fe: Sintered material; specific gravity 6.1–6.5 **UTS: 180**
SAE 865 A	0.8% C 4.0% Cu Fe: Sintered material: porosity 18%; specific gravity 5.7–6.1 **UTS: 160**
SAE 865 B	0.8% C 4.0% Cu Fe: Sintered material; specific gravity 6.1–6.5 **UTS: 210**
SAE 866 A	0.8% C 8.5% Cu Fe: Sintered material; porosity 18%; specific gravity 5.7–6.1 **UTS: 160**
SAE 866 B	0.8% C 8.5% Cu Fe: Sintered material; specific gravity 6.1–6.5 **UTS: 230**
SAE 867 A	0.8% C 20% Cu Fe: Sintered material; porosity 18%; specific gravity 5.7–6.1 **UTS: 330**
SAE 867 B	0.8% C 20% Cu Fe: Sintered material; specific gravity 6.1–6.5 **UTS: 210**
SAE 1044	0.47% C 0.4% Mn 0.04% S and P steel: Mechanical properties not quoted
SAE 1047	0.47% C 1.5% Mn 0.04% S and P steel: Mechanical properties not quoted
SAE 1048	0.48% C 1.3% Mn 0.04% S and P steel: Mechanical properties not quoted
SAE 1049	0.49% C 0.8% Mn 0.05% S 0.04% P steel: Bar; hot rolled **DPN: 179 UTS: 630 Elon: 15% Proof: 330**
SAE 1049	0.49% C 0.8% Mn 0.05% S 0.04% P steel: Bar; cold drawn **DPN: 197 UTS: 740 Elon: 10% Proof: 450**
SAE 1050	0.51% C 0.8% Mn 0.05% S 0.04% P steel: Bar; hot rolled **DPN: 179 UTS: 660 Elon: 15% Proof: 340**
SAE 1050	0.51% C 0.8% Mn 0.05% S 0.04% P steel: Bar; cold drawn **DPN: 197 UTS: 740 Elon: 10% Proof: 610**
SAE 1051	0.51% C 1.0% Mn 0.05% S 0.04% P steel: Bar, etc.
SAE 1052	0.51% C 1.2% Mn 0.05% S 0.04% P steel: Bar; hot rolled **DPN: 217 UTS: 770 Elon: 12% Proof: 420**
SAE 1053	0.53% C 0.8% Mn 0.05% S 0.04% P steel: Bar, etc.
SAE 1055	0.55% C 0.3% Mn 0.05% S 0.04% P steel: Bar; hot rolled **DPN: 192 UTS: 530 Elon: 12% Proof: 360**
SAE 1060	0.6% C 0.8% Mn 0.05% S 0.04% P steel: Bar; hot rolled **DPN: 201 UTS: 560 Elon: 12% Proof: 370**
SAE 1061	0.61% C 0.9% Mn 0.05% S 0.04% P steel: Bar, etc.
SAE 1064	0.63% C 0.6% Mn 0.05% S 0.04% P steel: Bar; hot rolled **DPN: 201 UTS: 560 Elon: 12% Proof: 370**
SAE 1065	0.65% C 0.8% Mn 0.05% S 0.04% P steel: Bar; hot rolled **DPN: 207 UTS: 730 Elon: 12% Proof: 390**
SAE 1066	0.66% C 1.0% Mn 0.05% S 0.04% P steel: Bar, etc.
SAE 1069	0.69% C 0.6% Mn 0.05% S 0.04% P steel: Bar, etc.
SAE 1070	0.7% C 0.8% Mn 0.05% S 0.04% P steel: Bar, hot rolled **DPN: 212 UTS: 750 Elon: 12% Proof: 390**
SAE 1072	0.72% C 1.2% Mn 0.05% S 0.04% P steel: Bar
SAE 1074	0.75% C 0.6% Mn 0.05% S 0.04% P steel: Bar; hot rolled **DPN: 217 UTS: 770 Elon: 12% Proof: 400**
SAE 1078	0.78% C 0.5% Mn 0.05% S 0.04% P steel: Bar; hot rolled **DPN: 207 UTS: 730 Elon: 12% Proof: 390**
SAE 1080	0.8% C 0.8% Mn 0.05% S 0.04% P steel: Bar; hot rolled **DPN: 229 UTS: 820 Elon: 10% Proof: 440**
SAE 1151	0.51% C 0.9% Mn 0.1% S 0.04% P steel: Free cutting; hot rolled **DPN: 187 UTS: 670 Elon: 15% Proof: 340**

Note. The following abbreviations and units are used in the tables:

DPN	Hardness, diamond pyramid number
UTS	Ultimate tensile strength, N/mm^2
Elon	Elongation, %
Proof	0.1% proof strength, N/mm^2

1 N/mm^2=0.1 hbar=0.102 kgf/mm^2=0.06475 $tonf/in.^2$=145.04 $lbf/in.^2$=1 MPa

See Appendix II for other abbreviations and conversion tables.

Symbol	Nominal analysis, supplier, condition and remarks.
SAE 1151	0.51% C 0.9% Mn 0.1% S 0.04% P steel: Free cutting; cold drawn **DPN: 207 UTS: 750 Elon: 10% Proof: 630**
SAE 1548	0.48% C 1.2% Mn 0.05% S 0.04% P steel: Bar, etc.
SAE 1551	0.51% C 1.0% Mn 0.05% S 0.04% P steel: Bar, etc.
SAE 1552	0.52% C 1.35% Mn 0.05% S 0.04% P steel: Bar, etc.
SAE 1561	0.61% C 0.9% Mn 0.05% S 0.04% P steel: Bar, etc.
SAE 1566	0.66% C 1.0% Mn 0.05% S 0.04% P steel: Bar, etc.
SAE 1572	0.72% C 1.15% Mn 0.05% S 0.04% P steel: Bar, etc.
SAE J 172	0.67% C 0.7% Mn steel: Wire for springs **UTS: 1700**
SAE J 271	0.85% C 0.8% Mn steel: Wire for springs **UTS: 1900**
SANBOLD 9	0.53% C 0.6% Mn 0.3% Si steel: Sanderson for BS alloy En 9
SANBOLD 42	0.8% C 0.6% Mn 0.2% Si steel: Sanderson for BS alloy En 42
SANBOLD 43	0.5% C 0.7% Mn 0.3% Si steel: Sanderson for BS alloy En 43
SANBOLD 45	0.55% C 0.9% Mn 1.7% Si steel: Sanderson for BS alloy En 45
SANBOLD 50	0.55% C 0.6% Mn 0.3% Si steel: Sanderson; replaced by Sanbold 9
SANBOLD 59	0.5% C 0.7% Mn 0.3% Si steel: Sanderson; replaced by Sanbold 43
SANBOLD 60	0.8% C 0.6% Mn 0.2% Si steel: Sanderson; replaced by Sanbold 42
SANBOLD 8058	0.55% C 0.9% Mn 1.7% Si steel: Sanderson; replaced by Sanbold 45
SH 4	0.48% C 0.7% Mn 0.35% Si steel: Pompey
SH 5	0.55% C 0.8% Mn 0.35% Si steel: Pompey
Si Mn	0.57% C 0.8% Mn 1.8% Si steel: For springs; ESC for BS alloy En 45 and En 45A

For SIS specifications see SS

Symbol	Nominal analysis, supplier, condition and remarks.
SK 6	0.75% C 0.5% Mn (max) 0.2% Cr and Ni (max) steel: Japanese Standards designation
SK 7	0.65% C 0.5% Mn 0.2% Cr and Ni (max) steel: Japanese Standards designation
SKC 3	0.77% C 0.5% Mn (max) 0.2% Cr (max) 0.25% V (max) steel: For hollow drills; Japanese Standards designation
SKF 511	0.7% C 0.9% Mn steel: SKF
SKO 67	0.7% C steel: SKF
SMN	0.55% C 0.8% Mn 1.8% Si steel: Punches, chisels, etc.; Huntsman **DPN: 610**
SOUDOKAY 089.0	3.5% C 0.3% Mn 2.0% Si steel: Welding electrode; Soudometal **DPN: 350**
SOUDOKAY GS-0	3.1% C 0.3% Mn 3.0% Si steel: Welding electrode; Soudometal **DPN: 450**
SPEAR 50	0.52% C steel: Spear and Jackson; hardened and tempered **DPN: 240 UTS: 780**
SPEAR 75	0.7% C steel: For press tools; Spear and Jackson
SPECIAL	C not specified 0.25% Mn steel: For tools; Braeburn
SPRASTEEL 80	0.8% C steel powder coating: Metco **DPN: 380**
SS 1606	0.5% C 0.6% Mn 0.04% S and P steel: Annealed; Swedish Standard **UTS: 630 Elon: 10% Proof: 300**
SS 1655	0.55% C 0.65% Mn 0.05% S and P steel: Swedish Standard
SS 1655/00	0.55% C steel: Swedish Standard; as rolled or forged; suffix indicates condition **DPN: 225 UTS: 700 Elon: 12% Proof: 340**
SS 1665	0.55% C 0.4% Mn 0.03% S and P steel: Strip; annealed; Swedish Standard **DPN: 165 UTS: 400**
SS 1665	0.55% C 0.4% Mn 0.3% S and P steel: Strip; cold rolled; Swedish Standard **DPN: 260 UTS: 780**
SS 1672	0.47% C steel: For flame and induction hardening; Swedish Standard; as rolled **DPN: 200 UTS: 650 Elon: 16% Proof: 300**
SS 1674	0.52% C steel: Sheet and plate for hardening and tempering; Swedish Standard **UTS: 700 Elon: 15% Proof: 440**
SS 1674	0.52% C steel: For induction or flame hardening; Swedish Standard
SS 1678	0.61% C steel: Swedish Standard; hardened and tempered **UTS: 850 Elon: 11% Proof: 580**
SS 1678	0.61% C 0.75% Mn 0.2% Si 0.035% S and P steel: Swedish Standard
SS 1770	0.7% C 0.6% Mn 0.035% S and P steel: Hardened and tempered; Swedish Standard **UTS: 1320**
SS 1774	0.65% C 0.5% Mn 0.25% Si 0.035% S and P spring steel
SS 1778	0.7% C 0.5% Mn 0.03% S and P steel: Strip; annealed; Swedish Standard **DPN: 170 UTS: 600**
SS 1778	0.7% C 0.5% Mn 0.03% S and P steel: Cold rolled; Swedish Standard **DPN: 270 UTS: 820**
SS 1778	0.75% C 0.65% Mn 0.03% S and P steel: Strip; cold rolled **DPN: 210 UTS: 650 Elon: 20% Proof: 440**
SS 1973	0.5% C 1% Mn 0.15% S 0.05% P steel: Free machining; normalized; Swedish Standard **UTS: 630**
SS 2090	0.56% C 0.8% Mn 0.3% Cr (max) 1.75% Si steel: For springs **DPN: 400 UTS: 1300**
SS 8011	0.75% C sintered steel: Swedish Standard
SSM	0.54% C 0.65% Mn 0.3% Si steel: Pompey
St 70/2	0.5% C 0.05% S and P steel: German Standard
St 70/202	0.5% C 0.05% S and P steel: German Standard
St 70/203	0.5% C 0.05% S and P steel: German Standard
ST BRAND	0.7% C steel: Carr
STA 3	0.75% C steel: Wire; hard drawn; replaced by BS alloy En 49D
STA 4	0.7% C steel: Wire; hard drawn; replaced by BS alloy En 49C
STA 5 V4B	0.52% C steel: Replaced by BS alloy En 43A
STA 5 V4B/1	0.52% C steel: Replaced by BS alloy En 43B
STA 5 V4B/2	0.52% C steel: Replaced by BS alloy En 43C
STA 5 V5	0.55% C steel: Replaced by BS alloy En 9K
STA 5 V5A	0.5% C steel: Replaced by BS alloy En 43D
STA 5 V5A/1	0.5% C steel: Replaced by BS alloy En 43D
STA 5 V5A/2	0.5% C steel: Replaced by BS alloy En 43E
STA 5 V22B	0.5% C steel: Replaced by BS alloy En 43
STA 5 V22B/1	0.5% C steel: Replaced by BS alloy En 43
STA 5 V22B/2	0.5% C steel: Replaced by BS alloy En 43
STA 5 V23	0.6% C 0.8% Mn 1.8% Si steel: Replaced by BS alloy En 45
STA 5 V23A	0.6% C 0.8% Mn 1.8% Si steel: Replaced by BS alloy En 45
STA 41	0.5% C steel: Strip; hardened and tempered; for springs
STANDARD	C not specified 0.25% Mn steel: For tools; Braeburn
STC 5	0.5% C 0.6% Mn 0.2% Si 0.06% S and P steel: Spencer; normalized **DPN: 240 UTS: 720 Elon: 18%**
STC 6	0.55% C 0.5% Mn 0.2% Si 0.05% S and P steel: Spencer; normalized **DPN: 240 UTS: 690 Elon: 18%**
T 11	0.72% C 0.2% Mn 0.2% Si steel: Jessop
T 12	0.81% C 0.2% Mn 0.2% Si steel: Jessop

Symbol	Nominal analysis, supplier, condition and remarks.
T 21	0.72% C 0.25% Mn 0.25% Si steel: Jessop
T 22	0.8% C 0.25% Mn 0.25% Si steel: Jessop
T 31	0.72% C 0.27% Mn 0.27% Si steel: Jessop
T 32	0.8% C 0.27% Mn 0.27% Si steel: Jessop
T 750	0.5% C 1.0% Mn 0.25% Si 0.15% S 0.05% P steel: Fagersta; free machining
TEMPER No.2	0.6% C steel: Scissors, knives, auger bits, etc.; Sanderson
TEMPER No.2½	0.7% C steel: Crowbars, hammers, stone bits, etc.; Sanderson
TEMPER No.3	0.8% C steel: Punches, shear blades, stone bits, etc.; Sanderson
TOLEDO 55	0.55% C 0.6% Mn 0.06% S and P steel: Toledo for BS alloy En 9
U 7	0.7% C 0.3% Mn 0.25% Ni (max) 0.2% Cr (max) steel: Russian Standards designation
U 7A	0.7% C 0.25% Mn 0.2% Ni (max) 0.15% Cr (max) steel: Russian Standards designation
U 8A	0.8% C 0.25% Mn 0.2% Ni (max) 0.15% Cr (max) steel: Russian Standards designation
UHB 11	0.5% C 0.5% Mn 0.2% Si 0.035% S steel: Uddelholm
UMS 75	0.53% C 0.65% Mn 0.25% Si 0.045% S and P steel: Krupp
UMS 80	0.6% C 0.65% Mn 0.25% Si 0.045% S and P steel: Krupp
V 953	0.53% C 0.6% Mn steel: VEW
V 960	0.61% C 0.8% Mn steel: VEW
VSMD	0.65% C 0.3% Mn 0.2% Si steel: Hammers, rock drills, etc.; Vulcan
VSS	0.66% C 0.5% Mn 0.2% Si steel: Springs; Vulcan
VSSM	0.57% C 1% Mn 1.5% Si steel: Heavy duty vehicle springs; Vulcan **DPN: 460**
W 6	C not specified 0.4% S and P steel: Galvanized wire; designation used by British Standards
W 21	0.6% C steel: Welding electrode; Cu covered; Armco; for hard surfacing
WLM BEST	0.7% C 0.35% S and P steel: Temper grade varies with carbon content; W Marrison
WTS 0.75%	0.75% C 0.35% Mn 0.15% Si steel: Centres, vice jaws etc.; Marsh Brothers
XBP	0.6% C 1% Mn 0.3% Si 0.04% S and P steel: Toledo for BS alloy En 49
XC 65	0.67% C 0.7% Mn steel: AFNOR
Y 65	0.65% C 0.25% Mn 0.025% S (max) 0.035% P (max) steel: French Standards designation
Y 65V	0.65% C 0.02% S and P (max) steel: French Standards designation
Y 75	0.75% C 0.25% Mn 0.025% S (max) 0.035% P (max) steel: French Standards designation
Y 75V	0.75% C 0.25% Mn 0.02% S and P (max) steel: French Standards designation

Note. The following abbreviations and units are used in the tables:

DPN	Hardness, diamond pyramid number
UTS	Ultimate tensile strength, N/mm^2
Elon	Elongation, %
Proof	0.1% proof strength, N/mm^2

1 N/mm^2=0.1 hbar=0.102 kgf/mm^2=0.06475 tonf/in.^{2}=145.04 lbf/in.^{2}=1 MPa

See Appendix II for other abbreviations and conversion tables.

44A4 steel – plain carbon 0.8 per cent minimum

Specific gravity	7.86
Density	7860 kg/m^3
Solidus/liquidus	1430–1480 °C
Thermal conductivity	41.9 W/m °C
Coefficient of linear expansion	11 × 10^{-6}/ °C
Electrical conductivity	7–10% IACS (copper 100%)
Specific resistance	180–280 microhm mm
Young's modulus of elasticity	–
Impact	–
Fatigue strength (no. of cycles)	–
Hot strength	–

The above properties are typical of the following group, and may not apply exactly to any one specification. It is possible that with certain specifications some of the values may not be applicable.

General metallurgical characteristics

The steels in this group are all hyper-eutectic, i.e. there is no free ferrite present at any time. This means that very little cold work can be accomplished without cracking or fracturing the parts and this limits the forms available to forging and relatively heavy section sheet and bar. Cold drawn wire is available when the carbon is at the lower end of the range.

The steels are relatively easily hardened and only in very special cases are they used in anything except the hardened and tempered or cold drawn condition.

There may be residual elements present, very small quantities of which have considerable effect on the hardenability. Where air hardening is undesirable it is advisable that the supplier be informed of this fact when ordering.

All these steels can be thermally treated by exposure to a temperature between 750 and 800 °C.

Where the sections are thin, i.e. less than 10 mm, or where there is any change in section, these steels should not be water quenched.

It is essential that these steels are obtained in the normalized condition and all rough machining is carried out prior to any thermal treatment.

Welding of these materials is not advised and if it is essential then expert advice must be obtained. Where welding is to be carried out it is essential that a test certificate is obtained to identify the carbon equivalent CE. The formula for this is

$$CE = C + \frac{Mn}{6} + \frac{Cr + Mo + V}{5} + \frac{Ni + Cu}{15}$$

A carbon equivalent above 0.7% gives problems during welding.

The materials in this section have no protection against corrosion pitting. The fact that many of these steels are used as springs or spring-like structures means that they are subjected to fatigue and fatigue failure commonly commences from stress raisers which could be corrosion pitting.

It must also be appreciated that the majority of these materials will have tensile strength above that which it is desirable to electroplate. This could result in hydrogen embrittlement and specialist advice should be obtained prior to any electroplating.

Many of the materials in this group will have competition from low alloy steels or from steels in Section 44M1, i.e. 12% chromium stainless steel.

Uses for the materials in this group include cutting tools. Many press tools and fixtures not subjected to impact or heat can also be produced in these materials.

Hard-wearing parts not subjected to shock, such as shear blades, scissors, many springs and knives, are included in this section. In all cases, however, it must be appreciated that where any corrosion is involved, as must occur in the case of many knives, then the materials in this group would not be suitable.

Symbol	Nominal analysis, supplier, condition and remarks.
0.0783	0.84% C 0.4% Mn 0.2% Si 0.05% S and P steel: German Standard
0.1274	1.0% C 0.4% Mn 0.2% Si 0.035% S and P steel: German Standard
1.1264	0.8% C 0.3% Mn 0.2% Si 0.025% S and P steel: German Standard
1.1274	1.0% C 0.4% Mn 0.2% Si 0.035% S and P steel: German Standard
1.1525	0.08% C 0.17% Mn 0.02% S and P (max) steel: German Standards designation
1.1545	1.05% C 0.17% Mn 0.02% S and P (max) steel: German Standards designation
1.1830	0.85% C 0.6% Mn 0.025% S and P (max) steel: German Standards designation
1.2002	1.0% C 0.5% Mn steel: German Standard
02	0.9% C 1.6% Mn steel: Designation used by AISI
16	0.8% C steel: Fagersta
17	1% C 0.45% Mn 0.25% Si steel: Sandvik for SS 1870 and SS 1880
17 AP	1% C 0.4% Mn 0.05% S 0.2% Pb steel: Sandvik; annealed; free machining **UTS: 660**
20	1.0% C steel: Fagersta
20	1.15% C 0.35% Mn 0.25% Si steel: Sandvik for SS 1885
20 P	1.15% C 0.35% Mn 0.2% Pb steel: Sandvik; free machining
24	1.2% C steel: Fagersta
40 D 35	1.0% C 0.45% Mn 0.25% Si steel: Fagersta
44 D 35	1.1% C 0.5% Mn 0.3% Si steel: Fagersta
48 D 35	1.2% C 0.5% Mn 0.3% Si steel: Fagersta **DPN: 400**
050 A 86	0.80% C 0.5% Mn steel: BS 970; obsolete
060 A 86	0.86% C 0.6% Mn steel: BS 970; obsolete
060 A 99	0.99% C 0.6% Mn steel: BS 970; obsolete
080 A 83	0.83% C 0.8% Mn steel: BS 970; obsolete
080 A 86	0.86% C 0.8% Mn steel: BS 970; obsolete

Note. The following abbreviations and units are used in the tables:

DPN	Hardness, diamond pyramid number
UTS	Ultimate tensile strength, N/mm^2
Elon	Elongation, %
Proof	0.1% proof strength, N/mm^2

1 N/mm^2=0.1 hbar=0.102 kgf/mm^2=0.06475 tonf/in.^{2}=145.04 lbf/in.^{2}=1 MPa

See Appendix II for other abbreviations and conversion tables.

Symbol	Nominal analysis, supplier, condition and remarks.
1075	0.75% C 0.04% S and P (max) steel: Designation used in the UK and USA
1078	0.78% C 0.04% S and P (max) steel: Designation used in the UK and USA
1080	0.80% C 0.04% S and P (max) steel: Designation used in the UK and USA
1084	0.84% C 0.04% S and P (max) steel: Designation used in the UK and USA
1085	0.85% C 0.04% S and P (max) steel: Designation used in the UK and USA
1086	0.86% C 0.04% S and P (max) steel: Designation used in the UK and USA
1090	0.9% C 0.04% S and P (max) steel: Designation used in the UK and USA
1095	0.95% C 0.04% S and P (max) steel: Designation used in the UK and USA
1101	1.2% C 0.12% Mn 0.02% S and P steel: French Standards designation
1102	1.05% C 0.02% S and P steel: French Standards designation
1103	0.9% C 0.02% S and P (max) steel: French Standards designation
1161	1.2% C 0.2% Mn 0.02% S and P (max) steel: French Standards designation
1162	1.05% C 0.2% Mn 0.02% S and P (max) steel: French Standards designation
1163	0.9% C 0.2% Mn 0.02% S and P (max) steel: French Standards designation
1200	1.35% C 0.25% Mn 0.025% S and P (max) steel: French Standards designation
1201	1.2% C 0.25% Mn 0.025% S and P (max) steel: French Standards designation
1202	1.05% C 0.25% Mn 0.025% S and P (max) steel: French Standards designation
1203	0.9% C 0.25% Mn 0.025% S and P (max) steel: French Standards designation
1303	0.9% C 0.25% Mn 0.2% S 0.3% P steel: French Standards designation
1590	0.9% C 1% Mn steel: Designation used in the UK and USA
AISI 1084	0.86% C 0.8% Mn 0.05% S 0.04% P steel
AISI 1086	0.86% C 0.4% Mn 0.05% S 0.04% P steel
AISI 1090	0.91% C 0.8% Mn 0.05% S 0.04% P steel
AISI 1095	0.95% C 0.4% Mn 0.05% S 0.04% P steel
AMS 5112E	0.9% C steel: Music and spring wire; best quality; AMS for SAE 1090
AMS 5121B	0.95% C steel: Spring and strip; annealed; AMS for SAE 1095
AMS 5122B	0.95% C steel: Spring and strip; rolled; AMS for SAE 1095

Symbol	Nominal analysis, supplier, condition and remarks.
AMS 5132C	0.95% C steel: Bar; AMS for SAE 1095
AMS 7304	1% C 0.4% Mn steel
AN QQ A/666/1	1.0% C 0.4% Mn 0.04% S and P steel: Strip; US Service
AN QQ W 441/3	0.85% C 0.4% Mn 0.2% Si 0.03% S and P steel: Wire; US Service; music wire **UTS: 2000**
AN S5	1.0% C 0.4% Mn 0.05% S and P steel: US Service
AQUATOUGH 100	0.95% C 0.3% Mn steel: Clyde Alloy **DPN: 770**
ASTM A68	1.0% C 0.4% Mn 0.2% Si steel: Bars for springs
ASTM A679	0.8% C 0.7% Mn steel: Wire; hard drawn for springs **UTS: 2000**
ASTM B770	Sand castings; copper beryllium alloy; general specification; designation by UNS
ATLAS ALPHA 8	0.8% C 0.25% Mn 0.2% Si steel: Atlas; as AISI type W1
ATLAS X10	1.05% C 0.2% Mn 0.2% Si steel: Atlas; as AISI type W1
ATLAS X12	1.2% C 0.25% Mn 0.20% Si steel: Atlas; as AISI type W1
ATLAS XX-95	0.95% C 0.3% Mn 0.3% Si steel: Atlas; as AISI type W1
B 15T	0.75% C steel: Bofors; obsolete **DPN: 820**
B 15V	0.8% C steel: Bofors; obsolete **DPN: 820**
B 20V	1.0% C steel: Bofors **DPN: 820**
B 24V	1.2% C steel: Bofors **DPN: 850**
B 28V	1.4% C steel: Bofors **DPN: 850**
BC	1.0% C tool steel: Origin unknown
BELL B and P	0.6%–1.2% C steel: John Vessey; a range of steels based on C content
BEST WARRANTED	0.9% C tool sheet: Balfour Darwin
BLACK LABEL	1.0% C tool steel: Jessop
BLUE LABEL	1.0% C steel: Origin unknown
BLUE LABEL	0.98% C 0.25% Mn 0.25% Si tool sheet: Huntsman **DPN: 630**
BS 24/3Ba	1.0% C 0.6% Mn 0.05% S and P steel: For springs; as BS alloy En 44
BS 970	See designation for details
BS 1407	1.1% C 0.3% Mn 0.5% Cr (optional) 0.3% Si steel: Supplied ground; 'Silver Steel'
BS 1423	1.0% C 0.6% Mn 0.5% Cr (optional) 0.3% Si steel: For springs; hardened and tempered **DPN: 780**
BS 1429 En 44B	0.95% C 0.6% Mn 0.05% S and P steel: Bar
BS 1449 HS90	0.9% C 0.5% Mn 0.2% Si 0.04% S and P steel: Sheet
BS 1449 HS100	1.0% C 0.4% Mn 0.2% Si 0.04% S and P steel: Sheet
BS 3141 C9A1	0.8% C 0.5% Mn 0.05% S and P steel: Wire; classification system used by BS; discontinued
BS 3141 C9B1	0.8% C 0.8% Mn 0.04% S and P steel: Wire; classification system used by BS; discontinued
BS 3141 C9C1	0.85% C 0.4% Mn 0.05% S and P steel: Classification system used by BS; discontinued
BS 3141 C9D1	0.85% C 0.7% Mn 0.05% S and P steel: Wire; classification system used by BS; discontinued
BS 3141 C11A1	1% C 0.5% Mn 0.05% S and P steel: Wire; classification system used by BS; discontinued
BS 3141 C11B1	1.0% C 0.5% Mn 0.04% S and P steel: Wire; classification system used by BS; discontinued
BS 3141 C12A1	1.1% C 0.5% Mn 0.05% S and P steel: Wire; classification system used by BS; discontinued
BS 3141 C13A1	1.1% C 0.3% Mn 0.04% S and P steel: Wire; classification system used by BS; discontinued
C 80W1	0.8% C 0.17% Mn 0.02% S and P (max) steel: German Standards designation
C 85 W	0.86% C 0.6% Mn 0.025% S and P (max): German Standard
C 85WS	0.85% C 0.6% Mn 0.025% S and P (max) steel: German Standards designation
C 100	1.0% C 0.42% Mn steel: Italian Standards designation **DPN: 270 UTS: 1450 Elon: 4% Proof: 110**
C 105W1	1.05% C 0.17% Mn 0.02% S and P (max) steel: German Standards designation
C 1105	1.05% C 0.4% Mn 0.25% Si 0.02% S and P steel: Drills, etc.; SHD
CARBON	1.05% C 0.2% Mn 0.08% Cr 0.2% Si steel: Latrobe
CARP RED LABEL	0.9% C steel: For tools; Carp
CARP YELLOW LABEL	1.15% C steel: For tools; Carp
CELFOR	1.0% C steel: Shear blades, dies, etc.; Sanderson **DPN: 950**
CHISEL STEEL	0.9% C steel: For cold chisels and punches; origin unknown
CHQ	Plain C die steel: Analysis vanes; Firth Sterling
COH	0.9% C 1.7% Mn 0.25% Si steel: Spencer; collets, gauges, etc. **DPN: 750**
COLUMBIA ELECTREX	1.06% C steel: Columbia; AISI type W1
COLUMBIA EXTRA	1.06% C steel: Columbia; AISI type W1
COLUMBIA EXTRA HEADERDIE	0.95% C steel: Columbia; AISI type W1
COLUMBIA STANDARD	1.06% C steel: Columbia; AISI type W1
CONQUEROR TEMP 1-6	1.0% C steel: C content and hardness varies; origin unknown
CRESCENT	1.1% C steel: Spencer
CRESCENT 1	1.3% C steel: Spencer
CRESCENT 3	1.0% C steel: Spencer
CRESCENT 4	0.85% C steel: Spencer
CRESCENT 5	0.75% C steel: Spencer
CRUSCA	0.8% C 0.3% Mn 0.15% Si steel: Crucible Steel Co.; for drills
DC	0.8% C 0.35% Mn 0.3% Si steel: Pompey
DIN 17222 M85	0.85% C 0.4% Mn 0.045% S and P steel: For springs
DIN 17222 MK101	1.0% C 0.4% Mn 0.045% S and P steel: For springs
DOH	0.9% C 1.7% Mn 0.25% Si steel: Spencer; collets, gauges, etc. **DPN: 750**
DTD 5 B	0.8% C 0.5% Mn 0.2% Si steel: Wire; cold drawn and tempered **UTS: 1550**
DTD 488	1.0% C 1.0% Mn 0.05% S 0.05% P steel: Strip; hardened and tempered **UTS: 1840 Proof: 1620**
E 11C	1.2% C 0.45% Mn steel: Pompey
ECLIPSE CHD	1.0% C steel: Origin unknown
En 42D	0.85% C 0.05% S and P steel: For springs; designation used in BS 970; obsolete

Note. The following abbreviations and units are used in the tables:

DPN	Hardness, diamond pyramid number
UTS	Ultimate tensile strength, N/mm^2
Elon	Elongation, %
Proof	0.1% proof strength, N/mm^2

1 N/mm^2=0.1 hbar=0.102 kgf/mm^2=0.06475 $tonf/in.^2$=145.04 $lbf/in.^2$=1 MPa

See Appendix II for other abbreviations and conversion tables.

Symbol	Nominal analysis, supplier, condition and remarks.
En 44	1% C 0.05% S and P steel: For springs; designation used in BS 970
En 44A	1.1% C 0.05% S and P steel: For springs; designation used in BS 970
En 44C	1.1% C 0.05% S and P steel: For springs; designation used in BS 970
EXTRA	1.0% C tool steel: Origin unknown
FAGERSTA 16	0.8% C 0.3% Mn 0.2% Si steel: Fagersta
FAGERSTA 20	1.0% C 0.3% Mn 0.2% Si steel: Fagersta **DPN: 730**
FAGERSTA 24	1.2% C 0.3% Mn 0.2% Si steel: Fagersta **DPN: 800**
FAGERSTA 1870	1.0% C 0.5% Mn steel: For springs; Fagersta
FAGERSTA 1884	1.2% C 0.3% Mn steel: For springs and cutting tools; Fagersta
FILE	1.2% C 0.3% Mn 0.18% Si 0.05% S and P steel: Also produced with up to 3% Cr additions; Toledo
G 10950	UNS designation for type 1095 steel
GAUGE STEEL	1.0% C steel: Hardened and ground; general trade term
GENUINE STUBS	1.1% C 0.35% Mn 0.2% Si 0.045% S and P steel: For BS 1407; Peter Stubs
GOST 1050/85	0.85% C 0.65% Mn 0.04% S and P (max) steel: For structures; Russian Standard
GOST 2052/60 S2	0.61% C 0.85% Mn 0.4% Ni (max) 0.3% Cr (max) 0.04% S and P (max) 0.25% Cu (max) steel: For springs; Russian Standard
GOST 2052/85	0.85% C 0.65% Mn 0.13% Cr and Ni (max) 0.04% S and P (max) 0.25% Cu (max) steel: For springs; Russian Standard
GRAPH A1	1.5% C 0.3% Mn 0.2% Si 0.15% Al steel: Free graphite; Timken; mandrels, dies, etc. **DPN: 850**
GREEN LABEL	1.2% C 0.2% Mn 0.2% Si steel: Carpenter; for drill rod
GREEN LABEL	0.8% C 0.25% Mn tool steel: Huntsman **DPN: 650**
GS 6	0.9% C 2% Mn steel: Cutting tools; Osborn; obsolete
H9 DOUBLE HEADER	0.9% C 0.4% Mn 0.4% Si steel: Carpenter for AISI type W1
HECLA 34	1.0% C 0.6% Mn steel: Hadfield for BS alloys En 44B, C, D and E
HM 1	0.9% C 1.6% Mn tool steel: Origin unknown
HYFORM	1.0% C steel: Origin unknown
ICS	1.0% C steel: Origin unknown
INMALLOY	0.6% C 1.1% Cr 0.25% Mo steel: Inman; shock resistant
ISI	0.45% C 1.25% Cr 2% W steel: Inman; shock resistant
JEM No. 1	1.4% C steel: Jonas
JEM No. 2	1.2% C steel: Jonas
JEM No. 3	1.1% C steel: Jonas
JEM No. 4	0.9% C steel: Jonas
JIS G4801 54P3	0.82% C 0.45% Mn 0.035% S and P (max) steel: For springs; Japanese Standard
JIS G4801 84P4	1.0% C 0.45% Mn 0.035% S and P (max) steel: Japanese Standard
K 980	0.8% C 0.2% Mn 0.02% S and P steel: VEW
K 985	0.86% C 0.6% Mn 0.025% S and P steel: VEW
K 990	1.05% C 0.2% Mn 0.02% S and P steel: VEW
KE 1006	1.0% C 0.2% Mn 0.2% Si steel: Coining dies, etc.; Kayser Ellison **DPN: 910**
KEA 108	1.0% C 0.5% Mn steel: Sanderson Kayser
KP 9	0.86% C (min) steel: Kiveton Park
L 11	1.1% C steel: Pompey
L 12	1.2% C steel: Pompey
L 13	1.3% C steel: Pompey
LC 100	0.9% C steel: Hardened and tempered; water quenched; Low Moor; obsolete **DPN: 800**
LC 110	1.1% C steel: Hardened and tempered; water quenched; Low Moor; obsolete **DPN: 800**
LC 120	1.2% C steel: Hardened and tempered; water quenched; Low Moor; obsolete **DPN: 800**
LC 140	1.4% C steel: Hardened and tempered; water quenched; Low Moor; obsolete **DPN: 800**
LT 20	0.9% C 2.0% Mn steel: Non-shrinking tool steel; Low Moor **DPN: 800**
MIL W 8957	0.87% C 0.32% Mn 0.02% S and P (max) steel: Casting; designation used by UNS
MOTOR BRAND	1.0% C steel: Supplied as ground flat stock, Carrs
MT	1.7% C 8.0% Al steel: Casting or forging for magnets; Japanese alloy listed by PMA
MUSIC WIRE	1% C steel: Cold drawn; common name; also piano wire
No. 1 HARDENITE	1.0% C steel: For tools; origin unknown
No. 2 HARDENITE	0.9% C tool steel: Origin unknown
No. 3 HARDENITE	0.9% C tool steel: Origin unknown
No. 9 TEMPER	0.9% C 0.2% Mn steel: Reamers and dies; ESC **DPN: 750**
No. 10 TEMPER	1.0% C 0.2% Mn steel: Hand chisels, axes, etc.; ESC **DPN: 750**
No. 11 SPECIAL	1.05% C 0.2% Mn steel: Carpenter
No. 11 TEMPER	1.1% C 0.2% Mn steel: Knives, reamers, etc.; ESC **DPN: 750**
No. 12 TEMPER	1.2% C 0.2% Mn steel: Woodmaking tools, razors, etc.; ESC **DPN: 750**
No. 14 TEMPER	1.4% C 0.2% Mn steel: Razors and surgical instruments; ESC **DPN: 750**
NOLL SPECIAL	1.05% C 0.2% Mn 0.2% Si steel: As AISI W1; Carpenter
NONVAR	0.92% C 1.75% Mn steel: Firth Brown
NSS 3	1% C 2% Mn steel: Press tools, jig brushes, etc.; Balfour **DPN: 660**
OTTOWA	0.8% C 0.2% Mn 0.15% Si 0.08% S and P steel: Tube; Atlas; for hollow rock drills
P 1	10.1% C steel: For moulds; designation used by AISI
PIANO WIRE	1% C steel: Cold drawn; common name; also music wire
QQ W 428	High carbon steel: Wire for springs; US Federal specification
QQ W 432	Carbon steel: Wire; US Federal specification
QQ W 470	0.85% C 0.4% Mn 0.2% Si 0.03% S and P steel: Wire; US Federal specification; cold drawn; music wire **UTS: 2280**
RCS	1.0% C tool steel: Origin unknown
RED LABEL	0.9% C 0.25% Mn 0.25% Si tool steel: Huntsman **DPN: 750**
REMOUNT	1.0% C tool steel: Origin unknown
S 90	0.9% C 0.5% Mn 0.3% Si steel: Pompey
SAE 852	0.8% C Fe: Sinter for mechanical parts; density 5900 kg/m^3
SAE 855	0.8% C Fe: Sinter for mechanical parts; density 6300 kg/m^3
SAE 872	0.8% C 20% Cu Fe alloy: Sinter; density 7100 kg/m^3
SAE 1084	0.86% C 0.8% Mn 0.05% S 0.04% P steel: Bar; hot rolled **DPN: 241 UTS: 900 Elon: 10% Proof: 460**
SAE 1085	0.86% C 0.9% Mn 0.05% S 0.04% P steel: Bar; hot rolled **DPN: 248 UTS: 910 Elon: 10% Proof: 480**
SAE 1086	0.86% C 0.4% Mn 0.05% S 0.04% P steel: Bar; hot rolled **DPN: 229 UTS: 830 Elon: 10% Proof: 440**

Symbol	Nominal analysis, supplier, condition and remarks.
SAE 1090	0.91% C 0.8% Mn 0.05% S 0.04% P steel: Bar; hot rolled
	DPN: 248 UTS: 910 Elon: 10% Proof: 480
SAE 1095	0.95% C 0.4% Mn 0.05% S 0.04% P steel: Bar; hot rolled
	DPN: 248 UTS: 900 Elon: 10% Proof: 460
SAE W108	0.8% C tool steel
SAE W109	0.9% C tool steel
SAE W110	1.0% C tool steel
SAE W112	1.2% C tool steel
SANBOLD 44	1.0% C 0.6% Mn 0.3% Si steel: For BS alloy En 44; Sanderson
SILVER STEEL	1.1% C 0.35% Mn 0.2% Si 0.04% S and P steel: Common name; as rolled; supplied centreless ground; BS 1407
SK 1	1.45% C 0.5% Mn (max) 0.2% Cr and Ni (max) steel: Japanese Standards designation
SK 2	1.2% C 0.5% Mn (max) 0.2% Cr and Ni (max) steel: Japanese Standards designation
SK 3	1.05% C 0.5% Mn (max) 0.2% Cr and Ni (max) steel: Japanese Standards designation
SK 4	0.95% C 0.5% Mn (max) 0.2% Cr and Ni (max) steel: Japanese Standards designation
SK 5	0.85% C 0.5% Mn (max) 0.2% Cr and Ni (max) steel: Japanese Standards designation
SKF 715	0.85% C 0.8% Mn steel: SKF
SOD	0.9% C 1.6% Mn steel: Braeburn; AISI type 02
SPEAR MD4	1.0% C 0.2% Mn 0.2% Si 0.03% S and P steel: Spear and Jackson; for coin dies
SPEAR No. 2	1.0% C steel: Spear and Jackson
SPEAR No. 3	0.85% C steel: Spear and Jackson
SPRASTEEL 80	0.8% C 0.7% Mn 0.04% S and P steel: Wire for spraying; Metco; as sprayed
	DPN: 352 UTS: 180
SS 1774	0.8% C 0.6% Mn 0.2% Si steel: Swedish Standard; included in MNC 870
SS 1870	1.0% C 0.45% Mn 0.03% S and P steel: Strip; cold rolled; Swedish Standard
	DPN: 620 UTS: 2000
SS 1870	1.0% C 0.45% Mn 0.03% S and P steel: Strip; cold rolled; hardened; Swedish Standard
	DPN: 600 UTS: 2000
SS 1870	0.95% C 0.5% Mn 0.03% S and P steel: Strip; annealed; Swedish Standard
	DPN: 185 UTS: 630
SS 1870	0.95% C 0.5% Mn 0.03% S and P steel: Strip; cold rolled; Swedish Standard
	DPN: 280 UTS: 850
SS 1880	1.0% C 0.3% Mn 0.025% S and P steel: Bar and forging; annealed; Swedish Standard
	DPN: 195
SS 1885	1.2% C 0.3% Mn steel: Annealed; Swedish Standard
	DPN: 195
STA 5/V22A	0.8% C steel: Replaced by BS alloy En 42
STA 5/V22A/1	0.8% C steel: Replaced by BS alloy En 42
STA 5/V22A/2	0.8% C steel: Replaced by BS alloy En 42
STA 27	1.0% C steel: Free cutting; for pinions, etc.
STENTOR	0.9% C 1.6% Mn 0.2% Si steel: Carpenter for AISI 02
STERLING	1.0% C steel: Firth Sterling
STUBS	1.1% C 0.35% Mn 0.2% Si 0.45% S and P steel: As rolled; supplied centreless ground; Peter Stubs
	DPN: 220 UTS: 760 Elon: 32% Proof: 580
T 13	0.9% C 0.2% Mn 0.2% Si steel: Jessop
T 14	1% C 0.2% Mn 0.2% Si steel: Jessop
T 15	1.15% C 0.2% Mn 0.2% Si steel: Jessop
T 16	1.3% C 0.2% Mn 0.2% Si steel: Jessop
T 23	0.9% C 0.25% Mn 0.25% Si steel: Jessop
T 24	1% C 0.25% Mn 0.25% Si steel: Jessop
T 25	1.15% C 0.25% Mn 0.25% Si steel: Jessop
T 26	1.3% C 0.25% Mn 0.25% Si steel: Jessop
T 27	1.4% C 0.25% Mn 0.25% Si steel: Jessop
T 33	0.9% C 0.27% Mn 0.27% Si steel: Jessop
T 34	1.0% C 0.27% Mn 0.27% Si steel: Jessop
T 35	1.15% C 0.27% Mn 0.27% Si steel: Jessop
T 36	1.3% C 0.27% Mn 0.27% Si steel: Jessop
T 37	1.4% C 0.27% Mn 0.27% Si steel: Jessop
T QUALITY	1.0% C tool steel: Origin unknown
TD	1.0% C tool steel: Origin unknown
TEMPER No. 3½	0.9% C steel: Shear blades, knives, chisels, etc.; Sanderson
TEMPER No. 4	1.0% C steel: Knives, wood tools, chisels, etc.; Sanderson
TEMPER No. 4½	1.1% C steel: Knives, chisels, wood tools, etc.; Sanderson
TEMPER No. 5	1.2% C steel: Chisels, knives, wood tools, etc.; Sanderson
TEMPER No. 5½	1.3% C steel: Knives, granite drills, etc.; Sanderson
TEMPER No. 6	1.4% C steel: Scrapers, surgical tools, etc.; Sanderson
TM	1.0% C tool steel: Origin unknown
TMS	0.92% C 1.75% Mn steel: Firth Brown
TREBLE EXTRA	1.0% C steel: Origin unknown
TREBLE EXTRA	0.7%–1.4% C range of steels: Firth Brown
U 8	0.8% C 0.25% Mn 0.25% Ni (max) 0.2% Cr (max) steel: Russian Standards designation
U 9	0.9% C 0.25% Mn 0.25% Ni (max) 0.2% Cr (max) steel: Russian Standards designation
U 9A	0.9% C 0.22% Mn 0.2% Ni (max) 0.15% Cr (max) steel: Russian Standards designation
U 10	1.0% C 0.25% Mn 0.25% Ni (max) 0.2% Cr (max) steel: Russian Standards designation
U 10A	1.0% C 0.22% Mn 0.2% Ni (max) 0.15% Cr (max) steel: Russian Standards designation
U 11	1.1% C 0.25% Mn 0.25% Ni (max) 0.2% Cr (max) steel: Russian Standards designation
U 11A	1.1% C 0.22% Mn 0.2% Ni (max) 0.15% Cr (max) steel: Russian Standards designation
U 12	1.2% C 0.25% Mn 0.25% Ni (max) 0.2% Cr (max) steel: Russian Standards designation
U 12A	1.2% C 0.22% Mn 0.2% Ni (max) 0.15% Cr (max) steel: Russian Standards designation
U 13	1.3% C 0.25% Mn 0.25% Ni (max) 0.2% Cr (max) steel: Russian Standards designation
U 13A	1.3% C 0.22% Mn 0.2% Ni (max) 0.15% Cr (max) steel: Russian Standards designation
U 85	0.85% C 0.5% Mn 0.25% Ni (max) 0.2% Cr (max) steel: Russian Standards designation
U BRAND	0.7%–1.7% C range of steel: Firth Brown
UGA	0.85% C 0.52% Mn 0.2% Ni (max) 0.15% Cr (max) steel: Russian Standards designation
UHB	1.0% C steel: For springs; Uddelholm
UHB 20	1.04% C 0.25% Mn 0.2% Si steel: Uddelholm; for type W1
VEDAS	0.8%–1.2% C steel: Available in four grades depending on C content; Hall and Pickles; obsolete
VEDAS	1.0% C steel: Origin unknown
VHRD	1.0% C tool steel: Origin unknown
VIZOR	0.8%–1.2% C steel: Available in four grades depending on C content; Hall and Pickles; obsolete

Note. The following abbreviations and units are used in the tables:

DPN	Hardness, diamond pyramid number
UTS	Ultimate tensile strength, N/mm^2
Elon	Elongation, %
Proof	0.1% proof strength, N/mm^2

1 N/mm^2=0.1 hbar=0.102 kgf/mm^2=0.06475 tonf/in.^{2}=145.04 lbf/in.^{2}=1 MPa

See Appendix II for other abbreviations and conversion tables.

Symbol	Nominal analysis, supplier, condition and remarks.
VIZOR	1.0% C steel: Origin unknown
W 1	1.0% C steel: C varies over wide limits; designation used by AISI
W 1	1.0% C steel: Osborn
W 110	1.0% C steel: For tools; designation used by SAE
W 112	1.2% C steel: For tools; designation used by SAE
WLM 9015	0.9% C 1.8% Mn steel: Press tools and shear blades; W Marrison **DPN: 720**
WLM SPECIAL	0.7%–1.3% C 0.03% S and P steel: Temper grade varies with C content; W Marrison
WORTLE PLATE	2.5% C steel: Used in the manufacture of wire drawing dies; origin unknown
WTS 1%	1.0% C 0.35% Mn 0.15% Si steel: Blanking tools and shear blades; Marsh Brothers
Y 90	0.9% C 0.25% Mn 0.02% S 0.03% P steel: French Standards designation
Y 90V	0.9% C 0.25% Mn 0.025% S and P (max) steel: French Standards designation
Y 105	1.05% C 0.02% S and P (max) steel: French Standards designation
Y 105	1.05% C 0.25% Mn 0.025% S and P (max) steel: French Standards designation
Y 120	1.2% C 0.02% S and P (max) steel: French Standards designation
Y 120	1.2% C 0.25% Mn 0.025% S and P (max) steel: French Standards designation
Y 135	1.35% C 0.25% Mn 0.025% S and P (max) steel: French Standards designation
YELLOW LABEL	0.9% C 0.03% S and P steel: T Turton

Note. The following abbreviations and units are used in the tables:

DPN	Hardness, diamond pyramid number
UTS	Ultimate tensile strength, N/mm^2
Elon	Elongation, %
Proof	0.1% proof strength, N/mm^2

1 N/mm^2=0.1 hbar=0.102 kgf/mm^2=0.06475 tonf/in.^{2}=145.04 lbf/in.^{2}=1 MPa

See Appendix II for other abbreviations and conversion tables.

44B1 Steel – low carbon with silicon 0.05% carbon – 3.5% silicon

Specific gravity	7.5–7.7
Density	7500–7700 kg/m^3
Solidus/liquidus	1470–1500 °C
Thermal conductivity	50.2 W/m °C
Coefficient of linear expansion	10 × 10^{-6}/ °C
Electrical conductivity	8–10% IACS (copper 100%)
Specific resistance	250 microhm mm
Young's modulus of elasticity	–
Impact	–
Fatigue strength (no. of cycles)	–
Hot strength	–

The above properties are typical of the following group, and may not apply exactly to any one specification. It is possible that with certain specifications some of the values may not be applicable.

General metallurgical characteristics

This is a completely ferritic material sometimes known as silicon iron and would be more correctly shown with the pure irons. The materials, however, are generally known as 'silicon steels' and are conveniently listed here.

There is insufficient carbon present for any thermal hardening, but the silicon gives some increase in strength with a tendency to brittleness and causes the material to work harden. Silicon also improves the corrosion resistance. Very often Swedish iron is used to manufacture the steels listed as this has a minimum of residual elements and is inherently low in sulphur and phosphorus.

The materials have peculiar electrical and magnetic properties which are the reason for their existence, the only useful form being thin sheet. This is to allow the grains to be preferentially oriented, thus improving the magnetic properties in one direction.

Materials in this group are almost always sheet and many are used for specific magnetic purposes. Annealing is at a temperature just below 1000 °C and under most conditions requires firstly a controlled atmosphere and secondly highly controlled heating and cooling curves in order to obtain the necessary properties. Specialist advice is essential when any of these materials is thermally treated.

No problems exist with welding provided normal protection is taken although with the higher silicon materials specialist advice should be sought.

Low silicon does not give any corrosion protection and the materials must therefore be oiled or protected by lacquer.

Where the materials are machined there is a problem with the abrasive properties of silicon oxide (silica), which is present in these materials, and tool wear will be found. These materials are seldom machined, however.

The most common use for these materials is laminations in all types of transformers including those in loudspeakers.

Symbol	Nominal analysis, supplier, condition and remarks.
1S 15	0.05% C (max) 1.4% Si 0.1% Cu steel: Sandvik
1S 35	0.05% C (max) 0.15% Mn 3.5% Si steel: Sandvik
AMS 7711	1.5% Si steel: Sheet and strip for electrical purposes
AMS 7712	2.5% Si steel: Sheet and strip for electrical purposes
AMS 7714	3% Si steel: Sheet and strip for electrical purposes
AMS 7715	4.25% Si steel: Sheet and strip for electrical purposes
BS 933/3	Si steel: Sheet for magnetic purposes; composition not part of specification
LOSIL 17	2.3% Si hot rolled dynamo steel: Electrical resistivity 400 microhm mm; Steel Co. of Wales; BS 601/1 **DPN: 175 UTS: 460 Elon: 19% Proof: 320**
LOSIL 17C	2.3% Si hot rolled and annealed dynamo steel: Electrical resistivity 400 microhm mm; Steel Co. of Wales **DPN: 175 UTS: 460 Elon: 22% Proof: 320**
LOSIL 19	1.7% Si hot rolled dynamo steel: Electrical resistivity 330 microhm mm; Steel Co. of Wales **DPN: 150 UTS: 420 Elon: 21% Proof: 270**
LOSIL 19C	1.7% Si hot rolled and annealed dynamo steel: Electrical resistivity 330 microhm mm; Steel Co. of Wales **DPN: 150 UTS: 420 Elon: 25% Proof: 270**
LOSIL 22	1.3% Si hot rolled dynamo steel: Electrical resistivity 280 microhm mm; Steel Co. of Wales; BS 601/1 **DPN: 130 UTS: 390 Elon: 22% Proof: 240**
LOSIL 22C	1.3% Si hot rolled and annealed dynamo steel: Electrical resistivity 220 microhm mm; Steel Co. of Wales **DPN: 105 UTS: 340 Elon: 28% Proof: 200**
LOSIL 25	0.4% Si hot rolled dynamo steel: Electrical resistivity 170 microhm mm; Steel Co. of Wales; BS 601/1 **DPN: 90 UTS: 320 Elon: 25% Proof: 170**
LOSIL 25C	0.4% Si hot rolled and annealed dynamo steel: Electrical resistivity 170 microhm mm; Steel Co. of Wales **DPN: 90 UTS: 320 Elon: 31% Proof: 170**
M 6W	Low C Si steel: Strip; grain oriented; Armco
M 7W	Low C Si steel: Strip; grain oriented; Armco
M 7X	Low C Si steel: Strip; grain oriented; Armco
M 8X	Low C Si steel: Strip; grain oriented; Armco
SILICON CORE IRON A	0.03% C 0.15% Mn 1% Si Fe alloy: Carpenter
SILICON CORE IRON B	0.03% C 2.5% Si Fe alloy: Carpenter; for magnetic purposes
SILICON CORE IRON C	0.03% C 4% Si Fe alloy: Carpenter; for magnetic purposes
TRANSIL 74	4.2% Si hot rolled steel: For transformers; electrical resistivity 620 microhm mm; Steel Co. of Wales; BS 601/1 **DPN: 265 UTS: 560 Elon: 4% Proof: 460**
TRANSIL 80	4.1% Si hot rolled steel: For transformers; electrical resistivity 610 microhm mm; Steel Co. of Wales; BS 601/1 **DPN: 260 UTS: 560 Elon: 6% Proof: 460**
TRANSIL 86	4% Si hot rolled steel: For transformers; electrical resistivity 590 microhm mm; Steel Co. of Wales; BS 601/1 **DPN: 255 UTS: 560 Elon: 9% Proof: 450**
TRANSIL 92	3.9% Si hot rolled steel: For transformers; electrical resistivity 580 microhm mm; Steel Co. of Wales; BS 601/1 **DPN: 250 UTS: 560 Elon: 11% Proof: 450**
TRANSIL 100	3.6% Si hot rolled transformer steel: Electrical resistivity 550 microhm mm; Steel Co. of Wales; BS 601/1 **DPN: 235 UTS: 560 Elon: 13% Proof: 420**
TRANSIL 107	3.4% Si hot rolled transformer steel: Electrical resistivity 520 microhm mm; Steel Co. of Wales; BS 601/1 **DPN: 225 UTS: 540 Elon: 14% Proof: 400**
TRANSIL 115	3.2% Si hot rolled transformer steel: Electrical resistivity 500 microhm mm; Steel Co. of Wales; BS 601/1 **DPN: 215 UTS: 530 Elon: 15% Proof: 390**
UNISIL 46	3.1% Si cold reduced steel: For transformers; electrical resistivity 480 microhm mm; Steel Co. of Wales; BS 601/11 **DPN: 165 UTS: 360 Elon: 13% Proof: 300**
UNISIL 51	3.1% Si cold reduced steel: For transformers; electrical resistivity 480 microhm mm; Steel Co. of Wales; BS 601/11 **DPN: 165 UTS: 360 Elon: 13% Proof: 300**
UNISIL 56	3.1% Si cold reduced steel: For transformers; electrical resistivity 480 microhm mm; Steel Co. of Wales; BS 601/11 **DPN: 165 UTS: 360 Elon: 13% Proof: 300**
UNISIL 62	3.1% Si cold reduced steel: For transformers; electrical resistivity 480 microhm mm; Steel Co. of Wales; BS 601/11 **DPN: 165 UTS: 360 Elon: 13% Proof: 300**

Note. The following abbreviations and units are used in the tables:

DPN	Hardness, diamond pyramid number
UTS	Ultimate tensile strength, N/mm^2
Elon	Elongation, %
Proof	0.1% proof strength, N/mm^2

1 N/mm^2=0.1 hbar=0.102 kgf/mm^2=0.06475 $tonf/in.^2$=145.04 $lbf/in.^2$=1 MPa

See Appendix II for other abbreviations and conversion tables.

44B2 Steel – medium carbon with silicon

Specific gravity	7.86
Density	7860 kg/m^3
Solidus/liquidus	1460–1500 °C
Thermal conductivity	41.9 W/m °C
Coefficient of linear expansion	10×10^{-6}/ °C
Electrical conductivity	10% IACS (copper 100%)
Specific resistance	250 microhm mm
Young's modulus of elasticity	–
Impact	40 J
Fatigue strength	–
Hot strength	–

The above properties are typical of the following group, and may not apply exactly to any one specification. It is possible that with certain specifications some of the values may not be applicable.

General metallurgical characteristics

Silicon raises the temperatures required for all thermal treatments and is a graphite former, thus reducing the carbon available for hardening and aiding de-carburization. Practically without exception all steels have some silicon added as it is a very efficient and cheap de-oxidizer, the product of which has little effect on ductility if trapped in the steel.

When less than 0.5% silicon is present it is not classified as an alloying element as it has no appreciable effect on the mechanical properties. Above this figure there is some increase in corrosion resistance and fatigue strength although not to such a marked degree as with other alloying elements.

There is a process analogous to carburizing which absorbs silicon into the steel surface, converting all the carbon to graphite and giving a high silicon layer which has excellent resistance to many acids.

The silicon steels all have fairly good fatigue strength with the very great advantage that notches for any reason have a minimum effect on reducing this fatigue strength. It is seldom used as the sole alloying element, generally being present with chromium and very often vanadium.

When thermal treatments are required such as hardening, normalizing or annealing, then temperatures very little below 1000 °C are required.

These steels tend to de-carburize very readily and thus care must be taken to use controlled atmospheres, salt baths or machining following thermal treatment.

Prior to any welding or brazing it is strongly advised that technical advice is obtained.

The increase in corrosion resistance given by the silicon content of these steels is not sufficient to allow them to be classified as stainless but, provided they are lightly oiled or greased, satisfactory results are generally obtained.

Use of these steels is confined generally to low duty springs of the leaf type and some tool steels.

It must be appreciated that tramp elements such as chromium, vanadium and molybdenum will dramatically increase the hardenability, and where any critical components are used, test certificates should be obtained to ensure that these elements are at an acceptable level.

The majority of high silicon steels have other elements such as chromium, molybdenum or vanadium added as part of the specification and thus appear in the 44K series of steels.

Symbol	Nominal analysis, supplier, condition and remarks.
AISI 9250	0.5% C 2% Si steel: Obsolete
AISI 9255	0.55% C 2.0% Si steel
AISI 9260	0.6% C 2.0% Si steel
ASTM A59	0.6% C 0.8% Mn 2% Si steel: Bars for springs; obsolete; see A552
CRD SILICO	0.55% C 2% Si 0.8% Mn steel: Shear blades; C Denton **DPN: 540**
S 4	0.55% C 0.8% Mn 2.0% Si steel: Designation used by AISI
SAE 9250	0.5% C 2% Si steel: Obsolete
SAE 9255	0.55% C 0.8% Mn 2% Si steel: Mechanical properties not specified
SAE 9260	0.6% C 0.8% Mn 2% Si steel: Mechanical properties not specified
SIL F	0.4% C 2.2% Cr 3.9% Si steel: For valves; origin unknown

Note. The following abbreviations and units are used in the tables:

DPN	Hardness, diamond pyramid number
UTS	Ultimate tensile strength, N/mm^2
Elon	Elongation, %
Proof	0.1% proof strength, N/mm^2

1 N/mm^2=0.1 hbar=0.102 kgf/mm^2=0.06475 tonf/in.^{2}=145.04 lbf/in.^{2}=1 MPa

See Appendix II for other abbreviations and conversion tables.

44C Steel – low and medium carbon with boron

Specific gravity	7.86
Density	7860 kg/m^3
Solidus/liquidus	1490–1520 °C
Thermal conductivity	50.2 W/m °C
Coefficient of linear expansion	11 × 10^{-6}/ °C
Electrical conductivity	8–10% IACS (copper 100%)
Specific resistance	180–250 microhm mm
Young's modulus of elasticity	90 × 10^9 N/m^2
Impact	54 J
Fatigue strength (no. of cycles)	–
Hot strength	

Temperature °C	Tensile strength N/mm^2	Elongation %
100	420	34
200	450	27
300	470	23
400	370	37
500	300	38
600	200	48
700	100	56
800	60	65
900	35	75

The properties above are typical of the following group and may not apply exactly to any one specification. It is possible that with certain specifications some of the values may not be applicable.

General metallurgical characteristics

Boron in very small amounts has a startling affect on the hardenability of steel. For example low carbon sheet material which would not air harden in plain carbon steel will be fully air hardenable with the addition of 0.0015% boron. Similar differences are found with thicker sections which can be successfully oil hardened where previously water quenching only achieved a degree of surface hardening. Boron also enhances the hardenability of other alloying elements and, in the United States of America at least, the element is being used as a very economical substitute for some of the more expensive elements.

Larger quantities of boron result in brittle, unworkable steels.

The beneficial effects of boron are only apparent with low and medium carbon steels, there being no real increase in hardenability above 0.6% carbon. There are no steels recorded with boron as the sole alloying element but it is found in conjunction with vanadium, molybdenum and chromium. There is no improvement in corrosion resistance but the reduction of alloying elements often improves the machinability.

The weldability of boron steels is one of the principal reasons for their use in this country at present.

The presence of this alloying element is often not noticed or is ignored because of the very small amounts present in steel. Boron has an influence on the metallurgy of steel much greater than might be expected from the very low percentages of boron which are normally present. It is generally present at less than 0.01% and very often in the region of 0.001% and still has some influence on the hardenability.

Boron has no appreciable affect on the critical change points for thermal treatment. These steels are therefore thermally treated at the same temperature as if no boron were present.

The steels are generally specified where welding is being carried out and controlled hardenability in the weld is necessary. These steels can be welded using all the normal production welding techniques.

Very often the heat input is carefully controlled to give the desired degree of hardenability. These steels are quite commonly tempered following welding to obtain the necessary ductility in the heat affected zone.

Boron is found where high quality welding is required on sheet metal components. Welded structures such as pipes, much of the canware on aircraft engines, and other high duty components have controlled quantities of boron.

The majority of boron steels have molybdenum present and very often other alloying elements. They will therefore, be found in Sections 44G1 and 44K1.

When boron is present in greater quantities it can act as a fluxing agent but this will not be found with low carbon steels.

Symbol	Nominal analysis, supplier, condition and remarks.
15 B 21	0.2% C 1% Mn steel: With B; designation used in the UK and USA
15 B 35	0.35% C 1% Mn steel: With B; designation used in the UK and USA
15 B 37	0.37% C 1.2% Mn steel: With B; designation used in the UK and USA
15 B 41	0.41% C 1.5% Mn steel: With B; designation used in the UK and USA
15 B 48	0.48% C 1.2% Mn steel: With B; designation used in the UK and USA
15 B 62	0.6% C 1.2% Mn steel: With B; designation used in the UK and USA
170 H 15	0.15% C 1% Mn 0.003% B steel: BS 970
170 H 20	0.2% C 1% Mn 0.003% B steel: BS 970
170 H 36	0.36% C 1% Mn 0.003% B steel: BS 970
170 H 41	0.41% C 1% Mn 0.003% B steel: BS 970
170 H 36	0.36% C 1% Mn 0.003% B steel: BS 970
170 H 41	0.41% C 1% Mn 0.003% B steel: BS 970
173 H 16	0.16% C 1.2% Mn 0.003% B steel: BS 970
174 H 20	0.2% C 1.3% Mn 0.003% B steel: BS 970
175 H 23	0.23% C 1.4% Mn 0.003% B steel: BS 970
185 H 40	0.4% C 0.25% Cr 0.12% Mo 0.003% B steel: BS 970

Note. The following abbreviations and units are used in the tables:

DPN	Hardness, diamond pyramid number
UTS	Ultimate tensile strength, N/mm^2
Elon	Elongation, %
Proof	0.1% proof strength, N/mm^2

1 N/mm^2=0.1 hbar=0.102 kgf/mm^2=0.06475 tonf/in.^{2}=145.04 ibf/in.^{2}=1 MPa

See Appendix II for other abbreviations and conversion tables.

44D Steel–manganese alloys

Specific gravity	7.7
Density	7700 kg/m^3
Solidus/liquidus	1430–1520 °C
Thermal conductivity	50.2 W/m °C
Coefficient of linear expansion	Variable
Electrical conductivity	8–10% IACS (copper 100%)
Specific resistance	200–250 microhm mm
Young's modulus of elasticity	–
Impact	–
Fatigue strength	–
Hot strength	–

The above properties are typical of the following group, and may not apply exactly to any one specification. It is possible that with certain specifications some of the values may not be applicable.

General metallurgical characteristics

Manganese is present in almost all steels but is not acknowledged as an alloying element unless present above approximately 2%. It is essentially a de-oxidizer and desulphurizer which is popular because any manganese oxide or sulphide trapped in the steel instead of the slag is more ductile than the steel itself. This means that all hot and cold working operations can take place without refractory oxides or sulphides tearing cavities in the more ductile steel.

Manganese also lowers the critical thermal treatment temperatures. With more than about 15% manganese the critical thermal treatment temperature is below room temperature, thus forming an austenitic, non-thermally-treatable steel.

Manganese also strengthens steel and this is particularly the case with higher carbon steels where 2–5% manganese gives a steel which is toughened and hardened by cold work. The presence of silicon aids this toughening process.

The steels in this group are very prone to cold working. That is, their mechanical properties are increased by a limited amount of cold work with a resultant drop in ductility. This property is very often used to advantage but where it is a disadvantage then annealing will be required. This will be at a temperature in the region of 750 °C.

The low manganese steels can accept thermal treatment in the normal manner with manganese lowering the critical temperatures. Welding should not be attempted with these materials without obtaining technical advice.

The properties of steels with manganese content above approximately 15% cannot be improved by thermal treatments. These steels will be non-magnetic.

Because of their tendency to cold work the materials in the group can be difficult to machine. As with other materials of this type it is essential that the cutting tool stays below the surface and that rubbing does not occur.

The materials in this group are plain carbon manganese steels and are used for chisels, picks, low cost drills, taps, etc., and occasionally shear blades and for other duties where cold working improves the mechanical strength. Items such as rail crossings subjected to considerable wear are commonly made from these steels. It must always be noted that, as the mechanical strength increases, the ductility becomes less; thus these steels will not accept impact.

Other manganese steels of a more complex nature will be found in Section 44P.

Symbol	Nominal analysis, supplier, condition and remarks.
1.3401	1.2% C 12.5% Mn steel: German Standard
1.3402	1.1% C 14.0% Mn 0.4% Si steel: German Standard
1.3405	0.9% C 17.5% Mn 0.8% Si steel: German Standard
1.3802	1.2% C 12.5% Mn steel: German Standard
13 Mn R	0.9% C 13.0% Mn steel: Electrode; Metrode; for hard facing; work hardens from 200 to 500 DPN
17 Mn Mo R	0.9% C 17.0% Mn 1.0% Mo steel: Electrode; Metrode; for hard facing; work hardens from 200 to 500 DPN
ABRADUR M14	1.2% C 13% Mn 0.35% Si steel: Pompey; wear resistant; annealed **DPN: 200 UTS: 1070 Elon: 55% Proof: 360**
ABRADUR M14N	1.2% C 13.5% Mn 2.7% Ni 0.7% Si steel: Pompey; wear resistant
AM	1.25% Mn austenitic steel: Casting; Wokingham **DPN: 229**
ASTM A128 A	1.2% C 11.0% Mn steel: Casting
ASTM A128 B1	1.0% C 12.5% Mn steel: Casting
ASTM A128 B2	1.1% C 12.5% Mn steel: Casting
ASTM A128 B3	1.2% C 12.5% Mn steel: Casting
ASTM A128 B4	1.28% C 12.5% Mn steel: Casting
ASTM A128 C	1.2% C 12.5% Mn 2.0% Cr steel: Casting
ASTM A128 D	1.0% C 12.5% Mn 1.0% Mo steel: Casting
ASTM A128 E1	1.0% C 12.5% Mn 1.0% Mo steel: Casting
ASTM A128 E2	1.2% C 12.5% Mn 2.0% Mo steel: Casting
AURIGA X	1.2% C 12% Mn steel: Casting; D Brown; austenitic; cold working
AURIGA XI	0.8% C 12% Mn 3.5% Ni steel: Casting; D Brown; normalized; austenitic; cold working
AURIGA XII	1.2% C 12% Mn 1.7% Cr steel: Casting; D Brown; austenitic; cold working
BS 1457	1.2% C 11.0% Mn steel: Casting; annealed; contained in BS 3100
CODE 630	1.3% C 14% Mn 1.6% Cr steel: Casting; Edgar Allen; 'Duraloy'
CODE 631	1.1% C 12% Mn steel: Casting; Edgar Allen; 'Ord Mn 5'
CODE 633	1.5% C 17.5% Mn steel: Casting; Edgar Allen; 'Mang 18'
CODE 634	1% C 12% Mn steel: Casting; Edgar Allen; 'Tumblers'
CRUCIBLE 13% Mn	1.2% C 13.0% Mn 0.2% Si steel: Annealed; Crucible Steel **DPN: 180 UTS: 940 Elon: 55% Proof: 360**
Cr-X 120 Mn 12	1.2% C 12.5% Mn steel: German Standard
DURALOY	1.3% C 14% Mn 1.6% Cr steel: Casting; Edgar Allen; Code 630
EPOK	13% Mn steel: Casting; annealed; Firth Brown **DPN: 220**
HARDTRODE 86.08	1.10% C 13.0% Mn steel: For electrodes; Esab
IMPERIAL MANGANESE	1.15% C 12.5% Mn steel: Chisels; Edgar Allen
JS 14	1.2% C 11.0% Mn steel: Casting; Jopling; for wear resistance **DPN: 230**
MANG 18	1.5% C 17.5% Mn steel: Casting; Edgar Allen; Code 633
NN 3	1.1% C 13% Mn steel: Bofors
ORD Mn 5	1.1% C 12% Mn steel: Casting; Edgar Allen; Code 631
RED DIAMOND	0.45% C 1.75% Mn steel: Plate for wear; Redheugh **DPN: 270 UTS: 940 Elon: 15% Proof: 550**
SS 2183	1.2% C 12.5% Mn steel: Casting; Swedish Standard
TITAN	1.2% C 13% Mn steel: Chisel steel; Osborn **DPN: 210 UTS: 940 Elon: 50% Proof: 340**
TITAN MANGANESE	1.2% C 13.0% Mn steel: Cold work; Osborn
TRM 45	0.4% C 15% Mn 5% Ni steel: Colt Ind.; hot rolled **DPN: 171 UTS: 830 Elon: 70% Proof: 260**
TUBROD 15.60	1.0% C 13% Mn steel: Tubular weld rod; Esab; for wear resistance; cold worked **DPN: 520**
TUMBLERS	1% C 12% Mn steel: Casting; Edgar Allen; Code 634
VS MANG	1.15% C 12% Mn 0.3% Si steel: Cold working steel; Vulcan **DPN: 200**
X 90 Mn 18	0.9% C 17.5% Mn 0.8% Si steel: Designation used by German Standards
X 110 Mn 14	1.1% C 14.0% Mn 0.45% Si steel: Designation used by German Standards
X 120 Mn 12	1.2% C 12.5% Mn steel: Designation used by German Standards

Note. The following abbreviations and units are used in the tables:

DPN	Hardness, diamond pyramid number
UTS	Ultimate tensile strength, N/mm^2
Elon	Elongation, %
Proof	0.1% proof strength, N/mm^2

$1\ N/mm^2 = 0.1\ hbar = 0.102\ kgf/mm^2 = 0.06475\ tonf/in.^2 = 145.04\ lbf/in.^2 = 1\ MPa$

See Appendix II for other abbreviations and conversion tables.

44E1 Steel – low carbon with chromium 0.1–0.35% carbon, up to 9% chromium

Specific gravity	7.84
Density	7840 kg/m^3
Solidus/liquidus	1470–1510 °C
Thermal conductivity	50.2 W/m °C
Coefficient of linear expansion	11×10^{-6}/ °C
Electrical conductivity	8–10% IACS (copper 100%)
Specific resistance	180–250 microhm mm
Young's modulus of elasticity	–
Impact	40–60 J (can be much lower)
Fatigue strength (no. of cycles)	–
Hot strength	

Temperature °C	Tensile strength N/mm^2	Elongation %
200	480	38
425	400	36
480	340	38
625	210	55

The above properties are typical of the following group, and may not apply exactly to any one specification. It is possible that with certain specifications some of the values may not be applicable.

General metallurgical characteristics

Chromium forms stable carbides with available carbon, the carbon combining with chromium in preference to iron. These carbides are harder than iron carbide and more sluggish in their metallurgical reactions; thus a longer time at temperature is necessary to allow solution to take place and the quenching rate can be slower.

As they are inherently harder the unhardened steel is harder than an equivalent plain carbon steel. Chromium in fact is the most efficient of the common hardening elements.

Chromium also imparts corrosion resistance as any chromium at the steel surface becomes oxidized and this improves the resistance to rusting. Unlike rust it is a stable very adherent oxide which inhibits further oxidation. Corrosion resistance is proportional to the chromium content but none of the steels listed can be classed as stainless or corrosion resistant although they do not rust as readily as plain carbon steel.

Being a carbide former, chromium aids carburization and some of the lower carbon steels listed were developed for carburizing steels. The case formed tends to be brittle, and as the core is relatively soft these steels should not be used for high duty purposes.

Many chrome steels become brittle and notch sensitive at sub-zero temperatures.

Some of the steels have small amounts of copper additions. These improve the corrosion resistance and reportedly help the adhesion of paint.

Larger quantities of chromium are present in all stainless steels which are described in Sections 44M and 44N.

Most steels in this group will cold work more readily than plain carbon steels and thus will require inter-stage annealing on many occasions. This will be at a temperature in the region of 850 °C.

Chromium raises the critical temperatures for thermal treatment and increases the hardenability by quite a marked degree. Chromium steels are susceptible to the problem of 'temper brittleness' and should never be tempered or heated at a temperature between approximately 300 and 600 °C. Within this temperature range carbide precipitation takes place at the grain boundary with resultant brittleness.

These steels should not be welded without particular care and advice is essential. The steels have an appreciable amount of hardenability and in addition to this there is the problem of temper brittleness.

The carbon equivalent (CE) should always be identified. The formula is

$$CE = C + \frac{Mn}{6} + \frac{Cr + Mo + V}{5} + \frac{Ni + Cu}{15}$$

A figure above 0.7% indicates that problems may exist.

The lower chromium steels are used for pressure vessels, structural steels, and low duty engineering purposes. In general, however, chromium carbon steels have been replaced by alloy steels where nickel and molybdenum are present along with the chromium.

The higher chromium steels are often used for plastic moulds and dies and for some fixtures where they are used under controlled conditions and no impact is involved.

Symbol	Nominal analysis, supplier, condition and remarks.
0.7033	0.33% C 1.0% Cr steel: German Standard
1.0345	0.16% C (max) 0.3% Cr steel: German Standard
1.0408	0.25% C (max) 0.3% Cr steel: German Standard
1.0425	0.2% C (max) 0.3% Cr steel: German Standard
1.0435	0.22% C (max) 0.3% Cr steel: German Standard
1.0445	0.26% C (max) 0.3% Cr steel: German Standard
1.0501	0.36% C 0.5% Cr steel: German Standard
1.0844	0.17% C 1.0% Mn 0.3% Cr steel: For boiler plate; German Standard
1.0845	0.2% C 1.2% Mn 0.3% Cr steel: Boiler plate; German Standard
1.0916	0.17% C 1.0% Mn 0.3% Cr steel: For boiler plate; German Standard
1.0935	0.2% C 1.2% Mn 0.3% Cr steel: Boiler plate; German Standard
1.2162	0.21% C 1.2% Cr tool steel: German Standards designation
1.4712	0.12% C (max) 6.0% Cr 2.2% Si steel: Heat resistant; German Standard
1.4713	0.12% C (max) 6.5% Cr 0.8% Al steel: Heat resistant; German Standard
1.4716	0.1% C (max) 9% Cr steel: Heat resistant; German Standard
1.4903	0.1% C (max) 8% Cr 5.5% Al steel: High thermal conductivity; German Standard
1.5053	0.2% C 1.2% Mn 0.3% Cr steel: Forging; German Standard
1.7015	0.15% C 0.7% Cr steel: For carburizing; German Standard
1.7033	0.35% C 1.1% Cr steel: Cold drawn for bolts; German Standard
1.7034	0.38% C 1.1% Cr steel: German Standard
1.7071	0.18% C 2.4% Cr steel: For high pressure; German Standard
1.7083	0.1% C 2.8% Cr steel: For high pressure; German Standard
1.7147	0.2% C 1.3% Mn 1.2% Cr steel: For carburizing; German Standard
1.8401	0.3% C 1.1% Mn 0.9% Cr 0.25% Ti 0.1% Al steel: Welding rod; German Standard
1.8504	0.33% C 1.4% Cr 1.0% Al steel: For nitriding; German Standard
1.8506	0.33% C 1.2% Cr 1.0% Al steel: For nitriding; German Standard
3 LC 2	0.15% C 0.5% Mn 0.65% Cr steel: Sandvik
3 MC 2	0.15% C 1.2% Mn 1% Cr steel: Sandvik
6 LMC	0.3% C 1.4% Mn 0.035% Cr steel: Sandvik
10 Cr 11	0.1% C 0.4% Mn 2.9% Cr steel: Designation used by DIN
16 Cr 9	0.16% C 0.4% Mn 2.4% Cr steel: Designation used by DIN
16 MC 5	0.16% C 0.95% Cr steel: For carburizing; AFNOR
17 Mn Cr 5	0.16% C 1% Cr steel: For carburizing; German Standard
19 Mn 5	0.2% C 1.2% Mn 0.3% Cr steel: Designation used by DIN
20 Mn 5	0.2% C 1.1% Mn 0.3% Cr steel: Designation used by DIN

Note. The following abbreviations and units are used in the tables:

DPN	Hardness, diamond pyramid number
UTS	Ultimate tensile strength, N/mm^2
Elon	Elongation, %
Proof	0.1% proof strength, N/mm^2

1 N/mm^2=0.1 hbar=0.102 kgf/mm^2=0.06475 $tonf/in.^2$=145.04 $lbf/in.^2$=1 MPa

See Appendix II for other abbreviations and conversion tables.

Symbol	Nominal analysis, supplier, condition and remarks.
21 Mn Cr 5	0.2% C 1.2% Cr steel: For tools; German Standards designation
30 Mn Cr Ti 4	0.3% C 1.1% Mn 0.9% Cr 0.25% Ti steel: Designation used by DIN
34 Cr Al 6	0.33% C 0.8% Mn 1.4% Cr 1.0% Al steel: Designation used by DIN
523 A14	0.14% C 0.45% Cr steel: BS 970; obsolete
523 M15	0.15% C 0.45% Cr steel: For case hardening; BS 970
527 A17	0.17% C 0.85% Cr steel: For case hardening; BS 970
527 M17	0.17% C 0.85% Cr steel: For case hardening; BS 970
527 M20	0.2% C 0.75% Cr steel: BS 970; obsolete
530 A30	0.3% C 1% Cr steel: BS 970
530 A32	0.32% C 1% Cr steel: BS 970
530 A36	0.36% C 1% Cr steel: BS 970
530 A40	0.4% C 1% Cr steel: BS 970
530 H30	0.3% C 1% Cr steel: BS 970; obsolete
530 H32	0.32% C 1% Cr steel: BS 970
530 H36	0.36% C 1% Cr steel: BS 970
530 H40	0.4% C 1% Cr steel: BS 970
590 A15	0.15% C 1.05% Cr steel: For case hardening; BS 970
590 H17	0.17% C 1% Cr steel: BS 970
590 M17	0.17% C 0.95% Cr steel: For case hardening; BS 970
605 M36	0.36% C 0.27% Cr steel: BS 970
608 M38	0.38% C 0.47% Mo steel: BS 970; obsolete
0708 Si Al	0.07% C 6.5% Cr 0.8% Si 0.8% Al steel: NYBY; annealed **DPN: 185 UTS: 280 Elon: 20% Proof: 160**
5015	0.15% C 0.4% Cr steel: Designation used in the UK and USA
5115	0.15% C 0.8% Mn 0.8% Cr steel: Designation used by AISI
5115	0.15% C 0.8% Cr steel: Designation used in the UK and USA
5117	0.17% C 0.8% Cr steel: Designation used in the UK and USA
5120	0.2% C 0.8% Cr steel: Designation used in the UK and USA
5130	0.3% C 1% Cr steel: Designation used in the UK and USA
5132	0.32% C 0.9% Cr steel: Designation used in the UK and USA
5135	0.35% C 1% Cr steel: Designation used in the UK and USA
A 202	0.2% C 1.2% Mn 0.45% Cr steel: Plate; Armco; for pressure tanks **UTS: 630 Elon: 20% Proof: 300**
ACM	Cr steel: For magnets; Holzer; further information from Permanent Magnet Association
AISI 5015	0.15% C 0.4% Cr steel
AISI 5117	0.17% C 0.8% Cr steel: Obsolete
AISI 5120	0.20% C 0.8% Cr steel
AISI 5130	0.3% C 0.8% Mn 0.9% Cr steel
AISI 5132	0.32% C 0.7% Mn 0.9% Cr steel
AISI 5135	0.35% C 0.7% Mn 0.9% Cr steel
APS 10	0.12% C 2.0% Cr 0.9% Al steel: Pompey **UTS: 460 Elon: 20% Proof: 300**
APS 10C	0.12% C 2.5% Cr 0.4% Al steel: Pompey **UTS: 460 Elon: 20% Proof: 300**
APS 20	0.12% C 4.0% Cr 1.0% Al steel: Pompey **UTS: 460 Elon: 20% Proof: 300**
APS 20C	0.12% C 4.0% Cr 0.4% Al steel: Pompey **UTS: 460 Elon: 20% Proof: 300**
APS 30	0.1% C 6.5% Cr 1.3% Al steel: Pompey **UTS: 460 Elon: 20% Proof: 300**
ASN 151	0.22% C 1.5% Mn 0.5% Cr 0.5% Cu steel: Australian Iron & Steel Co.; normalized
ASTM A202 A	0.17% C (max) 1.2% Mn 0.45% Cr 0.8% Si steel: For pressure vessels
ASTM A202 B	0.25% C (max) 1.2% Mn 0.5% Cr 0.8% Si steel: For pressure vessels

Symbol	Nominal analysis, supplier, condition and remarks.
ASTM A588 Cr	0.2% C (max) 1.2% Mn (max) 0.7% Cr 0.4% Cu steel: For structures **UTS: 440 Elon: 21% Proof: 320**
ASTM A588 D	0.15% C 0.7% Cr 0.1% Zr steel: For structures **UTS: 440 Elon: 21% Proof: 320**
ATLANTES HT	0.2% C 0.9% Mn 0.65% Cr 0.5% Cu steel: South Durham
BS 681	Carbon chromium steel: Replaced by BS 970
BS 968	0.2% C 1.5% Mn 0.5% Cr steel: Normalized **UTS: 530 Elon: 23% Proof: 340**
BS 970	See designation for details
BS 1956 A	0.5% C 0.7% Mn 1.0% Cr steel: Casting; abrasion resisting; normalized or hardened and tempered; contained in BS 3100 **DPN: 207 UTS: 690 Elon: 10%**
BS 1956 B	0.5% C 0.7% Mn 1.0% Cr steel: Casting; abrasion resisting; normalized or hardened and tempered; contained in BS 3100 **DPN: 293**
BS 3551/A	0.35% C 1.5% Mn 0.25% Cr steel: For shackles; hardened and tempered **DPN: 275**
BS S19	Chromium steel: Forgings for valves; replaced
BS S115	0.4% C 1.0% Cr steel: Wire for forged bolts; hardened and tempered after forging **UTS: 910 Elon: 18% Proof: 660**
BS S117	0.4% C 1.0% Cr steel: Bar and forging; hardened and tempered **DPN: 280 UTS: 910 Elon: 18% Proof: 660**
C 1K	0.12% C 0.9% Cr steel: Pompey
C 1MK	0.16% C 1.2% Mn 0.9% Cr steel: Pompey **UTS: 1150 Elon: 37% Proof: 770**
C 2K	0.16% C 0.9% Cr steel: Pompey **UTS: 980 Elon: 8% Proof: 770**
C 2MK	0.2% C 1.3% Cr steel: Pompey
C 35	0.36% C 0.5% Cr steel: German Standard
CA 30	0.34% C 1.35% Cr 0.95% Al steel: For nitriding; Krupp
CA 30 A	0.34% C 1.15% Cr 0.95% Al steel: For nitriding; Krupp
CBV	0.15% C 0.65% Cr 0.25% Si steel: Krupp
CBV 1	0.34% C 1.05% Cr steel: Krupp
CHROMADOR	0.3% C 0.8% Mn 0.8% Cr 0.3% Cu steel: Dorman Long Ltd; as rolled; resists corrosion **DPN: 180 UTS: 660 Elon: 17% Proof: 360**
CMBV/h	0.2% C 1.15% Cr 0.25% Si steel: Krupp
CMBV/w	0.16% C 0.95% Cr 0.25% Si steel: Krupp
CONICRO	0.23% C (max) 0.95% Mn 0.6% Cr 0.4% Cu steel: Consett
COR-TEN A	0.12% C (max) 1% Cr 0.49% Cu steel: Appleby; Frodingham; weldable **UTS: 570 Elon: 22% Proof: 330**
COR-TEN B	0.14% C 0.5% Cr 0.06% V 0.3% Cu steel: Appleby; Frodingham **UTS: 300 Elon: 21% Proof: 210**
D 3K	0.32% C 0.4% Cr steel: Pompey
DIN 1654 34 Cr 4	0.34% C 1.1% Cr steel: For cold forged bolts
DIN 8555 El 400	Mn Cr steel: Welding electrode
DIN 17110 17 Mn 4	0.17% C 1.0% Mn 0.3% Cr steel: For boiler plate
DIN 17155 17 Mn 4	0.17% C 1.0% Mn 0.3% Cr steel: For boiler plate
DIN 17155 19 Mn 4	0.2% C 1.2% Mn 0.3% Cr steel: Boiler plate
DIN 17155 H	0.2% C (max) 0.3% Cr steel: For boiler plate
DIN 17155 H 1	0.16% C (max) 0.4% Mn 0.3% Cr steel: Plate
DIN 17155 H 11a	0.2% C (max) 0.3% Cr steel: For boiler plate
DIN 17155 H 111	0.22% C 0.3% Cr steel: Seamless tube
DIN 17200 15 Cr 3	0.15% C 0.55% Mn 0.55% Cr steel: German Standard
DIN 17200 16 Mn Cr 5	0.16% C 1.15% Mn 0.9% Cr steel: German Standard
DIN 17200 16 Mn Cr S 5	0.16% C 1.15% Mn 0.95% Cr steel: German Standard
DIN 17200 20 Mn Cr 5	0.2% C 1.2% Mn 1.15% Cr steel: German Standard
DIN 17200 20 Mn Cr S 5	0.2% C 1.2% Mn 1.15% Cr steel: German Standard
DIN 17200 34 Cr 4	0.39% C 0.75% Mn 1.05% Cr steel: German Standard
DIN 17200 34 Cr S 4	0.34% C 0.75% Mn 1.05% Cr steel: German Standard
DIN 17210 15 Cr 3	0.15% C 0.6% Cr steel: For carburizing
DIN 17210 16 Mn Cr 5	0.16% C 1.1% Mn 1% Cr steel: For carburizing
DIN 17210 20 Mn Cr 5	0.2% C 1.2% Mn 1.1% Cr steel: For carburizing
DIN 17470 Cr A18/5	0.1% C (max) 8.0% Cr 5.5% Al steel: For conducting heat
DK 3	0.32% C 1.0% Cr steel: Pompey **UTS: 980 Elon: 9% Proof: 820**
DK 3E	0.32% C 1.0% Cr steel: Pompey
DK 35E	0.35% C 1.0% Cr steel: Pompey
DUK TEN	0.15% C (max) 0.5% Mn 3.5% Ni steel: German proprietor's specification
DURAPSO	0.19% C 0.4% Cr 0.4% Cu steel: Pompey
DURAPSO 3	0.2% C 0.4% Cr 0.5% Cu steel: Pompey
E 401	0.2% C 1.2% Cr steel: VEW
E 410	0.17% C 1.2% Mn 0.9% Cr steel: VEW
E 525	0.15% C 0.5% Mn 0.6% Cr steel: VEW
E 4111	0.17% C 1.2% Mn 1.0% Cr steel: VEW
E Fe 1	0.12% C 1.5% Cr steel: Weld electrode; designation used by AWS
E Fe 2	0.2% C 2.8% Cr steel: Weld electrode; designation used by AWS
EC Ni 1	0.12% C (max) 0.9% Ni steel: Weld electrode; designation used by AWS
EC Ni 1N	0.12% C (max) 0.9% Ni steel: Weld electrode; designation used by AWS
En 18 A	0.3% C 1% Cr steel: Designation used in BS 970; replaced by 530 A30
En 18 B	0.32% C 1% Cr steel: Designation used in BS 970; replaced by 530 A32
En 18 C	0.36% C 1% Cr steel: Designation used in BS 970; replaced by 530 A36
En 206	0.15% C 0.4% Cr steel: For carburizing; designation used in BS 970; replaced by 523 A14
En 207	0.18% C 0.7% Cr steel: For carburizing; designation used in BS 970; replaced by 527 A19
EXX T8 KG	0.15% C (max) 0.8% Ni steel: Weld electrode; designation used by AWS
F128	0.6% C 0.9% Mn 0.3% Cr steel: VEW
FDK 3	0.33% C 0.9% Cr steel: Pompey **UTS: 950 Elon: 11% Proof: 740**
GOST 550 KH5	0.15% C (max) 0.5% Mn (max) 5.0% Cr steel: Pipe; Russian Standard **UTS: 400 Proof: 220**
GOST 4543/14 KH Cr	0.14% C 1.1% Mn 0.3% Ni (max) 0.65% Cr 0.3% Cu (max) steel: Russian Standard
GOST 4543/15 KH	0.15% C 0.55% Mn 0.85% Cr steel: Russian Standard
GOST 4543/15 KHA	0.15% C 0.55% Mn 0.85% Cr steel: Russian Standard
GOST 4543/15 KHR	0.15% C 0.55% Mn 0.85% Cr 0.0035% B steel: Russian Standard
GOST 4543/15 KHRA	0.15% C 0.55% Mn 0.85% Cr 0.0035% B: Russian Standard
GOST 4543/18 KHG	0.18% C 1.1% Mn 1.05% Cr steel: Russian Standard
GOST 4543/18 KHGT	0.2% C 0.9% Mn 1.15% Cr 0.09% Ti steel: Russian Standard
GOST 4543/20 KH	0.2% C 0.65% Mn 0.85% Cr steel: Russian Standard
GOST 4543/20 KHG2C	0.22% C 1.7% Mn 0.3% Ni (max) 1.1% Cr 0.3% Cu (max) 0.1% Zr steel: Russian Standard
GOST 4543/20 KHGR	0.2% C 0.9% Mn 0.95% Cr steel: Russian Standard
GOST 4543/20 KHGSA	0.2% C 0.95% Mn 0.95% Cr steel: Russian Standard
GOST 4543/25 KHGSA	0.25% C 0.95% Mn 0.95% Cr steel: Russian Standard

Symbol	Nominal analysis, supplier, condition and remarks.
GOST 4543/30 KH	0.3% C 0.65% Mn 0.95% Cr steel: Russian Standard
GOST 4543/30 KHGS	0.3% C 0.95% Mn 0.95% Cr steel: Russian Standard
GOST 4543/30 KHGT	0.28% C 0.95% Mn 1.15% Cr steel: Russian Standard
GOST 4543/33 KHS	0.33% C 0.45% Mn 1.45% Cr steel: Russian Standard
GOST 4543/35 KH	0.35% C 0.65% Mn 0.95% Cr steel: Russian Standard
GOST 4543/35 KHRA	0.35% C 0.65% Mn 0.95% Cr steel: Russian Standard
GOST 5632 KH5	0.15% C (max) 0.5% Mn (max) 5.25% Cr steel: Plate; Russian Standard; annealed **UTS: 400** **Proof: 170**
GOST 5632 KH5	0.15% C (max) 5.2% Cr steel: Russian Standard
GOST 5632 KH6 S14	0.15% C (max) 6.2% Cr 0.9% Al steel: Russian Standard
GOST 8731/20 KH	0.2% C 0.65% Mn 0.2% Ni (max) 0.85% Cr steel: Pipe; Russian Standard; seamless **UTS: 440**
GOST 8733/15 KH	0.16% C 0.95% Mn 0.85% Cr steel: Pipe; Russian Standard; seamless **UTS: 420**
GOST 8733/20 KH	0.2% C 0.65% Mn 0.2% Ni (max) 0.85% Cr steel: Pipe; Russian Standard; seamless **UTS: 440**
GOST 12132/15 KH	0.15% C 0.55% Mn 0.85% Cr steel: Pipe; Russian Standard; welded **UTS: 420**
GOST 12132/30 KHGSA	0.3% C 0.95% Mn 0.25% Ni (max) 0.95% Cr steel: Pipe; Russian Standard; welded **UTS: 500**
H160	0.12% C (max) 6.8% Cr 0.75% Al steel: VEW; heat resistant
HARDTRODE 83.28	0.10% C 0.750% Mn 3.50% Cr steel: For electrodes; Esab **DPN: 320**
HECLA 207	0.18% C 0.7% Mn 0.7% Cr steel: Case hardening; Hadfields for BS alloy En 207
HECLA 260	0.15% C 0.4% Mn 0.4% Cr steel: Case hardening; Hadfields for BS alloy En 206
HEDEX 25	0.18% C 0.7% Cr steel: Wire for forging; Kiveton Park for BS alloy En 207
HSCR	0.12% C (max) 0.35% Mn 0.8% Cr 0.4% Cu steel: Patent Shop Steel Co.
HT2	0.09% C 0.35% Mn 5% Cr 2% Si steel: Sandvik
HTS	0.2% C (max) 1.5% Mn 0.5% Cr 0.5% Cu steel: Patent Shop Steel Co.
HYPLUS 1	0.2% C (max) 1.5% Mn 0.5% Cr 1.6% Mn + Cr (max) steel: Appleby-Frodingham
IDROTUB 52	0.18% C (max) 1.2% Mn 0.3% Cr steel: Italian proprietor's specification
JIS G4106 S Cr 2	0.3% C 0.7% Mn 1.05% Cr steel: Japanese Standard
JIS G4106 S Cr 3	0.35% C 0.7% Mn 1.05% Cr steel: Japanese Standard
JIS G4106 S Cr 21	0.15% C 0.7% Mn 1.05% Cr steel: Japanese Standard
JIS G4106 S Cr 22	0.22% C 0.7% Mn 1.05% Cr steel: Japanese Standard
JIS G4106 S Mn C 21	0.2% C 1.3% Mn 0.5% Cr steel: Japanese Standard
KF 18	0.3% C 1% Cr steel: Kirkstall for BS alloy En 18
KF 18 A	0.32% C 1% Cr steel: Kirkstall for BS alloy En 18
KF 18 B	0.37% C 1% Cr steel: Kirkstall for BS alloy En 18
KF 30 A	0.18% C 0.7% Cr steel: Case hardening; Kirkstall for BS alloy En 207
LS 55	0.12% C 6% Cr 0.55% Mo: Low Moor; obsolete **DPN: 220** **UTS: 480** **Elon: 23%** **Proof: 400**
M100	0.2% C 1.2% Mn 1.1% Cr steel: VEW
MAFERITE 6	C not quoted 6% Cr steel: Pompey **DPN: 200**
MAFERITE 8	C not quoted 8% Cr steel: Pompey **DPN: 200**
MANOIR APS10	0.12% C 2.0% Cr 0.65% Al steel: Pompey; normalized **UTS: 550** **Elon: 14%** **Proof: 340**
MANOIR APS20	0.12% C 4.0% Cr 0.65% Al steel: Pompey; normalized **UTS: 550** **Elon: 14%** **Proof: 340**
MANOIR APS25	0.12% C 0.9% Ni 4.0% Cr 0.9% Al steel: Pompey; normalized **UTS: 870** **Elon: 8%** **Proof: 670**
MANOIR APS30	0.12% C 7.0% Cr 1.0% Si 1.5% Al steel: Pompey; normalized **UTS: 600** **Elon: 10%** **Proof: 300**
MIL S12505.4	0.14% C (max) 0.7% Mn 0.5% Cr 0.1% Zr steel: Bethlehem
Mn Cr	0.17% C 0.3% Cr steel: Pompey
NF A35 551/16MC5	0.16% C 1.15% Mn 0.95% Cr steel: French Standard
NF A35 551/20MCS	0.2% C 1.2% Mn 1.15% Cr steel: French Standard
NF A35 551/32C	0.32% C 0.75% Mn 1.0% Cr steel: French Standard
P 5	0.1% C 2.25% Cr steel: Designation used by AISI
PLMA/2	0.15% C 0.5% Mn 1% Cr steel: For case hardening; ESC for plastic moulds
PM 12	0.18% C 1.2% Mn 0.15% Cr steel: Pompey
PS 58	0.18% C 0.55% Cr steel: SAE; potential specification
PS 59	0.2% C 0.8% Cr steel: SAE; potential specification
PS 61	0.25% C 0.8% Cr steel: SAE; potential specification
PS 63	0.34% C 0.55% Cr 0.0004% B steel: SAE; potential specification
PS 64	0.18% C 0.8% Cr steel: SAE; potential specification
PS 65	0.23% C 0.8% Cr steel: SAE; potential specification
SAE 5015	0.15% C 0.4% Cr steel
SAE 5115	0.15% C 0.8% Mn 0.8% Cr steel: Mechanical properties not specified
SAE 5117	0.17% C 0.8% Mn 0.8% Cr steel: Mechanical properties not specified
SAE 5120	0.2% C 0.8% Mn 0.8% Cr steel: Mechanical properties not specified
SAE 5130	0.3% C 0.8% Mn 0.9% Cr steel: Mechanical properties not specified
SAE 5132	0.32% C 0.7% Mn 0.9% Cr steel: Mechanical properties not specified
SAE 5135	0.35% C 0.7% Mn 0.9% Cr steel: Mechanical properties not specified
SAMSON EXTRA	0.1% C 2.3% Cr steel: Carpenter; for AISI type P5
For SIS specifications see SS	
SKF 234	0.16% C 0.95% Cr steel: For carburizing; SKF
SOUDER GA 250	0.3% C 1.0% Cr 0.2% Ti steel: Welding electrode; Soudometal; for hard surfacing **DPN: 245**
SOUDODER 400	0.25% C 2.7% Cr steel: Welding electrode; Soudometal; for build up of rollers, etc. **DPN: 400**
SOUDOKAY 242.0	0.14% C 2.5% Cr steel: Welding electrode; Soudometal **DPN: 400**
SOUDOKAY 252.0	0.2% C 1.8% Mn 3.5% Cr steel: Welding electrode; Soudometal **DPN: 470**
SS 1225	0.08% C 0.1% Cr 0.2% Cu 0.009% N steel: For chains; Swedish Standard; normalized **UTS: 340** **Elon: 30%**
SS 1232	0.13% C 0.5% Mn 0.25% Cr 0.3% Cu steel: Swedish Standard; for pressure vessels **UTS: 370** **Elon: 25%** **Proof: 200**

Note. The following abbreviations and units are used in the tables:

DPN	Hardness, diamond pyramid number
UTS	Ultimate tensile strength, N/mm^2
Elon	Elongation, %
Proof	0.1% proof strength, N/mm^2

1 N/mm^2=0.1 hbar=0.102 kgf/mm^2=0.06475 tonf/in.^{2}=145.04 ibf/in.^{2}=1 MPa

See Appendix II for other abbreviations and conversion tables.

Symbol	Nominal analysis, supplier, condition and remarks.
SS 1232E	0.13% C 0.5% Mn 0.25% Cr 0.3% Cu steel: Swedish Standard; for pressure vessels **UTS: 390 Elon: 18% Proof: 290**
SS 1233	0.17% C 0.2% Cr 0.3% Cu 0.009% N steel: Bar; Swedish Standard; as rolled **UTS: 370 Elon: 25%**
SS 1234	0.17% C 0.2% Cr 0.3% Cu 0.009% N steel: Bar; Swedish Standard; as rolled **UTS: 370 Elon: 25%**
SS 1306	0.18% C (max) 1% Mn 0.3% Cr 0.3% Cu steel: Swedish Standard; annealed **UTS: 390 Elon: 25% Proof: 210**
SS 1330	0.12% C 0.5% Mn 0.25% Cr 0.4% Cu 0.009% N steel: For pressure vessels; Swedish Standard; normalized **UTS: 390 Elon: 26% Proof: 220**
SS 1330E	0.15% C 0.6% Mn 0.25% Cr 0.3% Cu steel: Swedish Standard; for pressure vessels **UTS: 420 Elon: 26% Proof: 210**
SS 1331	0.11% C 0.7% Mn 0.25% Cr 0.4% Cu 0.009% N steel: For pressure vessels; Swedish Standard; normalized **UTS: 390 Elon: 26% Proof: 220**
SS 1332	0.1% C 0.7% Mn 0.25% Cr 0.4% Cu 0.009% N steel: For pressure vessels; Swedish Standard; normalized **UTS: 390 Elon: 26% Proof: 220**
SS 1411	0.2% C 0.9% Mn 0.3% Cr 0.4% Cu structural steel: Swedish Standard; as rolled **UTS: 450 Elon: 20% Proof: 240**
SS 1430	0.15% C 0.9% Mn 0.25% Cr 0.4% Cu 0.009% N steel: For pressure vessels; Swedish Standard; normalized **UTS: 450 Elon: 24% Proof: 240**
SS 1430E	0.18% C 0.7% Mn 0.25% Cr 0.30% Cu (max) steel: Swedish Standard; for pressure vessels **UTS: 470 Elon: 24% Proof: 260**
SS 1431	0.12% C 1% Mn 0.25% Cr 0.4% Cu 0.009% N steel: For pressure vessels; Swedish Standard; normalized **UTS: 450 Elon: 24% Proof: 240**
SS 1432	0.12% C 1% Mn 0.25% Cr 0.4% Cu 0.009% N steel: For pressure vessels; Swedish Standard; normalized **UTS: 450 Elon: 24% Proof: 240**
SS 1432E	0.14% C 1.1% Mn 0.25% Cr (max) 0.3% Cu steel: Swedish Standard; for pressure vessels **UTS: 470 Elon: 24% Proof: 250**
SS 1434	0.22% C (max) 0.2% Cr 0.3% Cu 0.009% N steel: For pressure vessels; Swedish Standard; hot rolled **DPN: 150 UTS: 450 Elon: 26% Proof: 240**
SS 1435	0.22% C (max) 0.2% Cr 0.3% Cu 0.009% N steel: For pressure vessels; Swedish Standard; hot rolled **UTS: 450 Elon: 21% Proof: 240**
SS 1435E	0.22% C 0.8% Mn 0.25% Cr (max) 0.30% Cu (max) steel: Swedish Standard; for pressure vessels **UTS: 500 Elon: 21% Proof: 260**
SS 2101	0.18% C 1% Mn 0.25% Cr 0.4% Cu 0.009% N steel: For pressure vessels; Swedish Standard; normalized **UTS: 510 Elon: 21% Proof: 320**
SS 2101E	0.2% C 1.2% Mn 0.25% Cr (max) 0.3% Cu (max) steel: Swedish Standard; for pressure vessels **UTS: 550 Elon: 21% Proof: 300**
SS 2102	0.14% C 1.6% Mn 0.25% Cr 0.4% Cu 0.009% N steel: For pressure vessels; Swedish Standard; normalized **UTS: 510 Elon: 18% Proof: 320**
SS 2103	0.14% C 1.6% Mn 0.25% Cr 0.4% Cu 0.009% N steel: For pressure vessels; Swedish Standard; normalized **UTS: 510 Elon: 18% Proof: 320**
SS 2103E	0.15% C 1.3% Mn 0.25% Cr 0.3% Cu steel: Swedish Standard; for pressure vessels; normalized **UTS: 550 Elon: 21% Proof: 300**
SS 2108	0.2% C 1.4% Mn 0.1% Cr 0.2% Cu 0.009% N steel: For chains; Swedish Standard; hardened and tempered **DPN: 180 UTS: 660 Elon: 18%**
SS 2127	0.16% C 1% Cr steel: For carburizing; Swedish Standard
SS 2130	0.27% C 0.2% Cr 0.002% B steel: Swedish Standard
SS 2131	0.27% C 0.4% Cr 0.002% B steel: Swedish Standard
SS 2172	0.2% C 1.3% Mn 0.3% Cr 0.4% Cu 0.009% N structural steel: Swedish Standard; as rolled **UTS: 530 Elon: 21% Proof: 320**
SS 2173	0.14% C 1.2% Mn 0.3% Cr 0.4% Cu 0.009% N structural steel: Swedish Standard; normalized **UTS: 530 Elon: 18% Proof: 320**
SS 2174	0.14% C 1.2% Mn 0.2% Cr 0.3% Cu 0.009% N structural steel: Swedish Standard; normalized **UTS: 530 Elon: 18% Proof: 320**
STUBS 710	C Cr steel: Welding electrode; Stubs; for hard facing **DPN: 650**
STUBS 711	C Cr Si steel: Welding electrode; Stubs; for hard facing **DPN: 600**
T 91	0.1% C 8.7% Cr 1% Mo 0.08% Nb steel: US designation
THIRTY OAK	0.22% C 1.0% Mn 0.5% Cr 0.2% Si 1.6% Mn + Cr (max) steel: Round Oak Steel; weldable; may have 0.5% Cu **UTS: 630 Elon: 17% Proof: 450**
THIRTY OAK 1	0.22% C (max) 1.5% Mn 0.5% Cr 0.08% Nb steel: Round Oak Steel
THIRTY OAK 2	0.22% C (max) 1.5% Mn 0.5% Cr 0.08% Nb steel: Round Oak Steel
TUBROD 15.40	0.09% C 2.0% Mn 3.8% Cr steel: Tubular welding rod; Esab; for use with submerged arc; properties vary with flux chased; as welded **DPN: 380**
WLM 4060	0.3% C 1% Cr steel: 600–900 N/m^2 tensile; W Marrison; analysis varies over wide range
X 7 AL	0.12% C (max) 7% Cr 0.75% Al steel: Valbruna
X 8 Cr 9	0.1% C (max) 1.5% Mn 9.0% Cr 1.5% Si steel: Designation used by German Standards
X 10 Cr Al 7	0.12% C (max) 6.5% Cr 0.7% Al steel: Designation used by German Standards
X 10 Cr Si 6	0.12% C (max) 6.0% Cr 2.2% Si steel: Designation used by German Standards

Note. The following abbreviations and units are used in the tables:

DPN	Hardness, diamond pyramid number
UTS	Ultimate tensile strength, N/mm^2
Elon	Elongation, %
Proof	0.1% proof strength, N/mm^2

1 N/mm^2=0.1 hbar=0.102 kgf/mm^2=0.06475 tonf/in.^{2}=145.04 lbf/in.^{2}=1 MPa

See Appendix II for other abbreviations and conversion tables.

44E2 Steel – medium carbon with chromium 0.4–0.65% carbon, up to 9% chromium with silicon

Specific gravity	7.84
Density	7840 kg/m^3
Solidus/liquidus	1400–1470 °C
Thermal conductivity	50 W/m °C
Coefficient of linear expansion	11 × 10^{-6}/ °C
Electrical conductivity	8–10% IACS (copper 100%)
Specific resistance	180–230 microhm mm
Young's modulus of elasticity	–
Impact	41–54 J (can be much lower)
Fatigue strength (no. of cycles)	–
Hot strength	–

The above properties are typical of the following group, and may not apply exactly to any one specification. It is possible that with certain specifications some of the values may not be applicable.

General metallurgical characteristics

Chromium has a twin effect on steel, the principal effect, with the steels listed, being to act as a carbide former. This improves the hardenability, thus allowing deeper hardening with less drastic quenching media. The carbides slow down the metallurgical reactions and retain their hardness at higher temperatures, the steels also being harder even in the annealed condition as the chromium carbides are harder than iron carbide.

The second effect of chromium on steel is to improve the corrosion resistance. This is achieved by the formation of an adherent stable oxide on the surface. There is never sufficient chromium present to make the steels listed stainless or even corrosion resistant, but they have better resistance than plain carbon steels, in proportion to the chromium present.

These steels are capable of a very high hardness, but like many plain carbon single alloy steels with a carbide former they are notch sensitive with a tendency to brittleness, particularly at low temperatures.

Some of the listed steels have up to 3.0% silicon additions which improve the fatigue and corrosion resistance, thus giving economical spring steels.

The steels become brittle when subjected to comparatively little cold work.

Steels in this group require considerable skill and care in any thermal treatments. Because of the carbon and chromium content, the steels have considerable hardenability and very little ductility. Cracking can readily occur during both the heating and cooling cycles. Chromium raises the temperature for critical thermal treatment and thus there is the considerable danger of grain growth which increases the brittleness; thus quench cracking can occur.

The defect known as 'temper brittleness' exists with these steels and they must not be tempered or heated for any appreciable time between 300 and 600 °C. This results in precipitation of chromium carbides at grain boundaries with resultant brittleness. They should therefore be tempered above 600 °C and quenched after tempering, or tempered below 300 °C.

The steels must not be welded without obtaining competent technical advice. The heat affected zone will be prone to cracking and in addition there is the danger of temper brittleness in the heat affected zone and the adjacent material.

These materials are used where abrasion resistance is required, but brittleness can be accepted. Thus cutting tools, dies for plastic and metal forming, shear blades, knives and needles are made from these steels. All these uses can be fulfilled with medium carbon alloy steels where nickel and molybdenum are present. These increase ductility and eliminate temper brittleness.

Symbol	Nominal analysis, supplier, condition and remarks.
0.7035	0.41% C 1.0% Cr steel: German Standard
0.7103	0.67% C 0.5% Cr 1.3% Si steel: For springs; German Standard
1.0503	0.46% C 0.5% Cr steel: German Standard
1.0958	0.6% C 0.3% Cr 1.7% Si spring steel: German Standard
1.0960	0.6% C 0.3% Cr 1.7% Si spring steel: German Standard
1.0961	0.6% C 0.85% Mn 0.3% Cr steel: For springs; German Standard
1.2101	0.62% C 0.6% Cr steel: German Standard
1.2721	0.5% C 1.1% Cr tool steel: German Standards designation
1.4704	0.45% C 2.7% Cr 4.0% Si steel: For valves; German Standard
1.4718	0.45% C 9% Cr 3.0% Si steel: For valves; German Standard
1.7035	0.41% C 1.1% Cr steel: For cold forged bolts; German Standard
1.8403	0.5% C 1.1% Mn 0.9% Cr 0.2% Ti 0.1% Al steel: Welding rod; German Standard
1.8404	0.6% C 0.9% Mn 0.9% Cr 0.2% Ti 0.1% Al steel: Welding rod; German Standard
4KHS	0.4% C 1.45% Cr steel: Russian Standards designation
6KHS	0.65% C 1.15% Cr steel: Russian Standards designation
7KH3	0.67% C 3.5% Cr steel: Russian Standards designation

Note. The following abbreviations and units are used in the tables:

DPN	Hardness, diamond pyramid number
UTS	Ultimate tensile strength, N/mm^2
Elon	Elongation, %
Proof	0.1% proof strength, N/mm^2

1 N/mm^2=0.1 hbar=0.102 kgf/mm^2=0.06475 tonf/in.^{2}=145.04 lbf/in.^{2}=1 MPa

See Appendix II for other abbreviations and conversion tables.

Symbol	Nominal analysis, supplier, condition and remarks.
5 LC	0.5% C 0.12% Cr 1.7% Si steel: Pompey
6 LC	0.55% C 0.25% Cr 1.8% Si steel: Pompey
6 LCK	0.6% C 0.35% Cr 1.85% Si steel: Pompey
12 C	0.65% C 0.15% Cr steel: Sandvik
34 Cr Al S5	0.33% C 1.2% Cr 1.0% Al steel: Designation used by German Standards
35 Cr Mn 5	0.35% C 0.95% Mn 1.15% Cr steel: Italian Standards designation; as rolled **DPN: 230**
37 Cr 4	0.38% C 1.1% Cr steel: Designation used by German Standards
40 Cr 4	0.4% C 0.65% Mn 1.05% Cr steel: Italian Standards designation; as rolled **DPN: 230**
41 Cr 4	0.41% C 1.1% Cr steel: Designation used by German Standards
50 B40	0.4% C 0.5% Cr 0.004% B steel: Designation used in the UK and USA
50 B44	0.44% C 0.5% Cr 0.004% B steel: Designation used in the UK and USA
50 B46	0.46% C 0.25% Cr 0.004% B steel: Designation used in the UK and USA
50 B50	0.5% C 0.5% Cr 0.004% B steel: Designation used in the UK and USA
50 B60	0.6% C 0.5% Cr 0.004% B steel: Designation used in the UK and USA
50 Mn Cr Ti 4	0.5% C 1.1% Mn 0.9% Cr 0.25% Ti steel: Designation used by German Standards
50 Ni Cr 13	0.5% C 1.1% Cr steel: For tools; German Standards designation
51 B60	0.6% C 0.8% Cr 0.0005% B (min) steel: Designation used in the UK and USA
54 Si Cr 6	0.5% C 0.6% Cr 1.5% Si steel: German Standard
60 Mn Cr Ti 4	0.6% C 0.9% Cr 0.25% Ti steel: Designation used by German Standards
60 SC7	0.55% C 0.65% Cr steel: AFNOR
60 Si Cr 7	0.6% C 0.3% Cr 1.7% Si spring steel: German Standard
60 Si Cr 8	0.6% C 0.85% Mn 0.32% Cr steel: Italian Standards designation **DPN: 250 UTS: 1600 Elon: 5% Proof: 1350**
65 Si Cr 7	0.6% C 0.3% Cr 1.7% Si spring steel; German Standard
67 Si Cr 5	0.67% C 0.5% Cr 1.3% Si steel: Designation used by German Standards
526 M60	0.6% C 0.65% Cr steel: BS 970; obsolete
527 A60	0.6% C 0.8% Cr steel: BS 970; obsolete
530 H32	0.32% C 1% Cr steel: BS 970
530 H36	0.36% C 1% Cr steel: BS 970
530 H40	0.4% C 1% Cr steel: BS 970
530 M40	0.4% C 1% Cr steel: BS 970
606 M36	0.36% C 0.27% Cr steel: BS 970
685 A57	0.57% C 0.75% Cr steel: For springs; BS 970
685 H57	0.57% C 0.75% Cr steel: For springs; BS 970
1234	0.75% C 0.25% Mn 0.32% Cr steel: French Standards designation
2323	0.6% C 0.6% Cr steel: French Standards designation
5060	0.6% C 0.5% Cr steel: Designation used in the UK and USA
5140	0.4% C 0.8% Cr steel: Designation used in the UK and USA
5145	0.45% C 0.8% Cr steel: Designation used in the UK and USA
5147	0.49% C 1% Cr steel: Designation used in the UK and USA
5150	0.5% C 0.8% Cr steel: Designation used in the UK and USA
5155	0.55% C 0.8% Cr steel: Designation used in the UK and USA
5160	0.6% C 0.8% Cr steel: Designation used in the UK and USA
9254	0.55% C 0.7% Cr 1.5% Si steel: Designation used by German Standards
9254	0.54% C 0.7% Cr 1.4% Si steel: Designation used in the UK and USA
AB 75	0.6% C 0.25% Cr 2.0% Si tool steel: Balfour; chisels, dies, etc. **DPN: 540**
ABRADUR SK	0.64% C 0.8% Cr 1.7% Si steel: Pompey **DPN: 660**
ABRASODUR II	0.5% C 9.5% Cr 2.0% Si: Welding electrode; Soudometal; for hard facing **DPN: 520**
ADAMANT	0.52% C Cr steel: Casting; Firth Brown; normalized and tempered **DPN: 210**
AISI 50 B44	0.44% C 0.5% Cr 0.0005% B steel
AISI 50 B46	0.46% C 0.3% Cr 0.0005% B steel
AISI 50 B50	0.5% C 0.5% Cr 0.0005% B steel
AISI 50 B60	0.6% C 0.5% Cr 0.0005% B steel
AISI 51 B60	0.6% C 0.8% Cr 0.0005% B steel
AISI 5045	0.45% C 0.6% Cr steel: Obsolete
AISI 5046	0.46% C 0.3% Cr steel
AISI 5140	0.4% C 0.8% Cr steel
AISI 5145	0.45% C 0.8% Cr steel
AISI 5147	0.47% C 1% Cr steel
AISI 5150	0.5% C 0.8% Cr steel
AISI 5152	0.52% C 1% Cr steel
AISI 5155	0.55% C 0.8% Cr steel
AISI 5160	0.6% C 0.8% Cr steel
AISI 9261	0.6% C 0.2% Cr 2% Si steel
AISI 9262	0.6% C 0.3% Cr 2.0% Si steel
ANTICHOC 2	0.48% C 0.12% Cr 1.7% Si steel: Pompey
ASTM A401	0.55% C 0.6% Cr 1.5% Si steel: For springs; cold drawn and tempered
AURIGA VII	0.5% C 0.7% Mn 1% Cr 0.4% S and P steel: Casting; D Brown; normalized and tempered **UTS: 680 Elon: 10% Proof: 370**
B 904	0.85% C 0.2% Cr steel: VEW
B 908	0.75% C 0.2% Cr steel: VEW
BCTA	0.6% C 0.65% Cr steel: ESC; hot stamping dies **DPN: 460**
BCTB	0.6% C 0.6% Cr steel: ESC; BS 970 En 11
BS 970	See designation for details
BS 1453 A5	0.4% C 1% Mn 1% Cr 0.5% Si: Weld filler rod for the repair of wearing surfaces; railway points
BS 3100 BW2	0.5% C 0.7% Mn 1.0% Cr steel: Casting for abrasion resistance
BS 3100 BW3	0.5% C 0.7% Mn 1.0% Cr steel: Casting for abrasion resistance
BS 3111/3	0.4% C 1.0% Cr steel: Wire; annealed and drawn; for cold forged bolts **UTS: 680**
BS 3146 CLA12A	0.5% C 1% Cr steel: Investment casting; hardened and tempered; for abrasion resistance **DPN: 207 UTS: 680 Elon: 10%**
BS 3146 CLA12B	0.5% C 1% Cr steel: Investment casting; hardened and tempered; for abrasion resistance **DPN: 293**
C 35	0.36% C 0.5% Cr steel: German Standard
CA 30	0.34% C 1.35% Cr 0.95% Al steel: For nitriding; Krupp
CA 30 A	0.34% C 1.15% Cr 0.95% Al steel: For nitriding; Krupp
CBV 2	0.37% C 1.05% Cr steel: Krupp
CBV 2h	0.36% C 1.5% Cr steel: Krupp
CBV/Z	0.41% C 1.05% Cr steel: Krupp
CCR 2	0.6% C 0.6% Cr steel: Huntsman; picks, axes, chisels, etc. **DPN: 600**
CCR 350	0.7% C 3.5% Cr steel: Origin unknown
CHISEL STEEL	0.35% C 1.5% Cr steel: Origin unknown

Symbol	Nominal analysis, supplier, condition and remarks.
CL 60	0.6% C 0.7% Mn 0.6% Cr steel: Origin unknown
CMN	0.65% C 12% Mn 0.4% Ni 2.5% Cr steel: American proprietary alloy listed in SAE Yearbook
CODE 620	0.5% C 0.8% Cr steel: Casting; Edgar Allen; as Cromax F
CP1	0.43% C 1.4% Cr 1% Si tool steel: Osborn; chisels; shock resisting; obsolete **DPN: 600**
CPG	0.5% C 1.25% Cr 1% Si steel: Jonas; chisels, hammers, etc. **DPN: 600**
CR	0.4% C 1% Cr steel: ESC for BS alloy En 18
CROMANSIL	0.2% C 1.2% Mn 0.5% Cr 0.7% Si steel: Origin unknown
CROMAX F	0.5% C 0.9% Cr steel: Casting; Edgar Allen; for wear resistance; Code 620 **DPN: 250**
DEF 13 B 5A	Alloy steel: Hardened and tempered; covers En 18; also includes En 15, 21, 22, 100, etc. **UTS: 760 Elon: 20% Proof: 550**
DIN 1654 41 Cr 4	0.41% C 1.1% Cr steel: For cold forged bolts
DIN 8555 E6-60	Medium C Cr steel: Welding electrode
DIN 8555 E10-60	Medium C Cr steel: Welding electrode
DIN 8555 SG6-60	0.45% C 9.5% Cr steel: Welding electrode
DIN 17200 37 Cr 4	0.37% C 0.75% Mn 1.05% Cr steel: German Standard
DIN 17200 41 Cr 4	0.41% C 1.1% Cr steel: Forging; German Standard
DIN 17200 46 Cr 2	0.46% C 0.65% Mn 0.27% Cr 0.0005% B (max) steel: German Standard
DIN 17200 34 Cr Al S 5	0.34% C 0.75% Mn 1.15% Cr 1.0% Al steel: German Standard
DIN 17200 37 Cr S 4	0.37% C 0.75% Mn 1.05% Cr steel: German Standard
DIN 17200 38 Cr 2	0.38% C 0.6% Mn 0.5% Cr steel: German Standard
DIN 17221 55 Cr 3	0.55% C 0.85% Mn 0.75% Cr 0.3% Cu (max) steel: For springs; German Standard
DIN 17221 60 Si Cr 7	0.6% C 0.85% Mn 0.3% Cr steel: For springs; German Standard
DIN 17221 67 Si Cr 5	0.67% C 0.5% Cr 1.3% Si steel: For springs; German Standard
DIN 17222 67 Si Cr 5	0.67% C 0.5% Cr 1.3% Si steel: For springs; German Standard
DIN 17225 67 Si Cr 5	0.67% C 0.5% Cr 1.3% Si steel: For springs; German Standard
DK 4	0.38% C 1.0% Cr steel: Pompey **UTS: 1070 Elon: 8% Proof: 870**
DK 6	0.6% C 0.75% Cr steel: Pompey
DUNELT 33	0.4% C 1.0% Cr steel: Dunford for BS alloy En 18
DUNELT 83	0.45% C 8% Cr 3.5% Si steel: For valves; Dunford for BS alloy En 52
En 11	0.6% C 0.7% Cr steel: Designation used in BS 970; replaced by 526 NM60
En 18	0.4% C 1% Cr steel: Designation used in BS 970; replaced by 530 M40
En 18D	0.4% C 1% Cr steel: Designation used in BS 970; replaced by 530 M40

Note. The following abbreviations and units are used in the tables:

DPN	Hardness, diamond pyramid number
UTS	Ultimate tensile strength, N/mm^2
Elon	Elongation, %
Proof	0.1% proof strength, N/mm^2

1 N/mm^2=0.1 hbar=0.102 kgf/mm^2=0.06475 tonf/in.^{2}=145.04 lbf/in.^{2}=1 MPa

See Appendix II for other abbreviations and conversion tables.

Symbol	Nominal analysis, supplier, condition and remarks.
En 48	0.5% C 1.2% Cr steel: Designation used in BS 970
En 52	0.45% C 8% Cr steel: For valves; designation used in BS 970; replaced by 401 S45
En 53	0.6% C 6.2% Cr steel: For valves; designation used in BS 970; obsolete
F 300	0.56% C 0.9% Mn 0.8% Cr steel: VEW
F 550	0.5% C 1.1% Cr steel: VEW
FAGERSTA B126	0.6% C 1.7% Mn 0.2% Cr steel: For diaphragms; Fagersta
FAGERSTA C182	0.9% C 0.4% Mn 0.5% Cr steel: For wood saws; Fagersta
FAGERSTA C261	0.4% C 0.5% Mn 0.5% Cr steel: Bar; Fagersta
FAGERSTA C525	0.75% C 0.6% Mn 0.3% Cr steel: For springs; Fagersta
FDK 4	0.38% C 0.9% Cr steel: Pompey **UTS: 1070 Elon: 10% Proof: 760**
FF 4H	0.48% C 0.17% Cr steel: Pompey **UTS: 870 Elon: 8% Proof: 610**
FIRTHAG	0.45% C 0.95% Cr steel: Firth Brown for BS alloy En 18
GOST 2052/60 S2KHA	0.6% C 0.55% Mn 0.85% Cr steel: Bar; Russian Standard
GOST 2052/70 S2KHA	0.7% C 0.5% Mn 0.3% Cr steel: Bar; Russian Standard
GOST 4543/30 KHGSA	0.3% C 0.95% Mn 0.95% Cr steel: Russian Standard
GOST 4543/35 KHG2	0.36% C 1.7% Mn 0.55% Cr steel: Russian Standard
GOST 4543/35 KHGSA	0.35% C 0.95% Mn 1.25% Cr steel: Russian Standard
GOST 4543/38 KHA	0.38% C 0.65% Mn 0.95% Cr steel: Bar; Russian Standard
GOST 4543/38 KHIU	0.38% C 0.35% Mn 1.65% Cr 0.65% Al steel: Russian Standard
GOST 4543/38 KHS	0.38% C 0.45% Mn 1.45% Cr steel: Russian Standard
GOST 4543/40 KH	0.4% C 0.65% Mn 0.95% Cr steel: Russian Standard
GOST 4543/40 KHG	0.4% C 1.1% Mn 1.0% Cr steel: Russian Standard
GOST 4543/40 KHGR	0.4% C 0.85% Mn 0.95% Cr 0.0035% B steel: Russian Standard
GOST 4543/40 KHR	0.4% C 0.65% Mn 0.95% Cr 0.0035% B steel: Russian Standard
GOST 4543/40 KHS	0.4% C 0.45% Mn 1.45% Cr steel: Russian Standard
GOST 4543/45 KH	0.45% C 0.65% Mn 0.95% Cr steel: Russian Standard
GOST 4543/45 KHC	0.46% C 0.65% Mn 0.95% Cr 0.5% Al (max) 0.2% Zr steel: Russian Standard
GOST 4543/50 KH	0.5% C 0.65% Mn 0.95% Cr steel: Russian Standard
GOST 5632/4 KH9S2	0.4% C 9.0% Cr steel: Russian Standard
GOST 5632/4 KH952	0.4% C 9.0% Cr steel: Plate; Russian Standard
GOST 7909/30 KHGS	0.31% C 0.25% Ni (max) 0.95% Cr steel: Pipe; Russian Standard **UTS: 1100 Proof: 850**
GOST 7909/40 KH	0.4% C 0.65% Mn 0.95% Cr steel: Pipe; Russian Standard **UTS: 1000 Proof: 800**
GOST 8467/30 KHGS	0.31% C 0.95% Mn 0.25% Ni (max) 0.95% Cr steel: Pipe; Russian Standard **UTS: 1100 Proof: 850**
GOST 8467/40 KH	0.4% C 0.65% Mn 0.25% Ni (max) 0.95% Cr steel: Pipe; Russian Standard **UTS: 1000 Proof: 800**
GOST 8731/30 KHGSA	0.31% C 0.9% Mn 0.2% Ni (max) 0.9% Cr steel: Pipe; Russian Standard; seamless **DPN: 700**
GOST 8731/40 KH	0.4% C 0.65% Mn 0.2% Ni (max) 0.9% Cr steel: Pipe; Russian Standard; seamless **UTS: 670**

Symbol	Nominal analysis, supplier, condition and remarks.
GOST 8733/30 KHGSA	0.31% C 0.95% Mn 0.95% Cr steel: Pipe; Russian Standard; seamless **UTS: 500**
GOST 8733/40 KH	0.4% C 0.05% Mn 0.95% Cr steel: Pipe; Russian Standard; seamless **UTS: 630**
H 3	0.6% C 2% Cr 1.5% Si steel: Jessop for BS alloy En 53
H 18	0.45% C 8.5% Cr 3.3% Si steel: Jessop specification for BS alloy En 52
H 23	0.48% C 1.3% Cr steel: Jessop
H 28	0.4% C 1% Cr steel: Jessop for BS alloy En 18
H 30	0.6% C 0.65% Cr steel: Jessop
H 700	0.46% C 9.0% Cr 3.0% Si steel: VEW; for valves
HECLA 104	0.6% C 0.65% Cr steel: Hadfields for BS alloy En 11
HECLA 105	0.5% C 1.2% Cr steel: Hadfields for BS alloy En 48
HECLA 120	0.45% C 8% Cr 3.5% Si steel: Hadfields for BS alloy En 52
HECLA 148	0.65% C 3.5% Cr steel: Origin unknown
HECLA 166	0.6% C 6.2% Cr 1.5% Si steel: Hadfields for BS 970 En 53
HECLA 198	0.55% C 0.75% Cr 1.5% Si steel: Hadfields for BS alloy En 48A
HEDEX 4	0.4% C 1% Cr steel: Wire for forging; Kiveton Park for BS alloy En 18
HEDEX 29	0.4% C 0.8% Cr steel: Wire for forging; Kiveton Park; annealed **UTS: 610**
HIV	0.26% C (max) 0.3% Cr steel: German Standard
HNV 2	0.4% C 2.2% Cr 3.9% Si steel: For inlet valves; designation used by SAE
HNV 3	0.45% C 8.5% Cr 3.3% Si steel: For valves; designation used by SAE
JIS G4106 S Cr 4	0.4% C 0.7% Mn 1.02% Cr steel: Japanese Standard
JIS G4106 S Cr 5	0.45% C 0.7% Mn 1.05% Cr steel: Japanese Standard
JIS G4106 S Mn C 3	0.43% C 1.5% Mn 0.5% Cr steel: Japanese Standard
JIS G4106 SUP10	0.5% C 0.85% Mn 0.95% Cr 0.15% V (max) steel: Japanese Standard
JIS G4801 SUP9	0.55% C 0.8% Mn 0.8% Cr steel: For springs; Japanese Standard
JIS G4801 SUP11	0.55% C 0.8% Mn 0.8% Cr steel: For springs; Japanese Standard
K 200	1.0% C 1.5% Cr steel: VEW
K 240	0.92% C 1.2% Cr steel: VEW
K 245	0.62% C 1.1% Mn 0.6% Cr steel: VEW
KF 18C	0.4% C 1% Cr steel: Kirkstall for BS alloy En 18
KF 18D	0.45% C 1% Cr steel: Kirkstall for BS alloy En 18
KP 32	0.4% C 1% Cr steel: Kiveton Park for BS alloy En 18
L 5	0.42% C 1.45% Mn 0.85% Cr steel: Casting; Jessop
LT 36	0.45% C 1.5% Cr steel: Low Moor; obsolete **DPN: 540**
LV 52	0.45% C 8% Cr 3.5% Si steel: For valves; Low Moor; obsolete **DPN: 320**
MONARCH GENERAL UTILITY	0.6% C 0.7% Mn 0.7% Cr steel: For tools; origin unknown
MSC 40	0.43% C 0.5% Cr steel: Krupp
MSC 50	0.5% C 0.7% Cr steel: Krupp
MY 0	0.55% C 0.95% Cr steel: Origin unknown
NF A35 551/38C2	0.38% C 0.75% Mn 0.45% Cr steel: French Standard
NF A35 551/38C4	0.38% C 0.75% Mn 1.0% Cr steel: French Standard
NF A35 551/42C2	0.43% C 0.75% Mn 0.45% Cr steel: French Standard
NF A35 551/42C4	0.42% C 0.75% Mn 1.0% Cr steel: French Standard
NF A35 551/10006	0.98% C 0.3% Mn 0.3% Ni (max) 1.5% Cr 0.1% Mo (max) steel: French Standard
NF A35 571/45C4	0.45% C 0.75% Mn 1.0% Cr steel: Bar; French Standard
NF A35 571/60SC7	0.6% C 0.75% Mn 0.52% Cr steel: For bars; French Standard
NITA	0.53% C 1.5% Cr 0.2% Al 1.1% Ac steel: Origin unknown

Symbol	Nominal analysis, supplier, condition and remarks.
No. 6 HARDENITE	1.0% C 1.0% Mn 1.35% Cr steel: For tools; origin unknown
NV 6	0.5% C 0.8% Cr steel: For valves; SAE 5150
NVG	0.5% C 0.8% Cr steel: For inlet valves; designation used by SAE
OCM 9	0.4% C 1% Cr steel: W Marrison for BS alloy En 18
P 609	0.4% C 0.8% Mn 1% Cr steel: Carrs; pressure dies for casting Zn Pb Sn base alloys **DPN: 400**
PHILIPS 600	0.65% C 2.5% Cr steel: Welding electrode; Philips **DPN: 600**
PN	0.42% C 0.45% Mn 0.45% Cr steel: Origin unknown
RMK	0.45% C 1.0% Cr steel: Pompey
RMKS	0.45% C 1.5% Cr steel: Pompey
RS 3K	0.6% C 0.4% Cr 1.8% Si steel: Pompey
RS 4K	0.65% C 0.8% Cr 1.7% Si steel: Pompey
RSMK	0.6% C 0.5% Cr steel: Pompey
S 4K	0.45% C 0.5% Cr steel: Pompey
SAE 50 B40	0.4% C 0.5% Cr steel with B
SAE 50 B44	0.44% C 0.5% Cr 0.0005% B steel
SAE 50 B46	0.46% C 0.3% Cr 0.0005% B steel
SAE 50 B50	0.5% C 0.5% Cr 0.0005% B steel
SAE 50 B60	0.6% C 0.8% Cr 0.0005% B steel
SAE 51 B60	0.6% C 0.8% Cr 0.0005% B steel
SAE 5045	0.45% C 0.7% Cr steel: Obsolete
SAE 5046	0.46% C 0.3% Cr steel: Mechanical properties not specified
SAE 5060	0.6% C 0.5% Cr steel
SAE 5140	0.4% C 0.8% Cr steel: Mechanical properties not specified
SAE 5145	0.45% C 0.8% Cr steel: Mechanical properties not specified
SAE 5147	0.47% C 1% Cr steel: Mechanical properties not specified
SAE 5150	0.5% C 0.8% Cr steel: Mechanical properties not specified
SAE 5152	0.52% C 1% Cr steel: Mechanical properties not specified
SAE 5155	0.55% C 0.8% Cr steel: Mechanical properties not specified
SAE 5160	0.6% C 0.8% Cr steel: Mechanical properties not specified
SAE 9254	0.55% C 0.7% Cr 1.5% Si steel: Mechanical properties not specified
SAE 9261	0.6% C 0.2% Cr 2.0% Si steel: Mechanical properties not specified
SAE 9262	0.6% C 0.3% Cr 2.0% Si steel: Mechanical properties not specified
SAE HNV 2	0.4% C 2.2% Cr 3.9% Si steel: For inlet valves
SAE HNV 3	0.45% C 8.5% Cr 3.3% Si steel: For valves **UTS: 980 Elon: 22% Proof: 740**
SAE J 157	0.55% C 0.7% Cr 1.5% Si steel: Wire for springs **DPN: 600 UTS: 2000**
SAE NVG	0.5% C 0.8% Cr steel: For inlet valves as SAE 5150
SANBOLD 11	0.6% C 0.65% Cr steel: Sanderson for BS alloy En 11
SANBOLD 18	0.4% C 1% Cr steel: Sanderson for BS alloy En 18
SANBOLD 54	0.4% C 1% Cr steel: Sanderson; replaced by Sanbold 18
SCRI	0.47% C 1.4% Cr 1% Si steel: Huntsman; shear blades, punches, etc. **DPN: 600**
SIL 1	0.45% C 8.5% Cr 3.3% Si steel: For valves; origin unknown
SIL 2	0.55% C 8% Cr 0.7% Mo 1.5% Si steel: For valves; origin unknown
SIL No. 1 VALVE STEEL	0.4% C 8.5% Cr 3.2% Si steel: Carpenter **DPN: 270 UTS: 920 Elon: 22% Proof: 560**

For SIS specifications see SS

Symbol	Nominal analysis, supplier, condition and remarks.
SKF 606	0.65% C 0.8% Cr steel: SKF
SKF 677	0.55% C 0.65% Cr steel: SKF

Symbol	Nominal analysis, supplier, condition and remarks.
SLV	0.45% C 8.3% Cr 3.4% Si steel: Firth Brown for BS alloy En 52 **DPN: 285 UTS: 1070 Elon: 15%**
SOUDODUR 600	0.65% C 2.0% Cr steel: Welding electrode; Soudometal; for hard facing wear resistant parts **DPN: 650**
SOUDOKAY A 12-0	0.45% C 9% Cr 2.7% Si steel: Welding electrode; Soudometal **DPN: 600**
SOUDOR GA 350	0.7% C 2.0% Mn 1.0% Cr 0.2% Ti steel: Welding electrode; Soudometal; for hard surfaces **DPN: 350**
SOUDOR GA 500	1.0% C 2.0% Mn 2.0% Cr 0.2% Ti steel: Welding electrode; Soudometal; for hard facings **DPN: 520**
SOUDOR GA 600	0.45% C 9.0% Cr 3.0% Si steel: Welding electrode; Soudometal; for hard facings **DPN: 600**
SPKS	0.45% C 9.0% Cr steel: Pompey
SS 2085	0.55% C 0.3% Cr 1.7% Si steel: Bar and forging; Swedish Standard; hardened and tempered **DPN: 480 UTS: 1380 Elon: 5% Proof: 1260**
SS 2090	0.55% C 0.3% Cr 1.7% Si steel: Bar and forging; Swedish Standard; hardened and tempered **DPN: 480 UTS: 1380 Elon: 5% Proof: 1260**
SS 2230	0.51% C 1.0% Cr 0.15% V steel: For springs; Swedish Standard
SS 2245	0.41% C 1.0% Cr steel: Swedish Standard **DPN: 255 UTS: 920**
SS 2253	0.55% C 0.75% Cr steel: For springs; Swedish Standard
SS 2254	0.6% C 0.75% Cr steel: For springs; Swedish Standard
STA 5 V 24	0.45% C 8.5% Cr 3.5% Si steel: Replaced by BS alloy En 52
STA 5 V9C	0.4% C 1.0% Cr steel: Replaced by BS alloy En 18
STA 5 V9C/2	0.4% C 1.0% Cr steel: Replaced by BS alloy En 18C
STA 5 V9C/3	0.4% C 1.0% Cr steel: Replaced by BS alloy En 18D
STA 5 V9C/4	0.4% C 1.0% Cr steel: Replaced by BS alloy En 18A
STA 5 V9C/a	0.4% C 1.0% Cr steel: Replaced by BS alloy En 18B
STUBS 67S	C Si Cr steel: Welding electrode; Stubs; for hard facings **DPN: 600**
STUBS 400	Mn Cr Si Fe alloy: Rod; Stubs; for hard surfaces **DPN: 450**
STUBS 670	C Cr Si steel: Welding electrode; Stubs; for hard facings **DPN: 600**
SUH 1	0.45% C 8.5% Cr 32% Si steel: Heat resisting; Japanese Standards designation
TR 3KD	0.38% C 0.9% Cr steel: Pompey
TR 4	0.45% C 0.15% Cr steel: Pompey
TR 5	0.52% C 0.15% Cr steel: Pompey
TRIPLE SIX	0.6% C 0.6% Cr steel: John Vessey **DPN: 700**
TRK	0.42% C 0.15% Cr steel: Pompey
V 500	0.4% C 1.1% Cr steel: VEW
VAL 5	0.45% C 9% Cr 3% Si steel: Valbruna
VALVEX 518D	0.45% C 8% Cr 3.5% Si steel: For valves; Swift Levick for BS alloy En 52
VIM-VAR 52100	1% C 1.4% Cr steel: Carpenters **DPN: 450 UTS: 1630 Proof: 1560**
VIPER 666	0.6% C 0.6% Cr steel: Hall and Pickles; lathe centres, leather cutters, etc.; obsolete **DPN: 600**
VS 1C	0.4% C 1% Cr steel: Vulcan **UTS: 1070**
VS 50	0.45% C 8.5% Cr 3.5% Si steel: Darwins for BS alloy En 52
VSCS	0.55% C 0.8% Cr steel: Vulcan; heavy duty vehicle springs **DPN: 440**
Y 60SC7	0.6% C 0.8% Mn 0.6% Cr steel: French Standards designation
Y 75V	0.75% C 0.25% Mn 0.32% Cr steel: French Standards designation
Z 45CS9	0.45% C 0.8% Mn (max) 8.0% Cr steel: French Standards designation

Note. The following abbreviations and units are used in the tables:

DPN	Hardness, diamond pyramid number
UTS	Ultimate tensile strength, N/mm^2
Elon	Elongation, %
Proof	0.1% proof strength, N/mm^2

1 N/mm^2=0.1 hbar=0.102 kgf/mm^2=0.06475 tonf/in.^{2}=145.04 lbf/in.^{2}=1 MPa

See Appendix II for other abbreviations and conversion tables.

44E3 Steel – high carbon with chromium 0.7% carbon minimum – 5% chromium maximum

Specific gravity	7.8
Density	7800 kg/m^3
Solidus/liquidus	1400–1450 °C
Thermal conductivity	60 W/m °C
Coefficient of linear expansion	11×10^{-6}/ °C
Electrical conductivity	8–10% IACS (copper 100%)
Specific resistance	170–230 microhm mm
Young's modulus of elasticity	200×10^9 N/m^2
Impact	Can be expected to be low
Fatigue strength	±570 N/mm^2
Hot strength	–

The above properties are typical of the following group, and may not apply exactly to any one specification. It is possible that with certain specifications some of the values may not be applicable.

General metallurgical characteristics

Chromium forms very hard stable carbides with the carbon present in steel. These are harder than iron carbides and retain their hardness at higher temperatures. Being stable the carbides are very sluggish in their metallurgical reactions and care must always be taken that the correct times and temperatures are used. As these steels are all high carbon there is considerable carbon available to form chromium carbides and even in the annealed condition they have considerable hardness. Chromium imparts some corrosion resistance by virtue of the tenacious uncorrodable oxide that it forms, but none of the steels listed can be classed as corrosion resistant. The steels are all air hardening, being the most readily hardened of any of the low alloy carbon steels.

In the hardened or tempered condition the steels have very little ductility and are notoriously notch sensitive. They will always be used in the hardened and tempered condition. Larger quantities of chromium (over 12%) are present in the stainless steels, which are described in Sections 44M and 44N. Small quantities of titanium or aluminium are sometimes added to control grain growth.

The steels in this group cannot accept cold work without becoming extremely brittle and cracking. Annealing, hardening and normalizing operations are carried out at a temperature in the region of 900–1000 °C depending on the alloy involved.

These steels have considerable hardenability and many will harden with air cooling. Larger sections will require oil quenching but great care is necessary to prevent cracking in the quenching operation.

None of these steels should be welded under any circumstances. If it is essential, then expert advice must be obtained and rigidly followed.

The relatively low chromium content of these materials will slightly enhance their corrosion resistance but under no circumstances can they be classified as stainless, and the very high notch sensitive characteristics of these steels means that any corrosion pitting which occurs could have a very serious effect on their service life.

As with other chromium steels, these materials are prone to temper brittleness when heated within the range of 300–600 °C for any appreciable time. This is the result of carbide precipitation at the grain boundaries and will make the materials even more brittle than they are in the normal state.

These steels will normally be used in the hardened condition; thus tempering will be carried out below 300 °C.

Machining of these steels will almost invariably be by final grinding and great care must be taken to prevent grinding abuse which can result in surface cracking.

Plating of these components should not be carried out without very careful control as hydrogen embrittlement can result in fracture of even relatively large sections during or immediately after the plating process.

The uses for these materials are in the range of shear blades, ball roller bearings, dies where impact is at a low level, and other uses where excellent abrasion resistance is required.

It will be found that many of the steels in the higher chromium, high carbon section, 44M2, are substitutes for these materials and generally will give better service because of the corrosion resistance in this group.

Symbol	Nominal analysis, supplier, condition and remarks.
1.2002	1.2% C 0.4% Cr steel: German Standard
1.2067	1.15% C 1.5% Cr steel: German Standards designation
1.2067	1% C 1.4% Cr steel: German Standard
1.3501	1.0% C 0.5% Cr steel: For bearings; German Standard
1.3503	1.02% C 0.35% Mn 1.5% Cr steel: For bearings; German Standard
1.3503	1.0% C 1.0% Cr steel: For ball bearings; German Standard
1.3505	1.0% C 1.5% Cr steel: For ball bearings; German Standard
1.3520	1.0% C 1.5% Cr 1.1% Mn steel: For ball races; German Standard
1.7131	1.6% C 1.1% Mn 1.0% Cr steel: German Standard
1.7176	0.55% C 0.85% Mn 0.85% Cr 0.3% Cu (max) steel: For springs; German Standard
1.7305	0.41% C 0.7% Mn 1.05% Cr steel:For bearings; German Standard
1.8405	0.7% C 2.0% Mn 1% Cr 0.2% Ti 0.1% Al steel: Welding rod; German Standard
1.8425	1.1% C 2.0% Mn 1.8% Cr 0.2% Ti 0.1% Al steel: Welding rod; German Standard
8KH3	0.8% C 3.5% Cr steel: Russian Standards designation
9KHS	0.9% C 1.1% Cr steel: Russian Standards designation

Note. The following abbreviations and units are used in the tables:

DPN	Hardness, diamond pyramid number
UTS	Ultimate tensile strength, N/mm^2
Elon	Elongation, %
Proof	0.1% proof strength, N/mm^2

1 N/mm^2=0.1 hbar=0.102 kgf/mm^2=0.06475 tonf/in.^{2}=145.04 lbf/in.^{2}=1 MPa

See Appendix II for other abbreviations and conversion tables.

Symbol	Nominal analysis, supplier, condition and remarks.
10 K 2	1.0% C 1.45% Cr steel: Pompey
11KH	1.1% C 0.55% Cr steel: Russian Standards designation
13KH	1.32% C 0.45% Mn 0.55% Cr steel: Russian Standards designation
14 Si C1	0.75% C 0.35% Cr steel: Sandvik
17C	1.0% C 0.15% Cr steel: Sandvik; annealed **UTS: 660**
17C1	1.0% C 0.5% Cr steel: Sandvik
21C	1.25% C 0.15% Cr steel: Sandvik; annealed **UTS: 690**
22C	1.3% C 0.15% Cr steel: Sandvik
22C1	1.25% C 0.45% Cr steel: Sandvik; annealed **UTS: 720**
41 Cr 4	0.41% C 0.7% Mn 1.05% Cr steel: For bearings; German Standard
67 Si Cr 5	0.67% C 0.5% Cr 1.3% Si steel: Designation used by German Standards
70 Mn Cr Ti 8	0.7% C 1.1% Cr 0.2% Ti steel: Designation used by German Standards
73/3.5% Cr	1.0% C 3.5% Cr steel: For magnets; Simonds
90 Cr Si 5	0.9% C 1.2% Cr steel: German Standard
100 C6	1% C 1.5% Cr steel: AFNOR
100 Cr 6	1.15% C 1.5% Cr steel: German Standards designation
100 Cr 6 (W3)	1.0% C 1.55% Cr steel: Designation used by German Standards
100 Cr Mn 6 (W4)	1.0% C 1.5% Cr steel: Designation used by German Standards
105 Cr 2 (W1)	1.05% C 0.5% Cr steel: Designation used by German Standards
105 Cr 4 (W2)	1.05% C 1.1% Cr steel: Designation used by German Standards
105 Cr 6	1.02% C 0.3% Mn 1.5% Cr steel: For bearings; German Standard
110 Mn Cr Ti 8	1.1% C 1.8% Cr 0.25% Ti steel: Designation used by German Standards

Symbol	Nominal analysis, supplier, condition and remarks.
150	1.0% C 1.45% Cr steel: F Parkin; ball races, collets, etc. **DPN: 870**
180	1.0% C 0.5% Cr steel: F Parkin; reamers, taps, etc. **DPN: 820**
534 A99	0.95% C 1.4% Cr steel: BS 970; obsolete
535 A99	0.99% C 1.4% Cr steel: BS 970
1230	1.35% C 0.3% Mn 0.35% Cr steel: French Standards designation
1231	1.2% C 0.2% Mn 0.32% Cr steel: French Standards designation
1232	1.05% C 0.2% Mn 0.32% Cr steel: French Standards designation
1233	0.9% C 0.25% Mn 0.32% Cr steel: French Standards designation
50100	1% C 0.5% Cr steel: Designation used in the UK and USA
51100	1% C 1.4% Cr steel: Designation used in the UK and USA
52100	1% C 1.4% Cr steel: Designation used in the UK and USA
ADAMANT GR	0.77% C Cr steel: Casting; Firth Brown; normalized **DPN: 300**
AFNOR 100 Cb	1.0% C 1.5% Cr steel: French Standard
AGS	1.05% C 1.3% Cr steel: Sheet; Balfour; templates, bushes, etc. **DPN: 770**
AISI E50100	1.0% C 0.5% Cr steel
AISI E51100	1.0% C 1.00% Cr steel
AISI E52100	1.0% C 1.5% Cr steel
AMS 6440D	1.0% C 1.5% Cr steel: Bar and forging; AMS for SAE 52100
AMS 6441B	1.0% C 1.5% Cr steel: Seamless tube; AMS for SAE 52100
AMS 6442B	1.0% C 0.5% Cr steel: Bar and forging; AMS for SAE 50100
AMS 6443B	1.0% C 1.0% Cr steel: For ball bearings; vacuum melted; AMS for SAE 51100
AMS 6444B	1.0% C 1.5% Cr steel: For ball bearings; vacuum melted; AMS for SAE 52100
AMS 6445A	1.0% C 1.0% Cr steel: For ball bearings; vacuum melted; AMS for SAE 51100
AMS 6446	1.0% C 1.0% Cr steel: Bar and forging; AMS for SAE 51100
AMS 6447	1.0% C 1.5% Cr steel: Bar and forging; AMS for SAE 52100
AMS 6449	1% C 1% Cr steel
ASTM A485/1	1.0% C 1.0% Mn 1.0% Cr 0.5% Si steel: High hardenability
ASTM A485/2	0.97% C 1.3% Mn 1.6% Cr 0.7% Si steel: High hardenability
ASTM A732/15 A	1.0% C 1.45% Cr steel: Investment casting; annealed **DPN: 240**
ATLAS Q	1.2% C 0.5% Cr steel: Atlas; as AISI type W5
B 4 CR	1.0% C 1.4% Cr steel: ESC; bushes, taps, etc. **DPN: 850**
BRI	1.0% C 1.0% Cr steel: Jonas; press tools **DPN: 720**
BRUNSWICK V 981	1.0% C 1.4% Cr steel: John Vessey **DPN: 800**
BS 970	See designation for details
C 46	1.3% C 0.5% Cr steel: Fagersta **DPN: 910**
C 71	0.9% C 1.0% Cr 1.5% Si steel: Fagersta **DPN: 750**
C 75 W	0.75% C 0.2% Cr steel: German Standard
C 87 WS	0.85% C 0.2% Cr steel: German Standard
C 182	0.9% C 0.5% Cr steel: Fagersta **DPN: 800**
C 525	0.75% C 0.3% Cr 1.3% Si steel: Fagersta
CH 4	1.25% C 0.7% Cr steel: Forez
CH 5	1.0% C 0.5% Cr steel: Forez
CNS NON SHRINK	1.0% C 1.0% Cr steel: T Turton
CODE 623	0.72% C 2.0% Cr steel: Casting; Edgar Allen; Cromax H
Cr 030	0.9% C 3.3% Cr steel: Forging; for magnets; German Standard included in DIN 17410
Cr 035	0.9% C 1.0% Mn 5.0% Cr steel: Forging; for magnets; German Standard included in DIN 17410
CRC	1.0% C 1.45% Cr steel: ESC for BS alloy En 31
CROMAX N	0.73% C 2.0% Cr steel: Casting; Edgar Allen; for wear resistant parts **DPN: 250**
DCCM	1.05% C 1.3% Cr tool steel: For cold work; Balfour Darwin
DTD 5022	1.0% C 1.5% Cr steel **DPN: 750**
DUNELT 90	1.0% C 1.2% Cr steel: Dunford for BS alloy En 31
E 10	1.0% C 0.13% Cr steel: Pompey
E Fe 3	0.65% C 6% Cr steel: Weld electrode; designation used by AWS
E Fe 4	1.5% C 4% Cr steel: Weld electrode; designation used by ANS
EH 4	1.0% C 4.0% Cr 1.0% Si steel: Forging; for magnets; Polish alloy listed by PMA
EKh	1.0% C 3.3% Cr steel: For magnets; Russian alloy listed by PMA
En 31	1% C 1.3% Cr steel: For bearings; designation used in BS 970; replaced by 535 A99
FAGERSTA C250	1.2% C 0.35% Mn 0.17% Cr steel: Bar; Fagersta
FAGERSTA N164	0.8% C 0.4% Mn 2.0% Cr steel: For wood saws; Fagersta
GOST 801 SHKH6	1.1% C 0.3% Mn 0.3% Ni (max) 0.55% Cr steel: For bearings; Russian Standard
GOST 801 SHKH9	1.05% C 0.3% Mn 0.3% Ni (max) 1.05% Cr steel: For bearings; Russian Standard
GOST 801 SHKH15	1.0% C 0.3% Mn 0.3% Ni (max) 1.5% Cr steel: For bearings; Russian Standard
GOST 801 SHKH15SG	1.0% C 1.05% Mn 0.3% Ni (max) 1.5% Cr steel: For bearings; Russian Standard
GOST 4543	0.3% C 0.65% Mn 1.15% Cr 0.0035% B steel: Russian Standard
H 2	1.0% C 1.5% Cr steel: Jessops **DPN: 800**
H 17	1.0% C 6.0% Cr steel: For magnets; Jessops
H 44	1.0% C 1.5% Cr steel: Jessops **DPN: 850**
H 63	1.0% C 0.5% Cr steel: Jessops
H 64	0.97% C 1.0% Cr steel: Jessops
HAO	1.0% C 6.0% Cr steel: For magnets; Belgian alloy listed by PMA
HECLA 108	1.0% C 1.3% Cr steel: Hadfields for BS alloy En 31
HEDEX 18	1.0% C 1.3% Cr steel: Wire for forgings; Kiveton Park for BS alloy En 31
HEDEX 41	1.0% C 0.32% Cr steel: Wire for forgings; Kiveton Park; annealed **UTS: 630**
HRS	1.0% C 1.3% Cr steel: Jonas; press tools **DPN: 680**

Note. The following abbreviations and units are used in the tables:

DPN	Hardness, diamond pyramid number
UTS	Ultimate tensile strength, N/mm^2
Elon	Elongation, %
Proof	0.1% proof strength, N/mm^2

$1\ N/mm^2 = 0.1\ hbar = 0.102\ kgf/mm^2 = 0.06475\ tonf/in.^2 = 145.04\ lbf/in.^2 = 1\ MPa$

See Appendix II for other abbreviations and conversion tables.

Symbol	Nominal analysis, supplier, condition and remarks.
HRS	1.0% C 1.4% Cr steel: Osborn; dies, balls, rollers, races; obsolete
	DPN: 720
ID 2	0.9% C 1.7% Mn 0.25% Cr steel: Inman
JCNS	0.9% C 1.3% Mn 0.4% Cr steel: Jonas
JIS G4805 SUJ1	1.0% C 0.3% Mn 1.05% Cr 0.08% Mo (max) steel: For bearings; Japanese Standard
JIS G4805 SUJ2	1.05% C 0.5% Mn (max) 1.45% Cr 0.08% Mo (max) steel: For bearings; Japanese Standard
JIS G4805 SUJ3	1.02% C 1.0% Mn 1.1% Cr 0.08% Mo (max) steel: For bearings; Japanese Standard
KE 4L	1.0% C 0.2% Cr steel: Kayser Ellison; for ball bearings; needles, etc.
KE 160	1.2% C 0.4% Mn 0.5% Cr steel: Kayser Ellison; taps, punches, etc.
KE 839	1.0% C 1.5% Cr steel: Kayser Ellison; taps, dies, thread rolls, etc.
	DPN: 870
KH	1.02% C 1.45% Cr steel: Russian Standards designation
KP 24	1.0% C 1.5% Cr steel: Kiveton Park for BS alloy En 31
KR 6	1.0% C 1.5% Cr steel: Pompey
L 1	1.0% C 1.25% Cr steel: Designation used by AISI
LESCALLOY 52100	1.0% C 1.5% Cr steel: For bearings; Latrobe; as AISI 52100
LK	1.4% C 0.7% Cr steel: Pompey
LT 2	1.0% C 0.5% Cr steel: Low Moor; obsolete
	DPN 870
LT 22	1.0% C 1.5% Cr steel: Low Moor; obsolete
	DPN 770
M CHROME REGENT	1.0% C 0.15% Ni 1.45% Cr 0.1% Mo 0.15% Cu steel: Latrobe
MARSH MN	0.9% C 0.25% Cr steel: Marsh Brothers; gauges, taps, etc.
MIC	0.95% C 0.35% Cr steel: ESC; symbol for 'Microlim'
	DPN: 770
MIC 2	0.95% C 0.35% Cr steel: Origin unknown
MICROLIM	0.95% C 0.3% Cr steel: ESC; punches, rivets, dies, etc.
	DPN: 770
MIL S 980	1% C 1.45% Cr steel: US military specification
MN	0.9% C 0.25% Cr steel: Marsh Brothers; gauges, taps, etc.
Mn Cr	0.6% C 0.8% Cr steel: ESC; springs
NF A35-565/100C2	1.02% C 0.3% Mn 0.5% Cr steel: For bearings; French Standard
NF A35-565/100C6	1.02% C 0.3% Mn 1.5% Cr 0.1% Mo (max) steel: For bearings; French Standard
NSC	1.0% C 1.4% Cr steel: Huntsman; centres, small dies, reamers, etc.
	DPN: 800
NSCD	0.9% C 0.3% Cr steel: Huntsman; slitters, gauges, reamers, etc.
	DPN: 720
NSCT	1.0% C 1.4% Cr steel: Huntsman; ball races, press tools, taps, etc.
	DPN: 800
P 280	0.6% C 0.6% Cr steel: Carrs; hot press, dies, etc.
	DPN: 450
P 704	0.95% C 0.5% Cr steel: Carrs; vice jaws, punches, etc.
	DPN: 630
P 720	1.0% C 1.3% Cr steel: Carrs; centres, ball bearings, etc.
	DPN: 730
PRESTO	1.0% C 1.4% Cr steel: Carpenter; for AISI type L1
R100	1.0% C 0.3% Mn 1.5% Cr steel: VEW
RED DIAMOND II	1.1% C 0.6% Mn 1.5% Cr steel: For wear; Redheugh
	DPN: 295 UTS: 1020 Elon: 10% Proof: 870
REGENT VAC-ARC	1.0% C 1.5% Cr steel: Latrobe; for SAE 52100
	DPN: 800
S 135	1% C 1.4% Cr steel: BS designation for aerospace steel
S 136	1% C 1.4% Cr steel: BS designation for aerospace steel

Symbol	Nominal analysis, supplier, condition and remarks.
SAE 50100	1.0% C 0.5% Cr steel: Mechanical properties not specified
SAE 51100	1.0% C 1.0% Cr steel: Mechanical properties not specified
SAE 52100	1.0% C 1.5% Cr steel: Mechanical properties not specified
SANBOLD 31	1.1% C 1.3% Cr steel: Sanderson for BS alloy En 31
SANBOLD 57	1.1% C 1.3% Cr steel: Sanderson; replaced by Sanbold 31
For SIS specifications see SS	
SK SILVER STEEL	1.2% C 0.4% Mn 0.4% Cr steel: Sanderson Kayser
SK SILVER STEEL	1.2% C 0.4% Cr steel: Sanderson Kayser
SKC 11	0.97% C 0.5% Mn (max) 1.2% Cr 0.4% Mo (max) 0.25% V (max) steel: For roller drills; Japanese Standard designation
SKF 1	1% C 1% Cr steel
SKF 2	0.9% C 1.5% Cr steel
SKF 3	1.0% C 1.5% Cr steel: Bar for bearings; Japanese Standard designation
SKF 707	1.05% C 0.5% Cr steel: SKF
SKF 801	0.97% C 0.5% Cr steel: SKF
SKF 802	1.05% C 0.85% Cr steel: SKF
SKF 831	1% C 1% Cr steel: SKF; as SKF 1
SKF 832	0.9% C 1.5% Cr steel: SKF; as SKF 2
SKF 837	1% C 1.55% Cr steel: SKF
SKS 8	1.45% C 0.35% Cr steel: Japanese Standards designation
SKS 93	1.05% C 0.95% Mn 0.4% Cr steel: Japanese Standards designation
SKS 94	0.95% C 0.9% Mn 0.4% Cr steel: Japanese Standards designation
SKS 95	0.85% C 0.9% Mn 0.4% Cr steel: Japanese Standards designation
SL 934	1.1% C 1.25% Cr steel: Swift Levick; press, dies, punches, etc.
SOUDODUR 600R	0.9% C 2.0% Cr steel: Welding electrode; Soudometal; for hard facing; wear resistant parts
	DPN: 650
SOUDOKAY 258 Ti Cr O	1.7% C 7% Cr 1% Mo 5% Ti steel: Welding electrode; Soudometal
	DPN: 600
SOUDOKAY 795.0	2.2% C 9.5% Cr 0.9% Mo steel: Welding electrode; Soudometal
	DPN: 460
SPEAR CRH	0.75% C 0.4% Cr steel: Origin unknown
SPEAR GFS	0.95% C 0.5% Cr steel: Origin unknown
SR 1855	1.0% C 1.0% Cr 1.5% Si steel: Bofors
	DPN: 850
SS 2092	1.0% C 1.0% Cr steel: Swedish Standard; annealed
	DPN: 240
SS 2258	1.0% C 1.5% Cr steel: Swedish Standard; annealed
	DPN: 210
SS 2912 E	0.6% C 0.65% Mn 0.5% Cr (max) 0.3% Cu steel: Swedish Standard; for pressure vessels
	UTS: 490 Elon: 23% Proof: 270
STA 5 V14	1.1% C 1.3% Cr steel: Replaced by BS alloy En 31
SUJ 2	1.0% C 1.45% Cr steel: Japanese Standards designation
T 1040	1.0% C 0.25% Cr steel: Origin unknown
TERFLING 2	0.6% C 0.7% Cr steel: Uddeholm; for type S4
TOLEDO BR	1.1% C 1.3% Cr steel: Toledo for BS alloy En 31
VAMPIRE	0.9% C 0.3% Cr steel: Hall and Pickles; press tools, gauges, etc.; obsolete
	DPN: 800
VANSTAR	1.0% C 5.0% Cr steel: Pompey
VIPER	1.0% C 1.0% Cr steel: Hall and Pickles; gauges, ball bearings, etc.; obsolete
	DPN: 800
VSCM	0.9% C 1.0% Cr steel: Vulcan; press tools; milling cutters, etc.
	DPN: 720

Symbol	Nominal analysis, supplier, condition and remarks.
W 3	1.0% C 0.5% Cr steel: Designation used by AISI; obsolete
W 4	1.0% C 0.25% Cr steel: Carbon varies over wide limits; designation used by AISI
W 5	1.1% C 0.5% Cr steel: Designation used by AISI
WATERDIE EXTRA	1.0% C 0.5% Cr steel: Columbia; AISI type W5
WATERDIE STANDARD	1.0% C 0.5% Cr steel: Columbia; AISI type W5
WLM 1015	1.0% C 1.4% Cr steel: W Marrison; for ball races
X 45 Cr Si 9	0.45% C 9.0% Cr 3.0% Si steel: Designation used by German Standards
X 45 Cr Si 9/3	0.45% C 9.0% Cr 3.0% Si steel: Designation used by German Standards
X 45 Si Cr 4	0.45% C 2.7% Cr 4.0% Si steel: Designation used by German Standards
XX 9CR	0.9% C 0.5% Cr steel: ESC; mandrels, punches, etc. **DPN: 750**
Y 90V	0.9% C 0.25% Mn 0.32% Cr steel: French Standards designation
Y 100 C6	1.05% C 1.4% Cr steel: French Standard
Y 105V	1.05% C 0.2% Mn 0.32% Cr steel: French Standards designation
Y 120V	1.20% C 0.2% Mn 0.32% Cr steel: French Standards designation
Y 135C	1.35% C 0.3% Mn 0.35% Cr steel: French Standards designation
Y 135V	1.35% C 0.2% Mn 0.32% Cr steel: French Standards designation
Y 2120 C	1.2% C 0.4% Cr steel: French Standard

Note. The following abbreviations and units are used in the tables:

DPN	Hardness, diamond pyramid number
UTS	Ultimate tensile strength, N/mm^2
Elon	Elongation, %
Proof	0.1% proof strength, N/mm^2

1 N/mm^2=0.1 hbar=0.102 kgf/mm^2=0.06475 $tonf/in.^2$=145.04 $lbf/in.^2$=1 MPa

See Appendix II for other abbreviations and conversion tables.

44F1 Steel – low carbon with nickel 0.35% carbon maximum

Specific gravity	7.8
Density	7800 kg/m^3
Solidus/liquidus	1470–1510 °C
Thermal conductivity	50 W/m °C
Coefficient of linear expansion	11×10^{-6}/ °C
Electrical conductivity	6–9% IACS (copper 100%)
Specific resistance	230–270 microhm mm
Young's modulus of elasticity	–
Impact	54–105 J
Fatigue strength	–
Hot strength	–

The above properties are typical of the following group, and may not apply exactly to any one specification. It is possible that with certain specifications some of the values may not be applicable.

General metallurgical characteristics

Nickel is a ferrite strengthener and toughener which also reduces the critical temperatures and widens the range of temperatures at which optimum hardening is obtained. The increased strength and toughness are apparent in the normalized or softened condition, even without the slight air hardening properties of nickel steels.

One important property of some nickel steels is that they retain their strength and ductility at very low temperatures when other materials become embrittled. The ductility and notch resistance is roughly proportional to the nickel content and inversely proportional to the carbon content.

With the higher nickel steels there is some corrosion resistance, but under 4% nickel they are little better than carbon steels.

The good ductility, toughness and flexible heat treatment of these low carbon nickel steels make them good case hardening materials.

The relatively large amounts of nickel required, of high cost and scarcity, led to research on alternative materials, which has been largely successful. Similar or superior properties to those obtainable from 3–5% straight nickel steels, with the exception of low temperature impact values, are now available with low nickel, chromium and molybdenum steels.

These steels on the whole can accept a considerable amount of cold work without appreciably increasing hardness but when annealing is required this will be at the same temperature as for hardening or normalizing, about 850 °C. Softening can be accomplished at lower temperatures when necessary.

Some of the materials in this group can be carburized to give excellent toughness in the core.

Welding will not present any serious problems but advice should be obtained regarding pre-heating, etc. It must be noted that welding of a carburized surface will result in serious problems.

These materials are noted for their toughness and, when appreciable amounts of nickel are present, for their good impact properties at low temperatures.

The steels are used for gears, shafts, etc., very often in the carburized condition. The majority of steels in this section, however, can be replaced by materials in Section 44K1 which contain lesser amounts of nickel, but also contain chromium and molybdenum, and possibly vanadium.

Symbol	Nominal analysis, supplier, condition and remarks.
1.5613	0.24% C 1.2% Ni steel: Forging; German Standard
1.5622	0.14% C 1.4% Ni steel: For low temperature use; German Standard
1.5633	0.24% C 2.1% Ni steel: Forging; German Standard
1.5680	0.2% C (max) 4.7% Ni steel: German Standard
1.6215	0.09% C 1.1% Mn 0.4% Ni steel: Welding rod; German Standard
1.6216	0.17% C 0.8% Ni steel: Welding rod; German Standard
1Ni	0.1% C 1% Mn 0.9% Ni steel: Electrode; Metrode
2Ni	0.1% C 1% Mn 2.3% Ni steel: Electrode; Metrode
2NiB	0.07% C 2.4% Ni steel: Electrode; Metrode; for low temperature toughness
2NiB	0.07% C 2.5% Ni steel: Electrode: Metrode; for low temperature use
2.2D8	0.2% C 0.75% Mn 2.2% Ni steel: Creusot-Loire
3NiB	0.07% C 3.3% Ni steel: Electrode; Metrode; for low temperature toughness
3NiB	0.05% C 2.3% Ni steel: Electrode; Metrode; for low temperature use
3NS	0.3% C 3% Ni steel: ESC for BS alloy En 21
3.5D8	0.19% C 0.75% Mn 3.5% Ni steel: Creusot-Loire
5NSR	0.15% C 5% Ni steel: For case hardening; Jonas; core properties **DPN: 321 UTS: 1000 Elon: 13%**
8CN2	0.08% C 2.0% Ni steel: Pompey
9 Mn Ni 4	0.09% C 1.1% Mn 0.45% Ni steel: Designation used by German Standards
10 N9	0.1% C 9.0% Ni steel: Pompey **UTS: 610 Elon: 22% Proof: 390**
12 Ni19	0.2% C (max) 0.45% Ni steel: Designation used by German Standards
14 N3	0.15% C 3.45% Ni steel: Pompey **UTS: 460 Elon: 25% Proof: 270**
14 Ni6	0.14% C 1.4% Ni steel: Designation used by German Standards
17 Mn Ni4	0.17% C 0.7% Ni steel: Designation used by German Standards
150 M19	0.25% C 0.4% Ni steel: Designation used in BS970
A 203A	0.2% C 2.2% Ni steel: Plate; Armco; for low temperature use **UTS: 670 Elon: 24% Proof: 220**
A 203B	0.24% C 2.2% Ni steel: Plate; Armco; for low temperature use **UTS: 610 Elon: 22% Proof: 270**
A 203D	0.2% C 3.5% Ni steel: Plate; Armco; for low temperature use **UTS: 500 Elon: 24% Proof: 220**
A 203E	0.23% C 3.5% Ni steel: Plate; Armco; for low temperature use **UTS: 580 Elon: 23% Proof: 270**
A 440	0.28% C 1.4% Mn 0.4% Ni 0.2% Cu steel: Armco **UTS: 480 Elon: 20% Proof: 300**
AFNOR 81.340 E 146 3 Ni B26T	0.03% C 3.5% Ni steel: Welding electrode; Federal specification
AISI 2317	0.17% C 3% Ni steel
AISI 2330	0.3% C 3.5% Ni steel: Obsolete
AISI 2512	0.12% C 5% Ni steel: Obsolete
AISI 2515	0.15% C 5% Ni steel
AISI 2517	0.17% C 5% Ni steel
ALFRIG Ni3	0.15% C 1.5% Mn 3.5% Ni steel: Austrian specification
ALLOY 36	0.03% C (max) 36% Ni steel: VDM
AMS 6242C	0.17% C 5% Ni steel: Bar and forging; AMS for SAE 2517
AN QQ S 676/1	0.25% C 1.75% Mn 0.2% Ni steel: US Service **UTS: 460 Elon: 15% Proof: 370**
AN QQ S 689/1	0.3% C 3.5% Ni steel: US Service; normalized and tempered **DPN: 230 UTS: 840 Elon: 17% Proof: 740**
ARMCO HS No. 1	0.15% C (max) 0.5% Mn 0.75% Ni 0.6% Cu steel: Armco
ARMCO HS No. 2	0.15% C (max) 0.6% Mn 0.75% Ni 0.6% Cu steel: Armco
ASTM A203 A	0.2% C 0.75% Mn 2.25% Ni steel: Plate for pressure vessels
ASTM A203 B	0.25% C 0.75% Mn 2.25% Ni steel: Plate for pressure vessels
ASTM A203 D	0.2% C 0.75% Mn 3.5% Ni steel: Plate for pressure vessels
ASTM A203 E	0.21% C 0.75% Mn 3.5% Ni steel: Plate for pressure vessels
ASTM A316 E8016C	0.12% C Ni steel: Welding electrodes; code varies with analysis
ASTM A316 E8016C1	0.12% C 2.4% Ni steel: Electrodes for arc welding
ASTM A316 E8016C2	0.12% C 3.4% Ni steel: Electrodes for arc welding
ASTM A316 E8018C1	0.12% C 2.4% Ni steel: Electrodes for arc welding
ASTM A316 E8018C2	0.12% C 3.4% Ni steel: Electrodes for arc welding
ASTM A320 L10	0.18% C 3.5% Ni steel: Normalized **UTS: 460 Elon: 25% Proof: 270**
ASTM A333/3	0.18% C 3.5% Ni steel: Pipe for low temperature use
ASTM A333/7	0.19% C 2.2% Ni steel: Pipe for low temperature use
ASTM A333/8	0.13% C 9.0% Ni steel: Pipe for low temperature use
ASTM A350 LF3	0.2% C (max) 3.5% Ni steel: For low temperature use **UTS: 460 Elon: 25% Proof: 270**
ASTM A352 IC3	0.15% C 3.5% Ni steel: Normalized and tempered for low temperature service **UTS: 440 Elon: 24% Proof: 270**
ASTM A352 LC2	0.25% C 2.5% Ni steel: Normalized and tempered for low temperature service **UTS: 440 Elon: 24% Proof: 270**
ASTM A352 LC3	0.15% C (max) 0.65% Mn 3.5% Ni steel: Casting for low temperature use
ASTM A353 A and B	0.13% C 9.0% Ni steel: Normalized and tempered **UTS: 630 Elon: 21% Proof: 420**

Note. The following abbreviations and units are used in the tables:

DPN	Hardness, diamond pyramid number
UTS	Ultimate tensile strength, N/mm^2
Elon	Elongation, %
Proof	0.1% proof strength, N/mm^2

1 N/mm^2=0.1 hbar=0.102 kgf/mm^2=0.06475 $tonf/in.^2$=145.04 $lbf/in.^2$=1 MPa

See Appendix II for other abbreviations and conversion tables.

Symbol	Nominal analysis, supplier, condition and remarks.
ASTM A453/660	0.08% C (max) 25.5% Ni 1.45% Cr 1.2% Mo 2.1% Ti 0.3% Al steel: For bolts
ASTM A522	0.13% C (max) 9.0% Ni steel: For valve fitting; low temperature service
ASTM A553	0.13% C (max) 9.0% Ni steel: Plate for pressure vessels **UTS: 750 Elon: 20% Proof: 580**
ASTM A588	0.2% C (max) 0.8% Mn 0.6% Ni steel: For structures **UTS: 440 Elon: 21% Proof: 320**
ASTM A658	0.1% C (max) 0.5% Mn (max) 36.0% Ni 0.5% Cr (max) steel: Plate for pressure vessels **UTS: 500 Elon: 30% Proof: 240**
ASTM A690	0.22% C (max) 0.7% Mn 0.52% Ni steel: For marine piles **UTS: 485 Elon: 18% Proof: 345**
ASTM A707 L7/1	0.2% C (max) 1.0% Mn (max) 3.4% Ni steel: Flanges; for low temperature use; impact 41 J at −73 °C **UTS: 290 Elon: 22%**
ASTM A707 L7/2	0.2% C (max) 1.0% Mn (max) 3.4% Ni steel: Flanges for low temperature use; impact 54 J at −73 °C **UTS: 360 Elon: 25%**
ASTM A707 L7/3	0.2% C (max) 1.0% Mn (max) 3.4% Ni steel: Flanges for low temperature use; impact 68 J at −73 °C **UTS: 415 Elon: 23%**
ASTM A707 L7/4	0.2% C (max) 1.0% Mn (max) 3.4% Ni steel: Flanges for low temperature use; impact 68 J at −73 °C **UTS: 515 Elon: 20%**
ASTM A714/5 E or S	0.2% C (max) 1.4% Mn 1.9% Ni 1.0% Cu steel: Pipe; welded or seamless **UTS: 448 Proof: 317**
ASTM A714/5 F	0.2% C (max) 1.4% Mn 1.9% Ni 1.0% Cu steel: Pipe; welded or seamless **UTS: 379 Elon: 27%**
ASTM A732/11 Q	0.2% C 1.8% Ni 0.25% Mo steel: Investment casting; hardened and tempered **UTS: 830 Elon: 10% Proof: 690**
ASTM A757-B2N	0.25% C 2.5% Ni 1% Cr + Mo + V + Cu (max) steel: Casting; normalized and tempered; impact 20 J at −75 °C **UTS: 485 Elon: 25% Proof: 275**
ASTM A757-B2Q	0.25% C 2.5% Ni 1% Cr + Mo + V + Cu (max) steel: Casting; quench and temper; impact 20 J at −75 °C **UTS: 485 Elon: 25% Proof: 275**
ASTM A757-B3N	0.25% C 3.5% Ni 1% Cr + Mo + V + Cu (max) steel: Casting; normalized and tempered; impact 20 J at −100 °C **UTS: 485 Elon: 25% Proof: 275**
ASTM A757-B3Q	0.25% C 3.5% Ni 1% Cr + Mo + V + Cu (max) steel: Casting; quench and temper; impact 20 J at −100 °C **UTS: 485 Elon: 25% Proof: 275**
ASTM A757-B4N	0.15% C 4.5% Ni 1% Cr + Mo + V + Cu (max) steel: Casting; normalized and tempered; impact 20 J at −115 °C **UTS: 485 Elon: 24% Proof: 275**

Note. The following abbreviations and units are used in the tables:

DPN	Hardness, diamond pyramid number
UTS	Ultimate tensile strength, N/mm^2
Elon	Elongation, %
Proof	0.1% proof strength, N/mm^2

1 N/mm^2=0.1 hbar=0.102 kgf/mm^2=0.06475 tonf/in.^{2}=145.04 lbf/in.^{2}=1 MPa

See Appendix II for other abbreviations and conversion tables.

Symbol	Nominal analysis, supplier, condition and remarks.
ASTM A757-B4Q	0.15% C 4.5% Ni 1% Cr + Mo + V + Cu (max) steel: Casting; quench and temper; impact 20 J at −115 °C **UTS: 485 Elon: 24% Proof: 275**
ASTM A844	0.13% C (max) 9% Ni steel: Plate for pressure vessel; impact 27 J at −200 °C **UTS: 720 Elon: 20% Proof: 585**
AURIGA XVII	0.15% C 3.5% Ni steel: Casting; D Brown; normalized and tempered for sub-zero use **UTS: 440 Elon: 25% Proof: 270**
AUSTEN 56	0.23% C 1.6% Mn 0.6% Ni steel: Australian Iron and Steel Co.
B and W Ni 57	0.2% C 3.5% Ni steel: Tube; Babcock and Wilcox (USA)
BS 970	See designation for details
BS 1501/503	0.15% C 3.50% Ni steel: Plate and bar; normalized **UTS: 440 Elon: 20% Proof: 260**
BS 1501/509	0.1% C (max) 9.0% Ni 0.2% Al (min) steel: Plate; hardened and tempered for pressure vessels; Charpy impact 47 J at 160 °C **UTS: 700 Elon: 18% Proof: 530**
BS 1501/510	0.1% C (max) 9.2% Ni 0.02% Al (min) steel: Plate; hardened and tempered for pressure vessels; Charpy impact 47 J at 160 °C **UTS: 700 Elon: 18% Proof: 600**
BS 1503/503	0.15% C 3.5% Ni steel: Forging; normalized and tempered **UTS: 460 Elon: 20% Proof: 270**
BS 1504/503	0.15% C (max) 3.5% Ni steel: Casting; normalized and tempered **UTS: 440 Elon: 25% Proof: 240**
BS 2493/1 Ni B	0.1% C (max) 1% Ni steel: For welding electrodes
BS 2493/1 Ni C	0.15% C (max) 1% Ni steel: For welding electrodes
BS 2493/1 Ni LB	0.07% C (max) 1% Ni steel: For welding electrodes
BS 2493/2 Ni B	0.1% C (max) 2.5 % Ni steel: For welding electrodes
BS 2493/2 Ni C	0.15% C (max) 2.5% Ni steel: For welding electrodes
BS 2493/2 Ni LB	0.07% C (max) 1% Ni steel: For welding electrodes
BS 2493/3 Ni B	0.1% C (max) 3.5% Ni steel: For welding electrodes
BS 2493/3 Ni LB	0.07% C (max) 3.5% Ni steel: For welding electrodes
BS 2493/Mn Ni B	0.1% C (max) 1.5% Ni steel: For welding electrodes
BS 3100 BL2	0.12% C (max) 3.5% Ni steel: Casting for low temperature use; impact 20 J at −60 °C **UTS: 460 Elon: 20% Proof: 280**
BS 3146 CIA10	0.15% C 3% Ni steel: Investment casting; for case hardening; refined and hardened core properties **UTS: 670 Elon: 18%**
BSN 3	0.3% C 3% Ni steel: Toledo for BS alloy En 21
CASALLOY	0.1% C 1.0% Ni steel: For case hardening; Clyde Alloy; annealed **DPN: 180**
CH 3N	0.13% C 3.1% Ni steel: For case hardening; ESC for BS alloy En 33
CMN 1	0.25% C 1.5% Mn 0.25% Si steel: Firth Brown for BS alloy En 14A and B
CRYONIC 5	0.13% C 5% Ni 0.27% Mo 0.1% V steel: Armco Steel; for low temperature use **UTS: 69 Elon: 20% Proof: 45**
DOFASCOLOY	0.18% C 1.0% Mn 0.9% Ni 0.6% Cu steel: Dominion
DOFASCOLOY 2	0.15% C 1.0% Mn 0.9% Ni 0.6% Cu steel: Dominion
DUCOL QT131	0.2% C 1.6% Mn 0.5% Ni steel: Colvilles; Al killed
E 0	0.12% C 3% Ni steel: Jessop for BS alloy En 33
E 1A	0.3% C 3% Ni steel: Jessop for BS alloy En 21
E 4	0.11% C 4.8% Ni steel: Jessop for BS alloy En 37
E 7 XT8 Ni 2	0.12% C (max) 2.2% Ni 1.8% Al (max) steel: Weld electrode; designation used by AWS
E 8 XT1 Ni 1	0.12% C (max) 1% Ni 1.8% Al steel: Weld electrode; designation used by AWS
E 8 XT5 Ni 1	0.12% C (max) 1% Ni 1.8% Al steel: Weld electrode; designation used by AWS
E 8 XT5 Ni 2	0.12% C (max) 2.2% Ni steel: Weld electrode; designation used by AWS

Symbol	Nominal analysis, supplier, condition and remarks.
E 10 XT1 K7	0.15% C (max) 2.3% Ni steel: Weld electrode; designation used by AWS
E 80 C Ni 1	0.12% C (max) 1% Ni steel: Weld electrode; designation used by AWS
E 80 C Ni 3	0.12% C (max) 3.3% Ni steel: Weld electrode
E 7015 C1L	0.05% C (max) 2.3% Ni steel: Weld electrode; designation used by AWS
E 7015 C2L	0.05% C (max) 3.3% Ni steel: Weld electrode; designation used by AWS
E 7016 C1L	0.05% C (max) 2.3% Ni steel: Weld electrode; designation used by AWS
E 7016 C2L	0.05% C (max) 3.3% Ni steel: Weld electrode; designation used by AWS
E 7018 C1L	0.05% C (max) 2.3% Ni steel: Weld electrode; designation used by AWS
E 7018 C2L	0.05% C (max) 3.3% Ni steel: Weld electrode; designation used by AWS
E 8015	0.12% C (max) 1% Ni 0.4% Si steel: Weld electrode; designation used by AWS
E 8016 C1	0.12% C (max) 2.3% Ni steel: Weld electrode; designation used by AWS
E 8016 C2	0.12% C (max) 3.3% Ni steel: Weld electrode; designation used by AWS
E 8016 C3	0.12% C (max) 1% Ni steel: Weld electrode; designation used by AWS
E 8018 C1	0.12% C (max) 2.3% Ni steel: Weld electrode; designation used by AWS
E 8018 C2	0.12% C (max) 3.3% Ni steel: Weld electrode; designation used by AWS
E 8018 C3	0.12% C (max) 1% Ni steel: Weld electrode; designation used by AWS
E G XX T3	0.1% C (max) 0.9% Ni steel: Weld electrode; designation used by AWS
E Ni 1	0.12% C (max) 1% Ni steel: Weld electrode; designation used by AWS
E Ni 1K	0.18% C (max) 1% Ni steel: Weld electrode; designation used by AWS
E Ni 1KN	0.12% C (max) 1% Ni steel: Weld electrode; designation used by AWS
E Ni 2	0.12% C (max) 2.5% Ni steel: Weld electrode; designation used by AWS
E Ni 2 N	0.12% C (max) 2.5% Ni steel: Weld electrode; designation used by AWS
E Ni N	0.12% C (max) 1% Ni steel: Weld electrode; designation used by AWS
E XX T1 K2	0.15% C (max) 1.5% Ni steel: Weld electrode; designation used by AWS
E XX T1 N2	0.12% C (max) 2.2% Ni steel: Weld electrode; designation used by AWS
E XX T5 K2	0.15% C (max) 1.5% Ni steel: Weld electrode; designation used by AWS
E XX T5 Ni 3	0.12% C (max) 3.2% Ni steel: Weld electrode; designation used by AWS
E7 XX T4 K2	0.15% C (max) 1.5% Ni 1.8% Al steel: Weld electrode; designation used by AWS
EC F1	0.12% C (max) 1.3% Ni steel: Weld electrode; designation used by AWS
EC F1 N	0.12% C (max) 1.3% Ni steel: Weld electrode; designation used by AWS
EC F4	0.17% C (max) 0.6% Ni steel: Weld electrode; designation used by AWS
EC F4 N	0.17% C (max) 0.6% Ni steel: Weld electrode; designation used by AWS
EC Ni 2	0.12% C 2.5% Ni steel: Weld electrode; designation used by AWS
EC Ni 2N	0.12% C 2.5% Ni steel: Weld electrode; designation used by AWS
EC Ni 3	0.12% C (max) 3.2% Ni steel: Weld electrode; designation used by AWS
EC Ni 3N	0.12% C (max) 3.2% Ni steel: Weld electrode; designation used by AWS

Symbol	Nominal analysis, supplier, condition and remarks.
ELGA P62 MR	0.06% C 1.2% Mn 0.9% Ni steel: Electrode; Metrode
En 14B	0.25% C 0.4% Ni steel: Designation used in BS 970; replaced by 150 M19
ER 80 S Ni 1	0.12% C (max) 1% Ni 0.6% Si steel: Weld electrode; designation used by AWS
ER 80 S Ni 2	0.12% C (max) 2.3% Ni steel: Weld electrode; designation used by AWS
ER 80 S Ni 3	0.12% C (max) 3.3% Ni steel: Weld electrode; designation used by AWS
ER 100 S-2	0.12% C (max) 1% Ni 0.5% Cu steel: Weld electrode; designation used by AWS
FLT 2C	0.1% C 0.56% Mn 2.5% Ni 0.1% Cu steel: Fuji
FLT 2D	0.07% C 0.5% Mn 3.5% Ni 0.14% Cu steel: Fuji
FN 30	0.3% C 3% Ni steel: Firth Brown **UTS: 840 Elon: 20%**
HARDY NICKEL IRON	0.08% C 2.0% Ni 0.2% Si steel: Atlas; for low duty sub-zero parts **DPN: 120 UTS: 340 Elon: 30% Proof: 210**
IN 787B	0.06% C (max) 1.3% Ni 0.02% Nb (min) 1.15% Cu steel: International Nickel Co.; solution treated and aged; impact 170 J at −60 °C **UTS: 630 Elon: 26% Proof: 550**
JIS G3460	0.18% C (max) 0.45% Mn 3.6% Ni 0.02% Cu (max) steel: Pipe; Japanese Standard **UTS: 460 Proof: 250**
JIS G3464 STBL46	0.18% C (max) 0.5% Mn 3.6% Ni steel: Pipe; Japanese Standard **UTS: 460 Proof: 250**
K 11268	0.12% C (max) 0.75% Ni 0.17% Al 1% Cu steel: Designation used by UNS
KF 13	0.32% C 3% Ni steel: Kirkstall for BS alloy En 21
KF 15/3	0.12% C 3.2% Ni case hardening steel: Kirkstall for BS alloy En 33
KOERZIT 120	27% Ni 13% Al steel: For magnets; German alloy listed by PMA
KP 11	Low C 2% Ni steel: For case hardening; Kiveton Park
LT 3N	0.12% C 3.5% Ni steel: For low temperature use; ESC; normalized and tempered **UTS: 480 Elon: 24% Proof: 270**
LT 9N	0.1% C 9% Ni steel: For low temperature use; ESC; normalized and tempered **UTS: 670 Elon: 22% Proof: 450**
MAXELOY 4	0.15% C (max) 1.2% Mn 0.5% Ni 0.2% Cu steel: Crucible Steel Co.
MIL S 12505.1	0.23% C (max) 1.25% Mn 0.75% Ni 0.45% Cu steel: Bethlehem Steel Co.
N 1/57	0.2% C 3.5% Ni steel: Tube; Babcock and Wilcox (USA)
N 3CH	0.14% C 3.1% Ni steel: For carburizing; Firth Brown for BS 970 En 33
N 5CH	0.1% C 5.0% Ni steel: For carburizing; Firth Brown for BS 970 En 37
NBV 1	0.24% C 1.15% Ni steel: Krupp
NBV 1/h	0.34% C 1.35% Ni steel: Krupp
NBV 2/w	0.14% C 1.5% Ni steel: Krupp
NBV 5/w	0.12% C 5.0% Ni steel: Krupp
NF A36 208/2.25Ni	0.15% C (max) 0.8% Mn (max) 2.0% Ni steel: Plate; French Standard; normalized **UTS: 500 Proof: 270**
NF A36 208/3.5Ni	0.15% C (max) 0.8% Mn (max) 3.5% Ni steel: Plate; French Standard **UTS: 500 Proof: 270**
NF A36 208/9Ni	0.1% C (max) 0.8% Mn (max) 9.0% Ni steel: Plate; French Standard **UTS: 700 Proof: 600**
Ni Cu Cb	0.06% C (max) 0.5% Mn 0.85% Ni 0.02% Nb (min) 1.15% Cu steel: International Nickel Co.; age hardened **Elon: 29%**
Ni Cu Ti	0.15% C (max) 1.0% Mn 0.7% Ni 0.05% Ti 0.2% Cu steel: Jones and Laughlin

Symbol	Nominal analysis, supplier, condition and remarks.
NICLOY	0.2% C 3.5% Ni steel: Tube; Babcock and Wilcox (USA)
NICLOY 5	0.14% C 5.0% Ni steel: Tube; Babcock and Wilcox (USA)
NICLOY 9	0.12% C 9% Ni 0.3% Cu steel: Tube; Babcock and Wilcox (USA)
NICUAGE	0.06% C 0.5% Mn 0.85% Ni 0.02% Nb 1.15% Cu steel: Strip, bar, section and tube, International Nickel Co.; normalized; aged **DPN: 215 UTS: 640 Elon: 22% Proof: 590**
NICUAGE 1	0.06% C 0.85% Ni 0.02% Nb 1.1% Cu steel: Inco; age hardened **DPN: 235 UTS: 690 Elon: 23% Proof: 640**
NICUAGE TYLER	0.06% C (max) 0.5% Mn 0.85% Ni 0.02% Nb (min) 1.2% Cu steel: Information from Inco; age hardened; impact 50 J at −20 °C **DPN: 230 UTS: 690 Elon: 23% Proof: 620**
NKR	0.3% C 3% Ni steel: Jonas **UTS: 920 Elon: 20% Proof: 760**
NMS	0.23% C 1.6% Mn 0.6% Ni steel: Australian Iron and Steel Co.
NSR	0.15% C 3% Ni steel: For case hardening; Jonas; core properties **DPN: 207 UTS: 690 Elon: 18%**
NV 20.0	0.1% C (max) 0.55% Mn 3.25% Ni (min) steel: Designation used by DNV **UTS: 450 Elon: 21% Proof: 350**
NV 20.00	0.1% C (max) 0.55% Mn 2.25% Ni (min) steel: Designation used by DNV **UTS: 400 Elon: 21% Proof: 320**
NV 20.1	0.1% C (max) 0.55% Mn 4.75% Ni (min) steel: Designation used by DNV **UTS: 500 Elon: 25% Proof: 400**
NV 20.2	0.08% C (max) 0.55% Mn 9.0% Ni (min) steel: Designation used by DNV **UTS: 650 Elon: 20% Proof: 450**
NVR 20.0	0.15% C (max) 0.45% Mn 3.25% Ni (min) steel: Designation used by DNV
NVR 20.00	0.15% C (max) 0.45% Mn 2.25% Ni (min) steel: Designation used by DNV
NVR 20.1	0.1% C (max) 0.55% Mn 4.75% Ni (min) steel: Designation used by DNV
NVR 20.2	0.08% C (max) 0.6% Mn 8.6% Ni (min) steel: Designation used by DNV
P 153	0.12% C 3% Ni steel: For case hardening; Carrs
PERNIFER 36	0.03% C 36% Ni steel: VDM
PHILIPS 75	0.06% C 0.5% Mn 2.2% Ni steel: Welding electrode; Philips; low hydrogen rod **UTS: 580 Elon: 24% Proof: 440**
PITT-TEN 1	0.12% C (max) 0.8% Mn 0.75% Ni 0.8% Cu steel: Pittsburgh Steel
PPT	0.2% C 4.0% Ni 1.2% Al steel: Origin unknown
S4	Nickel steel: Sheet; obsolete
S8	3.0% Ni steel: Hardened and tempered; replaced
S9	3.0% Ni steel: Forging; hardened and tempered; replaced
S10	3.0% Ni steel: Bar and forging; hardened and tempered; replaced
S15	0.12% C 3.0% Ni steel: For carburizing; blank carburized; hardened and tempered **UTS: 670 Elon: 18%**
S17	5% Ni steel: For carburizing; replaced
S46	Nickel alloy steel: Strip; obsolete
S67	Low carbon 5% Ni steel: For carburizing; obsolete
S83	Low carbon 5% Ni steel: For carburizing; obsolete
SAE 2317	0.17% C 3.5% Ni steel: Obsolete
SAE 2330	0.3% C 3.5% Ni steel: Obsolete
SAE 2512	0.12% C 5% Ni steel: Obsolete
SAE 2515	0.15% C 5% Ni steel: Mechanical properties not specified
SAE 2517	0.17% C 0.5% Mn 5% Ni steel: Mechanical properties not specified
SANBOLD 14	0.25% C 1.5% Mn 0.4% Ni steel: Sanderson for En 14B
SANBOLD 51	0.25% C 1.5% Mn 0.4% Ni steel: Sanderson; replaced by Sanbold 14
SHEF-TEN	0.28% C 1.4% Mn 0.4% Ni 0.2% Cu steel: Armco **UTS: 480 Elon: 20% Proof: 300**
SL 2N26	0.17% C (max) 0.7% Mn (max) 2.2% Ni steel: Plate; Japanese Standard; draft specification **UTS: 500 Proof: 260**
SL 3N26	0.15% C (max) 0.7% Mn (max) 3.5% Ni steel: Plate; Japanese Standard; draft specification **UTS: 500 Proof: 260**
SL 3N45	0.15% C (max) 0.7% Mn (max) 3.5% Ni steel: Plate; French Standard; draft; hardened and tempered **UTS: 620 Proof: 450**
SL 9N53	0.12% C (max) 9.0% Ni steel: Plate; Japanese Standard; draft; normalized; tempered **UTS: 770 Proof: 530**
SL 9N60	0.12% C (max) 0.9% Mn (max) 9.0% Ni steel: Plate; Japanese Standard; draft; hardened and tempered **UTS: 770 Proof: 600**
SN 5	Nickel steel sintered material: Sintered Products Ltd; 10–15% porosity **UTS: 400 Elon: 5%**
SN 10	Nickel steel sintered material: Sintered Products Ltd; 10–15% porosity **UTS: 630 Elon: 3%**
SN 15	Nickel steel sintered material: Sintered Products Ltd; 10–15% porosity **UTS: 840 Elon: 2%**
SOUDER GWT	0.1% C 0.9% Ni 0.4% Cu steel: Welding electrode; Soudometal; for inert gas welding **UTS: 570 Elon: 24% Proof: 440**
SPEAR 433	0.12% C 3.0% Ni steel: Spear and Jackson as BS alloy En 33
SPEAR 521	0.3% C 3.2% Ni steel: Spear and Jackson as BS alloy En 21
SS 8110	0.2% C (max) 3.0% Ni 1.0% Cu steel: Sintered
SUPER TOUGH B25	0.25% C 1.5% Mn 0.4% Ni steel: ESC for BS alloy En 14
TENASOUDO Ni 3 S	0.04% C 0.7% Mn 3.5% Ni: Welding electrode; Soudometal; welding low alloy steel **UTS: 570 Elon: 26% Proof: 480**
TUBROD 14.18	0.1% C 0.8% Mn 1.5% Ni 0.6% Al steel: Welding wire; Esab; obsolete
TUBROD 15.18	0.08% C 1.2% Mn 0.5% Ni 0.4% Si steel: Tubular weld rod; Esab; for hard facing; as welded **UTS: 625 Elon: 25% Proof: 470**
VS 3N	0.1% C 3% Ni steel: For carburizing; Vulcan **UTS: 690**
VS 5N	0.1% C 5% Ni steel: For carburizing; Vulcan **UTS: 1070**
VS 15M	0.25% C 1.5% Mn 0.4% Ni steel: Vulcan
VS 35N	0.35% C 3.5% Ni steel: Vulcan **UTS: 760**
YND 37	0.14% C (max) 1.5% Mn 0.7% Ni steel: Yamata
YOLOY S	0.2% C (max) 1.0% Mn 1.9% Ni 1.0% Cu steel: Youngstown Steel Co.

Note. The following abbreviations and units are used in the tables:

DPN	Hardness, diamond pyramid number
UTS	Ultimate tensile strength, N/mm^2
Elon	Elongation, %
Proof	0.1% proof strength, N/mm^2

1 N/mm^2=0.1 hbar=0.102 kgf/mm^2=0.06475 tonf/in.^{2}=145.04 lbf/in.^{2}=1 MPa

See Appendix II for other abbreviations and conversion tables.

44F2 Steel – medium carbon with nickel 0.4% carbon minimum

Specific gravity	7.8
Density	7800 kg/m^3
Solidus/liquidus	1450–1500 °C
Thermal conductivity	50 W/m °C
Coefficient of linear expansion	11×10^{-6}/ °C
Electrical conductivity	6–9% IACS (copper 100%)
Specific resistance	230–270 microhm mm
Young's modulus of elasticity	–
Impact	–102 J
Fatigue strength	±539 N/mm^2
Hot strength	

Temperature °C	Tensile strength N/mm^2	Elongation %
200	750	25.5
400	730	21.5
600	270	40
800	100	71
1000	40	72.5

The above properties are typical of the following group, and may not apply exactly to any one specification. It is possible that with certain specifications some of the values may not be applicable.

General metallurgical characteristics

Nickel additions in steel lower the critical change points, in proportion to the nickel content. This means that parts do not require to be taken to as high a temperature as with plain carbon steels. Nickel also toughens the ferrite, resulting in better mechanical properties even in the normalized or annealed conditions. There is some increase in corrosion resistance, again proportional to the nickel content, but this is never sufficient in the steels listed to be noticeable.

The one time scarcity and high cost of nickel resulted in research that culminated in the use of low nickel, chromium and molybdenum steels in lieu of the 4% nickel series and this use has remained long after the scarcity disappeared.

There is still considerable use of the low nickel steels which show a degree of toughness in the normalized condition, and also nickel bearing steels for good impact properties at low temperatures.

These steels with the higher carbon content become brittle when cold worked to any great extent and therefore require to be annealed to remove this embrittling effect. Full annealing is at the same temperature as hardening and normalizing, in the region of 800 °C. The majority of the steels in this group have considerable tolerance to the thermal treatment temperatures.

Welding presents no serious problems with steels with lower amounts of carbon but as the amount of carbon increases then problems can be encountered.

The carbon equivalent (CE) formula can supply useful information. This is as follows:

$$CE = C + \frac{Mn}{6} + \frac{Cr + Mo + V}{5} + \frac{Ni + Cu}{15}$$

Values above 0.7% can indicate problems.

Nickel has little effect on the hardenability of steel but with a higher proportion of nickel, above 3%, some effect does exist and specialist advice should be sought.

Some of the steels with lower amounts of carbon are used for structural steels but in the main they are used for engineering purposes, shafts and components such as pneumatic tools which are subjected to impact loading.

These steels have excellent impact resistance and toughness with excellent low temperature impact properties when an appreciable amount of nickel is present.

Note. The following abbreviations and units are used in the tables:

DPN	Hardness, diamond pyramid number
UTS	Ultimate tensile strength, N/mm^2
Elon	Elongation, %
Proof	0.1% proof strength, N/mm^2

1 N/mm^2=0.1 hbar=0.102 kgf/mm^2=0.06475 tonf/in.^{2}=145.04 lbf/in.^{2}=1 MPa
See Appendix II for other abbreviations and conversion tables.

Symbol	Nominal analysis, supplier, condition and remarks.
14N 3	0.75% C 2.6% Ni steel: Sandvik
503 A37	0.37% C 0.85% Ni steel: BS 970; obsolete
503 A42	0.42% C 0.85% Ni steel: BS 970; obsolete
503 H37	0.37% C 0.85% Ni steel: BS 970; obsolete
503 H42	0.42% C 0.85% Ni steel: BS 970; obsolete
503 M40	0.4% C 0.85% Ni steel: BS 970; obsolete
13728	2% Co 27% Ni 12% Al 3% Cu steel: For magnets; German Standards alloy for Alni 120 in DIN 17410
AISI A2340	0.4% C 3.5% Ni steel: Obsolete
AISI A2345	0.45% C 3.5% Ni steel: Obsolete
Al Ni	22% Ni 12% Al steel: Casting; German Standard

Symbol	Nominal analysis, supplier, condition and remarks.
Al Ni 120	27.5% Ni 13% Al 2.5% Cu steel: Casting for magnets; German Standard
ALNI	28% Ni 0.5% Ti 12% Al 0.4% Cu steel: Casting for magnets; PMA
ALNI 1	24.5% Ni 12.5% Al 6% Cu steel: For magnets; casting; Jessops
ALNICO 3	26% Ni 12% Al 2.0% Cu steel: Casting for magnets; American alloy; information from PMA
AM 4 Ni	0.4% C 1.5% Mn 0.8% Ni steel: Jessop for BS alloy En 12
ASTM A320 L9	0.4% C 3.5% Ni steel: Hardened and tempered **UTS: 880 Elon: 16% Proof: 720**
ASTM A352 LC2	0.25% C (max) 0.65% Mn 2.5% Ni steel: Casting for low temperature use
BS 970	See designation for details
BS 980 CDS 14	0.4% C 3% Ni steel: Cold drawn tube; hardened and tempered **UTS: 690 Proof: 580**
BSN 1	0.4% C 1.5% Mn 0.8% Ni steel: Toledo for BS alloy En 12
BSN 35	0.4% C 3.5% Ni steel: Toledo
CHISEL STEEL	0.4% C 3% Ni steel: Origin unknown
CL 15	0.4% C 0.5% Mn 3.25% Ni steel: Origin unknown
DIN 17200 41 Cr 4	0.41% C 0.65% Mn 1.05% Cr steel: German Standard
DIN 17200 41 Cr S 4	0.41% C 0.65% Mn 0.05% Cr steel: German Standard
DMA 3	26% Ni 12% Al 2.0% Cu steel: For magnets; Japanese alloy listed by PMA
DUNELT 26	0.4% C 0.7% Ni steel: Dunford for BS alloy En 12
E 1	0.4% C 3.5% Ni steel: Jessop
En 10	0.55% C 0.7% Ni steel: Designation used in BS 970; obsolete
En 12	0.4% C 0.8% Ni steel: Designation used in BS 970; obsolete
En 12 B	0.37% C 0.8% Ni steel: Designation used in BS 970; replaced by 503 A37
En 12 C	0.42% C 0.8% Ni steel: Designation used in BS 970; replaced by 503 A42
EXPANDAL	0.6% C 5.7% Mn 10% Ni steel: Latrobe; high expansion steel
GOST 2052/60 S2N2A	0.6% C 0.55% Mn 1.55% Ni 0.3% Cr (max) steel: Russian Standard
HDB 2	0.6% C 1.25% Ni steel: Huntsman; large press dies
HECLA 10	0.55% C 0.65% Ni steel: Hadfields for BS alloy En 10
HECLA 115	0.4% C 1.5% Mn 0.8% Ni steel: Hadfields for BS alloy En 12
HEDEX 24	0.4% C 0.8% Ni steel: Wire for forgings; Kiveton Park
HTCN	0.4% C 0.75% Ni steel: ESC for BS alloy En 12
KF 12A	0.4% C 1.5% Mn 0.8% Ni steel: Kirkstall for BS alloy En 12
KOERZIT 90	21% Ni 12% Al steel: For magnets; German alloy listed by PMA
KP 10	0.4% C 1.5% Mn 0.8% Ni steel: Kiveton Park for BS alloy En 12
LE	0.4% C 3.20% Ni tool steel: Osborn; chisel steel; obsolete **DPN: 600**
LT 226	0.4% C 3.2% Ni steel: Low Moor; obsolete **DPN: 600**
MANOIR 15N4	0.15% C 3.8% Ni steel: Pompey **UTS: 550 Elon: 18% Proof: 320**
MARSH N3	0.4% C 3.2% Ni steel: Marsh Brothers; chisels
MAX MITH	0.4% C 3.25% Ni steel: Edgar Allen; blacksmiths' tools, chisels, etc.
N 3	0.4% C 3.2% Ni steel: Marsh Brothers; chisels
N 10	Ni steel sinter: Firth Cleveland; specific gravity 6.6–7.0; hardened and tempered; file hard **UTS: 700 Elon: 0.8%**
N 20	Ni steel sinter: Firth Cleveland; specific gravity 6.8–7.2; hardened and tempered; file hard **UTS: 1010 Elon: 2%**
N 30	Ni steel sinter: Firth Cleveland; specific gravity 6.7–7.1; hardened and tempered; file hard **UTS: 890 Elon: 1%**
Ni Al 0	22% Ni 10% Al steel: Casting for magnets; Belgian alloy listed by PMA
Ni Al 1	30% Ni 12% Al Fe alloy: Casting for magnets; Belgian alloy listed PMA
NIPERMAG	30.0% Ni 0.4% Ti 12.0% Al Fe alloy: For magnets; origin unknown
ON 1	0.4% C 1.5% Mn 0.8% Ni steel: W Marrison for BS alloy En 12
S 4	Nickel steel: Sheet; obsolete; BS designation for S4
S 50	Ni Mn steel sinter: Firth Cleveland; specific gravity 6.4–6.8; hardened and tempered; file hard **UTS: 700 Elon: 0.8%**
SAE 2340	0.4% C 3.5% Ni steel: Obsolete
SAE 2345	0.45% C 3.5% Ni steel: Obsolete
SANBOLD 12	0.4% C 1.5% Mn 1% Ni steel: Sanderson for BS alloy En 12
SANBOLD 39	0.4% C 1.5% Mn 1.0% Ni steel: Sanderson; replaced by Sanbold 12
SPEAR D3	0.4% C 0.5% Mn 3.25% Ni steel: Origin unknown
STA 5/V/5/1	0.55% C 0.65% Ni steel: Replaced by BS alloy En 10
T 3N	0.4% C 3.2% Ni steel: ESC; punches, etc.

Note. The following abbreviations and units are used in the tables:

DPN	Hardness, diamond pyramid number
UTS	Ultimate tensile strength, N/mm^2
Elon	Elongation, %
Proof	0.1% proof strength, N/mm^2

1 N/mm^2=0.1 hbar=0.102 kgf/mm^2=0.06475 $tonf/in.^2$=145.04 $lbf/in.^2$=1 MPa

See Appendix II for other abbreviations and conversion tables.

44F3 Steel – low carbon, high nickel – maraging with molybdenum, cobalt, aluminium and titanium

Specific gravity		8.05
Density		8050 kg/m^3
Solidus/liquidus		–
Thermal conductivity		–
Coefficient of linear expansion		10×10^{-6}/ °C
Electrical conductivity	aged	4% IACS (copper 100%)
Specific resistance	aged	400 microhm mm
Young's modulus of elasticity		186×10^9 N/m^2
Impact		–
Fatigue strength		±900 N/mm^2
Hot strength		

Temperature °C	Tensile strength N/mm^2	Elongation %
400	1400	18
500	1300	20
550	1000	22

The above properties are typical of the following group, and may not apply exactly to any one specification. It is possible that with certain specifications some of the values may not be applicable.

General metallurgical characteristics

These steels have over 15% nickel, with titanium, generally some cobalt, molybdenum and aluminium. The carbon, phosphorus and sulphur are all held at low levels.

They are modern steels developed by the International Nickel Company and are not yet in general use.

The materials have very high tensile and proof strength, of the order of 2000 N/mm^2 with good elongation at 10–15% and unique notch resistance for any material of comparable strength.

The metallurgy depends on a complex precipitation reaction caused by intermetallic compounds of nickel, cobalt, titanium or aluminium hardening a fully martensitic structure. This martensite is very ductile owing to the low carbon and high nickel. The steels are martensitic in the annealed condition but can accept considerable cold work which can be used to enhance the final mechanical properties by subsequent ageing. They have no outstanding corrosion resistance, being only slightly better than normal alloy steels. Their excellent notch resistance, however, considerably reduces the danger normally associated with rust in high strength steels.

These steels are still at the development stage but the high strength, excellent notch strength and weldability indicate a large number of potential uses.

Because of the very low carbon content in the steels in this group they can generally accept a high degree of cold work without any serious increase in brittleness.

The thermal treatment of these steels requires specialist advice. They are expensive materials and have been formulated to give the highest tensile strength to ductility ratio of any steels. They must be treated correctly, however, in order to achieve the correct properties.

The same comments apply to welding. No technical problems exist regarding the production of a weld but this weld could seriously affect the mechanical properties in the heat affected zone unless the welding is very carefully controlled. Wherever possible welding should be carried out prior to the thermal treatment.

These steels are difficult to machine because of their toughness in the soft condition and after full ageing have high hardness and therefore will cause severe tool wear.

It is possible to age partially for best machinability results and then to age fully after machining.

These steels were developed because of their high strength to ductility ratio and, where rotating parts occur, uses are being found.

To date, however, there is no indication that these steels are finding common use in general engineering.

Note. The following abbreviations and units are used in the tables:

DPN	Hardness, diamond pyramid number
UTS	Ultimate tensile strength, N/mm^2
Elon	Elongation, %
Proof	0.1% proof strength, N/mm^2

1 N/mm^2=0.1 hbar=0.102 kgf/mm^2=0.06475 tonf/in.^{2}=145.04 lbf/in.^{2}=1 MPa

See Appendix II for other abbreviations and conversion tables.

Symbol	Nominal analysis, supplier, condition and remarks.
1.2706	0.03% C 18.5% Ni 5.3% Mo 9% Co 0.6% Ti 0.1% Al steel: German Standard
1.6351	0.03% C 17.0% Ni 4.7% Mo 10.2% Co 0.3% Ti steel: Casting; maraging; German Standard
1.6354	0.03% C 18.0% Ni 4.9% Mo 8.7% Co 0.75% Ti steel: Maraging; German Standard
1.6359	0.03% C 18.0% Ni 4.9% Mo 7.7% Co 0.45% Ti steel: Maraging; German Standard
17 Ni 1600	17.0% Ni 4.6% Mo 10.0% Co 0.3% Ti 0.05% Al steel: Castings; maraging; Inco **Proof: 1600**

Symbol	Nominal analysis, supplier, condition and remarks.
18 Ni 1400	18.0% Ni 3.0% Mo 8.5% Co 0.2% Ti 0.1% Al steel: Maraging; Inco **Proof: 1400**
18 Ni 1700	18.0% Ni 5.0% Mo 8.0% Co 0.4% Ti 0.01% Al steel: Maraging; Inco **Proof: 1700**
18 Ni 1900	18.0% Ni 5.0% Mo 9.0% Co 0.6% Ti 0.01% Al steel: Maraging; Inco **Proof: 1900**
18 Ni 2400	17.5% Ni 3.75% Mo 12.5% Co 1.8% Ti 0.15% Al steel: Maraging; Inco **Proof: 2400**
18 Ni MA	Low C 18.0% Ni 4.0% Mo 8.0% Co 0.25% Ti steel: Electrode; Metrode; maraging; aged **DPN: 570**
18 Ni MA	0.02% C 18% Ni 4% Mo 8% Co Ti steel: Electrode; Metrode; for maraging steel **DPN: 390**
18 Ni MP	0.02% C 18% Ni 4% Mo 8% Co Ti steel: Electrode; Metrode; for maraging steel **DPN: 390**
250 0.02% C	18.3% Ni 5.0% Mo 8.0% Co 0.4% Ti 0.1% Al steel: Maraging; information from Climax Molybdenum
300 0.02% C	18.3% Ni 5.0% Mo 9.0% Co 0.65% Ti 0.1% Al steel: Maraging; information from Climax Molybdenum
648A	0.02% C 18% Ni 3.2% Mo 8.5% Co 0.2% Ti 0.1% Al steel: Firth Brown; maraging **DPN: 138**
648B	0.02% C 18% Ni 4.8% Mo 7.7% Co 0.4% Ti 0.1% Al steel: Firth Brown; maraging **UTS: 1700 Elon: 12% Proof: 1600**
648C	0.02% C 18% Ni 5.0% Mo 9.0% Co 0.6% Ti 0.1% Al steel: Firth Brown; maraging **UTS: 2000 Elon: 12% Proof: 1900**
AL18 NiCoMo (250)	0.02% C 18% Ni 5% Mo 7% Co 0.4% Ti 0.1% Al steel: Allegheny Ludlum; solution treated and aged; maraging **DPN: 500 UTS: 1700 Elon: 12% Proof: 1600**
AL18 NiCoMo (300)	0.02% C 18% Ni 5% Mo 9% Co 0.6% Ti 0.1% Al steel: Allegheny Ludlum; solution treated and aged; maraging **UTS: 2200 Elon: 14% Proof: 2100**
AL20 Ni (250)	0.02% C 20% Ni 1.4% Ti 0.5% Nb 0.2% Al steel: Allegheny Ludlum; solution treated and aged; maraging **DPN: 630 UTS: 2300 Elon: 1.5% Proof: 2200**
AL25 Ni (250)	0.02% C 25% Ni 1.4% Ti 0.5% Nb 0.2% Al steel: Allegheny Ludlum; solution treated and aged; maraging **UTS: 2400 Elon: 2% Proof: 2200**
ALMAR 18/200	0.03% C 18% Ni 32% Mo 8.5% Co 0.2% Ti 0.02% Zr steel: Allegheny Ludlum; solution treated and aged; maraging **UTS: 1550 Elon: 15% Proof: 1500**
ALMAR 18/250	0.03% C 18% Ni 5% Mo 8% Co 0.4% Ti 0.1% Al 0.02% Zr steel: Allegheny Ludlum; solution treated and aged; maraging **DPN: 580 UTS: 2000 Elon: 12% Proof: 1850**
ALMAR 18/300	0.03% C 18.5% Ni 5% Mo 9% Co 0.7% Ti 0.1% Al 0.02% Zr steel: Allegheny Ludlum; solution treated and aged; maraging **DPN: 610 UTS: 2200 Elon: 11% Proof: 2100**
ALMAR 20	0.03% C 19.5% Ni 1.5% Ti 0.5% Nb 0.2% Al 0.02% Zr steel: Allegheny Ludlum; solution treated and aged; maraging **UTS: 1900 Elon: 8% Proof: 1700**
ALMAR 25	0.03% C 25.5% Ni 1.5% Ti 0.5% Nb 0.2% Al 0.02% Zr steel: Allegheny Ludlum; solution treated and aged; maraging **UTS: 2000 Elon: 8% Proof: 1850**
AMS 5337	0.03% C (max) 18.5% Ni 5% Mo 9% Co 0.65% Ti 0.1% Al steel: Casting
AMS 5339	0.03% C (max) 16.8% Ni 4.6% Mo 10.2% Co 0.35% Ti steel: Casting
AMS 6501	0.03% C 18% Ni 4.8% Mo 7.7% Co 0.45% Ti 0.1% Al steel
AMS 6512	0.03% C 18% Ni 4.8% Mo 7.7% Co 0.45% Ti 0.1% Al steel
AMS 6514	0.03% C (max) 18.5% Ni 4.9% Mo 8.7% Co 0.65% Ti 0.1% Al steel
AMS 6520	0.03% C (max) 18% Ni 4.8% Mo 7.7% Co 0.45% Ti 0.1% Al steel
AMS 6521	0.03% C (max) 18.5% Ni 4.9% Mo 8.7% Co 0.65% Ti 0.1% Al steel
ASTM A538 A	0.03% C 18% Ni 4.2% Mo 7.7% Co 0.2% Ti steel: Maraging **UTS: 1440 Elon: 8% Proof: 1380**
ASTM A538 B	0.03% C 18% Ni 4.8% Mo 7.7% Co 0.4% Ti steel: Maraging **UTS: 1650 Elon: 6% Proof: 1600**
ASTM A538 C	0.03% C 18.5% Ni 4.9% Mo 8.7% Co 0.7% Ti steel: Maraging **UTS: 1880 Elon: 6% Proof: 1700**
ASTM A579/75	0.03% C 12% Ni 5.0% Cr 3.0% Mo 0.17% Ti 0.4% Al steel: Forging; maraging **UTS: 1300 Elon: 14% Proof: 1250**
CFS	1.1% C 0.1% Ni 0.5% Cr steel: Latrobe
DTD 5212	18% Ni Co Mo steel: Maraging; double vacuum melted **UTS: 1800**
DTD 5232	18% Ni Co Mo steel: Maraging; vacuum melted **UTS: 1800**
E Fe 7	2.2% C 6% Cr steel: Weld electrode; designation used by AWS
FE PA95	0.03% C 18.0% Ni 4.9% Mo 8.0% Co 0.4% Ti steel: Maraging; origin unknown
G 140	0.02% C 18.0% Ni 4.8% Mo 8.0% Co 0.4% Ti steel: Jessop
INCO MS 250	0.03% C (max) 19% Ni 3% Mo 1.4% Ti 0.1% Al Fe alloy: Inco; solution test and aged **UTS: 1830 Elon: 12% Proof: 1750**
MAR 100	0.02% C 16.75% Ni 4.65% Mo 10.3% Co 0.37% Ti 0.11% Al Fe alloy: Union Carbide
MARAGE 200	0.03% C 18.0% Ni 3.2% Mo 8.5% Co 0.1% Ti steel
MARAGE 250	0.03% C (max) 18.0% Ni 4.9% Mo 7.7% Co 0.4% Ti 0.08% Al steel
MARAGE 300	0.03% C (max) 18.5% Ni 5.0% Mo 9.0% Co 0.6% Ti 0.08% Al steel
MARVAC 18	0.03% C 18% Ni 5.0% Mo 7.5% Co 0.4% Ti 0.1% Al steel: Latrobe Steel; maraged **UTS: 1900 Elon: 11% Proof: 1700**
MARVAC 18A	0.03% C 18% Ni 5.0% Mo 9.0% Co 0.6% Ti 0.1% Al steel: Latrobe Steel; maraged **UTS: 2100 Elon: 10% Proof: 2000**
MARVAC 18LT	0.03% C 18% Ni 4.5% Mo 8.0% Co 0.2% Ti 0.1% Al steel: Latrobe Steel; maraged **UTS: 11% Proof: 1650**
MARVAC 200	0.02% C 18% Ni 4.25% Mo 7.5% Co 0.2% Ti 0.1% Al steel: Latrobe; maraging

Note. The following abbreviations and units are used in the tables:

DPN	Hardness, diamond pyramid number
UTS	Ultimate tensile strength, N/mm^2
Elon	Elongation, %
Proof	0.1% proof strength, N/mm^2

1 N/mm^2=0.1 hbar=0.102 kgf/mm^2=0.06475 $tonf/in.^2$=145.04 $lbf/in.^2$=1 MPa

See Appendix II for other abbreviations and conversion tables.

Symbol	Nominal analysis, supplier, condition and remarks.
MARVAC 250	0.02% C 18% Ni 5.0% Mo 8.0% Co 0.4% Ti 0.1% Al steel: Latrobe; maraging
MARVAC 300	0.02% C 18% Ni 5.0% Mo 9.0% Co 0.6% Ti 0.1% Al steel: Latrobe; maraging
Ni MARK 1	0.3% C (max) 18.5% Ni 4.9% Mo 7.5% Co 0.4% Ti 0.1% Al steel: Carpenter; maraging
Ni MARK 250	0.03% C 4.8% Mo 7.5% Co 0.4% Ti 0.003% B 0.1% Al steel: Carpenter **DPN: 510 UTS: 1760 Elon: 12% Proof: 1720**
Ni MARK 300	0.03% C (max) 18.5% Ni 4.9% Mo 8.5% Co 0.7% Ti 0.1% Al steel: Carpenter; maraging
PYROMET ALLOY CTX-1	0.03% C 37.5% Ni 16% Co 1.7% Ti 3% Nb + Ta steel: Carpenter; solution treated and aged **UTS: 1500 Elon: 16% Proof: 1300**
PYROMET ALLOY CTX-3	0.06% C (max) 38% Ni 13% Co 1.5% Ti 4.8% Nb + Ta steel: Carpenter; solution treated and aged **UTS: 1180 Elon: 14% Proof: 830**
TELMAR 90	0.01% C 18% Ni 3.2% Mo 8.5% Co 0.2% Ti steel: Maraging; Telcon **DPN: 450 UTS: 1500 Elon: 11% Proof: 1420**
TELMAR 110	0.01% C 18% Ni 4.8% Mo 7.7% Co 0.4% Ti steel: Maraging; Telcon **DPN: 520 UTS: 1800 Elon: 12% Proof: 1700**
TELMAR 125	0.01% C 18.5% Ni 4.9% Mo 9.0% Co 0.6% Ti steel: Maraging; Telcon **DPN: 560 UTS: 1900 Elon: 12% Proof: 1850**
THYROTHERM 2799	0.02% C (max) 12% Ni 8% Mo 8% Co 0.5% Ti 0.05% Al steel: Thyssen **DPN: 600 Proof: 2413**
UCAR MARAGING 110	0.01% C 18.0% Ni 5.0% Mo 8.0% Co 0.5% Ti 0.1% Al maraging steel: Union Carbide
ULTIMATE 110	0.01% C 17.5% Ni 4.8% Mo 8.0% V with Ti and Al steel: For die casting aluminium; Jessop; maraging **DPN: 600**
VASCOMAX 250	0.03% C (max) 18% Ni 5% Mo 7% Co 4% Ti 1% Al steel: Vanadium Alloys; solution treated and aged; maraging steel **DPN: 540 UTS: 1900 Elon: 10% Proof: 1700**
VASCOMAX 300 CVM	0.03% C 18% Ni 5% Mo 9% Co 6% Ti 1% Al steel: Vanadium Alloys; solution treated and aged; maraged **DPN: 610 UTS: 2200 Elon: 11% Proof: 2000**
VMA-B	10% Co 18% Ni 4.5% Mo steel: Investment casting; BS designation; obsolete
VMAI-A	10% Co 18% Ni 4.5% Mo steel: Investment casting; BS designation; obsolete
X 3 Ni Co Mo 18-8-5	0.03% C 18.5% Ni 5.3% Mo 9% Co 0.6% Ti 0.1% Al steel: German Standard

Note. The following abbreviations and units are used in the tables:

DPN	Hardness, diamond pyramid number
UTS	Ultimate tensile strength, N/mm^2
Elon	Elongation, %
Proof	0.1% proof strength, N/mm^2

1 N/mm^2=0.1 hbar=0.102 kgf/mm^2=0.06475 $tonf/in.^2$=145.04 $ibf/in.^2$=1 MPa

See Appendix II for other abbreviations and conversion tables.

44G1 Steel – low carbon with molybdenum 0.05–0.35% carbon, 0.7% molybdenum maximum

Specific gravity	7.85
Density	7850 kg/m^3
Solidus/liquidus	1470–1510 °C
Thermal conductivity	50 W/m °C
Coefficient of linear expansion	11×10^{-6}/ °C
Electrical conductivity	8–10% IACS (copper 100%)
Specific resistance	180–250 microhm mm
Young's modulus of elasticity	–
Impact	54–108 J
Fatigue strength (no. of cycles)	–
Hot strength	–

The above properties are typical of the following group, and may not apply exactly to any one specification. It is possible that with certain specifications some of the values may not be applicable.

General metallurgical characteristics

Molybdenum forms very stable carbides which are harder than, and formed in preference to, iron carbides. Because of this molybdenum steels have better creep strength, maintain hardness at higher temperature and have increased hardenability over plain carbon steels. It also acts as a grain refiner, preventing excessive grain growth. Molybdenum has many similar effects to tungsten, which is generally cheaper but requires larger quantities to achieve the same purpose.

None of the steels listed has more than 0.7% molybdenum and very few have as much as 0.5%. Traces of boron are sometimes added in addition to the molybdenum. This has a similar effect and is additive to that of molybdenum as far as hardenability is concerned.

These steels are seldom used in the cold worked condition.

The small amount of molybdenum present in the steels of this group has little effect on the temperatures used during thermal treatment. It also has little effect on the cold working properties of the steels. The molybdenum, however, does affect the hardenability and these steels will generally require oil quench rather than water quench.

Welding will present no problem provided that it is appreciated that the hardenability will be better than for plain carbon steels. These materials should be treated in a similar manner to steels with a similar amount of chromium. For high integrity parts, specialist advice should be obtained.

It must be appreciated that there could be 'tramp' elements present, i.e. traces of chromium or nickel which will increase the hardenability and thus could present problems at welding.

The machinability of these steels is generally good regarding surface finish but, because of the abrasive nature of the carbides formed, tool wear can be expected.

In the annealed condition the steels will tend to tear rather than cut.

These steels find uses for boiler tubes and equipment used in steam-raising plant where the good creep properties of the steel increase the length of life available.

They will seldom be used where the other mechanical properties alone are required, because of the high cost of molybdenum.

Symbol	Nominal analysis, supplier, condition and remarks.
1.2 Mo6/7	0.22% C 1.1% Mn 0.5% Mo steel: Creusot-Loire
1.3 Mo6/7	0.22% C 1.25% Mn 0.5% Mo steel: Creusot-Loire
1.5415	0.16% C 0.3% Mo steel: For boiler plate; German Standard
1.5416	0.15% C 0.3% Mo steel: German Standard
1.5424	0.12% C 0.5% Mo 0.2% Cu steel: Welding rod; German Standard
1.5425	0.1% C 0.5% Mo steel: Welding rod: German Standard
1.5426	0.15% C 0.5% Mo steel: Welding rod; German Standard
1.5427	0.13% C 2.0% Mn 0.5% Mo steel: Welding rod; German Standard
1.5428	0.15% C 3.0% Mn 0.5% Mo steel: Welding rod; German Standard
3 Mo	0.15% C 0.35% Mo steel: Sandvik
3 Mo 1	0.13% C 0.45% Mo steel: Sandvik
9 Mn Mo 4/5	0.09% C 1.0% Mn 0.5% Mo 0.22% Cu steel: Designation used by German Standards
13 Mn Mo 3/5	0.12% C 0.5% Mo 0.22% Cu steel: Designation used by German Standards
13 Mn Mo 6/5	0.14% C 1.5% Mn 0.5% Mo steel: Designation used by German Standards
13 Mn Mo 8/5	0.13% C 2.0% Mn 0.5% Mo steel: Designation used by German Standards
13 Mn Mo 12/5	0.14% C 3.0% Mn 0.5% Mo steel: Designation used by German Standards
15 Mo 3	0.15% C 0.3% Mo steel: Designation used by German Standards
18 MM	0.18% C 1.3% Mn 0.2% Mo steel: Electrode; Metrode
20 Mn Mo 4	0.2% C 1.1% Mn 0.25% Mo steel: Designation used by German Standards
20 Mo 3	0.15% C 0.3% Mo steel: Designation used by German Standards
30 MM	0.3% C 1.3% Mn 0.25% Mo steel: Electrode; Metrode
605 A 32	0.32% C 0.27% Mo steel: BS 970
605 A 37	0.37% C 0.27% Mo steel: BS 970
605 H 32	0.32% C 0.27% Mo steel: BS 970
605 H 37	0.37% C 0.27% Mo steel: BS 970
605 M 30	0.3% C 0.27% Mo steel: BS 970; obsolete
605 M 36	0.36% C 0.27% Mo steel: BS 970
4012	0.12% C 0.2% Mo steel: Designation used in the UK and USA
4023	0.23% C 0.25% Mo steel: Designation used in the UK and USA
4024	0.24% C 0.25% Mo steel: Designation used in the UK and USA
4027	0.27% C 0.25% Mo steel: Designation used in the UK and USA
4028	0.28% C 0.25% Mo steel: Designation used in the UK and USA
4032	0.32% C 0.25% Mo steel: Designation used in the UK and USA
4037	0.37% C 0.25% Mo steel: Designation used in the UK and USA
4042	0.42% C 0.25% Mo steel: Designation used in the UK and USA
4047	0.47% C 0.25% Mo steel: Designation used in the UK and USA
4063	0.63% C 0.25% Mo steel: Designation used in the UK and USA
4419	0.19% C 0.5% Mo steel: Designation used in the UK and USA
4422	0.22% C 0.4% Mo steel: Designation used in the UK and USA
4427	0.27% C 0.4% Mo steel: Designation used in the UK and USA
4520	0.2% C 0.5% Mo steel: Designation used in the UK and USA
10013	0.1% C 1% Mo 0.45% Si steel: Weld electrode; USA designation; military specification
A 30	0.12% C (max) 0.55% Mo steel: For welding electrodes; designation for BS 2901
A 31	0.14% C (max) 0.5% Mo steel: For welding electrodes; designation used for BS 2901
A 204 A	0.23% C 0.5% Mo steel: Plate; Armco; pressure vessels used at high temperature **UTS: 480 Elon: 24% Proof: 240**
A 204 B	0.25% C 0.5% Mo steel: Plate; Armco; pressure vessels used at high temperature **UTS: 530 Elon: 23% Proof: 270**
A 204 C	0.26% C 0.5% Mo steel: Plate; Armco; pressure vessels used at high temperature **UTS: 640 Elon: 22% Proof: 300**
A Mn Mo	0.25% C Mn Mo steel: Casting; Firth Brown; normalized and tempered **DPN: 200 UTS: 700 Elon: 17% Proof: 450**
ABS/H	0.22% C (max) 0.9% Mn (max) 0.5% Mo steel: Plate; American Bureau of Shipping **UTS: 500 Elon: 19% Proof: 260**
ABS/I	0.23% C (max) 0.9% Mn (max) 0.52% Mo steel: Plate; American Bureau of Shipping **UTS: 550 Elon: 17% Proof: 280**
ABS/J	0.25% C (max) 0.9% Mn 0.52% Mo steel: Plate; American Bureau of Shipping **UTS: 580 Elon: 16% Proof: 305**
ABS/K	0.15% C 0.55% Mn 0.55% Mo steel: Tube; American Bureau of Shipping
ABS/L	0.2% C 0.6% Mn 0.55% Mo steel: Tube; American Bureau of Shipping
ABS/M	0.14% C (max) 0.55% Mn 0.55% Mo steel: Tube; American Bureau of Shipping
ABS/N	0.15% C (max) 0.5% Mn 0.55% Mo steel: Tube; American Bureau of Shipping
ABS/O	0.15% C (max) 0.5% Mn 0.55% Mo steel: Tube; American Bureau of Shipping

Note. The following abbreviations and units are used in the tables:

DPN	Hardness, diamond pyramid number
UTS	Ultimate tensile strength, N/mm^2
Elon	Elongation, %
Proof	0.1% proof strength, N/mm^2

1 N/mm^2=0.1 hbar=0.102 kgf/mm^2=0.06475 tonf/in.^{2}=145.04 lbf/in.^{2}=1 MPa

See Appendix II for other abbreviations and conversion tables.

Symbol	Nominal analysis, supplier, condition and remarks.
ABS/P	0.15% C (max) 0.5% Mn 1.0% Mo steel: Tube; American Bureau of Shipping
AFNOR 81-345 EC Mo B20	0.06% C 0.5% Mo steel: Welding electrode; French specification
AFNOR 81-345 EC Mo R22	0.08% C 0.5% Mo steel: Welding electrode; French specification
AISI 4012	0.12% C 0.2% Mo steel
AISI 4023	0.23% C 0.25% Mo steel
AISI 4024	0.22% C 0.25% Mo steel
AISI 4027	0.27% C 0.25% Mo steel
AISI 4028	0.27% C 0.25% Mo steel
AISI 4032	0.32% C 0.8% Mn 0.25% Mo steel
AISI 4037	0.37% C 0.25% Mo steel
AISI 4419	0.2% C 0.5% Mo steel
AISI 4422	0.22% C 0.4% Mo steel
AISI 4427	0.27% C 0.4% Mo steel
AISI 4520	0.2% C 0.5% Mo steel: Obsolete
AMS 6300	0.37% C 0.25% Mo steel: Bar and forging; AMS for SAE 4037
AMS 6386	0.16% C 0.62% Mo 0.003% B steel
AN S 9/1	0.37% C 0.25% Mo steel: US Service; hardened and tempered **DPN: 230 UTS: 840 Elon: 17% Proof: 740**
ASTM A161T1	0.15% C 0.5% Mn 0.5% Mo 0.05% S and P steel: Tube
ASTM A204 A	0.22% C 0.9% Mn 0.5% Mo steel: Plate for pressure vessels
ASTM A204 B	0.23% C 0.9% Mn 0.5% Mo steel: Plate for pressure vessels
ASTM A204 C	0.24% C 0.9% Mn 0.5% Mo steel: Plate for pressure vessels
ASTM A209 T1	0.15% C 0.5% Mn 0.55% Mo steel: Seamless tube
ASTM A209 T1A	0.2% C 0.5% Mn 0.55% Mo steel: Seamless tube
ASTM A209 T1B	0.14% C (max) 0.5% Mn 0.55% Mo steel: Seamless tube
ASTM A250 T1	0.15% C 0.5% Mn 0.55% Mo steel: Electric welded tube
ASTM A250 T1A	0.2% C 0.5% Mn 0.55% Mo steel: Electric welded tube
ASTM A250 T1B	0.14% C (max) 0.5% Mn 0.55% Mo steel: Electric welded tube
ASTM A295 F1	0.25% C 0.5% Mo steel: Seamless drum forgings; normalized and tempered **UTS: 500 Elon: 20% Proof: 270**
ASTM A316 E7010A1	0.12% C 0.6% Mn 0.52% Mo 0.4% Si steel: Electrodes for arc welding
ASTM A316 E7011A1	0.12% C 0.6% Mn 0.52% Mo 0.4% Si steel: Electrodes for arc welding
ASTM A316 E7015A1	0.12% C 0.9% Mn 0.52% Mo 0.6% Si steel: Electrodes for arc welding
ASTM A316 E7016A1	0.12% C 0.9% Mn 0.52% Mo 0.6% Si steel: Electrodes for arc welding
ASTM A316 E7018A1	0.12% C 0.9% Mn 0.52% Mo 0.8% Si steel: Electrodes for arc welding
ASTM A316 E7020A1	0.12% C 0.6% Mn 0.52% Mo 0.4% Si steel: Electrodes for arc welding
ASTM A316 E7027A1	0.12% C 0.52% Mo steel: Electrodes for arc welding
ASTM A316 E9015/8D1	0.12% C 1.5% Mn 0.3% Mo steel: Welding electrode; code varies with analysis
ASTM A316 E9015D1	0.12% C 0.35% Mo steel: Electrodes for arc welding
ASTM A316 E9018D1	0.12% C 0.35% Mo steel: Electrodes for arc welding
ASTM A316 E10015/8D2	0.15% C 1.8% Mn 0.3% Mo steel: Welding electrodes; code varies with analysis
ASTM A316 E10015D2	0.15% C 0.35% Mo steel: Electrodes for arc welding
ASTM A316 E10016D2	0.15% C 0.35% Mo steel: Electrodes for arc welding
ASTM A316 E10018D2	0.15% C 0.35% Mo steel: Electrodes for arc welding
ASTM A335 P1	0.15% C 0.5% Mo steel: Seamless pipe

Symbol	Nominal analysis, supplier, condition and remarks.
ASTM A335 P15	0.15% C (max) 0.5% Mo 1.4% Si steel: Seamless pipe
ASTM A336 F1	0.25% C 0.5% Mo: Seamless drum forging **UTS: 460 Elon: 20% Proof: 270**
ASTM A352 LC1	0.25% C 0.55% Mo steel: Normalized and tempered for low temperature service **UTS: 440 Elon: 24% Proof: 220**
ASTM A356/2	0.25% C 0.5% Mo steel: Casting; normalized and tempered; for turbines **UTS: 440 Elon: 22% Proof: 220**
ASTM A356/3	0.25% C 1.0% Mo steel: Casting; normalized and tempered; for turbines **UTS: 530 Elon: 18% Proof: 330**
ASTM A369 FP1	0.15% C 0.5% Mo steel: For forged and bored pipe
ASTM A426 CP1	0.15% C 0.5% Mo steel: Cast pipe
ASTM A426 CP15	0.15% C (max) 0.5% Mo steel: Cast pipe
ASTM A487/2 N	0.3% C (max) 1.2% Mn 0.2% Mo steel: Casting for pressure; normalized; 2Q grade is quenched and tempered
ASTM A514 J	0.17% C 0.56% Mo 0.003% B steel: Plate
ASTM A514 K	0.15% C 0.50% Mo 0.003% B steel: Plate
ASTM A517 C	0.15% C 0.25% Mo 0.003% B steel: Boiler plate
ASTM A517 J	0.17% C 0.58% Mo 0.003% B steel: Boiler plate
ASTM A517 K	0.15% C 0.5% Mo 0.003% B steel: Boiler plate
ASTM A533 A	0.25% C (max) 1.25% Mn 0.52% Mo steel: Plate
ASTM A588 E	0.15% C (max) 1.2% Mn (max) 0.17% Mo 0.65% Cu steel: For structures **UTS: 440 Elon: 21% Proof: 320**
ASTM A692	0.22% C 0.7% Mn 0.5% Mo steel: Pipe; seamless for pressure use **UTS: 500 Elon: 20% Proof: 290**
ASTM A699/1	0.06% C (max) 1.7% Mn 0.3% Mo 0.06% Nb steel **UTS: 700 Elon: 12% Proof: 485**
ASTM A699/2	0.06% C (max) 1.7% Mn 0.3% Mo 0.06% Nb steel **UTS: 700 Elon: 12% Proof: 515**
ASTM A699/3	0.06% C (max) 1.7% Mn 0.3% Mo 0.06% Nb steel **UTS: 680 Elon: 12% Proof: 485**
ASTM A699/4	0.06% C (max) 1.7% Mn 0.3% Mo 0.06% Nb steel **UTS: 680 Elon: 12% Proof: 515**
ASTM A707 L6/1	0.09% C (max) 1.8% Mn 0.3% Mo 0.07% Nb steel: Flanges for low temperature use; impact 41 J at −62 °C **UTS: 290 Elon: 22%**
ASTM A707 L6/2	0.09% C (max) 1.8% Mn 0.3% Mo 0.07% Nb steel: Flanges for low temperature use; impact 54 J at −62 °C **UTS: 360 Elon: 25%**
ASTM A707 L6/2	0.09% C (max) 1.8% Mn 0.3% Mo 0.07% Nb steel: Flanges for low temperature use; impact 68 J at −62 °C **UTS: 415 Elon: 23%**
ASTM A707 L6/3	0.09% C (max) 1.8% Mn 0.3% Mo 0.07% Nb steel: Flanges for low temperature use; impact 68 J at −62 °C **UTS: 515 Elon: 20%**
ASTM A732/6 N	0.35% C (max) 1.5% Mn 0.4% Mo steel: Investment casting; normalized and tempered **UTS: 620 Elon: 20% Proof: 415**
ASTM A735	0.08% C 1.5% Mn 0.35% Mo 0.08% Nb steel: Plate for pressure vessels
ASTM A737 C	0.24% C (max) 1.3% Mn 0.1% V steel: Plate for pressure vessels
AURIGA VIII LT	0.18% C 0.5% Mo steel: Casting; D Brown; normalized and tempered for sub-zero use **UTS: 440 Elon: 25% Proof: 240**
AUTROD 12.24	0.1% C 1.0% Mn 0.5% Mo steel: Welding wire; Esab; copper coated; used for submerged arc welding
BEARCOMO	0.16% C 1.5% Mn 0.25% Mo 0.5% Cu steel: Plate; Consett; normalized; suitable for welding; creep resisting **UTS: 460 Elon: 22% Proof: 300**
BS 806M	0.17% C 0.55% Mo steel: Seamless pipe; cold drawn and normalized **UTS: 480**
BS 970	See designation for details

Symbol	Nominal analysis, supplier, condition and remarks.
BS 980 CDS11	0.25% C 1.5% Mn 0.2% Mo steel: Cold drawn tube; as drawn and tempered; can be welded then heat treated **UTS: 700 Proof: 580**
BS 1398	0.2% C 0.5% Mo steel: Casting; normalized; contained in BS 3100 **UTS: 450 Elon: 20% Proof: 240**
BS 1453 A6	0.15% C (max) 0.5% Mo steel: Weld filler rod
BS 1501/240	0.15% C 0.5% Mo steel: Plate and bar; normalized **UTS: 400 Elon: 22% Proof: 210**
BS 1501/261	0.13% C 0.5% Mo 0.003% B steel: Plate; normalized; for pressure vessels **UTS: 600 Elon: 16% Proof: 430**
BS 1503/240 A	0.2% C 0.5% Mo steel: Forging; normalized **UTS: 450 Proof: 210**
BS 1503/240 B	0.2% C 0.5% Mo steel: Forging; normalized **UTS: 480 Elon: 20% Proof: 270**
BS 1504/240	0.2% C 0.6% Mo steel: Casting; normalized **UTS: 450 Elon: 20% Proof: 240**
BS 1506/240	0.3% C 0.5% Mo steel: Bar; hardened and tempered **UTS: 610 Elon: 22% Proof: 370**
BS 1507/240	0.20% C 0.5% Mo steel: Pipe; normalized **UTS: 460 Elon: 20%**
BS 1507/240 A	0.2% C 0.5% Mo steel: Seamless pipe; normalized **UTS: 450 Elon: 25%**
BS 1507/240 B	0.25% C 0.5% Mo steel: Seamless pipe; normalized **UTS: 460 Elon: 25%**
BS 1508/240	0.17% C 0.6% Mo steel: Tube; normalized **UTS: 450 Elon: 22%**
BS 1652	0.5% Mo steel: Tube; replaced by BS 3059
BS 1717 CDS109	0.26% C 0.2% Mo steel: Seamless tube; as drawn **UTS: 690 Proof: 610**
BS 1717 CDS109	0.26% C 0.2% Mo steel: Seamless tube; after welding, brazing or annealing **UTS: 550 Proof: 370**
BS 2493 A	0.12% C 0.5% Mo steel: Welding electrode
BS 2943/2 Mn Mo B	0.1% C 0.35% Mo steel: For welding electrodes
BS 2943/Mn Mo B	0.1% C 0.35% Mo steel: For welding electrodes
BS 2943/Mo B	0.1% C 0.6% Mo steel: For welding electrodes
BS 2943/Mo C	0.1% C 0.6% Mo steel: For welding electrodes
BS 3100 B1	0.2% C (max) 0.7% Mn 0.55% Mo 0.8% Cr + Ni + Cu (max) steel: Casting for high temperature use **UTS: 460 Elon: 18% Proof: 260**
BS 3100 BL1	0.2% C (max) 0.55% Mo steel: Casting for low temperature use; impact 20 J at −50 °C **UTS: 460 Elon: 18% Proof: 260**
BS S102	0.35% C 0.3% Mo steel: Hire for forged bolts; hardened and tempered after forging **UTS: 920 Elon: 18% Proof: 660**
BS S114	0.34% C 0.3% Mo steel: Hardened and tempered **DPN: 280 UTS: 920 Elon: 18% Proof: 660**
C Mo	0.1% C 1% Mn 0.5% Mo steel: Electrode; Metrode
C Mo 15	0.15% C 0.6% Mo steel: ESC; normalized **UTS: 400 Elon: 20% Proof: 210**
C Mo 25	0.25% C 0.55% Mo steel: ESC **UTS: 610 Elon: 22% Proof: 370**
CELLOCITO PL S2	0.13% C 0.6% Mn 0.55% Mo steel: Welding electrode; Soudometal; for welding alloy steel **UTS: 625 Elon: 20% Proof: 535**
CELLOCITO PL S2	0.13% C 0.60% Mn 0.55% Mo 0.20% Si steel: Soudometal; welding of high strength steel **UTS: 625 Elon: 20% Proof: 535**
CODE 629	0.21% C 0.25% Mo steel: Casting; Edgar Allen; MT Mo 'P'
CODE 641	0.3% C 0.25% Mo steel: Casting; Edgar Allen; MT Mo 'R'
CODE 644	0.25% C 0.6% Mo steel: Casting; Edgar Allen
CS 5	0.57% C 0.5% Mo tool steel: ASTM designation
DIN 17155 15 Mo 3	0.16% C 0.6% Mn 0.3% Mo steel: Plate; German Standard **UTS: 500 Proof: 270**
DIN 17155 15 Mo 3	0.15% C 0.3% Mo steel: For boiler plate
DIN 17175 15 Mo 3	0.15% C 0.7% Mn 0.3% Mo steel: For boiler plate
DTD 740	0.13% C 0.5% Mn 0.5% Mo 0.003% B steel: Tube; normalized and tempered; weldable **UTS: 610 Proof: 450**
DTD 5062	0.13% C 0.5% Mo 0.003% B steel: Sheet; normalized; weldable **UTS: 610 Elon: 12% Proof: 420**
DTD 5202	0.5% Mo B steel: Wrought **UTS: 600**
DUNELT 50	0.32% C 1.6% Mn 0.3% Mo steel: Dunford for BS alloy En 16
DUNELT 51	0.35% C 1.5% Mn 0.45% Mo steel: Dunford for BS alloy En 17
E 7X T5 A1	0.12% C (max) 0.52% Mo steel: Weld electrode; designation used by AWS
E 8X T1 A1	0.12% C (max) 0.52% Mo steel: Weld electrode; designation used by AWS
E 8X T5 K1	0.15% C (max) 0.4% Mo steel: Weld electrode; designation used by AWS
E 9X T1 D1	0.12% C (max) 0.4% Mo steel: Weld electrode; designation used by AWS
E 9X T1 D3	0.12% C (max) 0.55% Mo steel: Weld electrode; designation used by AWS
E 9X T5 D2	0.15% C (max) 0.4% Mo steel: Weld electrode; designation used by AWS
E 100 T5 D2	0.15% C (max) 0.35% Mo steel: Weld electrode; designation used by AWS
E 7010 Al	0.12% C (max) 0.52% Mo steel: Weld electrode; designation used by AWS
E 7011 Al	0.12% C (max) 0.52% Mo steel: Weld electrode; designation used by AWS
E 7015 Al	0.12% C (max) 0.52% Mo steel: Weld electrode; designation used by AWS
E 7016 Al	0.12% C (max) 0.52% Mo steel: Weld electrode; designation used by AWS
E 7018 Al	0.12% C (max) 0.52% Mo steel: Weld electrode; designation used by AWS
E 7020 Al	0.12% C (max) 0.52% Mo steel: Weld electrode; designation used by AWS
E 7027 Al	0.12% C (max) 0.52% Mo steel: Weld electrode; designation used by AWS
E 8016 D3	0.12% C (max) 0.55% Mo steel: Weld electrode; designation used by AWS
E 8018 D3	0.12% C (max) 0.55% Mo steel: Weld electrode; designation used by AWS
E 9015 D1	0.12% C (max) 0.25% Mo steel: Weld electrode; designation used by AWS
E 9016 A	0.12% C (max) 0.45% Mo 0.3% Si steel: Weld electrode; designation used by AWS
E 9018 Dl	0.12% C (max) 0.3% Mo steel: Weld electrode; designation used by AWS
E 10015 D2	0.15% C (max) 0.35% Mo steel: Weld electrode; designation used by AWS
E 10016 D2	0.15% C (max) 0.35% Mo steel: Weld electrode; designation used by AWS
E 10018 D2	0.15% C (max) 0.35% Mo steel: Weld electrode; designation used by AWS

Note. The following abbreviations and units are used in the tables:

DPN	Hardness, diamond pyramid number
UTS	Ultimate tensile strength, N/mm^2
Elon	Elongation, %
Proof	0.1% proof strength, N/mm^2

$1 N/mm^2 = 0.1 hbar = 0.102 kgf/mm^2 = 0.06475 tonf/in.^2 = 145.04 lbf/in.^2 = 1 MPa$

See Appendix II for other abbreviations and conversion tables.

Symbol	Nominal analysis, supplier, condition and remarks.
EA 1	0.12% C 0.55% Mo steel: Weld electrode; designation used by AWS
EA 1N	0.12% C 0.55% Mo steel: Weld electrode; designation used by AWS
EA 2	0.12% C 0.55% Mo steel: Weld electrode; designation used by AWS
EA 2N	0.12% C 0.55% Mo steel: Weld electrode; designation used by AWS
EA 3	0.12% C 0.55% Mo steel: Weld electrode; designation used by AWS
EA 3K	0.1% C 0.5% Mo steel: Weld electrode; designation used by AWS
EA 3K N	0.1% C 0.5% Mo steel: Weld electrode; designation used by AWS
EA 3N	0.12% C 0.55% Mo steel: Weld electrode; designation used by AWS
EA 4	0.12% C 0.55% Mo steel: Weld electrode; designation used by AWS
EA 4N	0.12% C 0.55% Mo steel: Weld electrode; designation used by AWS
EB 2	0.12% C 0.55% Mo 0.2% Si steel: Weld electrode; designation used by AWS
EC A1	0.12% C (max) 0.52% Mo steel: Weld electrode; designation used by AWS
EC A1 N	0.12% C (max) 0.52% Mo steel: Weld electrode; designation used by AWS
EC A2	0.12% C (max) 0.52% Mo steel: Weld electrode; designation used by AWS
EC A2 N	0.12% C (max) 0.52% Mo steel: Weld electrode; designation used by AWS
EC A3	0.15% C (max) 0.52% Mo steel: Weld electrode; designation used by AWS
EC A3 N	0.15% C (max) 0.52% Mo steel: Weld electrode; designation used by AWS
EC A4	0.15% C (max) 0.52% Mo steel: Weld electrode; designation used by AWS
EC A4 N	0.15% C (max) 0.52% Mo steel: Weld electrode; designation used by AWS
EH 10 Mo EW	0.1% C 0.5% Mo steel: Weld electrode; designation used by AWS
En 16	0.3% C 0.3% Mo steel: Designation used in BS 970; replaced by 605 M36
En 17	0.35% C 0.4% Mo steel: Designation used in BS 970; replaced by 608 M38
ERA 59	0.35% C 1.5% Mn 0.3% Mo steel: Hadfields for BS alloy En 16
ERA 156	0.35% C 1.5% Mn 0.45% Mo steel: Hadfields for BS alloy En 17
ESSHETE D 4C	0.12% C 0.5% Mo steel: Samuel Fox; weldable **UTS: 450**
FASTEX 100M	0.07% C 0.5% Mo steel: Welding electrode; Murex **UTS: 450 Elon: 30% Proof: 360**
FORTIWELD	0.16% C (max) 0.6% Mo 0.005% B steel: United Steel Co.; normalized; suitable for welding **UTS: 600 Elon: 16% Proof: 440**
GOST 4543 38KHGN	0.38% C 0.95% Mn 0.85% Ni 0.65% Cr steel: Russian Standard
GOST 4543 38KHM1UA	0.38% C 0.55% Mn 1.5% Cr 0.2% Mo 1.0% Al steel: Russian Standard
GOST 4543 38KHVF1UA	0.38% C 0.3% Mn 1.65% Cr 0.15% V 0.3% W 0.5% Al steel: Russian Standard
GOST 4543 38KHVFU	0.38% C 0.3% Mn 1.65% Cr 0.15% V 0.3% W 0.5% Al steel: Russian Standard
JALLOY AR 280	0.28% C 1.5% Mn 0.15% Mo 0.2% Cu 0.0005% B steel: Jones and Laughlin
JALLOY AR 320	0.28% C 1.5% Mn 0.15% Mo 0.2% Cu 0.0005% B steel: Jones and Laughlin
JALLOY AR 360	0.28% C 1.5% Mn 0.15% Mo 0.2% Cu 0.0005% B steel: Jones and Laughlin
JALLOY AR 400	0.28% C 1.5% Mn 0.15% Mo 0.2% Cu 0.0005% B steel: Jones and Laughlin
JALLOY No. 1	0.16% C 1.15% Mn 0.15% Mo 0.2% Cu steel: Jones and Laughlin
JALLOY No. 3	0.28% C 1.5% Mo steel: Jones and Laughlin
JALLOY S 90	0.15% C 1.25% Mn 0.25% Mo 0.003% B steel: Jones and Laughlin
JALLOY S 100	0.15% C 1.25% Mn 0.25% Mo 0.003% B steel: Jones and Laughlin
JALLOY S 110	0.15% C 1.35% Mn 0.25% Mo steel: Jones and Laughlin
JALLOY S 340	0.15% C 1.3% Mn 0.25% Mo steel: Jones and Laughlin
JALTEN 1	0.15% C (max) 1.3% Mn 0.05% V 0.3% Cu steel: Jones and Laughlin
JIS G3103 SB46M	0.25% C (max) 0.9% Mn (max) 0.5% Mo steel: Plate; Japanese Standard **UTS: 500 Proof: 250**
JIS G3103 SB49M	0.27% C (max) 0.9% Mn 0.55% Mo steel: Plate; Japanese Standard **UTS: 540 Proof: 270**
JIS G3103 SB56M	0.27% C (max) 1.2% Mn 0.5% Mo steel: Plate; Japanese Standard **UTS: 630 Proof: 350**
JIS G3119 SBV1A	0.25% C (max) 1.1% Mn 0.5% Mo steel: Plate: Japanese Standard **UTS: 600 Proof: 320**
JIS G3119 SBV1B	0.25% C (max) 1.2% Mn 0.5% Mo steel: Plate: Japanese Standard **UTS: 630 Proof: 350**
JIS G3120 SQV1A	0.25% C (max) 1.2% Mn 0.5% Mo steel: Plate; Japanese Standard **UTS: 630 Proof: 350**
JIS G3120 SQV1B	0.25% C (max) 1.2% Mn 0.5% Mo steel: Plate; Japanese Standard **UTS: 720 Proof: 490**
JIS G3458 STPA12	0.15% C 0.55% Mn 0.55% Mo steel: Pipe; Japanese Standard **UTS: 390 Proof: 210**
JIS G3462 STBA12	0.15% C 0.55% Mn 0.55% Mo steel: Pipe; Japanese Standard **UTS: 390 Proof: 210**
JIS G3462 STBA13	0.2% C 0.55% Mn 0.55% Mo steel: Pipe; Japanese Standard **UTS: 420 Proof: 210**
JS 6	0.2% C 0.5% Mo steel: Casting; Jopling; for use up to 425 °C **DPN: 180 UTS: 450 Elon: 20% Proof: 240**
JS 81	0.32% C 1.5% Mn 0.3% Mo steel: Casting; Jopling **DPN: 230 UTS: 760 Elon: 15% Proof: 480**
K 10614	0.06% C (max) 0.3% Mo 0.06% Nb steel: Casting; designation used by UNS
K 12220	0.22% C (max) 0.5% Mo steel: Designation used by UNS
K 13020	0.3% C (max) 0.5% Mo steel: Designation used by UNS
KAISALOY 70MB	0.15% C (max) 0.6% Mn 0.6% Mo 0.001% B steel: Kaiser Steel Corp.
KF 46	0.35% C 1.5% Mn 0.3% Mo steel: Kirkstall for BS alloy En 16
KF 46A	0.34% C 1.5% Mn 0.45% Mo steel: Kirkstall for BS alloy En 17
KP 14	0.35% C 1.5% Mn 0.3% Mo steel: Kiveton Park for BS alloy En 16
M 1	0.35% C 1.5% Mn 0.27% Mo steel: Jessop for BS alloy En 16
M 2	0.35% C 1.5% Mn 0.45% Mo steel: Jessop for BS alloy En 17
MANOIR PFO	0.15% C 0.5% Mo steel: Pompey; hardened and tempered **UTS: 450 Elon: 24% Proof: 260**
MAXEL 1B	0.18% C 0.1% Mo 0.08% S 0.04% P steel: Free machining; Crucible Steel Co.; may be carburized **DPN: 156 UTS: 480 Elon: 30% Proof: 330**

Symbol	Nominal analysis, supplier, condition and remarks.
MIL S 12504 Mn Mo	0.72% C 1% Mo steel: US military specification
MM 01	0.32% C 1.6% Mn 0.27% Mo steel: Firth Brown for BS alloy En 16
MM 02	0.33% C 1.6% Mn 0.4% Mo steel: Firth Brown for BS alloy En 17
Mn Mo	0.1% C 1.8% Mn 0.5% Mo steel: Electrode; Metrode
Mo B	0.07% C 0.8% Mn 0.5% Mo steel: Electrode; Metrode; for BS 2493
Mo B	0.05% C 0.5% Mo steel: Electrode; Metrode
MOLEX	0.06% C 0.5% Mo steel: Welding electrode; Murex **UTS: 460 Elon: 33% Proof: 440**
MOLYTERM 5	0.06% C 0.8% Mn 0.5% Mo: Welding electrode; Soudametal; welding electrode **UTS: 540 Elon: 26% Proof: 450**
MOLYTERM Ti 5	0.08% C 0.6% Mn 0.5% Mo: Welding electrode; Soudametal; for welding creep strength steels **UTS: 570 Elon: 24% Proof: 470**
MOLYTRODE 74.56	0.08% C 0.06% Mn 0.6% Mo steel: For electrodes; Esab **UTS: 556 Elon: 28% Proof: 463**
MT 'R'	0.3% C 1.4% Mn 0.25% Mo steel: Casting; Edgar Allen; code 641
MT Mo 'P'	0.21% C 1.4% Mn 0.25% Mo steel: Casting; Edgar Allen; code 629
NV 7-1	0.2% C (max) 0.6% Mn 0.35% Mo steel: Designation used by DNV **UTS: 440 Elon: 22% Proof: 280**
NV 7-2	0.18% C (max) 0.6% Mn 0.55% Mo steel: Designation used by DNV **UTS: 440 Elon: 22% Proof: 300**
NVR 7-1	0.2% C (max) 0.6% Mn 0.32% Mo steel: Designation used by DNV
OCM 7	0.35% C 1.6% Mn 0.3% Mo steel: W Marrison for BS alloy En 16
OX 1002	0.2% C (max) 1.7% Mn 0.7% Mo 0.005% B steel: Plate; Svenskt Stal; graded by impact **UTS: 980 Elon: 14% Proof: 890**
PFO	0.24% C 0.5% Mo steel: Pompey
PFO A	0.12% C 0.5% Mo steel: Pompey
PHILIPS KV 2	0.07% C 0.5% Mo steel: Welding electrode; Philips **UTS: 600 Elon: 26% Proof: 500**
REYNOLDS 531	0.26% C 1.5% Mn 0.2% Mo steel: Tube; Reynolds; suitable for brazing **UTS: 690 Elon: 10% Proof: 610**
REYNOLDS 753	0.26% C 0.2% Mo steel: Origin unknown; for bicycle frames
REYNOLDS 5313L	0.26% C 0.2% Mo steel: Origin unknown; for bicycle frames
SAE 4012	0.12% C 0.2% Mo steel
SAE 4023	0.23% C 0.25% Mo steel: Mechanical properties not specified
SAE 4024	0.22% C 0.25% Mo steel: Mechanical properties not specified
SAE 4027	0.27% C 0.25% Mo steel: Mechanical properties not specified
SAE 4028	0.28% C 0.25% Mo steel: Mechanical properties not specified
SAE 4032	0.32% C 0.25% Mo steel: Mechanical properties not specified
SAE 4037	0.37% C 0.25% Mo steel
SAE 4419	0.2% C 0.5% Mo steel: Mechanical properties not quoted
SAE 4427	0.27% C 0.4% Mo steel: Mechanical properties not quoted
SANBOLD 16	0.35% C 1.5% Mn 0.3% Mo steel: Sanderson for BS alloy En 16
SANBOLD 17	0.35% C 1.5% Mn 0.4% Mo steel: Sanderson for BS alloy En 17
SANBOLD 53	0.35% C 1.5% Mn 0.4% Mo steel: Sanderson; replaced by Sanbold 17
SANBOLD 74	0.35% C 1.5% Mn 0.3% Mo steel: Sanderson; replaced by Sanbold 16
SMM	0.35% C 1.5% Mn 0.5% Mo steel: Spencer for BS alloy En 17
SOUDOR G Mo	0.1% C 1.2% Mn 0.5% Mo steel: Welding electrode; Soudometal; for inert gas welding **UTS: 580 Elon: 22% Proof: 430**
SOUDOTIG Mo	0.1% C 1.1% Mn 0.45% Mo steel: Welding electrode; Soudometal; for inert gas welding **UTS: 630 Elon: 22% Proof: 470**
SOUDOTIG Mo	0.1% C 0.45% Mo steel: Welding electrode; Soudometal **UTS: 700 Elon: 22% Proof: 470**
SOUPERSOUDO	0.1% C 0.4% Mn 0.35% Mo steel: Welding electrode; Soudometal; build up of worn steel **DPN: 200**
SPEAR 416	0.37% C 1.5% Mn 0.25% Mo steel: Spear and Jackson for BS alloy En 16
SS 2912	0.2% C 0.35% Mo steel: For pressure vessel tubes; Swedish Standard; as drawn **UTS: 480 Elon: 22% Proof: 190**
STA 5 V9A	0.32% C 1.5% Mn 0.3% Mo steel: Replaced by BS alloy En 16
STA 5 V9A/1	0.32% C 1.5% Mn 0.27% Mo steel: Replaced by BS alloy En 16A
STA 5 V9B	0.35% C 1.5% Mn 0.45% Mo steel: Replaced by BS alloy En 17
STA 5 V9B/1	0.35% C 1.5% Mn 0.45% Mo steel: Obsolete
STUBS 62	C Si Mn Mo steel: Welding electrode; Stubs **DPN: 200 UTS: 685 Elon: 33% Proof: 490**
SUPERTOUGH C40	0.35% C 1.6% Mn 0.28% Mo steel: ESC for BS alloy En 16
SUPERTOUGH C40M	0.35% C 1.6% Mn 0.45% Mo steel: ESC for BS alloy En 17
UHB STATO 21	0.18% C 0.3% Mo steel: Tube; Uddelholm; normalized **UTS: 480 Elon: 30% Proof: 270**
VSMM	0.35% C 1.5% Mn 0.35% Mo steel: Vulcan **UTS: 840**
W 7	0.11% C 0.6% Mo steel: Welding electrode; Cu covered; Armco; as welded **UTS: 680 Elon: 22% Proof: 580**

Note. The following abbreviations and units are used in the tables:

DPN	Hardness, diamond pyramid number
UTS	Ultimate tensile strength, N/mm^2
Elon	Elongation, %
Proof	0.1% proof strength, N/mm^2

1 N/mm^2=0.1 hbar=0.102 kgf/mm^2=0.06475 $tonf/in.^2$=145.04 $lbf/in.^2$=1 MPa

See Appendix II for other abbreviations and conversion tables.

44G2 Steel – medium and high carbon with molybdenum
0.4% carbon minimum, 0.7% molybdenum maximum

Specific gravity	7.8
Density	7800 kg/m^3
Solidus/liquidus	1470–1510 °C
Thermal conductivity	56 W/M °C
Coefficient of linear expansion	11×10^{-6}/ °C
Electrical conductivity	8–10% IACS (copper 100%)
Specific resistance	180–250 microhm mm
Young's modulus of elasticity	–
Impact	50 J – lower impact figures can be expected
Fatigue strength (no. of cycles)	–
Hot strength	

Temperature °C	Tensile strength N/mm^2	Elongation %
200	460	35
425	400	35
480	330	40
650	210	50

The above properties are typical of the following group and may not apply exactly to any one specification. It is possible that with certain specifications some of the values may not be applicable.

General metallurgical characteristics

Carbon will always combine with molybdenum in preference to iron; thus the molybdenum content of these steels is always present as the carbide. This increases the creep strength and helps the steels to retain their hardness at higher temperatures. The stability of the carbides increases the hardenability and gives an inherent fine grained structure which improves the toughness of the correctly thermally treated parts. There is no apparent difference in corrosion resistance between these and plain carbon steels. In many respects molybdenum has the same effect as tungsten but less is required to achieve the same result.

Molybdenum steels are noted for their lack of notch sensitivity in the hardened condition, being superior to chromium steels in this respect.

These materials are almost invariably used hardened and tempered, seldom or never in the cold worked condition.

Because of the medium to high carbon content of these steels they do not have the capability of accepting cold work without becoming brittle. Where they are cold worked therefore, annealing may be required.

Thermal treatments to fully anneal, harden and normalize will be carried out in the region of 800 °C.

While molybdenum increases the hardenability of steel there is seldom sufficient molybdenum present in the steels listed to cause problems either at hardening or at welding.

The carbon equivalent (CE) formula gives a useful indication of the hardenability. The formula is:

$$CE = C + \frac{Mn}{6} + \frac{Cr + Mo + V}{5} + \frac{Ni + Cr}{15}$$

A value above 0.7% indicates appreciable hardenability and thus possible weld problems.

The fact that molybdenum produces very stable carbides, which do not diffuse, gives toughness and inhibits any tendency to brittleness from carbide precipitation.

Where high integrity components are involved, advice should be sought prior to any thermal treatment to increase the mechanical properties or before any welding operation.

The steels in this section are used for cutting tools, press tools and fixtures. Shear blades and similar components are also manufactured. Taps, dies, drills and cutting tools, where relatively low production is required, are economically manufactured.

All these steels are competing with the steels found in Sections 44K2 and 44K3, i.e. where alloy steels containing molybdenum plus chromium, nickel and vanadium exist.

Note. The following abbreviations and units are used in the tables:

DPN	Hardness, diamond pyramid number
UTS	Ultimate tensile strength, N/mm^2
Elon	Elongation, %
Proof	0.1% proof strength, N/mm^2

1 N/mm^2=0.1 hbar=0.102 kgf/mm^2=0.06475 $tonf/in.^2$=145.04 $lbf/in.^2$=1 MPa

See Appendix II for other abbreviations and conversion tables.

Symbol	Nominal analysis, supplier, condition and remarks.
06	1.45% C 0.25% Mo 1.0% Si steel: Designation used by AISI
19F	1.5% C 0.25% Mo 1.0% Si steel: Carrs; press and bending tools, etc. **DPN: 730**
608 H 37	0.37% C 0.47% Mo steel: BS 970; obsolete
608 M 38	0.38% C 0.47% Mo steel: BS 970; obsolete
2121	1.4% C 1.0% Mn 0.3% Mo steel: French Standards designation

Symbol	Nominal analysis, supplier, condition and remarks.
AISI 4042	0.42% C 0.25% Mo steel
AISI 4047	0.47% C 0.25% Mo steel
AISI 4053	0.53% C 0.25% Mo steel
AISI 4063	0.63% C 0.25% Mo steel
AISI 4068	0.68% C 0.25% Mo steel
AISI A4068	0.68% C 0.25% Mo steel: Obsolete
ASTM A194/4	0.45% C 0.8% Mn 0.25% Mo steel: For bolts
ASTM A194/7	0.42% C 0.9% Mn 0.2% Mo steel: For bolts
ATLAS CM	0.4% C 0.15% Mo steel: Atlas; hardened and tempered **DPN: 230 UTS: 760 Elon: 23% Proof: 610**
ATLAS SHAFTING	0.4% C 0.15% Mo steel: Bar; Atlas; normalized; centreless ground **UTS: 660**
BS 3111/2	0.4% C 0.3% Mo steel: Wire; annealed and drawn **UTS: 670**
CRD 515	0.5% C 0.5% Mo 1.0% Si steel: C Denton; shear blades; obsolete **DPN: 540**
E 356	1.5% C 0.45% Mo steel: Fagersta
GRAPH Mo	1.45% C 1.0% Mn 0.25% Mo 1.25% Si steel: Free graphite; Timken; tools for wear resistance **DPN: 425 UTS: 1200 Elon: 10% Proof: 1100**
GRAY DIAMOND	1.45% C 0.25% Mo 1.1% Si steel: Latrobe; free graphite
HEDEX 6	0.4% C 0.3% Mo steel: Wire for forging; Kiveton Park; annealed **UTS: 580**
HEDEX 22	0.42% C 0.25% Mo steel: Wire for forging; Kiveton Park; annealed **UTS: 640**
JALLOY No. 3AR	0.55% C 1.5% Mn 0.15% Mo steel: Jones and Laughlin
JALLOY No. 7	0.55% C 1.5% Mn 0.15% Mo steel: Jones and Laughlin
KF 57	0.38% C 1.6% Mn 0.3% Mo 0.2% S 0.05% P steel: Free machining; Kirkstall **UTS: 920 Elon: 16% Proof: 660**
LT 38	0.6% C 0.25% Mo 2.0% Si steel: Low Moor; obsolete **DPN: 670**
LTG	1.45% C 0.25% Mo 1.0% Si steel: Low Moor; obsolete
MIC 8	1.4% C 0.25% Mo 1.0% Si steel: Taps, punches, rolls, etc., ESC **DPN: 600**
M11	0.63% C 0.25% Mo steel: Round Oak
M12	0.67% C 0.25% Mo steel: Round Oak
MAXEL 2B	0.4% C 0.1% Mo 0.08% S steel: Free machining; Crucible Steel Co. **DPN: 220 UTS: 760 Elon: 18% Proof: 390**
MONARK 1	0.5% C 0.45% Mo steel: Atlas; as AISI type S2
S 60	Ni Mo steel sinter: Firth Cleveland; specific gravity 6.4-6.8; hardened and tempered; file hard **UTS: 780 Elon: 0.5%**
SAE 06	1.45% C 0.25% Mo steel: Oil hardening
SAE 06	1.5% C 0.25% Mo tool steel
SAE 4042	0.42% C 0.25% Mo steel: Mechanical properties not specified
SAE 4047	0.47% C 0.25% Mo steel: Mechanical properties not specified
SAE 4053	0.53% C 0.25% Mo steel: Mechanical properties not specified
SAE 4063	0.63% C 0.25% Mo steel: Mechanical properties not specified
SAE 4068	0.68% C 0.25% Mo steel: Mechanical properties not specified
SIMONDS 864	0.9% C 0.25% Mo steel: Simonds; AISI type 02
SMO	0.55% C 0.35% Mo 1.7% Si steel: Huntsman; chisels, shear blades, collets, etc. **DPN: 670**
SOLAR	0.5% C 0.4% Mn 0.5% Mo steel: Carpenter
SOLAR	0.5% C 0.4% Mn 0.5% Mo 1.0% Si steel: Carpenter; for AISI type S2
TG	1.5% C 0.3% Mo 1.0% Si graphitic tool steel: Osborn; brushes, gauges, etc.; self-lubricating steel; obsolete **DPN: 600**
TRITON	0.5% C 0.6% Mo steel: Braeburn; AISI type S2

Note. The following abbreviations and units are used in the tables:

DPN	Hardness, diamond pyramid number
UTS	Ultimate tensile strength, N/mm^2
Elon	Elongation, %
Proof	0.1% proof strength, N/mm^2

1 N/mm^2=0.1 hbar=0.102 kgf/mm^2=0.06475 tonf/in.^{2}=145.04 lbf/in.^{2}=1 MPa

See Appendix II for other abbreviations and conversion tables.

44H Steel – all carbon contents with tungsten

Specific gravity	7.8
Density	7800 kg/m^3
Solidus/liquidus	1470–1510 °C
Thermal conductivity	50 N/m °C
Coefficient of linear expansion	11×10^{-6}/ °C
Electrical conductivity	8–10% IACS (copper 100%)
Specific resistance	180–250 microhm mm
Young's modulus of elasticity	–
Impact	60 J – lower impact figures can be expected
Fatigue strength	–
Hot strength	–

The above properties are typical of the following group and may not apply exactly to any one specification. It is possible that with certain specifications some of the values may not be applicable.

General metallurgical characteristics

Tungsten combines with the carbon in steel to form very hard carbides which retain their hardness above red heat. These tungsten carbides form in preference to iron carbide; thus all available tungsten exists as the carbide. This inhibits grain growth and by slowing down the metallurgical reactions increases the hardenability.

Fine grained steels tend to be less notch sensitive and less brittle than course grained, very hard steels. Tungsten is therefore useful in that while it increases the hardenability by almost the same amount as chromium, it has a controlling effect on the grain size which tends to make the steels tougher and less notch sensitive. The intrinsically hard carbide particles ensure that these steels are never very soft and it is their hot hardness properties which find the greatest use. There are comparatively few steels containing tungsten only as it is generally present with chromium, molybdenum or vanadium.

Where any appreciable amount of tungsten is present, none of these steels can accept cold work without becoming extremely brittle. The thermal treatments such as full anneal, hardening and normalizing will require temperatures in the region of 750–800 °C.

These steels all have high hardenability and will be extremely brittle in the 'as quenched' condition.

Advice should be sought prior to any thermal treatment on these materials where critical parts are involved.

The same comments apply to welding where it could be expected that cracking in the heat affected zone would take place unless very careful precautions are taken. Advice must always be sought prior to any welding of components which have important uses in service.

As will be seen, there are very few steels in this category, i.e. with tungsten as the only alloying element. They are used for cutting tools, press tools and fixtures, and also shear blades.

Tungsten is mostly used with other alloying elements.

Symbol	Nominal analysis, supplier, condition and remarks.
AK	0.7% C 7% W steel: For tools; Osborn; obsolete **DPN: 950**
BRUNSWICK V104	1.2% C 2.0% W steel: John Vessey; chisels, punches **DPN: 540**
BRUNSWICK V105	0.3% C 9% W steel: John Vessey; hot work dies **DPN: 620**
BRUNSWICK V414	0.4% C 14% W steel: John Vessey **DPN: 540**
BS S68	High tungsten steel: Obsolete
F 1	1.0% C 1.25% W steel: Designation used by AISI
F 2	1.25% C 3.5% W steel: Designation used by AISI
FCW 5	1.35% C 5% W steel: Firth Brown
HALLAMITE	C not specified 14% W steel: For tools; Hallamshire
HALLAMITIER	C not specified 18% W steel: For tools; Hallamshire
HALLAMITIEST	C not specified 22% W steel: For tools; Hallamshire
INTRA	1.2% C 1.5% W steel: Jonas; taps, reamers, etc. **DPN: 670**
JO	1.15% C 1.4% W tool steel: Osborn; taps, reamers, small tools; obsolete **DPN: 850**
KE 621	1.05% C 0.4% W steel: Kayser Ellison; taps, punches, etc.
KE 637	1.0% C 1.7% Mn 0.25% W steel: Kayser Ellison; gauges, etc. **DPN: 850**
KAO	1.3% C 3.0% W steel: For tools; Huntsman; cold drawing dies **DPN: 790**
KAOX	1.3% C 5% W steel: Huntsman; cold drawing dies **DPN: 790**
K-W	1.3% C 0.3% Mn 3.5% W steel: Carpenter
LT5	1.1% C 1.5% W steel: For tools; Low Moor; obsolete **DPN: 850**
LT 13	1.7% C 7% W steel: For tools; Low Moor; obsolete **DPN: 950**
MEL-TROL K-W	1.3% C 3.5% W steel: Carpenter for AISI type F3
SAE HNV 7	0.5% C 14% W 3.5% Si steel: For inlet valves
SPEAR No. 1	1.0% C W steel: Spear and Jackson; for hand tools
SUPERDRAW	1.3% C 2.25% W steel: Edgar Allen; cold drawing dies **DPN: 950**
TDB	1.4% C 5% W steel: Jonas; dies **DPN: 670**
T 60601	UNS designation for type F1 tool steel
T 60602	UNS designation for type F2 tool steel
TDC	1.2% C 1.25% W steel: Firth Brown

Note. The following abbreviations and units are used in the tables:

DPN	Hardness, diamond pyramid number
UTS	Ultimate tensile strength, N/mm^2
Elon	Elongation, %
Proof	0.1% proof strength, N/mm^2

1 N/mm^2=0.1 hbar=0.102 kgf/mm^2=0.06475 tonf/in.^{2}=145.04 lbf/in.^{2}=1 MPa

See Appendix II for other abbreviations and conversion tables.

44J Steel – all carbon contents with vanadium

Specific gravity	7.8
Density	7800 kg/m^3
Solidus/liquidus	1470–1510 °C
Thermal conductivity	50 W/m °C
Coefficient of linear expansion	11 × 10^{-6}/ °C
Electrical conductivity	8–10% IACS (copper 100%)
Specific resistance	180–250 microhm mm
Young's modulus of elasticity	–
Impact	70 J – lower impact figures can be expected
Fatigue strength	–
Hot strength	–

The above properties are typical of the following group and may not apply exactly to any one specification. It is possible that with certain specifications some of the values may not be applicable.

General metallurgical characteristics

Vanadium is an excellent de-oxidizer, carbide former and grain refiner which is very expensive and relatively scarce. Vanadium is seldom used purely as a de-oxidizer and is not often found as the sole alloying element, being generally present with chromium. Its grain refining properties are associated with the carbide forming. These carbides, being stable, do not readily dissolve in the iron matrix and prevent the formation of large carbide masses. Carbon combines with vanadium in preference to iron; thus all the vanadium exists as the carbide, the remaining carbon forming iron carbide. Vanadium carbide is very hard and stable; consequently even annealed steel is hard.

Vanadium improves the fatigue strength on the one hand, but increases the notch sensitivity on the other, and has no appreciable effect on the corrosion resistance.

Most of the steels listed are of the high carbon type (0.8% carbon and upwards) which can be made glass hard by thermal treatment.

These steels, particularly the high carbon ones, cannot accept any cold work without becoming very hard and brittle and very often shattering.

Any thermal treatment carried out for full annealing, normalizing or hardening requires a temperature in the region of 900 °C.

The steels in this group will always be very hard and brittle in the as quenched condition and specialist advice should be obtained before carrying out any form of thermal treatment where quenching is involved.

The same comments apply to welding as it could be expected that these materials will crack in the heat affected zone unless great care is taken.

These steels will always be fine grained which gives better flexibility. Where heat treatment is correctly controlled, this will result in very fine grained steels with good impact and fatigue strength.

The steels in this section are used for cutting tools, press tools and shear blades. Some of the lower carbon steels may be used for structural steels under specific conditions.

In general these steels are being replaced by alloy steels where nickel, chromium, molybdenum are present along with the vanadium.

Symbol	Nominal analysis, supplier, condition and remarks.
1.2833	1.05% C 1.15% V steel: German Standard
1.2838	1.4% C 3.4% V steel: German Standard
1.5213	0.2% C 1.4% Mn 0.1% V steel: For seamless drums; German Standard
1.5223	0.41% C 1.7% Mn 0.1% V steel: Forging; German Standard
02	0.9% C 1.7% Mn 0.15% V steel: For cold work; Osborn
06S	0.95% C 0.25% V steel: Carrs; centres, cold heading dies, etc. **DPN: 750**

Note. The following abbreviations and units are used in the tables:

DPN	Hardness, diamond pyramid number
UTS	Ultimate tensile strength, N/mm^2
Elon	Elongation, %
Proof	0.1% proof strength, N/mm^2

1 N/mm^2=0.1 hbar=0.102 kgf/mm^2=0.06475 tonf/in.^{2}=145.04 lbf/in.^{2}=1 MPa

See Appendix II for other abbreviations and conversion tables.

Symbol	Nominal analysis, supplier, condition and remarks.
12W1	1.1% C 0.2% V steel: Pompey
15VDT	0.8% C 0.1% V steel: Sandvik
17VDT	1.0% C 0.1% V steel: Sandvik
20 Mn V6	0.2% C 1.5% Mn 0.12% V steel: Designation used by German Standards
42 Mn V7	0.41% C 1.7% Mn 0.1% V steel: Designation used by German Standards
90 MV8	0.9% C 2.0% Mn 0.2% V steel: French Standards designation
145 V33	1.45% C 3.3% V steel: German Standards designation
280 M 01	0.42% C 0.12% V 0.035% Al (max) steel: BS 970
1162	1.05% C 0.15% V steel: French Standards designation
1163	0.9% C 0.15% V steel: French Standards designation
2211	0.9% C 2.0% Mn 0.2% V steel: French Standards designation
9257	1.0% C 0.25% V tool steel: Origin unknown
A 441	0.22% C 1.2% Mn 0.2% V 0.2% Cu steel: Armco **UTS: 480 Elon: 22% Proof: 320**
A 540	0.8% C 0.1% V steel: Fagersta **DPN: 630**
A 575	0.73% C 0.18% V steel: SHD; drills
ARMCO HS No. 3	0.1% C 0.6% Mn 0.02% V 0.2% Cu steel: Armco

Symbol	Nominal analysis, supplier, condition and remarks.
ARMCO HS No. 5	0.22% C (max) 1.25% Mn 0.2% V 0.02% Nb 0.2% Cu steel: Armco
ARMCO HS No. 6	0.17% C 0.7% Mn 0.02% Nb steel: Armco
ASTM A225 A	0.18% C 1.45% Mn 0.12% V steel: Normalized **UTS: 510 Elon: 22%**
ASTM A225 B	0.2% C 1.45% Mn 0.12% V steel: Normalized **UTS: 550 Elon: 21%**
ASTM A292 1	0.45% C 0.1% V steel: For generator motors; hardened and tempered; obsolete; see A 469 **UTS: 530 Elon: 24% Proof: 270**
ASTM A293/1	0.45% C 0.1% V steel: For turbine rotors; hardened and tempered **UTS: 530 Elon: 24% Proof: 270**
ASTM A356/1	0.35% C 0.03% V steel: Casting; normalized and tempered **UTS: 460 Elon: 20% Proof: 240**
ASTM A469/1	0.45% C 0.2% V steel: For rotors; vacuum treated
ASTM A487 IN	0.3% C (max) 0.1% V steel: Casting for pressure; normalized; IQ grade is quenched and tempered
ASTM A508/1	0.35% C (max) 0.6% Mn 0.05% V (max) steel: For pressure vessels; vacuum treated
ASTM A618/11	0.26% C (max) 1.3% Mn (max) 0.01% V (min) 0.18% Cu steel: Tube; heat formed welded or seamless **UTS: 483 Elon: 22% Proof: 345**
ASTM A618/111	0.27% C (max) 1.4% Mn (max) 0.01% V (min) steel: Seamless and welded tube for structures **UTS: 448 Elon: 19% Proof: 345**
ASTM A633 E	0.22% C (max) 1.3% Mn 0.08% V 0.02% N steel: For structures; normalized **UTS: 590 Elon: 20% Proof: 415**
ASTM A656/1	0.18% C (max) 1.6% Mn (max) 0.1% V 0.02% Al (min) 0.01% N steel: For structures **UTS: 720 Elon: 12% Proof: 550**
ASTM A707 L3/1	0.25% C (max) 1.3% Mn 0.1% V 0.18% Cu (min) steel: Flanges for low temperature use; impact 41 J at −46 °C **UTS: 290 Elon: 22%**
ASTM A707 L3/2	0.25% C (max) 1.3% Mn 0.1% V 0.18% Cu (min) steel: Flanges for low temperature use; impact 54 J at −46 °C **UTS: 360 Elon: 25%**
ASTM A707 L3/3	0.25% C (max) 1.3% Mn 0.1% V 0.18% Cu (min) steel: Flanges for low temperature use; impact 68 J at −46 °C **UTS: 415 Elon: 23%**
ASTM A707 L3/4	0.25% C (max) 1.3% Mn 0.1% V 0.18% Cu (min) steel: Flanges for low temperature use; impact 68 J at −46 °C **UTS: 515 Elon: 20%**
ASTM A714/2	0.26% C (max) 1.3% Mn 0.01% V 0.18% Cu (min) steel: Pipe; welded or seamless **UTS: 463 Elon: 22% Proof: 345**
ASTM A714/3	0.27% C (max) 1.4% Mn (max) 0.01% V 0.18% Cu (min) steel: Pipe; welded or seamless **UTS: 448 Elon: 20% Proof: 345**
ASTM A732 SN	0.3% C (max) 0.8% Mn 0.1% V steel: Investment casting; normalized and tempered **UTS: 590 Elon: 22% Proof: 380**
ASTM A737 A	0.22% C (max) 1.2% Mn 0.1% V (max) steel: Plate for pressure vessels
ATLAS SPECIAL 10	1.05% C 0.2% V steel: Atlas; as AISI type W2
AW 441	0.22% C (max) 1.25% Mn 0.02% V 0.2% Cu steel: Alan Wood
AWV 50	0.22% C (max) 1.25% Mn 0.02% V 0.015% B steel: Alan Wood
AWV 55	0.22% C (max) 1.15% Mn 0.02% V 0.015% B steel: Alan Wood
AWV 60	0.22% C (max) 1.15% Mn 0.02% V 0.015% B steel: Alan Wood

Symbol	Nominal analysis, supplier, condition and remarks.
AWV 65	0.22% C (max) 1.6% Mn 0.02% V 0.015% B steel: Alan Wood
B 188	0.8% C 2.1% Mn 0.1% V steel: Fagersta **DPN: 790**
BC 3V	1.4% C 3.5% V steel: Origin unknown
BC 8V	0.82% C 0.18% V steel: ESC; rock drills, punches, etc. **DPN: 730**
BC 10V	1.0% C 0.2% V steel: ESC: twist drills, dental burrs **DPN: 730**
BH 36S	0.16% C (max) 1.3% Mn 0.1% V 0.025% S and P (max) steel: Rheinstahl
BH 39	0.17% C (max) 1.3% Mn 0.17% V steel: Rheinstahl
BH 395	0.16% C (max) 1.3% Mn 0.17% V steel: Rheinstahl
BLUE and WHITE LABEL	0.97% C 0.5% V steel: For tools; Huntsman **DPN: 750**
C 110	0.2% C 0.1% V steel
CB/V45	0.2% C 1.2% Mn 0.01% V 0.01% Nb steel: Algoma
CB/V50	0.2% C 1.2% Mn 0.01% V 0.01% Nb steel: Algoma
CB/V55	0.2% C 1.2% Mn 0.01% V 0.01% Nb steel: Algoma
CLAY-LOY	0.22% C 1.25% Mn 0.02% V 0.02% Nb steel: Phoenix Steel Corp.
CN/V60 022/C	0.2% C 1.25% Mn 0.01% V 0.01% Nb steel: Algoma
COLDIE	0.87% C 0.25% Mn 0.2% V steel: Braeburn
CONSUMET ELECTRICAL IRON	0.04% C 0.1% V 0.25% Si Fe alloy: Carpenter; for magnetic purposes
CRD 90C	0.9% C 0.25% V steel: C Denton; shear blades; obsolete **DPN: 690**
CRD 100C	1.0% C 0.25% V steel: C Denton; shear blades; obsolete **DPN: 690**
CRD 120C	1.15% C 0.25% V steel: C Denton; shear blades **DPN: 690**
DIAMOND 1	1.3% C 0.25% V steel: Spencer
DIAMOND 2	1.1% C 0.25% V steel: Spencer
DIAMOND 3	1.0% C 0.25% V steel: Spencer
DIAMOND 4	0.85% C 0.25% V steel: Spencer
DIAMOND 5	0.75% C 0.25% V steel: Spencer
DIAMOND 6	0.65% C 0.25% V steel: Spencer
DIN 17200 42 Mn V7	0.41% C 1.8% Mn 0.1% V steel: Forging
DOFASCOLOY MY	0.22% C 1.2% Mn 0.02% V 0.2% Cu steel: Dominion
DOUBLE EXTRA	1.0% C 0.25% V steel: Origin unknown
DOUBLE CONQUEROR VANADIUM	1.0% C 0.25% V tool steel: Origin unknown
DP 5	1.0% C 0.25% V tool steel: Osborn; cutting tools, cold forming dies; obsolete **DPN: 750**
EX-TEN 45	0.22% C 1.25% Mn 0.01% V or 0.01% Nb steel: US Steel Corp.
EX-TEN 50	0.22% C 1.35% Mn 0.01% V or 0.01% Nb steel: US Steel Corp.
EX-TEN 55	0.24% C 1.35% Mn 0.01% V or 0.01% Nb steel: US Steel Corp.
EX-TEN 60	0.25% C 1.35% Mn 0.02% V or 0.01% Nb steel: US Steel Corp.
EX-TEN 65	0.26% C 1.35% Mn 0.02% V or 0.01% Nb steel: US Steel Corp.
EX-TEN 70	0.26% C 1.35% Mn 0.02% V or 0.01% Nb steel: US Steel Corp.
EXTRA TOUGH HARD V	0.9% C 0.15% V steel: Schoeller-Bleckmann **DPN: 730**
F	1.0% C 0.3% V steel: Russian Standards designation
FB 50AK	0.2% C (max) 1.3% Mn 0.1% V 0.03% S and P (max) 0.015% N steel: German proprietor's specification; A1 killed
FMP 035	1.05% C 0.35% V steel: F Parkin; as tool steel W2
FTW 60	0.15% C (max) 1.23% Mn 0.08% V 0.005% S 0.01% P steel: Fuji

Symbol	Nominal analysis, supplier, condition and remarks.
GLS 441	0.22% C 1.25% Mn 0.2% Cu 0.02% N steel: National Steel Co.
GOST 5058/15 GF	0.15% C 1.0% Mn 0.3% Cr Ni Cu (max) 0.075% V steel: Plate; Russian Standard **UTS: 520** **Proof: 360**
GS 10 Mn 7	0.08% C 1.7% Mn 0.06% V 0.04% Nb cast steel: Thyssen; precipitation hardened
HECLA 28	1.0% C 0.25% V steel: Origin unknown
HI STRENGTH B	0.22% C 1.2% Mn 0.2% V 0.2% Cu steel: Armco **UTS: 480** **Elon: 22%** **Proof: 320**
HI YIELD 42	0.21% C 0.9% Mn 0.02% V steel: Granitility Steel Co.
IH 65	0.22% C 1.65% Mn 0.01% V or 0.01% Nb 0.2% Cu steel: I H Wisconsin Steel
IHX 45	0.2% C 1.0% Mn 0.01% V or 0.01% Nb steel: I H Wisconsin Steel
IHX 50	0.22% C 1.1% Mn 0.01% V or 0.01% Nb steel: I H Wisconsin Steel
IHX 55	0.24% C 1.4% Mn 0.01% V or 0.01% Nb steel: I H Wisconsin
IHX 60	0.26% C 1.55% Mn 0.01% V or 0.01% Nb steel: I H Wisconsin
IHX 65	0.22% C 1.6% Mn 0.01% V or 0.01% Nb steel: I H Wisconsin
IHX 70	0.26% C 1.65% Mn 0.01% V or 0.01% Nb steel: I H Wisconsin
INX 42	0.2% C 0.9% Mn 0.01% V or 0.01% Nb steel: Inland Steel Co.
INX 45	0.2% C 1.0% Mn 0.01% V or 0.01% Nb steel: Inland Steel Co.
INX 50	0.22% C 1.25% Mn 0.01% V or 0.01% Nb steel: Inland Steel Co.
INX 55	0.24% C 1.4% Mn 0.01% V or 0.01% Nb steel: Inland Steel Co.
INX 60	0.26% C 1.65% Mn 0.01% V or 0.01% Nb steel: Inland Steel Co.
INX 65	0.26% C 1.65% Mn 0.01% V or 0.01% Nb steel: Inland Steel Co.
INX 70	0.26% C 1.65% Mn 0.01% V or 0.01% Nb steel: Inland Steel Co.
J 4V	0.9% C 0.25% V steel: Jonas; cold swaging dies **DPN: 570**
JLX 42	0.2% C (max) 1.0% Mn 0.01% V or 0.01% Nb steel: Jones & Laughlan
JLX 65	0.26% C (max) 1.5% Mn 0.01% V or 0.01% Nb steel: Jones & Laughlan
JLX 70	0.26% C (max) 1.65% Mn 0.01% V or 0.01% Nb steel: Jones & Laughlan
K 11805	0.18% C (max) 0.02% V (min) steel: Designation used by UNS
KAISALOY 42CV	0.2% C (max) 0.9% Mn 0.01% V or 0.01% Nb steel: Kaiser Steel Corp.
KAISALOY 50MV	0.22% C (max) 1.25% Mn 0.02% V 0.2% Cu steel: Kaiser Steel Corp.
KE 1006	1% C 0.15% V steel: For coining dies; Kayser Ellison **DPN: 910**
KE A205	1.4% C 3.4% V steel: Kayser Ellison; dies, cold heading punches **DPN: 910**

Note. The following abbreviations and units are used in the tables:

DPN Hardness, diamond pyramid number
UTS Ultimate tensile strength, N/mm^2
Elon Elongation, %
Proof 0.1% proof strength, N/mm^2

1 N/mm^2=0.1 hbar=0.102 kgf/mm^2=0.06475 tonf/in.^{2}=145.04 lbf/in.^{2}=1 MPa
See Appendix II for other abbreviations and conversion tables.

Symbol	Nominal analysis, supplier, condition and remarks.
LT 1	1.0% C 0.25% V steel: Low Moor; obsolete **DPN: 850**
MEL-TROL MIRROMOLD	0.1% C 0.2% Mn 0.1% V steel: Carpenter for AISI type P1
MIL S 12505/5	0.23% C (max) 1.25% Mn 0.02% V 0.4% Cu steel: Bethlehem
MIL S 16113-1	0.18% C (max) 0.02% V (min) 0.005% Ti (min) steel: US military specification
MIL S 16113-11	0.18% C (max) 0.08% V (min) steel: US military specification
MIL S 24412	0.12% C (max) 0.02% V (min) steel: US military specification
MLX 45	0.15% C (max) 1.0% Mn 0.02% V or 0.005% Nb steel: McLouth
MLX 50	0.2% C (max) 1.0% Mn 0.02% V or 0.005% Nb steel: McLouth
MLX 55	0.24% C (max) 1.2% Mn 0.02% V or 0.005% Nb steel: McLouth
MLX 60	0.26% C (max) 1.5% Mn 0.2% V or 0.005% Nb steel: McLouth
MNV	0.22% C (max) 1.25% Mn 0.02% V 0.2% Cu steel: Lukens
MNV A441	0.22% C (max) 1.25% Mn 0.02% V 0.2% Cu steel: Bethlehem
NORESCO EXTRA TOUGH HARD V	0.9% C 0.15% V steel: Schoeller-Bleckmann **DPN: 730**
ORELLOY 441	0.22% C 1.25% Mn 0.02% V 0.2% Cu steel: Oregon
P121	0.6% C 0.1% V steel: Bofors **DPN: 800**
P171	0.85% C 0.1% V steel: Bofors **DPN: 800**
P181	0.9% C 0.1% V steel: Bofors; obsolete **DPN: 800**
P211	1.05% C 0.1% V steel: Obsolete; Bofors **DPN: 850**
PINK LABEL	0.9% C 0.2% V steel: T Turton
PM 15	0.14% C 1.5% Mn 0.07% V steel: Pompey
PM 15 E	0.14% C 1.5% Mn 0.07% V steel: Pompey
RENOWN	1.0% C 0.2% V steel: Latrobe
REPUBLIC A441	0.22% C (max) 1.25% Mn 0.02% V 0.2% Cu steel: Republic Steel Co.
REPUBLIC X42W	0.22% C (max) 1.1% Mn 0.01% V or 0.01% Nb steel: Republic Steel Co.
REPUBLIC X45W	0.2% C 0.75% Mn 0.01% V or 0.01% Nb steel: Republic Steel Co.
REPUBLIC X50W	0.2% C 0.75% Mn 0.01% V or 0.01% Nb steel: Republic Steel Co.
REPUBLIC X60W	0.25% C 1.5% Mn 0.01% V or 0.01% Nb steel: Republic Steel Co.
REPUBLIC X65W	0.25% C 1.5% Mn 0.01% V or 0.01% Nb steel: Republic Steel Co.
REPUBLIC X65W	0.26% C 1.5% Mn 0.01% V or 0.01% Nb steel: Republic Steel Co.
REPUBLIC X70W	0.26% C 1.65% Mn 0.01% V or 0.01% Nb steel: Republic Steel Co.
SAE W209	0.9% C 0.2% V steel: For tools
SAE W210	1.0% C 0.2% V steel: For tools
SAE W310	1.0% C 0.4% V steel: For tools
SB 39F	0.16% C 1.3% Mn 0.03% V 0.03% Ti 0.035% S and P (max) steel: German proprietor's specification
SB 42F	0.16% C 1.3% Mn 0.1% V 0.035% S and P (max) steel: German proprietor's specification; Al killed
SHEFFIELD HIGH STRENGTH B	0.22% C (max) 1.20% Mn 0.02% V 0.2% Cu steel: Armco
SILVAN STAR	1.0% C with V steel: Firth Sterling
SIX STAR VANADIUM	1.0% C 0.2% Mn 0.15% V 0.2% Si steel: For tools; Jessop
SKA 45	0.22% C (max) 1.25% Mn 0.02% V steel: Phoenix
SKA 50	0.26% C (max) 1.3% Mn 0.02% V steel: Phoenix
SKS 43	1.05% C 0.17% V steel: Japanese Standards designation

Symbol	Nominal analysis, supplier, condition and remarks.
SKS 44	0.85% C 0.17% V steel: Japanese Standards designation
SPECIAL CONQUEROR VANADIUM	1.0% C 0.25% V steel: For tools; origin unknown
SPECIAL V	C as specified 0.25% Mn 0.2% V steel: For tools; Braeburn
SS 2900	0.8% C 0.1% V steel: Swedish Standard; annealed **DPN: 195**
STELCO VANADIUM	0.22% C 1.2% Mn 0.02% V 0.2% Nb 0.2% Cu steel: Steel Co. of Canada
STERLING V	1.0% C with V steel: Firth Sterling
TCR	0.8% C 0.3% V steel: Swift Levick; chisels, punches
TOWER 60	0.33% C 1.65% Mn 0.05% V steel: Algoma
TRI-STEEL	0.2% C 1.25% Mn 0.02% Cu steel: Inland Steel
TRI-TEN	0.22% C (max) 1.25% Mn 0.02% V 0.02% Cu steel: US Steel Co.
UHB 19 Va	0.9% C 0.07% V steel: Uddelholm for type W2
V 42	0.22% C (max) 1.25% Mn 0.02% V steel: Bethlehem Steel
V 45	0.22% C (max) 1.25% Mn 0.02% V steel: Bethlehem
V 50	0.22% C (max) 1.25% Mn 0.02% V 0.015% N steel: Bethlehem
V 55	0.22% C (max) 1.25% Mn 0.02% V 0.015% N steel: Bethlehem
V 60	0.22% C (max) 1.25% Mn 0.02% V 0.015% N steel: Bethlehem Steel
V 65	0.22% C (max) 1.25% Mn 0.02% V steel: Bethlehem Steel
V 90V	0.9% C 0.15% V steel: French Standards designation
VANADIUM EXTRA	1.06% C 0.2% V steel: Columbia; AISI type W2
VANADIUM STANDARD	1.06% C 0.2% V steel: Columbia; AISI type W2
VANQUISH	0.95% C 0.25% V steel: Hall and Pickles; obsolete **DPN: 670**
VS 4	1.05% C 0.25% V steel: Edgar Allen; dies, hammers, etc.
W 2	1.0% C 0.25% V steel: Carbon varies over wide limits; designation used by AISI
W 2	1.0% C 0.2% V steel: Osborn
Y 1 105V	1.05% C 0.15% V steel: French Standard
Y 105V	1.05% C 0.15% V steel: French Standards designation
YEO 36	0.18% C 1.3% Mn 0.1% V steel: Yamata
YOLOY A 242	0.22% C (max) 1.25% Mn 0.02% V 0.2% Cu steel: Youngstown Steel

Note. The following abbreviations and units are used in the tables:

DPN	Hardness, diamond pyramid number
UTS	Ultimate tensile strength, N/mm^2
Elon	Elongation, %
Proof	0.1% proof strength, N/mm^2

1 N/mm^2=0.1 hbar=0.102 kgf/mm^2=0.06475 tonf/in.^{2}=145.04 lbf/in.^{2}=1 MPa

See Appendix II for other abbreviations and conversion tables.

44K1 Steel – low carbon alloys
0.05–0.35% carbon with two or more of chromium, nickel, molybdenum, tungsten or vanadium

Specific gravity	7.86
Density	7860 kg/m^3
Solidus/liquidus	1450–1510 °C
Thermal conductivity	37.7 W/m °C
Coefficient of linear expansion	11 × 10^{-6}/ °C
Electrical conductivity	7–9% IACS (copper 100%)
Specific resistance	230–270 microhm mm
Young's modulus of elasticity	206 × 10^9 N/m^2
Impact	54–108 J (can be lower)
Fatigue strength	–
Hot strength	

Temperature °C	Tensile strength N/mm^2	Elongation %
100	800	24
200	800	23
300	800	23
400	750	25
500	640	24
600	500	26
650	370	29

The properties above are typical of the following group and may not apply exactly to any one specification. It is possible that with certain specifications some of the values may not be applicable.

General metallurgical characteristics

These steels are tougher, more hardenable and have better fatigue strength and, to some extent, corrosion resistance than the plain carbon or single alloy steels.

At least two of the alloying elements are present in all steels listed, while some steels have all the elements present, the majority having some chromium. Some of the steels are designed specially for carburizing and all the steels can be carburized. Tungsten and vanadium should not be present, however, as there is no technical advantage and with the very hard carbide formed there is a danger of the case being too hard and brittle.

One characteristic of alloy steel is that it is inherently fine grained, particularly with vanadium present. This improves the fatigue strength and allows a greater laxity of temperatures during thermal treatment.

Very briefly the effect of each alloying element is as follows.

Chromium. This is a carbide former which aids carburization and slows down the metallurgical reactions, thus increasing hardenability. Chromium improves the properties of steel at high temperatures, including oxidation resistance. It has a tendency to promote grain growth, and as it raises the critical temperatures, there is a tendency for chromium to produce large grained brittle steels.

Nickel. This strengthens and toughens the ferritic matrix of steel, improving the impact strength, particularly at low temperatures. Some 10% nickel steels have been specially developed for very low temperature application.

Nickel added to carburized steels gives a tough ductile core supporting a hard but not brittle case.

Nickel lowers the temperatures required for heat treatment and widens the range of suitable thermal treatment temperatures, thus giving more flexibility.

Molybdenum. This is a carbide former, the resultant carbides being well distributed, small and very stable. Thus it gives fine grained steels, improves the hot strength and creep strength, increases the depth of hardness achieved and improves the fatigue strength.

In addition molybdenum added to nickel and nickel chromium steels eliminates temper brittleness. This is a brittle condition obtained when some steels are tempered in the range 250–450 °C and is caused by carbide precipitation. The molybdenum carbides inhibit this tendency by stabilizing the carbides.

Tungsten. This forms very hard stable carbides and acts in many ways as molybdenum does. More tungsten is required to give the same effect as very little molybdenum. The use is generally confined to tool steels where large quantities of very hard carbides are required.

Vanadium. This is a powerful de-oxidizer in addition to being a carbide former almost as powerful as molybdenum. The carbides are always very small and evenly distributed throughout the mass, making vanadium the best grain refining element. This prevents agglomeration of these and other carbides during long term thermal treatment. Vanadium also forms nitrides and is present in most nitriding steels. It is commonly used with chromium where it considerably improves the fatigue strength because of its grain refining property, thus overcoming the tendency of chromium to give large grains.

Copper, boron and aluminium. These are sometimes also present in these steels. The copper enhances the corrosion resistance and there are claims that its presence improves the adhesion of paint films. Boron improves the hardenability to a much greater extent than any other element and also increases this function in other alloying elements. It reduces the range of temperatures for successful heat treatment. Aluminium is a powerful de-oxidizer, but the refractory nature of the oxide formed reduces the usefulness for this purpose. Small amounts of aluminium added to the de-oxidizer steel increases the ability of these steels to nitride, and has some grain refining properties.

All steels listed will almost always be used in the hardened and tempered or surface hardened condition.

The majority of these steels cannot accept any appreciable cold work without becoming brittle.

The temperatures for thermal treatment, either full anneal, hardening or normalizing, are in the region of 850–900 °C.

These steels will all have an appreciable hardenability and the reader is advised to examine the carbon equivalent before having any sophisticated thermal treatment or welding of high integrity components carried out.

The carbon equivalent (CE) formula is:

$$\text{CE} = \text{Carbon} + \frac{\text{Mn}}{6} + \frac{\text{Cr} + \text{Mo} + \text{V}}{15} + \frac{\text{Ni} + \text{Cu}}{15}$$

Any value above 0.7% indicates an appreciable hardenability and thus possible weld cracking unless precautions are taken. The higher the carbon equivalent, the greater is the hardenability of the steel. This means that the high carbon equivalent steels will respond well to thermal treatment and a decision can be made, taking into account the ruling section and the general size of the component, whether or not water, oil or even air hardening would be possible.

The same comments apply to welding where the carbon equivalent can be used to decide the amount of pre-heating and the heat input during the welding operation.

Specialist advice should always be requested where high integrity components are involved, either for thermal treatment or welding.

The steels in this group find a wide variety of uses including hot working tools, forming dies and equipment used in casting. Punches, shears, drills, cutting tools for non-ferrous parts, and high duty structural steel will all be found in this group, as will gears and shafts which require carburizing. There are a limited number of steels in this section which can be nitrided. The vast majority of high tensile studs, bolts, nuts, etc., used in automobiles and aircraft engineering will also be found here.

Symbol	Nominal analysis, supplier, condition and remarks.
0.5% FO	0.2% C (max) 0.7% Mn 0.7% Cr 0.5% Mo steel: Creusot-Loire
0.5467	0.3% C 2.4% Cr 0.6% V 4.2% W steel: For springs; German Standard
1 Cr Mo	0.1% C 1.3% Cr 0.5% Mo steel: Electrode; Metrode
1 Cr Mo B	0.07% C 1.3% Cr 0.5% Mo steel: Electrode; Metrode for BS 2493
1 Ni B	0.05% C 1% Ni 0.2% Mo steel: Electrode; Metrode
1 Ni Cu B	0.06% C 0.7% Ni 0.5% Cr 0.6% Cu steel: Electrode; Metrode
1 Ni Mo B	0.09% C 1% Ni 0.4% Mo steel: Electrode; Metrode
1.1% FO	0.17% C (max) 0.5% Mn 1.0% Cr 0.5% Mo steel: Creusot-Loire
1.2% FO	0.17% C (max) 0.5% Mn 1.25% Cr 0.5% Mo steel: Creusot-Loire
1.2312	0.33% C 1.6% Cr 0.4% Mo steel: German Standard
1.2330	0.35% C 1.6% Cr 0.5% Mo steel: German Standard
1.2344	0.38% C 1.3% Ni 5.25% Cr 1% V steel: German Standard
1.2365	0.3% C 3% Cr 2.7% Mo 0.5% V steel: German Standard
1.2567	0.3% C 2.4% Cr 0.6% V 4.3% W steel: German Standards designation
1.2581	0.3% C 2.7% Cr 0.4% V 8.5% W steel: German Standard
1.2735	0.14% C 3.5% Ni 0.7% Cr steel: German Standard
1.2764	0.19% C 4.1% Ni 1.3% Cr 0.2% Mo 0.4% W tool steel: German Standards designation
1.2766	0.32% C 4.1% Ni 1.2% Cr 0.2% Mo steel: German Standard
1.3344	1.2% C 4.1% Cr 5.0% Mo 3.0% V 6.3% W steel: For tools; German Standards designation
1.5404	0.21% C 0.3% Ni 0.3% Cr 0.5% Mo 0.3% V steel: Forging; German Standard
1.5406	0.18% C 0.3% Ni 0.3% Cr 0.9% Mo 0.35% V steel: Forging; German Standard
1.5419	0.21% C 0.5% Mn 0.3% Cr 0.35% Mo steel: Forging; German Standard
1.5620	0.34% C 1.3% Ni 0.6% Cr steel: Forging; German Standard
1.5919	0.17% C 1.6% Ni 1.6% Cr steel: For carburizing; German Standard
1.5920	0.18% C 2.0% Ni 2.0% Cr steel: For carburizing; German Standard
1.5924	0.15% C 1.5% Ni 1.5% Cr steel: For carburizing; German Standard
1.5934	0.18% C 2.0% Ni 2.0% Cr steel: For carburizing; German Standard
1.6001	0.24% C 0.7% Ni 2.2% Cr steel: For pressure vessels; German Standard
1.6511	0.36% C 1.1% Ni 1.1% Cr 0.20% Mo steel: Forging; German Standard
1.6513	0.28% C 1.2% Ni 1.2% Cr 0.25% Mo steel: Forging; German Standard
1.6582	0.34% C 1.5% Ni 1.5% Cr 0.2% Mo steel: Forging; German Standard
1.6590	0.3% C 2.0% Ni 2.0% Cr 0.3% Mo steel: Forging; German Standard
1.6592	0.28% C 2.0% Ni 1.2% Cr 0.35% Mo 0.1% V steel: Forging; German Standard
1.6604	0.3% C 2.1% Ni 2.1% Cr 0.35% Mo steel: German Standard
1.6761	0.28% C 1.2% Ni 1.2% Cr 0.45% Mo steel: Forging; German Standard
1.7205	0.14% C 0.8% Cr 0.15% Mo steel: Bar; German Standard
1.7207	0.18% C 0.7% Cr 0.15% Mo steel: German Standard
1.7214	0.25% C 1.1% Cr 0.2% Mo steel: Forging; German Standard
1.7218	0.25% C 1.1% Cr 0.2% Mo steel: Forging; German Standard
1.7220	0.34% C 0.6% Ni 1.1% Cr 0.2% Mo steel: Forging; German Standard
1.7242	0.17% C 0.4% Ni 1.1% Cr 0.25% Mo steel: Forging; German Standard
1.7251	0.23% C 0.8% Cr 0.25% Mo steel: For seamless drums; German Standard
1.7254	0.25% C 1.0% Cr 0.25% Mo steel: German Standard
1.7258	0.24% C 0.6% Ni 1.1% Cr 0.25% Mo steel: For high temperature fasteners; German Standard
1.7273	0.24% C 0.8% Ni 2.4% Cr 0.25% Mo steel: For high pressures; German Standard
1.7281	0.16% C 2.2% Cr 0.35% Mo steel: For high pressures; German Standard
1.7283	0.2% C 0.8% Ni 2.2% Cr 0.3% Mo steel: For high pressures; German Standard
1.7307	0.17% C 4.0% Cr 1.0% Mo 0.1% V steel: Welding rod; German Standard
1.7324	0.2% C 0.4% Cr 0.45% Mo steel: Welding rod; German Standard
1.7334	0.2% C 0.5% Cr 0.4% Mo steel: For carburizing; German Standard
1.7335	0.14% C 0.8% Cr 0.45% Mo steel: For boiler plate; German Standard
1.7337	0.16% C 0.4% Ni 1.1% Cr 0.45% Mo steel: Forging; German Standard
1.7345	0.1% C 1.2% Cr 0.5% Mo steel: Welding rod; German Standard
1.7346	0.12% C 1.1% Cr 0.5% Mo steel: Welding rod; German Standard
1.7350	0.22% C 0.6% Ni 1.1% Cr 0.45% Mo steel: Forging; German Standard
1.7356	0.1% C (max) 1.7% Cr 0.5% Mo steel: Welding rod; German Standard
1.7361	0.32% C 0.6% Ni 3.0% Cr 0.4% Mo steel: German Standard
1.7362	0.15% C 5% Cr 0.55% Mo steel: For high pressure; German Standard
1.7373	0.15% C (max) 6.0% Cr 0.7% Mo steel: Welding rod; German Standard
1.7380	0.15% C (max) 2.2% Cr 1.1% Mo steel: German Standard
1.7384	0.1% C (max) 3.0% Cr 1.0% Mo steel: Welding rod; German Standard
1.7394	0.1% C (max) 9.5% Cr 1.0% Mo steel: Welding rod; German Standard
1.7513	0.22% C 1.1% Cr 0.2% V steel: German Standard
1.7704	0.3% C 2.5% Cr 0.2% Mo 0.15% V steel: Forging; German Standard
1.7707	0.3% C 0.6% Ni 2.5% Cr 0.2% Mo 0.2% V steel: Forging; German Standard
1.7708	0.16% C 1.1% Cr 0.3% Mo 0.2% V steel: German Standard
1.7709	0.14% C 1.1% Cr 0.25% Mo 0.2% V steel: German Standard
1.7710	0.24% C 1.3% Cr 0.22% Mo 0.25% V steel: German Standard
1.7733	0.24% C 0.6% Ni 1.4% Cr 0.55% Mo 0.2% V steel: For high temperature fasteners; German Standard

Note. The following abbreviations and units are used in the tables:

DPN	Hardness, diamond pyramid number
UTS	Ultimate tensile strength, N/mm^2
Elon	Elongation, %
Proof	0.1% proof strength, N/mm^2

$1\ N/mm^2 = 0.1\ hbar = 0.102\ kgf/mm^2 = 0.06475\ tonf/in.^2 = 145.04\ lbf/in.^2 = 1\ MPa$

See Appendix II for other abbreviations and conversion tables.

Symbol	Nominal analysis, supplier, condition and remarks.
1.7766	0.17% C 2.7% Cr 0.25% Mo 0.15% V steel: For high pressures; German Standard
1.7779	0.2% C 3.2% Cr 0.55% Mo 0.5% V steel: For high pressures; German Standard
1.8070	0.21% C 0.6% Ni 1.3% Cr 1.1% Mo 0.3% V steel: For high temperature fasteners; German Standard
1.8212	0.21% C 2.8% Cr 0.4% Mo 0.8% V 0.4% W steel: For pressure vessels; German Standard
1.8507	0.33% C 1.1% Cr 0.2% Mo 1.0% Al steel: For nitriding; German Standard
1.8514	0.3% C 2.5% Cr 0.2% Mo 0.15% V steel: For nitriding; German Standard
1.8519	0.3% C 2.5% Cr 0.2% Mo 0.15% V steel: For nitriding; German Standard
1.8544	0.32% C 1.1% Cr 0.2% Mo 1.1% Al steel: For nitriding; German Standard
1.8550	0.33% C 1.0% Ni 1.7% Cr 0.2% Mo 1.0% Al steel: For nitriding; German Standard
2 Cr Mo	0.1% C 2.5% Cr 1% Mo steel: Electrode; Metrode
2 Cr Mo B	0.07% C 2.2% Cr 1.0% Mo steel: Electrode; Metrode
2 Cr Mo BG	0.06% C 2.2% Cr 1% Mo steel: Electrode; Metrode
2 Cr Mo BX	0.06% C 2.2% Cr 1% Mo steel: Electrode; Metrode
2 Cr Mo C B	0.1% C 2.2% Cr 1.0% Mo steel: Electrode; Metrode
2 N3C1	0.12% C 0.55% Mn 3% Ni 0.7% Cr steel: Sandvik for SS 2514
2 Ni Mo B	0.05% C 2.8% Ni 0.6% Mo steel: Electrode; Metrode
2.2% FO	0.15% C (max) 0.45% Mn 2.25% Cr 1.0% Mo steel: Creusot-Loire
3% FO	0.15% C (max) 0.45% Mn 3.0% Cr 1.0% Mo steel: Creusot-Loire
3 G Mo B	0.07% C 3.0% Cr 1.0% Mo steel: Electrode; Metrode
3 KH2V8F	0.35% C 2.35% Cr 0.35% V 8.2% W steel: Russian Standards designation
03 NCMB	0.03% C 0.5% Ni 0.5% Cr 0.16% Mo steel: Electrode; Metrode
3 Ni Cr Mo B	0.04% C 4% Ni 0.9% Cr 0.7% Mo steel: Electrode; Metrode
3 Ni Mo B	0.04% C 3.3% Ni 0.6% Mo steel: Electrode; Metrode
3 Ni Mo LB	0.04% C 3.2% Ni 0.5% Mo steel: Electrode; Metrode; for low temperature use
4 KH2V5FM	0.35% C 2.5% Cr 0.8% Mo 0.8% V 5.0% W steel: Russian Standards designation
4 NHD	0.35% C 4% Ni 1.5% Cr 0.35% V 5.7% W steel: For tools; Huntsman; hot stamping dies **DPN: 530**
4 WHD	0.35% C 1.2% Cr 0.25% V 4% W steel: Huntsman; pressure die casting dies, etc. **DPN: 500**
4.5% DO	0.15% C (max) 0.5% Mn (max) 4.0% Ni 0.5% Cr 0.4% Mo steel: Creusot-Loire
5 C 2 Mo	0.25% C 1.1% Cr 0.2% Mo steel: Sandvik for SS 2225
5 Cr Mo	0.1% C 5.5% Cr 0.5% Mo steel: Electrode; Metrode
5 Cr Mo 10	0.1% C 5% Cr 0.55% Mo steel: ESC; annealed for BS 1501/625 **UTS: 400 Elon: 20% Proof: 210**
5 Cr Mo 22	0.22% C 0.45% Mn 5% Cr 0.55% Mo steel: ESC for BS 1506/625 **UTS: 670 Elon: 19% Proof: 580**
5 Cr Mo B	0.07% C 5.0% Cr 0.5% Mo steel: Metrode
6 F4	0.2% C 3.1% Ni 0.15% Cr (max) 3.4% Mo 0.1% V steel: Carpenter
6 N 3 C2 Mo	0.3% C 3.3% Ni 1% Cr 0.25% Mo steel: Sandvik for SS 2534
6 N 4 C 2	0.3% C 4.3% Ni 1.3% Cr steel: Sandvik
7 C 2 Mo	0.35% C 1.1% Cr 0.2% Mo steel: Sandvik
7 Cr Mo 7/5	0.1% C (max) 1.7% Cr 0.5% Mo steel: Designation used by German Standards
7 Cr Mo 12/10	0.1% C (max) 3.0% Cr 1.0% Mo steel: Designation used by German Standards
7 Cr Mo B	0.07% C 7.0% Cr 0.5% Mo steel: Electrode; Metrode
7 Cr Mo B	0.05% C 7% Cr 0.6% Mo steel: Electrode
07 NCMB	0.07% C 0.5% Ni 0.5% Cr 0.16% Mo steel: Electrode; Metrode
9 Cr Mo 4/5	0.1% C 1.2% Cr 0.5% Mo steel: Designation used by German Standards
9 Cr Mo B	0.07% C 9.0% Cr 1.0% Mo steel: Electrode; Metrode
9 Cr Mo L B	0.05% C 9.0% Cr 1.0% Mo steel: Electrode; Metrode
9 G Mo	0.08% C 9% Cr 1% Mo steel: Electrode; Metrode
10 C 2 M	0.1% C 2.2% Cr 1% Mo steel: Electrode; Metrode
10 CNK 1	0.09% C 1.4% Ni 1.0% Cr 1.0% Mo steel: Pompey
10 CNK 3	0.09% C 2.7% Ni 0.75% Cr steel: Pompey
10 CNK 32	0.09% C 3.0% Ni 0.75% Cr steel: Pompey
10 Cr Mo 9/10	0.15% C (max) 2.2% Cr 1.0% Mo steel: Designation used by German Standards
10 NC12	0.1% C 3.0% Ni 0.8% Cr steel: French Standards designation
10 NCMB	0.1% C 0.5% Ni 0.5% Cr 0.16% Mo steel: Electrode; Metrode
11 Cr Mo 4/5	0.12% C 1.1% Cr 0.5% Mo steel: Designation used by German Standards
12 CMV	0.12% C 0.5% Cr 0.5% Mo 0.25% V steel: Electrode; Metrode
12 CNK 3	0.14% C 2.7% Ni 0.75% Cr steel: Pompey
12 Cr Mo 19/5	0.15% C (max) 0.5% Cr 0.55% Mo steel: Designation used by German Standards
12 Cr Mo V B	0.07% C 12.0% Cr 0.7% Mo 0.15% V steel: Electrode; Metrode
12 KH1MF	0.11% C 0.55% Mn 1.05% Cr 0.3% Mo 0.22% V steel: Russian Standards designation
12 MKH	0.13% C 0.55% Mn 0.5% Cr 0.5% Mo steel: Russian Standards designation
12 NCD 6	0.12% C 1.4% Ni 0.9% Cr 0.2% V steel: Pompey
12 NCMB	0.12% C 0.5% Ni 0.5% Cr 0.16% Mo steel: Metrode
12 S	0.3% C 3% Cr 0.25% V 8.5% W steel: Carrs; hot work, punches, dies, etc. **DPN: 460**
13 Cr Mo 4/4	0.14% C 0.8% Cr 0.45% Mo steel: Designation used by German Standards
13 Cr Mo V 4/2	0.14% C 1% Cr 0.25% Mo 0.2% V steel: Designation used by German Standards
14 CNK 3	0.14% C 3.45% Ni 0.75% Cr steel: Pompey
14 CNK 4	0.14% C 4.25% Ni 0.65% Cr steel: Pompey
14 CNK 32	0.13% C 3.0% Ni 0.75% Cr steel: Pompey
14 NKD 2	0.14% C 1.4% Ni 1.0% Cr steel: Pompey
15 Cr Mo 3	0.14% C 0.8% Cr 0.15% Mo steel: Designation used by German Standards
15 Cr Mo 6	0.15% C 1.5% Cr 1.5% Si steel: For carburizing; designation used by German Standards
15 Cr Ni 6	0.2% C 1% Ni 0.8% Cr steel: Carburizing; German Standard
15 N 3 CM	0.15% C 3% Ni 1.5% Cr 0.4% Mo steel: Electrode; Metrode
15 N 4 CM	0.15% C 4% Ni 1.3% Cr 0.2% Mo steel: Electrode; Metrode
15 NCMB	0.15% C 1.0% Ni 0.5% Cr 0.16% Mo steel: Electrode; Metrode
15 Ni Cr 14	0.15% C 3.5% Ni 0.7% Cr steel: German Standard
15 NKD	0.14% C 0.55% Ni 0.5% Cr 0.2% Mo steel: Pompey
16 CNK 1	0.16% C 1.4% Ni 1.0% Cr steel: Pompey

Note. The following abbreviations and units are used in the tables:

DPN	Hardness, diamond pyramid number
UTS	Ultimate tensile strength, N/mm^2
Elon	Elongation, %
Proof	0.1% proof strength, N/mm^2

1 N/mm^2=0.1 hbar=0.102 kgf/mm^2=0.06475 tonf/in.^{2}=145.04 lbf/in.^{2}=1 MPa

See Appendix II for other abbreviations and conversion tables.

Symbol	Nominal analysis, supplier, condition and remarks.
16 Cr Mo 4	0.16% C 0.4% Ni 1.1% Cr 0.25% Mo steel: Designation used by German Standards
16 Cr Mo 4/4	0.17% C 0.4% Ni 1.1% Cr 0.45% Mo steel: Designation used by German Standards
16 Cr Mo 9/3	0.16% C 2.2% Cr 0.35% Mo steel: Designation used by German Standards
16 Cr Mo V 4/2	0.16% C 1.1% Cr 0.3% Mo 0.2% V steel: Designation used by German Standards
16 NC 6	0.16% C 1% Ni 0.8% Cr steel: Carburizing; AFNOR
16 NCK 3	0.16% C 2.75% Ni 0.8% Cr steel: Pompey
17/22 AS	0.30% C 1.2% Cr 0.5% Mo 0.25% V steel: Timken; hardened and tempered; for high temperature use **DPN: 415 UTS: 1550 Elon: 15% Proof: 1300**
17/22 AV	0.27% C 1.2% Cr 0.5% Mo 0.8% V steel: Timken; hardened and tempered; for high temperature use **DPN: 444 UTS: 1550 Elon: 14% Proof: 1380**
17 CNK	0.17% C 0.6% Ni 0.6% Cr steel: Pompey
17 Cr Mo V 10	0.18% C 2.7% Cr 0.25% Mo 0.15% V steel: Designation used by German Standards
17 Cr Mo V 16/10	0.17% C 1.0% Ni 4.0% Cr 0.12% V steel: Designation used by German Standards
17 ND 2	0.17% C 2.0% Ni 0.2% Mo steel: Pompey
17 ND 3	0.17% C 3.5% Ni 0.2% Mo steel: Pompey
18 Cr Mo 3	0.18% C 0.7% Cr 0.15% Mo steel: Designation used by German Standards
18 Cr Ni 8	0.18% C 2.0% Ni 2.0% Cr carburizing steel: Designation used by German Standards
18 KH3MV	0.18% C 0.4% Mn 2.75% Cr 0.6% Mo 0.07% V 0.65% W steel: Russian Standards designation
18 Mo V 8/4	0.18% C 0.3% Ni 0.3% Cr 0.9% Mo 0.35% V steel: Designation used by German Standards
18 NCD 4	0.2% C 1% Ni 0.4% Cr 0.2% Mo steel: Carburizing; AFNOR
18 NCD 14	0.17% C 3.5% Ni 1.5% Cr 0.2% Mo steel: Carburizing; AFNOR
18 Ni Cr Mo 14	0.17% C 3.5% Ni 1.5% Cr 0.2% Mo steel: Carburizing; German Standard
20 CD 2	0.2% C 0.5% Cr 0.1% Mo steel: Carburizing; AFNOR
20 Cr Mo 9	0.2% C 0.8% Ni 2.3% Cr 0.3% Mo steel: Designation used by German Standards
20 Cr Mo V 13/5	0.2% C 3.2% Cr 0.55% Mo 0.5% V steel: Designation used by German Standards
20 DN32.12	0.2% C 3.0% Ni 3.2% Mo steel: French Standards designation
20 KH3MVF	0.2% C 0.4% Mn 3.0% Cr 0.45% Mo 0.7% V 0.45% W steel: Russian Standards designation
20 NC 6	0.21% C 1% Ni 0.8% Cr steel: Carburizing; AFNOR
20 NCD 2	0.2% C 0.5% Ni 0.5% Cr 0.2% Mo steel: Carburizing; AFNOR
20 NCD 7	0.2% C 1.8% Ni 0.5% Cr 0.25% Mo steel: Carburizing; AFNOR
20 ND 2K	0.19% C 1.7% Ni 0.3% Cr 0.25% Mo steel: Pompey
20 Ni Cr Mo 2	0.2% C 0.6% Ni 0.5% Cr 0.2% Mo steel: Carburizing; German Standard
20 Ni Cr Mo 4	0.2% C 1% Ni 0.5% Cr 0.2% Mo steel: Carburizing; German Standard
20 Ni Cr Mo 7	0.2% C 1.8% Ni 0.5% Cr 0.25% Mo steel: Carburizing; German Standard
20 NK 1	0.2% C 1.4% Ni 1.0% Cr steel: Pompey
20 NK 3	0.2% C 2.75% Ni 0.75% Cr steel: Pompey
20 NXD	0.2% C 0.55% Ni 0.5% Cr 0.2% Mo steel: Pompey
21 Cr Mo 3	0.22% C 0.9% Cr 0.25% Mo steel: Designation used by German Standards
21 Cr Mo V 5/11	0.21% C 0.6% Ni 1.3% Cr 1.1% Mo 0.3% V steel: Designation used by German Standards
21 Cr V Mo W 12	0.22% C 2.9% Cr 0.4% Mo 0.8% V 0.4% W steel: Designation used by German Standards
21 Mo V 5/3	0.21% C 0.3% Ni 0.3% Cr 0.5% Mo 0.3% V steel: Designation used by German Standards
22 Cr Mo 4/4	0.23% C 0.6% Ni 1.1% Cr 0.45% Mo steel: Designation used by German Standards

Symbol	Nominal analysis, supplier, condition and remarks.
22 Cr V 4	0.22% C 1.1% Cr 0.2% V steel: Designation used by German Standards
22 Mo 4	0.22% C 0.3% Cr 0.3% Mo steel: Designation used by German Standards
22 NCM	0.2% C 0.7% Ni 0.7% Cr 0.2% Mo steel: Electrode; Metrode
22 NKD	0.22% C 0.55% Ni 0.5% Cr 0.2% Mo steel: Pompey
24 C 3 M	0.24% C 3.2% Cr 0.5% Mo steel: Electrode; Metrode
24 Cr Mo 5	0.24% C 0.6% Ni 1.1% Cr 0.25% Mo steel: Designation used by German Standards
24 Cr Mo 10	0.24% C 0.8% Ni 2.5% Cr 0.25% Mo steel: Designation used by German Standards
24 Cr Mo V 5/2	0.24% C 1.3% Cr 0.22% Mo 0.25% V steel: Designation used by German Standards
24 Cr Ni 9	0.24% C 0.7% Ni 2.2% Cr steel: Designation used by German Standards
24 Ni 4	0.24% C 1.2% Ni 0.3% Cr steel: Designation used by German Standards
24 Ni 8	0.24% C 2.1% Ni 0.3% Cr steel: Designation used by German Standards
24S	0.24% C 2.5% Ni 3% Cr 8.5% W steel: Carrs; hot extrusion and press dies, etc. **DPN: 500**
25 Cr Mo 4	0.25% C 1.1% Cr 0.2% Mo steel: Designation used by German Standards
25 Cr Mo 4	0.24% C 0.65% Mn 0.15% Cr 0.2% Mo steel: Italian Standards designation; as rolled **DPN: 210**
25 KH1MF	0.18% C 0.55% Mn 1.65% Cr 0.3% Mo 0.22% V steel: Russian Standards designation
25 KH2MF	0.25% C 0.55% Mn 2.3% Cr 1.05% Mo 0.32% V steel: Russian Standards designation
25 NCD11	0.26% C 0.5% Mn 0.27% Ni 0.75% Cr steel: Origin unknown
25 NK1	0.25% C 1.55% Ni 1.0% Cr steel: Pompey
25 S	0.25% C 4.25% Ni 1.5% Cr 6.7% W steel: Carrs; hot extrusion and press dies, etc. **DPN: 525**
27 Cr Mo Ni	0.27% C 0.6% Ni 1.1% Cr 0.3% Mo steel: Electrode; Metrode
27 Cr Mo Ni 1	0.27% C 1.0% Ni 1.1% Cr 0.3% Mo steel: Electrode; Metrode
27 Cr Mo Ni 2	0.27% C 2.0% Ni 1.1% Cr 0.3% Mo steel: Electrode; Metrode
27 NCM	0.27% C 0.6% Ni 1.1% Cr 0.25% Mo steel: Electrode; Metrode
28 C 3 M	0.28% C 3% Cr 0.45% Mo steel: Electrode; Metrode
28 F	0.1% C 0.3% Mn 5% Cr 1% Mo 0.3% V steel: For case hardening; Carrs
28 Ni Cr Mo 4	0.28% C 1.2% Ni 1.2% Cr 0.25% Mo steel: Designation used by German Standards
28 Ni Cr Mo 4/4	0.28% C 1.2% Ni 1.2% Cr 0.45% Mo steel: Designation used by German Standards
28 Ni Cr Mo 7/4	0.28% C 2.0% Ni 1.3% Cr 0.35% Mo 0.11% V steel: Designation used by German Standards
28 NKD 3	0.3% C 3.2% Ni 1.55% Cr 0.55% Mo steel: Pompey
30 CM	0.3% C 0.5% Cr 0.5% Mo steel: Electrode; Metrode
30 Cr Mo V 9	0.3% C 0.6% Ni 2.5% Cr 0.2% Mo 0.15% V steel: Forging; designation used by German Standards
30 Cr Ni Mo 8	0.3% C 2.0% Ni 2.0% Cr steel: Designation used by German Standards
30 DCV28	0.3% C 2.8% Cr 2.8% Mo 0.5% V steel: French Standards designation
30 DKCV28	0.3% C 2.8% Cr 2.8% Mo 0.5% V 2.5% Co steel: French Standards designation
30 N 4 CM	0.3% C 4% Ni 1.3% Cr 0.3% Mo steel: Electrode; Metrode
30 NCD2	0.3% C 0.8% Mn 0.5% Ni 0.4% Cr steel: Origin unknown

Symbol	Nominal analysis, supplier, condition and remarks.
30 NCD16	0.3% C 0.45% Mn 4.0% Ni 1.8% Cr 0.4% Mo steel: For low temperature use; origin unknown **UTS: 1200 Elon: 11% Proof: 900**
30 NK 3	0.3% C 2.7% Ni 0.75% Cr steel: Pompey
30 NKD 2	0.3% C 2.0% Ni 2.0% Cr 0.4% Mo steel: Pompey
30 NKD 3	0.3% C 3.0% Ni 0.9% Cr 0.35% Mo steel: Pompey
30 NKDE	0.3% C 0.6% Ni 0.6% Cr 0.2% Mo steel: Pompey
30 W Cr V 17/9	0.3% C 2.3% Cr 0.6% V 4.2% W steel: Designation used by German Standards
31 CM	0.3% C 1% Cr 0.2% Mo steel: Electrode; Metrode
31 Cr Mo V 9	0.3% C 2.5% Cr 0.2% Mo 0.15% V steel: Designation used by German Standards
32 Cr Mo 12	0.32% C 0.6% Ni 3.0% Cr 0.4% Mo steel: Designation used by German Standards
32 Cr Mo B	0.32% C 1.2% Cr 0.3% Mo steel: Electrode; Metrode
32 NDC18.12	0.32% C 4.5% Ni 0.5% Cr 1.2% Mo steel: French Standards designation
34 Cr Al Mo 5	0.33% C 1.2% Cr 0.2% Mo 1.0% Al steel: Designation used by German Standards
34 Cr Al Ni 7	0.33% C 1.0% Ni 1.7% Cr 0.2% Mo 1.0% Al steel: Designation used by German Standards
34 Cr Mo 4	0.34% C 0.6% Ni 1.1% Cr 0.2% Mo steel: Designation used by German Standards
34 Cr Ni Mo 6	0.34% C 1.5% Ni 1.5% Cr 0.2% Mo steel: Designation used by German Standards
34 Ni 5	0.34% C 1.4% Ni 0.6% Cr (max) steel: Designation used by German Standards
35 CP	0.35% C 1.0% Cr 0.25% Mo steel: Pompey
35 NK 1	0.35% C 1.45% Ni 1.0% Cr steel: Pompey
35 NKD 1	0.35% C 1.15% Ni 0.55% Cr 0.15% Mo steel: Pompey
35 NKD 2	0.35% C 1.4% Ni 1.0% Cr 0.15% Mo steel: Pompey
36 Cr Ni Mo 4	0.36% C 1.1% Ni 1.1% Cr 0.2% Mo steel: Designation used by German Standards
36 Cr Ni Mo 6	0.34% C 1.6% Ni 1.6% Cr 0.2% Mo steel: Designation used by German Standards
40 FK	0.16% C (max) 1.0% Mn 0.04% S and P (max) steel: HOAG; Al killed
43 BV 12	0.12% C 1.8% Ni 0.5% Cr 0.25% Mo 0.03% V steel: Designation used by SAE; obsolete
43 BV 14	0.14% C 1.8% Ni 0.5% Cr 0.1% Mo 0.03% V steel: Designation used by SAE; obsolete
46 B 12	0.12% C 0.6% Mn 1.8% Ni 0.25% Mo 0.3% Si B steel: Designation used by SAE
58 S	0.35% C 5% Cr 2% Mo 0.25% V 1.2% W 1% Si steel: Carrs; hot extrusion dies, shears, etc. **DPN: 550**
74 S	0.3% C 2.3% Cr 0.3% Mo 0.6% V 4.25% W steel: Carrs; hot extrusion dies, hot shears. etc. **DPN: 560**
75.75	0.050% C 1.40% Mn 1.80% Ni 0.350% Cr 0.40% Mo steel: For electrode: Esab **UTS: 800 Elon: 18% Proof: 730**
80 HLES	0.15% C (max) 0.5% Mn (max) 4.0% Ni 0.5% Cr 0.4% Mo 0.1% V steel: Creusot-Loire
94 B 15	0.15% C 0.4% Ni 0.4% Cr 0.1% Mo 0.0005% B (min) steel: Designation used in the UK and USA
94 B 17	0.17% C 0.4% Ni 0.4% Cr 0.1% Mo 0.004% B steel: Designation used in the UK and USA

Note. The following abbreviations and units are used in the tables:

DPN	Hardness, diamond pyramid number
UTS	Ultimate tensile strength, N/mm^2
Elon	Elongation, %
Proof	0.1% proof strength, N/mm^2

1 N/mm^2=0.1 hbar=0.102 kgf/mm^2=0.06475 tonf/in.^{2}=145.04 lbf/in.^{2}=1 MPa

See Appendix II for other abbreviations and conversion tables.

Symbol	Nominal analysis, supplier, condition and remarks.
94 B 30	0.3% C 0.4% Ni 0.4% Cr 0.1% Mo 0.004% B steel: Designation used in the UK and USA
94 B 40	0.4% C 0.4% Ni 0.4% Cr 0.1% Mo 0.004% B steel: Designation used in the UK and USA
158	0.1% C 3.5% Ni 1.5% Cr steel: Carpenter
227	0.3% C 3.2% Cr 0.35% V 9% W steel: Balfour; hot dies for forging and casting **DPN: 520**
293	0.36% C 3.25% Cr 0.35% V 9% W steel: For hot work; Balfour; for tools; non-ferrous casting die punches **DPN: 570**
505	0.32% C 0.3% Mn 2.25% Cr 0.5% Mo 0.5% V 9% W steel: F Parkin; hot forming; extrusion and forging dies **DPN: 540**
635 A 14	0.14% C 0.95% Ni 0.6% Cr 0.1% Mo (max) steel: For case hardening; BS 970
635 H 15	0.15% C 1% Ni 0.6% Cr steel: For case hardening; BS 970
635 M 15	0.15% C 0.9% Ni 0.6% Cr steel: For case hardening; BS 970
637 A 16	0.16% C 1% Ni 0.85% Cr 0.1% Mo (max) steel: For case hardening; BS 970
637 H 17	0.17% C 1% Ni 0.8% Cr steel: For case hardening; BS 970
637 M 17	0.17% C 1% Ni 0.75% Cr steel: For case hardening; BS 970
640 A 35	0.35% C 1.3% Ni 0.65% Cr steel: BS 970; obsolete
640 H 35	0.35% C 1.3% Ni 0.65% Cr steel: BS 970; obsolete
653 M 31	0.31% C 3% Ni 1% Cr steel: BS 970; obsolete
655 A 12	0.12% C 3.3% Ni 0.85% Cr steel: BS 970; obsolete
655 H 13	0.13% C 3.3% Ni 0.85% Cr steel: For case hardening; BS 970
655 M 13	0.13% C 3.3% Ni 0.85% Cr steel: For case hardening; BS 970
659 A 15	0.15% C 4.1% Ni 1.2% Cr steel: BS 970; obsolete
659 H 15	0.15% C 4.1% Ni 1.2% Cr steel: BS 970; obsolete
659 M 15	0.15% C 4.1% Ni 1.2% Cr steel: BS 970; obsolete
665 A 17	0.17% C 1.7% Ni 0.25% Mo steel: BS 970; obsolete
665 A 19	0.19% C 1.7% Ni 0.25% Mo steel: BS 970; obsolete
665 A 22	0.22% C 1.7% Ni 0.25% Mo steel: BS 970; obsolete
665 A 22	0.22% C 1.8% Ni 0.25% Mo steel: BS 970; obsolete
665 A 24	0.24% C 1.7% Ni 0.25% Mo steel: BS 970; obsolete
665 A 24	0.24% C 1.8% Ni 0.25% Mo steel: BS 970; obsolete
665 H 17	0.17% C 1.7% Ni 0.25% Mo steel: For case hardening; BS 970
665 H 20	0.2% C 1.7% Ni 0.25% Mo steel: For case hardening; BS 970
665 H 23	0.23% C 1.7% Ni 0.25% Mo steel: For case hardening; BS 970
665 M 17	0.17% C 1.75% Ni 0.25% Mo steel: For case hardening; BS 970
665 M 20	0.2% C 1.75% Ni 0.25% Mo steel: For case hardening; BS 970
665 M 23	0.2% C 1.75% Ni 0.25% Mo steel: For case hardening; BS 970
708 A 25	0.25% C 1% Cr 0.2% Mo steel: BS 970
708 A 30	0.3% C 1% Cr 0.2% Mo steel: BS 970
708 A 37	0.37% C 1% Cr 0.2% Mo steel: BS 970
708 A 40	0.4% C 1% Cr 0.2% Mo steel: BS 970
708 H 20	0.2% C 1% Cr 0.2% Mo steel: For case hardening
708 H 37	0.37% C 1% Cr 0.2% Mo steel: BS 970
709 A 37	0.37% C 1% Cr 0.3% Mo steel: BS 970
709 A 40	0.4% C 1% Cr 0.3% Mo steel: BS 970
720 M 32	0.32% C 3% Cr 0.5% Mo steel: BS 970
722 M 24	0.24% C 3.2% Cr 0.55% Mo steel: BS 970
785 M 19	0.19% C 0.55% Ni 0.25% Mo steel: BS 970; obsolete
796	0.3% C 1.5% Cr 0.3% V 6.3% W steel: F Parkin; hot punches and dies **DPN: 500**

Symbol	Nominal analysis, supplier, condition and remarks.
805 A 15	0.15% C 0.6% Ni 0.5% Cr 0.2% Mo steel: BS 970; obsolete
805 A 17	0.17% C 0.5% Ni 0.5% Cr 0.2% Mo steel: For case hardening; BS 970
805 A 20	0.2% C 0.5% Ni 0.5% Cr 0.2% Mo steel: For case hardening; BS 970
805 A 22	0.22% C 0.5% Ni 0.5% Cr 0.2% Mo steel: For case hardening; BS 970
805 A 24	0.24% C 0.6% Ni 0.5% Cr 0.2% Mo steel: BS 970; obsolete
805 H 17	0.17% C 0.5% Ni 0.55% Cr 0.2% Mo steel: For carburizing; BS 970
805 H 20	0.2% C 0.5% Ni 0.55% Cr 0.2% Mo steel: For carburizing; BS 970
805 H 22	0.22% C 0.5% Ni 0.55% Cr 0.2% Mo steel: For carburizing; BS 970
805 H 25	0.25% C 0.6% Ni 0.5% Cr 0.2% Mo steel: BS 970; obsolete
805 M 17	0.17% C 0.5% Ni 0.45% Cr 0.2% Mo steel: For case hardening; BS 970
805 M 20	0.2% C 0.5% Ni 0.5% Cr 0.2% Mo steel: For case hardening; BS 970
805 M 22	0.22% C 0.5% Ni 0.5% Cr 0.2% Mo steel: For case hardening; BS 970
805 M 25	0.25% C 0.5% Cr 0.2% Mo steel: BS 970; obsolete
808 H 17	0.17% C 0.5% Ni 0.55% Cr 0.35% Mo steel: For case hardening; BS 970
815 A 16	0.16% C 1.5% Ni 1% Cr 0.15% Mo steel: BS 970; obsolete
815 H 17	0.17% C 1.5% Ni 1% Cr 0.15% Mo steel: For case hardening; BS 970
815 M 17	0.17% C 1.5% Ni 1% Cr 0.15% Mo steel: For case hardening; BS 970
820 A 16	0.16% C 1.75% Ni 1% Cr 0.15% Mo steel: BS 970; obsolete
820 H 17	0.17% C 1.7% Ni 1% Cr 0.15% Mo steel: For case hardening; BS 970
822 A 17	0.17% C 2% Ni 1.5% Cr 0.2% Mo steel: BS 970; obsolete
822 H 17	0.17% C 2% Ni 1.5% Cr 0.2% Mo steel: For case hardening; BS 970
822 M 17	0.17% C 2% Ni 1.5% Cr 0.2% Mo steel: For case hardening; BS 970
823 M 30	0.31% C 2% Ni 2% Cr 0.4% Mo steel: BS 970; obsolete
826 M 31	0.31% C 2.5% Ni 0.65% Cr 0.5% Mo steel: BS 970
830 M 31	0.31% C 3% Ni 1.1% Cr 0.3% Mo steel: BS 970; obsolete
832 H 13	0.13% C 3.3% Ni 0.85% Cr 0.17% Mo steel: For case hardening; BS 970
832 M 13	0.13% C 3.3% Ni 0.85% Cr 0.2% Mo steel: For case hardening; BS 970
835 A 15	0.15% C 4.2% Ni 1.2% Cr 0.27% Mo steel
835 H 15	0.15% C 4.2% Ni 1.2% Cr 0.2% Mo steel: For case hardening; BS 970
835 M 15	0.15% C 4.1% Ni 1.2% Cr 0.22% Mo steel: For case hardening; BS 970
905 M 31	0.31% C 1.6% Cr 0.2% Mo 1.1% Al steel: For nitriding; BS 970; obsolete
0908 Mo	0.07% C 9.0% Cr 1.0% Mo steel: NYBY annealed **DPN: 190 UTS: 280 Elon: 20% Proof: 160**
942 X	0.21% C 1.3% Mn steel: With Nb or W; SAE high strength steel
945 A	0.15% C 1% Mn steel: Low alloy; SAE high strength steel
945 C	0.23% C 1.4% Mn steel: Low alloy; SAE high strength steel
945 M 38	0.38% C 0.75% Ni 0.5% Cr 0.2% Mo steel: BS 970
945 X	0.22% C 1.3% Mn steel: Low alloy; SAE high strength steel
950 A	0.15% C 1.3% Mn steel: Low alloy; SAE high strength steel
950 B	0.22% C 1.3% Mn steel: Low alloy; SAE high strength steel
950 C	0.25% C 1.6% Mn steel: Low alloy; SAE high strength steel
950 D	0.15% C 1% Mn steel: Low alloy; SAE high strength steel
950 X	0.23% C 1.3% Mn steel: Low alloy; SAE high strength steel
955 X	0.25% C 1.3% Mn steel: With Nb or V; SAE high strength steel
960 X	0.26% C 1.45% Mn steel: With Nb or V; SAE high strength steel
965 X	0.26% C 1.4% Mn steel: With Nb or V; SAE high strength steel
970 X	0.26% C 1.6% Mn steel: With Nb or V; SAE high strength steel
980 X	0.26% C 1.6% Mn steel: With Nb or V; SAE high strength steel
1418	0.3% C 4% Ni 1.25% Cr 0.3% Mo steel: For tools; Osborn; plastic dies **DPN: 570 UTS: 155**
1433	0.15% C 4.2% Ni 1.2% Cr steel: Case hardening; Osborn **DPN: 340**
2504	0.15% C (max) 0.5% Mn (max) 4.0% Ni 0.5% Cr 0.4% Mo 0.1% V steel: Creusot-Loire
2831	0.08% C 5.0% Cr 1.0% Mo 0.3% V steel: French Standards designation
2881	0.1% C 1.5% Ni 1.0% Cr steel: French Standards designation
2882	0.1% C 3.0% Ni 0.8% Cr steel: French Standards designation
3115	0.15% C 1.2% Ni 0.6% Cr steel: Designation used by AISI; obsolete
3120	0.2% C 1.2% Ni 0.6% Cr steel: Designation used by AISI
3130	0.3% C 1.2% Ni 0.6% Cr steel: Designation used by AISI
3215	0.15% C 1.7% Ni 1% Cr steel: Designation used by AISI; obsolete
3220	0.2% C 1.7% Ni 1% Cr steel: Designation used by AISI; obsolete
3230	0.30% C 1.7% Ni 1% Cr steel: Designation used by AISI; obsolete
3310	0.1% C 3% Ni 1.5% Cr steel: Designation used by AISI
3310	0.1% C 3.5% Ni 1.5% Cr steel: Designation used in the UK and USA
3312	0.1% C 3.5% Ni 1.5% Cr steel: Designation used by AISI; obsolete
3316	0.16% C 3.5% Ni 1.5% Cr steel: Designation used by AISI
3325	0.25% C 3.5% Ni 1.5% Cr steel: Designation used by AISI; obsolete
3335	0.35% C 3.5% Ni 1.5% Cr steel: Designation used by AISI; obsolete
3382	0.35% C 4.0% Ni 1.8% Cr 0.4% Mo 0.1% V steel: French Standards designation
3383	0.32% C 4.5% Ni 0.5% Cr 1.2% Mo steel: French Standards designation
3415	0.15% C 3% Ni 0.8% Cr steel: Designation used by AISI; obsolete
3436	0.35% C 3% Ni 0.8% Cr steel: Designation used by AISI; obsolete
3451	0.3% C 2.8% Cr 2.8% Mo 0.5% V steel: French Standards designation
3455	0.2% C 3.0% Ni 3.2% Mo steel: French Standards designation

Symbol	Nominal analysis, supplier, condition and remarks.
3543	0.3% C 3.0% Cr 0.4% V 9.0% W steel: French Standards designation
4118	0.2% C 0.5% Cr 0.1% Mo steel: Designation used in the UK and USA
4119	0.19% C 0.5% Cr 0.25% Mo steel: Designation used by AISI
4120	0.2% C 0.5% Cr 0.2% Mo steel: Designation used in the UK and USA
4121	0.2% C 0.55% Cr 0.2% Mo steel: Designation used in the UK and USA
4125	0.25% C 0.5% Cr 0.25% Mo steel: Designation used by AISI; obsolete
4130	0.3% C 1% Cr 0.2% Mo steel: Designation used in the UK and USA
4135	0.35% C 1% Cr 0.2% Mo steel: Designation used in the UK and USA
4320	0.2% C 1.8% Ni 0.5% Cr 0.25% Mo steel: Designation used in the UK and USA
4330	0.3% C 1.8% Ni 0.82% Cr 0.41% Mo 0.07% V steel
4335	0.35% C 1.8% Ni 0.72% Cr 0.35% Mo 0.2% V steel
4337	0.37% C 1.8% Ni 0.8% Cr 0.25% Mo steel: Designation used in the UK and USA
4615	0.15% C 1.8% Ni 0.25% Mo steel: Designation used in the UK and USA
4617	0.17% C 1.8% Ni 0.25% Mo steel: Designation used by AISI
4617	0.17% C 1.8% Ni 0.25% Mo steel: Designation used in the UK and USA
4620	0.2% C 1.8% Ni 0.25% Mo steel: Designation used in the UK and USA
4621	0.21% C 1.8% Ni 0.25% Mo steel: Designation used in the UK and USA
4626	0.26% C 0.9% Ni 0.2% Mo steel: Designation used in the UK and USA
4715	0.15% C 0.9% Ni 0.5% Cr 0.5% Mo steel: Designation used in the UK and USA
4718	0.18% C 1% Ni 0.45% Cr 0.2% Mo steel: Designation used in the UK and USA
4720	0.2% C 1% Ni 0.45% Cr 0.2% Mo steel: Designation used in the UK and USA
4812	0.12% C 3.5% Ni 0.25% Mo steel: Designation used by AISI
4815	0.15% C 3.5% Ni 0.25% Mo steel: Designation used in the UK and USA
4817	0.17% C 3.5% Ni 0.25% Mo steel: Designation used in the UK and USA
4820	0.2% C 3.5% Ni 0.25% Mo steel: Designation used in the UK and USA
6115	0.15% C 1% Cr 0.15% V steel: Designation used by AISI; obsolete
6118	0.18% C 0.6% Cr 0.12% V steel: Designation used in the UK and USA
6120	0.2% C 0.8% Cr 0.1% V (min) steel: Designation used in the UK and USA
6125	0.25% C 1% Cr 0.15% V steel: Designation used by AISI; obsolete
6130	0.3% C 1% Cr 0.15% V steel: Designation used by AISI; obsolete

Note. The following abbreviations and units are used in the tables:

DPN	Hardness, diamond pyramid number
UTS	Ultimate tensile strength, N/mm^2
Elon	Elongation, %
Proof	0.1% proof strength, N/mm^2

1 N/mm^2=0.1 hbar=0.102 kgf/mm^2=0.06475 tonf/in.^{2}=145.04 ibf/in.^{2}=1 MPa

See Appendix II for other abbreviations and conversion tables.

Symbol	Nominal analysis, supplier, condition and remarks.
6135	0.35% C 1% Cr 0.15% V steel: Designation used by AISI; obsolete
8115	0.15% C 0.3% Ni 0.4% Cr 0.1% Mo steel: Designation used in the UK and USA
8615	0.15% C 0.5% Ni 0.5% Cr 0.2% Mo steel: Designation used in the UK and USA
8617	0.17% C 0.5% Ni 0.5% Cr 0.2% Mo steel: Designation used in the UK and USA
8620	0.2% C 0.5% Ni 0.5% Cr 0.2% Mo steel: Designation used in the UK and USA
8622	0.22% C 0.5% Ni 0.5% Cr 0.2% Mo steel: Designation used in the UK and USA
8625	0.25% C 0.5% Ni 0.5% Cr 0.2% Mo steel: Designation used in the UK and USA
8720	0.2% C 0.5% Ni 0.5% Cr 0.25% Mo steel: Designation used in the UK and USA
8735	0.35% C 0.5% Ni 0.5% Cr 0.25% Mo steel: Designation used in the UK and USA
8740	0.4% C 0.5% Ni 0.5% Cr 0.25% Mo steel: Designation used in the UK and USA
8742	0.42% C 0.5% Ni 0.5% Cr 0.25% Mo steel: Designation used in the UK and USA
8822	0.22% C 0.5% Ni 0.5% Cr 0.35% Mo steel: Designation used in the UK and USA
9310	0.1% C 3.2% Ni 1.2% Cr 0.1% Mo steel: Designation used in the UK and USA
10018 Ni	0.15% C (max) 1.7% Ni 1% Cr 0.5% Mo 0.5% Si steel: Weld electrode; designation used by AWS
11015	0.10% C (max) 2% Ni 1.2% Cr 0.3% Mo 0.3% Si steel: Weld electrode; designation used by AWS
11015	0.1% C (max) 2% Ni 1.2% Cr 0.3% Mo 0.3% Si steel: Weld electrode; designation used by AWS
14018 HT	0.1% C (max) 8% Ni 0.4% Cr 0.5% Mo steel: Weld electrode; designation used by AWS
14018 MI	0.1% C (max) 3.4% Ni 0.9% Cr 0.7% Mo steel: Weld electrode; designation used by ANS
A 3	0.28% C 0.25% Ni 0.25% Cr steel: Designation for BS 1453
A 4	0.3% C 3% Ni 3% Cr steel: Designation for BS 1453
A 6	0.15% C (max) 0.2% Ni 0.2% Cr 0.55% Mo steel: Designation for BS 1453
A 32	0.12% C (max) 1.3% Cr 0.55% Mo steel: Designation for BS 1453
A 32	0.12% C (max) 1.3% Cr 0.55% Mo steel: For welding electrodes; designation for BS 2901
A 33	0.12% C (max) 2.4% Cr 1.0% Mo steel: For welding electrodes; designation used by BS
A 34	0.12% C (max) 5.5% Cr 0.55% Mo steel: For welding electrodes; designation for BS 2901
A 35	0.1% C (max) 0.5% Ni (max) 9.2% Cr 1.0% Mo steel: For welding electrodes, designation for BS 2901
A 242	0.12% C 0.7% Ni 0.6% Cr 0.1% Mo 0.4% Cu steel: Armco **UTS: 500 Elon: 22% Proof: 330**
A 325	0.25% C 3.2% Cr 0.5% Mo steel: SHD; drills, etc.
A 330	0.28% C 0.3% Ni 2.1% Cr 0.3% Mo steel: SHD; drills, etc.
A 387 A	0.2% C 0.7% Cr 0.55% Mo steel: Plate; Armco; for high temperature and slightly corrosive use **UTS: 520 Elon: 24% Proof: 270**
A 387 B	0.17% C 1.0% Cr 0.55% Mo steel: Plate; Armco; for high temperature and slightly corrosive use **UTS: 500 Elon: 24% Proof: 240**
A 387 C	0.17% C 1.25% Cr 0.55% Mo steel: Plate; Armco; for high temperature and slightly corrosive use **UTS: 500 Elon: 24% Proof: 240**
A 387 D	0.15% C 2.2% Cr 1.0% Mo steel: Plate; Armco; for high temperature and slightly corrosive use **UTS: 520 Elon: 20% Proof: 200**

Symbol	Nominal analysis, supplier, condition and remarks.
A 387 E	0.15% C 3.0% Cr 1.0% Mo steel: Plate; Armco; for high temperature and slightly corrosive use **UTS: 520 Elon: 28% Proof: 200**
A 420	0.2% C 1.75% Ni 0.5% Cr 0.25% Mo steel: SHD; drills
A 1000	0.32% C 2.5% Ni 0.65% Cr 0.5% Mo steel: Edgar Allen; hardened and tempered **UTS: 2000 Elon: 20% Proof: 1380**
A W DYNALLOY 1	0.15% C (max) 0.8% Mn 0.6% Ni 0.1% Mo 0.45% Cu steel: Alan Wood
A W DYNALLOY 50	0.15% C (max) 1.0% Mn 0.55% Ni 0.15% Mo 0.45% Cu steel: Alan Wood
AB 213	0.3% C 0.3% Ni 1% Cr 1.75% W steel: Balfour; chisels, punches, etc. **DPN: 570**
ADIC	0.37% C 5% Cr 1.3% Mo 1.0% V 1.0% Si steel: For tools; hot work; Balfour; Al and Mg die casting moulds, etc. **DPN: 570**
ADS	0.35% C 5% Cr 1.5% Mo 0.5% V steel: Inman; for aluminium die casting
AEI/M	0.12% C 4.2% Ni 1.2% Cr 0.3% Mo steel: For case hardening; Swift Levick for BS alloy En 39B
AF 1410	0.15% C 10% Ni 2% Cr 1% Mo steel
AFC 77	0.15% C 0.5% Ni 14% Cr 5% Mo 0.2% V steel
AFNOR 81-340 EY 50.1 Ni Mo B 11026 BN	0.05% C 1% Ni 0.1% Cr 0.1% Mo steel: Welding electrode; French Standard
AFNOR 81-340 EY 55.1 Ni Cr Mo B 11026 TBN	0.05% C 1% Ni 0.15% Cr 0.26% Mo steel: Welding electrode; French Standard
AFNOR 81-340 EY 69 2 Mn 2 Ni Cr Mo B 11026 TBH	0.09% C 2% Ni 0.3% Cr 0.4% Mo steel: Welding electrode; French Standard
AFNOR 81-345 EC 0.5 Cr Mo B 20	0.05% C 0.5% Cr 0.5% Mo steel: Welding electrode; French Standard
AFNOR 81-345 EC 1 Cr Mo B 20	0.06% C 1% Cr 0.5% Mo steel: Welding electrode; French Standard
AFNOR 81-345 EC 1 Cr Mo R 22	0.08% C 1% Cr 0.5% Mo steel: Welding electrode; French Standard
AFNOR 81-345 EC 2 Cr Mo B 11020	0.09% C 2.4% Cr 1% Mo steel: Welding electrode; French Standard
AFNOR 81-345 EC 5 Cr Mo B 20	0.04% C 5% Cr 0.5% Mo steel: Welding electrode; French Standard
AFNOR 81-345 EC 9 Cr Mo B 20	0.1% C 9.5% Cr 1% Mo steel: Welding electrode; French Standard
AGT VAC-ARC	0.1% C 3.2% Ni 1.2% Cr 0.12% Mo steel: For carburizing; Latrobe Steel
AHT 28	0.3% C 4.0% Ni 1.4% Cr 0.2% Mo steel: Atlas; hardened and tempered **DPN: 375 UTS: 1380 Elon: 12% Proof: 1180**
AIS	0.2% C 1.5% Mn 0.7% Cr 0.3% Mo steel: Australian Iron and Steel Co.
AISI 94 B17	0.17% C 0.45% Ni 0.4% Cr 0.1% Mo 0.0005% B steel
AISI 94 B30	0.3% C 0.45% Ni 0.4% Cr 0.1% Mo 0.0005% B (min) steel
AISI 501	0.1% C (min) 5% Cr 0.5% Mo steel
AISI 502	0.1% C 5% Cr 0.5% Mo steel
AISI 602	0.3% C 1.2% Cr 0.5% Mo 0.25% V steel: Hardened and tempered **DPN: 350 UTS: 1210 Elon: 18% Proof: 1000**
AISI 603	0.27% C 1.25% Cr 0.5% Mo 0.85% V steel: Hardened and tempered **DPN: 444 UTS: 1550 Elon: 14% Proof: 1300**
AISI 604	0.2% C 1% Cr 1% Mo 0.1% V steel: Normalized and tempered **UTS: 1000 Proof: 840**
AISI 3115	0.15% C 1.2% Ni 0.6% Cr steel: Obsolete
AISI 3120	0.2% C 1.2% Ni 0.6% Cr steel: Obsolete
AISI 3130	0.3% C 1.2% Ni 0.6% Cr steel: Obsolete
AISI 3135	0.35% C 1.2% Ni 0.6% Cr steel
AISI 3316	0.16% C 3.5% Ni 1.5% Cr steel: Obsolete
AISI 4118	0.2% C 0.5% Cr 0.12% Mo steel
AISI 4119	0.2% C 0.5% Cr 0.25% Mo steel: Obsolete
AISI 4125	0.25% C 0.5% Cr 0.25% Mo steel: Obsolete
AISI 4130	0.3% C 1% Cr 0.2% Mo steel
AISI 4137	0.37% C 1% Mn 0.2% Mo steel
AISI 4317	0.17% C 1.8% Ni 0.5% Cr 0.25% Mo steel: Obsolete
AISI 4320	0.2% C 1.8% Ni 0.5% Cr 0.25% Mo steel
AISI 4330	0.3% C 1.8% Ni 0.8% Cr 0.4% Mo 0.07% V steel
AISI 4335	0.35% C 1.8% Ni 0.72% Cr 0.35% Mo 0.2% V steel
AISI 4337	0.37% C 1.8% Ni 0.8% Cr 0.25% Mo steel
AISI 4608	0.08% C 1.5% Cr 0.2% Mo steel: Obsolete
AISI 4615	0.15% C 1.8% Ni 0.25% Mo steel
AISI 4620	0.2% C 0.5% Mn 1.8% Ni 0.25% Mo steel
AISI 4620	0.2% C 1.8% Ni 0.25% Mo steel: Obsolete
AISI 4621	0.21% C 0.8% Mn 1.8% Ni 0.25% Mo steel
AISI 4626	0.26% C 0.8% Ni 0.2% Mo steel
AISI 4718	0.18% C 1% Ni 0.45% Cr 0.35% Mo steel
AISI 4720	0.2% C 1.0% Ni 0.45% Cr 0.2% Mo steel
AISI 4815	0.15% C 3.5% Ni 0.25% Mo steel
AISI 4817	0.12% C 3.5% Ni 0.25% Mo steel: Obsolete
AISI 4820	0.2% C 3.5% Ni 0.25% Mo steel
AISI 6117	0.17% C 0.8% Cr 0.1% V steel
AISI 6118	0.18% C 0.6% Cr 0.12% V steel
AISI 6120	0.2% C 0.8% Cr 0.1% V steel
AISI 8615	0.15% C 0.5% Ni 0.5% Cr 0.2% Mo steel
AISI 8617	0.17% C 0.6% Ni 0.5% Cr 0.2% Mo steel
AISI 8620	0.2% C 0.6% Ni 0.5% Cr 0.2% Mo steel
AISI 8622	0.22% C 0.6% Ni 0.5% Cr 0.2% Mo steel
AISI 8625	0.25% C 0.6% Ni 0.5% Cr 0.2% Mo steel
AISI 8627	0.27% C 0.6% Ni 0.5% Cr 0.2% Mo steel
AISI 8630	0.3% C 0.6% Ni 0.5% Cr 0.2% Mo steel
AISI 8632	0.32% C 0.5% Ni 0.5% Cr 0.2% Mo steel: Obsolete
AISI 8635	0.35% C 0.6% Ni 0.5% Cr 0.2% Mo steel
AISI 8637	0.37% C 0.6% Ni 0.5% Cr 0.2% Mo steel
AISI 8715	0.15% C 0.6% Ni 0.5% Cr 0.25% Mo steel: Obsolete
AISI 8717	0.17% C 0.6% Ni 0.5% Cr 0.25% Mo steel: Obsolete
AISI 8719	0.2% C 0.6% Ni 0.5% Cr 0.25% Mo steel: Obsolete
AISI 8720	0.2% C 0.6% Ni 0.5% Cr 0.25% Mo steel
AISI 8735	0.35% C 0.6% Ni 0.5% Cr 0.25% Mo steel: Obsolete
AISI 8822	0.22% C 0.6% Ni 0.5% Cr 0.35% Mo steel
AISI 9310	0.1% C 3.2% Ni 1.2% Cr 0.1% Mo steel
AISI 9315	0.15% C 3.2% Ni 1.2% Cr 0.1% Mo steel: Obsolete
AISI 9317	0.17% C 3.2% Ni 1.2% Cr 0.1% Mo steel: Obsolete
AISI 9437	0.37% C 0.4% Ni 0.4% Cr 0.1% Mo steel: Obsolete
ALCODIE	0.36% C 5.0% Cr 1.5% Mo 0.3% V 1.3% W steel: Columbia; AISI type H12
ALDECOR	0.15% C (max) 2% Ni 1% Cr 0.2% Mo 1% Cu steel: Source unknown
ALGO-COR-TEN	0.14% C 1.1% Mn 0.5% Cr 0.15% V 0.3% Cu steel: Algoma
ALGO-LOY 1317	0.22% C 1.6% Mn 0.7% Cr 0.05% Mo 0.2% Cu steel: Algoma
ALGOMA	0.12% C 1.6% Mn 3.4% Ni 0.4% Mo 0.02% V steel: Algoma
ALGOMA 90	0.12% C 1.6% Mn 3.3% Cr 0.4% Mo 0.02% V steel: Algoma
ALGOTUF 50	0.2% C 1.1% Mn 0.5% Ni 0.02% V steel: Algoma
ALLOY 53	0.1% C 2% Ni 1% Cr 3.25% Mo 0.1% V 2% Cu steel: Carpenter
ALLOY MOULD	0.32% C 4.1% Ni 1.2% Cr 0.2% Mo steel: Sanderson Kayser
AMI	0.35% C 5% Cr 1.5% Mo 0.5% V 1.35% W steel: Edgar Allen; brass forging dies and inserts **DPN: 550**
AMI SPECIAL	0.35% C 5% Cr 1.5% Mo 0.5% V 1.35% W steel: Edgar Allen; also known as AMI
AMS 5027	0.3% C 0.65% Ni 1% Cr 1% Mo 0.08% V steel: Welding wire

Symbol	Nominal analysis, supplier, condition and remarks.
AMS 5028	0.37% C 0.65% Ni 1% Cr 1% Mo 0.08% V steel: Welding wire
AMS 5029	0.35% C 1.8% Ni 0.75% Cr 0.35% Mo 0.2% V steel: Welding wire
AMS 5333	0.15% C 0.5% Ni 0.5% Cr 0.2% Mo steel: Casting
AMS 5502 A	0.1% C 5% Cr 0.5% Mo steel: Sheet and strip; AMS for SAE 51501 + Mo
AMS 5602 A	0.1% C 5% Cr 0.5% Mo steel: Bar and forging; AMS for SAE 51501
AMS 5680	0.07% C (max) 11% Ni 18% Cr 0.8% Nb 0.8% Si steel: Weld electrode
AMS 5790	0.08% C (max) 10% Ni 20% Cr 0.8% Nb 0.4% Si steel: Weld electrode
AMS 6231 B	0.3% C 0.55% Ni 0.5% Cr 0.2% Mo steel: Tube; AMS for SAE 8630
AMS 6250 E	0.1% C 3.5% Ni 1.5% Cr steel: Bar and forging, etc.; AMS for SAE 3310
AMS 6255	0.19% C 1.4% Cr 1% Mo 0.08% Al steel
AMS 6256	0.13% C 3% Ni 1% Cr 4.5% Mo 0.4% V 0.8% Al steel
AMS 6260 F	0.1% C 3.25% Ni 1.2% Cr 0.1% Mo steel: Bar, forging, etc.; AMS for SAE 9310
AMS 6263 C	0.15% C 3.25% Ni 1.2% Cr 0.1% Mo steel: Bar, forging, etc.; AMS for SAE 9315
AMS 6264 C	0.17% C 3.25% Ni 1.2% Cr 0.1% Mo steel: Bar, forging, etc.; AMS for SAE 9317
AMS 6265 B	0.1% C 3.2% Ni 1.2% Cr 0.1% Mo steel: Bar, forging, etc.; AMS for SAE 9310; premium quality; vacuum melted
AMS 6266 B	0.11% C 1.8% Ni 0.5% Cr 0.25% Mo B steel: Bar and forging; AMS for SAE 43BV12
AMS 6267	0.1% C 3.2% Ni 1.2% Cr 0.1% Mo steel: Bar and forging; AMS for SAE 9310; premium quality
AMS 6270 F	0.15% C 0.5% Ni 0.5% Cr 0.2% Mo steel: Bar and forging; AMS for SAE 8615
AMS 6272 D	0.17% C 0.55% Ni 0.5% Cr 0.2% Mo steel: Bar and forging; AMS for SAE 8617
AMS 6274 F	0.2% C 0.55% Ni 0.5% Cr 0.2% Mo steel: Bar and forging; AMS for SAE 8620
AMS 6275 A	0.17% C 0.45% Ni 0.4% Cr 0.1% Mo B steel: Bar and forging; AMS for SAE 94B17
AMS 6276 B	0.2% C 0.55% Ni 0.5% Cr 0.2% Mo steel: Bar and forging; AMS for SAE 8620; premium quality
AMS 6277	0.2% C 0.55% Ni 0.5% Cr 0.2% Mo steel: Bar and forging; AMS for SAE 8620
AMS 6280 D	0.3% C 0.55% Ni 0.5% Cr 0.2% Mo steel: Bar and forging; AMS for SAE 8630
AMS 6281	0.3% C 0.55% Ni 0.5% Cr 0.2% Mo steel: Type 8630
AMS 6282 C	0.35% C 0.55% Ni 0.5% Cr 0.25% Mo steel: Tube; AMS for SAE 8735
AMS 6290 C	0.15% C 1.8% Ni 0.25% Mo steel: Bar and forging; AMS for SAE 4615
AMS 6292 C	0.17% C 1.8% Ni 0.25% Mo steel: Bar and forging; AMS for SAE 4617
AMS 6294 C	0.2% C 1.8% Ni 0.25% Mo steel: Bar and forging; AMS for SAE 4620
AMS 6299	0.2% C 1.7% Ni 0.5% Cr 0.25% Mo steel: Bar and forging; AMS for SAE 4320
AMS 6302	0.3% C 1.2% Cr 0.5% Mo 0.25% V steel: Forgings

Note. The following abbreviations and units are used in the tables:

DPN	Hardness, diamond pyramid number
UTS	Ultimate tensile strength, N/mm^2
Elon	Elongation, %
Proof	0.1% proof strength, N/mm^2

1 N/mm^2=0.1 hbar=0.102 kgf/mm^2=0.06475 $tonf/in.^2$=145.04 $lbf/in.^2$=1 MPa
See Appendix II for other abbreviations and conversion tables.

Symbol	Nominal analysis, supplier, condition and remarks.
AMS 6302 A	0.3% C 1.25% Cr 0.5% Mo 0.25% V steel: Bar and forging
AMS 6303	0.27% C 1.25% Cr 0.5% Mo 0.85% V steel: Bar and forging
AMS 6308	0.1% C 2% Ni 1% Cr 3.2% Mo 0.1% V 2% Cu steel
AMS 6320 E	0.35% C 0.5% Ni 0.5% Cr 0.25% Mo steel: Bar and forging; AMS for SAE 8735
AMS 6330 A	0.35% C 1.25% Ni 0.6% Cr steel: Bar and forging; AMS for SAE 3135
AMS 6348	0.3% C 1% Cr 0.2% Mo steel: Type 4130
AMS 6349	0.4% C 1% Cr 0.2% Mo steel: Type 4140
AMS 6350 C	0.3% C 1% Cr 0.2% Mo steel: Sheet and strip; annealed; AMS for SAE 4130
AMS 6352 B	0.35% C 1% Cr 0.2% Mo steel: Sheet and strip; annealed; AMS for SAE 4135
AMS 6354	0.15% C 0.6% Cr 0.2% Mo 0.1% Zr steel: Sheet and strip
AMS 6355 F	0.3% C 0.5% Ni 0.5% Cr 0.2% Mo steel: Sheet and strip; AMS for SAE 8630
AMS 6356	0.32% C 1% Cr 0.2% Mo steel: Sheet and strip
AMS 6357 C	0.35% C 0.5% Ni 0.5% Cr 0.25% Mo steel: Sheet and strip; AMS for SAE 8735
AMS 6360 D	0.3% C 1% Cr 0.2% Mo steel: Seamless tube; normalized; AMS for SAE 4130
AMS 6361	0.3% C 1% Cr 0.2% Mo steel: Seamless tube; hardened and tempered **UTS: 920**
AMS 6362	0.3% C 1% Cr 0.2% Mo steel: Seamless tube; hardened and tempered **UTS: 1100**
AMS 6365 C	0.35% C 1% Cr 0.2% Mo steel: Seamless tube; normalized; AMS for SAE 4135
AMS 6370 E	0.3% C 1% Cr 0.2% Mo steel: Bar and forging; AMS for SAE 4130
AMS 6371 C	0.3% C 1% Cr 0.2% Mo steel: Tube; AMS for SAE 4130
AMS 6372 C	0.35% C 1% Cr 0.2% Mo steel: Tube; AMS for SAE 4135
AMS 6373	0.3% C 1% Cr 0.2% Mo steel: Type 4130
AMS 6374	0.3% C 1% Cr 0.2% Mo steel: Tyre 4130
AMS 6385	0.3% C 1.25% Cr 0.5% Mo 0.25% V steel: Sheet and strip
AMS 6386	0.18% C 0.65% Cr 0.22% Mo 0.1% Zr steel
AMS 6395	0.4% C 1% Cr 0.2% Mo steel: Type 4140
AMS 6407 A	0.3% C 2% Ni 1.2% Cr 0.45% Mo steel: Bar and forging
AMS 6409	0.4% C 1.8% Ni 0.8% Cr 0.25% Mo steel: Type 4340
AMS 6411	0.3% C 1.8% Ni 0.9% Cr 0.42% Mo 0.08% V steel
AMS 6412 E	0.37% C 1.8% Ni 0.8% Cr 0.25% Mo steel: Bar and forging; AMS for SAE 4337
AMS 6413 C	0.37% C 1.8% Ni 0.8% Cr 0.25% Mo steel: Tube; AMS for SAE 4337
AMS 6418 B	0.25% C 1.8% Ni 0.3% Cr 0.4% Mo 1.5% Si steel: Bar and forging
AMS 6421	0.37% C 0.8% Ni 0.8% Cr 0.2% Mo B steel: Bar and forging; AMS for SAE 98B37
AMS 6427 C	0.3% C 1.8% Ni 0.8% Cr 0.4% Mo 0.07% V steel: Bar and forging
AMS 6428	0.35% C 1.8% Ni 0.8% Cr 0.35% Mo 0.2% V steel: Bar and forging
AMS 6429	0.35% C 1.8% Ni 0.8% Cr 0.35% Mo 0.2% V steel: Bar and forging; vacuum melted
AMS 6430	0.35% C 1.8% Ni 0.8% Cr 0.35% Mo 0.2% V steel: Bar and forging; special grade
AMS 6433	0.35% C 1.8% Ni 0.8% Cr 0.35% Mo 0.2% V steel: Sheet and strip; special grade
AMS 6434	0.35% C 1.8% Ni 0.8% Cr 0.35% Mo 0.2% V steel: Sheet and strip
AMS 6435	0.35% C 1.8% Ni 0.8% Cr 0.35% Mo 0.2% V steel: Sheet and strip; vacuum melted

Symbol	Nominal analysis, supplier, condition and remarks.
AMS 6436	0.22% C 1.25% Cr 0.5% Mo 0.8% V steel: Sheet and strip
AMS 6437	0.35% C 5% Cr 1.5% Mo 0.4% V steel: Sheet
AMS 6457	0.3% C 1% Cr 0.2% Mo steel: Welding wire
AMS 6458	0.3% C 1.2% Cr 0.5% Mo 0.2% V steel: Wire
AMS 6458 B	0.3% C 1.2% Cr 0.5% Mo 0.3% V steel: Welding wire; vacuum melted
AMS 6459	0.2% C 1% Cr 1% Mo 0.1% V steel
AMS 6460	0.15% C 0.6% Cr 0.2% Mo 0.7% Si 0.1% Zr steel: Welding wire
AMS 6461 B	0.3% C 1% Cr 0.2% V steel: Welding wire; AMS for SAE 6130
AMS 6462 B	0.3% C 1% Cr 0.2% V steel: Welding wire; AMS for SAE 6130
AMS 6464	0.1% C 1% Mo 0.2% V steel: Coated welding electrode
AMS 6466	0.1% C (max) 5% Cr 0.5% Mo steel: Welding wire; cold drawn
AMS 6467	0.1% C (max) 5% Cr 0.5% Mo steel: Coated welding electrode
AMS 6475 B	0.25% C 3.5% Ni 1.2% Cr 0.2% Mo 1.25% Al steel: Bar and forging; nitriding steel
AMS 6485	0.35% C 5% Cr 1.5% Mo 0.4% V steel: Forgings
AMS 6528	0.3% C 1% Cr 0.2% Mo steel
AMS 6530 D	0.3% C 0.5% Ni 0.5% Cr 0.2% Mo steel: Seamless tube; normalized; AMS for SAE 8630
AMS 6535 C	0.35% C 0.5% Ni 0.5% Cr 0.25% Mo steel: Seamless tube; normalized; AMS for SAE 8735
AMS 6546	0.27% C 8% Ni 0.42% Cr 0.42% Mo 0.08% V 4% Co steel
AMS 6550 D	0.3% C 0.5% Ni 0.5% Cr 0.2% Mo steel: Welded tube; normalized; AMS for SAE 8630
AMS 7464	0.38% C 3.3% Cr 2.5% Mo 0.5% V steel: Type H 10
AMS 7496	0.25% C 1.8% Ni 0.3% Cr 0.4% Mo steel
AN 1 A	0.3% C 4.1% Ni 1.3% Cr steel: W Marrison for BS alloy En 30A
AN 1 B	0.3% C 4.1% Ni 1.3% Cr 0.3% Mo steel: W Marrison for BS alloy En 30B
AN QQ S684a	0.3% C 1.0% Cr 0.2% Mo steel: US Service; normalized and tempered **DPN: 230 UTS: 840 Elon: 18% Proof: 740**
AN QQ S685/4	0.3% C 1.0% Cr 0.2% Mo steel: Strip and plate; US Service **UTS: 690 Elon: 18% Proof: 480**
AN QQ S686/1	0.35% C 1.0% Cr 0.2% Mo steel: Strip and plate; US Service; annealed **UTS: 620**
AN S 12/2	0.3% C 0.5% Ni 0.5% Cr 0.2% Mo steel: Strip; US Service; annealed **UTS: 620**
AN S 13a	0.2% C 0.5% Ni 0.5% Cr 0.2% Mo steel: US Service
AN S 14a	0.3% C 0.5% Ni 0.5% Cr 0.2% Mo steel: US Service; normalized and tempered **DPN: 230 UTS: 840 Elon: 16% Proof: 740**
AN S 15a	0.35% C 0.6% Ni 0.5% Cr 0.25% Mo steel: US Service; hardened and tempered **DPN: 230 UTS: 840 Elon: 16% Proof: 740**
AN S 22/1	0.35% C 0.6% Ni 0.5% Cr 0.25% Mo steel: Strip; US Service; annealed **UTS: 620**
AN T 3	0.3% C 1.0% Cr 0.2% Mo steel: Tube; welded; US Service; annealed **UTS: 640**
AN T 15/2	0.3% C 0.6% Ni 0.5% Cr 0.2% Mo steel: Seamless tube; US Service; normalized **UTS: 630 Elon: 12% Proof: 460**
AN T 22/2	0.35% C 0.5% Ni 0.5% Cr 0.25% Mo steel: Seamless tube; US Service; annealed **UTS: 749**
AN T 33/1	0.3% C 0.6% Ni 0.5% Cr 0.2% Mo steel: Welded tube: US Service; annealed **UTS: 630**
AN WW T 850/1	0.3% C 1.0% Cr 0.2% Mo steel: Seamless tube; US Service; normalized **UTS: 630 Elon: 12% Proof: 460**
AN WW T 852	0.35% C 1.0% Cr 0.2% Mo steel: Seamless tube; US Service **UTS: 700**
ANC 1	0.3% C 4.25% Ni 1.25% Cr steel: Firth Brown for BS alloy En 30A
ANCM	0.3% C 4.25% Ni 1.25% Cr 0.25% Mo steel: Firth Brown for BS alloy En 30B
APS 10 M	0.12% C 2.0% Cr 0.1% Mo 1.0% Al steel: Pompey **UTS: 460 Elon: 20% Proof: 300**
APS 10 M4	0.12% C 2.0% Cr 0.35% Mo 0.35% Al steel: Pompey **UTS: 460 Elon: 20% Proof: 300**
APS 20 M	0.12% C 4.0% Cr 0.12% Mo 0.9% Al steel: Pompey **UTS: 460 Elon: 20% Proof: 300**
APS 25	0.15% C 0.7% Ni 4.0% Cr 0.7% Al steel: Pompey **UTS: 820 Elon: 15% Proof: 580**
AR COL 360	0.13% C 1.35% Mn 0.8% Cr 0.4% Mo steel: Colvilles; abrasion resisting **DPN: 320**
ARMOUR PLATE	0.3% C 2.5% Ni 2.75% Cr 0.25% Mo steel: Plate; common name
ARO 75	0.4% C 1.7% Cr 0.3% Mo 1.1% Al steel: For nitriding; Bofors; hardened and tempered **DPN: 280 UTS: 920 Elon: 15% Proof: 620**
ASTM A54/2	0.27% C (max) 0.7% Ni 0.35% Cr 0.62% Mo 0.06% V steel: Forging
ASTM A182 F1	0.3% C 0.6% Mn 0.5% Mo steel: For pipe fittings
ASTM A182 F5	0.15% C (max) 0.45% Mn 5% Cr 0.5% Mo steel: For pipe fittings
ASTM A182 F5a	0.25% C (max) 0.6% Mn 0.5% Cr 0.5% Mo steel: For pipe fittings
ASTM A182 F7	0.15% C (max) 0.45% Mn 7% Cr 0.5% Mo 0.8% Si steel: For pipe fittings
ASTM A182 F9	0.15% C (max) 0.45% Mn 9% Cr 1.0% Mo 0.7% Si steel: For pipe fittings
ASTM A182 F11	0.15% C 0.5% Mn 1.2% Cr 0.5% Mo 0.7% Si steel: For pipe fittings
ASTM A182 F12	0.15% C 0.5% Mn 1% Cr 0.5% Mo 0.4% Si steel: For pipe fittings
ASTM A182 F21	0.15% C (max) 0.4% Mn 3% Cr 0.9% Mo steel: For pipe fittings
ASTM A182 F22	0.15% C (max) 0.45% Mn 2.2% Cr 0.95% Mo steel: For pipe fittings
ASTM A193 B5	0.1% C 1% Mn 5% Cr 0.5% Mo steel: For bolts, etc.; hardened and tempered **UTS: 730 Elon: 16% Proof: 560**
ASTM A194/3	0.1% C (min) 0.5% Mo 5% Cr steel: For bolts
ASTM A199 T3b	0.15% C (max) 2.0% Cr 0.5% Mo steel: Condenser tube
ASTM A199 T4	0.15% C (max) 2.5% Cr 0.5% Mo 0.7% Si steel: Condenser tube
ASTM A199 T5	0.15% C (max) 5.0% Cr 0.5% Mo steel: Condenser tube
ASTM A199 T7	0.15% C (max) 7.0% Cr 0.5% Mo steel: Condenser tube
ASTM A199 T9	0.15% C (max) 9% Cr 1.0% Mo steel: Condenser tube
ASTM A199 T11	0.15% C (max) 1.2% Cr 0.5% Mo steel: Condenser tube
ASTM A199 T21	0.15% C 3.0% Cr 0.95% Mo steel: Condenser tube
ASTM A199 T22	0.15% C 2.2% Cr 1.0% Mo steel: Condenser tube
ASTM A200	Alloy steel still tube; grades as in ASTM A199
ASTM A217 C5	0.2% C (max) 0.5% Mn 5.2% Cr 0.55% Mo steel: Casting
ASTM A217 C12	0.2% C (max) 0.5% Mn 9% Cr 1.1% Mo steel: Casting
ASTM A217 WC4	0.2% C (max) 0.65% Mn 0.9% Ni 0.65% Cr 0.55% Mo steel: Casting

Symbol	Nominal analysis, supplier, condition and remarks.
ASTM A217 WC5	0.2% C (max) 0.55% Mn 0.8% Ni 0.7% Cr 1.1% Mo steel: Castings
ASTM A217 WC6	0.2% C (max) 0.65% Mn 1.2% Cr 0.55% Mo steel: Castings
ASTM A217 WC9	0.18% C (max) 0.5% Mn 2.3% Cr 1.0% Mo steel: Castings
ASTM A290 D	0.35% C 1% Cr 0.2% Mo steel: Forging; hardened and tempered **DPN: 223 UTS: 760 Elon: 18% Proof: 530**
ASTM A290 D2	0.35% C 1% Cr 0.2% Mo steel: Forging; hardened and tempered **DPN: 245 UTS: 840 Elon: 15% Proof: 760**
ASTM A290 E	0.35% C 1.8% Ni 0.8% Cr 0.25% Mo steel: Forging; hardened and tempered **DPN: 302 UTS: 980 Elon: 16% Proof: 760**
ASTM A290 F	0.35% C 1.8% Ni 0.8% Cr 0.25% Mo steel: Forging: hardened and tempered **DPN: 302 UTS: 1070 Elon: 13% Proof: 840**
ASTM A290 G	0.35% C 1.8% Ni 0.8% Cr 0.25% Mo steel: Forging; hardened and tempered **DPN: 341 UTS: 1210 Elon: 11% Proof: 1000**
ASTM A292/2	0.3% C 2% Ni 0.5% Cr 0.2% Mo 0.1% V steel: For generator rotors; hardened and tempered; obsolete **UTS: 530 Elon: 22% Proof: 370**
ASTM A292/3	0.3% C 2% Ni 0.5% Cr 0.2% Mo 0.1% V steel: For generator rotors; hardened and tempered; obsolete **UTS: 640 Elon: 20% Proof: 500**
ASTM A292/4	0.35% C 2.5% Ni 0.7% Cr 0.2% Mo 0.1% V steel: For generator rotors; hardened and tempered; obsolete **UTS: 760 Elon: 18% Proof: 580**
ASTM A292/5	0.35% C 2.5% Ni 0.7% Cr 0.2% Mo 0.1% V steel: For generator rotors; hardened and tempered; obsolete **UTS: 840 Elon: 16% Proof: 640**
ASTM A293/2	0.3% C 2% Ni 0.7% Cr 0.25% Mo 0.1% V steel: For turbine rotors; hardened and tempered **UTS: 580 Elon: 22% Proof: 370**
ASTM A293/3	0.3% C 2% Ni 0.7% Cr 0.25% Mo 0.1% V steel: For turbine rotors; hardened and tempered **UTS: 700 Elon: 20% Proof: 500**
ASTM A293/4	0.35% C 2.5% Ni 0.7% Cr 0.25% Mo 0.1% V steel: For turbine rotors; hardened and tempered **UTS: 760 Elon: 18% Proof: 580**
ASTM A293/5	0.35% C 2.5% Ni 1.25% Cr 0.25% Mo 0.03% V steel: For turbine rotors; hardened and tempered **UTS: 880 Elon: 17% Proof: 700**
ASTM A293/6	0.37% C 1% Mn 0.05% Ni 1% Cr 1.2% Mo 0.25% V steel: For turbine rotors; hardened and tempered **UTS: 780 Elon: 16% Proof: 620**
ASTM A294 C	0.35% C 2.5% Ni 0.7% Cr 0.2% Mo 0.1% V steel: For turbine wheels
ASTM A295 F2	0.15% C 1% Cr 0.5% Mo steel: Seamless drum forging; normalized and tempered **UTS: 500 Elon: 18% Proof: 270**
ASTM A295 F5	0.15% C 5% Cr 0.5% Mo steel: Seamless drum forging; normalized and tempered **UTS: 500 Elon: 18% Proof: 270**
ASTM A295 F5A	0.25% C 5% Cr 0.5% Mo steel: Seamless drum forging; normalized and tempered **UTS: 580 Elon: 19% Proof: 340**

Symbol	Nominal analysis, supplier, condition and remarks.
ASTM A295 F22	0.15% C 2.2% Cr 1% Mo steel: Seamless drum forging; normalized and tempered **UTS: 500 Elon: 18% Proof: 270**
ASTM A295 F31	0.35% C 2.5% Ni 0.3% Mo 0.15% V steel: Seamless drum forging; normalized and tempered **UTS: 700 Elon: 18% Proof: 500**
ASTM A295 F32	0.35% C 0.7% Ni 3.3% Cr 0.4% Mo 0.1% V steel: Seamless drum forging; normalized and tempered **UTS: 740 Elon: 18% Proof: 420**
ASTM A298 E502	0.1% C 5% Cr 0.5% Mo steel: Welding rod annealed after welding, for plain carbon steels **UTS: 400 Elon: 20%**
ASTM A302 A	0.21% C 1.1% Mn 0.5% Mo steel: Plate for pressure vessels
ASTM A302 B	0.22% C 1.3% Mn 0.5% Mo steel: Plate for pressure vessels
ASTM A302 C	0.22% C 1.3% Mn 0.5% Ni 0.5% Mo steel: Plate for pressure vessels
ASTM A302 D	0.22% C 1.3% Mn 0.8% Ni 0.5% Mo steel: Plate for pressure vessels
ASTM A316 E7000A1	0.12% C 0.7% Mn 0.05% Mo steel: Welding electrodes; code varies with analysis
ASTM A316 E7020G	1.0% Mn (min) 0.05% Ni (min) 0.3% Cr (min) 0.2% Mo (min) 0.1% V (min) 0.8% Si (min) steel: Electrodes for arc welding
ASTM A316 E8015/18B	0.1% C and Mo steel: Electrodes; code varies with analysis
ASTM A316 E8015B2L	0.05% C 1.25% Cr 0.52% Mo steel: Electrodes for arc welding
ASTM A316 E8015B4L	0.05% C 2.0% Cr 0.52% Mo steel: Electrodes for arc welding
ASTM A316 E8016B1	0.12% C 0.52% Cr 0.52% Mo steel: Electrodes for arc welding
ASTM A316 E8016B2	0.12% C 1.25% Cr 0.52% Mo steel: Electrodes for arc welding
ASTM A316 E8016C3	0.12% C 0.95% Ni 0.15% Cr 0.35% Mo 0.05% V steel: Electrodes for arc welding
ASTM A316 E8018B2	0.12% C 1.25% Cr 0.52% Mo steel: Electrodes for arc welding
ASTM A316 E8018C3	0.12% C 0.95% Ni 0.15% Cr 0.35% Mo 0.05% V steel: Electrodes for arc welding
ASTM A316 E9015B3	0.12% C 2.25% Cr 1.05% Mo steel: Electrodes for arc welding
ASTM A316 E9015B3L	0.05% C 2.25% Cr 1.05% Mo steel: Electrodes for arc welding
ASTM A316 E9016/8B3	0.1% C 2.2% Cr Mo steel: Welding electrodes; code varies with analysis
ASTM A316 E9016B3	0.12% C 2.25% Cr 1.05% Mo steel: Electrodes for arc welding
ASTM A316 E9018B3	0.12% C 0.25% Cr 1.05% Mo steel: Electrodes for arc welding
ASTM A316 E9018M	0.10% C 1.6% Ni 0.15% Cr 0.35% Mo 0.05% V steel: Electrodes for arc welding
ASTM A316 E9018M	0.1% C 1% Mn 1.5% Ni 0.3% Mo steel: Welding electrodes
ASTM A316 E10018M	0.10% C 1.75% Ni 0.35% Cr 0.37% Mo 0.05% V steel: Electrodes for arc welding
ASTM A316 E10018M	0.1% C 1.2% Mn 1.8% Ni 0.3% Mo steel: Welding electrodes
ASTM A316 E11018M	0.10% C 1.9% Ni 0.40% Cr 0.42% Mo 0.05% V steel: Electrodes for arc welding
ASTM A316 E12018M	0.10% C 2.0% Ni 0.9% Cr 0.42% Mo 0.05% V steel: Electrodes for arc welding
ASTM A316 EXX10G	C not specified 1.0% Mn (min) 0.5% Ni (min) 0.3% Cr (min) 0.2% Mo (min) 0.1% V (min) 0.8% Si (min) steel: Electrodes for arc welding
ASTM A316 EXX11G	C not specified 1.0% Mn (min) 0.5% Ni (min) 0.3% Cr (min) 0.2% Mo (min) 0.1% V (min) 0.8% Si (min) steel: Electrodes for arc welding

Note. The following abbreviations and units are used in the tables:

DPN	Hardness, diamond pyramid number
UTS	Ultimate tensile strength, N/mm^2
Elon	Elongation, %
Proof	0.1% proof strength, N/mm^2

1 N/mm^2=0.1 hbar=0.102 kgf/mm^2=0.06475 tonf/in.^{2}=145.04 lbf/in.^{2}=1 MPa

See Appendix II for other abbreviations and conversion tables.

Symbol	Nominal analysis, supplier, condition and remarks.
ASTM A316 EXX13G	C not specified 1.0% Mn (min) 0.5% Ni (min) 0.3% Cr (min) 0.2% Mo (min) 0.1% V (min) 0.8% Si (min) steel: Electrodes for arc welding
ASTM A316 EXX15G	C not specified 1.0% Mn (min) 0.5% Ni (min) 0.3% Cr (min) 0.2% Mo (min) 0.1% V (min) 0.8% Si (min) steel: Electrodes for arc welding
ASTM A316 EXX16G	C not specified 1.0% Mn (min) 0.5% Ni (min) 0.3% Cr (min) 0.2% Mo (min) 0.1% V (min) 0.8% Si (min) steel: Electrodes for arc welding
ASTM A316 EXX18G	C not specified 1.0% Mn (min) 0.5% Ni (min) 0.3% Cr (min) 0.2% Mo (min) 0.1% V (min) 0.8% Si (min) steel: Electrodes for arc welding
ASTM A333/4	0.1% C 0.7% Mn 0.7% Ni 0.8% Cr 0.2% Al 0.6% Cu steel: Pipe for low temperature use
ASTM A335 P2	0.15% C 0.7% Cr 0.5% Mo steel: Seamless pipe
ASTM A335 P5	0.15% C (max) 5.0% Cr 0.5% Mo steel: Seamless pipe
ASTM A335 P5b	0.15% C (max) 5.0% Cr 0.5% Mo steel: Seamless pipe
ASTM A335 P5c	0.12% C (max) 5.0% Cr 0.5% Mo steel: Seamless pipe
ASTM A335 P7	0.15% C (max) 7.0% Cr 0.5% Mo steel: Seamless pipe
ASTM A335 P9	0.15% C (max) 9.0% Cr 1.0% Mo steel: Seamless pipe
ASTM A335 P11	0.15% C (max) 1.2% Cr 0.5% Mo steel: Seamless pipe
ASTM A335 P12	0.15% C (max) 1.0% Cr 0.5% Mo steel: Seamless pipe
ASTM A335 P21	0.15% C (max) 3.0% Cr 0.95% Mo steel: Seamless pipe
ASTM A335 P22	0.15% C (max) 2.3% Cr 1.0% Mo steel: Seamless pipe
ASTM A336 F2	0.15% C 1.0% Cr 0.5% Mo steel: Seamless drum forging **UTS: 460 Elon: 18% Proof: 270**
ASTM A336 F5	0.15% C (max) 5.0% Cr 0.5% Mo steel: Seamless drum forging **UTS: 400 Elon: 19% Proof: 240**
ASTM A336 F5a	0.25% C (max) 5.0% Cr 0.5% Mo steel: Seamless drum forging **UTS: 540 Elon: 19% Proof: 330**
ASTM A336 F22	0.15% C (max) 2.2% Cr 1.0% Mo steel: Seamless drum forging **UTS: 460 Elon: 18% Proof: 270**
ASTM A336 F31	0.35% C (max) 2.5% Ni 0.4% Mo 0.15% V (max) steel: Seamless drum forging **UTS: 630 Elon: 18% Proof: 330**
ASTM A336 F32	0.35% C (max) 3.3% Ni 3.3% Cr 0.4% Mo 0.1% V steel: Seamless drum forging **UTS: 670 Elon: 18% Proof: 400**
ASTM A350 LF4	0.12% C (max) 0.7% Ni 0.7% Cr 0.2% Al 0.5% Cu steel: For low temperature use **UTS: 400 Elon: 25% Proof: 200**
ASTM A355 C	0.23% C 3.5% Ni 1.2% Cr 0.2% Mo 1.0% Al steel: Hardened and tempered **DPN: 250**
ASTM A356/4	0.2% C 1.0% Mo 0.2% V steel: Casting; normalized and tempered **UTS: 620 Elon: 16% Proof: 400**
ASTM A356/5	0.25% C 0.5% Cr 0.5% Mo steel: Casting; normalized and tempered **UTS: 460 Elon: 22% Proof: 270**
ASTM A356/6	0.2% C 1.2% Cr 0.5% Mo steel: Casting; normalized and tempered **UTS: 460 Elon: 22% Proof: 300**
ASTM A356/7	0.2% C 1.2% Cr 0.5% Mo 0.2% V steel: Casting; normalized and tempered **UTS: 460 Elon: 22% Proof: 270**
ASTM A356/8	0.25% C 1.2% Cr 1.0% Mo steel: Casting; normalized and tempered **UTS: 530 Elon: 18% Proof: 330**
ASTM A356/9	0.2% C 1.2% Cr 1.0% Mo 0.2% V steel: Casting; normalized and tempered **UTS: 640 Elon: 15% Proof: 400**
ASTM A356/10	0.2% C 2.4% Cr 1.0% Mo steel: Casting; normalized and tempered **UTS: 580 Elon: 20% Proof: 360**
ASTM A357	0.15% C 5.0% Cr 0.5% Mo steel: Plate; annealed **UTS: 460 Elon: 20% Proof: 200**
ASTM A369 FP1	0.15% C 0.7% Cr 0.5% Mo steel: For forged or bored pipe
ASTM A369 FP3b	0.15% C (max) 2.0% Cr 0.5% Mo steel: For forged or bored pipe
ASTM A369 FP5	0.15% C (max) 5.0% Cr 0.5% Mo steel: For forged or bored pipe
ASTM A369 FP7	0.15% C (max) 7.0% Cr 0.5% Mo steel: For forged or bored pipe
ASTM A369 FP9	0.15% C (max) 9.0% Cr 1.0% Mo steel: For forged or bored pipe
ASTM A369 FP11	0.15% C (max) 1.2% Cr 0.5% Mo steel: For forged or bored pipe
ASTM A369 FP12	0.15% C (max) 1.1% Cr 0.5% Mo steel: For forged or bored pipe
ASTM A369 FP21	0.15% C (max) 3.0% Cr 1.0% Mo steel: For forged or bored pipe
ASTM A369 FP22	0.15% C (max) 2.3% Cr 1.0% Mo steel: For forged or bored pipe
ASTM A372 VI	0.18% C (max) 0.30% Mn 2.7% Ni 1.4% Cr 0.30% Mo 0.25% Si steel: Forging for pressure vessels
ASTM A387 A	0.21% C (max) 0.65% Cr 0.52% Mo steel: Plate for pressure vessels; normalized and tempered **UTS: 550 Elon: 20% Proof: 300**
ASTM A387 B	0.17% C (max) 1.0% Cr 0.5% Mo steel: Plate for pressure vessels; normalized and tempered **UTS: 510 Elon: 20% Proof: 270**
ASTM A387 C	0.17% C (max) 1.2% Cr 0.5% Mo steel: Plate for pressure vessels; normalized and tempered
ASTM A387 D	0.15% C (max) 2.2% Cr 1.0% Mo steel: Plate for pressure vessels; normalized and tempered **UTS: 600 Elon: 18% Proof: 300**
ASTM A387 E	0.15% C (max) 3.0% Cr 1.0% Mo steel: Plate for pressure vessels; normalized and tempered
ASTM A389 C23	0.2% C (max) 1.2% Cr 0.5% Mo 0.2% V steel: Casting for high temperature use **UTS: 460 Elon: 18% Proof: 270**
ASTM A389 C24	0.2% C (max) 1.0% Cr 1.0% Mo 0.2% V steel: Casting for high temperature use **UTS: 540 Elon: 15% Proof: 330**
ASTM A404 F24	0.2% C (max) 1.0% Cr 1.0% Mo 0.2% V steel: For high temperature use; flanges, etc.
ASTM A405 P24	0.15% C (max) 1.0% Cr 1.0% Mo 0.2% V steel: For high temperature use; seamless pipe
ASTM A410	0.12% C (max) 0.7% Ni 0.8% Cr 0.2% Al 0.5% Cu steel: Plate for pressure vessels
ASTM A423/1	0.15% C (max) 0.4% Ni 0.65% Cr steel: Welded tube
ASTM A423/2	0.15% C (max) 0.75% Ni 0.1% Mo steel: Welded tube
ASTM A426 CP2	0.15% C 0.65% Cr 0.5% Mo steel: Cast pipe
ASTM A426 CP5	0.15% C (max) 5.0% Cr 0.5% Mo steel: Cast pipe
ASTM A426 CP5b	0.15% C (max) 5.0% Cr 0.5% Mo steel: Cast pipe
ASTM A426 CP7	0.15% C (max) 7.0% Cr 0.5% Mo steel: Cast pipe
ASTM A426 CP9	0.15% C (max) 9.0% Cr 1.0% Mo steel: Cast pipe
ASTM A426 CP11	0.15% C (max) 1.2% Cr 0.5% Mo steel: Cast pipe
ASTM A426 CP12	0.15% C (max) 1.0% Cr 0.5% Mo steel: Cast pipe
ASTM A426 CP21	0.15% C (max) 3.0% Cr 0.95% Mo steel: Cast pipe
ASTM A426 CP22	0.15% C (max) 2.3% Cr 1.0% Mo steel: Cast pipe
ASTM A469/2	0.25% C 2.5% Ni 0.5% Cr (max) 0.3% Mo 0.03% V steel: For rotors; vacuum treated
ASTM A487/4 N	0.3% C (max) 0.6% Ni 0.6% Cr 0.2% Mo steel: Casting for pressure; normalized; 4Q grade is hardened and tempered
ASTM A487/5 N	0.3% C (max) 1.2% Mn 0.6% Ni 0.6% Cr 0.2% Mo steel: Casting for pressure; normalized; 5Q grade is hardened and tempered
ASTM A487/7 Q	0.2% C (max) 0.9% Ni 0.6% Cr 0.5% Mo 0.08% V 0.004% B steel: Casting for pressure; hardened and tempered

Symbol	Nominal analysis, supplier, condition and remarks.
ASTM A487/8 N	0.2% C (max) 2.5% Cr 1.0% Mo steel: Casting for pressure; normalized; 8Q grade is hardened and tempered
ASTM A487/9 N	0.33% C (max) 0.9% Cr 0.25% Mo steel: Casting for pressure; normalized; 9Q grade is hardened and tempered
ASTM A487/10 N	0.3% C (max) 1.7% Ni 0.75% Cr 0.3% Mo steel: Casting for pressure; normalized; 10Q grade is hardened and tempered
ASTM A508/2	0.27% C (max) 0.7% Mn 0.7% Ni 0.35% Cr 0.6% Mo 0.05% V steel: For pressure vessels; vacuum treated
ASTM A508/3	0.2% C (max) 1.3% Mn 0.6% Ni 0.52% Mo 0.05% V (max) steel: For pressure vessels; vacuum treated
ASTM A508/4	0.23% C (max) 3.3% Ni 1.7% Cr 0.5% Mo 0.03% V (max) steel: For pressure vessels; vacuum treated
ASTM A508/5	0.23% C (max) 3.3% Ni 1.7% Cr 0.05% Mo 0.1% V steel: For pressure vessels; vacuum treated
ASTM A514 A	0.17% C 0.65% Cr 0.22% Mo 0.0025% B (max) 0.1% Zr steel: For plate
ASTM A514 B	0.14% C 0.52% Cr 0.2% Mo 0.05% V 0.02% Ti 0.001% B steel: For plate
ASTM A514 C	0.15% C 0.25% Mo 0.003% B steel: Plate
ASTM A514 D	0.17% C 1.0% Cr 0.2% Mo 0.08% Ti 0.03% Cu 0.003% B steel: Plate
ASTM A514 E	0.16% C 1.7% Cr 0.5% Mo 0.07% Ti 0.3% Cu 0.003% B steel: Plate
ASTM A514 F	0.15% C 0.85% Ni 0.5% Cr 0.5% Mo 0.005% V 0.3% Cu 0.004% B steel: Plate
ASTM A514 G	0.18% C 0.7% Cr 0.5% Mo 0.0025% B (max) 0.1% Zr steel: Plate
ASTM A514 H	0.17% C 0.5% Ni 0.48% Cr 0.25% Mo 0.05% V 0.001% B steel: Plate
ASTM A517 A	0.18% C 0.65% Cr 0.23% Mo 0.0025% B (max) 0.1% Zr steel: Boiler plate
ASTM A517 B	0.15% C 0.5% Cr 0.2% Mo 0.005% V 0.02% Ti 0.001% B steel: Boiler plate
ASTM A517 D	0.17% C 1.0% Cr 0.2% Mo 0.07% Ti 0.3% Cu 0.003% B steel: Boiler plate
ASTM A517 E	0.16% C 1.7% Cr 0.5% Mo 0.07% Ti 0.3% Cu 0.003% B steel: Boiler plate
ASTM A517 F	0.15% C 0.9% Ni 0.5% Cr 0.5% Mo 0.05% V 0.4% Cu 0.004% B steel: Boiler plate
ASTM A517 G	0.18% C 0.7% Cr 0.5% Mo 0.0025% B 0.1% Zr steel: Boiler plate
ASTM A517 H	0.17% C 0.5% Ni 0.5% Cr 0.25% Mo 0.05% V 0.0005% B steel: Boiler plate
ASTM A533 B	0.25% C (max) 1.25% Mn 0.55% Ni 0.52% Mo steel: Plate
ASTM A533 C	0.25% C (max) 1.25% Mn 0.85% Ni 0.52% Mo steel: Plate
ASTM A534	carburizing steels; graded by AISI system; see also Section 44A1
ASTM A535	steels for ball and roller bearings; graded by AISI system; see Sections 44K2 and 44K3
ASTM A541/3	0.2% C 1.3% Mn 0.6% Ni 0.52% Mo 0.06% V steel: Forging
ASTM A541/4	0.18% C (max) 0.25% Ni (max) 0.15% Cr (max) 0.1% V steel: Forging

Note. The following abbreviations and units are used in the tables:

DPN	Hardness, diamond pyramid number
UTS	Ultimate tensile strength, N/mm^2
Elon	Elongation, %
Proof	0.1% proof strength, N/mm^2

1 N/mm^2=0.1 hbar=0.102 kgf/mm^2=0.06475 $tonf/in.^2$=145.04 $lbf/in.^2$=1 MPa
See Appendix II for other abbreviations and conversion tables.

Symbol	Nominal analysis, supplier, condition and remarks.
ASTM A541/5	0.15% C 1.25% Cr 0.52% Mo steel: Forging
ASTM A541/6	0.15% C (max) 2.2% Cr 1.0% Mo steel: Forging
ASTM A541/7	0.23% C (max) 3.2% Ni 1.5% Cr 0.5% Mo 0.03% V steel: Forging
ASTM A541/8	0.23% C (max) 3.2% Ni 1.5% Cr 0.5% Mo 0.1% V steel: Forging
ASTM A542	0.15% C (max) 2.25% Cr 1.0% Mo steel: Plate
ASTM A543	0.23% C (max) 3.2% Ni 1.75% Cr 0.5% Mo steel: Plate; Ni content varies with thickness **UTS: 780 Elon: 14% Proof: 650**
ASTM A574	0.33% C (min) 0.04% S and P steel: For cap screws; may contain Cr Ni Mo or V to ensure mechanical properties are obtained
ASTM A579/11	0.25% C 3.0% Ni 1.5% Cr 0.9% Mo 0.05% Nb steel: Forging
ASTM A579/12	0.12% C (max) 5.0% Ni 0.55% Cr 0.5% Mo 0.07% V steel: Forging
ASTM A579/13	0.3% C 0.95% Cr 0.2% Mo 0.07% V steel: Forging
ASTM A579/31	0.25% C 1.8% Ni 0.3% Cr 0.4% Mo steel: Forging
ASTM A588 A	0.14% C 1.1% Mn 0.52% Cr 0.06% V 0.32% Cu steel: For structures **UTS: 440 Elon: 21% Proof: 320**
ASTM A588 B	0.2% C (max) 1.0% Mn 0.32% Ni 0.55% Cr 0.05% V 0.3% Cu steel: For structures **UTS: 440 Elon: 320**
ASTM A588 C	0.15% C (max) 1.1% Mn 0.4% Cr 0.06% V 0.35% Cu steel: For structures **UTS: 440 Elon: 21% Proof: 320**
ASTM A588 F	0.15% C 0.75% Mn 0.7% Ni 0.15% Mo 0.05% V 0.7% Cu steel: For structures **UTS: 440 Elon: 21% Proof: 320**
ASTM A588 H	0.2% C (max) 1.25% Mn (max) 0.45% Ni 0.16% Cr 0.06% V 0.27% Cu steel: For structures **UTS: 440 Elon: 21% Proof: 320**
ASTM A592 A	0.18% C 1.0% Mn 0.65% Cr 0.2% Mo 0.1% Zr steel: For pressure vessels **UTS: 850 Elon: 18% Proof: 650**
ASTM A592 E	0.17% C 0.5% Mn 1.7% Cr 0.5% Mo 0.07% Ti 0.25% Cu 0.003% B steel: For pressure vessels **UTS: 850 Elon: 18% Proof: 650**
ASTM A592 F	0.15% C 1.0% Mn 0.9% Ni 0.6% Cr 0.5% Mo 0.05% V steel: For pressure vessels **UTS: 850 Elon: 18% Proof: 650**
ASTM A595 C	0.15% C (max) 0.35% Mn 0.7% Ni (max) 1.0% Cr 0.4% Cu steel: Tapered tube **UTS: 480 Proof: 410**
ASTM A597 CH13	0.36% C 5.2% Cr 1.5% Mo 1.0% V steel: Cast; for tools
ASTM A643 B	0.25% C (max) 1.3% Mn 0.7% Ni 0.5% Mo steel: Casting for pressure vessels **UTS: 650 Elon: 22% Proof: 345**
ASTM A643 C1	0.2% C (max) 2.3% Cr 1.0% Mo steel: Casting for pressure vessels **UTS: 690 Elon: 20% Proof: 380**
ASTM A643 C2	0.2% C (max) 2.3% Cr 1.0% Mo steel: Casting for pressure vessels **UTS: 750 Elon: 18% Proof: 515**
ASTM A643 C3	0.2% C (max) 2.3% Cr 1.0% Mo steel: Casting for pressure vessels **UTS: 830 Elon: 15% Proof: 585**
ASTM A643 D1	0.2% C (max) 3.2% Ni 1.7% Cr 0.5% Mo steel: Casting for pressure vessels **UTS: 830 Elon: 15% Proof: 585**
ASTM A643 D2	0.2% C (max) 3.2% Ni 1.7% Cr 0.5% Mo steel: Casting for pressure vessels **UTS: 900 Elon: 13% Proof: 690**
ASTM A645	0.15% C (max) 0.4% Mn 5.0% Ni 0.25% Mo 0.1% Al steel: For pressure vessels; impact 22 J at −170 °C **UTS: 720 Elon: 20% Proof: 450**
ASTM A646/1111	0.4% C 5.0% Cr 1.3% Mo 0.5% V steel

Symbol	Nominal analysis, supplier, condition and remarks.
ASTM A649/3	0.35% C (max) 0.55% Mn 1.0% Cr 0.2% Mo steel: for rollers; corrugated
ASTM A687/1	0.36% C 1.1% Mn 0.22% Ni 0.3% Cr 0.08% Mo steel: For bolts; studs; impact 20 J at −20 °C **UTS: 1034 Elon: 15% Proof: 724**
ASTM A687/11	0.4% C 0.9% Mn 0.9% Cr 0.2% Mo steel: For bolts and studs; impact 20 J at −30 °C **UTS: 1034 Elon: 15% Proof: 724**
ASTM A706	0.3% C (max) 1.5% Mn (max) Cr + Ni + V + Mo at controlled (max) to give C equivalent (CE) 0.55% (max) **UTS: 550 Elon: 14% Proof: 500**
ASTM A707 44/2	0.2% C (max) 0.5% Mn 1.8% Ni 0.21% Mo steel: Flanges for low temperature use; impact 41 J at −62 °C **UTS: 360 Elon: 25%**
ASTM A707 L4/1	0.2% C (max) 0.5% Mn 1.8% Ni 0.21% Mo steel: Flanges for low temperature use; impact 41 J at −62 °C **UTS: 290 Elon: 22%**
ASTM A707 L4/3	0.2% C (max) 0.5% Mn 1.8% Ni 0.21% Mo steel: Flanges for low temperature use; impact 68 J at −62 °C **UTS: 415 Elon: 23%**
ASTM A707 L4/4	0.2% C (max) 0.5% Mn 1.8% Ni 0.21% Mo steel: Flanges for low temperature use; impact 68 J at −62 °C **UTS: 515 Elon: 20%**
ASTM A707 L5/1	0.09% C (max) 0.8% Ni 0.75% Cr 0.2% Mo 0.02% Nb (min) 1.1% Cu steel: Flanges for low temperature use; impact 41 J at −62 °C **UTS: 290 Elon: 22%**
ASTM A707 L5/2	0.09% C (max) 0.8% Ni 0.75% Cr 0.2% Mo 0.2% Nb (min) 1.1% Cu steel: Flanges for low temperature use; impact 64 J at −62 °C **UTS: 360 Elon: 25%**
ASTM A707 L5/3	0.09% C (max) 0.8% Ni 0.75% Cr 0.2% Mo 0.02% Nb (min) 1.1% Cu steel: Flanges for low temperature use; impact 68 J at −62 °C **UTS: 415 Elon: 23%**
ASTM A707 L5/4	0.09% C (max) 0.8% Ni 0.75% Cr 0.2% Mo 0.02% Nb (min) 1.1% Cu steel: Flanges for low temperature use; impact 68 J at −62 °C **UTS: 515 Elon: 20%**
ASTM A707 L8/1	0.22% C (max) 0.3% Mn 3.2% Ni 1.8% Cr 0.5% Mo 0.04% V steel: Flanges for low temperature use; impact 41 J at −73 °C **UTS: 290 Elon: 22%**
ASTM A707 L8/2	0.22% C (max) 0.3% Mn 3.2% Ni 1.8% Cr 0.5% Mo 0.04% V steel: Flanges for low temperature use; impact 54 J at −73 °C **UTS: 360 Elon: 25%**
ASTM A707 L8/3	0.22% C (max) 0.3% Mn 3.2% Ni 1.8% Cr 0.5% Mo 0.04% V steel: Flanges for low temperature use; impact 68 J at −73 °C **UTS: 415 Elon: 23%**
ASTM A707 L8/4	0.22% C (max) 0.3% Mn 3.2% Ni 1.8% Cr 0.5% Mo 0.04% V steel: Flanges for low temperature use; impact 68 J at −73 °C **UTS: 515 Elon: 20%**
ASTM A710 A	0.07% C (max) 0.85% Ni 0.75% Cr 0.2% Mo 0.02% Nb (max) steel
ASTM A710 B	0.06% C (max) 1.3% Ni 0.02% Nb (min) 1.1% Cu steel
ASTM A714/6 E & S	0.18% C (max) 0.8% Mn 0.8% Ni 0.3% Cr (max) 0.15% Mo 0.6% Cu steel: Pipe; welded or seamless **UTS: 448 Proof: 317**
ASTM A714 4	0.13% C (max) 0.35% Ni 1.0% Cr 0.3% Cu steel: Pipe; welded or seamless **UTS: 400 Proof: 248**
ASTM A723/1	0.35% C (max) 1.4% Cr 0.3% Mo 0.2% V (max) steel: For pressure applications
ASTM A723/2	0.4% C (max) 1.4% Cr 0.4% Mo 0.2% V (max) steel: For pressure applications
ASTM A723/3	0.4% C (max) 1.4% Cr 0.6% Mo 0.2% V (max) steel: For pressure application
ASTM A732/7 Q	0.3% C 0.5% Mn 1.0% Cr 0.2% Mo steel: Investment casting; hardened and tempered **UTS: 1000 Elon: 7% Proof: 800**
ASTM A732/8 Q	0.4% C 0.8% Mn 1.0% Cr 0.2% Mo steel: Investment casting; hardened and tempered **UTS: 1240 Elon: 5% Proof: 1000**
ASTM A732/9 Q	0.3% C 1.8% Ni 0.8% Cr 0.25% Mo steel: Investment casting; hardened and tempered **UTS: 1030 Elon: 7% Proof: 795**
ASTM A732/10 Q	0.4% C 1.8% Ni 0.8% Cr 0.25% Mo steel: Investment casting; hardened and tempered **UTS: 1240 Elon: 5% Proof: 1000**
ASTM A732/13 Q	0.2% C 0.5% Ni 0.5% Cr 0.2% Mo steel: Investment casting; hardened and tempered **UTS: 720 Elon: 10% Proof: 590**
ASTM A732/14 Q	0.3% C 0.5% Ni 0.5% Cr 0.2% Mo steel: Investment casting; hardened and tempered **UTS: 1030 Elon: 7% Proof: 790**
ASTM A734 A	0.14% C (max) 0.6% Mn 1.1% Ni 1.0% Cr 0.3% Mo steel: Plate for pressure vessels; hardened and tempered **UTS: 600 Elon: 20% Proof: 450**
ASTM A734 B	0.19% C (max) 1.6% Mn (max) 0.3% Cr (max) 0.13% V steel: Plate for pressure vessels; hardened and tempered **UTS: 600 Elon: 20% Proof: 450**
ASTM A736	0.09% C (max) 0.55% Mn 0.8% Ni 0.75% Cr 0.2% Mo 0.02% Nb (min) 1.1% Cu steel: Plate for pressure use
ASTM A739-B 11	0.2% C (max) 1.2% Cr 0.5% Mo steel: Bar **UTS: 520 Elon: 18% Proof: 310**
ASTM A739-B 22	0.15% C (max) 2.2% Cr 1% Mo steel: Bar **UTS: 630 Elon: 18% Proof: 310**
ASTM A757 C1 Q	0.25% C 1.75% Ni 0.2% Mo 1% Cr + V + Cu (max) steel: Casting; quench and temper; impact 20 J at −50 °C **UTS: 515 Elon: 22% Proof: 380**
ASTM A757 D1 N1	0.2% C 1% Ni + V + Cu (max) 2.3% Cr 1% Mo steel: Casting; normalize and temper; impact not part of specification **UTS: 585 Elon: 20% Proof: 380**
ASTM A757 D1 N1	0.2% C 2.3% Cr 1% Mo steel: Casting; for low temperature use
ASTM A757 D1 N2	0.2% C 2.3% Cr 1% Mo steel: Casting; for low temperature use
ASTM A757 D1 N2	0.2% C 1% Ni + V + Cu (max) 2.3% Cr 1% Mo steel: Casting; normalize and temper; impact not part of specification **UTS: 655 Elon: 18% Proof: 520**
ASTM A757 D1 N3	0.2% C 2.3% Cr 1% Mo steel: Casting; for low temperature use
ASTM A757 D1 N3	0.2% C 1% Ni + V + Cu (max) 2.3% Cr 1% Mo steel: Casting; normalize and temper; impact not part of specification **UTS: 725 Elon: 15% Proof: 590**
ASTM A757 D1 Q1	0.2% C 1% Ni + V + Cu (max) 2.3% Cr 1% Mo steel: Casting; quench and temper; impact not part of specification **UTS: 585 Elon: 20% Proof: 380**
ASTM A757 D1 Q1	0.2% C 2.3% Cr 1% Mo steel: Casting; for low temperature use
ASTM A757 D1 Q2	0.2% C 1% Ni + V + Cu (max) 2.3% Cr 1% Mo steel: Casting; quench and temper; impact not part of specification **UTS: 655 Elon: 18% Proof: 520**
ASTM A757 D1 Q2	0.2% C 2.3% Cr 1% Mo steel: Casting; for low temperature use

Symbol	Nominal analysis, supplier, condition and remarks.
ASTM A757 D1 Q3	0.2% C 1% Ni + V + Cu (max) 2.3% Cr 1% Mo steel: Casting; quench and temper; impact not part of specification **UTS: 725 Elon: 15% Proof: 590**
ASTM A757 D1 Q3	0.2% C 2.3% Cr 1% Mo steel: Casting; for low temperature use
ASTM A757 E1 Q	0.22% C 3% Ni 1.5% Cr 0.5% Mo 0.7% V + Cu (max) steel: Casting; quench and temper; impact 40 J at −75 °C **UTS: 620 Elon: 22% Proof: 450**
ASTM A757 E2 N	0.2% C 3.5% Ni 1.7% Cr 0.5% Mo 0.7% V + Cu (max) steel: Casting; quench and temper; impact 40 J at −75 °C **UTS: 725 Elon: 15% Proof: 590**
ASTM A757 E2 N	0.2% C 3.1% Ni 1.7% Cr 0.5% Mo steel: Casting; for low temperature use
ASTM A757 E2 Q	0.2% C 3.5% Ni 1.7% Cr 0.5% Mo 0.7% V + Cu (max) steel: Casting; quench and temper; impact 40 J at −75 °C **UTS: 620 Elon: 18% Proof: 485**
ASTM A757 E3 N	0.2% C 3.1% Ni 1.7% Cr 0.5% Mo steel: Casting; for low temperature use
ASTM A757 EQ 1	0.22% C 3% Ni 1.5% Cr 0.5% Mo steel: Casting; for low temperature use
ASTM A757 EQ 2	0.2% C 3.1% Ni 1.7% Cr 0.5% Mo steel: Casting; for low temperature use
ASTM A765-1	0.3% C (max) 0.5% Ni (max) 0.4% Cr (max) 0.2% Mo (max) steel: For pressure vessels **UTS: 500 Elon: 25% Proof: 200**
ASTM A765-11	0.3% C (max) 0.5% Ni (max) 0.4% Cr (max) 0.2% Mo (max) steel: For pressure vessels **UTS: 530 Elon: 22% Proof: 250**
ASTM A765-111	0.3% C (max) 3.5% Ni 0.2% Cr (max) 0.06% Mo steel: For pressure vessels; impact 20 J (min) at −100 °C **UTS: 530 Elon: 22% Proof: 260**
ASTM A782-1	0.22% C 0.7% Cr 0.4% Mo 0.1% Zr steel: Plate; for pressure vessels **UTS: 740 Elon: 18% Proof: 550**
ASTM A782-2	0.22% C 0.7% Cr 0.4% Mo 0.1% Zr steel: Plate; for pressure vessels **UTS: 810 Elon: 17% Proof: 620**
ASTM A782-3	0.22% C 0.7% Cr 0.4% Mo 0.1% Zr steel: Plate; for pressure vessels **UTS: 840 Elon: 16% Proof: 690**
ASTM A816 E8018B1	0.12% C 0.52% Cr 0.52% Mo steel: Electrodes for arc welding
ASTM A832	0.12% C 3% Cr 1% Mo 0.22% V 0.03% Ti 0.002% B steel: Plate; for pressure vessels **UTS: 670 Elon: 18% Proof: 415**
ASTM A832	0.13% C 3% Cr 1% Mo 0.21% V steel: For pressure vessels **UTS: 620 Elon: 18% Proof: 415**
ASTM A841	0.2% C (max) 0.2% Ni (max) 0.2% Cr (max) steel: For pressure vessels **UTS: 520 Elon: 18% Proof: 330**
ASTM A852	0.2% C (max) 0.5% Ni (max) 0.5% Cr 0.08% V 0.3% Cu steel: Plate **UTS: 700 Elon: 19% Proof: 485**

Note. The following abbreviations and units are used in the tables:

DPN	Hardness, diamond pyramid number
UTS	Ultimate tensile strength, N/mm^2
Elon	Elongation, %
Proof	0.1% proof strength, N/mm^2

1 N/mm^2=0.1 hbar=0.102 kgf/mm^2=0.06475 tonf/in.^{2}=145.04 lbf/in.^{2}=1 MPa

See Appendix II for other abbreviations and conversion tables.

Symbol	Nominal analysis, supplier, condition and remarks.
ASTM A859/1	0.07% C (max) 0.8% Ni 0.7% Cr 0.2% Mo 1.1% Cu steel: For pressure vessels **UTS: 510 Elon: 20% Proof: 380**
ASTM A859/2	0.07% C (max) 0.8% Ni 0.7% Cr 0.2% Mo 1.1% Cu steel: For pressure vessels **UTS: 580 Elon: 20% Proof: 450**
ASTM A871	0.2% C (max) 1.25% Ni 0.9% Cr (max) 0.2% Mo 1% Cu steel: Plate
ASTM B432	Mild or low alloy steel plate clad with Cu or copper alloy
ATLANTES MT	0.12% C (max) 0.5% Mn 0.5% Ni 0.8% Cr 0.06% Cu steel: South Durham
AURIGA VIII	0.2% C 0.25% Cr 0.5% Mo 0.4% Si steel: Casting; D Brown; normalized and tempered for high temperature use **UTS: 450 Elon: 20% Proof: 240**
AURIGA VIIIB	0.18% C 0.25% Cr 0.5% Mo 0.3% V 0.3% Si steel: Casting; D Brown; normalized and tempered for high temperature use **UTS: 530 Elon: 20% Proof: 300**
AURIGA VIIIC	0.15% C 2.25% Cr 1% Mo 0.3% Si steel: Casting; D Brown; normalized and tempered for high temperature use **UTS: 460 Elon: 20% Proof: 270**
AURIGA IX	0.32% C 0.8% Mn 1% Ni 1.2% Cr 0.3% Mo steel: Casting; D Brown; hardened and tempered **UTS: 920 Elon: 12% Proof: 700**
AURIGA IXB	0.32% C 0.7% Mn 1.2% Ni 1.2% Cr 0.4% Mo steel: Casting; D Brown; hardened and tempered **UTS: 1110 Elon: 10% Proof: 780**
AURIGA XV	0.25% C 3% Cr 0.5% Mo 0.4% Si steel: Casting; D Brown; hardened and tempered for high temperature use **UTS: 620 Elon: 18% Proof: 360**
AURIGA XVI	0.3% C 0.7% Mn 2% Ni 0.8% Cr 0.4% Mo steel: Casting; D Brown; hardened and tempered **UTS: 1110 Elon: 10% Proof: 780**
AURIGA XXI	0.15% C 5% Cr 0.5% Mo 0.5% Si steel: Casting; D Brown; hardened and tempered for high temperature use **UTS: 620 Elon: 18% Proof: 400**
AURIGA XXII	0.15% C 9% Cr 1% Mo 0.8% Si steel: Casting; D Brown; hardened and tempered for high temperature use **UTS: 620 Elon: 18% Proof: 400**
AURIGA XXXIV	0.3% C 1.5% Mn 0.8% Ni 0.6% Cr 0.4% Mo steel: Casting; D Brown; hardened and tempered **UTS: 1110 Elon: 10% Proof: 780**
AUTROD 13/12	0.1% C 1.3% Mn 1.0% Cr 0.5% Mo 0.6% Si steel: Welding wire; copper coated; Esab; for gas shielded welding; as welded **UTS: 630 Elon: 26% Proof: 530**
AZ	0.3% C 3.2% Cr 0.25% V 8.5% W steel: For hot dies; S Osborn; tools; extrusion dies, rivet dies; now H31
AZN	0.25% C 2.2% Ni 3% Cr 0.25% V 8.5% W steel: For hot dies; tools; S Osborn; now H21N
B and W WL2	0.15% C 9.0% Cr 0.5% Mo steel: Tube; Babcock and Wilcox (USA)
B 6L	0.05% C (max) 5% Cr 0.5% Mo steel: Weld electrode; designation used by AWS
B 45CH	0.15% C 0.6% Ni 0.65% Cr 0.12% Mo steel: For carburizing; Firth Brown for BS 970 En 361
B 55CH	0.2% C 0.6% Ni 0.65% Cr 0.12% Mo steel: For carburizing; Firth Brown for BS 970 En 362
B 65CH	0.24% C 0.6% Ni 0.65% Cr 0.12% Mo steel: For carburizing; Firth Brown for BS 970 En 363
BCD 37	0.37% C 1.6% Cr 0.3% V 2.0% W steel: Carrs; chisels
BCH	0.18% C (max) 3.5% Ni 0.8% Cr steel: Toledo for BS alloy En 36
BCMF	0.35% C 2.5% Ni 0.6% Cr 0.5% Mo steel: Toledo for BS alloys En 25 and 26

Symbol	Nominal analysis, supplier, condition and remarks.
BCMO	0.3% C 3.25% Ni 0.9% Cr 0.5% Mo steel: Toledo for BS alloy En 27
BCN	0.3% C 3.25% Ni 0.7% Cr steel: Toledo for BS alloy En 23
BCN (Mo)	0.3% C 3.25% Ni 0.7% Cr 0.06% Mo steel: Toledo for BS alloy En 23
BENUM	0.3% C 4.1% Ni 1.25% Cr 0.3% Mo steel: Origin unknown
BH 43	0.18% C (max) 1.3% Mn 0.55% Ni 0.18% V steel: Rheinstahl
BH 43S	0.17% C (max) 1.3% Mn 0.55% Ni 0.18% V steel: Rheinstahl
BH 47	0.19% C (max) 1.3% Mn 0.55% Ni 0.2% V steel: Rheinstahl
BH 47S	0.18% C (max) 1.3% Mn 0.55% Ni 0.2% V steel: Rheinstahl
BH 51	0.2% C (max) 1.5% Mn 0.55% Ni 0.22% V steel: Rheinstahl
BH 51	0.2% C (max) 1.5% Mn 0.65% Ni 0.16% V steel: German proprietor's steel
BH 51S	0.2% C (max) 1.5% Mn 0.55% Ni 0.22% V steel: Rheinstahl
BH 57V	0.16% C 1.3% Mn 0.55% Ni 0.16% V steel: Rheinstahl; Al killed
BH 57VT	0.16% C 1.3% Mn 0.55% Ni 0.16% V steel: Rheinstahl; Al killed
BH 65V	0.18% C 1.5% Mn 0.55% Ni 0.18% V steel: Rheinstahl; Al killed
BH 65VT	0.18% C 1.5% Mn 0.55% Ni 0.18% V steel: Rheinstahl; Al killed
BL 30	0.25% C 1.5% Mn 0.8% Ni 0.7% Mo 0.2% V steel: Pompey
BRUNSWICK V108	0.35% C 5% Cr + Mo + V steel: John Vessey; hot work tools **DPN: 425**
BRUNSWICK V/110/CNA	0.3% C 4% Ni 1.2% Cr 0.3% Mo steel: John Vessey; stamping dies, shear blades, etc. **DPN: 400**
BS 24/5/614	0.23% C (max) 1.5% Mn Ni Cr steel: Plate and bar for railway use
BS 682	3% Ni Cr case hardening steel: Obsolete
BS 806 P	0.15% C 0.8% Cr 0.5% Mo steel: Seamless pipe; cold drawn and normalized **UTS: 500**
BS 806 Q	0.15% C 0.8% Cr 0.5% Mo steel: Seamless pipe; hot drawn and normalized **UTS: 500**
BS 970	See designation for details
BS 980 CDS12	0.25% C 1% Cr 0.2% Mo steel: Cold drawn tube; as drawn and tempered; can be welded and heat treated **UTS: 700 Proof: 580**
BS 980 CDS13	0.35% C 1% Cr 0.2% Mo steel: Cold drawn tube; hardened and tempered **UTS: 700 Proof: 580**
BS 980 CDS16	0.25% C 4% Ni 1.2% Cr 0.2% Mo steel: Cold drawn tube; hardened and tempered **UTS: 1210 Elon: 7% Proof: 1070**
BS 980 CDS17	0.3% C 4% Ni 1.2% Cr 0.2% Mo steel: Cold drawn tube; hardened and tempered **UTS: 1380 Elon: 5% Proof: 1140**
BS 1453 A4	0.3% C 3% Ni 0.3% Cr steel: Weld filler rod; for welds to be hardened and tempered
BS 1461	0.25% C (max) 3.0% Cr 0.45% Mo steel: Casting; hardened and tempered; contained in BS 3100 **UTS: 700 Elon: 18% Proof: 400**
BS 1462	0.2% C (max) 5.0% Cr 0.55% Mo steel: Casting; hardened and tempered; contained in BS 3100 **UTS: 610 Elon: 18% Proof: 400**
BS 1463	0.2% C (max) 9.0% Cr 1.1% Mo steel: Casting; hardened and tempered; contained in BS 3100 **UTS: 610 Elon: 18% Proof: 400**
BS 1501/271	0.14% C 1.2% Mn 0.55% Cr 0.23% Mo 0.1% V steel: Plate; normalized and tempered; for pressure vessels **UTS: 650 Elon: 16% Proof: 400**
BS 1501/281	0.13% C 1.1% Mn 0.9% Ni 0.55% Cr 0.24% Mo 0.1% V steel: Plate; normalized and tempered; for pressure vessels **UTS: 650 Elon: 16% Proof: 400**
BS 1501/282	0.14% C 1.1% Mn 1.5% Ni 0.5% Cr 0.35% Mo 0.1% V steel: Plate; normalized and tempered; for pressure vessels **UTS: 660 Elon: 18% Proof: 420**
BS 1501/620/27	0.12% C 1.0% Cr 0.5% Mo steel: Plate; normalized and tempered; for pressure vessels **UTS: 480 Elon: 21% Proof: 300**
BS 1501/620/31	0.15% C 1.0% Cr 0.5% Mo steel: Plate; normalized and tempered; for pressure vessels **UTS: 530 Elon: 18% Proof: 330**
BS 1501/620A	0.15% C 0.65% Cr 0.6% Mo steel: Plate and bar; normalized **UTS: 420 Elon: 20% Proof: 220**
BS 1501/620B	0.15% C 0.9% Cr 0.6% Mo steel: Plate and bar; normalized **UTS: 440 Elon: 20% Proof: 220**
BS 1501/621	0.12% C 1.2% Cr 0.55% Mo steel: Plate **UTS: 520 Elon: 18% Proof: 330**
BS 1501/622/31	0.13% C 2.2% Cr 1.0% Mo steel: Plate; normalized and tempered; for pressure vessels **UTS: 550 Elon: 18% Proof: 290**
BS 1501/622/45	0.15% C 2.2% Cr 1.0% Mo steel: Plate: normalized and tempered; for pressure vessels **UTS: 760 Elon: 15% Proof: 570**
BS 1501/625	0.15% C 5.0% Cr 0.55% Mo steel: Plate and bar; annealed **UTS: 400 Elon: 20% Proof: 210**
BS 1503/620	0.20% C 1.0% Cr 0.5% Mo steel: Forging; normalized and tempered **UTS: 460 Elon: 20% Proof: 270**
BS 1503/622	0.15% C 2.3% Cr 1.0% Mo steel: Forging; normalized and tempered **UTS: 460 Elon: 20% Proof: 270**
BS 1503/623	0.2% C 3% Cr 0.4% Mo steel: Normalized and tempered **UTS: 670 Elon: 23% Proof: 420**
BS 1503/625	0.25% C 5.0% Cr 0.5% Mo steel: Forgings; hardened and tempered **UTS: 610 Elon: 20% Proof: 440**
BS 1504/621	0.2% C (max) 1.2% Cr 0.5% Mo steel: Casting; normalized and tempered **UTS: 460 Elon: 20% Proof: 270**
BS 1504/623	0.25% C (max) 3.0% Cr 0.5% Mo steel: Casting; hardened and tempered **UTS: 700 Elon: 18% Proof: 360**
BS 1504/625	0.2% C (max) 5% Cr 0.5% Mo steel: Casting; hardened and tempered **UTS: 610 Elon: 18% Proof: 400**
BS 1504/629	0.2% C (max) 9% Cr 1.0% Mo steel: Casting; hardened and tempered **UTS: 610 Elon: 18% Proof: 400**
BS 1505/622	0.18% C (max) 2.5% Cr 1.0% Mo steel: Casting; normalized and tempered **UTS: 460 Elon: 20% Proof: 270**
BS 1506/625	0.3% C (max) 5.0% Cr 0.5% Mo steel: Bar; hardened and tempered **UTS: 700 Elon: 19% Proof: 580**
BS 1506/661	0.3% C 1.0% Cr 0.6% Mo 0.25% V steel: Bar; normalized and tempered **UTS: 840 Elon: 16% Proof: 720**
BS 1507/621	0.15% C 1% Cr 0.5% Mo steel: Seamless pipe; normalized **UTS: 450 Elon: 24%**

Symbol	Nominal analysis, supplier, condition and remarks.
BS 1507/623	0.15% C 3% Cr 0.5% Mo steel: Pipe; normalized **UTS: 450 Elon: 22%**
BS 1507/625	0.15% C 5% Cr 0.5% Mo steel: Pipe; normalized **UTS: 450 Elon: 20%**
BS 1508/621	0.15% C 1% Cr 0.5% Mo steel: Tube; normalized **UTS: 450 Elon: 22%**
BS 1508/622	0.15% C 2.5% Cr 1% Mo steel: Tube; normalized **UTS: 400 Elon: 20%**
BS 1508/625	0.15% C 5% Cr 0.5% Mo steel: Tube; normalized **UTS: 450 Elon: 18%**
BS 1628 A	0.15% C 2.5% Cr 1% Mo steel: Cold drawn tube; annealed **UTS: 400 Elon: 27%**
BS 1628 B	0.15% C 5% Cr 0.5% Mo steel: Cold drawn tube; annealed **UTS: 400 Elon: 27%**
BS 1653	Cr Mo steel: Tube; replaced by BS 3059
BS 1717 CDS110	0.26% C 1% Cr 0.25% Mo steel: Seamless tube; as drawn **UTS: 700 Proof: 370**
BS 1717 CDS110	0.26% C 1% Cr 0.25% Mo steel: Seamless tube; after welding, brazing or annealing **UTS: 550 Proof: 370**
BS 2493/1 Cr Mo B	0.1% C (max) 1.3% Cr 0.6% Mo steel: For welding electrodes
BS 2493/1 Cr Mo LB	0.05% C (max) 1.5% Cr 0.6% Mo steel: For welding electrodes
BS 2493/1 Cr Mo R	0.1% C (max) 1.3% Cr 0.6% Mo steel: For welding electrodes
BS 2493/1 Ni Mo B	0.1% C (max) 1.5% Ni 0.35% Mo steel: For welding electrodes
BS 2493/1 Ni Mo C	0.18% C (max) 1.5% Ni 0.4% Mo steel: For welding electrodes
BS 2493/2 Cr Mo B	0.1% C (max) 2.5% Cr 1% Mo steel: For welding electrodes
BS 2493/2 Cr Mo LB	0.05% C (max) 2.5% Cr 1% Mo steel: For welding electrodes
BS 2493/2 Cr Mo R	0.1% C (max) 2.5% Cr 1% Mo steel: For welding electrodes
BS 2493/2 Ni Cr Mo B	0.1% C (max) 2.0% Ni 1.3% Cr 0.35% Mo steel: For welding electrodes
BS 2493/2 Ni Mo B	0.1% C (max) 2.0% Ni 0.35% Mo steel: For welding electrodes
BS 2493/5 Cr Mo B	0.1% C 5% Cr 0.6% Mo steel: For welding electrodes
BS 2493/7 Cr Mo B	0.1% C 7% Cr 0.6% Mo steel: For welding electrodes
BS 2493/9 Cr Mo B	0.1% C (max) 9% Cr 1% Mo steel: For welding electrodes
BS 2493 B	0.1% C 1% Cr 0.5% Mo steel: Welding electrode
BS 2493 C	0.1% C 2.2% Cr 1% Mo steel: Welding electrode
BS 2493 D	0.1% C 5% Cr 0.5% Mo steel: Welding electrode
BS 3059/7	0.15% C (max) 0.3% Ni 0.5% Mo 0.3% Cu steel: Tube; seamless; hot finished and normalized **UTS: 440**
BS 3059/8	0.15% C (max) 0.3% Ni 0.5% Mo 0.3% Cu steel: Tube; cold drawn; seamless; normalized **UTS: 440**
BS 3059/9	0.12% C (max) 0.3% Ni 0.8% Cr 0.5% Mo 0.3% Cu steel: Tube; hot finished; seamless; normalized **UTS: 440**
BS 3059/10	0.12% C 0.3% Ni 0.8% Cr 0.5% Mo 0.3% Cu steel: Tube; cold drawn; seamless; normalized **UTS: 440**
BS 3059/11	0.1% C 0.3% Ni 2.2% Cr 1.0% Mo steel: Tube; seamless; hot finished; normalized **UTS: 460**
BS 3059/12	0.1% C 0.3% Ni 2.2% Cr 1.1% Mo steel: Tube; seamless; cold drawn and normalized **UTS: 460**
BS 3100 B2	0.2% C (max) 0.7% Mn 1.2% Cr 0.5% Mo steel: Casting for high temperature use **UTS: 480 Elon: 17% Proof: 280**
BS 3100 B3	0.18% C (max) 0.6% Mn 2.3% Cr 1.0% Mo steel: Casting for high temperature use **UTS: 540 Elon: 17% Proof: 325**
BS 3100 B4	0.25% C (max) 0.5% Mn 3.0% Cr 0.5% Mo steel: Casting for high temperature use **UTS: 620 Elon: 13% Proof: 370**
BS 3100 B5	0.2% C (max) 0.5% Mn 5.0% Cr 0.5% Mo steel: Casting for high temperature use **UTS: 620 Elon: 13% Proof: 420**
BS 3100 B6	0.2% C (max) 0.5% Mn 9.0% Cr 1.0% Mo steel: Casting for high temperature use **UTS: 620 Elon: 13% Proof: 420**
BS 3100 B7	0.12% C 0.5% Mn 0.4% Cr 0.5% Mo 0.25% V steel: Casting for high temperature use **UTS: 510 Elon: 17% Proof: 300**
BS 3100 BW1	0.16% C 0.45% Mn 3.3% Ni 0.85% Cr 0.2% Mo steel: Casting for case hardening **UTS: 1000 Elon: 7%**
BS 3100 BW4	0.6% C 0.7% Mn 1.1% Cr 0.3% Mo steel: Casting for abrasion resistance
BS 3111/4	0.35% C 1.2% Ni 0.6% Cr steel: Wire; annealed and drawn; for cold forged bolts **UTS: 700**
BS 3127/6	0.3% C 1% Cr 0.2% Mo steel: Tube; annealed; for bourdon tubes **DPN: 165**
BS 3146 CLA1A	0.25% C 1% Mn low Ni Cr Mo Cu steel: Investment casting; hardened and tempered **DPN: 150 UTS: 420 Elon: 22% Proof: 210**
BS 3146 CLA1B	0.35% C 1% Mn low Ni Cr Mo Cu steel: Investment casting; hardened and tempered **DPN: 160 UTS: 480 Elon: 20% Proof: 210**
BS 3146 CLA2A	0.21% C 1.5% Mn low Ni Cr Mo Cu steel: Investment casting; hardened and tempered **DPN: 175 UTS: 530 Elon: 18%**
BS 3146 CLA2B	0.29% C 1.5% Mn low Ni Cr Mo Cu steel: Investment casting; hardened and tempered **DPN: 220 UTS: 700 Elon: 15%**
BS 3146 CLA6	0.2% C 0.6% Mo low Ni Cr steel: Investment casting; normalized **DPN: 150 UTS: 450 Elon: 20% Proof: 240**
BS 3146 CLA7	0.25% C 3% Cr 0.45% Mo steel: Investment casting; hardened and tempered **DPN: 200 UTS: 610 Elon: 18% Proof: 360**
BS 3146 CLA9	0.15% C low Ni Cr Mo Cu steel: Investment casting for case hardening; refined and hardened core **UTS: 480 Elon: 20%**
BS 3146 CLA11	0.25% C 3.2% Cr 0.6% Mo steel: Investment casting; hardened and tempered; for nitriding **DPN: 260 UTS: 840 Elon: 12% Proof: 580**
BS 3551 C	0.2% C 0.6% Ni 0.6% Cr 0.2% Mo steel: For shackles; hardened and tempered **DPN: 270**
BS S16	Nickel chrome steel: For carburizing; replaced
BS S28	0.3% C 4.2% Ni 1.2% Cr 0.3% Mo steel: Bar and forging; hardened and tempered **DPN: 444 UTS: 1550 Elon: 12% Proof: 1030**

Note. The following abbreviations and units are used in the tables:

DPN	Hardness, diamond pyramid number
UTS	Ultimate tensile strength, N/mm^2
Elon	Elongation, %
Proof	0.1% proof strength, N/mm^2

1 N/mm^2=0.1 hbar=0.102 kgf/mm^2=0.06475 tonf/in.^{2}=145.04 lbf/in.^{2}=1 MPa

See Appendix II for other abbreviations and conversion tables.

Symbol	Nominal analysis, supplier, condition and remarks.
BS S82	0.15% C 4.1% Ni 1.2% Cr 0.2% Mo steel: For carburizing; blank carburized; hardened and tempered **UTS: 1300 Elon: 10%**
BS S90	Low C 5% Ni case hardening steel: Obsolete; see BS S107
BS S96	0.32% C 2.5% Ni 0.7% Cr 0.55% Mo steel: Hardened and tempered **DPN: 280 UTS: 920 Elon: 18% Proof: 670**
BS S97	0.32% C 2.5% Ni 0.7% Cr 0.55% Mo steel: Hardened and tempered **DPN: 320 UTS: 1070 Elon: 16% Proof: 820**
BS S103	0.35% C 1.2% Ni 0.7% Cr steel: Wire for forged bolts; hardened and tempered after forging **UTS: 920 Elon: 18% Proof: 670**
BS S106	0.24% C 3.2% Cr 0.55% Mo steel: For nitriding; hardened and tempered **DPN: 290 UTS: 1000 Elon: 17% Proof: 700**
BS S107	0.13% C 3.3% Ni 0.8% Cr 0.15% Mo steel: For carburizing; blank carburized; hardened and tempered **UTS: 1070 Elon: 12%**
BS S120	0.32% C 2.5% Ni 0.6% Cr 0.55% Mo steel: Hardened and tempered **DPN: 450 UTS: 1550 Elon: 12% Proof: 1030**
BS S122	0.35% C 1.2% Ni 0.6% Cr steel: Bar; hardened and tempered **UTS: 920 Elon: 18% Proof: 650**
BS S518	0.25% C (max) 1.0% Cr 0.2% Mo steel: Sheet; hardened and tempered; weldable **UTS: 840 Elon: 12% Proof: 610**
BS S519	0.25% C (max) 1.0% Cr 0.2% Mo steel: Sheet; hardened and tempered; weldable **UTS: 1210 Elon: 5% Proof: 1000**
BS STA 5 V20	0.3% C 1.5% Cr 0.2% Mo 1% Al steel: Replaced by BS 970 En 41
BS T2	0.25% C 4.0% Ni 1.0% Cr 0.2% Mo steel: Tube; hardened and tempered **UTS: 1450 Elon: 7% Proof: 1210**
BS T53	0.25% C (max) 1.0% Cr 0.2% Mo steel: Tube; hardened and tempered; weldable **UTS: 700**
BS T56	0.25% C (max) 1.0% Cr 0.2% Mo steel: Tube; hardened and tempered; weldable **UTS: 530 Proof: 450**
BS T57	0.25% C 4.0% Ni 1.0% Cr 0.2% Mo steel: Tube; hardened and tempered **UTS: 1210 Elon: 10% Proof: 1000**
BS T59	0.25% C 1.0% Cr 0.2% Mo steel: Tube; hardened and tempered; weldable **UTS: 760 Proof: 700**
BS T60	0.25% C (max) 1.0% Cr 0.2% Mo steel: Tube; hardened and tempered; weldable **UTS: 1210 Elon: 10% Proof: 1120**
BSS 1462	0.2% C 5% Cr 0.22% Mo steel: Jessop
BVR 20h/W	0.28% C 1.15% Ni 1.15% Cr 0.25% Mo steel: Krupp
BVR 20T	0.28% C 1.15% Ni 1.15% Cr 0.45% Mo steel: Krupp
BYR 1	0.28% C 1.95% Ni 1.25% Cr 0.35% Mo steel: Krupp
C 5	0.35% C 5% Cr 1.5% Mo 1% W steel: Turton Bros
C 642	0.25% C 1.0% Cr 0.2% Mo steel: Fagersta
C 643	0.34% C 1.0% Cr 0.2% Mo steel: Fagersta
CA 2	1% C 5.2% Cr 1.2% Mo tool steel: ASTM designation
CANO 30	0.34% C 1.0% Ni 1.65% Cr 0.2% Mo 0.95% Al steel: For nitriding; Krupp
CAO 30	0.34% C 1.15% Cr 0.2% Mo 0.95% Al steel: For nitriding; Krupp
CARPENTER 345	0.35% C 5.0% Cr 1.5% Mo 1.25% W steel: Carpenter for AISI type H12
CARPENTER 345	0.35% C 5.0% Cr 1.5% Mo 1.25% W tool steel: Carpenter for AISI type H12
CASCADE	0.2% C 4.1% Ni 0.25% Cr 0.2% V 1.2% Al steel: Latrobe; solution treated and aged; type P21 **DPN: 230 UTS: 1300 Elon: 16% Proof: 1180**
CBS 600	0.22% C 1.5% Cr 1.0% Mo steel: For case hardening; Timkin; for bearings at high temperature **DPN: 330 UTS: 1210 Elon: 17% Proof: 920**
CBVO	0.34% C 1.05% Cr 0.2% Mo steel: Krupp
CBVO/3R	0.32% C 3.0% Cr 0.4% Mo steel: Krupp
CBVO/M	0.25% C 1.05% Cr 0.2% Mo steel: Krupp
CBVO/TM	0.24% C 1.05% Cr 0.25% Mo steel: Krupp
CBVOV	0.3% C 2.5% Cr 0.2% Mo 0.15% V steel: Krupp
CBVOV/N	0.31% C 0.2% Ni 2.5% Cr 0.15% V steel: For nitriding; Krupp
CH 5	0.37% C 5% Cr 1.4% Mo 0.6% V steel: Marsh Brothers; hot forging dies, etc.
CH 12	0.35% C 5.2% Cr 1.5% Mo 0.4% Co tool steel: ASTM designation
CH 13	0.36% C 5.2% Cr 1.5% Mo 1% V tool steel: ASTM designation
CH 45D	0.15% C 0.8% Ni 0.6% Cr steel: For case hardening; ESC for BS alloy En 351
CH 55B	0.16% C 1% Ni 0.8% Cr steel: For case hardening; ESC for BS alloy En 352
CH 65B	0.17% C 1.3% Ni 1% Cr steel: For case hardening; ESC for BS alloy En 353
CH 65J	0.16% C 1.5% Ni 0.85% Cr 0.2% Mo steel: For case hardening; ESC **UTS: 1000 Elon 13%**
CH 75B	0.18% C 1.8% Ni 1% Cr 0.15% Mo steel: For case hardening; ESC for BS alloy En 354
CH 85A	0.18% C 2% Ni 1.5% Cr 0.2% Mo steel: For case hardening; ESC for BS alloy En 355
CH 361	0.15% C 0.6% Ni 0.7% Cr 0.1% Mo steel: For case hardening; ESC for BS alloy En 361
CH 362	0.2% C 0.6% Ni 0.7% Cr 0.1% Mo steel: For case hardening; ESC for BS alloy En 362
CH 363	0.25% C 0.6% Ni 0.7% Cr 0.1% Mo steel: For case hardening; ESC for BS alloy En 363
CHD	0.33% C 5.0% Cr 1.5% Mo 0.3% V 1.5% W tool steel: W Spencer
CHD 3	0.34% C 5% Cr 1.6% Mo 0.4% V 1.3% W 1% Si tool steel: Huntsman; hot punches, shear, etc. **DPN: 500**
CHIPPEWA	0.3% C 3.0% Ni 0.4% Cr 0.25% Mo steel: Tube; Atlas; for detachable rock drill bits
CHNC 55	0.15% C 3.2% Ni 0.8% Cr 0.17% Mo steel: For case hardening; ESC for BS alloy En 36A
CHNC 65	0.15% C 3.2% Ni 0.8% Cr 0.17% Mo steel: For case hardening; ESC for BS alloy En 36 B and C
CHNM 45	0.17% C 1.8% Ni 0.25% Mo steel: For case hardening; ESC for BS alloy En 34
CHNM 55	0.22% C 1.8% Ni 0.25% Mo steel: For case hardening; ESC for BS alloy En 35
CHROMET 1	0.06% C 1.2% Cr 0.5% Mo steel: Electrode; Metrode; for creep resistance
CHROMET 1L	0.04% C 1.2% Cr 0.5% Mo steel: Electrode; Metrode; for creep resistance
CHROMET 2	0.06% C 2.2% Cr 1% Mo steel: Electrode; Metrode
CHROMET 2L	0.04% C 2.2% Cr 1% Mo steel: Electrode; Metrode
CHROMET 5	0.05% C 5% Cr 0.5% Mo steel: Electrode; Metrode
CHROMET 9	0.05% C 9% Cr 1% Mo steel: Electrode; Metrode
CHROMET 9 MV	0.09% C 8.5% Cr 1% Mo 0.2% V Nb steel: Electrode; Metrode
CHROMOLOY	0.2% C 1% Cr 1% Mo 0.1% V steel: American proprietary alloy listed in SAE Yearbook
CHROMOTRODE 76/12	0.07% C 0.8% Mn 1.0% Cr 0.5% Mo steel: For electrodes; Esab **UTS: 710 Elon: 26% Proof: 602**
CHROMOTRODE 76/19	0.07% C 0.9% Mn 1.25% Cr 0.5% Mo steel: For electrodes; Esab **UTS: 726 Elon: 21% Proof: 618**
CHROMOTRODE 76/22	0.07% C 0.9% Mn 2.25% Cr 1.0% Mo steel: For electrodes; Esab **UTS: 726 Elon: 22% Proof: 618**

Symbol	Nominal analysis, supplier, condition and remarks.
CHROMOTRODE 76/28	0.07% C 0.9% Mn 2.25% Cr 1.0% Mo steel: For electrodes; Esab standard
	UTS: 772 Elon: 22% Proof: 679
CHRO-MOW	0.3% C 5% Cr 1.35% Mo 1.25% W 1% Si steel: Used for die holders; forging tools and extrusion punches; origin unknown
CHROMVA-W	0.18% C 2.7% Cr 0.5% Mo 0.7% V 0.05% W steel: Firth Brown; for use at temperatures up to 580 °C
	UTS: 920 Elon: 15% Proof: 700
CKD 1	0.12% C 1.0% Ni 0.25% Cr steel: Pompey
CKD 2	0.18% C 1.0% Cr 0.25% Mo steel: Pompey
	UTS: 1150 Elon: 7% Proof: 860
CL 222	0.35% C 0.8% Cr 0.5% Mo 0.5% W steel: For tools; origin unknown
CL 444	0.35% C 5.0% Cr 1.5% Mo 0.3% V 1.5% W steel: Origin unknown
CLW	0.3% C 3.3% Cr 0.5% V 9% W steel: Casting; Latrobe; for precision casting of forging dies
	DPN: 572
CM 1	0.07% C 5% Cr 1% Mo 0.3% V steel: Huntsman; for case hardened dies for Zn casting
CM 70	0.18% C 1.2% Mn 0.6% Ni + Cr + Mo steel: For carburizing; Steel Peech and Tozer; core properties; hardened and tempered
	UTS: 1070 Elon: 22% Proof: 760
CM 80F	0.2% C 1.2% Mn 0.8% Cr + Mo + Cu 0.004% B steel: Steel Peech and Tozer; can be carburized
	UTS: 1070 Elon: 10% Proof: 920
CM 90F	0.22% C 1.5% Mn 0.8% Ni + Cr + Mo 0.004% B steel: Steel Peech and Tozer
CMV	0.35% C 5% Cr 1.5% Mo 0.9% V 1% Si steel: Sanderson; forging dies, etc.
	DPN: 570
CMW	0.35% C 0.3% Mn 5% Cr 1.2% W 1% Si steel: Sanderson; forging dies
	DPN: 560
CN 1	0.15% C 3.1% Ni 0.3% Cr steel: For case hardening; W Marrison for BS alloy En 33
CN 2	0.18% C 3.4% Ni 0.9% Cr steel: For case hardening; W Marrison for BS alloy En 36
CN 3A	0.15% C 4.2% Ni 1.2% Cr steel: For case hardening; W Marrison for BS alloy En 39A
CN 3B	0.15% C 4.2% Ni 1.2% Cr 0.3% Mo steel: For case hardening; W Marrison for BS alloy En 39B
CN 5	0.18% C 2.0% Ni 2.0% Cr 0.2% Mo steel: For case hardening; W Marrison for BS alloy En 320
CN 9	0.2% C 0.8% Ni 0.6% Cr 0.1% Mo steel: For case hardening; W Marrison for BS alloy En 351
CN 10	0.22% C 1.1% Ni 0.8% Cr 0.1% Mo steel: For case hardening; W Marrison for BS alloy En 352
CN 11	0.22% C 1.3% Ni 1% Cr 0.15% Mo steel: For case hardening; W Marrison for BS alloy En 353
CN 12	0.22% C 1.8% Ni 1% Cr 0.2% Mo steel: For case hardening; W Marrison for BS alloy En 354
CN 13	0.2% C 2.0% Ni 1.6% Cr 0.2% Mo steel: For case hardening; W Marrison for BS alloy En 355
CNBN 1/W	0.15% C 1.5% Ni 1.5% Cr steel: Krupp
CNBV 2/W	0.18% C 2.0% Ni 2.0% Cr steel: Krupp
CNBV 2V/W	0.18% C 1.8% Ni 2.0% Cr 0.1% V steel: Krupp
CNBV 3/W	0.14% C 2.5% Ni 0.75% Cr steel: Krupp
CNBV 4/W	0.14% C 3.5% Ni 0.75% Cr steel: Krupp
CNBV 5/W	0.14% C 4.5% Ni 1.1% Cr steel: Krupp
CNBVO	0.36% C 1.05% Ni 1.05% Cr 0.2% Mo steel: Krupp
CNBVO 1	0.34% C 1.55% Ni 1.55% Cr 0.25% Mo steel: Krupp
CNBVO 1/1	0.17% C 1.55% Ni 1.65% Cr 0.35% Mo steel: Krupp
CNBVO 2	0.3% C 1.95% Ni 1.95% Cr 0.3% Mo steel: Krupp
CND	0.1% C 1.0% Ni 0.15% Mo steel: Pompey
CNKD1	0.16% C 1.2% Ni 0.6% Cr 0.15% Mo steel: Pompey; hardened and tempered
	UTS: 1030 Elon: 9% Proof: 740
CNKD2	0.18% C 1.3% Ni 1.0% Cr 0.2% Mo steel: Pompey; hardened and tempered
	UTS: 1360 Elon: 7% Proof: 920
CNKD3	0.16% C 3.2% Ni 1.0% Cr 0.2% Mo steel: Pompey; hardened and tempered
	UTS: 1360 Elon: 8% Proof: 990
CO 1	0.92% C 0.7% Cr 0.5% W tool steel: ASTM designation
CODE 107	0.3% C 2.5% Ni 0.7% Cr 0.5% Mo steel: Casting; Edgar Allen; 'A 100'
CODE 617	0.2% C (max) 5% Cr 0.5% Mo steel: Casting; Edgar Allen
CODE 625	0.22% C 3.1% Cr 0.5% Mo steel: Casting; Edgar Allen; 'Cr Mo 329'
COLUMBUS RECORD	0.26% C 1% Cr 0.2% Mo steel: Origin unknown; for bicycle frames
COMET J 56 LH	0.05% C 0.9% Mn 1.0% Ni 0.1% Cr 0.08% Mo: Welding electrode; Soudometal; for welding low alloy steel
	UTS: 600 Elon: 28% Proof: 510
COMET J 66 ELH	0.05% C 1.2% Mn 1.0% Ni 0.15% Cr 0.26% Mo: Welding electrode; Soudometal; for welding low alloy steel
	UTS: 655 Elon: 25% Proof: 580
COMET J 76 ELH	0.04% C 1.4% Mn 2.1% Ni 0.35% Cr 0.4% Mo: Welding electrode; Soudometal; for welding low alloy steel
	UTS: 770 Elon: 20% Proof: 690
COMPAX	0.13% C 3.7% Ni 1.4% Cr steel: Uddelholm for type PG
CORTEN 50	0.12% C (max) 0.35% Mn 0.65% Ni 0.8% Cr 0.5% Cu steel
CP 30	0.3% C 0.7% Ni 1.6% Cr 0.4% Mo steel: ESC
	UTS: 1000 Elon: 16% Proof: 800
Cr Mo 18	0.18% C 0.4% Ni 1.2% Cr 0.5% Mo 0.3% Cu steel: Casting; Wolsingham
	UTS: 460 Elon: 20% Proof: 270
Cr Mo 23	0.22% C 0.4% Ni 3% Cr 0.5% Mo 0.3% Cu steel: Casting; Wolsingham
	DPN: 190 UTS: 700 Elon: 18% Proof: 360
Cr Mo 33	0.32% C 0.4% Ni 1.2% Cr 0.5% Mo 0.3% Cu steel: Casting; Wolsingham
	UTS: 610
Cr Mo 329	0.22% C 3.2% Cr 0.5% Mo steel: Casting; Edgar Allen; Code 625
CRD HYKROM	0.35% C 5.0% Cr 1.5% Mo 0.4% V steel: Origin unknown
CRD HYKROM B	0.35% C 3.0% Cr 0.5% Mo 0.1% V steel: Origin unknown
CRD SPECIAL HOT DIE	0.3% C 5% Cr 1.5% Mo 1% V 1% Si steel: C Denton; hot shear blades; obsolete
	DPN: 460
CRD 9% W	0.3% C 3.5% Cr 0.1% V 9% W steel: C Denton; hot shear blades; obsolete
	DPN: 460
CREUSABRO 32	0.2% C 1.2% Mn 1.3% Cr 0.25% Mo 0.2% Cu steel: Creusot-Loire
CREUSABRO 32	0.2% C 1.2% Mn 1.3% Cr 0.25% Mo 0.2% Cu steel: For abrasion resistance; Creusot-Loire

Note. The following abbreviations and units are used in the tables:

DPN	Hardness, diamond pyramid number
UTS	Ultimate tensile strength, N/mm^2
Elon	Elongation, %
Proof	0.1% proof strength, N/mm^2

1 N/mm^2=0.1 hbar=0.102 kgf/mm^2=0.06475 $tonf/in.^2$=145.04 $lbf/in.^2$=1 MPa

See Appendix II for other abbreviations and conversion tables.

Symbol	Nominal analysis, supplier, condition and remarks.
CREUSELSO 32	0.2% C 1.2% Mn 1.3% Cr 1.2% Mo steel: Creusot-Loire
CREUSELSO 42	0.2% C 1.6% Mn 0.7% Cr + Ni + Mo + Cu steel: Creusot-Loire
CREUSELSO 47	0.18% C (max) 1.7% Mn (max) 0.5% Ni 0.07% V steel: Creusot-Loire
CRM 3	0.3% C 3% Cr 0.5% Mo steel: Firth Brown for BS alloy En 29B
CRO 42	0.2% C 0.6% Ni 0.5% Cr 0.2% Mo steel: For carburizing; Bofors; hardened and tempered; core properties
	DPN: 300 UTS: 1000 Elon: 12% Proof: 500
CRO 53	0.2% C 1.2% Ni 1.0% Cr 0.1% Mo steel: For carburizing; Bofors; hardened and tempered; core properties
	DPN: 360 UTS: 1200 Elon: 11% Proof: 600
CRO 684 BOVAC	0.3% C 0.6% Ni 1.7% Cr 0.4% Mo steel: Bofors; vacuum melted
CRO 861	0.35% C 1.4% Ni 1.4% Cr 0.2% Mo steel: Bofors; hardened and tempered
	DPN: 300 UTS: 1000 Elon: 14% Proof: 700
CRO 7146	0.32% C 0.8% Ni 2.7% Cr 0.6% Mo steel: Bofors
CRODI	0.35% C 5.0% Cr 1.4% Mo 0.3% V 1.2% W steel: Atlas as AISI type H12
CROLOY ½	0.15% C 0.7% Cr 0.5% Mo steel: Tube; Babcock and Wilcox (USA)
CROLOY 1	0.15% C 1.0% Cr 0.5% Mo steel: Tube; Babcock and Wilcox (USA)
CROLOY 1¼	0.15% C 1.25% Cr 0.5% Mo steel: Tube; Babcock and Wilcox (USA)
CROLOY 1¾	0.15% C 1.75% Cr 0.7% Mo steel: Tube; Babcock and Wilcox (USA)
CROLOY 2	0.15% C 2.0% Cr 0.5% Mo steel: Tube; Babcock and Wilcox (USA)
CROLOY 2¼	0.15% C 2.5% Cr 0.5% Mo steel: Tube; Babcock and Wilcox (USA)
CROLOY 2½	0.15% C 2.2% Cr 1.0% Mo steel: Tube; Babcock and Wilcox (USA)
CROLOY 3	0.15% C 3.0% Cr 1.0% Mo steel: Tube; Babcock and Wilcox (USA)
CROLOY 5	0.15% C 5.0% Cr 0.5% Mo steel: Tube; Babcock and Wilcox (USA)
CROLOY 5Si	0.15% C 5.0% Cr 0.5% Mo 1.5% Si steel: Tube; Babcock and Wilcox (USA)
CROLOY 5Ti	0.12% C 5.0% Cr 0.5% Mo 0.7% Ti steel: Tube; Babcock and Wilcox (USA)
CROLOY 7	0.15% C 7.0% Cr 0.5% Mo steel: Tube; Babcock and Wilcox (USA)
CROLOY 9M	0.15% C 9.0% Cr 1.0% Mo steel: Tube; Babcock and Wilcox (USA)
CROLOY 9½ Mo	0.15% C 9.0% Cr 0.5% Mo steel: Tube; Babcock and Wilcox (USA)
CROMODIE	0.35% C 5% Cr 1.5% Mo 1.0% V 1.0% Si steel: Firth Brown
CROMODIE W	0.31% C 5% Cr 1.7% Mo 0.2% V 1.1% W 0.9% Si steel: Firth Brown
CROVAN	0.35% C 5.0% Cr 1.4% Mo 0.9% V steel: Atlas; as AISI H13
CTHV	0.3% C 3.4% Cr 0.35% V 8.5% W steel: Marsh Brothers; hot forging dies, etc.
CTM	0.35% C 2.5% Ni 2.5% Cr 0.5% Mo 0.5% V 2.5% W steel: Swift Levick; hot dies
CTU	0.37% C 5% Cr 1.3% Mo 1.0% V 1.7% W 1.0% Si steel: For tools; Marsh Brothers; hot work dies
Cu Ni Mo	0.21% C (max) 0.95% Mn 1.4% Ni 0.25% Mo 0.65% Cu steel: US Steel Corp.
CUP-TEN	0.12% C (max) 0.6% Mn 0.6% Cr 0.22% Mo 0.35% Cu steel: Nippon-Kokan
CUP-TEN 6	0.12% C (max) 0.6% Mn 0.8% Cr 0.35% Mo 0.1% V 0.35% Cu steel: Nippon-Kokan

Symbol	Nominal analysis, supplier, condition and remarks.
CUP-TEN 50	0.18% C (max) 1.2% Mn 0.8% Cr 0.35% Mo 0.1% V 0.35% Cu steel: Nippon-Kokan
CUP-TEN 53	0.18% C (max) 1.2% Mn 0.8% Cr 0.35% Mo 0.1% V 0.35% Cu steel: Nippon-Kokan
CVM	0.36% C 5% Cr 1.3% Mo 0.3% V 1.0% Si steel: ESC; Extrusion, forging and die casting dies
	DPN: 500
CVM 2	0.36% C 5.0% Cr 1.3% Mo 0.3% V 1.4% W 2.0% Si steel: ESC; hot working dies
	DPN: 460
CVM 5	0.32% C 5.0% Cr 1.6% Mo 0.3% V 1.5% W steel: Origin unknown
D 66	0.3% C 3.0% Cr 0.3% V 9.0% W steel: Fagersta
	DPN: 550
D 230	0.23% C 1.3% Cr 0.8% Mo 0.3% V steel: VEW
D 320	0.1% C 2.4% Cr 1.0% Mo steel: VEW
D 330	0.14% C 1.0% Cr 0.6% Mo steel: VEW
DEF 13 B4	Weldable structural steel covers BS 968; BS alloy En 14A, etc.
	UTS: 480 Elon: 14% Proof: 290
DEF 13 B6C	Alloy case hardening steel covers BS alloy En 34; also includes En 351, etc.
	UTS: 700 Elon: 18%
DEF 13 B6D	Alloy case hardening steel covers BS alloy En 35; also includes En 36, En 325, etc.
	UTS: 840 Elon: 15%
DEF 13 B6E	Alloy case hardening steel covers BS alloy En 36; also includes En 353, etc.
	UTS: 1000 Elon: 12%
DEF 13 B6F	Alloy case hardening steel covers BS alloy En 354
	UTS: 1140 Elon: 12%
DEF 13 B6G	Alloy case hardening steel covers BS alloy En 39A, En 39B; also includes En 355, etc.
	UTS: 1300 Elon: 12%
DIE TOUGH 10 MS	0.1% C 0.5% Ni 0.2% Mo steel: Electrode; Metrode
DIE TOUGH 15 C	0.1% C 1.4% Ni 0.3% Cr 0.3% Mo steel: Electrode; Metrode
DIE TOUGH 25 C	0.1% C 1.6% Ni 1% Cr 0.6% Mo steel: Electrode; Metrode
DIE TOUGH 35 C	0.1% C 2.5% Ni 1.5% Cr 1% Mo steel: Electrode; Metrode
DIE TOUGH 40 C	0.15% C 2.5% Ni 2% Cr 1.5% Mo steel: Electrode; Metrode
DIN 8555 E7-250K	Ni Cr steel: Welding electrode
DIN 8556 E 18.8 Mn 6 R 26	Ni Cr steel: Welding electrode
DIN 8556 E 22.12.3 R 26.120	Ni Cr Mo steel: Welding electrode
DIN 8556 E 29.9 MPR 26.160	Ni Cr Mo: Welding electrode
DIN 17155 13 Cr Mo 4/4	0.14% C 0.6% Mn 0.8% Cr 0.45% Mo steel: Plate; German Standard; normalized and tempered
	UTS: 510 Proof: 300
DIN 17155 13 Cr Mo 4/4	0.14% C 0.8% Cr 0.45% Mo steel: For boiler plate; German Standard
DIN 17175 10 Cr Mo 9/10	0.15% C (max) 2.2% Cr 1.0% Mo steel: German Standard
DIN 17175 13 Cr Mo 4/4	0.14% C 0.8% Cr 0.45% Mo steel: For boiler plate; German Standard
DIN 17200 15 Cr Ni 6	0.15% C 0.5% Mn 1.6% Ni 1.6% Cr steel: German Standard
DIN 17200 17 Cr Mo 6	0.17% C 0.5% Mn 1.55% Ni 1.65% Cr 0.3% Mo steel: German Standard
DIN 17200 18 Cr Ni 8	0.18% C 0.5% Mn 1.95% Ni 1.95% Cr steel: German Standard
DIN 17200 20 Mo Cr 4	0.2% C 0.75% Mn 0.4% Cr 0.45% Mo steel: German Standard
DIN 17200 20 Mo Cr S4	0.2% C 0.75% Mn 0.41% Cr 0.45% Mo steel: German Standard

Symbol	Nominal analysis, supplier, condition and remarks.
DIN 17200 25 Cr Mo 4	0.25% C 0.65% Mn 1.05% Cr 0.22% Mo steel: German Standard
DIN 17200 25 Cr Mo 4	0.25% C 1.1% Cr 0.2% Mo steel: Forging; German Standard
DIN 17200 25 Mo Cr 4	0.25% C 0.75% Mn 0.5% Cr 0.45% Mo steel: German Standard
DIN 17200 25 Mo Cr S4	0.25% C 0.75% Mn 0.5% Cr 0.45% Mo steel: German Standard
DIN 17200 30 Cr Mo V9	0.3% C 0.6% Ni 2.5% Cr 0.2% Mo 0.15% V steel: Forging; German Standard
DIN 17200 30 Cr Ni Mo 8	0.3% C 2.0% Ni 2.0% Cr 0.3% Mo steel: Forging; German Standard
DIN 17200 30 Cr Ni Mo 8	0.3% C 0.45% Mn 2.0% Ni 2.0% Cr 0.45% Mo steel: German Standard
DIN 17200 31 Cr Mo 12	0.31% C 0.55% Mn 3.0% Cr 0.45% Mo 0.2% V steel: German Standard
DIN 17200 31 Cr Mo 12	0.32% C 0.55% Mn 3.0% Cr 0.45% Mo steel: German Standard
DIN 17200 34 Cr Al Mo 5	0.34% C 0.6% Mn 1.15% Cr 0.2% Mo 1.0% Al steel: German Standard
DIN 17200 34 Cr Al Ni 7	0.34% C 0.55% Mn 1.0% Ni 1.65% Cr 0.2% Mo 1.0% Al steel: German Standard
DIN 17200 34 Cr Mo 4	0.33% C 0.65% Mn 1.05% Cr 0.22% Mo steel: German Standard
DIN 17200 34 Cr Mo 4	0.34% C 0.4% Ni 1.1% Cr 0.2% Mo steel: Forging; German Standard
DIN 17200 34 Cr Mo S4	0.33% C 0.65% Mn 1.05% Cr 0.22% Mo steel: German Standard
DIN 17200 34 Cr Ni Mo 6	0.34% C 0.55% Mn 1.5% Ni 1.5% Cr 0.22% Mo steel: German Standard
DIN 17200 34 Cr Ni Mo 6	0.34% C 1.5% Ni 1.5% Cr 0.2% Mo steel: Forging; German Standard
DIN 17200 36 Cr Ni Mo 4	0.36% C 1.1% Ni 1.1% Cr 0.2% Mo steel: German Standard
DIN 17200 39 Cr Mo V	0.39% C 0.55% Mn 3.25% Cr 0.95% Mo 0.2% V steel: German Standard
DIN 17210 15 Cr Ni 6	0.15% C 1.5% Ni 1.5% Cr steel: For carburizing
DIN 17210 18 Cr Ni 8	0.17% C 2.0% Ni 2.0% Cr steel: For carburizing
DIN 17225 30 W Cr V 1/7/9	0.3% C 2.3% Cr 0.6% V 4.2% W steel: For springs
DIN 17240 21 Cr Mo V 5/11	0.21% C 0.6% Ni 1.3% Cr 1.1% Mo 0.3% V steel: For high temperature fasteners
DIN 17240 24 Cr Mo 5	0.24% C 0.6% Ni 1.1% Cr 0.25% Mo steel: For high temperature fasteners
DIN 17240 24 Cr Mo V 5/5	0.24% C 0.6% Ni 1.3% Cr 0.55% Mo 0.2% V steel: For high temperature fasteners
DNV	0.35% C 3.25% Cr 0.4% V 9.5% W steel: Simonds; AISI type H21
DOFASCOLOY P	0.16% C 0.6% Mn 0.9% Ni 0.6% Cr 0.6% Cu steel: Dominion
DOUBLE GRIFFIN	0.35% C 3.5% Ni 1.7% Cr steel: Balfour; chisels, wedges, etc. **DPN: 400**
DR 34	0.15% C 1.5% Ni 0.8% Cr steel: For case hardening; Bofors; annealed **DPN: 200**
DR 44	0.2% C 1.5% Ni 0.8% Cr steel: For case hardening; Bofors; annealed **DPN: 220**
DRO 86 BOVAC	0.35% C 1.5% Ni 1.5% Cr 0.2% Mo steel: Bofors; vacuum melted
DTD 713	0.3% C 2.5% Ni 0.6% Cr 0.5% Mo steel: Tube; hardened and tempered **UTS: 1140 Elon: 14%**
DTD 723	0.3% C 2.5% Ni 0.7% Cr 0.5% Mo steel: Tube; hardened and tempered **UTS: 1380 Elon: 12%**
DTD 730	0.35% C 3% Cr 1% Mo 0.2% V steel: For nitriding; hardened and tempered **DPN: 400 UTS: 1380 Elon: 12% Proof: 1070**
DTD 5002	0.15% C 0.8% Ni 0.6% Cr steel: For case hardening; blank carburized **UTS: 700 Elon: 18%**
DTD 5082	0.22% C 0.2% Mn 1% Cr steel: Bar and forging; hardened and tempered; weldable **DPN: 360 UTS: 1140 Elon: 10% Proof: 840**
DTD 5112	0.32% C 0.9% Cr 0.2% Mo steel: Sheet; hardened and tempered **UTS: 1210 Elon: 4% Proof: 1040**
DTD 5123	0.32% C 1% Cr 0.2% Mo steel: Forging and bar; hardened and tempered **UTS: 1140 Elon: 4% Proof: 920**
DTD 5132	0.32% C 1% Cr 0.2% Mo steel: Tube; hardened and tempered **UTS: 1210 Elon: 4%**
DTD 5142	0.32% C 1% Cr 0.2% Mo steel: Tube; hardened and tempered **UTS: 1210 Elon: 4%**
DTD 5152	0.32% C 1% Cr 0.2% Mo steel: Wire for welding filler; hardened and tempered; supplied annealed and copper coated **DPN: 380**
DTD 5219	1% Cr Mo steel: Casting; investment cast **UTS: 800**
DTD 5229	3% Cr Mo steel: Casting; investment cast **UTS: 710**
DTD 5239	3% Ni steel: Casting for case hardening; investment cast **UTS: 700**
DTD 5249	3% Cr Mo steel: Casting for nitriding; investment cast **UTS: 920**
DUCOL QT28	0.17% C (max) 0.5% Ni 0.3% Cr 0.3% Mo steel: Colvilles
DUCOL QT35	0.15% C (max) 1.2% Mn 1.2% Ni 1.0% Cr 0.5% Mo 0.12% V steel: Colvilles
DUCOL QT342	0.17% C (max) 1.5% Mn 0.3% Ni 0.7% Cr 0.25% Mo 0.1% V steel: Colvilles
DUCOL QT445A	0.17% C (max) 1.2% Mn 0.5% Cr 0.25% Mo 1.04% V steel: Colvilles
DUCOL W21	0.23% C (max) 1.6% Mn 0.25% Ni 0.5% Cr steel: Plate and section; Colvilles; Cu may be present up to 0.5% **UTS: 510 Elon: 23% Proof: 330**
DUCOL W23	0.2% C (max) 1.5% Mn 0.25% Ni 0.5% Cr steel: Plate and section; Colvilles alloy for BS 968; 0.5% Cu optional; 1.6% Mn + Cr (max) **UTS: 480 Elon: 23% Proof: 340**
DUCOL W23D	0.2% C (max) 1.5% Mn 0.25% Ni 0.5% Cr 0.5% Cu: With grain refining elements; Colvilles
DUCOL W24	0.22% C (max) 1.5% Mn 0.25% Ni 0.5% Cr steel: Plate and section; Colvilles alloy for BS 968; 0.5% Cu optional; 1.6% Cr + Mn (max) **UTS: 480 Elon: 22% Proof: 330**
DUCOL W25	0.2% C (max) 1.5% Mn 0.25% Ni 0.5% Cr steel: Plate; Colvilles; 0.5% Cu optional; 1.6% Mn + Cr (max) **UTS: 510 Elon: 23% Proof: 360**

Note. The following abbreviations and units are used in the tables:

DPN	Hardness, diamond pyramid number
UTS	Ultimate tensile strength, N/mm^2
Elon	Elongation, %
Proof	0.1% proof strength, N/mm^2

$1\ N/mm^2 = 0.1\ hbar = 0.102\ kgf/mm^2 = 0.06475\ tonf/in.^2 = 145.04\ lbf/in.^2 = 1\ MPa$

See Appendix II for other abbreviations and conversion tables.

Symbol	Nominal analysis, supplier, condition and remarks.
DUCOL W25D	0.2% C (max) 1.5% Mn 0.25% Ni 0.5% Cr: With grain refining elements; Colvilles
DUCOL W30	0.17% C (max) 1.5% Mn 0.7% Cr 0.28% Mo 0.1% V steel: Plate; Colvilles; weldable **UTS: 610 Elon: 17% Proof: 450**
DUNELT 10	0.32% C 3% Ni 0.3% Cr steel: Dunford for BS alloy En 21
DUNELT 15	0.12% C 3.2% Ni 0.3% Cr steel: For case hardening; Dunford for BS alloy En 38
DUNELT 16	0.12% C 3.2% Ni 1% Cr steel: For carburizing; Dunford for BS alloy En 36
DUNELT 17	0.12% C 5.0% Ni 0.3% Cr steel: For case hardening; Dunford for BS alloy En 37
DUNELT 18	0.13% C 5.0% Ni 0.3% Cr 0.2% Mo steel: For case hardening; Dunford for BS alloy En 38
DUNELT 20	0.18% C 1.7% Ni 0.25% Mo steel: For case hardening; Dunford for BS alloy En 34
DUNELT 55	0.3% C 3.2% Ni 0.7% Cr steel: Dunford for BS alloy En 23
DUNELT 60	0.34% C 3.3% Ni 0.7% Cr steel: Dunford; hardened and tempered **DPN: 300 UTS: 1040 Elon: 17%**
DUNELT 64A	0.32% C 2.5% Ni 0.7% Cr 0.5% Mo steel: Dunford; hardened and tempered **DPN: 350 UTS: 1000 Elon: 14%**
DUNELT 64A	0.4% C 2.5% Ni 0.7% Cr 0.5% Mo steel: Dunford; hardened and tempered **DPN: 375 UTS: 1070 Elon: 12%**
DUNELT 65	0.27% C 3.2% Ni 1.2% Cr 0.6% Mo 0.2% V steel: Dunford for BS alloy En 27
DUNELT 67	0.25% C 0.4% Ni (max) 3.2% Cr 0.5% Mo steel: For nitriding; Dunford for BS alloy En 40B
DUNELT 82M	0.16% C 4.2% Ni 1.2% Cr 0.25% Mo steel: For case hardening; Dunford for BS alloy En S82
DUNELT 128	0.3% C 4.2% Ni 1.3% Cr steel: Dunford for BS alloy En 30A
DUNELT 129	0.3% C 4.2% Ni 1.2% Cr 0.3% Mo steel: Dunford for BS alloy En 30B
DUREHETE 1050	0.2% C 1% Cr 1% Mo 0.7% V steel: For use at high temperature; Samuel Fox; capable of operating up to 500 °C **UTS: 860 Elon: 22% Proof: 770**
DUREHETE 1055	0.2% C 1.1% Cr 0.95% Mo 0.7% V 0.1% Ti steel: Samuel Fox; tempered **UTS: 840 Elon: 13% Proof: 720**
DUROTERM 8 R	0.15% C 2.7% Cr 0.7% V 4.3% W steel: Welding electrode; Soudometal; for cladding **DPN: 500**
E 4 Mo	0.12% C 5% Ni 0.2% Mo steel: Jessop for BS alloy En 38
E 6	0.17% C 1.7% Ni 0.25% Mo steel: Jessop for BS alloy En 34
E 9	0.24% C 1.7% Ni 0.25% Mo steel: Jessop for BS alloy En 35
E 11	0.19% C 1.8% Ni 0.25% Mo steel: Jessop
E 200	0.14% C 0.5% Mn 3.5% Ni 0.7% Cr steel: VEW
E 204	0.13% C 4.5% Ni 1.1% Cr steel: VEW
E 220	0.18% C 2.0% Ni 2.0% Cr steel: VEW
E 230	0.15% C 1.6% Ni 1.5% Cr steel: VEW
E 502	0.05% C (max) 5% Cr 0.5% Mo steel: Weld electrode; designation used by AWS
E 502 L	0.05% C (max) 5% Cr 0.5% Mo steel: Weld electrode; designation used by AWS
E 502 T	0.1% C (max) 5% Cr 0.5% Mo steel: Weld electrode; designation used by AWS
E 505	0.05% C (max) 9% Cr 1% Mo steel: Weld electrode; designation used by AWS
E 505 T	0.1% C (max) 9% Cr 1% Mo steel: Weld electrode; designation used by AWS
E 7018 W	0.12% C (max) 0.3% Ni 0.22% Cr 0.5% Si 0.45% Cu steel: Weld electrode; designation used by AWS

Symbol	Nominal analysis, supplier, condition and remarks.
E 8015 B2 L	0.05% C (max) 1.2% Cr 0.5% Mo steel: Weld electrode; designation used by AWS
E 8015 B2 L	0.05% C (max) 1.2% Cr 0.5% Mo steel: Weld electrode; designation used by AWS
E 8015 B4 L	0.05% C (max) 2% Cr 0.5% Mo steel: Weld electrode; designation used by AWS
E 8016 B1	0.1% C 0.5% Cr 0.5% Mo steel: Weld electrode; designation used by AWS
E 8016 B2	0.1% C 1.2% Cr 0.5% Mo steel: Weld electrode; designation used by AWS
E 8016 B5	0.1% C 0.5% Cr 1.2% Mo 0.45% Si steel: Weld electrode; designation used by AWS
E 8018 B1	0.1% C 0.5% Cr 0.5% Mo steel: Weld electrode; designation used by AWS
E 8018 B2	0.1% C 1.2% Cr 0.5% Mo steel: Weld electrode; designation used by AWS
E 8018 B2 L	0.05% C (max) 1.2% Cr 0.5% Mo steel: Weld electrode; designation used by AWS
E 8018 NM	0.1% C (max) 1% Ni 0.5% Mo steel: Weld electrode; designation used by AWS
E 8018 W	0.12% C (max) 0.6% Ni 0.52% Cr 0.6% Si 0.5% Cu steel: Weld electrode; designation used by AWS
E 9015 B3	0.1% C 2.2% Cr 1% Mo steel: Weld electrode; designation used by AWS
E 9015 B3L	0.05% C (max) 2.2% Cr 1.1% Mo steel: Weld electrode; designation used by AWS
E 9016 B3	0.1% C 2.2% Cr 1.1% Mo steel: Weld electrode; designation used by AWS
E 9018 B3	0.1% C 2.2% Cr 1.1% Mo steel: Weld electrode; designation used by AWS
E 9018 B3L	0.05% C (max) 2.2% Cr 1.1% Mo steel: Weld electrode; designation used by AWS
E 11018 M	0.1% C (max) 2% Ni 0.4% Mo steel: Weld electrode; designation used by AWS
E 12018 M	0.1% C (max) 2.2% Ni 1% Cr 0.4% Mo steel: Weld electrode; designation used by AWS
E 12018 M1	0.1% C (max) 3.4% Ni 0.25% Mo steel: Weld electrode; designation used by AWS
E Ni 4	0.15% C 1.8% Ni 0.2% Mo 0.2% Si steel: Weld electrode; designation used by AWS
E Ni 4N	0.15% C 1.8% Ni 0.2% Mo 0.2% Si steel: Weld electrode; designation used by AWS
E xxx T1 B3	0.05% C (max) 2.2% Cr 1.1% Mo steel: Weld electrode; designation used by AWS
E xxx T1 K3	0.15% C (max) 1.8% Ni 0.45% Mo steel: Weld electrode; designation used by AWS
E xxx T5 K3	0.15% C (max) 1.8% Ni 0.45% Mo steel: Weld electrode; designation used by AWS
E1 xx T5 K4	0.15% C (max) 2.2% Ni 0.4% Cr 0.4% Mo steel: Weld electrode; designation used by AWS
E6 x T5 B 8L	0.05% C (max) 9% Cr 1% Mo steel: Weld electrode; designation used by AWS
E7 Cr	0.05% C (max) 7% Cr 0.5% Mo steel: Weld electrode; designation used by AWS
E8 x T1 B1	0.1% C 0.5% Cr 0.5% Mo steel: Weld electrode; designation used by AWS
E8 x T1 B1 L	0.05% C (max) 0.5% Cr 0.5% Mo steel: Weld electrode; designation used by AWS
E8 x T1 B2	0.1% C 1.2% Cr 0.5% Mo steel: Weld electrode; designation used by AWS
E8 x T1 B2 H	0.12% C 1.2% Cr 0.5% Mo steel: Weld electrode; designation used by AWS
E8 x T1 B2 L	0.05% C (max) 1.2% Cr 0.5% Mo steel: Weld electrode; designation used by AWS
E8 x T1 W	0.12% C (max) 0.6% Ni 0.52% Cr 0.5% Cu steel: Weld electrode; designation used by AWS
E8 x T5 B2	0.1% C 1.2% Cr 0.5% Mo steel: Weld electrode; designation used by AWS
E8 x T5 B2 L	0.05% C (max) 1.2% Cr 0.5% Mo steel: Weld electrode; designation used by AWS

Symbol	Nominal analysis, supplier, condition and remarks.
E9 x T1 B3 H	0.12% C 2.2% Cr 1.1% Mo steel: Weld electrode; designation used by AWS
E9 x T1 B3 L	0.05% C (max) 2.2% Cr 1.1% Mo steel: Weld electrode; designation used by AWS
E9 x T5 B3	0.1% C 2.2% Cr 1.1% Mo steel: Weld electrode; designation used by AWS
E11 x T1 K4	0.15% C (max) 2.2% Ni 0.4% Cr 0.4% Mo steel: Weld electrode; designation used by AWS
E12 x T1 K5	0.17% C 1.2% Ni 0.5% Cr 0.35% Mo steel: Weld electrode; designation used by AWS
E80 B2 L	0.05% C (max) 1.2% Cr 0.5% Mo steel: Weld electrode; designation used by AWS
E80 C B2	0.1% C 1.2% Cr 0.5% Mo steel: Weld electrode; designation used by AWS
E90 C B3	0.1% C 2.2% Cr 1.1% Mo steel: Weld electrode; designation used by AWS
EB 1	0.1% C (max) 0.6% Cr 0.5% Mo steel: Weld electrode; designation used by AWS
EB 1N	0.1% C (max) 0.6% Cr 0.5% Mo steel: Weld electrode; designation used by AWS
EB 2H	0.3% C 1.25% Cr 0.52% Mo 0.25% V steel: Weld electrode; designation used by AWS
EB 2H N	0.3% C 1.25% Cr 0.52% Mo 0.25% V steel: Weld electrode; designation used by AWS
EB 2N	0.1% C 1.3% Cr 0.5% Mo steel: Weld electrode; designation used by AWS
EB 3	0.1% C 2.5% Cr 1% Mo steel: Weld electrode; designation used by AWS
EB 3N	0.1% C 2.5% Cr 1% Mo steel: Weld electrode; designation used by AWS
EB 5	0.2% C 0.55% Cr 1% Mo steel: Weld electrode; designation used by AWS
EB 5N	0.2% C 0.55% Cr 1% Mo steel: Weld electrode; designation used by AWS
EB 6H	0.35% C 5.5% Cr 0.45% Mo steel: Weld electrode; designation used by AWS
EB 6H N	0.35% C 5.5% Cr 0.45% Mo steel: Weld electrode; designation used by AWS
EB 8N	0.1% C (max) 9.2% Cr 1% Mo steel: Weld electrode; designation used by AWS
EB 82	0.12% C (max) 1% Ni 0.4% Mo steel: Weld electrode; designation used by AWS
EC 502	0.12% C (max) 5.2% Cr 0.5% Mo steel: Weld electrode; designation used by AWS
EC 505	0.1% C 9% Cr 1% Mo steel: Weld electrode
EC F2	0.17% C (max) 1.75% Mn 0.6% Ni 0.52% Mo steel: Weld electrode; designation used by AWS
EC F2 N	0.17% C (max) 1.75% Mn 0.6% Ni 0.5% Mo steel: Weld electrode; designation used by AWS
EC F3	0.17% C (max) 0.9% Ni 0.5% Mo steel: Weld electrode; designation used by AWS
EC F3 N	0.17% C (max) 0.9% Ni 0.52% Mo steel: Weld electrode; designation used by AWS
EC F5	0.17% C (max) 2.4% Ni 0.5% Mo steel: Weld electrode; designation used by AWS
EC F6	0.14% C (max) 1.8% Ni 0.3% Mo steel: Weld electrode; designation used by AWS
EC F6 N	0.14% C (max) 1.8% Ni 0.3% Mo steel: Weld electrode; designation used by AWS

Note. The following abbreviations and units are used in the tables:

DPN	Hardness, diamond pyramid number
UTS	Ultimate tensile strength, N/mm^2
Elon	Elongation, %
Proof	0.1% proof strength, N/mm^2

1 N/mm^2=0.1 hbar=0.102 kgf/mm^2=0.06475 $tonf/in.^2$=145.04 $lbf/in.^2$=1 MPa

See Appendix II for other abbreviations and conversion tables.

Symbol	Nominal analysis, supplier, condition and remarks.
EC M2	0.1% C (max) 1.7% Ni 0.45% Mo steel: Weld electrode; designation used by AWS
EC M2 N	0.1% C (max) 1.7% Ni 0.45% Mo steel: Weld electrode; designation used by AWS
EC M3	0.1% C (max) 2.2% Ni 0.5% Mo steel: Weld electrode; designation used by AWS
EC M3 N	0.1% C (max) 2.2% Ni 0.5% Mo steel: Weld electrode; designation used by AWS
EC M4	0.1% C (max) 2.4% Ni 0.5% Mo steel: Weld electrode; designation used by AWS
EC M4 N	0.1% C (max) 2.4% Ni 0.5% Mo steel: Weld electrode; designation used by AWS
EC XX T4	0.12% C (max) 1.7% Ni 0.5% Mo steel: Weld electrode; designation used by AWS
ECB 1	0.12% C (max) 0.5% Cr 0.5% Mo steel: Weld electrode; designation used by AWS
ECB 1N	0.12% C (max) 0.5% Cr 0.5% Mo steel: Weld electrode; designation used by AWS
ECB 2	0.15% C 1.2% Cr 0.5% Mo steel: Weld electrode; designation used by AWS
ECB 2H	0.2% C 1.2% Cr 0.5% Mo steel: Weld electrode; designation used by AWS
ECB 2H N	0.2% C 1.2% Cr 0.5% Mo steel: Weld electrode; designation used by AWS
ECB 2N	0.15% C 1.2% Cr 0.5% Mo steel: Weld electrode; designation used by AWS
ECB 3	0.15% C (max) 2.2% Cr 1.1% Mo steel: Weld electrode; designation used by AWS
ECB 3N	0.15% C (max) 2.2% Cr 1.1% Mo steel: Weld electrode; designation used by AWS
ECB 4	0.12% C (max) 2% Cr 0.5% Mo steel: Weld electrode; designation used by AWS
ECB 4N	0.12% C (max) 2% Cr 0.5% Mo steel: Weld electrode; designation used by AWS
ECB 5	0.18% C (max) 0.5% Cr 1% Mo steel: Weld electrode; designation used by AWS
ECB 5N	0.18% C (max) 0.5% Cr 1% Mo steel: Weld electrode; designation used by AWS
ECB 6H	0.1% C (max) 5.2% Cr 0.5% Mo steel: Weld electrode; designation used by AWS
ECB 6H N	0.1% C (max) 5.2% Cr 0.5% Mo steel: Weld electrode; designation used by AWS
ECB 6N	0.12% C (max) 5.2% Cr 0.5% Mo steel: Weld electrode; designation used by AWS
ECLIPSE LDC	0.35% C 5.0% Cr 1.5% Mo 0.4% V steel: Origin unknown
ECO/m	0.23% C 1.0% Cr 0.2% Mo steel: Krupp
EEEE	0.28% C 0.3% Ni 2.15% Cr 0.3% Mo steel: Crucible Steel Co. for drills
EF 1	0.12% C 1.3% Ni 0.4% Mo steel: Weld electrode; designation used by AWS
EF 1N	0.1% C 1.2% Ni 0.35% Mo steel: Weld electrode; designation used by AWS
EF 2	0.14% C 0.6% Ni 0.5% Mo steel: Weld electrode; designation used by AWS
EF 2N	0.14% C 0.6% Ni 0.5% Mo steel: Weld electrode; designation used by AWS
EF 3	0.14% C 0.8% Ni 0.5% Mo steel: Weld electrode; designation used by AWS
EF 3N	0.14% C 0.8% Ni 0.5% Mo steel: Weld electrode; designation used by AWS
EF 4	0.2% C 0.6% Ni 0.5% Cr 0.22% Mo steel: Weld electrode; designation used by AWS
EF 4N	0.2% C 0.6% Ni 0.5% Cr 0.22% Mo steel: Weld electrode; designation used by AWS
EF 5	0.13% C 2.5% Ni 0.35% Cr 0.55% Mo steel: Weld electrode; designation used by AWS
EF 6	0.11% C 2% Ni 0.32% Cr 0.5% Mo steel: Weld electrode; designation used by AWS
EF 6N	0.11% C 2% Ni 0.32% Cr 0.5% Mo steel: Weld electrode; designation used by AWS

Symbol	Nominal analysis, supplier, condition and remarks.
EHW 1	0.25% C 4.0% Cr 0.5% V 15% W steel: Latrobe; type H25
EM 3	0.1% C (max) 2.3% Ni 0.45% Mo steel: Weld electrode
EM 3N	0.1% C (max) 2.3% Ni 0.45% Mo steel: Weld electrode; designation used by AWS
EM 4	0.1% C (max) 2.4% Ni 0.42% Mo steel: Weld electrode
EM 4N	0.1% C (max) 2.4% Ni 0.42% Mo steel: Weld electrode; designation used by AWS
EM 70 S-5	0.13% C 0.62% Cr 0.2% Mo steel: Weld electrode; designation used by AWS
EM 8440	0.4% C 0.25% Ni 2% Cr 0.1% Mo 0.25% Cu steel: For valves; origin unknown
En 13	0.2% C 0.5% Ni 0.2% Mo steel: Designation used in BS 970
En 14A	0.2% C 0.4% Ni 0.25% Cr steel: Designation used in BS 970; replaced by 150 M19
En 20	0.35% C 1% Cr 0.6% Mo steel: Designation used in BS 970; obsolete
En 21	0.3% C 3% Ni 0.3% Cr steel: Designation used in BS 970; obsolete
En 23	0.3% C 3% Ni 0.75% Cr steel: Designation used in BS 970; replaced by 635 M31
En 25	0.3% C 2.5% Ni 0.7% Cr 0.5% Mo steel: Designation used in BS 970; replaced by 826 M31
En 27	0.3% C 3.5% Ni 1% Cr 0.5% Mo steel: Designation used in BS 970; replaced by 830 M31
En 28	0.3% C 3.5% Ni 1% Cr 0.4% Mo steel: Designation used in BS 970; obsolete
En 29	0.25% C 0.4% Ni 3% Cr 0.5% Mo steel: Designation used in BS 970
En 30A	0.3% C 4.25% Ni 1.2% Cr steel: Designation used in BS 970; see 835 M30
En 30B	0.3% C 4.25% Ni 1.2% Cr 0.3% Mo steel: Designation used in BS 970; replaced by 835 M30
En 33	0.12% C 3% Ni 0.3% Cr steel: For carburizing; designation used in BS 970; obsolete
En 34	0.17% C 1.75% Ni 0.25% Mo steel: For carburizing; designation used in BS 970; replaced by 665 M17
En 35	0.25% C 1.75% Ni 0.25% Mo steel: For carburizing; designation used in BS 970; replaced by 665 M23
En 35A	0.22% C 1.75% Ni 0.25% Mo steel: For carburizing; designation used in BS 970; replaced by 665 A22
En 35B	0.25% C 1.75% Ni 0.25% Mo steel: For carburizing; designation used in BS 970; replaced by 665 A24
En 36	0.18% C (max) 3.5% Ni 0.9% Cr steel: For carburizing; designation used in BS 970; replaced by 655 M13
En 37	0.16% C (max) 5% Ni 0.3% Cr steel: For carburizing; designation used in BS 970; obsolete
En 38	0.16% C (max) 5% Ni 0.2% Mo steel: For carburizing; designation used in BS 970; obsolete
En 39A	0.15% C 4.25% Ni 1.2% Cr steel: For carburizing; designation used in BS 970; replaced by 659 M15
En 39B	0.15% C 4.25% Ni 1.2% Cr 0.2% Mo steel: For carburizing; designation used in BS 970; replaced by 835 M15
En 40A	0.15% C 0.4% Ni 3% Cr 0.6% Mo steel: For nitriding; designation used in BS 970; obsolete
En 40B	0.25% C 0.4% Ni 3% Cr 0.6% Mo steel: For nitriding; designation used in BS 970; obsolete
En 41	0.25% C 0.4% Ni 1.5% Cr 0.2% Mo 1% Al steel: For nitriding; designation used in BS 970; obsolete
En 51	0.3% C 3% Ni 0.3% Cr steel: For valves; designation used in BS 970
En 100A	0.27% C 0.7% Ni 0.4% Cr 0.2% Mo steel: Designation used in BS 970; obsolete
En 100B	0.32% C 0.7% Ni 0.4% Cr 0.2% Mo steel: Designation used in BS 970; obsolete

Symbol	Nominal analysis, supplier, condition and remarks.
En 100C	0.37% C 0.7% Ni 0.4% Cr 0.2% Mo steel: Designation used in BS 970; replaced by 945 A40
En 111	0.35% C 1.25% Ni 0.6% Cr steel: Designation used in BS 970; replaced by 640 M40
En 111A	0.35% C 1.25% Ni 0.6% Cr steel: Designation used in BS 970; replaced by 640 A35
En 320	0.17% C 2% Ni 2% Cr 0.2% Mo steel: For carburizing; designation used in BS 970; obsolete
En 325	0.2% C 1.75% Ni 0.5% Cr 0.25% Mo steel: For carburizing; designation used in BS 970; obsolete
EOC/m	0.25% C 0.45% Cr 0.45% Mo steel: Krupp
EOC/m	0.2% C 0.4% Cr 0.45% Mo steel: Krupp
ER 1	0.32% C 3% Cr 2.8% Mo 0.5% V tool steel: Balfour; die casting brass, etc. **DPN: 550**
ER 70 S-2	0.07% C (max) 0.1% Ti 0.1% Al 0.1% Zr steel: Weld electrode; designation used by AWS
ER 80 S B2	0.1% C 1.35% Cr 0.5% Mo steel: Weld electrode; designation used by AWS
ER 80 S B2 L	0.05% C (max) 1.35% Cr 0.5% Mo steel: Weld electrode; designation used by AWS
ER 90 S B3	0.1% C 2.5% Cr 1% Mo steel: Weld electrode; designation used by AWS
ER 90 S B3 L	0.05% C (max) 2.5% Cr 1% Mo steel: Weld electrode; designation used by AWS
ER 100 S-1	0.08% C (max) 1.6% Ni 0.4% Mo steel: Weld electrode; designation used by AWS
ER 110 S-1	0.1% C (max) 2.3% Ni 0.42% Mo steel: Weld electrode; designation used by AWS
ER 120 S-1	0.1% C (max) 2.4% Ni 0.42% Mo steel: Weld electrode; designation used by AWS
ERA 164	0.2% C 1.6% Mn 0.25% Cr 0.4% Mo steel: Hadfields for BS alloy En 14A and B
ERA 165	0.2% C 1.6% Mn 0.6% Ni 0.25% Mo steel: Hadfields for BS alloy En 13
ESSHETE CML	0.12% C 1% Cr 0.5% Mo steel: Creep resisting; Samuel Fox; weldable **UTS: 510**
ESSHETE CRM 2	0.15% C (max) 2% Cr 1% Mo steel: Samuel Fox; normalized and tempered; high temperature use **UTS: 630 Elon: 30% Proof: 550**
ESSHETE CRM 5	0.12% C (max) 5% Cr 0.5% Mo steel: Samuel Fox; annealed; not easily welded **UTS: 450 Elon: 41%**
ESSHETE MV	0.15% C 0.3% Cr 0.5% Mo 0.3% V steel: Creep resisting; Samuel Fox; weldable **UTS: 610**
ETAD 1	0.3% C 3.8% Ni 1.35% Cr 0.5% V steel: Pompey; hardened and tempered **UTS: 1610 Elon: 6% Proof: 1300**
ETAD 3	0.35% C 3.85% Ni 1.85% Cr 0.5% Mo steel: Pompey; hardened and tempered **UTS: 1630 Elon: 6% Proof: 1300**
ETANO	0.35% C 3.75% Ni 1.7% Cr steel: Pompey; hardened and tempered **UTS: 1680 Elon: 6% Proof: 1300**
EW	0.12% C (max) 0.6% Ni 0.65% Cr 0.27% Si 0.5% Cu steel: Weld electrode; designation used by AWS
EWS	0.12% C (max) 0.6% Ni 0.65% Cr 0.27% Si 0.5% Cu steel: Weld electrode; designation used by AWS
EWT 2	0.12% C (max) 0.6% Ni 0.52% Cr 0.5% Cu steel: Weld electrode; designation used by AWS
EWT 3	0.12% C (max) 2% Ni 0.5% Mo 0.3% Si steel: Weld electrode; designation used by AWS
EWT 4	0.12% C (max) 0.6% Ni 0.52% Cr 0.5% Cu steel: Weld electrode; designation used by AWS
EX 22	0.15% C 0.9% Mn 0.5% Cr 0.25% Mo steel: For carburizing; SAE designation replacing 8615
EX 23	0.17% C 0.9% Mn 0.5% Cr 0.25% Mo steel: For carburizing; SAE designation replacing 8617

Symbol	Nominal analysis, supplier, condition and remarks.
EX 24	0.2% C 0.9% Mn 0.5% Cr 0.25% Mo steel: For carburizing; SAE designation replacing 8620
EX 25	0.22% C 0.9% Mn 0.5% Cr 0.25% Mo steel: For carburizing; SAE designation replacing 8622
EX 26	0.25% C 0.9% Mn 0.5% Cr 0.25% Mo steel: For carburizing; SAE designation for 8625
EX 27	0.27% C 0.9% Mn 0.5% Cr 0.25% Mo steel: For carburizing; SAE designation for 8627
EX 28	0.18% C 0.9% Mn 0.5% Ni 0.5% Cr 0.35% Mo steel: For carburizing; SAE designation replacing 4718
EX 29	0.2% C 0.9% Mn 0.5% Ni 0.5% Cr 0.35% Mo steel: For carburizing; SAE designation replacing 4320
EX 30	0.15% C 0.8% Mn 0.85% Ni 0.5% Cr 0.52% Mo steel: For carburizing; SAE designation replacing 4865
EX 31	0.17% C 0.8% Mn 0.85% Ni 0.5% Cr 0.52% Mo steel: For carburizing; SAE designation replacing 4817
EX 32	0.2% C 0.8% Mn 0.85% Ni 0.5% Cr 0.52% Mo steel: For carburizing; SAE designation replacing 4820
EX 55	0.17% C 0.9% Mn 1.85% Ni 0.5% Cr 0.72% Mo steel: For carburizing; SAE temporary designation
EX 56	0.1% C 0.9% Mn 1.8% Ni 0.5% Cr 0.72% Mo steel: For carburizing; SAE temporary designation
EXD 1	0.3% C 3.7% Ni 1.5% Cr 5.5% Mo 5.5% W steel: For tools; Balfour; hot extrusion dies for non-ferrous metal, etc. **DPN: 480**
F 9	0.2% C 2% Cr 0.25% V 2.4% W steel: Jessop
F 45CH	0.15% C 0.8% Ni 0.6% Cr steel: For carburizing; Firth Brown for BS 970 En 351
F 55CH	0.18% C 1.0% Ni 0.8% Cr steel: For carburizing; Firth Brown for BS 970 En 352
F 65CH	0.18% C 1.3% Ni 1.0% Mo steel: For carburizing; Firth Brown for BS 970 En 353
F 75CH	0.18% C 1.75% Ni 1.0% Cr 0.15% Mo steel: For carburizing; Firth Brown for BS 970 En 354
F 85CH	0.18% C 2.0% Ni 1.6% Cr 0.2% Mo steel: For carburizing; Firth Brown for BS 970 En 355
F 543	0.09% C 4.8% Ni 3.9% Cr 3.0% Mo steel: Firth Brown
FAGERSTA K353	0.15% C 0.8% Mn 1.5% Ni 0.8% Cr steel: For case hardening; Fagersta
FAGERSTA N104	0.1% C 0.4% Mn 2.1% Ni steel: For case hardening; Fagersta
FB 60	0.18% C (max) 1.4% Mn 0.7% Ni 0.18% V 0.015% N steel: German proprietor's standard; Al killed
FB 60AK	0.18% C (max) 1.4% Mn 0.7% Ni 0.18% V 0.015% N steel: German proprietor's specification; Al killed
FB 70	0.2% C (max) 1.5% Mn 0.7% Ni 0.18% V 0.015% N steel: Mannesmann; Al killed
FB 70AK	0.2% C (max) 1.5% Mn 0.7% Ni 0.18% V 0.015% N steel: Mannesmann; Al killed
FB 80	0.22% C (max) 1.6% Mn 0.7% Ni 0.18% V 0.02% N steel: Mannesmann; Al killed
FB 80AK	0.22% C (max) 1.6% Mn 0.7% Ni 0.22% V 0.02% N steel: Mannesmann; Al killed
FIRTH OB	0.09% C 1.25% Ni 0.5% Cr steel: Firth Brown for plastic moulds
FKD 3	0.35% C 1.0% Cr 0.25% Mo steel: Pompey **UTS: 1140 Elon; 10% Proof: 950**
fmp 505	0.32% C 3.2% Cr 0.5% Mo 9.2% W steel: F Parkin; as tool steel H 21
fmp 507	0.3% C 2.5% Ni 3.0% Cr 0.5% Mo 0.3% V 9.0% W steel: F Parkin
fmp 513	0.35% C 5.25% Cr 1.5% Mo 0.35% V 1.25% W steel: F Parkin; as tool steel H 12
FNCR	0.35% C 1.25% Ni 0.6% Cr steel: Firth Brown for BS alloy En 111
FOREMOST A	0.3% C 3.0% Cr 0.4% V 9.0% W steel: Swift Levick; hot dies
FORMITE 2	0.33% C 3.3% Cr 0.5% V 9.2% W steel: Columbia; AISI type H21
FORTREX 45	0.06% C 1.5% Ni 0.4% Mo steel: Welding electrode; Murex **UTS: 680 Elon: 27%**
FORTREX 55	0.07% C 1.1% Mn 2% Ni 0.6% Cr 0.4% Mo steel: Welding electrode; Murex **UTS: 820 Elon: 24% Proof: 670**
G 8	0.27% C 3% Ni 1.2% Cr 0.4% Mo steel: Jessop
G 11	0.3% C 2.5% Ni 0.6% Cr 0.5% Mo steel: Jessop for BS alloy En 25
G 15	0.12% C 3.2% Ni 0.9% Cr steel: Jessop for BS alloy En 36A
G 16	0.15% C 4.2% Ni 1.2% Cr 0.25% Mo steel: Jessop for BS alloy En 39B
G 27	0.37% C 0.4% Ni 3% Cr 1% Mo 0.2% V steel: Jessop for BS alloy En 40C
G 27	0.30% C 3.3% Ni 1.0% Cr 0.35% Mo steel: Jessop for BS alloy En 27 and En 28
G 28	0.12% C 1.3% Ni 0.5% Cr steel: Jessop
G 36	0.2% C 1.7% Ni 0.3% Cr 0.23% Mo steel: Jessop
G 45	0.35% C 1% Ni 0.6% Cr steel: Jessop for BS alloy En 111
G 46	0.15% C 0.8% Ni 0.6% Cr steel: Jessop for BS alloy En 351
G 47	0.2% C 1% Ni 0.8% Cr steel: Jessop for BS alloy En 352
G 48	0.2% C 1.25% Ni 1% Cr 0.11% Mo steel: Jessop for BS alloy En 353
G 49	0.2% C 1.7% Ni 1% Cr 0.15% Mo steel: Jessop for BS alloy En 354
G 50	0.2% C 1.5% Cr 2% Mo steel: Jessop for BS alloy En 355
G 51	0.15% C 0.6% Ni 0.7% Cr 0.11% Mo steel: Jessop for BS alloy En 361
G 52	0.2% C 0.6% Ni 0.7% Cr 0.11% Mo steel: Jessop for BS alloy En 362
G 53	0.24% C 0.6% Ni 0.7% Cr 0.11% Mo steel: Jessop for BS alloy En 363
G 86	0.10% C 6.0% Cr 2.0% Mo 11.0% W 1.5% Nb 8.0% Al: Jessop; scale or creep resistant
G 110	0.3% C 4.1% Ni 1.2% Cr 0.3% Mo steel: Toledo for BS alloy En 30A (no Mo) and En 30B
GEARITE	0.31% C 4.2% Ni 1.2% Cr steel: Swift Levick for BS alloy En 30
GEARMAC	0.31% C 2.5% Ni 0.7% Cr 0.6% Mo steel: Swift Levick for BS alloy En 25
GI SPECIAL	0.3% C 4% Ni 1.3% Cr 0.3% Mo steel: Jessop
GK 3	0.35% C 2.0% Cr 0.25% Mo 0.15% V steel: For carburizing; Firth Brown
GK 5	0.25% C 2.0% Cr 0.25% Mo 0.15% V steel: For carburizing; Firth Brown
GK 7	0.18% C 2.0% Cr 0.25% Mo 0.15% V steel: For carburizing; Firth Brown
GO	0.3% C 3% Ni 0.7% Cr steel: Jessop for BS alloy En 23
GOST 550/12 KHM	0.16% C (max) 0.55% Mn 0.95% Cr 0.47% Mo steel: Pipe; Russian Standard **UTS: 420 Proof: 250**
GOST 550 KH5M	0.15% C (max) 0.5% Mn (max) 5.0% Cr 0.55% Mo steel: Pipe; Russian Standard **UTS: 400 Proof: 220**

Note. The following abbreviations and units are used in the tables:

DPN	Hardness, diamond pyramid number
UTS	Ultimate tensile strength, N/mm^2
Elon	Elongation, %
Proof	0.1% proof strength, N/mm^2

1 N/mm^2=0.1 hbar=0.102 kgf/mm^2=0.06475 tonf/in.^{2}=145.04 ibf/in.^{2}=1 MPa

See Appendix II for other abbreviations and conversion tables.

Symbol	Nominal analysis, supplier, condition and remarks.
GOST 550 KH5YF	0.15% C (max) 0.5% Mn (max) 5.0% Cr 0.5% V 0.55% W steel: Pipe; Russian Standard **UTS: 400** **Proof: 220**
GOST 4543/12 KH2NGA	0.12% C 0.45% Mn 3.4% Ni 1.4% Cr steel: Russian Standard
GOST 4543/12 KHN2	0.12% C 0.45% Mn 1.7% Ni 0.75% Cr steel: Russian Standard
GOST 4543/12 KHN3	0.12% C 0.45% Mn 3.0% Ni 0.75% Cr steel: Russian Standard
GOST 4543/13 N2KHA	0.12% C 0.45% Mn 1.9% Ni 0.35% Cr steel: Russian Standard
GOST 4543/15 KHF	0.15% C 0.5% Mn 0.95% Cr 0.15% V steel: Russian Standard
GOST 4543/15 KHM	0.15% C 0.55% Mn 0.95% Cr 0.47% Mo steel: Russian Standard
GOST 4543/15 NM	0.15% C 0.55% Mn 1.7% Ni 0.3% Cr (max) 0.25% Mo steel: Russian Standard
GOST 4543/15 KH2GN2T	0.15% C 0.85% Mn 1.6% Ni 1.6% Cr 0.1% W steel: Russian Standard
GOST 4543/15 KH2GN2TA	0.15% C 0.85% Mn 1.6% Ni 1.6% Cr steel: Russian Standard
GOST 4543/15 KH2GN2TRA	0.15% C 0.85% Mn 1.6% Ni 1.6% Cr 0.003% V 0.1% W steel: Russian Standard
GOST 4543/15 KHGNT	0.15% C 0.85% Mn 1.6% Ni 0.85% Cr 0.1% W steel: Russian Standard
GOST 4543/15 KHGNTA	0.15% C 0.85% Mn 1.6% Ni 0.85% Cr 0.1% W steel: Russian Standard
GOST 4543/16 KHSN	0.16% C 0.45% Mn 0.75% Ni 0.95% Cr steel: Russian Standard
GOST 4543/18 KH2N4VA	0.17% C 0.4% Mn 4.2% Ni 1.5% Cr 1.0% W steel: Russian Standard
GOST 4543/18 KHGN	0.18% C 0.95% Mn 0.55% Ni 0.55% Cr steel: Russian Standard
GOST 4543/18 SNRA	0.18% C 0.75% Mn 0.95% Ni 0.95% Cr 0.0035% V steel: Russian Standard
GOST 4543/20 KHF	0.2% C 0.65% Mn 0.95% Cr 0.15% V steel: Russian Standard
GOST 4543/20 KHN	0.2% C 0.6% Mn 1.2% Ni 0.6% Cr steel: Russian Standard
GOST 4543/20 NM	0.21% C 0.55% Mn 1.7% Ni 0.3% Cr (max) 0.25% Mo steel: Russian Standard
GOST 4543/20 KH2NGA	0.19% C 0.45% Mn 3.4% Ni 1.4% Cr steel: Russian Standard
GOST 4543/20 KHGNR	0.2% C 0.85% Mn 0.95% Ni 0.85% Cr 0.0035% V steel: Russian Standard
GOST 4543/20 KHN3A	0.2% C 0.45% Mn 3.0% Ni 0.75% Cr steel: Russian Standard
GOST 4543/20 KHN4FA	0.2% C 0.4% Mn 4.0% Ni 0.95% Cr 0.22% B steel: Russian Standard
GOST 4543/20 KHNR	0.2% C 0.75% Mn 0.95% Ni 0.9% Cr 0.0025% B steel: Russian Standard
GOST 4543/25 KH2GNTA	0.25% C 0.95% Mn 1.1% Ni 1.5% Cr 0.1% W steel: Russian Standard
GOST 4543/25 KH2N4VA	0.25% C 0.4% Mn 4.2% Ni 1.5% Cr 1.0% W steel: Russian Standard
GOST 4543/30 KHM	0.3% C 0.5% Mn 0.95% Cr 0.2% Mo steel: Russian Standard
GOST 4543/30 KH2GN2	0.3% C 0.95% Mn 1.6% Ni 1.55% Cr steel: Russian Standard
GOST 4543/30 KH2NVA	0.30% C 0.45% Mn 1.6% Ni 1.8% Cr 1.4% W steel: Russian Standard
GOST 4543/30 KH2NVFA	0.3% C 0.45% Mn 1.6% Ni 1.8% Cr 1.4% W 0.21% B steel: Russian Standard
GOST 4543/30 KHGNA	0.3% C 0.75% Mn 0.45% Ni 1.1% Cr steel: Russian Standard
GOST 4543/30 KHGSNA	0.3% C 1.15% Mn 1.2% Ni 1.05% Cr steel: Russian Standard
GOST 4543/30 KHMA	0.3% C 0.5% Mn 0.95% Cr 0.2% Mo steel: Russian Standard
GOST 4543/30 KHN2VFA	0.3% C 0.45% Mn 2.2% Ni 0.75% Cr 0.65% W 0.2% B steel: Russian Standard

Symbol	Nominal analysis, supplier, condition and remarks.
GOST 4543/30 KHN3A	0.3% C 0.45% Mn 3.0% Ni 0.75% Cr steel: Russian Standard
GOST 4543/30 KHNVA	0.3% C 0.45% Mn 1.45% Ni 0.75% Cr 0.65% W steel: Russian Standard
GOST 4543/35 KHM	0.36% C 0.55% Mn 0.95% Cr 0.20% Mo steel: Russian Standard
GOST 4543/38 KHNVA	0.38% C 0.45% Mn 1.45% Ni 1.5% Cr 0.65% W steel: Russian Standard
GOST 4543/38 KHVA	0.38% C 0.4% Mn 1.1% Cr 0.65% W steel: Russian Standard
GOST 4543/45 KHN	0.45% C 0.65% Mn 1.2% Ni 0.6% Cr steel: Russian Standard
GOST 5058/10 KHSND	0.12% C (max) 0.65% Mn 0.75% Ni 0.8% Cr 0.5% Cu steel: Plate; Russian Standard; hardened and tempered **UTS: 540** **Proof: 400**
GOST 5058/15 KHSND	0.15% C 0.6% Mn 0.75% Ni 0.75% Cr 0.3% Cu steel: Plate; Russian Standard; hardened and tempered **UTS: 600** **Proof: 500**
GOST 5632/1 KH11 MF	0.15% C 10.7% Cr 0.7% Mo 0.3% V steel: Plate; Russian Standard
GOST 5632/1 KH8VF	0.12% C 7.7% Cr 0.4% V 0.8% W steel: Russian Standard
GOST 5632 KH5M	0.15% C (max) 5.2% Cr 0.52% Mo steel: Russian Standard
GOST 5632 KH5VF	0.15% C (max) 5.2% Cr 0.5% V 0.5% W steel: Russian Standard
GOST 5632 KH6SIU	0.15% C (max) 6.2% Cr 0.9% Al steel: Plate; Russian Standard
GOST 5632 KH6SM	0.15% C (max) 0.7% Mn (max) 5.75% Cr 0.55% Mo steel: Plate; Russian Standard; annealed **UTS: 580** **Proof: 310**
GOST 5632 KH17 G9AN4	0.14% C (max) 4.0% Ni 1.7% Cr 0.2% N steel: Russian Standard
GOST 5632 KHCSM	0.15% C (max) 6.2% Cr 0.52% Mo steel: Russian Standard
GOST 8731/12 KHN2	0.13% C 0.45% Mn 1.7% Ni 0.75% Cr steel: Pipe; Russian Standard; seamless **UTS: 500** **Proof: 400**
GOST 8731/15 KHM	0.14% C 0.5% Mn 0.9% Cr 0.5% Mo steel: Pipe; Russian Standard; seamless **UTS: 440** **Proof: 230**
GOST 8731/30 KHMA	0.3% C 0.5% Mn 0.9% Cr 0.2% Mo steel: Pipe; Russian Standard; seamless **UTS: 600** **Proof: 400**
GOST 8733/15 KH	0.14% C 0.55% Mn 0.95% Cr 0.5% Mo steel: Pipe; Russian Standard; seamless **UTS: 440** **Proof: 230**
GOST 10802/12 K2 MFB	0.1% C 0.55% Mn 2.3% Cr 0.6% Mo 0.27% V 0.25% Cu (max) steel: Pipe; Russian Standard; seamless **UTS: 420** **Proof: 210**
GOST 10802/12 KH1MF	0.11% C 0.55% Mn 0.95% Cr 0.3% Mo 0.22% V steel: Pipe; Russian Standard; seamless **UTS: 500** **Proof: 260**
GOST 10802/12 KH2 MFSR	0.11% C 0.55% Mn 1.75% Cr 0.61% Mo 0.27% V 0.65% Nb 0.25% Cu (max) steel: Pipe; Russian Standard; seamless **UTS: 480** **Proof: 260**
GOST 10802/15 KH1ME	0.13% C 0.55% Mn 1.25% Cr 1.0% Mo 0.27% V 0.2% Cu (max) steel: Pipe; Russian Standard; seamless **UTS: 620** **Proof: 350**
GOST 10802/15 KHM	0.14% C 0.55% Mn 0.95% Cr 0.47% Mo steel: Pipe; Russian Standard; seamless **UTS: 450** **Proof: 240**
GOST 12132/30 KHMA	0.3% C 0.5% Mn 0.25% Ni (max) 0.95% Cr 0.2% Mo steel: Pipe; Russian Standard; welded **UTS: 600**
GPS 1	0.15% C 4% Ni 1.5% Cr 0.25% Mo steel: For case hardening; Jonas; core properties **DPN: 415** **UTS: 1300** **Elon: 12%**

Symbol	Nominal analysis, supplier, condition and remarks.
GPS 2	0.15% C 3% Ni 1% Cr steel: For case hardening; Jonas; core properties **DPN: 341 UTS: 1070 Elon: 13%**
GPS 3	0.15% C 2.75% Ni 0.6% Cr steel: For case hardening; Jonas; core properties **DPN: 269 UTS: 840 Elon: 18%**
GS 10 Mn Mo7/4	0.08% C 1.7% Mn 0.4% Mo 0.06% V 0.04% Nb steel: Cast; Thyssen; precipitation hardened **DPN: 155 UTS: 530 Elon: 13% Proof: 405**
H 5DM	0.15% C 0.5% Mn 4.0% Ni 0.5% Cr 0.4% Mo 0.1% V steel: Creusot-Loire
H 11	0.35% C 5% Cr 1.5% Mo 0.4% V steel: For tools; designation used by AISI
H 12	0.35% C 5% Cr 1.5% Mo 0.3% V 1.5% W steel: For tools; designation used by AISI
H 12	0.35% C 5.0% Cr 1.6% Mo 0.25% V 1.4% W steel: For hot work; Osborn
H 13	0.35% C 5% Cr 1.5% Mo 1% V steel: For tools; designation used by AISI
H 14	0.28% C 1% Cr 0.2% Mo steel: Jessop
H 19	0.2% C 5% Cr 0.55% Mo steel: Jessop
H 20	0.35% C 2.0% Cr 9.0% W steel: For tools; designation used by AISI
H 21	0.35% C 3.5% Cr 0.4% V 9% W steel: For tools; designation used by AISI
H 21	0.3% C 3.0% Cr 0.3% V 8.8% W steel: For hot work; Osborn
H 21N	0.25% C 2.3% Ni 3.0% Cr 0.3% V 8.8% W steel: For hot work; Osborn
H 22	0.35% C 2.0% Cr 11.0% W steel: For tools; designation used by AISI
H 22	0.2% C 1.5% Cr 0.5% Mo steel: Jessop
H 25	0.25% C 4.0% Cr 15.0% W steel: For tools; designation used by AISI
H 32	0.24% C 0.3% Ni 3.2% Cr 0.5% Mo steel: Jessop for BS alloy En 40B
H 40	0.20% C 0.3% Ni 2.5% Cr 0.5% Mo 0.75% V steel: Jessop; creep resisting **DPN: 300 UTS: 990 Elon: 20% Proof: 840**
H 50	0.37% C 5% Cr 1.3% Mo 1.1% V 1.1% Si steel: Jessop **DPN: 450**
H 51	0.2% C 0.6% Mo 0.25% V steel: Jessop
H 57	0.15% C 2.2% Cr 1% Mo steel: Jessop
H 65	0.12% C 1.02% Cr 0.55% Mo steel: Jessop
H 600/2	0.3% C 4.5% Cr 1.0% Mo 0.3% V 1.3% Si steel: Electrode; Metrode; hardenable deposit **DPN: 600**
HA	0.2% C 9% Cr 1% Mo steel: Casting; Huntingdon; aged **UTS: 770 Elon: 21% Proof: 570**
HARDTRADE 85/58	0.35% C 10% Mn 1.5% Cr 0.8% V 8.0% W 2.0% Co steel: For electrodes; Esab **DPN: 480**
HCM 5	0.3% C 3.0% Cr 0.4% Mo steel: For nitriding; Firth Brown for BS 970 En 40B
HCM 7	0.2% C 3.0% Cr 0.4% Mo steel: For nitriding; Firth Brown for BS 970 En 40A
HCRS	0.25% C 0.75% Cr 0.6% Mo steel: ESC for BS alloy En 20A

Note. The following abbreviations and units are used in the tables:

DPN	Hardness, diamond pyramid number
UTS	Ultimate tensile strength, N/mm^2
Elon	Elongation, %
Proof	0.1% proof strength, N/mm^2

1 N/mm^2=0.1 hbar=0.102 kgf/mm^2=0.06475 tonf/in.^{2}=145.04 lbf/in.^{2}=1 MPa

See Appendix II for other abbreviations and conversion tables.

Symbol	Nominal analysis, supplier, condition and remarks.
HCRS 3	0.13% C 0.9% Cr 0.5% Mo steel: ESC; normalized and tempered BS alloy 620 **UTS: 440 Elon: 19% Proof: 220**
HCRS 4	0.13% C 2.25% Cr 1% Mo steel: ESC; good creep resistance
HD 3	0.32% C 3.5% Cr 0.3% V 9% W steel: T Turton
HD 3M	0.3% C 3.0% Cr 2.8% Mo 0.5% V steel: Origin unknown
HD 10	0.35% C 3.0% Cr 0.5% Mo 4.2% W steel: Spencer; forging dies, casting dies, etc. **DPN: 400**
HD 12	0.35% C 2% Ni 2.3% Cr 0.25% V 10% W steel: Turton Bros
HDS	0.28% C 3.2% Cr 0.3% V 9.5% W steel: Jessop; die for casting Cu **DPN: 400**
HDZ	0.3% C 3.4% Cr 0.35% V 8.4% W steel: Origin unknown
HECLA 66	0.12% C 3.2% Ni 0.3% Cr steel: Case hardening; Hadfields for BS alloy En 33
HECLA 67	0.3% C 4.1% Ni 1.2% Cr steel: Hadfields for BS alloy En 30A
HECLA 67B	0.3% C 4.1% Ni 1.2% Cr 0.3% Mo steel: Hadfields for BS alloy En 30B
HECLA 73	0.3% C 3% Ni 0.3% Cr steel: Hadfields for BS alloy En 21
HECLA 78	0.16% C 5% Ni 0.3% Cr case hardening steel: Hadfields for BS alloy En 37
HECLA 98	0.35% C 1.2% Ni 0.6% Cr steel: Hadfields for BS alloy En 111 and En 111A
HECLA 116	0.3% C 3.2% Ni 0.7% Cr 0.6% Mo steel: Hadfields for BS alloy En 23 and En 27
HECLA 138	0.3% C 2.5% Ni 0.6% Cr 0.5% Mo steel: Hadfields for BS alloy En 25
HECLA 142	0.3% C 3.5% Ni 1.2% Cr 0.45% Mo steel: Hadfields for BS alloy En 28
HECLA 143	0.15% C 3.5% Ni 0.8% Cr steel: Case hardening; Hadfields for BS alloy En 36A
HECLA 143B	0.15% C 3.5% Ni 0.8% Cr 0.2% Mo steel: Case hardening; Hadfields for BS alloy En 36C
HECLA 146	0.16% C 4% Ni 1.2% Cr steel: Case hardening; Hadfields for BS alloy En 39A
HECLA 146B	0.16% C 4% Ni 1.2% Cr 0.25% Mo steel: Case hardening; Hadfields
HECLA 149C	0.35% C 3.5% Cr 0.4% V 9.0% W steel: For tools; origin unknown
HECLA 151	0.16% C 1.75% Ni 0.25% Mo steel: Case hardening; Hadfields for BS alloy En 34
HECLA 151B	0.24% C 1.75% Ni 0.25% Mo steel: Case hardening; Hadfields for BS alloy En 35
HECLA 153	0.25% C 0.7% Cr 0.7% Mo steel: Hadfields for BS alloy En 20A
HECLA 163	0.16% C 5% Ni 0.3% Cr 0.25% Mo steel: Case hardening; Hadfields for BS alloy En 38
HECLA 174	0.35% C 5.0% Mo 0.4% V steel: Origin unknown
HECLA 177	0.35% C 5.0% Cr 1.5% Mo 0.3% V steel: Origin unknown
HECLA 181	0.2% C 0.8% Ni 0.6% Cr 0.1% Mo steel: Case hardening; Hadfields for BS alloy En 351
HECLA 182	0.2% C 1% Ni 0.8% Cr 0.1% Mo steel: Case hardening; Hadfields for BS alloy En 352
HECLA 183	0.2% C 1.25% Ni 1% Cr 0.12% Mo steel: Case hardening; Hadfields for BS alloy En 353
HECLA 184	0.2% C 1.75% Ni 1% Cr 0.15% Mo steel: Case hardening; Hadfields for BS alloy En 354
HECLA 185	0.2% C 2% Ni 1.5% Cr 0.2% Mo steel: Case hardening; Hadfields for BS alloy En 355
HECLA 191	0.15% C 0.6% Ni 0.7% Cr 0.1% Mo steel: Case hardening; Hadfields
HECLA 192	0.2% C 0.6% Ni 0.7% Cr 0.1% Mo steel: Case hardening; Hadfields for BS alloy En 362

Symbol	Nominal analysis, supplier, condition and remarks.
HECLA 193	0.24% C 0.6% Ni 0.7% Cr 0.1% Mo steel: Case hardening; Hadfields for BS alloy En 363
HECLA 196	0.3% C 0.4% Ni 1.6% Cr 0.2% Mo 1.1% Al steel: Nitriding; Hadfields for BS alloy En 41 A and En 41 B
HECLA 197	0.22% C 1.7% Ni 0.5% Cr 0.25% Mo steel: Case hardening; Hadfields for BS alloy En 325
HECLA 317	0.15% C 0.4% Ni 3.2% Cr 0.5% Mo steel: Nitriding; Hadfields for BS alloy En 40 A, En 40 B, En 29 A and En 29 B
HEDEX 7	0.3% C 1.2% Ni 0.55% Cr steel: Wire for forging; Kiveton Park for BS alloy En 111
HEDEX 16	0.12% C 3% Ni 0.3% Cr (max) steel: Wire for forging; Kiveton Park for BS alloy En 33
HEDEX 17	0.2% C 0.8% Ni 0.6% Cr 0.1% Mo steel: Wire for forging; Kiveton Park for BS alloy En 351
HEDEX 19	0.2% C 0.55% Ni 0.5% Cr 9.25% Mo steel: Wire for forging; Kiveton Park; annealed **UTS: 580**
HEDEX 20	0.3% C 4.2% Ni 1.2% Cr steel: Wire for forging; Kiveton Park for BS alloy En 30A
HEDEX 21	0.35% C 0.6% Ni 0.5% Cr 0.25% Mo steel: Wire for forging; Kiveton Park; annealed **UTS: 610**
HEDEX 23	0.27% C 0.5% Ni 0.5% Cr 0.25% Mo steel: Wire for forging; Kiveton Park; annealed **UTS: 610**
HEDEX 27	0.2% C 0.55% Ni 0.5% Cr 0.2% Mo steel: Wire for forging; Kiveton Park; annealed **UTS: 580**
HEDEX 28	0.17% C 0.55% Ni 0.5% Cr 0.2% Mo steel: Wire for forging; Kiveton Park; annealed **UTS: 540**
HEDEX 30	0.17% C 1.7% Ni 0.3% Cr (max) 0.25% Mo steel: Wire for forging; Kiveton Park for BS alloy En 34
HEDEX 31	0.12% C 3.5% Ni 0.8% Cr steel: Wire for forging; Kiveton Park for BS alloy En 36B
HEDEX 32	0.3% C 0.9% Cr 0.2% Mo steel: Wire for forging; Kiveton Park; annealed **UTS: 580**
HEDEX 34	0.31% C 2.5% Ni 0.7% Cr 0.6% Mo steel: Wire for forging; Kiveton Park for BS alloy En 25
HEDEX 38	0.16% C 4.2% Ni 0.2% Cr 0.2% Mo steel: Wire for forging; Kiveton Park for BS alloy En 39B
HEDEX 39	0.18% C (max) 3.5% Ni 0.8% Cr steel: Wire for forging; Kiveton Park for BS alloy En 36
HEDEX 40	0.27% C 0.55% Ni 0.5% Cr 0.2% Mo steel: Wire for forging; Kiveton Park; annealed **UTS: 580**
HEDEX 43	0.15% C 0.55% Ni 0.5% Cr 0.2% Mo steel: Wire for forging; Kiveton Park; annealed **UTS: 540**
HEDEX 46	0.37% C 0.9% Cr 0.2% Mo steel: Wire for forging; Kiveton Park; annealed **UTS: 580**
HI 440	0.12% C 0.7% Mn 0.5% Ni 0.18% Mo 0.9% Cu steel: Inland Steel Co.
HI-STEEL	0.12% C 0.75% Mn 0.5% Ni 0.18% Mo 0.9% Cu steel: Inland Steel Co.
HI STRENGTH A	0.12% C 0.7% Ni 0.6% Cr 0.1% Mo 0.4% Cu steel: Armco; weldable **UTS: 500 Elon: 22% Proof: 330**
HI TUFMOLD	0.35% C 1.2% Cr 0.4% Mo steel: Jessop
HI Z80	0.18% C (max) 0.9% Mn 0.85% Ni 0.6% Cr 0.5% Mo 0.07% V 0.2% Cu 0.004% B steel: Fuji
HK 5	0.25% C 2.0% Cr 1.0% Mo 0.5% V 0.6% Al steel: Firth Brown for nitriding
HOT DIE No 5	0.32% C 3.25% Cr 0.35% V 9.5% W steel: Edgar Allen; hot extrusion dies, etc. **DPN: 500**
HOTFORM No 1	0.35% C 5% Cr 1.4% Mo 0.45% V 1.4% W steel: Vanadium Alloys Ltd **DPN: 600**
HOTFORM No 2	0.35% C 5% Cr 1.3% Mo 0.45% V steel: Vanadium Alloys Ltd **DPN: 600**
HOV	0.3% C 5% Cr 1.5% Mo 1% W 1% Si steel: Jonas; gravity dies for Al casting **DPN: 550**
HOWARD A	0.35% C 5.0% Cr 1.5% Mo 0.4% V steel: Simmonds; AISI type H 11
HOWARD B	0.35% C 5.0% Cr 1.5% Mo 0.4% V 1.5% W steel: Simmonds; AISI type H 12
HOWARD C	0.35% C 5.0% Cr 1.5% Mo 1.0% V steel: Simmonds; AISI type H 13
HP	0.32% C 3.2% Cr 9% W 0.45% V steel: For tools; Hunstman; for hot work **DPN: 560**
HR 33	0.12% C 3.0% Ni 0.7% Cr steel: For case hardening; Bofors; annealed **DPN: 200**
HSB 51	0.18% C (max) 1.2% Mn 0.45% Ni 0.1% V 0.35% Cu steel: Phoenix Rheinrohr
HSB 51S	0.18% C (max) 1.2% Mn 0.45% Ni 0.1% V 0.35% Cu steel
HSB 52	0.18% C (max) 1.2% Mn 0.55% Ni 0.2% V 0.45% Cu steel: Phoenix Rheinrohr
HSB 52S	0.18% C (max) 1.2% Mn 0.55% Ni 0.12% V 0.45% Cu steel
HSB 55C	0.12% C 1.2% Mn 0.65% Ni 0.15% V 0.6% Cu steel: Phoenix Rheinrohr
HSB 77V	0.15% C (max) 1.2% Mn 1.2% Ni 0.5% Cr 0.3% Mo 0.1% V steel: Phoenix Rheinrohr
HSM/W9A	0.3% C 2.8% Cr 0.35% V 9.5% W die steel: ESC; for hot work **DPN: 460**
HT 1	0.09% C 5% Cr 0.5% Mo 1.5% Si steel: Sandvik
HT 3	0.1% C 5% Cr 0.5% Mo steel: Sandvik
HT 5	0.13% C 1% Cr 0.45% Mo steel: Sandvik for SS 2216
HT 7	0.1% C 9% Cr 1% Mo steel: Sandvik
HT 8	0.1% C 2.3% Cr 1% Mo steel: Sandvik for SS 2218
HT 51	0.12% C 1.2% Cr 0.5% Mo steel: Sandvik
HW 1	0.3% C 3.5% Cr 0.5% V 9% W steel: For tools; Darwin; dies, die inserts **DPN: 480**
HW 2N	0.28% C 2.25% Ni 0.85% Cr 0.5% Mo 0.3% V 2.1% W steel: Origin unknown
HW 4	0.3% C 3% Ni 2.8% Cr 0.25% V 9% W steel: For tools; Darwin; dies punches **DPN: 480**
HW 5	0.3% C 5% Cr 2% Mo 0.25% V 1% W 1% Si steel: For tools; Darwin; dies **DPN: 480**
HW 6 NV	0.33% C 3.7% Ni 1.5% Cr 0.27% V 5.5% W steel: ESC; hot forging dies, punches, etc. **DPN: 470**
HWD 1	0.35% C 5.0% Cr 1.5% Mo 0.3% V 1.4% H steel: Firth Sterling
HWT 7	0.3% C 3.7% Ni 1.5% Cr 5.5% W steel: For tools; origin unknown
HY 80 ARMCO	0.18% C 0.25% Mn 2.7% Ni 1.4% Cr 0.6% Mo 0.03% V 0.25% Cu steel: Armco
HY 100	0.2% C 0.25% Mn 2.9% Ni 1.4% Cr 0.6% Mo steel: Armco
HY 140	0.12% C (max) 5.0% Ni 0.55% Cr 0.5% Mo 0.07% V 0.15% Cu (max) steel: Origin unknown
HY 150	0.18% C 0.5% Mn 3.75% Ni 1.5% Cr 0.4% Mo 0.1% V steel: Armco
HYDRA E	0.3% C 3.5% Cr 0.35% V 8.5% W steel: Hall and Pickles; hot extrusion of copper alloys; obsolete **DPN: 550**

Symbol	Nominal analysis, supplier, condition and remarks.
HYDRA M	0.35% C 4% Ni 1.5% Cr 0.5% Mo 4% W steel: Hall and Pickles; die casting dies for Al, etc.; obsolete **DPN: 500**
HYDRA Z	0.26% C 2.5% Ni 3% Cr 0.5% Mo 0.25% V 8.5% W steel: Hall and Pickles; forging dies; obsolete **DPN: 500**
HYKRO	0.28% C 3.2% Cr 0.5% Mo steel: ESC; nitriding steel; BS alloy En 40 A and En 40 B **UTS: 1430 Elon: 17% Proof: 1140**
HYKROM A	0.35% C 5% Cr 1.5% Mo 0.3% V steel: C Denton; extrusion dies, casting dies, etc.; obsolete **DPN: 610**
HYKROM B	0.35% C 3% Cr 0.5% Mo 0.1% V steel: C Denton; drop forging dies, etc.; obsolete **DPN: 550**
HY-TUF	0.25% C 1.3% Mn 1.8% Ni 0.3% Cr 0.4% Mo 1.5% Si steel: Crucible Steel Co. **UTS: 1680 Elon: 14% Proof: 1380**
IBD	0.32% C 4.2% Ni 1.2% Cr 0.35% Mo steel: Inman; can be nitrided
IDROTUB 56	0.2% C (max) 1.4% Mn 0.3% Ni 0.3% Cr steel: Italian proprietory specification
IDROTUB 58T	0.18% C (max) 1.3% Mn 0.35% Ni 0.3% Cr 0.22% Mo steel: Italian proprietory specification
IDROTUB 62T	0.16% C (max) 1.3% Mn 0.25% Ni 0.3% Cr 0.22% Mo 0.11% V steel: Italian proprietory specification
IMPACTO	0.16% C 1.7% Ni 0.25% Mo steel: For carburizing; Atlas
IN 12	0.14% C (max) 0.8% Mn 1.0% Ni 0.55% Cr 0.35% Mo 0.07% V steel: Yawata
IN 787	0.04% C 0.5% Mn 1.2% Ni 0.6% Cr 0.2% Mo 0.05% Nb 1.2% Cu steel: Solution treated and aged; information from Climax Machy **UTS: 615 Elon: 30% Proof: 540**
IN 787A	0.07% C (max) 0.85% Ni 0.75% Cr 0.2% Mo 0.02% Nb (min) 1.2% Cu steel: International Nickel Co.; solution treated and aged; impact 176 J at −60 °C **UTS: 630 Elon: 26% Proof: 550**
INKOMO	0.3% C 3% Ni 0.7% Cr 0.5% Mo steel: Sanderson; hardened and tempered **DPN: 320 UTS: 1000 Elon: 14% Proof: 780**
IR 34	0.13% C 3.5% Ni 0.7% Cr steel: For case hardening; Bofors; annealed; obsolete **DPN: 220**
IR 74	0.3% C 3.5% Ni 0.8% Cr steel: Bofors; hardened and tempered; obsolete **DPN: 300 UTS: 1000 Elon: 15% Proof: 530**
IRO 743	0.35% C 3.4% Ni 1.1% Cr 0.3% Mo steel: Bofors **DPN: 475 EUT: 1500 Elon: 7%**
ISHIURATA	0.3% C 1% Cr 0.2% Mo steel: Origin unknown for bicycle frames
J 7	0.35% C 3.5% Cr 0.25% V 2.5% W steel: Jessop
J 12	0.26% C 3% Ni 2.7% Cr 0.5% Mo 0.3% V 10% W steel: Jessop; dies for casting Cu **DPN: 400**
J 21	0.21% C 3.2% Cr 4.2% Mo 1.1% W steel: Jessop
J 23	0.35% C 2.5% Cr 4.2% Mo 1% V 1.8% W steel: Jessop
J12545	0.25% C (max) 0.7% Ni 0.5% Mo steel: Castings; UNS designation
JAVELIN	0.26% C 0.15% Ni 3.2% Cr 0.55% Mo steel: Tube; Atlas; for detachable rock drill bits
JC 20	0.3% C 2.75% Cr 0.4% V 10% W steel: Jonas; extrusion dies for non-ferrous metals **DPN: 500**
JC 20N	0.3% C 2.25% Ni 2.5% Cr 0.25% V 10% W steel: Jonas; extrusion dies for non-ferrous metals **DPN: 560**
JH 2	0.3% C 3% Cr 0.55% Mo nitriding steel: S Osborn; obsolete **UTS: 920**
JIS G3114 SMA50A, B and C	0.19% C (max) 1.4% Mn (max) 0.8% Cr 0.5% Cu with Ni Mo Nb or V steel: Plate; Japanese Standard **UTS: 560 Proof: 360**
JIS G3114 SMA58	0.19% C (max) 1.4% Mn (max) 0.75% Cr 0.5% Cu with Ni Mo V etc. steel: Plate; Japanese Standard; hardened and tempered **UTS: 650 Proof: 460**
JIS G3119 SBV3	0.25% C (max) 1.2% Mn 0.9% Ni 0.5% Mo steel: Plate; Japanese Standard; normalized **UTS: 630 Proof: 350**
JIS G3120 SQV2A	0.25% C (max) 1.25% Mn 0.6% Ni 0.55% Mo steel: Plate; Japanese Standard; hardened and tempered **UTS: 630 Proof: 350**
JIS G3120 SQV2B	0.25% C (max) 1.2% Mn 0.6% Ni 0.5% Mo steel: Plate; Japanese Standard; hardened and tempered **UTS: 720 Proof: 490**
JIS G3120 SQV3A	0.25% C (max) 1.25% Mn 0.85% Ni 0.5% Mo steel: Plate; Japanese Standard; hardened and tempered **UTS: 630 Proof: 350**
JIS G3120 SQV3B	0.25% C (max) 1.2% Mn 0.9% Ni 0.5% Mo steel: Plate; Japanese Standard; hardened and tempered **UTS: 730 Proof: 490**
JIS G3125 SPAH	0.12% C (max) 0.35% Mn 0.8% Cr 0.5% Cu steel: Plate; Japanese Standard **UTS: 500 Proof: 360**
JIS G3179 SBV2	0.25% C (max) 1.2% Mn 0.6% Ni 0.5% Mo steel: Plate; Japanese Standard; normalized **UTS: 630 Proof: 350**
JIS G3441 STK3A	0.35% C 0.6% Mn 1.0% Cr 0.2% Mo steel: Pipe; Japanese Standard **UTS: 700**
JIS G3441 STK3B	0.35% C 0.6% Mn 1.0% Cr 0.2% Mo steel: Pipe; Japanese Standard **UTS: 560 Proof: 400**
JIS G3441 STK3C	0.35% C 0.6% Mn 1.0% Cr 0.2% Mo steel: Pipe; Japanese Standard **UTS: 670 Proof: 500**
JIS G3441 STK3D	0.35% C 0.6% Mn 1.0% Cr 0.2% Mo steel: Pipe; Japanese Standard **UTS: 880 Proof: 700**
JIS G3441 STK3E	0.35% C 0.6% Mn 1.0% Cr 0.2% Mo steel: Pipe; Japanese Standard **UTS: 1050 Proof: 950**
JIS G3441 STKS1A	0.3% C 0.6% Mn 1.0% Cr 0.2% Mo steel: Pipe; Japanese Standard **UTS: 670**
JIS G3441 STKS1B	0.3% C 0.6% Mn 1.0% Cr 0.2% Mo steel: Pipe; Japanese Standard **UTS: 560 Proof: 400**
JIS G3441 STKS1C	0.3% C 0.6% Mn 1.0% Cr 0.2% Mo steel: Pipe; Japanese Standard **UTS: 630 Proof: 500**
JIS G3441 STKS1D	0.3% C 0.6% Mn 1.0% Cr 0.2% Mo steel: Pipe; Japanese Standard **UTS: 880 Proof: 700**
JIS G3441 STKS1E	0.3% C 0.6% Mn 1.0% Cr 0.2% Mo steel: Pipe; Japanese Standard **UTS: 1050 Proof: 950**

Note. The following abbreviations and units are used in the tables:

DPN	Hardness, diamond pyramid number
UTS	Ultimate tensile strength, N/mm^2
Elon	Elongation, %
Proof	0.1% proof strength, N/mm^2

1 N/mm^2=0.1 hbar=0.102 kgf/mm^2=0.06475 $tonf/in.^2$=145.04 $lbf/in.^2$=1 MPa

See Appendix II for other abbreviations and conversion tables.

Symbol	Nominal analysis, supplier, condition and remarks.
JIS G3441 STKS2A	0.3% C 0.75% Mn 0.55% Ni 0.5% Cr 0.2% Mo steel: Pipe; Japanese Standard **UTS: 670**
JIS G3441 STKS2B	0.3% C 0.75% Mn 0.55% Ni 0.5% Cr 0.2% Mo steel: Pipe; Japanese Standard **UTS: 560** **Proof: 400**
JIS G3441 STKS2C	0.3% C 0.75% Mn 0.55% Ni 0.5% Cr 0.2% Mo steel: Pipe; Japanese Standard **UTS: 630** **Proof: 500**
JIS G3441 STKS2D	0.3% C 0.75% Mn 0.55% Ni 0.5% Cr 0.2% Mo steel: Pipe; Japanese Standard **UTS: 880** **Proof: 700**
JIS G3441 STKS2E	0.3% C 0.75% Mn 0.55% Ni 0.5% Cr 0.2% Mo steel: Pipe; Japanese Standard **UTS: 1050** **Proof: 950**
JIS G3441 STKS4A	0.34% C 0.85% Mn 0.5% Ni 0.55% Cr 0.25% Mo steel: Pipe; Japanese Standard **UTS: 700**
JIS G3441 STKS4B	0.34% C 0.85% Mn 0.5% Ni 0.55% Cr 0.25% Mo steel: Pipe; Japanese Standard **UTS: 560** **Proof: 400**
JIS G3441 STKS4C	0.34% C 0.85% Mn 0.5% Ni 0.55% Cr 0.25% Mo steel: Pipe; Japanese Standard **UTS: 670** **Proof: 500**
JIS G3441 STKS4D	0.34% C 0.85% Mn 0.5% Ni 0.55% Cr 0.25% Mo steel: Pipe; Japanese Standard **UTS: 880** **Proof: 700**
JIS G3441 STKS4E	0.34% C 0.85% Mn 0.5% Ni 0.55% Cr 0.25% Mo steel: Pipe; Japanese Standard **UTS: 1050** **Proof: 95**
JIS G3458 STPA22	0.15% C (max) 0.45% Mn 1.0% Cr 0.55% Cu 0.035% S and P (max) steel: Plate; Japanese Standard **UTS: 420** **Proof: 210**
JIS G3458 STPA23	0.15% C (max) 0.45% Mn 1.25% Cr 0.55% Mo 0.035% S and P (max) steel: Pipe; Japanese Standard **UTS: 420** **Proof: 210**
JIS G3458 STPA24	0.15% C (max) 0.45% Mn 2.25% Cr 1.0% Mo 0.03% S and P (max) steel: Pipe; Japanese Standard **UTS: 420** **Proof: 210**
JIS G3458 STPA25	0.15% C (max) 0.45% Mn 5% Cr 0.55% Mo steel: Pipe; Japanese Standard **UTS: 420** **Proof: 210**
JIS G3458 STPA26	0.15% C (max) 0.45% Mn 9% Cr 1% Mo steel: Pipe; Japanese Standard **UTS: 420** **Proof: 210**
JIS G3462 STBA22	0.15% C (max) 0.45% Mn 1.0% Cr 0.55% Mo steel: Pipe; Japanese Standard **UTS: 420** **Proof: 210**
JIS G3462 STBA23	0.15% C (max) 0.45% Mn 1.25% Cr 0.55% Mo steel: Pipe; Japanese Standard **UTS: 420** **Proof: 210**
JIS G3462 STBA24	0.15% C (max) 0.45% Mn 2.3% Cr 1.0% Mo steel: Pipe; Japanese Standard **UTS: 420** **Proof: 210**
JIS G3462 STBA25	0.15% C 0.45% Mn 0.65% Cr 0.52% Mo steel: Pipe; Japanese Standard **UTS: 420** **Proof: 210**
JIS G3462 STBA25	0.15% C (max) 0.45% Mn 5.0% Cr 0.55% Mo steel: Pipe; Japanese Standard **UTS: 420** **Proof: 210**
JIS G3462 STBA26	0.15% C (max) 9.0% Cr 1.0% Mo steel: Pipe; Japanese Standard **UTS: 420** **Proof: 210**
JIS G4106 SCM1	0.32% C 0.45% Mn 1.2% Cr 0.22% Mo steel: Japanese Standard
JIS G4106 SCM2	0.31% C 0.7% Mn 1.05% Cr 0.22% Mo steel: Japanese Standard
JIS G4106 SCM3	0.36% C 0.7% Mn 1.05% Cr 0.22% Mo steel: Japanese Standard
JIS G4106 SCM21	0.16% C 0.7% Mn 1.05% Cr 0.22% Mo steel: Japanese Standard

Symbol	Nominal analysis, supplier, condition and remarks.
JIS G4106 SCM22	0.21% C 0.7% Mn 1.05% Cr 0.22% Mo steel: Japanese Standard
JIS G4106 SCM23	0.2% C 0.85% Mn 1.05% Cr 0.22% Mo steel: Japanese Standard
JIS G4106 SCM24	0.23% C 0.7% Mn 1.05% Cr 0.4% Mo steel: Japanese Standard
JIS G4106 SNC2	0.31% C 0.5% Mn 2.75% Ni 0.8% Cr steel: Japanese Standard
JIS G4106 SNC21	0.15% C 0.5% Mn 2.25% Ni 0.35% Cr steel: Japanese Standard
JIS G4106 SNC22	0.15% C 0.5% Mn 3.25% Ni 1.85% Cr steel: Japanese Standard
JIS G4106 SNCM1	0.31% C 0.75% Mn 1.8% Ni 0.8% Cr 0.22% Mo steel: Japanese Standard
JIS G4106 SNCM2	0.25% C 0.5% Mn 3.25% Ni 1.25% Cr 0.22% Mo steel: Japanese Standard
JIS G4106 SNCM5	0.3% C 0.5% Mn 3.0% Ni 3.0% Cr 0.6% Mo steel: Japanese Standard
JIS G4106 SNCM21	0.21% C 0.85% Mn 0.5% Ni 0.5% Cr 0.22% Mo steel: Japanese Standard
JIS G4106 SNCM22	0.15% C 0.55% Mn 1.8% Ni 0.5% Cr 0.22% Mo steel: Japanese Standard
JIS G4106 SNCM25	0.15% C 0.45% Mn 4.25% Ni 0.85% Cr 0.22% Mo steel: Japanese Standard
JIS G4106 SNCM26	0.17% C 1.0% Mn 3.0% Ni 1.6% Cr 0.5% Mo steel: Japanese Standard
JIS G4109 SCMV1	0.21% C (max) 0.65% Mn 0.65% Cr 0.5% Mo steel: Plate; Japanese Standard; normalized and tempered **UTS: 560** **Proof: 320**
JIS G4109 SCMV2	0.17% C (max) 0.5% Mn 1.0% Ni 0.5% Mo steel: Plate; Japanese Standard; normalized and tempered **UTS: 530** **Proof: 280**
JIS G4109 SCMV3	0.17% C (max) 0.5% Mn 1.25% Cr 0.55% Mo steel: Plate; Japanese Standard; normalized and tempered **UTS: 600** **Proof: 320**
JIS G4109 SCMV4	0.15% C (max) 0.45% Mn 2.25% Cr 1.0% Mo steel: Plate; Japanese Standard; normalized and tempered **UTS: 610** **Proof: 320**
JIS G4109 SCMV5	0.15% C (max) 3.0% Cr 1.0% Mo steel: Plate; Japanese Standard; normalized and tempered **UTS: 610** **Proof: 320**
JIS G4109 SCMV6	0.15% C (max) 0.45% Mn 5.0% Cr 0.5% Mo steel: Plate; Japanese Standard; normalized and tempered **UTS: 610** **Proof: 320**
JO	0.37% C 3.2% Cr 0.3% V 9.5% W steel: Jessop
JO Sp1	0.27% C 3.2% Cr 0.3% V 9.5% W steel: Jessop
JS 1	0.25% C (max) 1.0% Mn 1% Cu + Ni + Cr + Mo (max) 0.6% Si steel: Casting; Jopling; weldable **DPN: 130** **UTS: 420** **Elon: 22%** **Proof: 210**
JS 1B	0.25% C (max) 0.4% Ni 0.25% Cr 0.1% Mo 0.4% Cu steel: Casting; Jopling; for magnetic properties **DPN: 140** **UTS: 440** **Elon: 20%** **Proof: 200**
JS 4	0.15% C (max) 0.4% Ni 0.2% Cr 0.1% Mo 0.4% Cu steel: Casting; Jopling; for magnetic properties **DPN: 110** **UTS: 420** **Elon: 22%**
JS 9	0.35% C (max) 1.0% Ni (max) 1.0% Cr (max) 0.6% Mo (max) steel: Casting; Jopling **DPN: 230** **UTS: 760** **Elon: 15%** **Proof: 480**
JS 10	0.35% C (max) 1.0% Ni (max) 1.0% Cr (max) 0.6% Mo (max) steel: Casting; Jopling **DPN: 275** **UTS: 920** **Elon: 12%** **Proof: 580**
JS 40	0.2% C (max) 0.4% Ni 1.2% Cr 0.5% Mo 0.4% Cu (max) steel: Casting; Jopling; for use up to 54 °C **DPN: 160** **UTS: 460** **Elon: 20%** **Proof: 270**
K 18	0.32% C 5% Cr 1.7% Mo 1.4% W steel: Jessop **DPN: 440**
K 291	0.13% C 3.0% Ni 0.7% Cr steel: Fagersta
K 336	0.18% C 3.0% Ni 0.7% Cr steel: Fagersta
K 353	0.15% C 1.5% Ni 0.8% Cr steel: Fagersta
K 354	0.2% C 1.5% Ni 0.8% Cr steel: Fagersta
K 825	0.36% C 1.4% Ni 1.4% Cr 0.2% Mo steel: Fagersta

Symbol	Nominal analysis, supplier, condition and remarks.
K12103	0.25% C 0.35% Cr 0.1% Mo steel: Designation used by UNS
K12125	0.23% C 0.5% Cr 0.17% Mo steel: B treated
K13262	0.32% C (max) 0.7% Ni (min) 0.4% Cr (min) 0.15% Mo (min) steel: Designation used by UNS
K22094	0.2% C (max) 1.25% Cr 0.55% Mo steel: Designation used by UNS
K23505	0.35% C 0.62% Cr 0.6% Mo 0.6% Cu steel: Designation used by UNS
K23578	0.35% C (max) 2.5% Ni (min) 0.25% Mo (min) 0.1% V steel: Designation used by UNS
K23579	0.35% C (max) 2.5% Ni (min) 0.25% Mo (min) 0.03% V (min) steel: Designation used by UNS
K27705	0.37% C (max) 1% Cr 1.25% Mo 0.25% V steel: Designation used by UNS
K42570	0.25% C 3.5% Ni 1.5% Cr 0.25% Mo steel: Designation used by UNS
K43170	0.31% C 3.5% Ni 0.25% Mo steel: Designation used by UNS
K91890	0.03% C (max) 12% Ni 5% Cr 3% Mo 0.27% Ti steel: Designation used by UNS
KAISALOY 45FG	0.12% C (max) 0.6% Mn 0.6% Ni 0.25% Cr 0.1% Mo 0.02% V 0.005% Ti 0.3% Cu steel: Kaiser Steel Corp.
KAISALOY 50CR	0.2% C (max) 1.25% Mn 0.6% Ni 0.25% Cr 0.15% Mo 0.02% V 0.005% Ti 0.35% Cu steel: Kaiser Steel Corp.
KAISALOY No. 1	0.2% C (max) 1.25% Mn 0.6% Ni 0.25% Cr 0.15% Mo 0.02% V 0.25% Al 0.35% Cu steel: Kaiser Steel Corp.
KAISALOY No. 2	0.12% C (max) 0.6% Mn 0.6% Ni 0.25% Cr 0.1% Mo 0.02% V 0.005% Ti 0.3% Cu steel: Kaiser Steel Co.
KD 2	0.25% C 1.0% Cr 0.25% Mo steel: Pompey **UTS: 1070 Elon: 10% Proof: 870**
KD 2SE	0.25% C 1.0% Cr 0.25% Mo steel: Pompey **UTS: 990 Elon: 11% Proof: 760**
KD 3	0.35% C 1.0% Cr 0.25% Mo steel: Pompey **UTS: 1150 Elon: 9% Proof: 990**
KE 41	0.12% C 4.2% Ni 1.35% Cr 0.2% Mo steel: For case hardening; Kayser Ellison; hardened and tempered; core properties **DPN: 375 UTS: 1300 Elon: 15%**
KE 169	0.12% C 3.2% Ni 0.9% Cr 0.2% Mo steel: For case hardening; Kayser Ellison; hardened and tempered; core properties **DPN: 302 UTS: 1000 Elon: 16%**
KE 287	0.12% C 3.2% Ni 0.3% Cr 0.2% Mo steel: For case hardening; Kayser Ellison; hardened and tempered; core properties **DPN: 235 UTS: 790 Elon: 22%**
KE 339	0.25% C 2.2% Ni 3% Cr 0.4% Mo 0.5% V 9.5% W 0.4% Co 0.7% Si steel: Kayser Ellison; for hot working copper **DPN: 470**
KE 339	0.28% C 2.2% Ni 2.5% Cr 0.15% V 9.5% W steel: Sanderson Kayser
KE 660	0.12% C 4.7% Ni 0.3% Cr 0.2% Mo steel: For case hardening; Kayser Ellison; hardened and tempered; core properties **DPN: 311 UTS: 1020 Elon: 16%**
KE 805	0.35% C 1.5% Ni 1% Cr 0.2% Mo steel: Kayser Ellison; hardened and tempered **DPN: 350 UTS: 1150 Elon: 17% Proof: 1030**
KE 897	0.3% C 4.2% Ni 1.2% Cr 0.2% Mo steel: Kayser Ellison; hardened and tempered **DPN: 470 UTS: 1680 Elon: 12%**
KE A145	0.35% C 5.5% Cr 1.3% Mo 1% V 1% Si steel: Kayser Ellison; for plastic extrusion dies **DPN: 402**
KE A220	0.35% C 1.6% Cr 0.5% Mo steel: Sanderson Kayser
KF 15A	0.14% C 3.4% Ni 1% Cr steel: Case hardening; Kirkstall for BS alloy En 36A
KF 15B	0.16% C 3.5% Ni 1% Cr 0.25% Mo steel: Case hardening; Kirkstall for BS alloy En 36C
KF 15C	0.16% C 4.2% Ni 1.2% Cr 0.3% Mo steel: Case hardening; Kirkstall for BS alloy En 39B
KF 16	0.32% C 3.2% Ni 0.9% Cr steel: Kirkstall for BS alloy En 23
KF 19	0.3% C 4.1% Ni 1.3% Cr 0.3% Mo steel: Kirkstall for BS alloy En 30B
KF 28	0.32% C 3.2% Ni 0.8% Cr 0.6% Mo steel: Kirkstall for BS alloy En 27
KF 28B	0.32% C 2.5% Ni 0.6% Cr 0.6% Mo steel: Kirkstall for BS alloy En 25
KF 31B	0.19% C 1.7% Ni 0.5% Cr 0.25% Mo steel: Case hardening; Kirkstall for BS alloy En 325
KF 38A	0.35% C 1.2% Ni 0.6% Cr steel: Kirkstall for BS alloy En 111
KF 42	0.17% C 1.7% Ni 0.25% Mo steel: Case hardening; Kirkstall for BS alloy En 34
KF 42A	0.22% C 1.7% Ni 0.25% Mo steel: Case hardening; Kirkstall for BS alloy En 35
KF 50A	0.2% C 1.6% Mn 0.6% Ni 0.25% Mo steel: Kirkstall for BS alloy En 13
KF 70	0.2% C (max) 0.8% Ni 0.6% Cr 0.1% Mo steel: Case hardening; Kirkstall for BS alloy En 351
KF 71	0.2% C 1% Ni 0.8% Cr 0.1% Mo steel: Case hardening; Kirkstall for BS alloy En 352
KF 72	0.2% C 1.2% Ni 1% Cr 0.11% Mo steel: Case hardening; Kirkstall for BS alloy En 353
KF 73	0.2% C (max) 1.7% Ni 1% Cr 0.15% Mo steel: Case hardening; Kirkstall for BS alloy En 354
KF 74	0.2% C (max) 2% Ni 1.5% Cr 0.2% Mo steel: Case hardening; Kirkstall for BS alloy En 355
KF 75	0.15% C 0.6% Ni 0.7% Cr 0.11% Mo steel: Case hardening; Kirkstall for BS alloy En 361
KF 76	0.2% C 0.5% Ni 0.7% Cr 0.11% Mo steel: Case hardening; Kirkstall for BS alloy En 362
KF 77	0.24% C 0.5% Ni 0.7% Cr 0.11% Mo steel: Case hardening; Kirkstall for BS alloy En 363
KH 5M	0.15% C (max) 5.2% Cr 0.52% Mo steel: Russian Standards designation
KH 5VF	0.15% C (max) 5.2% Cr 0.5% V 0.5% W steel: Russian Standards designation
KO	0.15% C 0.8% Mn 0.85% Ni 0.6% Cr 0.5% Mo 0.07% V 0.3% Cu steel: Kawasaki Iron and Steel Co.
KP 12	0.12% C 3% Ni 0.3% Cr steel: Case hardening; Kiveton Park for BS alloy En 33
KP 13	0.3% C 0.5% Mn 3% Ni 0.3% Cr steel: Kiveton Park for BS alloy En 21
KP 17	0.3% C 3.25% Ni 0.8% Cr 0.6% Mo steel: Kiveton Park for BS alloy En 23
KP 23	0.3% C 3.5% Ni 1% Cr 0.4% Mo steel: Kiveton Park for BS alloy En 27
KP 31	0.18% C (max) 3.5% Ni 0.8% Cr steel: For case hardening; Kiveton Park for BS alloy En 36
KP 35	0.16% C 4% Ni 1.2% Cr steel: For case hardening; Kiveton Park for BS alloy En 39
KP 36	0.3% C 4.22% Ni 1.2% Cr 0.3% Mo steel: Kiveton Park for BS alloy En 30B
KP 37	0.2% C 1.7% Ni 2.25% Mo steel: For case hardening; Kiveton Park for BS alloy En 34

Note. The following abbreviations and units are used in the tables:

DPN	Hardness, diamond pyramid number
UTS	Ultimate tensile strength, N/mm^2
Elon	Elongation, %
Proof	0.1% proof strength, N/mm^2

1 N/mm^2=0.1 hbar=0.102 kgf/mm^2=0.06475 tonf/in.^{2}=145.04 lbf/in.^{2}=1 MPa

See Appendix II for other abbreviations and conversion tables.

Symbol	Nominal analysis, supplier, condition and remarks.
KR 35	0.13% C 4.2% Ni 1.2% Cr steel: For case hardening; Bofors; annealed **DPN: 240**
KR 75	0.35% C 4.2% Ni 1.2% Cr steel: Bofors; hardened and tempered **DPN: 340 UTS: 1160 Elon: 12% Proof: 870**
KV 1	See Philips KV 1
KV 3	See Philips KV 3
KV 4	See Philips KV 4
KV 5	See Philips KV 5
KV 7	See Philips KV 7
L 2N	0.17% C 1.8% Ni 0.25% Mo steel: For case hardening; Swift Levick for BS alloy En 34
L 197	0.36% C 1.4% Ni 1.4% Cr 0.2% Mo steel: Fagersta; hardened and tempered **DPN: 400 UTS: 1360 Elon: 5% Proof: 1160**
L 536	0.3% C 3.2% Ni 1.0% Cr 0.25% Mo steel: Fagersta
LARPORT	0.3% C 4.25% Ni 1.25% Cr 0.3% Mo steel: Hall and Pickles; plastic dies and moulds; hardened and tempered; obsolete **UTS: 1140 Elon: 20% Proof: 1020**
LCHD	0.35% C 3.5% Cr 9.0% W 0.4% Co steel: Origin unknown
LESCALLOY 4330 +V	0.3% C 1.8% Ni 0.85% Cr 0.4% Mo 0.07% V steel: Latrobe
LESCALLOY 4335 +V	0.35% C 1.85% Ni 0.8% Cr 0.35% Mo 0.2% V steel: Latrobe
LESCALLOY 9310	0.1% C 3.2% Ni 1.2% Cr 0.12% Mo steel: Latrobe; for carburizing
LESCALLOY BG31	0.12% C 2.8% Ni 1.5% Cr 5.0% Mo steel: Latrobe; for carburizing
LESCALLOY UT19	0.16% C 4.25% Ni 1.2% Cr 0.25% Mo steel: Latrobe; for carburizing
LESCO HW114	0.3% C 1.65% Ni 2.75% Cr 9.75% W steel: Latrobe; for hot working
LHG	0.12% C 3.2% Ni 0.8% Cr steel: For case hardening; Swift Levick for BS alloy En 36
LK 5	0.3% C 1.6% Cr 0.2% Mo 1.1% Al steel: For nitriding; Firth Brown for En 41A
LK 7	0.2% C 1.6% Cr 0.2% Mo 1.1% Al steel: For nitriding; Firth Brown
LKP	0.24% C 3.5% Ni 1.3% Cr 0.3% Mo 1.0% Al steel: Firth Brown; for nitriding
LOYCON N	0.15% C (max) 1.2% Mn 1.6% Ni 0.6% Cr 0.3% Mo 0.12% V steel: Consett; normalized **UTS: 610 Elon: 16% Proof: 360**
LOYCON QT	0.15% C (max) 1.2% Mn 1.6% Ni 0.6% Cr 0.3% Mo 0.12% V steel: Consett; hardened and tempered **UTS: 770 Elon: 18% Proof: 710**
LPD	0.35% C 5% Cr 1.6% Mo 0.3% V 1.3% W 1% Si steel: Casting; Latrobe; steel for precision casting of forging dies **DPN: 550**
LT 39	0.35% C 5.0% Cr 1.3% Mo 1.0% V steel: For tools; Low Moor
LT 41W	0.35% C 5% Cr 1.6% Mo 0.25% V 1.5% W 1.0% Si steel: Low Moor; for hot work tools; obsolete **DPN: 550**
LT 43	0.3% C 3.2% Cr 0.2% V 9% W steel: Low Moor; for hot work tools; obsolete **DPN: 620**
LT 45	0.25% C 2.5% Ni 3% Cr 0.25% V 8.5% W steel: Low Moor; for hot work tools; obsolete **DPN: 630**
LT 52	0.35% C 4.25% Ni 1.25% Cr 0.3% Mo steel: Low Moor; die steel; obsolete **DPN: 580**
LT 53	0.12% C 4% Ni 1.25% Cr steel: Low Moor; carburizing die steel; obsolete
LT FORGING	0.33% C 3.5% Cr 0.5% V 9.5% W steel: Firth Sterling
M Mo 30	0.3% C 1.5% Mn 0.4% Ni 0.25% Cr 0.3% Mo steel: Casting; Wolsingham **DPN: 225 UTS: 760 Elon: 15% Proof: 480**
M 120	0.14% C 3.5% Ni 0.7% Cr steel: VEW
M 130	0.19% C 4.1% Ni 1.3% Cr 0.2% Mo steel: VEW
M 188	0.06% C (max) 1.7% Ni 0.2% Cr 0.2% Cu steel: Weld electrode; designation used by AWS
M 210	0.33% C 1.6% Cr 0.4% Mo steel: VEW
MANOIR PF1	0.15% C 0.5% Cr 0.5% Mo steel: Pompey **UTS: 500 Elon: 18% Proof: 280**
MANOIR PF2	0.15% C 1.0% Cr 0.5% Mo steel: Pompey **UTS: 550 Elon: 16% Proof: 320**
MANOIR PF5	0.15% C 5.0% Cr 0.5% Mo steel: Pompey **UTS: 600 Elon: 18% Proof: 260**
MANOIR PF6	0.15% C 2.2% Cr 1.0% Mo steel: Pompey **UTS: 500 Elon: 15% Proof: 240**
MANOIR PF15	0.2% C 0.8% Cr 0.4% Mo steel: Pompey **UTS: 650 Elon: 16% Proof: 400**
MANOIR PFV55	0.15% C 2.0% Cr 0.35% Mo + V + Al steel: Pompey **UTS: 550 Elon: 15% Proof: 280**
MANOIR PM17	0.12% C with Mo and V steel: Pompey **UTS: 600 Elon: 16% Proof: 420**
MANOIR PM35	0.17% C with Cr Mo and V steel: Pompey **UTS: 1100 Elon: 8% Proof: 800**
MANOIR ABRADIER 600	Cr V W steel: For carburizing; Pompey
MARSH CH 5	0.37% C 5% Cr 1.4% Mo 0.6% V steel: Marsh Brothers; hot forging dies, etc.
MARSH CTHY	0.3% C 3.4% Cr 0.35% V 8.5% W steel: Marsh Brothers; hot forging dies, etc.
MAX HIGH TENSILE	0.15% C (max) 0.75% Mn 0.55% Cr 0.2% Mo 0.2% Cu 0.07% Zr steel: National Steel
MAXINIUM	0.32% C 5.25% Cr 0.5% Mo 5% W steel: Edgar Allen; die casting moulds **DPN: 570**
MAXNAP	0.3% C 1.05% Cr 0.2% V steel: Edgar Allen; rivet hammers, lathe centres, etc. **DPN: 540**
MAXTRESS	0.35% C 3.5% Ni 1% Cr 0.5% Mo steel: Jonas **UTS: 1380 Elon: 13% Proof: 1300**
MAYARI R	0.12% C (max) 0.75% Mn 1.0% Ni 0.7% Cr 0.1% Zr steel: Bethlehem Steel Co.
MCHT	0.25% C 1.7% Ni 0.4% Cr 0.4% Mo steel: Pompey
MCL	0.3% C 3.7% Cr 6.25% Mo 0.75% V 1.0% W steel: Latrobe; hot work **DPN: 450**
MCN	0.33% C 3% Ni 0.7% Cr 0.3% Mo steel: For hot work tools; Osborn; stamping dies; casting dies, etc.; obsolete **DPN: 350**
MCV 24	0.27% C 0.75% Cr 0.1% V steel: Krupp
MEL-TROL SUPER SAMSON	0.1% C 5.0% Cr 0.9% Mo 0.25% V steel: Carpenter; for AISI type P4
MEL-TROL TK	0.35% C 3.5% Cr 0.4% V 9.0% W steel: Carpenter
MET-HARD 250	1.7% Cr 1% Mo steel: Electrode; Metrode **DPN: 250**
MET-HARD 350	0.15% C 3% Cr 0.5% Mo 0.3% V steel: Electrode; Metrode **DPN: 350**
MET-HARD 400HW	0.1% C 4.5% Cr 1.2% Mo 0.3% V 1.2% W steel: Electrode; Metrode
MET-HARD 450	0.2% C 3% Cr 1% Mo + V + Nb steel: Electrode, Metrode **DPN: 450**
MET-HARD 450HW	0.3% C 5% Cr 1.5% Mo 1.8% W steel: Electrode; Metrode
MET-HARD 500HW	0.25% C 2.5% Cr 1.5% Mo 0.6% V 4.5% W steel: Electrode; Metrode
MET-HARD 530HW	0.3% C 1.5% Cr 8% W 2% Co 0.8% Nb steel: Electrode; Metrode

Symbol	Nominal analysis, supplier, condition and remarks.
MET-HARD 550HW	0.25% C 4.5% Cr 0.2% Mo 0.3% V 3% W steel: Electrode; Metrode
MET-HARD H13	0.35% C 5% Cr 1.3% Mo 1% V steel: Electrode; Metrode **DPN: 600**
MIL-S 12504 Mn Cr Mo	0.57% C 1% Cr 0.2% Mo steel: US military specification
MIL-S 12505/2	0.12% C (max) 1.0% Mn 0.7% Ni 0.75% Cr 0.35% Cu steel: Bethlehem Steel Co.
MIL-S 15083-100-70	0.2% C (max) 2.8% Ni 1.5% Cr 0.4% Mo steel: Casting; US Federal specification
MIL-S 15464-1	0.2% C (max) 1.3% Cr 0.5% Mo steel: Casting
MIL-S 15464-2	0.18% C (max) 2.3% Cr 1% Mo steel: Casting; US Federal specification
MIL-S 15464-3	0.18% C (max) 1.3% Cr 0.5% Mo 0.15% V steel: Casting; US Federal specification
MIL-S 21925	0.2% C (max) 3% Ni 1.4% Cr 0.4% Mo steel: US military specification
MIL-S 46047	0.42% C 1.1% Cr 0.62% Mo 0.25% V steel: US military specification
MK 2	0.3% C 3.0% Cr 0.4% V 8.8% W steel: Origin unknown
ML 50	0.13% C 0.48% Mn 0.61% Ni 0.4% Cr 0.01% Mo 0.2% Cu steel: Origin unknown
ML 60	0.13% C 0.9% Mn 0.68% Ni 0.51% Cr 0.01% Mo 0.3% Cu steel: Origin unknown
ML 70	0.16% C 0.95% Mn 0.73% Ni 0.55% Cr 0.01% Mo 0.3% Cu steel: Origin unknown
MNC 831	General standard for SS range of pressure vessel steel tubes; Swedish Standard
MNC 832	General standard for SS range of seamless gas pressure vessels; Swedish Standard
MNC 853	General standard for SS range of alloyed steels intended for heat treatment; Swedish Standard
Mo 18	0.18% C 0.4% Ni 0.25% Cr 0.55% Mo 0.3% Cu steel: Casting; Wolsingham **UTS: 450 Elon: 20% Proof: 270**
MODAL 5	0.3% C 2.1% Ni 2.1% Cr 0.4% Mo steel: Pompey
MOHD	0.36% C 3.5% Cr 6% Mo 0.75% V 1% W tool steel : Huntsman; hot work dies **DPN: 600**
MOHD	0.36% C 3.5% Cr 6.0% Mo 0.75% V 1.0% W steel: Origin unknown
MOLYCROM 05	0.05% C 0.8% Mn 0.55% Cr 0.55% Mo: Welding electrode; Soudometal; for welding creep strength steels **UTS: 610 Elon: 24% Proof: 520**
MOLYCROM 15	0.06% C 0.8% Mn 1.0% Cr 0.5% Mo: Welding electrode; Soudometal; for welding creep strength steels **UTS: 610 Elon: 22% Proof: 520**
MOLYCROM 21	0.06% C 0.8% Mn 2.4% Cr 1.0% Mo: Welding rod; Soudometal; for welding creep strength steels **UTS: 670 Elon: 22% Proof: 570**
MOLYCROM 21 STC	0.09% C 0.6% Mn 2.4% Cr 1.0% Mo: Welding electrode; Soudometal; for welding creep strength steels **UTS: 600 Elon: 24% Proof: 510**
MOLYCROM 55	0.04% C 0.7% Mn 5.0% Cr 0.5% Mo: Welding electrode; Soudometal; for welding creep strength steels **UTS: 480 Elon: 30% Proof: 250**
MOLYCROM 91	0.09% C 0.8% Mn 9.4% Cr 1.0% Mo: Welding electrode; Soudometal; for welding creep strength oxidation resistant steels **UTS: 600 Elon: 35% Proof: 325**
MOLYCROM TI 15	0.08% C 0.6% Mn 1.0% Cr 0.5% Mo: Welding electrode; Soudometal; for welding creep strength steels **UTS: 620 Elon: 23% Proof: 530**
MONARCH DW8	0.3% C 2.75% Cr 0.45% V 7.7% W tool steel: Origin unknown
MONARCH NCG	0.35% C 4.0% Ni 1.5% Cr 0.3% Mo steel: Origin unknown
MONARCH SPECIAL TAN	0.3% C 2.5% Ni 0.75% Cr 0.6% Mo tool steel: Origin unknown
MOSTAR	0.32% C 5.0% Cr 1.15% Mo 0.35% V 1.2% W steel: Pompey
MOULD STEEL	0.35% C 1.65% Cr 0.5% Mo steel: Sanderson Kayser
MOV	0.18% C 0.6% Mo 0.3% V steel: ESC; creep resisting; for steam turbines
MOVA	0.2% C 0.3% Ni 0.8% Cr 0.7% Mo 0.25% V steel: Firth Brown; for use at temperatures up to 550 °C **UTS: 840 Elon: 15% Proof: 610**
MTA 2	0.35% C 3.75% Ni 1.7% Cr steel: Pompey
MTA 3	0.35% C 4.0% Ni 1.7% Cr 0.5% Mo steel: Pompey
MTR	0.13% C 0.8% Mn 1.4% Ni 0.55% Cr 0.5% Mo 0.08% V 0.25% Cu steel: Imphy
MVW	0.35% C 1.25% Ni 1.75% Cr 0.5% Mo 0.5% V 0.5% W steel: Swift Levick; for hot dies
N 1	0.2% C 0.5% Ni 0.5% Cr 0.2% Mo steel: Jessop
N 2	0.22% C 0.55% Ni 0.5% Cr 0.2% Mo steel: Jessop
N 2MCH	0.17% C 1.8% Ni 0.25% Mo steel: For carburizing; Firth Brown for BS 970 En 34
N 3	0.17% C 0.55% Ni 0.5% Cr 0.2% Mo steel: Jessop
N 3CCH	0.12% C 3.3% Ni 0.8% Cr steel: For carburizing; Firth Brown for BS 970 En 36A and En 36B
N 3CMCH	0.12% C 3.3% Ni 0.8% Cr 0.2% Mo steel: For carburizing; Firth Brown for BS 970 En 36
N 3MCH	0.24% C 1.8% Ni 0.25% Mo steel: For carburizing; Firth Brown for BS 970 En 35
N 4CCH	0.14% C 4.25% Ni 1.25% Cr steel: For carburizing; Firth Brown for BS 970 En 39A
N 4CMCH	0.14% C 4.25% Ni 1.25% Cr 0.25% Mo steel: For carburizing; Firth Brown for BS 970 En 39B
N 5MCH	0.12% C 5.1% Ni 0.2% Mo steel: For carburizing; Firth Brown for BS 970 En 38
N 12	0.12% C 3.5% Ni 0.2% Cr 0.15% Mo 0.3% Cu steel: Casting; Wolsingham **UTS: 440 Elon: 25% Proof: 240**
N A XTRA	0.15% C 0.85% Mn 0.6% Cr 0.3% Mo 0.0025% B steel: National Steel
N A XTRA 55	0.2% C (max) 1.0% Mn 0.7% Cr 0.25% Mo 0.07% Zr steel: Hoag
N A XTRA 60	0.2% C (max) 1.0% Mn 0.7% Cr 0.25% Mo 0.07% Zr steel: Hoag
N A XTRA 65	0.2% C (max) 1.0% Mn 0.7% Cr 0.25% Mo 0.07 Zr steel: Hoag
N A XTRA 70	0.2% C (max) 1.0% Mn 0.7% Cr 0.25% Mo 0.07% Zr steel: Hoag
N A XTRA 75	0.2% C (max) 1.0% Mn 0.7% Cr 0.25% Mo 0.07% Zr steel: Hoag
N A XTRA 80	0.2% C (max) 0.85% Mn 0.6% Cr 0.3% Mo 0.05% Zr steel: National Steel
N A XTRA 93	0.15% C 0.8% Mn 0.6% Cr 0.3% Mo steel: National Steel
N A XTRA 100	0.15% C 0.85% Mn 0.6% Cr 0.3% Mo steel: National Steel

Note. The following abbreviations and units are used in the tables:

DPN	Hardness, diamond pyramid number
UTS	Ultimate tensile strength, N/mm^2
Elon	Elongation, %
Proof	0.1% proof strength, N/mm^2

1 N/mm^2=0.1 hbar=0.102 kgf/mm^2=0.06475 $tonf/in.^2$=145.04 $lbf/in.^2$=1 MPa

See Appendix II for other abbreviations and conversion tables.

Symbol	Nominal analysis, supplier, condition and remarks.
NATIONALLOY 14	0.35% C 3.5% Ni 1.0% Cr 0.5% Mo 0.12% V steel: Source unknown
NAX 9115-AC	0.13% C 0.6% Cr 0.2% Mo steel
NBVOC 1/W	0.17% C 1.6% Ni 0.5% Cr 0.45% Mo steel: Krupp
NCM 1	0.34% C 1.5% Ni 1.1% Cr 0.2% Mo steel: Huntsman **DPN: 600**
NCM 2	0.35% C 2.5% Ni 0.6% Cr 0.6% Mo steel: Huntsman, punches, collets, moulds, etc. **DPN: 560**
NCM 3	0.35% C 3.5% Ni 1.2% Cr 0.4% Mo steel: Huntsman, punches, collets, moulds, etc. **DPN: 550**
NCM 3	0.3% C 2.4% Ni 0.6% Cr 0.45% Mo steel: Firth Brown for BS alloy En 25
NCM 4	0.35% C 4% Ni 1.3% Cr 0.3% Mo steel: Huntsman, dies, punches, moulds, etc. **DPN: 550**
NCM 6	0.3% C 3.25% Ni 0.8% Cr 0.3% Mo steel: Firth Brown for BS alloy En 27
NCM 30	0.3% C 1.7% Ni 1% Cr 0.35% Mo 0.3% Cu steel: Casting; Wolsingham **DPN: 230 UTS: 760 Elon: 15% Proof: 480**
NCM 40	0.4% C 1.7% Ni 1% Cr 0.35% Mo 0.3% Cu steel: Casting; Wolsingham
NCM SPECIAL	0.32% C 4.2% Ni 1.3% Cr 0.3% Mo steel: Swift Levick for BS alloy En 30B
NCMo	0.25% C Cr Mo steel: Casting; Firth Brown; hardened and tempered **DPN: 250 UTS: 840 Elon: 17% Proof: 610**
NCR 2	0.3% C 3.25% Ni 0.75% Cr steel: Firth Brown for BS alloy En 23
NE 8150	0.5% C 0.3% Ni 0.4% Cr 0.15% Mo steel: For valves; origin unknown
NF A35 551/16N06	0.15% C 0.75% Mn 1.4% Ni 1.0% Cr steel: French Standard
NF A35 551/ 18NCDG	0.18% C 0.75% Mn 1.4% Ni 1.0% Cr 0.2% Mo steel: French Standard
NF A35 551/20NC6	0.18% C 0.7% Mn 1.4% Ni 1.0% Cr steel: French Standard
NF A35 551/20NCD2	0.2% C 0.75% Mn 0.55% Ni 0.5% Cr 0.22% Mo steel: French Standard
NF A35 551/25CD4	0.26% C 0.75% Mn 1.0% Cr 0.22% Mo steel: French Standard
NF A35 551/ 30CAD6/12	0.3% C 0.65% Mn 1.65% Cr 0.32% Mo 1.15% Al steel; French Standard
NF A35 551/30CD4	0.31% C 0.75% Mn 0.95% Cr 0.22% Mo steel: French Standard
NF A35 551/30CD12	0.31% C 0.55% Mn 3.0% Cr 0.45% Mo steel: French Standard
NF A35 551/30CND8	0.3% C 0.45% Mn 2.0% Ni 2.0% Cr 0.45% Mo steel: French Standard
NF A35 551/35CD4	0.36% C 0.75% Mn 0.95% Cr 0.22% Mo steel: French Standard
NF A35 551/35NC6	0.35% C 0.75% Mn 1.4% Ni 1.0% Cr steel: French Standard
NF A35 551/35NCD6	0.34% C 0.75% Mn 1.4% Ni 1.0% Cr 0.22% Mo steel: French Standard
NF A35 565/ 16NCD13	0.15% C 0.5% Mn 3.25% Ni 1.0% Cr 0.22% Mo (max) steel: For bearings: French Standard
NF A35 565/18NCD4	0.101% C 0.65% Mn 1.1% Ni 0.42% Cr 0.22% Mo steel: For bearings; French Standard
NF A35 565/20NCD2	0.2% C 0.8% Mn 0.55% Ni 0.5% Cr 0.22% Mo steel: For bearings; French Standard
NF A35 565/20NCD7	0.19% C 0.55% Mn 1.8% Ni 0.4% Cr 0.25% Mo steel: For bearings; French Standard
NF A36 206/ 10CD910	0.15% C (max) 0.6% Mn 2.25% Cr 1.0% Mo 0.04% V (max) steel: Plate; French Standard; normalized and tempered **UTS: 570 Proof: 300**
NF A36 206/ 15CD205	0.18% C (max) 0.7% Mn 0.5% Cr 0.5% Mo 0.04% V (max) steel: Plate; French Standard; normalized and tempered **UTS: 510 Proof: 270**
NF A36 206/ 15CD405	0.18% C (max) 1.0% Mn 1.0% Cr 0.5% Mo 0.04% V (max) steel: Plate; French Standard; normalized and tempered **UTS: 510 Proof: 300**
NF A36 206/15D3	0.18% C (max) 0.65% Mn 0.3% Mo 0.04% V (max) steel: Plate; French Standard **UTS: 490 Proof: 260**
NF A36 206/ 15MDV405	0.18% C (max) 1.2% Mn 0.3% Cr 0.5% Mo 0.08% V (max) steel: Plate; French Standard; normalized and tempered **UTS: 570 Proof: 350**
NF A36 206/ 18MD405	0.2% C (max) 1.2% Mn 0.3% Cr (max) 0.5% Mo 0.04% V (max) steel: Plate; French Standard; normalized and tempered **UTS: 570 Proof: 350**
NF A36 208/0.5NiA	0.14% C (max) 0.5% Ni 0.2% Cr (max) 0.1% Mo (max) 0.05% V (max) steel: Plate; French Standard **UTS: 460 Proof: 270**
NF A36 208/1.5Ni	0.18% C (max) 1.5% Mn (max) 1.5% Ni 0.2% Cr (max) 0.1% Mo (max) 0.05% V (max) steel: Plate; French Standard **UTS: 550 Proof: 340**
NF A36 208/5Ni	0.1% C (max) 0.8% Mn (max) 5.0% Ni 0.2% Cr (max) 0.1% Mo (max) 0.05% V (max) steel: Plate; French Standard **UTS: 550 Proof: 380**
NF A36 208/A52FP2	0.2% C (max) 0.1% Mn (max) 0.3% Ni (max) 0.2% Cr (max) 0.1% Mo (max) 0.05% V (max) steel: Plate; French Standard; normalized **UTS: 570 Proof: 340**
NH	0.3% C 4.1% Ni 1.3% Cr 0.3% Mo steel: Obsolete; S Osborn; replaced by RABI
NHP	0.26% C 2.5% Ni 3% Cr 0.5% Mo 0.3% V 9% W steel: For tools; Huntsman; forging dies; for nimonic alloys **DPN: 460**
NI MO	0.26% C (max) 0.3% Mn 3.0% Ni 0.6% Mo steel: Isaacson
NiCr 322	0.3% C 3.2% Ni 0.75% Cr 0.3% Mo steel: Wrought; Belgian national specification
NiCr 342	0.3% C 4.2% Ni 1.2% Cr steel: Wrought; Belgian national specification
NiCrMo 335	0.3% C 3.3% Ni 0.9% Cr 0.4% Mo steel: Wrought: Belgian national specification
NiCrMo 342	0.3% C 4.2% Ni 1.3% Cr 0.3% Mo steel: Wrought; Belgian national specification
NiCrMo 415	0.3% C 1.5% Ni 1.2% Cr 0.28% Mo steel: Wrought; Belgian national specification
NICUAGE 1	0.06% C 0.85% Ni 0.02% Nb 1.15% Cu steel: International Nickel Co.; aged weldable **DPN: 230 UTS: 700 Elon: 23% Proof: 640**
NICUAGE 1	0.04% C 0.8% Ni 0.02% Nb (min) 1.1% Cu steel: Age hardening; for use in high strength structural work; International Nickel Co. **DPN: 230 UTS: 680 Elon: 23% Proof: 650**
NIT-KD	0.25% C 3.2% Cr 0.3% Mo steel: Pompey
NITRAL 4	0.34% C 1.65% Cr 0.2% Mo 0.9% Al steel: Pompey
NITRALLOY 5	0.3% C 1.5% Cr 0.2% Mo 1.1% Al steel: ESC; BS alloy En 41A; nitriding steel
NITRALLOY LK7	0.2% C 1.6% Cr 0.2% Mo 1.1% Al steel: For nitriding; Firth Brown
NK HITEN 60A	0.18% C (max) 1.5% Mn 0.3% Mo 0.1% V steel: Nippon Kokan
NK HITEN 60B	0.16% C (max) 1.35% Mn 0.6% Ni 0.4% Cr 0.3% Mo 0.15% Cu steel: Nippon Kokan

Symbol	Nominal analysis, supplier, condition and remarks.
NK HITEN 70	0.18% C (max) 1.2% Mn 1.0% Ni 0.8% Cr 0.6% Mo 0.22% Cu steel: Nippon Kokan
NK HITEN 80	0.18% C (max) 1.0% Mn 1.0% Ni 0.8% Cr 0.6% Mo 0.1% V 0.3% Cu 0.006% B steel: Nippon Kokan
NK HITEN 100 HW90	0.18% C (max) 0.9% Mn 0.9% Ni 0.6% Cr 0.5% Mo 0.15% V 0.3% Cu steel: Nippon Kokan
NO. 158	0.1% C 3.5% Ni 1.5% Cr steel: Carpenter for AISI type P6
NO158	0.1% C 3.5% Ni 1.5% Cr steel: Carpenter
NO345	0.35% C 5% Cr 1.5% Mo 1.25% W steel: Carpenter
NORESCO NSW	0.3% C 1.3% Cr 0.35% Mo 4.25% W steel: Schoeller-Bleckmann **UTS: 1530**
NORESCO PARFORCE SPECIAL 10	0.15% C 3.6% Ni 0.75% Cr steel: For case hardening; Schoeller-Bleckmann; for plastic moulds
NORESCO SG	0.26% C 2.75% Cr 0.15% V 9% W steel: Schoeller-Bleckmann **UTS: 1500**
NORESCO SGW	0.23% C 1.5% Ni 2.5% Cr 0.15% Mo 0.15% V 9.5% W steel: Schoeller-Bleckmann **UTS: 1500**
NORESCO WCD	0.35% C 5% Cr 1.2% Mo 0.35% V 1.2% W 1% Si steel: Schoeller-Bleckmann **UTS: 1550**
NORESCO WCD 2	0.33% C 5% Cr 1.1% Mo 0.5% V 1% Si steel: Schoeller-Bleckmann **UTS: 1500**
NSW	0.3% C 1.3% Cr 0.35% Mo 4.25% W steel: Schoeller-Bleckmann **UTS: 1500**
N-TUP CR196	0.13% C 5.5% Ni 0.2% Mo steel: Nippon Steel Corp.; information from Climax Molybdenum **UTS: 760 Elon: 22% Proof: 600**
NV 2	0.35% C 5% Cr 1.6% Mo 0.2% V 1.5% W 1% Si tool steel: For hot working; S Osborn; dies, shears, blades, punches, etc.; obsolete **DPN: 460**
NV 7-3	0.18% C (max) 0.45% Mn 2.25% Cr 1.0% Mo steel: Designation used by DNV **UTS: 450 Elon: 20% Proof: 280**
NVR 7-2	0.15% C (max) 0.6% Mn 0.9% Cr 0.55% Mo steel: Designation used by DNV
NVR 7-3	0.15% C (max) 0.45% Mn 2.25% Cr 1.0% Mo steel: Designation used by DNV
O 35 XHK	0.25% C steel sheet with alloy elements; SAE J1392 grade; strength given by figures; type by letters
O 40 XHK	0.25% C steel sheet with alloy elements: Coated; SAE J 1392 grade; strength given by figures; type by letters
O 45 XHK	0.25% C steel sheet with alloy elements: Coated; SAE J 1392 grade; strength given by figures; type by letters
O 50 XHK	0.25% C steel sheet with alloy elements: Coated; SAE J 1392 grade; strength given by figures; type by letters
O 60 XHK	0.25% C steel sheet with alloy elements: Coated; SAE J 1392 grade; strength given by figures; type by letters
O 70 XHK	0.25% C steel sheet with alloy elements: Coated; SAE J 1392 grade; strength given by figures; type by letters
O 80 XHK	0.25% C steel sheet with alloy elements: Coated; SAE J 1392 grade; strength given by figures; type by letters

Note. The following abbreviations and units are used in the tables:

DPN	Hardness, diamond pyramid number
UTS	Ultimate tensile strength, N/mm^2
Elon	Elongation, %
Proof	0.1% proof strength, N/mm^2

1 N/mm^2=0.1 hbar=0.102 kgf/mm^2=0.06475 tonf/in.^{2}=145.04 lbf/in.^{2}=1 MPa

See Appendix II for other abbreviations and conversion tables.

Symbol	Nominal analysis, supplier, condition and remarks.
OCM 5	0.2% C. 1.5% Mn 0.4% Ni 0.25% Cr steel: W Marrison for BS alloy En 14
ON 3	0.3% C 3.1% Ni 0.3% Cr steel: W Marrison for BS alloy En 21
ON 5	0.3% C 3.1% Ni 0.8% Cr 0.6% Mo steel: W Marrison for BS alloy En 23
ON 7	0.3% C 2.6% Ni 0.7% Cr 0.6% Mo steel: W Marrison for BS alloy En 25
ON 9	0.3% C 3.4% Ni 0.9% Cr 0.5% Mo steel: W Marrison for BS alloy En 27
ON 10	0.35% C 4% Ni 1.1% Cr 0.5% Mo steel: W Marrison for BS alloy En 28
ON 12	0.35% C 1.3% Ni 0.6% Cr steel: W Marrison for BS alloy En 111
ORELLOY 242	0.22% C 0.5% Mn 0.45% Ni 0.86% Cr 0.4% Cu steel: Oregon
OSBORN 501	0.2% C 5% Cr 0.5% Mo steel: Osborn; previously PY grade
OX602	0.2% C (max) 1.7% Mn 0.09% V steel: Plate; Svenskt Stal; graded by impact **UTS: 650 Elon: 18% Proof: 490**
OX702	0.2% C (max) 1.5% Mn 0.45% Cr + Mo + V (max) 0.005% B steel: Plate; Svenskt Stal; graded by impact **UTS: 770 Elon: 18% Proof: 600**
OX812	0.2% C (max) 1.7% Mn 0.37% Cr + Mo + V 0.005% B steel: Plate; Svenskt Stal; graded by impact **UTS: 810 Elon: 16% Proof: 620**
OX1222BM	0.25% C (max) 1.7% Mn 1.5% Ni 1% Mo 0.1% V 0.01% Al 0.005% B steel: Plate; Svenskt Stal **UTS: 1200 Elon: 10% Proof: 1100**
P 2	0.07% C 2.0% Cr 0.2% Mo 0.5% Co steel: Designation used by AISI
P 3	0.11% C (max) 1.25% Ni 0.55% Cr tool steel: Designation used by AISI
P 4	0.07% C 5.0% Cr 0.75% Mo steel: Designation used by AISI
P 20	0.35% C 1.25% Cr 0.4% Mo steel: Designation used by AISI
P 21	0.2% C 4.1% Ni 0.25% Cr 0.2% V steel: For tools; listed in ASTM A681 1977
P 21	0.2% C 4.1% Ni 0.25% Cr 0.2% V steel: For tools; designation used by AISI
P 155	0.15% C 3.5% Ni 1% Cr steel: For case hardening; Carrs
P 157	0.15% C 4.25% Ni 1.2% Cr steel: For case hardening; Carrs
P 158	0.15% C 4.2% Ni 1.2% Cr 0.3% Mo steel: Carpenter
P 552	0.3% C 4.25% Ni 1.25% Cr 0.3% Mo steel: Carrs; gears, shafts, bolts, etc.; can be case hardened **UTS: 1300**
P 558	0.3% C 2.5% Ni 0.7% Cr 0.5% Mo steel: Carrs; crankshafts, gears, spindles, etc. **DPN: 350 UTS: 1070**
P 564	0.3% C 3% Ni 0.75% Cr steel: Carrs; crankshafts, bolts, spindles, etc. **DPN: 350 UTS: 1000**
P 612	0.3% C 1.6% Cr 0.2% Mo 1% Al steel: For nitriding; Carrs **UTS: 610**
P 615	0.3% C 0.3% Ni 3% Cr 0.4% Mo steel: For nitriding; Carrs **UTS: 1000**
PARFORCE SPECIAL 10	0.15% C 3.6% Ni 0.75% Cr steel: For case hardening; Schoeller-Bleckmann; for plastic moulds
PCS	0.35% C 5% Cr 1.5% Mo 0.25% V 1.4% W 1% Si steel: T Turton
PDX	0.28% C 2.25% Ni 0.85% Cr 0.5% Mo 0.3% V 2.25% H steel: For tools; Balfour Darwin
PF 1	0.15% C 0.5% Cr 0.5% Mo steel: Pompey
PF 2	0.12% C 1.1% Cr 0.5% Mo steel: Pompey
PF 2M	0.1% C 1.1% Cr 0.5% Mo steel: Pompey

Symbol	Nominal analysis, supplier, condition and remarks.
PF 5	0.12% C 4.4% Cr 0.5% Mo steel: Pompey
PF 5K	0.08% C 5.0% Cr 0.6% Mo steel: Pompey
PF 6	0.12% C 2.25% Cr 1.0% Mo steel: Pompey
PF 15	0.15% C 0.8% Cr 0.4% Mo steel: Pompey
PFDV	0.13% C 1.35% Cr 0.8% Mo 0.25% V steel: Pompey
PFV 55	0.12% C 2.0% Cr 0.35% Mo 0.08% V 0.4% Al steel: Pompey
PHILIPS 35S	0.17% C 0.44% Ni 0.1% Cr 0.02% Mo 0.01% V steel: Welding electrode; Philips
PHILIPS 88	0.12% C (max) 0.7% Ni 0.1% Cr (max) 0.4% Mo steel: Welding electrode; Philips **UTS: 670 Elon: 23% Proof: 500**
PHILIPS 98	0.1% C (max) 1.6% Ni 0.15% Cr (max) 0.35% Mo (max) steel: Welding electrode; Philips **UTS: 670 Elon: 26% Proof: 580**
PHILIPS 108	0.1% C (max) 1.8% Ni 0.35% Cr (max) 0.35% Mo (max) steel: Welding electrode; Philips **UTS: 720 Elon: 24% Proof: 660**
PHILIPS 118	0.1% C (max) 2.0% Ni 0.4% Cr (max) 0.3% Mo steel: Welding electrode; Philips **UTS: 790 Elon: 20% Proof: 710**
PHILIPS 400	0.12% C 2.6% Cr 0.9% Mo steel: Welding electrode; Philips
PHILIPS KV1	0.07% C 0.8% Cr 0.5% Mo steel: Welding electrode; Philips **UTS: 600 Elon: 22% Proof: 500**
PHILIPS KV3	0.05% C (max) 22% Cr 1.0% Mo steel: Welding electrode: Philips **UTS: 640 Elon: 23% Proof: 520**
PHILIPS KV4	0.05% C (max) 5% Cr 0.5% Mo steel: Welding electrode: Philips **UTS: 640 Elon: 22% Proof: 540**
PHILIPS KV5	0.05% C (max) 1.2% Cr 0.5% Mo steel: Welding electrode; Philips **UTS: 590 Elon: 25% Proof: 490**
PHILIPS KV7	0.06% C 9.0% Cr 1.0% Mo steel: Welding electrode; Philips **UTS: 590 Elon: 27% Proof: 440**
PLASMOLD	0.35% C 4.3% Ni 1.3% Cr 0.3% Mo steel: Firth Brown; for plastic dies
PLASTIFORM	0.3% C 4% Ni 1.3% Cr 0.25% Mo steel: Clyde Alloy **DPN: 550**
PLMA/5	0.1% C 5% Cr 0.7% Mo 0.2% V steel: ESC for plastic moulds; case hardened
PLMC/3	0.16% C 4.2% Ni 1.1% Cr 0.25% Mo steel: ESC for plastic moulds; case hardened
PM 17	0.14% C 1.55% Mn 0.12% Mo 0.07 V steel: Pompey
PM 18	0.18% C 1.5% Mn 0.4% Ni 0.25% Cr 0.15% Mo 0.3% Cu steel: Casting; Wolsingham **UTS: 480 Elon: 20%**
PM 23	0.23% C 1.4% Mn 0.4% Ni 0.25% Cr 0.15% Mo 0.3% Cu steel: Casting; Wolsingham **UTS: 610 Elon: 20%**
PM 24	0.19% C 0.4% Mo 0.1% V steel: Pompey
PM 30	0.3% C 1.5% Mn 0.4% Ni 0.25% Cr 0.15% Mo 0.3% Cu steel: Casting; Wolsingham **UTS: 760 Elon: 17%**
PM 35	0.35% C 1.5% Mn 0.4% Ni 0.25% Cr 0.15% Mo 0.3% Cu steel: Casting; Wolsingham **UTS: 610 Elon: 17%**
PM 35	0.18% C 1.5% Cr 0.2% Mo 0.07% V steel: Pompey
PM 38	0.28% C 0.8% Ni 1.7% Cr 0.5% Mo 0.1% V steel: Pompey
PNUSNAP WH	0.35% C 1% Cr 1.8% W steel: Firth Brown; punches, shears
PREMO	0.05% C 3.9% Cr 0.5% Mo 0.1% V steel: Uddelholm for type P 4
PRESSURDIE/2	0.35% C 5.0% Cr 1.45% Mo 0.3% V 1.2% W steel: Braeburn; AISI type H12
PREXI	0.15% C 1.1% Cr 0.25% Mo steel: Uddelholm

Symbol	Nominal analysis, supplier, condition and remarks.
Pro No 1	0.3% C 0.85% Cr 0.15% Mo steel: Origin unknown; for bicycle frames
Pro No 2	0.3% C 0.85% Cr 0.15% Mo steel: Origin unknown; for bicycle frames
Pro No 3	0.3% C 0.85% Cr 0.15% Mo steel: Origin unknown; for bicycle frames
PS10	0.21% C 0.3% Ni 0.3% Cr 0.07% Mo steel: SAE potential specification
PS15	0.2% C 0.5% Cr 0.17% Mo steel: SAE potential specification
PS16	0.22% C 0.5% Cr 0.17% Mo steel: SAE potential specification
PS17	0.25% C 0.5% Cr 0.17% Mo steel: SAE potential specification
PS18	0.27% C 0.5% Cr 0.17% Mo steel: SAE potential specification
PS19	0.2% C 0.5% Cr 0.11% Mo 0.0004% B steel: SAE potential specification
PS20	0.15% C 0.5% Cr 0.17% Mo steel: SAE potential specification
PS20	0.17% C 0.5% Cr 0.17% Mo steel: SAE potential specification
PS24	0.2% C 0.55% Cr 0.25% Mo steel: SAE potential specification
PS30	0.15% C 0.8% Ni 0.55% Cr 0.52% Mo steel: SAE potential specification
PS31	0.17% C 0.8% Ni 0.55% Cr 0.52% Mo steel: SAE potential specification
PS32	0.2% C 0.8% Ni 0.55% Cr 0.52% Mo steel: SAE potential specification
PS33	0.2% C 0.2% Ni (min) 0.2% Cr (min) 0.05% Mo (min) steel: SAE potential specification
PS34	0.3% C 0.5% Cr 0.17% Mo steel: SAE potential specification
PS54	0.22% C 0.55% Cr 0.05% Mo (min) steel: SAE potential specification
PS55	0.17% C 1.8% Ni 0.55% Cr 0.7% Mo steel: SAE potential specification
PS56	0.1% C 1.8% Ni 0.55% Cr 0.7% Mo steel: SAE potential specification
PS66	0.18% C 1.8% Ni 0.6% Cr 0.1% Mo steel: SAE potential specification
PX 80 PLUS	0.2% 0.7% Cr 0.9% Mn 0.3% Mo 0.0025% B steel: Phoenix
PX 90 PHIS	0.2% C 0.9% Mn 0.7% Cr 0.3% Mo 0.0025% B steel: Phoenix
PX 100 PHIS	0.2% C 0.9% Mn 0.7% Cr 0.3% Mo 0.0025% B steel: Phoenix
PX 110 PHIS	0.2% C 0.9% Mn 0.7% Cr 0.3% Mo 0.0025% B steel: Phoenix
PX 360 PHIS	0.2% C 0.9% Mn 0.7% Cr 0.3% Mo 0.0025% B steel: Phoenix
PY 8720	0.2% C 0.5% Ni 0.5% Cr 0.2% Mo steel: Pompey
PYROTOUGH	0.35% C 2.75% Cr 0.25% V 0.5% W steel: Clyde Alloy **DPN: 460**
PYROWEAR ALLOY 53	0.1% C 2% Ni 1% Cr 3.2% Mo 0.1% V 2% Cu steel: Carpenter; for carburizing
PZ 6042	0.12% C 1.2% Cr 0.5% Mo steel wire: For welding; Philips **UTS: 600 Elon: 21% Proof: 510**
PZ 6043	0.1% C 2.4% Cr 1% Mo steel wire: For welding; Philips **UTS: 650 Elon: 23% Proof: 510**
PZ 6044	0.1% C 5.9% Cr 0.5% Mo steel wire: For welding; Philips **UTS: 600 Elon: 22% Proof: 540**
PZ 6047	0.12% C 0.5% Ni 0.5% Cr 0.2% Mo steel wire: For welding; Philips **UTS: 740 Elon: 20% Proof: 620**

Symbol	Nominal analysis, supplier, condition and remarks.
R 9030	0.35% C 2.7% Ni 0.7% Cr 0.5% Mo steel: For hot work tools; Balfour; moulding dies; piercing punches **DPN: 550**
R100	0.2% C 0.55% Ni 0.5% Cr 0.2% Mo steel: Weld electrode; designation used by AWS
RABI	0.3% C 4.1% Ni 1.3% Cr 0.3% Mo steel: S Osborn; mould steel; hardened and tempered **DPN: 500**
RBD	0.3% C 3% Cr 0.5% V 10% W steel: Firth Brown
RBD-E	0.22% C 2.2% Ni 2.2% Cr 0.45% Mo 0.2% V 10% W steel: Firth Brown
RED DIAMOND 20H	0.25% C 1.0% Cr 0.2% Mo steel: For wear; Redheugh **DPN: 260 UTS: 820 Elon: 18% Proof: 630**
RED DIAMOND 20S	0.25% C 1.0% Cr 0.2% Mo steel: For wear; Redheugh **DPN: 220 UTS: 710 Elon: 25% Proof: 600**
RED INDIAN	0.35% C 5.0% Cr 0.3% Mo 0.3% V 4.5% W 0.5% Co steel: Atlas; as AISI type H14
REPUBLIC 50	0.15% C (max) 0.7% Mn 0.7% Ni 0.3% Cr 0.1% Mo 0.7% Cu steel: Republic Steel
REPUBLIC 65	0.15% C 1.0% Mn 1.2% Ni 0.25% Mo 1.1% Cu steel: Republic Steel
REPUBLIC 70	0.2% C (max) 1.0% Mn 1.5% Ni 0.25% Mo 1.2% Cu steel: Republic Steel
RHB 65	0.2% C (max) 1.4% Mn 0.7% Ni 0.18% V steel: German proprietor's specification
RIVER ACE 70	0.18% C (max) 1.2% Mn 1.0% Ni 0.4% Cr 0.4% Mo 0.08% V 0.4% Cu 0.005% B steel: Kawasaki Iron and Steel Co.
RIVER ACE KO	0.18% C (max) 1.0% Mn 1.5% Ni 0.8% Cr 0.6% Mo 0.08% V 0.5% Cu 0.006% B steel: Kawasaki Iron and Steel Co.
RO 211	0.12% C 2.5% Cr 1.1% Mo steel: Bofors; normalized and tempered **DPN: 190 UTS: 610 Elon: 26% Proof: 460**
RO 346	0.16% C 1.0% Cr 0.6% Mo steel: Bofors; hardened and tempered **DPN: 240 UTS: 770 Elon: 20% Proof: 680**
RO 651	0.3% C 1.0% Cr 0.2% Mo steel: Bofors **DPN: 325 UTS: 1100 Elon: 11%**
RO 653	0.25% C 1.1% Cr 0.2% Mo steel: Bofors; hardened and tempered **DPN: 300 UTS: 990 Elon: 13% Proof: 700**
RO 663	0.28% C 1.2% Cr 0.3% Mo steel: Bofors
RO 752	0.35% C 1.1% Cr 0.2% Mo steel: Bofors; hardened and tempered **DPN: 280 UTS: 820 Elon: 14% Proof: 580**
RO 4154	0.25% C 3.0% Cr 0.5% Mo steel: Bofors; hardened and tempered **DPN: 260 UTS: 840 Elon: 20% Proof: 710**
RO 7155	0.35% C 3.0% Cr 0.5% Mo steel: Bofors; hardened and tempered **DPN: 350 UTS: 1140 Elon: 12% Proof: 890**
ROP 10	0.08% C 3.0% Cr 0.8% Mo 0.2% V steel: For carburizing; Bofors
ROP 63	0.3% C 0.5% Ni 2.5% Cr 0.3% Mo 0.3% V steel: For nitriding; Bofors; hardened and tempered; obsolete **DPN: 320 UTS: 1030 Elon: 12% Proof: 820**
ROP 5462	0.24% C 1.3% Cr 0.5% Mo steel: Bofors; hardened and tempered **DPN: 260 UTS: 820 Elon: 20% Proof: 700**
ROPT	0.2% C 3.0% Cr 0.6% Mo 0.8% V 0.5% W steel: Bofors; obsolete
RS 2	0.17% C 1.8% Ni 0.25% Mo steel: Toledo for BS alloy En 34 and En 35
RS 3	0.12% C 3% Ni 0.3% Cr steel: Toledo for BS alloy En 33
RT 27	0.35% C 3.5% Ni 1.0% Cr 0.2% V 5.5% W steel: Bofors; obsolete **DPN: 350**
RT 45	0.3% C 3.0% Cr 0.3% V 10% W steel: Bofors; obsolete **DPN: 450**
S133	0.16% C 0.9% Ni 0.6% Cr steel for carburizing: BS designation for aerospace steel
S134	0.4% C 3.2% Cr 1% Mo 0.2% V steel: BS designation for aerospace steel
S138	0.4% C 3.2% Cr 1% Mo steel: BS designation for aerospace steel
S139	0.4% C 1.5% Ni 1.3% Cr 0.27% Mo steel: BS designation for aerospace steel
S140	0.32% C 2.5% Ni 0.6% Cr 0.5% Mo steel: BS designation for aerospace steel
S142	0.25% C 1% Cr 0.2% Mo steel: BS designation for aerospace steel
S146	0.38% C 4% Ni 1.8% Cr 0.5% Mo steel: Vacuum melted; BS designation for aerospace steel
S147	0.4% C 0.5% Ni 0.5% Cr 0.25% Mo steel: For bolts; BS designation for aerospace steel
S148	0.38% C 1.2% Ni 0.6% Cr steel: For bolts; BS designation for aerospace steel
S149	0.4% C 1.8% Ni 0.8% Cr 0.25% Mo steel: For bolts; BS designation for aerospace steel
S153	0.3% C 2.5% Ni 0.6% Cr 0.5% Mo steel: BS designation for aerospace steel
S154	0.3% C 2.5% Ni 0.6% Cr 0.5% Mo steel: BS designation for aerospace steel
S155	0.41% C 1.8% Ni 0.8% Cr 0.4% Mo 0.08% V 1.7% Si steel: BS designation for aerospace steel
S156	0.16% C 4% Ni 1.2% Cr 0.25% Mo steel: Carburizing; BS designation for aerospace steel
S157	0.15% C 3.2% Ni 1% Cr 0.25% Mo steel: Carburizing; BS designation for aerospace steel
S158	0.25% C 1% Cr 0.2% Mo steel: For bolts; BS designation for aerospace steel
S534	0.25% C 1% Cr 0.2% Mo steel: Weldable
S535	0.25% C 1% Cr 0.2% Mo steel: Weldable
SABEX	0.3% C 1.7% Ni 2.2% Cr 0.1% V steel: Sanderson; hot shear blades, casting dies **DPN: 610**
SAE 43 BV12	0.12% C 1.8% Ni 0.5% Cr 0.25% Mo 0.03% V steel: Obsolete
SAE 43 BV14	0.14% C 1.8% Ni 0.5% Cr 0.1% Mo 0.03% V steel: Obsolete
SAE 46 B12	0.12% C 1.85% Ni 0.25% Mo steel
SAE 94 B15	0.15% C 0.5% Ni 0.35% Cr 0.11% Mo steel with B
SAE 94 B17	0.17% C 0.45% Ni 0.4% Cr 0.1% Mo 0.0005% B steel
SAE 94 B30	0.3% C 0.45% Ni 0.4% Cr 0.1% Mo 0.0005% B (min) steel
SAE 3115	0.15% C 1.2% Ni 0.6% Cr steel: Obsolete
SAE 3120	0.2% C 1.2% Ni 0.6% Cr steel: Obsolete
SAE 3130	0.3% C 1.2% Ni 0.6% Cr steel
SAE 3135	0.35% C 1.2% Ni 0.6% Cr steel
SAE 3215	0.15% C 1.7% Ni 1% Cr steel: Obsolete
SAE 3220	0.2% C 1.7% Ni 1% Cr steel: Obsolete
SAE 3230	0.3% C 1.7% Ni 1% Cr steel: Obsolete
SAE 3240	0.4% C 1.7% Ni 1% Cr steel: Obsolete
SAE 3310	0.1% C 3% Ni 1.5% Cr steel: As AISI 3310
SAE 3312	0.1% C 3.5% Ni 1.5% Cr steel: Obsolete

Note. The following abbreviations and units are used in the tables:

DPN	Hardness, diamond pyramid number
UTS	Ultimate tensile strength, N/mm^2
Elon	Elongation, %
Proof	0.1% proof strength, N/mm^2

1 N/mm^2=0.1 hbar=0.102 kgf/mm^2=0.06475 $tonf/in.^2$=145.04 $lbf/in.^2$=1 MPa

See Appendix II for other abbreviations and conversion tables.

Symbol	Nominal analysis, supplier, condition and remarks.
SAE 3316	0.16% C 3.5% Ni 1.6% Cr steel: Obsolete
SAE 3325	0.25% C 3.5% Ni 1.5% Cr steel: Obsolete
SAE 3335	10.35% C 3.5% Ni 1.5% Cr steel: Obsolete
SAE 3415	0.15% C 3% Ni 0.8% Cr steel: Obsolete
SAE 3435	0.35% C 3.0% Ni 0.8% Cr steel
SAE 3436	0.35% C 3% Ni 0.8% Cr steel: Obsolete
SAE 4118	0.2% C 0.5% Cr 0.12% Mo steel
SAE 4119	0.19% C 0.5% Cr 0.25% Mo steel
SAE 4125	0.25% C 0.5% Cr 0.25% Mo steel: Obsolete
SAE 4130	0.3% C 1% Cr 0.2% Mo steel
SAE 4135	0.35% C 1.0% Cr 0.2% Mo steel
SAE 4137	0.37% C 1% Cr 0.2% Mo steel
SAE 4317	0.17% C 1.8% Ni 0.5% Cr 0.25% Mo steel: Obsolete
SAE 4320	0.2% C 1.8% Ni 0.5% Cr 0.25% Mo steel
SAE 4337	0.37% C 1.8% Ni 0.8% Cr 0.25% Mo steel
SAE 4608	0.08% C 1.5% Ni 0.2% Mo steel
SAE 4615	0.15% C 1.8% Ni 0.25% Mo steel
SAE 4617	0.17% C 1.8% Ni 0.25% Mo steel
SAE 4620	0.2% C 1.8% Ni 0.25% Mo steel
SAE 4621	0.21% C 1.8% Ni 0.25% Mo steel
SAE 4626	0.26% C 0.8% Ni 0.2% Mo steel
SAE 4718	0.18% C 1% Ni 0.45% Cr 0.35% Mo steel
SAE 4720	0.2% C 0.1% Ni 0.45% Cr 0.2% Mo steel
SAE 4812	0.12% C 3.5% Ni 0.25% Mo steel
SAE 4815	0.15% C 3.5% Ni 0.25% Mo steel
SAE 4817	0.17% C 3.5% Ni 0.25% Mo steel
SAE 4820	0.2% C 3.5% Ni 0.25% Mo steel
SAE 6115	0.15% C 1% Cr 0.15% V steel: Obsolete
SAE 6117	0.17% C 0.8% Cr 0.1% V steel
SAE 6118	0.18% C 0.6% Cr 0.12% V steel
SAE 6120	0.2% C 0.8% Cr 0.1% V steel
SAE 6125	0.25% C 1% Cr 0.15% V steel: Obsolete
SAE 6130	0.3% C 1% Cr 0.15% V steel: Obsolete
SAE 6135	0.35% C 1% Cr 0.15% V steel: Obsolete
SAE 8115	0.15% C 0.3% Ni 0.4% Cr 0.11% Mo steel
SAE 8615	0.15% C 0.5% Ni 0.5% Cr 0.2% Mo steel
SAE 8617	0.17% C 0.6% Ni 0.5% Cr 0.2% Mo steel
SAE 8620	0.2% C 0.6% Ni 0.5% Cr 0.2% Mo steel
SAE 8622	0.22% C 0.6% Ni 0.5% Cr 0.2% Mo steel
SAE 8625	0.25% C 0.6% Ni 0.5% Cr 0.2% Mo steel
SAE 8627	0.27% C 0.6% Ni 0.5% Cr 0.2% Mo steel
SAE 8630	0.3% C 0.6% Ni 0.5% Cr 0.2% Mo steel
SAE 8632	0.32% C 0.6% Ni 0.5% Cr 0.2% Mo steel: Obsolete
SAE 8635	0.35% C 0.6% Ni 0.5% Cr 0.2% Mo steel
SAE 8637	0.37% C 0.6% Ni 0.5% Cr 0.2% Mo steel
SAE 8715	0.15% C 0.6% Ni 0.5% Cr 0.25% Mo steel
SAE 8717	0.17% C 0.6% Ni 0.5% Cr 0.25% Mo steel: Obsolete
SAE 8719	0.2% C 0.6% Ni 0.5% Cr 0.25% Mo steel: Obsolete
SAE 8720	0.2% C 0.6% Ni 0.5% Cr 0.25% Mo steel
SAE 8735	0.35% C 0.6% Ni 0.5% Cr 0.25% Mo steel: Obsolete
SAE 8822	0.22% C 0.6% Ni 0.5% Cr 0.35% Mo steel
SAE 9310	0.1% C 3.2% Ni 1.2% Cr 0.1% V steel
SAE 9315	0.15% C 3.2% Ni 1.2% Cr 0.1% Mo steel
SAE 9317	0.17% C 3.2% Ni 1.2% Cr 0.1% Mo steel
SAE 9437	0.37% C 0.4% Ni 0.4% Cr 0.1% Mo steel: Obsolete
SAE 51501	0.1% C (min) 5% Cr 0.5% Mo steel
SAE 51502	0.1% C 5% Cr 0.5% Mo steel
SAE 60502	0.25% C 5% Cr 0.5% Mo steel: Casting
SAE 70502	0.25% C (max) 5.5% Cr 0.5% Mo steel: Casting
SAE H11	0.35% C 5% Cr 1.5% Mo 0.4% V steel: For tools
SAE H12	0.35% C 5% Cr 1.5% Mo 0.3% V 1.5% H steel: For tools
SAE H13	0.35% C 5% Cr 1.5% Mo 1% V steel: For tools
SAE H21	0.35% C 3.5% Cr 0.4% V 9% W steel: For tools
SAE J1392	Specification of steel sheet: Coated; listed by grades
SAE X4620	0.2% C 1.8% Ni 0.25% Mo steel
SANBOLD 21	0.3% C 3% Ni 0.3% Cr steel: Sanderson for BS alloy En 21
SANBOLD 23	0.3% C 3% Ni 0.8% Cr 0.5% Mo steel: Sanderson for BS alloy En 23; Mo optional

Symbol	Nominal analysis, supplier, condition and remarks.
SANBOLD 25	0.3% C 2.5% Ni 0.7% Cr 0.5% Mo steel: Sanderson for BS alloy En 25
SANBOLD 30	0.32% C 4.2% Ni 1.3% Cr 0.3% Mo steel: Sanderson for BS alloy En 30
SANBOLD 33	0.12% C 3.25% Ni 0.3% Cr steel: Sanderson for BS alloy En 33
SANBOLD 34	0.18% C 1.7% Ni 0.25% Mo steel: Sanderson for BS alloy En 34
SANBOLD 36	0.16% C 3.5% Ni 0.8% Cr steel: Sanderson for BS alloy En 36A, B or C; C varied to suit
SANBOLD 37	0.16% C 5% Ni 0.3% Cr steel: Sanderson for BS alloy En 37
SANBOLD 38	0.16% C 5% Ni 0.3% Cr 0.2% Mo steel: Sanderson for BS alloy En 38
SANBOLD 39	0.16% C 4.2% Ni 1.2% Cr 0.2% Mo steel: Sanderson for BS alloy En 39
SANBOLD 56	0.3% C 2.5% Ni 0.7% Cr 0.5% Mo steel: Sanderson, replaced by Sanbold 25
SANBOLD 58	0.18% C 1.7% Ni 0.25% Mo steel: Sanderson; replaced by Sanbold 34
SANBOLD 68	0.16% C 4.2% Ni 1.2% Cr 0.2% Mo steel: Sanderson; replaced by Sanbold 39
SANBOLD 69	0.16% C 5% Ni 0.3% Cr 0.2% Mo steel: Sanderson; replaced by Sanbold 38
SANBOLD 77	0.32% C 4.2% Ni 1.3% Cr 0.3% Mo steel: Sanderson; replaced by Sanbold 30
SANBOLD 101	0.16% C 3.5% Ni 0.8% Cr steel: Sanderson; replaced by Sanbold 36
SANBOLD N 2832	0.12% C 3.25% Ni 0.3% Cr steel: Sanderson; replaced by Sanbold 33
SANBOLD N 3603	0.16% C 5% Ni 0.3% Cr steel: Sanderson; replaced by Sanbold 37
SANBOLD N 5366	0.3% C 3% Ni 0.3% Cr steel: Sanderson; replaced by Sanbold 21
SANBOLD NK 8055	0.3% C 3% Ni 0.8% Cr 0.5% Mo steel: Sanderson; replaced by Sanbold 23
SB 46F	0.15% C 1.2% Mn 0.45% Ni 0.2% V 0.35% Cu steel: German proprietor's specification
SB 46FT	0.15% C 1.2% Mn 0.45% Ni 0.2% V 0.35% Cu steel: German proprietor's specification
SCORPION	0.3% C 3.4% Ni 1.3% Cr steel: T Turton
SD 20	0.35% C 1.7% Ni 1.2% Cr steel: Balfour; collets, gears, etc. **DPN: 550**
SENECA	0.35% C 3.2% Cr 0.4% V 9.5% W steel: Atlas; as AISI type H 21
SG	0.26% C 2.75% Cr 0.15% V 9% W steel: Schoeller-Bleckmann **UTS: 1420**
SGW	0.23% C 1.5% Ni 2.5% Cr 0.15% Mo 0.15% V 9.5% W steel: Schoeller-Bleckmann **UTS: 1420**
SGW 1	0.35% C 1.5% Ni 1% Cr 0.25% Mo steel: Jonas **UTS: 1060 Elon: 16% Proof: 1000**
SGW 2	0.3% C 3% Ni 0.75% Cr steel: Jonas; hardened and tempered **UTS: 1300 Elon: 13% Proof 1140**
SGW 3	0.3% C 3% Ni 0.6% Cr steel: Jonas; hardened and tempered **UTS: 1210 Elon: 13% Proof: 1070**
SHEFFIELD HIGH STRENGTH A	0.12% C 0.4% Mn 0.7% Ni 0.6% Cr 0.1% Mo 0.4% Cu 0.03% Ti steel: Armco
SHEF-LO-TEMP	0.18% C 0.2% Ni 0.2% Cr 0.07% Mo 0.2% Cu steel: Armco; for low temperature use; hardened and tempered **UTS: 580 Elon: 23% Proof: 340**
SHEF-SUPER-LO-TEMP	0.16% C 0.23% Ni 0.18% Cr 0.07 Mo 0.2% Cu steel: Armco; for low temperature use; hardened and tempered **UTS: 670 Elon: 23% Proof: 390**

Symbol	Nominal analysis, supplier, condition and remarks.
SHNC	0.3% C 4.25% Ni 1.25% Cr steel: ESC; for BS alloy En 30A
For SIS specifications see SS	
SKC 31	0.18% C 0.9% Mn 3.0% Ni 1.6% Cr 0.6% Mo steel: For hollow drills; Japanese Standards designation
SKD 2	0.18% C 1.0% Cr 0.2% Mo steel: Pompey
SKD 4	0.3% C 2.5% Cr 0.4% V 5.5% W steel: Japanese Standards designation
SKD 5	0.3% C 2.5% Cr 0.4% V 9.5% W steel: Japanese Standards designation
SKF 126	0.2% C 0.1% Mn 0.5% Cr steel: Carburizing; SKF
SKF 145	0.16% C 1% Ni 0.8% Cr steel: Carburizing; SKF
SKF 146	0.21% C 1% Ni 0.8% Cr steel: Carburizing; SKF
SKF 152	0.2% C 0.5% Ni 0.5% Cr 0.2% Mo steel: Carburizing; SKF
SKF 153	0.2% C 1% Ni 0.45% Cr 0.2% Mo steel: Carburizing; SKF
SKF 157	0.2% C 1.8% Ni 0.5% Cr 0.25% Mo steel: Carburizing; SKF
SKF 255	0.17% C 3.5% Ni 1.5% Cr 0.2% Mo steel: Carburizing; SKF
SL 503 M	0.25% C 2.5% Ni 2.7% Cr 0.5% Mo 0.3% V 9.0% W steel: Swift Levick; hot dies
SL 598	0.25% C 4.0% Ni 1.5% Cr 0.3% V 6.0% W steel: Swift Levick; hot dies
SL 853	0.35% C 5.25% Cr 4.5% W 0.5% Co steel: Swift Levick; for hot dies
SL 942	0.35% C 2.5% Cr 4.3% Mo 1.0% V 2.0% W steel: Swift Levick; for hot dies
SMA 58	See JIS G3114 SMA58
SN 3	0.3% C 3.2% Ni 0.2% Cr steel: Spencer for BS alloy En 21
SN 5/½	sintered Ni alloy steel: High impact strength; can be hardened; Durasint; specific gravity 6.6; 16% porosity **DPN: 160 UTS: 470 Elon: 6%**
SNC 1	0.3% C 0.8% Cr 0.5% Mo steel: Spencer for BS alloy En 23
SNC 3G	0.3% C 4.2% Ni 1.2% Cr 0.3% Mo (optional) steel: Spencer for BS alloy En 30
SNCH 6	0.16% C 3.5% Ni 0.9% Cr steel: For case hardening; Spencer for BS alloy En 36
SNCH 9	0.16% C 4.2% Ni 1.2% Cr 0.2% Mo (optional) steel: Spencer for BS alloy En 39
SNCM	0.3% C 2.5% Ni 0.7% Cr 0.5% Mo steel: Spencer for BS alloy En 25
SNCM 26	0.32% C 3.7% Ni 1.2% Cr 0.5% Mo steel: Spencer for BS alloy En 27 and En 28
SNH 3	0.12% C 3.2% Ni 0.2% Cr steel: For case hardening; Spencer for BS alloy En 33
SNH 5	0.14% C 4.8% Ni 0.2% Cr steel: For carburizing; Spencer for BS alloy En 37
SOMER SUPAMOLD	0.35% C 1.5% Cr 0.4% Mo steel: Somers
SOMPLAS 30	0.32% C 4.3% Ni 1.2% Cr 0.3% Mo steel: Somers
SOUDODUR MR	0.03% C 18.0% Ni 4.2% Mo 11.3% Co 0.2% Ti steel welding electrode: Soudometal; can be heat treated **DPN: 400**
SOUDOKAY 258 LO	0.3% C 6% Cr 1.5% Mo 1.5% W steel: Welding electrode; Soudometal **DPN: 470**

Note. The following abbreviations and units are used in the tables:

DPN	Hardness, diamond pyramid number
UTS	Ultimate tensile strength, N/mm^2
Elon	Elongation, %
Proof	0.1% proof strength, N/mm^2

1 N/mm^2=0.1 hbar=0.102 kgf/mm^2=0.06475 $tonf/in.^2$=145.04 $ibf/in.^2$=1 MPa

See Appendix II for other abbreviations and conversion tables.

Symbol	Nominal analysis, supplier, condition and remarks.
SOUDOKAY BVO	0.15% C 0.45% Cr 0.35% Mo steel: Welding electrode; Soudometal **DPN: 300**
SOUDOR G65	0.06% C 1.3% Ni 0.3% Cr 0.2% Mo 0.1% V steel: Welding electrode; Soudometal; for inert gas welding **UTS: 750 Elon: 19% Proof: 700**
SOUDOR G Cr Mo 1	0.1% C 1.2% Cr 0.5% Mo steel: Welding electrode; Soudometal; for inert gas welding **UTS: 520 Elon: 25% Proof: 350**
SOUDOR G Cr Mo 2	0.08% C 2.7% Cr 1.0% Mo steel: Welding electrode; Soudometal; for inert gas welding **UTS: 560 Elon: 22% Proof: 300**
SOUDOTIG Cr Mo 1	0.1% C 1.1% Cr 0.45% Mo steel: Welding electrode; Soudometal; for inert gas welding **UTS: 650 Elon: 22% Proof: 510**
SOUDOTIG Cr Mo 2	0.07% C 2.7% Cr 1.0% Mo steel: Welding electrode; Soudometal; for inert gas welding **UTS: 610 Elon: 20% Proof: 450**
SOUDOTIG Cr Mo 55	0.06% C 5.8% Cr 0.6% Mo steel: Welding electrode; Soudometal; for inert gas welding **UTS: 600 Elon: 18% Proof: 400**
SP 10	Sintered alloy steel: Good wear resistance; specific gravity 6.0; 22% porosity; Durasint **DPN: 190 UTS: 440 Elon: 1%**
SP 20	Sintered Ni alloy steel: Machinable; specific gravity 6.7; porosity 15%; Durasint **DPN: 260 UTS: 520 Elon: 1%**
SP 22	Sintered Ni Mo steel: Specific gravity 6.9; Durasint **DPN: 300 UTS: 720 Elon: 1%**
SP 24	Sintered Ni alloy steel: Good wear resistance; can be carbo-nitrided; specific gravity 6.8; 13% porosity **DPN: 125 UTS: 390 Elon: 4%**
SPEAR 451	0.14% C 0.75% Ni 0.5% Cr steel: Spear and Jackson; as BS alloy En 351
SPEAR 524	0.32% C 1.5% Ni 1.1% Cr 0.2% Mo steel: Spear and Jackson for BS alloy En 24
SPEAR 525	0.3% C 2.5% Ni 0.6% Cr 0.6% Mo steel: Spear and Jackson; as BS alloy En 25
SPEAR 536	0.12% C 3.0% Ni 1.0% Cr steel: Spear and Jackson; as BS alloy En 36
SPEAR 552	0.17% C 1.0% Ni 0.8% Cr steel: Spear and Jackson; as BS alloy En 352
SPEAR 653	0.18% C 1.25% Ni 0.9% Cr 0.11% Mo steel: Spear and Jackson; as BS alloy En 353
SPEAR 855	0.18% C 2.0% Ni 1.6% Cr 0.2% Mo steel: Spear and Jackson; as BS alloy En 355
SPEAR 1030	0.32% C 4.1% Ni 1.2% Cr 0.25% Mo steel: Spear and Jackson; as BS alloy 30B
SPEAR B1	0.32% C 4.2% Ni 1.2% Cr 0.3% Mo steel: Spear and Jackson
SPEAR B4	0.32% C 0.6% Mn 2.1% Cr 0.2% Mo 0.15% V steel: Origin unknown
SPEAR D5	0.35% C 5.7% Cr 1.5% Mo 1.0% V steel: Spear and Jackson; for aluminium die casting moulds
SPEAR D7	0.32% C 5.0% Cr 1.5% Mo 0.4% V 1.5% W steel: Spear and Jackson; for extrusion dies
SPEAR D9	0.3% C 3.2% Cr 0.3% V 9.5% W steel: Spear and Jackson; for hot work
SPEAR DX	0.28% C 2.2% Ni 0.85% Cr 0.5% Mo 0.25% V 2.2% W steel: Origin unknown
SPECIAL BB	0.35% C 3.5% Cr 0.4% V 9.0% W steel: For tools; origin unknown
SPECIAL BB	0.28% C 3.2% Cr 0.3% V 9.5% W steel: Jessop; dies for casting Cu **DPN: 400**
SPECIAL HDS	0.35% C 3.5% Cr 0.4% V 9.0% W steel: For tools; origin unknown
SPENARD	0.32% C 4.1% Ni 1.25% Cr 0.3% Mo steel: Spencer
SPRASTEEL LS	0.04% C 2% Mn 4% Ni 1.5% Cr 1.3% Mo steel: Wire for spraying; Metco; as sprayed **DPN: 261 UTS: 220**

Symbol	Nominal analysis, supplier, condition and remarks.
SS 100	0.16% C 1.7% Cr 0.5% Mo 0.07% Ti 0.3% Cu 0.003% B steel: Armco; hardened and tempered **UTS: 840 Elon: 18% Proof: 740**
SS 2092	1.0% C 0.75% Mn 1.0% Cr steel: Swedish Standard; cold water tool steel **DPN: 240**
SS 2203	0.1% C 9% Cr 1.0% Mo steel: Fagersta
SS 2203E	0.15% C 0.45% Mn 0.85% Cr 1.0% Mo 0.25% Cu steel: Swedish Standard; for pressure vessels; normalized and annealed **UTS: 650 Elon: 18% Proof: 390**
SS 2216	0.15% C 0.9% Cr 0.6% Mo steel: For pressure vessels; Swedish Standard; as forged **UTS: 500 Elon: 22% Proof: 300**
SS 2218	0.15% C 2.2% Cr 1% Mo steel: For pressure vessels; Swedish Standard; as forged **UTS: 500 Elon: 20% Proof: 250**
SS 2218E	0.11% C 0.5% Mn 2.2% Cr 1.0% Mo 0.25% Cu steel: Swedish Standard; for pressure vessels; normalized and annealed **UTS: 550 Elon: 18% Proof: 265**
SS 2223	0.18% C 1% Cr 0.6% Mo steel: Casting; Swedish Standard; normalized and annealed **DPN: 160 UTS: 480 Elon: 20% Proof: 270**
SS 2224	0.18% C 2.2% Cr 1% Mo steel: Casting; Swedish Standard; normalized and annealed **DPN: 180 UTS: 480 Elon: 20% Proof: 270**
SS 2225	0.25% C 0.25% Ni 1.1% Cr 0.2% Mo steel: Bar, plate and forging; Swedish Standard; hardened and tempered **DPN: 350 UTS: 1070 Elon: 8% Proof 860**
SS 2225	0.25% C 1.05% Cr 0.2% Mo steel: Swedish Standard; hardened and tempered **UTS: 780 Elon: 14% Proof: 600**
SS 2233	0.3% C 1% Cr 0.25% Mo steel: For gas cylinders; Swedish Standard; hardened and tempered **DPN: 280 UTS: 930 Elon: 13% Proof: 720**
SS 2234	0.33% C 1.1% Cr 0.2% Mo steel: Bar and forging; Swedish Standard; hardened and tempered **DPN: 280 UTS: 840 Elon: 12% Proof: 700**
SS 2240	0.3% C 0.3% Ni 3% Cr 0.5% Mo steel: For nitriding; Swedish Standard; hardened and tempered **DPN: 370 UTS: 1210 Elon: 12% Proof: 880**
SS 2310	1.5% C 0.45% Mn 12% Cr 0.8% Mo 0.85% V steel: Swedish Standard; cold water tool steel **DPN: 260**
SS 2317	0.23% C 0.5% Ni 11.7% Cr 1.2% Mo 0.3% V 0.045% B steel: Swedish Standard **DPN: 260 UTS: 910 Elon: 16% Proof: 640**
SS 2506	0.2% C 0.55% Ni 0.5% Cr 0.2% Mo steel: For construction; Swedish Standard **DPN: 320 UTS: 100 Elon: 10% Proof: 550**
SS 2511	0.15% C 1.5% Ni 0.9% Cr steel: Bar and forging; Swedish Standard
SS 2512	0.2% C 1.5% Ni 0.9% Cr steel: Bar and forging; Swedish Standard
SS 2514	0.12% C 3% Ni 0.8% Cr steel: Bar and forging; Swedish Standard
SS 2515	0.17% C 3% Ni 0.8% Cr steel: Bar and forging; Swedish Standard
SS 2523	0.2% C 1.2% Ni 1.0% Cr 0.12% Mo steel: Swedish Standard; annealed **DPN: 220 UTS: 110 Elon: 8% Proof: 75**
SS 2534	0.31% C 3.2% Ni 1% Cr 0.25% Mo steel: Bar and forging; Swedish Standard
SS 2730	0.3% C 3% Cr 0.3% V 9% W steel: Swedish Standard; annealed **DPN: 250**
SS 2782	1% C 0.3% Mn 4.0% Cr 8.5% Mo 2% V 2.7% W 0.03% S and P steel: Swedish Standard; high speed tool steel **DPN: 260**

Symbol	Nominal analysis, supplier, condition and remarks.
SS 8100	0.2% C (max) 1.75% Ni 0.5% Mo 1.5% Cu steel: Sintered
SSG	0.3% C 4% Ni 1.5% Cr steel: Jonas; hardened and tempered **UTS: 1430 Elon: 15% Proof: 1300**
SSS 100	0.16% C 0.55% Mn 1.7% Cr 0.5% Mo 0.06% V 0.07% Ti 0.3% Cu steel: Armco
SSS 100A	0.17% C 1.0% Cr 0.2% Mo 0.07% Ti 0.3% Cu 0.003% B steel: Armco; hardened and tempered **UTS: 840 Elon: 18% Proof: 740**
SSS 100A	0.17% C 0.55% Mn 1.0% Cr 0.2% Mo 0.06% V 0.06% Ti 0.3% Cu steel: Armco
SSS 110	0.16% C 0.55% Mn 1.8% Cr 0.5% Mo 0.06% V 0.06% Ti 0.3% Cu steel; Armco
SSS 321	0.19% C 0.55% Mn 1.4% Cr 0.4% Mo 0.1% Ti 0.4% Cu 0.0015% B steel: Armco
SSS 360	0.19% C 0.5% Mn 1.4% Cr 0.4% Mo 0.1% Ti 0.3% Cu 0.0015% B steel: Armco
STA 5 V6	0.2% C 0.3% Ni 0.2% Cr steel: Replaced by BS alloy En 14
STA 5 V6A	0.2% C 0.3% Ni 0.2% Cr steel: Replaced by BS alloy En 14A
STA 5 V6B	0.2% C 0.3% Ni 0.2% Cr steel: Obsolete
STA 5 V8	0.20% C 0.6% Ni 0.25% Mo steel: Replaced by BS alloy En 13
STA 5 V9E	0.35% C 1.25% Ni 0.6% Cr steel: Replaced by BS alloy En 111
STA 5 V9E/1	0.35% C 1.25% Ni 0.6% Cr steel: Replaced by BS alloy En 111
STA 5 V11	0.31% C 0.7% Cr 0.6% Mo 2.5% Ni steel: Replaced by BS alloy En 25
STA 5 V13	0.3% C 4.2% Ni 1.2% Cr steel: Replaced by BS alloy En 30A
STA 5 V13A	0.3% C 4.2% Ni 1.2% Cr steel: Replaced by BS alloy En 30B
STA 5 V16A	0.17% C 1.75% Ni 0.25% Mo steel: Replaced by BS alloy En 34
STA 5 V16A/1	0.17% C 1.7% Ni 0.2% Mo steel: Obsolete
STA 5 V16B	0.24% C 1.75% Ni 0.25% Mo steel: Replaced by BS alloy En 35
STA 5 V16B/1	0.24% C 1.75% Ni 0.25% Mo steel: Replaced by BS alloy En 35A
STA 5 V16B/2	0.24% C 1.75% Ni 0.25% Mo steel: Replaced by BS alloy En 35B
STA 5 V17	0.12% C 3.2% Ni 0.25% Cr steel: Replaced by BS alloy En 33
STA 5 V17/1	0.12% C 3.2% Ni 0.25% Cr steel: Obsolete
STA 5 V18	0.16% C 3.5% Ni 0.8% Cr steel: Replaced by BS alloy En 36A
STA 5 V18A	0.16% C 3.5% Ni 0.8% Cr steel: Replaced by BS alloy En 36B
STA 5 V19	0.15% C 4.2% Ni 1.2% Cr steel: Replaced by BS alloy En 39A
STA 5 V19/1	0.15% C 4.2% Ni 1.2% Cr steel: Obsolete
STA 5 V20A	0.3% C 0.3% Ni 1.6% Cr 0.2% Mo 1.0% Al steel: Replaced by BS alloy En 41
STA 5 V20B	0.3% C 0.3% Ni 1.6% Cr 0.2% Mo 1.0% Al steel: Replaced by BS alloy En 41
STA 5 V20C	0.3% C 0.3% Ni 1.6% Cr 0.2% Mo 1.0% Al steel: Replaced by BS alloy En 41
STA 5 V20D	0.3% C 0.3% Ni 1.6% Cr 0.2% Mo 1.0% Al steel: Obsolete
STA 39	0.25% C 0.4% Ni 0.2% Mo steel: Suitable for welding
STA 40	0.2% C 0.6% Ni 0.25% Mo steel: Weldable
STAGMOLD	0.32% C 4.1% Ni 1.3% Cr 0.3% Mo steel: For tools; for cold working; Edgar Allen
STA V18/1	0.16% C 3.5% Ni 0.8% Cr steel: Obsolete
STUBS 81	Low alloy structured steel: Welding electrode; Stubs; for welding poor quality cast iron **DPN: 330 UTS: 360**

Symbol	Nominal analysis, supplier, condition and remarks.
STUBS 651	C Cr Ni Mo steel: Welding electrode; Stubs; for alloy steel welds **DPN: 245 UTS: 741 Elon: 22%**
SUPER AW 23	0.3% C 4.1% Ni 1.2% Cr 0.2% Mo steel: ESC; punches, etc. **DPN: 500**
SUPER CHNC	0.15% C 4.2% Ni 1.2% Cr 0.25% Mo steel: For case hardening; ESC for BS alloy En 39B
SUPERELSO	0.18% C (max) 0.9% Mn 1.5% Ni 1.1% Cr 0.25% Mo steel: Creusot-Loire
SUPERELSO 70	0.18% C (max) 1.1% Mn (max) 1.7% Ni 1.0% Cr 0.4% Mo 0.01% V steel: Creusot-Loire
SUPERELSO 600	0.09% C 1.1% Mn 0.8% Ni 1.7% Cr 0.2% Mo steel: Creusot-Loire
SUPERMOLD	0.3% C 1.7% Cr 0.4% Mo steel: Latrobe; dies and mould; type P20 **DPN: 340**
SUPER NITRALLOY	0.23% C 0.5% Cr 0.25% Mo 0.1% V 2.0% Al steel: Latrobe; for nitriding
SUPER NOVEX	0.3% C 3% Ni 0.8% Cr 0.25% Mo steel: Jonas **DPN: 320 UTS: 1050 Elon: 18% Proof: 920**
SUPER SAMSON	0.1% C 5.0% Cr 0.9% Mo 0.25% V steel: Carpenter for AISI type P4
SUPER SHNC	0.3% C 4.25% Ni 1.25% Cr 0.25% Mo steel: ESC for BS alloy En 30B
SUPERTOUGH C20	0.18% C 1.6% Mn 0.6% Ni 0.25% Mo steel: ESC for BS alloy En 13
SUPERVITAC	Free machining grade of steel by Creusot-Loire
SUPERVITUS 980	0.2% C 0.15% Cr 0.1% Mo 0.15% N steel: Origin unknown; for bicycle frames
SUPER VNCA	0.3% C 3% Ni 0.75% Cr 0.25% Mo steel: ESC for BS alloy En 23
SUPREX A	0.08% C 0.9% Mn 1.25% Cr 0.5% Mo steel: Welding electrode; Murex **UTS: 750 Elon: 24% Proof: 640**
SUPREX B	0.08% C 2.2% Cr 1% Mo steel: Welding electrode; Murex **UTS: 750 Elon: 22% Proof: 660**
SUPREX C	0.08% C 4.5% Cr 0.6% Mo steel: Welding electrode; Murex **UTS: 750 Elon: 23% Proof: 660**
SUPREX M	0.07% C 0.05% Cr 0.05% Mo steel: Welding electrode; Murex **UTS: 640 Elon: 30% Proof: 560**
SUPRIMPACTO	0.12% C 3.7% Ni 1.5% Cr steel: For carburizing; Atlas
SV 10KH5M	0.12% C (max) 5.0% Cr 0.5% Mo steel: For welding wire; Russian Standards designation
SVL	0.3% C 4% Ni 1.3% Cr 0.3% Mo steel: Jessop
T ALLOY A	0.33% C 3.5% Cr 0.5% V 9.2% W steel: Braeburn; AISI type H21
T ALLOY C	0.25% C 4.0% Cr 0.5% V 15% W steel: Braeburn
T91	0.1% C 8.7% Cr 1% Mo 0.2% V 0.08% Nb steel: US specification; steel for nuclear use
TANGE CHAMPION	0.3% C 0.8% Cr 0.15% Mo steel: Origin unknown; for bicycle frames
TENASOUDO 75	0.08% C 1.5% Mn 0.7% Ni 0.2% Cr 0.4% Mo steel: Welding electrode; Soudometal; for tough hard facing **DPN: 230**

Note. The following abbreviations and units are used in the tables:

DPN	Hardness, diamond pyramid number
UTS	Ultimate tensile strength, N/mm^2
Elon	Elongation, %
Proof	0.1% proof strength, N/mm^2

1 N/mm^2=0.1 hbar=0.102 kgf/mm^2=0.06475 tonf/in.^{2}=145.04 lbf/in.^{2}=1 MPa

See Appendix II for other abbreviations and conversion tables.

Symbol	Nominal analysis, supplier, condition and remarks.
TENASOUDO 105	0.08% C 0.8% Mn 2.8% Ni 0.9% Cr 0.8% Mo steel: Welding electrode; Soudometal; for buttering or high impact building **DPN: 300**
TENSITRODE 75.55	0.07% C 0.6% Mn 1.6% Ni 0.5% Mo 0.1% V steel: For electrodes; Esab standard **UTS: 726 Elon: 23% Proof: 618**
TENSITRODE 75.65	0.07% C 0.9% Mn 1.6% Ni 0.75% Mo 0.25% V steel: For electrodes; Esab **UTS: 726 Elon: 23% Proof: 618**
THERMENOL	0.05% C 3.3% Mo 0.3% V 16% Al steel: American proprietary alloy listed in SAE Yearbook
TI	0.15% C 0.8% Mn 0.8% Ni 0.6% Cr 0.5% Mo 0.06% V 0.35% Cu 0.004% B steel: Rheinstahl
TI	0.15% C 0.8% Mn 0.85% Ni 0.5% Cr 0.5% Mo 0.07% V 0.3% Cu steel: Lukens Steel Co.
TI TYPE A	0.17% C 0.8% Mn 0.6% Cr 0.2% Mo 0.06% V 0.02% Ti 0.003% B steel: Rheinstahl
TI TYPE A	0.15% C 0.85% Mn 0.5% Cr 0.3% Mo 0.05% V 0.02% Ti steel: Lukens Steel Co.
TI TYPE B	0.16% C 1.1% Mn 0.55% Ni 0.5% Cr 0.25% Mo 0.05% V 0.3% Cu steel: Lukens Steel Co.
T-K	0.35% C 3.5% Cr 0.4% V 9% W steel: Carpenter
TOBA	0.37% C 1% Cr 0.2% V steel: T Turton
TUBROD 15.42	0.15% C 2.0% Mn 0.5% Ni 4.5% Cr 0.5% Mo steel: Tubular weld rod; Esab for hard facing; as welded **DPN: 430**
TUBROD 15.50	0.3% C 2.5% Mn 5.0% Cr 0.8% Mo steel: Tubular weld rod; Esab; for hard facing; as welded **DPN: 520**
TUREX 9	0.3% C 2.4% Cr 0.1% V steel: Pompey
TY	0.28% C 4% Ni 1.5% Cr 0.25% V 5% W steel: For tools; S Osborn; for hot extrusions dies, punches, etc. **DPN: 500**
UHB STATO 23	0.15% C 0.9% Cr 0.45% Mo steel: Tube; Uddelholm; normalized **UTS: 500 Elon: 28% Proof: 290**
UHB STATO 28	0.15% C 2.25% Cr 1.0% Mo steel: Tube; Uddelholm; normalized **UTS: 480 Elon: 28% Proof: 250**
UHB UDDCO 2	0.13% C 5% Cr 0.5% Mo steel: Uddelholm; annealed **DPN: 150 UTS: 480 Elon: 20% Proof: 240**
UHB UDDCO 3	0.06% C 3% Cr 0.5% Mo steel: Uddelholm; annealed **DPN: 150 UTS: 440 Elon: 20% Proof: 240**
UHB UDDCO 9	0.1% C 9.0% Cr 1.0% Mo steel: Tube; Uddelholm; normalized **UTS: 610 Elon: 25% Proof: 370**
UNI 5331 12 Ni Cr 3	0.12% C 0.6% Ni 0.5% Cr steel: For carburizing; Italian Standard
UNI 5331 16 Cr Ni 4	0.15% C 1.0% Ni 1.0% Cr steel: For carburizing; Italian Standard
UNI 5331 16 Ni Cr 11	0.14% C 2.7% Ni 0.7% Cr steel: For carburizing; Italian Standard
UNI 5331 16 Ni Cr Mo 2	0.15% C 0.5% Ni 0.5% Cr 0.2% Mo steel: For carburizing; Italian Standard
UNI 5331 16 Ni Cr Mo 12	0.16% C 3.0% Ni 1.0% Cr 0.35% Mo steel: For carburizing; Italian Standard
UNI 5331 18 Ni Cr Mo 5	0.18% C 1.3% Ni 0.8% Cr 0.2% Mo steel: For carburizing; Italian Standard
UNI 5331 18 Ni Cr Mo 7	0.18% C 1.6% Ni 0.6% Cr 0.25% Mo steel: For carburizing; Italian Standard
UNI 5331 20 Cr Ni 4	0.2% C 1.0% Ni 1.0% Cr steel: For carburizing; Italian Standard
UNI 5331 20 Ni Cr Mo 2	0.2% C 0.5% Ni 0.5% Cr 0.2% Mo steel: For carburizing; Italian Standard
V 110	0.32% C 3.3% Ni 1.1% Cr 0.3% Mo steel: VEW
V 145	0.3% C 2.0% Ni 2.0% Cr 0.4% Mo steel: VEW
V 155	0.34% C 1.5% Ni 1.5% Cr 0.2% Mo steel: VEW
V 165	0.36% C 1.1% Ni 1.1% Cr 0.2% Mo steel: VEW
V 214	0.31% C 3.5% Ni 0.8% Cr steel: VEW
V 330	0.34% C 1.1% Cr 0.2% Mo steel: VEW

Symbol	Nominal analysis, supplier, condition and remarks.
V 340	0.26% C 1.1% Cr 0.3% Mo steel: VEW
V 350	0.3% C 2.5% Cr 0.2% Mo 0.2% V steel: VEW
V 810	0.34% C 1.2% Cr 0.2% Mo 0.95% Al steel: VEW
V 820	0.34% C 1.0% Ni 1.7% Cr 0.2% Mo 0.95% Al steel: VEW
VALAND 1	0.28% C 3.0% Cr 0.3% V 9.4% W steel: Uddelholm for type H21
VANACHROM	0.35% C 5.5% Cr 1.35% Mo 1.0% V steel: Swift Levick; for hot dies
VCN 15	0.35% C 1.25% Ni 0.5% Cr steel: Wrought; Austrian national specification
VCN 25w(g)	0.3% C 3.2% Ni 0.7% Cr 0.3% Mo steel: Wrought; Austrian national specification
VCN 35h	0.3% C 3.2% Ni 0.7% Cr 0.25% Mo steel: Wrought; Austrian national specification
VCN 35w(e)	0.3% C 3.2% Ni 0.7% Cr 0.3% Mo steel: Wrought; Austrian national specification
VCN 45(d)	0.3% C 4.2% Ni 1.2% Cr steel: Wrought; Austrian national specification
VIBRAC V30	0.3% C 2.5% Ni 0.6% Cr 0.5% Mo steel: ESC for BS alloy En 25
VITUS 181	0.2% C 0.15% Cr 0.1% Mo 0.15% N steel: Origin unknown; for bicycle frames
VMC	0.35% C 5.0% Cr 1.5% Mo 1.0% V steel: Origin unknown
VMC(H)	0.33% C 3.0% Cr 0.5% Mo 0.15% V steel: Origin unknown
VNCA	0.23% C 3% Ni 0.75% Cr steel: ESC for BS alloy En 23
VS 3 NC	0.35% C 3% Ni 1% Cr 0.35% Mo steel: Vulcan **UTS: 920**
VS 4 NC	0.3% C 4.1% Ni 1.3% Cr 0.35% Mo steel: Vulcan **UTS: 920**
VS 351	0.1% C 3.5% Ni 1% Cr steel: For carburizing; Vulcan; hardened and tempered; core properties **UTS: 1140**
VS INC	0.35% C 1.25% Ni 1% Cr 0.3% Mo steel: Vulcan **UTS: 920**
VSHD	0.3% C 3% Cr 0.5% V 10% W steel: Vulcan; hot dies **DPN: 440**
VWMC	0.35% C 5.0% Cr 1.5% Mo 1.5% V steel: Origin unknown
W 9B	0.22% C 2.7% Ni 2.5% Cr 0.45% Mo 0.3% V 9.5% W steel: Origin unknown
W 100	0.29% C 2.7% Cr 0.4% V 8.5% W steel: VEW
W 105	0.32% C 2.4% Cr 0.6% V 4.3% W steel: VEW
W 320	0.31% C 2.9% Cr 2.8% Mo 0.5% V steel: VEW
W 53116	0.05% C 2.2% Cr 1% Mo steel: Weld electrode; UNS designation
W 53315	0.1% C (max) 2% Cr 0.5% Mo steel: Weld electrode; designation used by UNS
W 53315	0.1% C (max) 2% Cr 0.5% Mo steel: Weld electrode; UNS designation
W 53316	0.1% C (max) 2% Cr 0.5% Mo steel: Weld electrode; designation used by UNS
W 53416	0.05% C (max) 2% Cr 0.5% Mo steel: Weld electrode; UNS designation
WC 10	0.1% C 0.4% Ni 0.25% Cr 0.15% Mo 0.3% Cu (all max) steel: Casting; Wolsingham **UTS: 390 Elon: 22% Proof: 180**
WC 18	0.18% C 0.4% Ni 0.25% Cr 0.15% Mo 0.3% Cu (all max) steel: Casting; Wolsingham **UTS: 450 Elon: 22% Proof: 210**
WC 23	0.23% C 0.4% Ni 0.25% Cr 0.15% Mo 0.3% Cu (all max) steel: Casting; Wolsingham **UTS: 480 Elon: 20% Proof: 220**
WC 28	0.28% C 0.4% Ni 0.25% Cr 0.15% Mo 0.3% Cu (all max) steel: Casting; Wolsingham **UTS: 520 Elon: 20% Proof: 240**

Symbol	Nominal analysis, supplier, condition and remarks.
WC 33	0.34% C 0.4% Ni 0.25% Cr 0.15% Mo 0.3% Cu (all max) steel: Casting; Wolsingham **UTS: 550 Elon: 20% Proof: 250**
WC 181	0.18% C 0.4% Ni 0.25% Cr 0.15% Mo 0.3% Cu (all max) steel: Casting; Wolsingham **UTS: 450 Elon: 30% Proof: 200**
WC 182	0.18% C 0.4% Ni 0.25% Cr 0.15% Mo 0.3% Cu (all max) steel: Casting; Wolsingham **UTS: 440 Elon: 24% Proof: 240**
WCD	0.35% C 5% Cr 1.2% Mo 0.35% V 1.2% W 1% Si steel: Schoeller-Bleckmann **UTS: 1550**
WCD 2	0.33% C 5% Cr 1.1% Mo 0.5% V 1% Si steel: Schoeller-Bleckmann **UTS: 1420**
WELCON 2H.S	0.12% C 0.9% Mn 1.0% Ni 0.5% Cr 0.5% Mo steel: Japan Steel Works
WELCON 2H.U	0.12% C 0.9% Mn 0.5% Ni 0.8% Cr 0.7% Mo 0.32% Cu steel: Japan Steel Works
WEL-MONIX	0.15% C (max) 1.4% Mn 0.3% Mo 0.14% V steel: German proprietory specification
WEL-TEN 60	0.16% C (max) 1.2% Mn 0.6% Ni 0.4% Cr 0.15% V steel
WEL-TEN 60H	0.18% C (max) 1.2% Mn 1.0% Ni 0.15% Ti with V steel: Yawata
WEL-TEN 62	0.18% C (max) 1.2% Mn 0.6% Ni 0.3% Cr 0.12% V steel: Yawata
WEL-TEN 80	0.18% C (max) 0.9% Mn 1.5% Ni 0.6% Cr 0.6% Mo 0.1% V 0.3% Cu 0.006% B steel: Yawata
WEL-TEN 100W	0.18% C (max) 0.9% Mn 1.5% Ni 0.6% Cr 0.6% Mo 0.1% V 0.3% Cu steel: Yawata
WL 2	0.15% C 9.0% Cr 0.5% Mo steel: Tube; Babcock and Wilcox (USA)
WLM 77	0.35% C 0.8% Cr 0.7% Mo 0.2% V steel: W Marrison; chisels, shears
WLM 1030	0.3% C 3.5% Cr 10% W steel: W Marrison; hot dies **DPN: 550**
WLM 1530	0.18% C 3% Ni 1% Cr steel: For case hardening; W Marrison; analysis varies over wide limits
WLM 3033	0.3% C 3.4% Ni 0.9% Cr 0.5% Mo steel: W Marrison; hardened and tempered **DPN: 320 UTS: 1000 Elon: 16% Proof: 800**
WLM 3042	0.3% C 4.1% Ni 1.3% Cr steel: Mo optional; W Marrison; hardened and tempered **DPN: 444 UTS: 1550 Elon: 10% Proof: 1300**
WLM 4053	0.35% C 0.5% Cr 0.5% W steel: W Marrison; chisels, punches
X 7 Cr Mo 6/1	0.15% C (max) 6.0% Cr 0.7% Mo steel: Designation used by German Standards
X 7 Cr Mo 10/1	0.1% C (max) 9.5% Cr 1.0% Mo steel: Designation used by German Standards
X 19 Ni Cr Mo4	0.19% C 4.1% Ni 1.3% Cr 0.2% Mo 0.4% W steel: For tools; German Standards designation
X 30 W Cr V 53	0.3% C 2.4% Cr 0.6% V 4.3% W steel: German Standards designation
X 32 Cr Mo V 33	0.32% C 3.0% Cr 2.8% Mo 0.5% V steel: German Standards designation
X 4620	0.2% C 1.8% Ni 0.25% Mo steel: Designation used by SAE; obsolete
X30W Cr V 93	0.3% C 2.7% Cr 0.4% V 8.6% W steel: German Standard
XAR 15	0.15% C 0.9% Mn 0.6% Cr 0.22% Mo 0.1% Zr steel: Republic Steel
XAR 30	0.3% C (max) 0.8% Mn 0.6% Cr 0.21% Mo 0.1% Zr steel: Republic Steel
Y 10 N66	0.1% C 1.5% Ni 1.0% Cr steel: French Standards designation
Y 35 NCD16	0.35% C 4.0% Ni 1.8% Cr 0.4% Mo 0.1% V steel: French Standards designation
Y 35 NCD16 (3)	0.32% C 4.1% Ni 1.2% Cr 0.2% Mo steel: French Standard

Symbol	Nominal analysis, supplier, condition and remarks.
YD1	0.25% C 2.25% Cr 9.5% W steel: Jonas; hot extrusion dies **DPN: 400**
YDC	0.3% C 2.25% Cr 0.35% V 8.5% W steel: Jonas; hot extrusion dies **DPN: 520**
YND 58	0.14% C (max) 0.7% Mn 2.3% Ni 0.5% Cr 0.5% Mo steel: Yamata
YOLOY	0.15% C (max) 0.75% Mn 1.7% Ni 0.1% V steel: Youngstoun Steel Co.
YOLOY EHS	0.18% C (max) 1.0% Mn 0.7% Ni 0.4% Cr 0.4% Mo 0.4% Cu steel: Youngstoun Steel Co.
YOLOY HS	0.15% C (max) 1.0% Mn 1.0% Ni 0.2% Mo 0.75% Cu steel: Youngstoun Steel Co.
YWA	0.28% C 3.4% Ni 1.3% Cr 0.25% V 5.8% W steel: Edgar Allen; extrusion dies, mandrels, etc. **DPN: 450**
Z 8CDV5	0.08% C 5.0% Cr 1.0% Mo 0.3% V steel: French Standard. designation
Z 30WCV9	0.3% C 3.0% Cr 0.4% V 9.0% W steel: French Standards designation

Note. The following abbreviations and units are used in the tables:

DPN	Hardness, diamond pyramid number
UTS	Ultimate tensile strength, N/mm^2
Elon	Elongation, %
Proof	0.1% proof strength, N/mm^2

1 N/mm^2=0.1 hbar=0.102 kgf/mm^2=0.06475 tonf/in.^{2}=145.04 lbf/in.^{2}=1 MPa

See Appendix II for other abbreviations and conversion tables.

44K2 Steel – medium carbon alloys 0.4–0.65% carbon with two or more of chromium, nickel, molybdenum, tungsten or vanadium

Specific gravity	7.84
Density	7840 kg/m^3
Solidus/liquidus	1450-1520 °C
Thermal conductivity	50 W/m °C
Coefficient of linear expansion	11 × 10^{-6}/ °C
Electrical conductivity	7–10% IACS (copper 100%)
Specific resistance	190–280 microhm mm
Young's modulus of elasticity	32.07 × 10^9 N/m^2
Impact	40–80 J (could be much lower)
Fatigue strength	–
Hot strength	

Temperature °C	Tensile strength N/mm^2	Elongation %
100	930	25
200	920	21
300	920	21
400	840	29
500	710	27
550	610	32
600	520	39

The above properties are typical of the following group and may not apply exactly to any one specification. It is possible that with certain specifications some of the values may not be applicable.

General metallurgical characteristics

These steels have a high enough carbon content to ensure a minimum strength of about 9000 N/mm^2 when fully tempered.

The top limit of the carbon present is around the eutectoid point; thus even in the fully hardened condition there should never be sufficient free cementite present to make the material too brittle. At least two of the alloying elements are present in all the steels with chromium present in the majority, while a few steels have all five alloying elements present.

Some of the lower carbon steels listed can be carburized but in general the core hardness is too high for general engineering applications. Many of these steels can be nitride hardened, but generally this requires a special formulation with some aluminium which speeds up the process.

The effect of each alloying element is described under the

section where the single alloy steels are listed. When two or more of these elements are present the combined effect is the sum total of the individual effects.

Briefly the elements affect the steels as follows.

Chromium. This is a carbide former, with the carbides being harder and more stable than iron carbides. Carbon will combine with chromium in preference to iron; thus all the chromium is generally present as carbide.

Chromium increases the hardenability of steel in that larger sections can be through hardened, or less drastic oil quenching can be used in place of water quenching, compared with plain carbon steels. There is a tendency, however, for chromium steels to be notch sensitive. Chromium also increases the corrosion resistance, although none of the steels listed can be classed as stainless. The higher chromium stainless steels are listed and described in Sections 44M and 44N. Chromium raises the critical temperatures for steel thermal treatment but slows down the speed at which the metallurgical reactions take place.

Nickel. This strengthens the ferritic matrix of the steel, imparting toughness rather than hardness and increasing the fatigue strength. Nickel lowers the critical heat treatment temperatures and widens the range required for successful treatment. It also improves the ductility of steel at low temperatures. A range of 10% nickel steels have been developed with good ductility at the temperatures required for liquid air and other low temperature uses.

The range of 4% nickel steels has largely been replaced with the more economical chromium, molybdenum, 1% nickel steels which have equal fatigue strength and ductility, with better tensile properties.

Molybdenum. This is a carbide former, the resultant molybdenum carbide being inherently small grained and evenly distributed. This ensures that the steels are fine grained, which improves the fatigue strength. These stable carbides slow down the metallurgical reactions, ensuring deeper hardening, greater hardenability and the elimination of temper brittleness. This is a defect caused by the undesirable precipitation of carbides during tempering between 250 and 500 °C which occurs with chromium/nickel steels. As carbon always combines with the molybdenum in preference to other carbide formers including chromium and stays in solution, this obviates carbide precipitation. Vanadium and tungsten have similar effects.

Tungsten. This is a carbide former which results in very hard, rather massive tungsten carbide particles. Generally tungsten is used with chromium and vanadium which aid the distribution and prevent agglomeration of the carbides. Tungsten has similar properties to molybdenum but larger quantities are usually required to achieve the same effect.

Vanadium. This forms very small carbides and also influences other carbide formers to produce tiny particles which in turn ensures a fine grain steel. Thus vanadium has a considerable effect on improving the fatigue strength and is very often used with chromium to achieve this purpose. The presence of vanadium aids the nitriding process by preventing the agglomeration of carbide particles. Vanadium has little effect on the temperatures at which the metallurgical changes take place, but slows down these reactions, thus improving the hardenability.

None of these elements in the quantities used has any appreciable effect on the corrosion resistance although these alloy steels will always have slightly better resistance than plain carbon steels.

The alloying elements actually used depend on whether maximum hardness or maximum ductility is required, with the economy of the final product the deciding factor.

These steels do not accept cold work without becoming hard and brittle. They therefore require annealing as an interstage operation.

Full annealing, hardening and normalizing require temperatures in the region of 750–850 °C depending on the alloy content.

Where high integrity parts are involved technical advice must be obtained. All these steels have considerable hardenability and thus can be brittle if not correctly treated. Any rapid change in section can result in quench cracking. Holding at temperature for too long, or taking to too high a temperature, will invariably result in a brittle material.

None of the materials in this group can be welded without very careful control. This requires that technical expertise should be obtained. Unless correctly treated cracking can occur during welding or more seriously brittleness resulting in failure during service.

These materials are all relatively expensive and should only be used when their specific properties are required. That is, they will have a controlled hardness with ductility and are therefore used for components such as gears, shafts, couplings, springs, high tensile studs, bolts and nuts which are required for engineering use. Some of the materials have sufficient alloy content and carbon to be used for cutting tools and many are used for press tools, jigs, fixtures, and hand-tools of all types.

These materials, in the correctly hardened condition, have good abrasion resistant properties.

Many of these steels can be nitrided, but this must always be stated on the order, as nitriding requires special conditions.

A small number of the lower carbon alloys can be carburized.

Note. The following abbreviations and units are used in the tables:

DPN	Hardness, diamond pyramid number
UTS	Ultimate tensile strength, N/mm^2
Elon	Elongation, %
Proof	0.1% proof strength, N/mm^2

1 N/mm^2=0.1 hbar=0.102 kgf/mm^2=0.06475 tonf/in.^{2}=145.04 lbf/in.^{2}=1 MPa

See Appendix II for other abbreviations and conversion tables.

Symbol	Nominal analysis, supplier, condition and remarks.
00	0.6% C 1.1% Cr 0.3% V 1.9% W tool steel: Balfour; punches, chisels, etc. **DPN: 710**
0.7225	0.42% C 0.6% Ni 1.1% Cr 0.2% Mo steel: Forging; German Standard
0.7561	0.42% C 1.5% Cr 0.1% V steel: For bolts; German Standard
0.7727	0.45% C 1.4% Cr 0.7% Mo 0.3% V steel: For springs; German Standard

Symbol	Nominal analysis, supplier, condition and remarks.
0.8159	0.5% C 1.1% Cr 0.1% V steel: German Standard
0.8161	0.58% C 1.0% Cr 0.1% V steel: For springs; German Standard
0.8239	0.65% C 3.7% Cr 0.85% Mo 0.7% V 8.5% W steel: For springs; German Standard
1.1232	0.38% C 5.3% Cr 1.1% Mo 0.4% V steel: German Standards designation
1.2241	0.5% C 1.1% Cr 0.2% V steel: German Standard
1.2323	0.45% C 1.4% Cr 0.7% Mo 0.3% V steel: German Standard
1.2343	0.53% C 8.3% Cr 1.2% Mo 1.2% W steel: German Standard
1.2344	0.38% C 5.2% Cr 1.35% Mo 1.0% V steel: German Standard
1.2344	0.4% C 5.3% Cr 1.4% Mo 1.0% V steel: German Standards designation
1.2345	0.5% C 5% Cr 1.4% Mo 1.4% V steel: German Standard
1.2367	0.4% C 5% Cr 3% Mo 0.5% V steel: German Standard
1.2541	0.35% C 1.1% Cr 0.2% V 2.0% W steel: German Standards designation
1.2542	0.45% C 1.1% Cr 0.2% V 2.0% W steel: German Standards designation
1.2547	0.5% C 1.5% Cr 0.2% V 2.25% W steel: German Standard
1.2550	0.6% C 1.1% Cr 0.2% V 2.0% W steel: German Standards designation
1.2631	0.53% C 8.3% Cr 1.2% Mo 1.2% W steel: German Standard
1.2711	0.54% C 1.7% Ni 0.7% Cr 0.3% Mo 0.1% V steel: For tools; German Standards designation
1.2713	0.55% C 1.7% Ni 0.7% Cr 0.3% Mo 0.1% V steel: Russian Standards designation
1.2714	0.55% C 1.7% Ni 1.0% Cr 0.5% Mo 0.1% V steel: German Standards designation
1.2721	0.50% C 3.3% Ni 1.1% Cr steel: For tools; German Standards designation
1.2767	0.45% C 4.1% Ni 1.4% Cr 0.3% Mo 0.5% W steel: For tools; German Standards designation
1.7225	0.42% C 0.6% Ni (max) 1.1% Cr 0.2% Mo steel: Forging; German Standard
1.7228	0.5% C 1.1% Cr 0.2% Mo steel: Forging; German Standard
1.7561	0.42% C 1.6% Cr 0.1% V steel: German Standard
1.7701	0.52% C 0.9% Mn 1.05% Cr 0.2% Mo 0.1% V steel: For springs; German Standard
1.8154	0.52% C 1.1% Cr 0.1% V steel: For springs; German Standard
1.8159	0.5% C 0.85% Mn 1.05% Cr 0.15% V steel: For springs; German Standard
2 CV	0.45% C 2% Cr 0.3% V 2% Si steel: T Turton
4KH8V2	0.4% C 7.5% Cr 2.5% W steel: Russian Standards designation
4KHV2S	0.4% C 1.15% Cr 2.25% W steel: Russian Standards designation
5/317	0.5% C 1.75% Ni 1.0% Cr steel: Carpenter
5KHNGM	0.55% C 0.75% Cr 0.22% Mo steel: Russian Standards designation

Note. The following abbreviations and units are used in the tables:

DPN	Hardness, diamond pyramid number
UTS	Ultimate tensile strength, N/mm^2
Elon	Elongation, %
Proof	0.1% proof strength, N/mm^2

$1\ N/mm^2 = 0.1\ hbar = 0.102\ kgf/mm^2 = 0.06475\ tonf/in.^2 = 145.04\ lbf/in.^2 = 1\ MPa$

See Appendix II for other abbreviations and conversion tables.

Symbol	Nominal analysis, supplier, condition and remarks.
5KHNM	0.55% C 1.6% Ni 0.65% Cr 0.22% Mo steel: Russian Standards designation
5KHNV	0.55% C 1.6% Ni 0.65% Cr 0.5% W steel: Russian Standards designation
5KHV2S	0.5% C 1.15% Cr 2.25% W steel: Russian Standards designation
7KHF	0.68% C 0.55% Cr 0.22% V steel: Russian Standards designation
8 MC 2 Mo	0.4% C 1% Cr 0.2% Mo steel: Sandvik for SS 2244
8 NIC 2	0.4% C 1.3% Ni 0.85% Cr steel: Sandvik for SS 2530
11.11	0.4% C 5.0% Cr 1.3% Mo 0.5% V steel: Origin unknown
14 HD	0.4% C 3.5% Cr 0.5% V 14% W steel: For tools; Huntsman; cutting tools; extrusion dies **DPN: 630**
17/22A	0.45% C 1.25% Cr 0.5% Mo 0.25% V steel: Timken; hardened and tempered for high temperature use **DPN: 415 UTS: 1550 Elon: 13% Proof: 1380**
18 D	0.5% C 4.0% Cr 1.0% V 18.0% W steel: For tools; origin unknown
18 HD	0.5% C 4% Cr 1% V 18% W steel: For tools; Huntsman; cutting tools; Cu extrusion dies **DPN: 700**
30 Ni Cr Mo 12	0.3% C 0.65% Mn 2.9% Ni 0.8% Cr 0.45% Mo steel: Italian Standards designation; as rolled **DPN: 260**
30 S	0.45% C 1.3% Cr 0.4% Mo 0.3% V 2.2% Si steel: Carrs **DPN: 520**
31 S	0.5% C 3.2% Cr 1.4% Mo steel: Carrs **DPN: 500**
34 Cr Ni Mo 8(e.f.)	0.4% C 1.5% Ni 1.2% Cr 0.28% Mo steel: Wrought; Austrian national specification
35 NC5 UNI 4365	0.36% C 1.2% Ni 0.9% Cr steel: Italian Standard; for bolts
35 NC9 UNI 4365	0.36% C 2.3% Ni 0.8% Cr steel: Italian Standard; for bolts
35 Ni Cr Mo 15	0.34% C 0.5% Mn 3.7% Ni 0.65% Cr 0.3% Mo steel: Italian Standards designation; as rolled **DPN: 275**
35 Ni Cr9	0.35% C 0.65% Mn 2.2% Ni 0.75% Cr steel: Italian Standards designation **DPN: 240**
35 W Cr V7	0.35% C 1.1% Cr 0.2% V 2.0% W steel: German Standards designation
37 S	0.4% C 4.2% Cr 0.3% Mo 2.2% V 4.2% W steel: Carrs
38 CD4 UNI 4365	0.38% C 1.0% Cr 0.2% Mo steel: Italian Standard; for bolts
38 CM	0.4% C 1.3% Cr 0.2% Mo steel: Electrode; Metrode
38 Ni Cr Mo 4	0.38% C 0.85% Mn 0.85% Ni 0.85% Cr 0.2% Mo steel: Italian Standards designation; as rolled **DPN: 240**
40 Cr Mo 4	0.4% C 0.85% Mn 1.05% Cr 0.2% Mo steel: Italian Standard designation; as rolled **DPN: 245**
40N2 CM	0.4% C 1.8% Ni 0.8% Cr 0.25% Mo steel: Electrode; Metrode
40 Ni Cr Mo 7	0.4% C 0.65% Mn 1.75% Ni 0.75% Cr 0.25% Mo steel: Italian Standards designation; as rolled **DPN: 250**
40 NKD	0.4% C 0.55% Ni 0.5% Cr 0.2% Mo steel: Pompey
42 Cr Mo 4	0.42% C 0.6% Ni (max) 1.1% Cr 0.2% Mo steel: Designation used by German Standards
42 Cr V 6	0.42% C 1.5% Cr 0.1% V steel: Designation used by German Standards
45 CM	0.45% C 1.1% Cr 0.2% Mo steel: Electrode; Metrode
45 CMV	0.45% C 1.1% Cr 0.3% Mo 0.25% V steel: Electrode; Metrode
45 CP	0.42% C 0.95% Cr 0.25% Mo steel: Pompey

Symbol	Nominal analysis, supplier, condition and remarks.
45 Cr 5 Ni 5	0.45% C 4.5% Ni 5.0% Cr steel: Electrode; Metrode; hardenable deposit **DPN: 400**
45 Cr Mo V 6-7	0.45% C 1.4% Cr 0.7% Mo 0.3% V steel: Designation used by German Standards
45 W Cr V 7	0.45% C 1.1% Cr 0.2% V 2.0% W steel: German Standards designation
48 Cr Mo V 67	0.45% C 1.4% Cr 0.7% Mo 0.3% V steel: German Standard
50 Cr Mo 4	0.5% C 1.1% Cr 0.2% Mo steel: Designation used by German Standards
50 CR V 4	0.5% C 1.1% Cr 0.1% V steel: Designation used by German Standards
50 Cr V4	0.5% C 0.8% Mn 1.0% Cr 0.15% V steel: Italian Standards designation **DPN: 250 UTS: 1550 Elon: 6% Proof: 150**
50 Ni Cr 13	0.5% C 3.3% Ni 1.1% Cr steel: For tools; German Standards designation
50 WCV20	0.5% C 1.5% Cr 0.2% V 2.2% W steel: French Standard
52 Si Cr Ni5	0.53% C 0.8% Mn 0.6% Ni 0.05% Cr 1.4% Si steel: Italian Standards designation **DPN: 260 UTS: 700 Elon: 5% Proof: 1350**
53 S	0.4% C 5% Cr 1.4% Mo 1% V 1% Si steel: Can be nitrided; Carrs; hot extrusion dies, mandrels, punches, etc. **DPN: 530**
54 Ni Cr Mo V6	0.54% C 1.7% Ni 0.7% Cr 0.3% Mo 0.1% V tool steel: German Standards designation
55 Ni Cr Mo V6	0.55% C 1.7% Ni 0.7% Cr 0.3% Mo 0.1% V steel: German Standards designation
55 Ni NCDV7	0.55% C 1.75% Ni 0.8% Cr 0.3% Mo 0.2% V steel: French Standards designation
55 WC20	0.55% C 1.0% Cr 2.0% W steel: French Standards designation
56 Ni Cr Mo V7	0.55% C 1.7% Ni 1.0% Cr 0.5% Mo 0.1% V steel: Germand Standards designation
58 Cr V 4	0.58% C 1.1% Cr 0.1% V steel: Designation used by German Standards
60 W Cr V7	0.6% C 1.1% Cr 0.2% V 2.0% W steel: German Standards designation
65 S	0.38% C 5% Cr 1.5% Mo 0.4% V steel: Carrs
65 W Mo 34–8	0.65% C 3.7% Cr 0.85% Mo 0.7% V 8.5% W steel: Designation used by German Standards
81 B40	0.4% C 0.35% Ni 0.4% Cr 0.1% Mo B steel: Designation used by AISI
81 B45	0.45% C 0.3% Ni 0.45% Cr 0.1% Mo steel: Designation used in the UK and USA
82 S	0.33% C 3% Cr 2.8% Mo 0.9% V 3% Co steel: Carrs
86 B30	0.3% C 0.55% Ni 0.5% Cr 0.2% Mo steel: Designation used in the UK and USA
86 B45	0.45% C 0.6% Ni 0.5% Cr 0.2% Mo 0.0005% B steel: Designation used by AISI
86 B45	0.45% C 0.5% Ni 0.5% Cr 0.2% Mo 0.004% B steel: Designation used in the UK and USA
185H40	0.4% C 0.2% Cr 0.12% Mo 0.003% B steel: BS 970
251A58	0.58% C 0.2% Cr 0.1% Mo (max) steel: For springs; BS 970
251A60	0.6% C 0.3% Cr 0.12% Mo (max) steel: For springs; BS 970
251H60	0.6% C 0.3% Cr 0.12% Mo steel: For springs; BS 970
300M	0.4% C 1.8% Ni 0.8% Cr 0.4% Mo 0.08% V steel: American proprietary alloy listed in SAE yearbook
328	0.4% C 5.5% Cr 0.8% Mo 0.5% V 1% Si steel: F Parkin; Al dies, cast moulds, etc. **DPN: 440**
348	0.4% C 4% Ni 1.6% Cr 0.3% Mo steel: F Parkin; shear blades, press dies **DPN: 700**

Symbol	Nominal analysis, supplier, condition and remarks.
361	0.52% C 1.1% Cr 0.3% V 1.9% W steel: For hot work tools; Balfour; forging dies, punches, etc. **DPN: 460**
399	0.5% C 1.3% Cr 0.2% V 2.3% W steel: F Parkin; trimming dies, chisels, etc. **DPN: 600**
525A58	0.58% C 0.8% Cr 0.1% Mo (max) steel: For springs; BS 970
525A60	0.6% C 0.9% Cr 0.06% Mo (max) steel: For springs; BS 970
525A61	0.61% C 0.95% Cr 1% Mo steel: For springs; BS 970
525H60	0.6% C 0.9% Cr 0.06% Mo (max) steel: For springs; BS 970
528	0.4% C 4.2% Cr 2.2% V 4.2% W 4.2% Co tool steel: Jessop; forging dies for copper alloys
640M40	0.4% C 1.3% Ni 0.65% Cr steel: BS 970; obsolete
704A60	0.6% C 0.9% Cr 0.2% Mo steel: For springs; BS 970
704H60	0.6% C 0.9% Cr 0.2% Mo steel: For springs; BS 970
705A60	0.6% C 0.9% Cr 0.3% Mo steel: For springs; BS 970
705H60	0.6% C 0.9% Cr 0.2% Mo steel: For springs; BS 970
708A42	0.42% C 1% Cr 0.2% Mo steel: BS 970
708A47	0.47% C 1% Cr 0.2% Mo steel: BS 970
708H37	0.37% C 1% Cr 0.2% Mo steel: BS 970
708H42	0.42% C 1% Cr 0.2% Mo steel: BS 970
708H45	0.45% C 1% Cr 0.2% Mo steel: BS 970
708M40	0.4% C 1% Cr 0.2% Mo steel: BS 970
709A42	0.42% C 1% Cr 0.3% Mo steel: BS 970
709M40	0.4% C 1% Cr 0.3% Mo steel: BS 970
735A51	0.51% C 1% Cr 0.15% V steel: For springs; BS 970
735A54	0.54% C 1% Cr 0.15% V steel: For springs; BS 970
735H51	0.51% C 1% Cr 0.15% V steel: For springs; BS 970
805H60	0.6% C 0.5% Ni 0.5% Cr 0.2% Mo steel: For springs; BS 970
816M40	0.4% C 1.5% Ni 1.2% Cr 0.15% Mo steel: BS 970; obsolete
817A37	0.37% C 1.5% Ni 1.2% Cr 0.3% Mo steel: BS 970
817A42	0.42% C 1.5% Ni 1.2% Cr 0.3% Mo steel: BS 970
817M40	0.4% C 1.5% Ni 1.2% Cr 0.27% Mo steel: BS 970
826M40	0.4% C 2.5% Ni 0.75% Cr 0.55% Mo steel: BS 970
925A60	0.6% C 0.3% Cr 0.25% Mo steel: For springs; BS 970
945A40	0.4% C 0.75% Ni 0.5% Cr 0.2% Mo steel: BS 970; obsolete
1213	0.45% C 0.6% Cr 0.2% Mo steel: French Standards designation
1413	0.38% C 1.6% Ni 1.2% Cr 0.25% Mo tool steel: S Osborn; plastic dies, zinc die cast dies; obsolete **DPN: 580**
1850	0.57% C 0.35% Mn 4% Cr 0.7% V 18% W steel: F Parkin; hot dies; can be used at higher hardness **DPN: 500**
2321	0.38% C 5.0% Cr 1.25% Mo 0.5% V steel: French Standards designation
2331	0.42% C 1.0% Cr 0.2% Mo steel: French Standards designation
2332	0.5% C 1.0% Cr 0.15% V steel: French Standards designation
2341	0.55% C 1.0% Cr 2.0% W steel: French Standards designation
3140	0.4% C 1.2% Ni 0.6% Cr steel: Designation used in the UK and USA
3145	0.40% C 1.2% Ni 0.8% Cr steel: Designation used by AISI; obsolete
3150	0.5% C 1.2% Ni 0.8% Cr steel: Designation used by AISI; obsolete
3240	0.4% C 1.7% Ni 1% Cr steel: Designation used by AISI; obsolete
3245	0.45% C 1.7% Ni 1% Cr steel: Designation used by AISI; obsolete
3250	0.5% C 1.7% Ni 1% Cr steel: Designation used by AISI; obsolete

Symbol	Nominal analysis, supplier, condition and remarks.
3340	0.4% C 3.5% Ni 1.5% Cr steel: Designation used by AISI; obsolete
3381	0.55% C 1.75% Ni 0.8% Cr 0.3% Mo 0.2% V steel: French Standards designation
3432	0.38% C 5.0% Cr 1.25% Mo 0.5% V 1.25% W steel: French Standards designation
3450	0.5% C 3% Ni 0.8% Cr steel: Designation used by AISI; obsolete
3541	0.4% C 4.0% Cr 0.5% Mo 0.5% W steel: French Standards designation
3548	0.65% C 4.0% Cr 5.0% Mo 2.0% V 6.0% W steel: French Standards designation
4137	0.37% C 1% Cr 0.2% Mo steel: Designation used in the UK and USA
4140	0.4% C 1% Cr 0.2% Mo steel: Designation used in the UK and USA
4142	0.42% C 1% Cr 0.2% Mo steel: Designation used in the UK and USA
4145	0.45% C 1% Cr 0.2% Mo steel: Designation used in the UK and USA
4147	0.47% C 1% Cr 0.2% Mo steel: Designation used in the UK and USA
4150	0.5% C 1% Cr 0.2% Mo steel: Designation used in the UK and USA
4161	0.61% C 0.8% Cr 0.3% Mo steel: Designation used in the UK and USA
4337	0.37% C 1.8% Ni 0.8% Cr 0.25% Mo steel: Designation used by AISI
4340	0.4% C 1.8% Ni 0.8% Cr 0.25% Mo steel: Designation used in the UK and USA
4620	0.4% C 1.8% Ni 0.25% Mo steel: Designation used by AISI; obsolete
4640	0.4% C 1.8% Ni 0.25% Mo steel: Designation used by AISI
6140	0.4% C 1% Cr 0.15% V steel: Designation used by AISI; obsolete
6150	0.5% C 1% Cr 0.15% V (min) steel: Designation used in the UK and USA
7260	0.6% C 0.8% Cr 1.8% W steel: Designation used by AISI; obsolete
8627	0.27% C 0.5% Ni 0.5% Cr 0.2% Mo steel: Designation used in the UK and USA
8630	0.3% C 0.5% Ni 0.5% Cr 0.2% Mo steel: Designation used in the UK and USA
8637	0.37% C 0.5% Ni 0.5% Cr 0.2% Mo steel: Designation used in the UK and USA
8640	0.4% C 0.5% Ni 0.5% Cr 0.2% Mo steel: Designation used in the UK and USA
8642	0.42% C 0.5% Ni 0.5% Cr 0.2% Mo steel: Designation used in the UK and USA
8645	0.45% C 0.5% Ni 0.5% Cr 0.2% Mo steel: Designation used in the UK and USA
8650	0.5% C 0.5% Ni 0.5% Cr 0.2% Mo steel: Designation used in the UK and USA
8655	0.55% C 0.5% Ni 0.5% Cr 0.2% Mo steel: Designation used in the UK and USA
8660	0.6% C 0.5% Ni 0.5% Cr 0.2% Mo steel: Designation used in the UK and USA

Note. The following abbreviations and units are used in the tables:

DPN	Hardness, diamond pyramid number
UTS	Ultimate tensile strength, N/mm^2
Elon	Elongation, %
Proof	0.1% proof strength, N/mm^2

1 N/mm^2=0.1 hbar=0.102 kgf/mm^2=0.06475 $tonf/in.^2$=145.04 $lbf/in.^2$=1 MPa

See Appendix II for other abbreviations and conversion tables.

Symbol	Nominal analysis, supplier, condition and remarks.
9763	0.63% C 0.5% Ni 0.2% Cr 0.2% Mo steel: Designation used by AISI; obsolete
9840	0.4% C 1% Ni 0.8% Cr 0.25% Mo steel: Designation used in the UK and USA
9850	0.5% C 1% Ni 0.8% Cr 0.25% Mo steel: Designation used in the UK and USA
71360	0.6% C 3.5% Cr 14% W steel: Designation used by AISI; obsolete
71660	0.6% C 3.5% Cr 16% W steel: Designation used by AISI; obsolete
A 5	0.4% C 1% Cr steel: Designation for BS 1453
A 6	0.7% C 2.0% Mn 1.0% Cr 1.0% Mo steel: For tools; designation used by AISI
A 8	0.55% C 5.0% Cr 1.25% Mo 1.25% W steel: For tools; designation used by AISI
A 9	0.5% C 1.5% Ni 5.0% Cr 1.4% Mo 1.0% V steel: For tools; designation used by AISI
A 13	0.4% C 1.5% Ni 1.15% Cr 0.3% Mo steel: Edgar Allen; hardened and tempered **DPN: 500 UTS: 1620 Elon: 11% Proof: 1500**
A 442	0.42% C 3% Ni 0.35% Cr 0.27% Mo steel: SHD; drills, etc.
ABRAZO	0.47% C 1.5% Mn 0.5% Cr 0.2% Mo steel: Plate; Colvilles; abrasion resisting **DPN: 230**
ACP	0.5% C 1% Cr 0.15% V steel: Toledo for BS alloy En 47
ADS HB	0.35% C 5.0% Cr 1.5% Mo 1.0% V tool steel: For hot working; Inman
AIRMO	0.7% C 1.0% Cr 1.35% Mo steel: Firth Sterling
AISI 81 B45	0.45% C 0.3% Ni 0.45% Cr 0.1% Mo 0.0005% B steel
AISI 86 B45	0.45% C 0.5% Ni 0.5% Cr 0.2% Mo steel
AISI 601	0.46% C 1% Cr 0.5% Mo 0.3% V steel: Hardened and tempered **UTS: 920 Elon: 29% Proof: 760**
AISI 610	0.4% C 5% Cr 1.3% Mo 0.5% V steel: Hardened and tempered **UTS: 2150 Elon: 7% Proof: 1920**
AISI 3140	0.4% C 1.2% Ni 0.6% Cr steel
AISI 3141	0.4% C 1.2% Ni 0.8% Cr steel: Obsolete
AISI 3145	0.45% C 1.2% Ni 0.8% Cr steel: Obsolete
AISI 3150	0.5% C 1.2% Ni 0.8% Cr steel: Obsolete
AISI 3240	0.4% C 1.7% Ni 1% Cr steel: Obsolete
AISI 4140	0.4% C 1% Cr 0.2% Mo steel
AISI 4142	0.42% C 1% Cr 0.2% Mo steel
AISI 4145	0.45% C 1% Cr 0.2% Mo steel
AISI 4147	0.47% C 1% Cr 0.2% Mo steel
AISI 4150	0.5% C 1% Cr 0.2% Mo steel
AISI 4161	0.61% C 0.8% Cr 0.3% Mo steel
AISI 4340	0.4% C 1.8% Ni 0.8% Cr 0.25% Mo steel
AISI 4340	0.40% C 1.8% Ni 0.8% Cr 0.25% Mo steel
AISI 4640	0.40% C 1.8% Ni 0.25% Mo steel: Obsolete
AISI 6145	0.45% C 1% Cr 0.15% V steel
AISI 6150	0.5% C 1% Cr 0.15% V steel
AISI 8640	0.4% C 0.6% Ni 0.5% Cr 0.2% Mo steel
AISI 8641	0.4% C 0.6% Ni 0.5% Cr 0.2% Mo steel
AISI 8642	0.42% C 0.6% Ni 0.5% Cr 0.2% Mo steel
AISI 8645	0.45% C 0.6% Ni 0.5% Cr 0.2% Mo steel
AISI 8647	0.47% C 0.6% Ni 0.5% Cr 0.2% Mo steel: Obsolete
AISI 8650	0.5% C 0.6% Ni 0.5% Cr 0.2% Mo steel
AISI 8653	0.53% C 0.6% Ni 0.7% Cr 0.2% Mo steel
AISI 8655	0.55% C 0.6% Ni 0.5% Cr 0.2% Mo steel
AISI 8660	0.6% C 0.6% Ni 0.5% Cr 0.2% Mo steel
AISI 8740	0.4% C 0.6% Ni 0.5% Cr 0.25% Mo steel
AISI 8742	0.42% C 0.6% Ni 0.5% Cr 0.25% Mo steel
AISI 8745	0.45% C 0.6% Ni 0.5% Cr 0.25% Mo steel: Obsolete
AISI 8750	0.5% C 0.6% Ni 0.5% Cr 0.25% Mo steel
AISI 9440	0.4% C 0.4% Ni 0.4% Cr 0.1% Mo steel: Obsolete
AISI 9442	0.42% C 0.4% Ni 0.5% Cr 0.1% Mo steel: Obsolete
AISI 9445	0.45% C 0.4% Ni 0.4% Cr 0.1% Mo steel: Obsolete
AISI 9447	0.47% C 0.4% Ni 0.4% Cr 0.1% Mo steel: Obsolete

Symbol	Nominal analysis, supplier, condition and remarks.
AISI 9747	0.47% C 0.5% Ni 0.2% Cr 0.2% Mo steel: Obsolete
AISI 9763	0.63% C 0.5% Ni 0.2% Cr 0.2% Mo steel: Obsolete
AISI 9840	0.4% C 1% Ni 0.8% Cr 0.25% Mo steel
AISI 9845	0.45% C 1% Ni 0.8% Cr 0.25% Mo steel: Obsolete
AISI 9850	0.5% C 1% Ni 0.8% Cr 0.25% Mo steel
AISI H11	0.4% C 5% Cr 1.3% Mo 0.5% V steel: Similar to AISI 610
ALHD	0.4% C 5% Cr 1.3% Mo 1% V 1% Si steel: Jonas; gravity dies for Al casting
	DPN: 620
ALLOY 10	0.6% C 0.4% Mo 0.25% V 1.8% Si steel: Braeburn
ALZ	0.4% C 5.0% Cr 1.4% Mo 1.0% V steel: Origin unknown
AM 3	0.4% C 5% Cr 1.35% Mo 1.1% V 1% Si steel: Edgar Allen; for Al die casting moulds
	DPN: 600
AMS 641	0.42% C 1.8% Ni 0.85% Cr 0.4% Mo steel
AMS 6304 B	0.45% C 1% Cr 0.5% Mo 0.3% V steel: Bar and forging; AMS for alloy 17/22A
AMS 6305	0.46% C 1% Cr 0.52% Mo 0.3% V steel
AMS 6307	0.45% C 1.2% Cr 0.5% Mo 0.3% V steel: Forgings
AMS 6312 A	0.4% C 1.8% Ni 0.25% Mo steel: Bar and forging; AMS for SAE 4640
AMS 6317 B	0.4% C 1.8% Ni 0.25% Mo steel: Bar and forging; hardened and tempered; AMS for SAE 4640
	UTS: 920
AMS 6321	0.4% C 0.3% Ni 0.4% Cr 0.1% Mo B steel: Bar and forging; AMS for SAE 81 B40
AMS 6322 E	0.4% C 0.5% Ni 0.5% Cr 0.25% Mo steel: Bar and forging; AMS for SAE 8740
AMS 6323 C	0.4% C 0.5% Cr 0.25% Mo steel: Tube; AMS for SAE 8740
AMS 6324 B	0.4% C 0.7% Ni 0.6% Cr 0.25% Mo steel: Bar and forging; AMS for SAE 8740
AMS 6325 C	0.4% C 0.5% Ni 0.5% Cr 0.25% Mo steel: Bar and forging; hardened and tempered; AMS for SAE 8740
	UTS: 760
AMS 6327 C	0.4% C 0.5% Ni 0.5% Cr 0.25% Mo steel: Bar and forging; hardened and tempered; AMS for SAE 8740
	UTS: 920
AMS 6328 D	0.5% C 0.5% Ni 0.5% Cr 0.25% Mo steel: Bar and forging; AMS for SAE 8750
AMS 6342 C	0.4% C 1% Ni 0.8% Cr 0.25% Mo steel: Bar and forging; AMS for SAE 9840
AMS 6351	0.4% C 1% Cr 0.2% Mo steel: Sheet and strip; spheroidized; AMS for SAE 4130
AMS 6358 A	0.4% C 0.5% Ni 0.5% Cr 0.25% Mo steel: Sheet and strip; AMS for SAE 8740
AMS 6359 A	0.4% C 1.8% Ni 0.8% Cr 0.25% Mo steel: Sheet and strip; AMS for SAE 4340
AMS 6378	0.4% C 1% Cr 0.2% Mo steel: Bar; drawn and tempered; AMS for SAE 4140
	Proof: 920
AMS 6379	0.5% C 1% Cr 0.2% Mo steel: Bar; drawn and tempered; AMS for SAE 4150
	Proof: 1210
AMS 6381 A	0.4% C 1% Cr 0.2% Mo steel: Tube; AMS for SAE 4140
AMS 6382 E	0.4% C 1% Cr 0.2% Mo steel: Bar and forging; AMS for SAE 4140
AMS 6390	0.4% C 1% Cr 0.2% Mo steel: Seamless tube; AMS for SAE 4140; special quality
AMS 6396	0.52% C 1.8% Ni 0.8% Cr 0.25% Mo steel
AMS 6406	0.44% C 2% Cr 5% Mo 0.05% V 1.6% Si steel: Sheet and strip
AMS 6408	0.4% C 5.2% Cr 1.5% Mo 1% V steel: Type H13
AMS 6414	0.4% C 1.8% Ni 0.8% Cr 0.25% Mo steel: Bar and forging; AMS for SAE 4340
AMS 6415 F	0.4% C 1.8% Ni 0.8% Cr 0.25% Mo steel: Bar and forging; AMS for SAE 4340
AMS 6416	0.45% C 1.8% Ni 0.8% Cr 0.4% Mo 0.07% V 1.6% Si steel: Bar and forging
AMS 6419	0.42% C 1.8% Ni 0.82% Cr 0.52% Mo steel
AMS 6422 B	0.4% C 0.8% Ni 0.8% Cr 0.2% Mo B steel: Bar and forging; AMS for SAE 98 B40
AMS 6423	0.42% C 0.7% Ni 0.9% Cr 0.5% Mo B steel: Bar and forging
AMS 6424	0.52% C 1.8% Ni 0.8% Cr 0.25% Mo steel
AMS 6431	0.45% C 0.5% Ni 1% Cr 1% Mo 0.1% V steel: Forgings
AMS 6432	0.47% C 0.55% Ni 1% Cr 1% Mo 0.1% V steel
AMS 6437 A	0.4% C 5% Cr 1.3% Mo 0.5% V steel: Sheet and strip
AMS 6438	0.47% C 0.55% Ni 1% Cr 1% Mo 0.1% V steel
AMS 6439	0.47% C 0.55% Ni 1% Cr 1% Mo 0.1% V steel
AMS 6445 B	0.5% C 1% Cr 0.15% V steel: Sheet and strip for springs; annealed; AMS for SAE 6150
AMS 6448 B	0.5% C 1% Cr 0.15% V steel: Bar and forging; AMS for SAE 6150
AMS 6450 B	0.5% C 1% Cr 0.2% V steel: Wire for springs; annealed; AMS for SAE 6150
AMS 6454	0.4% C 1.8% Ni 0.8% Cr 0.25% Mo steel
AMS 6455	0.5% C 1% Cr 0.15% V (min) steel
AMS 6470 E	0.4% C 1.6% Cr 0.35% Mo 1.1% Al steel: Bar and forging; nitriding steel
AMS 6471	0.4% C 1.6% Cr 0.35% Mo 1.1% Al steel: For nitriding
AMS 6472	0.4% C 1.6% Cr 0.35% Mo 1.1% Al steel: For nitriding
AMS 6485 A	0.4% C 5% Cr 1.3% Mo steel: Bar and forging
AMS 6487 B	0.4% C 5% Cr 1.3% Mo 0.5% V steel: Bar and forging; vacuum melted; premium quality
AMS 6488	0.4% C 5% Cr 1.3% Mo 0.5% V steel: Bar and forging; premium quality
AMS 6529	0.4% C 1% Cr 0.2% Mo steel
AMS 7301	0.5% C 1% Cr 0.15% V (min) steel
AMS 7452	0.4% C 0.55% Ni 0.5% Cr 0.25% Mo steel
AMS 7454	0.46% C 1% Cr 0.55% Mo 0.3% V steel
AMS 7455	0.45% C 0.55% Ni 1% Cr 0.3% V steel
AMS 7459	0.45% C 0.55% Ni 1% Cr 0.3% V steel
AN QQ S690/1	0.4% C 1.2% Ni 0.6% Cr steel: US Service; normalized and tempered
	DPN: 230
AN QQ S752a	0.38% C 1.0% Cr 0.2% Mo steel: US Service; hardened and tempered
	DPN: 230 UTS: 840 Elon: 18% Proof: 740
AN QQ S756a	0.4% C 1.8% Ni 0.75% Cr 0.25% Mo steel: US Service; hardened and tempered
	UTS: 1070 Elon: 16% Proof: 870
AN S 16a	0.41% C 0.6% Ni 0.5% Cr 0.25% Mo steel: US Service; hardened and tempered
	DPN: 230 UTS: 840 Elon: 16% Proof: 740
ANTICHOC H	0.45% C 1.1% Cr 0.15% V 1.9% W steel: Pompey
ANTICHOC IX	0.6% C 0.37% Cr 0.1% V 1.6% Si steel: Pompey
ARDHO 2	0.45% C 1.3% Cr 0.3% V 2.3% W steel: Spencer
ASJMA752	General requirements for wire rod and alloy covers from carbon and alloy steel
ASTM A60	0.5% C 1% Cr 0.1% V steel: Bar for springs; obsolete
ASTM A193 B7	0.4% C 1% Cr 0.2% Mo steel: For bolts, etc.; hardened and tempered
	UTS: 890 Elon: 16% Proof: 700
ASTM A193 B14	0.45% C 1% Cr 0.35% Mo 0.25% V steel: For bolts; hardened and tempered
	UTS: 920 Elon: 16% Proof: 760
ASTM A193 B16	0.4% C 1% Cr 0.5% Mo 0.3% V steel: For bolts; hardened and tempered
	UTS: 840 Elon: 17% Proof: 700
ASTM A231	0.5% C 1% Cr 0.15% V steel: Spring wire; supplied annealed or cold drawn
ASTM A232	0.5% C 1% Cr 0.2% V steel: Spring wire; cold drawn and tempered
	UTS: 1900

Symbol	Nominal analysis, supplier, condition and remarks.
ASTM A291/3	0.45% C 3% Ni 1.25% Cr 0.15% Mo 0.1% V steel: Forging; hardened and tempered
	UTS: 760 Elon: 19% Proof: 580
ASTM A291/4	0.4% C 1.6% Ni 0.6% Cr 0.2% Mo 0.1% V steel: Forging; hardened and tempered
	UTS: 840 Elon: 14% Proof: 700
ASTM A291/5	0.4% C 1.6% Ni 0.6% Cr 0.2% Mo 0.1% V steel: Forging; hardened and tempered
	UTS: 100 Elon: 14% Proof: 760
ASTM A291/6	0.4% C 1.6% Ni 0.6% Cr 0.2% Mo 0.1% V steel: Forging; hardened and tempered
	UTS: 1070 Elon: 13% Proof: 840
ASTM A291/7	0.45% C 1.6% Ni 0.6% Cr 0.2% Mo 0.1% V steel: Forging; hardened and tempered
	DPN: 80 UTS: 1000 Elon: 12%
ASTM A294 A	0.4% C 1.2% Cr 0.15% Mo steel: For turbine wheels
ASTM A294 B	0.45% C 2.5% Ni 1.2% Cr 0.2% Mo steel: For turbine wheels
ASTM A295 F30	0.45% C 0.5% Mo 0.2% V steel: Seamless drum forging; normalized and tempered
	UTS: 580 Elon: 21% Proof: 340
ASTM A320 L7	0.43% C 0.9% Cr 0.2% Mo steel: Hardened and tempered
	UTS: 860 Elon: 16% Proof: 730
ASTM A320 L43	0.4% C 1.8% Ni 0.8% Cr 0.2% Mo steel: Hardened and tempered
	UTS: 860 Elon: 16% Proof: 730
ASTM A331	Specification covering alloy steel bars; AISI designations are used to define the alloys
ASTM A332/8650	0.5% C 0.5% Ni 0.2% Mo steel: For springs; obsolete
ASTM A332/8655	0.55% C 0.6% Ni 0.5% Cr 0.2% Mo steel: For springs; obsolete
ASTM A332/8660	0.6% C 0.5% Ni 0.5% Cr 0.2% Mo steel: For springs; obsolete
ASTM A336 F30	0.45% C (max) 0.45% Cr 0.27% V steel: Seamless drum forgings
	UTS: 540 Elon: 21% Proof: 330
ASTM A354	Alloy steel with 0.04% S and P steel: For bolts and nuts; analysis not part of specification; grades vary with mechanical properties
ASTM A355 A	0.41% C 1.6% Cr 0.4% Mo 1.0% Al steel: Hardened and tempered
	DPN: 250
ASTM A355 B	0.35% C 1.2% Cr 0.2% Mo 0.2% Si 1.0% Al steel: Hardened and tempered
	DPN: 250
ASTM A355 D	0.35% C 1.1% Cr 0.2% Mo 1.0% Al steel: Hardened and tempered
	DPN: 250
ASTM A372 Class IV	0.45% C 1.60% Mn 0.22% Mo 0.27% Si steel: Forging for pressure vessels
ASTM A372 Class V Type A	0.30% C 0.55% Mn 1.0% Cr 0.27% Si 0.2% Mo steel: Forging for pressure vessels
ASTM A372 Class V Type B	0.35% C 0.85% Mn 1.0% Cr 0.2% Mo 0.27% Si steel: Forging for pressure vessels
ASTM A372 Class V Type C	0.30% C 0.55% Ni 0.52% Cr 0.20% Mo steel: For pressure vessels
ASTM A372 Class V Type D	0.35% C 0.55% Ni 0.52% Cr 0.20% Mo steel: Forging for pressure vessels

Note. The following abbreviations and units are used in the tables:

DPN	Hardness, diamond pyramid number
UTS	Ultimate tensile strength, N/mm^2
Elon	Elongation, %
Proof	0.1% proof strength, N/mm^2

1 N/mm^2=0.1 hbar=0.102 kgf/mm^2=0.06475 $tonf/in.^2$=145.04 $lbf/in.^2$=1 MPa
See Appendix II for other abbreviations and conversion tables.

Symbol	Nominal analysis, supplier, condition and remarks.
ASTM A372 Class V Type E	0.42% C 1.0% Cr 0.20% Mo steel: Forging for pressure vessels
ASTM A391	0.3% C (max) 0.04% S and P (max) steel: For chain; alloying elements will be present
ASTM A469/3	0.27% C 2.5% Ni 0.5% Cr (max) 0.35% Mo steel: For rotors, vacuum treated
ASTM A469/4	0.27% C 3.0% Ni 0.5% Cr (max) 0.3% Mo steel: For rotors; vacuum treated
ASTM A469/5	0.3% C 3.0% Ni 0.5% Cr (max) 0.45% Mo steel: For rotors, vacuum treated
ASTM A469/6	0.28% C 3.5% Ni 1.5% Cr 0.45% Mo 0.1% V steel: For rotors; vacuum treated
ASTM A470/2	0.25% C (max) 2.5% Ni 0.75% Cr (max) 0.25% Mo (min) steel: For turbine rotors; vacuum treated
ASTM A470/3 and 4	0.28% C (max) 2.5% Ni 0.75% Cr (max) 0.25% Mo (min) steel: For turbine rotors; vacuum treated
ASTM A470/5, 6 and 7	0.28% C 3.5% Ni 1.5% Cr 0.45% Mo 0.1% V steel: For turbine rotors; vacuum treated
ASTM A470/8	0.3% C 0.7% Ni 1.2% Cr 1.2% Mo 0.25% V steel: For turbine rotors; vacuum treated
ASTM A471	0.4% C (max) 3.0% Ni 2.0% Cr (max) 0.5% Mo 0.05% V steel: For turbine discs; vacuum treated; grades 1, 2, 3, 4, 5 and 6
ASTM A471/7	0.31% C 0.5% Ni (max) 1.1% Cr 1.2% Mo 0.25% V steel: For turbine discs; vacuum treated
ASTM A487 GN	0.38% C (max) 1.5% Mn 0.6% Ni 0.6% Cr 0.35% Mo steel: Castings for pressure; normalized; 6Q grade is hardened and tempered
ASTM A540 B21	0.4% C 1.0% Cr 0.55% Mo 0.3% V steel: For bolts
ASTM A540 B22	0.4% C 1.0% Cr 0.2% Mo steel: For bolts
ASTM A540 B23	0.4% C 1.7% Ni 0.8% Cr 0.25% Mo steel: For bolts
ASTM A540 B24	0.4% C 1.8% Ni 0.8% Cr 0.35% Mo steel: For bolts
ASTM A579/21	0.35% C 1.85% Ni 0.8% Cr 0.5% Mo 0.2% V steel: Forging
ASTM A579/22	0.4% C 1.8% Ni 0.8% Cr 0.5% Mo 0.07% V steel: Forging
ASTM A579/23	0.47% C 0.6% Ni 1.0% Cr 1.0% Mo 0.11% V steel: Forging
ASTM A579/32	0.42% C 1.85% Ni 0.8% Cr 0.42% Mo 1.6% Si steel: Forging
ASTM A579/38	0.43% C 2.1% Cr 0.5% Mo 0.05% V 1.5% Si steel: Forging
ASTM A579/41	0.4% C 5.0% Cr 1.3% Mo 0.5% V steel: Forging
ASTM A597 CS5	0.6% C 0.35% Cr (max) 0.5% Mo 0.35% V (max) 2.0% Si steel: Cast; for tools
ASTM A628	Steel plate for security purposes to resist drilling; composition by agreement
ASTM A646	See Nit 135
ASTM A646/98 BV40	0.43% C 0.75% Ni 0.9% Cr 0.5% Mo 0.03% V steel
ASTM A649/1 A	0.42% C 1.0% Cr 0.3% Mo steel: For rolls; corrugated
ASTM A649/1 B	0.4% C 1.8% Ni 0.8% Cr 0.25% Mo steel: For rolls; corrugated
ASTM A732/12 Q	0.5% C 0.95% Cr 0.15% V (min) steel: Investment casting; hardened and tempered
	UTS: 1300 Elon: 4% Proof: 1170
ASTRA	0.55% C 1.0% Cr 2.0% W steel: Origin unknown
AT	0.45% C 3.0% Ni 0.5% Cr steel: For cold work; Osborn
AUSTENITE	0.6% C 4% Cr 0.5% V 14% W steel: Inman; cutting tools
AWCW	0.63% C 4% Cr 0.5% Mo 0.7% V 14% W steel: For tools; ESC; shear blades, punches, etc.
	DPN: 790
BARTO	0.5% C 1.75% Ni 1% Cr steel: Carpenter
BCC	0.5% C 1.2% Cr 0.25% V 2.25% W 1% Si steel: Carrs; snaps, drifts, etc.
	DPN: 580
BCD 48	0.5% C 1.6% Cr 0.3% V 2% W steel: Carrs; chisels, snaps, etc.
	DPN: 600

Symbol	Nominal analysis, supplier, condition and remarks.
BESTEM	C not specified 3.0% Ni Cr Mo steel: Origin unknown
BKM	0.4% C 4.25% Ni 1.5% Cr steel: Jonas; bakelite moulds, etc. **DPN: 460**
BRAKE DIE	0.5% C 0.6% Cr 0.2% Mo steel: Atlas; hardened and tempered **DPN: 330 UTS: 1140 Elon: 13% Proof: 1070**
BRUNSWICK V 109	Medium C 1.5% Ni 0.6% Cr + Mo steel: John Vessey; stamping dies **DPN: 440**
BRUNSWICK V 114	5% Cr 1.5% Mo 1% W steel: John Vessey **DPN: 424**
BRUNSWICK V 130	Medium C 3% Cr 0.8% W steel: John Vessey **DPN: 820**
BS 224/2	0.6% C 1.2% Ni 0.3% Cr steel: For die blocks; hardened and tempered **DPN: 250**
BS 224/5	0.55% C 1.5% Ni 0.6% Cr 0.3% Mo steel: For die blocks; hardened and tempered **DPN: 300**
BS 970	See designation for details
BS 980 CDS15	0.55% C 1% Cr 0.2% Mo steel: Cold drawn tube; hardened and tempered **UTS: 1070 Elon: 10% Proof: 760**
BS 1506/621 A	0.4% C 1.1% Cr 0.3% Mo steel: Bar; hardened and tempered **UTS: 840 Elon: 18% Proof: 700**
BS 1506/621 B	0.4% C 1.3% Cr 0.7% Mo steel: Bar; hardened and tempered **UTS: 920 Elon: 17% Proof: 740**
BS 3111/5	0.4% C 1.2% Cr 0.3% Mo steel: Wire; annealed and drawn; for cold forged bolts **UTS: 730**
BS 3111/6	0.4% C 1.5% Ni 1.2% Cr 0.15% Mo steel: Wire; annealed and drawn; for cold forged bolts **UTS: 730**
BS 3146 CLA1C	0.45% C 1% Mn low Ni Cr Mo Cu steel: Investment casting; hardened and tempered **DPN: 180 UTS: 530 Elon: 15% Proof: 250**
BS 3146 CLA8	0.4% C low Cr Mo Cu steel: Investment casting; normalized and tempered
BS 3146 CLA12C	0.6% C 1.2% Cr 0.3% Mo steel: Investment casting; hardened and tempered; for abrasion resistance **DPN: 341**
BS 3551 B	0.4% C 1.3% Mn 0.7% Ni 0.5% Cr 0.2% Mo steel: For shackles; hardened and tempered **DPN: 270**
BS 5005	Wrought steels: For automobiles; withdrawn; replaced by BS 970
BS 5007	Sheet steels: For automobiles; withdrawn; replaced by BS 1449
BS 5008	Valve steels: For automobiles; withdrawn; replaced by BS 970
BS 9156/C	0.6% C 1.2% Cr 0.3% Mo steel: Casting; normalized or hardened and tempered; in BS 3100 **DPN: 341**
BS N35	0.4% C 3.5% Ni 0.3% Cr steel: Toledo for BS alloy En 22
BS S2	Medium carbon alloy steel: 850 N/mm^2 range; replaced by BS S94, S95 and S96
BS S11	Nickel chromium steel: 900 N/mm^2 range; replaced by BS S94, S95 and S96
BS S12	Alloy steel: Bar and forging; high tensile; replaced
BS S18	Alloy steel: 1300 N/mm^2 tensile range; replaced
BS S33	Medium carbon alloy steel: 900 N/mm^2 tensile; obsolete
BS S34	Medium carbon alloy steel: Bar; hardened and tempered; obsolete
BS S40	Medium carbon alloy steel: Strip; high tensile; obsolete
BS S43	Medium carbon alloy steel: Strip; obsolete
BS S65	Medium carbon nickel chrome steel: Replaced by BS/S97
BS S69	Medium carbon 3.5% Ni steel: Obsolete; see BS S94, S95 and S96
BS S81	Medium carbon chrome nickel steel: 1070 N/mm^2 tensile; obsolete; replaced by BS/S97
BS S86	Medium carbon chrome nickel steel: Sheet; obsolete; see BS S514
BS S87	Medium carbon chrome nickel steel: Sheet; obsolete
BS S88	Medium carbon chrome nickel steel: Sheet; obsolete; see BS S517
BS S94	0.4% C 0.7% Ni 0.5% Cr 0.2% Mo steel: Hardened and tempered **DPN: 280 UTS: 920 Elon: 18% Proof: 670**
BS S95	0.4% C 1.5% Ni 1.2% Cr 0.3% Mo steel: Hardened and tempered **DPN: 280 UTS: 920 Elon: 18% Proof: 670**
BS S98	0.4% C 2.5% Ni 0.7% Cr 0.55% Mo steel: Hardened and tempered **DPN: 360 UTS: 1210 Elon: 14% Proof: 1000**
BS S99	0.4% C 2.5% Ni 0.7% Cr 0.55% Mo steel: Hardened and tempered **DPN: 390 UTS: 1300 Elon: 14% Proof: 1070**
BS S118	0.4% C 1.5% Ni 1.2% Cr 0.15% Mo steel: Hardened and tempered **DPN: 280 UTS: 920 Elon: 18% Proof: 670**
BS S119	0.4% C 1.5% Ni 1.2% Cr 0.3% Mo steel: Hardened and tempered **DPN: 320 UTS: 1070 Elon: 16% Proof: 820**
BS T50	0.5% C (max) 1.0% Cr 0.2% Mo steel: Tube; hardened and tempered **UTS: 760 Proof: 700**
BUSTER ALLOY	0.52% C 1.35% Cr 0.25% V 2.25% W steel: Columbia; AISI type S1
B and W – WL 1	0.4% C 1.5% Ni 0.8% Cr 0.2% Mo steel: Tube; Babcock and Wilcox (USA)
C 5	5.0% Cr 1.5% Mo steel: For tools; hot forging; Turton B100; can be used cast
C 108	0.45% C 3.0% Cr 1.5% Mo steel: Fagersta **DPN: 620**
C 320	0.45% C 3.1% Cr 0.4% Mo 0.1% V steel: Fagersta
C 330	0.39% C 5.3% Cr 1.4% Mo 1.0% V 1.0% Si: Fagersta **DPN: 570**
C 345	0.42% C 1.1% Cr 0.2% Mo steel: Fagersta
C 424	0.5% C 1.0% Cr 0.15% V steel: Fagersta
CARPENTER 5/317	0.5% C 1.75% Ni 1.0% Cr steel: Carpenter
CARPENTER 481	0.55% C 0.8% Mn 0.25% Cr 0.4% Mo 1.9% Si steel: For tools; Carpenter for AISI S5
CARPENTER 883	0.37% C 5.25% Cr 1.3% Mo 1.0% V 1.0% Si steel: For tools; Carpenter for AISI H13
CBVO/h	0.5% C 1.05% Cr 0.2% Mo steel: Krupp
CBVO/z	0.42% C 1.05% Cr 0.2% Mo steel: Krupp
CCS	0.4% C 5.25% Cr 4.25% W 1.15% Si steel: For tools; origin unknown
CCV	0.4% C 1.25% Cr 0.15% V steel: Origin unknown
CCW	0.52% C 1.2% Cr 2.3% W steel: Firth Brown; chisels, punches
CDS	0.4% C 5% Cr 1.35% Mo 0.7% V 1% Si steel: T Turton
CDS 2	0.4% C 5% Cr 1.35% Mo 1.1% V 1% Si steel: T Turton
CDV	0.39% C 0.5% Cr 1.5% Mo 1.0% V steel: For tools
CEC SMOOTHCUT	0.58% C 0.3% Cr 0.25% V steel: Free cutting; Columbia; AISI type S4S
CHD 1	0.38% C 5% Cr 1.3% Mo 1% V 1% Si steel: Huntsman; for nitrided dies for Al and Mg casting **DPN: 550**
CHD 2	0.4% C 5% Cr 0.3% Mo 0.3% V 5% W 1% Si steel: Huntsman; dies for Al and brass casting **DPN: 540**
CHIMO	0.55% C 0.2% Cr 0.5% Mo 0.2% V 2.0% Si steel: Firth Sterling
CHISEL STEEL	Low alloy W Cr steel: Origin unknown

Symbol	Nominal analysis, supplier, condition and remarks.
CHROMETOUGH	0.55% C 0.5% Cr 0.2% V steel: Clyde Alloy **DPN: 600**
CHROME TUNGSTEN	0.55% C 0.16% Ni 7.5% Cr 7.4% W steel: Latrobe; for hot working
CHW	0.5% C 2.8% Cr 0.5% V 15.1% W steel: Latrobe; casting and forging dies for brass; type H24 **DPN: 570**
CL 40X	0.55% C 0.3% Cr 0.4% Mo 0.2% V steel: Origin unknown
CL 99	0.4% C 4.2% Ni 1.2% Cr 0.25% Mo steel: Origin unknown
CL 224	0.55% C 1.5% Ni 0.75% Cr 0.3% Mo steel: Origin unknown
CL 225	0.5% C 1.25% Ni 0.5% Cr 0.25% Mo steel: Origin unknown
CLARITE EW 50	0.53% C 4.0% Cr 1.0% V 18% W steel: Columbia; AISI type H26
CLOSELOY	Medium C Ni Cr Mo steel: Armstrong Whitworth; range of steels with varying C
CMN	0.52% C 1.5% Ni 0.6% Cr 0.22% Mo steel: Marsh Brothers; forging dies
CMVM	0.5% C 0.9% Cr 0.2% V steel: Huntsman; chisels, screwdrivers; dies, etc. **DPN: 500**
CODE 628	0.45% C 0.8% Cr 0.3% Mo steel: Casting; Edgar Allen; Cr Mo 'P' steel
CODE 638	0.48% C 4.2% Ni 0.6% Cr 0.25% Mo 1.7% Si steel: Casting; Edgar Allen; 'YCW STEM'
CODE 639	0.5% C 0.8% Cr 0.3% Mo steel: Casting; Edgar Allen; Cr Mo/F Mo steel
COLLET STEEL	0.6% C 0.4% Ni 0.8% Cr 0.2% Mo steel: Crucible Steel Co. **DPN: 570**
COLMONOY 155	0.5% C 1.5% Mn 2.5% Cr 0.4% Mo 0.1% Si iron alloy: Bronze metal; Wall Colomony; melting point 1450 °C
COMET 258S	0.3% C 5.5% Cr 1.5% Mo 1.4% W steel: Welding electrode; Soudometal; for abrasion resistance **DPN: 530**
COMMANDO	0.5% C 1.4% Cr 0.2% V 2.1% W steel: Simonds; AISI type S1
CONQUEROR LC	0.45% C 3.0% Cr 15% W steel: For tools; origin unknown
CONSUMET	0.5% C 0.1% Ni 4.0% Cr 4.5% Mo 1.0% V steel: Carpenter for bearings
CP 2	0.53% C 1.4% Cr 0.2% V 2% W steel; For tools; S Osborn; shock and fatigue resistant tools; obsolete **DPN: 570**
CP 30	C not specified 1.5% Cr 0.75% Ni 0.5% Mo steel: BSC Cyclop; armour plate
CP 40	0.4% C 0.7% Ni 1.6% Cr 0.5% Mo steel: ESC **UTS: 1070 Elon: 15% Proof: 880**
CP 50	C not specified 1.5% Cr 0.75% Ni 0.5% Mo steel: BSC Cyclop; armour plate
CR 43	0.43% C 0.4% Ni 1% Cr 0.15% Mo 0.3% Cu steel: Casting; Wolsingham **DPN: 200 UTS: 610 Elon: 12%**

Note. The following abbreviations and units are used in the tables:

DPN	Hardness, diamond pyramid number
UTS	Ultimate tensile strength, N/mm^2
Elon	Elongation, %
Proof	0.1% proof strength, N/mm^2

1 N/mm^2=0.1 hbar=0.102 kgf/mm^2=0.06475 $tonf/in.^2$=145.04 $lbf/in.^2$=1 MPa

See Appendix II for other abbreviations and conversion tables.

Symbol	Nominal analysis, supplier, condition and remarks.
CR 50	0.5% C 0.4% Ni 1% Cr 0.15% Mo 0.3% Cu steel: Casting; Wolsingham **DPN: 207 UTS: 700 Elon: 10%**
CR 83	0.4% C 1.3% Ni 0.8% Cr steel: Bofors; obsolete **DPN: 260 UTS: 840 Elon: 15% Proof: 580**
CR 501	0.5% C 0.4% Ni 1% Cr 0.15% Mo 0.3% Cu steel: Casting; Wolsingham **DPN: 293**
Cr Mo/F Mo	0.5% C 0.8% Cr 0.3% Mo steel: Casting; Edgar Allen; Code 639 steel
Cr Mo 'P'	0.45% C 0.8% Cr 0.3% Mo steel: Casting; Edgar Allen; Code 628 steel
CRD 23	0.4% C 1.5% Ni 1.2% Cr steel: Origin unknown
CRD 24	0.4% C 1.5% Ni 1.2% Cr 0.3% Mo steel: C Denton; shear blades; obsolete **DPN: 550**
CRD NICHROMA	0.45% C 3% Ni 0.65% Cr 1% Si steel: C Denton; shear blades; obsolete **DPN: 550**
CRD NTCS	0.5% C 1% Cr 0.25% V steel: C Denton; shear blades; obsolete **DPN: 550**
CRM 1	0.4% C 1.2% Cr 0.3% Mo steel: Firth Brown for BS alloy En 19 and 19A
CRM 2	0.4% C 1.3% Cr 0.8% Mo steel: Firth Brown for BS alloy En 20B
CrMo60	0.6% C 0.4% Ni 1.1% Cr 0.3% Mo 0.3% Cu steel: Casting; Wolsingham **DPN: 341**
CRO 684 BOVAC	0.3% C 0.6% Ni 1.7% Cr 0.4% Mo steel: Bofors; vacuum de-gassed for plastic moulds; hardened and tempered **DPN: 300 UTS: 900 Elon: 18% Proof: 750**
CROMAX Cr Mo F Mo	0.53% C 0.8% Cr 0.3% Mo steel: Casting; Edgar Allen; for wear resistance **DPN: 250**
CROMOL	0.4% C 1.1% Cr 0.3% Mo steel: ESC for BS alloy En 19
CROWN	0.4% C 1.3% Cr 0.15% V 2.3% W steel: Jessop **DPN: 620**
CRV	0.5% C 1% Cr 0.2% V spring steel: ESC for BS alloy En 47
CTC 1	0.45% C 1.25% Cr 2% W steel; Marsh Brothers; chisels, caulking irons, etc.
CTH	0.43% C 0.25% Cr 0.5% W steel: Marsh Brothers; hot forging dies, etc.
CV 4	0.45% C 2.3% Cr 0.3% V steel: Origin unknown
CV 40h	0.42% C 1.5% Cr 0.1% V steel: Krupp
CV 50	0.5% C 1.05% Cr 0.1% V steel: Krupp
CV 58	0.58% C 1.05% Cr 0.1% V steel: Krupp
CV Punch and Die	0.45% C 1% Cr 0.25% V steel: T Turton
CVHD	0.6% C 0.7% Cr 0.2% V steel; Huntsman; dies for zinc, casting, extrusion dies, etc. **DPN: 335**
CVM 3	0.4% C 5% Cr 1.3% Mo 1.1% V 1.0% Si die steel: ESC; forging dies, blanks, etc. **DPN: 550**
CVM 4	0.58% C 5.0% Cr 1.35% Mo 1.1% V steel: Origin unknown
CVM 6	0.38% C 5.0% Cr 1.75% Mo 0.95% V 2.0% W steel: Origin unkown
D 6A	0.46% C 0.5% Ni 1% Cr 1% Mo steel: American proprietary alloy listed in SAE yearbook
D 6AC	0.46% C 0.5% Ni 1.0% Cr 1.0% Mo steel: Origin unkown
D 6AC	0.5% C 0.6% Ni 1.05% Cr 1.01% Mo 0.1% V steel
D 249	0.48% C 1.2% Cr 0.25% Mo 0.15% V 2.2% W Fagersta **DPN: 550**
D 421	0.45% C 1.3% Cr 0.3% V 2.3% W steel: Spencer

Symbol	Nominal analysis, supplier, condition and remarks.
D 444	0.4% C 0.5% Ni 4.2% Cr 4.2% Mo 2.2% V steel: Fagersta RC51
DAMASCUS	0.55% C 0.25% Cr 0.3% V 2.0% Si steel: Latrobe
DART	0.41% C 3.3% Cr 2.4% Mo 0.35% V steel: Latrobe; for hot working
DEF 13B5B	Alloy steel: Hardened and tempered; covers En 19 and includes En 16, 18, 22, 100, etc. **UTS: 840 Elon: 18% Proof: 630**
DEF 13B5C	Alloy steel: Hardened and tempered; covers En 19 and includes En 16, 100, etc. **UTS: 920 Elon: 17% Proof: 710**
DEF 13B5D	Alloy steel: Hardened and tempered; covers En 19 and includes En 16, 17, 24, etc. **UTS: 1000 Elon: 16% Proof: 760**
DEF 13B5E	Alloy steel: Hardened and tempered; covers En 24, 110 and includes En 19, etc. **UTS: 1070 Elon: 14% Proof: 840**
DEF 13B5F	Alloy steel: Hardened and tempered; covers En 24, 25, etc. **UTS: 1140 Elon: 14% Proof: 900**
DEF 13B5G	Alloy steel: Hardened and tempered; covers En 24, 26, etc. **UTS: 121 Elon: 14% Proof: 99**
DIN 50CR V4	0.5% C 0.85% Mn 1.05% Cr 0.15% V steel: For springs; German Standard
DIN 1654/ 42 Cr Mo 4	0.42% C 0.6% Ni 1.1% Cr 0.2% Mo steel: Forging
DIN 1654/42 Cr V6	0.42% C 1.5% Cr 0.1% V steel: For bolts
DIN 8555 E3 40T	Cr Mo steel: Welding electrode
DIN 8555 E3 50T	Cr Mo steel: Welding electrode
DIN 8555 E3 55T	Cr Mo steel: Welding electrode
DIN 8555 E4 60ST	Cr Mo V W steel: Welding electrode
DIN 8555 E21 7022	Cr W steel: Welding electrode
DIN 8555 E23– 200 (45K) ZCKT	Cr Ni Mo W steel: Welding electrode
DIN 17200/ 36 Cr Ni Mo 4	0.36% C 0.65% Mn 1.05% Ni 1.05% Cr 0.22% Mo steel: German Standard
Din 17200/ 42 Cr Mo 4	0.42% C 0.6% Ni 1.1% Cr 0.2% Mo steel: Forging
DIN 17200/ 42 Cr Mo 4	0.41% C 0.65% Mn 1.05% Cr 0.22% Mo steel: German Standard
DIN 17200/ 42 Cr Mo S 4	0.42% C 0.65% Mn 1.05% Cr 0.22% Mo steel: German Standard
DIN 17200/ 50 Cr Mo 4	0.5% C 1.1% Cr 0.2% Mo steel: Forging
DIN 17200/ 50 Cr Mo 4	0.5% C 0.65% Mn 1.05% Cr 0.22% Mo steel: German Standard
DIN 17200/ 50 Cr V 4	0.51% C 0.9% Mn 0.95% Cr 0.15% V steel: German Standard
DIN 17221/ 50 Cr V 4	0.5% C 1.1% Cr 0.1% V steel: For springs
DIN 17221/ 51 Cr Mo V 4	0.52% C 0.9% Mn 1.05% Cr 0.2% Mo 0.1% V steel: For springs; German Standard
DIN 17221/ 58 Cr V 4	0.58% C 1.1% Cr 0.1% V steel: For springs; German Standard
DIN 17222/ 50 Cr V 4	0.5% C 1.1% Cr 0.1% V steel: For springs; German Standard
DIN 17222/ 58 Cr V 4	0.58% C 1.1% Cr 0.1% V steel: For springs; German Standard
DIN 17225/ 45 Cr Mo V 6/7	0.45% C 1.4% Cr 0.7% Mo 0.3% V steel: For springs; German Standard
DIN 17225/ 50 Cr V 4	0.5% C 1.1% Cr 0.1% V steel: For springs; German Standard
DIN 17225/ 65 W Mo 34/8	0.65% C 3.7% Cr 0.85% Mo 0.7% V 8.5% W steel: For springs; German Standard
DM3	0.42% C 1.5% Cr 0.35% V 3.5% W steel: T Turton
DOUBLE FLYGO	High C 18% W steel: For tools; high speed air hardening; Turton Bros
DOUBLE ZEBRA	0.4% C 3.5% Ni 0.3% Cr steel: Sanderson; chisels, drifts, etc. **DPN: 610**
DRO 86 BOVAC	0.35% C 1.5% Ni 1.5% Cr 0.2% Mo steel: Bofors; vacuum de-gassed; for plastic moulds; hardened and tempered **DPN: 300 UTS: 900 Elon: 18% Proof: 750**
DRO 1133	0.55% C 1.6% Ni 0.6% Cr 0.3% Mo steel: Bofors; dies **DPN: 400**
DTD 4 C	0.45% C 1.25% Cr 0.2% V steel: Springs for engine valves **UTS: 1550**
DTD 87 B	0.4% C 0.4% Ni 1.6% Cr 0.2% Mo 1% Al steel: For bar and forging; suitable for nitriding **DPN: 286 UTS: 920 Elon: 17% Proof: 640**
DTD 167 A	0.4% C 1.0% Cr 0.2% Mo steel: Tube; hardened and tempered **DPN: 200 UTS: 700 Proof: 610**
DTD 5012	0.4% C 3.2% Cr 1% Mo 0.2% V steel: Air hardened and tempered **DPN: 444 UTS: 1550 Elon: 12% Proof: 1040**
DTD 5042	0.42% C 1.75% Ni 1.25% Cr 1% Mo 0.2% V steel: Bar; hardened and tempered **DPN: 385 UTS: 1300 Elon: 12% Proof: 1070**
DTD 5052	0.42% C 1.75% Ni 1.25% Cr 1% Mo 0.2% V steel: Plate; hardened and tempered **DPN: 380 UTS: 1300 Elon: 12% Proof: 1070**
DTD 5162	High tensile steel: For bolts; high metallurgical quality **UTS: 900**
DTD 5172	Low alloy steel: Castings; hardened and tempered **DPN: 265 UTS: 920 Elon: 12% Proof: 670**
DTD 5192	Ni Cr Mo V steel: Wrought; vacuum remelted **UTS: 1900**
DTD 5222	5% Cr Mo V steel: Vacuum melted; steel for bolts **UTS: 1800**
DUNELT 34	0.38% C 1.3% Cr 0.3% Mo steel: Dunford for BS alloy En 19
DUNELT 36	0.42% C 1.2% Cr 0.2% V steel: Dunford for BS alloy En 50
DUNELT 69	0.4% C 0.3% Cr (max) 3.5% Ni steel: Dunford for BS alloy En 22
DUNELT 100	0.4% C 1.5% Ni 1.2% Cr steel: Dunford; hardened and tempered **DPN: 280 UTS: 920 Elon: 18%**
DUNELT 101	0.4% C 1.5% Ni 1.2% Cr 0.25% Mo steel: Dunford for BS alloy En 24
DUREHETE 900	0.4% C 1.25% Cr 0.7% Mo steel: Samuel Fox; for use up to 500 °C **UTS: 1010 Elon: 22% Proof: 930**
DUREHETE 950	0.4% C 0.7% Ni 1.25% Cr 0.25% V steel: Samuel Fox; for use up to 500 °C **UTS: 980 Elon: 25% Proof: 920**
DUROTERM 12R	0.35% C 8.5% Cr 2.7% Mo steel: Welding electrode; Soudometal; for build up of hot work tools **DPN: 580**
DUX 4	0.5% C 1.1% Cr 2.25% W steel: ESC; shear blades, punches, etc. **DPN: 600**
DYCAST 1	0.4% C 5.4% Cr 0.8% Mo 0.5% V steel: Latrobe; hot punches, etc.; type H 11 **DPN: 430**
DYNAFLEX VAC-ARC	0.4% C 5.0% Cr 1.3% Mo 0.5% V 0.9% Si steel: Latrobe steel; hardened and tempered **UTS: 2000 Elon: 8% Proof: 1550**
E	0.65% C 3.75% Cr 0.5% V 14% W steel: S Osborn; obsolete **DPN: 790**

Symbol	Nominal analysis, supplier, condition and remarks.
E 50	0.5% C 3% Cr 0.5% V 14% W tool steel: For hot working; S Osborn; for dies, inserts, covers, etc.; obsolete **DPN: 550**
E 4340	0.40% C 1.8% Ni 0.8% Cr 0.25% Mo steel: Designation used by AISI
E Fe 5	0.8% C (max) 2.2% Cr steel: Weld electrode; designation used by AWS
E Fe 5C	0.4% C 4% Cr 7% Mo 1% V 1.7% W steel: Weld electrode; obsolete
ECHO	0.65% C 4% Cr 0.6% V 14.5% W steel: John Vessey
ECONO	0.4% C 3.75% Cr 5.7% Mo 0.7% V 1.0% W steel: Braeburn
ELECTERN	0.55% C 1.5% Ni 0.75% Cr 0.28% Mo steel: Origin unknown
ELECTRITE 5	0.6% C 4.1% Cr 1.1% V 18% W steel: Latrobe; hot work; type H26
ELECTRITE 7	0.68% C 4.2% Cr 5.0% Mo 2.0% V 6.4% W steel: Latrobe; hot work; type H42
En 19	0.4% C 1.2% Cr 0.3% Mo steel: Designation used in BS 970; replaced by 709 M40
En 19A	0.4% C 1.1% Cr 0.25% Mo steel: Designation used in BS 970; replaced by 708 M40
En 19B	0.39% C 1.1% Cr 0.25% Mo steel: Designation used in BS 970; replaced by 708 A37
En 19C	0.42% C 1.1% Cr 0.25% Mo steel: Designation used in BS 970; replaced by 708 A42
En 22	0.4% C 3.5% Ni 0.3% Cr steel: Designation used in BS 970; obsolete
En 24	0.4% C 1.5% Ni 1% Cr 0.3% Mo steel: Designation used in BS 970; replaced by 817 M40
En 26	0.4% C 2.5% Ni 0.6% Cr 0.6% Mo steel: Designation used in BS 970; replaced by 826 M40
En 40C	0.4% C 0.4% Ni 3% Cr 1% Mo 0.2% V steel: For nitriding; designation used in BS 970; replaced by 897 M39
En 41A	0.31% C 0.4% Ni 1.5% Cr 0.2% Mo 1% Al steel: For nitriding; designation used in BS 970; replaced by 905 M31
En 41B	0.39% C 0.4% Ni 1.5% Cr 0.2% Mo 1% Al steel: For nitriding; designation used in BS 970; replaced by 905 M39
En 47	0.5% C 1% Cr 0.15% V steel: For springs; designation used in BS 970; replaced by 897 M39
En 50	0.45% C 1.25% Cr 0.15% V steel: For valve springs; designation used in BS 970; replaced by 897 M39
En 100	0.4% C 0.7% Ni 0.4% Cr 0.2% Mo steel: Designation used in BS 970; replaced by 945 M38
En 100D	0.42% C 0.7% Ni 0.4% Cr 0.2% Mo steel: Designation used in BS 970; obsolete
En 110	0.4% C 1.5% Ni 1.2% Cr 0.15% Mo steel: Designation used in BS 970; replaced by 816 M40
En 160	0.4% C 1.75% Ni 0.3% Mo steel: Designation used in BS 970; obsolete
En 160A	0.4% C 1.75% Ni 0.3% Mo steel: Obsolete
ERA 171	0.4% C 0.4% Ni 3% Cr 1% Mo 0.2% V nitriding steel: Hadfields for BS alloy En 40C
ETAD	0.4% C 4.6% Ni 2.5% Cr 0.5% Mo steel: Pompey
ETAD 2	0.35% C 4.0% Ni 1.3% Cr 0.5% Mo steel: Pompey

Note. The following abbreviations and units are used in the tables:

DPN	Hardness, diamond pyramid number
UTS	Ultimate tensile strength, N/mm^2
Elon	Elongation, %
Proof	0.1% proof strength, N/mm^2

1 N/mm^2=0.1 hbar=0.102 kgf/mm^2=0.06475 tonf/in.^{2}=145.04 lbf/in.^{2}=1 MPa

See Appendix II for other abbreviations and conversion tables.

Symbol	Nominal analysis, supplier, condition and remarks.
EXTENDO-DIE	0.4% C 6% Cr 1.9% Mo 0.8% V steel: Carpenter
F 1	0.44% C 1.25% Cr 0.2% V steel: Jessop
F 2	0.6% C 0.7% Cr 0.2% V steel: Jessop **DPN: 340**
F 5	0.44% C 1.3% Cr 0.12% V 2.3% W steel: Jessop
F 7	0.5% C 0.7% Ni 1.5% Cr 0.12% V 1.2% W steel: Jessop **DPN: 620**
F 10	0.55% C 0.35% Cr 0.3% Mo 0.2% V steel: Shock resistant; Jessop **DPN: 640**
F 500	0.52% C 1.0% Mn 1.1% Cr 0.2% Mo 0.1% V steel: VEW
F 550	0.51% C 1.0% Mn 1.1% Cr 0.1% V steel: VEW
F C V	0.4% C 2.5% V steel: Jonas; dies for casting Al **DPN: 441**
FAGERSTA D234	1.1% C 0.4% Mn 0.6% Cr 1.9% W steel: For saw blades; Fagersta
FAGERSTA K825	0.36% C 0.7% Mn 1.4% Ni 1.4% Cr 0.2% Mo steel: Fagersta; for valves
FALCON 4	0.45% C 1.5% Cr 0.25% V steel: Atlas; as AISI type S1
FALCON 6	0.55% C 1.5% Cr 0.25% V 2.0% W steel: Atlas; as AISI type S1
FIREDIE	0.4% C 5.0% Cr 1.4% Mo 0.5% V steel: Columbia; AISI type H11
FIREDIE 13	0.4% C 5.2% Cr 1.4% Mo 1.0% V steel: Columbia; AISI type H13
FIREDIE 13 SMOOTHCUT	0.4% C 5.2% Cr 1.4% Mo 1.0% V steel: Columbia; AISI type H13 S; free cutting
FIREX	0.4% C 4.25% Ni 1.2% Cr tool steel: Darwin; chisels, shear blades, etc. **DPN: 620**
FKD 4	0.42% C 1.0% Cr 0.25% Mo steel: Pompey **UTS: 1250 Elon: 9% Proof: 1040**
fmp 328	0.4% C 5.2% Cr 1.35% Mo 0.4% V steel: F Parkin; as tool steel H11
fmp 329	0.4% C 5.25% Cr 1.35% Mo 1.0% V steel: F Parkin; as tool steel H13
fmp 348	0.43% C 4.0% Ni 1.5% Cr 0.3% Mo steel: F Parkin
fmp 399	0.5% C 1.45% Cr 0.2% V 2.2% W steel: F Parkin; as tool steel S1
fmp 682	0.6% C 4.0% Cr 8.25% Mo 2.0% V steel; F Parkin; as tool steel H43
fmp 1850	0.55% C 4.1% Cr 1% Mo 0.7% V 18% W steel: F Parkin; as tool steel H26
FN 35	0.4% C 3.5% Ni 0.25% Cr steel: Firth Brown for BS alloy En 22
FOREMOST 257	Medium C Ni Cr Mo steel: Swift Levick for hot dies **DPN: 300**
FOREMOST PNEUMATIC	0.45% C 1.7% Cr 0.25% V 2% W steel: Swift Levick; punches, shears
FORMDIE SMOOTHCUT	0.51% C 1.5% Ni 5.2% Cr 1.4% Mo 1.05% V steel: Free cutting; Columbia; AISI type A9S
FORMITE 3	0.51% C 3.0% Cr 0.5% V 15% W steel: Columbia; AISI type H24
G 5 Spl	0.4% C 1.5% Ni 1.1% Cr 0.27% Mo steel: Jessop for BS alloy En 24
G 7	0.5% C 1.7% Ni 0.6% Cr 0.3% Mo steel: Jessop
G 12	0.4% C 2.5% Ni 0.6% Cr 0.55% Mo steel: Jessop for BS alloy En 26
G 14 A	0.42% C 3.7% Ni 1% Cr steel: Jessop
G 14 C	0.4% C 3.5% Ni 1.5% Cr 0.5% Mo steel: Jessop
G 20	0.4% C 1.3% Mn 0.75% Ni 0.5% Cr 0.2% Mo steel: Jessop for BS alloy En 100
G 26	0.4% C 1.4% Ni 1.1% Cr 0.15% Mo steel: Jessop for BS alloy En 110
G 33	0.4% C 1.37% Ni 0.7% Cr 0.22% Mo steel: Jessop
G 105	0.4% C 1.5% Ni 1.2% Cr 0.25% Mo steel: Toledo for BS alloy En 24; also produced with lower Mo to alloy En 110

Symbol	Nominal analysis, supplier, condition and remarks.
GEAROL	0.4% C 1.5% Ni 1.2% Cr 0.3% Mo steel: Swift Levick for BS alloy En 24
GJ 2	0.5% C 1% Cr 0.2% V steel: For tools; S Osborn; punches, spanners, tool holders; obsolete **DPN: 620**
GOST 2052/50 KHFA	0.5% C 0.75% Mn 0.95% Cr 0.15% V steel: Bar; Russian Standard
GOST 2052/50 KHFA	0.5% C 0.75% Mn 0.95% Cr 0.15% V 0.4% Cu (max): Russian Standard
GOST 2052/50 KHGFA	0.51% C 0.9% Mn 1.05% Cr 0.2% V steel: Bar; Russian Standard
GOST 2052/50 KHGFA	0.52% C 0.9% Mn 1.1% Cr 0.2% V 0.25% Cu (max) steel: For springs; Russian Standard
GOST 2052/60 S2KHFA	0.6% C 0.55% Mn 1.05% Cr 0.15% V steel: Bar; Russian Standard
GOST 2052/65 S2VA	0.65% C 0.85% Mn 0.3% Cr (max) 1.0% W steel: Bar; Russian Standard
GOST 4543/38 KHN3MFA	0.38% C 0.4% Mn 3.2% Ni 1.35% Cr 0.4% Mo 0.15% B steel: Russian Standard
GOST 4543/38 KHN3VA	0.38% C 0.35% Mn 3.0% Ni 1.0% Cr 0.65% W steel: Russian Standard
GOST 4543/38 KHN3VFA	0.38% C 0.4% Mn 3.2% Ni 1.2% Cr 0.65% W 0.15% B steel: Russian Standard
GOST 4543/40 KHN	0.4% C 0.65% Mn 1.2% Ni 0.6% Cr steel: Russian Standard
GOST 4543/40 KHFA	0.4% C 0.65% Mn 0.95% Cr 0.15% V steel: Russian Standard
GOST 4543/40 KHNMA	0.4% C 0.65% Mn 1.5% Ni 0.75% Cr 0.2% Mo steel: Russian Standard
GOST 4543/40 KHNVA	0.4% C 0.65% Mn 1.45% Ni 0.75% Cr 1.0% W steel: Russian Standard
GOST 4543/50 KHN	0.5% C 0.65% Mn 1.2% Ni 0.6% Cr steel: Russian Standard
GOST 5632/4 KH10S2M	0.4% C 9.7% Cr 0.8% Mo steel: Russian Standard
GOST 5632/4 KH10S2M	0.4% C 9.7% Cr 0.8% Mo steel: Plate; Russian Standard
GRANE 2	0.55% C 3.0% Ni 1.5% Cr steel: Uddelholm for type L6
GV	0.5% C 3.0% Ni 1% Cr 0.2% Mo 0.3% V steel: For tools; S Osborn; shears, punches, dies; obsolete **DPN: 630**
H 10	0.41% C 3.3% Cr 2.4% Mo 0.35% V steel: Designation used by AISI
H 13	0.39% C 5.0% Cr 1.4% Mo 1.0% V steel: For hot work; Osborn
H 14	0.4% C 5.0% Cr 5.0% W tool steel: Designation used by AISI
H 15	0.4% C 5.0% Cr 5.0% Mo steel: Designation used by AISI; obsolete
H 16	0.55% C 7.0% Cr 7.0% W steel: For tools; designation used by AISI
H 24	0.45% C 3.0% Cr 15.0% W steel: For tools; designation used by AISI
H 24	0.4% C 1.2% Cr 0.3% Mo steel: Jessop for BS alloy En 19
H 24	0.52% C 4.1% Cr 0.45% V 14.5% W steel: For hot work; Osborn
H 26	0.5% C 4.0% Cr 1.0% V 18.0% W steel: For tools, designation used by AISI
H 31	0.4% C 1.2% Cr 0.7% Mo steel: Jessop for BS alloy En 20B
H 41	0.65% C 4.0% Cr 8.0% Mo 1.0% V 1.5% W steel: For tools; designation used by AISI
H 42	0.6% C 4.0% Cr 5.0% Mo 2.0% V 6.0% W steel: For tools; designation used by AISI
H 43	0.55% C 40% Cr 8.0% Mo 2.0% V steel: For tools; designation used by AISI
H 55	0.55% C 1% Cr 0.5% Mo 0.07% V steel: Jessop
H 56	0.55% C 1% Cr 0.45% Mo 0.05% V with S steel: Jessop
H 600.1	0.4% C 1.5% Mn 3.0% Cr 3.0% Mo steel: Electrode; Metrode; hardenable deposit **DPN: 600**
H 750	0.6% C 1.5% Mn 5.0% Cr 5.0% Mo steel: Electrode; Metrode; hardenable deposit **DPN: 700**
HARDEES 350	0.3% C 2.7% Cr 1.2% Mo steel: Welding electrode; Philips; for hard facing **DPN: 570**
HARDEES 800	0.6% C 5% Cr 3.7% Mo 0.2% B steel: Welding electrode; Philips **DPN: 800**
HARDTEM	0.5% C 1% Cr 0.45% Mo steel: Somers; die steel
HAVOC	0.5% C 0.5% Cr 0.2% V steel: Simonds; AISI type S2
HCM 3	0.4% C 3.0% Cr 1.0% Mo 0.25% V steel: Firth Brown for BS 970 En 40C
HCRS 2	0.4% C 1.25% Cr 0.6% Mo steel: ESC for BS alloy En 20B
HCRS 5	0.4% C 1.25% Cr 0.7% Mo 0.25% V steel: ESC; hardened and tempered **UTS: 840 Elon: 16% Proof: 720**
HD 2	0.4% C 3.35% Cr 0.25% V 2.4% W steel: Spencer; forging dies and inserts **DPN: 400**
HD 3	W steel: For tools; hot forging; Turton Bros
HD 12	10% W steel: For tools; hot forging; Turton Bros
HDB 3	0.5% C 1.5% Ni 0.6% Cr steel: Huntsman; long run press dies
HDB 5	0.5% C 1.6% Ni 0.6% Cr 0.2% Mo steel: Huntsman; long run press dies
HECLA 70	0.4% C 3.2% Ni 0.3% Cr steel: Hadfields for BS alloy En 22
HECLA 76	0.5% C 1% Cr 0.15% V steel: Hadfields for BS alloy En 47, En 50
HECLA 100	0.4% C 1.2% Mn 0.7% Ni 0.4% Cr 0.2% Mo steel: Hadfields for BS alloy En 100, En 100 A, B, C and D
HECLA 110	0.4% C 1.5% Ni 1.2% Cr 0.15% Mo steel: Hadfields for BS alloy En 19 and En 19A
HECLA 135	0.6% C 2.0% Ni 2.0% Cr 0.5% Mo steel: Origin unknown
HECLA 139	0.55% C 1.75% Ni 0.7% Cr 0.25% Mo steel: Origin unknown
HECLA 157	0.4% C 1.2% Cr 0.3% Mo steel: Hadfields for BS alloy En 19 and En 19A
HECLA 160C	0.38% C 1.5% Cr 0.5% Mo 0.2% V 3.6% W steel: Origin unknown
HEDEX 8	0.4% C 1.5% Ni 1.2% Cr 0.15% Mo steel; Wire for forging; Kiveton Park for BS alloy En 110
HEDEX 9	0.4% C 1.2% Cr 0.3% Mo steel: Wire for forging; Kiveton Park for BS alloy En 19
HEDEX 10	0.4% C 0.55% Ni 0.5% Cr 0.25% Mo steel: Wire for forging; Kiveton Park; annealed **UTS: 640**
HEDEX 15	0.4% C 0.55% Ni 0.5% Cr 0.2% Mo steel; Wire for forging; Kiveton Park; annealed **UTS: 650**
HEDEX 33	0.45% C 1.2% Cr 0.15% V steel: Wire for forging; Kiveton Park for BS alloy En 50
HEDEX 47	0.4% C 1.5% Ni 1.2% Cr 0.3% Mo steel: Wire for forging; Kiveton Park for BS alloy En 24
HI SHOCK 60	0.7% C 0.5% Ni 1% Cr 1.1% Mo 0.15% V 2.5% Cu steel: Carpenter
HNV 1	0.55% C 0.8% Cr 0.75% Mo 1.5% Si steel: For inlet valves; designation used by SAE
HNV 4	0.45% C 1% Ni 7% Cr 3.3% Si steel: For inlet valves; designation used by SAE
HR 119	0.55% C 3.0% Ni 1.5% Cr steel: Bofors **DPN: 400**
HRO 1243	0.55% C 3.0% Ni 1.0% Cr 0.3% Mo steel: Bofors; dies **DPN: 440**

Symbol	Nominal analysis, supplier, condition and remarks.
HW 7	0.45% C 5.0% Cr 1.0% Mo 0.3% V 3.7% W 0.5% Co steel: Atlas; for hot working **DPN: 400**
HWD 2	0.38% C 5.25% Cr 1.35% Mo 0.5% V steel: Firth Sterling
HWD 3	0.4% C 5.25% Cr 1.25% Mo 1.05% V steel: Firth Sterling
HWD (Mod)	0.55% C 5.0% Cr 1.5% Mo 0.3% V 1.4% W steel: Firth Sterling
HYBLADE	0.55% C 0.3% Cr 0.4% Mo 0.2% V steel: Origin unknown
HYDIE	0.37% C 3% Cr 0.8% Mo 0.2% V steel: Somers; die steel
HYDRA HD	0.4% C 4% Cr 0.7% V 14% W steel: Hall and Pickles; forging dies; obsolete **DPN: 630**
HYDRA VK	0.4% C 4.75% Cr 1.2% Mo 1% V steel: Hall and Pickles; dies for casting Al; extrusion dies, etc.; obsolete **DPN: 550**
HYKRO V	0.4% C 3% Cr 0.9% Mo 0.2% V steel: ESC; nitriding steel **UTS: 1380 Elon: 10% Proof: 1070**
HY-TEN	0.5% C 1% Mn 0.7% Cr 0.2% Mo 0.08% S steel: Wheelock Lovejoy and Co; for machinability
HYTUF	0.37% C 5.3% Cr 1.4% Mo 1% V steel: Somers
IDEOR	0.4% C 1% Cr 0.15% V 2% W 1.0% Si tool steel: Shock resisting: Darwin; chisels, shear blades, etc. **DPN: 650**
IMPAX	0.36% C 1.4% Ni 1.4% Cr 0.2% Mo steel: Uddelholm
IVORESCO WM EXTRA	0.38% C 2.1% Cr 4.25% Mo 0.32% V 1.05% Si steel: Schoeller-Bleckman **UTS: 1550**
J 27	0.42% C 3.2% Cr 0.25% V 13% W steel: Jessop
JAYSEE	0.45% C 0.5% Cr 0.5% W steel: Jonas; chisels, hammers, etc. **DPN: 530**
JG 3	0.5% C 1% Mn 1% Cr 0.3% Mo steel: For tools; S Osborn; chisels; obsolete
JIS G4106 SCM4	0.4% C 0.75% Mn 0.95% Cr 0.22% Mo steel: Japanese Standard
JIS G4106 SCM5	0.45% C 0.72% Mn 1.05% Cr 0.22% Mo steel: Japanese Standard
JIS G4106 SNC1	0.36% C 0.65% Mn 1.25% Ni 0.7% Cr steel: Japanese Standard
JIS G4106 SNC3	0.36% C 0.5% Mn 3.25% Ni 0.8% Cr steel: Japanese Standard
JIS G4106 SNCM1	0.45% C 0.6% Mn (max) 0.3% Ni (max) 1.5% Cr 0.22% Mo 1.0% Al steel: Japanese Standard
JIS G4106 SNCM6	0.4% C 0.85% Mn 0.55% Ni 0.52% Cr 0.22% Mo steel: Japanese Standard
JIS G4106 SNCM7	0.45% C 0.85% Mn 0.55% Ni 0.52% Cr 0.22% Mo steel: Japanese Standard
JIS G4106 SNCM8	0.4% C 0.75% Mn 1.8% Ni 0.8% Cr 0.22% Mo steel: Japanese Standard
JIS G4801 SUP10	0.5% C 0.8% Mn 0.95% Cr 0.2% V steel: For springs; Japanese Standard
JL	0.4% C 5% Cr 1.4% Mo 0.4% V 1.0% Si tool steel: For tools; hot working; S Osborn; hot extrusion dies, etc.; obsolete **DPN: 460**
JS	0.5% C 1.4% Cr 0.25% V 2.5% W steel: Firth Sterling
JS 5	0.42% C 0.4% Ni 0.2% Cr 0.1% Mo 0.4% Cu steel casting: Jopling; for induction hardening **DPN: 200 UTS: 620 Elon: 12% Proof: 300**
JS 13	0.52% C 0.4% Ni 0.2% Cr 0.1% Mo 0.4% Cu steel casting: Jopling **DPN: 230 UTS: 700 Elon: 10% Proof: 330**
JS 26	0.5% C 0.4% Ni 0.2% Cr 0.1% Mo 0.4% Cu steel casting: Jopling **DPN: 200 UTS: 620 Elon: 12% Proof: 300**
K 19	0.4% C 1.5% Cr 0.55% Mo 0.3% V 3.5% W steel: Jessop
K 300	0.53% C 8.3% Cr 1.2% Mo 1.2% W steel: VEW
K 306	0.51% C 5.0% Cr 1.4% V steel: VEW
K 450	0.47% C 1.0% Cr 0.2% V 2.0% W steel: VEW
K 455	0.63% C 1.1% Cr 0.2% V 2.0% W steel: VEW
K 600	0.45% C 4.0% Ni 1.3% Cr 0.3% Mo steel: VEW
K 605	0.52% C 3.1% Ni 1.0% Cr 0.2% Mo steel: VEW
K 24535	0.45% C (max) 2.5% Ni 1% Cr 0.2% Mo (min) steel: Designation used by UNS
KAISALOY No.3	0.35% C 1.5% Mn 0.4% Ni 0.25% Cr 0.1% Mo 0.05% V 0.005% Ti 0.35% Cu steel: Kaiser Steel Co.
KD 4	0.4% C 1.0% Cr 0.25% Mo steel: Pompey **UTS: 1250 Elon: 8% Proof: 1070**
KE 241	0.55% C 8.5% Cr 2% W 3% Si steel: Kayser Ellison; valves
KE 275	0.4% C 3% Cr 0.4% Mo 0.5% V 9.5% W steel: Kayser Ellison for hot working copper **DPN: 530**
KE 355	0.43% C 4.2% Ni 1.5% Cr 0.2% Mo steel: Kayser Ellison; hardened and tempered for plastic dies **DPN: 460 UTS: 1680**
KE 396	0.6% C 1.5% Ni 1% Cr 0.2% Mo steel: Kayser Ellison; for punches, shear blades **DPN: 700**
KE 896	0.5% C 1% Cr 0.2% V steel: Kayser Ellison; hardened and tempered **DPN: 430 UTS: 1500 Elon: 10% Proof: 133**
KE 960	0.5% C 1.5% Cr 0.2% V 2.2% W steel: Kayser Ellison; shear blades, screwdriver bits **DPN: 670**
KE 1029	0.55% C 6.2% Cr 0.25% W 1.4% Si steel: Kayser Ellison; for valves
KEA 227	0.5% C 1.2% Mn 0.65% Cr 0.2% Mo steel: Sanderson Kayser
KELOCK	0.75% C 4.1% Cr 1.1% V 18% W steel: Sanderson Kayser
KF 26	0.4% C 1.5% Ni 1.2% Cr 0.25% Mo steel: Kirkstall for BS alloy En 24
KF 26A	0.4% C 1.4% Ni 1.2% Cr 0.15% Mo steel: Kirkstall for BS alloy En 110
KF 28C	0.4% C 2.5% Ni 0.6% Cr 0.6% Mo steel: Kirkstall for BS alloy En 26
KF 48	0.4% C 1.2% Cr 0.3% Mo steel: Kirkstall for BS alloy En 19
KF 53	0.42% C 0.2% Ni 1.2% Cr 0.7% Mo steel: Kirkstall for BS alloy En 20B
KF 56	0.42% C 1.4% Mn 0.7% Ni 0.45% Cr 0.2% Mo steel: Kirkstall for BS alloy En 100
KHNSV	0.55% C 1.0% Ni 1.5% Cr 0.55% W steel: Russian Standards designation
KLAH	0.55% C 1.75% Cr 1.75% W 1% Si steel: Jonas; chisels, punches, etc. **DPN: 700**
KP 16	0.4% C 3.5% Ni 0.3% Cr steel: Kiveton Park for BS alloy En 22

Note. The following abbreviations and units are used in the tables:

DPN	Hardness, diamond pyramid number
UTS	Ultimate tensile strength, N/mm^2
Elon	Elongation, %
Proof	0.1% proof strength, N/mm^2

1 N/mm^2=0.1 hbar=0.102 kgf/mm^2=0.06475 tonf/in.^{2}=145.04 ibf/in.^{2}=1 MPa

See Appendix II for other abbreviations and conversion tables.

Symbol	Nominal analysis, supplier, condition and remarks.
KP 22	0.4% C 1.5% Ni 1.2% Cr 0.3% Mo steel: Kiveton Park for BS alloy En 24
KP 26	0.5% C 1% Cr 0.15% V steel: Kiveton Park for BS alloy En 47
KP 33	0.4% C 1% Cr 0.3% Mo steel: Kiveton Park for BS alloy En 19
KP 34	0.4% C 1.4% Mn 0.8% Ni 0.5% Cr 0.2% Mo steel: Kiveton Park for BS alloy En 100
L 94	0.6% C 1.5% Ni 0.6% Cr 0.25% Mo steel: Fagersta **DPN: 730**
L 97	0.55% C 3.0% Ni 1.0% Cr 0.3% Mo steel: Fagersta **DPN: 600**
LANARK	0.55% C 0.16% Cr 1.3% Mo 0.28% V 1.9% Si steel: Latrobe; for coldwork
LCKD	0.6% C 0.12% Cr 0.25% Mo steel: Pompey
LESCALLOY HP 9-4-30	0.3% C 8.5% Ni 1.0% Cr 1.0% Mo 0.1% V 4.0% Co steel: Latrobe
LESCALLOY LADISH D6AC VAC-ARC	0.45% C 0.6% Ni 1.1% Cr 1.0% Mo 0.12% V steel: Latrobe steel; hardened and tempered **UTS: 1680 Elon: 7% Proof: 1300**
LESCALLOY 300M	0.42% C 1.8% Ni 0.8% Cr 0.4% Mo 0.07% V steel: Latrobe
LESCALLOY 4340	0.4% C 1.8% Ni 0.8% Cr 0.25% Mo steel: Latrobe
LESCALLOY UT18	0.4% C 3.2% Cr 0.95% Mo 0.2% V steel: Latrobe
LK 1	0.5% C 1.6% Cr 0.2% Mo 1.1% Al steel: For nitriding; Firth Brown
LK 3	0.4% C 1.5% Cr 0.2% Mo 1.1% Al steel: For nitriding; Firth Brown for En 41B
LNCM	0.4% C 1.3% Ni 1.2% Cr 0.15% Mo steel: ESC for BS alloy En 110
LOMINIUM	0.47% C 1.75% Cr 0.2% V steel: Edgar Allen; for Zn die casting moulds **DPN: 600**
LOW CARBON TATMO	0.6% C 3.75% Cr 8.7% Mo 1.0% V 1.7% W steel: Latrobe; for hot working
LOW CARBON TNW	0.6% C 4.0% Cr 8.0% Mo 2.0% V steel: Latrobe; for hot working
LS	0.5% C 3.2% Cr 1.7% Mo steel: Braeburn
LT 8	0.45% C 3.0% Ni 0.5% Cr steel: Low Moor
LT 9	0.5% C 1.0% Mn 1% Cr 0.3% Mo steel: Low Moor **DPN: 600**
LT 31	0.5% C 1.5% Cr 0.2% V 2% W 1.0% Si steel: Low Moor; for hot work tools; obsolete **DPN: 790**
LT 41	0.4% C 4.7% Cr 1.4% Mo 0.4% V 1.0% Si steel: Low Moor; for hot work tools; obsolete **DPN: 630**
LT 42	0.4% C 5% Cr 0.4% Mo 4% W 1.0% Si steel: Low Moor; for hot work and tools; obsolete **DPN: 550**
LT 47	0.5% C 3% Cr 0.5% V 14% W steel: Low Moor; for hot work tools; obsolete **DPN: 670**
LT 51	0.4% C 1.7% Ni 1.2% Cr 0.2% Mo steel: Low Moor; die steel; obsolete **DPN: 550**
LUNDIE	0.4% C 5.25% Cr 4.75% W steel: Latrobe; for hot working
LV 52 Mo	0.5% C 7.5% Mo 1.5% Si steel: Low Moor; valve steel; obsolete **DPN: 320**
M 200	0.4% C 1.5% Mn 1.9% Cr 0.2% Mo steel: VEW
M Mo 38	0.38% C 1.5% Mn 0.4% Ni 0.25% Cr 0.3% Mo 0.3% Cu steel: Casting; Wolsingham **DPN: 375 UTS: 920 Elon: 12% Proof: 580**
MALLOY	0.6% C 1.1% Cr 0.25% Mo steel: Origin unknown
MANOFORT 160	0.3% C 2% Ni 1.0% Cr 0.4% Mo steel: Pompey **UTS: 1600 Elon: 8% Proof: 1250**
MANOFORT 180	0.4% C 3.5% Ni 1.4% Cr 0.4% Mo steel: Pompey **UTS: 1700 Elon: 5% Proof: 1400**
MANOIR RS1	0.45% C 0.7% Mn 1.8% Si steel: Pompey **UTS: 1320 Elon: 4% Proof: 1150**
MANOIR ABRADUR 240	0.5% C 1.0% Mn 1.0% Cr steel: Pompey **UTS: 1200 Elon: 5% Proof: 1000**
MANOIR ABRADUR 300	0.6% C 0.8% Cr 1.6% Si steel: Pompey **UTS: 1500 Elon: 3% Proof: 1300**
MANOIR ABRADUR 400	0.3% C 1.8% Ni 1.0% Cr 35% Mo steel: Pompey **UTS: 1500 Elon: 5% Proof: 1150**
MANOIR ABRADUR 500	0.4% C 3.5% Ni 1.4% Cr 0.3% Mo steel: Pompey **UTS: 1600 Elon: 5% Proof: 1300**
MANOIR ABRADUR PM38	0.5% C 1.5% Mn 1.5% Cr + Mo + V + Ni steel: Pompey **UTS: 1650 Elon: 4% Proof: 1300**
MARSH CMN	0.52% C 1.5% Ni 0.6% Cr 0.22% Mo steel: Marsh Brothers; forging dies
MARSH CTC 1	0.45% C 1.25% Cr 2% W steel: Marsh Brothers; chisels, caulking irons, etc.
MARSH CTH	0.43% C 0.25% Cr 0.5% W steel: Marsh Brothers; hot forging dies, etc.
MARSH NC	0.45% C 3.7% Ni 1.1% Cr steel: Marsh Brothers; chisels
MARSH NC 1	0.4% C 1.5% Ni 1% Cr steel: Marsh Brothers; collets
MAS	0.52% C 2.0% Ni 1.0% Cr 0.4% Mo 0.2% V steel: Pompey
MAX CHIP	0.42% C 4.1% Ni 1.05% Cr 1.7% Si steel: Edgar Allen
MAXEL 3½	0.5% C 1.25% Mn 0.65% Cr 0.18% Mo 0.08% S steel: Free machining; Crucible Steel Co. **DPN: 250 UTS: 880 Elon: 22% Proof: 370**
MC	0.4% C 1% Ni 0.6% Cr steel: Jonas **UTS: 1000 Elon: 17% Proof: 840**
MCH	0.5% C 3.7% Cr 6.2% Mo 0.75% V 1.0% W steel: Latrobe; hot work **DPN: 550**
MCV	0.5% C 1.2% Mn 0.5% Ni 0.6% Cr 0.25% Mo steel: Sanderson; has additions of B, Ti, Zr and V **DPN: 255 UTS: 800 Elon: 23% Proof: 770**
MET-HARD 450HW	0.3% C 5% Cr 1.5% Mo 1.8% W steel: Electrode; Metrode **DPN: 520**
MET-HARD 650	0.5% C 8% Cr 1% Mo 0.7% V steel: Electrode; Metrode **DPN: 650**
MET-HARD 650M	0.5% C 4.5% Cr 6% Mo + Ni V steel: Electrode; Metrode **DPN: 650**
MGR	0.55% C 5.0% Cr 1.2% Mo 1.2% W steel: Latrobe; punch and dies; type A8 **DPN: 630**
MINERVA HC	0.53% C 1.8% Cr 0.2% V 1.9% W steel: Edgar Allen; punches, shear blades
MINERVA LC	0.43% C 1.8% Cr 0.2% V 1.9% W steel: Edgar Allen; chisels, punches, shear blades
MNC 851	General standard for SS range of case hardening steels; Swedish Standard
MOHD	0.36% C 3.5% Cr 6.0% Mo 0.75% V 1.0% W steel: For tools; Huntsman; extrusion dies **DPN: 600**
MOHICAN 6	0.62% C 3.7% Cr 8.7% Mo 1.0% V 1.7% W steel: Atlas; as AISI type H51
MOLITE HW60	0.63% C 4.0% Cr 8.2% Mo 1.9% V steel: Columbia; AISI type H43
MONARCH BLA	0.4% C 5.0% Cr 1.4% Mo 1.0% V steel: For tools; origin unknown
MONARCH NCV	0.4% C 4.2% Ni 1.5% Cr 0.2% Mo steel: Origin unknown
MONARK 2	0.6% C 0.3% Cr 0.2% Mo steel: Atlas; as AISI S5
MUSHET	0.65% C 3.7% Cr 0.5% V 14.0% W steel: For tools; origin unknown
N 4	0.42% C 1.7% Ni 1.2% Cr 1% Mo 0.2% V steel: Jessop

Symbol	Nominal analysis, supplier, condition and remarks.
NA	0.6% C 0.3% Mo 0.2% V 2% Si steel: For tools; S Osborn; punches, picks, shear blades; obsolete **DPN: 620**
NARVE	0.65% C 1.1% Cr 0.57% Mo steel: Uddelholm
NAT	0.32% C 5.0% Cr 1.4% Mo 0.4% V 1.0% Si tool steel: Balfour Darwin
NC	0.45% C 3.7% Ni 1.1% Cr steel: Marsh Brothers; chisels
NC 1	0.4% C 1.5% Ni 1% Cr steel: Marsh Brothers; collets, etc.
NC MO	0.38% C 1.4% Ni 1.2% Cr 0.15% Mo steel: Firth Brown for BS alloy EN 110
NCM 1	0.4% C 1.6% Ni 1.1% Cr 0.3% Mo steel: Firth Brown for BS alloy En 24
NCM 4	0.4% C 2.4% Ni 0.65% Cr 0.45% Mo steel: Firth Brown for BS alloy En 26
NCMS	0.6% C 0.6% Ni 0.6% Cr 0.15% Mo steel: For springs; ESC
NCMV	0.42% C 1.75% Ni 1.2% Cr 1% Mo 0.2% V steel: ESC; nitriding, fatigue resistant **UTS: 1840 Elon: 8% Proof: 1550**
NF	0.38% C 1.6% Ni 1.2% Cr 0.25% Mo steel: S Osborn; for moulds; hardened and tempered **DPN: 500**
NF A35 551/42CD4	0.42% C 0.75% Mn 0.95% Cr 0.22% Mo steel: French Standard
NF A35 551/50CV4	0.5% C 0.85% Mn 1.0% Cr 0.15% V steel: French Standard
NF A35 571/45SCD6	0.46% C 0.65% Mn 0.62% Cr 0.22% Mo steel: Bar; French Standard
NF A35 571/50CV4	0.5% C 0.85% Mn 1.0% Cr 0.15% V steel: For springs; French Standard
Ni Cr Si YCW STEM	0.48% C 4.2% Ni 0.6% Cr 0.25% Mo 1.7% Si steel: Casting; Edgar Allen; code 638 steel
NIT 135	0.41% C 1.55% Cr 0.35% Mo 1.1% Al steel
NITRAL	0.42% C 1.65% Cr 0.3% Mo steel: Pompey
NITRALLOY 3	0.4% C 1.5% Cr 0.2% Mo 1.1% Al steel: ESC for BS alloy En 41 B nitriding steel
NITRALLOY LK1	0.5% C 1.6% Cr 0.2% Mo 1.1% Al steel: For nitriding; Firth Brown
NITRALLOY LK3	0.4% C 1.6% Cr 0.2% Mo 1.1% Al steel: For nitriding; Firth Brown
NITRALLOY LK5	0.3% C 1.6% Cr 0.2% Mo 1.1% Al steel: For nitriding; Firth Brown
NITRALLOY N	0.25% C 3.5% Ni 1.25% Cr 0.25% Mo 1.1% Al steel: Latrobe; nitriding steel
NITRALLOY N135 Mod	0.4% C 1.6% Cr 0.35% Mo 1.1% Al steel: Latrobe; for nitriding
NITRIDING STEEL	0.4% C 1.6% Cr 0.35% Mo 1.1% Al steel: For nitriding; Crucible Steel Co. **DPN: 320 UTS: 1140 Elon: 17% Proof: 1000**
NMCM	0.4% C 1.3% Mn 0.75% Ni 0.4% Cr 0.2% Mo steel: Firth Brown for BS alloy En 100
No.1 MONARCH	0.65% C 4.0% Cr 0.6% V 14.0% W steel: For tools; origin unknown
No 5 ELECTERN	0.55% C 1.5% Ni 0.8% Cr 0.3% Mo steel: Somers
No 5-317	0.5% C 1.75% Ni 1% Cr steel: Carpenter **DPN: 500**
No 408	0.5% C 3% Ni 0.75% Cr steel: Carpenter
No 481	0.55% C 0.25% Cr 0.4% Mo steel: Carpenter; for AISI type S5
No 481	0.5% C 0.25% Cr 0.4% Mo steel: Carpenter
No 882	0.4% C 5% Cr 1.35% Mo 0.4% V steel: Carpenter
No 883	0.37% C 5.2% Cr 1.3% Mo 1% V steel: Carpenter
NORESCO SPK	0.45% C 2.35% Cr 0.3% V steel: Schoeller-Bleckmann **UTS: 1210**
NORESCO WAW	0.45% C 1.5% Cr 0.5% Mo 0.9% V 0.5% W steel: Schoeller-Bleckmann **UTS: 1380**
NORESCO PARFORCE SPECIAL 2	0.4% C 4.25% Ni 1.4% Cr 1.% W steel: Schoeller-Bleckmann; for large plastic dies **UTS: 1000**
NORESCO TYRANT EXTRA	0.45% C 1.05% Cr 2% W 1% Si steel: Schoeller-Bleckmann **DPN: 500**
NTC	0.4% C 0.4% Cr 0.5% W steel: For tools; Huntsman; punches, chisels, etc. **DPN: 550**
NUSHANK	0.43% C 3.0% Ni 0.4% Cr 0.25% Mo steel: Tube; Atlas; for detachable rock drill bits
NV 3	0.5% C 0.3% Ni 0.4% Cr 0.15% Mo steel: For inlet valves; designation used by SAE
NV 4	0.4% C 1.2% Ni 0.6% Cr steel: For inlet valves; designation used by SAE
NV 5	0.45% C 0.5% Ni 0.5% Cr 0.2% Mo steel: For inlet valves; designation used by SAE
NV 7	0.4% C 0.9% Cr 0.2% Mo steel: For valves; SAE 4140
NV 8	0.4% C 0.25% Ni 2.1% Cr 0.1% Mo steel: For valves; SAE designation
NYBLADE	0.48% C 3% Ni 0.65% Cr 1.1% Si steel: Sanderson; chisels, picks, shear blades, etc. **DPN: 600**
NZ 2	0.55% C 1.5% Ni 0.7% Cr 0.2% Mo steel: For hot work tools; S Osborn; forging dies, die holders, etc.; obsolete **DPN: 302**
OCM 10	0.4% C 1.2% Cr 0.3% Mo steel: W Marrison for BS alloy En 19
OK	0.4% C 1.3% Cr 0.15% V 2.3% W steel: Jessop **DPN: 620**
OK CROWN	0.5% C 1.5% Cr 0.4% Mo 0.2% V 2.0% W steel: Origin unknown
ON 2	0.4% C 0.8% Ni 0.5% Cr 0.2% Mo steel: W Marrison for BS alloy En 100
ON 4	0.4% C 3.5% Ni 0.3% Cr steel: W Marrison for BS alloy En 22
ON 6	0.4% C 1.6% Ni 1.2% Cr 0.3% Mo steel: W Marrison for BS alloy En 24
ON 8	0.4% C 2.6% Ni 0.7% Cr 0.6% Mo steel: W Marrison for BS alloy En 26
ON 11	0.4% C 1.4% Ni 1.2% Cr 0.2% Mo steel: W Marrison for BS alloy En 110
OOC	0.6% C 1.1% Cr 0.3% V 1.9% W steel: For tools; origin unknown
ORLEANS	0.55% C 0.25% Cr 0.35% Mo 0.2% V steel: Simonds; AISI type S5
ORVAR 2	0.37% C 5.3% Cr 1.4% Mo 1.0% V steel: Uddelholm; for type H13
P 553	0.4% C 1.5% Ni 1.1% Cr 0.3% Mo steel: Carrs; crankshafts, gears, etc. **DPN: 370 UTS: 920**
P 576	0.43% C 4.1% Ni 1.4% Cr 0.8% Mo steel: Carrs
P 602	0.4% C 1.1% Cr 0.3% Mo steel: Carrs; pressure dies for Zn Pb and Sn base castings **DPN: 460**
P 614	0.4% C 1.6% Cr 0.2% Mo 1% Al steel: For nitriding; Carrs **UTS: 760**

Note. The following abbreviations and units are used in the tables:

DPN	Hardness, diamond pyramid number
UTS	Ultimate tensile strength, N/mm^2
Elon	Elongation, %
Proof	0.1% proof strength, N/mm^2

$1\ N/mm^2 = 0.1\ hbar = 0.102\ kgf/mm^2 = 0.06475\ tonf/in.^2 = 145.04\ lbf/in.^2 = 1\ MPa$

See Appendix II for other abbreviations and conversion tables.

Symbol	Nominal analysis, supplier, condition and remarks.
P 618	0.4% C 0.3% Ni 3% Cr 1% Mo 0.25% V steel: For nitriding; Carrs **UTS: 1380**
PARFORCE SPECIAL 2	0.4% C 4.25% Ni 1.4% Cr 1% W steel: Schoeller-Bleckmann; for large plastic dies **UTS: 1000**
PAX 2	0.5% C 1.5% Cr 0.2% V 2.2% W steel: Sanderson; shear blades, punches **DPN: 700**
PAX NON-BREAK	0.4% C 1.5% Cr 0.2% V 2.2% W steel: Sanderson; chisels, dies, punches, etc. **DPN: 570**
PCSK	0.52% C 1.5% Cr 0.04% Mo 0.2% V 2.0% W steel: Origin unknown
PENUMO	0.45% C 1.25% Cr 2.0% W steel: For tools; origin unknown
PLM B/1	0.55% C 1.6% Ni 0.75% Cr 0.3% Mo steel: ESC; plastic moulds; short runs **UTS: 950**
PLM B/2	0.42% C 2.6% Ni 0.6% Cr 0.6% Mo steel: ESC plastic moulds; short runs **UTS: 1100**
PLM C/1	0.42% C 2.6% Ni 0.6% Cr 0.6% Mo steel: ESC; plastic moulds **DPN: 500**
PLM C/2	0.4% C 4.2% Ni 1.4% Cr 0.25% Mo steel: ESC; plastic moulds **DPN: 550**
PN 1	0.5% C 0.4% Mo 0.23% V 1.6% Si steel: ESC; rivet snaps, air hammer tools, etc. **DPN: 630**
PNEUTOUGH	0.5% C 1.5% Cr 0.25% V 2.25% W steel: Clyde Alloy **DPN: 570**
PNUSNAP OH	0.43% C 1% Cr 1.8% W 1% Si steel: Firth Brown; shear blades
PREGA	0.45% C 3.0% Cr 0.45% Mo steel: Uddelholm
PRESSURDIE	0.5% C 5.0% Cr 1.3% Mo 2.0% V steel: Braeburn
PRESSURDIE/1	0.38% C 5.0% Cr 0.25% Mo 0.2% V 5.0% W steel: Braeburn; AISI type H 14
PRESSURDIE/3	0.39% C 5.5% Cr 1.2% Mo 1.0% V steel: Braeburn; AISI type H 13
PRESSURDIE/3L	0.39% C 5.5% Cr 1.2% Mo 0.4% V steel: Braeburn; AISI type H 11
PRESSURDIE/5	0.38% C 3.5% Cr 1.0% Mo 1.0% V 1.2% W steel: Braeburn
PRESSURDIE/16	0.5% C 5.0% Cr 1.25% Mo 0.2% V 1.25% W steel: Braeburn; AISI type A8
PS 36	0.4% C 0.55% Cr 0.17% Mo steel: SAE potential specification
PS 38	0.45% C 0.55% Cr 0.17% Mo steel: SAE potential specification
PS 39	0.51% C 0.55% Cr 0.17% Mo steel: SAE potential specification
PS 40	0.55% C 0.55% Cr 0.17% Mo steel: SAE potential specification
PS 67	0.45% C 1% Cr 0.3% Mo steel: SAE potential specification
PW	0.4% C 5% Cr 0.4% Mo 4.0% W 1% Si steel: For hot working tools; S Osborn; hot dies, punches, mandrels, etc.; obsolete **DPN: 460**
PYRODIE	0.4% C 5% Cr 1.3% Mo 1% V 1% Si steel: Clyde Alloy **DPN: 500**
PYROMET 882	0.4% C 5.0% Cr 1.5% Mo 0.4% V steel: Carpenter for AISI type H 11
QQ W 405a	Alloy steel: Wire; US Federal Specification; covers a range of steels prefixed FS with AISI code following
QQ W 412	Alloy steel: Wire for springs; US Federal Specification; covers several analyses
RABI	0.3% C 4.1% Ni 1.3% Cr 0.3% Mo steel: Osborn
RED DIAMOND 21	0.4% C 1.1% Cr 0.25% Mo steel: For wear; Redheugh **DPN: 375 UTS: 1300 Elon: 10% Proof: 1100**
RED DIAMOND 22	0.4% C 1.5% Ni 1.25% Cr 0.3% Mo steel: Redheugh **DPN: 325 UTS: 1180 Elon: 15% Proof: 1070**
REGIN 3	0.5% C 1.2% Cr 0.2% V 2.5% W steel: Uddelholm for type S1
RKV	0.5% C 1.0% Cr 0.15% V steel: Pompey
RO 952	0.4% C 1.1% Cr 0.2% Mo steel: Bofors; hardened and tempered **DPN: 290 UTS: 990 Elon: 12% Proof: 700**
RO 8155	0.4% C 3.0% Cr 0.5% Mo steel: For dies; Bofors; annealed **DPN: 230**
ROP 19	0.4% C 5.3% Cr 1.4% Mo 1.0% V steel: Bofors **DPN: 450**
ROP 9653	0.45% C 1.5% Cr 0.7% Mo 0.3% V steel: Bofors **DPN: 500**
RP 1152	0.5% C 1.1% Cr 0.15% V steel: For springs; Bofors; hardened and tempered **DPN: 460 UTS: 1620 Elon: 7% Proof: 1210**
RSKD	0.45% C 0.8% Cr 0.25% Mo steel: Pompey
RTO 712	0.4% C 1.2% Cr 0.3% Mo 0.2% V 2.5% W steel: Bofors; obsolete **DPN: 640**
RTO 912	0.5% C 1.2% Cr 0.3% Mo 0.2% V 2.5% W steel: Bofors **DPN: 500**
S 1	0.52% C 1.4% Cr 0.2% V 2.2% W steel: For cold work; Osborn
S 1	0.5% C 1.5% Cr 0.4% Mo 0.2% V 2% W steel: Designation used by SAE
S 2	0.5% C 0.5% Mo 0.25% V 1.0% Si steel: For tools; designation used by SAE
S 3	0.5% C 0.75% Cr 1.0% W steel: Designation used by SAE; obsolete
S 4	0.55% C 0.25% Cr 0.2% V steel: Designation used by AISI
S 5	0.55% C 0.3% Cr 0.4% Mo 0.2% V 2.0% Si steel: For tools; designation used by SAE
S 6	0.45% C 1.4% Mn 1.5% Cr 0.4% Mo 2.2% Si steel: Designation used by SAE
S 7	0.5% C 3.2% Cr 1.4% Mo steel: Designation used by SAE
S 7	0.5% C 3.25% Cr 1.4% Mo steel: Carpenter
S 131	0.62% C 4.5% Mn 13% Ni 3.5% Cr steel: BS designation for aerospace steel
S 132	0.4% C 3.2% Cr 1% Mo 0.2% V steel: BS designation for aerospace steel
SAE 81 B45	0.45% C 0.3% Ni 0.45% Cr 0.1% Mo 0.0005% B steel
SAE 86 B45	0.45% C 0.6% Ni 0.5% Cr 0.2% Mo 0.0005% B steel
SAE 94 B40	0.40% C 0.45% Ni 0.4% Cr 0.11% Mo steel
SAE 3140	0.4% C 1.2% Ni 0.6% Cr steel
SAE 3145	0.45% C 1.2% Ni 0.8% Cr steel: Obsolete
SAE 3150	0.5% C 1.2% Ni 0.8% Cr steel: Obsolete
SAE 3245	0.45% C 1.7% Ni 1% Cr steel: Obsolete
SAE 3250	0.5% C 1.7% Ni 1% Cr steel: Obsolete
SAE 3340	0.40% C 3.5% Ni 1.5% Cr steel: Obsolete
SAE 3450	0.5% C 3% Ni 0.8% Cr steel: Obsolete
SAE 4140	0.4% C 1% Cr 0.2% Mo steel
SAE 4142	0.42% C 1% Cr 0.2% Mo steel
SAE 4145	0.45% C 1% Cr 0.2% Mo steel
SAE 4147	0.47% C 1% Cr 0.2% Mo steel
SAE 4150	0.5% C 1% Cr 0.2% Mo steel
SAE 4161	0.61% C 0.8% Cr 0.3% Mo steel
SAE 4340	0.4% C 1.8% Ni 0.8% Cr 0.25% Mo steel
SAE 4620	0.4% C 1.8% Ni 0.25% Mo steel: Obsolete
SAE 4640	0.4% C 1.8% Ni 0.25% Mo steel
SAE 6140	0.4% C 1% Cr 0.15% V steel: Obsolete
SAE 6145	0.45% C 1% Cr 0.15% V steel: Obsolete
SAE 6150	0.5% C 1% Cr 0.15% V steel
SAE 7260	0.6% C 0.8% Cr 1.8% W steel: Obsolete

Symbol	Nominal analysis, supplier, condition and remarks.
SAE 8640	0.4% C 0.6% Ni 0.5% Cr 0.2% Mo steel
SAE 8641	0.40% C 0.6% Ni 0.5% Cr 0.2% Mo steel (S higher than 8640)
SAE 8642	0.42% C 0.6% Ni 0.5% Cr 0.2% Mo steel
SAE 8645	0.45% C 0.6% Ni 0.5% Cr 0.2% Mo steel
SAE 8647	0.47% C 0.6% Ni 0.5% Cr 0.2% Mo steel: Obsolete
SAE 8650	0.5% C 0.6% Ni 0.5% Cr 0.2% Mo steel
SAE 8653	0.53% C 0.6% Ni 0.7% Cr 0.2% Mo steel
SAE 8655	0.55% C 0.6% Ni 0.5% Cr 0.2% Mo steel
SAE 8740	0.4% C 0.6% Ni 0.5% Cr 0.25% Mo steel
SAE 8742	0.42% C 0.6% Ni 0.5% Cr 0.25% Mo steel
SAE 8745	0.45% C 0.6% Ni 0.5% Cr 0.25% Mo steel: Obsolete
SAE 8750	0.5% C 0.6% Ni 0.5% Cr 0.25% Mo steel
SAE 9440	0.4% C 0.4% Ni 0.4% Cr 0.1% Mo steel: Obsolete
SAE 9442	0.42% C 0.4% Ni 0.4% Cr 0.1% Mo steel: Obsolete
SAE 9445	0.45% C 0.4% Ni 0.4% Cr 0.1% Mo steel: Obsolete
SAE 9447	0.47% C 0.5% Ni 0.4% Cr 0.1% Mo steel: Obsolete
SAE 9747	0.47% C 0.5% Ni 0.2% Cr 0.2% Mo steel: Obsolete
SAE 9763	0.63% C 0.5% Ni 0.2% Cr 0.2% Mo steel: Obsolete
SAE 9840	0.4% C 1% Ni 0.8% Cr 0.25% Mo steel
SAE 9845	0.45% C 1% Ni 0.3% Cr 0.25% Mo steel: Obsolete
SAE 9850	0.5% C 1% Ni 0.8% Cr 25% Mo steel
SAE 71360	0.6% C 3.5% Cr 14% W steel: Obsolete
SAE 71660	0.6% C 3.5% Cr 16% W steel: Obsolete
SAE E 4340	0.4% C 1.75% Ni 0.8% Cr 0.25% Mo steel
SAE HNV 1	0.55% C 0.8% Cr 0.75% Mo 1.5% Si steel: For inlet valves
SAE HNV 4	0.45% C 1% Ni 7% Cr 3.3% Si steel: For inlet valves
SAE J132	0.5% C 1% Cr 0.15% V (min) steel: For springs **DPN: 450**
SAE NV 3	0.5% C 0.3% Ni 0.4% Cr 0.15% Mo steel: For inlet valves
SAE NV 4	0.4% C 1.2% Ni 0.6% Cr steel: For inlet valves; as SAE 3140
SAE NV 5	0.45% C 0.5% Ni 0.5% Cr 0.2% Mo steel: For inlet valves; as SAE 8645
SAE S 1	0.5% C 1.5% Cr 0.4% Mo 0.2% V 2% W steel
SAE S 2	0.5% C 0.5% Mo 0.25% V steel: For tools
SAE S 5	0.5% C 0.3% Cr 0.4% Mo 0.2% V steel: For tools
SAE X 3140	0.4% C 1.2% Ni 0.8% Cr steel: Obsolete
SANBOLD 19	0.4% C 1.2% Cr 0.3% Mo steel: Sanderson for BS alloy En 19
SANBOLD 22	0.4% C 3.5% Ni 0.3% Cr steel: Sanderson for BS alloy En 22
SANBOLD 24	0.4% C 1.5% Ni 1.2% Cr 0.3% Mo steel: Sanderson for BS alloy En 24
SANBOLD 66	0.4% C 3.5% Ni 0.3% Cr steel: Sanderson; replaced by Sanbold 22
SANBOLD 155	0.4% C 1.2% Cr 0.3% Mo steel: Sanderson; replaced by Sanbold 19
SANBOLD CVS	0.5% C 1% Cr 0.15% V steel: Sanderson for BS alloy En 47
SANBOLD NK 2833	0.4% C 1.5% Ni 1.2% Cr 0.3% Mo steel: Sanderson; replaced by Sanbold 24
SBM	0.58% C 1.0% Cr 0.25% V steel: Origin unknown
SEARCHER	0.45% C 1% Cr 2% W steel: Hall and Pickles; chisels, wedges, shear blades, etc.; obsolete **DPN: 620**
SEARCHER A1	0.63% C 0.4% Cr 1.6% W steel: Origin unknown

Note. The following abbreviations and units are used in the tables:

DPN	Hardness, diamond pyramid number
UTS	Ultimate tensile strength, N/mm^2
Elon	Elongation, %
Proof	0.1% proof strength, N/mm^2

$1\ N/mm^2 = 0.1\ hbar = 0.102\ kgf/mm^2 = 0.06475\ tonf/in.^2 = 145.04\ lbf/in.^2 = 1\ MPa$

See Appendix II for other abbreviations and conversion tables.

Symbol	Nominal analysis, supplier, condition and remarks.
SHC 1	0.44% C 3% Ni 0.65% Cr 1% Si steel: Firth Brown: for chisels and punches
SHEARTOUGH	0.45% C 2.25% Ni 1.15% Cr 0.2% V steel: Clyde Alloy **DPN: 550**
SILICO ALLOY	0.58% C 0.3% Cr 0.5% Mo 1.95% Si steel: Columbia; AISI type S5
SILMO	0.55% C 0.35% Cr 0.3% Mo 0.2% V 2.0% Si steel: Jessop; shock resistant
For SIS specifications see SS	
SKC 24	0.38% C 0.65% Mn 3.0% Ni 0.5% Cr 0.22% Mo steel: For hollow drills; Japanese Standards designation
SKD 6	0.37% C 5.2% Cr 1.25% Mo 0.4% V steel: Japanese Standards designation
SKD 61	0.37% C 5.0% Cr 1.25% Mo 1.0% V steel: Japanese Standards designation
SKD 62	0.38% C 5.0% Cr 1.25% Mo 0.4% V 1.25% W steel: Japanese Standards designation
SKF 327A	0.4% C 1.2% Cr 0.3% Mo steel: Hardened and tempered **UTS: 930** **Proof: 720**
SKF 661	0.5% C 0.95% Cr 0.2% V steel: SKF
SKF 663	0.52% C 1% Cr 0.15% V steel: SKF
SKS 4	0.5% C 0.75% Cr 0.75% W steel: Japanese Standards designation
SKS 41	0.4% C 1.25% Cr 3.0% W steel: Japanese Standards designation
SKT 2	0.55% C 1.0% Mn 1.0% Cr 0.2% V steel: Japanese Standards designation
SKT 3	0.55% C 0.42% Ni 1.05% Cr 0.4% Mo 0.2% V (max) steel: Japanese Standards designation
SKT 4	0.55% C 0.8% Mn 1.7% Ni 0.85% Cr 0.35% Mo 0.2% V (max) steel: Japanese Standards designation
SKT 5	0.55% C 0.8% Mn 1.25% Cr 0.35% Mo 0.22% V steel: Japanese Standards designation
SKT 6	0.75% C 0.8% Mn 2.75% Ni 0.95% Cr 0.42% Mo 0.2% V (max) steel: Japanese Standards designation
SL 823	0.4% C 5.0% Cr 1.0% Mo 0.4% V steel: Swift Levick; for hot dies
SL 866	0.4% C 5.0% Cr 1.25% Mo 0.3% V 1.0% W steel: Swift Levick for hot dies
SL 962	0.4% C 1.5% Cr 0.5% Mo 0.35% V 3.5% W steel: Swift Levick for hot dies
SL 983	Medium C 2% Cr Mo steel: Swift Levick **DPN: 400**
SM	0.58% C 0.25% Cr 0.35% Mo 0.2% V steel: Forez
SN 3½	0.4% C 3.5% Ni 0.2% Cr steel: Spencer for BS alloy En 22
SNC4G	0.4% C 1.5% Ni 1.2% Cr 0.3% Mo steel: Spencer for BS alloy En 24
SOMDIE	0.48% C 1.0% Ni 1.2% Cr 0.5% Mo 0.07% V steel: origin unknown
SOMDIE SPECIAL	0.5% C 3.2% Cr 1.4% Mo steel: Origin unknown
SOUDOKAY 258.0	0.45% C 6% Cr 1.5% Mo 1.5% W steel: Welding electrode; Soudometal **DPN: 600**
SPARTAN 5	0.5% C 4.0% Cr 1.0% V 18% W steel: Atlas; as AISI type H26
SPEAR 419	0.4% C 1.2% Cr 0.3% Mo steel: Spear and Jackson for BS alloy En 19
SPEAR 500	0.4% C 0.45% Cr 0.2% Mo steel: Spear and Jackson for BS alloy En 100
SPEAR 510	0.4% C 1.5% Ni 1.2% Cr 0.15% Mo steel: Spear and Jackson for BS alloy En 110
SPEAR 626	0.4% C 2.5% Ni 0.6% Cr 0.6% Mo steel: Spear and Jackson; as BS alloy En 26
SPEAR AHC	0.5% C 1.5% Cr 2.0% W steel: Spear and Jackson
SPEAR CV	0.78% C 0.45% Cr 0.15% V steel: Origin unknown
SPEAR D1	0.5% C 0.8% Cr 0.25% V steel: For zinc die casting mould; Spear and Jackson; hardened and tempered **UTS: 870**

Symbol	Nominal analysis, supplier, condition and remarks.
SPEAR D2	0.5% C 2.5% Cr 0.2% V steel: Origin unknown
SPEAR EXM	0.73% C 1.5% Ni 0.2% Cr steel: Origin unknown
SPEAR LEAPFROG	0.65% C 3.75% Cr 0.6% V 14.0% W steel: For tools; origin unknown
SPECIAL ARDHO	0.4% C 3.0% Ni 0.8% Cr steel: Spencer; chisels, punches, etc.
SPECIAL HW No. 3	0.37% C 3.1% Cr 0.45% Mo 0.25% V 11.25% W steel: Origin unknown
SPECIAL WOLFRAM	0.5% C 1.2% Cr 2.2% W steel: Swift Levick; hot dies
SPECIFICATION 55	0.68% C 0.65% Ni 0.7% Cr 0.2% Mo steel: Crucible Steel Co.; for saw blades, etc.
SPEEDICUT 14	0.6% C 3.5% Cr 0.6% V 14% W steel: Firth Brown
SPK	0.45% C 2.35% Cr 0.3% V steel: Schoeller-Bleckmann **UTS: 1210**
SPKD	0.4% C 10.5% Cr 1.0% Mo steel: Pompey
SPS	0.5% C 0.9% Cr 0.2% V 2% W 0.9% Si steel: For tools; Huntsman; chisels, shear blades, punches, etc. **DPN: 600**
SPS 245	0.4% C 1.25% Ni 0.6% Cr 0.15% Mo steel: Atlas; hardened and tempered **DPN: 340 UTS: 1140 Elon: 16% Proof: 1070**
SRO	0.45% C 9% Cr 0.3% Mo 2.8% Si steel: Bofors; obsolete
SS 14-2216 E	0.4% C 0.6% Mn 1.0% Cr 0.5% Mo 0.25% Cu steel: For pressure vessels; Swedish Standard; normalized and annealed **UTS: 500 Elon: 19% Proof: 290**
SS 2230	0.52% C 0.9% Mn 1.1% Cr 0.15% V 0.22% Si steel: For springs **DPN: 400 UTS: 1300**
SS 2234	0.33% C 0.65% Mn 0.05% Cr 0.22% Mo steel: Swedish Standard; hardened and tempered **UTS: 1080 Elon: 10% Proof: 880**
SS 2242	0.38% C 5.5% Cr 1.4% Mo 1% V 1% Si steel: Swedish Standard; hardened and tempered **DPN: 410 UTS: 1380 Elon: 8% Proof: 1070**
SS 2244	0.42% C 1.1% Cr 0.2% Mo steel: Bar and forging; Swedish Standard; hardened and tempered **DPN: 350 UTS: 1140 Elon: 10% Proof: 880**
SS 2244	0.42% C 0.75% Mn 1.05% Cr 0.22% Mo steel: Swedish Standard; hardened and tempered **UTS: 1080 Elon: 10% Proof: 880**
SS 2541	0.35% C 1.4% Ni 1.4% Cr 0.2% Mo steel: Swedish Standard
SS 2550	0.55% C 3% Ni 1% Cr 0.3% Mo steel: Bar and forging; Swedish Standard; annealed **DPN: 260**
SS 2710	0.5% C 1.2% Cr 0.25% Mo 0.15% V 2.5% W steel: Swedish Standard; annealed **DPN: 245**
SS 2940	0.41% C 1.7% Cr 0.25% Mo 1.1% Al steel; For nitriding; Swedish Standard; hardened and tempered **DPN: 270 UTS: 920 Elon: 15% Proof: 630**
SS 8101	0.4% C 1.7% Ni 0.5% Mo 1.5% Cu steel: Sintered
SS 8102	0.6% C 1.75% Ni 0.5% Mo 1.5% Cu steel: Sintered
SSC	0.58% C 1.05% Cr 0.3% V steel: Origin unknown
STA 5/V9D	0.4% C 1.2% Cr 0.3% Mo steel: Replaced by BS alloy En 19
STA 5/V9D/1	0.4% C 1.2% Cr 0.3% Mo steel: Replaced by BS alloy En 19B
STA 5/V9D/2	0.4% C 1.2% Cr 0.3% Mo steel: Replaced by BS alloy En 19C
STA 5/V10	0.4% C 1.5% Ni 1.2% Cr 0.3% Mo steel: Replaced by BS alloy En 24
STA 5/V12	0.4% C 2.5% Ni 0.7% Cr 0.6% Mo steel: Replaced by BS alloy En 26
STA 5/V21A	0.4% C 0.3% Ni 3.0% Cr 1.0% Mo 0.2% V steel: Replaced by BS alloy En 40
STA 5/V21B	0.4% C 0.3% Ni 3.0% Cr 1.0% Mo 0.2% V steel: Replaced by BS alloy En 40

Symbol	Nominal analysis, supplier, condition and remarks.
STAG 14	0.67% C 3.5% Cr 0.37% V 15.0% W steel: For tools; origin unknown
STAG C	0.45% C 2.5% Cr 0.35% V steel: Edgar Allen
STAMINAL	0.55% C 2.7% Ni 0.4% Cr 0.45% Mo 0.13% V steel: Latrobe; dies, punches and stamps **DPN: 600**
STUBS 73 G2	C Mn Cr Mo Si steel: Electrode; Stubs; for welding tool steel; high temperature application **DPN: 500**
STUBS 73 G3	C Mn Cr Mo Si steel: Electrode; Stubs; for welding tool steel; high temperature application **DPN: 500**
STUBS 73 G4	C Mn Cr Mo Si steel: Electrode; Stubs; for welding tool steel; high temperature application **DPN: 450**
STUBS 700	C Ni Cr Mo W steel: Electrode; Stubs; for abrasion and impact resistance **DPN: 300 UTS: 620**
STUBS A675	0.45% C 9.5% Cr 3.0% Si steel: Electrode; Stubs; for hard facing **DPN: 600**
SUPER Si Mn	0.57% C 0.8% Mn 0.8% Cr 0.25% Mo 1.8% Si steel: For springs; ESC
SUPER TOUGH D	0.4% C 1.3% Mn 0.75% Ni 0.5% Cr 0.2% Mo steel: ESC for BS alloy En 100
T 20810	UNS designation for H10 type tool steel
T 20811	UNS designation for H11 type tool steel
T-ALLOY	0.38% C 3.5% Cr 0.4% V 10.5% W steel: Braeburn; AISI type H22
T-ALLOY-B	0.5% C 3.0% Cr 0.5% V 15% W steel: Braeburn; AISI type H 24
TH	0.38% C 2.1% Cr 0.15% V steel: For hot work tools; S Osborn; casting dies; obsolete **DPN: 400**
THERMODIE	0.55% C 2.1% Ni 0.9% Cr 0.8% Mo steel: Origin unknown
THERMOWEAR	0.6% C 4% Cr 2.5% Mo 1% V 0.1% Ti 1.5% Nb steel: Carpenter
TITANIC 141	0.6% C 0.3% Mo 0.2% V 2% Si steel: S Osborn; name for NA; obsolete
TM	0.4% C 5.0% Cr 1.2% Mo 0.5% V steel: Forez
TPC	0.45% C 1.2% Cr 0.15% V steel: Toledo for BS alloy En 50
TRKD	0.42% C 1.0% Cr 0.25% Mo steel: Pompey
TRNKD	0.4% C 0.5% Ni 0.5% Cr 0.2% Mo steel: Pompey
TYRANT EXTRA	0.45% C 1.05% Cr 2% W 1% Si steel: Schoeller-Bleckmann **DPN: 550**
UCX 2	0.4% C 1.1% Cr 0.25% Mo 0.15% V 1.0% Co steel: American proprietary alloy listed in SAE yearbook
ULTIMO 4	0.4% C 1.7% Ni 0.7% Cr 0.4% Mo steel: Atlas; hardened and tempered **DPN: 375 UTS: 1300 Elon: 15% Proof: 1250**
ULTIMO 6	0.55% C 1.6% Ni 1.0% Cr 0.75% Mo steel: Atlas; forging die inserts, etc. **DPN: 400**
V 130	0.4% C 1.4% Ni 0.8% Cr 0.4% Mo steel: VEW
V 310	0.5% C 1.1% Cr 0.2% Mo steel: VEW
V 320	0.4% C 1.1% Cr 0.2% Mo steel: VEW
VA	0.63% C 2% Ni 2% Cr 0.5% Mo steel: For hot work tools; S Osborn; obsolete **DPN: 400**
VAGS	0.4% C 1.5% Ni 1.1% Cr 0.3% Mo steel: ESC for BS alloy En 24
VAL 5M	0.4% C 10% Cr 1% Mo 2.5% Si steel: Valbruna
VDC	0.4% C 5% Cr 1.2% Mo 1% V 1% Si steel: Casting; Latrobe; for precision casting of forging dies **DPN: 550**
VERSATILE	0.4% C 1.5% Ni 1% Cr steel: Hall and Pickles; swaging tools, etc.; obsolete **UTS: 1150 Elon: 17% Proof: 1080**

Symbol	Nominal analysis, supplier, condition and remarks.
VIADUCT 15	0.4% C 1.1% Cr 0.3% V 1.9% W steel: Balfour; chisels, caulking tools, etc. **DPN: 550**
VIBRAC V 45	0.4% C 2.5% Ni 0.6% Cr 0.5% Mo steel: ESC for BS alloy En 26
VIBRO	0.5% C 1.4% Cr 0.25% V 2.1% W steel: Braeburn; AISI type S1
VINCO HOT WORK	0.5% C 4.0% Cr 1.0% V 18% W steel: Braeburn; AISI H 26
VISCOUNT 20	0.4% C 5.0% Cr 1.2% Mo 1.0% V steel: With sulphides; Latrobe; free machining form of VDC; type H13 **DPN: 650**
VISCOUNT 44	0.4% C 5.0% Cr 1.2% Mo 1.0% V steel: With sulphides; Latrobe; free machining form of VDC; type H13 **DPN: 650**
VS 35 V	0.45% C 1.5% Cr 0.35% V steel: Vulcan **UTS: 1070**
VSCV	0.55% C 1% Cr 0.25% V steel: Vulcan; vehicle springs **DPN: 460**
VSNT	0.45% C 0.4% Cr 0.6% W steel: Vulcan; chisels, punches, etc. **DPN: 560**
W 300	0.36% C 5.0% Cr 1.1% Mo 0.4% V steel: VEW
W 302	0.39% C 5.0% Cr 1.3% Mo 1.0% V steel: VEW
W 303	0.39% C 5.0% Cr 2.9% Mo 0.6% V steel: VEW
W 321	0.39% C 2.9% Cr 2.8% Mo 0.5% V steel: VEW
W 326	0.45% C 1.4% Cr 0.7% Mo 0.3% V steel: VEW
W 500	0.55% C 1.7% Ni 1.1% Cr 0.5% Mo 0.1% V steel
W 501	0.55% C 1.7% Ni 0.7% Cr 0.3% Mo 0.1% V steel: VEW
WAW	0.45% C 1.5% Cr 0.5% Mo 0.9% V 0.5% W steel: Schoeller-Bleckmann **UTS: 1380**
WC 38	0.38% C 0.4% Ni 0.25% Cr 0.15% Mo 0.3% Cu (all max) steel: Casting; Wolsingham **DPN: 163 UTS: 580 Elon: 15% Proof: 270**
WC 43	0.42% C 0.4% Ni 0.25% Cr 0.15% Mo 0.3% Cu (all max) steel: casting; Casting; Wolsingham **DPN: 175 UTS: 610 Elon: 12% Proof: 300**
WC 48	0.48% C 0.4% Ni 0.25% Cr 0.15% V 0.3% Cu (all max) steel: Casting; Wolsingham **DPN: 185 UTS: 630 Elon: 12% Proof: 300**
WC 53	0.53% C 0.4% Ni 0.25% Cr 0.15% V 0.3% Cu (all max) steel: Casting; Wolsingham **DPN: 190 UTS: 700 Elon: 10% Proof: 330**
WC 58	0.58% C 0.4% Ni 0.25% Cr 0.15% Mo 0.3% Cu (all max) steel: Casting; Wolsingham **DPN: 200 UTS: 700 Elon: 10% Proof: 330**
WC 63	0.63% C 0.4% Ni 0.25% Cr 0.15% Mo 0.3% Cu (all max) steel: Casting: Wolsingham **DPN: 210 UTS: 700 Elon: 10% Proof: 330**
WCPS	0.48% C 0.7% Ni 1.5% Cr 0.2% Mo 0.15% V 1.2% W tool steel: Huntsman; chisels, shear blades, punches, etc. **DPN: 670**
WL 1	0.4% C 1.8% Ni 1.0% Cr 0.25% Mo steel: Tube; Babcock and Wilcox (USA)
WLM 1214	0.4% C 1% Cr 2% W 1.3% Si steel: W Marrison; chisels, punches
WM EXTRA	0.38% C 2.1% Cr 4.25% Mo 0.32% V 1.05% Si steel: Schoeller-Bleckmann **UTS: 1550**
WZ	0.4% C 3.0% Cr 0.4% Mo 0.25% V 11.0% W steel: Origin unknown
X 38 Cr Mo V 51	0.38% C 5.3% Cr 1.1% Mo 0.4% V steel: German Standards designation
X 40 Cr Mo V 51	0.4% C 5.3% Cr 1.4% Mo 1.0% V steel: German Standards designation
X 45 Ni Cr Mo 4	0.45% C 4.1% Ni 1.4% Cr 0.3% Mo 0.5% W steel: For tools; German Standards designation
X 3140	0.4% C 1.2% Ni 0.8% Cr steel: Designation used by AISI; obsolete
X 40 Cr Mo V 53	0.4% C 5% Cr 3% Mo 0.5% V steel: German Standard
X 50 Cr Mo 51	0.5% C 5% Cr 1.4% Mo 1.4% V steel: German Standard
X 50 Cr Mo W 911	0.53% C 8.3% Cr 1.2% Mo 1.2% W steel: German Standard
XDH	0.55% C 4.0% Cr 1.0% V 18% W steel: Firth Sterling
XDL	0.48% C 3.5% Cr 0.75% V 14% W steel: Firth Sterling
XL CHISEL	0.55% C 1.35% Cr 0.1% Mo 0.25% V 2.0% W steel: Latrobe; tool steel
XP 180	0.5% C 5.0% Cr 1.3% Mo 2.0% V steel: Braeburn; as Pressurdie 7
Y 42CD4	0.42% C 1.0% Cr 0.2% Mo steel: French Standards designation
Y 45SCD6	0.45% C 0.6% Cr 0.2% Mo steel: French Standards designation
Y 50CV4	0.5% C 1.1% Cr 0.2% V steel: French Standard
Y 50CV4	0.5% C 1.0% Cr 0.15% V steel: French Standards designation
YCW STEM	0.48% C 4.2% Ni 0.6% Cr 0.25% Mo 1.7% Si steel: Casting; Edgar Allen; code 638 steel
Z 38CDV5	0.38% C 5.0% Cr 1.25% Mo 0.5% V steel: French Standards designation
Z 38CDV5	0.38% C 5.2% Cr 1.3% Mo 1.0% V steel: French Standard
Z 38CDWV5	0.38% C 5.0% Cr 1.25% Mo 0.5% V 1.25% W steel: French Standards designation
Z 40CSD10	0.4% C 0.8% Mn (max) 10.5% Cr 1.0% Mo steel: French Standards designation
Z 40WCV5	0.4% C 4.0% Cr 0.5% Mo 0.5% V 5.0% W steel: French Standards designation
Z 65WDCV605	0.65% C 4.0% Cr 5.0% Mo 2.0% V 6.0% W steel: French Standards designation

Note. The following abbreviations and units are used in the tables:

DPN	Hardness, diamond pyramid number
UTS	Ultimate tensile strength, N/mm^2
Elon	Elongation, %
Proof	0.1% proof strength, N/mm^2

1 N/mm^2=0.1 hbar=0.102 kgf/mm^2=0.06475 $tonf/in.^2$=145.04 $lbf/in.^2$=1 MPa
See Appendix II for other abbreviations and conversion tables.

44K3 Steel – high carbon alloys nickel, molybdenum, tungsten or vanadium

Specific gravity	7.87
Density	7870 kg/m^3
Solidus/liquidus	1450–1550 °C
Thermal conductivity	50 W/m °C
Coefficient of linear expansion	11 × 10^{-6}/ °C
Electrical conductivity	4–8% IACS (copper 100%)
Specific resistance	220–380 microhm mm
Young's modulus of elasticity	203.4 × 10^9 N/m^2
Impact	will be low
Fatigue strength	–
Hot strength	–

The above properties are typical of the following group and may not apply exactly to any one specification. It is possible that with certain specifications some of the values may not be applicable.

General metallurgical characteristics

These are all very hard brittle materials which have been developed for their high hardness. This is only achieved after hardening and tempering and they are thus only used in this condition. This hardness persists at fairly high temperatures depending on the amount of alloying carried out.

None of the alloying elements enhances the corrosion resistance to any great extent whereas the high hardness and low ductility reduce the notch sensitivity, thus making the presence of corrosion pitting particularly dangerous.

All the alloys have sufficient carbon present to ensure the presence at all times of free cementite which is inherently hard even in the annealed state.

Each steel has at least two of the listed alloying elements and many have all five present. The effect of each element is as described in the single alloy steel bearing that name, and when two or more elements are present the effect is additive.

Briefly the effects of the alloying elements are as follows.

Chromium. This is a carbide former, with the carbides being harder and more stable than iron carbides. Chromium increases the hardenability of steel in that larger sections can be through hardened, or less drastic oil or air quenching can be used in place of water, compared with plain or lower alloy carbon steels. Chromium also increases the corrosion resistance although none of the steels listed can be classed as stainless. The higher chromium stainless steels are listed and described in separate sections. Chromium raises the critical temperatures for steel thermal treatment and slows down the speed at which the metallurgical reactions take place it has a tendency to promote grain growth.

Nickel. This strengthens the ferrite matrix of the steel, imparting toughness rather than hardness, and increases the fatigue strength. Nickel lowers the critical thermal treatment temperatures and widens the range required for successful treatment. It also improves the ductility of steel at low temperatures. Nickel is seldom an important constituent in the steels listed as hardness and abrasion resistance are their prime requirements, and nickel has less effect on these high carbon steels than on the lower carbon varieties.

Molybdenum. This is a carbide former, the resultant molybdenum carbide being inherently small grained and evenly distributed. This ensures that the steels are fine grained, thus improving the fatigue strength. These stable carbides slow down the metallurgical reactions, ensuring deeper hardening, great hardenability and the elimination of temper brittleness. They also help to prevent agglomeration of the carbides and grain growth with prolonged high temperature treatments. As carbon always combines with the molybdenum in preference to other carbide formers, including iron, and stays in solution, this obviates carbide precipitation. Vanadium and tungsten have similar effects.

Tungsten. This is a carbide former which results in very hard, rather massive tungsten carbide particles. Generally tungsten is used with chromium and vanadium which aid the distribution and prevent agglomeration of the carbides. Molybdenum has similar properties to tungsten with generally less required to give the same results. Tungsten was the first element to be used in quantity in high carbon steels to improve the hot hardness.

Vanadium. This forms very small carbides and also influences other carbide formers to produce tiny particles which in turn ensure a fine grain steel. Thus vanadium has a considerable effect on improving the fatigue strength and is very often associated with chromium for this purpose. Vanadium has little effect on the temperatures at which the metallurgical changes take place, but slows down these reactions, thus improving the hardenability.

The steels listed here may have slightly better corrosion resistance than plain carbon steels but cannot be classified as stainless. Also the effect of corrosion pitting on the rather notch sensitive materials is much more serious than on more ductile steels.

The choice of elements in the steels is a complex process, balancing hardness, hot hardness, ductility and workability with the overriding factor that the final product must be economically saleable.

These steels will accept no cold work and therefore this should not be attempted.

The materials all have a high hardenability and many of the materials are hard with low ductility even in the annealed condition. The majority of the materials in this group will harden in air and certainly water quenching will seldom or never be required and will probably cause cracking.

With the variation in carbon content and alloy content of

the steels in this group it is necessary to obtain expert advice for each specification before attempting heat treatment. On all the parts before quenching it is essential that changes of section have good radii to prevent cracking even when air cooling is involved.

Because of their very high hardenability welding is not advised under any circumstances. Where it is essential then expert advice should be obtained and followed to the letter.

Brazing can be safely carried out on many of these materials but again advice should be obtained.

Many of these materials will only be used in the fully hardened condition and this will require grinding where a good finish and engineering tolerances are involved. Great care must be taken in the grinding operation as these materials are extremely prone to grinding cracking. The steels in this group will almost invariably have very low impact properties and low ductility; thus surface cracking could result in failure in service.

The materials will be used for cutting tools of all types, i.e. milling, drilling, tapping, broaching, etc. High quality hand-tools of all types, such as spanners, will also be found in this group, as will forging dies, both hot and cold work. The materials are used in certain engineering applications where wear and temperature use coincide, e.g. high duty ball and roller bearings.

Rock drills and chisels will also be found in this group but these very often compete with the carbide cutting tools. Some of the materials in this group are used as tips either mechanically joined to a tougher backing material or brazed. This can apply to cutting tools but also crushing jaws and such-like components.

Symbol	Nominal analysis, supplier, condition and remarks.
0 KX	0.9% C 0.8% Cr 0.12% V steel; Pompey
01	0.9% C 0.5% Cr 0.5% W steel: Designation used by AISI
02	0.9% C 1.4% Mn 0.3% Cr 0.3% Mo 0.2% V steel: For tools; designation used by SAE
06/05/04/02	0.85% C 4.0% Cr 5.0% Mo 2.0% V 6.0% W steel: For tools; French Standards designation
06/05/04/04	1.3% C 4.0% Cr 4.5% Mo 4.0% V 5.5% W steel: For tools; French Standards designation
07	1.2% C 0.75% Cr 1.75% W steel: Designation used by AISI
08/04/02/02	0.85% C 4.2% Cr 5.0% Mo 1.8% V 6.2% W steel: For tools; French Standards designation
09B	1% C 1% Mn 0.6% Cr 1% W steel: Carrs Ltd; shear blades, rivet dies, etc. **DPN: 750**
1.2063	1.45% C 1.4% Cr 0.1% V steel: German Standard
1.2210	1.15% C 0.7% Cr 0.1% V steel: German Standards designation
1.2235	0.8% C 0.5% Cr 0.2% V steel: German Standard
1.2327	0.88% C 2% Cr 0.5% Mo 0.1% V steel: German Standard
1.2363	1.0% C 5.25% Cr 1.0% Mo 0.2% V steel: German Standard
1.2510	1% C 0.5% Cr 0.2% V 0.5% W steel: German Standard
1.2516	1.2% C 0.2% Cr 0.1% V 1% W steel: German Standard
1.2552	0.8% C 1.1% Cr 0.3% V 1.2% W steel: German Standards designation
1.2703	0.7% C 0.4% Ni 0.3% Cr steel: German Standard
1.2770	0.85% C 0.8% Ni 0.1% V steel: German Standard
1.2842	0.9% C 2% Mn 0.4% Cr 0.1% V steel: German Standard
1.3333	0.99% C 4.0% Cr 2.65% Mo 2.35% V 2.85% W steel: For tools; German Standards designation
1.3342	1.0% C 4.1% Cr 5.0% Mo 2.0% V 6.3% W steel: For tools; German Standards designation
1.3343	0.88% C 4.1% Cr 5.0% Mo 1.85% V 6.3% W steel: For tools; German Standards designation
1.3346	0.8% C 3.9% Cr 8.5% Mo 1.1% V 1.5% W steel: For tools; German Standards designation
1.3348	1% C 3.8% Cr 8.7% Mo 2.1% V 1.8% W steel: German Standards designation
1.3355	0.75% C 4% Cr 1% V 18% W steel: German Standards designation
2AS	0.82% C 4.2% Cr 5.0% Mo 6.2% W steel: Forez
3AS	0.76% C 4.2% Cr 1.0% V 18% W steel: Forez
4AS	0.8% C 4.2% Cr 0.7% Mo 1.4% V 18% W steel: Forez
5/6/2 quality	0.83% C 4.2% Cr 5% Mo 1.9% V 6.4% W steel: T Turton
5CC	1.0% C 0.6% Mn 5.2% Cr 1% Mo 0.5% V steel: ESC; shear blades, punches, etc. **DPN: 750**
6/5/2	0.85% C 4% Cr 2% V 6% W steel: For tools; Darwins; reamers, punches, etc.
8KH4V4F1(P4)	0.8% C 4.5% Cr 1.2% V 4.5% W steel: Russian Standards designation
8KHF	0.75% C 0.55% Cr 0.22% V steel: Russian Standards designation
8N2	4% Cr 8% Mo 1% V 1.5% W steel: For tools; Vanadium Alloys Ltd; M1 type; AISI code M1A
9 KH	0.87% C 1.52% Cr 0.5% V 1.25% W steel: Russian Standards designation
9KH5F	0.92% C 5.0% Cr 0.22% V 1.0% W steel: Russian Standards designation
9KH5VF	0.9% C 0.25% Mn 5.0% Cr 0.22% V 1.0% W steel: Russian Standards designation
9KHF	0.85% C 0.55% Cr 0.22% V steel: Russian Standards designation
9KHVG	0.9% C 1.1% Mn 0.65% Cr 0.1% V 0.65% W steel: Russian Standards designation
10 KD	1.0% C 1.0% Cr 0.2% Mo steel: Pompey
12 KV	1.15% C 0.9% Cr 0.12% V steel: Pompey
12 W 12 C1	0.75% C 0.25% Mn 0.4% Cr 5.8% W steel: Sandvik
14 NC	0.8% C 0.35% Mn 0.5% Ni 0.15% Cr steel: Sandvik; annealed **UTS: 660**
16 C1 V	0.9% C 0.55% Mn 0.6% Cr 0.1% V steel: Sandvik
20 W 1V	1.15% C 0.25% Cr 0.1% V 0.6% W steel: Sandvik
21 T 10	1.25% C 0.35% Mn 0.3% Cr 0.1% V 1.7% W steel: Sandvik
21 T 10 P	1.25% C 0.3% Mn 0.25% Cr 0.1% V 1.7% W 0.2% Pb steel: Sandvik
32 B	1.0% C 0.7% Mn 5.25% Cr 1.15% Mo 0.3% V steel: Carrs; coining tools, swaging rolls, etc. **DPN: 680**
66 HS	4% Cr 5% Mo 2% V 6% W tool steel: Bethlehem Steel Co.; M2 type; AISI code M2E

Note. The following abbreviations and units are used in the tables:

DPN	Hardness, diamond pyramid number
UTS	Ultimate tensile strength, N/mm^2
Elon	Elongation, %
Proof	0.1% proof strength, N/mm^2

1 N/mm^2=0.1 hbar=0.102 kgf/mm^2=0.06475 tonf/in.^{2}=145.04 lbf/in.^{2}=1 MPa

See Appendix II for other abbreviations and conversion tables.

Symbol	Nominal analysis, supplier, condition and remarks.
67 S	1.4% C 0.5% Cr 3.25% W steel: Carrs; extrusion dies etc. **DPN: 800**
70 H	0.9% C 0.5% Cr 0.5% W steel: For tools; origin unknown
74 Ni Cr 2	0.7% C 0.4% Ni 0.3% Cr steel: German Standard
80 DCV 42.16	0.8% C 4.0% Cr 4.25% Mo 1.0% V steel: French Standards designation
80 W Cr V 8	0.8% C 1.1% Cr 0.3% V 1.2% W steel: German Standards designation
85 Ni V 4	0.85% C 2% Mn 0.4% Cr 0.1% V steel: German Standard
86 Cr Mo V 7	0.88% C 2% Cr 0.5% Mo 0.1% V steel: German Standard
90 MCW 5	0.9% C 1.25% Mn 0.5% Cr 0.5% W steel: French Standards designation
90 MCW V5	0.95% C 0.5% Cr 0.2% V 0.5% W steel: French Standard
90 Mn Cr V 8	0.9% C 2.0% Mn 0.4% Cr 0.1% V steel: German Standard
90 Mn V 8	0.9% C 2.0% Mn 0.3% Cr 0.1% V steel: German Standards designation
100 Mn Cr W 4	1.2% C 0.2% Cr 0.1% V 1% W steel: German Standard
100C3	1.0% C 0.75% Cr 0.15% V steel: French Standards designation
100CD7	0.97% C 1.8% Cr 0.2% Mo steel: AFNOR
100 Cr Mo 6	0.97% C 1.8% Cr 0.2% Mo steel: German Standard
100WC10	1.0% C 0.5% Cr 0.2% V 1.0% W steel: French Standards designation
110WC20	1.1% C 0.3% Mn 0.75% Cr 2.0% W steel: French Standards designation
115 Cr V 3	1.15% C 0.7% Cr 0.1% V steel: German Standards designation
120WV4	1.2% C 0.2% Cr 0.1% V 1% W steel: German Standard
140C3	1.4% C 0.75% Cr 0.15% V steel: French Standards designation
140SMD4	1.4% C 1.0% Mn 0.3% Mo steel: French Standards designation
145 Cr 6	1.45% C 1.4% Cr 0.1% V steel: German Standard
175	0.85% C 1.9% Mn 0.15% Cr 0.3% V steel: F Parkin; reamers, gauges, etc. **DPN: 830**
200	0.9% C 1.3% Mn 0.5% Cr 0.5% W steel: F Parkin; chisels, gauges, etc. **DPN: 850**
212	1.3% C 0.75% Cr 1% W steel: F Parkin; forming dies, gauges, etc. **DPN: 850**
300	1.5% C 0.3% Mn 0.8% Cr 5.5% W steel: F Parkin; cold draw dies, marble cutting tools, etc. **DPN: 800**
321	0.25% C 0.5% Mn 0.75% Cr 0.4% Mo 0.2% V 2.3% W steel: F Parkin; taps and dies, etc. **DPN: 870**
562	0.8% C 4.25% Cr 2% V 6.5% W steel: F Parkin; cutting tools, form tools **DPN: 800**
562	0.85% C 4.75% Cr 5.0% Mo 2.0% V 6.0% W steel: Spear and Jackson
599	0.68% C 4.2% Cr 0.7% V 14% W steel: F Parkin; taps, dies, etc. **DPN: 800**
622	0.75% C 4% Cr 1% V 18% W steel: F Parkin; cutting tools **DPN: 800**
2131	1.0% C 0.75% Cr 0.15% V steel: French Standards designation
2132	1.4% C 0.75% Cr 0.15% V steel: French Standards designation

Symbol	Nominal analysis, supplier, condition and remarks.
2133	0.8% C 1.0% Ni 0.35% Cr steel: French Standards designation
2141	1.0% C 0.5% Cr 0.2% V 1.0% W steel: French Standards designation
2142	1.1% C 0.3% Mn 0.75% Cr 2.0% W steel: French Standards designation
2212	0.9% C 0.5% Cr 0.5% W steel: French Standards designation
2231	1.0% C 5.0% Cr 1.0% Mo 0.3% V steel: German Standards designation
3551	0.8% C 4.0% Cr 4.25% Mo 1.0% V steel: French Standards designation
4151	0.8% C 4.0% Cr 2.0% Mo 2.0% V 12.0% W steel: For tools; French Standards designation
4161	0.8% C 4.0% Cr 3.5% V 12.0% W steel: For tools; French Standards designation
4201	0.8% C 4.0% Cr 1.0% V 18.0% W steel: For tools; French Standards designation
4203	0.85% C 4.0% Cr 18.0% Mo 2.0% V steel: For tools; French Standards designation
4301	0.85% C 4.0% Cr 5.0% Mo 2.0% V 6.0% W steel: For tools; French Standards deisgnation
4361	1.3% C 4.2% Cr 4.5% Mo 4.0% V 6.2% W steel: For tools; French Standards designation
4441	0.85% C 4.2% Cr 5.0% Mo 1.9% V 6.2% W steel: For tools; French Standards deisgnation
6542 MOLY	0.8% C 4.0% Cr 5.0% Mo 2.0% V 6.0% W steel: For tools; origin unknown
A 2	1.0% C 5% Cr 1.2% Mo 0.4% V steel: For tools; designation used by AISI
A 2	1.0% C 5.0% Cr 1.05% Mo 1.3% V steel: For cold work; Osborn
A 3	1.25% C 5.0% Cr 1.0% Mo 1.0% V steel: For tools; designation used by AISI
A 4	1.0% C 2.0% Mn 1.0% Cr 1.0% Mo steel: For tools; designation used by AISI
A 5	1.0% C 3.0% Mn 1.0% Cr 1.0% Mo steel: For tools; designation used by AISI
A 7	2.25% C 5.2% Cr 1.0% Mo 4.7% V 1.0% W steel: For tools; designation used by AISI
A 7	2.3% C 5.3% Cr 1.0% Mo 4.7% V 1.0% W steel: For cold work; Osborn
A 7W	2.25% C 5.25% Cr 1.0% Mo 4.7% V 1.0% W steel: Simonds; AISI type A7
A 10	1.35% C 1.8% Mn 1.8% Ni 1.5% Mo 1.2% Si steel: For tools; designation used by AISI
A 395	0.95% C 1.1% Cr 0.25% Mo steel: SHD; drills
ABC III	0.8% C 3.8% Cr 9% Mo 1.2% V 1.7% W steel: Origin unknown
ABRASODUR 16	4.3% C 9.7% Cr 1.8% Mo steel: Welding electrode; Soudometal; for hard facing **DPN: 650**
ADAMANT HP	0.85% C Cr Mo steel: Casting; Firth Brown; hardened and tempered **DPN: 370**
ADAMANT XL	1.25% C Cr Mo steel: Casting; Firth Brown; hardened and tempered **DPN: 400**
ADAMANT XTRA	0.77% C Cr Mo steel: Casting; Firth Brown: hardened and tempered **DPN: 330**
AH CHROME DIE	1.0% C 5.0% Cr 1.2% Mo 0.4% V steel: Origin unknown
AIRQUE	1.0% C 5.2% Cr 1.1% Mo 0.25% V steel: Braeburn; AISI type A2
AIRQUE 4	1.0% C 2.5% Mn 1.1% Cr 1.1% Mo steel: Braeburn
AIRQUE SPECIAL	0.8% C 5.25% Cr 1.1% Mo 0.25% V steel: Braeburn; AISI type A2
AIRQUE V	1.25% C 5.25% Cr 1.1% Mo 1.0% V steel: Braeburn; AISI type A3

Symbol	Nominal analysis, supplier, condition and remarks.
AIRTRUE	1.0% C 5.0% Cr 1.0% Mo 0.25% V steel: Simonds; AISI type A2
AIRVAN	1.0% C 5.25% Cr 1.15% Mo 0.25% V steel: Firth Sterling
AISI 611	0.84% C 4.2% Cr 5% Mo 1.9% V 6.3% W steel: Hardened and tempered: for cutting tools **UTS: 780 Elon: 21% Proof: 370**
AISI 612	0.87% C 4% Cr 8.2% Mo 1.9% V steel: Cutting tools **DPN: 910**
AISI 613	0.8% C 4% Cr 4.25% Mo 1% V steel: Hardened and tempered **DPN: 870 UTS: 2760 Elon: 2% Proof: 2420**
ALLOY B	1.25% C 0.3% Mn 1.1% Cr 0.25% V 4.5% W 0.3% Si steel: Jessop
AMS 5626A	4% Cr 1% V 18% W steel: Bar forging
AMS 6426	0.85% C 1% Cr 0.57% Mo steel
AMS 6490	0.8% C 4% Cr 4.2% Mo 1% V steel: Forgings
AMS 6490A	0.8% C 4% Cr 4.2% Mo 1% V steel: Bar and forging; vacuum melted; premium quality M50
AMS 6491	0.82% C 4% Cr 4.2% Mo 1% V steel: For bearings
AN 2	0.78% C 4.25% Cr 1% V 22% W steel: Osborn; obsolete **DPN: 872**
ARDENT	1.5% C 0.5% Cr 4% W steel: Hall and Pickles; cold drawing dies; obsolete **DPN: 750**
ARK SUPERIOR	0.8% C 4.25% Cr 1.3% V 18% W steel: Jessop **DPN: 750**
ARK SUPERIOR EXTRA	0.73% C 4.2% Cr 1.4% V 22.0% W steel: For tools; origin unknown
ARK TRIUMPH	0.67% C 0.3% Mn 3.7% Cr 0.6% V 13% W 0.3% Si steel: Jessop
ARK TRIUMPHANT	1.23% C 4.5% Cr 3.7% V 13% W steel: Jessop
ARLEY	1.4% C 3% Cr + W steel: John Vessey; extrusion dies **DPN: 870**
ARNE	0.9% C 0.5% Cr 0.1% V 0.5% W steel: Uddelholm; for type 01
ASP23	1.3% C 4.2% Cr 5% Mo 3.1% V 6.4% W steel: Speedsteel
ASTM A485/3	1.02% C 0.8% Mn 1.3% Cr 0.25% Mo 0.25% Si steel: High hardenability
ASTM A485/4	1.02% C 1.2% Mn 1.3% Cr 0.5% Mo 0.3% Si steel: High hardenability
ASTM A597	0.92% C 1.2% Mn 0.7% Cr 0.3% V (max) 0.5% W steel: For cast tools
ASTM A597 CA2	1.0% C 5.2% Mo 0.3% V steel: For cast tools
ASTM A600	High speed tool steel specification analysis given under grade AISI system
ATLAS M3/2	4% Cr 6% Mo 3% V 6% W steel: For tools; Atlas Steels Ltd; AISI code M3/2
ATLAS M4	4% Cr 4.5% Mo 4% V 5.5% W steel: For tools; Atlas Steels Ltd; AISI code M4Z
ATLAS XXX	1.3% C 0.35% Cr 3.7% W steel: Atlas; as AISI type F2
AW	0.7% C 4% Cr 1% V 14% W steel: ESC; cutting tools **DPN: 910**
AW 4 V	1.2% C 4.2% Cr 4% V 13.8% W steel: ESC; cutting, punches **DPN: 910**

Note. The following abbreviations and units are used in the tables:

DPN	Hardness, diamond pyramid number
UTS	Ultimate tensile strength, N/mm^2
Elon	Elongation, %
Proof	0.1% proof strength, N/mm^2

1 N/mm^2=0.1 hbar=0.102 kgf/mm^2=0.06475 $tonf/in.^2$=145.04 $lbf/in.^2$=1 MPa

See Appendix II for other abbreviations and conversion tables.

Symbol	Nominal analysis, supplier, condition and remarks.
AXL	0.7% C 4.5% Cr 1.2% V 18% W steel: Fagersta **DPN: 750**
B 1	1.12% C 0.27% Cr 0.2% V 1.0% W steel: Russian Standards designation
B 2/2 (3)	2.8% C 5.25% Cr 1.1% Mo 4.5% V steel: Latrobe
B 6	4% Cr 1% V 18% W steel: For tools; Universal Cyclops; AISI code T1D
B 9	4% Cr 2% V 18% W steel: For tools; Universal Cyclops; AISI code T2D
B 400	0.8% C 0.5% Cr 0.2% V steel: VEW
B 412	0.85% C 0.4% Ni 0.5% Cr 0.2% Mo steel: VEW
B 530	0.75% C 2.4% Ni 0.3% Cr steel: VEW
B 535	0.7% C 0.4% Ni 0.3% Cr steel: VEW
BADGER	0.94% C 0.5% Cr 0.5% W steel: Latrobe; punches and dies; type 01 **DPN: 740**
BESTEM	High C 3% Ni Mo steel: For dies; Somers
BFD	1.2% C 0.6% Cr 0.2% V 1.65% W steel: Simonds; AISI type 07
BLUE CHIP	0.73% C 4.0% Cr 1.15% V 18% W steel: Firth Sterling
BOLSTER STEEL	0.5% C 1.2% Mn 0.65% Cr 0.2% Mo steel: Sanderson Kayser
BP 4	1.25% C 0.15% V 3% W steel: Osborn; cold draw dies; burnishing tools; obsolete **DPN: 950**
BR 3	2.8% C 5.25% Cr 1.1% Mo 4.5% V steel: Latrobe
BR 3 DIE STEEL	2.8% C 5.25% Cr 1.1% Mo 4.5% V steel: Latrobe; abrasion resistant **DPN: 700**
BRAEBURN M7	4% Cr 8.7% Mo 2% V 1.7% W steel: For tools; Braeburn; AISI code M71
BRAEFOUR	1.27% C 4.5% Cr 4.5% Mo 4.0% V 5.5% W steel: Braeburn; AISI type M4
BRAEMOW	0.84% C 4.2% Cr 5.0% Mo 2.0% V 6.5% W steel: Braeburn; AISI type M2
BRAEMOW 2	4% Cr 5% Mo 2% V 6% W steel: For tools; Braeburn; AISI code M21
BRAEMOW SPECIAL	0.66% C 4.0% Cr 5.0% Mo 1.7% V 6.5% W steel: Braeburn
BRAEVAN	4% Cr 6% Mo 2.4% V 6% W steel: For tools; Braeburn; AISI code M31
BRAEVAN/1	1.02% C 4.0% Cr 5.7% Mo 2.5% V 6.2% W steel: Braeburn; AISI type M3
BRAEVAN/2	1.15% C 4.0% Cr 5.5% Mo 3.3% V 5.7% W steel: Braeburn; AISI type M3
BRUNSWICK V150	1.1% C 0.8% W + Cr steel: John Vessey **DPN: 770**
BRUNSWICK V888	1.0% C 5% Cr 1.0% Mo steel: John Vessey **DPN: 750**
BSS	0.8% C 4.2% Cr 9% Mo 1.3% V 2% W steel: Huntsman; thread rolling dies; cutting tools **DPN: 870**
C 253	1.0% C 1.0% Cr 0.25% Mo steel: Fagersta **DPN: 750**
C 254	1.0% C 1.2% Cr 0.35% Mo steel: Fagersta **DPN: 750**
C 550	1.0% C 5.2% Cr 1.1% Mo 0.21% V steel: Fagersta **DPN: 680**
CAPITAL 305	0.8% C 4.0% Cr 8.7% Mo 1.1% V 1.7% W steel: For tools; Balfour Darwin
CAPITAL 395	1.2% C 4.5% Cr 4.5% Mo 4.0% V 5.5% W steel: For tools; Balfour Darwin
CAPITAL 562	0.83% C 4.25% Cr 5% Mo 1.9% V 6.4% W steel: For tools; Balfour; reamers, twist drills, etc. **DPN: 850**
CARBON COLD HEADER	0.93% C 0.05% Cr 1.18% V steel: Latrobe; for cold work
CARPENTER 10 STAR	4% Cr 8% Mo 2% V steel: For tools; Carpenter Steel Co.; AISI code M10K

Symbol	Nominal analysis, supplier, condition and remarks.
CARPENTER SUPER SPEED STAR	4% Cr 6% Mo 2.4% V 6% W steel: For tools; Carpenter Steel Co.; AISI code M3K
CARRS BLUE LABEL	0.7% C steel: For tools; Carrs
CARVITE	4% Cr 4% V 18% W steel: For tools; Columbia Tool Steel Co.; AISI code T9H
CASTLE 6/6/2	0.85% C 4.0% Cr 5.2% Mo 2.0% V 6.0% W steel: For tools; origin unknown
CASTLE SUPERIOR	0.75% C 4.0% Cr 1.0% V 18.0% W steel: For tools; origin unknown
CDD	1.9% C 0.6% Cr 6% W steel: Firth Brown
CENTURY	1.0% C 4.0% Cr 8.0% Mo 2.0% V 0.8% W steel: Braeburn
CHD 4	0.6% C 0.3% Mn 3.7% Cr 0.6% Mo 0.4% V 0.3% Si steel: Huntsman; hot blanking, shears, etc. **DPN: 530**
CL 45	0.8% C 1.0% Cr 0.2% V steel: Origin unknown
CL 652	0.85% C 4.0% Cr 5.0% Mo 2.0% V 6.0% W steel: For tools; origin unknown
CLARITE	4% Cr 1% V 18% W steel: For tools; Columbia Tool Steel Co.; AISI code T1H
CLYDALL 18	0.8% C 4% Cr 1% V 18% W steel: Clyde Alloy **DPN: 800**
CLYDMO	0.83% C 4% Cr 5% Mo 1.8% V 6.2% W steel: Clyde Alloy **DPN: 800**
CMC	0.9% C 0.3% Cr 0.3% Mo 0.2% V steel: Origin unknown
COL-GRAPH	1.45% C 0.2% Cr 0.25% Mo steel: Columbia; AISI type 06
COLUMBIA SPECIAL	1.06% C 0.2% Cr 0.05% V steel: Columbia; AISI type W1
CONQUEROR 14	0.75% C 4.0% Cr 2.0% V 14.0% W steel: For tools; origin unknown
CORONA 2	0.8% C 0.25% Mn 4% Cr 5% Mo 2% V 6.5% W steel: Sandvik
CP 30	C not specified 1.5% Cr Mo steel: For armour plate; British Steel Armour Products
CP 50	C not specified 1.5% Cr Mo steel: For armour plate; British Steel Armour Products
CRD 18	0.7% C 4.2% Cr 0.75% V 18% W steel: C Denton; shear blades; obsolete **DPN: 580**
CRD 22	0.78% C 4.5% Cr 1.7% V 22% W steel: C Denton: drills, taps, etc.; obsolete
CRD 652	0.85% C 4.5% Cr 5.3% Mo 2% V 6% W steel: C Denton; punches, saws, etc.; obsolete
CRD DIAMOND YELLOW	0.9% C 1.2% Mn 0.5% Cr 0.1% V 0.5% W steel: C Denton; blanking dies, gauges, etc.; obsolete **DPN: 540**
CRM 1	1.0% C 0.6% Mn 5% Cr 1% Mo 0.25% V 0.2% Si steel: For tools; Huntsman; cold punches, shear blades, etc. **DPN: 770**
CRM 2	0.6% C 0.6% Mn 0.6% Cr 0.3% Mo 0.25% Si steel: Huntsman; rivet tools, arbors, wrenches, etc. **DPN: 650**
CROMODIE HC	1.0% C 5.25% Cr 1.1% Mo 0.25% V steel: Firth Brown
CROMODIE HCV	2.4% C 5.25% Cr 1.15% Mo 4.1% V steel: Firth Brown
CRO-MO-LOY	1.0% C 5.0% Cr 1.0% Mo 0.2% V steel: Atlas; as AISI type A2
CROWN(4)SUPERB	0.75% C 0.95% Cr 0.18% V steel: Latrobe
CRP	0.93% C 1.0% Cr 0.15% V steel: Origin unknown
CRV 14	0.68% C 3.75% Cr 0.65% V 14% W steel: Huntsman; cutting tools **DPN: 870**
CRV 1444	1.2% C 4.4% Cr 3.7% V 13.7% W steel: Huntsman; cutting tools **DPN: 850**
CV 18	0.7% C 4.2% Cr 1.3% V 18% W steel: Huntsman; cutting tools **DPN: 870**
CV 22	0.8% C 4.5% Cr 1.4% V 22% W steel: Huntsman; cutting tools **DPN: 910**
CV 1842	0.8% C 4.2% Cr 2% V 18% W steel: Huntsman; cutting tools **DPN: 870**
CY	0.78% C 4.5% Cr 0.7% Mo 1.25% V 18% W steel: For tools; Osborn; obsolete **DPN: 910**
CYC 18/4/1	0.75% C 4% Cr 1% V 18% W steel: Marsh Brothers; cutting tools
CYCLONE 4V	1.3% C 4.0% Cr 4.5% Mo 4.0% V 5.5% W steel: For tools; origin unknown
CYCLONE 56	0.8% C 4.2% Cr 15% Mo 2% V 6.5% W steel: For tools; ESC; cutting tools and drills **DPN: 870**
CYCLONE 56CW	0.78% C 4.2% Cr 5% Mo 2% V 6.5% W steel: For tools; ESC; blanking dies, thread rolling dies **DPN: 870**
CYCLONE 92	0.8% C 4.0% Cr 8.0% Mo 1.0% V 1.5% W steel: For tools; origin unknown
CYCLONE 92CW	0.8% C 4.0% Cr 8.0% Mo 1.0% V 1.5% W steel: For tools; origin unknown
D 61	0.9% C 1.2% Mn 0.5% Cr 0.1% V 0.5% W steel: Fagersta **DPN: 750**
D 416	1.4% C 1.6% Cr 0.5% V 5.5% W steel: Fagersta **DPN: 800**
D 522	0.9% C 4% Cr 4.5% Mo 1.8% V 1% W steel: Fagersta; obsolete
D 921	0.7% C 4% Cr 1% V 18% W steel: Fagersta; obsolete
D 941	0.85% C 4% Cr 5% Mo 2% V 6% W steel: Fagersta; obsolete
D 941	0.82% C 4.0% Cr 5.0% Mo 1.9% V 6.5% W steel: Fagersta **DPN: 870**
D 943	0.8% C 3.8% Cr 9.0% Mo 1.2% V 1.7% W steel: Fagersta **DPN: 750**
D 953	0.95% C 4.0% Cr 2.6% Mo 2.3% V 2.8% W steel: Fagersta **DPN: 680**
D 954	1.0% C 4.1% Cr 8.8% Mo 2.0% V 2.0% W steel: Fagersta **DPN: 750**
D 960	0.9% C 4% Cr 8% Mo 2% V steel: Fagersta; obsolete
DBL 2	4% Cr 5% Mo 2% V 6% W steel: For tools; Allegheny-Ludlum; AISI code M2G
DBL 2½	4% Cr 6% Mo 2.4% V 6% W steel: For tools; Allegheny-Ludlum; AISI code M3G
DBL 3	4% Cr 6% Mo 3% V 6% W steel: For tools; Allegheny-Ludlum; AISI code M3G
DCM	1.0% C 0.6% Mn 5% Cr 1% Mo 0.25% V steel: Jonas, shear, blades, etc. **DPN: 600**
DOUBLE DIAMOND	0.9% C 0.95% Cr 0.3% Mo steel: Crucible Steel Co.; for drills
DOUBLE FLYGO	0.73% C 4% Cr 1% V 18% W steel: Turton Bros
DOUBLE MUSHET	0.75% C 4.25% Cr 1.25% V 18.0% W steel: For tools; origin unknown
DOUBLE MUSKET	0.75% C 4.25% Cr 1.25% V 18% W steel: For tools; Osborn; obsolete
DOUBLE MUSKET ND	0.7% C 4.0% Cr 1.0% V 18.0% W steel: Origin unknown

Symbol	Nominal analysis, supplier, condition and remarks.
DOUBLE SPECIAL	1.4% C 0.8% Cr 5% W steel: Sanderson; turning tools for cast iron copper, etc. **DPN: 720**
DUREX	0.75% C 4.5% Cr 1% V 18.5% W tool steel: Darwin; cold heading dies, etc.
DUROTERM 20R	1.0% C 4.6% Cr 9.4% Mo 2.1% V 2.1% W steel: Welding electrode; Soudotherm; for hard facing; for use at high temperatures **DPN: 620**
DUX 12	1.2% C 0.75% Cr 1.4% W steel: ESC; drills, taps, dental burrs, etc. **DPN: 750**
E Fe 5A	0.8% C 4% Cr 5% Mo 1.7% V 6% W steel: Weld electrode; designation used by AWS; obsolete
E Fe 6	0.8% C 4% Cr 8% Mo 1% V 1% W steel: Weld electrode; designation used by AWS
ECHO MOLYBDENUM	0.82% C 4.1% Cr 5% Mo 1.9% V 6.4% W steel: John Vessey
ECLIPSE 103	0.7% C 4.0% Cr 1.0% V 18.0% W steel: Origin unknown
EF 10	1.0% C 0.1% Cr 0.03% Mo 0.03% V steel: Pompey
EGALIT	0.85% C 2.1% Mn 0.07% Cr 0.16% V steel: Pompey
EL	0.78% C 4.2% Cr 2% V 18% W steel: For tools; Osborn; obsolete **DPN: 850**
ELECTEM	1.5% Ni Cr Mo steel: For dies, Somers
ELECTRIDE CRUSADER XL	1.2% C 4.1% Cr 6.0% Mo 3.2% V 6.0% W steel: Latrobe; cutting tools; M3 type 2 **DPN: 850**
ELECTRITE 1	4% Cr 1% V 18% W steel: For tools; Latrobe Steel Co.; AISI code T1F
ELECTRITE 19	4% Cr 2% V 18% W steel: For tools; Latrobe Steel Co.; AISI code T2F
ELECTRITE CORSAIR	4% Cr 6% Mo 2.4% V 6% W steel: For tools; Latrobe Steel Co.; AISI code M3F
ELECTRITE CORSAIR XL	1.02% C 4.0% Cr 6.0% Mo 2.4% V 6.1% W steel: Latrobe; cutting tools; M3 Type 1 **DPN: 850**
ELECTRITE CRUSADER	4% Cr 6% Mo 3% V 6% W steel: For tools; Latrobe Steel Co.; AISI code M3F
ELECTRITE DOUBLE 6	4% Cr 5% Mo 2% V 6% W steel: For tools; Latrobe Steel Co.; AISI code M2F
ELECTRITE DOUBLE 6 M2XL	0.85% C 4.1% Cr 5.0% Mo 1.8% V 6.3% W steel: Latrobe; cutting tools **DPN: 800**
ELECTRITE MV1	0.8% C 4.1% Cr 4.5% Mo 1.8% V 0.15% W steel: Latrobe; for tools
ELECTRITE MV2	0.88% C 4.1% Cr 4.5% Mo 1.8% V 1.1% W steel: Latrobe; for tools
ELECTRITE No. 1XL	0.75% C 4.1% Cr 0.7% Mo 1.1% V 18% W steel: Latrobe; cutting tools; type T1 with sulphides **DPN: 800**
ELECTRITE No. 7	0.68% C 4.25% Cr 5.0% Mo 1.9% V 6.4% W steel: Latrobe; cutting tools; type H42
ELECTRITE No. 19	0.85% C 4.1% Cr 0.6% Mo 2.1% V 18% W steel: Latrobe; cutting tools; type T2

Note. The following abbreviations and units are used in the tables:

DPN	Hardness, diamond pyramid number
UTS	Ultimate tensile strength, N/mm^2
Elon	Elongation, %
Proof	0.1% proof strength, N/mm^2

1 N/mm^2=0.1 hbar=0.102 kgf/mm^2=0.06475 tonf/in.^{2}=145.04 ibf/in.^{2}=1 MPa

See Appendix II for other abbreviations and conversion tables.

Symbol	Nominal analysis, supplier, condition and remarks.
ELECTRITE STARK	1.28% C 4.5% Cr 4.5% Mo 4.0% V 5.5% W steel: Latrobe; cutting tools; type M4 **DPN: 850**
ELECTRITE TATMO	4% Cr 8% Mo 1% V 1.5% W steel: For tools; Latrobe Steel Co.; AISI code M1F
ELECTRITE TATMO V	1.0% C 3.7% Cr 8.7% Mo 2.0% V 1.7% W steel: Latrobe; cutting tools; type M7
ELECTRITE TATMO XL	0.8% C 4.0% Cr 8.5% Mo 1.0% V 1.5% W steel: Latrobe; cutting tools; type M1
ELECTRITE TNW	4% Cr 8% Mo 2% V steel: For tools; Latrobe Steel Co.; AISI code M10F
ELECTRITE TNWXL	0.87% C 4.0% Cr 8.0% Mo 1.9% V steel: Latrobe; cutting tools; type M10
ELECTRITE U	4% Cr 2% V 14% W steel: For tools; Latrobe Steel Co.; AISI code T7B
ELECTRITE VANADIUM	4% Cr 3% V 18% W steel: For tools; Latrobe Steel Co.; AISI code T3F
ER Fe C6	0.7% C 4% Cr 7% Mo 1% V 1.7% W steel: Weld electrode; designation used by AWS
ESA	1.45% C 0.5% Cr 0.25% V 4.0% W steel: Latrobe
ETA	0.9% C 0.15% V 0.5% W steel: Origin unknown
EV 5	0.75% C 0.5% Cr 6.0% W steel: For magnets; Russian alloy listed by PMA
EVM	4% Cr 2% V 18% W steel: For tools; Vanadium Alloys: AISI code T2A
EW 6	0.7% C 0.5% Cr 6.0% W steel: For magnets; Polish alloy listed by PMA
EXL-DIE	0.95% C 0.5% Cr 0.1% V steel: Columbia; AISI type 01
EXTRA DOUBLE MUSHET	0.76% C 4.2% Cr 0.3% Mo 0.8% V 18.2% W steel: For tools; origin unknown
EXTRA SUPER	0.79% C 4.5% Cr 1.4% V 22.0% W steel: For tools; origin unknown
EXTRA TRIPLE CONQUEROR	1.55% C 0.55% Cr 5.5% W steel: For tools; origin unknown
EXTRA TRIPLE GRIFFIN	1.7% C 0.2% Cr 6% W steel: For tools; Balfour; tube drawing, dies, etc. **DPN: 850**
EXTRA VANADIUM	0.7% C 4% Cr 1.25% V 18% W steel: For tools; Vulcan; cutting tools **DPN: 850**
E-Z-DIE V	2.23% C 5.25% Cr 1.1% Mo 4.8% V 1.1% W steel: Columbia; AISI type A7
E-Z-DIE SMOOTHCUT	1.0% C 5.25% Cr 1.1% Mo 0.2% V steel: Columbia: AISI type A2S; free cutting
F 1	1.23% C 0.45% Cr 0.25% Mo 0.2% V 1.35% W steel: Designation used by AISI
F 3	1.25% C 0.75% Cr 3.5% W steel: Designation used by AISI
F 4	0.95% C 1% Mn 0.5% Cr 0.2% V 0.2% Si steel: Jessop
FAGERSTA D941	0.82% C 0.3% Mn 4.0% Cr 5.0% Mo 1.9% V 6.5% W steel: For saw blades; Fagersta
FAGERSTA D953	0.95% C 0.3% Mn 4.0% Cr 2.6% Mo 2.3% V 2.8% W steel: Fagersta
FAVORIT	0.9% C 0.2% Cr 0.15% V steel: Schoeller-Bleckmann **DPN: 540**
FHS 18	0.75% C 4.0% Cr 1.0% V 18.0% W steel: For tools; origin unknown
FLASHKUT	0.95% C 3.7% Cr 0.25% Mo 0.18% V steel: Origin unknown
FMP	0.95% C 1.25% Mn 0.55% Cr 0.15% V 0.5% W steel: F Parkin **DPN: 800**
fmp 200	0.95% C 0.5% Cr 0.17% V 0.5% W steel: F Parkin: as tool steel 01
fmp 379	1.0% C 5.0% Cr 1.1% Mo 0.3% V steel: F Parkin; as tool steel A2
fmp 470	0.8% C 4.5% Cr 1.0% Mo 1.5% V 2.0% W steel: F Parkin
fmp 501	0.8% C 3.9% Cr 9.0% Mo 1.2% V 2.0% W steel: F Parkin; as tool steel M1

Symbol	Nominal analysis, supplier, condition and remarks.
fmp 504	1.3% C 4.4% Cr 4.6% Mo 4.0% V 5.7% W steel: F Parkin; as tool steel M4
fmp 562	0.82% C 4.1% Cr 5.0% Mo 2.0% V 6.2% W steel: F Parkin; as tool steel M2
fmp 563	1.0% C 4.0% Cr 5.0% Mo 2.3% V 6.0% W steel: F Parkin; as tool steel M3/1
fmp 599	0.7% C 4.2% Cr 0.8% V 14.0% W steel: F Parkin
fmp 842	0.84% C 4.2% Cr 1.0% Mo 2.2% V 18.5% W steel: F Parkin; as T2 tool steel
fmp 922	1.0% C 4.0% Cr 8.7% Mo 2.0% V 1.7% W steel: F Parkin
fmp 948	0.88% C 3.0% Cr 8.25% Mo 2.0% V steel: F Parkin; as tool steel M10
FOUR STAR	1.3% C 4.5% Cr 4.5% Mo 4% V 5.5% W steel: Carpenter
FRAMDIE	1.1% C 1.35% Cr 0.4% Mo steel: Columbia; AISI type L7
FS M2½	1.02% C 4.0% Cr 6.1% Mo 2.5% V 6.1% W steel: Firth Sterling
FS M10	0.87% C 4.0% Cr 8.2% Mo 2.0% V steel: Firth Sterling
FURIOUS	1.25% C 1.2% Cr 0.25% V 4.5% W steel: Hall and Pickles: broaches, taps, punches, etc.; obsolete **DPN: 750**
G 23	0.62% C 0.6% Mn 2% Ni 2% Cr 0.45% Mo 0.2% Si steel: Jessop
G 24	3.0% C 13% Ni 4.5% Cr 4.5% Cu steel: Casting; Jessop
G 25	0.7% C 0.3% Mn 0.45% Ni 0.2% Cr 0.2% Si steel: Jessop
GFS	0.9% C 0.5% Cr 0.5% W steel: Origin unknown
GOAHED	0.7% C 4.3% Cr 1.3% V 18% W steel: John Vessey
GRAPH-AIR	1.35% C 1.8% Ni 1.5% Mo steel: Free graphite; Timken; tools to withstand wear **UTS: 1950** **Proof: 1620**
GRAPH-TUNG	1.5% C 0.5% Mo 2.8% W steel: Free graphite; Timken; dies and tools to withstand wear **DPN: 800**
GS 8	0.9% C 1.2% Mn 0.6% Cr 0.5% W steel: For tools; Osborn; taps, dies, broaches, etc.; obsolete **DPN: 772**
H 4	0.9% C 1.8% Mn 0.5% Cr 0.5% W steel: Jessop **DPN: 800**
H 4 SPECIAL	0.9% C 1.4% Mn 0.3% Cr 0.3% Mo 0.2% V steel: Origin unknown
H 4 Spl	0.9% C 2% Mn 0.35% Cr 0.35% W 0.2% Si steel: Jessop
H 7	2.3% C 5.2% Cr 1.1% Mo 4.5% V steel: Jessop; refractory moulds **DPN: 730**
H 15	0.8% C 0.5% Mn 0.6% Cr 0.2% Mo 0.2% Si steel: Jessop
H 33	1.0% C 0.3% Mn 6.2% Cr 1% Mo 0.4% Si steel: Jessop **DPN: 750**
H 62	1.0% C 1.1% Mn 0.3% Ni 0.4% Cr 0.35% V 0.2% Si steel: Jessop
H M HIGH SPEED	4% Cr 8% Mo 1% V 1.5% W steel: For tools; Bethlehem Steel; AISI code M1E
H RV	1.0% C 1.5% Cr 0.25% V steel: Osborn; cutting tools **DPN: 850**
HARDTRODE 85.65	0.9% C 1.3% Mn 4.5% Cr 7.5% Mo 1.6% V 2.0% W steel: For electrodes; Esab **DPN: 700**
HECLA 15	0.9% C 0.3% Cr 0.3% Mo 0.2% V steel: Origin unknown
HECLA 114	0.7% C 1.5% Ni 0.8% Cr 0.2% Mo steel: Origin unknown
HECLA 175	1.0% C 5.0% Cr 1.2% Mo 0.4% V steel: For tools; origin unknown
HEDERVAN	1.4% C 0.15% Cr 0.1% Mo 3.5% V steel: Latrobe; punches, dies, etc. **DPN: 820**
HI WEAR 64	1.5% C 1.0% Cr 1.0% Mo 4.0% W steel: Carpenter **DPN: 870**
Hl-Mo	4% Cr 8% Mo 1% V 1.5% W steel: For tools; Firth Sterling Inc; AISI code M1B
HORSEMAN SUPER	0.75% C 4.0% Cr 1.0% V 18.0% W steel: For tools; origin unknown
HRW	1.0% C 0.6% Mn 1.5% Cr 0.5% W steel: Jonas; press tools **DPN: 720**
HS 22	0.75% C 5% Cr 2% V 22% W steel: For tools; Darwin; form cutters, etc.
HSP 41	0.7% C 3.7% Cr 0.75% V 14.5% W steel: For tools, origin unknown
HURCO EXCELLENT	0.75% C 4.0% Cr 1.0% V 18.0% W steel: For tools; origin unknown
HUSKY	0.8% C 4.5% Cr 1.5% V 22% W steel: Hall and Pickles; cutting tools; obsolete **DPN: 800**
HV Hi Mo	4% Cr 8.7% Mo 2% V 1.7% W steel: For tools; Firth Sterling Inc; AISI code M7B
HY BLUE CHIP	4% Cr 2% V 18% W steel: For tools; Firth Sterling Inc; AISI code T2B
HYDRA	0.65% C 4% Cr 0.5% V 15% W steel: Hall and Pickles, drills, taps, reamers, etc.; obsolete
HYDRA HUSKY	0.8% C 4.5% Cr 1.5% V 22% W steel: Hall and Pickles; cutting tools; obsolete **DPN: 800**
HYDRA VANTAGE	1.25% C 4.5% Cr 4% V 13.5% W steel: Hall and Pickles; taps, reamers, etc.; obsolete **DPN: 800**
HYKRO	C not specified 3% Cr Mo steel: For armour plate; British Steel Armour Products
HYPEAK 14	0.8% C 4% Cr 1% V 14% W steel: Swift Levick; cutting tools
HYPEAK 101	0.8% C 4% Cr 5% Mo 2% V 6.5% W steel: Swift Levick; cutting tools
HYPEAK 202	0.8% C 4.2% Cr 0.5% Mo 1.2% V 18% W steel: Swift Levick; cutting tools
IAS	1.2% C 0.75% Cr 1.75% W steel: Origin unknown
IDI	1.3% C 1.3% Mn 1.1% Cr 0.5% W steel: Inman
IMPERIAL	1.4% C 0.75% Cr 4.5% W steel: Edgar Allen; cutting tools
INMOLYTE M2	0.85% C 4.0% Cr 5.0% Mo 2.0% V 6.0% W steel: For tools; Inman
INVARD	0.9% C 0.5% Cr 0.2% V 0.5% W steel: Firth Sterling
INVINCIBLE 18	0.7% C 4.0% Cr 1.0% V 18.0% W steel: For tools; origin unknown
INVINCIBLE 22	0.75% C 4.0% Cr 1.25% V 22.0% W steel: For tools; origin unknown
J 1	0.67% C 0.3% Mn 3.7% Cr 0.6% V 13.2% W 0.3% Si steel: Jessop
J 13	1.23% C 4.5% Cr 3.7% V 13% W steel: Jessop
J 24	0.8% C 4% Cr 5% Mo 1.9% V 3.4% W with S steel: High speed; Jessop; cutting tools **DPN: 850**
J 30	0.75% C 0.3% Mn 4.2% Cr 1.1% V 18% W 0.3% Si steel: Jessop
J 34	0.8% C 4% Cr 5% Mo 1.9% V 6.5% W steel: Jessop **DPN: 750**
J 35	0.81% C 4% Cr 8.8% Mo 1.1% V 1.6% W steel: Jessop
J 38	0.8% C 4% Cr 0.75% Mo 2% V 18% W steel: Jessop
J 39	0.77% C 0.3% Mn 4.2% Cr 1.5% V 21% W 0.3% Si steel: Jessop
J 40	0.78% C 4.2% Cr 0.6% Mo 2.0% V 9.0% W steel: Jessop; high speed
J 41	0.85% C 4.5% Cr 0.8% Mo 3.0% V 11.5% W steel: Jessop; high speed

Symbol	Nominal analysis, supplier, condition and remarks.
JA 3	0.97% C 5% Cr 1% Mo 0.3% V steel: Osborn; dies, cold forming tools, etc.; obsolete **DPN: 770**
JIS G4805 SUJ4	1.02% C 0.5% Mn (max) 1.45% Cr 0.17% Mo steel: For bearings; Japanese Standard
JIS G4805 SUJ5	1.02% C 1.0% Mn 1.05% Cr 0.17% Mo steel: For bearings; Japanese Standard
K 4	0.9% C 1.2% Mn 0.5% Cr 0.5% W steel: Jessop **DPN: 800**
K 4 SPECIAL	0.9% C 0.5% Cr 0.5% W steel: Origin unknown
K 4 Spl	0.92% C 0.5% Cr 0.5% W steel: Jessop; carbon and alloy tools
K 6	1.25% C 0.3% Mn 0.3% Cr 1.3% W 0.2% Si steel: Jessop
K 6 Spl	1.4% C 0.3% Mn 0.3% Cr 1.5% W 0.2% Si steel: Jessop
K 8 HP	0.9% C 0.9% Mn 1.1% Cr 1.5% W 0.2% Si steel: Jessop
K 9	1.0% C 0.8% Mn 0.75% Cr 0.4% W steel: Edgar Allen; dies, gauges, etc. **DPN: 750**
K 10 Spl	1.6% C 0.3% Mn 0.6% Cr 5.5% W 0.3% Si steel: Jessop
K 15	1.25% C 0.3% Mn 1.1% Cr 0.25% V 4.5% W 0.3% Si steel: Jessop
K 16	0.9% C 1.2% Mn 0.5% Cr 0.2% V 0.5% W 0.2% Si steel: Jessop
K 17	0.7% C 0.2% Mn 0.4% Cr 6% W 0.5% Co 0.15% Si steel: For magnets; Jessop
K 20	1.5% C 1.1% Mn 0.6% Cr 0.7% W 0.2% Si steel: Jessop
K 21	1.05% C 0.5% Mn 1% Cr 0.6% W 0.2% Si steel: Jessop
K 305	0.98% C 5.1% Cr 1.0% Mo 0.2% V steel: VEW
K 310	0.88% C 2.0% Cr 0.5% Mo 0.1% V steel: VEW
K 405	1.2% C 0.2% Cr 0.1% V 1.0% W steel: VEW
K 451	0.8% C 0.7% Cr 0.4% Mo 0.1% V 2.6% W steel: VEW
K 460	0.95% C 0.5% Cr 0.1% V 0.5% W steel: VEW
K 465	1.05% C 1.0% Cr 1.1% W steel: VEW
K 505	1.45% C 1.4% Cr 0.1% V steel: VEW
K 510	1.18% C 0.7% Cr 0.1% V steel: VEW
K 614	0.68% C 0.7% Ni 0.5% Cr 0.2% Mo steel: VEW
K 630	0.85% C 0.8% Ni 0.1% V steel: VEW
K 720	0.9% C 0.4% Cr 0.1% V steel: VEW
K 942	0.9% C 4% Cr 3% Mo 2% V 6.3% W steel: Speedsteel
K 950	0.9% C 3.7% Cr 5% Mo 1.2% V 1.8% W steel: Speedsteel
KABC III	0.8% C 3.8% Cr 9% Mo 1.2% V 1.7% W steel: Speedsteel
KAOC	1.3% C 0.5% Mn 1% Cr 3% W 0.3% Si steel: For tools; Huntsman; taps, reamers, etc. **DPN: 750**
KBT 21	0.73% C 4% Cr 0.6% V 14.2% W steel: Speedsteel
KE 127	0.95% C 0.4% Mn 0.4% Cr 0.15% V steel: Kayser Ellison; ball and roller bearings
KE 226	1.0% C 0.5% Cr 2.1% W steel: Kayser Ellison; piercing punches, etc. **DPN: 950**

Note. The following abbreviations and units are used in the tables:

DPN	Hardness, diamond pyramid number
UTS	Ultimate tensile strength, N/mm^2
Elon	Elongation, %
Proof	0.1% proof strength, N/mm^2

1 N/mm^2=0.1 hbar=0.102 kgf/mm^2=0.06475 tonf/in.^{2}=145.04 lbf/in.^{2}=1 MPa
See Appendix II for other abbreviations and conversion tables.

Symbol	Nominal analysis, supplier, condition and remarks.
KE 595	1.25% C 0.8% Mn 1.2% Cr 1.2% W steel: Kayser Ellison; taps, dies, etc. **DPN: 910**
KE 672	1.1% C 1.5% Cr 0.5% W steel: Kayser Ellison; press tools **DPN: 870**
KE 708	1.4% C 0.35% Mn 0.6% Cr 1.5% W steel: Kayser Ellison; punches, taps, etc.
KE DIAMOND 10	1.5% C 0.4% Mn 0.6% Cr 5.7% W steel: Kayser Ellison; cutting tools **DPN: 960**
KEA 162	1.0% C 0.6% Mn 5.2% Cr 1% Mo 0.2% V steel: Kayser Ellison; cutting tools **DPN: 850**
KEEWATIN	0.9% C 0.5% Cr 0.2% V 0.5% W steel: Atlas; as AISI type 01
KELOCK 237	0.75% C 4.1% Cr 1.1% V 18.0% W steel: For tools; Sanderson Kayser
KELOCK 795	0.7% C 4% Cr 0.6% V 14% W steel: Kayser Ellison; cutting tools, shear blades
KELOCK 1014	0.75% C 6% Cr 0.75% Mo 1.4% V 14% W steel: Kayser Ellison; cutting tools
KELOCK A157	0.8% C 4% Cr 5% Mo 2% V 6% W steel: Kayser Ellison; cutting tools
KELOCK A182	0.8% C 3.9% Cr 8.5% Mo 1.15% W steel: For tools; Sanderson Kayser
KELOCK A229	1.05% C 3.7% Cr 9.5% Mo 1.1% V 1.5% W steel: For tools; Sanderson Kayser
KH 6VF	1.1% C 6.2% Cr 0.55% V 13.0% W steel: Russian Standards designation
KHV 5	1.35% C 0.55% Cr 0.22% V 0.45% W steel: Russian Standards designation
KHVG	1.0% C 1.0% Mn 1.05% Cr 1.4% W steel: Russian Standards designation
KHVSG	1.0% C 0.85% Cr 0.1% V 0.85% W steel: Russian Standards designation
KISKI	0.9% C 0.5% Cr 0.2% V 0.6% W steel: Braeburn; AISI type 01
KK	1.1% C 1.4% Cr 0.4% Mo steel: Atlas **DPN: 520**
KK 3	0.83% C 4.2% Cr 5% Mo 2% V 6.5% W steel: For tools; Osborn; obsolete **DPN: 850**
KM 1	0.8% C 4% Cr 8% Mo 1% V 1.5% W steel: Speedsteel
KM 2	0.85% C 4% Cr 5% Mo 2% V 6% W steel: Speedsteel
KM 3.1	1% C 4% Cr 5% Mo 2.4% V 6% W steel: Speedsteel
KM 3.2	1.2% C 4% Cr 5.6% Mo 3.2% V 6% W steel: Speedsteel
KM 4	1.3% C 4.3% Cr 4.5% Mo 4% V 6% W steel: Speedsteel
KM 7	1% C 4% Cr 8.7% Mo 2% V 1.7% W steel: Speedsteel
KM 10	0.9% C 4% Cr 8% Mo 2% V steel: Speedsteel
KM 50	0.8% C 4.1% Cr 4.2% Mo 1% V steel: Speedsteel
KM 52	0.9% C 4% Cr 4.5% Mo 1.8% V 1% W steel: Speedsteel
KONCOR	1.1% C 5.2% Cr 1.1% Mo 4% V 1% Si steel: Latrobe; precision casting of forging dies **DPN: 610**
KSA	0.9% C 1.8% Mn 0.5% Cr 0.5% W steel: Jessop **DPN: 800**
KSA	0.9% C 1.4% Mn 0.3% Cr 0.3% Mo 0.2% V steel: Origin unknown
KT 1	0.7% C 4% Cr 1% V 18% W steel: Speedsteel
L 2	0.8% C 1.0% Cr 0.2% V steel: Carbon varies over wide limits; designation used by AISI
L 3	1.0% C 1.5% Cr 0.2% V steel: Designation used by AISI
L 3	1.0% C 1.4% Cr 0.25% V steel: For cold work; Osborn
L 4	1.0% C 0.6% Mn 1.5% Cr 0.25% V steel: Designation used by AISI; obsolete

Symbol	Nominal analysis, supplier, condition and remarks.
L 5	1.0% C 1.0% Mn 1.0% Cr 0.25% Mo steel: Designation used by AISI; obsolete
L 6	0.7% C 1.5% Ni 0.8% Cr 0.2% Mo steel: Designation used by AISI
L 7	1.0% C 1.5% Cr 0.4% Mo steel: Designation used by AISI
LESCO HS29XL	0.08% C 4.1% Cr 5.05% Mo 1.8% V 6.3% W steel: Latrobe; tools; contains sulphides
LF	1.25% C 1.15% Cr 0.25% V 4.5% W steel: For tools; Osborn; broaches, reamers, taps, obsolete **DPN: 850**
LHS 14	0.7% C 3.5% Cr 0.5% V 14% W steel: For tools; Low Moor; obsolete **DPN: 850**
LHS 18	0.75% C 4% Cr 1.2% V 18% W steel: For tools; Low Moor; obsolete **DPN: 870**
LHS 22	0.75% C 4.5% Cr 1.2% V 22% W steel: For tools; Low Moor; obsolete **DPN: 910**
LHS 652	0.8% C 4% Cr 5% Mo 1.9% V 6% W steel: For tools; Low Moor; obsolete **DPN: 870**
LMW	High C 4% Cr 8% Mo 1% V 1.5% W steel: For tools; Allegheny Ludlum; AISI code M1G
LMW-V	High C 4% Cr 8.7% Mo 2% V 1.7% W steel: For tools; Allegheny Ludlum; AISI code M7G
LOCKPORT SPECIAL	High C 4% Cr 2% V 18% W tool steel: Simonds; AISI code T2N
LT 12	1.25% C 0.15% V 3% W steel: Low Moor; for tools; obsolete **DPN: 950**
LT 22V	1.0% C 1.5% Cr 0.2% V steel: Low Moor; non-shrinking; tools; obsolete **DPN: 800**
LT 23	0.9% C 1.25% Mn 0.5% Cr 5% W steel: Low Moor; non-shrinking; tools; obsolete **DPN: 800**
LT 24	1.0% C 1% Mn 0.8% Cr 0.7% W steel: Low Moor; non-shrinking; obsolete **DPN: 850**
LT 40	1.0% C 5% Cr 1% Mo 0.3% V steel: Low Moor; obsolete **DPN: 770**
LTAH	0.7% C 2% Mn 1% Cr 1.5% Mo 0.35% Si steel: Sanderson; forming dies, shear blades, etc. **DPN: 680**
LTTS	1.15% C 0.3% Mn 0.7% Cr 0.25% V 1.3% W 0.3% Si steel: For tools; Huntsman; centres, hacksaw blades, etc. **DPN: 720**
LXX	4% Cr 1% V 18% W steel: For tools; Allegheny Ludlum; AISI code T1G
M 1	0.8% C 4% Cr 8% Mo 1% V 1.5% W steel: For tools; designation used by AISI; the supplier is coded by a suffix letter
M 1	0.8% C 4.0% Cr 8.75% Mo 1.1% V 1.8% W steel: Osborn
M 2	0.85% C 4% Cr 5% Mo 2% V 6% W steel: For tools; designation used by AISI; the supplier is coded by a suffix letter
M 2	0.85% C 4.0% Cr 5.0% Mo 1.9% V 6.3% W steel: Osborn
M 3/1	1.05% C 4% Cr 6% Mo 2.4% V 6% W steel: For tools; designation used by AISI; the supplier is coded by a suffix letter
M 3/2	1.2% C 4% Cr 6% Mo 3% V 6% W steel: For tools; designation used by AISI; the supplier is coded by a suffix letter
M 4	1.3% C 4% Cr 4.5% Mo 4% V 5.5% W steel: For tools; designation used by AISI; the supplier is coded by a suffix letter
M 7	1.0% C 4% Cr 8.7% Mo 2% V 1.7% W steel: For tools; designation used by AISI; the supplier is coded by a suffix letter; (higher C than M1)
M 8	0.8% C 5.0% Cr 5.0% Mo 1.5% V 5.0% W 1.2% Nb steel: Designation used by AISI; obsolete
M 10	0.9% C 4% Cr 8% Mo 2% V steel: For tools; designation used by AISI; the supplier is coded by a suffix letter
M 50	0.8% C 4.1% Cr 4.25% Mo 1.1% V steel: Latrobe
M 52	0.88% C 4.1% Cr 4.5% Mo 1.8% V 1.1% W steel: Designation used by AISI; for tools
MANGDIE	0.95% C 1.25% Mn 0.5% Cr 0.2% V 0.5% W steel: Clyde Alloy **DPN: 750**
MANOIR ABRADUR 220	0.7% C 1.0% Mn 0.04% S and P steel: Pompey **UTS: 770 Elon: 7% Proof: 500**
MARSH CYC 1841	0.75% C 4% Cr 1% V 18% W steel: Marsh Brothers; cutting tools
MAXIMUM 18	0.75% C 4.2% Cr 1.1% V 18% W steel: Firth Brown
MC 2	0.8% C 3.75% Cr 8.7% Mo 1.1% V 1.7% W steel: For tools; Osborn; obsolete **DPN: 850**
MCMO	0.7% C 2% Mn 1% Cr 1.3% Mo steel: For tools; Huntsman; blanking dies, punches, etc. **DPN: 720**
MCT	0.95% C 1.1% Mn 0.5% Cr 0.5% W steel: Firth Brown
MEL-TROL SPEED STAR	0.8% C 4.2% Cr 5.0% Mo 1.9% V 6.2% W steel: Carpenter; for AISI type M2
MEL-TROL VEGA-FM	0.7% C 2.2% Mn 1.0% Cr 1.35% Mo steel: For tools; Carpenter; for AISI A6
MEL-TROL 484FM	1.0% C 5.2% Cr 1.1% Mo 0.2% V steel: Carpenter for AISI type A2; contains sulphides
MEL-TROL HI SHOCK 60	0.68% C 0.5% Ni 1.0% Cr 1.0% Mo 0.15% V 2.5% Cu steel: For tools; Carpenter
MEL-TROL HI SHOCK 60	0.7% C 0.5% Ni 1.0% Cr 1.0% Mo 0.15% V 2.5% Cu steel: Carpenter
MEL-TROL HI WEAR 64	1.5% C 0.9% Cr 1.0% Mo 4.0% W steel: Carpenter **DPN: 750**
MEL-TROL STAR ZENITH	0.7% C 4.0% Cr 1.1% V 18.2% W steel: Carpenter for AISI type T1
MEL-TROL VEGA	0.7% C 2.0% Mn 1.0% Cr 1.35% Mo 0.3% Si steel: For tools; Carpenter for AISI A6
MERMAID	0.75% C 4.0% Cr 1.4% V 18% W steel: Spear & Jackson
MET-HARD 750TS	0.7% C 4% Cr 8% Mo 1% V 2% W steel: Metrode; obsolete **DPN: 750**
MIC 3	0.95% C 1.5% Mn 0.5% Cr 0.5% W steel: ESC; taps, punches, reamers, etc. **DPN: 750**
MIC 4	0.95% C 1.25% Mn 0.5% Cr 0.2% V 0.5% W steel: ESC; taps, dies, reamers, etc. **DPN: 800**
MIC 9	1.0% C 0.8% Mn 0.7% Cr 0.4% W steel: ESC; small punches, form tools, etc. **DPN: 750**
ML	4% Cr 2% V 18% W steel: For tools; Allegheny Ludlum; AISI code T2G
MO CUT	4% Cr 8% Mo 1% V 1.5% W steel: For tools; Braeburn; AISI code M11
MOCARB	1.0% C 4.0% Cr 5.0% Mo 1.9% V 6.5% W steel: Braeburn
MOGUL	4% Cr 8% Mo 1% V 1.5% W steel: For tools; Jessop; AISI code M1W
MOHICAN 8	4% Cr 8% Mo 1% V 1.5% W steel: For tools; Atlas steel Ltd; AISI code M1Z
MOLITE	0.85% C 4% Cr 5% Mo 2% V 6% W steel: For tools; Columbia Tool Steel Co; AISI code M2H

Symbol	Nominal analysis, supplier, condition and remarks.
MOLITE 1	0.82% C 4.0% Cr 8.5% Mo 1.1% V 1.6% W steel: Columbia; AISI type M1
MOLITE 3	1.0% C 4% Cr 6% Mo 2.4% V 6% W steel: For tools; Columbia Tool Steel Co; AISI code M3H
MOLITE 3 SMOOTHCUT	1.03% C 4.0% Cr 6.2% Mo 2.5% V 6.2% W steel: Columbia; AISI type M3S/1; free machining
MOLITE 4	1.28% C 4.5% Cr 4.5% Mo 4.0% V 5.5% W steel: Columbia
MOLVA	0.82% C 4.0% Cr 8.0% Mo 1.9% V steel: Simonds; AISI type M10
MOLVA T	0.82% C 4.0% Cr 5.0% Mo 1.96% V 6.0% W steel: Simonds; AISI type M2
MOLVA TC	1.2% C 4.0% Cr 6.25% Mo 2.8% V 6.25% W steel: Simonds; AISI type M3
MOLYCUT 562	0.8% C 4.1% Cr 5% Mo 1.9% V 6.4% W steel: Firth Brown
MONARCH	0.75% C 7.5% Cr 1.3% V 22.0% W steel: For tools; origin unknown
MONARCH 652	0.85% C 4.0% Cr 5.0% Mo 2.0% V 6.0% W steel: For tools; origin unknown
MONARCH OHB	0.9% C 0.5% Cr 0.5% W steel: For tools; origin unknown
MONARCH PCS	0.8% C 0.5% Ni 1.25% Cr 1.9% W steel: For tools; origin unknown
MONARCH TAN	0.7% C 1.5% Ni 0.8% Cr 0.2% Mo steel: For tools; origin unknown
MOTEMP	0.88% C 4.0% Cr 8.0% Mo 2.0% V steel: Braeburn; AISI type M10
MO-TEMP M10	4% Cr 8% Mo 2% V steel: For tools; Braeburn; AISI code M10
MOTOR MAXIMUM	0.75% C 4.25% Cr 1.2% V 18% W steel: Carrs; boring tools; chisels **DPN: 910**
MOTOR MAGNUS	0.8% C 4.25% Cr 5% Mo 2% V 6.5% W steel: Carrs; form tools, gear cutters, etc. **DPN: 910**
MOTOR SPECIAL	0.7% C 4% Cr 0.25% V 14% W steel: Carrs; twist drills, reamers, etc. **DPN: 850**
MOTUF	1.0% C 3.7% Cr 8.7% Mo 2.1% V 1.7% W steel: Braeburn; AISI type M7
MOTUNG	4% Cr 8% Mo 1% V 1.5% W steel: For tools; Universal Cyclops; AISI code M1D
MOTUNG 652	4% Cr 5% Mo 2% V 6% W steel: For tools; Universal Cyclops; AISI code M2D
MOTUNG CV	4% Cr 8.7% Mo 2% V 1.7% W steel: For tools; Universal Cyclops; AISI code M7D
MOVAN	4% Cr 8% Mo 2% V steel: For tools; Universal Cyclops; AISI code M10D
MOW 562	0.83% C 4% Cr 5% Mo 2% V 6% W steel: Huntsman; cutting tools
MUSHET HIGH VANADIUM	1.25% C 4.6% Cr 3.7% V 13.0% W steel: For tools; origin unknown
MUSHET M	0.83% C 4.2% Cr 5.0% Mo 1.9% V 6.5% W steel: For tools; origin unknown
MUSHET MOLYB.	0.8% C 3.7% Cr 8.7% Mo 1.1% V 1.7% W steel: For tools; origin unknown
MUSKET MKK	0.85% C 4.0% Cr 5.0% Mo 2.0% V 6.0% W steel: For tools; origin unknown

Note. The following abbreviations and units are used in the tables:

DPN	Hardness, diamond pyramid number
UTS	Ultimate tensile strength, N/mm^2
Elon	Elongation, %
Proof	0.1% proof strength, N/mm^2

1 N/mm^2=0.1 hbar=0.102 kgf/mm^2=0.06475 tonf/in.^{2}=145.04 lbf/in.^{2}=1 MPa

See Appendix II for other abbreviations and conversion tables.

Symbol	Nominal analysis, supplier, condition and remarks.
MUSTANG	4% Cr 5% Mo 2% V 6% W steel: For tools; AISI code M2W
MUSTANG M3/1	4% Cr 6% Mo 2.4% V 6% W steel: For tools; Jessop; AISI code M3W
MUSTANG M3/2	4% Cr 6% Mo 3% V 6% W steel: For tools; Jessop; AISI code M3W
MV-1-VAC-ARC	0.8% C 4.1% Cr 4.25% Mo 1.1% V steel: Latrobe; ball races **DPN: 750**
NATRONA	0.88% C 4.0% Cr 4.0% Mo 1.75% V 0.5% W steel: Braeburn
NCM IHC	0.6% C 0.45% Mn 0.5% Ni 0.9% Cr 0.3% Mo 0.25% Si steel: Huntsman; thread rolling dies, forming tools, etc. **DPN: 720**
ND	0.75% C 4.25% Cr 1.25% V 18% W steel: For tools; Osborn 'Double Mushet'; obsolete **DPN: 850**
NDS	0.75% C 1.7% Ni 1.0% Cr steel: Latrobe
NEATRO	4% Cr 4.5% Mo 4% V 5.5% W steel: For tools; Vanadium Alloys; AISI code M14A
NEW CAPITAL	0.7% C 3.7% Cr 0.75% V 14.5% W steel: For tools; Balfour; punches, dies, etc. **DPN: 850**
NEWHALL	0.9% C 1.2% Mn 0.5% Cr 0.15% V 0.7% W steel: Sanderson; centres, collets, etc. **DPN: 750**
NF A35 565/100CD7	1.02% C 0.3% Mn 1.8% Cr 0.2% Mo steel: For bearings; French Standard
No. 4 HARDENITE VAN	1.4% C 0.4% Cr 0.4% Mo 3.6% V steel: For tools; origin unknown
No. 7 HARDENITE	0.9% C 0.5% Cr 0.5% W steel: For tools; origin unknown
No. 8 HARDENITE	0.85% C 1.35% Cr 0.2% V 1.0% W steel: For tools; origin unknown
No. 10 HARDENITE	1.2% C 3.0% Cr 1.0% W steel: For tools; origin unknown
No 484	1% C 5.2% Cr 1% Mo 0.2% V steel: Carpenter
No 610	1.5% C 12% Cr 0.8% Mo 0.9% V steel: Carpenter
NORESCO 88 HP	0.74% C 4.25% Cr 1% V 18% W steel: Schoeller-Bleckmann **DPN: 870**
NORESCO 99 HP	0.74% C 4.25% Cr 0.9% Mo 1.15% V 18.5% W steel: Schoeller-Bleckmann **DPN: 910**
NORESCO DHS	0.84% C 4.25% Cr 5% Mo 1.9% V 6.25% W steel: Schoeller-Bleckmann **DPN: 870**
NORESCO FAVORIT	0.9% C 0.2% Cr 0.15% V steel: Schoeller-Bleckmann **DPN: 540**
NORVAR	0.9% C 1.4% Mn 0.3% Cr 0.3% Mo 0.2% V steel: Origin unknown
NOVO	0.7% C 4% Cr 0.5% V 14% W steel: Jonas cutting tools **DPN: 850**
NOVO 6/6/2	0.85% C 4.0% Cr 5.0% Mo 2.0% V 6.0% W steel: For tools; origin unknown
NOVO 9/2	0.75% C 3.75% Cr 9.0% Mo 1.0% V 1.5% W tool steel: Jonas; type M1
NOVO C	0.75% C 6.2% Cr 0.7% Mo 1.25% V 13.5% W steel: Jonas; cutting or parting off tools **DPN: 870**
NOVO SUPERIOR	0.75% C 4% Cr 1% V 18% W steel: Jonas; cutting tools **DPN: 850**
NOVO SUPERIOR 6/6/2	0.85% C 4% Cr 5.25% Mo 2% V 6% W steel: Jonas; cutting tools **DPN: 910**
NOVO SUPERIOR SS	0.75% C 4.5% Cr 1.25% V 21% W steel: Jonas; cutting form tools **DPN: 910**

Symbol	Nominal analysis, supplier, condition and remarks.
NOVO TCV	1.5% C 5.0% Cr 5.0% V 11% W tool steel: Jonas
NRO 146	0.7% C 2.1% Mn 1.0% Cr 1.3% Mo 0.1% S steel: Bofors; annealed; for tools **DPN: 240**
NUTHERM	0.7% C 1.0% Cr 1.3% Mo steel: Atlas; AISI type A6
OHD	1.0% C 1.6% Cr 0.5% W steel: For tools; origin unknown
OI	0.93% C 0.5% Cr 0.2% V 0.5% W steel: For cold work; Osborn
OILDIE SMOOTHCUT	1.05% C 1.6% Cr 0.5% W steel: Free cutting; Columbia; AISI type 03S
P 9E5	1.45% C 4.1% Cr 4.6% V 9.7% W steel: For tools; Russian Standards designation
P 10	0.75% C 4.0% Cr 1.2% V 18% W steel: Bofors; cutting tools; obsolete **DPN: 850**
P 14F4	1.25% C 4.3% Cr 3.7% V 1.37% W steel: For tools; Russian Standards designation
P 86 OH	1.0% C 1% Mn 0.5% Cr 0.2% V 1% W steel: T Turton
PGT	1.15% C 0.8% Mn 0.75% Cr 1.5% W steel: Jonas; taps, dies, etc. **DPN: 630**
PILGER ROLL STEEL	1.1% C Cr Mo steel: Casting; Firth Brown; hardened and tempered **DPN: 340**
PITHO	0.9% C 1.2% Mn 0.5% Cr 0.15% V 0.7% W steel: Sanderson; alternative name to Newhall **DPN: 750**
PLUTOCRAT	0.8% C 4.25% Cr 0.5% Mo 1% V 22% W steel: Carrs; reamers, heavy duty drills, etc. **DPN: 910**
PO 2	0.85% C 4.0% Cr 5.0% Mo 2.0% V 6.5% W steel: Bofors; obsolete **DPN: 850**
PO 3	0.85% C 4.0% Cr 3.0% Mo 2.0% V 6.5% W steel: Bofors; obsolete **DPN: 850**
PO 7	1.0% C 3.8% Cr 9.0% Mo 2.0% V 1.8% W steel: Bofors; high speed steel as type M7
PP	1.0% C 1% Mn 0.8% Cr 0.8% W steel: Osborn; cutting tools; obsolete **DPN: 800**
PRN 2	1.05% C 1.3% Mn 1.3% Cr steel: Balfour; blanking tools, gauges, etc. **DPN: 660**
PYRO-VAN	0.75% C 5.2% Cr 1.1% Mo 2.5% V steel: Latrobe; hot work **DPN: 470**
QV	0.7% C 4.0% Cr 1.0% V 14.5% W steel: S Osborn; high speed steel
QV/E	0.7% C 4.0% Cr 1.0% V 14.5% W steel: Obsolete; S Osborn; replaced by QV
R 9	0.9% C 4.1% Cr 2.3% V 9.2% W steel: For tools; Russian Standards designation
R 12	0.85% C 3.4% Cr 1.7% V 12.5% W steel: For tools; Russian Standards designation
R 18	0.75% C 4.1% Cr 1.2% V 17.7% W steel: For tools; Russian Standards designation
R 18F2	0.9% C 4.1% Cr 2.1% V 17.7% W steel: For tools; Russian Standards designation
R Fe 5A	0.85% C 4% Cr 5% Mo 1.7% V 6% W steel: Weld rod; designation used by AWS
R Fe 5B	0.7% C 4% Cr 7.2% Mo 1.7% V 6% W steel: Weld rod; designation used by AWS
RDS	0.7% C 1.75% Ni 1.0% Cr steel: Carpenter; for AISI type L6
RED CUT SUPERIOR	4% Cr 1% V 18% W steel: For tools; Vanadium Alloys; AISI code T1A
RED STREAK	4% Cr 1% V 18% W steel: For tools; Simonds; AISI code T1N
REX 4V	4% Cr 4% V 18% W steel: For tools; Crucible Steel Co.; AISI code T9X
REX 939	4% Cr 3% V 18% W steel: For tools; Crucible Steel Co.; AISI code T3X
REX AA	4% Cr 1% V 18% W steel: For tools: Crucibile Steel Co.; AISI code T1X
REX CHAMPION	4% Cr 2% V 14% W steel: For tools; Crucible Steel Co.; AISI code T7X
REX M2	4% Cr 5% Mo 2% V 6% W steel: For tools; Crucible Steel Co.; AISI code M2X
REX M3/1	4% Cr 6% Mo 2.4% V 6% W steel: For tools; Crucible Steel Co.; AISI code M3X
REX M3/2	4% Cr 6% Mo 3% V 6% W steel: For tools; Crucible Steel Co.; AISI code M3X
REX M7	4% Cr 8.7% Mo 2% V 1.7% W steel: For tools; Crucible Steel Co.; AISI code M7X
REX SUPER VAN	4% Cr 2% V 18% W steel: For tools; Crucible Steel Co.; AISI code T2X
REX TMO	4% Cr 8% Mo 1% V 1.5% W steel: For tools; Crucible Steel Co.; AISI code M1X
REX VM	4% Cr 8% Mo 2% V steel: For tools; Crucible Steel Co.; AISI code M10X
RIGOR	1.0% C 5.3% Cr 1.1% Mo 0.2% V steel: Uddelholm for type A2
ROLLO	1.25% C 2.9% Cr 6.0% W steel: Origin unknown
ROP 21	1.0% C 5.5% Cr 1.1% Mo 0.2% V steel: Bofors **DPN: 800**
RSD	1% C 5.2% Cr 1.1% Mo 0.2% V steel: For tools; Balfour and Darwin
RT 1733	0.9% C 0.6% Cr 0.1% V 0.6% W steel: Bofors **DPN: 800**
RV	1.0% C 1.4% Cr 0.25% V steel: Obsolete; S Osborn; replaced by L3
RW 2	1.25% C 4.5% Cr 3.75% V 13% W steel: For tools; S Osborn; obsolete **DPN: 870**
S 2-9.1	0.83% C 3.8% Cr 8.8% Mo 1.3% V 1.8% W steel: German Standard
S 2-9.2	1% C 3.8% Cr 8.7% Mo 2.1% V 1.8% W steel: German Standard
S 3-3.2	1.0% C 4.2% Cr 2.65% Mo 2.3% V 2.85% W steel: For tools; German Standards designation
S 6-5.2	0.88% C 4.2% Cr 5.0% Mo 1.85% V 6.3% W steel: For tools; German Standards designation
S 6-5.3	1.22% C 4.2% Cr 5.0% Mo 2.0% V 6.3% W steel: For tools; German Standards designation
S 18-0.1	0.76% C 4.3% Cr 1% V 1.8% W steel: German Standard
S 200	0.75% C 4.3% Cr 1.1% V 18.0% W tool steel: VEW
S 400	1.0% C 3.8% Cr 8.7% Mo 2.1% V 1.8% W tool steel: VEW
S 401	0.83% C 3.8% Cr 8.8% Mo 1.3% V 1.8% W tool steel: VEW
S 600	0.9% C 4.3% Cr 5.0% Mo 1.9% V 6.4% W tool steel: VEW
S 607	1.2% C 4.1% Cr 5.0% Mo 2.9% V 6.4% W tool steel: VEW
S 610	1.0% C 4.3% Cr 2.7% Mo 2.4% V 2.9% W tool steel: VEW
SABEN 652	0.81% C 4.1% Cr 5% Mo 2% V 6.5% W steel: Sanderson; cutting tools **DPN: 770**
SABEN EXTRA	0.75% C 4% Cr 1% V 18% W steel: Sanderson; cutting tools **DPN: 770**
SABEN HC	1.25% C 4.5% Cr 0.3% Mo 3.7% V 13.5% W steel: Sanderson; cutting tools **DPN: 800**
SABEN KERAU	0.75% C 4.5% Cr 0.25% Mo 1.5% V 22% W steel: Sanderson; cutting tools **DPN: 800**

Symbol	Nominal analysis, supplier, condition and remarks.
SAE 01	0.9% C 1.2% Mn 0.5% Cr 0.2% V 0.5% W steel: For tools
SAE 02	0.9% C 1.4% Mn 0.3% Cr 0.3% Mo 0.2% V steel: For tools
SAE 16	0.7% C 0.8% Cr 0.2% Mo 0.2% V steel
SAE 17	1.0% C 1.5% Cr 0.4% Mo steel
SAE 6195	0.95% C 1% Cr 0.15% V steel: Obsolete
SAE 8660	0.6% C 0.6% Ni 0.5% Cr 0.2% Mo steel
SAE A2	1.0% C 5% Cr 1.2% Mo 0.4% V steel: For tools
SAE M1	0.8% C 4% Cr 9% Mo 1.1% V 1.7% W steel: For tools
SAE M2	0.85% C 4.2% Cr 5% Mo 2% V 6.2% W steel: For tools
SAE M3	1.1% C 4.2% Cr 6% Mo 2.7% V 6% W steel: For tools
SAE M4	1.2% C 4.5% Cr 5% Mo 4.2% V 6% W steel: For tools
SAE T1	0.7% C 4.2% Cr 1% V 18% W steel: For tools
SAE T2	0.8% C 4.1% Cr 0.8% Mo 2% V 18% W steel: For tools
SC 6-5-2	1.0% C 4.2% Cr 5.0% Mo 1.85% V 6.3% W steel: For tools; German Standards designation
SCV	1.0% C 1.0% Cr 0.2% V steel: Sanderson; cutting dies, blanking tools, etc. **DPN: 540**
SCV 3	0.95% C 1.1% Cr 0.25% V steel: Spencer; ball races, gauges, etc.
SCV 5	1.0% C 5.0% Cr 0.95% Mo 0.25% V steel: For tools; W Spencer
SE4AS	0.82% C 4.5% Cr 0.8% Mo 1.7% V 18.2% W steel: Forez
SEARCHER A1	0.63% C 0.4% Cr 1.6% W 1.12% Si steel: Hall and Pickles; wedges, shear blades, obsolete **DPN: 750**
SELECT BFM	1.0% C 5.25% Cr 1.1% Mo 0.25% V steel: Latrobe; contains sulphides; AISI type A2 die steel **DPN: 600**
SEVEN STAR	1.0% C 4.0% Cr 8.7% Mo 2% V 1.7% W steel: Carpenter
SHOCKTOUGH	High C 1.2% Cr + Ni 2% W steel: John Vessey; press tools, punches **DPN: 580**
SILVANITE	1.3% C 4.0% Cr 0.5% V 8.0% W steel: Columbia; AISI type F8

For SIS specifications see SS

Symbol	Nominal analysis, supplier, condition and remarks.
SIW	0.67% C 0.6% Cr 1.3% W 1.15% Si steel: Balfour; chisels, coal cutter blades **DPN: 580**
SIXIX	4% Cr 5% Mo 2% V 6% W steel: For tools; Atlas Steels Ltd; AISI code M2Z
SIXIX-fm	0.84% C 4.0% Cr 5.0% Mo 1.9% V 6.5% W 0.12% S steel: Atlas; as AISI type M2; free machining
SKD 12	1.0% C 0.5% Ni (max) 5.0% Cr 1.0% Mo 0.35% V steel: Japanese Standards designation
SKF 3	1.0% C 0.3% Mn 1.5% Ni 0.35% Mo steel: For bearings; SKF Ltd
SKF 24	1.0% C 1.8% Cr 0.2% Mo steel: Tube; SKF Ltd
SKF 25	1.0% C 1.8% Cr 0.35% Mo steel: Tube; SKF Ltd
SKF 26	0.97% C 1.8% Cr 0.5% Mo steel: SKF
SKF 27	0.97% C 1.95% Cr 0.5% Mo steel: SKF
SKF 28	0.97% C 1.95% Cr 0.55% Mo steel: SKF
SKF 562	0.68% C 0.5% Cr 0.18% V steel: SKF
SKF 822	1.05% C 1.3% Cr 0.25% Mo steel: SKF
SKH 2	0.77% C 4.1% Cr 1.0% V 18.0% W steel: For tools; Japanese Standards designation
SKH 9	0.85% C 4.2% Cr 5.0% Mo 1.9% V 6.1% W steel: For tools; Japanese Standards designation
SKH 52	1.05% C 4.2% Cr 5.5% Mo 2.4% V 6.2% W steel: For tools; Japanese Standards designation
SKH 53	1.2% C 4.3% Cr 5.5% Mo 3.1% V 6.2% W steel: For tools; Japanese Standards designation
SKH 54	1.32% C 4.2% Cr 5.0% Mo 4.2% V 5.9% W steel: For tools; Japanese Standards designation
SKS 1	1.35% C 0.75% Cr 0.2% V 4.5% W steel: For tools; Japanese Standards designation
SKS 2	1.05% C 0.8% Cr 0.2% V (max) 1.25% W steel: For tools; Japanese Standards designation
SKS 2	1.05% C 0.8% Cr 0.2% V (max) 1.25% W steel: Japanese Standards designation
SKS 3	0.95% C 0.75% Cr 0.75% W steel: Japanese Standards designation
SKS 5	0.8% C 1.0% Ni 0.32% Cr steel: Japanese Standards designation
SKS 7	1.15% C 0.35% Cr 0.2% V (max) 2.25% W steel: Japanese Standards designation
SKS 11	1.25% C 0.35% Cr 3.5% W steel: For tools; Japanese Standards designation
SKS 21	1.05% C 0.35% Cr 0.8% W steel: For tools; Japanese Standards designation
SKS 31	1.0% C 1.0% Mn 1.0% Cr 1.25% W steel: Japanese Standards designation
SKS 42	0.8% C 0.38% Cr 0.22% V 2.0% W steel: Japanese Standards designation
SKS 51	0.8% C 1.6% Ni 0.35% Cr steel: For tools; Japanese Standards designation
SL 173	1.0% C 1.2% Mn 0.5% Cr 0.5% W steel: Swift Levick; punches
SL 874	1.4% C 0.5% Cr 4.0% W steel: Swift Levick; shears, punches
SL 954	1.2% C 1.5% Cr 1.25% W steel: Swift Levick; press dies, punches
SNSC	0.9% C 1% Mn 0.5% Cr 0.2% V 0.5% W 0.28% Si steel: For tools; Huntsman; taps, shear blades, etc. **DPN: 720**
SOMDIE	1% Cr 1% Ni 0.5% Mo steel: For dies; Somers
SOUDOKAY D20.0	1.2% C 4.5% Cr 8% Mo 2.7% Nb steel: Welding electrode; Soudometal **DPN: 620**
SPARTAN 7	4% Cr 1% V 18% W steel: For tools; Atlas Steels Ltd; AISI code T1Z
SPEAR 35	0.8% C 2.5% Ni 0.2% Cr steel: Origin unknown
SPEAR 562	0.85% C 4.0% Cr 5.0% Mo 2.0% V 6.0% W steel: For tools; origin unknown
SPEAR MERMAID	0.7% C 4.0% Cr 1.0% V 18.0% W steel: Origin unknown
SPEAR PS	1.05% C 1.5% Cr 0.25% Mo steel: Spear & Jackson
SPECIAL ECHO	0.73% C 4.3% Cr 1.3% V 18.2% W steel: John Vessey
SPECIAL HS	4% Cr 1% V 18% W steel: For tools; Bethlehem Steel Co.; AISI code T1E
SPEED STAR	Carpenter for AISI type M2
SPEED STAR	4% Cr 5% Mo 2% V 6% W steel: For tools; Carpenter Steel Co.; AISI code M2K
SPEEDICUT MAX 22	0.73% C 4.2% Cr 1.4% V 22.0% W steel: For tools; origin unknown
SPEEDICUT MAXIMUM 18	0.75% C 4.2% Cr 1.1% V 18% W steel: Firth Brown
SPM	0.9% C 1.4% Mn 0.5% Cr 0.5% W steel: Jonas; punches; blanking and forming tools **DPN: 610**
SS 2140	0.9% C 1.2% Mn 0.5% Cr 0.1% V 0.5% W steel: Swedish Standard: annealed **DPN: 220**

Note. The following abbreviations and units are used in the tables:

DPN	Hardness, diamond pyramid number
UTS	Ultimate tensile strength, N/mm^2
Elon	Elongation, %
Proof	0.1% proof strength, N/mm^2

1 N/mm^2=0.1 hbar=0.102 kgf/mm^2=0.06475 tonf/in.^{2}=145.04 lbf/in.^{2}=1 MPa

See Appendix II for other abbreviations and conversion tables.

Symbol	Nominal analysis, supplier, condition and remarks.
SS 2260	1.0% C 5.2% Cr 1.1% Mo 0.2% V steel: Swedish Standard; annealed **DPN: 240**
SS 2722	0.82% C 4% Cr 5% Mo 2% V 6.5% W steel: Swedish Standard; annealed **DPN: 260**
SS 2724	0.88% C 4% Cr 3.2% Mo 2.1% V 6.5% W steel: Swedish Standard; annealed **DPN: 260**
STAG MO	0.85% C 4.75% Cr 5% Mo 2% V 6% W steel: Edgar Allen; cutting tools
STAG MO 562	0.85% C 4.75% Cr 5% Mo 2% V 6% W steel: Edgar Allen; also known as Stag Mo
STAG SPECIAL	0.7% C 5% Cr 1.5% V 19% W steel: Edgar Allen
STAR BLUE CHIP	4% Cr 2% V 14% W steel: For tools; Firth Sterling; AISI code T7X
STAR COLUMBRIUM	4% Cr 5% Mo 1.5% V 5% W 1.2% Nb steel: For tools; Carpenter Steel Co.; AISI code M8K
STAR MAX	4% Cr 8% Mo 1.0% V 1.5% W steel: For tools; Carpenter Steel Co.; AISI code M1K
STAR MO 68	1.02% C 4.0% Cr 5.0% Mo 2.0% V 6.4% W steel: Firth Sterling
STAR MO M2	4% Cr 5% Mo 2% V 6% W steel: For tools; Firth Sterling; AISI code M2B
STAR ZENITH	4% Cr 1.0% V 18% W steel: For tools; Carpenter Steel Co.; AISI code T1K
STM	4% Cr 8% Mo 1.0% V 1.5% W steel: For tools; Simonds; AISI code M1N
STUBS 69 C	Mn Cr Mo V W Si steel: Electrode; Stubs; for tool steel welding **DPN: 600**
STUBS 75	C Cr W steel: Electrode; Stubs; for severe abrasion resistance **DPN: 700**
STUBS 713	C Cr Mo V W Nb steel: Electrode; Stubs; for hard facing; high temperature application **DPN: 650**
STYR	0.78% C 4.5% Cr 0.5% Mo 1.4% V 21.5% W steel: For tools; ESC; symbol for SUPER TYR **DPN: 910**
SUPER AUSTENITE	0.7% C 4.5% Cr 0.5% Mo 1.0% V 18% W steel: Inman; cutting tools
SUPER AUSTENITE T1	0.7% C 4.0% Cr 1.0% V 18.0% W steel: For tools; Inman
SUPER DOUBLE MUSHET	0.78% C 4.25% Cr 1.0% V 22.5% W steel: For tools; origin unknown
SUPER ECLIPSE	1.0% C 1.0% Cr 0.25% Mo steel: Origin unknown
SUPER HS	4% Cr 2% V 18% W steel: For tools; Vulcan Crucible Steel Co.; AISI code T2P
SUPER HYDRA	0.75% C 4.25% Cr 1.1% V 18% W steel: For tools; Hall and Pickles **DPN: 770**
SUPER MONARCH	0.7% C 4.0% Cr 1.0% V 18.0% W steel: For tools; origin unknown
SUPER RAPID	0.8% C 4.0% Cr 1.2% V 19.0% W steel: For tools; origin unknown
SUPER RAPID	0.7% C 4.0% Cr 1.0% V 18.0% W steel: Origin unknown
SUPER Si Mn	0.57% C 0.8% Mn 0.3% Cr 0.25% Mo 1.8% Si steel: For springs; ESC
SUPER SPECIAL ECHO	0.7% C 4.4% Cr 1.4% V 21% W steel: John Vessey
SUPER TYR	0.78% C 4.5% Cr 0.5% Mo 1.4% V 21.5% W steel: ESC; cutting tools **DPN: 910**
SUPERIOR 5	1.35% C 6.2% Cr 1.1% Mo 6.2% V steel: Braeburn
SUPERIOR SPUR	0.72% C 4% Cr 0.3% Mo 1.0% V 18% W steel: T Turton
SUPREMUS	4% Cr 1.0% V 18% W steel: For tools; Jessop; AISI code T1W

Symbol	Nominal analysis, supplier, condition and remarks.
SUPREMUS EXTRA	4% Cr 2% V 18% W steel: For tools; Jessop; AISI code T2W
T 1	0.7% C 4% Cr 1.0% V 18% W steel: For tools; designation used by AISI; the supplier is coded by a suffix letter
T 1	0.75% C 4.0% Cr 1.1% V 18.0% W steel: Osborn
T 2	0.8% C 4% Cr 2% V 18% W steel: For tools; designation used by AISI; the supplier is coded by a suffix letter
T 3	1.05% C 4.0% Cr 3.0% V 18.0% W steel: Designation used by AISI; obsolete
T 7	0.75% C 4% Cr 2% V 14% W steel: For tools; designation used by AISI; the supplier is coded by a suffix letter
T 9	1.2% C 4% Cr 4% V 18% W steel: For tools; designation used by AISI; the supplier is coded by a suffix letter
T 10	0.9% C 0.5% V 2% W steel: Turton Bros
TBS 600	0.98% C 1.5% Cr 0.3% Mo steel: Timken; for bearings at high temperature **DPN: 800**
TBS 1000	0.8% C 1.1% Cr 5.0% Mo 1.1% V steel: Timken; bearings for high temperature use **DPN: 700**
TCA 2	0.65% C 1.5% Cr 0.25% V 2.5% W steel: T Turton
TDBC	1.4% C 0.3% Mn 0.7% Cr 5% W steel: Jonas; dies **DPN: 720**
TEENAX	0.9% C 0.5% Cr 0.15% V 0.5% W steel: Simonds; AISI type 01
TEN STAR	0.85% C 4% Cr 8% Mo 2% V steel: Carpenter
TM 6	4% Cr 5% Mo 2% V 6% W steel: For tools; Crucible Steel Co.; AISI code M2P
TMS	0.9% C 1.4% Mn 0.3% Mo 0.2% V steel: Origin unknown
TOH	0.9% C 1.3% Mn 0.4% Cr 0.5% W steel: Balfour; blanking dies, etc. **DPN: 650**
TPM	0.9% C 1.4% Mn 0.5% Cr 0.5% W steel: Jonas; fine grained grade of SPM **DPN: 660**
TRIPLE CONQUEROR	1.25% C 0.75% Cr 3.5% W steel: For tools; origin unknown
TRIPLE CRESCENT	1.25% C 1.15% Cr 0.3% V 4.25% W steel: Spencer; cutting tools, broaches, etc. **DPN: 780**
TRIPLE GRIFFIN	1.3% C 0.3% Cr 2.7% W steel: For tools; Balfour; tube drawing dies etc. **DPN: 850**
TRIUMPH SUPER	0.7% C 4.0% Cr 1.0% V 18.0% W steel: Origin unknown
TRIUMPH SUPERB	0.8% C 4.25% Cr 1.3% V 18% W steel: Jessop **DPN: 750**
TROJAN	4% Cr 2% V 18% W steel: For tools; Atlas Steels Ltd; AISI code T2Z
TRS	1.2% C 1.15% Cr 0.25% V 4.5% W steel: Jonas; taps; reamers, etc. **DPN: 650**
TUNGSTEN DIAMOND	1.25% C 0.75% Cr 3.5% W steel: For tools; origin unknown
TUNGSTEN SPUR	0.67% C 3.8% Cr 0.7% V 14% W steel: T Turton
TWIN VAN	4% Cr 2% V 18% W steel: For tools; Braeburn; AISI code T21
UHB 25	0.7% C 4% Cr 1% V 18% W steel: Speedsteel
UHB 29	0.85% C 4% Cr 5% Mo 2% V 6% W steel: Uddelholm; obsolete
UHB 431	0.8% C 4% Cr 8% Mo 1% V 1.5% W steel: Uddelholm; obsolete
UHB 432	0.9% C 4% Cr 8% Mo 2% V steel: Uddelholm; obsolete
UHB 433	1% C 4.2% Cr 8.7% Mo 2% V 1.7% W steel: Uddelholm; obsolete

Symbol	Nominal analysis, supplier, condition and remarks.
UHB 480	0.8% C 4% Cr 4.2% Mo 1% V steel: Uddelholm; obsolete
UHB 485	0.9% C 4% Cr 4.5% Mo 1.8% V 1% W steel: Uddelholm; obsolete
ULTRA CAPITAL	0.76% C 4.25% Cr 1.2% V 18.5% W steel: For tools; Balfour; reamers, twist drills, etc. **DPN: 870**
ULTRA CAPITAL 22	0.76% C 4.2% Cr 1.25% V 22% W steel: For tools; Balfour; reamers, twist drills, etc. **DPN: 850**
UNICUT	4% Cr 6% Mo 2.4% V 6% W steel: For tools; Universal Cyclops; AISI code M3D
UNI-DIE SMOOTHCUT	0.71% C 1.0% Cr 1.3% Mo steel: Columbia; AISI type A6S; free cutting
V 175	0.8% C 4.2% Cr 0.5% Mo 1.7% V 18% W steel: Spencer; cutting tools **DPN: 780**
V1M-VAR M 50	0.8% C 0.1% Ni 4% Cr 4.5% Mo 1% V steel: Carpenter; for bearings **DPN: 600**
VALIANT	1.5% C 0.5% Cr 4% W steel: Hall and Pickles; obsolete **DPN: 750**
VAN CHIP	4% Cr 6% Mo 3% V 6% W steel: For tools; Firth Sterling; AISI code M3B
VAN CUT 1	4% Cr 6% Mo 2.4% V 6% W steel: For tools; Vanadium Alloys, AISI code M3A
VAN CUT 2	4% Cr 6% Mo 3% V 6% W steel: For tools; Vanadium Alloys; AISI code M3A
VANGUARD	1.3% C 0.7% Mn 1.5% Cr 1% W steel: Hall and Pickles; taps, dies, gauges, etc.; obsolete **DPN: 850**
VANITE	4% Cr 2% V 18% W steel: For tools; Columbia Tool Steel Co.; AISI code T2H
VAN-LOM	4% Cr 8% Mo 2% V steel: For tools; Vanadium Alloys; AISI code M10A
VANTAGE	1.25% C 4.5% Cr 4% V 13.5% W steel: Hall & Pickles; taps, reamers, etc.; obsolete **DPN: 800**
VAP	0.75% C 4.1% Cr 1.2% V 18.2% W steel: For tools; ESC; cutting tools, taps and blades **DPN: 872**
VAP 2V	0.82% C 4.2% Cr 2.0% V 18.0% W steel: For tools; origin unknown
VAPCW	0.7% C 4% Cr 0.8% Mo 1.2% V 18% W steel: For tools; ESC; shear blades, chisels, etc. **DPN: 800**
VASCO M2	4% Cr 5% Mo 2% V 6% W steel: For tools; Vanadium Alloys; AISI code M2A
VASCO M7	4% Cr 8.7% Mo 2% V 1.7% W steel: For tools; Vanadium Alloys; AISI code M7A
VEGA	0.7% C 1% Cr 1.35% Mo steel: Carpenter
VELOS	0.85% C 4.2% Cr 5.2% Mo 1.9% V 6.4% W steel: For tools; W Spencer
VELOS UR	0.75% C 4.2% Cr 0.5% Mo 1.7% V 18% W steel: Spencer; cutting tools **DPN: 780**
VIBRESIST	0.9% C 1.0% Cr 0.25% Mo steel: Tube; Atlas; for hollow rockdrills
VICTOR	1.0% C 5.25% Cr 1.1% Mo 0.25% V steel: Hall and Pickles; press tools, gauges, etc.; obsolete **DPN: 680**
VIKING	1.1% C 1.35% Cr 0.45% Mo steel: Braeburn
VINCO	4% Cr 1% V 18% W steel: For tools; Braeburn; AISI code T11
VLM	4% Cr 8% Mo 2% V steel: For tools; Allegheny Ludlum; AISI code M10G
VMC	5% Cr 1.5% Mo 1.0% V steel: For dies; Somers
VS 1/2	1.2% C 0.5% Cr 2% W steel: Vulcan; twist drills, hacksaw blades, etc. **DPN: 720**
VS 6/6	0.8% C 4% Cr 6% Mo 2% V 6% W steel: Vulcan; cutting tools **DPN: 850**
VS 14	0.65% C 3.5% Cr 0.75% V 14% W steel: Vulcan; cutting tools **DPN: 770**
VS 16/15	0.75% C 4.5% Cr 1.5% Mo 1.0% V 16% W steel: Vulcan; cutting tools **DPN: 850**
VS 22	0.8% C 4% Cr 0.35% Mo 1.25% V 22% W steel: Vulcan; cutting tools **DPN: 870**
VSCW	1.2% C 0.75% Cr 0.15% V 2.25% W steel: Vulcan; dies, reamers **DPN: 750**
VSD	1.0% C 5% Cr 1.0% Mo 0.15% V steel: Vulcan; punching dies, shear blades **DPN: 720**
VSM	0.7% C 3.0% Cr 5.2% Mo steel: Carpenter; for high temperature use
VSMNC	0.8% C 4% Cr 2% Mo 2.4% V 7.5% W steel: Vulcan; cutting tools **DPN: 850**
VSR	1.2% C 0.75% Cr 4% W steel: Vulcan; reamers, etc. **DPN: 770**
VWMC	5% Cr 1.5% Mo 0.5% V 1.5% W steel: For dies, Somers
W TAP	1.23% C 0.45% Cr 0.25% Mo 0.2% V 1.35% W steel: Latrobe
W7	1.0% C 0.5% Cr 0.2% V steel: Designation used by AISI; obsolete
WEAREX	0.7% C 4% Cr 1% V 14.5% W steel: Darwin; shaping tools, hacksaw blades, etc.
WG	1.0% C 0.25% Cr 0.25% V steel: Designation used by AISI; obsolete
WKE	0.8% C 4.5% Cr 1.2% Mo 1.4% V 18.5% W 5.5% Co steel: Fagersta **DPN: 770**
WLM 1175	0.7% C 0.5% Cr 1.5% W 1% Si steel: W Marrison; shear blades
WLM 1261	1.25% C 2% Cr 6% W steel: W Marrison; cutting tools
WLM 1841	0.75% C 4% Cr 1% V 18% W steel: W Marrison, cutting tools
WLM 2241	0.75% C 4% Cr 1% V 22% W steel: W Marrison; cutting tools
WM	0.86% C 4.0% Cr 3.2% Mo 2.1% V 6.5% W steel: Fagersta **DPN: 850**
WOLFRAM	4% Cr 1% V 18% W steel: For tools; Crucible Steel Co.; AISI code T1P
WP 300	C not specified 1.5% Cr Mo steel: Wear resistant; British Steel Armour Products
WP 500	C not specified 1.5% Cr Mo steel: Wear resistant; British Steel Armour Products
WS	1.07% C 1.2% Mn 1% Cr 1.7% W 0.3% Si steel: Huntsman; turning tools, taps, etc. **DPN: 750**
WS 44	0.9% C 1.2% Mn 0.5% Cr 0.15% V 0.5% W steel: John Vessey; press tools, collets, etc. **DPN: 750**
WULCRO	12.3% C 1% Cr 0.15% V 4.5% W steel: Vulcan; extruding dies **DPN: 870**
X100 Cr Mo V 51	0.98% C 5% Cr 1% Mo 0.2% V steel: German Standard
XP 550	1.3% C 6.2% Cr 1.1% Mo 6.2% V steel: Braeburn; as Superior 5
Y 100C6	1.0% C 1.5% Cr 0.15% V steel: French Standards designation
Z 13.0WDCV	1.3% C 4.0% Cr 4.5% Mo 4.0% V 5.5% W steel: For tools; French Standards designation

Symbol	Nominal analysis, supplier, condition and remarks.
Z 80WCDV 12/0402/02	0.8% C 4.0% Cr 2.0% Mo 2.0% V 12.0% W steel: For tools; French Standards designation
Z 80WCV 18/04/01	0.8% C 4.0% Cr 1.0% V 18% W steel: For tools; French Standards designation
Z 85DCWV	0.85% C 4.2% Cr 5.0% Mo 1.8% V 6.2% W steel: For tools; French Standards designation
Z 85WCV 18/04/2	0.85% C 4.0% Cr 2.0% V 18.0% W steel: for tools; French Standards designation
Z 85WDCV	0.85% C 4.0% Cr 5.0% Mo 2.0% V 6.0% W steel: For tools; French Standards designation
Z 100CDV 5	1.0% C 5.0% Cr 1.0% Mo 0.3% V steel: German Standards designation
Z 100CDV 5	1.0% C 5.3% Cr 1.0% Mo 0.2% V steel: French Standard
Z 130WCV 12/04/04	0.8% C 4.0% Cr 3.5% V 12.0% W steel: For tools; French Standards designation

Note. The following abbreviations and units are used in the tables:

DPN	Hardness, diamond pyramid number
UTS	Ultimate tensile strength, N/mm^2
Elon	Elongation, %
Proof	0.1% proof strength, N/mm^2

1 N/mm^2=0.1 hbar=0.102 kgf/mm^2=0.06475 tonf/in.^{2}=145.04 lbf/in.^{2}=1 MPa

See Appendix II for other abbreviations and conversion tables.

44L Steel–cobalt alloys
High carbon with chromium, nickel, molybdenum tungsten or vanadium

Specific gravity	7.9
Density	7900 kg/m^3
Solidus, liquidus	1288–1354 °C
Thermal conductivity	61 W/m °C
Coefficient of linear expansion	17.5 × 10^{-6}/ °C
Electrical conductivity	2% IACS (copper 100%)
Specific resistance	930 microhm mm
Young's modulus of elasticity	–
Impact	Can vary from very good to very low
Fatigue strength	–
Hot strength	–

The above properties are typical of the following group, and may not apply exactly to any one specification. It is possible that with certain specifications some of the values may not be applicable.

General metallurgical characteristics

Cobalt is probably unique in that it is never present alone but always as an addition to already highly alloyed steels. Like nickel and silicon, it does not form carbides but dissolves in the ferritic matrix.

Nickel, iron and cobalt are the only common ferromagnetic elements and considerable advances have been made in recent years with complex cobalt alloys which have excellent magnetic, high temperature and permeability characteristics. These alloys are briefly discussed in Section 13A.

Additions of up to 30% cobalt to ferrous alloys also considerably improve certain magnetic properties and many of the materials listed have been developed for this reason. Among the valuable properties given by cobalt are the stability of magnets for use at high temperatures and under more arduous conditions than normal.

Apart from magnets, cobalt is added in quantities of up to 10% to alloy steels; it imbues them with unique hot strength properties and results in less galling and pick up under friction conditions. These are all comparatively new materials and much of the metallurgy remains to be explained.

Cobalt is similar in many respects to nickel, but as well as strengthening the ferrite, appears to stabilize the carbides and maintains their properties to much higher temperatures, while at the same time giving these high carbon steels a degree of toughness not obtainable by other means.

The other alloying elements found in these steels have the same effect as that described in Section 44K3.

The materials in this group cannot accept cold work without becoming brittle and hard. It is therefore essential that these materials are hot worked.

Where full annealing, hardening or normalizing is required this will be at a temperature above 900 °C and in some cases as high as 1200 °C. These operations should ideally be carried out in controlled atmospheres as cobalt is readily oxidized and lost from the surface.

Many of the materials in this group are thermally treated for magnetic purposes and it is essential that expert advice is obtained.

Welding either requires specialist equipment such as Flash Butt Weld or very specialist advice. It is very seldom that these materials will require welding but some of them can be applied as hard facings.

Specialist advice must always be obtained.

These materials are readily machined in the soft condition but after hardening will require grinding. This must be carefully carried out to prevent grinding abuse resulting in surface cracking.

They are used for cutting tools where arduous conditions exist such as interrupted cuts. They have the ability to accept impact loads much better than the steels without cobalt.

Many of the materials in this group have been developed for their magnetic properties.

Symbol	Nominal analysis, supplier, condition and remarks.
1.2886	0.16% C 10% Cr 5.1% Mo 0.5% V 10% Co steel: German Standard
1.3202	1.42% C 4.2% Cr 0.85% Mo 3.7% V 12.0% W 5.0% Co steel: For tools; German Standards designation
1.3207	1.27% C 4.1% Cr 4.0% Mo 3.25% V 10.2% W 10.5% Co steel: For tools; German Standards designation
1.3215	1.5% C 4.8% Cr 5% V 12.5% W 5% Co steel: German Standard
1.3243	0.92% C 4.1% Cr 5.0% Mo 1.85% V 6.3% W 5.0% Co steel: For tools; German Standards designation
1.3246	1.1% C 4.1% Cr 3.8% Mo 1.85% V 6.8% W 5.0% Co steel: For tools; German Standards designation
1.3247	1.05% C 3.7% Cr 9.5% Mo 1.1% V 1.5% W 8.0% Co steel: German Standard
1.3255	0.75% C 4.2% Cr 0.6% Mo 1.1% V 18.5% W 5.5% Co steel: For tools; German Standard
1.3265	0.75% C 5.0% Cr 0.6% Mo 1.7% V 18.5% W 9.5% Co steel: For tools; German Standard
2 Co 4 Cr	1.0% C 4.0% Cr 0.5% W 2.0% Co steel: For magnets; PMA
3% Co	1.0% C 0.5% Mn 9% Cr 1.2% Mo 3% Co 0.1% Si steel: For magnets; Jessop
03P	1.5% C 4.6% Cr 5% V 12.5% W 5% Co steel: Carrs **DPN: 650**
06/05/05/04/01	0.85% C 4.0% Cr 5.0% Mo 2.0% V 6.0% W 5.0% Co steel: For tools; French Standards designation
6% Co	1.0% C 0.5% Mn 9% Cr 1.2% Mo 6% Co 0.1% Si steel: For magnets; Jessop
07/05/05/05/04	1.5% C 4.0% Cr 5.0% Mo 5.0% V 6.5% W 5.0% Co steel: For tools; French Standards designation
8-N-2 COBALT	4% Cr 8% Mo 1.2% V 2% W 5% Co steel: For tools; Vanadium Alloys; AISI code M30A
08S	1.4% C 14% Cr 0.8% Mo 3% Co steel: Carrs; punches, dies, etc. **DPN: 710**
09/08/04/02/01	1.1% C 4.0% Cr 9.0% Mo 1.0% V 1.5% W 8.0% Co steel: For tools; French Standards designation
9% Co	1.0% C 0.5% Mn 9% Cr 1.2% Mo 9% Co 0.1% Si steel: For magnets; Jessop
10/07/05/05/04	1.75% C 4.0% Cr 5.0% Mo 5.0% V 6.5% W 10.0% Co steel: For tools; French Standards designation
15% Co	1.0% C 0.5% Mn 9% Cr 1.2% Mo 15% Co 0.1% Si steel: For magnets; Jessop
17 Co	0.7% C 2.5% Cr 8.0% W 17% Co steel: For magnets; American alloy listed by PMA
30 DCKY/28	0.32% C 3.0% Cr 2.8% Mo 0.35% V 3.0% Co steel: French Standard
35% Co	0.9% C 0.5% Mn 6% Cr 4% W 35% Co 0.1% Si steel: For magnets; Jessop
81/18.5% Co	0.75% C 3.7% Cr 5.0% W 18.5% Co steel: For magnets; Simonds
83/3% Co	1.0% C 4.0% Cr 3.2% Co steel: For magnets; Simonds

Note. The following abbreviations and units are used in the tables:

DPN	Hardness, diamond pyramid number
UTS	Ultimate tensile strength, N/mm^2
Elon	Elongation, %
Proof	0.1% proof strength, N/mm^2

1 N/mm^2=0.1 hbar=0.102 kgf/mm^2=0.06475 $tonf/in.^2$=145.04 $lbf/in.^2$=1 MPa

See Appendix II for other abbreviations and conversion tables.

Symbol	Nominal analysis, supplier, condition and remarks.
178 Co	0.95% C 0.6% Mn 5% Cr 4% W 35.5% Co steel: For magnets; Sandvik
333	1.55% C 4.7% Cr 5.0% V 12.5% W 5.0% Co steel: Forez
444	1.55% C 4.75% Cr 5.0% V 19.5% W 14.5% Co steel: Forez
455	0.78% C 4.2% Cr 0.4% Mo 1.3% V 18.5% W 5.5% Co steel: F Parkin; cutting tools **DPN: 820**
808	0.8% C 4.2% Cr 1% Mo 1.6% V 19% W 10.5% Co steel: F Parkin; cutting tools **DPN: 850**
3500/37% Co	0.75% C 3.8% Cr 5.0% W 38% Co steel: For magnets; Simonds
4171	1.3% C 4.0% Cr 5.0% V 12.0% W 5.0% Co steel: For tools; French Standards designation
4175	1.65% C 4.0% Cr 5.0% V 12.0% W 10.0% Co steel: For tools; French Standards designation
4271	0.8% C 4.0% Cr 1.0% V 18.0% W 5.0% Co steel: For tools; French Standards designation
4275	0.8% C 4.0% Cr 1.6% V 18.0% W 10.0% Co steel: For tools; French Standards designation
4371	0.85% C 4.0% Cr 5.0% Mo 2.0% V 6.0% W 5.0% Co steel: For tools; French Standards designation
4373	1.5% C 4.0% Cr 5.0% Mo 5.0% V 6.5% W 5.0% Co steel: For tools; French Standards designation
4375	1.75% C 4.0% Cr 5.0% Mo 5.0% V 6.5% W 10.0% Co steel: For tools; French Standards designation
4475	1.1% C 4.0% Cr 9.0% Mo 1.0% V 1.5% W 8.0% Co steel: For tools; French Standards designation
13743	24.5% Ni 11% Co 11.5% Al 3% Cu steel: For magnets; German Standards alloy for Alnico 160 listed in DIN 17410
13745	20% Ni 16% Co 11% Al 3% Cu steel: For magnets; German Standards alloy for Alnico 190 listed in DIN 17410
13756	15% Ni 27% Co 7% Ti 7% Al 4% Cu steel: For magnets; German Standards alloy for Alnico 220 listed in DIN 17410
13758	15% Ni 30% Co 5% Ti 7% Al 4% Cu steel: For magnets; German Standards alloy for Alnico 350 listed in DIN 17410
13760	15.5% Ni 24% Co 9% Al 4% Cu steel: For magnets; German Standard for Alnico 400 listed in DIN 17410
13761	15% Ni 24% Co 8.5% Al 3% Cu steel: For magnets; German Standard for Alnico 500 listed in DIN 17410
ACMITE	4% Cr 1% V 18% W 5% Co steel: For tools; Columbia Tool Steel Co.; AISI code T4H
AFC 77	0.15% C 14.5% Cr 5% Mo 0.35% V 13% Co steel: Crucible
AISI 661	0.12% C 19.8% Ni 20.7% Cr 2.9% Mo 2.3% W 19% Co 1.1% Nb Fe alloy: Solution treated and aged **UTS: 840 Elon: 43% Proof: 370**
AJ 2	0.76% C 4% Cr 0.3% Mo 0.8% V 18% W 2.25% Co steel: For tools; Osborn; obsolete **DPN: 820**
ALCOMAX I	11% Ni 25% Co 1.5% Ti 7.5% Al 3.0% Cu steel: Casting; PMA; anisotropic
ALCOMAX II	11.5% Ni 24% Co 8.0% Al 4.5% Cu steel: Casting; PMA; anisotropic
ALCOMAX II	11% Ni 21% Co 9% Al 4% Cu steel: For magnets (casting); Jessop; remanence 13 000 gauss; coercivity 580 oersted
ALCOMAX II Sc	11.5% Ni 24% Co 8.0% Al 4.5% Cu steel: Casting; PMA; anisotropic; semicolumnar

Symbol	Nominal analysis, supplier, condition and remarks.
ALCOMAX II Sc	11% Ni 21% Co 9% Al 4% Cu steel: Casting; Crystal oriented magnet; Jessops; remanence 13 700 gauss; coercivity 600 oersted
ALCOMAX III	13.5% Ni 24% Co 0.8% Nb 8.0% Al 3.0% Cu steel: Casting; PMA; anisotropic
ALCOMAX III	13.5% Ni 24% Co 0.7% Nb 8% Al 3% Cu steel: Cast magnets; Jessops; remanance 12 600 gauss; coercivity 650 oersted
ALCOMAX III Sc	13.5% Ni 24% Co 0.8% Nb 8.0% Al 3.0% Cu steel: Casting; PMA; anisotropic; semicolumnar
ALCOMAX III Sc	13.5% Ni 24% Co 0.7% Nb 8% Al 3% Cu steel: Cast crystal oriented magnet; Jessop; remanence 13 200 gauss; coercivity 700 oersted
ALCOMAX IV	1.35% Ni 24% Co 2.5% Nb 8.0% Al 3.0% Cu steel: Casting; PMA; anisotropic
ALCOMAX IV	13% Ni 24% Co 2.5% Nb 8% Al 3% Cu steel: Cast magnets; Jessops; remanence 11 500 gauss; coercivity 750 oersted
ALCOMAX IV Sc	13% Ni 24% Co 2.5% Nb 8% Al 3% Cu steel: Crystal oriented cast magnet; Jessops; remanence 12 200 gauss; coercivity 780 oersted
ALCONIT AN 16	Ni Co Ti Al Cu steel: Casting for magnets; Von Roll; further information from PMA
ALCONIT AN 0	Ni Co Ti Al Cu steel: Casting for magnets; Von Roll; further information from PMA
ALLOY 8	2.3% C 6% Ni 16% Cr 20% Co Fe alloy: Casting; Dewrance **DPN: 501 UTS: 800**
ALLOY 15	3.75% C 4% Ni 16% Cr 6.5% Mo 20% Co 1.5% Si Fe alloy: Casting; Dewrance **DPN: 680 UTS: 610**
ALLOY 18	2.5% C 15% Ni 25% Cr 8% Mo 25% Co Fe alloy: Casting; Dewrance **DPN: 400 UTS: 680**
ALNICO	18% Ni 12% Co 0.5% Ti 10% Al 5.0% Cu steel: Casting for magnets; PMA
ALNICO I	22% Ni 5.0% Co 0.35% Ti 12% Al steel: For magnets; Simonds
ALNICO I	18% Ni 12.5% Co 0.7% Ti 10% Al 6.5% Cu steel: Casting for magnets; Jessop; remanence 7250 gauss; coercivity 500 oersted
ALNICO 1A	22.5% Ni 5.0% Co 12.0% Al Fe steel: For magnets; origin unknown
ALNICO 1B	21.0% Ni 5.0% Co 12.0% Al 3.0% Cu Fe alloy: For magnets; origin unknown
ALNICO 1C	20.0% Ni 5.0% Co 12% Al 12.0% Cu Fe alloy: For magnets; origin unknown
ALNICO II	16% Ni 12% Co 0.7% Ti 9% Al 6% Cu steel: Casting for magnets; Jessops; remanence 6200 gauss; coercivity 480 oersted
ALNICO 2	17% Ni 12.5% Co 0.45% Ti 10% Al 6.0% Cu steel: For magnets; Simonds
ALNICO 2A	18.0% Ni 13.0% Co 10.0% Al 5.0% Cu Fe alloy: For magnets; origin unknown
ALNICO 2B	19.0% Ni 13.0% Co 10.0% Al 3.0% Cu Fe alloy: For magnets; origin unknown
ALNICO 2C	21% Ni 12.5% Co 0.3% Ti 10% Al 6% Cu steel: For magnets; Simonds
ALNICO 2H	19.0% Ni 14.5% Co 10.0% Al 3.0% Cu Fe alloy: For magnets; origin unknown
ALNICO III	20% Ni 12% Co 0.7% Ti 10% Al 6% Cu steel: Casting for magnets; Jessops; remanence 6500 gauss; coercivity 620 oersted
ALNICO 3A	26.0% Ni 12.0% Al 3.0% Cu Fe alloy: For magnets; origin unknown
ALNICO 3B	25.0% Ni 12.0% Al 3.0% Cu Fe alloy: For magnets; origin unknown
ALNICO 3C	24.0% Ni 12.0% Al 3.0% Cu Fe alloy: For magnets; origin unknown
ALNICO 4	25% Ni 5.0% Co 0.4% Ti 12% Al steel: For magnets; Simonds
ALNICO 4A	28.0% Ni 5.0% Co 12.0% Al Fe alloy: For magnets; origin unknown
ALNICO 4B	27.0% Ni 5.0% Co 12.0% Al Fe alloy: For magnets; origin unknown
ALNICO 5	14.5% Ni 24% Co 8% Al 3.0% Cu steel: For magnets; Simonds
ALNICO 5/7	1.45% Ni 24% Co 8% Al 3.0% Cu steel: For magnets; Simonds
ALNICO 5A	15.0% Ni 24.0% Co 8.0% Al 3.0% Cu Fe alloy: For magnets; origin unknown
ALNICO 5A,B,D and G	14.5% Ni 24.0% Co 8.0% Al 3.0% Cu Fe alloy: For magnets; origin unknown
ALNICO 5AB	14.5% Ni 24.0% Co 8.0% Al 3.0% Cu Fe alloy: For magnets; origin unknown
ALNICO 5B	14.0% Ni 24.0% Co 8.0% Al 3.0% Cu Fe alloy: For magnets; origin unknown
ALNICO 5B,D and G	14.0% Ni 24.0% Co 8.0% Al 3.0% Cu Fe alloy: For magnets; origin unknown
ALNICO 5DG	14% Ni 24.0% Co 8% Al 3.0% Cu steel: For magnets; columnar form; anisotropic; American Alloy; consult PMA
ALNICO 5E	15% Ni 22.5% Co 0.45% Ti 8% Al 3.6% Cu steel: For magnets; Simonds
ALNICO 6	15% Ni 24% Co 1.25% Ti 8% Al 3.4% Cu steel: For magnets; Simonds
ALNICO 6A	17.0% Ni 23.0% Co 8.0% Al 3.0% Cu Fe alloy: For magnets; origin unknown
ALNICO 6B	16.0% Ni 24.0% Co 1.0% Ti 8.0% Al 3.0% Cu Fe alloy: For magnets; origin unknown
ALNICO 7	20% Ni 23.5% Co 5.0% Ti 8.5% Al 3.0% Cu steel: For magnets; Simonds
ALNICO 7	18.0% Ni 24.0% Co 5.0% Ti 8.5% Al 3.25% Cu Fe alloy: For magnets; origin unknown
ALNICO 8	15% Ni 35% Co 5.0% Ti 7.5% Al 4.0% Cu steel: For magnets; Simonds
ALNICO 8A	14.0% Ni 38.0% Co 8.0% Ti 7.5% Al 3.0% Cu Fe alloy: For magnets; origin unknown
ALNICO 8B	15% Ni 35% Co 5% Ti 7% Al 4% Cu steel: For magnets; American alloy listed by PMA
ALNICO 9	15.0% Ni 35.0% Co 5.0% Ti 7.0% Al 4.0% Cu Fe alloy: For magnets; origin unknown
ALNICO 12	18% Ni 35% Co 8.0% Ti 6.0% Al steel: For magnets; American alloy; consult PMA; anisotropic
ALNICO 160	24.5% Ni 11% Co 11.5% Al 3.0% Cu steel: For magnets; German Standard
ALNICO 190	20% Ni 16% Co 11% Al 3% Cu steel: For magnets; German alloy listed by PMA
ALNICO 220	15% Ni 27% Co 7% Ti 7% Al 4% Cu steel: For magnets; German alloy listed by PMA
ALNICO 300	18% Ni 20% Co 10% Al 4% Cu steel: For magnets; German alloy listed by PMA
ALNICO 350	15% Ni 30% Co 5% Ti 7% Al 4% Cu steel: For magnets; German alloy listed by PMA
ALNICO 400	15.5% Ni 24% Co 9.0% Al 4.0% Cu steel: For magnets; German Standard; anisotropic
ALNICO 450	15% Ni 34% Co 5% Ti 7% Al 4% Cu steel: For magnets; German alloy listed by PMA
ALNICO 500	15% Ni 24% Co 8.5% Al 3% Cu steel: For magnets; German alloy listed by PMA
ALNICO 580	15% Ni 24% Co 8.5% Al 3% Cu steel: For magnets; German alloy listed by PMA
ALNICO 700	15% Ni 24% Co 8.5% Al 3% Cu steel: For magnets; German alloy listed by PMA
AM 367	0.02% C 3.4% Ni 15.2% Cr 2.0% Mo 15.5% Co 0.3% Ti steel: Allegheny Ludlum
AMS 5376 B	21% Cr 3% Mo 2.5% W 20% Co 1% Nb + Ti Fe alloy: Casting
AMS 5531	20% Ni 20% Cr 3% Mo 2% W 20% Co 0.7% Nb + Ti Fe alloy: Sheet
AMS 5532 B	20% Ni 20% Cr 3% Mo 2% W 20% Co 1% Nb Fe alloy: Sheet

Symbol	Nominal analysis, supplier, condition and remarks.
AMS 5533 A	20% Ni 20% Cr 4% Mo 4% W 20% Co 4% Nb Fe alloy: Sheet
AMS 5585	20% Ni 20% Cr 3% Mo 2% W 20% Co 1% Nb + Ti Fe alloy: Welded tube
AMS 5748	0.15% C 0.5% Ni 14% Cr 5% Mo 0.2% V 13.5% Co steel
AMS 5761	0.03% C (max) 15.2% Cr 2.75% Mo 20% Co steel
AMS 5768	0.15% C 20% Ni 21% Cr 3% Mo 2.5% W 20% Co 1% Nb steel: Forgings
AMS 5768 E	20% Ni 20% Cr 3% Mo 2% W 20% Co 1% Nb + Ti Fe alloy: Bar and forging; solution treated and aged
AMS 5769	20% Ni 20% Cr 3% Mo 2% W 20% Co 1% Nb + Ti Fe alloy: Bar; solution treated
AMS 5770	0.4% C 20% Ni 21% Cr 4% Mo 4% W 20% Co 4% Nb steel: Forgings
AMS 5770 B	20% Ni 20% Cr 4% Mo 4% W 20% Co 4% Nb + Ti Fe alloy: Bar; solution treated and aged
AMS 5794 A	20% Ni 20% Cr 3% Mo 2% W 20% Co 1% Nb + Ti Fe alloy: Welding wire
AMS 5795 B	20% Ni 20% Cr 3% Mo 2% W 20% Co 1% Nb + Ti Fe alloy: Coated electrode
AMS 5874	0.1% C 20% Ni 21% Cr 4% Mo 4% W 18.8% Co 0.07% La 0.2% N steel
AMS 6463	0.01% C (max) 18.5% Ni 5.2% Mo 8.5% Co 0.72% Ti steel: Weld electrode
AMS 6465	0.12% C 10% Ni 2% Cr 1% Mo 0.06% V 8% Co steel
AMS 6468	0.15% C 10% Ni 1% Cr 0.45% Mo 0.08% V 4% Co steel
AMS 6522	0.15% C 10% Ni 2% Cr 1% Mo 14% Co steel
AMS 6523	0.2% C 9% Ni 0.75% Cr 1% Mo 0.09% V 4.5% Co steel
AMS 6524	0.31% C 7.7% Ni 1% Cr 1% Mo 0.09% V 4.5% Co steel
AMS 6525	0.2% C 9% Ni 0.75% Cr 1% Mo 0.09% V 4.5% Co steel
AMS 6526	0.31% C 7.7% Ni 1% Cr 1% Mo 0.09% V 4.5% Co steel
AMS 6527	0.15% C 10% Ni 2% Cr 1% Mo 14% Co steel
AMS 6543	0.12% C 10% Ni 2% Cr 1% Mo 8% Co steel
AMS 6544	0.15% C 10% Ni 2% Cr 1% Mo 14% Co steel
ANSEMAX	14% Ni 24% Co 8% Al 3% Cu steel: For magnets; Italian alloy; listed by PMA
ARK SUPERLATIVE	0.75% C 4.0% Cr 1.0% V 18.0% W 5.0% Co steel: Origin unknown
ASP 30	1.3% C 4.2% Cr 5% Mo 3% V 6.4% W 8.5% Co steel: Speedsteel
ASP 60	2.3% C 4.2% Cr 7% Mo 6.5% V 6.5% W 10.5% Co steel: Speedsteel
ASTM A579/81	0.27% C 8.0% Ni 0.5% Cr 0.45% Mo 0.09% V 4.0% Co steel: Forging
ASTM A579/82	0.31% C 7.7% Ni 1.0% Cr 1.0% Mo 0.09% V 4.5% Co steel: Forging
ASTM A579/83	0.45% C 7.7% Ni 0.27% Cr 0.27% Mo 0.09% V 4.0% Co steel: Forging
ASTM A605	0.2% C 0.3% Mn 9.0% Ni 0.72% Cr 1.0% Mo 0.08% V 4.8% Co steel: For pressure vessels **UTS: 1430 Elon: 14% Proof: 1200**
ASTM A646 HP9/4/20	0.2% C 9.0% Ni 0.75% Cr 1.0% Mo 0.09% V 4.5% Co steel
ASTM A646 HP9/4/30	0.31% C 7.5% Ni 1.0% Cr 1.0% Mo 0.1% V 4.5% Co steel
B 7	4% Cr 1% V 18% W 5% Co steel: For tools; Universal Cyclops; AISI code T4D
B 8	4% Cr 2% V 14% W 5% Co steel: For tools; Universal Cyclops; AISI code T8D
B 10	4% Cr 2% V 18% W 8% Co steel: For tools; Universal Cyclops; AISI code T5D
BCC	1.4% C 0.35% Mn 12% Cr 0.8% Mo 0.3% V 2.7% Co 0.35% Si steel: For tools; Huntsman; shear blades; intricate punches, etc. **DPN: 750**
BONDED CARBIDE	4.5% Cr 1.5% V 20% W 12% Co steel: For tools; Braeburn; AISI code T61
BONDED CARBIDE Jr	4% Cr 2% V 18% W 8% Co steel: For tools; Braeburn; AISI code T51
BONDED CARBIDE Sr	0.75% C 4.25% Cr 0.7% Mo 2.0% V 18.5% W 7.6% Co steel: Braeburn
BRAEBURN M33	0.9% C 3.75% Cr 9.5% Mo 1.15% V 1.75% W 8.25% Co steel: Braeburn
BRAEBURN T15	1.5% C 4.5% Cr 0.5% Mo 4.7% V 13.0% W 4.5% Co steel: Braeburn
BRAECUT	1.5% C 4.25% Cr 6.25% Mo 2.2% V 5.2% W 12% Co steel: Braeburn
BRAEMAX	1.1% C 3.7% Cr 9.5% Mo 1.1% V 1.5% W 8.0% Co steel: Braeburn
BRAETUF	1.1% C 4.0% Cr 6.7% Mo 2.0% V 5.0% W 7.25% Co steel: Braeburn
BU 8	0.78% C 4.5% Cr 0.75% Mo 1.25% V 18% W 8.25% Co steel: For tools; Osborn; obsolete **DPN: 850**
BV 10	0.78% C 4.5% Cr 0.75% Mo 1.25% V 18% W 10% Co steel: For tools; Osborn; obsolete **DPN: 850**
C 445	0.3% C 2.8% Cr 2.8% Mo 0.5% V 2.8% Co steel: For tools; Fagersta **DPN: 460**
CALDUR	0.27% C 10.5% Cr 7.0% W 4.0% Co steel: Uddelholm
CALMAX	0.28% C 11.5% Cr 0.5% V 7.5% W 9.5% Co steel: Uddelholm
CAPITAL 398	0.82% C 4.15% Cr 5% Mo 1.9% V 6.4% W 5% Co steel: for tools; Balfour; reamers, toolbits, etc. **DPN: 950**
CCM 10	0.8% C 4.5% Cr 1.6% V 20% W 10.5% Co steel: Huntsman; cutting tools (used as tips) **DPN: 850**
CCMP 15	0.8% C 4.7% Cr 1.6% V 22% W 15% Co steel: Huntsman; cutting tools (used as tips) **DPN: 800**
CIRCLE C	0.77% C 4.5% Cr 1.0% Mo 2.0% V 18.5% W 9.0% Co steel: Firth Sterling
CIRCLE H	1.28% C 4.2% Cr 5.1% Mo 1.6% V 8.25% W 5.4% Co steel: Firth Sterling
CIRCLE M	4% Cr 5% Mo 6% W 8% Co tool steel: Firth Sterling: AISI code M36B
CIRCLE T 15	1.5% C 4.7% Cr 5.0% V 12.5% W 5.0% Co steel: Firth Sterling
CLYDALL 5 SPECIAL	0.78% C 4% Cr 1.3% V 18.5% W 5.5% Co steel: For tools; Clyde Alloy **DPN: 790**
CLYDALL 12 SPECIAL	0.78% C 4% Cr 1.35% V 22% W 12% Co steel: For tools; Clyde Alloy **DPN: 790**
CM 5	0.8% C 4.6% Cr 1.5% V 18% W 5.7% Co steel: Huntsman; cutting tools **DPN: 910**
CM 18	2.5% C 15% Ni 25% Cr 8% Mo 25% Co Fe alloy: For hardfacing gas welding; Dewrance
CM 32	1.25% C 23% Cr 8% Mo 10% Co Fe alloy: For hardfacing; Dewrance; gas welding

Note. The following abbreviations and units are used in the tables:

DPN	Hardness, diamond pyramid number
UTS	Ultimate tensile strength, N/mm^2
Elon	Elongation, %
Proof	0.1% proof strength, N/mm^2

$1\ N/mm^2 = 0.1\ hbar = 0.102\ kgf/mm^2 = 0.06475\ tonf/in.^2 = 145.04\ lbf/in.^2 = 1\ MPa$

See Appendix II for other abbreviations and conversion tables.

Symbol	Nominal analysis, supplier, condition and remarks.
CM 118	2.5% C 15% Ni 25% Cr 8% Mo 25% Co Fe alloy: For hardfacing; Dewrance; electric welding
CM 132	1.25% C 23% Cr 8% Mo 10% Co Fe alloy: For hardfacing; Dewrance; electric welding
CM 1255	1.5% C 4.7% Cr 5% V 12% W 5% Co steel: Huntsman; cutting tools **DPN: 910**
Co 040	1.0% C 4.0% Cr 0.7% W 2.0% Co steel: For magnets; German Standard; included in DIN 17410
Co 050	1.0% C 8.5% Cr 1.3% Mo 6.5% Co steel: For magnets; German Standard; included in DIN 17410
Co 060	1.0% C 8.5% Cr 1.6% Mo 11% Co steel: For magnets; German Standard; included in DIN 17410
Co 070	1.0% C 9% Cr 1.6% Mo 16% Co steel: For magnets; German Standard; included in DIN 17410
Co 090	0.9% C 5.0% Cr 0.4% Mo 5.0% W 31% Co steel: For magnets; German Standard; included in DIN 17410
COALNI	25.5% Ni 4% Co 11.5% Al 4% Cu steel: For magnets; Italian alloy listed by PMA
COALNIMAX	14% Ni 24% Co 8% Al 3% Cu steel: For magnets; Italian alloy listed by PMA
COBALT	0.74% C 4.0% Cr 0.8% Mo 1.0% V 18% W 5.0% Co steel: Braeburn; AISI type T4
COBITE	0.81% C 4.2% Cr 0.6% Mo 2.0% V 18.5% W 8.7% Co steel: Columbia; AISI type T5
CODE P	0.12% C 10.5% Cr 4.7% Mo 6.0% Co 1.25% Cu steel: American proprietary alloy
COLUMAX	13.5% Ni 24% Co 2.0% Nb 8% Al 3.0% Cu steel: For magnets; PMA; columnar crystal form of Alcomax III
COMO	0.82% C 4.0% Cr 8.5% Mo 1.2% V 1.5% W 5.0% Co steel: Braeburn; AISI type M30
COMO	12% Co 17% Mo steel (probably): Japanese alloy listed by PMA
COMOKUT	4% Cr 1% V 18% W 5% Co steel: For tools; Bethlehem Steel Co.; AISI code T4E
COMOL	12% Co 17% Mo steel: For magnets; American alloy listed by PMA
CONGO	0.78% C 4.0% Cr 5.0% Mo 1.4% V 4.0% W 12% Co steel: Braeburn; AISI type M6
CONGO HOT-WORK	0.1% C 4.0% Cr 5.0% Mo 0.75% V 4.0% W 23.5% Co steel: Braeburn
COVAN	1.5% C 5% Cr 5% V 12.5% W 5% Co steel: Hall and Pickles; cutting tools; obsolete **DPN: 820**
CRD 206	0.78% C 4.5% Cr 0.5% Mo 2% V 19% W 5% Co steel: C Denton; drills and cutters; obsolete
CRD 339	1.25% C 4.0% Cr 3.5% V 8.5% W 8.5% Co steel: For tools; origin unknown
CRD 2212	0.8% C 5.0% Cr 0.5% Mo 2.0% V 21% W 10% Co steel: C Denton; cutting tools; obsolete
CRD TVC	1.5% C 4.0% Cr 5.0% V 12.0% W 5.0% Co steel: For tools; origin unknown
CROMO CO	0.27% C 0.7% Ni 11.5% Cr 1.2% Mo 1% V 0.5% W 10% Co steel: Bethlehem Steel
CX	0.8% C 4.4% Cr 1.0% Mo 1.6% V 18.5% W 10% Co steel: For tools; ESC symbol for Cyclone Extra **DPN: 910**
CY	0.7% C 4.5% Cr 0.7% Mo 1.25% V 18% W steel: For tools; Osborn; obsolete **DPN: 910**
CYC 5% Co	0.7% C 4.5% Cr 1% Mo 1.5% V 18% W 5% Co steel: Marsh Brothers; cutting tools
CYC 275	0.8% C 4.5% Cr 1.2% Mo 1.2% V 20% W 10% Co steel: Marsh Brothers; cutting tools
CYCLONE EXTRA	0.8% C 4.4% Cr 1% Mo 1.6% V 18.5% W 10% Co steel: For tools; ESC; cutting tools **DPN: 910**
CYCLONE MC 6	4.0% Cr 5.0% Mo 2.0% V 6.0% W 5.0% Co steel: For tools; origin unknown
CYCLONE MC 6V	1.25% C 4.3% Cr 5.5% Mo 2.8% V 7% W 6% Co tool steel: ESC: cutting tools **DPN: 910**
CYCLONE MC 33	0.9% C 4.0% Cr 9.5% Mo 1.15% V 1.5% W 8.0% Co steel: For tools; origin unknown
CYCLONE MC 42	1.05% C 4.0% Cr 9.8% Mo 1.1% V 1.5% W 8.3% Co steel: For tools; origin unknown
CYCLONE MC 98	0.9% C 3.7% Cr 8.7% Mo 2% V 1.6% W 8.2% Co steel: ESC: cutting tools **DPN: 910**
CYCLONE SUPERCUT	1.5% C 4.7% Cr 5% V 12.5% W 5% Co steel: ESC; cutting tools **DPN: 910**
D 444	0.4% C 4.2% Cr 0.5% Mo 2.2% V 4.2% W 4.2% Co steel: Fagersta
D 946	4% Cr 5% Mo 2% V 6% W 5% Co steel: Fagersta; obsolete
D 948	1% C 4.1% Cr 9.8% Mo 1.1% V 1.5% W steel: Fagersta; obsolete
DARWIN 505	0.7% C 4.5% Cr 1% V 18.5% W 5% Co steel: For tools; Darwin; milling cutters, etc.
DARWIN 1366	0.73% C 4.5% Cr 2% V 20% W 12% Co steel: For tools; Darwin; hobs, borers, etc.
DG CYCLONE	0.8% C 4.5% Cr 1.4% V 18.5% W 5.7% Co steel: ESC; cutting tools, drills, etc. **DPN: 910**
DGC	0.8% C 4.5% Cr 1.4% V 18.5% W 5.7% Co steel: For tools; ESC symbol for DG cyclone **DPN: 910**
DMA 1	14% Ni 24% Co 8.0% Al 3.0% Cu steel: For magnets; Japanese alloy listed by PMA
DMA 2	14% Ni 24% Co 2.0% Ti 8.0% Al 3.0% Cu steel: For magnets; Japanese alloy listed by PMA
DMA 4	22% Ni 5.0% Co 12% Al steel: For magnets; Japanese alloy listed by PMA
DOUBLE ECHO COBALT	0.83% C 4.4% Cr 0.5% Mo 1.6% V 18% W 10% Co steel: John Vessey
DOUBLE RAPID	0.75% C 4.0% Cr 1.0% V 18.0% W 5.0% Co steel: Origin unknown
DOUBLE RAPID	0.8% C 4.75% Cr 0.5% Mo 1.45% V 19.0% W 6.25% Co steel: For tools; origin unknown
DOUBLE SEVEN	1.25% C 0.3% Mn 12.5% Cr 1.5% Mo 2.75% Co steel: Edgar Allen; punches, dies, etc. **DPN: 730**
DOUBLE SUPER HYDRA	0.8% C 4.5% Cr 1.7% V 18% W 5% Co steel: Hall and Pickles; cutting tools; obsolete **DPN: 820**
ECHO COBALT	0.83% C 4.7% Cr 0.5% Mo 1.4% V 18% W 5.2% Co steel: John Vessey
ECHO SPECIAL VANADIUM	1.5% C 5% Cr 13% W 5% Co steel: John Vessey
ECHO SUPERCUT	1.08% C 3.7% Cr 9.5% Mo 1.15% V 1.5% W 8% Co steel: John Vessey
ECLIPSE H5	0.8% C 4.0% Cr 2.0% V 18.0% W 8.0% Co steel: Origin unknown
EHK 3	1.0% C 6.0% Cr 2.0% Co steel: Forging for magnets; Polish alloy listed by PMA
EK 5	1.0% C 6.0% Cr 6.0% Co steel: For magnets; Russian alloy listed by PMA
EK 6	1.0% C 6.0% Cr 6.0% Co steel: Forging for magnets; Polish alloy listed by PMA
EK 15	1.0% C 8.0% Cr 1.5% Mo 15% Co steel: For magnets; Russian alloy listed by PMA
ELECTRITE Co	4% Cr 5% Mo 1.5% V 4% W 12% Co steel: For tools; Latrobe; AISI code M6F
ELECTRITE Co 9	0.9% C 4.1% Cr 5.2% Mo 1.9% V 5.7% W 9.0% Co steel: Latrobe; cutting tools; type M36
ELECTRITE COBALT	0.73% C 4.1% Cr 0.7% Mo 1.1% V 18% W 5.0% Co steel: Latrobe; cutting tools; type T4
ELECTRITE DYNACUT	1.26% C 3.7% Cr 8.7% Mo 2.0% V 1.8% W 8.2% Co steel: Latrobe; cutting tools; Type M43

Symbol	Nominal analysis, supplier, condition and remarks.
ELECTRITE DYNAVAN	4% Cr 12% W 5% Co steel: For tools; Latrobe; AISI code T15F
ELECTRITE DYNAVAN XL	4.7% Cr 0.5% Mo 4.7% V 12.5% W 5.0% Co 1.57% O steel: Latrobe; cutting tools; type T15
ELECTRITE KELVAN	0.87% C 3.7% Cr 9.5% Mo 1.1% V 1.7% W 5.0% Co steel: Latrobe
ELECTRITE LACOMO	4% Cr 8% Mo 1.2% V 2% W 5% Co steel: For tools; Latrobe; AISI code M30F
ELECTRITE SUPER	0.85% C 4.1% Cr 0.8% Mo 1.9% V 18.7% W 9.0% Co steel: Latrobe; cutting tools; type T5
ELECTRITE SUPER COBALT	4% Cr 2% V 18% W 8% Co steel: For tools; Latrobe; T5 type; AISI code TSF
ELECTRITE TATMO	0.9% C 4.0% Cr 8.7% Mo 2.0% V 1.5% W 8.5% Co steel: Latrobe; cutting tools; type M34
ELECTRITE TATMO COBALT	4% Cr 8% Mo 2% V 2% W 8% Co steel: For tools; Latrobe; M34; AISI code M34F
ELECTRITE UB	4% Cr 2% V 14% W 5% Co steel: For tools; Latrobe; AISI code T8F
ELECTRITE ULTRA	4.5% Cr 1.5% V 20% W 12% Co steel: For tools; Latrobe; AISI code T6F
ELECTRITE ULTRAVAN	1.5% C 4.2% Cr 5.0% Mo 4.7% V 6.3% W 5.0% Co steel: Latrobe; cutting tools; type M15
EMK 10	1.0% C 6.0% Cr 1.5% Mo 10% Co steel: For magnets; Polish alloy listed by PMA
EMK 15	1.0% C 9% Cr 1.5% Mo 15% Co steel: Polish alloy listed by PMA
ER Co 3	0.32% C 3% Cr 2.8% Mo 0.5% V 3.0% Co steel: For tools; Balfour Darwin
EWK 30	0.75% C 6.0% Cr 6.0% W 30% Co steel: For magnets; Polish alloy listed by PMA
EXTRA NAP SUPERIOR	0.8% C 4.5% Cr 0.5% Mo 1.25% V 20% W 10% Co steel: For tools; Vulcan; cutting tools **DPN: 910**
EXTRA SUPERIOR SPUR	0.72% C 4.5% Cr 1.5% V 18.5% W 3.7% Co steel: For tools; T Turton
EXTRA TRIPLE MUSHET	0.78% C 45% Cr 0.75% Mo 1.25% V 18.25% W 8.25% Co steel: For tools; origin unknown
F ALLOY	0.06% C 0.4% Mn 28% Ni 17.5% Co steel: Darwins
FHS 20/5	0.8% C 4.5% Cr 1.5% V 20.0% W 5.0% Co steel: For tools; origin unknown
FHS 20/10	0.83% C 4.5% Cr 1.5% V 20.0% W 10.0% Co steel: Origin unknown
fmp 455	0.8% C 4.2% Cr 0.7% Mo 1.3% V 18.5% W 5.0% Co steel: F Parkin; as tool steel T4
fmp 526	1.1% C 4.2% Cr 3.7% Mo 2.0% V 6.7% W 5.0% Co steel: F Parkin; as tool steel M41
fmp 530	0.8% C 4.0% Cr 8.2% Mo 1.25% V 2.0% W 5.0% Co steel: F Parkin; as tool steel M30
fmp 536	1.5% C 4.0% Cr 3.5% Mo 5.0% V 6.5% W 5.0% Co steel: F Parkin; as tool steel M15
fmp 542	1.05% C 3.7% Cr 9.5% Mo 1.1% V 1.5% W 8.0% Co steel: F Parkin; as tool steel M42
fmp 555C	1.5% C 4.7% Cr 1.0% Mo 5.0% V 12.7% W 5.0% Co steel: F Parkin; as T15 tool steel
fmp 808	0.8% C 4.2% Cr 1.0% Mo 1.6% V 18.5% W 10% Co steel: F Parkin; as tool steel T6
fmp 828	0.8% C 4.5% Cr 1.0% Mo 2.0% V 18.5% W 8.0% Co steel: F Parkin; as T5 tool steel
fmp 928	0.9% C 4.0% Cr 8.5% Mo 2.0% V 2.0% W 8.0% Co steel: F Parkin; as tool steel M34
fmp 929	1.25% C 3.7% Cr 8.7% Mo 2.0% V 1.8% W 8.2% Co steel: F Parkin; as tool steel M43
fmp 933	1.3% C 4.2% Cr 3.7% Mo 3.5% V 9.2% W 10.0% Co steel: F Parkin
FS 2.5	4% Cr 2% V 14% W 5% Co steel: For tools; Firth Sterling; AISI code T8
FS M34	0.9% C 4.0% Cr 8.0% Mo 2.0% V 2.0% W 8.0% Co steel: Firth Sterling
FV 535	0.07% C 0.3% Ni 10.5% Cr 0.7% Mo 0.2% V 6% Co 0.5% Nb steel: Firth Vickers
GECALLOY	General name for magnetic powders; Salford Electric
GOLD STAR	4% Cr 2% V 14% W 5% Co steel: For tools; Carpenter; AISI code T8K
GREY CUT COBALT	4.5% Cr 1.5% V 20% W 12% Co steel: For tools; Vanadium Alloys; AISI code T6A
GZ 6	0.8% C 4.8% Cr 0.75% Mo 1.25% V 18% W 5% Co steel: For tools; Osborn **DPN: 850**
H 10A	0.37% C 3.0% Cr 2.75% Mo 0.45% V 3.0% Co steel: Osborn
H 19	0.4% C 4.25% Cr 2.0% V 4.2% W 4.25% Co steel: For tools; designation used by AISI
H 19	0.39% C 4.3% Cr 0.4% Mo 2.2% V 4.25% W 4.3% Co steel: For hot work; Osborn
H 43	0.9% C 0.5% Mn 4% Cr 0.5% W 2% Co 0.15% Si steel: For magnets; Jessop
H 61	0.32% C 0.37% Mn 2.75% Cr 2.75% Mo 0.2% V 2.7% Co 0.35% Si steel: Jessop
HA 1	1.0% C 6.0% Cr 3.0% Co steel: For magnets; Belgian alloy listed by PMA
HA 2	1.0% C 8.0% Cr 12% Co steel: For magnets; Belgian alloy listed by PMA
HA 3	1.0% C 10% Cr 15% Co steel: For magnets; Belgian alloy listed by PMA
HA 4	0.7% C 6.0% Cr 35% Co steel: For magnets; Belgian alloy listed by PMA
HAYNES 56	0.25% C 13% Ni 20% Cr 4.5% Mo 1% W 11% Co 1% Nb Fe alloy: American proprietary alloy listed in SAE yearbook
HAYNES 93	3% C 17% Cr 16% Mo 6% Co steel: Osborn; abrasion and corrosion resistant **DPN: 850**
HD 3MX	0.33% C 3.0% Cr 2.8% Mo 0.9% V 1.0% W 3.0% Co steel: For tools; origin unknown
HEV 1	0.1% C 20% Ni 21.3% Cr 3% Mo 2.5% W 20% Co steel: Designation used by SAE; annealed **UTS: 800** **Elon: 63%** **Proof: 370**
HIPERCO 27	0.8% Mn 0.7% Ni 0.7% Cr 27% Co steel: For magnets; Westinghouse
HIPERCO 50	2% V 49% Co iron alloy: For magnets; Westinghouse
HP 9 4-25	0.27% C 8% Ni 0.52% Cr 0.5% Mo 0.1% V 4% Co steel
HP 9-4-30	See ASTM A646
HS 95	0.15% C 20% Ni 21% Cr 3% Mo 2.5% W 20% Co 1% Nb Fe alloy: Union Carbide
HSP 15/17	1.3% C 4.2% Cr 3.0% Mo 3.5% V 9.0% W 8.0% Co steel: For tools; origin unknown
HV 5	1.5% C 4.7% Cr 0.5% Mo 5.0% V 12.5% W 5.0% Co steel: Spencer
HYCOMAX I	21% Ni 20% Co 9% Al 1.6% Cu Fe alloy: Casting; PMA; anisotropic
HYCOMAX II	15% Ni 29% Co 4% Ti 2% Nb 7.0% Al 4% Cu Fe alloy: PMA; cast for magnets
HYCOMAX III	15% Ni 34% Co 5% Ti 7% Al 4% Cu Fe alloy: For magnets; PMA; cast and sintered
HYCOMAX IV	14% Ni 38% Co 8% Ti 7.5% Al 3% Cu steel: For magnets; cast PMA

Note. The following abbreviations and units are used in the tables:

DPN	Hardness, diamond pyramid number
UTS	Ultimate tensile strength, N/mm^2
Elon	Elongation, %
Proof	0.1% proof strength, N/mm^2

1 N/mm^2=0.1 hbar=0.102 kgf/mm^2=0.06475 $tonf/in.^2$=145.04 $lbf/in.^2$=1 MPa

See Appendix II for other abbreviations and conversion tables.

Symbol	Nominal analysis, supplier, condition and remarks.
HYDRA COVAN	1.5% C 5.0% Cr 12.5% W 5.0% Co steel: Hall and Pickles; cutting tools **DPN: 820**
HYDRA MULTICO	0.85% C 4.5% Cr 0.75% Mo 1.5% V 22% W 11% Co steel: Hall and Pickles; cutting tools **DPN: 790**
HYNICO II	20% Ni 20% Co 4.0% Ti 0.8% Nb 8% Al 4.0% Cu Fe alloy: Casting; PMA
HYPEAK 5 V 5	1.45% C 5% Cr 5% V 13% W 5% Co steel: Swift Levick; cutting tools
HYPEAK 303	0.75% C 4% Cr 0.25% Mo 1.25% V 18% W 5.5% Co steel: Swift Levick; cutting tools
INDALLOY	17% Mo 12% Co Fe alloy: Material for magnets; American material listed by PMA
INMANITE	0.8% C 4.5% Cr 1% Mo 1.5% V 20% W 5% Co steel: Inman; cutting tools
INMARITE T4	0.75% C 4.0% Cr 1.0% V 18.0% W 5.0% Co steel: For tools; Inman
J 5	0.82% C 0.3% Mn 4.2% Cr 1.6% V 18.5% W 5% Co 0.3% Si steel: Jessop; cutting tools
J 26	0.81% C 4.2% Cr 1.6% V 20% W 10% Co steel: For tools; Jessop
J 36	1.5% C 4.7% Cr 5% V 12.5% W 5% Co steel: For tools; Jessop **DPN: 720**
J 37	1.3% C 4.5% Cr 4% Mo 3.5% V 9.5% W 10% Co steel: For tools; Jessop **DPN: 720**
J 42	1.0% C 3.8% Cr 9.5% Mo 1.1% V 1.5% W 8% Co steel: Jessop; high speed cutting tools **DPN: 975**
JETHETE 7210	0.1% C 12% Cr 1.5% Mo 0.2% V 7% Co 0.015% B steel: S. Fox
K MAT 11	0.73% C 3.8% Cr 5% Mo 1% V 1% W 7% Co steel: Speedsteel
K 945	0.9% C 3.7% Cr 5.0% Mo 1.2% V 1.8% W 2.5% Co steel: Speedsteel
K 91313	0.32% C 7.5% Ni 1% Cr 1% Mo 0.1% V 4.5% Co steel: Designation used by UNS
KEA 218	0.32% C 3.0% Cr 2.8% Mo 0.35% V 3.0% Co steel: For tools; Sanderson Kayser
KELOCK 873	0.7% C 4.2% Cr 0.7% Mo 1% V 18.2% W 5% Co steel: Kayser Ellison; cutting tools
KELOCK 1021	0.7% C 4.2% Cr 0.7% Mo 1.4% V 18.2% W 10.5% Co steel: Kayser Ellison; cutting tools
KELOCK A 72	0.87% C 4.2% Cr 0.5% Mo 1% V 18.7% W 5.5% Co steel: Kayser Ellison; cutting tools
KERAU WUMDA	0.75% C 4% Cr 0.5% Mo 1.25% V 18% W 5.5% Co steel: Sanderson; now Saben Wunda **DPN: 790**
KING COBALT	4.5% Cr 1.5% V 20% W 12% Co steel: For tools; Jessop; AISI code T6W
KM 35	4% Cr 5% Mo 2% V 6% W 5% Co steel: Speedsteel
KM 42	1% C 4.1% Cr 9.8% Mo 1.1% V 1.5% W steel: Speedsteel
KOERZIT 130	25% Ni 5.0% Co 12% Al 4.0% Cu steel: Casting for magnets; German alloy listed by PMA
KOERZIT 160	21% Ni 14% Co 13% Al 3% Cu steel: For magnets; German alloy listed by PMA
KOERZIT 190	20% Ni 16% Co 9% Al 3.0% Cu steel: For magnets; German alloy listed by PMA
KOERZIT 250	17% Ni 26% Co 7.0% Ti 8.0% Al 4.0% Cu Fe alloy: Casting; German alloy listed by PMA
KOERZIT 300	17% Ni 20% Co 9% Al 3% Cu steel: For magnets; German alloy listed by PMA
KOERZIT 400	14% Ni 24% Co 8% Al 3% Cu steel: For magnets; German alloy listed by PMA
KOERZIT 580	14% Ni 24% Co 8% Al 3% Cu steel: For magnets; German alloy listed by PMA
KOERZIT 700	14% Ni 24% Co 8% Al 3% Cu steel: For magnets; German alloy listed by PMA

Symbol	Nominal analysis, supplier, condition and remarks.
KOERZIT VS 55	14% Ni 38% Co 8% Ti 7.5% Al 3% Cu steel: For magnets; German alloy listed by PMA
KOVA 57	1.3% C 4.2% Cr 3.0% Mo 3.5% V 9.0% W 8.5% Co steel: Origin unknown
KOVAR	0.4% Mn 29% Ni 17% Co Fe alloy: Westinghouse; low expansion alloy
LEDA	0.75% C 4% Mo 1.4% V 18% W 5.2% Co steel: For tools; Firth Brown
LHS 2C	0.75% C 4% Cr 0.3% Mo 1.8% V 18.25% W 2% Co steel: For tools; Low Moor; obsolete **DPN: 850**
LHS 5C	0.8% C 4.5% Cr 0.75% Mo 1.25% V 18.25% W 5% Co steel: For tools; Low Moor; obsolete **DPN: 910**
LHS 8C	0.75% C 4.25% Cr 0.75% Mo 18.25% W 1.25% Co steel: For tools; Low Moor; obsolete **DPN: 910**
LHS 10C	0.75% C 4.5% Cr 0.7% Mo 1.25% V 19% W 10% Co steel: For tools; Low Moor; obsolete **DPN: 950**
LHS 227	1.5% C 4.7% Cr 3% Mo 6.5% W 5% Co steel: For tools; Low Moor; obsolete **DPN: 975**
LHS 235	1.3% C 4% Cr 3.2% Mo 3% V 9% W 9.5% Co steel: For tools; Low Moor; obsolete **DPN: 975**
LT 49	0.4% C 4.2% Cr 0.4% Mo 2.2% V 4.2% W 4.2% Co steel: Low Moor; for hot work tools; obsolete **DPN: 572**
LU	0.75% C 4% Cr 0.7% Mo 1% V 20% W 10% Co steel: Osborn; obsolete **DPN: 850**
M 6	4% Cr 5% Mo 1.5% V 4% W 12% Co steel: For tools; designation used by AISI: the supplier is coded by a suffix letter
M 15	4% Cr 3.5% Mo 5% V 6.5% W 5% Co steel: For tools; designation used by AISI; the supplier is coded by a suffix letter
M 15	1.55% C 4.7% Cr 3.0% Mo 5.0% V 6.5% W 5.0% Co steel: Osborn
M 30	4% Cr 8% Mo 1.2% V 2% W 5% Co steel: For tools; designation used by AISI; the supplier is coded by a suffix letter
M 33	0.9% C 4.0% Cr 9.5% Mo 1.15% V 1.5% W 8.0% Co steel: For tools; designation used by AISI
M 34	4% Cr 8% Mo 2% V 8% Co steel: For tools; designation used by AISI; the supplier is coded by a suffix letter
M 35	4% Cr 5% Mo 2% V 6% W 5% Co steel: For tools; designation used by AISI; the supplier is coded by a suffix letter
M 36	4% Cr 5% Mo 2% V 6% W 8% Co steel: For tools; designation used by AISI; the supplier is coded by a suffix letter
M 40	4% Cr 8% Mo 1.5% V 8% Co steel with boron: For tools; designation used by AISI; the supplier is coded by a suffix letter
M 41	1.1% C 4.2% Cr 3.8% Mo 2.0% V 6.8% W 5.2% Co steel: For tools; AISI designation
M 42	1.07% C 4.1% Cr 9.8% Mo 1.1% V 1.5% W 8.3% Co steel: Osborn
M 43	1.2% C 3.7% Cr 8.0% Mo 1.6% V 2.7% W 8.25% Co steel: For tools; designation used by AISI
M 44	1.15% C 4.2% Cr 6.5% Mo 2.0% V 5.3% W 11.8% Co steel: For tools; AISI designation
M 46	1.28% C 4.0% Cr 8.2% Mo 3.2% V 2.0% W 8.2% Co steel: For tools; AISI designation
M 47	1.1% C 3.7% Cr 9.7% Mo 1.2% V 1.5% W 5.0% Co steel: For tools; AISI designation
MAGLOY 1	Ni Co Al Fe alloy: Casting for magnets; remanence 12 700 gauss; Preformations Ltd

Symbol	Nominal analysis, supplier, condition and remarks.
MAGLOY 5	Ni Co Al Fe alloy: Casting for magnets; remanence 7700 gauss; Preformations Ltd
MAGLOY 6	Ni Co Al Fe alloy: Casting for magnets; remanence 6200 gauss; Preformations Ltd
MAGLOY 7	Ni Co Al Fe alloy: Casting for magnets; remanence 9500 gauss; Preformations Ltd
MAGLOY 8A	Ni Co Al Fe alloy: Casting for magnets; remanence 9000 gauss; Preformations Ltd
MAGLOY 8B	Ni Co Al Fe alloy: Casting for magnets; remanence 8500 gauss; Preformations Ltd
MAGLOY 10	Ni Co Al Fe alloy: Casting for magnets; remanence 12 900 gauss; Preformations Ltd
MAGLOY 15	Ni Co Al Fe alloy: Casting for magnets; remanence 11 600 gauss; Preformations Ltd
MAGLOY 100	Ni Co Al Fe alloy: Casting for magnets; remanence 13 400 gauss; Preformations Ltd
MAGNACUT	0.75% C 4.5% Cr 1.25% V 18.0% W 10.0% Co steel: For tools; origin unknown
MAGNICO	14% Ni 24% Co 8.0% Al 3.0% Cu Fe alloy: Casting; Russian alloy listed by PMA
MARSH CYC 5% Co	0.7% C 4.5% Cr 1% Mo 1.5% V 18% W 5% Co steel: Marsh Brothers; cutting tools
MARSH CYC 275	0.8% C 4.5% Cr 1.2% Mo 1.2% V 20% W 10% Co steel: Marsh Brothers; cutting tools
MAXITE	4% Cr 2% V 14% W 5% Co steel: For tools; Columbia; AISI code T8H
MS 6	6.0% Co steel: For magnets; Swiss alloy listed by PMA
MS 15	15% Co steel: For magnets; Swiss alloy listed by PMA
MS 35	35% Co steel: For magnets; Swiss alloy listed by PMA
MULTICO	0.85% C 4.5% Cr 0.75% Mo 1.5% V 22% W 11% Co steel: Hall and Pickles; cutting tools **DPN: 790**
MULTIMET	0.1% C 20% Ni 21% Cr 3% Mo 2% W 20% Co 1.0% Nb iron alloy: Wrought; Osborn; solution treated; heat resistant alloy **UTS: 800 Elon: 49% Proof: 370**
MUSKET HIGH V Co	1.5% C 4.7% Cr 5.0% V 12.5% W 5.0% Co steel: For tools; origin unknown
MUSKET SPECIAL	1.5% C 4.7% Cr 3.0% Mo 5.0% V 6.5% W 5.0% Co steel: For tools; origin unknown
MUSKET SPECIAL VG	1.5% C 4.7% Cr 3.0% Mo 5.0% V 6.5% W 5.0% Co steel: For tools; origin unknown
N 153	0.3% C 15% Ni 17% Cr 2% W 12% Co 1% Nb Fe alloy: Union Carbide
N 155	0.15% C 20% Ni 21% Cr 3% Mo 2.5% W 20% Co 1% Nb Fe alloy: Union Carbide
N 156	0.3% C 33% Ni 17% Cr 3% Mo 2% W 24% Co 1% Nb Fe alloy: Union Carbide
NAP SUPERIOR	0.8% C 4.5% Cr 0.5% Mo 1.25% V 20% W 5% Co steel: For tools; Vulcan; cutting tools **DPN: 910**
NEW KS	17.7% Ni 27.2% Co 6.7% Ti 3.7% Al Fe alloy: For magnets; origin unknown
Ni Al Co 2	28% Ni 6.0% Co 10% Al Fe alloy: Casting for magnets; Belgian alloy listed by PMA
Ni Al Co 2B	26% Ni 8.0% Co 12% Al Fe alloy: Casting for magnets; Belgian alloy listed by PMA
Ni Al Co 6	20% Ni 12% Co 10% Al 6.0% Cu Fe alloy: Casting for magnets; Belgian alloy listed by PMA
NICOLLOY 15	0.4% C 1.25% Mn 10% Ni 19% Cr 1.5% Mo 1.5% V 1.5% W 15% Co steel: Swift Levick; for hot dies
NIPIGON	4% Cr 2% V 18% W 8% Co steel: For tools; Atlas; AISI code T5Z
NORESCO K5	0.74% C 4.25% Cr 0.65% Mo 1.4% V 18.5% W 4.5% Co steel: For tools; Schoeller-Bleckmann **DPN: 850**
NORESCO K10	0.68% C 4.25% Cr 0.6% Mo 18.5% W 10% Co steel: For tools; Schoeller-Bleckmann **DPN: 830**
NORESCO K12M	1.25% C 4.25% Cr 5.5% Mo 2.8% V 7% W 12.5% Co steel: For tools; Schoeller-Bleckmann **DPN: 910**
NORESCO WU	0.3% C 2.7% Cr 0.15% Mo 0.25% V 10% W 2.2% Co steel: For tools; Schoeller-Bleckmann **UTS: 1000**
NORESCO WVC	0.4% C 4.2% Cr 0.4% Mo 2.2% V 4.2% W 4.2% Co steel: For tools; Schoeller-Bleckmann **UTS: 1000**
NOVA MAX	0.75% C 4.5% Cr 1.25% V 18% W 5% Co steel: Jonas; cutting tools **DPN: 850**
NOVA V	1.45% C 4.75% Cr 5.0% V 14.0% W 5.0% Co steel: For tools; Jonas; type T15
NOVA VHC	1.05% C 3.75% Cr 9.5% Mo 1.2% V 1.75% W 8.0% Co steel: For tools; Jonas; type M42
NOVO ENORMOUS	0.75% C 4.5% Cr 1.25% V 18% W 10% Co steel: For tools; Jonas **DPN: 850**
NOVO SUPERB	1.25% C 4.25% Cr 3.75% Mo 3.2% V 10.5% W 10% Co steel: Jonas; cutting tools **DPN: 950**
NOX 1S4	15.0% Ni 12.0% Cr 3.0% Mo 2.0% W 10.0% Co Fe alloy: Origin unknown
OERSTIT	14% Ni 24% Co 8% Al 3% Cu Fe alloy: International Nickel Co.; magnetic alloy
OERSTIT 120R	22% Ni 3% Co 0.5% Ti 12% Al 3% Cu steel: For magnets; German alloy listed by PMA
OERSTIT 160	21% Ni 17% Co 1% Ti 10% Al 3% Cu steel: For magnets; German alloy listed by PMA
OERSTIT 190	21% Ni 17% Co 0.5% Ti 10% Al 3% Cu steel: For magnets; German alloy listed by PMA
OERSTIT 220	15% Ni 28% Co 8% Ti 7% Al 5% Cu steel: For magnets; German alloy listed by PMA
OERSTIT 300	17% Ni 22% Co 2% Ti 9% Al 3% Cu steel: For magnets; German alloy listed by PMA
OERSTIT 350	15% Ni 30% Co 5% Ti 7% Al 4% Cu steel: For magnets; German alloy listed by PMA
OERSTIT 400K	15% Ni 24% Co 1% Ti 9% Al 3% Cu steel: For magnets; German alloy listed by PMA
OERSTIT 400R	14% Ni 24% Co 9% Al 3% Cu steel: For magnets; German alloy listed by PMA
OERSTIT 450	15% Ni 32% Co 5% Ti 7% Al 4% Cu steel: For magnets; German alloy listed by PMA
OERSTIT 500	15% Ni 24% Co 8% Al 3% Cu steel: For magnets; German alloy listed by PMA
OERSTIT 600	15% Ni 24% Co 8% Al 3% Cu steel: For magnets; German alloy listed by PMA
OERSTIT 700	15% Ni 24% Co 8% Al 3% Cu steel: For magnets; German alloy listed by PMA
P 3	0.1% C 0.6% Cr 1.25% Co steel: For tools; designation used by AISI
P 6	0.1% C 1.5% Cr 3.5% Co steel: For tools; designation used by AISI
P 15	0.8% C 4.0% Cr 1.0% Mo 1.5% V 18% W 2.0% Co steel: Bofors; cutting tools **DPN: 840**
P 21	0.2% C 1.2% Ni 4.0% Co steel: For tools; designation used by AISI

Note. The following abbreviations and units are used in the tables:

DPN Hardness, diamond pyramid number
UTS Ultimate tensile strength, N/mm^2
Elon Elongation, %
Proof 0.1% proof strength, N/mm^2

1 N/mm^2=0.1 hbar=0.102 kgf/mm^2=0.06475 $tonf/in.^2$=145.04 $lbf/in.^2$=1 MPa

See Appendix II for other abbreviations and conversion tables.

Symbol	Nominal analysis, supplier, condition and remarks.
PANTHER 5	4% Cr 5% V 12% W 5% Co steel: For tools; Allegheny Ludlum; AISI code T15G
PANTHER SPECIAL	4% Cr 1% V 18% W 5% Co steel: For tools; Allegheny Ludlum; AISI code T4G
PB 2	1.5% C 4.75% Cr 5% V 12% W 5% Co steel: For tools; Osborn; obsolete **DPN: 910**
PBD	0.35% C 1% Cr 0.5% Mo 5% W 4% Co steel: Jonas; dies for casting Cu **DPN: 570**
PDCO	0.3% C 2.5% Cr 10% W 2% Co steel: Jonas; dies for casting Cu and also hot press dies **DPN: 547**
PERMANIT	Prefix for Austrian alloys using same grade numbers as Oerstit; listed by PMA
PERMET PF2	3.0% Co Fe alloy: For magnets; origin unknown
PLUTO PARAMOUNT	0.8% C 4.5% Cr 0.7% Mo 1.25% V 18% W 10% Co steel: Carrs; turning tools **DPN: 950**
PLUTO PEERLESS	1.3% C 4.1% Cr 3% Mo 3% V 9% W 9.5% Co steel: Carrs
PLUTO PERFECTUM	0.8% C 4.8% Cr 0.8% Mo 1.25% V 18% W 5% Co steel: For tools; Carrs; milling cutters, reamers, etc. **DPN: 820**
PLUTO PLUS	1.1% C 3.7% Cr 9.5% Mo 1.2% V 1.5% W 8% Co steel: Carrs
PLUTO PREMIER	1.5% C 4.5% Cr 3% Mo 5% V 6.3% W 5% Co steel: Carrs; cutting tools; broaches **DPN: 950**
PO 41	1.1% C 4.3% Cr 3.8% Mo 2.0% V 6.8% W 5.0% Co steel: Bofors
POWHATAN	4% Cr 1% V 18% W 5% Co steel: For tools; Atlas Steels Ltd; AISI code T4Z
PRESSURDIE C	0.38% C 4.2% Cr 0.5% Mo 2.2% V 4.2% W 4.2% Co steel: Braeburn; AISI type H19
PURPLE LABEL	4% Cr 1% V 18% W 5% Co steel: For tools; Jessop; AISI code T4W
PURPLE LABEL EXTRA	4% Cr 2% V 18% W 8% Co steel: For tools; Jessop; T5 type; AISI code T5W
PYROMET CTX-1	0.03% C 37.7% Ni 0.2% Cr (max) 16% Co 3% Nb + Ta steel: Carpenter **UTS: 1480 Elon: 16% Proof: 1300**
PYROMET X12	0.12% C 10.5% Cr 4.7% Mo 6.0% Co 1.2% Cu 0.08% N steel: Carpenter
PYROMET X15	0.03% C (max) 15.2% Cr 2.7% Mo 20% Co steel: Carpenter
Q 5	0.85% C 4.0% Cr 1.0% Mo 1.7% V 18% W 5.0% Co steel: Bofors; obsolete **DPN: 850**
Q 10	0.85% C 4.0% Cr 1.0% Mo 1.7% V 18% W 10% Co steel: Bofors; obsolete **DPN: 850**
QRO 45	0.3% C 2.8% Cr 2.8% Mo 0.5% V 2.8% Co steel: Bofors **DPN: 400**
QX 1	1.1% C 4.3% Cr 5.0% Mo 2.6% V 6.5% W 5.3% Co steel: Bofors; high speed steel similar to type M35
QX 2	1.1% C 4.0% Cr 5.0% Mo 2.5% V 6.5% W 10% Co steel: Bofors; cutting tools; obsolete **DPN: 850**
R 9K5	0.95% C 4.1% Cr 2.3% V 9.7% W 5.5% Co steel: For tools; Russian Standards designation
R 9K10	0.95% C 4.1% Cr 2.3% V 9.7% W 10.0% Co steel: For tools; Russian Standards designation
R 10K5F5	1.5% C 4.5% Cr 4.6% V 10.7% W 5.5% Co steel: For tools; Russian Standards designation
R 18K5F2	0.9% C 4.1% Cr 2.1% V 17.7% W 5.5% Co steel: For tools; Russian Standards designation
R 30155	0.1% C 20% Ni 22% Cr 3% Mo 2.5% W 20% Co 1% Nb + Ta Fe alloy: UNS designation used by ASTM

Symbol	Nominal analysis, supplier, condition and remarks.
RC 32	1.25% C 23% Cr 8.0% Mo 10% Co Fe alloy: Casting; Dewrance **DPN: 460 UTS: 810**
RECO 2A	20% Ni 20% Co 6.5% Ti 7.0% Al 7.0% Cu Fe alloy: Casting for magnets; Belgian alloy listed by PMA
RECO 3A	20% Ni 14% Co 0.5% Ti 9.5% Al 6.0% Cu Fe alloy: Casting for magnets; Belgian alloy listed by PMA
RED CHIP	4% Cr 1% V 18% W 5% Co steel: For tools; Firth Sterling; AISI code T4B
RED CUT COBALT	4% Cr 1% V 18% W 5% Co steel: For tools; Vanadium Alloys; AISI code T4A
RED CUT COBALT B	4% Cr 2% V 18% W 8% Co steel: For tools; Vanadium Alloys; AISI code T5A
RED SABRE	4% Cr 5% V 12% W 5% Co tool steel: Bethlehem Steel; AISI code T15E
REMALLOY	17% Mo 12% Co Fe alloy: For magnets; American alloy listed by PMA
REMALLOY 17	17% Mo 12% Co steel: For magnets; Simonds
REMALLOY 20	20% Mo 12% Co steel: For magnets; Simonds
REX 95	4% Cr 2% V 14% W 5% Co steel: For tools; Crucible Steel: AISI code T8X
REX 440	4.5% Cr 1.5% V 20% W 12% Co steel: For tools; Crucible Steel; AISI code T6X
REX AAA	4% Cr 1% V 18% W 5% Co steel: For tools; Crucible Steel; AISI code T4X
REX M25	4% Cr 5% Mo 2% V 6% W 5% Co steel: For tools; Crucible Steel; AISI code M35X
REX SUPER CUT	4% Cr 2% V 18% W 8% Co steel: For tools; Crucible Steel; AISI code T5X
REX T Mo 5	4% Cr 8% Mo 1.2% V 2% W 5% Co steel: For tools; Crucible Steel; AISI code M30X
S 6-5-2-5	0.92% C 4.2% Cr 5.0% Mo 1.85% V 6.3% W 4.7% Co steel: For tools; German Standards designation
S 7-4-2-5	1.10% C 4.2% Cr 3.8% Mo 1.85% V 6.8% W 5.0% Co steel: For tools; German Standards designation
S 10-4-3-10	1.28% C 4.2% Cr 4.0% Mo 3.2% V 10.2% W 10.5% Co steel: For tools; German Standards designation
S 12-1-4-5	1.4% C 4.2% Cr 0.85% Mo 3.7% V 12.0% W 5.0% Co steel: For tools; German Standards designation
S 18-1-2-5	0.8% C 4.2% Cr 0.65% Mo 1.6% V 17.8% W 5.0% Co steel: For tools; German Standards designation
S 497	4% C 20% Ni 14% Cr 4% Mo 4% W 19% Co 4% Nb Fe alloy: Allegheny Ludlum
S 590	0.4% C 20% Ni 20% Cr 4% W 20% Co 4% Nb Fe alloy: Allegheny Ludlum
S 2-10-1-8	1.1% C 4% Cr 9.5% Mo 1.2% V 1.5% W 7.8% Co steel: German Standard
S 12-0-5-5	1.5% C 4.8% Cr 5% V 12.5% W 5% Co steel: German Standard
S 18-1-2-10	0.76% C 4.3% Cr 0.6% Mo 1.5% V 18% W 9.5% Co steel: German Standard
S 300	0.76% C 4.3% Cr 0.6% Mo 1.5% V 18.0% W 9.5% Co tool steel: VEW
S 305	0.81% C 4.3% Cr 0.7% Mo 1.5% V 18.0% W 4.8% Co tool steel: VEW
S 307	1.5% C 4.8% Cr 5.0% V 12.5% W 5.0 Co tool steel: VEW
S 500	1.1% C 3.9% Cr 9.4% Mo 1.2% V 1.5% W 7.8% Co tool steel: VEW
S 700	1.25% C 4.3% Cr 3.8% Mo 3.2% V 9.8% W 10.4% Co tool steel: VEW
S 705	0.9% C 4.3% Cr 5.0% Mo 1.9% V 6.4% W 4.8% Co tool steel: VEW
SABEN TENCO	0.75% C 5% Cr 0.5% Mo 1.7% V 18% W 9.5% Co steel: Sanderson **DPN: 790**
SABEN WUNDA	0.75% C 4% Cr 0.5% Mo 1.25% V 18% W 5.5% Co steel: Sanderson; cutting tools **DPN: 790**
SABRE	4% Cr 5% V 12% W 5% Co steel: For tools; Atlas Steel Ltd; AISI code T15Z

Symbol	Nominal analysis, supplier, condition and remarks.
SAE HEV 1	0.1% C 20% Ni 21.3% Cr 3.0% Mo 2.5% W 20% Co steel: Annealed
SAE T4	0.75% C 4% Cr 0.8% Mo 1% V 18% W 5% Co steel: For tools
SAE T5	0.8% C 4% Cr 0.8% Mo 2.1% V 18.5% W 8% Co steel: For tools
SAE T8	0.8% C 4.2% Cr 0.8% Mo 2% V 13% W 5% Co steel: For tools
SC	0.8% C 4.5% Cr 1.5% V 21.5% W 12% Co tool steel: ESC; code for Super Cyclone **DPN: 910**
SERMALLOY A1	14% Ni 3.8% Co 8% Ti 7.5% Al 3% Cu steel: For magnets; French alloy listed by PMA
SERMALLOY A2	12% Ni 24% Co 0.2% Si 7.5% Al 2% Cu steel: For magnets; French alloy listed by PMA
SKH 3	0.77% C 4.2% Cr 1.0% V 18.0% W 5.0% Co steel: For tools; Japanese Standards designation
SKH 4A	0.77% C 4.1% Cr 1.25% V 18.0% W 9.5% Co steel: For tools; Japanese Standards designation
SKH 4B	0.77% C 4.2% Cr 1.25% V 19.0% W 15.0% Co steel: For tools; Japanese Standards designation
SKH 5	0.3% C 4.2% Cr 1.25% V 18.5% W 16.5% Co steel: For tools; Japanese Standards designation
SKH 10	1.52% C 12.5% Cr 4.2% Mo 5.0% V 12.5% W 4.7% Co steel: For tools; Japanese Standards designation
SKH 55	0.85% C 4.2% Cr 5.2% Mo 2.0% V 6.2% W 5.0% Co steel: For tools; Japanese Standards designation
SKH 56	0.85% C 4.1% Cr 5.5% Mo 2.0% V 6.1% W 8.0% Co steel: For tools; Japanese Standards designation
SKH 57	1.2% C 4.2% Cr 3.5% Mo 3.3% V 10.0% W 10.0% Co steel: For tools; Japanese Standards designation
SL 972	0.1% C 3.5% Cr 5.0% Mo 0.5% V 4.0% W 25% Co steel: Swift Levick; for hot dies
SL 977	0.25% C 12% Cr 0.5% V 8.0% W 10% Co steel: Swift Levick; for hot dies
SO 12/21	0.75% C 4.0% Cr 0.75% Mo 1.0% V 20.0% W 10.0% Co steel: Origin unknown
SOBV	0.78% C 4.5% Cr 0.75% Mo 1.25% V 18% W 10% Co steel: For tools; Osborn; obsolete
SOUDOKAY A46.0	4.5% C 20.7% Cr 5.2% Mo 8.8% Co steel: Welding electrode; Soudometal **DPN: 640**
SOVB	0.78% C 4.5% Cr 0.75% Mo 1.25% V 18.75% W 10.25% Co steel: For tools; origin unknown
SPEAR DOUBLE CENTURY	1.5% C 4.5% Cr 4.0% Mo 4.8% V 9.0% W 7.5% Co steel: Origin unknown
SPEAR SUPERIOR	0.8% C 4.5% Cr 1.5% V 20.0% W 12.0% Co steel: Origin unknown
SPEAR TRIPLE MERMAID	0.75% C 4.0% Cr 1.0% V 18.0% W 5.0% Co steel: Origin unknown
SPEEDICUT LEDA	0.75% C 4% Cr 0.7% Mo 1.4% V 18% W 5.2% Co steel: Firth Brown
SPEEDICUT SUPERLEDA	0.78% C 4.8% Cr 1% Mo 2% V 18.8% W 9.3% Co steel: Firth Brown
SPEEDICUT VANLEDA	1.5% C 4.5% Cr 5% V 12.2% W 4.7% Co steel: Firth Brown
SS 2733	0.88% C 0.3% Mn 4.0% Cr 5.0% Mo 1.9% V 6.5% W 5.0% Co steel: For cutting tools; Swedish Standard
SS 2750	0.7% C 4.5% Cr 0.5% Mo 1.2% V 18% W 0.6% Co steel: Swedish Standard; annealed **DPN: 265**
SS 2754	0.8% C 4.5% Cr 1.4% Mo 1.6% V 18% W 5.5% Co steel: Swedish Standard; annealed **DPN: 290**
SS 2756	0.8% C 4.5% Cr 1% Mo 1.6% V 18% W 10% Co steel: Swedish Standard; annealed **DPN: 310**
STAG EXTRA SPECIAL	0.8% C 5% Cr 0.5% Mo 1.5% V 19% W 6% Co steel: Edgar Allen
STAG MAJOR	0.8% C 5% Cr 0.5% Mo 1.5% V 21% W 11.5% Co steel: Edgar Allen
STAG VANCO	1.5% C 5% Cr 0.5% Mo 5% V 17.5% W 9% Co steel: Edgar Allen
STAR BORON	4% Cr 8% Mo 1.5% V 8% Co B steel: For tools; Carpenter; M40 type; AISI code M40K
STAR Mo M 2/5	4% Cr 5% Mo 2% V 6% W 5% Co steel: For tools; Firth Sterling; N 135 type AISI code M35B
SUPER CAPITAL	1.3% C 4.25% Cr 3% Mo 3.5% V 9% W 8.5% Co steel: For tools; Balfour, reamers, hobs **DPN: 975**
SUPER COBALT	4% Cr 2% V 18% W 8% Co steel: For tools; Simonds; T5 type AISI code T5N
SUPER CYCLONE	0.8% C 4.5% Cr 1.5% V 21.5% W 12% Co steel: For tools; ESC; cutting tools **DPN: 910**
SUPER DBL	4% Cr 5% Mo 2% V 6% W 8% Co steel: For tools; Allegheny Ludlum; M36 type AISI code M36G
SUPER HALLAMITE	18% W 5% Co steel (C not specified): For tools; Hallams
SUPER Hi-Mo	4% Cr 8% Mo 1.2% V 2% W 5% Co steel: For tools; Firth Sterling; M30 type AISI code M30B
SUPER INMANITE	0.8% C 4.5% Cr 1% Mo 1.5% V 22% W 10% Co steel: Inman; cutting tools
SUPER INVINCIBLE 5	0.75% C 4.0% Cr 1.0% V 18.0% W 5.0% Co steel: For tools; origin unknown
SUPER INVINCIBLE ADVANCE 10	0.8% C 4.0% Cr 2.0% V 18.0% W 8.0% Co steel: For tools; origin unknown
SUPER INVINCIBLE TB	0.75% C 4.0% Cr 2.0% V 14.0% W 5.0% Co steel: For tools; origin unknown
SUPERLEDA	0.78% C 4.8% Cr 1% Mo 18.8% W 2% V 9.3% Co steel: Firth Brown
SUPER LMW	4% Cr 8% Mo 1.2% V 2% W 5% Co steel: For tools; Allegheny Ludlum; M30 type AISI code M30G
SUPER Mo CHIP	4% Cr 8% Mo 1.5% V 8% Co B steel: For tools; Firth Sterling; M40 type AISI code M40B
SUPER MOTUNG	4% Cr 8% Mo 1.2% V 2% W 5% Co steel: For tools; Universal Cyclops; M30 type AISI code M30D
SUPER MOTUNG SPECIAL	4% Cr 8% Mo 2% V 2% W 8% Co steel: For tools; Universal Cyclops; M34 type AISI code M34D
SUPER PANTHER	4% Cr 2% V 18% W 8% Co steel: For tools; Allegheny Ludlum; T5 type AISI code T5G
SUPER STAR	1.08% C 3.7% Cr 9.5% Mo 1.1% V 1.5% W 8% Co steel: Carpenter
SUPER UNICUT	4% Cr 3.5% Mo 5% V 6.5% W 5% Co steel: For tools; Universal Cyclops; AISI code M15D
SUPER-SUPER MONARCH	0.73% C 4.0% Cr 0.8% Mo 1.1% V 18.0% W 4.7% Co steel: For tools; origin unknown
T 4	0.75% C 4% Cr 1% V 18% W 5% Co steel: For tools; designation used by AISI; the supplier is coded by a suffix letter
T 4	0.73% C 4.0% Cr 0.8% Mo 1.1% V 18.0% W 4.7% Co steel: Osborn
T 5	0.8% C 4% Cr 2% V 18% W 8% Co steel: For tools; designation used by AISI; the supplier is coded by a suffix letter
T 6	0.8% C 4.5% Cr 1.5% V 20% W 12% Co steel: For tools; designation used by AISI; the supplier is coded by a suffix letter

Note. The following abbreviations and units are used in the tables:

DPN	Hardness, diamond pyramid number
UTS	Ultimate tensile strength, N/mm^2
Elon	Elongation, %
Proof	0.1% proof strength, N/mm^2

1 N/mm^2=0.1 hbar=0.102 kgf/mm^2=0.06475 $tonf/in.^2$=145.04 $lbf/in.^2$=1 MPa

See Appendix II for other abbreviations and conversion tables.

Symbol	Nominal analysis, supplier, condition and remarks.
T 6	0.8% C 4.0% Cr 0.8% Mo 1.5% V 20.5% W 11.5% Co steel: Osborn
T 8	0.75% C 4% Cr 2% V 14% W 5% Co steel: For tools; designation used by AISI; the supplier is coded by a suffix letter
T 15	1.5% C 4% Cr 5% V 12% W 5% Co steel: For tools; designation used by AISI; the supplier is coded by a suffix letter
TEM 02C	1% C 4.2% Cr 4.8% Mo 2% V 6% W 2% Co steel: Speedsteel
THREE CASTLES	0.75% C 4.5% Cr 1.25% V 18.0% W 5.0% Co steel: For tools; origin unknown
TICONAL C	14% Ni 24% Co 1.0% Nb 8.0% Al 3.0% Cu steel: Casting; Mullard alloy listed by PMA
TICONAL D	14% Ni 24% Co 1.0% Ti 8.0% Al 3.0% Cu steel: Casting; Mullard alloy listed by PMA
TICONAL E	14% Ni 24% Co 1.5% Ti 8.0% Al 3.0% Cu steel: Casting; Mullard alloy listed by PMA
TICONAL F	14% Ni 24% Co 0.5% Ti 8.0% Al 3.0% Cu steel: Casting; Mullard alloy listed by PMA
TICONAL G	14% Ni 24% Co 8.0% Al 3.0% Cu steel: Casting; Mullard alloy listed by PMA
TICONAL GX	14% Ni 24% Co 8.0% Al 3.0% Cu steel: Casting Mullard alloy listed by PMA; columnar crystal form
TICONAL H	14% Ni 24% Co 2.5% Nb 8.0% Al steel: Casting; Mullard alloy listed by PMA
TICONAL L	14.5% Ni 19.5% Co 0.8% Si 7.0% Al 1.5% Cu steel: Casting; Mullard alloy listed by PMA
TICONAL S	14% Ni 24% Co 8.0% Al 3.0% Cu steel: Sintered; Mullard alloy listed by PMA
TICONAL X	15% Ni 34% Co 5.0% Ti 7.0% Al steel: Casting Mullard alloy listed by PMA
TREBLE SUPER MONARCH	0.8% C 4.5% Cr 1.5% V 20.0% W 12.0% Co steel: For tools; origin unknown
TRIPLE 5C	1.5% C 5% Cr 5% V 13% W 5% Co steel: For tools; F Parkin **DPN: 910**
TRIPLE 5 MONARCH	1.5% C 4.0% Cr 5.0% V 12.0% W 5.0% Co steel: For tools; origin unknown
TRIPLE FLYGO	0.73% C 18% Cr 1% Mo 1% V 5% Co steel: For tools; Turton Bros
TRIPLE MERMAID	0.8% C 4.5% Cr 1.5% V 18% W 5.5% Co steel: For tools; Spear and Jackson
TRIPLE MUSHET	0.8% C 4.8% Cr 0.75% Mo 1.25% V 18.2% W 5.0% Co steel: For tools; origin unknown
TRIPLE MUSKET GZ	0.75% C 4.0% Cr 1.0% V 18.0% W 5.0% Co steel: Origin unknown
TRIPLE SPUR	0.82% C 4.5% Cr 1% Mo 1.7% V 20% W 11% Co steel: For tools; T Turton
TRIPLE VELOS	0.8% C 4.2% Cr 0.5% Mo 1.15% V 18% W 5.0% Co steel: Spencer; cutting tools
TRIUMPH SUPERB 1000	0.8% C 4.7% Cr 1.6% V 18.5% W 5.7% Co steel: For tools; Jessop **DPN: 730**
TRIUMPH SUPERB DOUBLE	0.8% C 4.5% Cr 1.5% V 20.0% W 12.0% Co steel: Origin unknown
TRIUMPH SUPERB DOUBLE 1000	0.8% C 4.7% Cr 1.6% V 20% W 10% Co steel: For tools; Jessop **DPN: 730**
TTQ	0.8% C 4.7% Cr 1.6% V 20% W 10% Co steel: For tools; Jessop **DPN: 730**
TUNCO	4% Cr 1% V 18% W 5% Co steel: For tools; Simonds; AISI code T4N
TWO SPUR	0.78% C 4.5% Cr 0.7% Mo 1.7% V 21% W 6% Co steel: For tools; T Turton
UHB 424	4% Cr 5% Mo 2% V 6% W 5% Co steel: Uddelholm; obsolete
UHB 442	1% C 4% Cr 9.8% Mo 1% V 1.5% W steel: Uddelholm; obsolete

Symbol	Nominal analysis, supplier, condition and remarks.
ULTRA CAPITAL + 1	0.85% C 4.25% Cr 1.2% V 18.5% W 5% Co steel: For tools; Balfour; reamers, screwing, etc. **DPN: 910**
ULTRA CAPITAL + 2	0.76% C 4.25% Cr 1.5% V 20% W 10% Co steel: For tools; Balfour; cutting tools for hard material **DPN: 830**
VANCO 5	1.5% C 5% Cr 5% V 12% W 5% Co steel: For tools; Darwin; broaches
VANLEDA	1.5% C 4.5% Cr 5.0% V 12.2% W 4.7% Co steel: For tools; Firth Brown
VASCO SUPREME	4% Cr 5% V 12% W 5% Co steel: For tools; Vanadium Alloys; AISI code T15A
VASCO SUPREME A	4% Cr 3.5% Mo 5% V 6.5% W 5% Co steel: For tools; Vanadium Alloys; AISI code M15A
VC 12	0.8% C 4.2% Cr 0.5% Mo 1.8% V 20% W 12% Co steel: W Spencer; cutting tools **DPN: 800**
VELOS 42	1.05% C 4.0% Cr 9.5% Mo 1.2% V 1.6% W 8.5% Co steel: W Spencer
VF	0.4% C 4.2% Cr 0.4% Mo 2.2% V 4.2% W 4.2% Co steel: For tools; Osborn; for hot working; obsolete **DPN: 550**
VG	1.5% C 4.7% Cr 3% Mo 5% V 6.5% W 5% Co steel: For tools; Osborn; obsolete **DPN: 950**
VS 10	High C tool steel with 10.0% Co high speed: Turton Bros
VS 35	0.6% C 2% Cr 6% W 35% Co steel: For magnets; Vulcan
W 705	0.16% C 10.0% Cr 5.1% Mo 0.5% V 10.0% Co steel: VEW
W 720	0.03% C 18.5% Ni 5.3% Mo 9.0% Co 0.6% Ti 0.1% Al steel: VEW
W 725	0.03% C 18.0% Ni 4.2% Mo 12.3% Co 1.65% Ti 0.15% Al steel: VEW
WE	1.35% C 4% Cr 3% Mo 3% V 9% W 9% Co steel: For tools; Osborn; obsolete **DPN: 1000**
WKE	0.8% C 4.5% Cr 1.2% Mo 1.4% V 18.5% W 5.5% Co steel: Fagersta; milling cutters
WKE 4	1.25% C 4.1% Cr 3.1% Mo 3.1% V 9.0% W 9.0% Co steel: For tools; Fagersta **DPN: 670**
WKE 44	1.2% C 4.8% Cr 3.2% Mo 3.3% V 7.3% W 5.3% Co steel: Fagersta; parting off tools, milling cutters
WKE 45	1.4% C 4.2% Cr 3.5% Mo 3.9% V 9.0% W 12.5% Co steel: For tools; Fagersta **DPN: 850**
WKE 45	1.4% C 4.2% Cr 3.5% Mo 3.4% V 8.5% W 10.5% Co steel: Speedsteel
WKE 46	0.82% C 5.0% Mn 1.9% V 6.5% W 5.0% Co 4.0% Si steel: Fagersta **DPN: 790**
WKE 48	1% C 4% Cr 9.8% Mo 1.1% V 1.5% Co steel: Fagersta; obsolete
WKE Extra	0.8% C 4.5% Cr 1.0% Mo 1.4% V 18.5% W 10.0% Co steel: For tools; Fagersta **DPN: 671**
WLM 1020	0.8% C 4% Cr 1.25% V 20% W 10% Co steel: W Marrison; cutting tools
WLM 5018	0.8% C 4% Cr 1.25% V 18% W 5% Co steel: W Marrison; cutting tools
WOLFRAM COBALT	4% Cr 1% V 18% W 5% Co steel: For tools; Vulcan; AISI code T4P
WU	0.3% C 2.7% Cr 0.15% Mo 0.25% V 10% W 2.2% Co steel: For tools; Schoeller-Bleckmann **UTS: 1550**
WV Co	1.5% C 4.7% Cr 12.5% W 5% Co steel: For tools; ESC symbol for Cyclone Supercut **DPN: 910**

Symbol	Nominal analysis, supplier, condition and remarks.
WVC	0.4% C 4.2% Cr 0.4% Mo 2.2% V 4.2% W 4.2% Co steel: For tools; Schoeller-Bleckmann **UTS: 1550**
X15 Cr Co Mo 10.10.5	0.16% C 10% Cr 5% Mo 0.5% V 10% Co steel: German Standard
XERG 0.45	3% Cr 2.9% Mo 0.9% V 0.9% W 2.9% Co steel: W Somers
YCM 2B	15% Ni 24% Co 1.25% Ti 8% Al 3.4% Cu steel: For magnets; Japanese alloy listed by PMA
YMDC	0.35% C 3% Ni 1.5% Cr 5% W 0.5% Co 1% Si steel: For tools; Jonas; hot extrusion dies **DPN: 580**
Z 80WKCV 18-5/04-01	0.8% C 4.0% Cr 1.0% V 18.0% W 5.0% Co steel: For tools; French Standards designation
Z 80WKCV 18-10/04-02	0.8% C 4.0% Cr 1.6% V 18.0% W 10.0% Co steel: For tools; French Standards designation
Z 85WDKCV	0.85% C 4.0% Cr 5.0% Mo 2.0% V 6.0% W 5.0% Co steel: For tools; French Standards designation
Z 110DKCWV	1.1% C 3.8% Cr 9.5% Mo 1.15% V 1.5% W 8.0% Co steel: For tools; French Standards designation
Z 150WDKCV	1.5% C 4.0% Cr 5.0% Mo 5.0% V 6.5% W 5.0% Co steel: For tools; French Standards designation
Z 150WKCV 12-05/05-04	1.3% C 4.0% Cr 5.0% V 12.0% W 5.0% Co steel: For tools; French Standards designation
Z 165WKCV	1.65% C 4.0% Cr 5.0% V 12.0% W 10.0% Co steel: For tools; French Standards designation
Z 175KWDCV	1.75% C 4.0% Cr 5.0% Mo 5.0% V 6.5% W 10.0% Co steel: For tools; French Standards designation

Note. The following abbreviations and units are used in the tables:

DPN	Hardness, diamond pyramid number
UTS	Ultimate tensile strength, N/mm^2
Elon	Elongation, %
Proof	0.1% proof strength, N/mm^2

1 N/mm^2=0.1 hbar=0.102 kgf/mm^2=0.06475 tonf/in.^{2}=145.04 lbf/in.^{2}=1 MPa

See Appendix II for other abbreviations and conversion tables.

44M1 Steel – low and medium carbon, high chromium
0.8% carbon maximum, 10–30% chromium maximum with titanium; molybdenum and nickel

Specific gravity		7.7
Density		7700 kg/m^3
Solidus/liquidus		1470 °C
Thermal conductivity		18.8 W/m °C
Coefficient of linear expansion		12.3 × 10^{-6}/ °C
Electrical conductivity		2.4% IACS (copper 100%)
Specific resistance		720 microhm mm
Young's modulus of elasticity		207 × 10^9 N/m^2
Impact		7–9.5 J (can be higher)
Fatigue strength	endurance limit	±300 N/mm^2
Hot strength		

Temperature °C	Tensile strength N/mm^2	Elongation %
450	760	60
500	700	68
600	450	81
650	370	85

The above properties are typical of the following group and may not apply exactly to any one specification. It is possible that with certain specifications some of the values may not be applicable.

General metallurgical characteristics

These steels have a minimum of about 11% chromium, sufficient to ensure that the surface is always protected with a film of chromium oxide, which is firmly adherent and successfully prevents the formation of the loose iron oxides which normally form. None of the steels are truly stainless as they will corrode if used outdoors in damp climates.

As the carbon content is decreased the corrosion resistance improves, until with the carbon content held below that which allows hardening the corrosion resistance is excellent.

There are actually two distinct groups of steel under this heading as the low carbon or ferritic high chrome steels are different in many respects from the medium carbon martensitic high chrome steels. It was found impossible, however, to make the division on analysis alone as the same steel can be ferritic or martensitic depending on the thermal treatment.

The very low carbon steels are always ferritic, being non-heat treatable, but from 0.1% carbon the steels are almost always used in the hardened and tempered martensitic condition.

The lower carbon steels very seldom have any alloy content apart from chromium. The general effect of the alloying elements in the higher carbon steels is to improve the creep and other hot strength properties. The alloy elements are usually carbide formers which help to prevent grain growth, stabilize the structure and generally slow down all the metallurgical reactions. Nickel up to about 2% is sometimes present to improve the fatigue strength and low temperature properties which are poor in general with these steels.

The low carbon steels in this group will accept cold working to some extent but as the carbon increases the ability to accept cold work decreases and the steels become very hard and brittle with slight amounts of cold work.

Many of these steels are used in the annealed condition where corrosion resistance is of more importance than strength.

Full annealing, hardening and normalizing will all require temperatures close to, or above, 1000 °C. Most of the steels in this group will be oil quenched. Some of the higher carbon steels will give a satisfactory hardness with air quenching.

It must be noted that these steels are prone to temper brittleness. They should not therefore be held at temperatures between 300 and 600 °C for any appreciable length of time. This means that tempering should be carried out above 600 °C and this is one occasion where material should be quenched following the tempering operation. Temper brittleness is the result of precipitation of chromium carbides at the grain boundaries.

Expert advice should be obtained when these materials have to be thermally treated.

No technical problems exist when welding the lower carbon steels. With carbon above approximately 0.15% preheating is required. Advice should be obtained before welding. The use of inert gas welding is recommended. With the higher carbon steels there will be a great danger of cracking in the heat affected zone and specialist advice must always be obtained prior to welding these materials.

The corrosion resistance of these materials is not as good as those found in Sections 44N. The corrosion resistance in this instance is by the formation of the adherent chromium oxide which is self-healing. That is, as it is removed in a normal atmosphere it will re-form. It is thus necessary to design components so that all the surfaces are in contact with the atmosphere. Lap joints should not be used and the components should not be used in circumstances where the surfaces are covered and can be abraded.

Once corrosion commences the corrosion products will tend to trap the corroding material and thus corrosion can propagate rapidly. It is therefore advisable that components in these materials should be kept clean.

The uses for the materials in this group include cutlery, particularly knives, golf clubs, steam turbine blading, compressor blades in gas turbines, motor car trims, household ornaments of all types, steel tubing and sheet for medium corrosion conditions but not for marine or food uses, and architectural uses for inside trim or outside use in rural conditions but not marine or industrial.

These materials should always be identified as being magnetic stainless steels or basic 12% chromium steels to differentiate them from the 18/8 austenitic non-magnetic stainless steels found in Sections 44N1 to 44N2.

Symbol	Nominal analysis, supplier, condition and remarks.
0.000	0.25% C 3% Ni 29% Cr steel: Casting; D Brown **UTS: 370**
0.4021	0.19% C 13% Cr steel: German Standard
1.400	0.08% C (max) 13.0% Cr steel: German Standard
1.2082	0.25% C 3% Ni 29% Cr steel: Casting; D Brown **UTS: 370**
1.2083	0.43% C 13.5% Cr steel: German Standard
1.2316	0.38% C 0.8% Ni 1.6% Cr 1% Mo steel
1.2625	0.38% C 12.0% Cr 1.0% V 12.0% W steel: For tools; German Standard
1.3811	0.15% C (max) 25% Cr steel: Bar; German Standard
1.3827	0.1% C (max) 17% Cr 0.8% Ti steel: Forging; German Standard
1.4000	0.08% C (max) 13% Cr steel: German Standard
1.4001	0.08% C (max) 14% Cr steel: German Standard
1.4002	0.08% C (max) 13.0% Cr 0.2% Al steel: German Standard
1.4006	0.1% C 13% Cr steel: Forging; German Standard
1.4007	0.35% C 14% Cr steel: Welding rod; German Standard
1.4009	0.1% C (max) 14.5% Cr steel: Welding rod; German Standard
1.4015	0.1% C (max) 17.5% Cr steel: Forging; German Standard
1.4016	0.1% C (max) 16.5% Cr steel: Forging; German Standard
1.4021	0.19% C 13% Cr steel: For springs; German Standard
1.4024	0.15% C 13% Cr steel: Forging; German Standard
1.4034	0.41% C 13% Cr steel: For bearings; German Standard
1.4044	0.2% C 2% Ni 16% Cr steel: German Standard
1.4057	0.2% C 2% Ni 17% Cr steel: Forging; German Standard
1.4084	0.1% C (max) 28% Cr steel: German Standard
1.4104	0.13% C 16% Cr 0.25% Mo 0.2% S steel: German Standard; free machining
1.4113	0.07% C (max) 17.0% Cr 1.05% Mo steel: German Standard
1.4116	0.46% C 14.4% Cr 0.12% V steel: German Standard
1.4119	0.15% C 13% Cr 1.1% Mo steel: German Standard
1.4120	0.2% C 1.0% Ni 13% Cr 1.2% Mo steel: German Standard
1.4122	0.38% C 1.0% Ni 16% Cr 1.1% Mo steel: German Standard
1.4137	0.1% C (max) 2.0% Ni 30% Cr 2.1% Mo steel: German Standard
1.4502	0.1% C (max) 17.5% Cr 0.5% Ti steel: Forging; German Standard
1.4510	0.1% C (max) 17% Cr 0.8% Ti steel: Forging; German Standard
1.4511	0.1% C (max) 17.5% Cr 1.2% Nb steel: Forging; German Standard

Note. The following abbreviations and units are used in the tables:

DPN	Hardness, diamond pyramid number
UTS	Ultimate tensile strength, N/mm^2
Elon	Elongation, %
Proof	0.1% proof strength, N/mm^2

1 N/mm^2=0.1 hbar=0.102 kgf/mm^2=0.06475 tonf/in.^{2}=145.04 lbf/in.^{2}=1 MPa

See Appendix II for other abbreviations and conversion tables.

Symbol	Nominal analysis, supplier, condition and remarks.
1.4523	0.1% C (max) 17% Cr 1.8% Mo 0.8% Ti steel: German Standard
1.4525	0.1% C (max) 1.0% Ni 17.5% Cr 1.1% Mo 0.6% Ti steel: German Standard
1.4722	0.12% C (max) 13% Cr steel: Forging; German Standard
1.4723	0.12% C (max) 1.0% Ni 14.5% Cr steel: Forging; German Standard
1.4724	0.12% C (max) 13% Cr 1.0% Al steel: Forging; German Standard
1.4741	0.12% C (max) 18% Cr steel: Forging; German Standard
1.4742	0.12% C (max) 18% Cr 1.0% Al steel: Forging; German Standard
1.4749	0.18% C 27% Cr steel: German Standard
1.4762	0.12% C (max) 24% Cr 1.5% Al steel: Forging; German Standard
1.4765	25% Cr 5% Al steel: German Standard
1.4767	20% Cr 5% Al steel: German Standard
1.4772	0.12% C (max) 29% Cr steel: Forging; German Standard
1.4773	0.1% C (max) 30% Cr steel: Forging; German Standard
1.4774	0.1% C (max) 29% Cr 4.5% Al steel: Heat conductor; German Standard
1.4820	0.15% C (max) 4.5% Ni 26% Cr steel: Forging; German Standard
1.4821	0.2% C 4.5% Ni 25% Cr steel: Forging; German Standard
1.4905	0.1% C (max) 20% Cr 4.5% Al steel: Heat conductor; German Standard
1.4920	0.17% C 11.5% Cr 1.1% Mo 0.3% V steel: German Standard
1.4922	0.20% C 11.5% Cr 1.1% Mo 0.3% V steel: German Standard
1.4924	0.22% C 11% Cr 1.1% Mo 0.3% V steel: German Standard
1.4934	0.22% C 0.5% Ni 11% Cr 1.0% Mo 0.3% V 0.5% W steel: German Standard
1.4935	0.2% C 0.5% Ni 11.5% Cr 1.0% Mo 1.0% V 0.5% W steel: German Standard
1.4994	0.45% C 5.0% Ni 23.0% Cr 2.7% Mo steel: German Standard
1 C 27	0.08% C 13.5% Cr 1.0% Mo steel: Tube **DPN: 160 UTS: 520 Elon: 13% Proof: 350**
1C 36	0.12% C 17% Cr steel: Sandvik
1C 342	0.08% C 16.7% Cr 0.5% Ti steel: Sandvik
1CN	0.2% C 1% Ni 14% Cr steel: Inman; for plastic extrusion dies
1KH11MF	0.15% C 10.7% Cr 0.7% Mo 0.32% V steel: Russian Standards designation
1KH12N2VMF	0.13% C 1.65% Ni 11.5% Cr 0.4% Mo 0.22% V 1.8% W steel: Russian Standards designation
1KH12V2MF	0.13% C 12.0% Cr 0.75% Mo 0.22% V 2.0% W steel: Russian Standards designation
1KH12VNMF	0.15% C 0.6% Ni 12.0% Cr 0.6% Mo 0.2% V 0.9% W steel: Russian Standards designation
2 Cr 120 Mo 8 Co	0.2% C 12.3% Cr 0.8% Mo 0.3% V 2% Co steel: Origin unknown

Note. The following abbreviations and units are used in the tables:

DPN	Hardness, diamond pyramid number
UTS	Ultimate tensile strength, N/mm^2
Elon	Elongation, %
Proof	0.1% proof strength, N/mm^2

1 N/mm^2=0.1 hbar=0.102 kgf/mm^2=0.06475 tonf/in.^{2}=145.04 lbf/in.^{2}=1 MPa

See Appendix II for other abbreviations and conversion tables.

Symbol	Nominal analysis, supplier, condition and remarks.
2C 27	0.08% C 13.5% Cr steel: Sandvik
2C 34	0.12% C 17.2% Cr steel: Sandvik
2KH12VMBFR	0.18% C 12.0% Cr 0.5% Mo 0.2% V 0.5% W 0.3% Nb steel: Russian Standards designation
2R 27	0.12% C 13% Cr steel: Bofors; obsolete
2R 29	0.09% C 17.5% Cr steel: Bofors; obsolete
2R 47	0.21% C 13% Cr steel: Bofors
2R 57	0.3% C 0.5% Ni 13% Cr steel: Bofors; obsolete
2R 77	0.27% C 13% Cr steel: Bofors
2R 107	0.55% C 14% Cr steel: Bofors; hardened and tempered; obsolete **DPN: 325 UTS: 1070 Elon: 10% Proof: 870**
2RL 2	0.1% C 13.5% Cr 0.2% S steel: Free machining; Bofors; hardened and tempered; obsolete **DPN: 210 UTS: 700 Elon: 25% Proof: 530**
2RL 3	0.12% C 16.5% Cr 0.3% Mo 0.2% S steel: Free machining; Bofors; annealed **DPN: 185 UTS: 620 Elon: 20% Proof: 300**
2RL 14	0.2% C 1.0% Mn 13.5% Cr 0.2% S steel: Free machining; Bofors; hardened and tempered **DPN: 315 UTS: 750 Elon: 14% Proof: 680**
2RM 2	0.05% C 5.5% Ni 12.5% Cr steel: Bofors; hardened and tempered; suitable for welding **DPN: 265 UTS: 800 Elon: 15% Proof: 600**
2RMO	0.05% C 6% Ni 13% Cr 1.5% Mo steel: Weldable; Bofors; hardened and tempered **DPN: 290 UTS: 810 Elon: 15% Proof: 600**
2RO 26	0.1% C 0.5% Ni 12% Cr 0.5% Mo steel: Bofors; hardened and tempered **DPN: 220 UTS: 720 Elon: 22% Proof: 580**
2RO 27	0.1% C 13.5% Cr 1.2% Mo steel: Bofors
2RO 46	0.2% C 0.5% Ni 12% Cr 1.2% Mo steel: Bofors; obsolete
3C 27 Mo 2	0.1% C 13% Cr 1% Mo steel: Sandvik
3RE 60	0.03% C (max) 4.7% Ni 18.5% Cr 2.7% Mo steel: Tube; Sandvik; weldable **UTS: 836 Elon: 44% Proof: 420**
3RE 60	0.03% C (max) 4.7% Ni 18.5% Cr 2.7% Mo steel: Sandvik; ferritic; austenitic; weldable **DPN: 260 UTS: 820 Elon: 22% Proof: 500**
4C 27	0.2% C 12.5% Cr steel: Sandvik
4C 27A	0.22% C 1% Ni 13% Cr 1.3% Mo 0.2% S steel: Sandvik
4C 54	0.18% C 27% Cr steel: Sandvik
4N 2C 36	0.25% C 2% Ni 17% Cr steel: Sandvik
6C 27	0.32% C 14% Cr steel: Sandvik
6C 283	0.28% C 14% Cr 0.5% Mo 0.25% Cu steel: Sandvik
7C 27	0.27% C 13% Cr steel: Sandvik
7C 27 Mo2	0.35% C 13.5% Cr 1.3% Mo steel: Sandvik
10 RE 20	0.08% C 5% Ni 26.3% Cr steel: Sandik
10 RE 21	0.08% C (max) 5% Ni 26.5% Cr 1.5% Mo steel: Sandvik
10C 27	0.5% C 14% Cr steel: Sandvik
11 Cr 9 Mn B	0.4% C 9.0% Mn 11.0% Cr steel: Electrode; Metrode; work hardens from 200 to 400 DPN
12 Cr	0.1% C 12.5% Cr steel: Electrode; Metrode
12 Cr 152	0.12% C 2.5% Ni 12% Cr 1.8% Mo 0.3% V N steel: Electrode; Metrode
12 Cr Mo V	0.2% C 11.5% Cr 1% Mo 0.3% V 0.5% W steel: Electrode; Metrode
12 Cr Mo VB	0.2% C 0.5% Ni 12% Cr 1% Mo 0.3% V 0.5% W steel: Electrode; Metrode
12 Cr MOB	0.07% C 12.0% Cr 1.0% Mo steel: Electrode; Metrode
13.2R	0.06% C 2.0% Ni 13.0% Cr steel: Electrode; Metrode
13.2R	0.08% C 2% Ni 12.5% Cr steel: Electrode; Metrode
13.4 Mo LB	0.026% C 4.5% Ni 12% Cr 0.6% Mo steel: Electrode; Metrode
13.4 Mo LR	0.04% C 3.5% Ni 13.0% Cr 0.4% Mo steel: Electrode; Metrode
13.4 Mo LR	0.026% C 4.5% Ni 12% Cr 0.6% Mo steel: Electrode; Metrode

Symbol	Nominal analysis, supplier, condition and remarks.
13.4LB	0.04% C 3.5% Ni 13.0% Cr steel: Electrode; Metrode
13 BMP	0.06% C 12% Cr 0.2% Mo steel: Electrode; Metrode
13 RMP	0.06% C 12% Cr 0.2% Mo steel: Electrode; Metrode
13BMP	0.07% C 13.0% Cr steel: Electrode; Metrode
13RMP	0.07% C 13.0% Cr steel: Electrode; Metrode
15-15LC	0.03% C 15.3% Mn 1% Ni 16.3% Cr 1% Mo 0.5% Cu 0.4% N steel: Carpenter
	DPN: 210 UTS: 630 Elon: 50% Proof: 425
17 BMP	0.07% C 17.0% Cr steel: Electrode; Metrode
17 RMP	0.08% C 17% Cr steel: Electrode; Metrode
18.2	18.0% Cr 2.0% Mo steel: Resists chlorides; has good low temperature impact; Climax Molybdenum
18.2	0.025% C (max) 18.5% Cr 2.2% Mo 1% Ti steel: Designation used in the UK and USA
18 Cr 2 Mo	0.025% C 18.5% Cr 2% Mo 0.7% Nb + Ti steel: Designation used in the UK and USA
20RMP	0.07% C 20.0% Cr steel: Electrode; Metrode; resists scaling up to 950 °C
23C 27	0.12% C 13% Cr steel: Sandvik
25-4.4	0.025% C (max) 4% Ni 26.5% Cr 4% Mo 1% Ti steel: Designation used in the UK and USA
26.3.3	0.03% C (max) 2.2% Ni 26.5% Cr 3% Mo 0.2% Ti (min) steel: Designation used in the UK and USA
28RMP	0.07% C 28.0% Cr steel: Electrode; Metrode; resists scaling up to 1000 °C
29.4	0.01% C (max) 29% Cr 3.9% Mo steel: Designation used in the UK and USA
29.4.2	0.01% C (max) 2.1% Ni 29% Cr 3.9% Mo steel: Designation used in the UK and USA
29.4C	0.03% C (max) 29% Cr 4% Mo 0.2% Ti (min) steel: Designation used in the UK and USA
29 Cr 4 Mo	0.01% C (max) 29% Cr 4% Mo steel: Designation used in the UK and USA
29 Cr 4 Mo 2 Ni	0.01% C 2.2% Ni 29% Cr 3.8% Mo steel: Designation used in the UK and USA
40 Cr Mn Mo S 8.6	0.4% C 1.9% Cr 0.2% Mo steel: German Standard
45 Cr 13	0.45% C 13.0% Cr steel: Electrode; Metrode; hardenable corrosion resistant deposit
	DPN: 550
45 Cr 25	0.45% C 1.0% Mn 1.0% Ni 25.0% Cr steel: Electrode; Metrode; for repair welding of castings
	DPN: 300
56R	0.08% C 12% Cr 1% Mo 0.25% V 1% W steel: Origin unknown
56TS	0.2% C 11% Cr steel: Origin unknown
249H	0.25% C 2% Ni 17% Cr steel: Avesta
249T	0.12% C 17% Cr steel: Avesta
300 Cr 15 Mn 6	3.0% C 6.0% Mn 15.0% Cr 2.0% Si steel: Electrode; Metrode; for hard facing
	DPN: 550
393	0.12% C 13% Cr steel: Avesta
403 S17	0.08% C (max) 13% Cr steel: BS 970
409	0.08% C 11.1% Cr 0.46% Ti steel: Designation used in the UK and USA
410 S21	0.12% C 12.5% Cr steel: BS 970 designation
416 S21	0.12% C 12.5% Cr 0.6% Mo 0.2% S steel: For machining; BS 970 designation
416 S29	0.17% C 12.5% Cr 0.6% Mo 0.2% S steel: For machining; BS 970 designation
416 S37	0.24% C 13% Cr 0.6% Mo 0.2% S steel: For machining; BS 970 designation
416 S41	0.12% C 12.5% Cr 0.2% Se 0.6% Mo steel: For machining; BS 970 designation
419	0.25% C 11.5% Cr 0.5% Mo 2.5% W steel: American proprietary alloy listed in SAE yearbook
420 S29	0.18% C 12.5% Cr steel: BS 970 designation
420 S37	0.24% C 13% Cr steel: BS 970 designation
420 S45	0.32% C 13% Cr steel: BS 970; obsolete
422	0.2% C 12% Cr 1.0% Mo 0.3% V 1.0% W steel: Hardened and tempered; designation used by AISI
	DPN: 320 UTS: 1050 Elon: 18% Proof: 920
422M	0.85% C 12% Cr 2.25% Mo 0.5% V 1.7% W steel: American proprietary alloy listed in SAE yearbook
430 FR SOLENOID QUALITY	0.06% C 17.5% Cr 0.3% Mo 1.25% Si steel: For magnetic purposes in corrosive conditions; Carpenter
430 S17	0.08% C (max) 17% Cr: BS 970 designation
430 SOLENOID QUALITY	0.06% C 17.5% Cr 0.3% Mo 0.5% Si steel: For magnetic purposes in corrosive conditions; Carpenter
431 S29	0.16% C 2.5% Ni 16.5% Cr steel: BS 970 designation
434	0.12% C 17% Cr 1% Mo steel: Designation used in the UK and USA
436	0.12% C 17% Cr 1% Mo 0.6% Nb steel: Designation used in the UK and USA
439	0.07% C (max) 18% Cr 0.48% Ti steel: Designation used in the UK and USA
441 S29	0.16% C 2.5% Ni 16.5% Cr 0.6% Mo (max) steel: BS 970; obsolete
510	0.12% C 13% Cr steel: Soderfors
511	0.21% C 13% Cr steel: Soderfors
512	0.27% C 13% Cr steel: Soderfors
522	0.12% C 17% Cr steel: Soderfors
525	0.25% C 2% Ni 17% Cr steel: Soderfors
636	0.22% C 0.7% Ni 13% Cr 1.0% Mo 0.35% V 1.0% W steel: Carpenter for AISI type 422
739	0.21% C 13% Cr steel: Avesta
739H	0.27% C 13% Cr steel: Avesta
762	0.2% C 0.5% Ni 11% Cr 1% Mo 0.3% V 0.6% W steel: BS designation
831	0.12% C 21% Cr steel: Avesta
1408	0.07% C 13.0% Cr 0.2% Al steel: Nyby; annealed
	DPN: 190 UTS: 420 Elon: 20% Proof: 290
1410	0.12% C 13.0% Cr steel: Nyby
1410 Mo	0.08% C 14.0% Cr 1.0% Mo steel: Nyby; annealed
	DPN: 200 UTS: 480 Elon: 20% Proof: 290
1415	0.15% C 13.0% Cr steel: Nyby
1415 Mo	0.14% C 13.0% Cr 1.0% Mo steel: Nyby; annealed
	DPN: 190 UTS: 480 Elon: 20% Proof: 290
1435	0.27% C 13% Cr steel: Nyby
1706	0.05% C 17.0% Cr steel: Nyby; annealed
	DPN: 180 UTS: 420 Elon: 20% Proof: 290
1708	0.07% C 17.0% Cr steel: Nyby; annealed
	DPN: 200 UTS: 420 Elon: 20% Proof: 290
1708 Mo	0.07% C 17.0% Cr 1.0% Mo steel: Nyby; annealed
	DPN: 200 UTS: 420 Elon: 20% Proof: 290
1708 Mo Nb	0.07% C 17.0% Cr 1.0% Mo 0.9% Nb steel: Nyby; annealed
	DPN: 200 UTS: 480 Elon 20% Proof: 290
1708 Mo Ti	0.07% C 17.0% Cr 1.0% Mo 0.5% Ti steel: Nyby; annealed
	DPN: 200 UTS: 480 Elon: 20% Proof: 290
1708 Nb	0.07% C 17.0% Cr 0.9% Nb steel: Nyby; annealed
	DPN: 180 UTS: 420 Elon: 20% Proof: 290
1708 Ti	0.07% C 17.0% Cr 0.5% Ti steel: Nyby; annealed
	DPN: 180 UTS: 420 Elon: 20% Proof: 290
1710 Si	0.10% C 17.0% Cr 2.2% Si steel: Nyby; annealed
	DPN: 230 UTS: 530 Elon: 15% Proof: 330
2013	0.2% C 13.0% Cr steel: F Parkin; plastic moulds
2205	0.06% C (max) 4.2% Ni 14% Cr 2.3% Mo 0.2% Nb + Ta 3.2% Cu steel: Casting; designation used in the UK and USA
2520	0.20% C 0.15% Ni 25.0% Cr steel: Nyby; annealed
	DPN: 230 UTS: 480 Elon: 20% Proof: 330
2731	0.7% C 15.0% Cr steel: For tools; French Standards designation
2732	0.4% C 14.0% Cr steel: For tools; French Standards designation
2733	0.3% C 13.0% Cr steel: For tools; French Standards designation
4379	0.3% C 13.0% Cr steel: Sanderson; 'Non-Stain'
	DPN: 277 UTS: 940 Elon: 20% Proof: 800
AC 254	0.24% C 1% Ni 12% Cr 2.5% Mo 0.25% V 1% W 0.2% B steel: Origin unknown

Symbol	Nominal analysis, supplier, condition and remarks.
AFNOR 81 347 EF 2070 Ni Cr Fe Mo B 20 BH	0.03% C 22% Cr 9% Mo 3.5% Nb 1.6% Fe Ni alloy: Welding electrode; French Standard
AFNOR 81 347 EF 2070 Ni Cr Fe Mo B 150-26 BH	0.06% C 22% Cr 9% Mo 3.5% Nb 4.5% Fe Ni alloy: Welding electrode; French Standard
AH	0.2% C 19% Cr steel: Annealed; furnace equipment; Osborn **UTS: 530 Elon: 20% Proof: 450**
AISI 403	0.15% C 12% Cr steel: Hardened and tempered **DPN: 400 UTS: 1500 Elon: 2% Proof: 1350**
AISI 405	0.08% C 12.5% Cr 0.2% Al steel: Annealed **DPN: 180 UTS: 420 Elon: 20% Proof: 210**
AISI 410	0.15% C (max) 12% Cr steel: Hardened and tempered **DPN: 400 UTS: 1500 Elon: 2% Proof: 1350**
AISI 414	0.1% C 2% Ni 12.5% Cr steel: Hardened and tempered **DPN: 440 UTS: 1550 Elon: 2% Proof: 1350**
AISI 416	0.15% C (max) 13% Cr steel: Hardened and tempered **DPN: 280 UTS: 1030 Elon: 15% Proof: 840**
AISI 416 PLUS X	0.15% C (max) 13% Cr 0.15% S (min) steel
AISI 416 Se	0.15% C 13% Cr 0.15% Se steel: Free cutting
AISI 418	0.15% C 13% Cr 3% W steel: Hardened and tempered **DPN: 400 UTS: 1500 Elon: 10% Proof: 1350**
AISI 420	0.35% C 13% Cr steel **DPN: 550 UTS: 1840 Elon: 2% Proof: 1610**
AISI 420F	0.3% C 13% Cr steel: Hardened and tempered **DPN: 500 UTS: 1620 Elon: 5% Proof: 1350**
AISI 430	0.12% C (max) 16% Cr steel: Hardened and tempered **DPN: 500 UTS: 1100 Elon: 3% Proof: 880**
AISI 430 F	0.12% C (max) 16% Cr 0.15% S steel: Hardened and tempered **DPN: 300 UTS: 1100 Elon: 2% Proof: 880**
AISI 430 F Se	0.12% C 16% Cr 0.15% Se steel: Free machining
AISI 431	0.2% C (max) 2% Ni 16% Cr steel: Hardened and tempered **DPN: 440 UTS: 1580 Elon: 10% Proof: 1380**
AISI 434	0.12% C 17.0% Cr 1% Mo steel: Sheet, strip and wire **DPN: 164 UTS: 530 Elon: 23% Proof: 370**
AISI 440 A	0.7% C 17% Cr 0.7% Mo steel: Hardened and tempered **DPN: 550 UTS: 1920 Elon: 2% Proof: 1700**
AISI 440 B	0.8% C 17% Cr 0.7% Mo steel
AISI 442	0.2% C (max) 20% Cr steel: Annealed **DPN: 200 UTS: 580 Elon: 20% Proof: 300**
AISI 446	0.25% C (max) 0.25% Ni 26% Cr steel: Annealed **DPN: 200 UTS: 450 Elon: 20% Proof: 300**
AISI 614	0.12% C 12.2% Cr steel: Annealed **UTS: 620 Elon: 30% Proof: 200**
AISI 614	0.12% C 12.2% Cr steel: Hardened and tempered **DPN: 255 UTS: 880 Elon: 21% Proof: 760**
AISI 615	0.17% C 2% Ni 13% Cr 0.2% Mo 2.9% W steel: Hardened and tempered **UTS: 1140 Elon: 16% Proof: 1000**
AISI 616	0.23% C 12% Cr 1% Mo 0.25% V 1% W steel: Hardened and tempered **UTS: 1070 Elon: 20% Proof: 880**
AISI 619	0.3% C 11.4% Cr 2.7% Mo 0.25% V steel: Hardened and tempered **UTS: 1000 Elon: 18% Proof: 700**
AK 2MV	0.2% C 0.1% Ni 11.2% Cr 2.4% Mo 0.25% V steel: Origin unknown
AL 419	0.2% C 1.2% Ni 1.2% Cr 0.5% Mo 0.5% V 3% W steel: Hardened and tempered; designation used by ASTM **DPN: 330 UTS: 1210 Elon: 15% Proof: 1000**
AL29-4.2	Low C 29% Cr 2% Ni 4% Mo steel: Allegheny Ludlum **DPN: 220 UTS: 660 Elon: 22% Proof: 580**
AL29-4C	Low C 28.5% Cr 3.7% Mo Ti stabilized steel: Allegheny Ludlum **DPN: 200 UTS: 620 Elon: 25% Proof: 520**
ALCRESS	20% Cr 7% Al Fe alloy: Source unknown
ALKROTHAL	15% Cr 4.3% Al Fe alloy: Kanthal AB for heating element
ALUCHROM W	0.04% C 14.5% Cr 4.2% Al steel: VDM
ALUCROM 0	0.04% C 22.5% Cr 0.4% Si 5.2% Al steel: VDM
AMS 5340	4.2% Ni 14% Cr 2.5% Mo 0.25% Nb 3.2% Cu + Ti steel: Investment casting
AMS 5342	4% Ni 16% Cr 3.1% Cu steel: Investment casting; solution treated and aged **UTS: 950**
AMS 5343	4% Ni 16% Cr 3% Cu steel: Investment casting; solution treated and aged **UTS: 1110**
AMS 5344	4% Ni 16% Cr 3% Cu steel: Investment casting; solution treated and aged **UTS: 1350**
AMS 5349	13% Cr 0.25% S steel: Investment casting
AMS 5350	0.12% C 0.4% Ni 12.2% Cr 0.3% Mo steel
AMS 5350 D	12.5% Cr steel: Investment casting
AMS 5351	0.12% C 0.4% Ni 12.2% Cr 0.3% Mo steel: Casting
AMS 5351 B	0.15% C 12.5% Cr steel: Sand casting; AMS for SAE 60410
AMS 5353	1.8% Ni 16% Cr steel: Investment casting
AMS 5354	0.15% C 2% Ni 13% Cr 0.15% Mo 3% W 0.15% Cu steel: Castings
AMS 5354 B	2% Ni 13% Cr 3% W steel: Investment casting
AMS 5355 A	4% Ni 16% Cr 3% Cu steel: Investment casting
AMS 5359	4% Ni 15% Cr 2.3% Mo 0.1% N steel: Sand casting
AMS 5368	4% Ni 15% Cr 2.3% Mo 1.0% N steel: Investment casting
AMS 5372	2% Ni 15.8% Cr steel: Sand casting
AMS 5398 A	4% Ni 16% Cr 3% Cu Fe steel: Sand casting
AMS 5503	0.2% C 17% Cr steel: Sheet and strip; AMS for SAE 51431
AMS 5504	0.12% C 0.4% Ni 12.2% Cr 0.3% Mo steel: Sheet
AMS 5504 D	0.15% C 12.5% Cr steel: Sheet and strip; AMS for SAE 51410
AMS 5505	0.15% C 12.5% Cr steel: Sheet and strip; ferrite controlled; AMS for SAE 51410
AMS 5505	0.12% C 0.4% Ni 12.2% Cr 0.3% Mo steel: Sheet
AMS 5506 A	0.35% C 13% Cr steel: Sheet and strip; AMS for SAE 51420
AMS 5508	0.15% C 2% Ni 13% Cr 0.1% Mo 3% W steel: Sheet
AMS 5508 A	2% Ni 13% Cr 3% W steel: Sheet and strip
AMS 5573	0.05% C (max) 12.5% Ni 17% Cr 2.5% Mo steel: Tubing
AMS 5591	0.12% C 0.4% Ni 12.2% Cr 0.3% Mo steel: Tubing
AMS 5591 D	0.15% C 12.5% Cr steel: Seamless tube; AMS for SAE 51410
AMS 5609	0.15% C 12.5% Cr 0.12% Nb steel: Tubing; ferrite controlled; AMS for SAE 51410
AMS 5610 E	0.35% C 13% Cr steel: Bar and forgings; free machining; AMS for SAE 51416F
AMS 5611	0.15% C (max) 12.5% Cr 0.1% Nb steel
AMS 5612	0.15% C 12.5% Cr steel: Bar or tube; ferrite controlled; AMS for SAE 51410
AMS 5612	0.12% C 0.4% Ni 12.2% Cr 0.3% Mo steel: Forgings

Note. The following abbreviations and units are used in the tables:

DPN	Hardness, diamond pyramid number
UTS	Ultimate tensile strength, N/mm^2
Elon	Elongation, %
Proof	0.1% proof strength, N/mm^2

1 N/mm^2=0.1 hbar=0.102 kgf/mm^2=0.06475 $tonf/in.^2$=145.04 $lbf/in.^2$=1 MPa

See Appendix II for other abbreviations and conversion tables.

Symbol	Nominal analysis, supplier, condition and remarks.
AMS 5613	0.12% C 0.4% Ni 12.2% Cr 0.3% Mo steel: Forgings
AMS 5613 E	0.15% C 12.5% Cr steel: Bar or tube; AMS for SAE 51410
AMS 5614	0.15% C 12% Cr 0.5% Mo steel: Bar and forging; AMS for SAE 51410 + Mo
AMS 5615 B	0.1% C 2% Ni 12.5% Cr steel: Bar and forging; AMS for SAE 51414
AMS 5616	0.15% C 2% Ni 13% Cr 0.15% Mo 3% W steel: Forgings
AMS 5616 D	13% C 2% Ni 3% W steel: Bar and forging
AMS 5620 D	0.35% C 13% Cr steel: Bar and forging; free machining; AMS for SAE 51420F
AMS 5621	0.35% C 13% Cr steel: Bar and forging; AMS for SAE 51420
AMS 5627	0.2% C 17% Cr steel: Bar and forging
AMS 5628 B	0.2% C 2% Ni 16% Cr steel: Bar and forging; AMS for SAE 51431
AMS 5631	0.7% C 17% Cr 0.75% Mo steel: Bar and forging; AMS for SAE 51440A
AMS 5655	0.2% C 0.7% Ni 12.5% Cr 1% Mo 1% W steel: Forgings
AMS 5710 B	0.8% C 1.3% Ni 20% Cr 2.3% Si steel: Bar and forging
AMS 5719	0.12% C 2.5% Ni 11.7% Cr 1.7% Mo 0.32% V 0.03% N steel
AMS 5776	0.12% C 12.2% Cr 0.3% Mo steel: Wire; AMS for SAE 51410
AMS 5777	12.5% Cr steel: Coated electrode; AMS for SAE 51410
AMS 5817	0.15% C 2% Ni 13% Cr 0.15% Mo 3% W steel: Wire
AMS 5817 A	2% Ni 13% Cr 3% W steel: Welding wire
AMS 5821	0.12% C 0.4% Ni 12.2% Cr 0.3% Mo steel: Tubing
AMS 5821 A	12.5% Cr steel: Welding wire; special grade; AMS for SAE 51410
AMS 5822	0.12% C 2.7% Ni 11.7% Cr 1.7% Mo 0.32% V 1.7% Co steel: Weld electrode
AMS 5823	0.12% C 2.7% Ni 11.7% Cr 1.7% Mo 0.32% V 1.7% Co steel: Weld electrode
AMS 7207	0.15% C (min) 13% Cr steel: Type 420
AMS 7445	0.67% C 17% Cr steel: Type 440A
AMS 7470	0.17% C 2% Ni 13% Cr 3% W steel: Weld electrode
AMS 7493	0.15% C (max) 12.5% Cr steel
ANORINOX 35 AM	0.25% C 13% Cr 1.5% Mo steel: Anor
AN QQ S770/2/1	0.17% C 2.0% Ni 16.5% Cr steel: US Service; hardened and tempered **UTS: 1210 Elon: 13% Proof: 940**
AN QQ S770/2/11	0.17% C 2.0% Ni 16.5% Cr steel: US Service; hardened and tempered **UTS: 780 Elon: 15% Proof: 610**
ANTINIT KW 10M	0.1% C 13% Cr 1.0% Mo steel: Gebruder Bohler
ANTINIT KW 15M	0.15% C 13% Cr 1.0% Mo steel: Gebruder Bohler
ANTINIT KW 20M	0.2% C 13% Cr 1.0% Mo steel: Gebruder Bohler
ARD 3	0.38% C 0.6% Ni 17% Cr 1.2% Mo steel: Hardened and tempered; Schoeller-Bleckmann **DPN: 260 UTS: 870 Elon: 14% Proof: 580**
ARH	0.40% C 14.5% Cr steel: Hardened and tempered; Schoeller-Bleckmann **DPN: 400 UTS: 1280 Elon: 6% Proof: 1110**
ARH TOUGH	0.22% C 0.4% Ni 14% Cr steel: Hardened and tempered Schoeller-Bleckmann **DPN: 240 UTS: 800 Elon: 15% Proof: 460**
ARKS	0.13% C 0.5% Ni 17% Cr S steel: Free machining; hardened and tempered; Schoeller-Bleckmann **DPN: 230 UTS: 760 Elon: 12% Proof: 450**
ARL	0.22% C 1.4% Ni 17% Cr steel: Hardened and tempered; Scholler-Bleckmann **DPN: 260 UTS: 840 Elon: 14% Proof: 580**
ARMCO 12	0.15% C 12% Cr steel: Armco
ARMCO 12/2	0.15% C 2.0% Ni 12% Cr steel: Armco
ARMCO 12 Al	0.03% C 12% Cr 0.2% Al steel: Armco
ARMCO 12 FM	0.15% C 13% Cr 0.07% S or Se or Mo steel: Armco; free machining
ARMCO 12T	0.15% C 12% Cr steel: Armco
ARMCO 13-C-35	0.15% C 13% Cr 0.2% S steel: Armco; free machining
ARMCO 16/2	0.2% C 2.0% Ni 16% Cr steel: Armco
ARMCO 17	0.12% C 16% Cr steel: Armco
ARMCO 17-C-60	0.65% C 17% Cr 0.7% Mo steel: Armco
ARMCO 27	0.35% C 25% Cr 0.25% N steel: Armco
ARW	0.1% C 13.5% Cr steel: Bar and sheet; hardened and tempered; Schoeller-Bleckmann **DPN: 200 UTS: 640 Elon: 18% Proof: 440**
ARWA	0.08% C (max) 13% Cr Al steel: Annealed; Schoeller-Bleckmann **DPN: 150 UTS: 530 Elon: 20% Proof: 290**
ARWB	0.1% C 14% Cr steel: Annealed; Schoeller-Bleckmann **DPN: 160 UTS: 530 Elon: 20% Proof: 290**
ARWD	0.08% C (max) 17% Cr 1.7% Mo steel: For welding; annealed; Schoeller-Bleckmann **DPN: 160 UTS: 550 Elon: 20% Proof: 290**
ARWDN	0.08% C (max) 17% Cr 1.7% Mo Nb steel: Annealed; Schoeller-Bleckmann **DPN: 160 UTS: 550 Elon: 20% Proof: 290**
ARWDT	0.08% C (max) 17% Cr 1.7% Mo Ti steel: Annealed; Schoeller-Bleckmann **DPN: 160 UTS: 550 Elon: 20% Proof: 290**
ARWF	0.08% C (max) 12% Cr steel: Hardened and tempered; Schoeller-Bleckmann **DPN: 200 UTS: 650 Elon: 18% Proof: 440**
ARWK	0.08% C 16% Cr steel: Schoeller-Bleckmann **DPN: 150 UTS: 510 Elon: 20% Proof: 290**
ARWKIN	0.08% C (max) 18% Cr Nb steel: For welding; annealed; Schoeller-Bleckmann **DPN: 150 UTS: 510 Elon: 20% Proof: 290**
ARWKT	0.08% C (max) 18% Cr Ti steel: For welding; annealed; Schoeller-Bleckmann **DPN: 150 UTS: 510 Elon: 20% Proof: 290**
ARZ	0.2% C 13.5% Cr steel: Hardened and tempered; Schoeller-Bleckmann **DPN: 260 UTS: 840 Elon: 14% Proof: 530**
ASM 123	0.3% C 13% Cr 1.25% Mo steel: Sambre & Meuse
ASTM A176/403	0.15% C 12% Cr steel: Plate **DPN: 202 UTS: 480 Elon: 25% Proof: 200**
ASTM A176/405	0.08% C 13% Cr 1% Si steel: Plate **DPN: 202 UTS: 450 Elon: 22% Proof: 150**
ASTM A176/410	0.15% C 0.7% Ni 12% Cr steel: Plate **DPN: 202 UTS: 450 Elon: 22% Proof: 200**
ASTM A176/410 S	0.08% C 0.6% Ni 12% Cr steel: Plate **DPN: 202 UTS: 420 Elon: 22% Proof: 200**
ASTM A176/430	0.12% C 0.7% Ni 16% Cr steel: Plate **DPN: 202 UTS: 450 Elon: 22% Proof: 200**
ASTM A176/442	0.35% C 20% Cr steel: Plate **DPN: 202 UTS: 530 Elon: 20% Proof: 270**
ASTM A176/446	0.2% C 0.25% Ni 25% Cr steel: Plate **DPN: 217 UTS: 530 Elon: 20% Proof: 270**
ASTM A182 F6	0.12% C (max) 1.0% Mn 12% Cr steel: For pipe fittings
ASTM A193 B6	0.15% C 12% Cr steel: For bolts, etc.; hardened and tempered as AISI 410–416 **UTS: 800 Elon: 15% Proof: 610**
ASTM A194/6	0.15% C (max) 13% Cr steel: For bolts
ASTM A240/405	0.08% C 13% Cr 0.2% Al steel: For pressure vessels
ASTM A240/410	0.15% C 12.5% Cr steel: For pressure vessels
ASTM A240/410 S	0.08% C 12.5% Cr steel: For pressure vessels
ASTM A240/430 A	0.12% C 15% Cr steel: For pressure vessels
ASTM A240/430 B	0.12% C 17% Cr steel: For pressure vessels
ASTM A268 TP329	0.2% C 26% Cr 1.5% Mo steel: Tube
ASTM A268 TP405	0.08% C 12% Cr 0.2% Al steel: Tube
ASTM A268 TP410	0.15% C 12% Cr steel: Tube
ASTM A268 TP430	0.12% C 16% Cr steel: Tube
ASTM A268 TP433	0.2% C 21% Cr 1% Cu steel: Tube
ASTM A268 TP466	0.2% C 16% Cr 0.2% Ni steel: Tube
ASTM A276/403	0.15% C 12% Cr steel: Bar
ASTM A276/405	0.08% C 13% Cr 0.2% Al steel: Bar
ASTM A276/410	0.15% C 12% Cr steel: Bar

Symbol	Nominal analysis, supplier, condition and remarks.
ASTM A276/414	0.15% C 2% Ni 12% Cr steel: Bar
ASTM A276/416	0.15% C 13% Cr 0.6% Mo steel: Bar
ASTM A276/416 Se	0.15% C 13% Cr 0.15% Se steel: Bar; free machining
ASTM A276/420	0.15% C 13% Cr steel: Bar
ASTM A276/430	0.12% C 16% Cr steel: Bar
ASTM A276/430 F	0.12% C 16% Cr 0.6% Mo steel: Bar
ASTM A276/ 430 F Se	0.12% C 16% Cr 0.15% Se steel: Bar
ASTM A276/431	0.2% C 2% Ni 16% Cr steel: Bar
ASTM A276/440 A	0.7% C 7% Cr 0.7% Mo steel: Bar
ASTM A276/440 B	0.8% C 7% Cr 0.7% Mo steel: Bar
ASTM A276/446	0.2% C 0.2% Ni 25% Cr steel: Bar
ASTM A295 F6	0.12% C 1% Mn 12% Cr steel: Seamless drum forgings; normalized and tempered **UTS: 530 Elon: 18% Proof: 300**
ASTM A296 CA15	0.3% C 12% Cr steel: Casting; annealed **UTS: 610 Elon: 18% Proof: 440**
ASTM A296 CA40	0.2% C 12% Cr steel: Casting; annealed **UTS: 610 Elon: 18% Proof: 440**
ASTM A296 CB30	0.3% C 20% Cr steel: Casting; annealed **UTS: 440 Proof: 200**
ASTM A296 CC50	0.50% C 28% Cr steel: Casting; annealed **UTS: 360**
ASTM A297 HC	0.5% C 28% Cr steel: Casting; as cast **UTS: 360**
ASTM A297 HD	0.5% C 5% Ni 28% Cr steel: Casting; as cast **UTS: 500 Elon: 8% Proof: 240**
ASTM A298 E410	0.12% C 13% Cr steel: Welding rod; annealed after welding chromium stainless steels **UTS: 460 Elon: 20%**
ASTM A298 E430	0.1% C 17% Cr steel: Welding rod; annealed after welding chromium stainless steels **UTS: 460 Elon: 20%**
ASTM A336 F6	0.12% C (max) 12.0% Cr steel: Seamless drum forgings **UTS: 5710 Elon: 8% Proof: 360**
ASTM A351 CA15	0.15% C 1.0% Ni 13% Cr 0.5% Mo steel
ASTM A437	0.22% C 0.6% Ni 11.7% Cr 1.1% Mo 0.25% V 1.0% W steel: For high temperature bolts
ASTM A565	Martensitic steel: Forging; grades 615, 616 and 619 available
ASTM A579/51	0.15% C (max) 12% Cr steel: Forging
ASTM A579/52	0.2% C 1.0% Ni 12.2% Cr 1.0% Mo 0.25% V 1.0% W steel: Forging
ASTM A579/53	0.2% C (max) 2.0% Ni 16% Cr steel: Forging
ASTM A579/61	0.07% C (max) 4.0% Ni 16.5% Cr 0.3% V 4.0% Ti steel: Forging
ASTM A581/416	0.15% C (max) 13% Cr 0.15% S (min) steel: Wire; free machining
ASTM A581/416 Se	0.15% C (max) 13% Cr 0.15% Se (min) steel: Wire; free machining
ASTM A581/430 F	0.12% C (max) 16% Cr 0.15% S (min) steel: Wire; free machining
ASTM A581/430 F Se	0.12% C (max) 16% Cr 0.15% Se (min) steel: Wire; free machining
ASTM A581/XM6	0.15% C (max) 13% Cr 0.6% Mo 0.15% S (min) steel: Wire; free machining
ASTM A651 TP409	0.08% C (max) 11.0% Cr 0.75% Ti (max) steel: Tube for water
ASTM A651 TP430	0.12% C 17.0% Cr steel: Tube
ASTM A651 TP430 Ti	0.1% C 17.7% Cr 0.75% Ti (max) 0.15% Al (max) steel: Tube for water
ASTM A651 TP434	0.12% C 17.0% Cr 1.0% Mo steel: Tube for water
ASTM A651 TPXM8	0.07% C 18.0% Cr 1.1% Ti 0.15% Al (max) steel: Tube for water
ASTM A731 TPXM27	0.01% C (max) 26.2% Cr 1.2% Mo steel: Pipe; welded and seamless **UTS: 450 Elon: 20% Proof: 275**
ASTM A731 TPXM33	0.06% C (max) 26.0% Cr 1.2% Mo 0.6% Ti steel: Pipe; welded and seamless **UTS: 450 Elon: 20% Proof: 275**
ASTM A757 E3N	0.06% C 4% Ni 12.2% Cr 0.7% Mo steel: Casting; Normalized impact at −75 °C 27 J
ASTM A768	0.15% C (max) 0.6% Ni 12% Cr 0.5% Mo (max) steel: For turbine rotors and shafts; vacuum melted **UTS: 700 Elon: 16% Proof: 480**
ASTM A791	Ferritic stainless steel: Tube; see designation for alloys
ASTM A803	Ferritic stainless steel: Tube for heaters; see designation for alloys
ASTM A815 S41500	0.05% C (max) 4.5% Ni 12.7% Cr 0.7% Mo steel: Pipe fitting
ASTM A815 WP27	0.01% C (max) 26% Cr 1.2% Mo 0.1% Nb steel: Pipe fitting
ASTM A815 WP33	0.06% C (max) 26% Cr 1.2% Mo 0.8% Ti steel: Pipe fitting
ASTM A815 WP410	0.15% C (max) 12.5% Cr steel: Pipe fitting
ASTM A815 WP429	0.12% C (max) 15% Cr steel: Pipe fitting
ASTM A815 WP430	0.12% C (max) 17% Cr steel: Pipe fitting
ASTM A815 WP430 Ti	0.1% C (max) 17.5% Cr 0.6% Ti steel: Pipe fitting
ASTM A815 WP446	0.2% C (max) 26.5% Cr 0.2% N steel: Pipe fitting
AURIGA XXIII FCG	0.15% C 1% Ni 12% Cr 0.3% Mo 1% Si steel: Casting; hardened and tempered; D Brown **UTS: 700 Elon: 12% Proof: 440**
AUSTINOX F	0.1% C 0.5% Ni 18% Cr 0.2% Mo 0.2% S steel: Pompey **UTS: 580 Elon: 50% Proof: 240**
AVESTA 393 S	0.08% C 14.5% Cr 1.0% Mo steel: Avesta
AVESTA 739 C	0.14% C 14.5% Cr 1.0% Mo steel: Avesta
AVESTA 739 SG	0.12% C 1.1% Ni 13.0% Cr 1.1% Mo steel: Avesta; casting
AVESTA 739 SH	0.20% C 14.0% Cr 1.0% Mo steel: Avesta
B 3422C	0.10% C 17.0% Cr 2.2% Si steel: French Standard **DPN: 230 UTS: 530 Elon: 15% Proof: 330**
B 3423C	0.20% C 0.15% Ni 25.0% Cr steel: French Standard **DPN: 230 UTS: 480 Elon: 20% Proof: 330**
B2056 420S45	0.3% C 1% Ni 13% Cr steel: For springs
BREARLEY A	0.15% C (max) 12% Cr steel: Bar, billet, etc.; Brown Bayley for BS alloy En 56A
BREARLEY C	0.12% C (max) 17% Cr steel: Bar, billets, etc.; Brown Bayley for BS alloy En 60
BS 970	See designation for detail
BS 980 CDS 18	0.2% C 13% Cr steel: Tube; annealed; cold drawn **UTS: 420 Proof: 270**
BS 1449/403 S17	0.08% C (max) 13% Cr steel: Plate and sheet; annealed **DPN: 170 UTS: 430 Elon: 20% Proof: 250**
BS 1449/405 S17	0.08% C (max) 13% Cr 0.2% Al steel: Plate and sheet; annealed **DPN: 170 UTS: 430 Elon: 20% Proof: 250**
BS 1449/410 S21	0.1% C 12% Cr steel: Plate and sheet; hardened and tempered **DPN: 183 UTS: 620 Elon: 18% Proof: 380**
BS 1501/713	0.08% C 12% Cr 0.2% Al steel: Plate; annealed **UTS: 400 Elon: 21% Proof: 180**

Note. The following abbreviations and units are used in the tables:

DPN	Hardness, diamond pyramid number
UTS	Ultimate tensile strength, N/mm^2
Elon	Elongation, %
Proof	0.1% proof strength, N/mm^2

1 N/mm^2=0.1 hbar=0.102 kgf/mm^2=0.06475 $tonf/in.^2$=145.04 $lbf/in.^2$=1 MPa

See Appendix II for other abbreviations and conversion tables.

Symbol	Nominal analysis, supplier, condition and remarks.
BS 1502/713	0.08% C 12.5% Cr 0.2% Al steel: Forging; hardened and tempered **DPN: 260 UTS: 610 Elon: 20% Proof: 420**
BS 1504/713	0.18% C 12% Cr steel: Casting; hardened and tempered **UTS: 610 Elon: 18% Proof: 440**
BS 1506/713	0.12% C (max) 12% Cr steel: Bar; hardened and tempered **UTS: 580 Elon: 25% Proof: 370**
BS 1554 En 56B	0.15% C 13% Cr steel: Wire; hardened and tempered **UTS: 610**
BS 1554 En 57B	0.25% C 17.5% Cr steel: Wire; hardened and tempered **UTS: 920**
BS 1630 A	0.15% C (max) 12.5% Cr steel: Casting; hardened and tempered; contained in BS 3100 **DPN: 180 UTS: 530 Elon: 20% Proof: 360**
BS 1630 B	0.16% C 12.5% Cr steel: Casting; hardened and tempered; contained in BS 3100 **DPN: 200 UTS: 610 Elon: 18% Proof: 440**
BS 1630 C	0.25% C 12.5% Cr steel: Casting; hardened and tempered; contained in BS 3100 **DPN: 225 UTS: 700 Elon: 15% Proof: 450**
BS 1648 A	0.25% C (max) 14% Cr steel: Casting; as cast; contained in BS 3100
BS 2056 En 56A	0.12% C (max) 13% Cr steel: Wire for springs; as drawn **UTS: 700**
BS 2056 En 56B	0.16% C 13% Cr steel: Wire for springs; as drawn **UTS: 700**
BS 2056 En 56C	0.2% C 13% Cr steel: Wire for springs; as drawn **UTS: 760**
BS 2056 En 56D	0.3% C 13% Cr steel: Wire for springs; as drawn **UTS: 840**
BS 2056 En 57	0.25% C (max) 2% Ni 18% Cr steel: Wire for springs; as drawn **UTS: 840**
BS 2493/12 Cr Mo B	0.23% C (max) 0.5% Ni 12% Cr 1% Mo, 0.5% V steel: For welding electrodes
BS 2493/12 Cr Mo VB	0.23% C (max) 12% Cr 1% Mo 0.5% V steel: For welding electrodes
BS 2493/12 Cr Mo WVB	0.28% C (max) 0.5% Ni 12% Cr 1% Mo 0.5% V 1% W steel: For welding electrodes
BS 2926/13	0.08% C (max) 12% Cr steel: For welding electrodes
BS 2926/13/4 Mo	0.06% C (max) 4.5% Ni 12% Cr 0.5% Mo steel: For welding electrodes
BS 2926/17	0.1% C (max) 16% Cr steel: For welding electrodes
BS 2926/28	0.1% C (max) 29% Cr steel: For welding electrodes
BS 3100/410 C21	0.15% C (max) 1.25% Cr steel: Casting
BS 3100/420 C24	0.25% C (max) 14.0% Cr steel: Casting
BS 3100/420 C29	0.2% C (max) 12.5% Cr steel: Casting
BS 3100/425 C11	0.1% C 3.8% Ni 12.5% Cr steel: Casting
BS 3146 ANC1A	0.15% C 12% Cr steel: Investment casting; hardened and tempered **DPN: 180 UTS: 530 Elon: 20% Proof: 360**
BS 3146 ANC1B	0.18% C 12% Cr steel: Investment casting; hardened and tempered **DPN: 200 UTS: 610 Elon: 18% Proof: 440**
BS 3146 ANC1C	0.25% C 12% Cr steel: Investment casting; hardened and tempered **DPN: 220 UTS: 700 Elon: 15% Proof: 450**
BS 3146 ANC2	0.2% C 2% Ni 17% Cr steel: investment casting; hardened and tempered **DPN: 280 UTS: 840 Elon: 10% Proof: 700**
BS 6105 F1	0.12% C 0.2% Ni 16.5% Cr steel: For fasteners
BS 6323 LW12	0.06% C 12% Cr 0.2% Ti steel: Tube
BS 6323 LW19	0.08% C 11% Cr 0.3% Ti steel: Tube
BVT 130	0.2% C 12% Cr 1.2% Mo steel: Bochumer Verein
BVT 130V	0.2% C 0.4% Ni 12% Cr 1% Mo 0.3% V steel: Origin unknown
BVT 130V50	0.2% C 12% Cr 1% Mo 0.3% V 0.5% W steel: Origin unknown

Symbol	Nominal analysis, supplier, condition and remarks.
CA 6NM	0.06% C 4% Ni 13.0% Cr 0.7% Mo steel: Casting; hardened and tempered; information from Climax Molybdenum **UTS: 84 Elon: 24% Proof: 70**
CA 15	0.15% C (max) 1.0% Ni (max) 12.2% Cr 0.5% Mo (max) steel: Casting; information from Department of Mines, Canada
CA 15	0.3% C 12% Cr steel: Casting; designation used by ASTM
CA 15	0.15% C 1% Ni 12.5% Cr 0.5% Mo 1.5% Si steel: Casting; aged; Huntington **DPN: 390 UTS: 1420 Elon: 7% Proof: 1070**
CA 15M	0.15% C (max) 12.7% Cr 0.17% Mo steel: Casting; designation used in the UK and USA
CA 15M	0.15% C (max) 12.5% Cr 0.7% Mo steel: Casting; designation used in the UK and USA
CA 25 MWV	0.24% C 0.7% Ni 11.7% Cr 1.1% Mo 0.25% V 1% W steel: Casting; designation used in the UK and USA
CA 40	0.2% C 12% Cr steel: Casting; designation used by ASTM
CA 40	0.3% C 1% Ni 13% Cr 0.5% Mo steel: Casting; aged; Huntington **DPN: 470 UTS: 1630 Elon: 1% Proof: 1210**
CA 40F	0.3% C 12.7% Cr 0.3% S steel: Casting; designation used in the UK and USA; free machining
CA28 MWV	0.24% C 0.7% Ni 11.7% Cr 1% Mo 0.25% V 1% W steel: Casting; designation used in the UK and USA
CARPENTER 636	0.22% C 0.7% Ni 12% Cr 1% Mo 0.3% V 1% W steel: Carpenter
CARPENTER A46	0.17% C 0.5% Ni 12% Cr 0.7% Mo 0.3% V 0.4% Nb + Ta steel: Carpenter
CB 30	0.3% C 2% Ni 20% Cr steel: Casting; aged; Huntington **DPN: 195 UTS: 700 Elon: 15% Proof: 370**
CB 30	0.3% C 20% Cr steel: Casting; designation used by ASTM
CC 50	0.5% C 28% Cr steel: Casting; designation used by ASTM
CC 50	0.5% C 4% Ni 28% Cr steel: Casting; as cast; Huntington **DPN: 210 UTS: 720 Elon: 18% Proof: 420**
CD2	1.5% C 12% Cr 1% Mo tool steel: ASTM designation
CEKAS M151	0.15% C 13% Cr 1.0% Mo steel: Kuhbler and Son
CEKAS M152	0.2% C 13% Cr 1.0% Mo steel: Kuhbler and Son
CHROMEX 1	0.07% C 13.5% Cr steel: Welding electrode; Murex **UTS: 640 Elon: 32%**
CHROMEX 2	0.07% C 18% Cr steel: Welding electrode; Murex **UTS: 570 Elon: 25%**
CHROMEX 3	0.1% C 30% Cr steel: Welding electrode; Murex **UTS: 570 Elon: 25%**
CHROMIMPHY A2100	0.1% C 12% Cr 0.6% Mo 0.25% V 0.4% Nb steel: Imphy
CHROMIMPHY A2200	0.17% C 11.5% Cr 1% Mo 0.3% V steel: Imphy
CHROMIMPHY A3100	0.08% C 12% Cr 1% Mo 0.2% V 1% W steel: Imphy
CHROMIMPHY A3200	0.22% C 12% Cr 1% Mo 0.25% V 1% W steel: Imphy
CHROMODUR 22	0.2% C 0.4% Ni 12% Cr 1% Mo 0.3% V steel: Krupp
CHROMODUR 33	0.2% C 0.4% Ni 12% Cr 1% Mo 0.3% V 0.5% W steel: Krupp
CHROMODUR II	0.18% C 0.3% Ni 12% Cr 1.0% Mo steel: Krupp
CMVW 11	0.2% C 0.6% Ni 12.5% Cr 1.1% Mo 0.3% V 0.5% W steel: Origin unknown
CMW11	0.2% C 0.6% Ni 12.5% Cr 1.1% Mo 0.3% V steel: Origin unknown
COBALT ASCOLOY	0.2% C 12.2% Cr 0.2% V 3% W 5% Co steel: American proprietary alloy listed in SAE yearbook
COMET 410H	0.15% C 0.6% Ni 12.5% Cr welding electrode: Soudometal for welding 410 type steel **UTS: 680 Elon: 18% Proof: 490**

Symbol	Nominal analysis, supplier, condition and remarks.
COMET 410NM	0.06% C 4.5% Ni 12.0% Cr 0.5% Mo welding electrode: Soudometal for welding martensitic stainless steel **UTS: 960 Elon: 18% Proof: 800**
CONSUMET 355	0.12% C 4.5% Ni 15.5% Cr 2.9% Mo 0.1% N steel: Carpenter
CORNIX 1	0.2% C 13% Cr 15% Mo steel: Rochling
CORNIX 2	0.2% C 0.5% Ni 12% Cr 1.1% Mo 0.3% V steel: Origin unknown
CORRESIST 13 HMo	0.2% C 13% Cr 1.0% Mo steel: Pose-Marre
Cr Al 20/5	0.1% C 20% Cr 4.5% Al steel: Designation used by German Standards
Cr Al 25/5	25% Cr 5% Al steel: Designation used by German Standards
Cr Al 30/5	0.1% C (max) 30% Cr 4.5% Al steel: Designation used by German Standards
CRH 4/17	0.1% C 0.6% Ni 13% Cr 0.4% Mo steel: Tokushu Seiko Co.
CRO 13 Mo	0.15% C 0.6% Ni 13% Cr steel: Officine Metallurgiche
CROLOY 12	0.15% C 0.5% Ni 12% Cr steel: Babcock and Wilcox (USA)
CROLOY 12 Al	0.08% C 0.5% Ni 12% Cr 0.2% Al steel: Babcock and Wilcox (USA)
CROLOY 18	0.12% C 0.5% Ni 16% Cr steel: Babcock and Wilcox (USA)
CROLOY 22	0.2% C 0.5% Ni 21% Cr 1.0% Cu steel: Babcock and Wilcox (USA)
CROLOY 27	0.2% C 0.5% Ni 27% Cr steel: Babcock and Wilcox (USA)
CROLOY 27/4/1	0.2% C 4% Ni 25% Cr 1.5% Mo steel: Babcock and Wilcox (USA)
CROMFER 2803 Mo	0.015% C (max) 3.7% Ni 28% Cr 2.1% Mo steel: VDM
CROMIMPHY 1 bis Mo	0.8% C 0.5% Ni 13% Cr 0.5% Mo steel: Imphy
CROMIMPHY 33	0.3% C 1.0% Ni 12% Cr 0.3% Mo steel: Imphy
CROMIMPHY A8Mo	0.08% C 13.0% Cr 0.5% Mo steel: Imphy
CROMIMPHY A10Mo	0.15% C (max) 13.0% Cr 0.5% Mo steel: Imphy
CROMIMPHY A15Mo	0.18% C 13.0% Cr 0.5% Mo steel: Imphy
CROMIMPHY A18Mo	0.25% C (max) 13.0% Cr 0.5% Mo steel: Imphy
CROMIMPHY A20Mo	0.22% C 12.5% Cr 1.0% Mo steel: Imphy
CROMIMPHY A1007Mo	0.12% C (max) 13% Cr 0.7% Mo steel: Imphy
CROMO-N	0.23% C 0.7% Ni 10% Cr 1.2% Mo 1% V 0.4% W steel: Bethlehem Steel
CRUCIBLE 26-1	0.06% C (max) 0.5% Ni 26.0% Cr 1.2% Mo 0.6% Ti 0.2% Cu (max) steel: Crucible Steel Co. **UTS: 530 Elon: 30% Proof: 375**
CRUCIBLE 401	0.15% C 12% Cr steel: Hardened and tempered; Crucible Steel Co. **DPN: 250 UTS: 1070 Elon: 20% Proof: 920**
CRUCIBLE 403	0.15% C 12.5% Cr steel: Hardened and tempered; Crucible Steel Co. **DPN: 200 UTS: 1000 Elon: 20% Proof: 760**

Note. The following abbreviations and units are used in the tables:

DPN	Hardness, diamond pyramid number
UTS	Ultimate tensile strength, N/mm^2
Elon	Elongation, %
Proof	0.1% proof strength, N/mm^2

1 N/mm^2=0.1 hbar=0.102 kgf/mm^2=0.06475 tonf/in.^{2}=145.04 lbf/in.^{2}=1 MPa

See Appendix II for other abbreviations and conversion tables.

Symbol	Nominal analysis, supplier, condition and remarks.
CRUCIBLE 416	0.15% C 13% Cr 0.6% Mo or Zr steel: Hardened and tempered; Crucible Steel Co. **DPN: 300 UTS: 1070 Elon: 20% Proof: 920**
CRUCIBLE 420	0.15% C (min) 13% Cr steel: Hardened and tempered; Crucible Steel Co. **DPN: 350 UTS: 1170 Elon: 15% Proof: 1000**
CRUCIBLE 422	0.23% C 0.7% Ni 13% Cr 1% Mo 0.2% V 1% W steel: Crucible Steel Co.
CRUCIBLE 430	0.12% C 16% Cr steel: Annealed; Crucible Steel Co. **DPN: 160 UTS: 530 Elon: 30% Proof: 290**
CRUCIBLE 431	0.2% C 2% Ni 16% Cr steel: Hardened and tempered; Crucible Steel Co. **DPN: 400 UTS: 1380 Elon: 20% Proof: 1070**
CRUCIBLE 440A	0.7% C 17% Cr 0.7% Mo steel: Crucible Steel Co. **UTS: 610**
CRUCIBLE 440B	0.8% C 17% Cr 0.7% Mo steel: Crucible Steel Co. **DPN: 640**
CRUCIBLE 442	0.2% C 21% Cr steel: Annealed; Crucible Steel Co. **DPN: 175 UTS: 570 Elon: 22% Proof: 370**
CRUCIBLE 446	0.2% C 25% Cr 0.25% N steel: Annealed; Crucible Steel Co. **DPN: 180 UTS: 580 Elon: 20% Proof: 370**
CTX/D	0.08% C (max) 17% Cr 1.7% Mo 1% Nb steel: Valbruna
CUTLERY	0.15% C (min) 13% Cr steel: Bar; billet; Brown Bayley for BS alloy En 56D
D 12	Symbol of Durco CA15; Duriron Co.
D 8512	0.2% C 12% Cr 1.0% Mo steel: Grusstahlwerk
D8514	0.17% C 0.5% Ni 11% Cr 0.6% Mo 0.6% V steel: Origin unknown
D8518	0.22% C 0.5% Ni 11.7% Cr 1% Mo 0.3% V steel: Origin unknown
D8518W	0.2% C 0.5% Ni 12% Cr 1% Mo 0.3% V 0.5% W steel: Origin unknown
DARWIN 168	0.75% C 17% Cr steel: Surgical instruments, dies, rollers, etc.; Darwin **DPN: 640**
DAUPHINOX TPMo	0.25% C 13% Cr 0.8% Mo steel: Bonpertius **UTS: 275 Proof: 535**
DIEHARD LC	0.8% C 12% Cr 0.5% Mo 0.5% V steel: Firth Brown
DILVER O	25% Cr Fe alloy: Low expansion; Imphy
DILVER T	20% Cr Fe alloy: Low expansion; Imphy
DIN 17224 X20 Cr 13	0.19% C 13% Cr steel: For springs; German Standard
DIN 17440 X6 Cr Mo 17	0.07% C (max) 1.0% Mn (max) 17% Cr 1.0% Mo 0.05% S (max) 0.045% P (max) steel: Plate; German Standard; annealed
DIN 17440 X7 Cr 13	0.08% C (max) 1.30% Cr steel: Plate; German Standard; annealed **UTS: 610 Proof: 225**
DIN 17440 X7 Cr Al 13	0.08% C 1.0% Mn (max) 13% Cr 0.03% S (max) 0.045% P (max) 0.2% Al steel: Plate; German Standard; annealed **UTS: 560 Proof: 255**
DIN 17440 X8 Cr 17	0.10% C (max) 16.5% Cr steel: Plate; German Standard; annealed **UTS: 600 Proof: 275**
DIN 17440 X8 Cr Nb 17	0.1% C (max) 17.0% Cr 1.2% Nb steel: Plate; German Standard; annealed **UTS: 510 Proof: 275**
DIN 17440 X8 Cr Ti 17	0.1% C (max) 17.0% Cr 0.7% Ti steel: Plate; German Standard; annealed **UTS: 485 Proof: 275**
DIN 17440 X10 Cr 13	0.1% C 13.0% Cr steel: Plate; German Standard; annealed **UTS: 610 Proof: 300**
DIN 17440 X12 Cr Mo S17	0.13% C 16.5% Cr 0.25% Mo steel: Plate; German Standard; annealed **UTS: 700 Proof: 310**

Symbol	Nominal analysis, supplier, condition and remarks.
DIN 17440 X15 Cr 13	0.14% C 13.0% Cr steel: Plate; German Standard; annealed **UTS: 760**
DIN 17440 X20 Cr 13	0.2% C 13.0% Cr steel: Plate; German Standard; annealed **UTS: 760**
DIN 17440 X22 Cr Ni 17	0.19% C 1.0% Mn (max) 2% Ni 19% Cr 0.03% S (max) 0.045% P (max) steel: Plate; German Standard; annealed **DPN: 969**
DIN 17440 X40 Cr 13	0.45% C 13.0% Cr steel: Plate; German Standard; annealed **UTS: 810**
DIN 17440 X48 Cr Mo V15	0.46% C 14.4% Cr 0.5% Mo 0.12% V steel: Plate; German Standard; annealed **UTS: 920**
DIN 17470 Cr Al 20/5	0.1% C (max) 20% Cr 4.5% Al steel: Heat conductor; German Standard
DIN 17470 Cr Al 25/5	25% Cr 5% Al steel: Heat resisting; German Standard
DIN 17470 Cr Al 30/5	0.1% C (max) 29% Cr 4.5% Al steel: Heat conductor; German Standard
DN	0.18% C 1.5% Ni 17% Cr steel: Osborn; obsolete **UTS: 840 Elon: 20% Proof: 610**
DN 5	0.17% C 1.5% Ni 17% Cr steel: Osborn **UTS: 840 Elon: 20% Proof: 610**
DNH	0.22% C 1.5% Ni 17% Cr steel: Valve seats, etc.; Osborn; obsolete **UTS: 840 Elon: 20% Proof: 610**
DOMINIAL RM 13 Mo	0.2% C 13% Cr 1.0% Mo steel: Kind and Co
DOUBLE TWELVE	0.32% C 12% Cr 1% V 12% W steel: Pressure die casting moulds; Edgar Allen **DPN: 450**
DSC	0.35% C 13.5% Cr steel: Darwin **DPN: 610**
DTD 97B	0.15% C (max) 13% Cr steel: Tube; annealed **UTS 420 Proof: 270**
DTD 161A	0.12% C (max) 13% Cr steel: Annealed **UTS: 530**
DTD 203B	0.15% C 13% Cr steel: Tube; hardened and tempered **UTS: 920 Proof: 760**
DTD 271	0.3% C 13% Cr steel: Strip; hardened and tempered **UTS: 1550 Proof: 1210**
DTD 326A	0.3% C 13% Cr steel: Wire for springs; hardened and tempered; not engine valve springs **UTS: 1420**
DTD 525	0.3% C 12% Cr 0.7% S 0.5% P steel: Hardened and tempered; free machining **DPN: 240 UTS: 760 Elon: 15% Proof: 420**
DTD 715	0.25% C 2% Ni 18% Cr 0.3% S steel: Hardened and tempered; free cutting **DPN: 270 UTS: 920 Elon: 15%**
DTD 5046	12% Cr Mo V steel: Sheet and strip **UTS: 1020**
DTD 5065	12% Cr steel: For bolts, studs, etc.; heat resisting **UTS: 1100**
DUNELT 61	0.15% C 12% Cr steel: Dunford for BS alloy En 56A
DUNELT 62	0.22% C 12% Cr steel: Dunford for BS alloy En 56B
DUNELT 63	0.25% C 1% Ni 18% Cr steel: Dunford for BS alloy En 57
DURCO CA15	0.15% C (max) 1.0% Ni (max) 12.5% Cr 0.5% Mo (max) steel: Casting; Duriron Co. **UTS: 630 Elon: 18% Proof: 460**
E409T	0.1% C (max) 12% Cr 1% Ti steel: Weld electrode; designation used by AWS
E410	0.12% C (max) 12.2% Cr steel: Weld electrode; designation used by AWS
E410 Ni Mo	0.06% C (max) 4.5% Ni 11.7% Cr 0.5% Mo steel: Weld electrode; designation used by AWS
E410 Ni Mo T	0.06% C (max) 4.5% Ni 11.7% Cr 0.5% Mo steel: Weld electrode; designation used by AWS
E410 Ni Mo Ti T	0.03% C (max) 4% Ni 11.5% Cr 0.4% Ti steel: Weld electrode; designation used by AWS
E410T	0.12% C 11.7% Cr steel: Weld electrode; designation used by AWS
E420 16	0.2% C 13% Cr steel: Electrode; Metrode
E430	0.1% C (max) 16.5% Cr steel: Weld electrode; designation used by AWS
E430 T	0.1% C (max) 16.5% Cr steel: Weld electrode; designation used by AWS
E1747	0.15% C 0.6% Ni 11.2% Cr 0.6% Mo 0.3% V steel: Russian specification
E1755	0.13% C 0.3% Ni 10.9% Cr 0.7% Mo 0.1% V 2% W steel: Russian specification
E1756	0.12% C 0.7% Ni 11.2% Cr 0.7% Mo 0.25% V 2% W steel: Russian specification
E1757	0.12% C 0.7% Ni 11.2% Cr 0.7% Mo 0.25% V 4% W steel: Russian specification
E1802	0.15% C 0.7% Ni 12% Cr 0.5% Mo 0.2% V 0.9% W steel: Russian specification
E1961	0.13% C 1.6% Ni 11.2% Cr 0.4% Mo 0.2% V 1.7% W steel: Russian specification
E1993	0.2% C 0.1% Ni 12% Cr 0.4% Mo 0.3% V 0.6% W steel: Russian specification
EC26.1	0.01% C (max) 26.2% Cr 1% Mo steel: Weld electrode; designation used by AWS
EC410	0.12% C (max) 12.5% Cr steel: Weld electrode; designation used by AWS
EC410 Ni Mo	0.06% C (max) 4.5% Ni 11.7% Cr 0.5% Mo steel: Weld electrode; designation used by AWS
EC420	0.3% C 13% Cr steel: Weld electrode; designation used by AWS
EC430	0.1% C (max) 16.2% Cr steel: Weld electrode; designation used by AWS
ED12	0.2% C 11.5% Cr 1% Mo 0.3% V steel: Origin unknown
ED12G	0.18% C (max) 0.7% Ni 11% Cr 1% Mo 0.3% V steel: Origin unknown
ED12W	0.2% C 12% Cr 1% Mo 0.3% V 0.5% W steel: Origin unknown
EK	0.3% C 31% Cr steel: Annealed; static furnace equipment; Osborn **UTS: 500 Elon: 20% Proof: 300**
En 56A	0.12% C 13% Cr steel: Designation used in BS970; replaced by 410S21
En 56AM	0.12% C 13% Cr 0.12% S steel: Designation used in BS970; replaced by 416S21
En 56B	0.15% C 13% Cr steel: Designation used in BS970; replaced by 420S29
En 56BM	0.15% C 13% Cr 0.17% S steel: Designation used in BS970; replaced by 416S37
En 56C	0.22% C 13% Cr steel: Designation used in BS970; replaced by 420S37
En 56D	0.28% C 13% Cr steel: Designation used in BS970; replaced by 420S45
En 57	0.25% C 2% Ni 17% Cr steel: Designation used in BS970; replaced by 431S29
En 59	0.8% C 1.5% Ni 20% Cr steel: For valves; designation used in BS970; replaced by 443S65
En 60	0.12% C 17% Cr steel: Designation used in BS970; replaced by 430S15
En 61	0.12% C 21% Cr steel: Obsolete
ENDURO FC	0.13% C 0.4% Ni 12.5% Cr 0.5% Mo steel: Fiat
ENGINEERING	0.15% C (min) 13% Cr steel: Bar, billet, etc.; Brown Bayley for BS alloy En56C
EP 65	0.23% C 1% Ni 13% Cr 1% Mo 1% V 1% W steel: Russian specification
ER410 Ni Mo	0.04% C 4% Ni 13% Cr 0.5% Mo steel: Electrode; Metrode

Symbol	Nominal analysis, supplier, condition and remarks.
ER420	0.32% C 13% Cr steel: Weld electrode; designation used by AWS
ER446 LR	0.015% C (max) 26.2% Cr steel: Weld electrode; designation used by AWS
ER505	0.1% C (max) 9.2% Cr 1% Mo steel: Weld electrode; designation used by AWS
ERA HR4	29% Cr steel: Casting; Hadfields for BS 1648B
ERA HR4 HARD	High C 29% Cr steel: Casting; Hadfields for BS 1648C
EV 1	0.45% C 4.8% Ni 23% Cr 2.8% Mo steel: For exhaust valves; designation used by SAE **UTS: 920 Elon: 3%**
EV 2	0.4% C 3.8% Ni 24% Cr 1.4% Mo steel: For exhaust valves; designation used by SAE **UTS: 1070 Elon: 3%**
EXD 5	0.3% C 12% Cr 1.1% V 12% W steel: Extrusion dies for brass, etc.; Balfour **DPN: 395**
F 1	0.12% C 13% Cr steel: Ugine
F 15	0.12% C 13% Cr 0.75% S steel: Free machining; Ugine
F 17	0.12% C 17% Cr steel: Ugine
FAGERSTA R100	0.05% C 14.5% Cr steel: Fagersta
FAGERSTA R290	0.05% C 17.5% Cr steel: For oil burners; Fagersta
FAGERSTA R700	0.2% C 13.5% Cr steel: For cutting tools; Fagersta
FAGERSTA R710	0.3% C 13.5% Cr steel: For cutting tools; Fagersta
FAGERSTA R720	0.4% C 13.5% Cr steel: For cutting tools; Fagersta
FAGERSTA R730	0.5% C 13.5% Cr steel: For cutting tools; Fagersta
FAGERSTA R740	0.3% C 0.4% Ni 14.0% Cr 0.6% Mo steel: For cutting tools; Fagersta
FAL	0.1% C 13% Cr 4.2% Al steel: High electrical resistance; Firth Vickers **UTS: 610 Elon: 25% Proof: 450**
FAS	0.43% C 11.5% Cr steel: Firth Vickers **DPN: 240 UTS: 760 Elon: 20% Proof: 530**
FC1	0.1% C 13% Cr 0.4% Mo steel: Free machining; Firth Vickers **DPN: 180 UTS: 630 Elon: 28% Proof: 420**
FCS	0.2% C 0.4% Mn 13% Cr 0.23% S steel: Free machining; Firth Vickers **DPN: 220 UTS: 600 Elon: 25% Proof: 440**
FECRALLOY	0.03% C (max) 18.5% Cr 4.6% Al 0.2% Y steel: Resistalloy Ltd; oxidation resistant; annealed **UTS: 510 Elon: 23% Proof: 375**
FG	0.25% C 13% Cr steel: Hardened and tempered; Firth Vickers **DPN: 450 UTS: 1500 Elon: 18% Proof: 1350**
FG (L)	0.15% C 13% Cr steel: Hardened and tempered; Firth Vickers **DPN: 450 UTS: 1500 Elon: 18% Proof: 1350**
FH	0.25% C 13% Cr steel: Firth Vickers
FHM	0.8% C 16.5% Cr 0.5% Mo steel: Firth Vickers **DPN: 600**
FI	0.12% C 0.7% Ni 12% Cr steel: Hardened and tempered; Firth Vickers **DPN: 172 UTS: 570 Elon: 33% Proof: 360**
FI 17	0.08% C 2.5% Ni 17% Cr steel: Hardened and tempered; Firth Vickers; sheet **DPN: 175 UTS: 840 Elon: 28% Proof: 360**
FI 20	0.07% C 21% Cr steel: Firth Vickers **UTS: 480 Elon: 30%**
FI Ti	0.06% C (max) 1.0% Ni (max) 11.5% Cr 0.5% Ti steel: Firth Vickers
FLUGINOX 51	0.05% C 12.5% Cr 0.5% Mo 0.2% Al steel: Ugine
FLUGINOX 60	0.25% C 13% Cr 0.5% Mo steel: Ugine
FLUGINOX 62	0.2% C 0.8% Ni 12% Cr 1% Mo 0.2% V 1% W steel: Ugine
FLUGINOX 65	0.22% C 11% Cr 1% Mo steel: Ugine
FNZ	0.25% C 2% Ni 17% Cr steel: Firth Vickers **DPN: 380 UTS: 1300 Elon: 15% Proof: 1140**
FOREMOST 14 Cr	0.12–0.35% C 13% Cr range of steels: Swift Levick for BS alloys En 56 A, B and C
FOREMOST 467	0.25% C 1% Ni 18% Cr steel: Swift Levick for BS alloy En 57
FOREMOST HR 2	0.2% C 2% Ni 17% Cr steel: Casting; annealed; Swift Levick **DPN: 150 UTS: 530 Elon: 25% Proof: 340**
FOREMOST IRON	0.12% C 1% Ni 13% Cr steel: Hardened and tempered; Swift Levick **DPN: 170 UTS: 600 Elon: 30%**
FV 403	0.12% C 12% Cr steel: Hardened and tempered; Firth Vickers **DPN: 172 UTS: 570 Elon: 33% Proof: 360**
FV 406	0.1% C 13% Cr 4.25% Al steel: High electrical resistance; 1350 microhm mm at 600 °C; Firth Vickers **UTS: 610 Elon: 25% Proof: 450**
FV 410	0.1% C 13% Cr steel: hardened and tempered; Firth Vickers **DPN: 172 UTS: 570 Elon: 33% Proof: 360**
FV 420	0.25% C 13% Cr steel: Hardened and tempered; Firth Vickers **DPN: 450 UTS: 1500 Elon: 18% Proof: 1350**
FV 430	0.06% C 16.5% Cr steel: Firth Vickers **DPN: 170 UTS: 530 Elon: 32% Proof: 330**
FV 431	0.16% C 2.5% Ni 16% Cr steel: Firth Vickers **DPN: 380 UTS: 1300 Elon: 15% Proof: 1140**
FV 440B	0.8% C 16.5% Cr 0.5% Mo steel: Firth Vickers **DPN: 600**
FV 446	0.09% C 1.8% Ni 29% Cr steel: Firth Vickers **DPN: 180 UTS: 530 Elon: 25%**
FV 448	0.1% C 0.75% Ni 11% Cr 0.7% Mo steel: Firth Vickers **UTS: 940 Elon: 17% Proof: 800**
FV 507	0.12% C 10.5% Cr 0.8% Mo 0.15% V 0.4% Nb steel: Casting; Firth Vickers; ferrite stainless steel; hardened and tempered **DPN: 300 UTS: 970 Elon: 12% Proof: 820**
FV 535	0.07% C 0.3% Ni 10.5% Cr 0.75% Mo 6% Co steel: Creep resistant; Firth Vickers **UTS: 1020 Elon: 18% Proof: 880**
FV 566	0.12% C 2.3% Ni 11.5% Cr 1.4% Mo 0.15% V 0.3% Nb steel: Firth Vickers
FV 607	0.15% C 0.6% Ni 11% Cr 0.8% Mo 0.25% V steel: Hardened and tempered; Firth Vickers **UTS: 980 Elon: 23% Proof: 820**
FV 702	0.03% C 2.5% Ni 15.7% Cr 1.0% Mo 0.5% Nb steel: Firth Vickers
FV Fl	0.08% C 13% Cr steel: For turbine blades; Firth Vickers **UTS: 610 Elon: 33% Proof: 390**
FX	0.1% C 21% Cr steel: Annealed; furnace and domestic equipment; Osborn **UTS: 450 Elon: 30% Proof: 300**
G 85	0.06% C 20.0% Cr 4.5% Mo 13.5% W 3.0% Ti 1.5% Al steel: Jessop; scale or creep resistant
G 4301 Sec/1	0.15% C (max) 12% Cr steel: Japanese Standard
G 4301 Sec/4	0.15% C (min) 13% Cr steel: Japanese Standard
G 4301 Sec/5	0.12% C 16% Cr steel: Japanese Standard
GALA	0.1% C 13% Cr steel: Hadfields for BS alloy BS/S61

Note. The following abbreviations and units are used in the tables:

DPN	Hardness, diamond pyramid number
UTS	Ultimate tensile strength, N/mm^2
Elon	Elongation, %
Proof	0.1% proof strength, N/mm^2

1 N/mm^2=0.1 hbar=0.102 kgf/mm^2=0.06475 $tonf/in.^2$=145.04 $lbf/in.^2$=1 MPa

See Appendix II for other abbreviations and conversion tables.

Symbol	Nominal analysis, supplier, condition and remarks.
GALAHAD A	0.12% C 1% Ni 13% Cr steel: Wrought; Hadfields for BS alloy En 56A **UTS: 530 Elon: 20% Proof: 360**
GALAHAD AFC	0.12% C 1% Ni 13% Cr 0.6% Mo 0.75% S steel: Free machining; Hadfields for BS alloy En 56AM
GALAHAD B	0.16% C 1% Ni 13% Cr steel: Wrought; Hadfields for BS alloy En 56B **UTS: 610 Elon: 18% Proof: 440**
GALAHAD BFC	0.15% C 13% Cr 0.6% Mo 0.75% S steel: Free machining; Hadfields for BS alloy En 56BM
GALAHAD C	0.21% C 1% Ni 13% Cr steel: Wrought; Hadfields for BS alloy En 56C **UTS: 700 Elon: 15% Proof: 450**
GALAHAD CFC	0.21% C 1% Ni 13% Cr 0.6% Mo 0.75% S steel: Free machining; Hadfields for BS alloy En 56CM
GALAHAD D	0.3% C 1% Ni 13% Cr steel: Hadfields for BS alloy En 56D
GALAHAD DFC	0.3% C 1% Ni 13% Cr 0.8% Mo 0.75% S steel: Free machining; Hadfields for BS alloy En 56DM
GALAHAD E	0.25% C 2% Ni 17% Cr steel: Hadfields for BS alloy En 57
GALAHAD F	0.12% C 0.5% Ni 17% Cr steel: Hadfields for BS alloy En 60
GASC	0.5% C 14% Cr steel: Jonas; plastic moulds **DPN: 480**
GASD	0.35% C 14% Cr steel: Jonas; plastic moulds **DPN: 460**
GILBRUN 10	0.12% C (max) 13.0% Cr steel: Wire; Gilly Brunton; annealed **UTS: 500 Elon: 33% Proof: 310**
GILBRUN Cr 17	0.07% C 17.0% Cr steel: Wire; Gilly Brunton **UTS: 540 Elon: 33% Proof: 340**
GILBRUN Cr 20	0.07% C (max) 20.0% Cr steel: Wire; Gilly Brunton **UTS: 540 Elon: 33% Proof: 340**
GOST 5632/0 KH2 N5T	0.08% C (max) 5.2% Ni 21.0% Cr 0.45% Ti steel: Plate; Russian Standard
GOST 5632/0 KH13	0.08% C (max) 0.6% Mn (max) 12% Cr 0.025% S (max) 0.03% P (max): Russian Standard; annealed steel plate **UTS: 400**
GOST 5632/0 KH17T	0.08% C (max) 17.0% Cr 0.8% Ti steel: Plate; Russian Standard; annealed
GOST 5632/1 KH11 MF	0.16% C 10.7% Cr 0.7% Mo 0.32% W steel: Russian Standard
GOST 5632/1 KH12N2VMF	0.13% C 1.7% Ni 11.2% Cr 0.42% Mo 0.2% V 1.8% W steel: Plate; Russian Standard
GOST 5632/1 KH12S1U	0.1% C 13.0% Cr 1.4% Al steel: Russian Standard
GOST 5632/1 KH12S1U	0.9% C 13.0% Cr 1.4% Al steel: Plate; Russian Standard
GOST 5632/1 KH12V2MF	0.13% C 12.0% Cr 0.75% Mo 0.2% V 2.0% W steel: Plate; Russian Standard
GOST 5632/1 KH12VNMF	0.15% C 0.6% Ni 12.0% Cr 0.6% Mo 0.25% V 0.9% W steel: Plate; Russian Standard
GOST 5632/1 KH13N3	0.12% C 2.7% Ni 13.5% Cr steel: Russian Standard
GOST 5632/1 KH13	0.12% C 0.6% Mn (max) 13% Cr 0.025% S (max) 0.03% P (max) steel: Plate; Russian Standard; annealed **UTS: 550 Proof: 280**
GOST 5632/1 KH17N2	0.14% C 2.0% Ni 17.0% Cr steel: Russian Standard
GOST 5632/1 KH21N5T	0.12% C 5.2% Ni 21.0% Cr 0.7% Ti steel: Plate; Russian Standard
GOST 5632/2 KH12VMBMR	0.18% C 12.0% Cr 0.5% Mo 0.2% V 0.6% W 0.3% Nb 0.003% B steel: Plate; Russian Standard
GOST 5632/2 KH13	0.2% C 13.0% Cr steel: Plate; Russian Standard; annealed **UTS: 500 Proof: 250**
GOST 5632/2 KH13N4G9	0.22% C 4.2% Ni 13.0% Cr steel: Plate; Russian Standard
GOST 5632/2 KH17N2	0.25% C 0.8% Mn 2.0% Ni 17.0% Cr steel: Plate; Russian Standard
GOST 5632/3 KH13	0.3% C 13.0% Cr steel: Plate; Russian Standard; annealed **UTS: 550**
GOST 5632/3 KH13N7S2	0.3% C 7.2% Ni 13.0% Cr steel: Russian Standard
GOST 5632/4 KH13	0.4% C 13.0% Cr steel: Plate; Russian Standard; annealed **UTS: 520**
GOST 5632/15 KH12VMF	0.15% C 0.6% Ni 12.0% Cr 0.6% Mo 0.22% V 0.9% W steel: Plate; Russian Standard
GOST 5632 KH13N3	0.11% C 2.6% Ni 13.5% Cr steel: Plate; Russian Standard
GOST 5632 KH14	0.15% C (max) 0.7% Mn (max) 14.0% Cr steel: Plate; Russian Standard
GOST 5632 KH14G14N	0.12% C (max) 1.25% Ni 14.0% Cr steel: Plate; Russian Standard
GOST 5632 KH14G14N13T	0.1% C (max) 3.0% Ni 14.0% Cr 0.6% Ti steel: Russian Standard
GOST 5632 KH17	0.12% C (max) 17.0% Cr steel: Plate; Russian Standard; annealed **UTS: 450 Proof: 280**
GOST 5632 KH17AG14	0.15% C (max) 0.6% Ni (max) 17.0% Cr 0.45% N steel: Russian Standard
GOST 5632 KH17AG14	0.15% C (max) 17.0% Cr 0.35% N steel: Plate; Russian Standard
GOST 5632 KH17G9AN4	0.12% C (max) 4.0% Ni 17.0% Cr 0.2% N steel: Russian Standard
GOST 5632 KH17H2	0.14% C 0.8% Mn (max) 2% Ni 19% Cr 0.025% S (max) 0.03% P (max) steel: Plate; Russian Standard; solution treated **UTS: 1100 Proof: 800**
GOST 5632 KH18S1U	0.15% C (max) 0.5% Mn (max) 18.5% Cr 0.85% Al steel: Plate; Russian Standard; annealed **UTS: 500 Proof: 300**
GOST 5632 KH25T	0.15% C (max) 0.8% Mn (max) 25% Cr 0.8% Ti 0.025% S (max) 0.035% P steel: Plate; Russian Standard **UTS: 450 Proof: 300**
GOST 5632 KH28	0.15% C (max) 29.0% Cr steel: Plate; Russian Standard
GOST 5632 KH28AN	0.15% C 1.3% Ni 26.5% Cr 0.2% N steel: Plate; Russian Standard
GOST 9940/0 KH13	0.08% C (max) 12.0% Cr steel: Pipe; Russian Standard; seamless **UTS: 380**
GOST 9940/0 KH17T	0.08% C (max) 17.0% Cr 0.6% Ti steel: Pipe; Russian Standard; seamless
GOST 9940/1 KH13	0.12% C 0.6% Mn (max) 13.0% Cr steel: Pipe; Russian Standard; seamless **UTS: 400**
GOST 9940/1 KH17	0.12% C (max) 17.0% Cr steel: Pipe; Russian Standard; seamless **UTS: 450**
GOST 9940 KH28	0.15% C (max) 28.5% Cr steel: Pipe; Russian Standard; seamless **UTS: 450**
GOST 9940 KH28 T	0.15% C (max) 25.5% Cr 0.8% Ti steel: Pipe; Russian Standard; seamless **UTS: 450**
GOST 9941/0 KH13	0.08% C (max) 12.0% Cr steel: Pipe; Russian Standard; seamless **UTS: 380**
GOST 9941/0 KH17T	0.08% C (max) 17.0% Cr 0.8% Ti steel: Pipe; Russian Standard; seamless
GOST 9941/1 KH13	0.12% C 13.0% Cr steel: Pipe; Russian Standard; seamless **UTS: 408**

Symbol	Nominal analysis, supplier, condition and remarks.
GOST 9941 KH17	0.12% C (max) 17.0% Cr steel: Pipe; Russian Standard; seamless
	UTS: 450
GOST 9941 KH25 T	0.15% C (max) 25.5% Cr 0.8% Ti steel: Pipe; Russian Standard; seamless
	UTS: 450
GOST 10498/1 KH13S2M2	0.12% C 13% Cr 0.6% Mo steel: Pipe; Russian Standard; seamless
	UTS: 550
GOST 10802/1 KH11B2MF	0.13% C 0.65% Mn 11.0% Cr 0.75% Mo 0.2% V 2.0% W 0.32% Cu (max) steel: Pipe; Russian Standard; seamless
	UTS: 600 Proof: 400
GOST 14162/1 KH13	0.12% C 0.6% Mn (max) 13.0% Cr steel: Capillary tube; Russian Standard
GOST 14162/2 KH13	0.2% C 13.0% Cr steel: Capillary tube; Russian Standard
GQ	0.85% C 17% Cr steel: Osborn; ball races, gears, etc.; obsolete
	UTS: 760 Elon: 20% Proof: 370
GREEK ASCOLOY	0.17% C 2.0% Ni 13% Cr 3.0% W steel: Firth Sterling; hardened and tempered
	UTS: 1000 Elon: 20% Proof: 760
GREEK ASCOLOY	0.17% C 2% Ni 13% Cr 0.5% Mo 3% W steel: Carpenter
GX 5 Cr Ni 13.4	0.07% C 3.7% Ni 13.0% Cr 0.5% Mo steel: Casting; designation used by Swiss Standards
GX 5 Cr Ni 13.4	0.07% C 4.2% Ni 12.7% Cr 0.7% Mo (max) steel: Casting; designation used by German Standards
	UTS: 850 Elon: 10% Proof: 580
H 23	0.3% C 12.0% Cr 12.0% W steel: Designation used by AISI
H 29	0.8% C 0.4% Mn 1.4% Ni 20% Cr 2% Si steel: Jessop for BS alloy En 59
H 46	0.16% C 0.6% Ni 11.5% Cr 0.6% Mo 0.3% V steel: Jessop
H 53	0.08% C 0.7% Ni 10% Cr 0.8% Mo 0.4% V 0.5% W 6.5% Co 0.4% Nb steel: Jessop; high creep strength
H 59	0.07% C 3% Ni 11% Cr 1.5% Mo 0.4% V steel: Jessop
	UTS: 1070 Elon: 20% Proof: 1000
H 100	0.1% C (max) 23.8% Cr 1.45% Al steel: VEW; heat resistant
H 102	0.12% C (max) 26.8% Cr 0.1% N steel: VEW; heat resistant
H 120	0.12% C (max) 18% Cr 1.05% Al steel: VEW; heat resistant
H 140	0.12% C (max) 12.5% Cr 0.95% Al steel: VEW; heat resistant
H 300	0.13% C 4.0% Ni 24.8% Cr steel: VEW; heat resistant
H 304	0.2% C 4.0% Ni 25% Cr 0.1% N steel: VEW; heat resistant
HA1 Fe Cr Al 15	15% Cr 4.5% Al Fe alloy for resistance heating weld: Harrison Alloy
HA1 Fe Cr Al 20	15% Cr 5% Al Fe alloy for resistance heating weld: Harrison Alloy
HA1 Fe Cr Al 25	23% Cr 5.5% Al Fe alloy for resistance heating weld: Harrison Alloy
HARDINOX 3	0.3% C 13% Cr steel: Pompey
HARDINOX 4	0.4% C 13% Cr steel: Pompey
HARDTRODE 84.42	0.12% C 0.3% Mn 13.0% Cr steel: For electrodes; Esab
	DPN: 460
HARDTRODE 84.52	0.25% C 0.3% Mn 13.0% Cr steel: For electrodes; Esab
	DPN: 500
HC	0.5% C 4% Ni 28% Cr 0.5% Mo steel: Casting Huntington; aged
	UTS: 820 Elon: 18% Proof: 530
HC 30	0.3% C 4% Ni (max) 28% Cr steel: Casting; designation used in the UK and USA
HCA	0.3% C 12% Cr 0.7% V 12% W steel: Braeburn; AISI type H23
HD	0.5% C 5.5% Ni 28% Cr 0.5% Mo steel: Casting; Huntington; as cast
	UTS: 610 Elon: 16% Proof: 330
HD 50	0.5% C 5.5% Ni 28% Cr steel: Casting; designation used in the UK and USA
HECLA HGT4	0.16% C 11.2% Cr 0.6% Mo 0.2% V 0.02% B steel: Hadfields
HECLA SNS	0.8% C 1.5% Ni 19% Cr 2% Si steel: Hadfields for BS alloy En 59
HEDEX 5	0.12% C (max) 17% Cr steel: Wire for forging; Kiveton Park; annealed
	UTS: 630
HNV 6	0.8% C 1.3% Ni 20% Cr 2.3% Si steel: For valves; designation used by SAE
HNV8	0.23% C 0.8% Ni 12% Cr 1% Mo 0.25% V 1% W steel: For valves; SAE designation
HRM 2M1	0.15% C 13.0% Cr 1.15% Mo steel: Henricot
HRM 2M2	0.20% C 13.0% Cr 1.15% Mo steel: Henricot
HT 9	0.2% C 0.55% Ni 12.0% Cr 0.1% Mo 0.3% V 0.5% W steel: Sandvik
HWT AQUALLOY	0.07% C (max) 18% Cr 0.8% Ti steel: Origin unknown
HWX	0.36% C 5.8% Ni 13.5% Cr 0.65% V 2.8% W steel: for hot working; origin unknown
HYFLOW 420R	0.14% C 12% Cr steel: Wear resistant; British Steel Armour Products
HYFORM 409	Oxidation resistant stainless steel: British Steel Stainless
ICN	0.2% C 1.0% Mn 1.0% Ni 14.0% Cr steel: Origin unknown
IMMUNIT R10/13 Mo Ni	0.1% C 0.45% Ni 12.5% Cr 0.4% Mo steel: Bochum
IMMUNIT R12/13 Mo A	0.15% C 13.0% Cr 0.30% Mo steel: Bochum
IMMUNIT R15/13 Mo	0.15% C 13.0% Cr 1.0% Mo steel: Bochum
IMMUNIT R20/13 Mo	0.2% C 13.0% Cr 1.0% Mo steel: Bochum
IMMUNIT R2213	0.2% C 12% Cr 1% Mo 0.3% V 0.5% W steel: Origin unknown
IMPERIAL CT	0.3% C 13% Cr steel: Edgar Allen for BS alloy En 56D
IMPERIAL EQ	0.2% C 13% Cr steel: Edgar Allen for BS alloy En 56C
IMPERIAL R1	0.12% C (max) 13% Cr steel: Edgar Allen
IMPERIAL R10	0.06% C (max) 13% Cr steel: Edgar Allen
IMPERIAL S80	0.25% C 1.5% Ni 18% Cr steel: Edgar Allen for BS alloy En 57
INCOLOY HA 956	20% Cr 0.5% Mo 0.5% Ti 4.5% Al Fe alloy: Inco
INOFO 1	0.12% C 13% Cr steel: Forez
INOFO 2	0.3% C 13% Cr steel: Forez
INOFO 4	0.1% C 17% Cr steel: Forez
INOFO 4S	0.12% C 17% Cr steel: Forez
INOFO 5	0.18% C 1.5% Ni 16% Cr steel: Forez
INOX 1	0.1% C 13% Cr steel: Pompey
INOX 1F	0.12% C 13% Cr 0.2% Mo 0.2% S steel: Pompey; free machining
	UTS: 760 Elon: 15% Proof: 580
INOX 2	0.2% C 0.4% Ni 13% Cr steel: Pompey
INOX 3	0.3% C 0.4% Ni 13% Cr steel: Pompey

Note. The following abbreviations and units are used in the tables:

DPN	Hardness, diamond pyramid number
UTS	Ultimate tensile strength, N/mm^2
Elon	Elongation, %
Proof	0.1% proof strength, N/mm^2

1 N/mm^2=0.1 hbar=0.102 kgf/mm^2=0.06475 $tonf/in.^2$=145.04 $lbf/in.^2$=1 MPa

See Appendix II for other abbreviations and conversion tables.

Symbol	Nominal analysis, supplier, condition and remarks.
INOX 16	0.08% C 0.4% Ni 17% Cr steel: Pompey **UTS: 480 Elon: 17% Proof: 240**
INOX 17T	0.1% C 16% Cr steel: Pompey
INOX 430F	0.1% C 0.4% Ni 17% Cr steel: Pompey **UTS: 530 Elon: 21% Proof: 330**
INOX 431	0.15% C 2.25% Ni 16% Cr steel: Pompey; hardened and tempered **UTS: 870 Elon: 14% Proof: 730**
INOXESCO 13	0.17% C 0.5% Ni 13% Cr 0.6% Mo 0.2% V steel: Origin unknown
INVENTOR	0.15% C (min) 13% Cr steel: Bar, billet, etc.; Brown Hayley for BS alloy En 56B
IOX KM	0.11% C 12.75% Cr 0.50% Mo steel: Nazionale Cogne
IRRUBIGO 1 Mo	0.15% C 13.0% Cr 1.2% Mo steel: Schmidt and Clemens
IRRUBIGO 2 Mo	0.18% C 13.0% Cr 1.2% Mo steel: Schmidt and Clemens
J R ALLOY 1	0.15% C (max) 13% Cr 0.7% Ti 4% Al steel: Carpenter; for electrical resistance
J R ALLOY 2	0.15% C (max) 13% Cr 0.7% Ti 3.5% Al steel: Carpenter; for electrical resistance
J R ALLOY 3	0.15% C (max) 13% Cr 0.7% Ti 3% Al steel: Carpenter; for electrical resistance
JETHETE	0.14% C 2.4% Ni 12.0% Cr 1.8% Mo 0.35% V 0.05% N steel: Origin unknown
JETHETE M140	0.08% C 13.0% Cr 1.0% Mo steel: Samuel Fox
JETHETE M151	0.1% C 1.5% Ni 12% Cr 0.6% Mo 0.3% V steel: Samuel Fox; weldable **UTS: 930 Elon: 10% Proof: 780**
JETHETE M152	0.1% C 2.5% Ni 12% Cr 1.8% Mo 0.3% V steel: Samuel Fox; creep resisting; weldable **UTS: 1040 Elon: 20% Proof: 780**
JETHETE M153	0.15% C 2% Ni 12% Cr 1.5% Mo steel: Samuel Fox; creep resistant; weldable **UTS: 840 Elon: 24% Proof: 610**
JETHETE M154	0.1% C 2.5% Ni 12.0% Cr 17.5% Mo 0.37% V steel: Samuel Fox; hardened and tempered **UTS: 650 Elon: 11% Proof: 570**
JETHETE M160	0.2% C 1.25% Ni 12% Cr 0.7% Mo 0.4% V 0.7% Nb steel: Samuel Fox; creep resistant **UTS: 1110 Elon: 20% Proof: 870**
JETHETE M190	0.13% C 2.7% Ni 12% Cr 1.8% Mo 0.3% V 1.5% Co steel: Samuel Fox
JIS G3446 SUS410	0.15% C (max) 12.5% Cr steel: Tube for structural; Japanese Standard **UTS: 420 Proof: 210**
JIS G3446 SUS430	0.15% C (max) 17.0% Cr steel: Tube for structural; Japanese Standard **UTS: 420 Proof: 250**
JIS G3463 SUS410TB	0.15% C (max) 12.5% Cr steel: Pipe; Japanese Standard **UTS: 420 Proof: 210**
JIS G3463 SUS430TB	0.12% C (max) 17.0% Cr steel: Pipe; Japanese Standard **UTS: 420 Proof: 210**
JIS G4304 SUS403	0.15% C (max) 1.0% Mn (max) 0.6% Ni 12% Cr steel: Plate; Japanese Standard; annealed **UTS: 450 Proof: 210**
JIS G4304 SUS405	0.08% C (max) 1.0% Mn (max) 0.6% Ni (max) 13% Cr 0.03% S 0.04% P (max) 0.2% Al steel: Plate; Japanese Standard; annealed **UTS: 420 Proof: 180**
JIS G4304 SUS410	0.15% C 12.5% Cr steel: Plate; Japanese Standard; annealed **UTS: 450 Proof: 210**
JIS G4304 SUS429	0.12% C (max) 15.00% Cr steel: Plate; Japanese Standard; annealed **UTS: 460 Proof: 210**
JIS G4304 SUS430	0.12% C (max) 17.0% Cr steel: Plate; Japanese Standard; annealed **UTS: 460 Proof: 210**
JIS G4304 SUS434	0.12% C (max) 1.0% Mn (max) 0.6% Ni (max) 17% Cr 1.0% Mn 0.03% S (max) 0.04% P (max) steel: Plate; Japanese Standard; annealed **UTS: 460 Proof: 210**
JIS G4312 SUH446	0.2% C 1.5% Mn 2.5% Cr steel: Plate; Japanese Standard; solution treated **UTS: 570 Proof: 210**
K20L	0.2% C 13% Cr 1.0% Mo steel: Zapp
K92801	0.12% C (max) 28% Cr steel: Designation used by UNS
KALKOS	0.3% C 12% Cr 1.0% V 12% W steel: Latrobe; extrusion and forging dies for brass; type H23 **DPN: 390**
KANTHAL A1	0.06% C 23% Cr 2.0% Co 5.5% Al Fe alloy: Kanthal AB; for furnace elements **DPN: 230 UTS: 770 Elon: 20% Proof: 550**
KANTHAL AF	22% Cr 6% Al Fe alloy: Kanthal AB; for heating elements
KANTHAL APM	22% Cr 5.8% Al Fe alloy: Kanthal AB; for heating elements
KANTHAL D	0.1% C 22% Cr 2.0% Co 5.7% Al Fe alloy: Kanthal AB; for furnace elements **DPN: 200 UTS: 700 Elon: 20% Proof: 550**
KARONI 40/13 Mo	0.2% C 13% Cr 1.2% Mo steel: Kabel
KE 15	0.1% C 13% Cr steel: Kayser Ellison; hardened and tempered for plastic dies **UTS: 760**
KE 25	0.25% C 13% Cr steel: Kayser Ellison for BS alloy En 56C
KE 35	0.35% C 13% Cr steel: Kayser Ellison
KE 40A	0.1% C 13% Cr steel: For free cutting agent; Kayser Ellison for BS alloy En 56AM
KE 43	0.17% C 1.75% Ni 17% Cr steel: Kayser Ellison for BS alloy En 57
KEA 28	0.45% C 1% Ni 13.75% Cr steel: Kayser Ellison; hardened and tempered for plastic dies **DPN: 560**
KEA 138	0.35% C 12% Cr 1.05% V 12% W steel: Kayser Ellison for hot hobbing
KEA 203	0.85% C 17.5% Cr 0.55% Mo steel: Kayser Ellison
KEA 505	0.17% C 1.75% Ni 17.0% Cr with Se steel: Free machining; Kayser Ellison
KEA 508	0.25% C 13.0% Cr with Se: Free machining; Kayser Ellison
KK BREARLEY	0.15% C 13% Cr 0.15% Se steel: Bar, billet, etc,; Brown Bayley for BS alloy En 56AM
KK ENGINEERING	0.21% C 13% Cr with Se Zr or Mo steel: Bar, billet, etc.; Brown Bayley for BS En 56CM
KK TWOSCORE	0.25% C 2% Ni 17% Cr steel: Brown Bayley; free cutting form of BS alloy En 57
KORROSIL 2M	0.2% C 13% Cr 1.2% Mo steel: Klockner-Werke
KP 1	0.13% C 12.2% Cr 0.7% Mo steel: USSR
KP 27	0.25% C (max) 2% Ni 17% Cr steel: Kiveton Park for BS alloy En 57
KP 28	0.25% C 13% Cr steel: Kiveton Park for BS alloy En 56C
KP 29	0.12% C (max) 13% Cr steel: Kiveton Park for BS alloy En 56A
KV7	0.12% C 0.7% Ni 11.2% Cr 0.7% Mo 0.25% V 4% W steel: Russian specification
LANGALLOY 6V	0.08% C 0.6% Ni 12% Cr 0.2% Al steel: Langley Alloys
LANGALLOY 80V	0.2% C 2% Ni 17% Cr steel: Langley Alloys **UTS: 870 Elon: 16% Proof: 650**
LAPELLOY	0.3% C 0.3% Ni 11.5% Cr 2.7% Mo 0.25% V steel: Latrobe Steel: hardened and tempered **DPN: 231 UTS: 1070 Elon: 15% Proof: 840**
LAPELLOY C	0.22% C 11.5% Cr 2.75% Mo 2% Cu steel: Latrobe Steel

Symbol	Nominal analysis, supplier, condition and remarks.
LESCALLOY 5616	0.18% C 2.0% Ni 13% Cr 3.0% W steel: Latrobe Steel hardened and tempered **DPN: 321 UTS: 1030 Elon: 17% Proof: 880**
LINCO	0.3% C 0.3% Ni 11.5% Cr 2.7% Mo 0.25% V steel: Latrobe Steel; hardened and tempered **DPN: 331 UTS: 1070 Elon: 15% Proof: 840**
LOWSCORE	0.2% C 20% Cr steel: Bar, billets, etc.; Brown Bayley for BS alloy En 61
LS 1	0.08% C 13% Cr steel: Low Moor for BS/S61; annealed; obsolete
LS 16	0.1% C 17% Cr steel: Low Moor for BS alloy En 60; obsolete
LS 22	0.1% C 21% Cr steel: Low Moor for BS alloy En 61; obsolete
LS 61	0.12% C 13% Cr steel: Low Moor for BS/S61; obsolete
LS 62	0.2% C 13% Cr steel: Low Moor for BS/S62; obsolete
LS 80	0.18% C 2% Ni 17% Cr steel: Low Moor for BS S80; obsolete
LS CUT	0.3% C 13% Cr steel: Low Moor; for cutlery; obsolete **DPN: 480 UTS: 1550 Elon: 3% Proof: 1330**
LS CUT/HC	0.4% C 13% Cr steel: For cutlery; Low Moor; high carbon LS Cut; obsolete **UTS: 650**
LT 54	0.1% C (max) 13% Cr steel: Low Moor; can be carburized for die steel: obsolete
LT 55	0.25% C 13% Cr steel: Low Moor; die steel; obsolete **DPN: 480**
LV 19	0.85% C 19% Cr 2.5% Mo 0.5% V steel: Low Moor; valve steel; obsolete **DPN: 340**
LV 20	0.8% C 1.5% Ni 20% Cr 2.0% Si steel: Low Moor; valve steel; obsolete **DPN: 340**
LV 24	0.45% C 4.7% Ni 24% Cr 2.7% Mo steel: Low Moor; valve steel; obsolete
M10 FIRTH CLEVELAND	Martensitic stainless steel sinter: Firth Cleveland; specific gravity 6.4–6.8 hardened and tempered **UTS: 590 Elon: 1%**
M152	0.1% C 2.5% Ni 11.7% Cr 1.7% Mo 0.3% V steel
M300	0.38% C 0.8% Ni 16.0% Cr 1.0% Mo steel: VEW
M310	0.43% C 13.5% Cr steel: VEW
MAFERITE 14	14% Cr steel: Pompey **UTS: 2000**
MAFERITE 17	17% Cr steel: Pompey **UTS: 2400**
MAFERITE 20	20% Cr 1.0% Mo steel: Pompey **UTS: 2500**
MAFERITE 25	25% Cr 1.0% Mo steel: Pompey **UTS: 2500**
MAFERITE 25A	25% Cr 1.5% Si 1.5% Al steel: Pompey
MAFERITE 28N	4% Ni 28% Cr steel: Pompey
MAFERITE 30	30% Cr steel: Pompey **UTS: 2500**
MAFERITE 30P	30% Cr + Mo steel: Pompey
MALURECHER F12	0.2% C 0.4% Ni 12% Cr 1% Mo 0.3% V steel: Origin unknown
MANNESMANN-STAHL F12	0.2% C 0.4% Ni 1.2% Cr 1% Mo 0.3% V steel: Origin unknown
MARKER ED12G	0.18% C (max) 0.7% Ni 11% Cr 1% Mo 0.3% V steel: Origin unknown
MARKER ED12W	0.2% C 12% Cr 1% Mo 0.3% V 0.5% W steel: Origin unknown
MARKER EP12	0.2% C 11.5% Cr 1% Mo 0.3% V steel: Origin unknown
MARKER S14M	0.18% C 13% Cr 1.15% Mo steel: Schmidt and Clemens
MAXHETE 3	0.2% C (max) 28% Cr steel: Edgar Allen; furnace equipment
MB	0.08% C 15% Cr steel: Osborn; annealed for cutlery; obsolete **UTS: 530 Elon: 30% Proof: 450**
MB 2	0.1% C 17% Cr steel: Osborn; annealed; cutlery, motor trim; obsolete **UTS: 600 Elon: 32% Proof: 370**
MEL-TROL H46	0.17% C 0.5% Ni 12% Cr 0.7% Mo 0.3% V 0.06% N steel: Carpenter
MEL-TROL STAINLESS 2	0.35% C 13.0% Cr steel: Carpenter
METCO 42C	0.2% C 2% Ni 16% Cr steel: Wire for spraying; coarser grade than 42F; Metco **DPN: 360**
METCO 42F	0.2% C 2% Ni 16% Cr steel: Powder for spraying; Metco; finer grade than 42C **DPN: 420**
METCO 463	Cr Fe stainless steel: Spray deposit; Metco **DPN: 270**
METCO 465	Cr Fe stainless steel: Spray deposit; Metco **DPN: 270**
METCOLOY 2	0.3% C 13% Cr steel: Wire for spraying; Metco; as sprayed **DPN: 284 UTS: 270**
MET-MAX 25.5	0.1% C 5% Ni 25% Cr steel: Electrode; Metrode
MIDVAC 422	0.25% C 0.9% Ni 12% Cr 0.9% Mo 0.2% V 1% W steel: Origin unknown
MIL-S-16993	0.12% C 0.4% Ni 12.2% Cr 0.3% Mo steel
MIL-S-81591	0.15% C (max) 13% Cr steel: Casting; type 420; US Federal specification
MINOX 10	0.1% C 13% Cr steel: Pompey; hardened and tempered **DPN: 380 UTS: 1200 Elon: 6% Proof: 1000**
MINOX 15	0.2% C 13% Cr 1.0% Mo steel: Pompey; castings
MINOX 20	0.2% C 14% Cr steel: Pompey; hardened and tempered **UTS: 1500 Elon: 4% Proof: 1200**
MINOX 30	0.3% C 14% Cr steel: Pompey; hardened and tempered **UTS: 1800 Elon: 4% Proof: 1400**
MINOX 1610	0.1% C 16% Cr steel: Pompey **DPN: 170 UTS: 520 Proof: 280**
MINOX 1620	0.2% C 2% Ni 16% Cr steel: Pompey; hardened and tempered **UTS: 1300 Elon: 5% Proof: 1000**
MINOX 1820	0.15% C 1.0% Ni 19% Cr steel: Pompey; annealed **DPN: 190 UTS: 550 Proof: 300**
MNC 900	General standard for SS range of stainless steels; Swedish Standard
MOLY ASCOLOY	0.08% C 13% Cr 2% Mo steel: American proprietary alloy listed in SAE yearbook
MOSIT LOW C	4% Ni 25% Cr 4% Mo Ti stabilized steel: Origin unknown
MTS 1	0.18% C 11.5% Cr 1.0% Mo steel: Deutsche
MTS 5	0.2% C 0.4% Ni 11.5% Cr 1% Mo 0.3% V steel: Origin unknown
MTS 6	0.17% C 0.6% Ni 11.5% Cr 0.6% Mo 0.25% V steel: Origin unknown
N 23	0.32% C 14% Cr steel: Vulcan; hardened and tempered **UTS: 1140 Elon: 33%**
N100	0.1% C 13% Cr steel: VEW
N104	0.06% C (max) 12.5% Cr steel: VEW
N106	0.07% C (max) 14.3% Cr steel: VEW
N108	0.08% C (max) 12.5% Cr steel: VEW

Note. The following abbreviations and units are used in the tables:

DPN	Hardness, diamond pyramid number
UTS	Ultimate tensile strength, N/mm^2
Elon	Elongation, %
Proof	0.1% proof strength, N/mm^2

$1\ N/mm^2 = 0.1\ hbar = 0.102\ kgf/mm^2 = 0.06475\ tonf/in.^2 = 145.04\ lbf/in.^2 = 1\ MPa$

See Appendix II for other abbreviations and conversion tables.

Symbol	Nominal analysis, supplier, condition and remarks.
N200	0.08% C (max) 16.5% Cr 0.3% Mo steel: VEW
N310	0.13% C 16.5% Cr 0.3% Mo steel: VEW
N315	0.14% C 12.5% Cr steel: VEW
N320	0.2% C 13% Cr steel: VEW
N350	0.17% C 1.8% Ni 16.5% Cr steel: VEW
N400	0.05% C (max) 4.1% Ni 13.1% Cr 0.4% Mo steel: VEW
N404	0.05% C (max) 5.3% Ni 15.4% Cr 0.9% Mo steel: VEW
N530	0.33% C 13.5% Cr steel: VEW
N540	0.43% C 13.5% Cr steel: VEW
NC 2 (Mo)	0.85% C 17% Cr 0.7% Mo steel: Bar and forging; Brown Bayley
NEUTROTHERM KW20M	0.2% C 13.0% Cr 1.0% Mo steel: Bohler
NF A35 572/Z6C13	0.08% C (max) 0.5% Ni (max) 12.5% Cr steel: French Standard
NF A35 572/Z6CA13	0.08% C 12.5% Cr 0.2% Al steel: French Standard
NF A35 572/Z8C17	0.10% C (max) 0.5% Ni (max) 17.0% Cr steel: French Standard
NF A35 572/ Z8CD17/01	0.1% C (max) 0.5% Ni (max) 17.0% Cr 1.1% Mo steel: French Standard
NF A35 572/ Z10CF17	0.12% C (max) 0.5% Ni (max) 17.0% Cr 0.6% Mo (max) steel: French Standard
NF A35 572/Z12C13	0.11% C 0.5% Ni (max) 12.5% Cr steel: French Standard
NF A35 572/ Z12CF13	0.15% C (max) 13.0% Cr 0.6% Mo (max) steel: French Standard
NF A35 572/Z20C13	0.20% C 1.0% Ni (max) 13.0% Cr steel: French Standard
NF A35 572/Z30C13	0.3% C 1.0% Ni (max) 13.0% Cr steel: French Standard
NF A35 572/Z40C14	0.4% C 1.0% Ni (max) 13.5% Cr steel: French Standard
NF A35 572/ Z100CD17	1.05% C 0.5% Ni (max) 17.0% Cr 0.5% Mo (max) steel: French Standard
NF A35 573/Z6C13	0.08% C (max) 12.5% Cr steel: Plate; French Standard **UTS: 480 Proof: 270**
NF A35 573/Z6CA13	0.08% C 12.5% Cr 0.25% Al steel: Plate; French Standard
NF A35 573/Z8C17	0.1% C (max) 17.0% Cr steel: Plate; French Standard **UTS: 500 Proof: 250**
NF A35 573/ Z8CD17/01	0.1% C 1.0% Mn 0.5% Ni (max) 17% Cr 1.0% Mo 0.05% S 0.04% P (max) steel: Plate; French Standard **UTS: 600 Proof: 290**
NF A35 573/Z12C13	0.12% C 12.5% Cr steel: Plate; French Standard **UTS: 700 Proof: 430**
NICRAL S	25% C 2.0% Si steel: Imphy **DPN: 190 UTS: 600 Elon: 20% Proof: 400**
NIROSTA VK5M	0.20% C 13.0% Cr 1.2% Mo steel: Krupp
NIROSTA VK7M	0.15% C 13.0% Cr 1.2% Mo steel: Krupp
NON STAIN	0.3% C 1.0% Mn 0.5% Ni 13.0% Cr steel: Sanderson
NOVO 18/2	0.15% C 2% Ni 17% Cr steel: Jonas **DPN: 255 UTS: 840 Elon: 25% Proof: 630**
NOVO AS 1	0.1% C 14% Cr steel: Jonas; hardened and tempered **DPN: 200 UTS: 610 Elon: 30% Proof: 480**
NOVO AS 20	0.1% C 19% Cr steel: Jonas; annealed **DPN: 160 UTS: 570**
NOVO ASFC	0.1% C 14% Cr 0.35% S steel: Jonas; free cutting
NOVO GAS	0.3% C 14% Cr steel: Jonas **DPN: 207 UTS: 700 Elon: 23% Proof: 530**
NOVO GBS	0.2% C 14% Cr steel: Jonas **DPN: 217 UTS: 720 Elon: 26% Proof: 530**
NOVO NOX T131	0.2% C 13% Cr 1.2% Mo steel: Sudwestfalen
NOVO NOX TT131	0.15% C 13% Cr 1.2% Mo steel: Sudwestfalen
NOVO THERM	0.2% C 30% Cr 0.75% Mo steel: Jonas
NST M1	0.1% C 0.50% Ni 13.0% Cr 0.5% Mo steel: Japan Metal Industries
NST M2	0.15% C 0.50% Ni (max) 12.3% Cr 1.0% Mo steel: Japan Metal Industries
NVR 7-4	0.12% C 0.45% Mn 9.0% Cr 1.0% Mo steel: Designation used by DNV
NVR 7-5	0.22% C (max) 0.6% Mn 0.5% Ni 11.7% Cr 1.0% Mo steel: Designation used by DNV
NYBY 1410 Mo	0.08% C 14.0% Cr 1.0% Mo steel: Nyby
NYBY 1415 Mo	0.14% C 13.0% Cr 1.0% Mo steel: Nyby
OSBORN 405	0.08% C 14% Cr steel: Osborn; previously RPA grade
OSBORN 410	0.1% C 13% Cr steel: Osborn; previously RP grade
OSBORN 420C	0.18% C 13% Cr steel: Osborn; previously SA grade
OSBORN 420D	0.35% C 13% Cr steel: Osborn; previous PH grade
OSBORN 430AL	0.12% C (max) 18% Cr 1.0% Al steel: Osborn; previously MBA grade
OSBORN 431	0.18% C (max) 1.7% Ni 17% Cr steel: Osborn; previously DN grade
OSBORN 440B	0.6% C 17% Cr 0.5% Mo steel: Osborn; previously GQ grade
P 4	0.18% C 27.0% Cr steel: Sandvik
P 12	0.27% C 13% Cr steel: Ugine
P 1000	0.1% C 13% Cr steel: Carrs for BS alloy En 56
P 1001	0.16% C 13% Cr steel: Carrs for BS alloy En 56
P 1002	0.21% C 13% Cr steel: Carrs for BS alloy En 56
P 1003	0.3% C 13% Cr steel: Carrs for BS alloy En 56
P 1008	0.35% C 1% Ni 13% Cr steel: Plastic moulds and dies; Carrs
P 1009	0.18% C 2.1% Ni 17% Cr steel: Carrs for BS alloy En 56
P 557	0.08% C (max) 18% Cr 2% Mo steel: SAE potential specification
PANTEG 430	0.06% C 17% Cr steel: Sheet **UTS: 540 Elon: 25%**
PARALLOY M2	0.2% C (max) 1.3% Cr steel: APV Paramount
PARALLOY M3	0.3% C (max) 13% Cr steel: APV Paramount
PARALLOY M15	0.15% C (max) 13% Cr steel: APV Paramount
PARALLOY ML	2% Ni 17% Cr steel: APV Paramount
PARALLOY MPH	0.08% C (max) 4% Ni 17% Cr 2.5% Cu steel: APV Paramount
PARALLOY MPH2	0.08% C (max) 4% Ni 1.4% Cr 2% Mo 2.5% Cu steel: APV Paramount
PHEINROHR MV12	0.2% C 12% Cr 1% Mo 0.3% V 0.5% W steel: Origin unknown
PHEINROHR MVW12	0.2% C 0.4% Ni 11% Cr 1% Mo 0.2% V steel: Origin unknown
PHENIX HD 32	0.2% C 0.4% Ni 12.75% Cr 0.15% Mo steel: Schoeller-Bleckmann
PHENIX HD 301	0.15% C 0.5% Ni 12.75% Cr 1.15% Mo steel: Schoeller-Bleckmann
PHENIX HD40	0.2% C 0.5% Ni 12% Cr 1.1% Mo 0.3% V steel: Schoeller-Bleckmann
PHENIX HD50	0.5% C 0.6% Ni 12% Cr 1.1% Mo 0.3% V 0.5% W steel: Schoeller-Bleckmann
PKH	0.3% C 13% Cr steel: Bar; RTB for BS alloy En 56D
PKI	0.06% C 17% Cr steel: RTB for BS alloy En 60
PKL	0.1% C 13% Cr steel: Bar; RTB for BS alloy En 56A
PKM	0.2% C 13% Cr steel: Bar; RTB for BS alloy En 56C
PKN	0.15% C 2% Ni 17% Cr steel: Bar; RTB for BS alloy En 57
PROJECT 70-182FM	0.08% C (max) 18% Cr 2% Mo 0.2% S steel: Carpenter; free machining
PURO M Mo	0.2% C 13% Cr 1.0% Mo steel: Hagener
PURO M Mo2	0.15% C 13% Cr 1.0% Mo steel: Hagener
PYRISTA	0.09% C 1.8% Ni 29% Cr steel: Firth Vickers **DPN: 180 UTS: 530 Elon: 25%**
PYROMET 350	0.08% C 4.5% Ni 16.5% Cr 3% Mo 0.1% N steel: Carpenter
PYROMET 355	0.1% C 4.5% Ni 15.5% Cr 3% Mo 0.1% N steel: Carpenter
PYROMET ALLOY 80A	0.06% C 20% Cr 1% Co 2.3% Ti 1.25% Al steel: Carpenter
PYROMET ALLOY 751	0.04% C 15% Cr 2.5% Ti 1% Nb 1.2% Al steel: Carpenter

Symbol	Nominal analysis, supplier, condition and remarks.
QQ S 763/6	0.15% C 0.5% Ni 13% Cr 0.6% Mo S and Se steel: US Service specification **DPN: 200 UTS: 450 Elon: 25% Proof: 200**
QQS-763	0.12% C 0.4% Ni 12.2% Cr 0.3% Mo steel
R 1	0.1% C 13% Cr steel: Jessop **UTS: 720 Elon: 29% Proof: 510**
R 2	0.2% C 0.75% Ni 13% Cr steel: Jessop **UTS: 670 Elon: 21% Proof: 690**
R 4	0.07% C 4% Ni 25% Cr 1.5% Mo steel: Japanese steel
R 4	0.16% C 2.2% Ni 16.25% Cr steel: Jessop for BS alloy En 57; austenitic
R 12	0.1% C 1.25% Ni 27% Cr steel: Jessop
R 15	0.15% C 13% Cr 0.23% S steel: Free machining; Jessop for BS alloy En 56 BM
R 19	0.3% C 13% Cr 0.2% S steel: Free machining; Jessop
R 26	0.1% C 13% Cr 0.22% S steel: Free cutting; Jessop for BS alloy En 56AM
R 29	0.1% C 21% Cr steel: Jessop
R 34	0.3% C 13% Cr steel: Jessop for BS alloy En 56D
R 50	0.08% C 4.25% Ni 17.0% Cr 2.75% Mo steel: Jessop; corrosion resistant steel
R 55	0.10% C 4.25% Ni 15% Cr 2.5% Mo steel: Jessop; corrosion resistant steel
R 100	0.06% C 14.5% Cr steel: Annealed; Fagersta **UTS: 620 Elon: 25% Proof: 390**
R 110	0.1% C 13.5% Cr steel: Hardened and tempered; Fagersta **DPN: 220 UTS: 740 Elon: 14% Proof: 530**
R 120G	0.12% C 13.0% Cr steel: Casting; Fagersta
R 140	0.1% C 1.1% Ni 13.5% Cr steel: Hardened and tempered; Fagersta **DPN: 220 UTS: 740 Elon: 14% Proof: 530**
R 140G	0.12% C 13.0% Cr 1.0% Mo steel: Casting; Fagersta
R 200	0.09% C 17.5% Cr steel: Annealed; Fagersta **DPN: 200 UTS: 450 Elon: 23% Proof: 240**
R 220	0.17% C 2% Ni 17.5% Cr steel: Hardened and tempered; Fagersta **DPN: 260 UTS: 820 Elon: 14% Proof: 630**
R 290	0.06% C 17.5% Cr steel: Annealed; Fagersta **DPN: 200 UTS: 420 Elon: 23% Proof: 240**
R 600G	0.25% C (max) 24% Cr steel: Casting; scaling temperature 1070 °C; Fagersta
R 620	0.1% C 4.3% Ni 28% Cr steel: Annealed; Fagersta **DPN: 260 UTS: 620 Elon: 20% Proof: 370**
R 640	0.1% C 5.3% Ni 26% Cr 1.6% Mo steel: Annealed; Fagersta **DPN: 260 UTS: 620 Elon: 20% Proof: 400**
R 700	0.2% C 13.5% Cr steel: Hardened and tempered; Fagersta **DPN: 270 UTS: 870 Elon: 15% Proof: 690**
R 710	0.3% C 13.5% Cr steel: Hardened and tempered; Fagersta **DPN: 300 UTS: 920 Elon: 9% Proof: 740**
R 720	0.4% C 13.5% Cr steel: Hardened and tempered; Fagersta **DPN: 300 UTS: 920 Elon: 9% Proof: 740**
R 730	0.58% C 13.5% Cr steel: Fagersta **DPN: 630**

Note. The following abbreviations and units are used in the tables:

DPN	Hardness, diamond pyramid number
UTS	Ultimate tensile strength, N/mm^2
Elon	Elongation, %
Proof	0.1% proof strength, N/mm^2

1 N/mm^2=0.1 hbar=0.102 kgf/mm^2=0.06475 tonf/in.^{2}=145.04 lbf/in.^{2}=1 MPa

See Appendix II for other abbreviations and conversion tables.

Symbol	Nominal analysis, supplier, condition and remarks.
R 740	0.3% C 0.4% Ni 14% Cr 0.6% Mo steel: Hardened and tempered; Fagersta **DPN: 300 UTS: 920 Elon: 9% Proof: 740**
R THERM 2012 Mo	0.2% C 12% Cr 1.2% Mo steel: Bochum
R THERM 2012 Mo WV	0.2% C 0.4% Ni 12% Cr 1% Mo 0.3% V 0.5% W steel: Origin unknown
RCA 33	30% Cr 5.0% Al Fe alloy: Imphy
RCA 44	20% Cr 5.0% Al Fe alloy: Imphy
RE 39	0.2% C 2% Ni 17% Cr steel: Bofors; hardened and tempered **DPN: 290 UTS: 820 Elon: 14% Proof: 630**
REMANIT 1520 Mo	0.2% C 12.5% Cr 1.2% Mo steel: Deutsche
RESOURCE	0.3% C 13% Cr steel: Plastic moulds; Hall and Pickles; obsolete **DPN: 610**
RFK	0.20% C 13.0% Cr 1.0% Mo steel: Friedr Lohmann
RHM 5	0.30% C 13.0% Cr 0.4% Mo steel: Tokushu Seiko
RHM 7	0.17% C 0.6% Mn 0.7% Ni 12.0% Cr 1.10% Mo steel: Tokushu Seiko
RHM 8	0.14% C 0.5% Mn 0.6% Ni 12.0% Cr 1.1% Mo steel: Tokushu Seiko
RHM 9	0.2% C (max) 1.0% Ni 13.0% Cr 1.0% Mo steel: Tokushu Seiko
RHM 10	0.13% C 0.6% Ni (max) 13.0% Cr 0.45% Mo steel: Tokushu Seiko
RHM 11	0.15% C 0.5% Ni (max) 12.5% Cr 1.0% Mo steel: Tokusho Seiko
RHM 37	0.13% C 0.6% Ni (max) 12.25% Cr 0.45% Mo steel: Tokushu Seiko
RHMV4	0.22% C 0.6% Ni 11.7% Cr 1% Mo 0.3% V steel: Origin unknown
RHMW1	0.22% C 0.7% Ni 11.7% Cr 1% Mo 0.25% V 1% W steel: Origin unknown
RHMW2	0.2% C 0.5% Ni 11.5% Cr 1% Mo 0.3% V steel: Origin unknown
RNO Mo	0.15% C 1.0% Ni (max) 13.0% Cr 1.2% Mo steel: Rochling-Buderus
RNO Mo V	0.22% C 0.5% Ni 11.5% Cr 1% Mo 0.3% V steel: Origin unknown
RNO Mo WV	0.22% C 0.8% Ni 11.5% Cr 1% Mo 0.3% V 0.5% W steel: Origin unknown
RNOD	0.15% C 0.6% Ni 11% Cr 0.7% Mo 0.3% V steel: Origin unknown
RNOM	3.20% C 0.3% Ni (max) 13.0% Cr 1.2% Mo steel: Rochling-Buderus
ROP 43	0.2% C 0.6% Ni 12% Cr 1.2% Mo 0.3% V steel: Bofors; hardened and tempered **DPN: 280 UTS: 750 Elon: 15% Proof: 590**
ROP 46	0.2% C 12% Cr 0.5% Mo 0.4% V 0.2% Nb steel: Bofors; hardened and tempered **DPN: 290 UTS: 950 Elon: 16% Proof: 800**
ROSTODUR R3	0.20% C 13.5% Cr 1.0% Mo steel: Hoffmann
ROSTODUR R7	0.15% C 13.0% Cr 1.0% Mo steel: Hoffmann
RP	0.1% C 13% Cr steel: Plastic dies; Osborn
RP 1	0.12% C 13% Cr steel: Annealed; turbine blades etc.; Osborn **UTS: 530 Elon: 30% Proof: 450**
RPA	0.08% C 13% Cr steel: Annealed; golf clubs; rivets; Osborn **UTS: 530 Elon: 30% Proof: 450**
RRJ 11	0.12% C 13% Cr steel: Fagersta
RRM 20	0.12% C 17% Cr steel: Fagersta
RRM 22	0.25% C 2.0% Ni 17% Cr steel: Fagersta
RRS 70	0.21% C 13% Cr steel: Fagersta
RRS 71	0.27% C 13% Cr steel: Fagersta
RS (4)	0.42% C 13% Cr steel: Hard wearing parts with medium corrosion resistance; Osborn **UTS: 700 Elon: 30% Proof: 300**
RS (A)	0.17% C 13% Cr steel: Scissors and knives; Osborn **UTS: 700 Elon: 30% Proof: 580**

Symbol	Nominal analysis, supplier, condition and remarks.
RS (B)	0.22% C 13% Cr steel: Osborn **UTS: 740 Elon: 25% Proof: 610**
RS (C)	0.32% C 13% Cr steel: Osborn **UTS: 700 Elon: 30% Proof: 300**
RT 46	0.3% C 12% Cr 0.5% V 12% W steel: Bofors; obsolete **DPN: 380**
S 1R	0.1% C 25% Cr steel: Bofors; annealed
S61	0.12% C (max) 12% Cr steel: Hardened and tempered **DPN: 180 UTS: 610 Elon: 22% Proof: 330**
S62	0.22% C 13% Cr steel: Hardened and tempered **DPN: 220 UTS: 760 Elon: 17% Proof: 480**
S80	0.17% C 2% Ni 16.5% Cr steel: Hardened and tempered **DPN: 280 UTS: 920 Elon: 12% Proof: 620**
S124	0.2% C 13% Cr 0.3% S steel: Hardened and tempered: free machining **DPN: 230 UTS: 770 Elon: 15% Proof: 420**
S 137	0.16% C 2.5% Ni 17% Cr steel: BS Aerospace specification
S 141	0.1% C 13% Cr steel: BS Aerospace specification
S141	0.1% C 13% Cr steel
S150	0.1% C 1% Ni 10% Cr 0.6% Mo 0.2% V 0.25% Nb 0.05% N steel
S151	0.1% C 1% Ni 10% Cr 0.6% Mo 0.2% V 0.25% Nb 0.05% N steel
S152	0.08% C 0.6% Ni 10% Cr 0.7% Mo 0.2% V 6% Co 0.3% Nb 0.02% N steel
S159	0.1% C 2.5% Ni 11.7% Cr 1.7% Mo 0.3% V steel
S 159	0.1% C 2.5% Ni 11.7% Cr 1.7% Mo 0.3% V steel: BS Aerospace specification
S 410	Sintered martensitic stainless steel SG 6.9; Durasint **DPN: 170 UTS: 500 Elon: 7%**
S 538	0.1% C 2.5% Ni 11.7% Cr 1.7% Mo 0.3% V 0.2% N steel
S 41500	0.05% C (max) 4.5% Ni 12.7% Cr 0.7% Mo steel: UNS designation
S 42100	0.2% C 12% Cr 1% Mo 0.3% V 0.5% W steel: UNS designation
S 44100	0.03% C (max) 18.5% Cr 0.35% Ti 0.5% Nb steel: Designation used by UNS
S S1	0.12% C 13% Cr steel: Uddelholm
S S2	0.12% C 17% Cr steel: Uddelholm
S S6	0.27% C 13% Cr steel: Uddelholm
S S22	0.25% C 17% Cr Ni steel: Uddelholm
S S31	0.15% C 13% Cr steel: Uddelholm
SAE 51403	0.15% C 12% Cr steel: Mechanical properties not quoted
SAE 51405	0.08% C 12.5% Cr 0.2% Al steel: Mechanical properties not quoted
SAE 51409	0.08% C 11.25% Cr 0.75% Ti (max) steel: Non-hardenable
SAE 51410	0.15% C (max) 12% Cr steel: Mechanical properties not specified
SAE 51414	0.1% C 2% Ni 12.5% Cr steel: Mechanical properties not specified
SAE 51416	0.15% C (max) 13% Cr 0.6% Zr or Mo steel
SAE 51416 F	0.15% C (max) 13% Cr 0.07% S or Se steel: Free machining; mechanical properties not specified
SAE 51416 Se	0.15% C 13% Cr 0.15% Se steel: Free cutting; mechanical properties not quoted
SAE 51420	0.35% C 13% Cr steel: Mechanical properties not specified
SAE 51420 F	0.35% C 13% Cr 0.07% S or Se steel: Mechanical properties not specified
SAE 51420 F Se	0.35% C 13% Cr 0.15% Se steel: Mechanical properties not quoted
SAE 51429	0.12% C (max) 15% Cr steel: SAE for AISI 429
SAE 51430	0.12% C (max) 16% Cr steel: Mechanical properties not specified
SAE 51430 F	0.12% C (max) 16% Cr 0.07% S or Se steel: Free machining
SAE 51430 F Se	0.12% C 16% Cr 0.15% Se steel: Free machining; mechanical properties not quoted
SAE 51431	0.2% C (max) 2% Ni 16% Cr steel: Mechanical properties not specified
SAE 51434	0.12% C 16% Cr 1% Mo steel: Mechanical properties not quoted
SAE 51436	0.12% C 16% Cr 1% Mo 0.5% Nb steel: Mechanical properties not quoted
SAE 51439	0.07% C (max) 0.5% Ni 18% Cr steel: SAE for AISI 439
SAE 51439 LL	0.014% C (max) 0.5% Ni 18% Cr steel: SAE for AISI 439LL
SAE 51440 A	0.7% C 17% Cr 0.7% Mo steel: Mechanical properties not specified
SAE 51440 B	0.85% C 17% Cr 0.7% Mo steel: Mechanical properties not specified
SAE 51442	0.2% C (max) 20% Cr steel: Mechanical properties not specified
SAE 51446	0.25% C (max) 1% Ni 26% Cr steel: Mechanical properties not specified
SAE 51499	0.08% C 0.5% Ni 11% Cr 0.7% Nb steel: Mechanical properties not quoted
SAE 60326	0.07% C 4.2% Ni 16% Cr 2.8% Cu steel: Casting
SAE 60328	0.04% C 5.2% Ni 26% Cr 2% Mo 3% Cu steel: Casting
SAE 60410	0.15% C 1% Ni 13% Cr 0.5% Mo steel: Casting; ASTM alloy CA 15
SAE 60420	0.3% C 1% Ni 13% Cr 0.5% Mo steel: Casting; ASTM alloy CA 40
SAE 60442	0.3% C 2% Ni 20% Cr steel: Casting; ASTM alloy CB 30
SAE 60446	0.5% C 4% Ni 28% Cr steel: Casting; ASTM alloy CC 50
SAE 70446	0.5% C (max) 4% Ni 28% Cr 0.5% Mo steel: Casting; ASTM alloy HC
SAE EV 1	0.45% C 4.8% Ni 23% Cr 2.8% Mo steel: For exhaust valves **UTS: 920 Elon: 3%**
SAE EV 2	0.4% C 3.8% Ni 24% Cr 1.4% Mo steel: For exhaust valves **UTS: 1070 Elon: 3%**
SAE HNV 6	0.8% C 1.3% Ni 20% Cr 2.3% Si steel: For valves **UTS: 1140 Elon: 15% Proof: 800**
SANBRON BB	0.25% C 13% Cr steel: Sanderson for BS alloy En 56C or D
SANBRON DD	0.25% C 1.7% Ni 18% Cr steel: Sanderson for BS alloy En 57
SC 40	0.4% C 12% Cr steel: Valves, pump spindles; Balfour **DPN: 630**
SC 45	0.75% C 17% Cr steel: Plastic and rubber dies, etc.; Balfour **DPN: 660**
SC 140 Mo	0.14% C 14.5% Cr 1.0% Mo steel: Stavanger
SEA-CURE	Low C 27.5% Cr 1.2% Ni 3.5% Mo Ti stabilized steel: Origin unknown **DPN: 210 UTS: 620 Elon: 32% Proof: 520**
SEC 1	0.15% C (max) 12% Cr steel: Designation used by Japanese Standards
SEC 4	0.15% C (min) 13% Cr steel: Designation used by Japanese Standards
SEC 5	0.12% C 16% Cr steel: Designation used by Japanese Standards
SF 10	0.12% C 1% Ni 13% Cr steel: Annealed; Samuel Fox **UTS: 550 Elon: 25%**
SF 10 ELC	0.03% C (max) 1% Ni 13% Cr steel: Weldable; Samuel Fox
SF 11	0.15% C 1% Ni 13% Cr steel: hardened and tempered; Samuel Fox **UTS: 720 Elon: 30%**
SF 12	0.20% C 1% Ni 13% Cr steel: Hardened and tempered; Samuel Fox **UTS: 740 Elon: 28%**

Symbol	Nominal analysis, supplier, condition and remarks.
SF 13	0.3% C 1% Ni 13% Cr steel: Annealed; Samuel Fox **UTS: 540 Elon: 26%**
SF 14	0.4% C 1% Ni 14% Cr steel: Hardened and tempered; Samuel Fox **UTS: 750 Elon: 26%**
SF 15	0.7% C 1% Ni 14% Cr steel: Hardened and tempered; Samuel Fox **UTS: 780 Elon: 25%**
SF 17	0.1% C 17% Cr steel: Annealed; Samuel Fox **UTS: 510 Elon: 27%**
SF 18	0.2% C 2% Ni 17.5% Cr steel: Hardened and tempered; Samuel Fox **UTS: 900 Elon: 24%**
SF 610	0.12% C 1% Ni 13% Cr 0.2% Zr steel: Free machining; Samuel Fox **UTS: 540 Elon: 25% Proof: 360**
SF 611	0.15% C 1% Ni 13% Cr 0.2% Zr steel: Free machining; Samuel Fox **UTS: 730 Elon: 30% Proof: 580**
SF 612	0.21% C 1% Ni 13% Cr 0.2% Zr steel: Free machining; Samuel Fox **UTS: 740 Elon: 28% Proof: 600**
SF 613	0.27% C 1% Ni 13% Cr 0.2% Zr steel: Free machining; Samuel Fox **UTS: 540 Elon: 26% Proof: 270**
SFG	0.08% C 14% Cr steel: Samuel Fox **UTS: 530 Elon: 31%**
SIC HMo	0.12% C 13% Cr 1.1% Mo steel: Stavanger
SIL 746	0.7% C 6.3% Mn 2% Ni 21% Cr steel: For valves
For SIS specifications see SS	
SK 20	0.22% C 0.5% Ni 11.5% Cr 1% Mo 0.3% V steel: Origin unknown
SK 20W	0.2% C 0.4% Ni 12% Cr 1% Mo 0.3% V 0.5% W steel: Origin unknown
SL 440A	0.7% C 17% Cr 0.7% Mo steel: Hardened and tempered; Swift Levick **DPN: 600**
SL 440B	0.85% C 17% Cr 0.7% Mo steel: Hardened and tempered; Swift Levick **DPN: 600**
SOLEIL 4	0.14% C 14.5% Cr 1.0% Mo steel: Firminy
SOLEIL B2	0.08% C 0.6% Ni 12.5% Cr steel: Creusot-Loire
SOLEIL B3	0.08% C 0.6% Ni 13.0% Cr steel: Creusot-Loire
SOLEIL B4	0.12% C 0.75% Ni 17.2% Cr steel: Creusot-Loire
SOLEIL T1	0.2% C 0.5% Ni 11% Cr 0.7% Mo 0.3% V steel: Origin unknown
SOLEIL T2	0.2% C 0.8% Ni 11% Cr 1% Mo 0.3% V 0.6% W steel: Origin unknown
SOUDOR G 410	0.06% C 12.5% Cr steel: For welding electrode; Soudometal; for inert gas welding **UTS: 650 Elon: 15% Proof: 450**
SOUDOR G 430	0.065% C 17.7% Cr steel: Welding electrode; Soudometal; for inert gas welding **UTS: 540 Elon: 20% Proof: 340**
SPEAR B2	0.1% C 13.5% Cr steel: Spear and Jackson
SPECIAL GENCO	0.2% C 0.8% Ni 12% Cr 1% Mo 0.2% V 1% W steel: Latrobe
SPW	0.25% C 13.0% Cr steel: Origin unknown

Note. The following abbreviations and units are used in the tables:

DPN	Hardness, diamond pyramid number
UTS	Ultimate tensile strength, N/mm^2
Elon	Elongation, %
Proof	0.1% proof strength, N/mm^2

1 N/mm^2=0.1 hbar=0.102 kgf/mm^2=0.06475 tonf/in.^{2}=145.04 lbf/in.^{2}=1 MPa

See Appendix II for other abbreviations and conversion tables.

Symbol	Nominal analysis, supplier, condition and remarks.
SR 264	Low C 26% Cr 4% Mo Nb stabilized steel: Origin unknown
SRO 2	0.45% C 13% Cr 1.0% Mo 2.8% Si steel: Bofors; obsolete
S S1	0.12% C 13% Cr steel: Uddelholm
S S2	0.12% C 17% Cr steel: Uddelholm
S S6	0.27% C 13% Cr steel: Uddelholm
S S22	0.25% C 17% Cr Ni steel: Uddelholm
S S31	0.15% C 13% Cr steel: Uddelholm
SS 14-2383	0.14% C 1.5% Mn 0.5% Ni 17% Cr steel: Stainless; free cutting; Swedish Standard **DPN: 240 UTS: 735 Elon: 12% Proof: 440**
SS 2102	0.21% C 1.9% Ni 17.0% Cr steel: Swedish Standard **DPN: 270 UTS: 1000 Elon: 12% Proof: 640**
SS 2301	0.08% C 0.2% Ni 14% Cr steel: Wrought; annealed; Swedish Standard **DPN: 210 UTS: 420 Elon: 16% Proof: 210**
SS 2302	0.12% C 13% Cr steel: Swedish Standard; included in MNC 900 **DPN: 200**
SS 2302	0.12% C 1.0% Ni (max) 13.0% Cr steel: For construction; Swedish Standard; hardened and tempered **UTS: 700 Elon: 16% Proof: 410**
SS 2302	0.12% C 1.0% Ni (max) 13.0% Cr steel: Swedish Standard; hardened and tempered **UTS: 670 Elon: 16% Proof: 390**
SS 2303	0.2% C 12% Cr steel: Annealed; Swedish Standard **DPN: 230 UTS: 770**
SS 2303	0.2% C 12% Cr steel: Hardened and tempered; Swedish Standard **DPN: 300**
SS 2303	0.22% C 1.0% Ni (max) 13.0% Cr steel: Swedish Standard; hardened and tempered **UTS: 980 Elon: 14% Proof: 690**
SS 2303	0.22% C 1.0% Ni (max) 13.0% Cr steel: Swedish Standard **UTS: 1000 Elon: 14% Proof: 690**
SS 2304	0.35% C 13% Cr steel: Plate, bar, forging, etc.; annealed; Swedish Standard **DPN: 240 UTS: 820**
SS 2304	0.35% C 13% Cr steel: Plate, bar, forging, etc.; hardened and tempered; Swedish Standard **DPN: 550 UTS: 1500 Proof: 1070**
SS 2317	0.23% C 0.5% Ni 11.7% Cr 1.2% Mo 0.3% V 0.05% B steel: Bar or forging; Swedish Standard; hardened and tempered **UTS: 870 Elon: 16% Proof: 640**
SS 2320	0.1% C (max) 17% Cr steel: Swedish Standard included in MNC 900 **DPN: 200**
SS 2321	0.25% C 2% Ni 16% Cr steel: Annealed; Swedish Standard **DPN: 270**
SS 2321	0.21% C 1.9% Ni 17.0% Cr steel: For construction; Swedish Standard; hardened and tempered **UTS: 1000 Elon: 12% Proof: 640**
SS 2322	0.25% C 24% Cr steel: Annealed; Swedish Standard **DPN: 210 UTS: 820**
SS 2323	0.12% C 4% Ni 25% Cr steel: Plate, bar and forging; annealed; Swedish Standard **DPN: 260 UTS: 920**
SS 2324	0.1% C 4.5% Ni 25% Cr 1.3% Mo steel: Plate, bar, etc.; annealed; Swedish Standard **DPN: 260 UTS: 820 Elon: 20% Proof: 400**
SS 2325	0.08% C 0.2% Ni 17.5% Cr 1.8% Mo steel: Annealed **DPN: 180 UTS: 500 Elon: 25% Proof: 330**
SS 2326	0.025% C (max) 0.5% Ni (max) 18.0% Cr 2.2% Mo 0.06% Ti steel: Swedish Standard **DPN: 210 UTS: 550 Elon: 25% Proof: 340**

Symbol	Nominal analysis, supplier, condition and remarks.
SS 2326	0.02% C (max) 18% Cr 2.2% Mo 0.2% Ti (min) steel: Swedish Standard
SS 2380	0.12% C 13.0% Cr 0.22% S steel: Free machining; Swedish Standard; hardened and tempered **DPN: 223 UTS: 590 Elon: 12% Proof: 410**
SS 2382	0.03% C (max) 18% Cr 2.5% Mo 0.8% Ti steel: Swedish Standard
SS 2382	0.03% C 18% Cr 2.2% Mo 0.6% Ti 0.2% S 2% Cu steel: Swedish Standard; free machining
SS 2382	0.03% C + N 18% Cr 2.2% Mo steel: Swedish Standard
SS 2383	0.14% C 17% Cr 0.2% S steel: Free machining; Swedish Standard
STA 5 V25	0.2% C 13% Cr steel: Replaced by BS alloy En 56C
STA 5 V25A	0.3% C 13% Cr steel: Replaced by BS alloy En 56D
STA 5 V25B	0.12% C 13% Cr steel: Replaced by BS alloy En 56A
STA 5 V25B/1	0.12% C 13% Cr steel: Replaced by BS alloy En 56A
STA 5 V25B/2	0.16% C 13% Cr steel: Replaced by BS alloy En 56B
STA 5 V25M	0.25% C (average) 13% Cr steel: Free machining grade; replaced by BS alloy En 56 appropriate M
STA 5 V28	0.2% C 2% Ni 18% Cr steel: Replaced by En 57
STAINLESS 16	0.27% C 13.5% Cr 1.5% Mo steel: Uddelholm
STAINLESS 51	0.1% C 1.0% Ni 14% Cr steel: Uddelholm
STAINLESS 851	0.2% C 11.5% Cr 1% Mo 0.3% V steel: Uddelholm
STRAINTRODE 6815	0.05% C 0.5% Mn 13.5% Cr steel: For electrodes; Esab standard **UTS: 880**
STUBS 66	0.08% C 13.5% Cr steel: Welding electrode; Stubs; low hydrogen rod for heat resistant Cr steels **DPN: 350 UTS: 810 Elon: 2% Proof: 700**
SUH 3	0.4% C 11.0% Cr 1.0% Mo steel: Heat resisting; Japanese Standards designation
SUH 600	0.17% C 11.5% Cr 0.6% Mo steel: Heat resisting; Japanese Standard
SUH 616	0.22% C 12.0% Cr 1.0% Mo 0.25% V 1.0% W steel: Heat resisting; Japanese Standards designation
SV 06KH 14	0.08% C (max) 0.5% Mn 14.0% Cr steel: For welding; Russian Standards designation
SV 08KH 14GT	0.1% C (max) 1.1% Mn 14.0% Cr 0.8% Ti steel: For welding wire; Russian Standards designation
SV 10KH 11MFN	0.12% C 0.75% Ni 11.2% Cr 0.75% Mo 0.32% V steel: For welding wire; Russian Standards designation
SV 10KH 11VMFN	0.1% C 0.5% Mn 1.0% Ni 11.2% Cr 1.1% Mo 0.3% V 1.2% W steel: For welding wire; Russian Standards designation
SV 10KH 13	0.12% C 0.5% Mn 13.0% Cr steel: For welding wire; Russian Standards designation
SV 10KH 17T	0.12% C (max) 0.7% Mn (max) 17.0% Cr 0.5% Ti (max) steel: For welding wire; Russian Standards designation
SV 13KH 25T	0.15% C (max) 0.8% Mn (max) 24.5% Cr 0.5% Ti (max) steel: For welding wire; Russian Standards designation
SV 120KH 13	0.12% C 0.5% Mn 13.0% Cr steel: For welding wire; Russian Standards designation
T 550	0.23% C 0.7% Ni 11.8% Cr 1.1% Mo 0.3% V steel: VEW; creep resisting
T 552	0.11% C 2.5% Ni 11.4% Cr 1.7% Mo 0.3% V steel: VEW; creep resisting
T 558	0.2% C 0.7% Ni 12% Cr 1.0% Mo 0.3% V steel: VEW; creep resisting
T 560	0.17% C 0.5% Ni 11.1% Cr 0.6% Mo 0.3% V 0.28% Nb 0.005% B 0.05% N steel: VEW; creep resisting
T 608	0.13% C 0.5% Ni 11.9% Cr 0.5% Mo steel: VEW; creep resisting
T 650	0.21% C 0.4% Ni 12.3% Cr steel: VEW; creep resisting
T 655	0.14% C 0.4% Ni 11.6% Cr steel: VEW; creep resisting

Symbol	Nominal analysis, supplier, condition and remarks.
TEMPEST	0.16% C 1.8% Ni 17% Cr steel: Glass moulds and inserts; Hall & Pickles **UTS: 1130 Elon: 20% Proof: 920**
TENELON	0.12% C (max) 15.2% Mn 17.8% Cr steel
TRIM RITE	0.2% C 0.7% Ni 14% Cr 0.6% Mo steel: Carpenter
TRUBRITE B41	Low C 17% Cr steel: Arthur Lee
TRUBRITE B42	Low C 20% Cr steel: Arthur Lee
TRUBRITE C72	Low C 12% Cr steel: Hardenable; Arthur Lee
TRUBRITE C74	Low C 12% Cr steel: For cutlery; Arthur Lee
TRUBRITE C76	12% Cr steel: For cutlery; Arthur Lee
TUBROD 15.23	0.3% C 0.5% Mn 13.0% Cr steel: Tubular welding rod; Esab; for hard facing; as welded **DPN: 525**
TUBROD 15.70	0.06% C 0.5% Mn 13.0% Cr steel: Tubular weld rod for submerged arc welding; corrosion resistant; Esab **DPN: 330 UTS: 900**
TUBROD 15.73	0.3% C 0.4% Mn 13.0% Cr steel: Tubular welding rod for submerged arc welding; Esab; corrosion resistant; as welded **DPN: 520**
TURBOTHERM 15M	0.15% C 12% Cr 1.1% Mo steel: Bohler
TURBOTHERM 20M	0.2% C 12% Cr 1.2% Mo steel: Bohler
TURBOTHERM 20MV	0.2% C 12% Cr 1% Mo 0.3% V steel: Bohler
TURBOTHERM 20MV Nb	0.2% C 11.5% Cr 0.6% Mo 0.2% V steel: Bohler
TURBOTHERM 20MV W	0.2% C 12% Cr 1% Mo 0.3% V 0.3% W steel: Bohler
TWOSCORE	0.2% C 2% Ni 16% Cr steel: Bar, billet, etc.; Brown Bayley for BS alloy En 57
U 12	0.17% C 13% Cr steel: Ugine
UDDCO 12	0.2% C 12% Cr 1% Mo 0.3% V 0.5% W steel: Uddelholm
UHB STAINLESS 1	0.1% C 13.5% Cr steel: Uddelholm; annealed **DPN: 160 UTS: 480 Elon: 18% Proof: 290**
UHB STAINLESS 2	0.09% C 17.5% Cr steel: Uddelholm: annealed **DPN: 150 UTS: 490 Elon: 23% Proof: 330**
UHB STAINLESS 5	0.25% C 25% Cr steel: Uddelholm; annealed **DPN: 170 UTS: 530 Elon: 18% Proof: 330**
UHB STAINLESS 6	0.33% C 13.5% Cr steel: Uddelholm; hardened and tempered **DPN: 250 UTS: 820 Elon: 11% Proof: 630**
UHB STAINLESS 21	0.06% C 13.5% Cr steel: Uddelholm; annealed **DPN: 140 UTS: 440 Elon: 25% Proof: 250**
UHB STAINLESS 22	0.2% C 1.5% Ni 17% Cr steel: Uddelholm; hardened and tempered **DPN: 260 UTS: 820 Elon: 11% Proof: 580**
UHB STAINLESS 31	0.22% C 0.7% Ni 13.5% Cr steel: Uddelholm; hardened and tempered **DPN: 220 UTS: 710 Elon: 15% Proof: 540**
UHB STAINLESS 41	0.13% C 13.5% Cr steel: Free cutting; Uddelholm; hardened and tempered **DPN: 200 UTS: 630 Elon: 15% Proof: 480**
UHB STAINLESS 51	0.1% C 13.5% Cr 1.0% Mo steel: Uddelholm; hardened and tempered **DPN: 190 UTS: 650 Elon: 15% Proof: 500**
UHB STAINLESS 72	0.12% C 17% Cr 0.25% Mo steel: Free cutting; Uddelholm; annealed **DPN: 150 UTS: 530 Elon: 20% Proof: 290**
UHB STAINLESS 716	0.35% C 13.5% Cr 1.0% Mo steel: Uddelholm; hardened and tempered **DPN: 250 UTS: 820 Elon: 11% Proof: 630**
UHB STAINLESS 721	0.06% C 12% Cr steel: Uddelholm
UHB UDD Co 12	0.23% C 0.5% Ni 11.7% Cr 1.0% Mo 0.3% V 0.5% W steel: Tube; Uddelholm; hardened and tempered **UTS: 760 Elon: 20% Proof: 480**
UNI 3992 X43 CS8	0.42% C 8.7% Cr steel: For valves; Italian Standard
UNI 4047	0.16% C (max) 2.0% Ni 19% Cr steel: Italian Standard

Symbol	Nominal analysis, supplier, condition and remarks.
UNI 4047 X8 CA13	0.08% C (max) 12.5% Cr 0.2% Al steel: Italian Standard
UNI 4047 X12 C17	0.12% C (max) 17% Cr steel: Italian Standard
UNI 4047 X12 CA12	0.12% C (max) 12% Cr 1.2% Al steel: Italian Standard
UNI 4047 X12 CA23	0.12% C (max) 23% Cr 1.6% Al steel: Italian Standard
UNI 4047 X15 C13	0.15% C (max) 12.5% Cr steel: Italian Standard
UNI 4047 X15 CF13	0.15% C (max) 13% Cr 0.6% Mo (max) steel: Italian Standard
UNI 4047 X20 C13	0.2% C 13% Cr steel: Italian Standard
UNI 4047 X20 CN16	0.2% C (max) 2.0% Ni 16% Cr steel: Italian Standard
UNI 4047 X25 C 26	0.25% C (max) 25.5% Cr steel: Italian Standard
UNI 4047 X32 C13	0.3% C 13% Cr steel: Italian Standard
UNI 4047 X40 C14	0.4% C 14% Cr steel: Italian Standard
UNI S115	Replaced by UNI 3992 and UNI 4047: Italian Standard
UNILOY 1409TB	0.12% C 0.4% Ni 12.2% Cr 0.3% Mo steel: Cyclop
UNILOY 1420 WM	0.2% C 0.7% Ni 12.2% Cr 1% Mo 0.25% V 1% W steel: Origin unknown
UNILOY 1430 MV	0.3% C 0.3% Ni 11.5% Cr 2.7% Mo 0.25% V steel: Origin unknown
USS 12 Mo V	0.2% C 0.5% Ni 12% Cr 1% Mo 0.3% V steel: Origin unknown
V 3M	0.27% C 13% Cr steel: Krupps
V 5M	0.15% C 13% Cr steel: Krupps
V 13F	0.12% C 13% Cr steel: Krupps
V 13F Al	0.08% C 13% Cr 0.2% Al steel: Krupps
V 13F B	0.08% C 14% Cr steel: Krupps
V 13F Supra	0.08% C 13% Cr steel: Krupps
V 17F	0.1% C 16.5% Cr steel: Krupps
V 17F Extra	0.1% C 17% Cr 0.7% Ti steel: Krupps
V 17F X Extra	0.1% C 17% Cr 1.0% Nb steel: Krupps
V 274 M	0.1% C (max) 4.5% Ni 27.5% Cr 1.5% Mo steel: Valbruna
V2KG	0.2% C 11% Cr steel: Origin unknown
V254	0.2% C (max) 4% Ni 25% Cr steel: Valbruna
VACCUTHERM 5.32	0.15% C 11.5% Cr 1% Mo steel: Origin unknown
VACCUTHERM 5.32H	0.2% C 11.5% Cr 1% Mo steel: Origin unknown
VACCUTHERM 5.34	0.25% C 11.5% Cr 1% Mo 0.3% V steel: Origin unknown
VACCUTHERM 5.36	0.2% C 12% Cr 1% Mo 0.3% V 0.5% W steel: Origin unknown
VACCUTHERM 5.40H	0.18% C 0.6% Ni 11.5% Cr 0.6% Mo 0.2% V steel: Origin unknown
VAL 1A	0.12% C (max) 12.5% Cr steel: Valbruna
VAL 1AL	0.08% C (max) 13% Cr 0.2% Al steel: Valbruna
VAL 1B	0.15% C (max) 12.5% Cr 0.6% Mo steel: Valbruna
VAL 1MP	0.15% C (max) 12.5% Cr 0.5% Mo steel: Valbruna
VAL 1P	0.08% C (max) 12.5% Cr steel: Valbruna
VAL 1W	0.2% C 0.7% Ni 12.5% Cr 1.5% Mo 0.2% V 1% W steel: Origin unknown
VAL 2A	0.2% C 13% Cr steel: Valbruna
VAL 2B	0.3% C 13% Cr steel: Valbruna
VAL 2BZ	0.3% C 13% Cr steel: Valbruna
VAL 2C	0.4% C 14% Cr steel: Valbruna
VAL 2D	0.45% C 14% Cr 0.5% Mo 0.2% V steel: Valbruna
VAL 2MV	0.21% C 0.5% Ni 12% Cr 1% Mo 0.3% V steel: Valbruna

Note. The following abbreviations and units are used in the tables:

DPN	Hardness, diamond pyramid number
UTS	Ultimate tensile strength, N/mm^2
Elon	Elongation, %
Proof	0.1% proof strength, N/mm^2

1 N/mm^2=0.1 hbar=0.102 kgf/mm^2=0.06475 tonf/in.^{2}=145.04 lbf/in.^{2}=1 MPa

See Appendix II for other abbreviations and conversion tables.

Symbol	Nominal analysis, supplier, condition and remarks.
VAL 2W	0.2% C 12.5% Cr 1% Mo 0.25% V 1% W steel: Valbruna
VAL 3	0.38% C 16.5% Cr 1.1% Mo steel: Valbruna
VAL 4	0.2% C (max) 2% Ni 16% Cr steel: Valbruna
VAL 4S	0.15% C (max) 2% Ni 19% Cr steel: Valbruna
VAL 12	0.15% C (max) 13% Cr 0.15% S steel: Valbruna
VALVIC 857	0.8% C 1.5% Ni 20% Cr 2.1% Si Steel: For valves; Swift Levick **DPN: 500 UTS: 1620**
VH 217	0.21% C 13% Cr steel: Wykmanshyttan
VH 273	0.12% C 13% Cr steel: Wykmanshyttan
VH 417	0.27% C 13% Cr steel: Wykmanshyttan
VH 620	0.12% C 17% Cr steel: Wykmanshyttan
VH 626	0.25% C 2.0% Ni 17% Cr steel: Wykmanshyttan
VIRGO 2.0	0.2% C 13% Cr 1% Mo steel: Origin unknown
VIRGO 3	0.25% C (max) 13% Cr 0.5% Mo steel: Origin unknown
VIRGO 3S	0.15% C (max) 13% Cr 0.5% Mo steel: Origin unknown
VIRGO 4	0.15% C (max) 1% Ni 13% Cr 1% Mo steel: Origin unknown
VIRGO 6	0.12% C (max) 13% Cr 0.45% Mo steel: Origin unknown
VIRGO 7	0.22% C 0.5% Ni 11.5% Cr 1.1% Mo 0.3% V steel: Origin unknown
VIRGO 8	0.2% C 0.5% Ni 11% Cr 0.7% Mo 0.3% V 0.2% Nb steel: Origin unknown
VIRGO 8S	0.08% C (max) 1.2% Ni 11.5% Cr 0.5% Mo 0.1% V 0.2% Nb steel: Origin unknown
VIRGO 10	0.14% C 14% Cr 1% Mo steel: Origin unknown
VIS 11	0.25% C 0.6% Ni 10.8% Cr 1.1% Mo 0.4% V 0.4% W steel: Origin unknown
VK 5M	0.2% C 13% Cr 1.2% Mo steel: Origin unknown
VK 17F	0.07% C 17% Cr 1.0% Mo steel: Krupps
VS 17	0.1% C 17% Cr steel: Annealed; Vulcan **UTS: 840 Elon: 33%**
VS S61	0.1% C 13% Cr steel: Hardened and tempered; Vulcan **UTS: 950 Elon: 30%**
VS S80	0.2% C 1.5% Ni 16% Cr steel: Hardened and tempered; Vulcan **UTS: 1280 Elon: 24%**
W 12	0.12% C 13% Cr steel: Welding electrode; Cu coated; Armco
W IRONIT 720	0.2% C 13% Cr 1.2% Mo steel: Origin unknown
WCR 100	0.7% C 14% Cr 20% W 3.5% B steel: Rod for facing cutting tools, etc.; melting point 1200 °C; Wall Colmonoy **DPN: 950**
WEARTRODE 84.58	0.7% C 0.7% Mn 10.0% Cr steel: For electrodes; Esab **DPN: 635**
WESTFALIA RF 5	0.2% C 13% Cr 1.2% Mo steel: Origin unknown
WH 602	0.08% C 14.5% Cr 1% Mo steel: Origin unknown
WH 608	0.25% C 14% Cr 1% Mo steel: Origin unknown
X 3 Cr Mo Ti 17	0.1% C 17.0% Cr 1.75% Mo 0.8% Ti steel: Designation used by German Standards
X 6 Cr 13	0.08% C (max) 0.06% Ni 12.5% Cr steel: Italian Standard
X 6 Cr Mo 17	0.07% C (max) 17.0% Cr 1.05% Mo steel: German Standard
X 6 Cr Mo 17	0.07% C (max) 17.0% Cr 1.0% Mo steel: Designation used by German Standards
X 7 Cr 14	0.08% C (max) 14.0% Cr steel: Designation used by German Standards
X 7 Cr Al 13	0.08% C (max) 13.0% Cr 0.2% Al steel: German Standard
X 7 Cr Al 13	0.08% C (max) 13.0% Cr 0.2% Al steel: Designation used by German Standards
X 8 Cr 13	0.07% C 13% Cr 0.2% Al steel **DPN: 190 UTS: 420 Elon: 20% Proof: 290**
X 8 Cr 14	0.1% C (max) 14.5% Cr steel: Designation used by German Standards

Symbol	Nominal analysis, supplier, condition and remarks.
X 8 Cr 15	0.12% C (max) 14.5% Cr steel: Designation used by German Standards
X 8 Cr 17	0.1% C (max) 16.5% Cr steel: Designation used by German Standards
X 8 Cr 18	0.1% C (max) 17.5% Cr steel: Designation used by German Standards
X 8 Cr 28	0.1% C (max) 28.0% Cr steel: Designation used by German Standards
X 8 Cr 30	0.1% C (max) 2.0% Ni 30% Cr steel: Designation used by German Standards
X 8 Cr Mo 302	0.1% C (max) 2.0% Ni 30.0% Cr 2.25% Mo 2.0% Si 12% N steel: Designation used by German Standards
X 8 Cr Mo Ti 18	0.1% C (max) 17.5% Cr 1.2% Mo 0.6% Ti steel: Designation used by German Standards
X 8 Cr Nb 17	0.1% C (max) 17.5% Cr 1.3% Nb steel: Designation used by German Standards
X 8 Cr Ti 17	0.1% C (max) 17.0% Cr 1.0% Ti steel: Designation used by German Standards
X 8 Cr Ti 18	0.1% C (max) 17.5% Cr 0.6% Ti steel: Designation used by German Standards
X 10 C18	0.07% C 17.0% Cr steel: Italian Standard **DPN: 200 UTS: 180 Elon: 10% Proof: 290**
X 10 Cr 13	0.1% C 13.0% Cr steel: Designation used by German Standards
X 10 Cr 25	0.15% C (max) 25.0% Cr steel: Designation used by German Standards
X 10 Cr Al 13	0.12% C (max) 13.0% Cr 1.0% Al steel: Designation used by German Standards
X 10 Cr Al 18	0.12% C (max) 18.0% Cr 1.0% Al steel: Designation used by German Standards
X 10 Cr Al 24	0.12% C (max) 24.0% Cr 1.5% Al steel: Designation used by German Standards
X 10 Cr Si 13	0.12% C (max) 13.0% Cr 2.1% Si steel: Designation used by German Standards
X 10 Cr Si 18	0.12% C (max) 18.0% Cr 2.1% Si steel: Designation used by German Standards
X 10 Cr Si 29	0.12% C (max) 29.0% Cr steel: Designation used by German Standards
X 12 C 17	0.07% C 17.0% Cr steel: Italian Standard **DPN: 200 UTS: 180 Elon: 20% Proof: 290**
X 12 Cr Mo S 17	0.15% C 16.5% Cr 0.25% Mo 0.2% S steel: Designation used by German Standards
X 15 Cr 13	0.15% C 13.0% Cr steel: Designation used by German Standards
X 15 Cr Mo 12/1	0.15% C 11.5% Cr 1.1% Mo steel: Designation used by German Standards
X 15 Cr Mo 13	0.15% C 13.0% Cr 1.1% Mo steel: Designation used by German Standards
X 20 C 28	0.2% C 25.0% Cr 0.15% N steel: Italian Standard **DPN: 230 UTS: 480 Elon: 20% Proof: 330**
X 20 Cr 13	0.19% C 13% Cr steel: Designation used by German Standards
X 20 Cr Mo 13	0.2% C 13.0% Cr 1.1% Mo steel: Designation used by German Standards
X 20 Cr Mo W V 12/1	0.2% C 0.5% Ni 12% Cr 1.0% Mo 0.3% V 0.5% W steel: German Standard
X 20 Cr Ni Si 25/4	0.2% C 4.5% Ni 25.0% Cr steel: Designation used by German Standards
X 22 Cr Mo V 12/1	0.22% C 0.6% Ni 11.5% Cr 1.1% Mo 0.3% V steel: Designation used by German Standards
X 22 Cr Mo W V 12/1	0.2% C 0.5% Ni 12.0% Cr 1.0% Mo 0.3% V 0.5% W steel: Designation used by German Standards
X 22 Cr Ni 17	0.2% C 2.0% Ni 17.0% Cr steel: Designation used by German Standards
X 25 C26	0.2% C 25.0% Cr 0.15% N steel: Italian Standard **DPN: 230 UTS: 480 Elon: 20% Proof: 330**
X 35 Cr 14	0.35% C 14.0% Cr steel: Designation used by German Standards
X 35 Cr Mo 17	0.38% C 16.5% Cr 1.1% Mo steel: Designation used by German Standards
X 40 Cr 13	0.4% C 13.0% Cr steel: Designation used by German Standards
X 45 Cr Mo V 15	0.44% C 14.2% Cr 0.52% Mo 0.12% V steel: Designation used by German Standards
X 45 Cr Mo V 15	0.45% C 14.4% Cr 0.13% V steel: German Standard
X 45 Cr Ni Mo 23/5	0.45% C 5.0% Ni 23.0% Cr 2.7% Mo steel: Designation used by German Standards
X 80 Cr Ni Si 20	0.8% C 1.5% Ni 20.0% Cr 2.25% Si steel: Designation used by German Standards
X B	0.78% C 1.5% Ni 19% Cr 2% Si steel: Firth Vickers **DPN: 300 UTS: 970 Elon: 15%**
X B	0.8% C 17% Cr steel: Bar and forging; Brown Bayley for BS alloy En 59
X 13 RG	0.3% C 1% Ni 12% Cr 0.3% Mo steel: Origin unknown
X 17 AL	0.12% C (max) 18% Cr 1% Al steel: Valbruna
X 17 C	0.12% C (max) 17% Cr steel: Valbruna
X 17 Z	0.12% C (max) 17% Cr 0.15% S steel: Valbruna
X 36 Cr Mo 17	0.38% C 0.8% Ni 16% Cr 1% Mo steel: German Standard
X 42 Cr 13	0.43% C 13.5% Cr steel: German Standard
XM 8	0.07% C 0.5% Ni (max) 18.0% Cr 0.6% Nb (min) 0.15% Al (max) steel: Designation used by ASTM
XM 27	0.01% C (max) 26.0% Cr 0.1% Mo steel: Designation used by ASTM
XM 30	0.18% C 12.5% Cr 0.2% Nb steel: Designation used by ASTM
XM 33	0.06% C (max) 26% Cr 1% Mo 0.4% Ti steel: Designation used in the UK and USA
Z 6 C13	0.15% C 12% Cr steel: French National Standard
Z 6 C13	0.08% C (max) 0.6% Ni 12.5% Cr steel: French National Standard
Z 6 C13 Al	0.07% C 13% Cr 0.2% Al steel: French National Standard **DPN: 190 UTS: 420 Elon: 20% Proof: 290**
Z 6 CA13	0.08% C (max) 0.6% Ni 13.0% Cr steel: French National Standard
Z 6CND13/04M	0.08% C 3.0% Ni 12.5% Cr 1.0% Mo steel: Casting; designation used by French Standards
Z 6CNDNb17/12	0.08% C (max) 12.0% Ni 17.5% Cr 2.5% Mo steel: French National Standard
Z 8 C17	0.12% C 2% Ni 16% Cr steel: French National Standard **DPN: 200 UTS: 1130 Elon: 20% Proof: 290**
Z 8C17	0.12% C (max) 0.75% Ni 17.2% Cr steel: French National Standard
Z 8CD17/01	0.05% C 17% Cr 0.04% P (max) steel: Plate; French Standard **UTS: 600 Proof: 290**
Z 10 C13	0.12% C 13% Cr steel: French National Standard
Z 10 C17	0.07% C 17.0% Cr steel: French National Standard **DPN: 200 UTS: 180 Elon: 20% Proof: 290**
Z 10 C18	0.12% C 17% Cr steel: French National Standard **DPN: 200 UTS: 180 Elon: 20% Proof: 290**
Z 12 C13	0.15% C 12% Cr steel: French National Standard
Z 12 C18	0.12% C 2% Ni 16% Cr steel: French National Standard
Z 15 C27	0.2% C 25% Cr 0.2% N steel: French National Standard **DPN: 230 UTS: 480 Elon: 20% Proof: 330**
Z 15 CN16/2	0.2% C 2% Ni 16% Cr steel: French National Standard
Z 20 C13	0.12% C 13% Cr steel: French National Standard
Z 20 C25	0.2% C 25% Cr 0.15% N steel: French National Standard **DPN: 230 UTS: 480 Elon: 20% Proof: 330**
Z 20 C25	0.07% C 4% Ni 27% Cr steel: French Standard; annealed **DPN: 260 UTS: 630 Elon: 20% Proof: 370**
Z 20 CN25/05	0.07% C 4% Ni 27% Cr steel: French Standard; annealed **DPN: 260 UTS: 630 Elon: 20% Proof: 370**

Symbol	Nominal analysis, supplier, condition and remarks.
Z 30 C13	0.15% C 13% Cr steel: French National Standard
Z 30 C13	0.3% C 13.0% Cr steel: French Standards designation
Z 70 C14	0.4% C 14.0% Cr steel: French Standards designation
Z 70 C15	0.7% C 15.0% Cr steel: French Standards designation
Z 80 CSN20/02	0.8% C 0.8% Mn (max) 1.3% Ni 20.0% Cr steel: French Standards designation

Note. The following abbreviations and units are used in the tables:

DPN	Hardness, diamond pyramid number
UTS	Ultimate tensile strength, N/mm^2
Elon	Elongation, %
Proof	0.1% proof strength, N/mm^2

1 N/mm^2=0.1 hbar=0.102 kgf/mm^2=0.06475 tonf/in.^{2}=145.04 lbf/in.^{2}=1 MPa

See Appendix II for other abbreviations and conversion tables.

44M2 Steel – high carbon, high chromium
0.8% carbon minimum, 12% chromium, with titanium and molybdenum

Specific gravity	7.77
Density	7770 kg/m^3
Solidus/liquidus	1450 °C
Thermal conductivity	18.8 W/m °C
Coefficient of linear expansion	12×10^{-6}/ °C
Electrical conductivity	2% IACS (copper 100%)
Specific resistance	700 microhm mm
Young's modulus of elasticity	–
Impact	will generally be very low
Fatigue strength	–
Hot strength	–

The above properties are typical of the following group, and may not apply exactly to any one specification. It is possible that with certain specifications some of the values may not be applicable.

General metallurgical characteristics

These steels are very hard with a tendency to brittleness even in the annealed condition. The high chromium content ensures a degree of corrosion resistance but this does not equal that of the lower carbon high chrome steels owing to the presence of the carbides. Additions of titanium and molybdenum ensure grain refinement, improve the hot strength characteristics and raise the tempering temperature. The corrosion resistance of these steels is improved by hardening and tempering and they are almost always used in this condition.

These steels cannot accept cold work without becoming extremely hard and brittle. They are therefore unworkable in the cold.

Full annealing, hardening and normalizing will require temperatures just below or above 1000 °C.

Most of the materials in this group will air harden except where large masses or volumes are involved, where oil quenching will be necessary. These steels have extremely high hardenability and cracking or fracture will occur unless the hardening operation is correctly controlled. Any rapid change of section must be very carefully examined.

These steels are subject to temper brittleness unless specially formulated. This is the result of precipitation of chromium carbides at the grain boundaries when the components are held for any length of time at between 300 and 600 °C. Tempering should therefore be carried out either above 600 °C followed by water quenching or below 300 °C.

Welding on these steels should not be carried out but, where essential, specialist advice must be obtained and great care taken to follow this advice.

Brazing presents no problems apart from temper brittleness unless rapidly cooled after brazing.

These steels have excellent corrosion resistance in oxidizing conditions. The chromium oxide, which is adherent and forms at the surface in oxidizing conditions immediately the surface is abraded, prevents further corrosion. It is therefore essential that these materials are used in conditions where the oxide can re-form immediately.

Corrosion products, once formed, can absorb the corrosive media and corrosion can then be very rapid. These materials have a low notch sensitivity initially and any further notches can result in fracture.

The materials in this group are used for plastic dies and moulds, forging dies, shear blades or cutting equipment where corrosive material is involved. Many of the steels in this group can retain their hardness under high temperature conditions with corrosive materials.

They are also used for forging dies, surgical equipment, cutlery, particularly knives, and ball and roller bearings.

Symbol	Nominal analysis, supplier, condition and remarks.
1.2080	2.0% C 12.0% Cr steel: German Standards designation
1.2080	2.1% C 12.5% Cr 0.2% V steel: German Standards designation
1.2080	2% C 12.5% Cr 0.2% V steel: German Standards designation
1.2086	2.9% C 12% Cr steel: German Standards designation
1.2379	1.5% C 11.5% Cr 0.7% Mo 1% V steel: German Standards designation
1.2419	1.05% C 1.0% Cr 1.2% W steel: German Standards designation
1.2436	2.0% C 12.0% Cr 0.7% W steel: German Standards designation
1.2601	1.5% C 12% Cr 0.8% Mo 0.3% V steel: German Standards designation
1.4112	0.9% C 18% Cr 1.2% Mo 0.1% V steel: German Standards designation
1.4721	2.1% C 12% Cr steel: For valves; German Standards designation
1.4747	0.8% C 1.5% Ni 20% Cr steel: For valves; German Standards designation
1 CW	2.0% C 13% Cr steel: Inman
1 R3	2.3% C 11.75% Cr 0.25% V steel: Thread roll dies, punches, etc; Osborn **DPN: 540**
2 RO 189	1.0% C 17% Cr 0.6% Mo steel: Bofors; hardened and tempered **DPN: 350 UTS: 1070 Elon: 8% Proof: 820**
14 Cr 4 Mo	1% C 14.5% Cr 4.0% Mo 0.12% V steel: Proprietary alloy; information from SAE handbook
14S	2% C 0.5% Ni 14% Cr 0.3% Mo 0.3% V steel: Thread rolls, press tools, etc.; Carrs **DPN: 660**
18 C 283	1.05% C 14% Cr 0.5% Mo 0.5% Co 0.2% Cu steel: Sandvik
23S	2% C 13% Cr 0.3% Mo steel: Plastic moulds, shear blades, etc.; Carrs **DPN: 700**
25.51C	1.0% C 5.0% Ni 25.0% Cr steel: Electrode; Metrode; for cavitation resistance
69S	1.5% C 12% Cr 0.7% Mo 0.25% V steel: Press tools, thread rolls, etc.; Carrs **DPN: 660**
88S	1.05% C 17% Cr 0.5% Mo steel: Carrs **DPN: 600**
100 Cr 29	1.0% C 29.0% Cr steel: Electrode; Metrode; hard corrosion resistant deposit
105 W Cr 6	1.06% C 1.0% Cr 1.2% W steel: German Standards designation
120 Cr 25	1.2% C 1.0% Mn 1.0% Ni 25.0% Cr steel: Electrode; Metrode; for repair welding of castings **DPN: 630**
140 Cr 29	1.4% C 29.0% Cr steel: Electrode; Metrode; for hard facing **DPN: 640**
150 Cr 25	1.5% C 25.0% Cr steel: Electrode; Metrode; for hard facing **DPN: 640**

Note. The following abbreviations and units are used in the tables:

DPN	Hardness, diamond pyramid number
UTS	Ultimate tensile strength, N/mm^2
Elon	Elongation, %
Proof	0.1% proof strength, N/mm^2

1 N/mm^2=0.1 hbar=0.102 kgf/mm^2=0.06475 $tonf/in.^2$=145.04 $lbf/in.^2$=1 MPa
See Appendix II for other abbreviations and conversion tables.

Symbol	Nominal analysis, supplier, condition and remarks.
300 Cr 12 Mo W	3.0% C 12.0% Cr 2.0% Mo 2.0% W steel: Electrode; Metrode; for hard facing **DPN: 550**
336	1.5% C 0.3% Mn 12% Cr 0.6% Mo 0.6% V steel: F Parkin; thread roll dies, gauges, etc. **DPN: 630**
338	2.0% C 13% Cr steel: Plastic moulds, press tools, etc.; F Parkin **DPN: 800**
350 Cr 20	3.5% C 1.0% Mn 20.0% Cr steel: Electrode; Metrode; for hard facing **DPN: 620**
476	1.5% C 12% Cr 1% Mo 0.2% V steel: Blanking dies, plastic moulds; Sanderson
476 SPECIAL	2.2% C 12% Cr 1% Mo 0.2% V steel: Blanking dies and punches etc.; Sanderson **DPN: 750**
2233	2.0% C 12.0% Cr steel: French Standards designation
2234	2.0% C 12.0% Cr 0.8% Mo 0.2% V steel: French Standards designation
2235	1.6% C 12.0% Cr 0.8% Mo 0.4% V steel: French Standards designation
2236	1.8% C 12.0% Cr 0.8% Mo 0.5% V 3.0% Co steel: French Standards designation
2237	2.3% C 12.0% Cr 1.0% Mo 4.0% V steel: French Standards designation
A 6	1.5% C 11.7% Cr 0.5% Mo 1.3% V steel: T Turton
A 11	2.0% C 13% Cr 0.2% Mo 0.2% V steel: T Turton
A 22	1.5% C 0.6% Ni 14% Cr 1.5% Mo 0.2% V steel: T Turton
ABRASODUR 33	2.3% C 29.0% Cr 2.8% Mo steel: Welding electrode; Soudometal; for abrasion resistance **DPN: 450**
ABRASODUR 35	3.6% C 32.6% Cr steel: Welding electrode; Soudometal; for abrasion resistance **DPN: 600**
ABRASODUR 38	5.1% C 32.6% Cr steel: Welding electrode; Soudometal; for abrasion resistance **DPN: 600**
ABRASODUR 38AT	5.3% C 30% Cr steel: Welding electrode; Soudometal; for build up of high Mn steel **DPN: 620**
ABRASODUR 39	5.7% C 31.0% Cr steel: For welding electrode; Soudometal; for abrasion resistance **DPN: 640**
ABRASODUR 40	3.9% C 21.1% Cr 6.4% Nb steel: Welding electrode; Soudometal; for hard facing **DPN: 580**
ABRASODUR 43	6.8% C 24.0% Cr 7.0% Nb steel: Welding electrode; Soudometal; for hard facing **DPN: 580**
ABRASODUR 45	5.8% C 21.5% Cr 5.9% Mo 1.1% V 2.3% W 6.0% Nb steel: Welding electrode; Soudometal; for abrasion resistance at high temperature **DPN: 700**
ABRASODUR 45AT	6.5% C 20% Cr 6.0% Mo 1.0% V 2.5% W 6.0% Nb steel: Welding electrode; Soudometal; for abrasion resistance at high temperature **DPN: 700**
ABRASODUR 46	5.4% C 21.5% Cr 6.3% Mo 10.5% Co steel: Welding electrode; Soudometal; for abrasion resistance at high temperature **DPN: 690**
AISI 440 C	1.1% C 17% Cr 0.7% Mo steel **DPN: 600 UTS: 2000 Elon: 1% Proof: 1840**
AISI 440 F	1.0% C 17% Cr 0.07% Se steel: Free machining **DPN: 600 UTS: 2000 Elon: 1% Proof: 1840**
AISI 617	1.1% C 17.5% Cr 0.5% Mo steel: Annealed **DPN: 230 UTS: 760 Elon: 14% Proof: 450**

Symbol	Nominal analysis, supplier, condition and remarks.
AISI 617	1.1% C 17.5% Cr 0.5% Mo steel: Hardened and tempered **DPN: 580 UTS: 2000 Elon: 2% Proof: 1920**
AISI 618	1.05% C 14.5% Cr 4% Mo steel: Hardened and tempered for ball bearings, etc. **DPN: 820**
ALLOY 1	4% C 6% Ni 16% Cr 0.5% V Fe alloy: Casting; Dewrance **DPN: 500 UTS: 530**
ALLOY 4	4% C 6% Ni 16% Cr 0.5% V 4.5% Si Fe alloy: Casting; Dewrance **DPN: 750 UTS: 500**
ALLOY 10 Mod	4.2% C 2% Ni 16% Cr 8% Mo 1% V 1.5% Si Fe alloy: Casting; Dewrance **DPN: 870 UTS: 360**
ALLOY 19	1.25% C 32% Cr 2% Cu Fe alloy: Casting; Dewrance **DPN: 370 UTS: 800**
ALLOY C	2.3% C 13% Cr steel: Jessop **DPN: 870**
ALLOY CWPS	1.5% C 12.0% Cr 1.0% Mo steel: Origin unknown
AMS 5352	1.1% C 17.5% Cr 0.5% Mo steel: Castings
AMS 5352 A	1.0% C 17% Cr 0.5% Mo steel: Investment casting
AMS 5618	1% C 17% Cr steel: Type 440C
AMS 5630	1.1% C 17.5% Cr 0.5% Mo steel: Forgings
AMS 5630 C	1.1% C 17% Cr 0.5% Mo steel: Bar and forging; AMS for SAE 51440 C
AMS 5632	1% C 17% Cr 0.5% Mo 0.2% S steel: Type 440F
AMS 5632	1% C 17% Cr 0.15% Se (min) steel: Type 440 FSe
AMS 5632 B	1.0% C 17% Cr 0.5% Mo steel: Bar and forging; free machining; AMS for SAE 51440 F
AMS 5749	1.15% C 14.5% Cr 4% Mo 1.2% V steel: For bearings
AMS 5880	1% C 17% Cr steel: Type 440C
AMS 5900	1.1% C 14.2% Cr 2.1% Mo 1.2% V 0.3% Nb steel: For bearings
AMS 7445	0.85% C 17% Cr steel: Type 440 B
ARH 8	1.05% C 16% Cr 0.8% Mo Co steel: Hardened and tempered; Schoeller-Bleckmann **DPN: 770**
ARMCO 17 C 80	0.85% C 17% Cr 0.7% Mo steel: Armco
ARMCO 17 C 100	1.0% C 17% Cr 0.7% Mo steel: Armco
ARMCO 17 C 100 FM	1.0% C 17% Cr 30.2% Se steel: Armco
ARS	2.35% C 12% Cr 1.0% Mo 4.0% V steel: Simonds; AISI type B7
ASTM A 756	1% C 0.7% Ni (max) 17% Cr 0.5% Mo steel: For bearings
ASTM A276/440 C	1.1% C 17% Cr 0.7% Mo steel: Bar
ASTM A457/650	0.08% C (max) 25.5% Ni 16.2% Cr 6.2% Mo steel: Plate
ASTM A597 CD2	1.5% C 12.0% Cr 1.0% Mo 0.7% V 0.9% Co cast steel: For tools
ASTM A597 CD5	1.5% C 12.0% Cr 1.0% Mo 0.4% V 3.0% Co cast steel: For tools
AT 2	2.2% C 12% Cr 0.9% Mo 0.25% V steel: Braeburn; AISI type D4
ATMODIE	1.55% C 11.9% Cr 0.8% Mo 0.9% V steel: Columbia; AISI type D2
ATMODIE SMOOTHCUT	1.55% C 11.9% Cr 0.8% Mo 0.9% V steel: Columbia; AISI type D2S; free machining

Note. The following abbreviations and units are used in the tables:

DPN	Hardness, diamond pyramid number
UTS	Ultimate tensile strength, N/mm^2
Elon	Elongation, %
Proof	0.1% proof strength, N/mm^2

1 N/mm^2=0.1 hbar=0.102 kgf/mm^2=0.06475 tonf/in.^{2}=145.04 lbf/in.^{2}=1 MPa

See Appendix II for other abbreviations and conversion tables.

Symbol	Nominal analysis, supplier, condition and remarks.
AVS	0.8% C 1.5% Ni 20% Cr 2% Si steel: Valbruna
BCD	2% C 0.3% Mn 12% Cr 0.3% V 0.3% Si steel: For tools; Huntsman; plastic mould dies, shear blades, etc. **DPN: 770**
BCHV	1.8% C 12% Cr 0.8% Mo 0.25% V steel: For tools; light stamping dies, shear blades, etc.; Huntsman **DPN: 750**
BCRS	1.0% C 13% Cr 0.8% Mo 0.3% V steel: For tools; dies, shear blades, etc.; Huntsman **DPN: 800**
BCW	2.0% C 12% Cr 1% W steel: For tools; plastic mould dies; Huntsman **DPN: 770**
BKV	2.2% C 12.5% Cr 0.25% V steel: Origin unknown
BR 4 FM	2.3% C 12.5% Cr 1.1% Mo 4.0% V steel: Latrobe; die steel; type D7 **DPN: 830**
BRUNSWICK V106	2.0% C 13% Cr + Mo steel: Plastic moulds, shear blades; John Vessey **DPN: 720**
BS 648/B	1.0% C (max) 27% Cr steel: Casting; as cast; contained in BS 3100
BS 1648/C	1.5% C 27% Cr steel: Casting; as cast; contained in BS 3100
BS 3100/452 C11	1.0% C (max) 4.0% Ni (max) 27.0% Cr 1.5% Mo (max) steel: Casting
BS 3100/452 C12	1.5% C (max) 4.0% Ni (max) 27.0% Cr 1.5% Mo (max) steel: Casting
C 12	1.8% C 12.5% Cr 0.5% V steel: Origin unknown
C 265	1.6% C 12% Cr 0.8% Mo 0.8% V steel: Fagersta **DPN: 680**
CCM	1.75% C 13.5% Cr 0.7% Mo steel: For dies; Swift Levick
CCM	1.5% C 12% Cr 0.85% Mo 0.9% V steel: Simonds; AISI type D2
CD	2.1% C 12.5% Cr steel: Plastic moulding, dies, etc.; ESC **DPN: 870**
CDV 1	2.25% C 12.0% Cr 1.0% Mo steel: Origin unknown
CDV 2	1.5% C 12.0% Cr 1.0% Mo steel: Origin unknown
CDV 4	1.5% C 12.0% Cr 1.0% Mo 0.8% V steel: For tools; origin unknown
CDW	1.5% C 12.0% Cr 1.0% Mo steel: Origin unknown
CHIPPER KNIFE	0.81% C 11.5% Cr 0.4% Mo steel: Latrobe
CHROME FM	Latrobe; die steel; type D5 **DPN: 770**
CMD	2.0% C 12% Cr 0.25% V steel: Turton Bros
CMX-A/B/C	0.9% C 17% Cr 0.6% Mo steel: Valbruna
CMX-BM	0.9% C 18% Cr 1.1% Mo 0.1% V steel: Valbruna
COBALT	1.5% C 12.25% Cr 0.85% Mo 3.1% Co steel:
COBALTCROM	1.4% C 13% Cr 0.75% Mo 3% Co steel: Blanking dies, shear blades, etc.; Darwin **DPN: 680**
COLMONOY 1 SPECIAL	1.0% C 13% Cr 3% B steel: Rod for facing; abrasion resistant; melting point 1350 °C; Wall Colmonoy **DPN: 800**
CRB-7 ALLOY	1.1% C 14% Cr 2% Mo 1% V 0.2% Co steel: Carpenter **DPN: 700**
CRD DIAMOND	2.2% C 12.5% Cr steel: C Denton
CRD DIAMOND RED	1.5% C 12% Cr 0.8% Mo 0.2% V steel: Blanking tools, gauges; C Denton **DPN: 680**
CROMOVAN	1.5% C 12% Cr 1.0% Mo 1.0% V steel: Firth Sterling
CROSS PIPES	1.5% C 11% Cr 2% Co steel: Press tools and forging dies; John Vessey **DPN: 720**
CROSTAR	2.0% C 12.5% Cr 0.1% V steel: Pompey
CRUCIBLE 440 BM	1.0% C 18% Cr 0.6% Mo 0.15% V steel: Crucible Steel Co. **DPN: 660**

Symbol	Nominal analysis, supplier, condition and remarks.
CRUCIBLE 440 C	1.1% C 17% Cr 0.7% Mo steel: Crucible Steel Co. **DPN: 660**
C-XB VALVE STEEL	0.8% C 1.3% Ni 20% Cr steel: Carpenter **DPN: 295 UTS: 940 Elon: 8.5% Proof: 840**
D 1	1.0% C 12.0% Cr 1.0% Mo steel: Designation used by AISI
D 2	1.5% C 12.0% Cr 1.0% Mo steel: Designation used by AISI
D 2	1.5% C 12.0% Cr 0.9% Mo 0.9% V steel: For cold work; Osborn
D 3	2.25% C 12.0% Cr steel: Designation used by AISI
D 3	2.2% C 12.5% Cr 0.25% V steel: S Osborn; cold work steel; hardened and tempered **DPN: 700**
D 4	2.25% C 12.0% Cr 1.0% Mo steel: Designation used by AISI
D 5	1.5% C 12% Cr 1% Mo 0.8% V 3% Co steel: For tools; designation used by AISI
D 6	1.5% C 12% Cr 1.0% Mo steel: Designation used by AISI; obsolete
D 7	2.35% C 12% Cr 1.0% Mo 4.0% V steel: Designation used by AISI
D 20	1.5% C 12% Cr 1.0% Mo steel: John Vessey **DPN: 770**
D 65	2.0% C 13% Cr 1.2% W steel: Fagersta **DPN: 800**
D 165	2.1% C 13% Cr 4.0% V 1.4% W steel: Fagersta **DPN: 740**
DELCROME 50V	2.75% C 27% Cr 0.75% V Fe alloy: Melting range 1375–1400 °C; Delcro Stellite **DPN: 580**
DELCROME 450	0.2% C 12.0% Cr Fe alloy: Delcro Stellite **DPN: 450**
DELCROME C	3.7% C 21% Cr Fe alloy: Casting; Deloro **DPN: 610**
DELCROME R	3.0% C 6.0% Mn 30% Cr Fe alloy: Delcro Stellite **DPN: 580**
DELFER B	3.2% C 18% Cr 16% Mo 2.0% V 6.0% Co Fe alloy: Melting point 1178 °C; Delcro Stellite **DPN: 820**
DELFOR	2.5% C 13.5% Cr 9% Mo 5.5% W 16.5% Co 2.5% Fe alloy: Casting; Deloro **DPN: 610 UTS: 440 Elon: 1%**
DIEHARD HCD	2.1% C 13% Cr 0.5% Mo steel: Firth Brown
DIEHARD STANDARD	1.5% C 12% Cr 0.9% Mo 0.9% V steel: Firth Brown
DOMINATOR	1.9% C 12.5% Cr steel: Schoeller-Bleckmann **DPN: 640**
DOMINATOR VM	1.55% C 11.5% Cr 0.8% Mo 1% V steel: Schoeller-Bleckmann **DPN: 740**
DOMINATOR Z	1.9% C 11.5% Cr 2% W steel: Schoeller-Bleckmann **DPN: 740**
DOUBLE SIX	1.9% C 0.3% Mn 12.5% Cr 0.8% Mo 0.25% V steel: Punches, dies, etc.; Edgar Allen **DPN: 750**
DRS 16	0.8% C 1.5% Ni 21% Cr steel: Bofors; obsolete
DT	2.2% C 13% Cr steel: Forez
DTva	1.55% C 12% Cr 0.8% Mo 0.8% V 0.4% Co steel: Forez
E Fe Cr A1	4% C 22% Cr steel: Weld electrode; designation used by AWS
E Fe Cr A2	4% C 26% Cr 2% Mo steel: Weld electrode; designation used by AWS
E Fe Cr A3	3.5% C 17% Cr steel: Weld electrode; designation used by AWS
E Fe Cr A4	3.5% C 35% Cr steel: Weld electrode; designation used by AWS
E Fe Cr A5	3% C 8.2% Cr 1.6% Ti steel: Weld electrode; designation used by AWS

Symbol	Nominal analysis, supplier, condition and remarks.
E Fe Cr B1	2% C 28% Cr 4% Mo (max) steel: Weld electrode; designation used by AWS
E Fe Cr B2	3% C 27% Cr 1.2% Mo 1.7% Si steel: Weld electrode; designation used by AWS
E Fe Cr B3	4.2% C 27% Cr 3.2% Mo 1.7% Si steel: Weld electrode; designation used by AWS
E Fe Cr C1	5% C 25% Cr 6% Mo 1% V 5.5% Nb steel: Weld electrode; designation used by AWS
E Fe Cr C2	6% C 23% Cr 6% Mo 4% W 1.2% Si steel: Weld electrode; designation used by AWS
E Fe Cr C3	5% C 17% Cr 6% Mo steel: Weld electrode; designation used by AWS
E Fe Cr C4	5.7% C 13% Cr 5.5% Ti 1% Si steel: Weld electrode; designation used by AWS
ECLIPSE 91A	1.5% C 12.0% Cr 1.0% Mo steel: Origin unknown
FASTWORK	1.5% C 12.5% Cr 0.75% Mo 0.25% V tool steel: Balfour Darwin
fmp 336	1.5% C 12.0% Cr 0.8% Mo 0.9% V steel: F Parkin; as tool steel D2
fmp 338	2.05% C 13.0% Cr steel: F Parkin; as tool steel D3
FNS	1.5% C 12% Cr 0.8% Mo 0.85% V steel: Atlas; as AISI type D2
FNS fm	1.5% C 12% Cr 0.8% Mo 0.85% V 0.12% S steel: Atlas; as AISI type D2; free machining
GOST 5632/9 KH18	0.95% C 0.7% Mn (max) 18% Cr 0.025% S (max) 0.03% P (max) steel: Plate; Russian Standard; annealed **UTS: 700**
GSN	1.5% C 11.5% Cr 0.8% Mo 0.25% V steel: Broaches, ring gauges, etc.; S Osborn **DPN: 740**
GSN	2.1% C 13.0% Cr steel: Latrobe; die steel; type D3 **DPN: 830**
GSN + Mo	2.2% C 12.0% Cr 0.8% Mo 0.5% V steel: Latrobe
H 5	1.5% C 12% Cr 1% Mo 1% V steel: Jessop
H 9	2.1% C 13% Cr steel: Jessop
H 9 (alloy C)	2.1% C 12.7% Cr steel: Jessop; carbon and alloy tool steels
H 42	1.6% C 13% Cr 0.7% Mo 0.3% V steel: Jessop **DPN: 770**
H 142	1.6% C 13% Cr 0.7% Mo 0.3% V with S steel: For tools; Jessop **DPN: 730**
HAMPDEN	2.1% C 0.5% Ni 12% Cr steel: Carpenter
HAYNES 90	2.7% C 27% Cr steel: Abrasion and corrosion resistant; S. Osborn **DPN: 740**
HECLA 125	2.25% C 12.0% Cr steel: Origin unknown
HECLA 159	1.5% C 12.0% Cr 1.0% Mo steel: Origin unknown
HFC	2.1% C 12% Cr steel: Press tools, extruding dies, etc.; Marsh Bros
ICS	2.0% C 13.0% Cr steel: Inman
ICW	2.25% C 12.0% Cr steel: Origin unknown
ICW PLUS D3	1.5% C 12.0% Cr 1.0% Mo steel: For dies; Inman
IR	2.2% C 12.5% Cr 0.25% V steel: Obsolete; S Osborn; replaced by D3
IU	1.9% C 12.5% Cr 0.75% Mo 0.3% V steel: S Osborn; cold work steel; hardened and tempered **DPN: 700**
JJ	1.4% C 0.5% Ni 14% Cr 0.7% Mo 3.25% Co steel: Shear blades, etc.; S. Osborn **DPN: 740**
K 100	2.0% C 11.5% Cr steel: VEW
K 102	2.9% C 12.0% Cr steel: VEW
K 105	1.6% C 11.5% Cr 0.6% Mo 0.2% V 0.5% W steel: VEW
K 107	2.1% C 11.5% Cr 0.7% W steel: VEW
K 110	1.5% C 11.5% Cr 0.7% Mo 1.0% V steel: VEW
K 700	1.23% C 12.5% Cr steel: VEW
KE 200	1.4% C 13% Cr 0.6% Mo 3.5% Co steel: Press tools; Kayser Ellison

Symbol	Nominal analysis, supplier, condition and remarks.
KE 954	1.4% C 13% Cr 0.6% Mo 3.5% Co steel: For valves; Kayser Ellison
KE 961	1.5% C 12.5% Cr 0.5% W steel: For plastic dies; Kayser Ellison **DPN: 660**
KE 970	2.0% C 13.5% Cr 0.5% W steel: Press tools and dies; Kayser Ellison **DPN: 872**
KE 1036	0.9% C 13% Cr steel: Kayser Ellison
KEA 180	1.55% C 12% Cr 0.9% Mo 0.2% V steel: Hot shear blades; Kayser Ellison **DPN: 750**
KEA 180/476	1.55% C 12.0% Cr 0.85% Mo 0.3% V steel: Sanderson Kayser
KEA 207	1.05% C 17.0% Cr 0.6% Mo steel: Kayser Ellison
KEA 476	1.55% C 12% Cr 0.85% Mo 0.28% V steel: Sanderson Kayser
KH 12	12.2% C 12.2% Cr steel: French Standards designation
KH 12F1	1.32% C 11.7% Cr 0.8% V steel: Russian Standards designation
KH 12M	1.55% C 11.7% Cr 0.5% Mo 0.22% V steel: Russian Standards designation
KKK	2.0% C 0.8% Ni 12.5% Cr 0.6% Mo steel: Spencer; press tools; drawing dies **DPN: 640**
KMV	1.5% C 12.0% Cr 1.0% Mo steel: Origin unknown
LESCALLOY BG 42	1.15% C 14.5% Cr 4.0% Mo 1.2% V steel: Latrobe; for bearings
LESCALLOY BG 66	0.85% C 4.15% Cr 5.0% Mo 1.85% V 6.3% W steel: Latrobe
LESCO BC42 VAC ARC	1.15% C 14.5% Cr 4.0% Mo 1.2% V steel: Latrobe Steel; for bearings **DPN: 400**
LONGWAIR	2.2% C 13% Cr steel: For tools; BSC (Clyde Alloy); previously Nonwear **DPN: 550**
LT 25	2.0% C 12% Cr steel: Low Moor **DPN: 640**
LT 27	2.3% C 11.75% Cr 0.25% V steel: Low Moor
LT 33	1.5% C 12% Cr 0.75% Mo 0.25% V steel: Low Moor **DPN: 720**
LT 35	1.9% C 12.5% Cr 0.75% Mo 0.25% V steel: Low Moor
MAFERITE 30C	1.2% C 30% Cr steel: Pompey **DPN: 350**
MARSH HFC	2.1% C 12% Cr steel: Press tools, extruding dies, etc.; Marsh Bros
MCHD	2.0% C 13% Cr steel: Plastic moulds; John Vessey
MEL-TROL 610FM	1.5% C 12.0% Cr 0.8% Mo 0.9% V steel with sulphides: Carpenter for AISI type D 2
MEL-TROL HAMPDEN	2.1% C 0.5% Ni 12.5% Cr steel: Carpenter for type D3
MET-HARD 850	3% C 25% Cr 1.5% Mo + Nb + V + W steel: Electrode; Metrode **DPN: 800**
MET-HARD 950	4% C 34% Cr 2.5% Mo + Nb + V + W steel: Electrode; Metrode **DPN: 900**

Note. The following abbreviations and units are used in the tables:

DPN	Hardness, diamond pyramid number
UTS	Ultimate tensile strength, N/mm^2
Elon	Elongation, %
Proof	0.1% proof strength, N/mm^2

$1 N/mm^2 = 0.1 hbar = 0.102 kgf/mm^2 = 0.06475 tonf/in.^2 = 145.04 lbf/in.^2 = 1 MPa$

See Appendix II for other abbreviations and conversion tables.

Symbol	Nominal analysis, supplier, condition and remarks.
MET-HARD 1050	4.5% C 28% Cr 12 Mo + Nb + V + W steel: Electrode; Metrode **DPN: 1000**
MONARCH HCR	1.5% C 12.0% Cr 1.0% Mo steel: For tools; origin unknown
N 555	0.6% C 14.1% Cr 0.6% Mo steel: VEW
N 685	0.9% C 17.5% Cr 1.2% Mo steel: VEW
N 690	1.07% C 17.0% Cr 1.1% Mo steel: VEW
NEOR	2.0% C 13% Cr steel: Press tools, lathe centres, etc.; Darwin **DPN: 700**
NN	2.25% C 12% Cr 0.8% Mo 0.25% V steel: Atlas; as AISI type D4
NONWAIR	2.2% C 13% Cr steel: Clyde Alloy **DPN: 770**
NORESCO DOMINATOR	1.9% C 12.5% Cr steel: Schoeller-Bleckmann **DPN: 640**
NORESCO DOMINATOR VM	1.55% C 11.5% Cr 0.8% Mo 1% V steel: Schoeller-Bleckmann **DPN: 720**
NORESCO DOMINATOR Z	1.9% C 11.5% Cr 2% W steel: Schoeller-Bleckmann **DPN: 740**
NRM	1.7% C 11.5% Cr 0.8% Mo 0.25% V steel: Press tools; Jonas **DPN: 740**
NRW	2.2% C 12.5% Cr steel: Press tools; Jonas **DPN: 770**
OHIO DIE	1.5% C 12% Cr 0.8% Mo 0.8% V steel: Vanadium Alloys **DPN: 750**
OLYMPIC FM	1.5% C 12.0% Cr 0.75% Mo 1.0% V steel: Latrobe; die steel; type D2 **DPN: 800**
QQ S 763/10	1.0% C 0.5% Ni 16% Cr 0.5% Mo 1.5% Cu steel: US Federal; annealed **DPN: 240**
R 5	1.05% C 17% Cr steel: Jessop
R 680G	2.5% C 27.0% Cr steel: Casting; scaling temperature 1070 °C; Fagersta
R 770	0.95% C 13.5% Cr steel: Fagersta **DPN: 660**
REXWELD 54	4.0% C 4% Ni 30% Cr Fe alloy: For hard facing; Crucible Steel Co. **DPN: 540**
ROP 57	1.55% C 12% Cr 0.8% Mo 0.9% V steel: Bofors **DPN: 760**
RT 60	2.0% C 13% Cr 1.2% W steel: Bofors **DPN: 760**
RVM	1.55% C 11.5% Cr 0.7% Mo 0.25% V steel: Clyde Alloy **DPN: 750**
SAE 51440 C	1.1% C 17% Cr 0.7% Mo steel: Mechanical properties not specified
SAE 51440 F	1.1% C 17% Cr 0.7% Mo 0.07% S or Se steel: Mechanical properties not specified
SAE 51440 F Se	1.1% C 17% Cr 0.15% Se steel: Mechanical properties not specified; free machining
SAE D2	1.5% C 12% Cr 1% Mo 0.8% V 0.6% Co steel
SAE D3	2.2% C 12% Cr 0.8% Mo 0.8% V 0.7% W steel
SAE D5	1.5% C 12% Cr 1% Mo 0.8% V 3% Co steel
SAE D7	2.2% C 12% Cr 1% Mo 4% V steel
SC 13	2.0% C 13% Cr 0.2% V 0.3% Si steel: Blanking dies, hobs, etc.; Balfour **DPN: 740**
SC 25	1.6% C 13% Cr 0.7% Mo 0.2% V steel: Blanking dies, etc.; Balfour **DPN: 620**
SC 26	1.8% C 12% Cr 0.3% Si steel: Mandrels, forming punches, etc.; Balfour **DPN: 630**

Symbol	Nominal analysis, supplier, condition and remarks.
SC 38	1.5% C 12% Cr 0.7% Mo 1.1% V steel: Blanking dies, shear blades, etc.; Balfour **DPN: 660**
SIL XB	0.8% C 1.3% Ni 20% Cr 2.3% Si steel: For valves; origin unknown
SIMONDS 168	0.9% C 12% Cr 0.75% Mo steel: Simonds; AISI type D1
SIMONDS 12225	2.1% C 13% Cr steel: Simonds; AISI type D3
For SIS specifications see SS	
SKD 1	2.1% C 13.5% Cr 0.3% V (max) steel: Japanese Standards designation
SKD 2	2.0% C 0.5% Ni (max) 13.5% Cr 3.0% W steel: Japanese Standards designation
SKD 11	1.55% C 12.0% Cr 1.0% Mo 0.35% V steel: Japanese Standards designation
SL 440 B	0.85% C 17% Cr 0.7% Mo steel: Swift Levick; hardened and tempered **DPN: 600**
SL 440 C	1.1% C 17% Cr 0.7% Mo steel: Swift Levick; hardened and tempered **DPN: 600**
SL 951	High C high Cr Ni W Co steel: Casting; Swift Levick **DPN: 400**
SL 958	High C high Cr Ni W Co steel: Casting; Swift Levick **DPN: 400**
SOLAR ECLIPSE	2.25% C 12.0% Cr steel: Origin unknown
SONCOLD	1.5% C 12.0% Cr 1.0% Mo steel: Designation unknown
SOUDOKAY 240-0	3.5% C 15% Cr steel: Welding electrode; Soudometal **DPN: 510**
SOUDOKAY 255.0	4.5% C 27% Cr steel: Welding electrode; Soudometal **DPN: 650**
SOUDOKAY 410.0	0.085% C 13% Cr steel: Welding electrode; Soudometal **DPN: 460**
SOUDOKAY 420.0	0.5% C 13% Cr 0.25% Mo steel: Welding electrode; Soudometal **DPN: 580**
SOUDOKAY A40.0	4% C 21% Cr 6% Nb steel: Welding electrode; Soudometal **DPN: 600**
SOUDOKAY A43.0	5.5% C 22% Cr 7% Nb steel: Welding electrode; Soudometal **DPN: 660**
SOUDOKAY A45.0	5.5% C 22% Cr 7% Mo 2% W 7% Nb steel: Welding electrode; Soudometal **DPN: 660**
SPEAR D12	2.0% C 13% Cr steel: Spear & Jackson
SPEAR D13	1.9% C 13.5% Cr 1.3% Mo steel: Spear & Jackson
SPEAR D14	1.3% C 13% Cr 1.5% Mo 2.7% Co steel: Spear & Jackson
SPEAR D16	1.5% C 12.0% Cr 1.0% Mo steel: Origin unknown
SPECIAL K	2.25% C 12.0% Cr steel: Origin unknown
SS 2183	1.2% C 12.5% Cr steel: Swedish Standard
SS 2310	1.5% C 12% Cr 0.8% Mo 0.9% V steel: Swedish Standard; annealed **DPN: 260**
SS 2312	2.0% C 12.5% Cr 1.2% W steel: Swedish Standard; annealed **DPN: 280**
STA 5 V24A	0.8% C 1.4% Ni 19.5% Cr steel: Replaced by BS alloy En 59
SUH 4	0.8% C 1.4% Ni 19.7% Cr steel: Heat resisting; Japanese Standards designation
SUPER C12	1.5% C 12.0% Cr 1.0% Mo steel: Origin unknown
SUPER HICRO D5	1.5% C 12.0% Cr 1.0% Mo 3.0% Co steel: For dies; Inman
SUPERDIE	2.13% C 11.6% Cr 0.8% W steel: Columbia; AISI type D3
SUPERIOR 1	2.15% C 12% Cr 0.6% V steel: Braeburn; AISI type D3
SUPERIOR 2	1.4% C 0.5% Ni 12.5% Cr 0.8% Mo 3.5% Co steel: Braeburn; AISI type D5
SUPERIOR 3	1.5% C 12% Cr 0.8% Mo 0.8% V steel: Braeburn; AISI type D2
SUPERIOR 4	1.0% C 12% Cr 0.8% V 0.7% V steel: Braeburn
SUPERWEAR	2.0% C 13% Cr 0.5% Co steel: John Vessey **DPN: 750**
SVERKER 3	2.05% C 12.7% Cr 1.25% W steel: Uddelholm for type D6
SVERKER 21	1.55% C 11.8% Cr 0.8% Mo 0.9% V steel: Uddelholm for type D2
TRIPLE DIE	2.2% C 12% Cr 1.0% Mo 1.0% V steel: Firth Sterling
UHB STAINLESS AEB	0.95% C 13.5% Cr steel: Uddelholm; hardened and tempered **DPN: 390 UTS: 1250 Elon: 3% Proof: 950**
UNI 3992 X80 CSN20	0.8% C 19.5% Cr steel: For valves
VASCO 7152	1.7% C 17.4% Cr steel: Vanadium Alloys **DPN: 660**
VESUVIUS	1.0% C 0.3% Ni 29.0% Cr steel: Firth Vickers
VITAL	2.0% C 13% Cr steel: Blanking dies, thread rolls; Hall and Pickles **DPN: 740**
VITAL X	1.5% C 12% Cr 0.8% Mo 0.25% V steel: Press tools, punches, etc.; Hall and Pickles **DPN: 680**
VS 13 C	2.0% C 13% Cr 0.25% V steel: Rolling dies, gauges, etc.; Vulcan **DPN: 720**
WALLEX	2.5% C 1.7% Mn 12% Cr 6% Mo steel: Rod for facing; abrasion, impact, corrosion resistant; melting point 1180 °C; Wall Colmonoy **DPN: 720**
WCD	1.3% C 13% Cr 0.5% W steel: John Vessey **DPN: 800**
WCR 100	0.75% C 14% Cr 20% W 3.5% B Fe alloy: For hard facing; melting point 1200 °C; Wall Colmonoy
WEARCLAD 40	4.5% C 25% Cr 0.6% B steel: For armour plate; British Steel Armour products
WEARCLAD 43	5.5% C 23% Cr 7% Nb steel: For armour plate; British Steel Armour products
WEARCLAD 45	5.5% C 20% Cr 6.5% Mo 1% V 1.8% W 6% Nb steel: For armour plate; British Steel Armour products
WLM 2014	2.0% C 14% Cr steel: Press tools; W Marrison **DPN: 740**
WLM 2014 SPECIAL	2.0% C 14% Cr 1.5% W steel: Press tools; W Marrison **DPN: 740**
WLM 3141	1.4% C 14% Cr 0.7% Mo 3% Co steel: Plastic moulds and dies; W Marrison **DPN: 680**
WPS	2.3% C 0.4% Mn 13% Cr steel: Jessop **DPN: 870**
X 90 Cr Mo V 18	0.9% C 18.0% Cr 1.1% Mo 0.10% V steel: Designation used by German Standard
X 155 Cr V Mo 12.1	1.5% C 11.5% Cr 0.7% Mo 1% V steel: German Standard
X 165 Cr Mo V 12	1.6% C 11.5% Cr 0.6% Mo 0.2% V 0.5% W steel: German Standard
X 165 Cr Mo V 12	1.65% C 12.0% Cr 0.6% Mo 0.1% V 0.5% W steel: German Standards designation
X 210 Cr 12	2.1% C 11.5% Cr steel: Designation used by German Standard
X 210 Cr 12	2.0% C 12.0% Cr steel: German Standards designation
X 210 Cr W 12	2.0% C 12.0% Cr 0.7% W steel: German Standards designation
X 290 Cr 12	2.9% C 12% Cr steel: German Standards
Z 160CDV12	1.55% C 12.0% Cr 0.8% Mo 0.3% V steel: French Standards designation
Z 160CDV12	1.6% C 12.0% Cr 0.88% Mo 0.4% V steel: French Standards designation

Symbol	Nominal analysis, supplier, condition and remarks.
Z 180CKD1203	1.8% C 12.0% Cr 0.8% Mo 0.5% V 3.0% Co steel: French Standards designation
Z 200C12	2.1% C 12.5% Cr 0.2% V steel: French Standards designation
Z 200C12	2.0% C 12.0% Cr steel: French Standards designation
Z 200CD12	2.0% C 12.0% Cr 0.8% Mo 0.2% V steel: French Standards designation
Z 230CDV12-04	2.3% C 12.0% Cr 1.0% Mo 4.0% V steel: French Standards designation

Note. The following abbreviations and units are used in the tables:

DPN	Hardness, diamond pyramid number
UTS	Ultimate tensile strength, N/mm^2
Elon	Elongation, %
Proof	0.1% proof strength, N/mm^2

1 N/mm^2=0.1 hbar=0.102 kgf/mm^2=0.06475 tonf/in.^{2}=145.04 lbf/in.^{2}=1 MPa

See Appendix II for other abbreviations and conversion tables.

44N1 Steel – chromium–nickel – austenitic wrought and cast
With niobium, titanium, molybdenum and tungsten

Specific gravity		8.05
Density		8050 kg/m^3
Solidus/liquidus		1230–1280 °C
Thermal conductivity		16.3 W/m °C
Coefficient of linear expansion		18 × 10^{-6}/ °C
Electrical conductivity		3% IACS (copper 100%)
Specific resistance		700 microhm mm
Young's modulus of elasticity		207 × 10^9 N/m^2
Impact	cold worked	20 J
	annealed	108–136 J (can be higher)
Fatigue strength		–

Hot strength

Temperature °C	Tensile strength N/mm^2	Elongation %
650	580	40
700	450	45
800	330	45
900	170	58

The above properties are typical of the following group, and may not apply exactly to any one specification. It is possible that with certain specifications some of the values may not be applicable.

General metallurgical characteristics

When 18% chromium and 8% nickel are added to steel the effect is so to depress the critical change points that the austenitic or gamma phase is present at room temperature. This is a single phase solid solution and like all single phase materials has excellent corrosion resistance.

In this group are all the truly stainless as distinct from corrosion resistant steels. The fact that they are non-magnetic is a rapid but not very precise means of identification, as cold work can cause the steels to become slightly magnetic.

Since these materials are single phase, they cannot be hardened by thermal treatment, but their strength can be improved by cold working, when they can become magnetic as some austenite is transformed to martensite. The carbon content of these steels is important in that, the higher the carbon, the harder they can be made by cold work.

The presence of carbon, however, can cause the appearance of a second phase of martensite or chromium carbide, either of which can considerably reduce the corrosion resistance. The presence of martensite after cold working can generally be accepted and design allowance can be made for the reduction in corrosion resistance. The chromium carbides, however, contribute little to the mechanical properties and, as the corrosion resistance is drastically reduced at grain boundaries, it can have a serious effect on the fatigue strength as well. This is known as 'weld decay'.

By the addition of carbide stabilizing elements such as niobium or titanium which combine with carbon in preference to chromium and remain in solution, the problem of carbide precipitation at grain boundaries is eliminated. These steels are said to be 'weld stabilized' and are also suitable for use at high temperatures.

The low carbon steels generally have better ductility and corrosion resistance with lower tensile properties, but are all weldable as there are very few carbides present to be precipitated.

Many of the steels listed have considerably more chromium and nickel than the 18/8 steels. This generally improves their hot oxidation resistance and ability to withstand temperature shock without very much improving the creep strength.

Other elements are added to these alloys for specific purposes as follows:

Manganese. This is almost always present up to about 2.5% as a de-oxidizing and de-sulphurizing agent. No attempt is made to list this element in the alloy descriptions of this group, and when it is present in greater amounts the alloys are listed and discussed in Section 44P.

Molybdenum. This considerably enhances the acid resistance without affecting the weldability. It also confers some strength and can actually be used to prevent carbide precipitation, thus making the higher carbon versions weldable. Because of the high cost and scarcity of molybdenum its use is confined to improving the acid resistance at present.

Tungsten. This stabilizes the carbides, preventing their precipitation during welding or heat cycling. Tungsten also increases the strength and acid resistance. It is seldom found by itself, being generally associated with molybdenum.

Titanium. This forms very stable carbides which are taken into solution in the austenitic matrix and are not readily precipitated, thus preventing the formation of a second phase. Carbon combines with titanium in preference to iron or chromium, thus preventing the formation of these carbides.

Sufficient titanium must always be present to combine with all the carbon and most specifications are written calling for a minimum of eight times as much titanium as carbon. These are known as titanium stabilized weldable austenitic steels. There is some indication that these can present some difficulty when brazing or melt spraying is involved.

Niobium (columbium). This is the only element with two names and different symbols. Niobium is the common European form, and throughout this book use of this and the symbol Nb has been standardized. Columbium (Cb) is the American version.

The element acts in an analogous manner to titanium, forming stable carbides which are not readily precipitated from solid solutions. Tantalum and niobium are always found together and are difficult to separate. As tantalum acts in an identical manner to niobium the sum total of these elements is generally accepted. This should be ten times the carbon content to prevent carbide precipitation.

Tellurium, lead and sulphur are sometimes present to improve the machinability but there is some reduction in the corrosion resistance and mechanical properties.

The majority of these steels will accept appreciable amounts of cold work which will increase the tensile properties while reducing the ductility but in many cases this is an acceptable technique to increase the mechanical properties. This will reduce the corrosion resistance to some extent.

Steels with appreciable amounts of carbon and elements such as tungsten will become more brittle as the cold work increases.

The mechanical properties of these steels cannot be improved by any form of thermal treatment. Where cold work has to be removed for any reason, this requires a temperature of approximately 1050 °C and should be followed by a rapid cooling to prevent carbide precipitation.

These steels can be readily welded using the inert gas technique. It should be noted that the materials have a high coefficient of linear expansion and this can result in centreline cracking of the weld in some circumstances. Technical advice should be obtained where high integrity welds are required. Other forms of welding such as resistance welding and brazing present no problems.

The corrosion resistance of these steels results from the fact that they are single phase metals; thus no inherent galvanic corrosion takes place. They also have the self-healing adherent chromium oxide. They must always be used in oxidizing conditions. These steels are more prone to stress corrosion, crevice corrosion and weld decay corrosion.

Advice should be obtained at the design stage.

Many of these materials are difficult to machine because of their softness which results in tearing rather than cutting. High speed cutting is essential.

Grinding of these materials is very difficult as the grinding wheels become 'loaded' and this results in glazing of the surface.

Some of the alloys listed have additions of sulphur, tellurium, lead, etc. to improve the machinability. This will reduce the ductility to some extent and can result in problems during welding.

The materials in this group in many cases have replaced copper alloys where corrosion resistance is required.

Chemical plant, the food and drink industry, the aerospace and marine industries, all make use of these materials in large quantities. The transport of chemicals, food and drink is generally carried out in stainless steel tanks where the stainless steel is of the austenitic type. Laboratory equipment such as beakers, pans for balances, etc. and sinks are commonly manufactured in these materials. Household goods such as sinks, the trim on cookers, microwave ovens, etc. are very often austenitic stainless steel. The trim on cars is sometimes of this material but more often is the 12% chromium type found in Section 44M1.

A great amount of this material is now being used for ornamental purposes such as cutlery, and other equipment used in restaurants and as tableware. Knives and cutters etc. are now made from 12% chrome steels in the hardened condition.

The higher alloyed materials, i.e. is the higher nickel chromium materials found in this group, have excellent high temperature oxidation resistance and can be used for furnace equipment, etc.

These materials should always be referred to as austenitic stainless steels. They are non-magnetic except in the highly cold worked condition. This differentiates them from the materials in Sections 44M1 and 44M2.

Symbol	Nominal analysis, supplier, condition and remarks.
0.4401	0.07% C (max) 11% Ni 17% Cr 2.2% Mo steel: German Standard
1.0 W VAR	1.8% C 15.7% Ni 27.8% Cr 9.8% W Fe alloy: Origin unknown; for hard facing **DPN: 450**
1.2779	0.05% C 25% Ni 15% Cr 1.3% Mo 2% Ti steel: German Standard
1.4300	0.12% C (max) 9% Ni 18% Cr steel: German Standard
1.4301	0.07% C (max) 10% Ni 18% Cr steel: German Standard
1.4302	0.06% C (max) 9.5% Ni 19% Cr steel: Forging; German Standard
1.4303	0.07% C (max) 11.0% Ni 18.5% Cr steel: German Standard
1.4304	0.15% C (max) 9% Ni 18% Cr steel: German Standard
1.4305	0.15% C (max) 9% Ni 18% Cr 0.15% S steel: German Standard; free machining
1.4306	0.03% C (max) 9.5% Ni 19.0% Cr steel: German Standard
1.4307	0.1% C (max) 13% Ni 12% Cr steel: German Standard
1.4310	0.15% C (max) 7.5% Ni 17% Cr steel: German Standard
1.4311	0.03% C (max) 10.2% Ni 18.0% Cr 0.18% N steel: German Standard
1.4314	0.07% C (max) 10% Ni 18% Cr steel: German Standard
1.4324	0.15% C (max) 7.5% Ni 17% Cr steel: German Standard
1.4401	0.07% C (max) 11.5% Ni 17.5% Cr 2.3% Mo steel: German Standard
1.4402	0.06% C (max) 10.0% Ni 19.0% Cr 2.2% Mo steel: German Standard
1.4404	0.03% C (max) 12.0% Ni 17.0% Cr 2.2% Mo steel: German Standard
1.4406	0.03% C (max) 12.0% Ni 17.5% Cr 2.2% Mo 0.18% N steel: German Standard
1.4417	0.03% C (max) 13% Ni 17% Cr 2.8% Mo steel: German Standard
1.4429	0.03% C (max) 13.0% Ni 17.5% Cr 2.7% Mo 0.18% N steel: German Standard
1.4435	0.03% C (max) 12.0% Ni 17.5% Cr 2.7% Mo steel: German Standard
1.4436	0.07% C (max) 13.0% Ni 17.5% Cr 2.7% Mo steel: German Standard
1.4438	0.03% C (max) 13.0% Ni 19.0% Cr 3.5% Mo steel: German Standard
1.4447	0.06% C 13.5% Ni 18.0% Cr 4.7% Mo steel: German Standard
1.4449	0.07% C (max) 13.5% Ni 17.0% Cr 4.5% Mo steel: German Standard
1.4460	0.08% C 5% Ni 26% Cr 1.5% Mo steel: German Standard
1.4505	0.07% C (max) 20% Ni 17.5% Cr 2.2% Mo 0.7% Nb 2.0% Cu steel: German Standard
1.4507	0.08% C (max) 21.0% Ni 18.5% Cr 2.2% Mo 0.9% Nb 2.0% Cu steel: German Standard
1.4539	0.02% C (max) 25% Ni 20% Cr 4.5% Mo with Cu steel: German Standard
1.4541	0.1% C (max) 10% Ni 18% Cr 0.5% Ti steel: Forging; German Standard
1.4550	0.1% C (max) 10% Ni 18% Cr 1.0% Nb steel: Forging; German Standard
1.4551	0.1% C (max) 9% Ni 19% Cr 1.5% Nb steel: Forging; German Standard
1.4554	0.1% C (max) 9% Ni 19% Cr 1.5% Nb steel: Welding rod; German Standard
1.4558	0.03% C (max) 34% Ni 22% Cr with Ti and Al steel: German Standard
1.4570	0.1% C (max) 11.5% Ni 17.5% Cr 1.4% Mo steel: German Standard
1.4571	0.1% C (max) 11.5% Ni 17.5% Cr 2.2% Mo 0.6% Ti steel: German Standard
1.4573	0.1% C (max) 13.0% Ni 17.5% Cr 2.7% Mo 0.6% Ti steel: German Standard
1.4577	0.06% C (max) 25.0% Ni 25.0% Cr 2.2% Mo 0.4% Ti steel: German Standard
1.4580	0.1% C (max) 11.5% Ni 17.5% Cr 2.3% Mo 1.0% Nb steel: German Standard
1.4581	0.1% C (max) 10.5% Ni 19.0% Cr 2.2% Mo 1.3% Nb steel: German Standard
1.4583	0.1% C (max) 13.0% Ni 17.5% Cr 2.7% Mo 1.0% Nb steel: German Standard
1.4584	0.1% C (max) 11.5% Ni 19.0% Cr 2.7% Mo 1.3% Nb steel: German Standard
1.4587	0.1% C (max) 25.0% Ni 26.0% Cr 2.2% Mo 1.3% Nb steel: German Standard
1.4589	0.07% C (max) 13.5% Ni 18.0% Cr 4.7% Mo 1.0% Nb steel: German Standard
1.4828	0.2% C (max) 12% Ni 20% Cr steel: Forging; German Standard
1.4829	0.15% C (max) 11.0% Ni 22.0% Cr steel: German Standard
1.4833	0.08% C (max) 13.5% Ni 23.0% Cr steel: German Standard
1.4841	0.2% C (max) 20.0% Ni 25.0% Cr 2.0% Si steel: German Standard
1.4842	0.15% C (max) 20% Ni 25.0% Cr steel: German Standard
1.4843	0.2% C (max) 19.0% Ni 24.0% Cr steel: German Standard
1.4844	0.2% C (max) 20.0% Ni 25.0% Cr 2.0% Si steel: German Standard
1.4845	0.15% C (max) 21.0% Ni 25.0% Cr steel: German Standard
1.4854	0.15% C (max) 20% Ni 25.0% Cr steel: German Standard
1.4860	0.2% C (max) 30.0% Ni 21.0% Cr Fe steel: German Standard
1.4876	0.07% C 31% Ni 21% Cr with Ti and Al steel: German Standard
1.4878	0.15% C (max) 10% Ni 18% Cr 0.6% Ti steel: Forging; German Standard
1.4919	0.055% C (max) 13% Ni 17% Cr 2.3% Mo steel: German Standard
1.4943	0.45% C 9.0% Ni 18.0% Cr 1.0% W 2.5% Si steel: German Standard
1.4944	0.08% C (max) 25% Ni 15% Cr 1.2% Mo 0.3% V 2.1% Ti steel: German Standard
1.4954	0.08% C (max) 25% Ni 14% Cr 0.4% V 2.0% Ti steel: German Standard
1.4961	0.08% C (max) 13% Ni 16% Cr Nb steel: German Standard
1.4970	0.1% C 15% Ni 15% Cr 1.2% Mo with Ti and B steel: German Standard
1.4984	0.1% C (max) 16.5% Ni 16.5% Cr 1.7% Mo 1.0% Nb + Ta steel: German Standard
1.4986	0.07% C 16.7% Ni 16.8% Cr 1.8% Mo 0.8% Nb 0.7% B steel: German Standard

Note. The following abbreviations and units are used in the tables:

DPN	Hardness, diamond pyramid number
UTS	Ultimate tensile strength, N/mm^2
Elon	Elongation, %
Proof	0.1% proof strength, N/mm^2

$1 N/mm^2 = 0.1 hbar = 0.102 kgf/mm^2 = 0.06475 tonf/in.^2 = 145.04 lbf/in.^2 = 1 MPa$

See Appendix II for other abbreviations and conversion tables.

Symbol	Nominal analysis, supplier, condition and remarks.
1.6903	0.1% C (max) 10% Ni 18% Cr 0.5% Ti steel: For low temperature use; German Standard
1.6905	0.1% C (max) 10% Ni 18% Cr 1.0% Nb steel: For low temperature use; German Standard
1KH25N25TR	0.1% C 25.5% Ni 24.5% Cr 1.3% Ti 0.01% B (max) steel: Russian Standard designation
1R2	0.16% C (max) 9.0% Ni 18% Cr steel: Sandvik
2R	0.2% C 10% Ni 18% Cr 3.0% Mo steel: Sandvik
2R1	0.16% C (max) 13% Ni 13% Cr steel: Sandvik
2R2A	0.16% C 9% Ni 18% Cr 0.2% S steel: Sandvik; free machining
2R12	0.02% C (max) 11% Ni 18% Cr steel: Sandvik
2R16	0.02% C (max) 10.5% Ni 19.4% Cr steel: Sandvik
2RE10	0.02% C (max) 20% Ni 25% Cr steel: Sandvik
2RE69	0.02% C (max) 22% Ni 25% Cr 2.1% Mo steel: Sandvik
2RK65	0.02% C (max) 25% Ni 20% Cr 4.5% Mo steel: Sandvik
3KH19N9MVBT	0.32% C 9.0% Ni 19.0% Cr 1.25% Mo 1.25% W 0.35% Ti 0.35% Nb steel: Russian Standard designation
3R12	0.03% C (max) 11% Ni 18.5% Cr steel: Sandvik
3R16	0.03% C (max) 19.4% Cr 10.5% Mo steel: Sandvik; annealed **UTS: 700**
3R17	0.03% C (max) 9.7% Ni 20.6% Cr steel: Sandvik
3R19	0.03% C (max) 9% Ni 18% Cr steel: Sandvik
3R54	0.03% C (max) 10.5% Ni 18.5% Cr 1.5% Mo steel: Sandvik
3R60	0.03% C (max) 13.6% Ni 17% Cr 2.8% Mo steel: Sandvik
3R61	0.03% C (max) 10.5% Ni 18.5% Cr 2.5% Mo steel: Sandvik
3R62	0.03% C 13% Ni 19.5% Cr 2.3% Mo steel: Sandvik
3R63	0.03% C (max) 12.2% Ni 17.8% Cr 2.8% Mo steel: Sandvik
3R64	0.03% C 14.5% Ni 18.5% Cr 3.5% Mo steel: Sandvik
3R65	0.03% C (max) 14.5% Ni 18% Cr 4.5% Mo steel: Sandvik
4 Mo Var	1.7% C 9.7% Ni 30% Cr 4.3% Mo Fe alloy: Origin unknown; for hard facing **DPN: 420**
5R10	0.05% C (max) 9% Ni 18% Cr steel: Sandvik
5R60	0.05% C (max) 13% Ni 17% Cr 2.7% Mo steel: Sandvik
5RA50	0.05% C (max) 10% Ni 18% Cr 0.5% Mo steel: Sandvik; free machining with S
6R10	0.06% C (max) 9% Ni 18.5% Cr steel: Sandvik for SS 2333
6R40	0.06% C (max) 10% Ni 17% Cr steel: Sandvik; stabilized with Nb
6R55	0.06% C (max) 11% Ni 17.5% Cr 1.5% Mo steel: Sandvik for SS 2341
6R60	0.06% C (max) 13.5% Ni 17.5% Cr 2.8% Mo steel: Sandvik for SS 2343
6R64	0.06% C (max) 14% Ni 18.5% Cr 3.5% Mo steel: Sandvik
7 Mo Var	1.9% C 14% Ni 28% Cr 6.8% Mo Fe alloy: Origin unknown; for hard facing **DPN: 400**
7R60	0.05% C 13% Ni 17% Cr 2.3% Mo steel: Sandvik
7RE10	0.05% C 20% Ni 25% Cr steel: Sandvik
7RE12	0.07% C (max) 21% Ni 25.5% Cr steel: Sandvik
8R30	0.08% C 11.5% Ni 17.5% Cr 0.4% Ti steel: Sandvik for SS 2337
8R40	0.08% C (max) 11.5% Ni 17.5% Cr 0.8% Nb steel: Sandvik for SS 2338
8R41	0.08% C 13% Ni 16.5% Cr 0.8% Nb steel: Sandvik
8R42	0.08% C 9.5% Ni 18.5% Cr 0.8% Nb steel: Sandvik
8R70	0.08% C 12.4% Ni 17.5% Cr 2.5% Mo 0.4% Ti steel: Sandvik for SS 2344
8R80	0.08% C 13.5% Ni 17.5% Cr 2.5% Mo 0.8% Nb steel: Sandvik for SS 2365
8R81	0.08% C (max) 11% Ni 18.5% Cr 2.3% Mo 0.8% Nb steel: Sandvik
10R13	0.08% C 11% Ni 18% Cr steel: Sandvik
10R52	0.08% C 9% Ni 17.5% Cr 1.5% Mo steel: Sandvik for SS 2340
10R53	0.08% C 11% Ni 17% Cr 1.5% Mo steel: Sandvik
10RE51	0.08% C 5% Ni 26% Cr 1.5% Mo steel: Sandvik
11R50	0.09% C 8% Ni 17% Cr 0.7% Mo steel: Sandvik
11R51	0.09% C 8% Ni 17% Cr 0.7% Mo steel: Sandvik
12/12 EL	0.05% C 13% Ni 12% Cr steel: Nyby; annealed **DPN: 160 UTS: 480 Elon: 55% Proof: 180**
12R10	0.1% C 9% Ni 18% Cr steel: Sandvik for SS 2331
12R10 HV	0.1% C 9.0% Ni 18% Cr steel: Wire for springs; Sandvik; vacuum melted **UTS: 2400**
12R11	0.1% C 7.7% Ni 18% Cr steel: Sandvik for SS 2331
12R72	0.1% C 15% Ni 15% Cr 1.2% Mo steel: Sandvik; stabilized with Ti with B
15/35	0.1% C 35% Ni 15% Cr steel: Eastern Stainless for AISI 330
15.35.2 CB	0.2% C 35.0% Ni 15.0% Cr steel: Electrode; Metrode
15RA10	0.12% C 10% Ni 18% Cr 0.5% Mo 0.2% S steel: Free cutting; Sandvik for SS 2346
15RE10	0.12% C 22% Ni 23% Cr steel: Sandvik for SS 2361
15RE10	0.12% C 20% Ni 25% Cr steel: Sandvik
15RE11	0.12% C 21% Ni 25.5% Cr steel: Sandvik
15RE12	0.12% C 21% Ni 25.5% Cr steel: Sandvik
16/12 UMoR	0.03% C (max) 12.0% Ni 17.0% Cr 2.8% Mo steel: Nyby; annealed; nuclear reactor quality **DPN: 180 UTS: 480 Elon: 45% Proof: 200**
16/13 L Nb	0.07% C 13% Ni 16% Cr 0.7% Nb steel: Nyby; annealed **DPN: 190 UTS: 530 Elon: 35% Proof: 210**
16/14 L Mo Nb V	0.07% C 16% Ni 16% Cr 1.8% Mo 0.7% V 0.7% Nb steel: Nyby; annealed **DPN: 190 UTS: 530 Elon: 30% Proof: 220**
16/16 L Mo Nb	0.07% C 1.8% Mn 16% Ni 16% Cr 0.7% Nb steel: Nyby; annealed **DPN: 190 UTS: 530 Elon: 35% Proof: 220**
16/25/6	0.08% C 25% Ni 16% Cr 6% Mo 0.15% N steel: American proprietary alloy listed in SAE yearbook
17/12 UL	0.03% C (max) 12.0% Ni 17.0% Cr steel: Nyby; annealed **DPN: 160 UTS: 450 Elon: 45% Proof: 180**
17/12 ULR	0.03% C (max) 0.05% Ni 17.0% Cr 0.05% Co 0.25% Cu steel: Nyby; annealed; nuclear reactor quality **DPN: 160 UTS: 450 Elon: 45% Proof: 180**
17/14 Cu Mo	0.12% C 14% Ni 16% Cr 2.5% Mo 0.25% Ti 0.5% Nb 3% Cu steel: Armco
17.8.2 RCF	0.08% C 9.0% Ni 17.0% Cr 2.0% Mo steel: Electrode; Metrode; creep resisting; ferritic content 3–8%
17.8.2 RCF	0.07% C 9% Ni 17% Cr 2% Mo steel: Electrode; Metrode
17.12.2.4 MN	0.06% C 4.0% Mn 12.0% Ni 17.0% Cr 2.0% Mo steel: Electrode; Metrode
17.17.4 LR	0.04% C 17.0% Ni 17.0% Cr 4.0% Mo steel: Electrode; Metrode
17.38.4 CB	0.4% C 38.0% Ni 17.0% Cr steel: Electrode; Metrode
17.38.4 CR	0.4% C 38.0% Ni 17.0% Cr steel: Electrode; Metrode
18/8	0.15% C 9% Ni 18% Cr steel: Eastern Stainless; annealed **UTS: 700 Elon: 50% Proof: 300**
18/8	0.10% C 8.0% Ni 18.0% Cr steel: Nyby; annealed **DPN: 180 UTS: 480 Elon: 45% Proof: 180**
18/8 EL	0.05% C (max) 10.0% Ni 18.0% Cr steel: Nyby; annealed **DPN: 180 UTS: 480 Elon: 45% Proof: 180**

Symbol	Nominal analysis, supplier, condition and remarks.
18/8 EMo	0.05% C 10.0% Ni 17.0% Cr 1.6% Mo steel: Nyby; annealed **DPN: 180 UTS: 480 Elon: 45% Proof: 200**
18/8 FM	0.15% C 9% Ni 18% Cr 0.15% S or Se steel: Eastern Stainless; annealed; free machining **UTS: 630 Elon: 40% Proof: 270**
18/8 L	0.07% C 9.0% Ni 18.0% Cr steel: Nyby; annealed **DPN: 180 UTS: 480 Elon: 45% Proof: 180**
18/8 LNb	0.03% C (max) 10.0% Ni 18.0% Cr 0.3% Nb (max) steel: Nyby; annealed **DPN: 190 UTS: 480 Elon: 40% Proof: 210**
18/8 LNbR	0.03% C (max) 10.0% Ni 18.0% Cr 0.3% Nb (max) steel: Nyby; annealed; nuclear reactor quality **DPN: 190 UTS: 480 Elon: 40% Proof: 210**
18/8 LT	0.07% C 10.0% Ni 18.0% Cr 0.35% Ti steel: Nyby; annealed **DPN: 190 UTS: 480 Elon: 40% Proof: 180**
18/8 Si	0.15% C 9% Ni 18% Cr 2.5% Si steel: Eastern Stainless; annealed; for AISI 302B **UTS: 700 Elon: 45% Proof: 330**
18/8 UL	0.03% C (max) 11.0% Ni 19.0% Cr steel: Nyby; annealed **DPN: 180 UTS: 450 Elon: 45% Proof: 180**
18/8 ULR	0.03% C (max) 11.0% Ni 19.0% Cr steel: Nyby; annealed; nuclear reactor quality **DPN: 180 UTS: 450 Elon: 45% Proof: 180**
18/8 UMo	0.03% C 11.0% Ni 17.0% Cr 1.6% Mo steel: Nyby; annealed **DPN: 180 UTS: 450 Elon: 45% Proof: 180**
18/10 Cb	0.08% C 11% Ni 18% Cr 1% Nb + Ta steel: Eastern Stainless; annealed **UTS: 630 Elon: 45% Proof: 320**
18/10 EMo	0.05% C 11.0% Ni 17.0% Cr 2.3% Mo steel: Nyby; annealed **DPN: 180 UTS: 480 Elon: 45% Proof: 210**
18/10 LMoNb	0.07% C 12.0% Ni 17.0% Cr 2.3% Mo 0.7% (max) Nb steel: Nyby; annealed **DPN: 190 UTS: 480 Elon: 40% Proof: 220**
18/10 LMoT	0.07% C 12.0% Ni 17.0% Cr 2.3% Mo 0.35% Ti (max) steel: Nyby; annealed **DPN: 190 UTS: 480 Elon: 40% Proof: 210**
18/10 Ti	0.08% C 10.5% Ni 18% Cr 0.5% Ti steel: Eastern Stainless; annealed **UTS: 610 Elon: 48% Proof: 240**
18/10 UMo	0.03% C 11.0% Ni 17.0% Cr 2.3% Mo steel: Nyby; annealed **DPN: 180 UTS: 480 Elon: 45% Proof: 180**
18/11 EL	0.05% C 11% Ni 18% Cr steel: Nyby; annealed **DPN: 180 UTS: 480 Elon: 45% Proof: 180**
18/12	0.12% C 12% Ni 18% Cr steel: Eastern Stainless **UTS: 570 Elon: 47% Proof: 240**
18/12 EMo	0.05% C 12.0% Ni 17.0% Cr 2.8% Mo steel: Nyby; annealed **DPN: 180 UTS: 480 Elon: 45% Proof: 210**
18/12 LMo	0.07% C 11.0% Ni 17.0% Cr 2.8% Mo steel: Nyby; annealed **DPN: 180 UTS: 480 Elon: 45% Proof: 210**
18/12 LMoNB	0.07% C 13.0% Ni 17.0% Cr 2.8% Mo 0.7% (max) Nb steel: Nyby; annealed **DPN: 190 UTS: 480 Elon: 40% Proof: 220**
18/12 LMoT	0.07% C 13.0% Ni 17.0% Cr 2.8% Mo 0.35% Ti (max) steel: Nyby; annealed **DPN: 190 UTS: 480 Elon: 40% Proof: 210**
18/12 Mo	0.08% C 12% Ni 17% Cr 2.5% Mo steel: Eastern Stainless; annealed **UTS: 600 Elon: 49% Proof: 240**
18/12 MoL	0.03% C 12% Ni 17% Cr 2.5% Mo steel: Eastern Stainless; annealed **UTS: 570 Elon: 46% Proof: 240**
18/12 UMo	0.03% C (max) 12.0% Ni 17.0% Cr 2.8% Mo steel: Nyby; annealed **DPN: 180 UTS: 480 Elon: 45% Proof: 200**
18/13 EMo	0.05% C 14.0% Ni 17.0% Cr 2.8% Mo steel: Nyby; annealed **DPN: 190 UTS: 480 Elon: 45% Proof: 200**
18/14 EMo	0.05% C 14.0% Ni 19.0% Cr 3.5% Mo steel: Nyby; annealed **DPN: 190 UTS: 480 Elon: 45% Proof: 210**
18/14 UMo	0.03% C (max) 14.0% Ni 19.0% Cr 3.5% Mo steel: Nyby; annealed **DPN: 190 UTS: 480 Elon: 45% Proof: 200**
18/15 EMo	0.05% C 15.0% Ni 17.0% Cr 4.5% Mo steel: Nyby; annealed **DPN: 190 UTS: 480 Elon: 45% Proof: 210**
18/15 UMo	0.03% C (max) 15.0% Ni 17.0% Cr 4.5% Mo steel: Nyby; annealed **DPN: 190 UTS: 480 Elon: 45% Proof: 200**
18.18.3 Cu LR	0.04% C 18.0% Ni 18.0% Cr 3.0% Mo 1.0% Cu steel: Electrode; Metrode
18.20.2 Cu Nb	0.09% C 20.0% Ni 18.0% Cr 2.0% Mo 1.0% Nb 2.0% Cu steel: Electrode; Metrode
19/9	0.08% C 10% Ni 19% Cr steel: Eastern Stainless; annealed **UTS: 630 Elon: 50% Proof: 330**
19/9 DL	0.3% C 10% Ni 19% Cr 1.5% Mo 1.5% W 0.2% Ti 0.4% Nb + Ta 0.5% Cu steel: Eastern Stainless; annealed **UTS: 700 Elon: 30% Proof: 330**
19/9 DL	0.3% C 9% Ni 19% Cr 1.25% Mo 1.2% W 0.3% Ti 0.4% Nb steel: Universal Cyclops
19/9 DX	0.3% C 9% Ni 19.2% Cr 1.5% Mo 1.2% W 0.5% Ti steel: Universal Cyclops
19/9 L	0.03% C (max) 10% Ni 19% Cr steel: Eastern Stainless; annealed **UTS: 570 Elon: 52% Proof: 240**
19/9 WMo	0.1% C 9% Ni 19% Cr 0.4% Mo 1.3% W 0.4% Ti 0.4% Nb steel: Universal Cyclops
19/9 WX	0.1% C 8.5% Ni 20% Cr 0.5% Mo 1.5% W 0.2% Ti 1.3% Nb steel: American proprietary alloy listed in SAE yearbook
19/12 Mo	0.08% C 12% Ni 19% Cr 3.5% Mo steel: Eastern Stainless; annealed **UTS: 630 Elon: 45% Proof: 330**
19/12 MoL	0.03% C 12% Ni 19% Cr 3.5% Mo steel: Eastern Stainless; annealed **UTS: 610 Elon: 48% Proof: 290**
19.9.1 R	0.03% C 9.5% Ni 19.0% Cr steel: Electrode; Metrode
19.9 B	0.05% C 9% Ni 18.5% Cr 0.2% Mo steel: Electrode; Metrode
19.9 LB	0.03% C 9.5% Ni 19.0% Cr steel: Electrode; Metrode
19.9 L.HE	0.02% C 10% Ni 19% Cr steel: Electrode; Metrode
19.9 LRMP	0.03% C 9.5% Ni 19.0% Cr steel: Electrode; Metrode
19.9 NbB	0.06% C 9.5% Ni 19.0% Cr 0.8% Nb steel: Electrode; Metrode
19.9 NbLR	0.04% C (max) 9.5% Ni 19.0% Cr 0.7% Nb steel: Electrode; Metrode
19.9 NbR	0.06% C 9.5% Ni 19.0% Cr 0.8% Nb steel: Electrode; Metrode

Note. The following abbreviations and units are used in the tables:

DPN	Hardness, diamond pyramid number
UTS	Ultimate tensile strength, N/mm^2
Elon	Elongation, %
Proof	0.1% proof strength, N/mm^2

1 N/mm^2=0.1 hbar=0.102 kgf/mm^2=0.06475 tonf/in.^{2}=145.04 lbf/in.^{2}=1 MPa

See Appendix II for other abbreviations and conversion tables.

Symbol	Nominal analysis, supplier, condition and remarks.
19.9 R	0.05% C 9% Ni 18.5% Cr 0.2% Mo steel: Electrode; Metrode
19.12.3 B	0.06% C 11.0% Ni 19.0% Cr 3.0% Mo steel: Electrode; Metrode
19.12.3 ELRCF	0.03% C (max) 14.0% Ni 18.0% Cr 3.0% Mo steel: Electrode; Metrode; ferrite controlled
19.12.3 LR	0.03% C (max) 11.0% Ni 19.0% Cr 3.0% Mo steel: Electrode; Metrode; ferrite content 7–10%
19.12.3 LRCF	0.03% C (max) 14.0% Ni 18.0% Cr 3.0% Mo steel: Electrode; Metrode; ferrite content 1.5% (max)
19.12.3 LRCF2	0.03% C (max) 14.0% Ni 18.0% Cr 3.0% Mo steel: Electrode; Metrode; ferrite content 2% (max)
19.12.3 LRNF	0.03% C (max) 15.0% Ni 17.0% Cr 3.0% Mo steel: Electrode; Metrode; ferrite content 0.5% (max)
19.12.3 NbR	0.06% C 10.5% Ni 18.0% Cr 3.0% Mo 0.8% Nb steel: Electrode; Metrode
19.12.3 NLB	0.04% C 12.5% Ni 17.0% Cr 3.0% Mo 0.2% N steel: Electrode; Metrode
19.12.3 R	0.06% C 11.0% Ni 19.0% Cr 3.0% Mo steel: Electrode; Metrode
19.12.3 RMP	0.06% C 11.0% Ni 19.0% Cr 3.0% Mo steel: Electrode; Metrode
19.12.3 ULRCF	0.025% C (max) 14.0% Ni 18.0% Cr 3.0% Mo steel: Electrode; Metrode
19.13.4 ELRCF	0.03% C (max) 15.0% Ni 18.0% Cr 4.0% Mo steel: Electrode; Metrode; ferrite content 1–5%
19.13.4 LRCF	0.03% C 15.0% Ni 18.0% Cr 4.0% Mo steel: Electrode; Metrode; ferrite content 1–5%
19.13.4 R	0.06% C 12.0% Ni 19.0% Cr 4.0% Mo steel: Electrode; Metrode
19.13.4 LRCF	0.03% C 15% Ni 18.5% Cr 4% Mo steel: Electrode; Metrode
19.15.3 Mn RNF	0.03% C 16% Ni 18% Cr 3% Mo N steel: Electrode; Metrode
19.15.3 MnBNF	0.03% C 3.0% Mn 15.0% Ni 19.0% Cr 3.0% Mo steel: Electrode; Metrode; nil ferrite content
19.15.3 MnRNF	0.03% C 3.0% Mn 15.0% Ni 19.0% Cr 3.0% Mo steel: Electrode; Metrode; nil ferrite content
20/10	0.08% C 11% Ni 20% Cr steel: Eastern Stainless for AISI 308
20/12 L	0.07% C 11.0% Ni 20.0% Cr steel: Nyby; annealed **DPN: 180 UTS: 530 Elon: 50% Proof: 210**
20/12 Si	0.12% C 12.0% Ni 20.0% Cr 2.0% Si steel: Nyby; annealed **DPN: 190 UTS: 580 Elon: 40% Proof: 290**
20/25 N6	0.03% C 0.7% Mn 25% Ni 20.0% Cr 0.3% Ti 0.6% Si Fe alloy: Union Carbide
20/25 Ti	0.03% C 0.7% Mn 25% Ni 20.0% Cr 0.3% Ti 0.6% Si Fe alloy: Union Carbide
20.9.3 B	0.08% C 9.0% Ni 19.0% Cr 3.5% Mo steel: Electrode; Metrode; ferrite content 20%
20.9.3 R	0.08% C 9.0% Ni 19.0% Cr 3.5% Mo steel: Electrode; Metrode; ferrite content 20%
20.9.3 RMP	0.08% C 9.0% Ni 19.0% Cr 3.5% Mo steel: Electrode; Metrode; ferrite content 20%
20.10.3 CB	0.3% C 10.0% Ni 20.0% Cr steel: Electrode; Metrode; for welding stainless steel castings
20.15.3 CB	0.3% C 15.0% Ni 20.0% Cr steel: Electrode; Metrode
20.15.3 CR	0.3% C 15.0% Ni 20.0% Cr steel: Electrode; Metrode
20.23.3 CuNb	0.07% C 23.0% Ni 20.0% Cr 3.0% Mo 1.0% Nb 3.0% Cu steel: Electrode; Metrode
20.25.4 NbR	0.04% C 25.0% Ni 20.0% Cr 4.5% Mo 1.0% Nb steel: Electrode; Metrode
20.29.3 CuNbB	0.06% C 25.0% Ni 20.0% Cr 2.0% Mo 0.8% Nb 3.0% Cu steel: Electrode; Metrode
20.32 NbB	0.1% C 32.0% Ni 20.0% Cr 1.1% Nb steel: Electrode; Metrode; for welding Incoloy 800
20.34.3 CuNbB	0.06% C 34.0% Ni 20.0% Cr 3.0% Mo 0.8% Nb 3.0% Cu steel: Electrode; Metrode
20 Cb 3	0.06% C (max) 34% Ni 20% Cr 2.5% Mo 0.5% Nb 3.5% Cu and Ta steel: Carpenter
20 Mo 4	0.03% C (max) 37% Ni 23% Cr 4% Mo 0.2% Nb 1% Cu steel: Carpenter
20 Mo 6	0.03% C (max) 35% Ni 24% Cr 6% Mo 3% Cu steel: Carpenter **UTS: 640 Elon: 50% Proof: 310**
21.26.5 CuNbR	0.04% C 26.0% Ni 21.0% Cr 4.5% Mo 0.8% Nb 1.5% Cu steel: Electrode; Metrode
21-12 N Valve	0.2% C 11.5% Ni 21% Cr 0.2% N steel: Carpenter **DPN: 240 UTS: 850 Elon: 40% Proof: 520**
21-12 Valve Steel	0.2% C 11.5% Ni 21% Cr steel: Carpenter **DPN: 190 UTS: 670 Elon: 50% Proof: 360**
22 Cr 13 Ni 5 n	0.06% C (max) 5% Mn 12% Ni 22% Cr 2.2% Mo 0.2% V 0.2% Nb 0.3% N steel: Carpenter **UTS: 830 Elon: 45% Proof: 450**
22A	0.04% C 22% Ni 20% Cr 2% Mo 1.5% Cu steel: Japanese alloy; Japanese Metal Industry
23/14 L	0.07% C 13.0% Ni 23.0% Cr steel: Nyby; annealed **DPN: 190 UTS: 530 Elon: 40% Proof: 240**
23/14 LNb	0.07% C 14.0% Ni 23.0% Cr 0.7% (max) Nb steel: Nyby; annealed **DPN: 200 UTS: 530 Elon: 40% Proof: 240**
23.12.2 LR	0.03% C 12.0% Ni 24.0% Cr 2.5% Mo steel: Electrode; Metrode
23.12.2 NbR	0.06% C 12.0% Ni 24.0% Cr 2.5% Mo 0.8% Nb steel: Electrode; Metrode
23.12.2 R	0.06% C 12.0% Ni 24.0% Cr 2.5% Mo steel: Electrode; Metrode
23.12 LR	0.03% C 12.0% Ni 24.0% Cr steel: Electrode; Metrode
23.12 NbLR	0.03% C 12.0% Ni 24.0% Cr 0.8% Nb steel: Electrode; Metrode
23.12 NbR	0.06% C 12.0% Ni 24.0% Cr 0.8% Nb steel: Electrode; Metrode
23.12 R	0.06% C 12.0% Ni 24.0% Cr steel: Electrode; Metrode; ferrite content 15%
23.12 WR	0.1% C 12.0% Ni 24.0% Cr 3.0% W steel: Electrode; Metrode
23.35.4 CR	0.4% C 35.0% Ni 25.0% Cr steel: Electrode; Metrode
24.14.2 CWR	0.25% C 14.0% Ni 24.0% Cr 3.0% W steel: Electrode; Metrode
24.24.3 CNB	0.3% C 24.0% Ni 24.0% Cr 1.7% Nb steel: Electrode; Metrode
25/12	0.08% C 13% Ni 23% Cr steel: Eastern Stainless; annealed **UTS: 620 Elon: 45% Proof: 330**
25/20	0.08% C 21% Ni 25% Cr steel: Eastern Stainless; annealed **UTS: 620 Elon: 43% Proof: 330**
25/20 Si	0.08% C 21% Ni 25% Cr 2.5% Si steel: Eastern Stainless; for AISI 314
25/21 L	0.07% C 20.0% Ni 25.0% Cr steel: Nyby; annealed **DPN: 190 UTS: 530 Elon: 40% Proof: 240**
25/21 Si	0.12% C 21.0% Ni 25.0% Cr 2.0% Si steel: Nyby; annealed **DPN: 200 UTS: 580 Elon: 40% Proof: 240**
25/24 EMo	0.05% C 25.0% Ni 25.0% Cr 2.3% Mo steel: Nyby; annealed **DPN: 180 UTS: 530 Elon: 40% Proof: 240**
25.5.1 C	1.0% C 5.0% Ni 25.0% Cr steel: Electrode; Metrode; for resistance to cavitation **DPN: 380**
25.6.2 CuLR	0.03% C 6.0% Ni 25.0% Cr 2.0% Mo 3.0% Cu steel: Electrode; Metrode; for wear resistance
25.6.2 CuR	0.06% C 6.0% Ni 25.0% Cr 2.0% Mo 3.0% Cu steel: Electrode; Metrode; for corrosion and wear resistance **DPN: 250**
25.6.3 CuR	0.06% C 6.0% Ni 25.0% Cr 3.0% Mo 2.0% Cu steel: Electrode; Metrode
25.12.4 CR	0.4% C 12.0% Ni 25.0% Cr steel: Electrode; Metrode
25.20.2 CB	0.2% C 1.5% Mn 20.0% Ni 25.0% Cr steel: Electrode; Metrode

Symbol	Nominal analysis, supplier, condition and remarks.
25.20.3 CB	0.3% C 1.5% Mn 20.0% Ni 25.0% Cr steel: Electrode; Metrode
25.20 B	0.17% C 5.0% Mn 20.0% Ni 25.0% Cr steel: Electrode; Metrode
25.20 HB	0.4% C 1.5% Mn 20.0% Ni 25.0% Cr steel: Electrode; Metrode
25.20 HR	0.4% C 1.5% Mn 20.0% Ni 25.0% Cr steel: Electrode; Metrode
25.20 MoB	0.1% C 20.0% Ni 25.0% Cr 2.5% Mo steel: Electrode; Metrode
25.20 NbB	0.1% C 5.0% Mn 20.0% Ni 25.0% Cr 1.0% Nb steel: Electrode; Metrode
25.20 SUPER R	0.1% C 20% Ni 26% Cr steel: Electrode; Metrode
25.25.2 Nb	0.07% C 25.0% Ni 25.0% Cr 2.1% Mo 1.0% Nb steel: Electrode; Metrode
26.33.4 CNbR	0.4% C 33.0% Ni 26.0% Cr 1.5% Nb steel: Electrode; Metrode
27/4 L	0.07% C 4.0% Ni 27.0% Cr steel: Nyby; annealed
	DPN: 260 UTS: 580 Elon: 20% Proof: 370
27/5 LMo	0.07% C 5.0% Ni 27.0% Cr 1.6% Mo steel: Nyby; annealed
	DPN: 260 UTS: 630 Elon: 20% Proof: 400
28.6.4 CB	0.4% C 6.0% Ni 28.0% Cr steel: Electrode; Metrode
29.9 R	0.09% C 9.0% Ni 29.0% Cr steel: Electrode; Metrode; ferrite content 30–35%
29.9% SUPER R	0.12% C 9% Ni 30% Cr 1% Mo steel: Electrode; Metrode
	DPN: 275
30A	0.04% C 30% Ni 20% Cr 2.5% Mo 3.5% Cu steel: Japanese alloy
248 SV	0.03% C 5.0% Ni 16% Cr 1.0% Mo steel: Weldable; Avesta
254 EM	0.08% C 21% Ni 25% Cr steel: Avesta
	UTS: 650 Elon: 40% Proof: 300
254 SLX	0.02% C 25% Ni 20% Cr 4.5% Mo 1.5% Cu steel: Avesta
	UTS: 590 Elon: 40% Proof: 270
302 HQ-FM	0.06% C (max) 10% Ni 17.5% Cr 0.1% S 2% Cu steel: Carpenter
302 NILSTAIN	0.15% C 8% Ni 18% Cr steel for springs: Driver
302 S25	0.12% C (max) 9.5% Ni 18% Cr steel: BS 970; obsolete
302 S31	0.12% C (max) 9% Ni 18% Cr steel: BS 970
303 Cu	0.15% C (max) 9% Ni 18% Cr 3.2% Cu steel: Designation used in the UK and USA
303 MA	0.15% C (max) 9% Ni 18% Cr 0.6% Mo 0.2% S 0.17% Pb steel: Designation used in the UK and USA
303 Pb	0.15% C (max) 9% Ni 18% Cr 0.2% S 0.18% Pb steel: Designation used in the UK and USA
303 PLUS X	0.15% C (max) 8.5% Ni 18% Cr 0.15% S (min) steel: Designation used in the UK and USA
303 S21	0.12% C (max) 9.5% Ni 18% Cr 0.2% S steel: Free machining: BS 970; obsolete
303 S31	0.12% C (max) 9% Ni 18% Cr 1% Mo (max) 0.2% S steel: BS 970
303 S41	0.12% C (max) 9.5% Ni 18% Cr steel
303 S42	0.12% C (max) 9% Ni 18% Cr 1% Mo (max) 0.2% Se steel: BS 970
304 ELC NILSTAIN	0.03% C (max) 10% Ni 19% Cr steel: Driver

Note. The following abbreviations and units are used in the tables:

DPN	Hardness, diamond pyramid number
UTS	Ultimate tensile strength, N/mm^2
Elon	Elongation, %
Proof	0.1% proof strength, N/mm^2

$1\ N/mm^2 = 0.1\ hbar = 0.102\ kgf/mm^2 = 0.06475\ tonf/in.^2 = 145.04\ lbf/in.^2 = 1\ MPa$

See Appendix II for other abbreviations and conversion tables.

Symbol	Nominal analysis, supplier, condition and remarks.
304 LN	0.035% C (max) 9.5% Ni 19% Cr 0.13% N steel: Designation used in the UK and USA
304 NILSTAIN	0.08% C (max) 10% Ni 19% Cr steel: Driver
304 S11	0.03% C (max) 10.5% Ni 18% Cr steel: BS 970
304 S12	0.03% C (max) 10.5% Ni 18% Cr steel: BS 970; obsolete
304 S15	0.06% C (max) 9.5% Ni 18.7% Cr steel
304 S31	0.07% C (max) 9.5% Ni 18% Cr 0.4% Ti steel
305 NILSTAIN	0.12% C (max) 11.5% Ni 18% Cr steel: Driver
308 S92	0.03% C (max) 10% Ni 21% Cr steel: Welding electrode; designation for BS 2901
308 S93	0.03% C (max) 10% Ni 20% Cr steel: Welding electrode; designation for BS 2901
308 S96	0.08% C (max) 10% Ni 21% Cr steel: Welding electrode; designation for BS 2901
309 Cb	0.08% C (max) 14% Ni 23% Cr 0.8% Nb steel: Designation used in the UK and USA
309 NILSTAIN	0.2% C (max) 13.5% Ni 23% Cr steel: Driver
309 S	0.08% C (max) 14% Ni 23% Cr steel: Designation used in the UK and USA
309 S92	0.03% C (max) 13% Ni 24% Cr steel: Welding electrode; designation for BS 2901
309 S94	0.12% C (max) 13% Ni 24% Cr steel: Welding electrode; designation for BS 2901
310 NILSTAIN	0.25% C (max) 20.5% Ni 25% Cr steel: Driver
310 S24	0.15% C (max) 20.5% Ni 24.5% Cr steel: BS 970; obsolete
310 S31	0.15% C (max) 20.5% Ni 25% Cr steel: BS 970
310 S94	0.11% C 21% Ni 27% Cr steel: Welding electrode; designation for BS 2901
310 S98	0.4% C 21% Ni 27% Cr steel: Welding electrode; designation for BS 2901
311 S94	0.12% C (max) 13% Ni 24% Cr 1.2% Nb steel: Welding electrode; designation for BS 2901
312 S94	0.15% C (max) 9% Ni 30% Cr steel: Welding electrode; designation for BS 2901
313 S94	0.1% C 21% Ni 27% Cr 1.2% Nb steel: Welding electrode; designation for BS 2901
314 NILSTAIN	0.25% C (max) 20.5% Ni 24.5% Cr steel: Driver
315 S16	0.07% C (max) 11.5% Ni 17.5% Cr 2.7% Mo steel: BS 970; obsolete
316 ELC NILSTAIN	0.03% C (max) 12% Ni 17% Cr 2.5% Mo steel: Driver
316 LN	0.035% C (max) 12% Ni 17% Cr 2.5% Mo 0.13% N steel: Designation used in the UK and USA
316 NILSTAIN	0.08% C (max) 12% Ni 17% Cr 2.5% Mo steel: Driver
316 S11	0.03% C (max) 12.5% Ni 17.5% Cr 2.2% Mo steel: BS 970
316 S12	0.03% C (max) 12.5% Ni 17.5% Cr 2.7% Mo steel: BS 970; obsolete
316 S13	0.03% C (max) 12.5% Ni 17.5% Cr 2.7% Mo steel: BS 970
316 S16	0.07% C (max) 11.5% Ni 17.5% Cr 2.7% Mo steel: BS 970; obsolete
316 S31	0.07% C (max) 12.5% Ni 17.5% Cr 2.2% Mo steel: BS 970
316 S33	0.07% C (max) 12.5% Ni 17.5% Cr 2.7% Mo steel: BS 970
316 S92	0.03% C (max) 13% Ni 19% Cr 2.5% Mo steel: Welding electrode; designation for BS 2901
316 S93	0.03% C (max) 12% Ni 19% Cr 3% Mo steel: Welding electrode; designation for BS 2901
316 S96	0.08% C (max) 13% Ni 19% Cr 2.5% Mo steel: Welding electrode; designation for BS 2901
317 NILSTAIN	0.08% C 13% Ni 19% Cr 3.5% Mo steel: Driver
317 S12	0.03% C (max) 15.5% Ni 18.5% Cr 3.5% Mo steel: BS 970; obsolete
317 S16	0.08% C (max) 15.5% Ni 18.5% Cr 3.5% Mo steel: BS 970; obsolete
317 S96	0.08% C (max) 14% Ni 19.5% Cr 3% Mo steel: Welding electrode; designation for BS 2901

Symbol	Nominal analysis, supplier, condition and remarks.
318 S96	0.08% C (max) 13% Ni 19% Cr 2.5% Mo 1.0% Nb steel: Welding electrode; designation for BS 2901
320 S17	0.08% C (max) 12.5% Ni 17.5% Cr 2.7% Mo steel: BS 970; obsolete
320 S31	0.08% C (max) 12.5% Ni 17.5% Cr 2.2% Mo 0.6% Nb steel: BS 970
321 NILSTAIN	0.08% C (max) 10.5% Ni 18% Cr 0.4% Ti steel: Driver
321 S12	0.08% C (max) 10.5% Ni 18% Cr 0.6% Ti steel: BS 970; obsolete
321 S20	0.12% C (max) 9.5% Ni 18% Cr 0.6% Ti steel: BS 970; obsolete
325 S21	0.12% C (max) 9.5% Ni 18% Cr 0.6% Ti steel: BS 970; obsolete
325 S31	0.12% C (max) 9% Ni 18% Cr 0.7% Nb 0.2% S steel: BS 970
326 S36	0.12% C (max) 12.5% Ni 17.5% Cr 2.7% Mo steel: BS 970; obsolete
330 NILSTAIN	15% Cr 34% Ni steel: Driver
347 HC	0.1% C 10% Ni 19.5% Cr 0.8% Nb steel: Weld electrode; designation used by AWS
347 NILSTAIN	0.08% C (max) 11% Ni 18% Cr 0.8% Nb steel: Driver
347 S17	0.08% C (max) 10.5% Ni 18% Cr 0.8% Nb steel: BS 970; obsolete
347 S31	0.08% C (max) 9.5% Ni 18% Cr 0.9% Nb steel: BS 970
347 S96	0.08% C (max) 10% Ni 20.5% Cr 1.0% Nb steel: Rod for gas shielded arc welding; designation for BS 2901
384	0.08% C (max) 18% Ni 16% Cr steel: Designation used in the UK and USA
385	0.08% C (max) 15% Ni 12.5% Cr steel
430 NILSTAIN	0.12% C (max) 16% Cr steel: Driver
453S	0.1% C 5.0% Ni 26% Cr 1.5% Mo steel: Avesta **UTS: 650 Elon: 20% Proof: 500**
553	0.08% C 9% Ni 18% Cr steel: Soderfors
554	0.16% C (max) 9% Ni 18% Cr steel: Soderfors
556	0.16% C 9% Ni 18% Cr 0.2% S steel: Free machining; Soderfors
559	0.16% C (max) 13% Ni 13% Cr steel: Soderfors
564	0.12% C 10% Ni 18% Cr 3.0% Mo steel: Soderfors
832C	0.16% C 8% Ni 18% Cr 0.2% S steel: Free machining; Avesta
832K	0.12% C 10% Ni 18% Cr 3.0% Mo steel: Avesta
832M	0.1% C 9% Ni 18% Cr steel: Avesta
832MM	0.06% C 10.5% Ni 18.5% Cr steel: Avesta **UTS: 600 Elon: 70% Proof: 270**
832MP	0.07% C 8.5% Ni 17.5% Cr steel: Avesta **UTS: 660 Elon: 70% Proof: 300**
832MV	0.15% C 8% Ni 18% Cr steel: Avesta
832MVN	0.05% C 8.5% Ni 18.5% Cr 0.2% N steel: Avesta
832MVR	0.03% C 10.5% Ni 18.5% Cr steel: Avesta **UTS: 540 Elon: 60% Proof: 250**
832MVRN	0.03% C 10.5% Ni 18.5% Cr 0.2% N steel: Avesta
832MVT	0.08% C 10.5% Ni 17.5% Cr 0.4% Ti steel: Avesta **UTS: 600 Elon: 50% Proof: 280**
832P	0.16% C 13% Ni 13% Cr steel: Avesta
832SF	0.05% C 11% Ni 17% Cr 2.3% Mo steel: Avesta **UTS: 600 Elon: 55% Proof: 290**
832SFR	0.03% C 11.5% Ni 17% Cr 2.3% Mo steel: Avesta **UTS: 580 Elon: 55% Proof: 290**
832SFT	0.08% C 12% Ni 17% Cr 2.2% Mo 0.4% Ti steel: Avesta **UTS: 600 Elon: 50% Proof: 290**
832SK	0.05% C 11.5% Ni 17% Cr 2.7% Mo steel: Avesta **UTS: 600 Elon: 55% Proof: 290**
832SKR	0.03% C 13% Ni 17.5% Cr 2.7% Mo steel: Avesta **UTS: 580 Elon: 55% Proof: 290**
832SKRN	0.03% C 13% Ni 17.5% Cr 2.7% Mo 0.2% N steel: Avesta
832SN	0.05% C 14.5% Ni 18.5% Cr 3.5% Mo steel: Avesta **UTS: 600 Elon: 50% Proof: 300**
832SV	0.05% C 10% Ni 17% Cr 1.5% Mo steel: Avesta **UTS: 590 Elon: 60% Proof: 280**
832T	0.15% C 8% Ni 18% Cr 0.6% Ti steel: Avesta
832V	0.12% C 10% Ni 18% Cr 2.0% Mo steel: Avesta
2205 KS	0.03% C 10% Ni 24% Cr 3% Mo N steel: Electrode; Metrode
3632	0.12% C 20.0% Ni 25.0% Cr steel: French Standard designation
3636	0.12% C 37.0% Ni 18.0% Cr steel: French Standard designation
3682	0.06% C 25.0% Ni 15.0% Cr 1.25% Mo 0.3% V 2.0% Ti steel: French Standard designation
A 1/57	0.08% C 12% Ni 18% Cr 2.5% Si steel: Babcock and Wilcox (USA)
A 10	Austenitic stainless steel sinter: Firth Cleveland; specific gravity 6.4–6.8; as sintered **DPN: 100 UTS: 360 Elon: 6%**
A 16X	0.03% C (max) 24.0% Ni 20.0% Cr 6.5% Mo stainless steel: Allegheny Ludlum **UTS: 650 Elon: 45% Proof: 310**
A 100	0.05% C (max) 12.5% Ni 17% Cr 2.7% Mo steel: VEW
A 101	0.05% C (max) 12.0% Ni 17.8% Cr 2.7% Mo 0.16% S steel: VEW; face machining
A 120	0.05% C (max) 11% Ni 17% Cr 2.2% Mo steel: VEW
A 200	0.03% C (max) 11.5% Ni 17% Cr 2.2% Mo steel: VEW
A 205	0.03% C (max) 13% Ni 17.5% Cr 2.8% Mo steel: VEW
A 220	0.03% C (max) 14.5% Ni 17.5% Cr 2.7% Mo steel: VEW
A 300	0.06% C (max) 11% Ni 17% Cr 2.2% Mo 0.3% Ti (min) steel: VEW
A 400	0.04% C (max) 13.5% Ni 17% Cr 4.3% Mo 0.15% N steel: VEW
A 405	0.02% C (max) 22% Ni 25% Cr 2.2% Mo 0.12% N steel: VEW
A 410	0.03% C (max) 13.5% Ni 17.5% Cr 2.7% Mo 0.18% N steel: VEW
A 500	0.05% C (max) 9.5% Ni 18.5% Cr steel: VEW
A 506	0.09% C (max) 8.5% Ni 18% Cr 0.23% S steel: VEW; free machining
A 507	0.06% C 8.5% Ni 18.5% Cr steel: VEW
A 520	0.1% C 7.5% Ni 17.5% Cr steel: VEW
A 604	0.03% C (max) 9.5% Ni 18.5% Cr steel: VEW
A 610	0.02% C (max) 15% Ni 17.5% Cr 4.0% Si steel: VEW
A 700	0.06% C (max) 9.5% Ni 17.5% Cr 0.3% Ti (min) steel: VEW
A 700	0.12% C (max) 9.5% Ni 17.5% Cr 0.5% Ti (min) steel: VEW; heat resistant
A 750	0.06% C (max) 9.5% Ni 17.5% Cr 0.6% Nb (min) steel: VEW
A 900	0.1% C (max) 5.3% Ni 26.8% Cr 1.5% Mo steel: VEW
A 962	0.02% C (max) 25% Ni 20% Cr 4.5% Mo 1.5% Cu steel: VEW
AC1 HT-30	0.3% C 35% Ni 15% Cr Fe alloy: Origin unknown
AC1 HT-50	0.5% C 35% Ni 17% Cr Fe alloy: Origin unknown
AC1 HT-50C	0.5% C 35% Ni 15% Cr 1% Nb Fe alloy: Origin unknown
ACNW	14% Ni 2.5% W 0.5% C 14% Cr steel: Valhuna
AFN03 81-343 EZ 19.9LB 110-20	0.03% C 9.5% Ni 19% Cr steel: Welding electrode; French Standard
AFNOR 81-343 EZ 19.9LR23	0.02% C 9.5% Ni 19.5% Cr steel: Welding electrode; French Standard
AFNOR 81-343 EZ 19.9LR150-36	0.02% C 9.5% Ni 19.5% Cr steel: Welding electrode; French Standard
AFNOR 81-343 EZ 19.9LR 160-23X	0.03% C 9.5% Ni 19.5% Cr steel: Welding electrode; French Standard
AFNOR 81-343 EZ19.9 NbR23	0.02% C 9.5% Ni 18.5% Cr 0.25% Nb steel: Welding electrode; French Standard

Symbol	Nominal analysis, supplier, condition and remarks.
AFNOR 81-343 EZ19.12.3 LB 110-20	0.03% C 12% Ni 18.5% Cr 2.5% Mo steel: Welding electrode; French Standard
AFNOR 81-343 EZ19.12.3 LR23	0.02% C 12% Ni 18.5% Cr 2.6% Mo steel: Welding electrode; French Standard
AFNOR 81-343 EZ19.12.3 LR150-36	0.02% C 11.5% Ni 18.5% Cr 2.5% Mo steel: Welding electrode; French Standard
AFNOR 81-343 EZ19.12.3 LR170 23X	0.03% C 12% Ni 18.5% Cr 2.5% Mo steel: Welding electrode; French Standard
AFNOR 81-343 EZ19.12.3 NbR23	0.02% C 12% Ni 18% Cr 2.5% Mo 0.25% Nb steel: Welding electrode; French Standard
AFNOR 81-343 EZ19.13.4 R180-23X	0.04% C 13.5% Ni 19.5% Cr 3.3% Mo steel: Welding electrode; French Standard
AFNOR 81-343 EZ20.25.2 LCnRB20	0.03% C 25% Ni 22% Cr 5% Mo 2% Cu steel: Welding electrode; French Standard
AFNOR 81-343 EZ20.25.5 LCnR23	0.02% C 25% Ni 20% Cr 4.5% Mo 1.5% Cu steel: Welding electrode; French Standard
AFNOR 81-343 EZ23.12.2 R23	0.02% C 13% Ni 24% Cr 2.3% Mo steel: Welding electrode; French Standard
AFNOR 81-343 EZ23.12.2 R186-33X	0.04% C 13% Ni 23% Cr 2.3% Mo steel: Welding electrode; French Standard
AFNOR 81-343 EZ23.12 LR23	0.02% C 13% Ni 23% Cr steel: Welding electrode; French Standard
AFNOR 81-343 EZ23.12 LR150-36	0.02% C 12.5% Ni 23% Cr steel: Welding electrode; French Standard
AFNOR 81-343 EZ25.20 R23	0.1% C 21% Ni 26% Cr steel: Welding electrode; French Standard
AFNOR 81-343 EZ29.9 R23	0.12% C 9% Ni 30% Cr steel: Welding electrode; French Standard
AGT	0.16% C 13% Ni 13% Cr steel: Ugine
AIP	0.08% C (max) 11.5% Ni 18% Cr steel: Valhuna
AIS	0.08% C (max) 10% Ni 19% Cr steel: Valhuna
AISC	0.08% C (max) 11% Ni 18% Cr steel: Valhuna
AISI 302	0.1% C 8% Ni 18% Cr steel **DPN: 180 UTS: 580 Elon: 50% Proof: 200**
AISI 302B	0.15% C 9% Ni 18% Cr steel **DPN: 180 UTS: 580 Elon: 40% Proof: 180**
AISI 303	0.15% C (max) 9% Ni 18% Cr with S or Se steel **DPN: 160 UTS: 530 Elon: 40% Proof: 200**
AISI 303 Se	0.15% C 9% Ni 18% Cr 0.15% Se steel: Free cutting
AISI 304	0.08% C (max) 9% Ni 19% Cr steel **DPN: 180 UTS: 580 Elon: 50% Proof: 200**
AISI 304L	0.03% C 10% Ni 19% Cr steel **DPN: 180 UTS: 480 Elon: 40% Proof: 150**
AISI 305	0.12% C (max) 12% Ni 18% Cr steel **DPN: 180 UTS: 480 Elon: 50% Proof: 150**
AISI 308	0.08% C 11% Ni 20% Cr steel **UTS: 530 Elon: 50% Proof: 200**
AISI 309	0.2% C 13% Ni 23% Cr steel **DPN: 200 UTS: 580 Elon: 40% Proof: 200**
AISI 309S	0.08% C 14% Ni 23% Cr steel **DPN: 200 UTS: 530 Elon: 40% Proof: 200**
AISI 310	0.25% C (max) 21% Ni 25% Cr steel: Annealed **DPN: 180 UTS: 530 Elon: 40% Proof: 200**

Note. The following abbreviations and units are used in the tables:

DPN	Hardness, diamond pyramid number
UTS	Ultimate tensile strength, N/mm^2
Elon	Elongation, %
Proof	0.1% proof strength, N/mm^2

1 N/mm^2=0.1 hbar=0.102 kgf/mm^2=0.06475 tonf/in.^{2}=145.04 lbf/in.^{2}=1 MPa

See Appendix II for other abbreviations and conversion tables.

Symbol	Nominal analysis, supplier, condition and remarks.
AISI 310S	0.08% C 21% Ni 25% Cr steel
AISI 314	0.29% C 21% Ni 25% Cr steel: Annealed **DPN: 180 UTS: 530 Elon: 40% Proof: 200**
AISI 316	0.08% C (max) 12% Ni 17% Cr 2.5% Mo steel: Annealed **DPN: 200 UTS: 480 Elon: 40% Proof: 200**
AISI 316L	0.03% C 12% Ni 17% Cr 2.5% Mo steel: Annealed **DPN: 180 UTS: 480 Elon: 40% Proof: 180**
AISI 316N	0.08% C 12.0% Ni 17.0% Cr 2.5% Mo 0.13% N steel: Stainless
AISI 317	0.08% C (max) 13% Ni 19% Cr 3.5% Mo steel: Annealed **DPN: 200 Elon: 40% Proof: 200**
AISI 317 L	0.03% C 14% Ni 20% Cr 3.5% Mo steel
AISI 318	0.08% C 14% Ni 20% Cr 3.5% Mo steel: Annealed **DPN: 200 UTS: 530 Elon: 40% Proof: 180**
AISI 321	0.08% C (max) 10% Ni 18% Cr 0.4% Ti steel: Annealed **DPN: 200 UTS: 530 Elon: 40% Proof: 180**
AISI 321H	0.06% C 10.5% Ni 18.0% Cr 0.7% Ti (max) steel: Stainless
AISI 325	0.25% C (max) 21% Ni 9% Cr 1.2% Cu steel
AISI 327	0.1% C 4.8% Ni 25.5% Cr steel
AISI 329	0.09% C 5% Ni 26% Cr 1.5% Mo steel: Obsolete
AISI 330	0.1% C 35% Ni 15% Cr 1.2% Mo steel
AISI 347	0.08% C (max) 11% Ni 18% Cr 0.8% Nb steel: Annealed **DPN: 200 UTS: 580 Elon: 40% Proof: 200**
AISI 347 F	0.08% C 12% Ni 18% Cr 0.8% Nb 0.2% Se steel: Annealed **DPN: 200 UTS: 580 Elon: 40% Proof: 200**
AISI 347H	0.6% C 11.0% Ni 18.0% Cr 1.0% Nb (max) steel: Stainless
AISI 348	0.08% C 11.0% Ni 18.0% Cr 0.8% Nb + Ta (min) steel
AISI 348	0.08% C 11% Ni 18% Cr 0.8% Nb 0.1% Ta steel
AISI 348H	0.06% C 11.0% Ni 18.0% Cr 0.6% Nb (min) steel
AISL	0.03% C (max) 10.5% Ni 19% Cr steel: Valhuna
AIST	0.08% C (max) 10.5% Ni 18% Cr 0.6% Ti steel: Valhuna
AL-6X	0.035% C (max) 24.5% Ni 21% Cr 6.5% Mo Fe alloy: Allegheny Ludlum
AL-6XN	0.03% C (max) 24.5% Ni 21% Cr 6.5% Mo 0.2% N Fe alloy
ALLOY 20	0.05% C (max) 37% Ni 20% Cr 2.5% Mo 0.6% Nb 3.5% Cu Fe alloy: VDM; Nicrofer 3620 Nb
ALLOY 28	0.015% C (max) 31% Ni 27% Cr 3.5% Mo 1.2% Cu Fe alloy: VDM; Nicrofer 3127LC
ALLOY 306	0.015% C (max) 14.7% Ni 18% Cr 4.2% Si steel: Origin unknown
ALLOY 310S	0.08% C (max) 20% Ni 24.5% Cr steel: VDM
ALLOY 316LN	0.03% C (max) 15% Ni 17.5% Cr 2.7% Mo 0.17% N steel: Origin unknown
ALLOY 317LN	0.03% C (max) 13.5% Ni 17.7% Cr 4.2% Mo 0.2% N steel: Origin unknown
ALLOY 318LN	0.03% C (max) 5.7% Ni 22.5% Cr 3% Mo 0.15% N steel: VDM
ALLOY 330	0.15% C (max) 35.5% Ni 16.5% Cr 2% Si steel: VDM
ALLOY 800	0.1% C (max) 31% Ni 20% Cr steel: VDM
ALLOY 800H	0.08% C 31% Ni 20.5% Cr 0.3% Ti steel: VDM
ALLOY 800L	0.025% C (max) 33% Ni 21% Cr 0.4% Ti Fe alloy: VDM; Nicrofer 3220LC
ALLOY 904L	0.02% C (max) 25% Ni 20.5% Cr 4.7% Mo 0.07% N steel: Origin unknown
ALLOY 904LN	0.02% C (max) 25% Ni 20.5% Cr 4.7% Mo 0.18% N steel: Origin unknown
ALLOY 904RMo	0.025% C (max) 25% Ni 20.5% Cr 6.4% Mo steel: Origin unknown
ALLOY DS	0.15% C (max) 35.5% Ni 16.5% Cr 2% Si steel: VDM
AMS	0.08% C (max) 13% Ni 19% Cr 3.5% Mo steel: Valhuna

Symbol	Nominal analysis, supplier, condition and remarks.
AMS 5341	0.16% C (max) 11.5% Ni 19.5% Cr 0.2% S steel: Casting
AMS 5358	18% Cr 9% Ni steel: Investment casting
AMS 5360	0.05% C 12.5% Ni 17% Cr 2.5% Mo steel: Castings
AMS 5360 B	0.08% C 13% Ni 17% Cr 2% Mo steel: Investment casting; AMS for SAE 60316
AMS 5361	0.05% C 12.5% Ni 17% Cr 2.5% Mo steel: Castings
AMS 5361 B	0.08% C 13% Ni 18% Cr 2% Mo steel: Sand casting; AMS for SAE 60316
AMS 5362	0.05% C 11% Ni 18.5% Cr 0.7% Nb steel: Castings
AMS 5362 D	0.08% C 12% Ni 19% Cr Nb + Ta steel: Investment casting; AMS for SAE 60347
AMS 5363	0.05% C 11% Ni 18.5% Cr 0.7% Nb steel: Castings
AMS 5363 B	0.08% C 10.5% Ni 18% Cr Nb +Ta steel: Sand casting; AMS for SAE 60347
AMS 5364	0.1% C (max) 11% Ni 18.5% Cr 1.2% Nb + Ta steel: Castings
AMS 5365	0.12% C 20.5% Ni 25% Cr steel: Castings
AMS 5365 A	0.2% C 20% Ni 25% Cr steel: Sand casting; AMS for SAE 60310
AMS 5366	0.12% C 20.5% Ni 25% Cr steel: Castings
AMS 5366 A	0.2% C 20% Ni 25% Cr steel: Investment casting; AMS for SAE 60310
AMS 5368	0.08% C 9% Ni 18% Cr steel: Castings
AMS 5369 A	20% Cr 9% Ni 1.4% Mo 1.4% Nb Ti steel: Sand casting
AMS 5370	0.02% C (max) 10% Ni 19% Cr steel: Castings
AMS 5371	0.02% C (max) 10% Ni 19% Cr steel: Castings
AMS 5501	0.08% C (max) 9.7% Ni 19% Cr steel
AMS 5507	0.08% C 13% Ni 17% Cr 2.5% Mo steel: Sheet and strip; AMS for SAE 30316
AMS 5510	0.04% C (max) 11% Ni 18.5% Cr 0.4% Ti steel: Sheet
AMS 5510 H	0.08% C 10% Ni 18% Cr Ti steel: Sheet and strip; AMS for SAE 30321
AMS 5511	0.02% C (max) 10% Ni 19% Cr steel: Sheet
AMS 5511 A	Low C 18% Cr 8% Ni steel: Sheet and strip; AMS for SAE 30304
AMS 5512 B	0.08% C 11% Ni 18% Cr Nb + Ta steel: Sheet and strip; AMS for SAE 30347
AMS 5513	0.08% C 9% Ni 19% Cr steel: Sheet and strip; AMS for SAE 30304
AMS 5514 A	0.1% C 11% Ni 18% Cr steel: Sheet and strip for deep drawing; AMS for SAE 30305
AMS 5515	0.08% C 9% Ni 18% Cr steel: Sheet
AMS 5515 D	0.1% C 8% Ni 18% Cr steel: Sheet and strip for deep drawing; AMS for SAE 30302
AMS 5516	0.08% C 9% Ni 18% Cr steel: Sheet
AMS 5516 E	0.1% C 8% Ni 18% Cr steel: Sheet and strip; cold rolled; AMS for SAE 30302 **UTS: 530**
AMS 5517 D	0.1% C 8% Ni 18% Cr steel: Sheet and strip; cold rolled; AMS for SAE 30301 **UTS: 920**
AMS 5518 C	0.1% C 8% Ni 18% Cr steel: Sheet and strip; cold rolled; AMS for SAE 30301 **UTS: 1120**
AMS 5519 E	0.1% C 8% Ni 18% Cr steel: Sheet and strip; cold rolled; AMS for SAE 30301 **UTS: 1380**
AMS 5521	0.12% C 20% Ni 25% Cr steel: Sheet
AMS 5521 B	0.25% C 20% Ni 25% Cr steel: Sheet and strip for deep drawing; AMS for SAE 30310
AMS 5522	0.12% C 20.5% Ni 24% Cr steel: Sheet
AMS 5522 B	20% Ni 25% Cr 2% Si steel: Sheet and strip
AMS 5523	0.2% C 13% Ni 23% Cr steel: Sheet and strip; AMS for SAE 30309
AMS 5524	0.05% C (max) 12.5% Ni 17% Cr 2.5% Mo steel: Sheet
AMS 5524 B	0.08% C 13% Ni 18% Cr 2.5% Mo steel: Sheet and strip; AMS for SAE 30316
AMS 5527	0.3% C 9% Ni 18.5% Cr 1.4% Mo 1.3% W 0.2% Ti 0.4% Nb steel: Sheet
AMS 5531	0.15% C 20% Ni 21% Cr 3% Mo 2.5% W 20% Co 1% Nb steel: Sheet
AMS 5533	0.43% C 20% Ni 21% Cr 4% Mo 4% W 20% Co 4% Nb steel: Sheet
AMS 5538	0.3% C 9% Ni 18% Cr 1.6% Mo 1.3% W 0.5% Ti steel: Sheet
AMS 5543	0.03% C 26% Ni 13.5% Cr 1.7% Mo 3% Ti 0.15% Al steel: Sheet
AMS 5552	0.04% C 32% Ni 20% Cr 1% Ti 0.15% Cu steel: Sheet
AMS 5552 A	20% Cr 32% Ni 1% Ti Fe alloy: Sheet
AMS 5556 A	0.08% C 11% Ni 18% Cr Nb + Ta steel: Tube; AMS for SAE 30347
AMS 5557	0.12% C 20.5% Ni 25% Cr steel: Tubing
AMS 5557 A	0.08% C 11% Ni 18% Cr Ti steel: Tube; AMS for SAE 30321
AMS 5558	0.08% C 11% Ni 18% Cr Nb + Ta steel: Welded tube; AMS for SAE 30347
AMS 5559 A	0.08% C 10% Ni 18% Cr Ti steel: Welded tube; AMS for SAE 30321
AMS 5560	0.04% C (max) 10% Ni 19% Cr steel: Tubing
AMS 5560 D	0.08% C 10% Ni 19% Cr steel: Seamless tube; AMS for SAE 30304
AMS 5563	0.08% C (max) 9.7% Ni 19% Cr steel
AMS 5564	0.08% C (max) 9.7% Ni 19% Cr steel
AMS 5565	0.04% C (max) 10% Ni 19% Cr steel: Tubing
AMS 5565 D	0.08% C 9% Ni 19% Cr steel: Welded tube; AMS for SAE 30304
AMS 5566 D	0.08% C 10% Ni 19% Cr steel: Seamless or welded tube; AMS for SAE 30304
AMS 5567	0.08% C 10% Ni 19% Cr steel: Tube; AMS for SAE 30304
AMS 5570	0.04% C (max) 11% Ni 18.5% Cr 0.4% Ti steel: Tubing
AMS 5570 G	0.08% C 11% Ni 18% Cr Ti steel: Seamless tube; AMS for SAE 30321
AMS 5571	0.05% C (max) 11% Ni 18.5% Cr 0.7% Nb steel: Tubing
AMS 5571 B	0.08% C 11% Ni 18% Cr Nb + Ta steel: Seamless tube; AMS for SAE 30347
AMS 5572	0.12% C 20.5% Ni 25% Cr steel: Tubing
AMS 5572 B	0.25% C 20% Ni 25% Cr steel: Seamless tube; AMS for SAE 30310
AMS 5573 C	0.08% C 12% Ni 17% Cr 2.5% Mo steel: Seamless tube; AMS for SAE 30316
AMS 5574	0.2% C 13% Ni 23% Cr steel: Seamless tube; AMS for SAE 30309
AMS 5575	0.05% C (max) 11% Ni 18.5% Cr 0.7% Nb steel: Tubing
AMS 5575 F	0.08% C 11% Ni 18% Cr Nb + Ta steel: Welded tube; AMS for SAE 30347
AMS 5576	0.04% C (max) 11% Ni 18.5% Cr 0.4% Ti steel: Tubing
AMS 5576 C	0.08% C 10% Ni 18% Cr Ti steel: Welded tube; AMS for SAE 30321
AMS 5577 A	0.25% C 20% Ni 25% Cr steel: Welded tube; AMS for SAE 30310
AMS 5579	9% Ni 20% Cr 1.5% Mo 1.5% W 0.2% Ti 0.4% Nb + Ta steel: Welded tube
AMS 5592	0.06% C 35% Ni 19% Cr steel: Sheet
AMS 5600	0.15% C (max) 9% Ni 1.8% Cr steel: Type 302
AMS 5624 C	0.15% C 11% Ni 18% Cr Nb + Ta steel: Bar and forging; free machining; AMS for SAE 30303 F
AMS 5635	0.15% C 9% Ni 18% Cr steel: Bar and forging; free machining; AMS for SAE 30303 F
AMS 5636	0.08% C 9% Ni 18% Cr steel: Forgings
AMS 5636 A	0.1% C 8% Ni 18% Cr steel: Bar; cold drawn; AMS for SAE 30302 **UTS: 740**
AMS 5637	0.08% C 9% Ni 18% Cr steel: Forgings

Symbol	Nominal analysis, supplier, condition and remarks.
AMS 5637 A	0.1% C 8% Ni 18% Cr steel: Bar; cold drawn; AMS for SAE 30302 **UTS: 920**
AMS 5639	0.04% C (max) 10% Ni 19% Cr steel: Forgings
AMS 5639 A	0.08% C 9% Ni 19% Cr steel: Bar, forgings, etc.; AMS for SAE 30304
AMS 5640	0.15% C (max) 9% Ni 18% Cr 0.15% S (min) steel: Type 303
AMS 5640	0.15% C (max) 8.5% Ni 18% Cr 0.25% S (min) steel: Type 303
AMS 5640 F	0.15% C 9% Ni 18% Cr steel: Bar and forging; free machining; AMS for SAE 30303 F
AMS 5641 A	0.15% C 10% Ni 18.5% Cr steel: Bar and forging; free machining; for swaging; AMS for SAE 30303 F
AMS 5642	0.08% C (max) 10.5% Ni 18% Cr 1% Nb 0.2% Se steel
AMS 5645 G	0.08% C 10% Ni 18% Cr Ti steel: Bar, forging, etc.; AMS for SAE 30321
AMS 5646	0.05% C 11% Ni 18.5% Cr 0.7% Ti steel: Forgings
AMS 5646 E	0.08% C 11% Ni 18% Cr Nb + Ta steel: Bar, forging; AMS for SAE 30347
AMS 5647	0.02% C (max) 10% Ni 19% Cr steel: Forgings
AMS 5647 A	0.08% C 8% Ni 18% Cr steel: Bar and forging; AMS for SAE 30304
AMS 5648	0.05% C (max) 12.5% Ni 17% Cr 2.5% Mo steel: Forgings
AMS 5648 C	0.08% C 13% Ni 18% Cr 2.5% Mo steel: Bar and forging; AMS for SAE 30316
AMS 5649	0.08% C 13% Ni 18% Cr 2% Mo steel: Bar and forging; free machining; AMS for SAE 30316 L
AMS 5650	0.04% C (max) 13% Ni 23% Cr 0.25% Mo steel: Forgings
AMS 5650 A	0.2% C 13.5% Ni 23% Cr steel: Bar and forging; AMS for SAE 30309
AMS 5651	0.12% C 20% Ni 25% Cr steel: Forgings
AMS 5651 D	0.25% C 20% Ni 25% Cr steel: Bar and forging; AMS for SAE 30310
AMS 5652 B	25% Cr 20% Ni 2% Si steel: Bar and forging
AMS 5653	0.08% C 13% Ni 7% Cr 2.5% Mo steel: Bar and forging; AMS for SAE 30316
AMS 5654	0.08% C (max) 11% Ni 18% Cr 0.8% Nb steel
AMS 5674	0.08% C (max) 11% Ni 18% Cr 0.8% Nb steel
AMS 5680	0.05% C (max) 11% Ni 18.5% Cr 0.7% Nb steel: Wire
AMS 5680 B	0.08% C 11% Ni 18% Cr Nb + Ta steel: Welding wire; AMS for SAE 30347
AMS 5681 A	0.08% C 10% Ni 19% Cr Nb + Ta steel: Coated electrode; AMS for SAE 30347
AMS 5685 C	0.12% C 11% Ni 18% Cr steel: Wire; AMS for SAE 30305
AMS 5686 A	0.12% C 11% Ni 18% Cr steel: Wire for rivetting; AMS for SAE 30305
AMS 5688	0.08% C 9% Ni 18% Cr steel: Wire
AMS 5688 D	0.1% C 8% Ni 18% Cr steel: Wire; spring temper; AMS for SAE 30302
AMS 5689	0.08% C 9.5% Ni 18% Cr Ti steel: Screen wire; AMS for SAE 30321
AMS 5690	0.05% C (max) 12.5% Ni 17% Cr 2.5% Mo steel: Wire
AMS 5690 E	0.08% C 12% Ni 17% Cr 2.5% Mo steel: Screen wire; AMS for SAE 30316
AMS 5691	0.05% C (max) 12.5% Ni 17% Cr 2.5% Mo steel: Wire

Note. The following abbreviations and units are used in the tables:

DPN	Hardness, diamond pyramid number
UTS	Ultimate tensile strength, N/mm^2
Elon	Elongation, %
Proof	0.1% proof strength, N/mm^2

1 N/mm^2=0.1 hbar=0.102 kgf/mm^2=0.06475 $tonf/in.^2$=145.04 $lbf/in.^2$=1 MPa

See Appendix II for other abbreviations and conversion tables.

Symbol	Nominal analysis, supplier, condition and remarks.
AMS 5691 B	0.08% C 13% Ni 18% Cr 2% Mo steel: Coated electrode; AMS for SAE 30316
AMS 5692	0.08% C (max) 12.5% Ni 19% Cr 2.5% Mo steel: Type 316
AMS 5693	0.15% C (max) 9% Ni 18% Cr steel: Type 302
AMS 5694	0.12% C 20.5% Ni 25% Cr steel: Wire
AMS 5694 B	0.25% C 20% Ni 25% Cr steel: Welding wire; cold drawn; AMS for SAE 30310
AMS 5695 A	0.25% C 20% Ni 25% Cr steel: Coated electrode; AMS for SAE 30310
AMS 5696	0.08% C (max) 12% Ni 17% Cr 2.5% Mo steel: Type 316
AMS 5697	0.08% C 9% Ni 19% Cr steel: Wire; AMS for SAE 30304
AMS 5700 B	14% Ni 14% Cr 0.4% Mo 2.5% W steel: Bar and forging
AMS 5705 A	8% Ni 12.8% Cr 2.5% Si steel: Bar and forging
AMS 5716	0.08% C 35.5% Ni 18.5% Cr steel
AMS 5722	0.3% C 9% Ni 18% Cr 1.4% Mo 1.3% W 0.25% Ti 0.4% Nb steel: Forgings
AMS 5723	0.3% C 9% Ni 18% Cr 1.6% Mo 1.3% W 0.5% Ti steel: Forgings
AMS 5725 A	25% Ni 16% Cr 6% Mo Fe alloy: Bar; up to 40 mm in diameter
AMS 5727	0.5% C 25% Ni 16% Cr 6% Mo steel: Forgings
AMS 5727 B	25% Ni 16% Cr 6% Mo Fe alloy: Forging
AMS 5728 B	25% Ni 16% Cr 6% Mo Fe alloy: Forging; special casting process
AMS 5733	0.08% C 26% Ni 13.5% Cr 2.75% Mo 1.7% Ti steel: Forgings
AMS 5733 C	26% Ni 13.5% Cr 3% Mo 1.8% Ti Fe alloy: Bar and forging
AMS 5738	0.15% C 9% Ni 18% Cr steel: Bar; free machining; AMS for SAE 30303 F
AMS 5741 C	26% Ni 13.5% Cr 1.7% Mo 3% Ti Fe alloy: Bar and forging; vacuum melted
AMS 5742 A	32% Ni 20% Cr 1% Ti Fe alloy: Bar and forging
AMS 5764	0.06% C (max) 5% Mn 12.5% Ni 22% Cr 2.2% Mo 0.2% V 0.2% Nb 0.3% N steel
AMS 5766	0.1% C (max) 32% Ni 21% Cr 0.3% Al Fe alloy
AMS 5784	9% Ni 29% Cr steel: Wire
AMS 5785 B	9% Ni 29% Cr steel: Coated electrode
AMS 5871	0.1% C (max) 32.5% Ni 21% Cr 0.4% Al Fe alloy
AMS 7210	0.15% C (max) 9% Ni 18% Cr steel: Type 302
AMS 7211	0.08% C (max) 10.5% Ni 18% Cr 0.4% Ti steel: Type 321
AMS 7228	0.08% C 10.2% Ni 19% Cr steel: Type 304
AMS 7229	0.08% C (max) 11% Ni 18% Cr 0.8% Nb steel: Type 347
AMS 7236	0.1% C 10% Ni 20% Cr 15% W steel
AMS 7241	0.15% C (max) 9% Ni 18% Cr steel: Type 302
AMS 7245	0.08% C (max) 9.2% Ni 19% Cr steel: Type 304
AMS 7472	0.15% C (max) 9% Ni 18% Cr steel: Type 302
AMS 7490	0.08% C (max) 11% Ni 18% Cr 0.8% Nb steel
AMS 7490	0.08% C (max) 13.5% Ni 23% Cr steel: Type 309
AMS 7490	0.08% C (max) 20.5% Ni 25% Cr steel: Type 310
AMS 7490	0.25% C (max) 20.5% Ni 24.5% Cr steel: Type 314
AMS 7490	0.08% C (max) 12% Ni 17% Cr 2.5% Mo steel: Type 316
AMS 7490	0.08% C (max) 10.5% Ni 18% Cr 0.4% Ti steel: Type 321
AMSL	0.03% C (max) 13% Ni 19% Cr 3.5% Mo steel: Valhuna
AMST	0.08% C (max) 13% Ni 19% Cr 3.5% Mo 0.4% Ti steel: Valhuna
AN QQ S 772/MCR	0.1% C 10% Ni 17% Cr 2.0% Mo steel: US Service **UTS: 760 Elon: 40% Proof: 200**
AN QQ W 423/1	0.2% C 14% Ni 12% Cr steel: US Service; annealed **UTS: 760**
ANKA	0.15% C (max) 9% Ni 18% Cr steel: Bar, billets, etc.; Brown Bayley for BS alloy En 58A

Symbol	Nominal analysis, supplier, condition and remarks.
ANKA E	0.08% C (max) 10% Ni 19% Cr steel: Bar, billet, etc.; Brown Bayley for BS alloy En 58E
ANKA M	0.16% C 12% Ni 12% Cr steel: Bar, billet, etc.; Brown Bayley for BS alloy En 58D
APF1	0.25% C (max) 20.5% Ni 25% Cr steel: Valhuna
APF1S	0.08% C (max) 20.5% Ni 25% Cr steel: Valhuna
APFR	0.2% C (max) 13.25% Ni 23% Cr steel: Valhuna
APFRS	0.08% C (max) 13.25% Ni 23% Cr steel: Valhuna
APM	0.08% C (max) 12% Ni 17% Cr 2.25% Mo steel: Valhuna
APM/Sp	0.06% C (max) 12% Ni 17% Cr 2.75% Mo steel: Valhuna
APMC	0.08% C (max) 12% Ni 18% Cr 2.25% Mo steel: Valhuna
APML	0.03% C (max) 12% Ni 17% Cr 2.25% Mo steel: Valhuna
APML/Sp	0.03% C (max) 12% Ni 17% Cr 2.75% Mo steel
APMT	0.08% C (max) 12% Ni 17% Cr 2.5% Mo 0.4% Ti steel: Valhuna
APMZ	0.08% C (max) 12% Ni 17% Cr 2.25% Mo 0.15% S steel: Valhuna
ARC/098	14% Ni 25% Cr 1.25% Mo steel: Imphy
ARC 2233	10% Ni 18% Cr 2.4% Mo steel: Imphy
ARC 2266	10% Ni 18% Cr 3.0% Mo steel: Imphy
ARMARC	See Philips Armarc
ARMCO 18/8	0.12% C 9% Ni 18% Cr steel: Armco
ARMCO 18/8 Si	0.11% C 9% Ni 18% Cr 2.5% Si steel: Armco
ARMCO 18/10 Cb Ta	0.08% C 10% Ni 18% Cr 1.7% Nb + Ta steel: Armco
ARMCO 18/10 Ti	0.08% C 10% Ni 18% Cr 0.5% Ti steel: Armco
ARMCO 18/12 Mo	0.1% C 12% Ni 17% Cr 2.0% Mo steel: Armco
ARMCO 20/10	0.08% C 11% Ni 20% Cr steel: Armco
ARMCO 25/12	0.2% C 13% Ni 23% Cr steel: Armco
ARMCO 25/20	0.25% C 21% Ni 25% Cr steel: Armco
ARMET	9% Ni 20% Cr 3% Mo steel: Electrode; Metrode **DPN: 320**
ARMEX 2	0.08% C 8.5% Ni 20% Cr 3.5% Mo steel: Welding electrode; Murex **UTS: 760 Elon: 35%**
ARMEX 3	0.08% C 8.5% Ni 19.5% Cr 2.5% Mo steel: Welding electrode; Murex **UTS: 680 Elon: 37%**
ARMEX GT	0.07% C 8% Ni 17% Cr 2.0% Mo steel: Welding electrode; Murex **UTS: 610 Elon: 35%**
ASTM A167/302	0.15% C 9% Ni 18% Cr steel: Sheet and plate **DPN: 202 UTS: 530 Elon: 40% Proof: 200**
ASTM A167/302 B	0.15% C 9% Ni 18% Cr 2.5% Si steel: Plate and sheet **DPN: 217 UTS: 530 Elon: 40% Proof: 200**
ASTM A167/304	0.08% C 10% Ni 19% Cr steel: Sheet and plate **DPN: 202 UTS: 530 Elon: 40% Proof: 200**
ASTM A167/304 L	0.03% C 10% Ni 19% Cr steel: Sheet and plate **DPN: 202 UTS: 480 Elon: 40% Proof: 170**
ASTM A167/305	0.12% C 11% Ni 18% Cr steel: Sheet and plate **DPN: 202 UTS: 500 Elon: 40% Proof: 170**
ASTM A167/308	0.08% C 11% Ni 20% Cr steel: Sheet and plate **DPN: 202 UTS: 530 Elon: 40% Proof: 200**
ASTM A167/309	0.2% C 13% Ni 23% Cr steel: Sheet and plate **DPN: 217 UTS: 530 Elon: 40% Proof: 200**
ASTM A167/309 S	0.08% C 13% Ni 23% Cr steel: Sheet and plate **DPN: 217 UTS: 530 Elon: 40% Proof: 200**
ASTM A167/310	0.25% C 21% Ni 25% Cr steel: Sheet and plate **DPN: 217 UTS: 530 Elon: 40% Proof: 200**
ASTM A167/310 S	0.08% C 21% Ni 25% Cr steel: Sheet and plate **DPN: 217 UTS: 530 Elon: 40% Proof: 200**
ASTM A167/316	0.08% C 12% Ni 17% Cr 2.5% Mo steel: Sheet and plate **DPN: 217 UTS: 530 Elon: 40% Proof: 200**
ASTM A167/316 L	0.03% C 12% Ni 17% Cr 2.5% Mo steel: Sheet and plate **DPN: 217 UTS: 480 Elon: 40% Proof: 170**
ASTM A167/317	0.08% C 13% Ni 19% Cr 3.5% Mo steel: Sheet and plate **DPN: 217 UTS: 530 Elon: 35% Proof: 200**
ASTM A167/317L	0.03% C 14% Ni 19% Cr 3.5% Mo steel: Sheet and plate **DPN: 217 UTS: 530 Elon: 35% Proof: 200**
ASTM A167/321	0.08% C 10% Ni 18% Cr 0.4% Ti steel: Sheet and plate **DPN: 202 UTS: 530 Elon: 40% Proof: 200**
ASTM A167/347	0.08% C 11% Ni 18% Cr 1% Nb steel: Sheet and plate **DPN: 202 UTS: 530 Elon: 40% Proof: 200**
ASTM A167/348	0.08% C 11% Ni 18% Cr 1% Nb steel: Sheet and plate **DPN: 202 UTS: 530 Elon: 40% Proof: 200**
ASTM A182 F10	0.14% C 20.5% Ni 8% Cr steel: For pipe fittings
ASTM A182 F304	0.08% C (max) 9.5% Ni 19% Cr steel: For pipe fittings
ASTM A182 F304H	0.07% C 9.5% Ni 19% Cr steel: For pipe fittings
ASTM A182 F304L	0.035% C (max) 9.5% Ni 19% Cr steel: For pipe fittings
ASTM A182 F310	0.15% C (max) 20.5% Ni 25% Cr steel: For pipe fittings
ASTM A182 F316	0.08% C (max) 12% Ni 17% Cr 2.5% Mo steel: For pipe fittings
ASTM A182 F316H	0.07% C 12% Ni 17% Cr 2.5% Mo steel: For pipe fittings
ASTM A182 F316L	0.03% C (max) 12.5% Ni 17% Cr 2.5% Mo steel: For pipe fittings
ASTM A182 F321	0.08% C (max) 9% Ni (min) 17% Cr (min) steel: For pipe fittings
ASTM A182 F321H	0.07% C 9% Ni (min) 17% Cr (min) steel: For pipe fittings
ASTM A182 F347	0.08% C (max) 11% Ni 18.5% Cr steel: For pipe fittings
ASTM A182 F347H	0.07% C 11% Ni 18.5% Cr steel: For pipe fittings
ASTM A182 F348	0.08% C (max) 11% Ni 18.5% Cr steel: For pipe fittings
ASTM A182 F348H	0.07% C 11% Ni 18.5% Cr steel: For pipe fittings
ASTM A193 B8	0.08% C 10% Ni 19% Cr steel: Bar **UTS: 530 Elon: 35% Proof: 200**
ASTM A193 B8C	0.08% C 11% Ni 18% Cr 1% Nb steel: Bar **UTS: 530 Elon: 35% Proof: 200**
ASTM A193 B8M	0.08% C 12% Ni 17% Cr 2.5% Mo steel: Bar **UTS: 530 Elon: 35% Proof: 200**
ASTM A193 B8T	0.08% C 10% Ni 18% Cr 0.5% Ti steel: Bar **UTS: 530 Elon: 35% Proof: 200**
ASTM A194/8	0.08% C (max) 10% Ni 19% Cr steel: For bolts
ASTM A194/8 C	0.08% C (max) 11% Ni 18% V 0.08% Nb Cr steel: For bolts
ASTM A194/8 M	0.08% C (max) 12% Ni 17% Cr 2.5% Mo steel: For bolts
ASTM A194/8 T	0.08% C (max) 11% Ni 18% Cr 0.4% Ti steel: for bolts
ASTM A240/302	0.15% C 9% Ni 18% Cr steel: For pressure vessels
ASTM A240/304	0.08% C 10% Ni 19% Cr steel: For pressure vessels
ASTM A240/304 L	0.03% C 10% Ni 19% Cr steel: For pressure vessels
ASTM A240/305	0.12% C 12% Ni 18% Cr steel: For pressure vessels
ASTM A240/309S	0.08% C 13% Ni 23% Cr steel: For pressure vessels
ASTM A240/310S	0.08% C 21% Ni 25% Cr steel: For pressure vessels
ASTM A240/316	0.08% C 12% Ni 17% Cr 2.5% Mo steel: For pressure vessels
ASTM A240/316 L	0.03% C 12% Ni 17% Cr 2.5% Mo steel: for pressure vessels
ASTM A240/317	0.08% C 13% Ni 19% Cr 3.5% Mo steel: For pressure vessels
ASTM A240/317 L	0.03% C 13% Ni 19% Cr 3.5% Mo steel: For pressure vessels
ASTM A240/321	0.08% C 11% Ni 18% Cr 0.5% Ti steel: For pressure vessels
ASTM A240/347	0.08% C 11% Ni 18% Cr 1.0% Nb + Ta steel: For pressure vessels
ASTM A240/348	0.08% C 11% Ni 18% Cr 1.0% Nb + Ta steel: For pressure vessels
ASTM A249 TP304	0.08% C (max) 9.5% Ni 19% Cr steel: For boiler tube

Symbol	Nominal analysis, supplier, condition and remarks.
ASTM A249 TP304H	0.07% C 9.5% Ni 19% Cr steel: For boiler tube
ASTM A249 TP304L	0.035% C (max) 10% Ni 19% Cr steel: For boiler tube
ASTM A249 TP309	0.15% C (max) 13.5% Ni 23% Cr steel: For boiler tube
ASTM A249 TP310	0.15% C (max) 21% Ni 25% Cr steel: For boiler tube
ASTM A249 TP316	0.08% C (max) 12.5% Ni 17% Cr 2.5% Mo steel: For boiler tube
ASTM A249 TP316H	0.07% C 12.5% Ni 17% Cr 2.5% Mo steel: For boiler tube
ASTM A249 TP316L	0.035% C (max) 12.5% Ni 17% Cr 2.5% Mo steel: For boiler tube
ASTM A249 TP317	0.08% C (max) 12.5% Ni 19% Cr 3.5% Mo steel: For boiler tube
ASTM A249 TP321	0.08% C (max) 11% Ni 18.5% Cr steel: For boiler tube
ASTM A249 TP321H	0.07% C 11% Ni 18.5% Cr steel: For boiler tube
ASTM A249 TP347	0.08% C (max) 11% Ni 18.5% Cr steel: For boiler tube
ASTM A249 TP347H	0.07% C (max) 11% Ni 18.5% Cr steel: For boiler tube
ASTM A249 TP348	0.08% C (max) 11% Ni 18.5% Cr 0.1% Ta steel: For boiler tube
ASTM A249 TP348H	0.07% C 11% Ni 18.5% Cr 0.1% Ta steel: For boiler tube
ASTM A269 TP304	0.08% C 9% Ni 19% Cr steel: Tube
ASTM A269 TP304L	0.03% C 10% Ni 19% Cr steel: Tube
ASTM A269 TP316	0.08% C 12% Ni 17% Cr steel: Tube
ASTM A269 TP316L	0.03% C 12% Ni 17% Cr 2.5% Mo steel: Tube
ASTM A269 TP317	0.08% C 12% Ni 19% Cr 3.5% Mo steel: Tube
ASTM A269 TP321	0.08% C 11% Ni 18% Cr 0.4% Ti steel: Tube
ASTM A269 TP347	0.08% C 12% Ni 18% Cr 1.0% Nb + Ta steel: Tube
ASTM A269 TP348	0.08% C 11% Ni 18% Cr 1.0% Nb + Ta steel: Tube
ASTM A271	Stainless steel: Still tubes; see AISI range for grades
ASTM A276/302	0.15% C 9% Ni 18% Cr steel: Bar
ASTM A276/302B	0.15% C 9% Ni 18% Cr 2.5% Si steel: Bar
ASTM A276/303	0.15% C 9% Ni 18% Cr 0.6% Mo steel: Bar
ASTM A276/303 Se	0.15% C 9% Ni 18% Cr 0.15% Se steel: Bar
ASTM A276/304	0.08% C 10% Ni 19% Cr steel: Bar
ASTM A276/304 L	0.03% C 10% Ni 19% Cr steel: Bar
ASTM A276/308	0.08% C 11% Ni 20% Cr steel: Bar
ASTM A276/309	0.2% C 13% Ni 23% Cr steel: Bar
ASTM A276/309 S	0.08% C 13% Ni 23% Cr steel: Bar
ASTM A276/310	0.25% C 21% Ni 25% Cr steel: Bar
ASTM A276/310 S	0.08% C 21% Ni 25% Cr steel: Bar
ASTM A276/314	0.25% C 21% Ni 25% Cr steel; Bar
ASTM A276/316	0.08% C 12% Ni 17% Cr 2.5% Mo steel: Bar
ASTM A276/316 L	0.03% C 12% Ni 17% Cr 2.5% Mo steel: Bar
ASTM A276/321	0.08% C 11% Ni 18% Cr 0.4% Ti steel: Bar
ASTM A276/347	0.08% C 11% Ni 18% Cr 1.0% Nb + Ta steel: Bar
ASTM A276/348	0.08% C 11% Ni 18% Cr 1.0% Nb + Ta steel: Bar
ASTM A289 C	0.05% C 25% Ni 5% Cr steel: For non-magnetic retaining rings **UTS: 1140 Elon: 12% Proof: 840**
ASTM A295 F8	0.08% C 10% Ni 19% Cr steel: Seamless drum forging; annealed **UTS: 500 Elon: 30% Proof: 200**
ASTM A295 F8C	0.08% C 10% Ni 18% Cr 1.0% Nb steel: Seamless drum forging; annealed **UTS: 500 Elon: 30% Proof: 200**
ASTM A295 F8M	0.08% C 12% Ni 17% Cr 2.5% Mo steel: Seamless drum forging; annealed **UTS: 500 Elon: 30% Proof: 200**
ASTM A295 F8T	0.08% C 9% Ni 17% Cr 0.4% Ti steel: Seamless drum forging; annealed **UTS: 500 Elon: 30% Proof: 200**
ASTM A295 F10	0.15% C 21% Ni 8% Cr steel: Seamless drum forging; annealed **UTS: 580 Elon: 25% Proof: 200**
ASTM A295 F25	0.15% C 21% Ni 25% Cr steel: Seamless drum forging; annealed **UTS: 530 Elon: 30% Proof: 200**
ASTM A296 CE30	0.3% C 9% Ni 29% Cr steel: Casting; annealed **UTS: 540 Elon: 10% Proof: 240**
ASTM A296 CF3	0.03% C 9% Ni 19% Cr steel: Casting; annealed **UTS: 440 Elon: 35% Proof: 180**
ASTM A296 CF3M	0.03% C 10% Ni 19% Cr 3.0% Mo steel: Casting; annealed **UTS: 460 Elon: 30% Proof: 200**
ASTM A296 CF8	0.08% C 10% Ni 20% Cr steel: Casting; annealed **UTS: 450 Elon: 35% Proof: 180**
ASTM A296 CF8C	0.08% C 10% Ni 19% Cr 0.8% Nb steel: Casting; annealed **UTS: 460 Elon: 30% Proof: 200**
ASTM A296 CF8M	0.08% C 10% Ni 19% Cr Mo steel: Casting; annealed **UTS: 460 Elon: 30% Proof: 200**
ASTM A296 CF16F	0.16% C 9% Ni 19% Cr 1.5% Mo 0.3% Se cast steel: Annealed **UTS: 460 Elon: 25% Proof: 200**
ASTM A296 CF20	0.2% C 9% Ni 19% Cr steel: Casting; annealed **UTS: 460 Elon: 30% Proof: 200**
ASTM A296 CG8M	0.08% C 11% Ni 19% Cr 4.0% Mo steel: Casting; annealed **UTS: 500 Elon: 25% Proof: 240**
ASTM A296 CG12	0.12% C 12% Ni 22% Cr steel: Casting; annealed **UTS: 460 Elon: 35% Proof: 180**
ASTM A296 CH20	0.2% C 12% Ni 25% Cr steel: Casting; annealed **UTS: 460 Elon: 30% Proof: 200**
ASTM A296 CK20	0.2% C 20% Ni 25% Cr steel: Casting; annealed **UTS: 440 Elon: 30% Proof: 180**
ASTM A296 CN7M	0.07% C (max) 29.5% Ni 20.5% Cr 2.5% Mo 3.5% Cu steel: Casting
ASTM A297 HE	0.2% C 9% Ni 29% Cr steel: Casting; as cast **UTS: 580 Elon: 9% Proof: 270**
ASTM A297 HF	0.20% C 9% Ni 19% Cr steel: Casting; as cast **UTS: 460 Elon: 25% Proof: 240**
ASTM A297 HH	0.2% C 12% Ni 25% Cr steel: Casting; as cast **UTS: 500 Elon: 10% Proof: 240**
ASTM A297 HI	0.2% C 15% Ni 28% Cr steel: Casting; as cast **UTS: 460 Elon: 10% Proof: 240**
ASTM A297 HK	0.2% C 20% Ni 25% Cr steel: Casting; as cast **UTS: 440 Elon: 10% Proof: 240**
ASTM A297 HL	0.2% C 20% Ni 29% Cr steel: Casting; as cast **UTS: 440 Elon: 10% Proof: 240**
ASTM A297 HN	0.2% C 25% Ni 20% Cr steel: Casting; as cast **UTS: 420 Elon: 8%**
ASTM A297 HT	0.35% C 35% Ni 15% Cr steel: Casting; as cast **UTS: 440 Elon: 4%**
ASTM A297 HU	0.55% C 40% Ni 18% Cr Fe alloy: Casting
ASTM A297 HV	0.35% C 39% Ni 19% Cr steel: Casting; as cast **UTS: 440 Elon: 4%**
ASTM A298 E308	0.08% C 10% Ni 19% Cr steel: Welding rod; as welded **UTS: 540 Elon: 35%**
ASTM A298 E308ELC	0.04% C 10% Ni 19% Cr steel: Welding rod; as welded **UTS: 500 Elon: 35%**
ASTM A298 E309	0.15% C 13% Ni 24% Cr steel: Welding rod; as welded **UTS: 540 Elon: 35%**
ASTM A298 E309 Cb	0.12% C 13% Ni 24% Cr 0.8% Nb steel: Welding rod; as welded **UTS: 540 Elon: 30%**
ASTM A298 E309 Mo	0.12% C 13% Ni 24% Cr 2.0% Mo steel: Welding rod; as welded **UTS: 540 Elon: 35%**

Note. The following abbreviations and units are used in the tables:

DPN	Hardness, diamond pyramid number
UTS	Ultimate tensile strength, N/mm^2
Elon	Elongation, %
Proof	0.1% proof strength, N/mm^2

$1\ N/mm^2 = 0.1\ hbar = 0.102\ kgf/mm^2 = 0.06475\ tonf/in.^2 = 145.04\ lbf/in.^2 = 1\ MPa$

See Appendix II for other abbreviations and conversion tables.

Symbol	Nominal analysis, supplier, condition and remarks.
ASTM A298 E310	0.2% C 21% Ni 27% Cr steel: Welding rod; as welded **UTS: 540 Elon: 30%**
ASTM A298 E310 Cb	0.12% C 21% Ni 27% Cr 0.8% Nb steel: Welding rod; as welded **UTS: 540 Elon: 25%**
ASTM A298 E310 Mo	0.12% C 21% Ni 27% Cr 2.0% Mo steel: Welding rod; as welded **UTS: 540 Elon: 30%**
ASTM A298 E312	0.15% C 9% Ni 29% Cr steel: Welding rod; as welded **UTS: 580 Elon: 22%**
ASTM A298 E316	0.08% C 13% Ni 19% Cr 2.0% Mo steel: Welding rod; as welded **UTS: 540 Elon: 30%**
ASTM A298 E316ELC	0.04% C 13% Ni 19% Cr 2.0% Mo steel: Welding rod; as welded **UTS: 500 Elon: 30%**
ASTM A298 E317	0.08% C 13% Ni 20% Cr 3.0% Mo steel: Welding rod; as welded **UTS: 540 Elon: 30%**
ASTM A298 E318	0.08% C 13% Ni 19% Cr 2% Mo 1% Nb steel: Welding rod; as welded **UTS: 540 Elon: 25%**
ASTM A298 E330	0.25% C 33% Ni 16% Cr Fe alloy: Welding rod; as welded **UTS: 500 Elon: 30%**
ASTM A298 E347	0.08% C 10% Ni 20% Cr 1% Nb steel: Welding rod; as welded **UTS: 540 Elon: 30%**
ASTM A312	Austenitic steel pipe; graded as ASTM A249
ASTM A313	0.15% C 8.5% Ni 19% Cr steel: Spring wire; cold drawn **UTS: 2280**
ASTM A320 B8	0.08% C 10% Ni 19% Cr steel: Annealed **UTS: 500 Elon: 35% Proof: 200**
ASTM A320 B8C	0.08% C 11% Ni 18% Cr 0.8% Nb steel: Annealed **UTS: 500 Elon: 35% Proof: 200**
ASTM A320 B8D	0.08% C 11% Ni 18% Cr 0.8% Nb 0.1% Ta steel: Annealed **UTS: 500 Elon: 35% Proof: 200**
ASTM A320 B8F	0.15% C 9% Ni 18% Cr 0.6% Mo or 0.15% Se steel: Annealed **UTS: 500 Elon: 35% Proof: 200**
ASTM A320 B8T	0.08% C 10% Ni 18% Cr 0.4% Ti steel: Annealed **UTS: 500 Elon: 35% Proof: 200**
ASTM A336 F8	0.08% C 9.5% Ni 19% Cr steel: Seamless drum forging **UTS: 460 Elon: 30% Proof: 200**
ASTM A336 F8C	0.08% C 10.5% Ni 18% Cr steel: Seamless drum forging **UTS: 460 Elon: 30% Proof: 200**
ASTM A336 F8M	0.08% C 12% Ni 17% Cr 2.5% Mo steel: Seamless drum forging **UTS: 460 Elon: 30% Proof: 200**
ASTM A336 F8T	0.08% C (max) 9.0% Ni 17.0% Cr steel: Seamless drum forging **UTS: 460 Elon: 30% Proof: 200**
ASTM A336 F10	0.15% C 20% Ni 8% Cr steel: Seamless drum forging **UTS: 540 Elon: 25% Proof: 200**
ASTM A336 F25	0.15% C (max) 21% Ni 25% Cr steel: Seamless drum forging **UTS: 500 Elon: 30% Proof: 200**
ASTM A351 CF3	0.03% C 10% Ni 19.0% Cr steel: Annealed **UTS: 460 Elon: 35% Proof: 200**
ASTM A351 CF3M	0.03% C 11.0% Ni 19.0% Cr 2.0% Mo steel: Annealed **UTS: 460 Elon: 30% Proof: 200**
ASTM A351 CF8	0.08% C 9.0% Ni 19.0% Cr steel: Annealed **UTS: 460 Elon: 35% Proof: 200**
ASTM A351 CF8C	0.08% C 10.0% Ni 19.0% Cr 1.0% Nb steel: Annealed **UTS: 460 Elon: 30% Proof: 200**
ASTM A351 CF8M	0.08% C 10.0% Ni 19.0% Cr 2.0% Mo steel: Annealed **UTS: 460 Elon: 30% Proof: 200**
ASTM A351 CF10MC	0.1% C 14.0% Ni 16.0% Cr 2.0% Mo steel: Annealed **UTS: 460 Elon: 20% Proof: 200**
ASTM A351 CH8	0.08% C 13.0% Ni 24.0% Cr steel: Annealed **UTS: 440 Elon: 30% Proof: 180**
ASTM A351 CH10	0.10% C 13.0% Ni 24.0% Cr steel: Annealed **UTS: 460 Elon: 30% Proof: 200**
ASTM A351 CH20	0.20% C 13.0% Ni 24.0% Cr steel: Annealed **UTS: 460 Elon: 30% Proof: 200**
ASTM A351 CK45	0.35% C 20% Ni 24.0% Cr steel: Annealed **UTS: 440 Elon: 10% Proof: 220**
ASTM A351 CT35	0.35% C 34.0% Ni 14.0% Cr 0.5% Mo steel: Annealed **UTS: 440 Elon: 15% Proof: 180**
ASTM A362	Heat resistant tubular castings; analysis as ASTM A296 and A297
ASTM A371 ER308	0.08% C 10% Ni 20.25% Cr steel: Corrosion resistant for welding rods and electrodes
ASTM A371 ER308L	0.03% C 10% Ni 20.25% Cr steel: Corrosion resistant for welding rods and electrodes
ASTM A371 ER309	0.12% C 13.0% Ni 24% Cr steel: Corrosion resistant for welding rods and electrodes
ASTM A371 ER310	0.09% C 21.25% Ni 26.5% Cr steel: Corrosion resistant for welding rods and electrodes
ASTM A371 ER312	0.09% C 9.25% Ni 30.0% Cr steel: Corrosion resistant for welding rods and electrodes
ASTM A371 ER316	0.08% C 12.5% Ni 19.0% Cr 2.5% Mo steel: Corrosion resistant for welding rods and electrodes
ASTM A371-ER316L	0.03% C 12.5% Ni 19.0% Cr 2.5% Mo steel: Corrosion resistant for welding rods and electrodes
ASTM A371-ER317	0.08% C 14.0% Ni 19.5% Cr 3.5% Mo steel: Corrosion resistant for welding rods and electrodes
ASTM A371-ER318	0.08% C 12.5% Ni 19.0% Cr 2.5% Mo 0.8% Ta + Nb steel: Corrosion resistant for welding rods and electrodes
ASTM A376	Stainless steel: Pipe for high temperature service; see AISI range for grades
ASTM A403	Stainless steel: For welders fittings; see AISI range for grades
ASTM A409	Stainless steel: For pipes; see AISI for grades
ASTM A430	Stainless steel: Forged or bored pipe; see AISI range for grades
ASTM A451 CPF8	0.08% C (max) 9.5% Ni 19.5% Cr cast steel: Pipe
ASTM A451 CPF8M	0.08% C (max) 9.5% Ni 19.5% Cr 2.5% Mo cast steel: Pipe
ASTM A451 CPF10MC	0.1% C (max) 14.5% Ni 16.5% Cr 2.0% Mo 1.0% Nb cast steel: Pipe
ASTM A451 CPF87C	0.08% C (max) 10.5% Ni 19.5% Cr 0.9% Nb cast steel: Pipe
ASTM A451 CPH8	0.08% C 13.5% Ni 24% Cr cast steel: Pipe
ASTM A451 CPH10	0.2% C (max) 13.5% Ni 24% Cr cast steel: Pipe; alternative name for CPH 20
ASTM A451 CPH20	0.2% C (max) 13.5% Ni 24% Cr cast steel: Pipe; alternative name for CPH 10
ASTM A451 CPK20	0.2% C (max) 20.5% Ni 25% Cr cast steel: Pipe
ASTM A452	Austenitic steel: Cast pipe; subsequently cold worked and annealed; three grades as AISI standards
ASTM A473	Stainless steel: For forgings; graded by AISI system; includes steel in Section 44M1
ASTM A478	Stainless steel: Wire for wearing; graded by AISI system
ASTM A479	Stainless steel: Bar and shapes for boilers, etc.; graded by AISI system
ASTM A492	Stainless steel: For wire rope; graded by AISI system
ASTM A493	Stainless steel: Wire for cold heading; graded by AISI system
ASTM A511	Austenitic seamless steel: Mechanical tubing graded according to AISI system

Symbol	Nominal analysis, supplier, condition and remarks.
ASTM A554	Stainless steel: Welded tubing; AISI symbols used for different grades
ASTM A567/3	0.2% C (max) 1.5% Mn 20% Ni 21.2% Cr 3.0% Mo 2.5% W 19.7% Co 1.0% Nb + Ta iron: Casting as AISI 661
ASTM A567 HK40	0.4% C 20% Ni 26% Cr 0.1% Ni Fe alloy: Casting
ASTM A567 HK50	0.5% C 20% Ni 26% Cr 0.1% Ni Fe alloy: Casting
ASTM A580	Stainless steel: Wire classified by AISI system; see also Sections 44M1 and 44M2
ASTM A581/303	0.15% C (max) 9% Ni 18% Cr 0.15% S (min) steel: Wire, stainless; free machining
ASTM A581/303 Se	0.15% C (max) 9% Ni 18% Cr 0.15% Se (min) steel: Wire; stainless; free machining
ASTM A581 XM2	0.15% C (max) 9% Ni 18% Cr 0.5% Mo 0.13% S 0.8% Al steel: Wire; free machining
ASTM A581 XM3	0.15% C (max) 9% Ni 18% Cr 0.6% Mo 0.2% S 0.2% Pb steel: Wire; free machining
ASTM A582	Stainless steel: Wire, graded in same manner as ASTM A581; see also Sections 44M1 and 44P
ASTM A608 H135	0.35% C 16.0% Ni 28.0% Cr 0.5% Mo (max) steel: Tube; centrifugally cast
ASTM A608 HC30	28.0% Cr 0.7% Mn 4% Ni 0.5% Mo (max) steel: Tube; centrifugally cast; properties at 760 °C **UTS: 365** **Elon: 40%**
ASTM A608 HD50	0.5% C 5.5% Ni 28.0% Cr 0.5% Mo (max) steel: Tube; centrifugally cast; properties at 760 °C **UTS: 514**
ASTM A608 HE35	0.3% C 11.0% Ni 21.0% Cr 0.5% Mo (max) steel: Tube; centrifugally cast
ASTM A608 HF30	0.3% C 10.5% Ni 21.0% Cr 0.5% Mo (max) steel: Tube; centrifugally cast; properties at 760 °C **UTS: 179** **Elon: 7%**
ASTM A608 HH30	0.3% C 12.5% Ni 26.0% Cr 0.5% Mo (max) steel: Tube; centrifugally cast
ASTM A608 HH33	0.32% C 13% Ni 25% Cr 0.5% Mo (max) steel: Tube; centrifugally cast
ASTM A608 HK30	0.3% C 20.5% Ni 25.0% Cr 0.5% Mo (max) steel: Tube; centrifugally cast; properties at 760 °C **UTS: 179**
ASTM A608 HK40	0.4% C 20.5% Ni 25.0% Cr 0.5% Mo (max) steel: Tube; centrifugally cast; properties at 760 °C **UTS: 200** **Elon: 7%**
ASTM A608 HL30	0.3% C 20.0% Ni 30.0% Cr 0.5% Mo (max) steel: Tube; centrifugally cast
ASTM A608 HL40	0.4% C 20.0% Ni 30.0% Cr 0.5% Mo (max) steel: Tube; centrifugally cast
ASTM A608 HN40	0.4% C 25.0% Ni 21.0% Cr 0.5% Mo (max) steel: Tube; centrifugally cast
ASTM A608 HT50	0.5% C 35% Ni 17.0% Cr 0.5% Mo (max) steel: Tube; centrifugally cast
ASTM A608 HU50	0.5% C 39% Ni 19.0% Cr 0.5% Mo (max) steel: Tube; centrifugally cast
ASTM A632 TP304	0.08% C (max) 9.5% Ni 19.0% Cr steel: Tube
ASTM A632 TP304L	0.04% C (max) 10.5% Ni 19.0% Cr steel: Tube
ASTM A632 TP310	0.15% C 20.5% Ni 25.0% Cr steel: Tube
ASTM A632 TP316	0.08% C 12.5% Ni 17.0% Cr 2.5% Mo steel: Tube
ASTM A632 TP316L	0.04% C 15.5% Ni 17.0% Cr 2.5% Mo steel: Tube
ASTM A632 TP317	0.08% C 12.5% Ni 19.0% Cr 3.5% Mo steel: Tube
ASTM A632 TP321	0.08% C 11.0% Ni 18.5% Cr steel: Tube
ASTM A632 TP347	0.08% C 11.0% Ni 18.5% Cr steel: Tube
ASTM A632 TP348	0.08% C 11.0% Ni 18.5% Cr steel: Tube
ASTM A651 TP304	0.08% C 9.5% Ni 19.0% Cr steel: Tube for water
ASTM A651 TP316	0.08% C 12.5% Ni 17.0% Cr 2.5% Mo steel: Tube for water
ASTM A666	Austenitic stainless steel: For structures; graded by AISI system
ASTM A666	Stainless steel: Sheet, plate, bar, etc.; for structural purposes; graded by AISI system (see also Section 44P)
ASTM A688 TP304	0.08% C 9.5% Ni 19.0% Cr steel: Welded tube
ASTM A688 TP304L	0.035% C 10.5% Ni 19.0% Cr steel: Welded tube
ASTM A688 TP316	0.08% C 12.5% Ni 17.0% Cr 2.5% Mo steel: Welded tube
ASTM A688 TP316L	0.035% C 12.5% Ni 17.0% Cr 2.5% Mo steel: Welded tube
ASTM A733 A312	Welded or seamless pipe nipples
ASTM A744	General specification for stainless steel: Casting; see US designation
ASTM A757 B2N B2Q	0.25% C 2.5% Ni steel: Casting; for low temperature use
ASTM A757 B3N B3Q	0.15% C 3.5% Ni steel: Casting; for low temperature use
ASTM A757 B4N B4Q	0.15% C 4.5% Ni steel: Casting; for low temperature use
ASTM A771	Type 316 stainless steel: Tube; for nuclear reactor tube
ASTM A774	Austenitic stainless steel fittings: Type 304, 316, 317, 321 and 347
ASTM A778	As welded austenitic stainless steel tube: Type 304, 316, 317, 321 and 347
ASTM A793	0.08% C (max) 9% Ni 19% Cr steel: For floor plates: Type 304 steel
ASTM A814	General specification for cold worked austenitic steel pipe: Steel type shown as suffix
ASTM B639	0.1% C 20% Ni 21% Cr 3% Mo 2.5% W 20% Co 1% Nb + Ta Fe alloy
ASTRANIT N/Z	0.08% C 9.0% Ni 18% Cr steel: Krupp
ASTRANIT SST/Z	0.08% C 18% Cr 10% steel: Krupp
ASV	6% Cr 42% Ni Fe alloy: Low expansion; Imphy
AU 18/8	0.15% C (max) 9% Ni 18% Cr steel: Valhuna
AU 188/Z	0.15% C (max) 9% Ni 18% Cr 0.15% steel: Valhuna
AURIGA XXIVB	0.12% C 9% Ni 18% Cr 1.0% Nb steel: Casting; D Brown; annealed **UTS: 450** **Elon: 20%** **Proof: 200**
AURIGA XXIVB FCG	0.12% C 9% Ni 18% Cr 0.3% Mo 1% Nb steel: Casting; D Brown; annealed **UTS: 450** **Elon: 15%** **Proof: 200**
AURIGA XXIVC	0.12% C 10% Ni 17% Cr 2.5% Mo 1% Nb steel: Casting; D Brown; annealed **UTS: 450** **Elon: 15%** **Proof: 200**
AURIGA XXVII	0.3% C 13% Ni 24% Cr 3% W steel: Casting; D Brown; annealed **UTS: 530** **Elon: 10%** **Proof: 220**
AURIGA XXVIIA	0.3% C 12% Ni 25% Cr steel: Casting; D Brown; annealed **UTS: 530** **Elon: 10%** **Proof: 220**
AURIGA XXXIII	0.5% C 27% Ni 18% Cr Fe alloy: Casting; D Brown **UTS: 420**
AUSTINOX	0.06% C 10% Ni 18% Cr steel: Pompey; annealed **UTS: 540** **Elon: 48%** **Proof: 200**
AUSTINOX 3	0.02% C 10% Ni 18% Cr steel: Pompey **UTS: 460** **Elon: 45%** **Proof: 180**
AUSTINOX B	0.06% C 12% Ni 18% Cr steel: Pompey **UTS: 580** **Elon: 45%** **Proof: 210**
AUSTINOX BB	0.08% C 12% Ni 18.0% Cr 2.7% Mo steel: Pompey
AUSTINOX S	0.06% C 10% Ni 18% Cr 0.3% Ti steel: Pompey **UTS: 600** **Elon: 45%** **Proof: 210**
AUSTINOX SB	0.06% C 12% Ni 18% Cr 2.2% Mo 0.3% Ti steel: Pompey **UTS: 570** **Elon: 40%** **Proof: 210**
AUSTINOX SN	0.06% C 10% Ni 18% Cr 0.6% Nb steel: Pompey

Note. The following abbreviations and units are used in the tables:

DPN	Hardness, diamond pyramid number
UTS	Ultimate tensile strength, N/mm^2
Elon	Elongation, %
Proof	0.1% proof strength, N/mm^2

1 N/mm^2=0.1 hbar=0.102 kgf/mm^2=0.06475 tonf/in.^{2}=145.04 lbf/in.^{2}=1 MPa

See Appendix II for other abbreviations and conversion tables.

Symbol	Nominal analysis, supplier, condition and remarks.
AUTAAS 101	0.04% C (max) 9.5% Ni 19.2% Cr steel: Origin unknown
AUTOROD 16.10	0.025% C 9.0% Ni 20.0% Cr steel: Wire for welding; Esab; as welded **UTS: 580 Elon: 45% Proof: 390**
AUTOROD 16.12	0.025% C 10.0% Ni 20% Cr steel: For welding wire; Esab; as welded **UTS: 600 Elon: 45% Proof: 380**
AUTOROD 16.30	0.025% C 12.0% Ni 20% Cr 2.8% Mo steel: Welding wire; Esab; as welded **UTS: 620 Elon: 35% Proof: 340**
AUTOROD 16.32	0.025% C 12.0% Ni 20.0% Cr 2.8% Mo steel: Welding wire; Esab; as welded **UTS: 600 Elon: 40% Proof: 380**
AVW	0.45% C 9% Ni 19% Cr 1% W 2.5% Si steel: Valhuna
B and W A1	0.15% C 20% Ni 25% Cr steel: Tube; Babcock and Wilcox (USA)
B and W A1/57	0.08% C 12% Ni 18% Cr 2.5% Si steel: Babcock and Wilcox (USA)
B 3312a	0.05% C 10% Ni 17% Cr 1.6% Mo steel: French National Standard **DPN: 180 UTS: 480 Elon: 45% Proof: 180**
BB 0K	0.12% C 10% Ni 19% Cr 2% Mo steel: Bar, billets, etc.; Brown Bayley for BS alloy En 58H
BB 2K	0.08% C (max) 12% Ni 17% Cr 2.5% Mo steel: Bar, billets, etc.; Brown Bayley for BS alloy En 58H and En 58J
BB 4K	0.08% C (max) 14% Ni 19% Cr 3.5% Mo steel: Bar, billets, etc.; Brown Bayley for BS alloy En 58J
BBMK	18% Cr 18% Ni 2% Mo 2% Cu steel: Bar, billets, etc.; Brown Bayley
BLAW KNOX Mo-RE1	0.5% C (max) 30% Ni 22% Cr 1.3% W Fe alloy: Casting; for furnace fittings, etc.; Wellman
BRIGHTRAY F	0.15% C 37% Ni 18% Cr 2% Si Fe alloy: Wrought; H Wiggin; annealed; for use up to 1000 °C when carburizing **DPN: 186 UTS: 720**
BS	0.12% C 8% Ni 18% Cr steel: Osborn **UTS: 700 Elon: 50% Proof: 270**
BS 2	0.08% C 13% Ni 14% Cr steel: Osborn; for deep drawing **UTS: 610 Elon: 60% Proof: 220**
BS 970	See designation for details
BS 980 CDS 19	0.16% C (max) 8% Ni 18% Cr steel: Tube; cold drawn and annealed **UTS: 530 Proof: 220**
BS 980 CDS 20	0.16% C (max) 8% Ni 18% Cr Ti steel: Tube; cold drawn and annealed; weldable **UTS: 530 Proof: 220**
BS 1449/304 S12	0.03% C 10.5% Ni 18% Cr steel: Plate and sheet; weldable **DPN: 192 UTS: 500 Elon: 40% Proof: 200**
BS 1449/304 S15	0.06% C 9% Ni 18% Cr steel: Plate and sheet; annealed **DPN: 192 UTS: 520 Elon: 40% Proof: 210**
BS 1449/309 S24	0.15% C (max) 14.5% Ni 23.5% Cr steel: Plate and sheet; annealed **DPN: 207 UTS: 550 Elon: 40% Proof: 220**
BS 1449/310 S24	0.15% C (max) 20.5% Ni 24.5% Cr steel: Plate and sheet; annealed **DPN: 207 UTS: 550 Elon: 40% Proof: 220**
BS 1449/312 S24	0.15% C (max) 17.5% Ni 24.5% Cr steel: Plate and sheet; annealed **DPN: 207 UTS: 550 Elon: 40% Proof: 220**
BS 1449/316 S12	0.03% C (max) 12.5% Ni 17.5% Cr 2.7% Mo steel: Plate and sheet; annealed **DPN: 197 UTS: 500 Elon: 40% Proof: 200**
BS 1449/316 S16	0.07% C 11.5% Ni 17.5% Cr 2.7% Mo steel: Annealed **DPN: 207 UTS: 550 Elon: 40% Proof: 210**
BS 1449/317 S16	0.06% C (max) 13.5% Ni 18.5% Cr 3.5% Mo steel: Plate and sheet; annealed **DPN: 207 UTS: 550 Elon: 35% Proof: 210**
BS 1449/320 S17	0.08% C 12.5% Ni 17.5% Cr 2.7% Mo 0.5% Ti steel: Plate and sheet; annealed **DPN: 207 UTS: 550 Elon: 40% Proof: 210**
BS 1449/321 S12	0.08% C 10.5% Ni 18% Cr 0.6% Ti steel: Plate and sheet; weldable; annealed **DPN: 202 UTS: 520 Elon: 40% Proof: 210**
BS 1449/347 S17	0.08% C (max) 10.5% Ni 18% Cr 0.9% Nb steel: Plate and sheet; weldable; annealed **DPN: 202 UTS: 520 Elon: 40% Proof: 210**
BS 1453 A8 Nb	0.1% C (max) 9% Ni 18% Cr 1% Nb steel: Weld filler rod; for welding austenitic steel
BS 1453 A10	0.15% C (max) 12% Ni 25% Cr steel: Weld filler rod; for welding heat resistant stainless steel
BS 1453 A10 Nb	0.15% C (max) 12% Ni 25% Cr 1.2% Nb steel: Weld filler rod; for welding heat resistant stainless steel
BS 1453 A11	0.15% C (max) 21% Ni 25% Cr 1.5% Si steel: Weld filler rod; for welding heat resistant stainless steels
BS 1453 A11 Nb	0.15% C (max) 21% Ni 25% Cr 1.2% Nb 1.5% Si steel: Weld filler rod; for welding heat resistant stainless steels
BS 1453 A12	0.08% C (max) 11% Ni 17% Cr 3% Mo steel: Weld filler rod; for welding Mo bearing stainless steels
BS 1453 A12 Nb	0.08% C (max) 11% Ni 17% Cr 3% Mo 1.2% Nb steel: Weld filler rod; for welding Mo bearing stainless steels
BS 1501/801 B	0.08% C 10% Ni 19% Cr steel: Plate and bar; annealed **UTS: 530 Elon: 25% Proof: 210**
BS 1501/801 C	0.03% C 10.0% Ni 20% Cr steel: Plate and bar; annealed **UTS: 500 Elon: 25% Proof: 200**
BS 1501/821 Nb	0.08% C 9.0% Ni 20% Cr 0.8% Nb steel: Plate and bar; annealed **UTS: 530 Elon: 25% Proof: 200**
BS 1501/821 Ti	0.12% C 10% Ni 20% Cr 0.5% Ti steel: Plate and bar; annealed **UTS: 530 Elon: 25% Proof: 210**
BS 1501/845 B	0.08% C 10% Ni 18% Cr 2.5% Mo steel: Plate and bar; annealed **UTS: 530 Elon: 25% Proof: 210**
BS 1501/845 Ti	0.08% C 10% Ni 18% Cr 2.5% Mo 0.35% Ti steel: Plate and bar; annealed **UTS: 530 Elon: 25% Proof: 210**
BS 1501/846	0.08% C 12% Ni 20% Cr 3.5% Mo steel: Plate and bar; annealed **UTS: 530 Elon: 25% Proof: 210**
BS 1503/801	0.08% C 10.0% Ni 18.5% Cr steel: Forging; annealed **UTS: 530 Elon: 30% Proof: 210**
BS 1503/821 Nb	0.08% C 9.0% Ni 18% Cr 1.0% Nb steel: Forging; annealed **UTS: 530 Elon: 30% Proof: 210**
BS 1503/821 Ti	0.12% C 8% Ni 18% Cr 0.5% Ti steel: Forging; annealed **UTS: 530 Elon: 30% Proof: 210**
BS 1503/845 B	0.08% C 10% Ni 17.5% Cr 2.5% Mo steel: Forging; annealed **UTS: 530 Elon: 30% Proof: 200**
BS 1503/845 Ti	0.08% C 10% Ni 17.5% Cr 2.5% Mo 0.5% Ti steel: Annealed **UTS: 530 Elon: 30% Proof: 200**
BS 1503/846	0.08% C 13% Ni 19% Cr 3.5% Mo steel: Forging; annealed **UTS: 530 Elon: 30% Proof: 200**
BS 1504/801	0.12% C (max) 8% Ni 18% Cr steel: Casting; annealed **UTS: 450 Elon: 20% Proof: 200**
BS 1504/821 Nb	0.12% C (max) 9% Ni 18% Cr 1.0% Nb steel: Casting; annealed **UTS: 450 Elon: 20% Proof: 200**

Symbol	Nominal analysis, supplier, condition and remarks.
BS 1504/821 Ti	0.12% C (max) 8% Ni 18% Cr 0.5% Ti steel: Casting; annealed **UTS: 450 Elon: 20% Proof: 200**
BS 1504/845 B	0.08% C (max) 11% Ni 17% Cr 2.5% Mo steel: Casting; annealed **UTS: 450 Elon: 15% Proof: 200**
BS 1504/845 Nb	0.12% C (max) 11% Ni 17% Cr 2.5% Mo steel: Casting; annealed **UTS: 450 Elon: 15% Proof: 200**
BS 1504/846	0.08% C 12% Ni 19% Cr 3.5% Mo steel: Casting; annealed **UTS: 450 Elon: 15% Proof: 200**
BS 1506/801 A	0.16% C (max) 9% Ni 18% Cr steel: Bar; annealed **UTS: 530 Elon: 30% Proof: 200**
BS 1506/801 AM	0.16% C (max) 9% Ni 18% Cr 0.3% S steel: Bar; free machining; annealed **UTS: 530 Elon: 30% Proof: 200**
BS 1506/801 B	0.08% C (max) 9% Ni 18% Cr steel: Bar; annealed **UTS: 530 Elon: 30% Proof: 200**
BS 1506/801 C	0.03% C 11% Ni 18% Cr steel: Bar; annealed **UTS: 530 Elon: 30% Proof: 200**
BS 1506/821 Nb	0.08% C (max) 9% Ni 18% Cr 1.0% Nb steel: Bar; annealed **UTS: 530 Elon: 30% Proof: 200**
BS 1506/821 Ti	0.12% C 8% Ni 18% Cr 0.6% Ti steel: Bar; annealed **UTS: 530 Elon: 30% Proof: 200**
BS 1506/821 Ti M	0.12% C (max) 8% Ni 18% Cr 0.5% Ti 0.5% Zr steel: Bar; free machining; annealed **UTS: 530 Elon: 30% Proof: 200**
BS 1506/845	0.08% C (max) 10% Ni 17% Cr 2.7% Mo steel: Bar; annealed **UTS: 530 Elon: 30% Proof: 200**
BS 1507/821	0.15% C 2% Mn 10% Ni 18% Cr Ti or Nb stabilized seamless steel: Pipe; annealed **UTS: 530 Elon: 30%**
BS 1507/825	0.15% C 21% Ni 24% Cr 1% Ti seamless steel: Pipe; annealed **UTS: 530 Elon: 30%**
BS 1507/845	0.08% C 10% Ni 17% Cr 3% Mo seamless steel: Pipe; annealed; weldable **UTS: 530 Elon: 30%**
BS 1508/801	0.16% C 10% Ni 18% Cr steel: Tube; annealed **UTS: 530 Elon: 30%**
BS 1508/821	0.15% C 10% Ni 18% Cr 0.7% Ti steel: Tube; annealed **UTS: 530 Elon: 30%**
BS 1508/825	0.15% C 22% Ni 24% Cr 1.0% Ti steel: Tube; annealed **UTS: 530 Elon: 30%**
BS 1508/845	0.08% C 10% Ni 17% Cr 3.0% Mo steel: Tube; annealed **UTS: 530 Elon: 30%**
BS 1554 En 58A	0.16% C 9% Ni 18% Cr steel: Wire
BS 1554 En 58B	0.15% C 2% Mn 8% Ni 18% Cr steel: Wire
BS 1554 En 58C	0.15% C 11% Ni 18% Cr steel: Wire
BS 1554 En 58D	0.16% C 13% Ni 13% Cr steel: Wire
BS 1554 En 58E	0.08% C 10% Ni 18% Cr steel: Wire
BS 1554 En 58F	0.15% C 8% Ni 18% Cr steel: Wire

Note. The following abbreviations and units are used in the tables:

DPN	Hardness, diamond pyramid number
UTS	Ultimate tensile strength, N/mm^2
Elon	Elongation, %
Proof	0.1% proof strength, N/mm^2

1 N/mm^2=0.1 hbar=0.102 kgf/mm^2=0.06475 $tonf/in.^2$=145.04 $lbf/in.^2$=1 MPa

See Appendix II for other abbreviations and conversion tables.

Symbol	Nominal analysis, supplier, condition and remarks.
BS 1554 En 58G	0.15% C 11% Ni 18% Cr steel: Wire
BS 1554 En 58H	0.12% C 10% Ni 18% Cr steel: Wire
BS 1554 En 58J	0.12% C 10% Ni 18% Cr steel: Wire
BS 1631 A	0.12% C (max) 8% Ni 17% Cr steel: Casting; annealed; contained in BS 3100 **UTS: 450 Elon: 20% Proof: 200**
BS 1631 B Nb	0.12% C (max) 9% Ni 18% Cr 1.0% Nb steel: Casting; annealed; contained in BS 3100 **UTS: 450 Elon: 20% Proof: 200**
BS 1631 B Ti	0.12% C (max) 8% Ni 18% Cr 0.6% Ti steel: Casting; annealed; contained in BS 3100 **UTS: 450 Elon: 20% Proof: 200**
BS 1632 A	0.08% C (max) 12% Ni 19% Cr 3.5% Mo steel: Casting; annealed; contained in BS 3100 **UTS: 450 Elon: 15% Proof: 200**
BS 1632 B	0.08% C (max) 16% Ni 17% Cr 2.5% Mo steel: Casting; annealed; contained in BS 3100 **UTS: 450 Elon: 15% Proof: 200**
BS 1632 C Nb	0.12% C (max) 10% Ni 17% Cr 2.5% Mo 1.0% Nb steel: Casting; annealed; contained in BS 3100 **UTS: 450 Elon: 15% Proof: 200**
BS 1632 C Ti	0.12% C (max) 10% Ni 17% Cr 2.5% Mo 0.6% Ti steel: Casting; annealed; contained in BS 3100 **UTS: 450 Elon: 15% Proof: 200**
BS 1632 D	0.08% C (max) 8% Ni 17% Cr 2.5% Mo steel: Casting; annealed; contained in BS 3100 **UTS: 450 Elon: 15% Proof: 200**
BS 1648 D	0.4% C (max) 6% Ni 19% Cr steel: Casting; as cast; contained in BS 3100
BS 1648 E	0.5% C (max) 12% Ni 25% Cr steel: Casting; as cast; contained in BS 3100
BS 1648 F	0.5% C (max) 20% Ni 25% Cr steel: As cast; contained in BS 3100
BS 1648 G	0.5% C (max) 25% Ni 17% Cr steel: Casting; as cast; contained in BS 3100
BS 2014/2	0.15% C 8% Ni 18% Cr steel: Welded tube; annealed **UTS: 530 Elon: 45% Proof: 180**
BS 2056-301581	0.1% C 7% Ni 17% Cr 1% Al steel: For springs
BS 2056-302826	0.12% C 9% Ni 18% Cr steel: For springs
BS 2056-316542	0.07% C 12% Ni 17% Cr 2.2% Mo steel: For springs
BS 2901 A8 Nb	0.1% C (max) 9% Ni 18% Cr 1.0% Nb steel: Rod for all forms of welding
BS 2901 A8 Ti	0.1% C (max) 9% Ni 18% Cr 0.7% Ti steel: Rod for all forms of welding
BS 2901 A10	0.12% C 12% Ni 25% Cr steel: Rod for all forms of welding
BS 2901 A10 Nb	0.12% C 12% Ni 25% Cr 1.0% Nb steel: Rod for gas and electric welding
BS 2901 A10 Ti	0.12% C 12% Ni 25% Cr 0.7% Ti steel: Rod for all forms of welding
BS 2901 A11	0.12% C 21% Ni 25% Cr steel: Rod for all forms of welding
BS 2901 A11 Ti	0.12% C 21% Ni 25% Cr 0.7% Ti steel: For all forms of welding
BS 2901 A12	0.08% C 11% Ni 17% Cr 3% Mo steel: For all forms of welding
BS 2901 A12 Nb	0.08% C 11% Ni 17% Cr 3% Mo 1% Nb steel: Rod for gas and electric welding
BS 2901 A12 Ti	0.08% C 11% Ni 17% Cr 3% Mo 0.5% Ti steel: Rod for all forms of welding
BS 2926/15/35/H	0.4% C 36% Ni 17% Cr steel: Weld electrode
BS 2926/17/8/2	0.08% C 9% Ni 17.5% Cr 2% Mo steel: For welding electrodes
BS 2926/19/9	0.08% C (max) 10% Ni 20% Cr steel: For welding electrodes
BS 2926/19/9/3	0.1% C (max) 9% Ni 20% Cr 3% Mo steel: For welding electrodes
BS 2926/19/9 Nb	0.1% C 10% Ni 20% Cr 1.0% Nb steel: For welding electrodes
BS 2926/19/9L	0.04% C (max) 10% Ni 20% Cr steel: For welding electrodes

Symbol	Nominal analysis, supplier, condition and remarks.
BS 2926/19/12/2L	0.04% C (max) 13% Ni 19% Cr 2% Mo steel: For welding electrodes
BS 2926/19/12/3	0.08% C (max) 12% Ni 19% Cr 3.0% Mo steel: For welding electrodes
BS 2926/19/12/3 L	0.04% C (max) 12% Ni 19% Cr 3.0% Mo steel: For welding electrodes
BS 2926/19/12/3 Nb	0.1% C (max) 12% Ni 19% Cr 3.0% Mo 1.0% Nb steel: For welding electrodes
BS 2926/19/13/4	0.08% C (max) 13% Ni 19% Cr 4.5% Mo steel: For welding electrodes
BS 2926/19/13/4 L	0.04% C 13% Ni 19% Cr 4.5% Mo steel: For welding electrodes
BS 2926/19/13/4 Nb	0.1% C (max) 13% Ni 19% Cr 4.5% Mo 1.0% Nb steel: For welding electrodes
BS 2926/20/25/5 L Cu Nb	0.04% C (max) 26% Ni 21% Cr 5% Mo 0.5% Nb 2% Cu steel: For welding electrodes
BS 2926/20/34/2 Cu Nb	0.07% C (max) 34% Ni 20% Cr 2.5% Mo 1% Nb 3.5% Cu steel: For welding electrodes
BS 2926/23/12	0.15% C (max) 12.5% Ni 23.5% Cr steel: For welding electrodes
BS 2926/23/12/2	0.1% C (max) 12.5% Ni 23.5% Cr 2.5% Mo steel: For welding electrodes
BS 2926/23/12L	0.04% C (max) 13% Ni 24% Cr steel: For welding electrodes
BS 2926/23/12 Nb	0.1% C (max) 12.5% Ni 23.5% Cr 1.0% Nb steel: For welding electrodes
BS 2926/23/12 W	0.2% C (max) 12.5% Ni 23.5% Cr 3.0% W steel: For welding electrodes
BS 2926/25/6/2	0.06% C (max) 6% Ni 26% Cr 3% Mo steel: For welding electrodes
BS 2926/25/6/Cu	0.06% C (max) 6% Ni 26% Cr 3% Mo 2.5% Cu steel: For welding electrodes
BS 2926/25/20	0.2% C (max) 20% Ni 26% Cr steel: For welding electrodes
BS 2926/25/20/1	0.12% C (max) 20% Ni 27% Cr 1% Mo steel: For welding electrodes
BS 2926/25/20/2	0.12% C (max) 20% Ni 27% Cr 2.5% Mo steel: For welding electrodes
BS 2926/25/20 H	0.4% C 20% Ni 26% Cr steel: For welding electrodes
BS 2926/25/20 Nb	0.12% C (max) 20% Ni 26% Cr 1.0% Nb steel: For welding electrodes
BS 2926/25/35HNb	0.4% C 34% Ni 25% Cr steel: For welding electrodes
BS 2926/25/35M	0.4% C 34% Ni 25% Cr steel: For welding electrodes
BS 2926/29/9	0.15% C 10% Ni 30% Cr steel: For welding electrodes
BS 2926/29/9/1	0.15% C 9% Ni 50% Cr 1% Mo steel: For welding electrodes
BS 2926/29/9/2	0.15% C 9% Ni 30% Cr 2.5% Mo steel: For welding electrodes
BS 2926 A	0.08% C 10% Ni 19% Cr steel: Welding electrode; as welded **UTS: 540 Elon: 35%**
BS 2926 B	0.1% C 9% Ni 19% Cr 1.0% Nb steel: Welding electrode; as welded **UTS: 540 Elon: 30%**
BS 2926 C	0.08% C 12% Ni 19% Cr 3.5% Mo 1.0% Nb steel: Welding electrode; as welded **UTS: 540 Elon: 30%**
BS 2926 D	0.5% C 11% Ni 17% Cr 3% Mo 1.0% Nb steel: Welding electrode; as welded **UTS: 540 Elon: 25%**
BS 2926 E	0.6% C 9% Ni 19% Cr 3% Mo steel: Welding electrode; as welded **UTS: 540 Elon: 25%**
BS 2926 F	0.6% C 12% Ni 24% Cr steel: Welding electrode; as welded **UTS: 540 Elon: 35%**
BS 2926 G	1% C 12% Ni 25% Cr 3% W steel: Welding electrode; as welded **UTS: 540 Elon: 25%**
BS 3014/1	0.08% C 9% Ni 18% Cr steel: Welded tube; annealed **UTS: 530 Elon: 45% Proof: 180**
BS 3014/3	0.08% C 9% Ni 18% Cr 0.5% Ti steel: Welded tube; annealed **UTS: 530 Elon: 45% Proof: 180**
BS 3014/4	0.15% C 9% Ni 18% Cr 0.6% Ti steel: Welded tube; annealed **UTS: 530 Elon: 45% Proof: 180**
BS 3014/5	0.08% C 11% Ni 19% Cr 0.8% Nb: Welded tube; annealed **UTS: 530 Elon: 45% Proof: 180**
BS 3014/6	0.08% C 11% Ni 17.5% Cr 3.0% Mo steel: Welded tube; annealed **UTS: 530 Elon: 45% Proof: 180**
BS 3076 NA 15	32% Ni 21% Cr 0.4% Ti 0.4% Al Fe alloy: Rod
BS 3100/302 C25	0.12% C 8.0% Ni 19.0% Cr steel: Casting
BS 3100/302 C35	0.3% C (max) 8.0% Ni 19.5% Cr steel: Casting
BS 3100/304 C12	0.03% C 8.0% Ni 19.0% Cr steel: Casting
BS 3100/304 C15	0.08% C 8.0% Ni 19.0% Cr steel: Casting
BS 3100/309 C30	0.5% C (max) 12.0% Ni 23.5% Cr 1.0% Mo steel: Casting
BS 3100/309 C40	0.5% C (max) 10.0% Ni 27.5% Cr 1.0% Mo steel: Casting
BS 3100/310 C45	0.5% C (max) 19.5% Ni 24.5% Cr 1.0% Mo steel: Casting
BS 3100/315 C16	0.08% C 8.0% Ni 19.0% Cr 1.3% Mo steel: Casting
BS 3100/316 C12	0.03% C 10.0% Ni 19.0% Cr 2.5% Mo steel: Casting
BS 3100/316 C16	0.08% C 10.0% Ni 19.0% Cr 2.5% Mo steel: Casting
BS 3100/347 C17	0.08% C 8.5% Ni 19.0% Cr 1.0% Nb steel: Casting
BS 3127/7	0.08% C 12% Ni 17% Cr 2.5% Mo steel: Tube; lightly cold worked; for bourdon tubes **DPN: 220**
BS 3146 ANC 3A	0.1% C 8% Ni 17% Cr steel: Investment casting; annealed **UTS: 450 Elon: 25%**
BS 3146 ANC 3BNb	0.1% C 9% Ni 19% Cr 1.0% Nb steel: Investment casting; annealed **UTS: 450 Elon: 25% Proof: 210**
BS 3146 ANC 4A	0.08% C (max) 13% Ni 19% Cr 3.5% Mo steel: Investment casting; annealed **UTS: 450 Elon: 15% Proof: 210**
BS 3146 ANC 4B	0.08% C (max) 10% Ni 17% Cr 2.5% Mo steel: Investment casting; annealed **UTS: 450 Elon: 15% Proof: 210**
BS 3146 ANC 4CNb	0.12% C (max) 10% Ni 17% Cr 2.5% Mo 1.0% Nb steel: Investment casting; annealed **UTS: 450 Elon: 15% Proof: 210**
BS 3146 ANC 5A	0.5% C 20% Ni 25% Cr steel: Investment casting; as cast
BS 3146 ANC 5B	0.5% C 40% Ni 20% Cr steel: Investment casting; as cast
BS 3146 ANC 6A	0.2% C 12% Ni 22% Cr steel: Investment casting; as cast **DPN: 250 UTS: 450 Elon: 20%**
BS 3146 ANC 6B	0.22% C 12% Ni 22% Cr 3% W steel: Investment casting; as cast **DPN: 250 UTS: 450 Elon: 20%**
BS 3146 ANC 7	0.35% C 13% Ni 19% Cr 1.7% Mo 2.5% W 10% Co 3% Nb steel: Casting; as cast **UTS: 530 Proof: 300**
BS 3532/2	Austenitic stainless steel for implants and surgery tools classified into types 316 S16, 316 S12, 317 S10
BS 4534/1	0.08% C 12% Ni 18% Cr 0.5% Mo (max) cast steel: Tube; weldable; as cast **UTS: 450 Elon: 26%**
BS 4534/1 F	0.06% C 8% Ni 18.5% Cr cast steel: Tube; weldable; as cast **UTS: 480 Elon: 26%**
BS 4534/2	0.06% C 12.5% Ni 18% Cr 0.5% Mo (max) 1.0% Nb cast steel: Tube; weldable; as cast **UTS: 450 Elon: 22%**

Symbol	Nominal analysis, supplier, condition and remarks.
BS 4534/2 F	0.06% C 8% Ni 18.5% Cr 1.0% Nb cast steel: Tube; weldable; as cast **UTS: 480 Elon: 22%**
BS 4534/3	0.06% C 13.5% Ni 17% Cr 2.5% Mo cast steel: Tube **UTS: 450 Elon: 26%**
BS 4534/3 F	0.06% C 8% Ni 18.5% Cr 2.5% Mo cast steel: Tube; weldable; as cast **UTS: 480 Elon: 26%**
BS 4534/4	0.3% C 10% Ni 19.5% Cr 0.5% Mo (max) cast steel: Tube **UTS: 480 Elon: 22%**
BS 4534/5	0.3% C 12% Ni 25% Cr 0.5% Mo (max) cast steel: Tube weldable; as cast; can be aged **UTS: 510 Elon: 9%**
BS 4534/6	0.4% C 20% Ni 25% Cr 0.5% Mo (max) cast steel: Tube **UTS: 480 Elon: 9%**
BS 4534/7	0.15% C (max) 21% Ni 25% Cr 0.5% Mo (max) cast steel: Weldable tube; as cast **UTS: 450 Elon: 22%**
BS 4534/8	0.45% C 35% Ni 18% Cr 0.5% Mo (max) cast steel: Tube **UTS: 450 Elon: 7%**
BS 4534/9	0.5% C 35% Ni 18% Cr 0.5% Mo (max) cast steel: Weldable tube; as cast **UTS: 450 Elon: 7%**
BS 4534/10	0.15% C (max) 35% Ni 18% Cr 0.5% Mo (max) cast steel: Weldable tube; as cast **UTS: 420 Elon: 22%**
BS 4534/11	0.1% C (max) 32% Ni 20.5% Cr 0.5% Mo (max) cast steel: Weldable tube; as cast **UTS: 400 Elon: 26%**
BS 6105-A1	0.12% C 9% Ni 18% Cr 0.5% Mo steel: For fasteners
BS 6105-A2	0.08% C 10.5% Ni 18.5% Cr steel: For fasteners
BS 6105-A4	0.08% C 12% Ni 17.5% Cr 2.5% Mo steel: For fasteners
BS 6323 LW20	0.03% C 10.5% Ni 18% Cr steel: Tube
BS 6323 LW21	0.06% C 10.5% Ni 18% Cr steel: Tube
BS 6323 LW22	0.03% C 12% Ni 17.5% Cr 3% Mo steel: Tube
BS 6323 LW23	0.07% C 12.5% Ni 17.5% Cr 3% Mo steel: Tube
BS 6323 LW24	0.08% C 10.5% Ni 18% Cr steel: Tube
BS Cb	0.08% C 8% Ni 18% Cr steel: Osborn; weld stabilized **UTS: 700 Elon: 55% Proof: 270**
BS M	0.06% C 8% Ni 18% Cr steel: Osborn; weldable **UTS: 700 Elon: 60% Proof: 270**
BS T	0.08% C 8% Ni 18% Cr steel: Osborn; weld stabilized **UTS: 700 Elon: 55% Proof: 270**
BS W10	0.2% C 13% Ni 12% Cr steel: Wire
BS W11	0.2% C (max) 14% Ni 12% Cr steel: Wire
CA 6N	0.06% C (max) 7% Ni 11.5% Cr steel: Casting; designation used in the UK and USA
CALITE	40% Ni 5.5% Cr 4.5% Al Fe alloy: Heat resistant; origin unknown
CALMET	12% Ni 25% Cr steel: Casting; Calorizing Corp; heat resistant properties **DPN: 160 UTS: 540 Proof: 330**

Note. The following abbreviations and units are used in the tables:

DPN	Hardness, diamond pyramid number
UTS	Ultimate tensile strength, N/mm^2
Elon	Elongation, %
Proof	0.1% proof strength, N/mm^2

1 N/mm^2=0.1 hbar=0.102 kgf/mm^2=0.06475 $tonf/in.^2$=145.04 $lbf/in.^2$=1 MPa

See Appendix II for other abbreviations and conversion tables.

Symbol	Nominal analysis, supplier, condition and remarks.
CAMLOY	30% Ni 15% Cr Fe alloy: For marine valves; Cameron and Son; obsolete **UTS: 840 Elon: 35%**
CARPENTER 20	0.05% C 28.5% Ni 20% Cr 2.25% Mo 3.25% Cu steel: Jessops
CARPENTER 20C63	0.07% C (max) 34% Ni 20% Cr 2.5% Mo 3.5% Cu Fe alloy: Carpenter
CD 4M	Symbol for Durcornet 100; Duriron Co.
CD4 MCu	0.04% C (max) 5.2% Ni 25.5% Cr 2% Mo 3% Cu steel: Casting; designation used in the UK and USA
CE 30	0.3% C 9% Ni 28% Cr steel: Casting; ASTM designation; as cast **DPN: 170 UTS: 700 Elon: 15% Proof: 370**
CF 3	0.03% C 10% Ni 19% Cr steel: Casting; ASTM designation
CF 3M	0.03% C 11% Ni 19% Cr 2.5% Mo steel: Casting; ASTM designation
CF 3MN	0.03% C (max) 11% Ni 19% Cr 0.15% N steel: Casting; designation used in the UK and USA
CF 8	0.08% C 9% Ni 19% Cr steel: Casting; ASTM designation; as cast **UTS: 540 Elon: 55% Proof: 230**
CF 8 C	0.08% C 11% Ni 19% Cr 1.0% Nb steel: Casting; ASTM designation; as cast **UTS: 540 Elon: 9% Proof: 220**
CF 8 M	0.08% C 11% Ni 19% Cr 2.5% Mo steel: Casting; ASTM designation; as cast **UTS: 530 Elon: 50% Proof: 240**
CF 12 M	0.12% C 11% Ni 19% Cr 2.5% Mo steel: Casting; ASTM designation; as cast **UTS: 530 Elon: 50% Proof: 270**
CF 16 F	0.16% C 11% Ni 19% Cr 1.5% Mo 0.3% Se steel: Casting; ASTM designation; as cast; free machining **UTS: 540 Elon: 52% Proof: 240**
CF 20	0.2% C 9% Ni 19% Cr steel: Casting; ASTM designation; as cast **UTS: 540 Elon: 50% Proof: 210**
CF-10MC	0.1% C (max) 14.5% Ni 16.5% Cr 2% Mo steel: Casting; designation used in the UK and USA
CG 8 M	0.08% C 9% Ni 19% Cr 3% Mo steel: Casting; ASTM designation
CG6MMN	0.06% C (max) 5% Mn 12.5% Ni 21.5% Cr 2.2% Mo 0.3% V 0.2% Nb steel: Casting; designation used in the UK and USA
CG12	0.12% C (max) 11.5% Ni 21.5% Cr steel: Casting; designation used in the UK and USA
CG-12	0.12% C (max) 11.5% Ni 21.5% Cr steel: Casting; designation used in the UK and USA
CH 8	0.08% C (max) 13.5% Ni 24% Cr steel: Casting; designation used in the UK and USA
CH 10	0.1% C (max) 13.5% Ni 24% Cr steel: Casting; designation used in the UK and USA
CH 20	0.2% C 13% Ni 24% Cr steel: Casting; ASTM designation; as cast **UTS: 620 Elon: 38% Proof: 300**
CHROMAX	37% Ni 18% Cr Fe alloy: British Driver-Harris; furnace elements **UTS: 700**
CHRONIFER 2520	0.15% C (max) 20% Ni 24.5% Cr 1.7% Se steel: VDM
CHRONIFER 2520 NV	0.05% C (max) 20% Ni 24.5% Cr 0.5% Se steel: VDM
CK 20	0.2% C 21% Ni 25% Cr steel: Casting; Huntington; as cast **UTS: 520 Elon: 37% Proof: 220**
CK 20	0.2% C (max) 20.5% Ni 25% Cr steel: Casting; designation used in the UK and USA
CN 3 M	0.03% C (max) 25% Ni 21% Cr 5% Mo 3.5% Cu steel: Casting; designation used in the UK and USA
CN 7 M	0.07% C 29% Ni 20% Cr 2.5% Mo 3.5% Cu steel: Casting; Huntington; as cast **UTS: 450 Elon: 48% Proof: 180**

Symbol	Nominal analysis, supplier, condition and remarks.
CN 7 MS	0.07% C (max) 23.5% Ni 19% Cr 2.7% Mo 1.7% Cu steel: Casting; designation used in the UK and USA
CNS	0.35% C 8% Ni 13% Cr 0.5% Mo steel: For valves; origin unknown
COLMONOY C290	0.45% C 37.0% Ni 13.2% Cr 2.5% Si 1.5% B iron alloy: For brazing; Wall Colmonoy
COLMONOY C395	0.5% C 37.0% Ni 13.2% Cr 2.5% Si 1.5% B Fe alloy bronze metal: Wall Colmonoy
COMET 307	0.09% C 5.5% Mn 8.5% Ni 21.0% Cr 0.9% Mo steel: welding electrode; Soudometal; for hard facing work; hardness to 500 DPN **DPN: 200**
COMET 307R	0.05% C 6.0% Mn 8.5% Ni 21.0% Cr steel: Welding electrode; Soudometal; for hard facing work; hardness to 500 DPN **DPN: 200**
COMET 308L	0.03% C 9.5% Ni 19.5% Cr steel: Welding electrode; Soudometal; for welding austenitic stainless steel **UTS: 600 Elon: 46% Proof: 450**
COMET 309	0.04% C 13.0% Ni 23.0% Cr steel: Welding electrode; Soudometal; for buttering **UTS: 570 Elon: 40% Proof: 460**
COMET 309 Mo	0.04% C 13.0% Ni 23.0% Cr 2.3% Mo steel: Welding electrode; Soudometal; buttering of 316 type steel **UTS: 620 Elon: 30% Proof: 510**
COMET 316L	0.03% C 12.0% Ni 18.5% Cr 2.6% Mo steel: Welding electrode; Soudometal; for welding 316 type steel **UTS: 590 Elon: 42% Proof: 470**
COMET 317	0.07% C 1.2% Mn 13.5% Ni 19.5% Cr 3.3% Mo steel: Welding electrode; Soudometal; for welding 316 type steel **UTS: 610 Elon: 30% Proof: 470**
COMET SD	0.04% C 10.0% Ni 24.5% Cr steel: Welding electrode; Soudometal **UTS: 730 Elon: 30% Proof: 570**
CONTRACID	10% Ni Cr Mo Fe alloy: Corrosion resistant alloy; no further information
CORNES 310	21% Ni 25% Cr steel: Philip Cornes
Cr Ni 25-20	0.2% C (max) 19.0% Ni 24.0% Cr steel: Designation used by German Standards
CRM 6D	1.0% C 5.0% Mn 5.0% Ni 20% Cr 1.0% Mo 1.0% W Fe alloy: Information from SAE handbook
CROLOY 15/15N	0.15% C 2% Mn 15% Ni 16% Cr 1.5% Mo 1.4% W 1% Nb steel: Babcock and Wilcox (USA)
CROLOY 16/13/3	0.08% C 12% Ni 17% Cr 2.5% Mo steel: Tube; Babcock and Wilcox (USA)
CROLOY 16/13/3 ELC	0.035% C 14% Ni 17% Cr 2.5% Mo steel: Tube; Babcock and Wilcox (USA)
CROLOY 16/13/3 H	0.07% C 12% Ni 17% Cr 2.5% Mo steel: Tube; Babcock and Wilcox (USA)
CROLOY 18/8 Cb	0.08% C 10% Ni 18% Cr 1.0% Nb steel: Tube; Babcock and Wilcox (USA)
CROLOY 18/8 Cb H	0.6% C 10% Ni 18% Cr 1.0% Nb steel: Babcock and Wilcox (USA)
CROLOY 18/8 Cb Ta	0.08% C 10% Ni 18% Cr 1.0% Nb + Ta steel: Tube; Babcock and Wilcox (USA)
CROLOY 18/8 Cb Ta H	0.6% C 10% Ni 18% Cr 1.0% Nb + Ta steel: Babcock and Wilcox (USA)
CROLOY 18/8 ELC	0.035% C 12% Ni 19% Cr steel: Tube; Babcock and Wilcox (USA)
CROLOY 18/8 EM	0.15% C 10% Ni 18% Cr 0.1% Se steel: Tube; Babcock and Wilcox (USA); free machining
CROLOY 18/8 H	0.07% C 9% Ni 19% Cr steel: Tube; Babcock and Wilcox (USA)
CROLOY 18/8 HC	0.12% C 9% Ni 18% Cr steel: Babcock and Wilcox (USA)
CROLOY 18/8 S	0.08% C 9% Ni 19% Cr steel: Tube; Babcock and Wilcox (USA)
CROLOY 18/8 Si	0.08% C 12% Ni 18% Cr 2.5% Si steel: Tube; Babcock and Wilcox (USA)
CROLOY 18/8 S Ti H	0.7% C 10% Ni 18% Cr 0.6% Ti steel: Tube; Babcock and Wilcox (USA)
CROLOY 18/8 Ti	0.08% C 11% Ni 19% Cr 0.6% Ti steel: Tube; Babcock and Wilcox (USA)
CROLOY 18/12	0.12% C 12% Ni 18% Cr steel: Tube; Babcock and Wilcox (USA)
CROLOY 18/13/3	0.08% C 12% Ni 19% Cr 3.5% Mo steel: Tube; Babcock and Wilcox (USA)
CROLOY 25/12	0.15% C 13% Ni 23% Cr steel: Tube; Babcock and Wilcox (USA)
CROLOY 25/20	0.15% C 20% Ni 25% Cr steel: Tube; Babcock and Wilcox (USA)
CRONIFER 1925 LC	0.02% C (max) 25% Ni 20.5% Cr 4.7% Mo 1.6% Cu 0.18% N steel: VDM
CRONIFER S1925	0.02% C (max) 25% Ni 20.5% Cr 4.7% Mo Fe alloy: Weld rod; VDM
CRONIFER 1713 LCN	0.03% C (max) 13.5% Ni 17.7% Cr 4.2% Mo 0.15% N steel: VDM
CRONIFER 1812 LCN	0.03% C (max) 15% Ni 17.5% Cr 2.7% Mo 0.17% N steel: VDM
CRONIFER 1815 LC Si	0.015% C (max) 14.7% Ni 18% Cr 4% Si steel: VDM
CRONIFER 1925 H Mo	0.025% C (max) 25% Ni 20.5% Cr 6.4% Mo 0.9% Cu 0.19% N steel: VDM
CRONIFER 1925 LCN	0.02% C (max) 25% Ni 20.5% Cr 4.7% Mo 1.6% Cu 0.15% N steel: VDM
CRONIFER 2205 LCN	0.03% C (max) 5.7% Ni 22.5% Cr 3% Mo 0.15% N steel: VDM
CRONIFER 2328	0.03% C (max) 27% Ni 23% Cr 2.7% Mo 0.5% Ti 2.7% Cu steel: VDM
CRONIFER 2521 LC	0.02% C (max) 20.5% Ni 24.5% Cr steel: VDM
CRONIFER 2522 LCN	0.02% C (max) 22.5% Ni 24.7% Cr 2.2% Mo 0.13% N steel: VDM
CRONIFER 2525 Ti	0.03% C (max) 24.7% Ni 24.7% Cr 2.2% Mo 0.4% Ti steel: VDM
CRONITE 25/12	0.5% C (max) 1.0% Mn 12% Ni 25% Cr Fe alloy: Casting; heat resistant **UTS: 550 Elon: 10% Proof: 100**
CRONITE 25/20	0.5% C 20% Ni 24% Cr 0.7% W steel: Casting; Cronite; as cast **DPN: 170 UTS: 530 Elon: 18% Proof: 300**
CRONITE 37/18	0.3% C 36% Ni 19% Cr Fe alloy: Casting; heat resistant **UTS: 520 Elon: 8% Proof: 290**
CRONITE 37/18/Nb	0.3% C (max) 36% Ni 19% Cr 2.0% Nb Fe alloy: Casting; heat resistant **UTS: 570 Elon: 8% Proof: 330**
CRONITE 428	0.5% C 15% Ni 23% Cr 0.25% Mo 2.5% W steel: Casting; Cronite; as cast **DPN: 165 UTS: 570 Elon: 20% Proof: 300**
CROWN MAX	0.2% C 11.5% Ni 23% Cr 2.8% W steel: Firth Vickers for BS alloy En 55
CRUCIBLE 303	0.15% C 9% Ni 18% Cr 0.6% Mo or Zr steel: Crucible Steel Co.; annealed **UTS: 650 Elon: 50% Proof: 220**
CRUCIBLE 304	0.08% C 10% Ni 19% Cr steel: Crucible Steel Co.; annealed **DPN: 150 UTS: 620 Elon: 55% Proof: 240**
CRUCIBLE 309	0.2% C 13% Ni 23% Cr steel: Crucible Steel Co.; annealed **DPN: 170 UTS: 700 Elon: 50% Proof: 270**
CRUCIBLE 316	0.08% C 12% Ni 17% Cr 2.5% Mo steel: Crucible Steel Co.; annealed **DPN: 160 UTS: 630 Elon: 50% Proof: 270**
CRUCIBLE 321	0.08% C 11% Ni 18% Cr 0.5% Ti steel: Crucible Steel Co.; annealed **DPN: 160 UTS: 630 Elon: 50% Proof: 250**
CRUCIBLE 347	0.08% C 11% Ni 18% Cr 1% Nb + Ta steel: Crucible Steel Co.; annealed **DPN: 160 UTS: 690 Elon: 50% Proof: 170**

Symbol	Nominal analysis, supplier, condition and remarks.
CRUCIBLE 348	0.08% C 11% Ni 18% Cr 1% Nb + Ta (Ta 0.1% (max)) steel: Crucible Steel Co.; annealed **DPN: 160 UTS: 690 Elon: 50% Proof: 290**
CUSTOM FLO 302 HQ	0.08% C (max) 9% Ni 18% Cr 3.5% Cu steel: Carpenter
CUSTOM FLO 316 HQ	0.03% C (max) 12% Ni 17% Cr 2.5% Mo 3.5% Cu steel: Carpenter
D 2	Symbol for Durco CF 8; Duriron Co.
D 2L	Symbol for Durco CF 3; Duriron Co.
D 4	Symbol for Durco CF 8M; Duriron Co.
D 4L	Symbol for Durco FC 3M; Duriron Co.
D 20	Symbol for Durimet 20; Duriron Co.
DARWIN K	0.15% C 23% Ni 23% Cr 2% Mo 0.25% Cu Fe alloy: Casting; Darwin
DDQ	0.08% C 12.5% Ni 12.5% Cr steel: Firth Vickers **DPN: 160 UTS: 570 Elon: 50%**
DIN 1736 E1-Ni Cr 26 Mo	0.04% C 38% Ni 26% Cr 3.5% Mo 1.6% Cu steel: Welding electrode
DIN 8556 E19-9.2 MPR 23.180	0.03% C (max) 9.5% Ni 19% Cr 2% Mo steel: Welding electrode
DIN 8556 E19-9 Nb R 26	0.05% C (max) 10% Ni 19% Cr 0.4% Ta steel: Welding electrode
DIN 8556 E19-9 NC R 26	0.03% C (max) 10% Ni 19% Cr steel: Welding electrode
DIN 8556 E19-12.3 Nb R 26	0.05% C (max) 11.5% Ni 18% Cr 2.8% Mo 0.3% Nb steel: Welding electrode
DIN 8556 E25-20 R 26	0.12% C (max) 20% Ni 25% Cr steel: Welding electrode
DIN 17440 X 2 Cr Ni 18/9	0.03% C (max) 2.0% Mn (max) 11% Ni 18.5% Cr steel: Plate; German Standard **UTS: 586 Proof: 178**
DIN 17440 X 2 Cr Ni 18/10	0.03% C 2.0% Mn (max) 10.3% Ni 18% Cr steel: Plate; German Standard **UTS: 660 Proof: 275**
DIN 17440 X 2 Cr Ni Mo 18/10	0.03% C (max) 13.75% Ni 17.5% Cr 2.75% Mo steel: Plate; German Standard; solution treated **UTS: 590 Proof: 240**
DIN 17440 X 2 Cr Ni Mo 18/12	0.03% C (max) 13.75% Ni 17.5% Cr 2.75% Mo steel: Plate; German Standard; solution treated **UTS: 590 Proof: 240**
DIN 17440 X 2 Cr Ni Mo 18/16	0.03% C (max) 2.0% Mn (max) 16% Ni 18% Cr 3.5% Mo 0.045% P (max) 0.03% N (max) steel: Plate; German Standard; solution treated **UTS: 612 Proof: 199**
DIN 17440 X 2 Cr Ni Mo N 18/2	0.03% C (max) 12.0% Ni 17.5% Cr 2.25% Mo 0.18% N steel: Plate; German Standard; solution treated **UTS: 700 Proof: 325**
DIN 17440 X 2 Cr Ni Mo N 18/13	0.03% C (max) 13.25% Ni 17.5% Cr 2.75% Mo 0.18% N steel: Plate; German Standard; solution treated **UTS: 700 Proof: 350**
DIN 17440 X 5 Cr Ni 18/9	0.07% C (max) 9% Ni 18% Cr 2.0% Mo (max) steel: Plate; German Standard **UTS: 201 Proof: 189**
DIN 17440 X 5 Cr Ni 19/11	0.07% C (max) 11.2% Ni 18.5% Cr steel: Plate; German Standard; solution treated **UTS: 230 Proof: 190**
DIN 17440 X 5 Cr Ni Mo 18/10	0.07% C (max) 12.0% Ni 17.5% Cr 2.25% Mo steel: Plate; German Standard; solution treated **UTS: 600 Proof: 250**
DIN 17440 X 5 Cr Ni Mo 18/12	0.07% C (max) 12.75% Ni 17.5% Cr 2.75% Mo steel: Plate; German Standard; solution treated **UTS: 600 Proof: 250**
DIN 17440 X 10 Cr Ni Mo Nb 18/10	0.1% C (max) 12.0% Ni 17.5% Cr 2.25% Mo 0.8% Nb steel: Plate; German Standard; solution treated **UTS: 610 Proof: 230**
DIN 17440 X 10 Cr Ni Mo Ti 18/10	0.1% C (max) 12.2% Ni 17.5% Cr 2.25% Mo 0.5% Ti steel: Plate; German Standard; solution treated **UTS: 610 Proof: 270**
DIN 17440 X 10 Cr Ni Nb 18/9	0.1% C (max) 2.0% Mn (max) 10% Ni 18% Cr 0.8% Nb 0.03% S (max) 0.04% P (max) steel: Plate; German Standard; solution treated **UTS: 640 Proof: 209**
DIN 17440 X 10 Cr Ni Ti 18/9	0.1% C (max) 2.0% Mn (max) 10% Ni 18% Cr 0.5% Ti (min) 0.03% S (max) 0.045% P (max) steel: Plate; German Standard; solution treated **UTS: 640 Proof: 209**
DIN 17440 X 12 Cr Ni 5188	0.15% C (max) 9% Ni 18% Cr steel: Plate; German specification **UTS: 610 Proof: 219**
DIN 17470 Cr Ni 25/20	20% Ni 25% Cr Fe alloy
DIN 17470 Ni Cr 30/20	30% Ni 20% Cr Fe alloy
DISCALOY	26% Ni 13.5% Cr 2.8% Mo 1.8% Ti steel: Westinghouse; for high temperature use
DTD 189 A	0.15% C 8% Ni 18% Cr Ti or Nb steel: Wire; weldable **UTS: 530**
DTD 712 A	0.06% C (max) 10% Ni 18% Cr Ti or Nb steel: Sheet annealed; weldable **UTS: 450 Elon: 40% Proof: 150**
DTD 734	0.08% C 9% Ni 18% Cr: Obsolete
DTD 5006	0.15% C 8% Ni 18% Cr 0.6% Ti or 1.0% Nb steel: Cold drawn **UTS: 1550**
DTD 5016	0.1% C (max) 11% Ni 18% Cr Nb steel: Tube; annealed; weldable **UTS: 620 Elon: 40% Proof: 180**
DTD 5036	Low C Ni Cr steel: For rivets and split pins; weldable
DTD 5056	Ni Cr steel: For bolts, etc.; suitable for cold forming
DTD 5259	Ni/Cr steel: Casting; investment cast; not stabilized; for use up to 350 °C **UTS: 470**
DTD 5269	Ni/Cr steel: Casting; investment cast; Nb stabilized **UTS: 470**
DTD 5279	Ni/Cr 2.5% Mo steel: Casting; investment cast **UTS: 500**
DTD 5289	Ni/Cr 3.5% Mo steel: Casting; investment cast **UTS: 500**
DUNELT 84	0.43% C 10% Ni 14% Cr 0.5% Mo 3% W steel: Dunford for BS alloy En 54
DUNELT 88	0.2% C 9% Ni 18% Cr steel: Dunford; annealed; can be stabilized for welding **UTS: 530 Elon: 30%**
DURCO CF 3	0.03% C (max) 10.0% Ni 19.0% Cr steel: Casting; Duriron Co. **UTS: 455 Elon: 35% Proof: 195**
DURCO CF 3M	0.03% C (max) 11.0% Ni 19.0% Cr 2.5% Mo steel: Casting; Duriron Co. **UTS: 490 Elon: 30% Proof: 210**
DURCO CF 8	0.08% C (max) 9.0% Ni 19.5% Cr steel: Casting; Duriron Co. **UTS: 460 Elon: 35% Proof: 195**
DURCO CF 8M	0.08% C (max) 11.0% Ni 19.5% Cr 2.5% Mo steel: Casting; Duriron Co. **UTS: 490 Elon: 30% Proof: 211**
DURCOMET 5	0.025% C (max) 16% Ni 21% Cr 5% Si steel: Casting; designation used in the UK and USA

Note. The following abbreviations and units are used in the tables:

DPN	Hardness, diamond pyramid number
UTS	Ultimate tensile strength, N/mm^2
Elon	Elongation, %
Proof	0.1% proof strength, N/mm^2

$1\ N/mm^2 = 0.1\ hbar = 0.102\ kgf/mm^2 = 0.06475\ tonf/in.^2 = 145.04\ lbf/in.^2 = 1\ MPa$

See Appendix II for other abbreviations and conversion tables.

Symbol	Nominal analysis, supplier, condition and remarks.
DURCORNET 100	0.04% C (max) 5.2% Ni 25.5% Cr 3.0% Cu steel: Casting; Duriron Co. **UTS: 700 Elon: 20% Proof: 490**
DURIMET 20	0.07% C (max) 29.0% Ni 20.5% Cr 2.5% Mo steel: Casting; Duriron Co. **UTS: 500 Elon: 20% Proof: 323**
E 16-8.2	0.1% C (max) 8.5% Ni 15.5% Cr 1.5% Mo steel: Weld electrode; designation used by AWS
E 028L	0.03% C (max) 31.5% Ni 28.2% Cr 3.8% Mo 1% Cu steel: Weld electrode; designation used by AWS
E 71 T8 K2	0.15% C (max) 1.5% Ni 1.8% Al (max) steel: Weld electrode; designation used by AWS; obsolete
E 209	0.06% C (max) 10% Ni 22.2% Cr 2.2% Mo 0.2% N steel: Weld electrode; designation used by AWS
E 307	0.1% C 9.6% Ni 19.2% Cr 1% Mo steel: Weld electrode; designation used by AWS
E 307 T1	0.13% C (max) 9.7% Ni 19% Cr 1% Mo steel: Weld electrode; designation used by AWS
E 307 T3	0.1% C 9.2% Ni 20.2% Cr 1% Mo steel: Weld electrode; designation used by AWS
E 308	0.08% C (max) 10.5% Ni 19.5% Cr steel: Weld electrode; designation used by AWS
E 308 HC	0.1% C 10.5% Ni 19.5% Cr steel: Weld electrode; designation used by AWS
E 308 LT1	0.03% C (max) 10.5% Ni 19.5% Cr steel: Weld electrode; designation used by AWS
E 308 LT3	0.03% C (max) 10.5% Ni 21% Cr steel: Weld electrode; designation used by AWS
E 308 Mo	0.08% C (max) 9.5% Ni 19.5% Cr 2.5% Mo steel: Weld electrode; designation used by AWS
E 308 Mo LT1	0.03% C (max) 10.5% Ni 19.5% Cr 2.5% Mo steel: Weld electrode; designation used by AWS
E 308 MoL	0.04% C (max) 10.5% Ni 19.5% Cr 2.5% Mo steel: Weld electrode; designation used by AWS
E 308 MoT-1	0.08% C (max) 10.5% Ni 19.5% Cr 2.5% Mo steel: Weld electrode; designation used by AWS
E 308 T1	0.08% C (max) 10.5% Ni 19.5% Cr steel: Weld electrode; designation used by AWS
E 308 T-3	0.08% C (max) 10.5% Ni 19.5% Cr steel: Weld electrode; designation used by AWS
E 308L	0.04% C (max) 10.5% Ni 19.5% Cr steel: Weld electrode; designation used by AWS
E 309	0.15% C (max) 13% Ni 23.5% Cr steel: Weld electrode; designation used by AWS
E 309 Cb	0.12% C (max) 13% Ni 23.5% Cr 0.8% Nb steel: Weld electrode; designation used by AWS
E 309 Cb LT	0.03% C (max) 13% Ni 23.5% Cr 0.8% Nb steel: Weld electrode; designation used by AWS
E 309 L	0.04% C (max) 13% Ni 23.5% Cr steel: Weld electrode; designation used by AWS
E 309 LTX	0.03% C (max) 13% Ni 23.5% Cr steel: Weld electrode; designation used by AWS
E 309 Mo	0.12% C (max) 13% Ni 23.5% Cr 2.5% Mo steel: Weld electrode; designation used by AWS
E 309 MoL	0.04% C (max) 13% Ni 23.5% Cr 2.5% Mo steel: Weld electrode; designation used by AWS
E 309 T	0.1% C (max) 13% Ni 23.5% Cr steel: Weld electrode; designation used by AWS
E 310	0.14% C 21% Ni 26.5% Cr steel: Weld electrode; designation used by AWS
E 310 Cb	0.12% C (max) 21% Ni 26.5% Cr 0.8% Nb steel: Weld electrode; designation used by AWS
E 310 H	0.4% C 21% Ni 26.5% Cr steel: Weld electrode; designation used by AWS
E 310 Mo	0.12% C (max) 21% Ni 26.5% Cr 2.5% Mo steel: Weld electrode; designation used by AWS
E 310 T	0.2% C (max) 21% Ni 26.5% Cr steel: Weld electrode; designation used by AWS
E 312	0.15% C (max) 9.2% Ni 30% Cr steel: Weld electrode; designation used by AWS

Symbol	Nominal analysis, supplier, condition and remarks.
E 312 T	0.15% C (max) 9.2% Ni 30% Cr steel: Weld electrode; designation used by AWS
E 316	0.08% C (max) 12.5% Ni 18.5% Cr 2.5% Mo steel: Weld electrode; designation used by AWS
E 316 L	0.04% C (max) 12.5% Ni 18.5% Cr 2.5% Mo steel: Weld electrode; designation used by AWS
E 316 LT	0.04% C (max) 12.5% Ni 18.5% Cr 2.5% Mo steel: Weld electrode; designation used by AWS
E 316 T	0.08% C (max) 12.5% Ni 18.5% Cr 2.5% Mo steel: Weld electrode; designation used by AWS
E 317	0.08% C (max) 13% Ni 19.5% Cr 3.5% Mo steel: Weld electrode; designation used by AWS
E 317 L	0.04% C (max) 13% Ni 19.5% Cr 3.5% Mo steel: Weld electrode; designation used by AWS
E 317 LT	0.03% C (max) 13% Ni 19.5% Cr 3.5% Mo steel: Weld electrode; designation used by AWS
E 318	0.08% C (max) 12.5% Ni 18.5% Cr 2.2% Mo 0.5% Nb steel: Weld electrode; designation used by AWS
E 330	0.2% C 34% Ni 15.5% Cr steel: Weld electrode; designation used by AWS
E 330 H	0.4% C 35% Ni 15.5% Cr steel: Weld electrode; designation used by AWS
E 347	0.08% C (max) 10% Ni 19.5% Cr 0.5% Nb steel: Weld electrode; designation used by AWS
E 347 T	0.08% C (max) 10% Ni 19.5% Cr 0.5% Nb steel: Weld electrode; designation used by AWS
E 349	0.13% C (max) 9% Ni 19.5% Cr 0.45% Mo 1.5% W 1% Nb steel: Weld electrode; designation used by AWS
E 385 L	0.03% C (max) 20% Ni 25% Cr 4.7% Mo 1.8% Cu steel: Weld electrode; designation used by AWS
E 2209	0.04% C (max) 9.5% Ni 22.5% Cr 3% Mo 0.1% N steel: Weld electrode; designation used by AWS
E 2209 T	0.04% C (max) 8.5% Ni 22% Cr 3.5% Mo 0.1% N steel: Weld electrode; designation used by AWS
E 9015	0.1% C (max) 1.6% Ni 0.3% Si steel: Weld electrode; designation used by AWS
E 9018 M	0.1% C (max) 1.6% Ni steel: Weld electrode; designation used by AWS
EC 16.8.2	0.1% C (max) 8.5% Ni 15.5% Cr 1.5% Mo 0.4% Si steel: Weld electrode; designation used by AWS
EC 218	0.1% C (max) 8.5% Ni 17% Cr 4% Si 0.12% N steel: Weld electrode; designation used by AWS
EC 308 H	0.08% C (max) 10.5% Ni 21% Cr 0.5% Si steel: Weld electrode; designation used by AWS
EC 308 L	0.03% C (max) 10.5% Ni 21% Cr 0.5% Si steel: Weld electrode; designation used by AWS
EC 308 L Si	0.03% C (max) 10.5% Ni 21% Cr 0.8% Si steel: Weld electrode; designation used by AWS
EC 308 Mo	0.08% C (max) 10.5% Ni 20.5% Cr 2.5% Mo 0.4% Si steel: Weld electrode; designation used by AWS
EC 308 Mo L	0.04% C (max) 10.5% Ni 19.5% Cr 2.5% Mo 0.4% Si steel: Weld electrode; designation used by AWS
EC 308 Si	0.08% C (max) 10.5% Ni 21% Cr 0.8% Si steel: Weld electrode; designation used by AWS
EC 309	0.12% C (max) 13% Ni 24% Cr 0.4% Si steel: Weld electrode; designation used by AWS
EC 309 L	0.03% C (max) 13% Ni 24% Cr 0.4% Si steel: Weld electrode; designation used by AWS
EC 309 Si	0.12% C (max) 13% Ni 24% Cr 0.8% Si steel: Weld electrode; designation used by AWS
EC 310	0.12% C 21.7% Ni 26.5% Cr 0.4% Si steel: Weld electrode; designation used by AWS
EC 312	0.15% C (max) 9.2% Ni 30% Cr 0.4% Si steel: Weld electrode; designation used by AWS
EC 316	0.08% C (max) 12.5% Ni 19% Cr 2.5% Mo 0.4% Si steel: Weld electrode; designation used by AWS
EC 316 L	0.03% C (max) 12.5% Ni 20% Cr 2.5% Mo 0.4% Si steel: Weld electrode; designation used by AWS
EC 316 L Si	0.03% C (max) 12.5% Ni 19% Cr 2.5% Mo 0.8% Si steel: Weld electrode; designation used by AWS

Symbol	Nominal analysis, supplier, condition and remarks.
EC 316 Si	0.08% C (max) 12.5% Ni 19% Cr 2.5% Mo 0.8% Si steel: Weld electrode; designation used by AWS
EC 317	0.08% C (max) 14% Ni 19.5% Cr 3.5% Mo 0.4% Si steel: Weld electrode; designation used by AWS
EC 317 L	0.03% C (max) 14% Ni 19.5% Cr 3.5% Mo 0.4% Si steel: Weld electrode; designation used by AWS
EC 318	0.08% C (max) 12.5% Ni 19% Cr 2.5% Mo 0.5% Nb 0.4% Si steel: Weld electrode; designation used by AWS
EC 321	0.08% C (max) 9.7% Ni 19.5% Cr 0.5% Ti 0.4% Si steel: Weld electrode; designation used by AWS
EC 347	0.08% C (max) 10% Ni 19.5% Cr 0.8% Nb steel: Weld electrode; designation used by AWS
EC 347 Si	0.08% C (max) 10% Ni 19.5% Cr 0.8% Nb steel: Weld electrode; designation used by AWS
EC 349	0.1% C 8.7% Ni 20.5% Cr 0.5% Mo 1.2% Nb 0.4% Si steel: Weld electrode; designation used by AWS
EC M1	0.1% C (max) 1.7% Ni steel: Weld electrode; designation used by AWS
EC M1 N	0.1% C (max) 1.7% Ni steel: Weld electrode; designation used by AWS
EC Ni 4	0.14% C (max) 1.7% Ni steel: Weld electrode; designation used by AWS
EC Ni 4N	0.14% C (max) 1.7% Ni steel: Weld electrode; designation used by AWS
EM 2N	0.1% C (max) 1.7% Ni steel: Weld electrode; designation used by AWS
EMS	0.06% C 10% Ni 18.5% Cr steel: Sheet; Firth Vickers; weldable **DPN: 160 UTS: 620 Elon: 60%**
EMS 235	0.3% C 8% Ni 23% Cr steel: For valves; origin unknown
EMS 253	0.9% C 15% Ni 28% Cr 5.5% Mo iron alloy: Origin unknown; for hard facing **DPN: 460**
En 54	0.42% C 10% Ni 13% Cr 3% W steel: For valves; designation used in BS 970; replaced by 331 S40
En 55	0.3% C 9% Ni 17% Cr 3% W steel: For valves; designation used in BS 970; obsolete
En 58A	0.16% C (max) 9% Ni 18% Cr steel: Designation used in BS 970; replaced by 302 S25
En 58B	0.15% C (max) 9% Ni 18% Cr 0.6% Ti steel: Designation used in BS 970; replaced by 321 S12
En 58C	0.15% C (max) 11% Ni 19% Cr 0.6% Ti steel: Designation used in BS 970; replaced by 321 S12
En 58D	0.16% C (max) 13% Ni 18% Cr steel: Sheet for deep drawing; designation used in BS 970; obsolete
En 58E	0.08% C (max) 9% Ni 18% Cr steel: Designation used in BS 970; replaced by 304 S15
En 58F	0.15% C (max) 8% Ni 18% Cr Nb stabilized steel: Designation used in BS 970; replaced by 347 S17
En 58G	0.15% C (max) 10% Ni 18% Cr Nb stabilized steel: Designation used in BS 970; replaced by 347 S17
En 58H	0.12% C (max) 9% Ni 18% Cr 2% Mo steel: Designation used in BS 970; replaced by 315 S16
En 58J	0.12% C (max) 9% Ni 18% Cr 3% Mo steel: Designation used in BS 970; replaced by 316 S16

Note. The following abbreviations and units are used in the tables:

DPN	Hardness, diamond pyramid number
UTS	Ultimate tensile strength, N/mm^2
Elon	Elongation, %
Proof	0.1% proof strength, N/mm^2

1 N/mm^2=0.1 hbar=0.102 kgf/mm^2=0.06475 $tonf/in.^2$=145.04 $lbf/in.^2$=1 MPa

See Appendix II for other abbreviations and conversion tables.

Symbol	Nominal analysis, supplier, condition and remarks.
En 58M	0.12% C (max) 9% Ni 18% Cr 3% Mo steel: Free machining En 58; designation used in BS 970; replaced by 325 S21
ER 209	0.05% C (max) 11% Ni 22% Cr 2.2% Mo steel: Weld electrode; designation used by AWS
ER 307	0.1% C 4% Mn 9% Ni 20% Cr 1% Mo steel: Weld electrode; designation used by AWS
ER 308	0.08% C 10.0% Ni 20.5% Cr steel: Designation used by ASTM
ER 308	0.08% C (max) 10% Ni 20.5% Cr 0.4% Si steel: Weld electrode; designation used by AWS
ER 308 L	0.03% C (max) 10% Ni 20.5% Cr 0.4% Si steel: Weld electrode; designation used by AWS
ER 308 LSi	0.03% C (max) 10% Ni 20.5% Cr 0.8% Si steel: Weld electrode; designation used by AWS
ER 308 MoL	0.04% C (max) 10.5% Ni 19.5% Cr 1.7% Mo 0.4% Si steel: Weld electrode; designation used by AWS
ER 308 Si	0.08% C (max) 10% Ni 20.5% Cr 0.8% Si steel: Weld electrode; designation used by AWS
ER 309	0.12% C (max) 13% Ni 24% Cr 0.4% Si steel: Weld electrode; designation used by AWS
ER 309 L	0.03% C (max) 13% Ni 24% Cr 0.4% Si steel: Weld electrode; designation used by AWS
ER 309 Si	0.12% C (max) 13% Ni 24% Cr 0.8% Si steel: Weld electrode; designation used by AWS
ER 310	0.12% C 21% Ni 26.5% Cr 0.4% Si steel: Weld electrode; designation used by AWS
ER 312	0.15% C (max) 9.7% Ni 30% Cr 0.4% Si steel: Weld electrode; designation used by AWS
ER 316	0.08% C (max) 12.5% Ni 19% Cr 2.5% Mo steel: Weld electrode; designation used by AWS
ER 316 L	0.03% C (max) 12.5% Ni 19% Cr 2.5% Mo steel: Weld electrode; designation used by AWS
ER 316 LSi	0.03% C (max) 12.5% Ni 19% Cr 2.5% Mo 0.8% Si steel: Weld electrode; designation used by AWS
ER 316 Si	0.08% C (max) 12.5% Ni 19% Cr 2.5% Mo 0.8% Si steel: Weld electrode; designation used by AWS
ER 317	0.08% C (max) 14% Ni 19.5% Cr 3.5% Mo steel: Weld electrode; designation used by AWS
ER 317 L	0.03% C (max) 14% Ni 19.5% Cr 3.5% Mo steel: Weld electrode; designation used by AWS
ER 318	0.08% C (max) 12.5% Ni 19% Cr 2.5% Mo 0.8% Nb steel: Weld electrode; designation used by AWS
ER 320	0.07% C (max) 3.4% Ni 20% Cr 2.5% Mo 3.5% Cu steel: Weld electrode; designation used by AWS
ER 320	0.035% C (max) 34% Ni 20% Cr 2.5% Mo 3.5% Cu Fe alloy: Weld electrode; designation used by AWS
ER 320 LR	0.025% C 34% Ni 20% Cr 2.5% Mo 3.5% Cu steel: Weld electrode; designation used by AWS
ER 321	0.08% C (max) 9.7% Ni 19.5% Cr 0.8% Ti steel: Weld electrode; designation used by AWS
ER 330	0.2% C 35.5% Ni 16% Cr Fe alloy: Weld electrode; designation used by AWS
ERA 1414	0.4% C 10% Ni 14% Cr 3% W 0.2% Nb steel: Hadfields for BS alloy En 54 and 54A
ERA CR 1	0.1% C 8% Ni 18% Cr steel: Casting; Hadfields **DPN: 170 UTS: 580 Elon: 20% Proof: 270**
ERA CR 1	0.16% C 9% Ni 18% Cr steel: Wrought; Hadfields for BS alloy En 58A
ERA CR 2	0.3% C 9% Ni 21% Cr 4% W 2% Cu steel: Forging; Hadfield; not stabilized for welding **DPN: 235 UTS: 800 Elon: 31% Proof: 450**
ERA CR 2	0.3% C 9% Ni 21% Cr 4% W 1.5% Si 2% Cu steel: Casting; Hadfields **DPN: 229 UTS: 700 Elon: 15% Proof: 370**
ERA CR 4	12% Ni 18% Cr 3% Mo steel: Casting; Hadfields for BS 1632 A **UTS: 450 Elon: 15% Proof: 200**
ERA CR 4	0.1% C 9% Ni 19% Cr 3% Mo steel: Hadfields **DPN: 187 UTS: 710 Elon: 50% Proof: 330**

Symbol	Nominal analysis, supplier, condition and remarks.
ERA CR 4FC	10% Ni 18% Cr 2% Mo S steel: Hadfields for BS alloy En 58M; free machining
ERA CR 4S	0.1% C 9% Ni 19% Cr 3% Mo 0.6% Ti steel: Forging; Hadfields; stabilized for welding **DPN: 187 UTS: 710 Elon: 50% Proof: 300**
ERA CR 4S	0.12% C 9% Ni 19% Cr 3% Mo 0.6% Ti casting: Hadfields **DPN: 212 UTS: 690 Elon: 25% Proof: 300**
ERA CR 4S (Cb)	12% Ni 18% Cr 2.5% Nb steel: Casting; Hadfields for BS 1632 C **UTS: 450 Elon: 15% Proof: 200**
ERA CR 4SFC	10% Ni 18% Cr 2% Mo steel: Hadfields for BS alloy En 58M
ERA CRI FC	9% Ni 18% Cr S steel: Hadfields for BS alloy En 58M; free machining
ERA CRI SFC	9% Ni 18% Cr 0.6% Ti S steel: Hadfields for BS alloy En 58M; free machining
ERA CRIS	0.1% C 9% Ni 18% Cr 0.6% Ti steel: Casting; Hadfields **DPN: 187 UTS: 620 Elon: 18% Proof: 300**
ERA CRIS	0.1% C 9% Ni 18% Cr 0.6% Ti steel: Forging; Hadfields; stabilized for welding **DPN: 170 UTS: 580 Elon: 60% Proof: 300**
ERA CRIS (Cb)	8% Ni 18% Cr Nb stabilized steel: Casting; Hadfields for BS 1631B **UTS: 450 Elon: 20% Proof: 200**
ERA CRIS (Cb)	0.15% C 9% Ni 18% Cr 1.0% Nb steel: Wrought; Hadfields for BS alloy En 58F and 58G
ERA HR 1	0.3% C 9% Ni 17% Cr 3% W steel: Wrought; Hadfields for BS alloy En 55
ERA HR 1	7% Ni 21% Cr 4% W steel: Casting; Hadfields for BS 1648 D **UTS: 740 Elon: 15% Proof: 450**
ERA HR 1S	7% Ni 21% Cr 3.5% Si steel: Casting; Hadfields for BS 1648 D
ERA HR 2	7% Ni 21% Cr steel: Casting; Hadfields for BS 1648 D **UTS: 700 Elon: 15% Proof: 300**
ERA HR 2W	7% Ni 21% Cr 2% W steel: Casting; Hadfields for BS 1648 D
ERA HR 3	20% Ni 25% Cr steel: Casting; Hadfields for BS 1648 F **UTS: 600 Elon: 16%**
ERA HR 6	12% Ni 25% Cr steel: Casting; Hadfields for BS 1648 E **UTS: 450 Elon: 20%**
ERA HR 6W	12% Ni 25% Cr 3.5% W steel: Casting; Hadfields for BS 1648 E **UTS: 450 Elon: 20%**
ERA HR 7	15% Ni 15% Cr steel: Casting; Hadfields for BS 1648 H
ERA HR 8	25% Ni 25% Cr W steel: Casting; Hadfields for BS 1648 H
ESCALLOY 20	0.7% C 28% Ni 20% Cr 2.5% Mo 1% Nb 3.2% Cu steel: Eastern Stainless
ESSHETE 316	0.07% C 12.5% Ni 17.0% Cr 2.3% Mo steel: Samuel Fox; solution treated **UTS: 500 Elon: 30% Proof: 210**
ESSHETE 321	0.07% C 10.0% Ni 18.0% Cr 0.7% Ti (max) steel: Samuel Fox; solution treated **UTS: 500 Elon: 30% Proof: 210**
ESSHETE 347	0.07% C 11.0% Ni 18.0% Cr 0.7% Nb (min) steel: Samuel Fox; solution treated **UTS: 500 Elon: 30% Proof: 220**
EV 3	0.2% C 11.5% Ni 21% Cr steel: For exhaust valves; designation used by SAE **UTS: 840 Elon: 26% Proof: 450**
EV 10	1% C 14% Ni 14% Cr steel: Casting for exhaust valves; designation used by SAE
EV 15	0.32% C 7.5% Ni 23% Cr steel: For valves; SAE designation
EV 16	0.3% C 8% Ni 23% Cr steel: For valves; SAE designation
EV 17	0.1% C 9% Ni 18% Cr 3.5% Cu steel: For valves; SAE designation

Symbol	Nominal analysis, supplier, condition and remarks.
EVERSHYNE	Osborn trade name for austenitic steel; not weld stabilized
EW 2	0.08% C 10% Ni 18% Cr 1.5% Mo steel: Osborn; weldable; acid resistant **UTS: 740 Elon: 45% Proof: 300**
EW 3	0.08% C 10% Ni 18% Cr 2.5% Mo steel: Osborn; weldable; acid resistant **UTS: 760 Elon: 45% Proof: 330**
EW 4	0.06% C 10% Ni 18% Cr 4% Mo steel: Osborn; austenitic; weldable; acid resistant **UTS: 760 Elon: 45% Proof: 370**
EZEFORM 35	8% Ni 18% Cr steel: Atlas
F-10	0.15% C 21% Ni 8% Cr steel: Origin unknown
FAGERSTA R310	0.07% C 10.0% Ni 18.0% Cr steel: For deep drawing; Fagersta
FAGERSTA R320	0.06% C 9.0% Ni 18.0% Cr steel: For deep drawing; Fagersta
FAGERSTA R350	0.04% C 9.0% Ni 18.5% Cr steel: For deep drawing; Fagersta
FAGERSTA R358	0.04% C 9.5% Ni 18.0% Cr Nb: Stabilized steel; Fagersta; for heat resistance
FAGERSTA R359	0.04% C (max) 9.5% Ni 18.0% Cr Ti: Stabilized steel; Fagersta; for heat resistance
FAGERSTA R360	0.03% C 10.0% Ni 18.5% Cr steel: For tubes; Fagersta
FAGERSTA R390	0.05% C 11.5% Ni 18.0% Cr steel: For deep drawing and spinning; Fagersta
FAGERSTA R428	0.04% C 11.0% Ni 17.5% Cr 2.3% Mo with Nb steel: Fagersta
FAGERSTA R440	0.05% C 12.0% Ni 17.0% Cr 2.8% Mo steel: Acid resistant; Fagersta
FAGERSTA R450	0.04% C 10.5% Ni 18.0% Cr 1.5% Mo steel: For acid resistance; Fagersta
FAGERSTA R460	0.03% C 12.5% Ni 17.0% Cr 2.8% Mo steel: For acid resistance; Fagersta
FAGERSTA R590	0.06% C 12.0% Ni 13.5% Cr steel: For deep drawing: Fagersta
FAGERSTA R800	0.08% C 19.0% Ni 20.5% Cr steel: Tube for heat resistance; Fagersta
FAGERSTA R820	0.05% C 20.5% Ni 25.0% Cr steel: Tube for heat resistance; Fagersta
FCB	0.06% C 10% Ni 18% Cr 0.7% Nb steel: Firth Vickers **DPN: 175 UTS: 630 Elon: 58% Proof: 250**
FCB(T)	0.08% C 12% Ni 17.5% Cr 0.8% Nb steel: Firth Vickers **UTS: 580 Elon: 45%**
FDP	0.05% C 9% Ni 18% Cr 0.5% Ti steel: Firth Vickers **DPN: 160 UTS: 620 Elon: 50% Proof: 240**
FDP (L)	0.05% C 9% Ni 18% Cr 0.5% Ti steel: Firth Vickers
FMB	0.06% C 11% Ni 18% Cr 2.8% Mo steel: Firth Vickers
FMB Ti	0.06% C 12% Ni 18% Cr 2.8% Mo 0.3% Ti steel: Firth Vickers; weldable **DPN: 180 UTS: 620 Elon: 50% Proof: 270**
FMB (L)	0.03% C 12% Ni 17.5% Cr 2.8% Mo steel: Firth Vickers
FML	0.07% C 9.5% Ni 18% Cr 1.2% Mo steel: Firth Vickers; for acid resistance **DPN: 170 UTS: 620 Elon: 50% Proof: 250**
FNB	0.4% C 13.5% Ni 14% Cr 2.5% W steel: Darwins for BS alloy En 54
FOREMOST 18/8	0.15% C 9% Ni 18.5% Cr steel: Swift Levick for BS alloy En 58A
FOREMOST 18/8 Mo	0.12% C 9% Ni 18% Cr 2.7% Mo steel: Swift Levick for BS alloy En 58J
FOREMOST 18/8 Nb	0.1% C 9% Ni 18.5% Cr 1% Nb steel: Swift Levick for BS alloy En 58F
FOREMOST 18/8 Ti	0.1% C 9% Ni 18.5% Cr 0.5% Ti steel: Swift Levick for BS alloy En 58B
FOREMOST HR 1	0.3% C 20% Ni 25% Cr 3% W steel: Casting; Swift Levick **DPN: 220 UTS: 760 Elon: 27% Proof: 420**

Symbol	Nominal analysis, supplier, condition and remarks.
FOREMOST HR 5	0.3% C 8% Ni 20% Cr 4% W steel: Casting; Swift Levick; annealed **DPN: 220 UTS: 830 Elon: 49% Proof: 330**
FOREMOST HR 6	0.1% C 18% Ni 18% Cr 3.5% Mo 0.6% Ti 2.5% Cu steel: Casting; Swift Levick; annealed **DPN: 163 UTS: 390 Elon: 12% Proof: 180**
FOREMOST HR 7	0.2% C 12% Ni 25% Cr 3% W steel: Casting; Swift Levick; annealed **DPN: 220 UTS: 710 Elon: 32% Proof: 270**
FOREMOST HR 8	0.2% C 20% Ni 25% Cr steel: Casting; Swift Levick; annealed **DPN: 200 UTS: 630 Elon: 48%**
FORMWELL	0.11% C 9.5% Ni 17% Cr steel: Atlas; annealed **DPN: 130 UTS: 620 Elon: 63% Proof: 220**
FS 16/25/6	0.05% C 25% Ni 16% Cr 6.5% Mo steel: Firth Sterling **DPN: 201 UTS: 840 Elon: 35% Proof: 390**
FSL	0.06% C 10% Ni 18.5% Cr steel: Firth Vickers
FSL (L)	0.05% C 10% Ni 18.5% Cr steel: Sheet; Firth Vickers **UTS: 610 Elon: 60% Proof: 220**
FSQ	0.1% C 11% Ni 18% Cr steel: Firth Vickers
FST	0.08% C 9% Ni 18% Cr steel: Firth Vickers **DPN: 170 UTS: 610 Elon: 50%**
FST (L)	0.6% C 10% Ni 18.5% Cr steel: Firth Vickers
FV 302	0.08% C 9% Ni 18% Cr steel: Firth Vickers **DPN: 170 UTS: 610 Elon: 50%**
FV 303	0.1% C 9% Ni 18% Cr 0.3% Mo 0.6% Ti 0.25% S steel: Firth Vickers; free cutting; weldable **UTS: 600 Elon: 53% Proof: 240**
FV 304	0.06% C 10% Ni 18.5% Cr steel: Sheet; Firth Vickers; weldable **DPN: 160 UTS: 610 Elon: 60% Proof: 220**
FV 304 (L)	0.03% C 11% Ni 18% Cr steel: Firth Vickers; weldable **DPN: 160 UTS: 580 Elon: 60% Proof: 220**
FV 305	0.1% C 11% Ni 18% Cr steel: For deep drawing; Firth Vickers; annealed **UTS: 580 Elon: 60% Proof: 240**
FV 309	0.20% C 11.5% Ni 23% Cr 2.8% W steel: Firth Vickers **DPN: 220 UTS: 740 Elon: 28%**
FV 310	0.1% C 21% Ni 23% Cr steel: Firth Vickers **DPN: 190 UTS: 700 Elon: 35%**
FV 316	0.06% C 11% Ni 18% Cr 2.8% Mo steel: Firth Vickers; for acid resistance; sulphuric, chlorides, etc. **DPN: 180 UTS: 610 Elon: 50% Proof: 270**
FV 316 (L)	0.03% C 12% Ni 17.5% Cr 2.8% Mo steel: Firth Vickers; weldable **DPN: 160 UTS: 580 Elon: 50% Proof: 240**
FV 321	0.05% C 9% Ni 18% Cr 0.5% Ti steel: Firth Vickers; weldable **DPN: 160 UTS: 610 Elon: 50% Proof: 240**
FV 347	0.06% C 10% Ni 18% Cr 0.7% Nb steel: Firth Vickers; weldable **DPN: 175 UTS: 620 Elon: 58% Proof: 250**
FV 555	0.05% C 10.5% Ni 16.5% Cr 2.4% Mo steel: Firth Vickers; this steel is developed for its weldability **UTS: 540 Elon: 67% Proof: 200**

Symbol	Nominal analysis, supplier, condition and remarks.
FVS	0.4% C 14% Ni 13.6% Cr 2.6% W steel: Firth Vickers **DPN: 230 UTS: 760 Elon: 35%**
G 2	0.42% C 14% Ni 14% Cr 2.6% W steel: Wrought; Jessop; for BS alloy En 54
G 4	0.12% C 13% Ni 16% Cr 3% W 0.5% Ti steel: Jessop
G 9	0.12% C 11% Ni 16% Cr 1% W 0.4% Ti steel: Jessop
G 19	0.4% C 13% Ni 19% Cr 1.8% Mo 2.6% W 10% Co 3.1% Nb steel: Casting; Jessop; as cast; creep resisting **UTS: 580 Elon: 4% Proof: 300**
G 19	0.4% C 13% Ni 19% Cr 1.7% Mo 2.5% W 10% Co 3% Nb steel: Jessop; wrought
G 21	0.4% C 13% Ni 13% Cr 2.5% W 1.0% Nb steel: Jessop; creep resisting **UTS: 730 Elon: 47% Proof: 250**
G 38	0.17% C 12% Ni 16% Cr 2.5% Mo 1.25% V 1.3% W steel: Jessop
G 40	0.23% C 25% Ni 20% Cr 3% Mo 1% W 1% Ti steel: Jessop **UTS: 840 Elon: 18% Proof: 630**
G 71	0.10% C 37.0% Ni 18.0% Cr steel: Jessop; as cast; scale or creep resistant
G 72	0.10% C 32.0% Ni 21.0% Cr steel: Jessop; as cast; scale or creep resistant
G 125	0.02% C 18.5% Ni 9.0% Cr 5.0% Mo 0.65% Ti 0.9% Al steel: Jessop
G 3426 STC 52D	0.08% C 12% Ni 17% Cr 2.5% Mo steel: Japanese Standard
G 4301 SEC 7	0.15% C (max) 9% Ni 18% Cr steel: Japanese Standard
G 4301 SEC 8	0.08% C (max) 10% Ni 19% Cr steel: Japanese Standard
G 4301 SEC 9	0.03% C (max) 10% Ni 19% Cr steel: Japanese Standard
G 4301 SEC 10	0.08% C 10.5% Ni 18% Cr 0.5% Ti steel: Japanese Standard
G 4301 SEC 13	0.03% C 12% Ni 17% Cr 2.5% Mo steel: Japanese Standard
G 4302 SEH 5	0.25% C 20% Ni 25% Cr steel: Japanese Standard
GAMMA Cb	0.4% C 24.5% Ni 15.2% Cr 4.1% Mo 2.2% Nb Fe alloy: Origin unknown
GOST 5632/0 KH10N20T2	0.08% C 19.0% Ni 11.0% Cr 2.0% Ti 1.0% Al (max) steel: Russian Standard
GOST 5632/0 KH14N28V3T31UR	0.08% C (max) 27.5% Ni 14.0% Cr 3.2% W 2.8% Ti 0.8% Al 0.02% B steel: Russian Standard
GOST 5632/0 KH17N16M3T	0.08% C (max) 16.0% Ni 17.0% Cr 2.7% Mo 0.45% Ti steel: Plate; Russian Standard
GOST 5632/0 KH18N10	0.08% C (max) 10.0% Ni 18.0% Cr steel: Russian Standard
GOST 5632/00 KH18N10	0.04% C (max) 1.5% Mn 10% Ni 18% Cr steel: Plate; Russian Standard **UTS: 450 Proof: 160**
GOST 5632/0 KH18N10E	0.12% C (max) 10.0% Ni 18.0% Cr 0.25% Se Te steel: Russian Standard; free machining
GOST 5632/0 KH18N10T	0.08% C 1.5% Mn 10% Ni 18% Cr 0.5% Ti steel: Plate; Russian Standard; solution treated **UTS: 500 Proof: 200**
GOST 5632/0 KH18N11	0.06% C (max) 11.0% Ni 18.0% Cr steel: Plate; Russian Standard; solution treated **UTS: 520**
GOST 5632/0 KH18N12B	0.08% C (max) 1.5% Mn 12% Ni 18% Cr 0.9% Nb 0.02% S (max) 0.035% P (max) steel: Plate; Russian Standard; solution treated **UTS: 550 Proof: 200**
GOST 5632/0 KH18N12B	0.08% C (max) 12.5% Ni 18.0% Cr 1.2% Nb steel: Russian Standard
GOST 5632/0 KH18N12T	0.08% C (max) 1.5% Mn 12% Ni 18% Cr 0.5% Ti steel: Plate; Russian Standard; solution treated
GOST 5632/0 KH20N14S2	0.08% C (max) 13.5% Ni 20.5% Cr steel: Plate; Russian Standard
GOST 5632/0 KH21N5T	0.08% C (max) 0.8% Mn 5.2% Ni 21.0% Cr 0.45% Ti steel: Russian Standard

Note. The following abbreviations and units are used in the tables:

DPN	Hardness, diamond pyramid number
UTS	Ultimate tensile strength, N/mm^2
Elon	Elongation, %
Proof	0.1% proof strength, N/mm^2

1 N/mm^2=0.1 hbar=0.102 kgf/mm^2=0.06475 $tonf/in.^2$=145.04 $lbf/in.^2$=1 MPa

See Appendix II for other abbreviations and conversion tables.

Symbol	Nominal analysis, supplier, condition and remarks.
GOST 5632/0 KH21N6M2T	0.08% C (max) 6.0% Ni 21.0% Cr 2.1% Mo 0.3% Ti steel: Plate; Russian Standard
GOST 5632/0 KH23N18	0.1% C (max) 18.5% Ni 23.5% Cr steel: Russian Standard
GOST 5632/0 KH23N28M2T	0.06% C (max) 26.5% Ni 23.5% Cr 2.2% Mo 0.6% Ti steel: Plate; Russian Standard
GOST 5632/0 KH23N28M3D3T	0.06% C (max) 27.5% Ni 23.5% Cr 2.7% Mo 0.6% Ti steel: Plate; Russian Standard
GOST 5632/0 KH23N28M3DeT	0.06% C (max) 27.5% Ni 23.5% Cr 2.7% Mo 0.55% Ti 3.0% Cu steel; Russian Standard
GOST 5632/1 KH 13N3	0.12% C 22.5% Ni 13.5% Cr steel: Russian Standard
GOST 5632/1 KH14N16B	0.1% C 15.5% Ni 14.0% Cr 1.1% Nb steel: Plate; Russian Standard
GOST 5632/1 KH14N16BR	0.1% C (max) 15.5% Ni 14.0% Cr 1.1% Nb 0.005% B (max) 0.02% Zr (max) steel: Plate; Russian Standard
GOST 5632/1 KH14N18V2B	0.1% C 19.0% Ni 14.5% Cr 2.2% W 1.0% Nb steel: Plate; Russian Standard
GOST 5632/1 KH14N18V2BR	0.1% C 19.0% Ni 14.0% Cr 2.2% W 1.0% Nb 0.005% B (max) 0.02% Zr (max) steel: Plate; Russian Standard
GOST 5632/1 KH14N18V2BR1	0.1% C 19.0% Ni 14.0% Cr 2.2% W 1.0% Nb 0.025% B (max) 0.02% Zr (max) steel: Plate; Russian Standard
GOST 5632/1 KH16N13M2B	0.09% C 13.5% Ni 16.0% Cr 2.25% Mo 1.1% Nb steel: Plate; Russian Standard
GOST 5632/1 KH21N5T	0.12% C 0.8% Mn (max) 5.2% Ni 21.0% Cr 0.8% Ti steel: Russian Standard
GOST 5632/1 KH25N25TR	0.1% C 25.5% Ni 24.5% Cr 1.3% Ti 0.01% B steel: Russian Standard
GOST 5632/2 KH18N9	0.17% C 9% Ni 18% Cr 1.6% Mn steel: Plate; Russian Standard **UTS: 675** **Proof: 280**
GOST 5632/2 KH18N92	0.17% C 9.0% Ni 18.0% Cr steel: Russian Standard
GOST 5632/3 KH13N7S2	0.3% C 6.5% Ni 13.0% Cr steel: Plate; Russian Standard
GOST 5632/3 KH19N9MVBT	0.31% C 9.0% Ni 19.0% Cr 1.2% Mo 1.2% W 0.35% Ti 0.35% Nb steel
GOST 5632/4 KH12N8G8MFB	0.38% C 8.0% Ni 12.5% Cr 1.2% Mo 1.3% V 0.3% Ti steel: Plate; Russian Standard
GOST 5632/4 KH14N14V2M	0.45% C 14.0% Ni 14.0% Cr 0.32% Mo 2.2% W steel: Plate; Russian Standard
GOST 5632/4 KH18N2552	0.36% C 24.5% Ni 18.0% Cr steel: Plate; Russian Standard
GOST 5632/4 KH18N2582	0.36% C 24.5% Ni 18.0% Cr steel: Russian Standard
GOST 5632/610 KH18N10	0.08% C (max) 1.5% Mn 10% Ni 18% Cr 0.02% S (max) 0.035% P (max) steel: Plate; Russian Standard **UTS: 630** **Proof: 220**
GOST 5632 KH18N9	0.12% C 1.5% Mn 9.0% Ni 18% Cr steel: Plate; Russian Standard
GOST 5632 0KH18N10T	0.08% C (max) 10.0% Ni 18.0% Cr 0.6% Ti steel: Russian Standard
GOST 5632 0KH23N18	0.1% C (max) 18.5% Ni 23.5% Cr steel: Plate; Russian Standard
GOST 5632 KH12N20T3R	0.1% C (max) 19.5% Ni 11.2% Cr 2.9% Ti 0.8% Al 0.0014% B steel: Russian Standard
GOST 5632 KH12N22T3MR	0.1% C (max) 22.5% Ni 11.2% Cr 1.3% Mo 2.9% Ti 0.8% Al 0.02% B steel: Russian Standard
GOST 5632 KH15N91U	0.09% C (max) 8.1% Ni 15.0% Cr 1.0% Al steel: Plate; Russian Standard
GOST 5632 KH16N15M3B	0.09% C (max) 15.0% Ni 16.0% Cr 2.75% Mo 0.7% Nb steel: Plate; Russian Standard
GOST 5632 KH17N13M2T	0.1% C (max) 13.0% Ni 17.0% Cr 2.2% Mo 0.45% Ti steel: Plate; Russian Standard
GOST 5632 KH17N13M3T	0.1% C (max) 13.0% Ni 17.0% Cr 3.5% Mo 0.45% Ti steel: Plate; Russian Standard
GOST 5632 KH17N71U	0.09% C (max) 7.0% Ni 17% Cr 1.0% Al steel: Plate; Russian Standard
GOST 5632 KH18N9T	0.12% C (max) 1.5% Mn 8.5% Ni 19% Cr 0.7% Ti steel: Plate; Russian Standard; solution treated **UTS: 550** **Proof: 200**
GOST 5632 KH18N10E	0.12% C (max) 10.0% Ni 18.0% Cr 0.25% Te or Se steel: Plate; Russian Standard
GOST 5632 KH18N10T	0.12% C (max) 1.5% Mn 10% Ni 18% Cr 0.7% Ti steel: Plate; Russian Standard; solution treated **UTS: 520** **Proof: 200**
GOST 5632 KH18N12T	0.12% C (max) 1.5% Mn 12% Ni 18% Cr 0.7% Ti steel: Plate; Russian Standard; solution treated **UTS: 550** **Proof: 200**
GOST 5632 KH20N14S2	0.2% C (max) 1.5% Mn (max) 13.5% Ni 20.5% Cr steel: Russian Standard
GOST 5632 KH20N14S2	0.08% C (max) 13.5% Ni 20.5% Cr steel: Russian Standard
GOST 5632 KH23N13	0.2% C (max) 13.5% Ni 23.5% Cr steel: Plate; Russian Standard; solution treated **UTS: 500** **Proof: 300**
GOST 5632 KH23N18	0.2% C (max) 18.5% Ni 23.5% Cr steel: Plate; Russian Standard
GOST 5632 KH23NB	0.2% C (max) 2.0% Mn (max) 13.5% Ni 23.5% Cr steel: Russian Standard
GOST 5632 KH25H25TR	0.1% C 25.5% Ni 24.5% Cr 1.4% Ti 0.01% B (max) steel: Plate; Russian Standard
GOST 5632 KH25N16G7AR	0.12% C (max) 6.0% Mn 17.5% Ni 24.5% Cr 0.02% B 0.37% N steel: Russian Standard
GOST 5632 KH25N20S2	0.2% C (max) 19.5% Ni 25.5% Cr steel: Plate; Russian Standard; solution treated **UTS: 600** **Proof: 300**
GOST 9940/0 KH17N16M3T	0.08% C (max) 16.0% Ni 17.0% Cr 2.7% Mo 0.45% Ti steel: Pipe; Russian Standard; seamless **UTS: 500**
GOST 9940/00 KH18N10	0.04% C (max) 10.0% Ni 18.0% Cr steel: Pipe; Russian Standard; seamless **UTS: 450**
GOST 9940/0 KH18N10	0.08% C (max) 10.0% Ni 18.0% Cr steel: Pipe; Russian Standard; seamless **UTS: 520**
GOST 9940/0 KH18N10T	0.08% C (max) 10.0% Ni 18.0% Cr 0.6% Ti steel: Pipe; Russian Standard; seamless **UTS: 520**
GOST 9940/0 KH18N12B	0.08% C (max) 12.0% Ni 18.0% Cr 1.2% Nb steel: Pipe; Russian Standard; seamless **UTS: 520**
GOST 9940/0 KH18N12T	0.08% C (max) 12.0% Ni 18.0% Cr 0.6% Ti steel: Pipe; Russian Standard; seamless **UTS: 520**
GOST 9940/0 KH20N1452	0.08% C (max) 13.5% Ni 20.5% Cr steel: Pipe; Russian Standard; seamless
GOST 9940/0 KH21N5T	0.08% C (max) 5.2% Ni 21.0% Cr 0.45% Ti steel: Pipe; Russian Standard; seamless
GOST 9940/0 KH23N18	0.1% C (max) 18.5% Ni 23.5% Cr steel: Pipe; Russian Standard; seamless **UTS: 500**
GOST 9940/1 KH14N18V2BR	0.1% C 19.0% Ni 14.0% Cr 2.3% W 1.1% Nb steel: Plate; Russian Standard; seamless **UTS: 550**
GOST 9940/1 KH21N5T	0.12% C 5.2% Ni 21.0% Cr 0.45% Ti steel: Pipe; Russian Standard; seamless
GOST 9940/2 KH18N9	0.12% C (max) 9.0% Ni 18.0% Cr steel: Pipe; Russian Standard; seamless **UTS: 540**
GOST 9940 KH18N9	0.12% C (max) 9.0% Ni 18.0% Cr steel: Pipe; Russian Standard; seamless **UTS: 540**
GOST 9940 KH17NBM2T	0.1% C (max) 13.0% Ni 17.0% Cr 2.1% Mo 0.45% Ti steel: Pipe; Russian Standard; seamless **UTS: 540**

Symbol	Nominal analysis, supplier, condition and remarks.
GOST 9940 KH18N10T	0.12% C (max) 10.0% Ni 18.0% Cr 0.7% Ti steel: Pipe; Russian Standard; seamless **UTS: 540**
GOST 9940 KH18N12T	0.12% C (max) 12.0% Ni 18.0% Cr 0.7% Ti steel: Pipe; Russian Standard; seamless **UTS: 540**
GOST 9941/0 KH17N16M3T	0.08% C (max) 13.0% Ni 17.0% Cr 2.1% Mo 0.45% Ti steel: Pipe; Russian Standard; seamless **UTS: 540**
GOST 9941/0 KH18N10	0.08% C (max) 10.0% Ni 18.0% Cr steel: Pipe; Russian Standard; seamless **UTS: 450**
GOST 9941/00 KH18N10	0.04% C (max) 16.0% Ni 17.0% Cr 2.7% Mo 0.45% Ti: Russian Standard: seamless **UTS: 500**
GOST 9941/0 KH18N10T	0.08% C (max) 10.0% Ni 18.0% Cr 0.6% Ti steel: Pipe; Russian Standard; seamless **UTS: 520**
GOST 9941/0 KH18N12B	0.08% C (max) 12.0% Ni 18.0% Cr 0.7% Ti steel: Pipe; Russian Standard; seamless **UTS: 540**
GOST 9941/0 KH18N12T	0.08% C (max) 10.0% Ni 18.0% Cr 0.7% Ti steel: Pipe; Russian Standard; seamless **UTS: 540**
GOST 9941/0 KH20N1452	0.08% C (max) 13.5% Ni 20.5% Cr steel: Pipe; Russian Standard; seamless
GOST 9941/0 KH23N18	0.1% C (max) 18.5% Ni 23.5% Cr steel: Pipe; Russian Standard; seamless **UTS: 500**
GOST 9941/1 KH14N18V2BR	0.1% C 5.1% Ni 21.0% Cr 0.8% Ti steel: Pipe; Russian Standard; seamless
GOST 9941/1 KH18N9	0.17% C 9.0% Ni 18.0% Cr steel: Pipe; Russian Standard; seamless **UTS: 540**
GOST 9941/1 KH21N5T	0.12% C 5.2% Ni 21.0% Cr 0.45% Ti steel: Pipe; Russian Standard; seamless
GOST 9941 KH17N13M2T	0.1% C (max) 19.0% Ni 14.5% Cr 2.2% W 1.0% Nb steel: Pipe; Russian Standard; seamless **UTS: 550**
GOST 9941 KH18N9	0.12% C (max) 10.0% Ni 18.0% Cr steel: Pipe; Russian Standard; seamless **UTS: 520**
GOST 9941 KH18N12T	0.12% C (max) 12.0% Ni 18.0% Cr 0.6% Ti steel: Pipe; Russian Standard; seamless **UTS: 520**
GOST 9941 KH21N5T	0.08% C (max) 13.5% Ni 21.5% Cr steel: Pipe; Russian Standard; seamless
GOST 10498/0 KH16N15M3	0.03% C (max) 15.0% Ni 15.5% Cr 3.0% Mo steel: Pipe; Russian Standard; seamless **UTS: 560**
GOST 10498/00 KH16N15M3B	0.04% C (max) 15.0% Ni 16.0% Cr 2.7% Mo 0.5% Nb steel: Pipe; Russian Standard; seamless **UTS: 520**
GOST 10498/0 KH18N10T	0.05% C 10.0% Ni 18.0% Cr 0.6% Ti steel: Pipe; Russian Standard; seamless **UTS: 540**
GOST 10498/0 KH18N10T	0.85% C 10.0% Ni 18.0% Cr 0.8% Ti steel: Pipe; Russian Standard; seamless **UTS: 560**

Note. The following abbreviations and units are used in the tables:

DPN	Hardness, diamond pyramid number
UTS	Ultimate tensile strength, N/mm^2
Elon	Elongation, %
Proof	0.1% proof strength, N/mm^2

1 N/mm^2=0.1 hbar=0.102 kgf/mm^2=0.06475 $tonf/in.^2$=145.04 $lbf/in.^2$=1 MPa

See Appendix II for other abbreviations and conversion tables.

Symbol	Nominal analysis, supplier, condition and remarks.
GOST 10498/1 KH16N15M3B	0.06% C 15.0% Ni 16.0% Cr 3.0% Mo 0.6% Nb steel: Pipe; Russian Standard; seamless **UTS: 550**
GOST 10802/1 KH14N18V2BR	0.09% C 19.0% Ni 14.0% Cr 2.4% W 1.1% Nb steel: Pipe; Russian Standard; seamless **UTS: 550** **Proof: 240**
GOST 10802 KH16N14V2BR	0.09% C 14.0% Ni 16.5% Cr 2.4% W 1.1% Nb steel: Pipe; Russian Standard; seamless **UTS: 550** **Proof: 220**
GOST 10802 KH16N16V2MBR	0.08% C 16.0% Ni 16.0% Cr 0.65% Mo 2.5% W 0.8% Nb 0.3% Cu (max) steel: Pipe; Russian Standard; seamless **UTS: 550** **Proof: 240**
GOST 10802 KH18N12T	0.12% C (max) 12.0% Ni 18.0% Cr 0.7% Ti steel: Pipe; Russian Standard; seamless **UTS: 540** **Proof: 220**
GOST 11068/0 KH18N10T	0.08% C (max) 10.0% Ni 18.0% Cr 0.6% Ti steel: Pipe; Russian Standard; seamless **UTS: 540**
GOST 11068/00 KH18N10T	0.04% C (max) 10.0% Ni 18.0% Cr 0.5% Ti steel: Tube; Russian Standard; seamless **UTS: 500**
GOST 11068/0 KH18N12T	0.08% C (max) 12.0% Ni 18.5% Cr 0.6% Ti steel: Pipe; Russian Standard; seamless **UTS: 540**
GOST 11068/0 KH21N5T	0.08% C (max) 5.2% Ni 21.0% Cr steel: Pipe; Russian Standard; seamless
GOST 11068/0 KH23N28M3D3T	0.06% C (max) 27.5% Ni 23.5% Cr 3.0% Mo 0.5% Ti steel: Pipe; Russian Standard; seamless
GOST 11068/1 KH21N5T	0.11% C 5.2% Ni 21.0% Cr 0.8% Ti steel: Pipe; Russian Standard; seamless
GOST 110680 KH23N28M2T	0.06% C (max) 27.5% Ni 23.5% Cr 2.1% Mo 0.5% Ti steel: Pipe; Russian Standard; seamless
GOST 11068 KH17N13M2T	0.1% C (max) 13.0% Ni 17.0% Cr 2.1% Mo 0.45% Ti steel: Pipe; Russian Standard; seamless
GOST 11068 KH17N13M3T	0.1% C 13.0% Ni 17.0% Cr 3.5% Mo 0.45% Ti steel: Pipe; Russian Standard; seamless
GOST 11068 KH18N10T	0.12% C (max) 10.0% Ni 18.0% Cr 0.7% Ti steel: Pipe; Russian Standard; seamless **UTS: 560**
GOST 11068 KH18N12T	0.12% C (max) 12.0% Ni 18.0% Cr 0.7% Ti steel: Pipe; Russian Standard; seamless **UTS: 560**
GOST 14162/0 KH18N10T	0.08% C (max) 9.5% Ni 18.0% Cr 0.6% Ti steel: Capillary tube; Russian Standard **UTS: 540**
GOST 14162/0 KH18N12T	0.08% C (max) 12.0% Ni 18.0% Cr 0.6% Ti steel: Capillary tube; Russian Standard **UTS: 520**
GOST 14162 KH18N9	0.12% C (max) 9.0% Ni 18.0% Cr steel: Capillary tube; Russian Standard
GOST 14162 KH18N10T	0.12% C (max) 9.5% Ni 18.0% Cr 0.7% Ti steel: Capillary tube; Russian Standard **UTS: 560**
GOST 14162 KH18N12T	0.12% C (max) 12.0% Ni 18.0% Cr 0.7% Ti steel: Capillary tube; Russian Standard **UTS: 520**
GT 45	0.12% C 14% Ni 16% Cr 2.5% Mo 0.25% Ti 0.5% Nb 3% Cu steel: Armco
H 500	0.07% C 32.0% Ni 20.8% Cr 0.3% Ti 0.3% Al steel: VEW; heat resistant
H 508	0.4% C 32.5% Ni 20.5% Cr 0.6% Ti 0.5% Al steel: VEW; heat resistant
H 520	0.12% C (max) 35% Ni 15.8% Cr steel: VEW; heat resistant
H 522	0.15% C (max) 19.8% Ni 24.8% Cr steel: VEW; heat resistant
H 525	0.08% C 19.8% Ni 24.8% Cr steel: VEW; heat resistant
H 532	0.08% C (max) 20% Ni 25.3% Cr steel: VEW; heat resistant

Symbol	Nominal analysis, supplier, condition and remarks.
H 550	0.09% C 11.5% Ni 19.5% Cr steel: VEW; heat resistant
H 800	0.43% C 9.0% Ni 18% Cr 1.0% W 2.3% Si steel: VEW; for valves
HAI Ni Cr 35	35% Ni 20% Cr Fe alloy: For resistance heating weld; Harrison Alloy
HAI Ni Cr 35 Cb	35% Ni 20% Cr 1% Nb Fe alloy: Harrison Alloy
HAI Ni Cr 40	37% Ni 20% Cr Fe alloy: For resistance heating weld; Harrison Alloy
HAI Ni Cr 80 Cb	35% Ni 20% Cr 1% Nb Fe alloy: Harrison Alloy
HAYNES 20 MOD	0.05% C (max) 26% Ni 22% Cr 5% Mo 0.2% Ti Fe alloy
HE	0.35% C 9% Ni 28% Cr 0.5% Mo steel: Casting; Huntington; aged **UTS: 620 Elon: 10% Proof: 370**
HE-35	0.35% C 9.5% Ni 28% Cr steel: Casting; designation used in the UK and USA
HEADWELL	0.07% C 18% Ni 16% Cr steel: Atlas; annealed **DPN: 130 UTS: 530 Elon: 52% Proof: 200**
HEDEX 14	0.08% C 10% Ni 18% Cr steel: Wire for forging; Kiveton Park for BS alloy En 58E
HEDEX 35	0.08% C (max) 11% Ni 18% Cr steel: Wire for forging; Kiveton Park for BS alloy En 58E
HEDEX 36	0.12% C (max) 10% Ni 19% Cr 3% Mo steel: Wire for forging; Kiveton Park for BS alloy En 58J
HEDEX 37	0.05% C (max) 10% Ni (min) 19% Cr steel: Wire for forging; Kiveton Park for BS alloy En 58E
HEDEX 44	0.08% C (max) 14% Ni 18% Cr steel: Wire for forging; Kiveton Park; annealed **UTS: 690**
HEDEX 45	0.08% C (max) 16% Ni 18% Cr steel: Wire for forging; Kiveton Park; annealed **UTS: 690**
HEV7	0.08% C 26% Ni 14.75% Cr 1.2% Mo 0.3% V 2% Ti steel: For valves; SAE designation
HF	0.3% C 11% Ni 20% Cr 0.5% Mo steel: Casting; Huntington; aged **UTS: 710 Elon: 25% Proof: 330**
HF-30	0.3% C 10.5% Ni 21% Cr steel: Casting; designation used in the UK and USA
HH	0.3% C 12% Ni 26% Cr 0.5% Mo 0.2% N steel: Casting; Huntington; aged **UTS: 630 Elon: 8% Proof: 300**
HH-30	0.3% C 13.5% Ni 26% Cr steel: Casting; designation used in the UK and USA
HH-33	0.32% C 13% Ni 25% Cr steel: Casting; designation used in the UK and USA
HI	0.35% C 16% Ni 28% Cr 0.5% Mo steel: Casting; Huntington; aged **UTS: 630 Elon: 6% Proof: 450**
Hi Ni 1	3.7% C 8% Ni 16% Cr 0.5% V Fe alloy: Casting; Dewrance **DPN: 540 UTS: 580**
HI-35	0.35% C 16.5% Ni 28% Cr steel: Casting; designation used in the UK and USA
HI-PROOF 304	0.06% C (max) 9.5% Ni 19% Cr steel: S Fox **UTS: 580 Elon: 35% Proof: 300**
HI-PROOF 304L	0.03% C (max) 10% Ni 19% Cr steel: S Fox **UTS: 610 Elon: 35% Proof: 350**
HI-PROOF 316	0.07% C (max) 12% Ni 17% Cr 2.5% Mo steel: S Fox **UTS: 580 Elon: 35% Proof: 320**
HI-PROOF 316L	0.03% C (max) 12% Ni 17% Cr 2.5% Mo steel: S Fox **UTS: 580 Elon: 35% Proof: 320**
HI-PROOF 347	0.08% C (max) 10% Ni 18% Cr steel: S Fox **UTS: 610 Elon: 35% Proof: 350**
HK	0.4% C 20% Ni 26% Cr 0.5% Mo steel: Casting; Huntington; aged **UTS: 610 Elon: 10% Proof: 330**
HK-30	0.3% C 20.5% Ni 25% Cr steel: Casting; designation used in the UK and USA

Symbol	Nominal analysis, supplier, condition and remarks.
HK-40	0.4% C 20.5% Ni 25% Cr steel: Casting; designation used in the UK and USA
HL	0.4% C 20% Ni 30% Cr 0.5% Mo steel: Casting; Huntington; as cast **UTS: 570 Elon: 19% Proof: 330**
HL	0.4% C 19% Ni 30% Cr Fe alloy: Designation used by ASTM
HL 30	0.3% C 20% Ni 30% Cr Fe alloy: Designation used by ASTM
HL 40	0.4% C 20% Ni 30% Cr Fe alloy: Designation used by ASTM
HN	0.35% C 25% Ni 21% Cr 0.5% Mo steel: Casting; Huntington; as cast **UTS: 450 Elon: 17% Proof: 220**
HN-40	0.4% C 25% Ni 21% Cr steel: Casting; designation used in the UK and USA
HNM	0.3% C 9.5% Ni 18.5% Cr 0.23% P steel: American proprietary alloy listed in SAE year book
HNV 5	0.35% C 8% Ni 13% Cr 0.5% Mo steel: For valves; SAE designation
HOTSPUR A	0.4% C 10% Ni 14% Cr 3% W steel: Bar, billet, etc.; Brown Bayley for BS alloy En 54
HOTSPUR C	0.2% C 20% Ni 25% Cr steel: Bar, billet, etc.; Brown Bayley
HOTSPUR D	0.35% C 9% Ni 17% Cr 3% W steel: Bar, billet, etc.; Brown Bayley for BS alloy En 55
HOTSPUR F	0.2% C 12% Ni 25% Cr steel: Bar, billet, etc.; Brown Bayley
HP	0.55% C 36% Ni 26% Cr Fe alloy: Designation used by ASTM
HR CROWN 1	0.2% C 12.0% Ni 24.0% Cr 0.5% W steel: Casting; Firth Vickers; annealed **DPN: 220 UTS: 530 Elon: 32% Proof: 270**
HR CROWN MAX	0.2% C 11.5% Ni 23% Cr 2.8% W steel: Firth Vickers for BS alloy En 55 **DPN: 220 UTS: 740 Elon 28%**
HS 88	0.07% C 15% Ni 12.5% Cr 2% Mo 0.6% W 0.6% Ti 0.15% B steel: Haynes Stellite
HT	0.55% C 35% Ni 15% Cr 0.5% Mo steel: Casting; Huntington; aged **UTS: 540 Elon: 5% Proof: 300**
HT	0.4% C 19% Ni 30% Cr Fe alloy: Designation used by ASTM
HT 30	0.3% C 3.5% Ni 15% Cr Fe alloy: Designation used by ASTM
HU	0.55% C 39% Ni 19% Cr 0.5% Mo steel: Casting; Huntington; aged **UTS: 530 Elon: 5% Proof: 290**
HYDRA XL	0.25% C 11% Ni 23% Cr 3% W steel: Hall and Pickles; cores; die pins for non-ferrous casting **UTS: 700 Elon: 33% Proof: 390**
HYRESIST 94L	25% Ni 20% Cr 4.5% Mo 1.5% Cu steel: British Steel Corporation
HYRESIST 317LM	15% Ni 18% Cr 4.5% Mo steel: British Steel Corporation
ICL 164	0.08% C (max) 12.0% Ni 17.0% Cr 2.2% Mo steel: Creusot-Loire
ICL 164 BC	0.03% C (max) 12.0% Ni 17.0% Cr 2.2% Mo steel: Creusot-Loire
ICL 164 Nb	0.08% C (max) 12.0% Ni 17.5% Cr 2.5% Mo 0.4% Ti steel: Creusot-Loire
ICL 164 T	0.08% C (max) 12.0% Ni 17.5% Cr 2.5% Mo 0.9% Nb steel: Creusot-Loire
ICL 166	0.08% C 12.0% Ni 17.5% Cr 2.75% Mo steel: Creusot-Loire
ICL 166 BC	0.03% C 12.0% Ni 17.5% Cr 2.75% Mo steel: Creusot-Loire
ICL 167 CN	0.045% C (max) 12.0% Ni 17.6% Cr 2.4% Mo steel: Creusot-Loire
ICL 168	0.08% C (max) 13.0% Ni 19.0% Cr 3.5% Mo steel: Creusot-Loire

Symbol	Nominal analysis, supplier, condition and remarks.
ICL 168 BC	0.03% C (max) 13.0% Ni 19.0% Cr 3.5% Mo steel: Creusot-Loire
ICL 472	0.08% C (max) 9.2% Ni 19.0% Cr steel: Creusot-Loire
ICL 472 BC	0.03% C (max) 10.0% Ni 19.0% Cr steel: Creusot-Loire
ICL 472 Nb	0.08% C (max) 10.5% Ni 18.0% Cr 1.0% Nb + Ta steel: Creusot-Loire
ICL 472 T	0.08% C (max) 10.5% Ni 18.0% Cr 0.4% Ti steel: Creusot-Loire
ICL 473 BC	0.04% C (max) 9.5% Ni 19.2% Cr steel: Creusot-Loire
ICL 473 Nb	0.05% C (max) 9.5% Ni 19.0% Cr steel: Creusot-Loire
IMMACULATE	0.3% C 7.5% Ni 20% Cr 2.0% W steel: Wrought; Firth Vickers **DPN: 220 UTS: 830 Elon: 50% Proof: 380**
IMMACULATE 2W	0.25% C 1.0% Mn 9.0% Ni 21.0% Cr steel: Casting; Firth Vickers; annealed **DPN: 200 UTS: 530 Elon: 30% Proof: 250**
IMMACULATE 5	0.1% C 21% Ni 23% Cr steel: Firth Vickers **DPN: 180 UTS: 700 Elon: 35%**
IMMACULATE 5T	0.1% C 17% Ni 22% Cr 0.7% Ti steel: Firth Vickers **DPN: 180 UTS: 610 Elon: 45%**
INCO 020	0.07% C (max) 35% Ni 20% Cr 2.5% Mo 0.4% Nb 3.5% Cu steel: Inco; annealed **UTS: 620 Elon: 40% Proof: 300**
INCO 330	0.08% C (max) 35.5% Ni 18.5% Cr steel: Inco
INCO ALLOY 032	0.01% C 32% Ni 22% Cr 4.2% Mo Fe alloy
INCOLOY	0.1% C 32% Ni 20.5% Cr Fe alloy: Inco
INCOLOY	20% Cr 32% Ni Fe alloy: Heat resistant; H Wiggin; furnace equipment
INCOLOY 800	0.1% C 32% Ni 21% Cr 0.5% Cu Fe alloy: Wrought; H Wiggin; annealed **DPN: 180 UTS: 580**
INCOLOY 800H	32% Ni 20% Cr stainless steel: Wrought; H Wiggin
INCOLOY 800HT	0.08% C 32% Ni 21% Cr 0.4% Ti 0.4% Al Fe alloy: Inco
INCOLOY 801	0.04% C 32% Ni 20.5% Cr 1% Ti Fe alloy: Bar, sheet, etc.; Huntington
INCOLOY 802	0.3% C 32.5% Ni 21% Cr 1% Ti 0.7% Al Fe alloy
INCOLOY 810	0.25% C 32% Ni 21.0% Cr 0.5% Cu Fe alloy: Information from SAE handbook
INCOLOY 926	0.04% C (max) 28% Ni 16% Cr 3% Mo 2% Ti Fe alloy
INCOLOY D S	37% Ni 18% Cr 2% Se Fe alloy: Heat resistant; H Wiggin; furnace equipment; sulphur resistant
INCOLOY T	0.1% C 32% Ni 20.5% Cr 1.0% Ti Fe alloy: Inco
INOFO 10	0.16% C 9% Ni 18% Cr steel: Forez
INOFO 10B	0.07% C 10% Ni 18% Cr steel: Forez
INOFO 10Ti	0.12% C 10% Ni 18% Cr 0.5% Ti steel: Forez
INOFO 11	0.03% C 10% Ni 18% Cr steel: Forez
INOFO 12	0.07% C 11% Ni 18% Cr 2.5% Mo steel: Forez
INOFO 12B	0.04% C 11% Ni 18% Cr 2.5% Mo steel: Forez
INOFO 13S	0.12% C 10% Ni 18% Cr steel: Forez
IR 4	0.15% C (max) 9% Ni 18% Cr 0.6% Ti steel: Sandvik
IR 41	0.15% C (max) 9% Ni 18% Cr 1.0% Nb steel: Sandvik
JIS 3446 SU304	0.08% C (max) 9.5% Ni 19% Cr steel: Tube for structural use; Japanese Standard **UTS: 600 Proof: 210**
JIS G3446 SUS316	0.08% C (max) 12% Ni 17% Cr 2.5% Mo steel: Tube for structural use; Japanese Standard **UTS: 600 Proof: 210**
JIS G3446 SUS321	0.08% C (max) 11.0% Ni 18% Cr 0.4% Ti steel: Tube for structural use; Japanese Standard **UTS: 800 Proof: 530**
JIS G3446 SUS347	0.08% C (max) 11.0% Ni 18.0% Cr 0.8% Nb + Ta steel: Tube for structural use; Japanese Standard **UTS: 800 Proof: 530**
JIS G3447 SUS304LTBS	0.3% C (max) 11.0% Ni 19.0% Cr steel: Pipe; Japanese Standard **UTS: 490**
JIS G3447 SUS304TBS	0.08% C (max) 9.5% Ni 19.0% Cr steel: Pipe; Japanese Standard **UTS: 530**
JIS G3447 SUS316LTBS	0.03% C (max) 14.0% Ni 17.0% Cr 2.5% Mo steel: Pipe; Japanese Standard **UTS: 490**
JIS G3447 SUS316TBS	0.08% C (max) 12.0% Ni 17.0% Cr 2.5% Mo steel: Pipe; Japanese Standard **UTS: 530**
JIS G3459 SUS304HTP	0.07% C 9.5% Ni 19.0% Cr steel: Pipe; Japanese Standard **UTS: 530 Proof: 210**
JIS G3459 SUS304LTP	0.03% C (max) 11.0% Ni 19.5% Cr steel: Pipe; Japanese Standard **UTS: 490 Proof: 180**
JIS G3459 SUS304TP	0.08% C (max) 9.5% Ni 19.0% Cr steel: Pipe; Japanese Standard **UTS: 530 Proof: 210**
JIS G3459 SUS309TP	0.15% C 13.5% Ni 23.5% Cr steel: Pipe; Japanese Standard **UTS: 530 Proof: 210**
JIS G3459 SUS310TP	0.15% C (max) 20.5% Ni 25.0% Cr steel: Pipe; Japanese Standard **UTS: 530 Proof: 210**
JIS G3459 SUS316HTP	0.7% C 12.5% Ni 17.0% Cr 2.5% Mo steel: Pipe; Japanese Standard **UTS: 530 Proof: 210**
JIS G3459 SUS316LTP	0.03% C (max) 14.0% Ni 17.0% Cr 2.5% Mo steel: Pipe; Japanese Standard **UTS: 490 Proof: 210**
JIS G3459 SUS316TP	0.08% C (max) 12.0% Ni 17.0% Cr 2.5% Mo steel: Pipe; Japanese Standard **UTS: 530 Proof: 210**
JIS G3459 SUS321HTP	0.07% C 11.0% Ni 18.5% Cr 0.6% Ti steel: Pipe; Japanese Standard **UTS: 530 Proof: 210**
JIS G3459 SUS321TP	0.08% C (max) 11.0% Ni 18.0% Cr 0.4% Ti steel: Pipe; Japanese Standard **UTS: 530 Proof: 210**
JIS G3459 SUS347HTP	0.07% C 11.0% Ni 18.5% Cr 0.8% Nb + Ta steel: Pipe; Japanese Standard **UTS: 530 Proof: 210**
JIS G3459 SUS347TP	0.08% C (max) 11.0% Ni 18.0% Cr 0.8% Nb + Ta steel: Pipe; Japanese Standard **UTS: 530 Proof: 210**
JIS G3463 SUS304HTB	0.07% C 9.0% Ni 19.0% Cr steel: Pipe; Japanese Standard **UTS: 530 Proof: 210**
JIS G3463 SUS304LTB	0.03% C (max) 11.0% Ni 19.0% Cr steel: Pipe; Japanese Standard **UTS: 490 Proof: 180**
JIS G3463 SUS304TB	0.08% C (max) 19.0% Ni 19.0% Cr steel: Pipe; Japanese Standard **UTS: 530 Proof: 210**
JIS G3463 SUS309STB	0.15% C (max) 13.5% Ni 23.0% Cr steel: Pipe: Japanese Standard **UTS: 530 Proof: 210**
JIS G3463 SUS310STB	0.15% C (max) 20.5% Ni 25.0% Cr steel: Pipe; Japanese Standard **UTS: 530 Proof: 210**

Note. The following abbreviations and units are used in the tables:

DPN	Hardness, diamond pyramid number
UTS	Ultimate tensile strength, N/mm^2
Elon	Elongation, %
Proof	0.1% proof strength, N/mm^2

1 N/mm^2=0.1 hbar=0.102 kgf/mm^2=0.06475 $tonf/in.^2$=145.04 $lbf/in.^2$=1 MPa

See Appendix II for other abbreviations and conversion tables.

Symbol	Nominal analysis, supplier, condition and remarks.
JIS G3463 SUS316HTB	0.07% C 12.5% Ni 17.0% Cr 2.5% Mo steel: Pipe; Japanese Standard **UTS: 530** **Proof: 210**
JIS G3463 SUS 316LTB	0.03% C (max) 14.0% Ni 17.0% Cr 2.5% Mo steel: Pipe; Japanese Standard **UTS: 490** **Proof: 180**
JIS G3463 SUS316TB	0.08% C (max) 12.0% Ni 17.0% Cr 2.5% Mo steel: Pipe; Japanese Standard **UTS: 530** **Proof: 210**
JIS G3463 SUS321HTB	0.07% C 11.0% Ni 18.5% Cr 0.6% Ti steel: Pipe; Japanese Standard **UTS: 530** **Proof: 210**
JIS G3463 SUS321TB	0.08% C (max) 11.0% Ni 18.0% Cr 0.4% Ti steel: Pipe; Japanese Standard **UTS: 530** **Proof: 210**
JIS G3463 SUS329J1TB	0.08% C (max) 5.5% Ni 25.5% Cr 2.0% Mo steel: Pipe; Japanese Standard **UTS: 600** **Proof: 400**
JIS G3463 SUS347HTB	0.07% C 11.0% Ni 18.5% Cr 1.0% Nb + Ta steel: Pipe; Japanese Standard **UTS: 530** **Proof: 210**
JIS G3463 SUS347TB	0.08% C (max) 11.0% Ni 18.0% Cr 0.8% Nb + Tb steel: Pipe; Japanese Standard
JIS G4303 SUS	Suffix in agreement with AISI etc. steels
JIS G4304 SUS302	15% C (max) 2% Mn (max) 9% Ni 18% Cr steel: Plate; Japanese specification **UTS: 530** **Proof: 210**
JIS G4304 SUS304	0.3% C (max) 2% Mn (max) 10.5% Ni 19.0% Cr: Japanese specification **UTS: 530** **Proof: 210**
JIS G4304 SUS304L	0.08% C (max) 2.0% Mn (max) 9.5% Ni 19% Cr steel: Plate; Japanese Standard **UTS: 530** **Proof: 210**
JIS G4304 SUS305	0.12% C (max) 2.0% Mn (max) 11.5% Ni 18% Cr steel: Plate; Japanese specification; solution treated **Proof: 180**
JIS G4304 SUS309S	0.08% C (max) 13.5% Ni 23% Cr steel: Plate; Japanese Standard; solution treated **UTS: 530** **Proof: 210**
JIS G4304 SUS310S	0.08% C (max) 20.5% Ni 25.0% Cr steel: Plate; Japanese Standard; solution treated **UTS: 530** **Proof: 210**
JIS G4304 SUS316	0.08% C 12.0% Ni 17.0% Cr 2.5% Mo steel: Plate; Japanese Standard; solution treated **UTS: 530** **Proof: 210**
JIS G4304 SUS316JI	0.08% C (max) 12.0% Ni 18.0% Cr 2.0% Mo 1.2% Cu steel: Plate; Japanese Standard; solution treated **UTS: 530** **Proof: 210**
JIS G4304 SUS316JIL	0.03% C (max) 14.0% Ni 18.0% Cr 2.0% Mo 1.75% Cu steel: Plate; Japanese Standard; solution treated **UTS: 490** **Proof: 180**
JIS G4304 SUS316L	0.03% C (max) 13.5% Ni 17.0% Cr 2.5% Mo steel: Plate; Japanese Standard; solution treated **UTS: 490** **Proof: 180**
JIS G4304 SUS317	0.08% C (max) 2.0% Mn (max) 13% Ni 19% Cr 3.5% Mo 0.03% S (max) 0.04% P (max) steel: Plate; Japanese Standard; solution treated **UTS: 530** **Proof: 210**
JIS G4304 SUS317L	0.03% C (max) 2.0% Mn (max) 13% Ni 19% Cr 3.5% Mo 0.03% S (max) 0.04% P (max) steel: Plate; Japanese Standard; solution treated **UTS: 490** **Proof: 180**
JIS G4304 SUS321	0.08% C (max) 2.0% Mn 10.5% Ni 18% Cr 0.03% S (max) 0.04% P (max) 0.4% Ti (min) steel: Plate; Japanese Standard; solution treated **UTS: 530** **Proof: 210**
JIS G4304 SUS347	0.08% C (max) 2.0% Mn (max) 10% Ni 18% Cr 0.8% Nb 0.03% S (max) 0.04% P (max) Ta steel: Plate; Japanese Standard; solution treated **UTS: 530** **Proof: 210**
JIS G4312 SUH309	0.2% C (max) 13.5% Ni 23.0% Cr steel: Plate; Japanese Standard; solution treated **UTS: 570** **Proof: 210**
JIS G4312 SUH310	0.25% C (max) 20.5% Ni 25.0% Cr steel: Plate; Japanese Standard; solution treated **UTS: 600** **Proof: 210**
JIS G4312 SUH330	0.15% C (max) 35.0% Ni 15.5% Cr steel: Plate; Japanese Standard; solution treated **UTS: 570** **Proof: 210**
K 91800	0.08% C (max) 18% Cr 0.4% Ti steel: Designation used by UNS
KE 965	0.4% C 13.5% Ni 13% Cr 0.55% Mo 2.75% W steel: Kayser Ellison; for valves
KEA 23	0.1% C 9.0% Ni 18.5% Cr + Ti steel: Weldable; Kayser Ellison
KEA 23	0.07% C 9.0% Ni + Ti 18.0% Cr steel: Sanderson Kayser
KEA 221	0.07% C 11.5% Ni 18.0% Cr 2.6% Mo steel: Sanderson Kayser
KEA 221	0.09% C 10.0% Ni 18.5% Cr 2.7% Mo steel: Kayser Ellison
KEA 507	0.1% C 8.0% Ni 18.5% Cr with Se steel: Free machining; Kayser Ellison
KEA 507	0.09% C 8.5% Ni + Se 18.0% Cr steel: Sanderson Kayser
KEA 521	0.07% C 11.5% Ni 18.0% Cr 2.6% Mo steel: Sanderson Kayser
KEA 521	0.09% C 10.0% Ni 18.5% Cr 2.7% Mo + Se steel: Free machining; Kayser Ellison
KEH 22/A507	0.1% C 7.5% Ni 20% Cr + free cutting agent steel: Kayser Ellison; annealed **UTS: 530** **Elon: 30%** **Proof: 180**
KH 18N10T	0.12% C (max) 10.0% Ni 18.0% Cr 0.7% Ti steel: Russian Standard designation
KH 18N12T	0.12% C (max) 12.0% Ni 18.0% Cr 0.7% Ti steel: Russian Standard designation
KH 25N16G7AR	0.12% C (max) 16.5% Ni 24.5% Cr 0.02% B (max) 0.4% N steel: Russian Standard designation
KH 35VT	0.12% C (max) 36.0% Ni 15.0% Cr 3.2% W 1.3% Ti Fe alloy: Russian Standard designation
KH 35VT1U	0.08% C (max) 35.0% Ni 15.0% Cr 3.2% W 2.8% Ti 1.0% Al Fe alloy: Russian Standard designation
KH 35VTR	0.1% C (max) 36.5% Ni 15.0% Cr 4.5% W 1.3% Ti Fe alloy: Russian Standard designation
KH 38VT	0.09% C 37.0% Ni 21.5% Cr 3.2% V 1.0% Ti Fe alloy: Russian Standard designation
KHN 35VMT	0.12% C (max) 34.0% Ni 15.0% Cr 2.5% Mo 2.7% W 1.3% Ti Fe alloy: Russian Standard designation
KK WELDANKA	0.15% C (max) 9% Ni 18% Cr 0.15% Se steel: Bar, billets, etc.; Brown Bayley for BS alloy En 58M
LANGALLOY 1V	0.08% C 9% Ni 18% Cr 1.0% Nb steel: Langley for BS alloy 821
LANGALLOY 2V	0.08% C 10% Ni 18% Cr 2.5% Mo steel: Langley for BS alloy 845; free machining
LANGALLOY 3V	0.08% C 10% Ni 18% Cr 2.5% Mo steel: Langley for BS alloy 845
LANGALLOY 4V	12% Ni 25% Cr W steel: Langley **DPN: 170** **UTS: 580** **Elon: 15%** **Proof: 270**
LANGALLOY 5V	0.08% C 10% Ni 18% Cr 2.5% Mo steel: Langley for BS alloy 845 with low magnetic permeability; free machining
LANGALLOY 10V	0.08% C 10% Ni 19% Cr steel: Langley; for low temperature use **UTS: 480** **Elon: 40%** **Proof: 240**
LANGALLOY 12V	0.08% C 12% Ni 20% Cr 3.5% Mo steel: Langley for BS alloy 846; free machining
LANGALLOY 13V	0.08% C 12% Ni 20% Cr 3.5% Mo steel: Langley for BS alloy 846
LANGALLOY 15V	0.15% C 9% Ni 18% Cr steel: Langley **UTS: 480** **Elon: 40%** **Proof: 240**

Symbol	Nominal analysis, supplier, condition and remarks.
LO	0.07% C 10% Ni 18% Cr 1.0% Mo steel: Osborn; weldable; acid resistant **UTS: 710 Elon: 50% Proof: 270**
LS 12	0.1% C 13% Ni 13% Cr steel: Low Moor **DPN: 160 UTS: 530 Elon: 45% Proof: 220**
LS 18	0.1% C 9% Ni 18% Cr steel: Low Moor; not weldable **DPN: 150 UTS: 570 Elon: 50% Proof: 240**
LS 18/2	0.08% C 10% Ni 18% Cr 2.0% Mo steel: Low Moor; acid resisting; weldable **DPN: 160 UTS: 610 Elon: 48% Proof: 250**
LS 18/3	0.08% C 10% Ni 18% Cr 2.75% Mo steel: Low Moor; acid resistant; weldable **DPN: 165 UTS: 610 Elon: 48% Proof: 270**
LS 18/4	0.07% C 10% Ni 18% Cr 3.75% Mo steel: Low Moor; acid resistant; weldable **DPN: 160 UTS: 610 Elon: 48% Proof: 270**
LS 18L	0.06% C 9% Ni 18% Cr steel: Low Moor; corrosion resistance reduced by welding **DPN: 150 UTS: 570 Elon: 50% Proof: 240**
LS 20/12	0.15% C 13% Ni 20% Cr steel: Low Moor; heat resistant **DPN: 160 UTS: 610 Elon: 48% Proof: 250**
LS 25/12/3	0.2% C 11% Ni 24% Cr 3% W steel: Low Moor; heat resistant **DPN: 230 UTS: 840 Elon: 35% Proof: 440**
LS 25/20	0.15% C 18% Ni 25% Cr steel: Low Moor **DPN: 200 UTS: 630 Elon: 45% Proof: 300**
LS WF	0.08% C 9% Ni 18% Cr Ti or Nb steel: Low Moor **DPN: 160 UTS: 610 Elon: 50% Proof: 250**
LV 10	0.4% C 9% Ni 19% Cr 2.5% Si steel: Low Moor; valve steel
LV 21/12	0.25% C 12% Ni 21% Cr steel: Low Moor; valve steel
LV 21/12N	0.2% C 12% Ni 21% Cr 0.2% Nb steel: Low Moor; valve steel
LV 54	0.4% C 11% Ni 13% Cr 3% W steel: Low Moor; valve steel **DPN: 260**
LV 55	0.45% C 9% Ni 19% Cr 1.1% W 2.5% Si steel: Low Moor; valve steel
M 5	0.03% C 16% Ni 18% Cr 5% Mo 0.2% Cu steel: Japanese specification
M 7	0.06% C 17% Ni 10% Cr 7% Mo steel: Japanese specification
MA 1	0.12% C 9.5% Ni 18% Cr steel: Schoeller-Bleckmann; annealed **DPN: 175 UTS: 540 Elon: 50% Proof: 210**
MA 3	0.09% C (max) 13% Ni 12.5% Cr steel: Schoeller-Bleckmann; annealed **DPN: 170 UTS: 540 Elon: 55% Proof: 200**
MA 4	0.08% C (max) 8.5% Ni 18% Cr steel: Schoeller-Bleckmann; annealed **DPN: 175 UTS: 540 Elon: 50% Proof: 210**
MACDN	0.06% C 20% Ni 18% Cr 2.2% Mo Cu Nb steel: Schoeller-Bleckmann; annealed **DPN: 185 UTS: 610 Elon: 40% Proof: 220**
MACM	0.07% C 22% Ni 18% Cr 3% Mo Cu Nb steel: Schoeller-Bleckmann; annealed **DPN: 185 UTS: 610 Elon: 40% Proof: 220**
MAN	0.08% C 9.5% Ni 18% Cr Nb steel: Weldable; Schoeller-Bleckmann; annealed **DPN: 185 UTS: 570 Elon: 40% Proof: 240**
MANAURITE 8S	8% Ni 18% Cr steel: Pompey **UTS: 500 Elon: 20% Proof: 240**
MANAURITE 10	10% Ni 22% Cr steel: Pompey **UTS: 500 Elon: 18% Proof: 260**
MANAURITE 12	12% Ni 25% Cr steel: Pompey **UTS: 540 Elon: 18% Proof: 260**
MANAURITE 20	20% Ni 25% Cr steel: Pompey **UTS: 560 Elon: 20% Proof: 260**
MANAURITE 35	35% Ni 15% Cr steel: Pompey **UTS: 500 Elon: 20% Proof: 240**
MANGANESE P	0.1% C 2.0% Mn 8% Ni 20% Cr steel: Welding electrode; Murex; used as a buffer layer when hard facing **DPN: 200**
MANIFLEX	0.2% C 11.5% Ni 21% Cr steel: Carpenter
MANIFLEX 11 ALLOY	0.08% C 11.5% Ni 21% Cr steel: Carpenter **UTS: 600 Elon: 60% Proof: 270**
MANIFLEX FM ALLOY	0.2% C 11.5% Ni 21% Cr steel: Carpenter **DPN: 210 UTS: 710 Elon: 60% Proof: 400**
MAO	0.05% C (max) 9.5% Ni 18% Cr steel: Schoeller-Bleckmann; annealed **DPN: 175 UTS: 540 Elon: 50% Proof: 200**
MAO SUPERIOR	0.03% C (max) 9.5% Ni 18% Cr steel: Schoeller-Bleckmann; annealed **DPN: 175 UTS: 540 Elon: 50% Proof: 200**
MAOY	0.06% C (max) 9.5% Ni 18% Cr steel: Schoeller-Bleckmann; annealed **DPN: 175 UTS: 540 Elon: 50% Proof: 200**
MASBw	0.06% C 13% Ni 17% Cr 4.2% Mo steel: Schoeller-Bleckmann; annealed **DPN: 185 UTS: 610 Elon: 45% Proof: 220**
MASD	0.06% C (max) 9% Ni 17% Cr 1.5% Mo steel: Schoeller-Bleckmann; annealed **DPN: 180 UTS: 610 Elon: 45% Proof: 210**
MASN	0.08% C 11.5% Ni 18% Cr 2.2% Mo Nb steel: Schoeller-Bleckmann; annealed **DPN: 185 UTS: 570 Elon: 40% Proof: 240**
MASO	0.05% C 11% Ni 18% Cr 2.2% Mo steel: Schoeller-Bleckmann; annealed **DPN: 175 UTS: 540 Elon: 45% Proof: 200**
MASO SUPERIOR	0.03% C (max) 11% Ni 18% Cr 2.2% Mo steel: Schoeller-Bleckmann; annealed **DPN: 175 UTS: 540 Elon: 45% Proof: 200**
MASOY	0.05% C 12% Ni 18% Cr 2.6% Mo steel: Schoeller-Bleckmann; annealed **DPN: 175 UTS: 570 Elon: 45% Proof: 200**
MASOY SUPERIOR	0.03% C 12% Ni 18% Cr 2.6% Mo steel: Schoeller-Bleckmann; annealed **DPN: 175 UTS: 570 Elon: 45% Proof: 200**
MASW	0.07% C 11.5% Ni 18% Cr 2.2% Mo Ti steel: Schoeller-Bleckmann; annealed **DPN: 185 UTS: 570 Elon: 40% Proof: 240**
MASWY	0.07% C 12% Ni 17% Cr 2.7% Mo Ti steel: Schoeller-Bleckmann; annealed **DPN: 185 UTS: 610 Elon: 40% Proof: 240**
MAT	0.07% C 9.5% Ni 18% Cr Ti steel: Weldable; Schoeller-Bleckmann; annealed **DPN: 185 UTS: 570 Elon: 40% Proof: 240**
MAUSTINOX A	0.12% C (max) 9% Ni 18% Cr steel: Pompey **UTS: 500 Elon: 30% Proof: 240**
MAUSTINOX Abc	0.08% C (max) 9% Ni 18% Cr steel: Pompey **UTS: 500 Elon: 35% Proof: 240**
MAUSTINOX Atbc	0.05% C (max) 10% Ni 18% Cr steel: Pompey **UTS: 470 Elon: 40% Proof: 220**
MAUSTINOX B	0.12% C (max) 10% Ni 18% Cr 2.5% Mo steel: Pompey **UTS: 540 Elon: 30% Proof: 240**

Note. The following abbreviations and units are used in the tables:

DPN	Hardness, diamond pyramid number
UTS	Ultimate tensile strength, N/mm^2
Elon	Elongation, %
Proof	0.1% proof strength, N/mm^2

1 N/mm^2=0.1 hbar=0.102 kgf/mm^2=0.06475 $tonf/in.^2$=145.04 $lbf/in.^2$=1 MPa

See Appendix II for other abbreviations and conversion tables.

Symbol	Nominal analysis, supplier, condition and remarks.
MAUSTINOX Bbc	0.08% C 10% Ni 18% Cr 2.5% Mo steel: Pompey **UTS: 500 Elon: 35% Proof: 240**
MAUSTINOX C	0.12% C 8% Ni 20% Cr 2.5% Mo steel: Pompey **UTS: 550 Elon: 25% Proof: 310**
MAUSTINOX F	0.1% C 13% Ni 24% Cr steel: Pompey **UTS: 570 Elon: 30% Proof: 240**
MAUSTINOX H	0.1% C 20% Ni 25% Cr steel: Pompey **UTS: 520 Elon: 30% Proof: 240**
MAUSTINOX SA	0.08% C (max) 9% Ni + Nb + Ta 18% Cr steel: Pompey **UTS: 500 Elon: 35% Proof: 240**
MAUSTINOX SB	0.08% C 10% Ni 18% Cr 2.5% Mo + Nb + Ta steel: Pompey
MAUSTINOX X	0.1% C 30% Ni 15% Cr 4.5% Mo steel: Pompey **UTS: 470 Elon: 35% Proof: 230**
MAUSTINOX XD	0.1% C 30% Ni 15% Cr 6% Mo steel: Pompey **UTS: 470 Elon: 35% Proof: 230**
MAUSTINOX Y	0.1% C 20% Ni 20% Cr 4.5% Mo steel: Pompey
MAUSTINOX YD	0.07% C 20% Ni 20% Cr 6% Mo steel: Pompey **UTS: 480 Elon: 32% Proof: 260**
MAUSTINOX YN	0.07% C 30% Ni 20% Cr 5% Mo Cu steel: Pompey **UTS: 480 Elon: 32% Proof: 260**
MAXHETE 1	0.35% C 8.5% Ni 18.5% Cr 2.5% W steel: Edgar Allen
MAXHETE 1A	0.4% C 13% Ni 13% Cr steel: Edgar Allen; internal combustion engine valves
MAXHETE 2	0.25% C (max) 13.5% Ni 25% Cr steel: Edgar Allen; high temperature uses
MAXHETE 4	0.25% C (max) 20% Ni 25% Cr steel: Edgar Allen; furnace equipment
MAXILVRY	0.12% C (max) 8% Ni 18% Cr steel: Edgar Allen for BS alloy En 58A
MAXILVRY ADS	0.12% C (max) 12.5% Ni 12.5% Cr steel: Edgar Allen for BS alloy En 58D
MAXILVRY AWP	0.1% C (max) 8% Ni 18% Cr steel: Edgar Allen for BS alloy En 58B
MAXILVRY CB	0.1% C (max) 8% Ni 18% Cr steel: Edgar Allen for BS alloy En 58F
MAZ	0.14% C 8.5% Ni 18% Cr S steel: Free machining; Schoeller-Bleckmann; annealed **DPN: 175 UTS: 540 Elon: 45% Proof: 210**
MCS 6	0.05% C 13.5% Ni 17.5% Cr 2.5% Mo 1.2% Cu steel: Japanese specification
METCO 41C	0.1% C 12% Ni 18% Cr 2% Mo 1.0% Si steel: Powder for spraying; Metco **DPN: 189**
METCO 41F-NS	Type 316 powder coating; Metco; for thin deposits **DPN: 170**
METCO 43F-NS	Cr Ni alloy steel: Metco **DPN: 190**
METCO 402	Austenitic steel with Al and Ni composite wire: Metco; for wear and corrosion resistance **DPN: 300**
METCO 625	Austenitic stainless steel: Plastic composite powder coating; Metco
METCOLOY 1	0.08% C 9% Ni 19% Cr steel: Wire for spraying; Metco; as sprayed **DPN: 146 UTS: 200**
METCOLOY 4	0.08% C 12% Ni 17% Cr 2.5% Mo steel: Wire for spraying; Metco
MET-MAX 19.9L	0.02% C 10% Ni 19% Cr steel: Electrode; Metrode
MET-MAX 19.12.3L	0.03% C 12% Ni 18% Cr 2.5% Mo steel: Electrode; Metrode
MET-MAX 20.9.3	0.1% C 9% Ni 20% Cr 3% Mo steel: Electrode; Metrode **DPN: 320**
MET-MAX 23.12	0.06% C 12% Ni 24% Cr steel: Electrode; Metrode
MET-MAX 23.12.2	0.05% C 12% Ni 24% Cr 2.5% Mo steel: Electrode; Metrode
MET-MAX 29.9	0.12% C 9% Ni 30% Cr 1% Mo steel: Electrode; Metrode **DPN: 275**
MET-MAX 307R	0.12% C 10% Ni 19% Cr 1% Mo steel: Electrode; Metrode **DPN: 320**
MIL-S-16538	0.5% C 25% Ni 16% Cr 6% Mo steel
MIL-S-16598	0.15% C (max) 36% Ni 0.2% Se steel: US military specification
MIL-T-8506	0.04% C 10% Ni 19% Cr steel
MNC 900	Covers stainless steel: Swedish Standard; includes detailed specification
N 08028	0.03% C (max) 32% Ni 27% Cr 3.5% Mo 1% Cu steel: UNS designation
N 08028	0.03% C 32% Ni 27% Cr 3.5% Mo 1% Cu steel: UNS designation
N 08330	0.08% C (max) 35.5% Ni 18.5% Cr steel: UNS designation
N 08332	0.07% C 35.5% Ni 18.5% Cr steel: UNS designation
N 08700	0.04% C (max) 25% Ni 21% Cr 4.2% Mo 0.3% Nb steel: UNS designation
NAG 19.9L	0.015% C 10% Ni 20% Cr steel: Electrode; Metrode
NAG 19.9LR	0.02% C 10% Ni 19% Cr steel: Electrode; Metrode
NB	0.15% C (max) 9% Ni 18% Cr 1.0% Nb steel: Avesta
NCF 2	0.1% C (max) 32.5% Ni 21.0% Cr 0.35% Ti 0.35% Al Fe alloy: Japanese Standard designation; Incoloy 800
NEUTRO SORB	0.08% C 14% Ni 19% Cr 2% B (max) 0.1% N (max) steel: For nuclear industry; Carpenter
NEUTRO SORB PLUS	0.08% C (max) 13.5% Ni 19% Cr 2% B (max) 0.1% N steel: For nuclear industry; Carpenter
NF 25-20L	0.02% C 20% Ni 24% Cr steel: Electrode; Metrode
NF 25-20LR	0.02% C 5.5% Mn 21% Ni 25% Cr 0.2% Nb steel: Electrode; Metrode
NF A35 572/ Z2CN18/10	0.04% C (max) 10.0% Ni 18.0% Cr steel: French Standard
NF A35 572/ Z2CND17/11	0.07% C (max) 11.0% Ni 17.0% Cr 2.25% Mo steel: French Standard
NF A35 572/ Z2CND17/12	0.07% C (max) 12.0% Ni 17.0% Cr 2.7% Mo steel: French Standard
NF A35 572/ Z2CND17/13	0.03% C (max) 13.0% Ni 17.0% Cr 2.7% Mo steel: French Standard
NF A35 572/ Z2CND19/15	0.03% C (max) 15.0% Ni 19.0% Cr 3.5% Mo steel: French Standard
NF A35 572/ Z5CNDU21/08	0.06% C (max) 8.0% Ni 21.0% Cr 2.5% Mo steel: French Standard
NF A35 572/ Z6CN18/09	0.07% C (max) 9.0% Ni 18.0% Cr steel: French Standard
NF A35 572/ Z6CNDNb17/12	0.08% C (max) 12.0% Ni 17.0% Cr 2.2% Mo Nb + Ta 0.8% steel: French Standard
NF A35 572/ Z6CNDNb17/13	0.08% C (max) 13.0% Ni 17.0% Cr 2.7% Mo 0.8% Nb + Ta steel: French Standard
NF A35 572/ Z6CNNb18/11	0.08% C (max) 11.0% Ni 18.0% Cr 1.0% Nb + Ta steel: French Standard
NF A35 572/ Z6CNT18/11	0.08% C (max) 11.0% Ni 18.0% Cr 0.06% Ti steel: French Standard
NF A35 572/ Z8CN13/13	0.1% C (max) 13.0% Ni 13.0% Cr steel: French Standard
NF A35 572/ Z8CNDT17/12	0.1% C (max) 12.0% Ni 17.0% Cr 2.2% Mo 0.6% Ti steel: French Standard
NF A35 572/ Z10CN18/09	0.12% C 9.0% Ni 18.0% Cr steel: French Standard
NF A35 572/ Z10CNF18/09	0.12% C (max) 9.0% Ni 18.0% Cr steel: French Standard
NF A35 572/ Z10CNT18/11	0.12% C (max) 11.0% Ni 18.0% Cr 0.8% Ti steel: French Standard
NF A35 573/ Z2CN18/10	0.03% C (max) 10% Ni 18% Cr 2.0% Mo (max) 0.03% S (max) 0.04% P (max) steel: Plate; French Standard **UTS: 185 Proof: 570**
NF A35 573/ Z2CND17/12	0.03% C (max) 12.0% Ni 17.0% Cr 2.25% Mo steel: Plate; French Standard **UTS: 600 Proof: 240**
NF A35 573/ Z2CND17/13	0.03% C (max) 13.0% Ni 17.0% Cr 2.75% Mo steel: Plate; French Standard **UTS: 600 Proof: 240**

Symbol	Nominal analysis, supplier, condition and remarks.
NF A35 573/ Z2CND19/15	0.03% C (max) 2.0% Mn (max) 15% Ni 19% Cr 3.5% Mo 0.03% S 0.04% P steel: Plate; French Standard **UTS: 600 Proof: 210**
NF A35 573/ Z5CNUD21/08	0.06% C (max) 8.0% Ni 21.0% Cr 2.6% Mo 1.5% Cu steel: Plate; French Standard **UTS: 600 Proof: 350**
NF A35 573/ Z6CN18/09	0.07% C (max) 9% Ni 2.0% 18% Cr steel: Plate; French Standard **UTS: 620 Proof: 200**
NF A35 573/ Z6CND17/11	0.07% C (max) 11.0% Ni 17.0% Cr 2.25% Mo steel: Plate; French Standard **UTS: 620 Proof: 250**
NF A35 573/ Z6CNDNb17/12	0.08% C (max) 12.0% Ni 17.0% Cr 2.25% Mo 1.0% Nb + Ta steel: Plate; French Standard **UTS: 650 Proof: 260**
NF A35 573/ Z6CNNb18/11	0.08% C (max) 2.0% Mn (max) 11% Ni 18% Cr 0.9% Nb 0.03% S (max) 0.04% P (max) Ta steel: Plate; French Standard **UTS: 600 Proof: 210**
NF A35 573/ Z6CNT18/11	0.08% C (max) 2.0% Mn (max) 11% Ni 18% Cr 0.5% Ti 0.03% S (max) 0.04% P (max) steel: Plate; French Standard **UTS: 600 Proof: 210**
NF A35 573/ Z8CN13/13	0.1% C (max) 13% Ni 13.0% Cr 2.0% Cu steel: Plate; French Standard **UTS: 500 Proof: 240**
NF A35 573/ Z8CN18/12	0.1% C (max) 2.0% Mn (max) 12.0% Ni 18.0% Cr steel; Plate; French Standard **UTS: 600 Proof: 220**
NF A35 573/ Z8CNDT17/12	0.1% C (max) 12.0% Ni 17.0% Cr 2.2% Mo 0.6% Ti steel: Plate; French Standard **UTS: 650 Proof: 260**
NF A35 573/ Z8CNDT17/12	0.1% C (max) 12.0% Ni 17.0% Cr 2.2% Mo 0.6% Ti steel: Plate; French Standard **UTS: 650 Proof: 260**
NF A35 573/ Z10CN18/09	0.12% C (max) 2% Mn (max) 9% Ni 18% Cr steel: Plate; French Standard **UTS: 650 Proof: 225**
NF A35 573/ Z10CNT18/11	0.12% C (max) 2.0% Mn 11% Ni 18% Cr 0.15% Ti 0.03% S (max) 0.04% P (max) steel: Plate; French Standard **UTS: 650 Proof: 225**
NF A36 209/ Z2CN18/10	0.03% C (max) 2.0% Mn (max) 10% Ni 18% Cr 0.03% S (max) 0.04% P (max) steel: Plate; French Standard **UTS: 185 Proof: 570**
NF A36 209/ Z6CN18/09	0.01% C (max) 2.0% Mn (max) 9% Ni 18% Cr steel: Plate; French Standard **UTS: 620 Proof: 200**
NF A36 209/ Z6CNT18/11	0.08% C (max) 1.5% Mn 12% Ni 18% Cr 0.5% Ti 0.02% S (max) 0.035% P (max) steel: Plate; French Standard **UTS: 600 Proof: 210**
NF A35 572/ Z8CN18/12	0.10% C (max) 12.0% Ni 18.0% Cr steel: French Standard
Ni Cr 30/20	0.2% C (max) 30.0% Ni 21.0% Cr steel: Designation used by German Standards
NICRAL C	36% Ni 18% Cr steel: Imphy **DPN: 190 UTS: 650 Elon: 30% Proof: 350**
NICRAL C	0.08% C (max) 35.5% Ni 18.5% Cr steel: Creusot-Loire
NICRAL D	0.08% C (max) 20.5% Ni 25.0% Cr steel: Creusot-Loire
NICRAL D	20% Ni 25% Cr steel: Imphy **DPN: 175 UTS: 650 Elon: 35% Proof: 350**
NICRAL DM	20% Ni 25% Cr steel: Imphy **DPN: 175 UTS: 600 Elon: 15% Proof: 300**
NICRAL H	12% Ni 25% Cr steel: Imphy **DPN: 175 UTS: 650 Elon: 30% Proof: 300**
NICRAL H	0.08% C (max) 13.5% Ni 23.0% Cr steel: Creusot-Loire
NICRAL HR2	0.15% C 9.0% Ni 19.0% Cr steel: Creusot-Loire
NICRAL K	40% Ni 22% Cr Fe alloy: Imphy
NICRALT	0.12% C 13% Ni 16% Cr 3.0% W 0.5% Ti steel: Imphy; annealed **UTS: 840 Elon: 35% Proof: 220**
NICREX 1	0.1% C 20% Ni 26% Cr 0.4% Mo steel: Welding electrode; Murex **UTS: 630 Elon: 40%**
NICREX 1 Cb	0.1% C 20% Ni 26% Cr 0.4% Mo 1% Nb steel: Welding electrode; Murex; Nb also called Cb **UTS: 740 Elon: 35%**
NICREX 2	0.1% C 18% Ni 23% Cr 3.2% Mo 1% Nb steel: Welding electrode; Murex **UTS: 450 Elon: 27%**
NICREX AC	0.08% C 8% Ni 18% Cr 0.8% Nb steel: Welding electrode; Murex **UTS: 670 Elon: 40%**
NICROFER 3127LC	0.015% C (max) 31% Ni 27% Cr 3.5% Mo 1.2% Cu Fe alloy: VDM **UTS: 500 Elon: 35% Proof: 210**
NICROFER 3220	0.1% C (max) 31% Ni 20% Cr steel: VDM
NICROFER 3220 H	0.08% C 31% Ni 20.5% Cr 0.3% Ti steel: VDM
NICROFER 3220LC	0.025% C (max) 33% Ni 21% Cr 0.4% Ti Fe alloy: VDM **UTS: 450 Elon: 35% Proof: 180**
NICROFER 3620 Nb	0.05% C (max) 37% Ni 20% Cr 2.5% Mo 0.6% Nb 3.5% Cu Fe alloy: VDM **UTS: 590 Elon: 30% Proof: 275**
NICROFER 3718 So	0.15% C (max) 35.15% Ni 16.5% Cr 2% Si steel: VDM
NICROMAZ B	40% Ni 15% Cr steel: Imphy; resistant to hot concentrated sulphuric acid
NICROMAZ C	25% Ni 20% Cr 5% Mo 1.5% Cu steel: Imphy
NIKROTHAL 20 PLUS	20% Ni 25% Cr Fe alloy: For heating element; Kanthal AB
NIKROTHAL 40 PLUS	35% Ni 20% Cr Fe alloy: For heating element; Kanthal AB
NISIMAZ	21% Ni 2.5% Cr 7.0% Si steel: Imphy; resistant to hot acids
NITRONIC 20	0.3% C 7.5% Ni 23% Cr steel: For valves; origin unknown
NITRONIC 50	0.04% C 12.5% Ni 22% Cr 2.2% Mo 0.2% V 0.2% Nb 0.3% N steel
NOVO 18/3	0.1% C 8% Ni 18% Cr steel: Jonas; fully softened; not weld stabilized **DPN: 183 UTS: 460 Elon: 65% Proof: 320**
NOVO 18/8W	0.06% C 8% Ni 18% Cr steel: Jonas; annealed; stabilized with 0.35% Ti or 0.6% Nb **DPN: 183 UTS: 460 Elon: 65% Proof: 320**
NS 9S	0.15% C 9% Ni 19% Cr 1.0% Nb steel: Ugine
NS 20E	0.16% C 9% Ni 19% Cr steel: Ugine
NS 21A	0.08% C (max) 10% Ni 19% Cr steel: Ugine
NS 21S	0.15% C (max) 9% Ni 19% Cr 0.6% Ti steel: Ugine
NSCD	18% Cr 16% Ni 6% Mo 3% Cu steel: Ugine
NSM 21	0.12% C (max) 10% Ni 19% Cr 3.0% Mo steel: Ugine
NSMC	0.12% C (max) 10% Ni 19% Cr 3.0% Mo steel: Ugine
NSU	9% Ni 18% Cr steel: Ugine; free machining

Note. The following abbreviations and units are used in the tables:

DPN	Hardness, diamond pyramid number
UTS	Ultimate tensile strength, N/mm^2
Elon	Elongation, %
Proof	0.1% proof strength, N/mm^2

1 N/mm^2=0.1 hbar=0.102 kgf/mm^2=0.06475 tonf/in.^{2}=145.04 lbf/in.^{2}=1 MPa

See Appendix II for other abbreviations and conversion tables.

Symbol	Nominal analysis, supplier, condition and remarks.
NTK M7	0.06% C 17% Ni 10% Cr 7% Mo steel: For use with hydrochloric acid. Kawasaki; annealed **UTS: 840 Elon: 60% Proof: 370**
NV 25 1	0.05% C (max) 8.0% Ni 18.5% Cr steel: Designation used by DNV **UTS: 500 Elon: 40% Proof: 200**
NV 25 2	0.08% C (max) 8.5% Ni 18.5% Cr steel: Designation used by DNV **UTS: 500 Elon: 40% Proof: 200**
NV 25 3	0.05% C (max) 9.5% Ni 19.0% Cr 2.5% Mo steel: Designation used by DNV **UTS: 500 Elon: 40% Proof: 200**
NV 25 4	0.08% C (max) 10.0% Ni 19.0% Cr 2.5% Mo steel: Designation used by DNV **UTS: 500 Elon: 40% Proof: 200**
NVR 25 1	0.06% C (max) 10.0% Ni 18.5% Cr steel: Designation used by DNV
NVR 25 2	0.03% C (max) 11.0% Ni 18.5% Cr steel: Designation used by DNV
NVR 25 3	0.06% C (max) 11.0% Ni 18.2% Cr 1.8% Mo steel: Designation used by DNV
NVR 25 4	0.06% C 12.0% Ni 18.2% Cr 2.8% Mo steel: Designation used by DNV
NVR 25 5	0.03% C 13.7% Ni 18.2% Cr 2.8% Mo steel: Designation used by DNV
OR 2	0.08% C 9% Ni 18% Cr steel: Sandvik
OR 11	0.12% C 10% Ni 19% Cr 3.0% Mo steel: Sandvik
OSBORN 303	0.07% C 9.0% Ni 18% Cr steel: Osborn; previously BSMF grade
OSBORN 304	0.05% C 9.5% Ni 18% Cr steel: Osborn; previously BSM grade
OSBORN 309	0.08% C 14.5% Ni 22.5% Cr steel: Osborn; previously IK grade
OSBORN 310	0.1% C 20.0% Ni 24.5% Cr steel: Osborn; previously FD grade
OSBORN 316	0.05% C 11% Ni 18% Cr 2.7% Mo steel: Osborn; previously EW grade
OSBORN 317	0.05% C 13.5% Ni 18% Cr 3.7% Mo steel: Osborn; previously EWA grade
OSBORN 321	0.06% C 9.5% Ni 18% Cr steel: Osborn; previously BST grade
OSBORN 330	0.1% C (max) 37.5% Ni 18% Cr 2.0% Si steel: Osborn; previously FL grade
OSBORN 347	0.07% C 9.5% Ni 18% Cr 0.6% Nb steel: Osborn; previously BSC grade
P 1010	0.12% C 8.5% Ni 18.5% Cr steel: Carrs for BS alloy En 58A
P 1011	0.14% C 8% Ni 18% Cr 0.5% Ti steel: Carrs for BS alloy En 58B
P 1012	0.14% C 10% Ni 18% Cr 0.5% Ti steel: Carrs for BS alloy En 58C
P 1013	0.14% C 12% Ni 12% Cr steel: Carrs for BS alloy En 58
P 1014	0.07% C 10% Ni 18% Cr steel: Carrs for BS alloy En 58E
P 1015	0.14% C 8% Ni 18% Cr 1.0% Nb steel: Carrs for BS alloy En 58F
P 1016	0.14% C 10% Ni 18% Cr 1.0% Nb steel: Carrs for BS alloy En 58G
P 1017	0.1% C 10% Ni 18% Cr 2.0% Mo steel: Carrs for BS alloy En 58H
P 1018	0.1% C 10% Ni 18% Cr 3.0% Mo steel: Carrs for BS alloy En 58J
PANTEG 304	0.05% C 9.75% Ni 18% Cr steel: Sheet; Richard Thomas and Baldwin; air cooled **UTS: 600 Elon: 50%**
PANTEG 315	0.06% C 11.0% Ni 17.5% Cr 1.5% Mo steel: Sheet; Richard Thomas and Baldwin: air cooled **UTS: 650 Elon: 45%**
PANTEG 316	0.05% C 11.8% Ni 17.5% Cr 3.0% Mo steel: Sheet; Richard Thomas and Baldwin; air cooled **UTS: 660 Elon: 40%**
PANTEG 321	0.05% C 9.6% Ni 17.5% Cr 0.4% Ti steel: Sheet; Richard Thomas and Baldwin; air cooled **UTS: 640 Elon: 50%**
PARALLOY 0	8% Ni 18% Cr steel: APV Paramount
PARALLOY 0KW	10% Ni 19% Cr Nb: Stabilized APV Paramount
PARALLOY 0L	12% Ni 17% Cr steel (low magnetic permeability): APV Paramount
PARALLOY 0LC	0.03% C (max) 8% Ni 18% Cr steel: APV Paramount
PARALLOY 0LS	12% Ni 17% Cr steel: Free machining; low magnetic permeability; APV Paramount
PARALLOY 0S	8% Ni 18% Cr steel: Free machining; APV Paramount
PARALLOY 0SW	8% Ni 18% Cr Nb steel: Free machining; stabilized; APV Paramount
PARALLOY 0W	8% Ni 18% Cr Nb steel: Stabilized; APV Paramount
PARALLOY 3	8% Ni 18% Cr 3% Mo steel: APV Paramount
PARALLOY 3K	10% Ni 17% Cr 3% Mo steel: APV Paramount
PARALLOY 3KLC	0.03% C (max) 10% Ni 17% Cr 3% Mo steel: APV Paramount
PARALLOY 3L	14% Ni 18% Cr 3% Mo steel: Low magnetic permeability; APV Paramount
PARALLOY 3LS	14% Ni 18% Cr 3% Mo steel: Free machining; low magnetic permeability; APV Paramount
PARALLOY 3S	8% Ni 18% Cr 3% Mo steel: Free machining; APV Paramount
PARALLOY 3W	10% Ni 17% Cr 3% Mo Nb steel: Stabilized; APV Paramount
PARALLOY 3WS	10% Ni 17% Cr 3% Mo Nb steel: Stabilized; free machining: APV Paramount
PARALLOY 4K	12% Ni 19% Cr 4% Mo steel: APV Paramount
PARALLOY 4KLC	0.03% C (max) 12% Ni 19% Cr 4% Mo steel: APV Paramount
PEL	0.06% C 8.5% Ni 17.5% Cr steel: Sheet; RTB for BS alloy En 58A
PET	0.07% C 9% Ni 17.5% Cr 0.4% Ti steel: Sheet; RTB for BS alloy En 58B
PHILIPS ARMARC	0.07% C 9.3% Ni 18.1% Cr 3% Mo steel: Welding electrode; Philips **UTS: 725 Elon: 35%**
PHILIPS STAINARC 25/25 R	0.1% C 19.3% Ni 25.7% Cr steel: Welding electrode; Philips
PHILIPS STAINARC M	0.07% C 11.8% Ni 18.3% Cr 3.75% Mo steel: Welding electrode; Philips
PHILIPS STAINARC MN	0.08% C 10.8% Ni 17.2% Cr 2.8% Mo 0.7% Nb steel: Welding electrode; Philips
PHILIPS STAINARC N	0.065% C 9.5% Ni 18.8% Cr 0.8% Mo steel: Welding electrode; Philips
PMH	0.06% C 10.5% Ni 17.5% Cr 3.0% Mo steel: Sheet; RTB for BS alloy En 58J
PML	0.06% C 9.5% Ni 17.5% Cr 1.5% Mo steel: Sheet; RTB for BS alloy En 58H
PROJECT 70-303	0.12% C 9% Ni 18% Cr 0.25% S steel: Free machining; Carpenter
PROJECT 70-303 DQ	0.12% C (max) 9% Ni 18% Cr steel: Carpenter
PROJECT 70-304	0.08% C 9.2% Ni 19% Cr steel: Carpenter
PROJECT 70-316	0.08% C (max) 12% Ni 17% Cr 2.5% Mo steel: Carpenter
PYROMET 860	0.1% C 42% Ni 14% Cr 6% Mo 4% Co 3% Ti 1.0% Al 0.01% B Fe alloy: Carpenter
PYROTOOL A	0.04% C 25% Ni 14.5% Cr 1.5% Mo 2.2% Ti steel: Carpenter
PYROTOOL V	0.04% C 27% Ni 14.5% Cr 1.2% Mo 0.2% V 3% Ti steel: Carpenter
PZ 6072	0.08% C (max) 10% Ni 20% Cr steel: Wire for welding; Philips
PZ 6074	0.08% C (max) 12% Ni 19% Cr 2.5% Mo steel: Wire for welding; Philips

Symbol	Nominal analysis, supplier, condition and remarks.
QQ S 763/7	0.12% C 8% Ni 17% Cr 0.7% Mo 0.2% Se 0.7% Zr steel: US Federal; annealed **UTS: 630 Elon: 28% Proof: 200**
QQ S 766/1	0.08% C 8% Ni 18% Cr steel: US Federal; annealed **UTS: 530 Elon: 40%**
QQ W 423a	Corrosion resisting steel: Wire; US Federal specification
QQ-S-766	0.12% C 20.5% Ni 25% Cr steel
R 3	0.1% C 9% Ni 18% Cr steel: Jessop for BS alloy En 58A
R 9	0.08% C 9.5% Ni 18.5% Cr 2.0% Mo 0.3% Ti steel: Jessop for BS alloy En 58H
R 10	0.08% C 12% Ni 12% Cr steel: Jessop for BS alloy En 58D
R 11	0.25% C 14% Ni 23% Cr 1.2% Si steel: Jessop
R 16	0.08% C 9% Ni 18% Cr 0.3% Ti steel: Jessop for BS 970 En 58B
R 18	0.32% C 7.5% Ni 18% Cr 2.7% W steel: Jessop; for BS 970 En 55
R 20	0.1% C 12% Ni 19% Cr 1.25% Nb steel: Jessop **UTS: 580 Elon: 52% Proof: 200**
R 22	0.2% C 11% Ni 22% Cr 2.7% W steel: Casting; Jessop; as cast **UTS: 510 Elon: 28%**
R 22	0.17% C 11.5% Ni 22% Cr 3% W steel: Jessop
R 23	0.1% C 22% Ni 25% Cr steel: Jessop **UTS: 720 Elon: 45% Proof: 270**
R 24	0.08% C 9% Ni 18% Cr 0.22% S steel: Free machining; Jessop; for BS alloy En 58M
R 25	0.08% C 9.5% Ni 18% Cr 2.8% Mo steel: Jessop; for BS alloy En 58J
R 27	0.05% C 28.5% Ni 20% Cr 2.25% Mo 3.25% Cu steel: Jessop
R 27 Nb	0.05% C 28.5% Ni 20% Cr 2.25% Mo 3.25% Cu Nb steel: Jessop
R 35	0.03% C 25% Ni 20% Cr 0.3% Nb steel: Jessop
R 41	0.06% C 11% Ni 17% Cr 2.35% Mo steel: Jessop
R 42	0.06% C 9% Ni 19% Cr steel: Jessop
R 43	0.1% C 8% Ni 17% Cr 1.0% Nb steel: Jessop
R 45	0.12% C 17.5% Ni 21.5% Cr 0.5% Ti steel: Jessop
R 47	0.45% C 9% Ni 19% Cr 1.4% W 2.2% Si steel: Jessop
R 48	0.09% C 9% Ni 18% Cr 0.6% Nb 0.22% S steel: Jessop; free machining; weld stabilized
R 51	0.08% C 15.0% Ni 23.0% Cr 0.32% Ti steel: Jessop; corrosion resistant steel
R 52	0.08% C 15.0% Ni 23.0% Cr 0.04% Nb steel: Jessop; corrosion resistant steel
R 53	0.08% C 17.0% Ni 24.0% Cr 0.32% Ti steel: Jessop; corrosion resistant steel
R 54	0.08% C 17.0% Ni 24.0% Cr 0.64% Nb steel: Jessop; corrosion resistant steel
R 300	0.1% C 8.5% Ni 18% Cr steel: Fagersta; annealed **DPN: 200 UTS: 530 Elon: 45% Proof: 200**
R 310	0.07% C 10% Ni 18% Cr steel: Fagersta; annealed **DPN: 180 UTS: 530 Elon: 45% Proof: 180**
R 320	0.07% C 9% Ni 18% Cr steel: Fagersta; annealed **DPN: 180 UTS: 530 Elon: 55% Proof: 200**
R 350	0.04% C 9% Ni 18.5% Cr steel: Fagersta; annealed **DPN: 180 UTS: 530 Elon: 50% Proof: 200**

Note. The following abbreviations and units are used in the tables:

DPN	Hardness, diamond pyramid number
UTS	Ultimate tensile strength, N/mm^2
Elon	Elongation, %
Proof	0.1% proof strength, N/mm^2

1 N/mm^2=0.1 hbar=0.102 kgf/mm^2=0.06475 tonf/in.^{2}=145.04 lbf/in.^{2}=1 MPa

See Appendix II for other abbreviations and conversion tables.

Symbol	Nominal analysis, supplier, condition and remarks.
R 350G	0.05% C 8.5% Ni 18.0% Cr steel: Casting; Fagersta
R 358	0.05% C 9.5% Ni 18.0% Cr Nb steel: Wrought; stainless; type 347; Fagersta
R 358G	0.03% C 9.5% Ni 18.0% Cr 0.8% Nb steel: Casting; Fagersta
R 359	0.05% C 9.5% Ni 18.0% Cr Ti steel: Wrought; stainless; type 321; Fagersta
R 360	0.03% C 10% Ni 18% Cr steel: Fagersta; annealed **DPN: 180 UTS: 530 Elon: 50% Proof: 200**
R 360G	0.03% C (max) 10.0% Ni 18.5% Cr steel: Casting; Fagersta
R 380	0.1% C 8.5% Ni 18% Cr 0.15% Se steel: Fagersta; free machining **UTS: 520 Elon: 40%**
R 390	0.05% C 11.5% Ni 18.0% Cr steel: Wrought; stainless; type 304; Fagersta
R 400	0.09% C 9% Ni 18% Cr 1.5% Mo steel: Fagersta; annealed **DPN: 200 UTS: 530 Elon: 45% Proof: 200**
R 420	0.05% C 11% Ni 17.5% Cr 2.3% Mo steel: Fagersta; annealed **DPN: 180 UTS: 530 Elon: 45% Proof: 200**
R 428	0.05% C 11.0% Ni 17.5% Cr 2.3% Mo Nb steel: Wrought; stainless; Fagersta
R 429	0.05% C 11.0% Ni 17.0% Cr 2.3% Mo Ti stainless steel: Wrought; Fagersta
R 440	0.05% C 12% Ni 17% Cr 2.8% Mo steel: Fagersta; annealed **DPN: 180 UTS: 530 Elon: 45% Proof: 200**
R 440G	0.05% C 10.5% Ni 18.0% Cr 2.8% Mo steel: Casting; Fagersta
R 450	0.05% C 10.5% Ni 18% Cr 1.5% Mo steel: Fagersta; annealed **DPN: 180 UTS: 530 Elon: 45% Proof: 200**
R 460	0.03% C 12% Ni 17% Cr 2.8% Mo steel: Fagersta; annealed **DPN: 180 UTS: 530 Elon: 45% Proof: 20**
R 460G	0.03% C (max) 12.0% Ni 17.0% Cr 2.8% Mo steel: Casting; Fagersta
R 470	0.05% C 14.5% Ni 17% Cr 4.5% Mo steel: Fagersta; annealed **DPN: 180 UTS: 530 Elon: 45% Proof: 200**
R 470G	0.04% C 15.5% Ni 17.5% Cr 3.7% Mo steel: Casting; Fagersta
R 500	0.06% C 9.5% Ni 18% Cr 1% Nb steel: Fagersta; annealed **DPN: 200 UTS: 530 Elon: 45% Proof: 240**
R 510	0.06% C 9.5% Ni 18% Cr 0.5% Ti steel: Fagersta; annealed **DPN: 200 UTS: 530 Elon: 45% Proof: 240**
R 530	0.06% C 11% Ni 17.5% Cr 2.3% Mo + Ti steel: Fagersta; annealed **DPN: 180 UTS: 530 Elon: 40% Proof: 240**
R 590	0.06% C 12% Ni 13.5% Cr steel: Fagersta; annealed **DPN: 170 UTS: 480 Elon: 50% Proof: 180**
R 620G	0.1% C 5.0% Ni 24.0% Cr steel: Casting; scaling temperature 1070 °C; Fagersta
R 640G	0.1% C 5.0% Ni 24.0% Cr 1.5% Mo steel: Casting; scaling temperature 1070 °C; Fagersta
R 800	0.08% C 19% Ni 20.5% Cr steel: Fagersta; annealed **DPN: 220 UTS: 530 Elon: 40% Proof: 200**
R 800G	0.2% C 19.0% Ni 20.0% Cr 1.5% Si steel: Casting; scaling temperature 1050 °C; Fagersta
R 820	0.07% C 20% Ni 25% Cr steel: Fagersta; annealed **DPN: 230 UTS: 580 Elon: 40% Proof: 240**
R 830	0.1% C 23.5% Ni 22.5% Cr steel: Fagersta; annealed **DPN: 230 UTS: 580 Elon: 30% Proof: 240**
R 830G	0.2% C 22.0% Ni 23.0% Cr 1.5% Si steel: Casting; scaling temperature 1125 °C; Fagersta
R 850G	0.2% C 11.5% Ni 23.0% Cr 1.5% Si steel: Casting; scaling temperature 1125 °C; Fagersta

Symbol	Nominal analysis, supplier, condition and remarks.
R 860G	0.2% C 40% Ni 20.0% Cr steel: Casting; scaling temperature 1050 °C; Fagersta
R 890Gr	0.05% C 24.0% Ni 20.0% Cr 3.0% Mo 2.5% Cu steel: Casting
RA 310	0.05% C 20% Ni 25% Cr steel: Origin unknown
RA 330	0.08% C 35% Ni 18% Cr steel: Simonds
RA 330	0.08% C (max) 35.5% Ni 19% Cr Fe alloy: Origin unknown
RA 330 TX	0.3% C 35.5% Ni 18.5% Cr Fe alloy
RCK 3	0.15% C 13% Ni 25% Cr 0.2% N steel: Bofors; annealed
	DPN: 170 UTS: 630 Elon: 54% Proof: 340
RCK 4	0.2% C 12% Ni 26% Cr 0.2% N steel: Bofors; obsolete
RCT	0.45% C 12% Ni 15% Cr 3.0% W steel: Bofors; obsolete
RCT 3	0.2% C 13% Ni 21% Cr 3.0% W steel: Bofors; obsolete
REB 210	0.18% C 21% Ni 25% Cr steel: Bofors; obsolete
RIM 21	0.08% C 11% Ni 18% Cr 1.5% Mo steel: Bofors; obsolete
RIM 29	0.16% C 8% Ni 18% Cr steel: Bofors
RIM 210	0.05% C 11% Ni 18% Cr 1.5% Mo steel: Bofors; obsolete
RIM 213	0.07% C 11% Ni 18% Cr 2.3% Mo 0.4% Ti steel: Bofors; annealed
	DPN: 165 UTS: 580 Elon: 50% Proof: 200
RIM 215	0.12% C (max) 10% Ni 18% Cr 3.0% Mo steel: Bofors
RIM 217	0.02% C 13% Ni 18% Cr 2.8% Mo steel: Bofors
	DPN: 180 UTS: 500 Elon: 45% Proof: 200
RIM 290	0.02% C 10% Ni 18% Cr steel: Bofors; annealed; obsolete
	DPN: 140 UTS: 510 Elon: 60% Proof: 200
RIM 291	0.05% C 9% Ni 18% Cr steel: Bofors; annealed
	DPN: 155 UTS: 570 Elon: 65% Proof: 200
RIM 294	0.15% C (max) 9% Ni 18% Cr 0.6% Ti steel: Bofors
RIM 295	0.15% C 9% Ni 19% Cr 0.6% Ti steel: Bofors
RIO 214	0.09% C 5% Ni 26% Cr 1.5% Mo steel: Bofors; annealed
	DPN: 240 UTS: 710 Elon: 28% Proof: 530
RLH 2	9% Ni 18% Cr S steel: Free machining; Bofors
RNC 0	12% Ni 12% Cr Fe alloy: For electrical resistance; Imphy
RNC 1	35% Ni 12% Cr Fe alloy: For electrical resistance; Imphy
RRN J 30	0.15% C 9% Ni 18% Cr steel: Fagersta
RRN J 31	0.08% C (max) 9% Ni 19% Cr steel: Fagersta
RRN J 38	0.1% C 9% Ni 18% Cr S steel: Free machining; Fagersta
RRN J 42	0.12% C (max) 10% Ni 18% Cr 2.0% Mo steel: Fagersta
RRN J 44	0.12% C (max) 10% Ni 18% Cr 3.0% Mo steel: Fagersta
RRN J 50	0.15% C 9% Ni 18% Cr 1.0% Nb steel: Fagersta
RRN J 51	0.15% C 9% Ni 18% Cr 0.6% Ti steel: Fagersta
RRN J 59	9% Ni 18% Cr steel: For deep drawing; Fagersta
S 18/8	Stainless steel: Sintered material; Sintered Products Ltd; 15–20% porosity
	UTS: 400 Elon: 22%
S 18/8	Sintered 18/8 stainless steel: Specific gravity 6.9; Durasint
	DPN: 130 UTS: 440 Elon: 10%
S85	Austenitic steel: Sheet; obsolete
S108	0.2% C (max) 14% Ni 23% Cr 1.5% Nb steel: Annealed
	UTS: 530 Elon: 30% Proof: 200
S109	0.2% C (max) 18% Ni 23% Cr 1.5% Nb steel: Annealed
	UTS: 530 Elon: 30% Proof: 200
S110	0.16% C (max) 9% Ni 18% Cr Ti or Nb steel: Annealed; stabilized
	UTS: 530 Elon: 30% Proof: 200
S111	0.42% C 14% Ni 14% Cr 0.4% Mo 2.8% W steel: Annealed
	DPN: 269
S125	0.15% C (max) 15% Ni 23% Cr 0.7% Ti steel: Annealed
	UTS: 530 Elon: 30% Proof: 200
S126	0.15% C (max) 14% Ni 23% Cr 1.1% Nb steel: Annealed
	UTS: 530 Elon: 30% Proof: 200
S127	0.15% C (max) 17% Ni 25% Cr 0.7% Ti steel: Annealed
	UTS: 530 Elon: 30% Proof: 200
S128	0.15% C (max) 17% Ni 25% Cr 1.1% Nb steel: Annealed
	UTS: 530 Elon: 30% Proof: 200
S129	0.12% C (max) 10% Ni 18% Cr 0.7% Ti steel: Annealed
	UTS: 530 Elon: 30% Proof: 200
S130	0.08% C 10% Ni 18% Cr 1.0% Nb steel: Annealed
	UTS: 530 Elon: 30% Proof: 200
S 495	0.4% C 20% Ni 14% Cr 4% Mo 4% W 4% Nb steel: Allegheny Ludlum
S520	0.16% C (max) 10% Ni 18% Cr Ti or Nb steel: Sheet; stabilized; cold rolled and tempered
	UTS: 920 Elon: 15% Proof: 620
S521	0.16% C (max) 10% Ni 18% Cr Ti or Nb steel: Sheet; stabilized; annealed
	UTS: 530 Elon: 30% Proof: 200
S522	0.16% C (max) 14% Ni 23% Cr Ti or Nb steel: Sheet; stabilized; annealed
	UTS: 530 Elon: 30%
S523	0.16% C (max) 18% Ni 23% Cr Ti or Nb steel: Sheet; stabilized; annealed
	UTS: 530 Elon: 30%
S 524	0.08% C 10% Ni 18% Cr 0.4% Ti steel: BS designation
S 525	0.08% C 10% Ni 18% Cr 0.8% Nb steel: BS designation
S 526	0.08% C 10% Ni 18% Cr 0.4% Ti steel: BS designation
S 527	0.08% C 10% Ni 18% Cr 0.8% Nb steel: BS designation
S 528	0.12% C 14.5% Ni 23% Cr 0.6% Ti steel: BS designation
S 529	0.12% C 14.5% Ni 23% Cr 1% Nb steel: BS designation
S 530	0.12% C 17.5% Ni 24% Cr 0.6% Ti steel: BS designation
S 531	0.12% C 17.5% Ni 24% Cr 1% Nb steel: BS designation
S 536	0.03% C (max) 11% Ni 18% Cr steel: BS designation
S 537	0.03% C (max) 12.5% Ni 17.5% Cr 2.75% Mo steel: BS designation
S 588	0.4% C 20% Ni 18.5% Cr 4% Mo 4% W 4% Nb steel: Allegheny Ludlum
S 21000	0.1% C (max) 5.5% Mn 18% Ni 20% Cr 5% Mo steel: Designation used by UNS
S 30115	0.1% C 8% Ni 17% Cr 0.7% Mo steel: Designation used by UNS
S 30210	0.3% C 3.5% Mn 9% Ni 18% Cr 0.2% P steel: Designation used by UNS
S 30260	0.15% C (max) 11% Ni 17% Cr 0.3% P steel: Designation used by UNS
S 30415	0.05% C 9.5% Ni 18.5% Cr 0.16% N 0.05% Ce steel: Designation used by UNS
S 30431	0.06% C (max) 10% Ni 17.5% Cr 1.8% Cu steel: Designation used by UNS
S 30815	0.1% C (max) 11% Ni 21% Cr steel: UNS designation
S 34740	0.08% C (max) 10.5% Ni 18% Cr 0.8% Nb 0.2% S steel: Obsolete; designation used by UNS
S 34741	0.08% C (max) 10.5% Ni 18% Cr 0.8% Nb 0.2% Se steel: Obsolete; designation used by UNS

Symbol	Nominal analysis, supplier, condition and remarks.
SAE 30301	0.1% C 7% Ni 17% Cr steel: Mechanical properties not specified; AISI type 301
SAE 30302	0.1% C 9% Ni 18% Cr steel: Mechanical properties not specified; AISI type 302
SAE 30302 B	0.15% C 9% Ni 18% Cr steel: As SAE 30302 with different C range
SAE 30303	0.15% C (max) 9% Ni 18% Cr 0.6% Zr or Mo steel
SAE 30303 F	0.15% C (max) 9% Ni 18% Cr 0.6% Mo steel: Free machining; mechanical properties not specified; P or S or Se 0.07%; AISI type 303F
SAE 30303 Se	0.15% C 9% Ni 18% Cr 0.5% Se (min) steel: Free cutting; mechanical properties not specified
SAE 30304	0.08% C (max) 9% Ni 19% Cr steel: Mechanical properties not specified; AISI type 304
SAE 30304 L	0.03% C (max) 10% Ni 19% Cr steel: Mechanical properties not specified
SAE 30305	0.12% C (max) 12% Ni 18% Cr steel: Mechanical properties not specified; AISI type 305
SAE 30308	0.08% C 11% Ni 20% Cr steel
SAE 30309	0.2% C (max) 14% Ni 23% Cr steel: Mechanical properties not specified; AISI type 309
SAE 30309 S	0.08% C 14% Ni 23% Cr steel: Mechanical properties not specified
SAE 30310	0.25% C (max) 20% Ni 25% Cr steel: Mechanical properties not specified; AISI type 310
SAE 30310 S	0.08% C 21% Ni 25% Cr steel: Mechanical properties not specified
SAE 30314	0.25% C 21% Ni 24% Cr steel: Mechanical properties not specified
SAE 30316	0.08% C (max) 12% Ni 17% Cr 2.5% Mo steel: Mechanical properties not specified; AISI type 316
SAE 30316 L	0.03% C 12% Ni 17% Cr 2.5% Mo steel: Mechanical properties not specified
SAE 30317	0.08% C (max) 13% Ni 19% Cr 3.5% Mo steel: Mechanical properties not specified; AISI type 317
SAE 30321	0.08% C (max) 10% Ni 18% Cr 0.5% Ti steel: Mechanical properties not specified; AISI type 321
SAE 30325	0.25% C (max) 21% Ni 8% Cr 1.2% Cu steel: Mechanical properties not specified; AISI type 325
SAE 30329	0.09% C 5% Ni 26% Cr 1.5% Mo steel: Obsolete
SAE 30330	0.25% C (max) 35% Ni 15% Cr steel: Mechanical properties not specified
SAE 30330 A	0.45% C 35% Ni 15% Cr steel: Mechanical properties not specified
SAE 30347	0.08% C 10% Ni 18% Cr 1.0% Nb steel: Mechanical properties not specified; AISI type 347
SAE 30348	0.08% C 11% Ni 18% Cr 0.8% Nb 0.1% Ta steel: Mechanical properties not specified
SAE 60303	0.16% C 10% Ni 19% Cr 1.5% Mo 0.3% Se steel: Casting; ASTM alloy CF 16F
SAE 60304	0.08% C 9% Ni 19% Cr steel: Casting; ASTM alloy CF 8
SAE 60304 L	0.03% C 1.5% Mn 10% Ni 19% Cr steel: Casting; as cast; weldable
SAE 60309	0.2% C 13% Ni 24% Cr steel: Casting; ASTM alloy CH 20
SAE 60310	0.2% C 21% Ni 24% Cr steel: Casting; ASTM alloy CK 20

Note. The following abbreviations and units are used in the tables:

DPN	Hardness, diamond pyramid number
UTS	Ultimate tensile strength, N/mm^2
Elon	Elongation, %
Proof	0.1% proof strength, N/mm^2

1 N/mm^2=0.1 hbar=0.102 kgf/mm^2=0.06475 tonf/in.^{2}=145.04 lbf/in.^{2}=1 MPa

See Appendix II for other abbreviations and conversion tables.

Symbol	Nominal analysis, supplier, condition and remarks.
SAE 60312	0.3% C 9% Ni 28% Cr steel: Casting; ASTM alloy CE 30
SAE 60316	0.08% C 11% Ni 19% Cr 2.5% Mo steel: Casting; ASTM alloy CF 8 M
SAE 60316 L	0.03% C 11% Ni 19% Cr 2.5% Mo steel: Casting; as cast; weldable
SAE 60317	0.08% C 11% Ni 19% Cr 3.5% Mo steel: Casting; as cast
SAE 60347	0.08% C 10% Ni 19% Cr 0.7% Nb steel: Casting
SAE 70308	0.3% C 10% Ni 20% Cr 0.5% Mo steel: Casting; ASTM alloy HF
SAE 70309	0.35% C 12% Ni 26% Cr 0.5% Mo 0.2% N steel: Casting; ASTM alloy HH
SAE 70310	0.5% C 20% Ni 26% Cr 0.5% Mo steel: Casting; ASTM alloy HK
SAE 70310 A	0.5% C 20% Ni 30% Cr 0.5% Mo steel: Casting; ASTM alloy HL
SAE 70311	0.45% C 25% Ni 21% Cr 0.5% Mo steel: Casting
SAE 70312	0.35% C 10% Ni 28% Cr 0.5% Mo steel: Casting; ASTM alloy HE
SAE 70327	0.5% C (max) 6% Ni 28% Cr 0.5% Mo steel: Casting; ASTM alloy HD
SAE 70330	0.5% C 35% Ni 15% Cr 0.5% Mo steel: Casting; ASTM alloy HT
SAE 70331	0.5% C 39% Ni 19% Cr 0.5% Mo steel: Casting; ASTM alloy HU
SAE EV 3	0.2% C 11.5% Ni 21% Cr steel: For exhaust valves **UTS: 840 Elon: 26% Proof: 450**
SAE EV 10	1% C 14% Ni 14% Cr steel: Casting for exhaust valves
SAE HNV 5	0.35% C 8% Ni 13% Cr 0.5% Mo 2.5% Si steel: For inlet valves
SAE J230	0.15% C (max) 9% Ni 18% Cr steel: Cold drawn wire for springs **UTS: 2000**
SAF 2205	0.03% C (max) 5.5% Ni 22% Cr 3% Mo 0.14% N steel: Sandrik
SANBRON AA	0.16% C 8% Ni 19% Cr steel: Sanderson for BS alloy En 58A
SANICRO 28	0.02% C (max) 31% Ni 27% Cr 3.5% Mo 1% Cu steel: Sandvik
SANICRO 30	0.03% C (max) 34% Ni 22% Cr steel: With Ti and Al additions; Sandvik
SANICRO 31	0.04% C 31% Ni 21% Cr Fe alloy: Tube; Sandvik **DPN: 150 UTS: 490 Elon: 30% Proof: 170**
SANICRO 31H	0.07% C 31% Ni 21% Cr steel: With Ti and Al additions; Sandvik
SCF 19	0.03% C 18% Ni 20% Cr 5% Mo 0.3% N steel: Carpenter **UTS: 830 Elon: 52% Proof: 430**
SEC 7	0.15% C (max) 9% Ni 18% Cr steel: Designation used by Japanese Standard
SEC 8	0.08% C (max) 10% Ni 19% Cr steel: Designation used by Japanese Standard
SEC 9	0.03% C (max) 10% Ni 19% Cr steel: Designation used by Japanese Standard
SEC 10	0.08% C 10.5% Ni 18% Cr 0.5% Ti steel: Designation used by Japanese Standard
SEC 13	0.03% C 12% Ni 17% Cr 2.5% Mo steel: Designation used by Japanese Standard
SEH 5	0.25% C 20% Ni 25% Cr steel: Designation used by Japanese Standard
SF 24	0.08% C 13% Ni 13% Cr steel: Annealed; Samuel Fox **UTS: 540 Elon: 56%**
SF 25	0.1% C 9% Ni 18% Cr 2.0% Mo + Ti steel: Weldable; Samuel Fox **UTS: 690 Elon: 46%**
SF 50	0.12% C 10% Ni 17% Cr Mo Ti S steel: Weldable; free machining; Samuel Fox **UTS: 630 Elon: 56%**
SF 302	0.08% C 9% Ni 18% Cr steel: Annealed; Samuel Fox **UTS: 620 Elon: 57%**

Symbol	Nominal analysis, supplier, condition and remarks.
SF 304	0.06% C 10% Ni 19% Cr steel: Annealed; Samuel Fox **UTS: 600 Elon: 57%**
SF 304 L	0.03% C 10% Ni 19% Cr steel: Annealed; weldable; Samuel Fox **UTS: 600 Elon: 57%**
SF 305	0.12% C 12% Ni 18% Cr steel: Annealed; Samuel Fox **UTS: 580 Elon: 54%**
SF 316	0.08% C 12% Ni 17% Cr 2.5% Mo steel: Weldable; up to 10 mm; Samuel Fox **UTS: 610 Elon: 50%**
SF 316 ELC	0.03% C (max) 12% Ni 17% Cr 2.5% Mo steel: Annealed; weldable; Samuel Fox **UTS: 610 Elon: 50% Proof: 32**
SF 316 Ti	0.08% C 12% Ni 17% Cr 2.5% Mo + Ti steel: Weldable; Samuel Fox **UTS: 610 Elon: 50%**
SF 317	0.08% C 13% Ni 19% Cr 3.5% Mo steel: Weldable; Samuel Fox **UTS: 630 Elon: 50%**
SF 321	0.08% C 10% Ni 18% Cr Ti steel: Weldable; Samuel Fox **UTS: 610 Elon: 53%**
SF 347	0.08% C 10% Ni 18% Cr Nb steel: Weldable; Samuel Fox **UTS: 620 Elon: 54%**
SF 835	0.08% C 12% Ni 17% Cr 3.5% Mo + Ti steel: Weldable for use on check plates; annealed; Samuel Fox **UTS: 610 Elon: 50%**
SIL 10	0.38% C 8% Ni 19% Cr 3% Si steel: For valves; origin unknown
SIL 10 N	0.38% C 8% Ni 19% Cr 3% Si steel: For valves; origin unknown
SIRIUS 3	0.08% C (max) 20.5% Ni 25.0% Cr steel: Creusot-Loire
SIRIUS 35	0.08% C (max) 35.5% Ni 18.5% Cr steel: Creusot-Loire
SIRIUS 345	0.08% C (max) 13.5% Ni 23.0% Cr steel: Creusot-Loire
SIRIUS S12	0.15% C (max) 9.0% Ni 19.0% Cr steel: Creusot-Loire
For SIS specifications see SS	
SKF 304	As AISI 304 SKF
SKF 316	As AISI 316 SKF
SKF 321	As AISI 321 SKF
SOUDINOX 25-4	0.1% C 5.0% Ni 26.0% Cr steel: Welding electrode; Soudometal; for welding 25% Cr steels **UTS: 700 Elon: 16% Proof: 500**
SOUDINOX 25-20	0.1% C 20.0% Ni 25.0% Cr steel: Welding electrode; Soudometal; for welding and building of high alloy austenitic steel **UTS: 600 Elon: 30% Proof: 300**
SOUDINOX A8	0.11% C 6.5% Mn 8.0% Ni 17.0% Cr steel: Welding rod; Soudometal; for welding hardenable steels **UTS: 650 Elon: 40% Proof: 450**
SOUDINOX A308H	0.05% C 9.5% Ni 19.0% Cr steel: Welding electrode; Soudometal; for welding austenitic steels **UTS: 590 Elon: 53% Proof: 415**
SOUDINOX A308L	0.03% C 9.5% Ni 19.0% Cr steel: Welding electrode; Soudometal; for welding austenitic steels **UTS: 570 Elon: 47% Proof: 420**
SOUDINOX B316H	0.05% C 1.3% Mn 11.5% Ni 18.5% Cr 2.3% Mo steel: Welding electrode; Soudometal; for welding 316 type steel **UTS: 550 Elon: 35% Proof: 345**
SOUDINOX B316L	0.03% C 1.7% Mn 12.0% Ni 18.5% Cr 2.6% Mo steel: Welding electrode; Soudometal; for welding 316 type steel **UTS: 570 Elon: 40% Proof: 450**
SOUDINOX LF	0.03% C 22.0% Ni 25.0% Cr 2.0% Mo 0.15% N steel: Welding electrode; Soudometal; for corrosion resistant welding **UTS: 625 Elon: 35% Proof: 425**
SOUDINOX S65	0.03% C 19.0% Ni 25.0% Cr 0.25% Nb steel: Welding electrode; Soudometal; for corrosion resistant welding **UTS: 550 Elon: 35% Proof: 380**
SOUDOCROM 56	0.02% C 25.0% Ni 20.0% Cr 4.5% Mo 1.5% Cu steel: Welding electrode; Soudometal; for corrosion resistant welding **UTS: 560 Elon: 36% Proof: 375**
SOUDOCROM A	0.02% C 9.5% Ni 18.5% Cr steel: Welding electrode; Soudometal; for welding austenitic steel **UTS: 615 Elon: 48% Proof: 450**
SOUDOCROM AD	0.02% C 9.5% Ni 19.5% Cr steel: Welding electrode; Soudometal; for welding austenitic steels **UTS: 615 Elon: 48% Proof: 460**
SOUDOCROM B	0.02% C 11.5% Ni 18.0% Cr 2.1% Mo steel: Welding electrode; Soudometal; for welding type 316 steel **UTS: 590 Elon: 45% Proof: 475**
SOUDOCROM BD	0.02% C 11.5% Ni 18.5% Cr 2.6% Mo steel: Welding electrode; Soudometal; for welding type 316 type steel **UTS: 580 Elon: 45% Proof: 465**
SOUDOCROM D	0.12% C 9.0% Ni 30.0% Cr steel: Welding electrode; Soudometal **UTS: 810 Elon: 22% Proof: 670**
SOUDOCROM G	0.02% C 10.5% Ni 21.0% Cr 3.5% Mo steel: Welding electrode; Soudometal; for welding type 316 steel **UTS: 780 Elon: 30% Proof: 620**
SOUDOCROM HR 308L	0.02% C 9.5% Ni 19.5% Cr steel: Welding electrode; Soudometal; for welding austenitic steel **UTS: 600 Elon: 45% Proof: 450**
SOUDOCROM HR 309L	0.02% C 12.5% Ni 23.0% Cr steel: Welding electrode; Soudometal; for buttering of 3082 type steel **UTS: 550 Elon: 40% Proof: 450**
SOUDOCROM HR 316L	0.02% C 11.5% Ni 28.5% Cr 2.6% Mo steel: Welding electrode; Soudometal; for welding 316 type steel **UTS: 580 Elon: 45% Proof: 465**
SOUDOCROM L 309L	0.02% C 13.0% Ni 24.0% Cr steel: Welding electrode; Soudometal; for buttering of 308 type steel **UTS: 570 Elon: 40% Proof: 470**
SOUDOCROM A308L	0.02% C 9.5% Ni 19.5% Cr steel: Welding electrode; Soudometal; for welding austenitic steels **UTS: 615 Elon: 48% Proof: 460**
SOUDOCROM B316L	0.02% C 12.0% Ni 18.5% Cr 2.6% Mo steel: Welding electrode; Soudometal; for welding type 316 steel **UTS: 590 Elon: 42% Proof: 480**
SOUDOCROM C310	0.1% C 21.0% Ni 26.0% Cr steel: Welding rod: Soudometal; for welding high alloy austenitic steel **UTS: 590 Elon: 35% Proof: 420**
SOUDOCROM F347	0.02% C 9.5% Ni 18.5% Cr 0.25% Nb steel: Welding electrode; Soudometal; for welding 321 type steels **UTS: 610 Elon: 45% Proof: 470**
SOUDOCROM K318	0.02% C 12.0% Ni 18.0% Cr 2.6% Mo 0.25% Nb steel: Welding electrode; Soudometal; for welding 316 type steel **UTS: 620 Elon: 42% Proof: 495**
SOUDOCROM L309 Mo	0.02% C 13.0% Ni 24.0% Cr 2.3% Mo steel: Welding electrode; Soudometal; for buttering 316L type steel **UTS: 700 Elon: 32% Proof: 560**
SOUDOCROM M317L	0.02% C 0.8% Mn 13.0% Ni 18.5% Cr 3.3% Mo steel: Welding electrode; Soudometal; for welding 316 type steel **UTS: 520 Elon: 30%**
SOUDOCROM S4462	0.03% C 8.5% Ni 22.0% Cr 3.0% Mo 0.15% N steel: Welding rod; Soudometal; for welding Duplex stainless steel **UTS: 790 Elon: 28% Proof: 640**
SOUDOKAY 308L-0	0.02% C 10% Ni 21% Cr steel: Welding electrode; Soudometal **UTS: 660 Elon: 40% Proof: 490**

Symbol	Nominal analysis, supplier, condition and remarks.
SOUDOKAY 309L-0	0.02% C 13% Ni 24% Cr steel: Welding electrode; Soudometal **UTS: 640 Elon: 40% Proof: 490**
SOUDOKAY 316LO	0.02% C 12% Ni 20% Cr 2.2% Mo steel: Welding electrode; Soudometal **UTS: 625 Elon: 40% Proof: 480**
SOUDOKAY 402.0	0.12% C 6.5% Mn 8% Ni 20% Cr steel: Welding electrode; Soudometal **DPN: 155 UTS: 640 Elon: 30% Proof: 440**
SOUDOR G18.8.6	0.07% C 8.6% Ni 19% Cr steel: Welding electrode; Soudometal **UTS: 610 Elon: 40% Proof: 350**
SOUDOR G309	0.06% C 13% Ni 23.5% Cr steel: Welding electrode; Soudometal; for inert gas welding **UTS: 640 Elon: 28% Proof: 360**
SOUDOR G309 Si	0.08% C 13.3% Ni 23.6% Cr steel: Welding electrode; Soudometal; for inert gas welding **UTS: 640 Elon: 35% Proof: 400**
SOUDOR G310	0.1% C 20.5% Ni 25.7% Cr steel: Welding electrode; Soudometal; for inert gas welding **UTS: 590 Elon: 43% Proof: 390**
SOUDOR G316L Si	0.01% C 12.3% Ni 18.5% Cr 2.6% Mo steel: Welding electrode; Soudometal; for inert gas welding **UTS: 590 Elon: 34% Proof: 320**
SOUDOR G318	0.03% C 12% Ni 18.7% Cr 2.7% Mo 0.5% Nb steel: Welding electrode; Soudometal; for inert gas welding **UTS: 610 Elon: 36% Proof: 400**
SOUDOR G347	0.04% C 9.7% Ni 19.3% Cr 0.6% Nb steel: Welding electrode; Soudometal; for inert gas welding **UTS: 610 Elon: 42% Proof: 400**
SOUDOR G20.25.5L Cu	0.015% C 25% Ni 20% Cr 4.6% Mo 1.5% Cu steel: Welding electrode; Soudometal; for inert gas welding **UTS: 540 Elon: 37% Proof: 320**
SOUDOR G22.8.3L	0.02% C 8% Ni 22.8% Cr 3% Mo 0.15% N steel: Welding electrode; Soudometal; for inert gas welding **UTS: 540 Elon: 37% Proof: 320**
SOUDOR G308L Si	0.01% C 10% Ni 20% Cr steel: Welding electrode; Soudometal; for inert gas welding **UTS: 590 Elon: 34% Proof: 320**
SOUDOTIG 18.8.6	0.07% C 8.5% Ni 19% Cr steel: Welding electrode; Soudometal; for inert gas welding **UTS: 610 Elon: 40% Proof: 350**
SOUDOTIG 308L	0.02% C 10% Ni 20% Cr steel: Welding electrode; Soudometal; for inert gas welding **UTS: 590 Elon: 30% Proof: 320**
SOUDOTIG 309 Si	0.08% C 13.3% Ni 23.6% Cr 0.85% Si steel: Welding electrode; Soudometal; for inert gas welding **UTS: 640 Elon: 35% Proof: 400**
SOUDOTIG 310	0.1% C 20.5% Ni 25.7% Cr steel: Welding electrode; Soudometal **UTS: 590 Elon: 43% Proof: 390**
SOUDOTIG 316L	0.015% C 12% Ni 18.5% Cr 2.6% Mo steel: Welding electrode; Soudometal; for inert gas welding **UTS: 600 Elon: 31% Proof: 340**
SOUDOTIG 316L Si	0.1% C 12.3% Ni 18.5% Cr 2.6% Mo steel: Welding electrode; Soudometal; for inert gas welding **UTS: 590 Elon: 34% Proof: 320**
SOUDOTIG 318	0.03% C 12% Ni 18.7% Cr 2.7% Mo 0.5% Nb steel: Welding electrode; Soudometal; for inert gas welding **UTS: 620 Elon: 30% Proof: 350**
SOUDOTIG 347	0.04% C 9.7% Ni 19.3% Cr 0.6% Nb steel: Welding electrode; Soudometal; for inert gas welding **UTS: 630 Elon: 30% Proof: 350**
SOUDOTIG 20.25.2L Cu	0.15% C 25% Ni 20% Cr 4.6% Mo 1.5% Cu steel: Welding electrode; Soudometal; for inert gas welding **UTS: 540 Elon: 37% Proof: 320**
SOUDOTIG 22.8.3L	0.02% C 8% Ni 22.8% Cr 3% Mo 0.15% N steel: Welding electrode; Soudometal; for inert gas welding **UTS: 540 Elon: 37% Proof: 320**
SOUDOTIG 308L Si	0.015% C 10% Ni 20% Cr steel: Welding electrode; Soudometal; for inert gas welding **UTS: 590 Elon: 34% Proof: 320**
SPW	0.28% C 10% Ni 20% Cr 0.2% Mo steel: Pompey
S R ALLOY	0.6% C 37% Ni 20% Cr 0.4% Mo 2.0% W steel: Casting; Cronite; as cast **DPN: 170 UTS: 450 Elon: 9% Proof: 250**
S S 3M	0.16% C 9.0% Ni 18% Cr steel: Uddelholm
S S 3MM	0.08% C 9% Ni 19% Cr steel: Uddelholm
S S 4	0.12% C 10% Ni 19% Cr 2.0% Mo steel: Uddelholm
S S 24	0.12% C (max) 10% Ni 19% Cr 3.0% Mo steel: Uddelholm
S S 33	0.16% C 12% Ni 12% Cr steel: For deep drawing; Uddelholm
S S 43	9.0% Ni 18% Cr S steel: Free machining; Uddelholm; C content to be agreed
S S 53	0.15% C 9.0% Ni 18% Cr 0.6% Ti steel: Uddelholm
SS 2324	0.1% C 5% Ni 20% Cr 1.5% Mo steel: Swedish Standard; included in MNC 900 **DPN: 260**
SS 2332	0.08% C 10% Ni 18% Cr steel: Bar, forging, etc.; annealed; Swedish Standard **DPN: 180 UTS: 480 Elon: 45% Proof: 200**
SS 2333	0.06% C (max) 10% Ni 18% Cr steel: Bar, forging, etc.; annealed; Swedish Standard **DPN: 180 UTS: 480 Elon: 45% Proof: 200**
SS 2334	0.15% C (max) 9.0% Ni 18% Cr Ti or Nb steel: Swedish Standard
SS 2337	0.08% C 10% Ni 18% Cr 0.5% Ti steel: Bar, forging, etc.; annealed; Swedish Standard **DPN: 190 UTS: 480 Elon: 40% Proof: 200**
SS 2338	0.08% C 10% Ni 18% Cr 0.8% Nb + Ta steel: Bar, forging, etc.; annealed; Swedish Standard **DPN: 190 UTS: 500 Elon: 40% Proof: 200**
SS 2340	0.1% C (max) 2.0% Mn (max) 9% Ni 17% Cr 1.5% Mo steel: For reinforcing; Swedish Standard **Elon: 7% Proof: 780**
SS 2341	0.06% C 10% Ni 18% Cr 1.8% Mo steel: Bar, forging, etc.; annealed; Swedish Standard **DPN: 180 UTS: 500 Elon: 45% Proof: 200**
SS 2342	0.1% C 9% Ni 17% Cr 2.3% Mo steel: Bar, forging, etc.; annealed; Swedish Standard **DPN: 200 UTS: 580 Elon: 40% Proof: 200**
SS 2343	0.06% C 12% Ni 17% Cr 2.7% Mo steel: Bar, forging, etc.; annealed; Swedish Standard **DPN: 180 UTS: 530 Elon: 45% Proof: 210**
SS 2344	0.08% C 12% Ni 17% Cr 2.7% Mo 0.6% Ti steel: Bar, forging, etc.; Swedish Standard **DPN: 190 UTS: 540 Elon: 40% Proof: 200**
SS 2345	0.08% C 12% Ni 17% Cr 2.7% Mo 0.9% Nb + Ta steel: Bar, forging, etc.; Swedish Standard **DPN: 190 UTS: 540 Elon: 40% Proof: 200**
SS 2346	0.15% C 10% Ni 17% Cr 0.7% Mo steel: Plate and bar; annealed; Swedish Standard **DPN: 200 UTS: 530 Elon: 40% Proof: 200**
SS 2346	0.15% C 10% Ni 17% Cr 0.7% Mo steel: Plate and bar; cold rolled; Swedish Standard **DPN: 310 UTS: 680 Elon: 20% Proof: 480**

Note. The following abbreviations and units are used in the tables:

DPN	Hardness, diamond pyramid number
UTS	Ultimate tensile strength, N/mm^2
Elon	Elongation, %
Proof	0.1% proof strength, N/mm^2

1 N/mm^2=0.1 hbar=0.102 kgf/mm^2=0.06475 $tonf/in.^2$=145.04 $lbf/in.^2$=1 MPa

See Appendix II for other abbreviations and conversion tables.

Symbol	Nominal analysis, supplier, condition and remarks.
SS 2347	0.5% C (max) 13.0% Ni 17.0% Cr 2.0% Mo steel: Swedish Standard **DPN: 190 UTS: 600 Elon: 45% Proof: 210**
SS 2348	0.03% C (max) 12.5% Ni 17.0% Cr 2.2% Mo steel: Swedish Standard **DPN: 190 UTS: 603 Elon: 45% Proof: 200**
SS 2350	0.08% C 12.0% Ni 17.0% Cr 2.2% Mo steel: Swedish Standard **DPN: 200 UTS: 600 Elon: 40% Proof: 210**
SS 2352	0.03% C 10.5% Ni 18.5% Cr steel: Swedish Standard; included in MNC 900 **DPN: 180**
SS 2353	0.03% C 12% Ni 17% Cr 2.7% Mo steel: Swedish Standard; included in MNC 900 **DPN: 180**
SS 2360	0.12% C 18% Ni 19% Cr steel: Plate, bar, forging; annealed; Swedish Standard **DPN: 220 UTS: 820**
SS 2361	0.2% C 19% Ni 22% Cr steel: Bar, forging, etc.; annealed; Swedish Standard **DPN: 230 UTS: 770**
SS 2366	0.05% C 14.5% Ni 18.5% Cr 3.5% Mo steel: Swedish Standard **DPN: 190 UTS: 600 Elon: 40% Proof: 220**
SS 2367	0.03% C (max) 16.0% Ni 18.5% Cr 3.5% Mo steel: Swedish Standard **DPN: 190 UTS: 600 Elon: 48% Proof: 210**
SS 2370	0.05% C (max) 9.5% Ni 18.0% Cr steel: Swedish Standard **DPN: 195 UTS: 700 Elon: 40% Proof: 290**
SS 2371	0.03% C (max) 10.5% Ni 18.5% Cr steel: Swedish Standard
SS 2374	0.05% C (max) 12.0% Ni 17.5% Cr 3.0% Mo steel: Swedish Standard **DPN: 190 UTS: 700 Elon: 40% Proof: 300**
SS 2375	0.03% C (max) 12.5% Ni 17.5% Cr 2.7% Mo steel: Swedish Standard
SS 2384	0.05% C (max) 13% Ni 18% Cr 2.7% Mo 1% Ti 0.2% S 2% Cu steel: Swedish Standard; free machining
SS 2564	0.06% C (max) 25% Ni 20% Cr 4.5% Mo 3.2% Cu steel: Casting; Swedish Standard
SS 2570	0.08% C 25% Ni 15% Cr 1.2% Mo 0.2% V 2.1% Ti steel: Swedish Standard; included in MNC 870 **DPN: 300**
STA 5 V26	0.42% C 10% Ni 13% Cr 3% W steel: Replaced by BS alloy En 56
STA 5 V27	0.16% C (max) 9% Ni 18% Cr steel: Replaced by BS alloy En 58A
STA 5 V27A	0.15% C 10% Ni 18% Cr steel: Weld stabilized; replaced by BS alloy En 58; stabilized grade
STA 5 V27M	0.1% C 10% Ni 18% Cr steel: Free machining; replaced by BS alloy En 58M
STAYBRITE	18% Cr 8% Ni steel: Firth Vickers; general trade name covering stainless steels
STC 52D	0.08% C 12% Ni 17% Cr 2.5% Mo steel: Designation used by Japanese Standard
STRAINTRODE 61.81	0.06% C 1.5% Mn 9.5% Ni 20.0% Cr 0.7% Nb 0.8% Si steel: For electrodes; Esab; 8% ferrite **UTS: 700 Elon: 30% Proof: 570**
STRAINTRODE 62.30	0.03% C 0.6% Mn 10.0% Ni 18.5% Cr 1.7% Mo 0.7% Si steel: For electrodes; Esab; 8% ferrite **UTS: 570 Elon: 45% Proof: 460**
STRAINTRODE 63.30	0.03% C 0.6% Mn 12.5% Ni 18.5% Cr 2.8% Mo steel: For electrodes; Esab; 9% ferrite **UTS: 590 Elon: 35% Proof: 490**
STRAINTRODE 63.35	0.05% C 0.7% Mn 12.5% Ni 18.5% Cr 2.8% Mo 0.5% Si steel: For electrodes; Esab; 15% ferrite **UTS: 590 Elon: 35% Proof: 490**
STRAINTRODE 63.80	0.03% C 1.5% Mn 12.5% Ni 18.5% Cr 2.8% Mo 0.6% Nb steel: For electrodes; Esab; 10% ferrite **UTS: 664 Elon: 35% Proof: 455**
STRAINTRODE 64.30	0.03% C 0.6% Mn 10.0% Ni 19.0% Cr 0.8% Si steel: For electrodes; Esab; 6% ferrite **UTS: 564 Elon: 46% Proof: 420**
STRAINTRODE 64.30	0.03% C 0.7% Mn 13.5% Ni 19.0% Cr 4.0% Mo steel: For electrodes; Esab; 11% ferrite **UTS: 602 Elon: 35% Proof: 455**
STRAINTRODE 67.15	0.1% C 1.7% Mn 20.0% Ni 26.0% Cr steel: For electrodes; Esab Standard **UTS: 610 Elon: 35% Proof: 450**
STRAINTRODE 67.45	0.05% C 6.5% Mn 9.0% Ni 17.5% Cr steel: For electrodes; Esab Standard **UTS: 620 Elon: 42% Proof: 390**
STRAINTRODE 67.70	0.04% C 1.0% Mn 12.0% Ni 22.0% Cr 2.5% Mo steel: For electrodes; Esab Standard; 15% ferrite **UTS: 730 Elon: 30% Proof: 640**
STRAINTRODE 67.75	0.05% C 3.0% Mn 12.0% Ni 23.0% Cr 0.5% Mo steel: For electrodes; Esab; 17% ferrite **UTS: 630 Elon: 30% Proof: 510**
STRAINTRODE 68.60	0.1% C 1.5% Mn 5.0% Ni 26.0% Cr 1.5% Mo steel: For electrodes; Esab Standard **UTS: 650 Elon: 35%**
STUBS 63 C Mn Cr Ni	Austenitic steel: Welding electrode; Stubs; for ductile weld; impact resistance **DPN: 200 UTS: 685 Elon: 40% Proof: 390**
STUBS 65 XL C Si Cr Ni Mo	Austenitic steel: Welding electrode; Stubs **DPN: 220 UTS: 800 Elon: 20% Proof: 640**
STUBS 65 C Si Mn Cr Ni Mo	Austenitic steel: Welding electrode; Stubs **DPN: 235 UTS: 805 Elon: 25%**
STUBS 68	0.05% C (max) 10% Ni 20% Cr 0.4% Nb + Ta steel: Stubs **UTS: 590 Elon: 30% Proof: 375**
STUBS 68 LC	0.03% C (max) 10% Ni 20% Cr steel: Stubs **DPN: 180 UTS: 590 Elon: 35% Proof: 375**
STUBS 68 Mo	0.05% C (max) 11.5% Ni 18% Cr 2.8% Mo 0.4% Nb + Ta steel; Stubs **DPN: 200 UTS: 590 Elon: 30% Proof: 390**
STUBS 68 Mo LC	0.03% C (max) 12% Ni 18% Cr 2.8% Mo steel; Stubs **DPN: 180 UTS: 590 Elon: 35% Proof: 390**
STUBS 68 Ti Mo	0.03% C (max) 9.5% Ni 19% Cr 2.0% Mo steel: Stubs **DPN: 190 UTS: 640 Elon: 23% Proof: 390**
STUBS 68H	0.12% C (max) 20% Ni 25% Cr steel: Welding electrode; Stubs; for heat resistance **DPN: 200 UTS: 590 Elon: 30% Proof: 345**
STUBS 653 C Cr Ni Mo	Austenitic steel: Welding electrode; Stubs; can be work hardened **DPN: 240 UTS: 785**
STUBS 5820 LC	0.02% C 9.5% Ni 18.5% Cr steel: Welding electrode; Stubs; for welding type 304 stainless steel **DPN: 170 UTS: 590 Elon: 35% Proof: 400**
STUBS 5820 Mo LC	0.02% C 11.5% Ni 18.5% Cr 2.5% Mo steel: Welding electrode; Stubs; for welding type 316 steel **DPN: 170 UTS: 550 Elon: 35% Proof: 410**
STUBS 5820 Mo Nb	0.024% C 12% Ni 18.5% Cr 2.4% Mo 0.25% Nb steel: Welding electrode; Stubs; for welding type 318 steel **DPN: 170 UTS: 550 Elon: 35% Proof: 410**
STUBS 5820 Nb	0.02% C 10% Ni 18.5% Cr 0.2% Nb steel: Welding electrode; Stubs; for welding type 321 and 347 steels **DPN: 170 UTS: 590 Elon: 35% Proof: 400**
STUBS 5824 LC	0.02% C 12.7% Ni 22.8% Cr steel: Welding electrode; Stubs; for cladding mild and low alloy steel **DPN: 170 UTS: 540 Elon: 30% Proof: 390**
STUBS 5824 Mo LC	0.02% C 14% Ni 22.5% Cr 2.6% Mo steel: Welding electrode; Stubs **DPN: 170 UTS: 590 Elon: 30% Proof: 400**
STUBS 6820 LC	0.025% C 10% Ni 19.5% Cr steel: Welding electrode; Stubs; for welding type 304 steel **DPN: 170 UTS: 590 Elon: 35% Proof: 400**
STUBS 6820 Mo LC	0.03% C 11.5% Ni 19% Cr 2.4% Mo steel: Welding electrode; Stubs; for welding type 316 stainless steel **DPN: 170 UTS: 550 Elon: 35% Proof: 410**

Symbol	Nominal analysis, supplier, condition and remarks.
STUBS 6820 Mo Nb	0.028% C 12% Ni 18.5% Cr 2.2% Mo 0.3% Nb steel: Welding electrode; Stubs; for welding type 318 type steel **DPN: 170 UTS: 550 Elon: 35% Proof: 410**
STUBS 6824 LC	0.025% C 12.4% Ni 22.5% Cr steel: Welding electrode; Stubs; for cladding mild and low alloy steel **DPN: 170 UTS: 540 Elon: 30% Proof: 390**
STUBS 6824 Mo LC	0.03% C 12.5% Ni 22.5% Cr 2.3% Mo steel: Welding electrode; Stubs **DPN: 170 UTS: 590 Elon: 30% Proof: 400**
STUBS 6825	0.1% C 21.5% Ni 25.5% Cr steel: Welding electrode; Stubs; for heat resistance **DPN: 200 UTS: 590 Elon: 30% Proof: 345**
SUH 31	0.4% C 14.0% Ni 15.0% Cr 2.5% W steel: Heat resisting; Japanese Standard designation
SUPERCROM	0.02% C 9.0% Ni 30.0% Cr steel: Welding electrode; Soudometal **UTS: 800 Elon: 22% Proof: 660**
SUPERCROM 29.9	0.1% C 9.0% Ni 29.0% Cr steel: Welding electrode; Soudometal **UTS: 750 Elon: 25% Proof: 590**
SUPERCROM 308L	0.02% C 9.5% Ni 19.5% Cr steel: Welding electrode; Soudometal; for welding austenitic steel **UTS: 600 Elon: 48% Proof: 450**
SUPERCROM 318	0.02% C 12.0% Ni 18.0% Cr 2.6% Mo 0.25% Nb steel: Welding electrode; Soudometal; for welding 316 type steel **UTS: 600 Elon: 42% Proof: 495**
SUPERCROM 347	0.02% C 9.5% Ni 18.5% Cr 0.25% Nb steel: Welding electrode; Soudometal; for welding 321 type steel **UTS: 615 Elon: 45% Proof: 475**
SUPERCROM SD	0.1% C 9.0% Ni 29.0% Cr steel: Welding electrode; Soudometal **UTS: 750 Elon: 25% Proof: 590**
SUPERCROM 309L	0.02% C 13.0% Ni 23.0% Cr steel: Welding electrode; Soudometal; for buttering 308L type steel **UTS: 550 Elon: 40% Proof: 420**
SUPERCROM 316L	0.02% C 0.6% Mn 12.0% Ni 18.5% Cr 2.6% Mo steel: Welding electrode; Soudometal; for welding 316 type steel **UTS: 590 Elon: 45% Proof: 480**
SUPERMET 308L	0.02% C 10% Ni 19% Cr steel: Electrode; Metrode
SUPERMET 309L	0.02% C 13% Ni 24% Cr steel: Electrode; Metrode
SUPERMET 316L	0.03% C 12% Ni 18% Cr 2.5% Mo steel: Electrode; Metrode
SUPERMET 347	0.03% C 10% Ni 19% Cr 0.5% Nb steel: Electrode; Metrode
SUPERMET 309 Cb	0.02% C 12% Ni 24% Cr 0.5% Nb steel: Electrode; Metrode
SUPERMET 309 Mo	0.02% C 12% Ni 24% Cr 2.5% Mo steel: Electrode; Metrode
SUPERMET 318	0.03% C 11% Ni 18% Cr 2.2% Mo Nb steel: Electrode; Metrode
SUPERMET 2205	0.03% C 10% Ni 24% Cr 3% Mo steel: Electrode; Metrode
SV 02KH19N9	0.04% C (max) 9.0% Ni 19.5% Cr steel: For welding wire; Russian Standard designation
SV 04KH19N9	0.6% C (max) 9.0% Ni 19.0% Cr steel: For welding wire; Russian Standard designation

Note. The following abbreviations and units are used in the tables:

DPN	Hardness, diamond pyramid number
UTS	Ultimate tensile strength, N/mm^2
Elon	Elongation, %
Proof	0.1% proof strength, N/mm^2

1 N/mm^2=0.1 hbar=0.102 kgf/mm^2=0.06475 $tonf/in.^2$=145.04 $lbf/in.^2$=1 MPa
See Appendix II for other abbreviations and conversion tables.

Symbol	Nominal analysis, supplier, condition and remarks.
SV 04KH19N11M3	0.06% C (max) 11.0% Ni 19.0% Cr 2.5% Mo steel: For welding wire; Russian Standard designation
SV 04KH19N952	0.06% C (max) 9.0% Ni 19.0% Cr steel: For welding wire; Russian Standard designation
SV 05KH19N9F352	0.07% C (max) 9.0% Ni 19.0% Cr 2.5% V steel: For welding wire; Russian Standard
SV 06KH19N9T	0.08% C (max) 9.0% Ni 19.0% Cr 0.7% Ti steel: For welding wire; Russian Standard designation
SV 07KH18N9T1U	0.09% C (max) 9.0% Ni 18.0% Cr 1.2% Ti 0.7% Al steel: For welding wire; Russian Standard designation
SV 07KH25N13	0.09% C (max) 13.0% Ni 24.5% Cr steel: For welding wire; Russian Standard designation
SV 08KH19N9F252	0.1% C (max) 9.0% Ni 19.0% Cr 2.1% V steel: For welding wire; Russian Standard designation
SV 08KH19N10B	0.07% C 9.7% Ni 19.5% Cr 1.35% Nb steel: For welding wire; Russian Standard designation
SV 08KH19N12M3	0.08% C 12.2% Ni 19.5% Cr 2.6% Mo steel: For welding wire; Russian Standard designation
SV 08KH25N5TMF	0.1% C(max) 5.2% Ni 25.0% Cr 0.09% Mo 0.11% V 0.14% Ti 0.15% N steel: For welding wire; Russian Standard designation
SV 10KH16N25M6	0.1% C 25.5% Ni 16.2% Cr 6.2% Mo 0.15% N steel: For welding wire; Russian Standard designation
SV 10KH20N15	0.12% C (max) 15.0% Ni 20.5% Cr steel: For welding wire; Russian Standard designation
SV 13KH25N18	0.15% C (max) 18.5% Ni 25.5% Cr steel: For welding wire; Russian Standard designation
SV 25KH25N16G7	0.24% C 7.0% Mn 16.0% Ni 25.5% Cr steel: For welding wire; Russian Standard designation
SV 30KH15N35V3B 3T	0.08% C (max) 10.0% Ni 19.0% Cr 2.5% Mo 2.7% W 0.7% Ti steel: For welding wire; Russian Standard designation
SV 30KH15N35V3B 3T	0.29% C 35.0% Ni 15.0% Cr 0.8% Ti 3.0% Nb steel: For welding wire; Russian Standard designation
T55	0.16% C 10% Ni 18% Cr Ti or Nb stabilized steel: Tube; annealed **UTS: 530 Proof: 230**
T58	0.16% C (max) 10% Ni 18% Cr Ti or Nb steel: Tube; stabilized; cold drawn and tempered **UTS: 760 Proof: 700**
T61	0.16% C 17% Ni 23% Cr Ti or Nb steel: Tube; stabilized; annealed **UTS: 530 Proof: 220**
T 200	0.05% C 25% Ni 15% Cr 1.3% Mo 0.3% V 2.1% Nb 0.2% Al 0.005% B steel: Creep resisting; VEW
T 240	0.13% C 13.5% Ni 16% Cr 2.8% W 0.6% Ti 0.004% B steel: Creep resisting; VEW
T 275	0.06% C 12.5% Ni 16% Cr 0.6% Nb steel: Creep resisting; VEW
THERMALLOY 40 A2	0.5% C 15% Ni 26% Cr 1% Nb 0.13% N steel: Casting; American proprietary alloy listed in SAE yearbook.
THERMALLOY 40 E	0.9% C 12% Ni 26% Cr steel: Casting; American proprietary alloy listed in SAE yearbook
THERMALLOY 47	0.4% C 20% Ni 26% Cr 0.1% N steel: Casting; American proprietary alloy listed in SAE yearbook
THERMALLOY 50 CQ	0.5% C 35% Ni 15% Cr 1% Nb steel: Casting; American proprietary alloy listed in SAE yearbook
THERMALLOY D47	0.5% C 20% Ni 28% Cr 0.1% N steel: Casting; American proprietary alloy listed in SAE yearbook
THERMINOX	0.12% C 20% Ni 25% Cr steel: Pompey
THERMINOX 13	0.1% C 13% Ni 25% Cr steel: Pompey
THERMINOX 20	0.07% C 20% Ni 25% Cr steel: Pompey
THERMINOX 821	0.2% C 5.0% Ni 25% Cr steel: Pompey
TIMKEN 16/25/6	0.1% C 25% Ni 16% Cr 6% Mo 0.15% N steel: Timken
TOLOY 9 TYPE HE	0.35% C 9.5% Ni 28% Cr steel: Casting; resistant to sulphur bearing atmosphere; Wellman
TOLOY 10 TYPE HF	0.3% C 10.5% Ni 21% Cr steel: Casting for use at 600–850 °C; Wellman

Symbol	Nominal analysis, supplier, condition and remarks.
TOLOY 12	0.3% C 12% Ni 25% Cr steel: Casting; Thompson L'Hospied; annealed **DPN: 180 UTS: 630 Elon: 15% Proof: 340**
TOLOY 12 HH2	0.45% C 12% Ni 24.5% Cr steel: Casting for use between 925 and 1050 °C; Wellman
TOLOY 16	0.3% C 16% Ni 27% Cr steel: Casting; Thompson L'Hospied; annealed **DPN: 175 UTS: 610 Elon: 8% Proof: 480**
TOLOY 20	0.4% C 20% Ni 26% Cr steel: Casting; Thompson L'Hospied; annealed **DPN: 180 UTS: 530 Elon: 16% Proof: 340**
TOLOY 20 HK40	0.4% C 20% Ni 26% Cr steel: Casting; for use at temperatures above 1050 °C; Wellman
TOLOY 25 HN	0.37% C 25% Ni 21% Cr steel: Casting; good high temperature corrosion resistance; Wellman
TOLOY 37	0.55% C 37% Ni 19% Cr steel: Casting; Thompson L'Hospied; annealed **DPN: 165 UTS: 540 Elon: 12% Proof: 360**
TOLOY 37 HU	0.45% C 37.5% Ni 18.5% Cr Fe casting: Resists sulphur and carbon attack at high temperature; Wellman
TOPHET 12.12	0.16% C (max) 13.0% Ni 12.0% Cr steel: Wire; Getly-Brunton for weaving close mesh
TOPHET 18.8	0.16% C (max) 9.0% Ni 18.0% Cr steel: Getly-Brunton; can be hard drawn
TOPHET 18.8 Low C	0.08% C 9.5% Ni 18.0% Cr steel: Getly-Brunton; hard drawn **DPN: 160**
TOPHET 25.20	0.1% C 20.0% Ni 25.0% Cr steel: Wire; Getly-Brunton; annealed **UTS: 725 Elon: 55% Proof: 360**
TOPHET Cb	0.08% C (max) 9.2% Ni 18.0% Cr 0.6% Nb (min) steel: Getly-Brunton; annealed **UTS: 700 Elon: 55% Proof: 300**
TOPHET Mo	0.08% C (max) 10.0% Ni 18.0% Cr 2.85% Mo steel: Getly-Brunton; annealed **UTS: 720 Elon: 55% Proof: 360**
TOPHET Ti	0.08% C (max) 9.2% Ni 18.0% Cr 0.3% Ti (min) steel: Getly-Brunton; hard drawn **DPN: 165**
TP XM 15	See XM 15
TPA	0.45% C 14% Ni 14% Cr 0.35% Mo 2.4% W steel: For valves; origin unknown
TRUBRITE	8% Ni 18% Cr Ti stabilized steel: Arthur Lee
TRUBRITE A1	12% Ni 12% Cr steel: For deep drawing; Arthur Lee
TRUBRITE A3	8% Ni 18% Cr steel: Arthur Lee
TRUBRITE A6	Low C 10% Ni 19% Cr steel: Weldable; Arthur Lee
TURBALOY 13	0.13% C 23.5% Ni 17.5% Cr 2.5% Mo 1% W 1.5% Nb 1.5% Al Fe alloy: General Electric
TXCR	0.4% C 3.8% Ni 24% Cr 1.4% Mo steel: For valves; origin unknown
UCAR 11	0.05% C 39.0% Ni 18.0% Cr 5.2% Mo 0.5% Co 2.5% Ti Fe alloy: Casting; Union Carbide
UCAR 706	0.06% C (max) 41.5% Ni 16.0% Cr 1.0% Co 3.0% Mo 1.7% Nb Fe alloy: Union Carbide
UCAR 800	0.05% C 32.5% Ni 21.0% Cr 0.4% Ti 0.4% Al Fe alloy: Union Carbide
UHB STAINLESS 3	0.1% C 8.5% Ni 18% Cr steel: Uddelholm; annealed **DPN: 160 UTS: 630 Elon: 50% Proof: 220**
UHB STAINLESS 3L	0.03% C 10% Ni 18% Cr steel: Uddelholm; annealed **DPN: 160 UTS: 580 Elon: 50% Proof: 200**
UHB STAINLESS 3M	0.08% C 8.5% Ni 18% Cr steel: Uddelholm; annealed **DPN: 160 UTS: 630 Elon: 50% Proof: 220**
UHB STAINLESS 3MM	0.06% C 9% Ni 18% Cr steel: Uddelholm; annealed **DPN: 160 UTS: 630 Elon: 50% Proof: 220**
UHB STAINLESS 4	0.09% C 9.5% Ni 17.5% Cr 1.5% Mo steel: Uddelholm; annealed **DPN: 160 UTS: 630 Elon: 50% Proof: 240**
UHB STAINLESS 4MM	0.06% C 10.5% Ni 17.5% Cr 1.5% Mo steel: Uddelholm; annealed **DPN: 160 UTS: 630 Elon: 50% Proof: 240**
UHB STAINLESS 15	0.07% C 20.5% Ni 18.5% Cr steel: Uddelholm; annealed **DPN: 160 UTS: 580 Elon: 35% Proof: 250**
UHB STAINLESS 24	0.06% C 12% Ni 17.5% Cr 2.5% Mo steel: Uddelholm; annealed **DPN: 160 UTS: 630 Elon: 45% Proof: 240**
UHB STAINLESS 24L	0.03% C 13% Ni 17.5% Cr 2.8% Mo steel: Uddelholm; annealed **DPN: 160 UTS: 580 Elon: 50% Proof: 220**
UHB STAINLESS 25	0.07% C 21.5% Ni 23.5% Cr steel: Uddelholm; annealed **DPN: 160 UTS: 580 Elon: 35% Proof: 240**
UHB STAINLESS 33MM	0.06% C 12.5% Ni 12.5% Cr steel: Uddelholm; annealed **DPN: 130 UTS: 530 Elon: 55% Proof: 210**
UHB STAINLESS 34	0.05% C 14% Ni 17.5% Cr 4.3% Mo steel: Uddelholm; annealed **DPN: 170 UTS: 530 Elon: 40% Proof: 270**
UHB STAINLESS 34L	0.03% C 14.5% Ni 17.5% Cr 4.2% Mo steel: Uddelholm
UHB STAINLESS 43	0.08% C 9.5% Ni 17.5% Cr 0.5% Mo steel: Free cutting; Uddelholm; annealed **DPN: 160 UTS: 630 Elon: 45% Proof: 220**
UHB STAINLESS 44	0.1% C 5.3% Ni 25% Cr 1.5% Mo steel: Uddelholm; annealed **DPN: 230 UTS: 680 Elon: 20% Proof: 500**
UHB STAINLESS 45	0.1% C 4.8% Ni 25.5% Cr steel: Uddelholm; annealed **DPN: 220 UTS: 630 Elon: 25% Proof: 450**
UHB STAINLESS 53	0.06% C 9.5% Ni 18% Cr Ti steel: Uddelholm; annealed **DPN: 160 UTS: 630 Elon: 45% Proof: 220**
UHB STAINLESS 55	0.15% C 13% Ni 23% Cr steel: Uddelholm; annealed **DPN: 160 UTS: 630 Elon: 40% Proof: 290**
UHB STAINLESS 63	0.06% C 10% Ni 18% Cr Nb steel: Uddelholm; annealed **DPN: 160 UTS: 630 Elon: 40% Proof: 220**
UHB STAINLESS 63H	0.08% C 13% Ni 16% Cr 1.0% Nb steel: Tube; Uddelholm; annealed **DPN: 210 UTS: 610 Elon: 35% Proof: 210**
UHB STAINLESS 64	0.06% C 18% Ni 18% Cr 2.2% Mo 2.0% Cu + Nb steel: Uddelholm; annealed **DPN: 160 UTS: 530 Elon: 40% Proof: 240**
UHB STAINLESS 524	0.06% C 13% Ni 17.5% Cr 2.8% Mo Ti steel: Uddelholm; annealed **DPN: 160 UTS: 620 Elon: 50% Proof: 250**
UHB STAINLESS 624	0.06% C 13% Ni 17.5% Cr 2.8% Mo Nb steel: Uddelholm; annealed **DPN: 160 UTS: 620 Elon: 40% Proof: 250**
UHB STAINLESS 703	0.09% C 10.5% Ni 17.5% Cr steel: Uddelholm; annealed **DPN: 160 UTS: 620 Elon: 50% Proof: 220**
UHB STAINLESS 724	0.06% C 13.5% Ni 17% Cr 2.75% Mo steel: Tube; Uddelholm; annealed **DPN: 180 UTS: 580 Elon: 45% Proof: 210**
UHB STAINLESS 734L	0.03% C 13.5% Ni 17.5% Cr 3.4% Mo steel: Uddelholm
UHB STAINLESS 734Nb	0.06% C 13.5% Ni 17.5% Cr 3.4% Mo steel: Uddelholm
UHB STAINLESS 753	0.08% C 10% Ni 18% Cr 0.5% Ti steel: Tube; Uddelholm; annealed **DPN: 190 UTS: 630 Elon: 40% Proof: 200**
UNI 3992 X45 CNW 1909	0.45% C 9% Ni 19% Cr 1.0% W steel: For valves
UNI 3992 X50 CNW 1414	0.5% C 14% Ni 14% Cr 2.5% W steel: For valves
UNI 4047 X3 CN 1911	0.03% C (max) 10.5% Ni 19% Cr steel
UNI 4047 X6 CN 1911	0.06% C (max) 10.5% Ni 19% Cr steel

Symbol	Nominal analysis, supplier, condition and remarks.
UNI 4047 X8 CN 1910	0.08% C (max) 10.5% Ni 19% Cr steel
UNI 4047 X15 CN 1808	0.15% C (max) 9% Ni 18% Cr steel
UNI 4047 X15 CNF 1808	0.15% C (max) 9% Ni 18% Cr 0.6% Mo steel
URANUS 50	8% Ni 20% Cr 2.5% Mo 1.5% Cu steel: Holtzer-Loire
URANUS 65	0.03% C (max) 20.5% Ni 24.5% Cr with Nb steel: Creusot-Loire
URANUS A5	0.06% C (max) 17.5% Ni 2.25% Mo 2.0% Cu steel: Creusot-Loire
URANUS B6	Low C 25% Ni 20% Cr 4.5% Mo 1.5% Cu steel: Holtzer-Loire
URANUS S1	0.015% C (max) 1.5% Mn 14.2% Ni 17.5% Cr with Nb steel: Creusot-Loire
V 2A	0.16% C 9.0% Ni 18% Cr steel: Krupps
V 2A EXTRA	0.15% C 9.0% Ni 18% Cr 0.6% Ti steel: Krupps
V 2A FH	0.15% C 7.5% Ni 17% Cr steel: Krupps
V 2A NORMAL	0.12% C 9% Ni 18% Cr steel: Krupps
V 2A SUPRA	0.07% C 10% Ni 18% Cr steel: Krupps
V 2A SUPRA NK	0.03% C 11% Ni 18% Cr steel: Krupps
V 2A X EXTRA	0.1% C 10% Ni 18% Cr 1.0% Nb steel: Krupps
V 2A XET	0.08% C 10% Ni 18% Cr 0.2% Mo 1.0% Nb steel: Krupps
V 4A EXTRA	0.1% C 11.5% Ni 17.5% Cr 2.2% Mo 0.5% Ti steel: Krupps
V 4A SUPRA	0.07% C 11.5% Ni 17.5% Cr 2.2% Mo steel: Krupps
V 4A X EXTRA	0.1% C 11.5% Ni 17.5% Cr 2.2% Mo 1.0% Nb steel: Krupps
V 8A SUPRA	0.12% C 10% Ni 18% Cr 2.0% Mo steel: Krupps
V 12A SUPRA	0.16% C 12% Ni 12% Cr steel: Krupps
V 14A SUPRA	0.07% C 13.5% Ni 17% Cr 4.5% Mo steel: Krupps
V 16A EXTRA	0.07% C 20% Ni 17.5% Cr 2.2% Mo 0.3% Ti 2.0% Cu steel: Krupps
V 16A X EXTRA	0.07% C 20% Ni 17.5% Cr 2.2% Mo 0.5% Nb 2.0% Cu steel: Krupps
V 18A SUPRA NK	0.03% C 16.5% Ni 18.5% Cr 3.5% Mo steel: Krupps
V 24A EXTRA	0.06% C 25% Ni 25% Cr 2.2% Mo 0.3% Ti steel: Krupps
V 25AF X EXTRA	0.06% C 7% Ni 25% Cr 1.5% Mo 0.4% Nb steel: Krupps
V 44A EXTRA	0.1% C 13% Ni 17.5% Cr 2.7% Mo 0.5% Ti steel: Krupps
V 44A SUPRA	0.07% C 13% Ni 17.5% Cr 2.7% Mo steel: Krupps
V 44A SUPRA NK	0.03% C 12% Ni 17.5% Cr 2.2% Mo steel: Krupps
V 44A X EXTRA	0.1% C 13% Ni 17.5% Cr 2.7% Mo 1.0% Nb steel: Krupps
V 57	0.05% C 27.2% Ni 14.7% Cr 1.3% Mo 0.3% V 3.0% Ti 0.2% Al stainless steel: Information from SAE handbook
VALVO 829	0.42% C 13.5% Ni 13.5% Cr 2.7% W steel: For valves; Swift Levick for BS alloy En 54
VERSALLOY	0.11% C 8% Ni 17% Cr steel: Atlas; annealed **DPN: 150 UTS: 730 Elon: 55% Proof: 240**
VERTAMET 308L	0.02% C 10% Ni 19% Cr steel: Electrode
VERTAMET 316L	0.02% C 12% Ni 19% Cr 3% Mo steel: Electrode
VF 11	2.25% C 11% Ni 24% Cr 5.5% Mo steel: For valve facing; SAE designation
VH 274	0.16% C 9.0% Ni 18% Cr steel: Wykmanshyttan
VH 650	0.12% C 10% Ni 18% Cr 3.0% Mo steel: Wykmanshyttan
VIKRO 10	0.35% C 25.0% Ni 15.0% Cr steel: Casting; Firth Vickers; annealed **UTS: 500 Elon: 8% Proof: 280**
VITRESIST	0.06% C (max) 23% Ni 23% Cr 5.5% Mo 0.4% Nb 2% Si 1.5% Cu steel: D Brown; for use with hot sulphuric acid
VMS 585	2.2% C 11% Ni 24% Cr 5% Mo steel: For valve facing; origin unknown
VS 12/12	0.08% C 12.5% Ni 13% Cr steel: Vulcan; annealed **UTS: 920 Elon: 65%**
VS 18/8	0.09% C 8% Ni 18% Cr steel: Vulcan; annealed **UTS: 1070 Elon: 60%**
VS 18/8/2	0.08% C 8.5% Ni 18.5% Cr 2.5% Mo steel: Vulcan; annealed **UTS: 1140 Elon: 51%**
VS 18/8 Ti	0.06% C 8.5% Ni 18% Cr Ti steel: Vulcan; annealed **UTS: 1110 Elon: 56%**
W 18/8	8% Ni 18% Cr Nb steel: Welding electrode; Cu coated; Armco
W 19/9	0.03% C 9% Ni 19% Cr steel: Welding electrode; Cu coated; Armco
WELDANKA A	0.15% C (max) 9% Ni 18% Cr steel: Castings; Brown Bayley
WELDANKA B	0.08% C 10% Ni 18% Cr 0.4% Ti steel: Bar, billet, etc.; Brown Bayley for BS alloy En 58B and 58C
WELDANKA CB	0.08% C 12% Ni 18% Cr 0.8% Nb + Ta steel: Bar, billet, etc.; Brown Bayley for BS alloy En 58F and 58G
X 2 CN18/10	0.03% C 11.0% Ni 19.0% Cr steel: French Standard; annealed **DPN: 180 UTS: 450 Elon: 45% Proof: 170**
X 2 Cr Ni 18/9	0.03% C (max) 10% Ni 19.0% Cr steel: German Standard
X 2 Cr Ni 18/9	0.03% C (max) 11.2% Ni 18.5% Cr steel: Designation used by German Standards
X 2 Cr Ni 18/10	0.03% C (max) 10.2% Ni 18.0% Cr 0.17% N steel: German Standard
X 2 Cr Ni 18/11	0.03% C (max) 10.0% Ni 19.0% Cr steel: German Standard
X 2 Cr Ni 18/19	0.03% C (max) 11.2% Ni 18.5% Cr steel: German Standard
X 2 Cr Ni Mo 17/12	0.03% C (max) 12.0% Ni 17.0% Cr 2.2% Mo steel: German Standard
X 2 Cr Ni Mo 17/12	0.03% C (max) 12.0% Ni 17.5% Cr 2.7% Mo steel: German Standard
X 2 Cr Ni Mo 18/10	0.03% C (max) 12.0% Ni 17.0% Cr 2.2% Mo steel: German Standard
X 2 Cr Ni Mo 18/10	0.03% C (max) 12.5% Ni 17.5% Cr steel: Designation used by German Standards
X 2 Cr Ni Mo 18/10	0.03% C (max) 16.0% Ni 18.0% Cr 35% Mo steel: German Standard
X 2 Cr Ni Mo 18/10	0.03% C (max) 12.5% Ni 17.5% Cr 2.2% Mo steel: German Standard
X 2 Cr Ni Mo 18/12	0.03% C (max) 13.5% Ni 17.5% Cr 2.7% Mo steel: German Standard
X 2 Cr Ni Mo 18/12	0.03% C (max) 12.0% Ni 17.5% Cr 2.7% Mo steel: German Standard
X 2 Cr Ni Mo 18/12	0.03% C (max) 13.7% Ni 17.5% Cr steel: Designation used by German Standards
X 2 Cr Ni Mo 18/16	0.03% C (max) 13.0% Ni 19.0% Cr 3.5% Mo steel: German Standard
X 2 Cr Ni Mo 18/16	0.03% C 16.0% Ni 18.0% Cr 3.5% Mo steel: Designation used by German Standards
X 2 Cr Ni Mo N 18/12	0.03% C (max) 12.0% Ni 17.5% Cr 2.2% Mo 0.16% N steel: Designation used by German Standards
X 2 Cr Ni Mo N 18/13	0.03% C (max) 13.2% Ni 17.5% Cr 2.7% Mo 0.16% N steel: Designation used by German Standards
X 2 Cr Ni Mo Ni 18/12	0.03% C (max) 12.0% Ni 17.5% Cr 2.2% Mo 0.18% N steel: German Standard

Note. The following abbreviations and units are used in the tables:

DPN Hardness, diamond pyramid number
UTS Ultimate tensile strength, N/mm^2
Elon Elongation, %
Proof 0.1% proof strength, N/mm^2

1 N/mm^2=0.1 hbar=0.102 kgf/mm^2=0.06475 $tonf/in.^2$=145.04 $lbf/in.^2$=1 MPa
See Appendix II for other abbreviations and conversion tables.

Symbol	Nominal analysis, supplier, condition and remarks.
X 2 Cr Ni N 18/10	0.03% C (max) 10.2% Ni 18.0% Cr 0.18% N steel: Designation used by German Standards
X 3 CN 19 11	0.03% C 11% Ni 19% Cr steel: Italian Standard; annealed **DPN: 180 UTS: 450 Elon: 45% Proof: 170**
X 5 Cr Ni 18/9	0.07% C (max) 10.0% Ni 18.0% Cr steel: Designation used by German Standards
X 5 Cr Ni 18/10	0.08% C (max) 9.2% Ni 19.0% Cr steel: German Standard
X 5 Cr Ni 19/9	0.06% C (max) 9.5% Ni 19.0% Cr steel: Designation used by German Standards
X 5 Cr Ni 19/11	0.07% C (max) 11.2% Ni 18.5% Cr steel: Designation used by German Standards
X 5 Cr Ni Mo 17/12	0.08% C (max) 12.0% Ni 17.0% Cr 2.2% Mo steel: German Standard
X 5 Cr Ni Mo 17/13	0.07% C (max) 13.5% Ni 17.0% Cr 4.5% Mo steel: Designation used by German Standards
X 5 Cr Ni Mo 18/10	0.07% C (max) 11.5% Ni 17.5% Cr 2.2% Mo steel: Designation used by German Standards
X 5 Cr Ni Mo 18/12	0.07% C 13.0% Ni 17.5% Cr 2.7% Mo steel: Designation used by German Standards
X 5 Cr Ni Mo 18/13	0.06% C (max) 13.5% Ni 18.0% Cr 4.7% Mo steel: Designation used by German Standards
X 5 Cr Ni Mo 19/10	0.06% C (max) 10.0% Ni 19.0% Cr 2.2% Mo steel: Designation used by German Standards
X 5 Cr Ni Mo Cu Nb 18/18	0.07% C (max) 20.0% Ni 17.5% Cr 2.2% Mo 1.0% Nb 2.0% Cu steel: Designation used by German Standards
X 5 Cr Ni Mo Cu Nb 20/18	0.06% C (max) 20.5% Ni 17.5% Cr 2.2% Mo 0.5% Nb 2.0% Cu steel: German Standard
X 5 Cr Ni Mo Ti 25/25	0.06% C (max) 25% Ni 25% Cr 2.2% Mo 0.5% Ti steel: Designation used by German Standards
X 6 CN 1911	0.05% C 10.0% Ni 18.0% Cr steel: Italian Standard; annealed **DPN: 180 UTS: 480 Elon: 45% Proof: 180**
X 6 Cr Ni 24/18	0.08% C (max) 13.5% Ni 23.0% Cr steel: Italian Standard
X 6 Cr Ni 25/20	0.08% C (max) 20.5% Ni 25.0% Cr steel: Italian Standard
X 6 Cr Ni Mo Nb 17/12	0.08% C (max) 12.0% Ni 17.5% Cr 2.5% Mo steel: Italian Standard
X 6 Cr Ni Mo Ti 17/12	0.08% C (max) 12.0% Ni 17.5% Cr 2.5% Mo steel: Italian Standard
X 6 Cr Ni Nb 18/11	0.08% C (max) 10.5% Ni 18.0% Cr Nb steel: Italian Standard
X 6 Cr Ni Ti 18/11	0.08% C (max) 10.5% Ni 18.0% Cr Ti steel: Italian Standard
X 7 Cr Ni 23/14	0.08% C (max) 13.5% Ni 23.0% Cr steel: German Standard
X 8 CN 1910	0.07% C 9.0% Ni 18.0% Cr steel: Italian Standard; annealed **DPN: 180 UTS: 480 Elon: 45% Proof: 180**
X 8 CN 2520	0.07% C 20.0% Ni 25.0% Cr steel: Italian Standard; annealed **DPN: 190 UTS: 530 Elon: 40% Proof: 240**
X 8 CN Nb 1811	0.07% C 10.0% Ni 18.0% Cr 0.7% (max) Nb steel: Italian Standard; annealed **DPN: 190 UTS: 480 Elon: 40% Proof: 210**
X 8 CND 1712	0.05% C 11.0% Ni 17.0% Cr 2.3% Mo steel: Italian Standard; annealed **DPN: 180 UTS: 480 Elon: 45% Proof: 210**
X 8 CND 1712	0.05% C 12.0% Ni 17.0% Cr 2.8% Mo steel: Italian Standard; annealed **DPN: 190 UTS: 480 Elon: 45% Proof: 200**
X 8 CNT 1808	0.07% C 10.0% Ni 18.0% Cr 0.35% Ti (max) steel: Italian Standard; annealed **DPN: 190 UTS: 480 Elon: 40% Proof: 180**
X 8 CNT 1810	0.07% C 10.0% Ni 18.0% Cr 0.35% Ti (max) steel: Italian Standard; annealed **DPN: 190 UTS: 480 Elon: 40% Proof: 180**
X 8 Cr Ni 12/12	0.1% C (max) 12.5% Ni 13.0% Cr steel: Designation used by German Standards
X 8 Cr Ni Mo B Nb 16/16K	0.08% C 16.7% Ni 16.8% Cr 1.8% Mo 1.0% Nb 0.08% B steel: Designation used by German Standards
X 8 Cr Ni Mo Cu Nb 20/18	0.08% C (max) 21.0% Ni 18.5% Cr 2.2% Mo 0.8% Nb 2.0% Cu steel: Designation used by German Standards
X 8 Cr Ni Mo V Nb 16/13	0.1% C (max) 13% Ni 23% Cr 20% Mo 0.7% V 1.0% Nb steel: German Standard
X 8 Cr Ni Mo Nb 17/13	0.07% C (max) 13.5% Ni 18.0% Cr 4.7% Mo 1.0% Nb steel: Designation used by German Standards
X 8 Cr Ni Mo Nb 19/10	0.1% C (max) 10.5% Ni 19.0% Cr 2.2% Mo 1.3% Nb steel: Designation used by German Standards
X 8 Cr Ni Mo Nb 19/12	0.1% C (max) 11.5% Ni 19.0% Cr 2.7% Mo 1.2% Nb steel: Designation used by German Standards
X 8 Cr Ni Mo Nb 25/25	0.1% C (max) 25.0% Ni 26.0% Cr 2.2% Mo 1.3% Nb steel: Designation used by German Standards
X 8 Cr Ni Nb 19/9	0.1% C (max) 9.0% Ni 19.0% Cr 1.3% Nb steel: Designation used by German Standards
X 8 Ni Cr Mo Cu Nb	0.08% C 21% Ni 18.5% Cr 2.2% Mo 1.0% Nb 2.0% Cu steel: German Standard
X 10 CND 1808	0.05% C 10% Ni 17% Cr 1.6% Mo steel: Italian Standard; annealed **DPN: 180 UTS: 480 Elon: 45% Proof: 180**
X 10 CNDT 1808	0.07% C 12% Ni 17% Cr 2.3% Mo 0.3% Ti steel: Italian Standard; annealed **DPN: 120 UTS: 480 Elon: 40% Proof: 210**
X 10 Cr Ni Mo Nb 18/10	0.1% C (max) 11.5% Ni 17.5% Cr 2.2% Mo 0.8% Nb steel: Designation used by German Standards
X 10 Cr Ni Mo Nb 18/12	0.1% C (max) 13.0% Ni 17.5% Cr 2.7% Mo 0.8% Nb steel: Designation used by German Standards
X 10 Cr Ni Mo Ti 18/10	0.1% C (max) 11.5% Ni 17.5% Cr 2.2% Mo 0.6% Ti steel: Designation used by German Standards
X 10 Cr Ni Mo Ti 18/10/1	0.1% C (max) 11.5% Ni 17.5% Cr 1.4% Mo 0.5% Ti steel: Designation used by German Standards
X 10 Cr Ni Mo Ti 18/12	0.1% C (max) 13.0% Ni 17.5% Cr 2.7% Mo 0.6% Ti steel: Designation used by German Standards
X 10 Cr Ni Nb 1809	0.1% C 9% Ni 18% Cr 0.4% Nb steel: Italian Standard
X 10 Cr Ni Nb 18/9	0.1% C (max) 10.0% Ni 18.0% Cr 0.8% Nb steel: Designation used by German Standards
X 10 Cr Ni Nb 18/10	0.1% C (max) 10.0% Ni 18.0% Cr 0.8% Nb steel: Designation used by German Standards
X 10 Cr Ni S 1809	0.15% C (max) 9% Ni 18% Cr 0.15% S steel: Free machining; Italian Standard
X 10 Cr Ni Ti 18/9	0.1% C (max) 10.0% Ni 18.0% Cr 0.6% Ti steel: Designation used by German Standards
X 10 Cr Ni Ti 18/10	0.1% C (max) 10.0% Ni 18.0% Cr 0.6% Ti steel: Designation used by German Standards
X 12 Cr Ni 18/8	0.12% C (max) 9.0% Ni 18.0% Cr steel: Designation used by German Standards
X 12 Cr Ni 18/9	0.12% C (max) 9.0% Ni 18.0% Cr steel: Designation used by German Standards
X 12 Cr Ni 22/12	0.15% C (max) 11.0% Ni 22.0% Cr steel: Designation used by German Standards
X 12 Cr Ni 25/4	0.15% C (max) 4.5% Ni 26.0% Cr steel: Designation used by German Standards
X 12 Cr Ni 25/20	0.15% C (max) 20% Ni 25.0% Cr steel: Designation used by German Standards
X 12 Cr Ni 25/21	0.15% C (max) 20.0% Ni 25.0% Cr steel: Designation used by German Standards
X 12 Cr Ni S 18/8	0.15% C (max) 9.0% Ni 18.0% Cr 0.15% S steel: Designation used by German Standards
X 12 Cr Ni Si 36/16	0.08% C (max) 35.5% Ni 18.5% Cr steel: German Standard
X 12 Cr Ni Ti 18/9	0.15% C (max) 10.0% Ni 18.0% Cr 0.7% Ti steel: Designation used by German Standards
X 12 Ni Cr 36/18	0.2% C 38% Ni 18% Cr steel: German Standard
X 12 Ni Cr Si 36/16	0.15% C 36% Ni 16% Cr steel: German Standard
X 15 CN 1808	0.10% C 8.0% Ni 18.0% Cr steel: Italian Standard; annealed **DPN: 180 UTS: 480 Elon: 45% Proof: 180**
X 15 CN 2412	0.07% C 13.0% Ni 23.0% Cr steel: Italian Standard; annealed **DPN: 190 UTS: 530 Elon: 40% Proof: 240**

Symbol	Nominal analysis, supplier, condition and remarks.
X 15 CN 2420	0.07% C 20.0% Ni 25.0% Cr steel: Italian Standard; annealed **DPN: 190 UTS: 530 Elon: 40% Proof: 240**
X 15 Cr Ni 25/20	0.15% C 20% Ni 25% Cr steel: German Standard
X 15 Cr Ni Si 20/12	0.2% C (max) 12.0% Ni 20.0% Cr steel: Designation used by German Standards
X 15 Cr Ni Si 20/12	0.15% C (max) 9.0% Ni 19.0% Cr steel: German Standard
X 15 Cr Ni Si 25/20	0.2% C (max) 20.0% Ni 25.0% Cr steel: Designation used by German Standards
X 20 CN 245	0.07% C 4.0% Ni 27.0% Cr steel: Italian Standard; annealed **DPN: 260 UTS: 630 Elon: 20% Proof: 370**
X 20 CN 2412	0.07% C 13.0% Ni 23.0% Cr steel: Italian Standard; annealed **DPN: 190 UTS: 530 Elon: 40% Proof: 240**
X 20 T2	12.0% Ni 23.0% Cr 4.0% W Fe alloy: Origin unknown
X 25 CN 2520	0.12% C 21.0% Ni 25.0% Cr 2.0% Si steel: Italian Standard; annealed **DPN: 200 UTS: 580 Elon: 40% Proof: 290**
X 45 Cr Ni W 18/9	0.45% C 9.0% Ni 18.0% Cr 1.0% W steel: Designation used by German Standards
XCR	0.45% C 4.8% Ni 23.5% Cr 2.8% Mo steel: For valves; origin unknown
XG Ni Cr Ti 26.15	0.05% C 25% Ni 15% Cr 1.3% Mo 2% Ti steel: German Standard
XM 15	0.08% C (max) 18% Ni 18% Cr steel: Designation used in the UK and USA
XM 19	0.06% C 12.5% Ni 21.5% Cr 2.5% Mo 0.2% V 0.2% Nb steel: Designation used by ASTM
Z 1 NCDU25/20	0.02% C (max) 25.5% Ni 20.5% Cr 4.4% Mo 1.5% Cu steel: French National Standard
Z 2 CND17/12	0.03% C (max) 12.0% Ni 17.0% Cr 2.2% Mo steel: French National Standard
Z 2 CND17/13	0.03% C (max) 12.0% Ni 17.5% Cr 2.7% Mo 1.0% Nb steel: French National Standard
Z 2 CND18/10	0.03% C 12.0% Ni 17.0% Cr 2.8% Mo steel: French Standard; annealed **DPN: 180 UTS: 480 Elon: 45% Proof: 180**
Z 2 CND19/15	0.03% C (max) 13.0% Ni 19.0% Cr 3.5% Mo steel: French National Standard
Z 3 CN18/08	0.03% C (max) 9.0% Ni 19.0% Cr steel: French Standard; annealed **DPN: 180 UTS: 450 Elon: 45% Proof: 170**
Z 3 CN18/10	0.03% C 10% Ni 19% Cr steel: French National Standard; annealed **DPN: 180 UTS: 480 Elon: 45% Proof: 180**
Z 3 CND18/12	0.03% C 12% Ni 17% Cr 2.5% Mo steel: French National Standard; annealed **DPN: 180 UTS: 480 Elon: 45% Proof: 180**
Z 5 CN18/08	0.07% C 9% Ni 18% Cr steel: French National Standard; annealed **DPN: 180 UTS: 480 Elon: 45% Proof: 180**
Z 5 CND18/10	0.12% C 10% Ni 18% Cr 3.0% Mo steel: French National Standard; annealed **DPN: 180 UTS: 480 Elon: 45% Proof: 180**
Z 5 CNDU21/08	0.06% C (max) 7.5% Ni 21% Cr 2.5% Mo steel: French National Standard

Note. The following abbreviations and units are used in the tables:

DPN	Hardness, diamond pyramid number
UTS	Ultimate tensile strength, N/mm^2
Elon	Elongation, %
Proof	0.1% proof strength, N/mm^2

1 N/mm^2=0.1 hbar=0.102 kgf/mm^2=0.06475 tonf/in.^{2}=145.04 lbf/in.^{2}=1 MPa
See Appendix II for other abbreviations and conversion tables.

Symbol	Nominal analysis, supplier, condition and remarks.
Z 6 CN18/09	0.08% C (max) 9.2% Ni 19.0% Cr steel: French National Standard
Z 6 CN18/10	10% Ni 19% Cr steel: Philip Cornes
Z 6 CN18/10	0.08% C 10% Ni 19% Cr steel: French National Standard; annealed **DPN: 180 UTS: 480 Elon: 45% Proof: 180**
Z 6 CND17/11	0.08% C (max) 12.0% Ni 17.0% Cr 2.2% Mo steel: French National Standard
Z 6 CND17/12	0.08% C (max) 12.0% Ni 17.5% Cr 2.7% Mo 0.4% Ti steel: French National Standard
Z 6 CND18/12	0.08% C 12% Ni 17% Cr 2.5% Mo steel: French National Standard; annealed **DPN: 180 UTS: 480 Elon: 45% Proof: 210**
Z 6 CNDT17/12	0.08% C (max) 12.0% Ni 17.5% Cr 2.5% Mo steel: French National Standard
Z 6 CNT18/10	0.08% C (max) 10.5% Ni 18.0% Cr 0.9% Ti steel: French National Standard
Z 6 CNT25/15	0.06% C 25.0% Ni 15.0% Cr 1.25% Mo 0.3% V 2.0% Ti steel: French Standard designation
Z 7 CN18/10	0.07% C 9.0% Ni 18.0% Cr steel: French National Standard; annealed **DPN: 180 UTS: 480 Elon: 45% Proof: 180**
Z 8 CND17/12	12% Ni 17% Cr 2.5% Mo steel: Philip Cornes
Z 8 CNDNb18/12	0.07% C 13.0% Ni 17.0% Cr 2.8% Mo 0.7% Nb (max) steel: French National Standard; annealed **DPN: 190 UTS: 480 Elon: 40% Proof: 220**
Z 8 CNDT18/12	0.07% C 12.0% Ni 17.0% Cr 2.3% Mo 0.35 Ti (max) steel: French National Standard; annealed **DPN: 120 UTS: 480 Elon: 40% Proof: 210**
Z 10 CN11/12	0.16% C 12% Ni 12% Cr steel: For deep drawing; French National Standard
Z 10 CN12/12	0.05% C 13% Ni 12% Cr steel: French National Standard; annealed **DPN: 160 UTS: 480 Elon: 55% Proof: 200**
Z 10 CNDNb18/10	0.07% C 12% Ni 17% Cr 2.3% Mo 0.7% Nb steel: French National Standard; annealed **DPN: 190 UTS: 480 Elon: 40% Proof: 220**
Z 10 CNT18/10	10.5% Ni 18% Cr steel: French Standard
Z 10 CNDT18/12	0.07% C 12% Ni 17% Cr 2.3% Mo 0.3% Ti steel: French National Standard; annealed **DPN: 120 UTS: 480 Elon: 40% Proof: 210**
Z 10 CNNb18/10	0.08% C 12% Ni 18% Cr 1.0% Nb steel: French National Standard; annealed **DPN: 190 UTS: 480 Elon: 40% Proof: 210**
Z 10 CNS25/13	0.2% C 14% Ni 23% Cr steel: French National Standard; annealed **DPN: 190 UTS: 530 Elon: 40% Proof: 240**
Z 10 CNS25/20	0.25% C 21% Ni 25% Cr steel: French National Standard; annealed **DPN: 200 UTS: 580 Elon: 40% Proof: 290**
Z 10 CNT18/08	0.07% C 10% Ni 18% Cr 0.35% Ti steel: French National Standard; annealed **DPN: 190 UTS: 480 Elon: 40% Proof: 180**
Z 10 CNT18/10	0.08% C 11% Ni 18% Cr 0.4% Ti steel: French National Standard; annealed **DPN: 190 UTS: 480 Elon: 40% Proof: 180**
Z 10 NNb18/10	0.15% C (max) 9.0% Ni 18% Cr 1.2% Nb steel: French National Standard; annealed
Z 12 CN	0.15% C 9% Ni 18% Cr steel: French National Standard
Z 12 CN18/08	0.08% C 9.0% Ni 18% Cr steel: French National Standard; annealed **DPN: 180 UTS: 480 Elon: 45% Proof: 180**
Z 12 CN18/10	0.1% C 8% Ni 18% Cr steel: French National Standard; annealed **DPN: 180 UTS: 480 Elon: 45% Proof: 180**
Z 12 CNS25/20	0.08% C (max) 20.5% Ni 25.0% Cr steel: French National Standard
Z 12 CNS25/20	0.12% C 20.0% Ni 25.0% Cr steel: French National Standard

Symbol	Nominal analysis, supplier, condition and remarks.
Z 12 NCS37/18	0.08% C (max) 35.5% Ni 18.5% Cr steel: French National Standard
Z 12 NCS37/18	0.12% C 37.0% Ni 18.0% Cr steel: French National Standard
Z 15 CN25/13	0.08% C (max) 13.5% Ni 23.0% Cr steel: French National Standard
Z 15 CNS20/10	0.12% C 12% Ni 20% Cr 2.0% Si steel: French National Standard; annealed **DPN: 190 UTS: 580 Elon: 40% Proof: 290**
Z 15 CNS20/12	0.15% C (max) 9.0% Ni 19.0% Cr steel: French National Standard
Z 15 CNS25/13	0.12% C 12.0% Ni 20.0% Cr 2.0% Si steel: French National Standard; annealed **DPN: 190 UTS: 580 Elon: 40% Proof: 290**
Z 15 CNS25/20	0.25% C 20% Ni 25% Cr steel: French National Standard
Z 20 CNS24/13	0.07% C 13.0% Ni 23.0% Cr steel: French National Standard; annealed **DPN: 190 UTS: 530 Elon: 40% Proof: 240**
Z 20 CNS24/20	0.12% C 21.0% Ni 25.0% Cr 2.0% Si steel: French National Standard; annealed **DPN: 200 UTS: 580 Elon: 40% Proof: 290**
Z 25 CNWS20/09	0.25% C 9.0% Ni 20.0% Cr 2.0% W steel: French National Standard
Z 35 CNS14/14	0.35% C 14.0% Ni 14.0% Cr 2.5% W steel: French National Standard

Note. The following abbreviations and units are used in the tables:

DPN	Hardness, diamond pyramid number
UTS	Ultimate tensile strength, N/mm^2
Elon	Elongation, %
Proof	0.1% proof strength, N/mm^2

1 N/mm^2=0.1 hbar=0.102 kgf/mm^2=0.06475 tonf/in.^{2}=145.04 lbf/in.^{2}=1 MPa
See Appendix II for other abbreviations and conversion tables.

44N2 Steel – chromium–nickel – wrought and cast, age hardening With molybdenum, niobium, tungsten, titanium, aluminium or nitrogen

Specific gravity		8.05
Density		8050 kg/m^3
Solidus/liquidus		1230–1280 °C
Thermal conductivity		16.7 W/m °C
Coefficient of linear expansion		18×10^{-6}/ °C
Electrical conductivity		3% IACS (copper 100%)
Specific resistance		850 microhm mm
Young's modulus of elasticity		200×10^9 N/m^2
Impact	aged	20 J (can be higher)
	over aged	115 J
Fatigue strength (endurance limit)	aged	±480 N/mm^2
	over aged	±600 N/mm^2

Hot strength

Temperature °C	Tensile strength N/mm^2	Elongation %	
420	1150	22	
530	1130	21	These figures can
650	1000	20	be much lower
700	880	12	

The above properties are typical of the following group, and may not apply exactly to any one specification. It is possible that with certain specifications some of the values may not be applicable.

General metallurgical characteristics

The steels in this group are based on the 18/8 chromium nickel series discussed in Section 44N1. All the alloys listed are fully austenitic and do not contain any appreciable amounts of ferrite or martensite except under peculiar conditions.

The alloys, however, have additional elements added which makes them basically different from Section 44N1 in that their mechanical properties can be improved or altered by thermal treatment. In some cases the elements present are such that the mechanical properties will have been increased anyway by the alloying elements present in appreciable amounts such as tungsten and cobalt.

These materials tend to be rather complex and were developed in order to improve the mechanical properties but retaining the excellent corrosion resistance of austenitic steels.

Many of the materials have constituents present which allow them to be solution treated and aged in an identical manner to other non-ferrous alloys.

In the solution treated condition the steels are soft and ductile and in the aged condition they have improved tensile properties with some reduction in ductility.

In the annealed condition the materials in this group are

identical, or very similar, to those found in Section 44N1. There are many specifications appearing in this group which could equally have been listed in Section 44N1. There are possibly specifications in 44N1 which should technically appear in Section 44N2.

All the materials in this group can be considerably hardened by cold work. There is some indication that those materials which can be aged can have their properties improved by carrying out a cold working operation after the solution treatment. Ageing will then improve the mechanical properties with a minimum reduction in corrosion resistance and better ductility than without the cold work.

These materials can be welded and brazed without any problems in the same manner as those in Section 44N1. It must be appreciated, however, that where they have been either cold worked or aged the welding operation will affect the mechanical properties immediately adjacent to the weld or braze.

In general the corrosion resistant properties of these materials in any condition where the mechanical properties have been increased will have decreased. The same comments apply here as given in Section 44N1.

These materials were developed specifically for steam and gas turbine high duty rotating parts where high strength was necessary in a corrosive atmosphere.

Some chemical plant components and internal combustion engine components may also be made of these materials.

In general, however, it must be stated that the materials found in Section 44N3, i.e. the ferritic austenitic type stainless steels, have more or less replaced the need for the materials in this section.

Symbol	Nominal analysis, supplier, condition and remarks.
1.4529	0.03% C (max) 25% Ni 20% Cr 6.2% Mo 0.02% W 1.5% Cu steel: German Standard
1.4974	0.12% C 20% Ni 21% Cr 3.0% Mo 2.5% W 20% Co 1.0% Nb + Ta steel; German Standard
1KH14N16B	0.1% C 15.5% Ni 14.0% Cr 1.1% Nb steel: Russian Standard
1KH14N18V2B	0.1% C 19.0% Ni 14.0% Cr 2.3% W 1.1% Nb steel: Russian Standard
1KH14N18V2BR	0.1% C 19.0% Ni 14.0% Cr 2.3% W 1.1% Nb 0.005% B (max) 0.02% Ce (max) steel: Russian Standard
1KH14N18V2BR1	0.1% C 19.0% Ni 14.0% Cr 2.3% W 1.1% Nb 0.005% B (max) 0.02% Ce (max) steel: Russian Standard
4KH14N14V2M	0.45% C 14.0% Ni 14.0% Cr 0.32% Mo 2.3% W steel: Russian Standards designation
16/25/6	0.5% C 25% Ni 16% Cr 6.0% Mo 0.15% N steel: Information from SAE handbook
19/9 DL	0.3% C 9% Ni 19% Cr 1.25% Mo 1.2% W 0.4% Nb 0.3% Al steel: Age hardening; American proprietary alloy listed in SAE yearbook
19/9W Mo	0.1% C 9% Ni 19% Cr 0.4% Mo 1.3% W 0.4% Nb 0.4% Al steel: Age hardening; American proprietary alloy listed in SAE yearbook
254SMO	18% Ni 20% Cr 6% Mo 0.7% Cu steel: Origin unknown **DPN: 220 UTS: 650 Elon: 35% Proof: 300**
A 286	0.06% C 25% Ni 15% Cr 1.3% Mo 0.3% V 2.0% Ti steel: Bofors; solution treated and aged **DPN: 290 UTS: 1030 Elon: 24% Proof: 720**
A 286	0.08% C 26% Ni 15% Cr 1.25% Mo 2% Ti steel: Allegheny Ludlum; solution treated and aged **UTS: 1000 Elon: 25%**
A 950	0.04% C (max) 27.3% Ni 23.3% Cr 2.8% Mo 0.7% Nb 3.0% Cu steel: VEW
A 955	0.05% C (max) 22.3% Ni 17.5% Cr 3.2% Mo 0.4% Nb 1.8% Cu steel: VEW
A 963	0.03% C (max) 16% Ni 17% Cr 6.3% Mo 1.6% Cu 0.16% N steel: VEW

Note. The following abbreviations and units are used in the tables:

DPN	Hardness, diamond pyramid number
UTS	Ultimate tensile strength, N/mm^2
Elon	Elongation, %
Proof	0.1% proof strength, N/mm^2

1 N/mm^2=0.1 hbar=0.102 kgf/mm^2=0.06475 tonf/in.^{2}=145.04 lbf/in.^{2}=1 MPa

See Appendix II for other abbreviations and conversion tables.

Symbol	Nominal analysis, supplier, condition and remarks.
ACI CN-7M	0.07% C (max) 28.7% Ni 20.5% Cr 2.5% Mo 3.5% Cu Fe alloy: Origin unknown
AISI 650	0.05% C 25% Ni 16% Cr 6% Mo 0.15% N steel: Warm worked **UTS: 630 Elon: 23% Proof: 500**
AISI 651	0.32% C 9% Ni 18.5% Cr 1.4% Mo 1.4% W 0.2% Ti 0.4% Nb steel: Solution treated and aged **DPN: 189 UTS: 450 Elon: 36% Proof: 150**
AISI 652	0.3% C 9% Ni 18.5% Cr 1.6% Mo 1.35% W 0.55% Ti steel: Solution treated and aged **DPN: 189 UTS: 450 Elon: 36% Proof: 150**
AISI 653	0.12% C 14.1% Ni 15.9% Cr 2.5% Mo 0.25% Ti 0.45% Nb steel: Solution treated and aged **UTS: 610 Elon: 45% Proof: 220**
AISI 660	0.05% C 25.2% Ni 14.7% Cr 1.3% Mo 2.15% Ti 0.2% Al steel: Solution treated and aged **UTS: 1000 Elon: 25% Proof: 610**
AISI 662	0.04% C 26% Ni 13.5% Cr 2.7% Mo 1.7% Ti 0.07% Al 0.005% B Fe alloy: Solution treated and aged **UTS: 1040 Elon: 24% Proof: 650**
AISI 663	0.05% C 27.2% Ni 14.7% Cr 1.3% Mo 0.3% V 3% Ti 0.2% Al 0.01% B Fe alloy: Solution treated and aged **DPN: 341**
AISI 665	0.03% C 26% Ni 13.5% Cr 1.7% Mo 3% Ti 0.15% Al 0.02% B Fe alloy: Solution treated and aged **DPN: 350 UTS: 1210 Elon: 13% Proof: 970**
ALLOY 5X	0.07% C (max) 20% Ni 18% Cr 1% Mo 5.2% Si 2% Cu steel: Origin unknown
AMS 5221 A	42% Ni 5.2% Cr 2.3% Ti 0.5% Al Fe alloy: Strip; solution treated; AMS for SAE 902
AMS 5525 B	26% Ni 15% Cr 1.3% Mo 0.3% V 2% Ti steel: Sheet and strip
AMS 5526 C	9% Ni 20% Cr 1.5% Mo 1.5% W Nb Ti steel: Sheet and strip
AMS 5527 A	9% Ni 20% Cr 1.5% Mo 1.5% W 0.2% Ti 0.4% Nb + Ta steel: Sheet and strip; hot rolled and stress relieved **UTS: 92**
AMS 5538	9% Ni 20% Cr 1.6% Mo 1.4% W Ti steel: Sheet and strip
AMS 5539	9% Ni 20% Cr 1.6% Mo 1.4% W Ti steel: Sheet and strip; hot rolled and stress relieved **UTS: 920**
AMS 5543	26% Ni 13.5% Cr 1.75% Mo 3% Ti steel: Sheet; vacuum melted; solution treated
AMS 5721 B	9% Ni 20% Cr 1.5% Mo 1.5% W Nb Ti Ta steel: Bar; up to 25 mm in diameter
AMS 5722 A	9% Ni 20% Cr 1.5% Mo 1.5% W Nb Ti steel: Bar and forging
AMS 5723 A	9% Ni 20% Cr 1.5% Mo 1.5% W 0.6% Ti steel: Bar and forging etc.

Symbol	Nominal analysis, supplier, condition and remarks.
AMS 5724	9% Ni 20% Cr 1.5% Mo 1.5% W 0.6% Ti steel: Bar; up to 25 mm in diameter
AMS 5726	0.08% C (max) 25.5% Ni 14.75% Cr 1.25% Mo 0.3% V 2.1% Ti steel
AMS 5729	9% Ni 20% Cr 1.5% Mo 1.5% W 0.6% Ti steel: Bar; up to 40 mm in diameter
AMS 5731 A	26% Ni 15% Cr 1.3% Mo 0.3% V 2% Ti steel: Bar, forging, etc.; solution treated
AMS 5731-32	0.05% C 26% Ni 15% Cr 1.3% Mo 0.3% V 2.1% Ti 0.2% Al steel: Forgings
AMS 5732 A	26% Ni 15% Cr 1.3% Mo 0.3% V 2% Ti steel: Bar, forging, etc.; solution treated and aged
AMS 5734	26% Ni 15% Cr 1.3% Mo 0.3% V 2% Ti steel: Bar, forging, etc.; annealed
AMS 5735 E	26% Ni 15% Cr 1.5% Mo 0.3% V 2% Ti steel: Bar, forging, etc.
AMS 5736 B	26% Ni 15% Cr 1.5% Mo 0.3% V 2% Ti steel: Bar and forging; solution treated
AMS 5737 B	26% Ni 15% Cr 1.5% Mo 0.3% V 2% Ti steel: Bar, forging, etc.; annealed and aged
AMS 5741	0.03% C 26% Ni 13.5% Cr 1.75% Mo 3% Ti 0.15% Al 0.02% B steel: Forgings
AMS 5782 A	9% Ni 19% Cr 0.5% Mo 1.5% W 0.2% Ti 1% Nb + Ta steel: Wire for welding
AMS 5783 B	9% Ni 19% Cr 0.5% Mo 1.5% W 1% Nb + Ta steel: Coated electrode
AMS 5804 A	26% Ni 15% Cr 1.3% Mo 0.3% V 2.2% Ti steel: Welding wire
AMS 5805 A	26% Ni 15% Cr 1.3% Mo 0.3% V 2.2% Ti steel: Welding wire; vacuum melted
AMS 5853	0.08% C (max) 25.5% Ni 14.75% Cr 1.25% Mo 0.3% V 2.1% Ti steel
AMS 5858	0.08% C (max) 25.5% Ni 14.75% Cr 1.25% Mo 0.3% V 2.1% Ti steel
AMS 5895	0.08% C (max) 25.5% Ni 14.75% Cr 1.25% Mo 0.3% V 2.1% Ti steel
AMS 7235	0.08% C 25.5% Ni 15.2% Cr 1.25% Mo 0.3% V 2.1% Ti steel
AMS 7477	0.08% C (max) 25.5% Ni 15.2% Cr 1.25% Mo 0.35% W 2.1% Ti steel
AMS 7478	0.08% C (max) 25.5% Ni 15.2% Cr 1.25% Mo 0.35% W 2.1% Ti steel
AMS 7479	0.08% C (max) 25.5% Ni 15.2% Cr 1.25% Mo 0.35% W 2.1% Ti steel
AMS 7481	0.08% C (max) 25.5% Ni 15.2% Cr 1.25% Mo 0.35% W 2.1% Ti steel
ASTM A453/651	0.32% C 9.5% Ni 19.5% Cr 1.5% W 0.2% Ti 0.4% Nb steel: For high expansion bolts
ASTM A453/660	0.08% C (max) 25.5% Ni 14.7% Cr 1.2% Mo 2.1% Ti 0.005% B steel: For high expansion bolts
ASTM A453/662	0.8% C (max) 26% Ni 13.5% Cr 2.8% Mo 2.0% Ti 0.3% Al 0.005% B steel: For high expansion bolts
ASTM A453/665	0.08% C (max) 26% Ni 14% Cr 1.8% Mo 3.0% Ti 0.2% Al 0.04% B steel: For high expansion bolts
ASTM A457/651	0.32% C 9.5% Ni 19% Cr 1.3% Mo 1.3% W 0.2% Ti 0.45% Nb steel: Plate
ASTM A457/652	0.32% C 9.5% Ni 19% Cr 1.6% Mo 1.3% W 0.5% Ti steel: Plate
ASTM A458	Stainless steel: Bar; graded as ASTM A457
ASTM A461/660	0.08% C (max) 25.5% Ni 14.7% Cr 1.2% Mo 2.1% Ti 0.005% B steel: Bar
ASTM A461/661	0.12% C 20% Ni 21.5% Cr 3.0% Mo 2.5% W 20% Co 1.0% Nb + Ta Fe alloy: Bar
ASTM A461/662	0.08% C 26% Ni 13.5% Cr 3.2% Mo 1.8% Ti 0.005% B steel: Bar
ASTM A638/660	0.08% C (max) 26.0% Ni 15.0% Cr 1.2% Mo 0.25% V 2.1% Ti 0.005% B Fe alloy
ASTM A638/662	0.08% C (max) 26.0% Ni 13.5% Cr 3.0% Mo 1.8% Ti 0.005% B Fe alloy
ASTM A639/661	0.12% C 20.0% Ni 21.2% Cr 3.0% Mo 2.5% W 19.7% Co 1.0% Nb + Ta Fe alloy
ATGB	0.4% C 13% Ni 13% Cr 2.0% Mo 2.5% W 10% Co 3.0% Nb steel: Imphy; annealed **UTS: 760 Elon: 15% Proof: 500**
ATV S7	0.1% C 30% Ni 19% Cr 20% Co 2.0% Ti 1.0% Al Fe alloy: Imphy **DPN: 270 UTS: 890 Elon: 15% Proof: 480**
CARPENTER 20 Mo6	0.03% C (max) 35% Ni 24% Cr 5.7% Mo 3% Cu Fe alloy: Carpenter
CD 4M Cu	0.04% C (max) 5.3% Ni 25.5% Cr 2% Mo 3% Cu steel: Casting; US designation
CD 4M Cu	0.04% C (max) 5.3% Ni 25.5% Cr 2% Mo 3% Cu steel: Casting; US designation
CG 27	0.05% C 38% Ni 13% Cr 5.7% Mo 2.5% Ti 0.7% Nb 1.6% Al steel: Crucible Steel Co.; solution treated and aged **UTS: 1420 Elon: 20% Proof: 1000**
CONIFER 1925 HMO	25% Ni 20% Cr 6% Mo 1.5% Cu steel: Origin unknown **DPN: 210 UTS: 600 Elon: 35% Proof: 300**
CREUSOT UR SB 8	0.02% C (max) 25% Ni 25% Cr 5.2% Mo 1.5% Cu 0.2% N Fe alloy
CRUCIBLE 901	0.05% C 42% Ni 12.5% Cr 5.7% Mo 3% Ti 0.2% Al steel: Crucible Steel Co.; solution treated and aged **UTS: 1210 Elon: 15% Proof: 920**
CRUCIBLE A286	0.05% C 25% Ni 15% Cr 1.25% Mo 0.3% V 2.2% Ti 0.2% Al steel: Crucible Steel Co.; solution treated and aged **UTS: 1130 Elon: 25% Proof: 800**
CRUCIBLE CG27	0.05% C 38% Ni 13% Cr 5.7% Mo 2.5% Ti 0.7% Nb 1.6% Al steel: Crucible Steel Co.; solution treated and aged **UTS: 1420 Elon: 20% Proof: 1000**
CUSTOM 455	0.05% C 8% Ni 11.5% Cr 1.2% Ti 0.3% Nb + Ta 2% Cu steel: Carpenter; aged **DPN: 400 UTS: 1300 Elon: 14% Proof: 1200**
DISCOLOY 24	0.04% C 26% Ni 13% Cr 2.7% Mo 1.7% Ti 0.1% Al steel; Westinghouse
E2553	0.05% C (max) 7% Ni 25.5% Cr 3.5% Mo 2% Cu 0.2% N steel: Weld electrode; designation used by AWS
E2553T	0.04% C (max) 9% Ni 25.5% Cr 3.5% Mo 2% Cu 0.15% N steel: Weld electrode; designation used by AWS
EME	0.1% C 12% Ni 19% Cr 3.2% W 1.2% Nb 0.15% N steel: American proprietary alloy listed in SAE yearbook
ER 209	0.05% C (max) 5.5% Mn 11.2% Ni 22.2% Cr 2.2% Mo 0.2% N steel: Weld electrode; designation used by AWS
ER 218	0.1% C (max) 8% Mn 8.5% Ni 17% Cr 0.12% N steel: Weld electrode; designation used by AWS
ER 219	0.05% C (max) 9% Mn 6.2% Ni 20% Cr 0.2% N steel: Weld electrode; designation used by AWS
ER 240	0.05% C (max) 12% Mn 5% Ni 18% Cr 0.15% N steel: Weld electrode; designation used by AWS
ER 349	0.1% C 8.7% Ni 20.2% Cr 0.45% Mo 1.5% W 0.2% Ti 1% Nb + Ta steel: Weld electrode; designation used by AWS
EV 4	0.2% C 11.5% Ni 21% Cr 0.2% N steel: For exhaust valves; designation used by SAE
EV 5	0.38% C 8% Ni 19% Cr steel: For exhaust valves; designation used by SAE
EV 6	0.38% C 8% Ni 19% Cr 0.2% N steel: For exhaust valves; designation used by SAE
EV 8	0.53% C 9% Mn 3.7% Ni 21% Cr 0.42% N steel: For exhaust valves; designation used by SAE
EV 9	0.45% C 14% Ni 14% Cr 0.3% Mo 2.4% W steel; For exhaust valves; designation used by SAE
EV 11	0.7% C 6% Mn 2% Ni 21% Cr 0.2% N steel: For exhaust valves

Symbol	Nominal analysis, supplier, condition and remarks.
EV 12	0.55% C 8.3% Mn 2.2% Ni 21% Cr steel: For valves; SAE designation
EV 13	0.53% C 11.5% Mn 21% Cr 0.4% N steel: For valves; SAE designation
EV 14	0.2% C 6.5% Mn 5.5% Ni 21% Cr steel: For valves; SAE designation
FS A 286	0.05% C 26% Ni 15% Cr 1.25% Mo 2.1% Ti 0.2% Al steel: Firth Sterling; solution treated and aged **UTS: 1030 Elon 25% Proof: 760**
FV 326	0.25% C 18.5% Ni 17% Cr 2.5% Mo 7% Co 1.75% Nb steel: Firth Vickers; age hardened **UTS: 730 Elon: 44% Proof: 300**
G 18B	0.4% C 13% Ni 13% Cr 1.7% Mo 2.5% W 10% Co 3% Nb steel: Jessop; solution treated and aged **DPN: 216 UTS: 710 Elon: 35% Proof: 270**
G 19	0.4% C 13% Ni 19% Cr 1.8% Mo 2.6% W 10% Co 3.1% Nb steel: Casting; Jessop; as cast; creep resistant **UTS: 580 Elon: 4% Proof: 300**
G 40	0.23% C 25% Ni 20% Cr 3% Mo 1% W 1% Ti steel: Jessop **UTS: 840 Elon: 18% Proof: 630**
G 68	0.08% C 26% Ni 15% Cr 1.25% Mo 2% Ti steel: Jessop; solution treated and aged **UTS: 1000 Elon 25%**
GOST 5632/ 0KH14N28V 3T31UR	0.08% C (max) 27.5% Ni 14.0% Cr 3.1% W 2.8% Ti 0.9% Al 0.02% B (max) steel: Plate; Russian Standard
GOST 5632/ 3KH19N9MVBT	0.32% C 9.0% Ni 19.0% Cr 1.25% Mo 1.25% W 0.35% Ti 0.25% Nb steel: Plate; Russian Standard
GOST 5632 KH12H22T3MP	0.1% C (max) 23.5% Ni 15% Cr 1.3% Mo 3.0% Ti 0.8% Al (max) 0.02% B (max) steel: Plate; Russian Standard
GOST 5632 KH12N20T3R	0.1% C (max) 19.5% Ni 11.5% Cr 3.0% Ti 0.8% Al 0.012% B steel: Plate; Russian Standard
GOST 5632 KH25N16G7AR	0.12% C (max) 16.5% Ni 24.5% Cr 0.02% B (max) 0.4% N steel: Plate; Russian Standard
GX3 Cr Ni Mo Cu 246	0.04% C (max) 6% Ni 25.2% Cr 2.3% Mo 3.2% Cu steel: German specification
H 1	0.35% C 16% Ni 28% Cr 0.5% Mo steel: Casting; Huntington; age hardened **UTS: 410 Elon: 6% Proof: 300**
HR 51	0.08% C (max) 25.5% Ni 15% Cr 1.2% Mo 0.3% V 2% Ti 0.008% B Fe alloy: BS Aerospace designation
HR 52	0.06% C (max) 25.5% Ni 15% Cr 1.2% Mo 0.3% V 1.8% Ti 0.008% B Fe alloy: BS Aerospace designation
HS 556	0.1% C 21% Ni 22% Cr 3.2% Mo 2.7% W 1% Ta 0.3% Al 16% Co 0.08% La Fe alloy
J 1300	0.08% C 33% Ni 14% Cr 4% Mo 6.5% W 2% Ti 0.2% Al 0.2% Fe alloy: American proprietary alloy listed in SAE yearbook
JIS G4312 SUH 661	0.12% C 20% Ni 21.5% Cr 3.0% Mo 2.7% W 20% Co 1.0% Nb + Ta steel: Plate; Japanese Standard; solution treated **UTS: 700 Proof: 320**
KH 12N20T3R	0.1% C (max) 19.0% Ni 11.2% Cr 2.9% Ti 0.8% Al (max) 0.01% B steel: Russian Standard
KH 12N22T3MR	0.1% C (max) 23.5% Ni 11.2% Cr 1.3% Mo 2.9% Ti 0.8% Al (max) 0.2% B (max) steel: Russian Standard

Note. The following abbreviations and units are used in the tables:

DPN	Hardness, diamond pyramid number
UTS	Ultimate tensile strength, N/mm^2
Elon	Elongation, %
Proof	0.1% proof strength, N/mm^2

1 N/mm^2=0.1 hbar=0.102 kgf/mm^2=0.06475 tonf/in.^{2}=145.04 lbf/in.^{2}=1 MPa

See Appendix II for other abbreviations and conversion tables.

Symbol	Nominal analysis, supplier, condition and remarks.
LANGALLOY 20V	0.05% C 29% Ni 20% Cr 3% Mo 0.5% Nb 3.5% Cu Fe alloy: Langley **DPN: 140 UTS: 420 Elon: 25% Proof: 210**
LESCALLOY 901 VAC-ARC	0.1% C 42% Ni 12% Cr 6.0% Mo 1.0% Co 3.0% Ti 0.3% Al 0.015% B steel: Latrobe Steel; solution treated and aged **UTS: 1300 Elon: 18% Proof: 920**
LESCALLOY V57	0.05% C 27% Ni 14.7% Cr 1.25% Mo 0.5% V 3% Ti 0.25% Al steel: Latrobe; age hardening
LESCALLOY V57 VAC-ARC	0.08% C 26% Ni 15% Cr 1.2% Mo 0.5% V 3.0% Ti 2% Al 0.01% B steel: Latrobe Steel; solution treated and aged **UTS: 880 Elon: 12% Proof: 700**
LT 21/42	0.5% C 9.0% Mn 4.5% Ni 21% Cr 2.0% Nb 0.4% N steel: Low Moor
LV 21/69	0.08% C 9% Mn 6.0% Ni 21% Cr 0.4% N steel: Sheet and bar; Low Moor
M 30B	0.08% C 33% Ni 14% Cr 4% Mo 6.5% W 2% Ti 0.2% Al 0.2% Fe alloy: American proprietary alloy listed in SAE yearbook
M 813	0.08% C 35% Ni 18% Cr 4% Mo 2.2% Ti 1.4% Al Fe alloy: American proprietary alloy listed in SAE yearbook
N 08366	0.035% C (max) 24.5% Ni 21% Cr 6.5% Mo steel: UNS designation
N 08367	0.03% C (max) 24.5% Ni 21% Cr 6.5% Mo 0.2% N steel
N 08904	0.02% C (max) 25% Ni 21% Cr 4.5% Mo 1.5% Cu 1.5% Mo steel: UNS designation
N 08925	0.02% C (max) 25% Ni 20.5% Cr 6.5% Mo 1.2% Cu 0.15% N steel: UNS designation
NORIDUR 9.4460	0.04% C (max) 6% Ni 25.2% Cr 2.3% Mo 3.2% Cu steel: KSB; solution treated and aged **DPN: 230 UTS: 690 Elon: 25% Proof: 480**
NTK Mo	25% Ni 20% Cr 6% Mo 0.2% N steel: Origin unknown **DPN: 180 UTS: 790 Elon: 50% Proof: 415**
PH 13.8 Mo	0.05% C 8% Ni 13% Cr 2.2% Mo 1.1% Al 0.1% N steel: Carpenter
PYROMET 901	0.1% C (max) 42% Ni 13% Cr 6% Mo 3% Ti 0.015% B steel: Carpenter **UTS: 1200 Elon: 15% Proof: 860**
PYROMET ALLOY A286	0.08% C 25% Ni 15% Cr 1.5% Mo 0.3% V 2% Ti 0.005% B steel: Carpenter **UTS: 1000 Elon: 24% Proof: 650**
PYROMET ALLOY V57	0.08% C 26% Ni 15% Cr 1.2% Mo 0.5% V 3% Ti 0.01% B steel: Carpenter **UTS: 1200 Elon: 21% Proof: 860**
REX 559	0.05% C 25.5% Ni 15.5% Cr 1.3% Mo 0.3% V 2.1% Ti 0.2% Al 0.006% B steel: Firth Vickers; solution treated **UTS: 1090 Elon: 33% Proof: 800**
S 590	0.42% C 20.5% Ni 20.5% Cr 4% Mo 4% W 20% Co Fe alloy: Origin unknown
S 38600	0.4% C 15.5% Ni 13.5% Cr 2% Mo 0.2% Ti steel: UNS designation
SANICRO 28	0.03% C (max) 31% Ni 27% Cr 3.5% Mo 1% Cu Fe alloy: Origin unknown
For SIS specifications see SS	
SOUDINOX S28	0.03% C 31.0% Ni 28.0% Cr 4.2% Mo 1.0% Cu welding electrode: Soudometal; for corrosion resistant welding **UTS: 620 Elon: 36% Proof: 500**
SOUDOCROM B6	0.03% C 25.0% Ni 22.0% Cr 5.0% Mo 2.0% Co welding electrode: Soudometal; for corosion resistant welding **UTS: 565 Elon: 30% Proof: 375**
SOUDOCROM S17	0.02% C 25.0% Ni 21.0% Cr 4.0% Mo 0.35% Nb 4.0% Cu welding electrode: Soudometal; for corrosion resistant welding **UTS: 520 Elon: 30% Proof: 420**

Symbol	Nominal analysis, supplier, condition and remarks.
SOUDOCROM S50	0.02% C 7.5% Ni 21.0% Cr 2.5% Mo 1.7% Cu welding electrode: Soudometal; for corrosion resistant welding **UTS: 745 Elon: 28% Proof: 570**
SS 2570	0.08% C 25% Ni 14% Cr 1.2% Mo 0.4% V 2.1% Ti steel: For springs; Swedish Standard; solution treated cold worked and aged **DPN: 350 UTS: 1070 Elon: 8% Proof: 1000**
UNITEMP 212	0.8% C 20% Ni 16% Cr 4% Ti 0.05% Al 0.07% B 0.5% Cb 0.05% Zr steel: American proprietary alloy listed in SAE yearbook
URANUS 50	0.03% C (max) 5.5% Ni 22% Cr 3% Mo 1.5% Cu steel: Creusot-Loire; impact 100 J; Duplex steel **UTS: 690 Elon: 25% Proof: 450**
URANUS 50M	0.07% C (max) 7.2% Ni 21% Cr 2.5% Mo 1.5% Cu steel: Creusot-Loire; impact 80 J; Duplex steel
W 545	0.08% C 26% Ni 13.5% Cr 1.5% Mo 2.8% Ti 0.2% Al 0.08% B steel: American proprietary alloy listed in SAE yearbook
W 545	0.08% C (max) 26% Ni 13.5% Cr 1.7% Mo 3% Ti 0.06% B steel: Origin unknown
X2 Cr Ni Mo N225	0.03% C (max) 5.5% Ni 22% Cr 3% Mo 0.14% N steel: German Standard
XM 17	0.08% C 8.0% Mn 6.0% Ni 20.0% Cr 2.5% Mo 0.35% N steel: Designation used by ASTM
XM 18	0.03% C 6.0% Ni 20.0% Cr 2.5% Mo 0.35% N steel: Designation used by ASTM
XM 29	0.08% C 12.5% Mn 3.0% Ni 18.0% Cr 0.3% N steel: Designation used by ASTM

Note. The following abbreviations and units are used in the tables:

DPN	Hardness, diamond pyramid number
UTS	Ultimate tensile strength, N/mm^2
Elon	Elongation, %
Proof	0.1% proof strength, N/mm^2

1 N/mm^2=0.1 hbar=0.102 kgf/mm^2=0.06475 $tonf/in.^2$=145.04 $lbf/in.^2$=1 MPa

See Appendix II for other abbreviations and conversion tables.

44N3 Steel – chromium nickel – age hardening, ferritic – austenitic – (Duplex steels)

Specific gravity	7.8
Density	7800 kg/m^3
Solidus/liquidus	1430–1470 °C
Thermal conductivity	16.5 W/m °C
Coefficient of linear expansion	12×10^{-6}/ °C
Electrical conductivity	20% IACS
Specific resistance	800 microhm mm^2
Young's modulus of elasticity	215×10^9 N/m^2
Impact	60 J (can vary)
Fatigue strength	±500 N/mm^2
Hot strength	considerable variation

The above properties are typical of the following group and may not apply exactly to any one specification. It is possible that with some specifications some of the values may not be applicable.

General metallurgical characteristics

The steels in this group vary in analysis but are based on a chromium–nickel ratio which prevents the formation of the fully austenitic phase. The generic names for these were 17/7, 17/4 chrome nickel. There are now many additions made to these to strengthen the ferrite phase. Included in this section are the Duplex stainless steels.

The metallurgical structure of these materials will be ferritic with austenite always being present. They have been developed quite recently and many of these steels are known under the heading of Duplex stainless steels. This is to differentiate them from austenitic stainless steels and the 12% chromium martensitic/ferritic steels.

They were developed to have corrosion resistance but also significant mechanical strength.

The principal reason for the development was to find materials which would have better stress corrosion resistance. Austenitic steels have poor stress corrosion resistance, the 12% chromium steels, have good mechanical but less corrosion resistance although significant stress corrosion resistance.

All the materials in this group are capable of being solution treated and aged.

In the solution treated, solution annealed or fully annealed condition these materials will accept a considerable amount of cold work. If they have been solution treated this cold work could enhance the mechanical properties once the materials are aged. The full anneal, solution treat, or solution anneal temperatures are in the region of 1000–1150 °C.

All the materials in this group, in theory at least, are capable of being aged, or precipitation hardened. This must always follow the solution treat or solution quench process, and results in an increase in the tensile properties, and a reduction in ductility and impact properties. This reduction in impact properties in particular, can be quite dramatic, and many of the materials listed must not be aged, except under very carefully controlled conditions.

It must be appreciated that any welding operation could result in undesirable precipitation or age hardening. Thus, weld procedures must be prepared with this knowledge, and great care taken that the procedure is rigidly followed.

It is essential that those producing the weld procedure are aware of the problems which can exist with Duplex steels.

Precipitation hardening or ageing takes place relatively slowly between 350–600 °C, and thus care must be taken to pass rapidly through these temperatures at welding and cooling after welding. This is, therefore, quite different to the control of alloy steel welding.

All these materials are complex, and specialist advice must be obtained before specifying these materials as they are expensive and unless correctly treated there will be alternative material which could give the same result more economically.

The steels are readily weldable, but it must always be appreciated that if they are welded in the solution treated, annealed and fully aged condition, the welding will affect the properties at the weld itself and adjacent to the weld. It is common that these materials are welded in the solution treated or solution treated/part aged condition, and then aged. It is essential that specialist advice is obtained.

Where ageing is desirable, welding can result in overageing and therefore a reduction in tensile strength, but in general an improvement in ductility. The results on impact strength cannot always be predicted.

Where ageing is not desirable, then great care is required to obviate any tendency to age as this can reduce impact properties dramatically. This means not holding at temperatures of 350–600 °C for any appreciable time.

Many of these materials are difficult to machine in the solution treated or annealed condition as they tend to tear. In the fully aged condition they are hard and abrasion-resistant. It is generally found that it is most economical to machine these materials in the solution treated/part aged condition and then to age fully after machining. Duplex stainless alloys have the ferrite phase strengthened and are thus machinable in the solution annealed condition.

Provided care is taken during machining there will be a minimum of distortion at the final ageing operation.

These materials find considerable use in the petrochemical industry, in particular in the oil industry where stress corrosion cracking is a serious problem.

They should be the first choice where highly stressed components are required to have good corrosion resistance and, in particular, where stress corrosion problems have been identified. This applies to pumps, turbines, etc., used in atmospheres where hydrogen sulphide or moist carbon dioxide gases or similar substances could be present.

Symbol	Nominal analysis, supplier, condition and remarks.
0.4310	0.15% C (max) 7.5% Ni 17% Cr steel: German Standard
1.4462	0.03% C (max) 5.5% Ni 22% Cr 3% Mo 0.14% N steel: German Standard
1.4504	0.1% C (max) 7% Ni 17% Cr 1% Al steel: German Standard
1.4514	0.1% C 7% Ni 15% Cr 2.5% Mo 1% Al steel: German Standard
1.4568	0.1% C (max) 7.2% Ni 17% Cr 1.1% Al steel: German Standard
1.4574	0.1% C (max) 7% Ni 15% Cr 2.5% Mo 1% Al steel: German Standard
3 RE 60	0.03% C (max) 4.7% Ni 18.5% Cr 2.7% Mo steel: Tube, etc. Sandvik; solution treated **DPN: 260 UTS: 800 Elon: 22% Proof: 500**
7 Mo Plus Stainless	0.03% C (max) 4.2% Ni 27.5% Cr 1.7% Mo 0.2% N steel: Carpenter **DPN: 230 UTS: 760 Elon: 38% Proof: 570**
7 Mo Stainless	0.1% C 4.5% Ni 27% Cr 1.5% Mo steel: Carpenter; Duplex stainless steel **DPN: 230 UTS: 725 Elon: 25% Proof: 550**
13.4.2 Cu R	0.06% C 4% Ni 13% Cr 2% Mo 2% Cu steel: For weld electrodes; Metrode; precipitation hardening
14/4 PH	0.03% C 4.2% Ni 14.1% Cr 3.3% Mo 0.25% Nb 3.2% Cu steel: Common name; age hardenable
14.6.2B	0.07% C 6% Ni 14% Cr 1.5% Mo steel: Electrode; Metrode
14.6.2 R	0.05% C 5.8% Ni 13.5% Cr 1.7% Mo steel: Electrode; Metrode
15Cr5Ni	0.07% C (max) 4.2% Ni 14.7% Cr 0.25% Nb + Ta 3.2% Cu steel: Carpenter
15.4 Mn Si R	0.05% C 3.0% Mn 4.0% Ni 15% Cr 3.5% Si steel: Electrode; Metrode
15.5 PH	0.3% C 4.5% Ni 14.7% Cr 0.3% Nb 3.5% Cu steel: Origin unknown
17/4 PH	0.04% C 4% Ni 16.5% Cr 3.5% Cu 0.3% Cb steel: Age hardening; American proprietary alloy listed in SAE yearbook
17/7	0.1% C 7% Ni 17% Cr steel: Hyby; annealed **DPN: 180 UTS: 480 Elon 45% Proof: 180**
17/7	0.15% C 7% Ni 17% Cr steel: Eastern Stainless; annealed **UTS: 700 Elon 50% Proof: 250**
17/7	0.15% C 7% Ni 17% Cr steel: Eastern Stainless; aged **UTS: 1300 Elon: 8% Proof: 1000**
17/7 PH	0.07% C 7% Ni 17% Cr 1.1% Al steel: Age hardening; American proprietary alloy listed in SAE yearbook
17/14 Cu Mo	0.12% C 14% Ni 16% Cr 2.5% Mo 0.2% Ti 0.4% Nb 3% Cu steel: Age hardening; American proprietary alloy listed in SAE yearbook
17.4 Cu R	0.06% C 4.0% Ni 16.0% Cr 2.0% Cu steel: Electrodes; Metrode; precipitation hardening
17.4.CuR	0.06% C 4% Ni 15% Cr 2% Cu steel: Electrode; Metrode
17.4.IR	0.04% C 4% Ni 17% Cr 1.3% Mo steel: Electrode; Metrode
301	0.1% C 7% Ni 17% Cr steel: US designation
630	0.4% C 4.25% Ni 16% Cr 0.27% Nb 3.3% Cu steel: Solution treated and aged (aged 1 hour at 430 °C); US designation **DPN: 420 UTS: 1420 Elon: 14% Proof: 1540**
631	0.07% C 7.1% Ni 17% Cr 1.17% Al steel: Solution treated; deep frozen and aged **DPN: 430 UTS: 1420 Elon: 10% Proof: 1300**
632	0.07% C 7.1% Ni 15.1% Cr 2.25% Mo 1.17% Al steel: Solution treated; deep frozen and aged; US designation **UTS: 1750 Elon: 6% Proof: 1620**
632	0.07% C 7.1% Ni 15.1% Cr 2.25% Mo 1.17% Al steel: Annealed **UTS: 920 Elon: 30% Proof: 370**
635	0.06% C 7% Ni 17% Cr 0.8% Ti 0.2% Al steel: Solution treated and aged; US designation **DPN: 460**
AFNOR 81-343-E2 13.4 Mo B 120.26X	0.06% C 4.5% Ni 12% Cr 0.5% Mo steel: Welding electrode; French specification
AISI 633	0.1% C 4.25% Ni 16.5% Cr 2.75% Mo 0.09% N steel: Solution treated; deep frozen and aged; US designation **UTS: 1550 Elon: 12% Proof: 1300**

Note. The following abbreviations and units are used in the tables:

DPN	Hardness, diamond pyramid number
UTS	Ultimate tensile strength, N/mm^2
Elon	Elongation, %
Proof	0.1% proof strength, N/mm^2

1 N/mm^2=0.1 hbar=0.102 kgf/mm^2=0.06475 $tonf/in.^2$=145.04 $lbf/in.^2$=1 MPa

See Appendix II for other abbreviations and conversion tables.

Symbol	Nominal analysis, supplier, condition and remarks.
AISI 634	0.13% C 4.25% Ni 15.5% Cr 2.7% Mo 0.1% N steel: Solution treated; deep frozen and aged **UTS: 1580 Elon: 19% Proof: 1300**
AL 350	0.1% C 4.2% Ni 17% Cr 217% Mo steel: origin unknown **DPN: 400 UTS: 1550 Elon: 9% Proof: 1350**
ALLOY 255	0.08% C (max) 5.5% Ni 25.5% Cr 3% Mo 1.7% Cu 0.1% N steel: Origin unknown; solution treated and aged **UTS: 880 Elon: 28% Proof: 650**
ALLOY 288	0.08% C 7.5% Ni 27.5% Cr 2.5% Mo 1.2% Cu 0.1% N steel: Origin unknown; solution treated **UTS: 800 Elon: 25% Proof: 460**
Almar 362 S36200	0.1% C (max) 6.5% Ni 14.5% Cr 0.7% Ti steel: Allegheny Ludlum
AM 350	0.08% C 4.3% Ni 16.5% Cr 2.7% Mo 0.1% N steel: Allegheny Ludlum; solution treated and double aged **DPN: 400 UTS: 1380 Elon: 14% Proof: 1100**
AM 355	0.13% C 4.3% Ni 15.5% Cr 2.7% Mo 0.1% N steel: Allegheny Ludlum; solution treated and double aged **UTS: 1380 Elon: 13% Proof: 1070**
AMS 5340	0.03% C (max) 4.25% Ni 14% Cr 2.4% Mo 3.2% Cu 0.02% N steel: Castings
AMS 5346 J 92110	0.07% C (max) 4.8% Ni 14.7% Cr 0.25% Nb 2.8% Cu steel: Casting
AMS 5347	0.07% C (max) 4.8% Ni 14.7% Cr 0.25% Nb 2.8% Cu steel: Casting
AMS 5348	0.07% C (max) 4.8% Ni 14.7% Cr 0.25% Nb 2.8% Cu steel: Casting
AMS 5355	0.03% C (max) 4.25% Ni 14.1% Cr 2.4% Mo 0.2% Nb 3.2% Cu 0.02% N steel: Castings
AMS 5356	0.07% C (max) 4.8% Ni 14.7% Cr 0.25% Nb 2.8% Cu steel: Casting
AMS 5357	0.07% C (max) 4.8% Ni 14.7% Cr 0.25% Nb 2.8% Cu steel: Casting
AMS 5359	0.15% C 4.2% Ni 15.5% Cr 2.75% Mo 0.1% N steel: Castings
AMS 5400	0.04% C (max) 5% Ni 14.7% Cr 0.25% Nb 2.8% Cu steel: Casting
AMS 5520A	7% Ni 15% Cr 2.5% Mo 1% Al steel: Sheet and strip
AMS 5528	0.07% C (max) 7% Ni 17% Cr 1.2% Al steel: Sheet
AMS 5528A	7% Ni 17% Cr 1% Al steel: Sheet and strip
AMS 5529A	7% Ni 17% Cr 1% Al steel: Sheet and strip; Age hardened
AMS 5541A	4.5% Ni 15.5% Cr 3% Mo 0.1% N steel: Sheet and strip
AMS 5546	4.5% Ni 16.5% Cr 3% Mo 0.1% N steel: Sheet and strip; cold rolled and tempered
AMS 5546	0.1% C 4.25% Ni 16.5% Cr 2.7% Mo 0.1% N steel: Sheet
AMS 5547	0.15% C 4.25% Ni 15.5% Cr 2.7% Mo 0.1% N steel: Sheet
AMS 5547	0.1% C 4.5% Ni 16.5% Cr 2.8% Mo 0.1% N steel
AMS 5548A	4.5% Ni 16.5% Cr 3% Mo 0.1% N steel: Sheet, high temperature annealed
AMS 5549A	4.5% Ni 15.5% Cr 3% Mo 0.1% N steel: Plate; solution treated
AMS 5554	4.5% Ni 16.5% Cr 3% Mo 0.1% N steel: Seamless tube
AMS 5568	0.09% C (max) 7% Ni 17% Cr 1% Al steel: Precipitation hardening
AMS 5594	4.5% Ni 15.5% Cr 3% Mo 0.1% N steel: Plate
AMS 5601	0.05% C (max) 8.2% Ni 14.5% Cr 25% Mo 1.2% Al steel
AMS 5603	0.05% C (max) 8.2% Ni 14.5% Cr 2.5% Mo 1.2% Al steel
AMS 5604	0.07% C (max) 4% Ni 16.5% Cr 0.3% Nb steel
AMS 5617	0.05% C (max) 8.2% Ni 11.7% Cr 1.1% Ti 0.35% Nb steel
AMS 5622	0.07% C (max) 4% Ni 16.5% Cr 0.3% Nb steel
AMS 5629	0.05% C (max) 8% Ni 12.75% Cr 2.25% Mo 1.1% Al steel
AMS 5643F	4.0% Ni 17% Cr 4% Cu steel: Bar and forging
AMS 5644B	7% Ni 17% Cr 1% Al steel: Bar and forging
AMS 5657	7% Ni 15% Cr 2.5% Mo 1% Al steel: Bar and forging
AMS 5658 SI5500	0.07% C (max) 4.5% Ni 14.7% Cr 0.3% Nb 3.5% Cu steel
AMS 5659	0.07% C (max) 4.5% Ni 14.7% Cr 0.3% Nb 3.5% Cu steel
AMS 5672	0.05% C (max) 8.2% Ni 11.7% Cr 1.1% Ti 0.35% Nb steel
AMS 5673A	7% Ni 17% Cr 1% Al steel: Wire; aged
AMS 5678	0.09% C (max) 7.2% Ni 17% Cr 1.2% Al steel
AMS 5718	0.12% C 2.5% Ni 11.7% Cr 1.7% Mo 0.32% V 0.03% N steel
AMS 5727	0.03% C (max) 4.2% Ni 14.1% Cr 2.4% Mo 0.25% Nb 3.2% Cu 0.02% N steel: Wire
AMS 5739	0.05% C (max) 6.5% Ni 14.5% Cr 0.7% Ti steel
AMS 5740	0.05% C (max) 6.5% Ni 14.5% Cr 0.7% Ti steel
AMS 5743B	4.5% Ni 15.5% Cr 3% Mo 0.1% N steel: Bar and forging
AMS 5744	0.12% C 4.5% Ni 15.5% Cr 2.8% Mo 0.1% N steel
AMS 5745	4.5% Ni 16.5% Cr 3% Mo 0.1% N steel: Bar and forging
AMS 5762	0.08% C (max) 5.7% Mn 5.7% Ni 17% Cr 0.25% S steel
AMS 5763	0.05% C (max) 6% Ni 15% Cr 0.4% Nb 1.5% Cu steel
AMS 5773	0.05% C (max) 6% Ni 15% Cr 0.4% Nb 1.5% Cu steel
AMS 5774	4.5% Ni 16.5% Cr 3% Mo 0.1% N steel: Wire for welding
AMS 5774	0.1% C 4.2% Ni 16.5% Cr 2.7% Mo 0.1% N steel: Wire
AMS 5775	4.5% Ni 16.5% Cr 3% Mo 0.1% N steel: Wire coated electrode
AMS 5780	4.5% Ni 15.5% Cr 3% Mo 0.1% N steel: Welding wire
AMS 5780	0.1% C 4.5% Ni 16.5% Cr 3% Mo 0.1% N steel: Weld electrode
AMS 5781	4.5% Ni 15.5% Cr 3% Mo 0.1% N steel: Coated electrode
AMS 5781	0.12% C 4.5% Ni 15.5% Cr 2.8% Mo 0.1% N steel: Weld electrode
AMS 5812	0.07% C (max) 7% Ni 15% Cr 2.2% Mo 1.2% Al steel: Wire
AMS 5812B	7% Ni 15% Cr 2.5% Mo 1% Al steel: Welding wire; vacuum melted
AMS 5813A	7% Ni 15% Cr 2% Mo 1.0% Al steel: Welding wire
AMS 5824 S17780	0.09% C (max) 7% Ni 16.7% Cr 1% Al steel
AMS 5825A	5% Ni 17% Cr 0.2% Nb + Ta 3.5% Cu steel: Welding wire
AMS 5826	0.07% C (max) 4.5% Ni 14.7% Cr 0.3% Nb 3.5% Cu steel
AMS 5827	5% Ni 17% Cr 0.2% Nb + Ta 3.5% Cu steel: Welding electrode
AMS 5840	0.05% C (max) 8% Ni 12.75% Cr 2.25% Mo 1.1% Al steel
AMS 5859	0.05% C (max) 6% Ni 15% Cr 0.4% Nb 1.5% Cu steel
AMS 5860	0.05% C (max) 8.5% Ni 11.7% Cr 0.3% Nb 2% Cu steel
AMS 5862 515500	0.07% C (max) 4.5% Ni 14.7% Cr 0.3% Nb 3.5% Cu steel
AMS 5863	0.05% C (max) 6% Ni 15% Cr 0.4% Nb 1.5% Cu steel
AMS 5864	0.05% C (max) 8% Ni 12.75% Cr 2.25% Mo 1.1% Al steel
AMS 5870	0.15% C 4.2% Ni 15.5% Cr 2.7% Mo 0.1% N steel: Wire
AMS 7474	0.07% C (max) 4% Ni 16.5% Cr 3.5% Cu steel
AN QQ S 757/3	0.1% C 7% Ni 17% Cr 0.5% Ti 1% Nb steel: US military specification; annealed **UTS: 700 Elon: 40%**

Symbol	Nominal analysis, supplier, condition and remarks.
AN QQ S 771/4 FM	0.12% C 7% Ni 17% Cr 0.7% Mo steel: US military specification; annealed **UTS: 690 Elon: 35%**
AN QQ S 771/4 MCR	0.1% C 7% Ni 17% Cr 2% Mo 0.5% Cu steel: US military specification; cold drawn **UTS: 840 Elon: 12% Proof: 700**
AN QQ S 771/4G	0.12% C 7% Ni 17% Cr 0.5% Cu steel: US military specification; cold drawn **UTS: 840 Elon: 12% Proof: 700**
AN QQ S 772 G	0.15% C 6.5% Ni 17% Cr 0.5% Cu Ti or Nb steel: US military specification; annealed **UTS: 540 Elon: 40% Proof: 200**
AN T 14/1	0.1% C 7% Ni 17% Cr 0.5% Cu Nb or Ti steel: US military specification; flexible tube
AN T 43	0.12% C 6.5% Ni 17% Cr 0.5% Cu steel tube: US military specification; annealed **UTS: 740 Elon: 30% Proof: 200**
AN WWT 855/1	0.12% C 7% Ni 17% Cr 0.5% Cu steel: US military specification; annealed **UTS: 700 Elon: 30% Proof: 200**
AN WWT 858/1	0.1% C 7% Ni 17% Cr + Ti or Nb seamless steel tube: US military specification
AN WWT T 861/2	0.1% C 7% Ni 17% Cr 0.5% Cu Ti or Nb steel: Welded tube; US military specification **UTS: 740**
ARMCO 17/7	0.12% C 7% Ni 17% Cr steel: Armco
ASTM A 167/301	0.15% C 7% Ni 17% Cr steel: Sheet and plate; annealed **DPN: 202 UTS: 530 Elon: 40% Proof: 200**
ASTM A 693/630	0.07% C 4.0% Ni 16.5% Cr 4.0% Cu steel: Precipitation hardened **UTS: 1200 Elon: 5% Proof: 1050**
ASTM A 693/634	0.2% C 4.5% Ni 15.5% Cr 2.9% Mo steel: Precipitation hardened **UTS: 1250 Elon: 12% Proof: 1080**
ASTM A 747 CB 7 CU1 392180	0.07% C (max) 4.2% Ni 16.5% Cr 0.2% Nb 3% Cu steel: Casting
ASTM A 757 E3N	0.06% C (max) 4% Ni 12.7% Cr 0.6% Mo steel: Casting
ASTM A177	0.12% C 7% Ni 17% Cr sheet and strip: Hard **UTS: 1380 Elon: 4% Proof: 1030**
ASTM A461/630	0.07% C 4.0% Ni 16.5% Cr 0.3% Nb 4.0% Cu + Ta steel: Bar
ASTM A461/631	0.09% C (max) 7.2% Ni 17% Cr 1.2% Al steel: Bar
ASTM A461/632	0.09% C (max) 7.2% Ni 15% Cr 2.5% Mo 1.2% Al steel: Bar
ASTM A461/634	0.12% C 4.5% Ni 15.5% Cr 2.8% Mo 0.1% N steel: Bar
ASTM A579	0.12% C 4.5% Ni 15.5% Cr 2.8% Mo 0.1% N steel: Forging
ASTM A579	0.1% C (max) 7% Ni 14.8% Cr 2.3% Mo 1% Al steel: Forging
ASTM A669	0.03% C (max) 4.7% Ni 18% Cr 2.7% Mo steel: Tube **DPN: 290 UTS: 630 Elon: 30% Proof: 440**
ASTM A693/631	0.09% C 7.2% Ni 17% Cr 1.2% Al steel: Precipitation hardened **UTS: 1300 Elon: 5% Proof: 1200**
ASTM A693/632	0.09% C 7.2% Ni 15% Cr 25% Mo 1.2% Al steel: Precipitation hardened **UTS: 1480 Elon: 3% Proof: 1250**
ASTM A693/635	0.08% C 6.8% Ni 16.8% Cr 0.8% Ti 0.4% Al steel: Precipitation hardened **UTS: 1280 Elon: 7% Proof: 1050**
ASTM A693 XM9	0.05% C 6.5% Ni 14.2% Cr 0.3% Mo 0.75% Ti 0.1% Al steel: Precipitation hardened **UTS: 1200 Elon: 8% Proof: 1050**
ASTM A693 XM12	0.07% C 4.0% Ni 14.7% Cr 3.5% Cu steel: Precipitation hardened **UTS: 1200 Elon: 8% Proof: 1050**
ASTM A693 XM13	0.05% C 8.0% Ni 12.7% Cr 2.2% Mo 1.1% Al steel: Precipitation hardened **UTS: 1400 Elon: 8% Proof: 1350**
ASTM A693 XM16	0.06% C 8.5% Ni 11.7% Cr 0.5% Mo 1.1% Ti 1.0% Cu steel: Precipitation hardened **UTS: 1520 Elon: 3% Proof: 1400**
ASTM A693 XM25	0.05% C 6.0% Ni 15.0% Cr 0.7% Mo 1.5% Cu steel: Precipitation hardened **UTS: 1100 Elon: 6% Proof: 1000**
ASTM A747 CB7 Cu 1	0.07% C 4% Ni 16.5% Cr 0.2% Nb 2.8% Cu steel: Casting; precipitation hardened **DPN: 375 UTS: 1170 Elon: 5% Proof: 1000**
ASTM A747 CB7 Cu 2	0.07% C 5% Ni 14.7% Cr 0.2% Nb 2.8% Cu steel: Casting; precipitation hardened **DPN: 375 UTS: 1170 Elon: 5% Proof: 1000**
ASTM A789 S31200	0.03% C (max) 6% Ni 25% Cr 1.6% Mo 0.17% N steel: Tube; solution treated and aged **DPN: 280 UTS: 690 Elon: 25% Proof: 450**
ASTM A789 S31260	0.03% C (max) 6.5% Ni 25% Cr 3% Mo 0.2% N steel: Tube; solution treated and aged **DPN: 290 UTS: 630 Elon: 30% Proof: 440**
ASTM A789 S31500	0.03% C (max) 4.75% Ni 18.5% Cr 2.7% Mo 0.07% N steel: Tube; solution treated and aged **DPN: 290 UTS: 630 Elon: 30% Proof: 440**
ASTM A789 S31803	0.03% C (max) 5.5% Ni 22% Cr 3% Mo 0.14% N steel: Tube; solution treated and aged **DPN: 290 UTS: 620 Elon: 25% Proof: 450**
ASTM A789 S32304	0.03% C (max) 4.5% Ni 23% Cr 0.3% Mo 0.3% Cu 0.1% N steel: Tube; solution treated and aged **DPN: 290 UTS: 600 Elon: 25% Proof: 400**
ASTM A789 S32550	0.04% C 5.5% Ni 25.5% Cr 3.4% Mo 2% Cu 0.2% N steel: Tube; solution treated and aged **DPN: 300 UTS: 750 Elon: 15% Proof: 550**
ASTM A789 S32950	0.03% C (max) 4.2% Ni 27.5% Cr 1.7% Mo 0.2% N steel: Tube; solution treated and aged **DPN: 290 UTS: 620 Elon: 20% Proof: 480**
ASTM A815 S31803	0.03% C (max) 5.5% Ni 22% Cr 3% Mo 0.12% N steel: Pipe fitting
BS 3146 ANC 20A	0.07% C 4.5% Ni 13.5% Cr 1.7% Mo 0.5% Nb 2% Cu steel: Casting
BS 3146 ANC 20B	0.07% C 4.5% Ni 13.5% Cr 1.7% Mo 0.5% Nb 2% Cu steel: Casting
BS 3146 ANC 21	0.05% C 5% Ni 26% Cr 2% Mo 3% Cu 0.1% N steel: Casting
BS 3146 ANC 22 A, B and C	0.06% C 4% Ni 16% Cr 3% Cu 0.05% N steel
CA 6NM	0.06% C (max) 4.0% Ni 12.2% Cr 0.7% Mo steel: Casting; information from Department of Mines, Canada
CROLOY 16.6 PH	0.035% C 7.5% Ni 15.5% Cr 0.4% Ti 0.35% Al steel: Babcock and Wilcox, USA
CRUCIBLE 301	0.15% C 7% Ni 17% Cr steel: Crucible Steel Co.; hard **UTS: 1380 Elon: 8% Proof: 1000**
CUSTOM 450	0.05% C 6% Ni 15% Cr 0.7% Mo 0.4% Nb 1.5% Cu steel: Carpenter
CUSTOM 455	0.03% C 9.0% Ni 11.75% Cr 1.2% Ti 0.3% Nb 2.2% Cu + Ta steel: Carpenter; age hardening
CUSTOM 630	0.07% C 4% Ni 16.2% Cr 0.3% Nb + Ta 4% Cu steel: Carpenter

Note. The following abbreviations and units are used in the tables:

DPN	Hardness, diamond pyramid number
UTS	Ultimate tensile strength, N/mm^2
Elon	Elongation, %
Proof	0.1% proof strength, N/mm^2

$1\ N/mm^2 = 0.1\ hbar = 0.102\ kgf/mm^2 = 0.06475\ tonf/in.^2 = 145.04\ lbf/in.^2 = 1\ MPa$

See Appendix II for other abbreviations and conversion tables.

Symbol	Nominal analysis, supplier, condition and remarks.
D 51	0.24% C 6.0% Ni 12% Cr 0.6% Mo 1.0% V steel: ESC; solution treated and aged **DPN: 550 UTS: 1500 Elon: 17%**
D 70	0.03% C 4.2% Ni 12% Cr 4.2% Mo 14.5% Co 0.06% Ti steel: ESC; solution treated and aged **UTS: 1600**
DTD 5086	7% Ni 17% Cr steel with additions: Age hardening for rod, wire and springs
DTD 5299	Ni/Cr with Cu and Mo steel: Casting; investment cast; age hardenable **UTS: 950**
DTD 5309	Ni/Cr with Cu and Mo steel: Casting; investment cast; age hardenable
E 630	0.05% C (max) 4.7% Ni 16.4% Cr 0.2% Nb 3.7% Cu steel: Weld electrode; designation used by AWS
EC 219	0.05% C (max) 6.2% Ni 19.7% Cr 0.2% N steel: Weld electrode; designation used by AWS
EC 240	0.06% C (max) 5% Ni 18% Cr 0.15% N steel: Weld electrode
EC 630	0.05% C (max) 417% Ni 16.4% Cr 0.2% Nb 3.7% Cu steel: Weld electrode; designation used by AWS
ER 16.8.2	0.1% C (max) 8.5% Ni 15.5% Cr 1.5% Mo steel: Weld electrode; designation used by AWS
ER 2553	0.04% C (max) 5.5% Ni 25.5% Cr 3% Mo 2% Cu 0.2% N steel: Weld electrode; designation used by AWS
ESSHETE 1250	0.15% C 10% Ni 15% Cr 1% Mo 0.4% V 1.0% Nb steel: Samuel Fox; solution treated; weldable **UTS: 450 Elon: 67% Proof: 210**
EV7	0.2% C 4.5% Ni 21% Cr 0.3% N steel: For exhaust valves; designation used by SAE
FERRALIUM	5% Ni 25% Cr Mo Cu steel: Transformable; Langley Alloys; solution treated and aged **DPN: 320 UTS: 650 Elon: 20% Proof: 550**
FERRALIUM 255	Low C 5% Ni 26% Cr 3% Mo 2% Cu 0.15% N steel: Langley; Duplex steel **DPN: 230 UTS: 830 Elon: 30% Proof: 660**
FV 301	0.1% C 7.5% Ni 17.5% Cr steel: Firth Vickers; annealed **UTS: 740 Elon: 55%**
FV 301	0.1% C 7.5% Ni 17.5% Cr steel: Firth Vickers; hard **UTS: 1210 Elon: 10%**
FV 467	0.18% C 10% Ni 14% Cr 2% Mo 0.8% Ti 2.5% Cu steel: Firth Vickers; age hardened **UTS: 650 Elon: 44% Proof: 300**
FV 520(B)	0.07% C 5.4% Ni 14% Cr 1.8% Mo 0.7% Nb 1.9% Cu steel: Sheet; Firth Vickers; aged **DPN: 410 UTS: 1250 Elon: 10% Proof: 1070**
FV 520(S)	0.06% C 5.4% Ni 15.8% Cr 1.8% Mo 0.2% Ti 1.9% Cu steel: Sheet; Firth Vickers; aged **DPN: 410 UTS: 1250 Elon: 10% Proof: 1070**
FV 548	0.08% C 11.5% Ni 16.5% Cr 1.5% Mo 1.0% Nb steel: Firth Vickers; age hardened **UTS: 630 Elon: 52% Proof: 250**
G 38	0.17% C 12% Ni 16% Cr 2.5% Mo 1.2% V 1.3% W steel: Jessop; solution treated and warm worked **UTS: 820 Elon: 34% Proof: 710**
G 41	0.17% C 12% Ni 16% Cr 1.3% V 1.3% W 1.3% Nb steel: Jessop; annealed and warm worked **UTS: 730 Elon: 26% Proof: 530**
GOST 5632/4 KH 14N7G7FMS	0.42% C 7% Ni 15% Cr 0.8% Mo 1.7% V steel: Plate; Russian Standard
GOST 5632/4 KH 15N17G7F2MS	0.42% C 7% Ni 15% Cr 0.8% Mo 1.7% V steel: Plate; Russian Standard
GOST 5632 KH 17N71U	0.1% C (max) 7% Ni 17% Cr 1% Al steel: Plate; Russian Standard
HY RESIST 22/5 DUPLEX	5.5% Ni 22% Cr 3% Mo 0.15% N steel: British Steel Stainless
JIS G3459 SUS 329 JITP	0.08% C (max) 4.5% Ni 25% Cr 2% Mo steel: Plate; Japanese Standard **UTS: 600 Proof: 400**

Symbol	Nominal analysis, supplier, condition and remarks.
JIS G4304 SUS 329JI	0.08% C 4.5% Ni 25% Cr 2% Mo steel: Plate; Japanese Standard; solution treated **UTS: 600 Proof: 400**
JIS G4304 SUS631	0.1% C (max) 7% Ni 17% Cr 1.5% Al steel: Plate; Japanese Standard
KCR–D183 J93183	0.03% C (max) 5% Ni 2.5% Cr 3% Mo 1% Co 0.2% N steel: Casting; designation used in the UK and USA
KCR–D283 J93550	0.03% C (max) 22.5% Cr 6.5% Mo 1% Co 0.2% N steel: Casting
LANGALLOY 40V	5% Ni 25% Cr transformable steel: Langley; solution treated and aged **DPN: 320 UTS: 1020 Elon: 15% Proof: 760**
LANGALLOY 40V	5% Ni 25% Cr transformable steel: Langley; solution treated **DPN: 260 UTS: 730 Elon: 20% Proof: 630**
N 700	0.04% C 4.5% Ni 16.0% Cr 0.25% Nb 0.3% Cu steel: VEW
NF A 35 572/ Z12CN 17/08	0.1% C 7.5% Ni 17% Cr steel: French Standard
NF A35 572/ Z6CNU 17/04	4.0% Ni 16.5% Cr steel: French Standard
NF A35 572/ Z6CNUD 15/04	0.07% C (max) 4.0% Ni 1.3% Mo 14.5% Co steel: French Standard
PH 13.8 Mo	0.05% C (max) 8% Ni 12.8% Cr 2.2% Mo 1% Al steel: Origin unknown
PH 14.8 Mo	0.05% C (max) 8% Ni 14% Cr 2.5% Mo 1.2% Al steel: Origin unknown
PH 15.7 Mo	0.1% C (max) 7.2% Ni 14.7% Cr 2.3% Mo 1% Al steel: Origin unknown
PH 15.7 Mo	0.07% C 7% Ni 15% Cr 2.2% Mo 1.1% Al steel: Age hardening; American proprietary alloy listed in SAE yearbook
QQ S763/8	0.1% C 7% Ni 17% Cr 1% Nb or 0.5% Ti steel: US Federal; annealed **UTS 450 Elon: 35% Proof: 150**
R690G	0.05% C 6% Ni 13% Cr steel: Casting; Fagersta
S 143	0.07% C 5.4% Ni 14% Cr 1.8% Mo 1.8% Cu 0.8% N steel: BS Aerospace specification
S 144	0.07% C 5.4% Ni 14% Cr 1.8% Mo 1.8% Cu 0.8% N steel: BS Aerospace specification
S 145	0.07% C 5.4% Ni 14% Cr 1.8% Mo 1.8% Cu 0.8% N steel: BS Aerospace specification
S 532	0.055% C 5.4% Ni 15.8% Cr 1.8% Mo 0.1% Ti 2% Cu steel: BS Aerospace specification
S 533	0.055% C 5.4% Ni 15.8% Cr 1.8% Mo 0.1% Ti 2% Cu steel: BS Aerospace specification
S 31803	0.03% C (max) 5.5% Ni 22% Cr 3% Mo 0.12% N steel: UNS designation; solution treated and aged **DPN: 290 UTS: 620 Elon: 25% Proof: 450**
SAE J217	0.09% C (max) 7% Ni 17% Cr 1% Al steel: Cold drawn and aged; wire for springs **UTS: 2000**
SAF 2205	0.03% C (max) 5.5% Ni 22% Cr 3% Mo 0.14% N steel: Sandvik; cold worked **UTS: 965 Elon: 10% Proof: 895**
SAE 30301	0.1% C 7% Ni 17% Cr steel: AISI type 301
SF301	0.15% C 7% Ni 17% Cr steel: Samuel Fox; annealed **UTS: 920 Elon: 55%**
For SIS specifications see SS	
SS 2330	0.15% C 7% Ni 17% Cr steel: Plate, bar, etc.; Swedish Standard **DPN: 220 UTS: 770**
SS 2331	0.15% C 7% Ni 17% Cr steel: Sheet; Swedish Standard; cold rolled **UTS: 920 Elon: 3% Proof: 770**
SS 2376	0.03% C (max) 4.8% Ni 18.5% Cr 2.7% Nb 0.07% N steel: Swedish Standard
SS 2377	0.4% C (max) 5% Ni 22% Cr 3% Mo 0.07% N steel: Swedish Standard
SS 2385	0.1% C (max) 5.5% Ni 13% Cr steel: Swedish Standard

Symbol	Nominal analysis, supplier, condition and remarks.
SS 2385	0.05% C (max) 5% Ni 16% Cr 1.1% Mo steel: Swedish Standard
SS 2387	0.5% C (max) 4.5% Ni 16% Cr 1.2% Mo steel: Swedish Standard
SS 2387	0.05% C (max) 5% Ni 16% Cr 1.1% Mo steel: Swedish Standard
STAINLESS W	0.12% C 7% Ni 17% Cr 1% Ti 1% Al 0.2% N steel: Age hardening; American proprietary alloy listed in SAE yearbook
STAINLESS W	0.12% C 7% Ni 17% Cr 1% Ti 1% Al 0.2% N steel: Age hardening; American proprietary alloy listed in SAE yearbook
STAINLESS W	0.08% C (max) 6.7% Ni 16.2% Cr 0.8% Ti steel: Origin unknown
UHB STAINLESS 673	0.1% C 7.5% Ni 17% Cr steel: Uddelholm
V 174	0.07% C (max) 4% Ni 17% Cr 0.3% Nb + Ta 4% Cu steel: Valbuna
VK	0.07% C 7% Ni 17% Cr 1.0% Al steel: Osborn; age hardened **DPN: 380 UTS: 1210 Elon: 15% Proof: 760**
X 7 Cr Ni Al 17/7	0.1% C (max) 7.2% Ni 17% Cr 1.1% Al steel: German Standard
X 12 Cr Ni 17/7	0.15% C (max) 7.5% Ni 17% Cr steel: German Standard
X 15 CN 1707	0.1% C 7% Ni 17% Cr steel: Italian Standard; annealed **DPN: 180 UTS: 480 Elon: 45% Proof: 180**
XM9	0.05% C (max) 6.7% Ni 14.2% Cr 0.3% Mo 0.7% Ti 0.1% Al steel: Precipitation hardened **UTS: 1240 Elon: 10% Proof: 1205**
XM12	0.07% C (max) 4.5% Ni 14.2% Cr 3.5% Cu steel: Precipitation hardened **UTS: 1070 Elon: 12% Proof: 1000**
XM13	0.05% C (max) 8.0% Ni 12.7% Cr 2.2% Mo 1.1% Al steel: Precipitation hardened **UTS: 1415 Elon: 10% Proof: 1310**
XM16	0.05% C (max) 8.5% Ni 11.7% Cr 2.0% Cu 0.5% 1.1% Ti steel: Precipitation hardened **UTS: 1380 Elon: 10% Proof: 1275**
XM25	0.05% C (max) 6.0% Ni 15.0% Cr 0.7% Mo 1.5% Cu steel: Precipitation hardened **UTS: 1100 Elon: 12% Proof: 1035**
ZERON 100	0.03% C (max) 7.2% Ni 25.5% Cr 3.5% Mo 0.7% W 0.25% N steel: Casting; designation used in the UK and USA

Note. The following abbreviations and units are used in the tables:

DPN	Hardness, diamond pyramid number
UTS	Ultimate tensile strength, N/mm²
Elon	Elongation, %
Proof	0.1% proof strength, N/mm²

1 N/mm²=0.1 hbar=0.102 kgf/mm²=0.06475 tonf/in.²=145.04 lbf/in.²=1 MPa

See Appendix II for other abbreviations and conversion tables.

44P Steel – high manganese austenitic

Specific gravity	8.0
Density	8000 kg/m³
Solidus/liquidus	1470–1510 °C
Thermal conductivity	15 W/m °C
Coefficient of linear expansion	16 /°C – some materials have special expansion characteristics
Electrical conductivity	3% IACS (copper 100%)
Specific resistance	800 microhm mm
Young's modulus of elasticity	–
Impact	will be variable
Fatigue strength	–
Hot strength	–

The above properties are typical of the following group, and may not apply exactly to any one specification. It is possible that with certain specifications some of the values may not be applicable.

General metallurgical characteristics

Manganese alone has the effect of depressing the critical points of steel, and with more than 18% manganese steel becomes austenitic.

These high manganese steels are discussed in Section 44D. The steels in the following list are all austenitic and contain above approximately 5% manganese with varying amounts of chromium and nickel to ensure a stable austenitic structure. Manganese has a considerable influence on the cold working properties of steel and this increases in proportion to the manganese content which may be as high as 20%.

These high manganese steels are relatively unworkable and are generally used as castings or sprayed coatings for abrasion resistant purposes.

Chromium improves the corrosion resistant properties, particularly the hot oxidation resistance, and, when nickel is also present, the hot strength of the alloys.

There are certain specially formulated 4–6% manganese chromium nickel alloys which have a high coefficient of linear expansion which can be useful when steel is associated with aluminium.

These steels are all hardened by cold working and very little cold work will result in a dramatic increase in the hardness or tensile strength and reduction in ductility.

The steels are all fully austenitic and therefore their properties cannot be improved by any form of thermal treatment. If the cold work has to be removed this requires a temperature of 1050 °C. The materials, however, are very often chosen because of their ability to accept cold work so that their mechanical properties are dramatically increased.

Where welding is required the inert gas welding technique should be used. It will be appreciated that if any cold work is present this will be removed by the welding operation. In general expert advice should be obtained before attempting to weld these materials. It should be noted that these materials have a high coefficient of linear expansion and this can result in centre-line weld cracking if the correct procedure is not applied.

These materials are almost invariably difficult to machine either by normal conventional machining or by grinding.

The uses for the materials within this group include studs and inserts which require a high coefficient of expansion in order to be used with aluminium and its alloys.

Many of the materials used are castings where good mechanical strength is required in the 'as cast' condition and this applies with the materials in this group.

They are commonly used in items such as buckets on dredgers, etc. where the teeth are subjected to abrasion. These can be as-cast teeth or very often these materials are used as inserts or can be applied by welding or metal spraying.

They are first choice where abrasion and corrosive conditions exist.

Symbol	Nominal analysis, supplier, condition and remarks.
1.3817	0.4% C 18.0% Mn 4.0% Cr steel: German Standard
1.3960	0.45% C 13.0% Mn 6.5% Ni 5.5% Cr 0.9% V steel: German Standard
1.3962	0.12% C 6.0% Mn 10.0% Ni 11.5% Cr steel: German Standard
1.4370	0.15% C (max) 6.5% Mn 8.5% Ni 18.5% Cr steel: German Standard
1.4371	0.1% C (max) 8.5% Mn 5.5% Ni 18.0% Cr 0.15% N steel: German Standard
1.4451	0.15% C (max) 18.0% Mn 2.0% Ni 12.0% Cr 0.5% Mo steel: German Standard
1.5151	0.4% C 18.0% Mn 0.1% Ni 4.0% Cr steel: German Standard
1.5152	0.4% C 22.5% Mn 4.0% Cr steel: German Standard
1.5161	0.15% C (max) 18.0% Mn 2.0% Ni 12.0% Cr 0.5% Mo steel: German Standard
10.17.5.9 Mn	0.15% C 9.0% Mn 17.0% Ni 10.0% Cr 5.0% Mo steel: Electrode; Metrode
14 Cr 14 Mn R	0.35% C 14.0% Mn 2.0% Ni 14.0% Cr 0.7% Mo 0.3% V steel: Electrode; Metrode; for hard facing; work hardens from 250 to 450 DPN
17.12.1.4 Mn Nb	0.08% C 4.0% Mn 12.0% Ni 17.0% Cr 1.5% Mo 1.0% Nb steel: Electrode; Metrode
18/5/9	0.08% C 18% Mn 9.0% Ni 5.0% Cr steel: Nyby, annealed **DPN: 230 UTS: 580 Elon: 45% Proof: 270**
18.6.9 Mn Si Nb	0.06% C 9.0% Mn 6.0% Ni 18.0% Cr 1.0% Nb 2.0% Si steel: Electrode; Metrode
18.8.3 Mn Si Nb N	0.06% C 7.0% Mn 8.0% Ni 18.0% Cr 3.0% Mo 0.5% Nb 2.0% Si 0.2% N steel: Electrode; Metrode
18.18 PLUS	0.15% C 18% Mn 18% Cr 1% Mo 1% Cu 0.5% N steel: Carpenter **UTS: 820 Elon: 50% Proof: 450**
18 Cr 2 Ni 12 Mn	0.15% C 12% Mn 1.8% Ni 17% Cr 0.3% N steel: Carpenter
19.9.6 Mn	0.1% C 6% Mn 9% Ni 19% Cr 1% Mo steel: Electrode; Metrode **DPN: 320**
21/4N	0.55% C 9% Mn 4% Ni 21% Cr 0.4% N steel: Jessop
21 Cr 6 Ni 9 Mn	0.03% C 9% Mn 6% Ni 20% Cr 0.25% N steel: Carpenter
21 MN	0.7% C 6% Mn 1.6% Ni 21% Cr 0.2% N steel: Valhuna
22/4/9	0.5% C 8.5% Mn 3.5% Ni 20% Cr 0.4% N steel: Age hardening American proprietary alloy listed in SAE yearbook
65 Mn 12 Cr 5 Ni 3	0.65% C 12.0% Mn 3.0% Ni 4.5% Cr steel: Electrode; Metrode; work hardening deposit
203 E2	0.08% C (max) 5.7% Mn 5.7% Ni 17% Cr 0.2% S steel: Designation used in the UK and USA
214 MN	0.5% C 9% Mn 4% Ni 21% Cr 0.4% N steel: Valhuna
284 S 16	0.07% C 8.5% Mn (max) 5.2% Ni 17.5% Cr 0.2% N steel: Sheet and plate; designation in BS 1449
284 S 16	0.07% C (max) 8.5% Mn 5.2% Ni 17% Cr 0.2% N steel: Sheet and plate
664 M	0.06% C 6.5% Mn 5.5% Ni 17% Cr 0.1% N steel: Avesta
664 MV	0.05% C 8.0% Mn 5.5% Ni 18% Cr 0.2% N steel: Avesta
1031 ALLOY	0.3% C 6% Mn 13% Ni 2% Cu Fe alloy: Darwin
1250	0.1% C 6% Mn 10% Ni 15% Cr 1% Mo 0.3% V 1% Nb 0.006% B steel tube: Designation used by BS
AF 71	0.3% C 18% Mn 12.5% Cr 3% Mo 8% V 0.2% B 0.2% N steel: American proprietary alloy listed in SAE yearbook
AF 183	0.3% C 18% Mn 12.5% Cr 3% Mo 0.8% V 0.2% N steel: American proprietary alloy listed in SAE yearbook
AFNOR 81.343 EZ18.8 Mn B160 23X	0.1% C 5.5% Mn 8.5% Ni 21% Cr 0.85% Mo steel: Welding electrode; French Standard
AFNOR 81.343 EZ23.12R 180.33X	0.04% C 13% Ni 23% Cr steel: Welding electrode; French Standard
AFNOR 81-343 EZ18.8 Mn B23	0.1% C 6.5% Mn 8% Ni 17% Cr steel: Welding electrode; French Standard
AFNOR 81-343 EZ18.8 Mn R160 21X	0.05% C 6% Mn 8.5% Ni 21% Cr steel: Welding electrode; French Standard
AISI 201	0.15% C 6% Mn 4.5% Ni 17% Cr 0.25% N steel
AISI 202	0.15% C (max) 8.7% Mn 5.0% Ni 18.0% Cr steel
AKMC/Z	0.4% C 18% Mn 4.0% Cr 0.1% N steel: Krupps
AM 88	0.2% C 8.0% Mn 5.7% Ni 8.0% Cr 0.3% Si steel: Pompey
AMS 5561	0.08% C (max) 9% Mn 6.5% Ni 20% Cr 0.25% N steel
AMS 5562	0.04% C (max) 9% Mn 6.2% Ni 20% Cr 0.25% N steel
AMS 5595	0.04% C (max) 9% Mn 6.5% Ni 20.2% Cr 0.3% N steel
AMS 5623	0.6% C 5.5% Mn 9.5% Ni steel: Bar and forging; high expansion steel; annealed
AMS 5624 A	0.55% C 4.5% Mn 12.5% Ni 4% Cr steel: Bar; high expansion
AMS 5625 A	0.6% C 5.5% Mn 9.5% Ni steel: Bar; cold drawn; high expansion
AMS 5656	0.04% C (max) 9% Mn 6.5% Ni 20% Cr 0.3% N steel
AMS 5848	0.1% C (max) 8.2% Mn 8.5% Ni 17% Cr 2.5% Mo 0.35% N steel

Note. The following abbreviations and units are used in the tables:

DPN	Hardness, diamond pyramid number
UTS	Ultimate tensile strength, N/mm^2
Elon	Elongation, %
Proof	0.1% proof strength, N/mm^2

1 N/mm^2=0.1 hbar=0.102 kgf/mm^2=0.06475 $tonf/in.^2$=145.04 $lbf/in.^2$=1 MPa

See Appendix II for other abbreviations and conversion tables.

Symbol	Nominal analysis, supplier, condition and remarks.
ASTM A289 A	0.55% C 8% Mn 8% Ni 5% Cr steel: For non-magnetic retaining rings **UTS: 1070 Elon: 20% Proof: 920**
ASTM A289 B	0.5% C 18% Mn 2% Ni 5% Cr steel: For non-magnetic retaining rings **UTS: 1070 Elon: 20% Proof: 920**
ASTM A412/201	0.15% C (max) 6.5% Mn 4.5% Ni 27% Cr steel: Sheet, etc.
ASTM A412/202	0.15% C (max) 8.7% Mn 5% Ni 18% Cr steel: Sheet, etc.
ASTM A429/201	0.15% C (max) 6.5% Mn 4.5% Ni 17% Cr steel: Bar
ASTM A429/202	0.15% C (max) 8.7% Mn 5% Ni 18% Cr steel: Bar
ASTM A581 XM1	0.08% C 5.7% Mn 5.7% Ni 17% Cr 0.2% S steel: Wire; free machining
ASTM A666	Stainless steel: Sheet, plate, bar, etc.; for structural purposes; graded by AISI system (see also Section 44N1)
ASTM A688 TPXM29	0.06% C 13.0% Mn 3.0% Ni 18% Cr 0.3% N steel: Tube; welded **UTS: 690 Elon: 35% Proof: 380**
ASTRANIT M/Z	0.12% C 18% Mn 2.0% Ni 12% Cr 0.5% Mo steel: Krupps
AURIGA XXXI	0.27% C 8% Mn 8% Ni 8% Cr steel: Casting; D Brown; annealed **UTS: 540 Elon: 24% Proof: 240**
AV1	0.5% C 8.2% Mn 2.8% Ni 21% Cr 1.7% W steel: For rods; origin unknown
BS 2926/18/15/3 L Mn	0.04% C (max) 5% Mn 15% Ni 18% Cr 3% Mo steel: For welding electrodes
BS 2926/25/20/1 Nb	0.12% C (max) 4% Mn 20% Ni 26% Cr 1% Mo steel: For welding electrodes
BS 2926/25/21/2 L Mn	0.04% C (max) 5% Mn 21% Ni 25% Cr 2.5% Mo steel: For welding electrodes
BS 2926 H	0.6% C 6% Mn 19% Ni 26% Cr 1% Mo 2% Nb steel: Welding electrode; as welded **UTS: 540 Elon: 25%**
BS 3059-1250	0.1% C 6% Mn 10% Ni 15% Cr 1% Mo 0.3% V 1% Nb 0.006% B steel: Tube
CF 10S Mn N	0.1% C (max) 8% Mn 8.5% Ni 17% Cr steel: Casting; designation used in the UK and USA
CG6 MMN	0.06% C (max) 5% Mn 12.5% Ni 22% Cr 2.2% Mo 0.2% Nb steel: Casting; designation used in the UK and USA
COSSET MC	0.6% C 16.7% Mn 13.6% Cr steel: Welding electrode; Soudometal; for hard facing; work hardens to 500 DPN **DPN: 200**
COSSET 624S	1.2% C 17.5% Mn 9.5% Cr 2.8% Nb steel: Welding electrode; Soudometal; work hardens to 500 DPN **DPN: 250**
CRM 15D	1.0% C 5.0% Mn 5.0% Ni 20.0% Cr 2.0% Mo 2.0% Nb 2.0% Cu 0.2% N Fe alloy: Information from SAE handbook
CRUCIBLE 201	0.15% C 6% Mn 4.5% Ni 17% Cr 0.25% N steel: Crucible Steel Co.; annealed **UTS: 800 Elon: 55% Proof: 370**
CRUCIBLE 202	0.15% C 8.5% Mn 5% Ni 18% Cr 0.25% N steel: Crucible Steel Co.; annealed **UTS: 760 Elon: 55% Proof: 370**
CRYOGENIC TENDON	0.12% C (max) 15% Mn 5.5% Ni 18% Cr steel: Origin unknown
CSA	0.25% C 4% Mn 5% Ni 18% Cr 1.3% Mo 1.3% W 1% Nb steel: Crucible Steel Co.
DTD 247	0.7% C (max) 4.5% Mn 13% Ni 3% Cr steel: Bar and forging; as rolled or softened; high expansion steel **UTS: 620 Elon: 25%**
DV2A	0.5% C 11.5% Mn 20% Cr 2% W steel: For valves; origin unknown
DV2B	0.5% C 11.5% Mn 20% Cr 2% W steel: For valves; origin unknown
E Fe Mn E	0.8% C 17.5% Mn 4.5% Cr steel: Welding electrode; designation used by AWS
E Fe Mn F	1% C 19% Mn 4.5% Cr steel: Welding electrode; designation used by AWS
E219	0.06% C (max) 9% Mn 6.2% Ni 19.7% Cr steel: Welding electrode; designation used by AWS
ESSHERE 1250	0.1% C 6.2% Mn 10% Ni 15% Cr 1% Mo 0.25% V 1% Nb steel
EV 8	0.53% C 9% Mn 3.7% Ni 21% Cr 0.42% N steel: For exhaust valves; designation used by SAE
EV II	0.7% C 6% Mn 2% Ni 21% Cr 0.2% N steel: For exhaust valves
F Fe Mn A	0.8% C 16% Mn 3.2% Ni steel: Welding electrode; designation used by AWS
F Fe Mn B	0.8% C 14% Mn 1.2% Mo steel: Welding electrode; designation used by AWS
F Fe Mn C	0.8% C 14% Mn 4.2% Ni 4.2% Cr steel: Welding electrode; designation used by AWS
F Fe Mn Cr	0.8% C 16% Mn 15% Cr steel: Welding electrode; designation used by AWS
F Fe Mn D	0.8% C 17.5% Mn 6.2% Cr 0.8% V steel: Welding electrode; designation used by AWS
G 192	0.6% C 8.5% Mn 22% Cr steel: American proprietary alloy listed in SAE yearbook
GAMAN H	0.5% C 11% Mn 21% Cr steel: For valves; origin unknown
GOST 5632 ICHITE9AN4	12% C (max) 9% Mn 4% Ni 17% Cr 2% N steel: Russian specification **UTS: 700 Proof: 350**
GOST 5632/2 KH13N4G9	0.22% C 9.0% Mn 4.2% Ni 13.0% Cr steel: Russian Standard
GOST 5632 KH17G9AN4	0.12% C 9.2% Mn 4.0% Ni 17.0% Cr 0.2% N steel: Russian Standard
H 35	0.08% C 3.25% Mn 4.2% Ni 11.7% Cr 0.6% Nb steel: Jessop
H 850	0.52% C 9.0% Mn 3.9% Ni 20.8% Cr 0.45% N steel: For valves; VEW
HTX	0.04% C 8.5% Mn 8% Ni 21% Cr 1.5% Mo 0.23% P 0.2% N steel: American proprietary alloy listed in SAE yearbook
JIS G4SUS202	0.15% C (max) 8.5% Mn (max) 5% Ni 18% Cr 0.75% N steel: Plate; Japanese specification
JIS G4303 SUS201	0.15% C (max) 6.5% Mn 4.5% Ni 17.0% Cr steel: Japanese Standard
JIS G4303 SUS202	0.15% C 8.2% Mn 5.0% Ni 18.0% Cr steel: Japanese Standard
LT 21/42	0.5% C 9.0% Mn 4.5% Ni 21% Cr 2.0% Nb 0.4% N steel: Low Moor
LV 21/2 N	0.5% C 8.5% Mn 2.0% Ni 20% Cr steel: For valves; S Osborn
LV 21/4 N	0.5% C 9% Mn 4% Ni 21% Cr 0.4% N steel: Low Moor; for valves
LV 21/42	0.5% C 9.0% Mn 4.0% Ni 21.0% Cr 2.0% Nb steel: For valves; S Osborn
LV 21/43	0.5% C 9.0% Mn 4.0% Ni 21.0% Cr 1.5% Nb steel: For valves; S Osborn
LV 21/69	0.08% C 9.0% Mn 6.0% Ni 21% Cr 0.4% N steel: Sheet and bar; Low Moor
MANGANESE Ni	0.8% C 13% Mn 3.5% Ni steel: Welding electrode; Murex; for hard facing; work hardens **DPN: 300**

Note. The following abbreviations and units are used in the tables:

DPN	Hardness, diamond pyramid number
UTS	Ultimate tensile strength, N/mm^2
Elon	Elongation, %
Proof	0.1% proof strength, N/mm^2

1 N/mm^2=0.1 hbar=0.102 kgf/mm^2=0.06475 $tonf/in.^2$=145.04 $lbf/in.^2$=1 MPa

See Appendix II for other abbreviations and conversion tables.

Symbol	Nominal analysis, supplier, condition and remarks.
MET MAX 307R	19% Cr 5% Mn 9% Ni 1% Mo steel: Electrode, Metrode
METCOLOY 5	0.15% C 8% Mn 5% Ni 18% Cr steel: Wire for spraying; Metco
MIL AB259	1.3% C 13.5% Mn steel: US military specification
MIL S.17249	1.2% C 13% Mn steel: Casting; US Federal specification
NICOLLOY 4	0.52% C 9.0% Mn 4% Ni 21% Cr 0.4% N steel: Swift Levick for hot dies
NITRONIC 60	0.1% C (max) 8% Mn 8.5% Ni 17% Cr 4% Si 0.1% N steel: Origin unknown
NITRONIC 60W	0.1% C (max) 8% Mn 8.5% Ni 17% Cr 4% Si 0.1% N steel: Origin unknown
R 32	0.55% C 4.5% Mn 12.5% Ni 3.7% Cr steel: Jessop
R 46	0.55% C 9% Mn 4% Ni 21% Cr 0.4% N steel: Jessop
R 520	0.08% C 8% Mn 4.8% Ni 18% Cr steel: Fagersta; annealed **DPN: 240 UTS: 580 Elon: 50% Proof: 270**
RA330.04	0.22% C 5% Mn 35.5% Ni 18.5% Cr Fe alloy: Origin unknown
RED DIAMOND 14H	1.1% C 12.5% Mn steel: For wear; Redheugh **DPN: 360 UTS: 520 Elon: 8%**
RED DIAMOND 14S	1.1% C 12.5% Mn steel: For wear; Redheugh **DPN: 280 UTS: 630 Elon: 12%**
RNK 29	0.09% C 8.5% Mn 5% Ni 18% Cr 0.2% N steel: Bofors
S21300	0.25% C 16.5% Mn 18.5% Cr 2% Mo 1.2% Cu 0.4% N steel: Designation used by UNS
S31254	0.08% C (max) 13% Mn 3% Ni 18% Cr steel: UNS designation
SAE 30201	0.15% C 6% Mn 4.5% Ni 17% Cr 0.25% N steel: Mechanical properties not quoted
SAE 30202	0.15% C 8% Mn 5% Ni 18% Cr 0.25% N (max) steel: Mechanical properties not quoted
SOUDOKAY	1% C 20% Mn 5% Cr steel: Welding electrode; Soudometal; for high Mn steels **DPN: 210 UTS: 850 Elon: 30% Proof: 640**
SOUDOKAY 218.0	0.95% C 14% Mn 0.4% Ni 3.5% Cr steel: Welding electrode; Soudometal; for welding high Mn steel **DPN: 200 UTS: 840 Elon: 32% Proof: 560**
SOUDOKAY 624.0	1.2% C 16.5% Mn 9% Cr 3% Nb steel: Welding electrode; Soudometal **DPN: 250**
SOUDOKAY AP.0	0.5% C 16.5% Mn 13% Cr steel: Welding electrode; Soudometal; for welding high Mn steels **DPN: 210 UTS: 860 Elon: 37% Proof: 580**
SOUDO-MANGANESE CN	0.95% C 13.8% Mn 3.6% Ni 3.7% Cr 0.6% Mo steel: Welding electrode; Soudometal; for hard facing; work hardened to 520 DPN **DPN: 200**
SOUDO-MANGANESE	1.2% C 12.6% Mn 0.75% Mo steel: Welding electrode; Soudometal; for hard facing; work hardened to 500 DPN **DPN: 220**
SS 2357	0.12% C 8% Mn 5% Ni 18% Cr 0.2% N steel: Bar, forging etc.; Swedish Standard; annealed **DPN: 220 UTS: 580 Elon: 45% Proof: 270**
STUBS 7200	Mn Ni Cr steel: Electrode; Stubs; austenitic high manganese steel; can be work hardened **DPN: 250**
STUBS A63	0.14% C 7.0% Mn 8.5% Ni 19% Cr steel: Welding electrode; Stubs; for joining; also surfacing for impact resistance **UTS: 610 Elon: 45% Proof: 370**
SV08KH20N9G7T	0.1% C (max) 7% Mn 9% Ni 19.5% Cr 0.75% Ti: Weld work; Russian Standards designation
SV08KH20N10G6	0.1% C (max) 6% Mn 10% Ni 21% Cr steel: Weld work; Russian Standards designation
TPXM10	See XM10
TPXM11	See XM11
VALMAG 984	High Mn Ni Cr austenitic steel: For valves; Swift Levick; solution treated and aged **DPN: 350 UTS: 1110 Elon: 7%**
WALMANG 3	0.7% C 14% Mn 3.5% Ni steel: Rod for facing; Wall Colmonoy; for repairing high Mn alloys; melting point 1430 °C **DPN: 630**
WORK-HARD 11 Cr 9 Mn	0.3% C 10% Mn 11% Cr steel: Electrode; Metrode **DPN: 320**
WORK-HARD 12 Mn Cr Ni	0.8% C 13% Mn 3% Ni 4% Cr steel: Electrode; Metrode **DPN: 300**
WORK-HARD 13 Mn	0.8% C 13% Mn 1% Mo steel: Electrode; Metrode **DPN: 300**
WORK-HARD 14 Cr 14 Mn	0.4% C 14% Mn 2% Ni 14% Cr 1% Mo 0.3% V steel: Electrode; Metrode **DPN: 320**
X 8 Cr Mn Ni 18/9	0.1% C (max) 8.5% Mn 5.5% Ni 18.0% Cr 0.15% N steel: Designation used by German Standards
X 12 Mn Cr 18/10	0.15% C (max) 18.0% Mn 2.0% Ni 12.0% Cr 0.6% Mo steel: Designation used by German Standards
X 12 Mn Cr 18/11	0.15% C (max) 18.0% Mn 2.0% Ni 12.0% Cr 0.5% Mo steel: Designation used by German Standards
X 15 Cr Ni Mn 18/8	0.15% C (max) 6.5% Mn 8.5% Ni 19.0% Cr steel: Designation used by German Standards
X 20 Cr Ni Mn 12/9	0.12% C 6.0% Mn 10.0% Ni 11.5% Cr steel: Designation used by German Standards
X 40 Mn Cr 18	0.4% C 18.0% Mn 4.0% Cr steel: Designation used by German Standards
X 40 Mn Cr 22	0.4% C 23.0% Mn 4.0% Cr steel: Designation used by German Standards
X 40 Mn Cr N 18	0.4% C 18.0% Mn 4.0% Cr 0.1% N steel: Designation used by German Standards
X 55 Mn Ni Cr 14	0.45% C 13.0% Mn 6.5% Ni 5.5% Cr 0.9% V steel: Designation used by German Standards
XEV-A	0.5% C 11% Mn 20% Cr 0.4% V 2% W 1% Nb steel: For valves; SAE
XEV-B	0.5% C 11% Mn 20% Cr 0.4% V 2% W 1% Nb steel: For valves; SAE
XEV-D	0.5% C 8% Mn 2.8% Ni 21% Cr 0.3% V steel: For valves; SAE
XEV-E	0.4% C 6% Mn 8% Ni 19% Cr 0.75% V steel: For valves; SAE
XEV-F	0.6% C 9% Mn 0.3% Ni 21% Cr 1% Mo 0.7% V 1% Nb steel: For valves; SAE
XEV-G	0.1% C 8.5% Mn 0.4% Ni 27.8% Cr 2% Ti steel: For valves; SAE
XM 11	0.04% C (max) 9% Mn 6.5% Ni 18.2% Cr steel: Designation used in the UK and USA
XM 17	0.08% C 8.0% Mn 6% Ni 20% Cr 2.5% Mo 0.35% N steel: Designation used by ASTM
XM 29	0.08% C 12.5% Mn 3.0% Ni 18.0% Cr 0.3% N steel: Designation used by ASTM
XM 10	0.08% C (max) 9% Mn 6.5% Ni 18.2% Cr steel: Designation used in the UK and USA
Z 10 CMN 18/7	0.08% C 9.0% Mn 5.0% Ni 18% Cr steel: French National Standard; annealed **DPN: 230 UTS: 580 Elon: 45% Proof: 270**
Z 10 CMN 19/9	0.08% C 9.0% Mn 5.0% Ni 18% Cr steel: French National Standard; annealed **DPN: 230 UTS: 580 Elon: 45% Proof: 270**
Z 12 CMN 18/7	0.08% C 9.0% Mn 5.0% Ni 18% Cr steel: French National Standard; annealed **DPN: 230 UTS: 580 Elon: 45% Proof: 270**
Z52CMN21/09	0.5% C 9% Mn 3.7% Ni 21% Cr 0.45% N steel: French Standard

Note. The following abbreviations and units are used in the tables:

DPN	Hardness, diamond pyramid number
UTS	Ultimate tensile strength, N/mm^2
Elon	Elongation, %
Proof	0.1% proof strength, N/mm^2

1 N/mm^2=0.1 hbar=0.102 kgf/mm^2=0.06475 tonf/in.^{2}=145.04 lbf/in.^{2}=1 MPa

See Appendix II for other abbreviations and conversion tables.

45. Strontium Sr

Physical properties

Atomic number	38
Atomic weight	87.63
Crystal structure	Face-centred cubic
Colour	Yellowish white
Specific gravity	2.6
Density	2600 kg/m^3
Melting point	770 °C
Boiling point	1366 °C
Specific heat	0.3077 J/g °C
Thermal conductivity	–
Coefficient of liner expansion (20–100 °C)	100×10^{-6}/ °C
Latent heat of fusion	105 J/g
Latent heat of vaporization	1717 J/g
Thermal neutron absorption cross-section	1.16 barns/atom
Electrical conductivity	7.5% IACS (copper 100%)
Specific resistance	250 microhm mm
Temperature coefficient of electrical resistance	0.0038/ °C
Electrochemical equivalent	–
Electrode potential	−2.89 V
Magnetic susceptibility	-0.2×10^{-6}
Young's modulus of elasticity	–
Tensile strength	–
Hardness	–

45.1 General notes on strontium

Strontium is always found as a compound, the sulphate or the carbonate being the most common, generally associated with lead ores. It was first isolated in 1790 in the lead mines then worked near the village of Strontian in the north-west of Scotland.

The metal is yellowish white when freshly cut, but tarnishes rapidly and decomposes water, releasing hydrogen. Strontium resembles calcium and as it has no unique properties justifying extraction on a commercial scale, there are therefore no alloys based on the metal.

It exhibits photo-electric properties, but it cannot compete economically with selenium. A limited amount of strontium has been used to de-oxidize copper and some of its alloys, but the high cost of the metal makes this an uneconomical method. Lead is hardened when traces of strontium are added along with small amounts of tin.

Strontium compounds are used in small quantities in pharmacy and the electronic industry. They have been added to open hearth furnace brick work and it is claimed that lower sulphur steels are obtained. The strong red colour given off when compounds are ignited is used in fireworks and distress or identification flares. Strontium sulphate is a stable powder which can be ground to a very fine flour and finds some use as a filler material for plastics and rubber.

Strontium is an alkaline earth metal which has no real value to mankind, but it has a radioactive isotope which is a decay product of materials produced by nuclear fission and is in the fallout. As this particular isotope takes a long time to decompose (has a long half-life) and is absorbed into the bone and stored there, it constitutes a danger to health.

Note. The following abbreviations and units are used in the tables:

DPN	Hardness, diamond pyramid number
UTS	Ultimate tensile strength, N/mm^2
Elon	Elongation, %
Proof	0.1% proof strength, N/mm^2

1 N/mm^2=0.1 hbar=0.102 kgf/mm^2=0.06475 tonf/in.^{2}=145.04 lbf/in.^{2}=1 MPa

See Appendix II for other abbreviations and conversion tables.

Symbol	Nominal analysis, supplier, condition and remarks.
STRONTIUM	Pure metal: Blackwells

46. Tantalum Ta

Physical properties		
Atomic number		73
Atomic weight		181.0
Crystal structure		Body-centred cubic
Colour		Iron grey
Specific gravity		16.65
Density		16 650 kg/m^3
Melting point		2950 °C
Boiling point		5300 °C
Specific heat		0.153 J/g °C
Thermal conductivity		54.4 W/m °C
Coefficient of linear expansion (20–100 °C)		6.5 × 10^{-6}/ °C
Latent heat of fusion		159 J/g
Latent heat of vaporization		–
Thermal neutron absorption cross-section		21.3 barns/atom
Electrical conductivity		13.9% IACS (copper 100%)
Specific resistance		125 microhm mm
Temperature coefficient of electrical resistance		0.0031 °C
Electrochemical equivalent		–
Electrode potential		4.1 V
Magnetic susceptibility		0.90 × 10^{-6}
Young's modulus of elasticity	cold worked	186 × 10^9 N/m^2
Tensile strength	annealed	450 N/mm^2
	cold worked	900 N/mm^2
Hardness	annealed	100 DPN
	cold worked	200 DPN

46.1 General notes on tantalum

This occurs as an oxide in combination with iron, always associated with the niobium–iron oxide.

Metal tantalum of 99.9% purity is now commercially available but it is still an expensive material and will remain so as long as the present scarcity of workable ores exists.

The basic product is a powder which must be sintered and forged under vacuum. With modern techniques of vacuum casting and electron beam melting, a product capable of further hot reduction is available which may prove more economical than vacuum sintering.

Tantalum is a white colour as polished, but in air immediately forms an adherent oxide which gives the metal an iron grey coloration.

It is capable of accepting considerable cold work without hardening to any great extent, and thus does not require repeated anneals. As tantalum combines with most gases in contact with it from 500 °C upwards, all heating must be in vacuum. None of the inert gases combine with the metal but it is always necessary to produce a high vacuum to remove atmospheric gases which would contaminate the inert gases. Tantalum can be readily welded provided adequate flow of an inert gas (generally argon or helium) is available on both sides of the weld, as well as fore and aft the weld pool. Best results are obtained by vacuum welding or electron beam welding if this is possible. Tantalum castings can be produced, provided melting, casting and cooling take place under vacuum.

Tantalum has excellent resistance to chemical attack from all acids except hydrofluoric and hot concentrated sulphuric acids. It has reasonable electrical properties and maintains these at a high temperature. Oxidation resistance is very good up to 450 °C when it falls off rapidly, but above 500 °C tantalum must be used under vacuum or one of the inert gases.

This chemical resistance together with an excellent heat transfer and its workability makes tantalum a favourite metal in the chemical industry, there being no doubt that only the high cost of the metal restricts its use at present. One method of reducing the initial capital outlay is to use a thin coating of tantalum on a strong cheaper base metal. Austenitic stainless steel is often used and electroplating techniques are being developed for this purpose.

Tantalum has various uses in the electronic industry, partly for its high strength at elevated temperatures where it is used as anode and grid supports in valves. The ability to combine with atmospheric gases makes it a popular, if expensive, 'getter' to remove the last traces of these gases from the vacuum glass envelope.

Tantalum also has peculiar properties connected with its oxide film which make it a useful material in the manufacture of capacitors and rectifiers.

Vacuum furnace equipment sometimes uses tantalum for any shields or supports which require considerable shaping and welding during manufacture and high strength at the elevated working temperatures.

The good corrosion characteristics and proven compatibility with body fluids allows the use of tantalum implants in surgery.

Tantalum is similar in many respects to niobium, and like this metal has its use restricted by the high price and scarcity.

Alloyed with stainless steels and nickel alloys it prevents carbide precipitation (weld decay) but is generally second choice to niobium. A series of sintered carbide tool materials includes tantalum, tungsten and titanium carbides, generally bound in a nickel or cobalt matrix. Tantalum has been used as an alloying element in high speed steels, but has little to offer over tungsten and is more expensive.

Some tantalum compounds are used as catalysts in chemical engineering, particularly in the manufacture of certain synthetic rubbers.

Symbol	Nominal analysis, supplier, condition and remarks.
AMS 7846	0.01% C max 10% W Ta alloy
AMS 7847	0.01% C max 10% W Ta alloy
AMS 7848	10% W Ta alloy: Bar
AMS 7849	Ta: Sheet, strip or foil
ASTM B364	High purity Ta: Ingot and sheet; cold worked **UTS: 530 Elon: 2%**
ASTM B364	High purity Ta: Ingot and sheet; annealed **UTS: 220 Elon: 27%**
ASTM B364	High purity Ta: Ingot and sheet; stress relieved **UTS: 370 Elon: 8%**
ASTM B365	High purity Ta: Rod and wire; cold worked **UTS: 500 Elon: 1%**
ASTM B365	High purity Ta: Rod and wire; annealed **UTS: 270 Elon: 10%**
h Ta 16	99.98% Ta: Ingot; Light Ltd; high purity metal
h Ta 73	99.95% Ta: Single crystal 3 mm diameter; Light Ltd
METCO 62	Pure Ta: spray deposit; Metco
RO 5200	High purity Ta: UNS designation
RO 5240	40% Nb Ta alloy: UNS designation
RO 5252	2.7% W Ta alloy: UNS designation
RO 5255	10% W Ta alloy: UNS designation
RO 5400	High purity Ta: UNS designation
T 222	0.01% C 9.6% W 2.2% Ha Ta alloy: Westinghouse Electrical Corporation
TANTALUM	99.6% Ta: powder; Kennametal
TANTALUM	Ta metal: Blackwells; pure metal
TANTALUM	Ta: Ingot, rod, sheet and wire; Murex
TANTALUM	Ta: Rod, sheet and wire; Tungsten Manufacturing

Note. The following abbreviations and units are used in the tables:

DPN	Hardness, diamond pyramid number
UTS	Ultimate tensile strength, N/mm^2
Elon	Elongation, %
Proof	0.1% proof strength, N/mm^2

1 N/mm^2=0.1 hbar=0.102 kgf/mm^2=0.06475 tonf/in.^{2}=145.04 lbf/in.^{2}=1 MPa

See Appendix II for other abbreviations and conversion tables.

47. Tellurium Te

Physical properties	
Atomic number	52
Atomic weight	127.61
Crystal structure	–
Colour	Shining white
Specific gravity	6.24
Density	6240 kg/m^3
Melting point	452 °C
Boiling point	1007 °C
Specific heat	0.201 J/g °C
Thermal conductivity	1.3 W/m °C
Coefficient of linear expansion (20–100 °C)	16.8×10^{-6}/ °C
Latent heat of fusion	30.6 J/g
Latent heat of vaporization	666 J/g
Thermal neutron absorption cross-section	4.7 barns/atom
Electrical conductivity	1.0% IACS (copper 100%)
Specific resistance	1500 microhm mm
Temperature coefficient of electrical resistance	–
Electrochemical equivalent	0.99 g/A/h
Electrode potential	–
Magnetic susceptibility	-0.31×10^{-6}
Young's modulus of elasticity	–
Tensile strength	–
Hardness	–

47.1 General notes on tellurium

Natural tellurium occurs in small amounts and is fairly widespread in complex gold, silver, lead, antimony, bismuth and copper compounds. Tellurium is a metalloid very similar to selenium and like all metalloids has non-metallic characteristics, including poor heat and electrical conductivity. It has few uses in the pure state and there is no alloy system based on tellurium. Some use has been made of the material in the vapour form in lamps when it is claimed artificial sunlight can be obtained. In the finely divided form tellurium is used as an additive to rubber to improve the oxidation resistance.

When electroplated on steel, tellurium resists the attack of most mild acids and alkalis and as the plating technique appears to be relatively simple there is considerable potential use in this field.

In small quantities of less than 1% it considerably improves the machinability of copper and can be used for the same purpose in steel where the effect is identical to that of selenium.

Tellurium improves the mechanical properties and corrosion resistance of lead and has been added to lead and tin base bearing metals for the same reasons. In cast irons minute quantities of about 0.05% (maximum) tellurium inhibit graphite formation, thus increasing the hardness. It is often used in mould washes to cause chill hardening of the cast surfaces, as the surface carbon is then present as carbides instead of graphite.

Light has no effect on the electrical resistance of tellurium, thus unlike selenium it has no use in light meters. Some of its compounds however, particularly bismuth telluride, are finding use in the electronic industry as semiconductors.

Tellurium and most of its compounds are to a degree toxic, but to a lesser extent than selenium compounds. There is never sufficient tellurium present in metallic alloys in common use to present any health hazard whatever.

There are no alloy systems based on the metal.

Note. The following abbreviations and units are used in the tables:

DPN	Hardness, diamond pyramid number
UTS	Ultimate tensile strength, N/mm^2
Elon	Elongation, %
Proof	0.1% proof strength, N/mm^2

1 N/mm^2=0.1 hbar=0.102 kgf/mm^2=0.06475 $tonf/in.^2$=145.04 $lbf/in.^2$=1 MPa
See Appendix II for other abbreviations and conversion tables.

Symbol	Nominal analysis, supplier, condition and remarks.
h Te 6	99.999% Te metal: Lump form; Light Ltd; high purity metal
h Te 73	99.995% Te single crystals, 6 mm diameter: Light Ltd
TELLURIUM	High purity metal: impurities 5 p.p.m.; Johnson Matthey; supplied as cast bar

48. Thallium Tl

Physical properties	
Atomic number	81
Atomic weight	204.39
Crystal structure	Close-packed hexagonal to 262 °C Body-centred cubic above 262 °C
Colour	White
Specific gravity	11.85
Density	11 850 kg/m^3
Melting point	304 °C
Boiling point	1450 °C
Specific heat	0.136 J/g °C
Thermal conductivity	39.4 W/m °C
Coefficient of linear expansion (20–100 °C)	28×10^{-6}/ °C
Latent heat of fusion	20.9 J/g
Latent heat of vaporization	1231 J/g
Thermal neutron absorption cross-section	3.4 barns/atom
Electrical conductivity	10% IACS (copper 100%)
Specific resistance	150 microhm mm
Temperature coefficient of electrical resistance	0.0051/ °C
Electrochemical equivalent	2.5 g/A/h
Electrode potential	−0.336 V
Magnetic susceptibility	-0.24×10^{-6}
Young's modulus of elasticity	–
Tensile strength	7.5 N/mm^2
Hardness	2 DPN

48.1 General notes on thallium

Thallium has a high specific gravity and resembles lead in colour and hardness, being readily cut by a knife. It oxidizes if left in air and slowly decomposes water, and thus must be stored under paraffin or alcohol.

The use of thallium and many of its compounds is severely restricted by its toxic nature, added to the ease with which it oxidizes.

Alloys with silver and lead up to about 20% thallium have been developed which appear to have excellent corrosion resistance under certain conditions. Silver with 10–20% thallium does not blacken in contact with sulphide compounds as do most other silver alloys.

The lead–thallium alloys have higher melting points than other lead alloys and have been used for heat fuses, while thallium added to lead/tin bearing alloys considerably improves the corrosion resistance and increases the hardness. Claims have been made that a tin/thallium/lead based anode has considerably less corrosion attack than the antimony/lead anodes normally used for chromium plating.

Some thallium compounds are sensitive to infra-red light and find some uses in equipment analogous to photo-electric cells using selenium in normal light.

In general thallium is a little used metal whose potential, if any, is severely restricted by the toxic nature of the metal and its compounds.

There are no alloy systems based on thallium.

Note. The following abbreviations and units are used in the tables:

DPN	Hardness, diamond pyramid number
UTS	Ultimate tensile strength, N/mm^2
Elon	Elongation, %
Proof	0.1% proof strength, N/mm^2

1 N/mm^2=0.1 hbar=0.102 kgf/mm^2=0.06475 tonf/in.^{2}=145.04 lbf/in.^{2}=1 MPa

See Appendix II for other abbreviations and conversion tables.

Symbol	Nominal analysis, supplier, condition and remarks.
THALLIUM	High purity metal: Impurities 5 p.p.m.; Johnson Matthey; supplied as wire in sealed containers
THALLIUM	Pure metal: Blackwells
w Th Tl 18	99.9% Th Tl powder: Light Ltd

49. Thorium Th

Physical properties

Atomic number	90
Atomic weight	232.12
Crystal structure	Face-centred cubic
Colour	–
Specific gravity	11.3
Density	11 300 kg/m^3
Melting point	1800 °C
Boiling point	3500 °C
Specific heat	0.116 J/g °C
Thermal conductivity	37.7 W/m °C
Coefficient of linar expansion (20–100 °C)	11.1 × 10^{-6}/ °C
Latent heat of fusion	82.9 J/g
Latent heat of vaporization	2617 J/g
Thermal neutron absorption cross-section	–
Electrical conductivity	14% IACS (copper 100%)
Specific resistance	130 microhm mm
Temperature coefficient of electrical resistance	0.0038/ °C
Electrochemical equivalent	–
Electrode potential	−2.1 V
Magnetic susceptibility	0.11 × 10^{-6}
Young's modulus of elasticity	73.1 × 10^9 N/m^2
Tensile strength	200 N/mm^2
Hardness	35 DPN

49.1 General notes on thorium

Thorium does not exist in the native state but the oxide is fairly widespread and deposits of the silicate are also known. The sand monazite contains thorium oxide along with the rare earth metals and this is the principal source. Thoria (thorium oxide) crucibles are used in the extraction and working of many of the refractory metals as it withstands temperatures up to 2500 °C.

High purity thorium can now be produced and there are indications that this will result in cheaper metal of a purity capable of accepting cold work without becoming too brittle.

Thorium is greyish white and quite soft. Its freshly cut surfaces rapidly oxidize but this prevents further oxide formation. It does not ignite spontaneously but will continue burning in air, forming the very white powder thoria – thorium oxide. The fact that thorium is radioactive restricts its widespread use and its low mechanical strength reduces its value in nuclear engineering where most of the metal is used. It finds some use as a substitute for radium as it is considerably cheaper to produce, although not so radioactive. Thorium is also used in X-ray equipment and as a source of radioactivity for non-destructive flaw detection. Certain types of photo-electric cells use thorium, but the demand for the metal at present is very low.

There are no alloys based on thorium but a range of magnesium alloys exists with 2% thorium which have good ductility and improved high temperature properties while retaining the tensile strength of other magnesium alloys.

As previously stated the oxide thoria is a valuable refractory and considerable quantities at one time were used to make gas mantles. This was the highly purified form as the white crystalline structure gave the desired incandescent effect.

There are only limited uses for the metal thorium and these are not likely to increase much even with a reduction in the present high cost of the metal. The low mechanical properties coupled with the radioactivity account for this unpopularity.

Note. The following abbreviations and units are used in the tables:

DPN	Hardness, diamond pyramid number
UTS	Ultimate tensile strength, N/mm^2
Elon	Elongation, %
Proof	0.1% proof strength, N/mm^2

1 N/mm^2=0.1 hbar=0.102 kgf/mm^2=0.06475 tonf/in.^{2}=145.04 lbf/in.^{2}=1 MPa

See Appendix II for other abbreviations and conversion tables.

Symbol	Nominal analysis, supplier, condition and remarks.
THORIUM	Pure metal: Blackwells

50. Tin Sn

Physical properties		
Atomic number		50
Atomic weight		118.7
Crystal structure	alpha	Tetrahedral cubic – 18 °C
	beta	Body-centred tetragonal – 18–161 °C
	gamma	Close-packed hexagonal – 161 °C
Colour		Bright white
Specific gravity	alpha	5.81
	beta	7.29
	gamma	6.56
Density	alpha	5810 kg/m^3
	beta	7290 kg/m^3
	gamma	6560 kg/m^3
Melting point		231.84 °C
Boiling point		2270 °C
Specific heat		0.234 J/g °C
Thermal conductivity		63.2 W/m °C
Coefficient of linear expansion (20–100 °C)		23.8×10^{-6}/ °C
Latent heat of fusion		59.5 J/g
Latent heat of vaporization		2399 J/g
Thermal neutron absorption cross-section		0.65 barns/atom
Electrical conductivity		14% IACS (copper 100%)
Specific resistance		1150 microhm mm
Temperature coefficient of electrical resistance		0.0044/ °C
Electrochemical equivalent		–
Electrode potential		−0.14 V
Magnetic susceptibility		-0.25×10^{-6}
Young's modulus of elasticity		41.4×10^9 N/m^2
Tensile strength		220 N/mm^2
Hardness		–

50.1 General notes on tin

Tin is never found in the native state but deposits of the oxide ore – called tinstone – are widespread throughout the world. These deposits are seldom large and it is becoming progressively harder to win the ore.

Metal of 99.75–99.99% purity is commonly available, the most undesirable elements being lead, arsenic and bismuth, all of which embrittle tin.

The metal is silver-white and although it rapidly loses its original lustre it does not readily tarnish or discolour in air. It is a soft, ductile metal, being readily cut by a knife, and in the pure state is not subject to cold work. When block tin or tin pipes are bent a characteristic noise known as 'tin creak' is heard.

Tin is capable of being rolled into very thin sheets, but owing to its low strength cannot be drawn into thin wire.

The present high cost of tin prevents it being used for many purposes for which it is eminently suitable. The complete freedom from toxicity and its extreme ease of manipulation make it ideal as food and drink containers and at one time water pipes and 'silver' paper were made in pure tin.

The largest single use of tin is to coat steel sheet which is subsequently used to manufacture the well-known 'tins' used to store all types of food and drink. Although plastics and, to a greater extent, aluminium are now entering this field, tin plate remains an economical and one of the commonest means of permanently storing foods of all kinds.

Tin can be applied to sheet steel, cast iron, aluminium, copper and its alloys generally by dip tinning after cleaning and fluxing, but also by electroplating and metal spraying techniques.

The low melting point, high fluidity and good wetting properties ensure excellent coverage by dip tinning, while the position of tin in the electrochemical series ensures that it sacrifices itself to protect the base metal if there is any break in the coating. Tin has comparable corrosion resistance in most atmospheres to zinc and cadmium without any danger from toxicity.

Apart from use in food containers, tin is used to coat copper conductor wires to prevent corrosion and ease the soldering of connections. Pure tin is also used to coat bearing shells and moulds prior to white metalling, and electrical connections of all types before soldering.

A limited quantity of pure tin piping and foil is still used for piping beer and wrapping foods, such as certain cheeses which corrode aluminium foil, but in general stainless steel and plastics have taken over in these fields.

About 50% of all tin mined is used for these purposes, while the remainder goes to the many alloys which use tin. The tin based alloys are described in the section following, but in addition it is valuable as a minor alloying element, particularly in conjunction with copper. The copper–tin alloys take the generic name 'bronze' and are used where bearing properties, strength and corrosion resistance are required. The alloys are discussed in more detail under Section 14, particularly Section 14K.

Tin is also commonly found in lead alloys where it is used to strengthen. Small quantities increase the creep strength of lead while the lead–tin solders cover a wide range of analysis for a variety of purposes relating to melting temperatures, corrosion resistance and strength. In general the higher the tin content, the higher duty the alloys.

Tin is also added to lead based bearing alloys where it increases the load bearing properties and reduces the frictional coefficient.

It is of interest that tin added to lead almost always has a beneficial effect, whereas lead added to tin results in deleterious embrittlement and reduction in other properties.

Some zinc–tin and nickel–tin alloys are available which find most use as electrodeposits where an attractive silver-type lustre is desired. The nickel–tin deposit can have the analysis adjusted to give a bright plate which requires very little polishing. These deposits are relatively expensive.

There is no doubt that tin would find considerably more use if the price could be reduced. This is unlikely as tin is readily and economically obtained from its ores which are scarce and difficult to win. Tin has been used by mankind for centuries and the discovery of new easily obtained ore deposits is therefore highly improbable.

Following are listed the brand names and available information on pure tin.

Symbol	Nominal analysis, supplier, condition and remarks.
ASTM B339	99–99.8% Sn: Pig; primary metal; covers 5 grades
BANKA	0.03% Pb 0.006% Cu 0.014% As 99.935% Sn: Brand name; further information from Tin Research Institute
BS 3252 T1	99.9% Sn: Ingot; high purity tin
BS 3252 T2	99.75% Sn: Ingot; refined tin
BS 3252 T3	99.0% Sn: Ingot; common tin
CORNISH REFINED	0.065% Pb 0.34% Cu 0.03% As 99.82% Sn: Brand name; further information from Tin Research Institute
E S COY Ltd	0.03% Pb 0.012% Cu 0.031% As 99.899% Sn: Brand name; further information from Tin Research Institute
HAWTHORNE	0.028% Pb 0.028% Cu 0.029% As 99.891% Sn: Brand name; further information from Tin Research Institute
HERBERTON	0.18% Pb 0.03% Cu 0.037% As 99.715% Sn: Brand name; further information from Tin Research Institute
L 13002	99.9% Sn: UNS designation
L 13004	99.8% Sn: UNS designation
L 13006	99.8% Sn: UNS designation
L 13007	99.8% Sn: UNS designation
L 13008	99.8% Sn: UNS designation
L 13010	99.65% Sn: UNS designation
L 13012	99.5% Sn: UNS designation
L 13014	99% Sn: UNS designation
LAMB & FLAG	0.38% Pb 0.025% Cu 0.02% As 99.56% Sn: Brand name; further information from Tin Research Institute
LAMB & FLAG COMMER	0.763% Pb 0.03% Cu 0.029% As 99.128% Sn: Brand name; further information from Tin Research Institute
LONGHORN 3 STAR	0.04% Pb 0.03% Cu 0.024% As 99.86% Sn: Brand name; further information from Tin Research Institute
M & T ELECTROLYTIC	0.008% Pb 0.001% Cu 99.98% Sn: Brand name; further information from Tin Research Association
M & T No 1	0.01% Pb 0.001% Cu 0.016% As 99.96% Sn: Brand name; further information from Tin Research Institute
MELLANEAR GUAR 99.9	0.037% Pb 0.009% Cu 0.008% As 99.915% Sn: Brand name; further information from Tin Research Institute
MELLANEAR REFINED	0.065% Pb 0.022% Cu 0.025% As 99.827% Sn: Brand name; further information from Tin Research Institute
METCO TIN	Sn spray deposit: Metco; suitable for food use
PASS No 1	0.001% Pb 0.0002% Cu 0.003% As 99.975% Sn: Brand name; further information from Tin Research Institute
PYRMANT	0.048% Pb 0.022% Cu 0.043% As 99.85% Sn: Brand name; further information from Tin Research Institute
QQT 371	99.9% Sn: US Federal specification
REGIS	0.011% Pb 0.001% Cu 0.004% As 99.972% Sn: Brand name; further information from Tin Research Institute
ROSE	0.002% Pb 0.012% Cu 99.955% Sn: Brand name; further information from Tin Research Institute
TIN	Sn: Ingot, sheet, sponge, etc.; Blackwells; primary metal
TOYO	0.04% Pb 0.03% Cu 0.03% As 99.8% Sn: Brand name; further information from Tin Research Institute
TULIP	0.086% Pb 0.01% Cu 0.008% As 99.87% Sn: Brand name; further information from Tin Research Institute
UMHK	0.12% Pb 0.013% Cu 0.006% As 99.966% Sn: Brand name; further information from Tin Research Institute
ZTM	0.09% Pb 0.04% Cu 0.063% As 99.78% Sn: Brand name; further information from Tin Research Institute

Note. The following abbreviations and units are used in the tables:

DPN	Hardness, diamond pyramid number
UTS	Ultimate tensile strength, N/mm^2
Elon	Elongation, %
Proof	0.1% proof strength, N/mm^2

1 N/mm^2=0.1 hbar=0.102 kgf/mm^2=0.06475 $tonf/in.^2$=145.04 $lbf/in.^2$=1 MPa

See Appendix II for other abbreviations and conversion tables.

50A Tin alloys

Specific gravity	7.35–7.95
Density	7350–7950 kg/m^3
Solidus/liquidus	185–370 °C
Thermal conductivity	–
Coefficient of linear expansion	–
Electrical conductivity	–
Specific resistance	–
Young's modulus of elasticity	–
Impact	–
Hot strength	

Temperature °C	Tensile strength N/mm^2
20	31
100	16
150	9

The above properties are typical of the following group, and may not apply exactly to any one specification. It is possible that with certain specifications some of the values may not be applicable.

General metallurgical characteristics

Tin based alloys are expensive and thus are only used when their properties cannot be reproduced using cheaper metals or other methods. Both the principal uses of the tin alloys also employ lead based alloys which are much cheaper. The first and greatest use of tin in the alloy form is as solder. This always has some lead present and may have other elements added for specific purposes.

In general the high tin solders have better corrosion resistance and are stronger than the high lead varieties. Silver increases the strength at temperature. The temperature range over which solders melt is of considerable importance as this decides the workability of the metal. Where high strength, mass produced articles are joined by soldering, a short solidification range is generally desirable to prevent movement and give speedy joins. Where it is necessary to work the solder during solidification, as with plumbers' or electricians' wiped joints, a long slow solidification range is essential. The alloy content decides the liquidus/solidus points and hence the choice of alloy for any particular use.

The second largest use for tin alloys is as a bearing metal where it again has lead as a second best competitor. Tin bearing alloys must be low in lead to prevent embrittlement, and generally have under 0.5% lead with up to 10% antimony and 5% copper.

Tin has an excellent frictional coefficient, and even when alloyed is soft enough to accept grit particles, which become embedded and covered with tin, causing little or no scoring. There appears to be some affinity between tin and oil which considerably aids the tin bearing metals and allows them to run for longer periods in oil-starved conditions than most other bearing alloys.

There is also a considerable increase in temperature characteristics over the lead bearing alloys. Except for low load high speed duties, the tin bearing alloys must always be supported by a stronger metal shell on to which they are cast. The bond of tin alloy to steel or bronze must be excellent to prevent exfoliation and failure by plucking. Tin alloy bearings are now being replaced with new design bearings and alternative materials.

PTFE and PTFE impregnated bearings, sinter materials and nylon are all competing with tin alloy bearings.

It is not possible to improve the mechanical properties of any alloys listed by any form of thermal treatment.

Some of the alloys can accept considerable quantities of cold work without any hardening while others will become brittle with cold working.

Joining of these materials in general presents no problems as they can be soldered or brazed or even technically welded without any problem using low temperatures and correct fluxes.

All traces of flux must be removed prior to going into service.

These materials are difficult to machine because they are soft and tear rather than cut.

Use of the tin alloys at present is as solders. Many bearing metals are still tin-based, particularly where high duty is involved.

Some of the alloys are still used because of their non-toxic properties and, having a slightly higher strength than tin itself, for foodstuff.

The high cost of tin and its general scarcity precludes this as a material in common use.

Symbol	Nominal analysis, supplier, condition and remarks.
2.3820	11% Zn Sn alloy: Working temperature 210 °C; German Standard
2.3830	40% Zn Sn alloy: Working temperature 260 °C; German Standard
2.3852	45% Pb 10% Zn + Cd Sn alloy: Working temperature 220 °C; German Standard
A 85	Tin base white metal: For diesel engine bearings; Anti-Attrition Ltd
A 89	Tin base white metal: For marine petrol engine bearings; Anti-Attrition Ltd
A 127	Tin base white metal: For electric motor bearings; Anti-Attrition Ltd
A 129	Tin base white metal: For electric motor bearings; Anti-Attrition Ltd
A 139	Tin base white metal: For railway carriage bearings; Anti-Attrition Ltd
ADASTRAL 'A'	3% Cu 4.5% Sb Sn alloy: Bearing metal; Stone Manganese for BS 3332/1
AERO No 2	Sn base Pb free bearing metal: High load; high speed; Magnolia Anti-Friction Metal Co.
ALGIERS METAL	10% Sb 0.3% Cu Sn alloy: Source unknown
ALICIA	Sn base bearing metal: Medium duty; Magnolia Anti-Friction Metal Co.
AMS 4751	37% Pb Sn solder
AMS 4800 A	4.5% Cu 4.5% Sb Sn alloy: Bearing metal; 'Babbit'
ARGENTINE METAL	14.5% Sb Sn alloy: For toys; source unknown
ARGENTINE METAL	15% Sb Sn alloy
ARMA	Sb Cu Sn base bearing metal: For high duty loads; Phosphor Bronze Co.
ARSENIC-TIN	5% As Sn alloy: Blackwells; primary metal
ASHBERRY	16% Sb 3% Cb 2% Zn Sn alloy: White metal; information from Tin Research Institute
ASTM B23/1	4.5% Sb 4.5% Cu 0.3% Pb Sn alloy: Bearing metal; melting point 223 °C; compressive strength 90 N/mm^2 **DPN: 17**
ASTM B23/2	7.5% Sb 0.3% Pb 3.5% Cu Sn alloy: Bearing metal; melting point 241 °C; compressive strength 100 N/mm^2 **DPN: 24**
ASTM B23/3	8% Sb 0.3% Pb 8% Cu Sn alloy: Bearing metal; melting point 240 °C; compressive strength 120 N/mm^2 **DPN: 27**
ASTM B32/50 A	50% Pb Sn: Solder; melting range 183–216 °C
ASTM B32/50 B	50% Pb 0.3% Sb Sn: Solder; melting range 183–216 °C
ASTM B32/60 A	40% Pb Sn: Solder; melting range 183–190 °C
ASTM B32/60 B	40% Pb 0.3% Sb Sn: Solder; melting range 183–190 °C
ASTM B32/63 A	37% Pb Sn: Solder (eutectic solder); melting point 183 °C
ASTM B32/63 B	37% Pb 0.3% Sb Sn: Solder (eutectic solder); melting point 183 °C
ASTM B32/70 A	30% Pb Sn: Solder; melting range 183–192 °C
ASTM B32/70 B	30% Pb 0.3% Sb Sn: Solder; melting range 183–192 °C
ASTM B32/95 TA	5% Sb Sn: Solder; melting range 234–240 °C
ASTM B102 CY44A	4.5% Sb 4.5% Cu Sn alloy: Bearing metal
ASTM B102 PY1815A	15% Sb 18% Pb 2% Cu Sn alloy: Bearing metal

Note. The following abbreviations and units are used in the tables:

DPN	Hardness, diamond pyramid number
UTS	Ultimate tensile strength, N/mm^2
Elon	Elongation, %
Proof	0.1% proof strength, N/mm^2

1 N/mm^2=0.1 hbar=0.102 kgf/mm^2=0.06475 tonf/in.^{2}=145.04 lbf/in.^{2}=1 MPa

See Appendix II for other abbreviations and conversion tables.

Symbol	Nominal analysis, supplier, condition and remarks.
ASTM B102 YC135A	13% Sb 5% Cu Sn alloy: Bearing metal
ASTM B774–281/338	40% Bi Sn alloy: Melting range 138–170 °C
ASTM B774–291/325	14% Bi 42% Pb Sn alloy: Melting range 144–163 °C
ASTM B774–293	30% Pb 18% Cd Sn alloy: Melting point 145 °C
ASTM B774–307/323	18% Pb 12% In Sn alloy: Melting range 153–162 °C
ASTM F67	Ti-wrought pure metal for surgical implants: Graded by purity
ATLAS ADMIRALTY 'A'	Zn Sn base bearing alloy: For stem tube bearings; Eyre Smelting Co.; melting range 200–425 °C; compressive strength 60 N/mm^2 **DPN: 30 UTS: 75**
ATLAS ADMIRALTY 'B'	Sn base Pb free bearing metal: For steam turbines; Eyre Smelting Co.; melting range 240–325 °C; compressive strength 70 N/mm^2 **DPN: 31 UTS: 90**
ATLAS AMACOL	Pb Sn base bearing alloy: For reciprocating engines; Eyre Smelting Co.; melting range 190–370 °C; compressive strength 70 N/mm^2 **DPN: 33 UTS: 90**
ATLAS D8	Sn base bearing metal: For diesels; Eyre Smelting Co.; melting range 240–405 °C; compressive strength 70 N/mm^2 **DPN: 35 UTS: 90**
ATLAS DD	Cu Sn base Pb free bearing metal: For diesels; Eyre Smelting Co.; melting range 240–370 °C; compressive strength 70 N/mm^2 **DPN: 32 UTS: 80**
ATLAS DG	Sn base bearing metal: For diesels; Eyre Smelting Co.; melting range 230–376 °C; compressive strength 70 N/mm^2 **DPN: 33 UTS: 90**
ATLAS FDS	Sn base Pb free bearing metal: For steam turbines; Eyre Smelting Co.; melting range 240–330 °C; compressive strength 70 N/mm^2 **DPN: 32 UTS: 90**
ATLAS INFRANGA	Sn base Pb free bearing metal: For high speed diesels, Eyre Smelting Co.; melting range 240–330 °C; compressive strength 70 N/mm^2 **DPN: 32 UTS: 90**
ATLAS TENAXAS	Pb Sn base bearing alloy: For reciprocating engines; Eyre Smelting Co.; melting range 186–340 °C; compressive strength 70 N/mm^2 **DPN: 30 UTS: 80**
AUTO 'C'	4% Pb 11% Sb 4% Cu Sn alloy: Bearing metal; Stone Manganese for BS 3332/3
BABBITT	Sb Cu Pb Sn white metal bearing alloys: Information from Tin Research Institute
BABBITT No 1	Sn base bearing metal: For high speed use; Eyre Smelting Co.; melting point 376 °C **DPN: 33 UTS: 90 Elon: 5%**
BRITTANIA	7% Sb 2% Cu Sn alloy: White metal; information from Tin Research Institute
BS 2B 21	4% Cu 9% Sb Sn alloy: White metal; bearing metal
BS 2B 22	4% Cu 4% Sb 0.3% Ni Sn alloy: White metal; bearing metal
BS 219/95 A	5% Sb 0.07% Pb Sn alloy: Solder; melting range 236–243 °C
BS 219A	35% Pb 0.6% Sb Sn alloy: Solder; melting range 183–185 °C
BS 219B	47% Pb 2.7% Sb Sn alloy: Solder; melting range 185–204 °C
BS 219F	50% Pb 0.5% Sb (max) Sn alloy: Solder; melting range 183–212 °C
BS 219K	40% Pb 0.5% Sb (max) Sn alloy: Solder; melting range 183–188 °C
BS 3332/1	7% Sb 3% Cu Sn alloy: Bearing metal
BS 3332/2	9% Sb 4% Cu Sn alloy: Bearing metal
BS 3332/3	10% Sb 5% Cu 4% Pb Sn alloy: Bearing metal
BS 3332/4	12% Sb 3% Cu 10% Pb Sn alloy: Bearing metal
BS 3332/5	7% Sb 3% Cu 15% Pb Sn alloy: Bearing metal

Symbol	Nominal analysis, supplier, condition and remarks.
BS 3332/6	10% Sb 3% Cu 27% Pb Sn alloy: Bearing metal
BS 3332/9	1.5% Cu 30% Zn Sn alloy: Bearing metal
CERROBEND	Bi Cd Pb Sn alloy: Casting filler metal for tube bending: Mining & Chemical Products Ltd; melting point 70 °C; expands on solidification
CERROMATRIX	Bi Sb Pb Sn alloy: Casting for short run dies; Mining & Chemical Products Ltd; melting range 102–227 °C; expands on solidification
CERROSEAL - 35	50% Tn Sn alloy: Mining & Chemical Products Ltd; melting range 117–127 °C; for fusible links
CERROTRU	Bi Sn alloy: Casting for pattern metal; Mining & Chemical Products Ltd; melting point 137.5 °C; no volume changes on solidification
CONDENSER FOIL	15% Pb 2% Sb Sn alloy: Used as the dielectric in electronic industry; origin unknown
CY 44A	4.5% Sb 4.5% Cu Sn alloy: Designation used by ASTM
DE	Sn base Pb free bearing metal: High load; high speed; Magnolia Anti-Friction Metal Co.
DIN 1704	99.0% Sn: Supplied in four grades
DIN 1707 L Sn 50	46% Pb 3.3% Sb Sn alloy
DIN 1707 L Sn 60	35% Pb 3.5% Sb Sn alloy
DIN 1707 L Sn 90	8% Pb 1.3% Sb Sn alloy
DIN 1742 Sg Sn 50	4% Cu 33% Pb 13% Sb Sn alloy **DPN: 26 UTS: 45 Elon: 2%**
DIN 1742 Sg Sn 60	4% Cu 23% Pb 13% Sb Sn alloy **DPN: 28 UTS: 45 Elon: 1.5%**
DIN 1742 Sg Sn 70	5% Cu 10% Pb 15% Sb Sn alloy **DPN: 30 UTS: 100 Elon: 1%**
DIN 1742 Sg Sn 75	5% Cu 3% Pb 17% Sb Sn alloy **DPN: 30 UTS: 100 Elon: 2%**
DIN 1742 Sg Sn 78	4% Cu 1% Pb 17% Sb Sn alloy **DPN: 30 UTS: 180 Elon: 3%**
DIN 8512 L Sn Pb Zn	45% Pb 10% Zn + Cd Sn alloy: Working temperature 220 °C
DIN 8512 L Sn Zn 10	11% Zn Sn alloy: Working temperature 210 °C
DIN 8512 L Sn Zn 40	40% Zn Sn alloy: Working temperature 260 °C
DIN L Sn Ag 5	Ag Sn welders and brazers electrode
DSYL	7% Cu 9% Sb 0.3% Pb Sn alloy: Bearing metal; Stone Manganese for BS 3332/2
DTD 214	3.5% Cu 7% Sb 5% Cb Sn alloy: White metal; suitable for bearings
DTD 244	6% Cu 6.5% Sb 0.5% Ni Sn alloy: White metal; suitable for bearings
h Sn 1	99.9999% Sn: Bar; Light Ltd; high purity metal
h Sn 73	99.99% Sn: Single crystal; 25 mm × 25 mm; Light Ltd
HOYT 11D	Sb Cu Sn base bearing metal: Hoyt Ltd
HOYT 11R	Sn base Pb free bearing metal: Hoyt Ltd
HOYT 11Z3	Sn base Pb free bearing metal: Hoyt Ltd
HOYT 38	Sb Cu Sn base bearing metal: Hoyt Ltd
HOYT 71	Sb Cu Sn base bearing metal: Hoyt Ltd
HOYT 133C	Sn base bearing metal: Hoyt Ltd
HOYT 156B	Sn base bearing metal: Hoyt Ltd
HOYT 175	Sn base bearing metal: Hoyt Ltd **DPN: 37 UTS: 80 Elon: 1.5%**
HOYT BLOWPIPE SOLDER	35.5% Pb Sn: Solder; quick setting; melting range 183–185 °C; Hoyt
HOYT FIFTY	50% Pb Sn: Solder; may contain Sb; melting range 183–212 °C; Hoyt
HOYT MARINE A	Sn base bearing metal: Hoyt Ltd
HOYT SIXTY	40% Pb Sn: Solder; melting range 183–188 °C; Hoyt
HOYTICE	Sn base bearing metal: Hoyt Ltd **DPN: 27 UTS: 65 Elon: 6%**
L 13600	40% Pb Sn alloy: Solder; UNS designation
L 13601	40% Pb 0.35% Sb Sn alloy: Solder; UNS designation
L 13630	37% Pb Sn alloy: Solder; UNS designation
L 13631	37% Pb 0.3% Sb Sn alloy: Solder; UNS designation
L 13650	18% Pb 15% Sb 2% Cu Sn diecast alloy: UNS designation
L 13700	30% Pb 70% Sn alloy: Solder; UNS designation
L 13701	30% Pb 0.3% Sb Sn alloy: Solder; UNS designation
L 13820	5% Cu 13% Sb Sn alloy: Die casting; UNS designation
L 13840	8% Cu 8% Sb Sn alloy: Babbitt metal; UNS designation
L 13870	5.7% Cu 6.7% Sb Sn alloy: Babbitt metal; UNS designation
L 13890	3.5% Cu 7.5% Sb Sn alloy: Babbitt metal; UNS designation
L 13910	4.5% Cu 4.5% Sb Sn alloy: Babbitt metal; UNS designation
L 13911	1.5% Cu 7% Sb Sn alloy: Pewter; UNS designation
L 13912	2.2% Cu 6% Sb Sn alloy: Pewter; UNS designation
L 13913	4.5% Cu 4.5% Sb Sn alloy: Die casting; UNS designation
L 13940	5% Sb Sn alloy: Solder; UNS designation
L 13950	5% Sb Sn alloy: Solder; UNS designation
L 13960	0.3% Sb 4% Ag Sn alloy: Solder; UNS designation
L 13961	4% Ag Sn alloy: Solder; UNS designation
L 13963	2% Sb 1.5% Cu Sn alloy: Pewter; UNS designation
L 13965	0.3% Sb 3.5% Ag Sn alloy: Solder; UNS designation
L Sn 50	46% Pb 3.3% Sb Sn alloy: Designation used by German Standards
L Sn 60	45% Pb 3.2% Sb Sn alloy: Designation used by German Standards
L Sn 90	8% Pb 1.3% Sb Sn alloy: Designation used by German Standards
L Sn Pb Zn	45% Pb 10% Zn + Cd Sn alloy: Designation used by German Standards
L Sn Zn 10	11% Zn Sn alloy: Designation used by German Standards
L Sn Zn 40	40% Zn Sn alloy: Designation used by German Standards
LION BRAND	7% Cu 5% Sb Sn alloy: White metal; Blackwells; marine; main bearings
LM 10A	Ag Cu Sn alloy: Soft Solder; Johnson Matthey; electrical conductivity 13% IACS; melting range 214–275 °C **UTS: 70 Elon: 16%**
MCP 20	Ga Sn alloy: Mining & Chemical Products Ltd; melting point 20 °C
MCP 117	In Sn alloy: Mining & Chemical Products Ltd; melting point 117 °C
MCP 145	Pb Cd Sn alloy: Mining & Chemical Products Ltd; melting point 145 °C
MCP 176	Cd Sn alloy: Mining & Chemical Products Ltd; melting point 176 °C
MCP 183	Pb Sn alloy: Mining & Chemical Products Ltd; melting point 183 °C
MCP 200	Zn Sn alloy: Mining & Chemical Products Ltd; melting point 200 °C **DPN: 22 UTS: 6 Elon: 30% Proof: 4.5**
MCP 221	Ag Sn alloy: Mining & Chemical Products Ltd; melting point 221 °C
MCP 227	Cu Sn alloy: Mining & Chemical Products Ltd; melting point 227 °C
NAVY	4% Cu 5% Sb 15% Pb Sn alloy: Bearing metal; Stone Manganese for BS 3332/5

Note. The following abbreviations and units are used in the tables:

DPN Hardness, diamond pyramid number
UTS Ultimate tensile strength, N/mm^2
Elon Elongation, %
Proof 0.1% proof strength, N/mm^2

1 N/mm^2=0.1 hbar=0.102 kgf/mm^2=0.06475 tonf/in.^{2}=145.04 lbf/in.^{2}=1 MPa
See Appendix II for other abbreviations and conversion tables.

Symbol	Nominal analysis, supplier, condition and remarks.
No 75	Sn base bearing metal: High speed; medium load; Magnolia Anti-Friction Metal Co. **DPN: 28 UTS: 80**
OE 1 METAL	Sn base bearing metal: High speed; medium load; Magnolia Anti-Friction Metal Co.
OE 2 METAL	Sn base bearing metal: Medium duty; Magnolia Anti-Friction Metal Co.
PEWTER	Some modern alloys are lead free Sn alloy: White metal; information from Tin Research Institute
PLUMBSOL	Ag Sn base soft solder: Johnson Matthey; electrical conductivity 13.3% IACS; melting range 221–225 °C **UTS: 2.5 Elon: 60%**
PY 1815A	15% Sb 18% Pb 2% Cu Sn alloy: Designation used by ASTM range
QQ M 161/1	4.0% Sb 4.0% Cu Sn alloy: For bearings; US Federal specification
QQ M 161/2	7.5% Sb 4.0% Cu Sn alloy: For bearings; US Federal specification
QQ M 161/3	8.0% Sb 8.0% Cu 0.3% Pb Sn alloy: For bearings; US Federal specification
QQ M 161/4	13% Sb 5.5% Cu 0.2% Pb Sn alloy: For bearings; US Federal specification
QQ M 161/5	10% Sb 3.0% Cu 25% Pb Sn alloy: For bearings; US Federal specification
QQ S 571	40% Pb Sn alloy: Solder; US Federal specification
QQ S 571 A	50% Pb Sn alloy: Solder; US Federal specification
QQ S 571 Sb 5	5% Sb Sn alloy: Solder; US Federal specification
QQ S 571 Sn 63	37% Pb Sn alloy: US Federal specification
QQ S 571 Sn 70	30% Pb Sn alloy: US Federal specification
QQ S 571 Sn 96	4% Ag Sn alloy: Solder; US Federal specification
QQ S 571b Sn 60	0.45% Sb 40% Pb Sn: Solder; US Federal specification; melting range 182–189 °C
QQ S 571d Sb 5	5.0% Sb Sn: Solder; US Federal specification; melting range 232–240 °C
QQ S 571d Sn 50	0.45% Sb 50% Pb Sn: Solder; US Federal specification; melting range 182–216 °C
QQ S 571d Sn 62	0.45% Sb 38% Pb Sn: Solder; US Federal specification; melting range 177–189 °C
QQ S 571d Sn 63	0.45% Sb 37% Pb Sn: Solder; US Federal specification; melting point 182 °C
QQ S 571d Sn 70	0.45% Sb 29% Pb Sn: Solder; US Federal specification; melting range 182–193 °C
QQ S 571d Sn 96	0.1% Pb (max) 4.0% Ag Sn: Solder; US Federal specification; melting point 221 °C
QQ T 390	8% Cu 8% Sb Sn alloy: Babbitt; US Federal specification
RAILWAY C	5% Pb 11% Sb 4% Cu Sn alloy: Bearing metal; Stone Manganese for BS 3332/3
SAE 10	4.5% Cu 4.5% Sb Sn alloy: For bearings
SAE 11	5.7% Cu 6.7% Sb Sn alloy: For bearings
SAE 12	3.5% Cu 7.5% Sb Sn alloy: For bearings
SAE 783	1% Cu 19% Sb Sn alloy: For bearings
Sg Sn 50	4% Cu 33% Pb 13% Sb Sn alloy: Designation used by German Standards
Sg Sn 60	4% Cu 23% Pb 13% Sb Sn alloy: Designation used by German Standards
Sg Sn 70	5% Cu 10% Pb 15% Sb Sn alloy: Designation used by German Standards
Sg Sn 75	5% Cu 3% Pb 17% Sb Sn alloy: Designation used by German Standards
Sg Sn 78	4% Cu 1% Pb 17% Sb Sn alloy: Designation used by German Standards
SILVER BABBITT	Ag Cd Ni Sn base bearing metal: Highest duty; Magnolia Anti-Friction Metal Co. **DPN: 33.6**
SOVEREIGN 8TS	3.0% Cu 8.9% Sb 1.1% Cd Sn alloy: Bearing metal; Glacier Metals **DPN: 30 UTS: 98 Elon: 15% Proof: 84**
SOVEREIGN 8TS	8.8% Sb 3.0% Cu 1% Cd Sn: Bearing metal; Glacier Metals **DPN: 30 UTS: 100 Elon: 16% Proof: 80**
SPRABABBITT A	3.5% Cu 0.25% Pb 7.5% Sb Sn alloy: Wire for spraying; Metco
SPRABABBITT	Sn base Pb free spray deposit: For high speed high duty bearings; Metco
STA 7 TB1B	8% Sb 8% Cu 0.3% Pb Sn alloy: Bearing metal; obsolete
STA 7 TB2	Sn base bearing metal: Obsolete
STERN TUBE METAL	30% Zn 1.5% Cu Sn alloy: White metal; information from Tin Research Institute
SX 4	4% Ag Sn alloy: For soldering; Sheffield Smelting Co.; melting range 221–224 °C **DPN: 14 UTS: 50 Elon: 48% Proof: 45**
TANDEM DE	Sn base bearing metal: For reciprocating engines; Eyre Smelting Co.; melts at 315 °C **DPN: 27 UTS: 80 Elon: 24%**
TANDEM HP	Sn base bearing metal: For reciprocating engines; Eyre Smelting Co.; melts at 330 °C **DPN: 32 UTS: 80 Elon: 6%**
TANDEM ME	Sn base bearing metal: For high speed use; Eyre Smelting Co.; melts at 370 °C **DPN: 32 UTS: 90 Elon: 4%**
TANDEM ML	Sn base bearing metal: For high speed use; Eyre Smelting Co.; melts at 340 °C **DPN: 30 UTS: 80 Elon: 3%**
TANDEM PLUS	Sn base bearing metal: For reciprocating engines; Eyre Smelting Co.; melts at 330 ° C **DPN: 33 UTS: 90 Elon: 16%**
TC	Ag Cd Ni Sn base bearing metal: Highest duty; Magnolia Anti-Friction Metal Co. **UTS: 95**
TCS	Ag Cd Ni Sn base bearing metal: Highest duty; Magnolia Anti-Friction Metal Co.; silver 'Babbitt' **DPN: 33.6**
TENAXAS Al	Pb Sn base bearing metal; For reciprocating engines; Eyre Smelting Co.; melting range 186–340 °C **DPN: 30 UTS: 80 Elon: 3.5%**
TINMANS SOLDER	33.3% Pb Sn alloy: White metal: Information from Tin Research Institute
TURBEX	Sn base Pb free bearing metal: High speed; high load; Magnolia Anti-Friction Metal Co. **DPN: 25.5 UTS: 80**
UNDER-WATER	30% Zn 1.5% Cu 0.5% Pb 0.5% Sb Sn alloy: Bearing metal; Stone Manganese for BS 3332/9
VULCAN	Sn base bearing metal: For high duty loads; Phosphor Bronze Ltd
WAGNER'S ALLOY	10% Sb 1% Cu 3% Zn 0.8% Bi Sn alloy: Origin unknown
WAMATO METAL	4.5% Cu 4.5% Sb 1% Pb 1% Ni 1% Co (max) Sn alloy: Used for main bearings in internal combustion engines; origin unknown
WARNE'S METAL	26% Ni 26% Bi 11% Co Sn alloy: Used for jewellery; origin unknown
WELCH'S ALLOY	48% Ag Sn alloy: Used for dental purposes; origin unknown
YC 135A	13% Sb 5% Cb Sn alloy: Designation used by ASTM
ZINN	0.7% Pb 0.3% Cu Sn alloy: Used for hard service bearings; origin unknown

Note. The following abbreviations and units are used in the tables:

DPN	Hardness, diamond pyramid number
UTS	Ultimate tensile strength, N/mm^2
Elon	Elongation, %
Proof	0.1% proof strength, N/mm^2

1 N/mm^2=0.1 hbar=0.102 kgf/mm^2=0.06475 $tonf/in.^2$=145.04 $lbf/in.^2$=1 MPa
See Appendix II for other abbreviations and conversion tables.

51. Titanium Ti

Physical properties

Atomic number		22
Atomic weight		47.9
Crystal structure	alpha	Close-packed hexagonal up to 880 °C
	beta	Body-centred cubic above 880 °C
Colour		Dark grey
Specific gravity		4.5
Density		4500 kg/m^3
Melting point		1680 °C
Boiling point		2800 °C
Specific heat		0.528 J/g °C
Thermal conductivity		17 W/m °C
Coefficient of linear expansion (20–100 °C)		8.9×10^{-6} °C
Latent heat of fusion		435.4 J/g
Latent heat of vaporization		–
Thermal neutron absorption cross-section		5.6 barns/atom
Electrical conductivity		3% IACS (copper 100%)
Specific resistance		554 microhm mm
Temperature coefficient of electrical resistance		0.0026/ °C
Electrochemical equivalent		–
Electrode potential		+0.2 V
Magnetic susceptibility		1.25×10^{-6}
Young's modulus of elasticity		116×10^9 N/m^2
Tensile strength		220 N/mm^2
Hardness		60 DPN

51.1 General notes on titanium

Titanium is one of the more common metals in the earth and is found widespread in abundant quantities, but never as the native metal. Titanium oxide made from 'ilmenite' has been used for many years as a white pigment for paints which is not subject to yellowing, as is zinc oxide when in contact with sulphur or sulphide.

It is only in recent years that it has been possible to produce titanium in commercial quantities and it is still relatively expensive. Titanium is now available in all the normal forms such as castings, forging, bar, tube and extruded sections. There is also a well-defined series of alloys based on titanium which are discussed and listed in the following sections.

The metal is steel grey in appearance, having a density twice that of aluminium and two-thirds that of steel. The mechanical properties at room temperature are comparable with those of many steels, and at high temperature titanium and its alloys are superior to all except the special creep resistant steels.

Titanium metal resists the attack of almost all acids over a very wide range of temperatures and this with its high temperature properties makes it a favourite material with chemical, nuclear and aerospace engineers.

As the usage increases so the price drops and further uses for titanium and its alloys become an economical proposition.

Titanium was a laboratory curiosity until very recently, but already the quantity used annually can be measured in thousands of tons. The fact that it is available in commercial quantities of a purity to allow working with conventional equipment at easily obtainable temperatures accounts for this rapid increase.

In this book titanium and its alloys have been divided into three groups: Section 51A commercially pure, Section 51B alloys with no apparent increase in mechanical properties by thermal treatment and Section 51C alloys with some mechanical property improved by thermal treatment. It should be appreciated that there is some overlap with Sections 51B and 51C particularly for alloys in Section 51C which are used in the annealed condition in service. For the purposes of this book if the alloy is heat treatable it has been included in Section 51C.

51A Titanium – commercially pure

Specific gravity	4.51
Density	4500 kg/m^3
Solidus/liquidus	1725 °C
Thermal conductivity	15 W/m °C
Coefficient of linear expansion	9.1 × 10^{-6}
Electrical conductivity	3.0% IACS (copper 100%)
Specific resistance	550 microhm mm
Young's modulus of elasticity	(103–124) × 10^9 N/m^2
Impact	–
Fatigue strength – endurance limit	±220–300 N/mm^2
Hot strength	

Temperature °C	Tensile strength N/mm^2	Elongation %
100	450	25
200	340	28
300	270	28
400	220	28
500	180	28
600	140	150

The above properties are typical of the following group, and may not apply exactly to any one specification. It is possible that with certain specifications some of the values may not be applicable.

General metallurgical characteristics

The specifications listed do not contain any alloying metallic elements. Controlled amounts of oxygen however, up to 0.2%, affect the mechanical properties. It is important that carbon, iron and hydrogen are held to low limits, otherwise there will be a considerable drop in ductility. The materials listed have excellent corrosion resistance, can be formed and welded, and have reasonable mechanical strength.

Titanium is approximately two-thirds the weight of steel and twice the weight of aluminium, and depending on the purity and cold work can be the same strength as medium alloy steel. The higher the strength, the more difficult the material is to weld in general. There is some evidence that titanium is more notch sensitive than other materials of comparative strength; thus stress raisers must be absent.

Thermal treatment

None of the materials in this section can have their mechanical properties improved by any form of thermal treatment. They can be cold worked to improve their mechanical strength, however, but this will reduce their ductility at the same time. Where this cold work must be removed then an annealing operation is necessary at approximately 500 °C. The time at temperature must be held to the minimum and the temperature must not be exceeded as compounds will form and diffuse into the material to cause considerable embrittlement.

The problem is that above temperatures in the 400–600 °C range titanium has an affinity for oxygen, nitrogen, hydrogen, carbon and almost any other element and will form a brittle compound at the surface.

By the use of inert gases such as argon and helium, this effect will be reduced. It has been shown that it is quite difficult and expensive to eliminate air by replacing this with an inert gas without producing an initial vacuum. It is therefore very often more economical to anneal under vacuum than under inert gas.

Welding of pure titanium presents considerable problems in that it is essential that this is carried out either under vacuum or under inert gas.

Provided the correct controls are available, then excellent welds in titanium can be produced but expert advice is obviously necessary.

Machining of titanium can be difficult as it cold works readily. It must also be appreciated that titanium in fine form is readily oxidizable and thus will ignite and be explosive when it is produced in fine ribbon form or as dust. There are regulations which must be complied with when polishing, grinding or machining titanium.

The materials in this group find increasing uses in the chemical and food industry where acids are involved. Only fuming nitric, hot concentrated sulphuric and phosphoric acids attack commercially pure titanium.

While the materials have low mechanical strength, they have good ductility and reasonable impact properties provided they are not contaminated with oxide, nitride, hydrides, etc. They find some use in the aircraft industry as fire-retardant barriers: very thin sheets of titanium are used where flames could be produced during failures.

Symbol	Nominal analysis, supplier, condition and remarks.
3.7024	Commercially pure Ti: German Standard
3.7025	Commercially pure Ti: German Standard
3.7034	Commercially pure Ti: German Standard
3.7064	Commercially pure Ti: German Standard
12	Commercially pure Ti: Krupp
15	Commercially pure Ti: Krupp
18S	Commercially pure Ti: Krupp
30	Commercially pure Ti: Atlas
35	Commercially pure Ti: Ugine
40	Commercially pure Ti: Ugine
50	Commercially pure Ti: Ugine
A 40	Commercially pure Ti: Crucible Steel Co. **Proof: 270**
A 55	Commercially pure Ti: Crucible Steel Co. **Proof: 370**
A 70	Commercially pure Ti: Crucible Steel Co. **Proof: 480**
AIR 9/182 T35	0.1% C (max) 0.2% Fe Ti: Commercially pure; origin unknown
AMS 4900	0.4% C (max) 0.007% H 0.2% Fe Ti: Sheet
AMS 4900 B	Commercially pure Ti: Sheet and strip; annealed; AMS for alloy A 55 **Proof: 370**
AMS 4901 C	Commercially pure Ti: Sheet and strip; annealed; AMS for commercial alloy A 70 **Proof: 500**
AMS 4902 A	Commercially pure Ti: Sheet and strip; annealed; AMS for commercial alloy A 40
AMS 4921 A	Commercially pure Ti: Bar and forging; annealed; AMS for commercial alloy A 70 **Proof: 500**
AMS 4941	Commercially pure Ti: Welded tube; annealed; AMS for commercial alloy A 40 **Proof: 270**
AMS 4942	Commercially pure Ti: Seamless tube; annealed **Proof: 270**
AMS 4951	0.05% C (max) 0.12% O Ti: Weld metal
AMS 4951 A	Commercially pure Ti: Wire; annealed
ASTM B265/1	0.1% C 0.01% H 0.2% Fe Ti: Sheet; commercially pure titanium **UTS: 270 Elon: 22% Proof: 200**
ASTM B265/2	0.1% C 0.01% H 0.2% Fe Ti: Sheet; commercially pure titanium **UTS: 340 Elon: 20% Proof: 270**
ASTM B265/3	0.1% C 0.01% H 0.2% Fe Ti: Sheet; commercially pure titanium **UTS: 420 Elon: 18% Proof: 340**
ASTM B265/4	0.1% C 0.01% H 0.2% Fe Ti: Sheet; commercially pure titanium **UTS: 580 Elon: 15% Proof: 500**
ASTM B299 MD120	0.02% C 0.005% H 99.3% Ti: Sponge; primary metal **DPN: 120**
ASTM B299 MD160	0.05% C 0.005% H 99.1% Ti: Sponge; primary metal **DPN: 160**
ASTM B299 ML120	0.025% C 0.03% H 99.1% Ti: Sponge; primary metal **DPN: 120**
ASTM B299 ML140	0.06% C 0.05% H 99.1% Ti: Sponge; primary metal **DPN: 140**

Note. The following abbreviations and units are used in the tables:

DPN	Hardness, diamond pyramid number
UTS	Ultimate tensile strength, N/mm^2
Elon	Elongation, %
Proof	0.1% proof strength, N/mm^2

1 N/mm^2=0.1 hbar=0.102 kgf/mm^2=0.06475 tonf/in.^{2}=145.04 lbf/in.^{2}=1 MPa

See Appendix II for other abbreviations and conversion tables.

Symbol	Nominal analysis, supplier, condition and remarks.
ASTM B299 ML160	0.05% C 0.05% H 99.1% Ti: Sponge; primary metal **DPN: 160**
ASTM B299 SL120	0.025% C 0.0125% H 99.1% Ti: Sponge; primary metal **DPN: 120**
ASTM B299 SL140	0.06% C 0.015% H 99.1% Ti: Sponge; primary metal **DPN: 140**
ASTM B299 SL160	0.03% C 0.015% H 99.1% Ti: Sponge; primary metal **DPN: 160**
ASTM B337/1	0.1% C 0.015% H 0.05% N Ti: Tube; commercially pure titanium **UTS: 270 Elon: 22% Proof: 200**
ASTM B337/2	0.1% C 0.015% H 0.05% N Ti; Tube; commercially pure titanium **UTS: 340 Elon: 20% Proof: 270**
ASTM B337/3	0.1% C 0.015% H 0.07% N Ti; Tube; commercially pure titanium **UTS: 420 Elon: 18% Proof: 340**
ASTM B337/4	0.15% C 0.015% H 0.07% N Ti; Tube; commercially pure titanium **UTS: 580 Elon: 15% Proof: 500**
ASTM B338/1	0.1% C 0.25% O 0.015% H Ti: Tube; commercially pure; annealed **UTS: 270 Elon: 22%**
ASTM B338/2	0.1% C 0.25% O 0.015% H Ti: Tube; commercially pure; annealed **UTS: 340 Elon: 20%**
ASTM B338/3	0.1% C 0.35% O 0.015% H Ti: Tube; commercially pure; annealed **UTS: 420 Elon: 18%**
ASTM B338/4	0.15% C 0.45% O 0.015% H Ti: Tube; commercially pure; annealed **UTS: 580 Elon: 15%**
ASTM B348/1	0.1% C 0.2% O 0.012% H Ti: Bar; commercially pure; annealed **UTS: 270 Elon: 22% Proof: 200**
ASTM B348/2	0.1% C 0.2% O 0.001% H Ti: Bar; commercially pure; annealed **UTS: 270 Elon: 22% Proof: 200**
ASTM B348/3	0.1% C 0.3% O 0.012% H Ti: Bar; commercially pure; annealed **UTS: 420 Elon: 18% Proof: 340**
ASTM B348/4	0.15% C 0.4% O 0.012% H Ti: Bar; commercially pure; annealed **UTS: 580 Elon: 15% Proof: 500**
ASTM B367 C1	0.1% C 0.04% O 0.01% H Ti: Casting; annealed; commercially pure **UTS: 450 Elon: 12% Proof: 370**
ASTM B381 F1	0.1% C 0.2% O 0.01% H Ti: Forging; annealed; commercially pure **UTS: 270 Elon: 22% Proof: 200**
ASTM B381 F2	0.1% C 0.2% O 0.01% H Ti: Forging; annealed; commercially pure **UTS: 330 Elon: 20% Proof: 270**
ASTM B381 F3	0.1% C 0.3% O 0.012% H Ti: Forging; annealed; commercially pure **UTS: 420 Elon: 18% Proof: 330**
ASTM B381 F4	0.15% C 0.4% O 0.012% H Ti: Forging; annealed; commercially pure **UTS: 580 Elon: 15% Proof: 480**
ASTM B382 ER Ti	Commercially pure Ti: Welding rod
A Ti 24	Ti commercially pure: For deep drawing; Avesta **UTS: 350 Elon: 25% Proof: 220**
A Ti 30	Ti commercially pure: For lining vessels etc.; Avesta **UTS: 480 Elon: 16% Proof: 270**
A Ti 35	Ti commercially pure: Avesta **UTS: 530 Elon: 13% Proof: 330**
A Ti Pd	0.2% Pd Ti alloy: Avesta; for use with dilute acids **UTS: 360 Elon: 25% Proof: 220**
BS 3531/1.5	99.0% Ti: Wrought; for implants and surgery tools
BSTA 1	0.012% H (max) 0.2% Fe (max) Ti: Sheet and strip
BSTA 2	0.012% H (max) 0.2% Fe (max) Ti: Sheet and strip

Symbol	Nominal analysis, supplier, condition and remarks.
BSTA 6	0.012% H (max) 0.2% Fe (max) Ti: Sheet and strip
BSTA 7	0.012% H (max) 0.2% Fe (max) Ti: Bar
BSTA 8	0.012% H (max) 0.2% Fe (max) Ti: Forging
BSTA 9	0.012% H (max) 0.2% Fe (max) Ti: Forging
BSTA 21	0.01% H (max) 0.2% Fe (max) Ti: Sheet and strip
BSTA 22	0.01% H (max) 0.2% Fe (max) Ti: Bar
BSTA 23	0.01% H (max) 0.2% Fe (max) Ti: Forging
BSTA 24	0.01% H (max) 0.2% Fe (max) Ti: Forging
BSTA 52	0.01% H (max) 0.2% Fe (max) Ti: Sheet
BSTA 53	0.01% H (max) 0.2% Fe (max) Ti: Bar
BSTA 54	0.01% H (max) 0.2% Fe (max) Ti: Forging
BSTA 55	0.01% H (max) 0.2% Fe (max) Ti: Forging
BSTA 58	0.01% H (max) 0.2% Fe (max) Ti: Plate
CRUCIBLE A 40	Commercially pure Ti: Crucible Steel Co. **Proof: 270**
CRUCIBLE A 55	Commercially pure Ti: Crucible Steel Co. **Proof: 370**
CRUCIBLE A 70	Commercially pure Ti: Crucible Steel Co. **Proof: 480**
DTD 5003 B	99.7% Ti: Commercially pure bar; annealed **UTS: 530 Elon: 20% Proof: 300**
DTD 5013 B	0.013% H Ti: Commercially pure bar; annealed **UTS: 450 Elon: 25% Proof: 200**
DTD 5023 B	99.7% Ti: Commercially pure sheet; annealed; suitable for welding **UTS: 530 Elon: 20% Proof: 330**
DTD 5033 B	99.7% Ti: Commercially pure sheet; annealed **UTS: 450 Elon: 25% Proof: 200**
DTD 5063 A	99.7% Ti: Commercially pure sheet; annealed **UTS: 700 Elon: 15% Proof: 450**
DTD 5073	0.1% C 0.2% Fe Ti: Commercially pure Ti tube; annealed **UTS: 370 Proof: 270**
DTD 5183	Commercially pure Ti: Sheet and strip **UTS: 450 Elon: 22% Proof: 270**
DTD 5193	Commercially pure Ti: Sheet and strip **UTS: 610 Elon: 18% Proof: 370**
DTD 5273	0.012% H (max) 0.2% Fe (max) Ti: Bar
DTD 5283	0.012% H (max) 0.2 Fe (max) Ti: Forging
DTD 5293	0.012% H (max) 0.2% Fe (max) Ti: Forging
ER Ti 0.2 Pd	0.05% C (max) 0.2% Pd Ti: Weld electrode; designation used by AWS
ER Ti 1	0.03% C (max) Ti: Weld electrode; designation used by AWS
ER Ti 2	0.05% C (max) Ti: Weld electrode; designation used by AWS
ER Ti 3	0.05% C (max) 0.1% O Ti: Weld electrode; designation used by AWS
ER Ti 4	0.05% C (max) 0.2% O Ti: Weld electrode; designation used by AWS
h Ti 24	99.6% Ti: Sheet; Koch-Light Ltd
HYLITE 1	High purity Ti: Chemical grade; Jessops; obsolete
HYLITE 10	0.013% H (max) commercially pure Ti: Bar, forgings, sheet, etc.; Jessops; annealed; obsolete **UTS: 400 Elon: 30% Proof: 240**
HYLITE 15	0.013% H (max) commercially pure Ti: Bar and forging; Jessop; annealed; obsolete **UTS: 500 Elon: 25% Proof: 300**

Note. The following abbreviations and units are used in the tables:

DPN	Hardness, diamond pyramid number
UTS	Ultimate tensile strength, N/mm^2
Elon	Elongation, %
Proof	0.1% proof strength, N/mm^2

1 N/mm^2=0.1 hbar=0.102 kgf/mm^2=0.06475 tonf/in.^{2}=145.04 lbf/in.^{2}=1 MPa

See Appendix II for other abbreviations and conversion tables.

Symbol	Nominal analysis, supplier, condition and remarks.
HYLITE 15	0.013% H (max) commercially pure Ti: Sheet; Jessop; cold rolled and annealed; obsolete **UTS: 520 Elon: 18% Proof: 370**
HYLITE 15 H	Commercially pure Ti: Jessop; obsolete
ICI 115	Commercially pure Ti: ICI; now IMI see Ti 115
ICI 125	Commercially pure Ti: ICI; now IMI see Ti 125
ICI 130	Commercially pure Ti: ICI; now IMI see Ti 130
IMI 260	0.2% Pd Ti alloy: IMI
L 512	Commercially pure Ti: VEW annealed **UTS: 350 Elon: 25% Proof: 175**
L 515	Commercially pure Ti: VEW annealed **UTS: 400 Elon: 20% Proof: 250**
L 518	Commercially pure Ti: VEW annealed **UTS: 650 Elon: 15% Proof: 390**
METCO 102	Ti oxide: Spray deposit; Metco **DPN: 600**
METCO 110	Al and Ti oxides: Spray deposit; Metco **DPN: 550**
MIL-T-9046	0.04% C 0.007% H (max) 0.2% Fe Ti
MST 0.2 Pd	0.1% C 0.01% H 0.2% Fe Ti: Wrought; Reactive Metal Products Ltd
MST 30	Commercially pure Ti: Reactive Metal Products Ltd; annealed **UTS: 250 Elon: 22% Proof: 200**
MST 40	Commercially pure Ti: Reactive Metal Products Ltd; annealed **UTS: 330 Elon: 20% Proof: 250**
MST 55	Commercially pure Ti: Reactive Metal Products Ltd; annealed **UTS: 450 Elon: 18% Proof: 370**
MST 70	Commercially pure Ti: Reactive Metal Products Ltd; annealed **UTS: 580 Elon: 15% Proof: 480**
RM I 0.2% Pd	Commercially pure Ti: Reactive Metal Products Ltd; annealed **UTS: 340 Elon: 22% Proof: 250**
RM I 30	Commercially pure Ti: Reactive Metal Products Ltd; annealed **UTS: 250 Elon: 25% Proof: 200**
RM I 40	Commercially pure Ti: Reactive Metal Products Ltd; annealed **UTS: 340 Elon: 22% Proof: 250**
RM I 55	Commercially pure Ti: Reactive Metal Products Ltd; annealed **UTS: 530 Elon: 20% Proof: 370**
RM I 70	Commercially pure Ti: Reactive Metal Products Ltd; annealed **UTS: 570 Elon: 15% Proof: 480**
RT 12 Pd	0.08% C (max) 0.2% Fe (max) 0.2% Pb Ti: Krupp-Lockner-Titrutan **UTS: 350 Elon: 25% Proof: 180**
RT 15 Pd	0.08% C (max) 0.25% Fe (max) 0.2% Pd Ti: Krupp-Lockner-Titrutan **UTS: 450 Elon: 21% Proof: 250**
RT 18 Pd	0.1% C (max) 0.3% Fe (max) 0.2% Pd Ti: Krupp-Lockner-Titrutan **UTS: 520 Elon: 17% Proof: 320**
RT 20	0.1% C (max) 0.35% Fe (max) Ti: Krupp-Lockner-Titrutan **UTS: 640 Elon: 15% Proof: 390**
SB 265	See ASMT B265/1
T 35	Commercially pure Ti: French National Standard
T 40	Commercially pure Ti: French National Standard
T 50	Commercially pure Ti: French National Standard
T 60	Commercially pure Ti: French National Standard
Ti 0.15 Pd	0.08% C 0.015% H 0.25% Fe 0.05% N 0.15% Pd Ti: Sheet, bar, etc.; Titanium Metals Ltd; annealed for chemical industry **DPN: 200 UTS: 370 Elon: 22% Proof: 300**

Symbol	Nominal analysis, supplier, condition and remarks.
Ti 35 A	0.08% C 0.015% H 0.25% Fe 0.05% N Ti: Sheet, bar, etc.; Titanium Metals Ltd; annealed **DPN: 120 UTS: 240 Elon: 25% Proof: 170**
Ti 55 A	0.08% C 0.015% H 0.25% Fe 0.05% N Ti: Sheet, bar, etc.; Titanium Metals Ltd; annealed **DPN: 200 UTS: 360 Elon: 22% Proof: 300**
Ti 65 A	0.08% C 0.015% H 0.25% Fe 0.05% N Ti: Sheet, bar, etc.; Titanium Metals Ltd; annealed **DPN: 220 UTS: 440 Elon: 20% Proof: 360**
Ti 75 A	0.08% C 0.015% H 0.25% Fe 0.05% N Ti: Sheet, bar, etc.; Titanium Metals Ltd; annealed **DPN: 265 UTS: 530 Elon: 15% Proof: 480**
Ti 100 A	0.08% C 0.01% H 0.25% Fe 0.05% N Ti: Bar and wire; Titanium Metals Ltd; annealed **DPN: 295 UTS: 540 Elon: 15% Proof: 460**
Ti 115	Commercially pure Ti: IMI; annealed **UTS: 390 Elon: 30% Proof: 200**
Ti 120	Commercially pure Ti: Rod, sheet and tube; IMI; withdrawn
Ti 125	Commercially pure Ti: IMI; annealed **UTS: 450 Elon: 22% Proof: 270**
Ti 130	Commercially pure Ti: IMI; annealed **UTS: 530 Elon: 20% Proof: 330**
Ti 150	Commercially pure Ti: IMI; annealed **UTS: 610 Elon: 18% Proof: 370**
Ti 155	Commercially pure Ti: IMI; annealed **UTS: 650 Elon: 24% Proof: 550**
Ti 160	Commercially pure Ti: IMI; annealed **UTS: 700 Elon: 15% Proof: 450**
Ti 260	Commercially pure Ti with addition of Pd: IMI; for use with non-oxidizing acids **UTS: 360 Elon: 25% Proof: 200**
Ti Po 1	Commercially pure Ti: AICMA
Ti Po 2	Commercially pure Ti: AICMA
Ti Po 2	Commercially pure Ti: AICMA
TITANIUM	Ti 99–100% purity: Blackwells
TITRUTAN	Trade name for titanium and its alloys: Krupp-Lockner
U 99/100	0.03% Ag 99.95% Ti: Commercially pure Ti; origin unknown
VDM TITAN 993	0.1% C (max) 0.013% H (max) 0.05% N (max) Ti: VDM
VDM TITAN 994	0.08% C (max) 0.013% H (max) 0.03% N (max) Ti: VDM
VDM TITAN 994 Pd	0.08% C (max) 0.013% H (max) 0.03% N (max) 0.13% Pd Ti: VDM
VDM TITAN 995	0.08% C (max) 0.013% H (max) 0.03% N (max) Ti: VDM
VDM TITAN 995 Pd	0.08% C (max) 0.013% H (max) 0.03% N (max) 0.013 Pd Ti: VDM

Note. The following abbreviations and units are used in the tables:

DPN	Hardness, diamond pyramid number
UTS	Ultimate tensile strength, N/mm^2
Elon	Elongation, %
Proof	0.1% proof strength, N/mm^2

1 N/mm^2=0.1 hbar=0.102 kgf/mm^2=0.06475 tonf/in.^{2}=145.04 lbf/in.^{2}=1 MPa

See Appendix II for other abbreviations and conversion tables.

51B Titanium alloys – not solution treated

Specific gravity	4.5
Density	4314 kg/m^3
Solidus/liquidus	1510–1660 °C
Thermal conductivity	14.7 W/m °C
Coefficient of linear expansion	8.7×10^{-6}/ °C
Electrical conductivity	1.0% IACS (copper 100%)
Specific resistance	1500 microhm mm
Young's modulus of elasticity	$103–117 \times 10^9$ N/m^2
Impact	15 J
Fatigue strength – endurance limit	±450 N/mm^2

Hot strength

Temperature °C	Tensile strength N/mm^2	Elongation %
100	840	20
200	760	20
300	720	20
400	700	18
500	580	18
600	270	150

The above properties are typical of the following group, and may not apply exactly to any one specification. It is possible that with certain specifications some of the values may not be applicable.

General metallurgical characteristics

These materials have alloying additions of aluminium, manganese, tin or vanadium which strengthen the titanium matrix, particularly when cold worked. With the alloys listed, the alloying elements are dissolved in the titanium and remain in solution at all temperatures; thus these alloys cannot be strengthened by any form of thermal treatment.

Titanium alloys have a fatigue endurance limit below which failure by fatigue does not occur; they thus differ from other non-ferrous metals. They are rather notch sensitive, however, in that the fatigue strength falls by a disproportionate amount when stress raisers are present. Care must therefore be exercised not to design changes of section, joins or other stress raisers where there are stress situations.

Also listed in this section are several titanium carbide specifications. These are hard, unworkable, special purpose materials.

Thermal treatment

None of the materials in this group can be hardened by thermal treatment but many, or all, can be hardened by cold work and most will accept a considerable amount of deformation before they become too brittle for normal use.

Where the inter-stage anneal is required this should be for 2 hours at temperatures of approximately 500 °C. It must be appreciated that even at this relatively low temperature titanium will form oxides, carbides, nitrides and hydrides which can then diffuse into the material causing considerable embrittlement.

These materials are not normally welded although specific techniques are now available under highly controlled conditions where welding can be carried out. Very specialist advice and control are essential.

It is not possible to use the normal technique of a moving welding head but the welding must be carried out in a chamber which has been purged to remove all air and replace it with inert gas.

The machinability of these materials is similar to that of the higher alloyed austenitic stainless steels or the nickel chromium alloys.

Titanium in ribbon or powder form is extremely flammable or explosive and this material is subject to special precautions by law when grinding, machining or polishing.

The uses of these materials include aircraft structural parts where reasonable strength is required at temperatures up to 300 °C. This includes engine fireproof bulkheads, compressor blades, jet pipes, exhaust units, etc.

In chemical engineering these alloys are used for a similar application where pure titanium would be used but where greater strength is necessary. All the alloys have exceptionally good corrosion resistance against acids and alkalis. Cold working achieved either by machining or hammering, etc., or misuse during service can be removed, normally at 300 °C, but it may require temperatures as high as 450–600 °C.

Symbol	Nominal analysis, supplier, condition and remarks.
2 Cr 2 Fe 2 Mo	0.05% C 2.2% Fe 2.2% Mo 2.2% Cr Ti alloy: Alpha–beta grade; information from SAE handbook
3.7124	25% Cu Ti alloy: German Standard
4 Al 4 Mn	0.05% C 4.0% Al 4.0% Mn Ti alloy: Alpha–beta alloy; information from SAE handbook
5 Al 1.25 Fe 2.75 Cr	0.08% C 1.2% Fe 5.0% Al 2.7% Cr Ti alloy: Alpha–beta grade; information from SAE handbook
5 Al 2.5 Sn	0.15% C 5.0% Al 2.5% Sn Ti alloy: Alpha grade; information from SAE handbook
5 Al 5 Sn 5 Zr	0.02% C 5.0% Al 4.8% Sn 5.2% Zr Ti alloy: Alpha grade; information from SAE handbook
7 Al 2 Cb 1 Ta	0.04% C 7.0% Al 1.0% W 2.0% Nb Ti alloy: Alpha grade; information from SAE handbook
7 Al 4 Mo	0.05% C 6.9% Al 4.0% Mo Ti alloy: Alpha–beta grade; information from SAE handbook
7 Al 12 Zr	0.02% C 7.0% Al 12.0% Zr Ti alloy: Alpha grade; information from SAE handbook
8 Al 1 Mo 1 V	0.04% C 8.0% Al 1.0% Mo 1.0% V Ti alloy: Alpha grade; information from SAE handbook
8 Mn	0.1% C 8.0% Mn Ti alloy: Alpha–beta grade; information from SAE handbook
A 110 AT	5% Al 2.5% Sn Ti alloy: Crucible Steel Co.; annealed **UTS: 920 Elon: 15% Proof: 440**
AMS 4905	6.2% Al 4% V Ti alloy
AMS 4906	6.2% Al 4% V Ti alloy
AMS 4907	6% Al 4% V Ti alloy
AMS 4908	0.1% C (max) 8% Mn Ti alloy: Sheet
AMS 4908 A	8% Mn Ti alloy: Sheet and strip; annealed **Proof: 760**
AMS 4909	5% Al 2.5% Sn Ti alloy
AMS 4910	5% Al 2.5% Sn Ti alloy: Sheet and strip; annealed; AMS for commercial alloy A 110AT **Proof: 760**
AMS 4912	0.04% C 0.15% Fe 4.25% Al 3% Mo 1% V Ti alloy: Sheet
AMS 4915	0.04% C (max) 8% Al 1% Mo 1% V Ti alloy: Sheet
AMS 4916	8% Al 1% Mo 1% V Ti alloy
AMS 4918	5.5% Al 2% Sn 5.5% V Ti alloy
AMS 4919	6% Al 2% Mo 2% Sn 4% Zr Ti alloy
AMS 4920	6.2% Al 4% V Ti alloy
AMS 4923	2% Fe 2% Mo 2% Cr Ti alloy: Bar and forging; annealed **Proof: 880**
AMS 4924	5% Al 2.5% Sn Ti alloy
AMS 4925 A	4% Al 4% Mn Ti alloy: Bar and forging; annealed; AMS for commercial alloy C130 AM **Proof: 930**
AMS 4926	5% Al 2.5% Sn Ti alloy: Bar; annealed; AMS for commercial alloy A 110AT **Proof: 810**
AMS 4927	3% Al 5% Cr Ti alloy: Bar and forging
AMS 4928	0.02% C (max) 6% Al 3.8% V Ti alloy: Forgings
AMS 4929	0.04% C (max) 1.4% Fe 5.5% Al 1.2% Mo 1.4% Cr Ti alloy: Forgings
AMS 4930	6.2% Al 4% V Ti alloy
AMS 4931	6.2% Al 4% V Ti alloy
AMS 4933	8% Al 1% Mo 1% V Ti alloy
AMS 4934	6.2% Al 4% V Ti alloy
AMS 4936	5.5% Al 2% Sn 5.5% V Ti alloy
AMS 4943	3% Al 2.5% V Ti alloy
AMS 4944	3% Al 2.5% V Ti alloy
AMS 4952	6% Al 2% Mo 2% Sn 4% Zr Ti alloy
AMS 4953	5% Al 2.5% Sn Ti alloy: Welding wire; annealed; AMS for commercial alloy A 110AT
AMS 4955	0.04% C 8% Al 1% Mo 1% V Ti alloy: Wire
AMS 4959	3% Al 13% V 11% Cr Ti alloy
AMS 4965	6.2% Al 4% V Ti alloy

Note. The following abbreviations and units are used in the tables:

DPN	Hardness, diamond pyramid number
UTS	Ultimate tensile strength, N/mm^2
Elon	Elongation, %
Proof	0.1% proof strength, N/mm^2

1 N/mm^2=0.1 hbar=0.102 kgf/mm^2=0.06475 $tonf/in.^2$=145.04 $lbf/in.^2$=1 MPa

See Appendix II for other abbreviations and conversion tables.

Symbol	Nominal analysis, supplier, condition and remarks.
AMS 4966	5% Al 2.5% Sn Ti alloy: Forging; annealed; AMS for commercial alloy A 110AT **Proof: 800**
AMS 4967 A	2.25% Al 1% Mo 11% Sn 5% Zr 0.2% Si Ti alloy
AMS 4968	0.02% C (max) 5% Al 4.8% Sn 5.2% Zr: Forgings
AMS 4970	0.05% C (max) 0.1% Fe 7% Al 4% Mo Ti alloy: Forgings
AMS 4972	0.04% C (max) 8% Al 1% Mo 1% V Ti alloy: Forgings
AMS 4974	2.2% Al 4% Mo 11% Sn 40.2% Si Ti alloy
AMS 4982	45% Nb Ti alloy
AMS 4984	2% Fe 3% Al 10% V Ti alloy
AMS 4985	6% Al 4% V Ti alloy
AMS 4986	2% Fe 3% Al 10% V Ti alloy
AMS 4987	3% Al 10% V 2% Ti alloy
AMS 4991	6% Al 4% V Ti alloy
AMS 4993	6.2% Al 4% V Ti alloy
AMS 4996	6% Al 4% V Ti alloy
AMS 4998	6% Al 4% V Ti alloy
AMS 7460	6% Al 4% V Ti alloy
AMS 7461	6% Al 4% V Ti alloy
AMS 7496	0.27% C 0.5% Mo 0.85% V 1.2% Cr Ti: Sheet
ASTM B265/6	0.1% C 0.01% H 5% Al 2.5% Sn Ti alloy: Sheet **UTS: 840 Elon: 10% Proof: 800**
ASTM B265/7	0.15% C 0.01% H 7% Mn Ti alloy: Sheet **UTS: 880 Elon: 10% Proof: 780**
ASTM B348/6	0.1% C 0.02% H 5% Al 2.5% Sn Ti alloy: Bar; annealed **UTS: 840 Elon: 10% Proof: 780**
ASTM B348/7	0.1% C 0.01% H 4% Al 4% Mn Ti alloy: Bar; annealed **UTS: 1050 Elon: 10% Proof: 980**
ASTM B367 C3	0.1% C 0.01% H 5% Al 2.5% Sn Ti alloy: Casting; annealed **UTS: 840 Elon: 8% Proof: 760**
ASTM B381 F6	0.1% C 0.02% H 5% Al 2.5% Sn Ti alloy: forging; annealed **UTS: 820 Elon: 10% Proof: 760**
ASTM B381 F7	0.1% C 0.012% H 4% Al 4% Mn Ti alloy: Forging; annealed **UTS: 1030 Elon: 10% Proof: 920**
ASTM B382 ERTI 2.5 Al 16 V	2.5% Al 16% V Ti alloy: Welding rod
ASTM B382 ERTI 3 Al	3% Al Ti alloy: Welding rod
ASTM B382 ERTI 5 Al 2.5 Sn	5% Al 2.5% Sn Ti alloy: Welding rod
ASTM B382 ERTI 8 Al 2 Cb 1 Ta	8% Al 2% Nb 1% Ta Ti alloy: Welding rod
ASTM F 136	6% Al 4% V Ti alloy: For surgical implants **UTS: 870 Elon: 10% Proof: 800**
BSTA 10	0.0125% H (max) 6.1% Al 4% V Ti alloy: Sheet
BSTA 11	0.0125% H (max) 6.1% Al 4% V Ti alloy: Bar
BSTA 12	0.01% H (max) 6.1% Al 4% V Ti alloy: Forging
BSTA 13	0.015% H (max) 6.1% Al 4% V Ti alloy: Forging
BSTA 28	0.012% H (max) 6.1% Al 4% V Ti alloy: Forging
BSTA 38	0.012% H (max) 4% Al 4% Mo 0.5% Si Ti alloy: Bar
BSTA 39	0.012% H (max) 4% Al 4% Mo 0.5% Si Ti alloy: Forging
BSTA 40	0.012% H (max) 4% Al 4% Mo 0.5% Si Ti alloy: Bar
BSTA 41	0.012% H (max) 4% Al 4% Mo 0.5% Si Ti alloy: Forging
BSTA 42	0.012% H (max) 4% Al 4% Mo 0.5% Si Ti alloy: Forging
BSTA 43	0.006% H (max) 6% Al 0.5% Mo 0.3% Si Ti alloy: Forging
BSTA 44	0.006% H (max) 6% Al 0.5% Mo 0.3% Si Ti alloy: Forging
BSTA 45	0.012% H (max) 4% Al 4% Mo 0.5% Si Ti alloy: Bar
BSTA 46	0.012% H (max) 4% Al 4% Mo 0.5% Si Ti alloy: Bar
BSTA 47	0.012% H (max) 4% Al 4% Mo 0.5% Si Ti alloy: Forging
BSTA 48	0.012% H (max) 4% Al 4% Mo 0.5% Si Ti alloy: Forging
BSTA 49	0.012% H (max) 4% Al 4% Mo 0.5% Si Ti alloy: Bar
BSTA 50	0.01% H (max) 4% Al 4% Mo 0.5% Si Ti alloy: Forging
BSTA 51	0.01% H (max) 4% Al 4% Mo 0.5% Si Ti alloy: Forging
BSTA 56	0.01% H (max) 6.1% Al 4% V Ti: Plate
BSTA 57	0.01% H (max) 6.1% Al 4% V Ti: Plate
BSTA 59	0.012% H (max) 6.1% Al 4% V Ti alloy: Sheet
C130 AM	4% Al 4% Mn Ti alloy: Crucible Steel Co.; annealed **DPN: 1070 Elon: 12% Proof: 1000**
CRUCIBLE A110AT	5% Al 2.5% Sn Ti alloy: Crucible Steel Co.; annealed **UTS: 920 Elon: 15% Proof: 760**
CRUCIBLE C130AM	4% Al 4% Mn Ti alloy: Crucible Steel Co.; annealed **UTS: 1070 Elon: 12% Proof: 1000**
DTD 5043 B	1.5% Al 1.5% Mn Ti alloy: Bar; annealed **UTS: 730 Elon: 20% Proof: 450**
DTD 5053	4% Al 4% Mn Ti alloy: Bar; stress relieved **UTS: 950 Elon: 12% Proof: 870**
DTD 5083	5% Al 22.5% Sn Ti alloy: Bar; annealed **UTS: 760 Elon: 12% Proof: 700**
DTD 5093	5% Al 2.5% Sn Ti alloy: Sheet; annealed **UTS: 760 Elon: 12% Proof: 700**
DTD 5123	2.5% Cu Ti alloy: Bar; annealed **UTS: 530 Elon: 20% Proof: 370**
DTD 5133	2.5% Cu Ti alloy: Sheet; annealed **UTS: 580 Elon: 20% Proof: 440**
DTD 5143	4.0% Al 4.0% Mn Ti alloy: Forging; annealed **UTS: 930 Elon: 12% Proof: 870**
DTD 5203	Al Mo Sn Si Ti alloy: Bar and billet **UTS: 1150**
DTD 5213	11% Al Mo Sn Si Ti alloy: Bar and billet; annealed **UTS: 1210 Elon: 10% Proof: 1080**
DTD 5223	Al Mo Sn Si Ti alloy: Forging **UTS: 1150**
DTD 5233	2.5% Cu Ti alloy **UTS: 800**
DTD 5243	2.5% Cu Ti alloy **UTS: 800**
DTD 5253	2.5% Cu Ti alloy **UTS: 770**
DTD 5263	2.5% Cu Ti alloy **UTS: 770**
DTD 5303	0.012% H (max) 6.1% Al 4% V Ti alloy: Bar
DTD 5313	0.012% H (max) 6.1% Al 4% V Ti alloy: Forging
DTD 5323	0.012% H (max) 6.1% Al 4% V Ti alloy: Forging
DTD 5333	0.012% H (max) 4% Al 4% Mo 2% Sn 0.5% Si Ti alloy: Bar
DTD 5343	0.012% H (max) 4% Al 4% Mo 2% Sn 0.5% Si Ti alloy: Forging
DTD 5353	0.012% H (max) 4% Al 4% Mo 2% Sn 0.5% Si Ti alloy: Forging
DTD 5363	0.1% C (max) 0.15% H (max) 6.1% Al 4% V Ti alloy: Casting
DTD M159	11% Al Mo Sn Si Ti alloy: Draft specification **UTS: 1210 Elon: 10% Proof: 1050**

Note. The following abbreviations and units are used in the tables:

DPN	Hardness, diamond pyramid number
UTS	Ultimate tensile strength, N/mm^2
Elon	Elongation, %
Proof	0.1% proof strength, N/mm^2

1 N/mm^2=0.1 hbar=0.102 kgf/mm^2=0.06475 $tonf/in.^2$=145.04 $lbf/in.^2$=1 MPa

See Appendix II for other abbreviations and conversion tables.

Symbol	Nominal analysis, supplier, condition and remarks.
DTD M160	11% Al Mo Sn Si Ti alloy: Draft specification **UTS: 1210 Elon: 10% Proof: 1050**
ER Ti 3 Al 2.5 V	3% Al 2.5% V Ti alloy: Weld electrode; designation used by AWS
ER Ti 3 Al 2.5 V.1	3% Al 2.5% V Ti alloy: Weld electrode; designation used by AWS
ER Ti 5 Al 2.5 Sn	5% Al 2.5% Sn Ti alloy: Weld electrode; designation used by AWS
ER Ti 5 Al 2.5 Sn 1	5% Al 2.5% Sn Ti alloy: Weld electrode; designation used by AWS
ER Ti 6 Al 2 Cb 1Ta 1Mo	6% Al 2% Mo 1% Ta 2% Nb Ti alloy: Weld electrode; designation used by AWS
ER Ti 6 Al 4 V	6% Al 4% V Ti alloy: Weld electrode; designation used by AWS
ER Ti 6 Al 4 V.1	6% Al 4% V Ti alloy: Weld electrode; designation used by AWS
ER Ti 12	0.03% C (max) 3% Mo 0.75% Ni Ti alloy: Weld electrode; designation used by AWS
ER Ti 13 V 11 Cr 3 Al	11% Cr 3% Al 13% V Ti alloy: Weld electrode; designation used by AWS
FERRO-TITANIUM	0.1% C 40% Fe 8% Al 0.1% Cu 1% Si Ti alloy: Metal Alloys Ltd; primary metal
HYLITE 20	0.013% H (max) 5% Al 2.5% Sn Ti alloy: Annealed; Jessop; obsolete **UTS: 860 Elon: 11% Proof: 720**
HYLITE 25	2.5% Cu Ti alloy: Jessop; obsolete
HYLITE 55	3% Al 6% Sn 5% Zr 0.5% Si Ti alloy: Jessop; obsolete
ICI 314 A	4% Al 4% Mn Ti alloy: ICI; now IMI; see Ti 314 A
ICI 314 C	2% Al 2% Mn Ti alloy: ICI; now IMI; see Ti 314 C
ICI 317	5% Al 2.5% Sn Ti alloy: ICI; now IMI; see Ti 317
K 138 A	Ti C with Co binder: Kennametal; seal rings, bearings, etc.; in contact with Co bound WC
K 151 A	Ti C with Ni bonder: Kennametal; difficult to wet; used for brazing fixtures, etc.
K 162 B	Ti C with Mo Ni binder: Kennametal; seal rings, valve parts
K 163 B1	Ti C with Mo Ni binder: Kennametal; stronger than K 162B
K 164 B	Ti C with Mo Ni binder: Kennametal; stands thermal and mechanical shock
K 165	Ti C with Mo Ni binder: Kennametal; wear resistant
KENTANIUM	Ti C: Available in various grades; Kennametal
L 521	0.08% C (max) 5% Al 2.5% Sn Ti alloy: VEW; annealed **UTS: 780 Elon: 10% Proof: 735**
METCO 111	Cr and Ti oxides: Spray deposit; Metco **DPN: 580**
METCO 130	Al and Ti oxides: Spray deposit; Metco **DPN: 750**
METCO 130 SF	Al and Ti oxides: Spray coating; Metco **DPN: 1150**
METCO 131 VF	Al and Ti oxides: Spray deposit; Metco; fine, dense deposit **DPN: 750**
MIL A 18001	0.3% Al 0.1% Cd Zn anode: US military specification
MIL–F–83142	5% Al 5% Sn 5% Zr Ti alloy: US military specification
MIL–T–009046	0.1% C 8% Mn Ti alloy
MIL–T–9046	5% Al 1% Mo 6% Sn 2% Zr Ti alloy: US military specification
MIL–T–46035	0.02% C 5.5% Al 2% Sn 0.7% Cu 5.5% V Ti alloy
MST 3 Al 2.5 V	0.05% C 0.015% H 0.3% Fe 3% Al 2.5% V Ti alloy: Wrought; Reactive Metals Ltd
MST 4 Al 4 Mn	0.08% C 0.01% H 0.4% Fe 4% Al 4% Mn Ti alloy: Wrought; Reactive Metals Ltd
MST 5 Al 2.5 Sn	0.08% C 5% Al 2.5% Sn low N and H Ti alloy: Wrought; Reactive Metals Ltd; for low temperature use
MST 7 Al 2 Cb 1 Ta	0.08% O 7% Al 2% Nb 1% Ta Ti alloy: Wrought; Reactive Metals Ltd
MST 7 Al 12 Zr	0.04% C 7% Al 12% Zr Ti alloy: Wrought; Reactive Metals Ltd
MST 8 Mn	0.2% C 8% Mn low N and H Ti alloy: Wrought; Reactive Metals Ltd
R 54621	6% Al 2% Mo 2% Sn 4% Zr Ti alloy: Weld electrode; designation used by UNS
R 56440	4% Al 4% Mn Ti alloy: Obsolete; designation used by UNS
RMI 4 Al 4 Mn	0.08% C 0.01% H 0.4% Fe 4% Al 4% Mn Ti alloy: Wrought; Reactive Metals Ltd; annealed **UTS: 1000 Elon: 10% Proof: 920**
RMI 5 Al 2.5 Sn	0.08% C 0.015% H 5% Al 2.5% Sn Ti alloy: Reactive Metals Ltd; annealed **UTS: 870 Elon: 10% Proof: 780**
RMI 5 Al 2.5 Sn EL1	0.05% C 0.015% H 5% Al 2.5% Sn Ti alloy: Reactive Metals Ltd; annealed; low impurity grade **UTS: 730 Elon: 10% Proof: 630**
RMI 5 Al 6 Sn 2 Zr 1 Mo Si	0.05% C 5% Al 1.0% Mo 6% Sn 2% Zr 0.2% Si Ti alloy: Reactive Metals Ltd; alpha alloy **UTS: 650 Elon: 12% Proof: 560**
RMI 6 Al 2 Cb 1 Ta 1 Mo	0.05% C 6% Al 1.0% Mo 2% Nb 1.0% Ta Ti alloy: Bar, etc.; Reactive Metals Ltd
RMI 6 Al 2 Cb 1 Ta 1 Mo	0.05% C 6% Al 1.0% Mo 2% Nb 1.0% Ta Ti alloy: Reactive Metals Ltd; alpha alloy **UTS: 540 Elon: 10% Proof: 490**
RMI 7 Al 2 Cb 1 Ta	0.4% C 7% Al 2% Nb 1% Ta Ti alloy: Reactive Metals Ltd; annealed **UTS: 820 Elon: 10%**
RMI 7 Al 12 Zr	0.04% C 7% Al 12% Zr Ti alloy: Reactive Metals Ltd; annealed **UTS: 940 Elon: 10% Proof: 870**
RMI 8 Al 1 Mo 1 V	0.08% C 0.015% H 8% Al 1% Mo 1% V Ti alloy: Wrought; Reactive Metals Ltd; annealed **UTS: 990 Elon: 10% Proof: 880**
RMI 8 Mn	0.2% C 0.12% H 8% Mn Ti alloy: Reactive Metals Ltd; annealed **UTS: 950 Elon: 10% Proof: 780**
T 283	2.5% Al 15% V Ti alloy: Jessop; obsolete
TA 2M	2% Al 2% Mn Ti alloy: French National Standard
TA 4DE	4% Al 4% Mo 2% Sn 0.5% Si Ti alloy: French National Standard
TA 4M	4% Al 4% Mn Ti alloy: French National Standard **UTS: 950 Elon: 15% Proof: 870**
TA 5E	5.0% Al 2.5% Sn Ti alloy: French National Standard **UTS: 760 Elon: 12% Proof: 700**
TA 6Z-5D	Al Mo Zr Si Ti alloy: For creep resistance; French National Standard
TA 6ZW	5% Al 5% Zr 1% W 0.3% Si Ti alloy: For creep strength; French National Standard
TE11DA	2½% Al 4% Mo 11% Sn 0.2% Si Ti alloy: French National Standard
Ti 5 Al 2.5 Sn	0.08% C 4.6% Al 2.5% Sn low N and H Ti alloy: Sheet, bar, etc.; Titanium Metals Ltd; annealed **DPN: 352 UTS: 820 Elon: 10% Proof: 770**
Ti 5 Al 2.5 Sn EL1	0.08% C 5% Al 2.5% Sn low N and H Ti alloy: Sheet, bar, etc.; Titanium Metals Ltd; annealed; for low temperature use **DPN: 340 UTS: 690 Elon: 10% Proof: 620**

Note. The following abbreviations and units are used in the tables:

DPN	Hardness, diamond pyramid number
UTS	Ultimate tensile strength, N/mm^2
Elon	Elongation, %
Proof	0.1% proof strength, N/mm^2

1 N/mm^2=0.1 hbar=0.102 kgf/mm^2=0.06475 $tonf/in.^2$=145.04 $lbf/in.^2$=1 MPa

See Appendix II for other abbreviations and conversion tables.

Symbol	Nominal analysis, supplier, condition and remarks.
Ti 5 Al 6 Sn 5 Zr	0.04% C 5% Al 5% Sn 5% Zr Ti alloy: Sheet, bar, etc.; Titanium Metals Ltd; annealed; high creep strength **UTS: 800 Elon: 10% Proof: 750**
Ti 7 Al 12 Zr	0.04% C 7% Al 12% Zr Ti alloy: Sheet, bar, etc.; Titanium Metals Ltd; annealed; high creep strength **UTS: 900 Elon: 10% Proof: 820**
Ti 8 Mn	0.2% C 8% Mn low N and H Ti alloy: Sheet, bar, etc.; Titanium Metals Ltd; annealed **UTS: 820 Elon: 10% Proof: 750**
Ti 314 A	4% Al 4% Mn Ti alloy: Plate; IMI **UTS: 930 Elon: 15% Proof: 870**
Ti 314 C	2% Al 2% Mn Ti alloy: Plate; IMI **UTS: 740 Elon: 20% Proof: 450**
Ti 315	2.0% Al 2.0% Mn Ti alloy: IMI **UTS: 730 Elon: 20% Proof: 450**
Ti 317	5% Al 2.5% Sn Ti alloy: Bar and plate; IMI **UTS: 760 Elon: 12% Proof: 700**
Ti 550	4% Al 4% Mo 2% Sn 0.5% Si Ti alloy: Bar for creep use up to 400 °C; IMI **UTS: 1140 Elon: 9% Proof: 1000**
Ti 551	4% Al 4% Mo 4% Sn 0.5% Si Ti alloy: Bar; IMI **UTS: 1240 Elon: 8% Proof: 1100**
Ti 680	11% Al Mo Sn Ti alloy: IMI **UTS: 1210 Elon: 10% Proof: 1050**
Ti 684	6% Al 5% Zr 1% W 0.3% Si Ti alloy: Bar; IMI; for creep use up to 520 °C **UTS: 1000 Elon: 6% Proof: 880**
Ti 685	Al Mo Zr Ti: Bar; IMI; for creep use up to 550 °C **UTS: 1000 Elon: 6% Proof: 880**
Ti Cu 2	2.5% Cu Ti alloy: German specification
Ti P62	4.0% Al 4.0% Mn Ti alloy: AICMA **UTS: 930 Elon: 15% Proof: 870**
VR 65	Mo Ni Ti: Sinter material; VR/Wesson specific gravity 5.9; hardness 1870 DPN
XEV-J	5.8% Al 4% V Ti alloy: For cables; SAE

Note. The following abbreviations and units are used in the tables:

DPN	Hardness, diamond pyramid number
UTS	Ultimate tensile strength, N/mm^2
Elon	Elongation, %
Proof	0.1% proof strength, N/mm^2

1 N/mm^2=0.1 hbar=0.102 kgf/mm^2=0.06475 $tonf/in.^2$=145.04 $lbf/in.^2$=1 MPa

See Appendix II for other abbreviations and conversion tables.

51C Titanium alloys – solution treated

Specific gravity	4.42–4.96
Density	4420–4960 kg/m^3
Solidus/liquidus	1510–1650 °C
Thermal conductivity	6.3 W/m °C
Coefficient of linear expansion	8×10^{-6}/ °C
Electrical conductivity	1.0% IACS (copper 100%)
Specific resistance	1700 microhm mm
Young's modulus of elasticity	$103–117 \times 10^9$ N/m^2
Impact	16 J
Fatigue strength – endurance limit	±370–450 N/mm^2
Hot strength	

Temperature °C	Tensile strength N/mm^2
100	910
200	780
300	760
400	690
500	610

The above properties are typical of the following group, and may not apply exactly to any one specification. It is possible that with certain specifications some of the values may not be applicable.

General metallurgical characteristics

With these alloys use is made of the phase change which occurs in pure titanium at about 880 °C, from alpha to beta structure. This is caused by changing from a hexagonal close-packed crystal structure to a body-centred cubic crystal structure. This change is analogous to that which occurs in iron and is the basis for all thermal treatments of steel where one phase dissolves iron carbide while the other does not. To date no element analogous to carbon in steel has been found to give a similar response with titanium but for all the alloys listed their mechanical properties can be improved by some form of thermal treatment.

The alloys form two groups, the more common being hardened by a complex metallurgical process relying on the instability of the beta phase.

The second series of alloys is much more recent and not yet fully developed. With these the beta phase has been stabilized at room temperature instead of the 880 °C for pure titanium; this is achieved by alloying and is analogous to the 18/8 chromium/nickel stainless steels. Hardening is then achieved by precipitating intermetallic compounds from solution in a manner similar to the complex stainless steel alloys discussed in Section 44N2.

All the alloys have good workability in the annealed or solution treated condition. The corrosion resistance of titanium is seldom markedly decreased by the addition of alloying elements but little benefit is achieved from these elements, particularly regarding hot oxidation and hydrogen contamination; thus use is limited to below 500 °C for most purposes.

Unlike other non-ferrous metals, titanium alloys have an endurance limit below which failure does not occur by fatigue. They have rather poor notch sensitivity characteristics, however, and thus considerable care must be taken to remove potential stress raisers from areas of stress concentration. These include joints, rapid changes in section, damage marks or corrosion pits.

The mechanical properties of all the materials in this group can be enhanced by thermal treating. The thermal treatment consists of a solution treatment where the inter-metallic compounds are taken into solution and the material is in its softest, most ductile condition. This is carried out at temperatures in the region of 800 °C and at this temperature all these materials will have a very high tendency to form brittle compounds at the surface. These compounds then diffuse into the body giving a very brittle material. The compounds are formed in the atmosphere (oxides, nitrides, carbides, hydrides, etc.) and, therefore, the materials must be held at high temperature for as short a time as possible in highly controlled atmospheres, which must be either inert gas or vacuum.

There are techniques whereby the surface of the component is protected by other means such as molten glass, but these require very specialist techniques.

Following the solution treatment the materials are soft and are in their most readily cold worked condition. They can then be aged and any cold work which is applied to the materials between solution treating and ageing will improve the tensile properties with a minimum effect on the ductility.

These materials, correctly treated, have good impact properties and tensile strengths comparable with medium strength steels and are much lighter, thus giving one of the best strength to weight ratios of any series of alloys, along with excellent corrosion resistance.

If any welding is required specialist advice must be obtained; welding is carried out either under vacuum or in conditions where inert gas has been correctly applied.

Welding should be carried out before heat treatment where necessary, or very carefully controlled so that the reduction in mechanical strength is minimized.

The use of these materials is confined to high duty components in the aircraft industry and some petro-chemical and other chemical industries where high strength, low weight and excellent corrosion resistance are required, e.g. in rotating components such as compressor blades in gas turbines, shafts and other such components.

It should be noted that they cannot be used at temperatures above about 300 °C unless in controlled atmospheres.

Symbol	Nominal analysis, supplier, condition and remarks.
1 Al 8 V 5 Fe	0.05% C 5.0% Fe 1.0% Al 8.0% V Ti alloy: Beta grade; information from SAE handbook
2.5 Al 16 V	0.04% C 2.5% Al 16% V Ti alloy: Alpha–beta grade; information from SAE handbook
3 Al 12.5 V	0.02% C 3.0% Al 2.5% V Ti alloy: Alpha–beta grade; information from SAE handbook
3 Al 13 V 11 Cr	0.02% C 3.0% Al 13.5% V 11.0% Cr Ti alloy: Beta grade; information from SAE handbook
3.7144	6% Al 2% Mo 2% Sn 4% Zr Ti alloy: German Standard
3.7164	6.0% Al 4.0% V Ti alloy: German Standard **UTS: 95 Elon: 10% Proof: 700**
3.7174	0.7% Fe 5.5% Al 2% Sn 0.7% Cu 5.5% V Ti alloy: German Standard
3.7184	4% Al 4% Mo 2% Sn Ti alloy: German Standard
4 Al 3 Mo 1 V	0.04% C 4.25% Al 3.0% Mo 1.0% V Ti alloy: Alpha–beta alloy: information from SAE handbook
6 Al 4 V	0.02% C 6.1% Al 3.8% V Ti alloy: Alpha–beta grade; information from SAE handbook
6 Al 6 V 2 Sn	0.02% C 5.5% Al 2.0% Sn 5.5% V Ti alloy: Alpha–beta grade; information from SAE handbook
AMS 4911 A	6% Al 4% V Ti alloy: Sheet and strip; annealed; AMS for commercial alloy C 120 AV **Proof: 880**
AMS 4912	4% Al 3% Mo 1% V Ti alloy: Sheet and strip; solution treated
AMS 4913	4% Al 3% Mo 1% V Ti alloy: Sheet and strip; solution treated and aged

Note. The following abbreviations and units are used in the tables:

DPN	Hardness, diamond pyramid number
UTS	Ultimate tensile strength, N/mm^2
Elon	Elongation, %
Proof	0.1% proof strength, N/mm^2

1 N/mm^2=0.1 hbar=0.102 kgf/mm^2=0.06475 tonf/in.^{2}=145.04 lbf/in.^{2}=1 MPa

See Appendix II for other abbreviations and conversion tables.

Symbol	Nominal analysis, supplier, condition and remarks.
AMS 4917	3% Al 13.5% V 11% Cr Ti alloy: Sheet and strip; solution treated; AMS for commercial alloy B 120 VCA
AMS 4928 A	6% Al 4% V Ti alloy: Bar and forgings; annealed; AMS for commercial alloy C 120 AV **Proof: 880**
AMS 4929	1.3% Fe 5.4% Al 1.25% Mo 1.5% Cr Ti alloy: Bar; annealed; AMS for commercial alloy Ti 155A **Proof: 1000**
AMS 4935	6% Al 4% V Ti alloy: Bar; multiple vacuum melted; annealed; AMS for commercial alloy C 120 V
AMS 4954	0.05% C 4% Al 4% Mn Ti alloy: Wire
AMS 4954 A	6% Al 4% V Ti alloy: Welding wire; AMS for commercial alloy C 120 AV
AMS 4955	8% Al 1% Mo 1% V Ti alloy
AMS 4956	6.2% Al 4% V Ti alloy
AMS 4957	3% Al 4% Mo 4% Zr 8% V 6% Cr Ti alloy
AMS 4958	3% Al 4% Mo 4% Zr 8% V 6% Cr Ti alloy
AMS 4969	1.3% Fe 5.5% Al 1.25% Mo 1.5% Cr Ti alloy: Forging; annealed; AMS for commercial alloy Ti 155 A
AMS 4971	5.5% Al 2% Sn 5.5% V Ti alloy
AMS 4972	8% Al 1% Mo 1% V Ti alloy
AMS 4973	8% Al 1% Mo 1% V Ti alloy
AMS 4975	6% Al 2% Mo 2% Sn 4% Zr Ti alloy
AMS 4976	6% Al 2% Mo 2% Sn 4% Zr Ti alloy
AMS 4977	0.1% C (max) 11.5% Mo 4.2% Sn 5.2% Zr Ti alloy
AMS 4978	5.5% Al 2% Sn 5.5% V Ti alloy
AMS 4979	5.5% Al 2% Sn 5.5% V Ti alloy
AMS 4980	0.1% C (max) 11.5% Mo 4.2% Sn 5.2% Zr Ti alloy
AMS 4981	6% Al 6% Mo 2% Sn 4% Zr Ti alloy
AMS 4995	5% Al 4% Mo 2% Sn 2% Zr 4% Cr Ti alloy
AMS 4997	5% Al 4% Mo 2% Sn 2% Zr 4% Cr Ti alloy
ASTM B265/5	0.1% C 0.01% H 6.2% Al 4% V Ti alloy: Sheet **UTS: 950 Elon: 10% Proof: 88**
ASTM B348/5	0.1% C 0.012% H 6% Al 4% V Ti alloy: Bar; annealed **UTS: 960 Elon: 10% Proof: 880**
ASTM B367 C2	0.1% C 0.01% H 6% Al 4% V Ti alloy: Casting; annealed **UTS: 950 Elon: 6% Proof: 870**
ASTM B381 F5	0.1% C 0.012% H 6% Al 4% V Ti alloy: Forging; annealed **UTS: 920 Elon: 10% Proof: 840**

Symbol	Nominal analysis, supplier, condition and remarks.
ASTM B381 F8	0.1% C 0.012% H 6% Al 2% Sn 0.7% Cu 5.5% V Ti alloy: Forging; annealed **UTS: 1070 Elon: 12% Proof: 1000**
ASTM B381 F9	0.1% C 0.012% H 6.5% Al 4% Mo Ti alloy: Forging; annealed **UTS: 1070 Elon: 10% Proof: 1000**
ASTM B382 ER Ti 4 Al 3 Mo 1 V	4% Al 3% Mo 1% V Ti alloy: Welding rod
ASTM B382 ER Ti 4 Al 4 V	4% Al 4% V Ti alloy: Welding rod
ASTM B382 ER Ti 5 Al 4 Fe Cr	1% Fe 5% Al 2.8% Cr Ti alloy: Welding rod
ASTM B382 ER Ti 6 Al 4 V	6% Al 4% V Ti alloy: Welding rod
ASTM B382 ER Ti 13 V 11 Cr 3 Al	3% Al 13% V 11% Cr Ti alloy: Welding rod
B 120 VCA	3% Al 13% V 11% Cr Ti alloy: Crucible Steel Co.; solution treated and aged **UTS: 1520 Elon: 6% Proof: 1250**
C 120 AV	6% Al 4% V Ti alloy: Crucible Steel Co.; solution treated and aged **UTS: 1110 Elon: 7% Proof: 1050**
C 135 AMO	7% Al 4% Mo Ti alloy: Crucible Steel Co.; solution treated and aged **Elon: 6% Proof: 1140**
CRUCIBLE B120 VCA	3% Al 13% V 11% Cr Ti alloy: Crucible Steel Co.; solution treated and aged **UTS: 1410 Elon: 6% Proof: 1250**
CRUCIBLE C 120 AV	6% Al 4% V Ti alloy: Crucible Steel Co.; solution treated and aged **UTS: 1110 Elon: 7% Proof: 1050**
CRUCIBLE C 135 AMo	7% Al 4% Mo Ti alloy: Crucible Steel Co.; solution treated and aged **UTS: 1410 Elon: 6% Proof: 1140**
DTD 5103	4% Al 4% Mo 2% Sn 0.5% Si Ti alloy: Bar; solution treated and aged **UTS: 1070 Elon: 10% Proof: 880**
DTD 5113	2.2% Al 1% Mo 11% Sn 5% Zr 0.4% Si Ti alloy: Bar; solution treated and aged **UTS: 1030 Elon: 10% Proof: 850**
DTD 5153	4.0% Al 4.0% Mo 2.0% Sn 0.5% Si Ti alloy: Forging; solution treated and aged **UTS: 1070 Elon: 10% Proof: 880**
DTD 5163	6% Al 4% V Ti alloy: Sheet; solution treated and aged **UTS: 950 Elon: 8% Proof: 870**
DTD 5173	6% Al 4% V Ti alloy: Bar; annealed **UTS: 950 Elon: 10% Proof: 870**
DTD 5203	4% Al 4% Mo 2% Sn 0.5% Si Ti alloy
DTD 5223	4% Al 4% Mo 4% Sn 0.5% Si Ti alloy
DTD M200	6% Al 5.0% Zr 1.0% W 0.3% Si Ti alloy
DTD M201	6% Al 4.0% Mo 1.0% Cu 5.0% Zr 0.2% Si Ti alloy
EX 684	Al Zr Si Ti alloy: Bar and forging; IMI; weldable; beta stabilized **UTS: 1030**
HA 7146	0.08% C 7% Al 4% Mo Ti alloy: Harvey Aluminium; solution treated and aged **UTS: 920 Elon: 6% Proof: 750**
HA 8116	0.08% C 8.0% Al 1.0% Mo 1.0% V Ti alloy: Harvey Aluminium; annealed **DPN: 350 UTS: 580 Elon: 10% Proof: 530**
HYLITE 30	1.5% Al 1.5% Mn Ti alloy: Jessop; solution treated; obsolete **DPN: 230 UTS: 690 Elon: 20% Proof: 530**
HYLITE 40	4% Al 4% Mn Ti alloy: Jessop; solution treated; obsolete **DPN: 310 UTS: 950 Elon: 18% Proof: 880**
HYLITE 45	6% Al 4% V Ti alloy: Jessop; solution treated and aged; obsolete **UTS: 940 Elon: 22% Proof: 880**
HYLITE 50	4% Al 4% Mo 2% Sn 0.5% Si Ti alloy: Jessop; solution treated and aged; obsolete **DPN: 350 UTS: 1210 Elon: 15% Proof: 1080**
HYLITE 51	4% Al 4% Mo 4% Sn 0.5% Si Ti alloy: Jessop; solution treated and aged; obsolete **UTS: 1300 Elon: 14% Proof: 1160**
HYLITE 60	3% Al 2% Mo 6% Sn 5% Zr 0.5% Si Ti alloy: Jessop; solution treated and aged; obsolete **UTS: 1070 Elon: 14% Proof: 1050**
HYLITE 65	Al Mo Sn Zr Si alloy: Forging; Jessop; obsolete **UTS: 1070**
ICI 318A	6% Al 4% V Ti alloy: ICI; now IMI; see Ti 318 A
ICI 230	2% Cu Ti alloy: ICI; now IMI; see Ti 230
L 531	0.08% C (max) 6% Al 4% V Ti alloy: VEW; specification hardened **UTS: 1230 Elon: 15% Proof: 1070**
L 532	0.08% C (max) 7% Al 4% Mo Ti alloy: VEW; specification hardened **UTS: 1320**
L 535	4% Al 4% Mo 2% Sn 0.5% Si Ti alloy: VEW; specification hardened **UTS: 1200 Elon: 14% Proof: 1070**
LT 24	0.05% C (max) 5% Al 2% Mo 2% Sn 4% Zr Ti alloy: Krupp-Lockner; Titrutan **UTS: 900 Elon: 8% Proof: 830**
LT 25	0.1% C (max) 0.2% Fe (max) 2.5% Cu Ti alloy: Krupp-Lockner; aged condition; Titrutan **UTS: 650 Elon: 10% Proof: 540**
LT 31	0.08% C (max) 0.25% Fe (max) 6% Al 4% V Ti alloy: Krupp-Lockner; aged; Titrutan **UTS: 1140 Elon: 8% Proof: 1070**
LT 33	0.05% C 0.6% Fe 5.5% Al 2.0% Sn 0.6% Cu 5.5% V Ti alloy: Krupp-Lockner; aged; Titrutan **UTS: 1240 Elon: 6% Proof: 1170**
LT 34	0.08% C (max) 0.2% Fe (max) 4% Al 4% Mo 2% Sn Ti alloy: Krupp-Lockner; aged; Titrutan **UTS: 1100 Elon: 9% Proof: 870**
LT 35	0.08% C (max) 2.5% Fe 4% Al Ti alloy: Krupp-Lockner; aged; Titrutan **UTS: 860 Elon: 8% Proof: 780**
MST 8 Al 1 Mo 1 V	0.08% C 0.01% H 8% Al 1% Mo 1% V Ti alloy: Wrought; Reactive Metals Ltd
RMI 1 Al 8 V 5 Fe	0.05% C 0.012% H 5% Fe 1.2% Al 8% V Ti alloy: Wrought; Reactive Metals Ltd; solution treated and aged **UTS: 1550 Elon: 8% Proof: 1410**
RMI 3 Al 2.5 V	0.05% C 0.015% H 0.3% Fe 3% Al 2.5% V Ti alloy: Wrought; Reactive Metals Ltd; annealed **UTS: 610 Elon: 16% Proof: 530**
RMI 4 Al 3 Mo 1 V	0.08% C 0.015% H 0.2% Fe 4% Al 3% Mo 1% V Ti alloy: Wrought; Reactive Metals Ltd; annealed **UTS: 880 Elon: 10% Proof: 820**
RMI 4 Al 3 Mo 1 V	0.08% C 0.015% H 0.2% Fe 4% Al 3% Mo 1% V Ti alloy: Wrought; Reactive Metals Ltd; solution treated and aged **UTS: 1300 Elon: 4% Proof: 1120**
RMI 6 Al 2 Sn 4 Zr 2 Mo	6% Al 2% Mo 2% Sn 4% Zr Ti alloy: Wrought; Reactive Metals Ltd; alpha–beta alloy **UTS: 580 Elon: 10% Proof: 540**
RMI 6 Al 4 V	0.08% C 0.012% H 6.2% Al 4% V Ti alloy: Wrought; Reactive Metals Ltd; solution treated and aged **UTS: 1070 Elon: 7% Proof: 1000**
RMI 6 Al 4 V EL 1	0.09% C 0.012% H 6% Al 4% V Ti alloy: Reactive Metals Ltd; solution treated and aged; low impurities **UTS: 1070 Elon: 7% Proof: 1000**
RMI 6 Al 6 V 2 Sn	0.05% C 5.5% Al 2% Sn 0.6% Cu 5.5% V Ti alloy: Reactive Metals Ltd; solution treated and aged **UTS: 1250 Elon: 8% Proof: 1170**
RMI 7 Al 4 Mo	0.08% C 0.012% H 7% Al 4% Mo Ti alloy: Reactive Metals Ltd; solution treated and aged **UTS: 1160 Elon: 8% Proof: 1070**

Symbol	Nominal analysis, supplier, condition and remarks.
RMI 13 V 11 Cr 3 Al	0.05% C 0.02% H 3% Al 13% V 10.5% Cr Ti alloy: Wrought; Reactive Metals Ltd; annealed **UTS: 920 Elon: 10% Proof: 780**
RMI 13 V 11 Cr 3 Al	0.05% C 0.02% H 3% Al 13% V 10.5% Cr Ti alloy: Wrought; Reactive Metals Ltd; solution treated and aged **UTS: 1280 Elon: 3% Proof: 1180**
RMI 16 V 2.5 Al	0.06% C 0.15% H 0.3% Fe 2.5% Al 17% V Ti alloy: Wrought; Reactive Metals Ltd; solution treated and aged **UTS: 1180 Elon: 4% Proof: 1070**
RS 115	0.08% C 4% Al 3% Mo 1% V low N and H Ti alloy: Republic Steel Co.; aged
RS 135	0.08% C 7% Al 4% Mo Ti alloy: Republic Steel Co.; aged **UTS: 920 Elon: 6% Proof: 750**
RS 811 X	0.08% C 8.0% Al 1.0% Mo 1.0% V Ti alloy: Republic Steel Co.; annealed **DPN: 350 UTS: 580 Elon: 10% Proof: 530**
T 443	3% Al 13% V 11% Cr Ti alloy: Jessop; obsolete
T 713	2.2% Al 4% Mo 11% Sn 0.3% Si Ti alloy: Jessop; obsolete
TA 6V	6.0% Al 4.0% V Ti alloy: French National Standard **UTS: 950 Elon: 10% Proof: 1000**
Ti 4 Al 3 Mo 1 V	0.08% C 4% Al 3% Mo 1% V low N and H Ti alloy: Sheet; Titanium Metals Ltd; annealed **DPN: 340 UTS: 840 Elon: 10% Proof: 770**
Ti 4 Al 3 Mo 1 V	0.08% C 4% Al 3% Mo 1% V low N and H Ti alloy: Sheet; Titanium Metals Ltd; solution treated and aged **DPN: 426 UTS: 1210 Elon: 5% Proof: 1050**
Ti 5 Al 4 Fe Cr	0.1% C 1.5% Fe 5.25% Al 2.7% Cr low N and H Ti alloy: Sheet, bar, etc.; Titanium Metals Ltd; annealed **DPN: 340 UTS: 1050 Elon: 10% Proof: 970**
Ti 5 Al 4 Fe Cr	1.5% Fe 5.25% Al 0.1% Cr 2.7% Cr low N and H Ti alloy: Sheet, bar, etc.; Titanium Metals Ltd; solution treated and aged **DPN: 390 UTS: 1210 Elon: 4% Proof: 1050**
Ti 6 Al 4 V	0.8% C 6% Al 4% V low N and H Ti alloy: Sheet, bar, etc.; Titanium Metals Ltd; annealed **DPN: 352 UTS: 880 Elon: 10% Proof: 830**
Ti 6 Al 4 V	0.08% C 6% Al 4% V low N and H Ti alloy: Sheet, bar, etc.; Titanium Metals Ltd; solution treated and aged **UTS: 1080 Elon: 10% Proof: 1050**
Ti 6 Al 4 V EL 1	0.08% C 6% Al 4% V low N and H Ti alloy: Sheet, bar, etc.; Titanium Metals Ltd; annealed; for low temperature use **DPN: 340 UTS: 880 Elon: 10% Proof: 820**
Ti 6 Al 6 V 2 Sn	0.05% C 0.7% Fe 5.5% Al 2% Sn 0.7% Cu 5.5% V Ti alloy: Bar; Titanium Metals Ltd; annealed **UTS: 1050 Elon: 9% Proof: 950**
Ti 6 Al 6 V 2 Sn	0.05% C 0.7% Fe 5.5% Al 2% Sn 0.7% Cu 5.5% V Ti alloy: Bar; Titanium Metals Ltd; solution treated and aged **UTS: 1210 Elon: 7% Proof: 1080**
Ti 7 Al 4 Mo	0.08% C 7% Al 4% Mo Ti alloy: Bar and wire; Titanium Metals Ltd; annealed **DPN: 380 UTS: 990 Elon: 10% Proof: 920**
Ti 7 Al 4 Mo	0.08% C 7% Al 4% Mo Ti alloy: Bar and wire; Titanium Metals Ltd; solution treated and aged **UTS: 1160 Elon: 7% Proof: 1080**
Ti 8 Al 1 Mo 1 V	0.08% C 8% Al 1% Mo 1% V low N and H Ti alloy: Sheet, bar, etc.; Titanium Metals Ltd; annealed **DPN: 350 UTS: 880 Elon: 10% Proof: 820**
Ti 8 Al 1 Mo 1 V	0.08% C 8% Al 1% Mo 1% V low N and H Ti alloy: Sheet, bar, etc.; Titanium Metals Ltd; solution treated and aged **UTS: 920 Elon: 10% Proof: 840**
Ti 13 V 11 Cr 3 Al	0.05% C 3% Al 13% V 10.5% Cr low N and H Ti alloy: Sheet, bar, etc.; Titanium Metals Ltd; annealed **DPN: 340 UTS: 860 Elon: 10% Proof: 810**
Ti 13 V 11 Cr 3 Al	0.05% C 3% Al 13% V 10.5% Cr low N and H Ti alloy: Sheet, bar, etc.; Titanium Metals Ltd; solution treated and aged **UTS: 1160 Elon: 4% Proof: 1080**
Ti 140 A	2.0% Fe 2.0% Mo 2.0% Cr Ti alloy: Titanium Metals Ltd
Ti 155 A	0.08% C 1.5% Fe 5.5% Al 1.2% Mo 1.5% Cr low N and H Ti alloy: Bar; Titanium Metals Ltd; annealed **DPN: 385 UTS: 1000 Elon: 9% Proof: 920**
Ti 155 A	0.08% C 1.5% Fe 5.5% Al 1.2% Mo 1.5% Cr low N and H Ti alloy: Bar; Titanium Metals Ltd; solution treated and aged **UTS: 1180 Elon: 9% Proof: 1080**
Ti 205	15% Mo Ti alloy: IMI; solution treated and aged **UTS: 1060 Elon: 4% Proof: 900**
Ti 230	2% Cu Ti alloy: IMI; solution treated and aged **UTS: 760 Elon: 23% Proof: 580**
Ti 318 A	6% Al 4% V Ti alloy: IMI **UTS: 950 Elon: 10% Proof: 700**
Ti 550	4% Al 4% Mo 2% Sn 0.5% Si Ti alloy: Rod; IMI; solution treated and aged **UTS: 1230 Elon: 12% Proof: 1120**
Ti 551	4% Al 4% Mo 4% Sn 0.5% Si Ti alloy: Rod; IMI; air cooled and aged **UTS: 1350 Elon: 10% Proof: 1250**
Ti 679	Al Mo 11% Sn 5% Zr Si Ti alloy: IMI; solution treated and aged **UTS: 800 Elon: 10% Proof: 580**
Ti 684	6% Al 5% Zr 1% W 0.3% Si Ti alloy: Rod; IMI; solution treated and aged **UTS: 1030 Elon: 10% Proof: 930**
Ti 685	6% Al 0.5% Mo 5% Zr 0.5% Si Ti alloy: Rod; IMI; solution treated and aged **UTS: 1030 Elon: 10% Proof: 930**
Ti 700	6% Al 4% Mo 1% Cu 5% Zr Ti: Bar; IMI; for creep use up to 400 °C; solution treated and aged **UTS: 1350 Elon: 6% Proof: 1240**
Ti 700	6.0% Al 4.0% Mo 1.0% Cu 5.0% Zr 0.2% Si Ti alloy: Rod; IMI; oil quenched and aged **UTS: 1400 Elon: 10% Proof: 1300**
Ti Al 4 Mo 4 Sn 2 Si	3.5% Al 3.5% Mo 2% Sn Ti alloy: German specification
Ti Al 5 Fe 2.5	2.5% Fe 4% Al Ti alloy: German specification
Ti Al 6 Sn 2 Zr 4 Mo 2	6% Al 2% Mo 2% Sn 4% Zr Ti alloy: German specification
Ti Al 6 V 4	6% Al 4% V Ti alloy: German specification
Ti Al 6 V 6 Sn 2	0.7% Fe 5.5% Al 2% Sn 0.7% Cu 5.5% V Ti alloy: German specification
UTA 8 DV	0.08% C 8% Al 1.0% Mo 1.0% V Ti alloy: French National Standard; annealed **DPN: 350 UTS: 880 Elon: 10% Proof: 810**

Note. The following abbreviations and units are used in the tables:

DPN	Hardness, diamond pyramid number
UTS	Ultimate tensile strength, N/mm^2
Elon	Elongation, %
Proof	0.1% proof strength, N/mm^2

1 N/mm^2=0.1 hbar=0.102 kgf/mm^2=0.06475 tonf/in.^{2}=145.04 lbf/in.^{2}=1 MPa

See Appendix II for other abbreviations and conversion tables.

52. Tungsten W

Physical properties		
Atomic number		74
Atomic weight		184.0
Crystal structure		Body-centred cubic
Colour		Steel grey
Specific gravity		19.3
Density		19 300 kg/m^3
Melting point		3370 °C
Boiling point		5930 °C
Specific heat		0.134 J/g °C
Thermal conductivity		163.3 W/m °C
Coefficient of linear expansion (20–100 °C)		4.4×10^{-6}/ °C
Latent heat of fusion		184.2 J/g
Latent heat of vaporization		–
Thermal neutron absorption cross-section		19.2 barns/atom
Electrical conductivity		31% IACS (copper 100%)
Specific resistance		56.5 microhm mm
Temperature coefficient of electrical resistance		0.0048/ °C
Electrochemical equivalent		3.430 g/A/h
Electrode potential		4.5 V
Magnetic susceptibility		0.33×10^{-6}
Young's modulus of elasticity		345×10^9 N/m^2
Tensile strength	sintered	15
	drawn wire	450
Hardness	sintered	255 DPN
	drawn and annealed	480 DPN

52.1 General notes on tungsten

Tungsten exists in nature as wolframite, a complex iron manganese tungsten oxide, and scheelite, a calcium tungsten oxide. The metal is still commonly called wolfram and it is from this that the chemical symbol is derived. The ores are generally found in association with those of other refractory metals, such as molybdenum, and also with tin ores.

Tungsten of at least 99.7% purity is available, being steel grey in colour and resembling molybdenum in many respects. The tensile strength varies from 150 N/mm^2 in the fully annealed condition to over 4500 N/mm^2 for fine wire finally cold drawn using diamond dies.

Tungsten does not oxidize appreciably below about 500 °C and does not form a volatile oxide as readily as molybdenum. It has excellent resistance to the attack of most of the common acids, even at high temperature.

These properties make it a valuable element in nuclear engineering and there is little doubt that only the difficulty in working the pure metal precludes much greater use of tungsten in this field.

Pure tungsten wire is widely used in electric lamps when it generally has some thoria present to prevent sag at temperature.

The low thermal expansion of tungsten makes it useful when glass metal seals are required and this property is enhanced by the reasonably good electrical conductivity.

Tungsten is generally used as the non-consumable electrode in inert gas shielded arc welding, again because of the reasonably good electrical conductivity, coupled this time with the high melting point.

Because of the excellent refractory nature of the metal, considerable work is in progress to obtain a more ductile material. Using vacuum arc and electron beam melting and re-melting, reasonably ductile tungsten has been obtained, which would appear to indicate that if the undesirable impurities could be identified and eliminated a workable tungsten could be produced.

Probably the best known use of tungsten is in combination with carbon as sintered tungsten carbide for cutting tools and other uses, where abrasion at high temperature has to be overcome.

The largest quantity of tungsten is used as an alloying element in steel, where it combines with the carbon and gives the steel the ability to retain its hardness at temperatures up to red heat. Tungsten is seldom the sole alloying element, generally being present up to 18–20% with chromium and vanadium in lesser amounts. The degree of hot hardness is in direct proportion to the tungsten present. Tungsten is also used as an alloying element with other metals where its refractory qualities are useful. Many of the cobalt and nickel high temperature alloys include tungsten and it is also a valuable addition to some creep resistant steels.

Alloyed with copper, tungsten gives a hard material with good electrical conductivity, useful for resistance welding electrodes.

Some sintered alloys with about 90% tungsten have a specific gravity almost double that of lead. These are used for balance weights where space saving is important, for example high speed rotating parts and aircraft control surfaces.

These alloys are also used in nuclear engineering and medicine as X-ray barriers.

Tungsten is one of the modern metals which have valuable

properties regarding temperature and corrosion resistance. It is difficult to obtain in pure state, however, and present knowledge and techniques are such that very great difficulty exists in working the metal.

Following are the various grades of tungsten, including the available trade names for tungsten carbides.

Symbol	Nominal analysis, supplier, condition and remarks.
2A 3	11.0% Co WC: Sinter material; VR/Wesson; specific gravity 14.2; hardness 1470 DPN
2A 5	6.0% Co WC: Sinter material; VR/Wesson as GI; specific gravity 14.85; hardness 1860 DPN
2A 6	8.0% Co WC: Sinter material; VR/Wesson; specific gravity 147; hardness 1480 DPN
2A 7	4.0% Co WC: Sinter material; VR/Wesson; specific gravity 15.1; hardness 1875 DPN
2A 68	6.0% Co WC: Sinter material; VR/Wesson; specific gravity 14.85; hardness 1700 DPN
26	7.0% Co 10.0% TaC WC: Sinter material; VR/Wesson; specific gravity 12.4; hardness 1790 DPN
44 A	6.0% WC: For tool tips; General Electric Co. **UTS: 4500 Elon: 1.0% Proof: 1420**
55A	13% Co WC: For tool tips; General Electric Co. **UTS: 3700 Elon: 1.9% Proof: 530**
55B	16% Co WC: For tool tips; General Electric Co. **UTS: 2700 Elon: 2.7% Proof: 700**
78	8.0% Co 4.0% TaC 12% TiC WC: For tool tips; General Electric Co. **UTS: 4600 Elon: 1.0% Proof: 1700**
78B	9.0% Co 8.0% TiC WC: For tool tips; General Electric Co. **UTS: 3900 Elon: 2.0% Proof: 610**
90	10% Co WC: For tool tips; General Electric Co. **UTS: 4200 Elon: 1.9% Proof: 920**
120	12% Co WC: For tool tips; General Electric Co. **UTS: 3700 Elon: 3.5% Proof: 840**
190	25% Co WC: For tool tips; General Electric Co. **UTS: 3300 Elon: 3.5% Proof: 370**
330	12% Ni 2.0% Mo 9.5% TaC 30% TiC WC: For tool tips; General Electric Co. **UTS: 3700 Elon: 0.5% Proof: 1750**
350	4.5% Co 12.2% TaC 12.5% TiC WC: For tool tips; General Electric Co. **UTS: 4600 Elon: 0.9% Proof: 1200**
370	8.5% Co 11.5% TaC 8.0% TiC WC: For tool tips; General Electric Co. **UTS: 4600 Elon: 1.0% Proof: 1700**
779	9.0% Co WC: For tool tips; General Electric Co. **UTS: 4200 Elon: 1.5% Proof: 920**
860	5.0% Co 4.0% TaC WC: For tool tips; General Electric Co. **UTS: 4800 Elon: 1.1% Proof: 2000**
883	6.0% Co WC: For tool tips; General Electric Co. **UTS: 4500 Elon: 0.8% Proof: 2000**
900	3.0% Co 4.0% TaC WC: For tool tips; General Electric Co. **UTS: 4500 Elon: 0.7% Proof: 2500**
907	6.0% Co 20% TaC WC: For tool tips; General Electric Co. **UTS: 4800 Elon: 1.7% Proof: 1700**

Note. The following abbreviations and units are used in the tables:

DPN	Hardness, diamond pyramid number
UTS	Ultimate tensile strength, N/mm^2
Elon	Elongation, %
Proof	0.1% proof strength, N/mm^2

1 N/mm^2=0.1 hbar=0.102 kgf/mm^2=0.06475 tonf/in.^{2}=145.04 lbf/in.^{2}=1 MPa

See Appendix II for other abbreviations and conversion tables.

Symbol	Nominal analysis, supplier, condition and remarks.
999	3.0% Co WC: For tool tips; General Electric Co. **UTS: 4500 Elon: 0.5% Proof: 2400**
3047	WC: Tungsten carbide; Kennametal; high strength; high impact resistance
3109	WC: With Co binder; Kennametal; extrusion punches
3411	WC: Tungsten carbide; Kennametal; high hardness and impact value
AGATE	10% Co WC: Sinter; Dymet **DPN: 2400**
AMD 78 BU	W: Sintered forgings; Interim AMS
AMETHYST	10% Co 3% TiC 8% TaC WC: Sinter; Dymet **DPN: 2000**
AMS 7725	W base high density sintered alloy shapes
AMS 7897	99.95% W
AMS 7898	W: Sintered sheet, strip and foil
AMS 7899	W: Sheet, strip and foil
AQUAMARINE	6% Ni WC: Sinter; Dymet **DPN: 1700**
ASTM B297 EW P	99.5% W: Electrode for arc welding
ASTM B297 EW Th 1	1.0% Th W: Electrode for arc welding
ASTM B297 EW Th 2	2.0% Th W: Electrode for arc welding
ASTM B297 EW Zr	0.4% Zr W: Electrode for arc welding
ASTM B346	6% Ni 3% Cu W alloy: Powder compact; physical properties vary with weight of part **UTS: 58 Elon: 1% Proof: 450**
ASTM B760	0.01% C (max) W: Plate, sheet and foil
ASTM B772/3	95% W: Density 18 g/cc **DPN: 340**
ASTM B772/4	97% W: Density 18.5 g/cc **DPN: 350**
ASTM B777/1	90% W: Density 17 g/cc **DPN: 320**
ASTM B777/2	92.5% W: Density 17.5 g/cc **DPN: 330**
BERYL	0.3% Cr 15% Co WC: Sinter; Dymet **DPN: 2400**
BS 4276 BS1	5.0% Co W: Hard metal; grain size 0.5–1 μm **DPN: 1750 UTS: 1700**
BS 4276 BS2	5.5% Co W: Hard metal; grain size 0.5–1.5 μm **DPN: 1720 UTS: 1700**
BS 4276 BS3	5.5% Co W: Hard metal; grain size 0.5–2 μm **DPN: 1650 UTS: 1700**
BS 4276 BS4	6.0% Co W: Hard metal; grain size 1–3 μm **DPN: 1550 UTS: 2000**
BS 4276 BS5	8% Co W: Hard metal; grain size 1–5 μm **DPN: 1450 UTS: 2000**
BS 4276 BS6	10.5% Co W: Hard metal; grain size 1–5 μm **DPN: 1350 UTS: 2200**
BS 4276 BS7	12% Co W: Hard metal; grain size 1–5 μm **DPN: 1240 UTS: 2300**
BS 4276 BS8	15% Co W: Hard metal; grain size 1–5 μm **DPN: 1150 UTS: 2500**
CARBALOY	General Electric name for tungsten carbide grades covered by designation numbers
COLMONOY 705	2.4% C 2.0% Si 6.8% Cr 2.2% Fe 1.5% B 2.0% Co 39% Ni W alloy: Powder for metal spraying; melting point 1040 °C; Colmonoy
COPELMET	Cu + W or WC sintered material: Metro-Cutanit; range of materials
CORAL	0.3% CrC 6% Co 1% TaC WC: Sinter; Dymet **DPN: 1800**
CUTANIT	Co + Ti + Ta carbides with WC sintered materials: Metro-Cutanit; range of materials

Symbol	Nominal analysis, supplier, condition and remarks.
DIADUR 1C	WC welding electrode: Steel filler rod; Soudometal; tungsten carbide deposit; 0.7 mm grain size
DIADUR 2C	WC welding electrode: Steel filler rod; Soudometal; tungsten carbide deposit; 0.7 mm grain size
DIADUR 2000 DE 5.5	WC welding electrode: Wire coated; Soudometal; tungsten carbide deposit; by oxy-acetylene
DIADUR 2000 DE 8	WC welding electrode: Wire coated; Soudometal; tungsten carbide deposit; by oxy-acetylene
DIADUR AT	WC welding electrode: Steel tube filler; Soudometal; tungsten carbide deposit
DIADUR HR	WC welding electrode: Steel tube filler; Soudometal; tungsten carbide deposit
DIADUR OC	WC welding electrode: Steel filler rod; Soudometal; tungsten carbide deposit; 0.5 mm grain size
EMERALD	6% Co 1% TaC WC: Sinter; Dymet **DPN: 1900**
EWP	99.5% W: Weld electrode; designation used by AWS
EWTR1	1% Th W: Weld electrode; designation used by AWS
EWTR2	2% Th W: Weld electrode; designation used by AWS
EWTR3	0.45% Th W: Weld electrode; designation used by AWS
EWZr	0.32% Zr W: Weld electrode; designation used by AWS
F 26	WC: Tungsten carbide; Kennametal; high hardness and impact value
GARNET	20% Co WC: Sinter; Dymet **DPN: 2700**
GE 125	25% Rh W alloy: Sintered and wrought; billet and tube; General Electric Co.
GE HEVIMET	6.0% Ni 4.0% Cu W alloy: General Electric Co.; specific gravity 17.0 **UTS: 2000**
GI	6.0% Co WC: Sinter material; VR/Wesson now 2A5
HEAVY METAL	Cu Ni W alloy: Specific gravity 16–18; Tungsten Mfg Co.; sintered **DPN: 250**
HEVIMET	6.0% Ni 4.0% Cu W alloy: General Electric Co.; specific gravity 17.0 **UTS: 2000**
HR	9.0% Co 8.0% T C 10% Ta + Nb WC: Sinter material; VR/Wesson now VR R77
HV	3.0% Co 13.0% Ti C 1.0% Mo WC: Sinter material; VR/Wesson; specific gravity 11.9; hardness 1890 DPN
hW 11	99.99% W: Rod; Koch-Light Ltd
hW 72	99.999% W: Single crystal 3 mm diameter; Koch-Light Ltd
JADE	17% Co WC: Sinter; Dymet **DPN: 2600**
K 1	WC with Co binder: Kennametal; highest impact value
K 2S	WC with $WTiC_2$: Kennametal; heavy to medium loads and interrupted cuts
K 3H	WC with high percentage $WTiC_2$: Kennametal; medium heavy cuts on low C steel
K 4H	WC with $WTiC_2$: Kennametal; light load form tools
K 5H	WC with high percentage $WTiC_2$: Kennametal; for semi-finishing cuts
K 6	WC with Co binder: Kennametal; highest compressive strength
K 7H	WC with high percentage $WTiC_2$: Kennametal; high speed low load machining
K 8	WC with Co binder: Kennametal; best wear resistance
K 9	WC: Tungsten carbide; Kennametal; high wear resistance
K 11	WC with Co binder: Kennametal; hardest grade
K 21	WC with $WTiC_2$: Kennametal; medium loads with interrupted cuts
K 81	WC with $WTiC_2$ and Co binder: Kennametal; for cold punches
K 82	WC with $WTiC_2$ and Co binder: Kennametal; impact extrusion dies
K 84	WC with $WTiC_2$ and Co binder: Kennametal; tube sizing mandrels
K 86	WC with $WTiC_2$ and Co binder: Kennametal; draw dies, ball valves, etc.
K 90	WC with Co binder: Kennametal; blanking dies, thrust bearings
K 90A	WC with Co binder: Kennametal; heavy leading dies
K 91	WC with Co binder: Kennametal; medium blanking dies, crushing hammers
K 92	WC with Co binder: Kennametal; medium blanking dies, etc.
K 94	WC with Co binder: Kennametal; light blanking dies, etc.
K 95	WC with Co binder: Kennametal; light blanking dies, valve parts, etc.
K 96	WC with Co binder: Kennametal; compacting dies, seal rings, etc.
K 601	WC and TaC: Binder free; tantalum and tungsten carbide; Kennametal; corrosion and wear resistant
K 701	WC with Cr and Co binder: Kennametal; corrosion and wear resistant
K 801	WC with Ni binder: Kennametal; corrosion resistant and high hardness
KENNAMETAL	Tungsten carbide: Available in various grades; Kennametal
KENNERTIUM W2	W alloy: For powder metallurgy; Kennametal; high density; specific gravity 18.5 **UTS: 720 Elon: 3% Proof: 580**
KENNERTIUM W10	W alloy: For powder metallurgy; Kennametal; high density; specific gravity 17.0 **UTS: 920 Elon: 15% Proof: 640**
KM	WC with high percentage $WTiC_2$: Kennametal; for interrupted rough cuts
MATTHEY 1W3	40% Cu W alloy: Sinter for contacts; conductivity 41% IACS; Johnson Matthey **DPN: 140**
MATTHEY 3W3	32% Cu W alloy: Sinter for contacts; conductivity 33% IACS; Johnson Matthey **DPN: 160**
MATTHEY 20K3	33% Cu WC: For resisting weld electrode; Johnson Matthey; conductivity 20% IACS **DPN: 300 UTS: 680**
MATTHEY 30W3	22% Cu W: Sinter alloy for contacts; conductivity 28% IACS; Johnson Matthey **DPN: 240**
MATTHEY 35S	35% Ag W alloy: Sinter for contacts; conductivity 52% IACS; Johnson Matthey **DPN: 140**
MATTHEY 205	27% Ag W: Sintered alloy for contacts; conductivity 43% IACS; Johnson Matthey **DPN: 220**
MATTHEY 2265	35% Ag W: Sinter alloy for contacts; conductivity 50% IACS; Johnson Matthey **DPN: 135**
MATTHEY 2355	45% Ag W alloy: Sinter for contacts; conductivity 60% IACS; Johnson Matthey **DPN: 125**

Note. The following abbreviations and units are used in the tables:

DPN	Hardness, diamond pyramid number
UTS	Ultimate tensile strength, N/mm^2
Elon	Elongation, %
Proof	0.1% proof strength, N/mm^2

$1\ N/mm^2 = 0.1\ hbar = 0.102\ kgf/mm^2 = 0.06475\ tonf/in.^2 = 145.04\ lbf/in.^2 = 1\ MPa$

See Appendix II for other abbreviations and conversion tables.

Symbol	Nominal analysis, supplier, condition and remarks.
MATTHEY 2373	27% Ag W: Sinter alloy for contacts; conductivity 45% IACS; Johnson Matthey **DPN: 180**
MATTHEY G14	40% Ag WC: Sinter alloy for contacts; conductivity 36% IACS; Johnson Matthey **DPN: 200**
METCO 32C	Cr Ni 12% C WC: Spray deposit; Metco **DPN: 1500**
METCO 61	Pure W: Spray deposit; Metco
METCO 71 NS	12% Co WC: Composite spray deposit; Metco **DPN: 400**
METCO 71 VF NS	12% Co WC: Spray deposit; Metco; hard dense deposit **DPN: 580**
METCO 72 F NS	Co WC: Spray deposit; Metco **DPN: 600**
METCO 73 NS1	17% Co WC: Spray deposit; Metco **DPN: 720**
METCO 74 SF	12% Co WC: Spray deposit; Metco **DPN: 520**
METCO 75 F	18% Co WC: Composite deposit; Metco **DPN: 670**
METCO 1123	25% Ni Co WC: Spray deposit; Metco **DPN: 1500**
MUTCO 32C	0.1% C 0.8% Si 3.5% Cr 14% Ni 0.8% B WC alloy: Powder for spraying; Metco; tungsten carbide in stainless steel matrix **DPN: 865**
ONYX	25% Co WC: Sinter; Dymet **DPN: 2800**
OPAL	13% Co WC: Sinter; Dymet **DPN: 2500**
PEARL	0.7% CrC 5% Co 4% TiC 5% TaC WC: Sinter; Dymet **DPN: 1900**
PROLITE	Tungsten carbide: Murex Ltd
R 07004	0.01% C (max) 99.95% W: Obsolete; designation used by UNS
R 07030	96.99% W: Obsolete; designation used by UNS
R 07031	3% Re W alloy: Electrode grade; designation used by UNS
R 07050	94.96% W: Designation used by UNS
R 07080	91.94% W: Designation used by UNS
R 07100	89.9% W: Designation used by UNS
RUBY	0.5% CrC 10% Co WC: Sinter; Dymet **DPN: 2000**
SAPPHIRE	9% Co 9% TiC 4% TaC WC: Sinter; Dymet **DPN: 1700**
SILVELMET	Ag + W or WC sintered material: Metro-Cutanit; range of materials
TOPAZ	7% Co WC: Sinter; Dymet **DPN: 2200**
TUNGSTEN	W: Powder, sheet, wire and amorphous powder; Blackwells
VOLOMIT	4.5% C 2% Fe W alloy: Origin unknown
VR 13	10.0% Co WC: Sinter material; VR/Wesson; specific gravity 14.4 **DPN: 1350**
VR 14	11.5% Co WC: Sinter material; VR/Wesson; specific gravity 14.35 **DPN: 1250**
VR 15	13.0% Co WC: Sinter material; VR/Wesson; specific gravity 14.2 **DPN: 1180**
VR 52	4.0% Co WC: Sinter material for abrasion resistance; VR/Wesson; specific gravity 15.15 **DPN: 1870**
VR 54	8.0% Co WC: Sinter material; VR/Wesson; specific gravity 14.6 **DPN: 1860**
VR 71	6.0% Co 18.0% TiC 10.0% TaC + NbC WC: Sinter material; VR/Wesson; specific gravity 10.7 **DPN: 1875**
VR 73	6.5% Co 12.0% TiC 10.0% TaC + NbC WC: Sinter material; VR/Wesson; specific gravity 11.9 **DPN: 1860**
VR 75	7.5% Co 8.0% TiC 10.0% TaC + NbC WC: Sinter material; VR/Wesson; specific gravity 12.65 **DPN: 1790**
VR 77	9.0% Co 8.0% TiC 10.0% TaC + NbC: Sinter material; VR/Wesson as HR; specific gravity 12.55 **DPN: 1710**
VR 87	17.0% Co 28.0% TaC WC: Sinter material; VR/Wesson; specific gravity 13.5 **DPN: 894**
VR 89	11.0% Co 18.0% TaC WC: Sinter material; VR/Wesson; specific gravity 14.05 **DPN: 1480**
W 18	99.9% W: Powder; Koch-Light Ltd; high purity metal
W 367	99.9% W Fe + Mo 0.04% (max): Sintered and swaged; Tungsten Mfg Co.; suitable for brazing; for electrical contacts
W 408	99.9% W 0.02% Fe + Mo (max): Sintered and swaged; Tungsten Mfg Co.; suitable for brazing; for electrical contacts
W 410L	99.9% W: Sintered and swaged; Tungsten Mfg Co.; suitable for brazing; for electrical contacts
W 410M	99.9% W 0.02% Fe + Mo (max): Sintered and swaged; Tungsten Mfg Co.; suitable for brazing; for electrical contacts
WALLEX 55	3.5% C 1.4% Si 12.5% Cr 0.5% Fe 1.5% B 5.5% Ni 28% Co W alloy: Powder for metal spraying; melting point 1120 °C; Colmonoy
WALLEX 505	2.4% C 1.4% Si 12.5% Cr 0.5% Fe 1.5% B 5.6% Ni 24.0% Co W alloy: Powder for metal spraying; melting point 1120 °C; Colmonoy
WH	7.0% Co 13.0% TiC 2.0% TaC WC: Sinter material; VR/Wesson; specific gravity 11.75; hardness 1860 DPN
WM	10.0% Co 13.0% TiC 2.0% TaC + NbC WC: Sinter material; VR/Wesson; specific gravity 11.5 **DPN: 1710**
WOLFRAM	WO: Tungsten oxide; Blackwells; primary material
WS	13.0% Co 4.0% TiC 2.0% TaC + NbC WC: Sinter material; VR/Wesson; specific gravity 13.2 **DPN: 1400**
ZIRCON	7.5% Co 17% TiC 4% TaC WC: Sinter; Dymet **DPN: 1400**

Note. The following abbreviations and units are used in the tables:

DPN	Hardness, diamond pyramid number
UTS	Ultimate tensile strength, N/mm^2
Elon	Elongation, %
Proof	0.1% proof strength, N/mm^2

1 N/mm^2=0.1 hbar=0.102 kgf/mm^2=0.06475 tonf/in.^{2}=145.04 lbf/in.^{2}=1 MPa

See Appendix II for other abbreviations and conversion tables.

53. Uranium U

Physical properties

Atomic number		92
Atomic weight		238.07
Crystal structure		Body-centred cubic
Colour		Lustrous silvery white
Specific gravity		19.07
Density		19 070 kg/m^3
Melting point		1689 °C
Boiling point		3818 °C
Specific heat		0.1156 J/g °C
Thermal conductivity		26.8 W/m °C
Coefficient of linear expansion (20–100 °C)		19 × 10^{-6}/ °C
Latent heat of fusion		–
Latent heat of vaporization		–
Thermal neutron absorption cross-section		694 barns/atom
Electrical conductivity		6.0% IACS (copper 100%)
Specific resistance		300 microhm mm
Temperature coefficient of electrical resistance		0.0034/ °C
Electrochemical equivalent		–
Electrode potential		−1.4 V
Magnetic susceptibility		–
Young's modulus of elasticity		190 × 10^9 N/m^2
Tensile strength	cast and annealed	340 N/mm^2
Hardness		190 DPN

53.1 General notes on uranium

This is classed as a rare earth metal, belonging to the radioactive group. It is well known, however, as a source of atomic energy and as such is discussed in this section separately from the much lesser known metals classed as 'rare earths'.

Uranium is not widely distributed, being present it is estimated at about four parts per million of the earth's surface. It is thus a rare element in the true sense. Like radium it is found in pitchblende and also as carnotite.

Uranium is a lustrous silvery white metal. It is quite hard and with a specific gravity of almost 20 is one of the heaviest known elements.

Apart from its use in nuclear engineering, uranium has comparatively few applications. These are sufficient, however, for the metal to have been known and used before the atomic bomb was exploded. It has photo-electric properties useful in the ultra-violet range, and has been used in glow discharge lamps.

As an alloying element in certain fields it is claimed that uranium establishes the carbides and prevents agglomeration, thus maintaining an even distribution of fine carbide, ensuring good ductility and fatigue properties.

Some of the compounds of uranium are used in the glass and ceramic industries to impart special properties and colours.

With the development of nuclear fission uranium has become an important element, with a literature comparable in quantity to many of the older metals which have served mankind for centuries. The reason for this is that an isotope of uranium – uranium 235 – can have its nucleus broken fairly readily by bombarding it with neutrons and this releases energy. This breaking or splitting of the nuclei of elements has been known for some time, but until uranium 235 the energy input always exceeded the energy output. The products of the fission are other elements – generally radioactive – which themselves are unstable and change to other elements through a lengthy process culminating in a stable non-radioactive element such as lead.

The newly discovered elements with atomic numbers 93–102, which are discussed with the rare earths in Section 35, are all products of nuclear bombardment and generally have uranium as the starting element.

Uranium itself is radioactive, and is thus a health hazard along with some of its alloys and compounds. Not all the compounds exhibit this effect and many can be safely handled, but it is suggested that where there is any doubt, advice is obtained before handling uranium compounds and salts.

Symbol	Nominal analysis, supplier, condition and remarks.
AMS 7730	0.07% C (max) 99.00% U (min): Depleted uranium
COMMERCIAL U1	U with total impurities 1000 p.p.m.: British Nuclear Fuels
COMMERCIAL U2	U with total impurities 500 p.p.m.: British Nuclear Fuels
COMMERCIAL U3	U with total impurities 200 p.p.m.: British Nuclear Fuels
COMMERCIAL U4	High purity U: British Nuclear Fuels
h U 22	99.7% U: Chips; Koch-Light Ltd
URANIUM 1	Commercial metal: British Nuclear Fuels
URANIUM 2	Commercial as hexafluoride: British Nuclear Fuels
URANIUM 3	Commercial as dioxide: British Nuclear Fuels
URANIUM 4	Commercial as tetrafluoride: British Nuclear Fuels
URANIUM U	Commercial and high purity fo၊ms; Blackwells

Note. The following abbreviations and units are used in the tables:

DPN	Hardness, diamond pyramid number
UTS	Ultimate tensile strength, N/mm^2
Elon	Elongation, %
Proof	0.1% proof strength, N/mm^2

1 N/mm^2=0.1 hbar=0.102 kgf/mm^2=0.06475 $tonf/in.^2$=145.04 $lbf/in.^2$=1 MPa

See Appendix II for other abbreviations and conversion tables.

54. Vanadium V

Physical properties

Atomic number		23
Atomic weight		50.95
Crystal structure		Body-centred cubic
Colour		Brilliant white
Specific gravity		6.11
Density		6110 kg/m^3
Melting point		1735 °C
Boiling point		3000 °C
Specific heat		0.502 J/g °C
Thermal conductivity		31.0 W/m °C
Coefficient of linear expansion (20–100 °C)		8.33×10^{-6}/ °C
Latent heat of fusion		–
Latent heat of vaporization		–
Thermal neutron absorption cross-section		4.7 barns/atom
Electrical conductivity		7.0% IACS (copper 100%)
Specific resistance		248 microhm mm
Temperature coefficient of electrical resistance		0.0028/ °C
Electrochemical equivalent		–
Electrode potential		−1.5 V
Magnetic susceptibility		1.4×10^{-6}
Young's modulus of elasticity		125.5×10^9 N/m^2
Tensile strength	cold rolled	800 N/mm^2
Hardness		170 DPN

54.1 General notes on vanadium

This metal occurs sparingly, very often in conjunction with the rare earth metals, and is never found in the native state.

The concentration, extraction and purification are complex processes, with the added disadvantage that the pentoxide is toxic and this is usually present at one stage of the purification.

The metal is usually obtained as a powder requiring compacting and further purification to give a ductile product. The vast majority of vanadium is used in steel as an alloying element and for this purpose a high purity product is not normally required provided it is free from undesirable impurities such as sulphur.

The pure metal is a brilliant white colour, quite stable at room temperature but combining with oxygen and nitrogen when heated. The metal is also attacked by the strong oxidizing acids.

In the highly purified state, vanadium can be hot and cold worked, showing very little tendency to harden under the influence of excessive cold work – a reduction of over 80% being possible between anneals.

Because of the affinity for oxygen and the fact that nitrogen and hydrogen have deleterious effects, the metal can only be heated under vacuum or using argon.

Vanadium is machined very easily and can be welded using the inert gas shielded arc process.

The difficulty in extraction, together with the scarcity of the ore, prices the metal out of competition with materials of comparable properties.

By far the largest use for vanadium is as an alloy in steel, where it combines with carbon to form hard stable carbides. These are always minute and evenly dispersed and have a considerable influence on the grain size of the steel. It has also been shown that small quantities of vanadium inhibit the tendency of chromium carbides to agglomerate. These properties have a considerable effect on the ductility, fatigue strength and notch sensitivity of the high tensile chromium and tungsten–chromium steels which are notoriously lacking in these properties.

Vanadium is also added to cast iron to improve the hardenability and fatigue strength, while small additions to the age hardening aluminium casting alloys refine the grain, increase the response to thermal treatment and improve the fatigue strength.

Vanadium has considerable potential as a general purpose material which can be very easily worked and fabricated but has low resistance to atmospheric attack at higher temperatures, although the mechanical properties are reasonable. The scarcity and resultant high cost precludes the use of the material at present and there are no known alloys.

Symbol	Nominal analysis, supplier, condition and remarks.
h V 22	99.8% V: Granules; Koch-Light Ltd
h V 74	99.9% V: Single crystals 3 mm diameter; Koch-Light Ltd
VANADIUM	V metal: Blackwells; pure metal

Note. The following abbreviations and units are used in the tables:

DPN	Hardness, diamond pyramid number
UTS	Ultimate tensile strength, N/mm^2
Elon	Elongation, %
Proof	0.1% proof strength, N/mm^2

1 N/mm^2=0.1 hbar=0.102 kgf/mm^2=0.06475 tonf/in.^{2}=145.04 lbf/in.^{2}=1 MPa

See Appendix II for other abbreviations and conversion tables.

55. Zinc Zn

Physical properties	
Atomic number	30
Atomic weight	65.38
Crystal structure	Close-packed hexagonal
Colour	Bluish white
Specific gravity	7.1
Density	7100 kg/m^3
Melting point	419.5 °C
Boiling point	907 °C
Specific heat	0.3898 J/g °C
Thermal conductivity	112.2 W/m °C
Coefficient of linear expansion (20–100 °C)	31.2×10^{-6}/ °C
Latent heat of fusion	110 J/g
Latent heat of vaporization	1754 J/g
Thermal neutron absorption cross-section	1.06 barns/atom
Electrical conductivity	28% IACS (copper 100%)
Specific resistance	59 microhm mm
Temperature coefficient of electrical resistance	0.0041/ °C
Electrochemical equivalent	1.219 g/A/h
Electrode potential	−0.76 V
Magnetic susceptibility	-0.157×10^{-6}
Young's modulus of elasticity	96.5×10^9 N/m^3
Tensile strength	cast 37 N/mm^2
Hardness	30 DPN

55.1 General notes on zinc

Zinc ores occur in large deposits, generally easily mined and widespread throughout the world. Important deposits are worked in Australia and Canada.

Zinc is a bluish white metal which is comparatively brittle at normal temperatures, but soft and ductile at 100 °C. It has been known for many years as an excellent corrosion protective coating for steel. Galvanized corrugated steel sheeting was a favourite building material where economy had first consideration and appearance was not considered. The degree of protection afforded by the zinc coating is proportional to the coating thickness, but varies considerably depending on the atmospheric conditions. A coating of one ounce per square foot (300 g/m^2) of surface should have an indefinite life indoors (at least 60 years), twenty-five years in inland rural conditions, twenty years in rural maritime, and less than ten years in industrial outdoor conditions, these figures being halved for coatings of half an ounce per square foot (150 g/m^2). The method of application has little effect on the life as long as the correct pre-treatment is followed.

Zinc can be applied by hot dipping, by electroplating, by sherardizing, by metal spraying when it must be sealed, or as a zinc pigmented paint.

Considerable development has been carried out in recent years on the continuous production of a zinc coating on steel strip and the coating has ductility and adhesion sufficient to withstand relatively deep deformation on presses.

The advantage of zinc coating is that it is cheap to apply and sacrificial by nature. This means that if slightly damaged to bare the steel substrate, the zinc will corrode itself to protect the steel – this is the galvanic cell action, hence the name 'galvanizing'.

Considerable quantities of zinc are used as the can in 'dry' batteries where it acts as the sacrificial electrode in the production of electric current. This galvanic action is similar to that taking place when steel is protected by zinc and the coating is damaged.

Zinc of commercial purity is used in the form of sheet or strip for building applications such as roofing, flashings and rainwater goods. T-metal (see under Section 55A), an alloy which possesses better mechanical properties, particularly as regards creep strength, is also used for roofing.

Rolled zinc photo-engraving plates of controlled fine grain size are used by the printing industry for letterpress work and reproducing half-tone illustrations.

Rolled and cast zinc anodes are used to protect buried and immersed steelwork cathodically.

Zinc is also used as an alloying element with several distinct purposes as follows.

Copper alloys. Zinc is the main constituent along with copper in the series of alloys known as brass. These range from about 25 to 40% zinc and are the commonest of copper alloys.

In small quantities zinc is used to de-oxidize copper and as an economical addition which has little effect on the attractive properties of copper. These alloys are discussed in Sections 14B and 14C.

Zinc is always present in the nickel silvers where it is a much more economical addition than nickel when colour is more important than strength. These alloys are discussed in Section 14F.

Aluminium alloys. Zinc additions with copper, magnesium and chromium form the strongest series of wrought aluminium alloys. Zinc is usually present from 5% to 7%. These alloys are discussed in Section 1G.

Zinc is also added to some aluminium casting alloys where

its main function is economy. These alloys appear in Section 1M.

Magnesium alloys. The addition of zinc to magnesium, along with zirconium, gives a range of alloys with very fine grain which have excellent hot strength properties. These are listed in Section 23B.

Zinc also forms a series of alloys which are used for a variety of purposes. These alloys are discussed in Section 55A, following which are listed all the names and specifications of zinc and its alloys.

55A Zinc alloys

General metallurgical characteristics

The elements added to zinc all strengthen the alloys without in any way detracting from the excellent casting characteristics; in fact, by increasing the range of temperatures over which solidification occurs, the castability is improved for some purposes.

Some of the alloys have slightly improved tensile properties at higher temperatures while others have been developed to give specific casting shrinkage or growth. Zinc alloys are used for their economy when competing with brass, aluminium and other casting materials, but also in their own right for the ability to reproduce intricate shapes accurately. They can also compete with iron and steel in that they are much more readily melted; thus although not as cheap initially, they are more economical for short run production purposes, where they can be readily remelted and used again. This cannot be so easily carried out with cast iron or steel where more specialized foundry equipment is necessary.

A 20% Al zinc alloy, heated to 250 °C, can have ductility up to 1000% measured as ductility.

None of the alloys noted can have their mechanical properties improved to any great extent by cold working or by thermal treatment. Cold working gives some surface hardening and this property is occasionally used in the manufacture of dies whereby as the die is used the surface hardens. Excessive cold work, however, results in cracking. There are some alloys listed which have age hardening characteristics but this is over a long term and is of little significance as they do not improve the mechanical properties to any great extent.

Most of the alloys in the group can be welded using skill and normal gas welding. Very often, however, modern adhesives give satisfactory results.

These alloys, along with zinc itself, form 'white rust' as they corrode. They do have good corrosion resistance in their own right and in contact with iron or steel will sacrifice themselves to protect the steel.

The alloys are used for die casting materials where lightweight low duty components are involved such as door handles, covers and casings for all sorts of household and transport articles. Intricate toys such as model cars are manufactured from these alloys. Press tool bolsters and dies for short production runs are also made from some of these alloys.

Symbol	Nominal analysis, supplier, condition and remarks.
2Z 4	97.75% Zn: Ingot; designation used by French Standards
2Z 5	98.5% Zn: Ingot; designation used by French Standards
2Z 6	98.75% Zn: Ingot; designation used by French Standards
2.2360	40% Cd 2.0% Al Zn alloy: Working temperature 300 °C; German Standard
AC 41A	1.0% Cu 4% Al Zn alloy: Die casting; designation used by ASTM
AC 41A	4.2% Al 0.04% Mg Zn alloy: Castings; designation used in the UK and USA
AC 43A	4.2% Al 2.7% Cu 0.035% Mg Zn alloy: Ingot; designation used in the UK and USA
AG 40A	0.2% Cu 4% Al Zn alloy: Die casting; designation used by ASTM
AG 40A	4% Al Zn alloy: Castings; designation used in the UK and USA

Note. The following abbreviations and units are used in the tables:

DPN	Hardness, diamond pyramid number
UTS	Ultimate tensile strength, N/mm^2
Elon	Elongation, %
Proof	0.1% proof strength, N/mm^2

1 N/mm^2=0.1 hbar=0.102 kgf/mm^2=0.06475 tonf/in.^{2}=145.04 lbf/in.^{2}=1 MPa

See Appendix II for other abbreviations and conversion tables.

Symbol	Nominal analysis, supplier, condition and remarks.
AG 40B	4.2% Al 0.01% Ni Zn alloy: Castings; designation used in the UK and USA
ALBALOY	Electrodeposited bright Cu Sn Zn alloy: Origin unknown
ALMEN 305	30% Al 5% Cu Zn alloy: Casting as cast; for bearings; Hill Alzen **DPN: 100 UTS: 200 Elon: 2% Proof: 120**
ALZEN 305K	30% Al 5% Cu Zn alloy: Extrusion for bearings; Hill Alzen **DPN: 110 UTS: 250 Elon: 12% Proof: 150**
AMS 4803A	4% Al 0.04% Mg Zn alloy: Casting; as cast
ARSENIC-ZINC	40% As Zn alloy: Blackwells; primary metal
ASTM B6	98.5–99.99% Zn: Slab; spelter; primary metal; covers 5 grades
ASTM B69	Rolled zinc: Sheet and plate; gives information on commercially pure Zn
ASTM B86 AC41A	1.0% Cu 4% Al Zn alloy: Die casting **DPN: 91 UTS: 33 Elon: 7%**
ASTM B86 AG40A	0.2% Cu (max) 4% Al Zn alloy: Die casting **DPN: 82 UTS: 29 Elon: 10%**
ASTM B240 AC41A	4.0% Al 0.04% Mg Zn: Ingot for die casting; as cast
ASTM B240 AG40A	4.0% Al 0.04% Mg Zn: Ingot for die casting; as cast
AVONMOUTH	98.5% Zn ingot: Imperial Smelting Co. Ltd; brand name for BS 3436/4
BATT/Z	High purity Zn sheet for battery cases: London Zinc; Zincon H
BBS/Z	Zn alloy: London Zinc; Zincon/G
BIDENY METAL	88.5% Zn balance Pb and Cu: Used for domestic utensils; Indian origin

Symbol	Nominal analysis, supplier, condition and remarks.
BINDING METAL	3.7% Sb 28% Sn Zn alloy: Used for wire ropes; origin unknown
BIRMINGHAM PLATINA	25% Cu Zn alloy: Origin unknown
BISMUTH-ZINC	40% Bi Zn alloy: Blackwells; primary metal
BS 220	Fine zinc: Replaced by BS 3436
BS 221	Special zinc: Replaced by BS 3436
BS 222	Foundry zinc: Replaced by BS 3436
BS 849	1.2% Pb Zn metal: Sheet; for roofing purposes
BS 1003	Zn: High purity; replaced by BS 3436
BS 1004 A	4.1% Al 0.05% Mg Zn: Ingot and die casting; as cast **DPN: 83 UTS: 270 Elon: 15%**
BS 1004 B	4.0% Al 1.0% Cu 0.05% Mg Zn alloy: For die casting; after 1 year age **DPN: 74 UTS: 200 Elon: 11%**
BS 1141	Secondary zinc alloy: Die castings; withdrawn
BS 2656	0.003% Pb 0.003% Cd 0.004% Hg Zn anodes
BS 3436 Zn 1	99.99% Zn: Ingot; primary metal
BS 3436 Zn 2	99.95% Zn: Ingot; primary metal
BS 3436 Zn 3	99.5% Zn: Ingot; primary metal
BS 3436 Zn 4	98.5% Zn: Ingot; primary metal
DEF 17A	4.1% Al 0.05% Mg Zn: Die castings; defence specification covering BS 1004/A material
DICO	Zinc: Casting for press tools; melting point 390 °C; Platt Metal Co.
DI-METAL	4% Al 3% Cu Zn alloy: Used for die castings; origin unknown
DIN 1706	97.5–99.995% Zn (min) pure metal
DIN 1743 G Zn Al 4 Cu 1	4.0% Al 0.8% Cu 0.03% Mg Zn alloy: German Standard; for sand castings **DPN: 70 UTS: 180 Elon: 0.5%**
DIN 1743 G Zn Al 6 Cu 1	5.8% Al 1.4% Cu Zn alloy: Casting; German Standard; for sand castings **DPN: 80 UTS: 180 Elon: 1.0%**
DIN 1743 GD Zn Al 4	4.0% Al 0.3% Cu 0.03% Mg Zn alloy: Casting; German Standard
DIN 1743 GD Zn Al 4 Cu 1	4.0% Al 0.8% Cu 0.03% Mg Zn alloy: German National Standard; for pressure castings **DPN: 80 UTS: 270 Elon: 2%**
DIN 1743 GK Zn Al 4 Cu 1	4.0% Al 0.8% Cu 0.03% Mg Zn alloy: German Standard; for die castings **DPN: 70 UTS: 200 Elon: 1.0%**
DIN 1743 GK Zn Al 6 Cu 1	5.8% Al 1.4% Cu Zn alloy: German Standard; for die castings
DIN 1743 Z400	4.0% Al 0.3% Cu 0.03% Mg Zn alloy: Casting: German Standard **DPN: 70 UTS: 250 Elon: 1.5%**
DIN 1743 Z410	4.0% Al 0.8% Cu 0.03% Mg Zn alloy: Casting; German Standard
DIN 1743 Z610	5.8% Al 1.4% Cu Zn alloy: Casting; German Standard
DIN 8512 L Zn Cd 40	40% Cd 2.0% Al Zn alloy: Working temperature 300 °C
DOLER	4% Al 3% Cu Zn alloy: Used for die castings; origin unknown
EHRHARDT	3% Cu 6% Sn 2% Pb Zn alloy: Origin unknown
ELEC/Z	Zn alloy: London Zinc; now Zincon 1
ERAYDE	0.1% Ag 2% Cu Zn alloy: Used for radio shields; origin unknown
EZDA	Cu free Zn alloy: Die casting; Electrolytic Zinc Co.; further information from Zinc Development Corporation
FENTON	6% Cu 14% Sn Zn alloy: Used for bearings; origin unknown
FONTAIN MOREAU	6% Cu 1% Pb 1% Fe Zn alloy: Used for ornamental work; origin unknown
h Zn 1	99.9999% Zn: Rod; Koch-Light Ltd; high purity metal
h Zn 73a	99.99% Zn: Single crystals 6.3 mm × 25 mm; Koch-Light Ltd
ILZRO 12	12% Al 1% Cu 0.02% Mg Zn alloy: For casting; International Lead Zinc Research Organization; further information from Zinc Development Corporation **DPN: 115 UTS: 230 Elon: 6% Proof: 140**
ILZRO 14	0.02% Al 1.2% Cu 0.27% Ti Zn alloy: For casting; International Lead Zinc Research Organization; further information from Zinc Development Corporation
ILZRO 16	1.25% Cu 0.2% Ti 0.15% Cr Zn alloy: Origin unknown
KAYEM	Zinc: Rolled sheet; Imperial Smelting Co.; blanking dies; melting point 390 °C **DPN: 96 UTS: 320 Elon: 12%**
KAYEM	Zinc: Casting; Imperial Smelting Co.; melting point 390 °C; for press tools **DPN: 109 UTS: 220 Elon: 1.6%**
KAYEM 2	Zinc: Casting; Imperial Smelting Co.: melting point 358 °C; press tools **DPN: 140 UTS: 140**
KB 90	Iron copper zinc: Sintered material; Sintered Products Ltd; 3.8% porosity **UTS: 340 Elon: 12%**
KIRKSIRE I OR A	4% Al 3% Cu 0.07% Mg Zn alloy: Designation used by German Standards
KIRKSIRE II OR B	4.1% Al 2.7% Cu 0.03% Mg Zn alloy: Designation used by German Standards
KIRKSITE A	Zn base alloy: For press tools; Hoyt Ltd
L M 15	Ag Cd Zn base: Soft solder; Johnson Matthey; electrical conductivity 15% IACS; melting range 280–320 °C **UTS: 180 Elon: 5%**
L Zn Cd 40	40% Cd 2.0% Al Zn alloy: Designation used by German Standards
MAIN METAL	Al Zn alloy: Casting for bearings; Main Metal **DPN: 110 UTS: 220 Elon: 2% Proof: 150**
MAIN METAL	Al Zn alloy: Extrusion for bearings; Main Metal **DPN: 140 UTS: 310 Elon: 16% Proof: 260**
MAZAK 3	4.0% Al 0.05% Mg Zn alloy: For die casting; Imperial Smelting Co.; results after 1 year natural ageing **DPN: 70 UTS: 250 Elon: 25%**
MAZAK 5	4.0% Al 1% Cu 0.05% Mg Zn alloy: For die casting; Imperial Smelting Co.; results after 1 year natural ageing **DPN: 74 UTS: 300 Elon: 11%**
MAZAK 7	4% Al low Mg Cu Ni Zn alloy: For die casting; Imperial Smelting Co.
METCO ZINC	Zn: Metal spray deposit; Metco
METCO ZINC Al	15% Al Zn alloy: Spray deposit; Metco
MNC 71	General standard for SS range of zinc die casting alloys: Swedish Standard
NF A55 101/2Z4	99.75% Zn: Ingot; French National Standard
NF A55 101/2Z5	98.5% Zn: Ingot; French National Standard
NF A55 101/2Z6	98.75% Zn: Ingot; French National Standard
NF A55 101/Z5	98.0% Zn: Ingot; French National Standard
NF A55 101/Z6	98.5% Zn: Ingot; French National Standard
NF A55 101/Z7	99.5% Zn: Ingot; intermediate; French National Standard
NF A55 101/Z8	99.95% Zn: Ingot; fine; French National Standard
NF A55 101 Z9	99.993% Zn: Ingot; extra fine; French National Standard
NF A55 102 ZA4 U1G	4.0% Al 1.0% Cu 0.04% Mg Zn alloy: Ingot; French National Standard

Note. The following abbreviations and units are used in the tables:

DPN	Hardness, diamond pyramid number
UTS	Ultimate tensile strength, N/mm^2
Elon	Elongation, %
Proof	0.1% proof strength, N/mm^2

$1\ N/mm^2 = 0.1\ hbar = 0.102\ kgf/mm^2 = 0.06475\ tonf/in.^2 = 145.04\ lbf/in.^2 = 1\ MPa$

See Appendix II for other abbreviations and conversion tables.

Symbol	Nominal analysis, supplier, condition and remarks.
NF A55 102 ZA4 U3G	4.0% Al 3.0% Cu 0.04% Mg Zn alloy: Ingot; French National Standard
NF A55 102 ZA4G	4.0% Al 0.04% Mg Zn alloy: Ingot; French National Standard
OC/Z	Zn alloy: London Zinc; now Zincon F
QQ Z 35/a	Zinc: Slab or spelter; US Federal specification
QQ Z 100/a	Zinc: Alloy, sheet and strip; US Federal specification
QQ Z 285	Zinc: For anodes; US Federal specification
QQ Z 301/c	Zinc: Sheet and strip; US Federal specification
QQ Z 351/a	Zinc: Slab or spelter; US Federal specification
QQ Z 363/a	Zinc base alloys: For die castings; US Federal specification
SAE 903	4% Al Zn alloy: Casting; as cast **DPN: 82 UTS: 270 Elon: 10%**
SAE 925	4% Al 1% Cu Zn alloy: Casting; as cast **DPN: 91 UTS: 300 Elon: 7%**
SALGE	4% Cu 10% Sn 1% Sb 1% Pb Zn alloy: Used for novelties; origin unknown
SCHULZ	3% Al 6% Cu Zn alloy: Type metal; origin unknown
SEVERN	99.5% Zn: Ingot; Imperial Smelting Co.; brand name for BS 3436/3
For SIS specifications see SS	
SOREL	10% Cu 10% Fe Zn alloy: Ornamental use; origin unknown
SS 7020	4% Al Zn: Die casting; Swedish Standard
SS 7020	4.0% Al 0.04% Mg Zn alloy: Die casting; Swedish Standard
SS 7030	4.0% Al 1.0% Cu 0.04% Mg Zn alloy: Die casting; Swedish Standard
SS 7030	4% Al 1% Cu Zn: Die casting; Swedish Standard
STUBS 56	Cd Zn alloy: Brazing rod; Stubs; melting point 300 °C; for Zn Cu etc. joints **UTS: 139**
STZ 20	Cu Ti Zn alloy: Made by Stolberger Zinc AG; information from Zinc Development Corporation
STZ 30	Cu Ti Zn alloy: Made by Stolberger Zinc AG; information from Zinc Development Corporation
SWANSEA VALE	98.5% Zn: Ingot; Imperial Smelting Co.; brand name for BS 3436/4
T METAL	0.8% Cu 0.1% Ti 0.003% Mn 0.002% Cr Zn alloy: Wrought; London Zinc Mills; heat treated **UTS: 220 Elon: 45% Proof: 120**
TANDEM PTA	Zinc: Casting for press tools; melting point 390 °C; Fry Metal
UNI 3718 Gp Zn Al 4	4.0% Al 0.04% Mg Zn alloy: Italian Standard
UNI 3718 Gp Zn Al 4 Cu 1	4.1% Al 1.0% Cu 0.04% Mg Zn alloy: Italian Standard
UNI 3718 Gp Zn Al 4 Cu 3	4.1% Al 3.1% Cu 0.04% Mg Zn alloy: Italian Standard
VAUCHERS ALLOY	18% Sn 4.5% Pb 2.5% Sb Zn alloy: For bearings; origin unknown
W Zn 3	99.9999% Zn: Powder; Koch-Light Ltd
Z 5	98.0% Zn: Ingot; designation used by French Standards
Z 6	99.5% Zn: Ingot; designation used by French Standards
Z 7 INTERMEDIARE	99.5% Zn: Ingot; designation used by French Standards
Z 8 FIN	99.95% Zn: Ingot; designation used by French Standards
Z 9 EXTRA-FIN	99.993% Zn: Ingot; designation used by French Standards
Z 35630	11% Al 0.7% Cu 0.02% Mg Zn alloy: UNS designation
Z 35635	8.4% Al 1% Cu 0.02% Mg Zn alloy: UNS designation
Z 35840	26% Al 2.2% Cu 0.015% Mg Zn alloy: UNS designation

Symbol	Nominal analysis, supplier, condition and remarks.
ZA 4	Designation used by Italian Standards for UNI 3718 Gp Zn Al 4
ZA 4 C1	Designation used by Italian Standards for UNI 3718 Gp Zn Al 4 Cu 1
ZA 4 C3	Designation used by Italian Standards for UNI 3718 Gp Zn Al 4 Cu 3
ZA 4 G	4.0% Al 0.04% Mg Zn alloy: Ingot; designation used by French Standards
ZA 4 U1G	4.0% Al 1.0% Cu 0.04% Mg Zn alloy: Ingot; designation used by French Standards
ZA 4 U3G	4.0% Al 3.0% Cu 0.04% Mg Zn alloy: Ingot; designation used by French Standards
ZA 8	8.4% Al 1% Cu 0.02% Mg Zn alloy: Ingot; designation used in the UK and USA
ZA 12	10.7% Al 0.8% Cu 0.02% Mg Zn alloy: Castings; designation used in the UK and USA
ZA 27	26.5% Al 2.2% Cu 0.015% Mg Zn alloy: Ingot; designation used in the UK and USA
ZA 27	27% Al 2% Cu 0.02% Mg Zn alloy: Origin unknown **DPN: 115 UTS: 450 Elon: 1% Proof: 390**
ZAM METAL	Zinc base alloy containing Al and Hg: Used for Zn plating anodes; origin unknown
ZAMA	4% Al Zn alloy: Die casting; Italian alloy; further information from Zinc Development Corporation
ZAMAC	4% Al Zn alloy: For die castings; French alloy; further information from Zinc Development Corporation
ZAMAK	4% Al Zn alloy: Die casting; American alloy: further information from Zinc Development Corporation
ZELCO	15% Al 2% Cu Zn: Solder; origin unknown
ZILLAY	1% Cu 0.8% Co 0.1% Mg Zn alloy: Used for roofing; origin unknown
ZIMAL	4.15% Al 3.0% Cu Zn alloy: Origin unknown
ZINC	Zn: Ingot, sticks, powder, dust, etc.; Blackwells
ZINC TINSEL	40% Pb Zn alloy: Origin unknown
ZINCON F	0.7% Pb Zn alloy: Sheet and strip for roofing; London Zinc Mills; OC/Z
ZINCON G	0.7% Pb Zn alloy: Sheet; London Zinc Mills; high quality; BBS/Z
ZINCON H	0.6% Pb Zn alloy: Strip and calots; London Zinc Mills; for dry battery cases Batt/Z
ZINCON I	99.99% Zn: Sheet; high purity; London Zinc Mills; elec/Z
ZINCON J	0.7% Cu 0.08% Ti Zn alloy: Sheet and strip; London Zinc Mills; heat treatable; T metal **UTS: 150 Elon: 45% Proof: 80**
ZISKON	29% Al Zn alloy: Origin unknown
Zn Al 4	4.0% Al 0.04% Mg Zn alloy: For die casting; Swedish Standard designation
Zn Al Cu 1	4.0% Al 1.0% Cu 0.04% Mg Zn alloy: Die casting; Swedish Standard designation

Note. The following abbreviations and units are used in the tables:

DPN	Hardness, diamond pyramid number
UTS	Ultimate tensile strength, N/mm^2
Elon	Elongation, %
Proof	0.1% proof strength, N/mm^2

1 N/mm^2=0.1 hbar=0.102 kgf/mm^2=0.06475 $tonf/in.^2$=145.04 $lbf/in.^2$=1 MPa

See Appendix II for other abbreviations and conversion tables.

56. Zirconium Zr

Physical properties

Atomic number		40
Atomic weight		91.22
Crystal structure	alpha	Close-packed hexagonal up to 860 °C
	beta	Body-centred cubic – 860 °C to melting point
Colour		Silvery white
Specific gravity		6.53
Density		6530 kg/m^3
Melting point		1850 °C
Boiling point		3580 °C
Specific heat		0.285 J/g °C
Thermal conductivity		16.7 W/m °C
Coefficient of linear expansion (20–100 °C)		5.8×10^{-6}/ °C
Latent heat of fusion		251 J/g
Latent heat of vaporization		–
Thermal neutron absorption cross-section		0.18 barns/atom
Electrical conductivity		4% IACS (copper 100%)
Specific resistance		390 microhm mm
Temperature coefficient of electrical resistance		0.004/ °C
Electrochemical equivalent		0.844 g/A/h
Electrode potential		–
Magnetic susceptibility		1.32×10^{-6}
Young's modulus of elasticity		94.5×10^9 N/m^2
Tensile strength	annealed	330 N/mm^2
Hardness	annealed	150 DPN

56.1 General notes on zirconium

This is too chemically active an element ever to be found in the native state. It occurs in considerable quantities as zircon, which is zirconium silicate, and in various other less common forms. Although it is treated as a rare metal zirconium is in fact the twelfth most common element in order of abundance, the scarcity being solely caused by extreme difficulty in extraction.

The pure metal is soft, malleable and ductile, silvery white in colour, but becomes very brittle when impure. Zirconium cannot be treated above about 850 °C in air without becoming embrittled by absorbing the oxide and nitride which start to form on the surface at about 400 °C but which do not diffuse in until about 850 °C.

Satisfactory hot working can be carried out between 600 and 800 °C.

There are no known means of improving the mechanical properties of zirconium or its alloys by thermal treatment, but cold working can be used to increase the tensile strength without causing too rapid a drop in ductility. Annealing at 600 °C in air removes most of the effects of cold work but full annealing requires 800 °C for at least one hour which necessitates inert atmosphere or vacuum to prevent oxide and nitride diffusion.

The scale formed when the metal is heated in air is difficult to remove except by grit blasting and must be eliminated before machining as it is very abrasive.

Machining of zirconium presents no great difficulty as it does not harden rapidly with cold work. In many respects it resembles the austenitic stainless steels but note must be taken of the fire danger that exists when zirconium is in the finely divided state. There is no evidence that the metal or any of its compounds presents any other hazard to health.

Zirconium metal resists most acids and all alkalis and is not appreciably attacked by many molten metals, including the sodium–potassium alloy used in nuclear reactors. The acid resistance is not as good as that of tantalum, particularly when oxidizing acids such as nitric are involved.

The pure metal has been found to be very compatible with body fluids, and tissue adheres to metal inserts in a satisfactory manner. Zirconium thus finds use, with tantalum, for repairing bones and in some cases seized joints are replaced using these metals.

The largest use of zirconium is in the nuclear field where the excellent corrosion resistance joins with the low neutron absorption properties to make this the first choice for 'canning' the reactor material in certain types of nuclear power installations.

For this purpose the metal must be free from hafnium but is sometimes alloyed with small amounts of tin, chromium and nickel.

The excellent corrosion resistance of zirconium finds some uses in chemical engineering but these are neither as many nor as varied as with tantalum. Rayon spinnerets are now made from zirconium and the metal is also used as electric lamp filaments when the power required per candle power produced is lower than with the tungsten filament lamps.

The metal can be welded but requires either high vacuum or copious supplies of inert gas around the weld pool to prevent inclusions of the oxide or nitride embrittling the weld. Wherever possible welding should be in a gas tight chamber under inert atmosphere. Fusion and resistance welding are possible, and zirconium can be welded to other refractory materials such as tungsten and molybdenum using these methods.

Although it resists the attack of most acids zirconium is actually very chemically reactive and this becomes apparent when it is in a finely divided form. This has been put to use as a means of producing a smokeless white flash for photographic and other purposes and is also used as a primer for ammunition. It replaces the normal mercury fulminate or lead azide primers when smokeless ammunition is required.

Zirconium metal is also a popular 'getter' in vacuum apparatus as it can be made into anodes or grids which combine the last traces of gas in the vacuum and then has sufficient strength at high temperatures to prevent distortion and sagging.

This same activity becomes a serious disadvantage when zirconium dust or fine turnings are involved as they are prone to spontaneous combustion under certain conditions. Zirconium fires become explosive if water is added and continue fuming under carbon dioxide and many chlorinated compounds. Dry powder or inert material such as lime, fluorspar or graphite should be used first to confine the blaze and then to reduce the ignited material. Provided good housekeeping is practised and fine zirconium dust or turnings are not stored in quantity, machining and fabrication of zirconium should present no greater hazard than for magnesium. It is necessary, however, to comply with certain Factory Act requirements before handling zirconium.

There are very few alloys with zirconium as the base metal; additions are made in amounts of under 0.5% to about 40% zirconium.

Zirconium is now a popular metal for de-oxidizing high quality steels. Some sulphides and nitrides are removed at the same time, resulting in better cold working properties and increased ductility.

It is also used as an alloying element in certain stainless and tool steels when it is reputed to improve the wear resistance. It is added to heat resistant cast irons when there is an improvement in the machinability with no appreciable reduction in other properties. Higher zirconium contents with silicon cast iron improve the resistance to hot sulphuric acid. Zirconium added to copper acts as a de-oxidizer and residual amounts are sometimes found in the brasses and bronzes.

A new alloy of copper and zirconium has been developed which has certain age hardening characteristics and is similar in many respects to copper–chromium alloys.

Higher zirconium contents in copper give alloys which have mechanical properties after thermal treatment comparable to copper–beryllium.

Added to magnesium, zirconium acts as a grain refiner and gives improved response to age hardening, particularly in the magnesium–zinc range of alloys.

An alloy of zirconium and lead is used in the manufacture of flints for lighters.

The oxide, and purified silicate, is used as a refractory material—particularly the silicate, zircon, which is a popular lining material for electric furnaces.

Other uses for the oxide include abrasive polishing dust, refractory cement, ceramic and porcelain glaze, and as filler material for plastics and rubber.

Symbol	Nominal analysis, supplier, condition and remarks.
ASTM B349 R1	Zr: Sponge; high purity; for nuclear applications; primary metal **DPN: 150**
ASTM B350 R1	Zr: Ingot; high purity; for nuclear applications **DPN: 160**
ASTM B350 RA1	1.5% Sn 0.1% Cr 0.05% Ni Zr: Ingot; for nuclear applications **DPN: 200**
ASTM B351 R1	Zr: Bar; high purity for nuclear applications; annealed **DPN: 290 Elon: 18% Proof: 120**
ASTM B351 RA1	1.5% Sn 0.1% Cr 0.05% Ni Zr: Bars; for nuclear applications; annealed **UTS: 420 Elon: 14% Proof: 220**
ASTM B352 R1	Zr: Sheet and plate; high purity; for nuclear applications; annealed **UTS: 300 Elon: 18% Proof: 170**
ASTM B352 RA1	1.5% Sn 0.1% Cr 0.05% Ni Zr: Sheet and plate; for nuclear applications; annealed **UTS: 440 Elon: 14% Proof: 300**
ASTM B353 R1	Zr: Tube; high purity; for nuclear applications; annealed **UTS: 320 Elon: 18% Proof: 129**
ASTM B353 R1	Zr: Tube; high purity; for nuclear applications; $\frac{1}{4}$-hard **UTS: 420 Elon: 8% Proof: 300**
ASTM B353 R1	Zr: Tube; high purity; for nuclear applications; $\frac{1}{2}$-hard **UTS: 580 Elon: 5% Proof: 450**
ASTM B353 RA1	1.5% Sn 0.1% Cr 0.05% Ni Zr: Tube; for nuclear applications; annealed **UTS: 420 Elon: 14% Proof: 220**
ASTM B353 RA1	1.5% Sn 0.1% Cr 0.05% Ni Zr: Tube; for nuclear applications; $\frac{1}{4}$-hard **UTS: 530 Elon: 8% Proof: 450**
ASTM B353 RA1	1.5% Sn 0.1% Cr 0.5% Ni Zr: Tube; for nuclear applications; $\frac{1}{2}$-hard **UTS: 700 Elon: 3% Proof: 610**
ASTM B356 R1	Zr: Forging; high purity; for nuclear applications; annealed **UTS: 300 Elon: 18% Proof: 170**
ASTM B356 RA1	1.5% Sn 0.1% Cr 0.05% Ni Zr: Forging for nuclear application; annealed **UTS: 420 Elon: 14% Proof: 270**
ATR	0.58% Mo 0.56% Cn Zr alloy: Jessop
ER Zr 2	0.05% C (max) 99.2% Zr (min): Weld electrodes; designation used by AWS
ER Zr 3	0.05% C (max) 1.5% Sn Zr alloy: Weld electrodes; designation used by AWS
ER Zr 4	0.05% C (max) 2.5% Nb Zr alloy: Weld electrodes; designation used by AWS
h Zr 25	99.5% Zr: Wire 1 mm diameter; Koch-Light Ltd
h Zr 74	99.5% Zr: Single crystals 3 mm × 6 mm; Koch-Light Ltd
METCO	7.5% Y Zr oxide: Spray deposit; Metco **DPN: 300**
METCO 143	Y, Ti and Zr oxides: Spray deposit; Metco; ceramic coating **DPN: 910**
METCO 201	Ca Zr oxide: Spray coating; Metco **DPN: 260**
METCO 202 NS	Y and Zr oxides: Spray deposit; Metco **DPN: 395**
METCO 210 NS	Mg Zr oxide: Spray deposit; Metco **DPN: 310**
R 1	High purity Zr: Designation used by ASTM

Note. The following abbreviations and units are used in the tables:

DPN	Hardness, diamond pyramid number
UTS	Ultimate tensile strength, N/mm^2
Elon	Elongation, %
Proof	0.1% proof strength, N/mm^2

1 N/mm^2=0.1 hbar=0.102 kgf/mm^2=0.06475 tonf/in.^{2}=145.04 lbf/in.^{2}=1 MPa

See Appendix II for other abbreviations and conversion tables.

Symbol	Nominal analysis, supplier, condition and remarks.
R 60702	4.5% Hf Zr alloy: UNS designation
R 60704	4.5% Hf 1.5% Sn 0.3% Fe + Cr Zr alloy: UNS designation
R 60705	4.5% Hf 2.5% Nb Zr alloy: UNS designation
R 60902	0.03% C (max) 2.6% Nb 0.45% Cu 0.1% O Zr: reactor grade; designation used by UNS
RA 1	1.5% Sn 0.1% Cr 0.05% Ni Zr: Designation used by ASTM
VDM ZIRKONIUM 702	4.5% Hf 0.16% O 0.2% Fe + Cr (max) Zr alloy; VDM
VDM ZIRKONIUM 704	4.5% Hf 1.5% Sn 0.3% Fe + Cr Zr alloy; VDM
Z 1	Commercially pure Zr: Reactor grade; Jessop
Z 2	0.12% Fe 0.1% Cr 0.05% Ni 1.4% Sn Zr alloy: Jessop
Z 3	0.58% Mo 0.56% Cr Zr alloy: Jessop
Z 5	Commercially pure Zr: Non-reactor grade; Jessop
ZIRCALLOY 11	0.12% Fe 0.1% Cr 0.05% Ni 1.4% SN Zr alloy: Jessop; annealed **UTS: 500** **Elon: 28%** **Proof: 290**
ZIRCONIUM	Zr 99–100% purity: Blackwells
ZIRCONIUM 10	Pure Zr: For reactors; IMI **UTS: 250** **Elon: 25%** **Proof: 160**
ZIRCONIUM 20	1.5% Sn 0.12% Fe 0.1% Cr 0.05% Ni Zr alloy: IMI **UTS: 300** **Elon: 21%** **Proof: 220**
ZIRCONIUM 30	0.5% Cu 0.55% Mo Zr alloy: IMI **DPN: 150** **UTS: 320** **Elon: 21%** **Proof: 200**
ZIRCONIUM 40	1.5% Sn 0.2% Fe 0.1% Cr Zr alloy: IMI

Note. The following abbreviations and units are used in the tables:

DPN	Hardness, diamond pyramid number
UTS	Ultimate tensile strength, N/mm^2
Elon	Elongation, %
Proof	0.1% proof strength, N/mm^2

1 N/mm^2=0.1 hbar=0.102 kgf/mm^2=0.06475 $tonf/in.^2$=145.04 $lbf/in.^2$=1 MPa

See Appendix II for other abbreviations and conversion tables.

Appendix I

Names and addresses of firms and organizations

The following is a list of names and addresses of the firms and organizations whose materials are mentioned in this book.

It must be emphasized that although permission to use the information has been given by the relevant firms and organizations, they are in no way responsible for the accuracy of the extracts. It must also be emphasized that much of the information contained within this book will be covered by patents or trademarks, and that none of the information can be construed as giving permission to make use of subsequent materials such that these trademark rights are infringed upon.

AA *see* Aluminium Association of America
AALCO 31 Davies Street, London W1
ABS *see* American Bureau of Shipping
AEI Limited Trafford Park, Manchester 17
Aerospace Material Specification issued by SAE
AFNOR *see* Association Français de Normalisation
AICMA *see* Association des Constructeurs de Material Aerospatial
AISI *see* American Iron and Steel Institute
ALCAN (Germany) GMBH Postfach 85, Nurnberg, Germany
ALCAN Industries Ltd Bush House, Aldwych, London WC2
ALCOA *see* Aluminium Company of America
Allegheny Ludlum Steel Company Brackenbridge, Pennsylvania, USA
All-State Welding Alloys Co. Inc. White Plains, New York, USA
Aluminium Association of America 420 Lexington Avenue, New York
Aluminium Co. Ltd Norfolk House, St. James Square, London SW1
(now known as British Aluminium Co. Ltd)
Aluminium Company of America 1501 Alcoa Building, Pittsburgh PA 15219, USA
Aluminium Federation Broadway House, Calthorpe Road, Birmingham B15 1TN
Aluminium Français 23 rue Balzac, Paris 8, France
Aluminium Wire & Cable Co. Ltd Port Tenant, Swansea, Glamorgan
Aluminium Zentral EV 4 Dusseldorf 10, Postfach 10008, Germany
American Bureau of Shipping 45 Eisenhower Drive, P.O. Box 910, Peramis, New Jersey 70653–0910, USA
American Iron & Steel Institute 1133 15th Street, Suite 3000, Washington DC 20005, USA
American Petroleum Institute 1220 L ST NW, Washington DC 20005
American Society for Testing & Materials (ASTM) 1916 Race Street, Philadelphia, Pennsylvania, USA
AMPCO Metal Incorporated 1745 South 38th Street, P.O. Box 2004, Milwaukee W153201, USA
AMS *see* Aerospace Material Specification
Anglo-Swiss Aluminium Co. Mandor House, Mandor Centre, Wolverhampton WV1 3ND
Anor (Acieries & Forges d') *information from* Climax Molybdenum Company
Anti-Attrition Co. Ltd *no record*
API *see* American Petroleum Institute
Appleby-Frodingham Steel Company *no record*
ARMCO Ltd P.O. Box 11, Jubilee Road, Letchworth, Hertfordshire, SG6 1NQ
Armstrong Whitworth (Metal Industries) Ltd *no record*
N.C. Ashton *no record*
Association des Constructeurs de Material Aerospatiale Paris, France
Association Français de Normalisation (AFNOR) 23 rue Notre Dame des Victoires, Paris 2, France
ASTM *see* American Society for Testing and Materials
Atlas Steel Co. Inc. 2085 Lasalle Street, Chicago, IL 60604, USA
Atlas Steels Ltd Welland, Ontario, Canada
Austrian Specification Borsegasse 18, 1010 Vienna, Austria
(copy held by British Standards Institution)
Avesta Jernwerks Aktiebolas, Avesta, Sweden

Babcock & Wilcox (USA) 1010 Common Street, New Orleans LA 70112
Baird & Scottish Steel Co. Ltd *no longer exists*
Baldwins Ltd *see* Baldwins PLC, Windsford, Cheshire CW7 3QJ
Balfour & Darwins Ltd *no record*
Barker & Allen Ltd Nickel Silver Works, Springhill, Birmingham
Batterium Metal Ltd *no record*
Belgian Specification (Institute Belge Normalisation NBN) 29 avenue de la Brabanconne, Brussels 4, Belgium
Beryllium Corporation Reading, Pennsylvania, USA
Bethlehem Steel Company Martin Tower, Bethlehem, PA 18016
T.M. Birkett, Billington & Newton Ltd *no record*
Birmetals Ltd *no record*
Birmingham Aluminium Co. *no record*
Birmingham Battery & Metal Co. Ltd Selly Oak, Birmingham 29

BKL Alloys Ltd *no record*
Blackwells Metallurgical Ltd *no record*
Blaw-Knox Ltd Rochester, Kent
BNFL *see* British Nuclear Fuels
Bochumer Verein *information from* Climax Molybdenum Company
Bofors Aktiebolaget, Bofors, Sweden
Bofors Steel Ltd *no record*
Bofors Steels Incorporated 2 Henderson Drive, West Caldwell, New Jersey, USA
Gebruder Bohler & Co. AG *information from* Climax Molybdenum Company
G. Bohler & Company Aktiengesellschaft, Vienna, Austria
Bohler Bros. & Company *no record*
Thomas Bolton & Johnston Ltd Froghall, Stoke-on-Trent, Staffs.
Bonperithis (Forge & Acieries de) *information from* Climax Molybdenum
James Booth Aluminium Ltd *no record*
Bradley & Foster Ltd P.O. Box 4, Heath Road, Wednesbury WS10 8JL
Braeburn Steel Co. Braeburn, Pennsylvania, USA
Brandhurst Co. Ltd *no record*
Brimabright Ltd *no record*
Britannia Iron & Steel Works *no record*
British Aluminium Co. Ltd Norfolk House, St James Square, London SW1
British Driver-Harris Co. Ltd Pepper Road, Hazel Grove, Stockport, Cheshire SK7 5BP
British Metal Corporation Ltd *no record*
British Nuclear Fuels Ltd Risley, Warrington
British Rollery Mills Ltd Brymill Steel Works, P.O. Box 10, Tipton, Staffs.
British Rolling Mills Ltd Bloomfield Road, P.O. Box 10, Tipton, Staffs.
British Standards Institute (BS) 2 Park Street, London W1
British Steel Corporation 9 Albert Embankment, London SE1 7SN
Brown Bayley Steels Ltd *no record*
David Brown Industries Ltd Foundries Division, Penistone, Nr. Sheffield
Brush Beryllium Company 17876 St Clair Avenue, Cleveland, Ohio 44110, USA

Cameron & Sons *no longer exists*
Canadian Standards Association 235 Montreal Road, Ottawa 7, Canada
Carobronze *no record*
Carpenter Steel Company 135 West Barn Street, Reading, Pennsylvania, USA
R. Carr & Co. Ltd Pluto Works, Wadsley Bridge, Sheffield 6
CDA *see* Copper Development Association
CDAA *see* Copper Development Association of America
C. Clifford Ltd Dog Pool Mills, Birmingham 30
Climax Molybdenum 29 Gresham Street, London EC2V 7DA
Climax Molybdenum Corporation 1270 Avenue of Americas, New York 20, USA
Cobalt Information Centre Chichester House, 278/282 High Holborn, London WC1
Columbia Tool Steel Company 500 Lincoln Highway, Chicago Heights, Illinois, USA
Colvilles Ltd 195 West George Street, Glasgow G2, Scotland
COMALCO (Europe) Ltd *no record*
Consett Iron Co. Ltd *no record*
Consolidated Beryllium P.O. Box 5, Marblehall Road, Milford Haven, Dyfed, SA73 2PP
Constrictor Ltd *no longer exists*
Copper Development Association (CDA) Orchard House, Mutton Lane, Potters Bar, Hertfordshire
Copper Development Association of America (CDAA) P.O. Box 1840, Greenwich, CT 06836
Copperweld Steel Company 4000 Mahoning Avenue, NW Warren, OH 44482
Phillip Cornes & Co. Ltd Claybrook Drive, Washford, Redditch, Worcestershire
Creusot-Loire Steel Co. Ltd *no record*
Cronite Group PLC Monaco House, Bristol St, Birmingham B5 7AS
Crucible Steel Company (now Crucible) The Materials Research Centre, P.O. Box 88, Parle Way West, Route 80, Pittsburg PA 15230
Cutokumfu Helsinki, Finland

Dacral 164 rue Ambroise Croizat, 93204 St Denis, France
Danish Specification *see* Danish Standards Association
Danish Standards Association (Dansk Standardiseringsrad) Aurehojvej 12, 2900 Hellerup, Denmark
(copies held by British Standards Institution)
Darwins Ltd Tinsley, Sheffield, U.K.
DEF Ministry of Aviation, St Giles Court, 1–13 St Giles High Street, London WC2
Deloro Stellite Ltd Stratton St Margaret, Swindon, Wilts.
Delta Metal Company Dartmouth, Birmingham
C. Denton Steel & Tool Company *no record*
Det Norske Veritas Grenseveien 92, Oslo, Norway
Deutsche Edelstahlwerke AG *information from* Dewramet Ltd, York House, Mossland Road, Glasgow G52 44D, Scotland
Dewrance & Co. Ltd Special Alloy Division, Great Dover Street, London SE1
DIN *see* German Standards
Director of Materials Research & Development (DTD) Ministry of Aviation, St Giles Court, 1–13 St Giles High Street, London WC2
DNV *see* Det Norske Veritas
DOMAL *see* Dominion Magnesium Ltd
Dominion Magnesium Ltd Haley, Ontario, Canada
Dorman Long Steel Company *no record*
Driver-Harris 201 Middlesex Street, Harrison, New Jersey, USA
DTD *see* Director of Materials Research & Development
Dunford & Elliott *no record*
Dunford & Hadfield Ltd *no record*
Durasint Products Ltd *no record*
Duriron Company *no record*
Duriron Company Inc. (DURCO) 425 North Findlay Street, Dayton, OH 45401, USA

Eastern Stainless Steel Company Baltimore, P.O. Box 1975 MD, USA

Edgar & Allen & Co. Ltd P.O. Box 93, Imperial Steel Works, Sheffield 9
Electrolytic Zinc Company *no record*
Elm Engineering Ltd *no record*
Enfield Rolling Mills Ltd *no record*
Engelhard Industries *no record*
English Steel Corporation (ESC) *no record*
ESAB *no record*
ESC *see* English Steel Corporation
Esperance Longder 60 rue d'Harscamp, Liège, Belgium
Eyre Smelting Company *no record*

Fagersta Bruk AB Fagersta, Sweden
Fagersta Steels Ltd *no record*
Fansteel Metallurgical Corporation No. 1 Tantallen Place, North Chicago, USA
Fiat Spa Milan, Italy
Firminy (Forge de) *information from* Climax Molybdenum Company
Firth Brown *no record*
Firth Cleveland Sintered Products Ltd Treforest, Pontypridd, Glamorgan, Wales
Firth Sterling Incorporated Industrial Park, Interchange, City Lavertne, TN 37086
Firth Vickers Foundry Ltd P.O. Box 160, Garter Street, Sheffield S4 7QY
Forez (Acieres du) Saint Etienne, France
Samuel Fox & Co. Ltd *no record*
French Afnor Association Français de Normalisation, 23 rue Notre Dame des Victoires, Paris 2, France
French National Standard Association Français de Normalisation, 23 rue Notre Dame des Victoires, Paris 2, France
Friedr Lohmann *information from* Climax Molybdenum Company
S. Fry & Co. Ltd *no record*
Furnival Steel Co. Ltd Furnival Steel Works, Sheffield 4

William Gallimore & Sons Ltd *no record*
A.B. Galnol & Company *no record*
Gebruder Bohler Kapfenberg, Steiermark, Austria
Gebruder Bohler & Company AG *information from* Climax Molybdenum Company
General Alloys Company Boston, Massachusetts, USA
General Electric Company Detroit 32, Michigan, USA
General Electric Company Metallurgical Products Dept, Detroit, USA
General Steel Industries Incorporated Eddystone, PA, USA
General Tool & Die Company East Orange, New Jersey, USA
Georgsmarienwerke Selesiastahl GMBH Georgsmarienhutte, Germany
Gerhardi & Company Ladenscheid, Germany
German Standards 175 Uhlandstrasse, 1 Berlin 15, Germany
Gilby Brunton Ltd *no record*
Gilby-Foder Trefileries et Lamenoirsde Precision, Rueil-Malmaison, France
Gimo-Osterby Bruks AB Gimo, Sweden
Giulini Werke AG Rohrsbach, Germany
Glacier Metal Co. Ltd Argyle House, Joel Street, Northwood Hills, HA6 1CN
Gloucester Foundry *no record*
T.R. Goldschmidt AG, Abteilung Metalle Essen, Germany
Gorham Tool Company Detroit, Michigan, USA
GOST-Russian Specification (copies held by British Standards Institution)
Gottingen Aluminiumwerke GMBH Gottingen, Germany
Granite City Steel Company Granite City, Illinois, USA
Grant & West Ltd *no record*
Graylom Steel Company New York, USA
Great Lakes Steel Company Detroit, Michigan, USA
Great Western Steel Division Hayland Steel Company, Los Angeles, California, USA
Gruss Stahlwerk Witten AG, Germany
Guronitwerke Vervoot GMBH Dusseldorf, Germany
Gusstahl-Hondels GMBH Van Dohlen-Stahl, Wetter, Ruhr, Germany

Hadfields Ltd *no record*
Hagener Gusstahlwerke Remy, Germany
Hall & Pickles Port Street, Manchester 1 (now part of Osborne Steels)
Hallamshire Hardmetal Products 315–17 Coleford Road, Sheffield, S9 5NF
Hallamskie Steel Company (now part of Sheffield Rolling Mills)
Hanricot *information from* Climax Molybdenum Company
Harrison Fischer *no record*
Harvey Aluminium Ltd Torrance, California, USA
Haynes Stellite Samuel Osborn & Co. Ltd, P.O. Box 1, Clyde Steel Works, Sheffield 3
HDA *see* High Duty Alloys
A. Heckford Ltd Birmingham Metal Works, Frederick Street, Birmingham
High Duty Alloys Ltd (HDA) *no record*
Hill Alzen (Sales) Ltd *no record*
Hoffmann *information from* Climax Molybdenum Company
I. Holroyd & Co. Ltd P.O. Box 24, Holfus Works, Rochdale
Holtzer-Loire *no record*
Hopkinsons P.O. Box 27, Brittania Works, Huddersfield, Yorks.
Hoyt Metal Co. (Great Britain) Ltd *no record*
Huntingdon Alloy Products The International Nickel Company, 67 Wall Street, New York 5, USA
B. Huntsman *no record*

ICI Ltd *see* Imperial Metal Industries
IMI *see* Imperial Metal Industries
Imperial Aluminium Co. Ltd *no record*
Imperial Metal Industries Ltd *no record*
Imperial Smelting Processes Ltd P.O. Box 50, Castlemead Lower, Castle Street, Bristol BS99 7YR
Imphy Siege Social, 84 rue de Lille, Paris, France
Incanite Foundries Ltd *no record*
INCO *no record*
I. Inman & Co. Ltd *no record*
INTAL *see* International Alloys Ltd

International Alloys Ltd (INTAL) *no record*
International Lead Zinc Research Organisation Incorporated 2525 Meridian Parkway, P.O. Box 12036, Research Triangle Park, NC 27709, USA
International Meehanite Co. Ltd Meerion House, Albert Road North, Reigate, Surrey
International Nickel Co. (INCO) *no record*
Italian Standards Ente Nazionale Italiano de Unificazione Milano, Italy

Japan Metal Industry, Kawasaki, Japan
Jessop-Saville Ltd *no record*
Johnson Matthey Orchard Road, Royston, Hertfordshire, SG8 3ME
Jones & Colver Ltd *no record*
Jones & Rourke *no record*

Kawasaki Plant Japan Metal Industries Co. Ltd, Kawasaki, Japan
Kawecki Berylco Incorporated *no record*
E. & E. Kaye Ltd *no record*
Kayser-Ellison & Co. Ltd *no record*
Kennametal (GB) Ltd P.O. Box 29, Pensett Trading Estate, Brierley Hill, DY6 7NP
Kind & Company Edelstahlwerk *information from* Climax Molybdenum Company
Kirkstall Forge Engineering Ltd *no record*
Kiveton Park Steel Company Kiveton Park, Sheffield S31 8NQ
Klockner-Werge AG *information from* Climax Molybdenum Company
Krupps Steel Works Fried, Essen, Germany
C. Kuhbler & Son *information from* Climax Molybdenum Company

L'Aluminium Français 23 rue Balzac, Paris 8, France
Langley Alloys Ltd Langley, Slough, Bucks.
Latrobe Steel Company Suby of Timken Co., 2626 Ligonier Street, Latrobe, PA 15650, USA
Lead Development Association 34 Berkeley Square, London, W1
Arthur Lee & Sons Ltd P.O. Box 54, Sheffield
Light Alloys Ltd Dales Road, Ipswich IP1 4JR
Light Laboratories Ltd *no record*
London Zinc Mills Ltd *no record*
Low Moor Alloy Steelworks Ltd *no record*

McKechnie Metals Ltd Middlemore Lane, Aldridge, Near Walsall, Staffordshire
Magnesium Elektron Ltd Regal House, London Road, Twickenham, London TW1 3QA
Magnolia Anti-Friction Metal Company *no record*
Main Metal (Great Britain) Ltd *no record*
Mallory Metallurgical Products Ltd *no record*
Manganese Bronze Ltd Elton Park Works, P.O. Box 19, Hadleigh Road, Ipswich
W.L. Marrison Ltd *no record*
Marsh Bros & Co. Ltd *no record*
Martin Marietta Company *no record*
Martin Metals Company Division Martin Marietta Corporation, Wheeling, Illinois, USA
Meehanite (International Meehanite Co. Ltd) Murion House, Albert Road North, Reigate, Surrey
Meigh Casting Co. Ltd Werkinton Foundry, Nr. Swindon, Cheltenham, Gloucestershire
Mel-Magnesium Elektron Ltd *no record*
Metal Alloys Ltd Treforest Estate, Pontypridd, Wales
Metco Ltd Chobham, Woking, Surrey
Metro-Cutanit Ltd *no record*
Metrode Products Ltd Hanworth Lane, Chertsey, Surrey
William Mills & Co. Ltd *no record*
Mining & Chemical Products Ltd Alperton, Wembley, Middlesex
Miralite Co. Ltd *no longer exists*
Murex Ltd Rainham, Essex

Nazionale Cogne *information from* Climax Molybdenum Company
NBN *see* Belgian Specification
NELCO Metal Corporation Canaan, Connecticut, USA
Netherlands Standardization Institute (Nederlands Normalisatie Institut: Dutch Specification) Polaweg 5, Ruswuk, Netherlands
Northern Aluminium Co. Ltd *no record*
NYBY Stainless Steels *no record*

Officine Metallurgiche de Pont, St Martin, France
Samuel Osborne & Co. Ltd *no record*

Park Gate Iron & Steel Co. Ltd *no record*
F. Parkin, Wallsend Road North, Sheffield S6 1LL
Patent Shaft Steel *no record*
Permanent Magnet Association (PMA) *no record*
Phosphor Bronze Co. Ltd Temple Manor Works, Rochester, Kent
Platt Metal Ltd *no record*
PMA *no record*
Pompey Society of Acieres 61 rue de Monceau, Paris 8, France
Posse-Marre *information from* Climax Molybdenum Company
Kabel C. Pouplier Stahlwerk *information from* Climax Molybdenum Company
Preformations Ltd *no record*

Quebec Metallurgical Corporation, Quebec, Canada

Reactive Metals Incorporated Niles, Ohio, USA
Redheugh Iron & Steel Co. Ltd *no record*
Republic Steel Company Republic Building, Cleveland 1, Ohio, USA
Resistalloy Ltd Riverside Works, Weedon Street, Sheffield S9 2FT
Reynolds Tubes Company *no record*
Richards, Thomas & Baldwins Ltd *no record*
Rochling-Budenis *information from* Climax Molybdenum Company
Round Oak Steel *no record*
Russian Specifications (GOST) (copies held by British Standards Institution)

SAE *see* Society of Automotive Engineers
Salford Electric *no record*
Sambre & Meusse (Acieries de) *information from* Climax Molybdenum Company
Sanderson Brothers & Newbould Ltd (now Sanderson Kayser Ltd)
Sanderson Kayser Ltd *no record*

Sandvik Ltd Manor Way, Halesowen, BG2 8QZ
Schmidt & Clemens Edelstahlwerk *information from* Climax Molybdenum Company
Schoeller-Bleckmann Steels (Great Britain) Ltd *no record*
Sheffield Hollow Drill Steel Co. Ltd (SHD) *no record*
Sheffield Smelting Company *no record*
Simonds 110 Safety First Bank Buildings, Fitchburgh, Massachusetts, USA
Sintered Products Ltd *no record*
SIS *see* Swedish Standardizing Commission
Fred Smith & Company AEI Cables Ltd, Cretehall Road, Gravesend DA11 9AF
Society Metallurgique d'Imphy, Siege Social, 84 rue de Lille, Paris, France
Society of Automotive Engineers (SAE) 485 Lexington Avenue, New York 17, USA
Soderfors Steel Works Soderfors, Sweden
Walter Somers Haywood Forge, P.O. Box 7, Halesowen, Worcestershire
Southern Forge Ltd
(now Imperial Aluminium Co. Ltd, P.O. Box 216, Witton, Birmingham)
Spanish National Standardization Institute (Instituto Nationale de Racionalizacion del Trabajo) Sarrano 150, Madrid, Spain
Spear & Jackson Industrial Tool Steel Aetna Works, Saville Street, Sheffield 4
Walter Spenser & Co. Ltd *no record*
STA The Under Secretary of State, War Office (DS), First Avenue House, High Holborn, London WC1
Standard Telephones & Cables Ltd *no record*
Standards Association of Australia Standards House, 80–86 Arthur Street, North Sydney, NSW 2060, Australia
Stavanger Elektre Salverk Stavanger, Norway
STD Services Ltd *no record*
Steel Company of Wales *no record*
Steel Peech & Tozer Ltd P.O. Box 50, The Ickles, Rotherham, UK
(now British Steel Corporation)
J. Stone Ltd *no record*
Stone Manganese, J. Stone & Company *no record*
Peter Stubbs Scotland Road, Warrington, Cheshire
Sudwestfalen-Stahlwerk AG *information from* Climax Molybdenum Company
Suffolk Iron Co. Ltd *no record*
Svenskt Stal *no record*
Swedish Standardisering Kommission Metals Department, Box 3295, Stockholm, Sweden
Swift Levick & Sons Ltd Clarence Steel Works, Leveson Street, Sheffield 4

Telcon Metals Ltd P.O. Box 12, Manor Royal, Crawley, Sussex
Thomson L'Hospied & Co. Ltd *no record*
Thyssen Glesserei AG Mulheim, West Germany
Timken Roller Bearing Company Canton, Ohio, USA
Tin Research Institute *no record*
Titanium Metal Company of America 233 Broadway, New York, USA
Tokushu Seiko Co. Ltd *information from* Isfan Iron & Steel Federation, Tekko Building, Tokyo
Toledo Steel Works *no record*
Toshiba Tokyo Shibaura Electric Company, Tokyo, Japan
The Tungum Company Ltd The White House, Arle, Cheltenham, Gloucestershire
T. Turton & Sons Ltd *no longer exists*
Turton Bros & Matthews Ltd Burton Road, Sheffield S3 8DA

Uddelholm Aktiebolag Uddelholm, Sweden
Uddelholm Swedish Steels Ltd Kinwarton Farm Road, Alcester B49 6ET
Ugine (Acieries d') Paris, France
The Under Secretary of State (STA) War Office (DS), First Avenue House, High Holborn, London WC1
Union Carbide Ltd Shepley Street, Glossop, Derbyshire
United Steel Company *no record*
Universal Cyclops Steel Corporation Bridgeville, USA
US Federal Specifications
(copies held by British Standards Institution)
US Service Specifications US Government Printing Office, Washington, DC, USA

Valhuna *no record*
Vanadium Alloy Steel Company Latrobe, Pennsylvania, USA
John Vessey & Sons Ltd *no record*
Vickers Armstrong Ltd *no record*
The Vulcan Steel & Tool Co. Ltd *no record*

Wai-Met Alloys Company Division of Howe Sound Company, Dearborn, Michigan, USA
Wall Colmonoy (Canada) Ltd Pontardawe, Swansea SA8 4HL
Wellman plc Devon House, 12–15 Dartmouth Street, London SW1H 9BJ
V.R. Wesson Company 800 Market Street, Waukegan, Illinois, USA
Westinghouse Electric Corporation East Pittsburgh, PA, USA
Henry Wiggin Holmer Road, Hertford, UK
Wolsingham Steel Co. Ltd *no record*

Yorkshire Imperial Alloys Ltd P.O. Box 166, Leeds LS1 1RD, Yorks.

Zapp-Robt *information from* Climax Molybdenum Company
Zinc Development Association 34 Berkeley Square, London W1

Appendix II · Conversion tables

Appendix IIA

Temperature conversion table

To convert any temperature in Celsius or Fahrenheit degrees to the other, take the figure in the centre column, and if converting from Celsius to Fahrenheit, read off to the right-hand column; if converting from Fahrenheit to Celsius, read off to the left-hand column.

Example 720 °C = 1328 °F
1200 °F = 649 °C

Formula for conversion
Celsius to Fahrenheit = $°C \times 1.8 + 32 = °F$
Fahrenheit to Celsius = $°F - 32 \div 1.8 = °C$

°C		°F	°C		°F	°C		°F	°C		°F	°C		°F	°C		°F
−128.9	−200	−328	−46.1	−51	−59.8	−13.3	8	46.4	19.4	67	152.6	177	350	662	504	940	1724
−123.3	−190	−310	−45.5	−50	−58	−12.8	9	48.2	20.0	68	154.4	182	360	680	510	950	1742
−117.8	−180	−292	−45	−49	−56.2	−12.2	10	50.0	20.6	69	156.2	188	370	698	516	960	1760
−112.2	−170	−274	−44.4	−48	−54.4	−11.7	11	51.8	21.1	70	158.0	193	380	716	521	970	1778
−106.7	−160	−256	−43.9	−47	−52.6	−11.1	12	53.6	21.7	71	159.8	199	390	734	527	980	1796
−101.1	−150	−238	−43.3	−46	−50.8	−10.6	13	55.4	22.2	72	161.6	204	400	752	532	990	1814
−95.6	−140	−220	−42.8	−45	−49	−10.0	14	57.2	22.8	73	163.4	210	410	770	538	1000	1832
−90.0	−130	−202	−42.2	−44	−47.2	−9.44	15	59.0	23.3	74	165.2	216	420	788	543	1010	1850
−84.4	−120	−184	−41.7	−43	−45.4	−8.89	16	60.8	23.9	75	167.0	221	430	806	549	1020	1868
−78.9	−110	−166	−41.1	−42	−43.6	−8.33	17	62.6	24.4	76	168.8	227	440	824	554	1030	1886
−73.3	−100	−148	−40.6	−41	−41.8	−7.78	18	64.4	25.0	77	170.6	232	450	842	560	1040	1904
−72.8	−99	−146.2	−40	−40	−40	−7.22	19	66.2	25.6	78	172.4	238	460	860	566	1050	1922
−72.2	−98	−144.4	−39.4	−39	−38.2	−6.67	20	68.0	26.1	79	174.2	243	470	878	571	1060	1940
−71.7	−97	−142.6	−38.9	−38	−36.4	−6.11	21	69.8	26.7	80	176.0	249	480	896	577	1070	1958
−71.1	−96	−140.8	−38.3	−37	−34.6	−5.56	22	71.6	27.2	81	177.8	254	490	914	582	1080	1976
−70.6	−95	−139	−37.8	−36	−32.8	−5.00	23	73.4	27.8	82	179.6	260	500	932	588	1090	1994
−70.0	−94	−137.2	−37.2	−35	−31	−4.44	24	75.2	28.3	83	181.4	266	510	950	593	1100	2012
−69.4	−93	−135.4	−36.7	−34	−29.2	−3.89	25	77.0	28.9	84	183.2	271	520	968	599	1110	2030
−68.9	−92	−133.6	−36.1	−33	−27.4	−3.33	26	78.8	29.4	85	185.0	277	530	986	604	1120	2048
−68.3	−91	−131.8	−35.6	−32	−25.6	−2.78	27	80.6	30.0	86	186.8	282	540	1004	610	1130	2066
−67.8	−90	−130	−35	−31	−23.8	−2.22	28	82.4	30.6	87	188.6	288	550	1022	616	1140	2084
−67.2	−89	−128.2	−34.4	−30	−22	−1.67	29	84.2	31.1	88	190.4	293	560	1040	621	1150	2102
−66.7	−88	−126.4	−33.9	−29	−20.2	−1.11	30	86.0	31.7	89	192.2	299	570	1058	627	1160	2120
−66.1	−87	−124.6	−33.3	−28	−18.4	−0.56	31	87.8	32.2	90	194.0	304	580	1076	632	1170	2138
−65.6	−86	−122.8	−32.8	−27	−16.6	0	32	89.6	32.8	91	195.8	310	590	1094	638	1180	2156
−65.0	−85	−121	−32.2	−26	−14.8	0.56	33	91.4	33.3	92	197.6	316	600	1112	643	1190	2174
−64.4	−84	−119.2	−31.7	−25	−13	1.11	34	93.2	33.9	93	199.4	321	610	1130	649	1200	2192
−63.9	−83	−117.4	−31.1	−24	−11.2	1.67	35	95.0	34.4	94	201.2	327	620	1148	654	1210	2210
−63.3	−82	−115.6	−30.6	−23	−9.4	2.22	36	96.8	35.0	95	203.0	332	630	1166	660	1220	2228
−62.8	−81	−113.8	−30	−22	−7.6	2.78	37	98.6	35.6	96	204.8	338	640	1184	666	1230	2246
−62.2	−80	−112	−29.4	−21	−5.8	3.33	38	100.4	36.1	97	206.6	343	650	1202	671	1240	2264
−61.7	−79	−110.2	−28.9	−20	−4	3.89	39	102.2	36.7	98	208.4	349	660	1220	677	1250	2282
−61.1	−78	−108.4	−28.3	−19	−2.2	4.44	40	104.0	37.2	99	210.2	354	670	1238	682	1260	2300
−60.6	−77	−106.6	−27.8	−18	−0.4	5.00	41	105.8	38	100	212	360	680	1256	688	1270	2318
−60.0	−76	−104.8	−27.2	−17	1.4	5.56	42	107.6	43	110	230	366	690	1274	693	1280	2336
−59.4	−75	−103	−26.7	−16	3.2	6.11	43	109.4	49	120	248	371	700	1292	699	1290	2354
−58.9	−74	−101.2	−26.1	−15	5	6.67	44	111.2	54	130	266	377	710	1310	704	1300	2372
−58.3	−73	−99.4	−25.6	−14	6.8	7.22	45	113.0	60	140	284	382	720	1328	710	1310	2390
−57.8	−72	−97.6	−25	−13	8.6	7.78	46	114.8	66	150	302	388	730	1346	716	1320	2408
−57.2	−71	−95.8	−24.4	−12	10.4	8.33	47	116.6	71	160	320	393	740	1364	721	1330	2426
−56.7	−70	−94	−23.9	−11	12.2	8.89	48	118.4	77	170	338	399	750	1382	727	1340	2444
−56.1	−69	−92.2	−23.3	−10	14	9.44	49	120.2	82	180	356	404	760	1400	732	1350	2462
−55.6	−68	−90.4	−22.8	−9	15.8	10.0	50	122.0	88	190	374	410	770	1418	738	1360	2480
−55	−67	−88.6	−22.2	−8	17.6	10.6	51	123.8	93	200	392	416	780	1436	743	1370	2498
−54.4	−66	−86.8	−21.7	−7	19.4	11.1	52	125.6	99	210	410	421	790	1454	749	1380	2516
−53.9	−65	−85	−21.1	−6	21.2	11.7	53	127.4	100	212	414	427	800	1472	754	1390	2534
−53.3	−64	−83.2	−20.6	−5	23	12.2	54	129.2	104	220	428	432	810	1490	760	1400	2552
−52.8	−63	−81.4	−20	−4	24.8	12.8	55	131.0	110	230	446	438	820	1508	766	1410	2570
−52.2	−62	−79.6	−19.4	−3	26.6	13.3	56	132.8	116	240	464	443	830	1526	771	1420	2588
−51.7	−61	−77.8	−18.9	−2	28.4	13.9	57	134.6	121	250	482	449	840	1544	777	1430	2606
−51.1	−60	−76	−18.3	−1	30.2	14.4	58	136.4	127	260	500	454	850	1562	782	1440	2624
−50.6	−59	−74.2	−17.8	0	32	15.0	59	138.2	132	270	518	460	860	1580	788	1450	2642
−50	−58	−72.4	−17.2	1	33.8	15.6	60	140.0	138	280	536	466	870	1598	793	1460	2660
−49.4	−57	−70.6	−16.7	2	35.6	16.1	61	141.8	143	290	554	471	880	1616	799	1470	2678
−48.9	−56	−68.8	−16.1	3	37.4	16.7	62	143.6	149	300	572	477	890	1634	804	1480	2696
−48.3	−55	−67	−15.6	4	39.2	17.2	63	145.4	154	310	590	482	900	1652	810	1490	2714
−47.8	−54	−65.2	−15.0	5	41.0	17.8	64	147.2	160	320	608	488	910	1670	816	1500	2732
−47.2	−53	−63.4	−14.4	6	42.8	18.3	65	149.0	166	330	626	493	920	1688	821	1510	2750
−46.7	−52	−61.6	−13.9	7	44.6	18.9	66	150.8	171	340	644	499	930	1706	827	1520	2768

Appendix IIA—*cont.*

°C		°F	°C		°F	°C		°F	°C		°F	°C		°F	°C		°F
832	1530	2786	977	1790	3254	1121	2050	3722	1266	2310	4190	1410	2570	4658	1554	2830	5126
838	1540	2804	982	1800	3272	1127	2060	3740	1271	2320	4208	1416	2580	4676	1560	2840	5144
843	1550	2822	988	1810	3290	1132	2070	3758	1277	2330	4226	1421	2590	4694	1566	2850	5162
849	1560	2840	993	1820	3308	1138	2080	3776	1282	2340	4244	1427	2600	4712	1571	2860	5180
854	1570	2858	999	1830	3326	1143	2090	3794	1288	2350	4262	1432	2610	4730	1577	2870	5198
860	1580	2876	1004	1840	3344	1149	2100	3812	1293	2360	4280	1438	2620	4748	1582	2880	5216
866	1590	2894	1010	1850	3362	2254	2110	3830	1299	2370	4298	1443	2630	4766	1588	2890	5234
871	1600	2912	1016	1860	3380	1160	2120	3848	1304	2380	4316	1449	2640	4784	1593	2900	5252
877	1610	2930	1021	1870	3398	1166	2130	3866	1310	2390	4334	1454	2650	4802	1599	2910	5270
882	1620	2948	1027	1880	3416	1171	2140	3884	1316	2400	4352	1460	2660	4820	1604	2920	5288
888	1630	2966	1032	1890	3434	1177	2150	3902	1321	2410	4370	1466	2670	4838	1610	2930	5306
893	1640	2984	1038	1900	3452	1182	2160	3920	1327	2420	4388	1471	2680	4856	1616	2940	5324
899	1650	3002	1043	1910	3470	1188	2170	3938	1332	2430	4406	1477	2690	4874	1621	2950	5342
904	1660	3020	1049	1920	3488	1193	2180	3956	1338	2440	4424	1482	2700	4892	1627	2960	5360
910	1670	3038	1054	1930	3506	1199	2190	3974	1343	2450	4442	1488	2710	4910	1632	2970	5378
916	1680	3056	1060	1940	3524	1204	2200	3992	1349	2460	4460	1493	2720	4928	1638	2980	5396
921	1690	3074	1066	1950	3542	1210	2210	4010	1354	2470	4478	1499	2730	4946	1643	2990	5414
927	1700	3092	1071	1960	3560	1216	2220	4028	1360	2480	4496	1504	2740	4964	1649	3000	5432
932	1710	3110	1077	1970	3578	1221	2230	4046	1366	2490	4514	1510	2750	4982	1705	3100	5612
938	1720	3128	1082	1980	3596	1227	2240	4064	1371	2500	4532	1516	2760	5000	1760	3200	5792
943	1730	3146	1088	1990	3614	1232	2250	4082	1377	2510	4550	1521	2770	5018	1816	3300	5972
949	1740	3164	1093	2000	3632	1238	2260	4100	1382	2520	4568	1527	2780	5036	1871	3400	6152
954	1750	3182	1099	2010	3650	1243	2270	4118	1388	2530	4586	1532	2790	5054			
960	1760	3200	1104	2020	3668	1249	2280	4136	1393	2540	4604	1538	2800	5072			
966	1770	3218	1110	2030	3686	1254	2290	4154	1399	2550	4622	1543	2810	5090			
971	1780	3236	1116	2040	3704	1260	2300	4172	1404	2560	4640	1549	2820	5108			

Appendix IIB

Strength conversion table

hbar	N/mm² (MPa)	tonf/in²	lbf/in²	kgf/mm²	hbar	N/mm² (MPa)	tonf/in²	lbf/in²	kgf/mm²	hbar	N/mm² (MPa)	tonf/in²	lbf/in²	kgf/mm²
1	10	0.647	1 450	1.02	81	810	52.45	117 500	82.60	161	1610	104.2	233 500	164.2
2	20	1.295	2 900	2.04	82	820	53.09	118 900	83.62	162	1620	104.9	235 000	165.2
3	30	1.942	4 350	3.06	83	830	53.74	120 400	84.64	163	1630	105.5	236 400	166.2
4	40	2.590	5 800	4.08	84	840	54.39	121 800	85.65	164	1640	106.2	237 900	167.2
5	50	3.237	7 250	5.10	85	850	55.04	123 300	86.67	165	1650	106.8	239 300	168.3
6	60	3.885	8 700	6.12	86	860	55.68	124 700	87.69	166	1660	107.5	240 800	169.3
7	70	4.532	10 150	7.14	87	870	56.33	126 200	88.71	167	1670	108.1	242 200	170.3
8	80	5.180	11 600	8.16	88	880	56.98	127 600	89.73	168	1680	108.8	243 700	171.3
9	90	5.827	13 050	9.18	89	890	57.63	129 100	90.75	169	1690	109.4	245 100	172.3
10	100	6.475	14 500	10.20	90	900	58.27	130 500	91.77	170	1700	110.1	246 600	173.3
11	110	7.122	15 950	11.22	91	910	58.92	132 000	92.79	171	1710	110.7	248 000	174.4
12	120	7.770	17 400	12.24	92	920	59.57	133 400	93.81	172	1720	111.4	249 500	175.4
13	130	8.417	18 850	13.26	93	930	60.22	134 900	94.83	173	1730	112.0	250 900	176.4
14	140	9.065	20 300	14.28	94	940	60.86	136 300	95.85	174	1740	112.7	252 400	177.4
15	150	9.712	21 750	15.30	95	950	61.51	137 800	96.87	175	1750	113.3	253 800	178.4
16	160	10.36	23 200	16.32	96	960	62.16	139 200	97.89	176	1760	114.0	255 300	179.5
17	170	11.01	24 650	17.33	97	970	62.80	140 700	98.91	177	1770	114.6	256 700	180.5
18	180	11.65	26 100	18.35	98	980	63.45	142 100	99.93	178	1780	115.3	258 200	181.5
19	190	12.30	27 550	19.37	99	990	64.10	143 600	101.0	179	1790	115.9	259 600	182.5
20	200	12.95	29 000	20.39	100	1000	64.75	145 000	102.0	180	1800	116.5	261 100	183.5
21	210	13.60	30 450	21.41	101	1010	65.37	146 500	103.0	181	1810	117.2	262 500	184.6
22	220	14.24	31 900	22.43	102	1020	66.04	147 900	104.0	182	1820	117.8	264 000	185.6
23	230	14.89	33 350	23.45	103	1030	66.69	149 400	105.0	183	1830	118.5	265 400	186.6
24	240	15.54	34 800	24.47	104	1040	67.34	150 800	106.0	184	1840	119.1	266 900	187.6
25	250	16.19	36 250	25.49	105	1050	67.99	152 300	107.1	185	1850	119.8	268 300	188.6
26	260	16.83	37 700	26.51	106	1060	68.63	153 700	108.1	186	1860	120.4	269 800	189.7
27	270	17.48	39 150	27.53	107	1070	69.28	155 200	109.1	187	1870	121.1	271 200	190.7
28	280	18.13	40 600	28.55	108	1080	69.93	156 600	110.1	188	1880	121.7	272 700	191.7
29	290	18.78	42 050	29.57	109	1090	70.58	158 100	111.1	189	1890	122.4	274 100	192.7
30	300	19.42	43 500	30.59	110	1100	71.22	159 500	112.2	190	1900	123.0	275 600	193.7
31	310	20.07	44 950	31.61	111	1110	71.87	161 000	113.2	191	1910	123.7	277 000	194.8
32	320	20.72	46 400	32.63	112	1120	72.52	162 400	114.2	192	1920	124.3	278 500	195.8
33	330	21.37	47 850	33.65	113	1130	73.17	163 900	115.2	193	1930	125.0	279 900	196.8
34	340	22.01	49 300	34.67	114	1140	73.81	165 300	116.2	194	1940	125.6	281 400	197.8
35	350	22.66	50 750	36.69	115	1150	74.46	166 800	117.3	195	1950	126.3	282 800	198.8
36	360	23.31	52 200	36.71	116	1160	75.11	168 200	118.3	196	1960	126.9	184 300	199.9
37	370	23.96	53 650	37.73	117	1170	75.76	169 700	119.3	197	1970	127.6	285 700	200.9
38	380	24.60	55 100	38.75	118	1180	76.40	171 100	120.3	198	1980	128.2	287 200	201.9
39	390	25.25	56 550	39.77	119	1190	77.05	172 600	121.3	199	1990	128.9	288 600	202.9
40	400	25.90	58 000	40.79	120	1200	77.70	174 000	122.4	200	2000	129.5	290 100	203.9
41	410	26.55	59 450	41.81	121	1210	78.35	175 500	123.4	201	2010	130.1	291 500	205.0
42	420	27.19	60 900	42.83	122	1220	78.99	176 900	124.4	202	2020	130.8	293 000	206.0
43	430	27.84	62 350	43.85	123	1230	79.64	178 400	125.4	203	2030	131.4	294 400	207.0
44	440	28.49	63 800	44.87	124	1240	80.29	179 800	126.4	204	2040	132.1	295 900	208.0
45	450	29.14	65 250	45.89	125	1250	80.93	181 300	127.5	205	2050	132.7	297 300	209.0
46	460	29.78	66 700	46.91	126	1260	81.58	182 800	128.5	206	2060	133.4	298 800	210.1
47	470	30.43	68 150	47.93	127	1270	82.23	184 200	129.5	207	2070	134.0	300 200	211.1
48	480	31.08	69 600	48.95	128	1280	82.88	185 700	130.5	208	2080	134.7	301 700	212.1
49	490	31.73	71 050	49.97	129	1290	83.53	187 100	131.5	209	2090	135.3	303 100	213.1
50	500	32.37	72 500	50.99	130	1300	84.17	186 600	132.6	210	2100	136.0	304 600	214.1
51	510	33.02	73 950	52.00	131	1310	84.82	190 000	133.6	211	2110	136.6	306 000	215.2
52	520	33.67	75 400	53.02	132	1320	85.47	191 500	134.6	212	2120	137.3	307 500	216.2
53	530	34.32	76 850	54.04	133	1330	86.12	192 900	135.6	213	2130	137.9	308 900	217.2
54	540	34.96	78 300	55.06	134	1340	86.76	194 400	136.6	214	2140	138.6	310 400	218.2
55	550	35.61	79 750	56.08	135	1350	87.41	195 800	137.7	215	2150	139.2	311 800	219.2
56	560	36.26	81 200	57.10	136	1360	88.06	197 300	138.7	216	2160	139.9	313 300	220.3
57	570	36.91	82 650	58.12	137	1370	88.71	198 700	139.7	217	2170	140.5	314 700	221.3
58	580	37.55	84 100	59.14	138	1380	89.35	200 200	140.7	218	2180	141.2	316 200	222.3
59	590	38.20	85 550	60.16	139	1390	90.00	201 600	141.7	219	2190	141.8	317 600	223.3
60	600	38.85	87 000	61.18	140	1400	90.65	203 100	142.8	220	2200	142.4	319 100	224.3
61	610	39.50	88 450	62.20	141	1410	91.30	204 500	143.8	221	2210	143.1	320 500	225.4
62	620	40.14	89 900	63.22	142	1420	91.94	206 000	144.8	222	2220	143.7	322 000	226.4
63	630	40.79	91 350	64.24	143	1430	92.59	207 400	145.8	223	2230	144.4	323 400	227.4
64	640	41.44	92 800	65.26	144	1440	93.24	208 900	146.8	224	2240	145.0	324 900	228.4
65	650	42.09	94 250	66.28	145	1450	93.89	210 300	147.9	225	2250	145.7	326 300	229.4
66	660	42.74	95 700	67.30	146	1460	94.53	211 800	148.9	226	2260	146.3	327 800	230.5
67	670	43.38	97 150	68.32	147	1470	95.18	213 200	149.9	227	2270	147.0	329 200	231.5
68	680	44.02	98 600	69.34	148	1480	95.83	214 700	150.9	228	2280	147.6	330 700	232.5
69	690	44.68	100 050	70.36	149	1490	96.48	216 100	151.9	229	2290	148.3	332 100	233.5
70	700	45.32	101 500	71.38	150	1500	97.12	217 600	153.0	230	2300	148.9	333 600	234.5
71	710	45.97	103 000	72.40	151	1510	97.77	219 000	154.0	231	2310	149.6	335 000	235.6
72	720	46.62	104 400	73.42	152	1520	98.42	220 500	155.0	232	2320	150.2	336 500	236.6
73	730	47.27	105 900	74.44	153	1530	99.07	221 900	156.0	233	2330	150.9	337 900	237.6
74	740	47.91	107 300	75.46	154	1540	99.71	223 400	157.0	234	2340	151.5	339 400	238.6
75	750	48.56	108 800	76.48	155	1550	100.4	224 800	158.1	235	2350	152.2	340 800	239.6

Appendix IIB—*cont.*

hbar	N/mm² (MPa)	tonf/in²	lbf/in²	kgf/mm²	hbar	N/mm² (MPa)	tonf/in²	lbf/in²	kgf/mm²	hbar	N/mm² (MPa)	tonf/in²	lbf/in²	kgf/mm²
76	760	49.21	110 200	77.50	156	1560	101.0	226 300	159.1	236	2360	152.8	342 300	240.6
77	770	49.86	111 700	78.52	157	1570	101.7	227 700	160.1	237	2370	153.5	343 700	241.7
78	780	50.50	113 100	79.54	158	1580	102.3	229 200	161.1	238	2380	154.1	345 200	242.7
79	790	51.15	114 600	80.56	159	1590	103.0	230 600	162.1	239	2390	154.8	346 600	243.7
80	800	51.80	116 000	81.58	160	1600	103.6	232 100	163.2	240	2400	155.4	348 100	244.7

1 MPa = $1 N/mm^2 = 0.1/hbar = 0.102 kgf/mm^2 = 0.06475 tonf/in^2 = 145.04 lbf/in^2$

Appendix IIC

Impact data conversion table

joules	ft/lb	kg/m	joules	ft/lb	kg/m	joules	ft/lb	kg/m	joules	ft/lb	kg/m
1	1	0.14	35	26	3.60	69	51	7.05	103	76	10.51
3	2	0.28	37	27	3.73	71	52	7.19	104	77	10.65
4	3	0.42	38	28	3.97	72	53	7.33	106	78	10.78
5	4	0.55	39	29	4.01	73	54	7.47	107	79	10.92
7	5	0.69	41	30	4.15	75	55	7.60	108	80	11.06
8	6	0.83	42	31	4.29	76	56	7.74	110	81	11.20
9	7	0.97	43	32	4.42	77	57	7.88	111	82	11.34
11	8	1.11	45	33	4.56	78	58	8.02	113	83	11.48
12	9	1.24	46	34	4.70	80	59	8.16	114	84	11.61
14	10	1.38	47	35	4.84	81	60	8.30	115	85	11.75
15	11	1.52	49	36	4.98	83	61	8.43	117	86	11.89
16	12	1.66	50	37	5.12	84	62	8.57	118	87	12.03
18	13	1.80	52	38	5.25	85	63	8.71	119	88	12.17
19	14	1.94	53	39	5.39	87	64	8.85	121	89	12.31
20	15	2.07	54	40	5.53	88	65	8.99	122	90	12.44
22	16	2.21	56	41	5.67	89	66	9.13	123	91	12.58
23	17	2.35	57	42	5.81	91	67	9.26	124	92	12.72
24	18	2.49	58	43	5.95	92	68	9.40	126	93	12.86
26	19	2.63	60	44	6.08	94	69	9.54	127	94	13.00
27	20	2.77	61	45	6.22	95	70	9.68	129	95	13.13
28	21	2.90	62	46	6.36	96	71	9.82	130	96	13.27
30	22	3.04	64	47	6.50	98	72	9.95	132	97	13.41
31	23	3.18	65	48	6.64	99	73	10.09	133	98	13.55
33	24	3.32	66	49	6.78	100	74	10.23	134	99	13.69
34	25	3.46	68	50	6.91	102	75	10.37	136	100	13.83

Conversion ft/lb to joules multiply by 1.356
joules to ft/lb multiply by 0.737
joules to kg/m multiply by 0.1
kg/m to joules multiply by 10
kg/m to ft/lb multiply by 7.37

Appendix IID

Hardness and tensile values approximate conversion table for steel

Brinell				Diamond	Rockwell				
10 mm ball, 3000 kg load		Approximate equivalent tensile strength for steel		(HY, DPN or VPN)	120° cone			1/16 in. ball	Scleroscope
Diameter (mm)	HB HBN	(tonf/in^2)	N/mm^2 (MPa)		C scale (150 kg) (HRC)	D scale (100 kg) (HRD)	A scale (60 kg) (HRA)	B scale (100 kg) (HRB)	
2.00	945			1250	71	80	87	—	—
2.05	898			1150	70	79	87	—	—
2.10	856			1050	69	79	86	—	—
2.15	816			1000	68	78	86	—	—
2.20	781			975	67	78	85	—	106
2.25	745	163	2517	950	66	77	85	—	100
2.30	712	155	2394	910	65	76	84	—	95
2.35	683	150	2317	850	64	75	84	—	91
2.40	653	143	2209	790	62	73	83	—	87
2.45	627	137	2116	750	61	72	82	—	84
2.50	601	132	2039	715	59	71	81	—	81
2.55	578	127	1961	671	57	69	80	—	78
2.60	555	122	1884	633	56	68	79	—	75
2.65	534	117	1807	599	54	67	78	—	72
2.70	514	112	1730	572	52	65	77	—	70
2.75	495	108	1668	547	50	64	76	—	67
2.80	477	105	1622	523	49	63	75	—	65
2.85	461	101	1560	501	48	62	75	—	63
2.90	444	98	1514	479	47	61	74	—	61
2.95	429	95	1467	459	45	60	73	—	59
3.00	415	92	1421	441	44	59	73	—	57
3.05	401	88	1359	424	42	58	72	—	55
3.10	388	85	1313	409	41	57	71	—	54
3.15	375	82	1266	395	40	56	71	—	52
3.20	363	80	1236	382	39	55	70	—	51
3.25	352	77	1189	369	37	53	69	—	49
3.30	341	75	1158	358	36	52	68	—	48
3.35	331	73	1127	344	34	51	67	—	46
3.40	321	71	1097	332	33	50	67	—	45
3.45	311	68	1050	321	32	50	67	—	44
3.50	302	66	1019	310	31	49	66	—	43
3.55	293	64	988	299	30	49	66	—	42
3.60	285	63	973	290	29	48	65	—	41
3.65	277	61	942	282	27	46	64	—	40
3.70	269	59	911	274	26	45	64	—	39
3.75	262	58	896	267	25	45	63	—	38
3.80	255	56	865	260	24	44	63	—	37
3.85	248	55	849	253	23	43	62	—	36
3.90	241	53	819	246	22	42	62	—	35
3.95	235	51	788	240	21	41	61	100	34
4.00	229	50	772	234	20	41	61	99	33
4.05	223	49	757	228	19	40	60	98	32
4.10	217	48	741	222	18	—	60	97	31
4.15	212	46	710	217	·17	—	59	96	31
4.20	207	45	695	212	16	—	58	95	30
4.25	201	44	680	206	15	—	57	94	30
4.30	197	43	664	202	13	—	57	93	29
4.35	192	42	649	197	12	—	56	92	28
4.40	187	41	633	192	10	—	56	91	28
4.45	183	40	618	188	9	—	55	90	27
4.50	179	39	602	184	8	—	55	89	27
4.55	174	38	587	179	7	—	54	88	26
4.60	170	38	587	175	6	—	54	87	26
4.65	167	38	587	172	4	—	53	86	25
4.70	163	37	571	168	3	—	52	84	24
4.75	159	36	556	164	2	—	51	83	24
4.80	156	36	556	161	1	—	51	82	23
4.85	152	35	541	157	—	—	50	81	23
4.90	149	34	525	154	—	—	50	80	22
4.95	146	33	510	151	—	—	49	79	22
5.00	143	33	510	148	—	—	49	78	21
5.05	140	32	494	145	—	—	48	76	21
5.10	137	31	479	142	—	—	47	75	21
5.15	134	31	479	139	—	—	47	74	20
5.20	131	30	463	136	—	—	46	73	20
5.25	128	30	463	133	—	—	45	72	20
5.30	126	29	448	131	—	—	45	71	—
5.35	123	28	432	128	—	—	44	69	—
5.40	121	28	432	126	—	—	44	68	—
5.45	118	27	417	123	—	—	43	67	—

Appendix IID—*cont.*

Brinell				Diamond	Rockwell				
10 mm ball, 3000 kg load		Approximate equivalent tensile strength for steel		(HY, DPN or VPN)	120° cone			1/16 in. ball	Sclero-scope
Diameter (mm)	HB HBN	(tonf/in^2)	N/mm^2 (MPa)		C scale (150 kg) (HRC)	D scale (100 kg) (HRD)	A scale (60 kg) (HRA)	B scale (100 kg) (HRB)	
5.50	116	27	417	121	—	—	43	65	—
5.55	114	26	402	119	—	—	42	64	—
5.60	112	25	386	117	—	—	41	63	—
5.65	109	25	386	115	—	—	40	62	—
5.70	107	24	371	113	—	—	40	60	—
5.75	105	24	371	111	—	—	39	59	—
5.80	103	23	355	109	—	—	39	57	—
5.85	101	23	355	107	—	—	38	55	—
5.90	99	22	340	104	—	—	38	54	—
5.95	97	22	340	102	—	—	37	53	—
6.00	95	21	324	100	—	—	35	51	—

Appendix IIE

Thickness conversion table – approximate

Millimetres (mm)	Microns (μm)	Inches (in)
25	25,000	1.0
20	20,000	0.8
15	15,000	0.6
12.5	12,500	0.5
10	10,000	0.4
5	5,000	0.2
4	4.000	0.16
3	3,000	0.12
2	2,000	0.08
1	1,000	0.04
0.5	500	0.02
0.4	400	0.015
0.3	300	0.012
0.2	200	0.008
0.1	100	0.004
0.09	90	0.0035
0.08	80	0.0032
0.07	70	0.0028
0.06	60	0.0025
0.05	50	0.002
0.04	40	0.0015
0.03	30	0.0012
0.02	20	0.0008
0.01	10	0.0004
0.005	5	0.0002
0.004	4	0.00015
0.003	3	0.00012
0.002	2	0.00008
0.001	1	0.00004
0.0008	0.8	0.00003
0.0005	0.5	0.00002
0.0001	0.1	0.000004

Note

In verbal discussion there can be some confusion in that millimetres (mm) are often referred to as 'mils', meaning one thousandth of a metre while thousandths of an inch are also on occasion referred to as 'mils', meaning one thousandth of an inch.

With inches 0.001 inch will be called 'one thou', and 0.0001 inch will be called 'one tenth'.

It is thus clear that considerable confusion can be caused unless the basic unit being discussed is defined.

In writing, the term 'mil' will almost always refer to one thousandth of an inch; this applies in particular to US literature.

Appendix III

List of elements with their symbols

The section of the book under which each metal is discussed is shown alongside the element name.

	Element	Symbol		Element	Symbol
	Actinium	Ac	35	Mendelevium	Md
1	Aluminium	Al	25	Mercury	Hg
35	Americium	Am	26	Molybdenum	Mo
2	Antimony	Sb	35	Neodymium	Nd
	Argon	Ar		Neon	Ne
3	Arsenic	As	35	Neptunium	Np
	Astatine	At	27	Nickel	Ni
4	Barium	Ba	28	Niobium	Nb
35	Berkelium	Bk		(also called	
5	Beryllium	Be		Columbium, Cb)	
6	Bismuth	Bi		Nitrogen	N
7	Boron	B		Nobelium	No
	Bromine	Br	29	Osmium	Os
8	Cadmium	Cd		Oxygen	O
9	Caesium	Cs	30	Palladium	Pd
10	Calcium	Ca		Phosphorus	P
35	Californium	Cf	31	Platinum	Pt
	Carbon	C	32	Plutonium	Pu
11	Cerium	Ce		Polonium	Po
	Chlorine	Cl	33	Potassium	K
12	Chromium	Cr	35	Praseodymium	Pr
13	Cobalt	Co	35	Promethium	Pm
28	Columbium	Cb	35	Protectinium	Pa
	(also called		34	Radium	Ra
	Niobium, Nb)			Radon	Rn
14	Copper	Cu	36	Rhenium	Re
35	Curium	Cm	35	Rhodium	Rh
35	Dysprosium	Dy	38	Rubidium	Rb
35	Einsteinium	Es	39	Ruthenium	Ru
35	Erbium	Er	35	Samarium	Sm
35	Europium	Eu	35	Scandium	Sc
35	Fermium	Fm	40	Selenium	Se
	Fluorine	F	41	Silicon	Si
	Francium	Fr	42	Silver	Ag
35	Gadolinium	Gd	43	Sodium	Na
15	Gallium	Ga	45	Strontium	Sr
16	Germanium	Ge		Sulphur	S
17	Gold	Au	46	Tantalum	Ta
17A	Hafnium	Hf		Technetium	Tc
	Helium	He	47	Tellurium	Te
35	Holmium	Ho	35	Terbium	Tb
	Hydrogen	H	48	Thallium	Tl
18	Indium	In	49	Thorium	Th
	Iodine	I	38	Thulium	Tm
19	Iridium	Ir	50	Tin	Sn
20	Iron	Fe	51	Titanium	Ti
	Krypton	Kr	52	Tungsten	W
35	Lanthanum	La	53	Uranium	U
	Lawrencium	Lw	54	Vanadium	V
21	Lead	Pb		Xenon	Xe
22	Lithium	Li	35	Ytterbium	Yb
35	Lutetium	Lu	35	Yttrium	Y
23	Magnesium	Mg	53	Zinc	Zn
24	Manganese	Mn	56	Zirconium	Zr

Index

00 44K2
00 Iron 20A
0.000 44M1
0.0040 44A1
0.0041 44A1
0.0042 44A1
0.0043 44A1
0.0044 44A1
0.0073 44A3
0.0105 44A1
0.0115 44A1
0.0133 44A1
0.0402 44A1
0.0403 44A1
0.0404 44A1
0.0415 44A1
0.0433 44A1
0.0510 44A1
0.0511 44A1
0.0515 44A1
0.0516 44A1
0.0517 44A1
0.0518 44A1
0.0524 44A1
0.0525 44A1
0.0572 44A1
0.0612 44A1
0.0652 44A2
0.0722 44A3
0.0740 44A3
0.0761 44A3
0.0763 44A3
0.0783 44A4
0.0931 44A3
0.0967 44A2
0.0968 44A2
0.0969 44A3
0.0970 44A3
0.0971 44A3
0.1210 44A3
0.1221 44A3
0.1231 44A3
0.1248 44A3
0.1274 44A4
0.4021 44M1
0.4310 44N3
0.4401 44N1
0.5% FO 44K1
0.5028 44A3
0.5029 44A3
0.5467 44K1
0.70 Li Pja 44A3
0.7033 44E1
0.7035 44E2
0.7103 44E2
0.7225 44K2
0.7561 44K2
0.7727 44K2
0.8159 44K2
0.8161 44K2
0.8239 44K2
0 KX 44K3
0 STEEL 44A1
01
see SAE 01
01 44K3
1
see BS range 1A
1.0 W VAR 44N1
1.0022 44A1
1.0061 44A1
1.0100 44A1
1.0102 44A1
1.0103 44A1
1.0104 44A1
1.0106 44A1
1.0110 44A1
1.0112 44A1
1.0116 44A1
1.0122 44A1
1.0123 44A1
1.0130 44A2
1.0132 44A2
1.0136 44A2
1.0142 44A2
1.0143 44A2
1.0301 44A1
1.0305 44A1
1.0308 44A1
1.0309 44A1
1.0318 44A1
1.0322 44A1
1.0323 44A1
1.0325 44A1
1.0328 44A1
1.0329 44A1
1.0330 44A1
1.0331 44A1
1.0333 44A1
1.0334 44A1
1.0336 44A1
1.0337 44A1
1.0338 44A1
1.0345 44E1
1.0356 44A1
1.0401 44A1
1.0402 44A1
1.0405 44A1
1.0408 44E1
1.0418 44A1
1.0425 44E1
1.0435 44E1
1.0437 44A1
1.0439 44A1
1.0445 44E1
1.0448 44A1
1.0456 44A1
1.0501 44E1
1.0503 44E2
1.0505 44A3
1.0507 44A2
1.0509 44A2
1.0530 44A2
1.0532 44A2
1.0540 44A2
1.0542 44A2
1.0580 44A1
1.0581 44A1
1.0601 44A3
1.0603 44A3
1.0605 44A3
1.0611 44A1
1.0632 44A3
1.0651 44A2
1.0711 44A1
1.0713 44A1
1.0716 44A1
1.0719 44A1
1.0721 44A1
1.0723 44A1
1.0724 44A1
1.0726 44A2
1.0727 44A2
1.0728 44A3
1.0736 44A1
1.0737 44A1
1.0751 44A3
1.0831 44A1
1.0832 44A1
1.0833 44A1
1.0841 44A1
1.0844 44E1
1.0845 44E1
1.0902 44A3
1.0903 44A2
1.0904 44A3
1.0906 44A3
1.0909 44A3
1.0913 44A3
1.0916 44E1
1.0935 44E1
1.0958 44E2
1.0960 44E2
1.0961 44E2
1.0970 44A2
1.1% FO 44K1
1.1121 44A1
1.1141 44A1
1.1144 44A1
1.1151 44A1
1.1165 44A2
1.1174 44A2
1.1181 44A2
1.1183 44A2
1.1191 44A2
1.1193 44A2
1.1194 44A2
1.1210 44A3
1.1213 44A3
1.1221 44A3
1.1231 44A3
1.1232 44K2
1.1248 44A3
1.1249 44A3
1.1264 44A4
1.1274 44A4
1.1525 44A4
1.1545 44A4
1.1730 44A2

1.1740 44A3
1.1830 44A4
1.2% FO 44K1
1.2 Mo6/7 44G1
1.2002 44A4
1.2002 44E3
1.2063 44K3
1.2067 44E3
1.2080 44M2
1.2082 44M1
1.2083 44M1
1.2086 44M2
1.2101 44E2
1.2162 44E1
1.2210 44K3
1.2235 44K3
1.2241 44K2
1.2312 44K1
1.2316 44M1
1.2323 44K2
1.2327 44K3
1.2330 44K1
1.2343 44K2
1.2344 44K1
1.2344 44K2
1.2345 44K2
1.2363 44K3
1.2365 44K1
1.2367 44K2
1.2379 44M2
1.2419 44M2
1.2436 44M2
1.2510 44K3
1.2516 44K3
1.2541 44K2
1.2542 44K2
1.2547 44K2
1.2550 44K2
1.2552 44K3
1.2567 44K1
1.2581 44K1
1.2601 44M2
1.2625 44M1
1.2631 44K2
1.2703 44K3
1.2706 44F3
1.2711 44K2
1.2713 44K2
1.2714 44K2
1.2721 44E2
1.2721 44K2
1.2735 44K1
1.2764 44K1
1.2766 44K1
1.2767 44K2
1.2770 44K3
1.2779 44N1
1.2833 44J
1.2838 44J
1.2842 44K3
1.2886 44L
1.3 Mo6/7 44G1
1.3202 44L
1.3207 44L
1.3215 44L
1.3243 44L
1.3246 44L
1.3247 44L
1.3255 44L
1.3265 44L
1.3333 44K3
1.3342 44K3
1.3343 44K3
1.3344 44K1
1.3346 44K3
1.3348 44K3
1.3355 44K3
1.3401 44D
1.3402 44D
1.3405 44D
1.3501 44E3
1.3503 44E3
1.3505 44E3
1.3520 44E3
1.3802 44D
1.3811 44M1
1.3817 44P
1.3827 44M1
1.3921 20C
1.3926 20C
1.3927 20C
1.3960 44P
1.3962 44P
1.400 44M1
1.4000 44M1
1.4001 44M1
1.4002 44M1
1.4006 44M1
1.4007 44M1
1.4009 44M1
1.4015 44M1
1.4016 44M1
1.4021 44M1
1.4024 44M1
1.4034 44M1
1.4044 44M1
1.4057 44M1
1.4084 44M1
1.4104 44M1
1.4112 44M2
1.4113 44M1
1.4116 44M1
1.4119 44M1
1.4120 44M1
1.4122 44M1
1.4137 44M1
1.4300 44N1
1.4301 44N1
1.4302 44N1
1.4303 44N1
1.4304 44N1
1.4305 44N1
1.4306 44N1
1.4307 44N1
1.4310 44N1
1.4311 44N1
1.4314 44N1
1.4324 44N1
1.4370 44P
1.4371 44P
1.4401 44N1
1.4402 44N1
1.4404 44N1
1.4406 44N1
1.4417 44N1
1.4429 44N1
1.4435 44N1
1.4436 44N1
1.4438 44N1
1.4447 44N1
1.4449 44N1
1.4451 44P
1.4460 44N1
1.4462 44N3
1.4502 44M1
1.4504 44N3
1.4505 44N1
1.4507 44N1
1.4510 44M1
1.4511 44M1
1.4514 44N3
1.4523 44M1
1.4525 44M1
1.4529 44N2
1.4539 44N1
1.4541 44N1
1.4550 44N1
1.4551 44N1
1.4554 44N1
1.4558 44N1
1.4568 44N3
1.4570 44N1
1.4571 44N1
1.4573 44N1
1.4574 44N3
1.4577 44N1
1.4580 44N1
1.4581 44N1
1.4583 44N1
1.4584 44N1
1.4587 44N1
1.4589 44N1
1.4704 44E2
1.4712 44E1
1.4713 44E1
1.4716 44E1
1.4718 44E2
1.4721 44M2
1.4722 44M1
1.4723 44M1
1.4724 44M1
1.4741 44M1
1.4742 44M1
1.4747 44M2
1.4749 44M1
1.4762 44M1
1.4765 44M1
1.4767 44M1
1.4772 44M1
1.4773 44M1
1.4774 44M1
1.4820 44M1
1.4821 44M1
1.4828 44N1
1.4829 44N1
1.4833 44N1
1.4841 44N1
1.4842 44N1
1.4843 44N1
1.4844 44N1
1.4845 44N1

1.4854 44N1
1.4860 44N1
1.4876 27A
1.4876 44N1
1.4878 44N1
1.4903 44E1
1.4905 44M1
1.4919 44N1
1.4920 44M1
1.4922 44M1
1.4924 44M1
1.4934 44M1
1.4935 44M1
1.4943 44N1
1.4944 44N1
1.4954 44N1
1.4961 44N1
1.4970 44N1
1.4974 44N2
1.4984 44N1
1.4986 44N1
1.4994 44M1
1.5034 44A1
1.5035 44A1
1.5038 44A2
1.5053 44E1
1.5063 44A1
1.5064 44A1
1.5066 44A2
1.5067 44A2
1.5074 44A1
1.5083 44A1
1.5086 44A1
1.5098 44A1
1.5120 44A2
1.5121 44A2
1.5122 44A2
1.5151 44P
1.5152 44P
1.5161 44P
1.5213 44J
1.5223 44J
1.5331 44A1
1.5340 44A1
1.5404 44K1
1.5406 44K1
1.5415 44G1
1.5416 44G1
1.5417 44A1
1.5419 44K1
1.5424 44G1
1.5425 44G1
1.5426 44G1
1.5427 44G1
1.5428 44G1
1.5613 44F1
1.5620 44K1
1.5622 44F1
1.5633 44F1
1.5680 44F1
1.5919 44K1
1.5920 44K1
1.5924 44K1
1.5934 44K1
1.6001 44K1
1.6215 44F1
1.6216 44F1
1.6351 44F3
1.6354 44F3
1.6359 44F3
1.6511 44K1
1.6513 44K1
1.6582 44K1
1.6590 44K1
1.6592 44K1
1.6604 44K1
1.6761 44K1
1.6903 44N1
1.6905 44N1
1.7015 44E1
1.7033 44E1
1.7034 44E1
1.7035 44E2
1.7071 44E1
1.7083 44E1
1.7131 44E3
1.7147 44E1
1.7176 44E3
1.7205 44K1
1.7207 44K1
1.7214 44K1
1.7218 44K1
1.7220 44K1
1.7225 44K2
1.7228 44K2
1.7242 44K1
1.7251 44K1
1.7254 44K1
1.7258 44K1
1.7273 44K1
1.7281 44K1
1.7283 44K1
1.7305 44E3
1.7307 44K1
1.7324 44K1
1.7334 44K1
1.7335 44K1
1.7337 44K1
1.7345 44K1
1.7346 44K1
1.7350 44K1
1.7356 44K1
1.7361 44K1
1.7362 44K1
1.7373 44K1
1.7380 44K1
1.7384 44K1
1.7394 44K1
1.7513 44K1
1.7561 44K2
1.7701 44K2
1.7704 44K1
1.7707 44K1
1.7708 44K1
1.7709 44K1
1.7710 44K1
1.7733 44K1
1.7766 44K1
1.7779 44K1
1.8070 44K1
1.8154 44K2
1.8159 44K2
1.8212 44K1
1.8401 44E1
1.8403 44E2
1.8404 44E2
1.8405 44E3
1.8425 44E3
1.8504 44E1
1.8506 44E1
1.8507 44K1
1.8514 44K1
1.8519 44K1
1.8544 44K1
1.8550 44K1
1/360 27C
1 A
see BS range 1A
1A 1A
1A 14G
1 Al 8 V 5 Fe 51C
1 B
see BS range 1A
1B 1A
1B 44A1
1 C
see BS range 1A
1C 1A
1C 14A
1 C 27 44M1
1C 36 44M1
1C 342 44M1
1C 901 27C
1CN 44M1
1CR 44A1
1 Cr Mo 44K1
1 Cr Mo B 44K1
1CS 44A1
1 CW 44M2
1DT 44A1
1DTR 44A1
1 E
see BS range 1A
1E 1A
1HR 44A1
1HS 44A1
1Ni 44F1
1 Ni B 44K1
1 Ni Cu B 44K1
1 Ni Mo B 44K1
1R2 44N1
1 R3 44M2
1S 1A
1S 15 44B1
1S 35 44B1
02
see SAE 02 44K3
02 44A4
02 44J
02 44K3
2.0050 14A
2.0060 14A
2.0070 14A
2.0080 14A
2.0090 14A
2.0100 14A
2.0110 14A
2.0120 14A
2.0150 14A
2.0170 14A
2.0181 14A

2.0290 14B
2.0291 14B
2.0293 14B
2.0295 14B
2.0340 14C
2.0341 14C
2.0343 14C
2.0366 14C
2.0470 14B
2.0531 14C
2.0590 14C
2.0592 14C
2.0596 14C
2.0598 14C
2.0837 14E
2.0882 14E
2.0917 14G
2.0921 14K
2.0928 14G
2.0929 14G
2.0940 14G
2.0941 14G
2.0962 14G
2.0966 14G
2.0970 14G
2.0971 14G
2.0975 14G
2.0980 14G
2.1006 14A
2.1021 14K
2.1050 14K
2.1051 14K
2.1052 14K
2.1053 14K
2.1056 14K
2.1057 14K
2.1086 14K
2.1087 14K
2.1090 14K
2.1091 14K
2.1096 14K
2.1097 14K
2.1166 14K
2.1170 14K
2.1171 14K
2.1176 14K
2.1177 14K
2.1182 14K
2.1183 14K
2.1188 14K
2.1189 14K
2.1211 14A
2.1461 14M
2.2% FO 44K1
2.2D8 44F1
2.2360 55A
2.2480 8.1
2.3010 21.1
2.3020 21.1
2.3021 21.1
2.3030 21.1
2.3040 21.1
2.3075 21.1
2.3085 21.1
2.3131 21.1
2.3132 21.1
2.3137 21.1
2.3138 21.1
2.3139 21.1
2.3201 21.1
2.3202 21.1
2.3203 21.1
2.3205 21.1
2.3208 21.1
2.3212 21.1
2.3229 21.1
2.3299 21.1
2.3561 1J
2.3820 50A
2.3830 50A
2.3852 50A
2.4050 27A
2.4051 27A
2.4052 27A
2.4053 27A
2.4056 27A
2.4060 27A
2.4062 27A
2.4066 27A
2.4068 27A
2.4106 27A
2.4108 27F
2.4110 27F
2.4116 27F
2.4122 27F
2.4128 27F
2.4132 27F
2.4360 27D
2.4374 27D
2.4375 27D
2.4400–2.4419 27F
2.4472 20C
2.4480 20C
2.4486 20C
2.4500–2.4519 27E
2.4520–2.4539 27F
2.4537 27C
2.4540–2.4559 27E
2.4600 27F
2.4602 27C
2.4605 27C
2.4634 27C
2.4640 27B
2.4662 27C
2.4665 27C
2.4816 27B
2.4858 27B
2.4858 27C
2.4867 27B
2.4867 27E
2.4869 27B
2.4870 27B
2.4889 27C
2.4951 27B
2.4952 27C
2.4969 27C
2.5 Al 16 V 51C
2A 14F
2A 14G
2A 3 52.1
2A 5 52.1
2A 6 52.1
2A 7 52.1
2A 68 52.1
2AS 44K3
2 B8
see BS/B8 14K
2C 27 44M1
2C 34 44M1
2 Co 4 Cr 44L
2CR 44A1
2 Cr 2 Fe 2 Mo 51B
2 Cr 120 Mo 8 Co 44M1
2 Cr Mo 44K1
2 Cr Mo B 44K1
2 Cr Mo BG 44K1
2 Cr Mo BX 44K1
2 Cr Mo C B 44K1
2CS 44A1
2 CV 44K2
2EC 1E
2H 14L
2HR 44A1
2HS 44A1
2LS 44A1
2 N3C1 44K1
2Ni 44F1
2NiB 44F1
2 Ni Mo B 44K1
2R 44N1
2R1 44N1
2R2A 44N1
2R12 44N1
2R16 44N1
2R 27 44M1
2R 29 44M1
2R 47 44M1
2R 57 44M1
2R 77 44M1
2R 107 44M1
2RE10 44N1
2RE69 44N1
2RK65 44N1
2RL 2 44M1
2RL 3 44M1
2RL 14 44M1
2RM 2 44M1
2RMO 44M1
2RO 26 44M1
2RO 27 44M1
2RO 46 44M1
2 RO 189 44M2
2S 1A
2S 44A1
2SE 44A1
2Z 4 55A
2Z 5 55A
2Z 6 55A
03 NCMB 44K1
03P 44L
3 14D
3% Co 44L
3% FO 44K1
3.0001 1G
3.0002 1G
3.0120 1G
3.0185 1A
3.0200 1A
3.0205 1A
3.0250 1A
3.0255 1A

3.0256 1A
3.0257 1A
3.0270 1A
3.0275 1A
3.0280 1A
3.0285 1A
3.0305 1A
3.0385 1A
3.040 1A
3.0515 1B
3.0570 1B
3.0615 1E
3.1190 1F
3.1255 1F
3.1263 1L
3.1263 1M
3.1305 1F
3.1325 1F
3.1335 1F
3.1355 1F
3.1365 1F
3.1371 1L
3.1590 1F
3.1645 1F
3.1655 1F
3.1841 1L
3.2131 1C
3.2131 1F
3.2151 1L
3.2152 1C
3.2153 1C
3.2161 1L
3.2245 1C
3.2285 1C
3.2305 1E
3.2315 1E
3.2341 1C
3.2371 1C
3.2381 1C
3.2383 1C
3.2385 1C
3.2572 1C
3.2573 1C
3.2581 1C
3.2582 1C
3.2583 1C
3.2585 1C
3.2685 1N
3.3205 1E
3.3206 1E
3.3241 1E
3.3243 1E
3.3245 1E
3.3261 1J
3.3282 1J
3.3292 1J
3.3308 1D
3.3309 1D
3.3315 1D
3.3318 1D
3.3319 1D
3.3329 1D
3.3350 1D
3.3525 1D
3.3527 1D
3.3535 1D
3.3537 1D
3.3547 1D
3.3555 1D
3.3557 1D
3.3575 1D
3.3591 1J
3.4335 1G
3.4345 1G
3.4355 1G
3.4365 1G
3.4375 1G
3.4415 1A
3.5D8 44F1
3.5002 23A
3.5003 23A
3.5101 23B
3.5103 23B
3.5104 23B
3.5161 23B
3.5200 23B
3.5312 23B
3.5612 23B
3.5632 23B
3.5812 23B
3.5912 23B
3.5922 23B
3.6104 23B
3.6204 23B
3.7024 51A
3.7025 51A
3.7034 51A
3.7064 51A
3.7124 51B
3.7144 51C
3.7164 51C
3.7174 51C
3.7184 51C
3A 14G
3 Al 12.5 V 51C
3 Al 13 V 11 Cr 51C
3AS 44K3
3BS 44A1
3BSX 44A1
3C 27 Mo 2 44M1
3CR 44A1
3CS 44A1
3 G Mo B 44K1
3HR 44A1
3HS 44A1
3 KH2V8F 44K1
3L7 44A1
3 LC 2 44E1
3LS 44A1
3 MC 2 44E1
3 Mo 44G1
3 Mo 1 44G1
3NiB 44F1
3 Ni Cr Mo B 44K1
3 Ni Mo B 44K1
3 Ni Mo LB 44K1
3NS 44F1
3R12 44N1
3R16 44N1
3R17 44N1
3R19 44N1
3R54 44N1
3R60 44N1
3R61 44N1
3R62 44N1
3R63 44N1
3R64 44N1
3R65 44N1
3 RE 60 44N3
3RE 60 44M1
3S 1B
3S 1 1B
4 14C
4A 14G
4 Al 3 Mo 1 V 51C
4 Al 4 Mn 51B
4AS 44K3
4C 27 44M1
4C 27A 44M1
4C 54 44M1
4CR 44A1
4CS 44A1
4F 14E
4H 14L
4HR 44A1
4HS 44A1
4 KH2V5FM 44K1
4KH8V2 44K2
4KH14N14V2M 44N2
4KHS 44E2
4KHV2S 44K2
4L7 44A1
4LM 44A1
4LS 44A1
4M 14C
4 Mo Var 44N1
4N 2C 36 44M1
4 NHD 44K1
4S 1B
4VH 14C
4 WHD 44K1
4.5% DO 44K1
5 14C
5/6/2 quality 44K3
5/317 44K2
5 Al 1.25 Fe 2.75 Cr 51B
5 Al 2.5 Sn 51B
5 Al 5 Sn 5 Zr 51B
5 C 2 Mo 44K1
5CC 44K3
5 Cr Mo 44K1
5 Cr Mo 10 44K1
5 Cr Mo 22 44K1
5 Cr Mo B 44K1
5KHNGM 44K2
5KHNM 44K2
5KHNV 44K2
5KHV2S 44K2
5 LC 44E2
5NSR 44F1
5R10 44N1
5R60 44N1
5RA50 44N1
5V 14C
06
see SAE 06 44G2
06 44G2
06/05/04/02 44K3
06/05/04/04 44K3
06/05/05/04/01 44L
06S 44J

6% Co 44L
6/5/2 44K3
6A 14G
6 Al 4 V 51C
6 Al 6 V 2 Sn 51C
6C 27 44M1
6C 283 44M1
6 F4 44K1
6KHS 44E2
6 LC 44E2
6 LCK 44E2
6 LM 44A2
6 LMC 44E1
6 N 3 C2 Mo 44K1
6 N 4 C 2 44K1
6R10 44N1
6R40 44N1
6R55 44N1
6R60 44N1
6R64 44N1
07 44K3
07/05/05/05/04 44L
07 NCMB 44K1
7A 14G
7 Al 2 Cb 1 Ta 51B
7 Al 4 Mo 51B
7 Al 12 Zr 51B
7 C 2 Mo 44K1
7C 27 44M1
7C 27 Mo2 44M1
7 Cr Mo 7/5 44K1
7 Cr Mo 12/10 44K1
7 Cr Mo B 44K1
7KH3 44E2
7KHF 44K2
7 L 44A2
7M 14C
7 Mo Plus Stainless 44N3
7 Mo Stainless 44N3
7 Mo Var 44N1
7R60 44N1
7RE10 44N1
7RE12 44N1
08/04/02/02 44K3
08S 44L
8 Al 1 Mo 1 V 51B
8CN2 44F1
8KH3 44E3
8KH4V4F1(P4) 44K3
8KHF 44K3
8 LM 44A2
8M 14B
8 MC 2 Mo 44K2
8 Mn 51B
8N2 44K3
8-N-2 COBALT 44L
8 NIC 2 44K2
8R30 44N1
8R40 44N1
8R41 44N1
8R42 44N1
8R70 44N1
8R80 44N1
8R81 44N1
09/08/04/02/01 44L
09B 44K3
9% Co 44L
9 Cr Mo 4/5 44K1
9 Cr Mo B 44K1
9 Cr Mo L B 44K1
9ct No. 90 17.1
9ct No. 91 17.1
9ct No. 92 17.1
9ct No. 93 17.1
9ct No. 94 17.1
9ct No. 95 17.1
9ct No. 97 17.1
9ct No. 98 17.1
9ct No. 99 17.1
9ct No. 910 17.1
9ct No. 911 17.1
9ct No. 912 17.1
9ct S 91 (easy) 17.1
9ct S 92 (medium) 17.1
9ct S 93 (hard) 17.1
9ct W 91 17.1
9ct W 93 17.1
9ct W 94 17.1
9ct W 95 17.1
9ct W 96 17.1
9ct W 97 17.1
9ct W 98 17.1
9ct W 99 17.1
9 G Mo 44K1
9 KH 44K3
9KH5F 44K3
9KH5VF 44K3
9KHF 44K3
9KHS 44E3
9KHVG 44K3
9 L 44A2
9 L 7 44A2
9 LM 44A2
9M 14C
9 Mn Mo 4/5 44G1
9 Mn Ni 4 44F1
9 S 20 44A1
9 S 27 44A1
9 S Mn 23 44A1
9 S Mn 28 44A1
9 S Mn 36 44A1
9 S Mn Pb 23 44A1
9 S Mn Pb 28 44A1
9 S Mn Pb 36 44A1
9 S Pb 23 44A1
10 14C
10 14D
10.17.5.9 Mn 44P
10/07/05/05/04 44L
10A 14G
10C 14D
10 C 2 M 44K1
10C 27 44M1
10 CD9.10
see NF A36 206 44K1
10 CNK 1 44K1
10 CNK 3 44K1
10 CNK 32 44K1
10CR 14D
10 Cr 11 44E1
10 Cr Mo 9/10 44K1
10 K 2 44E3
10 KD 44K3
10 M1 44A1
10 N9 44F1
10 NC12 44K1
10 NCMB 44K1
10R13 44N1
10R52 44N1
10R53 44N1
10 RE 20 44M1
10 RE 21 44M1
10RE51 44N1
10 S 20 44A1
11 1H
11.11 44K2
11CR 14D
11Cr0.09CrTi
see 409 44M1
11 Cr 9 Mn B 44M1
11 Cr Mo 4/5 44K1
11KH 44E3
11M 14C
11 Mn 4 44A1
11 Mn 4 A1 44A1
11R50 44N1
11R51 44N1
11S 1F
12 1H
12 51A
12.40 44A1
12/12 EL 44N1
12A 14G
12 C 44E2
12 CMV 44K1
12 CNK 3 44K1
12 Cr 44M1
12 Cr 152 44M1
12 Cr Mo 19/5 44K1
12 Cr MOB 44M1
12 Cr Mo V 44M1
12 Cr Mo V B 44K1
12 Cr Mo VB 44M1
12 KH1MF 44K1
12 KV 44K3
12L13 44A1
12L14 44A1
12 MKH 44K1
12 Mn 6 44A1
12 Mn 6 A1 44A1
12 Mn 8 A1 44A1
12 NCD 6 44K1
12 NCMB 44K1
12 Ni19 44F1
12R10 44N1
12R10 HV 44N1
12R11 44N1
12R72 44N1
12 S 44K1
12W1 44J
12 W 12 C1 44K3
13 1C
13.2R 44M1
13.4.2 Cu R 44N3
13.4LB 44M1
13.4 Mo LB 44M1
13.4 Mo LR 44M1
13BMP 44M1
13 BMP 44M1
13 Cr Mo 4/4 44K1
13 Cr Mo V 4/2 44K1

13Cr0.32C
see 420 S45 44M1
13KH 44E3
13 Mn 12 44A1
13 Mn Mo 3/5 44G1
13 Mn Mo 6/5 44G1
13 Mn Mo 8/5 44G1
13 Mn Mo 12/5 44G1
13 Mn R 44D
13RMP 44M1
14.6.2B 44N3
14.6.2 R 44N3
14.75 Mn Nb 27B
14.75 Mo Nb 27B
14.75 Nb 27B
14/4 PH 44N3
14/450 17.1
14 CNK 3 44K1
14 CNK 4 44K1
14 CNK 32 44K1
14 Cr 4 Mo 44M2
14 Cr 14 Mn R 44P
14ct No. 144 17.1
14ct No. 145 17.1
14ct No. 146 17.1
14ct No. 147 17.1
14ct No. 148 17.1
14ct No. 149 17.1
14ct S 143 (medium) 17.1
14ct S 145 (hard) 17.1
14ct S 146 (easy) 17.1
14ct S 147 (easy) 17.1
14ct W 142 17.1
14ct W 143 17.1
14 HD 44K2
14HR 44A1
14HS 44A1
14 Mn 4 44A1
14 N3 44F1
14N 3 44F2
14 NC 44K3
14 Ni6 44F1
14 NKD 2 44K1
14S 1F
14S 44M2
14 Si C1 44E3
15 51A
15% Co 44L
15% MANGANESE SILVER 42.1
15.35.2 CB 44N1
15.4 Mn Si R 44N3
15.5 PH 44N3
15.60 Mn WNb 27B
15.60 Nb 27B
15/35 44N1
15-15LC 44M1
15 B 21 44C
15 B 35 44C
15 B 37 44C
15 B 41 44C
15 B 48 44C
15 B 62 44C
15 CD4.05
see NF A36 206 44K1
15 CD205
see NF A36 206 44K1
15 Cr 3
see DIN range 44E1
15 Cr Mo 3 44K1
15 Cr Mo 6 44K1
15 Cr Ni 6 44K1
15Cr5Ni 44N3
15ct S 151 (easy) 17.1
15HR 44A1
15HS 44A1
15MDV405
see NF A36 206 44K1
15 Mo 3 44G1
15 N 3 CM 44K1
15 N 4 CM 44K1
15 NCMB 44K1
15 Ni Cr 14 44K1
15 NKD 44K1
15RA10 44N1
15RE10 44N1
15RE11 44N1
15RE12 44N1
15 S 20 44A1
15VDT 44J
16 44A4
16/12 UMoR 44N1
16/13 L Nb 44N1
16/14 L Mo Nb V 44N1
16/16 L Mo Nb 44N1
16/25/6 44N1
16/25/6 44N2
16/550 17.1
16 C1 V 44K3
16 CNK 1 44K1
16 Cr 9 44E1
16 Cr Mo 4 44K1
16 Cr Mo 4/4 44K1
16 Cr Mo 9/3 44K1
16 Cr Mo V 4/2 44K1
16 MC 5 44E1
16 Mn Cr 5
see DIN range 44E1
16 NC 6 44K1
16 NCK 3 44K1
17 44A4
17.12.1.4 Mn Nb 44P
17.12.2.4 MN 44N1
17.17.4 LR 44N1
17.38.4 CB 44N1
17.38.4 CR 44N1
17.4 Cu R 44N3
17.4.IR 44N3
17.8.2 RCF 44N1
17/4 PH 44N3
17/7 44N3
17/7 PH 44N3
17/12 UL 44N1
17/12 ULR 44N1
17/14 Cu Mo 44N1
17/14 Cu Mo 44N3
17/22A 44K2
17/22 AS 44K1
17/22 AV 44K1
17A 14G
17 AP 44A4
17 BMP 44M1
17C 44E3
17C1 44E3
17 CNK 44K1
17 Co 44L
17Cr
see 430 S17 44M1
17 Cr Mo V 10 44K1
17 Cr Mo V 16/10 44K1
17Cr+Mo
see 434 44M1
17 Mn 4
see DIN range 44E1
17 Mn Cr 5 44E1
17 Mn Mo R 44D
17 Mn Ni4 44F1
17 Mn Ti 6 44A1
17 Mn Zr 4 44A1
17 ND 2 44K1
17 ND 3 44K1
17 Ni 1600 44F3
17 RMP 44M1
17S 1F
17VDT 44J
18.18.3 Cu LR 44N1
18.18 PLUS 44P
18.2 44M1
18.20.2 Cu Nb 44N1
18.6.9 Mn Si Nb 44P
18.8.3 Mn Si Nb N 44P
18/5/9 44P
18/8 44N1
18/8 EL 44N1
18/8 EMo 44N1
18/8 FM 44N1
18/8 L 44N1
18/8 LNb 44N1
18/8 LNbR 44N1
18/8 LT 44N1
18/8 Si 44N1
18/8 UL 44N1
18/8 ULR 44N1
18/8 UMo 44N1
18/10 Cb 44N1
18/10 EMo 44N1
18/10 LMoNb 44N1
18/10 LMoT 44N1
18/10 Ti 44N1
18/10 UMo 44N1
18/11 EL 44N1
18/12 44N1
18/12 EMo 44N1
18/12 LMo 44N1
18/12 LMoNB 44N1
18/12 LMoT 44N1
18/12 Mo 44N1
18/12 MoL 44N1
18/12 UMo 44N1
18/13 EMo 44N1
18/14 EMo 44N1
18/14 UMo 44N1
18/15 EMo 44N1
18/15 UMo 44N1
18/650 17.1
18A 14G
18 C 283 44M2
18 Cr 2 Mo 44M1
18 Cr 2 Ni 12 Mn 44P
18 Cr Mo 3 44K1
18 Cr Ni 8 44K1

18ct No. 185 17.1
18ct No. 186 17.1
18ct No. 187 17.1
18ct No. 188 17.1
18ct S 181 (hard) 17.1
18ct S 185 (easy) 17.1
18ct S 186 (medium) 17.1
18ct W 181 17.1
18ct W 182 17.1
18ct W 183 17.1
18ct W 184 17.1
18ct W 185 17.1
18 D 44K2
18 HD 44K2
18 KH3MV 44K1
18MD405
  see NF A36 206 44G1
18 MM 44G1
18 Mo V 8/4 44K1
18 NCD 4 44K1
18 NCD 14 44K1
18 Ni 1400 44F3
18 Ni 1700 44F3
18 Ni 1900 44F3
18 Ni 2400 44F3
18 Ni Cr Mo 14 44K1
18 Ni MA 44F3
18 Ni MP 44F3
18S 1F
18S 51A
19 13A
19.12.3 B 44N1
19.12.3 ELRCF 44N1
19.12.3 LR 44N1
19.12.3 LRCF 44N1
19.12.3 LRCF2 44N1
19.12.3 LRNF 44N1
19.12.3 NbR 44N1
19.12.3 NLB 44N1
19.12.3 R 44N1
19.12.3 RMP 44N1
19.12.3 ULRCF 44N1
19.13.4 ELRCF 44N1
19.13.4 LRCF 44N1
19.13.4 R 44N1
19.15.3 MnBNF 44N1
19.15.3 Mn RNF 44N1
19.9.1 R 44N1
19.9.6 Mn 44P
19.9 B 44N1
19.9 L.HE 44N1
19.9 LB 44N1
19.9 LRMP 44N1
19.9 NbB 44N1
19.9 NbLR 44N1
19.9 NbR 44N1
19.9 R 44N1
19/9 44N1
19/9 DL 44N1
19/9 DL 44N2
19/9 DX 44N1
19/9 L 44N1
19/9 WMo 44N1
19/9W Mo 44N2
19/9 WX 44N1
19/12 Mo 44N1
19/12 MoL 44N1
19F 44G2
19 Mn 4
  see DIN range 44E1
19 Mn 5 44E1
20 14B
20 44A4
20.10.3 CB 44N1
20.15.3 CB 44N1
20.15.3 CR 44N1
20.23.3 CuNb 44N1
20.25.4 NbR 44N1
20.29.3 CuNbB 44N1
20.32 NbB 44N1
20.34.3 CuNbB 44N1
20.55 4Si Cu W 27B
20.9.3 B 44N1
20.9.3 R 44N1
20.9.3 RMP 44N1
20/10 44N1
20/12 L 44N1
20/12 Si 44N1
20/25 N6 44N1
20/25 Ti 44N1
20/750 17.1
20C 14D
20 Cb 3 44N1
20 CD 2 44K1
20CR 14D
20 Cr Mo 9 44K1
20 Cr Mo V 13/5 44K1
20 DN32.12 44K1
20 KH3MVF 44K1
20 Mn 5 44E1
20 Mn Cr 5
  see DIN range 44E1
20 Mn Mo 4 44G1
20 Mn V6 44J
20 Mo 3 44G1
20 Mo 4 44N1
20 Mo 6 44N1
20 NC 6 44K1
20 NCD 2 44K1
20 NCD 7 44K1
20 ND 2K 44K1
20 Ni Cr Mo 2 44K1
20 Ni Cr Mo 4 44K1
20 Ni Cr Mo 7 44K1
20 NK 1 44K1
20 NK 3 44K1
20 NXD 44K1
20 P 44A4
20RMP 44M1
20 W 1V 44K3
21 1H
21 14B
21.26.5 CuNbR 44N1
21/4N 44P
21-12 N Valve 44N1
21-12 Valve Steel 44N1
21A 14G
21C 44E3
21 Cr 6 Ni 9 Mn 44P
21 Cr Mo 3 44K1
21 Cr Mo V 5/11 44K1
21 Cr V Mo W 12 44K1
21 MN 44P
21 Mn 6 44A1
21 Mn Cr 5 44E1
21 Mo V 5/3 44K1
21 T 10 44K3
21 T 10 P 44K3
22 1H
22.3 ALLOY 20C
22/4/9 44P
22/800 17.1
22A 44N1
22C 44E3
22C1 44E3
22 Cr 13 Ni 5 n 44N1
22 Cr Mo 4/4 44K1
22 Cr V 4 44K1
22ct No. 223 17.1
22H 27B
22 Mo 4 44K1
22 MS 14H
22 NCM 44K1
22 NKD 44K1
22 S 20 44A1
23 1H
23.12.2 LR 44N1
23.12.2 NbR 44N1
23.12.2 R 44N1
23.12 LR 44N1
23.12 NbLR 44N1
23.12 NbR 44N1
23.12 R 44N1
23.12 WR 44N1
23.35.4 CR 44N1
23/14 L 44N1
23/14 LNb 44N1
23C 27 44M1
23S 44M2
24 1H
24 44A4
24.14.2 CWR 44N1
24.24.3 CNB 44N1
24A 14G
24 C 3 M 44K1
24 Cr Mo 5 44K1
24 Cr Mo 10 44K1
24 Cr Mo V 5/2 44K1
24 Cr Ni 9 44K1
24 Ni 4 44K1
24 Ni 8 44K1
24S 1F
24S 44K1
25 14D
25.12.4 CR 44N1
25.20.2 CB 44N1
25.20.3 CB 44N1
25.20 B 44N1
25.20 HB 44N1
25.20 HR 44N1
25.20 MoB 44N1
25.20 NbB 44N1
25.20 SUPER R 44N1
25.25.2 Nb 44N1
25.35.4 CW Co 27F
25.5.1 C 44N1
25.50 COW Nb Zr 27F
25.51C 44M2
25.55.6 Cu W 27B
25.6.2 CuLR 44N1
25.6.2 CuR 44N1

25.6.3 CuR 44N1
25/12 44N1
25/20 44N1
25/20 Si 44N1
25/21 L 44N1
25/21 Si 44N1
25/24 EMo 44N1
25-4.4 44M1
25CB 44A1
25 Cr Mo 4 44K1
25 KH1MF 44K1
25 KH2MF 44K1
25MS 44A1
25 NCD11 44K1
25 Ni 13A
25 Ni & V 13A
25 NK1 44K1
25S 1F
25 S 44K1
26 52.1
26.3.3 44M1
26.33.4 CNbR 44N1
26.35 COW Nb 27F
26A 14G
27/4 L 44N1
27/5 LMo 44N1
27 Cr Mo Ni 44K1
27 Cr Mo Ni 1 44K1
27 Cr Mo Ni 2 44K1
27 NCM 44K1
28.37.4 Cu B 27F
28.6.4 CB 44N1
28 C 3 M 44K1
28 F 44K1
28 Ni Cr Mo 4 44K1
28 Ni Cr Mo 4/4 44K1
28 Ni Cr Mo 7/4 44K1
28 NKD 3 44K1
28RMP 44M1
29.4 44M1
29.4.2 44M1
29.4C 44M1
29.9% SUPER R 44N1
29.9 R 44N1
29 Cr 4 Mo 44M1
29 Cr 4 Mo 2 Ni 44M1
30 51A
30A 44N1
30 ALLOY 14E
30 CM 44K1
30CR 14D
30 Cr Mo V 9 44K1
30 Cr Ni Mo 8 44K1
30 DAK 44A1
30 DCKY/28 44L
30 DCV28 44K1
30 DKCV28 44K1
30 MM 44G1
30 Mn 5
see DIN range 44A2
30 Mn Cr Ti 4 44E1
30 N 4 CM 44K1
30 NCD 12 UNI 4365 44A2
30 NCD2 44K1
30 NCD16 44K1
30 Ni Cr Mo 12 44K2
30 NK 3 44K1
30 NKD 2 44K1
30 NKD 3 44K1
30 NKDE 44K1
30 S 44K2
30 W Cr V 17/9 44K1
31 14C
31 CM 44K1
31 Cr Mo V 9 44K1
31 S 44K2
32 B 44K3
32 Cr Mo 12 44K1
32 Cr Mo B 44K1
32 NDC18.12 44K1
32S 1C
34/20 44A1
34/20 CR 44A1
34/20 CS 44A1
34/20 HR 44A1
34/20 HS 44A1
34 Cr 4
see DIN range 44E1
34 Cr Al 6 44E1
34 Cr Al Mo 5 44K1
34 Cr Al Ni 7 44K1
34 Cr Al S5 44E2
34 Cr Mo 4 44K1
34 Cr Ni Mo 6 44K1
34 Cr Ni Mo 8(e.f.) 44K2
34 Ni 5 44K1
035 AHK 44A1
35 14D
35 51A
35% Co 44L
35C 14D
35 CP 44K1
35 Cr Mn 5 44E2
35 NC5 UNI 4365 44K2
35 NC9 UNI 4365 44K2
35 Ni Cr Mo 15 44K2
35 Ni Cr9 44K2
35 NK 1 44K1
35 NKD 1 44K1
35 NKD 2 44K1
35 S 20 44A2
35 W Cr V7 44K2
36A 14G
36 Cr Ni Mo 4 44K1
36 Cr Ni Mo 6 44K1
36 Mn 5 44A2
37 14C
37/20 44A1
37/23 CR 44A1
37/23 CS 44A1
37/23 HR 44A1
37/23 HS 44A1
37 CB 44A2
37 Cr 4 44E2
37 Mn Si 5
see DIN range 44A2
37 MS 44A2
37 S 44K2
38 CD4 UNI 4365 44K2
38 CM 44K2
38 Mn Si 4 44A2
38 NCD 4 UNI 4365 44A2
38 Ni Cr Mo 4 44K2
38 Si 6 44A2
38 Si 8 44A2
040 A 04 44A1
040 A 10 44A1
040 A 12 44A1
040 A 22 44A1
040 AHK 44A1
40 51A
40/30 44A1
40 Cr 4 44E2
40 Cr Mn Mo S 8.6 44M1
40 Cr Mo 4 44K2
40 D 35 44A4
40 E 1M
40 EE 44A1
40F30 44A1
40 FK 44K1
40 Mn 4
see DIN range 44A2
40N2 CM 44K2
40 NCD 7 UNI 4365 44A2
40 Ni Cr Mo 7 44K2
40 NKD 44K2
41A 14G
41 Cr 4 44E2
41 Cr 4 44E3
42A 14G
42 A 44A1
42 ALLOY 20C
42 Cr Mo 4 44K2
42 Cr V 6 44K2
42 Mn V 7 44A2
42 Mn V7 44J
43 1C
43/25 44A1
43/25 HR 44A1
43/25 HS 44A1
43/35 44A1
43 BV 12 44K1
43 BV 14 44K1
43 EE 44A1
43F35 44A1
43S 1C
44 A 52.1
44 D 35 44A4
44 K MONEL 27D
045 A10 44A1
045 AHK 44A1
045 M10 44A1
45 CM 44K2
45 CMV 44K2
45 CP 44K2
45 Cr 5 Ni 5 44K2
45 Cr 13 44M1
45 Cr 25 44M1
45 Cr Mo V 6-7 44K2
45FG 44A1
45 MS 44A2
45 S 20 44A2
45 W Cr V 7 44K2
46/40 44A1
46 B 12 44K1
46F40 44A1
46 Mn Si 4 44A3
46 Si 7 44A2
48.04 44A1
48.14 44A1
48 Cr Mo V 67 44K2

48 D 35 44A4
050 A04 44A1
050 A10 44A1
050 A12 44A1
050 A15 44A1
050 A20 44A1
050 A22 44A1
050 A 86 44A4
50 14D
50 51A
50.50 27B
50.50 Nb 27B
50.50 Nb 27F
50/35 44A1
50/35 HR 44A1
50/35 HS 44A1
50/45 44A2
50 A 44A1
50 B40 44E2
50B44
  see AISI 44E2
50 B44 44E2
50B46
  see AISI 44E2
50 B46 44E2
50B50
  see AISI 44E2
50 B50 44E2
50 B60 44E2
50C 14D
50CR 14D
50 Cr Mo 4 44K2
50 CR V 4 44K2
50 DD 44A1
50 EE 44A1
50 F 44A1
50F45 44A1
50FK 44A1
50 Mn 7 44A3
50 Mn Cr Ti 4 44E2
50 Mn Si 4 44A3
50 Ni Cr 13 44E2
50 Ni Cr 13 44K2
50P 14D
50S 1D
50 Si 7 44A2
50 W 44A1
50 WCV20 44K2
51 31.1
51A 14C
51B60
  see AISI 44E2
51 Si 7
  see DIN range 44A3
52 31.1
52S 1D
52 Si Cr Ni5 44K2
53 31.1
53.05 44A1
53.35 44A1
53 Mn Si 4 44A3
53 S 44K2
54 Ni Cr Mo V6 44K2
54 Si Cr 6 44E2
055 M15 44A1
55 31.1
55A 52.1
55B 52.1
55 EE 44A2
55 F 44A1
55FK 44A1
55HSW 27B
55 Ni Cr Mo V6 44K2
55 Ni NCDV7 44K2
55 S 7
  see DIN range 44A3
55 Si 7 44A3
55 Si 8 44A2
55 WC20 44K2
56 Ni Cr Mo V7 44K2
56R 44M1
56S 1D
56TS 44M1
58 Cr V 4 44K2
58 S 44K1
060 A 86 44A4
060 A 99 44A4
060 A10 44A1
060 A12 44A1
060 A15 44A1
060 A17 44A1
060 A20 44A1
060 A22 44A1
060A25 44A2
060A27 44A2
060A30 44A2
060A32 44A2
060A35 44A2
060A37 44A2
060A40 44A2
060A42 44A2
060A45 44A2
060A47 44A2
060A52 44A3
060A57 44A3
060A62 44A3
060A67 44A3
060A72 44A3
060A78 44A3
060A81 44A3
60/55 44A2
60 A 44A1
60 ALLOY 14E
60 F 44A1
60F55 44A1
60L 14C
60 Mn Cr Ti 4 44E2
60 S 20
  see DIN range 44A3
60 SC7 44E2
60 Si 7 44A3
60 Si Cr 7 44E2
60 Si Cr 8 44E2
60 Si Mn 5
  see DIN range 44A3
60 W Cr V7 44K2
62S 1E
63S 1E
64 K MONEL 27D
65 14B
65 Mn 12 Cr 5 Ni 3 44P
65 S 44K2
65 Si 7
  see DIN range 44A3
65 Si Cr 7 44E2
65 W Mo 34–8 44K2
66 HS 44K3
66 Si 7
  see DIN range 44A3
67 S 44K3
67 Si Cr 5 44E2
67 Si Cr 5 44E3
68F62 44A1
69 INCONEL X 27C
69S 44M2
070M20 44A2
070M26 44A2
070M55 44A3
70CR 14D
70 F 44A1
70 H 44K3
70 Mn Cr Ti 8 44E3
71 Si 7
  see DIN range 44A3
72S 1A
73.08 44A1
73.68 44A1
73/3.5% Cr 44E3
74 Ni Cr 2 44K3
74 S 44K1
75 14G
75.75 44K1
75F70 44A1
75S 1G
76S 1G
78 52.1
78B 52.1
080 A 83 44A4
080 A 86 44A4
080 A15 44A1
080 A17 44A1
080 A20 44A1
080 A22 44A1
080 A25 44A1
080A27 44A2
080A30 44A2
080A32 44A2
080A35 44A2
080A37 44A2
080A40 44A2
080A42 44A2
080A47 44A3
080A52 44A3
080A57 44A3
080A62 44A3
080A67 44A3
080A72 44A3
080A78 44A3
080H36 44A2
080H41 44A2
080 M15 44A1
080M30 44A2
080M36 44A2
080M40 44A2
080M46 44A3
080M50 44A3
080M55 44A3
80.20 ALLOY 27B
80 DCV 42.16 44K3
80 F 44A1
80 HLES 44K1

80 W Cr V 8	44K3
81/18.5% Co	44L
81 B40	44K2
81 B 45	
see AISI 81B45	44K2
82 S	44K2
83/3% Co	44L
85 Ni V 4	44K3
86 B30	44K2
86 B45	44K2
86 Cr Mo V 7	44K3
88S	44M2
90	52.1
90 ALLOY	14E
90 Cr Si 5	44E3
90 MCW 5	44K3
90 MCW V5	44K3
90 Mn Cr V 8	44K3
90 Mn V 8	44K3
90 MV8	44J
91E	1A
94 B 15	44K1
94 B 17	
see AISI 94B17	44K1
94 B 30	
see AISI 94 B	44K2
94 B 30	44K1
94 B 40	44K1
95 ALLOY	14E
99.8	1A
99 ALLOY	27A
100.1	1A
100C3	44K3
100 C6	44E3
100CD7	44K3
100 Cr 6	44E3
100 Cr 6 (W3)	44E3
100 Cr 29	44M2
100 Cr Mn 6 (W4)	44E3
100 Cr Mo 6	44K3
100 Mn Cr W 4	44K3
100WC10	44K3
101	
see BS range	44A1
101	
see BS range	23B
101	14A
102	
see SAE CA 102	14A
102	31.1
103	31.1
104	14A
105	14A
105 Cr 2 (W1)	44E3
105 Cr 4 (W2)	44E3
105 Cr 6	44E3
105 W Cr 6	44M2
107	14A
110	
see SAE CA 110	44A
110 Mn Cr Ti 8	44E3
110WC20	44K3
111	
see SAE CA 111	14A
111	
see BS range	23B
111 ALLOY	14E

113	
see SAE CA 113	14A
113	1L
114	
see SAE CA 114	14A
115	14C
115	31.1
115 Cr V 3	44K3
116	
see SAE CA 116	14A
117	1L
120	42.1
120	52.1
120 Cr 25	44M2
120 M19	44A1
120M28	44A2
120M36	44A2
120WV4	44K3
121	
see BS range	23B
121	14A
122	
see SAE CA 122	14A
122	1L
123	1C
123	14A
125	14A
125 A15	44A1
127	1L
127	14A
128	14A
130	14A
130.1	1A
130 M15	44A1
131	
see BS range	44A1
133 ALLOY	27F
134 K MONEL	27D
135	1K
135M44	44A2
138	1L
140C3	44K3
140 Cr 29	44M2
140SMD4	44K3
141	
see BS range	23B
141	14A
141 ALLOY	27E
142	1L
142	14A
142 ALLOY	20C
143	1L
145	
see SAE CA 145	14A
145 Cr 6	44K3
145 V33	44J
146 ALLOY	20C
150	
see SAE CA 150	14A
150	44E3
150.1	1A
150 Cr 25	44M2
150 M19	44A1
150 M19	44F1
150M28	44A2
150M36	44A2
150M40	44A2

150P	14D
151	
see BS range	44A1
151	
see BS range	23B
152	1L
152 ALLOY	20C
158	44K1
160	1C
160	14G
160.1	1A
160S	1C
161	
see BS range	44G1
161	
see BS range	23B
162	
see SAE CA 162	14A
162	1K
164	14A
164	14G
165	14A
165	14D
165C	14D
165CR	14D
170	
see SAE CA 170	14D
170.1	1A
170 H 15	44C
170 H 20	44C
170 H20	44A1
170 H 36	44C
170 H 41	44C
170M	44A1
171	
see BS range	44A1
172	
see SAE CA 172	14D
173 H 16	44C
174	14D
174 H 20	44C
175	
see SAE CA 175	14D
175	44K3
175 H 23	44C
178 Co	44L
179	42.1
180	44E3
180 ALLOY	14E
181	
see BS range	44A1
182	14L
184	
see SAE CA 184	14L
185	14L
185 H 40	44C
185H40	44K2
187	
see SAE CA 187	14A
189	14A
190	14D
190	14E
190	52.1
191	14E
193	14B
193 ALLOY	20C
195	1L

197 14G
200 44K3
200 ALLOY 27A
200C 27F
200P 14D
201
see AISI 201 44P
201.0 1L
201.2 1L
201 ALLOY 27A
202
see AISI 202 44P
202.0 1L
202.2 1L
203.0 1L
203.2 1L
203 E2 44P
204.0 1L
204.2 1L
205 14B
205 ALLOY 27A
206.0 1L
206.2 1L
208.0 1L
208.1 1L
208.2 1L
210
see SAE CA 210 14B
210 14C
210 A15 44A1
211 ALLOY 27F
212 44K3
212A37 44A2
212A42 44A2
212M36 44A2
212M44 44A2
213.0 1L
213.1 1L
214 A15 44A1
214 M15 44A1
214 MN 44P
216A42 44A2
216M28 44A2
216M36 44A2
216M44 44A2
218 14C
220
see SAE CA 220 14B
220 1C
220 1J
220.0 1L
220C 27F
220 M07 44A1
222.1 1L
224.0 1L
224.2 1L
225M36 44A2
226 1L
226 14B
226 ALLOY 30.1
226M44 44A2
227 44K1
230
see SAE CA 230 14B
230 M07 44A1
234 14B
238.0 1L
238.1 1L
238.2 1L
240
see BS range 44G1
240
see SAE CA 240 14B
240.0 1L
240.1 1L
240 M07 44A1
242.0 1L
242.1 1L
242.2 1L
243.0 1L
243.1 1L
245C 14D
245Cr 14D
248 SV 44N1
249.0 1M
249.2 1M
249H 44M1
249T 44M1
250 14B
250 0.02% C 44F3
251A58 44K2
251A60 44K2
251H60 44K2
254 EM 44N1
254 SLX 44N1
254SMO 44N2
260
see SAE CA 260 14B
260C 27F
261 14B
262 14B
267 30.1
268
see SAE CA 268 14B
270
see SAE CA 270 14B
274 14B
275C 14D
275Cr 14D
278 30.1
279 14H
280 14C
280 M 01 44J
284 S 16 44P
288 27F
290 14D
293 44K1
295.0 1L
295.1 1L
295.2 1L
296.0 1L
296.1 1L
296.2 1L
298 14C
300 44K3
300 0.02% C 44F3
300 Cr 12 Mo W 44M2
300 Cr 15 Mn 6 44M1
300M 44K2
301
see AISI 301 44N1
301 14C
301 44N3
302
see AISI 302 44N1
302 B
see AISI 302 B 44N1
302 HQ-FM 44N1
302 NILSTAIN 44N1
302 S25 44N1
302 S31 44N1
303
see AISI 303 44N1
303Cu
see AISI 303Cu 44N1
303 Cu 44N1
303MA
see AISI 303MA 44N1
303 MA 44N1
303 Pb 44N1
303 plus X
see AISI 303 plus X 44N1
303 PLUS X 44N1
303 S21 44N1
303 S31 44N1
303 S41 44N1
303 S42 44N1
304
see AISI 304 44N1
304 ELC NILSTAIN 44N1
304 L
see AISI 304 L 44N1
304 LN 44N1
304 NILSTAIN 44N1
304 S11 44N1
304 S12 44N1
304 S15 44N1
304 S31 44N1
305
see AISI 305 44N1
305 14B
305.0 1C
305.2 1C
305 NILSTAIN 44N1
306 14B
308
see AISI 308 44N1
308.0 1L
308.1 1L
308.2 1L
308 S92 44N1
308 S93 44N1
308 S96 44N1
309
see AISI 309 44N1
309 C
see AISI 309C 44N1
309 Cb 44N1
309 NILSTAIN 44N1
309 S
see AISI 309S 44N1
309 S92 44N1
309 S94 44N1
310
see AISI 310 44N1
310 14B
310 NILSTAIN 44N1
310 S24 44N1
310 S31 44N1
310 S94 44N1

310 S98 44N1
311 14B
311 S94 44N1
312 14C
312 S94 44N1
313 S94 44N1
314
see AISI 314 44N1
314 14B
314 NILSTAIN 44N1
315 S16 44N1
316
see AISI 316 44N1
316 14B
316 ELC NILSTAIN 44N1
316L
see AISI 316 L 44N1
316 LN 44N1
316 N
see AISI 316 N 44N1
316 NILSTAIN 44N1
316 S11 44N1
316 S12 44N1
316 S13 44N1
316 S16 44N1
316 S31 44N1
316 S33 44N1
316 S92 44N1
316 S93 44N1
316 S96 44N1
317
see AISI 317 44N1
317 L
see AISI 317 L 44N1
317 NILSTAIN 44N1
317 S12 44N1
317 S16 44N1
317 S96 44N1
318
see AISI 318 44N1
318 27F
318 S96 44N1
319 1L
319.0 1L
319.1 1L
319.2 1L
320 1J
320 14B
320.0 1L
320.1 1L
320 S17 44N1
320 S31 44N1
321
see AISI 321 44N1
321 44K3
321 H
see AISI 321 H 44N1
321 NILSTAIN 44N1
321 S12
see BS 970/321 44N1
321 S20 44N1
323 14B
324.0 1C
324.1 1C
324.2 1C
325
see AISI 325 44N1
325 14B
325 S21 44N1
325 S31 44N1
326 S36 44N1
327
see AISI 327 44N1
328 14B
328 44K2
328.0 1C
328.0 1L
328.1 1C
329
see AISI 329 44N1
330
see SAE CA 330 14B
330 52.1
330 NILSTAIN 44N1
331
see SAE CA 331 14B
332 14B
332.0 1L
332.1 1L
332.2 1L
333 44L
333.0 1L
333.1 1L
335 14B
336 44M2
336.0 1C
336.1 1C
336.2 1C
338 44A1
338 44M2
339.0 1C
339.0 1L
339.1 1C
340 1J
340 14B
342 14B
343.0 1C
343.1 1C
344 14B
345
see SAE CA 345 14B
347 14B
347 F
see AISI 347 F 44N1
347 H
see AISI 347 H 44N1
347 HC 44N1
347 NILSTAIN 44N1
347 S17 44N1
347 S31 44N1
347 S96 44N1
348
see AISI 348 44N1
348 14B
348 44K2
350
see SAE CA 350 14B
350 1J
350 52.1
350 Cr 20 44M2
353 14B
354 1K
354.0 1C
354.1 1C
355 1K
355.0 1C
355.0 1L
355.1 1C
355.2 1C
356 1K
356 14B
356.0 1C
356.0 1K
356.1 1C
356.2 1C
357.0 1C
357.0 1K
357.1 1K
358.0 1K
358.2 1K
359.0 1C
359.0 1K
359.2 1K
360
see SAE CA 360 14B
360 1K
360.0 1C
360.2 1C
361 44K2
361.0 1C
361.1 1C
362 14B
363.0 1C
363.1 1C
364.0 1C
364.2 1C
365 14B
366 14B
367 14B
368 14B
369.0 1C
369.1 1C
370 14B
370 14F
370 52.1
371 14B
377 14B
380 1L
380 14C
380.0 1C
380.2 1C
380–17 20B
383.0 1C
383.1 1C
383.2 1C
384 1L
384 44N1
384.0 1C
384.1 1C
384.2 1C
385 14C
385 44N1
385.0 1C
385.1 1C
390.0 1C
390.2 1C
392.0 1C
392.1 1C
393 44M1
393.0 1C
393.1 1C

393.2 1C
399 44K2
400 27D
400–12 20B
403
see AISI 403 44M1
403 S17 44M1
405
see AISI 405 44M1
405 14B
408 14B
408.2 1C
409 14B
409 44M1
409.2 1C
410
see AISI 410 44M1
410 14B
410 S21 44M1
411 14B
411.2 1C
413 14B
413.0 1C
413.2 1C
414
see AISI 414 44M1
415 14B
416
see AISI 416 44M1
416 plus X
see AISI 416 plus X 44M1
416 S21 44M1
416 S29 44M1
416 S37 44M1
416 S41 44M1
419 14K
419 44M1
420
see AISI 420 44M1
420 14B
420CR 27F
420 F
see AISI 420 F 44M1
420 S29 44M1
420 S37 44M1
420 S45 44M1
421 14B
422 13A
422 14B
422 44M1
422/19 13A
422M 44M1
424 17.1
425 14B
426 ALLOY 20C
429 17.1
430
see AISI 430 44M1
430 14B
430 F
see AISI 430 F 44M1
430 FR SOLENOID QUALITY 44M1
430 NILSTAIN 44N1
430 S17 44M1
430 SOLENOID QUALITY 44M1
431 S29 44M1
432 14B
434 14B
434 44M1
435 14B
435.2 1C
436 14B
436 44M1
438 14B
439 44M1
440 14B
440 A
see AISI 440 A 44M1
440 B
see AISI 440 B 44M1
440 C
see AISI 440 C 44M2
440 F
see AISI 440 F 44M2
441 14B
441 17.1
441 S29 44M1
442
see AISI 442 44M1
442 14B
443 14B
443.0 1C
443.1 1C
443.2 1C
444 14B
444 44L
444.0 1C
444.2 1C
445 14B
445.2 1C
446
see AISI 446 44M1
448 14B
450 31.1
450–5 20B
450 (fine) 17.1
452 14B
453S 44N1
455 44L
457 14C
462 14B
462 14C
463 14C
464
see SAE CA 464 14C
465 14C
466 14C
467 14C
470 14C
472 14C
473 14C
476 14B
476 44M2
476 SPECIAL 44M2
482 14B
485 14B
500–2 20B
501
see AISI 501 44K1
502
see AISI 502 44K1
502 14K
503
see BS range 44F1
503 A37 44F2
503 A42 44F2
503 H37 44F2
503 H42 44F2
503 M40 44F2
505 14K
505 44K1
507 14K
508 14K
509 14K
510
see SAE CA 510 14K
510 44M1
511 44M1
511.0 1J
511.1 1J
511.2 1J
512 44M1
512.0 1J
512.2 1J
513.0 1J
513.2 1J
514.0 1J
514.1 1J
514.2 1J
515.0 1J
515.2 1J
516.0 1J
516.1 1J
518 14K
518.0 1J
518.1 1J
518.2 1J
520.0 1J
520.2 1J
521
see SAE CA 521 14K
522 44M1
523 A14 44E1
523 M15 44E1
524 14K
525 44M1
525A58 44K2
525A60 44K2
525A61 44K2
525H60 44K2
526 M60 44E2
527 A17 44E1
527 A60 44E2
527 M17 44E1
527 M20 44E1
528 44K2
530 A30 44E1
530 A32 44E1
530 A36 44E1
530 A40 44E1
530 H30 44E1
530 H32 44E1
530 H32 44E2
530 H36 44E1
530 H36 44E2
530 H40 44E1
530 H40 44E2
530 M40 44E2
532 14K

534 14K
534 A99 44E3
535.0 1J
535.2 1J
535 A99 44E3
540 14C
544
  see SAE CA 544 14K
546 14K
550 (fine) 17.1
553 44N1
554 44N1
556 44N1
559 44N1
562 44K3
564 44N1
590 A15 44E1
590 H17 44E1
590 M17 44E1
599 44K3
600–2 20B
602
  see AISI 602 44K1
603
  see AISI 603 44K1
604
  see AISI 604 44K1
605 A32 44G1
605 A37 44G1
605 H 32 44G1
605 H 37 44G1
605 M 30 44G1
605 M 36 44G1
605 M36 44E1
606 14G
606 M36 44E2
607 14G
608 12.1
608 14G
608 H 37 44G2
608 M 38 44G2
608 M38 44E1
610 14G
612 14G
613 14G
614
  see SAE CA 614 14G
614
  see AISI 614 44M1
615
  see AISI 615 44M1
616
  see AISI 616 44M1
616 14G
617
  see AISI 617 44M2
617
  see SAE CA 617 14G
618
  see AISI 618 44M2
618 14G
619
  see AISI 619 44M1
620
  see BS range 44K1
620 14G
621
  see BS range 44K1
622
  see BS range 44K1
622 14G
622 44K3
623
  see BS range 44K1
623
  see SAE CA 623 14G
624
  see SAE CA 624 14G
625
  see BS range 44K1
625R 17.1
626 14G
628 14G
630
  see SAE CA 630 14G
630
  see AISI 630 44N2
630 44N3
631
  see AISI 631 44N2
631 44N3
632
  see AISI 632 44N2
632 44N3
633
  see AISI 633 44N2
634
  see AISI 634 44N2
634 14G
635
  see AISI 635 44N2
635 44N3
635 A 14 44K1
635 H 15 44K1
635 M 15 44K1
636 44M1
637
  see SAE CA 637 14G
637 A 16 44K1
637 H 17 44K1
637 M 17 44K1
639 14G
640 A 35 44K1
640 H 35 44K1
640 M40 44K2
642 14G
645 1L
647 14E
648A 44F3
648B 44F3
648C 44F3
650 (fine) 17.1
651
  see AISI 651 44N2
651 14M
652
  see AISI 652 44N2
653
  see AISI 653 44N1
653 14M
653 M 31 44K1
655
  see SAE CA 655 14M
655 A 12 44K1
655 H 13 44K1
655 M 13 44K1
656 14M
658 14M
659 A 15 44K1
659 H 15 44K1
659 M 15 44K1
660
  see AISI 660 44N2
661 14M
664 M 44P
664 MV 44P
665 14B
665 A 17 44K1
665 A 19 44K1
665 A 22 44K1
665 A 24 44K1
665 H 17 44K1
665 H 20 44K1
665 H 23 44K1
665 M 17 44K1
665 M 20 44K1
665 M 23 44K1
667 14B
670
  see SAE CA 670 14B
671 14C
673
  see SAE CA 673 14B
674 14B
675 14B
676 14B
677 14C
680 14C
681 14C
685 14B
685 A57 44E2
685 H57 44E2
687 14B
692 14M
694 14M
697 14M
700–2 20B
702 14E
703 14E
704 14E
704A60 44K2
704H60 44K2
705 14G
705.0 1M
705.2 1M
705A60 44K2
705H60 44K2
706
  see SAE CA 706 14E
0707 20B
707 14E
707.0 1M
707.1 1M
0708 20B
0708 Si Al 44E1
708 14E
708 A 25 44K1
708 A 30 44K1
708 A 37 44K1
708 A 40 44K1

708A42 44K2
708A47 44K2
708 H 20 44K1
708 H 37 44K1
708H37 44K2
708H42 44K2
708H45 44K2
708M40 44K2
709 14E
709 A 37 44K1
709 A 40 44K1
709A42 44K2
709M40 44K2
710
see SAE CA 710 14E
710.0 1M
710.1 1M
711.0 1M
711.1 1M
712.0 1M
712.2 1M
713
see BS range 44M1
713.0 1M
713.1 1M
713C 27C
713LC 27C
0715 20B
715
see SAE CA 715 14E
0716 20B
718 27C
718 42.1
719 42.1
720 14E
720 M 32 44K1
722 42.1
722 M 24 44K1
723 42.1
724 42.1
725 42.1
0727 20B
732 14F
735 14F
735A51 44K2
735A54 44K2
735H51 44K2
736 14F
738 27C
739 44M1
739H 44M1
740 14F
745 14F
748 42.1
750 (fine) 17.1
752
see SAE CA 752 14F
754 14F
757 14F
762 14F
762 44M1
764 14F
764 14G
766 14F
767 14F
768 14G
770
see SAE CA 770 14F
771.0 1M
771.2 1M
772.0 1M
772.2 1M
773 14F
774 14F
775 14G
776 14F
778 14G
779 52.1
781 14G
782 14F
784 14F
785 M 19 44K1
786 14F
788 14F
790 14F
792 14F
794 14F
796 14C
796 44K1
798 14C
800–2 20B
800 (fine) 17.1
801
see BS range 44N1
805 A 15 44K1
805 A 17 44K1
805 A 20 44K1
805 A 22 44K1
805 A 24 44K1
805 H 17 44K1
805 H 20 44K1
805 H 22 44K1
805 H 25 44K1
805H60 44K2
805 M 17 44K1
805 M 20 44K1
805 M 22 44K1
805 M 25 44K1
808 44L
808 H 17 44K1
0814
see SIS 14-0814 20B
0815
see SIS 14-0815 20B
815 A 16 44K1
815 H 17 44K1
815 M 17 44K1
816 M40 44K2
817A37 44K2
817A42 44K2
817M40 44K2
820 A 16 44K1
820 H 17 44K1
821
see BS range 44N1
822 A 17 44K1
822 H 17 44K1
822 M 17 44K1
823 M 30 44K1
825
see BS range 44N1
826 M 31 44K1
826M40 44K2
830 M 31 44K1
831 44M1
832C 44N1
832 H 13 44K1
832K 44N1
832M 44N1
832 M 13 44K1
832MM 44N1
832MP 44N1
832MV 44N1
832MVN 44N1
832MVR 44N1
832MVRN 44N1
832MVT 44N1
832P 44N1
832SF 44N1
832SFR 44N1
832SFT 44N1
832SK 44N1
832SKR 44N1
832SKRN 44N1
832SN 44N1
832SV 44N1
832T 44N1
832V 44N1
835 A 15 44K1
835 H 15 44K1
835 M 15 44K1
836 14K
838 14K
842 14K
844 14K
845
see BS range 44N1
846
see BS range 44N1
848 14K
850 14C
850.0 1N
850.1 1N
851.0 1N
851.1 1N
0852
see SIS 14-0852 20B
852.0 1N
852.1 1N
853.0 1N
853.2 1N
0854
see SIS 14-0854 20B
0856
see SIS 14-0856 20B
860 52.1
0862
see SIS 14-0862 20B
862 14B
863 14B
0864
see SIS 14-0864 20B
864 14B
883 52.1
897 M39
see BS 970 44K2
900 52.1
900–2 20B
901 27C
903 14K

905 14K
905 M 31 44K1
907 14K
907 52.1
0908 Mo 44K1
910 14K
912 14C
913 14K
915 14K
922 14K
923 14K
925 14K
925A60 44K2
927 14K
928 14K
932 14K
934 14K
935 14K
937 14K
938 14K
939 14K
940 14K
941 14K
942 X 44K1
943 14K
945 A 44K1
945A40 44K2
945 C 44K1
945 M 38 44K1
945 X 44K1
947 14K
948 14K
950 A 44K1
950 B 44K1
950 C 44K1
950 D 44K1
950 X 44K1
952 14G
953 14G
954 14G
955 14G
955 X 44K1
960 X 44K1
964 14E
965 X 44K1
970 X 44K1
980 X 44K1
990A 1A
995 A
see ASTM range 1A
996 A
see ASTM range 1A
999 52.1
1005 44A1
1006
see AISI C1006 44A1
1008
see AISI C1008 44A1
1009
see AISI C1009 44A1
1010
see AISI C1010 44A1
1011 44A1
1012
see AISI C1012 44A1
1013 44A1
1015
see AISI C1015 44A1
1016
see AISI C1016 44A1
1017
see AISI C1017 44A1
1018
see AISI C1018 44A1
1019
see AISI C1019 44A1
1020
see AISI C1020 44A1
1021
see AISI C1021 44A1
1022 44A1
1022 44A2
1023
see AISI C1023 44A2
1023 44A1
1024
see AISI C1024 44A2
1025
see AISI C1025 44A2
1026
see AISI C1026 44A2
1027
see AISI C1027 44A2
1029 44A2
1030
see AISI C1030 44A2
1030 1A
1031 ALLOY 44P
1033
see AISI C1033 44A2
1034 44A2
1035
see AISI C1035 44A2
1035 1A
1036
see AISI C1036 44A2
1037
see AISI C1037 44A2
1038
see AISI C1038 44A2
1039
see AISI C1039 44A2
1040
see AISI C1040 44A2
1040 1A
1041
see AISI C1041 44A2
1042
see AISI C1042 44A2
1043
see AISI C1042 44A2
1044 44A2
1045
see AISI C1045 44A2
1045 1A
1046 44A2
1046 44A3
1047 44A3
1048
see AISI C1048 44A3
1049
see AISI C1049 44A3
1049 44A2
1050
see AISI C1050 44A3
1050 1A
1050 44A2
1050A
see BS range 1A
1051
see AISI C1051 44A2
1052
see AISI C1052 44A3
1053
see AISI C1053 44A3
1055
see AISI C1055 44A2
1055 1A
1055 44A3
1059 44A3
1060
see AISI C1060 44A3
1060
see ASTM range 1A
1064
see AISI C1064 44A3
1065
see AISI C1065 44A3
1065 1A
1069 44A3
1070
see AISI C1070 44A3
1070 1A
1074
see AISI C1074 44A3
1075 1A
1075 44A4
1078
see AISI C1078 44A3
1078 44A4
1080
see AISI C1080 44A3
1080 1A
1080 44A4
1080A
see BS range 1A
1084
see AISI C1084 44A4
1085 1A
1085 44A4
1086
see AISI C1086 44A4
1090
see AISI C1090 44A4
1090 1A
1095
see AISI C1095 44A4
1095 1A
1100
see ASTM range 1A
1101 44A4
1102 44A4
1103 42.1
1103 44A4
1104 44A3
1105 44A2
1108
see AISI C1108 44A1
1109
see AISI B1109 44A1

1110
 see AISI 1110 44A1
1111
 see AISI B1111 44A1
1112
 see AISI B1112 44A1
1113
 see AISI B1113 44A1
1115
 see AISI B1115 44A1
1116
 see AISI B1116 44A1
1117
 see AISI B1117 44A1
1118
 see AISI B1118 44A1
1119
 see AISI C1119 44A1
1120
 see AISI C1120 44A1
1123 44A1
1126
 see AISI C1126 44A2
1130 1A
1132
 see AISI C1132 44A2
1135 1A
1137
 see AISI C1137 44A2
1138
 see AISI C1138 44A2
1139 44A2
1140
 see AISI C1140 44A2
1141
 see AISI C1141 44A2
1144
 see AISI C1144 44A2
1144 42.1
1145 1A
1145 44A2
1146 44A2
1151
 see AISI C1151 44A3
1151 44A2
1161 44A4
1162 44A4
1162 44J
1163 44A4
1163 44J
1164 44A3
1165 44A3
1170 1A
1175 1A
1180 1A
1185 1A
1188 1A
1193 1A
1199 1A
1200 1A
1200 44A4
1201 44A4
1202 44A4
1203 44A4
1204 44A3
1211
 see AISI 1211 44A1
1212
 see AISI 1212 44A1
1213
 see AISI 1213 44A1
1213 44K2
1215
 see AISI 1215 44A1
1230 1A
1230 44E3
1231 44E3
1232 44E3
1233 44E3
1234 44E2
1235 1A
1250 1A
1250 44P
1260 1A
1285 1A
1303 44A4
1304 44A3
1305 44A3
1306 44A2
1307 44A2
1308 44A2
1309 44A1
1320
 see AISI A1320 44A1
1322 42.1
1323 42.1
1325 42.1
1330
 see AISI 1330 44A2
1335
 see AISI 1335 44A2
1340
 see AISI 1340 44A2
1340 42.1
1345
 see AISI 1345 44A2
1345 1A
1350 1A
1379 42.1
1380 42.1
1386 42.1
1396 42.1
1408 44M1
1410 44M1
1410 Mo 44M1
1413 44K2
1415 44M1
1415 Mo 44M1
1418 44K1
1433 44K1
1435 1D
1435 44M1
1438 42.1
1439 42.1
1464 42.1
1465 42.1
1513 44A1
1518 44A1
1522 44A1
1524 44A1
1524 44A2
1525 44A2
1526 44A2
1527 44A2
1533 44A2
1534 44A2
1536 44A2
1541 44A2
1547 44A2
1548 44A2
1551 44A2
1552 44A2
1553 44A3
1561 44A3
1566 44A3
1570 44A3
1572 44A3
1580 44A3
1590 44A4
1706 44M1
1708 44M1
1708 Mo 44M1
1708 Mo Nb 44M1
1708 Mo Ti 44M1
1708 Nb 44M1
1708 Ti 44M1
1710 Si 44M1
1845/4 42.1
1845/5 42.1
1850 44K2
2004 1F
2008 1F
2011 1F
2013 44M1
2014
 see ASTM range 1F
2014A
 see BS range 1F
2017
 see ASTM range 1F
2018 1F
2020 1F
2021 1F
2024
 see ASTM range 1F
2025
 see ASTM range 1F
2034 1F
2036 1F
2037 1F
2038 1F
2048 1F
2090 1F
2091 1F
2117
 see ASTM range 1F
2121 44G2
2124 1F
2131 44K3
2132 44K3
2133 44K3
2141 44K3
2142 44K3
2205 44M1
2205 KS 44N1
2211 44J
2212 44K3
2214 1F
2218
 see ASTM range 1F

- 2219
  - see ASTM range 1F
- 2224 1F
- 2231 44K3
- 2233 44M2
- 2234 44M2
- 2235 44M2
- 2236 44M2
- 2237 44M2
- 2307 27F
- 2317
  - see AISI 2137 44F1
- 2319
  - see ASTM range 1F
- 2321 44A2
- 2321 44K2
- 2323 44E2
- 2324 1F
- 2330
  - see AISI A2330 44F1
- 2331 44K2
- 2332 44K2
- 2340
  - see AISI A2340 44F2
- 2341 44K2
- 2345
  - see AISI A2345 44F2
- 2419 1F
- 2504 44K1
- 2512
  - see AISI E2512 44F1
- 2515
  - see AISI 2515 44F1
- 2517
  - see AISI E2517 44F1
- 2519 1F
- 2520 44M1
- 2601 42.1
- 2602 42.1
- 2618 1F
- 2618A
  - see BS range
- 2731 44M1
- 2732 44M1
- 2733 44M1
- 2831 44K1
- 2881 44K1
- 2882 44K1
- 2949 42.1
- 3002 1B
- 3003
  - see ASTM range 1B
- 3004
  - see ASTM range 1B
- 3005 1B
- 3006 1B
- 3007 1B
- 3009 1B
- 3010 1B
- 3011 1B
- 3015 1B
- 3016 1B
- 3047 52.1
- 3102 1B
- 3103
  - see BS range 1B
- 3104 1B
- 3105 1B
- 3107 1B
- 3109 52.1
- 3115 44K1
- 3120 44K1
- 3130 44K1
- 3135
  - see AISI 3135 44K1
- 3140
  - see AISI 3140 44K2
- 3145 44K2
- 3150 44K2
- 3215 44K1
- 3220 44K1
- 3230 44K1
- 3240 44K2
- 3245 44K2
- 3250 44K2
- 3303 1B
- 3307 1B
- 3310 44K1
- 3312 44K1
- 3316 44K1
- 3325 44K1
- 3335 44K1
- 3340 44K2
- 3381 44K2
- 3382 44K1
- 3383 44K1
- 3411 52.1
- 3415 44K1
- 3432 44K2
- 3436 44K1
- 3450 44K2
- 3451 44K1
- 3455 44K1
- 3500/37% Co 44L
- 3541 44K2
- 3543 44K1
- 3548 44K2
- 3551 44K3
- 3632 44N1
- 3636 44N1
- 3682 44N1
- 4002 1N
- 4004 1N
- 4008 1C
- 4009 1E
- 4011 1E
- 4012
  - see AISI 4012 44G1
- 4013 1E
- 4023
  - see AISI 4023 44G1
- 4024
  - see AISI 4024 44G1
- 4027
  - see AISI 4027 44G1
- 4028
  - see AISI 4028 44G1
- 4032
  - see AA 4032 44G1
- 4032 1E
- 4037
  - see AISI 4037 44G1
- 4042
  - see AISI 4042 44G2
- 4042 44G1
- 4043 1C
- 4043A
  - see BS range 1C
- 4044 1C
- 4045 1C
- 4047
  - see AISI 4047 44G2
- 4047 1C
- 4047 44G1
- 4047A
  - see BS range 1C
- 4053
  - see AISI 4053 44G2
- 4063
  - see AISI 4063 44G2
- 4063 44G1
- 4068
  - see AISI 4068 44G2
- 4104 1C
- 4118
  - see AISI 4118 44K1
- 4119 44K1
- 4120 44K1
- 4121 44K1
- 4125 44K1
- 4130
  - see AISI 4130 44K1
- 4135 44K1
- 4137
  - see AISI 4137 44K1
- 4137 44K2
- 4140
  - see AISI 4140 44K2
- 4142
  - see AISI 4142 44K2
- 4145
  - see AISI 4145 44K2
- 4145 1C
- 4147
  - see AISI 4147 44K2
- 4150
  - see AISI 4150 44K2
- 4151 44K3
- 4161
  - see AISI 4161 44K2
- 4161 44K3
- 4171 44L
- 4175 44L
- 4201 44K3
- 4203 44K3
- 4271 44L
- 4275 44L
- 4301 44K3
- 4317
  - see AISI 4317 44K1
- 4320
  - see AISI 4320 44K1
- 4330 44K1
- 4335 44K1
- 4337 44K1
- 4337 44K2
- 4340
  - see AISI 4340 44K2
- 4343 1C
- 4361 44K3
- 4371 44L

4373 44L
4375 44L
4379 44M1
4419
see AISI 4419 44G1
4422
see AISI 4422 44G1
4427
see AISI 4427 44G1
4441 44K3
4475 44L
4520 44G1
4543 1C
4608
see AISI 4608 44K1
4615
see AISI 4615 44K1
4617 44K1
4620
see AISI 4620 44K1
4620 44K2
4621
see AISI 4621 44K1
4626
see AISI 4626 44K1
4640 44K2
4643 1C
4715 44K1
4718
see AISI 4718 44K1
4720
see AISI 4720 44K1
4812 44K1
4815
see AISI 4815 44K1
4817
see AISI 4817 44K1
4820
see AISI 4820 44K1
5005
see ASTM range 1D
5005
see BS range 1D
5005 1D
5006 1D
5010 1D
5015
see AISI 5015 44E1
5016 1D
5017 1D
5034 1D
5039 1D
5040 1D
5042 1D
5043 1D
5045
see AISI 5045 44E2
5046
see AISI 5046 44E2
5050
see ASTM range 1D
5051 1D
5052
see ASTM range 1D
5056
see ASTM range 1D
5056
see BS range 1D
5056A
see BS range 1D
5060 44E2
5082 1D
5083
see ASTM range 1D
5083
see BS range 1D
5083 1D
5086
see ASTM range 1D
5115 44E1
5117
see AISI 5117 44E1
5120
see AISI 5120 44E1
5130
see AISI 5130 44E1
5132
see AISI 5132 44E1
5135
see AISI 5135 44E1
5140
see AISI 5140 44E2
5145
see AISI 5145 44E2
5147
see AISI 5147 44E2
5150
see AISI 5150 44E2
5151 1D
5152
see AISI 5152 44E2
5154
see ASTM range 1D
5154
see BS range 1D
5155
see AISI 5155 44E2
5160
see AISI 5160 44E2
5182 1D
5183 1D
5205 1D
5250 1D
5251
see BS range 1D
5252
5254
see ASTM range 1D
5351 1D
5352 1D
5356
see ASTM range 1D
5357 1D
5451 1D
5454
see ASTM range 1D
5454
see BS range 1D
5456
see ASTM range 1D
5457 1D
5552 1D
5554
see ASTM range 1D
5554
see BS range 1D
5556
see ASTM range 1D
5556A
see BS range 1D
5557 1D
5652
see ASTM range 1D
5654 1D
5657 1D
6003
see ASTM range 1E
6004 1E
6005 1E
6006 1E
6007 1E
6009 1E
6010 1E
6011 1E
6013 1E
6017 1E
6053
see ASTM range 1E
6060 1E
6061
see ASTM range 1E
6061
see BS range 1E
6062
see ASTM range 1E
6063
see ASTM range 1E
6063
see BS range 1E
6063 1L
6066
see ASTM range 1E
6066 1L
6070 1L
6082
see BS range 1E
6101 1E
6101 1L
6101A
see BS range 1L
6105 1E
6110 1E
6111 1E
6115 44K1
6117
see AISI 6117 44K1
6118
see AISI 6118 44K1
6120
see AISI 6120 44K1
6125 44K1
6130 44K1
6135 44K1
6140 44K2
6145
see AISI 6145 44K2
6150
see AISI 6150 44K2

6151
see ASTM range 1E
6162 1E
6195
see SAE 6195 44K3
6201 1E
6205 1E
6253 1E
6261 1E
6262 1E
6301 1E
6351
see ASTM range 1E
6463
see BS range 1E
6542 MOLY 44K3
6563 1E
6763 1E
6961 1E
7001 1G
7004 1G
7005 1G
7008 1G
7010 1G
7011 1G
7013 1G
7016 1G
7020
see BS range 1G
7021 1G
7029 1G
7039 1G
7046 1G
7049 1G
7050 1G
7070 1G
7072
see ASTM range 1G
7075
see ASTM range 1G
7076 1G
7079
see ASTM range 1G
7090 1G
7091 1G
7104 1G
7108 1G
7116 1G
7129 1G
7146 1G
7149 1G
7150 1G
7175 1G
7178
see ASTM range 1G
7179 1G
7229 1G
7260 44K2
7277 1G
7472 1G
7475 1G
8001 1A
8006 1B
8007 1B
8010 1B
8013 1A
8014 1B
8017 1A
8020 1A
8040 1A
8076 1A
8077 1A
8081 1N
8111 1A
8112 1A
8115 44K1
8130 1A
8176 1A
8177 1A
8280 1N
8615
see AISI 8615 44K1
8617
see AISI 8617 44K1
8620
see AISI 8620 44K1
8622
see AISI 8622 44K1
8625
see AISI 8625 44K1
8627
see AISI 8627 44K1
8627 44K2
8630
see AISI 8630 44K1
8630 44K2
8632
see AISI 8632 44K1
8635
see AISI 8635 44K1
8637
see AISI 8637 44K1
8637 44K2
8640
see AISI 8640 44K2
8641
see AISI 8641 44K2
8642
see AISI 8642 44K2
8645
see AISI 8645 44K2
8647
see AISI 8647 44K2
8653
see AISI 8653 44K2
8655
see AISI 8655 44K2
8660
see AISI 8660 44K3
8660 44K2
8715
see AISI 8715 44K1
8717
see AISI 8717 44K1
8719
see AISI 8719 44K1
8720
see AISI 8720 44K1
8735
see AISI 8735 44K1
8740
see AISI 8740 44K2
8742
see AISI 8742 44K2
8742 44K1
8745
see AISI 8745 44K2
8750
see AISI 8750 44K2
8820
see AISI 8822 44K1
8822 44K1
9250
see AISI 9250 44B2
9254 44E2
9255
see AISI 9255 44B2
9255 44A2
9257 44J
9260
see AISI 9260 44B2
9260 44A2
9261
see AISI 9261 44E2
9262
see AISI 9262 44E2
9262 44A2
9310
see AISI E9310 44K1
9315
see AISI E9315 44K1
9317
see AISI E9317 44K1
9437
see AISI 9437 44K1
9440
see AISI 9440 44K2
9442
see AISI 9442 44K2
9445
see AISI 9445 44K2
9477
see AISI 9447 44K2
9747
see AISI 9747 44K2
9763 44K2
9840
see AISI 9840 44K2
9845
see AISI 9845 44K2
9850
see AISI 9850 44K2
10013 44G1
10018 Ni 44K1
11015 44K1
13728 44F2
13743 44L
13745 44L
13756 44L
13758 44L
13760 44L
13761 44L
14018 HT 44K1
14018 MI 44K1
30201
see SAE 30201 44P
30202
see SAE 30202 44P
30301
see SAE 30301 44N1

30302
see SAE 30302 44N1
30302 B
see SAE 30302 B 44N1
30303 F
see SAE 30303 F 44N1
30304
see SAE 30304 44N1
30304 L
see SAE 30304 L 44N1
30305
see SAE 30305 44N1
30308
see SAE 30308 44N1
30309
see SAE 30309 44N1
30309 S
see SAE 30309 S 44N1
30310
see SAE 30310 44N1
30310 S
see SAE 30310 S 44N1
30314
see SAE 30314 44N1
30316
see SAE 30316 44N1
30316 L
see SAE 30316 L 44N1
30317
see SAE 30317 44N1
30321
see SAE 30321 44N1
30325
see SAE 30325 44N1
30329
see SAE 30329 44N1
30330
see SAE 30330 44N1
30330 A
see 30330 A 44N1
30347
see SAE 30347 44N1
30348
see SAE 30348 44N1
50100
see AISI E51100 44E3
50502
see SAE 51502 44K1
51100
see AISI E51100 44E3
51403
see SAE 51403 44M1
51405
see SAE 51405 44M1
51410
see SAE 51410 44M1
51414
see SAE 51414 44M1
51416 F
see SAE 51416 F 44M1
51420
see SAE 51420 44M1
51420 F
see SAE 51420 F 44M1
51430
see SAE 51430 44M1
51430 F
see SAE 51430 F 44M1
51431
see SAE 51431 44M1
51434
see SAE 51434 44M1
51436
see SAE 51436 44M1
51440 A
see SAE 51440 A 44M1
51440 B
see SAE 51440 B 44M1
51440 C
see SAE 51440 C 44M1
51440 F
see SAE 51440 F 44M2
51442
see SAE 51442 44M1
51446
see SAE 51446 44M1
51449
see SAE 51499 44M1
51501
see SAE 51501 44K1
52100
see SAE 52100 44E3
60303
see SAE 60303 44N1
60303 A
see SAE 60303 A 44N1
60304
see SAE 60304 44N1
60304 L
see SAE 60304 L 44N1
60309
see SAE 60309 44N1
60310
see SAE 60310 44N1
60312
see SAE 60312 44N1
60316
see SAE 60316 44N1
60316 L
see SAE 60316 L 44N1
60317
see SAE 60317 44N1
60347
see SAE 60347 44N1
606608
UNS for AWS A5.20
(E6XT.8) obsolete 44A1
70308
see SAE 70308 44N1
70309
see SAE 70309 44N1
70310
see SAE 70310 44N1
70310A
see SAE 70310A 44N1
70311
see SAE 70311 44N1
70312
see SAE 70312 44N1
70327
see SAE 70327 44N1
70330
see SAE 70330 44N1
70331
see SAE 70331 44N1
71360 44K2
71660 44K2
A 0 14C
A 0 44A1
A 1 44A1
A 1 BABBITT 21.1
A 1/57 44N1
A1 Mg 7 1D
A1Q
See ASTMA757 44A2
A 2 1A
A 2 44A1
A 2 44A2
A 2 44K3
A2Q
See ASTMA757 44A2
A 3 1A
A 3 44A2
A 3 44K1
A 3 44K3
A 4 1A
A 4 44A2
A 4 44K1
A 4 44K3
A 4.5 1A
A 5 1A
A 5 21.1
A 5 44A2
A 5 44K2
A 5 44K3
A 5.3 (E91100) 1A
A5.3 (E93003) 1B
A5.3 (E94043) 1C
A5.8 (BA1 Si 2) 1C
A5.8 (BAl Si 3) 1C
A5.8 (BAl Si 4) 1C
A5.8 (BAl Si 5) 1C
A5.8 (BAl Si 7) 1C
A5.9 (BAl Si 11) 1C
A 5B 1A
A 5L 1A
A 6 44A2
A 6 44A3
A 6 44K1
A 6 44K2
A 6 44M2
A 7 1A
A 7 44A1
A 7 44K3
A 7W 44K3
A 8 1A
A 8 23B
A 8 44K2
A 8 Nb
see BS range 44N1
A 8 Ti
see BS range 44N1
A 9 1A
A 9 44K2
A 10 1J
A 10 44K3
A 10 44N1
A 10 Nb
see BS range 44N1

A 10 Ti
see BS range 44N1
A 11
see BS range 44N1
A 11 44M2
A 11 Ti
see BS range 44N1
A 12
see BS range 44N1
A 12 Nb
see BS range 44N1
A 12 Ti
see BS range 44N1
A 13 44K2
A15
Designation in BS 2901 44A1
A 15 44A1
A16
Designation in BS 2901 44A2
A 16 44A2
A 16X 44N1
A17
Designation in BS 2901 44A1
A 17 44A1
A 17S 1F
A18
Designation in BS 2901 44A1
A 18 44A1
A19
Designation in BS 2901 44A1
A 19 44A1
A 22 44M2
A 25 14K
A 25 21.1
A30
Designation in BS 2901 44G1
A 30 44G1
A31
Designation in BS 2901 44G1
A 31 44G1
A32
Designation in BS 2901 44K1
A 32 44K1
A 32 STEEL 44A3
A 32/101 20B
A 32/101 F40D 20B
A 32/201 20B
A33
Designation in BS 2901 44K1
A 33 44K1
A34
Designation in BS 2901 44K1
A 34 44K1
A35
Designation in BS 2901 44K1
A 35 44K1
A 36 21.1
A 36 44A1
A 37 C1
see NF A36 205 44A1
A 37 C2
see NF A36 205 44N1
A 37 P1
see NF A36 205 44A1
A 37 P2
see NF A36 205 44A1
A 40 51A
A 42 C1
see NF A36 205 44A1
A 42 C2
see NF A36 205 44A1
A 42 FP1
see NF A36 208 44A1
A 42 FP2
see NF A36 208 44A1
A 42 P1
see NF A36 205 44A1
A 42 P2
see NF A36 205 44A1
A 42-1
see NF A35 501 44A1
A 42-2
see NF A35 501 44A1
A 42-3
see NF A35 501 44A1
A 42-4
see NF A35 501 44A1
A 47/2
see NF A35 501 44A1
A 47/3
see NF A35 501 44A1
A 47/4
see NF A35 501 44A1
A 48 C1
see NF A36 205 44A1
A 48 C2
see NF A36 205 44A2
A 48 FP1
see NF A36 208 44A1
A 48 FP2
see NF A36 208 44K1
A 48 P1
see NF A36 205 44A1
A 48 P2
see NF A36 205 44A1
A 50/1 44A1
A 50/2 44A1
A 51S 1E
A 52 44A1
A 52 C1
see NF A36 205 44A1
A 52 C2
see NF A36 205 44A1
A 52 FP1
see NF A36 208 44A1
A 52 FP2
see NF A36 208 44K1
A 52 HS 44A1
A 52 P1
see NF A36 205 44A1
A 52 P2
see NF A36 205 44A1
A 55 51A
A 60-1
see NF A35 501 44A1
A 60-2
see NF A35 501 44A1
A 70 51A
A 71 NB 14C
A 85 50A
A 89 50A
A 100 44N1
A 101 44N1
A 105 21.1
A 108 1L
A 110 21.1
A 110 AT 51B
A 113 44A1
A 120 44N1
A 122 21.1
A 127 50A
A 129 50A
A 131 44A1
A 132 1K
A 139 50A
A 143 1L
A 200 44N1
A 201 A 44A2
A 201 B 44A2
A 202 44E1
A 203A 44F1
A 203B 44F1
A 203D 44F1
A 203E 44F1
A 204 A 44G1
A 204 B 44G1
A 204 C 44G1
A 205 21.1
A 205 44N1
A 210 21.1
A 212 A 44A2
A 212 B 44A2
A 220 44N1
A 242 44K1
A 283 44A1
A 285 44A2
A 286 44N2
A 299 44A1
A 300 44N1
A 325 44K1
A 330 44K1
A 356 1K
A 356.0 1K
A 360 1K
A 373 44A1
A 380 1L
A 387 A 44K1
A 387 B 44K1
A 387 C 44K1
A 387 D 44K1
A 387 E 44K1
A 395 44K3
A 400 44N1
A 405 44N1
A 410 44N1
A 420 44K1
A 440 44F1
A 441 44J
A 442 44A1
A 442 44K2
A 455 44A2
A 500 44N1
A 506 44N1
A 507 44N1
A 520 44N1
A 540 44J
A 575 44J
A 604 44N1
A 610 44N1
A 700 44N1
A 750 44N1

A 900 44N1
A 950 44N2
A 955 44N2
A 962 44N1
A 963 44N2
A 1000 44K1
A01001
UNS designation for 100.1 1A
A01301
UNS designation for 130.1 1A
A01501
UNS designation for 150.1 1A
A01601
UNS designation for 160.1 1A
A01701
UNS designation for 170.1 1A
A02010
UNS designation for 201.0 1L
A02012
UNS designation for 201.2 1L
A02020
UNS designation for 202.0 1L
A02022
UNS designation for 202.2 1L
A02030
UNS designation for 203.0 1L
A02032
UNS designation for 203.2 1L
A02042
UNS designation for 204.2 1L
A02060
UNS designation for 206.0 1L
A02062
UNS for 206.2 1L
A02080
UNS designation for 208.0 1L
A02081
UNS for 208.1 1L
A02082
UNS for 208.2 1L
A02130
UNS for 213.0 1L
A02131
UNS for 213.1 1L
A02220
UNS designation for 222.0 1L
A02221
UNS for 222.1 1L
A02240
UNS for 224.0 1L
A02242
UNS for 224.2 1L
A02380
UNS for 238.0 1L
A02381
UNS for 238.1 1L
A02382
UNS for 238.2 1L
A02400
UNS designation for 240.0 1L
A02401
UNS designation for 240.1 1L
A02420
UNS designation for 242.0 1L
A02421
UNS designation for 242.1 1L
A02422
UNS designation for 242.2 1L
A02430
UNS designation for 243.0 1L
A02431
UNS designation for 243.1 1L
A02490
UNS for 249.0 1M
A02492
UNS for 249.2 1M
A02591
UNS for 295.1 1L
A 2855 23B
A02950
UNS designation for 295.0 1L
A02952
UNS for 295.2 1L
A02960
UNS for 296.0 1L
A02961
UNS for 296.1 1L
A02962
UNS for 296.2 1L
A03050
UNS for 305.0 1C
A03052
UNS for 305.2 1C
A03080
UNS for 308.0 1L
A03081
UNS for 308.1 1L
A03082
UNS for 308.2 1L
A03191
UNS for 319.1 1L
A03192
UNS for 319.2 1L
A03200
UNS for 320.0 1L
A03201
UNS for 320.1 1L
A03240
UNS for 324.0 1C
A03241
UNS for 324.1 1C
A03242
UNS for 324.2 1C
A03280
UNS for 328.0 1C
A03281
UNS for 328.1 1C
A03320
UNS for designation 332.0 1L
A03321
UNS for designation 332.1 1L
A03322
UNS for designation 332.2 1L
A03330
UNS for designation 333.0 1L
A03331
UNS for designation 333.1 1L
A03360
UNS for designation 336.0 1C
A03361
UNS for designation 336.1 1C
A03362
UNS for designation 336.2 1C
A03390
UNS for designation 339.0 1L
A03391
UNS for designation 339.1 1L
A03430
UNS for designation 343.0 1C
A03431
UNS for designation 343.1 1C
A03540
UNS for designation 354.0 1C
A03541
UNS for designation 354.1 1C
A03551
UNS for designation 355.1 1C
A03552
UNS for designation 355.2 1C
A 03560 1K
A03561
UNS for designation 356.1 1C
A03562
UNS for designation 356.2 1C
A03570
UNS for designation 357.0 1K
A03571
UNS for designation 357.1 1K
A03580
UNS for designation 358.0 1K
A03582
UNS for designation 358.2 1K
A03590
UNS for designation 359.0 1C
A03592
UNS for designation 359.2 1K
A03600
UNS for designation 360.0 1K
A03602
UNS for designation 360.2 1C
A03610
UNS for designation 361.0 1C
A03611
UNS for designation 361.1 1C
A03630
UNS for designation 363.0 1C
A03631
UNS for designation 363.1 1C
A03640
UNS for designation 364.0 1C
A03642
UNS for designation 364.2 1C
A03690
UNS for designation 369.0 1C
A03691
UNS for designation 369.1 1C
A03800
UNS for designation 380.0 1C
A03802
UNS for designation 380.2 1C
A03830
UNS for designation 383.0 1C
A03831
UNS for designation 383.1 1C
A03832
UNS for designation 383.2 1C
A03840
UNS for designation 384.0 1C
A03841
UNS for designation 384.1 1C

A03842
UNS for designation 384.2 1C
A03850
UNS for designation 385.0 1C
A03851
UNS for designation 385.1 1C
A03900
UNS for designation 390.0 1C
A03902
UNS for designation 390.2 1C
A03920
UNS for designation 392.0 1C
A03921
UNS for designation 392.1 1C
A03930
UNS for designation 393.0 1C
A03931
UNS for designation 393.1 1C
A03932
UNS for designation 393.2 1C
A04082
UNS for designation 408.2 1C
A04092
UNS for designation 409.2 1C
A04112
UNS for designation 411.2 1C
A04130
UNS for designation 413.0 1C
A04132
UNS for designation 413.2 1C
A04352
UNS for 435.2 1C
A04431
UNS for 443.1 1C
A04432
UNS for 443.2 1C
A04440
UNS for 444.0 1C
A04442
UNS for 444.2 1C
A04452
UNS for 445.2 1C
A05110
UNS for 511.0 1J
A05111
UNS for 511.1 1J
A05112
UNS for 511.2 1J
A05120
UNS for 512.0 1J
A05122
UNS for 512.2 1J
A05130
UNS for 513.0 1J
A05132
UNS for 513.2 1J
A05140
UNS for 514.0 1J
A05141
UNS for 514.1 1J
A05142
UNS for 514.2 1J
A05150
UNS for 515.0 1J
A05152
UNS for 515.2 1J
A05160
UNS for 516.0 1J
A05161
UNS for 516.1 1J
A05180
UNS for 518.0 1J
A05181
UNS for 518.1 1J
A05182
UNS for 518.2 1J
A05202
UNS for 520.2 1J
A05352
UNS for 535.2 1J
A07051
UNS for 705.0 1M
A07071
UNS for 707.1 1M
A07100
UNS for 710.0 1M
A07101
UNS for 710.1 1M
A07110
UNS for 711.0 1M
A07111
UNS for 711.1 1M
A07120
UNS for 712.0 1M
A07122
UNS for 712.2 1M
A07131
UNS for 713.1 1M
A07712
UNS for 771.2 1M
A07720
UNS for 772.0 1M
A07722
UNS for 772.2 1M
A08501
UNS for 850.1 1N
A08510
UNS for 851.0 1N
A08511
UNS for 851.1 1N
A08520
UNS for 852.0 1N
A08521
UNS for 852.1 1N
A08530
UNS for 853.0 1N
A08532
UNS for 853.2 1N
A12010
UNS for 201.0 1L
A12011
UNS for 201.1 1L
A12012
UNS for 201.2 1L
A12060
UNS for 206.0 1L
A12062
UNS for 206.2 1L
A12400
UNS for 240.0 1L
A12401
UNS for 240.1 1L
A12402
UNS for 240.2 1L
A12421
UNS for 242.1 1L
A12422
UNS for 242.2 1L
A13050
UNS for 305.0 1C
A13051
UNS for 305.1 1C
A13052
UNS for 305.2 1C
A13190
UNS for 319.0 1L
A13191
UNS for 319.1 1L
A13320
UNS replaced by A03360
A13321
UNS replaced by A03361
A13322
UNS replaced by A03362
A13330
UNS for 333.0 1L
A13331
UNS for 333.1 1L
A13550
UNS for 355.0 1C
A13552
UNS for 355.2 1C
A 13560 1K
A13561
UNS for 356.1 1C
A13562
UNS for 356.2 1C
A13570
UNS for 357.0 1K
A13572
UNS for 357.2 1K
A13600
UNS for 360.0 1C
A13601
UNS for 360.1 1C
A13602
UNS for 360.2 1C
A13800
UNS for 380.0 1C
A13801
UNS for 380.1 1C
A13802
UNS for 380.2 1C
A13840
UNS for 384.0 1C
A13841
UNS for 384.1 1C
A13900
UNS for 390.0 1C
A13901
UNS for 390.1 1C
A14130
UNS for 413.0 1C
A14131
UNS for 413.1 1C
A14132
UNS for 413.2 1C
A14431
UNS for 443.1 1C

A14440
UNS for 444.0 1C
A14441
UNS for 444.1 1C
A14442
UNS for 444.2 1C
A15142
Replaced by A05132
A15350
UNS for 535.0 1J
A15351
UNS for 535.1 1J
A17120
Replaced by A070100
A17121
Replaced by A07101
A18501
Replaced by A08511
A22010
UNS for 201.0 1L
A22950
Replaced by A02960
A22951
Replaced by A02961
A22952
Replaced by A02962
A23190
UNS for 319.0 1L
A23191
UNS for 319.1 1L
A23560
UNS for 356.0 1C
A23562
UNS for 356.2 1C
A23570
UNS for 357.0 1K
A23572
UNS for 357.2 1K
A23580
Replaced by A03580
A23582
Replaced by A03582
A23800
UNS for 380.0 1C
A23801
UNS for 380.1 1C
A23840
Replaced by A03850
A23841
Replaced by A03851
A23900
UNS for 390.0 1C
A23901
UNS for 390.1 1C
A24130
UNS for 413.0 1C
A24131
UNS for 413.1 1C
A24431
UNS for 443.1 1C
A24442
Replaced by A04452
A25140
Replaced by A05120
A25142
Replaced by A05122
A25350
UNS for 535.0 1J
A25352
UNS for 535.2 1J
A27710
Replaced by A07720
A27712
Replaced by A07722
A28500
Replaced by A08520
A28501
Replaced by A08521
A 33550 1L
A33551
UNS for 355.1 1C
A33552
UNS for 355.2 1C
A33560
UNS for 356.0 1K
A33562
UNS for 356.2 1K
A33570
UNS for 357.0 1K
A33572
UNS for 357.2 1K
A34430
UNS for 443.0 1C
A34431
UNS for 443.1 1C
A34432
UNS for 443.2 1C
A37120
Replaced by A07110
A37121
Replaced by A07111
A43570
UNS for 357.0 1C
A47120
Replaced by A07120
A47122
Replaced by A07122
A63320
Replaced by A03320
A63321
Replaced by A03321
A63322
Replaced by A03322
A63560
UNS for 356.0 1K
A63562
UNS for 356.2 1K
A65140
Replaced by A05110
A65141
Replaced by A05111
A65142
Replaced by A05112
A82004
UNS for 2004 clad with A91070 1H
A82014
UNS for 2014 clad with A96003 1H
A82024
UNS for 2024 clad with A91230 1H
A82219
UNS for 2219 clad with A97072 1H
A83003
UNS for 3003 clad with A94343 1H
A86061
UNS for 6061 clad with A97072 1H
A86951
UNS for 6951 clad with A94343 1H
A87050
UNS for 7050 clad with A97072 1H
A87075
UNS for 7075 clad with A97072 or A97011 1H
A87079
UNS for 7079 clad with A97072 1H
A87178
UNS for 7178 clad with A97072 1H
A87475
UNS for 7475 clad with A97072 1H
A91030
UNS for 1030 obsolete 1A
A91035
UNS for 1035 1A
A91040
UNS for 1040 1A
A91045
UNS for 1045 1A
A91050
UNS for 1050 1A
A91055
UNS for 1055 - obsolete 1A
A91060
UNS for 1060 1A
A91065
UNS for 1065 1A
A91070
UNS for 1070 1A
A91075
UNS for 1075 - obsolete 1A
A91080
UNS for 1080 1A
A91085
UNS for 1085 1A
A91090
UNS for 1090 1A
A91095
UNS for 1095 - obsolete 1A
A91100
UNS for 1100 - obsolete 1A
A91135
UNS for 1135 - obsolete 1A
A91145
UNS for 1145 - obsolete 1A
A91170
UNS for 1170 1A
A91175
UNS for 1175 1A
A91180
UNS for 1180 1A

A91185
UNS for 1185 1A
A91188
UNS for 1188 1A
A91193
UNS for 1193 - obsolete 1A
A91199
UNS for 1199 1A
A91200
UNS for 1200 1A
A91230
UNS for 1230 1A
A91235
UNS for 1235 1A
A91250
UNS for 1250 - obsolete 1A
A91260
UNS for 1260 - obsolete 1A
A91285
UNS for 1285 1A
A91345
UNS for 1345 1A
A91350
UNS for 1350 1A
A91435
UNS for 1435 1D
A92004
UNS for 2004 1F
A92008
UNS for 2008 1F
A92011
UNS for 2011 1F
A92014
UNS for 2014 1F
A92017
UNS for 2017 1F
A92018
UNS for 2018 1F
A92020
UNS for 2020 - obsolete 1F
A92021
UNS for 2021 - obsolete 1F
A92024
UNS for 2024 1F
A92025
UNS for 2025 1F
A92034
UNS for 2034 1F
A92036
UNS for 2036 1F
A92037
UNS for 2037 1F
A92038
UNS for 2038 1F
A92048
UNS for 2048 1F
A92090
UNS for 2090 1F
A92091
UNS for 2091 1F
A92117
UNS for 2117 1F
A92124
UNS for 2124 1F
A92214
UNS for 2214 1F
A92218
UNS for 2218 1F
A92219
UNS for 2219 1F
A92224
UNS for 2224 1F
A92319
UNS for 2319 1F
A92324
UNS for 2324 1F
A92419
UNS for 2419 1F
A92519
UNS for 2519 1F
A92618
UNS for 2618 1F
A93002
UNS for 3002 1B
A93003
UNS for 3003 1B
A93004
UNS for 3004 1B
A93005
UNS for 3005 1B
A93006
UNS for 3006 1B
A93007
UNS for 3007 1B
A93009
UNS for 3009 1B
A93010
UNS for 3010 1B
A93011
UNS for 3011 1B
A93015
UNS for 3015 1B
A93016
UNS for 3016 1B
A93102
UNS for 3102 1B
A93104
UNS for 3104 1B
A93105
UNS for 3105 1B
A93107
UNS for 3107 1B
A93303
UNS for 3303 1B
A93307
UNS for 3307 1B
A94002
UNS for 4002 - obsolete 1N
A94004
UNS for 4004 1N
A94008
UNS for 4008 1C
A94009
UNS for 4009 1E
A94011
UNS for 4011 1E
A94013
UNS for 4013 1E
A94032
UNS for 4032 1E
A94043
UNS for 4043 1C
A94044
UNS for 4044 1C
A94047
UNS for 4047 1C
A94104
UNS for 4104 1C
A94145
UNS for 4145 1C
A94343
UNS for 4343 1C
A94543
UNS for 4543 1C
A94643
UNS for 4643 1C
A95005
UNS for 5005 1D
A95006
UNS for 5006 1D
A95010
UNS for 5010 1D
A95016
UNS for 5016 1D
A95017
UNS for 5017 1D
A95034
UNS for 5034 - obsolete 1D
A95039
UNS for 5039 - obsolete 1D
A95040
UNS for 5040 1D
A95042
UNS for 5042 1D
A95043
UNS for 5043 1D
A95050
UNS for 5050 1D
A95051
UNS for 5051 1D
A95052
UNS for 5052 1D
A95056
UNS for 5056 1D
A95082
UNS for 5082 1D
A95083
UNS for 5083 1D
A95086
UNS for 5086 1D
A95151
UNS for 5151 1D
A95154
UNS for 5154 1D
A95182
UNS for 5182 1D
A95183
UNS for 5183 1D
A95205
UNS for 5205 1D
A95250
UNS for 5250 1D
A95252
UNS for 5252 1D
A95254
UNS for 5254 1D
A95351
UNS for 5351 1D

A95352
UNS for 5352 1D
A95356
UNS for 5356 1D
A95357
UNS for 5357 1D
A95451
UNS for 5351 1D
A95454
UNS for 5454 1D
A95456
UNS for 5456 1D
A95457
UNS for 5457 1D
A95552
UNS for 5552 1D
A95554
UNS for 5554 1D
A95556
UNS for 5556 1D
A95557
UNS for 5557 1D
A95652
UNS for 5652 1D
A95654
UNS for 5654 1D
A95657
UNS for 5657 1D
A96003
UNS for 6003 1E
A96004
UNS for 6004 1E
A96005
UNS for 6005 1E
A96006
UNS for 6006 1E
A96007
UNS for 6007 1E
A96009
UNS for 6009 1E
A96010
UNS for 6010 1E
A96011
UNS for 6011 1E
A96013
UNS for 6013 1E
A96017
UNS for 6017 1E
A96053
UNS for 6053 1E
A96060
UNS for 6060 1E
A96061
UNS for 6061 1E
A96063
UNS for 6063 1E
A96066
UNS for 6060 1E
A96070
UNS for 6070 1E
A96101
UNS for 6101 1E
A96105
UNS for 6105 1E
A96110
UNS for 6110 1E
A96111
UNS for 6111 1E
A96151
UNS for 6151 1E
A96162
UNS for 6162 1E
A96201
UNS for 6201 1E
A96205
UNS for 6205 1E
A96253
UNS for 6253 1E
A96261
UNS for 6261 - obsolete 1E
A96262
UNS for 6262 1E
A96301
UNS for 6301 1E
A96351
UNS for 6351 1E
A96463
UNS for 6463 1E
A96763
UNS for 6763 1E
A96961
UNS for 6961 1E
A97001
UNS for 7001 1G
A97004
UNS for 7005 1G
A97008
UNS for 7008 1G
A97010
UNS for 7010 1G
A97011
UNS for 7011 - obsolete 1G
A97013
UNS for 7013 1G
A97016
UNS for 7016 1G
A97021
UNS for 7021 1G
A97029
UNS for 7029 1G
A97039
UNS for 7039 1G
A97046
UNS for 7046 1G
A97049
UNS for 7049 1G
A97050
UNS for 7050 1G
A97070
UNS for 7070 - obsolete 1G
A97072
UNS for 7072 1G
A97075
UNS for 7075 1G
A97076
UNS for 7076 1G
A97079
UNS for 7079 1G
A97090
UNS for 7090 1G
A97091
UNS for 7091 1G
A97104
UNS for 7104 - obsolete 1G
A97108
UNS for 7108 1G
A97116
UNS for 7116 1G
A97129
UNS for 7129 1G
A97146
UNS for 7146 1G
A97149
UNS for 7149 1G
A97150
UNS for 7150 1G
A97175
UNS for 7175 1G
A97178
UNS for 7178 1G
A97179
UNS for 7179 1G
A97229
UNS for 7229 1G
A97277
UNS for 7277 1G
A97472
UNS for 7472 1G
A97475
UNS for 7475 1G
A98001
UNS for 8001 1A
A98006
UNS for 8006 1B
A98007
UNS for 8007 1B
A98010
UNS for 8010 1B
A98013
UNS for 8013 - obsolete 1A
A98014
UNS for 8014 1B
A98017
UNS for 8017 1A
A98020
UNS for 8020 1A
A98030
UNS for 8030 1A
A98040
UNS for 8040 1A
A98076
UNS for 8076 1A
A98077
UNS for 8077 1A
A98079
UNS for 8079 1A
A98081
UNS for 8081 1N
A98111
UNS for 8111 1A
A98112
UNS for 8112 1A
A98130
UNS for 8130 1A
A98176
UNS for 8176 1A
A98177
UNS for 8177 1A

A98280
UNS for 8280 1N
AA
Aluminium Association - followed by designation: e.g. AA3003 will be found under 3003.
AAA
Aluminium Association of America - followed by designation
AAB
Aluminium Association of America - followed by designation
AAC
Aluminium Association of America - followed by designation
AAD
Aluminium Association of America - followed by designation
AAF
Aluminium Association of America - followed by designation
AAF 11078A/1 6.1
AAF 11078A/2 6.1
AAPHOS 14K
AB 14G
AB1
BS designation 14G
AB2
BS designation 14G
AB3
BS designation 14G
AB 49 TRI-FOIL 42.1
AB 75 44E2
AB 213 44K1
ABC III 44K3
ABRADUR 2 44A3
ABRADUR M14 44D
ABRADUR M14N 44D
ABRADUR SK 44E2
ABRASODUR 16 44K3
ABRASODUR 33 44M2
ABRASODUR 35 44M2
ABRASODUR 38 44M2
ABRASODUR 38AT 44M2
ABRASODUR 39 44M2
ABRASODUR 40 44M2
ABRASODUR 43 44M2
ABRASODUR 45 44M2
ABRASODUR 45AT 44M2
ABRASODUR 46 44M2
ABRASODUR 53 27F
ABRASODUR 83 13A
ABRASODUR II 44E2
ABRAZO 44K2
ABS 44A1
ABS/A 44A1
ABS/AH32 44A1
ABS/AH36 44A1
ABS/B 44A1
ABS/B42 14A
ABS/B43 14B
ABS/CS 44A1
ABS/D 44A1
ABS/DH32 44A1
ABS/DH36 44A1
ABS/DS 44A1
ABS/E 44A1
ABS/EH32 44A1
ABS/EH36 44A1
ABS/F 44A2
ABS/G 44A1
ABS/G 44A2
ABS/G2A 44A1
ABS/G2B 44A1
ABS/G3A 44A1
ABS/G3B 44A1
ABS/GA 44A1
ABS/GB 44A1
ABS/GC 44A1
ABS/GI 44A1
ABS/H 44A1
ABS/H 44G1
ABS/I 44G1
ABS/J 44A1
ABS/J 44G1
ABS/K 44A1
ABS/K 44G1
ABS/L 44A1
ABS/L 44G1
ABS/M 44A1
ABS/M 44G1
ABS/N 44A1
ABS/N 44G1
ABS/O 44G1
ABS/P 44G1
ABS TYPE 2 14B
ABS TYPE 3 14C
ABS TYPE 4 14G
ABS TYPE 5 14G
AC 1E
AC1 BAR 44A1
AC1 BEL 44A1
AC1 HT-30 44N1
AC1 HT-50 44N1
AC1 HT-50C 44N1
A C4 1N
AC 5 1C
AC 41A 55A
AC 43A 55A
AC 254 44M1
ACA
see ASTMA823 20B
ACB
see ASTMA823 20B
ACCOLAY 27B
AC HT 20C
ACIBOND 44A1
ACIBRADE 44A3
ACI CN-7M 44N2
ACID BRONZE 14K
ACI HU 27E
ACI HU 50 27E
ACI HW 27B
ACI HW 50 27E
ACI HX 27B
ACM 44E1
ACMITE 44L
ACNW 44N1
ACP 44K2
ACTIVE Ni 27A
AD 3 1E
AD 31 1E
ADAMANT 44E2
ADAMANT GR 44E3
ADAMANT HP 44K3
ADAMANT XL 44K3
ADAMANT XTRA 44K3
ADASTRAL 'A' 50A
ADIC 44K1
ADMIRALTY BRASS 14B
ADMIRALTY GUNMETAL 14K
ADNIC 14E
ADR 20C
ADS 44K1
ADS HB 44K2
ADVANCE 14E
AEI/M 44K1
AERAL 1L
AEROLITE 1L
AERON 1L
AERO No 2 50A
AETERNA 614 14G
AETERNA 630 14G
AF2 – 1DA 27C
Af 3 1L
AF 71 44P
AF 94 13A
AF 183 44P
AF 1410 44K1
AF 1753 27C
AFC 77 44K1
AFC 77 44L
A Fe 10 1N
AFN03 81-343 EZ 19.9LB 110-20 44N1
A FNDR 81.347 EF 2070 Ni Cr Mn Fe B 14020 BH 27B
AF No 3 81.309 E43 1/0 RR130 32 44A1
AF No 3 81.309 E51 3/2 AR160 31 44A1
AF No 3 E43 2/2 RII 44A1
AFNOR 23.15 NK 38 20C
AFNOR 81 347 EF 2070 Ni Cr Fe Mo B 20 BH 44M1
AFNOR 81 347 EF 2070 Ni Cr Fe Mo B 150-26 BH 44M1
AFNOR 81.309 E43 2/2 R12 44A1
AFNOR 81.309 E43 2/2 RR 150 32 44A1
AFNOR 81.309 E43 2/2 RR2 44A1
AFNOR 81.309 E43 3/2 C10 44A1
AFNOR 81.309 E43 3/3 AR150 32 44A1
AFNOR 81.340 E 146 3 Ni B26T 44F1
AFNOR 81.343 EZ18.8 Mn B160 23X 44P
AFNOR 81.343 EZ23.12R 180.33X 44P
AFNOR 81-340 EY 50.1 Ni Mo B 11026 BN 44K1
AFNOR 81-340 EY 55.1 Ni Cr Mo B 11026 TBN 44K1

AFNOR 81-340 EY 69 2 Mn 2 Ni Cr Mo B 11026 TBH 44K1
AFNOR 81-343 EZ 19.9LR 23 44N1
AFNOR 81-343 EZ 19.9LR 160-23X 44N1
AFNOR 81-343 EZ 19.9LR 150-36 44N1
AFNOR 81-343 EZ18.8 Mn B23 44P
AFNOR 81-343 EZ18.8 Mn R160 21X 44P
AFNOR 81-343 EZ19.12.3 LB 110-20 44N1
AFNOR 81-343 EZ19.12.3 LR23 44N1
AFNOR 81-343 EZ19.12.3 LR150-36 44N1
AFNOR 81-343 EZ19.12.3 LR170 23X 44N1
AFNOR 81-343 EZ19.12.3 NbR23 44N1
AFNOR 81-343 EZ19.13.4 R180-23X 44N1
AFNOR 81-343 EZ19.9 NbR23 44N1
AFNOR 81-343 EZ20.25.2 LCnRB20 44N1
AFNOR 81-343 EZ20.25.5 LCnR23 44N1
AFNOR 81-343 EZ23.12 LR23 44N1
AFNOR 81-343 EZ23.12 LR150-36 44N1
AFNOR 81-343 EZ23.12.2 R23 44N1
AFNOR 81-343 EZ23.12.2 R186-33X 44N1
AFNOR 81-343 EZ25.20 R23 44N1
AFNOR 81-343 EZ29.9 R23 44N1
AFNOR 81-343-E2 13.4 Mo B 120.26X 44N3
AFNOR 81-345 EC 0.5 Cr Mo B 20 44K1
AFNOR 81-345 EC 1 Cr Mo B 20 44K1
AFNOR 81-345 EC 1 Cr Mo R 22 44K1
AFNOR 81-345 EC 2 Cr Mo B 11020 44K1
AFNOR 81-345 EC 5 Cr Mo B 20 44K1
AFNOR 81-345 EC 9 Cr Mo B 20 44K1
AFNOR 81-345 EC Mo B20 44G1
AFNOR 81-345 EC Mo R22 44G1
AFNOR 81-347 EF2070 Ni Cr Fe B110 20BH 27F
AFNOR 100 Cb 44E3
Ag/Cu EUTECTIC 42.1
AG 0.5 1D
AG 0.6 1D
AG 1 1D
AG 2 1D
AG 2.5MC 1D
AG 2M 1D
AG 3 1D
A G3T 1J
AG 4 1D
AG 4 MC 1D
A G4Z 1J
AG 4.5 MC 1D
AG 5 1D
A G6 1J
AG 7 1D
A G10Y4 1J
AG 40A 55A
AG 40B 55A
AGATE 52.1
AGG 1 1A
AGG 2 1A
AGG 4 1D
AGG 50 1D
AGG 51 1E
AGG 54 1D
AGG 57 1D
AGP V 9 44A1
A GRADE 20A
AGS 1E
AGS 44E3
AGS/L 1E
AGS 50 1E
AGS 65 1E
AGSUC 1E
AGT 44N1
AGT VAC-ARC 44K1
AH 44M1
AH CHROME DIE 44K3
AHT 28 44K1
AICHS METAL 14B
AIP 44N1
AIR 9/182 T35 51A
AIRMO 44K2
AIRQUE 44K3
AIRQUE 4 44K3
AIRQUE SPECIAL 44K3
AIRQUE V 44K3
AIR RESIST 13 13A
AIR RESIST 213 13A
AIR RESIST 215 13A
AIRTRUE 44K3
AIRVAN 44K3
AIS 44K1
AIS 44N1
AISA 300 M see 300M 44K2
AISC 44N1
AISI 12 L14 44A1
AISI 50 B44 44E2
AISI 50 B46 44E2
AISI 50 B50 44E2
AISI 50 B60 44E2
AISI 51 B60 44E2
AISI 81 B45 44K2
AISI 86 B45 44K2
AISI 94 B17 44K1
AISI 94 B30 44K1
AISI 201 44P
AISI 202 44P
AISI 302 44N1
AISI 302B 44N1
AISI 303 44N1
AISI 303 Se 44N1
AISI 304 44N1
AISI 304L 44N1
AISI 305 44N1
AISI 308 44N1
AISI 309 44N1
AISI 309S 44N1
AISI 310 44N1
AISI 310S 44N1
AISI 314 44N1
AISI 316 44N1
AISI 316L 44N1
AISI 316N 44N1
AISI 317 44N1
AISI 317 L 44N1
AISI 318 44N1
AISI 321 44N1
AISI 321H 44N1
AISI 325 44N1
AISI 327 44N1
AISI 329 44N1
AISI 330 44N1
AISI 347 44N1
AISI 347 F 44N1
AISI 347H 44N1
AISI 348 44N1
AISI 348H 44N1
AISI 403 44M1
AISI 405 44M1
AISI 410 44M1
AISI 414 44M1
AISI 416 44M1
AISI 416 PLUS X 44M1
AISI 416 Se 44M1
AISI 418 44M1
AISI 420 44M1
AISI 420F 44M1
AISI 430 44M1
AISI 430 F 44M1
AISI 430 F Se 44M1
AISI 431 44M1
AISI 434 44M1
AISI 440 A 44M1
AISI 440 B 44M1
AISI 440 C 44M2
AISI 440 F 44M2
AISI 442 44M1
AISI 446 44M1
AISI 501 44K1
AISI 502 44K1
AISI 601 44K2
AISI 602 44K1
AISI 603 44K1
AISI 604 44K1
AISI 610 44K2
AISI 611 44K3
AISI 612 44K3
AISI 613 44K3
AISI 614 44M1
AISI 615 44M1
AISI 616 44M1
AISI 617 44M2
AISI 618 44M2
AISI 619 44M1
AISI 633 44N3
AISI 634 44N3
AISI 650 44N2
AISI 651 44N2
AISI 652 44N2
AISI 653 44N2

AISI 660 44N2
AISI 661 44L
AISI 662 44N2
AISI 663 44N2
AISI 664 27C
AISI 665 44N2
AISI 670 13A
AISI 671 13A
AISI 680 27C
AISI 681 27C
AISI 682 27C
AISI 683 27C
AISI 684 27C
AISI 685 27C
AISI 686 27C
AISI 687 27C
AISI 688 27C
AISI 689 27C
AISI 690 27C
AISI 1006 44A1
AISI 1008 44A1
AISI 1009 44A1
AISI 1010 44A1
AISI 1012 44A1
AISI 1015 44A1
AISI 1016 44A1
AISI 1017 44A1
AISI 1018 44A1
AISI 1019 44A1
AISI 1020 44A1
AISI 1021 44A1
AISI 1022 44A1
AISI 1023 44A2
AISI 1024 44A2
AISI 1025 44A2
AISI 1026 44A2
AISI 1027 44A2
AISI 1029 44A2
AISI 1030 44A2
AISI 1033 44A2
AISI 1035 44A2
AISI 1036 44A2
AISI 1037 44A2
AISI 1038 44A2
AISI 1039 44A2
AISI 1040 44A2
AISI 1041 44A2
AISI 1042 44A2
AISI 1043 44A2
AISI 1045 44A2
AISI 1046 44A2
AISI 1048 44A3
AISI 1049 44A3
AISI 1050 44A3
AISI 1051 44A3
AISI 1052 44A3
AISI 1052 44A3
AISI 1053 44A3
AISI 1055 44A3
AISI 1060 44A3
AISI 1064 44A3
AISI 1065 44A3
AISI 1070 44A3
AISI 1074 44A3
AISI 1078 44A3
AISI 1080 44A3
AISI 1084 44A4
AISI 1086 44A4
AISI 1090 44A4
AISI 1095 44A4
AISI 1108 44A1
AISI 1109 44A1
AISI 1110 44A1
AISI 1111 44A1
AISI 1112 44A1
AISI 1113 44A1
AISI 1115 44A1
AISI 1116 44A1
AISI 1117 44A1
AISI 1118 44A1
AISI 1119 44A1
AISI 1120 44A1
AISI 1126 44A2
AISI 1132 44A2
AISI 1137 44A2
AISI 1138 44A2
AISI 1139 44A2
AISI 1140 44A2
AISI 1141 44A2
AISI 1144 44A2
AISI 1145 44A2
AISI 1146 44A2
AISI 1151 44A3
AISI 1211 44A1
AISI 1212 44A1
AISI 1213 44A1
AISI 1215 44A1
AISI 1320 44A1
AISI 1330 44A2
AISI 1335 44A2
AISI 1340 44A2
AISI 1345 44A2
AISI 2317 44F1
AISI 2330 44F1
AISI 2512 44F1
AISI 2515 44F1
AISI 2517 44F1
AISI 3115 44K1
AISI 3120 44K1
AISI 3130 44K1
AISI 3135 44K1
AISI 3140 44K2
AISI 3141 44K2
AISI 3145 44K2
AISI 3150 44K2
AISI 3240 44K2
AISI 3310
see 3310 44K1
AISI 3316 44K1
AISI 4012 44G1
AISI 4023 44G1
AISI 4024 44G1
AISI 4027 44G1
AISI 4028 44G1
AISI 4032 44G1
AISI 4037 44G1
AISI 4042 44G2
AISI 4047 44G2
AISI 4053 44G2
AISI 4063 44G2
AISI 4068 44G2
AISI 4118 44K1
AISI 4119 44K1
AISI 4125 44K1
AISI 4130 44K1
AISI 4137 44K1
AISI 4140 44K2
AISI 4142 44K2
AISI 4145 44K2
AISI 4147 44K2
AISI 4150 44K2
AISI 4161 44K2
AISI 4317 44K1
AISI 4320 44K1
AISI 4330 44K1
AISI 4335 44K1
AISI 4337 44K1
AISI 4340 44K2
AISI 4419 44G1
AISI 4422 44G1
AISI 4427 44G1
AISI 4520 44G1
AISI 4608 44K1
AISI 4615 44K1
AISI 4620 44K1
AISI 4621 44K1
AISI 4626 44K1
AISI 4640 44K2
AISI 4718 44K1
AISI 4720 44K1
AISI 4815 44K1
AISI 4817 44K1
AISI 4820 44K1
AISI 5015 44E1
AISI 5045 44E2
AISI 5046 44E2
AISI 5117 44E1
AISI 5120 44E1
AISI 5130 44E1
AISI 5132 44E1
AISI 5135 44E1
AISI 5140 44E2
AISI 5145 44E2
AISI 5147 44E2
AISI 5150 44E2
AISI 5152 44E2
AISI 5155 44E2
AISI 5160 44E2
AISI 6117 44K1
AISI 6118 44K1
AISI 6120 44K1
AISI 6145 44K2
AISI 6150 44K2
AISI 8615 44K1
AISI 8617 44K1
AISI 8620 44K1
AISI 8622 44K1
AISI 8625 44K1
AISI 8627 44K1
AISI 8630 44K1
AISI 8632 44K1
AISI 8635 44K1
AISI 8637 44K1
AISI 8640 44K2
AISI 8641 44K2
AISI 8642 44K2
AISI 8645 44K2
AISI 8647 44K2
AISI 8650 44K2
AISI 8653 44K2
AISI 8655 44K2

AISI 8660 44K2
AISI 8715 44K1
AISI 8717 44K1
AISI 8719 44K1
AISI 8720 44K1
AISI 8735 44K1
AISI 8740 44K2
AISI 8742 44K2
AISI 8745 44K2
AISI 8750 44K2
AISI 8822 44K1
AISI 9250 44B2
AISI 9255 44B2
AISI 9260 44B2
AISI 9261 44E2
AISI 9262 44E2
AISI 9310 44K1
AISI 9315 44K1
AISI 9317 44K1
AISI 9437 44K1
AISI 9440 44K2
AISI 9442 44K2
AISI 9445 44K2
AISI 9447 44K2
AISI 9747 44K2
AISI 9763 44K2
AISI 9840 44K2
AISI 9845 44K2
AISI 9850 44K2
AISI A2340 44F2
AISI A2345 44F2
AISI A4068 44G2
AISI E50100 44E3
AISI E51100 44E3
AISI E52100 44E3
AISI H11 44K2
AISL 44N1
AIST 44N1
AITCH METAL 14C
AJ 2 44L
AK 44H
AK 2MV 44M1
AK 4-1 1F
AK 8 1F
AKMC/Z 44P
AL 0 1 14A
AL 1 14G
AL 2 1C
AL 3 1A
AL 4 1C
AL 5 1A
AL 5 1C
AL-6X 44N1
AL-6XN 44N1
AL 8 1D
AL 9 1C
Al 16 V 1L
Al 18 V 1L
AL18 NiCoMo (250) 44F3
AL18 NiCoMo (300) 44F3
AL20 Ni (250) 44F3
AL25 Ni (250) 44F3
AL 26 1C
AL 27 1D
Al 28 1J
AL29-4.2 44M1
AL29-4C 44M1
AL 30 1C
Al 75 1L
AL 99/99R 1A
Al 100 (79) 1M
AL 350 44N3
AL 419 44M1
ALAR 005 1C
ALAR 0012 1C
ALAR 308 1L
Al B42 1M
Al B133 1M
ALBALOY 55A
ALBATRA METAL 14F
AL BRONZE 184 14G
AL BRONZE 197 14G
Al Bz 5 14G
ALCAN 74S 1G
ALCAN 100
see ALCAN GB 100 1A
ALCAN 117
see ALCAN GB 117 1L
ALCAN 125
see ALCAN GB 125 1K
ALCAN 160
see ALCAN GB 160 1C
ALCAN 162
see ALCAN GB 162 1K
ALCAN 218
see ALCAN GB 218 1L
ALCAN 226
see ALCAN GB 226 1L
ALCAN 350
see ALCAN GB 350 1J
ALCAN B 116
see ALCAN GB B 116 1K
ALCAN B 320
see ALCAN GB B 320 1J
ALCAN C 117
see ALCAN GB C 117 1L
ALCAN DURALCOTE 1N
ALCAN GB 1S 1A
ALCAN GB 2S 1A
ALCAN GB 3S 1B
ALCAN GB 16S 1F
ALCAN GB 17S 1F
ALCAN GB 19S 1F
ALCAN GB 24S 1F
ALCAN GB 26S 1F
ALCAN GB 28S 1F
ALCAN GB 33S 1C
ALCAN GB 38S 1C
ALCAN GB 42S 1F
ALCAN GB 50S 1E
ALCAN GB 51S 1E
ALCAN GB 54S 1D
ALCAN GB 58S 1D
ALCAN GB 62S 1F
ALCAN GB 65S 1E
ALCAN GB 99.7 1A
ALCAN GB 99.8 1A
ALCAN GB 99.99 1A
ALCAN GB 100 1A
ALCAN GB 117 1L
ALCAN GB 125 1K
ALCAN GB 160 1C
ALCAN GB 162 1K
ALCAN GB 218 1L
ALCAN GB 226 1L
ALCAN GB 350 1J
ALCAN GB 730 1N
ALCAN GB A143 1L
ALCAN GB B1S 1A
ALCAN GB B19S 1F
ALCAN GB B26S 1F
ALCAN GB B33S 1C
ALCAN GB B51S 1E
ALCAN GB B75S 1G
ALCAN GB B116 1K
ALCAN GB B117 1L
ALCAN GB B320 1J
ALCAN GB C1S 1A
ALCAN GB C50S 1E
ALCAN GB C75S 1G
ALCAN GB C77S 1G
ALCAN GB C117 1L
ALCAN GB C125 1K
ALCAN GB D19S 1F
ALCAN GB D50S 1E
ALCAN GB D57S 1E
ALCAN GB D135 1K
ALCAN GB M75S 1G
ALCAN GB24S 1F
ALCAN GB24S ALCLAD 1H
ALCAN GB26S ALCLAD 1H
ALCAN GBA 56S 1D
ALCAN GBB 53S 1D
ALCAN GBD 54S 1D
ALCAN GBL 57S 1D
ALCAN GBM 57S 1D
ALCAN GBM75S ALCLAD 1H
ALCAN GBS 57S 1D
ALCHROME 750 20C
ALCHROME D 20C
ALCHROME DK 20C
ALCOA 050 1A
ALCOA 051 1A
ALCOA 102 1A
ALCOA 190 1B
ALCOA 460 1H
ALCOA 470 1H
ALCOA 510 1D
ALCOA 520 1D
ALCOA 540 1D
ALCOA 550 1D
ALCOA 660 1F
ALCOA 740 1G
ALCOA 760 1G
ALCOA 800 1F
ALCOA 910 1E
ALCOA 912 1E
ALCOA 918 1E
ALCOA 920 1E
ALCOA 940 1E
ALCOA 945 1E
ALCOA 946 1E
ALCODIE 44K1
ALCOMAX I 44L
ALCOMAX II 44L
ALCOMAX II Sc 44L
ALCOMAX III 44L
ALCOMAX III Sc 44L
ALCOMAX IV 44L
ALCOMAX IV Sc 44L
ALCONIT AN 0 44L

ALCONIT AN 16 44L
ALCRESS 44M1
Al Cu Mg 0.5 1F
Al Cu Mg 1 1F
Al Cu Mg 2 1F
Al Cu Mg Pb 1F
Al Cu Si Mn 1F
Al Cu 4 Mg 1.5 1F
Al Cu4 Si Mn 1F
Al Cu6 Bi Pb 1F
ALDAL 1L
ALDECOR 44K1
ALDREY 1E
ALDREY 051 1E
ALDUR 45/60 44A1
ALDUR 50 44A1
ALDUR 50/65 44A1
ALDUR 55 44A1
ALDUR 55/63 44A1
ALDUR 58 44A1
ALDUR 58/72 44A1
ALDURAL B 1H
ALDURAL G 1H
ALDURAL JJ 1H
ALDURAL K 1H
ALDURAL L 1H
ALDURAL Q 1F
ALDURAL S 1H
ALDURBRA 14B
ALFENIDE METAL 14F
ALFENOL 20A
ALFER 20A
ALFERIUM 1L
ALFORT 44A1
ALFREEZE I 44A1
ALFREEZE II 44A1
ALFRIG 5UV 44A1
ALFRIG Ni3 44F1
ALGIERS METAL 50A
ALGO-COR-TEN 44K1
ALGO-LOY 1315 44A1
ALGO-LOY 1317 44K1
ALGOMA 44K1
ALGOMA 90 44K1
ALGOTUF 50 44K1
ALHD 44K2
ALICIA 50A
ALKROTHAL 44M1
ALLEN'S METAL 14K
ALLOY 1 27F
ALLOY 1 44M2
ALLOY 1M 27F
ALLOY 2 27F
ALLOY 4 44M2
ALLOY 5X 44N2
ALLOY 6 13A
ALLOY 8 44L
ALLOY 9 13A
ALLOY 10 44K2
ALLOY 10 Mod 44M2
ALLOY 15 44L
ALLOY 18 44L
ALLOY 19 44M2
ALLOY 20 44N1
ALLOY 28 44N1
ALLOY 36 44F1
ALLOY 49 27C
ALLOY 53 44K1
ALLOY 63 13A
ALLOY 75 27B
ALLOY 80 27C
ALLOY 81 27C
ALLOY 82 27D
ALLOY 200 27A
ALLOY 201 27A
ALLOY 205 27A
ALLOY 255 44N3
ALLOY 288 44N3
ALLOY 306 44N1
ALLOY 310S 44N1
ALLOY 316LN 44N1
ALLOY 317LN 44N1
ALLOY 318LN 44N1
ALLOY 330 44N1
ALLOY 333 27C
ALLOY 400 14E
ALLOY 400 27D
ALLOY 600 27B
ALLOY 600 H 27B
ALLOY 600L 27F
ALLOY 601 27F
ALLOY 601 H 27B
ALLOY 617 27C
ALLOY 625 27F
ALLOY 625 H 27C
ALLOY 690 27B
ALLOY 690 27F
ALLOY 713 C 27C
ALLOY 718 27F
ALLOY 800 44N1
ALLOY 800H 44N1
ALLOY 800L 44N1
ALLOY 825 27F
ALLOY 825 h Mo 27F
ALLOY 904L 44N1
ALLOY 904LN 44N1
ALLOY 904RMo 44N1
ALLOY B 44K3
ALLOY B2 27F
ALLOY C 44M2
ALLOY C4 27F
ALLOY C22 27F
ALLOY Cu Ni 70/30 14E
ALLOY Cu Ni 90/10 14E
ALLOY CWPS 44M2
Alloy D 14G
ALLOY DS 44N1
Alloy E 14G
ALLOY G 27F
ALLOY G3 27F
ALLOY K 500 27D
ALLOY MOULD 44K1
ALLOY R 405 27D
ALLOY X 27C
ALL-STATE 8.60 27B
ALLULOY No. 1 21.1
ALLULOY No. 2 21.1
ALLVAC 500ZB 27C
ALLVAC 718 27C
ALLVAC GMR 235 27C
ALLVAC M 252 27C
ALLVAC R 41 27C
ALLVAC WASPALOY 27C
ALM 1B
ALMAR 18/200 44F3
ALMAR 18/250 44F3
ALMAR 18/300 44F3
ALMAR 20 44F3
ALMAR 25 44F3
Almar 362 S36200 44N3
ALMASILIUM 1E
ALMELEC 1K
ALMEN 305 55A
Al Mg 1 1D
Al Mg 2 1D
Al Mg 3 1D
Al Mg 3 Si 1D
Al Mg 5 1D
Al Mg Mn 1D
Al Mg Si 0.5 1E
Al Mg Si 1 1E
Al Mg Si Pb 1E
ALMINAL 2 1C
ALMINAL 3 1A
ALMINAL 4 1D
ALMINAL 5 1D
ALMINAL 6 1C
ALMINAL 6 1D
ALMINAL 8 1C
ALMINAL 9 1C
ALMINAL 9 1E
ALMINAL 10 1C
ALMINAL 10 1E
ALMINAL 11 1E
ALMINAL 11 1F
ALMINAL 15 1F
ALMINAL 20 1C
ALMINTRODE 96.0 1C
ALMINTRODE 96.10 1A
ALMINTRODE 96.20 1B
Al Mn 1B
ALNI 44F2
Al Ni 44F2
Al Ni 120 44F2
ALNI 1 44F2
ALNICO 44L
ALNICO 1A 44L
ALNICO 1B 44L
ALNICO 1C 44L
ALNICO 2 44L
ALNICO 2A 44L
ALNICO 2B 44L
ALNICO 2C 44L
ALNICO 2H 44L
ALNICO 3 20C
ALNICO 3 44F2
ALNICO 3A 44L
ALNICO 3B 44L
ALNICO 3C 44L
ALNICO 4 44L
ALNICO 4A 44L
ALNICO 4B 44L
ALNICO 5 44L
ALNICO 5/7 44L
ALNICO 5A 44L
ALNICO 5A,B,D and G 44L
ALNICO 5AB 44L
ALNICO 5B 44L
ALNICO 5B,D and G 44L
ALNICO 5DG 44L
ALNICO 5E 44L

ALNICO 6 44L
ALNICO 6A 44L
ALNICO 6B 44L
ALNICO 7 44L
ALNICO 7 44L
ALNICO 8 44L
ALNICO 8A 44L
ALNICO 8B 44L
ALNICO 9 44L
ALNICO 12 44L
ALNICO 160 44L
ALNICO 190 44L
ALNICO 220 44L
ALNICO 300 44L
ALNICO 350 44L
ALNICO 400 44L
ALNICO 450 44L
ALNICO 500 44L
ALNICO 580 44L
ALNICO 700 44L
ALNICO (high remanence) 20C
ALNICO I 44L
ALNICO I 44L
ALNICO II 44L
ALNICO III 44L
ALNI (high coercivity) 20C
ALOLINE 1A
ALPAX 1C
ALPAX BETA 1K
ALPAX GAMMA 1K
Al R Mg 0.5 1D
Al R Mg 1 1D
Al R Mg 2 1D
ALSIFER 20A
ALSIMIN 20A
ALTEMP S 816 13A
ALUCHROM W 44M1
ALUCROM 0 44M1
ALUFONT 3 1L
ALUFONT 42 1L
ALUFONT 47 1L
ALUMAGNESE 10 C 1D
ALUMAGNESE 20 V 1D
ALUMAGNESE 35 B 1D
ALUMAGNESE 45 1D
ALUMAGNESE 50J 1D
ALUMAGNESE CS 1D
ALUMAL 1B
ALUMAN 100 1B
ALUMBRO 14B
ALUMEL 27B
ALUMINAL 7 1L
ALUMINIUM BRASS 14B
ALUMOWELD 1A
ALUMOWELD 44A1
ALUM T 1A
ALZ 44K2
ALZEN 305K 55A
Al Zn 1 1A
Al Zn 4.5 Mg 1 1M
Al Zn Mg 3 1G
Al Zn Mg Cu 0.5 1G
Al Zn Mg Cu 1.5 1G
Al-Zr 1N
AM 14C
AM 44D
AM 1 1B
AM 3 44K2
A M4 1N
AM 4 44A2
AM 4 Ni 44F2
A M10 1N
AM 60 A 23B
AM 88 44P
AM 100A
see ASTM range 23B
AM 350 44N3
AM 355 44N3
AM 367 44L
AM 503 23B
AM 503 S 23B
AMALGAM 25.1
AMBRAC (1) 14F
AMBRAC (2) 14F
AMD 57 BX 27C
AMD 78 BU 52.1
AMERICAN BRASS 14B
AMERICAN GOLD 17.1
AMETHYST 52.1
AMG 3 1D
AMG05 1B
AMI 1B
AMI 44K1
AMIG 1B
AMI SPECIAL 44K1
A Mn 44A2
A Mn Mo 44G1
AMOLCTC 2 14K
AMPCO 8 14G
AMPCO 12 14G
AMPCO 15 14G
AMPCO 16 14G
AMPCO 18 14G
AMPCO 18/13 14G
AMPCO 18/22 14G
AMPCO 18/23 14G
AMPCO 20 14G
AMPCO 20/13 14G
AMPCO 21 14G
AMPCO 22 14G
AMPCO 24 14G
AMPCO 25 14G
AMPCO 501 27A
AMPCOLAY 45 14G
AMPCOLAY 90 14A
AMPCOLAY 405 14G
AMPCOLAY 483 14G
AMPCOLAY 495 14G
AMPCOLAY 502 27A
AMPCOLAY 531 27B
AMPCOLAY 551 27D
AMPCOLAY 552 27D
AMPCOLAY 553 27D
AMPCOLAY 557 27D
AMPCOLAY 558 27D
AMPCOLAY 559 27D
AMPCOLAY 561 27C
AMPCOLAY 570 14G
AMPCOLAY 900 14A
AMPCOLAY 901 14A
AMPCOLAY 910 14A
AMPCOLAY A1 14G
AMPCOLAY B2 14G
AMPCOLAY B2 (wrought) 14G
AMPCOLAY C3 14G
AMPCOLAY D4 14G
AMPCOLAY E5 14G
AMPCOLOY 31 14K
AMPCOLOY 32 14K
AMPCOLOY 35 14K
AMPCOLOY 38 14K
AMPCOLOY 54 14K
AMPCOLOY 62 14C
AMPCOLOY 64 14C
AMPCOLOY 66 14C
AMPCOLOY 71 14K
AMPCOLOY 72 14K
AMPCOLOY 74 14K
AMPCOLOY 79 14K
AMPCOLOY 83 14D
AMPCOLOY 84 14D
AMPCOLOY 96 14D
AMPCOLOY 521 14E
AMPCOLOY 522 14E
AMPCOLOY 525 14E
AMPCOLOY 526 14E
AMPCOLOY 666 14C
AMPCOLOY 711 14K
AMPCOLOY 712 14K
AMPCOLOY 715 14K
AMS 44N1
AMS 78 CD 26A
AMS 641 44K2
AMS 4000 B 1A
AMS 4001 B 1A
AMS 4003 B 1A
AMS 4004 1D
AMS 4005 1D
AMS 4006 B 1B
AMS 4007 1F
AMS 4008 C 1B
AMS 4009 1E
AMS 4010 1B
AMS 4011 1A
AMS 4014 1F
AMS 4015 E 1D
AMS 4016 E 1D
AMS 4017 E 1D
AMS 4018 1D
AMS 4019 1D
AMS 4020 1H
AMS 4021 B 1H
AMS 4022 C 1H
AMS 4023 C 1H
AMS 4024 A 1G
AMS 4025 D 1E
AMS 4026 D 1E
AMS 4027 E 1E
AMS 4028 A 1F
AMS 4029 A 1F
AMS 4031 1F
AMS 4033 A 1F
AMS 4034 A 1H
AMS 4035 E 1F
AMS 4036 1H
AMS 4037 F 1F
AMS 4038 1G
AMS 4039 1H
AMS 4040 E 1H
AMS 4041 G 1H
AMS 4042 E 1H

AMS 4043 1E
AMS 4044 C 1G
AMS 4045 C 1G
AMS 4046 1H
AMS 4047 B 1H
AMS 4048 C 1H
AMS 4049 C 1H
AMS 4050 1G
AMS 4051 A 1H
AMS 4052 A 1H
AMS 4053 1E
AMS 4054 1H
AMS 4055 1H
AMS 4056 B 1D
AMS 4057 A 1D
AMS 4058 A 1D
AMS 4059 B 1D
AMS 4062 C 1A
AMS 4063 1H
AMS 4064 1H
AMS 4065 B 1B
AMS 4066 1F
AMS 4067 B 1B
AMS 4069 1D
AMS 4070 F 1D
AMS 4071 F 1D
AMS 4077 1H
AMS 4078 1G
AMS 4079 1E
AMS 4080 E 1E
AMS 4081 A 1E
AMS 4082 E 1E
AMS 4083 1G
AMS 4083 D 1E
AMS 4084 1G
AMS 4086 F 1F
AMS 4087 C 1F
AMS 4088 E 1F
AMS 4089 1G
AMS 4090 1G
AMS 4091 1E
AMS 4092 1E
AMS 4093 1E
AMS 4094 1H
AMS 4095 1H
AMS 4096 1H
AMS 4100 1H
AMS 4101 1F
AMS 4102 A 1A
AMS 4107 1G
AMS 4108 1G
AMS 4111 1G
AMS 4112 1F
AMS 4113 1E
AMS 4114 B 1D
AMS 4115 1E
AMS 4116 A 1E
AMS 4117 A 1E
AMS 4118 C 1F
AMS 4119 A 1F
AMS 4120 F 1F
AMS 4121 C 1F
AMS 4122 C 1G
AMS 4123 A 1G
AMS 4124 1G
AMS 4125 E 1E
AMS 4126 1G
AMS 4127 B 1E
AMS 4128 1E
AMS 4130 G 1F
AMS 4131 1G
AMS 4132 A 1F
AMS 4133 1F
AMS 4134 A 1F
AMS 4135 J 1F
AMS 4136 1G
AMS 4137 A 1G
AMS 4138 1G
AMS 4139 F 1G
AMS 4140 D 1F
AMS 4141 1G
AMS 4142 D 1F
AMS 4143 1F
AMS 4145 E 1C
AMS 4146 1E
AMS 4147 1G
AMS 4148 1G
AMS 4149 1G
AMS 4150 C 1E
AMS 4152 G 1F
AMS 4153 B 1F
AMS 4154 F 1G
AMS 4155 A 1E
AMS 4156 C 1E
AMS 4158 A 1G
AMS 4160 1E
AMS 4161 1E
AMS 4162 1F
AMS 4163 1F
AMS 4164 C 1F
AMS 4165 C 1F
AMS 4166 1G
AMS 4167 1G
AMS 4168 A 1G
AMS 4169 B 1G
AMS 4170 1G
AMS 4171 A 1G
AMS 4172 1E
AMS 4173 1E
AMS 4174 1G
AMS 4175 1D
AMS 4176 1D
AMS 4177 1D
AMS 4178 1D
AMS 4179 1G
AMS 4180 B 1A
AMS 4181 1C
AMS 4182 A 1D
AMS 4184 B 1C
AMS 4185 A 1C
AMS 4186 1G
AMS 4187 1G
AMS 4189 1C
AMS 4190 A 1C
AMS 4191 A 1F
AMS 4192 1F
AMS 4193 1F
AMS 4194 1F
AMS 4195 1F
AMS 4196 1H
AMS 4197 1H
AMS 4198 1H
AMS 4199 1H
AMS 4200 1G
AMS 4201 1G
AMS 4202 1G
AMS 4203 1G
AMS 4204 1G
AMS 4205 1G
AMS 4207 1G
AMS 4208 1A
AMS 4209 1A
AMS 4210 F 1K
AMS 4212 E 1K
AMS 4214 D 1K
AMS 4215 B 1K
AMS 4217 D 1K
AMS 4218 B 1K
AMS 4219 1C
AMS 4220 D 1L
AMS 4221 1F
AMS 4222 D 1L
AMS 4223 1L
AMS 4224 1L
AMS 4225 1L
AMS 4226 1L
AMS 4227 A 1L
AMS 4228 1L
AMS 4229 1L
AMS 4230 C 1L
AMS 4231 C 1L
AMS 4235 1L
AMS 4236 1L
AMS 4237 1L
AMS 4238 A 1J
AMS 4239 1J
AMS 4240 C 1J
AMS 4241 1C
AMS 4242 1L
AMS 4243 1H
AMS 4247 1G
AMS 4248 1C
AMS 4260 1K
AMS 4261 1C
AMS 4275B 1N
AMS 4280 E 1K
AMS 4281 C 1K
AMS 4282 E 1L
AMS 4283 D 1L
AMS 4284 D 1K
AMS 4285 1K
AMS 4286 A 1K
AMS 4290 1K
AMS 4291 B 1L
AMS 4306 1G
AMS 4307 1G
AMS 4310 1G
AMS 4311 1G
AMS 4312 1E
AMS 4313 1F
AMS 4314 1F
AMS 4320 1G
AMS 4321 1G
AMS 4323 1G
AMS 4333 1G
AMS 4340 1G
AMS 4341 1G
AMS 4342 1G
AMS 4343 1G
AMS 4344 1G
AMS 4346 1F

AMS 4347 1E
AMS 4348 1D
AMS 4349 1D
AMS 4350 F 23B
AMS 4351 1F
AMS 4352 A 23B
AMS 4358 A 23B
AMS 4360 C 23B
AMS 4362 23B
AMS 4363 23B
AMS 4375 23B
AMS 4375 D 23B
AMS 4376 A 23B
AMS 4377 B 23B
AMS 4382 23B
AMS 4383 23B
AMS 4384 B 23B
AMS 4385 C 23B
AMS 4386 23B
AMS 4387 23B
AMS 4388 23B
AMS 4389 A 23B
AMS 4390 B 23B
AMS 4395 A 23B
AMS 4396 23B
AMS 4397 23B
AMS 4418 A 23B
AMS 4419 23B
AMS 4420 H 23B
AMS 4422 J 23B
AMS 4424 F 23B
AMS 4425 23B
AMS 4434 F 23B
AMS 4437 23B
AMS 4437 B 23B
AMS 4438 23B
AMS 4439 23B
AMS 4440 A 23B
AMS 4441 A 23B
AMS 4442 A 23B
AMS 4442 B 23B
AMS 4443 A 23B
AMS 4443 B 23B
AMS 4444 23B
AMS 4445 A 23B
AMS 4446 23B
AMS 4447 23B
AMS 4452 23B
AMS 4453 23B
AMS 4455 23B
AMS 4483 23B
AMS 4484 23B
AMS 4484 E 23B
AMS 4490 D 23B
AMS 4500 D 14A
AMS 4501 14A
AMS 4505 D 14B
AMS 4507 C 14B
AMS 4508 B 14B
AMS 4510 C 14K
AMS 4520 E 14K
AMS 4530 B 14D
AMS 4532 A 14D
AMS 4544 B 27D
AMS 4555 C 14B
AMS 4558 B 14B
AMS 4574 B 27D
AMS 4575 27D
AMS 4602 14A
AMS 4610 H 14B
AMS 4611 C 14B
AMS 4612 D 14B
AMS 4614 D 14B
AMS 4615 C 14M
AMS 4616 B 14M
AMS 4619 B 14C
AMS 4625 D 14K
AMS 4630 E 14G
AMS 4631 C 14G
AMS 4632 C 14G
AMS 4635 B 14G
AMS 4640 C 14G
AMS 4650 D 14D
AMS 4651 14D
AMS 4665 A 14M
AMS 4674 C 27D
AMS 4675 27D
AMS 4675 B 27D
AMS 4676 27D
AMS 4677 27D
AMS 4700 14A
AMS 4701 B 14A
AMS 4710 14B
AMS 4712 A 14B
AMS 4713 A 14B
AMS 4720 B 14K
AMS 4725 A 14D
AMS 4730C 27D
AMS 4731 27D
AMS 4732 14E
AMS 4750A 21.1
AMS 4751 50A
AMS 4755A 21.1
AMS 4756 21.1
AMS 4764 14N
AMS 4765 42.1
AMS 4766 A 42.1
AMS 4767 42.1
AMS 4768 42.1
AMS 4769 42.1
AMS 4770 B 42.1
AMS 4771 42.1
AMS 4772 A 42.1
AMS 4773 42.1
AMS 4774 42.1
AMS 4775 A 27B
AMS 4776 27B
AMS 4777 27B
AMS 4778 A 27F
AMS 4779 27F
AMS 4780 24.1
AMS 4782 27F
AMS 4783 13A
AMS 4784 17.1
AMS 4785 17.1
AMS 4786 17.1
AMS 4787 17.1
AMS 4800 A 50A
AMS 4803A 55A
AMS 4805 B 14J
AMS 4824 14K
AMS 4827 14K
AMS 4840 A 14K
AMS 4842 A 14K
AMS 4845 D 14K
AMS 4846 A 14K
AMS 4855 B 14K
AMS 4860 A 14C
AMS 4862 B 14B
AMS 4870 B 14G
AMS 4871 B 14G
AMS 4872 B 14G
AMS 4873 A 14G
AMS 4880 14G
AMS 4881 14G
AMS 4890 A 14D
AMS 4892 27D
AMS 4893 27D
AMS 4900 51A
AMS 4900 B 51A
AMS 4901 C 51A
AMS 4902 A 51A
AMS 4905 51B
AMS 4906 51B
AMS 4907 51B
AMS 4908 51B
AMS 4908 A 51B
AMS 4909 51B
AMS 4910 51B
AMS 4911 A 51C
AMS 4912 51B
AMS 4912 51C
AMS 4913 51C
AMS 4915 51B
AMS 4916 51B
AMS 4917 51C
AMS 4918 51B
AMS 4919 51B
AMS 4920 51B
AMS 4921 A 51A
AMS 4923 51B
AMS 4924 51B
AMS 4925 A 51B
AMS 4926 51B
AMS 4927 51B
AMS 4928 51B
AMS 4928 A 51C
AMS 4929 51B
AMS 4929 51C
AMS 4930 51B
AMS 4931 51B
AMS 4933 51B
AMS 4934 51B
AMS 4935 51C
AMS 4936 51B
AMS 4941 51A
AMS 4942 51A
AMS 4943 51B
AMS 4944 51B
AMS 4951 51A
AMS 4951 A 51A
AMS 4952 51B
AMS 4953 51B
AMS 4954 51C
AMS 4954 A 51C
AMS 4955 51B
AMS 4955 51C
AMS 4956 51C
AMS 4957 51C
AMS 4958 51C
AMS 4959 51B

AMS 4965 51B
AMS 4966 51B
AMS 4967 A 51B
AMS 4968 51B
AMS 4969 51C
AMS 4970 51B
AMS 4971 51C
AMS 4972 51B
AMS 4972 51C
AMS 4973 51C
AMS 4974 51B
AMS 4975 51C
AMS 4976 51C
AMS 4977 51C
AMS 4978 51C
AMS 4979 51C
AMS 4980 51C
AMS 4981 51C
AMS 4982 51B
AMS 4984 51B
AMS 4985 51B
AMS 4986 51B
AMS 4987 51B
AMS 4991 51B
AMS 4993 51B
AMS 4995 51C
AMS 4996 51B
AMS 4997 51C
AMS 4998 51B
AMS 5010 D 44A1
AMS 5020 44A2
AMS 5022 F 44A1
AMS 5024 C 44A2
AMS 5027 44K1
AMS 5028 44K1
AMS 5029 44K1
AMS 5030 A 44A1
AMS 5031 44A1
AMS 5032 A 44A1
AMS 5036 D 44A1
AMS 5040 E 44A1
AMS 5041 44A1
AMS 5042 E 44A1
AMS 5044 C 44A1
AMS 5045 B 44A1
AMS 5046 44A1
AMS 5047 44A1
AMS 5050 E 44A1
AMS 5053 B 44A1
AMS 5060 B 44A1
AMS 5061 A 44A1
AMS 5062 A 44A2
AMS 5069 44A1
AMS 5070 B 44A2
AMS 5075 A 44A2
AMS 5077 A 44A2
AMS 5080 A 44A2
AMS 5082 44A2
AMS 5085 44A3
AMS 5110 A 44A3
AMS 5112E 44A4
AMS 5115 B 44A3
AMS 5120 D 44A3
AMS 5121B 44A4
AMS 5122B 44A4
AMS 5132C 44A4
AMS 5221 20C
AMS 5221 A 44N2
AMS 5223 A 20C
AMS 5225 20C
AMS 5225 A 20C
AMS 5310 B 20B
AMS 5315 20B
AMS 5316 20B
AMS 5328 20B
AMS 5329 20B
AMS 5330 20B
AMS 5331 20B
AMS 5333 20B
AMS 5333 44K1
AMS 5334A 20B
AMS 5335A 20B
AMS 5336 20B
AMS 5337 44F3
AMS 5338 20B
AMS 5339 44F3
AMS 5340 44M1
AMS 5340 44N3
AMS 5341 44N1
AMS 5342 44M1
AMS 5343 44M1
AMS 5344 44M1
AMS 5346 J 92110 44N3
AMS 5347 44N3
AMS 5348 44N3
AMS 5349 44M1
AMS 5350 44M1
AMS 5350 D 44M1
AMS 5351 44M1
AMS 5351 B 44M1
AMS 5352 44M2
AMS 5352 A 44M2
AMS 5353 44M1
AMS 5354 44M1
AMS 5354 B 44M1
AMS 5355 44N3
AMS 5355 A 44M1
AMS 5356 44N3
AMS 5357 44N3
AMS 5358 44N1
AMS 5359 44M1
AMS 5359 44N3
AMS 5360 44N1
AMS 5360 B 44N1
AMS 5361 44N1
AMS 5361 B 44N1
AMS 5362 44N1
AMS 5362 D 44N1
AMS 5363 44N1
AMS 5363 B 44N1
AMS 5364 44N1
AMS 5365 44N1
AMS 5365 A 44N1
AMS 5366 44N1
AMS 5366 A 44N1
AMS 5368 44M1
AMS 5368 44N1
AMS 5369 A 44N1
AMS 5370 44N1
AMS 5371 44N1
AMS 5372 44M1
AMS 5373 13A
AMS 5375 B 13A
AMS 5376 B 44L
AMS 5378 B 13A
AMS 5380 C 13A
AMS 5382 13A
AMS 5383 27C
AMS 5384 27C
AMS 5385 13A
AMS 5385 C 13A
AMS 5387 13A
AMS 5388 27C
AMS 5389 27C
AMS 5390 27C
AMS 5391 27C
AMS 5392 G 20C
AMS 5393 B 20C
AMS 5394 20C
AMS 5395 20C
AMS 5396 27C
AMS 5397 27C
AMS 5398 A 44M1
AMS 5399 27C
AMS 5400 44N3
AMS 5401 27C
AMS 5402 27C
AMS 5404 27C
AMS 5405 27C
AMS 5406 27C
AMS 5407 27C
AMS 5501 44N1
AMS 5502 A 44K1
AMS 5503 44M1
AMS 5504 44M1
AMS 5504 D 44M1
AMS 5505 44M1
AMS 5506 A 44M1
AMS 5507 44N1
AMS 5508 44M1
AMS 5508 A 44M1
AMS 5509 27C
AMS 5510 44N1
AMS 5510 H 44N1
AMS 5511 44N1
AMS 5511 A 44N1
AMS 5512 B 44N1
AMS 5513 44N1
AMS 5514 A 44N1
AMS 5515 44N1
AMS 5515 D 44N1
AMS 5516 44N1
AMS 5516 E 44N1
AMS 5517 D 44N1
AMS 5518 C 44N1
AMS 5519 E 44N1
AMS 5520A 44N3
AMS 5521 44N1
AMS 5521 B 44N1
AMS 5522 44N1
AMS 5522 B 44N1
AMS 5523 44N1
AMS 5524 44N1
AMS 5524 B 44N1
AMS 5525 B 44N2
AMS 5526 C 44N2
AMS 5527 44N1
AMS 5527 A 44N2
AMS 5528 44N3
AMS 5528A 44N3
AMS 5529A 44N3

AMS 5530 27C
AMS 5530 C 13A
AMS 5531 44L
AMS 5531 44N1
AMS 5532 B 44L
AMS 5533 44N1
AMS 5533 A 44L
AMS 5534 13A
AMS 5536 27C
AMS 5536 C 27C
AMS 5537 13A
AMS 5538 44N1
AMS 5538 44N2
AMS 5539 44N2
AMS 5540 F 27B
AMS 5541 27C
AMS 5541 A 27C
AMS 5541A 44N3
AMS 5542 C 27C
AMS 5543 44N1
AMS 5543 44N2
AMS 5544 27C
AMS 5545 27C
AMS 5546 44N3
AMS 5547 44N3
AMS 5548A 44N3
AMS 5549A 44N3
AMS 5550 27C
AMS 5550A 27C
AMS 5551 27C
AMS 5552 44N1
AMS 5552 A 44N1
AMS 5554 44N3
AMS 5556 A 44N1
AMS 5557 44N1
AMS 5557 A 44N1
AMS 5558 44N1
AMS 5559 A 44N1
AMS 5560 44N1
AMS 5560 D 44N1
AMS 5561 44P
AMS 5562 44P
AMS 5563 44N1
AMS 5564 44N1
AMS 5565 44N1
AMS 5565 D 44N1
AMS 5566 D 44N1
AMS 5567 44N1
AMS 5568 44N3
AMS 5570 44N1
AMS 5570 G 44N1
AMS 5571 44N1
AMS 5571 B 44N1
AMS 5572 44N1
AMS 5572 B 44N1
AMS 5573 44M1
AMS 5573 C 44N1
AMS 5574 44N1
AMS 5575 44N1
AMS 5575 F 44N1
AMS 5576 44N1
AMS 5576 C 44N1
AMS 5577 A 44N1
AMS 5579 44N1
AMS 5580 C 27B
AMS 5581 27C
AMS 5582 27C
AMS 5583 27C
AMS 5585 44L
AMS 5586 27C
AMS 5587 27C
AMS 5588 27C
AMS 5589 27C
AMS 5590 27C
AMS 5591 44M1
AMS 5591 D 44M1
AMS 5592 44N1
AMS 5593 27C
AMS 5594 44N3
AMS 5595 44P
AMS 5596 27C
AMS 5597 27C
AMS 5598 27C
AMS 5599 27C
AMS 5600 44N1
AMS 5601 44N3
AMS 5602 A 44K1
AMS 5603 44N3
AMS 5604 44N3
AMS 5605 27E
AMS 5606 27E
AMS 5607 27F
AMS 5608 13A
AMS 5609 44M1
AMS 5610 E 44M1
AMS 5611 44M1
AMS 5612 44M1
AMS 5613 44M1
AMS 5613 E 44M1
AMS 5614 44M1
AMS 5615 B 44M1
AMS 5616 44M1
AMS 5616 D 44M1
AMS 5617 44N3
AMS 5618 44M2
AMS 5620 D 44M1
AMS 5621 44M1
AMS 5622 44N3
AMS 5623 44P
AMS 5624 A 44P
AMS 5624 C 44N1
AMS 5625 A 44P
AMS 5626A 44K3
AMS 5627 44M1
AMS 5628 B 44M1
AMS 5629 44N3
AMS 5630 44M2
AMS 5630 C 44M2
AMS 5631 44M1
AMS 5632 44M2
AMS 5632 B 44M2
AMS 5633 27E
AMS 5634 27E
AMS 5635 44N1
AMS 5636 44N1
AMS 5636 A 44N1
AMS 5637 44N1
AMS 5637 A 44N1
AMS 5639 44N1
AMS 5639 A 44N1
AMS 5640 44N1
AMS 5640 F 44N1
AMS 5641 A 44N1
AMS 5642 44N1
AMS 5643F 44N3
AMS 5644B 44N3
AMS 5645 G 44N1
AMS 5646 44N1
AMS 5646 E 44N1
AMS 5647 44N1
AMS 5647 A 44N1
AMS 5648 44N1
AMS 5648 C 44N1
AMS 5649 44N1
AMS 5650 44N1
AMS 5650 A 44N1
AMS 5651 44N1
AMS 5651 D 44N1
AMS 5652 B 44N1
AMS 5653 44N1
AMS 5654 44N1
AMS 5655 44M1
AMS 5656 44P
AMS 5657 44N3
AMS 5658 SI5500 44N3
AMS 5659 44N3
AMS 5660 27C
AMS 5660 A 27C
AMS 5661 27C
AMS 5662 B 26A
AMS 5663 B 26A
AMS 5664 A 26A
AMS 5665 F 27B
AMS 5666 27C
AMS 5667 27C
AMS 5668 D 27C
AMS 5669 27C
AMS 5671 27C
AMS 5672 44N3
AMS 5673A 44N3
AMS 5674 44N1
AMS 5675 27C
AMS 5676 27B
AMS 5677 27B
AMS 5678 44N3
AMS 5679 B 27B
AMS 5680 44K1
AMS 5680 44N1
AMS 5680 B 44N1
AMS 5681 44A1
AMS 5681 A 44N1
AMS 5682 A 27B
AMS 5683 B 27B
AMS 5684 B 27B
AMS 5685 C 44N1
AMS 5686 A 44N1
AMS 5687 C 27B
AMS 5688 44N1
AMS 5688 D 44N1
AMS 5689 44N1
AMS 5690 44N1
AMS 5690 E 44N1
AMS 5691 44N1
AMS 5691 B 44N1
AMS 5692 44N1
AMS 5693 44N1
AMS 5694 44N1
AMS 5694 B 44N1
AMS 5695 A 44N1
AMS 5696 44N1
AMS 5697 44N1

AMS 5698	27C
AMS 5699 B	27C
AMS 5700 B	44N1
AMS 5701	27E
AMS 5702	27E
AMS 5703	27E
AMS 5704	27C
AMS 5705 A	44N1
AMS 5706	27C
AMS 5707 A	27C
AMS 5708	27C
AMS 5709	27C
AMS 5710 B	44M1
AMS 5711	27C
AMS 5712	27C
AMS 5713	27C
AMS 5714	27C
AMS 5715	27C
AMS 5716	27B
AMS 5716	44N1
AMS 5717	27C
AMS 5718	44N3
AMS 5719	44M1
AMS 5721 B	44N2
AMS 5722	44N1
AMS 5722 A	44N2
AMS 5723	44N1
AMS 5723 A	44N2
AMS 5724	44N2
AMS 5725 A	44N1
AMS 5726	44N2
AMS 5727	44N1
AMS 5727	44N3
AMS 5727 B	44N1
AMS 5728 B	44N1
AMS 5729	44N2
AMS 5731 A	44N2
AMS 5731-32	44N2
AMS 5732 A	44N2
AMS 5733	44N1
AMS 5733 C	44N1
AMS 5734	44N2
AMS 5735 E	44N2
AMS 5736 B	44N2
AMS 5737 B	44N2
AMS 5738	44N1
AMS 5739	44N3
AMS 5740	44N3
AMS 5741	44N2
AMS 5741 C	44N1
AMS 5742 A	44N1
AMS 5743B	44N3
AMS 5744	44N3
AMS 5745	44N3
AMS 5746	27C
AMS 5747	27C
AMS 5748	44L
AMS 5749	44M2
AMS 5750	27C
AMS 5751	27C
AMS 5753	27C
AMS 5754D	27C
AMS 5755	27C
AMS 5756	27C
AMS 5757	27C
AMS 5758	13A
AMS 5759	13A
AMS 5759 B	13A
AMS 5761	44L
AMS 5762	44N3
AMS 5763	44N3
AMS 5764	44N1
AMS 5765	13A
AMS 5766	44N1
AMS 5768	44L
AMS 5768 E	44L
AMS 5769	44L
AMS 5770	44L
AMS 5770 B	44L
AMS 5771	27F
AMS 5772	13A
AMS 5773	44N3
AMS 5774	44N3
AMS 5775	44N3
AMS 5776	44M1
AMS 5777	44M1
AMS 5778	27C
AMS 5779	27C
AMS 5780	44N3
AMS 5781	44N3
AMS 5782 A	44N2
AMS 5783 B	44N2
AMS 5784	44N1
AMS 5785 B	44N1
AMS 5786	27C
AMS 5787	27C
AMS 5788	13A
AMS 5789	13A
AMS 5790	44K1
AMS 5794 A	44L
AMS 5795 B	44L
AMS 5796	13A
AMS 5797	13A
AMS 5798	27C
AMS 5799	27C
AMS 5800 A	27C
AMS 5801	13A
AMS 5804 A	44N2
AMS 5805 A	44N2
AMS 5812	44N3
AMS 5812B	44N3
AMS 5813A	44N3
AMS 5817	44M1
AMS 5817 A	44M1
AMS 5821	44M1
AMS 5821 A	44M1
AMS 5822	44M1
AMS 5823	44M1
AMS 5824 S17780	44N3
AMS 5825A	44N3
AMS 5826	44N3
AMS 5827	44N3
AMS 5828	27C
AMS 5829	27C
AMS 5832	27C
AMS 5833	13A
AMS 5834	13A
AMS 5837	27C
AMS 5838	27C
AMS 5840	44N3
AMS 5841	13A
AMS 5842	13A
AMS 5843	13A
AMS 5844	13A
AMS 5845	13A
AMS 5846	27C
AMS 5848	44P
AMS 5851	27C
AMS 5852	27C
AMS 5853	44N2
AMS 5855	27C
AMS 5856	27C
AMS 5858	44N2
AMS 5859	44N3
AMS 5860	44N3
AMS 5862 515500	44N3
AMS 5863	44N3
AMS 5864	44N3
AMS 5865	27A
AMS 5870	27B
AMS 5870	44N3
AMS 5871	44N1
AMS 5872	27C
AMS 5873	27F
AMS 5874	44L
AMS 5875	13A
AMS 5876	13A
AMS 5880	44M2
AMS 5881	27C
AMS 5882	27C
AMS 5886	27C
AMS 5890	27A
AMS 5895	44N2
AMS 5900	44M2
AMS 6231 B	44K1
AMS 6242C	44F1
AMS 6250 E	44K1
AMS 6255	44K1
AMS 6256	44K1
AMS 6260 F	44K1
AMS 6263 C	44K1
AMS 6264 C	44K1
AMS 6265 B	44K1
AMS 6266 B	44K1
AMS 6267	44K1
AMS 6270 F	44K1
AMS 6272 D	44K1
AMS 6274 F	44K1
AMS 6275 A	44K1
AMS 6276 B	44K1
AMS 6277	44K1
AMS 6280 D	44K1
AMS 6281	44K1
AMS 6282 C	44K1
AMS 6290 C	44K1
AMS 6292 C	44K1
AMS 6294 C	44K1
AMS 6299	44K1
AMS 6300	44G1
AMS 6302	44K1
AMS 6302 A	44K1
AMS 6303	44K1
AMS 6304 B	44K2
AMS 6305	44K2
AMS 6307	44K2
AMS 6308	44K1
AMS 6312 A	44K2
AMS 6317 B	44K2
AMS 6320 E	44K1
AMS 6321	44K2
AMS 6322 E	44K2

AMS 6323 C	44K2	AMS 6439	44K2	AMS 7234	27D
AMS 6324 B	44K2	AMS 6440D	44E3	AMS 7235	44N2
AMS 6325 C	44K2	AMS 6441B	44E3	AMS 7236	44N1
AMS 6327 C	44K2	AMS 6442B	44E3	AMS 7237	27C
AMS 6328 D	44K2	AMS 6443B	44E3	AMS 7238	13A
AMS 6330 A	44K1	AMS 6444B	44E3	AMS 7240	44A3
AMS 6342 C	44K2	AMS 6445 B	44K2	AMS 7241	44N1
AMS 6348	44K1	AMS 6445A	44E3	AMS 7245	44N1
AMS 6349	44K1	AMS 6446	44E3	AMS 7246	27C
AMS 6350 C	44K1	AMS 6447	44E3	AMS 7301	44K2
AMS 6351	44K2	AMS 6448 B	44K2	AMS 7304	44A4
AMS 6352 B	44K1	AMS 6449	44E3	AMS 7320	14K
AMS 6354	44K1	AMS 6450 B	44K2	AMS 7322	14K
AMS 6355 F	44K1	AMS 6454	44K2	AMS 7445	44M1
AMS 6356	44K1	AMS 6455	44K2	AMS 7445	44M2
AMS 6357 C	44K1	AMS 6457	44K1	AMS 7452	44K2
AMS 6358 A	44K2	AMS 6458	44K1	AMS 7454	44K2
AMS 6359 A	44K2	AMS 6458 B	44K1	AMS 7455	44K2
AMS 6360 D	44K1	AMS 6459	44K1	AMS 7459	44K2
AMS 6361	44K1	AMS 6460	44K1	AMS 7460	51B
AMS 6362	44K1	AMS 6461 B	44K1	AMS 7461	51B
AMS 6365 C	44K1	AMS 6462 B	44K1	AMS 7464	44K1
AMS 6370 E	44K1	AMS 6463	44L	AMS 7468	13A
AMS 6371 C	44K1	AMS 6464	44K1	AMS 7469	27C
AMS 6372 C	44K1	AMS 6465	44L	AMS 7470	44M1
AMS 6373	44K1	AMS 6466	44K1	AMS 7471	27C
AMS 6374	44K1	AMS 6467	44K1	AMS 7472	44N1
AMS 6378	44K2	AMS 6468	44L	AMS 7474	44N3
AMS 6379	44K2	AMS 6470 E	44K2	AMS 7475	13A
AMS 6381 A	44K2	AMS 6471	44K2	AMS 7477	44N2
AMS 6382 E	44K2	AMS 6472	44K2	AMS 7478	44N2
AMS 6385	44K1	AMS 6475 B	44K1	AMS 7479	44N2
AMS 6386	44G1	AMS 6485	44K1	AMS 7481	44N2
AMS 6386	44K1	AMS 6485 A	44K2	AMS 7490	27C
AMS 6390	44K2	AMS 6487 B	44K2	AMS 7490	44N1
AMS 6395	44K1	AMS 6488	44K2	AMS 7493	44M1
AMS 6396	44K2	AMS 6490	44K3	AMS 7496	44A1
AMS 6406	44K2	AMS 6490A	44K3	AMS 7496	44K1
AMS 6407 A	44K1	AMS 6491	44K3	AMS 7496	51B
AMS 6408	44K2	AMS 6501	44F3	AMS 7701	20C
AMS 6409	44K1	AMS 6512	44F3	AMS 7702	20C
AMS 6411	44K1	AMS 6514	44F3	AMS 7705	20C
AMS 6412 E	44K1	AMS 6520	44F3	AMS 7706	20A
AMS 6413 C	44K1	AMS 6521	44F3	AMS 7707	20A
AMS 6414	44K2	AMS 6522	44L	AMS 7711	44B1
AMS 6415 F	44K2	AMS 6523	44L	AMS 7712	44B1
AMS 6416	44K2	AMS 6524	44L	AMS 7714	44B1
AMS 6418 B	44K1	AMS 6525	44L	AMS 7715	44B1
AMS 6419	44K2	AMS 6526	44L	AMS 7717	20C
AMS 6421	44K1	AMS 6527	44L	AMS 7718	20C
AMS 6422 B	44K2	AMS 6528	44K1	AMS 7719	20C
AMS 6423	44K2	AMS 6529	44K2	AMS 7720	21.1
AMS 6424	44K2	AMS 6530 D	44K1	AMS 7721	21.1
AMS 6426	44K3	AMS 6535 C	44K1	AMS 7725	52.1
AMS 6427 C	44K1	AMS 6543	44L	AMS 7726 A	20C
AMS 6428	44K1	AMS 6544	44L	AMS 7727	20C
AMS 6429	44K1	AMS 6546	44K1	AMS 7728 A	20C
AMS 6430	44K1	AMS 6550 D	44K1	AMS 7730	53.1
AMS 6431	44K2	AMS 7207	44M1	AMS 7731	17.1
AMS 6432	44K2	AMS 7210	44N1	AMS 7734	20C
AMS 6433	44K1	AMS 7211	44N1	AMS 7735	30.1
AMS 6434	44K1	AMS 7225	44A1	AMS 7800	26A
AMS 6435	44K1	AMS 7228	44N1	AMS 7801	26A
AMS 6436	44K1	AMS 7229	44N1	AMS 7805	26A
AMS 6437	44K1	AMS 7232	27B	AMS 7806	26A
AMS 6437 A	44K2	AMS 7232	27F	AMS 7807	26A
AMS 6438	44K2	AMS 7233	27D	AMS 7811 A	26A

AMS 7813 26A
AMS 7817 26A
AMS 7819 26A
AMS 7846 46.1
AMS 7847 46.1
AMS 7848 46.1
AMS 7849 46.1
AMS 7850 28.1
AMS 7851 28.1
AMS 7852 28.1
AMS 7855 28.1
AMS 7857 28.1
AMS 7897 52.1
AMS 7898 52.1
AMS 7899 52.1
AMS 7900 5.1
AMS 7901 5.1
AMS 7902 5.1
AMSL 44N1
AMST 44N1
AMTROD 18.01 1A
AN 1 27B
AN 1 A 44K1
AN 1 B 44K1
AN 2 27B
AN 2 44K3
A N20 1N
AN A4 1L
AN A5 1L
ANA 8 1F
ANA 12.1 1F
ANA 13 1F
AN A17 1M
ANACOS 14A
ANACOS CADMIUM 14A
ANACOS SILVER 14A
ANATOMICAL ALLOY 6.1
AN B 16 14G
ANC 1 44K1
ANC 1A
see BS 3146 44M1
ANC 1B
see BS 3146 44M1
ANC 1C
see BS 3146 44M1
ANC 2
see BS 3146 44M1
ANC 3A
see BS range 44N1
ANC 3B Nb
see BS range 44N1
ANC 4A
see BS range 44N1
ANC 4B
see BS range 44N1
ANC 4C Nb
see BS range 44N1
ANC 5A
see BS range 44N1
ANC 5B
see BS range 44N1
ANC 5C 27B
ANC 6A
see BS range 44N1
ANC 6B
see BS range 44N1
ANC 7
see BS range 44N1
ANC 8 27B
ANC 9
see BS range 27C
ANC 10
see BS range 27C
ANC 11
see BS range 27C
ANC 12
see BS range 27C
ANC 13 13A
ANC 14 13A
ANC 15
see BS range 27F
ANC 16
see BS range 27C
ANC 17
see BS range 27F
ANC 18A
see BS range 27D
ANC 18B
see BS range 27D
ANC 18C
see BS range 27D
ANCM 44K1
A Nickel 27A
ANKA 44N1
ANKA E 44N1
ANKA M 44N1
AN M 16 23B
AN N4 27B
ANORINOX 35 AM 44M1
ANORMAT 17.1
AN QQ A 366 1J
AN QQ A 376 1K
AN QQ A 379 1L
AN QQ A 386 1K
AN QQ A 392 1J
AN QQ A 394 1K
AN QQ A 402 1J
AN QQ A/666/1 44A4
AN QQ A383 1L
AN QQ A390 1L
AN QQ A397 1L
AN QQ A399 1L
ANQQ A-405 1C
AN QQ B 646 14C
AN QQ B 672 14G
AN QQ M 56/1A 23B
AN QQ M 56/1C 23B
AN QQ N 268 27B
AN QQ N 271 27B
AN QQ S 646 1 44A1
AN QQ S 676/1 44F1
AN QQ S684a 44K1
AN QQ S685/4 44K1
AN QQ S686/1 44K1
AN QQ S 689/1 44F1
AN QQ S690/1 44K2
AN QQ S752a 44K2
AN QQ S756a 44K2
AN QQ S 757/3 44N3
AN QQ S770/2/1 44M1
AN QQ S770/2/11 44M1
AN QQ S 771/4 FM 44N3
AN QQ S 771/4G 44N3
AN QQ S 771/4 MCR 44N3
AN QQ S 772 G 44N3
AN QQ S 772/MCR 44N1
AN QQ W 298/2 1F
AN QQ W 423/1 44N1
AN QQ W 429/1 44A2
AN QQ W 435/3 44A1
AN QQ W 441/3 44A4
AN S 4 44A2
AN S5 44A4
AN S 9/1 44G1
AN S 11 44A1
AN S 12/2 44K1
AN S 13a 44K1
AN S 14a 44K1
AN S 15a 44K1
AN S 16a 44K2
AN S 22/1 44K1
ANS A5 15 E Ni Ci 27A
ANSCOL 50 44A1
ANSEMAX 44L
ANSI B125.1
see ASTMA53 44A1
ANSI B125.16
see ASTMA312 44N1
ANSI B125.17
see ASTMA333 44K1
ANSI B125.24
see ASTMA335 44K1
ANSI B125.25
see ASTMA376 44N1
ANSI B125.26
see ASTMA405 44K1
ANSI B125.27
see ASTMA369 44K1
ANSI B125.3
see ASTMA135 44A1
ANSI B125.30
see ASTMA106 44A2
ANSI B125.31
see ASTMA139 44A2
ANSI B125.35
see ASTMA381 44A2
ANSI B125.37
see ASTMA524 44A1
ANSI G62.5
see ASTMA523 44A1
AN T 3 44K1
AN T 4/1 44A2
AN T 14/1 44N3
AN T 15/2 44K1
AN T 22/2 44K1
AN T 33/1 44K1
AN T 43 44N3
ANTICHOC 2 44E2
ANTICHOC H 44K2
ANTICHOC IX 44K2
ANTICORODAL 1E
ANTICORODAL 1K
ANTICORODAL 5S1 1C
ANTICORODAL 34 1C
ANTICORODAL 041 1E
ANTICORODAL 70 1K
ANTICORODAL 71 1K
ANTICORODAL 090 1E
ANTICORODAL 110 1E
ANTICORODAL 850 1E

ANTICORODAL 990 1E
ANTICORODAL Pb 108 1E
ANTIMONIAL LEAD 21.1
ANTIMONY 2.1
ANTIMONY BRONZE 14E
ANTINIT KW 10M 44M1
ANTINIT KW 15M 44M1
ANTINIT KW 20M 44M1
AN WW C 561a/1 1D
AN WW T 831 27B
AN WW T 833 27B
AN WW T 846/1 44A2
AN WW T 850/1 44K1
AN WW T 852 44K1
AN WWT 855/1 44N3
AN WWT 858/1 44N3
AN WWT T 861/2 44N3
AP 303 1L
AP 309 1L
AP 403 1C
AP 501 1J
AP 505 1J
AP 701 1M
APF1 44N1
APF1S 44N1
APFR 44N1
APFRS 44N1
API 5A C75/1 44A2
API 5A C75/2 44A2
API 5A C75/3 44A2
API 5A C95 44A2
API 5A H40 44A1
API 5A J55 44A1
API 5A K55 44A1
API 5A N80 44A1
API 5L A 44A1
API 5L A25/1 44A1
API 5L A25/2 44A1
API 5L B 44A1
API 5L S 44A2
API 5L SA 44A1
API 5L SB 44A1
API 5L X42 44A2
API 5L X46 44A2
API 5L X52 44A2
API 5L X56 44A2
API 5L X60 44A2
API 5L X65 44A2
API 5L X70 44A1
API 5LU X80 44A2
API 5LU X100 44A2
APM 44N1
APM/Sp 44N1
APMC 44N1
APML 44N1
APML/Sp 44N1
APMT 44N1
APMZ 44N1
APS 10 44E1
APS 10C 44E1
APS 10 M 44K1
APS 10 M4 44K1
APS 20 44E1
APS 20C 44E1
APS 20 M 44K1
APS 25 44K1
APS 30 44E1
AQUACIDOX 44A1
AQUAMARINE 52.1
AQUATOUGH 70 44A3
AQUATOUGH 100 44A4
ARC/098 44N1
ARC 164 27F
ARC 2233 44N1
ARC 2266 44N1
ARCHITECTURAL
BRONZE C94 14C
AR COL 360 44K1
ARCTIC A 44A1
ARCTIC B 44A1
ARCTIC C 44A1
ARCTIC D 44A1
ARD 3 44M1
ARDAL 1F
ARDENT 44K3
ARDHO 2 44K2
ARGENTINE METAL 50A
ARGO-BOND 14H
ARGO BRAZE 25 14H
ARGO BRAZE 27 14H
ARGO-BRAZE 40 42.1
ARGO-BRAZE 49 H 42.1
ARGO-BRAZE 50 42.1
ARGO-BRAZE 56 42.1
ARGOFIL 14M
ARGO-FLO 42.1
ARGO-SWIFT 14H
ARH 44M1
ARH 8 44M2
ARH TOUGH 44M1
ARIEL BRONZE 14K
ARISTALOY 25.1
ARKS 44M1
ARK SUPERIOR 44K3
ARK SUPERIOR EXTRA 44K3
ARK SUPERLATIVE 44L
ARK TRIUMPH 44K3
ARK TRIUMPHANT 44K3
ARL 44M1
ARLEY 44K3
ARMARC 44N1
ARMCO 12 44M1
ARMCO 12 Al 44M1
ARMCO 12 FM 44M1
ARMCO 12/2 44M1
ARMCO 12T 44M1
ARMCO 13-C-35 44M1
ARMCO 16/2 44M1
ARMCO 17 44M1
ARMCO 17-C-60 44M1
ARMCO 17 C 80 44M2
ARMCO 17 C 100 44M2
ARMCO 17 C 100 FM 44M2
ARMCO 17/7 44N3
ARMCO 18/8 44N1
ARMCO 18/8 Si 44N1
ARMCO 18/10 Cb Ta 44N1
ARMCO 18/10 Ti 44N1
ARMCO 18/12 Mo 44N1
ARMCO 20/10 44N1
ARMCO 25/12 44N1
ARMCO 25/20 44N1
ARMCO 27 44M1
ARMCO HS No 4R 44A1
ARMCO HS No 4S 44A1
ARMCO HS No 7 44A1
ARMCO HS No. 1 44F1
ARMCO HS No. 2 44F1
ARMCO HS No. 3 44J
ARMCO HS No. 5 44J
ARMCO HS No. 6 44J
ARMCO IRON 20A
ARMET 44N1
ARMEX 2 44N1
ARMEX 3 44N1
ARMEX GT 44N1
ARMICO 20A
ARMOUR PLATE 44K1
ARNAVAR 13A
ARNE 44K3
ARO 75 44K1
AR Q 40.1
ARS 44M2
ARSENIC 3.1
ARSENIC-ANTIMONY 2.1
ARSENIC COPPER 3.1
ARSENIC-IRON 3.1
ARSENIC-LEAD 21.1
ARSENIC-NICKEL 27F
ARSENIC-TIN 50A
ARSENIC-ZINC 55A
ARTIC BRONZE 14K
ARW 44M1
ARWA 44M1
ARWB 44M1
ARWD 44M1
ARWDN 44M1
ARWDT 44M1
ARWF 44M1
ARWK 44M1
ARWKIN 44M1
ARWKT 44M1
ARZ 44M1
A S2U 1L
A-S4G 1C
A-S5 1C
A S5U 1L
A S5U2 1L
A S5U3 1L
A-S7G 1C
A S9K7 1N
A-S9U3Y4 1C
A S9U3Y4 1L
A-S10G 1C
A-S10UG 1C
A-S12 1C
A-S12N2G 1C
A-S12UN 1C
A-S13 1C
A-S22 1C
A-S22UNK 1C
AS 25 13A
AS 41 A 23B
AS 307 1L
AS 313 1L
AS 401 1C
AS 601 1C
ASA
see ASTMA823 20B
ASB
see ASTMA823 20B

ASC
see ASTMA823 20B
ASC 20C
ASC A 20C
A SG 1E
A SGM 1E
A SGM 0.7 1E
ASHBERRY 50A
ASJMA752 44K2
ASM 123 44M1
ASM A27/65/35 44A2
ASN 151 44E1
ASP23 44K3
ASP 30 44L
ASP 60 44L
ASS
see ASTMA823 20B
A St 2 44A1
A St 3 44A1
A St 4 44A1
A St 35/8 44A1
A St 45/8 44A1
A STEEL 44A3
ASTM 122 44A2
ASTM A 167/301 44N3
ASTM A 560/50/50 12.1
ASTM A 693/630 44N3
ASTM A 693/634 44N3
ASTM A 747 CB 7 CU1
392180 44N3
ASTM A 756 44M2
ASTM A 757 E3N 44N3
ASTM A1 44A3
ASTM A21 44A3
ASTM A25 Class A 44A3
ASTM A25 Class B 44A3
ASTM A25 Class C 44A3
ASTM A25 Class V 44A3
ASTM A27 N1 44A2
ASTM A27 N2 44A2
ASTM A27 U60/30 44A2
ASTM A27/60/30 44A2
ASTM A27/70/36 44A2
ASTM A27/70/40 44A2
ASTM A31 A 44A2
ASTM A31 B 44A2
ASTM A36 44A2
ASTM A42 20A
ASTM A43 20B
ASTM A47 20B
ASTM A48 20B
ASTM A53 E GA 44A1
ASTM A53 E GB 44A1
ASTM A53 F 44A1
ASTM A53 S GA 44A1
ASTM A53 S GB 44A1
ASTM A53 T/T 60/40/18 20B
ASTM A53 T/T 65/45/12 20B
ASTM A53 T/T 80/55/06 20B
ASTM A53 T/T 100/70/03 20B
ASTM A53 T/T 120/90/02 20B
ASTM A54/2 44K1
ASTM A59 44B2
ASTM A60 44K2
ASTM A67 44A1
ASTM A68 44A4
ASTM A84 20A
ASTM A105/1 44A2
ASTM A106 A 44A2
ASTM A106 B 44A2
ASTM A106 C 44A2
ASTM A108 44A2
ASTM A109/1 44A2
ASTM A109/2 44A2
ASTM A109/3 44A2
ASTM A109/4 44A1
ASTM A109/5 44A1
ASTM A113 44A1
ASTM A120 44A1
ASTM A126 A 20B
ASTM A126 B 20B
ASTM A128 A 44D
ASTM A128 B1 44D
ASTM A128 B2 44D
ASTM A128 B3 44D
ASTM A128 B4 44D
ASTM A128 C 44D
ASTM A128 D 44D
ASTM A128 E1 44D
ASTM A128 E2 44D
ASTM A131 A 44A2
ASTM A131 B 44A2
ASTM A131 C 44A2
ASTM A135 44A1
ASTM A139 A 44A2
ASTM A139 B 44A2
ASTM A148 44A2
ASTM A159 20B
ASTM A161 (low carbon) 44A1
ASTM A161T1 44G1
ASTM A167/302 44N1
ASTM A167/302 B 44N1
ASTM A167/304 44N1
ASTM A167/304 L 44N1
ASTM A167/305 44N1
ASTM A167/308 44N1
ASTM A167/309 44N1
ASTM A167/309 S 44N1
ASTM A167/310 44N1
ASTM A167/310 S 44N1
ASTM A167/316 44N1
ASTM A167/316 L 44N1
ASTM A167/317 44N1
ASTM A167/317L 44N1
ASTM A167/321 44N1
ASTM A167/347 44N1
ASTM A167/348 44N1
ASTM A176/403 44M1
ASTM A176/405 44M1
ASTM A176/410 44M1
ASTM A176/410 S 44M1
ASTM A176/430 44M1
ASTM A176/442 44M1
ASTM A176/446 44M1
ASTM A177 44N3
ASTM A178 A 44A1
ASTM A178 C 44A2
ASTM A179 44A1
ASTM A181/1 44A2
ASTM A182 F1 44K1
ASTM A182 F5 44K1
ASTM A182 F5a 44K1
ASTM A182 F6 44M1
ASTM A182 F7 44K1
ASTM A182 F9 44K1
ASTM A182 F10 44N1
ASTM A182 F11 44K1
ASTM A182 F12 44K1
ASTM A182 F21 44K1
ASTM A182 F22 44K1
ASTM A182 F304 44N1
ASTM A182 F304H 44N1
ASTM A182 F304L 44N1
ASTM A182 F310 44N1
ASTM A182 F316 44N1
ASTM A182 F316H 44N1
ASTM A182 F316L 44N1
ASTM A182 F321 44N1
ASTM A182 F321H 44N1
ASTM A182 F347 44N1
ASTM A182 F347H 44N1
ASTM A182 F348 44N1
ASTM A182 F348H 44N1
ASTM A183 44A2
ASTM A192 44A1
ASTM A193 B5 44K1
ASTM A193 B6 44M1
ASTM A193 B7 44K2
ASTM A193 B8 44N1
ASTM A193 B8C 44N1
ASTM A193 B8M 44N1
ASTM A193 B8T 44N1
ASTM A193 B14 44K2
ASTM A193 B16 44K2
ASTM A194/1 44A1
ASTM A194/2 44A2
ASTM A194/3 44K1
ASTM A194/4 44G2
ASTM A194/6 44M1
ASTM A194/7 44G2
ASTM A194/8 44N1
ASTM A194/8 C 44N1
ASTM A194/8 M 44N1
ASTM A194/8 T 44N1
ASTM A197 20B
ASTM A199 T3b 44K1
ASTM A199 T4 44K1
ASTM A199 T5 44K1
ASTM A199 T7 44K1
ASTM A199 T9 44K1
ASTM A199 T11 44K1
ASTM A199 T21 44K1
ASTM A199 T22 44K1
ASTM A200 44K1
ASTM A202 A 44E1
ASTM A202 B 44E1
ASTM A203 A 44F1
ASTM A203 B 44F1
ASTM A203 D 44F1
ASTM A203 E 44F1
ASTM A204 A 44G1
ASTM A204 B 44G1
ASTM A204 C 44G1
ASTM A207 20A
ASTM A209 T1 44G1
ASTM A209 T1A 44G1
ASTM A209 T1B 44G1
ASTM A210 A1 44A2
ASTM A210 C 44A2

ASTM A213 BOILER
see ASTM A182 for different grades
ASTM A214 44A1
ASTM A216 WCA 44A2
ASTM A216 WCB 44A2
ASTM A216 WCC 44A2
ASTM A217 C5 44K1
ASTM A217 C12 44K1
ASTM A217 WC1 44A2
ASTM A217 WC4 44K1
ASTM A217 WC5 44K1
ASTM A217 WC6 44K1
ASTM A217 WC9 44K1
ASTM A220 20B
ASTM A225 A 44J
ASTM A225 B 44J
ASTM A226 44A1
ASTM A227 44A3
ASTM A228 44A3
ASTM A229 A 44A3
ASTM A229B 44A3
ASTM A230 44A3
ASTM A231 44K2
ASTM A232 44K2
ASTM A233 44A1
ASTM A235 44A1
ASTM A237 44A1
ASTM A240/302 44N1
ASTM A240/304 44N1
ASTM A240/304 L 44N1
ASTM A240/305 44N1
ASTM A240/309S 44N1
ASTM A240/310S 44N1
ASTM A240/316 44N1
ASTM A240/316 L 44N1
ASTM A240/317 44N1
ASTM A240/317 L 44N1
ASTM A240/321 44N1
ASTM A240/347 44N1
ASTM A240/348 44N1
ASTM A240/405 44M1
ASTM A240/410 44M1
ASTM A240/410 S 44M1
ASTM A240/430 A 44M1
ASTM A240/430 B 44M1
ASTM A241 44A3
ASTM A242 44A2
ASTM A249 TP304 44N1
ASTM A249 TP304H 44N1
ASTM A249 TP304L 44N1
ASTM A249 TP309 44N1
ASTM A249 TP310 44N1
ASTM A249 TP316 44N1
ASTM A249 TP316H 44N1
ASTM A249 TP316L 44N1
ASTM A249 TP317 44N1
ASTM A249 TP321 44N1
ASTM A249 TP321H 44N1
ASTM A249 TP347 44N1
ASTM A249 TP347H 44N1
ASTM A249 TP348 44N1
ASTM A249 TP348H 44N1
ASTM A250 T1 44G1
ASTM A250 T1A 44G1
ASTM A250 T1B 44G1
ASTM A251 44A1
ASTM A254 44A1
ASTM A263 44A1
ASTM A264 44A1
ASTM A265 44A1
ASTM A266/1 and 2 44A2
ASTM A266/3 44A2
ASTM A268 TP329 44M1
ASTM A268 TP405 44M1
ASTM A268 TP410 44M1
ASTM A268 TP430 44M1
ASTM A268 TP433 44M1
ASTM A268 TP466 44M1
ASTM A269 TP304 44N1
ASTM A269 TP304L 44N1
ASTM A269 TP316 44N1
ASTM A269 TP316L 44N1
ASTM A269 TP317 44N1
ASTM A269 TP321 44N1
ASTM A269 TP347 44N1
ASTM A269 TP348 44N1
ASTM A271 44N1
ASTM A276 20B
ASTM A276/ 430 F Se 44M1
ASTM A276/302 44N1
ASTM A276/302B 44N1
ASTM A276/303 44N1
ASTM A276/303 Se 44N1
ASTM A276/304 44N1
ASTM A276/304 L 44N1
ASTM A276/308 44N1
ASTM A276/309 44N1
ASTM A276/309 S 44N1
ASTM A276/310 44N1
ASTM A276/310 S 44N1
ASTM A276/314 44N1
ASTM A276/316 44N1
ASTM A276/316 L 44N1
ASTM A276/321 44N1
ASTM A276/347 44N1
ASTM A276/348 44N1
ASTM A276/403 44M1
ASTM A276/405 44M1
ASTM A276/410 44M1
ASTM A276/414 44M1
ASTM A276/416 44M1
ASTM A276/416 Se 44M1
ASTM A276/420 44M1
ASTM A276/430 44M1
ASTM A276/430 F 44M1
ASTM A276/431 44M1
ASTM A276/440 A 44M1
ASTM A276/440 B 44M1
ASTM A276/440 C 44M2
ASTM A276/446 44M1
ASTM A278 20B
ASTM A283 A 44A1
ASTM A283 B 44A1
ASTM A283 C 44A1
ASTM A283 D 44A1
ASTM A284 A 44A2
ASTM A284 B 44A2
ASTM A284 C 44A2
ASTM A284 D 44A2
ASTM A285 A 44A1
ASTM A285 B 44A1
ASTM A285 C 44A1
ASTM A289 A 44P
ASTM A289 B 44P
ASTM A289 C 44N1
ASTM A290 A 44A2
ASTM A290 B 44A2
ASTM A290 C1 44A2
ASTM A290 C2 44A3
ASTM A290 D 44K1
ASTM A290 D2 44K1
ASTM A290 E 44K1
ASTM A290 F 44K1
ASTM A290 G 44K1
ASTM A291/1 44A2
ASTM A291/2 44A2
ASTM A291/3 44K2
ASTM A291/4 44K2
ASTM A291/5 44K2
ASTM A291/6 44K2
ASTM A291/7 44K2
ASTM A292 1 44J
ASTM A292/2 44K1
ASTM A292/3 44K1
ASTM A292/4 44K1
ASTM A292/5 44K1
ASTM A293/1 44J
ASTM A293/2 44K1
ASTM A293/3 44K1
ASTM A293/4 44K1
ASTM A293/5 44K1
ASTM A293/6 44K1
ASTM A294 A 44K2
ASTM A294 B 44K2
ASTM A294 C 44K1
ASTM A295 F1 44G1
ASTM A295 F2 44K1
ASTM A295 F5 44K1
ASTM A295 F5A 44K1
ASTM A295 F6 44M1
ASTM A295 F8 44N1
ASTM A295 F8C 44N1
ASTM A295 F8M 44N1
ASTM A295 F8T 44N1
ASTM A295 F10 44N1
ASTM A295 F22 44K1
ASTM A295 F25 44N1
ASTM A295 F30 44K2
ASTM A295 F31 44K1
ASTM A295 F32 44K1
ASTM A296 CA15 44M1
ASTM A296 CA40 44M1
ASTM A296 CB30 44M1
ASTM A296 CC50 44M1
ASTM A296 CE30 44N1
ASTM A296 CF3 44N1
ASTM A296 CF3M 44N1
ASTM A296 CF8 44N1
ASTM A296 CF8C 44N1
ASTM A296 CF8M 44N1
ASTM A296 CF16F 44N1
ASTM A296 CF20 44N1
ASTM A296 CG8M 44N1
ASTM A296 CG12 44N1
ASTM A296 CH20 44N1
ASTM A296 CK20 44N1
ASTM A296 CN7M 44N1
ASTM A296 CW 12M 27B
ASTM A296 CY40 27B
ASTM A297 HC 44M1

ASTM A297 HD 44M1
ASTM A297 HE 44N1
ASTM A297 HF 44N1
ASTM A297 HH 44N1
ASTM A297 HI 44N1
ASTM A297 HK 44N1
ASTM A297 HL 44N1
ASTM A297 HN 44N1
ASTM A297 HT 44N1
ASTM A297 HU 44N1
ASTM A297 HV 44N1
ASTM A297 HW 27B
ASTM A297 HX 27B
ASTM A298 E308 44N1
ASTM A298 E308ELC 44N1
ASTM A298 E309 44N1
ASTM A298 E309 Cb 44N1
ASTM A298 E309 Mo 44N1
ASTM A298 E310 44N1
ASTM A298 E310 Cb 44N1
ASTM A298 E310 Mo 44N1
ASTM A298 E312 44N1
ASTM A298 E316 44N1
ASTM A298 E316ELC 44N1
ASTM A298 E317 44N1
ASTM A298 E318 44N1
ASTM A298 E330 44N1
ASTM A298 E347 44N1
ASTM A298 E410 44M1
ASTM A298 E430 44M1
ASTM A298 E502 44K1
ASTM A299 44A2
ASTM A302 A 44K1
ASTM A302 B 44K1
ASTM A302 C 44K1
ASTM A302 D 44K1
ASTM A306/45 44A2
ASTM A306/50 44A2
ASTM A306/55 44A2
ASTM A306/60 44A2
ASTM A306/65 44A2
ASTM A306/70 44A3
ASTM A306/75 44A3
ASTM A306/80 44A3
ASTM A312 44N1
ASTM A313 44N1
ASTM A316 E7000A1 44K1
ASTM A316 E7010A1 44G1
ASTM A316 E7011A1 44G1
ASTM A316 E7015A1 44G1
ASTM A316 E7016A1 44G1
ASTM A316 E7018A1 44G1
ASTM A316 E7020A1 44G1
ASTM A316 E7020G 44K1
ASTM A316 E7027A1 44G1
ASTM A316 E8015/18B 44K1
ASTM A316 E8015B2L 44K1
ASTM A316 E8015B4L 44K1
ASTM A316 E8016B1 44K1
ASTM A316 E8016B2 44K1
ASTM A316 E8016C 44F1
ASTM A316 E8016C1 44F1
ASTM A316 E8016C2 44F1
ASTM A316 E8016C3 44K1
ASTM A316 E8018B2 44K1
ASTM A316 E8018C1 44F1
ASTM A316 E8018C2 44F1
ASTM A316 E8018C3 44K1
ASTM A316 E9015/8D1 44G1
ASTM A316 E9015B3 44K1
ASTM A316 E9015B3L 44K1
ASTM A316 E9015D1 44G1
ASTM A316 E9016/8B3 44K1
ASTM A316 E9016B3 44K1
ASTM A316 E9018B3 44K1
ASTM A316 E9018D1 44G1
ASTM A316 E9018M 44K1
ASTM A316 E10015/8D2 44G1
ASTM A316 E10015D2 44G1
ASTM A316 E10016D2 44G1
ASTM A316 E10018D2 44G1
ASTM A316 E10018M 44K1
ASTM A316 E11018M 44K1
ASTM A316 E12018M 44K1
ASTM A316 EXX10G 44K1
ASTM A316 EXX11G 44K1
ASTM A316 EXX13G 44K1
ASTM A316 EXX15G 44K1
ASTM A316 EXX16G 44K1
ASTM A316 EXX18G 44K1
ASTM A319/1 20B
ASTM A319/11 20B
ASTM A319/111 20B
ASTM A320 B8 44N1
ASTM A320 B8C 44N1
ASTM A320 B8D 44N1
ASTM A320 B8F 44N1
ASTM A320 B8T 44N1
ASTM A320 L7 44K2
ASTM A320 L9 44F2
ASTM A320 L10 44F1
ASTM A320 L43 44K2
ASTM A321 44A3
ASTM A325 44A2
ASTM A331 44K2
ASTM A332/8650 44K2
ASTM A332/8655 44K2
ASTM A332/8660 44K2
ASTM A333/1 44A2
ASTM A333/3 44F1
ASTM A333/4 44K1
ASTM A333/6 44A2
ASTM A333/7 44F1
ASTM A333/8 44F1
ASTM A334 Steel tube for low temperature see ASTM A333
ASTM A335 P1 44G1
ASTM A335 P2 44K1
ASTM A335 P5 44K1
ASTM A335 P5b 44K1
ASTM A335 P5c 44K1
ASTM A335 P7 44K1
ASTM A335 P9 44K1
ASTM A335 P11 44K1
ASTM A335 P12 44K1
ASTM A335 P15 44G1
ASTM A335 P21 44K1
ASTM A335 P22 44K1
ASTM A336 F1 44G1
ASTM A336 F2 44K1
ASTM A336 F5 44K1
ASTM A336 F5a 44K1
ASTM A336 F6 44M1
ASTM A336 F8 44N1
ASTM A336 F8C 44N1
ASTM A336 F8M 44N1
ASTM A336 F8T 44N1
ASTM A336 F10 44N1
ASTM A336 F22 44K1
ASTM A336 F25 44N1
ASTM A336 F30 44K2
ASTM A336 F31 44K1
ASTM A336 F32 44K1
ASTM A350 LF1 44A2
ASTM A350 LF2 44A2
ASTM A350 LF3 44F1
ASTM A350 LF4 44K1
ASTM A351 CA15 44M1
ASTM A351 CF3 44N1
ASTM A351 CF3M 44N1
ASTM A351 CF8 44N1
ASTM A351 CF8C 44N1
ASTM A351 CF8M 44N1
ASTM A351 CF10MC 44N1
ASTM A351 CH8 44N1
ASTM A351 CH10 44N1
ASTM A351 CH20 44N1
ASTM A351 CK45 44N1
ASTM A351 CT35 44N1
ASTM A352 IC3 44F1
ASTM A352 LC1 44G1
ASTM A352 LC2 44F1
ASTM A352 LC2 44F2
ASTM A352 LC3 44F1
ASTM A352 LCB 44A2
ASTM A353 A and B 44F1
ASTM A354 44K2
ASTM A355 A 44K2
ASTM A355 B 44K2
ASTM A355 C 44K1
ASTM A355 D 44K2
ASTM A356/1 44A2
ASTM A356/1 44J
ASTM A356/2 44G1
ASTM A356/3 44G1
ASTM A356/4 44K1
ASTM A356/5 44K1
ASTM A356/6 44K1
ASTM A356/7 44K1
ASTM A356/8 44K1
ASTM A356/9 44K1
ASTM A356/10 44K1
ASTM A357 44K1
ASTM A362 44N1
ASTM A366 44A1
ASTM A369 FP1 44G1
ASTM A369 FP1 44K1
ASTM A369 FP3b 44K1
ASTM A369 FP5 44K1
ASTM A369 FP7 44K1
ASTM A369 FP9 44K1
ASTM A369 FP11 44K1
ASTM A369 FP12 44K1
ASTM A369 FP21 44K1
ASTM A369 FP22 44K1
ASTM A371 ER308 44N1
ASTM A371 ER308L 44N1
ASTM A371 ER309 44N1
ASTM A371 ER310 44N1
ASTM A371 ER312 44N1

ASTM A371 ER316 44N1
ASTM A371-ER316L 44N1
ASTM A371-ER317 44N1
ASTM A371-ER318 44N1
ASTM A372 Class I 44A2
ASTM A372 Class II 44A2
ASTM A372 Class III 44A2
ASTM A372 Class IV 44K2
ASTM A372 Class V Type A 44K2
ASTM A372 Class V Type B 44K2
ASTM A372 Class V Type C 44K2
ASTM A372 Class V Type D 44K2
ASTM A372 Class V Type E 44K2
ASTM A372 Class VII 44A2
ASTM A372 VI 44K1
ASTM A374 44A2
ASTM A375 44A2
ASTM A376 44N1
ASTM A381 44A2
ASTM A382 20A
ASTM A387 A 44K1
ASTM A387 B 44K1
ASTM A387 C 44K1
ASTM A387 D 44K1
ASTM A387 E 44K1
ASTM A389 C23 44K1
ASTM A389 C24 44K1
ASTM A391 44K2
ASTM A398 20B
ASTM A401 44E2
ASTM A403 44N1
ASTM A404 F24 44K1
ASTM A405 P24 44K1
ASTM A407 44A3
ASTM A409 44N1
ASTM A410 44K1
ASTM A412/201 44P
ASTM A412/202 44P
ASTM A413 44A2
ASTM A414 44A2
ASTM A415 44A1
ASTM A417 44A3
ASTM A421 44A3
ASTM A423/1 44K1
ASTM A423/2 44K1
ASTM A424 44A1
ASTM A425 44A1
ASTM A426 CP1 44G1
ASTM A426 CP2 44K1
ASTM A426 CP5 44K1
ASTM A426 CP5b 44K1
ASTM A426 CP7 44K1
ASTM A426 CP9 44K1
ASTM A426 CP11 44K1
ASTM A426 CP12 44K1
ASTM A426 CP15 44G1
ASTM A426 CP21 44K1
ASTM A426 CP22 44K1
ASTM A429/201 44P
ASTM A429/202 44P
ASTM A430 44N1
ASTM A433 44A2
ASTM A436/1 20B
ASTM A436/1b 20B
ASTM A436/2 20B
ASTM A436/2b 20B
ASTM A436/3 20B
ASTM A436/4 20B
ASTM A436/5 20B
ASTM A436/6 20B
ASTM A437 44M1
ASTM A439 D2 20B
ASTM A439 D2B 20B
ASTM A439 D2C 20B
ASTM A439 D3 20B
ASTM A439 D3A 20B
ASTM A439 D4 20B
ASTM A439 D5 20B
ASTM A439 D5B 20B
ASTM A440 44A2
ASTM A441 44A1
ASTM A442 44A1
ASTM A445 20B
ASTM A447 20B
ASTM A448 20B
ASTM A449 44A2
ASTM A451 CPF8 44N1
ASTM A451 CPF8M 44N1
ASTM A451 CPF10MC 44N1
ASTM A451 CPF87C 44N1
ASTM A451 CPH8 44N1
ASTM A451 CPH10 44N1
ASTM A451 CPH20 44N1
ASTM A451 CPK20 44N1
ASTM A452 44N1
ASTM A453/651 44N2
ASTM A453/660 44F1
ASTM A453/660 44N2
ASTM A453/662 44N2
ASTM A453/665 44N2
ASTM A454 44A2
ASTM A455 44A2
ASTM A457/650 44M2
ASTM A457/651 44N2
ASTM A457/652 44N2
ASTM A458 44N2
ASTM A461/630 44N3
ASTM A461/631 44N3
ASTM A461/632 44N3
ASTM A461/634 44N3
ASTM A461/660 44N2
ASTM A461/661 44N2
ASTM A461/662 44N2
ASTM A461/671 13A
ASTM A461/684 27C
ASTM A461/685 27C
ASTM A461/688 27C
ASTM A461/689 27C
ASTM A465 LI 44A2
ASTM A465 LII 44A2
ASTM A465 LIII 44A2
ASTM A465 LIV 44A2
ASTM A466 44A2
ASTM A467 44A2
ASTM A469/1 44J
ASTM A469/2 44K1
ASTM A469/3 44K2
ASTM A469/4 44K2
ASTM A469/5 44K2
ASTM A469/6 44K2
ASTM A470/1 44A2
ASTM A470/2 44K2
ASTM A470/3 and 4 44K2
ASTM A470/5, 6 and 7 44K2
ASTM A470/8 44K2
ASTM A471 44K2
ASTM A471/7 44K2
ASTM A473 44N1
ASTM A476 20B
ASTMA477
see designation for details
ASTM A478 44N1
ASTM A479 44N1
ASTM A485/1 44E3
ASTM A485/2 44E3
ASTM A485/3 44K3
ASTM A485/4 44K3
ASTM A486 44A2
ASTM A487 GN 44K2
ASTM A487 IN 44J
ASTM A487/2 N 44G1
ASTM A487/3 N 44A2
ASTM A487/4 N 44K1
ASTM A487/5 N 44K1
ASTM A487/7 Q 44K1
ASTM A487/8 N 44K1
ASTM A487/9 N 44K1
ASTM A487/10 N 44K1
ASTM A489 44A1
ASTM A490 44A2
ASTM A492 44N1
ASTM A493 44N1
ASTM A494 CW12M2B 27B
ASTM A494 N12M2B 27B
ASTM A494 Ni Mo 27F
ASTM A494 Ni Mo Cr 27F
ASTM A500 44A2
ASTM A501 44A2
ASTM A502/1 44A1
ASTM A502/2 44A1
ASTM A504 44A3
ASTM A508/1 44J
ASTM A508/2 44K1
ASTM A508/3 44K1
ASTM A508/4 44K1
ASTM A508/5 44K1
ASTM A511 44N1
ASTM A512 44A1
ASTM A513 44A1
ASTM A514 A 44K1
ASTM A514 B 44K1
ASTM A514 C 44K1
ASTM A514 D 44K1
ASTM A514 E 44K1
ASTM A514 F 44K1
ASTM A514 G 44K1
ASTM A514 H 44K1
ASTM A514 J 44G1
ASTM A514 K 44G1
ASTM A515 44A2
ASTM A516 44A2
ASTM A517 A 44K1
ASTM A517 B 44K1
ASTM A517 C 44G1
ASTM A517 D 44K1
ASTM A517 E 44K1
ASTM A517 F 44K1
ASTM A517 G 44K1
ASTM A517 H 44K1
ASTM A517 J 44G1
ASTM A517 K 44G1

ASTM A518 20B
ASTM A519 44A2
ASTM A522 44F1
ASTM A523 A 44A1
ASTM A523 B 44A2
ASTM A524 44A1
ASTM A529 44A1
ASTM A532/1 20B
ASTM A532/11 20B
ASTM A532/111 20B
ASTM A532(1D) 20B
ASTM A532(11A) 20B
ASTM A532(11B) 20B
ASTM A532(11C) 20B
ASTM A532(11D) 20B
ASTM A532(11E) 20B
ASTM A532(111A) 20B
ASTM A533 A 44G1
ASTM A533 B 44K1
ASTM A533 C 44K1
ASTM A534 44K1
ASTM A535 44K1
ASTM A536 60/40/18 20B
ASTM A536 65/45/12 20B
ASTM A536 80/55/06 20B
ASTM A536 80/60/03 20B
ASTM A536 100/70/03 20B
ASTM A536 120/90/02 20B
ASTM A537 44A1
ASTM A538 A 44F3
ASTM A538 B 44F3
ASTM A538 C 44F3
ASTM A539 44A1
ASTM A540 B21 44K2
ASTM A540 B22 44K2
ASTM A540 B23 44K2
ASTM A540 B24 44K2
ASTM A541/1 44A2
ASTM A541/3 44K1
ASTM A541/4 44K1
ASTM A541/5 44K1
ASTM A541/6 44K1
ASTM A541/7 44K1
ASTM A541/8 44K1
ASTM A542 44K1
ASTM A543 44K1
ASTM A548 44A1
ASTM A549 44A1
ASTM A551 44A3
ASTM A553 44F1
ASTM A554 44N1
ASTM A556 A2 44A1
ASTM A556 B2 44A2
ASTM A556 C2 44A2
ASTM A557 44A2
ASTM A558 EL8 44A1
ASTM A558 EL8K 44A1
ASTM A558 EL12 44A1
ASTM A558 EM5K 44A1
ASTM A558 EM12 44A1
ASTM A558 EM12K 44A1
ASTM A558 EM13K 44A1
ASTM A558 EM15K 44A1
ASTM A558 EN14 44A1
ASTM A559 E60S1 44A1
ASTM A559 E60S2 44A1
ASTM A559 E60S3 44A1
ASTM A559 E70S4 44A1
ASTM A559 E70S5 44A1
ASTM A559 E70S6 44A1
ASTM A559 E70T 44A1
ASTM A560/60/40 12.1
ASTM A562 44A1
ASTM A563 44A2
ASTMA564
see designation for details
ASTM A565 44M1
ASTM A567 HH90 20B
ASTM A567 HI50C 20B
ASTM A567 HK40 44N1
ASTM A567 HK50 44N1
ASTM A567 HT50C 20B
ASTM A567/1 13A
ASTM A567/2 13A
ASTM A567/3 44N1
ASTM A567/4 27C
ASTM A567/5 27C
ASTM A567/6 V 27C
ASTM A567/7 V 27C
ASTM A567/8 27C
ASTM A567/9 V 27C
ASTM A567/10 V 27C
ASTM A569 44A1
ASTM A570 44A2
ASTM A571 20B
ASTM A572 44A1
ASTM A573 44A1
ASTM A573/70 44A2
ASTM A574 44K1
ASTM A575 44A1
ASTM A576 44A1
ASTM A579 44N3
ASTM A579/11 44K1
ASTM A579/12 44K1
ASTM A579/13 44K1
ASTM A579/21 44K2
ASTM A579/22 44K2
ASTM A579/23 44K2
ASTM A579/31 44K1
ASTM A579/32 44K2
ASTM A579/38 44K2
ASTM A579/41 44K2
ASTM A579/51 44M1
ASTM A579/52 44M1
ASTM A579/53 44M1
ASTM A579/61 44M1
ASTM A579/75 44F3
ASTM A579/81 44L
ASTM A579/82 44L
ASTM A579/83 44L
ASTM A580 44N1
ASTM A581 XM1 44P
ASTM A581 XM2 44N1
ASTM A581 XM3 44N1
ASTM A581/XM6 44M1
ASTM A581/303 44N1
ASTM A581/303 Se 44N1
ASTM A581/416 44M1
ASTM A581/416 Se 44M1
ASTM A581/430 F 44M1
ASTM A581/430 F Se 44M1
ASTM A582 44N1
ASTM A583 44A3
ASTM A587 44A1
ASTM A588 44F1
ASTM A588 A 44K1
ASTM A588 B 44K1
ASTM A588 C 44K1
ASTM A588 Cr 44E1
ASTM A588 D 44E1
ASTM A588 E 44G1
ASTM A588 F 44K1
ASTM A588 H 44K1
ASTM A589 44A1
ASTM A589 B 44A1
ASTM A591 44A1
ASTM A592 A 44K1
ASTM A592 E 44K1
ASTM A592 F 44K1
ASTM A594, 594/1 44A1
ASTM A594/2 44A1
ASTM A594/3 44A1
ASTM A594/4 44A1
ASTM A595 A 44A1
ASTM A595 B 44A1
ASTM A595 C 44K1
ASTM A597 44K3
ASTM A597 CA2 44K3
ASTM A597 CD2 44M2
ASTM A597 CD5 44M2
ASTM A597 CH13 44K1
ASTM A597 CS5 44K2
ASTM A599 44A1
ASTM A600 44K3
ASTM A601 A 24.1
ASTM A601 B 24.1
ASTM A601 C 24.1
ASTM A601 D 24.1
ASTM A601 E 24.1
ASTM A601 F 24.1
ASTM A602 M3210 20B
ASTM A602 M4504 20B
ASTM A602 M5003 20B
ASTM A602 M5503 20B
ASTM A602 M7002 20B
ASTM A602 M8501 20B
ASTM A603 44A1
ASTM A605 44L
ASTM A606/2 44A1
ASTM A606/4 44A1
ASTM A607/45 44A1
ASTM A607/50 44A1
ASTM A607/55 44A1
ASTM A607/60 44A1
ASTM A607/65 44A1
ASTM A607/70 44A1
ASTM A608 H135 44N1
ASTM A608 HC30 44N1
ASTM A608 HD50 44N1
ASTM A608 HE35 44N1
ASTM A608 HF30 44N1
ASTM A608 HH30 44N1
ASTM A608 HH33 44N1
ASTM A608 HK30 44N1
ASTM A608 HK40 44N1
ASTM A608 HL30 44N1
ASTM A608 HL40 44N1
ASTM A608 HN40 44N1
ASTM A608 HT50 44N1
ASTM A608 HU50 44N1
ASTM A608 HW50 27B

ASTM A608 HX50 27B
ASTM A611 A 44A1
ASTM A611 A, B, C and E 44A1
ASTM A611 B 44A1
ASTM A611 C 44A1
ASTM A611 D 44A1
ASTM A611 E 44A1
ASTM A612 44A2
ASTM A615/40 44A1
ASTM A615/60 44A1
ASTM A618/1 44A2
ASTM A618/11 44J
ASTM A618/111 44J
ASTM A619 44A1
ASTM A620 44A1
ASTM A621 44A1
ASTM A622 44A1
ASTM A623 D 44A1
ASTM A623 L 44A1
ASTM A623 MC 44A1
ASTM A623 MR 44A1
ASTM A628 44K2
ASTM A631 44A3
ASTM A632 TP304 44N1
ASTM A632 TP304L 44N1
ASTM A632 TP310 44N1
ASTM A632 TP316 44N1
ASTM A632 TP316L 44N1
ASTM A632 TP317 44N1
ASTM A632 TP321 44N1
ASTM A632 TP347 44N1
ASTM A632 TP348 44N1
ASTM A633 A 44A1
ASTM A633 B 44A1
ASTM A633 C 44A1
ASTM A633 D 44A1
ASTM A633 E 44J
ASTM A635 44A1
ASTM A636 27A
ASTM A638/660 44N2
ASTM A638/662 44N2
ASTM A639/661 44N2
ASTM A639/671 13A
ASTM A640 44A1
ASTM A641 44A1
ASTM A643 A 44A1
ASTM A643 B 44K1
ASTM A643 C1 44K1
ASTM A643 C2 44K1
ASTM A643 C3 44K1
ASTM A643 D1 44K1
ASTM A643 D2 44K1
ASTM A645 44K1
ASTM A646 44K2
ASTM A646 D64C
see D 64C 44K2
ASTM A646 HP9/4/20 44L
ASTM A646 HP9/4/30 44L
ASTM A646 MARAGE
see MARAGE 200 44F3
ASTM A646 MARAGE
see MARAGE 250 44F3
ASTM A646 MARAGE
see MARAGE 300 44F3
ASTM A646/98 BV40 44K2
ASTM A646/1111 44K1
ASTM A648/1 44A3
ASTM A648/11 44A3
ASTM A648/111 44A3
ASTM A649/1 A 44K2
ASTM A649/1 B 44K2
ASTM A649/2 44A2
ASTM A649/3 44K1
ASTM A649/4 44A2
ASTM A650 44A1
ASTM A651 TP304 44N1
ASTM A651 TP316 44N1
ASTM A651 TP409 44M1
ASTM A651 TP430 44M1
ASTM A651 TP430 Ti 44M1
ASTM A651 TP434 44M1
ASTM A651 TPXM8 44M1
ASTM A656/1 44J
ASTM A656/2 44A1
ASTM A657 44A1
ASTM A658 44F1
ASTM A659 44A1
ASTM A660 WCA 44A1
ASTM A660 WCB 44A1
ASTM A660 WCC 44A1
ASTM A662 A 44A1
ASTM A662 B 44A1
ASTM A663 44A1
ASTM A666 44N1
ASTM A666 44P
ASTM A667 20B
ASTM A668 A-F 44A1
ASTM A668 AH-FH 44A1
ASTM A668 GH-NH 44A1
ASTM A668 G-N 44A1
ASTM A669 44N3
ASTM A670 27C
ASTM A675 44A1
ASTM A678 A 44A1
ASTM A678 B 44A1
ASTM A678 C 44A1
ASTM A679 44A3
ASTM A679 44A4
ASTM A680 44A2
ASTM A680 44A3
ASTM A682 44A3
ASTM A684 44A3
ASTM A687/1 44K1
ASTM A687/11 44K1
ASTM A688 TP304 44N1
ASTM A688 TP304L 44N1
ASTM A688 TP316 44N1
ASTM A688 TP316L 44N1
ASTM A688 TPXM29 44P
ASTM A690 44F1
ASTM A692 44G1
ASTM A693 XM9 44N3
ASTM A693 XM12 44N3
ASTM A693 XM13 44N3
ASTM A693 XM16 44N3
ASTM A693 XM25 44N3
ASTM A693/631 44N3
ASTM A693/632 44N3
ASTM A693/635 44N3
ASTM A694 44A2
ASTM A695 A 44A2
ASTM A695 B 44A2
ASTM A695 C 44A2
ASTM A695 D 44A2
ASTM A696 B 44A2
ASTM A696 C 44A2
ASTM A699/1 44G1
ASTM A699/2 44G1
ASTM A699/3 44G1
ASTM A699/4 44G1
ASTM A701 24.1
ASTMA705
see designation for details
ASTM A706 44K1
ASTM A707 44/2 44K1
ASTM A707 L1/1 44A1
ASTM A707 L1/2 44A1
ASTM A707 L1/3 44A1
ASTM A707 L1/4 44A1
ASTM A707 L2/1 44A2
ASTM A707 L2/2 44A2
ASTM A707 L2/3 44A2
ASTM A707 L2/4 44A2
ASTM A707 L3/1 44J
ASTM A707 L3/2 44J
ASTM A707 L3/3 44J
ASTM A707 L3/4 44J
ASTM A707 L4/1 44K1
ASTM A707 L4/3 44K1
ASTM A707 L4/4 44K1
ASTM A707 L5/1 44K1
ASTM A707 L5/2 44K1
ASTM A707 L5/3 44K1
ASTM A707 L5/4 44K1
ASTM A707 L6/1 44G1
ASTM A707 L6/2 44G1
ASTM A707 L6/2 44G1
ASTM A707 L6/3 44G1
ASTM A707 L7/1 44F1
ASTM A707 L7/2 44F1
ASTM A707 L7/3 44F1
ASTM A707 L7/4 44F1
ASTM A707 L8/1 44K1
ASTM A707 L8/2 44K1
ASTM A707 L8/3 44K1
ASTM A707 L8/4 44K1
ASTM A709/36 44A2
ASTM A709/50 44A2
ASTM A709/50 W 44A1
ASTM A709/100 44A1
ASTM A709/100 W 44A1
ASTM A710 A 44K1
ASTM A710 B 44K1
ASTM A711 44A2
ASTM A713 44A3
ASTM A714/1 44A1
ASTM A714/2 44J
ASTM A714/3 44J
ASTM A714 4 44K1
ASTM A714/5 E or S 44F1
ASTM A714/5 F 44F1
ASTM A714/6 E & S 44K1
ASTM A715 44A1
ASTM A716 20B
ASTM A722 44A2
ASTM A723/1 44K1
ASTM A723/2 44K1
ASTM A723/3 44K1
ASTM A724 A 44A1
ASTM A727 44A1
ASTM A729 44A3

ASTM A730 A 44A1
ASTM A730 B 44A1
ASTM A730 C, D and E 44A3
ASTM A730 F 44A3
ASTM A730 G and H 44A3
ASTM A731 TPXM27 44M1
ASTM A731 TPXM33 44M1
ASTM A732 SN 44J
ASTM A732/1 A 44A1
ASTM A732/2 A 44A2
ASTM A732/2 Q 44A2
ASTM A732/3 A 44A2
ASTM A732/3 Q 44A2
ASTM A732/4 A 44A3
ASTM A732/4 Q 44A3
ASTM A732/6 N 44G1
ASTM A732/7 Q 44K1
ASTM A732/8 Q 44K1
ASTM A732/9 Q 44K1
ASTM A732/10 Q 44K1
ASTM A732/11 Q 44F1
ASTM A732/12 Q 44K2
ASTM A732/13 Q 44K1
ASTM A732/14 Q 44K1
ASTM A732/15 A 44E3
ASTM A733 A53 44A1
ASTM A733 A120 44A1
ASTM A733 A312 44N1
ASTM A734 A 44K1
ASTM A734 B 44K1
ASTM A735 44G1
ASTM A736 44K1
ASTM A737 A 44J
ASTM A737 B 44A1
ASTM A737 C 44G1
ASTM A738 44A2
ASTM A739-B 11 44K1
ASTM A739-B 22 44K1
ASTM A741 44A1
ASTMA743
stainless steel castings - see designation for steel type
ASTMA744
general specification for stainless steel castings - see designation for steel type
ASTM A744 44N1
ASTM A746 20B
ASTM A747 CB7 Cu 1 44N3
ASTM A747 CB7 Cu 2 44N3
ASTM A748 20B
ASTM A749 44A1
ASTM A757-A1Q 44A2
ASTM A757-A2Q 44A2
ASTM A757-B2N 44F1
ASTM A757 B2N B2Q 44N1
ASTM A757-B2Q 44F1
ASTM A757-B3N 44F1
ASTM A757 B3N B3Q 44N1
ASTM A757-B3Q 44F1
ASTM A757-B4N 44F1
ASTM A757 B4N B4Q 44N1
ASTM A757-B4Q 44F1
ASTM A757 C1 Q 44K1
ASTM A757 D1 N1 44K1
ASTM A757 D1 N2 44K1
ASTM A757 D1 N3 44K1
ASTM A757 D1 Q1 44K1
ASTM A757 D1 Q2 44K1
ASTM A757 D1 Q3 44K1
ASTM A757 E1 Q 44K1
ASTM A757 E2 N 44K1
ASTM A757 E2 Q 44K1
ASTM A757 E3 N 44K1
ASTM A757 E3N 44M1
ASTM A757 EQ 1 44K1
ASTM A757 EQ 2 44K1
ASTM A759 44A2
ASTM A759 44A3
ASTM A764 44A3
ASTM A765-1 44K1
ASTM A765-11 44K1
ASTM A765-111 44K1
ASTM A766 44A1
ASTM A767 44A2
ASTM A768 44M1
ASTM A769 44A1
ASTM A769-36 44A1
ASTM A769-40 44A1
ASTM A769-45 44A1
ASTM A769-45 W 44A1
ASTM A769-50 44A1
ASTM A769-50 W 44A1
ASTM A769-60 44A1
ASTM A769-80 44A1
ASTM A771 44N1
ASTM A774 44N1
ASTM A775 44A2
ASTM A778 44N1
ASTM A779 44A2
ASTM A782/1 44A1
ASTM A782/2 44A1
ASTM A782/3 44A1
ASTM A782-1 44K1
ASTM A782-2 44K1
ASTM A782-3 44K1
ASTM A787 MT 1010 44A1
ASTM A787 MT 1015 44A1
ASTM A787 MT 1020 44A1
ASTM A787 MTX 1015 44A1
ASTM A787 MTX 1020 44A1
ASTM A789 S31200 44N3
ASTM A789 S31260 44N3
ASTM A789 S31500 44N3
ASTM A789 S31803 44N3
ASTM A789 S32304 44N3
ASTM A789 S32550 44N3
ASTM A789 S32950 44N3
ASTM A791 44M1
ASTM A793 44N1
ASTM A794 44A1
ASTM A795 A 44A1
ASTM A795 B 44A1
ASTM A795 E 44A1
ASTM A795 F 44A1
ASTM A795 S 44A1
ASTM A803 44M1
ASTM A805 44A1
ASTM A808 44A1
ASTM A812 G 65 44A1
ASTM A812 G 80 44A1
ASTM A812/65 44A1
ASTM A812/80 44A1
ASTMA813
general specification for welded austentic stainless steel pipe - see designation for details
ASTM A814 44N1
ASTM A815 S31803 44N3
ASTM A815 S41500 44M1
ASTM A815 WP27 44M1
ASTM A815 WP33 44M1
ASTM A815 WP410 44M1
ASTM A815 WP429 44M1
ASTM A815 WP430 44M1
ASTM A815 WP430 Ti 44M1
ASTM A815 WP446 44M1
ASTM A816 E8018B1 44K1
ASTM A821 44A2
ASTM A822 44A1
ASTM A823 ACA 20B
ASTM A823 ACB 20B
ASTM A823 ACC 20B
ASTM A823 ASA 20B
ASTM A823 ASB 20B
ASTM A823 ASC 20B
ASTM A823 ASS 20B
ASTM A823 N-CA 20B
ASTM A823 N-CB 20B
ASTM A823 N-CC 20B
ASTM A823 N-SA 20B
ASTM A823 N-SB 20B
ASTM A823 N-SC 20B
ASTM A823 N-SS 20B
ASTMA826
steel for breeder reactor tubes - see designation for type
ASTM A827 44A1
ASTMA829
steel plate for structures operated by AISI designation
ASTMA830
carbon steel plate for structure graded by AISI designation
ASTM A832 44K1
ASTM A836 44A1
ASTM A841 44A1
ASTM A841 44K1
ASTM A844 44F1
ASTM A847 44A1
ASTM A850-1 44A1
ASTM A850-2 44A1
ASTM A851 44A1
ASTM A852 44K1
ASTM A853 44A1
ASTM A857 44A2
ASTM A858 44A1
ASTM A859/1 44K1
ASTM A859/2 44K1
ASTM A860 44A1
ASTM A865 44A1
ASTM A871 44K1
ASTM B 30/12 A 14M

Note Designation follows ASTM number

ASTM B 30/13 A 14M
ASTM B 30/13 B 14M
ASTM B 291 14B
ASTM B 359/230 14B
ASTM B 359/442 14B
ASTM B 359/443 14B
ASTM B 359/444 14B
ASTM B 359/445 14B
ASTM B 642 14B
ASTM B 706 14B
ASTM B1 14A
ASTM B2 14A
ASTM B3 14A
ASTM B4 14A
ASTM B5 14A
ASTM B6 55A
ASTM B11 ATP 14A
ASTM B11 DHP 14A
ASTM B11 DPA 14A
ASTM B11 ETP 14A
ASTM B11 FRTP 14A
ASTM B11/110 14A
ASTM B11/122 14A
ASTM B11/125 14A
ASTM B11/141 14A
ASTM B11/142 14A
ASTM B12 14A
ASTM B12/102 14A
ASTM B12/110 14A
ASTM B12/120 14A
ASTM B12/122 14A
ASTM B12/125 14A
ASTM B12/141 14A
ASTM B12/142 14A
ASTM B16 14B
ASTM B19 14B
ASTM B21 A 14C
ASTM B21 B 14C
ASTM B21 C 14C
ASTM B22 A 14K
ASTM B22 B 14K
ASTM B22 C 14K
ASTM B22 D 14K
ASTM B22 E 14B
ASTM B23/1 50A
ASTM B23/2 50A
ASTM B23/3 50A
ASTM B23/7 21.1
ASTM B23/8 21.1
ASTM B23/13 21.1
ASTM B23/15 21.1
ASTM B26 C4A 1L
ASTM B26 CG100A 1L
ASTM B26 CN42A 1L
ASTM B26 CS43A 1L
ASTM B26 CS72A 1L
ASTM B26 CS74A 1L
ASTM B26 G4A 1J
ASTM B26 G10A 1J
ASTM B26 GM70B 1J
ASTM B26 GS42A 1J
ASTM B26 S5A 1C
ASTM B26 S5B 1C
ASTM B26 SC8 1L
ASTM B26 SC51A 1L
ASTM B26 SC64D 1L
ASTM B26 SC82A 1K
ASTM B26 SG70A 1K
ASTM B26 ZC32A 1M
ASTM B26 ZC81A 1M
ASTM B26 ZG42A 1M
ASTM B26 ZG61A 1M
ASTM B26 ZG61B 1M
ASTM B29 21.1
ASTM B30/1 A 14K
ASTM B30/1 B 14K
ASTM B30/2 A 14K
ASTM B30/2 B 14K
ASTM B30/3 A 14K
ASTM B30/3 B 14K
ASTM B30/3 C 14K
ASTM B30/3 D 14K
ASTM B30/4 A 14K
ASTM B30/4 B 14K
ASTM B30/5 A 14K
ASTM B30/5 B 14K
ASTM B30/6 A 14K
ASTM B30/6 B 14K
ASTM B30/6 C 14C
ASTM B30/7 A 14C
ASTM B30/8 A 14C
ASTM B30/8 B 14C
ASTM B30/8 C 14C
ASTM B30/9 A 14G
ASTM B30/9 B 14G
ASTM B30/9 C 14G
ASTM B30/9 D 14G
ASTM B30/10 A 14F
ASTM B30/11 A 14F
ASTM B30/11 B 14F
ASTM B32/1/5 S 21.1
ASTM B32/2 B 21.1
ASTM B32/2/5 S 21.1
ASTM B32/2A 21.1
ASTM B32/5 A 21.1
ASTM B32/5 B 21.1
ASTM B32/10 B 21.1
ASTM B32/15 B 21.1
ASTM B32/20 B 21.1
ASTM B32/20 C 21.1
ASTM B32/25 A 21.1
ASTM B32/25 B 21.1
ASTM B32/25 C 21.1
ASTM B32/30 A 21.1
ASTM B32/30 B 21.1
ASTM B32/30 C 21.1
ASTM B32/35 A 21.1
ASTM B32/35 B 21.1
ASTM B32/35 C 21.1
ASTM B32/40 A 21.1
ASTM B32/40 B 21.1
ASTM B32/40 C 21.1
ASTM B32/45 A 21.1
ASTM B32/45 B 21.1
ASTM B32/50 A 50A
ASTM B32/50 B 50A
ASTM B32/60 A 50A
ASTM B32/60 B 50A
ASTM B32/63 A 50A
ASTM B32/63 B 50A
ASTM B32/70 A 50A
ASTM B32/70 B 50A
ASTM B32/95 TA 50A
ASTM B36/1 14B
ASTM B36/2 14B
ASTM B36/3 14B
ASTM B36/4 14B
ASTM B36/6 14B
ASTM B36/8 14B
ASTM B37 1A
ASTM B39 A Shot 27A
ASTM B39 Electrolytic 27A
ASTM B39 Ingot 27A
ASTM B39 X Shot 27A
ASTM B42 14A
ASTM B42/102 14A
ASTM B42/120 14A
ASTM B42/122 14A
ASTM B43 14B
ASTM B47 14A
ASTM B48 14A
ASTM B49 14A
ASTM B52A 14A
ASTM B52B 14A
ASTM B53 A 14M
ASTM B53 B 14M
ASTM B53 C 14M
ASTM B61 14K
ASTM B62 14K
ASTM B66/MEDIUM BRONZE 14K
ASTM B66/HARD BRONZE 14K
ASTM B66/PHOS BRONZE 14K
ASTM B66/SOFT BRONZE 14K
ASTM B68 DHP 14A
ASTM B68 DLP 14A
ASTM B68 OF 14A
ASTM B69 55A
ASTM B72 A 14A
ASTM B72 B 14A
ASTM B75 DHP 14A
ASTM B75 DLP 14A
ASTM B75 DPA 14A
ASTM B75 OF 14A
ASTM B75/102 14A
ASTM B75/120 14A
ASTM B75/122 14A
ASTM B75/142 14A
ASTM B80 AM 100 A 23B
ASTM B80 AZ 63 A 23B
ASTM B80 AZ 81 A 23B
ASTM B80 AZ 91 C 23B
ASTM B80 AZ 92 A 23B
ASTM B80 EK 30 A 23B
ASTM B80 EK 41 A 23B
ASTM B80 EZ 33 A 23B
ASTM B80 HK 31 A 23B
ASTM B80 HZ 32 A 23B
ASTM B80 K 1 A 23B
ASTM B80 QE 22 A 23B
ASTM B80 ZE 41 A 23B
ASTM B80 ZH 62 A 23B
ASTM B80 ZK 51 A 23B
ASTM B80 ZK 61 A 23B
ASTM B85 G8A 1J
ASTM B85 S5C 1C
ASTM B85 S12A 1C
ASTM B85 S12B 1C
ASTM B85 SC84A 1L
ASTM B85 SC84B 1L
ASTM B85 SC114A 1L

ASTM B85 SG100A 1K
ASTM B85 SG100B 1K
ASTM B86 AC41A 55A
ASTM B86 AG40A 55A
ASTM B88 14A
ASTM B90 AZ 31 B 23B
ASTM B90 HK 31 A 23B
ASTM B90 HM 21 A 23B
ASTM B90 ZE 10 A 23B
ASTM B91 AZ 61 A 23B
ASTM B91 AZ 80 A 23B
ASTM B91 TA 54 A 23B
ASTM B91 ZK 60 A 23B
ASTM B92/9980 A 23A
ASTM B92/9990 A 23A
ASTM B92/9990 B 23A
ASTM B92/9995 A 23A
ASTM B93 AM 100 A 23B
ASTM B93 AZ 63 A 23B
ASTM B93 AZ 81 A 23B
ASTM B93 AZ 91 A 23B
ASTM B93 AZ 91 B 23B
ASTM B93 AZ 91 C 23B
ASTM B93 AZ 92 A 23B
ASTM B94 AZ 91 A 23B
ASTM B94 AZ 91 B 23B
ASTM B96 A 14M
ASTM B96 C 14M
ASTM B97 14M
ASTM B97 B 14M
ASTM B97 C 14M
ASTM B98 A 14M
ASTM B98 B 14M
ASTM B98 D 14M
ASTM B98/651 14M
ASTM B98/655 14M
ASTM B98/661 14M
ASTM B99 A 14M
ASTM B99 B 14M
ASTM B100/1 14K
ASTM B102 CY44A 50A
ASTM B102 PY1815A 50A
ASTM B102 Y10A 21.1
ASTM B102 YC135A 50A
ASTM B102 YT155A 21.1
ASTM B103 A 14K
ASTM B103 B 14K
ASTM B103 B1 14K
ASTM B103 C 14K
ASTM B103 D 14K
ASTM B105 14B
ASTM B107 AZ 31 B 23B
ASTM B107 AZ 31 C 23B
ASTM B107 AZ 61 A 23B
ASTM B107 AZ 80 A 23B
ASTM B107 M 1 A 23B
ASTM B107 ZK 60 A 23B
ASTM B108 CG100A 1L
ASTM B108 CN42A 1L
ASTM B108 CS42A 1L
ASTM B108 CS66A 1L
ASTM B108 CS72A 1L
ASTM B108 CS104A 1L
ASTM B108 GM42A 1J
ASTM B108 GM70B 1J
ASTM B108 GZ42A 1J
ASTM B108 S5A 1C
ASTM B108 S5B 1C
ASTM B108 SC9IAI 1C
ASTM B108 SC41 1C
ASTM B108 SC51A 1L
ASTM B108 SC64D 1L
ASTM B108 SC92A 1L
ASTM B108 SC94A 1L
ASTM B108 SC103A 1L
ASTM B108 SC122A 1L
ASTM B108 SG51B 1K
ASTM B108 SG70A 1K
ASTM B108 SG70B 1K
ASTM B108 SN122A 1K
ASTM B108 ZC60A 1M
ASTM B108 ZC81B 1M
ASTM B108 ZG32A 1M
ASTM B108 ZG42A 1M
ASTM B111 14A
ASTM B111 70/30 Cu Ni 14E
ASTM B111 80/20 Cu Ni 14E
ASTM B111 90/10 Cu Ni 14E
ASTM B111 A 14B
ASTM B111 A1 BRASS 14B
ASTM B111 A1 BRONZE 14G
ASTM B111 B 14B
ASTM B111 C 14B
ASTM B111 D 14B
ASTM B111 MUNTZ 14B
ASTM B111 RED BRASS 14B
ASTM B111/102 14A
ASTM B111/120 14A
ASTM B111/122 14A
ASTM B111/142 14A
ASTM B111/230 14B
ASTM B111/280 14C
ASTM B111/442 14B
ASTM B111/443 14B
ASTM B111/444 14B
ASTM B111/445 14B
ASTM B111/608 14G
ASTM B111/687 14B
ASTM B111/687 14G
ASTM B111/704 14E
ASTM B111/706 14E
ASTM B111/710 14E
ASTM B111/715 14E
ASTM B111/716 14E
ASTM B111/720 14E
ASTM B115 14A
ASTM B119 14B
ASTM B121 14B
ASTM B121/1 14B
ASTM B121/2 14B
ASTM B121/3 14B
ASTM B121/4 14B
ASTM B121/5 14B
ASTM B122/1 14F
ASTM B122/2 14F
ASTM B122/3 14F
ASTM B122/4 14F
ASTM B122/5 14E
ASTM B122/6 14E
ASTM B122/7 14F
ASTM B122/8 14F
ASTM B122/9 14F
ASTM B124/2 14C
ASTM B124/3 14C
ASTM B124/4 14C
ASTM B124/7 14M
ASTM B124/11 A 14G
ASTM B124/11 B 14G
ASTM B124/12 14A
ASTM B124/13 14B
ASTM B124/14 14F
ASTM B127 27D
ASTM B129 14B
ASTM B130 14B
ASTM B131 14B
ASTM B132 A 14C
ASTM B132 B 14C
ASTM B134/1 14B
ASTM B134/2 14B
ASTM B134/3 14B
ASTM B134/4 14B
ASTM B134/6 14B
ASTM B134/7 14B
ASTM B134/8 14B
ASTM B135/1 14B
ASTM B135/2 14B
ASTM B135/3 14B
ASTM B135/4 14B
ASTM B135/5 14C
ASTM B135/6 14C
ASTM B135/7 14B
ASTM B138 A 14B
ASTM B138 B 14B
ASTM B139 A 14K
ASTM B139 B1 14K
ASTM B139 B2 14K
ASTM B139 C 14K
ASTM B139 D 14K
ASTM B140 A 14B
ASTM B140 B 14B
ASTM B143/1 A 14K
ASTM B143/1 B 14K
ASTM B143/2 A 14K
ASTM B143/2 B 14K
ASTM B144/3 A 14K
ASTM B144/3 B 14K
ASTM B144/3 C 14K
ASTM B144/3 D 14K
ASTM B144/3 E 14K
ASTM B145/4 A 14K
ASTM B145/4 B 14K
ASTM B145/5 A 14K
ASTM B145/5 B 14K
ASTM B146/6 A 14K
ASTM B146/6 B 14K
ASTM B146/6 C 14K
ASTM B147/7 A 14K
ASTM B147/8 A 14C
ASTM B147/8 B 14C
ASTM B147/8 C 14C
ASTM B148/9 A 14G
ASTM B148/9 B 14G
ASTM B148/9 C 14G
ASTM B148/9 D 14G
ASTM B149/10 A 14F
ASTM B149/11 A 14F
ASTM B149/11 B 14F
ASTM B150/1 14G
ASTM B150/2 14G
ASTM B150/3 14G
ASTM B150/614 14G

ASTM B150/630 14G
ASTM B150/642 14G
ASTM B151 A 14F
ASTM B151 B 14F
ASTM B151 B1 14F
ASTM B151 C 14F
ASTM B151 D 14F
ASTM B151 E 14F
ASTM B152 ATP 14A
ASTM B152 DHP 14A
ASTM B152 DPS 14A
ASTM B152 ETP 14A
ASTM B152 FRTP 14A
ASTM B152 OF 14A
ASTM B152 OFS 14A
ASTM B152 STP 14A
ASTM B152/102 14A
ASTM B152/104 14A
ASTM B152/105 14A
ASTM B152/107 14A
ASTM B152/122 14A
ASTM B152/123 14A
ASTM B159 A 14K
ASTM B159 C 14K
ASTM B159 D 14K
ASTM B160 27A
ASTM B161 27A
ASTM B162 27A
ASTM B164 A 27D
ASTM B164 B 27D
ASTM B165 27D
ASTM B166 27B
ASTM B167 27B
ASTM B168 27B
ASTM B169 A 14G
ASTM B169 C 14G
ASTM B169 D 14G
ASTM B169/612 14G
ASTM B171 14B
ASTM B171/464 14G
ASTM B171/614 14G
ASTM B171/628 14G
ASTM B171/706 14E
ASTM B171/715 14E
ASTM B176 Z 30 A 14K
ASTM B176 ZS 144 A 14M
ASTM B176 ZS331A 14B
ASTM B179 57A 1C
ASTM B179 C4A 1L
ASTM B179 CG100A 1K
ASTM B179 CG100A 1L
ASTM B179 CN42A 1L
ASTM B179 CS43A 1L
ASTM B179 CS66A 1L
ASTM B179 CS72A 1L
ASTM B179 CS76A 1L
ASTM B179 CS104A 1L
ASTM B179 G4A 1J
ASTM B179 G8A 1J
ASTM B179 G10A 1J
ASTM B179 GM70B 1J
ASTM B179 GS42A 1J
ASTM B179 S5A 1C
ASTM B179 S5B 1C
ASTM B179 S5C 1C
ASTM B179 S12A–B 1C
ASTM B179 SC51A 1L
ASTM B179 SC51B 1K
ASTM B179 SC64C 1L
ASTM B179 SC71A 1C
ASTM B179 SC82A 1K
ASTM B179 SC84A–B 1L
ASTM B179 SC91A 1C
ASTM B179 SC94A 1C
ASTM B179 SC103A 1L
ASTM B179 SC114A 1L
ASTM B179 SC122A 1L
ASTM B179 SC649 1L
ASTM B179 SG70A 1K
ASTM B179 SG70B 1K
ASTM B179 SG100A-B 1K
ASTM B179 ZC60A 1M
ASTM B179 ZC81A 1M
ASTM B179 ZC81B 1M
ASTM B179 ZG32A 1M
ASTM B179 ZG42A 1M
ASTM B179 ZG61A 1M
ASTM B179 ZG61B 1M
ASTM B179/995 A 1A
ASTM B184/A12 1A
ASTM B184 A143 1C
ASTM B187 14A
ASTM B188 14A
ASTM B194 14D
ASTM B196 14D
ASTM B197 14D
ASTM B198/12 A 14M
ASTM B198/13 A 14M
ASTM B198/13 B 14M
ASTM B199 AM 91 C 23B
ASTM B199 AM 92 A 23B
ASTM B199 AM 100 A 23B
ASTM B199 AZ81A 1D
ASTM B199 EK 41 A 23B
ASTM B199 EZ 33 A 23B
ASTM B199 HK31A 1N
ASTM B199 QE22A 1N
ASTM B206 A 14F
ASTM B206 B 14F
ASTM B206 B1 14F
ASTM B206 C 14F
ASTM B206 D 14F
ASTM B206 E 14F
ASTMB209
see designation for details
ASTM B209/1060 1A
ASTM B209/1100 1A
ASTM B209/1230 1A
ASTM B209/2014 CLAD 1H
ASTM B209/2024 1F
ASTM B209/2024 CLAD 1H
ASTM B209/2219 1F
ASTM B209/3003 1B
ASTM B209/3003 CLAD 1H
ASTM B209/3004 1B
ASTM B209/3004 CLAD 1H
ASTM B209/3005 1B
ASTM B209/5005 1D
ASTM B209/5050 1D
ASTM B209/5052 1D
ASTM B209/5083 1D
ASTM B209/5086 1D
ASTM B209/5154 1D
ASTM B209/5155 1D
ASTM B209/5155 CLAD 1H
ASTM B209/5252 1D
ASTM B209/5254 1D
ASTM B209/5257 1D
ASTM B209/5454 1D
ASTM B209/5456 1D
ASTM B209/5457 1D
ASTM B209/5557 1D
ASTM B209/5652 1D
ASTM B209/5657 1D
ASTM B209/6003 1E
ASTM B209/6061 1E
ASTM B209/6061 CLAD 1H
ASTM B209/7039 1G
ASTM B209/7072 1A
ASTM B209/7075 1G
ASTM B209/7075 1H
ASTMB210
see designation for details
ASTM B210/1060 1A
ASTM B210/1100 1A
ASTM B210/2024 1F
ASTM B210/3003 1B
ASTM B210/3003 CLAD 1H
ASTM B210/3004 1B
ASTM B210/5050 1D
ASTM B210/5052 1D
ASTM B210/5062 1E
ASTM B210/5083 1D
ASTM B210/5086 1D
ASTM B210/5154 1D
ASTM B210/5254 1D
ASTM B210/5652 1D
ASTM B210/6061 1E
ASTM B210/6063 1E
ASTM B210/7072 1A
ASTM B210/7075 1G
ASTMB211
see designation for details
ASTM B211/1060 1A
ASTM B211/1100 1A
ASTM B211/2011 1F
ASTM B211/2014 1F
ASTM B211/2017 1F
ASTM B211/2024 1F
ASTM B211/2219 1F
ASTM B211/3003 1B
ASTM B211/5052 1D
ASTM B211/5056 1D
ASTM B211/5056 CLAD 1H
ASTM B211/5154 1D
ASTM B211/5254 1D
ASTM B211/5652 1D
ASTM B211/6061 1E
ASTM B211/6262 1E
ASTM B211/7075 1G
ASTM B217 AZ 31 B 23B
ASTM B217 AZ 31 C 23B
ASTM B217 AZ 61 A 23B
ASTM B217 M 1 A 23B
ASTM B217 ZK 60 A 23B
ASTM B221/1060 1A
ASTM B221/1100 1A
ASTM B221/2014 1F
ASTM B221/2024 1F
ASTM B221/2219 1F
ASTM B221/3003 1B

ASTM B221/3004 1B
ASTM B221/5052 1D
ASTM B221/5083 1D
ASTM B221/5086 1D
ASTM B221/5154 1D
ASTM B221/5454 1D
ASTM B221/5456 1D
ASTM B221/5652 1D
ASTM B221/6061 1E
ASTM B221/6062 1E
ASTM B221/6063 1E
ASTM B221/6351 1E
ASTM B221/7072 1A
ASTM B221/7075 1G
ASTM B221/7079 1G
ASTM B221/7178 1G
ASTM B224 ATP 14A
ASTM B224 CAST 14A
ASTM B224 CATH 14A
ASTM B224 DHP 14A
ASTM B224 DLP 14A
ASTM B224 DPA 14A
ASTM B224 DPS 14A
ASTM B224 DPTE 14A
ASTM B224 ETP 14A
ASTM B224 FRHC 14A
ASTM B224 FRTP 14A
ASTM B224 OF 14A
ASTM B224 OFP 14A
ASTM B224 OFPTE 14A
ASTM B224 OFS 14A
ASTM B224 OFTE 14A
ASTM B224 SATP 14A
ASTM B224 STP 14A
ASTM B225 E Cu 14K
ASTM B225 E Cu Al A 1 14G
ASTM B225 E Cu Al A 2 14G
ASTM B225 E Cu Al B 14G
ASTM B225 E Cu Ni 14E
ASTM B225 E Cu Si 14K
ASTM B225 E Cu Sn A 14K
ASTM B225 E Cu Sn C 14K
ASTM B227 44A1
ASTM B228 44A1
ASTM B229 14A
ASTM B233 1350 1A
ASTM B234/1060 1A
ASTM B234/3003 1B
ASTM B234/3003 ALCLAD 1H
ASTM B234/5052 1D
ASTM B234/5454 1D
ASTM B234/6061 1E
ASTM B234/6062 1E
ASTM B234/7072 1A
ASTM B235/1060 1A
ASTM B235/1100 1A
ASTM B235/2014 1F
ASTM B235/2024 1F
ASTM B235/3003 1B
ASTM B235/3003 ALCLAD 1H
ASTM B235/3004 1B
ASTM B235/5052 1D
ASTM B235/5154 1D
ASTM B235/5254 1D
ASTM B235/5454 1D
ASTM B235/5456 1D
ASTM B235/5652 1D
ASTM B235/6061 1E
ASTM B235/6062 1E
ASTM B235/6063 1E
ASTM B235/6351 1E
ASTM B235/7075 1G
ASTM B237 B 2.1
ASTM B237A 2.1
ASTM B240 AC41A 55A
ASTM B240 AG40A 55A
ASTM B241/1060 1A
ASTM B241/1100 1A
ASTM B241/2014 1F
ASTM B241/2024 1F
ASTM B241/2219 1F
ASTM B241/3003 1B
ASTM B241/3003 ALCLAD 1H
ASTM B241/5052 1D
ASTM B241/5083 1D
ASTM B241/5154 1D
ASTM B241/5254 1D
ASTM B241/5454 1D
ASTM B241/5456 1D
ASTM B241/5652 1D
ASTM B241/6061 1E
ASTM B241/6062 1E
ASTM B241/6063 1E
ASTM B241/6351 1E
ASTM B241/7072 1A
ASTM B241/7075 1G
ASTM B241/7079 1G
ASTM B241/7178 1G
ASTM B246 14A
ASTM B247 4032 1C
ASTM B247/1100 1A
ASTM B247/2014 1F
ASTM B247/2018 1F
ASTM B247/2025 1F
ASTM B247/2218 1F
ASTM B247/3003 1B
ASTM B247/5083 1D
ASTM B247/5456 1D
ASTM B247/6053 1E
ASTM B247/6061 1E
ASTM B247/6151 1E
ASTM B247/7075 1G
ASTM B259 R Cu 14K
ASTM B259 R Cu Al A 2 14G
ASTM B259 R Cu Al B 14G
ASTM B259 R Cu Ni 14E
ASTM B259 R Cu Si A 14M
ASTM B259 R Cu Si B 14M
ASTM B259 R Cu Sn A 14K
ASTM B259 RB Cu Zn A 14C
ASTM B259 RB Cu Zn B 14C
ASTM B259 RB Cu Zn C 14C
ASTM B259 RB Cu Zn D 14C
ASTM B260 Ag 7 42.1
ASTM B260 Ag 18 42.1
ASTM B260 Ag 19 42.1
ASTM B260 B Ag 1 42.1
ASTM B260 B Ag 1 A 42.1
ASTM B260 B Ag 2 42.1
ASTM B260 B Ag 3 42.1
ASTM B260 B Ag 4 42.1
ASTM B260 B Ag 5 42.1
ASTM B260 B Ag 6 42.1
ASTM B260 B Ag 8 42.1
ASTM B260 B Ag 8 A 42.1
ASTM B260 B Ag 9 42.1
ASTM B260 B Ag 10 42.1
ASTM B260 B Ag 11 42.1
ASTM B260 B Ag 13 42.1
ASTM B260 B Ag Mn 42.1
ASTM B260 B Al Si 1 1C
ASTM B260 B Al Si 2 1C
ASTM B260 B Al Si 3 1C
ASTM B260 B Al Si 4 1C
ASTM B260 B Al Si 5 1C
ASTM B260 B Au 1 14H
ASTM B260 B Au 3 14H
ASTM B260 B Cu 14A
ASTM B260 B Cu 1 14A
ASTM B260 B Cu 1 14H
ASTM B260 B Cu 1 A 14H
ASTM B260 B Cu 1a 14A
ASTM B260 B Cu 2 14H
ASTM B260 B Cu Au 1 14H
ASTM B260 B Cu Au2 17.1
ASTM B260 B Cu P 1 14H
ASTM B260 B Cu P 2 14H
ASTM B260 B Cu P 3 14H
ASTM B260 B Cu P 4 14H
ASTM B260 B Cu P 5 14H
ASTM B260 B Mg 1 23B
ASTM B260 B Mg 2 23B
ASTM B260 B Mg 2a 23B
ASTM B260 B Ni 1 27B
ASTM B260 B Ni 2 27B
ASTM B260 B Ni 3 27F
ASTM B260 B Ni 4 27F
ASTM B260 B Ni 5 27B
ASTM B260 B Ni 6 27B
ASTM B260 B Ni 6 27F
ASTM B260 B Ni 7 27B
ASTM B260 B Ni 7 27C
ASTM B260 B Ni Cr 27B
ASTM B260 RB Cu Zn A 14C
ASTM B260 RB Cu Zn D 14F
ASTM B260/B Au2 17.1
ASTM B260/B Au4 17.1
ASTM B265/1 51A
ASTM B265/2 51A
ASTM B265/3 51A
ASTM B265/4 51A
ASTM B265/5 51C
ASTM B265/6 51B
ASTM B265/7 51B
ASTM B280 14A
ASTM B280 DHP 14A
ASTM B280 DLP 14A
ASTM B283 14B
ASTM B285 ER1060 1A
ASTM B285 ER1100 1A
ASTM B285 ER1260 1A
ASTM B285 ER2014 1F
ASTM B285 ER3004 1B
ASTM B285 ER4043 1C
ASTM B285 ER5050 1D
ASTM B285 ER5052 1D
ASTM B285 ER5154 1D
ASTM B285 ER5183 1D
ASTM B285 ER5254 1D
ASTM B285 ER5356 1D
ASTM B285 ER5554 1D

ASTM B285 ER5556 1D
ASTM B285 ER5652 1D
ASTM B285 RC4A 1F
ASTM B285 RCN42A 1F
ASTM B285 RS5B 1C
ASTM B285 RSC51A 1C
ASTM B285 RSG70A 1C
ASTM B285/R ZG61A 1G
ASTM B292 A 14F
ASTM B292 B 14F
ASTM B295 E 3 N 12 27B
ASTM B295 E 4 N 12 27B
ASTM B295 E Ni 1 27F
ASTM B295 E Ni Cr 1 27B
ASTM B295 E Ni Cr Fe 1 27B
ASTM B295 E Ni Cr Fe 2 27B
ASTM B295 E Ni Cr Fe 3 27B
ASTM B295 E Ni Cu 1 27D
ASTM B295 E Ni Cu 2 27D
ASTM B295 E Ni Cu 4 27D
ASTM B295 E Ni Mo 1 27F
ASTM B295 E Ni Mo 2 27F
ASTM B295 E Ni Mo 3 27F
ASTM B295 E3N1B 27C
ASTM B295 E3N1C 27C
ASTM B295 E3N10 27D
ASTM B295 E3N11 27F
ASTM B295 E3N14 27D
ASTM B295 E3N19 27C
ASTM B295 E4N1B 27C
ASTM B295 E4N1C 27C
ASTM B295 E4N10 27D
ASTM B295 E4N11 27F
ASTM B297 EW P 52.1
ASTM B297 EW Th 1 52.1
ASTM B297 EW Th 2 52.1
ASTM B297 EW Zr 52.1
ASTM B299 MD120 51A
ASTM B299 MD160 51A
ASTM B299 ML120 51A
ASTM B299 ML140 51A
ASTM B299 ML160 51A
ASTM B299 SL120 51A
ASTM B299 SL140 51A
ASTM B299 SL160 51A
ASTM B301 14A
ASTM B304 ER Ni 3 27A
ASTM B304 ER Ni Cr 3 27B
ASTM B304 ER Ni Cr Fe 5 27B
ASTM B304 ER Ni Cr Fe 6 27B
ASTM B304 ER Ni Cu 7 27D
ASTM B304 ER Ni Mo 4 27F
ASTM B304 ER Ni Mo 5 27F
ASTM B304 ERN 6N 27B
ASTM B304 ERN 62 27B
ASTM B304 ERN7B 27C
ASTM B304 ERN7C 27C
ASTM B304 ERN7W 27C
ASTM B304 ERN60 27D
ASTM B304 ERN61 27F
ASTM B304 ERN64 27D
ASTM B304 ERN69 27C
ASTM B304 R Ni 2 27A
ASTM B304 R Ni Cr Fe 4 27B
ASTM B304 R Ni Cu 5 27D
ASTM B304 RN 42 27B
ASTM B304 RN40 27D
ASTM B304 RN41 27F
ASTM B304 RN43 27D
ASTM B307/1100 1A
ASTM B307/3003 1B
ASTM B307/3003 ALCLAD 1H
ASTM B307/3004 1B
ASTM B307/5005 1B
ASTM B307/5050 1D
ASTM B307/5050 ALCLAD 1H
ASTM B307/5052 1D
ASTM B308/2014 1F
ASTM B308/5083 1D
ASTM B308/5086 1D
ASTM B308/5454 1D
ASTM B308/5456 1D
ASTM B308/6061 1E
ASTM B308/6062 1E
ASTM B308/6066 1E
ASTM B308/6351 1E
ASTM B313/1100 1A
ASTM B313/3003 1B
ASTM B313/3004 ALCLAD 1H
ASTM B313/5050 1D
ASTM B313/5052 1D
ASTM B313/5086 1D
ASTM B313/5154 1D
ASTM B313/6061 1E
ASTM B315 A 14M
ASTM B315 A7 14M
ASTM B315/615 14M
ASTM B315/655 14M
ASTM B315/658 14M
ASTM B316/1100 1A
ASTM B316/2017 1F
ASTM B316/2024 1F
ASTM B316/2117 1F
ASTM B316/5052 1D
ASTM B316/5056 1D
ASTM B316/5652 1D
ASTM B316/6053 1E
ASTM B316/6061 1E
ASTM B316/7075 1G
ASTM B317 1E
ASTM B317/6101 1E
ASTM B318/3003 1B
ASTM B318/3004 1B
ASTM B318/5005 1D
ASTM B318/5050 1D
ASTM B318/5052 1D
ASTM B322 Ni Mo 27C
ASTM B322 Ni Mo Cr 27C
ASTM B325 21.1
ASTM B327 CG181A 1N
ASTM B327 G1C 1A
ASTM B327 ZG71A 1G
ASTM B333 27C
ASTM B334 27C
ASTM B335 27C
ASTM B336 27C
ASTM B337/1 51A
ASTM B337/2 51A
ASTM B337/3 51A
ASTM B337/4 51A
ASTM B338/1 51A
ASTM B338/2 51A
ASTM B338/3 51A
ASTM B338/4 51A
ASTM B339 50.1
ASTM B344 27B
ASTM B344/1 27B
ASTM B344/2 27B
ASTM B345/1060 1A
ASTM B345/3003 1B
ASTM B345/3003 ALCLAD 1H
ASTM B345/3004 1B
ASTM B345/5050 1D
ASTM B345/5052 1D
ASTM B345/5083 1D
ASTM B345/5086 1D
ASTM B345/5154 1D
ASTM B345/5456 1D
ASTM B345/6061 1E
ASTM B345/6061 ALCLAD 1H
ASTM B345/6062 1E
ASTM B345/6063 1E
ASTM B345/6351 1E
ASTM B346 52.1
ASTM B348/1 51A
ASTM B348/2 51A
ASTM B348/3 51A
ASTM B348/4 51A
ASTM B348/5 51C
ASTM B348/6 51B
ASTM B348/7 51B
ASTM B349 R1 56.1
ASTM B350 R1 56.1
ASTM B350 RA1 56.1
ASTM B351 R1 56.1
ASTM B351 RA1 56.1
ASTM B352 R1 56.1
ASTM B352 RA1 56.1
ASTM B353 R1 56.1
ASTM B353 RA1 56.1
ASTM B356 R1 56.1
ASTM B356 RA1 56.1
ASTM B357 22.1
ASTM B358 A 14E
ASTM B358 B 14E
ASTM B359/102 14A
ASTM B359/120 14A
ASTM B359/122 14A
ASTM B359/142 14A
ASTM B359/608 14G
ASTM B359/687 14G
ASTM B360 14A
ASTM B360 B 14E
ASTM B361 1A
ASTM B364 46.1
ASTM B365 46.1
ASTM B367 C1 51A
ASTM B367 C2 51C
ASTM B367 C3 51B
ASTM B369 A 14E
ASTM B370 14A
ASTM B371 A 14M
ASTM B371 B 14M
ASTM B373/1145 1A
ASTM B373/1235 1A
ASTM B379 14A
ASTM B381 F1 51A
ASTM B381 F2 51A
ASTM B381 F3 51A
ASTM B381 F4 51A
ASTM B381 F5 51C

ASTM B381 F6 51B
ASTM B381 F7 51B
ASTM B381 F8 51C
ASTM B381 F9 51C
ASTM B382 ER Ti 51A
ASTM B382 ER Ti 4 Al 3 Mo 1 V 51C
ASTM B382 ER Ti 4 Al 4 V 51C
ASTM B382 ER Ti 5 Al 4 Fe Cr 51C
ASTM B382 ER Ti 6 Al 4 V 51C
ASTM B382 ER Ti 13 V 11 Cr 3 Al 51C
ASTM B382 ERTI 2.5 Al 16 V 51B
ASTM B382 ERTI 3 Al 51B
ASTM B382 ERTI 5 Al 2.5 Sn 51B
ASTM B382 ERTI 8 Al 2 Cb 1 Ta 51B
ASTM B396/5805 1D
ASTM B398/6201 1E
ASTM B402/70/30 14E
ASTM B402/90/10 14E
ASTM B403 AM 100 A 23B
ASTM B403 AZ 81 A 23B
ASTM B403 AZ 91 C 23B
ASTM B403 AZ 92 A 23B
ASTM B403 EZ 33 A 23B
ASTM B403 HK 31 A 23B
ASTM B403 K 1 A 23B
ASTM B403 QE 22 A 23B
ASTM B403 ZK 61 A 23B
ASTM B404/1100 1A
ASTM B404/3003 1B
ASTM B404/3003 ALCLAD 1H
ASTM B404/5052 1D
ASTM B404/5454 1D
ASTM B404/6061 1E
ASTM B411 14E
ASTM B412 14E
ASTM B422 14E
ASTM B427/908 14K
ASTM B427/915 14K
ASTM B427/916 14K
ASTM B427/917 14K
ASTM B429/6061 1E
ASTM B429/6063 1E
ASTM B432 44K1
ASTM B433/70/30 14E
ASTM B433/90/10 14E
ASTM B441 14D
ASTM B442 14A
ASTM B447 14A
ASTM B451 14A
ASTM B453 14B
ASTM B455/380 14C
ASTM B455/385 14C
ASTM B465 14A
ASTM B466 14E
ASTM B467 14E
ASTM B469 14A
ASTM B483 Al alloy
See designation for grades
ASTM B491 A1 alloy
See designation for grades
ASTM B492 14E
ASTM B505 14B
ASTM B508/411 14B
ASTM B508/505 14K
ASTM B531 1D
ASTM B544 1A
ASTM B547 1A
ASTM B566 1A
ASTM B606 1A
ASTM B606 44A3
ASTMB608
see designation for details
ASTM B616 37.1
ASTM B618 A1 alloy investment castings;
see designation for details
ASTMB619
see N or R or UNS designation for details
ASTMB620
Cr Ni Mo Fe alloy plate sheet - see N or UNS designation for detail
ASTMB621
Cr Fe Mo Fe alloy rod - see N or UNS designation for details
ASTMB622
Ni + Ni Co alloy pipe and tube - see N or UNS designation for detail
ASTM B623 14A
ASTMB625
Cr Ni Mo Cu Fe - low C alloy plate sheet - see N or UNS designation for details
ASTMB626
Ni + Ni Co tube - welder - see N or UNS designation for details
ASTMB637
Cr Mo Co Ti etc Ni alloy bar and forger - see N and UNS designation for details
ASTM B637 27C
ASTM B637/80 A 27C
ASTM B637/684 27C
ASTM B637/685 27C
ASTM B637/688 27C
ASTM B637/689 27C
ASTM B637/718 27C
ASTM B639 44N1
ASTM B640 14A
ASTM B641 14A
ASTM B643 14D
ASTMB652
Hf Nb alloy ingot - see R and UNS designation
ASTMB653
Zr + Zr alloy welding fittings graded by R and UNS designation
ASTMB654
specification for Nb-Hf alloy sheet etc graded by R and UNS designation
ASTMB655
Hf Nb alloy bar rod - see R and UNS designation
ASTMB658
Zr + Zr alloy pipe - see R and UNS designation
ASTMB668
seamless tubes - see N and UNS designation
ASTMB669
Zn alloy ingots - see Z and UNS designation
ASTMB670
specification for age hardened Ni alloy plate etc - see N and UNS designation
ASTM B670 27C
ASTM B671 – 99.8 19.1
ASTM B671 – 99.9 19.1
ASTMB672
Cr Ni Mo Nb stabilized steel wire and bar - see N and UNS designation
ASTMB673
Austenitic steel welded pipe - see N and UNS designation
ASTMB674
Austenitic steel welded tube - see N and UNS designation
ASTMB675
Austenitic steel welded pipe - see N and + UNS designation
ASTMB676
Austenitic steel welded tube - see N and UNS designation
ASTMB677
Austenitic steel seamless pipe and tube - see N and UNS designation
ASTMB686
Specification for high strength Al castings - see designation
ASTMB688
Austenitic steel plate sheet etc - see N and UNS designation
ASTMB690
Austenitic steel pipe and tube - seamless - see N and UNS designation
ASTMB691
Austenitic steel rod bar etc - see N and UNS designation
ASTMB704
Ni alloy welded tube - see N and UNS designation

Note Designation follows ASTM number

ASTMB705
Ni alloy welded pipe - see N and UNS designation
ASTM B707 14A
ASTMB708
Ta + Ta alloy plate sheet etc - see R and UNS designation
ASTMB709
High Ni Cr steel plate sheet - see R and UNS designation
ASTMB710
High Ni Cr steel welded pipe - see N and UNS designation
ASTM B716 14A
ASTM B717 99.8 39.1
ASTM B717 99.9 39.1
ASTMB718
Cr Mo Fe Ni alloy plate sheet etc - see N and UNS designation
ASTMB719
Cr Mo Fe Ni alloy bar - see N and UNS designation
ASTMB720
Austenitic steel tube - see N and UNS designation
ASTMB722
Cr Mo Fe Ni alloy seamless pipe and tube - see N and UNS designation
ASTMB723
Cr Mo Fe Ni alloy welded pipe - see N and UNS designation
ASTMB725
Specification for welded Ni pipe - see N and UNS designation
ASTMB726
Cr Mo Fe Ni alloy welded tube - see N and UNS designation
ASTMB729
Cr Mo Co Fe Ni alloy pipe tube - see N and UNS designation
ASTMB730
Ni tube - welded - see N and UNS designation
ASTM B737 R1 17A.1
ASTM B737 R2 17A.1
ASTM B737 R3 17A.1
ASTMB739
Austenitic steel - welded tube - see N and UNS designation
ASTM B740 Cu 9 Ni 6 Sn 14E
ASTM B740 Cu 15 Ni 8 Sn 14E
ASTM B743 14A
ASTM B747 C15100 14A
ASTMB749
Pb + Pb alloy strip etc - see L and UNS designation
ASTMB750
Zinc alloy inject - for hot dif coolers - see Z and UNS designation
ASTMB755
Cr Mo W Ni alloy plate sheet etc - see N and UNS designation
ASTMB756
Cr Mo W Ni alloy rod and bar - see N and UNS designation
ASTMB757
Cr Mo W Ni alloy welded pipe - see N and UNS designation
ASTMB758
Cr Mo W Ni alloy welded tube - see N and UNS designation
ASTMB759
Cr Mo W Ni alloy pipe and tube - see N and UNS designation
ASTM B760 52.1
ASTMB763
Sand castings - copper alloys - see C and UNS designation
ASTM B768 C 17400 14D
ASTM B768 C 17410 14D
ASTM B770 44A4
ASTM B772/3 52.1
ASTM B772/4 52.1
ASTM B774-117 6.1
ASTM B774-136 6.1
ASTM B774-158 6.1
ASTM B774-158/190 6.1
ASTM B774-174 6.1
ASTM B774-203 6.1
ASTM B774 244 18.1
ASTM B774-255 6.1
ASTM B774-281 6.1
ASTM B774–281/338 50A
ASTM B774–291/325 50A
ASTM B774–293 50A
ASTM B774 296 18.1
ASTM B774 300/302 18.1
ASTM B774–307/323 50A
ASTM B774 320/345 18.1
ASTM B776 R1 17A.1
ASTM B776 R2 17A.1
ASTM B776 R3 17A.1
ASTM B777/1 52.1
ASTM B777/2 52.1
ASTM F67 50A
ASTM F106 TB Ag 42.1
ASTM F 136 51B
ASTMF467
Non-ferrous nuts for general use graded by UNS system
ASTMF468
Non-ferrous bolts - cap screws etc for general use graded by UNS system
ASTRA 44K2
ASTRANIT M/Z 44P
ASTRANIT N/Z 44N1
ASTRANIT SST/Z 44N1
ASTROLOY 27C
ASTROLOY M 27C
ASV 44N1
AT 44K2
AT 2 44M2
A T4 1N
ATERCLIFFE 44A3
ATGB 44N2
ATGF 27B
ATG R 27C
ATG S3 27C
ATG S4 27C
ATG Z 13A
A Ti 24 51A
A Ti 30 51A
A Ti 35 51A
A Ti Pd 51A
ATLANTES HT 44E1
ATLANTES MT 44K1
ATLAS ADMIRALTY 'A' 50A
ATLAS ADMIRALTY 'B' 50A
ATLAS ALPHA 8 44A4
ATLAS AMACOL 50A
ATLAS CM 44G2
ATLAS CTC 1 14K
ATLAS D8 50A
ATLAS DD 50A
ATLAS DG 50A
ATLAS FDS 50A
ATLAS INFRANGA 50A
ATLAS M3/2 44K3
ATLAS M4 44K3
ATLAS Q 44E3
ATLAS SHAFTING 44G2
ATLAS SPECIAL 10 44J
ATLAS TENAXAS 50A
ATLAS X10 44A4
ATLAS X12 44A4
ATLAS XX-95 44A4
ATLAS XXX 44K3
ATMODIE 44M2
ATMODIE SMOOTHCUT 44M2
AT NICKEL 27A
ATP
see ASTM range 14A
ATR 56.1
ATV S7 44N2
A U2G 1F
A U2GN 1F
A U2N 1F
A U4G 1F
A U4G1 1F
A U4N 1F
A U4NT 1L
A U4PB 1F
A U4SG 1F
A U5GT 1L
A U5 Pb Bi 1F
A U10G 1L
AU 18/8 44N1
A U40T5 1L
A U50 1L
AU 188/Z 44N1
A U1054 1L

AUBEL 17.1
AUDENWIRE 17.1
AUDIOLLOY 20C
AUER 11.1
AUMET 17.1
AURIGA 1A 44A1
AURIGA 1B 44A1
AURIGA I 44A2
AURIGA III 44A2
AURIGA IIIA 44A2
AURIGA IX 44K1
AURIGA IXB 44K1
AURIGA V 44A3
AURIGA V1A 44A1
AURIGA VIB 44A2
AURIGA VII 44E2
AURIGA VIII 44K1
AURIGA VIII LT 44G1
AURIGA VIIIB 44K1
AURIGA VIIIC 44K1
AURIGA X 44D
AURIGA XI 44D
AURIGA XII 44D
AURIGA XV 44K1
AURIGA XVI 44K1
AURIGA XVII 44F1
AURIGA XXI 44K1
AURIGA XXII 44K1
AURIGA XXIII FCG 44M1
AURIGA XXIVB 44N1
AURIGA XXIVB FCG 44N1
AURIGA XXIVC 44N1
AURIGA XXVII 44N1
AURIGA XXVIIA 44N1
AURIGA XXXI 44P
AURIGA XXXII 27B
AURIGA XXXIII 44N1
AURIGA XXXIV 44K1
AUS 101 A
see BS range 20B
AUS 101 B
see BS range 20B
AUS 102 A
see BS range 20B
AUS 102 B
see BS range 20B
AUS 104
see BS range 20B
AUS 105
see BS range 20B
AUS 202 A
see BS range 20B
AUS 202 B
see BS range 20B
AUS 203
see BS range 20B
AUS 204
see BS range 20B
AUS 205
see BS range 20B
AUSTEN 56 44F1
AUSTENITE 44K2
AUSTINOX 44N1
AUSTINOX 3 44N1
AUSTINOX B 44N1
AUSTINOX BB 44N1
AUSTINOX F 44M1
AUSTINOX S 44N1
AUSTINOX SB 44N1
AUSTINOX SN 44N1
AUTAAS 101 44N1
AUTO 'C' 50A
AUTO A 21.1
AUTOROD 16.10 44N1
AUTOROD 16.12 44N1
AUTOROD 16.30 44N1
AUTOROD 16.32 44N1
AUTROD 12.10 44A1
AUTROD 12.24 44G1
AUTROD 12.30 44A1
AUTROD 12.51 44A1
AUTROD 13/12 44K1
AUTROD 13.42 44A1
AV1 44P
Av 22 1F
Av 24 1F
AVESTA 393 S 44M1
AVESTA 739 C 44M1
AVESTA 739 SG 44M1
AVESTA 739 SH 44M1
AVIONAL 22 1F
AVIONAL 24 1F
AVIONAL 050 1F
AVIONAL 100 1F
AVIONAL 150 1F
AVIONAL 660 1F
AVIONAL 662 1F
AVIONAL Pb 118 1F
AVONMOUTH 55A
AVS 44M2
AVW 44N1
AW 44K3
A W DYNALLOY 1 44K1
A W DYNALLOY 50 44K1
AW 4 V 44K3
AW 5 44A3
AW 8 44A3
AW 10 44A1
AW 55 44A1
AW 441 44J
AWA 14F
AW AC 44A1
AWCO 07 1D
AWCO 21 1D
AWCO 22 1E
AWCO 24 1E
AWCO 25 1E
AWCO 27 1D
AWCO 28 1D
AWCO 31 1F
AWCO 40 1C
AWCO 45 1C
AWCO 46 1C
AWCO 60 1B
AWCO 99.5 1A
AWCO 99.8 1A
AWCO 282 1D
AWCO 283 1D
AWCO 284 1D
AWCO 285 1D
AWCO 301 1F
AWCO 304 1F
AWCO 305 1F
AWCO 308 1F
AWCO 309 1F
AWCO 701 1G
AWCO EP 1A
AWCO SILMALEC 1E
AWCO SP 1A
AWCW 44K2
Note AWS specifications have the letter A followed by 5.1 to 5.30+. The alloy is identified by a code or designation: e.g. AWS A5.15-RC1 has the designation RC1 – found in Section 20B; AWS A5.20 E6XT5 has the designation E6XT5 – found in Section 44A1.
AWV 50 44J
AWV 55 44J
AWV 60 44J
AWV 65 44J
AWX 45 44A1
AWX 50 44A1
AWX 55 44A1
AXL 44K3
AZ 44K1
A Z3G2 1G
A Z4G 1G
A Z5G 1G
AZ 5G 1M
A Z5GU 1G
A Z6G 1G
A Z8GU 1G
AZ 21 X 23B
AZ 31 23B
AZ 31B
see ASTM range 23B
AZ 31C
see ASTM range 23B
AZ 31 X 23B
AZ 61 23B
AZ 61A
see ASTM range 23B
AZ 61 X 23B
AZ 63 23B
AZ 63A
see ASTM range 23B
AZ 80A
see ASTM range 23B
AZ 80 X 23B
AZ 81 23B
AZ 81A
see ASTM range 23B
AZ 91 23B
AZ 91A
see ASTM range 23B
AZ 91B
see ASTM range 23B
AZ 91C
see ASTM range 23B
AZ 91 E 23B
AZ 91 X 23B
AZ 92 23B
AZ 92A
see ASTM range 23B
AZ 855 23B
AZM 23B

AZN 44K1
A Zr5 1N
B 0 44A1
B 1 14K
B 1 44A1
B 1 44A3
B 1 44K3
B 2 14K
B 2 44A1
B 2/2 (3) 44K3
B2N
  See ASTMA757 44N1
B2Q
  See ASTMA757 44N1
B 3 44A2
B 3 Mn 44A2
B3N
  See ASTMA757 44N1
B3Q
  See ASTMA757 44N1
B 4 44A1
B 4 44A2
B 4 CR 44E3
B4N
  See ASTMA757 44N1
B4Q
  See ASTMA757 44N1
B 4 V 44A1
B 5 44A2
B 6 14H
B 6 14K
B 6 44A3
B 6 44K3
B 6L 44K1
B 7 44A2
B 7 44L
B 8
  see BS B8 14K
B 8 44L
B 9 14K
B 9 44K3
B 9 S 44A2
B 10 14J
B 10 44A3
B 10 44L
B 11
  see BS/B11 14B
B 12 44A3
B 13 14K
B 14 14K
B 14 44A3
B 15 14J
B 15T 44A4
B 15V 44A4
B 18 14K
B 18/6
  see BS range 20B
B 18S 1F
B 20 14J
B 20 14K
B 20/10
  see BS range 20B
B 20V 44A4
B 21
  see BS 2B21 50A
B 22
  see BS 2B22 50A
B 22/14
  see BS range 20B
B 24V 44A4
B 28V 44A4
B 29 14G
B 31 14C
B 32 14K
B 37 14A
B 38 14K
B 39 14K
B 40 14K
B 41 14C
B 42F 14C
B 42 HC 14B
B 42HS 14C
B 45 14K
B 45CH 44K1
B 47 14E
B 50 14C
B 55CH 44K1
B 60 14K
B 65 14K
B 65CH 44K1
B 68 14E
B 70 14K
B 76 14B
B 80 14K
B 80/1 14K
B 85 14K
B 85/P 14K
B 90/P 14K
B 103 44A2
B 120 VCA 51C
B 127
  see ASTM range 27D
B 132 1L
B 143 1L
B 164
  see ASTM range 27D
B 165
  see ASTM range 27D
B 188 44J
B 195 1L
B 282
  see HASTELLOY 27C
B 303 44A1
B 400 44K3
B 412 44K3
B 443.0 1C
B 514.0 1J
B 530 44K3
B 535 44K3
B 805 44A2
B 850.0 1N
B 904 44E2
B 908 44E2
B 1111
  see AISI B1111 44A1
B 1112
  see AISI B1112 44A1
B 1113
  see AISI B1113 44A1
B2056 420S45 44M1
B 3312a 44N1
B 3422C 44M1
B 3423C 44M1
BA 1A
B A Ag 27 42.1
BA 21 1D
BA 22 1E
BA 24 1E
BA 25 1E
BA 27 1D
BA 28 1D
BA 28 1J
BA 29 1J
BA 32 1L
BA 33 1L
BA 35 1F
BA 40 1C
BA 41 1K
BA 42 1K
BA 45 1C
BA 60 1B
BA 99 1A
BA 99.5 1A
BA 99.8 1A
BA 177 1A
BA 211 1D
BA 212 1D
BA 213 1D
BA 218 1D
BA 226 1D
BA 227 1D
BA 241 1E
BA 251 1E
BA 271 1D
BA 281 1D
BA 284 1D
BA 301 1F
BA 303 1F
BA 305 1F
BA 306 1F
BA 307 1F
BA 308 1F
BA 309 1F
BA 351 1H
BA 352 1F
BA 353 1F
BA 355 1H
BA 359 1H
BA 451 1K
BA 703 1G
BA 704 1G
BA 705 1G
BA 706 1G
BA 707 1G
BA 751 1H
BA 757 1H
BA 3059/6 44A2
BA 5052 1D
BABBITT 21.1
BABBITT 50A
BABBITT No 1 50A
BABBITT No. 6 21.1
BA COMMERCIAL
  QUALITY 1A
BADGER 44K3
BA ELECTRICAL PURITY 1A
B Ag 1 42.1
B Ag 1a 42.1
B Ag 2 42.1
B Ag 2a 42.1
B Ag 3 42.1
B Ag 4 42.1
B Ag 5 42.1
B Ag 6 42.1
B Ag 7 42.1
B Ag 8 42.1
B Ag 8a 42.1

B Ag 9 42.1
B Ag 10 42.1
B Ag 13 42.1
B Ag 18 42.1
B Ag 19 42.1
B Ag 20 42.1
B Ag 21 42.1
B Ag 22 42.1
B Ag 23 42.1
B Ag 24 42.1
B Ag 25 42.1
B Ag 26 42.1
B Ag 28 42.1
B Ag 33 42.1
B Ag 34 42.1
BAHN METAL 21.1
BAKER 4 17.1
BALCO 27E
BALLAST NICKEL 27A
B and W A1 44N1
B and W A1/57 44N1
B and W Ni 57 44F1
B and W – WL 1 44K2
B and W WL2 44K1
BANKA 50.1
BARIUM 4.1
BARRONIA 14B
BARTO 44K2
BA SILMALEC 1E
BASIS BRASS 14B
BASIS QUALITY BRASS 14B
BA SP11 1D
BA SP12 1D
BA SP 16 1E
BA SUPER PURITY 1A
BATALBRA 14B
BATH METAL 14C
BATNAVAL 14B
BATNICKON 5 14E
BATNICKON 10 14E
BATNICKON 30 14E
BATT/Z 55A
BATTERIUM 14G
BATURNAL 14B
BAZAR METAL 42.1
BB 0K 44N1
BB 1 1D
BB 1 - X 1D
BB 2 1D
BB 2K 44N1
BB 3 1D
BB 4 1D
BB 4K 44N1
BB 5 1D
BB 5 1J
BB 5 - X 1D
BB 7 1D
BB 17 1D
BB 019 1E
BB 127 1D
BBMK 44N1
B-BRONZE 14H
BBS/Z 55A
BBZ 36 1G
BC 44A4
BC 3V 44J
BC 8V 44J
BC 10V 44J
BCC 44K2
BCC 44L
BCD 44M2
BCD 37 44K1
BCD 48 44K2
BCH 44K1
BCHV 44M2
BCMF 44K1
BCMO 44K1
BCN 44K1
BCN (Mo) 44K1
BCO 1 13A
BCRS 44M2
BCTA 44E2
BCTB 44E2
BCW 44M2
BD3605
See designation for details
BEARCOMO 44G1
BEARING BRONZE 14K
BELL B and P 44A4
BELL METAL 14K
BENUM 44K1
BERALOY A 14D
BERYDUR 14D
BERYL 52.1
BERYLCO 14D
BERYLCO 25 14D
BERYLCO 33/25 14D
BERYLCO 50 14D
BERYLCO 165 14D
BERYLLIUM 5.1
BERYLLIUM BRONZE 14D
BERYLLOY 14D
BESTEM 44K2
BESTEM 44K3
BEST WARRANTED 44A4
BEST YORKSHIRE 20A
BE T9 1A
BETA BRASS 14C
BFD 44K3
B GRADE 20A
BH 36 44A1
BH 36S 44J
BH 39 44J
BH 43 44K1
BH 43S 44K1
BH 47 44K1
BH 47S 44K1
BH 51 44K1
BH 51S 44K1
BH 57V 44K1
BH 57VT 44K1
BH 65V 44K1
BH 65VT 44K1
BH 395 44J
BIDENY METAL 55A
BINDING BRASS 14B
BINDING METAL 55A
BINORMAT 17.1
BIRMABRIGHT 1J
BIRMAG 23B
BIRMALITE 1L
BIRMASIL SPEC 1C
BIRMASTIC 1C
BIRMETAL 1 1A
BIRMETAL 2M 1A
BIRMETAL 3 1B
BIRMETAL 005 1C
BIRMETAL 016 1E
BIRMETAL 055 1E
BIRMETAL 065 1E
BIRMETAL 069 1E
BIRMETAL 071 1E
BIRMETAL 161 1F
BIRMETAL 212 1G
BIRMETAL 230 1F
BIRMETAL 477 1F
BIRMETAL 477 CLAD 1H
BIRMETAL 478 1F
BIRMETAL 478 CLAD 1H
BIRMETAL Z36 1G
BIRMETAL Z36 CLAD 1H
BIRMIDAL 1K
BIRMINGHAM PLATINA 55A
BISMANOL 6.1
BISMUTH 6.1
BISMUTH-ZINC 55A
BKL 305 1L
BKL 308 1L
BKM 44K2
BKST 3
see GOST 10705 44A1
BKST 4
see GOST 10705 44A1
BKV 44M2
BL 30 44K1
BLACK LABEL 44A4
BLAW KNOX 22H 27B
BLAW KNOX Mo-RE1 44N1
BLUE and WHITE LABEL 44J
BLUE CHIP 44K3
BLUE LABEL 44A4
BM 78 14J
BMB 473 1F
BMB 551 1F
BMB 761 1F
BMB 1306 1G
BMB 2308 1G
B METAL 44A2
B Mg 1 23B
BMST 4SP
see GOST 8731 44A1
BMST 5SP
see GOST 8731 44A2
B Ni 1 27F
B Ni 1A 27F
B Ni 2 27F
B Ni 3 27F
B Ni 4 27F
B Ni 5 27F
B Ni 6 27F
B Ni 7 27F
B Ni 8 27F
B Ni 9 27F
B Ni 10 27F
B Ni 11 27F
BOBIERE'S METAL 14B
BO BRONZE 14K
Bohn L4 1L
B O H T BRONZE 14K
BOLSTER STEEL 44K3
BOLTOMET 10 14K
BOLTOMET 11 14K
BOLTOMET 12 14K
BOLTOMET 15 14K
BOLTOMET 16 14K
BOLTOMET 58 14K
BOLTOMET 103 14A

BOLTOMET 105	14A
BOLTOMET 107	14A
BOLTOMET 108	14A
BOLTOMET 112	14A
BOLTOMET 113	14A
BOLTOMET 115	14A
BOLTOMET 117	14A
BOLTOMET 121	14A
BOLTOMET 123	14A
BOLTOMET 152	14A
BOLTOMET 154	14A
BOLTOMET 156	14A
BOLTOMET 160	14A
BOLTOMET 162	14A
BOLTOMET 170	14A
BOLTOMET 175	14A
BOLTOMET 206	14A
BOLTOMET 208	14A
BOLTOMET 210	14A
BOLTOMET 302	14A
BOLTOMET 304	14A
BOLTOMET 305	14K
BOLTOMET 307	14K
BOLTOMET 309	14K
BOLTOMET 312	14K
BOLTOMET 317	14K
BOLTOMET 317	14K
BOLTOMET 317	14K
BOLTOMET 319	14K
BOLTOMET 320	14K
BOLTOMET 327	14K
BOLTOMET 338	14K
BOLTOMET 349	14K
BOLTOMET 356	14K
BOLTOMET 366	14A
BOLTOMET 445	14K
BOLTOMET 506	14B
BOLTOMET 510	14B
BOLTOMET 514	14B
BOLTOMET 516	14B
BOLTOMET 518	14B
BOLTOMET 520	14B
BOLTOMET 522	14B
BOLTOMET 524	14C
BOLTOMET 570	14B
BOLTOMET 607	14B
BOLTOMET 611	14C
BOLTOMET 613	14C
BOLTOMET 619	14C
BOLTOMET 710	14B
BOLTOMET 715	14C
BOLTOMET 721	14C
BOLTOMET 731	14B
BOLTOMET 803	14G
BOLTOMET 807	14G
BOLTOMET 814	14L
BOLTOMET 818	14L
BOLTOMET 820	14A
BOLTOMET 909	14A
BOLTOMET 952	14A
BOLTOMET 968	14M
BONDED CARBIDE	44L
BONDED CARBIDE Jr	44L
BONDED CARBIDE Sr	44L
BORAWIRE	17.1
BOROFIL	14A
BORON	7.1
BORON CARBIDE	7.1
BP 4	44K3
BP 87	27C
BR 3	44K3
BR 3 DIE STEEL	44K3
BR 4 FM	44M2
BRAEBURN M7	44K3
BRAEBURN M33	44L
BRAEBURN T15	44L
BRAECUT	44L
BRAEFOUR	44K3
BRAEMAX	44L
BRAEMOW	44K3
BRAEMOW 2	44K3
BRAEMOW SPECIAL	44K3
BRAETUF	44L
BRAEVAN	44K3
BRAEVAN/1	44K3
BRAEVAN/2	44K3
BRAKE DIE	44K2
BREARLEY A	44M1
BREARLEY C	44M1
BRI	44E3
BRIGHTRAY 35	27B
BRIGHTRAY B	27B
BRIGHTRAY C	27B
BRIGHTRAY F	44N1
BRIGHTRAY H	27B
BRIGHTRAY S	27B
BRILLUM	1F
BRIOUDE	2.1
BRITTANIA	50A
BRONZE 44	14K
BRONZE 66	14K
BRONZETRODE 94.15	14K
BROWN METAL	14B
BRUNSWICK V 109	44K2
BRUNSWICK V 114	44K2
BRUNSWICK V 130	44K2
BRUNSWICK V 981	44E3
BRUNSWICK V/110/CNA	44K1
BRUNSWICK V104	44H
BRUNSWICK V105	44H
BRUNSWICK V106	44M2
BRUNSWICK V108	44K1
BRUNSWICK V150	44K3
BRUNSWICK V414	44H
BRUNSWICK V888	44K3
BRUSH 190	14D
BS	44N1
BS 2	44N1
BS 2B 21	50A
BS 2B 22	50A
BS 9	44A3
BS 11	44A3
BS 14	44A1
BS 15/1	44A2
BS 15/2	44A2
BS 15/3	44A2
BS 23	14A
BS 24/1/1	44A1
BS 24/1/2	44A1
BS 24/1/3	44A1
BS 24/2/4	44A2
BS 24/2/5	44A2
BS 24/3 A 9	44A1
BS 24/3 A 10	44A1
BS 24/3 A 11	44A1
BS 24/3 A1	44A3
BS 24/3 A2	44A3
BS 24/3 A5	44A3
BS 24/3 A6	44A3
BS 24/3 A7	44A3
BS 24/3 A8	44A3
BS 24/3 BB	44A3
BS 24/3 BC	44A3
BS 24/3A/3	44A2
BS 24/3A/4	44A2
BS 24/3Ba	44A4
BS 24/4 A	44A1
BS 24/4 B	44A1
BS 24/4 E	44A1
BS 24/4 F	44A1
BS 24/4C	44A2
BS 24/4D	44A2
BS 24/5	14A
BS 24/5/611	44A1
BS 24/5/612	44A1
BS 24/5/613	44A1
BS 24/5/614	44K1
BS 29/22/26	44A1
BS 29/28/32	44A1
BS 29/32/36	44A1
BS 29/36/40	44A1
BS 31	14C
BS 32/1	44A2
BS 32/2	44A2
BS 32/3	44A1
BS 32/4	44A1
BS 32/5	44A1
BS 47A	44A2
BS 47B	44A2
BS 48	20A
BS 51 A	20A
BS 51 B	20A
BS 51 C	20A
BS 53	44A2
BS 75	44A1
BS 99	14K
BS 125	14A
BS 165	44A2
BS 198	14A
BS 200	14A
BS 201	14A
BS 202	14A
BS 203	14A
BS 207	14C
BS 208	14C
BS 215	1A
BS 218	14C
BS 219A	50A
BS 219B	50A
BS 219 C	21.1
BS 219 D	21.1
BS 219F	50A
BS 219 G	21.1
BS 219 H	21.1
BS 219 J	21.1
BS 219K	50A
BS 219 L	21.1
BS 219 M	21.1
BS 219 N	21.1
BS 219 R	21.1
BS 219 V	21.1
BS 219/1S	21.1
BS 219/5S	21.1
BS 219/95 A	50A

BS 220 55A
BS 221 55A
BS 222 55A
BS 224/1 44A3
BS 224/2 44K2
BS 224/5 44K2
BS 249 CZ121 14C
BS 250 CZ114 14C
BS 250 CZ122 14C
BS 251 CZ112 14B
BS 252 CZ113 14C
BS 264 14C
BS 265 CZ108 14B
BS 266 CZ107 14B
BS 267 CZ106 14B
BS 309 W22/4 20B
BS 309 W24/8 20B
BS 310 B18/6 20B
BS 310 B20/10 20B
BS 310 B22/14 20B
BS 321 A 20B
BS 321 C 20B
BS 334 A 21.1
BS 334 B 21.1
BS 335/2 21.1
BS 335/3 21.1
BS 335/4 21.1
BS 335/5 21.1
BS 352 14K
BS 359 1A
BS 360 1A
BS 361 1L
BS 362 1L
BS 363 1L
BS 369 PB102 14K
BS 374 CN103 14E
BS 374 CN104 14E
BS 374 CN105 14E
BS 374 CN106 14E
BS 375 A 27A
BS 378 C106 14A
BS 378 C107 14A
BS 378 CA102 14G
BS 378 CN102 14E
BS 378 CN107 14E
BS 378 CZ105 14B
BS 378 CZ110 14B
BS 378 CZ111 14B
BS 382/3 14K
BS 384 14K
BS 385 1A
BS 386 1A
BS 395 1F
BS 396 1F
BS 399 44A2
BS 400 44A1
BS 401 44A1
BS 407 PB101 14K
BS 407 PB102 14K
BS 407 PB103 14K
BS 409 14C
BS 414 1E
BS 421 14K
BS 477 1D
BS 478 1F
BS 494 44A2
BS 518 14A
BS 532 1D
BS 533 1F
BS 548 44A2
BS 558 27A
BS 592 A 44A2
BS 592 B 44A2
BS 592 C 44A2
BS 601/1 20A
BS 601/2 20A
BS 602/1 21.1
BS 602/2 21.1
BS 602/3 21.1
BS 640/1 44A2
BS 640/2 44A1
BS 640/3 44A1
BS 643 21.1
BS 648/B 44M2
BS 659 14A
BS 672 14A
BS 681 44E1
BS 682 44K1
BS 702 1L
BS 703 1L
BS 711 CZ103 14B
BS 712 CZ102 14B
BS 713 CZ101 14B
BS 725 44A1
BS 762 20A
BS 782 44A1
BS 785 44A2
BS 790 NS103 14F
BS 790 NS104 14F
BS 790 NS105 14F
BS 790 NS106 14F
BS 790 NS108 14F
BS 790 NS109 14F
BS 790 NS110 14F
BS 801 21.1
BS 801 B 21.1
BS 801 D 21.1
BS 801 E 21.1
BS 806 A 44A1
BS 806 B 44A1
BS 806 C 44A1
BS 806 D 44A1
BS 806 E 44A1
BS 806 F 44A1
BS 806 P 44K1
BS 806 Q 44K1
BS 806M 44G1
BS 821 (High) 20B
BS 821 (Medium) 20B
BS 821 (Ordinary) 20B
BS 847 44A1
BS 849 55A
BS 858 20A
BS 885 CZ105 14B
BS 885 CZ110 14B
BS 897 14K
BS 898 14K
BS 899 C101 14A
BS 899 C102 14A
BS 899 C103 14A
BS 899 C104 14A
BS 899 C105 14A
BS 899 C106 14A
BS 899 C107 14A
BS 918 1G
BS 920 14B
BS 932 14C
BS 933/3 44B1
BS 933/4 20C
BS 944/1 14C
BS 960 14K
BS 961 14K
BS 962 14K
BS 963 14K
BS 964 14K
BS 965 14K
BS 968 44E1
BS 970
See designation for details: e.g. BS 970-En 24, see En24 – Section 44K2; BS 970-230-M07, see 230 M07 – Section 44A1.
BS 971 44A1
BS 980 CDS1 44A1
BS 980 CDS2 44A1
BS 980 CDS3 44A1
BS 980 CDS4 44A1
BS 980 CDS5 44A2
BS 980 CDS6 44A2
BS 980 CDS7 44A2
BS 980 CDS 8 44A3
BS 980 CDS9 44A2
BS 980 CDS10 44A2
BS 980 CDS11 44G1
BS 980 CDS12 44K1
BS 980 CDS13 44K1
BS 980 CDS 14 44F2
BS 980 CDS15 44K2
BS 980 CDS16 44K1
BS 980 CDS17 44K1
BS 980 CDS 18 44M1
BS 980 CDS 19 44N1
BS 980 CDS 20 44N1
BS 980 CEW3 44A2
BS 980 CEW4 44A2
BS 980 ERW2 44A2
BS 980 ERW3 44A2
BS 1001 CZ115 14C
BS 1002 14C
BS 1003 55A
BS 1004 A 55A
BS 1004 B 55A
BS 1021 14K
BS 1022 14K
BS 1023 14K
BS 1024 14K
BS 1025 14K
BS 1026 14K
BS 1027 14K
BS 1028 14K
BS 1029 14M
BS 1030 14M
BS 1031 14G
BS 1032 14G
BS 1035 14A
BS 1036 14A
BS 1037 14A
BS 1038 14A
BS 1039 14A
BS 1040 14A

BS 1045 44A2
BS 1052 44A1
BS 1058 14K
BS 1059 14K
BS 1060 14K
BS 1061 14K
BS 1072 14G
BS 1073 14G
BS 1080 1F
BS 1085 21.1
BS 1110 14A
BS 1141 55A
BS 1144 44A2
BS 1158 14K
BS 1159 14K
BS 1172 14A
BS 1173 14A
BS 1174 14A
BS 1178 21.1
BS 1272/80 23B
BS 1287 44A2
BS 1354 NG2 1C
BS 1387 44A1
BS 1398 44G1
BS 1400 AB1 C 14G
BS 1400 AB2 1 14G
BS 1400 AB2 C 14G
BS 1400 AB3 14G
BS 1400 CMA1 14G
BS 1400 CMA1C 14G
BS 1400 CMA2 1 14G
BS 1400 CMA2 C 14G
BS 1400 CN1 14E
BS 1400 CN2 14E
BS 1400 DCB1 C 14C
BS 1400 DCB1/1 14C
BS 1400 DCB2 C 14C
BS 1400 DCB2/1 14C
BS 1400 DCB3 C 14C
BS 1400 DCB3/1 14C
BS 1400 G1 C 14K
BS 1400 G1/1 14K
BS 1400 G2 C 14K
BS 1400 G2/1 14K
BS 1400 G3 C 14K
BS 1400 G3 WP 14K
BS 1400 G3/1 14K
BS 1400 HCC1C 14A
BS 1400 HCC2C 14A
BS 1400 HTB1 C 14C
BS 1400 HTB1/1 14C
BS 1400 HTB2 C 14C
BS 1400 HTB2/1 14C
BS 1400 HTB3 C 14C
BS 1400 HTB3/1 14C
BS 1400 LB1 C 14K
BS 1400 LB1/1 14K
BS 1400 LB2 C 14K
BS 1400 LB2/1 14K
BS 1400 LB3 C 14K
BS 1400 LB3/1 14K
BS 1400 LB4 C 14K
BS 1400 LB4/1 14K
BS 1400 LB5 C 14K
BS 1400 LB5/1 14K
BS 1400 LG1 C 14K
BS 1400 LG1/1 14K
BS 1400 LG2 C 14K
BS 1400 LG2/1 14K
BS 1400 LG3 C 14K
BS 1400 LG3/1 14K
BS 1400 LG4/1 14K
BS 1400 LPB1 C 14K
BS 1400 LPB1/1 14K
BS 1400 PB1 C 14K
BS 1400 PB1/1 14K
BS 1400 PB2 C 14K
BS 1400 PB2/1 14K
BS 1400 PB3 C 14K
BS 1400 PB3/1 14K
BS 1400 PB4 C 14K
BS 1400 PB4/1 14K
BS 1400 SCB1 C 14K
BS 1400 SCB1-C 14B
BS 1400 SCB1/1 14K
BS 1400 SCB2-1 14B
BS 1400 SCB2-C 14B
BS 1400 SCB3-1 14B
BS 1400 SCB3-C 14B
BS 1400 SCB4 C 14K
BS 1400 SCB4/1 14K
BS 1400 SCB5 C 14K
BS 1400 SCB5-C 14B
BS 1400 SCB5/1 14K
BS 1400 SCB6-1 14B
BS 1400 SCB6-C 14B
BS 1401 14A
BS 1402 14B
BS 1403 14B
BS 1407 44A4
BS 1408 44A3
BS 1414B 44A2
BS 1423 44A4
BS 1429 44A3
BS 1429 En 42C 44A3
BS 1429 En 44B 44A4
BS 1432 C101 14A
BS 1432 C102 14A
BS 1432 C103 14A
BS 1433 14A
BS 1434 14A
BS1449
See designation for details
BS 1449 CS1 44A1
BS 1449 CS2 44A1
BS 1449 CS3 44A1
BS 1449 CS4 44A1
BS 1449 CS12 44A1
BS 1449 CS17 44A1
BS 1449 CS22 44A1
BS 1449 HR1 44A1
BS 1449 HR2 44A1
BS 1449 HR3 44A1
BS 1449 HR4 44A1
BS 1449 HR5A 44A2
BS 1449 HR5B 44A2
BS 1449 HR5C 44A2
BS 1449 HR6A 44A2
BS 1449 HR6B 44A2
BS 1449 HR6C 44A2
BS 1449 HR11 44A1
BS 1449 HR12 44A1
BS 1449 HR13 44A1
BS 1449 HR14 44A1
BS 1449 HR15 44A1
BS 1449 HR16A 44A2
BS 1449 HR16B 44A2
BS 1449 HR16C 44A2
BS 1449 HR17A 44A2
BS 1449 HR17B 44A2
BS 1449 HR17C 44A2
BS 1449 HS1 44A1
BS 1449 HS2 44A1
BS 1449 HS3 44A1
BS 1449 HS4A 44A1
BS 1449 HS4B 44A1
BS 1449 HS12 44A1
BS 1449 HS17 44A1
BS 1449 HS20 44A1
BS 1449 HS22 44A1
BS 1449 HS23 44A1
BS 1449 HS30 44A2
BS 1449 HS40 44A2
BS 1449 HS50 44A3
BS 1449 HS60 44A3
BS 1449 HS70 44A3
BS 1449 HS80 44A3
BS 1449 HS90 44A4
BS 1449 HS100 44A4
BS 1449 NHR12 44A1
BS 1449 NHR13 44A1
BS 1449 NHR14 44A1
BS 1449 NHR15 44A1
BS 1449 NHR21 44A1
BS 1449 NHR22 44A1
BS 1449 NHR23 44A1
BS 1449 NHR24 44A2
BS 1449 NHR25 44A1
BS 1449-10 44A2
BS 1449-20 44A2
BS 1449-30 44A2
BS 1449-40 44A3
BS 1449-50 44A3
BS 1449-60 44A3
BS 1449-70 44A3
BS 1449-80 44A3
BS 1449/304 S12 44N1
BS 1449/304 S15 44N1
BS 1449/309 S24 44N1
BS 1449/310 S24 44N1
BS 1449/312 S24 44N1
BS 1449/316 S12 44N1
BS 1449/316 S16 44N1
BS 1449/317 S16 44N1
BS 1449/320 S17 44N1
BS 1449/321 S12 44N1
BS 1449/347 S17 44N1
BS 1449/403 S17 44M1
BS 1449/405 S17 44M1
BS 1449/410 S21 44M1
BS 1452/10 20B
BS 1452/12 20B
BS 1452/14 20B
BS 1452/17 20B
BS 1452/20 20B
BS 1452/23 20B
BS 1452/26 20B
BS1453
See designation for details
BS 1453 A1 44A1
BS 1453 A2 44A1

BS 1453 A3 44A2
BS 1453 A4 44K1
BS 1453 A5 44E2
BS 1453 A6 44G1
BS 1453 A8 Nb 44N1
BS 1453 A10 44N1
BS 1453 A10 Nb 44N1
BS 1453 A11 44N1
BS 1453 A11 Nb 44N1
BS 1453 A12 44N1
BS 1453 A12 Nb 44N1
BS 1453 B1 20B
BS 1453 B2 20B
BS 1453 B3 20B
BS 1453 C1 14A
BS 1453 C2 14C
BS 1453 C4 14C
BS 1453 C5 14F
BS 1453 C6 14F
BS 1453 D1 23B
BS 1453 D2 23B
BS 1453 G1 1A
BS 1453 G1A 1A
BS 1453 G1B 1A
BS 1453 G1C 1A
BS 1453 N3 1B
BS 1453 NG5 1D
BS 1453 NG6 1D
BS 1453 NG21 1C
BS 1456 A 44A1
BS 1456 B 44A2
BS 1457 44D
BS 1458 44A2
BS 1459 44A2
BS 1461 44K1
BS 1462 44K1
BS 1463 44K1
BS 1464 CA 102 14G
BS 1464 CN107 14E
BS 1464 CZ105 14B
BS 1464 CZ110 14B
BS 1464 CZ111 14B
BS1470
Al alloy wrought plate etc - see designation for details
BS 1470 1050A 1A
BS 1470 1080A 1A
BS 1470 1200 1A
BS 1470 2014A Clad 1F
BS 1470 3103 1B
BS 1470 5083 1D
BS 1470 5154A 1D
BS 1470 5251 1D
BS 1470 6082 1E
BS 1470 HC14 1H
BS 1470 HC15 1H
BS 1470 HS 20 1E
BS 1470 HS 30 1E
BS 1470 HS14 1F
BS 1470 HS15 1F
BS 1470 NS3 1B
BS 1470 NS4 1D
BS 1470 NS5 1D
BS 1470 NS6 1D
BS 1470 S1 1A
BS 1470 S1A 1A
BS 1470 S1B 1A
BS 1470 S1C 1A
BS1471
Al alloy drawn tube - see designation for details
BS 1471 1050A 1A
BS 1471 1200 1A
BS 1471 2014A 1F
BS 1471 5083 1B
BS 1471 5154A 1D
BS 1471 6061 1E
BS 1471 6063 1E
BS 1471 6082 1E
BS 1471 F1A 1A
BS 1471 HT 9 1E
BS 1471 HT14 1F
BS 1471 HT15 1F
BS 1471 HT19 1E
BS 1471 HT20 1E
BS 1471 HT30 1E
BS 1471 NT4 1D
BS 1471 NT5 1D
BS 1471 NT6 1D
BS 1471 NT8 1D
BS 1471 T1A 1A
BS 1471 T1B 1A
BS 1471 T1C 1A
BS1472
Al alloy forged - see designation for details
BS 1472 1050A 1A
BS 1472 2014A 1F
BS 1472 2031 1E
BS 1472 2618A 1F
BS 1472 5083 1F
BS 1472 5154A 1D
BS 1472 5251 1D
BS 1472 6063 1E
BS 1472 6082 1E
BS 1472 F1B 1A
BS 1472 F1C 1A
BS 1472 HF9 1E
BS 1472 HF11 1F
BS 1472 HF12 1F
BS 1472 HF14 1F
BS 1472 HF15 1F
BS 1472 HF16 1F
BS 1472 HF17 1F
BS 1472 HF18 1F
BS 1472 HF30 1E
BS 1472 NF3 1B
BS 1472 NF4 1D
BS 1472 NF5 1D
BS 1472 NF6 1D
BS 1472 NF7 1D
BS 1472 NF8 1D
BS1473
Al alloy rivet bolts etc - see designation for details
BS 1473 1050A 1A
BS 1473 2014A 1F
BS 1473 5056A 1D
BS 1473 5154A 1D
BS 1473 6061 1E
BS 1473 6082 1E
BS 1473 HB15 1F
BS 1473 HB30 1E
BS 1473 HR15 1F
BS 1473 HR30 1E
BS 1473 NB6 1D
BS 1473 NG7 1D
BS 1473 NR5 1D
BS 1473 NR6 1D
BS 1473 R1B 1A
BS 1473 R1C 1A
BS1474
Al alloy bar extruded tube - see designation for details
BS 1474 1050A 1A
BS 1474 1200 1A
BS 1474 2014A 1F
BS 1474 5083 1D
BS 1474 5154A 1D
BS 1474 5251 1D
BS 1474 6061 1E
BS 1474 6063 1E
BS 1474 6063A 1E
BS 1474 6082 1E
BS 1474 HV9 1E
BS 1474 HV11 1F
BS 1474 HV14 1F
BS 1474 HV15 1F
BS 1474 HV19 1E
BS 1474 HV30 1E
BS 1474 NV4 1D
BS 1474 NV5 1D
BS 1474 NV6 1D
BS 1474 NV8 1D
BS 1474 V1A 1A
BS 1474 V1B 1A
BS 1474 V1C 1A
BS1475
Al alloy wire - see designation for details
BS 1475 1080A 1A
BS 1475 2014A 1F
BS 1475 3103 1B
BS 1475 4043A 1C
BS 1475 4047A 1C
BS 1475 5056 1D
BS 1475 5154A 1D
BS 1475 5251 1D
BS 1475 5556A 1D
BS 1475 6061 1E
BS 1475 6063 1E
BS 1475 G1 1A
BS 1475 G1A 1A
BS 1475 G1B 1A
BS 1475 G1C 1A
BS 1475 HG9 1E
BS 1475 HG15 1F
BS 1475 HG20 1E
BS 1475 HG30 1E
BS 1475 NG2 1C
BS 1475 NG3 1B
BS 1475 NG4 1D
BS 1475 NG5 1D
BS 1475 NG6 1D
BS 1475 NG21 1C
BS 1476 E1A 1A
BS 1476 E1B 1A
BS 1476 E1C 1A
BS 1476 HE9 1E
BS 1476 HE11 1F
BS 1476 HE14 1F

BS 1476 HE15 1F
BS 1476 HE19 1E
BS 1476 HE20 1E
BS 1476 HE30 1E
BS 1476 NE4 1D
BS 1476 NE5 1D
BS 1476 NE6 1D
BS 1476 NE8 1D
BS 1477 HP14 1F
BS 1477 HP15 1F
BS 1477 HP20 1E
BS 1477 HP30 1E
BS 1477 HPC14 1H
BS 1477 HPC15 1H
BS 1477 NP3 1B
BS 1477 NP4 1D
BS 1477 NP6 1D
BS 1477 NP8 1D
BS 1477 P1A 1A
BS 1477 P1B 1A
BS 1477 P1C 1A
BS1490
  Al alloy ingot and castings
  - see designation for details
BS 1490 LM1 1L
BS 1490 LM1M 1L
BS 1490 LM2 1C
BS 1490 LM2M 1C
BS 1490 LM3 1L
BS 1490 LM3M 1L
BS 1490 LM4 1L
BS 1490 LM4M 1L
BS 1490 LM4WP 1L
BS 1490 LM5 1J
BS 1490 LM5M 1J
BS 1490 LM6 1C
BS 1490 LM6M 1C
BS 1490 LM7 1L
BS 1490 LM8 1K
BS 1490 LM8M 1K
BS 1490 LM8P 1K
BS 1490 LM8W 1K
BS 1490 LM8WP 1K
BS 1490 LM9 1K
BS 1490 LM9WP 1K
BS 1490 LM10 1J
BS 1490 LM10W 1J
BS 1490 LM11 1L
BS 1490 LM11WP 1L
BS 1490 LM12 1L
BS 1490 LM12WP 1L
BS 1490 LM13 1K
BS 1490 LM13WP 1K
BS 1490 LM14 1L
BS 1490 LM14WP 1L
BS 1490 LM15 1L
BS 1490 LM16 1K
BS 1490 LM16WP 1K
BS 1490 LM17 1C
BS 1490 LM18 1C
BS 1490 LM18M 1C
BS 1490 LM20 1C
BS 1490 LM20M 1C
BS 1490 LM21 1L
BS 1490 LM21M 1L
BS 1490 LM22 1L
BS 1490 LM22W 1L
BS 1490 LM23 1L
BS 1490 LM23P 1L
BS 1490 LM24 1L
BS 1490 LM24M 1L
BS 1490 LM25 1C
BS 1490 LM26 1L
BS 1490 LM27 1L
BS 1490 LM28 1L
BS 1490 LM29 1L
BS 1490 LM30 1L
BS 1490 LMO 1A
BS 1490 LMOM 1A
BS1501.3
  See designation for details
BS 1501/101 44A1
BS 1501/151 44A1
BS 1501/151 A 44A1
BS 1501/151 B 44A1
BS 1501/161 A 44A1
BS 1501/161 B 44A1
BS 1501/161 C 44A1
BS 1501/221 44A1
BS 1501/240 44G1
BS 1501/261 44G1
BS 1501/271 44K1
BS 1501/281 44K1
BS 1501/282 44K1
BS 1501/503 44F1
BS 1501/509 44F1
BS 1501/510 44F1
BS 1501/620/27 44K1
BS 1501/620/31 44K1
BS 1501/620A 44K1
BS 1501/620B 44K1
BS 1501/621 44K1
BS 1501/622/31 44K1
BS 1501/622/45 44K1
BS 1501/625 44K1
BS 1501/713 44M1
BS 1501/801 B 44N1
BS 1501/801 C 44N1
BS 1501/821 Nb 44N1
BS 1501/821 Ti 44N1
BS 1501/845 B 44N1
BS 1501/845 Ti 44N1
BS 1501/846 44N1
BS1502
  For stainless, see
  designation for details
BS 1502/713 44M1
BS 1503/161 A and B 44A2
BS 1503/161 C 44A2
BS 1503/221 44A2
BS 1503/240 A 44G1
BS 1503/240 B 44G1
BS 1503/503 44F1
BS 1503/620 44K1
BS 1503/622 44K1
BS 1503/623 44K1
BS 1503/625 44K1
BS 1503/801 44N1
BS 1503/821 Nb 44N1
BS 1503/821 Ti 44N1
BS 1503/845 B 44N1
BS 1503/845 Ti 44N1
BS 1503/846 44N1
BS 1504/101 A 44A2
BS 1504/101 B 44A2
BS 1504/101 C 44A2
BS 1504/161 A 44A2
BS 1504/161 B 44A2
BS 1504/240 44G1
BS 1504/503 44F1
BS 1504/621 44K1
BS 1504/623 44K1
BS 1504/625 44K1
BS 1504/629 44K1
BS 1504/713 44M1
BS 1504/801 44N1
BS 1504/821 Nb 44N1
BS 1504/821 Ti 44N1
BS 1504/845 B 44N1
BS 1504/845 Nb 44N1
BS 1504/846 44N1
BS 1505/622 44K1
BS 1506/111 44A1
BS 1506/162 44A3
BS 1506/240 44G1
BS 1506/621 A 44K2
BS 1506/621 B 44K2
BS 1506/625 44K1
BS 1506/661 44K1
BS 1506/713 44M1
BS 1506/801 A 44N1
BS 1506/801 AM 44N1
BS 1506/801 B 44N1
BS 1506/801 C 44N1
BS 1506/821 Nb 44N1
BS 1506/821 Ti 44N1
BS 1506/821 Ti M 44N1
BS 1506/845 44N1
BS 1507/101 44A1
BS 1507/131 44A1
BS 1507/151 44A1
BS 1507/171 44A1
BS 1507/181 44A1
BS 1507/240 44G1
BS 1507/240 A 44G1
BS 1507/240 B 44G1
BS 1507/621 44K1
BS 1507/623 44K1
BS 1507/625 44K1
BS 1507/821 44N1
BS 1507/825 44N1
BS 1507/845 44N1
BS 1508/151 44A1
BS 1508/171 44A1
BS 1508/240 44G1
BS 1508/621 44K1
BS 1508/622 44K1
BS 1508/625 44K1
BS 1508/801 44N1
BS 1508/821 44N1
BS 1508/825 44N1
BS 1508/845 44N1
BS 1525 27A
BS 1526 27A
BS 1527 27A
BS 1528 27A
BS 1529 27A
BS 1530 27A
BS 1531 27A
BS 1532 27A
BS 1533 27A

BS 1534 27A
BS 1535 27A
BS 1536 27A
BS 1537 27D
BS 1541 C106 14A
BS 1541 CA102 14G
BS 1541 CA105 14G
BS 1541 CA106 14G
BS 1541 CN101 14E
BS 1541 CN102 14E
BS 1541 CN104 14E
BS 1541 CN106 14E
BS 1541 CN107 14E
BS 1541 CS101 14M
BS 1541 CZ110 14B
BS 1541 CZ112 14B
BS 1541 CZ123 14C
BS 1554 En 56B 44M1
BS 1554 En 57B 44M1
BS 1554 En 58A 44N1
BS 1554 En 58B 44N1
BS 1554 En 58C 44N1
BS 1554 En 58D 44N1
BS 1554 En 58E 44N1
BS 1554 En 58F 44N1
BS 1554 En 58G 44N1
BS 1554 En 58H 44N1
BS 1554 En 58J 44N1
BS 1560 B 44A2
BS 1561 42.1
BS 1570 B 44A2
BS 1591 20B
BS 1616 A 1A
BS 1616 B 1C
BS 1616 C 1D
BS 1617 A 44A1
BS 1617 B 44A1
BS 1627 44A1
BS 1628 A 44K1
BS 1628 B 44K1
BS 1630 A 44M1
BS 1630 B 44M1
BS 1630 C 44M1
BS 1631 A 44N1
BS 1631 B Nb 44N1
BS 1631 B Ti 44N1
BS 1632 A 44N1
BS 1632 B 44N1
BS 1632 C Nb 44N1
BS 1632 C Ti 44N1
BS 1632 D 44N1
BS 1633 A 44A2
BS 1633 B 44A2
BS 1633 C 44A2
BS 1633 D 44A2
BS 1633 E 44A2
BS 1633 L 44A1
BS 1648 A 44M1
BS 1648 D 44N1
BS 1648 E 44N1
BS 1648 F 44N1
BS 1648 G 44N1
BS 1648 H 27B
BS 1648 K 27B
BS 1648/C 44M2
BS 1652 44G1
BS 1653 44K1
BS 1654 44A1
BS 1663 44A2
BS 1678 44A1
BS 1683 1A
BS 1717 CDS 101 44A2
BS 1717 CDS102 44A2
BS 1717 CDS103 44A2
BS 1717 CDS104 44A2
BS 1717 CDS105 44A2
BS 1717 CDS106 44A2
BS 1717 CDS107 44A2
BS 1717 CDS108 44A2
BS 1717 CDS109 44G1
BS 1717 CDS110 44K1
BS 1717 CEW101 44A2
BS 1717 CEW102 44A2
BS 1717 CEW103 44A2
BS 1717 CEW104 44A2
BS 1717 ERW101 44A2
BS 1717 ERW102 44A2
BS 1717 ERW103 44A2
BS 1730 44A1
BS 1760 A 44A3
BS 1760 B 44A3
BS 1775 CDS11 44A1
BS 1775 CDS13 44A1
BS 1775 CDS16 44A1
BS 1775 CDS20 44A1
BS 1775 CDS23 44A1
BS 1775 CDS24 44A1
BS 1775 CDS28 44A1
BS 1775 CDS35 44A1
BS 1775 CEW11 44A1
BS 1775 CEW16 44A1
BS 1775 CEW23 44A1
BS 1775 CEW24 44A1
BS 1775 CEW28 44A1
BS 1775 EFW16 44A1
BS 1775 ERW11 44A1
BS 1775 ERW16 44A1
BS 1775 ERW20 44A1
BS 1775 ERW23 44A1
BS 1775 HFS11 44A1
BS 1775 HFS13 44A1
BS 1775 HFS16 44A1
BS 1775 HFS20 44A1
BS 1775 HFS23 44A1
BS 1775 HFW13 44A1
BS 1775 HFW16 44A1
BS 1775 HFW23 44A1
BS 1775 HLW16 44A1
BS 1775 OAW11 44A1
BS 1824 NS104 14F
BS 1824 NS107 14F
BS 1845 Ag 1 42.1
BS 1845 Ag 2 42.1
BS 1845 Ag 3 42.1
BS 1845 Ag 4 42.1
BS 1845 Ag 5 42.1
BS 1845 Ag 6 42.1
BS 1845 Ag 7 42.1
BS 1845 Ag 8 42.1
BS 1845 Au 1 17.1
BS 1845 Au 2 17.1
BS 1845 Au 3 14H
BS 1845 Au 4 14H
BS 1845 Au 5 17.1
BS 1845 Au 6 17.1
BS 1845 C28 14E
BS 1845 Cp 1 14H
BS 1845 Cp 2 14H
BS 1845 Cp 3 14H
BS 1845 CU1 14A
BS 1845 CU2 14A
BS 1845 CU3 14A
BS 1845 CU4 14A
BS 1845 CU5 14A
BS 1845 CU6 14A
BS 1845 CZ1 14B
BS 1845 CZ2 14B
BS 1845 CZ3 14B
BS 1845 CZ4 14B
BS 1845 CZ5 14B
BS 1845 CZ6 14B
BS 1845 CZ7 14B
BS 1845 N11 27F
BS 1845 N12 27F
BS 1845 N13 27B
BS 1845 N14 27F
BS 1845 N15 27F
BS 1845 N16 27B
BS 1845 N17 27B
BS 1845 N18 27B
BS 1845 Pd 1 42.1
BS 1845 Pd 2 42.1
BS 1845 Pd 3 42.1
BS 1845 Pd 4 42.1
BS 1845 Pd 5 42.1
BS 1845 Pd 6 42.1
BS 1845 Pd 7 42.1
BS 1845 Pd 8 14H
BS 1845 Pd 9 42.1
BS 1845 Pd 10 42.1
BS 1845 Pd 12 14H
BS 1845 Pd 13 14H
BS 1845 PD11 27F
BS 1845 PD14 30.1
BS 1845/3 42.1
BS 1845/4 42.1
BS 1845/5 42.1
BS 1845/6 14H
BS 1845/7 14H
BS 1845/8 14C
BS 1845/9 14C
BS 1845/10 14C
BS 1845/11 14C
BS 1845/12 14C
BS 1861 14A
BS 1866 14M
BS 1867 14G
BS 1882 44A1
BS 1942/1 1C
BS 1942/2 1C
BS 1942/3 1C
BS 1942/4 1C
BS 1948 CS101 14M
BS 1949 CZ109 14C
BS 1949 CZ123 14C
BS 1956 A 44E1
BS 1956 B 44E1
BS 1977 C101 14A
BS 1977 C102 14A
BS 2014/2 44N1
BS 2027 14A

BS 2032 CA103 14G
BS 2033 CA104 14G
BS 2056 En 56A 44M1
BS 2056 En 56B 44M1
BS 2056 En 56C 44M1
BS 2056 En 56D 44M1
BS 2056 En 57 44M1
BS 2056-301581 44N1
BS 2056-302826 44N1
BS 2056-316542 44N1
BS 2061 14K
BS 2453 44A3
BS 2493 A 44G1
BS 2493 B 44K1
BS 2493 C 44K1
BS 2493 D 44K1
BS 2493/1 Cr Mo B 44K1
BS 2493/1 Cr Mo LB 44K1
BS 2493/1 Cr Mo R 44K1
BS 2493/1 Ni B 44F1
BS 2493/1 Ni C 44F1
BS 2493/1 Ni LB 44F1
BS 2493/1 Ni Mo B 44K1
BS 2493/1 Ni Mo C 44K1
BS 2493/2 Cr Mo B 44K1
BS 2493/2 Cr Mo LB 44K1
BS 2493/2 Cr Mo R 44K1
BS 2493/Mn Ni B 44F1
BS 2493/2 Ni B 44F1
BS 2493/2 Ni C 44F1
BS 2493/2 Ni Cr Mo B 44K1
BS 2493/2 Ni LB 44F1
BS 2493/2 Ni Mo B 44K1
BS 2493/3 Ni B 44F1
BS 2493/3 Ni LB 44F1
BS 2493/5 Cr Mo B 44K1
BS 2493/7 Cr Mo B 44K1
BS 2493/9 Cr Mo B 44K1
BS 2493/12 Cr Mo B 44M1
BS 2493/12 Cr Mo VB 44M1
BS 2493/12 Cr Mo WVB 44M1
BS 2627 G1E 1A
BS 2656 55A
BS 2686/1 8.1
BS 2691 44A3
BS 2755 a 14A
BS 2755/2 14A
BS 2762 NDI 44A1
BS 2762 NDII 44A1
BS 2762 NDIII 44A1
BS 2762 NDIV 44A1
BS 2763 44A2
BS 2772/2 44A2
BS 2772/3 44A1
BS 2785 CZ118 14B
BS 2785 CZ119 14B
BS 2785 CZ120 14B
BS 2786 CZ107 14B
BS 2789 2 A 20B
BS 2789 SNG24/17 20B
BS 2789 SNG27/12 20B
BS 2789 SNG32/7 20B
BS 2789 SNG37/2 20B
BS 2789 SNG42/2 20B
BS 2789 SNG47/2 20B
BS 2789/1 20B
BS 2789/2 B 20B
BS 2791 1A
BS 2803 44A3
BS 2857 20C
BS 2857 A 27E
BS 2857 B 20C
BS 2857 C 20C
BS 2857 D 20C
BS 2858 44A1
BS 2870 C101 14A
BS 2870 C102 14A
BS 2870 C103 14A
BS 2870 C104 14A
BS 2870 C105 14A
BS 2870 C105 14A
BS 2870 C106 14A
BS 2870 C107 14A
BS 2870 C108 14A
BS 2870 C109 14A
BS 2870 CA101 14G
BS 2870 CA102 14G
BS 2870 CA103 14G
BS 2870 CA104 14G
BS 2870 CA105 14G
BS 2870 CA106 14G
BS 2870 CB101 14D
BS 2870 CN101 14E
BS 2870 CN102 14E
BS 2870 CN103 14E
BS 2870 CN104 14E
BS 2870 CN105 14E
BS 2870 CN106 14E
BS 2870 CN107 14E
BS 2870 CS101 14M
BS 2870 CZ101 14B
BS 2870 CZ102 14B
BS 2870 CZ103 14B
BS 2870 CZ104 14B
BS 2870 CZ105 14B
BS 2870 CZ106 14B
BS 2870 CZ107 14B
BS 2870 CZ108 14B
BS 2870 CZ109 14C
BS 2870 CZ110 14B
BS 2870 CZ111 14B
BS 2870 CZ112 14B
BS 2870 CZ113 14C
BS 2870 CZ114 14C
BS 2870 CZ115 14C
BS 2870 CZ116 14B
BS 2870 CZ118 14B
BS 2870 CZ119 14B
BS 2870 CZ120 14C
BS 2870 CZ121 14C
BS 2870 CZ122 14C
BS 2870 CZ123 14C
BS 2870 CZ124 14B
BS 2870 CZ125 14B
BS 2870 NS101 14F
BS 2870 NS102 14F
BS 2870 NS103 14F
BS 2870 NS104 14F
BS 2870 NS105 14F
BS 2870 NS106 14F
BS 2870 NS107 14F
BS 2870 NS108 14F
BS 2870 NS109 14F
BS 2870 NS110 14F
BS 2870 NS111 14F
BS 2870 NS112 14F
BS 2870 NS113 14F
BS 2870 PB101 14K
BS 2870 PB102 14K
BS 2870 PB102 14K
BS 2870 PB103 14K
BS 2870 PB104 14K
BS 2871 C101 14A
BS 2871 C102 14A
BS 2871 C103 14A
BS 2871 C104 14A
BS 2871 C105 14A
BS 2871 C106 14A
BS 2871 C107 14A
BS 2871 C108 14A
BS 2871 C109 14A
BS 2871 CA101 14G
BS 2871 CA102 14G
BS 2871 CA103 14G
BS 2871 CA104 14G
BS 2871 CN101 14E
BS 2871 CN102 14E
BS 2871 CN103 14E
BS 2871 CN104 14E
BS 2871 CN105 14E
BS 2871 CN106 14E
BS 2871 CN107 14E
BS 2871 CN108 14E
BS 2871 CS101 14M
BS 2871 CZ101 14B
BS 2871 CZ102 14B
BS 2871 CZ103 14B
BS 2871 CZ104 14B
BS 2871 CZ105 14B
BS 2871 CZ106 14B
BS 2871 CZ107 14B
BS 2871 CZ108 14B
BS 2871 CZ109 14C
BS 2871 CZ110 14B
BS 2871 CZ111 14B
BS 2871 CZ112 14C
BS 2871 CZ113 14C
BS 2871 CZ114 14C
BS 2871 CZ115 14C
BS 2871 CZ116 14B
BS 2871 CZ117 14C
BS 2871 CZ118 14B
BS 2871 CZ119 14B
BS 2871 CZ120 14C
BS 2871 CZ121 14C
BS 2871 CZ122 14C
BS 2871 CZ123 14C
BS 2871 CZ124 14B
BS 2871 CZ126 14B
BS 2871 CZ127 14B
BS 2871 NS101 14F
BS 2871 NS102 14F
BS 2871 NS103 14F
BS 2871 NS104 14F
BS 2871 NS105 14F
BS 2871 NS106 14F
BS 2871 NS107 14F
BS 2871 NS108 14F
BS 2871 NS109 14F
BS 2871 NS110 14F
BS 2871 PB101 14K

BS 2871 PB102 14K
BS 2871 PB103 14K
BS 2871 PB104 14K
BS 2872 C101 14A
BS 2872 C102 14A
BS 2872 C103 14A
BS 2872 C104 14A
BS 2872 C105 14A
BS 2872 C106 14A
BS 2872 C107 14A
BS 2872 C108 14A
BS 2872 C109 14A
BS 2872 CA101 14G
BS 2872 CA102 14G
BS 2872 CA103 14G
BS 2872 CA104 14G
BS 2872 CA105 14G
BS 2872 CA106 14G
BS 2872 CB101 14D
BS 2872 CN101 14E
BS 2872 CN102 14E
BS 2872 CN103 14E
BS 2872 CN104 14E
BS 2872 CN105 14E
BS 2872 CN106 14E
BS 2872 CN107 14E
BS 2872 CS101 14M
BS 2872 CZ101 14B
BS 2872 CZ102 14B
BS 2872 CZ103 14B
BS 2872 CZ104 14B
BS 2872 CZ105 14B
BS 2872 CZ106 14B
BS 2872 CZ107 14B
BS 2872 CZ108 14B
BS 2872 CZ109 14C
BS 2872 CZ110 14B
BS 2872 CZ111 14B
BS 2872 CZ112 14B
BS 2872 CZ113 14C
BS 2872 CZ114 14C
BS 2872 CZ115 14C
BS 2872 CZ116 14B
BS 2872 CZ118 14B
BS 2872 CZ119 14B
BS 2872 CZ120 14C
BS 2872 CZ121 14C
BS 2872 CZ122 14C
BS 2872 CZ123 14C
BS 2872 CZ124 14B
BS 2872 CZ128 14C
BS 2872 CZ129 14C
BS 2872 NS101 14F
BS 2872 NS102 14F
BS 2872 NS103 14F
BS 2872 NS104 14F
BS 2872 NS105 14F
BS 2872 NS106 14F
BS 2872 NS107 14F
BS 2872 NS108 14F
BS 2872 NS109 14F
BS 2872 NS110 14F
BS 2872 NS111 14F
BS 2872 NS112 14F
BS 2872 NS113 14F
BS 2872 PB101 14K
BS 2872 PB102 14K
BS 2872 PB103 14K
BS 2872 PB104 14K
BS 2873 C101 14A
BS 2873 C102 14A
BS 2873 C103 14A
BS 2873 C104 14A
BS 2873 C105 14A
BS 2873 C106 14A
BS 2873 C107 14A
BS 2873 C108 14A
BS 2873 C109 14A
BS 2873 CA101 14G
BS 2873 CA102 14G
BS 2873 CA103 14G
BS 2873 CA104 14G
BS 2873 CA105 14G
BS 2873 CA106 14G
BS 2873 CB 101 14D
BS 2873 CN101 14E
BS 2873 CN102 14E
BS 2873 CN103 14E
BS 2873 CN104 14E
BS 2873 CN105 14E
BS 2873 CN106 14E
BS 2873 CN107 14E
BS 2873 CS101 14M
BS 2873 CZ101 14B
BS 2873 CZ102 14B
BS 2873 CZ103 14B
BS 2873 CZ104 14B
BS 2873 CZ105 14B
BS 2873 CZ106 14B
BS 2873 CZ107 14B
BS 2873 CZ108 14B
BS 2873 CZ109 14C
BS 2873 CZ110 14B
BS 2873 CZ111 14B
BS 2873 CZ112 14B
BS 2873 CZ113 14C
BS 2873 CZ114 14C
BS 2873 CZ115 14C
BS 2873 CZ116 14B
BS 2873 CZ118 14B
BS 2873 CZ119 14B
BS 2873 CZ120 14C
BS 2873 CZ121 14C
BS 2873 CZ122 14C
BS 2873 CZ123 14C
BS 2873 CZ124 14B
BS 2873 NS101 14F
BS 2873 NS102 14F
BS 2873 NS103 14F
BS 2873 NS104 14F
BS 2873 NS105 14F
BS 2873 NS106 14F
BS 2873 NS107 14F
BS 2873 NS108 14F
BS 2873 NS109 14F
BS 2873 NS110 14F
BS 2873 NS111 14F
BS 2873 NS112 14F
BS 2873 NS113 14F
BS 2873 PB101 14K
BS 2873 PB102 14K
BS 2873 PB103 14K
BS 2873 PB104 14K
BS 2874 C101 14A
BS 2874 C102 14A
BS 2874 C103 14A
BS 2874 C104 14A
BS 2874 C105 14A
BS 2874 C106 14A
BS 2874 C107 14A
BS 2874 C108 14A
BS 2874 C109 14A
BS 2874 CA101 14G
BS 2874 CA102 14G
BS 2874 CA103 14G
BS 2874 CA104 14G
BS 2874 CA105 14G
BS 2874 CA106 14G
BS 2874 CA107 14G
BS 2874 CB101 14D
BS 2874 CN101 14E
BS 2874 CN102 14E
BS 2874 CN103 14E
BS 2874 CN104 14E
BS 2874 CN105 14E
BS 2874 CN106 14E
BS 2874 CN107 14E
BS 2874 CS101 14M
BS 2874 CZ101 14B
BS 2874 CZ102 14B
BS 2874 CZ103 14B
BS 2874 CZ104 14B
BS 2874 CZ105 14B
BS 2874 CZ106 14B
BS 2874 CZ107 14B
BS 2874 CZ108 14B
BS 2874 CZ109 14C
BS 2874 CZ110 14B
BS 2874 CZ111 14B
BS 2874 CZ112 14C
BS 2874 CZ113 14C
BS 2874 CZ114 14C
BS 2874 CZ115 14C
BS 2874 CZ116 14B
BS 2874 CZ118 14B
BS 2874 CZ119 14B
BS 2874 CZ120 14C
BS 2874 CZ121 14C
BS 2874 CZ122 14C
BS 2874 CZ123 14C
BS 2874 CZ124 14B
BS 2874 CZ124 14B
BS 2874 CZ128 14C
BS 2874 CZ129 14C
BS 2874 CZ130 14C
BS 2874 CZ131 14C
BS 2874 CZ132 14B
BS 2874 CZ133 14C
BS 2874 CZ134 14C
BS 2874 CZ135 14C
BS 2874 CZ136 14C
BS 2874 CZ137 14C
BS 2874 NS101 14F
BS 2874 NS102 14F
BS 2874 NS103 14F
BS 2874 NS104 14F
BS 2874 NS105 14F
BS 2874 NS106 14F
BS 2874 NS107 14F
BS 2874 NS108 14F
BS 2874 NS109 14F

BS 2874 NS110 14F
BS 2874 NS111 14F
BS 2874 NS112 14F
BS 2874 NS113 14F
BS 2874 PB101 14K
BS 2874 PB102 14K
BS 2874 PB103 14K
BS 2874 PB104 14K
BS 2875 C101 14A
BS 2875 C102 14A
BS 2875 C103 14A
BS 2875 C104 14A
BS 2875 C105 14A
BS 2875 C106 14A
BS 2875 C107 14A
BS 2875 C108 14A
BS 2875 C109 14A
BS 2875 CA101 14G
BS 2875 CA102 14G
BS 2875 CA103 14G
BS 2875 CA104 14G
BS 2875 CA105 14G
BS 2875 CA106 14G
BS 2875 CB101 14D
BS 2875 CN101 14E
BS 2875 CN102 14E
BS 2875 CN103 14E
BS 2875 CN104 14E
BS 2875 CN105 14E
BS 2875 CN106 14E
BS 2875 CN107 14E
BS 2875 CS101 14M
BS 2875 CZ101 14B
BS 2875 CZ102 14B
BS 2875 CZ103 14B
BS 2875 CZ104 14B
BS 2875 CZ105 14B
BS 2875 CZ106 14B
BS 2875 CZ107 14B
BS 2875 CZ108 14B
BS 2875 CZ109 14C
BS 2875 CZ110 14B
BS 2875 CZ111 14B
BS 2875 CZ112 14B
BS 2875 CZ113 14C
BS 2875 CZ114 14C
BS 2875 CZ115 14C
BS 2875 CZ116 14B
BS 2875 CZ121 14C
BS 2875 CZ122 14C
BS 2875 CZ123 14C
BS 2875 CZ124 14B
BS 2875 NS101 14F
BS 2875 NS102 14F
BS 2875 NS103 14F
BS 2875 NS104 14F
BS 2875 NS105 14F
BS 2875 NS106 14F
BS 2875 NS107 14F
BS 2875 NS108 14F
BS 2875 NS109 14F
BS 2875 NS110 14F
BS 2875 NS111 14F
BS 2875 NS112 14F
BS 2875 NS113 14F
BS 2875 PB101 14K
BS 2875 PB101 14K
BS 2875 PB102 14K
BS 2875 PB102 14K
BS 2875 PB103 14K
BS 2875 PB104 14K
BS 2897 1350 1A
BS 2897 D1E 1A
BS 2898 1350 1A
BS 2898 6101A 1E
BS 2898 E1E 1A
BS 2898 E91E 1A
BS 2901 A8 Nb 44N1
BS 2901 A8 Ti 44N1
BS 2901 A10 44N1
BS 2901 A10 Nb 44N1
BS 2901 A10 Ti 44N1
BS 2901 A11 44N1
BS 2901 A11 Ti 44N1
BS 2901 A12 44N1
BS 2901 A12 Nb 44N1
BS 2901 A12 Ti 44N1
BS 2901 A15 44A1
BS 2901 A 16 44A2
BS 2901 A17 44A1
BS 2901 C4 14C
BS 2901 C7 14A
BS 2901 C8 14A
BS 2901 C9 14M
BS 2901 C10 14K
BS 2901 C11 14K
BS 2901 C12 14G
BS 2901 C13 14G
BS 2901 C14 14B
BS 2901 C15 14B
BS 2901 C16 14E
BS 2901 C17 14E
BS 2901 C18 14E
BS 2901 C19 14E
BS 2901 C20 14G
BS 2901 D1 23B
BS 2901 D2 23B
BS 2901 D3 23B
BS 2901 D4 23B
BS 2901 D5 23B
BS 2901 D6 23B
BS 2901 D7 23B
BS 2901 D8 23B
BS 2901 G1 1A
BS 2901 G1A 1A
BS 2901 G1B 1A
BS 2901 G1C 1A
BS 2901 NA32 27A
BS 2901 NA33 27D
BS 2901 NA34 27B
BS 2901 NA35 27B
BS 2901 NA 36 27C
BS 2901 NA 37 27C
BS 2901 NA 38 27C
BS 2901 NA 39 27C
BS 2901 NA40 27B
BS 2901 NA41 27E
BS 2901 NA 42 27C
BS 2901 NA43 27B
BS 2901 NA44 27F
BS 2901 NA45 27B
BS 2901 NG2 1C
BS 2901 NG3 1B
BS 2901 NG4 1D
BS 2901 NG5 1D
BS 2901 NG6 1D
BS 2901 NG7 1D
BS 2901 NG21 1C
BS 2926/13 44M1
BS 2926/13/4 Mo 44M1
BS 2926/15/35/H 44N1
BS 2926/17 44M1
BS 2926/17/8/2 44N1
BS 2926/18/15/3 L Mn 44P
BS 2926/19/9 44N1
BS 2926/19/9 Nb 44N1
BS 2926/19/9/3 44N1
BS 2926/19/9L 44N1
BS 2926/19/12/2L 44N1
BS 2926/19/12/3 44N1
BS 2926/19/12/3 L 44N1
BS 2926/19/12/3 Nb 44N1
BS 2926/19/13/4 44N1
BS 2926/19/13/4 L 44N1
BS 2926/19/13/4 Nb 44N1
BS 2926/20/25/5 L Cu Nb 44N1
BS 2926/20/34/2 Cu Nb 44N1
BS 2926/23/12 44N1
BS 2926/23/12 Nb 44N1
BS 2926/23/12 W 44N1
BS 2926/23/12/2 44N1
BS 2926/23/12L 44N1
BS 2926/25/6/2 44N1
BS 2926/25/6/Cu 44N1
BS 2926/25/20 44N1
BS 2926/25/20 H 44N1
BS 2926/25/20 Nb 44N1
BS 2926/25/20/1 44N1
BS 2926/25/20/1 Nb 44P
BS 2926/25/20/2 44N1
BS 2926/25/21/2 L Mn 44P
BS 2926/25/35HNb 44N1
BS 2926/25/35M 44N1
BS 2926/28 44M1
BS 2926/29/9 44N1
BS 2926/29/9/1 44N1
BS 2926/29/9/2 44N1
BS 2926 A 44N1
BS 2926 B 44N1
BS 2926 C 44N1
BS 2926 D 44N1
BS 2926 E 44N1
BS 2926 F 44N1
BS 2926 G 44N1
BS 2926 H 44P
BS 2943/2 Mn Mo B 44G1
BS 2943/Mn Mo B 44G1
BS 2943/Mo B 44G1
BS 2943/Mo C 44G1
BS 2970 MAG 1 M 23B
BS 2970 MAG 1 W 23B
BS 2970 MAG 2 M 23B
BS 2970 MAG 2 W 23B
BS 2970 MAG 3 M 23B
BS 2970 MAG 3 W 23B
BS 2970 MAG 3 WP 23B
BS 2970 MAG 4 M 23B
BS 2970 MAG 4 P 23B
BS 2970 MAG 5 M 23B
BS 2970 MAG 5 P 23B
BS 2970 MAG 6 23B

BS 2970 MAG 7 M 23B
BS 2970 MAG 7 W 23B
BS 2970 MAG 7 WP 23B
BS 3014/1 44N1
BS 3014/3 44N1
BS 3014/4 44N1
BS 3014/5 44N1
BS 3014/6 44N1
BS 3059/1 44A1
BS 3059/2 44A1
BS 3059/3 44A1
BS 3059/4 44A1
BS 3059/5 44A2
BS 3059/7 44K1
BS 3059/8 44K1
BS 3059/9 44K1
BS 3059/10 44K1
BS 3059/11 44K1
BS 3059/12 44K1
BS 3059-1250 44P
BS 3071 NA1 27D
BS 3071 NA2 27D
BS 3071 NA3 27D
BS 3072 NA11 27A
BS 3072 NA12 27A
BS 3072 NA13 27D
BS 3072 NA14 27B
BS 3073 NA11 27A
BS 3073 NA13 27D
BS 3073 NA14 27B
BS 3074 NA11 27A
BS 3074 NA12 27A
BS 3074 NA13 27D
BS 3074 NA14 27B
BS 3075 NA11 27A
BS 3075 NA13 27D
BS 3075 NA14 27B
BS 3076 NA11 27A
BS 3076 NA12 27A
BS 3076 NA13 27D
BS 3076 NA14 27B
BS 3076 NA 15 44N1
BS 3076 NA 16 27C
BS 3076 NA17 27B
BS 3076 NA18 27D
BS 3083 44A2
BS 3100 44A2
BS 3100 A1 44A2
BS 3100 A2 44A2
BS 3100 A3 44A2
BS 3100 A4 44A1
BS 3100 A5 44A2
BS 3100 A6 44A2
BS 3100 AL1 44A1
BS 3100 AM1 44A1
BS 3100 AM2 44A1
BS 3100 AW1 44A1
BS 3100 AW2 44A2
BS 3100 AW3 44A2
BS 3100 B1 44G1
BS 3100 B2 44K1
BS 3100 B3 44K1
BS 3100 B4 44K1
BS 3100 B5 44K1
BS 3100 B6 44K1
BS 3100 B7 44K1
BS 3100 BL1 44G1
BS 3100 BL2 44F1
BS 3100 BT1 44A2
BS 3100 BT2 44A2
BS 3100 BT3 44A2
BS 3100 BW1 44K1
BS 3100 BW2 44E2
BS 3100 BW3 44E2
BS 3100 BW4 44K1
BS 3100/302 C25 44N1
BS 3100/302 C35 44N1
BS 3100/304 C12 44N1
BS 3100/304 C15 44N1
BS 3100/309 C30 44N1
BS 3100/309 C40 44N1
BS 3100/310 C45 44N1
BS 3100/315 C16 44N1
BS 3100/316 C12 44N1
BS 3100/316 C16 44N1
BS 3100/347 C17 44N1
BS 3100/410 C21 44M1
BS 3100/420 C24 44M1
BS 3100/420 C29 44M1
BS 3100/425 C11 44M1
BS 3100/452 C11 44M2
BS 3100/452 C12 44M2
BS 3111/1 44A3
BS 3111/2 44G2
BS 3111/3 44E2
BS 3111/4 44K1
BS 3111/5 44K2
BS 3111/6 44K2
BS 3127/2 14D
BS 3127/2 14D
BS 3127/3 14B
BS 3127/4 27D
BS 3127/5 14K
BS 3127/6 44K1
BS 3127/7 44N1
BS 3127/8 44A2
BS 3127/9 20C
BS 3141 C1A1 44A1
BS 3141 C2A1 44A1
BS 3141 C2B1 44A1
BS 3141 C2E1 44A1
BS 3141 C2F1 44A1
BS 3141 C2G1 44A1
BS 3141 C2K1 44A1
BS 3141 C2L1 44A1
BS 3141 C3A1 44A1
BS 3141 C3B1 44A1
BS 3141 C3C1 44A2
BS 3141 C3E1 44A2
BS 3141 C3G1 44A2
BS 3141 C3H1 44A2
BS 3141 C4A1 44A2
BS 3141 C4B1 44A2
BS 3141 C4C1 44A2
BS 3141 C4D1 44A2
BS 3141 C4E1 44A2
BS 3141 C4F1 44A2
BS 3141 C4G1 44A2
BS 3141 C4J1 44A2
BS 3141 C4K1 44A2
BS 3141 C4L1 44A2
BS 3141 C5A1 44A2
BS 3141 C5C1 44A2
BS 3141 C5D1 44A2
BS 3141 C5F1 44A2
BS 3141 C5G1 44A2
BS 3141 C5J1 44A2
BS 3141 C5K1 44A2
BS 3141 C5L1 44A2
BS 3141 C6A1 44A3
BS 3141 C6B1 44A3
BS 3141 C6C1 44A3
BS 3141 C6E1 44A3
BS 3141 C6F1 44A3
BS 3141 C6G1 44A3
BS 3141 C6H1 44A3
BS 3141 C7A1 44A3
BS 3141 C7B1 44A3
BS 3141 C7C1 44A3
BS 3141 C9A1 44A4
BS 3141 C9B1 44A4
BS 3141 C9C1 44A4
BS 3141 C9D1 44A4
BS 3141 C11A1 44A4
BS 3141 C11B1 44A4
BS 3141 C12A1 44A4
BS 3141 C13A1 44A4
BS 3141 CF2C1 44A1
BS 3141 CF2D1 44A1
BS 3141 CF2H1 44A1
BS 3141 CF2J1 44A1
BS 3141 CF3D1 44A1
BS 3141 CF3E1 44A1
BS 3141 CF4H1 44A2
BS 3141 CF4M1 44A2
BS 3141 CF5B1 44A2
BS 3141 CF5E1 44A2
BS 3141 CF5H1 44A2
BS 3146 ANC1A 44M1
BS 3146 ANC1B 44M1
BS 3146 ANC1C 44M1
BS 3146 ANC2 44M1
BS 3146 ANC 3A 44N1
BS 3146 ANC 3BNb 44N1
BS 3146 ANC 4A 44N1
BS 3146 ANC 4B 44N1
BS 3146 ANC 4CNb 44N1
BS 3146 ANC 5A 44N1
BS 3146 ANC 5B 44N1
BS 3146 ANC5C 27B
BS 3146 ANC 6A 44N1
BS 3146 ANC 6B 44N1
BS 3146 ANC 7 44N1
BS 3146 ANC8 27B
BS 3146 ANC9 27C
BS 3146 ANC10 27C
BS 3146 ANC11 27C
BS 3146 ANC12 27C
BS 3146 ANC13 13A
BS 3146 ANC14 13A
BS 3146 ANC 15 27F
BS 3146 ANC16 27C
BS 3146 ANC 17 27F
BS 3146 ANC18A 27D
BS 3146 ANC18B 27D
BS 3146 ANC18C 27D
BS 3146 ANC 20A 44N3
BS 3146 ANC 20B 44N3
BS 3146 ANC 21 44N3
BS 3146 ANC 22 A, B and C 44N3
BS 3146 CIA10 44F1

BS 3146 CLA1A 44K1
BS 3146 CLA1B 44K1
BS 3146 CLA1C 44K2
BS 3146 CLA2A 44K1
BS 3146 CLA2B 44K1
BS 3146 CLA3 44A2
BS 3146 CLA4 44A2
BS 3146 CLA5A 44A2
BS 3146 CLA5B 44A2
BS 3146 CLA6 44K1
BS 3146 CLA7 44K1
BS 3146 CLA8 44K2
BS 3146 CLA9 44K1
BS 3146 CLA11 44K1
BS 3146 CLA12A 44E2
BS 3146 CLA12B 44E2
BS 3146 CLA12C 44K2
BS 3242 1E
BS 3252 T1 50.1
BS 3252 T2 50.1
BS 3252 T3 50.1
BS 3313 NS3 1B
BS 3313 S1C 1A
BS 3332/1 50A
BS 3332/2 50A
BS 3332/3 50A
BS 3332/4 50A
BS 3332/5 50A
BS 3332/6 50A
BS 3332/7 21.1
BS 3332/8 21.1
BS 3332/9 50A
BS 3333 P28/6 20B
BS 3333 P33/4 20B
BS 3370 MAG 131 23B
BS 3370 MAG S101 23B
BS 3370 MAG S111 23B
BS 3370 MAG S141 23B
BS 3370 MAG S151 23B
BS 3371 MAG T101 23B
BS 3371 MAG T111 23B
BS 3371 MAG T121 23B
BS 3371 MAG T141 23B
BS 3372 MAG F101 23B
BS 3372 MAG F121 23B
BS 3372 MAG F151 23B
BS 3373 MAG E101 23B
BS 3373 MAG E111 23B
BS 3373 MAG E121 23B
BS 3373 MAG E141 23B
BS 3373 MAG E151 23B
BS 3373 MAG E161 23B
BS 3374 MAG P101 23B
BS 3374 MAG P111 23B
BS 3374 MAG P141 23B
BS 3374 MAG P151 23B
BS 3384 A 17.1
BS 3384 B 17.1
BS 3436 Zn 1 55A
BS 3436 Zn 2 55A
BS 3436 Zn 3 55A
BS 3436 Zn 4 55A
BS 3458 44A1
BS 3468 AUS101 A 20B
BS 3468 AUS101 B 20B
BS 3468 AUS102 A 20B
BS 3468 AUS102 B 20B
BS 3468 AUS104 20B
BS 3468 AUS105 20B
BS 3468 AUS202 A 20B
BS 3468 AUS202 B 20B
BS 3468 AUS203 20B
BS 3468 AUS204 20B
BS 3468 AUS205 20B
BS 3531/1.5 51A
BS 3531/1–4 13A
BS 3531/13 13A
BS 3532/2 44N1
BS 3551/A 44E1
BS 3551 B 44K2
BS 3551 C 44K1
BS 3601 BW 44A1
BS 3601 CDS 44A1
BS 3601 EFW 44A1
BS 3601 ERW 44A1
BS 3601 HFS 44A1
BS 3601 HLW 44A1
BS 3601 SFW 44A1
BS 3602 CDS 44A1
BS 3602 EFW 44A1
BS 3602 ERW 44A1
BS 3602 HFS 44A1
BS 3706/1 44A2
BS 3706/2 44A2
BS 3706/3 44A2
BS 3740 44A2
BS 3975 NA11 27A
BS 3976 NA19 27C
BS 3988 C1E 1A
BS 4109 C101 14A
BS 4109 C102 14A
BS 4276 BS1 52.1
BS 4276 BS2 52.1
BS 4276 BS3 52.1
BS 4276 BS4 52.1
BS 4276 BS5 52.1
BS 4276 BS6 52.1
BS 4276 BS7 52.1
BS 4276 BS8 52.1
BS 4300/1 5251 1D
BS 4300/1 NJ3 1B
BS 4300/1 NJ4 1D
BS 4300/1 NJ5 1D
BS 4300/2 BTRS1 1D
BS 4300/2 BTRS2 1D
BS 4300/3 1E
BS 4300/4 1E
BS 4300/4 6463 1E
BS 4300/5 201 1F
BS 4300/5 FC 1 1F
BS 4300/6 3105 1D
BS 4300/6 NS31 1B
BS 4300/7 5005 1D
BS 4300/7 NS41 1A
BS 4300/7 NS41 1D
BS 4300/8 5454 1D
BS 4300/8 NS51 1D
BS 4300/9 NG41 1D
BS 4300/10 NT51 1D
BS 4300/11 5454 1D
BS 4300/11 NF51 1D
BS 4300/12 5454 1D
BS 4300/12 NE51 1D
BS 4300/13 5554 1D
BS 4300/13 NG52 1D
BS 4300/14 HS17 1G
BS 4300/14–7020 1G
BS 4300/15 HE17 1G
BS 4300/15.7020 1G
BS 4300/16-8011 1C
BS 4360/40 A 44A1
BS 4360/40 B 44A1
BS 4360/40 C 44A1
BS 4360/40 D 44A1
BS 4360/40 E 44A1
BS 4360/43 A 44A2
BS 4360/43A1 44A2
BS 4360/43 B 44A2
BS 4360/43 C 44A2
BS 4360/43 D 44A1
BS 4360/43 E 44A1
BS 4360/50 A 44A2
BS 4360/50 B 44A2
BS 4360/50 C 44A2
BS 4360/50 D 44A2
BS 4360/55 C 44A2
BS 4360/55 E 44A2
BS 4393 14A
BS 4425 17.1
BS 4449 44A1
BS 4461 44A1
BS 4534/1 44N1
BS 4534/1 F 44N1
BS 4534/2 44N1
BS 4534/2 F 44N1
BS 4534/3 44N1
BS 4534/3 F 44N1
BS 4534/4 44N1
BS 4534/5 44N1
BS 4534/6 44N1
BS 4534/7 44N1
BS 4534/8 44N1
BS 4534/9 44N1
BS 4534/10 44N1
BS 4534/11 44N1
BS 4577 A1/1 14A
BS 4577 A1/2 14A
BS 4577 A1/3 14A
BS 4577 A2/1 14L
BS 4577 A2/2 14L
BS 4577 A3/1 14D
BS 4577 A3/2 14E
BS 4577 A4/1 14E
BS 4577 A4/2 14D
BS 4577 A4/3 14A
BS 4577 A4/4 14G
BS 4608 14A
BS4882-6
See BS1506-713 44M1
BS4882-8
See BS1506-801 44N1
BS4882-B6
See BS1506-713 44M1
BS4882-B8
See BS1506-801B 44N1
BS4882B8C
See BS1506-821Nb 44N1
BS4882-B8M
See BS1506-845 44N1
BS4882B8TX
See BS1506-821Ft 44N1

BS4882-B8X
See BS1506-801B 44N1
BS4882-L8
See BS1506-801B 44N1
BS4882L8C
See BS1506-821Nb 44N1
BS4882L8M
See BS1506-845 44N1
BS4882L8MX
See BS1506-845 44N1
BS4882L8TX
See BS1506-821Ft 44N1
BS4882LBX
See BS1506-801B 44N1
BS 5001 20B
BS 5005 44K2
BS 5006 44A2
BS 5007 44K2
BS 5008 44K2
BS 5009 44A2
BS 5010 44A2
BS 5022 20B
BS 5024 20B
BS 5025 20B
BS 5026 20B
BS 5028 44A2
BS5770
See designation for details
BS 6105-A1 44N1
BS 6105-A2 44N1
BS 6105-A4 44N1
BS 6105 F1 44M1
BS 6323 LW12 44M1
BS 6323 LW19 44M1
BS 6323 LW20 44N1
BS 6323 LW21 44N1
BS 6323 LW22 44N1
BS 6323 LW23 44N1
BS 6323 LW24 44N1
BS 6360 14A
BS 6791 1A
BS 9156/C 44K2
BS 43002 BTRS1 1A
BS 43009 NG41 1A
BS48828MX
See BS1506-845 44M1
BS B1 14C
BS B8 14K
BS B11 14B
BS B12 14C
BS B13 14C
BS B20 14C
BSB649
Cr Ni Mo Cu Fe alloy - see N and UNS designation
BS Cb 44N1
BS K6 20B
BS K11 20B
BS L 23B
BS L120 23A
BS M 44N1
BSN 1 44F2
BSN 3 44F1
BS N35 44K2
BSN 35 44F2
BSS 44K3
BS S2 44K2
BS S3 44A1
BS S11 44K2
BS S12 44K2
BS S13 44A1
BS S14 44A1
BS S16 44K1
BS S18 44K2
BS S19 44E1
BS S25 44A3
BS S28 44K1
BS S32 44A1
BS S33 44K2
BS S34 44K2
BS S40 44K2
BS S43 44K2
BS S65 44K2
BS S68 44H
BS S69 44K2
BS S70 44A3
BS S79 44A3
BS S81 44K2
BS S82 44K1
BS S84 44A1
BS S86 44K2
BS S87 44K2
BS S88 44K2
BS S90 44K1
BS S91 44A1
BS S92 44A1
BS S94 44K2
BS S95 44K2
BS S96 44K1
BS S97 44K1
BS S98 44K2
BS S99 44K2
BS S102 44G1
BS S103 44K1
BS S106 44K1
BS S107 44K1
BS S112 44A1
BS S114 44G1
BS S115 44E1
BS S117 44E1
BS S118 44K2
BS S119 44K2
BS S120 44K1
BS S122 44K1
BS S510 44A1
BS S511 44A1
BS S512 44A1
BS S513 44A3
BS S514 44A1
BS S515 44A1
BS S517 44A3
BS S518 44K1
BS S519 44K1
BSS 1462 44K1
BS STA 5 V20 44K1
BS T 44N1
BST 0
see GOST 380 44A1
BST 1KP
see GOST 380 44A1
BST 1KP2
see GOST 380 44A1
BST 1PS2
see GOST 380 44A1
BST 1SP
see GOST 380 44A1
BST 1SP2
see GOST 380 44A1
BS T2 44K1
BST 2KP
see GOST 380 44A1
BST 2KP2
see GOST 380 44A1
BST 2PS
see GOST 380 44A1
BST 2PS2
see GOST 380 44A1
BST 2SP
see GOST 380 44A1
BST 2SP2
see GOST 380 44A1
BST 3 44F
BST 3
see GOST 10705 44A1
BS T3 14F
BST 3GPS
see GOST 380 44A1
BST 3GPS2
see GOST 380 44A1
BST 3KP2
see GOST 380 44A1
BST 3PS
see GOST 380 44A1
BST 3PS2
see GOST 380 44A1
BST 3SP
see GOST 380 44A1
BST 3SP2
see GOST 380 44A1
BST 4 1F
BST 4
see GOST 10705 44A1
BS T4 1F
BST 4GPS
see GOST 380 44A1
BST 4KP
see GOST 380 44A1
BST 4KP2
see GOST 380 44A1
BST 4KP2 44A1
BST 4PS
see GOST 380 44A1
BST 4PS2
see GOST 380 44A1
BST 4SP2
see GOST 380 44A1
BST 5GPS2
see GOST 380 44A1
BST 5PS
see GOST 380 44A2
BST 5PS2
see GOST 380 44A2
BST 5SP
see GOST 380 44A2
BST 5SP2
see GOST 380 44A2
BS T6 44A1
BST 6P2
see GOST 380 44A2
BST 6PS2
see GOST 380 44A2

BST 6SP
see GOST 380 44A2
BST 6SP2
see GOST 380 44A2
BS T26 44A1
BS T50 44K2
BS T52 14K
BS T53 44K1
BS T56 44K1
BS T57 44K1
BS T59 44K1
BS T60 44K1
BSTA 1 51A
BSTA 2 51A
BSTA 6 51A
BSTA 7 51A
BSTA 8 51A
BSTA 9 51A
BSTA 10 51B
BSTA 11 51B
BSTA 12 51B
BSTA 13 51B
BSTA 21 51A
BSTA 22 51A
BSTA 23 51A
BSTA 24 51A
BSTA 28 51B
BSTA 38 51B
BSTA 39 51B
BSTA 40 51B
BSTA 41 51B
BSTA 42 51B
BSTA 43 51B
BSTA 44 51B
BSTA 45 51B
BSTA 46 51B
BSTA 47 51B
BSTA 48 51B
BSTA 49 51B
BSTA 50 51B
BSTA 51 51B
BSTA 52 51A
BSTA 53 51A
BSTA 54 51A
BSTA 55 51A
BSTA 56 51B
BSTA 57 51B
BSTA 58 51A
BSTA 59 51B
B STAR H 14C
BS W6 44A3
BS W10 44N1
BS W11 44N1
BT 1 1D
BT 2 1D
BT 3 1E
BT 4 1D
BT 5 1D
BT 6 1E
BTRE 6
see BS 4300/4 1E
BTRF 6
see BS 4300/3 1E
BTRS 1
see BS 4300/2 1A
BTRS 2
see BS 4300/2 1D
BU 8 44L
BULLET ALLOY 21.1
BULL HEAD RAIL 44A3
BULLS HEAD 44A3
BUSTER ALLOY 44K2
B V Ag 0 42.1
B V Ag 8 42.1
B V Ag 8a 42.1
B V Ag 18 42.1
B V Ag 29 42.1
B V Ag 30 42.1
B V Ag 31 42.1
B V Ag 32 42.1
B V Ag C 42.1
BV 10 44L
BV Pd 1 30.1
BVR 20h/W 44K1
BVR 20T 44K1
BVT 130 44M1
BVT 130V 44M1
BVT 130V50 44M1
BWB 3-7024
see 3-7024 51A
BWB 3-7034
see 3-7034 51A
BWB 3-7064
see 3-7034 51A
BWB 3-7164
see 3-7164 51C
BYR 1 44K1
C 23B
C 1 44A1
C 1A1
see BS range 44A1
C 1K 44E1
C1 MATCH COMP Fe C 20B
C1 MATCH Sg 20 Ni 20B
C 1MK 44E1
C1Q
See ASTMA757
C 2 44A1
C 2A1
see BS range 44A1
C 2B 14C
C 2B1
see BS range 44A1
C 2C 14C
C 2E1
see BS range 44A1
C 2F1
see BS range 44A1
C 2G1
see BS range 44A1
C 2K 44E1
C 2K1
see BS range 44A1
C 2L1
see BS range 44A1
C 2MK 44E1
C 3 44A2
C 3A1
see BS range 44A1
C 3B1
see BS range 44A1
C 3C1
see BS range 44A2
C 3E1
see BS range 44A2
C 3G1
see BS range 44A2
C 3H1
see BS range 44A2
C 4 14C
C 4 14H
C 4 44A2
C 4A
see ASTM range 1L
C 4A1
see BS range 44A2
C 4B1
see BS range 44A2
C 4C1
see BS range 44A2
C 4D1
see BS range 44A2
C 4E1
see BS range 44A2
C 4F1
see BS range 44A2
C 4G1
see BS range 44A2
C 4J1
see BS range 44A2
C 4K1
see BS range 44A2
C 4L1
see BS range 44A2
C 5 44A2
C 5 44K1
C 5 44K2
C 5A1
see BS range 44A2
C 5C1
see BS range 44A2
C 5D1
see BS range 44A2
C 5F1
see BS range 44A2
C 5G1
see BS range 44A2
C 5J1
see BS range 44A2
C 5K1
see BS range 44A2
C 5L1
see BS range 44A2
C 6 44A3
C 6A1
see BS range 44A3
C 6B1
see BS range 44A3
C 6C1
see BS range 44A3
C 6D1
see BS range 44A3
C 6E1
see BS range 44A3
C 6F1
see BS range 44A3
C 6G1
see BS range 44A3
C 6H1
see BS range 44A3

C 7
see BS range 14A
C 7A1
see BS range 44A3
C 7B1
see BS range 44A3
C 8
see BS range 14A
C 8 UNI 4365 44A1
C 9
see BS range 14M
C9 14M
C 9/15 44A1
C 9/22 44A1
C 9/35
see BS range 44A2
C 9/45
see BS range 44A2
C 9/A1
see BS range 44A4
C 9/B1
see BS range 44A4
C 9/C1
see BS range 44A4
C 9/D1
see BS range 44A4
C 10
see BS range 14K
C10 14K
C 10 14J
C 10 44A1
C 10 UNI 4365 44A1
C 11
see BS range 14K
C11 14K
C 11/A1
see BS range 44A1
C 11/B1
see BS range 44A4
C 12
see BS range 14G
C12 14G
C 12 44M2
C 12/A1
see BS range 44A4
C12 Fe 14G
C 13
see BS range 14G
C13 14G
C 13/A1
see BS range 44A4
C 14 14B
C 15 14B
C 15 44A1
C 15 UNI 4365 44A1
C 16
see BS range 14E
C 18 14E
C 20
see BS range 14G
C20 14G
C 20 44A1
C 20 UNI 4365 44A1
C 21 UNI 4365 44A1
C22 14G
C 22 44A1
C23 14G
C 23 44A1
C 24 14A
C25 14K
C26 14G
C 30 44A2
C 30 UNI 4365 44A2
C 35 44A2
C 35 44E1
C 35 44E2
C 35 UNI 4365 44A2
C 40 44A2
C 40 UNI 4365 44A2
C 45
see DIN range 44A2
C45W 44A3
C 45W3 44A2
C 46 44E3
C 50 14C
C 50 44A2
C 53 44A3
C 60
see DIN range 44A3
C 60 W 44A3
C 60 W3 44A3
C 63 14F
C 67
see DIN range 44A3
C 70 44A3
C 71 44E3
C 75
see DIN range 44A3
C 75 W 44E3
C 80 14A
C 80W1 44A4
C 85 W 44A4
C 85WS 44A4
C 87 WS 44E3
C 90 14C
C 90 44A3
C 92 14C
C 94 14C
C 100 44A4
C 101
see BS range 14A
C 102
see BS range 14A
C 103
see BS range 14A
C 104
see BS range 14A
C 105
see BS range 14A
C 105W1 44A4
C 106
see BS range 14A
C 107
see BS range 14A
C 108
see BS range 14A
C 108 44K2
C 109
see BS range 14A
C 110 44J
C 120 AV 51C
C130 AM 51A
C 135 AMO 51C
C 140 44A2
C 175 44A3
C 182 44E3
C 242 27C
C 253 44K3
C 254 44K3
C 255 44A3
C 263 27C
C 265 44M2
C 276
see HASTELLOY C276 27F
C 276 27C
C 320 44K2
C 330 44K2
C 345 44K2
C 355.0 1L
C 424 44K2
C 445 44L
C 525 44E3
C 550 44K3
C 612 1M
C 642 44K1
C 643 44K1
C 1023 27C
C 1105 44A4
C10100 14A
C10200
UNS for SAECA102 14A
C10300 14A
C10400 14A
C10500 14A
C10700 14A
C10800 14A
C10920 14A
C10930 14A
C10940 14A
C11000
UNS for SAECA110 14A
C11000 14A
C11010 14A
C11020 14A
C11030 14A
C11100
UNS for SAECA111 14A
C11300
UNS for SAECA113 14A
C11400
UNS for SAECA114 14A
C11500
UNS for SAECA115 14A
C11600
UNS for SAECA116 14A
C11700 14A
C11904 14A
C11905 14A
C11906 14A
C11907 14A
C12000
UNS for SAECA120 14A
C12100 14A
C12200
UNS for SAECA122 14A
C12210 14A
C12220 14A
C12300 14A
C12500 14A
C12700 14A
C12800 14A

C12900 14A
C13000 14A
C14100 14A
C14180 14A
C14181 14A
C14200 14A
C14210 14A
C14300 14A
C14310 14A
C14400 14A
C14410 14A
C14420 14A
C14430 14A
C14440 14A
C14500
UNS for SAECA145 14A
C14510 14A
C14520 14A
C14530 14A
C14700
UNS for SAECA147 14A
C14700 14A
C14710 14A
C14720 14A
C14730 14A
C15000
UNS for SAECA150 14A
C15100 14A
C15150 14A
C15500 14A
C15600 14A
C15710 14A
C15715 14A
C15720 14A
C15725 14A
C15735 14A
C15760 14A
C16200
UNS for SAECA162 14A
C16210 14A
C16400 14A
C16500 14A
C 17000 14D
C17200
UNS for SAECA172 14D
C 17200 14D
C 17300 14D
C 17400 14D
C 17410 14D
C 17420 14D
C17500
UNS for SAECA175 14D
C 17500 14D
C 17510 14D
C 17520 14D
C17600
UNS for SAECA176 14D
C 17600 14D
C 17700 14D
C18000 14L
C18030 14L
C18040 14L
C18050 14L
C18070 14L
C18090 14L
C18100 14L
C18135 14L
C18150 14L
C18200 14L
C18400 14L
C18500 14L
C18550 14L
C18700
UNS for SAECA187 14A
C18900 14A
C18980 14A
C18990 14K
C 19000 14E
C 19010 14E
C 19020 14E
C 19100 14E
C19200
UNS for SAECA192 14A
C19210 14A
C19220 14A
C19250 14A
C19260 14A
C19280 14A
C19400 14A
C19410 14A
C19450 14A
C19500 14A
C19520 14A
C19600 14A
C19700 14A
C19750 14A
C 20500 14B
C21000
UNS for SAECA210 14B
C 21000 14B
C22000
UNS for SAECA220 14B
C 22000 14B
C 22600 14B
C23000
UNS for SAECA230 14B
C 23000 14B
C 23030 14B
C 23400 14B
C24000
UNS for SAECA240 14B
C 24000 14B
C 24080 14B
C 25000 14B
C26000
UNS for SAECA260 14B
C 26000 14B
C 26100 14B
C 26130 14B
C 26200 14B
C 26380 14B
C26800
UNS for SAECA268 14B
C 26800 14B
C27000
UNS for SAECA270 14B
C 27000 14B
C 27200 14B
C 27400 14B
C 28000 14B
C 28200 14C
C 28580 14C
C 29800 14B
C 31000 14B
C 31200 14B
C 31400 14B
C 31600 14B
C 32000 14B
C 32500 14B
C 32510 14B
C33000
UNS for SAECA330 14B
C 33000 14B
C33100
UNS for SAECA331 14B
C 33100 14B
C 33200 14B
C 33500 14B
C 33530 14B
C 34000 14B
C34200
UNS for SAECA342 14B
C 34200 14B
C 34400 14B
C34500
UNS for SAECA345 14A
C 34500 14B
C 34700 14B
C 34800 14B
C 34900 14B
C35000
UNS for SAECA350 14B
C 35000 14B
C 35300 14B
C 35330 14B
C 35340 14B
C 35600 14B
C36000
UNS for SAECA360 14B
C 36000 14B
C 36200 14B
C 36500 14B
C 36600 14B
C 36700 14B
C 36800 14B
C 37000 14B
C 37100 14B
C37700
UNS for SAECA377 14C
C 37700 14B
C 37710 14C
C 37800 14C
C 38000 14C
C 38010 14C
C 38500 14C
C 38510 14C
C 38590 14C
C 38600 14C
C40400 14K
C40500 14K
C40800 14K
C 40900 14B
C41000 14K
C41100 14K
C41300 14K
C41500 14K
C41900 14K
C42000 14K
C42100 14K
C42200 14K
C42500 14K

C43000 14K
C43200 14K
C43400 14K
C43500 14K
C43600 14K
C 43800 14B
C44300 14K
C44400 14K
C44500 14K
C 45450 14B
C 46200 14B
C 46210 14B
C46400
UNS for SAECA464 14C
C 46400 14C
C 46420 14C
C46500
UNS for SAECA465 14C
C 46500 14C
C46600
UNS for SAECA466 14C
C 46600 14C
C46700
UNS for SAECA467 14C
C 46700 14C
C 47000 14C
C 47200 14B
C 47600 14B
C 47940 14B
C 48200 14C
C 48500 14C
C 48510 14B
C 48600 14C
C 49080 14C
C50100 14K
C50200 14K
C50500 14K
C50700 14K
C50710 14K
C50715 14K
C50800 14K
C50900 14K
C51000
UNS for SAECA510 14K
C51100
UNS for SAECA511 14K
C51800 14K
C51900 14K
C52100
UNS for SAECA521 14K
C52400
UNS for SAECA524 14K
C52400 14K
C52600 14K
C52900 14K
C53200 14K
C53400 14K
C54400
UNS for SAECA544 14K
C54600 14K
C54800 14K
C55180 14A
C55181 14A
C55280 14A
C55281 14A
C55282 14A
C55283 14A
C55284 14A
C60600 14G
C60700 14G
C60800
UNS for SAECA608 14G
C61000 14G
C61300 14G
C61400
UNS for SAECA614 14G
C61500 14G
C61550 14G
C61800
UNS for SAECA618 14G
C61900 14G
C62200 14G
C62300
UNS for SAECA623 14G
C62400
UNS for SAECA624 14G
C62500 14G
C62580 14G
C62581 14G
C62582 14G
C62730 14G
C63000
UNS for SAECA630 14G
C63010 14G
C63020 14G
C63200 14G
C63230 14G
C63280 14G
C63300 14G
C63380 14G
C63400 14G
C63600 14G
C63800 14G
C64110 14G
C64200
UNS for SAECA642 14G
C64210 14G
C64250 14G
C64400 14G
C 64700 14E
C 64710 14E
C 64720 14E
C64900 14K
C65100 14M
C65300 14M
C65400 14M
C65500
UNS for SAECA655 14M
C65600 14M
C65620 14M
C65800 14M
C66100 14M
C 66400 14B
C 66410 14B
C 66700 14B
C 66800 14B
C 66900 14B
C67000
UNS for SAECA670 14B
C 67000 14C
C 67130 14C
C67300
UNS for SAECA673 14B
C 67300 14C
C67400
UNS for SAECA674 14C
C 67400 14C
C 67410 14C
C67500
UNS for SAECA675 14C
C 67500 14C
C 67620 14C
C 67700 14C
C 67800 14C
C 67810 14C
C 67820 14C
C 68000 14C
C 68100 14C
C 68200 14C
C 68600 14C
C 68700 14B
C 68800 14B
C 69000 14B
C 69100 14B
C69400 14M
C69430 14M
C69440 14M
C69450 14M
C69700 14M
C69710 14M
C69720 14M
C69730 14M
C 69800 14B
C69900 14N
C69910 14N
C69950 14N
C 70100 14E
C 70200 14E
C 70250 14E
C 70300 14E
C 70320 14E
C 70400 14E
C 70440 14E
C 70500 14E
C70600
UNS for SAECA706 14E
C 70600 14E
C 70610 14E
C 70690 14E
C 70700 14E
C 70800 14E
C 70900 14E
C71000
UNS for SAECA710 14E
C 71000 14E
C 71100 14E
C 71110 14E
C 71300 14E
C71500
UNS for SAECA715 14E
C 71500 14E
C 71580 14E
C 71581 14E
C 71590 14E
C 71630 14E
C 71640 14E
C 71700 14E
C 71900 14E
C 72150 14E
C 72200 14E
C72400 14N

C 72400 14E
C72420 14N
C 72420 14E
C72500 14N
C 72500 14E
C72600 14N
C 72600 14E
C 72700 14E
C 72800 14E
C 72900 14E
C 72950 14E
C73150 14F
C73200 14F
C73500 14F
C73800 14F
C74000 14F
C74300 14F
C74500 14F
C75200
UNS for SAECA752 14F
C75400 14F
C75700 14F
C75720 14F
C75900 14F
C76000 14F
C76100 14F
C76200 14F
C76300 14F
C76390 14F
C76400 14F
C76600 14F
C76700 14F
C77000
UNS for SAECA770 14F
C77010 14F
C77300 14F
C77310 14F
C77400 14F
C77600 14F
C78200 14F
C78800 14F
C79000 14F
C79200 14F
C79300 14F
C79600 14F
C79620 14F
C79800 14F
C79810 14F
C 79820 14C
C79900 14F
C80100 14A
C80300 14A
C80500 14A
C80700 14A
C80900 14A
C81100 14A
C81200 14A
C 81300 14D
C 81400 14D
C81500 14L
C81540 14A
C 81700 14D
C 81800 14D
C 82000 14D
C 82100 14D
C 82200 14D
C 82400 14D
C 82500 14D
C 82510 14D
C 82600 14D
C 82700 14D
C 82800 14D
C 83300 14B
C 83400 14B
C 83410 14B
C 83420 14B
C 83450 14B
C83500 14K
C83520 14K
C83600
UNS for SAECA836 14K
C83800
UNS for SAECA838 14K
C83810 14K
C84200 14K
C84400 14K
C84410 14K
C84500 14K
C84800 14K
C85200
UNS for SAECA852 14B
C 85200 14B
C 85210 14B
C 85300 14B
C 85310 14B
C85400
UNS for SAECA854 14B
C 85400 14B
C 85500 14B
C 85600 14B
C 85700 14B
C 85710 14B
C85800
UNS for SAECA858 14B
C 85800 14B
C 86100 14B
C86200
UNS for SAECA862 14B
C 86200 14B
C86300
UNS for SAECA863 14B
C 86300 14B
C 86400 14B
C86500
UNS for SAECA865 14B
C 86500 14B
C 86700 14B
C 86800 14B
C87200
UNS for SAECA872 14B
C87200 14M
C87300 14M
C87400
UNS for SAECA874 14B
C87400 14M
C87410 14M
C87420 14M
C87430 14M
C87500
UNS for SAECA875 14B
C87500 14M
C87510 14M
C87520 14M
C87530 14M
C87600 14M
C87610 14M
C87800
UNS for SAECA878 14B
C87800 14M
C87900
UNS for SAECA879 14B
C 87900 14B
C90200 14K
C90250 14K
C90300
UNS for SAECA903 14K
C90500
UNS for SAECA905 14K
C90700 14K
C90710 14K
C90800 14K
C90810 14K
C90900 14K
C91000 14K
C91100 14K
C91300 14K
C91600 14K
C91700 14K
C92200
UNS for SAECA922 14K
C92300
UNS for SAECA923 14K
C92310 14K
C92400 14K
C92410 14K
C92500
UNS for SAECA925 14K
C92600 14K
C92610 14K
C92700
UNS for SAECA927 14K
C92710 14K
C92800 14K
C92810 14K
C92900
UNS for SAECA929 14K
C93100 14K
C93200
UNS for SAECA932 14K
C93400 14K
C93500
UNS for SAECA935 14K
C93600 14K
C93700
UNS for SAECA937 14K
C93720 14K
C93800
UNS for SAECA938 14K
C93900 14K
C94000 14K
C94100 14K
C94200 14K
C94300
UNS for SAECA943 14K
C94310 14K
C94320 14K
C94330 14K
C94400 14K
C94500 14K
C94700
UNS for SAECA947 14K

C94800
UNS for SAECA948 14K
C94900 14K
C95200
UNS for SAECA952 14G
C95210 14G
C95220 14G
C95300
UNS for SAECA953 14G
C95400
UNS for SAECA954 14G
C95410 14G
C95420 14G
C95500
UNS for SAECA955 14G
C95510 14G
C95520 14G
C95600 14G
C95700 14G
C95710 14G
C95800
UNS for SAECA958 14G
C95810 14G
C95900 14G
C96200
UNS for SAECA962 14A
C 96200 14E
C 96300 14E
C 96400 14E
C 96600 14E
C 96700 14E
C 96800 14E
C97300 14F
C97400 14F
C97600 14F
C97800 14F
C98200 14K
C98400 14K
C98600 14K
C98800 14K
C98820 14K
C98840 14K
C99300 14G
C99350 14G
C 99400 14E
C 99500 14E
C99600 14N
C99700 14N
C99750 14N
CA 2 44K1
CA 6N 44N1
CA 6NM 44M1
CA 6NM 44N3
CA 15 44M1
CA 15M 44M1
CA 25 MWV 44M1
CA28 MWV 44M1
CA 30 44E1
CA 30 44E2
CA 30 A 44E1
CA 30 A 44E2
CA 40 44M1
CA 40F 44M1
CA 101
see BS range 14G
CA 102
see BS range 14A
CA 102
see BS range 14G
CA 103
see BS range 14G
CA 104
see BS range 14G
CA 105
see BS range 14G
CA107
BS designation 14G
CA 110
see SAE range 14A
CA 111
see SAE range 14A
CA 113
see SAE range 14A
CA 114
see SAE range 14A
CA 116
see SAE range 14A
CA 120
see SAE range 14A
CA 122
see SAE range 14A
CA 145
see SAE range 14A
CA 147
see SAE range 14A
CA 150
see SAE range 14A
CA 162
see SAE range 14A
CA 170
see SAE range 14D
CA 172
see SAE range 14D
CA 175
see SAE range 14D
CA 176
see SAE range 14D
CA 184
see SAE range 14L
CA 187
see SAE range 14A
CA 210
see SAE range 14B
CA 230
see SAE range 14B
CA 240
see SAE range 14B
CA 260
see SAE range 14B
CA 268
see SAE range 14B
CA 270
see SAE range 14B
CA 330
see SAE range 14B
CA 331
see SAE range 14B
CA 342
see SAE range 14B
CA 345
see SAE range 14B
CA 350
see SAE range 14B
CA 360
see SAE range 14B
CA 377
see SAE range 14C
CA 464
see SAE range 14C
CA 510
see SAE range 14K
CA 521
see SAE range 14K
CA 544
see SAE range 14K
CA 614
see SAE range 14G
CA 617
see SAE range 14G
CA 623
see SAE range 14G
CA 624
see SAE range 14G
CA 630
see SAE range 14G
CA 637
see SAE range 14G
CA 655
see SAE range 14M
CA 670
see SAE range 14B
CA 673
see SAE range 14B
CA 674
see SAE range 14C
CA 675
see SAE range 14C
CA 706
see SAE range 14E
CA 710
see SAE range 14E
CA 715
see SAE range 14E
CA 752
see SAE range 14F
CA 770
see SAE range 14F
CA836
See SAECA836 14K
CA838
See SAECA838 14K
CA852
See SAECA852 14B
CA854
See SAECA854 14B
CA858
See SAECA858 14B
CA862
See SAECA862 14B
CA863
See SAECA863 14B
CA865
See SAECA865 14B
CA872
See SAECA872 14B
CA874
See SAECA874 14B
CA875
See SAECA875 14B

CA878
  See SAECA878 14B
CA879
  See SAECA879 14B
CA903
  See SAECA903 14K
CA905
  See SAECA905 14K
CA922
  See SAECA922 14K
CA923
  See SAECA923 14K
CA925
  See SAECA925 14K
CA927
  See SAECA927 14K
CA929
  See SAECA929 14K
CA932
  See SAECA932 14K
CA935
  See SAECA935 14K
CA937
  See SAECA937 14K
CA938
  See SAECA938 14K
CA943
  See SAECA943 14K
CA947
  See SAECA947 14K
CA948
  See SAECA948 14K
CA952
  See SAECA952 14G
CA953
  See SAECA953 14G
CA954
  See SAECA954 14G
CA955
  See SAECA955 14G
CA958
  See SAECA958 14G
CA962
  See SAECA962 14A
CABRA 170 14D
CAC
  see ASTMA823 20B
CADMIUM 8.1
CADMIUM COPPER 14A
CALCIUM 10.1
CALDUR 44L
CALITE 44N1
CALMALLOY 27D
CALMAX 44L
CALMET 44N1
CALOMIC 27B
CALORITE 27B
CAMLOY 44N1
CA NICKEL 27F
CANO 30 44K1
CAO 30 44K1
CAP COPPER 14B
CAPITAL 305 44K3
CAPITAL 395 44K3
CAPITAL 398 44L
CAPITAL 562 44K3
CAPSULE METAL 21.1
CARBALOY 52.1
CARBON 44A4
CARBON COLD HEADER 44K3
CARBONIZED NICKEL 27A
CAROBRONZE 14K
CARO-BUHLER 14B
CARPENTER 5/317 44K2
CARPENTER 10 STAR 44K3
CARPENTER 20 44N1
CARPENTER 20 Mo6 44N2
CARPENTER 20C63 44N1
CARPENTER 345 44K1
CARPENTER 481 44K2
CARPENTER 636 44M1
CARPENTER 883 44K2
CARPENTER A46 44M1
CARPENTER SUPER
  SPEED STAR 44K3
CARP RED LABEL 44A4
CARP YELLOW LABEL 44A4
CARRS BLUE LABEL 44K3
CARTRIDGE BRASS 14B
CARVITE 44K3
CASALLOY 44F1
CASCADE 44K1
CASE 1280 44A2
CASONA 44A1
CASTEES M 27D
CASTEES N 27A
CASTEES SG 20C
CASTLE 6/6/2 44K3
CASTLE SUPERIOR 44K3
CATH
  see ASTM range 14A
CATHODE ALLOY 27F
CATHODE COPPER 14A
CATHODE NICKEL 27A
CAUSAL METAL 20B
CAZIN 8.1
CB 20B
CB/V45 44J
CB/V50 44J
CB/V55 44J
CB 3 20B
CB 30 44M1
CB 60A 1F
CB 101
  see BS range 14D
CB CUFRON 14E
C BRONZE 14H
CBS 600 44K1
CBV 44E1
CBV/Z 44E2
CBV 1 44E1
CBV 2 44E2
CBV 2h 44E2
CBVO 44K1
CBVO/3R 44K1
CBVO/h 44K2
CBVO/M 44K1
CBVO/TM 44K1
CBVO/z 44K2
CBVOV 44K1
CBVOV/N 44K1
CBX CUFRON 14E
CC 20B
CC 50 44M1
CCM 44M2
CCM 10 44L
CCMP 15 44L
CCR 2 44E2
CCR 350 44E2
CCS 44K2
CCU 44A2
CCV 44K2
CCW 44K2
CD 44M2
CD2 44M1
CD 4M 44N1
CD 4M Cu 44N2
CD4 MCu 44N1
CD 28 44A1
CD 32 44A1
CD 40 44A1
CD 60 44A3
CDA 170 14D
CDA 172 14D
CDA 175 14D
CDA 176 14D
CDD 44K3
CDS 44K2
CDS 2 44K2
CDV 44K2
CDV 1 44M2
CDV 2 44M2
CDV 4 44M2
CDW 44M2
CE 30 44N1
CEAX 44A2
CEC SMOOTHCUT 44K2
CEKAS M151 44M1
CEKAS M152 44M1
CELFOR 44A4
CELLOCITO PL S2 44G1
CELLOCITO PL2 44A1
CENTURY 44K3
CERALUMIN ASM 1L
CERALUMIN B 1L
CERALUMIN C 1L
CERAMIC SEALING
  ALLOY 20C
CERIUM 11.1
CERROBASE 6.1
CERROBASE 21.1
CERROBEND 50A
CERRO-CAST 6.1
CERRO-LOW 117 6.1
CERRO-LOW 136 6.1
CERRO-LOW 147 6.1
CERROMATRIX 50A
CERROSAFE 6.1
CERROSEAL - 35 50A
CERROTRIC 6.1
CERROTRU 6.1
CERROTRU 50A
CEW 3
  see BS range 44A2
CEW 4
  see BS range 44A2
CF 2C1
  see BS range 44A1
CF 2D1
  see BS range 44A1

CF 2H1
see BS range 44A1
CF 2J1
see BS range 44A1
CF 3 44N1
CF 3D1
see BS range 44A1
CF 3E1
see BS range 44A1
CF 3M 44N1
CF 3MN 44N1
CF 4H1
see BS range 44A2
CF 4M1
see BS range 44A2
CF 5H1
see BS range 44N1
CF 8 44N1
CF 8 C 44N1
CF 8 M 44N1
CF-10MC 44N1
CF 10S Mn N 44P
CF 12 M 44N1
CF 16 F 44N1
CF 20 44N1
Cf 35 44A2
Cf 43 44A2
Cf 53 44A2
Cf 70 44A3
CFS 44F3
CG6MMN 44N1
CG6 MMN 44P
CG 8 M 44N1
CG12 44N1
CG 27 27C
CG 27 44N2
CG 30A 1F
CG 42A 1F
CG 100A
see ASTM range 1L
CH 3N 44F1
CH 4 44E3
CH 5 44E3
CH 5 44K1
CH 8 44N1
CH 10 44N1
CH 12 44K1
CH 13 44K1
CH 20 44N1
CH 45D 44K1
CH 55B 44K1
CH 65B 44K1
CH 65J 44K1
CH 75B 44K1
CH 85A 44K1
CH 361 44K1
CH 362 44K1
CH 363 44K1
CHANNEL BRONZE 14C
CHAR-PAC 44A1
CHD 44K1
CHD 1 44K2
CHD 2 44K2
CHD 3 44K1
CHD 4 44K3
CHICAGO 4 17.1
CHILI BAR 14A
CHIMO 44K2
CHINESE ART METAL 14B
CHINESE BRONZE 14K
CHIPPER KNIFE 44M2
CHIPPEWA 44K1
CHISEL STEEL 44A4
CHISEL STEEL 44E2
CHISEL STEEL 44F2
CHISEL STEEL 44K2
CHLORIMET 2 27B
CHLORIMET 3 27B
CHMS 44A1
CHNC 55 44K1
CHNC 65 44K1
CHNM 45 44K1
CHNM 55 44K1
CHQ 44A4
CHROMADOR 44E1
CHROMAX 44N1
CHROMAX BRONZE 14F
CHROME FM 44M2
CHROMEL A 27B
CHROMEL P 27B
CHROMET 1C
CHROMET 1 44K1
CHROMET 1L 44K1
CHROMET 2 44K1
CHROMET 2L 44K1
CHROMET 5 44K1
CHROMET 9 44K1
CHROMET 9 MV 44K1
CHROMETOUGH 44K2
CHROME TUNGSTEN 44K2
CHROMEX 1 44M1
CHROMEX 2 44M1
CHROMEX 3 44M1
CHROMIMPHY A2100 44M1
CHROMIMPHY A2200 44M1
CHROMIMPHY A3100 44M1
CHROMIMPHY A3200 44M1
CHROMIUM 12.1
CHROMIUM 99 12.1
CHROMIUM 99 LOW GAS 12.1
CHROMIUM COPPER 14L
CHROMODUR 22 44M1
CHROMODUR 33 44M1
CHROMODUR II 44M1
CHROMOLOY 44K1
CHROMOTRODE 76/12 44K1
CHROMOTRODE 76/19 44K1
CHROMOTRODE 76/22 44K1
CHROMOTRODE 76/28 44K1
CHRO-MOW 44K1
CHROMVA-W 44K1
CHRONIFER 2520 44N1
CHRONIFER 2520 NV 44N1
CHRONITE 27B
CHW 44K2
CI CAVITY FILL Ni 27A
CI LOW CON Ni 27A
CI MET Ni Fe 27E
CINDAL 1A
CINEX 27F
CI-NI 27A
CI Ni Fe 27E
CIRCLE C 44L
CIRCLE H 44L
CIRCLE M 44L
CIRCLE T 15 44L
CI SOFT FLOW Ni 27A
CI SOFT FLOW Ni Cu 27D
CITOBRONZE A5 14K
CITOBRONZE A8 14K
CITOBRONZE AA 14K
CK 10
see DIN range 44A1
CK 15
see DIN range 44A1
CK 20 44N1
CK 22
see DIN range 44A1
CK 35
see DIN range 44A2
CK 45
see DIN range 44A2
CK 53 44A3
CK 60 44A3
CK 67 44A3
CKD 1 44K1
CKD 2 44K1
CL 15 44F2
CL 40X 44K2
CL 45 44K3
CL 60 44E2
CL 99 44K2
CL 222 44K1
CL 224 44K2
CL 225 44K2
CL 444 44K1
CL 652 44K3
CLA 1A
see BS range 44K1
CLA 1B
see BS range 44K2
CLA 1C
see BS range 44K2
CLA 2A
see BS range 44K1
CLA 2B
see BS range 44K1
CLA 3
see BS range 44A2
CLA 4
see BS range 44A2
CLA 5A
see BS range 44A2
CLA 5B
see BS range 44A2
CLA 6
see BS range 44A2
CLA 7
see BS range 44K1
CLA 8
see BS range 44K1
CLA 9
see BS range 44K1
CLA 11
see BS range 44K1
CLA 12C
see BS range 44K2
CLAD CG 42A 1H
CLAD CS 41A 1H
CLAD GM 50A 1H
CLAD GS 11A 1H

CLAD MG 11A 1H
CLAD MIA 1H
CLAD ZG 62A 1H
CLARITE 44K3
CLARITE EW 50 44K2
CLAY-LOY 44J
CLOCK BRASS 14B
CLOSELOY 44K2
CLW 44K1
CLYDALL 5 SPECIAL 44L
CLYDALL 12 SPECIAL 44L
CLYDALL 18 44K3
CLYDMO 44K3
CM 1 44K1
CM 2 14C
CM 5 44L
CM 6 13A
CM 9 13A
CM 10 13A
CM 18 44L
CM 32 44L
CM 40 27C
CM 41A 1F
CM 52 27F
CM 53 27B
CM 55 27B
CM 56 27B
CM 63 13A
CM 70 44K1
CM 80F 44K1
CM 90F 44K1
CM 106 13A
CM 110 13A
CM 118 44L
CM 132 44L
CM 1255 44L
CMA 1
see BS range 14G
CMA 1-C
see BS range 14G
CMA 2-1
see BS range 14G
CMA 2-C
see BS range 14G
CMBV/h 44E1
CMBV/w 44E1
CMC 44K3
CMD 44M2
C METAL 44A2
CMN 44E2
CMN 44K2
CMN 1 44A2
CMN 1 44F1
CMN 2 44A2
C Mo 44G1
C Mo 15 44G1
C Mo 25 44G1
CMSZ 41.1
CMV 44K1
CMVM 44K2
CMVW 11 44M1
CMW 44K1
CMW11 44M1
CMX-A/B/C 44M2
CMX-BM 44M2
CMX FB 30W 1 26A
CMX FB TZM 1 26A
CMX S 1 26A
CMX S TZM 1 26A
CMX WB LC 1 26A
CMX WB T 1 26A
CMX WB TZM 1 26A
CMX WB TZM 2 26A
CN/V60 022/C 44J
CN1
BS designation 14E
CN 1 44K1
CN2
BS designation 14E
CN 2 44K1
CN 3A 44K1
CN 3B 44K1
CN 3 M 44N1
CN 5 44K1
CN 7 M 44N1
CN 7 MS 44N1
CN 9 44K1
CN 10 44K1
CN 11 44K1
CN 12 44K1
CN 13 44K1
CN 42A
see ASTM range 1L
CN 101
see BS range 14E
CN 102
see BS range 14E
CN 103
see BS range 14E
CN 104
see BS range 14E
CN 105
see BS range 14E
CN 106
see BS range 14E
CN 107
see BS range 14E
CN108
BS designation 14E
CNBN 1/W 44K1
CNBV 2/W 44K1
CNBV 2V/W 44K1
CNBV 3/W 44K1
CNBV 4/W 44K1
CNBV 5/W 44K1
CNBVO 44K1
CNBVO 1 44K1
CNBVO 1/1 44K1
CNBVO 2 44K1
CND 44K1
C Ni 98.5
see DIN range 27A
C Ni 99.5
see DIN range 27A
C Ni 99.8
see DIN range 27A
CNKD1 44K1
CNKD2 44K1
CNKD3 44K1
CNS 44N1
CNS NON SHRINK 44E3
CO 1 44K1
Co 040 44L
Co 050 44L
Co 060 44L
Co 070 44L
Co 090 44L
COALNI 44L
COALNIMAX 44L
COBALT 13.1
COBALT 44L
COBALT 44M2
COBALT ASCOLOY 44M1
COBALTCROM 44M2
COBANIC 27F
COBITE 44L
COBSTEL 1 13A
COBSTEL 6 13A
COBSTEL 8 13A
COBSTEL 12 13A
COBSTEL 25 13A
COBSTEL 50 13A
CODE 107 44K1
CODE 602 44A1
CODE 603 44A2
CODE 604 44A2
CODE 605 44A1
CODE 606 44A2
CODE 609 44A2
CODE 610 44A2
CODE 611 44A2
CODE 617 44K1
CODE 620 44E2
CODE 623 44E3
CODE 625 44K1
CODE 626 44A2
CODE 628 44K2
CODE 629 44G1
CODE 630 44D
CODE 631 44D
CODE 633 44D
CODE 634 44D
CODE 638 44K2
CODE 639 44K2
CODE 641 44G1
CODE 642 44A2
CODE 644 44G1
CODE 687 20A
CODE P 44L
CO-ELINVAR 13A
COGWHEEL 11 14K
COGWHEEL VII 14K
COGWHEEL VIII 14K
COGWHEEL XI 14K
COH 44A4
COILEX 44A1
COILEX F 44A1
COINAGE BRONZE 14K
COINAGE GOLD 17.1
COIN SILVER 42.1
COLCLAD 12 Cr 44A1
COLCLAD 13/Cr/Al 44A1
COLCLAD 18/8 44A1
COLCLAD 18/8 ELC 44A1
COLCLAD 18/8 Nb 44A1
COLCLAD 18/8 Ti 44A1
COLCLAD 18/10/2 ELC 44A1
COLCLAD 18/10/2 Nb 44A1
COLCLAD 18/10/2 Ti 44A1
COLCLAD 18/10/3 ELC 44A1
COLCLAD 18/10/3 Nb 44A1

COLCLAD INCONEL 44A1
COLCLAD NICKEL 44A1
COLDIE 44J
COL-GRAPH 44K3
COLLET STEEL 44K2
COLMONOY 1 SPECIAL 44M2
COLMONOY 4 27B
COLMONOY 5 27B
COLMONOY 6 27B
COLMONOY 8 27B
COLMONOY 15 14H
COLMONOY 20 27B
COLMONOY 21 27B
COLMONOY 22 27F
COLMONOY 23 27F
COLMONOY 23A 27F
COLMONOY 24 27F
COLMONOY 25 27F
COLMONOY 26 27F
COLMONOY 27 27F
COLMONOY 28 27F
COLMONOY 29 27B
COLMONOY 41 27B
COLMONOY 42 27F
COLMONOY 43 27F
COLMONOY 44 27F
COLMONOY 45 27B
COLMONOY 46 27F
COLMONOY 52 27F
COLMONOY 53 27F
COLMONOY 56 27B
COLMONOY 56 27B
COLMONOY 60 27B
COLMONOY 61 27B
COLMONOY 62 27B
COLMONOY 62 27F
COLMONOY 63 27F
COLMONOY 69 27F
COLMONOY 70 27C
COLMONOY 72 27B
COLMONOY 75 27B
COLMONOY 155 44K2
COLMONOY 705 52.1
COLMONOY C290 44N1
COLMONOY C395 44N1
COLORADO 44A3
COLORCOAT 44A1
COLTUF 26 44A1
COLTUF 28 44A1
COLTUF 32 44A1
COLUMAX 44L
COLUMBIA ELECTREX 44A4
COLUMBIA EXTRA 44A4
COLUMBIA EXTRA HEADERDIE 44A4
COLUMBIA SPECIAL 44K3
COLUMBIA STANDARD 44A4
COLUMBIUM 28.1
COLUMBUS RECORD 44K1
COLUMNAR HYCOMAX 11 20C
COMALCO 117 1L
COMALCO 127 1L
COMALCO 135 1K
COMALCO 138 1L
COMALCO 143 1L
COMALCO 160 1C
COMALCO 160S 1C
COMALCO 162 1K
COMALCO 360 1K
COMALCO A143 1L
COMALCO B143 1L
COMBARLOY 14A
COMET 20C
COMET 95 27F
COMET 95 Co 27F
COMET 97 27F
COMET 258S 44K2
COMET 307 44N1
COMET 307R 44N1
COMET 308L 44N1
COMET 309 44N1
COMET 309 Mo 44N1
COMET 316L 44N1
COMET 317 44N1
COMET 410H 44M1
COMET 410NM 44M1
COMET B 44A1
COMET J 50 44A1
COMET J 50 N 44A1
COMET J 56 LH 44K1
COMET J 66 ELH 44K1
COMET J 76 ELH 44K1
COMET R 12 44A1
COMET SD 44N1
COMET T 44A1
COMET V 44A1
COMET W 44A1
COMMANDO 44K2
COMMERCIAL BRONZE 14B
COMMERCIAL U1 53.1
COMMERCIAL U2 53.1
COMMERCIAL U3 53.1
COMMERCIAL U4 53.1
COMMON HARDENING 44A3
COMO 44L
COMOKUT 44L
COMOL 44L
COMPAX 44K1
COMSOL 21.1
CON 14B
CONDENSER FOIL 50A
CONGO 44L
CONGO HOT-WORK 44L
CONICRO 44E1
CONIFER 1925 HMO 44N2
CONLO 1 44A1
CONLO 11 44A1
CONLOY 1F
CON-PAC 44A1
CONQUEROR 14 44K3
CONQUEROR LC 44K2
CONQUEROR TEMP 1-6 44A4
CONSTANTAN 14E
CONSUMET 44K2
CONSUMET 355 44M1
CONSUMET ELECTRICAL IRON 44J
CONTRACID 44N1
COOKSON'S C 2.1
COPEL 27D
COPELMET 52.1
COPPER 14A
COPPER-FLO 14H
COPPERWELD 44A1
COPPERWELD HM 14A
COPPERWELD LC 14A
CORAL 52.1
CORNES 11 27A
CORNES 12 27A
CORNES 13 27D
CORNES 14 27B
CORNES 15 H 27B
CORNES 16 27B
CORNES 17 27B
CORNES 18 27D
CORNES 80/20 27B
CORNES 310 44N1
CORNES 601 27B
CORNES 625 27B
CORNES C276 27C
CORNISH REFINED 50.1
CORNIX 1 44M1
CORNIX 2 44M1
CORODENT 17.1
CORONA 2 44K3
CORRESIST 13 HMo 44M1
CORRONEL 27D
CORRONEL 220 27C
CORRONEL 230 27B
CORRONEL 230 27C
CORRONIUM 14B
CORROSTITE 44A1
CORTEN 50 44K1
COR-TEN A 44E1
COR-TEN B 44E1
COSSET 624S 44P
COSSET MC 44P
COVAN 44L
CP1 44E2
CP 2 44K2
CP 30 44K1
CP 30 44K2
CP 30 44K3
CP 40 44K2
CP 50 44K2
CP 50 44K3
CP 601 1C
CPG 44E2
CPNM 2 14H
CPV 44A2
Cq 15 44A1
Cq 22 44A1
Cq 35 44A2
Cq 45 44A2
CQ 51A 1L
CR 44E2
CR1 20B
CR1 27F
CR2 20B
CR 2 27F
CR3 20B
CR4 20B
CR5 20B
CR6 20B
CR7 20B
CR8 20B
CR9 20B
Cr 030 44E3
Cr 035 44E3
CR 43 44K2

CR 50 44K2
CR 83 44K2
CR 501 44K2
Cr A1 8/5
see DIN range 44E1
Cr Al 20/5 44M1
Cr Al 25/5 44M1
Cr Al 30/5 44M1
CRB-7 ALLOY 44M2
CRC 44E3
CRD 9% W 44K1
CRD 18 44K3
CRD 22 44K3
CRD 23 44K2
CRD 24 44K2
CRD 90C 44J
CRD 100C 44J
CRD 120C 44J
CRD 206 44L
CRD 339 44L
CRD 515 44G2
CRD 652 44K3
CRD 2212 44L
CRD DIAMOND 44M2
CRD DIAMOND RED 44M2
CRD DIAMOND YELLOW 44K3
CRD HYKROM 44K1
CRD HYKROM B 44K1
CRD NICHROMA 44K2
CRD NTCS 44K2
CRD SILICO 44B2
CRD SPECIAL HOT DIE 44K1
CRD TVC 44L
CRESCENT 44A4
CRESCENT 1 44A4
CRESCENT 3 44A4
CRESCENT 4 44A4
CRESCENT 5 44A4
CRESCENT 6 44A3
CREUSABRO 32 44K1
CREUSELSO 32 44K1
CREUSELSO 38 44A1
CREUSELSO 42 44A1
CREUSELSO 42 44K1
CREUSELSO 47 44K1
CREUSOT UR SB 8 44N2
CRH 4/17 44M1
CRM 1 44K2
CRM 1 44K3
CRM 2 44K2
CRM 2 44K3
CRM 3 44K1
CRM 6D 44N1
CRM 15D 44P
Cr Mo ‘P’ 44K2
Cr Mo 18 44K1
Cr Mo 23 44K1
Cr Mo 33 44K1
Cr Mo 329 44K1
Cr Mo/F Mo 44K2
CrMo60 44K2
Cr Ni 17/7
See 301 S21 44N1
Cr Ni 18/9 0.07C
See 304 S31 44N1
Cr Ni 18/9/Nb
See 347 S31 44N1
Cr Ni 18/9/Ti
See 321 S31 44N1
Cr Ni 18/10 0.03C
See 304 S11 44N1
Cr Ni 18/10 0.06C
See 304 S16 44N1
Cr Ni 18/11 0.10C
See 305 S19 44N1
Cr Ni 25-20 44N1
Cr Ni Mn17/5/8
See 284 SI6 44P
Cr Ni Mo 17/10/1 1/2
See 315 S16 44N1
Cr Ni Mo 17/11/2 1/4 0.07C
See 316 S31 44N1
Cr Ni Mo 17/11/2 3/4 0.07C
See 316 S33 44N1
Cr Ni Mo 17/12/2 1/4 0.030C
See 316 S11 44N1
Cr Ni Mo 17/12/2 1/4 TT
See 320 S31 44N1
Cr Ni Mo 17/12/2 3/4 Ti
See 320 S33 44N1
Cr Ni Mo 17/12/3 3/4 0.030C
See 316 S13 44N1
CRO 13 Mo 44M1
CRO 42 44A1
CRO 42 44K1
CRO 53 44A1
CRO 53 44K1
CRO 684 BOVAC 44K1
CRO 684 BOVAC 44K2
CRO 861 44K1
CRO 7146 44K1
CRODI 44K1
CROFORM 13A
CROLOY 44K1
CROLOY 1 44K1
CROLOY 2 44K1
CROLOY 3 44K1
CROLOY 5 44K1
CROLOY 5Si 44K1
CROLOY 5Ti 44K1
CROLOY 7 44K1
CROLOY 9 Mo 44K1
CROLOY 9M 44K1
CROLOY 12 44M1
CROLOY 12 Al 44M1
CROLOY 15/15N 44N1
CROLOY 16.6 PH 44N3
CROLOY 16/13/3 44N1
CROLOY 16/13/3 ELC 44N1
CROLOY 16/13/3 H 44N1
CROLOY 18 44M1
CROLOY 18/8 Cb 44N1
CROLOY 18/8 Cb H 44N1
CROLOY 18/8 Cb Ta 44N1
CROLOY 18/8 Cb Ta H 44N1
CROLOY 18/8 ELC 44N1
CROLOY 18/8 EM 44N1
CROLOY 18/8 H 44N1
CROLOY 18/8 HC 44N1
CROLOY 18/8 S 44N1
CROLOY 18/8 S Ti H 44N1
CROLOY 18/8 Si 44N1
CROLOY 18/8 Ti 44N1
CROLOY 18/12 44N1
CROLOY 18/13/3 44N1
CROLOY 22 44M1
CROLOY 25/12 44N1
CROLOY 25/20 44N1
CROLOY 27 44M1
CROLOY 27/4/1 44M1
CROMANSIL 44E2
CROMAX Cr Mo F Mo 44K2
CROMAX F 44E2
CROMAX N 44E3
CROMFER 2803 Mo 44M1
CROMIMPHY 1 bis Mo 44M1
CROMIMPHY 33 44M1
CROMIMPHY A8Mo 44M1
CROMIMPHY A10Mo 44M1
CROMIMPHY A15Mo 44M1
CROMIMPHY A18Mo 44M1
CROMIMPHY A20Mo 44M1
CROMIMPHY A1007Mo 44M1
CROMO CO 44L
CROMODIE 44K1
CROMODIE HC 44K3
CROMODIE HCV 44K3
CROMODIE W 44K1
CROMOL 44K2
CRO-MO-LOY 44K3
CROMO-N 44M1
CROMOVAN 44M2
CRONIFER 1713 LCN 44N1
CRONIFER 1812 LCN 44N1
CRONIFER 1815 LC Si 44N1
CRONIFER 1925 H Mo 44N1
CRONIFER 1925 LC 44N1
CRONIFER 1925 LCN 44N1
CRONIFER 2205 LCN 44N1
CRONIFER 2328 44N1
CRONIFER 2521 LC 44N1
CRONIFER 2522 LCN 44N1
CRONIFER 2525 Ti 44N1
CRONIFER S1925 44N1
CRONITE 27C
CRONITE 25/12 44N1
CRONITE 25/20 44N1
CRONITE 37/18 44N1
CRONITE 37/18/Nb 44N1
CRONITE 50/50 12.1
CRONITE 60/40 12.1
CRONITE 428 44N1
CRONIX 70 E 27B
CRONIX 80 E 27B
CROSS PIPES 44M2
CROSTAR 44M2
CROTORITE 14G
CROTORITE IV 14G
CROTORITE V 14G
CROTORITE Z 14G
CROVAN 44K1
CROWN 44K2
CROWN(4)SUPERB 44K3
CROWN MAX 44N1
CRP 44K3
CRS 20B
CRS 1 20B
CRS 2 20B
CRS 3 20B
CRS 4 20B
CRS 5 20B

CRS 6 20B
CRS 7 20B
CRS 8 20B
CRS 9 20B
CRS 10 20B
CRS 11 20B
CRUCIBLE 13% Mn 44D
CRUCIBLE 26-1 44M1
CRUCIBLE 201 44P
CRUCIBLE 202 44P
CRUCIBLE 301 44N3
CRUCIBLE 303 44N1
CRUCIBLE 304 44N1
CRUCIBLE 309 44N1
CRUCIBLE 316 44N1
CRUCIBLE 321 44N1
CRUCIBLE 347 44N1
CRUCIBLE 348 44N1
CRUCIBLE 401 44M1
CRUCIBLE 403 44M1
CRUCIBLE 416 44M1
CRUCIBLE 420 44M1
CRUCIBLE 422 44M1
CRUCIBLE 430 44M1
CRUCIBLE 431 44M1
CRUCIBLE 440 BM 44M2
CRUCIBLE 440 C 44M2
CRUCIBLE 440A 44M1
CRUCIBLE 440B 44M1
CRUCIBLE 442 44M1
CRUCIBLE 446 44M1
CRUCIBLE 718 27C
CRUCIBLE 901 44N2
CRUCIBLE A 40 51A
CRUCIBLE A 55 51A
CRUCIBLE A 70 51A
CRUCIBLE A11OAT 51B
CRUCIBLE A286 44N2
CRUCIBLE B120 VCA 51C
CRUCIBLE C 120 AV 51C
CRUCIBLE C130AM 51B
CRUCIBLE C 135 AMo 51C
CRUCIBLE CG27 44N2
CRUCIBLE M252 27C
CRUCIBLE X750 27C
CRUSCA 44A4
CRUSHER 21.1
CRV 44K2
CRV 14 44K3
CRV 1444 44K3
Cr-X 120 Mn 12 44D
CRYOGENIC TENDON 44P
CRYONIC 5 44F1
CS 5 44G1
CS 41A 1F
CS 42A
see ASTM range 1L
CS 43A
see ASTM range 1L
CS 66A
see ASTM range 1L
CS 72A
see ASTM range 1L
CS 101
see ASTM range 14M
CS 104A
see ASTM range 1L
CSA 44P
CSC STAR 14B
CT 0010-N 14J
CT 0010-R 14J
CTC 1 44K2
CTH 44K2
CTHV 44K1
CTM 44K1
CTU 44K1
CTX/D 44M1
Cu/a1 14A
Cu/a2 14A
Cu/a3 14A
Cu/b 14A
Cu/c1 14A
Cu/c2 14A
Cu/d 14A
Cu 1 44A1
Cu Al 5 14G
Cu Al 8 14G
Cu Al 8 Fe 14G
Cu Al 9 Mn 14G
Cu Al 10 Ni 14G
Cu Al 10 Ni 5 Fe 4 14G
Cu Al 11 Ni 14G
Cu Al 15 14G
Cu Al 17 14G
Cu Al 18 14G
CU BE 50 14D
CU BE 250 14D
CU BE 275 14D
CUBELLOY 14D
Cu Ni 10 Fe
see DIN range 14E
Cu Ni 10 Fe Mn 14E
Cu Ni 20 Fe
see DIN range 14E
Cu Ni 30 Fe
see DIN range 14E
Cu Ni 30 Fe 2 Mn 2 14E
Cu Ni 30 Fe Mn 14E
Cu Ni Mo 44K1
CUNICO 1 14E
CUNICO 11 13A
CUNIFE 1 14E
CUNIFE 11 14E
CUNIFER 10 14E
CUNIFER 30 14E
CUNIFER 302 14E
CUNIFER B 7030 14E
CUNIFER S 7030 14E
CUNIFER S 9010 14E
CUPALOY 14L
CUPRALITH 14A
CUPREX 44A1
CUPROCHROM 14L
CUPROMET 100 14A
CUPRON 14E
CUPRONET N10 14E
CUPRONET N30 14E
CUPRO NICKEL 14E
CUP-TEN 44K1
CUP-TEN 6 44K1
CUP-TEN 50 44K1
CUP-TEN 53 44K1
CUSIL 42.1
CUSILAY 14K
CUSIL DECARBONIZED 42.1
CUSIL VT 42.1
CUSTOM 450 44N3
CUSTOM 455 44N2
CUSTOM 455 44N3
CUSTOM 630 44N3
CUSTOM AGE 625 27B
CUSTOM FLO 302 HQ 44N1
CUSTOM FLO 316 HQ 44N1
CUTANIT 52.1
CUTLERY 44M1
Cu-Zr MASTER ALLOY 14A
CV 4 44K2
CV 18 44K3
CV 22 44K3
CV 40h 44K2
CV 50 44K2
CV 58 44K2
CV 1842 44K3
CVHD 44K2
CVM 44K1
CVM 2 44K1
CVM 3 44K2
CVM 4 44K2
CVM 5 44K1
CVM 6 44K2
CV Punch and Die 44K2
CW 12 M 27F
CX 44L
CX 1 14C
C-XB VALVE STEEL 44M2
CY 44K3
CY 44L
CY 40 27B
CY 44A 50A
CYC 5% Co 44L
CYC 18/4/1 44K3
CYC 275 44L
CYCLONE 4V 44K3
CYCLONE 56 44K3
CYCLONE 56CW 44K3
CYCLONE 92 44K3
CYCLONE 92CW 44K3
CYCLONE EXTRA 44L
CYCLONE MC 6 44L
CYCLONE MC 6V 44L
CYCLONE MC 33 44L
CYCLONE MC 42 44L
CYCLONE MC 98 44L
CYCLONE SUPERCUT 44L
CZ 100 27A
CZ 101
see BS range 14B
CZ 102
see BS range 14B
CZ 103
see BS range 14B
CZ 104
see BS range 14B
CZ 105
see BS range 14B
CZ 106
see BS range 14B
CZ 107
see BS range 14B
CZ 108
see BS range 14B

CZ 109
  see BS range 14C
CZ 110
  see BS range 14B
CZ 111
  see BS range 14B
CZ 112
  see BS range 14B
CZ 112
  see BS range 14C
CZ 114
  see BS range 14C
CZ 115
  see BS range 14C
CZ 116
  see BS range 14B
CZ 117
  see BS range 14C
CZ 118
  see BS range 14B
CZ 119
  see BS range 14B
CZ 120
  see BS range 14C
CZ 120
  see BS range 14B
CZ 121
  see BS range 14C
CZ 122
  see BS range 14C
CZ 123
  see BS range 14C
CZ 124
  see BS range 14B
CZ 125
  see BS range 14B
D 20B
D'ARCET 6.1
D 1
  see BS range 23B
D 1 14C
D 1 44A1
D 1 44M2
D1N1
  See ASTMA757 44K1
D1N2
  See ASTMA757 44K1
D1N3
  See ASTMA757 44K1
D1Q1
  ASTMA757 44K1
D1Q2
  See ASTMA757 44K1
D1Q3
  See ASTMA757 44K1
D 1S 44A1
D 1SS 44A1
D2
  See SAED2 44M2
D 2 44A1
D 2 44M2
D 2 44N1
D 2K 44A1
D 2L 44N1
D 2Mn 44A1
D 2S 44A1
D 2SS 44A1
D 3
  see BS range 23B
D3
  See SAED3 44M2
D 3 14H
D 3 44A2
D 3 44M2
D 3K 44E1
D 3S 44A2
D 4
  see BS range 23B
D 4 44A2
D 4 44M2
D 4 44N1
D 4L 44N1
D 4Mn 44A2
D 5
  see BS range 23B
D5
  See SAED5 44M2
D 5 44A3
D 5 44M2
D 6
  see BS range 23B
D 6 44A3
D 6 44M2
D 6A 44K2
D 6AC 44K2
D 7
  see BS range 23B
D7
  See SAED7 44M2
D 7 1L
D 7 14C
D 7 44A3
D 7 44M2
D 8
  see BS range 23B
D 8 1L
D 8 44A3
D 9 14C
D 12 1K
D 12 44M1
D 16 1F
D 18 1F
D 20 44M2
D 20 44N1
D 48 44A2
D 51 20B
D 51 44N3
D 61 44K3
D 65 44M2
D 66 44K1
D 70 44N3
D 165 44M2
D 230 44K1
D 249 44K2
D 320 44K1
D 330 44K1
D 416 44K3
D 421 44K2
D 444 44K2
D 444 44L
D 522 44K3
D 712.0 1M
D 921 44K3
D 941 44K3
D 943 44K3
D 946 44L
D 948 44L
D 953 44K3
D 954 44K3
D 960 44K3
D 979 27C
D 979 27F
D 4018 20B
D 4512 20B
D 5506 20B
D 7003 20B
D 8512 44M1
D8514 44M1
D8518 44M1
D8518W 44M1
DAMASCUS 44K2
DART 44K2
DARWIN 55 27C
DARWIN 168 44M1
DARWIN 505 44L
DARWIN 654A 27C
DARWIN 655B 27C
DARWIN 656C 27C
DARWIN 1366 44L
DARWIN K 44N1
DAUPHINOX TPMo 44M1
DBL 2 44K3
DBL 3 44K3
D BRONZE 14H
DC 44A4
DC 1 44A1
DCB 1
  see BS range 14C
DCB 2
  see BS range 14C
DCB 3
  see BS range 14C
DCCM 44E3
DCM 27C
DCM 44K3
DDQ 44N1
DD Q 40.1
DE 50A
DECOLTAL 500 1F
DEEPEX 44A1
DEF 13 B1 44A1
DEF 13 B2 44A1
DEF 13 B3 44A2
DEF 13 B3A 44A2
DEF 13 B3B 44A2
DEF 13 B4 44K1
DEF 13 B 5A 44E2
DEF 13B5B 44K2
DEF 13B5C 44K2
DEF 13B5D 44K2
DEF 13B5E 44K2
DEF 13B5F 44K2
DEF 13B5G 44K2
DEF 13 B6 44A1
DEF 13 B6 B 44A1
DEF 13 B6C 44K1
DEF 13 B6D 44K1
DEF 13 B6E 44K1
DEF 13 B6F 44K1
DEF 13 B6G 44K1
DEF 17A 55A

DEF 30 1C
DEF 5192 27E
DELCRO 1300K 13A
DELCROME 50V 44M2
DELCROME 450 44M2
DELCROME C 44M2
DELCROME R 44M2
DELFER B 44M2
DELFOR 44M2
DELORO 40G 27F
DELORO 45 27F
DELORO 50 27F
DELORO 60 27F
DELORO B 27C
DELORO C 27C
DELORO PW22 27F
DELORO SF 40 27B
DELORO SF 50 27B
DELORO SF 60 27B
DENTECON 17.1
DENTORMAT 17.1
DGC 44L
DG CYCLONE 44L
DGS 129 14G
DGS 320 14E
DGS 348 14G
DGS 357 14G
DGS 361A 14G
DGS 1043 14G
DGS 1044 14G
DGS 8451 14G
DGS 8452A 14G
DGS 8453 14G
DGS 8454 14E
DHP 14A
DIADUR 1C 52.1
DIADUR 2C 52.1
DIADUR 2000 DE 5.5 52.1
DIADUR 2000 DE 8 52.1
DIADUR AT 52.1
DIADUR HR 52.1
DIADUR OC 52.1
DIAMOND 1 44J
DIAMOND 2 44J
DIAMOND 3 44J
DIAMOND 4 44J
DIAMOND 5 44J
DIAMOND 6 44J
DICO 55A
DIEHARD HCD 44M2
DIEHARD LC 44M1
DIEHARD STANDARD 44M2
DIE TOUGH 10 MS 44K1
DIE TOUGH 15 C 44K1
DIE TOUGH 25 C 44K1
DIE TOUGH 35 C 44K1
DIE TOUGH 40 C 44K1
DIL 14C
DILLINAL 54 44A1
DILLINAL 54T 44A1
DILLINAL 56 44A1
DILLINAL 56T 44A1
DILVER O 44M1
DILVER P0 20C
DILVER P1 20C
DILVER T 44M1
DI-METAL 55A
DIMPALLOY 42.1
DIN 50CR V4 44K2
DIN 1613 St 35 13K 44A1
DIN 1624 St 0 44A1
DIN 1624 St 1 44A1
DIN 1624 St 2 44A1
DIN 1624 St 3 44A1
DIN 1624 St 4 44A1
DIN 1626 42/2 44A1
DIN 1626 52/3 44A1
DIN 1626 St 33 44A1
DIN 1626 St 34/2 44A1
DIN 1626 St 37 44A1
DIN 1626 St 37/2 44A1
DIN 1626 St 42 44A1
DIN 1629 St 00 44A1
DIN 1629 St 35 44A1
DIN 1629 St 35/4 44A1
DIN 1629 St 45 44A1
DIN 1629 St 45/4 44A1
DIN 1629 St 52 44A1
DIN 1629 St 52/4 44A1
DIN 1629 St 55 44A2
DIN 1629 St 55/4 44A2
DIN 1651 9 S 20 44A1
DIN 1651 9 S 27 44A1
DIN 1651 9 S Mn 23 44A1
DIN 1651 9 S Mn 28 44A1
DIN 1651 9 S Mn 36 44A1
DIN 1651 9 S Mn Pb 23 44A1
DIN 1651 9 S Mn Pb 28 44A1
DIN 1651 9 S Mn Pb 36 44A1
DIN 1651 10 S 20 44A1
DIN 1651 15 S 20 44A1
DIN 1651 22 S 20 44A1
DIN 1651 35 S 20 44A2
DIN 1651 45 S 20 44A2
DIN 1651 60 S 20 44A3
DIN 1654 34 Cr 4 44E1
DIN 1654 41 Cr 4 44E2
DIN 1654 C 9 15 44A1
DIN 1654 C 9 22 44A1
DIN 1654 C9/35 44A2
DIN 1654 C9/45 44A2
DIN 1654/ 42 Cr Mo 4 44K2
DIN 1654/42 Cr V6 44K2
DIN 1691 20B
DIN 1691 GG40 20B
DIN 1693 20B
DIN 1693 GGG38 20B
DIN 1693 GGG42 20B
DIN 1693 GGG45 20B
DIN 1693 GGG50 20B
DIN 1693 GGG60 20B
DIN 1693 GGG70 20B
DIN 1701 C Ni 98.5 27A
DIN 1701 C Ni 99.5 27A
DIN 1701 C Ni 99.8 27A
DIN 1701 E Ni 99.5 27A
DIN 1701 E Ni 99.8 27A
DIN 1701 M Ni 99.5 27A
DIN 1701 W Ni 99 27A
DIN 1702 27A
DIN 1704 50A
DIN 1705 G Sn Bz 10 14K
DIN 1705 G Sn Bz 14 14K
DIN 1705 Rg 5 14K
DIN 1705 Rg 7 14K
DIN 1705 Rg 10 14K
DIN 1706 55A
DIN 1707 L Sn 50 50A
DIN 1707 L Sn 60 50A
DIN 1707 L Sn 90 50A
DIN 1707 LPB 98.5 21.1
DIN 1707 LSn 8 21.1
DIN 1707 LSn 25 21.1
DIN 1707 LSn 30 21.1
DIN 1707 LSn 33 21.1
DIN 1707 LSn 35 21.1
DIN 1707 LSn 40 21.1
DIN 1708 14A
DIN 1709 G Ms 65 14B
DIN 1709 G So Ms F 30 14C
DIN 1709 G So Ms F 45 14C
DIN 1709 G So Ms F 60 14C
DIN 1709 G So Ms F 75 14C
DIN 1709 GD Ms 60 14C
DIN 1712 1A
DIN 1714 Cu Al 10 Fe 14G
DIN 1714 Cu Al 10 Ni 14G
DIN 1714 G Al Bz 9 14G
DIN 1714 G Fe Al Bz F 50 14G
DIN 1714 G Mn Al Bz F 42 14G
DIN 1714 G Ni Al Bz F 50 14G
DIN 1714 G Ni Al Bz F 60 14G
DIN 1714 G Ni Al Bz F 68 14G
DIN 1714 GZ Ni Al Bz F 70 14G
DIN 1716 G Pb Bz 25 14K
DIN 1716 G Sn Pb Bz 5 14K
DIN 1716 G Sn Pb Bz 10 14K
DIN 1716 G Sn Pb Bz 15 14K
DIN 1716 G Sn Pb Bz 20 14K
DIN 1719 21.1
DIN 1725 1A
DIN 1729 Mg Al 3 Zn 23B
DIN 1729 Mg Al 6 Zn 23B
DIN 1729 Mg Al 8 Zn 23B
DIN 1729 Mg Mn 2 23B
DIN 1729 Mg Zn 6 Zr 23B
DIN 1729G Mg Al 6 Zn 3 23B
DIN 1729G Mg Al 8 Zn 1 23B
DIN 1729G Mg Al 9 Zn 1 23B
DIN 1729G Mg Al 9 Zn 2 23B
DIN 1732 S Al Mn 1B
DIN 1732 S Al Si12 1C
DIN 1733 El Cu Ni 10 Mn 14E
DIN 1733 Ms 60 14C
DIN 1733 S Al Bz 6 14G
DIN 1733 S Al Bz 8 14G
DIN 1733 S Cu Ag 14A
DIN 1733 S Cu Al 8 14K
DIN 1733 S Cu Ni 30 Fe 14E
DIN 1733 S Cu Si 14M
DIN 1733 S Cu Sn 14A
DIN 1733 S Cu Sn 13 14K
DIN 1733 S Sn Bz 6 14K
DIN 1733 S Sn Bz 12 14K
DIN 1733 S So Ms 14C
DIN 1734 L Ag 8 14H
DIN 1734 L Ag 12 14H
DIN 1734 L Ag 12 Cd 14H
DIN 1734 L Ag 15 14H
DIN 1734 L Ag 15 P 14H
DIN 1734 L Ag 20 14H

DIN 1734 L Ag 25 14H
DIN 1734 L Ag 25 Cd 14H
DIN 1734 L Ag 27 14H
DIN 1734 L Ag 30 Cd 5 14H
DIN 1734 L Ag 30 Cd 12 14H
DIN 1734 L Ag 38 14H
DIN 1734 L Ag 44 42.1
DIN 1734 L Ag 45 42.1
DIN 1734 L Ag 49 42.1
DIN 1734 L Ag 50 42.1
DIN 1735 L Ag 50 Cd 42.1
DIN 1735 L Ag 60 42.1
DIN 1735 L Ag 60 Cd 42.1
DIN 1735 L Ag 67 42.1
DIN 1735 L Ag 67 Cd 42.1
DIN 1735 L Ag 75 42.1
DIN 1735 L Ag 83 42.1
DIN 1736 E1-Ni Cr 26 Mo 44N1
DIN 1736 S Ni Cr 19 ND 27B
DIN 1741 Sb Pb 46 21.1
DIN 1741 Sb Pb 59 21.1
DIN 1741 Sb Pb 85 21.1
DIN 1741 Sb Pb 87 21.1
DIN 1741 Sb Pb 97 21.1
DIN 1742 Sg Sn 50 50A
DIN 1742 Sg Sn 60 50A
DIN 1742 Sg Sn 70 50A
DIN 1742 Sg Sn 75 50A
DIN 1742 Sg Sn 78 50A
DIN 1743 G Zn Al 4 Cu 1 55A
DIN 1743 G Zn Al 6 Cu 1 55A
DIN 1743 GD Zn Al 4 55A
DIN 1743 GD Zn Al 4 Cu 1 55A
DIN 1743 GK Zn Al 4 Cu 1 55A
DIN 1743 GK Zn Al 6 Cu 1 55A
DIN 1743 Z400 55A
DIN 1743 Z410 55A
DIN 1743 Z610 55A
DIN 1785 Al Bz 5 14G
DIN 1785 Cu Ni 10 Fe 14E
DIN 1785 Cu Ni 20 Fe 14E
DIN 1785 Cu Ni 30 Fe 14E
DIN 1785 K Ms 63 14B
DIN 1785 K Ms 72 14B
DIN 1785 K Sn Bz 2 14K
DIN 1785 S B Cu 14A
DIN 1785 S D Cu 14A
DIN 1785 So Ms 71 14B
DIN 1785 So Ms 76 14B
DIN 2440 44A1
DIN 2441 44A1
DIN 2442 44A1
DIN 8512 L Cd Zn 20 8.1
DIN 8512 L Sn Pb Zn 50A
DIN 8512 L Sn Zn 10 50A
DIN 8512 L Sn Zn 40 50A
DIN 8512 L Zn Cd 40 55A
DIN 8513 L Cu Zn 40 14B
DIN 8513 L L Cu Ni 10 Zn 42 14F
DIN 8555 El 400 44E1
DIN 8555 E3 40T 44K2
DIN 8555 E3 50T 44K2
DIN 8555 E3 55T 44K2
DIN 8555 E4 60ST 44K2
DIN 8555 E6-60 44E2
DIN 8555 E7-250K 44K1
DIN 8555 E10-60 44E2
DIN 8555 E21 7022 44K2
DIN 8555 E23–200(45K) ZCKT 44K2
DIN 8555 E31 300 14G
DIN 8555 SG6-60 44E2
DIN 8556 E 18.8 Mn 6 R 26 44K1
DIN 8556 E19-9 Nb R 26 44N1
DIN 8556 E19-9 NC R 26 44N1
DIN 8556 E19-9.2 MPR 23.180 44N1
DIN 8556 E19-12.3 Nb R 26 44N1
DIN 8556 E 22.12.3 R 26.120 44K1
DIN 8556 E 29.9 MPR 26.160 44K1
DIN 8556 E25-20 R 26 44N1
DIN 16512 Pb Sn 3 Sb 4 21.1
DIN 16512 Pb Sn 3 Sb 12 21.1
DIN 16512 Pb Sn 4 Sb 15 21.1
DIN 16512 Pb Sn 5 Sb 12 21.1
DIN 16512 Pb Sn 5 Sb 28 21.1
DIN 16512 Pb Sn 9 Sb 17 21.1
DIN 16512 Pb Sn 15 Sb 4 21.1
DIN 16512 V Pb Sn 5 Sb 28 21.1
DIN 16512 V Pb Sn 30 Sb 6 21.1
DIN 17100 RST 34/1 44A1
DIN 17100 RST 34/2 44A1
DIN 17100 RST 37/1 44A1
DIN 17100 RST 37/2 44A1
DIN 17100 RSt 42/1 44A2
DIN 17100 RSt 42/2 44A2
DIN 17100 RST 46/7 44A1
DIN 17100 St 33/1 44A1
DIN 17100 St 33/2 44A1
DIN 17100 St 37/3 44A1
DIN 17100 St 46/3 44A1
DIN 17100 St 50/1 44A1
DIN 17100 St 50/2 44A2
DIN 17100 St 52/3 44A1
DIN 17100 St 60/1 44A2
DIN 17100 St 60/2 44A2
DIN 17100 St 70/2 44A3
DIN 17100 USt 34/1 44A1
DIN 17100 USt 34/2 44A1
DIN 17100 USt 37/1 44A1
DIN 17100 USt 37/2 44A1
DIN 17100 USt 42/1 44A1
DIN 17100 UST 42/2 44A2
DIN 17110 17 Mn 4 44E1
DIN 17110 MR St 34 44A1
DIN 17110 MR St 44 44A1
DIN 17110 MU St 34 44A1
DIN 17110 Q St 34 44A1
DIN 17110 TU St 34 44A1
DIN 17155 13 Cr Mo 4/4 44K1
DIN 17155 15 Mo 3 44G1
DIN 17155 17 Mn 4 44A1
DIN 17155 17 Mn 4 44E1
DIN 17155 19 Mn 4 44E1
DIN 17155 19 Mn 5 44A1
DIN 17155 Gd 1 44A1
DIN 17155 Gd 11 44A1
DIN 17155 H 44E1
DIN 17155 H 1 44E1
DIN 17155 H 11a 44E1
DIN 17155 H 111 44E1
DIN 17155 H1V 44A2
DIN 17155 H1VA 44A2
DIN 17155 H1VL 44A2
DIN 17155 H11 44A1
DIN 17155 HI 44A1
DIN 17155 HM 44A1
DIN 17172 RRSt 34/7 44A1
DIN 17172 RRSt 38/7 44A1
DIN 17172 RSt 34/7 44A1
DIN 17172 RSt 38/7 44A1
DIN 17172 St 43/7 44A1
DIN 17172 St 47/7 44A1
DIN 17172 St 53/7 44A1
DIN 17172 USt 34/7 44A1
DIN 17172 USt 38/7 44A1
DIN 17175 10 Cr Mo 9/10 44K1
DIN 17175 13 Cr Mo 4/4 44K1
DIN 17175 14 Mn 4 44A1
DIN 17175 15 Mo 3 44G1
DIN 17175 St 35/8 44A1
DIN 17175 St 45/8 44A1
DIN 17200 15 Cr 3 44E1
DIN 17200 15 Cr Ni 6 44K1
DIN 17200 16 Mn Cr 5 44E1
DIN 17200 16 Mn Cr S 5 44E1
DIN 17200 17 Cr Mo 6 44K1
DIN 17200 18 Cr Ni 8 44K1
DIN 17200 20 Mn Cr 5 44E1
DIN 17200 20 Mn Cr S 5 44E1
DIN 17200 20 Mo Cr 4 44K1
DIN 17200 20 Mo Cr S4 44K1
DIN 17200 25 Cr Mo 4 44K1
DIN 17200 25 Cr Mo 4 44K1
DIN 17200 25 Mo Cr S4 44K1
DIN 17200 28 Mn 6 44A2
DIN 17200 30 Cr Mo V9 44K1
DIN 17200 30 Cr Ni Mo 8 44K1
DIN 17200 30 Mn 5 44A2
DIN 17200 31 Cr Mo 12 44K1
DIN 17200 34 Cr 4 44E1
DIN 17200 34 Cr Al Mo 5 44K1
DIN 17200 34 Cr Al Ni 7 44K1
DIN 17200 34 Cr Al S 5 44E2
DIN 17200 34 Cr Mo 4 44K1
DIN 17200 34 Cr Mo S4 44K1
DIN 17200 34 Cr Ni Mo 6 44K1
DIN 17200 34 Cr S 4 44E1
DIN 17200 36 Cr Ni Mo 4 44K1
DIN 17200 37 Cr 4 44E2
DIN 17200 37 Cr S 4 44E2
DIN 17200 37 Mn Si 5 44A2
DIN 17200 38 Cr 2 44E2
DIN 17200 39 Cr Mo V 44K1
DIN 17200 40 Mn 4 44A2
DIN 17200 41 Cr 4 44E2
DIN 17200 41 Cr 4 44F2
DIN 17200 41 Cr S 4 44F2
DIN 17200 42 Mn V7 44J
DIN 17200 46 Cr 2 44E2
DIN 17200 C 35 44A2
DIN 17200 C 45 44A2
DIN 17200 C 55 44A3
DIN 17200 C 60 44A3
DIN 17200 C10 44A1
DIN 17200 C15 44A1
DIN 17200 C22 44A1
DIN 17200 C60 44A3
DIN 17200 CK 10 44A1
DIN 17200 CK 15 44A1
DIN 17200 CK 22 44A1

DIN 17200 CK 35 44A2
DIN 17200 CK 45 44A2
DIN 17200 CK 55 44A3
DIN 17200 CK 60 44A3
DIN 17200 CM 15 44A1
DIN 17200 CM 35 44A2
DIN 17200 CM 45 44A2
DIN 17200 CM 55 44A3
DIN 17200 CM 60 44A3
DIN 17200/ 36 Cr Ni Mo 4 44K2
DIN 17200/ 42 Cr Mo 4 44K2
DIN 17200/ 42 Cr Mo S 4 44K2
DIN 17200/ 50 Cr Mo 4 44K2
DIN 17200/ 50 Cr V 4 44K2
DIN 17210 15 Cr 3 44E1
DIN 17210 15 Cr Ni 6 44K1
DIN 17210 16 Mn Cr 5 44E1
DIN 17210 18 Cr Ni 8 44K1
DIN 17210 20 Mn Cr 5 44E1
DIN 17210 C 10 44A1
DIN 17210 C 15 44A1
DIN 17210 CK 10 44A1
DIN 17210 CK 15 44A1
DIN 17220 C 45 44A2
DIN 17221 38 Si 6 44A2
DIN 17221 38 Si 7 44A2
DIN 17221 46 Si 7 44A2
DIN 17221 51 S 7 44A3
DIN 17221 55 Cr 3 44E2
DIN 17221 55 S 7 44A3
DIN 17221 60 Si Cr 7 44E2
DIN 17221 60 Si Mn 5 44A3
DIN 17221 65 Si 7 44A3
DIN 17221 66 Si 7 44A3
DIN 17221 67 Si Cr 5 44E2
DIN 17221/ 50 Cr V 4 44K2
DIN 17221/ 51 Cr Mo V 4 44K2
DIN 17221/ 58 Cr V 4 44K2
DIN 17222 60 Si Mn 5 44A3
DIN 17222 65 Si 7 44A3
DIN 17222 66 Si 7 44A3
DIN 17222 67 Si Cr 5 44E2
DIN 17222 71 Si 7 44A3
DIN 17222 C67 44A3
DIN 17222 C75 44A3
DIN 17222 M75 44A3
DIN 17222 M85 44A4
DIN 17222 MK75 44A3
DIN 17222 MK101 44A4
DIN 17222/ 50 Cr V 4 44K2
DIN 17222/ 58 Cr V 4 44K2
DIN 17224 X20 Cr 13 44M1
DIN 17225 30 W Cr V 1/7/9 44K1
DIN 17225 67 Si Cr 5 44E2
DIN 17225/ 45 Cr Mo V 6/7 44K2
DIN 17225/ 50 Cr V 4 44K2
DIN 17225/ 65 W Mo 34/8 44K2
DIN 17240 21 Cr Mo V 5/11 44K1
DIN 17240 24 Cr Mo 5 44K1
DIN 17240 24 Cr Mo V 5/5 44K1
DIN 17410 13A
DIN 17440 X 2 Cr Ni 18/9 44N1
DIN 17440 X 2 Cr Ni 18/10 44N1
DIN 17440 X 2 Cr Ni Mo 18/10 44N1
DIN 17440 X 2 Cr Ni Mo 18/12 44N1
DIN 17440 X 2 Cr Ni Mo 18/16 44N1
DIN 17440 X 2 Cr Ni Mo N 18/2 44N1
DIN 17440 X 2 Cr Ni Mo N 18/13 44N1
DIN 17440 X 5 Cr Ni 18/9 44N1
DIN 17440 X 5 Cr Ni 19/11 44N1
DIN 17440 X 5 Cr Ni Mo 18/10 44N1
DIN 17440 X 5 Cr Ni Mo 18/12 44N1
DIN 17440 X 10 Cr Ni Mo Nb 18/10 44N1
DIN 17440 X 10 Cr Ni Mo Ti 18/10 44N1
DIN 17440 X 10 Cr Ni Nb 18/9 44N1
DIN 17440 X 10 Cr Ni Ti 18/9 44N1
DIN 17440 X 12 Cr Ni 5188 44N1
DIN 17440 X6 Cr Mo 17 44M1
DIN 17440 X7 Cr 13 44M1
DIN 17440 X7 Cr Al 13 44M1
DIN 17440 X8 Cr 17 44M1
DIN 17440 X8 Cr Nb 17 44M1
DIN 17440 X8 Cr Ti 17 44M1
DIN 17440 X10 Cr 13 44M1
DIN 17440 X12 Cr Mo S17 44M1
DIN 17440 X15 Cr 13 44M1
DIN 17440 X20 Cr 13 44M1
DIN 17440 X22 Cr Ni 17 44M1
DIN 17440 X40 Cr 13 44M1
DIN 17440 X48 Cr Mo V15 44M1
DIN 17470 Cr A18/5 44E1
DIN 17470 Cr Al 20/5 44M1
DIN 17470 Cr Al 25/5 44M1
DIN 17470 Cr Al 30/5 44M1
DIN 17470 Cr Ni 25/20 44N1
DIN 17470 Ni Cr 30/20 44N1
DIN 17470 Ni Cr 60/15 27E
DIN 17470 Ni Cr 80/20 27B
DIN 17640 21.1
DIN 17641 21.1
DIN 17656 GB Al Bz 9 14G
DIN 17656 GB Fe Al Bz 14G
DIN 17656 GB Ms 60 A 14C
DIN 17656 GBMs 65 A 14B
DIN 17656 GB Ni Al Bz 14G
DIN 17656 GB Rg 10 14K
DIN 17656 GB Sn Bz 14 14K
DIN 17656 GB Sn Pb Bz 5 14K
DIN 17656 GB Sn Pb Bz 10 14K
DIN 17660 14B
DIN 17661 14B
DIN 17665 Cu Al 5 14G
DIN 17665 Cu Al 8 14G
DIN 17665 Cu Al 8 Fe 14G
DIN 17665 Cu Al 9 Mn 14G
DIN 17665 Cu Al 10 Fe 14G
DIN 17665 Cu Al 10 Ni 14G
DIN 17665 Cu Al 11 Ni 14G
DIN 17740 LC Ni 99 27A
DIN 17740 Ni 99.8 27A
DIN 17741 Ni Al 4 Ti 27F
DIN 17741 Ni Mn 1 27A
DIN 17741 Ni Mn 3 Al 27F
DIN 17742 Ni Cr 10 27B
DIN 17742 Ni Cr 15 Fe Mo 27B
DIN 17742 Ni Cr 22 Mo 27C
DIN 17742 Ni Mo 16 Cr 27C
DIN 17743 Ni Cu 30 Al 27D
DIN 17743 Ni Cu 30 Fe 27D
DIN 17745 Ni 48 20C
DIN 17745 Ni 49 20C
DIN 17745 Ni Cu 14 Fe Mo 27F
DIN 17745 Ni Fe 15 Mo 27E
DIN 17745 Ni Fe 16 Cu Cr 27E
DIN 17745 Ni Fe 16 Cu Mo 27E
DIN 17745 Ni Fe 45 20C
DIN 17745 Ni Fe 46 Cr 27E
DIN 17745 Ni Fe 48 Cr 20C
DIN 17800 H Mg 99.8 23A
DIN 17800 H Mg 99.95 23A
DIN ARGO-FLO 42.1
DIN L Sn Ag 5 50A
DISCALOY 44N1
DISCOLOY 24 44N2
DISCUS 44A1
DISPLAY MONOTYPE ALLOY 21.1
DISQUE E 44A2
DISQUE H 44A2
DK 3 44E1
DK 3E 44E1
DK 4 44E2
DK 6 44E2
DK 35E 44E1
DLP
see ASTM range 14A
DM3 44K2
DMA 1 44L
DMA 2 44L
DMA 3 44F2
DMA 4 44L
D METAL 44A2
DMS 44A2
DN 44M1
DN 5 44M1
DNC M 38 14G
DNH 44M1
D NICKEL 27F
DNV 44K1
DO 14A
DOFASCOLOY 44F1
DOFASCOLOY 2 44F1
DOFASCOLOY 45 W 44A1
DOFASCOLOY 50 W 44A1
DOFASCOLOY 55 W 44A1
DOFASCOLOY 60 W 44A1
DOFASCOLOY 100 W 44A1
DOFASCOLOY M 44A1
DOFASCOLOY MY 44J
DOFASCOLOY P 44K1
DOH 44A4
DOLER 55A
DOMAL 2 10.1
DOMAL 3 10.1
DOMAL 4 10.1
DOMAL AZ 21X 23B
DOMAL AZ 31X 23B
DOMAL AZ 61X 23B
DOMAL AZ 80X 23B
DOMAL K 1 23B
DOMINATOR 44M2
DOMINATOR VM 44M2

DOMINATOR Z 44M2
DOMINIAL RM 13 Mo 44M1
DONA 14A
DORDENT 17.1
DOUBLE CONQUEROR
VANADIUM 44J
DOUBLE DIAMOND 44K3
DOUBLE ECHO COBALT 44L
DOUBLE EXTRA 44A2
DOUBLE EXTRA 44J
DOUBLE FLYGO 44K2
DOUBLE FLYGO 44K3
DOUBLE GRIFFIN 44K1
DOUBLE MUSKET 44K3
DOUBLE MUSKET ND 44K3
DOUBLE RAPID 44L
DOUBLE SEVEN 44L
DOUBLE SIX 44M2
DOUBLE SPECIAL 44K3
DOUBLE SUPER HYDRA 44L
DOUBLE TWELVE 44M1
DOUBLE ZEBRA 44K2
DOXA 14A
DP 5 44J
DPA
see ASTM range 14A
DPS
see ASTM range 14A
DPTE
see ASTM range 14A
DQ & T 20B
DR 34 44K1
DR 44 44K1
DRAGONITE 44A1
DRICO 20B
DRO 86 BOVAC 44K1
DRO 86 BOVAC 44K2
DRO 1133 44K2
DRS 16 44M2
DS 11/301 20B
DS 11/301 GG40 20B
DS 11/303 20B
DS 15000 1C
DS 15000 Al 101 1L
DS 15000 Al 511 1L
DS 15000 Al 512 1L
DS 15000 Al501 1C
DSC 44M1
DSI 5000 A1252 1J
DSI 5000 A1621 1M
DSYL 50A
DT 44M2
DTD 4 C 44K2
DTD 5 A 44A3
DTD 5 B 44A4
DTD 10 B 27D
DTD 87 B 44K2
DTD 88 C 23B
DTD 97B 44M1
DTD 118 B 23B
DTD 130 B 1F
DTD 131 1L
DTD 133C 1L
DTD 135 1C
DTD 140 C 23B
DTD 142 A 23B
DTD 147 1F
DTD 150 A 1F
DTD 160 14G
DTD 161A 44M1
DTD 164 A 14G
DTD 165 A 1J
DTD 167 A 44K2
DTD 174 A 14G
DTD 175 1D
DTD 180B 1D
DTD 182B 1D
DTD 186B 1D
DTD 189 A 44N1
DTD 190 1D
DTD 192 27D
DTD 196 27D
DTD 197 A 14G
DTD 200 A 27D
DTD 203B 44M1
DTD 204 A 27D
DTD 213 1B
DTD 214 50A
DTD 215 A 44A3
DTD 232 27D
DTD 233 A 20B
DTD 237 27D
DTD 239 A 44A3
DTD 244 50A
DTD 245 A 1K
DTD 246 B 1F
DTD 247 44P
DTD 253 A 14B
DTD 259 A 23B
DTD 263 14M
DTD 265 A 14K
DTD 267 14M
DTD 268 27D
DTD 271 44M1
DTD 273 1F
DTD 276A 1L
DTD 283 A 14B
DTD 287 1L
DTD 297A 1D
DTD 298B 1L
DTD 300 A 1J
DTD 303 1D
DTD 304B 1L
DTD 307 14B
DTD 310 1D
DTD 318 A 14B
DTD 318 A 14K
DTD 319 14B
DTD 323 B 14B
DTD 324A 1C
DTD 326A 44M1
DTD 327 1F
DTD 328 27B
DTD 330 20A
DTD 337 14B
DTD 346 A 1E
DTD 348 A 23B
DTD 354 14L
DTD 355 14M
DTD 361B 1L
DTD 363 A 1G
DTD 364 1F
DTD 372 B 1E
DTD 387 14B
DTD 390 1H
DTD 410 1F
DTD 412 14G
DTD 413 20B
DTD 423 C 1F
DTD 424A 1L
DTD 440 1D
DTD 443 A 1F
DTD 450 A 1F
DTD 459 14K
DTD 460 A 1F
DTD 462 20B
DTD 464 1F
DTD 477 27D
DTD 485 A 20B
DTD 487 27D
DTD 488 44A4
DTD 498 14E
DTD 503 A 44A2
DTD 504 14E
DTD 520 1F
DTD 525 44M1
DTD 546 B 1F
DTD 603 B 1F
DTD 604 14B
DTD 606A 1D
DTD 607 14A
DTD 610 1H
DTD 614 20B
DTD 619 23B
DTD 622 A 23B
DTD 626 B 23B
DTD 627 14B
DTD 631 14A
DTD 634A 1D
DTD 646 B 1F
DTD 653 1B
DTD 666 44A2
DTD 683 A 1G
DTD 684 23B
DTD 685 21.1
DTD 687B 1H
DTD 690 23B
DTD 693 1G
DTD 703B 27B
DTD 705 44A2
DTD 706 1F
DTD 708 23B
DTD 710B 1H
DTD 711 A 23B
DTD 712 A 44N1
DTD 713 44K1
DTD 715 44M1
DTD 716 A 1K
DTD 717 1F
DTD 718 23B
DTD 719 20B
DTD 720A 44A1
DTD 721 A 23B
DTD 722 A 1K
DTD 723 44K1
DTD 724 1F
DTD 725 27C
DTD 727 A 1K
DTD 728 23B
DTD 729 23B
DTD 730 44K1

DTD 731 A 1F
DTD 732 A 23B
DTD 734 44N1
DTD 735 A 1K
DTD 736 27C
DTD 737 23B
DTD 738 23B
DTD 740 44G1
DTD 741A 1L
DTD 742 A 23B
DTD 745 1F
DTD 746 1F
DTD 747A 27C
DTD 748 23B
DTD 749 23B
DTD 5001 A 23B
DTD 5002 44K1
DTD 5003 B 51A
DTD 5004 1F
DTD 5005 23B
DTD 5006 44N1
DTD 5007 27C
DTD 5008A 1M
DTD 5009 14B
DTD 5010 1F
DTD 5011 23B
DTD 5012 44K2
DTD 5013 B 51A
DTD 5014 1F
DTD 5015 23B
DTD 5016 44N1
DTD 5017 27C
DTD 5018 1J
DTD 5018 1L
DTD 5019 14B
DTD 5020 A 1F
DTD 5021 23B
DTD 5022 44E3
DTD 5023 B 51A
DTD 5024 1G
DTD 5025 23B
DTD 5027 27C
DTD 5028 1C
DTD 5028 1K
DTD 5030 1F
DTD 5031 23B
DTD 5032 44A2
DTD 5033 B 51A
DTD 5034 1G
DTD 5035 23B
DTD 5036 44N1
DTD 5040 1H
DTD 5041 23B
DTD 5042 44K2
DTD 5043 B 51B
DTD 5044 1G
DTD 5046 44M1
DTD 5047 27B
DTD 5050 B 1G
DTD 5051 23B
DTD 5052 44K2
DTD 5053 51B
DTD 5054 1G
DTD 5055 23B
DTD 5056 44N1
DTD 5057 27B
DTD 5060 1H
DTD 5061 23B
DTD 5062 44G1
DTD 5063 A 51A
DTD 5064 1G
DTD 5065 44M1
DTD 5067 27C
DTD 5070A 1H
DTD 5071 23B
DTD 5072 44A2
DTD 5073 51A
DTD 5074 1G
DTD 5077 27C
DTD 5080 1E
DTD 5081 23B
DTD 5082 44K1
DTD 5083 51B
DTD 5084 1F
DTD 5086 44N3
DTD 5087 27C
DTD 5090 1F
DTD 5091 23B
DTD 5092 20A
DTD 5093 51B
DTD 5094 1G
DTD 5100 1H
DTD 5101 23B
DTD 5102 20A
DTD 5103 51C
DTD 5104 A 1G
DTD 5110 1G
DTD 5111 23B
DTD 5112 44K1
DTD 5113 51C
DTD 5114 1G
DTD 5120 B 1G
DTD 5123 44K1
DTD 5123 51B
DTD 5124 1G
DTD 5130 1G
DTD 5132 44K1
DTD 5133 51B
DTD 5142 44K1
DTD 5143 51B
DTD 5152 44K1
DTD 5153 51C
DTD 5162 44K2
DTD 5163 51C
DTD 5172 44K2
DTD 5173 51C
DTD 5183 51A
DTD 5192 44K2
DTD 5193 51A
DTD 5199 44A1
DTD 5202 44G1
DTD 5203 51B
DTD 5203 51C
DTD 5209 44A1
DTD 5212 44F3
DTD 5213 51B
DTD 5219 44K1
DTD 5222 44K2
DTD 5223 51B
DTD 5223 51C
DTD 5229 44K1
DTD 5232 44F3
DTD 5233 51B
DTD 5239 44K1
DTD 5243 51B
DTD 5249 44K1
DTD 5253 51B
DTD 5259 44N1
DTD 5263 51A
DTD 5269 44N1
DTD 5273 51A
DTD 5279 44N1
DTD 5283 51A
DTD 5289 44N1
DTD 5293 51A
DTD 5299 44N3
DTD 5303 51BA
DTD 5309 44N3
DTD 5313 51B
DTD 5323 51B
DTD 5333 51B
DTD 5343 51B
DTD 5353 51B
DTD 5363 51B
DTD 5636 1G
DTD M159 51B
DTD M160 51B
DTD M200 51C
DTD M201 51C
DTva 44M2
DU BRAND 1F
DUCOL HT 44A2
DUCOL QT 122 44A1
DUCOL QT28 44K1
DUCOL QT35 44K1
DUCOL QT131 44F1
DUCOL QT342 44K1
DUCOL QT445A 44K1
DUCOL W21 44K1
DUCOL W23 44K1
DUCOL W23D 44K1
DUCOL W24 44K1
DUCOL W25 44K1
DUCOL W25D 44K1
DUCOL W30 44K1
DUK TEN 44E1
DUMET 20C
DUNELT 44A2
DUNELT 3 44A1
DUNELT 4 44A1
DUNELT 5 44A1
DUNELT 7 44A1
DUNELT 10 44K1
DUNELT 14 44A1
DUNELT 15 44K1
DUNELT 16 44K1
DUNELT 17 44K1
DUNELT 18 44K1
DUNELT 20 44K1
DUNELT 21 44A1
DUNELT 26 44F2
DUNELT 33 44E2
DUNELT 34 44K2
DUNELT 36 44K2
DUNELT 39 44A1
DUNELT 40 44A2
DUNELT 41 44A2
DUNELT 50 44G1
DUNELT 51 44G1
DUNELT 53 44A3
DUNELT 55 44K1

DUNELT 60 44K1
DUNELT 61 44M1
DUNELT 62 44M1
DUNELT 63 44M1
DUNELT 64A 44K1
DUNELT 65 44K1
DUNELT 67 44K1
DUNELT 69 44K2
DUNELT 82M 44K1
DUNELT 83 44E2
DUNELT 84 44N1
DUNELT 88 44N1
DUNELT 90 44E3
DUNELT 100 44K2
DUNELT 101 44K2
DUNELT 128 44K1
DUNELT 129 44K1
DURAL B 1F
DURALBRITE 1 1A
DURALBRITE 2 1A
DURALBRITE 3 1A
DURALBRITE 4 1D
DURALBRITE 6 1E
DURALCOTE 1A
DURAL E 1F
DURAL F 1E
DURAL G 1F
DURAL H 1E
DURAL J 1F
DURAL JJ 1F
DURAL K 1G
DURAL L 1G
DURAL M 1F
DURALOY 44D
DURAL Q 1F
DURAL S 1F
DURAL T 1F
DURALUMIN KC 1G
DURALUMIN LC 1G
DURAL X 1E
DURANICKEL 27A
DURANICKEL 27F
DURANICKEL 301 27F
DURAPSO 44E1
DURAPSO 3 44E1
DURASINT 14J
DURASTIC A 1 21.1
DURATHERM 477 13A
DURATHERM 600 13A
DURATHERM 700 13A
DURATHERM 2602 13A
DURATLAS GM 20B
DURCILIUM E 1E
DURCILIUM Q 1E
DURCILIUM R 1E
DURCILIUM S 1E
DURCILIUM T 1A
DURCILIUM W 1E
DURCO CA15 44M1
DURCO CF 3 44N1
DURCO CF 3M 44N1
DURCO CF 8 44N1
DURCO CF 8M 44N1
DURCO CY40 27B
DURCOMET 5 44N1
DURCORNET 100 44N1
DUREHETE 900 44K2
DUREHETE 950 44K2
DUREHETE 1050 44K1
DUREHETE 1055 44K1
DUREX 44K3
DURICHLOR 51 20B
DURICILIUM F 1F
DURICILIUM H 1F
DURICILIUM K 1F
DURICILIUM M 1F
DURICILIUM N 1G
DURICILIUM P 1G
DURICILIUM X 1G
DURICILIUM Z 1G
DURIMET 20 44N1
DURIRON 20B
DUROTERM 8 R 44K1
DUROTERM 12R 44K2
DUROTERM 20R 44K3
DUX 4 44K2
DUX 12 44K3
DV2A 44P
DV2B 44P
DYCAST 1 44K2
DYNAFLEX VAC-ARC 44K2
DYNAMET 13A
DYNAVAR 13A
E 14C
E 44K2
E 00000 35
E 0 44F1
E 1 44F2
E 1A 44F1
E1Q
 See ASTMA757 44K1
E1 xx T5 K4 44K1
E2N
 See ASTMA757 44K1
E2Q
 See ASTMA757 44K1
E3N
 See ASTMA757 44K1
E 3 Ni 1
 see ASTM range 27B
E 3 Ni 2
 see ASTM range 27B
E 3 Ni 4
 see ASTM range 27D
E 3 Ni 9
 see ASTM range 27C
E 3 Ni B
 see ASTM range 27C
E 3 Ni C
 see ASTM range 27C
E 3 Ni O
 see ASTM range 27D
E 4 44F1
E 4 Mo 44K1
E 4 Ni 1
 see ASTM range 27F
E 4 Ni 2
 see ASTM range 27B
E 4 Ni B
 see ASTM range 27C
E 4 Ni C
 see ASTM range 27C
E 5M 44A2
E 6 44K1
E 6 X T1
 see AWS 44A1
E 6 X T4
 see AWS 44A1
E 6 X T5
 see AWS 44A1
E6 x T5 B 8L 44K1
E 6 X T6
 see AWS 44A1
E 6 X T7
 see AWS 44A1
E 6 X T8
 see AWS 44A1
E 6 X T11
 see AWS 44A1
E 6 X T13
 see AWS 44A1
E 7 44A3
E7 Cr 44K1
E 7 X T1 44A1
E 7 X T2 44A1
E 7 X T3 44A1
E 7 X T4 44A1
E 7 X T5 44A1
E 7X T5 A1 44G1
E 7 X T6 44A1
E 7 X T7 44A1
E 7 X T8 44A1
E 7 X T8 Ni 1 44A1
E 7 XT8 Ni 2 44F1
E 7 X T9 44A1
E 7 X T10 44A1
E 7 X T11 44A1
E 7 X T12 44A1
E 7 X T14 44A1
E7 XX T4 K2 44F1
E 8 FX 44A3
E 8 XT1 Ni 1 44F1
E 8X T1 A1 44G1
E8 x T1 B1 44K1
E8 x T1 B1 L 44K1
E8 x T1 B2 44K1
E8 x T1 B2 H 44K1
E8 x T1 B2 L 44K1
E8 x T1 W 44K1
E8 x T5 B2 44K1
E8 x T5 B2 L 44K1
E 8X T5 K1 44G1
E 8 XT5 Ni 1 44F1
E 8 XT5 Ni 2 44F1
E 9 44K1
E9 x T1 B3 H 44K1
E9 x T1 B3 L 44K1
E9 x T5 B3 44K1
E 9X T1 D1 44G1
E 9X T1 D3 44G1
E 9X T5 D2 44G1
E 10 20A
E 10 20C
E 10 44E3
E 10 XT1 K7 44F1
E 11 44K1
E 11C 44A4
E11 x T1 K4 44K1
E12 x T1 K5 44K1
E 16-8.2 44N1
E 20 20A

E 22 14C
E 26-2
  see NF A35 501 44A1
E 26-3
  see NF A35 501 44A1
E 26-4
  see NF A35 501 44A1
E 028L 44N1
E 30-2
  see NF A35 501 44A1
E 30-3
  see NF A35 501 44A1
E 30-4
  see NF A35 501 44A1
E 48 14B
E 50 44K2
E 56 14B
E 71 T8 K2 44N1
E 78 14B
E80 B2 L 44K1
E80 C B2 44K1
E 80 C Ni 1 44F1
E 80 C Ni 3 44F1
E 86 14G
E90 C B3 44K1
E 100 T5 D2 44G1
E 200 44K1
E 204 44K1
E 209 44N1
E219 44P
E 220 44K1
E 230 44K1
E 307 44N1
E 307 T1 44N1
E 307 T3 44N1
E 308 44N1
E 308 HC 44N1
E 308 LT1 44N1
E 308 LT3 44N1
E 308 Mo 44N1
E 308 Mo LT1 44N1
E 308 MoL 44N1
E 308 MoT-1 44N1
E 308 T1 44N1
E 308 T-3 44N1
E 308L 44N1
E 309 44N1
E 309 Cb 44N1
E 309 Cb LT 44N1
E 309 L 44N1
E 309 LTX 44N1
E 309 Mo 44N1
E 309 MoL 44N1
E 309 T 44N1
E 310 44N1
E 310 Cb 44N1
E 310 H 44N1
E 310 Mo 44N1
E 310 T 44N1
E 312 44N1
E 312 T 44N1
E 316 44N1
E 316 L 44N1
E 316 LT 44N1
E 316 T 44N1
E 317 44N1
E 317 L 44N1
E 317 LT 44N1
E 318 44N1
E 320 27B
E 320 LR 27B
E 330 44N1
E 330 H 44N1
E 347 44N1
E 347 T 44N1
E 349 44N1
E 356 44G2
E 385 L 44N1
E 401 44E1
E409T 44M1
E410 44M1
E 410 44E1
E410 Ni Mo 44M1
E410 Ni Mo T 44M1
E410 Ni Mo Ti T 44M1
E410T 44M1
E420 16 44M1
E430 44M1
E430 T 44M1
E 502 44K1
E 502 L 44K1
E 502 T 44K1
E 505 44K1
E 505 T 44K1
E 525 44E1
E 630 44N3
E 900 44A1
E 920 44A1
E 01000 35
E 01401 35
E1747 44M1
E1755 44M1
E1756 44M1
E1757 44M1
E1802 44M1
E1961 44M1
E1993 44M1
E 2209 44N1
E 2209 T 44N1
E2553 44N2
E2553T 44N2
E 4111 44E1
E 4340 44K2
E 7010 Al 44G1
E 7011 Al 44G1
E 7014 44A1
E 7015 44A1
E 7015 Al 44G1
E 7015 C1L 44F1
E 7015 C2L 44F1
E 7016 44A1
E 7016 Al 44G1
E 7016 C1L 44F1
E 7016 C2L 44F1
E 7018 44A1
E 7018 Al 44G1
E 7018 C1L 44F1
E 7018 C2L 44F1
E 7018 W 44K1
E 7020 Al 44G1
E 7024 44A1
E 7027 44A1
E 7027 Al 44G1
E 7028 44A1
E 7048 44A1
E 8015 44F1
E 8015 B2 L 44K1
E 8015 B4 L 44K1
E 8016 B1 44K1
E 8016 B2 44K1
E 8016 B5 44K1
E 8016 C1 44F1
E 8016 C2 44F1
E 8016 C3 44F1
E 8016 D3 44G1
E 8018 B1 44K1
E 8018 B2 44K1
E 8018 B2 L 44K1
E 8018 C1 44F1
E 8018 C2 44F1
E 8018 C3 44F1
E 8018 D3 44G1
E 8018 NM 44K1
E 8018 W 44K1
E 9015 44N1
E 9015 B3 44K1
E 9015 B3L 44K1
E 9015 D1 44G1
E 9016 A 44G1
E 9016 B3 44K1
E 9018 B3 44K1
E 9018 B3L 44K1
E 9018 Dl 44G1
E 9018 M 44N1
E 10015 D2 44G1
E 10016 D2 44G1
E 10018 D2 44G1
E 11018 M 44K1
E 12018 M 44K1
E 12018 M1 44K1
E 21000 35
E 21100 35
E 21101 35
E 21102 35
E 21103 35
E 21200 35
E 21201 35
E 21300 35
E 21301 35
E 21302 35
E 21500 35
E 23130 35
E 23310 35
E 23311 35
E 23312 35
E 28531 35
E 31000 35
E 46006 35
E 48000 35
E 52000 35
E 56000 35
E 58000 35
E 58401 35
E 68000 35
E 69006 35
E 74000 35
E 74401 35
E 78000 35
E 79000 35
E 79401 35
E 83000 35

E 85000 35
E 87000 35
E 88000 35
E 90000 35
E 90401 35
E 90501 35
EA 1 44G1
EA 1N 44G1
EA 2 44G1
EA 2N 44G1
EA 3 44G1
EA 3K 44G1
EA 3K N 44G1
EA 3N 44G1
EA 4 44G1
EA 4N 44G1
E Al 1A
E Al 99.5 1A
E Al Mg Si 1E
EASY-FLO 42.1
EASY-FLO 1 42.1
EASY-FLO 2 42.1
EASY-FLO 3 42.1
EASY-FLO TRI-FOIL C 42.1
EATONITE 27F
EATONITE 3 27F
EATONITE 5 27F
EB 1 44K1
EB 1N 44K1
EB 2 44G1
EB 2H 44K1
EB 2H N 44K1
EB 2N 44K1
EB 3 44K1
EB 3N 44K1
EB 5 44K1
EB 5N 44K1
EB 6H 44K1
EB 6H N 44K1
EB 8N 44K1
EB 82 44K1
EC 14A
EC 16.8.2 44N1
EC26.1 44M1
EC 218 44N1
EC 219 44N3
EC 240 44N3
EC 308 H 44N1
EC 308 L 44N1
EC 308 L Si 44N1
EC 308 Mo 44N1
EC 308 Mo L 44N1
EC 308 Si 44N1
EC 309 44N1
EC 309 L 44N1
EC 309 Si 44N1
EC 310 44N1
EC 312 44N1
EC 316 44N1
EC 316 L 44N1
EC 316 L Si 44N1
EC 316 Si 44N1
EC 317 44N1
EC 317 L 44N1
EC 318 44N1
EC 321 44N1
EC 347 44N1
EC 347 Si 44N1
EC 349 44N1
EC410 44M1
EC410 Ni Mo 44M1
EC420 44M1
EC430 44M1
EC 502 44K1
EC 505 44K1
EC 630 44N3
EC A1 44G1
EC A1 N 44G1
EC A2 44G1
EC A2 N 44G1
EC A3 44G1
EC A3 N 44G1
EC A4 44G1
EC A4 N 44G1
ECB 1 44K1
ECB 1N 44K1
ECB 2 44K1
ECB 2H 44K1
ECB 2H N 44K1
ECB 2N 44K1
ECB 3 44K1
ECB 3N 44K1
ECB 4 44K1
ECB 4N 44K1
ECB 5 44K1
ECB 5N 44K1
ECB 6H 44K1
ECB 6H N 44K1
ECB 6N 44K1
EC F1 44F1
EC F1 N 44F1
EC F2 44K1
EC F2 N 44K1
EC F3 44K1
EC F3 N 44K1
EC F4 44F1
EC F4 N 44F1
EC F5 44K1
EC F6 44K1
EC F6 N 44K1
ECHO 44K2
ECHO COBALT 44L
ECHO MOLYBDENUM 44K3
ECHO SPECIAL VANADIUM 44L
ECHO SUPERCUT 44L
ECLIPSE 91A 44M2
ECLIPSE 103 44K3
ECLIPSE CHD 44A4
ECLIPSE H5 44L
ECLIPSE LDC 44K1
EC M1 44N1
EC M1 N 44N1
EC M2 44K1
EC M2 N 44K1
EC M3 44K1
EC M3 N 44K1
EC M4 44K1
EC M4 N 44K1
EC Ni 1 44E1
EC Ni 1N 44E1
EC Ni 2 44F1
EC Ni 2N 44F1
EC Ni 3 44F1
EC Ni 3N 44F1
EC Ni 4 44N1
EC Ni 4N 44N1
E Co Cr A 13A
E Co Cr B 13A
E Co Cr C 13A
E Co Cr E 13A
ECO/m 44K1
ECONO 44K2
E Cu 14A
E Cu Al A2 14G
E Cu Al C 14G
E Cu Al D 14G
E Cu Al E 14G
E Cu Mn Ni Al 14G
E Cu Ni 14E
E Cu Ni Al 14G
E Cu Si 14M
E Cu Sn A 14K
E Cu Sn C 14K
EC XX T4 44K1
ED 1E
E D 14C
ED12 44M1
ED12G 44M1
ED12W 44M1
EEEE 44K1
EF 1 44K1
EF 1N 44K1
EF 2 44K1
EF 2N 44K1
EF 3 44K1
EF 3N 44K1
EF 4 44K1
EF 4N 44K1
EF 5 44K1
EF 6 44K1
EF 6N 44K1
EF 10 44K3
EF 206 44A1
E Fe 1 44E1
E Fe 2 44E1
E Fe 3 44E3
E Fe 4 44E3
E Fe 5 44K2
E Fe 5A 44K3
E Fe 5C 44K2
E Fe 6 44K3
E Fe 7 44F3
E Fe Cr A1 44M2
E Fe Cr A2 44M2
E Fe Cr A3 44M2
E Fe Cr A4 44M2
E Fe Cr A5 44M2
E Fe Cr B1 44M2
E Fe Cr B2 44M2
E Fe Cr B3 44M2
E Fe Cr C1 44M2
E Fe Cr C2 44M2
E Fe Cr C3 44M2
E Fe Cr C4 44M2
E Fe Mn E 44P
E Fe Mn F 44P
EG 1 27D
EG 2 27D
EG 7 X T1 44A1
EG 7 X T2 44A1

EGALIT 44K3
EG XX S5 44A1
E G XX T3 44F1
EH 4 44E3
EH 10 K EW 44A1
EH 10 Mo EW 44G1
EH 14 44A1
EHC ALLOY 20C
EHK 3 44L
EHRHARDT 55A
EHW 1 44K1
E in C No. C 106 14G
EK 44M1
EK 5 44L
EK 6 44L
EK 15 44L
EK 30A
see ASTM range 23B
EK 41A
see ASTM range 23B
EKh 44E3
EL 44K3
EL 8 44A1
EL 8K 44A1
EL 12 44A1
EL 12N 44A1
ELEC/Z 55A
ELECTEM 44K3
ELECTERN 44K2
ELECTRIDE CRUSADER XL 44K3
ELECTRITE 1 44K3
ELECTRITE 5 44K2
ELECTRITE 7 44K2
ELECTRITE 19 44K3
ELECTRITE Co 44L
ELECTRITE Co 9 44L
ELECTRITE COBALT 44L
ELECTRITE CORSAIR 44K3
ELECTRITE CORSAIR XL 44K3
ELECTRITE CRUSADER 44K3
ELECTRITE DOUBLE 6 44K3
ELECTRITE DOUBLE 6 M2XL 44K3
ELECTRITE DYNACUT 44L
ELECTRITE DYNAVAN 44L
ELECTRITE DYNAVAN XL 44L
ELECTRITE KELVAN 44L
ELECTRITE LACOMO 44L
ELECTRITE MV1 44K3
ELECTRITE MV2 44K3
ELECTRITE No. 1XL 44K3
ELECTRITE No. 7 44K3
ELECTRITE No. 19 44K3
ELECTRITE STARK 44K3
ELECTRITE SUPER 44L
ELECTRITE SUPER COBALT 44L
ELECTRITE TATMO 44K3
ELECTRITE TATMO 44L
ELECTRITE TATMO COBALT 44L
ELECTRITE TATMO V 44K3
ELECTRITE TATMO XL 44K3
ELECTRITE TNW 44K3
ELECTRITE TNWXL 44K3
ELECTRITE U 44K3
ELECTRITE UB 44L
ELECTRITE ULTRA 44L
ELECTRITE ULTRAVAN 44L
ELECTRITE VANADIUM 44K3
ELECTROID IRON 44A1
ELECTRO TYPE 21.1
ELEKTRON MCZ 23B
ELEKTRON Z5Z 23B
ELEKTRON ZCM 23B
ELEKTRON ZZ 23B
EL GA 31 44A1
EL GA MAYETA II 44A1
EL GA P 43 44A1
EL GA P 44 44A1
EL GA P 47 44A1
EL GA P 51 44A1
ELGA P62 MR 44F1
ELGILOY 13A
ELINVAR 20C
ELKALOY A 14A
ELKONITE 1 W 3 14J
ELKONITE 3 W 3 14J
ELKONITE 10 W 3 14J
ELKONITE 20 K 3 14J
ELKONITE 20 W 3 14J
ELKONITE 20S 42.1
ELKONITE 30 W 3 14J
ELKONITE 35S 42.1
ELKONITE 50S 42.1
ELKONITE 100 M 14J
ELKONITE 100 M 26A
ELKONITE 100 W 14J
ELKONITE D54 42.1
ELKONITE D54L 42.1
ELKONITE D54X 42.1
ELKONITE D55X 42.1
ELKONITE D56 42.1
ELKONITE D58/1 42.1
ELKONITE D58/2 42.1
ELKONITE D510 42.1
ELKONITE D520 42.1
ELKONITE G13 42.1
ELKONITE G14 42.1
ELKONITE G17 42.1
ELKONITE G18 42.1
ELM 8 1D
ELM 10 1D
ELM 15 1B
ELMET SILNO 8.1
ELVERITE CG 20B
ELVERITE I 20B
ELVERITE K 20B
EM 2N 44N1
EM 3 44K1
EM 3N 44K1
EM 4 44K1
EM 4N 44K1
EM 12 EW 44A1
EM 12 K EW 44A1
EM 12 KN 44A1
EM 13 K EW 44A1
EM 14 K 44A1
EM 15 K EW 44A1
EM 70 S-5 44K1
EM 8440 44K1
EMBEESH 14C
EME 44N2
EMERALD 52.1
EMK 10 44L
EMK 15 44L
EMS 44N1
EMS 235 44N1
EMS 242 13A
EMS 253 44N1
En 1A 44A1
En 1B 44A1
En 2 44A1
En 2A 44A1
En 2A/1 44A1
En 2B 44A1
En 2C 44A1
En 2D 44A1
En 3 44A2
En 3A 44A2
En 3B 44A2
En 4 44A2
En 4A 44A2
En 5 44A2
En 5A 44A2
En 5C 44A2
En 5D 44A2
En 5K 44A2
En 6 44A2
En 6A 44A2
En 6B 44A2
En 6K 44A2
En 7 44A1
En 8 44A2
En 8A 44A2
En 8B 44A2
En 8C 44A2
En 8D 44A2
En 8E 44A2
En 8K 44A2
En 8M 44A2
En 9 44A3
En 9K 44A3
En 10 44F2
En 11 44E2
En 12 44F2
En 12 B 44F2
En 12 C 44F2
En 13 44K1
En 14A 44K1
En 14B 44F1
En 15 44A2
En 15A 44A2
En 15B 44A2
En 16 44G1
En 16A 44A2
En 16B 44A2
En 16C 44A2
En 17 44G1
En 18 44E2
En 18 A 44E1
En 18 B 44E1
En 18 C 44E1
En 18D 44E2
En 19 44K2
En 19A 44K2
En 19B 44K2
En 19C 44K2
En 20 44K1

En 21 44K1
En 22 44K2
En 23 44K1
En 24 44K2
En 25 44K1
En 26 44K2
En 27 44K1
En 28 44K1
En 29 44K1
En 30A 44K1
En 30B 44K1
En 31 44E3
En 32A 44A1
En 32B 44A1
En 32C 44A1
En 32M 44A1
En 33 44K1
En 34 44K1
En 35 44K1
En 35A 44K1
En 35B 44K1
En 36 44K1
En 37 44K1
En 38 44K1
En 39A 44K1
En 39B 44K1
En 40A 44K1
En 40B 44K1
En 40C 44K2
En 41 44K1
En 41A 44K2
En 41B 44K2
En 42 44A3
En 42A 44A3
En 42B 44A3
En 42D 44A4
En 43 44A3
En 43A 44A3
En 43B 44A3
En 43C 44A3
En 43D 44A3
En 43E 44A3
En 43F 44A3
En 44 44A4
En 44A 44A4
En 44C 44A4
En 45 44A3
En 45A 44A3
En 46 44A2
En 47 44K2
En 48 44E2
En 49 44A3
En 49A 44A3
En 49B 44A3
En 49C 44A3
En 49D 44A3
En 50 44K2
En 51 44K1
En 52 44E2
En 53 44E2
En 54 44N1
En 55 44N1
En 56A 44M1
En 56AM 44M1
En 56B 44M1
En 56BM 44M1
En 56C 44M1
En 56D 44M1
En 57 44M1
En 58A 44N1
En 58B 44N1
En 58C 44N1
En 58D 44N1
En 58E 44N1
En 58F 44N1
En 58G 44N1
En 58H 44N1
En 58J 44N1
En 58M 44N1
En 59 44M1
En 60 44M1
En 61 44M1
En 100 44K2
En 100A 44K1
En 100B 44K1
En 100C 44K1
En 100D 44K2
En 110 44K2
En 111 44K1
En 111A 44K1
En 160 44K2
En 160A 44K2
En 201 44A1
En 202 44A1
En 206 44E1
En 207 44E1
En 320 44K1
En 325 44K1
ENCON 14K
ENDURO FC 44M1
ENELEC SK 44A1
ENGINEERING 44M1
E Ni 1 27A
E Ni 1 44F1
E Ni 1K 44F1
E Ni 1KN 44F1
E Ni 2 44F1
E Ni 2 N 44F1
E Ni 4 44K1
E Ni 4N 44K1
E Ni CI 27F
E Ni CI A 27F
E Ni Cr 3T 27B
E Ni Cr A 27B
E Ni Cr B 27B
E Ni Cr C 27B
E Ni Cr Fe 1 27B
E Ni Cr Fe 2 27B
E Ni Cr Fe 3 27B
E Ni Cr Fe 4 27B
E Ni Cr Mo 1 27B
E Ni Cr Mo 1 27C
E Ni Cr Mo 2 27C
E Ni Cr Mo 4 27B
E Ni Cr Mo 6 27B
E Ni Cr Mo 7 27B
E Ni Cr Mo 9 27B
E Ni Cr Mo 12 27B
E Ni Cr Mo B 27E
E Ni Cu 7 27D
E Ni Cu A 27D
E Ni Cu B
 see AWS 27D
E Ni Cu B 27D
E Ni Fe CI 27E
E Ni Fe CI A 27E
E Ni Fe Cl
 see AWS 27E
E Ni Fe Cr 2T 27B
E Ni Fe Mn CI 27E
E Ni Fe T3 CI 27E
E Ni Mo 3 27B
E Ni Mo 7 27F
E Ni N 44F1
E NICKEL 27F
ENROX 44A1
EOC/m 44K1
EP 65 44M1
EPK 24 27C
EPOK 44D
E QQ A 601/13 1L
E QQ A 601/14 1L
E QQ S 571/R 21.1
ER 1 44K1
ER 16.8.2 44N3
ER 70 S4 44A1
ER 70 S6 44A1
ER 70 S7 44A1
ER 70 S-2 44K1
ER 80 S B2 44K1
ER 80 S B2 L 44K1
ER 80 S Ni 1 44F1
ER 80 S Ni 2 44F1
ER 80 S Ni 3 44F1
ER 90 S B3 44K1
ER 90 S B3 L 44K1
ER 100 S-1 44K1
ER 100 S-2 44F1
ER 110 S-1 44K1
ER 120 S-1 44K1
ER 209 44N1
ER 209 44N2
ER 218 44N2
ER 219 44N2
ER 240 44N2
ER 307 44N1
ER 308 44N1
ER 308 L 44N1
ER 308 LSi 44N1
ER 308 MoL 44N1
ER 308 Si 44N1
ER 309 44N1
ER 309 L 44N1
ER 309 Si 44N1
ER 310 44N1
ER 312 44N1
ER 316 44N1
ER 316 L 44N1
ER 316 LSi 44N1
ER 316 Si 44N1
ER 317 44N1
ER 317 L 44N1
ER 318 44N1
ER 320 44N1
ER 320 LR 44N1
ER 321 44N1
ER 330 44N1
ER 349 44N2
ER410 Ni Mo 44M1
ER420 44M1
ER446 LR 44M1

ER505 44M1
ER 2553 44N3
ERA 51 44A2
ERA 53 44A2
ERA 59 44G1
ERA 60 44A1
ERA 60M 44A1
ERA 156 44G1
ERA 164 44K1
ERA 165 44K1
ERA 171 44K2
ERA 1414 44N1
ERA CR 1 44N1
ERA CR 2 44N1
ERA CR 4 44N1
ERA CR 4FC 44N1
ERA CR 4S 44N1
ERA CR 4S (Cb) 44N1
ERA CR 4SFC 44N1
ERA CRI FC 44N1
ERA CRI SFC 44N1
ERA CRIS 44N1
ERA CRIS (Cb) 44N1
ERA HR 1 44N1
ERA HR 1S 44N1
ERA HR 2 44N1
ERA HR 2W 44N1
ERA HR 3 44N1
ERA HR4 44M1
ERA HR4 HARD 44M1
ERA HR 5 27B
ERA HR 6 44N1
ERA HR 6W 44N1
ERA HR 7 44N1
ERA HR 8 44N1
ERAYDE 55A
ER AZ61A 23B
ER Co 3 44L
ER Fe C6 44K3
ERM 3A 14E
ERMAL 99.0 1A
ERMAL 99.5 1A
ERMAL 99.7 1A
ERMAL NS4 1D
ERM ALW 14A
ERM CCS 14L
ERM CCS/Mg 14L
ERM CCS/Z 14L
ERM HSM 14A
ERM N S 14D
ERM SC 65 14H
ERN 6N
see ASTM range 27B
ERN 7B
see ASTM range 27C
ERN 7C
see ASTM range 27C
ERN 7W
see ASTM range 27C
ERN 60
see ASTM range 27D
ERN 61
see ASTM range 27F
ERN 62
see ASTM range 27B
ERN 64
see ASTM range 27D
ERN 69
see ASTM range 27C
ER Ni 1 27B
ER Ni C 1 27B
ER Ni Cr 3 27B
ER Ni Cr Fe 5 27C
ER Ni Cr Fe 6 27B
ER Ni Cr Fe 7 27B
ER Ni Cr Mo 1 27B
ER Ni Cr Mo 2 27C
ER Ni Cr Mo 3 27B
ER Ni Cr Mo 7 27B
ER Ni Cr Mo 8 27B
ER Ni Cr Mo 11 27C
ER Ni Cu 7 27D
ER Ni Fe Cr 1 27B
ER Ni Fe Cr 2 27B
ER Ni Fe Mn Ci 20C
ER Ni Mo 1 27F
ER Ni Mo 2 27F
ER Ni Mo 3 27F
ER Ni Mo 7 27F
ER Ti 0.2 Pd 51A
ER Ti 1 51A
ER Ti 2 51A
ER Ti 3 51A
ER Ti 3 Al 2.5 V 51B
ER Ti 3 Al 2.5 V.1 51B
ER Ti 4 51A
ER Ti 5 Al 2.5 Sn 51B
ER Ti 5 Al 2.5 Sn 1 51B
ER Ti 6 Al 2 Cb 1Ta 1Mo 51B
ER Ti 6 Al 4 V 51B
ER Ti 6 Al 4 V.1 51B
ER Ti 12 51B
ER Ti 13 V 11 Cr 3 Al 51B
ERW 2
see BS range 44A2
ERW 3
see BS range 44A2
ER Zr 2 56.1
ER Zr 3 56.1
ER Zr 4 56.1
E S COY Ltd 50.1
ESA 44K3
ESCALLOY 20 44N1
ESCO ALLOY 72 13A
ESCO TCC1 14K
ESCO TCC2 14K
ESD 32 21.1
ESD 42 21.1
ESSCALLOY L 605 13A
ESSHERE 1250 44P
ESSHETE 316 44N1
ESSHETE 321 44N1
ESSHETE 347 44N1
ESSHETE 1250 44N3
ESSHETE CML 44K1
ESSHETE CRM 2 44K1
ESSHETE CRM 5 44K1
ESSHETE D 4C 44G1
ESSHETE MV 44K1
ETA 44K3
ETAD 44K2
ETAD 1 44K1
ETAD 2 44K2
ETAD 3 44K1
ETANO 44K1
ETG 44A2
ETG88 44A2
ETP
See ASTM range 14A
ETP 14A
ETS 14C
ETW 14C
EUTECTIC ALLOY 42.1
EV 1 44M1
EV 2 44M1
EV 3 44N1
EV 4 44N2
EV 5 44K3
EV 5 44N2
EV 6 44N2
EV7 44N3
EV 8 44N2
EV 8 44P
EV 9 44N2
EV 10 44N1
EV 11 44N2
EV 12 44N2
EV 13 44N2
EV 14 44N2
EV 15 44N1
EV 16 44N1
EV 17 44N1
EVANOHM 27C
EVERDUR 14M
EVERDUR A 14M
EVERDUR D 14M
EVERSHYNE 44N1
EV II 44P
EVM 44K3
EW 44K1
EW 2 44N1
EW 3 44N1
EW 4 44N1
EW 6 44K3
EWK 30 44L
EWP 52.1
EWS 44K1
EWT 2 44K1
EWT 3 44K1
EWT 4 44K1
EWTR1 52.1
EWTR2 52.1
EWTR3 52.1
EWZr 52.1
EX 22 44K1
EX 23 44K1
EX 24 44K1
EX 25 44K1
EX 26 44K1
EX 27 44K1
EX 28 44K1
EX 29 44K1
EX 30 44K1
EX 31 44K1
EX 32 44K1
EX 55 44K1
EX 56 44K1
EX 684 51C
EXD 1 44K1
EXD 5 44M1
EXL-DIE 44K3

EXPANDAL 44F2
EX-TEN 42 44A1
EX-TEN 45 44J
EX-TEN 50 44J
EX-TEN 55 44J
EX-TEN 60 44J
EX-TEN 65 44J
EX-TEN 70 44J
EXTENDO-DIE 44K2
EXTRA 44A3
EXTRA 44A4
EXTRA DOUBLE MUSHET 44K3
EXTRA NAP SUPERIOR 44L
EXTRA SUPER 44K3
EXTRA SUPERIOR SPUR 44L
EXTRA TOUGH HARD V 44J
EXTRA TRIPLE CONQUEROR 44K3
EXTRA TRIPLE GRIFFIN 44K3
EXTRA TRIPLE MUSHET 44L
EXTRA VANADIUM 44K3
EXTRUDAL 1E
EXTRUDAL 050 1E
E XX T1 K2 44F1
E XX T1 N2 44F1
E XX T5 K2 44F1
E XX T5 Ni 3 44F1
EXX T8 KG 44E1
E xxx T1 B3 44K1
E xxx T1 K3 44K1
E xxx T5 K3 44K1
EZ 33A
  see ASTM range 23B
EZDA 55A
E-Z-DIE SMOOTHCUT 44K3
E-Z-DIE V 44K3
EZEFORM 35 44N1
F 44J
F 1 44H
F 1 44K2
F 1 44K3
F 1 44M1
F 2 44H
F 2 44K2
F3 23B
F 3 44K3
F 4 44K3
F 5 44K2
F 7 44K2
F 7S 14B
F 7S STAR 14B
F 9 44K1
F10 23B
F 10 20A
F 10 44K2
F-10 44N1
F 15 44M1
F 17 44M1
F 20 20A
F 26 52.1
F 30 20A
F 40 20A
F 45CH 44K1
F 50 20A
F 55CH 44K1
F 58 14C
F 65CH 44K1
F 70 20A
F 75CH 44K1
F 80 20A
F 85CH 44K1
F100 44A3
F105 44A3
F108 44A3
F110 44A3
F114 44A2
F128 44E1
F180 44A3
F200 44A3
F 300 44E2
F 484 13A
F 500 44K2
F 543 44K1
F 550 44E2
F 550 44K2
F1001
  UNS for ASTMA319(i) 20B
F1002
  UNS for ASTMA319(ii) 20B
F1003
  UNS for ASTMA319(iii) 20B
F1004
  UNS for ASTMA159 20B
F1005
  UNS for ASTMA159 20B
F1006
  UNS for ASTMA159 20B
F1007
  UNS for G3500 cast iron 20B
F1008
  UNS for G4000 cast iron 20B
F1009
  UNS for G2500A cast iron 20B
F2003
  UNS for ASTMA602 (M5603) 20B
F2005
  UNS for ASTMA602 (M8501) 20B
F10010
  UNS for G3500B cast iron 20B
F10011
  UNS for G3500C cast iron 20B
F10012
  UNS for G3500D cast iron 20B
F10090
  UNS for AWSA5.15 (RCI) 20B
F10091
  UNS for AWSA5.15 (RCIA) 20B
F10092
  UNS for AWSA5.15 (RCIB) 20B
F11401
  UNS for ASTMA48.20 20B
F11501
  UNS for ASTMA126A 20B
F11701
  UNS for ASTMA48.25 20B
F12102
  UNS for ASTMA126B 20B
F12401
  UNS for ASTMA48(35) 20B
F12801
  UNS for ASTMA48.40 20B
F12802
  UNS for ASTMA126C 20B
F12803
  UNS for ASTMA278(40) 20B
F13101
  UNS for ASTMA48(45) 20B
F13102
  UNS for ASTMA278(45) 20B
F13501
  UNS for ASTMA48(50) 20B
F13502
  UNS for ASTMA278(50) 20B
F13801
  UNS for ASTMA48(55) 20B
F13802
  UNS for ASTMA278(55) 20B
F14101
  UNS for ASTMA48(60) 20B
F14102
  UNS for ASTMA278(60) 20B
F14801
  UNS for ASTMA278(70) 20B
F15501
  UNS for ASTMA278(80) 20B
F20000
  UNS for ASTMA602 (M3210) 20B
F20001
  UNS for ASTMA602 (M4504) 20B
F20002
  UNS for ASTMA602 (M5003) 20B
F20004
  UNS for ASTMA602 (M7002) 20B
F22000
  UNS for ASTMA197 20B
F22200
  UNS for ASTMA47 20B
F22400
  UNS for ASTMA47 20B
F22830
  UNS for ASTMA220 20B
F23130
  UNS for ASTMA220 20B
F23131
  UNS for ASTMA220 20B
F23330
  UNS for AMS5310 20B
F23530
  UNS for ASTMA220 20B
F24130
  UNS for ASTMA220 20B
F24830
  UNS for ASTMA220 20B
F25530
  UNS for ASTMA220 20B
F26230
  UNS for ASTMA220 20B
F 30000 20B
F32800
  UNS for ASTMA536 20B
F32900
  UNS for ASTMA716 20B
F33100
  UNS for ASTMA536 20B
F33101
  UNS for AMS5315 20B

F33800
UNS for ASTMA536 20B
F34100
UNS for ASTMA476 20B
F34800
UNS for ASTMA536 20B
F36200
UNS for ASTMA536 20B
F41000
UNS for ASTMA436 20B
F41001
UNS for ASTMA436(1) 20B
F41002
UNS for ASTMA436(2) 20B
F41003
UNS for ASTMA436(2b) 20B
F41004
UNS for ASTMA436(3) 20B
F41005
UNS for ASTMA436(4) 20B
F41006
UNS for ASTMA436(5) 20B
F41007
UNS for ASTMA436(6) 20B
F43000
UNS for ASTMA439(D2) 20B
F43001
UNS for ASTMA439(D2B) 20B
F43002
UNS for ASTMA439(D2C) 20B
F43003
UNS for ASTMA439(D3) 20B
F43004
UNS for ASTMA439 (D3A) 20B
F43005
UNS for ASTMA439(D4) 20B
F43006
UNS for ASTMA439(D5) 20B
F43007
UNS for ASTMA439(D5B) 20B
F43010
UNS for ASTMA571 20B
F43020
UNS for MIL Spec 1-24137(B) 20B
f43021
UNS for MIL Spec 1-24137(C) 20B
F43030
UNS for AMS5395 20B
F45000
UNS for ASTMA532(1) 20B
F45001
UNS for ASTMA532(11) 20B
F45002
UNS for ASTMA532(111) 20B
F45003
UNS for ASTMA532(1D) 20B
F45004
UNS for ASTMA532(11A) 20B
F45005
UNS for ASTMA532(11B) 20B
F45006
UNS for ASTMA532(11C) 20B
F45007
UNS for ASTMA532(11D) 20B
F45008
UNS for ASTMA532(11E) 20B
F45009
UNS for ASTMA532 (111A) 20B
F45100
UNS for AWSA5.13(R Fe Cr Al)
F 47001 20B
F 47002 20B
F47004
UNS for AMS5392 20C
F47005
UNS for AMS5393 20C
F47006
UNS for AMS5394 20C
FABIS 44A3
FAGERSTA 16 44A4
FAGERSTA 20 44A4
FAGERSTA 24 44A4
FAGERSTA 1150 44A1
FAGERSTA 1160 44A1
FAGERSTA 1265 44A1
FAGERSTA 1357 44A1
FAGERSTA 1660 44A2
FAGERSTA 1680 44A3
FAGERSTA 1773 44A3
FAGERSTA 1778 44A3
FAGERSTA 1779 44A3
FAGERSTA 1870 44A4
FAGERSTA 1884 44A4
FAGERSTA B126 44E2
FAGERSTA B805 44A2
FAGERSTA C182 44E2
FAGERSTA C250 44E3
FAGERSTA C261 44E2
FAGERSTA C525 44E2
FAGERSTA D234 44K2
FAGERSTA D941 44K3
FAGERSTA D953 44K3
FAGERSTA K353 44K1
FAGERSTA K825 44K2
FAGERSTA N104 44K1
FAGERSTA N164 44E3
FAGERSTA R100 44M1
FAGERSTA R290 44M1
FAGERSTA R310 44N1
FAGERSTA R320 44N1
FAGERSTA R350 44N1
FAGERSTA R358 44N1
FAGERSTA R359 44N1
FAGERSTA R360 44N1
FAGERSTA R390 44N1
FAGERSTA R428 44N1
FAGERSTA R440 44N1
FAGERSTA R450 44N1
FAGERSTA R460 44N1
FAGERSTA R590 44N1
FAGERSTA R700 44M1
FAGERSTA R710 44M1
FAGERSTA R720 44M1
FAGERSTA R730 44M1
FAGERSTA R740 44M1
FAGERSTA R800 44N1
FAGERSTA R820 44N1
FAL 44M1
FALCON 4 44K2
FALCON 6 44K2
F ALLOY 44L
FAS 44M1
FASTEX 5 44A1
FASTEX 7 44A1
FASTEX 100 44A1
FASTEX 100M 44G1
FASTWORK 44M2
FAVORIT 44K3
FB 50 44A1
FB 50AK 44J
FB 60 44K1
FB 60AK 44K1
FB 70 44K1
FB 70AK 44K1
FB 80 44K1
FB 80AK 44K1
F BRONZE 14H
FC1 44M1
FC 3/75 14J
FC 10 GRADE 1 20B
FC 15 20B
FC 15 GRADE 2 20B
FC 20 20B
FC 20 GRADE 3 20B
FC 25 20B
FC 25 GRADE 4 20B
FC 30 20B
FC 30 GRADE 5 20B
FC 35 20B
FC 35 GRADE 6 20B
FC 40 20B
FC 71/1 14J
FC 75 14J
FC 80 14J
FC 85/1 14J
FC 150 20B
FC 200 20B
FC 250 20B
FC 275 20B
FC 300 20B
FC 350 20B
FC 400 20B
FCB 44N1
FCB(T) 44N1
FCCH 44A1
FCD 40 20B
FCD 45 20B
FCD 50 20B
FCD 60 20B
FCD 70 20B
FCS 44M1
F C V 44K2
FCW 5 44H
FDK 3 44E1
FDK 4 44E2
FDP 44N1
FDP (L) 44N1
FE 15 Si 20B
Fe 33 44A1
Fe 34A 44A1
Fe 34B 44A1
Fe 34C 44A1
Fe 37A 44A1
Fe 37B 44A1
Fe 37C 44A1
Fe 37D 44A1
Fe 42A 44A1
Fe 42B 44A1

Fe 42C 44A1
Fe 42D 44A1
Fe 44A 44A1
Fe 44B 44A1
Fe 44C 44A1
Fe 44D 44A1
Fe 50 44A1
Fe 52B 44A1
Fe 52C 44A1
Fe 52D 44A1
Fe 60 44A1
Fe 70 44A1
FE 85 20A
FE 90 20A
FEC 90 14J
FECRALLOY 44M1
FEMAX 33.80 44A1
FENTON 55A
FE PA95 44F3
FERALSIN 52 44A1
FERALSIN 58 44A1
FERRALIUM 44N3
FERRALIUM 255 44N3
FERRO ALUMINIUM 1N
FERROARC R155 44A1
FERROARC R160 44A1
FERROCOR 216 20A
FERROCOR 253 20A
FERROCOR 320 20A
FERRO-MANGANESE 24.1
FERROSIL 100 20A
FERROSIL 107 20A
FERROSIL 146 20A
FERROSIL 170 20A
FERROSIL 187 20A
FERROSIL 216 20A
FERROSIL CR253 20A
FERRO-TITANIUM 51B
FERROVAC E 20A
FERRY 14E
FF 3 44A2
FF 3H 44A2
FF 4 44A2
FF 4H 44E2
FF 38 44A2
F Fe Mn A 44P
F Fe Mn B 44P
F Fe Mn C 44P
F Fe Mn Cr 44P
F Fe Mn D 44P
FG 44M1
FG 10 20B
FG 15 20B
FG 15D 20B
FG 20 20B
FG 20D 20B
FG 25 20B
FG 25D 20B
FG 30 20B
FG 30D 20B
FG 35 20B
FG 35D 20B
FG 40D 20B
FGC 75 14J
FGC 85 14J
FGE 38.17 20B
FGE 42.12 20B
FGE 50.7 20B
FGE 60.2 20B
FGE 70.2 20B
FGE 80.2 20B
FGE 370.17 20B
FGE 400.12 20B
FGE 500.7 20B
FGE 600.3 20B
FGE 700.2 20B
FGE 800.2 20B
FGG 20B
FGG 10 20B
FGG 15 20B
FGG 20 20B
FGG 25 20B
FGG 30 20B
FGG 35 20B
FGG 40 20B
FG (L) 44M1
FGN 70.3 20B
FGS 38.15 20B
FGS 42.12 20B
FGS 50.7 20B
FGS 60.2 20B
FGS 70.2 20B
FGS 370.17 20B
FGS 400.12 20B
FGS 500.7 20B
FGS 600.3 20B
FGS 700.2 20B
FGS 800.2 20B
FH 44M1
FHM 44M1
FHS 18 44K3
FHS 20/5 44L
FHS 20/10 44L
FI 44M1
FI 17 44M1
FI 20 44M1
FILE 44A4
FILLETRODE 43.33 44A1
FIREDIE 44K2
FIREDIE 13 44K2
FIREDIE 13 SMOOTHCUT 44K2
FIREX 44K2
FIRTHAG 44E2
FIRTH OB 44K1
FISSIUM 39.1
FI Ti 44M1
FIVEOHM 14E
FIXAMPER 27E
FKD 3 44K1
FKD 4 44K2
FLASHKUT 44K3
FLAT BOTTOM RAILS 44A3
FLOWER BRAND 21.1
FLT 2A 44A1
FLT 2B 44A1
FLT 2C 44F1
FLT 2D 44F1
FLUGINOX 51 44M1
FLUGINOX 60 44M1
FLUGINOX 62 44M1
FLUGINOX 65 44M1
FMA 1F
FMB 44N1
FMB (L) 44N1
FMB Ti 44N1
FML 44N1
FMP 44K3
FMP 035 44J
fmp 200 44K3
fmp 328 44K2
fmp 329 44K2
fmp 336 44M2
fmp 338 44M2
fmp 348 44K2
fmp 379 44K3
fmp 399 44K2
fmp 455 44L
fmp 470 44K3
fmp 501 44K3
fmp 504 44K3
fmp 505 44K1
fmp 507 44K1
fmp 513 44K1
fmp 526 44L
fmp 530 44L
fmp 536 44L
fmp 542 44L
fmp 555C 44L
fmp 562 44K3
fmp 563 44K3
fmp 599 44K3
fmp 682 44K2
fmp 808 44L
fmp 828 44L
fmp 842 44K3
fmp 922 44K3
fmp 928 44L
fmp 929 44L
fmp 933 44L
fmp 948 44K3
fmp 1850 44K2
FN 30 44F1
FN 35 44K2
FNB 44N1
FNB 36A 44A1
FNB 36B 44A1
FNB 40A 44A1
FNB 40B 44A1
FNCR 44K1
FNG 1 20B
FNG 38.17 20B
FNG 42.12 20B
FNG 50.7 20B
FNG 60.2 20B
FNG 70.2 20B
FNG 80.2 20B
F NICKEL 27F
FNS 44M2
FNS fm 44M2
FNZ 44M1
FONTAIN MOREAU 55A
FOREMOST 14 Cr 44M1
FOREMOST 18/8 44N1
FOREMOST 18/8 Mo 44N1
FOREMOST 18/8 Nb 44N1
FOREMOST 18/8 Ti 44N1
FOREMOST 257 44K2
FOREMOST 467 44M1
FOREMOST A 44K1
FOREMOST HR 1 44N1
FOREMOST HR 2 44M1

FOREMOST HR 5 44N1
FOREMOST HR 6 44N1
FOREMOST HR 7 44N1
FOREMOST HR 8 44N1
FOREMOST HR3 27B
FOREMOST HR4 27B
FOREMOST IRON 44M1
FOREMOST PNEUMATIC 44K2
FORMDIE SMOOTHCUT 44K2
FORMITE 2 44K1
FORMITE 3 44K2
FORMWELL 44N1
FORTIWELD 44G1
FORTREX 30 44A1
FORTREX 35 44A1
FORTREX 35A 44A1
FORTREX 45 44K1
FORTREX 55 44K1
FOUR STAR 44K3
FRAMDIE 44K3
FRARY METAL 21.1
f Rb 16 38.1
FRHC
  see ASTM range 14A
FRONTIER 40E 1M
FRSTP 14A
FRTP 14A
Fs 39.1
FS 2.5 44L
FS 16/25/6 44N1
FS 718 27C
FS 901 27C
FS A 286 44N2
FSL 44N1
FSL (L) 44N1
FS M10 44K3
FS M21/2 44K3
FS M34 44L
FSQ 44N1
FST 44N1
FST (L) 44N1
FS X 750 27C
FT 23B
FT 15 D 20B
FT 20 D 20B
FT 25 D 20B
FT 30 D 20B
FT 35 D 20B
FT 40 D 20B
FTW 52 44A1
FTW 55 44A1
FTW 58 44A1
FTW 60 44J
FU E10-58 14J
FU E10-62 14J
FUJI CON SUPER 44A2
FURIOUS 44K3
FV 301 44N3
FV 302 44N1
FV 303 44N1
FV 304 44N1
FV 304 (L) 44N1
FV 305 44N1
FV 309 44N1
FV 310 44N1
FV 316 44N1
FV 316 (L) 44N1
FV 321 44N1
FV 326 44N2
FV 347 44N1
FV 403 44M1
FV 406 44M1
FV 410 44M1
FV 420 44M1
FV 430 44M1
FV 431 44M1
FV 440B 44M1
FV 446 44M1
FV 448 44M1
FV 467 44N3
FV 507 44M1
FV 520(B) 44N3
FV 520(S) 44N3
FV 535 44L
FV 535 44M1
FV 548 44N3
FV 555 44N1
FV 566 44M1
FV 607 44M1
FV 702 44M1
FV Fl 44M1
FVS 44N1
FX 44M1
FX 10 14J
FXY 14C
G
  see GOST 4543 44A1
G
  see GOST 5058 44A1
G
  see GOST 8731 44A2
G 1
  see BS range 14K
G 1A 1D
G 1B 1D
G2
  see GOST 5058 44A1
G2
  see GOST 4543 44A1
G2
  see GOST 5058 44A1
G 2
  see BS range 14K
G 2 44N1
G2S
  see GOST 4543 44A1
G 2S
  see GOST 5058 44A1
G2S
  see GOST 8731 44A2
G2SD
  see GOST 4543 44A1
G2SR
  see GOST 4543 44A1
G 3
  see BS range 14K
G4 42.1
G 4 44N1
G 4A
  see ASTM range 1J
G 5 Sp1 44K2
G 6 Cu Zn 42.1
G7 42.1
G 7 44K2
G 8 44K1
G 8A
  see ASTM range 1J
G9 42.1
G 9 44N1
G 10 20B
G 10A
  see ASTM range 1J
G 11 44K1
G 12 44K2
G 14 A 44K2
G 14 C 44K2
G 15 20B
G 15 44K1
G 16 44K1
G 18B 44N2
G 19 44N1
G 19 44N1
G 19 44N2
G 20 20B
G 20 44K2
G 21 44N1
G 23 44K3
G 24 44K3
G 25 20B
G 25 44K3
G 26 44K2
G 27 44K1
G 28 44K1
G 30 20B
G 32 13A
G 33 44K2
G 34 13A
G 34C 13A
G 35 20B
G 36 44K1
G 38 44N1
G 38 44N3
G 39 27C
G 40 44N1
G 40 44N2
G 41 44N3
G 44 27C
G 45 44K1
G 46 44K1
G 47 44K1
G 48 44K1
G 49 44K1
G 50 44K1
G 51 44K1
G 52 44K1
G 53 44K1
G 54 27C
G 55 27C
G 60 13A
G 62 27C
G 63 27B
G 64 27C
G 65 13A
G 66 27C
G 67 27C
G 68 44N2
G 69 27C
G 70 27C
G 71 44N1
G 72 44N1
G 73 27C

G 74 27C
G 75 13A
G 76 27C
G 77 27C
G 79 27C
G 80 27C
G 81 27C
G 82 27C
G 83 27C
G 84 27C
G 85 44M1
G 86 44K1
G 87 13A
G 88 13A
G 94 27C
G 95 27C
G 100 27C
G 101 27C
G 103 27C
G 105 44K2
G 110 44K1
G 125 44N1
G 140 44F3
G 157 27C
G 192 44P
G 267 27C
G 1800 20B
G 2000 20B
G 2500 20B
G 3000 20B
G 3101/73/5534 44A1
G 3426 STC 52D 44N1
G 3500 20B
G 4000 20B
G 4301 SEC 7 44N1
G 4301 SEC 8 44N1
G 4301 SEC 9 44N1
G 4301 SEC 10 44N1
G 4301 SEC 13 44N1
G 4301 Sec/1 44M1
G 4301 Sec/4 44M1
G 4301 Sec/5 44M1
G 4302 SEH 5 44N1
G 4404 Japanese standard for tool steels; see designation for details
G10050
  UNS for AISI1005 44A1
G10060
  UNS for 1006 44A1
G10080
  UNS for AISI1008 44A1
G10100
  UNS for AISI1010 44A1
G10110
  UNS for 1011 44A1
G10120
  UNS for AISI1012 44A1
G10130
  UNS for AISI1013 44A1
G10130
  UNS for 1013 44A1
G10150
  UNS for AISI1015 44A1
G10160
  UNS for AISI1016 44A1
G10170
  UNS for AISI1017 44A1
G10180
  UNS for AISI1018 44A1
G10190
  UNS for AISI1019 44A1
G10200
  UNS for AISI1020 44A1
G10210
  UNS for AISI1021 44A1
G10220
  UNS for AISI1022 44A1
G10230
  UNS for AISI1023 44A2
G10250
  UNS for 1025 44A2
G10260
  UNS for 1026 44A2
G10290
  UNS for AISI1029 44A2
G10300
  UNS for AISI1030 44A2
G10350
  UNS for AISI1035 44A2
G10370
  UNS for AISI1037 44A2
G10380
  UNS for AISI1038 44A2
G10390
  UNS for AISI1039 44A2
G10400
  UNS for AISI1040 44A2
G10420
  UNS for AISI1042 44A2
G10430
  UNS for AISI1043 44A2
G10440
  UNS for AISI1044 44A2
G10450
  UNS for AISI1045 44A2
G10460
  UNS for AISI1046 44A2
G10490
  UNS for AISI1049 44A2
G10500
  UNS for AISI1050 44A3
G10530
  UNS for AISI1053 44A3
G10550
  UNS for AISI1055 44A3
G10590
  UNS for AISI1059 44A3
G10600
  UNS for AISI1060 44A3
G10640
  UNS for AISI1064 44A3
G10650
  UNS for AISI1065 44A3
G10690
  UNS for AISI1069 44A3
G10700
  UNS for AISI1070 44A3
G10740
  UNS for AISI1074 44A3
G10750
  UNS for AISI1075 44A3
G10780
  UNS for AISI1078 44A3
G10800
  UNS for AISI1080 44A3
G10840
  UNS for AISI1084 44A4
G10850
  UNS for AISI1085 44A4
G10860
  UNS for AISI1086 44A4
G10900
  UNS for AISI1090 44A4
G10950
  UNS for AISI1095 44A4
G 10950 44A4
G11080
  UNS for AISI1108 44A1
G11090
  UNS for 1109 44A1
G11100
  UNS for AISI1110 44A1
G11160
  UNS for 1116 44A1
G11170
  UNS for AISI1117 44A1
G11180
  UNS for AISI1118 44A1
G11180
  UNS for 1118 44A1
G11190
  UNS for 1119 44A1
G11230
  UNS for 1123 44A1
G11320
  UNS for 1132 44A2
G11370
  UNS for AISI1137 44A2
G11374
  AMS5020 44A2
G11390
  UNS designation for AISI1139 44A2
G11400
  UNS designation for AISI1140 44A2
G11406
  UNS for 1140 44A2
G11410
  UNS for 1141 44A2
G11440
  UNS for 1144 44A3
G11450
  UNS for 1145 44A3
G11460
  UNS for 1146 44A3
G11510
  UNS for 1151 44A3
G12110
  UNS for 1211 44A1
G12120
  UNS for 1212 44A1
G12130
  UNS for 1213 44A1
G12134
  UNS for 12L13 44A1
G12144
  UNS for 12L14 44A1

G12150
UNS for 1215 44A1
G13300
UNS for 1330 44A2
G13350
UNS for 1335 44A2
G13400
UNS for 1340 44A2
G13450
UNS for 1345 44A2
G15116
UNS now G51986
G15130
UNS for 1513 44A1
G15180
UNS for 1518 44A1
G15216
UNS now G52986
G15220
UNS for 1522 44A1
G15240
UNS for 1524 44A1
G15250
UNS for 1525 44A2
G15260
UNS for 1526 44A2
G15270
UNS for 1527 44A2
G15330
UNS for 1533 44A2
G15340
UNS for 1534 44A2
G15360
UNS for 1536 44A2
G15410
UNS for 1541 44A2
G15450
UNS for 5145 44A2
G15470
UNS for 1547 44A2
G15480
UNS for 1548 44A2
G15510
UNS for 1551 44A2
G15520
UNS for 1552 44A2
G15530
UNS for 1553 44A3
G15610
UNS for 1561 44A3
G15660
UNS for 1566 44A3
G15700
UNS for 1570 44A3
G15720
UNS for 1572 44A3
G15800
UNS for 1580 44A3
G15900
UNS for 1590 44A4
G31400
UNS for 3140 44K2
G33106
UNS for 3310 44K1
G40120
UNS for 4012 44G1
G40230
UNS for 4023 44G1
G40240
UNS for 4024 44G1
G40270
UNS for 4027 44G1
G40280
UNS for 4028 44G1
G40320
UNS for 4032 44G1
G40370
UNS for 4037 44G1
G40420
UNS for 4042 44G1
G40470
UNS for 4047 44G2
G40630
UNS for 4063 44G2
G41200
UNS for 4120 44K1
G41210
UNS for 4121 44K1
G41300
UNS for 4130 44K1
G41350
UNS for 4135 44K1
G41370
UNS for 4137 44K1
G41400
UNS for 4140 44K2
G41420
UNS for 4142 44K2
G41450
UNS for 4145 44K2
G41470
UNS for 4147 44K2
G41500
UNS for 4150 44K2
G41610
UNS for 4161 44K3
G43200
UNS for 4320 44K1
G43370
UNS for 4337 44K2
G43376
UNS for 4337 44K2
G43400
UNS for 4340 44K2
G43406
UNS for 4340 44K2
G44190
UNS for 4419 44G1
G44220
UNS for 4422 44G1
G44270
UNS for 4427 44G1
G45200
UNS for 4520 44G1
G46150
UNS for 4615 44G1
G46170
UNS for 4617 44K1
G46200
UNS for 4620 44K1
G46210
UNS for 4621 44K1
G46260
UNS for 4626 44K1
G47150
UNS for 4715 44K1
G47180
UNS for 4718 44K1
G47200
UNS for 4720 44K1
G48150
UNS for 4815 44K1
G48170
UNS for 4817 44K1
G48200
UNS for 4820 44K1
G50150
UNS for 5015 44E1
G50401
UNS for 50B40 44E2
G50441
UNS for 50B44 44E2
G50460
UNS for 5046 44E2
G50461
UNS for 50B46 44E2
G50501
UNS for 5060 44E2
G50600
UNS for 5060 44E2
G50601
UNS for 50B60 44E2
G50986
UNS for 50100 44E3
G51150
UNS for 5115 44E1
G51170
UNS for 5117 44E1
G51200
UNS for 5120 44E1
G51300
UNS for 5130 44E1
G51320
UNS for 5132 44E1
G51350
UNS for 5135 44E1
G51400
UNS for 5140 44E2
G51470
UNS for 5147 44E2
G51500
UNS for 5150 44E2
G51550
UNS for 5155 44E2
G51600
UNS for 5160 44E2
G51601
UNS for 51B60 44E2
G51986
UNS for AISI E51100 44E3
G52986
UNS for SAE 52100 44E3
G61180
UNS for 6118 44K1
G61200
UNS for 6120 44K1
G61500
UNS for 6150 44K2

G71406
UNS for ASTMA355A 44K2
G81150
UNS for 8115 44K1
G81451
UNS for 81B45 44K2
G86150
UNS for 8615 44K1
G86170
UNS for 8617 44K1
G86200
UNS for 8620 44K1
G86220
UNS for 8622 44K1
G86250
UNS for 8625 44K1
G86270
UNS for 8627 44K2
G86300
UNS for 8630 44K2
G86370
UNS for 8637 44K2
G86400
UNS for 8640 44K2
G86420
UNS for 8642 44K2
G86450
UNS for 8645 44K2
G86451
UNS for 86B45 44K2
G86500
UNS for 8650 44K2
G86550
UNS for 8655 44K2
G86600
UNS for 8660 44K2
G87200
UNS for 8720 44K1
G87350
UNS for 8735 44K1
G87400
UNS for 8740 44K1
G87420
UNS for 8742 44K1
G88220
UNS for 8822 44K1
G92540
UNS for 9254 44E2
G92550
UNS for 9255 44B2
G92600
UNS for 9260 44A2
G92620
UNS for 9262 44E2
G93106
UNS for 9310 44K1
G94151
UNS for 94B15 44K1
G94171
UNS for AISI 94B17 44K1
G94301
UNS for AISI 94B30 44K2
G94401
UNS for 94B40 44K2
G98400
UNS for 9840 44K2
G98500
UNS for 9850 44K2
GA 20B
G A 3 Z1 23B
G A 6 Z1 23B
G A 9 23B
G A 9 Z1 23B
G A1 Bx 9
see DIN range 14G
GA 350 20B
G Ag 25 TR 23B
G Al Cu 4 Ti 1L
G Al Cu 4 Ti Mg 1L
G Al Cu 5 Si 3 1L
G Al Mg 3 1J
G Al Mg 3(Cu) 1J
G Al Mg 5 1J
G Al Mg 10 1J
G Al Mg(Cu) 1J
G Al Si 5 Cu 1 1C
G Al Si 5 Mg 1C
G Al Si 6 Cu 4 1C
G Al Si 6 Cu 6 1C
G Al Si 7 Cu 3 1C
G Al Si 9 (Cu) 1C
G Al Si 10 Mg 1C
G Al Si 10 Mg (Cu) 1C
G Al Si 12 1C
G Al Si 12 (Cu) 1C
GALA 44M1
GALAHAD A 44M1
GALAHAD AFC 44M1
GALAHAD B 44M1
GALAHAD BFC 44M1
GALAHAD C 44M1
GALAHAD CFC 44M1
GALAHAD D 44M1
GALAHAD DFC 44M1
GALAHAD E 44M1
GALAHAD F 44M1
GALLIUM 15.1
GALVALUME 1N
GALVALUNE SHEET 44A1
GALVAMATT 44A1
GAMAN H 44P
GAMMA Cb 44N1
GARNET 52.1
GASC 44M1
GASD 44M1
GAUGE STEEL 44A4
GAUSSIT 180 13A
GB 14C
GB 20B
GB 300 20B
GB A1 Bz 9
see DIN range 14G
GB Cu 60 Zn 14C
GB Cu 65 Zn 14B
GB Cu Al 9 14G
GB Cu Al 9 Ni 14G
GB Cu Al 10 Fe 14G
GB Cu Pb 5 Sn 14K
GB Cu Pb 10 Sn 14K
GB Cu Pb 15 Sn 14K
GB Cu Pb 20 Sn 14K
GB Cu Sn 5 Zn Pb 14K
GB Cu Sn 7 Zn Pb 14K
GB Cu Sn 10 14K
GB Cu Sn 10 Zn 14K
GB Cu Sn 12 14K
GB Cu Sn 14 14K
GB Fe A1 Bz
see DIN range 14G
GB Ms 60A 14C
GB Ms 60B 14C
GB Ms 65 A 14B
GB Ms 65 B 14B
GB Ms 65 C 14B
GB Ni A1 Bz
see DIN range 14G
GB Rg 5 14K
GB Rg 7 14K
GB Rg 10 14K
G BRONZE 14H
GB Sn Bz 10 14K
GB Sn Bz 12 14K
GB Sn Bz 14 14K
GB Sn Pb Bz 5 14K
GB Sn Pb Bz 10 14K
GB Sn Pb Bz 15 14K
GB Sn Pb Bz 20 14K
GC 20B
GC15 42.1
GC 275 20B
GC Cu Sn 5 Zn Pb 14K
GC Cu Sn 7 Zn Pb 14K
GC Cu Sn 10 Zn 14K
GC Cu Sn 12 14K
GC Rg 5 14K
GC Rg 7 14K
GC Rg 10 14K
GC Sn Bz 12 14K
G Cu 55 Zn Al 14C
G Cu 55 Zn Al 2 14C
G Cu 55 Zn Al 4 14C
G Cu 55 Zn Mn 14C
G Cu 65 Zn 14B
G Cu Al 8 Mn 14G
G Cu Al 9 14G
G Cu Al 9 Ni 14G
G Cu Al 10 Fe 14G
G Cu Al 10 Ni 14G
G Cu Al 11 Ni 14G
G Cu Pb 5 Sn 14K
G Cu Pb 10 Sn 14K
G Cu Pb 15 Sn 14K
G Cu Pb 20 Sn 14K
G Cu Pb 25 14K
G Cu Sn 5 Zn Pb 14K
G Cu Sn 7 Zn Pb 14K
G Cu Sn 10 14K
G Cu Sn 10 Zn 14K
G Cu Sn 12 14K
G Cu Sn 14 14K
GD 20B
GD 1
see DIN range 44A1
GD10 42.1
GD 11
see DIN range 44A1
GD25 42.1
GD35 42.1
GD 250 20B
GD Al Mg 8(Cu) 1J

GD Al Mg 9 1J
GD Al Si 6 Cu 3 1C
GD Al Si 10 (Cu) 1C
GD Al Si 12 (Cu) 1C
GD Al Si 12. 1C
GD Cu 60 Zn 14C
GD Mg Al 9 Zn 1 23B
GD Mg Al 9 Zn 2 23B
GD Mg Al Zn 1 23B
GD Ms 60 14C
GD Zn A1 4
 see German standard range 55A
GD Zn A1 4 Cu 1
 see German standard range 55A
GE 20B
GE 125 52.1
GE 200 20B
GE 225 20B
GEARITE 44K1
GEARMAC 44K1
GEAROL 44K2
GECALLOY 44L
GE HEVIMET 52.1
GENEX 44A1
GENUINE STUBS 44A4
GERMANIUM 16.1
GERMANIUM I 16.1
GERMANIUM II 16.1
GF 20B
GF 150 20B
GFA NICKEL 27A
G Fe Al Bz F 50 14G
GFS 44K3
GG 10 20B
GG 15 20B
GG 20 20B
GG 25 20B
GG 30 20B
GG 35 20B
GG 40 20B
GG Ft 15 20B
GG Ft 20 20B
GG Ft 25 20B
GG Ft 30 20B
GGG 35.3 20B
GGG 38 20B
GGG 40 20B
GGG 40.3 20B
GGG 42 20B
GGG 45 20B
GGG 50 20B
GGG 60 20B
GGG 70 20B
GGG 80 20B
GGG 400 20B
GGG 400 K 20B
GGG 500 20B
GGG 600 20B
GGG 700 20B
GGG FGS 38 20B
GGG FGS 42 20B
GGG FGS 50 20B
GGG FGS 60 20B
GGG FGS 70 20B
GGL 20B
GI 52.1
G I BRONZE 14K
GILBRUN 10 44M1
GILBRUN Cr 17 44M1
GILBRUN Cr 20 44M1
GILDING METAL 14B
GILGRID 20C
GI SPECIAL 44K1
GJ 2 44K2
GK 3 44K1
GK 5 44K1
GK 7 44K1
GK Cu 60 Zn 14C
GK Mg Al 8 Zn 1 23B
GK Mg Al 9 Zn 1 23B
GK Mg Al 9 Zn 2 23B
GK Ms 60 14C
GK Zn A1 4 Cu 1
 see German standard range 55A
GK Zn A1 6 Cu 1
 see German standard range 55A
GL 1 27B
GLANCE 6.1
GLASS SEALING 27 20C
GLASS SEALING 42 20C
GLASS SEALING 42.6 20C
GLASS SEALING 52 27E
GLS 441 44J
GLX 45W 44A1
GLX 50W 44A1
GLX 55 44A1
GLX 60W 44A1
GLX 65 44A1
GLX 70 44A1
GM 20B
G M2 23B
GM 31A 1D
GM 40A 1D
GM 41A 1D
GM 50A 1D
GM 51A 1D
GM 70B
 see ASTM range 1J
GM 400 20B
G Mg A1 8 Zn 1
 see DIN range 23B
G Mg A1 9 Zn 1
 see DIN range 23B
G Mg A1 9 Zn 2
 see DIN range 23B
G Mg Al 6 Zn 3 23B
G Mn A1 Bz F 42
 see DIN range 14G
GMNA 100 1L
GMR 235 27C
GMR 235D 27C
G Ms 65 14B
GN1 42.1
GN 38 20B
GN 42 20B
GN 50 20B
GN 60 20B
GN 70 20B
G Ni A1 Bz F 50
 see DIN range 14G
G Ni A1 Bz F 60
 see DIN range 14G
G Ni A1 Bz F 68
 see DIN range 14G
GO 44K1
GOAHED 44K3
GOLD 17.1
GOLD STAR 44L
GOST 380 2SP2 44A1
GOST 380 BST 25 P 44A1
GOST 380 BST2 PS 44A1
GOST 380 BST2 PS2 44A1
GOST 380 BST3 GPS 44A1
GOST 380 BST3 GPS2 44A1
GOST 380 BST3 KP2 44A1
GOST 380 BST3 PS 44A1
GOST 380 BST3 SP 44A1
GOST 380 BST3 SP2 44A1
GOST 380 BST4 KP 44A1
GOST 380 BST4 PS 44A1
GOST 380 BST4 PS2 44A1
GOST 380 BST4 SP2 44A1
GOST 380 BST5 GPS 44A1
GOST 380 BST5 PS 44A2
GOST 380 BST5 SP 44A2
GOST 380 BST5 SP2 44A2
GOST 380 BST6 PS 44A2
GOST 380 BST6 PS2 44A2
GOST 380 BST6 SP 44A2
GOST 380 BST6 SP2 44A2
GOST 380 ST0 44A1
GOST 380 ST1 KP 44A1
GOST 380 ST1 KP2 44A1
GOST 380 ST1 PS 44A1
GOST 380 ST1 PS2 44A1
GOST 380 ST1 SP 44A1
GOST 380 ST2 KP 44A1
GOST 380 ST2 KP2 44A1
GOST 380 VST2 KP 44A1
GOST 380 VST2 PS 44A1
GOST 380 VST2 PS2 44A1
GOST 380 VST2 SP2 44A1
GOST 380 VST3 44A1
GOST 380 VST3 CPS3 44A1
GOST 380 VST3 EPS4 44A1
GOST 380 VST3 EPS5 44A1
GOST 380 VST3 EPS6 44A1
GOST 380 VST3 PS 44A1
GOST 380 VST3 PS2 44A1
GOST 380 VST3 PS3 44A1
GOST 380 VST3 PS4 44A1
GOST 380 VST3 PS5 44A1
GOST 380 VST3 PS6 44A1
GOST 380 VST3 SP 44A1
GOST 380 VST3 SP2 44A1
GOST 380 VST3 SP3 44A1
GOST 380 VST3 SP4 44A1
GOST 380 VST3 SP5 44A1
GOST 380 VST3 SP6 44A1
GOST 380 VST4 PS2 44A1
GOST 380 VST4 PS3 44A1
GOST 380 VST4 SP 44A1
GOST 380 VST4 SP3 44A1
GOST 380 VST5 GPS 44A2
GOST 380 VST5 GPS2 44A2
GOST 380 VST5 PS 44A2
GOST 380 VST5 PS2 44A2

GOST 380 VST5 SP 44A2
GOST 380 VST5 SP2 44A2
GOST 380 VST25 P 44A1
GOST 550 10 44A1
GOST 550 10G2 44A1
GOST 550 20 44A1
GOST 550 KH5 44E1
GOST 550 KH5M 44K1
GOST 550 KH5YF 44K1
GOST 550/12 KHM 44K1
GOST 613 E 44A1
GOST 631 D 44A1
GOST 631 K 44A1
GOST 631 L 44A1
GOST 631 M 44A1
GOST 632 S 44A1
GOST 633 D 44A1
GOST 801 SHKH6 44E3
GOST 801 SHKH9 44E3
GOST 801 SHKH15 44E3
GOST 801 SHKH15SG 44E3
GOST 1050/05 KP 44A1
GOST 1050/08 44A1
GOST 1050/08 KP 44A1
GOST 1050/10 44A1
GOST 1050/10 KP 44A1
GOST 1050/15 44A1
GOST 1050/15 G 44A1
GOST 1050/15 KP 44A1
GOST 1050/20 44A1
GOST 1050/20 G 44A1
GOST 1050/20 KP 44A1
GOST 1050/25 44A2
GOST 1050/25 G 44A2
GOST 1050/30 44A2
GOST 1050/30 G 44A2
GOST 1050/35 44A2
GOST 1050/35 G 44A2
GOST 1050/40 44A2
GOST 1050/40 G 44A2
GOST 1050/45 44A2
GOST 1050/45 G 44A2
GOST 1050/50 44A3
GOST 1050/50 G 44A3
GOST 1050/55 44A3
GOST 1050/60 44A3
GOST 1050/60 G 44A3
GOST 1050/65 44A3
GOST 1050/65 G 44A3
GOST 1050/70 44A3
GOST 1050/70 G 44A3
GOST 1050/75 44A3
GOST 1050/80 44A3
GOST 1050/85 44A4
GOST 1060/10 44A1
GOST 1412 20B
GOST 2052/50 KHFA 44K2
GOST 2052/50 KHGFA 44K2
GOST 2052/50 S2 44A2
GOST 2052/55 S2 44A3
GOST 2052/55 SG 44A2
GOST 2052/60 S2 44A4
GOST 2052/60 S2A 44A3
GOST 2052/60 S2KHA 44E2
GOST 2052/60 S2KHFA 44K2
GOST 2052/60 S2N2A 44F2
GOST 2052/60 SG 44A3
GOST 2052/60 SGA 44A3
GOST 2052/62 S2A 44A3
GOST 2052/65 44A3
GOST 2052/65 G 44A3
GOST 2052/65 S2VA 44K2
GOST 2052/70 44A3
GOST 2052/70 S2KHA 44E2
GOST 2052/70 S3A 44A3
GOST 2052/75 44A3
GOST 2052/85 44A4
GOST 2446
Russian specification for welding wire; see appropriate designation
GOST 2685/63 Al 1 1L
GOST 2685/63 Al 3 1L
GOST 2685/63 Al 3B 1L
GOST 2685/63 Al 6 1L
GOST 2685/63 Al 7 1L
GOST 2685/63 Al 10B 1L
GOST 2685/63 Al 12 1L
GOST 2685/63 Al 14B 1L
GOST 2685/63 Al 15 1L
GOST 2685/63 Al 16B 1L
GOST 2685/63 Al 18B 1L
GOST 2685/63 Al 19 1L
GOST 2685/63 Al 20B 1L
GOST 2685/63 Al 21 1L
GOST 2685/63 Al7B 1L
GOST 2685-63 Al 30 1C
GOST 2685-63 Al2 1C
GOST 2685-63 Al4 1C
GOST 2685-63 Al4B 1C
GOST 2685-63 Al8 1C
GOST 2685-63 Al9 1C
GOST 2685-63 Al9B 1C
GOST 2685-63 AL13 1J
GOST 2685/63 A124 1M
GOST 2685-63 Al25 1C
GOST 2685-63 Al27 1C
GOST 3262 44A1
GOST 4543 44E3
GOST 4543 38KHGN 44G1
GOST 4543 38KHM1UA 44G1
GOST 4543 38KHVF1UA 44G1
GOST 4543 38KHVFU 44G1
GOST 4543/09 C2S 44A1
GOST 4543/09 G2 44A1
GOST 4543/10 G2S1 44A1
GOST 4543/12 GS 44A1
GOST 4543/12 KH2NGA 44K1
GOST 4543/12 KHN2 44K1
GOST 4543/12 KHN3 44K1
GOST 4543/13 N2KHA 44K1
GOST 4543/14 G 44A1
GOST 4543/14 G2 44A1
GOST 4543/14 KH Cr 44E1
GOST 4543/15 GF 44A1
GOST 4543/15 KH 44E1
GOST 4543/15 KH2GN2T 44K1
GOST 4543/15 KH2GN2TA 44K1
GOST 4543/15 KH2GN2TRA 44K1
GOST 4543/15 KHA 44E1
GOST 4543/15 KHF 44K1
GOST 4543/15 KHGNT 44K1
GOST 4543/15 KHGNTA 44K1
GOST 4543/15 KHM 44K1
GOST 4543/15 KHR 44E1
GOST 4543/15 KHRA 44E1
GOST 4543/15 NM 44K1
GOST 4543/16 ES 44A1
GOST 4543/16 KHSN 44K1
GOST 4543/17 GS 44A1
GOST 4543/18 G2 44A1
GOST 4543/18 G2S 44A1
GOST 4543/18 KH2N4VA 44K1
GOST 4543/18 KHG 44E1
GOST 4543/18 KHGN 44K1
GOST 4543/18 KHGT 44E1
GOST 4543/18 SNRA 44K1
GOST 4543/19 G 44A1
GOST 4543/20 KH 44E1
GOST 4543/20 KH2NGA 44K1
GOST 4543/20 KHF 44K1
GOST 4543/20 KHG2C 44E1
GOST 4543/20 KHGNR 44K1
GOST 4543/20 KHGR 44E1
GOST 4543/20 KHGSA 44E1
GOST 4543/20 KHN 44K1
GOST 4543/20 KHN3A 44K1
GOST 4543/20 KHN4FA 44K1
GOST 4543/20 KHNR 44K1
GOST 4543/20 NM 44K1
GOST 4543/25 G2S 44A2
GOST 4543/25 KH2GNTA 44K1
GOST 4543/25 KH2N4VA 44K1
GOST 4543/25 KHGSA 44E1
GOST 4543/27 SG 44A1
GOST 4543/30 KH 44E1
GOST 4543/30 KH2GN2 44K1
GOST 4543/30 KH2NVA 44K1
GOST 4543/30 KH2NVFA 44K1
GOST 4543/30 KHGNA 44K1
GOST 4543/30 KHGS 44E1
GOST 4543/30 KHGSA 44E2
GOST 4543/30 KHGSNA 44K1
GOST 4543/30 KHGT 44E1
GOST 4543/30 KHM 44K1
GOST 4543/30 KHMA 44K1
GOST 4543/30 KHN2VFA 44K1
GOST 4543/30 KHN3A 44K1
GOST 4543/30 KHNVA 44K1
GOST 4543/33 KHS 44E1
GOST 4543/35 G2 44A2
GOST 4543/35 GS 44A2
GOST 4543/35 KH 44E1
GOST 4543/35 KHG2 44E2
GOST 4543/35 KHGSA 44E2
GOST 4543/35 KHM 44K1
GOST 4543/35 KHRA 44E1
GOST 4543/35 SG 44A2
GOST 4543/36 G2S 44A2
GOST 4543/38 KHA 44E2
GOST 4543/38 KHIU 44E2
GOST 4543/38 KHN3MFA 44K2
GOST 4543/38 KHN3VA 44K2
GOST 4543/38 KHN3VFA 44K2
GOST 4543/38 KHNVA 44K1
GOST 4543/38 KHS 44E2
GOST 4543/38 KHVA 44K1
GOST 4543/40 G2 44A2
GOST 4543/40 KH 44E2
GOST 4543/40 KHFA 44K2

GOST 4543/40 KHG 44E2
GOST 4543/40 KHGR 44E2
GOST 4543/40 KHN 44K2
GOST 4543/40 KHNMA 44K2
GOST 4543/40 KHNVA 44K2
GOST 4543/40 KHR 44E2
GOST 4543/40 KHS 44E2
GOST 4543/45 G2 44A2
GOST 4543/45 KH 44E2
GOST 4543/45 KHC 44E2
GOST 4543/45 KHN 44K1
GOST 4543/50 G2 44A3
GOST 4543/50 KH 44E2
GOST 4543/50 KHN 44K2
GOST 5005/15 44A1
GOST 5005/20 44A1
GOST 5058/09 G2 44A1
GOST 5058/09 G2S 44A1
GOST 5058/10 G2S1 44A1
GOST 5058/10 KHSND 44K1
GOST 5058/12 GS 44A1
GOST 5058/14 G 44A1
GOST 5058/14 G2 44A1
GOST 5058/14 KHGS 44A1
GOST 5058/15 GF 44J
GOST 5058/15 KHSND 44K1
GOST 5058/16 GS 44A1
GOST 5058/17 GS 44A1
GOST 5058/18 G2 44A1
GOST 5058/19 G 44A1
GOST 5520/12 K 44A1
GOST 5520/15 K 44A1
GOST 5520/18 K 44A1
GOST 5520/69/16 K 44A1
GOST 5520/69/20 K 44A1
GOST 5632 0KH18N10T 44N1
GOST 5632 0KH23N18 44N1
GOST 5632 ICHITE9AN4 44P
GOST 5632 KH 17N71U 44N3
GOST 5632 KH5 44E1
GOST 5632 KH5M 44K1
GOST 5632 KH5VF 44K1
GOST 5632 KH6 S14 44E1
GOST 5632 KH6SIU 44K1
GOST 5632 KH6SM 44K1
GOST 5632 KH12H22T3MP 44N2
GOST 5632 KH12N20T3R 44N1
GOST 5632 KH12N20T3R 44N2
GOST 5632 KH12N22T3MR 44N1
GOST 5632 KH13N3 44M1
GOST 5632 KH14 44M1
GOST 5632 KH14G14N 44M1
GOST 5632 KH14G14N13T 44M1
GOST 5632 KH15N91U 44N1
GOST 5632 KH16N15M3B 44N1
GOST 5632 KH17 44M1
GOST 5632 KH17 G9AN4 44K1
GOST 5632 KH17AG14 44M1
GOST 5632 KH17G9AN4 44M1
GOST 5632 KH17G9AN4 44P
GOST 5632 KH17H2 44M1
GOST 5632 KH17N13M2T 44N1
GOST 5632 KH17N13M3T 44N1
GOST 5632 KH17N71U 44N1
GOST 5632 KH18N9 44N1
GOST 5632 KH18N9T 44N1
GOST 5632 KH18N10E 44N1
GOST 5632 KH18N10T 44N1
GOST 5632 KH18N12T 44N1
GOST 5632 KH18S1U 44M1
GOST 5632 KH20N14S2 44N1
GOST 5632 KH23N13 44N1
GOST 5632 KH23N18 44N1
GOST 5632 KH23NB 44N1
GOST 5632 KH25H25TR 44N1
GOST 5632 KH25N16G7AR 44N1
GOST 5632 KH25N16G7AR 44N2
GOST 5632 KH25N20S2 44N1
GOST 5632 KH25T 44M1
GOST 5632 KH28 44M1
GOST 5632 KH28AN 44M1
GOST 5632 KHCSM 44K1
GOST 5632/ 0KH14N28V
3T31UR 44N2
GOST 5632/ 3KH19N9MVBT 44N2
GOST 5632/00 KH18N10 44N1
GOST 5632/0 KH2 N5T 44M1
GOST 5632/0 KH10N20T2 44N1
GOST 5632/0 KH13 44M1
GOST 5632/0 KH14N28
V3T31UR 44N1
GOST 5632/0 KH17N16M3T 44N1
GOST 5632/0 KH17T 44M1
GOST 5632/0 KH18N10 44N1
GOST 5632/0 KH18N10E 44N1
GOST 5632/0 KH18N10T 44N1
GOST 5632/0 KH18N11 44N1
GOST 5632/0 KH18N12B 44N1
GOST 5632/0 KH18N12B 44N1
GOST 5632/0 KH18N12T 44N1
GOST 5632/0 KH20N14S2 44N1
GOST 5632/0 KH21N5T 44N1
GOST 5632/0 KH21N6M2T 44N1
GOST 5632/0 KH23N18 44N1
GOST 5632/0 KH23N28M2T 44N1
GOST 5632/0 KH23N28
M3D3T 44N1
GOST 5632/0 KH23N28
M3DeT 44N1
GOST 5632/1 KH 13N3 44N1
GOST 5632/1 KH8VF 44K1
GOST 5632/1 KH11 MF 44M1
GOST 5632/1 KH12N2VMF 44M1
GOST 5632/1 KH12S1U 44M1
GOST 5632/1 KH12V2MF 44M1
GOST 5632/1 KH12VNMF 44M1
GOST 5632/1 KH13 44M1
GOST 5632/1 KH13N3 44M1
GOST 5632/1 KH14N16B 44N1
GOST 5632/1 KH14N16BR 44N1
GOST 5632/1 KH14N18
V2B 44N1
GOST 5632/1 KH14N18V2BR 44N1
GOST 5632/1 KH14N18
V2BR1 44N1
GOST 5632/1 KH16N13M2B 44N1
GOST 5632/1 KH17N2 44M1
GOST 5632/1 KH21N5T 44M1
GOST 5632/1 KH21N5T 44N1
GOST 5632/1 KH25N25TR 44N1
GOST 5632/2 KH12VMBMR 44M1
GOST 5632/2 KH13 44M1
GOST 5632/2 KH13N4G9 44M1
GOST 5632/2 KH13N4G9 44P
GOST 5632/2 KH17N2 44M1
GOST 5632/2 KH18N9 44N1
GOST 5632/2 KH18N92 44N1
GOST 5632/3 KH13 44M1
GOST 5632/3 KH13N7S2 44M1
GOST 5632/3 KH13N7S2 44N1
GOST 5632/3 KH19N9MVBT 44N1
GOST 5632/4 KH
14N7G7FMS 44N3
GOST 5632/4 KH 15N17
G7F2MS 44N3
GOST 5632/4 KH9S2 44E2
GOST 5632/4 KH10S2M 44K2
GOST 5632/4 KH10S2M 44K2
GOST 5632/4
KH12N8G8MFB 44N1
GOST 5632/4 KH13 44M1
GOST 5632/4 KH14N14V2M 44N1
GOST 5632/4 KH18N2552 44N1
GOST 5632/4 KH18N2582 44N1
GOST 5632/4 KH952 44E2
GOST 5632/9 KH18 44M2
GOST 5632/15 KH12VMF 44M1
GOST 5632/610 KH18N10 44N1
GOST 5654/10 44A1
GOST 5654/20 44A1
GOST 7293/54 20B
GOST 7909/30 KHGS 44E2
GOST 7909/36 Cr 2S 44A2
GOST 7909/40 KH 44E2
GOST 8467 D 44A1
GOST 8467/30 KHGS 44E2
GOST 8467/36 G2S 44A2
GOST 8467/40 KH 44E2
GOST 8696 MST2KP 44A1
GOST 8696 MST3 44A1
GOST 8696 MST3KP 44A1
GOST 8696/10/G2SD 44A1
GOST 8731 BMST4 SP 44A1
GOST 8731 BMST5 SP 44A2
GOST 8731/10 44A1
GOST 8731/10/G2 44A1
GOST 8731/12 KHN2 44K1
GOST 8731/15 KHM 44K1
GOST 8731/20 44A1
GOST 8731/20 KH 44E1
GOST 8731/30 KHGSA 44E2
GOST 8731/30 KHMA 44K1
GOST 8731/35 44A2
GOST 8731/40 KH 44E2
GOST 8731/45 44A2
GOST 8733/10 44A1
GOST 8733/10 G2 44A1
GOST 8733/15 KH 44E1
GOST 8733/15 KH 44K1
GOST 8733/20 44A1
GOST 8733/20 KH 44E1
GOST 8733/30 KHGSA 44E2
GOST 8733/35 44A2
GOST 8733/40 KH 44E2
GOST 8733/45 44A2
GOST 9373 60 Russian
standard for tool steels;
see designation for details
GOST 9940 KH17NBM2T 44N1
GOST 9940 KH18N9 44N1
GOST 9940 KH18N10T 44N1

GOST 9940 KH18N12T 44N1
GOST 9940 KH28 44M1
GOST 9940 KH28 T 44M1
GOST 9940/00 KH18N10 44N1
GOST 9940/0 KH13 44M1
GOST 9940/0 KH17N16M3T 44N1
GOST 9940/0 KH17T 44M1
GOST 9940/0 KH18N10 44N1
GOST 9940/0 KH18N10T 44N1
GOST 9940/0 KH18N12B 44N1
GOST 9940/0 KH18N12T 44N1
GOST 9940/0 KH20N1452 44N1
GOST 9940/0 KH21N5T 44N1
GOST 9940/0 KH23N18 44N1
GOST 9940/1 KH13 44M1
GOST 9940/1 KH14N18V2BR 44N1
GOST 9940/1 KH17 44M1
GOST 9940/1 KH21N5T 44N1
GOST 9940/2 KH18N9 44N1
GOST 9941 KH17 44M1
GOST 9941 KH17N13M2T 44N1
GOST 9941 KH18N9 44N1
GOST 9941 KH18N12T 44N1
GOST 9941 KH21N5T 44N1
GOST 9941 KH25 T 44M1
GOST 9941/00 KH18N10 44N1
GOST 9941/0
see designation
KH17N16M3T 44N1
GOST 9941/0 KH13 44M1
GOST 9941/0 KH17N16M3T 44N1
GOST 9941/0 KH17T 44M1
GOST 9941/0 KH18N10 44N1
GOST 9941/0 KH18N10T 44N1
GOST 9941/0 KH18N12B 44N1
GOST 9941/0 KH18N12T 44N1
GOST 9941/0 KH20N1452 44N1
GOST 9941/0 KH23N18 44N1
GOST 9941/1 KH13 44M1
GOST 9941/1 KH14N18V2BR 44N1
GOST 9941/1 KH18N9 44N1
GOST 9941/1 KH21N5T 44N1
GOST 10498/00 KH16N15
M3B 44N1
GOST 10498/0 KH16N15M3 44N1
GOST 10498/0 KH18N10T 44N1
GOST 10498/1 KH13S2M2 44M1
GOST 10498/1
GOST 10705 BKST3 44A1
GOST 10705 BKST4 44A1
GOST 10705 BST3 44A1
GOST 10705 BST4 44A1
GOST 10705 MST2 44A1
GOST 10705 MST3 44A1
GOST 10705 MST4 44A1
GOST 10705/08 44A1
GOST 10705/10 44A1
GOST 10705/15 44A1
GOST 10705/20 44A1
GOST 10706/09 G2 44A1
GOST 10706/14 G 44A1
GOST 10706/14 G2 44A1
GOST 10706/15 G2S 44A1
GOST 10706/18 G2S 44A1
GOST 10706/19 G 44A1
GOST 10706/24 G 44A2
GOST 10706/25 G2S 44A2
GOST 10706/102 SR 44A1
GOST 10802 KH16N14
V2BR 44N1
GOST 10802
KH16N16V2MBR 44N1
GOST 10802 KH18N12T 44N1
GOST 10802/1 KH11B2MF 44M1
GOST 10802/1 KH14N18
V2BR 44N1
GOST 10802/12 K2 MFB 44K1
GOST 10802/12 KH1MF 44K1
GOST 10802/12 KH2 MFSR 44K1
GOST 10802/15 GS 44A1
GOST 10802/15 KH1ME 44K1
GOST 10802/15 KHM 44K1
GOST 10802/20 44A1
GOST 11017/20 44A1
GOST 110680 KH23N28M2T 44N1
GOST 11068 KH17N13M2T 44N1
GOST 11068 KH17N13M3T 44N1
GOST 11068 KH18N10T 44N1
GOST 11068 KH18N12T 44N1
GOST 11068/00 KH18N10T 44N1
GOST 11068/0 KH18N10T 44N1
GOST 11068/0 KH18N12T 44N1
GOST 11068/0 KH21N5T 44N1
GOST 11068/0 KH23N28
M3D3T 44N1
GOST 11068/1 KH21N5T 44N1
GOST 12132/08 44A1
GOST 12132/10 44A1
GOST 12132/15 KH 44E1
GOST 12132/20 44A1
GOST 12132/30 KHGSA 44E1
GOST 12132/30 KHMA 44K1
GOST 12132/35 44A2
GOST 12132/45 44A2
GOST 14162 KH18N9 44N1
GOST 14162 KH18N10T 44N1
GOST 14162 KH18N12T 44N1
GOST 14162/0 KH18N10T 44N1
GOST 14162/0 KH18N12T 44N1
GOST 14162/1 KH13 44M1
GOST 14162/2 KH13 44M1
GOST 14162/08 44A1
GOST 14162/10 44A1
GOST VST4 PS 44A1
GOST VST4 SP2 44A1
GOV 38 20B
GOV 40 20B
GOV 45 20B
GOV 50 20B
GOV 60 20B
GOV 70 20B
G Pb Bz 25 14K
GPS 1 44K1
GPS 2 44K1
GPS 3 44K1
GQ 44M1
GR 20A 1D
GR 20B 1D
GR 40A 1D
GR 40B 1D
GRADE A 27A
GRAIN 21.80 20B
GRAIN 21.81 20B
GRANE 2 44K2
GRAPH A1 44A4
GRAPH-AIR 44K3
GRAPH Mo 44G2
GRAPH-TUNG 44K3
GRAY DIAMOND 44G2
GREEK ASCOLOY 44M1
GREEN LABEL 44A4
GREY CUT COBALT 44L
GRINATAL 1C
GRINATAL 350 1C
GRP 38 20B
GRP 40 20B
GRP 50 20B
GRP 60 20B
GRP 70 20B
GRP 340 20B
GRP 370 20B
GRP 400 20B
GRP 500 20B
GRP 600 20B
GRP 700 20B
GRP 800 20B
GRS 10 20B
GRS 15 20B
GRS 20 20B
GRS 25 20B
GRS 30 20B
GRS 35 20B
GRS 40 20B
GS 6 44A4
GS 8 44K3
GS 10 Mn 7 44J
GS 10 Mn Mo7/4 44K1
GS 10A 1E
GS 11A 1E
GS 11B 1E
GS 11C 1E
GS 42A
see ASTM range 1J
GS 370.17 20B
GS 400.12 20B
GS 500.7 20B
GS 600.2 20B
GS 700.2 20B
GS 800 20B
GSN 44M2
G Sn Bz 10 14K
G Sn Bz 12 14K
G Sn Bz 14 14K
GSN + Mo 44M2
G Sn Pb Bz 5 14K
G Sn Pb Bz 10 14K
G Sn Pb Bz 15 14K
G Sn Pb Bz 20 14K
G So Ms F 30 14C
G So Ms F 45 14C
G So Ms F 60 14C
G So Ms F 75 14C
GT 45 44N1
GV 44K2
GX3 Cr Ni Mo Cu 246 44N2
GX 5 Cr Ni 13.4 44M1
G Z 4 TR 23B
G Z 5 Zr 23B
GZ 6 44L
GZ 42A
see ASTM range 1J
GZ Cu Al 10 Ni 14G

GZ Cu Sn 5 Zn Pb 14K
GZ Cu Sn 10 Zn 14K
GZ Cu Sn 12 14K
G Zn A1 4 Cu 1
see German standard range 55A
G Zn A1 6 Cu 1
see German standard range 55A
GZ Ni A1 Bz F 70
see DIN range 14G
GZ Rg 5 14K
GZ Rg 7 14K
GZ Rg 10 14K
GZ Sn Bz 12 14K
H 1 44A1
H 1 44N2
H 1A
see DIN range 44E1
H1 V 44A1
H 2 44E3
H 3 44E2
H 4 44K3
H 4 SPECIAL 44K3
H 4 Spl 44K3
H 5 44M2
H 5DM 44K1
H 7 44K3
H 9
see BS range 1E
H 9 1E
H 9 44M2
H 9 (alloy C) 44M2
H9 DOUBLE HEADER 44A4
H 10 44K2
H 10A 44L
H 11
see BS range 1F
H11
See SAE H11 44K1
H 11 1F
H 11 44K1
H 11–52 20B
H 11–52 GRS40 20B
H 11A
see DIN range 44E1
H 12
see BS range 1F
H12
See SAE H12 44K1
H 12 1F
H 12 14H
H 12 42.1
H 12 44K1
H13
See SAE H13 44K1
H 13 44K1
H 13 44K2
H 14
see BS range 1F
H 14 1F
H 14 44K1
H 14 44K2
H 15
see BS range 1F
H 15
see BS range 1H
H 15 1F
H 15 14C
H 15 44K2
H 15 44K3
H 16 1F
H 16 44K2
H 17
see BS range 1F
H 17 1F
H 17 44E3
H 18
see BS range 1F
H 18 1F
H 18 44E2
H 19
see BS range 1E
H 19 1E
H 19 44K1
H 19 44L
H 20
see BS range 1E
H 20 1E
H 20 44K1
H21
See SAE H21 44K1
H 21 44K1
H 21N 44K1
H 22 44K1
H 23 44E2
H 23 44M1
H 24 44K2
H 25 44K1
H 26 44K2
H 28 44E2
H 29 44M1
H 30
see BS range 1E
H 30 1E
H 30 44E2
H 31 44K2
H 32 44K1
H 33 44K3
H 35 44P
H 40 44K1
H 41 44K2
H 42 44K2
H 42 44M2
H 43 44K2
H 43 44L
H 44 44E3
H 46 44M1
H 49–1 1L
H 49-2 1C
H 49–3 1L
H 49 4 1J
H 49-5 1C
H 49-6 1C
H 49-7 1C
H 49 8 1J
H 49–9 1L
H 49–10 1L
H 49-11 1C
H 49–12 1L
H 49–13 1L
H 49-14 1C
H 49-15 1C
H 49–16 1L
H 49-17 1M
H 50 44K1
H 51 44K1
H 53 44M1
H 55 44K2
H 56 44K2
H 57 44K1
H 59 44M1
H 61 44L
H 62 44K3
H 63 44E3
H 64 44E3
H 65 44K1
H 75 14J
H 100 44M1
H 102 44M1
H 111
see DIN range 44E1
H 120 44M1
H 140 44M1
H 142 44M2
H160 44E1
H 300 44M1
H 304 44M1
H 500 44N1
H 508 44N1
H 520 44N1
H 522 44N1
H 525 44N1
H 532 44N1
H 550 44N1
H 600.1 44K2
H 600/2 44K1
H 700 44E2
H 750 44K2
H 800 44N1
H 850 44P
H10380
UNS for 1038 44A2
H10450
UNS for 1045 44A2
H13300
UNS for 1330 44A2
H13350
UNS for 1335 44A2
H13400
UNS for 1340 44A2
H13450
UNS for 1345 44A2
H15211
UNS for 15B21 44C
H15220
UNS for 1522 44A1
H15240
UNS for 1524 44A1
H15260
UNS for 1526 44A2
H15351
UNS for 15B35 44C
H15371
UNS for 15B37 44C
H15410
UNS for 1541 44A2
H15411
UNS for 15B41 44C
H15481
UNS for 15B48 44C

H15621
UNS for 15B62 44C
H40270
UNS for 4027 44G1
H40280
UNS for 4028 44G1
H40320
UNS for 4032 44G1
H40370
UNS for 4037 44G1
H40420
UNS for 4042 44G1
H40470
UNS for 4047 44G1
H41180
UNS for 4118 44K1
H41300
UNS for 4130 44K1
H41350
UNS for 4135 44K1
H41370
UNS for 4137 44K2
H41400
UNS for 4140 44K2
H41420
UNS for 4142 44K2
H41450
UNS for 4145 44K2
H41470
UNS for 4147 44K2
H41500
UNS for 4150 44K2
H41610
UNS for 4161 44K2
H43200
UNS for 4320 44K2
H43400
UNS for 4340 44K2
H43406
UNS for 4340 44K2
H44190
UNS for 4419 44G1
H46200
UNS for 4620 44K1
H46210
UNS for 4621 44K1
H46260
UNS for 4626 44K1
H47180
UNS for 4718 44K1
H47200
UNS for 4720 44K1
H48150
UNS for 4815 44K1
H48170
UNS for 4817 44K1
H48200
UNS for 4820 44K1
H50401
UNS for 50B40 44E2
H50441
UNS for 50B44 44E2
H50460
UNS for 5046 44E2
H50461
UNS for 50B46 44E2
H50501
UNS for 50B50 44E2
H50601
UNS for 50B60 44E2
H51200
UNS for 5120 44E1
H51300
UNS for 5130 44E1
H51320
UNS for 5132 44E1
H51350
UNS for 5135 44E1
H51400
UNS for 5140 44E2
H51450
UNS for 5145 44E2
H51470
UNS for 5147 44E2
H51500
UNS for 5150 44E2
H51550
UNS for 5155 44E2
H51600
UNS for 5160 44E2
H51601
UNS for 51B60 44E2
H61180
UNS for 6118 44K1
H61500
UNS for 6150 44K2
H81451
UNS for 81B45 44K2
H86170
UNS for 8617 44K1
H86200
UNS for 8620 44K1
H86220
UNS for 8622 44K1
H86250
UNS for 8625 44K1
H86270
UNS for 8627 44K2
H86300
UNS for 8630 44K2
H86301
UNS for 86B30 44K2
H86370
UNS for 8637 44K2
H86400
UNS for 8640 44K2
H86420
UNS for 8642 44K2
H86450
UNS for 8645 44K2
H86451
UNS for 86B45 44K2
H86500
UNS for 8650 44K2
H86550
UNS for 8655 44K2
H86600
UNS for 8660 44K2
H87200
UNS for 8720 44K1
H87400
UNS for 8740 44K1
H88220
UNS for 8822 44K1
H92600
UNS for AISI 9260 44B2
H93100
UNS for 9310 44K1
H94151
UNS for 94B15 44K1
H94171
UNS for 94B17 44K1
H94301
UNS for 94B30 44K1
HA 20B
HA 44K1
HA 1 44L
HA1 Fe Cr Al 15 44M1
HA1 Fe Cr Al 20 44M1
HA1 Fe Cr Al 25 44M1
HA 2 44L
HA 2.9900 1A
HA 2.9950 1A
HA 2.9970 1A
HA 2.9980 1A
HA 2.9990 1A
HA 2.9999 1A
HA 3 13A
HA 3 44L
HA 3 C4 1L
HA 3 CG50 1L
HA 3 CS42 1L
HA 3 CS72 1L
HA 3 G8 1J
HA 3 G10 1J
HA 3 GS40 1J
HA 3 S5 1C
HA 3 S12N 1C
HA 3 S12P 1C
HA 3 SC51N 1L
HA 3 SC51P 1L
HA 3 SC53 1L
HA 3 SC84N 1L
HA 3 SC84P 1L
HA 3 SC84R 1L
HA 3 SG70N 1C
HA 3 SG70P 1C
HA 3 SG71 1C
HA 3 SN122 1C
HA 3 ZG61N 1M
HA 3 ZG61P 1M
HA 4 13A
HA 4 44L
HA 4 CG42 1F
HA 4 CG42 ALCLAD 1H
HA 4 CS41N ALCLAD 1H
HA 4 GM 31N 1D
HA 4 GM41 1D
HA 4 GR20 1D
HA 4 GS11N 1E
HA 4 ZG62 1G
HA4 MC10 1B
HA4 ZG62 (ALCLAD) 1G
HA 4.990 1A
HA 5 CB60 1F
HA 5 CG42 1F
HA 5 CS41N 1F
HA 5 GM41 1D
HA 5 GR20 1D

HA 5 GS10 1E
HA 5 GS11N 1E
HA 5 GS11R 1E
HA 5 GS11T 1E
HA 5 ZG62 1G
HA 5.990C 1A
HA 6 13A
HA 6 CG30 1F
HA 6 CM41 1F
HA 6 GM50R 1D
HA 6 GS11N 1E
HA 6 GS11P 1D
HA 6.990 1A
HA 7 CG42 1F
HA 7 GR20 1D
HA 7 GS10 1E
HA 7 GS11N 1E
HA 7 GS11R 1E
HA 7 GS11T 1E
HA 7 ZG62 1G
HA7 MC10 1B
HA 7.995 1A
HA 8 CN42 1F
HA 8 CS41N 1F
HA 8 CS41P 1F
HA 8 GS11N 1E
HA 8 GS11P 1E
HA 8 SG121 1C
HA 8 ZG62 1F
HA 9 1L
HA 9 CG50 1L
HA 9 CS42 1L
HA 9 CS72 1L
HA 9 G10 1J
HA 9 GS40 1J
HA 9 S5 1C
HA 9 S12N 1C
HA 9 SC51N 1K
HA 9 SC51P 1K
HA 9 SC53 1L
HA 9 SG70N 1C
HA 9 SG71 1C
HA 9 ZG61N 1L
HA 9 ZG61P 1L
HA 10 CS42 1L
HA 10 S5 1C
HA 10 SC51N 1K
HA 10 SC51P 1K
HA 10 SC53 1L
HA 10 SG70N 1K
HA 10 SG70P 1K
HA 10 SG71 1K
HA 10 SN122 1K
HA 12 13A
HA 19 13A
HA 25 13A
HA 36 13A
HA 98 M2 13A
HA 150 13A
HA 151 13A
HA 152 13A
HA 302 13A
HA 7146 51C
HA 8116 51C
h Ag 1 42.1
h Ag 96 42.1
HAI 30 ALLOY 14E
HAI 36 ALLOY 20C
HAI 42 ALLOY 20C
HAI 49 ALLOY 20C
HAI 52 ALLOY 27E
HAI 60 ALLOY 14E
HAI 63 ALLOY 27F
HAI 64 ALLOY 27B
HAI 66 ALLOY 27B
HAI 90 ALLOY 14E
HAI 180 ALLOY 14E
HAI 373 ALLOY 20C
HAI 380 ALLOY 27E
HAI 400 ALLOY 27D
HAI 600 ALLOY 27B
HAI Cu Ni 102 27D
HAI Cu Ni 103 27D
HAI Cu Ni 104 27D
HAI-EN 14E
HAI EP 27B
HAI-II ALLOY 14E
HAI-JN 14E
HAI – JP 20A
HAI-KN 27F
HAI KP 27B
HAI Ni 200 27A
HAI Ni 201 27A
HAI Ni 205 27A
HAI Ni 211 27A
HAI Ni 233 27A
HAI Ni 233 C 27A
HAI Ni 499 C 27A
HAI Ni 634 27F
HAI Ni Cr 35 44N1
HAI Ni Cr 35 Cb 44N1
HAI Ni Cr 40 44N1
HAI Ni Cr 60 27B
HAI Ni Cr 70 27B
HAI Ni Cr 80 27B
HAI Ni Cr 80 Cb 44N1
HAI-NN 27F
HAI NP 27B
HAI P6 ALLOY 20C
HAI-TN 14E
H Al 1 1A
H Al 6A 1A
H Al 9 1A
HALLAMITE 44H
HALLAMITIER 44H
HALLAMITIEST 44H
HALLETTS 2.1
HAMPDEN 44M2
HAO 44E3
HARDEES 350 44K2
HARDEES 800 44K2
HARDFLEX 9 44A2
HARDFLEX 13M 44A3
HARDFLEX II 44A2
HARDINOX 3 44M1
HARDINOX 4 44M1
HARDSILVER 42.1
HARDTEM 44K2
HARDTRADE 85/58 44K1
HARDTRODE 83.28 44E1
HARDTRODE 84.42 44M1
HARDTRODE 84.52 44M1
HARDTRODE 85.65 44K3
HARDTRODE 86.08 44D
HARDWARE BRONZE 14B
HARDY NICKEL IRON 44F1
HASB 27F
HAS B 27B
HAS B 2L 27B
HASC 27F
HAS C 27C
HAS C 276 27B
HASC4 27F
HAS CH 27B
HAS CH 27F
HASD 27F
HASF 27F
HASG 27F
HAS G 27B
HAS G 27C
HAS G3 27B
HAS G-3 27C
HASN 27F
HAS N 27B
HASS 27F
HASTELLOY 5 27F
HASTELLOY A 27C
HASTELLOY B 27C
HASTELLOY B 27C
HASTELLOY B282 27C
HASTELLOY C 27C
HASTELLOY C4 27F
HASTELLOY C276 27C
HASTELLOY C276 27F
HASTELLOY D 27B
HASTELLOY F 27C
HASTELLOY G-2 27F
HASTELLOY G-3 27F
HASTELLOY N 27C
HASTELLOY R235 27C
HASTELLOY W 27C
HASTELLOY X 27C
HASW 27F
HAS W 27B
HAS X 27B
h Au 3 17.1
h Au 8 17.1
HAVAR 13A
HAVOC 44K2
HAWTHORNE 50.1
HAYES COMPACTED IRON 20B
HAYES HYDRAULIC IRON 20B
HAYES HYDRAULIC NODULAR 20B
HAYNES 3 13A
HAYNES 4 13A
HAYNES 6 13A
HAYNES 6B 13A
HAYNES 6K 13A
HAYNES 12 13A
HAYNES 19 13A
HAYNES 20 MOD 44N1
HAYNES 25 13A
HAYNES 36 13A
HAYNES 56 44L
HAYNES 90 44M2
HAYNES 93 44L
HAYNES 98 M2 13A
HAYNES 150 13A

HAYNES 151 13A
HAYNES 152 13A
HAYNES 302 13A
HAYNES STAR J 13A
HB 20B
HB 6 7.1
hBi1 6.1
H BRONZE 14H
HC 14A
HC 20B
HC 44M1
HC 14
see BS range 1H
HC 30 44M1
HCA 44M1
HCC 1C
see BS range 14A
HCC 2C
see BS range 14A
h Cd 1 8.1
h Ce 16 11.1
h Ce 19 11.1
h Ce 20a 11.1
h Ce 96 11.1
HCM 3 44K2
HCM 5 44K1
HCM 7 44K1
HCOKOF 14A
HCR 20C
h Cr 7a 12.1
h Cr 7b 12.1
h Cr 8 12.1
h Cr 12 12.1
HCRS 44K1
HCRS 2 44K2
HCRS 3 44K1
HCRS 4 44K1
HCRS 5 44K2
h Cu 6 a 14A
HD 20B
HD 44M1
HD 2 44K2
HD 3 44K1
HD 3 44K2
HD 3M 44K1
HD 3MX 44L
HD 10 44K1
HD 12 44K1
HD 12 44K2
HD 50 44M1
HDA 8151 13A
HDB 1 44A3
HDB 2 44F2
HDB 3 44K2
HDB 5 44K2
HDS 44K1
h Dy 16 35
h Dy 20a 35
h Dy 74 35
HDZ 44K1
HE 20B
HE 44N1
HE-35 44N1
HE 1049 13A
HEADWELL 44N1
HEATING ELEMENTS 27B
HEAVY METAL 52.1
HECLA 10 44F2
HECLA 15 44K3
HECLA 18 44A3
HECLA 26 44A1
HECLA 28 44J
HECLA 34 44A4
HECLA 35 44A1
HECLA 36 44A3
HECLA 37 44A2
HECLA 37M 44A2
HECLA 40 44A2
HECLA 41 44A3
HECLA 42 44A3
HECLA 46 44A1
HECLA 66 44K1
HECLA 67 44K1
HECLA 67B 44K1
HECLA 70 44K2
HECLA 73 44K1
HECLA 76 44K2
HECLA 78 44K1
HECLA 98 44K1
HECLA 100 44K2
HECLA 104 44E2
HECLA 105 44E2
HECLA 108 44E3
HECLA 110 44K2
HECLA 114 44K3
HECLA 115 44F2
HECLA 116 44K1
HECLA 120 44E2
HECLA 125 44M2
HECLA 135 44K2
HECLA 138 44K1
HECLA 139 44K2
HECLA 142 44K1
HECLA 143 44K1
HECLA 143B 44K1
HECLA 146 44K1
HECLA 146B 44K1
HECLA 148 44E2
HECLA 149C 44K1
HECLA 151 44K1
HECLA 151B 44K1
HECLA 153 44K1
HECLA 157 44K2
HECLA 159 44M2
HECLA 160C 44K2
HECLA 163 44K1
HECLA 166 44E2
HECLA 174 44K1
HECLA 175 44K3
HECLA 177 44K1
HECLA 181 44K1
HECLA 182 44K1
HECLA 183 44K1
HECLA 184 44K1
HECLA 185 44K1
HECLA 191 44K1
HECLA 192 44K1
HECLA 193 44K1
HECLA 196 44K1
HECLA 197 44K1
HECLA 198 44E2
HECLA 207 44E1
HECLA 260 44E1
HECLA 317 44K1
HECLA CH 31 44A1
HECLA D17 44A3
HECLA HGT4 44M1
HECLA S55 44A3
HECLA SNS 44M1
HECNUM 14E
HECORROS 76 14B
HECORROS 76 14C
HEDERVAN 44K3
HEDEX 1 44A2
HEDEX 1A 44A2
HEDEX 1B 44A2
HEDEX 2 44A2
HEDEX 3 44A2
HEDEX 4 44E2
HEDEX 5 44M1
HEDEX 6 44G2
HEDEX 7 44K1
HEDEX 8 44K2
HEDEX 9 44K2
HEDEX 10 44K2
HEDEX 11 44A3
HEDEX 12 44A2
HEDEX 13 44A2
HEDEX 14 44N1
HEDEX 15 44K2
HEDEX 16 44K1
HEDEX 17 44K1
HEDEX 18 44E3
HEDEX 19 44K1
HEDEX 20 44K1
HEDEX 21 44K1
HEDEX 22 44G2
HEDEX 23 44K1
HEDEX 24 44F2
HEDEX 25 44E1
HEDEX 26 44A3
HEDEX 27 44K1
HEDEX 28 44K1
HEDEX 29 44E2
HEDEX 30 44K1
HEDEX 31 44K1
HEDEX 32 44K1
HEDEX 33 44K2
HEDEX 34 44K1
HEDEX 35 44N1
HEDEX 36 44N1
HEDEX 37 44N1
HEDEX 38 44K1
HEDEX 39 44K1
HEDEX 40 44K1
HEDEX 41 44E3
HEDEX 43 44K1
HEDEX 44 44N1
HEDEX 45 44N1
HEDEX 46 44K1
HEDEX 47 44K2
HEDEX O 44A1
h Er 16 35
h Er 20a 35
HERBERTON 50.1
HERMEES 44A1
h Eu 16 35
h Eu 20b 35
HEUSLER ALLOY 14G
HEV 1 44L
HEV 2 27C

HEV 3 27C
HEV 4 13A
HEV 5 27C
HEV 6 27C
HEV7 44N1
HEV 8 27C
HEVIMET 52.1
HF 44N1
HF-30 44N1
HFC 44M2
h Fe 11a 20A
HG 2-9990 23A
HG 2-9995 23A
HG 2-9997 23A
HG 2-9999 23A
h Ga 1 15.1
h Gd 16 35
h Gd 20a 35
h Gd 74 35
h Ge 6 16.1
h Ge 8 16.1
HG NICKEL 27A
HH 44N1
HH-30 44N1
HH-33 44N1
HI 44N1
HI-35 44N1
HI 440 44K1
HIBIL 36 44A1
HIBIL TYPE A 44A1
HIBIL TYPE B 44A1
HIBIL TYPE C 44A1
HICON 44A1
HICON 36 44A1
HICON 40 44A1
HIDUMINIUM 00 1C
HIDUMINIUM 01 1F
HIDUMINIUM 1A 1A
HIDUMINIUM 1B 1A
HIDUMINIUM 1C 1A
HIDUMINIUM 02 1F
HIDUMINIUM 03 1F
HIDUMINIUM 04 1H
HIDUMINIUM 05 1D
HIDUMINIUM 07 1D
HIDUMINIUM 08 1C
HIDUMINIUM 10 1C
HIDUMINIUM 11 1B
HIDUMINIUM 11 1B
HIDUMINIUM 12 1D
HIDUMINIUM 14 1D
HIDUMINIUM 16 1D
HIDUMINIUM 17 1D
HIDUMINIUM 18 1E
HIDUMINIUM 20 1L
HIDUMINIUM 22 1D
HIDUMINIUM 24 1D
HIDUMINIUM 29 1N
HIDUMINIUM 33 1D
HIDUMINIUM 35 1D
HIDUMINIUM 40 1K
HIDUMINIUM 42 1E
HIDUMINIUM 43 1E
HIDUMINIUM 44 1E
HIDUMINIUM 46 1E
HIDUMINIUM 48 1G
HIDUMINIUM 55 1F
HIDUMINIUM 66 1F
HIDUMINIUM 72 1F
HIDUMINIUM 78 1G
HIDUMINIUM 80 1L
HIDUMINIUM 89 1G
HIDUMINIUM 90 1J
HIDUMINIUM 100 1A
HIDUMINIUM 512 1C
HIDURAL 5 14E
HIDURAL 6 14L
HIDURAL 7 14G
HIDURAL 640 14L
HIDURAX 1 14G
HIDURAX 2 14G
HIDURAX 3 14G
HIDURAX 4 14G
HIDURAX 5 14G
HIDURAX 6 14G
HIDURAX 7 14G
HIDURAX SPECIAL 13A 14E
HIDURAX SPECIAL 19A 14E
HIDUREL 6 14M
HIDUREL 640 14M
HIDURIT 10 14C
HIDURIT 11 14C
HIDURIT 12 14C
HIDURIT 15 14C
HIDURON 102 14E
HIDURON 107 14E
HIDURON 130 14G
HIDURON 191 14G
HIDURON 501 14E
HI-F 44A1
HIGH PERMEABILITY 49 20C
HIGH TENSILE BRASS 14C
HI-KILLED 44A1
HI-MAN 44A2
h In 1 18.1
h In 96 18.1
Hi Ni 1 44N1
HIPERCO 27 44L
HIPERCO 50 13A
HIPERCO 50 44L
HIPERM (SUPER) 20A
HIPERNIK 20C
HIPERNOM 27E
HIPERNOM ALLOY 27E
HI-PROOF 304 44N1
HI-PROOF 304L 44N1
HI-PROOF 316 44N1
HI-PROOF 316L 44N1
HI-PROOF 347 44N1
h Ir 8 19.1
HI SHOCK 60 44K2
HI-STEEL 44K1
HI STRENGTH A 44K1
HI STRENGTH B 44J
HI STRENGTH C 44A2
HI-STRESS 44A2
HITENSPEED 45 44A2
HITENSPEED 45A 44A2
HI TUFMOLD 44K1
HIV 44E2
HI WEAR 64 44K3
HI YIELD 42 44J
HI-YIELD 45 44A1
HI-YIELD 50 44A1
HI-YIELD 55 44A1
HI-YIELD 60 44A1
HI Z80 44K1
HK 44N1
HK 5 44K1
HK 31 23B
HK-30 44N1
HK-40 44N1
HL 44N1
HL 30 44N1
HL 40 44N1
HI-Mo 44K3
h Lu 16 35
H M HIGH SPEED 44K3
HM 1 44A4
H Mg 11a 23A
H Mg 99.8
see DIN range 23A
H Mg 99.95
see DIN range 23A
h Mn 11 24.1
H MONEL 27D
HMS 35 44A1
HMS 35/Z 44A1
HMS 40 44A1
HMS 41/Z 44A1
HMS 45 44A1
HMS 45/Z 44A1
HMS 55 44A2
HMS 65 44A2
HMS 75 44A3
HMS 80 44A3
HN 44N1
HN-40 44N1
h Nd 16 35
h Nd 20 35
h Ni 8 27A
h Ni 15 27A
HNM 44N1
HNV 1 44K2
HNV 2 44E2
HNV 3 44E2
HNV 4 44K2
HNV 5 44N1
HNV 6 44M1
HNV 7
see SAE range 44H
HNV8 44M1
HOLFOS AB1 14G
HOLFOS AB2 14G
HOLFOS BO 14K
HOLFOS BOHT 14K
HOLFOS G1 14K
HOLFOS G2 14K
HOLFOS G3 14K
HOLFOS G3WP 14K
HOLFOS HTB1 14G
HOLFOS HTB2 14G
HOLFOS HTB3 14G
HOLFOS JH17 14K
HOLFOS JHR42 14M
HOLFOS LB1 14K
HOLFOS LB2 14K
HOLFOS LB3 14K
HOLFOS LB4 14K
HOLFOS LB5 14K
HOLFOS LG1 14K

HOLFOS LG2 14K
HOLFOS LG3 14K
HOLFOS LG4 14K
HOLFOS LG773 14K
HOLFOS LPB1 14K
HOLFOS PB1 14K
HOLFOS PB2 14K
HOLFOS PB3 14K
HOLFOS SPUNCAST 14K
HOLFOS WW 14K
HONIAL 2 13A
HONIAL 5 13A
HONIAL 12 13A
HORSEMAN SUPER 44K3
h Os 18 29.1
HOT DIE No 5 44K1
HOTFORM No 1 44K1
HOTFORM No 2 44K1
HOTSPUR A 44N1
HOTSPUR C 44N1
HOTSPUR D 44N1
HOTSPUR F 44N1
HOV 44K1
HOWARD A 44K1
HOWARD B 44K1
HOWARD C 44K1
HOYT 3 M 21.1
HOYT 4 A 21.1
HOYT 11D 50A
HOYT 11R 50A
HOYT 11Z3 50A
HOYT 30 21.1
HOYT 32 21.1
HOYT 35 21.1
HOYT 38 50A
HOYT 40 21.1
HOYT 71 50A
HOYT 133C 50A
HOYT 142 21.1
HOYT 155 21.1
HOYT 156B 50A
HOYT 175 50A
HOYT BLOWPIPE
 SOLDER 50A
HOYT FIFTY 50A
HOYTICE 50A
HOYT MARINE A 50A
HOYT SIXTY 50A
HOYT STAR 21.1
HP 44K1
HP 44N1
HP 9 4.20
 see ASTM A646 44L
HP 9 4-25 44L
HP 9-4-30 44L
h Pb 1 21.1
h Pb 2 21.1
h Pb 4b 21.1
h Pb 73a 21.1
h Pd 15 30.1
h Pd 73 30.1
HPM 27A
h Pr 16 35
h Pr 20 35
h Pt 10 31.1
h Pt 71 31.1
HPW NICKEL 27F
HR 20B
HR 52.1
HR 1 20B
HR 1 27C
HR 2 27C
HR 3 27C
HR 4 27C
HR 5 27B
HR 6 27C
HR 10 27C
HR 33 44K1
HR 40 13A
HR 51 44N2
HR 52 44N2
HR 53 27C
HR 55 27C
HR 119 44K2
HR201
 see HR1 20B
HR202
 see HR2 27C
HR203
 see HR5 27B
HR206
 see HR10 27C
HR207
 see HR55 27C
HR240
 see HR40 13A
HR251
 see HR52 44N2
HR401
 see HR1 20B
HR402
 see HR2 27C
HR403
 see HR5 27B
HR404
 see HR10 27C
HR501
 see HR2 27C
HR502
 see HR2 27C
HR503
 see HR2 27C
HR504
 see HR5 27B
HR601
 see HR1 20B
HR650
 see HR51 44N2
HR CROWN 1 44N1
HR CROWN MAX 44N1
HRM 2M1 44M1
HRM 2M2 44M1
HRO 1243 44K2
HRS 44E3
H RV 44K3
HRW 44K3
HS 20B
HS 21 13A
HS 22 44K3
HS 23 13A
HS 25 13A
HS 27 13A
HS 28W 27B
HS 30 13A
HS 31 13A
HS 36 13A
HS 88 44N1
HS 95 44L
HS 188 13A
HS 556 44N2
HSA 1 44A1
HSB 44A1
h Sb 1 2.1
HSB 40 44A1
HSB 40S 44A1
HSB 45 44A1
HSB 45S 44A1
HSB 50 44A1
HSB 50S 44A1
HSB 51 44K1
HSB 51S 44K1
HSB 52 44K1
HSB 52S 44K1
HSB 55C 44K1
HSB 77V 44K1
h Sb 96 2.1
h Sb 97 2.1
h Sc 16 35
h Sc 20 35
HSCR 44E1
h Se 2 40.1
h Sm 16 35
h Sm 19 35
HSM/S 14A
HSM/W9A 44K1
h Sn 1 50A
h Sn 73 50A
HSP 15/17 44L
HSP 41 44K3
HSV 20B
HT 44N1
HT 1 44K1
HT2 44E1
HT 3 44K1
HT 5 8.1
HT 5 44K1
HT 7 44K1
HT 8 44K1
HT 9 44M1
HT 30 44N1
HT 51 44K1
h Ta 16 46.1
h Ta 73 46.1
h Tb 16 35
h Tb 74 35
HTB 1
 see BS range 14C
HTB 2
 see BS range 14C
HTB 3
 see BS range 14C
HTC 44A2
HTCN 44F2
h Te 6 47.1
h Te 73 47.1
h Ti 24 51A
h Tm 16 35
h Tm 20 35
HTP 52W 44A1
HTP 57W 44A1
HTS 44E1

HTX 44P
HU 44N1
h U 22 53.1
HUBER 11.1
HURCO EXCELLENT 44K3
HUSKY 44K3
HV 52.1
h V 22 54.1
h V 74 54.1
HV 5 44L
HV Hi Mo 44K3
HW 27B
HW 1 44K1
HW 2N 44K1
HW 4 44K1
HW 5 44K1
HW 6 NV 44K1
HW 7 44K2
HW 10C 42.1
hW 11 52.1
hW 72 52.1
HWD 1 44K1
HWD 2 44K2
HWD 3 44K2
HWD (Mod) 44K2
HWT 7 44K1
HWT AQUALLOY 44M1
HWX 44M1
HX 27B
HX 27C
h Y 16 35
h Y 20 35
HY 80 ARMCO 44K1
HY 100 44K1
HY 140 44K1
HY 150 44K1
h Yb 16 35
h Yb 20 35
HYBLADE 44K2
HY BLUE CHIP 44K3
HYC 27C
HYCOMAX I 44L
HYCOMAX II 44L
HYCOMAX III 44L
HYCOMAX IV 44L
HYDIE 44K2
HYDRA 44K3
HYDRA COVAN 44L
HYDRA E 44K1
HYDRA HD 44K2
HYDRA HUSKY 44K3
HYDRA M 44K1
HYDRA MULTICO 44L
HYDRA VANTAGE 44K3
HYDRA VK 44K2
HYDRA XL 44N1
HYDRA Z 44K1
HYFLOW 420R 44M1
HYFORM 44A4
HYFORM 409 44M1
HYKRO 44K1
HYKRO 44K3
HYKROM A 44K1
HYKROM B 44K1
HYKRO V 44K2
HYLITE 1 51A
HYLITE 10 51A
HYLITE 15 51A
HYLITE 15 H 51A
HYLITE 20 51B
HYLITE 25 51B
HYLITE 30 51C
HYLITE 40 51C
HYLITE 45 51C
HYLITE 50 51C
HYLITE 51 51C
HYLITE 55 51B
HYLITE 60 51C
HYLITE 65 51C
HY Mu 80 27E
HY Mu 800 27E
HY Mu 800A 27E
HYNICO II 44L
HYPEAK 5 V 5 44L
HYPEAK 14 44K3
HYPEAK 101 44K3
HYPEAK 202 44K3
HYPEAK 303 44L
HYPLUS 44A1
HYPLUS 1 44E1
HYPLUS 23 44A1
HYPLUS 29 44A1
HYPLUS 111 44A1
HYPRESS 26-32 44A1
HYREM RADIOMETAL 20C
HY RESIST 22/5 DUPLEX 44N3
HYRESIST 94L 44N1
HYRESIST 317LM 44N1
HYRHO RADIOMETAL 20C
HYTEMCO 27E
HY-TEN 44K2
HY-TUF 44K1
HYTUF 44K2
HZ 32 23B
HZ 32 A 23B
HZ 32A
see ASTM range 23B
h Zn 1 55A
h Zn 73a 55A
h Zr 25 56.1
h Zr 74 56.1
I 336 13A
I 400 5.1
IAS 44K3
IAT 14C
IBD 44K1
IBIS 21.1
IC 901 27C
ICI 115 51B
ICI 125 51B
ICI 130 51B
ICI 230 51C
ICI 314 A 51A
ICI 314 C 51A
ICI 317 51A
ICI 318A 51C
ICL 164 44N1
ICL 164 BC 44N1
ICL 164 Nb 44N1
ICL 164 T 44N1
ICL 166 44N1
ICL 166 BC 44N1
ICL 167 CN 44N1
ICL 168 44N1
ICL 168 BC 44N1
ICL 472 44N1
ICL 472 BC 44N1
ICL 472 Nb 44N1
ICL 472 T 44N1
ICL 473 BC 44N1
ICL 473 Nb 44N1
ICN 44M1
ICS 44A4
ICS 44M2
ICSW 1 44A3
ICW 44M2
ICW PLUS D3 44M2
ID 2 44E3
IDEOR 44K2
IDI 44K3
IDROTUB 52 44E1
IDROTUB 56 44K1
IDROTUB 58T 44K1
IDROTUB 62T 44K1
IH 50 44A1
IH 65 44J
IHX 45 44J
IHX 50 44J
IHX 55 44J
IHX 60 44J
IHX 65 44J
IHX 70 44J
ILLIUM 98 27C
ILLIUM B 27C
ILLIUM D 13A
ILLIUM G 27C
ILLIUM H 13A
ILLIUM R 27C
ILLIUM X 13A
ILZRO 12 55A
ILZRO 14 55A
ILZRO 16 55A
IMI 100 14A
IMI 103 14A
IMI 121 14A
IMI 131 14A
IMI 134 14A
IMI 138 14A
IMI 146 14A
IMI 153 14A
IMI 161 14A
IMI 166 14A
IMI 171 14L
IMI 176 14M
IMI 181 14A
IMI 185 14D
IMI 210 14B
IMI 212 14B
IMI 215 14B
IMI 220 14B
IMI 230 14B
IMI 237 14B
IMI 239 14B
IMI 246 14C
IMI 260 51A
IMI 276 14B
IMI 303 14B
IMI 312 14C
IMI 330 14C
IMI 333 14C
IMI 334 14C

IMI 345 14B
IMI 360 14C
IMI 365 14C
IMI 366 14C
IMI 388 14C
IMI 432 14B
IMI 433 14B
IMI 442 14C
IMI 443 14B
IMI 452 14C
IMI 453 14C
IMI 456 14C
IMI 467 14C
IMI 469 14C
IMI 511 14F
IMI 512 14F
IMI 513 14F
IMI 514 14F
IMI 515 14F
IMI 525 14F
IMI 530 14F
IMI 551 14F
IMI 581 14G
IMI 651 14A
IMI 657 14K
IMI 659 14K
IMI 660 14K
IMI 661 14K
IMI 705 14M
IMI 756 14G
IMI 757 14G
IMI 764 14G
IMI 842 14E
IMI 849 14E
IMI 981 14A
IMI Ti 115
see Ti 115 51A
IMI Ti 120
see Ti 120 51A
IMI Ti 125
see Ti 125 51A
IMI Ti 130
see Ti 130 51A
IMI Ti 150
see Ti 150 51A
IMI Ti 155
see Ti 155 51A
IMI Ti 160
see Ti 160 51A
IMI Ti 205
see Ti 205 51A
IMI Ti 230
see Ti 230 51A
IMI Ti 260
see Ti 260 51A
IMI Ti 314 A
see Ti 314 A 51B
IMI Ti 314 C
see Ti 314 C 51B
IMI Ti 315
see Ti 315 51B
IMI Ti 317
see Ti 317 51B
IMI Ti 318 A
see Ti 318 A 51C
IMI Ti 550
see Ti 550 51C
IMI Ti 551
see Ti 551 51C
IMI Ti 679
see Ti 679 51C
IMI Ti 680
see Ti 680 51B
IMI Ti 684
see Ti 684 51C
IMI Ti 685
see Ti 685 51C
IMI Ti 700
see Ti 700 51C
IMI Zr 10
see ZIRCONIUM 10 56
IMI Zr 20
see ZIRCONIUM 20 56
IMI Zr 30
see ZIRCONIUM 30 56
IMI Zr 40
see ZIRCONIUM 40 56
IMMACULATE 44N1
IMMACULATE 2W 44N1
IMMACULATE 5 44N1
IMMACULATE 5T 44N1
IMMADIUM II 14B
IMMADIUM IV 14C
IMMADIUM V 14B
IMMADIUM VI 14B
IMMUNIT R10/13 Mo Ni 44M1
IMMUNIT R12/13 Mo
see Ti 115 51A
IMMUNIT R12/13 Mo A 44M1
IMMUNIT R15/13 Mo 44M1
IMMUNIT R20/13 Mo 44M1
IMMUNIT R2213 44M1
IMPACTO 44K1
IMPALCO 030 1A
IMPALCO 050 1A
IMPALCO 051 1A
IMPALCO 102 1A
IMPALCO 190 1B
IMPALCO 450 1C
IMPALCO 460 1C
IMPALCO 470 1C
IMPALCO 510 1D
IMPALCO 520 1D
IMPALCO 530 1D
IMPALCO 531 1D
IMPALCO 540 1D
IMPALCO 550 1D
IMPALCO 560 1D
IMPALCO 570 1D
IMPALCO 660 1F
IMPALCO 690 1F
IMPALCO 700 1A
IMPALCO 740 1G
IMPALCO 750 1G
IMPALCO 760 1G
IMPALCO 770 1G
IMPALCO 800 1F
IMPALCO 810 1F
IMPALCO 830 1F
IMPALCO 840 1F
IMPALCO 860 1F
IMPALCO 900 1D
IMPALCO 901 1D
IMPALCO 910 1E
IMPALCO 912 1E
IMPALCO 918 1E
IMPALCO 920 1E
IMPALCO 940 1E
IMPALCO 945 1E
IMPALCO 946 1E
IMPALCO 950 1E
IMPALCO 2024 1F
IMPALCO C66 1F
IMPALCO C66A 1H
IMPALCO C69 1F
IMPALCO C80 1F
IMPALCO M31 1D
IMPALCO M32 1D
IMPALCO M32X 1D
IMPALCO M34 1D
IMPALCO M35/1 1D
IMPALCO M35/2 1D
IMPALCO M36 1D
IMPALCO M38 1E
IMPALCO M39/1 1E
IMPALCO M39/2 1E
IMPALCO M40 1E
IMPALCO M41 1E
IMPALCO M42 1E
IMPALCO P3 1A
IMPALCO P5 1A
IMPALCO P5E 1A
IMPALCO P10 1A
IMPALCO PA15 1C
IMPALCO PA16 1C
IMPALCO PA17 1C
IMPALCO PA19 1B
IMPALCO SUPERSPEED
946 1E
IMPAX 44K2
IMPERIAL 44K3
IMPERIAL CT 44M1
IMPERIAL EQ 44M1
IMPERIAL MANGANESE 44D
IMPERIAL R1 44M1
IMPERIAL R10 44M1
IMPERIAL S80 44M1
IN 1 44A1
IN 4 44A1
IN 5 44A1
IN 6 44A1
IN 10 44A1
IN 12 44K1
IN 15 42.1
IN 100 27C
IN 102 27C
IN 162 27C
IN 504 27F
IN 568 20C
IN 713 27C
IN 722 27C
IN 732X 14F
IN 738 27C
IN 787 44K1
IN 787A 44K1
IN 787B 44F1
IN A 44A1
IN B 44A1
INCANITE 1 20B
INCANITE 2 20B
INCANITE 3 20B

INCANITE 4 20B
INCANITE 5 20B
INCANITE 6 20B
INCANITE HRA 20B
INCANITE HRB 20B
INCO 27C
INCO 020 44N1
INCO 032 44N1
INCO 276 27C
INCO 330 44N1
INCO 700 27C
INCO 739 27C
INCO 901 27C
INCOCAL 10 27F
INCO G3 27C
INCO HX 27C
INCOLOY 27C
INCOLOY 44N1
INCOLOY 65 27B
INCOLOY 135 27B
INCOLOY 800 44N1
INCOLOY 800H 44N1
INCOLOY 800HT 44N1
INCOLOY 801 44N1
INCOLOY 802 44N1
INCOLOY 804 27C
INCOLOY 805 20C
INCOLOY 807 27C
INCOLOY 810 44N1
INCOLOY 825 27C
INCOLOY 901 27C
INCOLOY 901Mod 27C
INCOLOY 903 20C
INCOLOY 904 27F
INCOLOY 907 20C
INCOLOY 909 20C
INCOLOY 925 27C
INCOLOY 926 44N1
INCOLOY D S 44N1
INCOLOY FM65 27F
INCOLOY HA 956 44M1
INCOLOY T 44N1
INCOMAG 1 27F
INCOMAG 1LC 27F
INCOMAG 1M 27F
INCOMAG 2M 27F
INCOMAG 3 27F
INCOMAG 4 27F
INCOMAG Z 27F
INCO MS 250 44F3
INCONEL 27B
INCONEL 42 27B
INCONEL 62 27B
INCONEL 69 27C
INCONEL 82 27C
INCONEL 92 27C
INCONEL 112 27B
INCONEL 132 27B
INCONEL 182 27B
INCONEL 600 27B
INCONEL 601 27B
INCONEL 602 27B
INCONEL 604 27B
INCONEL 610 27B
INCONEL 617 27C
INCONEL 625 27C
INCONEL 690 27B
INCONEL 700 27C
INCONEL 702 27C
INCONEL 705 27B
INCONEL 706 27B
INCONEL 718 27C
INCONEL 721 27C
INCONEL 722 27C
INCONEL 751 27C
INCONEL ALLOY 725 27C
INCONEL FILLER 82 27B
INCONEL FILLER 92 27B
INCONEL FM62 27B
INCONEL FM69 27C
INCONEL FM82 27B
INCONEL FM92 27B
INCONEL M 27C
INCONEL MA 754 27C
INCONEL MA754 27F
INCONEL W 27C
INCONEL X 27C
INCONEL X 27C
INCONEL X550 27C
INCONEL X750 27C
INCO WELD A 27C
INCRAMET 800 14G
INCRAMUTE 1 14N
INDALLOY 44L
INDIUM 18.1
INKOMO 44K1
INMALLOY 44A4
INMANITE 44L
INMARITE T4 44L
INMOLYTE M2 44K3
INOFO 1 44M1
INOFO 2 44M1
INOFO 4 44M1
INOFO 4S 44M1
INOFO 5 44M1
INOFO 10 44N1
INOFO 10B 44N1
INOFO 10Ti 44N1
INOFO 11 44N1
INOFO 12 44N1
INOFO 12B 44N1
INOFO 13S 44N1
INOR 8 27F
INOX 1 44M1
INOX 1F 44M1
INOX 2 44M1
INOX 3 44M1
INOX 16 44M1
INOX 17T 44M1
INOX 430F 44M1
INOX 431 44M1
INOXESCO 13 44M1
INOXYDA 14G
INTAL 7Q5 1L
INTERTYPE METAL 21.1
INTRA 44H
INVAR 20C
INVAR 36 20C
INVAR 42 20C
INVARD 44K3
INVENTOR 44M1
INVINCIBLE 18 44K3
INVINCIBLE 22 44K3
INX 42 44J
INX 45 44J
INX 50 44J
INX 55 44J
INX 60 44J
INX 65 44J
INX 70 44J
IOX KM 44M1
IR 44M2
IR 4 44N1
IR 34 44K1
IR 41 44N1
IR 74 44K1
IRCUSIL 10 42.1
IRCUSIL 15 42.1
IRIDIUM 19.1
IRIDOSIUM 19.1
IRO 743 44K1
IRON 20A
IRONEX 2 44A1
IRONEX 5 44A1
IRONEX 7 44A1
IRRU 31.1
IRRUBIGO 1 Mo 44M1
IRRUBIGO 2 Mo 44M1
IRWAR FM 20C
IRWAR STANDARD 20C
IS 2283NS10 14E
ISHIURATA 44K1
ISI 44A4
ITA 14C
IU 44M2
IVORESCO WM EXTRA 44K2
J 1 44K3
J 4V 44J
J 5 44L
J 7 44K1
J 12 44K1
J 13 44K3
J 21 44K1
J 23 44K1
J 24 44K3
J 26 44L
J 27 44K2
J 30 44K3
J 34 44K3
J 35 44K3
J 36 44L
J 37 44L
J 38 44K3
J 39 44K3
J 40 44K3
J 41 44K3
J 42 44L
J0200
UNS for casting 1020 44A1
J 1300 44N2
J 1500 27C
J 1570 13A
J 1650 13A
J01700
UNS for SAE J 435 44A1
J02001
UNS for casting 1020 44A1
J02002
UNS for ASTM A732.1A 44A1
J02500
UNS for ASTM A27/60.30 44A2

J02501
UNS for ASTM A27/70.40 44A2
J02502
UNS for ASTM A216 WCA 44A2
J02503
UNS for ASTM A216 WCC 44A2
J02504
UNS for ASTM A352 LCB 44A2
J02505
UNS for ASTM A660 WCC 44A1
J02506
replaced by J22500 44F1
J02507
UNS for SAE J435-0025 44A1
J02605
UNS for ASTM A297-HC 44M1
J02610
UNS for 304 casting 44N1
J03000
UNS for ASTM A27-60-30 44A2
J03001
UNS for ASTM A27-65-35 44A2
J03002
UNS for ASTM A216 WCB 44A2
J03003
UNS for ASTM A352 LCB 44A2
J03003
UNS for ASTM A139-B 44A2
J03004
replaced by J13002
J03005
UNS for casting 1030 44A2
J03006
UNS for casting 1030 44A2
J03007
UNS for casting 1030 44A2
J03008
UNS for casting 1030 44A2
J03009
UNS for casting 1030 44A2
J03010
UNS for casting SAE J435.0050 44A2
J03011
UNS for ASTM A732-2K-2Q 44A2
J03401
UNS for CH10 44N1
J03402
UNS for CH20 44N1
J03500
UNS for ASTM A27 N2 44A2
J03501
UNS for ASTM A27-70.36 44A2
J03502
UNS for ASTM A356-1 44A2
J03503
UNS for ASTM A486-70 44A2
J03504
UNS for Mil S 15083-70.36 44A2
J03513
UNS for HH30 44N1
J04000
UNS for Mil S 81591-1050 44A2
J04001
UNS for Mil S 22141-1050 44A2
J04002
UNS for ASTM A732-3K-3Q 44A2
J04302
UNS for 310 casting 44N1
J04500
UNS for ASTM A487 GN 44K2
J04500
UNS for Mil S 81591 1050 44A2
J04501
UNS for SAE J 435 0050 44A2
J05001
UNS for Mil S 22141 1050 44A2
J05002
UNS for Mil S 15083-80-40 44A2
J05003
UNS for ASTM A487 44K2
J05405
now N08004
J11442
UNS for AMS 5333 44M1
J11522
UNS for ASTM A426-CP15 44G1
J11547
UNS for ASTM A426CP2 44K1
J11562
UNS for ASTM A426 CP12 44K1
J11697
UNS for ASTM A356-8 44K1
J11872
UNS for ASTM A217-WC1 44A2
J11875
UNS for Mil S15464-3 44K1
J12047
UNS for 8620-casting 44K1
J12048
UNS for ASTM A732.13Q 44K1
J12070
UNS for Mil S 15464-1 44K1
J12072
UNS for ASTM A217-WC6 44K1
J12073
UNS for ASTM A356-6 44K1
J12080
UNS for ASTM A389-C23 44K1
J12082
UNS for ASTM A217-WC4 44K1
J12084
UNS for ASTM A487-7Q 44K1
J12092
UNS for ASTM A389-C24 44K1
J12093
UNS for 4620 casting 44K1
J12094
UNS for ASTM A732-11Q 44F1
J12520
UNS for ASTM A217-WC1 44A2
J12521
UNS for ASTM A426-CP1 44G1
J12522
UNS for ASTM A352-LC1 44G1
J12523
UNS for A356-2 44G1
J12524
UNS for ASTM A217-WC1 44A2
J12540
UNS for ASTM A356-5 44K1
J12545 44K1
J12582
UNS for ASTM A757-C1Q 44K1
J13002
UNS for ASTM A487-1N 44J
J13005
UNS for ASTM A487-2N 44G1
J13042
UNS for AMS 5334 20B
J13045
UNS for ASTM A732-7Q 44K1
J13046
UNS for AMS 5356 44N3
J13047
UNS for ASTM A487-4N 44K1
J13048
UNS for 4130 casting 44K1
J13049
UNS for 8630 casting 44K1
J13050
UNS for AMS 5335 44N3
J13051
UNS for ASTM A732-14Q 44K1
J13052
UNS for ASTM A732-5N 44G1
J13080
UNS for ASTM A487-3N 44A2
J13345
UNS for ASTM A487 9N 44K1
J13432
UNS for 4335 casting 44K1
J13442
UNS for 8735 casting 44K1
J13512
UNS for ASTM A732-6N 44G1
J13855
UNS for ASTM A487-6N 44K2
J14046
UNS for AMS 5338 20B
J14047
UNS for 4140 casting 44K1
J14048
UNS 8640 castings 44K2
J14049
UNS for ASTM A732-8Q 44K1
J15047
UNS for 6150 casting 44K2
J15048
UNS for ASTM A732-12Q 44K2
J15580
UNS for ASTM A487-14Q 44K1
J19965
UNS for 52100 casting 44E3
J19966
UNS for ASTM A732-15A 44E3
J21610
UNS for ASTM A356-9 44K1
J21880
UNS for Mil S.15464-2 44K1
J21890
UNS for ASTM A217-WC9 44K1
J22000
UNS for ASTM A217-WC5 44K1
J22055
UNS for ASTM A732-9Q 44K1

J22090
UNS for A356-10 44K1
J22091
UNS for ASTM A487-8N 44K1
J22092
UNS for ASTM A643-C 44K1
J22500
UNS for ASTM A352-LC2 44F1
J22501
UNS for ASTM A757-B2N 44F1
J23015
UNS for ASTM A487-10N 44K1
J23260
UNS for AMS 5328 20B
J24054
UNS for ASTM A732-10Q 44K1
J24055
UNS for 4340 casting 44K1
J24056
UNS for Nitralloy 3 44K2
J24060
UNS for AMS 5330 20B
J31500
UNS for ASTM A757-B3N 44N1
J31545
UNS for ASTM A426-CP21 44K1
J31550
UNS for ASTM A352-LC3 44F1
J32075
UNS for Mil S 15083-100-70 44A2
J41501
UNS for ASTM A757-B4N 44F1
J42015
UNS for HY80 44K1
J42045
UNS for ASTM A217-C5 44K1
J42065
UNS for ASTM A757-E2N 44K1
J42215
UNS for ASTM A352-LC2-1 44F2
J42220
UNS for ASTM A757-E1Q 44K1
J42240
UNS for HY100 44K1
J51545
UNS for ASTM A426-CP5b 44K1
J61594
UNS for ASTM A426-CP7 44K1
J82090
UNS for ASTM A217-C12 44K1
J91109
UNS for ASTM A128A 44D
J91119
UNS for ASTM A128-B1 44D
J91129
UNS for ASTM A128-B2 44D
J91139
UNS for ASTM A128-B3 44D
J91149
UNS for ASTM A128-B4 44D
J91150
UNS for AMS 5351 44M1
J91151
UNS for CA15M 44M1
J91152
UNS for AMS 5350 44M1
J91153
UNS for CA40 44M1
J91161
UNS for AMS 5349 44M1
J91201
UNS for AISI 420 casting 44M1
J91209
UNS for Mil S 17249 44P
J91249
UNS for ASTM A128-E1 44D
J91261
UNS for Mil S 16993-2 44M1
J91309
UNS for ASTM A128-C 44D
J91339
UNS for ASTM A128E-2 44D
J91459
UNS for ASTM A128D 44D
J91540
UNS for CA6NM 44M1
J91550
UNS for ASTM A757-E3N 44M1
J91601
UNS for AMS 5372 44M1
J91606
UNS for 440A
J91631
UNS for AMS 5354 44M1
J91639
UNS for AMS 5352 44M2
J91650
UNS for CA6N 44M1
J91651
UNS for AISI 431 44M1
J91803
UNS for CB30 44M1
J92001
UNS for AMS 5359 44M1
J92110
UNS for AMS 5346 44N3
J92130
see J92110
J92150
UNS for 17.4 casting 44N3
J92180
UNS for ASTM A747-CB7Cu1 44N3
J92200
UNS for AMS 5342 44M1
J92205
UNS for 2205 44M1
J92240
UNS for AMS 5340 44N3
J92500
UNS for ASTM A351-CF3 44N1
J92501
UNS for AISI 302 casting 44N1
J92502
UNS for CF3 44N1
J92511
UNS for AISI 303 casting 44N1
J92530
UNS for AISI 321 casting 44N1
J92590
UNS for AISI 304 casting 44N1
J92600
UNS for ASTM A351CF8 44N1
J92603
UNS for ASTM A297-HF 44N1
J92613
UNS for ASTM A608 HC30 44N1
J92615
UNS for CC50 44M1
J92620
UNS for AMS 5370 44N1
J92640
UNS for AISI 347 casting 44N1
J92641
UNS for ASTM 5363 44N1
J92650
UNS for AISI 347H 44N1
J92660
UNS for AISI 317H 44N1
J92692
UNS for AMS 5358 44N1
J92700
UNS for CF3 44N1
J92701
UNS for CF16F 44N1
J92710
UNS for CF8C 44N1
J92711
UNS for AMS 5341 44N1
J92720
UNS for AMS 5341 44N1
J92730
UNS for AMS 5341 44N1
J92740
UNS for Durcomet 101 44N1
J92800
UNS for CF3M 44N1
J92801
UNS for CF3M 44N1
J92803
UNS for HF30 44N1
J92810
UNS for 316 casting 44N1
J92811
UNS for AMS 5362 44N1
J92843
UNS for AMS 5369 44N1
J92900
UNS for CF8M 44N1
J92910
UNS for CF8M 44N1
J92920
UNS for 316H 44N1
J92951
UNS for AMS 5360 44N1
J92971
UNS for ASTM A351 CF10MC 44N1
J93000
UNS for CG8M 44N1
J93001
UNS for CG12 44N1
J93005
UNS for ASTM A297HD 44N1
J93010
UNS for AMS 5339 44F3
J93015
UNS for HD50
J93072
UNS for AMS 5361 44N1

J93150
UNS for AMS 5337 44F3
J93183
UNS for KCR-D183 44N3
J93303
UNS for ASTM A447 20B
J93370
UNS for ASTM 744-CD4MCU 44N1
J93380
UNS for ASTM A351 44N1
J93400
UNS for CH8 44N1
J93403
UNS for HE 44N1
J93404
UNS for Alloy CH8 44N1
J93413
UNS for HE35 44N1
J93423
UNS for CE30 44N1
J93503
UNS for HH 44N1
J93550
UNS for KCR-D283 44N3
J93633
UNS for HH33 44N1
J93790
UNS for ASTM 351-CF10MC 44N1
J93900
UNS for DURCOMET5 44N1
J94013
UNS for ASTM A608 H135 44N1
J94202
UNS for CK20 44N1
J94203
UNS for HK30 44N1
J94204
UNS for ASTM A608HK40 44N1
J94211
UNS for AMS 5365 44N1
J94213
UNS for HN 44N1
J94214
UNS for HN40 44N1
J94224
UNS for HK 44N1
J94603
now N08030
J94604
now N06604
J94605
now N08002
J94613
now N08613
J94614
now N08614
J94650
UNS for CN7MS 44N1
J94805
now N08050
J95150
now N08007
J95151
now N08151
J95404
now N08005
J95705
now N08705
JA 3 44K3
JADE 52.1
JAE 27D
JALLOY AR 280 44G1
JALLOY AR 320 44G1
JALLOY AR 360 44G1
JALLOY AR 400 44G1
JALLOY No. 1 44G1
JALLOY No. 3 44G1
JALLOY No. 3AR 44G2
JALLOY No. 7 44G2
JALLOY S 90 44G1
JALLOY S 100 44G1
JALLOY S 110 44G1
JALLOY S 340 44G1
JALTEN 1 44G1
JALTEN 2 44A1
JALTEN 3 44A1
JALTEN 3R 44A1
JALTEN 3S 44A1
JAVELIN 44K1
JAYSEE 44K2
JB 33 A 14M
JC 20 44K1
JC 20N 44K1
JCNS 44E3
JEM No. 1 44A4
JEM No. 2 44A4
JEM No. 3 44A4
JEM No. 4 44A4
JEM No.5 44A3
JEM No.6 44A3
JETALLOY 209 13A
JETALLOY 249 13A
JETALLOY 1570 13A
JETHETE 44M1
JETHETE 7210 44L
JETHETE M140 44M1
JETHETE M151 44M1
JETHETE M152 44M1
JETHETE M153 44M1
JETHETE M154 44M1
JETHETE M160 44M1
JETHETE M190 44M1
JEWELRY BRONZE 14B
JG 3 44K2
JH 2 44K1
JH 17 14K
JHR 42 14M
JIS 202
see JIS G4304 44P
JIS 3446 SU304 44N1
JIS 4051/28 C 44A2
JIS 4403 Japanese standard high speed tool steels; see designation for details
JIS 4404 Japanese standard for tool steels; see appropriate designation for details
JIS B2351 STPS1 44A1
JIS B2351 STPS2 44A1
JIS C8305 44A1
JIS G4SUS202 44P
JIS G405 S48C 44A3
JIS G3101 SS41 44A1
JIS G3101 SS50 44A1
JIS G3101 SS55 44A2
JIS G3103 SB35 44A1
JIS G3103 SB42 44A1
JIS G3103 SB46 44A1
JIS G3103 SB46M 44G1
JIS G3103 SB49M 44G1
JIS G3103 SB56M 44G1
JIS G3103 SG30 44A1
JIS G3103 SP49 44A1
JIS G3106 B 44A1
JIS G3106 SM41A 44A1
JIS G3106 SM41B 44A1
JIS G3106 SM41C 44A1
JIS G3106 SM50A 44A2
JIS G3106 SM50C 44A1
JIS G3106 SM50Y A and B 44A1
JIS G3106 SM53 B and C 44A1
JIS G3106 SM58 44A1
JIS G3114 SMA41 A, B and C 44A1
JIS G3114 SMA50A, B and C 44K1
JIS G3114 SMA58 44K1
JIS G3115 SPV24 44A1
JIS G3115 SPV32 44A1
JIS G3115 SPV36 44A1
JIS G3115 SPV46 44A1
JIS G3115 SPV50 44A1
JIS G3116 SG26 44A1
JIS G3116 SG30 44A1
JIS G3116 SG33 44A1
JIS G3116 SG37 44A1
JIS G3118 SGV42 44A1
JIS G3118 SGV46 44A1
JIS G3118 SGV49 44A1
JIS G3119 SBV1A 44G1
JIS G3119 SBV1B 44G1
JIS G3119 SBV3 44K1
JIS G3120 SQV1A 44G1
JIS G3120 SQV1B 44G1
JIS G3120 SQV2A 44K1
JIS G3120 SQV2B 44K1
JIS G3120 SQV3A 44K1
JIS G3120 SQV3B 44K1
JIS G3125 SPAH 44K1
JIS G3126 SLA24A 44A1
JIS G3126 SLA24B 44A1
JIS G3126 SLA33A 44A1
JIS G3126 SLA33B 44A1
JIS G3126 SLA37 44A1
JIS G3179 SBV2 44K1
JIS G3441 STK3A 44K1
JIS G3441 STK3B 44K1
JIS G3441 STK3C 44K1
JIS G3441 STK3D 44K1
JIS G3441 STK3E 44K1
JIS G3441 STKS1A 44K1
JIS G3441 STKS1B 44K1
JIS G3441 STKS1C 44K1
JIS G3441 STKS1D 44K1
JIS G3441 STKS1E 44K1
JIS G3441 STKS2A 44K1
JIS G3441 STKS2B 44K1
JIS G3441 STKS2C 44K1
JIS G3441 STKS2D 44K1
JIS G3441 STKS2E 44K1

JIS G3441 STKS4A 44K1
JIS G3441 STKS4B 44K1
JIS G3441 STKS4C 44K1
JIS G3441 STKS4D 44K1
JIS G3441 STKS4E 44K1
JIS G3444 STK30 44A1
JIS G3444 STK41 44A1
JIS G3444 STK50 44A1
JIS G3444 STK51 44A1
JIS G3444 STK55 44A1
JIS G3445 STKM 17 C 44A3
JIS G3445 STKM 17A 44A3
JIS G3445 STKM11A 44A1
JIS G3445 STKM12A 44A1
JIS G3445 STKM12B 44A1
JIS G3445 STKM12C 44A1
JIS G3445 STKM13A 44A1
JIS G3445 STKM13B 44A1
JIS G3445 STKM13C 44A1
JIS G3445 STKM14A 44A2
JIS G3445 STKM14B 44A2
JIS G3445 STKM14C 44A2
JIS G3445 STKM15A 44A2
JIS G3445 STKM15C 44A2
JIS G3445 STKM16A 44A2
JIS G3445 STKM16C 44A2
JIS G3445 STKM18A 44A1
JIS G3445 STKM18B 44A1
JIS G3445 STKM18C 44A1
JIS G3446 SUS316 44N1
JIS G3446 SUS321 44N1
JIS G3446 SUS347 44N1
JIS G3446 SUS410 44M1
JIS G3446 SUS430 44M1
JIS G3447 SUS304LTBS 44N1
JIS G3447 SUS304TBS 44N1
JIS G3447 SUS316LTBS 44N1
JIS G3447 SUS316TBS 44N1
JIS G3452 SGP 44A1
JIS G3454 STPG38 44A1
JIS G3454 STPG42 44A2
JIS G3455 STS35 44A1
JIS G3455 STS38 44A1
JIS G3455 STS42 44A2
JIS G3455 STS49 44A2
JIS G3456 STPT38 44A1
JIS G3456 STPT42 44A2
JIS G3456 STPT49 44A2
JIS G3457 STPY41 44A1
JIS G3458 STPA12 44G1
JIS G3458 STPA22 44K1
JIS G3458 STPA23 44K1
JIS G3458 STPA24 44K1
JIS G3458 STPA25 44K1
JIS G3458 STPA26 44K1
JIS G3459 SUS 329 JITP 44N3
JIS G3459 SUS304HTP 44N1
JIS G3459 SUS304LTP 44N1
JIS G3459 SUS304TP 44N1
JIS G3459 SUS309TP 44N1
JIS G3459 SUS310TP 44N1
JIS G3459 SUS316HTP 44N1
JIS G3459 SUS316LTP 44N1
JIS G3459 SUS316TP 44N1
JIS G3459 SUS321HTP 44N1
JIS G3459 SUS321TP 44N1
JIS G3459 SUS347HTP 44N1
JIS G3459 SUS347TP 44N1
JIS G3460 44F1
JIS G3460 STPL39 44A1
JIS G3461 STB30 44A1
JIS G3461 STB33 44A1
JIS G3461 STB35 44A1
JIS G3461 STB42 44A1
JIS G3462 STBA12 44G1
JIS G3462 STBA13 44G1
JIS G3462 STBA22 44K1
JIS G3462 STBA23 44K1
JIS G3462 STBA24 44K1
JIS G3462 STBA25 44K1
JIS G3462 STBA26 44K1
JIS G3463 SUS 316LTB 44N1
JIS G3463 SUS304HTB 44N1
JIS G3463 SUS304LTB 44N1
JIS G3463 SUS304TB 44N1
JIS G3463 SUS309STB 44N1
JIS G3463 SUS310STB 44N1
JIS G3463 SUS316HTB 44N1
JIS G3463 SUS316TB 44N1
JIS G3463 SUS321HTB 44N1
JIS G3463 SUS321TB 44N1
JIS G3463 SUS329J1TB 44N1
JIS G3463 SUS347HTB 44N1
JIS G3463 SUS347TB 44N1
JIS G3463 SUS410TB 44M1
JIS G3463 SUS430TB 44M1
JIS G3464 STBL39 44A1
JIS G3464 STBL46 44F1
JIS G3465 STMC55 44A1
JIS G3465 STMC65 44A1
JIS G3465 STMR60 44A1
JIS G3465 STMR70 44A1
JIS G3465 STMR80 44A1
JIS G3466 STKR41 44A1
JIS G3466 STKR50 44A1
JIS G4051 S9CK 44A1
JIS G4051 S10C 44A1
JIS G4051 S12C 44A1
JIS G4051 S15C 44A1
JIS G4051 S17C 44A1
JIS G4051 S22C 44A2
JIS G4051 S25C 44A2
JIS G4051 S30C 44A2
JIS G4051 S33C 44A2
JIS G4051 S35C 44A2
JIS G4051 S38C 44A2
JIS G4051 S40C 44A2
JIS G4051 S43C 44A2
JIS G4051 S45C 44A2
JIS G4051 S50C 44A3
JIS G4051 S53G 44A3
JIS G4051 S55C 44A3
JIS G4051 S58C 44A3
JIS G4106 S Cr 2 44E1
JIS G4106 S Cr 3 44E1
JIS G4106 S Cr 4 44E2
JIS G4106 S Cr 5 44E2
JIS G4106 S Cr 21 44E1
JIS G4106 S Cr 22 44E1
JIS G4106 S Mn 1 44A2
JIS G4106 S Mn 2 44A2
JIS G4106 S Mn 3 44A2
JIS G4106 S Mn C 3 44E2
JIS G4106 S Mn C 21 44E1
JIS G4106 SCM1 44K1
JIS G4106 SCM2 44K1
JIS G4106 SCM3 44K1
JIS G4106 SCM4 44K2
JIS G4106 SCM5 44K2
JIS G4106 SCM21 44K1
JIS G4106 SCM22 44K1
JIS G4106 SCM23 44K1
JIS G4106 SCM24 44K1
JIS G4106 SNC1 44K2
JIS G4106 SNC2 44K1
JIS G4106 SNC3 44K2
JIS G4106 SNC21 44K1
JIS G4106 SNC22 44K1
JIS G4106 SNCM1 44K1
JIS G4106 SNCM1 44K2
JIS G4106 SNCM2 44K1
JIS G4106 SNCM5 44K1
JIS G4106 SNCM6 44K2
JIS G4106 SNCM7 44K2
JIS G4106 SNCM8 44K2
JIS G4106 SNCM21 44K1
JIS G4106 SNCM22 44K1
JIS G4106 SNCM25 44K1
JIS G4106 SNCM26 44K1
JIS G4106 SUP10 44E2
JIS G4109 SCMV1 44K1
JIS G4109 SCMV2 44K1
JIS G4109 SCMV3 44K1
JIS G4109 SCMV4 44K1
JIS G4109 SCMV5 44K1
JIS G4109 SCMV6 44K1
JIS G4303 SUS 44N1
JIS G4303 SUS201 44P
JIS G4303 SUS202 44P
JIS G4304 SUS 329JI 44N3
JIS G4304 SUS302 44N1
JIS G4304 SUS304 44N1
JIS G4304 SUS304L 44N1
JIS G4304 SUS305 44N1
JIS G4304 SUS309S 44N1
JIS G4304 SUS310S 44N1
JIS G4304 SUS316 44N1
JIS G4304 SUS316JI 44N1
JIS G4304 SUS316JIL 44N1
JIS G4304 SUS316L 44N1
JIS G4304 SUS317 44N1
JIS G4304 SUS317L 44N1
JIS G4304 SUS321 44N1
JIS G4304 SUS347 44N1
JIS G4304 SUS403 44M1
JIS G4304 SUS405 44M1
JIS G4304 SUS410 44M1

JIS G4304 SUS429 44M1
JIS G4304 SUS430 44M1
JIS G4304 SUS434 44M1
JIS G4304 SUS631 44N3
JIS G4311 Japanese specification for heat resisting steels; see appropriate designation
JIS G4312 SUH309 44N1
JIS G4312 SUH310 44N1
JIS G4312 SUH330 44N1
JIS G4312 SUH446 44M1
JIS G4312 SUH661 44N2
JIS G4410 Japanese specification for heat resisting steels; see appropriate designation
JIS G4801 54P3 44A4
JIS G4801 84P4 44A4
JIS G4801 SUP7 44A3
JIS G4801 SUP9 44E2
JIS G4801 SUP10 44K2
JIS G4801 SUP11 44E2
JIS G4805 SUJ1 44E3
JIS G4805 SUJ2 44E3
JIS G4805 SUJ3 44E3
JIS G4805 SUJ4 44K3
JIS G4805 SUJ5 44K3
JIS G5501 20B
JIS G5502 20B
JJ 44M2
JL 44K2
JLX 42 44J
JLX 65 44J
JLX 70 44J
JLX W 45 44A1
JLX W 50 44A1
JLX W 55 44A1
JLX W 60 44A1
JMC 77 30.1
JMC 625 17.1
JMC 625 R 17.1
JMC 1715 Mg Ni 42.1
JMM 77 30.1
JMM 77 42.1
JMM 625 17.1
JMM 1715 42.1
JMMB BRONZE 14H
JO 44H
JO 44K1
JO Spl 44K1
JOUVENCEL 44A1
J R ALLOY 1 44M1
J R ALLOY 2 44M1
J R ALLOY 3 44M1
JS 44K2
JS 1 44K1
JS 1B 44K1
JS 2 44A1
JS 3 44A2
JS 4 44K1
JS 5 44K2
JS 6 44G1
JS 9 44K1
JS 10 44K1
JS 12 44A2
JS 13 44K2
JS 14 44D
JS 16 44A2
JS 26 44K2
JS 40 44K1
JS 81 44G1
J SG4051 S20C 44A1
J SG4106 S Mn 21 44A1
JUS CJ 2020 20B
JUS CJ 2022 20B
K 14C
K 1 23B
K 1 52.1
K 2S 52.1
K 3H 52.1
K 4 44K3
K 4 SPECIAL 44K3
K 4 Spl 44K3
K 4H 52.1
K 5H 52.1
K 6
    see BS range 20B
K 6 44K3
K 6 52.1
K 6 Spl 44K3
K 7H 52.1
K 8 52.1
K 8 HP 44K3
K 9 44K3
K 9 52.1
K 10 Spl 44K3
K 11
    see BS range 20B
K 11 52.1
K 15 44K3
K 16 44K3
K 17 44K3
K 18 44K1
K 19 44K2
K 20 44K3
K20L 44M1
K 21 44K3
K 21 52.1
K00040
    UNS for ASTM A620 44A1
K 42B 27C
K00045
    UNS for AWS designation RG45 44A1
K00060
    UNS for AWS designation RG60 44A1
K00065
    UNS for AWS designation RG65 44A1
K 81 52.1
K 82 52.1
K 84 52.1
K 86 52.1
K 90 52.1
K 90A 52.1
K 91 52.1
K 92 52.1
K 94 52.1
K00095
    UNS for AMS7706 20A
K 95 52.1
K 96 52.1
K00100
    UNS for ASTM A424-1 44A1
K 100 44M2
K 102 44M2
K 105 44M2
K 107 44M2
K 110 44M2
K 138 A 51B
K 151 A 51B
k00158
    UNS for AWS designation ENiK 44F1
K 162 B 51B
K 163 B1 51B
K 164 B 51B
K 165 51B
K 200 44E2
K 240 44E2
K 245 44E2
K 291 44K1
K 300 44K2
K 305 44K3
K 306 44K2
K 310 44K3
K 336 44K1
K 353 44K1
K 354 44K1
K00400
    UNS for ASTM A424-11A 44A1
K 405 44K3
K 450 44K2
K 451 44K3
K 455 44K2
K 460 44K3
K 465 44K3
K 500 27D
K 505 44K3
K 510 44K3
K00600
    UNS for ASTM A594-4 44A1
K 600 44K2
K 601 52.1
K 605 44K2
K00606
    UNS for AMS 5030 44A1
K 614 44K3
K 630 44K3
K 700 44M2
K 701 52.1
K 720 44K3
K00800
    UNS for ASTM A594-3 44A1
K00801
    UNS for ASTM A424-11B 44A1
K 801 52.1
K00802
    UNS for AMS 5061 44A1
K 825 44K1
K00912
    UNS for AWS designation EL12N 44A1
K 942 44K3
K945 44A2
K 945 44L
K950 44A3
K 950 44K3

K960 44A3
K 980 44A4
K 985 44A4
K 990 44A4
K01000
UNS for ASTM A594-2 44A1
K01001
UNS for ASTM A254 44A1
K01008
UNS for AWS designation EL8 44A1
K01009
UNS for AWS designation EL8K 44A1
K01010
UNS for AWS designation EH10K-EW 44A1
K01012
UNS for AWS designation EL12 44A1
K01013
UNS for AWS designation EM12KN 44A1
K01112
UNS for AWS designation EM12-EW 44A1
K01113
UNS for AWS designation EM12KEW 44A1
K01200
UNS for ASTM A178-A 44A1
K01201
UNS for ASTM A192 44A1
K01313
UNS for AWS designation EM13KEW 44A1
K01314
UNS for AWS designation EM14K 44A1
K01500
UNS for ASTM A594-1 44A1
K01502
UNS for ASTM A730-A 44A1
K01503
UNS for ASTM A194-1 44A1
K01504
UNS for ASTM A161-LCST 44A1
K01505
UNS for ASTM A67 44A1
K01506
UNS for ASTM A539 44A1
K01507
UNS for ASTM A109-4.5 44A1
K01515
UNS AWS designation EM15KEW 44A1
K01520
UNS AWS designation EM15KEW 44A1
K01591
UNS for ASTM A414-A 44A2
K01600
UNS for ASTM A678-A 44A1
K01601
UNS for ASTM A131-CS 44A2
K01602
UNS for Mil S 645 44A1
K01700
UNS for ASTM A285 44A1
K01701
UNS for ASTM A662 44A1
K01800
UNS for ASTM A516 44A2
K01801
UNS for ASTM A131-E
K01801
UNS for ASTM A633-A 44A1
K01803
UNS for ASTM A633 44A1
K01804
UNS for ASTM A284 44A2
K01807
UNS for ASTM A214 44A1
K01900
UNS for ASTM A502-1 44A1
K01907
UNS for C110 44J
K02000
UNS for ASTM A730B 44A1
K02001
UNS for ASTM A284 44A2
K02002
UNS for ASTM A678-B 44A1
K02003
replaced by K12037
K02004
UNS for ASTM A595-A 44A1
K02005
UNS for ASTM A595B 44A1
K 02006 44A1
K02007
UNS for ASTM A662 44A1
K02100
UNS for ASTM A516 44A2
K02101
UNS for ASTM A131-C 44A2
K02102
UNS for ASTM A131-B 44A2
K02104
UNS for ASTM A524 44A1
K02200
UNS for ASTM A285 44A1
K02201
UNS for ASTM A414-B 44A2
K02202
UNS for ASTM A442 44A1
K02202
UNS for ASTM A442 44A1
K02203
UNS for ASTM A662 44A1
K02204
UNS for ASTM A678.C 44A1
K02300
UNS for ASTM A131-A 44A2
K02301
UNS for ASTM A573 44A1
K02302
UNS for ASTM A707-L1 44A1
K02303
UNS for ASTM A572-50-1 44A1
K02304
UNS for ASTM A572-50-2 44A1
K02305
UNS for ASTM A572-50-3 44A1
K02306
UNS for ASTM A472-50-4 44A1
K02400
UNS for ASTM A537 44A1
K02401
UNS for ASTM A284-C 44A2
K02402
UNS for ASTM A442 44A1
K02403
UNS for ASTM A516 44A2
K02404
UNS for ASTM A573 44A1
K02405
UNS for ASTM A502-2 44A1
K02500
UNS for ASTM A109-1 44A2
K02501
UNS for ASTM A106-A 44A2
K02502
UNS for ASTM A570-30 44A2
K02503
UNS for ASTM A414-C 44A2
K02504
UNS for ASTM A53-EA 44A1
K02505
UNS for ASTM A414-D 44A2
K02506
UNS for ASTM A727 44A1
K02507
UNS for ASTM A570-45 44A2
K02508
UNS for AMS 5062 44A2
K02600
UNS for ASTM A36 44A2
K02601
UNS for ASTM A618-1 44A2
K02603 44A2
K02700
UNS for ASTM A516 44A2
K02701
UNS for ASTM A573 44A2
K02702
UNS for ASTM A284-D 44A2
K02703
UNS for ASTM A529 44A1
K02704
UNS for ASTM A414-E 44A2
K02705
UNS for ASTM A500-C 44A2
K02706
UNS for ASTM A325-1 44A2
K02707
UNS for ASTM A210-A1 44A2
K02708
UNS for Mil S 22698C 44A2
K02741
UNS for ASTM A210-A1 44A2
K02800
UNS for ASTM A515 44A2
K02801
UNS for ASTM A285 44A1
K 02802 44A1
K02803
UNS for ASTM A299 44A2

K02900
UNS for ASTM A612 44A2
K02901
UNS for Mil S 3289 44A2
K03000
UNS for ASTM A500-A 44A2
K03002
UNS for ASTM A372-1 44A2
K03004
UNS for ASTM A139-C 44A2
K03005
UNS for ASTM A53-EB 44A1
K03006
UNS for ASTM A106-B 44A2
K03007
UNS for ASTM A557-B2 44A2
K03008
UNS for ASTM A333-1 44A2
K03009
UNS for ASTM A350-LF1 44A2
K03010
UNS for ASTM A139B 44A2
K03011
UNS for ASTM A350-LF2 44A2
K03012
UNS for ASTM A139B 44A2
K03013
UNS for ASTM A381 44A2
K03014
UNS for ASTM A694 44A2
K03015
UNS for ASTM A183 44A2
K 03016 44A1
K03017
UNS for ASTM A266-3 44A2
K03046
UNS for ASTM A765-1 44K1
K03047
UNS for ASTM A765-11 44K1
K03100
UNS for ASTM A31-B 44A2
K03101
UNS for ASTM A515 44A2
K03102
UNS for ASTM A414-F 44A2
K03103
UNS for ASTM A414-G 44A2
K03104
UNS for ASTM A574 44K1
K03200
UNS for ASTM A696-B&C 44A2
K03201
UNS for Mil S-43 44A2
K03300
UNS for ASTM A455-1 44A2
K03301
UNS for ASTM A707-L2 44A1
K03500 44A2
K03501
UNS for ASTM A106-C 44A2
K03502
UNS for ASTM A181 44A2
K03503
UNS for ASTM A178-C 44A2
K03504
UNS for ASTM A105 44A2
K03505
UNS for ASTM A557-C2 44A2
K03506
UNS for ASTM A541-1 44A2
K03700
UNS for ASTM A413 44A2
K03800
UNS for ASTM A563-DH 44A2
K03810
UNS for AMS 6321 44K2
K03900
UNS for ASTM A490 44A2
K04000
UNS for ASTM A290-A&B 44A2
K04001
UNS for ASTM A372-2 44A2
K04002
UNS for ASTM A194-2 44A2
K04100
UNS for ASTM A354 44K2
K04200
UNS for ASTM A449 44A2
K04500
UNS for ASTM A290-C&D 44K1
K04500
UNS for ASTM A290-C&D 44K1
K04600
UNS for ASTM A489 44A3
K04700
UNS for ASTM A21-U 44A3
K04701
UNS for ASTM A21 44A3
K04800
UNS for ASTM A489 44A1
K04801
UNS for ASTM A372-3 44A2
K05000
UNS for ASTM A291-2 44A2
K05001
UNS for ASTM A266-3 44A2
K05002
UNS for ASTM A266-3 44A2
K05003 44A3
K05200
UNS for ASTM A21-F 44A3
K05500
UNS for ASTM A291-1 44A2
K05501
UNS for ASTM A321 44A3
K05700
UNS for ASTM A25-A 44A3
K05701
UNS for ASTM A551-A 44A3
K05800 44A3
K05801
UNS for ASTM A563-D 44A2
K05802
UNS for ASTM A563-0 44A2
K06000
UNS for ASTM A6481 44A3
K06001
UNS for ASTM A729 44A3
K06002
UNS for ASTM A67-2 44A1
K06100
UNS for ASTM A1 44A3
K06200
UNS for ASTM A25-B 44A3
K06201
UNS for ASTM A417 44A3
K06500
UNS for Mil W 21425-1&111 44A3
K06501
UNS for ASTM A227 44A3
K06700
UNS for ASTM A648-11 44A3
K06701
UNS for ASTM A230 44A3
K06702
UNS for ASTM A551-B 44A3
K06703
UNS for ASTM A228 44A3
K07000
UNS for ASTM A1 44A3
K07001
UNS for ASTM A229 44A3
K07100
UNS for ASTM A648-111 44A3
K07200
UNS for ASTM A25-U 44A3
K07201
UNS for ASTM A25-C 44A3
K07301
UNS for ASTM A1-91-120 44A3
K07500
UNS for ASTM A1-121 44A3
K07700
UNS for ASTM A228 44A3
K08200
UNS for ASTM A679 44A3
K08201
UNS for ASTM A228 44A3
K08500
UNS for ASTM A228 44A3
K08700
UNS for Mil W 8957 44A4
K 10614 44G1
K10623
UNS for ASTM A735 44G1
K10726
UNS for AWS designation ER7082 44K1
K10882
UNS for AWS designation ER100S-1 44K1
K10940
UNS for AWS designation ENi1N
K10943
UNS for AWS designation EB1N 44K1
K10945
UNS for AWS designation EH10MoEW 44G1
K10958
UNS for AWS designation ENi1KN 44F1
K10982
UNS for AWS designation EM2N 44K1

K11022
UNS for AWS designation ER70S-2 44K1
K11040
UNS for AWS designation ENi1 44F1
K11043
UNS for AWS designation EB1 44K1
K11060
UNS for AWS designation EF1N 44K1
K11072
UNS for AWS designation EB2N 44K1
K11122
UNS for AWS designation EA1N 44G1
K11123
UNS for AWS designation EA2N 44G1
K11125
UNS for AWS designation ER70S-7 44A1
K11132
UNS for AWS designation ER70S.4 44A1
K11140
UNS for AWS designation ER70S.6 44A1
K11160
UNS for AWS designation EF1 44A1
K11172
UNS for AWS designation EB2 44K1
K11201
UNS for Mil S 24412 44J
K11210
UNS for ASTM A65
K11222
UNS for AWS designation EA1 44G1
K11223
UNS for AWS designation EA2 44G1
K11224
UNS for ASTM A562 44A1
K11245
UNS for AWS designation EWS 44K1
K11250
UNS for AWS designation ER100S-2 44F1
K11260
UNS for AWS designation ER80S Nil 44F1
K11267
UNS for ASTM A333-4 44K1
K 11268 44F1
K11323
UNS for AWS designation EA3N 44G1
K11324
UNS for AWS designation EA4N 44G1
K11325
UNS for AWS designation EGxxS.5
K11356
UNS for ASTM A714-4 44K1
K11357
UNS for AWS designation ER7086 44A1
K11365
UNS for AMS 6460 44K1
K11385
UNS for AWS designation ENi4N 44K1
K11422
UNS for ASTM A209-T1b 44G1
K11423
UNS for AWS designation EA3 44G1
K11424
UNS for AWS designation EA4 44G1
K11430
UNS for ASTM A502-3A 44A1
K11485
UNS for AWS designation ENi4 44K1
K11500
UNS for ASTM A587 44A1
k11501
UNS for ASTM A715-1 44A1
K11502
UNS for ASTM A715-2 44A1
K11502
now K11509
K11504
UNS for ASTM A715-4 44A1
K11505
UNS for ASTM A715-5 44A1
K11506
UNS for ASTM A715-6 44A1
K11507
UNS for ASTM A715-7 44A1
K11508
UNS for ASTM A715-8 44A1
K11509
UNS for ASTM A656-2 44A1
K11510
UNS for ASTM A242-1 44A2
K11511
UNS for AMS 6386 44K1
K11522
UNS for ASTM A161-T1 44G1
K11523
UNS for ASTM A514-K 44G1
K11526
UNS for ASTM A595 C 44K1
K11535
UNS for ASTM A423-1 44K1
K11538
UNS for ASTM A588-C 44K1
K11540
UNS for ASTM A423-2 44K1
K11541
UNS for ASTM A588-F 44K1
K11542
UNS for AMS 6378 44K2
K11546
UNS for AMS 6379 44K2
K11547
UNS for ASTM A369-FP3 44K1
K11552
UNS for ASTM A588-D 44E1
K11562
UNS for ASTM A369-FP3 44K1
K11564
UNS for ASTM A182-F12 44K1
K11567
UNS for ASTM A588E 44G1
K11572
UNS for ASTM A182-F11 44K1
K11576
UNS for ASTM A514F 44K1
K11578
UNS for ASTM 335-P15 44G1
K11585
UNS for AWS designation EH14 44A1
K11591
UNS for ASTM A405-P24 44K1
K11597
UNS for ASTM A199-T11 44K1
K11598
UNS for ASTM A541-5 44K1
K11625
UNS for AMS 6386 44G1
K11630
UNS for AMS 6386 44G1
K11640
UNS for AMS 6324 44K2
K11646
UNS for ASTM A514-H 44K1
K11662
UNS for AMS 6386 44G1
K11682
UNS for ASTM A514-D 44K1
K11683
UNS for ASTM A514-D 44K1
K11695
UNS for ASTM A592-E 44K1
K11711
UNS for ASTM A766 44A1
K11720
UNS for ASTM A734-A 44K1
K11742
UNS for ASTM A202-A 44E1
K11757
UNS for ASTM A387 44K1
K11789
UNS for ASTM A387 44K1
K11797
UNS for ASTM A739-B11 44K1
K11800
UNS for ASTM A541-4 44K1
K11801
UNS for Mil S 16113-1 44J
K11802
UNS for ASTM A225 44J
K11804
UNS for ASTM A656-1 44J
K 11805 44J
K11820
UNS for ASTM A204-A 44G1

K11831
UNS for ASTM A724-A 44A1
K11835
UNS for ASTM A714-VI 44K1
K11846
UNS for ASTM A131 44A2
K11847
UNS for ASTM A514-N 44A2
K11852
UNS for ASTM A131 44A2
K11856
UNS for AMS 6386 44K1
K11872
UNS for ASTM A514-G 44K1
K11900
UNS for ASTM A325-2 44A2
K11914
UNS for AMS 6354 44K1
K11918
UNS obsolete
K11940
UNS for AMS 6422 44K2
K11948
UNS for AWS designation EF4N 44K1
K12000
UNS for ASTM A633C 44A1
K12001
UNS for ASTM A737-B 44A1
K12003
UNS for ASTM A225 44J
K12010
UNS for ASTM A242-2 44A2
K12020
UNS for ASTM A204-B 44G1
K12021
UNS for ASTM A302-A 44K1
K12022
UNS for ASTM A302-B 44K1
K12023
UNS for ASTM A209-TiA 44G1
K12031
UNS for ASTM A724-B 44A1
K12032
UNS for ASTM A588-H 44K1
K12033
UNS for ASTM A325-C 44A2
K12037
UNS for ASTM A633-D 44A1
K12039
UNS for ASTM A302-C 44K1
K12040
UNS for ASTM A588G 44E1
K12042
UNS for ASTM A508-3 44K1
K12043
UNS for ASTM A588B 44K1
K12044
UNS for ASTM A588B 44K1
K12045
UNS for ASTM A541-3 44K1
K12047
UNS for ASTM A182 F12 44K1
K12048
UNS for AWS designation EF4 44K1
K12054
UNS for ASTM A302-D 44K1
K12059
UNS for ASTM A325-D 44A2
K12087
UNS for AWS designation EB5N 44K1
K12089
UNS for ASTM A707-44 44K1
K12103 44K1
K12121
UNS for ASTM A692 44G1
K12122
UNS for ASTM A182-F1 44K1
K12125 44K1
K12143
UNS for ASTM A387-2 44K1
K12147
UNS for AWS designation R100 44M1
K12148
UNS for AMS 6461 44K1
K12149
UNS for AMS 6462 44K1
K12187
UNS for AWS designation EB5 44K1
K12202
UNS for ASTM A633-E 44J
K12211
UNS for ASTM A441 44A1
K 12220 44G1
K12238
UNS for ASTM A325 44A2
K12244
UNS for ASTM A5022 44A1
K12249
UNS for ASTM A690 44F1
k12254
UNS for ASTM A325 44A2
K12320
UNS for ASTM A204-C 44G1
K12437
UNS for ASTM A537 44A1
K12447
UNS for ASTM A738 44A2
K12510
UNS for ASTM A707-L3 44J
K12520
UNS for ASTM A336F1 44G1
K12521
UNS for ASTM A533A 44A1
K12524
UNS for ASTM A225 44J
K12529
UNS for ASTM A533C 44K1
K12539
UNS for ASTM A533B 44K1
K12542
UNS for ASTM A292 44K1
K12554
UNS for ASTM A533C 44K1
K12608
UNS for ASTM A714-1 44A1
K12609
UNS for ASTM A618-11 44J
K12700
UNS for ASTM A618-111 44J
K12709
UNS for ASTM A714-111 44J
K12765
UNS for ASTM A541-2 44K1
K12766
UNS for ASTM A508-2 44K1
K12810 44A2
K12821
UNS for ASTM A182F1 44K1
K12822
UNS for ASTM A182-F1 44K1
K 13020 44G1
K13047
UNS for ASTM A372-VA 44K2
K13048
UNS for AMS 6550 44K1
K13049
UNS for ASTM A372 44A2
K13051
UNS for ASTM A579-13 44K1
K13147
UNS for AMS 6457 44K1
K13247
UNS for AMS 6356 44K1
K13262 44K1
K13502
UNS for ASTM A508-1 44J
K13521
UNS for ASTM A687-1 44K1
K13547
UNS for ASTM A372 44A2
K13548
UNS for ASTM A372 44A2
K13550
UNS for AMS 6328 44K2
K13586
UNS for ASTM A325A 44A2
K13643
UNS for ASTM A325-A 44A2
K13650
UNS for ASTM A563 44A2
K14044
UNS for ASTM A687-11 44K1
K14047
UNS for ASTM A4140 44K2
K14048
UNS for ASTM A290-E&F 44K1
K14072
UNS for ASTM A193-B16 44K2
K14073
UNS for ASTM A540-B21 44K2
K14185
UNS for Mil S46047 44K1
K14247
UNS for ASTM A649-1A 44K2
K14248
UNS for ASTM A372 5E&8 44K2
K14358
UNS for ASTM A325-B 44A2
K14394
UNS for ASTM A579-32 44K2
K14501
UNS for ASTM A469-1 44J

K14507
UNS for ASTM A291-3 44K2
K14508
UNS for A372-4 44K2
K14510
UNS for ASTM A194-4 44G2
K14520
UNS for ASTM A336-F30 44K2
K14542
UNS for 4145 44K2
K14557
UNS for ASTM 291-3A 44K2
K14675
UNS for AMS 6304 44K2
K15047
UNS for ASTM A232 44K2
K15048
UNS for ASTM A231 44K2
K15590
UNS for ASTM A401 44E2
K15747
UNS for Mil S 12504 MnCrMo 44K1
K17145
UNS for Mil S 12504 Mn Mo 44K1
K18597
UNS for AMS 6426 44K3
K19195
UNS for ASTM A485-2 44E3
K19667
UNS for Mil S980 44E3
K19965
UNS for ASTM A485-3 44K3
K19990
UNS for ASTM 485-4 44K3
K20500
UNS for AWS designation ER80SB2L 44K1
K20622
UNS for ASTM A710B 44K1
K20747
UNS for ASTM A710-A 44K1
K20900
UNS for AWS designation ER80SB2 44K1
K20902
UNS for ASTM A707-L6 44G1
K20910
UNS for ASW designation ENi2N 44F1
K20915
UNS for AWS designation EM3N 44K1
K20930
UNS for AWS designation EM4N 44K1
K20934
UNS for ASTM A707-L5 44K1
K21010
UNS for AWS designation ENi2 44F1
K21015
UNS for AWS designation EM3 44K1
K21028
UNS for AMS 6266
K21030
UNS for AWS designation EM4 44K1
K21135
UNS for AWS designation EF6N 44K1
K21205
UNS for ASTM A724A 44A1
K21240
UNS for AWS designation ER80SNi2 44F1
K21350
UNS for AWS designation EF2N 44K1
K21351
UNS for AWS designation EA3KN 44G1
K21385
UNS for AWS designation EF3N 44K1
K21390
UNS for ASTM A739-B22 44K1
K21450
UNS for AWS designation EF2 44K1
K21451
UNS for AWS designation EA3K 44G1
K21485
UNS for AWS designation EF3 44K1
K21509
UNS for ASTM A199-T3 44K1
K21590
UNS for ASTM A182-F22 44K1
K21604
UNS for ASTM A514-E 44K1
K21650
UNS for ASTM A514-E 44K1
K21730
UNS for ASTM A203 44F1
K21903
UNS for ASTM A333-7 44F1
K21940
UNS for AMS 6255 44K1
K22033
UNS for AMS 6330 44K1
K22035
UNS for ASTM A182-F5 44K1
K22036
UNS for ASTM A350-LF4 44K1
K22094 44K1
K22097
UNS for AMS 6445 44K2
K22103
UNS for ASTM A203-B 44F1
K22375
UNS for ASTM 508 4-4a&b 44K1
K22440
UNS for AMS 6312 44K2
K22573
UNS for ASTM A469-2 44K1
K22578
UNS for ASTM A470-2 44K2
K22720
UNS for AMS 6459 44K1
K22770
UNS for AMS 6303 44K1
K22773
UNS FOR ASTM A469-3 44K2
K22878
UNS for ASTM A370-3&4
K22925
UNS for AMS 5027 44K1
K22950
UNS for AMS 6396 44K2
K23010
UNS for ASTM A470-8 44K2
K23015
UNS for AMS 6302 44K1
K23016
UNS for AWS designation EB2H 44K1
K23080
UNS for AMS 6411 44K1
K23116
UNS for AWS designation EB2HN 44K1
K23205
UNS for ASTM A471 44K2
K23477
UNS for AMS 6428 44K1
K23505 44K1
K23510
UNS for ASTM A355-D 44K2
K23545
UNS for ASTM A336-F31 44K1
K23577
UNS for AMS 5029 44K1
K23578 44K1
K23579 44K1
K23725
UNS for AMS 5028 44K1
K23745
UNS for ASTM A355-B 44K2
K24040
UNS for ASTM A649-1B 44K2
K24045
UNS for ASTM A290-G 44K1
K24055
UNS for ASTM A372-7 44A2
K24064
UNS for ASTM A540-B24 44K2
K24065
UNS for AMS 6470 44K2
K24070
UNS for ASTM A540-B24V 44K2
K24245
UNS for ASTM A291-4 44K2
K24336
UNS for AMS 6423 44K2
K 24535 44K2
K24562
UNS for 4340 44K2
K24628
now combined with K24728
K24728
UNS for AMS 6431 44K2
K24729
now combined with K24728

K25550
UNS for ASTM A723-1 44K1
K27705 44K1
K30560
UNS for AWS designation ER90SB3L 44K1
K30960
UNS for AWS designation ER90SB3 44K1
K31015
UNS for AWS designation EB3N 44K1
K31115
UNS for AWS designation EB3 44K1
K31210
UNS for AWS designation Eni3N
K31240
UNS for AWS designation ER80SNi3 44F1
K31310
UNS for AWS designation ENi3
K31509
UNS for ASTM a199-T4 44K1
K31545
UNS for ASTM A182-F21 44K1
K31718
UNS for ASTM A203D 44F1
K31820
UNS for ASTM A765-11 44K1
K31830
UNS for ASTM A182 F2.2 44K1
K31918
UNS for ASTM A333-3 44F1
K32025
UNS for ASTM A350-LF3 44F1
K32026
UNS for ASTM A765-111 44K1
K32045
UNS for HY100 44K1
K32218
UNS for ASTM A707-L7 44F1
K32550
UNS for AMS 6418 44K1
K32723
UNS for ASTM A469-4 44K2
K32800
UNS for ASTM 471-1 44K2
K33020
UNS for AMS 6407 44K1
K33125
UNS for ASTM A469-5 44K2
K33370
UNS for 4335 44K2
K33517
UNS for AMS 6429 44K1
K33585
UNS for ASTM A336-F32 44K1
K34035
UNS for ASTM A723-2 44K1
K34378
UNS for AMS 6406 44K2
K41245
UNS for ASTM A182-F5 44K1
K41370
UNS for AWS designation EF5 44K1
K41545
UNS for ASTM A199-T5 44K1
K41583
UNS for ASTM A645 44K1
K42247
UNS for ASTM A707-L8 44K1
K42338
UNS for ASTM A543-A 44K1
K42339
UNS for ASTM A543-B 44K1
K42343
UNS for ASTM A541-7 44K1
K42348
UNS for ASTM A541-8 44K1
K42365
UNS for ASTM A508-5 44K1
K42544
UNS for ASTM A182-F5a 44K1
K42570 44K1
K42598
UNS for ASTM A579-11 44K1
K42885
UNS for ASTM A469-6 44K2
K43170 44K1
K44045
UNS for ASTM A723-3 44K1
K44210
replaced by K44220
K44220
UNS for AMS 6419 44K2
K44315
now K44220
K44414
UNS for AMS 6263 44K1
K44910
UNS for AMS 6250 44K1
K51210
UNS for ASTM A534 44K1
K51255
UNS for ASTM A579-12 44K1
K51545
UNS for ASTM A213-T5b
K52355
UNS for AMS 6457 44K1
K52440
UNS for ASTM A355-C 44K1
K61595
replaced by S50300
K63001
UNS for EV11 44N2
K63005
replaced by S63005
K63007
replaced by S63007
K63008
replaced by S63008
K63008
UNS for EV8 44P
K63009
UNS for EV9 44N2
K63011
replaced by S63011
K63012
replaced by S63012
K63012
UNS for EV12 44N2
K63013
replaced by S63013
K63014
replaced by S63014
K63014
UNS for EV5 44N2
K63015
replaced by S63015
K63016
replaced by S63016
K63017
replaced by S63017
K63017
UNS for EV4 44N2
K63198
replaced by S63198
K63199
replaced by S63199
K64005
replaced by S64005
K64006
replaced by S64006
K64006
UNS for HEV2 27C
K64152
replaced by S64152
K64299
replaced by S64299
K65006
replaced by S65006
K65150
replaced by S65159
K65770
replaced by S65770
K66220
replaced by S66220
K66286
UNS for A286 44N2
K66288
replaced by S66286
K66545
replaced by S66545
K66979
replaced by N09979
K70640
replaced by N09902
K71040
UNS for AMS 6308 44K1
K71340
UNS for ASTM A522 44F1
K71350
UNS for AMS 6256 44K1
K74015
replaced by T20811
K81340
UNS for ASTM A333-8 44F1
K81590
replaced by S50400
K88165
replaced by T11350
K90941
UNS for ASTM A182-F9 44K1
K90987
UNS for AWS designation RFe5B 44K3

K91094
UNS for ASTM A579-83 44L
K91122
UNS for ASM 6546 44K1
K91151
replaced by S41000
K91161
replaced by S41001
K91209
UNS for Mil A 13259
K91283
UNS for AMS 6524 44L
K91308
UNS for AWS designation RFE5A 44K3
K 91313 44L
K91342
replaced by S42201
K91352
UNS for ASTM A437-B4B 44M1
K91360
now combined with K93120
K91401
UNS for ASTM A605 44L
K91456
UNS for AMS 5623 44P
K91461
UNS for AMS 6468 44L
K91472
UNS for AMS 6523 44L
K91505
UNS for AMS 5624 44N1
K91555
UNS for ASTM A289-A 44P
K 91800 44N1
K91890 44K1
K91930
UNS for ASTM A579-74 44F3
K91940
UNS for ASTM A579-75 44F3
K91955
UNS for ASTM A289-B 44P
K91970
UNS for AMS 6543 44L
K91971
UNS for AMS 6465 44L
K92571
UNS for AMS 6522 44L
K92801 44M1
K92810
UNS for ASTM A538-A 44F3
K92820
UNS for ASTM A579-71 44F3
K92850
UNS for ASTM B388
K92890
UNS for AMS 6501 44F3
K92940
UNS for ASTM A579-72 44F3
K93120
UNS for AMS 6514 44F3
K93130
UNS for AMS 6463 44L
K93601
UNS for ASTM A658 44F1
K 94100 20C
K94101
UNS for AMS 7734 20C
K 94600 20C
K94610
UNS for AMS 7726 20C
K 94760 20C
K 94800 20C
K95000
UNS for AMS 7717 20C
K95050
replaced by N14052
K95100
replaced by R30005
KABC III 44K3
KAISALOY 42CV 44J
KAISALOY 45FG 44K1
KAISALOY 50CR 44K1
KAISALOY 50MV 44J
KAISALOY 70MB 44G1
KAISALOY No. 1 44K1
KAISALOY No. 2 44K1
KAISALOY No.3 44K2
KAISERALOY 45CB 44A1
KAISERALOY 50MM 44A1
KAISERALOY 55CB 44A1
KAISERALOY 60CB 44A1
KALKOS 44M1
KANTHAL A1 44M1
KANTHAL AF 44M1
KANTHAL APM 44M1
KANTHAL D 44M1
KAO 44H
KAOC 44K3
KAOX 44H
KARMA 27C
KARONI 40/13 Mo 44M1
KAYEM 55A
KAYEM 2 55A
KB 90 20A
KB 90 55A
KB 90/3/4 20A
Kb Pb 21.1
Kb Pb Sb 21.1
Kb Pb Sn 2.5 21.1
Kb Pb Te 0.04 21.1
KBT 21 44K3
KC 20B
KCR–D183 44N3
KCR–D283 44N3
KD 2 44K1
KD 2SE 44K1
KD 3 44K1
KD 4 44K2
KE 4L 44E3
KE 15 44M1
KE 25 44M1
KE 35 44M1
KE 40A 44M1
KE 41 44K1
KE 43 44M1
KE 127 44K3
KE 160 44E3
KE 169 44K1
KE 200 44M2
KE 226 44K3
KE 241 44K2
KE 275 44K2
KE 287 44K1
KE 339 44K1
KE 355 44K2
KE 396 44K2
KE 595 44K3
KE 621 44H
KE 637 44H
KE 660 44K1
KE 672 44K3
KE 708 44K3
KE 805 44K1
KE 839 44E3
KE 896 44K2
KE 897 44K1
KE 954 44M2
KE 960 44K2
KE 961 44M2
KE 965 44N1
KE 970 44M2
KE 1006 44A4
KE 1006 44J
KE 1029 44K2
KE 1036 44M2
KEA 23 44N1
KEA 28 44M1
KEA 108 44A4
KEA 138 44M1
KE A145 44K1
KEA 162 44K3
KEA 180 44M2
KEA 180/476 44M2
KEA 203 44M1
KE A205 44J
KEA 207 44M2
KEA 218 44L
KE A220 44K1
KEA 221 44N1
KEA 227 44K2
KEA 476 44M2
KEA 505 44M1
KEA 507 44N1
KEA 508 44M1
KEA 521 44N1
KEA 581 44A1
KE DIAMOND 10 44K3
KEEWATIN 44K3
KEH 22/A507 44N1
KELOCK 44K2
KELOCK 237 44K3
KELOCK 795 44K3
KELOCK 873 44L
KELOCK 1014 44K3
KELOCK 1021 44L
KELOCK A 72 44L
KELOCK A157 44K3
KELOCK A182 44K3
KELOCK A229 44K3
KENNAMETAL 52.1
KENNERTIUM W2 52.1
KENNERTIUM W10 52.1
KENTANIUM 51B
KERAU WUMDA 44L
KF 44A1
KF 1 44A1
KF 1A 44A1
KF 3 44A1
KF 4 44A1

KF 4A 44A2
KF 5 44A2
KF 8 44A2
KF 9 44A2
KF 10 44A2
KF 10A 44A2
KF 10B 44A3
KF 10C 44A3
KF 10D 44A3
KF 12A 44F2
KF 13 44F1
KF 15/3 44F1
KF 15A 44K1
KF 15B 44K1
KF 15C 44K1
KF 16 44K1
KF 18 44E1
KF 18 A 44E1
KF 18 B 44E1
KF 18C 44E2
KF 18D 44E2
KF 19 44K1
KF 23 44A1
KF 26 44K2
KF 26A 44K2
KF 28 44K1
KF 28B 44K1
KF 28C 44K2
KF 30 A 44E1
KF 31B 44K1
KF 38A 44K1
KF 39 44A1
KF 39C 44A1
KF 39E 44A1
KF 39E/1 44A1
KF 40 44A2
KF 40A 44A2
KF 42 44K1
KF 42A 44K1
KF 46 44G1
KF 46A 44G1
KF 48 44K2
KF 50A 44K1
KF 51 44A2
KF 53 44K2
KF 56 44K2
KF 57 44G2
KF 59 44A2
KF 60 44A1
KF 63 44A2
KF 63A 44A2
KF 64 44A2
KF 70 44K1
KF 71 44K1
KF 72 44K1
KF 73 44K1
KF 74 44K1
KF 75 44K1
KF 76 44K1
KF 77 44K1
KH
see GOST 4543 44E1
KH
see GOST 4543 44K1
KH
see GOST 8731 44E1
KH
see GOST 8731 44E2
KH 44E3
KH 5
see GOST 5632 44K1
KH 5M
see GOST 5632 44K1
KH 5M 44K1
KH 5VF
see GOST 5632 44K1
KH 5VF 44K1
KH 6S1U
see GOST 5632 44K1
KH 6SM
see GOST 5632 44K1
KH 6VF 44K3
KH 8VF
see GOST 5632 44K1
KH 9S2
see GOST 5632 44E2
KH 10N20T2
see GOST 5632 44N1
KH 10S2M
see GOST 5632 44K2
KH 11MF
see GOST 5632 44M1
KH11MF 44M1
KH 12 44M2
KH 12F1 44M2
KH 12H22T3MP
see GOST 5632 44N2
KH 12M 44M2
KH 12N2VFM
see GOST 5632 44M1
KH12N2VMF 44M1
KH 12N8G8MFB
see GOST 5632 44M1
KH 12N20T3R
see GOST 5632 44N2
KH 12N20T3R 44N2
KH 12N22T3MR 44N2
KH 12S1U
see GOST 5632 44M1
KH 12V2MF
see GOST 5632 44M1
KH12V2MF 44M1
KH12VMBFR 44M1
KH 12VMBMR
see GOST 5632 44M1
KH12VMF
see GOST 5632 44M1
KH 12VNMF
see GOST 5632 44M1
KH12VNMF 44M1
KH 13
see GOST 5632 44M1
KH 13
see GOST 5632 44M1
KH 13
see GOST 5632 44M1
KH 13
see GOST 5632 44M1
KH 13
see GOST 5632 44M1
KH 13N3
see GOST 5632 44M1
KH 13N4G9
see GOST 5632 44M1
KH 13N7S2
see GOST 5632 44M1
KH 14
see GOST 5632 44M1
KH 14G14N
see GOST 5632 44M1
KH 14G14N35
see GOST 5632 44M1
KH 14N14V2M
see GOST 5632 44N1
KH 14N16B
see GOST 5632 44N1
KH14N16B 44N2
KH 14N16BR
see GOST 5632 44N1
KH 14N18V2B
see GOST 5632 44N1
KH14N18V2B 44N2
KH 14N18V2BR
see GOST 5632 44N1
KH14N18V2BR 44N2
KH 14N18V2BR1
see GOST 5632 44N1
KH14N18V2BR1 44N2
KH 14N28V3T31UR
see GOST 5632 44N2
KH 15N7G7F2MS
see GOST 5632 44N2
KH 15N17G7F2MS
see GOST 5632 44N1
KH 15N91U
see GOST 5632 44M1
KH 16N13M2B
see GOST 5632 44N1
KH 16N15M3B
see GOST 5632 44N1
KH 17
see GOST 5632 44M1
KH 17AG14
see GOST 5632 44M1
KH 17E9AN4
see GOST 5632 44K1
KH 17H2
see GOST 5632 44M1
KH 17N2
see GOST 5632 44M1
KH 17N13M2T
see GOST 5632 44N1
KH 17N13M3T
see GOST 5632 44N1
KH 17N16M3T
see GOST 5632 44N1
KH 17N71U
see GOST 5632 44N1
KH 17NBM2T
see GOST 5632 44N1
KH 17T
see GOST 5632 44M1
KH 18
see GOST 5632 44M2
KH 18N9
see GOST 5632 44N1
KH 18N9T
see GOST 5632 44N1

KH 18N10
see GOST 5632 44N1
KH 18N10
see GOST 5632 44N1
KH 18N10
see GOST 5632 44N1
KH 18N10E
see GOST 5632 44N1
KH 18N10T
see GOST 5632 44N1
KH 18N10T
see GOST 5632 44N1
KH 18N10T 44N1
KH 18N11
see GOST 5632 44N1
KH 18N12B
see GOST 5632 44N1
KH 18N12T
see GOST 5632 44N1
KH 18N12T
see GOST 5632 44N1
KH 18N12T 44N1
KH 18N25S2
see GOST 5632 44N1
KH 18S1U
see GOST 5632 44M1
KH 19N9MVBT
see GOST 5632 44N2
KH19N9MVBT 44N1
KH 20N14S2
see GOST 5632 44N1
KH 20N14S2
see GOST 5632 44N1
KH 21N5T
see GOST 5632 44M1
KH 21N5T
see GOST 5632 44N1
KH 21N5T
see GOST 5632 44M1
KH 21N6M2T
see GOST 5632 44N1
KH 23N13
see GOST 5632 44N1
KH 23N18
see GOST 5632 44N1
KH 23N18
see GOST 5632 44N1
KH 23N28M2T
see GOST 5632 44N1
KH 23N28M3D3T
see GOST 5632 44N1
KH 25H25TR
see GOST 5632 44N1
KH 25N16G7AR
see GOST 5632 44N2
KH 25N16G7AR 44N1
KH 25N20S2
see GOST 5632 44N1
KH25N25TR 44N1
KH 25T
see GOST 5632 44M1
KH 28
see GOST 5632 44M1
KH 28AN
see GOST 5632 44M1
KH 35VT 44N1
KH 35VT1U 44N1
KH 35VTR 44N1
KH 38VT 44N1
KH 70 27B
KH 70MVTIUB 27C
KH 70VMTIU 27C
KH 75MBTIU 27F
KH 77TIUR 27C
KH 78T 27F
KH 80TBIU 27C
KH 601U 27F
KH 701U 27B
KHB 36 44A1
KH GSA
see GOST 8731 44E2
KHN 35VMT 44N1
KHN 60V 27C
KHNSV 44K2
KHV 5 44K3
KHVG 44K3
KHVSG 44K3
KIA 23B
KING COBALT 44L
KIRKASE 44A1
KIRKSIRE I OR A 55A
KIRKSIRE II OR B 55A
KIRKSITE A 55A
KISKI 44K3
KK 44K3
KK 3 44K3
KK BREARLEY 44M1
KK ENGINEERING 44M1
KKK 44M2
KK TWOSCORE 44M1
KK WELDANKA 44N1
KLAH 44K2
KM 52.1
KM 1 44K3
KM 2 44K3
KM 3.1 44K3
KM 3.2 44K3
KM 4 44K3
KM 7 44K3
KM 10 44K3
KM 35 44L
KM 42 44L
KM 50 44K3
KM 52 44K3
K MAT 11 44L
K MONEL 27D
K Ms 63 14B
K Ms 72 14B
KMV 44M2
KO 44K1
KOERFLEX 30 13A
KOERFLEX 200 13A
KOERFLEX 300 13A
KOERZIT 90 44F2
KOERZIT 120 44F1
KOERZIT 120K 20C
KOERZIT 120R 20C
KOERZIT 130 44L
KOERZIT 160 20C
KOERZIT 160 44L
KOERZIT 190 20C
KOERZIT 190 44L
KOERZIT 220 20C
KOERZIT 250 44L
KOERZIT 300 44L
KOERZIT 350 20C
KOERZIT 400 44L
KOERZIT 400K 20C
KOERZIT 450 20C
KOERZIT 500 20C
KOERZIT 580 44L
KOERZIT 600 20C
KOERZIT 700 44L
KOERZIT H 13A
KOERZIT T 13A
KOERZIT VS 55 44L
KONCOR 44K3
KONEL 27E
KORA 2 44A1
KORROSIL 2M 44M1
KOVA 57 44L
KOVAR 44L
KP 14C
KP 1 44M1
KP 2 44A1
KP 3 44A1
KP 4 44A2
KP 5 44A2
KP 6 44A2
KP 7 44A3
KP 8 44A3
KP 9 44A4
KP 10 44F2
KP 11 44F1
KP 12 44K1
KP 13 44K1
KP 14 44G1
KP 16 44K2
KP 17 44K1
KP 19 44A1
KP 20 44A2
KP 21 44A2
KP 22 44K2
KP 23 44K1
KP 24 44E3
KP 25 44A3
KP 26 44K2
KP 27 44M1
KP 28 44M1
KP 29 44M1
KP 31 44K1
KP 32 44E2
KP 33 44K2
KP 34 44K2
KP 35 44K1
KP 36 44K1
KP 37 44K1
KR 6 44E3
KR 35 44K1
KR 75 44K1
KR MONEL 27D
KROMORE 27B
KS 14C
KSA 44K3
K Sn Bz 2 14K
KT 14C
KT 1 44K3
KUFIL 14A
KUMANAL 14G

KUMIUM 14L
KUNIAL BRASS 14F
KUNIFER 5 14E
KUNIFER 5T 14E
KUNIFER 10 14E
KUNIFER 10T 14E
KUNIFER 30 14E
KUNIFER 30A 14E
KUNIFER 30T 14E
KUNIFORM 14B
KUTERN 14A
KUTHERM 41 14G
KUTTWELL 44A1
KV 1 44K1
KV 3 44K1
KV 4 44K1
KV 5 44K1
KV7 44M1
KV 7 44K1
K-W 44H
KYCUBE 25 14D
KYCUBE 25/1 14D
KYNAL 90 1F
KYNAL C65 1F
KYNAL C66 1F
KYNAL C67 1F
KYNAL C69 1E
KYNAL-CORE C65A 1H
KYNAL-CORE C66A 1H
KYNAL-CORE C67A 1H
KYNAL-CORE C68A 1H
KYNAL-CORE Z93A 1H
KYNAL M35/1 1D
KYNAL M35/2 1D
KYNAL M36 1D
KYNAL M37 1D
KYNAL M39/1 1E
KYNAL M39/2 1E
KYNAL P5 1A
KYNAL P10 1A
KYNAL PA15 1C
KYNAL PA16 1C
KYNAL PA17 1C
KYNAL PA19 1B
KYNAL Y88 1F
KYNAL Y92 1F
KYNAL Z93 1G
KZ 14C
L 1
see BS range 1F
L1 1F
L 1 44A1
L 1 44E3
L 2 44A2
L 2 44K3
L 2N 44K1
L 3
see BS range 1F
L3 1F
L 3 14H
L 3 44A2
L 3 44K3
L 4
see BS range 1A
L4 1A
L 4 1L
L 4 44K3
L 5
see BS range 1M
L5 1M
L 5 44E2
L 5 44K3
L 6 44K3
L 7 14H
L 7 44A3
L 7 44K3
L 8
see BS range 1C
L8 1C
L11 1N
L 11 44A4
L 12 44A4
L 13 44A4
L 16
see BS range 1A
L16 1A
L 17
see BS range 1A
L17 1A
L 24
see BS range 1F
L24 1F
L 25
see BS range 1F
L25 1F
L 27
see BS range 1N
L27 1N
L 28
see BS range 1N
L28 1N
L 29
see BS range 1N
L29 1N
L30 1A
L31 1A
L 32
see BS range 1A
L32 1A
L 33
see BS range 1C
L33 1C
L 34
see BS range 1A
L34 1A
L 35
see BS range 1L
L35 1L
L 36
see BS range 1A
L36 1A
L 37
see BS range 1F
L37 1F
L 38
see BS range 1H
L38 1H
L 39
see BS range 1F
L39 1F
L 40
see BS range 1F
L40 1F
L 42
see BS range 1F
L42 1F
L 43
see BS range 1F
L43 1F
L 44
see BS range 1D
L44 1D
L 45
see BS range 1F
L45 1F
L 46
see BS range 1D
L46 1D
L 47
see BS range 1H
L47 1H
L48 1A
L49 1A
L 50
see BS range 1L
L50 1L
L 51
see BS range 1L
L51 1L
L 52
see BS range 1L
L52 1L
L 53
see BS range 1J
L53 1J
L 54
see BS range 1A
L54 1A
L 55
see BS range 1D
L55 1D
L 56
see BS range 1D
L56 1D
L 57
see BS range 1F
L57 1F
L 58
see BS range 1D
L58 1D
L 59
see BS range 1B
L59 1B
L 60
see BS range 1B
L60 1B
L 61
see BS range 1B
L61 1B
L 62
see BS range 1F
L62 1F
L 63
see BS range 1F
L63 1F
L 64
see BS range 1F
L64 1F
L 65
see BS range 1F

L65	1F
L 67	
see BS range	1A
L67	1A
L 69	
see BS range	1F
L69	1F
L 70	
see BS range	1F
L70	1F
L 71	
see BS range	1F
L71	1F
L 72	
see BS range	1H
L72	1H
L 73	
see BS range	1F
L73	1H
L 74	
see BS range	1H
L 75	
see BS range	1K
L75	1K
L 76	
see BS range	1F
L76	1F
L 77	
see BS range	1F
L77	1F
L 78	
see BS range	1L
L78	1L
L 79	
see BS range	1L
L79	1L
L 80	
see BS range	1D
L80	1D
L 81	
see BS range	1D
L81	1D
L 82	
see BS range	1D
L82	1D
L 83	
see BS range	1F
L83	1F
L 84	
see BS range	1F
L84	1F
L 85	
see BS range	1F
L85	1F
L 86	
see BS range	1F
L86	1F
L 87	
see BSL87	1F
L87	1F
L 88	1G
L 91	
see BSL91	1L
L91	1L
L 92	
see BSL92	1L
L92	1L
L 94	
see BS range	44K2
L 94	44K2
L 97	
see BS range	44K2
L 97	44K2
L 99	
see BSL 99	1A
L99	1C
L102	1F
L103	1F
L105	1F
L109	1F
L110	1F
L111	1E
L112	1E
L113	1E
L114	1E
L116	1A
L117	1E
L118	1E
L119	1F
L 120	
see BS range	23A
L 121	
see BS range	23B
L121	23B
L 122	
see BS range	23B
L122	23B
L 123	
see BS range	23B
L123	23B
L 124	
see BS range	23B
L124	23B
L 125	
see BS range	23B
L125	23B
L126	23B
L127	23B
L128	23B
L154	1F
L155	1F
L156	1F
L157	1F
L158	1F
L159	1F
L 160	1G
L 161	1G
L 162	1G
L163	1F
L164	1F
L165	1F
L166	1F
L167	1F
L168	1F
L 197	44K1
L 202	13A
L 210	13A
L 211	1L
L 212	1L
L 213	1L
L 214	1L
L 215	1L
L 215	13A
L 216	1L
L 216	13A
L 218	13A
L 219	13A
L 225	13A
L231	1J
L 232	1L
L 251	13A
L 252	1C
L 253	1C
L 254	1C
L 255	1C
L 256	1C
L 257	1C
L 258	1L
L 306	27C
L307	27F
L311	27F
L317	27F
L326	27F
L 335	27C
L 411	1L
L 451	1L
L 452	1L
L 453	1L
L503	23B
L504	23B
L505	23B
L508	23B
L512	23B
L 512	51A
L513	23B
L514	23B
L515	23B
L 515	51A
L 518	51A
L 521	51B
L 531	51C
L 532	51C
L 535	51C
L 536	44K1
L 605	13A
L 01900	8.1
L 01950	8.1
L05020	
replaced by L54210	21.1
L05021	
replaced by L54211	21.1
L05025	
replaced by L51050	
L05050	
replaced by L54320	21.1
L05051	
replaced by L54321	21.1
L05055	
replaced by L50180	21.1
L05070	
replaced by L53131	
L05090	
replaced by L53340	21.1
L05100	
replaced by L54520	21.1
L05110	
replaced by L50151	21.1
L 05120	21.1
L05150	
replaced by L54560	21.1

L05153
  replaced by L53346 21.1
L05155
  replaced by L53620 21.1
L05180
  replaced by L53560 21.1
L05188
  replaced by L53565 21.1
L05201
  replaced by L54711 21.1
L05202
  replaced by L54712 21.1
L05237
  replaced by L53585 21.1
L05250
  replaced by L54720 21.1
L05251
  replaced by L54721 21.1
L05252
  replaced by L54722 21.1
L05300
  replaced by L54820 21.1
L05301
  replaced by L54821 21.1
L05302
  replaced by L54822 21.1
L05350
  replaced by L54850 21.1
L05351
  replaced by L54851 21.1
L05400
  replaced by L54915 21.1
L05401
  replaced by L54916 21.1
L05402
  replaced by L54918 21.1
L05450
  replaced by L54950 21.1
L05451
  replaced by L54951 21.1
L05500
  replaced by L55020
L05501
  replaced by L55031 21.1
L 06990 22.1
L 13002 50.1
L 13004 50.1
L 13006 50.1
L 13007 50.1
L 13008 50.1
L 13010 50.1
L 13012 50.1
L 13014 50.1
L 13600 50A
L 13601 50A
L 13630 50A
L 13631 50A
L 13650 50A
L 13700 50A
L 13701 50A
L 13820 50A
L 13840 50A
L 13870 50A
L 13890 50A
L 13910 50A
L 13911 50A
L 13912 50A
L 13913 50A
L 13940 50A
L 13950 50A
L 13960 50A
L 13961 50A
L 13963 50A
L 13965 50A
L 50001 21.1
L 50005 21.1
L 50010 21.1
L 50020 21.1
L 50025 21.1
L 50035 21.1
L 50040 21.1
L 50042 21.1
L 50045 21.1
L 50050 21.1
L 50060 21.1
L 50065 21.1
L 50070 21.1
L 50080 21.1
L 50090 21.1
L 50101 21.1
L 50110 21.1
L 50113 21.1
L 50115 21.1
L 50120 21.1
L 50121 21.1
L 50122 21.1
L 50131 21.1
L 50132 21.1
L 50134 21.1
L 50140 21.1
L 50150 21.1
L 50151 21.1
L 50152 21.1
L 50170 21.1
L 50172 21.1
L 50180 21.1
L 50310 21.1
L 50510 21.1
L 50520 21.1
L 50521 21.1
L 50522 21.1
L 50530 21.1
L 50535 21.1
L 50540 21.1
L 50541 21.1
L 50542 21.1
L 50543 21.1
L 50710 21.1
L 50712 21.1
L 50713 21.1
L 50720 21.1
L 50722 21.1
L 50725 21.1
L 50728 21.1
L 50730 21.1
L 50735 21.1
L 50736 21.1
L 50737 21.1
L 50740 21.1
L 50745 21.1
L 50750 21.1
L 50755 21.1
L 50760 21.1
L 50765 21.1
L 50770 21.1
L 50775 21.1
L 50780 21.1
L 50790 21.1
L 50795 21.1
L 50800 21.1
L 50810 21.1
L 50820 21.1
L 50840 21.1
L 50850 21.1
L 50880 21.1
L 50940 21.1
L 51110 21.1
L 51121 21.1
L 51123 21.1
L 51124 21.1
L 51125 21.1
L 51180 21.1
L 51510 21.1
L 51511 21.1
L 51512 21.1
L 51530 21.1
L 51532 21.1
L 51535 21.1
L 51540 21.1
L 51550 21.1
L 51705 21.1
L 51708 21.1
L 51710 21.1
L 51720 21.1
L 51740 21.1
L 51748 21.1
L 52515 21.1
L 52525 21.1
L 52540 21.1
L 52550 21.1
L 52560 21.1
L 52605 21.1
L 52615 21.1
L 52620 21.1
L 52625 21.1
L 52725 21.1
L 52760 21.1
L 52765 21.1
L 52805 21.1
L 52810 21.1
L 52860 21.1
L 52901 21.1
L 52910 21.1
L 52915 21.1
L 53020 21.1
L 53105 21.1
L 53120 21.1
L 53122 21.1
L 53125 21.1
L 53130 21.1
L 53235 21.1
L 53238 21.1
L 53260 21.1
L 53265 21.1
L 53305 21.1
L 53320 21.1
L 53340 21.1
L 53343 21.1
L 53345 21.1
L 53346 21.1
L 53405 21.1

L 53420 21.1
L 53425 21.1
L 53452 21.1
L 53455 21.1
L 53456 21.1
L 53460 21.1
L 53465 21.1
L 53470 21.1
L 53480 21.1
L 53505 21.1
L 53510 21.1
L 53530 21.1
L 53550 21.1
L 53555 21.1
L 53558 21.1
L 53560 21.1
L 53565 21.1
L 53570 21.1
L 53575 21.1
L 53580 21.1
L 53585 21.1
L 53620 21.1
L 53650 21.1
L 53655 21.1
L 53685 21.1
L 53710 21.1
L 53730 21.1
L 53740 21.1
L 53750 21.1
L 53780 21.1
L 53790 21.1
L 53795 21.1
L 54030 21.1
L 54050 21.1
L 54210 21.1
L 54211 21.1
L 54250 21.1
L 54280 21.1
L 54310 21.1
L 54320 21.1
L 54321 21.1
L 54322 21.1
L 54360 21.1
L 54370 21.1
L 54410 21.1
L 54510 21.1
L 54520 21.1
L 54525 21.1
L 54530 21.1
L 54540 21.1
L 54555 21.1
L 54560 21.1
L 54570 21.1
L 54580 21.1
L 54610 21.1
L 54710 21.1
L 54711 21.1
L 54712 21.1
L 54713 21.1
L 54715 21.1
L 54720 21.1
L 54721 21.1
L 54722 21.1
L 54727 21.1
L 54750 21.1
L 54755 21.1
L 54805 21.1
L 54810 21.1
L 54815 21.1
L 54820 21.1
L 54821 21.1
L 54822 21.1
L 54827 21.1
L 54829 21.1
L 54830 21.1
L 54832 21.1
L 54833 21.1
L 54835 21.1
L 54840 21.1
L 54850 21.1
L 54851 21.1
L 54852 21.1
L 54855 21.1
L 54860 21.1
L 54865 21.1
L 54905 21.1
L 54910 21.1
L 54915 21.1
L 54916 21.1
L 54918 21.1
L 54925 21.1
L 54930 21.1
L 54935 21.1
L 54940 21.1
L 54945 21.1
L 54950 21.1
L 54951 21.1
L 54955 21.1
L 55005 21.1
L 55030 21.1
L 55031 21.1
L 55210 21.1
L 55260 21.1
L 55290 21.1
LAC 10 1L
LAC 112 A 1C
LAC 113B 1L
L Ag 8
  see DIN range 14H
L Ag 12
  see DIN range 14H
L Ag 12 Cd
  see DIN range 14H
L Ag 15
  see DIN range 14H
L Ag 15P
  see DIN range 14H
L Ag 20
  see DIN range 14H
L Ag 25
  see DIN range 14H
L Ag 25 Cd
  see DIN range 14H
L Ag 27
  see DIN range 14H
L Ag 30 Cd 5
  see DIN range 14H
L Ag 30 Cd 12
  see DIN range 14H
L Ag 38
  see DIN range 14H
L Ag 44
  see DIN range 42
L Ag 45
  see DIN range 42
L Ag 49
  see DIN range 42
L Ag 50
  see DIN range 42
L Ag 50 Cd
  see DIN range 42
L Ag 60
  see DIN range 42
L Ag 60 Cd
  see DIN range 42
L Ag 67
  see DIN range 42
L Ag 67 Cd
  see DIN range 42
L Ag 75
  see DIN range 42
L Ag 83
  see DIN range 42
LAKE COPPER 14A
L Al Si 12 1C
L Al Si Sn 1N
LA LUCETTE 2.1
LAMB & FLAG 50.1
LAMB & FLAG COMMER 50.1
LANARK 44K2
LANGALLOY 1N 27D
LANGALLOY 1V 44N1
LANGALLOY 2N 27D
LANGALLOY 2R 27F
LANGALLOY 2V 44N1
LANGALLOY 3V 44N1
LANGALLOY 4R 27C
LANGALLOY 4V 44N1
LANGALLOY 5N 27D
LANGALLOY 5R 27C
LANGALLOY 5V 44N1
LANGALLOY 6V 44M1
LANGALLOY 7R 27C
LANGALLOY 10V 44N1
LANGALLOY 12V 44N1
LANGALLOY 13V 44N1
LANGALLOY 15V 44N1
LANGALLOY 20V 44N2
LANGALLOY 40V 44N3
LANGALLOY 80V 44M1
LANSTON STANDARD 21.1
LANTERN 44A1
LAPELLOY 44M1
LAPELLOY C 44M1
LARPORT 44K1
LB 1
  see BS range 14K
LB 2
  see BS range 14K
LB 3
  see BS range 14K
LB 4
  see BS range 14K
LB 5
  see BS range 14K
LBZ 14B
LC 60 44A3
LC 80 44A3
LC 100 44A4
LC 110 44A4

LC 120 44A4
LC 140 44A4
L Cd Zn 20 8.1
LCHD 44K1
LCKD 44K2
LC MONEL 27D
LC Ni 99
see DIN range 27A
L Co 8 13.1
L Co 11 13.1
L Co 73 13.1
LE 44F2
LEAD 21.1
LEAD 6 A 21.1
LEDA 44L
LED-O-LOY 14B
LEGA 42 20C
LEGA 46 20C
LEGA 52 27E
LESCALLOY 300M 44K2
LESCALLOY 600 27B
LESCALLOY 718 VAC–ARC 27C
LESCALLOY 901 VAC-ARC 44N2
LESCALLOY 4330 +V 44K1
LESCALLOY 4335 +V 44K1
LESCALLOY 4340 44K2
LESCALLOY 5616 44M1
LESCALLOY 9310 44K1
LESCALLOY 52100 44E3
LESCALLOY BG 42 44M2
LESCALLOY BG 66 44M2
LESCALLOY BG31 44K1
LESCALLOY D979 27C
LESCALLOY HP 9-4-30 44K2
LESCALLOY LADISH D6AC VAC-ARC 44K2
LESCALLOY UT18 44K2
LESCALLOY UT19 44K1
LESCALLOY V57 44N2
LESCALLOY V57 VAC-ARC 44N2
LESCALLOY X750 27C
LESCO BC42 VAC ARC 44M2
LESCO HS29XL 44K3
LESCO HW114 44K1
LF 44K3
LG 1
see BS range 14K
LG 2
see BS range 14K
LG 3
see BS range 14K
LG 4
see BS range 14K
LHG 44K1
LHS 2C 44L
LHS 5C 44L
LHS 8C 44L
LHS 10C 44L
LHS 14 44K3
LHS 18 44K3
LHS 22 44K3
LHS 227 44L
LHS 235 44L
LHS 652 44K3
LICHTENBERG 6.1
LINCO 44M1
LINOTYPE METAL 21.1
LION BRAND 50A
LIPOWITZ 6.1
LITHIUM 22.1
LK 44E3
LK 1 44K2
LK 3 44K2
LK 5 44K1
LK 7 44K1
LKP 44K1
LM 0
see BS range 1A
LM 1 1L
LM2 1C
LM 2 1C
LM 2 1L
LM 3 1L
LM 4 1L
LM 5 1K
LM 5 8.1
LM 6 1C
LM 7 1L
LM 8 1K
LM 9 1K
LM 10 1J
LM 10 1K
LM 10A 50A
LM 11 1L
LM 12 1L
LM 13 1K
LM 14 1L
LM 15 1L
L M 15 55A
LM 16 1K
LM 17 1C
LM 18 1C
LM 20 1C
LM 21 1L
LM 22 1L
LM 23 1L
LM 24 1L
LM 25 1C
LM 26 1L
LM 27 1L
LM 28 1L
LM 29 1L
LM 30 1L
L MOLLA 26A
LMW 44K3
LMW-V 44K3
L NC 20.2 20B
L NC 30.3 20B
LNCM 44K2
L NUC 15.6.2 20B
LO 44N1
Lo5352
replaced by L54852
LOCKPORT SPECIAL 44K3
LOCO 21.1
LODEX 13A
LO-EX 1K
LOHM 14E
LOMINIUM 44K2
LONESTAR 2.1
LONGHORN 3 STAR 50.1
LONGWAIR 44M2
LOSIL 17 44B1
LOSIL 17C 44B1
LOSIL 19 44B1
LOSIL 19C 44B1
LOSIL 22 44B1
LOSIL 22C 44B1
LOSIL 25 44B1
LOSIL 25C 44B1
LOW BRASS 14B
LOW CARBON TATMO 44K2
LOW CARBON TNW 44K2
LOWHEES 44A1
LOWSCORE 44M1
LOYCON N 44K1
LOYCON QT 44K1
L Pb 98.5
see DIN range 21
LPB 1
see BS range 14K
LPD 44K1
LRT 44A3
LS 44K2
LS 1 44M1
LS 12 44N1
LS 16 44M1
LS 18 44N1
LS 18/2 44N1
LS 18/3 44N1
LS 18/4 44N1
LS 18L 44N1
LS 20/12 44N1
LS 22 44M1
LS 25/12/3 44N1
LS 25/20 44N1
LS 55 44E1
LS 61 44M1
LS 62 44M1
LS 80 44M1
LS CUT 44M1
LS CUT/HC 44M1
L Si 2 41.1
L Si 71 41.1
L Sn 8
see DIN range 21
L Sn 25
see DIN range 21
L Sn 30
see DIN range 21
L Sn 33
see DIN range 21
L Sn 35
see DIN range 21
L Sn 40
see DIN range 21
L Sn 50 50A
L Sn 60 50A
L Sn 90 50A
L Sn Pb Zn 50A
L Sn Zn 10 50A
L Sn Zn 40 50A
L St 45/8 44A1
LS WF 44N1
LT 1 44J
LT 2 44E3
LT 3N 44F1
LT5 44H

LT 8 44K2
LT 9 44K2
LT 9N 44F1
LT 12 44K3
LT 13 44H
LT 20 44A4
LT 21/42 44N2
LT 21/42 44P
LT 22 44E3
LT 22V 44K3
LT 23 44K3
LT 24 44K3
LT 24 51C
LT 25 44M2
LT 25 51C
LT 27 44M2
LT 31 44K2
LT 31 51C
LT 33 44M2
LT 33 51C
LT 34 51C
LT 35 44M2
LT 35 51C
LT 36 44E2
LT 38 44G2
LT 39 44K1
LT 40 44K3
LT 41 44K2
LT 41W 44K1
LT 42 44K2
LT 43 44K1
LT 45 44K1
LT 47 44K2
LT 49 44L
LT 51 44K2
LT 52 44K1
LT 53 44K1
LT 54 44M1
LT 55 44M1
LT 75N 44A1
LT 75QT 44A1
LT 226 44F2
LTAH 44K3
LT FORGING 44K1
LTG 44G2
LTTS 44K3
LU 44L
LUKENS 45 44A1
LUKENS 50 44A1
LUKENS 55 44A1
LUKENS 60 44A1
LUKENS 80 44A1
LUKENS 440 44A1
LUNDIE 44K2
LV 10 44N1
LV 19 44M1
LV 20 44M1
LV 21/2 N 44P
LV 21/4 N 44P
LV 21/12 44N1
LV 21/12N 44N1
LV 21/42 44P
LV 21/43 44P
LV 21/69 44N2
LV 21/69 44P
LV 24 44M1
LV 52 44E2
LV 52 Mo 44K2
LV 54 44N1
LV 55 44N1
LX 13 14H
LX 18 14H
LXX 44K3
LYMANS ALLOY 21.1
L Zn Cd 40 55A
M & T ELECTROLYTIC 50.1
M & T No 1 50.1
M 1 14C
M 1 14H
M 1 23B
M 1 44G1
M 1 44K3
M 1A
see BS range 23B
M1A 1B
M1-F 44A1
M 2 23B
M 2 44G1
M 2 44K3
M 3 44A2
M 3/1 44K3
M 3/2 44K3
M 4 44A1
M 4 44K3
M 5 44N1
M 6 44L
M 6W 44B1
M 7 44K3
M 7 44N1
M 7W 44B1
M 7X 44B1
M 8 44K3
M 8X 44B1
M 10 44K3
M10 FIRTH CLEVELAND 44M1
M11 44G2
M12 44G2
M 15 44L
M21 27F
M 21 27C
M22 27F
M 22 27C
M 22B 27C
M 25 14D
M 30 44L
M 30B 44N2
M 33 44L
M 34 44L
M 35 44L
M 35–1 27D
M 35–2 27D
M 36 44L
M 40 44L
M 41 44L
M 42 44L
M 43 44L
M 44 44L
M 46 44L
M 47 44L
M 50 44K3
M 52 44K3
M 75 44A3
M 85
see DIN range 44A4
M100 44E1
M 113 44A1
M 120 44K1
M 130 44K1
M152 44M1
M 188 44K1
M 200 44K2
M 203 13A
M 204 13A
M 205 13A
M 210 44K1
M 252 27C
M300 44M1
M310 44M1
M 313 27C
M 600 27C
M 813 44N2
M00995
UNS for Mil A 10841 2.1
M00998
UNS for Mil A 10841 2.1
M1008
0.1'/C MAX steel - SAE - merchant quality 44A1
M1010
0.1'/C steel - SAE - merchant quality 44A1
M1012
0.12'/C steel - SAE - merchant quality 44A1
M1017
0.17'/C steel - SAE - merchant quality 44A1
M1020
0.2'/C steel - SAE - merchant quality 44A1
M1023
0.23'/C steel - SAE - merchant quality 44A1
M1025
0.25'/C steel - SAE - merchant quality 44A2
M1031
0.3'/C steel - SAE - merchant quality 44A2
M1044
0.4'/C steel - SAE - merchant quality 44A2
M1628 27F
M 3191 20B
M 3191 GG35 20B
M 3193 20B
M 3210 20B
M 4504 20B
M 5003 20B
M 5503 20B
M 05995 32.1
M 7002 20B
M 8501 20B
M10030
UNS for ASTM B275-A3A 23B
M10100
UNS for AMS 4455 23B
M10102
UNS for ASTM B80-AM100B 23B

M10410
UNS for AS41A 23B
M10600
UNS for ASTM B93-AM60A 23B
M10602
UNS for ASTM B94 23B
M10603
UNS for ASTM B93 23B
M10800
UNS for ASTM B275 AM80H 23B
M10900
UNS for ASTM B275 AM90A 23B
M11100
UNS for ASTM B275 AZ10A 23B
M11101
UNS for ASTM B275 AZ101A 23B
M11125
UNS for ASTM B275 AZ125A 23B
M11210
UNS for ASTM B275 AZ21A 23B
M11310
UNS for ASTM B90 AZ31H 23B
M11311
UNS for AMS 4375 23B
M11312
UNS for ASTM B90 AZ31C 23B
M11610
UNS for AMS 4350 23B
M11611
UNS for designation AWS ERAZ61A 23B
M11630
UNS for AMS 4420 23B
M11800
UNS for AMS 4360 23B
M11810
UNS for ASTM B80AZ81A 23B
M11900
UNS for ASTM B275 AZ90A 23B
M11910
UNS for AMS 4490 23B
M11912
UNS for ASTM B93AZ91B 23B
M11914
UNS for AMS 4437 23B
M11916
UNS for ASTM B94 23B
M11917
UNS for ASTM B93 23B
M11919
UNS for AMS 4446 23B
M11920
UNS for AMS 4434 23B
M11922
UNS for AMS 4395 23B
M12230
UNS for AMS 4442 23B
M12300
UNS for ASTM B80EK30A 23B
M12331
UNS for AMS 4396 23B
M12350
UNS for AMS 4419 23B
M12410
UNS for AMS 4440 23B
M13210
UNS for AMS 4363 23B
M13310
UNS for AMS 4384 23B
M13312
UNS for AMS 4388 23B
M13320
UNS for AMS 4447 23B
M14141
UNS for AMS 4386 23B
M14142
UNS for Mil M46130 23B
M14145
UNS for Mil M46130 23B
M15100
UNS for ASTM B107 MIA 23B
M15101
UNS for ASTM B93 M1B 23B
M15102
UNS for ASTM B91 MIC 23B
M16100
UNS for ASTM M90ZE10A 23B
M16210
UNS for AMS 4387 23B
M16400
UNS for ASTM B107ZK40A 23B
M16410
UNS for AMS 4439 23B
M16510
UNS for AMS 4443 23B
M16600
UNS for AMS 4352 23B
M16601
UNS for ASTM B275 ZK60B 23B
M16610
UNS for AMS 4444 23B
M16620
UNS for AMS 4438 23B
M16630
UNS for AMS 4425 23B
M16631
UNS for ASTM B80ZC63 23B
M18010
UNS for ASTM B80KIA 23B
M18210
UNS for ASTM B80 23B
M18220
UNS for AMS 4418 23B
M18330
UNS for ASTM B80 23B
M18410
UNS for ASTM B80 23B
M19980
UNS for ASTM B92-9980A 23A
M19990
UNS for ASTM B92-9990A 23A
M19991
UNS for ASTM B92-9990B 23A
M19995
UNS for ASTM B92-9995A 23A
M19998
UNS for ASTM B92-9998A 23A
M26800
UNS for AMS 4780 24.1
M29350
UNS for ASTM A601-E 24
M29450
UNS for ASTM A601D 24
M29951
UNS for ASTM A601-A 24
M29952
UNS for ASTM A601-B 24
M29953
UNS for ASTM A601-C 24
M29954
UNS for ASTM A601-F 24
MA 1 44N1
MA 3 44N1
MA 4 44N1
MACDN 44N1
MACHINERY 30 44A2
MACHINERY 40 44A2
MACHINETRODE 92.78 27C
MACM 44N1
MAFERITE 6 44E1
MAFERITE 8 44E1
MAFERITE 14 44M1
MAFERITE 17 44M1
MAFERITE 20 44M1
MAFERITE 25 44M1
MAFERITE 25A 44M1
MAFERITE 28N 44M1
MAFERITE 30 44M1
MAFERITE 30C 44M2
MAFERITE 30P 44M1
MAG 1 23B
MAG 2
see BS range 23B
MAG 3 23B
MAG 4
see BS range 23B
MAG 5 23B
MAG 6 23B
MAG 7
see BS range 23B
MAG 8 23B
MAG 9 23B
MAG 101
see BS range 23B
MAG 111 23B
MAG 121
see BS range 23B
MAG131
BS designation
MAG 141 23B
MAG 151
see BS range 23B
MAG 161
see BS range 23A
MAGLOY 1 20C
MAGLOY 1 44L

MAGLOY 2	20C
MAGLOY 5	20C
MAGLOY 5	44L
MAGLOY 6	44L
MAGLOY 7	20C
MAGLOY 7	44L
MAGLOY 8	20C
MAGLOY 8A	44L
MAGLOY 8B	44L
MAGLOY 10	20C
MAGLOY 10	44L
MAGLOY 15	44L
MAGLOY 100	20C
MAGLOY 100	44L
MAGNACUT	44L
MAGNESIUM	23A
MAGNETOFLEX 1 2	14E
MAGNETOFLEX 20	13A
MAGNICO	44L
MAGNO	27F
MAGNOLIA METAL	21.1
MAGNOX A 12	23B
MAGNUMINIUM 133	23B
MAGNUMINIUM 133 X	23B
MAGNUMINIUM 266	23B
MAIN METAL	55A
MALCALLOY	13A
MALLORY 3	14L
MALLORY 53	14M
MALLORY 53B	14D
MALLORY 73	14D
MALLORY 100	14D
MALLORY 150	14D
MALLORY 328	14L
MALLORY 1000	14J
MALLORY NO-CHAT	14J
MALLOY	44K2
MALOTTE	6.1
MALTEX X No 1	21.1
MALTEX X No 1 A	21.1
MALTEX X No 7	21.1
MALURECHER F12	44M1
MAN	44N1
MANAURITE 8S	44N1
MANAURITE 10	44N1
MANAURITE 12	44N1
MANAURITE 20	44N1
MANAURITE 35	44N1
MANAURITE 50W	27B
MANAURITE 60	27B
MANG 18	44D
MANGALAL	1B
MANGANESE	24.1
MANGANESE 94–95	24.1
MANGANESE 96–97	24.1
MANGANESE BRONZE	14B
MANGANESE BRONZE DI	14C
MANGANESE BRONZE DIL	14C
MANGANESE BRONZE IAT	14C
MANGANESE BRONZE IX	14C
MANGANESE Ni	44P
MANGANESE P	44N1
MANGANIN	14A
MANGANIN	14E
MANGDIE	44K3
MANGEAR	44A1
MANGNOL	27F
MANGONIC 2	27F
MANGONIC 3	27F
MANGONIC 5	27F
MANIFLEX	44N1
MANIFLEX 11 ALLOY	44N1
MANIFLEX FM ALLOY	44N1
MANITOBA	44A1
MANNESMANN-STAHL F12	44M1
MANOFORT 160	44K2
MANOFORT 180	44K2
MANOIR 15N4	44F2
MANOIR ABRADIER 600	44K1
MANOIR ABRADUR 220	44K3
MANOIR ABRADUR 240	44K2
MANOIR ABRADUR 300	44K2
MANOIR ABRADUR 400	44K2
MANOIR ABRADUR 500	44K2
MANOIR ABRADUR PM38	44K2
MANOIR APS10	44E1
MANOIR APS20	44E1
MANOIR APS25	44E1
MANOIR APS30	44E1
MANOIR PF1	44K1
MANOIR PF2	44K1
MANOIR PF5	44K1
MANOIR PF6	44K1
MANOIR PF15	44K1
MANOIR PFO	44G1
MANOIR PFV55	44K1
MANOIR PM17	44K1
MANOIR PM35	44K1
MANOIR RS1	44K2
MAN-TEN	44A1
MAN-TEN A440	44A1
MAO	44N1
MAO SUPERIOR	44N1
MAOY	44N1
MAR – M 918	13A
MAR 100	44F3
MARAGE 200	44F3
MARAGE 250	44F3
MARAGE 300	44F3
MARINE I	21.1
MARINE II	21.1
Marinel	14G
MARINER	44A1
MARKER ED12G	44M1
MARKER ED12W	44M1
MARKER EP12	44M1
MARKER S14M	44M1
MAR M 200	27C
MAR M 246	27C
MAR M 421	27C
MAR M432	27F
MAR-M-ALLOY Hf Modified	27F
MARSH CH 5	44K1
MARSH CMN	44K2
MARSH CTC 1	44K2
MARSH CTH	44K2
MARSH CTHY	44K1
MARSH CYC 5% Co	44L
MARSH CYC 275	44L
MARSH CYC 1841	44K3
MARSH HFC	44M2
MARSH MN	44E3
MARSH N3	44F2
MARSH NC	44K2
MARSH NC 1	44K2
MARVAC 18	44F3
MARVAC 18A	44F3
MARVAC 18LT	44F3
MARVAC 200	44F3
MARVAC 250	44F3
MARVAC 300	44F3
MAS	44K2
MASBw	44N1
MASCOT	21.1
MASD	44N1
MASN	44N1
MASO	44N1
MASO SUPERIOR	44N1
MASOY	44N1
MASOY SUPERIOR	44N1
MASW	44N1
MASWY	44N1
MAT	44N1
MATTABRAZE 12	14H
MATTABRAZE 45	14H
MATTE LADLE	44A2
MATTHEY 1W3	52.1
MATTHEY 3	14L
MATTHEY 3W3	52.1
MATTHEY 20K3	52.1
MATTHEY 30W3	52.1
MATTHEY 35S	52.1
MATTHEY 50S	42.1
MATTHEY 55X	42.1
MATTHEY 100	14D
MATTHEY 205	52.1
MATTHEY 328	14L
MATTHEY 2265	52.1
MATTHEY 2355	52.1
MATTHEY 2373	52.1
MATTHEY 3045	42.1
MATTHEY D54	42.1
MATTHEY D54X	42.1
MATTHEY D55	42.1
MATTHEY D56	42.1
MATTHEY D58	42.1
MATTHEY D510	42.1
MATTHEY D520	42.1
MATTHEY G13	42.1
MATTHEY G14	52.1
MATTIBRAZE 34	42.1
MAUSTINOX A	44N1
MAUSTINOX Abc	44N1
MAUSTINOX Atbc	44N1
MAUSTINOX B	44N1
MAUSTINOX Bbc	44N1
MAUSTINOX C	44N1
MAUSTINOX F	44N1
MAUSTINOX H	44N1
MAUSTINOX SA	44N1
MAUSTINOX SB	44N1
MAUSTINOX X	44N1
MAUSTINOX XD	44N1
MAUSTINOX Y	44N1

MAUSTINOX YD 44N1
MAUSTINOX YN 44N1
MAX CHIP 44K2
MAXEL 1B 44G1
MAXEL 2B 44G2
MAXEL 3 44K2
MAXELOY 44A1
MAXELOY 4 44F1
MAX FINE GRAIN 44A1
MAXHETE 1 44N1
MAXHETE 1A 44N1
MAXHETE 2 44N1
MAXHETE 3 44M1
MAXHETE 4 44N1
MAXHETE 5 27B
MAXHETE 8 27B
MAX HIGH MARG 44A1
MAX HIGH TENSILE 44K1
MAXILVRY 44N1
MAXILVRY AWP 44N1
MAXILVRY CB 44N1
MAXIMUM 18 44K3
MAXINIUM 44K1
MAXITE 44L
MAX MITH 44F2
MAXNAP 44K1
MAXTRESS 44K1
MAYARI R 44K1
MAZ 44N1
MAZAK 3 55A
MAZAK 5 55A
MAZAK 7 55A
MB 44M1
MB 2 44M1
MB 7 1N
MBK 8 44A1
MBK 8A1 44A1
MBK 10 44A1
MBK 13 44A1
M BOL-1 14J
M BOL-2 14J
MBV 44A2
MBV/m 44A2
MBV/w 44A2
MC 44K2
MC 2 44K3
MC 7 44A3
MC 102 27C
MCH 44K2
MCHD 44M2
M CHROME REGENT 44E3
MCHT 44K1
MCL 44K1
MCMO 44K3
MCN 44K1
MCP 20 50A
MCP 47 6.1
MCP 58 6.1
MCP 70 6.1
MCP 78 6.1
MCP 92 6.1
MCP 96 6.1
MCP 100 6.1
MCP 103 6.1
MCP 117 50A
MCP 124 6.1
MCP 137 6.1
MCP 145 50A
MCP 176 50A
MCP 183 50A
MCP 200 50A
MCP 221 50A
MCP 227 50A
MCS 6 44N1
MCT 44K3
MCV 44K2
MCV 24 44K1
MED Mn A440 44A1
MEIGH METAL 14G
MELLANEAR GUAR 99.9 50.1
MELLANEAR REFINED 50.1
MELPURE 23A
MELT-ESI MX 12 42.1
MELT-ESI MX 18 42.1
MELT-ESI MX 18 PLUS 42.1
MELT-ESI No 2 42.1
MEL-TROL 484FM 44K3
MEL-TROL 610FM 44M2
MEL-TROL H46 44M1
MEL-TROL HAMPDEN 44M2
MEL-TROL HI SHOCK 60 44K3
MEL-TROL HI WEAR 64 44K3
MEL-TROL K-W 44H
MEL-TROL MIRROMOLD 44J
MEL-TROL SPEED STAR 44K3
MEL-TROL STAINLESS 2 44M1
MEL-TROL STAR ZENITH 44K3
MEL-TROL SUPER SAMSON 44K1
MEL-TROL TGS 44A2
MEL-TROL TK 44K1
MEL-TROL VEGA 44K3
MEL-TROL VEGA-FM 44K3
MERMAID 44K3
METALFLO 14H
METASIL 14H
Met Bronze A6MN 14G
Met Bronze A8N 14G
MET BRONZE PT8 14K
METCO 56.1
METCO 12C 27B
METCO 14E 27B
METCO 15C 27B
METCO 15E 27B
METCO 15F 27B
METCO 16C 27C
METCO 17F 27C
METCO 18C 13A
METCO 19E 27B
METCO 31 CNS 27F
METCO 31C 27B
METCO 32C 52.1
METCO 36 C 27F
METCO 41C 44N1
METCO 41F-NS 44N1
METCO 42C 44M1
METCO 42F 44M1
METCO 43C 27B
METCO 43F-NS 44N1
METCO 44 27B
METCO 45C NS 13A
METCO 45VF NS 13A
Metco 51 14G
Metco 51F NS 14G
METCO 52C NS 1C
Metco 54NS 1A
METCO 55 14A
METCO 56 C NS 27A
METCO 56 F NS 27A
METCO 57 NS 27D
METCO 58 NS 27D
METCO 61 52.1
METCO 62 46.1
METCO 63 NS 26A
METCO 64 26A
METCO 68F NSI 13A
METCO 71 NS 52.1
METCO 71 VF NS 52.1
METCO 72 F NS 52.1
METCO 73 NS1 52.1
METCO 74 SF 52.1
METCO 75 F 52.1
METCO 81NS 12.1
METCO 83VF 12.1
METCO 91 44A1
METCO 101B NS 1N
METCO 101 NS 1N
METCO 101 SF 1N
METCO 102 51A
Metco 105 1A
Metco 105 NS 1A
Metco 105 NS.1 1A
Metco 105 SF 1A
METCO 106 12.1
METCO 110 51A
METCO 111 51B
METCO 130 51B
METCO 130 SF 51B
METCO 131 VF 51B
METCO 136CP 12.1
METCO 136F 12.1
METCO 143 56.1
METCO 201 56.1
METCO 202 NS 56.1
METCO 210 NS 56.1
METCO 350 26A
METCO 350 C 26A
METCO 402 44N1
METCO 404 NS 27F
METCO 405 27F
METCO 405 NS 27F
METCO 430 NS 12.1
METCO 439 13A
METCO 440 27C
METCO 443 NS 27F
METCO 451 27F
METCO 463 44M1
METCO 465 44M1
METCO 470 AW 27F
METCO 501 26A
METCO 505 26A
METCO 601 NS 1C
Metco 605 NS 14G
Metco 610 NS 14G
METCO 625 44N1
METCO 1123 52.1
Metco Aluminium 1A
METCO COPPER 14A
METCOLOY 1 44N1
METCOLOY 2 44M1

METCOLOY 4 44N1
METCOLOY 5 44P
METCOLOY 33 27B
METCO MOREL 27D
METCO NICKEL 27A
METCO SF Al 1C
Metco Sprabronze AA 14G
METCO TIN 50.1
METCO XP 1150 27B
METCO ZINC 55A
METCO ZINC Al 55A
METER BRONZE 14B
MET-HARD 250 44K1
MET-HARD 350 44K1
MET-HARD 400HW 44K1
MET-HARD 450 44K1
MET-HARD 450HW 44K1
MET-HARD 450HW 44K2
MET-HARD 500HW 44K1
MET-HARD 530HW 44K1
MET-HARD 550HW 44K1
MET-HARD 650 44K2
MET-HARD 650M 44K2
MET-HARD 750TS 44K3
MET-HARD 850 44M2
MET-HARD 950 44M2
MET-HARD 1050 44M2
MET-HARD H13 44K1
MET MAX 307R 44P
MET-MAX 19.12.3L 44N1
MET-MAX 19.9L 44N1
MET-MAX 20.9.3 44N1
MET-MAX 23.12 44N1
MET-MAX 23.12.2 44N1
MET-MAX 25.5 44M1
MET-MAX 29.9 44N1
MET-MAX 307R 44N1
MG 2 1D
MG 3 1D
MG 5 1D
MG 5S 1D
MG 7 1D
MG 7 1J
MG11A 1A
MG 11A 1B
Mg A1 3 Zn
see DIN range 23B
Mg A1 6 Zn
see DIN range 23B
Mg A1 8 Zn
see DIN range 23B
MG C 42 23B
MG C 43 23B
MG C 51 23B
MG C 61 23B
MG C 91 23B
Mg Mn 2
see DIN range 23B
MG P 43 23B
MG P 62 23B
MG P 63 23B
MGR 44K2
Mg Zn 6 Zr
see BS range 23B
MIC 44E3
MIC 2 44E3
MIC 3 44K3
MIC 4 44K3
MIC 8 44G2
MIC 9 44K3
MICROBRAZ 5025 14H
MICROBRAZ 5027 14H
MICROLIM 44E3
MIDOHM 14E
MIDVAC 422 44M1
MIL A 10841 2.1
MIL A 18001 51B
MIL AB259 44P
MILDEES 44A1
MILDTRODE 46.00 44A1
MIL–F–83142 51B
MIL M 8916 23B
MIL M 8917 23B
MIL M 26075 23B
MIL M 45207 23B
MIL M 46037 23B
MIL M 46062 23B
MIL M 46062 B 23B
MIL-N-18088 27C
MIL-N-6840 27B
MIL-N-7786 27C
MIL R 6944 23B
MIL-R-5031 27C
Mil S 43 44A2
MIL S 645 44A1
MIL S 780.9 44A1
MIL S 980 44E3
Mil S 3289 44A2
MIL-S 12504 Mn Cr Mo 44K1
MIL S 12504 Mn Mo 44G1
MIL S 12505.1 44F1
MIL-S 12505/2 44K1
MIL S 12505.3 44A1
MIL S12505.4 44E1
MIL S 12505/5 44J
MIL S 12505.6 44A1
Mil S 15083-70-36 44A2
Mil S 15083-80-40 44A2
MIL-S 15083- 100-70 44K1
MIL-S 15464-1 44K1
MIL-S 15464-2 44K1
MIL-S 15464-3 44K1
MIL S 16113-1 44J
MIL S 16113-11 44J
MIL-S-16538 44N1
MIL-S-16598 44N1
MIL-S-16993 44M1
MIL S.17249 44P
MIL-S 21925 44K1
Mil S 18591-1040 44A2
Mil S 22141-1040 44A2
Mil S 22141-1050 44A2
Mil S 22698(C) 44A2
MIL S 24412 44J
MIL-S 46047 44K1
MIL-S-81591 44M1
Mil S 81591-1050 44A2
MIL Spec 1·24137(B) 20B
MIL Spec 1·24137(C) 20B
MIL–T–009046 51B
MIL–T–9046 51B
MIL-T-9046 51A
MIL-T-8506 44N1
MIL–T–46035 51B
MIL W 8957 44A4
MIL W 21425-1 44A3
MINERVA HC 44K2
MINERVA LC 44K2
MINOX 10 44M1
MINOX 15 44M1
MINOX 20 44M1
MINOX 30 44M1
MINOX 1610 44M1
MINOX 1620 44M1
MINOX 1820 44M1
MIRALITE 1L
MISCHMETALL 11.1
MISPICKEL 3.1
MK 2 44K1
MK 75 44A3
MK 101
see DIN range 44A4
MKDC 1 14C
ML 44K3
ML 2 1F
ML 50 44K1
ML 60 44K1
ML 70 44K1
ML 1700 13A
MLX 45 44J
MLX 50 44J
MLX 55 44J
MLX 60 44J
MM 01 44G1
MM 02 44G1
M Mo 30 44K1
M Mo 38 44K2
MN 44E3
MNC 46E 23B
MNC 71 55A
MNC 71E
Swedish standard for zinc die castings; see designation (SS) for details
MNC705
General specification for grey cast iron; see SS designation
MNC 706 20B
MNC 707 E 20B
MNC 720 44A2
MNC 810 44A1
MNC 811 44A1
MNC 815 44A1
MNC 815E 44A1
MNC830
General specification - steel for pressure vessels; see SS designation
MNC 831 44A1
MNC 831 44K1
MNC 831E
Swedish standard for pressure vessels; see designation (SS) for details
MNC 832 44A1
MNC 832 44K1
MNC 840 44A1
MNC 845 44A1
MNC 851 44A1
MNC 851 44K2

MNC 851
Swedish standard for case hardening steels; see designation (SS) for details
MNC 852 44A2
MNC 853 44K1
MNC 854 44A2
MNC 870 44A2
MNC 900 44M1
MNC 900 44N1
MNC 900E Swedish standard; summary of stainless steel; see designation (SS) for details
MNC901
General specification for stainless steel for pressure vessel; see SS designation
MNC 915 44A1
MNC 916 44A1
MNC 920 44A1
MNC 1205 20B
Mn Cr 44E1
Mn Cr 44E3
Mn Mo 44G1
MN Ni 99.5
see DIN range 27A
MNS NICKEL 27A
MNV 44J
MNV A441 44J
Mo 18 44K1
Mo B 44G1
MOCARB 44K3
MO CUT 44K3
MODAL 5 44K1
MODIFIED HILO 27F
MOGUL 44K3
MOHD 44K1
MOHD 44K2
MOHICAN 6 44K2
MOHICAN 8 44K3
MOLEX 44G1
MOLITE 44K3
MOLITE 1 44K3
MOLITE 3 44K3
MOLITE 3 SMOOTHCUT 44K3
MOLITE 3 steel;
see SMOOTHCUT 44K3
MOLITE 4 44K3
MOLITE HW60 44K2
MOLVA 44K3
MOLVA T 44K3
MOLVA TC 44K3
MOLY ASCOLOY 44M1
MOLYBDENUM 26A
MOLYCROM 05 44K1
MOLYCROM 15 44K1
MOLYCROM 21 44K1
MOLYCROM 21 STC 44K1
MOLYCROM 55 44K1
MOLYCROM 91 44K1
MOLYCROM TI 15 44K1
MOLYCUT 562 44K3
MOLYTERM 5 44G1
MOLYTERM Ti 5 44G1
MOLYTRODE 74.56 44G1
MONARC 44A1
MONARCH 44K3
MONARCH 652 44K3
MONARCH BLA 44K2
MONARCH DW8 44K1
MONARCH GENERAL
UTILITY 44E2
MONARCH HCR 44M2
MONARCH NCG 44K1
MONARCH NCV 44K2
MONARCH OHB 44K3
MONARCH PCS 44K3
MONARCH SPECIAL TAN 44K1
MONARCH TAN 44K3
MONARK 1 44G2
MONARK 2 44K2
MONEL 40 27D
MONEL 43 27D
MONEL 44 27D
MONEL 60 27D
MONEL 64 27D
MONEL 130 27D
MONEL 134 27D
MONEL 140 27D
MONEL 180 27D
MONEL 400 27D
MONEL 401 27D
MONEL 402 27D
MONEL 403 27D
MONEL 404 27D
MONEL 406 27D
MONEL 414 27D
MONEL 450 14E
MONEL 501 27D
MONEL 502 27D
MONEL 505 27D
MONEL 506 27D
MONEL C and C 27D
MONEL FM60 27D
MONEL K405 27D
MONEL K500 27D
MONEL R405 27C
MONEMEL 1 44A1
MONEMEL 2 44A1
MONOTYPE ALLOY 21.1
MONOTYPE METAL 21.1
MOSIT LOW C 44M1
MOSTAR 44K1
MOTEMP 44K3
MO-TEMP M10 44K3
MOTOR BRAND 44A4
MOTOR MAGNUS 44K3
MOTOR MAXIMUM 44K3
MOTOR SPECIAL 44K3
MOTUF 44K3
MOTUNG 44K3
MOTUNG 652 44K3
MOTUNG CV 44K3
MOULD STEEL 44K1
MOV 44K1
MOVA 44K1
MOVAN 44K3
MOW 562 44K3
MP 35N 27C
MP 35N ALLOY 13A
MP 35N MULTIPHASE 27F
MP–159 27F
MR 6 44A3
MR St 1 44A1
MR St 2 44A1
MR St 3 44A1
MR St 4 44A1
MR St 34/2 44A1
MR St 37/2 44A1
MR St 42/2 44A1
MR St 44 44A1
MS 6 44L
MS 15 44L
MS 34 44A2
MS 35 44L
MS 43 44A2
MS 45 44A3
MS 50 44A3
Ms 56 14B
Ms 58 14B
Ms 60 14B
Ms 60 Pb 14B
Ms 63 14B
Ms 63 Pb 14B
Ms 67 14B
Ms 72 14B
Ms 80 14B
Ms 85 14B
Ms 90 14B
MS BV 44A2
MSC 40 44E2
MSC 50 44E2
MSR A 23B
MSR B 23B
M St 0 44A1
M St 34/2 44A1
M St 34/3 44A1
M St 37/2 44A1
M St 37/3 44A1
M St 42/2 44A1
M St 42/3 44A1
M St 50/2 44A2
M St 52/3 44A1
M St 60/2 44A2
MST 0.2 Pd 51A
MST 2
see GOST range 44A1
MST 2KP
see GOST range 44A1
MST 3
see GOST range 44A1
MST 3 Al 2.5 V 51B
MST 3KP
see GOST range 44A1
MST 4
see GOST range 44A1
MST 4 Al 4 Mn 51B
MST 5 Al 2.5 Sn 51B
MST 7 Al 2 Cb 1 Ta 51B
MST 7 Al 12 Zr 51B
MST 8 Al 1 Mo 1 V 51C
MST 8 Mn 51B
MST 30 51A
MST 40 51A
MST 55 51A
MST 70 51A
MSV 44A1
MS V3 44A2
MSVB/W 44A1
MSVD 44A1

MSZ 8277 20B
MSZ 8280 66 C 81 20B
MT 44A4
MT 'P' 44A2
MT 'R' 44A2
MT 'R' 44G1
MTA 2 44K1
MTA 3 44K1
MTC 26A
MT Mo 'P' 44G1
MTR 44K1
MTS 1 44M1
MTS 5 44M1
MTS 6 44M1
MTZ 23B
MUK 8 44A1
MUK 10 44A1
MULTICO 44L
MULTIMET 44L
MUMETAL 27C
MUMETAL 40 27C
MUMETAL 40CT 27C
MUMETAL 60 27C
MUNTZ 14B
MURAWIRE W1 44A1
MURAWIRE W2 44A1
MUSHET 44K2
MUSHET HIGH
 VANADIUM 44K3
MUSHET M 44K3
MUSHET MOLYB. 44K3
MUSIC WIRE 44A4
MUSKET HIGH V Co 44L
MUSKET MKK 44K3
MUSKET SPECIAL 44L
MUSKET SPECIAL VG 44L
MU St 1 44A1
MU St 2 44A1
MU St 3 44A1
MU St 4 44A1
MU St 34 44A1
MU St 34/2 44A1
MU St 37/2 44A1
MU St 42/2 44A1
MUSTANG 44K3
MUSTANG M3/1 44K3
MUSTANG M3/2 44K3
MUTCO 32C 52.1
MV-1-VAC-ARC 44K3
MVC 1C
MVW 44K1
MX 0 42.1
MX 4 14H
MX 5 42.1
MX 8 42.1
MX 12 42.1
MX 18 42.1
MX 18 PLUS 42.1
MY 0 44E2
N 1 44K1
N 1/57 44F1
N 2 1C
N 2 44K1
N 2MCH 44K1
N 3 1B
N 3 20B
N 3 44F2
N 3 44K1
N 3CCH 44K1
N 3CH 44F1
N 3CMCH 44K1
N 3MCH 44K1
N4 1D
N 4 44K2
N 4CCH 44K1
N 4CMCH 44K1
N 5
 see BS range 1D
N5 1D
N 5CH 44F1
N 5MCH 44K1
N 6
 see BS range 1D
N6 1D
N7 1D
N 8
 see BS range 1D
N8 1D
N 10 44F2
N 12 44K1
N 12M 27F
N 20 44F2
N 21 1C
N 23 44M1
N 30 44F2
N 31
 see BS range 1D
N41 1D
N 42 20C
N 50A 5.1
N 51
 see BS range 1D
N51 1D
N52 1D
N 52 27E
N54 44A1
N 54 27E
N 58 27E
N 75 27B
N 82 44A2
N 91 44A2
N100 44M1
N 100A 5.1
N104 44M1
N106 44M1
N108 44M1
N 153 44L
N 155 44L
N 156 44L
N200 44M1
N 200A 5.1
N 220 C 27A
N310 44M1
N315 44M1
N320 44M1
N350 44M1
N400 44M1
N404 44M1
N 501 20C
N 501 27E
N530 44M1
N540 44M1
N 555 44M2
N 685 44M2
N 690 44M2
N 700 44N3
N02016
 UNS for Alumel 27B
N02061
 UNS for AWS designation
 ERNi1 27B
N02100
 UNS for CZ100 27A
N02200
 UNS for Nickel 200 27A
N02201
 UNS for Nickel 201 27A
N02205
 UNS for Nickel 205 27A
N02211
 UNS for Nickel 211 27F
N02215
 UNS for AWS designation
 ERNiCl 27B
N02216
 UNS for AWS designation
 ERNiFeMnCl 20C
N02220
 UNS for Nickel 220 27A
N02225
 UNS for Nickel 225 27A
N02230
 UNS for Nickel 230 27A
N02233
 UNS for Nickel 233 27A
N02290
 UNS for Commercially
 pure nickel 27A
N03220
 UNS for 220C 27F
N03260
 UNS for AMS5865 27A
N03300
 UNS for Permanickel 300 27A
N03301
 UNS for Duranickel 301 27F
N04019
 UNS for AMS4892 27D
N04020
 UNS for QQ N281 27D
N04060
 UNS for AWS designation
 ERNiCu7 27D
N04400
 UNS for Monel 400 27D
N04401
 UNS for Monel 401 27D
N04404
 UNS for Monel 404 27D
N04405
 UNS for Monel R405 27D
N05500
 UNS for Monel K500 27D
N05502
 UNS for Monel 502 27D
N06001
 UNS for Hastelloy F 27C
N06002
 UNS for Hastelloy X 27C
N06003
 UNS for AMS5676 27B

N06004
UNS for nichrome 27B
N06005
UNS for Eatonite 27F
N06006
UNS for ASTM A297 44N1
N06007
UNS for Hastelloy G 27F
N 06008 27B
N 06009 27B
N06010
UNS for Chromel 27B
N06017
replaced by N06985
N06022
UNS for Hastelloy C 27C
N06030
UNS for Hastelloy G 27F
N06040
UNS for CY40 27B
N06050
UNS for AC1 HX50
N06062
UNS for AMS 5679 27B
N06075
UNS for Nimonic 75 27B
N06082
UNS for Inconel FM82 27B
N06102
UNS for IN102 27C
N06132
UNS for Irconel WE132 27B
N06333
UNS for AMS 5593 27C
N 06333 27B
N06455
UNS for Hastelloy C4 27F
N06600
UNS for AMS 5540 27B
N06601
UNS for AMS 5715 27C
N06617
UNS for AWS designation for ERNiCrMo1 27B
N 06621 27B
N06625
UNS for AMS 5401 27C
N 06625 27B
N06635
UNS for AMS 5401 27C
N06690
UNS for Inconel 690 27B
N06782
UNS for VF4 27C
N06804
UNS for Incoloy 804 27C
N06975
UNS for AWS designation ERNiCrMo8 27B
N07002
UNS for HEV2 27C
N07012
UNS for AMS 5855 27C
N07013
UNS for AMS 5404 27C
N07031
UNS for Pyromet 31 27C
N07041
UNS for Rene 41 27C
N07069
UNS for AWS designation ERniCrFe7 27B
N07080
UNS for Nimonic 80A 27C
N07090
UNS for Nimonic 90 27C
N07092
UNS for AWS designation ERNiCrFe6 27B
N07252
UNS for AMS 5551 27C
N07263
UNS for AMS 5872 27C
N07500
UNS for AMS 5384 27C
N07626
UNS for NACE MR01-75 27B
N07702
UNS for AMS 5556 44N1
N07713
UNS for IN713 27C
N07716
UNS for ASTM B637 27C
N07718
UNS for AMS 5383 27C
N 07718 27C
N07721
UNS for Inconel 721 27C
N07722
UNS for AMS 5541 27C
N07725
UNS for Inconel 721 27C
N07750
UNS for AMS 5542 27C
N07751
UNS for Inconel 751 27C
N07754
UNS for Inconel MA754 27C
N08001
UNS for ASTM A297 27C
N08002
UNS for ASTM A297 27C
N08004
UNS for ASTM A297 27C
N08005
UNS for ASTM A608 27B
N08006
UNS for ASTM A608 27B
N08007
UNS for ASTM A351
N08008
UNS for HT 44N1
N08020
UNS for Carpenter 20CD3 44N1
N08021
UNS for AWS designation ER320 44N1
N08022
UNS for AWS designation ER320LR 44N1
N08024
UNS for Carpenter 20MO6 44N2
N08026
UNS for Carpenter 20MO6 44N2
N08028
UNS for Sanicro 28 44N2
N 08028 44N1
N08030
UNS for ASTM A351 44N1
N08032
UNS for INCO-032 44N1
N08050
UNS for ASTM A608 44N1
N08065
UNS for AWS designation ERNiFECr1 27B
N08151
UNS for ASTM A351 44N1
N08221
UNS for Carpenter 20Cb3 44N1
N08245
UNS for ARMCO20-10 Obsolete 44N1
N08320
UNS for Carpenter 20Cb3 44N1
N08321
UNS for AWS designation ER320 44N1
N08330
UNS for AMS 5592 44N1
N 08330 44N1
N08331
UNS for AWS designation ER330 44N1
N08332
UNS for Carpenter 20Cb3 44N1
N 08332 44N1
N08334
UNS for RA 330 44N1
N 08366 44N2
N 08367 44N2
N 08421 27B
N08603
UNS for HT30 44N1
N08604
UNS for ASTM A297-HL 44N1
N08605
UNS for ASTM A297-HT 44N1
N08613
UNS for ASTM A608HL30 44N1
N08614
UNS for ASTM A608 HL40 44N1
N 08700 44N1
N08705
UNS for HP 44N1
N08800
UNS for AMS 5766 44N1
N08801
UNS for AMS 5552 44N1
N08801
UNS for Incoloy 800H 44N1
N08802
UNS for Incoloy 802 44N1
N08811
UNS for Incoloy 800HT 44N1
N08825
UNS for Incoloy 825 27C
N 08825 27B
N 08904 44N2

N 08925 44N2
N08932
UNS for Creusot URSB8 44N2
N09027
UNS for AMS 5633 27E
N09706
UNS for AMS 5605 27E
N09901
UNS for AMS 5660 27C
N09902
UNS for AMS 5221 20C
N 09911 27B
N 09925 27B
N09926
UNS for Incoloy 926 44N1
N09979
UNS for AMS 5509 27C
N10001
UNS for AWS designation ERNiMol 27F
N10002
UNS for AMS 5388 27C
N10003
UNS for AMS 5607 27F
N10004
UNS for AWS designation ERNiMo3 27F
N10276
UNS for Hastelloy 276 27F
N10665
UNS for AWS designation ERNiMo7 27B
N13009
UNS for AMS 5407 27C
N13010
UNS for AMS 5405 27C
N13017
UNS for AMS 5851 27C
N13020
UNS for AMS 5846 27C
N13100
UNS for AMS 5397 27C
N13246
UNS for Mar M-Alloy - Hf modified 27F
N19903
UNS for Incoloy 903 20C
N 19907 27F
N 19909 27F
N96612
UNS for AWS designation BN.9 27F
N96644
UNS for AWS designation RNiCrA 27F
N99600
UNS for AMS 4175 1D
N99610
UNS for AWS designation BNila 27F
N99620
UNS for AWS designation BNi2 27F
N99622
UNS for AWS designation BNi10 27F
N99624
UNS for AWS designation BNi11 27F
N99630
UNS for AWS designation BNi3 27F
N99640
UNS for AWS designation BNi4 27F
N99645
UNS for AWS designation RNiCrB 27F
N99646
UNS for AWS designation RNiCRB 27F
N99646
UNS for AWS designation RNiCrC 27F
N99650
UNS for AWS designation BNi5 27F
N99700
UNS for BNi6 27F
N99710
UNS for BNi7 27F
N99800
UNS for BNi8 27F
NA 44K2
N A XTRA 44K1
N A XTRA 55 44K1
N A XTRA 60 44K1
N A XTRA 65 44K1
N A XTRA 70 44K1
N A XTRA 75 44K1
N A XTRA 80 44K1
N A XTRA 93 44K1
N A XTRA 100 44K1
NA 1
see BS range 27D
NA 2
see BS range 27D
NA 3
see BS range 27D
NA 11
see BS range 27A
NA 12
see BS range 27A
NA 13
see BS range 27C
NA 14
see BS range 27B
NA 16
see BS range 27C
NA 17
see BS range 27B
NA 18
see BS range 27B
NA 19
see BS range 27C
NA 19 27C
NA 20 27C
NA 21 27C
Na 22 H 27B
NA 22H (c) 27C
NA 32
see BS range 27A
NA 33
see BS range 27D
NA 34
see BS range 27B
NACE MR 01.75 27B
NAG 19.9L 44N1
NAG 19.9LR 44N1
NAP SUPERIOR 44L
NARITE 14G
NARVE 44K2
NAT 44K2
NATIONALLOY 14 44K1
NATRONA 44K3
NAVAL BRASS 14C
NAVY 50A
NAX 9115-AC 44K1
NB 27C
NB 44N1
NBN 267/21/22 Br Sn 5 Zn 5 Pb 5 14K
NBN 267/21/22 Br Sn 7 Zn 3 Pb 4 14K
NBN 267/21/22 Br Sn 10 14K
NBN 267/21/22 Br Sn 10 Zn 2 14K
NBN 267/21/22 Br Sn 12 14K
NBN 267/21/23 Br Sn 9 P 14K
NBN 267/23 Br Pb 15 Sn 8 14K
NBN 267/23 Br Pb 24 Sn 5 14K
NBN 267/23 Br Sn 10 Pb 10 14K
NBN 436
see BS range AlCu4Ni2Mg 1L
NBN 436 Al Mg 6 1J
NBN 436 Al Mg 11 1J
NBN 436 AlCu2NiMg 1L
NBN 436 AlCu4Ni2Mg 1L
NBN 436 AlCu5MgTi 1L
NBN 436 AlSi2Cu 1L
NBN 436 AlSi5Cu 1L
NBN436 Al Si 5 Mg 1C
NBN 436 AlSi9Cu3 1L
NBN 436 Al Si 10 Mg 1C
NBN 436 Al Si 12 1C
NBN 830.01 20B
NBN 830.01 FGG40 20B
NBN 830.02 20B
NBV 1 44F1
NBV 1/h 44F1
NBV 2/w 44F1
NBV 5/w 44F1
NBVOC 1/W 44K1
NC 27C
NC 44K2
NC 1 44K2
NC 2 (Mo) 44M1
NC 20T 27B
NCA
see ASTM A823 20B
NCB
see ASTM A823 20B
NCC
see ASTM A823 20B
NCC PIG 27D
NCF 2 44N1
NCFI 27B
NCM 1 44K1
NCM 1 44K2
NCM 2 44K1

NCM 3 44K1
NCM 4 44K1
NCM 4 44K2
NCM 6 44K1
NCM 30 44K1
NCM 40 44K1
NCM IHC 44K3
NCMo 44K1
NC MO 44K2
NCMS 44K2
NCM SPECIAL 44K1
NCMV 44K2
NCR 2 44K1
ND 44K3
NDS 44K3
NE 8150 44K1
NEATRO 44K3
NEBALOY 14B
NELCO 10.1
NEN 6002 A 20B
NEN 6002 D 20B
NEN 6033/Cu Ni 10 14E
NEODYMIUM 35
NEOR 44M2
NEUTRO SORB 44N1
NEUTRO SORB PLUS 44N1
NEUTROTHERM KW20M 44M1
NEW CAPITAL 44K3
NEWHALL 44K3
NEW KS 44L
NEWLOY 14E
NEWTON 6.1
NF 44K2
NF 2 14C
NF 25-20L 44N1
NF 25-20LR 44N1
NF A 35 572/ Z12CN 17/08 44N3
NFA 32.101 20B
NFA 32.201 20B
NF A35 351/XC55 44A3
NF A35 501/A33 44A1
NF A35 501/A33/2 44A1
NF A35 501/A34/1 44A1
NF A35 501/A34/2 44A1
NF A35 501/A37/1 44A1
NF A35 501/A37/2 44A1
NF A35 501/A37/3 44A1
NF A35 501/A37/4 44A1
NF A35 501/A42/1 44A1
NF A35 501/A42/2 44A1
NF A35 501/A42/3 44A1
NF A35 501/A42/4 44A1
NF A35 501/A47/2 44A1
NF A35 501/A47/3 44A1
NF A35 501/A47/4 44A1
NF A35 501/A52/2 44A1
NF A35 501/A52/3 44A1
NF A35 501/A52/4 44A1
NF A35 501/A60/1 44A1
NF A35 501/A60/2 44A1
NF A35 501/A70/2 44A3
NF A35 501/E24/1 44A1
NF A35 501/E24/2 44A1
NF A35 501/E24/3 44A1
NF A35 501/E24/4 44A1
NF A35 501/E26/1 44A1
NF A35 501/E26/2 44A1
NF A35 501/E26/3 44A1
NF A35 501/E26/4 44A1
NF A35 501/E30/2 44A1
NF A35 501/E30/3 44A1
NF A35 501/E30/4 44A1
NF A35 501/E36/2 44A1
NF A35 501/E36/3 44A1
NF A35 501/E36/4 44A1
NF A35 501/E37/2 44A1
NF A35 551/ 18NCDG 44K1
NF A35 551/ 30CAD6/12 44K1
NF A35 551/16MC5 44E1
NF A35 551/16N06 44K1
NF A35 551/20M5 44A1
NF A35 551/20MCS 44E1
NF A35 551/20NC6 44K1
NF A35 551/20NCD2 44K1
NF A35 551/25CD4 44K1
NF A35 551/30CD4 44K1
NF A35 551/30CD12 44K1
NF A35 551/30CND8 44K1
NF A35 551/32C 44E1
NF A35 551/35 M5 44A2
NF A35 551/35CD4 44K1
NF A35 551/35NC6 44K1
NF A35 551/35NCD6 44K1
NF A35 551/38C2 44E2
NF A35 551/38C4 44E2
NF A35 551/42C2 44E2
NF A35 551/42C4 44E2
NF A35 551/42CD4 44K2
NF A35 551/50CV4 44K2
NF A35 551/10006 44E2
NF A35 551/X12 44A1
NF A35 551/XC10 44A1
NF A35 551/XC18 44A1
NF A35 551/XC25 44A2
NF A35 551/XC32 44A2
NF A35 551/XC38 44A2
NF A35 551/XC42 44A2
NF A35 551/XC48 44A3
NF A35 551/XC65 44A3
NF A35 551/XC70 44A3
NF A35 551/XC80 44A3
NF A35 561/12MF4 44A1
NF A35 561/S250 44A1
NF A35 561/S250 Pb 44A1
NF A35 561/S300 44A1
NF A35 561/S300 Pb 44A1
NF A35 562/10 Pb F2 44A1
NF A35 562/10F2 44A1
NF A35 562/20F2 44A1
NF A35 562/35MF4 44A1
NF A35 565/ 16NCD13 44K1
NF A35 565/18NCD4 44K1
NF A35 565/20NCD2 44K1
NF A35 565/20NCD7 44K1
NF A35-565/100C2 44E3
NF A35-565/100C6 44E3
NF A35 565/100CD7 44K3
NF A35 571/41S7 44A2
NF A35 571/45C4 44E2
NF A35 571/45SCD6 44K2
NF A35 571/46S7 44A2
NF A35 571/50CV4 44K2
NF A35 571/50S7 44A2
NF A35 571/51S7 44A2
NF A35 571/56S7 44A3
NF A35 571/60S7 44A3
NF A35 571/60SC7 44E2
NF A35 571/61S7 44A3
NF A35 572/ Z2CN18/10 44N1
NF A35 572/ Z2CND17/11 44N1
NF A35 572/ Z2CND17/12 44N1
NF A35 572/ Z2CND17/13 44N1
NF A35 572/ Z2CND19/15 44N1
NF A35 572/ Z5CNDU21/08 44N1
NF A35 572/ Z6CN18/09 44N1
NF A35 572/ Z6CNDNb17/12 44N1
NF A35 572/ Z6CNDNb17/13 44N1
NF A35 572/ Z6CNNb18/11 44N1
NF A35 572/ Z6CNT18/11 44N1
NF A35 572/ Z6CNU 17/04 44N3
NF A35 572/ Z6CNUD 15/04 44N3
NF A35 572/ Z8CD17/01 44M1
NF A35 572/ Z8CN13/13 44N1
NF A35 572/ Z8CN18/12 44N1
NF A35 572/ Z8CNDT17/12 44N1
NF A35 572/ Z10CF17 44M1
NF A35 572/ Z10CN18/09 44N1
NF A35 572/ Z10CNF18/09 44N1
NF A35 572/ Z10CNT18/11 44N1
NF A35 572/ Z12CF13 44M1
NF A35 572/ Z100CD17 44M1
NF A35 572/Z6C13 44M1
NF A35 572/Z6CA13 44M1
NF A35 572/Z8C17 44M1
NF A35 572/Z12C13 44M1
NF A35 572/Z20C13 44M1
NF A35 572/Z30C13 44M1
NF A35 572/Z40C14 44M1
NF A35 573/ Z2CN18/10 44N1
NF A35 573/ Z2CND17/12 44N1
NF A35 573/ Z2CND17/13 44N1
NF A35 573/ Z2CND19/15 44N1
NF A35 573/ Z5CNUD21/08 44N1
NF A35 573/ Z6CN18/09 44N1
NF A35 573/ Z6CND17/11 44N1
NF A35 573/ Z6CNDNb17/12 44N1
NF A35 573/ Z6CNNb18/11 44N1
NF A35 573/ Z6CNT18/11 44N1
NF A35 573/ Z8CD17/01 44M1
NF A35 573/ Z8CN13/13 44N1
NF A35 573/ Z8CN18/12 44N1
NF A35 573/ Z8CNDT17/12 44N1
NF A35 573/ Z10CN18/09 44N1
NF A35 573/ Z10CNT18/11 44N1
NF A35 573/Z6C13 44M1
NF A35 573/Z6CA13 44M1
NF A35 573/Z8C17 44M1
NF A35 573/Z12C13 44M1
NF A36 205/A37C1 44A1
NF A36 205/A37C2 44A1
NF A36 205/A37P1 44A1
NF A36 205/A37P2 44A1
NF A36 205/A42C1 44A1
NF A36 205/A42C2 44A1
NF A36 205/A42P1 44A1
NF A36 205/A42P2 44A1
NF A36 205/A48C1 44A1
NF A36 205/A48C2 44A1
NF A36 205/A48P1 44A1
NF A36 205/A48P2 44A1
NF A36 205/A52C1 44A1

NF A36 205/A52C2 44A1
NF A36 205/A52P1 44A1
NF A36 205/A52P2 44A1
NF A36 206/ 10CD910 44K1
NF A36 206/ 15CD205 44K1
NF A36 206/ 15CD405 44K1
NF A36 206/ 15MDV405 44K1
NF A36 206/ 18MD405 44K1
NF A36 206/15D3 44K1
NF A36 208/0.5NiA 44K1
NF A36 208/1.5Ni 44K1
NF A36 208/2.25Ni 44F1
NF A36 208/3.5Ni 44F1
NF A36 208/5Ni 44K1
NF A36 208/9Ni 44F1
NF A36 208/A42FP1 44A1
NF A36 208/A42FP2 44A1
NF A36 208/A48FP1 44A1
NF A36 208/A48FP2 44A1
NF A36 208/A52FP1 44A1
NF A36 208/A52FP2 44K1
NF A36 209/ Z2CN18/10 44N1
NF A36 209/ Z6CN18/09 44N1
NF A36 209/ Z6CNT18/11 44N1
NF A51 102 Cu Ni 10 Fe 1 Mn 14E
NF A51 102 Cu Ni 30 Mn 1 Fe 14E
NF A53 012 14K
NF A53 607 14K
NF A53 707 14K
NFA 53-100 14A
NF A55 101 Z9 55A
NF A55 101/2Z4 55A
NF A55 101/2Z5 55A
NF A55 101/2Z6 55A
NF A55 101/Z5 55A
NF A55 101/Z6 55A
NF A55 101/Z7 55A
NF A55 101/Z8 55A
NF A55 102 ZA4 U1G 55A
NF A55 102 ZA4 U3G 55A
NF A55 102 ZA4G 55A
NF A57 350 1N
NF A57 650 1N
NF A57-101 1A
NF A57-101-A4 1A
NF A57-101-A5 1A
NF A57-101-A7 1A
NF A57-101-A8 1A
NFW 1 13A
NFW 2 13A
NFW 3 13A
NFW 4 13A
NFW 5 13A
NFW 6 13A
NG 2 1C
NG 3 1B
NG 21 1C
NH 27B
NH 44K1
NHP 44K1
NHR 12
see BS 1449 44A1
NHR 13
see BS 1449 44A1
NHR 14
see BS 1449 44A1
NHR 15
see BS 1449 44A1
NHR 21
see BS 1449 44A1
NHR 22
see BS 1449 44A1
NHR 23
see BS 1449 44A1
NHR 24
see BS 1449 44A2
NHR 25
see BS 1449 44A1
Ni 48
see DIN range 20C
Ni 49
see DIN range 20C
Ni 99.8 Mg 27A
NI 3021 27C
Ni Al 0 44F2
Ni Al 1 44F2
Ni Al Co 2 44L
Ni Al Co 2B 44L
Ni Al Co 6 44L
NIALCO I 20C
NIALCO IF 20C
NIALCO II 20C
NIALCO III 20C
NIALCO IIIF 20C
NIALCO IV 20C
NIALCO IVF 20C
NIALCO V 20C
NICHROME 27B
NICHROME II 27B
NICHROME III 27B
NICHROME IV 27B
NICHROME V 27B
NICKEL 27A
NICKEL 2 27A
NICKEL 41 27A
NICKEL 61 27A
NICKEL 131 27A
NICKEL 141 27A
NICKEL 200 27A
NICKEL 201 27A
NICKEL 204 27F
NICKEL 205 27A
NICKEL 211 27F
NICKEL 212 27F
NICKEL 213 27A
NICKEL 215 27F
NICKEL 220 27A
NICKEL 222 27A
NICKEL 225 27A
NICKEL 229 27F
NICKEL 230 27A
NICKEL 233 27A
NICKEL 270 27A
NICKEL BRASS 14F
NICKEL FILLER 61 27B
NICKELOID 14F
NICKEL SILVER 14F
NICKELVAC 700 27C
NICKELVAC 901 27C
NICKELVAC L. 605 13A
NICKELVAC N 27B
NICKELVAC W 27C
NICKELVAC X 27C
NICLOY 44F1
NICLOY 5 44F1
NICLOY 9 44F1
NICOLLOY 4 44P
NICOLLOY 15 44L
NICORO 14H
NICORO 80 17.1
NICORROS 27D
NICORROS Al 27D
NICORROS B 6530 27D
NICORROS S 27D
NICORROS S 6530 27D
NICOSEAL 20C
NICOSEL 20C
Ni Cr 15 Fe 27B
Ni Cr 15 Fe Mo
see DIN range 27B
Ni Cr 22 Mo
see DIN range 27C
Ni Cr 30/20 44N1
Ni Cr 60/15
see DIN range 27E
Ni Cr 60/15 27B
Ni Cr 70/30 27B
Ni Cr 80/20
see DIN range 27B
Ni Cr 80/20 27B
Ni Cr Mo 4 27C
Ni Cr Si YCW STEM 44K2
NiCr 322 44K1
NiCr 342 44K1
NICRAL C 44N1
NICRAL D 44N1
NICRAL DM 44N1
NICRAL H 44N1
NICRAL HR2 44N1
NICRAL K 44N1
NICRAL S 44M1
NICRALT 44N1
NICRAL Z 27B
NICREX 1 44N1
NICREX 1 Cb 44N1
NICREX 2 44N1
NICREX 3 27B
NICREX 4 27B
NICREX AC 44N1
NiCrMo 335 44K1
NiCrMo 342 44K1
NiCrMo 415 44K1
NICROBODY 5007 27B
NICROBRAZ 10 27F
NICROBRAZ 30 27B
NICROBRAZ 35 27B
NICROBRAZ 50 27B
NICROBRAZ 51 27B
NICROBRAZ 60 27F
NICROBRAZ 65 27F
NICROBRAZ 120 27B
NICROBRAZ 125 27B
NICROBRAZ 130 27F
NICROBRAZ 135 27F
NICROBRAZ 150 27B
NICROBRAZ 160 27B
NICROBRAZ 170 27B
NICROBRAZ 171 27F
NICROBRAZ 180 27B
NICROBRAZ 200 27B

NICROBRAZ 210 13A
NICROBRAZ 220 27F
NICROBRAZ 230 27F
NICROBRAZ 300 13A
NICROBRAZ 1351 27F
NICROBRAZ 3001 27B
NICROBRAZ 3002 27B
NICROBRAZ 3003 27B
NICROBRAZ 3004 27B
NICROBRAZ 3005 27B
NICROBRAZ 5007 27B
NICROBRAZ 5040 27F
NICROBRAZ 5060 27F
NICROBRAZ 5075 27F
NICROBRAZ LC 27B
NICROBRAZ LM 27B
NICROBRAZ LMO1 27B
NICROBRAZ LMO2 27B
NICROBRAZ LMO3 27B
NICROBRAZ LMO4 27B
NICROBRAZ LMO5 27B
NICROBRAZ LMO6 27B
NICROBRAZ LMO7 27B
NICROBRAZ LMO8 27B
NICROBRAZ
(STANDARD) 27B
NICROBRAZ WG 27B
NICROCOAT 27B
NICROCOAT 1 27B
NICROCOAT 2 27F
NICROCOAT 3 27B
NICROCOAT 4 27B
NICROCOAT 6 27B
NICROCOAT 7 27F
NICROCOAT 8 27B
NICROCOAT 9 27B
NICROCOAT 610 27B
NICROCOAT 620 27B
NICROCOAT 621 27B
NICROCOAT 630 27B
NICROCOAT 700 27B
NICROFER 3127LC 44N1
NICROFER 3220 44N1
NICROFER 3220 H 27B
NICROFER 3220 H 44N1
NICROFER 3220LC 44N1
NICROFER 3620 Nb 44N1
NICROFER 3718 27B
NICROFER 3718 So 44N1
NICROFER 4221 27B
NICROFER 4221 27F
NICROFER 4221 h Mo 27F
NICROFER 4520 h Mo 27F
NICROFER 4626 Mo W 27C
NICROFER 4722 Co 27C
NICROFER 4823 h Mo 27F
NICROFER 5120 Co Ti 27C
NICROFER 5219 Nb 27C
NICROFER 5219 Nb 27F
NICROFER 5520 Co 27C
NICROFER 5621 h Mo W 27F
NICROFER 5716 h Mo W 27C
NICROFER 5716 h Mo W 27F
NICROFER 6020 h Mo 27B
NICROFER 6020 h Mo 27F
NICROFER 6022 h Mo 27C
NICROFER 6023 27B
NICROFER 6023 27F
NICROFER 6023 H 27B
NICROFER 6030 27B
NICROFER 6030 27F
NICROFER 6616 h Mo 27F
NICROFER 7016 Ti Nb 27C
NICROFER 7216 27B
NICROFER 7216 H 27B
NICROFER 7216 LC 27F
NICROFER 7520 27B
NICROFER 7520 Ti 27C
NICROFER B 6020 27C
NICROFER B 7020 27B
NICROFER S 4225 27C
NICROFER S 4722 27C
NICROFER S 5520 27C
NICROFER S 5621 27C
NICROFER S 5716 27C
NICROFER S 6020 27C
NICROFER S 6616 27C
NICROFER S 7020 27B
NICROMAZ B 44N1
NICROMAZ C 44N1
NICROTUNG 27B
NICROTUNG 27C
Ni Cu 14 Fe Mo
see DIN range 27F
Ni Cu 30 A1
see DIN range 27D
Ni Cu 30 Fe
see DIN range 27D
Ni Cu Cb 44F1
Ni Cu Ti 44F1
NICUAGE 44F1
NICUAGE 1 44F1
NICUAGE 1 44K1
NICUAGE TYLER 44F1
NICUSIL 3 42.1
Ni Fe 15 Mo
see DIN range 27E
Ni Fe 16 Cu Cr
see DIN range 27E
Ni Fe 16 Cu Mo
see DIN range 27E
Ni Fe 45
see DIN range 20C
Ni HARD 20B
Ni HARD 1 20B
Ni HARD 2 20B
Ni HARD 3 20B
Ni HARD 4 20B
NIKALIUM 14G
NIKO 20C
NIKROTHAL 20 PLUS 44N1
NIKROTHAL 40 PLUS 44N1
NIKROTHAL 60 27B
NIKROTHAL 80 27B
NIL 14C
NILGRO 36 27E
NILGRO 42 20C
NILO 36 20C
NILO 40 20C
NILO 42 20C
NILO 48 20C
NILO 50 20C
NILO 51 27E
NILO 55 27E
NILO 475 20C
NILO K 20C
NILO K45 20C
NILOMAG 77 27E
NILOMAG 471 20C
NILOMAG 771 27F
NILO P59 20C
NILSIL 44A1
NILVAR 20C
Ni MARK 1 44F3
Ni MARK 250 44F3
Ni MARK 300 44F3
NIMAX C 27F
NIMAX CH 27F
Ni Mn 3 A1
see DIN range 27F
Ni Mo
see ASTM range 27C
NI MO 44K1
Ni Mo 16 Cr
see DIN range 27C
Ni Mo 30 27F
Ni Mo Cr
see ASTM range 27C
NIMOCAST 75 27B
NIMOCAST 80 27C
NIMOCAST 90 27C
NIMOCAST 242 27F
NIMOCAST 257 27B
NIMOCAST 258 27B
NIMOCAST 258 27C
NIMOCAST 263 27F
NIMOCAST 713 27C
NIMOCAST 713LC 27C
NIMOCAST PD16 27B
NIMOCAST PE10 27B
NIMOCAST PE10 27C
NIMOCAST PK24 27C
NIMOFER 27F
NIMOFER 6928 27F
NIMOLOY PK37 27C
NIMONIC 75 27B
NIMONIC 80 A 27C
NIMONIC 81 27C
NIMONIC 90 27C
NIMONIC 93 27C
NIMONIC 95 27C
NIMONIC 100 27C
NIMONIC 105 27C
NIMONIC 115 27C
NIMONIC 118 27C
NIMONIC 120 27C
NIMONIC 242 27C
NIMONIC 263 27C
NIMONIC 901 27C
NIMONIC 942 27C
NIMONIC AP1 27C
NIMONIC PE 11 27C
NIMONIC PE 13 27C
NIMONIC PE 16 27C
NIMONIC PK 25 27C
NIMONIC PK 31 27C
NIMONIC PK 33 27C
NIMROD 182 27B
NIMROD 182 KS 27B
NIMROD 625 27B
NIMROD 625 27F

NIMROD 625 KS 27B
NIMROD A KS 27B
NIMROD AB 27B
NI O NEL 65 27C
NI O NEL 135 27C
NI O NEL 825 27C
NIOBIUM 28.1
NIORO 17.1
NIORO 17.1
NIPERMAG 20C
NIPERMAG 44F2
NIPIGON 44L
NIPPLE 14B
Ni-RESIST 1 20B
Ni-RESIST 1B 20B
Ni-RESIST 2 20B
Ni-RESIST 2B 20B
Ni-RESIST 3 20B
Ni-RESIST 4 20B
Ni-RESIST 5 20B
Ni-RESIST D 2 20B
Ni-RESIST D 2B 20B
Ni-RESIST D 2C 20B
Ni-RESIST D 2M 20B
Ni-RESIST D 3 20B
Ni-RESIST D 3A 20B
Ni-RESIST D 4 20B
Ni-RESIST D 5 20B
Ni-RESIST D 5B 20B
NI-ROD 27A
NI-ROD 55 20C
NIROMET 42 20C
NIROMET 46 20C
NIRON 52 27E
NIROSTA VK5M 44M1
NIROSTA VK7M 44M1
NISIMAZ 44N1
NI-SPAN-C 20C
NI-SPAN-C 902 20C
NIT 135 44K2
NITA 44E2
NIT-KD 44K1
NITRAL 44K2
NITRAL 4 44K1
NITRALLOY 3 44K2
NITRALLOY 5 44K1
NITRALLOY LK1 44K2
NITRALLOY LK3 44K2
NITRALLOY LK5 44K2
NITRALLOY LK7 44K1
NITRALLOY N 44K2
NITRALLOY N135 Mod 44K2
NITRIDING STEEL 44K2
NITROFIL 14A
NITRON 20B
NITRONIC 20 44N1
NITRONIC 50 44N1
NITRONIC 60 44P
NITRONIC 60W 44P
NIVAC P 27A
NIVAFLEX 13A
NIVCO 10 13A
Ni-Zr MASTER ALLOY 27F
NJ 3 1B
NK Hiten 50 44A1
NK Hiten 55 44A1
NK HITEN 60A 44K1
NK HITEN 60B 44K1
NK HITEN 70 44K1
NK HITEN 80 44K1
NK HITEN 100 HW90 44K1
NKR 44F1
NL 38 20B
NL 42 20B
NL 50 20B
NL 60 20B
NL 70 20B
NMCM 44K2
NMP 27F
NMS 44F1
NMX 20B
NN 44M2
NN 3 44D
n Nb 16 28.1
n Nb 73 28.1
No. 1 GUNMETAL 14K
No. 1 HARDENITE 44A4
No. 1 HIGH TENSILE BRASS 14C
No. 1 LEADED BRONZE 14K
No. 1 PHOSPHOR BRONZE 14K
No. 2 GUNMETAL 14K
No. 2 HARDENITE 44A4
No. 2 LEADED BRONZE 14K
No. 2 PHOSPHOR BRONZE 14K
No. 3 GUNMETAL 14K
No. 3 HARDENITE 44A4
No. 3 HIGH TENSILE BRASS 14C
No. 3 PHOSPHOR BRONZE 14K
No. 4 ALLOY 20C
No. 4 ALUMINIUM BRONZE 14G
No. 4 HARDENITE VAN 44K3
No. 4 HIGH TENSILE BRASS 14C
No. 4 LEADED BRONZE 14K
No. 4 PHOSPHOR BRONZE 14K
No. 6 ALUMINIUM BRONZE 14G
No. 6 HARDENITE 44E2
No. 6 HIGH TENSILE BRASS 14C
No. 7 HARDENITE 44K3
No. 8 HARDENITE 44K3
No. 9 TEMPER 44A4
No. 10 HARDENITE 44K3
No. 10 TEMPER 44A4
No. 11 SPECIAL 44A4
No. 11 TEMPER 44A4
No. 12 TEMPER 44A4
No. 14 TEMPER 44A4
NO. 158 44K1
No.1 MONARCH 44K2
No.7 TEMPER 44A3
No.8 TEMPER 44A3
No 5 ELECTERN 44K2
No 5-317 44K2
No 75 50A
NO158 44K1
NO345 44K1
No 408 44K2
No 481 44K2
No 484 44K3
No 610 44K3
No 882 44K2
No 883 44K2
NO2270
UNS for Nickel 270 27A
NO 4402 27D
NO6985
UNS for Hastelloy G3 27F
NO7001
UNS for Waspaloy 27C
NO-CHAT 14J
NOIL 14K
NOLL SPECIAL 44A4
NON STAIN 44M1
NONVAR 44A4
NONWAIR 44M2
NORAL 3S 1B
NORAL 16S 1F
NORAL 17S 1F
NORAL 19S 1F
NORAL 24S 1F
NORAL 26S 1F
NORAL 33S 1C
NORAL 38S 1C
NORAL 42S 1F
NORAL 50S 1E
NORAL 51S 1E
NORAL 54S 1D
NORAL 58S 1D
NORAL 62S 1E
NORAL 117 1L
NORAL 123 1C
NORAL 125 1K
NORAL 158 1C
NORAL 160 1C
NORAL 161 1K
NORAL 162 1K
NORAL 218 1L
NORAL 226 1L
NORAL 237 1L
NORAL 252 1L
NORAL 350 1J
NORAL A56S 1D
NORAL A111 1L
NORAL C77S 1G
NORAL M57S 1D
NORAL M75S 1G
NORAL S 1A
NORESCO 88 HP 44K3
NORESCO 99 HP 44K3
NORESCO DHS 44K3
NORESCO DOMINATOR 44M2
NORESCO DOMINATOR VM 44M2
NORESCO DOMINATOR Z 44M2
NORESCO EXTRA TOUGH HARD V 44J
NORESCO FAVORIT 44K3
NORESCO K5 44L
NORESCO K10 44L
NORESCO K12M 44L
NORESCO NSW 44K1
NORESCO PARFORCE SPECIAL 2 44K2
NORESCO PARFORCE SPECIAL 10 44K1
NORESCO SG 44K1

NORESCO SGW 44K1
NORESCO SPK 44K2
NORESCO TYRANT EXTRA 44K2
NORESCO WAW 44K2
NORESCO WCD 44K1
NORESCO WCD 2 44K1
NORESCO WU 44L
NORESCO WVC 44L
NORIDUR 9.4460 44N2
NORMAL Ni 27A
NORVAR 44K3
NOVA MAX 44L
NOVA V 44L
NOVA VHC 44L
NOVO 44K3
NOVO 6/6/2 44K3
NOVO 9/2 44K3
NOVO 18/2 44M1
NOVO 18/3 44N1
NOVO 18/8W 44N1
NOVO AS 1 44M1
NOVO AS 20 44M1
NOVO ASFC 44M1
NOVO C 44K3
NOVO ENORMOUS 44L
NOVO GAS 44M1
NOVO GBS 44M1
NOVO NOX T131 44M1
NOVO NOX TT131 44M1
NOVOSTON 14G
NOVO SUPERB 44L
NOVO SUPERIOR 44K3
NOVO SUPERIOR 6/6/2 44K3
NOVO SUPERIOR SS 44K3
NOVO TCV 44K3
NOVO THERM 44M1
NOX 1S4 44L
NP 3 1B
NP 1759 20B
NRM 44M2
NRO 146 44K3
NRW 44M2
NS 14C
NS 3 1B
NS 9S 44N1
NS 20E 44N1
NS 21A 44N1
NS 21S 44N1
NS 31 1B
NS 36 20C
NS 42 20C
NS 49 20C
NS 101
see BS range 14F
NS 102
see BS range 14F
NS 103
see BS range 14F
NS 104
see BS range 14F
NS 105
see BS range 14F
NS 106
see BS range 14F
NS 107
see BS range 14F
NS 108
see BS range 14F
NS 109
see BS range 14F
NS 110
see BS range 14F
NS 111
see BS range 14F
NS 112
see BS range 14F
NS 113
see BS range 14F
NS 722 20B
NS 11301 20B
NS 11338 20B
NS 11342 20B
NS 11350 20B
NS 11360 20B
NS 11370 20B
NSC 44E3
NSCD 44E3
NSCD 44N1
NSCT 44E3
NSM 21 44N1
NSMC 44N1
NSR 44F1
NSS 3 44A4
NST M1 44M1
NST M2 44M1
NSU 44N1
NSW 44K1
NTC 44K2
NTK M7 44N1
NTK Mo 44N2
N-TUP CR196 44K1
NURAL 25 1K
NURAL 1761 1C
NURAL 1761P 1C
NURAL 1762 1C
NURAL 2361 1C
NURAL 3210 1C
NUSHANK 44K2
NUTHERM 44K3
NV 1
see SAE NV1 44A2
NV 1-0 44A1
NV 1-1 44A1
NV 1-2 44A1
NV 2
see SAe NV2 44A2
NV 2 44K1
NV 2-0 44A1
NV 2-1 44A1
NV 2-2 44A1
NV 2-4 44A1
NV 3 44K2
NV 3-0 44A1
NV 3-1 44A1
NV 3-2 44A1
NV 4 44K2
NV 4-0 44A1
NV 4-1 44A1
NV 4-2 44A1
NV 4-3 44A1
NV 4-4 44A1
NV 5 44K2
NV 5-0 44A1
NV 5-1 44A1
NV 5-2 44A1
NV 6 44E2
NV 6-0 44A1
NV 6-1 44A1
NV 6-2 44A1
NV 7 44K2
NV 7-1 44G1
NV 7-2 44G1
NV 7-3 44K1
NV 8 44K2
NV9 44A2
NV 20.0 44F1
NV 20.00 44F1
NV 20.1 44F1
NV 20.2 44F1
NV 25 1 44N1
NV 25 2 44N1
NV 25 3 44N1
NV 25 4 44N1
NVA 44A1
NVA 27 44A1
NVA 32 44A1
NVA 36 44A1
NVA 40 44A1
NVC 44A1
NVD 44A1
NVD 27 44A1
NVD 32 44A1
NVD 36 44A1
NVD 40 44A1
NVE 44A1
NVE 27 44A1
NVE 32 44A1
NVE 36 44A1
NVE 40 44A1
NVG 44E2
NVK1 44A1
NVK2 44A1
NVK3 44A1
NVR 1-1 44A1
NVR 1-2 44A1
NVR 1-3 44A1
NVR 1-4 44A1
NVR 7-1 44G1
NVR 7-2 44K1
NVR 7-3 44K1
NVR 7-4 44M1
NVR 7-5 44M1
NVR 20.0 44F1
NVR 20.00 44F1
NVR 20.1 44F1
NVR 20.2 44F1
NVR 25 1 44N1
NVR 25 2 44N1
NVR 25 3 44N1
NVR 25 4 44N1
NVR 25 5 44N1
NVW 44A1
NYBLADE 44K2
NYBY 1410 Mo 44M1
NYBY 1415 Mo 44M1
NY–RA 80 27E
NZ 2 44K2
O 35 XHK 44K1
O 40 XHK 44K1
O 45 XHK 44K1

O 50 XHK 44K1
O 60 XHK 44K1
O 70 XHK 44K1
O 80 XHK 44K1
OC/Z 55A
OCM 5 44K1
OCM 6 44A2
OCM 7 44G1
OCM 9 44E2
OCM 10 44K2
OE 1 METAL 50A
OE 2 METAL 50A
OERSTIT 44L
OERSTIT 120R 20C
OERSTIT 120R 44L
OERSTIT 160 44L
OERSTIT 190 44L
OERSTIT 220 44L
OERSTIT 300 44L
OERSTIT 350 44L
OERSTIT 400K 44L
OERSTIT 400R 44L
OERSTIT 450 44L
OERSTIT 500 44L
OERSTIT 600 44L
OERSTIT 700 44L
OERSTIT 900 CP 31.1
OF
  see ASTM range 14A
OF
  Oxygen free 14A
OFE 14A
OFHC
  see ASTM range 14A
OFLP 14A
OFP
  see ASTM range 14A
OFPTE
  see ASTM range 14A
OFS
  see ASTM range 14A
OFTE
  see ASTM range 14A
OFXLP 14A
OHD 44K3
OHIO DIE 44M2
OI 44K3
OILDIE SMOOTHCUT 44K3
OK 44K2
OK CROWN 44K2
OLYMPIC FM 44M2
ON 1 44F2
ON 2 44K2
ON 3 44K1
ON 4 44K2
ON 5 44K1
ON 6 44K2
ON 7 44K1
ON 8 44K2
ON 9 44K1
ON 10 44K1
ON 11 44K2
ON 12 44K1
ONERAL M 47 13A
ONERAL S 90 13A
ONION 6.1
ONYX 52.1
OOC 44K2
OPAL 52.1
OR 2 44N1
OR 11 44N1
ORD Mn 5 44D
ORELLOY 242 44K1
ORELLOY 440 44A1
ORELLOY 441 44J
ORLEANS 44K2
ORM 14C
ORO BRAZE 17.1
ORO BRAZE 910 17.1
ORO BRAZE 940 17.1
ORO BRAZE 950 17.1
ORO BRAZE 990 17.1
ORO BRAZE 1040 17.1
OROBRAZE 998 14H
OROBRAZE 1018 14H
ORO CAST 17.1
ORPIMENT 3.1
ORTHOMUMETAL 20C
ORVAR 2 44K2
OSBORN 303 44N1
OSBORN 304 44N1
OSBORN 309 44N1
OSBORN 310 44N1
OSBORN 316 44N1
OSBORN 317 44N1
OSBORN 321 44N1
OSBORN 330 44N1
OSBORN 347 44N1
OSBORN 405 44M1
OSBORN 410 44M1
OSBORN 420C 44M1
OSBORN 420D 44M1
OSBORN 430AL 44M1
OSBORN 431 44M1
OSBORN 440B 44M1
OSBORN 501 44K1
OSMIRIDIUM 19.1
OSMIRIDIUM 31.1
OSMIUM 29.1
OTTOWA 44A4
OUNCE METAL 14K
OV 15 20B
OV 20 20B
OV 25 20B
OV 30 20B
OV 35 20B
OV 40 20B
OX 520A 44A1
OX 522 44A1
OX 522A 44A1
OX 525B 44A1
OX 542 44A1
OX602 44K1
OX702 44K1
OX812 44K1
OX 1002 44G1
OX1222BM 44K1
P 1 44A4
P 2 44K1
P 3 44K1
P 3 44L
P 3G 1D
P 4 44K1
P 4 44M1
P 5 44E1
P 5G 1D
P 6 13A
P 6 44L
P 9E5 44K3
P 9G 1D
P 10 44K3
P 12 44M1
P 14F4 44K3
P00015
  UNS for AMS 7731 17.1
P 15 44L
P00020
  UNS for AMS 7731 17.1
P 20 44K1
P 21 44K1
P 21 44L
P00025
  UNS for ASTM B562-99.5
P 28/6
  see BS range 20B
P 30 44A1
P 33/4
  see BS range 20B
P 86 OH 44K3
P121 44J
P 151 44A1
P 153 44F1
P 155 44K1
P 157 44K1
P 158 44K1
P171 44J
P181 44J
P211 44J
P 256 44A3
P 280 44E3
P 282 44A1
P00300
  UNS for AMS 4785 17.1
P00500
  UNS for AMS 4784 17.1
P 552 44K1
P 553 44K2
P 557 44M1
P 558 44K1
P 564 44K1
P 576 44K2
P 602 44K2
P 609 44E2
P 612 44K1
P 614 44K2
P 615 44K1
P 618 44K2
P00700
  UNS for AMS 4786 17.1
P 704 44E3
P 720 44E3
P 1000 44M1
P 1001 44M1
P 1002 44M1
P 1003 44M1
P 1008 44M1
P 1009 44M1
P 1010 44N1
P 1011 44N1
P 1012 44N1
P 1013 44N1

P 1014 44N1
P 1015 44N1
P 1016 44N1
P 1017 44N1
P 1018 44N1
P03300
UNS for AMS 7735 30.1
P03350
UNS for 226 Alloy 30.1
P 03590 30.1
P03980
UNS for 99.8Pd 30.1
P 03990 30.1
P03995
UNS for Refined Pd 30.1
P 04840 31.1
P04980
UNS for Refined Pt 31
P04995
UNS for Refined Pt 31
P05980
UNS for ASTM B616-99.8 37
P05981
UNS for ASTM B616-99.9 37
P05982
UNS for ASTM B616-99.95 37
P05990
UNS for ASTM B616-99.9 37
P05995
UNS for ASTM B616-99.95 37
P06100
UNS for ASTM B671-99.8 19.1
P07010
UNS for 99.99 Ag 42
P07015
UNS for 99.95 Ag 42
P07016
UNS for 99.95 Ag 42
P07020
UNS for 99.9 Ag 42
P 07505 42.1
P 07507 42.1
P 07540 42.1
P 07547 42.1
P 07560 42.1
P 07563 42.1
P 07587 42.1
P 07600 42.1
P 07607 42.1
P 07627 42.1
P 07630 42.1
P 07650 42.1
P 07687 42.1
P 07700 42.1
P 07720 42.1
P 07723 42.1
P 07727 42.1
P 07728 42.1
P 07850 42.1
P 07900 42.1
P 07925 42.1
P 07931 42.1
P 07932 42.1
PA 10 27F
PA 22 NICKEL 27A
PA 23 NICKEL 27A
PAC 44A1
PAC 1 44A1
PAC 2 44A1
PALAURAL 30.1
PALCO 30.1
PALCUSIL 5 42.1
PALCUSIL 10 42.1
PALCUSIL 15 42.1
PALCUSIL 25 42.1
PALINEY No. 7 30.1
PALLABRAZE 14H
PALLABRAZE 810 14H
PALLABRAZE 840 14H
PALLABRAZE 850 14H
PALLABRAZE 880 14H
PALLABRAZE 900 14H
PALLABRAZE 950 14H
PALLABRAZE 1010 42.1
PALLABRAZE 1090 14H
PALLABRAZE 1225 42.1
PALLABRAZE 1237 31.1
PALLACAST 30.1
PALLADIUM 30.1
PALNIRO 1 17.1
PALORO 17.1
PANTEG 304 44N1
PANTEG 315 44N1
PANTEG 316 44N1
PANTEG 321 44N1
PANTEG 430 44M1
PANTHER 5 44L
PANTHER SPECIAL 44L
PARALLOY 0 44N1
PARALLOY 0KW 44N1
PARALLOY 0L 44N1
PARALLOY 0LC 44N1
PARALLOY 0LS 44N1
PARALLOY 0S 44N1
PARALLOY 0SW 44N1
PARALLOY 0W 44N1
PARALLOY 3 44N1
PARALLOY 3K 44N1
PARALLOY 3KLC 44N1
PARALLOY 3L 44N1
PARALLOY 3LS 44N1
PARALLOY 3S 44N1
PARALLOY 3W 44N1
PARALLOY 3WS 44N1
PARALLOY 4K 44N1
PARALLOY 4KLC 44N1
PARALLOY M2 44M1
PARALLOY M3 44M1
PARALLOY M15 44M1
PARALLOY ML 44M1
PARALLOY MPH 44M1
PARALLOY MPH2 44M1
PARFORCE SPECIAL 2 44K2
PARFORCE SPECIAL 10 44K1
PARSONS C 63 14F
PARSONS C 90 14C
PARSONS C 92 14C
PARSONS ITA 14C
PARSONS No 1 STANDARD 14C
PARSONS SB 6 14C
PARSONS SS 15 14F
PAR-TEN 44A1
P ASLA 3.1
PASSIVE Ni 27A
PASS No 1 50.1
PAX 2 44K2
PAX NON-BREAK 44K2
PB 1
see BS range 14K
PB 2
see BS range 14K
PB 2 44L
PB 3
see BS range 14K
PB 4
see BS range 14K
Pb 5 21.1
PB 101
see BS range 14K
PB 102
see BS range 14K
PB 103
see BS range 14K
PB 104
see BS range 14K
PBD 44L
P Be 12 5.1
Pb Sb 5 21.1
Pb Sb 8 21.1
Pb Sb 9 21.1
Pb Sb 9X 21.1
Pb Sb 12 21.1
Pb Sn 3 Sb 4
see DIN range 21
Pb Sn 3 Sb 12
see DIN range 21
Pb Sn 4 Sb 15
see DIN range 21
Pb Sn 5 Sb 12
see DIN range 21
Pb Sn 5 Sb 28
see DIN range 21
Pb Sn 9 Sb 17
see DIN range 21
Pb Sn 15 Sb 4
see DIN range 21
PC 400 20B
p Cr 23 12.1
PCS 44K1
PCSK 44K2
PDCO 44L
PDX 44K1
PE 10 27C
PE 11 27C
PE 13 27C
Pe 15 1D
PE 16 27C
Pe 30 1D
Pe 40 1D
Pe 50 1D
PEARL 52.1
PEGASE S 44A3
PEL 44N1
PENETRODE 29.10 44A1
PENUMO 44K2
PER 1 27B
PER 2 27B
PER 2B 27B

PER 2U 27B
PER 2V 27B
PER 13 27B
PERALUMAN 3G 1D
PERALUMAN 5G 1D
PERALUMAN 9G 1D
PERALUMAN 15 1D
PERALUMAN 30 1D
PERALUMAN 34 1J
PERALUMAN 40 1D
PERALUMAN 50 1D
PERALUMAN 75 1J
PERALUMAN 100 1D
PERALUMAN 150 1D
PERALUMAN 200 1D
PERALUMAN 260 1D
PERALUMAN 300 1D
PERALUMAN 350 1D
PERALUMAN 400 1D
PERALUMAN 460 1D
PERALUMAN 500 1D
PERMAGRID 27F
PERMALLOY 'B' 20C
PERMALLOY 'D' 20C
PERMALLOY C 27C
PERMALLOY F 27E
PERMANICKEL 27A
PERMANICKEL 300 27A
PERMANIT 44L
PERMANIT 900 CP 31.1
PERMENDUR 24 13A
PERMENDUR V 13A
PERMET 14E
PERMET PF1 20A
PERMET PF2 44L
PERNIFER 36 44F1
PERUNAL 1G
PERUNAL 215 1G
PET 44N1
PEWTER 50A
PF 1 44K1
PF 2 44K1
PF 2M 44K1
PF 5 44K1
PF 5K 44K1
PF 6 44K1
PF 15 44K1
PFDV 44K1
PFO 44G1
PFO A 44G1
PFV 55 44K1
PG 605 44A1
PGF 1 44A1
PGF 4 44A1
PGF 5 44A1
PGS 31.1
PGT 44K3
PH 13.8 Mo 44N2
PH 13.8 Mo 44N3
PH 14.8 Mo 44N3
PH 15.7 Mo 44N3
PHEINROHR MV12 44M1
PHEINROHR MVW12 44M1
PHENIX HD 32 44M1
PHENIX HD 301 44M1
PHENIX HD40 44M1
PHENIX HD50 44M1
p Hg 1 25.1
PHILIPS 23C 44A1
PHILIPS 27 44A1
PHILIPS 27P 44A1
PHILIPS 35S 44A1
PHILIPS 36S 44A1
PHILIPS 46 44A1
PHILIPS 48 44A1
PHILIPS 55C (55) 44A1
PHILIPS 56S 44A1
PHILIPS 68 44A1
PHILIPS 75 44F1
PHILIPS 78 44A1
PHILIPS 88 44K1
PHILIPS 98 44K1
PHILIPS 108 44K1
PHILIPS 118 44K1
Philips 250 44A2
PHILIPS 400 44K1
PHILIPS 600 44E2
PHILIPS ARMARC 44N1
PHILIPS C16 44A1
PHILIPS C33 M 44A1
PHILIPS ENROX 44A1
PHILIPS HERMEES 44A1
PHILIPS KV 2 44G1
PHILIPS KV1 44K1
PHILIPS KV3 44K1
PHILIPS KV4 44K1
PHILIPS KV5 44K1
PHILIPS KV7 44K1
PHILIPS LOWHEES 44A1
PHILIPS MILDEES 44A1
PHILIPS MONARC 44A1
PHILIPS QUEENARC 44A1
PHILIPS STAINARC 25/25 R 44N1
PHILIPS STAINARC M 44N1
PHILIPS STAINARC MN 44N1
PHILIPS STAINARC N 44N1
PHOENIX 1 44A1
PHOENIX 2 44A1
PHOENIX 4 44A1
PHOENIX MANGEAR 44A1
PHOSPHALLOY No 1 14H
PHOSPHALLOY No 2 14H
PHOSPHOR BRONZE 14K
PIANO WIRE 44A4
PILGER ROLL STEEL 44K3
PINK LABEL 44J
PIREKS 60 27B
PIREKS 60/13 27B
PITHO 44K3
PITT-TEN 1 44F1
PITT-TEN 2 44A1
PITT-TEN X45W 44A1
PITT-TEN X50W 44A1
PITT-TEN X55W 44A1
PITT-TEN X60W 44A1
PK 24 27C
PK 31 27C
PK 33 27C
PK 42 27C
PKH 44M1
PKI 44M1
PKL 44M1
PKM 44M1
PKN 44M1
PL 14C
PLACO 31.1
PLACOVAR 31.1
PLANEMEL 44A1
PLASMOLD 44K1
PLASTIC METAL 21.1
PLASTICS HOBBING 44A1
PLASTIFORM 44K1
PLATINAX 11 31.1
PLATINIRIDIUM 19.1
PLATINITE 27E
PLATINUM 31.1
PLATINUM COLOUR SOLDER 17.1
PLATINUM SUBSTITUTE 28.1
PLATNAM 27D
PLMA/2 44E1
PLMA/5 44K1
PLM B/1 44K2
PLM B/2 44K2
PLM C/1 44K2
PLM C/2 44K2
PLMC/3 44K1
PLUMBERS SOLDER 21.1
PLUMBSOL 50A
PLUTOCRAT 44K3
PLUTO PARAMOUNT 44L
PLUTO PEERLESS 44L
PLUTO PERFECTUM 44L
PLUTO PLUS 44L
PLUTO PREMIER 44L
P M 21.1
PM 5 44A1
PM 6 44A1
PM 12 44E1
PM 15 44J
PM 15 E 44J
PM 17 44K1
PM 18 44K1
PM 23 44K1
PM 24 44K1
PM 30 44K1
PM 35 44K1
PM 38 44K1
PMG 14M
PMH 44N1
PML 44N1
PN 44E2
PN 1 30.1
PN 1 44K2
PNEUTOUGH 44K2
PNH 82123 20B
PNH 83101 20B
PNUSNAP OH 44K2
PNUSNAP WH 44K1
PO 2 44K3
PO 3 44K3
PO 7 44K3
PO 41 44L
PO Bronze 14A
POO560 17.1
POO580 17.1
POROSINT 14J
POSITRODE 46.28 44A1
POWHATAN 44L
PP 44K3
PPT 44F1

PREGA 44K2
PREMO 44K1
PRESSURDIE 44K2
PRESSURDIE/1 44K2
PRESSURDIE/2 44K1
PRESSURDIE/3 44K2
PRESSURDIE/3L 44K2
PRESSURDIE/5 44K2
PRESSURDIE/16 44K2
PRESSURDIE C 44L
PRESTO 44E3
PREXI 44K1
PRIMAFIXE 44A1
PRIMEX 44A1
PRN 2 44K3
PROJECT 70- 182FM 44M1
PROJECT 70-303 44N1
PROJECT 70-303 DQ 44N1
PROJECT 70-304 44N1
PROJECT 70-316 44N1
PROLITE 52.1
PROMET 6 14K
Pro No 1 44K1
Pro No 2 44K1
Pro No 3 44K1
PROPELLER BRASS 14C
PS10 44K1
PS15 44K1
PS16 44K1
PS17 44K1
PS18 44K1
PS19 44K1
PS20 44K1
PS24 44K1
PS30 44K1
PS31 44K1
PS32 44K1
PS33 44K1
PS34 44K1
PS 36 44K2
PS 38 44K2
PS 39 44K2
PS 40 44K2
PS54 44K1
PS55 44K1
PS56 44K1
PS 58 44E1
PS 59 44E1
PS 61 44E1
PS 63 44E1
PS 64 44E1
PS 65 44E1
PS66 44K1
PS 67 44K2
Pu 1G
PURO M Mo 44M1
PURO M Mo2 44M1
PURPLE LABEL 44L
PURPLE LABEL EXTRA 44L
PW 14C
PW 44K2
PX 80 PLUS 0.2% 44K1
PX 90 PHIS 44K1
PX 100 PHIS 44K1
PX 110 PHIS 44K1
PX 360 PHIS 44K1
PY 1815A 50A
PY 8720 44K1
PYRISTA 44M1
PYRMANT 50.1
PYRODIE 44K2
PYROMET 31 27C
PYROMET 80A 27C
PYROMET 88 27B
PYROMET 350 44M1
PYROMET 355 44M1
PYROMET 600 27B
PYROMET 751 27C
PYROMET 860 44N1
PYROMET 882 44K2
PYROMET 901 44N2
PYROMET ALLOY 31 27C
PYROMET ALLOY 41 27C
PYROMET ALLOY 80A 44M1
PYROMET ALLOY 88 27C
PYROMET ALLOY 90 27C
PYROMET ALLOY 102 27F
PYROMET ALLOY 625 27F
PYROMET ALLOY 680 27F
PYROMET ALLOY 718 27F
PYROMET ALLOY 751 44M1
PYROMET ALLOY 860 27C
PYROMET ALLOY A286 44N2
PYROMET ALLOY CTX-1 44F3
PYROMET ALLOY CTX-3 44F3
PYROMET ALLOY M252 27C
PYROMET ALLOY V57 44N2
PYROMET ALLOY X 750 27B
PYROMET CTX-1 44L
PYROMET X12 44L
PYROMET X15 44L
PYROMIC 27B
PYROTOOL 7 27C
PYROTOOL A 44N1
PYROTOOL EX 27C
PYROTOOL M 27C
PYROTOOL V 44N1
PYROTOOL W 27C
PYROTOUGH 44K1
PYRO-VAN 44K3
PYROWEAR ALLOY 53 44K1
PZ 6000 44A1
PZ 6008 44A1
PZ 6042 44K1
PZ 6043 44K1
PZ 6044 44K1
PZ 6047 44K1
PZ 6072 44N1
PZ 6074 44N1
PZ 6101S 44A1
PZ 6120S 44A1
Q 5 44L
Q 9 100 A 44A1
Q 9 100 B 44A1
Q 9 110 A 44A1
Q 10 44L
Q 980 44A1
Q 980 A 44A1
Q 980 B 44A1
Q 990 A 44A1
Q 990 B 44A1
QA Wire 17.1
Q BRAND 44A2
QE 22 23B
QE 22A 23B
QH 21 23B
QH 21A 23B
QMV 5.1
QQ 561/4 42.1
QQ A 35lb 1F
QQ A 59/7 1J
QQ A 315 1D
QQ A 318 1D
QQ A 327 1E
QQ A331B 1E
QQ A 334 1E
QQ A 353a 1F
QQ A 354a 1F
QQ A 355A 1F
QQ A 356b 1B
QQ A 359a 1B
QQ A 361 1F
QQ A 362/1 1F
QQ A 367b/1 1F
QQ A 411b 1A
QQ A 561 1A
QQ A 591/1 1C
QQ A 591/2 1C
QQ A 591/3 1C
QQ A 591/4 1L
QQ A 591/5 1L
QQ A 591/8 1C
QQ A 591/9 1J
QQ A 591/10 1L
QQ A 591/11 1L
QQ A 596/1 1L
QQ A 596/3 1L
QQ A 596/4 1L
QQ A 596/5 1L
QQ A 596/6 1K
QQ A 596/7 1C
QQ A 596/8 1K
QQ A 596/9 1K
QQ A 596/10 1K
QQ A 601/2 1C
QQ A 601/3 1K
QQ A 601/4 1L
QQ A 601/5 1J
QQ A 601/6 1L
QQ A 601/8 1L
QQ A 601/9 1L
QQ A 601/10 1K
QQ A 601/11 1K
QQ A 601/12 1K
QQ A671 8.1
QQ B 601/A 14B
QQ B 601/B 14B
QQ B 611a/A 14C
QQ B 611a/B 14C
QQ B 611a/E 14B
QQ B 621/B 14B
QQ B 621/C 14B
QQ B 636 14C
QQ B 666 A 14G
QQ B 666 B 14G
QQ B 671 A 14G
QQ B 671 B 14G
QQ B 671 C 14G
QQ B 671 D 14G
QQ B 691/1 14K
QQ B 691/2 14K

QQ B 691/3	14K
QQ B 691/6	14K
QQ B 691/11	14K
QQ B 691b/5	14K
QQ B 721a/A	14C
QQ B 721a/B	14B
QQ B 726/B	14B
QQ B 726/C	14B
QQ B 746/A	14K
QQ B 746/A	14M
QQ C 40	21.1
QQ C 501	14A
QQ C 591/A	14B
QQ C 591a/B	14K
QQ C 591a/D	14B
QQ C 593	14M
QQ C.551	27B
QQ L 171	21.1
QQ L 171 (Grade C)	21.1
QQ L 201	21.1
QQ L 201 A	21.1
QQ L 201 B	21.1
QQ M 31	23B
QQ M 31 B	23B
QQ M 38	23B
QQ M 40	23B
QQ M 40 B	23B
QQ M 44	23B
QQ M 54	23B
QQ M 55	23B
QQ M 56	23B
QQ M 56 B	23B
QQ M 161/1	50A
QQ M 161/2	50A
QQ M 161/3	50A
QQ M 161/4	50A
QQ M 161/5	50A
QQ N 281	27D
QQ N 321/A	14F
QQ N 321/B	14F
QQ S 551/A	14C
QQ S 551/B	14C
QQ S 551/C	14B
QQ S 551/D	14B
QQ S 561/3	14H
QQ S 571	50A
QQ S 571 A	50A
QQ S 571 (Ag 2.5)	21.1
QQ S 571 (Ag 5.5)	21.1
QQ S 571 B	21.1
QQ S 571 D	21.1
QQ S 571 D Ag 1.5	21.1
QQ S 571 D Ag 2.5	21.1
QQ S 571 D Ag 5.5	21.1
QQ S 571 D Pb 65	21.1
QQ S 571 D Pb 80	21.1
QQ S 571 D Sn 5	21.1
QQ S 571 D Sn 10	21.1
QQ S 571 D Sn 20	21.1
QQ S 571 D Sn 30	21.1
QQ S 571 D Sn 35	21.1
QQ S 571 (Pb 65)	21.1
QQ S 571 (Pb 70)	21.1
QQ S 571 (Pb 80)	21.1
QQ S 571 Sb 5	50A
QQ S 571 (Sn 5)	21.1
QQ S 571 (Sn 20)	21.1
QQ S 571 (Sn 30)	21.1
QQ S 571 (Sn 35)	21.1
QQ S 571 (Sn 40)	21.1
QQ S 571 (Sn 50)	21.1
QQ S 571 Sn 63	50A
QQ S 571 Sn 70	50A
QQ S 571 Sn 96	50A
QQ S 571b Sn 60	50A
QQ S 571d Pb 70	21.1
QQ S 571d Sb 5	50A
QQ S 571d Sn 50	50A
QQ S 571d Sn 62	50A
QQ S 571d Sn 63	50A
QQ S 571d Sn 70	50A
QQ S 571d Sn 96	50A
QQS-763	44M1
QQ S 763/6	44M1
QQ S 763/7	44N1
QQ S763/8	44N3
QQ S 763/10	44M2
QQ S 766/1	44N1
QQ-S-766	44N1
QQT 371	50.1
QQ T 390	21.1
QQ T 390	50A
QQ W 321/A	14B
QQ W 321/C	14B
QQ W 336	14A
QQ W 341	14A
QQ W 401	14K
QQ W 405a	44K2
QQ W 409a	44A2
QQ W 412	44K2
QQ W 414a	44A2
QQ W 418a	44A2
QQ W 423a	44N1
QQ W 428	44A4
QQ W 432	44A4
QQ W 461f	44A3
QQ W 470	44A4
QQ Z 35/a	55A
QQ Z 100/a	55A
QQ Z 285	55A
QQ Z 301/c	55A
QQ Z 351/a	55A
QQ Z 363/a	55A
QRO 45	44L
Q St 34	44A1
Q St 34/2	44A1
Q St 34/3	44A1
Q St 37/2	44A1
Q St 37/3	44A1
Q St 42/2	44A1
Q St 42/3	44A1
Q St 52/3	44A1
QUEENARC	44A1
QUICKSILVER	25.1
QV	44K3
QV/E	44K3
QX 1	44L
QX 2	44L
R 001	20B
R 1	44M1
R 1	56.1
R 2	44M1
R 3	44N1
R 4	44M1
R 5	44M2
R 9	44K3
R 9	44N1
R 9K5	44L
R 9K10	44L
R 10	44N1
R 10K5F5	44L
R 11	44N1
R 12	44K3
R 12	44M1
R 15	44M1
R 16	44N1
R 18	44K3
R 18	44N1
R 18F2	44K3
R 18K5F2	44L
R 19	44M1
R 20	44N1
R 22	44N1
R 23	44N1
R 24	44N1
R 25	44N1
R 26	44M1
R 27	44N1
R 27 Nb	44N1
R 29	44M1
R 32	44P
R 34	44M1
R 35	44N1
R 41	44N1
R 42	44N1
R 43	44N1
R 45	44N1
R 46	44P
R 47	44N1
R 48	44N1
R 50	44M1
R 51	44N1
R 52	44N1
R 53	44N1
R 54	44N1
R 55	44M1
R 63	27F
R100	44E3
R100	44K1
R 100	44M1
R 110	44M1
R 120G	44M1
R 140	44M1
R 140G	44M1
R 200	44M1
R 220	44M1
R 235	27C
R 290	44M1
R 300	44N1
R 310	44N1
R 320	44N1
R 350	44N1
R 350G	44N1
R 358	44N1
R 358G	44N1
R 359	44N1
R 360	44N1
R 360G	44N1
R 380	44N1
R 390	44N1
R 400	44N1

R 420 44N1
R 428 44N1
R 429 44N1
R 440 44N1
R 440G 44N1
R 450 44N1
R 460 44N1
R 460G 44N1
R 470 44N1
R 470G 44N1
R 500 44N1
R 510 44N1
R 520 44P
R 530 44N1
R 590 44N1
R 600G 44M1
R 620 44M1
R 620G 44N1
R 640 44M1
R 640G 44N1
R 680G 44M2
R690G 44N3
R 700 44M1
R 710 44M1
R 720 44M1
R 730 44M1
R 740 44M1
R 770 44M2
R 800 44N1
R 800G 44N1
R 820 44N1
R 830 44N1
R 830G 44N1
R 850G 44N1
R 860G 44N1
R 870G 27B
R 890Gr 44N1
R 2799 27E
R 2800 20C
R03600
UNS for Pure Mo 26A
R03601 26A
R03602 26A
R03605
UNS for AMS 7801 26A
R03606
UNS for AMS 7805 26A
R03610
UNS for AMS 7800 26A
R03620
UNS for 0.5% TiMo 26A
R03630
UNS for AMS 7817 26A
R03640
UNS for AMS 7811 26A
R03650
UNS for Low Mo 26A
R04206
UNS for Niobium 28
R04210
UNS for Niobium 28
R04211
UNS for AMS 7850 28
R04251
UNS for 1% Zr Nb 28
R04261
UNS for 1% Zr Nb 28
R04271
UNS for AMS 7851 28.1
R04295
UNS for AMS 7852 28.1
R04295 28.1
R05200
UNS for ASTM B364 46.1
R05210
UNS for AMS 7849 46.1
R05240
UNS for ASTM B364 46.1
R05255
UNS for AMS 7846 46.1
R05400
UNS for ASTM B364 46.1
R 07004 52.1
R07005
UNS for AMS 7897 52.1
R07006
UNS for AMS 7898 52.1
R 07030 52.1
R 07031 52.1
R 07050 52.1
R 07080 52.1
R 07100 52.1
R07900
UNS for AWS designation EWP 52.1
R07911
UNS for AWS designation EWTR1 52.1
R07912
UNS for AWS designation EWTR2 52.1
R07913
UNS for AWS designation EWTR3 52.1
R07920
UNS for AWS designation EWZr 52.1
R 9030 44K1
R19800
UNS for AMS 7901 5.1
R19801
UNS for AMS 7902 5.1
R19920
UNS for AMS 7900 5.1
R20500
UNS for ASTM A560-50Ni 12.1
R20600
UNS for ASTM A560-60Ni 12.1
R20990
UNS for 99% Cr 12.1
R20994
UNS for 99.4% Cr 12.1
R30001
UNS for Stellite 1 13A
R30002
UNS for Stellite F 13A
R30003
UNS for AMS 5833 13A
R 30004 13A
R30006
UNS for AMS 5373 13A
R30012
UNS for Stellite 12 13A
R30021
UNS for AMS 5385 13A
R30023
UNS for AMS 5357 44N3
R30027
UNS for AMS 5378 13A
R30030
UNS for AMS 5380 13A
R30031
UNS for AMS 5382 27C
R30035
UNS for AMS 5844 13A
R30036
UNS for Haynes 36 13A
R30040
UNS for AMS 4783 13A
R30155
UNS for AMS 5376 44L
R 30155 44L
R30159
UNS for AMS 5841 13A
R30188
UNS for AMS 5608 13A
R30260
UNS for Duratherm 2602 13A
R30477
UNS for Duratherm 477 13A
R30556
UNS for AMS 5874 13A
R30600
UNS for AMS 5770 44L
R30600
UNS for Duratherm 600 13A
R30605
UNS for AMS 5537 44N1
R30700
UNS for Duratherm 700
R30816
UNS for AMS 5534 27C
R 30816 13A
R39001
UNS for AWS designation BCo1 13A
R50100
UNS for AWS ERTi1 51A
R50120
UNS for ERTi2 51A
R50125
UNS for AWS ERTi3 51A
R50130
UNS for AWS ERTi4 51A
R50400
UNS for AMS 4902 51A
R50550
UNS for AMS 4900 51A
R50700
UNS for AMS 4901 51A
R52250
UNS for ASTM B265 51A
R52400
UNS for ASTM B265-7 51A
R52401
UNS for AWS ERTiOPd 51A
R52550
UNS for ASTM B367 51A
R54520
UNS for AMS 4910 51A

R54521
UNS for AMS 4909 51A
R54522
UNS for AMS 4953 51A
R54523
UNS for AWS ERTi5A-2.5Sn-1 51A
R54560
UNS for Mil T9046 51A
R54620
UNS for AMS 4919 51A
R 54621 51B
R54790
UNS for AMS 4974 51A
R54810
UNS for AMS 4915 51A
R56080
UNS for AMS 4908 51A
R56210
UNS for AWS ERTi6Al-2Cb-1Ta1Mo 51A
R56260
UNS for AMS 4981 51C
R56320
UNS for AMS 4943 51A
R56321
UNS for AWS designation ERTi3Al2.5V-1 51A
R56400
UNS for AMS 4905 51A
R56401
UNS for AMS 4907 51A
R56402
UNS for AMS 4956 51C
R56410
UNS for AMS 4984 51A
R56430
UNS for AMS 4912 51A
R 56440 51B
R56620
UNS for AMS 4918 51A
R56740
UNS for AMS 4970 51A
R58010
UNS for AMS 4917 51A
R58030
UNS for AMS 4977 51C
R58450
UNS for AMS 4982 51A
R58640
UNS for AMS 4957 51C
R58650
UNS for AMS 4995 51C
R58820
UNS for Mil T9046 51A
R60001
UNS for ASTM B349 56.1
R60701
UNS for Pure Zr 56.1
R60702
UNS for AWS designation ERZr2 56.1
R 60702 56.1
R60704
UNS for AWS designation ERZr3 56.1
R 60704 56.1
R 60705 56.1
R60706
as R60705 56.1
R60707
UNS for AWS designation ERZr4 56.1
R60802
UNS for ASTM B350 56.1
R60804
UNS for ASTM B350 56.1
R60901
UNS for ASTM B350 56.1
R 60902 56.1
RA 1 56.1
RA 310 44N1
RA 330 44N1
RA 330 TX 44N1
RA330.04 44P
RA 333 27B
RA 333 27F
RABI 44K1
RABI 44K2
RADIO METAL 36 20C
RADIO METAL 50 20C
RAFFINAL 990 1A
RAILWAY A 21.1
RAILWAY C 50A
RBD 44K1
RBD-E 44K1
RC 1
see AWS 20B
RC 1A
see AWS 20B
RC 1B
see AWS 20B
RC 32 44L
RCA 33 44M1
RCA 44 44M1
RCK 3 44N1
RCK 4 44N1
R Co Cr A 13A
R Co Cr B 13A
R Co Cr C 13A
RCS 44A4
RCT 44N1
RCT 3 44N1
RDS 44K3
RE 39 44M1
REALGAR 3.1
REB 210 44N1
RECO 2A 44L
RECO 3A 44L
RECO 100 20C
RECO 120 20C
RECO 140 20C
RECO 160 20C
RECO 170 20C
RECO 220 20C
RED BRASS 14B
RED CHIP 44L
RED CUT COBALT 44L
RED CUT COBALT B 44L
RED CUT SUPERIOR 44K3
RED DIAMOND 44D
RED DIAMOND 14H 44P
RED DIAMOND 14S 44P
RED DIAMOND 20H 44K1
RED DIAMOND 20S 44K1
RED DIAMOND 21 44K2
RED DIAMOND 22 44K2
RED DIAMOND II 44E3
RED INDIAN 44K1
RED LABEL 44A4
RED SABRE 44L
RED STREAK 44K3
REFLECTAL 050 1D
REFLECTAL 100 1D
REFRACTALOY 26 27C
REFRACTALOY 70 13A
REGENT 14E
REGENT VAC-ARC 44E3
REGIN 3 44K2
REGIS 50:1
REGULUS 2.1
REGULUS METAL 21.1
REINALUMINIUM 99.0 1A
REINALUMINIUM 99.5 1A
REMALLOY 44L
REMALLOY 17 44L
REMALLOY 20 44L
REMANIT 1520 Mo 44M1
REMENDUR 13A
REMOUNT 44A4
RENE 41 27C
RENOWN 44J
REPUBLIC 50 44K1
REPUBLIC 65 44K1
REPUBLIC 70 44K1
REPUBLIC A441 44J
REPUBLIC M1 44A1
REPUBLIC M2 44A1
REPUBLIC X42W 44J
REPUBLIC X45W 44J
REPUBLIC X50W 44J
REPUBLIC X60W 44J
REPUBLIC X65W 44J
REPUBLIC X70W 44J
RESISCO 14G
RESISTCO 44A1
RESOURCE 44M1
RETORT 44A2
REX 4V 44K3
REX 95 44L
REX 440 44L
REX 559 44N2
REX 939 44K3
REX AA 44K3
REX AAA 44L
REX CHAMPION 44K3
REX M2 44K3
REX M3/1 44K3
REX M3/2 44K3
REX M7 44K3
REX M25 44L
REX SUPER CUT 44L
REX SUPER VAN 44K3
REX T Mo 5 44L
REX TMO 44K3
REX VM 44K3
REXWELD 33W 13A
REXWELD 54 44M2
REXWELD 64 27F
REXWELD 66 27C
REXWELD A 13A

REXWELD B 13A
REXWELD C 13A
REXWELD VT 13A
REYNOLDS 531 44G1
REYNOLDS 753 44G1
REYNOLDS 5313L 44G1
R Fe 5A 44K3
R Fe 5B 44K3
RFK 44M1
Rg 5
see DIN range 14K
Rg 7
see DIN range 14K
Rg 10
see DIN range 14K
RG 45 44A1
RG 60 44A1
RG 65 44A1
RHB 65 44K1
RHENIUM 36.1
RHM 5 44M1
RHM 7 44M1
RHM 8 44M1
RHM 9 44M1
RHM 10 44M1
RHM 11 44M1
RHM 37 44M1
RHMV4 44M1
RHMW1 44M1
RHMW2 44M1
RHODIUM 37.1
RIGIDMESH 14J
RIGOR 44K3
RIM 21 44N1
RIM 29 44N1
RIM 210 44N1
RIM 213 44N1
RIM 215 44N1
RIM 217 44N1
RIM 290 44N1
RIM 291 44N1
RIM 294 44N1
RIM 295 44N1
RIO 214 44A1
RIO 214 44N1
RIVER ACE 60 44A1
RIVER ACE 70 44K1
RIVER ACE KO 44K1
RKV 44K2
RLH 2 44N1
RM 1 METAL 21.1
RM1015
see SAE 1015 44A1
RM I 0.2% Pd 51A
RM I 30 51A
RM I 40 51A
RM I 55 51A
RM I 70 51A
RMI 1 Al 8 V 5 Fe 51C
RMI 3 Al 2.5 V 51C
RMI 4 Al 3 Mo 1 V 51C
RMI 4 Al 4 Mn 51B
RMI 5 Al 2.5 Sn 51B
RMI 5 Al 2.5 Sn EL1 51B
RMI 5 Al 6 Sn 2 Zr 1 Mo Si 51B
RMI 6 Al 2 Cb 1 Ta 1 Mo 51B
RMI 6 Al 2 Sn 4 Zr 2 Mo 51C
RMI 6 Al 4 V 51C
RMI 6 Al 4 V EL 1 51C
RMI 6 Al 6 V 2 Sn 51C
RMI 7 Al 2 Cb 1 Ta 51B
RMI 7 Al 4 Mo 51C
RMI 7 Al 12 Zr 51B
RMI 8 Al 1 Mo 1 V 51B
RMI 8 Mn 51B
RMI 13 V 11 Cr 3 Al 51C
RMI 16 V 2.5 Al 51C
RMK 44E2
RMKS 44E2
RMM 2.1
R MONEL 27D
RN 40
see ASTM range 27D
RN 41
see ASTM range 27F
RN 42
see ASTM range 27B
RN 43
see ASTM range 27D
RNC 0 44N1
RNC 1 44N1
RNC 30 27B
RNC-CARBIMPHY 27B
RNC-SUPERIMPHY 27B
R Ni Cr A 27F
R Ni Cr B 27F
R Ni Cr C 27F
RNK 29 44P
RNOD 44M1
RNOM 44M1
RNO Mo 44M1
RNO Mo V 44M1
RNO Mo WV 44M1
RO 211 44K1
RO 346 44K1
RO 651 44K1
RO 653 44K1
RO 663 44K1
RO 752 44K1
RO 952 44K2
RO 4154 44K1
RO 5200 46.1
RO 5240 46.1
RO 5252 46.1
RO 5255 46.1
RO 5400 46.1
RO 7155 44K1
RO 8155 44K2
RODAR 20C
ROLLO 44K3
ROP 10 44K1
ROP 19 44K2
ROP 21 44K3
ROP 43 44M1
ROP 46 44M1
ROP 57 44M2
ROP 63 44K1
ROP 5462 44K1
ROP 9653 44K2
ROPT 44K1
ROSE 6.1
ROSE 50.1
ROSTODUR R3 44M1
ROSTODUR R7 44M1
ROTOR ALUMINIUM 04 1A
RP 44M1
RP 1 44M1
RP 1152 44K2
RPA 44M1
R Pb 21.1
RR 50 1L
RR 53B 1L
RR 56 1F
RR 57 1F
RR 58 1F
RR 59 1F
RR 77 1G
RR 88 1G
RR 250 1L
RR 257 1F
RR 350 1L
RRAC 9 A 1L
RRJ 11 44M1
RRM 20 44M1
RRM 22 44M1
RRN J 30 44N1
RRN J 31 44N1
RRN J 38 44N1
RRN J 42 44N1
RRN J 44 44N1
RRN J 50 44N1
RRN J 51 44N1
RRN J 59 44N1
RR S R 1F
RRS 70 44M1
RRS 71 44M1
RS 44A2
RS 1 44A2
RS 2 44A3
RS 2 44K1
RS 3 44A3
RS 3 44K1
RS 3K 44E2
RS 4 44A3
RS (4) 44M1
RS 4K 44E2
RS 115 51C
RS 135 51C
RS 811 X 51C
RS (A) 44M1
RS (B) 44M1
RS (C) 44M1
RSD 44K3
RSKD 44K2
RSMK 44E2
R St 13 44A1
R St 13/03 44A1
R St 13/04 44A1
R St 13/05 44A1
R St 14 44A1
R St 14/04 44A1
R St 14/05 44A1
R St 34 44A1
R St 34/2 44A1
R St 37 44A1
R St 37/2 44A1
R St 37/202 44A1
R St 37/203 44A1
R St 37/204 44A1
R St 37/205 44A1
R St 42 44A1

R St 42/2	44A1
R St 42/202	44A1
R St 42/203	44A1
R St 42/204	44A1
R St 42/205	44A1
R St34.1	
see DIN 17100	44A1
R St34-2	
see DIN 17100	44A
R St37.1	
see DIN 17100	44A1
R St37.2	
see DIN 17100	44A1
R St38.7	
see DIN 17172	44A1
RT 12 Pd	51A
RT 15 Pd	51A
RT 18 Pd	51A
RT 20	51A
RT 27	44K1
RT 45	44K1
RT 46	44M1
RT 60	44M2
RT 1733	44K3
R THERM 2012 Mo	44M1
R THERM 2012 Mo WV	44M1
RTO 712	44K2
RTO 912	44K2
RUBIDIUM	38.1
RUBY	52.1
RULES MONOTYPE	
ALLOY	21.1
RUTHENIUM	39.1
RV	44K3
RVM	44M2
RW 2	44K3
RWMA Class 1	14A
RWMA Class 2	14L
RWMA Class 3	14D
RWMA Class 4	14D
RWMA CLASS 12	14J
RWMA CLASS 13	14J
RZ 5	23B
S	14C
S 1	44K2
S 1A	
see BS range	44A2
S1 A	44A2
S 1B	
see BS range	44A2
S1B	14B
S1 B	44A2
S 1C	
see BS range	44A2
S1 C	44A2
S 1R	44M1
S 2	
see BS range	44K2
S2	
see SAE S2	44K2
S 2	44K2
S 2-9.1	44K3
S 2-9.2	44K3
S 2-10-1-8	44L
S 3	
see BS range	44A1
S 3	44K2
S 3-3.2	44K3
S 3F	44A2
S4	44F1
S 4	44B2
S 4	44F2
S 4	44K2
S 4F	44A2
S 4K	44E2
S 4M	44A2
S 4S	44A2
S5	
see SAE S5	
S 5	44K2
S 5A	
see ASTM range	1C
S 5B	
see ASTM range	1C
S 5C	
see ASTM range	1C
S 6	
see BS range	44A2
S6	44A2
S 6	44K2
S 6-5.2	44K3
S 6-5.3	44K3
S 6-5-2-5	44L
S 6F	44A3
S 7	44K2
S 7-4-2-5	44L
S 7F	44A3
S 8	
see BS range	44F1
S8	44F1
S 8	44A3
S 9	
see BS range	44F1
S9	44F1
S 10	
see BS range	44F1
S10	44F1
S 10	20A
S 10-4-3-10	44L
S 11	
see BS range	44K2
S 12	
see BS range	44K2
S 12-0-5-5	44L
S 12-1-4-5	44L
S 12A	
see ASTM range	1C
S 12B	
see ASTM range	1C
S 13	
see BS range	44A1
S 14	
see BS range	44A1
S 15	
see BS range	44F1
S15	44F1
S 16	
see BS range	44K1
S 17	
see BS range	44F1
S17	44F1
S 18	
see BS range	44K2
S 18/8	44N1
S 18-0.1	44K3
S 18-1-2-5	44L
S 18-1-2-10	44L
S 19	
see BS range	44E1
S 20	
see BS range	44A2
S20	44A2
S 20	20A
S 21	
see BS range	44A2
S21	44A2
S 23	
see BS range	44A2
S23	44A2
S 24	
see BS range	44A2
S24	44A2
S 25	
see BS range	44A3
S 25	44A3
S 26	
see BS range	44A2
S26	44A2
S 27	
see BS range	44A2
S27	44A2
S 28	
see BS range	44K1
S 30	20A
S 31	
see BS range	44A2
S31	44A2
S 32	
see BS range	44A1
S 33	
see BS range	44K2
S 34	
see BS range	44K2
S 40	
see BS range	44K2
S 40	20A
S 43	
see BS range	44K2
S 46	
see BS range	44F1
S46	44F1
S 50	44F2
S 60	14J
S 60	44G2
S 61	
see BS range	44M1
S61	44M1
S 62	
see BS range	44M1
S62	44M1
S 65	44A3
S 67	
see BS range	44F1
S67	44F1
S 68	
see BS range	44H
S 69	
see BS range	44K2
S 70	
see BS range	44A3
S 70	44A1

S 70 44A3
S 71
see BS range 44A2
S71 44A2
S 75 44A3
S 76
see BS range 44A2
S76 44A2
S 77
see BS range 44A2
S77 44A2
S 79
see BS range 44A3
S 79 44A3
S 80
see BS range 44M1
S80 44M1
S 81
see BS range 44K2
S 82
see BS range 44K1
S 83
see BS range 44K1
S83 44F1
S 84
see BS range 44A1
S 85
see BS range 44N1
S85 44N1
S 86
see BS range 44K2
S 87
see BS range 44K2
S 88
see BS range 44K2
S 90
see BS range 44K1
S 90 44A4
S 91
see BS range 44A1
S 92
see BS range 44A1
S 93
see BS range 44A2
S93 44A2
S 94
see BS range 44K2
S 95
see BS range 44K2
S 96
see BS range 44K1
S 97
see BS range 44K1
S 98
see BS range 44K2
S 99
see BS range 44K2
S 100C 5.1
S 102
see BS range 44G1
S 103
see BS range 44K1
S 105
see BS range 44A2
S105 44A2
S 106
see BS range 44K1
S 107
see BS range 44K1
S 108
see BS range 44N1
S108 44N1
S 109
see BS range 44N1
S109 44N1
S 110
see BS range 44N1
S110 44N1
S 111
see BS range 44N1
S111 44N1
S 112
see BS range 44A1
S 113
see BS range 44A2
S113 44A2
S 114
see BS range 44G1
S 115
see BS range 44E1
S 116
see BS range 44A2
S116 44A2
S 117
see BS range 44E1
S 118
see BS range 44K2
S 119
see BS range 44K2
S 120
see BS range 44K1
S 122
see BS range 44K1
S 124
see BS range 44M1
S124 44M1
S 125
see BS range 44N1
S125 44N1
S 126
see BS range 44N1
S126 44N1
S 127
see BS range 44N1
S127 44N1
S 128
see BS range 44N1
S128 44N1
S 129
see BS range 44N1
S129 44N1
S 130
see BS range 44N1
S130 44N1
S131 44K2
S132 44K2
S133 44K1
S134 44K1
S 135 44E3
S 136 44E3
S137 44M1
S138 44K1
S139 44K1
S140 44K1
S141 44M1
S142 44K1
S143 44N3
S144 44N3
S145 44N3
S 145 44A3
S146 44K1
S147 44K1
S148 44K1
S149 44K1
S150 44M1
S151 44M1
S152 44M1
S153 44K1
S154 44K1
S155 44K1
S156 44K1
S157 44K1
S158 44K1
S159 44M1
S 159 44M1
S 200 44K3
S 200C 5.1
S 300 44L
S 300C 5.1
S 305 44L
S 307 44L
S 310 44A3
S 400 44K3
S 401 44K3
S 410 44M1
S 495 44N1
S 497 44L
S 500 44L
S 510
see BS range 44A1
S 511
see BS range 44A1
S 512
see BS range 44A1
S 513
see BS range 44A3
S 513 44A3
S 514
see BS range 44A1
S 515
see BS range 44A1
S 516
see BS range 44A2
S516 44A2
S 517
see BS range 44A3
S 517 44A3
S 518
see BS range 44K1
S 519
see BS range 44K1
S 520
see BS range 44N1
S520 44N1
S 521
see BS range 44N1
S521 44N1
S 522
see BS range 44N1
S522 44N1

S 523
see BS range 44N1
S523 44N1
S 524 44N1
S 525 44N1
S 526 44N1
S 527 44N1
S 528 44N1
S 529 44N1
S 530 44N1
S 531 44N1
S 532 44N3
S 533 44N3
S534 44K1
S535 44K1
S 536 44N1
S 537 44N1
S 538 44M1
S 588 44N1
S 590 44L
S 590 44N2
S 600 44K3
S 607 44K3
S 610 44K3
S 700 44L
S 705 44L
S–816 13A
S 816 13A
S 844 13A
S13800
UNS for AMS 5629 44N3
S13889
UNS for AMS 5840 44N3
S14800
UNS for AMS 5601 44N3
S15500
UNS for AMS 5658 44N3
S15700
UNS for AMS 5520 44N3
S15780
UNS for AMS 5813 44N3
S15789
UNS for AMS 5812 44N3
S16600
UNS for Croloy 16.6 PH 44N3
S16800
UNS for ASTM A376 44N1
S16880
UNS for ER16-8-2 44N3
S17400
UNS for AMS 5604 44N3
S17480
UNS for AMS 5825 44N3
S17600
UNS for ASTM A693 (635) 44N3
S17700
UNS for AMS 5528 44N3
S17780
UNS for AMS 5824 44N3
S18200
UNS for ASTM A581 (XM6) 44M1
S20100
UNS for AISI 201 44P
S20161
UNS for ASTM A167 44N1
S20200
UNS for AISI 202 44P
S20300
UNS for AMS 5762 44N3
S20910
UNS for AMS 5764 44N1
S20980
UNS for ER209 44N2
S 21000 44N1
S21300 44P
S21400
UNS for TENELON 44M1
S21460
UNS for CRYOGENIC TENELON 44P
S21500
UNS for ESSHETE 1250 44P
S21600
UNS for XM-17 44P
S21603
UNS for XM18 44N2
S21800
UNS for NITRONIC 60 44P
S21880
UNS for ER218 44N2
S21900
UNS for AMS 5561 44P
S21904
UNS for AMS 5562 44P
S21980
UNS for ER219 44N2
S23980
UNS for ER240 44N2
S24000
UNS for XM29 44P
S24100
UNS for ASTM A580 44N1
S28200
UNS for ASTM A276 44N1
S30100
UNS for AISI 301 44N1
S 30115 44N1
S30200
UNS for AISI 302 44N1
S30205
UNS for AISI 302B 44N1
S 30210 44N1
S 30260 44N1
S30300
UNS for AISI 303 44N1
S30310
UNS for 303 plus X 44N1
S30323
UNS for AISI 303Se 44N1
S30330
UNS for 303Cu 44N1
S30345
UNS for 303MA 44N1
S30360
UNS for 303Pb 44N1
S30400
UNS for AISI 304 44N1
S30403
UNS for AISI 304L 44N1
S30409
UNS for AISI 304L 44N1
S 30415 44N1
S 30431 44N1
S30451
UNS for AISI 304N 44N1
S30453
UNS for 304LN 44N1
S30500
UNS for AISI 305 44N1
S30600
UNS for ASTM A479 44N1
S30780
UNS for ER307 44N1
S30800
UNS for AISI 308 44N1
S30815
UNS for ASTM A479 44N1
S 30815 44N1
S30880
UNS for ER308 44N1
S30881
UNS for ER308Si 44N1
S30882
UNS for ER308Mo 44N1
S30883
UNS for ER308L 44N1
S30886
UNS for ER308MoL 44N1
S30888
UNS for ER308LSi 44N1
S30900
UNS for 309 44N1
S30908
UNS for 309Si 44N1
S30909
UNS for 309H 44N1
S30940
UNS for 309 44N1
S30941
UNS for 309 44N1
S30980
UNS for ER309 44N1
S30981
UNS for ER309Si 44N1
S30983
UNS for ER309L 44N1
S31000
UNS for 310 44N1
S31008
UNS for 310 44N1
S31009
UNS for 310 44N1
S31040
UNS for 310 44N1
S31041
UNS for 310 44N1
S31050
UNS for ASTM A240 44N1
S31080
UNS for ER310 44N1
S31100
UNS for ASTM A276-XM26 44N1
S31254 44P
S31260
UNS for ASTM A789 44N3
S31380
UNS for ER312 44N1

S31400
UNS for AISI 314 44N1
S31500
UNS for ASTM A669 44N3
S31600
UNS for AISI 316 44N1
S31600
UNS for AISI 316L 44N1
S31600
UNS for 316 44N1
S31620
UNS for AMS 5649 44N1
S31635
UNS for 316 44N1
S31640
UNS for 316 44N1
S31651
UNS for AISI 316N 44N1
S31653
UNS for 316LN 44N1
S31654
UNS for 316L 44N1
S31680
UNS for ER316 44N1
S31681
UNS for ER316Si 44N1
S31683
UNS for ER316L 44N1
S31688
UNS for ER316LSi 44N1
S31700
UNS for AISI 317 44N1
S31703
UNS for AISI 317L 44N1
S31725
UNS for AISI 317L 44N1
S31726
UNS for AISI 317L 44N1
S31753
UNS for AISI 317L 44N1
S31780
UNS for ER317 44N1
S31783
UNS for ER317L 44N1
S31803
UNS for 2205 44M1
S 31803 44N3
S31980
UNS for ER318 44N1
S32100
UNS for ASTM A182-F50 44N1
S32100
UNS for AISI 321 44N1
S32109
UNS for AISI 321H 44N1
S32180
UNS for ER321 44N1
S32304
UNS for ASTM A789 44N3
S32404
UNS for Uranus 50 44N1
S32550
UNS for ER2553 44N3
S32615
UNS for ASTM A479 44N1
S32750
UNS for ASTM A789 44N3
S32760
UNS for ASTM A815 44M1
S32900
UNS for AISI 329 44N1
S32950
UNS for ASTM A789 44N3
S33100
UNS for ASTM A182F10 44N1
S33600
UNS for AISI 446 44M1
S34700
UNS for AISI 347 44N1
S34709
UNS for AISI 347H 44N1
S34720
UNS for AMS 5642 44N1
S34723
UNS for AMS 5642 44N1
S 34740 44N1
S 34741 44N1
S34780
UNS for 347 44M1
S34781
UNS for AMS 5680 44N1
S34788
UNS for 347Si 44M1
S34800
UNS for AISI 348 44N1
S34809
UNS for AISI 348 44N1
S35000
UNS for AM350 44N3
S35080
UNS for AMS 5774 44N3
S35500
UNS for AM355 44N3
S35580
UNS for AM355 weld rod 44N3
S36200
UNS for AMS 5739 44N3
S38100
UNS for ASTM A167 44N1
S38400
UNS for 384 44N1
S38500
UNS for 385 - Obsolete 44N1
S 38600 44N2
S38660
UNS for ASTM A771 44N1
S40300
UNS for ASTM 403 44N1
S40500
UNS for AISI 405 44M1
S40800
UNS for ASTM A268 44M1
S40900
UNS for 409 44M1
S40940
UNS for ASTM A493 44N1
S41000
UNS for AISI 410 44M1
S41001
UNS for ASTM A579/51 44M1
S41008
UNS for AISI 410 44M1
S41025
UNS for AMS 5614 44M1
S41026
UNS for ASTM A182 (F6B) 44M1
S41040
UNS for AMS 5609 44M1
S41050
UNS for ASTM A176 44M1
S41080
UNS for AMS 5778 27C
S41081
UNS for AMS 5821 44M1
S41086
UNS for 410NiMo 44M1
S41400
UNS for AISI 414 44M1
S41500
UNS for ASTM A182 F6 44M1
S 41500 44M1
S41600
UNS for AISI 416 44M1
S41610
UNS for AISI 416-X 44M1
S41623
UNS for AISI 416Se 44M1
S41780
UNS for AMS 5822 44M1
S41800
UNS for AMS 5508 44M1
S41880
UNS for AMS 5817 44M1
S42000
UNS for AISI 420 44M1
S42010
UNS for ASTM A276/403 44M1
S42020
UNS for AISI 420F 44M1
S42023
UNS for AISI 420FSe 44M1
S42080
UNS for ER420 44M1
S 42100 44M1
S42200
UNS for 422 44M1
S42201
UNS for ASTM A579-52 44M1
S42300
UNS for ASTM A565-619 44M1
S42400
UNS for ASTM A182-F6NM 44M1
S42700
UNS for AMS 5749 44M2
S42800
UNS for AMS 5900 44M2
S43000
UNS for AISI 430 44M1
S43020
UNS for AISI 430F 44M1
S43023
UNS for AISI 430FSe 44M1
S43035
UNS for 439 44M1
S43036
UNS for ASTM A268-430T1 44M1
S43080
UNS for AISI 430 44M1

S43100
  UNS for 431 44M1
S43400
  UNS for 434 44M1
S43600
  UNS for 436 44M1
S44002
  UNS for AISI 440A 44M1
S44003
  UNS for AISI 440B 44M1
S44004
  UNS for AISI 440C 44M1
S44020
  UNS for AMS 5632 44M2
S44023
  UNS for AMS 5632 44M2
S 44100 44M1
S44200
  UNS for AISI 442 44M1
S44300
  UNS for AISI 442 44M1
S44400
  UNS for ASTM A176 44M1
S44625
  UNS for ASTM A176XM27 - obsolete 44M1
S44626
  UNS for ASTM A176XM33 44M1
S44627
  UNS for ASTM A176 44M1
S44635
  UNS for ASTM A176 44M1
S44660
  UNS for ASTM A176-SC-1 44M1
S44700
  UNS for ASTM A176-29-4 44M1
S44735
  UNS for ASTM A176-29-4C 44M1
S44736
  UNS for ASTM A268-28-4C+b 44M1
S44800
  UNS for ASTM A176-29-42 44M1
S45000
  UNS for AMS 5763 44N3
S45500
  UNS for AMS 5617 44N3
S50100
  UNS for AISI 501 44K1
S50180
  UNS for EB6H 44K1
S50181
  UNS for EB6HN 44K1
S50200
  UNS for AISI 502 44K1
S50280
  UNS for AMS 6466 44K1
S50281
  UNS for EB6H 44K1
S50300
  UNS for ASTM A182-F7 44K1
S50400
  UNS for ASTM A199 T9 44K1
S50480
  UNS for EB8 44K1
S50481
  UNS for EB8N 44K1
S63005
  UNS for AMS 5705 44N1
S63007
  UNS for AMS 5705 44N1
S63008
  UNS for EV8 44P
S63011
  UNS for EV11 44N2
S63012
  UNS for EV12 44N2
S63013
  UNS for EV13 44N2
S63014
  UNS for EV5 44K3
S63015
  UNS for EV6 44N2
S63016
  UNS for EV3 44N1
S63017
  UNS for EV4 44N2
S63018
  UNS for EV16 44N1
S63019
  UNS for XEVF 44P
S63197
  UNS for ER349 44N2
S63198
  UNS for AMS 5526 44N2
S63199
  UNS for AMS 5782 44N2
S64005
  UNS for HNV-1 44K2
S64006
  UNS for HNV2 44E2
S64152
  UNS for AMS 5718 44N3
S64299
  UNS for AMS 5784 44N1
S65006
  UNS for AMS 5710 44M1
S65007
  UNS for HNV3 44E2
S65150
  UNS for AMS 5761 44L
S65770
  UNS for AMS 5748 44L
S66009
  UNS for AMS 5700 44N1
S66220
  UNS for AMS 5733 44N1
S66286
  UNS for AMS 5525 44N2
S66545
  UNS for AMS 5543 44N1
SA
  see ASTM A specifications
SA 4 14M
SA 31
  see ASTM A31 44A2
SA 47
  see ASTM A47 20B
SA 84
  see ASTM A84 20A
SA 105
  see ASTM A105 44A2
SA 106
  see ASTM A106 44A2
SA 106A
  see ASTM A106 A 44A2
SA 106B
  see ASTM A106 B 44A2
SA 106C
  see ASTM A106 C 44A2
SA 113
  see ASTM A113 44A1
SA 135
  see ASTM A135 44A1
SA 178 A
  see ASTM A178 44A1
SA 178 C
  see ASTM A178 C 44A2
SA 179
  see ASTM A179 44A1
SA 181/1
  see A181/1 44A2
SA 181/11
  see 181/11 44A2
SA 182 F1
  see ASTM A182 44K1
SA 182 F5
  see ASTM A182 44K1
SA 182 F5a
  see ASTM A182 44K1
SA 182 F6
  see ASTM A182 44M1
SA 182 F7
  see ASTM A182 44K1
SA 182 F9
  see ASTM A182 44K1
SA 182 F10
  see ASTM A182 44N1
SA 182 F11
  see ASTM A182 44K1
SA 182 F12
  see ASTM A182 44K1
SA 182 F21
  see ASTM A182 44K1
SA 182 F22
  see ASTM A182 44K1
SA 182 F304
  see ASTM 182 44N1
SA 182 F304 H
  see ASTM A182 44N1
SA 182 F304 L
  see ASTM F304 L 44N1
SA 182 F310
  see ASTM A182 44N1
SA 182 F316
  see ASTM A182 44N1
SA 182 F316H
  see ASTM A182 44N1
SA 182 F316L
  see ASTM A182 44N1
SA 182 F321
  see ASTM A182 44N1
SA 182 F321H
  see ASTM A182 44N1
SA 182 F347
  see ASTM A182 44N1

SA 182 F347H
see ASTM A182 44N1
SA 182 F348
see ASTM A182 44N1
SA 182 F348H
see ASTM A182 44N1
SA 192
see ASTM A192 44A1
SA 193 B5
see ASTM A193 44K1
SA 193 B6
see ASTM A193 44M1
SA 193 B8
see ASTM A193 44N1
SA 193 B8C
see ASTM A193 44N1
SA 193 B8M
see ASTM A193 44N1
SA 193 B8T
see ASTM A193 44N1
SA 193 B16
see ASTM A193 44K2
SA 194/1
see ASTM A194/1 44A1
SA 194/2
see ASTM A194/2 44A2
SA 194/3
see ASTM A194/3 44K1
SA 194/4
see ASTM A194/4 44G2
SA 194/6
see ASTM A194/6 44M1
SA 194/7
see ASTM A194/7 44G2
SA 194/8
see ASTM A194/8 44N1
SA 194/8C
see ASTM A194/8 44N1
SA 194/8M
see ASTM A194/8 44N1
SA 194/8T
see ASTM A194/8 44N1
SA 197
see ASTM A197 20B
SA 202A
see ASTM A202A 44E1
SA 202B
see ASTM A202B 44E1
SA 203 B
see ASTM A203 B 44F1
SA 203 D
see ASTM A203 D 44F1
SA 203 E
see ASTM A203 E 44F1
SA 203A
see ASTM A203A 44F1
SA 204 A
see ASTM A204 A 44G1
SA 204 B
see ASTM A204 B 44G1
SA 204 C
see ASTM A204 C 44G1
SA 209 T1
see ASTM A209 44G1
SA 209 Tla
see ASTM A209 44G1
SA 209 Tlb
see ASTM A209 44G1
SA 210 Al
see ASTM A210 44A2
SA 210 C
see ASTM A210 C 44A2
SA 213
see SA 182 for different grades
SA 214
see ASTM A214 44A1
SA 216 WCA
see ASTM A216 44A2
SA 216 WCB
see ASTM A216 44A2
SA 216 WCC
see ASTM A216 44A2
SA 217 C5
see ASTM A217 44K1
SA 217 C12
see ASTM A217 44A2
SA 217 WC1
see ASTM A217 44K1
SA 217 WC4
see ASTM A217 44K1
SA 217 WC5
see ASTM A217 44K1
SA 217 WC6
see ASTM A217 44K1
SA 217 WC9
see ASTM A217 44K1
SA 225
see ASTM A225 44J
SA 226
see ASTM A226 44A1
SA 233
see ASTM A233 44A1
SA 240 TP321 H
see ASTM A249 44N1
SA 249 TP304
see ASTM A249 44N1
SA 249 TP304 H
see ASTM A249 44N1
SA 249 TP304 L
see ASTM A249 44N1
SA 249 TP309
see ASTM A249 44N1
SA 249 TP310
see ASTM A249 44N1
SA 249 TP316
see ASTM A249 44N1
SA 249 TP316 H
see ASTM A249 44N1
SA 249 TP316 L
see ASTM A249 44N1
SA 249 TP317
see ASTM A249 44N1
SA 249 TP321
see ASTM A249 44N1
SA 249 TP347
see ASTM A249 44N1
SA 249 TP347 H
see ASTM A249 44N1
SA 249 TP348
see ASTM A249 44N1
SA 249 TP348 H
see ASTM A249 44N1
SA 250 Tl
see ASTM A250 44G1
SA 250 Tla
see ASTM A250 44G1
SA 250 Tlb
see ASTM A250 44G1
SA 251
see ASTM A251 44A1
SA 263
see ASTM A263 44A1
SA 264
see ASTM A264 44A1
SA 265
see ASTM A265 44A1
SA 266/1 & 2
see ASTM A266/1 & 2 44A2
SA 266/3
see ASTM A266/3 44A2
SA 268
see ASTM A268 44M1
SA 278
see ASTM A278 20B
SA 283
see ASTM A283 44A1
SA 285 A
see ASTM A285 A 44A1
SA 285 B
see ASTM A285 B 44A1
SA 285 C
see ASTM A285 C 44A1
SA 298
see ASTM A298 44N1
SA 299
see ASTM A299 44A2
SA 302 A
see ASTM A302 A 44K1
SA 302 B
see ASTM A302 B 44K1
SA 302 C
see ASTM A302 C 44K1
SA 302 D
see ASTM A302 D 44K1
SA 312
see ASTM A312 44N1
SA 316 E7000 Al
see ASTM A316 44K1
SA 316 E8015/18 B
see ASTM A316 44K1
SA 316 E8016 C
see ASTM A316 44F1
SA 316 E9015/8 D1
see ASTM A316 44G1
SA 316 E9016/8 B3
see ASTM A316 44K1
SA 316 E9018 M
see ASTM A316 44K1
SA 316 E10018 M
see ASTM A316 44K1
SA 316 E12018 M
see ASTM A316 44K1
SA 320
see ASTM A320 44N1
SA 325
see ASTM A325 44A2
SA 333/1
see ASTM A333/1 44A2

SA 333/3
see ASTM A333/3 44A1
SA 333/4
see ASTM A333/4 44A1
SA 333/6
see ASTM A333/6 44A2
SA 333/7
see ASTM A333/7 44A1
SA 333/8
see ASTM A333/8 44A1
SA 334
see ASTM A334 44A1
SA 336
see ASTM A336 44G1
SA 350
see ASTM A350 44A2
SA 351
see ASTM A351 44N1
SA 352
see ASTM A352 44F1
SA 353
see ASTM A357 44K1
SA 369
see ASTM A369 44G1
SA 371
see ASTM A371 44N1
SA 372
see ASTM A372 44A2
SA 376
see ASTM A376 44N1
SA 387
see ASTM A387 44K1
SA 410
see ASTM A410 44K1
SA 414
see ASTM A414 44A2
SA 423
see ASTM A423 44K1
SA 430
see ASTM A430 44N1
SA 433
see ASTM A433 44A2
SA 442
see ASTM A442 44A1
SA 452
see ASTM A452 44N1
SA 453
see ASTM A453 44N2
SA 455
see ASTM A455 44A2
SA 476
see ASTM A476 20B
SA 479
see ASTM A479 44N1
SA 487
see ASTM A508 44J
SA 508
see ASTM A508 44J
SA 515
see ASTM A515 44A2
SA 516
see ASTM A516 44A2
SA 517
see ASTM A517 44K1
SA 522
see ASTM A522 44F1
SA 533
see ASTM A533 44G1
SA 537
see ASTM A537 44A1
SA 540
see ASTM A540 44K2
SA 541
see ASTM A541 44A2
SA 553
see ASTM A553 44F1
SA 556
see ASTM A556 44A1
SA 557
see ASTM A557 44A2
SA 558
see ASTM A558 44A1
SA 559
see ASTM A559 44A1
SABEN 652 44K3
SABEN EXTRA 44K3
SABEN HC 44K3
SABEN KERAU 44K3
SABEN TENCO 44L
SABEN WUNDA 44L
SABEX 44K1
SABRE 44L
SAE 01 44K3
SAE 1 A 21.1
SAE 1 B 21.1
SAE 02 44K3
SAE 2 A 21.1
SAE 2 B 21.1
SAE 003 C 44A2
SAE 3 A 21.1
SAE 3 B 21.1
SAE 4 A 21.1
SAE 4 B 21.1
SAE 5 A 21.1
SAE 5 B 21.1
SAE 06 44G2
SAE 6 A 21.1
SAE 6 B 21.1
SAE 7 A 21.1
SAE 8 A 21.1
SAE 9 B 21.1
SAE 10 50A
SAE 11 50A
SAE 12 50A
SAE 12L14 44A1
SAE 13 21.1
SAE 14 21.1
SAE 15 21.1
SAE 16 21.1
SAE 16 44K3
SAE 17 44K3
SAE 18 8.1
SAE 19 21.1
SAE 20 1B
SAE 0022 44A1
SAE 24 1F
SAE 0025 44A2
SAE 25 1A
SAE 26 1F
SAE 28 1A
SAE 29 1B
SAE 33 1L
SAE 34 1L
SAE 0035 44A2
SAE 35 1C
SAE 38 1L
SAE 39 1L
SAE 40 14K
SAE 41 14K
SAE 43 14C
SAE 43 BV12 44K1
SAE 43 BV14 44K1
SAE 46 B12 44K1
SAE 48 14K
SAE 49 14K
SAE 0050 44A2
SAE 50 23B
SAE 50 B40 44E2
SAE 50 B44 44E2
SAE 50 B46 44E2
SAE 50 B50 44E2
SAE 50 B60 44E2
SAE 51 23B
SAE 51 B60 44E2
SAE 52 23B
SAE 53 23B
SAE 62 14K
SAE 63 14K
SAE 64 14K
SAE 65 14K
SAE 66 14K
SAE 67 14K
SAE 68 A 14G
SAE 68 B 14G
SAE 70A 14B
SAE 70B 14B
SAE 70C 14B
SAE 71 14A
SAE 71D 14G
SAE 72 14B
SAE 73 14C
SAE 74 A 14C
SAE 74B 14B
SAE 74C 14B
SAE 74D 14B
SAE 75 14A
SAE 77 A 14K
SAE 77 C 14K
SAE 79A 14B
SAE 79B 14B
SAE 080 44A1
SAE 80A 14B
SAE 80B 14B
SAE 81 14K
SAE 81 B45 44K2
SAE 83 14A
SAE 86 B45 44K2
SAE 88 14C
SAE 090 44A2
SAE 94 B15 44K1
SAE 94 B17 44K1
SAE 94 B30 44K1
SAE 94 B40 44K2
SAE 0105 44A2
SAE 110 20B
SAE 111 20B
SAE 113 20B
SAE 114 20B
SAE 0120 44A2
SAE 120 20B

SAE 121 20B
SAE 122 20B
SAE 0150 44A2
SAE 0175 44A2
SAE 180 8.1
SAE 190 21.1
SAE 201 1D
SAE 202 1F
SAE 203 1F
SAE 204 1F
SAE 205 1C
SAE 206 1C
SAE 207 1D
SAE 208 1D
SAE 209 1D
SAE 210 1E
SAE 211 1E
SAE 212 1E
SAE 213 1E
SAE 214 1A
SAE 215 1G
SAE 240 1H
SAE 242 1H
SAE 244 1H
SAE 245 1H
SAE 246 1H
SAE 247 1H
SAE 248 1H
SAE 249 1H
SAE 250 1H
SAE 251 1D
SAE 252 1D
SAE 253 1E
SAE 260 1F
SAE 270 1F
SAE 280 1E
SAE 281 1E
SAE 282 1E
SAE 290 1C
SAE 300 1L
SAE 303 1L
SAE 304 1C
SAE 305 1C
SAE 306 1L
SAE 307 1L
SAE 308 1L
SAE 309 1K
SAE 310 1M
SAE 311 1M
SAE 312 1M
SAE 313 1M
SAE 314 1M
SAE 315 1M
SAE 320 1J
SAE 321 1K
SAE 322 1K
SAE 323 1K
SAE 324 1J
SAE 326 1L
SAE 327 1K
SAE 328 1K
SAE 329 1L
SAE 330 1L
SAE 331 1L
SAE 332 1L
SAE 334 1L
SAE 335 1L
SAE 336 1K
SAE 380 1L
SAE 382 1L
SAE 383 1L
SAE 430 A 14C
SAE 430 B 14C
SAE 480 14K
SAE 481 14K
SAE 482 14K
SAE 484 14K
SAE 485 21.1
SAE 500 23B
SAE 501 23B
SAE 501A 23B
SAE 502 23B
SAE 503 23B
SAE 504 23B
SAE 505 23B
SAE 506 23B
SAE 507 23B
SAE 508 23B
SAE 509 23B
SAE 510 23B
SAE 513 23B
SAE 520 23B
SAE 522 23B
SAE 523 23B
SAE 524 23B
SAE 531 23B
SAE 532 23B
SAE 533 23B
SAE 534 23B
SAE 620 14K
SAE 621 14K
SAE 622 14K
SAE 640 14K
SAE 660 14K
SAE 701 A 14G
SAE 701 B 14G
SAE 701 C 14G
SAE 770 1N
SAE 780 1N
SAE 781 1N
SAE 782 1N
SAE 783 50A
SAE 791 14K
SAE 792 14K
SAE 793 14K
SAE 794 14K
SAE 795 14B
SAE 797 14K
SAE 798 14K
SAE 799 14K
SAE 840 14K
SAE 841 14K
SAE 842 14K
SAE 843 14K
SAE 850 44A1
SAE 851 44A2
SAE 852 44A4
SAE 853 44A1
SAE 855 44A4
SAE 861 44A2
SAE 862 44A2
SAE 863 44A2
SAE 864 A 44A3
SAE 864 B 44A3
SAE 865 A 44A3
SAE 865 B 44A3
SAE 866 A 44A3
SAE 866 B 44A3
SAE 867 A 44A3
SAE 867 B 44A3
SAE 870 44A1
SAE 872 44A4
SAE 890 14B
SAE 890 14J
SAE 891 14B
SAE 891 14J
SAE 903 55A
SAE 925 55A
SAE 950 44A1
SAE 1005 44A1
SAE 1006 44A1
SAE 1008 44A1
SAE 1009 44A1
SAE 1010 44A1
SAE 1012 44A1
SAE 1013 44A1
SAE 1015 44A1
SAE 1016 44A1
SAE 1017 44A1
SAE 1018 44A1
SAE 1019 44A1
SAE 1020 44A1
SAE 1021 44A1
SAE 1022 44A1
SAE 1023 44A1
SAE 1024 44A2
SAE 1025 44A2
SAE 1026 44A2
SAE 1027 44A2
SAE 1029 44A2
SAE 1030 44A2
SAE 1033 44A2
SAE 1035 44A2
SAE 1036 44A2
SAE 1037 44A2
SAE 1038 44A2
SAE 1039 44A2
SAE 1040 44A2
SAE 1041 44A2
SAE 1042 44A2
SAE 1043 44A2
SAE 1044 44A3
SAE 1045 44A2
SAE 1046 44A2
SAE 1047 44A3
SAE 1048 44A3
SAE 1049 44A3
SAE 1050 44A3
SAE 1051 44A3
SAE 1052 44A3
SAE 1053 44A3
SAE 1055 44A3
SAE 1060 44A3
SAE 1061 44A3
SAE 1064 44A3
SAE 1065 44A3
SAE 1066 44A3
SAE 1069 44A3
SAE 1070 44A3
SAE 1072 44A3
SAE 1074 44A3

SAE 1078 44A3
SAE 1080 44A3
SAE 1084 44A4
SAE 1085 44A4
SAE 1086 44A4
SAE 1090 44A4
SAE 1095 44A4
SAE 1108 44A1
SAE 1109 44A1
SAE 1110 44A1
SAE 1111 44A1
SAE 1112 44A1
SAE 1113 44A1
SAE 1115 44A1
SAE 1116 44A1
SAE 1117 44A1
SAE 1118 44A1
SAE 1119 44A1
SAE 1120 44A1
SAE 1126 44A2
SAE 1132 44A2
SAE 1137 44A2
SAE 1138 44A2
SAE 1139 44A2
SAE 1140 44A2
SAE 1141 44A2
SAE 1144 44A2
SAE 1145 44A2
SAE 1146 44A2
SAE 1151 44A3
SAE 1215 44A1
SAE 1320 44A1
SAE 1330 44A2
SAE 1335 44A2
SAE 1340 44A2
SAE 1345 44A2
SAE 1513 44A1
SAE 1518 44A1
SAE 1522 44A1
SAE 1524 44A2
SAE 1525 44A2
SAE 1527 44A2
SAE 1541 44A2
SAE 1548 44A3
SAE 1551 44A3
SAE 1552 44A3
SAE 1561 44A3
SAE 1566 44A3
SAE 1572 44A3
SAE 2317 44F1
SAE 2330 44F1
SAE 2340 44F2
SAE 2345 44F2
SAE 2512 44F1
SAE 2515 44F1
SAE 2517 44F1
SAE 3115 44K1
SAE 3120 44K1
SAE 3130 44K1
SAE 3135 44K1
SAE 3140 44K2
SAE 3145 44K2
SAE 3150 44K2
SAE 3215 44K1
SAE 3220 44K1
SAE 3230 44K1
SAE 3240 44K1
SAE 3245 44K2
SAE 3250 44K2
SAE 3310 44K1
SAE 3312 44K1
SAE 3316 44K1
SAE 3325 44K1
SAE 3335 44K1
SAE 3340 44K2
SAE 3415 44K1
SAE 3435 44K1
SAE 3436 44K1
SAE 3450 44K2
SAE 4012 44G1
SAE 4023 44G1
SAE 4024 44G1
SAE 4027 44G1
SAE 4028 44G1
SAE 4032 44G1
SAE 4037 44G1
SAE 4042 44G2
SAE 4047 44G2
SAE 4053 44G2
SAE 4063 44G2
SAE 4068 44G2
SAE 4118 44K1
SAE 4119 44K1
SAE 4125 44K1
SAE 4130 44K1
SAE 4135 44K1
SAE 4137 44K1
SAE 4140 44K2
SAE 4142 44K2
SAE 4145 44K2
SAE 4147 44K2
SAE 4150 44K2
SAE 4161 44K2
SAE 4317 44K1
SAE 4320 44K1
SAE 4337 44K1
SAE 4340 44K2
SAE 4419 44G1
SAE 4427 44G1
SAE 4608 44K1
SAE 4615 44K1
SAE 4617 44K1
SAE 4620 44K1
SAE 4620 44K2
SAE 4621 44K1
SAE 4626 44K1
SAE 4640 44K2
SAE 4718 44K1
SAE 4720 44K1
SAE 4812 44K1
SAE 4815 44K1
SAE 4817 44K1
SAE 4820 44K1
SAE 5015 44E1
SAE 5045 44E2
SAE 5046 44E2
SAE 5060 44E2
SAE 5115 44E1
SAE 5117 44E1
SAE 5120 44E1
SAE 5130 44E1
SAE 5132 44E1
SAE 5135 44E1
SAE 5140 44E2
SAE 5145 44E2
SAE 5147 44E2
SAE 5150 44E2
SAE 5152 44E2
SAE 5155 44E2
SAE 5160 44E2
SAE 6115 44K1
SAE 6117 44K1
SAE 6118 44K1
SAE 6120 44K1
SAE 6125 44K1
SAE 6130 44K1
SAE 6135 44K1
SAE 6140 44K2
SAE 6145 44K2
SAE 6150 44K2
SAE 6195 44K3
SAE 7260 44K2
SAE 8115 44K1
SAE 8615 44K1
SAE 8617 44K1
SAE 8620 44K1
SAE 8622 44K1
SAE 8625 44K1
SAE 8627 44K1
SAE 8630 44K1
SAE 8632 44K1
SAE 8635 44K1
SAE 8637 44K1
SAE 8640 44K2
SAE 8641 44K2
SAE 8642 44K2
SAE 8645 44K2
SAE 8647 44K2
SAE 8650 44K2
SAE 8653 44K2
SAE 8655 44K2
SAE 8660 44K3
SAE 8715 44K1
SAE 8717 44K1
SAE 8719 44K1
SAE 8720 44K1
SAE 8735 44K1
SAE 8740 44K2
SAE 8742 44K2
SAE 8745 44K2
SAE 8750 44K2
SAE 8822 44K1
SAE 9250 44B2
SAE 9254 44E2
SAE 9255 44B2
SAE 9260 44B2
SAE 9261 44E2
SAE 9262 44E2
SAE 9310 44K1
SAE 9315 44K1
SAE 9317 44K1
SAE 9437 44K1
SAE 9440 44K2
SAE 9442 44K2
SAE 9445 44K2
SAE 9447 44K2
SAE 9747 44K2
SAE 9763 44K2
SAE 9840 44K2
SAE 9845 44K2
SAE 9850 44K2

SAE 12413 44A1
SAE 30201 44P
SAE 30202 44P
SAE 30301 44N1
SAE 30301 44N3
SAE 30302 44N1
SAE 30302 B 44N1
SAE 30303 44N1
SAE 30303 F 44N1
SAE 30303 Se 44N1
SAE 30304 44N1
SAE 30304 L 44N1
SAE 30305 44N1
SAE 30308 44N1
SAE 30309 44N1
SAE 30309 S 44N1
SAE 30310 44N1
SAE 30310 S 44N1
SAE 30314 44N1
SAE 30316 44N1
SAE 30316 L 44N1
SAE 30317 44N1
SAE 30321 44N1
SAE 30325 44N1
SAE 30329 44N1
SAE 30330 44N1
SAE 30330 A 44N1
SAE 30347 44N1
SAE 30348 44N1
SAE 32510 20B
SAE 35013 20B
SAE 35018 20B
SAE 43010 20B
SAE 48005 20B
SAE 50100 44E3
SAE 51100 44E3
SAE 51403 44M1
SAE 51405 44M1
SAE 51409 44M1
SAE 51410 44M1
SAE 51414 44M1
SAE 51416 44M1
SAE 51416 F 44M1
SAE 51416 Se 44M1
SAE 51420 44M1
SAE 51420 F 44M1
SAE 51420 F Se 44M1
SAE 51429 44M1
SAE 51430 44M1
SAE 51430 F 44M1
SAE 51430 F Se 44M1
SAE 51431 44M1
SAE 51434 44M1
SAE 51436 44M1
SAE 51439 44M1
SAE 51439 LL 44M1
SAE 51440 A 44M1
SAE 51440 B 44M1
SAE 51440 C 44M2
SAE 51440 F 44M2
SAE 51440 F Se 44M2
SAE 51442 44M1
SAE 51446 44M1
SAE 51499 44M1
SAE 51501 44K1
SAE 51502 44K1
SAE 52100 44E3
SAE 53004 20B
SAE 60003 20B
SAE 60303 44N1
SAE 60304 44N1
SAE 60304 L 44N1
SAE 60309 44N1
SAE 60310 44N1
SAE 60312 44N1
SAE 60316 44N1
SAE 60316 L 44N1
SAE 60317 44N1
SAE 60326 44M1
SAE 60328 44M1
SAE 60347 44N1
SAE 60410 44M1
SAE 60420 44M1
SAE 60442 44M1
SAE 60446 44M1
SAE 60502 44K1
SAE 70002 20B
SAE 70308 44N1
SAE 70309 44N1
SAE 70310 44N1
SAE 70310 A 44N1
SAE 70311 44N1
SAE 70312 44N1
SAE 70327 44N1
SAE 70330 44N1
SAE 70331 44N1
SAE 70334 27B
SAE 70335 27B
SAE 70446 44M1
SAE 70502 44K1
SAE 71360 44K2
SAE 71660 44K2
SAE A2 44K3
SAE CA 102 14A
SAE CA 110 14A
SAE CA 111 14A
SAE CA 113 14A
SAE CA 114 14A
SAE CA 115 14A
SAE CA 116 14A
SAE CA 120 14A
SAE CA 122 14A
SAE CA 145 14A
SAE CA 147 14A
SAE CA 150 14A
SAE CA 162 14A
SAE CA170 14D
SAE CA 172 14D
SAE CA 175 14D
SAE CA 176 14D
SAE CA184 14L
SAE CA 187 14A
SAE CA 192 14A
SAE CA 210 14B
SAE CA 220 14B
SAE CA 230 14B
SAE CA 240 14B
SAE CA 260 14B
SAE CA 268 14B
SAE CA 270 14B
SAE CA 330 14B
SAE CA 331 14B
SAE CA 342 14B
SAE CA 345 14B
SAE CA 350 14B
SAE CA 360 14B
SAE CA 377 14C
SAE CA 464 14C
SAE CA 465 14C
SAE CA 466 14C
SAE CA 467 14C
SAE CA 510 14K
SAE CA 511 14K
SAE CA 521 14K
SAE CA 524 14K
SAE CA 544 14K
SAE CA 608 14G
SAE CA 614 14G
SAE CA 617 14G
SAE CA 618 14G
SAE CA 623 14G
SAE CA 624 14G
SAE CA 630 14G
SAE CA 637 14G
SAE CA 642 14G
SAE CA 655 14M
SAE CA655 14M
SAE CA 670 14B
SAE CA 673 14B
SAE CA 674 14C
SAE CA 675 14C
SAE CA 706 14E
SAE CA 710 14E
SAE CA 715 14E
SAE CA 752 14F
SAE CA 770 14F
SAE CA 836 14K
SAE CA 838 14K
SAE CA 852 14B
SAE CA 854 14B
SAE CA 858 14B
SAE CA 862 14B
SAE CA 863 14B
SAE CA 865 14B
SAE CA 872 14B
SAE CA 874 14B
SAE CA 875 14B
SAE CA 878 14B
SAE CA 879 14B
SAE CA 903 14K
SAE CA 905 14K
SAE CA 909 14K
SAE CA 922 14K
SAE CA 923 14K
SAE CA 925 14K
SAE CA 927 14K
SAE CA 929 14K
SAE CA 932 14K
SAE CA 935 14K
SAE CA 937 14K
SAE CA 938 14K
SAE CA 943 14K
SAE CA 947 14K
SAE CA 948 14K
SAE CA 952 14G
SAE CA 953 14G
SAE CA 954 14G
SAE CA 955 14G
SAE CA 958 14G
SAE CA 962 14A
SAE D2 44M2

SAE D3 44M2
SAE D5 44M2
SAE D7 44M2
SAE D4018 20B
SAE D4512 20B
SAE D5506 20B
SAE D7003 20B
SAE E 4340 44K2
SAE EV
see designation for detail; e.g. EV2 – 44M1
SAE EV 1 44M1
SAE EV 2 44M1
SAE EV 3 44N1
SAE EV 10 44N1
SAE G 2000 20B
SAE G 3000 20B
SAE G 3500 20B
SAE G 4000 20B
SAE G 4500 20B
SAE H11 44K1
SAE H12 44K1
SAE H13 44K1
SAE H21 44K1
SAE HEV 1 44L
SAE HEV 2 27C
SAE HEV 3 27C
SAE HEV 5 27C
SAE HEV 6 27C
SAE HNV 1 44K2
SAE HNV 2 44E2
SAE HNV 3 44E2
SAE HNV 4 44K2
SAE HNV 5 44N1
SAE HNV 6 44M1
SAE HNV 7 44H
SAE J 113-1 44A2
SAE J 113-2 44A2
SAE J 157 44E2
SAE J 172 44A3
SAE J 178 44A1
SAE J 271 44A3
SAE J 435-0022 44A1
SAE J 435-0025 44A1
SAE J 435-0050 44A2
SAE J132 44K2
SAE J158 20B
SAE J217 44N3
SAE J230 44N1
SAE J368a
specification for quench and tempered steels listed by grades
SAE J434 20B
SAE J434 B 20B
SAE J859 20B
SAE J1392
specification for steel sheet-coated listed by grades
SAE J1392 44K1
SAE J1442
specification for high strength steel plate bar etc hot rolled listed by grades
SAE M1 44K3
SAE M2 44K3
SAE M3 44K3
SAE M4 44K3
SAE NV 3 44K2
SAE NV 4 44K2
SAE NV 5 44K2
SAE NV1 44A2
SAE NV2 44A2
SAE NVG 44E2
SAE S 1 44K2
SAE S 2 44K2
SAE S 5 44K2
SAE T1 44K3
SAE T2 44K3
SAE T4 44L
SAE T5 44L
SAE T8 44L
SAE VF 1 27B
SAE VF 3 27C
SAE VF 4 27C
SAE VF 5 13A
SAE W108 44A4
SAE W109 44A4
SAE W110 44A4
SAE W112 44A4
SAE W209 44J
SAE W210 44J
SAE W310 44J
SAE X 3140 44K2
SAE X4620 44K1
SAF 2205 44N1
SAF 2205 44N3
S Al Bz 6 14G
S Al Bz 8 14G
S Al Si 5 1C
S Al Si 12 1C
SALGE 55A
SAMSON EXTRA 44E1
SAN BOLD 3 44A1
SANBOLD 8 44A2
SANBOLD 9 44A3
SANBOLD 11 44E2
SANBOLD 12 44F2
SANBOLD 14 44F1
SANBOLD 15 44A2
SANBOLD 16 44G1
SANBOLD 17 44G1
SANBOLD 18 44E2
SANBOLD 19 44K2
SANBOLD 21 44K1
SANBOLD 22 44K2
SANBOLD 23 44K1
SANBOLD 24 44K2
SANBOLD 25 44K1
SANBOLD 30 44K1
SANBOLD 31 44E3
SANBOLD 33 44K1
SANBOLD 34 44K1
SANBOLD 36 44K1
SANBOLD 37 44K1
SANBOLD 38 44K1
SANBOLD 39 44F2
SANBOLD 39 44K1
SANBOLD 42 44A3
SANBOLD 43 44A3
SANBOLD 44 44A4
SANBOLD 45 44A3
SANBOLD 50 44A3
SANBOLD 51 44F1
SANBOLD 52 44A2
SANBOLD 53 44G1
SANBOLD 54 44E2
SANBOLD 56 44K1
SANBOLD 57 44E3
SANBOLD 58 44K1
SANBOLD 59 44A3
SANBOLD 60 44A3
SANBOLD 66 44K2
SANBOLD 68 44K1
SANBOLD 69 44K1
SANBOLD 74 44G1
SANBOLD 77 44K1
SANBOLD 101 44K1
SANBOLD 155 44K2
SANBOLD 3890 44A2
SANBOLD 8058 44A3
SANBOLD CVS 44K2
SANBOLD N 2832 44K1
SANBOLD N 3603 44K1
SANBOLD N 5366 44K1
SANBOLD NA 35 20C
SANBOLD NA 47 20C
SANBOLD NA76 27E
SANBOLD NK 2833 44K2
SANBOLD NK 8055 44K1
SANBRON AA 44N1
SANBRON BB 44M1
SANBRON DD 44M1
SANDHOLME GA 20B
SANICRO 28 44N1
SANICRO 28 44N2
SANICRO 30 44N1
SANICRO 31 44N1
SANICRO 31H 44N1
SANICRO 70 27B
SANICRO 71 27B
SAPPHIRE 52.1
SATMUMETAL 20C
SATP
see ASTM range 14A
SB
See ASTM B specifications
SB 14C
SB 6 14C
SB 26 S5A
see ASTM B26 S5 1C
SB 36F 44A1
SB 36FT 44A1
SB 38F 44A1
SB 38FT 44A1
SB 39F 44J
SB 39FT 44A1
SB 40F 44A1
SB 40FT 44A1
SB 42F 44J
SB 42FT 44A1
SB 43
see ASTM B43 14K
SB 46F 44K1
SB 46FT 44K1
SB 61
see ASTM B61 14K
SB 62
see ASTM B62 14K

SB 96 C
see ASTM 96 C 14M
SB 96A
see ASTM B96 A 14M
SB 127
see ASTM B127 27D
SB 143
see ASTM B143 14K
SB 144
see ASTM B144 14K
SB 149
see ASTM B149 14F
SB 164
see ASTM B164 B 27D
SB 164 A
see ASTM B164 A 27D
SB 260 B Cu 2 14H
SB 265 51A
SB 304
see ASTM B304 14A
SB 308
see ASTM B308 1E
SB 308
see ASTM B308 1D
SB 315
see ASTM B315 14M
SB 333
see ASTM B333 27C
SB 334
see ASTM B334 27C
SB 335
see ASTM B335 27C
SB 336
see ASTM B336 27C
SB 337
see ASTM B337 51A
SB 338
see ASTM B338 51A
SB 348
see ASTM B348 51A
SB 359
see ASTM B359 14A
SB 381
see ASTM B381 51A
SB 382
see ASTM B382 51A
SB 402
see ASTM B402 14E
SB Cu
see DIN range 14A
SBM 44K2
SBV 1A
see JIS G3119 44G1
SBV 1B
see JIS G3119 44K1
SBV 2
see JIS G3179 44K1
SBV 3
see JIS G3119 44K1
SC 14A
SC 20B
SC 44L
SC 6-5-2 44K3
SC 12–28 20B
SC 13 44M2
SC 15–32 20B
SC 21–40 20B
SC 25 44M2
SC 26 44M2
SC 28–48 20B
SC 32–52 20B
SC 36–56 20B
SC 38 44M2
SC 40 44M1
SC 40–60 20B
SC 45 44M1
SC 51 A
see ASTM range 1L
SC 51 B
see ASTM range 1K
SC 64 C
see ASTM range 1L
SC 64 D
see ASTM range 1L
SC 82 A
see ASTM range 1L
SC 84 A
see ASTM range 1L
SC 84 B
see ASTM range 1L
SC 92A 1C
SC 103 A
see ASTM range 1L
SC 114 A
see ASTM range 1L
SC 122 A
see ASTM range 1L
SC 140 Mo 44M1
SCB 1
see BS range 14K
SCB 2
see BS range 14B
SCB 3
see BS range 14B
SCB 4
see BS range 14C
SCB 4 14C
SCB 5
see BS range 14B
SCB 6
see BS range 14B
SCF 19 44N1
SCH 1 44A1
SCHULZ 55A
SCIMITARS 14F
SCMV 1
see JIS G4109 44K1
SCMV 2
see JIS G4109 44K1
SCMV 3
see JIS G4109 44K1
SCMV 4
see JIS G4109 44K1
SCMV 5
see JIS G4109 44K1
SCMV 6
see JIS G4109 44K1
SCORPION 44K1
SCP 1 42.1
SCP 2 42.1
SCP 3 42.1
SCP 4 42.1
SCP 5 42.1
SCP 6 42.1
SCP 7 42.1
SCRI 44E2
S Cu
see DIN range 14A
S Cu Ag
see DIN range 14A
S Cu Ni 30 Fe
see DIN range 14E
S Cu Si 14M
S Cu Sn
see DIN range 14A
SCV 44K3
SCV 3 44K3
SCV 5 44K3
SD 20 44K1
SD 51 20B
SD Cu
see DIN range 14A
SE4AS 44K3
SE 90 44A1
SEA-CURE 44M1
SEARCHER 44K2
SEARCHER A1 44K2
SEARCHER A1 44K3
SEC 1 44M1
SEC 4 44M1
SEC 5 44M1
SEC 7 44N1
SEC 8 44N1
SEC 9 44N1
SEC 10 44N1
SEC 13 44N1
SEH 5 44N1
SEL 1 27C
SELCO 53 44A1
SELCO 53V 44A1
SELECT BFM 44K3
SELENIUM 40.1
SENECA 44K1
SENSITIVE 11.1
SERMALLOY A1 44L
SERMALLOY A2 44L
SEVA 14B
SEVEN STAR 44K3
SEVERN 55A
SF 20B
Sf 1 1C
Sf 2 1C
Sf 3 1C
Sf 4 1C
Sf 5 1C
Sf 6 1C
Sf 7 1C
SF 10 44M1
SF 10 ELC 44M1
SF 11 44M1
SF 12 44M1
SF 13 44M1
SF 14 44M1
SF 15 44M1
SF 17 44M1
SF 18 44M1
SF 24 44N1
SF 25 44N1
SF 40 27B
SF 50 27B
SF 50 44N1

SF 60 27B
SF301 44N3
SF 302 44N1
SF 304 44N1
SF 304 L 44N1
SF 305 44N1
SF 316 44N1
SF 316 ELC 44N1
SF 316 Ti 44N1
SF 317 44N1
SF 321 44N1
SF 347 44N1
SF 400 20B
SF 420.12 20B
SF 610 44M1
SF 611 44M1
SF 612 44M1
SF 613 44M1
SF 835 44N1
SFA 420–12 20B
SF Aluminium 1C
SFF 20B
SFF 350 20B
SFF 400 20B
SFG 14K
SFG 44M1
SFP 500 20B
SFSH 1152 20B
SFT 20B
SG 44K1
SG 11 B 1E
SG 30
see JIS G3116 44A1
SG 33
see JIS G3116 44A1
SG 37
see JIS G3116 44A1
SG 38 20B
SG 42 20B
SG 50 20B
SG 60 20B
SG 70 20B
SG 70 A
see ASTM range 1K
SG 70 B
see ASTM range 1K
SG 100 A
see ASTM range 1K
SG 100 B
see ASTM range 1K
SGP
see JIS G3452 44A1
Sg Pb 46
see DIN range 21
Sg Pb 59
see DIN range 21
Sg Pb 85
see DIN range 21
Sg Pb 87
see DIN range 21
Sg Pb 97
see DIN range 21
Sg Sn 50 50A
Sg Sn 60 50A
Sg Sn 70 50A
Sg Sn 75 50A
Sg Sn 78 50A
SGV 46
see JIS G3118 44A1
SGV 49
see JIS G3118 44A1
SGW 44K1
SGW 1 44K1
SGW 2 44K1
SGW 3 44K1
SH 20B
SH 4 44A3
SH 5 44A3
SH 13-54 44A1
SH 13-60 44A1
SH 800 20B
SH 1000 20B
SHC 1 44K2
SHEARTOUGH 44K2
SHEFFIELD HIGH
STRENGTH A 44K1
SHEFFIELD HIGH
STRENGTH B 44J
SHEF-LO-TEMP 44K1
SHEF-LO TEMP 44A1
SHEF-SUPER LO TEMP 44A1
SHEF-SUPER-LO-T EMP 44K1
SHEF-TEN 44F1
SHH 20B
SH N 20B
SHNC 44K1
SHOCKTOUGH 44K3
SHOT ALLOY 21.1
SIC HMo 44M1
SIFBRONZE 14C
SIL 1 44E2
SIL 2 44E2
SIL 10 44N1
SIL 10 N 44N1
SIL 746 44M1
SILAFONT 1 1C
SILAFONT 2 1C
SILAFONT 3 1C
SILAFONT 4 1C
SILAFONT 5 1C
SILAFONT 6 1C
SILAFONT 7 1C
SILAFONT 14 1C
SILAFONT 30 1K
SILAFONT 54 1K
SILAFONT 64 1K
SILAFONT 66 1K
SILAFONT 74 1C
SILAFONT 84 1C
SILBERLOT 8 14H
SILBERLOT 12 14H
SILBERLOT 12 Cd 14H
SILBERLOT 15 14H
SILBERLOT 15 P 14H
SILBERLOT 20 14H
SILBERLOT 25 14H
SILBERLOT 25 Cd 14H
SILBERLOT 27 14H
SILBERLOT 30 Cd 5 14H
SILBERLOT 30 Cd 12 14H
SILBERLOT 38 14H
SILBERLOT 44 42.1
SILBERLOT 45 42.1
SILBERLOT 49 42.1
SILBERLOT 50 42.1
SILBERLOT 50 Cd 42.1
SILBERLOT 60 42.1
SILBERLOT 60 Cd 42.1
SILBERLOT 67 42.1
SILBERLOT 67 Cd 42.1
SILBERLOT 75 42.1
SILBERLOT 83 42.1
SILBRALLOY 14H
SILBRAZE 14H
SILCORO 60 17.1
SILCORO 75 17.1
SIL F 44B2
SILFLO 0 14H
SILFLO 5 14H
SILFLO 15 14H
SIL-FOS 14H
SIL-FOS 5 14H
SIL-FOS 6 14H
SILICO ALLOY 44K2
SILICON BRASS 14C
SILICON BRONZE SA 4 14M
SILICON CORE IRON A 44A1
SILICON CORE IRON A 44B1
SILICON CORE IRON
A-FM 44A1
SILICON CORE IRON B 44B1
SILICON CORE IRON C 44B1
SILMANAL 42.1
SILMET 14F
SILMO 44K2
SIL No. 1 VALVE STEEL 44E2
SILVA BRONZE 14G
SILVAISE 41.1
SILVANITE 44K3
SILVAN STAR 44J
SILVELMET 52.1
SILVER 42.1
SILVER BABBITT 50A
SILVER COPPER
EUTECTIC 42.1
SILVER-FLO 5 14H
SILVER-FLO 12 14H
SILVER-FLO 20 14H
SILVER-FLO 25 14H
SILVER FLO 33 42.1
SILVER FLO 34 42.1
SILVER FLO 38 42.1
SILVER FLO 40 42.1
SILVER FLO 43 42.1
SILVER FLO 44 42.1
SILVER FLO 55 42.1
SILVER FLO 56 42.1
SILVER FLO 60 42.1
SILVER-FLO 302 14J
SILVER FLO 452 42.1
SILVERSMITHS G RANGE 42.1
SILVER SOLDER 42.1
SILVER STEEL 44A4
SIL XB 44M2
SIMALEC 1E
SIMALLOY 600 27B
SIMALLOY 750 27C
SIMANCON 44A2
SIMGAL 1E
Si Mn 44A3
SIMONDS 168 44M2

SIMONDS 864 44G2
SIMONDS 12225 44M2
SINT A50 14J
SIRIUS 3 44N1
SIRIUS 35 44N1
SIRIUS 345 44N1
SIRIUS S12 44N1
SIS
Swedish standard - now SS
SIS 14 range Swedish standard
for steels;
see SS range for details
SIW 44K3
SIX EIGHTY ALLOY 42.1
SIXIX 44K3
SIXIX-fm 44K3
SIX STAR VANADIUM 44J
SJ G15 20B
SJ G20 20B
SJ G25 20B
SJ G30 20B
SJ G35 20B
SJ G40 20B
SK 1 44A4
SK 2 44A4
SK 3 44A4
SK 4 44A4
SK 5 44A4
SK 6 44A3
SK 7 44A3
SK 20 44M1
SK 20W 44M1
SK 42 44A2
SKA 45 44J
SKA 50 44J
SKC 3 44A3
SKC 11 44E3
SKC 24 44K2
SKC 31 44K1
SKD 1 44M2
SKD 2 44K1
SKD 2 44M2
SKD 4 44K1
SKD 5 44K1
SKD 6 44K2
SKD 11 44M2
SKD 12 44K3
SKD 61 44K2
SKD 62 44K2
SKF 1 44E3
SKF 2 44E3
SKF 3 44E3
SKF 3 44K3
SKF 017 44A1
SKF 21T2 44A1
SKF 24 44K3
SKF 25 44K3
SKF 26 44K3
SKF 27 44K3
SKF 28 44K3
SKF 126 44K1
SKF 145 44K1
SKF 146 44K1
SKF 152 44K1
SKF 153 44K1
SKF 157 44K1
SKF 234 44E1
SKF 255 44K1
SKF 304 44N1
SKF 316 44N1
SKF 321 44N1
SKF 327A 44K2
SKF 511 44A3
SKF 562 44K3
SKF 606 44E2
SKF 661 44K2
SKF 663 44K2
SKF 677 44E2
SKF 707 44E3
SKF 715 44A4
SKF 801 44E3
SKF 802 44E3
SKF 822 44K3
SKF 831 44E3
SKF 832 44E3
SKF 837 44E3
SKF CEAX 44A1
SKF REMKO 20A
SKF TERMEK 44A1
SKH 2 44K3
SKH 3 44L
SKH 4A 44L
SKH 4B 44L
SKH 5 44L
SKH 9 44K3
SKH 10 44L
SKH 52 44K3
SKH 53 44K3
SKH 54 44K3
SKH 55 44L
SKH 56 44L
SKH 57 44L
SKO 67 44A3
SKS 1 44K3
SKS 2 44K3
SKS 3 44K3
SKS 4 44K2
SKS 5 44K3
SKS 7 44K3
SKS 8 44E3
SKS 11 44K3
SKS 21 44K3
SKS 31 44K3
SKS 41 44K2
SKS 42 44K3
SKS 43 44J
SKS 44 44J
SKS 51 44K3
SKS 93 44E3
SKS 94 44E3
SKS 95 44E3
SK SILVER STEEL 44E3
SKT 2 44K2
SKT 3 44K2
SKT 4 44K2
SKT 5 44K2
SKT 6 44K2
SL 2N26 44F1
SL 3N26 44F1
SL 3N45 44F1
SL 9N53 44F1
SL 9N60 44F1
SL 15 20B
SL 20 20B
SL 25 20B
SL 30 20B
SL 35 20B
SL 40 20B
SL 173 44K3
SL 440 B 44M2
SL 440 C 44M2
SL 440A 44M1
SL 440B 44M1
SL 503 M 44K1
SL 598 44K1
SL 823 44K2
SL 853 44K1
SL 866 44K2
SL 874 44K3
SL 934 44E3
SL 942 44K1
SL 951 44M2
SL 954 44K3
SL 958 44M2
SL 959 13A
SL 962 44K2
SL 970 13A
SL 972 44L
SL 977 44L
SL 983 44K2
SLA 24A
see JIS G3126 44A1
SLA 24B
see JIS G3126 44A1
SLA 33A
see JIS G3126 44A1
SLA 33B
see JIS G3126 44A1
SLA 37
see JIS G3126 44A1
SLIDEX 14F
SLV 44E2
SM 44K2
SM 41A,B & C
see JIS G3106 44A1
SM 50
see JIS G3106 44A1
SM 58
see JIS G3106 44A1
SM 302 13A
SM 322 13A
SMA 41A,B, & C
see JIS G3114 44K1
SMA 58
see JIS G3114 44K1
SMA 58 44K1
SMM 44G1
SMN 44A3
SMO 44G2
S Mo 1000 26A
S MONEL 27D
SMOOTHTRODE 43.23 44A1
SMR St 2 44A1
SMR St 3 44A1
SMR St 4 44A1
S Ms 60 14C
SMU St 2 44A1
SMU St 3 44A1
SMU St 4 44A1
SN 3 44K1
SN 5 44F1

SN 5/ 44K1
SN 10 44F1
SN 15 44F1
SN 22 20B
SN 31/2 44K2
SN 122 A
see ASTM range 1K
S NC 20.2 20B
S NC 30.3 20B
SNC 1 44K1
SNC 3G 44K1
SNC4G 44K2
SNCH 6 44K1
SNCH 9 44K1
SNCM 44K1
SNCM 26 44K1
SNG 24/17
see BS range 20B
SNG 27/12
see BS range 20B
SNG 32/7
see BS range 20B
SNG 37/2
see BS range 20B
SNG 42/2
see BS range 20B
SNG 47/2
see BS range 20B
SNH 3 44K1
SNH 5 44K1
S NM 23.4 20B
SNSC 44K3
SO 12/21 44L
SOBV 44L
SOD 44A4
SODIUM 43.1
SOLAR 44G2
SOLAR ECLIPSE 44M2
SOLEIL 4 44M1
SOLEIL B2 44M1
SOLEIL B3 44M1
SOLEIL B4 44M1
SOLEIL T1 44M1
SOLEIL T2 44M1
SOMDIE 44K2
SOMDIE 44K3
SOMDIE SPECIAL 44K2
SOMER SUPAMOLD 44K1
SOMPLAS 30 44K1
So Ms 57 Al 2 14B
So Ms 58 14B
So Ms 58 Al 1 14B
So Ms 58 Pb 14B
So Ms 59 14B
So Ms 60 14B
So Ms 71 14B
So Ms 76 14B
SONCOLD 44M2
SONDALU Si 5 1C
SONDALU Si 12 1C
SONDOR Cr Al Si 5 1C
SONDOR G Al Mg 1D
SONDOR G Al Mg 5 1D
SONDOTIG Al Si 5 1C
SONOSTON 24.1
SOREL 55A
SOUDAIR 44A1
Soudalu Al 100 1A
SOUDALU Mn 2 1B
SOUDEN BE 55 14H
SOUDER G625 27F
SOUDER GA 250 44E1
SOUDER GWT 44F1
SOUDINOX 25-4 44N1
SOUDINOX 25-20 44N1
SOUDINOX A8 44N1
SOUDINOX A308H 44N1
SOUDINOX A308L 44N1
SOUDINOX B316H 44N1
SOUDINOX B316L 44N1
SOUDINOX LF 44N1
SOUDINOX S28 44N2
SOUDINOX S65 44N1
SOUDITIG G1 44A1
SOUDITIG G3 44A1
Soudobronze 101 14G
Soudobronze 201 14G
Soudobronze 301 14G
Soudobronze MnS 14G
SOUDOCROM 56 44N1
SOUDOCROM A 44N1
SOUDOCROM A308L 44N1
SOUDOCROM AD 44N1
SOUDOCROM B 44N1
SOUDOCROM B6 44N2
SOUDOCROM B316L 44N1
SOUDOCROM BD 44N1
SOUDOCROM C310 44N1
SOUDOCROM D 44N1
SOUDOCROM F347 44N1
SOUDOCROM G 44N1
SOUDOCROM HR 308L 44N1
SOUDOCROM HR 309L 44N1
SOUDOCROM HR 316L 44N1
SOUDOCROM K318 44N1
SOUDOCROM L 309L 44N1
SOUDOCROM L309 Mo 44N1
SOUDOCROM M317L 44N1
SOUDOCROM S17 44N2
SOUDOCROM S50 44N2
SOUDOCROM S4462 44N1
SOUDOCUIVRE 14A
SOUDODER 400 44E1
SOUDODUR 600 44E2
SOUDODUR 600R 44E3
SOUDODUR MR 44K1
SOUDOFIL 1 44A1
SOUDOFIL 2 44A1
SOUDOFONTE B1 27A
SOUDOFONTE B12 27A
SOUDOFONTE B24 27A
SOUDOFONTE C 27E
SOUDOFONTE D 27E
SOUDOFONTE F 27E
SOUDOFONTE GS 20B
SOUDOFONTE K 27D
SOUDOFONTE Mo 27D
SOUDOGEN BE 1 42.1
SOUDOGEN BE 2 42.1
SOUDOGEN BE 3 42.1
SOUDOGEN BE 5 14H
SOUDOGEN BE 15 42.1
SOUDOGEN BE 100 42.1
SOUDOGEN BE 200 42.1
SOUDOGEN BE 300 14H
SOUDOGEN BE50 14A
SOUDOKAY 44A1
SOUDOKAY 44P
SOUDOKAY 089.0 44A3
SOUDOKAY 218.0 44P
SOUDOKAY 240-0 44M2
SOUDOKAY 242.0 44E1
SOUDOKAY 252.0 44E1
SOUDOKAY 255.0 44M2
SOUDOKAY 258 LO 44K1
SOUDOKAY 258 Ti Cr O 44E3
SOUDOKAY 258.0 44K2
SOUDOKAY 308L-0 44N1
SOUDOKAY 309L-0 44N1
SOUDOKAY 316LO 44N1
SOUDOKAY 402.0 44N1
SOUDOKAY 410.0 44M2
SOUDOKAY 420.0 44M2
SOUDOKAY 624.0 44P
SOUDOKAY 795.0 44E3
SOUDOKAY A 12-0 44E2
SOUDOKAY A40.0 44M2
SOUDOKAY A43.0 44M2
SOUDOKAY A45.0 44M2
SOUDOKAY A46.0 44L
SOUDOKAY AP.0 44P
SOUDOKAY BU-CI 44A1
SOUDOKAY BVO 44K1
SOUDOKAY D20.0 44K3
SOUDOKAY GS-0 44A3
SOUDOKAY STELKAY 1-G 13A
SOUDOKAY STELKAY 6-G 13A
SOUDOKAY STELKAY
12-G 13A
SOUDOKAY STELKAY
21-G 13A
SOUDOKAY TOOL
ALLOY CO 27F
SOUDO-MANGANESE 44P
SOUDO-MANGANESE CN 44P
SOUDONEL 27F
SOUDONEL 625 27F
SOUDONEL BS 27B
SOUDONEL D 27F
SOUDONEL F 27D
SOUDONEL G 27F
SOUDONEL S4648 27F
SOUDOR Cr Ni Cu 7 27D
SOUDORECORD 44A1
SOUDOR G 410 44M1
SOUDOR G 430 44M1
Soudor G Al 99.5 1A
SOUDOR G Cr Mo 1 44K1
SOUDOR G Cr Mo 2 44K1
SOUDOR G Cu 14A
Soudor G Cu Al 8 14G
SOUDOR G Cu Sn 6 14K
SOUDOR G Mo 44G1
SOUDOR G Ni Cr 3 27B
SOUDOR G Ni Ti 27A
SOUDOR G18.8.6 44N1
SOUDOR G20.25.5L Cu 44N1
SOUDOR G22.8.3L 44N1
SOUDOR G65 44K1
SOUDOR G308L Si 44N1
SOUDOR G309 44N1

SOUDOR G309 Si 44N1
SOUDOR G310 44N1
SOUDOR G316L Si 44N1
SOUDOR G318 44N1
SOUDOR G347 44N1
SOUDOR GA 350 44E2
SOUDOR GA 500 44E2
SOUDOR GA 600 44E2
SOUDOSPRAY 162-5 13A
SOUDOSPRAY 166-1 13A
SOUDOSPRAY 166-2 13A
SOUDOSPRAY 166-3 13A
SOUDOSPRAY 166-5 13A
SOUDOSPRAY 166-6 13A
SOUDOSPRAY 167-0 13A
SOUDOSPRAY 167-1 13A
SOUDOSPRAY 167-2 13A
SOUDOSPRAY 167-3 13A
SOUDOSPRAY 167-5 13A
SOUDOSPRAY 167-6 13A
SOUDOSPRAY 168-5 13A
SOUDOSPRAY 327-0 13A
SOUDOSPRAY 327-1 13A
SOUDOSPRAY 327-2 13A
SOUDOSPRAY 327-3 13A
SOUDOSPRAY 327-5 13A
SOUDOSPRAY 327-6 13A
SOUDOSPRAY 328-0 13A
SOUDOSPRAY 328-1 13A
SOUDOSPRAY 328-2 13A
SOUDOSPRAY 328-3 13A
SOUDOSPRAY 328-5 13A
SOUDOSPRAY 328-6 13A
SOUDOSTEL 1 AP 13A
SOUDOSTEL 1 C 13A
SOUDOSTEL 6 AP 13A
SOUDOSTEL 6 C 13A
SOUDOSTEL 12 AP 13A
SOUDOSTEL 12 C 13A
SOUDOSTEL 12 HC 13A
SOUDOSTEL 21 AP 13A
SOUDOSTEL 21 C 13A
SOUDOSTEL 60 27F
SOUDOSTEL HR 1 13A
SOUDOSTEL HR 6 13A
SOUDOSTEL HR 12 13A
SOUDOSTEL HR 21 13A
SOUDOSTEL HR 25 13A
Soudoteg Al 99.5 1A
SOUDOTEG Al Mg 1D
SOUDOTEG Al Mg 5 1D
Soudoteg Cu Al 8 14G
SOUDOTEG Cu Sn 6 14K
SOUDOTIG 18.8.6 44N1
SOUDOTIG 20.25.2L Cu 44N1
SOUDOTIG 22.8.3L 44N1
SOUDOTIG 308L 44N1
SOUDOTIG 308L Si 44N1
SOUDOTIG 309 Si 44N1
SOUDOTIG 310 44N1
SOUDOTIG 316L 44N1
SOUDOTIG 316L Si 44N1
SOUDOTIG 318 44N1
SOUDOTIG 347 44N1
SOUDOTIG 625 27B
SOUDOTIG Cr Mo 1 44K1
SOUDOTIG Cr Mo 2 44K1
SOUDOTIG Cr Mo 55 44K1
SOUDOTIG Cu 14A
SOUDOTIG Mo 44G1
SOUDOTIG Ni Cr 3 27B
SOUDOTIG Ni Cu 7 27D
SOUDOTIG Ni Ti 27A
SOUNDOTENAX 52 44A1
SOUNDOTENAX 56 44A1
SOUNDOTENAX 66 44A1
SOUPERSOUDO 44G1
SOVB 44L
SOVEREIGN 8TS 50A
SP 20B
SP 10 44K1
SP 20 44K1
SP 22 44K1
SP 24 44K1
SP 25 44A1
SP 26 44A1
SP 100C 5.1
SP 200C 5.1
SP 300C 5.1
SP 700 20B
SPARKOMITE 10 14J
SPARTAN 5 44K2
SPARTAN 7 44K3
SPEAR 35 44K3
SPEAR 50 44A3
SPEAR 75 44A3
SPEAR 412 44A2
SPEAR 416 44G1
SPEAR 419 44K2
SPEAR 433 44F1
SPEAR 451 44K1
SPEAR 500 44K2
SPEAR 510 44K2
SPEAR 521 44F1
SPEAR 524 44K1
SPEAR 525 44K1
SPEAR 536 44K1
SPEAR 552 44K1
SPEAR 562 44K3
SPEAR 626 44K2
SPEAR 653 44K1
SPEAR 855 44K1
SPEAR 1030 44K1
SPEAR AHC 44K2
SPEAR B1 44K1
SPEAR B2 44M1
SPEAR B4 44K1
SPEAR B8 44A1
SPEAR CRH 44E3
SPEAR CV 44K2
SPEAR D1 44K2
SPEAR D2 44K2
SPEAR D3 44F2
SPEAR D5 44K1
SPEAR D7 44K1
SPEAR D9 44K1
SPEAR D12 44M2
SPEAR D13 44M2
SPEAR D14 44M2
SPEAR D16 44M2
SPEAR DOUBLE
CENTURY 44L
SPEAR DX 44K1
SPEAR EXM 44K2
SPEAR GFS 44E3
SPEAR LEAPFROG 44K2
SPEAR MD4 44A4
SPEAR MERMAID 44K3
SPEAR No. 1 44H
SPEAR No. 2 44A4
SPEAR No. 3 44A4
SPEAR PS 44K3
SPEAR SUPERIOR 44L
SPEAR TRIPLE MERMAID 44L
SPECIAL 44A3
SPECIAL ADVANCE 14E
SPECIAL ARDHO 44K2
SPECIAL BB 44K1
SPECIAL CONQUEROR
VANADIUM 44J
SPECIAL ECHO 44K3
SPECIAL GENCO 44M1
SPECIAL HDS 44K1
SPECIAL HS 44K3
SPECIAL HW No. 3 44K2
SPECIAL K 44M2
SPECIAL V 44J
SPECIAL WOLFRAM 44K2
SPECIFICATION 55 44K2
SPECULUM 14K
SPEDEX 14F
SPEEDICUT 14 44K2
SPEEDICUT LEDA 44L
SPEEDICUT MAX 22 44K3
SPEEDICUT MAXIMUM 18 44K3
SPEEDICUT SUPERLEDA 44L
SPEEDICUT VANLEDA 44L
SPEED STAR 44K3
SPELTAFAST 44A1
SPENARD 44K1
SPF 20B
SPF 600 20B
SPINNING BRASS 14B
SPK 44K2
SPKD 44K2
SPKS 44E2
SPM 44K3
SPM 1 42.1
SPM 2 42.1
SPOOLARC GRADE C 14K
SPRABABBITT 50A
SPRABABBITT A 50A
SPRABABBITT L 21.1
SPRABOND 26A
SPRABRASSY 14B
SPRABRONZE AA 14G
SPRABRONZE C 14B
SPRABRONZE P 14K
SPRABRONZE TM 14C
SPRASTEEL 10 44A1
SPRASTEEL 25 44A1
SPRASTEEL 80 44A3
SPRASTEEL 80 44A4
Sprasteel LS 44A2
SPRASTEEL LS 44K1
SPS 44K2
SPS 245 44K2
SPV 32
see JIS G3115 44A1
SPV 36
see JIS G3115 44A1

SPV 46
  see JIS G3115 44A1
SPV 50
  see JIS G3115 44A1
SPW 44M1
SPW 44N1
SQV 1A
  see JIS G3120 44G1
SQV 1B
  see JIS G3120 44G1
SQV 2A
  see JIS G3120 44K1
SQV 2B
  see JIS G3120 44K1
SQV 3A
  see JIS G3120 44K1
SQV 3B
  see JIS G3120 44K1
S R ALLOY 44N1
SR 200 5.1
SR 264 44M1
SR 1855 44E3
SRO 44K2
SRO 2 44M1
SS 14C
S S1 44M1
SS 1 14C
S S2 44M1
S S 3M 44N1
S S 3MM 44N1
S S 4 44N1
S S6 44M1
SS 08–1002 20B
SS 08–1400 20B
SS 08–1500 20B
SS 08–5200 20B
SS 08–5400 20B
SS 08–5403 20B
SS 08–5600 20B
SS 08–5603 20B
SS 08–6203 20B
SS 08–6403 20B
SS 14–0814 20B
SS 14–0815 20B
SS 14–0852 20B
SS 14–0854 20B
SS 14–0856 20B
SS 14–0862 20B
SS 14–0864 20B
SS 14-2216 E 44K2
SS 14-2383 44M1
SS 15 14F
S S22 44M1
S S 24 44N1
S S31 44M1
S S 33 44N1
SS 40-20 1A
SS 40-21 1A
SS 40-22 1A
SS 40-24 1A
SS 41
  see JIS G3101 44A1
S S 43 44N1
S S 53 44N1
SS 0100 20B
SS 100 44K1
SS 0110 20B
SS 0115 20B
SS 0120 20B
SS 0125 20B
SS 0130 20B
SS 0135 20B
SS 0140 20B
SS 0212 20B
SS 0215 20B
SS 0217 20B
SS 0219 20B
SS 0221 20B
SS 0223 20B
SS 0707 20B
SS 0717 20B
SS 0727 20B
SS 0732 20B
SS 0737 20B
SS 0810 20B
SS 0854 20B
SS 0856 20B
SS 1111 44A1
SS 1140 44A1
SS 1142 44A1
SS 1145 44A1
SS 1146 44A1
SS 1147 44A1
SS 1148 44A1
SS 1150 44A1
SS 1151 44A1
SS 1152 44A1
SS 1157 44A1
SS 1160 44A1
SS 1211 44A1
SS 1225 44E1
SS 1232 44A1
SS 1232 44E1
SS 1232E 44E1
SS 1233 44E1
SS 1234 44E1
SS 1250 44A1
SS 1265 44A1
SS 1270 44A1
SS 1305 44A1
SS 1306 44A1
SS 1306 44E1
SS 1311 44A1
SS 1312 44A1
SS 1313 44A1
SS 1316 44A1
SS 1330 44E1
SS 1330E 44E1
SS 1331 44A1
SS 1331 44E1
SS 1332 44E1
SS 1350 44A1
SS 1350–6 44A1
SS 1357 44A1
SS 1360 44A1
SS 1370 44A1
SS 1386 44A1
SS 1387 44A1
SS 1410 44A2
SS 1411 44A1
SS 1411 44E1
SS 1412 44A1
SS 1412 44A2
SS 1413 44A1
SS 1414 44A1
SS 1426 44A1
SS 1430 44A1
SS 1430 44E1
SS 1430E 44E1
SS 1431 44E1
SS 1432 44E1
SS 1432E 44E1
SS 1434 44A1
SS 1434 44E1
SS 1435 44A1
SS 1435 44E1
SS 1435E 44E1
SS 1450 44A1
SS 1450 44A2
SS 1450-0 44A2
SS 1505 44A2
SS 1510 44A2
SS 1550 44A2
SS 1550-06 44A2
SS 1572 44A2
SS 1606 44A2
SS 1606 44A3
SS 1650 44A2
SS 1655 44A2
SS 1655 44A3
SS 1655/00 44A3
SS 1660 44A2
SS 1665 44A3
SS 1672 44A2
SS 1672 44A3
SS 1674 44A2
SS 1674 44A3
SS 1678 44A3
SS 1757 44A2
SS 1770 44A3
SS 1774 44A3
SS 1774 44A4
SS 1778 44A3
SS 1870 44A4
SS 1880 44A4
SS 1885 44A4
SS 1912 44A1
SS 1914 44A1
SS 1922 44A1
SS 1926 44A1
SS 1940 44A2
SS 1957 44A2
SS 1973 44A3
SS 2032 44A1
SS 2085 44E2
SS 2090 44A3
SS 2090 44E2
SS 2092 44E3
SS 2092 44K1
SS 2101 44A1
SS 2101 44E1
SS 2101E 44E1
SS 2102 44E1
SS 2102 44M1
SS 2103 44E1
SS 2103E 44E1
SS 2106 44A1
SS 2107 44A1
SS 2108 44E1
SS 2116 44A1
SS 2117 44A1

SS 2120 44A2
SS 2121 44A1
SS 2122 44A1
SS 2127 44A1
SS 2127 44E1
SS 2128 44A2
SS 2130 44E1
SS 2131 44E1
SS 2132 44A1
SS 2133 44A1
SS 2134 44A1
SS 2135 44A1
SS 2136 44A1
SS 2137 44A1
SS 2140 44K3
SS 2142 44A1
SS 2143 44A1
SS 2144 44A1
SS 2145 44A1
SS 2146 44A1
SS 2164 44A1
SS 2165 44A1
SS 2165 44A2
SS 2167 44A1
SS 2168 44A1
SS 2168 44A2
SS 2172 44E1
SS 2172-04 44A1
SS 2172-21 44A1
SS 2173 44E1
SS 2174 44E1
SS 2183 44D
SS 2183 44M2
SS 2203 44K1
SS 2203E 44K1
SS 2216 44K1
SS 2218 44K1
SS 2218E 44K1
SS 2223 44K1
SS 2224 44K1
SS 2225 44K1
SS 2230 44E2
SS 2230 44K2
SS 2233 44K1
SS 2234 44K1
SS 2234 44K2
SS 2240 44K1
SS 2242 44K2
SS 2244 44K2
SS 2245 44E2
SS 2253 44E2
SS 2254 44E2
SS 2258 44E3
SS 2260 44K3
SS 2301 44M1
SS 2302 44M1
SS 2303 44M1
SS 2304 44M1
SS 2310 44K1
SS 2310 44M2
SS 2312 44M2
SS 2317 44K1
SS 2317 44M1
SS 2320 44M1
SS 2321 44M1
SS 2322 44M1
SS 2323 44M1
SS 2324 44M1
SS 2324 44N1
SS 2325 44M1
SS 2326 44M1
SS 2330 44N3
SS 2331 44N3
SS 2332 44N1
SS 2333 44N1
SS 2334 44N1
SS 2337 44N1
SS 2338 44N1
SS 2340 44N1
SS 2341 44N1
SS 2342 44N1
SS 2343 44N1
SS 2344 44N1
SS 2345 44N1
SS 2346 44N1
SS 2347 44N1
SS 2348 44N1
SS 2350 44N1
SS 2352 44N1
SS 2353 44N1
SS 2357 44P
SS 2360 44N1
SS 2361 44N1
SS 2366 44N1
SS 2367 44N1
SS 2370 44N1
SS 2371 44N1
SS 2374 44N1
SS 2375 44N1
SS 2376 44N3
SS 2377 44N3
SS 2380 44M1
SS 2382 44M1
SS 2383 44M1
SS 2384 44N1
SS 2385 44N3
SS 2387 44N3
SS 2506 44K1
SS 2511 44K1
SS 2512 44K1
SS 2514 44K1
SS 2515 44K1
SS 2523 44K1
SS 2534 44K1
SS 2541 44K2
SS 2550 44K2
SS 2564 44N1
SS 2570 44N1
SS 2570 44N2
SS 2710 44K2
SS 2722 44K3
SS 2724 44K3
SS 2730 44K1
SS 2733 44L
SS 2750 44L
SS 2754 44L
SS 2756 44L
SS 2782 44K1
SS 2900 44J
SS 2912 44G1
SS 2912 E 44E3
SS 2940 44K2
SS 3401 44A1
SS 3402 44A1
SS 3403 44A1
SS 3404 44A1
SS 3405 44A1
SS 3406 44A1
SS 3421 44A1
SS 3422 44A1
SS 3423 44A1
SS 3424 44A1
SS 3425 44A1
SS 3426 44A1
SS 4005 1A
SS 4007 1A
SS 4008 1A
SS 4020 1A
SS 4021 1A
SS 4022 1A
SS 4023 1A
SS 4024 1A
SS 4054 1B
SS 4104 1E
SS 4106 1D
SS 4120 1D
SS 4163 1D
SS 4212 1E
SS 4230 1L
SS 4231 1L
SS 4244 1C
SS 4247 1C
SS 4251 1L
SS 4252 1L
SS 4253 1C
SS 4254 1L
SS 4255 1C
SS 4260 1C
SS 4261 1C
SS 4335 1F
SS 4338 1F
SS 4438 1M
SS 4602 23A
SS 4604 23A
SS 4635 23B
SS 4637 23B
SS 4640 23B
SS 5010 14A
SS 5011 14A
SS 5013 14A
SS 5015 14A
SS 5030 14A
SS 5053 14A
SS 5112 14B
SS 5114 14B
SS 5122 14B
SS 5124 14B
SS 5140 14B
SS 5144 14B
SS 5150 14B
SS 5163 14C
SS 5165 14C
SS 5168 14C
SS 5170 14C
SS 5173 14C
SS 5202 14B
SS 5203 14K
SS 5204 14K
SS 5217 14B
SS 5220 14B
SS 5234 14B

SS 5236 14B
SS 5238 14B
SS 5243 14F
SS 5243 14F
SS 5246 14F
SS 5272 14C
SS 5420 14K
SS 5428 14K
SS 5431 14K
SS 5443 14K
SS 5444 14K
SS 5465 14K
SS 5475 14K
SS 5640 14K
SS 5667 14E
SS 5682 14E
SS 7020 55A
SS 7030 55A
SS 8001 44A1
SS 8010 44A2
SS 8011 44A3
SS 8020 44A1
SS 8051 44A2
SS 8100 44K1
SS 8101 44K2
SS 8102 44K2
SS 8110 44F1
SS 8300 14K
SS 8320 14K
SS 8321 14K
SS 8330 14K
SSC 44K2
SSG 44K1
SSM 44A3
S Sn Bz 6
see DIN range 14K
S Sn Bz 12
see DIN range 14K
S So Ms 14C
SSS 100 44K1
SSS 100A 44K1
SSS 110 44K1
SSS 321 44K1
SSS 360 44K1
S St 2 44A1
S St 3 44A1
S St 4 44A1
St 00
see DIN 1629 44A1
St 0
see DIN range 44A1
St 1
see DIN range 44A1
ST 2
see GOST 380 44A1
St 2
see DIN range 44A1
ST 3
see GOST 380 44A1
St 3
see DIN range 44A1
St 4
see DIN range 44A1
St 10 44A1
St 10/01 44A1
St 10/02 44A1
St 10/03 44A1
St 33
see DIN 1626 44A1
St 34
see DIN range 44A1
St 34/2 44A1
St 34/3 44A1
St 35 44A1
St 35/4 44A1
St 35/8
see DIN range 44A1
St 35/13 K 44A1
St 37
see DIN 1626 44A1
St 37 44A1
St 37/2 44A1
St 37/3 44A1
St 42 44A1
St 42/2 44A1
St 42/3 44A1
St 43.7
see DIN range 44A1
St 45 44A1
St 45/4 44A1
St 45/8
see DIN range 44A1
St 47.7
see DIN range 44A1
St 50 44A2
St 50/2 44A2
St 50/202 44A2
St 50/203 44A2
St 52 44A1
St 52/3 44A1
St 52/4 44A1
St 52/302 44A1
St 53.7
see DIN range 44A1
St 55 44A2
St 55/4 44A2
St 60 44A2
St 60.1
see DIN 17100 44A2
St 60.2
see DIN 17100 44A2
St 60/2 44A2
St 60/202 44A2
St 60/203 44A2
St 70.2
see DIN 17100 44A3
St 70/2 44A3
St 70/202 44A3
St 70/203 44A3
STA 3 44A3
STA 4 44A3
STA 5 V 24 44E2
STA 5 V1A 44A1
STA 5 V1B 44A1
STA 5 V2 44A1
STA 5 V2A 44A1
STA 5 V2A/1 44A1
STA 5 V2B 44A1
STA 5 V2C 44A1
STA 5 V2D 44A1
STA 5 V3 44A1
STA 5 V4 44A2
STA 5 V4/1 44A2
STA 5 V4/2 44A2
STA 5 V4/3 44A2
STA 5 V4/4 44A2
STA 5 V4A 44A2
STA 5 V4A/1 44A2
STA 5 V4A/2 44A2
STA 5 V4A/3 44A2
STA 5 V4AM 44A2
STA 5 V4B 44A3
STA 5 V4B/1 44A3
STA 5 V4B/2 44A3
STA 5 V5 44A3
STA 5 V5A 44A3
STA 5 V5A/1 44A3
STA 5 V5A/2 44A3
STA 5 V6 44K1
STA 5 V6A 44K1
STA 5 V6B 44K1
STA 5 V7 44A2
STA 5 V8 44K1
STA 5 V9A 44G1
STA 5 V9A/1 44G1
STA 5 V9B 44G1
STA 5 V9B/1 44G1
STA 5 V9C 44E2
STA 5 V9C/2 44E2
STA 5 V9C/3 44E2
STA 5 V9C/4 44E2
STA 5 V9C/a 44E2
STA 5 V9E 44K1
STA 5 V9E/1 44K1
STA 5 V11 44K1
STA 5 V13 44K1
STA 5 V13A 44K1
STA 5 V14 44E3
STA 5 V15 44A1
STA 5 V15/1 44A1
STA 5 V15/A 44A1
STA 5 V15A/1 44A1
STA 5 V15AM 44A1
STA 5 V16A 44K1
STA 5 V16A/1 44K1
STA 5 V16B 44K1
STA 5 V16B/1 44K1
STA 5 V16B/2 44K1
STA 5 V17 44K1
STA 5 V17/1 44K1
STA 5 V18 44K1
STA 5 V18A 44K1
STA 5 V19 44K1
STA 5 V19/1 44K1
STA 5 V20A 44K1
STA 5 V20B 44K1
STA 5 V20C 44K1
STA 5 V20D 44K1
STA 5 V22B 44A3
STA 5 V22B/1 44A3
STA 5 V22B/2 44A3
STA 5 V23 44A3
STA 5 V23A 44A3
STA 5 V24A 44M2
STA 5 V25 44M1
STA 5 V25A 44M1
STA 5 V25B 44M1
STA 5 V25B/1 44M1
STA 5 V25B/2 44M1
STA 5 V25M 44M1
STA 5 V26 44N1

STA 5 V27 44N1
STA 5 V27A 44N1
STA 5 V27M 44N1
STA 5 V28 44M1
STA 5/V/5/1 44F2
STA 5/V9D 44K2
STA 5/V9D/1 44K2
STA 5/V9D/2 44K2
STA 5/V10 44K2
STA 5/V12 44K2
STA 5/V21A 44K2
STA 5/V21B 44K2
STA 5/V22A 44A4
STA 5/V22A/1 44A4
STA 5/V22A/2 44A4
STA 7 A2 1A
STA 7 A3 1A
STA 7 A4B 1A
STA 7 A4C 1A
STA 7 A4D 1A
STA 7 A4E 1A
STA 7 AC1 1L
STA 7 AC2 1C
STA 7 AC3 1L
STA 7 AC4 1L
STA 7 AC5 1J
STA 7 AC6 1C
STA 7 AC7 1L
STA 7 AC8 1K
STA 7 AC9 1K
STA 7 AC10 1J
STA 7 AC11 1L
STA 7 AC12 1L
STA 7 AC13 1K
STA 7 AC14 1L
STA 7 AC15 1L
STA 7 AW 10B 1E
STA 7 AW1 1C
STA 7 AW2C 1C
STA 7 AW4A 1D
STA 7 AW4B 1D
STA 7 AW4C 1D
STA 7 AW5A 1D
STA 7 AW5B 1D
STA 7 AW5C 1D
STA 7 AW6A 1D
STA 7 AW6B 1D
STA 7 AW6C 1D
STA 7 AW6D 1D
STA 7 AW7A 1D
STA 7 AW7B 1D
STA 7 AW7C 1D
STA 7 AW8 1F
STA 7 AW9A 1E
STA 7 AW9B 1E
STA 7 AW10A 1E
STA 7 AW10C 1E
STA 7 AW10D 1E
STA 7 AW10E 1E
STA 7 AW10F 1E
STA 7 AW11A 1F
STA 7 AW12 1F
STA 7 AW13 1F
STA 7 AW14 1F
STA 7 AW15A 1F
STA 7 AW15B 1F
STA 7 AW15C 1F
STA 7 AW15D 1F
STA 7 AW15E 1F
STA 7 AW15G 1F
STA 7 AW15H 1F
STA 7 AW16A 1G
STA 7 AW17B 1F
STA 7 AW18 1F
STA 7 C1 14A
STA 7 C2 14A
STA 7 C3 14A
STA 7 C4 14A
STA 7 C5 14A
STA 7 C6 14A
STA 7 C7A 14A
STA 7 C7B 14A
STA 7 C8A 14A
STA 7 C8B 14A
STA 7 C8C 14A
STA 7 C9 14A
STA 7 C10 14A
STA 7 C11 14A
STA 7 CA1 14G
STA 7 CA2 14G
STA 7 CA3 14G
STA 7 CA4 14G
STA 7 CA5 14G
STA 7 CG1 14K
STA 7 CG2 14K
STA 7 CG3 14K
STA 7 CG4 14K
STA 7 CG5 14K
STA 7 CG6 14K
STA 7 CM1 14C
STA 7 CM2 14C
STA 7 CM3 14C
STA 7 CM4A 14C
STA 7 CM4B 14B
STA 7 CM4C 14B
STA 7 CM5 14C
STA 7 CM6 A and B 14C
STA 7 CM7 A and B 14C
STA 7 CP1 14K
STA 7 CP2 14K
STA 7 CP3 14K
STA 7 CP4 14K
STA 7 CP5 14K
STA 7 CP6 14K
STA 7 CS1A 14B
STA 7 CS1B 14B
STA 7 CS1D 14B
STA 7 CS2 14B
STA 7 CSIC 14M
STA 7 CX1 14B
STA 7 CX2 14B
STA 7 CX3 14C
STA 7 CX4 14C
STA 7 CX4 14C
STA 7 CX5 14C
STA 7 CX6 14C
STA 7 CX7 14K
STA 7 CX8 14K
STA 7 CX9A 14K
STA 7 CX9B 14K
STA 7 CX10 14G
STA 7 CX11 14E
STA 7 CX12 14K
STA 7 CX13 14C
STA 7 CZ1 14B
STA 7 CZ2 14B
STA 7 CZ3 14B
STA 7 CZ4A 14B
STA 7 CZ4B 14B
STA 7 CZ5 14B
STA 7 CZ6 14B
STA 7 CZ7 14B
STA 7 CZ8A 14B
STA 7 CZ8B 14B
STA 7 CZ9A 14C
STA 7 CZ9B 14C
STA 7 CZ9B1 14C
STA 7 CZ9C 14C
STA 7 CZ9D 14C
STA 7 CZ9E 14C
STA 7 CZ10 14C
STA 7 CZ11A 14B
STA 7 CZ11B 14C
STA 7 CZ12 14K
STA 7 CZ13 14K
STA 7 CZ14 14C
STA 7 CZ15 14C
STA 7 LB1 21.1
STA 7 LB2 21.1
STA 7 TB1B 50A
STA 7 TB2 50A
STA 7 TB2A 21.1
STA7 AW 3C 1B
STA 8 20B
STA 9 20B
STA 13 C1 14A
STA 13 C2 14A
STA 17 14B
STA 19 14B
STA 25 20B
STA 27 44A4
STA 28 44A2
STA 30 44A2
STA 36 A 44A2
STA 38 B 44A1
STA 39 44K1
STA 40 44K1
STA 41 44A3
STA 51 14B
STA AW11B 1F
STABILOHM 43 14E
STA CX11 14E
STAG 14 44K2
STAG C 44K2
STAG EXTRA SPECIAL 44L
STAG MAJOR 44L
STAG MO 44K3
STAG MO 562 44K3
STAGMOLD 44K1
STAG SPECIAL 44K3
STAG VANCO 44L
Stainarc 25/20R
  See Phillips Stainarc 25/20R 44N1
Stainarc M
  See Phillips Stainarc M 44N1
Stainarc MN
  See Phillips Stainarc MN 44N1
Stainarc N
  See Phillips Stainarc 44N1
STAINLESS 16 44M1

STAINLESS 51 44M1
STAINLESS 851 44M1
STAINLESS W 44N3
STAMINAL 44K2
STANDARD 44A3
STANDARD SILVER 42.1
STAN-FOS 14H
Star 99.2 1A
Star 99.5 1A
Star 99.5E 1A
Star 99.8 1A
STAR BLUE CHIP 44K3
STAR BORON 44L
STAR COLUMBRIUM 44K3
STAR J 13A
STAR MAX 44K3
STAR MO 68 44K3
STAR Mo M 2/5 44L
STAR MO M2 44K3
STAR Si 050 1C
STAR Si 120 1C
STAR ZENITH 44K3
STAS 568 20B
STAS 6071 20B
STA V18/1 44K1
STAYBRITE 44N1
STB 12
see JIS G3462 STBA 12 44G1
STB 13
see JIS G3462 STBA 13 44G1
STB 20
see JIS G3462 STBA 20 44K1
STB 22
see JIS G3462 STBA 22 44K1
STB 23
see JIS G3462 STBA 23 44K1
STB 24
see JIS G3462 STBA 24 44K1
STB 25
see JIS G3462 STBA 25 44K1
STB 26
see JIS G3462 STBA 26 44K1
STB 30
see JIS G3461 44A1
STB 33
see JIS G3461 44A1
STB 35
see JIS G3461 44A1
STB 39
see JIS G3462 STBL 39 44A1
STB 42
see JIS G3461 44A1
ST BRAND 44A3
STC 1 44A1
STC 2 44A2
STC 3 44A2
STC 4 44A2
STC 5 44A3
STC 6 44A3
STC 52D 44N1
STELCO C845 44A1
STELCO C850 44A1
STELCO C855 44A1
STELCO C860 44A1
STELCOLOY G 44A1
STELCOLOY S 44A1
STELCO VANADIUM 44J
STELLITE 1 13A
STELLITE 3 13A
STELLITE 4 13A
STELLITE 6 13A
STELLITE 7 13A
STELLITE 8 13A
STELLITE 12 13A
STELLITE 12P 13A
STELLITE 20 13A
STELLITE 21 13A
STELLITE 23 13A
STELLITE 27 13A
STELLITE 30 13A
STELLITE 31 13A
STELLITE 100 13A
STELLITE 208 13A
STELLITE 306 13A
STELLITE 506 13A
STELLITE F 13A
STELLITE X 40 13A
STELVETITE G 44A1
STELVETITE R 44A1
STELVETITE Z 44A1
STENTOR 44A4
STEREO 14C
STEREOTYPE 21.1
STEREOTYPE ALLOY 21.1
STERLING 42.1
STERLING 44A4
STERLING V 44J
STERN TUBE METAL 50A
STK 30
see JIS G3444 44A1
STK 41
see JIS G3444 44A1
STK 50
see JIS G3444 44A1
STK 51
see JIS G3444 44A1
STK 55
see JIS G3444 44A1
STKM 11A
see JIS G3445 44A1
STKM 12A
see JIS G3445 44A1
STKM 12C
see JIS G3445 44A1
STKM 13A
see JIS G3445 44A1
STKM 13B
see JIS G3445 44A1
STKM 13C
see JIS G3445 44A1
STKM 14A
see JIS G3445 44A2
STKM 14B
see JIS G3445 44A2
STKM 14C
see JIS G3445 44A2
STKM 15A
see JIS G3445 44A2
STKM 15C
see JIS G3445 44A2
STKM 16A
see JIS G3445 44A2
STKM 16C
see JIS G3445 44A2
STKM 17A
see JIS G3445 44A3
STKM 17C
see JIS G3445 44A3
STKM 18A
see JIS G3445 44A1
STKM 18B
see JIS G3445 44A1
STKR 18C
see JIS G3445 44A1
STKS 1A-1E
see JIS G3441 44K1
STKS 2A-2E
see JIS G3441 44K1
STKS 3A-3E
see JIS G3441 44K1
STKS 4A-4E
see JIS G3441 44K1
STM 44K3
STP 14A
STPA 12
see JIS G3458 44G1
STPA 22
see JIS G3458 44K1
STPA 23
see JIS G3458 44K1
STPA 24
see JIS G3458 44K1
STPA 25
see JIS G3458 44K1
STPA 26
see JIS G3458 44K1
STPG 38
see JIS 3454 44A1
STPG 42
see JIS G3454 44A2
STPL 39
see JIS 3460 44A1
STPS 1
see JIS B2351 44A1
STPS 2
see JIS B2351 44A1
STPT 38
see JIS G3456 44A1
STPT 42
see JIS G3456 44A2
STPT 49
see JIS G3456 44A2
STPY 41
see JIS G3457 44A1
STRAINTRODE 61.81 44N1
STRAINTRODE 62.30 44N1
STRAINTRODE 63.30 44N1
STRAINTRODE 63.35 44N1
STRAINTRODE 63.80 44N1
STRAINTRODE 64.30 44N1
STRAINTRODE 67.15 44N1
STRAINTRODE 67.45 44N1
STRAINTRODE 67.70 44N1
STRAINTRODE 67.75 44N1
STRAINTRODE 68.60 44N1
STRAINTRODE 6815 44M1
STRENLITE 50 44A2
STRENLITE 440 44A2
STRONTIUM 45.1
STS 35
see JIS G3455 44A1

STS 38
see JIS G3455 44A1
STS 42
see JIS G3455 44A2
STS 49
see JIS G3455 44A2
Stub G1B 1A
STUBS 44A4
STUBS 1MS 14B
STUBS 1S 14B
STUBS 2H 14B
STUBS 2H M 14B
STUBS 3 42.1
STUBS 3M 42.1
STUBS 4 1C
STUBS 5 44A1
STUBS 6 14H
STUBS 6M 14H
STUBS 7 14H
STUBS 7M 14H
STUBS 8 27A
STUBS 8 FNN 27E
STUBS 8 NC 27A
STUBS 8 Nifer 27E
STUBS 8 S 27E
STUBS 22 14H
STUBS 22M 14H
STUBS 31 42.1
STUBS 31M 42.1
STUBS 32 14K
STUBS 32W 14K
Stubs 34 14G
Stubs 34 N 14G
STUBS 35R 14H
STUBS 35S 14H
STUBS 36 14H
STUBS 38 14H
STUBS 39 14K
STUBS 39W 14K
STUBS 48 1C
STUBS 49 1B
STUBS 56 55A
STUBS 62 44G1
STUBS 63 C Mn Cr Ni 44N1
STUBS 65 C Si Mn Cr Ni Mo 44N1
STUBS 65 XL C Si Cr Ni Mo 44N1
STUBS 66 44M1
STUBS 67S 44E2
STUBS 068 HH 27B
STUBS 68 44N1
STUBS 68 LC 44N1
STUBS 68 Mo 44N1
STUBS 68 Mo LC 44N1
STUBS 68 Ti Mo 44N1
STUBS 68H 44N1
STUBS 69 C 44K3
STUBS 73 G2 44K2
STUBS 73 G3 44K2
STUBS 73 G4 44K2
STUBS 75 44K3
STUBS 80 MON 27D
STUBS 80 Ni 27A
STUBS 81 44K1
STUBS 86 FN 27E
STUBS 88 27E
STUBS 306 42.1
STUBS 306M 42.1
STUBS 320 14K
Stubs 343 14G
STUBS 387 14E
STUBS 389 14E
STUBS 400 44E2
STUBS 570 K 42.1
STUBS 610 44A1
STUBS 611 44A1
STUBS 612 44A1
STUBS 614 44A1
STUBS 651 44K1
STUBS 653 C Cr Ni Mo 44N1
STUBS 670 44E2
STUBS 700 44K2
STUBS 701 13A
STUBS 701 A 13A
STUBS 706 13A
STUBS 706 A 13A
STUBS 710 44E1
STUBS 711 44E1
STUBS 712 13A
STUBS 712 A 13A
STUBS 713 44K3
STUBS 3033 42.1
STUBS 3033 M 42.1
STUBS 3040 42.1
STUBS 3040 M 42.1
STUBS 4221 27B
STUBS 5820 LC 44N1
STUBS 5820 Mo LC 44N1
STUBS 5820 Mo Nb 44N1
STUBS 5820 Nb 44N1
STUBS 5824 LC 44N1
STUBS 5824 Mo LC 44N1
STUBS 6222 Mo 27B
STUBS 6820 LC 44N1
STUBS 6820 Mo LC 44N1
STUBS 6820 Mo Nb 44N1
STUBS 6824 LC 44N1
STUBS 6824 Mo LC 44N1
STUBS 6825 44N1
STUBS 7015 27B
STUBS 7015 Mo 27B
STUBS 7200 44P
STUBS A1 44A1
STUBS A15 44A1
STUBS A18 44A1
Stubs A34 14G
STUBS A63 44P
STUBS A 320 14K
STUBS A675 44K2
Stubs A3444 14G
STUBS NG 6 1D
STUBS NG 21 1C
STUBS NG 61 1D
STYR 44K3
STZ 20 55A
STZ 30 55A
SU 13 28.1
SU 16 28.1
SU 31 28.1
SUH 1 44E2
SUH 3 44M1
SUH 4 44M2
SUH 31 44N1
SUH 309
see JIS G4312 44N1
SUH 310
see JIS G4312 44N1
SUH 330
see JIS 4312 44N1
SUH 446
see JIS G4312 44M1
SUH 600 44M1
SUH 616 44M1
SUH 661
see JIS 4312 44N2
SUJ 2 44E3
SUPAVIT 44A1
SUPER AUSTENITE 44K3
SUPER AUSTENITE T1 44K3
SUPER AW 23 44K1
SUPER C12 44M2
SUPER CAPITAL 44L
SUPERCHLOR 20B
SUPER CHNC 44K1
SUPER COBALT 44L
SUPERCOUPE 44A1
SUPERCOUPE A 44A1
SUPERCROM 44N1
SUPERCROM 29.9 44N1
SUPERCROM 308L 44N1
SUPERCROM 309L 44N1
SUPERCROM 316L 44N1
SUPERCROM 318 44N1
SUPERCROM 347 44N1
SUPERCROM SD 44N1
SUPER CYCLONE 44L
SUPER DBL 44L
SUPERDIE 44M2
SUPER DOUBLE MUSHET 44K3
SUPERDRAW 44H
SUPER ECLIPSE 44K3
SUPERELSO 44K1
SUPERELSO 60 44A1
SUPERELSO 70 44K1
SUPERELSO 600 44K1
SUPER HALLAMITE 44L
SUPER HICRO D5 44M2
SUPER Hi-Mo 44L
SUPER-HOLFOS WW 14K
SUPER HS 44K3
SUPER HYDRA 44K3
SUPER INMANITE 44L
SUPER INVINCIBLE 5 44L
SUPER INVINCIBLE
ADVANCE 10 44L
SUPER INVINCIBLE TB 44L
SUPERIOR 1 44M2
SUPERIOR 2 44M2
SUPERIOR 3 44M2
SUPERIOR 4 44M2
SUPERIOR 5 44K3
SUPERIOR SPUR 44K3
SUPERLEDA 44L
SUPER LMW 44L
SUPERMENDUR 13A
SUPERMET 308L 44N1
SUPERMET 309 Cb 44N1
SUPERMET 309 Mo 44N1
SUPERMET 309L 44N1
SUPERMET 316L 44N1
SUPERMET 318 44N1
SUPERMET 347 44N1

SUPERMET 2205 44N1
SUPER Mo CHIP 44L
SUPERMOLD 44K1
SUPER MONARCH 44K3
SUPER MOTUNG 44L
SUPER MOTUNG SPECIAL 44L
SUPER MUMETAL 50 27C
SUPER MUMETAL 100 27C
SUPER MUMETAL Co 27C
SUPER NITRALLOY 44K1
SUPER NOVEX 44K1
SUPER ORALIUM 17.1
SUPER PANTHER 44L
SUPER RADIO METAL 50 20C
SUPER RAPID 44K3
SUPER SAMSON 44K1
SUPER SHNC 44K1
SUPER Si Mn 44K2
SUPER Si Mn 44K3
SUPER SPECIAL ECHO 44K3
SUPER STAR 44L
SUPERSTON 14G
SUPERSTON 10 14G
SUPERSTON 40 14G
SUPERSTON 40 14G
SUPERSTON 50 14C
SUPERSTON 60 14G
SUPERSTON 70 14G
SUPER-SUPER MONARCH 44L
SUPER TOUGH B25 44F1
SUPER TOUGH D 44K2
SUPERTOUGH B 35 44A2
SUPERTOUGH B20 44A1
SUPERTOUGH C20 44K1
SUPERTOUGH C40 44G1
SUPERTOUGH C40M 44G1
SUPERTRODE 14G
SUPER TUFCOR 44A1
SUPER TYR 44K3
SUPER UGIMAX 600 20C
SUPER UGIMAX 800 20C
SUPER UNICUT 44L
SUPERVITAC 44K1
SUPERVITUS 980 44K1
SUPER VNCA 44K1
SUPERWEAR 44M2
SUPREMUS 44K3
SUPREMUS EXTRA 44K3
SUPREX A 44K1
SUPREX B 44K1
SUPREX C 44K1
SUPREX M 44K1
SUPRIMPACTO 44K1
SUS 302
see JIS G4304 44N1
SUS 304
see JIS G4304 44N1
SUS 304 L
see JIS G4304 44N1
SUS 305
see JIS G4304 44N1
SUS 309 S
see JIS G4304 44N1
SUS 310 S
see JIS G4304 44N1
SUS 316
see JIS G4304 44N1
SUS 316 J1
see JIS G4304 44N1
SUS 316 J1L
see JIS G4304 44N1
SUS 316 L
see JIS G4304 44N1
SUS 317
see JIS G4304 44N1
SUS 317 L
see JIS G4304 44N1
SUS 321
see JIS G4304 44N1
SUS 329 J1
see JIS G4304 44N3
SUS 347
see JIS G4304 44N1
SUS 403
see JIS G4304 44M1
SUS 405
see JIS G4304 44M1
SUS 410
see JIS G4304 44M1
SUS 429
see JIS G4304 44M1
SUS 430
see JIS G4304 44M1
SUS 434
see JIS G4304 44M1
SUS 631
see JIS G4304 44N3
SV 02KH19N9 44N1
SV 04KH19N9 44N1
SV 04KH19N11M3 44N1
SV 04KH19N952 44N1
SV 05KH19N9F352 44N1
SV 06KH 14 44M1
SV 06KH19N9T 44N1
SV 07KH18N9T1U 44N1
SV 07KH25N13 44N1
SV 08KH 14GT 44M1
SV 08KH19N9F252 44N1
SV 08KH19N10B 44N1
SV 08KH19N12M3 44N1
SV08KH20N9G7T 44P
SV08KH20N10G6 44P
SV 08KH25N5TMF 44N1
SV 10KH 11MFN 44M1
SV 10KH 11VMFN 44M1
SV 10KH 13 44M1
SV 10KH 17T 44M1
SV 10KH5M 44K1
SV 10KH16N25M6 44N1
SV 10KH20N15 44N1
SV 13KH 25T 44M1
SV 13KH25N18 44N1
SV 25KH25N16G7 44N1
SV 30KH15N35V3B 3T 44N1
SV 120KH 13 44M1
SVERKER 3 44M2
SVERKER 21 44M2
SVL 44K1
SWANSEA VALE 55A
SWM 14F
SX 4 50A
SX 20 14J
SX 25 21.1
SYLVALOY 27F
T 44A1
T 1
see BS T 1 44A2
T1 44A2
T 1 27B
T 1 44K3
T 2
see BS T 2 44K1
T2 23B
T 2 27F
T 2 44K3
T 3
see BS T 3 14F
T 3 44K3
T 3N 44F2
T 4 44L
T 5
see BS T 5 44A2
T5 44A2
T 5 44L
T 6
see BS T 6 44A1
T 6 44L
T 7
see BS T 7 14A
T7 14A
T 7 44K3
T 8
see BS T 8 14B
T8 14B
T 8 44L
T 9 1A
T 9 44K3
T 10 44K3
T 11 44A3
T 12 44A3
T 13 44A4
T 14
see BS T 14 44A2
T14 44A2
T 14 44A4
T 15 44A4
T 15 44L
T 16 44A4
T 18
see BS T 18 44A2
T18 14B
T21 44A2
T 21 44A3
T 22 44A3
T 23 44A4
T 24 44A4
T 25 44A4
T 26 44A4
T 27 44A4
T 31 44A3
T 32 44A3
T 33 44A4
T 34 44A4
T 35
see BS T 35 44A2
T35 44A2
T 35 44A4
T 35 51A
T 36 44A4
T 37 44A4
T 40 51A

T 45
see BS T 45 44A2
T45 44A2
T 47
see BS T 47 14B
T47 14B
T 48
see BS T 48 14B
T48 14B
T 50
see BS T 50 44K2
T 50 51A
T 51
see BS T 51 14A
T51 14A
T 52 14K
T 53
see BS T 53 44K1
T 54
see BS T 54 44A2
T54 44A2
T 55
see BS T 55 44N1
T55 44N1
T 56
see BS T 56 44K1
T 57
see BS T 57 44K1
T 58
see BS T 58 44N1
T58 44N1
T 59
see BS T 59 44K1
T 60
see BS T 60 44K1
T 60 51A
T 61
see BS T 61 44N1
T61 44N1
T91 44K1
T 91 44E1
T 200 44N1
T 222 46.1
T 240 44N1
T 275 44N1
T 283 51B
T 443 51C
T 550 44M1
T 552 44M1
T 558 44M1
T 560 44M1
T 608 44M1
T 650 44M1
T 655 44M1
T 708 44A1
T 709 44A1
T 713 51C
T 715 44A1
T 725 44A1
T 735 44A2
T 750 44A3
T 1040 44E3
T11301
UNS for M1 Tool steel 44K3
T11302
UNS for M2 Tool steel 44K3
T11304
UNS for M4 Tool steel 44K3
T11306
UNS for M6 Tool steel 44L
T11307
UNS for M7 Tool steel 44K3
T11310
UNS for M10 Tool steel 44K3
T11313
UNS for M3-C1.1 Tool steel 44K3
T11323
UNS for M3 C1.2 Tool steel 44K3
T11330
UNS for M30 Tool steel 44L
T11333
UNS for M33 Tool steel 44L
T11334
UNS for M34 Tool Steel 44L
T11336
UNS for M36 Tool steel 44L
T11341
UNS for M41 Tool steel 44L
T11342
UNS for M42 Tool steel 44L
T11343
UNS for M43 obsolete 44L
T11344
UNS for M44 obsolete 44L
T11346
UNS for M46 Tool steel 44L
T11347
UNS for M47 obsolete 44L
T11350
UNS for AMS 6490 44K3
T11352
UNS for M52 Tool steel 44K3
T12001
UNS for T1 Tool steel 44K3
T12002
UNS for T2 obsolete 44K3
T12004
UNS for T4 Tool steel 44L
T12005
UNS for T5 Tool steel 44L
T12006
UNS for T6 Tool steel 44L
T12008
UNS for T8 Tool steel 44L
T12015
UNS for T15 Tool steel 44L
T20806
UNS for H26 Tool steel 44K2
T20810
UNS for H10 Tool steel 44K2
T 20810 44K2
T20811
UNS for H11 Tool steel 44K1
T 20811 44K2
T20812
UNS for H12 Tool steel 44K1
T20813
UNS for H13 Tool steel 44K1
T20814
UNS for H14 Tool steel 44K1
T20819
UNS for H19 Tool steel 44L
T20821
UNS for H21 Tool steel 44K1
T20822
UNS for H22 Tool steel 44K1
T20823
UNS for H23 Tool steel 44M1
T20824
UNS for H24 Tool steel 44K2
T20825
UNS for H25 obsolete 44K1
T20841
UNS for H41 obsolete 44K2
T20842
UNS for H42 Tool steel 44K2
T20843
UNS for H43 obsolete 44K2
T30102
UNS for A2 Tool steel 44K3
T30103
UNS for A3 obsolete 44K3
T30104
UNS for A4 Tool steel 44K3
T30105
UNS for A5 obsolete 44K3
T30106
UNS for A6 Tool steel 44K2
T30107
UNS for A7 Tool steel 44K3
T30108
UNS for A8 tool steel 44K2
T30109
UNS for A9 tool steel 44K2
T30110
UNS for A10 tool steel 44K3
T30111
UNS for A11 tool steel 44M2
T30402
UNS for D2 tool steel 44M2
T30403
UNS for D3 tool steel 44M2
T30404
UNS for D4 tool steel 44M2
T30405
UNS for D5 tool steel 44M2
T30407
UNS for D7 tool steel 44M2
T31501
UNS for 01 Tool steel 44K3
T31502
UNS for 02 Tool steel 44K3
T31506
UNS for 06 Tool steel 44G2
T31507
UNS for 07 Tool steel 44K3
T41901
UNS for S1 Tool steel 44K2
T41902
UNS for S2 Tool steel 44K2
T41904
UNS for S4 Tool steel 44B2
T41905
UNS for S5 Tool steel 44K2
T41906
UNS for S6 Tool steel 44K2

T41907
UNS for S7 Tool steel 44K2
T51602
UNS for P2 Tool steel obsolete 44K1
T51603
UNS for P3 Tool steel obsolete 44K1
T51604
UNS for P4 Tool steel obsolete 44K1
T51605
UNS for P5 Tool steel obsolete 44E1
T51606
UNS for P6 Tool steel 44L
T51620
UNS for P20 Tool steel 44K1
T51621
UNS for P21 Tool steel 44K1
T60601
UNS for F1 Tool steel obsolete 44H
T 60601 44H
T60602
UNS for F2 tool steel obsolete 44H
T 60602 44H
T61202
UNS for L2 tool steel obsolete 44K3
T61206
UNS for L7 tool steel 44K3
T72301
UNS for W1 tool steel 44A4
T72302
UNS for W2 tool steel 44J
T72305
UNS for W5 tool steel 44E3
T90102
UNS for CA2 tool steel 44K1
T90402
UNS for CD2 tool steel 44M1
T90812
UNS for CH12 tool steel 44K1
T90813
UNS for CH13 tool steel 44K1
T91501
UNS for CO1 tool steel 44K1
T91905
UNS for CS5 tool steel 44G1
TA1
Designation used by BS 51A
TA2
Designation used by BS 51A
TA 2M 51B
TA 4DE 51B
TA 4M 51B
TA 5E 51B
TA6
Designation used by BS 51A
TA 6V 51C
TA 6Z-5D 51B
TA 6ZW 51B
TA7
Designation used by BS 51A
TA8
Designation used by BS 51A
TA9
Designation used by BS 51A
TA10
Designation used by BS 51B
TA11
Designation used by BS 51B
TA12
Designation used by BS 51B
TA13
Designation used by BS 51B
TA21
Designation used by BS 51A
TA22
Designation used by BS 51A
TA23
Designation used by BS 51A
TA24
Designation used by BS 51A
TA28
Designation used by BS 51B
TA38
Designation used by BS 51B
TA39
Designation used by BS 51B
TA40
Designation used by BS 51B
TA41
Designation used by BS 51B
TA42
Designation used by BS 51B
TA43
Designation used by BS 51B
TA44
Designation used by BS 51B
TA45
Designation used by BS 51B
TA46
Designation used by BS 51B
TA47
Designation used by BS 51B
TA48
Designation used by BS 51B
TA49
Designation used by BS 51B
TA50
Designation used by BS 51B
TA51
Designation used by BS 51B
TA52
Designation used by BS 51A
TA53
Designation used by BS 51A
TA54
Designation used by BS 51A
TA 54A
see ASTM range 23B
TA55
Designation used by BS 51A
TA56
Designation used by BS 51B
TA57
Designation used by BS 51B
TA58
Designation used by BS 51A
TA59
Designation used by BS 51B
T-ALLOY 44K2
T ALLOY A 44K1
T ALLOY C 44K1
T-ALLOY-B 44K2
TANDEM DE 50A
TANDEM HDL 21.1
TANDEM HP 50A
TANDEM ME 50A
TANDEM ML 50A
TANDEM PLUS 50A
TANDEM PTA 55A
TANDEM RM 21.1
TANDEM SC 21.1
TANGE CHAMPION 44K1
TANTALUM 46.1
TANTUNG 13A
TANTUNG G 13A
TAURUS 14E
TAURUS 14K
TAURUS I 14K
TAURUS II 14K
TAURUS III 14K
TAURUS IV 14K
TAURUS IX 14K
TAURUS VI 14K
TAURUS VII 14K
TAURUS VIII 14K
TAURUS XI 14K
TAURUS XIV 14K
TAURUS XVI 14K
TAURUS XX 14G
TAURUS XXI 14C
TAURUS XXII 14B
TAURUS XXIII 14B
TAURUS XXIV 14B
TAURUS XXVI 14E
TAURUS XXX 14B
TAURUS XXXI 14C
TAURUS XXXIII 14K
TBS 600 44K3
TBS 1000 44K3
TC 14A
TC 50A
TCA 2 44K3
TCR 44J
TCS 50A
TD 44A4
TDB 44H
TDBC 44K3
TDC 44H
TE11DA 51B
TEENAX 44K3
TELCALLOY 1 14E
TELCALLOY 1.5 14E
TELCALLOY 2 14E
TELCALLOY 3 14E
TELCALLOY 4 14E
TELCONSTAN 14E
TELCOSEAL I 20C
TELCOSEAL II 20C
TELCOSEAL III 20C
TELCOSEAL V 20C
TELCOSEAL VI 20C
TELFOS 14K
TELLURIUM 47.1

TELMAR 90 44F3
TELMAR 110 44F3
TELMAR 125 44F3
TEM 02C 44L
TEMPER No. 4 44A4
TEMPER No. 5 44A4
TEMPER No. 6 44A4
TEMPER No. 31/2 44A4
TEMPER No. 41/2 44A4
TEMPER No. 51/2 44A4
TEMPER No.2 44A3
TEMPER No.3 44A3
TEMPEST 44M1
TENAPSO 44A1
TENASOUDO 50 44A1
TENASOUDO 75 44K1
TENASOUDO 105 44K1
TENASOUDO Ni 3 S 44F1
TENAXAS Al 50A
TENELON 44M1
TENOHM 14E
TENSITRODE 48.30 44A1
TENSITRODE 55.00 44A1
TENSITRODE 75.55 44K1
TENSITRODE 75.65 44K1
TENSOLA I 20B
TENSOLA II 20B
TENSOLA III 20B
TEN STAR 44K3
TENZALOY 1M
TERFLING 2 44E3
TERNALLOY 5 1M
TERNALLOY 7 1M
TG 44G2
TGS 44A1
TH 44K2
THALLIUM 48.1
THERLO 20C
THERMALLOY 40 A2 44N1
THERMALLOY 40 E 44N1
THERMALLOY 47 44N1
THERMALLOY 50 CQ 44N1
THERMALLOY D47 44N1
THERMELAST 2602 13A
THERMENOL 44K1
THERMINOX 44N1
THERMINOX 13 44N1
THERMINOX 20 44N1
THERMINOX 821 44N1
THERMODIE 44K2
THERMOFIXE 44A1
THERMOWEAR 44K2
THESSCO 14H
THESSCO D 42.1
THESSCO M 42.1
THESSCO MX12 42.1
THESSCO MX18 42.1
THESSCO MX18 PLUS 42.1
THIRTY OAK 44E1
THIRTY OAK 1 44E1
THIRTY OAK 2 44E1
THORIUM 49.1
THREE CASTLES 44L
THYROTHERM 2799 44F3
TI 44K1
Ti 0.15 Pd 51A
TI 01 1F
TI 03 1E
TI 04 1F
Ti 4 Al 3 Mo 1 V 51C
TI 05 1D
Ti 5 Al 2.5 Sn 51B
Ti 5 Al 2.5 Sn EL1 51B
Ti 5 Al 4 Fe Cr 51C
Ti 5 Al 6 Sn 5 Zr 51B
Ti 6 Al 4 V 51C
Ti 6 Al 4 V EL 1 51C
Ti 6 Al 6 V 2 Sn 51C
TI 07 1D
Ti 7 Al 4 Mo 51C
Ti 7 Al 12 Zr 51B
Ti 8 Al 1 Mo 1 V 51C
Ti 8 Mn 51B
TI 11 1B
Ti 13 V 11 Cr 3 Al 51C
TI 22 1D
TI 33 1D
Ti 35 A 51A
TI 40 1E
TI 44 1E
TI 53 1F
TI 55 1F
Ti 55 A 51A
TI 56 1F
Ti 65 A 51A
TI 66 1F
Ti 75 A 51A
TI 77 1G
TI 88 1G
Ti 100 A 51A
Ti 115 51A
Ti 120 51A
Ti 125 51A
Ti 130 51A
Ti 140 A 51C
Ti 150 51A
Ti 155 51A
Ti 155 A 51C
Ti 160 51A
Ti 205 51C
Ti 230 51C
Ti 260 51A
Ti 314 A 51B
Ti 314 C 51B
Ti 315 51B
Ti 317 51B
Ti 318 A 51C
Ti 550 51B
Ti 550 51C
Ti 551 51B
Ti 551 51C
Ti 679 51C
Ti 680 51B
Ti 684 51B
Ti 684 51C
Ti 685 51B
Ti 685 51C
Ti 700 51C
Ti Al 4 Mo 4 Sn 2 Si 51C
Ti Al 5 Fe 2.5 51C
Ti Al 6 Sn 2 Zr 4 Mo 2 51C
Ti Al 6 V 4 51C
Ti Al 6 V 6 Sn 2 51C
TICONAL 190 20C
TICONAL 360 20C
TICONAL 400 20C
TICONAL 450 20C
TICONAL 500 20C
TICONAL 600 20C
TICONAL 650 20C
TICONAL 700 20C
TICONAL 750 20C
TICONAL 750 20C
TICONAL 800 20C
TICONAL 1500 20C
TICONAL 1500F 20C
TICONAL 1800 20C
TICONAL C 20C
TICONAL C 44L
TICONAL D 44L
TICONAL E 44L
TICONAL F 44L
TICONAL FRITTE 20C
TICONAL G 44L
TICONAL GX 44L
TICONAL H 44L
TICONAL L 44L
TICONAL S 44L
TICONAL X 44L
TICONIUM 27C
Ti Cu 2 51B
TIMKEN 16/25/6 44N1
TIN 50.1
TINMANS SOLDER 50A
Ti P62 51B
Ti Po 1 51A
Ti Po 2 51A
TITAN 44D
TITANIC 141 44K2
TITANIUM 51A
TITAN MANGANESE 44D
TITRUTAN 51A
TI TYPE A 44K1
TI TYPE B 44K1
T-K 44K1
TM 44A4
TM 44K2
TM 6 44K3
T METAL 55A
TMS 44A4
TMS 44K3
TOBA 44K1
TOBIN BRASS 14C
TOBIN BRONZE 14C
TOH 44K3
TOLEDO 15 44A1
TOLEDO 20 44A1
TOLEDO 30 44A2
TOLEDO 40 44A2
TOLEDO 55 44A3
TOLEDO BR 44E3
TOLOY 9 TYPE HE 44N1
TOLOY 10 TYPE HF 44N1
TOLOY 12 44N1
TOLOY 12 HH2 44N1
TOLOY 16 44N1
TOLOY 20 44N1
TOLOY 20 HK40 44N1
TOLOY 25 HN 44N1
TOLOY 37 44N1
TOLOY 37 HU 44N1

TOLOY 50 27B
TOLOY 50/50 27B
TOLOY 60/40 12.1
TOLOY 65 27B
TOLOY 65 HX 27B
TONCAN 44A1
TONUM 14C
TOPAZ 52.1
TOPHET 12.12 44N1
TOPHET 18.8 44N1
TOPHET 18.8 Low C 44N1
TOPHET 25.20 44N1
TOPHET 30 27B
TOPHET A 27B
TOPHET C 27B
TOPHET Cb 44N1
TOPHET D 20C
TOPHET Mo 44N1
TOPHET Ti 44N1
TOUGH PITCH 14A
TOWER 50 44A2
TOWER 55 44A2
TOWER 60 44J
TOYO 50.1
T P W 44A1
TP409
  see 409
TP439
  see 439
TPA 44N1
TPC 44K2
TPM 27B
TPM 44K3
TP XM 15 44N1
TPXM10 44P
TPXM11 44P
TPXM33
  see XM33
T QUALITY 44A4
TR 3KD 44E2
TR 4 44E2
TR 5 44E2
TRAN-COR A 5 20A
TRAN-COR A 6 20A
TRAN-COR M 14 20A
TRAN-COR M 15 20A
TRAN-COR M 17 20A
TRAN-COR M 19 20A
TRAN-COR M 22 20A
TRAN-COR M 27 20A
TRAN-COR M 36 20A
TRAN-COR M 43 20A
TRANSIL 74 44B1
TRANSIL 80 44B1
TRANSIL 86 44B1
TRANSIL 92 44B1
TRANSIL 100 44B1
TRANSIL 107 44B1
TRANSIL 115 44B1
TREBLE EXTRA 44A4
TREBLE SUPER
  MONARCH 44L
TRIBALOY 100 13A
TRIBALOY 400 13A
TRIBALOY 700 27F
TRIBALOY 800 13A
TRIBALOY T400 13A
TRIM RITE 44M1
TRIPLE 5 MONARCH 44L
TRIPLE 5C 44L
TRIPLE CONQUEROR 44K3
TRIPLE CRESCENT 44K3
TRIPLE DIE 44M2
TRIPLE FLYGO 44L
TRIPLE GRIFFIN 44K3
TRIPLE MERMAID 44L
TRIPLE MUSHET 44L
TRIPLE MUSKET GZ 44L
TRIPLE SIX 44E2
TRIPLE SPUR 44L
TRIPLE VELOS 44L
TRI-STEEL 44J
TRI-TEN 44J
TRITON 44G2
TRIUMPH SUPER 44K3
TRIUMPH SUPERB 44K3
TRIUMPH SUPERB 1000 44L
TRIUMPH SUPERB
  DOUBLE 44L
TRIUMPH SUPERB
  DOUBLE 1000 44L
TRK 44E2
TRKD 44K2
TRM 45 44D
TRNKD 44K2
TROJAN 44K3
TRS 44K3
TRUBRITE 44N1
TRUBRITE A1 44N1
TRUBRITE A3 44N1
TRUBRITE A6 44N1
TRUBRITE B41 44M1
TRUBRITE B42 44M1
TRUBRITE C72 44M1
TRUBRITE C74 44M1
TRUBRITE C76 44M1
TRUCAST HARD 17.1
TRUCAST MEDIUM 17.1
TRUCAST SOFT 17.1
TRW 1800 27C
T St 0 44A1
T St 10 44A1
T St 10/01 44A1
T St 10/02 44A1
T St 10/03 44A1
TTQ 44L
TT St 35 44A1
TT St 41 44A1
TT St 45 44A1
TUBROD 14.18 44F1
TUBROD 15.10 44A1
TUBROD 15.18 44F1
TUBROD 15.23 44M1
TUBROD 15.40 44E1
TUBROD 15.42 44K1
TUBROD 15.50 44K1
TUBROD 15.60 44D
TUBROD 15.70 44M1
TUBROD 15.73 44M1
TUFCOR 44A1
TULIP 50.1
TUMBLERS 44D
TUNCO 44L
TUNGSTEN 52.1
TUNGSTEN DIAMOND 44K3
TUNGSTEN SPUR 44K3
TUNGUM 14B
TURBALOY 13 44N1
TURBEX 50A
TURBISTON 3 14C
TURBISTON M 14C
TURBOTHERM 15M 44M1
TURBOTHERM 20M 44M1
TURBOTHERM 20MV 44M1
TURBOTHERM 20MV Nb 44M1
TURBOTHERM 20MV W 44M1
TUREX 9 44K1
TU St 1 44A1
TU St 2 44A1
TU St 34 44A1
TU St 37 44A1
TU St 37/02 44A1
TU St 37/03 44A1
TU St 37/04 44A1
TU St 37/05 44A1
TWIN VAN 44K3
TWOSCORE 44M1
TWO SPUR 44L
TXCR 44N1
TY 44K1
TYNE 2.1
TYPE M 44A1
TYPE XX 44A1
TYRANT EXTRA 44K2
TZ 6 23B
TZM 1 26A
U 3 14J
U 4 14J
U 7 44A3
U 7A 44A3
U 8 14J
U 8 44A4
U 8A 44A3
U 9 44A4
U 9A 44A4
U 10 44A4
U 10A 44A4
U 11 44A4
U 11A 44A4
U 12 44A4
U 12 44M1
U 12A 44A4
U 13 44A4
U 13A 44A4
U15 1M
U 85 44A4
U 99/100 51A
U-A 10NM 14G
U BRAND 44A4
UCAR 11 44N1
UCAR 16 27C
UCAR 25 13A
UCAR 75 27B
UCAR 80 27C
UCAR 90 27C
UCAR 600 27B
UCAR 601 27B
UCAR 625 27C
UCAR 700 27C
UCAR 702 27C
UCAR 706 44N1

UCAR 713C 27C
UCAR 713LC 27C
UCAR 718 27C
UCAR 800 44N1
UCAR 901 27C
UCAR C130 27C
UCAR C1023 27C
UCAR FSX414 13A
UCAR GMR 235 27C
UCAR GMR 235D 27C
UCAR HASTELLOY R 235 27C
UCAR IN100 27C
UCAR IN731 27C
UCAR IN792 27C
UCAR M21 27C
UCAR M509 13A
UCAR MARAGING 110 44F3
UCAR R 41 27C
UCAR U500 27C
UCAR U520 27C
UCAR U700 27C
UCAR U710 27C
UCAR U722 27C
UCAR WASPALLOY 27C
UCAR X 27C
UCAR X45 13A
UCX 2 44K2
UDDCO 12 44M1
UDIMET 200 27C
UDIMET 500 27C
UDIMET 600 27C
UDIMET 700 27C
UE 3Z9 14K
UE 5P 14K
UE 5Pb 5Z5 14K
UE 5Z4 14K
UE 7Z5Pb4 14K
UE 8Z2 14K
UE 9P 14K
UE 10Z1 14K
UE 12P 14K
UE 12Z1 14K
UE 14 14K
UE 16 14K
UE 18 14K
UE Pb Z 14K
UGA 44A4
UGIMAX 600 20C
UGIMAX 800 20C
UHB 44A4
UHB 3 44A1
UHB 4M10 44A1
UHB 11 44A3
UHB 19 Va 44J
UHB 20 44A4
UHB 25 44K3
UHB 29 44K3
UHB 424 44L
UHB 431 44K3
UHB 432 44K3
UHB 433 44K3
UHB 442 44L
UHB 480 44K3
UHB 485 44K3
UHB STAINLESS 1 44M1
UHB STAINLESS 2 44M1
UHB STAINLESS 3 44N1
UHB STAINLESS 3L 44N1
UHB STAINLESS 3M 44N1
UHB STAINLESS 3MM 44N1
UHB STAINLESS 4 44N1
UHB STAINLESS 4MM 44N1
UHB STAINLESS 5 44M1
UHB STAINLESS 6 44M1
UHB STAINLESS 15 44N1
UHB STAINLESS 21 44M1
UHB STAINLESS 22 44M1
UHB STAINLESS 24 44N1
UHB STAINLESS 24L 44N1
UHB STAINLESS 25 44N1
UHB STAINLESS 31 44M1
UHB STAINLESS 33MM 44N1
UHB STAINLESS 34 44N1
UHB STAINLESS 34L 44N1
UHB STAINLESS 41 44M1
UHB STAINLESS 43 44N1
UHB STAINLESS 44 44N1
UHB STAINLESS 45 44N1
UHB STAINLESS 51 44M1
UHB STAINLESS 53 44N1
UHB STAINLESS 55 44N1
UHB STAINLESS 63 44N1
UHB STAINLESS 63H 44N1
UHB STAINLESS 64 44N1
UHB STAINLESS 72 44M1
UHB STAINLESS 524 44N1
UHB STAINLESS 624 44N1
UHB STAINLESS 673 44N3
UHB STAINLESS 703 44N1
UHB STAINLESS 716 44M1
UHB STAINLESS 721 44M1
UHB STAINLESS 724 44N1
UHB STAINLESS 734L 44N1
UHB STAINLESS 734Nb 44N1
UHB STAINLESS 753 44N1
UHB STAINLESS AEB 44M2
UHB STATO 21 44G1
UHB STATO 23 44K1
UHB STATO 28 44K1
UHB UDD Co 12 44M1
UHB UDDCO 2 44K1
UHB UDDCO 3 44K1
UHB UDDCO 9 44K1
ULTIMATE 110 44F3
ULTIMO 4 44K2
ULTIMO 6 44K2
ULTRA CAPITAL 44K3
ULTRA CAPITAL + 1 44L
ULTRA CAPITAL + 2 44L
ULTRA CAPITAL 22 44K3
ULTRAMILD 44A1
UM Co 50 13A
UM Co 51 13A
UMHK 50.1
UMS 45 44A1
UMS 55 44A2
UMS 65 44A2
UMS 75 44A3
UMS 80 44A3
UN 30 14E
UNBREAKABLE 44A1
UNDER-WATER 50A
UNE 36–111 20B
UNE 36–118–73 20B
UNI 743 44A1
UNI 815 44A1
UNI 820 1A
UNI 2952 44A1
UNI 2953 44A1
UNI 3021 APR04 1A
UNI 3021 APT04 1A
UNI 3022 MA625 1C
UNI 3022 MAC33 1F
UNI 3022 MAF5 1N
UNI 3022 MAF10 1N
UNI 3022 MAK10 1N
UNI 3022 MAM10 1N
UNI 3022 MAN 25 1N
UNI 3022 MAR4 1N
UNI 3022 MAS6 1C
UNI 3022 MAT3 1N
UNI 3040 1L
UNI 3041 1L
UNI 3042 1L
UNI 3043 1L
UNI 3044 1L
UNI 3045 1L
UNI 3046 1L
UNI 3048 1C
UNI 3049 1C
UNI 3050 1L
UNI 3051 1K
UNI 3052 1L
UNI 3054 1K
UNI 3055 1K
UNI 3056 1J
UNI 3057 1J
UNI 3058 1J
UNI 3059 1J
UNI 3501 1E
UNI 3567/66 1A
UNI 3568 1B
UNI 3569 1E
UNI 3570 1E
UNI 3571 1E
UNI 3572 1C
UNI 3573 1D
UNI 3574 1D
UNI 3575 1D
UNI 3576 1D
UNI 3577 1F
UNI 3578 1F
UNI 3579 1F
UNI 3580 1H
UNI 3581 1F
UNI 3582 1H
UNI 3583 1F
UNI 3599 1K
UNI 3600 1K
UNI 3601 1L
UNI 3602 1L
UNI 3718 Gp Zn Al 4 55A
UNI 3718 Gp Zn Al 4 Cu 1 55A
UNI 3718 Gp Zn Al 4 Cu 3 55A
UNI 3735 1G
UNI 3736 1H
UNI 3737 1G
UNI 3738 1G
UNI 3950 AP0 1A
UNI 3950 AP3 1A
UNI 3950 AP5 1A

UNI 3950 AP7 1A
UNI 3950 AP8 1A
UNI 3992 X43 CS8 44M1
UNI 3992 X45 CNW 1909 44N1
UNI 3992 X50 CNW 1414 44N1
UNI 3992 X80 CSN20 44M2
UNI 4047 44M1
UNI 4047 X3 CN 1911 44N1
UNI 4047 X6 CN 1911 44N1
UNI 4047 X8 CA13 44M1
UNI 4047 X8 CN 1910 44N1
UNI 4047 X12 C17 44M1
UNI 4047 X12 CA12 44M1
UNI 4047 X12 CA23 44M1
UNI 4047 X15 C13 44M1
UNI 4047 X15 CF13 44M1
UNI 4047 X15 CN 1808 44N1
UNI 4047 X15 CNF 1808 44N1
UNI 4047 X20 C13 44M1
UNI 4047 X20 CN16 44M1
UNI 4047 X25 C 26 44M1
UNI 4047 X32 C13 44M1
UNI 4047 X40 C14 44M1
UNI 4365 44A1
UNI 4507 1A
UNI 4508 1A
UNI 4509 1A
UNI 4510 1D
UNI 4511 1D
UNI 4512 1D
UNI 4513 1C
UNI 4514 1C
UNI 5007 20B
UNI 5074 1C
UNI 5075 1L
UNI 5076 1C
UNI 5077 1C
UNI 5078 1C
UNI 5079 1C
UNI 5080 1J
UNI 5331 12 Ni Cr 3 44K1
UNI 5331 16 Cr Ni 4 44K1
UNI 5331 16 Ni Cr 11 44K1
UNI 5331 16 Ni Cr Mo 2 44K1
UNI 5331 16 Ni Cr Mo 12 44K1
UNI 5331 18 Ni Cr Mo 5 44K1
UNI 5331 18 Ni Cr Mo 7 44K1
UNI 5331 20 Cr Ni 4 44K1
UNI 5331 20 Ni Cr Mo 2 44K1
UNI 5331 C 10 44A1
UNI 5331 C 16 44A1
UNI 5334 Fe 33 44A1
UNI 5334 Fe 34 44A1
UNI 5334 Fe 37 44A1
UNI 5334 Fe 42 44A1
UNI 5334 Fe 50 44A1
UNI 5335 Fe 33 44A1
UNI 5335 Fe 34 44A1
UNI 5335 Fe 37 44A1
UNI 5335 Fe 42 44A1
UNI 5335 Fe 50 44A1
UNI 5452 1D
UNI 5764 1D
UNI 5784 1D
UNI 6170 1E
UNI 6250 1C
UNI 6251 1C
UNI 6252 1N
UNI 6253 1N
UNI 6359 1E
UNI 6360 1D
UNI 6361 1B
UNI 6362 1F
UNI 6785 14E
UNI 6786 14E
UNI 6896 14C
UNI 7256 1F
UNI 7257 1C
UNI 7789 1D
UNI 7790 1D
UNI 7791 1G
UNI 14544 20B
UNICUT 44K3
UNI-DIE SMOOTHCUT 44K3
UNIDOR 080 1G
UNIDOR 100 1G
UNIFONT 5 1M
UNIFONT 54 1M
UNIGOLD 17.1
UNILOY 1409TB 44M1
UNILOY 1420 WM 44M1
UNILOY 1430 MV 44M1
UNION 32 44A1
UNION 36 44A1
UNION 40 44A1
UNION 45 44A1
UNION HB36 44A1
UNION Q38 44A1
UNION Q42 44A1
UNI S115 44M1
UNISIL 46 44B1
UNISIL 51 44B1
UNISIL 56 44B1
UNISIL 62 44B1
UNITEMP 212 44N2
UNITEMP AF 1753 27C
UNITEMP L 605 13A
UNITEMP S 816 13A
UNITRODE 48.00 44A1
UNS
  unified numbering system
UNS A
  Aluminium and alloys, see A; e.g. AO3802 – UNS for 380.2 – see Section 1C
UNS C
  copper and alloy, see C; e.g. C10200 – UNS for SAE CA102 – see Section 14A
UNS D
  steel with mechanical properties specified
UNS E
  rare earth, etc., see E
UNS F
  cast iron, see F
UNS G
  AISI & SAE steels, see G
UNS H
  AISI H steel, see H
UNS J
  cast steels, see J
UNS K
  miscellaneous steels, see K
UNS L
  low melting point alloys see, L
UNS L05001–L05999 20C
UNS L5001–L5999 21.1
UNS M
  miscellaneous non ferrous metal, see M
UNS N
  nickel alloys, see N
UNS N07718 27C
UNS P
  precious metal alloy, see P
UNS R
  refractory metals and alloys, see R
UNS R3001–R3999 26A
UNS R30001-R39999 13A
UNS S
  stainless steels, see S
UNS T
  tool steels, see T
UNS W
  weld filler metals, see W
UNS Z
  zinc and alloy, see Z
URAL 1 1G
URAL III 1G
URAL IV 1G
URANIUM 1 53.1
URANIUM 2 53.1
URANIUM 3 53.1
URANIUM 4 53.1
URANIUM U 53.1
URANUS 50 44N1
URANUS 50 44N2
URANUS 50M 44N2
URANUS 65 44N1
URANUS A5 44N1
URANUS B6 44N1
URANUS S1 44N1
USS 12 Mo V 44M1
U St 12 44A1
U St 12/03 44A1
U St 12/04 44A1
U St 12/05 44A1
U St 13 44A1
U St 13/03 44A1
U St 13/04 44A1
U St 13/05 44A1
U St 14 44A1
U St 14/04 44A1
U St 14/05 44A1
U St 34/2 44A1
U St 37 44A1
U St 37/2 44A1
U St 37/202 44A1
U St 37/203 44A1
U St 37/204 44A1
U St 37/205 44A1
U St 42 44A1
U St 42/2 44A1
U St 42/202 44A1
U St 42/203 44A1
U St 42/204 44A1

U St 42/205 44A1
UST 34/1
see DIN 17100 44A1
UST 34/2
see DIN 17100 44A1
UST 37/1
see DIN 17100 44A1
UST 37/2
see DIN 17100 44A1
UST 42/1
see DIN 17100 44A1
UTA 8 DV 51C
UV Al Mg 1D
UZ 29E1 14B
V1M-VAR M 50 44K3
V 2A 44N1
V 2A EXTRA 44N1
V 2A FH 44N1
V 2A NORMAL 44N1
V 2A SUPRA 44N1
V 2A SUPRA NK 44N1
V 2A X EXTRA 44N1
V 2A XET 44N1
V2KG 44M1
V 3M 44M1
V 4A EXTRA 44N1
V 4A SUPRA 44N1
V 4A X EXTRA 44N1
V 5M 44M1
V 8A SUPRA 44N1
V 12A SUPRA 44N1
V 13F 44M1
V 13F Al 44M1
V 13F B 44M1
V 13F Supra 44M1
V 14A SUPRA 44N1
V 16A EXTRA 44N1
V 16A X EXTRA 44N1
V 17F 44M1
V 17F Extra 44M1
V 17F X Extra 44M1
V 18A SUPRA NK 44N1
V 24A EXTRA 44N1
V 25AF X EXTRA 44N1
V 36 13A
V 42 44J
V 44A EXTRA 44N1
V 44A SUPRA 44N1
V 44A SUPRA NK 44N1
V 44A X EXTRA 44N1
V 45 44J
V 50 44J
V 55 44J
V 57 44N1
V 60 44J
V 65 44J
V 90V 44J
V 95 1G
V 110 44K1
V 130 44K2
V 145 44K1
V 155 44K1
V 165 44K1
V 174 44N3
V 175 44K3
V 214 44K1
V254 44M1
V 274 M 44M1
V 310 44K2
V 320 44K2
V 330 44K1
V 340 44K1
V 350 44K1
V 500 44E2
V 732 44A2
V 762 44A2
V 810 44K1
V 820 44K1
V 930 44A2
V 935 44A2
V 945 44A2
V 953 44A3
V 960 44A3
V 1036 A/B BIX 1C
V 1036 A/B BVI 1C
V 1036 A/B BVII 1C
V 1036 A/B BXI 1C
V 1036 A/B BXIII 1L
V 1036 A/B BXV 1L
V 1036 A/B BXVI 1L
V1036 A/B BXIV 1J
VA 44K2
VACCUTHERM 5.32 44M1
VACCUTHERM 5.32H 44M1
VACCUTHERM 5.34 44M1
VACCUTHERM 5.36 44M1
VACCUTHERM 5.40H 44M1
VACODIL 36 20C
VACODIL 42 20C
VACODIL 46 20C
VACON 10 20C
VACON 12 20C
VACON 20 20C
VACON 70 20C
VACOVIT 485 20C
VACOVIT 501 20C
VACOVIT 511 27E
VACOVIT 540 27E
VACUMELTROL 41 27C
VACUMET HY Mu 800 27E
VAGS 44K2
V Al Cu 1F
V Al Cu Mn 1 1F
V Al Mn 1 1B
VAL 1A 44M1
VAL 1AL 44M1
VAL 1B 44M1
VAL 1MP 44M1
VAL 1P 44M1
VAL 1W 44M1
VAL 2A 44M1
VAL 2B 44M1
VAL 2BZ 44M1
VAL 2C 44M1
VAL 2D 44M1
VAL 2MV 44M1
VAL 2W 44M1
VAL 3 44M1
VAL 4 44M1
VAL 4S 44M1
VAL 5 44E2
VAL 5M 44K2
VAL 12 44M1
VALAND 1 44K1
VALIANT 44K3
VALMAG 984 44P
VALRAY 1 27B
VALVE BRONZE 14K
VALVE METAL 14K
VALVEX 518D 44E2
VALVIC 857 44M1
VALVIT 14K
VALVO 829 44N1
VAM 20 20B
VAM 27 20B
VAMPIRE 44E3
VANACHROM 44K1
VANADIUM 54.1
VANADIUM EXTRA 44J
VANADIUM STANDARD 44J
VANALUIM 1L
VAN CHIP 44K3
VANCO 5 44L
VAN CUT 1 44K3
VAN CUT 2 44K3
VANGUARD 44K3
VANITE 44K3
VANLEDA 44L
VAN-LOM 44K3
VANQUISH 44J
VANSTAR 44E3
VANTAGE 44K3
VAP 44K3
VAP 2V 44K3
VAPCW 44K3
VASCO 7152 44M2
VASCO M2 44K3
VASCO M7 44K3
VASCOMAX 250 44F3
VASCOMAX 300 CVM 44F3
VASCO SUPREME 44L
VASCO SUPREME A 44L
VAUCHERS ALLOY 55A
VC 12 44L
VCh 15 20B
VCh 20 20B
VCh 25 20B
VCh 30 20B
VCh 35 20B
VCh 38.17 20B
VCh 42.12 20B
VCh 45.0 20B
VCh 45.5 20B
VCh 50.1.5 20B
VCh 50.2 20B
VCh 50.7 20B
VCh 60.2 20B
VCh 70.2 20B
VCh 80.2 20B
VCh 100.2 20B
VCN 15 44K1
VCN 25w(g) 44K1
VCN 35h 44K1
VCN 35w(e) 44K1
VCN 45(d) 44K1
VCR 40–10 20B
VDC 44K2
VDM LC NICKEL 99.2 27A
VDM LC NICKEL 99.6 27A
VDM Mn Bz 2 14A
VDM NAVAL BRASS 14B

VDM NICKEL 99.2 27A
VDM NICKEL B 9604 27A
VDM NICKEL S 9604 27A
VDM TITAN 993 51A
VDM TITAN 994 51A
VDM TITAN 994 Pd 51A
VDM TITAN 995 51A
VDM TITAN 995 Pd 51A
VDM ZIRKONIUM 702 56.1
VDM ZIRKONIUM 704 56.1
VEDAS 44A4
VEGA 44K3
VELINVAR 13A
VELOS 44K3
VELOS 42 44L
VELOS UR 44K3
VELVARC TWO 44A1
VERILITE 14L
VERSALLOY 44N1
VERSATILE 44K2
VERSITRODE 46.58 44A1
VERTAMET 308L 44N1
VERTAMET 316L 44N1
VESUVIUS 44M2
VF 44L
VF 1 27B
VF 2 13A
VF 3 27C
VF 4 27C
VF 5 13A
VF 6 13A
VF 7 13A
VF 8 13A
VF 9 27F
VF 10 27F
VF 11 44N1
VG 44L
VH 217 44M1
VH 273 44M1
VH 274 44N1
VH 417 44M1
VH 620 44M1
VH 626 44M1
VH 650 44N1
VHRD 44A4
VIADUCT 15 44K2
VIALBRA 14B
VIBRAC V 45 44K2
VIBRAC V30 44K1
VIBRESIST 44K3
VIBRO 44K2
VICALLOY 13A
VICALLOY 1 13A
VICALLOY 11 13A
VICTOR 44K3
VICTOR BRONZE 14C
VICTOR METAL 14F
VIGILANT 44A2
VIKING 44K3
VIKRO 1 27B
VIKRO 10 44N1
VIKRO 11 27B
VIMETAL 13A
VIM-VAR 52100 44E2
VINCO 44K3
VINCO HOT WORK 44K2
VIPER 44E3
VIPER 666 44E2
VIRGO 2.0 44M1
VIRGO 3 44M1
VIRGO 3S 44M1
VIRGO 4 44M1
VIRGO 6 44M1
VIRGO 7 44M1
VIRGO 8 44M1
VIRGO 8S 44M1
VIRGO 10 44M1
VIS 11 44M1
VISCOUNT 20 44K2
VISCOUNT 44 44K2
VITAL 1L
VITAL 44M2
VITALLIUM 13A
VITAL X 44M2
VITRESIST 44N1
VITUS 181 44K1
VIVAL 1E
VIZOR 44A4
VK 44N3
VK 5M 44M1
VK 17F 44M1
VLM 44K3
VMA 2 27C
VMA 3 27C
VMA 4 13A
VMA 5 27B
VMA 6A 27B
VMA 6B 27B
VMA 6C 27B
VMA 7A 27B
VMA 7B 27B
VMA 8 27B
VMA 9 27B
VMA 10 27B
VMA 11 27B
VMA 12 27B
VMA 13 27B
VMA 14 27B
VMA 15 27B
VMA 16A 27B
VMA 16B 27B
VMA-B 44F3
VMAI-A 44F3
VMC 44K1
VMC 44K3
VMC(H) 44K1
VMS 585 44N1
VNCA 44K1
VODEX 44A1
VOLOMIT 52.1
V Pb Sn 5 Sb 28
see DIN range 21
V Pb Sn 5 Sb 28 21.1
V Pb Sn 30 Sb 6
see DIN range 21
V Pb Sn 30 Sb 6 21.1
VR 13 52.1
VR 14 52.1
VR 15 52.1
VR 52 52.1
VR 54 52.1
VR 65 51B
VR 71 52.1
VR 73 52.1
VR 75 52.1
VR 77 52.1
VR 87 52.1
VR 89 52.1
VS 1/2 44K3
VS 1C 44E2
VS 3 NC 44K1
VS 3N 44F1
VS 4 44J
VS 4 NC 44K1
VS 5N 44F1
VS 6/6 44K3
VS 10 44L
VS 12/12 44N1
VS 13 C 44M2
VS 14 44K3
VS 15M 44F1
VS 16/15 44K3
VS 17 44M1
VS 18/8 44N1
VS 18/8 Ti 44N1
VS 18/8/2 44N1
VS 22 44K3
VS 35 44L
VS 35 V 44K2
VS 35N 44F1
VS 50 44E2
VS 351 44K1
VSCM 44E3
VSCS 44E2
VSCV 44K2
VSCW 44K3
VSD 44K3
VSHD 44K1
VS INC 44K1
VSM 44K3
VSM 10691 20B
VSM 10693 20B
VSM 10803/Cu Ni 10 Fe Mn 14E
VSM 10803/Cu Ni 30 Fe Mn 14E
VS MANG 44D
VSMD 44A3
VSMM 44G1
VSMNC 44K3
VSNT 44K2
VSR 44K3
VSS 44A3
VS S61 44M1
VS S80 44M1
VSSM 44A3
VST2 P2 25P2
see GOST 380 44A1
VST2 P5VST 25P
see GOST 380 44A1
VST 2PS
see GOST 380 44A1
VST 2PS2
see GOST 380 44A1
VST 2SP
see GOST 380 44A1
VST 2SP2
see GOST 380 44A1
VST 3
see GOST 380 44A1
VST 3CPS3
see GOST 380 44A1

VST 3PS
see GOST 380 44A1
VST 3PS2
see GOST 380 44A1
VST 3PS3
see GOST 380 44A1
VST 3PS4
see GOST 380 44A1
VST 3PS5
see GOST 380 44A1
VST 3PS6
see GOST 380 44A1
VST 3SP
see GOST 380 44A1
VST 3SP2
see GOST 380 44A1
VST 3SP3
see GOST 380 44A1
VST 3SP4
see GOST 380 44A1
VST 3SP5
see GOST 380 44A1
VST 3SP6
see GOST 380 44A1
VST 4PS
see GOST 380 44A1
VST 4PS2
see GOST 380 44A1
VST 4PS3
see GOST 380 44A1
VST 4SP
see GOST 380 44A1
VST 4SP2
see GOST 380 44A1
VST 4SP3
see GOST 380 44A1
VST 5GPS
see GOST 380 44A2
VST 5GPS2
see GOST 380 44A2
VST 5PS
see GOST 380 44A2
VST 5PS2
see GOST 380 44A2
VST 5SP
see GOST 380 44A2
VST 5SP2
see GOST 380 44A2
VULCAN 50A
VULCAN METAL 14G
VWMC 44K1
VWMC 44K3
W 14C
W 1
see BS W 1 44A2
W1 44A2
W 1 44A4
W 2
see BS W 2 44A2
W2 44A2
W 2 44J
W3 44A2
W 3 1B
W 3 44E3
W 4 1D
W 4 44E3
W 5 1D
W 5 27F
W 5 44E3
W 6 1D
W 6 27F
W 6 44A1
W 6 44A3
W7 44K3
W 7 27F
W 7 44G1
W 8
see BS W 3 44A2
W8 44A2
W 9
see BS W 9 44A2
W9 44A2
W 9 1E
W 9 27F
W 9B 44K1
W 010 1D
W 10
see BS W 10 44N1
W 10 1E
W 011 1D
W 11
see BS W 11 44N1
W 11 1F
W 012 1D
W 12 1F
W 12 44M1
W 16 1G
W 17 1F
W 18 1F
W 18 52.1
W 18/8 44N1
W 19/9 44N1
W 20 1E
W 21 44A3
W 22/4 20B
W 24/8 20B
W 26 1G
W 30 1E
W 61 44A1
W 100 44K1
W 105 44K1
W 108
see SAE W 108 44A4
W 109
see SAE W 109 44A4
W 110
see SAE W 110 44A4
W 110 44A4
W 112
see SAE W 112 44A4
W 112 44A4
W 150 1F
W 152 1F
W 160 1G
W 209
see SAE W 209 44J
W 210
see SAE W 210 44J
W 260 1G
W 300 44K2
W 302 44K2
W 303 44K2
W 310
see SAE W 310 44J
W 320 44K1
W 321 44K2
W 326 44K2
W 367 52.1
W 408 52.1
W 410L 52.1
W 410M 52.1
W 500 44K2
W 501 44K2
W 545 44N2
W 705 44L
W 720 44L
W 725 44L
W06601
UNS for AWS
A5.20(E6XT.1) obsolete 44A1
W06604
UNS for AWS
A5.20(E6XT.4) obsolete 44A1
W06605
UNS for AWS
A5.20(E6XT.5) obsolete 44A1
W06606
UNS for AWS
A5.20(E6XT.6) obsolete 44A1
W06607
UNS for AWS
A5.20(E6XT.7) obsolete 44A1
W06611
UNS for E6xT11 obsolete 44A1
W06613
UNS for E6xT13 44A1
W06614
UNS for E7014 44A1
W07015
UNS for E7015 44A1
W07016
UNS for E7016 44A1
W07018
UNS for E7018 44A1
W07024
UNS for E7024 44A1
W07027
UNS for E7027 44A1
W07028
UNS for E7028 44A1
W07048
UNS for E7048 44A1
W07061
UNS for E7xT1 44A1
W07075
UNS for E7xT5 44A1
W07116
UNS for E7016 44A1
W07118
UNS for E7018 44A1
W07124
UNS for E7024 44A1
W07301
UNS for EG7xT1 44A1
W07302
UNS for EG7xT2 44A1
W07602
UNS for E7xT2 44A1
W07603
UNS for E7xT3 44A1

W07604
UNS for E7xT4 44A1
W07606
UNS for E7xT6 44A1
W07607
UNS for E7xT7 44A1
W07608
UNS for E7xT8 44A1
W07609
UNS for E7xT9 44A1
W07610
UNS for E7xT10 44A1
W07611
UNS for E7xT11 44A1
W07612
UNS for E7xT12 44A1
W07614
UNS for E7xT14 44A1
W10013
UNS for 10013 44G1
W10015
UNS for E10015-D2 44G1
W10016
UNS for E10016-D2 44G1
W10018
UNS for E10018-D2 44G1
W10235
UNS for E10015-D2 44G1
W17010
UNS for E7010-A1 44G1
W17011
UNS for E7011-A1 44G1
W17015
UNS for E7015-A1 44G1
W17016
UNS for E7016 A1 44G1
W17020
UNS for E7020 A1 44G1
W17027
UNS for E7027-A1 44G1
W17031
UNS for E8xT1A1 44G1
W17035
UNS for E7xT5A1 44G1
W17041
UNS for ECA1 44G1
W17042
UNS for ECA2 44G1
W17043
UNS for ECA3 44G1
W17044
UNS for ECA4 44G1
W17108
UNS for E7018-A1 44G1
W17141
UNS for ECA1N 44G1
W17142
UNS for ECA2N 44G1
W17143
UNS for ECA3N 44G1
W17144
UNS for ECA4N 44G1
W18016
UNS for E8016 D3 44G1
W18018
UNS for E8018 D3 44G1
W19015
UNS for E9015 D1 44G1
W19016
UNS for E9016 A 44G1
W19018
UNS for E9018 D1 44G1
W19131
UNS for E9xT1D1 44G1
W19235
UNS for E9xt5D2 44G1
W19331
UNS for E9xT1D3 44G1
W20018
UNS for E7018-W 44K1
W20118
UNS for E8018W 44K1
W20131
UNS for E8xT1W 44K1
W20140
UNS for EWT2 44K1
W20240
UNS for ECF2 44F1
W20241
UNS for ECF2N 44K1
W20440
UNS for ECF4 44F1
W20441
UNS for ECF4N 44F1
W21015
UNS for E8015 44F1
W21016
UNS for E8016-C3 44F1
W21018
UNS for E8018-C3 44F1
W21030
UNS for E80CNi1 44F1
W21031
UNS for E8xT1Ni1 44F1
W21033
UNS for EGxxT3 44A1
W21038
UNS for E7xT8Ni1 44A1
W21040
UNS for ECNi1 44E2
W21041
UNS for ECNi1N 44A1
W21118
UNS for E8018NM 44K1
W21135
UNS for E8xT5K1 44G1
W21140
UNS for ECF3 44K1
W21141
UNS for ECF3N 44K1
W21150
UNS for ECF1 44F1
W21151
UNS for ECF1N 44F1
W21215
UNS for E9015 44N1
W21216
UNS for E9016 44G1
W21218
UNS for E9018M 44N1
W21231
UNS for EXXT1K2 44K1
W21234
UNS for E7xT4K2 44K1
W21235
UNS for EXXT5K2 44K1
W21235
UNS for E71T8K2 44N1
W21240
UNS for ECM1 44N1
W21241
UNS for ECM1N 44N1
W21250
UNS for ECNi4 44N1
W21251
UNS for ECNi4N 44N1
W21318
UNS for E10018M 44K1
W21331
UNS for ExxxT1K3 44K1
W21335
UNS for ExxxT5K3 44K1
W21340
UNS for ECM2 44K1
W21341
UNS for ECM2N 44K1
W21418
UNS for E11018M 44K1
W21631
UNS for E12xT1K5 44K1
W21780
UNS for EB82 44K1
W22016
UNS for E8016C1 44F1
W22018
UNS for E8018C1 44F1
W22030
UNS for E80CNi2 44F1
W22031
UNS for ExxT1Ni2 44F1
W22035
UNS for E8xT5Ni2 44F1
W22040
UNS for ECNi2 44F1
W22041
UNS for ECNi2N 44F1
W22115
UNS for E7015C1L 44F1
W22116
UNS for E7016C1L 44F1
W22118
UNS for E12018M 44K1
W22118
UNS for E7018CIL 44F1
W22218
UNS for E12018M 44K1
W22231
UNS for E11xT1K4 44K1
W22235
UNS for E1xxT5K4 44K1
W22240
UNS for ECM3 44K1
W22241
UNS for ECM3N 44K1
W22340
UNS for EWT3 44K1
W22440
UNS for ECM4 44K1

W22540
UNS for ECF5 44K1
W22640
UNS for ECF6 44K1
W22641
UNS for ECF6N 44K1
W22718
UNS for 10018N1 44K1
W22815
UNS for 11015 44K1
W22816
UNS for 11016 44K1
W23016
UNS for E8016C2 44F1
W23018
UNS for E8018C2 44F1
W23030
UNS for E80CNi3 44F1
W23035
UNS for ExxT5Ni3 44F1
W23040
UNS for ECNi3N 44F1
W23115
UNS for E7015C2L 44F1
W23116
UNS for E7016C2L 44F1
W23118
UNS for E7018C2L 44F1
W23218
UNS for E12018M1 44K1
W23318
UNS for 14018M1 44K1
W26018
UNS for 14018HT 44K1
W30710
UNS for E307 44N1
W30731
UNS for E307T1 44N1
W30733
UNS for E307T3 44N1
W30737
UNS for E308LTx 44N1
W30740
UNS for ER307 44N1
W30810
UNS for E308 44N1
W30813
UNS for E308L 44N1
W30815
UNS for 308HC 44N1
W30816
UNS for 308HC 44N1
W30820
UNS for E308Mo 44N1
W30821
UNS for type 308 electrode 44N1
W30823
UNS for E308MoL 44N1
W30831
UNS for E308T1 44N1
W30832
UNS for E308MoTx 44N1
W30833
UNS for E308T3 44N1
W30835
UNS for E308LTx 44N1
W30837
UNS for E308LT3 44N1
W30838
UNS for E308LT2 44N1
W30840
UNS for EC308 44N1
W30841
UNS for EC308Si 44N1
W30843
UNS for EC308L 44N1
W30848
UNS for EC308LSi 44N1
W30850
UNS for EC308Mo 44N1
W30853
UNS for EC308MoL 44N1
W30910
UNS for EC309 44N1
W30913
UNS for E309L 44N1
W30917
UNS for E309Cb 44N1
W30920
UNS for E309Mo 44N1
W30923
UNS for E309MoL 44N1
W30931
UNS for E309T2 44N1
W30932
UNS for E309CbLT 44N1
W30933
UNS for E309T 44N1
W30934
UNS for E309CbLT 44N1
W30935
UNS for E309LT 44N1
W30937
UNS for E309LT 44N1
W30940
UNS for EC309 44N1
W30941
UNS for EC309Si 44N1
W30943
UNS for EC309L 44N1
W31010
UNS for E310 44N1
W31015
UNS for E310H 44N1
W31017
UNS for E210Cb 44N1
W31020
UNS for E310Mo 44N1
W31031
UNS for E310T 44N1
W31040
UNS for E310 44N1
W31310
UNS for E312 44N1
W31331
UNS for E312T 44N1
W31340
UNS for EC312 44N1
W31610
UNS for E316 44N1
W31613
UNS for E316L 44N1
W31631
UNS for E316T 44N1
W31633
UNS for E316T 44N1
W31635
UNS for E316LT 44N1
W31637
UNS for E316LT 44N1
W31637
UNS for E316LT 44N1
W31640
UNS for EC316H 44N1
W31641
UNS for EC316Si 44N1
w31643
UNS for EC316L 44N1
W31648
UNS for EC316LSi 44N1
W31710
UNS for E317 44N1
W31713
UNS for E317L 44N1
W31735
UNS for E317L 44N1
W31735
UNS for E317LT 44N1
W31737
UNS for E317LT 44N1
W31740
UNS for EC317 44N1
W31743
UNS for EC317L 44N1
W31810
UNS for E2209 44N1
W31831
UNS for E2209T 44N1
W31910
UNS for E318 44N1
W31940
UNS for EC318 44N1
W32140
UNS for EC321 44N1
W32210
UNS for E209 44N1
W32240
UNS for ER209 44N1
W32310
UNS for E219 44P
W32340
UNS for EC219 44N3
W32410
UNS for EC240 44N3
W32440
UNS for EC240 44N3
W32540
UNS for EC218 44N1
W34710
UNS for E347 44N1
W34712
UNS for 347HC 44N1
W34730
UNS for EC347 44N1
W34731
UNS for E347T1 44N1
W34733
UNS for E347T3 44N1

W34748
UNS for EC347Si 44N1
W34910
UNS for E349 44N1
W35010
UNS for AMS 5775 44N3
W35510
UNS for AMS 5781 44N3
W36840
UNS for EC16-8.2 44N1
W37410
UNS for E630 44N3
W37440
UNS for EC630 44N3
W40931
UNS for E409T 44N1
W41010
UNS for E410 44M1
W41016
UNS for E410NiMo 44M1
W41031
UNS for E410T 44M1
W41036
UNS for E410NiMoT 44M1
W41038
UNS for E410NiTiT 44M1
W41040
UNS for EC410 44M1
W41046
UNS for EC410NiMo 44M1
W42040
UNS for EC420 44M1
W43010
UNS for E430 44M1
W43031
UNS for E430T 44M1
W43040
UNS for EC430 44M1
W50210
UNS for E502 44K1
W50216
UNS for E8016B2 44K1
W50231
UNS for E502T 44K1
W50301
UNS for E7Cr 44K1
W50410
UNS for E505 44K1
W50430
UNS for E6xT5B8L 44K1
W50431
UNS for E505T 44K1
W50440
UNS for EC505 44K1
W51016
UNS for E8016B1 44K1
W51018
UNS for E8018B1 44K1
W51316
UNS for E8016B5 44K1
W52015
UNS for E8015B2 44K1
W52018
UNS for E8018B2 44K2
W52030
UNS for E80CB2 44K1
W52031
UNS for E8xT1B2 44K1
W52035
UNS for E8xT5B2 44K1
W52040
UNS for ECB2 44K1
W52041
UNS for ECB2N 44K1
W52115
UNS for E8015B2L 44K1
W52118
UNS for E8018B2L 44K1
W52130
UNS for E80CB2L 44K1
W52131
UNS for E8xT1B2L 44K1
W52135
UNS for E8xT5B2L 44K1
W52231
UNS for E8xT1B2H 44K1
W52240
UNS for ECB2H 44K1
W52241
UNS for ECB2HN 44K1
W53015
UNS for E9015B3 44K1
W53016
UNS for 9016B3 44K1
W53018
UNS for E9018B3 44K1
W53030
UNS for E90CB3 44K1
W53031
UNS for ExxxT1B3 44K1
W53035
UNS for E9xT5B3 44K1
W53040
UNS for ECB3 44K1
W53041
UNS for ECB3N 44K1
W53115
UNS for E9015B3L 44K1
W 53116 44K1
W53130
UNS for E90CB3L 44K1
W53131
UNS for E9xT1B3L 44K1
W53231
UNS for E9xT1B3H 44K1
W 53315 44K1
W 53316 44K1
W53340
UNS for ECB4 44K1
W53341
UNS for ECB4N 44K1
W53415
UNS for E8015-B4L 44K1
W 53416 44K1
W60189
UNS for ECu 14A
W60518
UNS for ECuSnA 14K
W60521
UNS for ECuAlA2 14G
W60614
UNS for ECuAlA2 14G
W60619
UNS for ECuAlB 14G
W60625
UNS for ECuAlC 14G
W60632
UNS for ECuNiAl 14G
W60633
UNS for ECuMnNiAl 14G
W60656
UNS for ECuSi 14M
W60715
UNS for ECuNi 14E
W61625
UNS for ECuAlD 14G
W62625
UNS for ECuAlE 14G
W73001
UNS for ECoCrC 13A
W73006
UNS for ECoCrA 13A
W73012
UNS for ECoCrB 13A
W73021
UNS for ECoCrE 13A
W73155
UNS for AMS 5795 44L
W73605
UNS for AMS 5797 13A
W74001
UNS for EFe1 44E1
W74002
UNS for EFe2 44E1
W74003
UNS for EFe3 44E3
W74004
UNS for EFe4 44E3
W74011
UNS for EFeCrAl 44M2
W74012
UNS for EFeCrA2 44M2
W74013
UNS for EFeCrA3 44M2
W74014
UNS for EFeCrA4 44M2
W74015
UNS for EFeCrA5 44M2
W74111
UNS for EFeCrB1 44M2
W74112
UNS for EFeCrB2 44M2
W74113
UNS for EFeCrB3 44M2
W74212
UNS for EFeCr2 44M2
W74213
UNS for EFeCrC3 44M2
W74214
UNS for EFeCrC4 44M2
W74510
UNS for EFeCrA1 obsolete 44M2
W77310
UNS for EFe5C obsolete 44K2
W77510
UNS for EFe6 44K3

W77530
UNS for ERFeC6 44K3
W77610
UNS for EFe7 44F3
W77710
UNS for EFe5A obsolete 44K3
W79510
UNS for EFeMnE 44P
W79610
UNS for EFeMnF 44P
W80001
UNS for ERNiMo1 27F
W80004
UNS for ENiMo3 27B
W80276
UNS for ENiCrMo4 27B
W80665
UNS for ENiMo7 27B
W82001
UNS for ENiCl 27F
W82002
UNS for ENiFeCl 27E
W82003
UNS for ENiClA 27F
W82004
UNS for ENiFeClA 27E
W82006
UNS for ENiFeMnCl 27E
W82032
UNS for ENiFeT3Cl 27E
W82141
UNS for ENil 44F1
W83002
UNS for ENiCrMoB 27E
W84001
UNS for ENiCuA 27D
W84002
UNS for ENiCuB 27D
W84190
UNS for ENiCu7 27D
W86002
UNS for ENiCrMo2 27C
W86007
UNS for ENiCrMo1 27B
W86040
UNS for ENiCrMo12 27B
W86132
UNS for ENiCrFe1 27B
W86133
UNS for ENiCrFe2 27B
W86134
UNS for ENiCrFe4 27B
W86182
UNS for ENiCrFe3 27B
W86333
UNS for RA333 27F
W86455
UNS for ENiCrMo7 27B
W86620
UNS for ENiCrMo6 27B
W86985
UNS for ENiCrMo9 27B
W88021
UNS for E230 44K1
W88022
UNS for E320LR 27B
W88331
UNS for E330 44N1
W88334
UNS for RA330-04 44N1
W88335
UNS for E330H 44N1
W88338
UNS for RA330-80 44N1
W89604
UNS for ENiCrA 27A
W89605
UNS for ENiCrB 27B
W89606
UNS for ENiCrC 27C
WA 20B
WAGNER'S ALLOY 50A
WAI-MET 50 13A
W Al 8 1A
WALLEX 44M2
WALLEX 1 13A
WALLEX 1NE 13A
WALLEX 4 13A
WALLEX 6 13A
WALLEX 6NE 13A
WALLEX 7 13A
WALLEX 7NE 13A
WALLEX 12 13A
WALLEX 12NE 13A
WALLEX 42 13A
WALLEX 50 13A
WALLEX 55 52.1
WALLEX 505 52.1
WALLEX F 13A
W ALLOY 30.1
WALMANG 3 44P
WAMATO METAL 50A
WARNE'S METAL 50A
WAS 14B
WASPALOY 27C
WASPALOY MOD 27C
WATERDIE EXTRA 44E3
WATERDIE STANDARD 44E3
WAUKESHA 88 27B
WAW 44K2
WB 20B
w Ba 21 4.1
WBC 20B
WC 10 44K1
WC 18 44K1
WC 23 44K1
WC 28 44K1
WC 33 44K1
WC 38 44K2
WC 43 44K2
WC 48 44K2
WC 53 44K2
WC 58 44K2
WC 63 44K2
WC 181 44K1
WC 182 44K1
W Ca 17 10.1
WCC 2.1
WCD 44K1
WCD 44M2
WCD 2 44K1
W Co 23 13.1
WCPS 44K2
WCR 100 44M1
WCR 100 44M2
WE 44L
W E WATSON BRAND 21.1
WE 132 27B
WEARCLAD 40 44M2
WEARCLAD 43 44M2
WEARCLAD 45 44M2
WEAREX 44K3
WEARTRODE 84.58 44M1
WELCH'S ALLOY 50A
WELCON 2H 44A1
WELCON 2H.S 44K1
WELCON 2H.U 44K1
WELCON 50 44A1
WELDANKA A 44N1
WELDANKA B 44N1
WELDANKA CB 44N1
WEL-MONIX 44K1
WELSBACH 11.1
WEL-TEN 50 44A1
WEL-TEN 55 44A1
WEL-TEN 60 44K1
WEL-TEN 60H 44K1
WEL-TEN 62 44K1
WEL-TEN 80 44K1
WEL-TEN 100W 44K1
WESSELS ALLOY 14F
WESTFALIA RF 5 44M1
WF 11 13A
WF 31 13A
W Fe 18 20A
WG 44K3
WH 20B
WH 52.1
WH 602 44M1
WH 608 44M1
WHC 44A2
WHEEL BRASS 14B
WHITE ARSENIC 3.1
WHITE BENEDICT METAL 14F
WHITE GOLD SOLDER 17.1
WHITE PINE LAKE COPPER 14A
WI 52 13A
WILMIL 1C
WILMIL M 1K
WINNS BRONZE 14F
WIPTAM 13A
W IRONIT 720 44M1
WK 16 33.1
WKE 44K3
WKE 44L
WKE 4 44L
WKE 44 44L
WKE 45 44L
WKE 46 44L
WKE 48 44L
WKE Extra 44L
WL 1 44K2
WL 2 44K1
W Li 16 22.1
W Li 20 22.1
WLM 77 44K1
WLM 1015 44E3
WLM 1020 44L

WLM 1030 44K1
WLM 1175 44K3
WLM 1214 44K2
WLM 1261 44K3
WLM 1530 44K1
WLM 1841 44K3
WLM 2014 44M2
WLM 2014 SPECIAL 44M2
WLM 2241 44K3
WLM 3033 44K1
WLM 3042 44K1
WLM 3141 44M2
WLM 4053 44K1
WLM 4060 44E1
WLM 5018 44L
WLM 9015 44A4
WLM BEST 44A3
WLM SPECIAL 44A4
WLM WARRANTED 44A2
WM 14F
WM 44K3
WM 52.1
WM 11 14F
WM EXTRA 44K2
WMW 14F
w Na 11 43.1
w Na 16 43.1
W Nb 18 28.1
WO 90 1E
WOLFRAM 44K3
WOLFRAM 52.1
WOLFRAM BRASS 14F
WOLFRAM BRONZE 14J
WOLFRAM COBALT 44L
WOLFRAMIUM 1N
WOODS 6.1
WORK-HARD 11 Cr 9 Mn 44P
WORK-HARD 12 Mn Cr Ni 44P
WORK-HARD 13 Mn 44P
WORK-HARD 14 Cr 14 Mn 44P
WORTLE PLATE 44A4
WP 300 44K3
WP 500 44K3
WPS 44M2
WS 14C
WS 44K3
WS 52.1
WS 44 44K3
WSH 20B
WSH 1 20B
WSH 2 20B
W TAP 44K3
w Th Tl 18 48.1
WTS 0.75% 44A3
WTS 1% 44A4
WU 44L
WULCRO 44K3
WU St 12 44A1
WU St 12/03 44A1
WU St 12/04 44A1
WU St 12/05 44A1
WU St 37/2 44A1
WU St 37/202 44A1
WU St 37/203 44A1
WU St 37/204 44A1
WU St 37/205 44A1
WVC 44L
WV Co 44L
WW T 783a 1A
WW T 785 1F
WW T 786a 1F
WW T 788A 1B
WW T 791/1 14B
WW T 791/2 14B
WW T 799a 14A
WW T 799a/N 14A
WW T 825 23B
WW T 825 B 23B
WWT 787 1D
WWT 789 1E
WWT 790 1E
WYMDALOY 14E
WZ 44K2
W Zn 3 55A
X 2 CN18/10 44N1
X 2 Cr Ni 18/9
see DIN 17440 44N1
X 2 Cr Ni 18/9 44N1
X 2 Cr Ni 18/10 44N1
X 2 Cr Ni 18/11 44N1
X 2 Cr Ni 18/19 44N1
X 2 Cr Ni 18-10
see DIN 17440 44N1
X 2 Cr Ni Mo 17/12 44N1
X 2 Cr Ni Mo 18/10 44N1
X 2 Cr Ni Mo 18/12 44N1
X 2 Cr Ni Mo 18/16 44N1
X 2 Cr Ni Mo 18-10
see DIN 17440 X 44N1
X 2 Cr Ni Mo 18-12
see DIN 17440 X 44N1
X 2 Cr Ni Mo N 18/12 44N1
X 2 Cr Ni Mo N 18/13 44N1
X 2 Cr Ni Mo N 18-9
see DIN 17440 44N1
X 2 Cr Ni Mo N 18-10
see DIN 17440 44N1
X 2 Cr Ni Mo Ni 18/12 44N1
X 2 Cr Ni N 18/10 44N1
X2 Cr Ni Mo N225 44N2
X 3 CN 19 11 44N1
X 3 Cr Mo Ti 17 44M1
X 3 Ni Co Mo 18-8-5 44F3
X 5 Cr Ni 18/9
see DIN 17440 44N1
X 5 Cr Ni 18/9 44N1
X 5 Cr Ni 18/10 44N1
X 5 Cr Ni 19/9 44N1
X 5 Cr Ni 19/11 44N1
X 5 Cr Ni 19-11
see DIN 17440 44N1
X 5 Cr Ni Mo 17/
see DIN 17440 44N1
X 5 Cr Ni Mo 17/12 44N1
X 5 Cr Ni Mo 17/13 44N1
X 5 Cr Ni Mo 18/10
see DIN 17440 44N1
X 5 Cr Ni Mo 18/10 44N1
X 5 Cr Ni Mo 18/12 44N1
X 5 Cr Ni Mo 18/13 44N1
X 5 Cr Ni Mo 19/10 44N1
X 5 Cr Ni Mo Cu Nb 18/18 44N1
X 5 Cr Ni Mo Cu Nb 20/18 44N1
X 5 Cr Ni Mo Ti
see DIN 17440 44N1
X 5 Cr Ni Mo Ti 25/25 44N1
X 6 CN 1911 44N1
X 6 Cr 13 44M1
X 6 Cr Mo 17
see DIN 17440 44M1
X 6 Cr Mo 17 44M1
X 6 Cr Ni 24/18 44N1
X 6 Cr Ni 25/20 44N1
X 6 Cr Ni Mo Nb 17/12 44N1
X 6 Cr Ni Mo Ti 17/12 44N1
X 6 Cr Ni Nb 18/11 44N1
X 6 Cr Ni Ti 18/11 44N1
X 7 AL 44E1
X 7 Cr 13
see DIN 17440 44M1
X 7 Cr 14 44M1
X 7 Cr Al 13
see DIN 17440 44M1
X 7 Cr Al 13 44M1
X 7 Cr Mo 6/1 44K1
X 7 Cr Mo 10/1 44K1
X 7 Cr Ni 23/14 44N1
X 7 Cr Ni Al 17/7 44N3
X 8 CN 1910 44N1
X 8 CN 2520 44N1
X 8 CN Nb 1811 44N1
X 8 CND 1712 44N1
X 8 CNT 1808 44N1
X 8 CNT 1810 44N1
X 8 Cr 9 44E1
X 8 Cr 13 44M1
X 8 Cr 14 44M1
X 8 Cr 15 44M1
X 8 Cr 17
see DIN 17440 44M1
X 8 Cr 17 44M1
X 8 Cr 18 44M1
X 8 Cr 28 44M1
X 8 Cr 30 44M1
X 8 Cr Mn Ni 18/9 44P
X 8 Cr Mo 302 44M1
X 8 Cr Mo Ti 18 44M1
X 8 Cr Nb 17
see DIN 17440 44M1
X 8 Cr Nb 17 44M1
X 8 Cr Ni 12/12 44N1
X 8 Cr Ni Mo B Nb 16/16K 44N1
X 8 Cr Ni Mo Cu Nb 20/18 44N1
X 8 Cr Ni Mo Nb
see DIN 17440 44M1
X 8 Cr Ni Mo Nb 17/13 44N1
X 8 Cr Ni Mo Nb 19/10 44N1
X 8 Cr Ni Mo Nb 19/12 44N1
X 8 Cr Ni Mo Nb 25/25 44N1
X 8 Cr Ni Mo V Nb
see DIN 17440 44N1
X 8 Cr Ni Mo V Nb 16/13 44N1
X 8 Cr Ni Nb 19/9 44N1
X 8 Cr Ti 17
see DIN 17440 44M1
X 8 Cr Ti 17 44M1
X 8 Cr Ti 18 44M1
X 8 Ni Cr Mo Cu Nb 44N1
X 10 C18 44M1
X 10 CND 1808 44N1

X 10 CNDT 1808 44N1
X 10 Cr 13
see DIN 17440 44M1
X 10 Cr 13 44M1
X 10 Cr 25 44M1
X 10 Cr Al 7 44E1
X 10 Cr Al 13 44M1
X 10 Cr Al 18 44M1
X 10 Cr Al 24 44M1
X 10 Cr Ni Mo Nb
see DIN 17440 44N1
X 10 Cr Ni Mo Nb 18/10 44N1
X 10 Cr Ni Mo Nb 18/12 44N1
X 10 Cr Ni Mo Ti
see DIN 17440 44N1
X 10 Cr Ni Mo Ti 18/10 44N1
X 10 Cr Ni Mo Ti 18/10/1 44N1
X 10 Cr Ni Mo Ti 18/12 44N1
X 10 Cr Ni Nb 18/9
see DIN 17440 44N1
X 10 Cr Ni Nb 18/9 44N1
X 10 Cr Ni Nb 18/10 44N1
X 10 Cr Ni Nb 1809 44N1
X 10 Cr Ni S 1809 44N1
X 10 Cr Ni Ti 18/9
see DIN 17440 44N1
X 10 Cr Ni Ti 18/9 44N1
X 10 Cr Ni Ti 18/10 44N1
X 10 Cr Si 6 44E1
X 10 Cr Si 13 44M1
X 10 Cr Si 18 44M1
X 10 Cr Si 29 44M1
X 10 Ni Cr 27A
X 12 C 17 44M1
X 12 Cr Mo S 17
see DIN 17440 44M1
X 12 Cr Mo S 17 44M1
X 12 Cr Ni 17/7 44N3
X 12 Cr Ni 18/8 44N1
X 12 Cr Ni 18/9 44N1
X 12 Cr Ni 22/12 44N1
X 12 Cr Ni 25/4 44N1
X 12 Cr Ni 25/20 44N1
X 12 Cr Ni 25/21 44N1
X 12 Cr Ni S 18/8
see DIN 17440 44N1
X 12 Cr Ni S 18/8 44N1
X 12 Cr Ni Si 36/16
see DIN 17440 44N1
X 12 Cr Ni Si 36/16 44N1
X 12 Cr Ni Ti 18/9 44N1
X 12 Mn Cr 18/10 44P
X 12 Mn Cr 18/11 44P
X 12 Ni Cr 36/18 44N1
X 12 Ni Cr Si 36/16
see DIN 17440 X 44N1
X 12 Ni Cr Si 36/16 44N1
X 13 RG 44M1
X 15 CN 1707 44N3
X 15 CN 1808 44N1
X 15 CN 2412 44N1
X 15 CN 2420 44N1
X 15 Cr 13
see DIN 17440 44M1
X 15 Cr 13 44M1
X 15 Cr Mo 12/1 44M1
X 15 Cr Mo 13 44M1
X 15 Cr Ni 25/20 44N1
X 15 Cr Ni Mn 18/
see DIN 17440 44N1
X 15 Cr Ni Mn 18/8 44P
X 15 Cr Ni Si 20/12 44N1
X 15 Cr Ni Si 25/20 44N1
X15 Cr Co Mo 10.10.5 44L
X 17 AL 44M1
X 17 C 44M1
X 17 Z 44M1
X 19 Ni Cr Mo4 44K1
X 20 C 28 44M1
X 20 CN 245 44N1
X 20 CN 2412 44N1
X 20 Cr 13
see DIN 17440 44M1
X 20 Cr 13 44M1
X 20 Cr Mo 13 44M1
X 20 Cr Mo W V 12/1 44M1
X 20 Cr Ni Mn 12/9 44P
X 20 Cr Ni Si 25/4 44M1
X 20 T2 44N1
X 22 Cr Mo V 12/1 44M1
X 22 Cr Mo W V 12/1 44M1
X 22 Cr Ni 17
see DIN 17440 44M1
X 22 Cr Ni 17 44M1
X 25 C26 44M1
X 25 CN 2520 44N1
X 30 W Cr V 53 44K1
X30W Cr V 93 44K1
X 32 Cr Mo V 33 44K1
X 35 Cr 14 44M1
X 35 Cr Mo 17 44M1
X 38 Cr Mo V 51 44K2
X 40 13A
X 40 Cr 13
see DIN 17440 44M1
X 40 Cr 13 44M1
X 40 Cr Mo V 51 44K2
X 40 Cr Mo V 53 44K2
X 40 Mn Cr 18 44P
X 40 Mn Cr 22 44P
X 40 Mn Cr N 18 44P
X 42 Cr 13 44M1
X 45 Cr Mo V 15 44M1
X 45 Cr Ni Mo 23/5 44M1
X 45 Cr Ni W 18/9 44N1
X 45 Cr Si 9 44E3
X 45 Cr Si 9/3 44E3
X 45 Ni Cr Mo 4 44K2
X 45 Si Cr 4 44E3
X 50 13A
X 50 Cr Mo 51 44K2
X 50 Cr Mo W 911 44K2
X 55 Mn Ni Cr 14 44P
X 63 13A
X 80 Cr Ni Si 20 44M1
X 90 Cr Mo V 18 44M2
X 90 Mn 18 44D
X100 Cr Mo V 51 44K3
X 110 Mn 14 44D
X 120 Mn 12 44D
X 155 Cr V Mo 12.1 44M2
X 165 Cr Mo V 12 44M2
X 210 Cr 12 44M2
X 210 Cr W 12 44M2
X 290 Cr 12 44M2
X 750 27C
X 3140 44K2
X 4620 44K1
X-782 27F
X 8081 1N
X-ALLOY 1L
XANTAL 14G
XAR 15 44K1
XAR 30 44K1
X B 44M1
XBP 44A3
XC 18 S 44A1
XC 65 44A3
XCR 44N1
XDH 44K2
XDL 44K2
XERG 0.45 44L
XEV-A 44P
XEV-B 44P
XEV-D 44P
XEV-E 44P
XEV-F 44P
XEV-G 44P
XEV H 27F
XEV I 27F
XEV-J 51B
XG Ni Cr Ti 26.15 44N1
XL CHISEL 44K2
XL CUT 44A1
XL CUT Pb 44A1
XL CUT SPb 44A1
XM 8 44M1
XM9 44N3
XM 10 44P
XM 11 44P
XM12 44N3
XM13 44N3
XM 15 44N1
XM16 44N3
XM 17 44N2
XM 17 44P
XM 18 44N2
XM 19 44N1
XM25 44N3
XM 27 44M1
XM 29 44N2
XM 29 44P
XM 30 44M1
XM 33 44M1
XP 180 44K2
XP 550 44K3
XP 1150
see METCO XP
XSH 13A
XX 14C
XX 9CR 44E3
Y 1 105V 44J
Y 10 N66 44K1
Y 10A
see ASTM range 21
Y 12 44A1
Y 35 44A2
Y 35 NCD16 44K1
Y 35 NCD16 (3) 44K1
Y 42CD4 44K2
Y 45 44A2

Y 45S7 44A2
Y 45SCD6 44K2
Y 50CV4 44K2
Y 55 44A2
Y 55S7 44A2
Y 60SC7 44E2
Y 65 44A3
Y 65V 44A3
Y 75 44A3
Y 75V 44A3
Y 75V 44E2
Y 90 44A4
Y 90V 44A4
Y 90V 44E3
Y 100 C6 44E3
Y 100C6 44K3
Y 105 44A4
Y 105V 44E3
Y 105V 44J
Y 120 44A4
Y 120V 44E3
Y 135 44A4
Y 135C 44E3
Y 135V 44E3
Y 2120 C 44E3
YALE BRONZE 14B
Y-ALLOY 1L
Y ALLOY 1F
YAW-TEN 50 44A1
YB TEN 44A1
YC 135A 50A
YCM 2B 44L
YCW STEM 44K2
YD1 44K1
YDC 44K1
YELLOW BRASS 14B
YELLOW INGOT METAL 14B
YELLOW LABEL 44A4
YEO 36 44J
YEO 40 44A1
YMDC 44L
YND 33 44A1
YND 37 44F1
YND 58 44K1
YOLOY 44K1
YOLOY 50W 44A1
YOLOY A 242 44J
YOLOY EHS 44K1
YOLOY HS 44K1
YOLOY M, G and A 44A1
YOLOY S 44F1
YO-MAN B 44A1
YORCALBRO 14B
YORCASTON 14K
YORCORON 14E
YORCORON 3/3 14E
YR 14B
YS 14B
YT 155A
see ASTM range 21
YWA 44K1
Z - ALLOY 1N
Z 1 56.1
Z 1 NCDU25/20 44N1
Z 2 56.1
Z 2 CND17/12 44N1
Z 2 CND17/13 44N1
Z 2 CND18/10 44N1
Z 2 CND19/15 44N1
Z 2CN 18-10
see NF A36 209 44N1
Z 2CND 17-12
see NF A35 573 44N1
Z 2CND 17-13
see NF A35 573 44N1
Z 2CND 19-15
see NF A35 573 44N1
Z 3 1L
Z 3 56.1
Z 3 CN18/08 44N1
Z 3 CN18/10 44N1
Z 3 CND18/12 44N1
Z 5 55A
Z 5 56.1
Z 5 CN18/08 44N1
Z 5 CND18/10 44N1
Z 5 CNDU21/08 44N1
Z 5CNUD 21-08
see NF A35 573 44N1
Z 6 55A
Z 6 C13 44M1
Z 6 C13 Al 44M1
Z 6 CA13 44M1
Z 6 CN18/09 44N1
Z 6 CN18/10 44N1
Z 6 CND17/11 44N1
Z 6 CND17/12 44N1
Z 6 CND18/12 44N1
Z 6 CNDT17/12 44N1
Z 6 CNT18/10 44N1
Z 6 CNT25/15 44N1
Z 6C 13
see NF A35 573 44M1
Z 6CA 13
see NF A35 573 44M1
Z 6CN 18-09
see NF A36 209 44N1
Z 6CND 17-11
see NF A35 573 44N1
Z 6CND13/04M 44M1
Z 6CNDNb 17-12
see NF A35 573 44N1
Z 6CNDNb17/12 44M1
Z 6CNNb 18-11
see NF A35 573 44N1
Z 6CNT 18-11
see NF A36 209 44N1
Z 7 CN18/10 44N1
Z 7 INTERMEDIARE 55A
Z 8 C17 44M1
Z 8 CND17/12 44N1
Z 8 CNDNb18/12 44N1
Z 8 CNDT18/12 44N1
Z 8 FIN 55A
Z 8C 17
see NF A35 573 44M1
Z 8C17 44M1
Z 8CD 17-01
see NF A35 573 44M1
Z 8CD17/01 44M1
Z 8CDV5 44K1
Z 8CN 13-13
see NF A35 573 44N1
Z 8CN 18-12
see NF A35 573 44N1
Z 8CNDT 17-12
see NF A35 573 44N1
Z 9 EXTRA-FIN 55A
Z 10 14J
Z 10 C13 44M1
Z 10 C17 44M1
Z 10 C18 44M1
Z 10 CMN 18/7 44P
Z 10 CMN 19/9 44P
Z 10 CN11/12 44N1
Z 10 CN12/12 44N1
Z 10 CNDNb18/10 44N1
Z 10 CNDT18/12 44N1
Z 10 CNNb18/10 44N1
Z 10 CNS25/13 44N1
Z 10 CNS25/20 44N1
Z 10 CNT18/08 44N1
Z 10 CNT18/10 44N1
Z 10 NNb18/10 44N1
Z 10CN 18-09
see NF A35 572 44N1
Z 10CNT 18-11
see NF A35 573 44N1
Z 12 C13 44M1
Z 12 C18 44M1
Z 12 CMN 18/7 44P
Z 12 CN 44N1
Z 12 CN18/08 44N1
Z 12 CN18/10 44N1
Z 12 CNS25/20 44N1
Z 12 NCS37/18 44N1
Z 12C 13
see NF A35 573 44M1
Z 12CN 17-08
see NF A35 572 44N1
Z 13.0WDCV 44K3
Z 15 C27 44M1
Z 15 CN16/2 44M1
Z 15 CN25/13 44N1
Z 15 CNS 25-20 27C
Z 15 CNS20/10 44N1
Z 15 CNS20/12 44N1
Z 15 CNS25/13 44N1
Z 15 CNS25/20 44N1
Z 20 C13 44M1
Z 20 C25 44M1
Z 20 CN25/05 44M1
Z 20 CNS24/13 44N1
Z 20 CNS24/20 44N1
Z 25 CNWS20/09 44N1
Z 30 C13 44M1
Z 30WCV9 44K1
Z 35 CNS14/14 44N1
Z 38CDV5 44K2
Z 38CDWV5 44K2
Z 40CSD10 44K2
Z 40WCV5 44K2
Z 45CS9 44E2
Z 52 23B
Z52CMN21/09 44P
Z 65WDCV605 44K2
Z 70 C14 44M1
Z 70 C15 44M1
Z 80 CSN20/02 44M1
Z 80WCDV 12/0402/02 44K3

Z 80WCV 18/04/01 44K3
Z 80WKCV 18-5/04-01 44L
Z 80WKCV 18-10/04-02 44L
Z 85DCWV 44K3
Z 85WCV 18/04/2 44K3
Z 85WDCV 44K3
Z 85WDKCV 44L
Z 100CDV 5 44K3
Z 110DKCWV 44L
Z 115 20B
Z 120 20B
Z 125 20B
Z 130 20B
Z 130WCV 12/04/04 44K3
Z 135 20B
Z 140 20B
Z 150WDKCV 44L
Z 150WKCV 12-05/05-04 44L
Z 160CDV12 44M2
Z 165WKCV 44L
Z 175KWDCV 44L
Z 180CKD1203 44M2
Z 200C12 44M2
Z 200CD12 44M2
Z 230CDV12-04 44M2
Z 908 44A1
Z 908 44A2
Z 912 44A2
Z 952 44A1
Z 952 44A2
Z 986 44A2
Z19001
  UNS for ASTM B6 Prime western
Z 35630 55A
Z 35635 55A
Z 35840 55A
ZA 23B
ZA 4 55A
ZA 4 C1 55A
ZA 4 C3 55A
ZA 4 G 55A
ZA 4 U1G 55A
ZA 4 U3G 55A
ZA 8 55A
ZA 12 55A
ZA 27 55A
ZAMA 55A
ZAMAC 55A
ZAMAK 55A
ZAM METAL 55A
ZC 60A
  see ASTM range 1M
ZC 63 23B
ZC 81A
  see ASTM range 1M
ZC 81B
  see ASTM range 1M
ZCM 630 23B
ZE 10A
  see ASTM range 23B
ZE 41 A 23B
ZE 41A
  see ASTM range 23B
ZE 63 A 23B
ZELCO 55A
ZERALOY 20C
ZERON 100 44N3
ZG 32A
  see ASTM range 1M
ZG 42A
  see ASTM range 1M
ZG 61A
  see ASTM range 1M
ZG 61B
  see ASTM range 1M
ZG 62A 1G
ZG 71B 1M
ZG 81A 1M
ZH 62 23B
ZH 62A
  see ASTM range 23B
ZICRAL 1N
ZILLAY 55A
ZIMAL 55A
ZIMALUIM 1M
ZINC 55A
ZINCGRIP 44A1
ZINCON F 55A
ZINCON G 55A
ZINCON H 55A
ZINCON I 55A
ZINCON J 55A
ZINCROMETAL 44A1
ZINC TINSEL 55A
ZINN 50A
ZINTEX 44A1
ZIRCALLOY 11 56.1
ZIRCON 52.1
ZIRCONIUM 56.1
ZIRCONIUM 10 56.1
ZIRCONIUM 20 56.1
ZIRCONIUM 30 56.1
ZIRCONIUM 40 56.1
ZIRKONAL 1F
ZISIUM 1G
ZISKON 55A
ZK 51A
  see ASTM range 23B
ZK 60 23B
ZK 60A
  see ASTM range 23B
ZK 61 23B
ZK 61A
  see ASTM range 23B
ZN 1
  see BS range 55A
ZN 2
  see BS range 55A
ZN 3
  see BS range 55A
ZN 4
  see BS range 55A
Zn Al 4 55A
Zn Al Cu 1 55A
Z-NICKEL 27A
ZODIAC 14F
ZRE 1 23B
Zs 3817 20B
Zs 4012 20B
Zs 4505 20B
Zs 5002 20B
Zs 6002 20B
Zs 8002 20B
Zs 9002 20B
ZT 1 23B
ZTM 50.1
ZTY 23B
ZW 1 23B
ZW 3 23B
ZW 6 23B